THE PRINCIPAL FUNCTIONAL GROUPS OF ORGANIC CHEMISTRY

	Example	*Acceptable Name(s) of Example*	*Characteristic Reaction Type*
Carboxylic acid derivatives			
Acyl halides	$\overset{\overset{\displaystyle O}{\|\|}}{CH_3C}Cl$	Ethanoyl chloride or acetyl chloride	Nucleophilic acyl substitution
Acid anhydrides	$\overset{\overset{\displaystyle O \quad O}{\|\| \quad \|\|}}{CH_3COCCH_3}$	Ethanoic anhydride or acetic anhydride	Nucleophilic acyl substitution
Esters	$\overset{\overset{\displaystyle O}{\|\|}}{CH_3COCH_2CH_3}$	Ethyl ethanoate or ethyl acetate	Nucleophilic acyl substitution
Amides	$\overset{\overset{\displaystyle O}{\|\|}}{CH_3CNHCH_3}$	N-Methylethanamide or N-methylacetamide	Nucleophilic acyl substitution
Nitrogen-containing organic compounds			
Amines	$CH_3CH_2NH_2$	Ethanamine or Ethylamine	Nitrogen acts as a base or as a nucleophile
Nitriles	$CH_3C \equiv N$	Ethanenitrile or acetonitrile	Nucleophilic addition to carbon-nitrogen triple bond
Nitro compounds	$C_6H_5NO_2$	Nitrobenzene	Reduction of nitro group to amine
Sulfur-containing organic compounds			
Thiols	CH_3CH_2SH	Ethanethiol	Oxidation to a sulfenic, sulfinic, or sulfonic acid or to a disulfide
Sulfides	$CH_3CH_2SCH_2CH_3$	Diethyl sulfide	Alkylation to a sulfonium salt; oxidation to a sulfoxide or sulfone

IMPORTANT:

HERE IS YOUR REGISTRATION CODE TO ACCESS

YOUR PREMIUM McGRAW-HILL ONLINE RESOURCES.

For key premium online resources you need THIS CODE to gain access. Once the code is entered, you will be able to use the Web resources for the length of your course.

If your course is using **WebCT** or **Blackboard**, you'll be able to use this code to access the McGraw-Hill content within your instructor's online course.

Access is provided if you have purchased a new book. If the registration code is missing from this book, the registration screen on our Website, and within your WebCT or Blackboard course, will tell you how to obtain your new code.

Registering for McGraw-Hill Online Resources

To gain access to your McGraw-Hill web resources simply follow the steps below:

1. USE YOUR WEB BROWSER TO GO TO: **http://www.mhhe.com/carey5/**

2. CLICK ON **FIRST TIME USER**.

3. ENTER THE REGISTRATION CODE* PRINTED ON THE TEAR-OFF BOOKMARK ON THE RIGHT.

4. AFTER YOU HAVE ENTERED YOUR REGISTRATION CODE, CLICK **REGISTER**.

5. FOLLOW THE INSTRUCTIONS TO SET-UP YOUR PERSONAL UserID AND PASSWORD.

6. WRITE YOUR UserID AND PASSWORD DOWN FOR FUTURE REFERENCE. KEEP IT IN A SAFE PLACE.

TO GAIN ACCESS to the McGraw-Hill content in your instructor's **WebCT** or **Blackboard** course simply log in to the course with the UserID and Password provided by your instructor. Enter the registration code exactly as it appears in the box to the right when prompted by the system. You will only need to use the code the first time you click on McGraw-Hill content.

Thank you, and welcome to your McGraw-Hill online Resources!

0-07-291913-2 CAREY: ORGANIC CHEMISTRY, 5E

MCGRAW-HILL
ONLINE RESOURCES

REGISTRATION CODE

guaranteed-61020507

ORGANIC CHEMISTRY

fifth edition

Francis A. Carey
University of Virginia

Boston   Burr Ridge, IL   Dubuque, IA   Madison, WI   New York   San Francisco   St. Louis
Bangkok   Bogotá   Caracas   Kuala Lumpur   Lisbon   London   Madrid   Mexico City
Milan   Montreal   New Delhi   Santiago   Seoul   Singapore   Sydney   Taipei   Toronto

McGraw-Hill Higher Education

A Division of The **McGraw-Hill** *Companies*

ORGANIC CHEMISTRY, FIFTH EDITION

International   2 3 4 5 6 7 8 9 0 VNH/VNH 0 9 8 7 6 5 4 3
Domestic   3 4 5 6 7 8 9 0 VNH/VNH 0 9 8 7 6 5 4 3

ISBN 0-07-242458-3
ISBN 0-07-115148-6 (ISE)

Publisher: *Kent A. Peterson*
Developmental editor: *Shirley R. Oberbroeckling*
Marketing manager: *Thomas D. Timp*
Lead project manager: *Peggy J. Selle*
Lead production supervisor: *Sandra Hahn*
Designer: *K. Wayne Harms*
Cover/interior designer: *Jamie E. O'Neal*
Cover photo credit: *NASA Ames Home Page*
Senior photo research coordinator: *John C. Leland*
Media project manager: *Sandra M. Schnee*
Supplement producer: *Brenda A. Ernzen*
Media technology lead producer: *Steve Metz*
Compositor: *GTS Graphics, Inc.*
Typeface: *10/12 Times Roman*
Printer: *Von Hoffmann Press, Inc.*

The credits section for this book begins on page C-1 and is considered an extension of the copyright page.

Library of Congress Cataloging-in-Publication Data

Carey, Francis A., 1937–
   Organic chemistry / Francis A. Carey.—5th ed.
     p.   cm.
   Includes index.
   ISBN 0-07-242458-3 (acid-free paper)—ISBN 0-07-115148-6 (ISE : acid-free paper)
     1. Chemistry, Organic.   I. Title.

QD251.3 .C37   2003
547—dc21

2002016637
CIP

INTERNATIONAL EDITION ISBN 0-07-115148-6
Copyright © 2003. Exclusive rights by The McGraw-Hill Companies, Inc., for manufacture and export. This book cannot be re-exported from the country to which it is sold by McGraw-Hill. The International Edition is not available in North America.

www.mhhe.com

ABOUT THE AUTHOR

Francis A. Carey is a native of Pennsylvania, educated in the public schools of Philadelphia, at Drexel University (B.S. in chemistry, 1959), and at Penn State (Ph.D. 1963). Following postdoctoral work at Harvard and military service, he was appointed to the chemistry faculty of the University of Virginia in 1966. Prior to retiring in 2000, he regularly taught the two-semester lecture courses in general chemistry and organic chemistry.

With his students, Professor Carey has published over 40 research papers in synthetic and mechanistic organic chemistry. In addition to this text, he is coauthor (with Robert C. Atkins) of *Organic Chemistry: A Brief Course* and (with Richard J. Sundberg) of *Advanced Organic Chemistry,* a two-volume treatment designed for graduate students and advanced undergraduates. He was a member of the Committee of Examiners of the Graduate Record Examination in Chemistry from 1993–2000.

Frank and his wife Jill, who is a teacher/director of a preschool and a church organist, are the parents of Andy, Bob, and Bill and the grandparents of Riyad and Ava.

ABOUT THE COVER

The cover shows a buckminsterfullerene (C_{60}) molecule traveling through interstellar space. A helium atom is trapped within the fullerene cage. Where the fullerene and helium atom came from, where they are going and when, combine to give the fascinating story of scientific detective work told on page 437.

   We thank Dr. Luann Becker, a geochemist at the University of Hawaii and one of the lead investigators who discovered these "spacebuckys," for providing the striking graphic displayed on the cover.

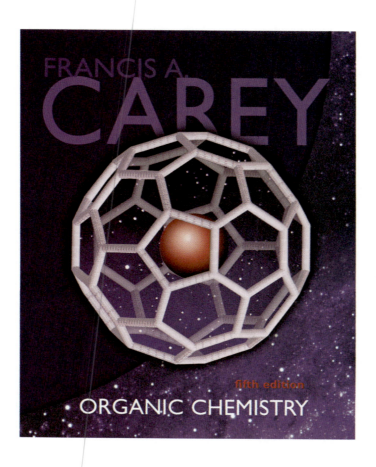

This edition is dedicated
with affection to my wife Jill.

BRIEF CONTENTS

CONTENTS

5.17 The E1 Mechanism of Dehydrohalogenation of Alkyl Halides 217
5.18 SUMMARY 220
      Problems 223

CHAPTER 6

REACTIONS OF ALKENES: ADDITION REACTIONS   **230**

6.1   Hydrogenation of Alkenes 230
6.2   Heats of Hydrogenation 231
6.3   Stereochemistry of Alkene Hydrogenation 234
6.4   Electrophilic Addition of Hydrogen Halides to Alkenes 235
6.5   Regioselectivity of Hydrogen Halide Addition: Markovnikov's Rule 236
6.6   Mechanistic Basis for Markovnikov's Rule 238
      Rules, Laws, Theories, and the Scientific Method 239
6.7   Carbocation Rearrangements in Hydrogen Halide Addition to Alkenes 241
6.8   Free-Radical Addition of Hydrogen Bromide to Alkenes 242
6.9   Addition of Sulfuric Acid to Alkenes 245
6.10 Acid-Catalyzed Hydration of Alkenes 247
6.11 Hydroboration–Oxidation of Alkenes 250
6.12 Stereochemistry of Hydroboration–Oxidation 252
6.13 Mechanism of Hydroboration–Oxidation 252
6.14 Addition of Halogens to Alkenes 254
6.15 Stereochemistry of Halogen Addition 256
6.16 Mechanism of Halogen Addition to Alkenes: Halonium Ions 256
6.17 Conversion of Alkenes to Vicinal Halohydrins 259
6.18 Epoxidation of Alkenes 260
6.19 Ozonolysis of Alkenes 262
6.20 Introduction to Organic Chemical Synthesis 265
6.21 Reactions of Alkenes with Alkenes: Polymerization 266
      **Ethylene and Propene: The Most Important Industrial
      Organic Chemicals 269**
6.22 SUMMARY 271
      Problems 274

CHAPTER 7

STEREOCHEMISTRY   **281**

7.1   Molecular Chirality: Enantiomers 281
7.2   The Chirality Center 282
7.3   Symmetry in Achiral Structures 286
7.4   Optical Activity 287
7.5   Absolute and Relative Configuration 289
7.6   The Cahn–Ingold–Prelog R–S Notational System 290
7.7   Fischer Projections 293
7.8   Properties of Enantiomers 295
      Chiral Drugs 296
7.9   Reactions That Create a Chirality Center 297
7.10 Chiral Molecules with Two Chirality Centers 300
7.11 Achiral Molecules with Two Chirality Centers 303
      Chirality of Disubstituted Cyclohexanes 305
7.12 Molecules with Multiple Chirality Centers 306

CHAPTER 8

NUCLEOPHILIC SUBSTITUTION   326

CHAPTER 9

ALKYNES   363

CHAPTER 23

ARYL HALIDES   971

CHAPTER 24

PHENOLS   993

CHAPTER 25

CARBOHYDRATES   **1026**

CHAPTER 26

LIPIDS   **1069**

LIST OF BOXED ESSAYS

PREFACE

From the first edition through this, its fifth, *Organic Chemistry* has been designed to meet the needs of the "mainstream," two-semester undergraduate organic chemistry course. From the beginning and with each new edition, we have remained grounded in some fundamental notions. These include important issues concerning the intended audience. Does the topic appropriately take into consideration their interests, aspirations, and experience? Just as important is the need to present an accurate picture of the present state of organic chemistry. How do we know what we know? What makes organic chemistry worth knowing? Where are we now? Where are we headed?

The central message of chemistry is that the properties of a substance come from its structure. What is less obvious, but very powerful, is the corollary. Someone with training in chemistry can look at the structure of a substance and tell you a lot about its properties. Organic chemistry has always been, and continues to be, the branch of chemistry that best connects structure with properties. Our objective has been to emphasize the connection between structure and properties, using the tools best suited to make that connection.

One tool is organizational. The time-honored *functional group approach* focuses attention on the *structural* units within a molecule that are most closely identified with its *properties*. The text is organized according to functional groups, but *emphasizes mechanisms* and encourages students to look for similarities in mechanisms among different functional groups.

Another tool relates to presentation. We decided to emphasize molecular modeling in the third edition, expanded its usefulness by adding Spartan electrostatic potential maps in the fourth, and continue this trend in the fifth. Molecular models, *and the software to make their own models*, not only make organic chemistry more accessible to students who are "visual learners," they enrich the educational experience for all.

WHAT'S NEW FOR THE FIFTH EDITION?

Organization

Key changes appear at the beginning and at the end. The changes in Chapter 1 are major and send ripples through the book. Chapter 28 is new.

- **New!** Chapter 1 has been retitled "Structure Determines Properties" to better reflect its purpose and has been rewritten to feature a detailed treatment of *acids and bases.* Rather than a review of what students learned about acids and bases in general chemistry, Sections 1.12–1.17 discuss acids and bases from an organic chemistry perspective.

- To accommodate the new material on acids and bases in Chapter 1, the orbital hybridization model of bonding in organic compounds has been rewritten and placed in Chapter 2. In keeping with its expanded role, Chapter 2 in now titled "Hydrocarbon Frameworks. Alkanes."

- Chapter 13, "Spectroscopy," has been supplemented by an expanded discussion of ^1H and ^{13}C chemical shifts and a new section on 2D NMR. A **new** box, *Spectra by the Thousands,* points the way to websites that feature libraries of spectra and spectroscopic problems of every range of difficulty.

- Chapter 27 has been shortened by removing material related to nucleic acids and its title changed to "Amino Acids, Peptides, and Proteins."

- **New!** Chapter 28 "Nucleosides, Nucleotides, and Nucleic Acids" is new. Its presence testifies to the importance of these topics and the explosive growth of our knowledge of the molecular basis of genetics.

Pedagogy

- **New!** The continuing positive response to the generous use of tables in *Organic Chemistry* has encouraged us to create new ones. The new tables are

Table 1.6 VSEPR and Molecular Geometry
Table 1.7 Dissociation Constants (pK_a) of Acids
Table 2.5 Oxidation Numbers in Compounds with More
    Than One Carbon
Table 28.2 The Major Nucleosides in DNA and/or RNA

- **New!** The number of boxed essays has been increased to 42 with the addition of

Electrostatic Potential Maps
Curved Arrows
Ring Currents: Aromatic and Antiaromatic
Spectra by the Thousands
Nonsteroidal Antiinflammatory Drugs (NSAIDs) and
    COX-2 Inhibitors
Oh NO! It's Inorganic!
"It Has Not Escaped Our Notice . . ."
RNA World

- The *Learning By Modeling* CD-ROM developed by Wavefunction, Inc. in connection with the fourth edition of this text accompanies the fifth as well. We were careful to incorporate Spartan so it would work with the textbook—from the Spartan images used in the text to the icons directing the student to opportunities to build models of their own or examine those in a collection of more than 250 already prepared ones.

- A number of **new** in-chapter and end-of-chapter problems have been added.

Art Program

- Instead of limiting molecular models to figures, "bonus" models have been integrated into the body of the text in places where they reveal key features more clearly than words or structural formulas alone can. (See page 175.)

- **New!** Attention is paid to the nodal properties of orbitals throughout the text in order to foster an appreciation for this important aspect of bonding theory. (See Figure 2.16 on page 90.)

Media

- *Learning By Modeling* for building, examining, and evaluating molecular models specific to organic chemistry

- **New!** Essential Study Partner (ESP) interactive student tutorial

- Improved Online Learning Center website for instructors and students

- **New!** Mechanism Animations CD-ROM for the instructor

INSTRUCTOR RESOURCES

McGraw-Hill offers various tools and technology products to support the fifth edition of *Organic Chemistry*. Instructors can obtain teaching aids by calling the Customer Service Department at 800-338-3987 or contacting your local McGraw-Hill sales representative.

Test Bank

Written by Bruce Osterby (University of Wisconsin-LaCrosse), this manual contains over 1,000 multiple-choice questions. The Test Bank is available under the Instructor Center on the Online Learning Center at www.mhhe.com/carey.

Computerized Test Bank

Written by Bruce Osterby (University of Wisconsin-LaCrosse), the Test Bank is formatted for easy integration into the following course management systems: PageOut, WebCT, and Blackboard.

Digital Content Manager

This presentation CD-ROM contains a multimedia collection of visual resources allowing instructors to use artwork from the text in multiple formats to create customized classroom presentations, visually-based tests and quizzes, dynamic course website content, or attractive printed support materials. The Digital Content Manager is a cross-platform CD containing an image library, a tables library, and a PowerPoint presentation.

Online Learning Center

The comprehensive website (www.mhhe.com/carey) is book-specific and offers excellent tools for both the instructor and the student. Instructors can create an interactive course with the integration of this site, and a secure Instructor Center stores your essential course materials to save you prep time before class. This center offers PowerPoint images, a PowerPoint lecture outline, mechanism animations, and more.

Learning By Modeling CD-ROM

In collaboration with Wavefunction, we have created a cross-function CD-ROM that contains an electronic model-building kit and a rich collection of molecular models that reveal the interplay between electronic structure and reactivity in organic chemistry.

Solutions Manual

Written by Robert Atkins (James Madison University) and Francis Carey, this manual provides complete solutions to all of the problems in the text.

Overhead Transparencies

These full-color transparencies of illustrations from the text include reproductions of spectra, orbital diagrams, key tables, computer-generated molecular models, and step-by-step reaction mechanisms.

Organic Chemistry Animations Library

Created by Rainer Glaser (University of Missouri), the animations are basic mechanisms that can be presented at full-screen size in your classroom. The animations on the CD can be played directly from the CD or can be imported easily into your own lecture presentation.

Course-Specific PageOut

Designed specifically to help you with your individual course needs, PageOut will assist you in integrating your syllabus with *Organic Chemistry* and with state-of-the-art new media tools. At the heart of PageOut you will find integrated multimedia and a full-scale Online Learning Center. You can upload your original test questions and create your own custom designs. More than 60,000 professors have chosen PageOut to create customized course websites.

WebCT, Blackboard

The Test Bank and Online Learning Center with self-assessments quizzes and review aids are available in the various course management systems. Please ask your sales representative for details if you are interested.

STUDENT RESOURCES

McGraw-Hill offers various tools and technology to support the fifth edition of *Organic Chemistry*. Students can order supplemental study materials by contacting the McGraw-Hill Customer Service Department at 800-338-3987.

Solutions Manual

Written by Robert C. Atkins and Francis A. Carey, the solutions manual provides step-by-step solutions guiding the student through the reasoning behind each problem in the text. There is also a self-test at the end of each chapter designed to assess the student's mastery of the material.

Learning By Modeling CD-ROM

In collaboration with Wavefunction, we have created a cross-function CD-ROM that contains an electronic model-building kit and a rich collection of molecular models that reveal the interplay between electronic structure and reactivity in organic chemistry. Icons in the text point the way to where you can use this state-of-art molecular modeling application to expand your understanding and sharpen your conceptual skills.

OLC (Online Learning Center)

The Online Learning Center is a comprehensive, exclusive website that provides a wealth of electronic resources for instructors and students alike. For students, the OLC features tutorial, problem-solving strategies and assessment exercises for every chapter in the book that were developed by Ian Hunt and Rick Spinney from the University of Calgary. You can also access the Essential Student Partner from the OLC. Log on at www.mhhe.com/carey.

Schaum's Outline of Organic Chemistry

This helpful study aid provides students with hundreds of solved and supplementary problems for the organic chemistry course.

ACKNOWLEDGMENTS

No textbook is solely the creative work of the person whose name is on the cover. Left to their own devices, authors would include too much, address the wrong audience, and, worst of all, never finish. The indispensable people, the folks who keep us focused, are our editors. At McGraw-Hill, Shirley Oberbroeckling combines all the qualities an author appreciates in an editor. In developing "Carey 5/e," Shirley never lost sight of the big picture, yet stayed on top of every detail. She was a consistent source of ideas and encouragement. Likewise, Kent Peterson, now publisher for the physical sciences, participated actively in shaping this edition, including identifying the content areas to emphasize. To use a sports metaphor, Kent was the coach of the team, Shirley was the captain. It was a pleasure to be on the field with them.

I also thank Linda Davoli, who has edited copy for most of our recent writing projects. In addition to being both professional and thorough, Linda makes a special effort to go beyond the words themselves to ensure that the ideas get through. As she did in the fourth edition, Peggy Selle oversaw the production of the text with skill and enthusiasm. Warren Hehre of Wavefunction, Inc., continues to support our efforts to expand the role of molecular modeling in the introductory organic chemistry course. He is generous with his time and a rich source of ideas.

I appreciate the help of Thomas Gallaher of the Department of Chemistry at James Madison University in obtaining the 2D NMR spectra that are new to this edition. As in every previous edition of this text, the contributions of Dr. Robert C. Atkins of James Madison University are beyond measure. I can't think of anything connected with *Organic Chemistry* that hasn't benefited from Bob's efforts and insight.

The comments offered by an unusually large number of teachers of organic chemistry were especially helpful in developing this edition. I appreciate their help. A special thank you to the following participants in a McGraw-Hill organic chemistry summit held during the summer of 2000 for their insight into the needs of organic chemistry professors:

Carl Bumgardner, *North Carolina State University*
Rob Coleman, *Ohio State University*
Rainer Glaser, *University of Missouri*
Gary Gray, *University of Minnesota*
Kenn Harding, *Texas A & M*
J. Howard Hargis, *Auburn University*
Charles Kingsbury, *University of Nebraska*
Cary Morrow, *University of New Mexico*
Anne Padias, *University of Arizona*
Bradley Smith, *University of Notre Dame*

The reviewers listed below have reviewed for the fifth edition in various ways: sharing ideas on improving this edition from the fourth edition, helping fine-tune the manuscript of the fifth edition, giving direction on new art, and reviewing media including the *Learning By Modeling* CD and the Online Learning Center.

REVIEWERS FOR THE FIFTH EDITION

Chris Abelt, *William & Mary*
David Adams, *University of Massachusetts—Amherst*
Sheila Adanus, *Providence College*
Rigoberto C. Advincula, *University of Alabama—Birmingham*
Carol P. Anderson, *University of Connecticut—Groton*
D. Timothy Anstine, *Northwest Nazarene University*
Cindy Applegate, *University of Colorado—Colorado Springs*
Mark Arant, *University of Louisiana—Monroe*
Lisa R. Arnold, *South Georgia College*
Jeffrey B. Arterburn, *New Mexico State University—Las Cruces*
William F. Bailey, *University of Connecticut*
Satinder Bains, *Arkansas State University—Beebe*
Dave Baker, *Delta College*
George C. Bandik, *University of Pittsburgh*
John Barbaro, *University of Alabama—Birmingham*
John Barbas, *Valdosta State University*
Bal Barot, *Lake Michigan College*
John Bellefeuille, *Texas A & M University—College Station*
John Benson, *Ridgewater College*
David E. Bergbreiter, *Texas A & M University—College Station*
Thomas Berke, *Brookdale Community College*
K. Darrell Berlin, *Oklahoma State University—Stillwater*
Chris Bernhardt, *Greensboro College*
Narayan G. Bhat, *University of Texas—Pan American*
Pradip K. Bhowmik, *University of Nevada—Las Vegas*
Joseph F. Bieron, *Canisius College*
T. Howard Black, *Eastern Illinois University*
Salah M. Blaih, *Kent State University—Trumbull Campus*
Erich C. Blossey, *Rollins College—Winter Park*

John Bonte, *Clinton Community College*
Eric Bosch, *Southwest Missouri State*
Peter H. Boyle, *Trinity College*
R. Bozak, *California State University—Hayward*
Lynn Bradley, *The College of New Jersey*
Andrew R. Bressette, *Berry College*
Patricia A. Brletic, *Washington & Jefferson College*
David Brook, *University of Detroit Mercy*
Edward Brown, *Lee University*
Paul Buonora, *University of Scranton*
Kevin Burgess, *Texas A & M University—College Station*
Sybil K. Burgess, *University of North Carolina—Wilmington*
Magnus Campbell, *Lansing Community College*
Suzanne Carpenter, *Armstrong Atlantic State University*
Jin K. Cha, *University of Alabama—Tuscaloosa*
Jeff Charonnat, *California Sate University—Northridge*
Dana Chatellier, *University of Delaware*
Clair J. Cheer, *San Jose State University*
John C. Cochran, *Colgate University*
Don Colborn, *Missouri Baptist College*
Wheeler Conover, *Southeast Community College*
Wayne B. Counts, *Georgia Southwestern State University*
Brian E. Cox, *Cochise College—Sierra Vista*
Thomas D. Crute, *Augusta State University*
Timothy P. Curran, *College of the Holy Cross*
S. Todd Deal, *Georgia Southern University*
Gary DeBoer, *Le Tourneau University*
Max Deinzer, *Oregon State University*
Leslie A. Dendy, *University of New Mexico—Los Alamos*
Ray Di Martini, *NYC Technical College*
Charles Dickson, *Catawba Valley Community College*
Edward Dix, *Assumption College*
Ajit S. Dixit, *A B Technical Community College*
Marvin Doerr, *Clemson University*
Charles M. Dougherty, *Lehman College*
Veljko Dragojlovic, *MST Nova Southeastern University*
Karen Walte Dunlap, *Sierra College*
Stella Elakovich, *University of Southern Mississippi—Hattiesburg*
Janice Ems-Wilson, *Valencia Community College—West Campus*
Dorothy N. Eseonu, *Virginia Union University*
Michael F. Falcetta, *Anderson University*
Mitchell Fedak, *Community College Allegheny County*
K. Thomas Finley, *SUNY—Brockport*
Jan M. Fleischer, *The College of New Jersey*
Malcolm Forbes, *University of North Carolina*
Laura Lowe Furge, *Kalamazoo College*
Ana M. Gaillat, *Greenfield Community College*
Roy Garvey, *North Dakota State University—Fargo*
Edwin Geels, *Dordt College*
Graeme C. Gerrans, *University of Virginia*
Mary S. Gin, *University of Illinois*
Ted Gish, *St. Mary's College*
Brett Goldston, *North Seattle Community College*
Manuel Grande, *Universidad de Salamanca*
Gary Gray, *University of Minnesota*
Gamini Gunawardena, *Utah Valley State College—Orem*
Sapna Gupta, *Park University—Parkville*
Frank Guziec, *Southwestern University*

Kevin P. Gwaltney, *Bucknell University*
Christopher Hadad, *Ohio State University*
Midge Hall, *Clark State Community College*
Philip D. Hampton, *University of New Mexico—Albuquerque*
Carol Handy, *Portland Community College—Sylvania*
Ryan Harden, *Central Lakes College*
Steve Hardinger, *University of California—Los Angeles*
Sidney Hecht, *University of Virginia*
Carl Heltzel, *Transylvania University*
Alvan Hengge, *Utah State University—Logan*
Reza S. Herati, *Southwest Missouri State University*
James R. Hermanson, *Wittenberg University*
Paul Higgs, *Barry University*
Brian Hill, *Bryan College*
Richard W. Holder, *University of New Mexico—Albuquerque*
Robert W. Holman, *Western Kentucky University*
Gary D. Holmes, *Butler County Community College*
Roger House, *Moraine Valley College*
William C. Hoyt, *St. Joseph's College*
Jeffrey Hugdahl, *Mercer University—Macon*
Lyle Isaacs, *University of Maryland*
T. Jackson, *University of South Alabama—Mobile*
George F. Jackson, *University of Tampa*
Richard Jarosch, *University of Wisconsin—Sheboygan*
Ronald M. Jarret, *College of the Holy Cross*
George T. Javor, *Loma Linda University School of Medicine*
John Jefferson, *Luther College*
Anton Jensen, *Central Michigan University*
Teresa Johnson, *Bevill State Community College—Sumiton*
Todd M. Johnson, *Weber State University*
Graham Jones, *Northeastern University*
Taylor B. Jones, *The Masters College*
Robert O. Kalbach, *Finger Lakes Community College*
John W. Keller, *University of Alaska—Fairbanks*
Floyd W. Kelly, *Casper College*
Dennis Kevill, *Northern Illinois University*
Tony Kiessling, *Wilkes University*
Lynn M. Kirms, *Southern Oregon University*
Susan Klein, *Manchester College*
Beata Knoedler, *Springfield College in Illinois*
Albert C. Kovelesky, *Limestone College*
Lawrence J. Krebaum, *Missouri Valley College*
Paul J. Kropp, *University of North Carolina*
Mahesh K. Lakshman, *University of North Dakota*
Andrew Langrehr, *Jefferson College*
Elizabeth M. Larson, *Grand Canyon University*
Melanie J. Lesko, *Texas A & M University—Galveston*
Nicholas Leventis, *University of Missouri—Rolla*
Brian Love, *East Carolina University*
John Lowbridge, *Madisonville Community College*
Rachel Lum, *Regis University*
Raymond Lutz, *Portland State University*
Elahe Mahdavian, *South Carolina State University*
William Mancini, *Paradise Valley Community College*
Alan P. Marchand, *University of North Texas*
David F. Maynard, *California State University—San Bernardino*
John McCormick, *University of Missouri—Columbia*
Alison McCurdy, *Harvey Mudd College*
Nancy McDonald, *Athens State University*

Michael B. McGinnis, *Georgia College and State University*
Harold T. McKone, *St. Joseph College*
Ed McNelis, *New York Univesity*
Tyler D. McQuade, *Cornell University*
Jerrold Meinwald, *Cornell University*
David Mendenhall, *Michigan Technology University*
K. Troy Milliken, *Broward Community College—North*
Qui-chee Mir, *Pierce College—Puyallup*
Anthony A. Molinero, *State University College—Potsdam SUNY*
Gary Mong, *Columbia Basin College*
Mel Mosher, *Missouri Southern College*
Richard Musgrave, *St. Petersburg Junior College—St. Petersburg*
Jennifer Muzyka, *Centre College of Kentucky*
David Myers, *Simons Rock College*
Richard Narske, *Augustana College*
Thomas A. Newton, *University of Southern Maine—Portland*
Maria Ngu-Schwe, *Southern University—Baton Rouge*
Olivier Nicaise, *St. Louis University*
Bradley Norwood, *Coastal Carolina University*
George O'Doherty, *University of Minnesota*
Thomas O'Hare, *Willamette University*
Kimberly A. Opperman, *Erskine College*
Oladayo Oyelola, *Lane College*
Michael J. Panigot, *Arkansas State University*
E. P. Papadopoulos, *University of New Mexico—Albuquerque*
Edward J. Parish, *Auburn University—Auburn*
Cyril Parkanyi, *Florida Atlantic University—Boca Raton*
G. Anthony Parker, *Seton Hill/St. Vincent's College*
Abby Parrill, *University of Memphis*
Gitendra Paul, *Ohio University—Athens*
Donald R. Paulson, *California State University—Los Angeles*
David Peitz, *Wayne State College*
Thomas C. Pentecost, *Aims Community College—Main Campus*
Phillip J. Persichini III, *University of Minnesota—Duluth*
Gary O. Pierson, *Central State University*
Suzette Polson, *Kentucky State University*
Steve Pruett, *Jefferson Community College*
Ann Randolph, *Rosemont College*
Casey Raymond, *Kent State University*
Preston Reeves, *Texas Lutheran College*
Jill Rehmann, *St. Joseph's College*
Thomas Reitz, *Bradford College*
Mitchell A. Rhea, *Pensacola Junior College*
Stacia Rink, *Pacific Lutheran University*
Randall E. Robinson, *Luther College*
Harold R. Rogers, *California State University—Fullerton*
Ellen Roskes, *Villa Julie College*
Thomas Russo, *Jacksonville University*
Thomas Ruttledge, *Ursinus College*
Vyacheslav V. Samoshin, *University of the Pacific*
Susan M. Schelble, *University of Colorado—Denver*
Judith E. Schneider, *SUNY—Oswego*
Keith Schray, *Lehigh University*
Allen J. Scism, *Central Missouri State University*
John Searle, *College of San Mateo*
Alexander Seed, *Kent State University*
William R. Shay, *University of North Dakota*

Dale F. Shellhamer, *Point Loma University*
Angela Rap Sherman, *College of Notre Dame of Maryland*
Barry R. Sickles, *Durham Technical Community College*
Jack Sidler, *Mansfield University*
Steven B. Silbering, *CUNY York College*
Phil Silberman, *Scottsdale Community College*
Eric E. Simanek, *Texas A & M University—College Station*
Carmen M. Simone, *Casper College*
Cynthia Somers, *Red Rocks Community College*
Melinda M. Sorenson, *University of Louisiana—Lafayette*
Tami Spector, *University of San Francisco*
Gary O. Spessard, *St. Olaf College*
Richard Steiner, *University of Utah—Salt Lake City*
James D. Stickler, *Allegheny College of Maryland*
Heinz Stucki, *Kent State University*
Sam Subramania, *Miles College*
Sam Sun, *Norfolk State University*
Paris Svoronos, *Queensborough Community College*
Erach R. Talaty, *Wichita State University*
Jamal Tayh, *Scott Community College—Bettendorf*
Richard T. Taylor, *Miami University—Oxford*
Stephen S. Templin, *Cardinal Stritch University*
Patricia C. Thorstenson, *University of the District of Columbia*
Tammy Tiner, *Texas A & M University*
Tony Toste, *Southwest Missouri State University*
Denise Tridle, *Highland Community College*
Eric Trump, *Emporia State University*
F. Warren Villaescusa, *Our Lady of the Lake University*
Paul Vorndam, *University of Southern Colorado*
J. S. Walia, *Loyola University—New Orleans*
George H. Wahl, *North Carolina State University*
William A. Wallace, *Barton College*
Chad Wallace, *Asbury College*
Carl C. Wamser, *Portland State University*
Steven P. Wathan, *Siena Heights University*
Paul L. Weber, *Briar Cliff College*
Sonia C. Weeks, *Shaw University*
Lyle D. Wescott, *Christian Brothers University*
Craig A. Wesolowski, *University of Central Arkansas*
Kraig Wheeler, *Delaware State University*
David Wiedenfeld, *University of North Texas*
Larry Wiginton, *Clarendon College*
Carrie Wolfe, *Union College*
Thomas G. Wood, *College Misericordia*
Mali Yin, *Sarah Lawrence College*
Viktor Zhdankin, *University of Minnesota—Duluth*
Hans Zimmer, *University of Cincinnati—Cincinnati*

My wife Jill, our sons Andy, Bob, and Bill, and daughter-in-law Tasneem, provide encouragement. Our grandchildren Riyad and Ava provide inspiration. It all adds up.

Comments, suggestions, and questions—especially from students—are welcome. I invite you to contact me at fac6q@unix.mail.virginia.edu.

—**Francis A. Carey**

A GUIDE TO USING THIS TEXT

The following two pages walk you through some of the key features of this text. This book was designed with you, the student in mind—to help you succeed in organic chemistry. My main goal in writing this text was to make the material you study as accessible and appealing as possible. I hope you enjoy your study of organic chemistry.

Francis A. Carey

TEXT

Art Program

- Molecular models make organic chemistry more accessible and enrich the education experience for all.
- Molecular models are integrated into the content revealing key features more clearly than words or structural formulas alone.
- Nodal properties of orbitals are included to show this important aspect of bonding theory.

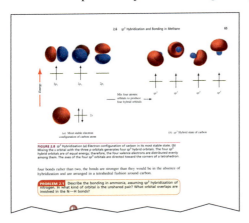

Tables and Summaries

- Summary tables allow the student easy access to a wealth of information in an easy to use format while reviewing information from previous chapters.
- End-of-Chapter Summaries highlight and consolidate all of the important concepts and reactions within a chapter.

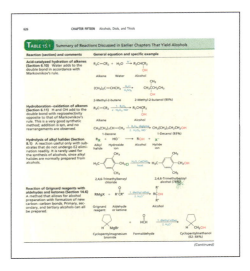

Biological Applications

- New chapter on *Nucleosides, Nucleotides, and Nucleic Acids* testifies to the explosive growth of the molecular basis of genetics.
- Numerous boxed essays throughout the text highlight biological applications of organic chemistry.

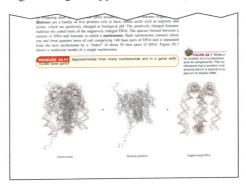

Problem-Solving

- Problem-solving strategies and skills are emphasized throughout. Understanding is continually reinforced by problems that appear within topic sections. For many problems, sample solutions are given.
- Every chapter ends with a comprehensive set of problems that give students liberal opportunity to master skills by working problems and integrating the use of Spartan modeling.

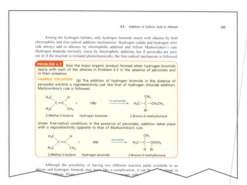

MEDIA

Instructor Media

- Online Learning Center (OLC) is a secure, book-specific website. The OLC is the doorway to a library of resources for instructors.
- PowerPoint Presentation—is organized by chapter and ready for the classroom, or the instructor can customize the lecture to reflect their own teaching style.
- Computerized Test Bank—contains over 1,000 multiple-choice questions.
- Course Management Systems—PageOut, WebCT, and Blackboard. All of the following tools are available on the Online Learning Center or in a cartridge for your course delivery system:
  1. Computerized Test Bank
  2. Images from the text
  3. Tables from the text
  4. Essential Study Partner links to home page

Create a custom course Website with **PageOut**, free with every McGraw-Hill textbook.

To learn more, contact your McGraw-Hill publisher's representative or visit www.mhhe.com/solutions.

- Organic Chemistry Mechanism Animation Library—CD-ROM set includes 8 important and standard mechanisms that can be used directly from the CD or imported into your own lecture presentation.

- Digital Content Manager—CD-ROM set includes electronic files of all full-color images in the text. Import the images into your own presentation, or use the PowerPoint presentation provided for each chapter.

Student Media

- Online Learning Center—is the doorway to access almost all of the media for *Organic Chemistry* Fifth Edition. The OLC includes self-assessment quizzes, resources.
- NetTutor—is a tool for the student when they need extra help with an end-of-chapter problem. Live sessions are scheduled or you can e-mail a tutor at any time.

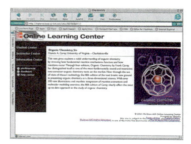

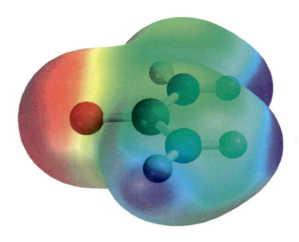

INTRODUCTION

At the root of all science is our own unquenchable curiosity about ourselves and our world. We marvel, as our ancestors did thousands of years ago, when fireflies light up a summer evening. The colors and smells of nature bring subtle messages of infinite variety. Blindfolded, we know whether we are in a pine forest or near the seashore. We marvel. And we wonder. How does the firefly produce light? What are the substances that characterize the fragrance of the pine forest? What happens when the green leaves of summer are replaced by the red, orange, and gold of fall?

THE ORIGINS OF ORGANIC CHEMISTRY

As one of the tools that fostered an increased understanding of our world, the science of chemistry—the study of matter and the changes it undergoes—developed slowly until near the end of the eighteenth century. About that time, in connection with his studies of combustion, the French nobleman Antoine Laurent Lavoisier provided the clues that showed how chemical compositions could be determined by identifying and measuring the amounts of water, carbon dioxide, and other materials produced when various substances were burned in air. By the time of Lavoisier's studies, two branches of chemistry were becoming recognized. One branch was concerned with matter obtained from natural or living sources and was called *organic chemistry*. The other branch dealt with substances derived from nonliving matter—minerals and the like. It was called *inorganic chemistry*. Combustion analysis soon established that the compounds derived from natural sources contained carbon, and eventually a new definition of organic chemistry emerged: **organic chemistry is the study of carbon compounds.** This is the definition we still use today.

BERZELIUS, WÖHLER, AND VITALISM

As the eighteenth century gave way to the nineteenth, Jöns Jacob Berzelius emerged as one of the leading scientists of his generation. Berzelius, whose training was in medicine, had wide-ranging interests and made numerous contributions in diverse areas of chemistry. It was he who in 1807 coined the term "organic chemistry" for the study of compounds derived from natural sources. Berzelius, like almost everyone else at the time, subscribed to the doctrine known as vitalism. **Vitalism** held that living systems possessed a "vital force" that was absent in nonliving systems. Compounds derived from natural sources (organic) were thought to be fundamentally different from inorganic compounds; it was believed inorganic compounds could be synthesized in the laboratory, but organic compounds could not—at least not from inorganic materials.

In 1823, Friedrich Wöhler, fresh from completing his medical studies in Germany, traveled to Stockholm to study under Berzelius. A year later Wöhler accepted a position teaching chemistry and conducting research in Berlin. He went on to have a distinguished career, spending most of it at the University of Göttingen, but is best remembered for a brief paper he published in 1828. Wöhler noted that when he evaporated an aqueous solution of ammonium cyanate, he obtained "colorless, clear crystals often more than an inch long," which were not ammonium cyanate but were instead urea.

$$NH_4^{+\,-}OCN \longrightarrow O{=}C(NH_2)_2$$

Ammonium cyanate   Urea
(an inorganic compound)   (an organic compound)

The transformation observed by Wöhler was one in which an *inorganic* salt, ammonium cyanate, was converted to urea, a known *organic* substance earlier isolated from urine. This experiment is now recognized as a scientific milestone, the first step toward overturning the philosophy of vitalism. Although Wöhler's synthesis of an organic compound in the laboratory from inorganic starting materials struck at the foundation of vitalist dogma, vitalism was not displaced overnight. Wöhler made no extravagant claims concerning the relationship of his discovery to vitalist theory, but the die was cast, and over the next generation organic chemistry outgrew vitalism.

What particularly seemed to excite Wöhler and his mentor Berzelius about this experiment had very little to do with vitalism. Berzelius was interested in cases in which two clearly different materials had the same elemental composition, and he invented the term **isomerism** to define it. The fact that an inorganic compound (ammonium cyanate) of molecular formula CH_4N_2O could be transformed into an organic compound (urea) of the same molecular formula had an important bearing on the concept of isomerism.

The article "Wöhler and the Vital Force" in the March 1957 issue of the *Journal of Chemical Education* (pp. 141–142) describes how Wöhler's experiment affected the doctrine of vitalism. A more recent account of the significance of Wöhler's work appears in the September 1996 issue of the same journal (pp. 883–886).

Lavoisier as portrayed on a 1943 French postage stamp.

A 1979 Swedish stamp honoring Berzelius.

This German stamp depicts a molecular model of urea and was issued in 1982 to commemorate the hundredth anniversary of Wöhler's death. The computer graphic that opened this introductory chapter is also a model of urea.

THE STRUCTURAL THEORY

From the concept of isomerism we can trace the origins of the **structural theory**—the idea that a precise arrangement of atoms uniquely defines a substance. Ammonium cyanate and urea are different compounds because they have different structures. To some degree the structural theory was an idea whose time had come. Three scientists stand out, however, for independently proposing the elements of the structural theory: August Kekulé, Archibald S. Couper, and Alexander M. Butlerov.

It is somehow fitting that August Kekulé's early training at the university in Giessen was as a student of architecture. Kekulé's contribution to chemistry lies in his description of the architecture of molecules. Two themes recur throughout Kekulé's work: critical evaluation of experimental information and a gift for visualizing molecules as particular assemblies of atoms. The essential features of Kekulé's theory, developed and presented while he taught at Heidelberg in 1858, were that carbon normally formed four bonds and had the capacity to bond to other carbons so as to form long chains. Isomers were possible because the same elemental composition (say, the CH_4N_2O molecular formula common to both ammonium cyanate and urea) accommodates more than one pattern of atoms and bonds.

Shortly thereafter, but independently of Kekulé, Archibald S. Couper, a Scot working in the laboratory of Charles-Adolphe Wurtz at the École de Medicine in Paris, and Alexander Butlerov, a Russian chemist at the University of Kazan, proposed similar theories.

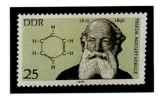

A 1968 German stamp combines a drawing of the structure of benzene with a portrait of Kekulé.

ELECTRONIC THEORIES OF STRUCTURE AND REACTIVITY

In the late nineteenth and early twentieth centuries, major discoveries about the nature of atoms placed theories of molecular structure and bonding on a more secure foundation. Structural ideas progressed from simply identifying atomic connections to attempting to understand the bonding forces. In 1916, Gilbert N. Lewis of the University of California at Berkeley described covalent bonding in terms of shared electron pairs. Linus Pauling at the California Institute of Technology subsequently elaborated a more sophisticated bonding scheme based on Lewis' ideas and a concept called **resonance,** which he borrowed from the quantum mechanical treatments of theoretical physics.

Once chemists gained an appreciation of the fundamental principles of bonding, the next logical step became understanding how chemical reactions occurred. Most notable among the early workers in this area were two British organic chemists: Sir Robert Robinson and Sir Christopher Ingold. Both held a number of teaching positions, with Robinson spending most of his career at Oxford while Ingold was at University College, London.

Robinson, who was primarily interested in the chemistry of natural products, had a keen mind and a penetrating grasp of theory. He was able to take the basic elements of Lewis' structural theories and apply them to chemical transformations by suggesting that chemical change can be understood by focusing on electrons. In effect, Robinson analyzed organic reactions by looking at the electrons and understood that atoms moved because they were carried along by the transfer of electrons. Ingold applied the quantitative methods of physical chemistry to the study of organic reactions so as to better understand the sequence of events, the **mechanism,** by which an organic substance is converted to a product under a given set of conditions.

Our current understanding of elementary reaction mechanisms is quite good. Most of the fundamental reactions of organic chemistry have been scrutinized to the degree that we have a relatively clear picture of the intermediates that occur during the passage

The University of Kazan was home to a number of prominent nineteenth-century organic chemists. Their contributions are recognized in two articles published in the January and February 1994 issues of the *Journal of Chemical Education* (pp. 39–42 and 93–98).

Linus Pauling is portrayed on this 1977 Volta stamp. The chemical formulas depict the two resonance forms of benzene, and the explosion in the background symbolizes Pauling's efforts to limit the testing of nuclear weapons.

The discoverer of penicillin, Sir Alexander Fleming, has appeared on two stamps. This 1981 Hungarian issue includes both a likeness of Fleming and a structural formula for penicillin.

For more on Tyrian purple, see the article "Indigo and Tyrian Purple—In Nature and in the Lab" in the November 2001 issue of the *Journal of Chemical Education*, pp. 1442–1443.

of starting materials to products. Extending of the principles of mechanism to reactions that occur in living systems, on the other hand, is an area in which a large number of important questions remain to be answered.

THE INFLUENCE OF ORGANIC CHEMISTRY

Many organic compounds were known to and used by ancient cultures. Almost every known human society has manufactured and used beverages containing ethyl alcohol and has observed the formation of acetic acid when wine was transformed into vinegar. Early Chinese civilizations (2500–3000 BCE) extensively used natural materials for treating illnesses and prepared a drug known as *ma huang* from herbal extracts. This drug was a stimulant and elevated blood pressure. We now know that it contains ephedrine, an organic compound similar in structure and physiological activity to adrenaline, a hormone secreted by the adrenal gland. Almost all drugs prescribed today for the treatment of disease are organic compounds—some are derived from natural sources; many others are the products of synthetic organic chemistry.

As early as 2500 BCE in India, indigo was used to dye cloth a deep blue. The early Phoenicians discovered that a purple dye of great value, Tyrian purple, could be extracted from a Mediterranean sea snail. The beauty of the color and its scarcity made purple the color of royalty. The availability of dyestuffs underwent an abrupt change in 1856 when William Henry Perkin, an 18-year-old student, accidentally discovered a simple way to prepare a deep-purple dye, which he called *mauveine,* from extracts of coal tar. This led to a search for other synthetic dyes and forged a permanent link between industry and chemical research.

The synthetic fiber industry as we know it began in 1928 when E. I. Du Pont de Nemours & Company lured Professor Wallace H. Carothers from Harvard University to direct their research department. In a few years Carothers and his associates had produced *nylon,* the first synthetic fiber, and *neoprene,* a rubber substitute. Synthetic fibers and elastomers are both products of important contemporary industries, with an economic influence far beyond anything imaginable in the middle 1920s.

COMPUTERS AND ORGANIC CHEMISTRY

A familiar arrangement of the sciences places chemistry between physics, which is highly mathematical, and biology, which is highly descriptive. Among chemistry's subdisciplines, organic chemistry is less mathematical than descriptive in that it emphasizes the qualitative aspects of molecular structure, reactions, and synthesis. The earliest applications of computers to chemistry took advantage of the "number crunching" power of mainframes to analyze data and to perform calculations concerned with the more quantitative aspects of bonding theory. More recently, organic chemists have found the graphics capabilities of minicomputers, workstations, and personal computers to be well suited to visualizing a molecule as a three-dimensional object and assessing its ability to interact with another molecule. Given a biomolecule of known structure, a protein, for example, and a drug that acts on it, molecular-modeling software can evaluate the various ways in which the two may fit together. Such studies can provide information on the mechanism of drug action and guide the development of new drugs of greater efficacy.

The influence of computers on the practice of organic chemistry is a significant recent development and will be revisited numerous times in the chapters that follow.

Many countries have celebrated their chemical industry on postage stamps. The stamp shown was issued in 1971 by Argentina.

CHALLENGES AND OPPORTUNITIES

A major contributor to the growth of organic chemistry during this century has been the accessibility of cheap starting materials. Petroleum and natural gas provide the building blocks for the construction of larger molecules. From petrochemicals comes a dazzling array of materials that enrich our lives: many drugs, plastics, synthetic fibers, films, and elastomers are made from the organic chemicals obtained from petroleum. As we enter an age of inadequate and shrinking supplies, the use to which we put petroleum looms large in determining the kind of society we will have. Alternative sources of energy, especially for transportation, will allow a greater fraction of the limited petroleum available to be converted to petrochemicals instead of being burned in automobile engines. At a more fundamental level, scientists in the chemical industry are trying to devise ways to use carbon dioxide as a carbon source in the production of building-block molecules.

A DNA double helix as pictured on a 1964 postage stamp issued by Israel.

Many of the most important processes in the chemical industry are carried out in the presence of **catalysts.** Catalysts increase the rate of a particular chemical reaction but are not consumed during it. In searching for new catalysts, we can learn a great deal from **biochemistry,** the study of the chemical reactions that take place in living organisms. All these fundamental reactions are catalyzed by enzymes. Rate enhancements of several millionfold are common when one compares an enzyme-catalyzed reaction with the same reaction performed in its absence. Many diseases are the result of specific enzyme deficiencies that interfere with normal metabolism. In the final analysis, effective treatment of diseases requires an understanding of biological processes at the molecular level—what the substrate is, what the product is, and the mechanism by which substrate is transformed to product. Enormous advances have been made in understanding biological processes. Because of the complexity of living systems, however, we have only scratched the surface of this fascinating field of study.

Spectacular strides have been made in genetics during the past few years. Although generally considered a branch of biology, genetics is increasingly being studied at the molecular level by scientists trained as chemists. Gene-splicing techniques and methods for determining the precise molecular structure of DNA are just two of the tools driving the next scientific revolution.

You are studying organic chemistry at a time of its greatest influence on our daily lives, at a time when it can be considered a mature science, when the challenging questions to which this knowledge can be applied have never been more important.

WHERE DID THE CARBON COME FROM?

According to the "big-bang" theory, the universe began expanding about 12 billion years ago when an incredibly dense (10^{96} g·cm^{-3}), incredibly hot (10^{32} K) ball containing all the matter in the universe exploded. No particles more massive than protons or neutrons existed until about 100 s after the big bang. By then, the temperature had dropped to about 10^9 K, low enough to permit the protons and neutrons to combine to form helium nuclei.

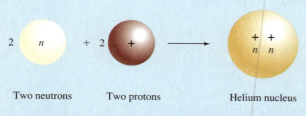

Two neutrons   Two protons   Helium nucleus

Conditions favorable for the formation of helium nuclei lasted for only a few hours, and the universe continued to expand without much "chemistry" taking place for approximately a million years.

As the universe expanded, it cooled, and the positively charged protons and helium nuclei combined with electrons to give hydrogen and helium atoms. Together, hydrogen and helium account for 99% of the mass of the universe and 99.9% of its atoms. Hydrogen is the most abundant element; 88.6% of the atoms in the universe are hydrogen, and 11.3% are helium.

Some regions of space have higher concentrations of matter than others, high enough so that the expansion and cooling that followed the big bang is locally reversed. Gravitational attraction causes the "matter clouds" to collapse and their temperature to increase. After the big bang, the nuclear fusion of hydrogen to helium took place when the temperature dropped to 10^9 K. The same nuclear fusion begins when gravitational attraction heats matter clouds to 10^7 K and the ball of gas becomes a star. The star expands, reaching a more or less steady state at which hydrogen is consumed and heat is evolved. The size of the star remains relatively constant, but its core becomes enriched in helium. After about 10% of the hydrogen is consumed, the amount of heat produced is insufficient to maintain the star's size, and it begins to contract. As the star contracts the temperature of the helium-rich core increases, and helium nuclei fuse to form carbon.

Three helium nuclei   Nucleus of ^{12}C

Fusion of a nucleus of ^{12}C with one of helium gives ^{16}O. Eventually the helium, too, becomes depleted, and gravitational attraction causes the core to contract and its temperature to increase to the point at which various fusion reactions give yet heavier nuclei.

Sometimes a star explodes in a supernova, casting debris into interstellar space. This debris includes the elements formed during the life of the star, and these elements find their way into new stars formed when a cloud of matter collapses in on itself. Our own sun is believed to be a "second generation" star, one formed not only from hydrogen and helium, but containing the elements formed in earlier stars as well.

According to one theory, earth and the other planets were formed almost 5 billion years ago from the gas (the solar nebula) that trailed behind the sun as it rotated. Being remote from the sun's core, the matter in the nebula was cooler than that in the interior and therefore it contracted, accumulating heavier elements and becoming the series of planets that now circle the sun.

Oxygen is the most abundant element on earth. The earth's crust is rich in carbonate and silicate rocks, the oceans are almost entirely water, and oxygen constitutes almost one fifth of the air we breathe. Carbon ranks only fourteenth among the elements in natural abundance, but trails only hydrogen and oxygen in its abundance in the human body. It is the chemical properties of carbon that make it uniquely suitable as the raw material for the building blocks of life. Let's find out more about those chemical properties.

STRUCTURE DETERMINES PROPERTIES

Structure* is the key to everything in chemistry. The properties of a substance depend on the atoms it contains and the way the atoms are connected. What is less obvious, but very powerful, is the idea that someone who is trained in chemistry can look at a structural formula of a substance and tell you a lot about its properties. This chapter begins your training toward understanding the relationship between structure and properties in organic compounds. It reviews some fundamental principles of the Lewis approach to molecular structure and bonding. By applying these principles, you will learn to recognize structural patterns that are more stable than others and develop skills in communicating structural information that will be used throughout your study of organic chemistry. A key relationship between structure and properties will be introduced by examining the fundamentals of acid–base chemistry from a structural perspective.

1.1 ATOMS, ELECTRONS, AND ORBITALS

Before discussing structure and bonding in *molecules,* let's first review some fundamentals of *atomic* structure. Each element is characterized by a unique **atomic number Z,** which is equal to the number of protons in its nucleus. A neutral atom has equal numbers of protons, which are positively charged, and electrons, which are negatively charged.

Electrons were believed to be particles from the time of their discovery in 1897 until 1924, when the French physicist Louis de Broglie suggested that they have wavelike properties as well. Two years later Erwin Schrödinger took the next step and calculated the energy of an electron in a hydrogen atom by using equations that treated the electron as if it were a wave. Instead of a single energy, Schrödinger obtained a series of **energy levels,** each of which corresponded to a different mathematical description of the electron wave. These mathematical descriptions are called **wave functions** and are symbolized by the Greek letter ψ (psi).

*A glossary of important terms may be found immediately before the index at the back of the book.

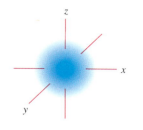

FIGURE 1.1 Probability distribution (ψ^2) for an electron in a 1s orbital.

According to the Heisenberg uncertainty principle, we can't tell exactly where an electron is, but we can tell where it is most likely to be. The probability of finding an electron at a particular spot relative to an atom's nucleus is given by the square of the wave function (ψ^2) at that point. Figure 1.1 illustrates the probability of finding an electron at various points in the lowest energy (most stable) state of a hydrogen atom. The darker the color in a region, the higher the probability. The probability of finding an electron at a particular point is greatest near the nucleus, and decreases with increasing distance from the nucleus but never becomes zero. We commonly describe Figure 1.1 as an "electron cloud" to call attention to the spread-out nature of the electron probability. Be careful, though. The "electron cloud" of a hydrogen atom, although drawn as a collection of many dots, represents only one electron.

Wave functions are also called **orbitals.** For convenience, chemists use the term "orbital" in several different ways. A drawing such as Figure 1.1 is often said to represent an orbital. We will see other kinds of drawings in this chapter, and use the word "orbital" to describe them too.

Orbitals are described by specifying their size, shape, and directional properties. Spherically symmetrical ones such as shown in Figure 1.1 are called *s orbitals*. The letter *s* is preceded by the **principal quantum number** n ($n = 1, 2, 3,$ etc.) which specifies the **shell** and is related to the energy of the orbital. An electron in a 1s orbital is likely to be found closer to the nucleus, is lower in energy, and is more strongly held than an electron in a 2s orbital.

Instead of probability distributions, it is more common to represent orbitals by their **boundary surfaces,** as shown in Figure 1.2 for the 1s and 2s orbitals. The boundary surface encloses the region where the probability of finding an electron is high—on the order of 90–95%. Like the probability distribution plot from which it is derived, a picture of a boundary surface is usually described as a drawing of an orbital.

A hydrogen atom ($Z = 1$) has one electron; a helium atom ($Z = 2$) has two. The single electron of hydrogen occupies a 1s orbital, as do the two electrons of helium. We write their electron configurations as:

Hydrogen: $1s^1$ Helium: $1s^2$

In addition to being negatively charged, electrons possess the property of **spin.** The **spin quantum number** of an electron can have a value of either $+\frac{1}{2}$ or $-\frac{1}{2}$. According to the **Pauli exclusion principle,** two electrons may occupy the same orbital only when

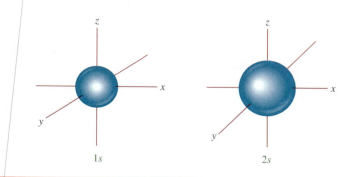

FIGURE 1.2 Boundary surfaces of a 1s orbital and a 2s orbital. The boundary surfaces enclose the volume where there is a 90–95% probability of finding an electron.

they have opposite, or "paired," spins. For this reason, no orbital can contain more than two electrons. Because two electrons fill the $1s$ orbital, the third electron in lithium ($Z = 3$) must occupy an orbital of higher energy. After $1s$, the next higher energy orbital is $2s$. The third electron in lithium therefore occupies the $2s$ orbital, and the electron configuration of lithium is

Lithium: $1s^2 2s^1$

The **period** (or **row**) of the periodic table in which an element appears corresponds to the principal quantum number of the highest numbered occupied orbital ($n = 1$ in the case of hydrogen and helium). Hydrogen and helium are first-row elements; lithium ($n = 2$) is a second-row element.

> A complete periodic table of the elements is presented on the inside back cover.

With beryllium ($Z = 4$), the $2s$ level becomes filled, and the next orbitals to be occupied in it and the remaining second-row elements are the $2p_x$, $2p_y$, and $2p_z$ orbitals. These three orbitals (Figure 1.3) are of equal energy and are characterized by boundary surfaces that are usually described as "dumbell-shaped." Each orbital consists of two "lobes," represented in Figure 1.3 by regions of different colors. Regions of a single orbital, in this case, each $2p$ orbital, may be separated by **nodal surfaces** where the wave function changes sign and the probability of finding an electron is zero.

> Other methods are also used to contrast the regions of an orbital where the signs of the wave function are different. Some mark one lobe of a p orbital $+$ and the other $-$. Others shade one lobe and leave the other blank. When this level of detail isn't necessary, no differentiation is made between the two lobes.

The electron configurations of the first 12 elements, hydrogen through magnesium, are given in Table 1.1. In filling the $2p$ orbitals, notice that each is singly occupied before any one is doubly occupied. This general principle for orbitals of equal energy is known as **Hund's rule.** *Of particular importance in Table 1.1 are hydrogen, carbon, nitrogen, and oxygen.* Countless organic compounds contain nitrogen, oxygen, or both in addition to carbon, the essential element of organic chemistry. Most of them also contain hydrogen.

It is often convenient to speak of the **valence electrons** of an atom. These are the outermost electrons, the ones most likely to be involved in chemical bonding and reactions. For second-row elements these are the $2s$ and $2p$ electrons. Because four orbitals ($2s$, $2p_x$, $2p_y$, $2p_z$) are involved, the maximum number of electrons in the **valence shell** of any second-row element is 8. Neon, with all its $2s$ and $2p$ orbitals doubly occupied, has eight valence electrons and completes the second row of the periodic table.

PROBLEM 1.1   How many valence electrons does carbon have?

> Answers to all problems that appear within the body of a chapter are found in Appendix 2. A brief discussion of the problem and advice on how to do problems of the same type are offered in the Solutions Manual.

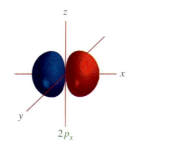
$2p_x$

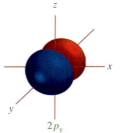
$2p_y$

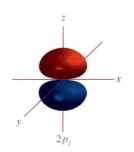
$2p_z$

FIGURE 1.3 Boundary surfaces of the $2p$ orbitals. The wave function changes sign at the nucleus. The two halves of each orbital are indicated by different colors. The *yz*-plane is a nodal surface for the $2p_x$ orbital. The probability of finding a $2p_x$ electron in the *yz*-plane is zero. Analogously, the *xz*-plane is a nodal surface for the $2p_y$ orbital, and the *xy*-plane is a nodal surface for the $2p_z$ orbital. You may examine different presentations of a $2p$ orbital on *Learning By Modeling*.

		Number of electrons in indicated orbital					
Element	Atomic number Z	$1s$	$2s$	$2p_x$	$2p_y$	$2p_z$	$3s$
Hydrogen	1	1					
Helium	2	2					
Lithium	3	2	1				
Beryllium	4	2	2				
Boron	5	2	2	1			
Carbon	6	2	2	1	1		
Nitrogen	7	2	2	1	1	1	
Oxygen	8	2	2	2	1	1	
Fluorine	9	2	2	2	2	1	
Neon	10	2	2	2	2	2	
Sodium	11	2	2	2	2	2	1
Magnesium	12	2	2	2	2	2	2

TABLE 1.1 Electron Configurations of the First Twelve Elements of the Periodic Table

Once the $2s$ and $2p$ orbitals are filled, the next level is the $3s$, followed by the $3p_x$, $3p_y$, and $3p_z$ orbitals. Electrons in these orbitals are farther from the nucleus than those in the $2s$ and $2p$ orbitals and are of higher energy.

In-chapter problems that contain multiple parts are accompanied by a sample solution to part (a). Answers to the other parts of the problem are found in Appendix 2, and detailed solutions are presented in the Solutions Manual.

PROBLEM 1.2 Referring to the periodic table as needed, write electron configurations for all the elements in the third period.

SAMPLE SOLUTION The third period begins with sodium and ends with argon. The atomic number Z of sodium is 11, and so a sodium atom has 11 electrons. The maximum number of electrons in the $1s$, $2s$, and $2p$ orbitals is ten, and so the eleventh electron of sodium occupies a $3s$ orbital. The electron configuration of sodium is $1s^2 2s^2 2p_x^2 2p_y^2 2p_z^2 3s^1$.

Neon, in the second period, and argon, in the third, possess eight electrons in their valence shell; they are said to have a complete **octet** of electrons. Helium, neon, and argon belong to the class of elements known as **noble gases** or **rare gases.** The noble gases are characterized by an extremely stable "closed-shell" electron configuration and are very unreactive.

Structure determines properties and the properties of atoms depend on atomic structure. All of an element's protons are in its nucleus, but the element's electrons are distributed among orbitals of varying energy and distance from the nucleus. More than anything else, we look at its electron configuration when we wish to understand how an element behaves. The next section illustrates this with a brief review of ionic bonding.

1.2   IONIC BONDS

Atoms combine with one another to give **compounds** having properties different from the atoms they contain. The attractive force between atoms in a compound is a **chemical bond.** One type of chemical bond, called an **ionic bond,** is the force of attraction between oppositely charged species **(ions)** (Figure 1.4). Ions that are positively charged are referred to as **cations;** those that are negatively charged are **anions.**

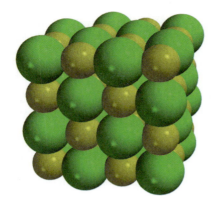

FIGURE 1.4 An ionic bond is the force of attraction between oppositely charged ions. Each Na^+ ion (yellow) in the crystal lattice of solid NaCl is involved in ionic bonding to each of six surrounding Cl^- ions (green) and vice versa.

Whether an element is the source of the cation or anion in an ionic bond depends on several factors, for which the periodic table can serve as a guide. In forming ionic compounds, elements at the left of the periodic table typically lose electrons, giving a cation that has the same electron configuration as the nearest noble gas. Loss of an electron from sodium, for example, yields Na^+, which has the same electron configuration as neon.

$$Na(g) \longrightarrow Na^+(g) + e^-$$

Sodium atom   Sodium ion   Electron
$1s^22s^22p^63s^1$ $1s^22s^22p^6$

[The (g) indicates that the species is present in the gas phase.]

A large amount of energy, called the **ionization energy,** must be added to any atom to dislodge one of its electrons. The ionization energy of sodium, for example, is 496 kJ/mol (119 kcal/mol). Processes that absorb energy are said to be **endothermic.** Compared with other elements, sodium and its relatives in group IA have relatively low ionization energies. In general, ionization energy increases across a row in the periodic table.

Elements at the right of the periodic table tend to gain electrons to reach the electron configuration of the next higher noble gas. Adding an electron to chlorine, for example, gives the anion Cl^-, which has the same closed-shell electron configuration as the noble gas argon.

$$Cl(g) + e^- \longrightarrow Cl^-(g)$$

Chlorine atom   Electron   Chloride ion
$1s^22s^22p^63s^23p^5$ $1s^22s^22p^63s^23p^6$

Energy is released when a chlorine atom captures an electron. Energy-releasing reactions are described as **exothermic,** and the energy change for an exothermic process has a negative sign. The energy change for addition of an electron to an atom is referred to as its **electron affinity** and is -349 kJ/mol (-83.4 kcal/mol) for chlorine.

The SI (*Système International d'Unites*) unit of energy is the *joule* (J). An older unit is the *calorie* (cal). Most organic chemists still express energy changes in units of kilocalories per mole (1 kcal/mol = 4.184 kJ/mol).

PROBLEM 1.3 Which of the following ions possess a noble gas electron configuration?

(a) K^+

(b) He^+

(c) H^-

(d) O^-

(e) F^-

(f) Ca^{2+}

Transfer of an electron from a sodium atom to a chlorine atom yields a sodium cation and a chloride anion, both of which have a noble gas electron configuration:

$$Na(g) \quad + \quad Cl(g) \quad \longrightarrow \quad Na^+Cl^-(g)$$

Sodium atom   Chlorine atom   Sodium chloride

Were we to simply add the ionization energy of sodium (496 kJ/mol) and the electron affinity of chlorine (-349 kJ/mol), we would conclude that the overall process is endothermic with $\Delta H° = +147$ kJ/mol. The energy liberated by adding an electron to chlorine is insufficient to override the energy required to remove an electron from sodium. This analysis, however, fails to consider the force of attraction between the oppositely charged ions Na^+ and Cl^-, which exceeds 500 kJ/mol and is more than sufficient to make the overall process exothermic. Attractive forces between oppositely charged particles are termed **electrostatic,** or **coulombic, attractions** and are what we mean by an **ionic bond** between two atoms.

PROBLEM 1.4 What is the electron configuration of C^+? Of C^-? Does either one of these ions have a noble gas (closed-shell) electron configuration?

Ionic bonds are very common in *inorganic* compounds, but rare in *organic* ones. The ionization energy of carbon is too large and the electron affinity too small for carbon to realistically form a C^{4+} or C^{4-} ion. What kinds of bonds, then, link carbon to other elements in millions of organic compounds? Instead of losing or gaining electrons, carbon *shares* electrons with other elements (including other carbon atoms) to give what are called covalent bonds.

1.3   COVALENT BONDS AND THE OCTET RULE

The **covalent, or shared electron pair,** model of chemical bonding was first suggested by G. N. Lewis of the University of California in 1916. Lewis proposed that a *sharing* of two electrons by two hydrogen atoms permits each one to have a stable closed-shell electron configuration analogous to helium.

H·   ·H   H:H

Two hydrogen atoms,   Hydrogen molecule:
each with a single   covalent bonding by way of
electron   a shared electron pair

Structural formulas of this type in which electrons are represented as dots are called **Lewis structures.**

The amount of energy required to dissociate a hydrogen molecule H_2 to two separate hydrogen atoms is called its **bond dissociation energy** (or **bond energy**). For H_2 it is quite large, amounting to 435 kJ/mol (104 kcal/mol). The main contributor to the strength of the covalent bond in H_2 is the increased binding force exerted on its two electrons. Each electron in H_2 "feels" the attractive force of two nuclei, rather than one as it would in an isolated hydrogen atom.

Ionic bonding was proposed by the German physicist Walther Kossel in 1916, in order to explain the ability of substances such as molten sodium chloride to conduct an electric current. He was the son of Albrecht Kossel, winner of the 1910 Nobel Prize in physiology or medicine for early studies in nucleic acids.

Gilbert Newton Lewis (born Weymouth, Massachusetts, 1875; died Berkeley, California, 1946) has been called the greatest American chemist. The January 1984 issue of the *Journal of Chemical Education* contains five articles describing Lewis's life and contributions to chemistry.

Covalent bonding in F_2 gives each fluorine eight electrons in its valence shell and a stable electron configuration equivalent to that of the noble gas neon:

:F·   ·F:| :F:F:

Two fluorine atoms, each
with seven electrons in
its valence shell

Fluorine molecule:
covalent bonding by way of
a shared electron pair

PROBLEM 1.5 Hydrogen is bonded to fluorine in hydrogen fluoride by a covalent bond. Write a Lewis formula for hydrogen fluoride.

The Lewis model limits second-row elements (Li, Be, B, C, N, O, F, Ne) to a total of eight electrons (shared plus unshared) in their valence shells. Hydrogen is limited to two. Most of the elements that we'll encounter in this text obey the **octet rule:** *in forming compounds they gain, lose, or share electrons to give a stable electron configuration characterized by eight valence electrons.* When the octet rule is satisfied for carbon, nitrogen, oxygen, and fluorine, they have an electron configuration analogous to the noble gas neon.

Now let's apply the Lewis model to the organic compounds methane and carbon tetrafluoride.

Combine ·C· and four H·   to write a Lewis structure for methane

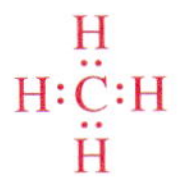

H:C:H with H above and below

Combine ·C· and four ·F: to write a Lewis structure for carbon tetrafluoride

:F:C:F: with :F: above and below

Carbon has eight electrons in its valence shell in both methane and carbon tetrafluoride. By forming covalent bonds to four other atoms, carbon achieves a stable electron configuration analogous to neon. Each covalent bond in methane and carbon tetrafluoride is quite strong—comparable to the bond between hydrogens in H_2 in bond dissociation energy.

PROBLEM 1.6 Given the information that it has a carbon–carbon bond, write a satisfactory Lewis structure for C_2H_6 (ethane).

Representing a two-electron covalent bond by a dash (—), the Lewis structures for hydrogen fluoride, fluorine, methane, and carbon tetrafluoride become:

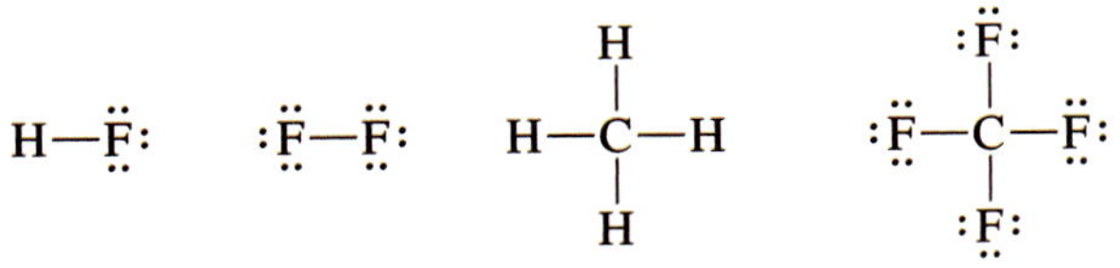

H—F: :F—F: H—C—H (with H above and below)   :F—C—F: (with :F: above and below)

Hydrogen   Fluorine   Methane   Carbon tetrafluoride
fluoride

1.4   DOUBLE BONDS AND TRIPLE BONDS

Lewis's concept of shared electron pair bonds allows for four-electron **double bonds** and six-electron **triple bonds.** Carbon dioxide (CO_2) has two carbon–oxygen double bonds, and the octet rule is satisfied for both carbon and oxygen. Similarly, the most stable Lewis structure for hydrogen cyanide (HCN) has a carbon–nitrogen triple bond.

Carbon dioxide: $:\ddot{O}::C::\ddot{O}:$ or $:\ddot{O}=C=\ddot{O}:$

Hydrogen cyanide: $H:C:::N:$ or $H-C\equiv N:$

Multiple bonds are very common in organic chemistry. Ethylene (C_2H_4) contains a carbon–carbon double bond in its most stable Lewis structure, and each carbon has a completed octet. The most stable Lewis structure for acetylene (C_2H_2) contains a carbon–carbon triple bond. Here again, the octet rule is satisfied.

Ethylene: or

Acetylene: $H:C:::C:H$ or $H-C\equiv C-H$

PROBLEM 1.7   Write the most stable Lewis structure for each of the following compounds:

(a) Formaldehyde, CH_2O. Both hydrogens are bonded to carbon. (A solution of formaldehyde in water was formerly used to preserve biological specimens.)

(b) Tetrafluoroethylene, C_2F_4. (The starting material for the preparation of Teflon.)

(c) Acrylonitrile, C_3H_3N. The atoms are connected in the order CCCN, and all hydrogens are bonded to carbon. (The starting material for the preparation of acrylic fibers such as Orlon and Acrilan.)

SAMPLE SOLUTION   (a) Each hydrogen contributes 1 valence electron, carbon contributes 4, and oxygen 6 for a total of 12 valence electrons. We are told that both hydrogens are bonded to carbon. Because carbon forms four bonds in its stable compounds, join carbon and oxygen by a double bond. The partial structure so generated accounts for 8 of the 12 electrons. Add the remaining four electrons to oxygen as unshared pairs to complete the structure of formaldehyde.

Partial structure showing
covalent bonds

Complete Lewis structure
of formaldehyde

1.5   POLAR COVALENT BONDS AND ELECTRONEGATIVITY

Electrons in covalent bonds are not necessarily shared equally by the two atoms that they connect. If one atom has a greater tendency to attract electrons toward itself than the other, we say the electron distribution is *polarized,* and the bond is referred to as a **polar covalent bond.** Hydrogen fluoride, for example, has a polar covalent bond. Because fluorine attracts electrons more strongly than hydrogen, the electrons in the H—F bond are pulled toward fluorine, giving it a partial negative charge, and away from hydrogen giving it a partial positive charge. This polarization of electron density is represented in various ways.

$$\overset{\delta+}{H}\!-\!\overset{\delta-}{F}$$

(The symbols $\delta+$ and $\delta-$
indicate partial positive
and partial negative
charge, respectively)

$$\overset{\longrightarrow}{H\!-\!F}$$

(The symbol $\longleftarrow\!\!\!\longrightarrow$ represents
the direction of polarization
of electrons in the H—F bond)

The tendency of an atom to draw the electrons in a covalent bond toward itself is referred to as its **electronegativity.** An **electronegative** element attracts electrons; an **electropositive** one donates them. Electronegativity increases across a row in the periodic table. The most electronegative of the second-row elements is fluorine; the most electropositive is lithium. Electronegativity decreases in going down a column. Fluorine is more electronegative than chlorine. The most commonly cited electronegativity scale was devised by Linus Pauling and is presented in Table 1.2.

PROBLEM 1.8   Examples of carbon-containing compounds include methane (CH_4), chloromethane (CH_3Cl), and methyllithium (CH_3Li). In which one does carbon bear the greatest partial positive charge? The greatest partial negative charge?

Another way to show the polarization of electron distribution in a molecule is with an electrostatic potential map, which uses the colors of the rainbow to show the charge distribution. Red through blue tracks regions of greater negative charge to greater positive charge, respectively. (For more details, see the boxed essay *Electrostatic Potential Maps* in this section.) The following electrostatic potential maps of hydrogen fluoride illustrate this polarization with blue being the dominant color in the region near the positively polarized hydrogen, and red dominating near the negatively polarized fluorine. The map on the left shows the electrostatic potential map as a solid surface; the one on the right is made transparent so that the atoms and bonds are also visible.

Linus Pauling (1901–1994) was born in Portland, Oregon, and was educated at Oregon State University and at the California Institute of Technology, where he earned a Ph.D. in chemistry in 1925. In addition to research in bonding theory, Pauling studied the structure of proteins and was awarded the Nobel Prize in chemistry for that work in 1954. Pauling won a second Nobel Prize (the Peace Prize) in 1962 for his efforts to limit the testing of nuclear weapons. He was one of only four scientists to have won two Nobel Prizes. The first double winner was a woman. Can you name her?

TABLE 1.2	Selected Values from the Pauling Electronegativity Scale

	Group number						
Period	**I**	**II**	**III**	**IV**	**V**	**VI**	**VII**
1	H 2.1						
2	Li 1.0	Be 1.5	B 2.0	C 2.5	N 3.0	O 3.5	F 4.0
3	Na 0.9	Mg 1.2	Al 1.5	Si 1.8	P 2.1	S 2.5	Cl 3.0
4	K 0.8	Ca 1.0					Br 2.8
5							I 2.5

ELECTROSTATIC POTENTIAL MAPS

All of the material in this text, and most of chemistry generally, can be understood on the basis of what physicists call the *electromagnetic force*. Its major principle is that opposite charges attract and like charges repel. As you learn organic chemistry, a good way to start to connect structure to properties such as chemical reactivity is to find the positive part of one molecule and the negative part of another. Most of the time, these will be the reactive sites.

Imagine that you bring a positive charge toward a molecule. The interaction between that positive charge and some point in the molecule will be attractive if the point is negatively charged, repulsive if it is positively charged, and the strength of the interaction will depend on the magnitude of the charge. Computational methods make it possible to calculate and map these interactions. It is convenient to display this map using the colors of the rainbow from red to blue. Red is the negative (electron-rich) end and blue is the positive (electron-poor) end.

The electrostatic potential map of hydrogen fluoride (HF) was shown in the preceding section and is repeated here. Compare it to the electrostatic potential map of lithium hydride (LiH).

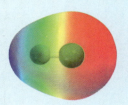 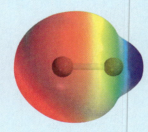

The H—F bond is polarized so that hydrogen is partially positive (blue) and fluorine partially negative

(red). Because hydrogen is more electronegative than lithium, the H—Li bond is polarized in the opposite sense, making hydrogen partially negative (red) and lithium partially positive (blue).

We will use electrostatic potential maps often to illustrate charge distribution in both organic and inorganic molecules. However, we need to offer one cautionary note. Electrostatic potential mapping within a single molecule is fine, but we need to be careful when comparing maps of different molecules. The reason for this is that the entire red-to-blue palette is used to map the electrostatic potential regardless of whether the charge difference is large or small. This is apparent in the H—F and H—Li electrostatic potential maps just shown. If, as shown in the following map, we use the same range for H—F that was used for H—Li we see that H is less blue than before and F is less red.

Thus, electrostatic potential maps can give an exaggerated picture of the charge distribution when the entire palette is used. In most cases, that won't matter to us inasmuch as we are mostly concerned with the distribution within a single molecule. In those few cases where we want to compare trends in a series of molecules, we'll use a common scale and will point that out.

Centers of positive and negative charge that are separated from each other constitute a **dipole**. The **dipole moment** μ of a molecule is equal to the charge e (either the positive or the negative charge, because they must be equal) multiplied by the distance between the centers of charge:

$$\mu = e \times d$$

Because the charge on an electron is 4.80×10^{-10} electrostatic units (esu) and the distances within a molecule typically fall in the 10^{-8} cm range, molecular dipole moments

TABLE 1.3	Selected Bond Dipole Moments		
Bond*	Dipole moment, D	Bond*	Dipole moment, D
H—F	1.7	C—F	1.4
H—Cl	1.1	C—O	0.7
H—Br	0.8	C—N	0.4
H—I	0.4	C=O	2.4
H—C	0.3	C=N	1.4
H—N	1.3	C≡N	3.6
H—O	1.5		

*The direction of the dipole moment is toward the more electronegative atom. In the listed examples hydrogen and carbon are the positive ends of the dipoles. Carbon is the negative end of the dipole associated with the C—H bond.

are on the order of 10^{-18} esu·cm. To simplify the reporting of dipole moments this value of 10^{-18} esu·cm is defined as a **debye, D.** Thus the experimentally determined dipole moment of hydrogen fluoride, 1.7×10^{-18} esu·cm is stated as 1.7 D.

Table 1.3 lists the dipole moments of various bond types. For H—F, H—Cl, H—Br, and H—I these "bond dipoles" are really molecular dipole moments. A **polar** molecule has a dipole moment, a **nonpolar** one does not. Thus, all of the hydrogen halides are polar molecules. To be polar, a molecule must have polar bonds, but can't have a shape that causes all the individual bond dipoles to cancel. We will have more to say about this in Section 1.11 after we have developed a feeling for the three-dimensional shapes of molecules.

The bond dipoles in Table 1.3 depend on the difference in electronegativity of the bonded atoms and on the bond distance. The polarity of a C—H bond is relatively low; substantially less than a C—O bond, for example. Don't lose sight of an even more important difference between a C—H bond and a C—O bond, and that is the *direction of the dipole moment*. In a C—H bond the electrons are drawn away from H, toward C. In a C—O bond, electrons are drawn from C toward O. As we'll see in later chapters, the kinds of reactions that a substance undergoes can often be related to the size and direction of key bond dipoles.

The debye unit is named in honor of Peter Debye, a Dutch scientist who did important work in many areas of chemistry and physics and was awarded the Nobel Prize in chemistry in 1936.

1.6   FORMAL CHARGE

Lewis structures frequently contain atoms that bear a positive or negative charge. If the molecule as a whole is neutral, the sum of its positive charges must equal the sum of its negative charges. An example is nitric acid, HNO_3:

$$H-\overset{\cdot\cdot}{\underset{\cdot\cdot}{O}}-\overset{+}{N}\overset{\displaystyle\overset{\cdot\cdot}{O}:}{\underset{\displaystyle :\overset{\cdot\cdot}{\underset{\cdot\cdot}{O}}:^{-}}{\Big\langle}}$$

As written, the structural formula for nitric acid depicts different bonding patterns for its three oxygens. One oxygen is doubly bonded to nitrogen, another is singly bonded to both nitrogen and hydrogen, and the third has a single bond to nitrogen and a negative charge. Nitrogen is positively charged. The positive and negative charges are called

formal charges, and the Lewis structure of nitric acid would be incomplete were they to be omitted.

We calculate formal charges by counting the number of electrons "owned" by each atom in a Lewis structure and comparing this **electron count** with that of the neutral atom. Figure 1.5 illustrates how electrons are counted for each atom in nitric acid. Counting electrons for the purpose of computing the formal charge differs from counting electrons to see if the octet rule is satisfied. A second-row element has a filled valence shell if the sum of all the electrons, shared and unshared, is eight. Electrons that connect two atoms by a covalent bond count toward filling the valence shell of both atoms. When calculating the formal charge, however, only half the number of electrons in covalent bonds can be considered to be "owned" by an atom.

To illustrate, let's start with the hydrogen of nitric acid. As shown in Figure 1.5, hydrogen is associated with only two electrons: those in its covalent bond to oxygen. It shares those two electrons with oxygen, and so we say that the electron count of each hydrogen is $\frac{1}{2}(2) = 1$. Because this is the same as the number of electrons in a neutral hydrogen atom, the hydrogen in nitric acid has no formal charge.

Moving now to nitrogen, we see that it has four covalent bonds (two single bonds + one double bond), and so its electron count is $\frac{1}{2}(8) = 4$. A neutral nitrogen has five electrons in its valence shell. The electron count for nitrogen in nitric acid is one less than that of a neutral nitrogen atom, so its formal charge is $+1$.

Electrons in covalent bonds are counted as if they are shared equally by the atoms they connect, but unshared electrons belong to a single atom. Thus, the oxygen that is doubly bonded to nitrogen has an electron count of six (four electrons as two unshared pairs + two electrons from the double bond). Because this is the same as a neutral oxygen atom, its formal charge is 0. Similarly, the OH oxygen has two bonds plus two unshared electron pairs, giving it an electron count of six and no formal charge.

The green oxygen in Figure 1.5 owns three unshared pairs (six electrons) and shares two electrons with nitrogen to give it an electron count of seven. This is one more than the number of electrons in the valence shell of an oxygen atom, and so its formal charge is -1.

The method described for calculating formal charge has been one of reasoning through a series of logical steps. It can be reduced to the following equation:

$$\text{Formal charge} = \frac{\text{group number in}}{\text{periodic table}} - \text{number of bonds} - \frac{\text{number of}}{\text{unshared electrons}}$$

The number of valence electrons in an atom of a main-group element such as nitrogen is equal to its group number. In the case of nitrogen this is five.

It will always be true that a covalently bonded hydrogen has no formal charge (formal charge = 0).

It will always be true that a nitrogen with four covalent bonds has a formal charge of +1. (A nitrogen with four covalent bonds cannot have unshared pairs, because of the octet rule.)

It will always be true that an oxygen with two covalent bonds and two unshared pairs has no formal charge.

It will always be true that an oxygen with one covalent bond and three unshared pairs has a formal charge of −1.

Electron count (O) $= \frac{1}{2}(4) + 4 = 6$

Electron count (H) $= \frac{1}{2}(2) = 1$

Electron count (N) $= \frac{1}{2}(8) = 4$

Electron count (O) $= \frac{1}{2}(4) + 4 = 6$

Electron count (O) $= \frac{1}{2}(2) + 6 = 7$

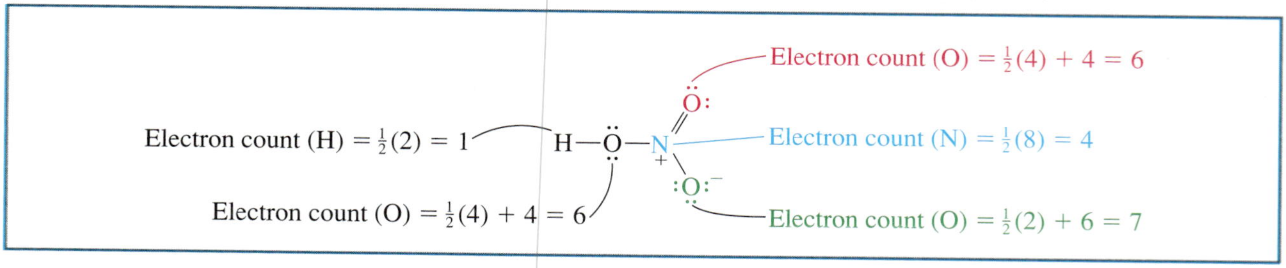

FIGURE 1.5 Counting electrons in nitric acid. The electron count of each atom is equal to half the number of electrons it shares in covalent bonds plus the number of electrons in its own unshared pairs.

PROBLEM 1.9 Like nitric acid, each of the following inorganic compounds will be frequently encountered in this text. Calculate the formal charge on each of the atoms in the Lewis structures given.

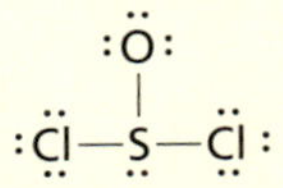

$$:\ddot{O}:$$
$$|$$
$$:\ddot{C}l-S-\ddot{C}l:\qquad H-\ddot{O}-S-\ddot{O}-H\qquad H-\ddot{O}-\ddot{N}=\ddot{O}:$$
$$|$$
$$:\ddot{O}:$$

(a) Thionyl chloride 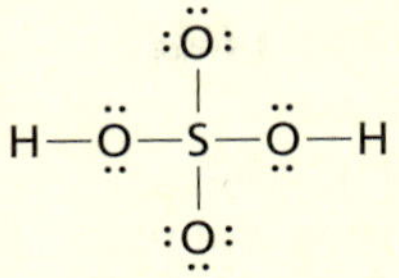   (b) Sulfuric acid 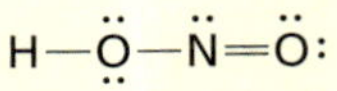   (c) Nitrous acid

SAMPLE SOLUTION (a) The formal charge is the difference between the number of valence electrons in the neutral atom and the electron count in the Lewis structure. (The number of valence electrons is the same as the group number in the periodic table for the main-group elements.)

	Valence electrons of neutral atom	Electron count	Formal charge
Sulfur:	6	$\frac{1}{2}(6) + 2 = 5$	+1
Oxygen:	6	$\frac{1}{2}(2) + 6 = 7$	−1
Chlorine:	7	$\frac{1}{2}(2) + 6 = 7$	0

The formal charges are shown in the Lewis structure of thionyl chloride

$$:\ddot{O}:^{-}$$
$$|$$
as $:\ddot{C}l-\overset{+}{S}-\ddot{C}l:$

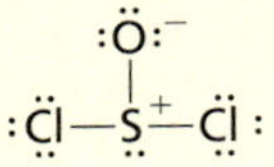

So far we've only considered neutral molecules—those in which the sums of the positive and negative formal charges were equal. With ions, of course, these sums will not be equal. Ammonium cation and borohydride anion, for example, are ions with net charges of +1 and −1, respectively. Nitrogen has a formal charge of +1 in ammonium ion, and boron has a formal charge of −1 in borohydride. None of the hydrogens in the Lewis structures shown for these ions bears a formal charge.

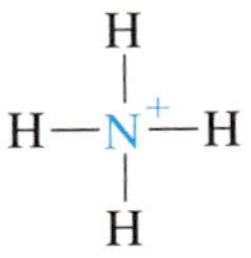

$$\begin{array}{cc} H & H \\ | & | \\ H-\overset{+}{N}-H & H-\overset{-}{B}-H \\ | & | \\ H & H \end{array}$$

Ammonium ion 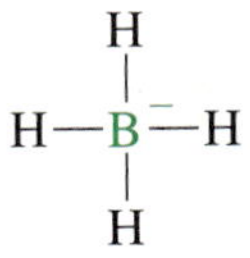   Borohydride ion

PROBLEM 1.10 Verify that the formal charges on nitrogen in ammonium ion and boron in borohydride ion are as shown.

Formal charges are based on Lewis structures in which electrons are considered to be shared equally between covalently bonded atoms. Actually, polarization of N—H bonds in ammonium ion and of B—H bonds in borohydride leads to some transfer of positive and negative charge, respectively, to the hydrogens.

PROBLEM 1.11 Use δ+ and δ− notation to show the dispersal of charge to the hydrogens in NH_4^+ and BH_4^-.

TABLE 1.4   How to Write Lewis Structures

Step	Illustration
1. The molecular formula and the connectivity are determined experimentally and are included among the information given in the statement of the problem.	Methyl nitrite has the molecular formula CH_3NO_2. All hydrogens are bonded to carbon, and the order of atomic connections is CONO.
2. Count the number of valence electrons available. For a neutral molecule this is equal to the sum of the valence electrons of the constituent atoms.	Each hydrogen contributes 1 valence electron, carbon contributes 4, nitrogen contributes 5, and each oxygen contributes 6 for a total of 24 in CH_3NO_2.
3. Connect bonded atoms by a shared electron pair bond (:) represented by a dash (—).	For methyl nitrite we write the partial structure $$\begin{array}{c} H \\ | \\ H-C-O-N-O \\ | \\ H \end{array}$$
4. Count the number of electrons in shared electron pair bonds (twice the number of bonds), and subtract this from the total number of electrons to give the number of electrons to be added to complete the structure.	The partial structure in step 3 contains 6 bonds equivalent to 12 electrons. Because CH_3NO_2 contains 24 electrons, 12 more electrons need to be added.
5. Add electrons in pairs so that as many atoms as possible have eight electrons. (Hydrogen is limited to two electrons.) When the number of electrons is insufficient to provide an octet for all atoms, assign electrons to atoms in order of decreasing electronegativity.	With four bonds, carbon already has eight electrons. The remaining 12 electrons are added as indicated. Both oxygens have eight electrons, but nitrogen (less electronegative than oxygen) has only six. $$\begin{array}{c} H \\ | \\ H-C-\ddot{O}-\ddot{N}-\ddot{O}: \\ | \\ H \end{array}$$
6. If one or more atoms have fewer than eight electrons, use unshared pairs on an adjacent atom to form a double (or triple) bond to complete the octet.	An electron pair on the terminal oxygen is shared with nitrogen to give a double bond. $$\begin{array}{c} H \\ | \\ H-C-\ddot{O}-\ddot{N}=\ddot{O}: \\ | \\ H \end{array}$$ The structure shown is the best (most stable) Lewis structure for methyl nitrite. All atoms except hydrogen have eight electrons (shared + unshared) in their valence shell.
7. Calculate formal charges.	None of the atoms in the Lewis structure shown in step 6 possesses a formal charge. An alternative Lewis structure for methyl nitrite, $$\begin{array}{c} H \\ | \\ H-C-\overset{+}{\ddot{O}}=\ddot{N}-\ddot{O}:^{-} \\ | \\ H \end{array}$$ although it satisfies the octet rule, is less stable than the one shown in step 6 because it has a separation of positive charge from negative charge.

Determining formal charges on individual atoms of Lewis structures is an important element in good "electron bookkeeping." So much of organic chemistry can be made more understandable by keeping track of electrons that it is worth taking some time at the beginning to become proficient at the seemingly simple task of counting them.

Table 1.4 summarizes the procedure we have developed for writing Lewis structures. Notice that the process depends on knowing not only the molecular formula, but also the order in which the atoms are attached to one another. This order of attachment is called the **constitution,** or **connectivity,** of the molecule and is determined by experiment. Only rarely is it possible to deduce the constitution of a molecule from its molecular formula.

1.7   STRUCTURAL FORMULAS OF ORGANIC MOLECULES

Organic chemists have devised a number of shortcuts to speed the writing of structural formulas. Sometimes we leave out unshared electron pairs, but only when we are sure enough in our ability to count electrons to know when they are present and when they're not. We've already mentioned representing covalent bonds by dashes. In **condensed structural formulas** we leave out some, many, or all of the covalent bonds and use subscripts to indicate the number of identical groups attached to a particular atom. These successive levels of simplification are illustrated as shown for isopropyl alcohol ("rubbing alcohol").

$$
\begin{array}{c}
\overset{\displaystyle H}{\underset{\displaystyle H}{|}}\;\; \overset{\displaystyle H}{\underset{\displaystyle \ddot{O}:}{|}}\;\; \overset{\displaystyle H}{|}
\end{array}
$$

H—C—C—C—H   written as CH_3CHCH_3 or condensed even further to
$\;\;\;\;\;\;\;\;\;|$ $\;$OH $(CH_3)_2CHOH$
H :O: H
$\;\;\;\;\;\;|$
$\;\;\;\;\;\;$H

PROBLEM 1.12 Expand the following condensed formulas so as to show all the bonds and unshared electron pairs.

(a) $HOCH_2CH_2NH_2$

(b) $(CH_3)_3CH$

(c) $ClCH_2CH_2Cl$

(d) CH_3CHCl_2

(e) $CH_3NHCH_2CH_3$

(f) $(CH_3)_2CHCH{=}O$

SAMPLE SOLUTION   (a) The molecule contains two carbon atoms, which are bonded to each other. Both carbons bear two hydrogens. One carbon bears the group HO—; the other is attached to —NH_2.

$$
\text{H—O—}\overset{\displaystyle H}{\underset{\displaystyle H}{\overset{|}{\underset{|}{C}}}}\text{—}\overset{\displaystyle H}{\underset{\displaystyle H}{\overset{|}{\underset{|}{C}}}}\text{—}\overset{\displaystyle H}{\overset{|}{N}}\text{—H}
$$

When writing the constitution of a molecule, it is not necessary to concern yourself with the spatial orientation of the atoms. There are many other correct ways to represent the constitution shown. What is important is to show the connectivity OCCN (or its equivalent NCCO) and to have the correct number of hydrogens on each atom.

To locate unshared electron pairs, first count the total number of valence electrons brought to the molecule by its component atoms. Each hydrogen contributes 1, each carbon 4, nitrogen 5, and oxygen 6, for a total of 26. Ten bonds are shown, accounting for 20 electrons; therefore 6 electrons must be contained in unshared pairs. Add pairs of electrons to oxygen and nitrogen so that their octets are complete, two unshared pairs to oxygen and one to nitrogen.

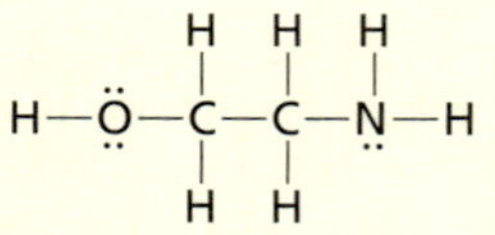

$$H\!-\!\overset{..}{\underset{..}{O}}\!-\!\overset{\overset{\displaystyle H}{|}}{\underset{\underset{\displaystyle H}{|}}{C}}\!-\!\overset{\overset{\displaystyle H}{|}}{\underset{\underset{\displaystyle H}{|}}{C}}\!-\!\overset{\overset{\displaystyle H}{|}}{\underset{..}{N}}\!-\!H$$

As you practice, you will begin to remember patterns of electron distribution. A neutral oxygen with two bonds has two unshared electron pairs. A neutral nitrogen with three bonds has one unshared pair.

With practice, writing structural formulas for organic molecules soon becomes routine and can be simplified even more. For example, a chain of carbon atoms can be represented by drawing all of the C—C bonds while omitting individual carbons. The resulting structural drawings can be simplified still more by stripping away the hydrogens.

$CH_3CH_2CH_2CH_3$ becomes $\underset{\underset{\text{H H}}{}}{\overset{\overset{\text{H H}}{}}{}}$ H ... H   simplified to   (bond-line formula)

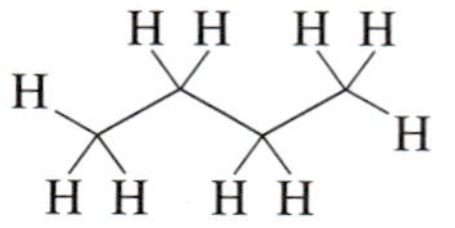

In these simplified representations, called **bond-line formulas** or **carbon skeleton diagrams,** the only atoms specifically written in are those that are neither carbon nor hydrogen bound to carbon. Hydrogens bound to these *heteroatoms* are shown, however.

$CH_3CH_2CH_2CH_2OH$ becomes   (bond-line formula) OH

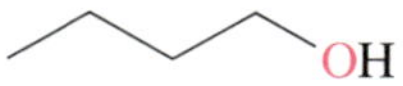

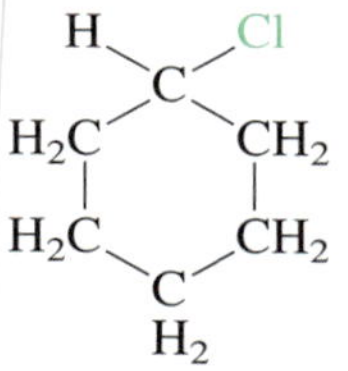

(structure)   becomes   (structure) Cl

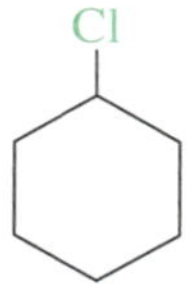

PROBLEM 1.13 Expand the following bond-line representations to show all the atoms including carbon and hydrogen.

(a)  (bond-line formula)

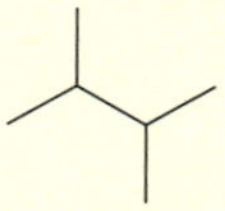

(c) 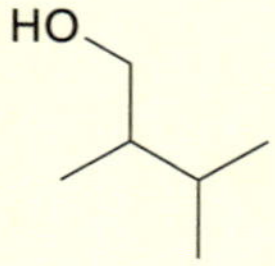 HO (bond-line formula)

(b) (bond-line formula)

(d) 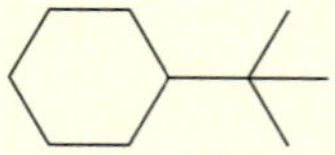 (bond-line formula)

SAMPLE SOLUTION (a) There is a carbon at each bend in the chain and at the ends of the chain. Each of the ten carbon atoms bears the appropriate number of hydrogens to give it four bonds.

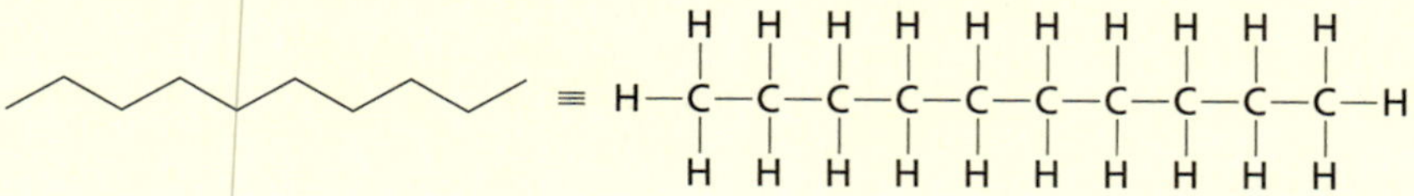

(bond-line formula) $\equiv$ $H\!-\!C\!-\!C\!-\!C\!-\!C\!-\!C\!-\!C\!-\!C\!-\!C\!-\!C\!-\!C\!-\!H$ (with H's above and below each carbon)

Alternatively, the structure could be written as
$CH_3CH_2CH_2CH_2CH_2CH_2CH_2CH_2CH_2CH_3$ or in condensed form as $CH_3(CH_2)_8CH_3$.

1.8   CONSTITUTIONAL ISOMERS

In the introduction we noted that both Berzelius and Wöhler were fascinated by the fact that two different compounds with different properties, ammonium cyanate and urea, possessed exactly the same molecular formula, CH_4N_2O. Berzelius had studied examples of similar phenomena earlier and invented the word **isomer** to describe *different compounds that have the same molecular formula.*

We can illustrate isomerism by referring to two different compounds, *nitromethane* and *methyl nitrite*, both of which have the molecular formula CH_3NO_2. Nitromethane,

Nitromethane   Methyl nitrite

> The suffix *-mer* in the word "isomer" is derived from the Greek word *meros*, meaning "part," "share," or "portion." The prefix *iso-* is also from Greek (*isos*, meaning "the same"). Thus isomers are different molecules that have the same parts (elemental composition).

used to power race cars, is a liquid with a boiling point of 101°C. Methyl nitrite is a gas boiling at −12°C, which when inhaled causes dilation of blood vessels. Isomers that differ in the order in which their atoms are bonded are often referred to as **structural isomers.** A more modern term is **constitutional isomer.** As noted in the previous section, the order of atomic connections that defines a molecule is termed its *constitution*, and we say that two compounds are *constitutional isomers* if they have the same molecular formula but differ in the order in which their atoms are connected.

PROBLEM 1.14 The formula CH_3NO_2 fits more isomers than just nitromethane and methyl nitrite. Some, such as *carbamic acid,* an intermediate in the commercial preparation of urea for use as a fertilizer, are too unstable to isolate. Given the information that the nitrogen and both oxygens of carbamic acid are bonded to carbon and that one of the carbon–oxygen bonds is a double bond, write a Lewis structure for carbamic acid.

PROBLEM 1.15 Write structural formulas for all the constitutionally isomeric compounds having the given molecular formula.
(a) C_2H_6O (b) C_3H_8O (c) $C_4H_{10}O$

SAMPLE SOLUTION (a) Begin by considering the ways in which two carbons and one oxygen may be bonded. There are two possibilities: C—C—O and C—O—C. Add the six hydrogens so that each carbon has four bonds and each oxygen two. There are two constitutional isomers: ethyl alcohol and dimethyl ether.

Ethyl alcohol   Dimethyl ether

In Chapter 3 another type of isomerism, called **stereoisomerism,** will be introduced. Stereoisomers have the same constitution but differ in the arrangement of atoms in space.

1.9 RESONANCE

When writing a Lewis structure, we restrict a molecule's electrons to certain well-defined locations, either linking two atoms by a covalent bond or as unshared electrons on a single atom. Sometimes more than one Lewis structure can be written for a molecule, especially those that contain multiple bonds. An example often cited in introductory chemistry courses is ozone (O_3). Ozone occurs naturally in large quantities in the upper atmosphere, where it screens the surface of the earth from much of the sun's ultraviolet rays. Were it not for this ozone layer, most forms of surface life on earth would be damaged or even destroyed by the rays of the sun. The following Lewis structure for ozone satisfies the octet rule; all three oxygens have eight electrons in their valence shell.

$$:\overset{..}{O}\overset{\overset{+}{\overset{..}{O}}}{\diagup\diagdown}\overset{..}{O}:_{-}$$

This Lewis structure, however, doesn't accurately portray the bonding in ozone, because the two terminal oxygens are bonded differently to the central oxygen. The central oxygen is depicted as doubly bonded to one and singly bonded to the other. Because it is generally true that double bonds are shorter than single bonds, we would expect ozone to exhibit two different O—O bond lengths, one of them characteristic of the O—O single bond distance (147 pm in hydrogen peroxide, H—O—O—H) and the other one characteristic of the O=O double bond distance (121 pm in O_2). Such is not the case. Both bond distances in ozone are exactly the same (128 pm)—somewhat shorter than the single bond distance and somewhat longer than the double bond distance. The structure of ozone requires that the *central oxygen must be identically bonded to both terminal oxygens.*

An electrostatic potential map shows the equivalence of the two terminal oxygens. Notice, too, that the central oxygen is blue (positively charged) and both terminal oxygens are red (negatively charged).

Bond distances in organic compounds are usually 1–2 Å ($1Å = 10^{-10}$m). Because the angstrom (Å) is not an SI unit, we will express bond distances in picometers (1 pm $= 10^{-12}$m). Thus, 128 pm $= 1.28$ Å.

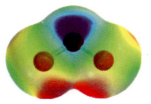

To deal with circumstances such as the bonding in ozone, the notion of **resonance** between Lewis structures was developed. According to the resonance concept, when more than one Lewis structure may be written for a molecule, a single structure is insufficient to describe it. Rather, the true structure has an electron distribution that is a "hybrid" of all the possible Lewis structures that can be written for the molecule. In the case of ozone, two equivalent Lewis structures may be written. We use a double-headed arrow to represent resonance between these two Lewis structures.

$$:\overset{..}{O}\overset{\overset{+}{\overset{..}{O}}}{\diagup\diagdown}\overset{..}{O}:_{-} \longleftrightarrow _{-}:\overset{..}{O}\overset{\overset{+}{\overset{..}{O}}}{\diagup\diagdown}\overset{..}{O}:$$

It is important to remember that the double-headed resonance arrow does *not* indicate a *process* in which the two Lewis structures interconvert. Ozone, for example, has a *single* structure; it does not oscillate back and forth between two Lewis structures, rather its true structure is not adequately represented by any single Lewis structure.

Resonance attempts to correct a fundamental defect in Lewis formulas. Lewis formulas show electrons as being **localized;** they either are shared between two atoms in a covalent bond or are unshared electrons belonging to a single atom. In reality, electrons distribute themselves in the way that leads to their most stable arrangement. This sometimes means that a pair of electrons is **delocalized,** or shared by several nuclei. What we try to show by the resonance description of ozone is the delocalization of an unshared pair of electrons of one oxygen and the electrons in the double bond over the three atoms of the molecule. Organic chemists often use curved arrows to show this electron delocalization. Alternatively, an average of two Lewis structures is sometimes drawn using a dashed line to represent a "partial" bond. In the dashed-line notation the central oxygen is linked to the other two by bonds that are halfway between a single bond and a double bond, and the terminal oxygens each bear one half of a unit negative charge.

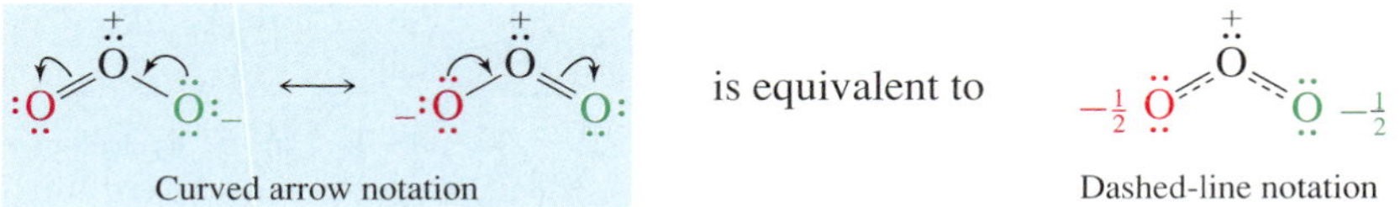

Curved arrow notation   is equivalent to   Dashed-line notation

Electron delocalization in ozone

The rules to be followed when writing resonance structures are summarized in Table 1.5.

PROBLEM 1.16 Electron delocalization can be important in ions as well as in neutral molecules. Using curved arrows, show how an equally stable resonance structure can be generated for each of the following anions:

(a)

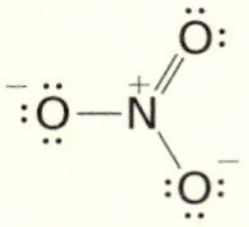

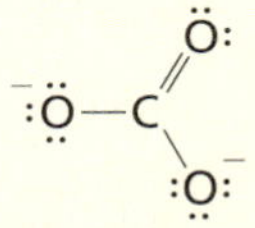

(b)

(c)

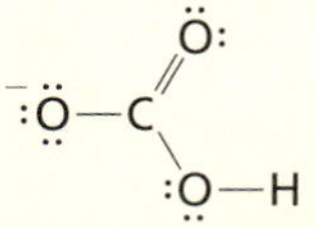

(d)

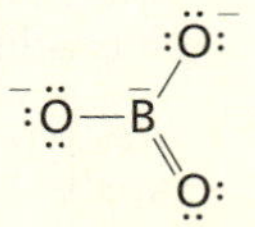

SAMPLE SOLUTION (a) When using curved arrows to represent the delocalization of electrons, begin at a site of high electron density, preferably an atom that is negatively charged. Move electron pairs until a proper Lewis structure results. For nitrate ion, this can be accomplished in two ways:

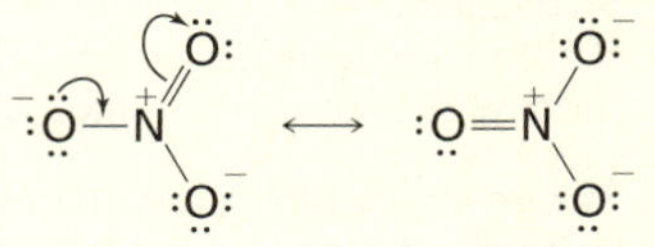

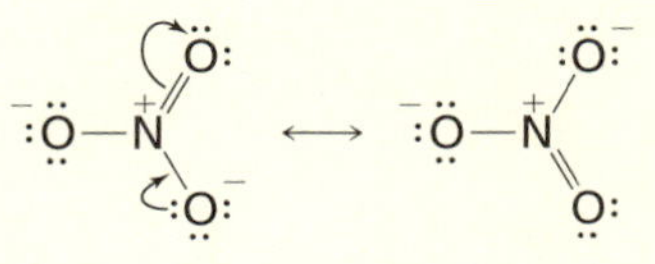

Three equally stable Lewis structures are possible for nitrate ion. The negative charge in nitrate is shared equally by all three oxygens.

It is good chemical practice to represent molecules by their most stable Lewis structure. The ability to write alternative resonance forms and to compare their relative stabilities, however, can provide insight into both molecular structure and chemical behavior. This will become particularly apparent in the last two thirds of this text, where the resonance concept will be used regularly.

TABLE 1.5   Introduction to the Rules of Resonance*

Rule	Illustration
1. Atomic positions (connectivity) must be the same in all resonance structures; only the electron positions may vary among the various contributing structures.	The structural formulas $CH_3-\overset{+}{N}\overset{\displaystyle \ddot{O}:}{\underset{\displaystyle \ddot{O}:^-}{}}$ and $CH_3-\ddot{O}-\ddot{N}=\ddot{O}:$ A B represent different compounds, not different resonance forms of the same compound. A is a Lewis structure for *nitromethane*; B is *methyl nitrite*.
2. Lewis structures in which second-row elements own or share more than eight valence electrons are especially unstable and make no contribution to the true structure. (The octet rule may be exceeded for elements beyond the second row.)	Structural formula C, $CH_3-N\overset{\displaystyle \ddot{O}:}{\underset{\displaystyle \ddot{O}:}{}}$ C has ten electrons around nitrogen. It is not a permissible Lewis structure for nitromethane and so cannot be a valid resonance form.
3. When two or more structures satisfy the octet rule, the most stable one is the one with the smallest separation of oppositely charged atoms.	The two Lewis structures D and E of methyl nitrite satisfy the octet rule: $CH_3-\ddot{O}-\ddot{N}=\ddot{O}: \longleftrightarrow CH_3-\overset{+}{\ddot{O}}=\ddot{N}-\ddot{O}:^-$ D E Structure D has no separation of charge and is more stable than E, which does. The true structure of methyl nitrite is more like D than E.
4. Among structural formulas in which the octet rule is satisfied for all atoms and one or more of these atoms bears a formal charge, the most stable resonance form is the one in which negative charge resides on the most electronegative atom.	The most stable Lewis structure for cyanate ion is F because the negative charge is on its oxygen. $:N{\equiv}C-\ddot{O}:^- \longleftrightarrow :\ddot{N}=C=\ddot{O}:$ F G In G the negative charge is on nitrogen. Oxygen is more electronegative than nitrogen and can better support a negative charge.

(Continued)

TABLE 1.5   Introduction to the Rules of Resonance* *(Continued)*

Rule	Illustration
5. Each contributing Lewis structure must have the same number of electrons and the same *net* charge although the formal charges of individual atoms may vary among the various Lewis structures.	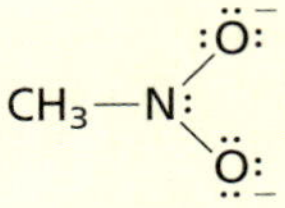 The Lewis structures 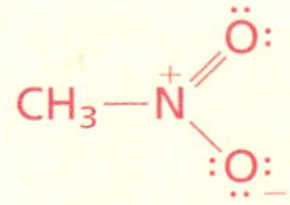 $CH_3-N \begin{smallmatrix} O \\ O \end{smallmatrix}$ (H) and $CH_3-N \begin{smallmatrix} O \\ O \end{smallmatrix}$ (I) are *not* resonance forms of one another. Structure H has 24 valence electrons and a net charge of 0; I has 26 valence electrons and a net charge of -2.
6. Each contributing Lewis structure must have the same number of *unpaired* electrons.	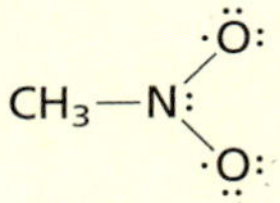 Structural formula J is a Lewis structure of nitromethane; K is not, even though it has the same atomic positions and the same number of electrons. 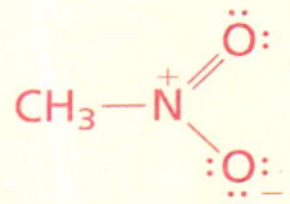 (J) and (K) Structure K has two unpaired electrons. Structure J has all its electrons paired and is a more stable structure.
7. Electron delocalization stabilizes a molecule. A molecule in which electrons are delocalized is more stable than implied by any of the individual Lewis structures that may be written for it. The degree of stabilization is greatest when the contributing Lewis structures are of equal stability.	Nitromethane is stabilized by electron delocalization more than methyl nitrite is. *The two most stable resonance forms of nitromethane are equivalent to each other.* 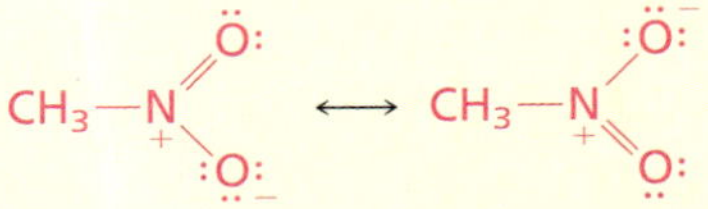 $CH_3-N \begin{smallmatrix} O \\ O \end{smallmatrix} \longleftrightarrow CH_3-N \begin{smallmatrix} O \\ O \end{smallmatrix}$ *The two most stable resonance forms of methyl nitrite are not equivalent.*  $CH_3-\ddot{O}-\ddot{N}=\ddot{O}: \longleftrightarrow CH_3-\overset{+}{\ddot{O}}=\ddot{N}-\ddot{O}:^-$

*These are the most important rules to be concerned with at present. Additional aspects of electron delocalization, as well as additional rules for its depiction by way of resonance structures, will be developed as needed in subsequent chapters.

LEARNING BY MODELING

As early as the nineteenth century many chemists built scale models to better understand molecular structure. We can gain a clearer idea about the features that affect structure and reactivity when we examine the three-dimensional shape of a molecule. Several types of molecular models are shown for methane in Figure 1.6. Probably the most familiar are ball-and-stick models (Figure 1.6*b*), which direct approximately equal attention to the atoms and the bonds that connect them. Framework models (Figure 1.6*a*) and space-filling models (Figure 1.6*c*) represent opposite extremes. Framework models emphasize the pattern of bonds of a molecule while ignoring the sizes of the atoms. Space-filling models emphasize the volume occupied by individual atoms at the cost of a clear depiction of the bonds; they are most useful in cases in which one wishes to examine the overall molecular shape and to assess how closely two non-bonded atoms approach each other.

The earliest ball-and-stick models were exactly that: wooden balls in which holes were drilled to accommodate dowels that connected the atoms. Plastic versions, including relatively inexpensive student sets, became available in the 1960s and proved to be a valuable learning aid. Precisely scaled stainless steel framework and plastic space-filling models, although relatively expensive, were standard equipment in most research laboratories.

Computer graphics-based representations are rapidly replacing classical molecular models. Indeed, the term "molecular modeling" as now used in organic chemistry implies computer generation of models. The methane models shown in Figure 1.6 were all drawn on a personal computer using software that possesses the feature of displaying and printing the same molecule in framework, ball-and-stick, and space-filling formats. In addition to permitting models to be constructed rapidly, even the simplest software allows the model to be turned and viewed from a variety of perspectives.

As useful as molecular models are, they are limited in that they only show the location of the atoms and the space they occupy. Another important dimension to molecular structure is its electron distribution. We introduced electrostatic potential maps in Section 1.5 as a way of illustrating charge distribution and will continue to use them throughout the text. Figure 1.6(*d*) shows the electrostatic potential map of methane. Its overall shape is similar to the volume occupied by the space-filling model. The most electron-rich regions are closer to carbon and the most electron-poor ones are closer to the hydrogens.

Organic chemistry is a very visual science and computer modeling is making it even more so.

—*Cont.*

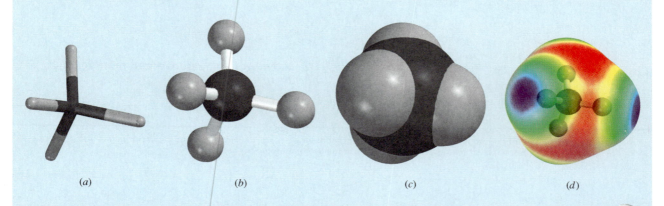

(*a*)   (*b*)   (*c*)   (*d*)

FIGURE 1.6 Molecular models of methane (CH_4). (*a*) Framework (tube) models show the bonds connecting the atoms, but not the atoms themselves. (*b*) Ball-and-stick (ball-and-spoke) models show the atoms as balls and the bonds as rods. (*c*) Space-filling models portray overall molecular size; the radius of each sphere approximates the van der Waals radius of the atom. (*d*) An electrostatic potential map of methane.

Accompanying this text is a CD entitled *Learning By Modeling*. As its name implies, it is a learning tool, designed to help you better understand molecular structure and properties, and contains two major components:

- *SpartanBuild* software that you can use to build molecular models of various types include tube, ball-and-spoke, and space-filling. This text includes a number of modeling exercises for you to do, but don't limit yourself to them. You can learn a lot by simply experimenting with *SpartanBuild* to see what you can make.

- *SpartanView* software with which you can browse through an archive of already-prepared models on the *Learning By Modeling* CD. These models include many of the same substances that appear in this text. *SpartanView* is the tool you will use to view

electrostatic potential maps and a collection of already prepared models.

All of the models, those you make yourself and those already provided on *Learning By Modeling,* can be viewed in different formats and rotated in three dimensions.

Immediately preceding the Glossary at the back of this text is a tutorial showing you how to use *SpartanBuild* and *SpartanView* and describing some additional features.

As you go through this text, you will see two different modeling icons. The *SpartanBuild* icon alerts you to a model-building opportunity, the *SpartanView* icon indicates that the *Learning By Modeling* CD includes a related model or animation.

SpartanBuild icon   *SpartanView* icon

1.10   THE SHAPES OF SOME SIMPLE MOLECULES

So far we have emphasized structure in terms of "electron bookkeeping." We now turn our attention to molecular geometry and will see how we can begin to connect the three-dimensional shape of a molecule to its Lewis formula. Table 1.6 lists some simple compounds illustrating the geometries that will be seen most often in our study of organic chemistry.

Methane is a tetrahedral molecule; its four hydrogens occupy the corners of a tetrahedron with carbon at its center. We often show three-dimensionality in structural formulas by using a solid wedge (━◀━) to depict a bond projecting from the paper toward you and a dashed wedge (⋯⋯) for one receding away from you. A simple line (—) represents a bond that lies in the plane of the paper.

The tetrahedral geometry of methane is often explained with the **valence shell electron-pair repulsion (VSEPR) model.** The VSEPR model rests on the idea that an electron pair, either a bonded pair or an unshared pair, associated with a particular atom will be as far away from the atom's other electron pairs as possible. Thus, a tetrahedral geometry permits the four bonds of methane to be maximally separated and is characterized by H—C—H angles of 109.5°, a value referred to as the **tetrahedral angle.**

Water, ammonia, and methane share the common feature of an approximately tetrahedral arrangement of four electron pairs. Because we describe the shape of a molecule according to the positions of its atoms rather than the disposition of its electron pairs, however, water is said to be *bent,* and ammonia is *trigonal pyramidal.*

The H—O—H angle in water (105°) and the H—N—H angles in ammonia (107°) are slightly smaller than the tetrahedral angle. These bond-angle contractions are easily accommodated by VSEPR by reasoning that electron pairs in bonds take up less space than an unshared pair. The electron pair in a covalent bond feels the attractive force of

Although reservations have been expressed concerning VSEPR as an *explanation* for molecular geometries, it remains a useful *tool* for predicting the shapes of organic compounds.

TABLE 1.6   VSEPR and Molecular Geometry

Compound	Structural Formula	Repulsive Electron Pairs	Arrangement of Electron Pairs	Molecular Shape	Molecular Model
Methane (CH_4)		Carbon has four bonded pairs	Tetrahedral	**Tetrahedral**	
Water (H_2O)		Oxygen has two bonded pairs + two unshared pairs	Tetrahedral	**Bent**	
Ammonia (NH_3)		Nitrogen has three bonded pairs + one unshared pair	Tetrahedral	**Trigonal pyramidal**	
Boron trifluoride (BF_3)		Boron has three bonded pairs	Trigonal planar	**Trigonal planar**	
Formaldehyde (H_2CO)		Carbon has two bonded pairs + one double bond, which is counted as one bonded pair	Trigonal planar	**Trigonal planar**	
Carbon dioxide (CO_2)		Carbon has two double bonds, which are counted as two bonded pairs	Linear	**Linear**	

two nuclei and is held more tightly than an unshared pair localized on a single atom. Thus, repulsive forces increase in the order:

Increasing force of repulsion between electron pairs

Bonded pair–bonded pair   >   Unshared pair–bonded pair   >   Unshared pair–unshared pair
Least repulsive   Most repulsive

Repulsions among the four bonded pairs of methane gives the normal tetrahedral angle of 109.5°. Repulsions among the unshared pair of nitrogen in ammonia and the three bonded pairs causes the bonded pair-bonded pair H—N—H angles to be smaller than 109.5°. In water, the largest repulsive force involves the two unshared pairs of oxygen. As the distance between the two unshared pairs increases, the H—O—H angle decreases.

Boron trifluoride is a *trigonal planar* molecule. There are six electrons, two for each B—F bond, associated with the valence shell of boron. These three bonded pairs are farthest apart when they are coplanar, with F—B—F bond angles of 120°.

PROBLEM 1.17 The salt sodium borohydride, $NaBH_4$, has an ionic bond between Na^+ and the anion BH_4^-. What are the H—B—H angles in the borohydride anion?

Multiple bonds are treated as a single unit in the VSEPR model. Formaldehyde is a trigonal planar molecule in which the electrons of the double bond and those of the two single bonds are maximally separated. A linear arrangement of atoms in carbon dioxide allows the electrons in one double bond to be as far away as possible from the electrons in the other double bond.

PROBLEM 1.18 Specify the shape of the following:

(a) H—C≡N: (Hydrogen cyanide)

(b) H_4N^+ (Ammonium ion)

(c) :N̈=N⁺=N̈⁻ (Azide ion)

(d) CO_3^{2-} (Carbonate ion)

SAMPLE SOLUTION (a) The structure shown accounts for all the electrons in hydrogen cyanide. No unshared electron pairs are associated with carbon, and so the structure is determined by maximizing the separation between its single bond to hydrogen and the triple bond to nitrogen. Hydrogen cyanide is a *linear* molecule.

1.11 MOLECULAR DIPOLE MOMENTS

We can combine our knowledge of molecular geometry with a feel for the polarity of chemical bonds to predict whether a molecule has a dipole moment or not. The molecular dipole moment is the resultant of all of the individual bond dipole moments of a substance. Some molecules, such as carbon dioxide, have polar bonds, but lack a dipole moment because their geometry causes the individual C=O bond dipoles to cancel.

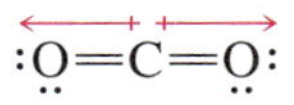

$$:\ddot{O}=C=\ddot{O}: \qquad \text{Dipole moment} = 0 \text{ D}$$

Carbon dioxide

Carbon tetrachloride, with four polar C—Cl bonds and a tetrahedral shape, has no net dipole moment, because the result of the four bond dipoles, as shown in Figure 1.7, is zero. Dichloromethane, on the other hand, has a dipole moment of 1.62 D. The C—H bond dipoles reinforce the C—Cl bond dipoles.

PROBLEM 1.19 Which of the following compounds would you expect to have a dipole moment? If the molecule has a dipole moment, specify its direction.

(a) BF_3

(b) H_2O

(c) CH_4

(d) CH_3Cl

(e) CH_2O

(f) HCN

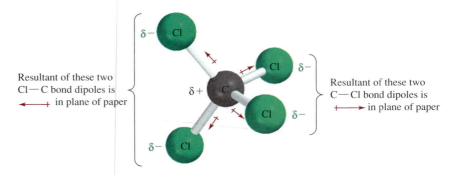

FIGURE 1.7 Contribution of individual bond dipole moments to the molecular dipole moments of (*a*) carbon tetrachloride (CCl_4) and (*b*) dichloromethane (CH_2Cl_2).

Resultant of these two
Cl—C bond dipoles is
in plane of paper
Resultant of these two
C—Cl bond dipoles is
in plane of paper
δ−
δ+
Cl
C

(*a*) There is a mutual cancellation of individual bond dipoles in carbon tetrachloride. It has no dipole moment.

Resultant of these two
H—C bond dipoles is
in plane of paper
Resultant of these two
C—Cl bond dipoles is
in plane of paper
δ+
δ−
H
C
Cl

(*b*) The H—C bond dipoles reinforce the C—Cl bond moment in dichloromethane. The molecule has a dipole moment of 1.62 D.

SAMPLE SOLUTION (a) Boron trifluoride is planar with 120° bond angles. Although each boron–fluorine bond is polar, their combined effects cancel and the molecule has no dipole moment.

F
B
F
F

$\mu = 0$ D

The opening paragraph of this chapter emphasized that the connection between structure and properties is what chemistry is all about. We have just seen one such connection. From the Lewis structure of a molecule, we can use electronegativity to tell us about the polarity of bonds and combine that with VSEPR to predict whether the molecule has a dipole moment. In the next several sections we'll see a connection between structure and *chemical reactivity* as we review acids and bases.

1.12 ACIDS AND BASES: THE ARRHENIUS VIEW

Acids and bases are a big part of organic chemistry, but the emphasis is much different from what you may be familiar with from your general chemistry course. Most of the attention in general chemistry is given to numerical calculations: pH, percent ionization, buffer problems, and so on. Some of this returns in organic chemistry, but mostly we are concerned with the roles that acids and bases play as reactants, products, and catalysts in chemical reactions. We'll start by reviewing some general ideas about acids and bases.

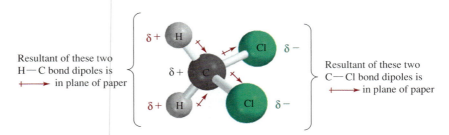

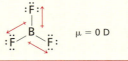

According to the theory proposed by Svante Arrhenius, a Swedish chemist and winner of the 1903 Nobel Prize in chemistry, an acid is a substance that ionizes to give protons when dissolved in water.

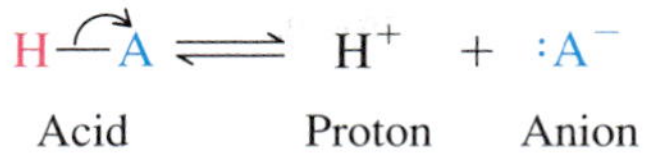

$$H{-}A \rightleftharpoons H^+ \;+\; :A^-$$

Acid   Proton   Anion

A base ionizes to give hydroxide ions.

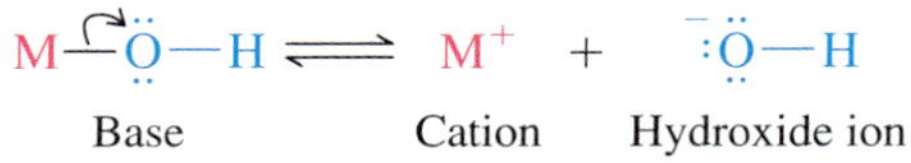

$$M{-}\ddot{O}{-}H \rightleftharpoons M^+ \;+\; :\ddot{O}{-}H$$

Base   Cation   Hydroxide ion

Acids differ in the degree to which they ionize. Those that ionize completely are called **strong acids;** those that do not are **weak acids.** Likewise, **strong bases** ionize completely; **weak bases** do not.

The strength of a weak acid is measured by its **acid dissociation constant,** which is the equilibrium constant K_a for its ionization in aqueous solution.

$$K_a = \frac{[H^+][:A^-]}{[HA]}$$

A convenient way to express the strength of an acid is by its pK_a, defined as:

$$pK_a = -\log_{10}K_a$$

Thus, acetic acid with $K_a = 1.8 \times 10^{-5}$ has a pK_a of 4.7. The advantage of pK_a over K_a is that it avoids exponentials. You are probably more familiar with K_a, but most organic chemists and biochemists use pK_a. It is a good idea to be comfortable with both systems, so you should practice converting K_a to pK_a and vice versa.

PROBLEM 1.20   Salicylic acid, the starting material for the preparation of aspirin, has a K_a of 1.06×10^{-3}. What is its pK_a?

PROBLEM 1.21   Hydrogen cyanide (HCN) has a pK_a of 9.1. What is its K_a?

1.13   ACIDS AND BASES: THE BRØNSTED–LOWRY VIEW

A more general theory of acids and bases was devised independently by Johannes Brønsted (Denmark) and Thomas M. Lowry (England) in 1923. In the Brønsted–Lowry approach, an acid is a **proton donor,** and a base is a **proton acceptor.** The reaction that occurs between an acid and a base is **proton transfer.**

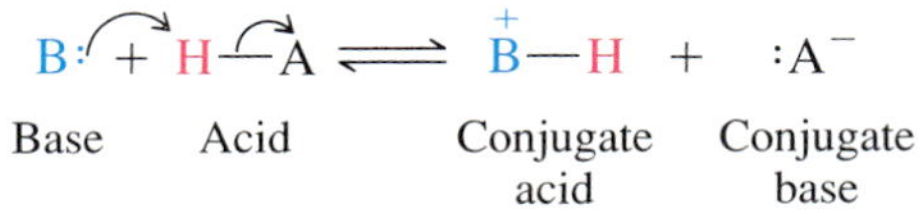

$$B: \;+\; H{-}A \rightleftharpoons \overset{+}{B}{-}H \;+\; :A^-$$

Base   Acid   Conjugate   Conjugate
acid   base

In the equation shown, the base uses an unshared pair of electrons to remove a proton from an acid. The base is converted to its **conjugate acid,** and the acid is converted to

CURVED ARROWS

"Follow the electrons; the atoms can't be far behind."

In Section 1.9 we introduced curved arrows as a tool to systematically generate resonance structures by moving electrons. The main use of curved arrows, however, is to show the bonding changes that take place in chemical reactions. The acid–base reactions to be discussed in Sections 1.12–1.17 furnish numerous examples of this and deserve some preliminary comment.

Consider the reaction of sodium hydroxide with hydrogen fluoride in aqueous solution.

$$\text{NaOH} \ + \ \text{HF} \ \longrightarrow \ \text{H}_2\text{O} \ + \ \text{NaF}$$

| Sodium hydroxide | Hydrogen fluoride | Water | Sodium fluoride |

Writing the equation in the usual way directs too much attention to the atoms and not enough to the electrons. We can remedy that by deleting any spectator ions and by showing the unshared electron pairs and covalent bonds that are made and broken. Both sodium hydroxide and sodium fluoride are completely ionized in water; therefore Na^+, which appears on both sides of the equation, is a spectator ion. Hydrogen fluoride is a weak acid and exists as undissociated HF molecules in water.

$$\text{H}\ddot{\text{O}}\text{:}^- \ + \ \text{H}—\ddot{\text{F}}\text{:} \ \longrightarrow \ \text{H}\ddot{\text{O}}—\text{H} \ + \ \text{:}\ddot{\text{F}}\text{:}^-$$

| Hydroxide ion | Hydrogen fluoride | Water | Fluoride ion |

Now use curved arrows to track electron flow. Move electrons from sites of high electron density (unshared pairs, negative charge) toward sites of lower electron density.

$$\text{H}\ddot{\text{O}}\text{:}^- \ +\ \text{H}—\ddot{\text{F}}\text{:} \ \longrightarrow \ \text{H}\ddot{\text{O}}—\text{H} \ + \ \text{:}\ddot{\text{F}}\text{:}^-$$

One of the unshared pairs of the hydroxide oxygen is used to form a covalent bond to the positively polarized proton of hydrogen fluoride. The covalent bond betwen H and F in hydrogen fluoride breaks, with the pair of electrons in this bond becoming an unshared pair of fluoride ion.

Curved arrows originate at electron pairs—in this case an electron pair of the hydroxide oxygen, and the shared pair in the covalent bond of HF. Curved arrows terminate at an atom or between two atoms.

Resist the temptation to use curved arrows to show the movement of atoms. Not only is this contrary to general practice, but it is also less reasonable. Electrons are much more mobile than atoms, so it makes sense to focus on them.

Showing electron flow is not just one more aspect of organic chemistry to learn, it is a genuinely useful aid to understanding what happens in a particular reaction.

---

its **conjugate base.** A base and its conjugate acid always differ by a single proton. Likewise, an acid and its conjugate base always differ by a single proton.

In the Brønsted–Lowry view, an acid doesn't dissociate in water; it transfers a proton to water. Water acts as a base.

$$\underset{\substack{\text{Water}\\\text{(base)}}}{\text{H}—\ddot{\text{O}}\text{:}—\text{H}} \ + \ \underset{\text{Acid}}{\text{H}—\text{A}} \ \rightleftharpoons \ \underset{\substack{\text{Conjugate}\\\text{acid of water}}}{\text{H}—\overset{+}{\ddot{\text{O}}}—\text{H}} \ + \ \underset{\substack{\text{Conjugate}\\\text{base}}}{\text{:}\text{A}^-}$$

The systematic name for the conjugate acid of water (H_3O^+) is **oxonium ion.** Its common name is **hydronium ion.**

Write an equation for proton transfer from hydrogen chloride (HCl) to

(a) Ammonia (:NH$_3$)

(b) Trimethylamine [(CH$_3$)$_3$N:]

Use curved arrows to track electron movement and identify the acid, base, conjugate acid, and conjugate base.

SAMPLE SOLUTION   We are told that a proton is transferred from HCl to :NH$_3$. Therefore, HCl is the Brønsted acid and :NH$_3$ is the Brønsted base.

$$H_3N: \quad + \quad H-\overset{..}{\underset{..}{Cl}}: \quad \rightleftharpoons \quad H_3\overset{+}{N}-H \quad + \quad :\overset{..}{\underset{..}{Cl}}:^-$$

Ammonia	Hydrogen chloride	Ammonium ion	Chloride ion
(base)	(acid)	(conjugate acid)	(conjugate base)

The acid dissociation constant K_a has the same form in Brønsted–Lowry as in the Arrhenius approach, but is expressed in the concentration of H_3O^+ rather than H^+. The concentration terms $[H_3O^+]$ and $[H^+]$ are considered equivalent quantities in equilibrium constant expressions.

$$K_a = \frac{[H_3O^+][:A^-]}{[HA]}$$

Even though water is a reactant (a Brønsted base), its concentration does not appear in the expression for K_a because it is the solvent. The convention for equilibrium constant expressions is to omit concentration terms for pure solids, liquids, and solvents.

Water can also be a Brønsted acid, donating a proton to a base. Sodium amide (NaNH$_2$), for example, is a source of the strongly basic amide ion, which reacts with water to give ammonia.

$$:\overset{H}{\underset{H}{N}}:\quad + \quad H-\overset{H}{\underset{..}{O}}: \quad \rightleftharpoons \quad :\overset{H}{\underset{H}{N}}-H \quad + \quad :\overset{H}{\underset{..}{O}}:^-$$

Amide ion	Water	Ammonia	Hydroxide ion
(base)	(acid)	(conjugate acid)	(conjugate base)

Potassium hydride (KH) is a source of the strongly basic hydride ion (:H$^-$). Using curved arrows to track electron movement, write an equation for the reaction of hydride ion with water. What is the conjugate acid of hydride ion?

Table 1.7 lists a number of acids, their dissociation constants, and their conjugate bases. The list is more extensive than we need at this point, but we will return to it repeatedly throughout the text as new aspects of acid–base behavior are introduced. The table is organized so that acid strength decreases from top to bottom. Conversely, the strength of the conjugate base increases from top to bottom. Thus, *the stronger the acid, the weaker its conjugate base. The stronger the base, the weaker its conjugate acid.*

TABLE 1.7 Dissociation Constants (pK_a) of Acids

Acid	pK_a	Formula	Conjugate Base	Discussed in Section
Hydrogen iodide	−10.4	HI	I$^-$	1.15
Hydrogen bromide	−5.8	HBr	Br$^-$	1.15
Sulfuric acid	−4.8	$HOSO_2OH$	$HOSO_2O^-$	1.16
Hydrogen chloride	−3.9	HCl	Cl$^-$	1.15
Hydronium ion	−1.7	H_3O^+	H_2O	1.16
Nitric acid	−1.4	$HONO_2$	$^-ONO_2$	1.15
Hydrogen sulfate ion	2.0	$HOSO_2O^-$	$^-OSO_2O^-$	1.16
Hydrogen fluoride	3.1	HF	F$^-$	1.15
Anilinium ion	4.6	$C_6H_5\overset{+}{N}H_3$	$C_6H_5NH_2$	22.5
Acetic acid	4.7	$CH_3\overset{O}{\overset{\|}{C}}OH$	$CH_3\overset{O}{\overset{\|}{C}}O^-$	1.15; 19.6
Pyridinium ion	5.2	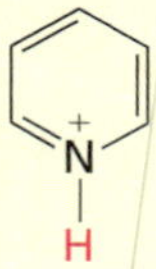 (pyridinium ion structure)	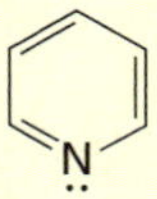 (pyridine structure)	1.14; 22.5
Carbonic acid	6.4	H_2CO_3	HCO_3^-	19.9
2,4-Pentanedione	9	$CH_3\overset{O}{\overset{\|}{C}}CH_2\overset{O}{\overset{\|}{C}}CH_3$	$CH_3\overset{O}{\overset{\|}{C}}\overset{..}{C}HCCH_3$	18.6
Hydrogen cyanide	9.1	HCN	CN$^-$	
Ammonium ion	9.3	NH_4^+	NH_3	1.14; 22.5
Glycine	9.6	$H_3\overset{+}{N}CH_2\overset{O}{\overset{\|}{C}}O^-$	$H_2NCH_2\overset{O}{\overset{\|}{C}}O^-$	27.3
Phenol	10	C_6H_5OH	$C_6H_5O^-$	1.16; 24.4
Hydrogen carbonate ion	10.2	HCO_3^-	CO_3^{2-}	19.9
Methanethiol	10.7	CH_3SH	CH_3S^-	15.4
Dimethylammonium ion	10.7	$(CH_3)_2\overset{+}{N}H_2$	$(CH_3)_2NH$	22.5
Ethyl acetoacetate	11	$CH_3\overset{O}{\overset{\|}{C}}CH_2\overset{O}{\overset{\|}{C}}OCH_2CH_3$	$CH_3\overset{O}{\overset{\|}{C}}\overset{..}{C}HCOCH_2CH_3$	21.1
Diethyl malonate	13	$CH_3CH_2O\overset{O}{\overset{\|}{C}}CH_2\overset{O}{\overset{\|}{C}}OCH_2CH_3$	$CH_3CH_2O\overset{O}{\overset{\|}{C}}\overset{..}{C}HCOCH_2CH_3$	21.7
Methanol	15.2	CH_3OH	CH_3O^-	1.15
2-Methylpropanal	15.5	$(CH_3)_2CH\overset{O}{\overset{\|}{C}}H$	$(CH_3)_2\overset{..}{C}\overset{O}{\overset{\|}{C}}H$	18.6
Water	15.7	H_2O	HO$^-$	1.15

(Continued)

TABLE 1.7	Dissociation Constants (pK_a) of Acids *(Continued)*			
Acid	**pK_a**	**Formula**	**Conjugate Base**	**Discussed in Section**
Ethanol	16	CH_3CH_2OH	$CH_3CH_2O^-$	1.15
Cyclopentadiene	16			11.20
Isopropyl alcohol	17	$(CH_3)_2CHOH$	$(CH_3)_2CHO^-$	1.15
tert-Butyl alcohol	18	$(CH_3)_3COH$	$(CH_3)_3CO^-$	1.15
Acetone	19	$CH_3\overset{O}{\overset{\|}{C}}CH_3$	$CH_3\overset{O}{\overset{\|}{C}}\ddot{C}H_2$	18.6
Methyl butanoate	22	$CH_3CH_2CH_2\overset{O}{\overset{\|}{C}}OCH_3$	$CH_3CH_2\overset{..}{C}H\overset{O}{\overset{\|}{C}}OCH_3$	21.10
Acetylene	26	$HC{\equiv}CH$	$HC{\equiv}C{:}^-$	9.5
Ammonia	36	NH_3	H_2N^-	1.15; 9.5
Diisopropylamine	36	$[(CH_3)_2CH]_2NH$	$[(CH_3)_2CH]_2N^-$	21.10
Benzene	43			14.5
Ethylene	45	$H_2C{=}CH_2$	$H_2C{=}\ddot{C}H^-$	9.4
Methane	60	CH_4	${:}\bar{C}H_3$	1.15; 14.5
Ethane	62	CH_3CH_3	$CH_3\ddot{C}H_2{}^-$	14.5

1.14   WHAT HAPPENED TO pK_b?

The Brønsted–Lowry approach involving conjugate relationships between acids and bases makes a separate basicity constant K_b unnecessary. Rather than having separate tables listing K_a for acids and K_b for bases, the usual practice is to give only K_a or pK_a as was done in Table 1.7. Assessing relative basicities requires only that we remember that the weaker the acid, the stronger the conjugate base and find the appropriate acid–base pair in the table.

Suppose, for example, we wished to compare the basicities of ammonia and pyridine.

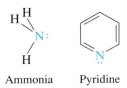

Ammonia   Pyridine

The stronger base is derived from the weaker conjugate acid. Therefore, add a proton to ammonia to give its conjugate acid (ammonium ion) and a proton to pyridine to give its conjugate acid (pyridinium ion), then look up the pK_a values for each.

	Ammonium ion	Pyridinium ion
pK_a	9.3	5.2
	weaker acid	stronger acid

Examine the models of ammonia and pyridine on your *Learning By Modeling* CD. Are the calculated charges on nitrogen consistent with their relative basicities? What about their electrostatic potential maps?

Ammonium ion is a weaker acid than pyridinium ion; therefore, ammonia is a stronger base than pyridine.

The conjugate bases listed in Table 1.7 that are anions are commonly encountered as their sodium or potassium salts. Thus, sodium methoxide ($NaOCH_3$), for example, is a source of methoxide ion (CH_3O^-), which is the conjugate base of methanol.

PROBLEM 1.24 Which is the stronger base in each of the following pairs? (*Note:* This information will prove useful when you get to Chapter 9.)
(a) Sodium ethoxide ($NaOCH_2CH_3$) or sodium amide ($NaNH_2$)
(b) Sodium acetylide ($NaC{\equiv}CH$) or sodium amide ($NaNH_2$)
(c) Sodium acetylide ($NaC{\equiv}CH$) or sodium ethoxide ($NaOCH_2CH_3$)

SAMPLE SOLUTION (a) $NaOCH_2CH_3$ contains the ions Na^+ and $CH_3CH_2O^-$. $NaNH_2$ contains the ions Na^+ and H_2N^-. $CH_3CH_2O^-$ is the conjugate base of ethanol; H_2N^- is the conjugate base of ammonia.

Base	$CH_3CH_2O^-$	H_2N^-
Conjugate acid	CH_3CH_2OH	NH_3
pK_a of conjugate acid	16	36

The conjugate acid of $CH_3CH_2O^-$ is stronger than the conjugate acid of H_2N^-. Therefore, H_2N^- is a stronger base than $CH_3CH_2O^-$.

1.15 HOW STRUCTURE AFFECTS ACID STRENGTH

The acids in Table 1.7 span a range of more than 70 pK_a units (10^{70} in K_a). In this section we'll introduce some generalizations that will permit us to connect molecular structure with acidity, at least insofar as trends in related compounds are concerned. The main ways in which structure affects acidity depend on:

1. The strength of the bond to the atom from which the proton is lost
2. The electronegativity of the atom from which the proton is lost
3. Changes in electron delocalization on ionization

Bond Strength.   The effect of bond strength is easy to see by comparing the acidities of the hydrogen halides.

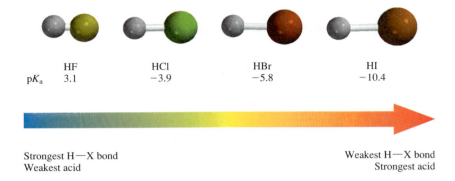

	HF	HCl	HBr	HI
pK_a	3.1	−3.9	−5.8	−10.4

Strongest H—X bond   Weakest H—X bond
Weakest acid   Strongest acid

For a discussion of the pK_a's of HF, HCl, and HI, see the January 2001 issue of the *Journal of Chemical Education*, pp. 116–117.

In general, bond strength decreases going down a group in the periodic table. As the H—X bond becomes longer, it also becomes weaker, so breaking it becomes energetically less costly, and acid strength increases.

Because of the conjugate relationship between acidity and basicity, the strongest acid (HI) has the weakest conjugate base (I⁻), and the weakest acid (HF) has the strongest conjugate base (F⁻).

PROBLEM 1.25   Which is the stronger acid, H_2O or H_2S? Which is the stronger base, HO⁻ or HS⁻? Check your predictions against the data in Table 1.7

Electronegativity.   The effect of electronegativity on acidity is evident in the following series involving bonds between hydrogen and the second-row elements C, N, O, and F.

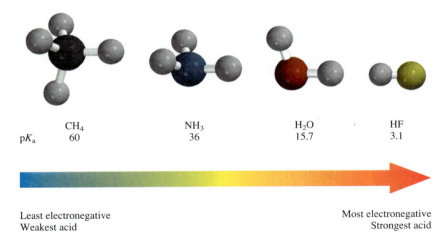

	CH_4	NH_3	H_2O	HF
pK_a	60	36	15.7	3.1

Least electronegative   Most electronegative
Weakest acid   Strongest acid

As the atom (A) to which H is bonded becomes more electronegative, the polarization $^{\delta+}$H—A$^{\delta-}$ becomes more pronounced and H is more easily lost as H⁺. An alternative approach to the same conclusion is based on the equation for proton transfer, especially with regard to the flow of electrons as shown by curved arrows.

Here we see that when the H—A bond breaks, both electrons in the bond are retained by A. The more electronegative atom A is, the easier it becomes for the electrons to flow in its direction.

When the two effects—bond strength and electronegativity—conflict, bond strength is more important when comparing elements in the same group of the periodic table as the pK_a's for the hydrogen halides show. Fluorine is the most electronegative and iodine the least electronegative of the halogens, but HF is the weakest acid while HI is the strongest. Electronegativity is the more important factor when comparing elements in the same row of the periodic table.

PROBLEM 1.26 Try to do this problem without consulting Table 1.7.

(a) Which is the stronger acid: $(CH_3)_3\overset{+}{N}H$ or $(CH_3)_2\overset{+}{O}H$?

(b) Which is the stronger base: $(CH_3)_3N\colon$ or $(CH_3)_2\overset{..}{O}\colon$?

SAMPLE SOLUTION (a) The ionizable proton is bonded to N in $(CH_3)_3\overset{+}{N}H$ and to O in $(CH_3)_2\overset{+}{O}H$.

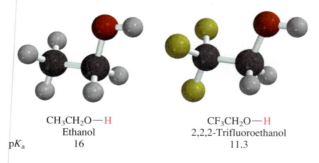

Nitrogen and oxygen are in the same row of the periodic table, so their relative electronegativities are the determining factor. Oxygen is more electronegative than nitrogen; therefore $(CH_3)_2\overset{+}{O}H$ is a stronger acid than $(CH_3)_3\overset{+}{N}H$.

The relative acidities of water and methanol are compared in an article in the November 2001 issue of the *Journal of Chemical Education*, pp. 1496–1498.

In many acids the acidic proton is bonded to oxygen. Such compounds can be considered as derivatives of water. Among organic compounds, the ones most closely related to water are alcohols. Most alcohols are somewhat weaker acids than water; methanol is slightly stronger.

	HO—H	CH_3O—H	CH_3CH_2O—H	$(CH_3)_2CHO$—H	$(CH_3)_3CO$—H
	Water	Methanol	Ethanol	Isopropyl alcohol	*tert*-Butyl alcohol
pK_a	15.7	15.2	16	17	18

Electronegative substituents in a molecule can affect acidity even when they are not directly bonded to the ionizable proton. Compare ethanol (CH_3CH_2OH) with a related compound in which a CF_3 group replaces the CH_3 group.

	CH_3CH_2O—H	CF_3CH_2O—H
	Ethanol	2,2,2-Trifluoroethanol
pK_a	16	11.3

We see that the substitution of C—H bonds by C—F increases the acidity of the O—H proton by 4.7 pK_a units, which corresponds to a difference of $10^{4.7}$ in K_a. The simplest explanation for this enhanced acidity is that the fluorine atoms attract electrons toward themselves and that this attraction is transmitted through the bonds, increasing the positive character of the O—H proton.

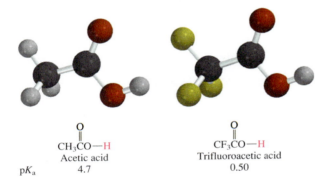

The greater positive character, hence the increased acidity, of the O—H proton of 2,2,2-trifluoroethanol can be seen in the electrostatic potential maps displayed in Figure 1.8.

Structural effects such as this that are transmitted through bonds are called **inductive effects.** A substituent *induces* a polarization in the bonds between it and some remote site. A similar inductive effect is evident when comparing acetic acid and its trifluoro derivative. Trifluoroacetic acid is more than 4 pK_a units stronger than acetic acid.

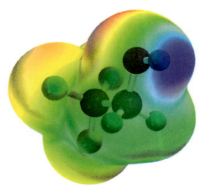

$$\overset{\overset{\displaystyle O}{\|}}{CH_3C}O{-}H$$
Acetic acid

$$\overset{\overset{\displaystyle O}{\|}}{CF_3C}O{-}H$$
Trifluoroacetic acid

pK_a	4.7	0.50

Inductive effects depend on the electronegativity of the substituent and the number of bonds between it and the affected site. As the number of bonds increases, the inductive effect decreases.

Ethanol (CH_3CH_2OH)

2,2,2-Trifluoroethanol (CF_3CH_2OH)

FIGURE 1.8 Electrostatic potential maps of ethanol and 2,2,2-trifluoroethanol. As indicated by the more blue, less green color in the region near the OH proton in 2,2,2-trifluoroethanol, this proton bears a greater degree of positive charge and is more acidic than the OH proton in ethanol.

Electron Delocalization in the Conjugate Base.   With a pK_a of -1.4, nitric acid is almost completely ionized in water. If we look at the Lewis structure of nitric acid in light of what we have said about inductive effects, we can see why. The N atom in nitric acid is not only electronegative in its own right, but bears a formal charge of $+1$, which enhances its ability to attract electrons away from the —OH group.

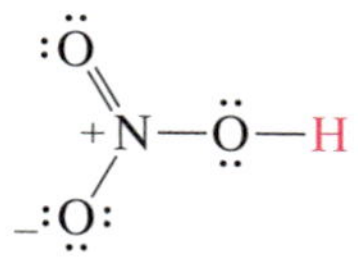

But inductive effects are only part of the story. When nitric acid transfers its proton to water, nitrate ion is produced.

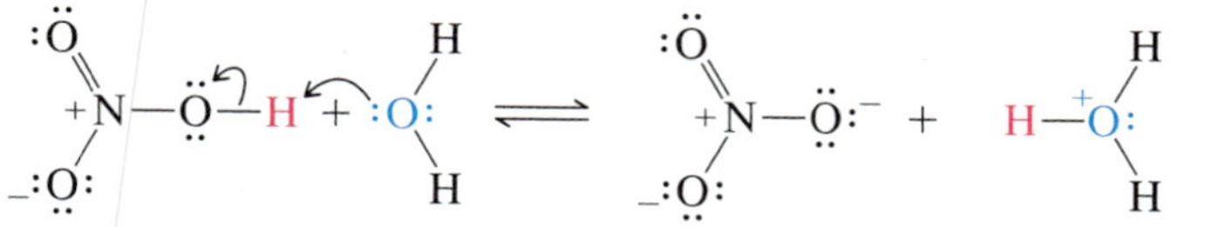

Nitric acid	Water	Nitrate ion	Hydronium ion
(acid)	(base)	(conjugate base)	(conjugate acid)

Nitrate ion is stabilized by electron delocalization, which we can represent in terms of resonance between three equivalent Lewis structures:

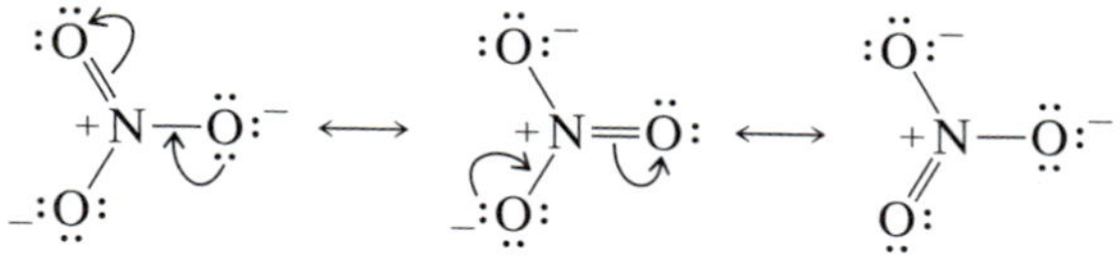

The negative charge is shared equally by all three oxygens. Stabilization of nitrate ion by electron delocalization increases the equilibrium constant for its formation.

PROBLEM 1.27   What is the average formal charge on each oxygen in nitrate ion?

A similar electron delocalization stabilizes acetate ion and related species.

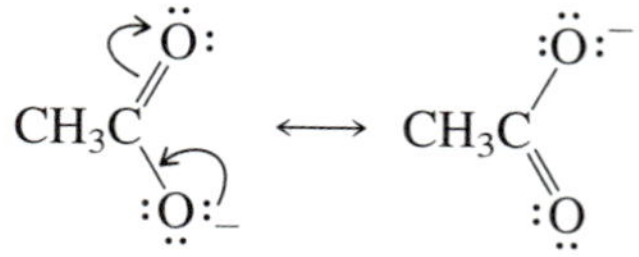

Both oxygens of acetate share the negative charge equally, which translates into a K_a for acetic acid that is greater than it would be if the charge were confined to a single oxygen.

PROBLEM 1.28   Show by writing appropriate resonance structures that the two compounds shown form the same conjugate base on ionization. Which atom, O or S, bears the greater share of negative charge?

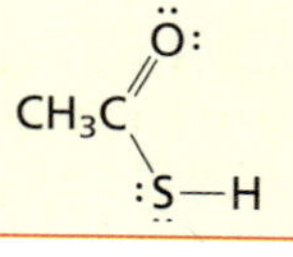

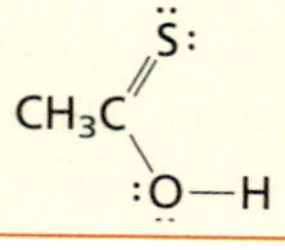

Organic chemistry involves a good bit of reasoning by analogy and looking for trends. The kind of reasoning we carried out in this section will become increasingly familiar as we learn more about the connection between structure and properties.

1.16   ACID–BASE EQUILIBRIA

In any proton-transfer reaction:

$$\text{Acid} + \text{Base} \rightleftharpoons \text{Conjugate acid} + \text{Conjugate base}$$

we are concerned with the question of whether the position of equilibrium lies to the side of products or reactants. There is an easy way to determine this. The reaction proceeds in the direction that converts the stronger acid and the stronger base to the weaker acid and the weaker base.

$$\text{Stronger acid} + \text{Stronger base} \xrightarrow{K > 1} \text{Weaker acid} + \text{Weaker base}$$

This generalization can be stated even more simply. *The reaction will be favorable when the stronger acid is on the left and the weaker acid is on the right.* This means that the equilibrium lies to the side of the acid that holds the proton more tightly.

Consider first the case of adding a strong acid such as HBr to water. The equation for the Brønsted acid–base reaction that occurs between them is:

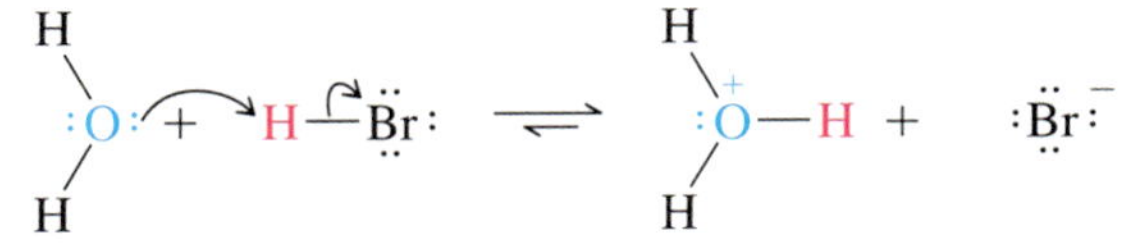

Water	Hydrogen bromide	Hydronium ion	Bromide ion
	$pK_a = -5.8$	$pK_a = -1.7$	
	stronger acid	weaker acid	

We identify the acid on the left and the acid on the right and compare their pK_a's to decide which is stronger. (Remember, the smaller the pK_a, the stronger the acid.) The acid on the left is HBr, which has a pK_a of -5.8. The acid on the right is H_3O^+, which has a pK_a of -1.7. The stronger acid (HBr) is on the left and the weaker acid (H_3O^+) is on the right, so the position of equilibrium lies to the right. Moreover, the difference in pK_a of more than 4 pK_a units is large enough to ensure that the equilibrium lies almost entirely to the side of products.

Compare the reaction of HBr with water to that of acetic acid with water.

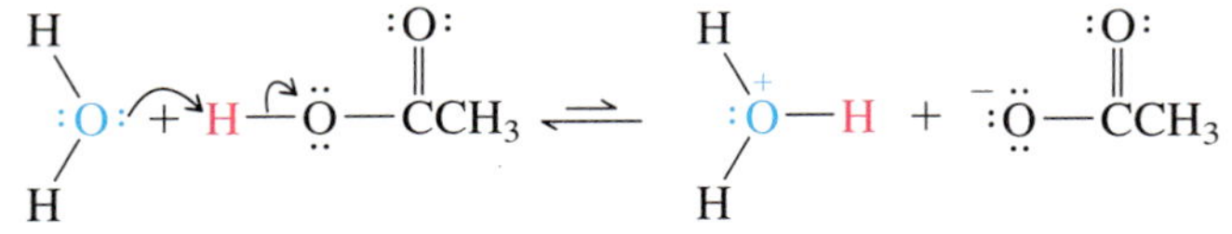

Water	Acetic acid	Hydronium ion	Acetate ion
	$pK_a = 4.7$	$pK_a = -1.7$	
	weaker acid	stronger acid	

Here, the weaker acid (acetic acid) is on the left and the stronger acid (hydronium ion) is on the right. The equilibrium constant is less than 1, and the position of equilibrium lies to the left.

For which of the following Brønsted bases is the reaction with acetic acid characterized by an equilibrium constant greater than 1?

(a) Ammonia   (b) Fluoride ion

SAMPLE SOLUTION   (a) Always start with an equation for an acid–base reaction. Ammonia is a Brønsted base and accepts a proton from the —OH group of acetic acid. Ammonia is converted to its conjugate acid, and acetic acid to its conjugate base.

| Ammonia | Acetic acid
$pK_a = 4.7$
stronger acid | Ammonium ion
$pK_a = 9.2$
weaker acid | Acetate ion |

From their respective pK_a's, we see that acetic acid is a stronger acid than ammonium ion. Therefore, the equilibrium lies to the right; K is greater than 1.

Two important points come from using relative pK_a's to analyze acid–base equilibria:

1. They permit clear-cut distinctions between strong and weak acids and bases. *A strong acid is one that is stronger than H_3O^+.* Conversely, a weak acid is one that is weaker than H_3O^+.

   Example: The pK_a's for the first and second ionizations of sulfuric acid are -4.8 and 2.0, respectively. Sulfuric acid ($HOSO_2OH$) is a strong acid; hydrogen sulfate ion ($HOSO_2O^-$) is a weak acid.

   Likewise, a *strong base is one that is stronger than HO^-.*

   Example: A common misconception is that the conjugate base of a weak acid is strong. This is sometimes, but not always, true. It is true, for example, for ammonia, which is a very weak acid (pK_a 36). Its conjugate base amide ion (H_2N^-) is a much stronger base than HO^-. It is not true, however, for acetic acid; both acetic acid and its conjugate base acetate ion are weak. The conjugate base of a weak acid will be strong only when the acid is a weaker acid than water.

2. The strongest acid present in significant amounts at equilibrium after a strong acid is dissolved in water is H_3O^+. The strongest acid present in significant amounts when a weak acid is dissolved in water is the weak acid itself.

   Example: $[H_3O^+] = 1.0$ M in a 1.0 M aqueous solution of HBr. The concentration of undissociated HBr molecules is near zero. $[H_3O^+] = 0.004$ M in a 1.0 M aqueous solution of acetic acid. The concentration of undissociated acetic acid molecules is near 1.0 M. Likewise, HO^- is the strongest base that can be present in significant quantities in aqueous solution.

Rank the following in order of decreasing concentration in a solution prepared by dissolving 1.0 mol of sulfuric acid in enough water to give 1.0 L of solution. (It is not necessary to do any calculations.)

$$H_2SO_4, \ HSO_4^-, \ SO_4^{2-}, \ H_3O^+$$

Analyzing acid–base reactions according to the Brønsted–Lowry picture provides yet another benefit. Table 1.7, which lists acids according to their strength in descending order along with their conjugate bases, can be used to predict the direction of proton transfer. Acid–base reactions in which a proton is transferred from an acid to a base that lies below it in the table have favorable equilibrium constants. Proton transfers from an acid to a base that lies above it in the table are unfavorable. Thus the equilibrium constant for proton transfer from phenol to hydroxide ion is greater than 1, but that for proton transfer from phenol to hydrogen carbonate ion is less than 1.

$$C_6H_5-\overset{..}{\underset{..}{O}}-H \; + \; :\overset{..}{\underset{..}{O}}-H \; \underset{}{\overset{K>1}{\rightleftharpoons}} \; C_6H_5-\overset{..}{\underset{..}{O}}:^- \; + \; H-\overset{..}{\underset{..}{O}}-H$$

| Phenol | Hydroxide ion | Phenoxide ion | Water |

$$C_6H_5-\overset{..}{\underset{..}{O}}-H \; + \; :\overset{..}{\underset{..}{O}}-\overset{\overset{\textstyle O}{\|}}{C}OH \; \underset{}{\overset{K<1}{\rightleftharpoons}} \; C_6H_5-\overset{..}{\underset{..}{O}}:^- \; + \; H-\overset{..}{\underset{..}{O}}-\overset{\overset{\textstyle O}{\|}}{C}OH$$

| Phenol | Hydrogen carbonate ion | Phenoxide ion | Carbonic acid |

Hydroxide ion lies below phenol in Table 1.7; hydrogen carbonate ion lies above phenol. The practical consequence of the reactions shown is that NaOH is a strong enough base to convert phenol to phenoxide ion, but $NaHCO_3$ is not.

> **PROBLEM 1.31** Verify that the position of equilibrium for the reaction between phenol and hydroxide ion lies to the right by comparing the pK_a of the acid on the left to the acid on the right. Which acid is stronger? Do the same for the reaction of phenol with hydrogen carbonate ion.

1.17 LEWIS ACIDS AND LEWIS BASES

The same G. N. Lewis who gave us electron-dot formulas also suggested a way of thinking about acids and bases that is more general than the Brønsted–Lowry approach. Where Brønsted and Lowry viewed acids and bases as donors and acceptors of protons (positively charged), Lewis took the opposite view and focused on electron pairs (negatively charged). According to Lewis *an acid is an electron-pair acceptor, and a base is an electron-pair donor.*

If we apply Lewis's definitions narrowly, we can write an equation for the reaction between a Lewis acid and a Lewis base as:

$$A^+ \; + \; :B^- \; \rightleftharpoons \; A-B$$

| Lewis acid | Lewis base |

An unshared pair of electrons from the Lewis base is used to form a covalent bond between the Lewis acid and the Lewis base. The Lewis acid and the Lewis base are shown as ions in the equation, but they need not be. If both are neutral molecules, the analogous equation becomes:

$$A \; + \; :B \; \rightleftharpoons \; \overset{-}{A}-\overset{+}{B}$$

| Lewis acid | Lewis base |

We can illustrate this latter case by the reaction:

Verify that the formal charges on boron and oxygen in "boron trifluoride etherate" are correct.

$$F_3B \quad + \quad :O: \qquad \rightleftharpoons \qquad F_3\overset{-}{B}-\overset{+}{O}:$$

Boron trifluoride	Diethyl ether	"Boron trifluoride etherate"
(Lewis acid)	(Lewis base)	(Lewis acid-Lewis base complex)

The product of this reaction, a **Lewis acid-Lewis base complex** called informally "boron trifluoride etherate," may look unusual but it is a stable species with properties different from those of the reactants. Its boiling point (126°C) for example, is much higher than that of boron trifluoride—a gas with a boiling point of −100°C—and diethyl ether, a liquid that boils at 34°C.

PROBLEM 1.32 Write an equation for the Lewis acid-Lewis base reaction between boron trifluoride and dimethyl sulfide [$(CH_3)_2S$]. Use curved arrows to track the flow of electrons and show formal charges if present.

The Lewis acid-Lewis base idea also includes certain **substitution** reactions in which one atom or group replaces another.

$$H\overset{..}{O}: ^- \quad + \quad H_3C-\overset{..}{\underset{..}{B}r}: \quad \rightleftharpoons \quad H\overset{..}{O}-CH_3 \quad + \quad :\overset{..}{\underset{..}{B}r}:^-$$

Hydroxide ion	Bromomethane	Methanol	Bromide ion
(Lewis base)	(Lewis acid)		

The carbon atom in bromomethane can accept an electron pair if its covalent bond with bromine breaks with both electrons in that bond becoming an unshared pair of bromide ion. Thus, bromomethane acts as a Lewis acid in this reaction.

Notice the similarity of the preceding reaction to one that is more familiar to us.

$$H\overset{..}{O}: ^- \quad + \quad H-\overset{..}{\underset{..}{B}r}: \quad \rightleftharpoons \quad H\overset{..}{O}-H \quad + \quad :\overset{..}{\underset{..}{B}r}:^-$$

Hydroxide ion	Hydrogen bromide	Water	Bromide ion
(Lewis base)	(Lewis acid)		

Clearly, the two reactions are analogous and demonstrate that the reaction between hydroxide ion and hydrogen bromide is simultaneously a Brønsted acid–base reaction and a Lewis acid-Lewis base reaction. *Brønsted acid–base reactions constitute a subcategory of Lewis acid-Lewis base reactions.*

Many important biochemical reactions involve Lewis acid-Lewis base chemistry. Carbon dioxide is rapidly converted to hydrogen carbonate ion in the presence of the enzyme *carbonic anhydrase*.

$$H\overset{..}{O}: ^- \quad + \quad \underset{:\overset{..}{O}:}{\overset{\overset{..}{O}:}{\|}{C}\|} \quad \underset{\text{carbonic anhydrase}}{\rightleftharpoons} \quad H\overset{..}{O}-\underset{:\overset{..}{O}:^-}{\overset{\overset{..}{O}:}{\|}C}$$

Hydroxide ion	Carbon dioxide	Hydrogen carbonate ion
(Lewis base)	(Lewis acid)	

Recall that the carbon atom of carbon dioxide bears a partial positive charge because of the electron-attracting power of its attached oxygens. When hydroxide ion (the Lewis base) bonds to this positively polarized carbon, a pair of electrons in the carbon–oxygen double bond leaves carbon to become an unshared pair of oxygen.

Lewis bases use an unshared pair to form a bond to some other atom and are also referred to as **nucleophiles** ("nucleus seekers"). Conversely, Lewis acids are **electrophiles** ("electron seekers"). We will use these terms hundreds of times throughout the remaining chapters.

Examine the table of contents. What chapters include terms related to "nucleophile" or "electrophile" in their title?

1.18 SUMMARY

This chapter sets the stage for all of the others by reminding us that the relationship between structure and properties is what chemistry is all about. It begins with a review of Lewis structures, moves to a discussion of the Arrhenius, Brønsted–Lowry, and Lewis pictures of acids and bases, and the effects of structure on acidity and basicity.

Section 1.1   A review of some fundamental knowledge about atoms and electrons leads to a discussion of **wave functions, orbitals,** and the **electron configurations** of atoms. Neutral atoms have as many electrons as the number of protons in the nucleus. These electrons occupy orbitals in order of increasing energy, with no more than two electrons in any one orbital. The most frequently encountered atomic orbitals in this text are s orbitals (spherically symmetrical) and p orbitals ("dumbbell"-shaped).

Boundary surface of a carbon $2s$ orbital

Boundary surface of a carbon $2p$ orbital

Section 1.2   An **ionic bond** is the force of electrostatic attraction between two oppositely charged ions. Atoms at the upper right of the periodic table, especially fluorine and oxygen, tend to gain electrons to form anions. Elements toward the left of the periodic table, especially metals such as sodium, tend to lose electrons to form cations. Ionic bonds in which carbon is the cation or anion are rare.

Section 1.3   The most common kind of bonding involving carbon is **covalent bonding.** A covalent bond is the sharing of a pair of electrons between two atoms. **Lewis structures** are written on the basis of the **octet rule,** which limits second-row elements to no more than eight electrons in their valence shells. In most of its compounds, carbon has four bonds.

$$H - \overset{\displaystyle H}{\underset{\displaystyle H}{\overset{|}{\underset{|}{C}}}} - \overset{\displaystyle H}{\underset{\displaystyle H}{\overset{|}{\underset{|}{C}}}} - \overset{..}{\underset{..}{O}} - H$$

Each carbon has four bonds in ethyl alcohol;
oxygen and each carbon are surrounded by eight electrons.

Section 1.4   Many organic compounds have **double** or **triple bonds** to carbon. Four electrons are involved in a double bond; six in a triple bond.

$$\underset{H}{\overset{H}{\diagdown}}C=C\underset{H}{\overset{H}{\diagup}} \qquad H-C\equiv C-H$$

Ethylene has a carbon–carbon double bond;
acetylene has a carbon–carbon triple bond.

Section 1.5   When two atoms that differ in **electronegativity** are covalently bonded, the electrons in the bond are drawn toward the more electronegative element.

$$\delta+\ \overset{}{C}-F\ \delta-$$

The electrons in a carbon–fluorine bond are
drawn away from carbon, toward fluorine.

Section 1.6   Counting electrons and assessing charge distribution in molecules is essential to understanding how structure affects properties. A particular atom in a Lewis structure may be neutral, positively charged, or negatively charged. The **formal charge** of an atom in the Lewis structure of a molecule can be calculated by comparing its electron count with that of the neutral atom itself.

Formal charge = (number of electrons in neutral atom)
      − (number of electrons in unshared pairs)
      − $\frac{1}{2}$ (number of electrons in covalent bonds)

Section 1.7   Table 1.4 in this section sets forth the procedure to be followed in writing Lewis structures for organic molecules. It begins with experimentally determined information: the **molecular formula** and the **constitution** (order in which the atoms are connected).

$$\underset{\overset{|}{H}}{\overset{H}{\underset{|}{H}}}\ H-\overset{\overset{H}{|}}{\underset{\underset{|}{H}}{C}}-\overset{:O:}{\overset{\|}{C}}-\overset{..}{\underset{..}{O}}-H$$

The Lewis structure of acetic acid

Section 1.8   Different compounds that have the same molecular formula are called **isomers.** If they are different because their atoms are connected in a different order, they are called **constitutional isomers.**

$$\underset{H}{\overset{:\overset{..}{O}}{\diagdown}}C-N\overset{H}{\underset{H}{\diagup}} \qquad \underset{H}{\overset{H}{\diagdown}}C=N\underset{:\overset{..}{O}-H}{\diagup}$$

Formamide (*left*) and formaldoxime (*right*) are constitutional isomers; both have the same molecular formula (CH_3NO), but the atoms are connected in a different order.

Section 1.9   Many molecules can be represented by two or more Lewis structures that differ only in the placement of electrons. In such cases the electrons are delocalized, and the real electron distribution is a hybrid of the contributing Lewis structures, each of which is called a **resonance** form. The rules for resonance are summarized in Table 1.5.

Two Lewis structures (resonance forms) of formamide; the atoms are connected in the same order, but the arrangment of the electrons is different.

Section 1.10   The shapes of molecules can often be predicted on the basis of **valence shell electron-pair repulsions.** A tetrahedral arrangement gives the maximum separation of four electron pairs (*left*); a trigonal planar arrangement is best for three electron pairs (*center*), and a linear arrangement for two electron pairs (*right*).

Section 1.11   Knowing the shape of a molecule and the polarity of its various bonds allows the presence or absence of a **molecular dipole moment** and its direction to be predicted.

$$H \overset{\overset{\displaystyle O}{|}}{} H \qquad O{=}C{=}O$$

Both water and carbon dioxide have polar bonds, but water is a polar molecule and carbon dioxide is not.

Section 1.12   According to the **Arrhenius** definitions, an acid ionizes in water to produce protons (H^+) and a base produces hydroxide ions (HO^-). The strength of an acid is given by its equilibrium constant K_a for ionization in aqueous solution:

$$K_a = \frac{[H^+][:A^-]}{[HA]}$$

or more conveniently by its pK_a:

$$pK_a = -\log_{10} K_a$$

Section 1.13   According to the **Brønsted–Lowry** definitions, an acid is a proton donor and a base is a proton acceptor.

$$B: + H{-}A \rightleftharpoons \overset{+}{B}{-}H \ + \ :A^-$$

Base   Acid   Conjugate   Conjugate
                                  acid   base

The Brønsted–Lowry approach to acids and bases is more generally useful than the Arrhenius approach.

Section 1.14   **Basicity constants** are not necessary in the Brønsted–Lowry approach. Basicity is measured according to the pK_a of the conjugate acid. The weaker the conjugate acid, the stronger the base.

Section 1.15   The strength of an acid depends on the atom to which the proton is bonded. The two main factors are the strength of the H—X bond and the electronegativity of X. Bond strength is more important for atoms in the same group of the periodic table, electronegativity is more important for atoms in the same row. Electronegative atoms elsewhere in the molecule can increase the acidity by **inductive effects.**

Electron-delocalization in the conjugate base, usually expressed via resonance between Lewis structures, increases acidity.

Section 1.16   The position of equilibrium in an acid–base reaction lies to the side of the weaker acid.

$$\text{Stronger acid + Stronger base} \xrightleftharpoons{K > 1} \text{Weaker acid + Weaker base}$$

This is a very useful relationship. You should practice writing equations according to the Brønsted–Lowry definitions of acids and bases and familiarize yourself with Table 1.7 which gives the pK_a's of various Brønsted acids.

Section 1.17   The Lewis definitions of acids and bases provide for a more general view of acid–base reactions than either the Arrhenius or Brønsted–Lowry picture. A **Lewis acid** is an electron-pair acceptor. A **Lewis base** is an electron-pair donor. The Lewis approach incorporates the Brønsted–Lowry approach as a subcategory in which the atom that accepts the electron pair in the Lewis acid is a proton.

PROBLEMS

1.33   Each of the following species will be encountered at some point in this text. They all have the same number of electrons binding the same number of atoms and the same arrangement of bonds; they are *isoelectronic*. Specify which atoms, if any, bear a formal charge in the Lewis structure given and the net charge for each species.

(a) :N≡N:

(b) :C≡N:

(c) :C≡C:

(d) :N≡O:

(e) :C≡O:

1.34   You will meet all the following isoelectronic species in this text. Repeat the previous problem for these three structures.

(a) :Ö=C=Ö:

(b) :N̈=N=N̈:

(c) :Ö=N=Ö:

1.35   All the following compounds are characterized by ionic bonding between a group I metal cation and a tetrahedral anion. Write an appropriate Lewis structure for each anion, remembering to specify formal charges where they exist.

(a) $NaBF_4$

(b) $LiAlH_4$

(c) K_2SO_4

(d) Na_3PO_4

1.36 Determine the formal charge at all the atoms in each of the following species and the net charge on the species as a whole.

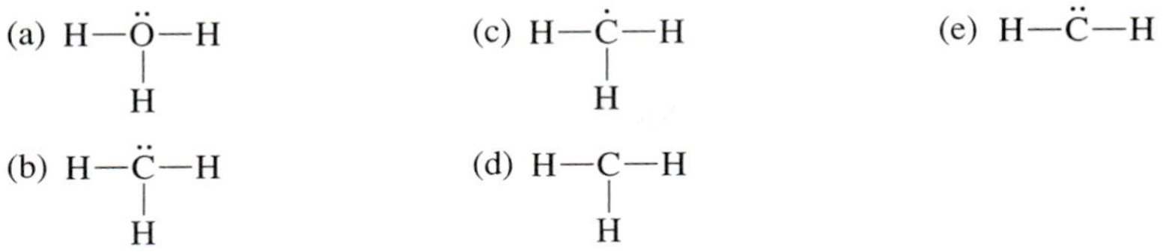

(a) $H - \overset{..}{\underset{H}{O}} - H$

(c) $H - \overset{.}{\underset{H}{C}} - H$

(e) $H - \overset{..}{C} - H$

(b) $H - \overset{..}{\underset{H}{C}} - H$

(d) $H - \underset{H}{C} - H$

1.37 What is the formal charge of oxygen in each of the following Lewis structures?

(a) $CH_3\overset{..}{\underset{..}{O}}:$

(b) $(CH_3)_2\overset{..}{\underset{..}{O}}:$

(c) $(CH_3)_3O:$

1.38 Write a Lewis structure for each of the following organic molecules:

(a) C_2H_5Cl (ethyl chloride: sprayed from aerosol cans onto skin to relieve pain)

(b) C_2H_3Cl [vinyl chloride: starting material for the preparation of poly(vinyl chloride), or PVC, plastics]

(c) $C_2HBrClF_3$ (halothane: a nonflammable inhalation anesthetic; all three fluorines are bonded to the same carbon)

(d) $C_2Cl_2F_4$ (Freon 114: formerly used as a refrigerant and as an aerosol propellant; each carbon bears one chlorine)

1.39 Write a structural formula for the CH_3NO isomer characterized by the structural unit indicated. None of the atoms in the final structure should have a formal charge.

(a) $C - N = O$

(c) $O - C = N$

(b) $C = N - O$

(d) $O = C - N$

1.40 Consider structural formulas A, B, and C:

$$H_2\overset{..}{C} - N \equiv N: \qquad H_2C = N = \overset{..}{N}: \qquad H_2C - \overset{..}{N} = \overset{..}{N}:$$

$$A \qquad\qquad\qquad B \qquad\qquad\qquad C$$

(a) Are A, B, and C constitutional isomers, or are they resonance forms?

(b) Which structures have a negatively charged carbon?

(c) Which structures have a positively charged carbon?

(d) Which structures have a positively charged nitrogen?

(e) Which structures have a negatively charged nitrogen?

(f) What is the net charge on each structure?

(g) Which is a more stable structure, A or B? Why?

(h) Which is a more stable structure, B or C? Why?

(i) What is the CNN geometry in each structure according to VSEPR?

1.41 Consider structural formulas A, B, C, and D:

$$H-\overset{..}{C}=N=\overset{..}{O}: \qquad H-C\equiv N-\overset{..}{\underset{..}{O}}: \qquad H-C\equiv N=\overset{..}{O}: \qquad H-\overset{..}{C}=\overset{..}{N}-\overset{..}{\underset{..}{O}}:$$

$$\text{A} \qquad\qquad\qquad \text{B} \qquad\qquad\qquad \text{C} \qquad\qquad\qquad \text{D}$$

(a) Which structures contain a positively charged carbon?

(b) Which structures contain a positively charged nitrogen?

(c) Which structures contain a positively charged oxygen?

(d) Which structures contain a negatively charged carbon?

(e) Which structures contain a negatively charged nitrogen?

(f) Which structures contain a negatively charged oxygen?

(g) Which structures are electrically neutral (contain equal numbers of positive and negative charges)? Are any of them cations? Anions?

(h) Which structure is the most stable?

(i) Which structure is the least stable?

1.42 In each of the following pairs, determine whether the two represent resonance forms of a single species or depict different substances. If two structures are not resonance forms, explain why.

(a) $:\overset{..}{\underset{..}{N}}-N\equiv N:$ and $:N=N=\overset{..}{\underset{..}{N}}:$

(b) $:\overset{..}{\underset{..}{N}}-N\equiv N:$ and $:\overset{..}{N}-N=\overset{..}{N}:$

(c) $:\overset{..}{\underset{..}{N}}-N\equiv N:$ and $:\overset{..}{N}-\overset{..}{N}-\overset{..}{\underset{..}{N}}:$

1.43 Among the following four structures, one is *not* a permissible resonance form. Identify the wrong structure. Why is it incorrect?

$$\overset{+}{C}H_2-\overset{..}{N}-\overset{..}{\underset{..}{O}}:^{-} \qquad CH_2=\overset{+}{N}-\overset{..}{\underset{..}{O}}:^{-} \qquad CH_2=N=\overset{..}{O}: \qquad :\overset{-}{C}H_2-\overset{+}{N}=\overset{..}{O}:$$
$$\underset{CH_3}{|} \qquad\qquad \underset{CH_3}{|} \qquad\qquad \underset{CH_3}{|} \qquad\qquad \underset{CH_3}{|}$$

$$\text{A} \qquad\qquad\qquad \text{B} \qquad\qquad\qquad \text{C} \qquad\qquad\qquad \text{D}$$

1.44 Keeping the same atomic connections and moving only electrons, write a more stable Lewis structure for each of the following. Be sure to specify formal charges, if any, in the new structure.

(a) $H-\overset{\overset{\displaystyle H}{|}}{\underset{\underset{\displaystyle H}{|}}{C}}-\overset{..}{N}=\overset{+}{N}:$

(b) $H-C\overset{\displaystyle \overset{..}{\underset{..}{O}}:^{-}}{\underset{\displaystyle \overset{+}{\underset{..}{O}}-H}{\big\backslash}}$

(c) $\underset{\displaystyle H}{\overset{\displaystyle H}{}}\overset{+}{C}-\overset{-}{C}\underset{\displaystyle H}{\overset{\displaystyle H}{}}$

(d) $\underset{\displaystyle H}{\overset{\displaystyle H}{}}C-\overset{+}{C}=C-\overset{-}{C}\underset{\displaystyle H}{\overset{\displaystyle H}{}}$ (with H, H on middle carbons)

(e) $\underset{\displaystyle H}{\overset{\displaystyle H}{}}\overset{+}{C}-C=C-\overset{..}{\underset{..}{O}}:^{-}$ (with H, H on middle carbons)

(f) $\underset{\displaystyle H}{\overset{\displaystyle H}{}}\overset{-}{C}-C\overset{\displaystyle \overset{..}{O}:}{\underset{\displaystyle H}{\big\backslash}}$

(g) $H-\overset{-}{\underset{..}{C}}=\overset{..}{O}:$

(h) $\underset{\displaystyle H}{\overset{\displaystyle H}{}}\overset{+}{C}-\overset{..}{\underset{..}{O}}H$

(i) $\underset{\displaystyle H}{\overset{\displaystyle H}{}}:\overset{-}{C}-\overset{..}{N}=\overset{+}{N}H_2$

1.45 (a) Write a Lewis structure for sulfur dioxide in which the octet rule is satisfied for all three atoms. Show all electron pairs and include any formal charges. The atoms are connected in the order OSO.

    (b) The octet rule may be violated for elements beyond the second period of the periodic table. Write a Lewis structure for sulfur dioxide in which each oxygen is connected to sulfur by a double bond. Show all electron pairs and formal charges.

1.46 Write structural formulas for all the constitutionally isomeric compounds having the given molecular formula.

    (a) C_4H_{10} (c) $C_2H_4Cl_2$ (e) C_3H_9N

    (b) C_5H_{12} (d) C_4H_9Br

1.47 Write structural formulas for all the constitutional isomers of

    (a) C_3H_8 (b) C_3H_6 (c) C_3H_4

1.48 Write structural formulas for all the constitutional isomers of molecular formula C_3H_6O that contain

    (a) Only single bonds   (b) One double bond

1.49 For each of the following molecules that contain polar covalent bonds, indicate the positive and negative ends of the dipole, using the symbol ↦. Refer to Table 1.2 as needed.

    (a) HCl   (c) HI   (e) HOCl

    (b) ICl   (d) H_2O

1.50 The compounds FCl and ICl have dipole moments μ that are similar in magnitude (0.9 and 0.7 D, respectively) but opposite in direction. In one compound, chlorine is the positive end of the dipole; in the other it is the negative end. Specify the direction of the dipole moment in each compound, and explain your reasoning.

1.51 Which compound in each of the following pairs would you expect to have the greater dipole moment μ? Why?

    (a) NaCl or HCl   (e) $CHCl_3$ or CCl_3F

    (b) HF or HCl   (f) CH_3NH_2 or CH_3OH

    (c) HF or BF_3 (g) CH_3NH_2 or CH_3NO_2

    (d) $(CH_3)_3CH$ or $(CH_3)_3CCl$

1.52 Apply the VSEPR method to deduce the geometry around carbon in each of the following species:

    (a) $:\overset{-}{C}H_3$ (b) $\overset{+}{C}H_3$ (c) $:CH_2$

1.53 Expand the following structural representations so as to more clearly show all the atoms and any unshared electron pairs.

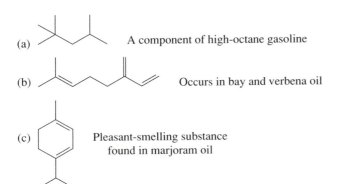

(a) A component of high-octane gasoline

(b) Occurs in bay and verbena oil

(c) Pleasant-smelling substance found in marjoram oil

(d) 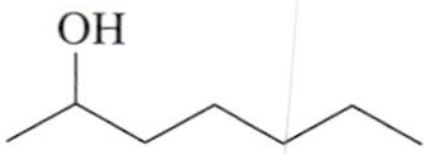 [structure: OH on a six-carbon chain]   Present in oil of cloves

(e) 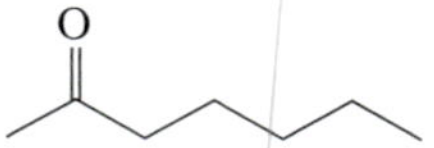 [structure: ketone on carbon chain]   Found in Roquefort cheese

(f)  [benzene ring structure]   Benzene: parent compound of a large family of organic substances

(g) 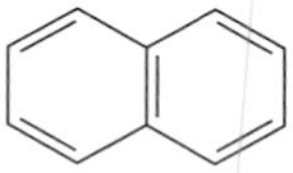 [naphthalene structure]   Naphthalene: sometimes used as a moth repellent

(h) 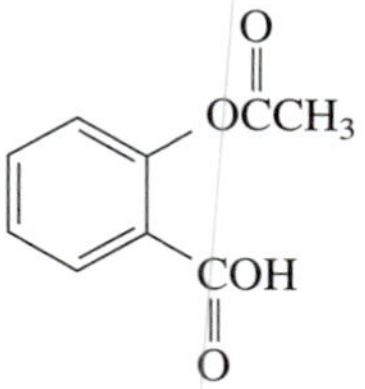 [aspirin structure with OCCH₃ and COH groups]   Aspirin

(i) 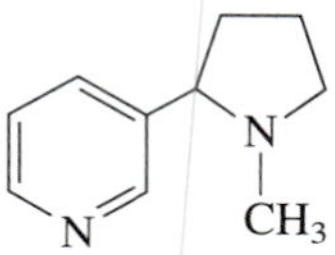 [nicotine structure with CH₃]   Nicotine: a toxic substance present in tobacco

(j) 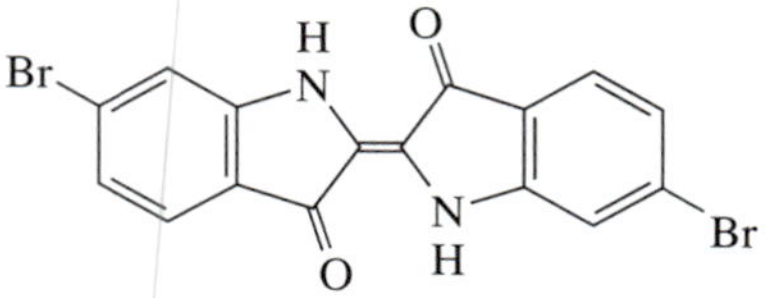 [Tyrian purple structure with Br substituents]   Tyrian purple: a purple dye extracted from a species of Mediterranean sea snail

(k) 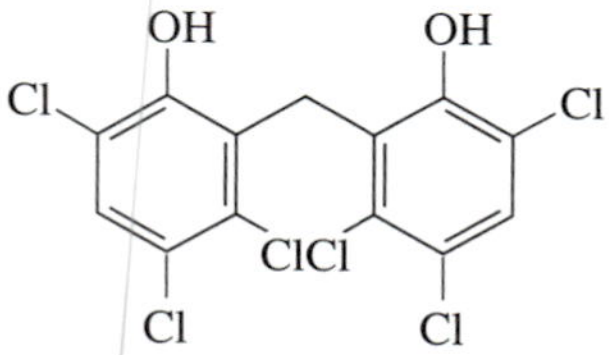 [hexachlorophene structure with OH, Cl substituents]   Hexachlorophene: an antiseptic

1.54 Molecular formulas of organic compounds are customarily presented in the fashion $C_2H_5BrO_2$. The number of carbon and hydrogen atoms are presented first, followed by the other atoms in alphabetical order. Give the molecular formulas corresponding to each of the compounds in the preceding problem. Are any of them isomers?

1.55 Calculate K_a for each of the following acids, given its pK_a. Rank the compounds in order of decreasing acidity.

(a) Aspirin: $pK_a = 3.48$

(b) Vitamin C (ascorbic acid): $pK_a = 4.17$

(c) Formic acid (present in sting of ants): $pK_a = 3.75$

(d) Oxalic acid (poisonous substance found in certain berries): $pK_a = 1.19$

1.56 Rank the following in order of decreasing acidity. Although none of these specific structures appear in Table 1.7, you can use analogous structures in the table to guide your reasoning.

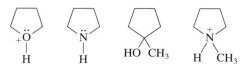

1.57 Rank the following in order of decreasing basicity. As in the preceding problem, Table 1.7 should prove helpful.

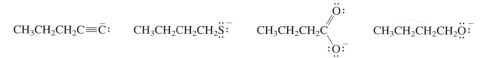

$CH_3CH_2CH_2C\equiv\bar{C}$: $CH_3CH_2CH_2CH_2\ddot{\underset{..}{S}}$:⁻ $CH_3CH_2CH_2C\overset{\ddot{O}:}{\underset{:\overset{..}{O}:^-}{\diagup}}$ $CH_3CH_2CH_2CH_2\ddot{\underset{..}{O}}$:⁻

1.58 Only one of the following statements is true. Use Table 1.7 to determine the true statement.

(a) HF is a weak acid and its conjugate base is strong.

(b) F⁻ is a weak base and its conjugate acid is strong.

(c) NH_3 is a weak acid and its conjugate base is strong.

(d) NH_3 is a weak base and its conjugate acid is strong.

1.59 Consider 1.0 M aqueous solutions of each of the following. Which solution is more basic?

(a) Sodium cyanide (NaCN) or sodium fluoride (NaF)

(b) Sodium carbonate (Na_2CO_3) or sodium acetate ($CH_3\overset{O}{\overset{\|}{C}}ONa$)

(c) Sodium sulfate (Na_2SO_4) or sodium methanethiolate ($NaSCH_3$)

1.60 (a) Which is the stronger acid: $(CH_3)_3NH^+$ or $(CH_3)_3PH^+$?

(b) Which is the stronger base: $(CH_3)_3N$: or $(CH_3)_3P$:?

1.61 Write an equation for the Brønsted acid–base reaction that occurs when each of the following acids reacts with water. Show all unshared electron pairs and formal charges, and use curved arrows to track electron movement.

(a) H—C≡N:

(b) [structure: CH₃CH₂—N⁺(H)(H)—CH₂CH₃]

(c) $H_3C—C\overset{:\overset{+}{O}—H}{\underset{:\underset{..}{O}—CH_3}{\diagup}}$

1.62 Write an equation for the Brønsted acid–base reaction that occurs when each of the following bases reacts with water. Show all unshared electron pairs and formal charges, and use curved arrows to track electron movement.

(a) $H_3C—C\equiv\bar{C}$:

(b) [structure: CH₃CH₂—N̈⁻—CH₂CH₃]

(c) $H_3C—C\overset{\ddot{N}—H}{\underset{H}{\diagup}}$

1.63 Each of the following pairs of compounds undergoes a Brønsted acid–base reaction for which the equilibrium lies to the right. Give the products of each reaction, and identify the acid, the base, the conjugate acid, and the conjugate base. Show the flow of electrons using curved arrows.

(a)

$$CH_3\overset{\overset{\displaystyle CH_3}{|}}{\underset{\underset{\displaystyle CH_3}{|}}{C}}-\overset{..}{\underset{..}{O}}:^- \quad + \quad \left\langle\!\!\!\bigcirc\!\!\!\right\rangle\!-\overset{..}{\underset{..}{S}}H \quad \rightleftharpoons$$

(b)

$$CH_3CH_2\overset{\overset{\displaystyle \overset{..}{O}:}{\|}}{C}\underset{:\underset{..}{O}H}{} \quad + \quad \overset{\overset{\displaystyle :\overset{..}{O}:}{\|}}{\underset{-:\overset{..}{\underset{..}{O}} \quad \overset{..}{\underset{..}{O}}:^-}{C}} \quad \rightleftharpoons$$

(c) $CH_3CH_2-\overset{..}{\underset{..}{N}}{}^--CH_2CH_3 \; + \; CH_3\overset{\overset{\displaystyle }{|}}{\underset{\underset{\displaystyle :\overset{..}{O}H}{|}}{C}}HCH_3 \rightleftharpoons$

1.64 One mol of each of the following is dissolved in water; the final volume in each case is 1.0 L. Identify the species that is present in greatest concentration in the respective solutions. Detailed calculations are not necessary.

(a) NH_3

(b) NH_4Cl (major species other than Cl^-)

(c) $NaNH_2$ (major species other than Na^+)

1.65 Use Table 1.7 to decide if NaOH is a strong enough base to remove a proton from acetylene ($HC\equiv CH$). Is $NaNH_2$ strong enough?

1.66 Practice working with your *Learning By Modeling* software. Construct molecular models of ethane, ethylene, and acetylene, and compare them with respect to their geometry, bond angles, and C—H and C—C bond distances.

1.67 How many different structures (isomers) can you make that have the formula (a) CH_2Cl_2; (b) $Cl_2C\!=\!CH_2$; and (c) $ClCH\!=\!CHCl$?

1.68 Examine the molecular models of H_2, HF, CH_4, CH_3F, and CF_4. Find the calculated dipole moment of each compound, and examine their electrostatic potential maps.

1.69 (a) Find the models of I—Br and Cl—F, and compare their calculated dipole moments. Which is more important, the difference in electronegativity between the bonded halogens or the length of the bond between them? [Remember that the dipole moment depends on both charge and distance ($\mu = e \times d$).]

(b) Compare the electrostatic potential maps of IBr and ClF. How do they correspond to the information provided by the dipole moment calculations?

1.70 Compare the dipole moments of cyanogen bromide ($BrC\equiv N$) and cyanogen chloride ($ClC\equiv N$). Which is larger? Why? What does this tell you about the electronegativity of the CN group?

1.71 Problem 1.8 concerned the charge distribution in methane (CH_4), chloromethane (CH_3Cl), and methyllithium (CH_3Li). Inspect molecular models of each of these compounds, and compare them with respect to how charge is distributed among the various atoms (carbon, hydrogen, chlorine, and lithium). Compare their electrostatic potential maps.

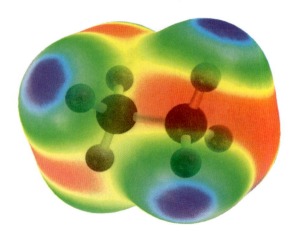

CHAPTER 2

HYDROCARBON FRAMEWORKS. ALKANES

This chapter continues the connection between structure and properties begun in Chapter 1. In it we focus on the simplest organic compounds—those that contain only carbon and hydrogen, called *hydrocarbons.* These compounds occupy a key position in the organic chemical landscape. Their framework of carbon–carbon bonds provides the scaffolding on which more reactive groups, called *functional groups,* are attached. We'll have more to say about functional groups beginning in Chapter 4; for now, we'll explore aspects of structure and bonding in hydrocarbons, especially alkanes.

We'll expand our picture of bonding by introducing two approaches that grew out of the idea that electrons can be described as waves—the *valence bond* and *molecular orbital* models. In particular, one aspect of the valence bond model, called **orbital hybridization,** will be emphasized.

A major portion of this chapter deals with how we name organic compounds. The system used throughout the world is based on a set of rules for naming hydrocarbons, then extending these rules to encompass other families of organic compounds.

2.1 CLASSES OF HYDROCARBONS

Hydrocarbons are divided into two main classes: **aliphatic** and **aromatic.** This classification dates from the nineteenth century, when organic chemistry was devoted almost entirely to the study of materials from natural sources, and terms were coined that reflected a substance's origin. Two sources were fats and oils, and the word *aliphatic* was derived from the Greek word *aleiphar* meaning ("fat"). Aromatic hydrocarbons, irrespective of their own odor, were typically obtained by chemical treatment of pleasant-smelling plant extracts.

Aliphatic hydrocarbons include three major groups: *alkanes, alkenes,* and *alkynes.* **Alkanes** are hydrocarbons in which all the bonds are single bonds, **alkenes** contain at least one carbon–carbon double bond, and **alkynes** contain at least one carbon–carbon triple bond. Examples of the three classes of aliphatic hydrocarbons are the two-carbon compounds *ethane, ethylene,* and *acetylene.*

Ethane	Ethylene	Acetylene
(alkane)	(alkene)	(alkyne)

Another name for aromatic hydrocarbons is **arenes.** Arenes have properties that are much different from alkanes, alkenes, and alkynes. The most important aromatic hydrocarbon is *benzene.*

Benzene
(arene)

We'll begin our discussion of hydrocarbons by introducing two additional theories of covalent bonding: the valence bond model and the molecular orbital model.

2.2   ELECTRON WAVES AND CHEMICAL BONDS

De Broglie's and Schrödinger's contributions to our present understanding of electrons were described in Section 1.1.

G. N. Lewis proposed his shared electron-pair model of bonding in 1916, almost a decade before Louis de Broglie's theory of wave–particle duality. De Broglie's radically different view of an electron, and Erwin Schrödinger's success in using wave equations to calculate the energy of an electron in a hydrogen *atom,* encouraged the belief that bonding in *molecules* could be explained on the basis of interactions between electron waves. This thinking produced two widely used theories of chemical bonding; one is called the **valence-bond model,** the other the **molecular orbital model.**

All of the forces in chemistry, except for nuclear chemistry, are electrical. Opposite charges attract; like charges repel. This simple fact can take you a long way.

Before we describe these theories, let's first think about bonding between two hydrogen atoms in the most fundamental terms. We'll begin with two hydrogen atoms that are far apart and see what happens as the distance between them decreases. The forces involved are electron–electron ($--$) repulsions, nucleus–nucleus ($++$) repulsions, and electron–nucleus ($-+$) attractions. All of these forces *increase* as the distance between the two hydrogens *decreases.* Because the electrons are so mobile, however, they can choreograph their motions so as to minimize their mutual repulsion while maximizing their attractive forces with the protons. Thus, as shown in Figure 2.1, a net, albeit weak, attractive force exists between the two hydrogens even when the atoms are far apart. This interaction becomes stronger as the two atoms approach each other—the electron of each hydrogen increasingly feels the attractive force of two protons rather than one, the total energy decreases, and the system becomes more stable. A potential energy minimum is reached when the separation between the nuclei reaches 74 pm, which corresponds to the H—H bond length in H_2. At distances shorter than this, the nucleus–nucleus and electron–electron repulsions dominate, and the system becomes less stable.

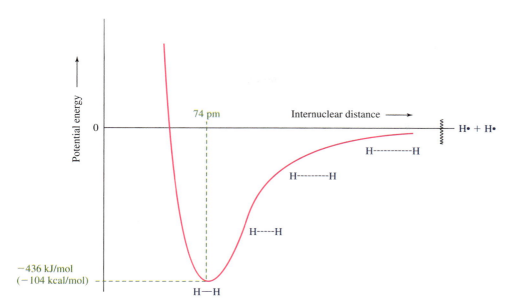

FIGURE 2.1 Plot of potential energy versus distance for two hydrogen atoms. At long distances, there is a weak attractive force. As the distance decreases, the potential energy decreases, and the system becomes more stable because each electron now "feels" the attractive force of two protons rather than one. The lowest energy state corresponds to a separation of 74 pm, which is the normal bond distance in H_2. At shorter distances, nucleus–nucleus and electron–electron repulsions are greater than electron–nucleus attractions, and the system becomes less stable.

Valence bond and molecular orbital theory both incorporate the wave description of an atom's electrons into this picture of H_2, but in somewhat different ways. Both assume that electron waves behave like more familiar waves, such as sound and light waves. One important property of waves is called interference in physics. *Constructive interference* occurs when two waves combine so as to reinforce each other (in phase); *destructive interference* occurs when they oppose each other (out of phase) (Figure 2.2). Recall from Section 1.1 that electron waves in atoms are characterized by their wave function, which is the same as an orbital. For an electron in the most stable state of a hydrogen atom, for example, this state is defined by the $1s$ wave function and is often called the $1s$ orbital. The *valence bond* model bases the connection between two atoms on the overlap between half-filled orbitals of the two atoms. The *molecular orbital* model assembles a set of molecular orbitals by combining the atomic orbitals of *all* of the atoms in the molecule.

For a molecule as simple as H_2, valence bond and molecular orbital theory produce very similar pictures. The next two sections describe these two approaches.

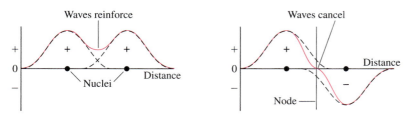

(a) Amplitudes of wave functions added

(b) Amplitudes of wave functions subtracted

FIGURE 2.2 Interference between waves. (a) Constructive interference occurs when two waves combine in phase with each other. The amplitude of the resulting wave at each point is the sum of the amplitudes of the original waves. (b) Destructive interference decreases the amplitude when two waves are out of phase with each other.

2.3   BONDING IN H_2: THE VALENCE BOND MODEL

The characteristic feature of valence bond theory is that it pictures a covalent bond between two atoms in terms of an in-phase overlap of a half-filled orbital of one atom with a half-filled orbital of the other, illustrated for the case of H_2 in Figure 2.3. Two hydrogen atoms, each containing an electron in a 1s orbital, combine so that their orbitals overlap to give a new orbital associated with both of them. In-phase orbital overlap (constructive interference) increases the probability of finding an electron in the region between the two nuclei where it feels the attractive force of both of them.

Figure 2.4 uses electrostatic potential maps to show this build-up of electron density in the region between two hydrogen atoms as they approach each other closely enough for their orbitals to overlap.

A bond in which the orbitals overlap along a line connecting the atoms (the *internuclear axis*) is called a **sigma (σ) bond.** The electron distribution in a σ bond is cylindrically symmetric; were we to slice through a σ bond perpendicular to the internuclear axis, its cross-section would appear as a circle. Another way to see the shape of the electron distribution is to view the molecule end-on.

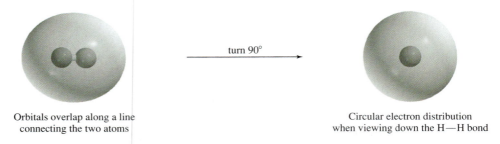

Orbitals overlap along a line
connecting the two atoms

turn 90°

Circular electron distribution
when viewing down the H—H bond

We will use the valence bond approach extensively in our discussion of organic molecules and expand on it shortly. First though, let's introduce the molecular orbital method to see how it uses the 1s orbitals of two hydrogen atoms to generate the orbitals of an H_2 molecule.

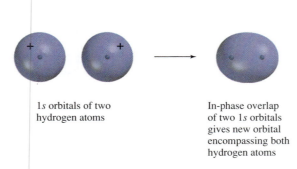

1s orbitals of two
hydrogen atoms

In-phase overlap
of two 1s orbitals
gives new orbital
encompassing both
hydrogen atoms

FIGURE 2.3 Valence bond picture of bonding in H_2. Overlap of half-filled 1s orbitals of two hydrogen atoms gives a new orbital that encompasses both atoms and contains both electrons. The electron density (electron probability) is highest in the region between the two atoms. The nuclei are shown by black dots. When the wave functions are of the same sign, (both + in this case), constructive interference increases the probability of finding an electron in the region where the two orbitals overlap.

(*a*) The 1*s* orbitals of two separated hydrogen atoms, sufficiently far apart so that essentially no interaction takes place between them. Each electron is associated with only a single proton.

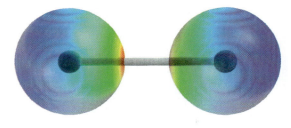

FIGURE 2.4 Valence bond picture of bonding in H₂ as illustrated by electrostatic potential maps. The 1*s* orbitals of two hydrogen atoms overlap to give an orbital that contains both electrons of an H₂ molecule.

(*b*) As the hydrogen atoms approach each other, their 1*s* orbitals begin to overlap and each electron begins to feel the attractive force of both protons.

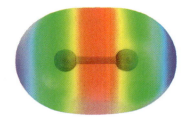

(*c*) The hydrogen atoms are close enough so that appreciable overlap of the two 1*s* orbitals occurs. The concentration of electron density in the region between the two protons is more readily apparent.

(*d*) A molecule of H₂. The center-to-center distance between the hydrogen atoms is 74 pm. The two individual 1*s* orbitals have been replaced by a new orbital that encompasses both hydrogens and contains both electrons. The electron density is greatest in the region between the two hydrogens.

2.4   BONDING IN H₂: THE MOLECULAR ORBITAL MODEL

The molecular orbital approach to chemical bonding rests on the notion that, as electrons in atoms occupy *atomic orbitals,* electrons in molecules occupy *molecular orbitals.* Just as our first task in writing the electron configuration of an atom is to identify the atomic orbitals that are available to it, so too must we first describe the orbitals available to a molecule. In the molecular orbital method this is done by representing molecular orbitals as combinations of atomic orbitals, the *linear combination of atomic orbitals-molecular orbital* (LCAO-MO) method.

Two molecular orbitals (MOs) of H₂ are generated by combining the 1*s* atomic orbitals (AOs) of two hydrogen atoms. In one combination, the two wave functions are added; in the other they are subtracted. The two new orbitals that are produced are portrayed in Figure 2.5. The additive combination generates a **bonding orbital;** the subtractive combination generates an **antibonding orbital.** Both the bonding and antibonding orbitals have σ symmetry. The two are differentiated by calling the bonding orbital

(*a*) Add the 1s wave functions of two hydrogen atoms to
generate a bonding molecular orbital (σ) of H_2. There is a
high probability of finding both electrons in the region
between the two nuclei.

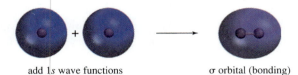

add 1s wave functions   σ orbital (bonding)

(*b*) Subtract the 1s wave function of one hydrogen atom from the
other to generate an antibonding molecular orbital (σ*) of H_2.
There is a nodal surface where there is a zero probability of
finding the electrons in the region between the two nuclei.

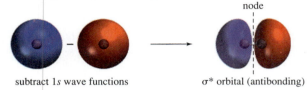

subtract 1s wave functions   σ* orbital (antibonding)

σ and the antibonding orbital σ*("sigma star"). The bonding orbital is characterized by
a region of high electron probability between the two atoms while the antibonding orbital
has a nodal surface between them.

A molecular orbital diagram for H_2 is shown in Figure 2.6. The customary format
shows the starting AOs at the left and right sides and the MOs in the middle. It must
always be true that *the number of MOs is the same as the number of AOs that combine
to produce them.* Thus, when the 1s AOs of two hydrogen atoms combine, two MOs
result. The bonding MO (σ) is lower in energy and the antibonding MO (σ*) higher in
energy than either of the original 1s orbitals.

When assigning electrons to MOs, the same rules apply as for writing electron configurations of atoms. Electrons fill the MOs in order of increasing orbital energy, and the

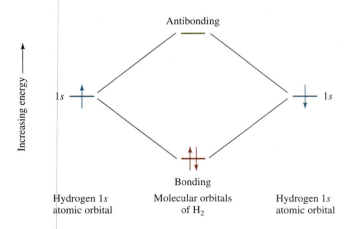

FIGURE 2.6 Two molecular orbitals (MOs) are generated by combining two hydrogen 1s atomic
orbitals (AOs). The bonding MO is lower in energy than either of the AOs that combine to produce it. The antibonding MO is of higher energy than either AO. Each arrow indicates one electron, and the electron spins are opposite in sign. Both electrons of H_2 occupy the bonding MO.

maximum number of electrons in any orbital is two. Both electrons of H_2 occupy the bonding orbital, have opposite spins, and both are held more strongly than they would be in separated hydrogen atoms. There are no electrons in the antibonding orbital.

For a molecule as simple as H_2, it is hard to see much difference between the valence bond and molecular orbital methods. The most important differences appear in molecules with more than two atoms. In those cases, the valence bond method continues to view a molecule as a collection of bonds between connected atoms. The molecular orbital method, however, leads to a picture in which the same electron can be associated with many, or even all, of the atoms in a molecule. We'll have more to say about the similarities and differences in valence bond and molecular orbital theory as we continue to develop their principles, beginning with the simplest alkanes: methane, ethane, and propane.

2.5   INTRODUCTION TO ALKANES: METHANE, ETHANE, AND PROPANE

Alkanes have the general molecular formula C_nH_{2n+2}. The simplest one, **methane** (CH_4), is also the most abundant. Large amounts are present in our atmosphere, in the ground, and in the oceans. Methane has been found on Jupiter, Saturn, Uranus, Neptune, and Pluto, and even on Halley's Comet.

Ethane (C_2H_6: CH_3CH_3) and **propane** (C_3H_8: $CH_3CH_2CH_3$) are second and third, respectively, to methane in many ways. Ethane is the alkane next to methane in structural simplicity, followed by propane. Ethane ($\approx 10\%$) is the second and propane ($\approx 5\%$) the third most abundant component of natural gas, which is $\approx 75\%$ methane. Natural gas is colorless and nearly odorless, as are methane, ethane, and propane. The characteristic odor of natural gas we use for heating our homes and cooking comes from trace amounts of unpleasant-smelling sulfur-containing compounds that are deliberately added to it to warn us of potentially dangerous leaks.

Methane is the lowest boiling alkane, followed by ethane, then propane.

	CH_4	CH_3CH_3	$CH_3CH_2CH_3$
	Methane	Ethane	Propane
Boiling point:	$-160°C$	$-89°C$	$-42°C$

> Boiling points cited in this text are at 1 atm (760 mm Hg) unless otherwise stated.

This will generally be true as we proceed to look at other alkanes; as the number of carbon atoms increases, so does the boiling point. All the alkanes with four carbons or less are gases at room temperature and atmospheric pressure. With the highest boiling point of the three, propane is the easiest one to liquefy. We are all familiar with "propane tanks." These are steel containers in which a propane-rich mixture of hydrocarbons called *liquefied petroleum gas* (LPG) is maintained in a liquid state under high pressure as a convenient clean-burning fuel.

The structural features of methane, ethane, and propane are summarized in Figure 2.7. All of the carbon atoms have four bonds, all of the bonds are single bonds, and the bond angles are close to tetrahedral. In the next section we'll see how to adapt the valence bond model to accommodate the observed structures.

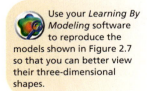 Use your *Learning By Modeling* software to reproduce the models shown in Figure 2.7 so that you can better view their three-dimensional shapes.

2.6   *sp³* HYBRIDIZATION AND BONDING IN METHANE

Before we describe the bonding in methane, it is worth pointing out that bonding theories attempt to describe a molecule on the basis of its component atoms; bonding theories do not attempt to explain *how* bonds form. Thus, the world's methane does not come

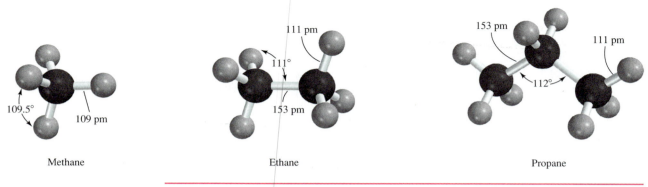

FIGURE 2.7 Structures of methane, ethane, and propane showing bond distances and bond angles.

from the reaction of carbon atoms with hydrogen atoms; it comes from biological processes. The boxed essay *Methane and the Biosphere* tells you more about the origins of methane and other organic compounds.

We *begin* with the experimentally determined three-dimensional structure of a molecule, then propose bonding models that are consistent with the structure. We do not claim that the observed structure is a result of the bonding model. Indeed, there may be two or more equally satisfactory models. Structures are facts; bonding models are theories that we use to try to understand the facts.

A vexing puzzle in the early days of valence bond theory concerned the fact that methane is CH_4 and that the four bonds to carbon are directed toward the corners of a tetrahedron. Valence bond theory is based on the overlap of half-filled orbitals of the connected atoms, but with an electron configuration of $1s^2 2s^2 2p_x{}^1 2p_y{}^1$ carbon has only two half-filled orbitals (Figure 2.8*a*). How can it have bonds to four hydrogens?

In the 1930s Linus Pauling offered an ingenious solution to this puzzle. He suggested that the electron configuration of a carbon bonded to other atoms need not be the same as a free carbon atom. By mixing ("hybridizing") the $2s$, $2p_x$, $2p_y$, and $2p_z$ orbitals, four new orbitals are obtained (Figure 2.8*b*). These four new orbitals are called **sp^3 hybrid orbitals** because they come from one s orbital and three p orbitals. Each sp^3 hybrid orbital has 25% s character and 75% p character. Among their most important features are the following:

1. *All four sp^3 orbitals are of equal energy.* Therefore, according to Hund's rule (Section 1.1) the four valence electrons of carbon are distributed equally among them, making four half-filled orbitals available for bonding.

2. *The axes of the sp^3 orbitals point toward the corners of a tetrahedron.* Therefore, sp^3 hybridization of carbon is consistent with the tetrahedral structure of methane. Each C—H bond is a σ bond in which a half-filled $1s$ orbital of hydrogen overlaps with a half-filled sp^3 orbital of carbon along a line drawn between them.

3. *σ Bonds involving sp^3 hybrid orbitals of carbon are stronger than those involving unhybridized 2s or 2p orbitals.* Each sp^3 hybrid orbital has two lobes of unequal size, making the electron density greater on one side of the nucleus than the other. In a C—H σ bond, it is the larger lobe of a carbon sp^3 orbital that overlaps with a hydrogen $1s$ orbital. This concentrates the electron density in the region between the two atoms.

As illustrated in Figure 2.9, the orbital hybridization model accounts for carbon having

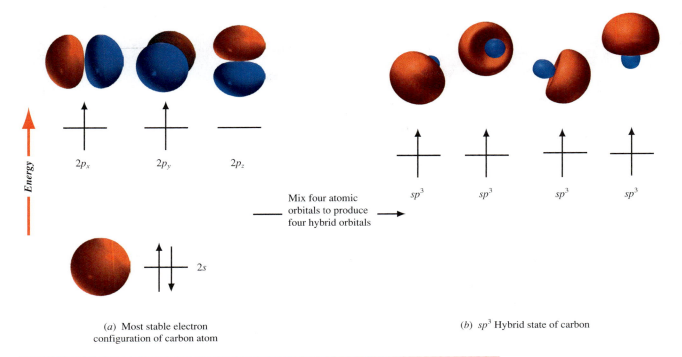

(*a*) Most stable electron configuration of carbon atom

Mix four atomic orbitals to produce four hybrid orbitals

(*b*) *sp*³ Hybrid state of carbon

FIGURE 2.8 *sp*³ Hybridization (*a*) Electron configuration of carbon in its most stable state. (*b*) Mixing the *s* orbital with the three *p* orbitals generates four *sp*³ hybrid orbitals. The four *sp*³ hybrid orbitals are of equal energy; therefore, the four valence electrons are distributed evenly among them. The axes of the four *sp*³ orbitals are directed toward the corners of a tetrahedron.

four bonds rather than two, the bonds are stronger than they would be in the absence of hybridization and are arranged in a tetrahedral fashion around carbon.

PROBLEM 2.1   Describe the bonding in ammonia, assuming *sp*³ hybridization of nitrogen. In what kind of orbital is the unshared pair? What orbital overlaps are involved in the N—H bonds?

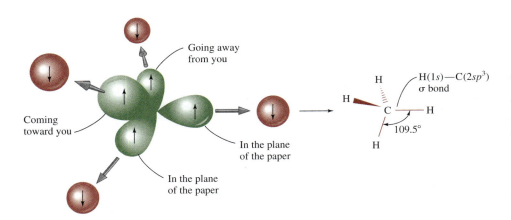

Going away from you

Coming toward you

In the plane of the paper

In the plane of the paper

H(1s)—C(2*sp*³) σ bond

109.5°

FIGURE 2.9 Each half-filled *sp*³ orbital overlaps with a half-filled hydrogen 1*s* orbital along a line between them giving a tetrahedral arrangement of four σ bonds. Only the major lobe of each *sp*³ orbital is shown. Each orbital contains a smaller back lobe, which has been omitted for clarity.

METHANE AND THE BIOSPHERE*

One of the things that environmental scientists do is to keep track of important elements in the biosphere—in what form do these elements normally occur, to what are they transformed, and how are they returned to their normal state? Careful studies have given clear, although complicated, pictures of the "nitrogen cycle," the "sulfur cycle," and the "phosphorus cycle," for example. The "carbon cycle," begins and ends with atmospheric carbon dioxide. It can be represented in an abbreviated form as:

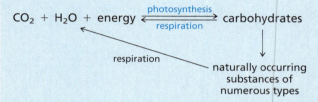

Methane is one of literally millions of compounds in the carbon cycle, but one of the most abundant. It is formed when carbon-containing compounds decompose in the absence of air (anaerobic conditions). The organisms that bring this about are called *methanoarchaea.* Cells can be divided into three types: *archaea, bacteria,* and *eukarya.* Methanoarchaea are one kind of archaea and may rank among the oldest living things on earth. They can convert a number of carbon-containing compounds, including carbon dioxide and acetic acid, to methane.

Virtually anywhere water contacts organic matter in the absence of air is a suitable place for methanoarchaea to thrive—at the bottom of ponds, bogs, and rice fields, for example. *Marsh gas (swamp gas)* is mostly methane. Methanoarchaea live inside termites and grass-eating animals. One source quotes 20 L/day as the methane output of a large cow.

The scale on which methanoarchaea churn out methane, estimated to be 10^{11}–10^{12} lb/year, is enormous. About 10% of this amount makes its way into the atmosphere, but most of the rest simply ends up completing the carbon cycle. It exits the anaerobic environment where it was formed and enters the aerobic world where it is eventually converted to carbon dioxide by a variety of processes.

When we consider sources of methane we have to add "old" methane, methane that was formed millions of years ago but became trapped beneath the earth's surface, to the "new" methane just described. *Firedamp,* an explosion hazard to miners, occurs in layers of coal and is mostly methane. Petroleum deposits, formed by microbial decomposition of plant material under anaerobic conditions, are always accompanied by pockets of natural gas, which is mostly methane.

An interesting thing happens when trapped methane leaks from sites under the deep-ocean floor. If the pressure is high enough (50 atm) and the water cold enough (4°C), the methane doesn't simply bubble to the surface. Individual methane molecules become trapped inside clusters of 6–18 water molecules forming *methane clathrates* or *methane hydrates.* Aggregates of these clathrates stay at the bottom of the ocean in what looks like a lump of dirty ice. Ice that burns. Far from being mere curiosities, methane clathrates are potential sources of energy on a scale greater than that of all known oil reserves combined. At present, it is not economically practical to extract the methane, however.

Methane clathrates have received recent attention from a different segment of the scientific community. While diving in the Gulf of Mexico in 1997, a research team of biologists and environmental scientists were surprised to find a new species of worm grazing on the mound of a methane clathrate. What were these worms feeding on? Methane? Bacteria that live on the methane? A host of questions having to do with deep-ocean ecosystems suddenly emerged. Stay tuned.

*The biosphere is the part of the earth where life is; it includes the surface, the oceans, and the lower atmosphere.

2.7   BONDING IN ETHANE

The orbital hybridization model of covalent bonding is readily extended to carbon–carbon bonds. As Figure 2.10 illustrates, ethane is described in terms of a carbon–carbon σ bond joining two CH_3 (**methyl**) groups. Each methyl group consists of an sp^3-hybridized carbon attached to three hydrogens by sp^3–$1s$ σ bonds. Overlap of the remaining half-filled orbital of one carbon with that of the other generates a σ bond between them. Here is a third kind of σ bond, one that has as its basis the overlap of two sp^3-hybridized orbitals. *In general, you can expect that carbon will be sp^3-hybridized when it is directly bonded to four atoms.*

> **PROBLEM 2.2**   Describe the bonding in propane according to the orbital hybridization model.

We will return to the orbital hybridization model to discuss bonding in other aliphatic hydrocarbons—alkenes and alkynes—later in the chapter. At this point, however, we'll turn our attention to alkanes to examine them as a class in more detail.

2.8   ISOMERIC ALKANES: THE BUTANES

Methane is the only alkane of molecular formula CH_4, ethane the only one that is C_2H_6, and propane the only one that is C_3H_8. Beginning with C_4H_{10}, however, constitutional isomers (Section 1.8) are possible; two alkanes have this particular molecular formula. In one, called **n-butane,** four carbons are joined in a continuous chain. The *n* in *n*-butane stands for "normal" and means that the carbon chain is unbranched. The second isomer has a branched carbon chain and is called **isobutane**.

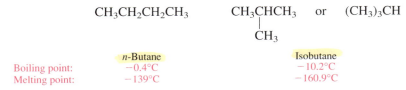

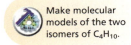

Make molecular models of the two isomers of C_4H_{10}.

	n-Butane	Isobutane
Boiling point:	−0.4°C	−10.2°C
Melting point:	−139°C	−160.9°C

As just noted (Section 2.7), CH_3 is called a *methyl* group. In addition to having methyl groups at both ends, *n*-butane contains two CH_2, or **methylene** groups. Isobutane contains three methyl groups bonded to a CH unit. The CH unit is called a **methine** group.

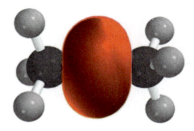

FIGURE 2.10 The C—C σ bond in ethane, pictured as an overlap of a half-filled sp^3 orbital of one carbon with a half-filled sp^3 hybrid orbital of the other.

n-Butane and isobutane have the same molecular formula but differ in the order in which their atoms are connected. They are *constitutional isomers* of each other (Section 1.8). Because they are different in structure, they can have different properties. Both are gases at room temperature, but *n*-butane boils almost 10°C higher than isobutane and has a melting point that is over 20°C higher.

Bonding in *n*-butane and isobutane continues the theme begun with methane, ethane, and propane. All of the carbon atoms are sp^3-hybridized, all of the bonds are σ bonds, and the bond angles at carbon are close to tetrahedral. This generalization holds for all alkanes regardless of the number of carbons they have.

2.9   HIGHER *n*-ALKANES

n-Alkanes are alkanes that have an unbranched carbon chain. ***n*-Pentane** and ***n*-hexane** are *n*-alkanes possessing five and six carbon atoms, respectively.

$$CH_3CH_2CH_2CH_2CH_3 \qquad CH_3CH_2CH_2CH_2CH_2CH_3$$

n-Pentane   *n*-Hexane

These condensed formulas can be abbreviated even more by indicating within parentheses the number of methylene groups in the chain. Thus, *n*-pentane may be written as $CH_3(CH_2)_3CH_3$ and *n*-hexane as $CH_3(CH_2)_4CH_3$. This shortcut is especially convenient with longer-chain alkanes. The laboratory synthesis of the "ultralong" alkane $CH_3(CH_2)_{388}CH_3$ was achieved in 1985; imagine trying to write a structural formula for this compound in anything other than an abbreviated way!

PROBLEM 2.3 An *n*-alkane of molecular formula $C_{28}H_{58}$ has been isolated from a certain fossil plant. Write a condensed structural formula for this alkane.

n-Alkanes have the general formula $CH_3(CH_2)_xCH_3$ and are said to belong to a **homologous series** of compounds. A homologous series is one in which successive members differ by a —CH_2— group.

Unbranched alkanes are sometimes referred to as "straight-chain alkanes," but, as we'll see in Chapter 3, their chains are not straight but instead tend to adopt the "zigzag" shape portrayed in the bond-line formulas introduced in Section 1.7.

Bond-line formula of *n*-pentane   Bond-line formula of *n*-hexane

PROBLEM 2.4 Much of the communication between insects involves chemical messengers called *pheromones*. A species of cockroach secretes a substance from its mandibular glands that alerts other cockroaches to its presence and causes them to congregate. One of the principal components of this *aggregation pheromone* is the alkane shown in the bond-line formula that follows. Give the molecular formula of this substance, and represent it by a condensed formula.

2.10 THE C$_5$H$_{12}$ ISOMERS

Three isomeric alkanes have the molecular formula C$_5$H$_{12}$. The unbranched isomer is, as we have seen, *n*-pentane. The isomer with a single methyl branch is called **isopentane.** The third isomer has a three-carbon chain with two methyl branches. It is called **neopentane.**

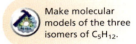

Make molecular models of the three isomers of C$_5$H$_{12}$.

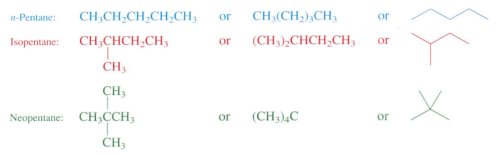

n-Pentane: $CH_3CH_2CH_2CH_2CH_3$ or $CH_3(CH_2)_3CH_3$ or

Isopentane: $CH_3CHCH_2CH_3$ or $(CH_3)_2CHCH_2CH_3$ or
$\quad\quad\quad\quad\; |$
$\quad\quad\quad\; CH_3$

Neopentane: CH_3CCH_3 or $(CH_3)_4C$ or
$\quad\quad\quad\quad\;\; |$
$\quad\quad\quad\quad CH_3$ (with CH$_3$ above)

Table 2.1 presents the number of possible alkane isomers as a function of the number of carbon atoms they contain. As the table shows, the number of isomers increases enormously with the number of carbon atoms and raises two important questions:

1. How can we tell when we have written all the possible isomers corresponding to a particular molecular formula?

2. How can we name alkanes so that each one has a unique name?

The answer to the first question is that you cannot easily calculate the number of isomers. The data in Table 2.1 were determined by a mathematician who concluded that no simple expression can calculate the number of isomers. The best way to ensure that you have written all the isomers of a particular molecular formula is to work systematically, beginning with the unbranched chain and then shortening it while adding branches one by one. It is essential that you be able to recognize when two different-looking structural formulas are actually the same molecule written in different ways. The key point is the *connectivity* of the carbon chain. For example, the following group of structural

The number of C$_n$H$_{2n+2}$ isomers has been calculated for values of *n* from 1 to 400 and the comment made that the number of isomers of C$_{167}$H$_{336}$ exceeds the number of particles in the known universe (10^{80}). These observations and the historical background of isomer calculation are described in a paper in the April 1989 issue of the *Journal of Chemical Education* (pp. 278–281).

TABLE 2.1	The Number of Constitutionally Isomeric Alkanes of Particular Molecular Formulas
Molecular formula	**Number of constitutional isomers**
CH$_4$	1
C$_2$H$_6$	1
C$_3$H$_8$	1
C$_4$H$_{10}$	2
C$_5$H$_{12}$	3
C$_6$H$_{14}$	5
C$_7$H$_{16}$	9
C$_8$H$_{18}$	18
C$_9$H$_{20}$	35
C$_{10}$H$_{22}$	75
C$_{15}$H$_{32}$	4,347
C$_{20}$H$_{42}$	366,319
C$_{40}$H$_{82}$	62,491,178,805,831

formulas do *not* represent different compounds; they are just a portion of the many ways we could write a structural formula for isopentane. Each one has a continuous chain of four carbons with a methyl branch located one carbon from the end of the chain.

The fact that all of these structural formulas represent the same substance can be clearly seen by making molecular models.

$$CH_3CHCH_2CH_3 \qquad CH_3CHCH_2CH_3 \qquad CH_3CH_2CHCH_3$$
$$| \qquad\qquad\qquad | \qquad\qquad\qquad\qquad\qquad |$$
$$CH_3 \qquad\qquad\quad CH_3 \qquad\qquad\qquad\qquad\qquad CH_3$$

$$CH_3CH_2CHCH_3 \qquad CHCH_2CH_3$$
$$| \qquad\qquad\qquad\qquad | $$
$$CH_3 \qquad\qquad\qquad CH_3$$

All same

PROBLEM 2.5 Write condensed and bond-line formulas for the five isomeric C_6H_{14} alkanes.

SAMPLE SOLUTION When writing isomeric alkanes, it is best to begin with the unbranched isomer.

$$CH_3CH_2CH_2CH_2CH_2CH_3 \qquad or$$

Next, remove a carbon from the chain and use it as a one-carbon (methyl) branch at the carbon atom next to the end of the chain.

$$CH_3CHCH_2CH_2CH_3 \qquad or$$
$$|$$
$$CH_3$$

Now, write structural formulas for the remaining three isomers. Be sure that each one is a unique compound and not simply a different representation of one written previously.

The answer to the second question—how to provide a name that is unique to a particular structure—is presented in the following section. It is worth noting, however, that being able to name compounds in a *systematic* way is a great help in deciding whether two structural formulas represent isomeric substances or are the same compound represented in two different ways. By following a precise set of rules, you will always get the same systematic name for a compound, regardless of how it is written. Conversely, two different compounds will always have different names.

2.11 IUPAC NOMENCLATURE OF UNBRANCHED ALKANES

A more detailed account of the history of organic nomenclature may be found in the article "The Centennial of Systematic Organic Nomenclature" in the November 1992 issue of the *Journal of Chemical Education* (pp. 863–865).

Nomenclature in organic chemistry is of two types: **common** (or "trivial") and **systematic.** Some common names existed long before organic chemistry became an organized branch of chemical science. Methane, ethane, propane, *n*-butane, isobutane, *n*-pentane, isopentane, and neopentane are common names. One simply memorizes the name that goes with a compound in just the same way that one matches names with faces. So long as there are only a few names and a few compounds, the task is manageable. But there are millions of organic compounds already known, and the list continues to grow! A system built on common names is not adequate to the task of communicating structural information. Beginning in 1892, chemists developed a set of rules for naming organic compounds based on their structures. We call these the **IUPAC rules;** *IUPAC* stands for

TABLE 2.2	IUPAC Names of Unbranched Alkanes				
Number of carbon atoms	Name	Number of carbon atoms	Name	Number of carbon atoms	Name
1	Methane	11	Undecane	21	Henicosane
2	Ethane	12	Dodecane	22	Docosane
3	Propane	13	Tridecane	23	Tricosane
4	Butane	14	Tetradecane	24	Tetracosane
5	Pentane	15	Pentadecane	30	Triacontane
6	Hexane	16	Hexadecane	31	Hentriacontane
7	Heptane	17	Heptadecane	32	Dotriacontane
8	Octane	18	Octadecane	40	Tetracontane
9	Nonane	19	Nonadecane	50	Pentacontane
10	Decane	20	Icosane*	100	Hectane

*Spelled "eicosane" prior to 1979 version of IUPAC rules.

the "International Union of Pure and Applied Chemistry." (See the boxed essay, *A Brief History of Systematic Organic Nomenclature.*)

The IUPAC rules assign names to unbranched alkanes as shown in Table 2.2. Methane, ethane, propane, and butane are retained for CH_4, CH_3CH_3, $CH_3CH_2CH_3$, and $CH_3CH_2CH_2CH_3$, respectively. Thereafter, the number of carbon atoms in the chain is specified by a Latin or Greek prefix preceding the suffix *-ane,* which identifies the compound as a member of the alkane family. Notice that the prefix *n-* is not part of the IUPAC system. The IUPAC name for $CH_3CH_2CH_2CH_3$ is butane, not *n*-butane.

PROBLEM 2.6 Refer to Table 2.2 as needed to answer the following questions:

(a) Beeswax contains 8–9% hentriacontane. Write a condensed structural formula for hentriacontane.

(b) Octacosane has been found to be present in a certain fossil plant. Write a condensed structural formula for octacosane.

(c) What is the IUPAC name of the alkane described in Problem 2.4 as a component of the cockroach aggregation pheromone?

SAMPLE SOLUTION (a) Note in Table 2.2 that hentriacontane has 31 carbon atoms. All the alkanes in Table 2.2 have unbranched carbon chains. Hentriacontane has the condensed structural formula $CH_3(CH_2)_{29}CH_3$.

In Problem 2.5 you were asked to write structural formulas for the five isomeric alkanes of molecular formula C_6H_{14}. In the next section you will see how the IUPAC rules generate a unique name for each isomer.

2.12 APPLYING THE IUPAC RULES: THE NAMES OF THE C_6H_{14} ISOMERS

We can present and illustrate the most important of the IUPAC rules for alkane nomenclature by naming the five C_6H_{14} isomers. By definition (Table 2.2), the unbranched C_6H_{14} isomer is hexane.

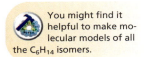
You might find it helpful to make molecular models of all the C_6H_{14} isomers.

$$CH_3CH_2CH_2CH_2CH_2CH_3$$

IUPAC name: hexane
(common name: *n*-hexane)

The IUPAC rules name branched alkanes as *substituted derivatives* of the unbranched alkanes listed in Table 2.2. Consider the C_6H_{14} isomer represented by the structure

$$\begin{array}{l} CH_3\underset{|}{C}HCH_2CH_2CH_3 \\ \quad\;\; CH_3 \end{array}$$

Step 1
Pick out the *longest continuous carbon chain,* and find the IUPAC name in Table 2.2 that corresponds to the unbranched alkane having that number of carbons. This is the parent alkane from which the IUPAC name is to be derived.

In this case, the longest continuous chain has *five* carbon atoms; the compound is named as a derivative of pentane. The key word here is *continuous.* It does not matter whether the carbon skeleton is drawn in an extended straight-chain form or in one with many bends and turns. All that matters is the number of carbons linked together in an uninterrupted sequence.

Step 2
Identify the substituent groups attached to the parent chain.

The parent pentane chain bears a methyl (CH_3) group as a substituent.

Step 3
Number the longest continuous chain in the direction that gives the lowest number to the substituent group at the first point of branching.

The numbering scheme

$$\begin{array}{l} \overset{1}{C}H_3\overset{2}{\underset{|}{C}}H\overset{3}{C}H_2\overset{4}{C}H_2\overset{5}{C}H_3 \\ \quad\;\; CH_3 \end{array} \quad \text{is equivalent to} \quad \begin{array}{l} CH_3\overset{2}{\underset{|}{C}}H\overset{3}{C}H_2\overset{4}{C}H_2\overset{5}{C}H_3 \\ \quad\;\; \overset{1}{C}H_3 \end{array}$$

Both schemes count five carbon atoms in their longest continuous chain and bear a methyl group as a substituent at the second carbon. An alternative numbering sequence that begins at the other end of the chain is incorrect:

$$\begin{array}{l} \overset{5}{C}H_3\overset{4}{\underset{|}{C}}H\overset{3}{C}H_2\overset{2}{C}H_2\overset{1}{C}H_3 \\ \quad\;\; CH_3 \end{array} \quad \text{(methyl group attached to C-4)}$$

Step 4
Write the name of the compound. The parent alkane is the last part of the name and is preceded by the names of the substituent groups and their numerical locations **(locants).** Hyphens separate the locants from the words.

$$\begin{array}{l} CH_3\underset{|}{C}HCH_2CH_2CH_3 \\ \quad\;\; CH_3 \end{array}$$

IUPAC name: 2-methylpentane

The same sequence of four steps gives the IUPAC name for the isomer that has its methyl group attached to the middle carbon of the five-carbon chain.

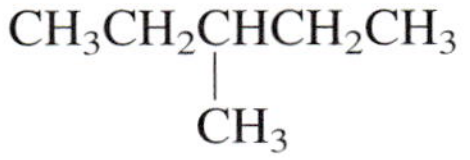

$$CH_3CH_2CHCH_2CH_3 \qquad \text{IUPAC name: 3-methylpentane}$$
$$| \\ CH_3$$

Both remaining C_6H_{14} isomers have two methyl groups as substituents on a four-carbon chain. Thus the parent chain is butane. When the same substituent appears more than once, use the multiplying prefixes *di-*, *tri-*, *tetra-*, and so on. A separate locant is used for each substituent, and the locants are separated from each other by commas and from the words by hyphens.

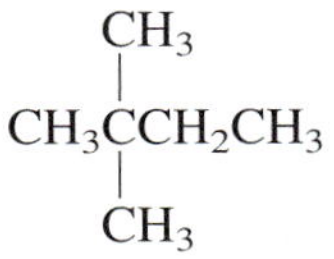

$$\begin{array}{cc} CH_3 & CH_3 \\ | & | \\ CH_3CCH_2CH_3 & CH_3CHCHCH_3 \\ | & | \\ CH_3 & CH_3 \end{array}$$

IUPAC name: 2,2-dimethylbutane 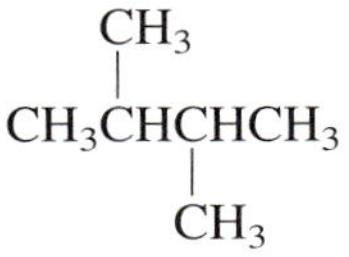   **IUPAC name: 2,3-dimethylbutane**

PROBLEM 2.7 Phytane is a naturally occurring alkane produced by the alga *Spirogyra* and is a constituent of petroleum. The IUPAC name for phytane is 2,6,10,14-tetramethylhexadecane. Write a structural formula for phytane.

PROBLEM 2.8 Derive the IUPAC names for

(a) The isomers of C_4H_{10} (c) $(CH_3)_3CCH_2CH(CH_3)_2$

(b) The isomers of C_5H_{12} (d) $(CH_3)_3CC(CH_3)_3$

SAMPLE SOLUTION (a) There are two C_4H_{10} isomers. Butane (see Table 2.2) is the IUPAC name for the isomer that has an unbranched carbon chain. The other isomer has three carbons in its longest continuous chain with a methyl branch at the central carbon; its IUPAC name is 2-methylpropane.

$$CH_3CH_2CH_2CH_3 \qquad\qquad CH_3CHCH_3 \\ | \\ CH_3$$

IUPAC name: butane   **IUPAC name: 2-methylpropane**
(common name: *n*-butane)   (common name: isobutane)

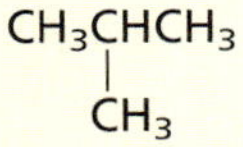

So far, the only branched alkanes that we've named have methyl groups attached to the main chain. What about groups other than CH_3? What do we call these groups, and how do we name alkanes that contain them?

2.13 ALKYL GROUPS

An alkyl group lacks one of the hydrogens of an alkane. A methyl group (CH_3—) is an alkyl group derived from methane (CH_4). Unbranched alkyl groups in which the point of attachment is at the end of the chain are named in IUPAC nomenclature by replacing the *-ane* endings of Table 2.2 by *-yl*.

$$CH_3CH_2— \qquad\qquad CH_3(CH_2)_5CH_2— \qquad\qquad CH_3(CH_2)_{16}CH_2—$$

Ethyl group   **Heptyl** group   **Octadecyl** group

The dash at the end of the chain represents a potential point of attachment for some other atom or group.

Carbon atoms are classified according to their degree of substitution by other carbons. A **primary** carbon is *directly* attached to one other carbon. Similarly, a **secondary** carbon is directly attached to two other carbons, a **tertiary** carbon to three, and a **quaternary** carbon to four. Alkyl groups are designated as primary, secondary, or tertiary according to the degree of substitution of the carbon at the potential point of attachment.

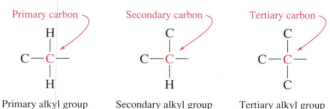

Primary alkyl group   Secondary alkyl group   Tertiary alkyl group

Ethyl (CH_3CH_2—), heptyl [$CH_3(CH_2)_5CH_2$—], and octadecyl [$CH_3(CH_2)_{16}CH_2$—] are examples of primary alkyl groups.

Branched alkyl groups are named by using the longest continuous chain that begins at the point of attachment as the base name. Thus, the systematic names of the two C_3H_7 alkyl groups are propyl and 1-methylethyl. Both are better known by their common names, *n*-propyl and isopropyl, respectively.

$$CH_3CH_2CH_2— \qquad \underset{2}{CH_3}\underset{1}{\overset{\overset{\displaystyle CH_3}{|}}{CH}}— \quad or \quad (CH_3)_2CH—$$

Propyl group
(common name: *n*-propyl)

1-Methylethyl group
(common name: isopropyl)

An isopropyl group is a *secondary* alkyl group. Its point of attachment is to a secondary carbon atom, one that is directly bonded to two other carbons.

The C_4H_9 alkyl groups may be derived either from the unbranched carbon skeleton of butane or from the branched carbon skeleton of isobutane. Those derived from butane are the butyl (*n*-butyl) group and the 1-methylpropyl (*sec*-butyl) group.

$$CH_3CH_2CH_2CH_2— \qquad \underset{3}{CH_3}\underset{2}{CH_2}\underset{1}{\overset{\overset{\displaystyle CH_3}{|}}{CH}}—$$

Butyl group
(common name: *n*-butyl)

1-Methylpropyl group
(common name: *sec*-butyl)

Those derived from isobutane are the 2-methylpropyl (isobutyl) group and the 1,1-dimethylethyl (*tert*-butyl) group. Isobutyl is a primary alkyl group because its potential point of attachment is to a primary carbon. *tert*-Butyl is a tertiary alkyl group because its potential point of attachment is to a tertiary carbon.

$$\underset{3}{CH_3}\underset{2}{\overset{\overset{\displaystyle CH_3}{|}}{CH}}\underset{1}{CH_2}— \quad or \quad (CH_3)_2CHCH_2— \qquad\qquad \underset{2}{CH_3}\underset{1}{\overset{\overset{\displaystyle CH_3}{|}}{\underset{\underset{\displaystyle CH_3}{|}}{C}}}— \quad or \quad (CH_3)_3C—$$

2-Methylpropyl group
(common name: isobutyl)

1,1-Dimethylethyl group
(common name: *tert*-butyl)

PROBLEM 2.9 Give the structures and IUPAC names of all the C_5H_{11} alkyl groups, and identify them as primary, secondary, or tertiary alkyl groups, as appropriate.

SAMPLE SOLUTION Consider the alkyl group having the same carbon skeleton as $(CH_3)_4C$. All the hydrogens are equivalent; replacing any one of them by a potential point of attachment is the same as replacing any of the others.

$$\begin{array}{c} CH_3 \\ | \\ \overset{3}{CH_3}-\overset{2}{C}-\overset{1}{CH_2}- \\ | \\ CH_3 \end{array} \quad \text{or} \quad (CH_3)_3CCH_2-$$

Numbering always begins at the point of attachment and continues through the longest continuous chain. In this case the chain is three carbons and there are two methyl groups at C-2. The IUPAC name of this alkyl group is **2,2-dimethylpropyl.** (The common name for this group is *neopentyl*.) It is a *primary* alkyl group because the carbon that bears the potential point of attachment (C-1) is itself directly bonded to one other carbon.

In addition to methyl and ethyl groups, *n*-propyl, isopropyl, *n*-butyl, *sec*-butyl, isobutyl, *tert*-butyl, and neopentyl groups will appear often throughout this text. Although these are common names, they have been integrated into the IUPAC system and are an acceptable adjunct to systematic nomenclature. You should be able to recognize these groups on sight and to give their structures when needed.

The names and structures of the most frequently encountered alkyl groups are given on the inside back cover.

2.14   IUPAC NAMES OF HIGHLY BRANCHED ALKANES

By combining the basic principles of IUPAC notation with the names of the various alkyl groups, we can develop systematic names for highly branched alkanes. We'll start with the following alkane, name it, then increase its complexity by successively adding methyl groups at various positions.

$$\begin{array}{c} CH_2CH_3 \\ | \\ \underset{1}{CH_3}\underset{2}{CH_2}\underset{3}{CH_2}\underset{4}{CH}\underset{5}{CH_2}\underset{6}{CH_2}\underset{7}{CH_2}\underset{8}{CH_3} \end{array}$$

As numbered on the structural formula, the longest continuous chain contains eight carbons, and so the compound is named as a derivative of octane. Numbering begins at the end nearest the branch, and so the ethyl substituent is located at C-4, and the name of the alkane is **4-ethyloctane.**

What happens to the IUPAC name when a methyl replaces one of the hydrogens at C-3?

$$\begin{array}{c} CH_2CH_3 \\ | \\ \underset{1}{CH_3}\underset{2}{CH_2}\underset{3}{CH}\underset{}{CH}\underset{4}{CH_2}\underset{5}{CH_2}\underset{6}{CH_2}\underset{7}{CH_2}\underset{8}{CH_3} \\ | \\ CH_3 \end{array}$$

The compound becomes an octane derivative that bears a C-3 methyl group and a C-4 ethyl group. *When two or more different substituents are present, they are listed in alphabetical order in the name.* The IUPAC name for this compound is **4-ethyl-3-methyloctane.**

Replicating prefixes such as *di-, tri-,* and *tetra-* (see Section 2.12) are used as needed but are ignored when alphabetizing. Adding a second methyl group to the original structure, at C-5, for example, converts it to **4-ethyl-3,5-dimethyloctane.**

$$\underset{\substack{4 \quad |5\; 6 \qquad 7 \quad 8 \\ \text{CH}_3 \quad \text{CH}_3}}{\overset{\substack{\text{CH}_2\text{CH}_3 \\ | \\ 1 \quad 2 \quad 3 \; |}}{\text{CH}_3\text{CH}_2\text{CHCHCHCH}_2\text{CH}_2\text{CH}_3}}$$

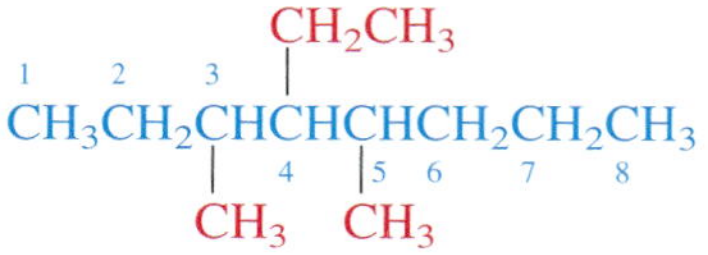

Italicized prefixes such as *sec-* and *tert-* are ignored when alphabetizing except when they are compared with each other. *tert-*Butyl precedes isobutyl, and *sec-*butyl precedes *tert-*butyl.

PROBLEM 2.10   Give an acceptable IUPAC name for each of the following alkanes:

(a) 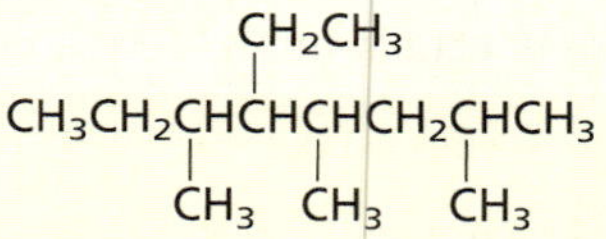 $$\underset{\substack{\text{CH}_3 \quad \text{CH}_3 \quad \text{CH}_3}}{\overset{\text{CH}_2\text{CH}_3}{\text{CH}_3\text{CH}_2\text{CHCHCHCH}_2\text{CHCH}_3}}$$

(b) $(\text{CH}_3\text{CH}_2)_2\text{CHCH}_2\text{CH}(\text{CH}_3)_2$

(c) 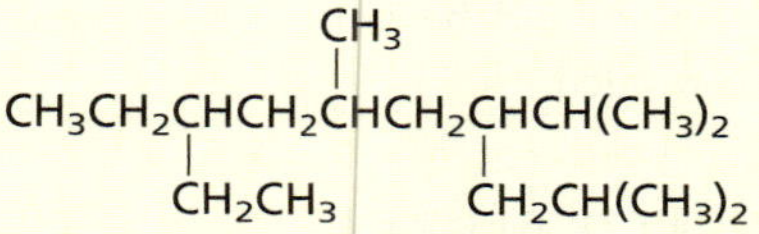 $$\underset{\substack{\text{CH}_2\text{CH}_3 \qquad\qquad \text{CH}_2\text{CH}(\text{CH}_3)_2}}{\overset{\text{CH}_3}{\text{CH}_3\text{CH}_2\text{CHCH}_2\text{CHCH}_2\text{CHCH}(\text{CH}_3)_2}}$$

SAMPLE SOLUTION   (a) This problem extends the preceding discussion by adding a third methyl group to 4-ethyl-3,5-dimethyloctane, the compound just described. It is, therefore, an *ethyltrimethyloctane*. Notice, however, that the numbering sequence needs to be changed in order to adhere to the rule of numbering from the end of the chain nearest the first branch. When numbered properly, this compound has a methyl group at C-2 as its first-appearing substituent.

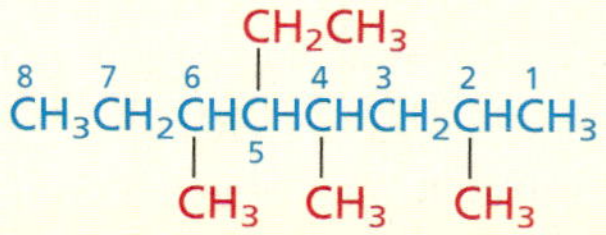

$$\underset{\substack{5 \qquad | \qquad | \\ \text{CH}_3 \quad \text{CH}_3 \quad \text{CH}_3}}{\overset{\substack{\text{CH}_2\text{CH}_3 \\ 8 \quad 7 \quad 6 \; | \; 4 \quad 3 \quad 2 \; 1}}{\text{CH}_3\text{CH}_2\text{CHCHCHCH}_2\text{CHCH}_3}}$$

5-Ethyl-2,4,6-trimethyloctane

An additional feature of IUPAC nomenclature that concerns the direction of numbering is the "first point of difference" rule. Consider the two directions in which the following alkane may be numbered:

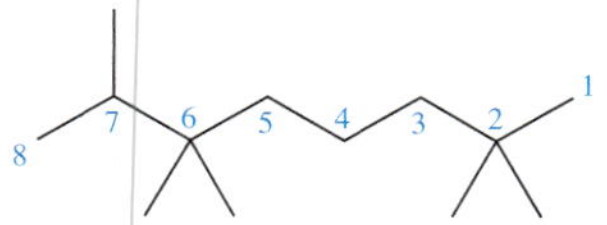

2,2,6,6,7-Pentamethyloctane   2,3,3,7,7-Pentamethyloctane
         (correct)   (incorrect!)

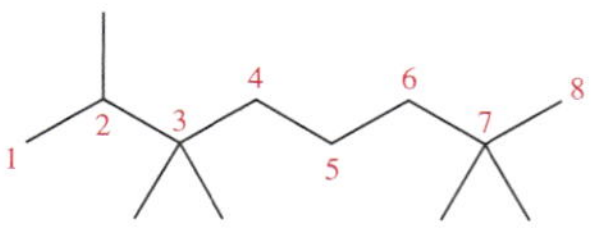

When deciding on the proper direction, a point of difference occurs when one order gives a lower locant than another. Thus, although 2 is the first locant in both numbering schemes, the tie is broken at the second locant, and the rule favors 2,2,6,6,7, which has

2 as its second locant, whereas 3 is the second locant in 2,3,3,7,7. Notice that locants are *not* added together, but examined one by one.

Finally, when equal locants are generated from two different numbering directions, choose the direction that gives the lower number to the substituent that appears first in the name. (Remember, substituents are listed alphabetically.)

The IUPAC nomenclature system is inherently logical and incorporates healthy elements of common sense into its rules. Granted, some long, funny-looking, hard-to-pronounce names are generated. Once one knows the code (rules of grammar) though, it becomes a simple matter to convert those long names to unique structural formulas.

> Tabular summaries of the IUPAC rules for alkane and alkyl group nomenclature appear on pages 96–98.

2.15   CYCLOALKANE NOMENCLATURE

Cycloalkanes are alkanes that contain a ring of three or more carbons. They are frequently encountered in organic chemistry and are characterized by the molecular formula C_nH_{2n}. Some examples include:

> Cycloalkanes are one class of *alicyclic* (*aliphatic cyclic*) hydrocarbons.

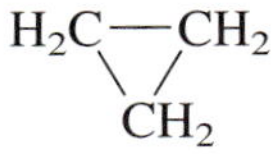

Cyclopropane

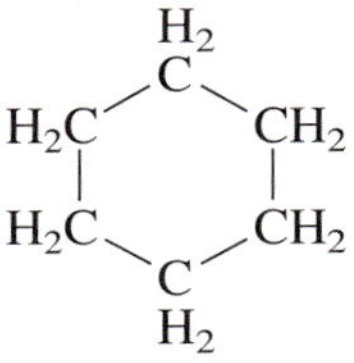

> If you make a molecular model of cyclohexane, you will find its shape to be very different from a planar hexagon. We'll discuss the reasons why in Chapter 3.

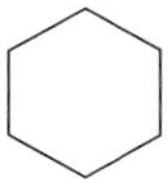

Cyclohexane

As you can see, cycloalkanes are named, under the IUPAC system, by adding the prefix *cyclo-* to the name of the unbranched alkane with the same number of carbons as the ring. Substituent groups are identified in the usual way. Their positions are specified by numbering the carbon atoms of the ring in the direction that gives the lowest number to the substituents at the first point of difference.

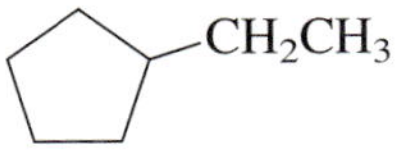

Ethylcyclopentane

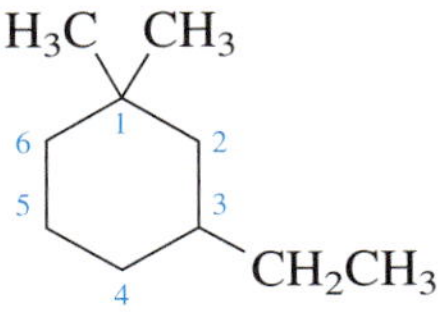

3-Ethyl-1,1-dimethylcyclohexane
(not 1-ethyl-3,3-dimethylcyclohexane, because first point of difference rule requires 1,1,3 substitution pattern rather than 1,3,3)

When the ring contains fewer carbon atoms than an alkyl group attached to it, the compound is named as an alkane, and the ring is treated as a cycloalkyl substituent:

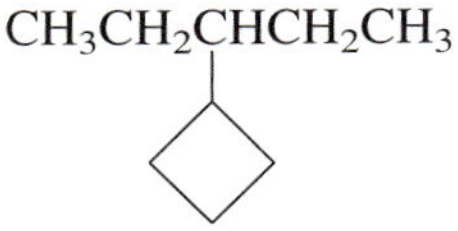

$$CH_3CH_2CHCH_2CH_3$$

3-Cyclobutylpentane

A BRIEF HISTORY OF SYSTEMATIC ORGANIC NOMENCLATURE

The first successful formal system of chemical nomenclature was advanced in France in 1787 to replace the babel of common names which then plagued the science. Hydrogen (instead of "inflammable air") and oxygen (instead of "vital air") are just two of the substances that owe their modern names to the proposals described in the *Méthode de nomenclature chimique*. It was then that important compounds such as sulfuric, phosphoric, and carbonic acid and their salts were named. The guidelines were more appropriate to inorganic compounds; it was not until the 1830s that names reflecting chemical composition began to appear in organic chemistry.

In 1889, a group with the imposing title of the International Commission for the Reform of Chemical Nomenclature was organized, and this group, in turn, sponsored a meeting of 34 prominent European chemists in Switzerland in 1892. Out of this meeting arose a system of organic nomenclature known as the **Geneva rules.** The principles on which the Geneva rules were based are the forerunners of our present system.

A second international conference was held in 1911, but the intrusion of World War I prevented any substantive revisions of the Geneva rules. The International Union of Chemistry was established in 1930 and undertook the necessary revision leading to publication in 1930 of what came to be known as the **Liège rules.**

After World War II, the International Union of Chemistry became the International Union of Pure and Applied Chemistry (known in the chemical community as the *IUPAC*). Since 1949, the IUPAC has issued reports on chemical nomenclature on a regular basis. The most recent **IUPAC rules** for organic chemistry were published in 1993. The IUPAC rules often offer several different ways to name a single compound. Thus although it is true that no two compounds can have the same name, it is incorrect to believe that there is only a single IUPAC name for a particular compound.

The 1993 IUPAC recommendations and their more widely used 1979 predecessors may both be accessed at the same web site:

www.acdlabs.com/iupac/nomenclature

The IUPAC rules are not the only nomenclature system in use today. Chemical Abstracts Service surveys all the world's leading scientific journals that publish papers relating to chemistry and publishes brief abstracts of those papers. The publication *Chemical Abstracts* and its indexes are absolutely essential to the practice of chemistry. For many years *Chemical Abstracts* nomenclature was very similar to IUPAC nomenclature, but the tremendous explosion of chemical knowledge has required *Chemical Abstracts* to modify its nomenclature so that its indexes are better adapted to computerized searching. This means that whenever feasible, a compound has a single *Chemical Abstracts* name. Unfortunately, this *Chemical Abstracts* name may be different from any of the several IUPAC names. In general, it is easier to make the mental connection between a chemical structure and its IUPAC name than its *Chemical Abstracts* name.

The **generic name** of a drug is not directly derived from systematic nomenclature. Furthermore, different pharmaceutical companies will call the same drug by their own trade name, which is different from its generic name. Generic names are invented on request (for a fee) by the U.S. Adopted Names Council, a private organization founded by the American Medical Association, the American Pharmaceutical Association, and the U.S. Pharmacopeial Convention.

PROBLEM 2.11 Name each of the following compounds:

(a) —C(CH$_3$)$_3$

(c)

(b) (CH$_3$)$_2$CH— H$_3$C   CH$_3$

2.16   SOURCES OF ALKANES AND CYCLOALKANES

As noted earlier, natural gas is especially rich in methane and also contains ethane and propane, along with smaller amounts of other low-molecular-weight alkanes. Natural gas is often found associated with petroleum deposits. Petroleum is a liquid mixture containing hundreds of substances, including approximately 150 hydrocarbons, roughly half of which are alkanes or cycloalkanes. Distillation of crude oil gives a number of fractions, which by custom are described by the names given in Figure 2.11. High-boiling fractions such as kerosene and gas oil find wide use as fuels for diesel engines and furnaces, and the nonvolatile residue can be processed to give lubricating oil, greases, petroleum jelly, paraffin wax, and asphalt.

The word *petroleum* is derived from the Latin words for "rock" (*petra*) and "oil" (*oleum*).

Although both are closely linked in our minds and by our own experience, the petroleum industry predated the automobile industry by half a century. The first oil well, drilled in Titusville, Pennsylvania, by Edwin Drake in 1859, provided "rock oil," as it was then called, on a large scale. This was quickly followed by the development of a process to "refine" it so as to produce kerosene. As a fuel for oil lamps, kerosene burned with a bright, clean flame and soon replaced the more expensive whale oil then in use. Other oil fields were discovered, and uses for other petroleum products were found—illuminating city streets with gas lights, heating homes with oil, and powering locomotives. There were oil refineries long before there were automobiles. By the time the first Model T rolled off Henry Ford's assembly line in 1908, John D. Rockefeller's Standard Oil holdings had already made him one of the half-dozen wealthiest people in the world.

Modern petroleum refining involves more than distillation, however, and includes two major additional operations:

1. **Cracking.** The more volatile, lower-molecular-weight hydrocarbons are useful as automotive fuels and as a source of petrochemicals. Cracking increases the proportion of these hydrocarbons at the expense of higher-molecular-weight ones by processes that involve the cleavage of carbon–carbon bonds induced by heat *(thermal cracking)* or with the aid of certain catalysts *(catalytic cracking)*.

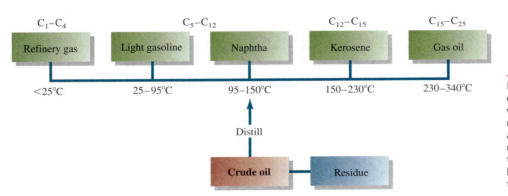

FIGURE 2.11 Distillation of crude oil yields a series of volatile fractions having the names indicated, along wih a nonvolatile residue. The number of carbon atoms that characterize the hydrocarbons in each fraction is approximate.

2. Reforming. The physical properties of the crude oil fractions known as *light gasoline* and *naphtha* (Figure 2.11) are appropriate for use as a motor fuel, but their ignition characteristics in high-compression automobile engines are poor and give rise to preignition, or "knocking." Reforming converts the hydrocarbons in petroleum to aromatic hydrocarbons and highly branched alkanes, both of which show less tendency for knocking than unbranched alkanes and cycloalkanes.

The leaves and fruit of many plants bear a waxy coating made up of alkanes that prevents loss of water. In addition to being present in beeswax (see Problem 2.6), hentriacontane, $CH_3(CH_2)_{29}CH_3$, is a component of the wax of tobacco leaves.

Cyclopentane and cyclohexane are present in petroleum, but as a rule, unsubstituted cycloalkanes are rarely found in natural sources. Compounds that contain rings of various types, however, are quite abundant.

The tendency of a gasoline to cause "knocking" in an engine is given by its octane number. The lower the octane number, the greater the tendency. The two standards are heptane (assigned a value of 0) and "isooctane" (2,2,4-trimethylpentane, which is assigned a value of 100). The octane number of a gasoline is equal to the percentage of isooctane in a mixture of isooctane and heptane that has the same tendency to cause knocking as that sample of gasoline.

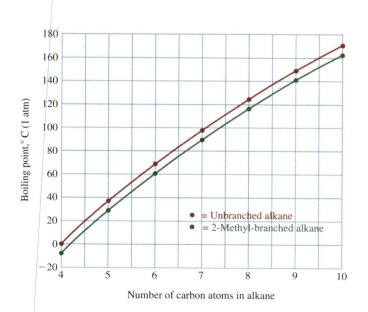

Limonene
(present in lemons and oranges)

Muscone
(responsible for odor of musk; used in perfumery)

Chrysanthemic acid
(obtained from chrysanthemum flowers)

2.17 PHYSICAL PROPERTIES OF ALKANES AND CYCLOALKANES

Appendix 1 lists selected physical properties for representative alkanes as well as members of other families of organic compounds.

Boiling Point. As we have seen earlier in this chapter, methane, ethane, propane, and butane are gases at room temperature. The unbranched alkanes pentane (C_5H_{12}) through heptadecane ($C_{17}H_{36}$) are liquids, whereas higher homologs are solids. As shown in Figure 2.12, the boiling points of unbranched alkanes increase with the number of carbon

FIGURE 2.12 Boiling points of unbranched alkanes and their 2-methyl-branched isomers. (Temperatures in this text are expressed in degrees Celsius, °C. The SI unit of temperature is the kelvin, K. To convert degrees Celsius to kelvins add 273.15.)

● = Unbranched alkane
● = 2-Methyl-branched alkane

Boiling point, °C (1 atm) — Number of carbon atoms in alkane atoms. Figure 2.12 also shows that the boiling points for 2-methyl-branched alkanes are lower than those of the unbranched isomer. By exploring at the molecular level the reasons for the increase in boiling point with the number of carbons and the difference in boiling point between branched and unbranched alkanes, we can begin to connect structure with properties.

A substance exists as a liquid rather than a gas because attractive forces between molecules (**intermolecular attractive forces**) are greater in the liquid than in the gas phase. Attractive forces between neutral species (atoms or molecules, but not ions) are referred to as **van der Waals forces** and may be of three types:

1. dipole–dipole
2. dipole/induced-dipole
3. induced-dipole/induced-dipole

Van der Waals forces involving induced dipoles are often called *London forces*, or *dispersion forces*.

These forces are electrical in nature, and in order to vaporize a substance, enough energy must be added to overcome them. Most alkanes have no measurable dipole moment, and therefore the only van der Waals force to be considered is the induced-dipole/induced-dipole attractive force.

It might seem that two nearby molecules A and B of the same nonpolar substance would be unaffected by each other.

The electric field of a molecule, however, is not static, but fluctuates rapidly. Although, on average, the centers of positive and negative charge of an alkane nearly coincide, at any instant they may not, and molecule A can be considered to have a temporary dipole moment.

The neighboring molecule B "feels" the dipolar electric field of A and undergoes a spontaneous adjustment in its electron positions, giving it a temporary dipole moment that is complementary to that of A.

The electric fields of both A and B fluctuate, but always in a way that results in a weak attraction between them.

Extended assemblies of induced-dipole/induced-dipole attractions can accumulate to give substantial intermolecular attractive forces. An alkane with a higher molecular weight has more atoms and electrons and, therefore, more opportunities for intermolecular attractions and a higher boiling point than one with a lower molecular weight.

As noted earlier in this section, branched alkanes have lower boiling points than their unbranched isomers. Isomers have, of course, the same number of atoms and electrons, but a molecule of a branched alkane has a smaller surface area than an unbranched one. The extended shape of an unbranched alkane permits more points of contact for intermolecular associations. Compare the boiling points of pentane and its isomers:

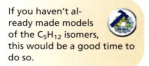

If you haven't already made models of the C$_5$H$_{12}$ isomers, this would be a good time to do so.

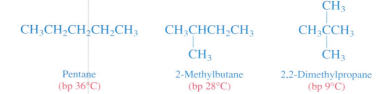

$$CH_3CH_2CH_2CH_2CH_3 \qquad CH_3CHCH_2CH_3 \qquad CH_3CCH_3$$

Pentane
(bp 36°C)

2-Methylbutane
(bp 28°C)

2,2-Dimethylpropane
(bp 9°C)

The shapes of these isomers are clearly evident in the space-filling models depicted in Figure 2.13. Pentane has the most extended structure and the largest surface area available for "sticking" to other molecules by way of induced-dipole/induced-dipole attractive forces; it has the highest boiling point. 2,2-Dimethylpropane has the most compact structure, engages in the fewest induced-dipole/induced-dipole attractions, and has the lowest boiling point.

Induced-dipole/induced-dipole attractions are very weak forces individually, but a typical organic substance can participate in so many of them that they are collectively the most important of all the contributors to intermolecular attraction in the liquid state. They are the only forces of attraction possible between nonpolar molecules such as alkanes.

> **PROBLEM 2.12** Match the boiling points with the appropriate alkanes.
> *Alkanes:* octane, 2-methylheptane, 2,2,3,3-tetramethylbutane, nonane
> *Boiling points* (°C, 1 atm): 106, 116, 126, 151

Melting Point. Solid alkanes are soft, generally low-melting materials. The forces responsible for holding the crystal together are the same induced-dipole/induced-dipole interactions that operate between molecules in the liquid, but the degree of organization is greater in the solid phase. By measuring the distances between the atoms of one molecule and its neighbor in the crystal, it is possible to specify a distance of closest approach characteristic of an atom called its **van der Waals radius.** In space-filling molecular models, such as those of pentane, 2-methylbutane, and 2,2-dimethylpropane shown in Figure 2.13, the radius of each sphere corresponds to the van der Waals radius of the atom it represents. The van der Waals radius for hydrogen is 120 pm. When two alkane molecules are brought together so that a hydrogen of one molecule is within 240 pm of a hydrogen of the other, the balance between electron–nucleus attractions versus electron–electron and nucleus–nucleus repulsions is most favorable. Closer approach is resisted by a strong increase in repulsive forces.

Solubility in Water. A familiar physical property of alkanes is contained in the adage "oil and water don't mix." Alkanes—indeed all hydrocarbons—are virtually insoluble in water. In order for a hydrocarbon to dissolve in water, the framework of hydrogen bonds between water molecules would become more ordered in the region around each molecule of the dissolved hydrocarbon. This increase in order, which corresponds to a decrease in entropy, signals a process that can be favorable only if it is reasonably

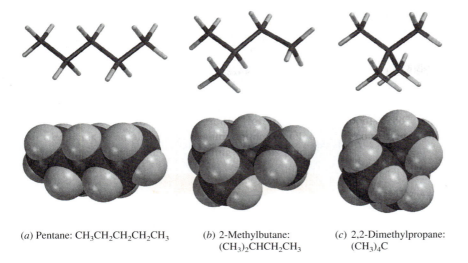

(a) Pentane: $CH_3CH_2CH_2CH_2CH_3$

(b) 2-Methylbutane: $(CH_3)_2CHCH_2CH_3$

(c) 2,2-Dimethylpropane: $(CH_3)_4C$

FIGURE 2.13 Tube (*top*) and space-filling (*bottom*) models of (*a*) pentane, (*b*) 2-methylbutane, and (*c*) 2,2-dimethylpropane. The most branched isomer, 2,2-dimethylpropane, has the most compact, most spherical three-dimensional shape.

exothermic. Such is not the case here. Being insoluble, and with densities in the 0.6–0.8 g/mL range, alkanes float on the surface of water. The exclusion of nonpolar molecules, such as alkanes, from water is called the **hydrophobic effect.** We will encounter it again at several points later in the text.

2.18 CHEMICAL PROPERTIES. COMBUSTION OF ALKANES

An older name for alkanes is **paraffin hydrocarbons.** *Paraffin* is derived from the Latin words *parum affinis* ("with little affinity") and testifies to the low level of reactivity of alkanes.

Table 1.7 shows that hydrocarbons are extremely weak acids. Among the classes of hydrocarbons, acetylene is a stronger acid than methane, ethane, ethylene, or benzene, but even its K_a is 10^{10} smaller than that of water.

> Alkanes are so unreactive that George A. Olah of the University of Southern California was awarded the 1994 Nobel Prize in chemistry in part for developing novel substances that do react with alkanes.

$HC{\equiv}CH$ H—⟨benzene⟩—H $H_2C{=}CH_2$ CH_4 CH_3CH_3

pK_a: 26   43   45   60   62

Although essentially inert in acid–base reactions, alkanes do participate in oxidation–reduction reactions as the compound that undergoes oxidation. Burning in air (**combustion**) is the best known and most important example. Combustion of hydrocarbons is exothermic and gives carbon dioxide and water as the products.

$$CH_4 \ + \ 2O_2 \ \longrightarrow \ CO_2 \ + \ 2H_2O \qquad \Delta H° = -890 \text{ kJ } (-212.8 \text{ kcal})$$

Methane   Oxygen   Carbon   Water
                                   dioxide

$$(CH_3)_2CHCH_2CH_3 + 8O_2 \longrightarrow 5CO_2 + 6H_2O \qquad \Delta H° = -3529 \text{ kJ } (-843.4 \text{ kcal})$$

2-Methylbutane   Oxygen   Carbon   Water
                                    dioxide

> **PROBLEM 2.13** Write a balanced chemical equation for the combustion of cyclohexane.

The heat released on combustion of a substance is called its **heat of combustion.** The heat of combustion is equal to $-\Delta H°$ for the reaction written in the direction shown. By convention

$$\Delta H° = H°_{products} - H°_{reactants}$$

where $H°$ is the heat content, or **enthalpy,** of a compound in its standard state, that is, the gas, pure liquid, or crystalline solid at a pressure of 1 atm. In an exothermic process the enthalpy of the products is less than that of the starting materials, and $\Delta H°$ is a negative number.

Table 2.3 lists the heats of combustion of several alkanes. Unbranched alkanes have slightly higher heats of combustion than their 2-methyl-branched isomers, but the most important factor is the number of carbons. The unbranched alkanes and the 2-methyl-branched alkanes constitute two separate *homologous series* (see Section 2.9) in which there is a regular increase of about 653 kJ/mol (156 kcal/mol) in the heat of combustion for each additional CH_2 group.

> **PROBLEM 2.14** Using the data in Table 2.3, estimate the heat of combustion of
> (a) 2-Methylnonane (in kcal/mol)   (b) Icosane (in kJ/mol)

TABLE 2.3   Heats of Combustion ($-\Delta H°$) of Representative Alkanes

Compound	Formula	$-\Delta H°$ kJ/mol	$-\Delta H°$ kcal/mol
Unbranched alkanes			
Hexane	$CH_3(CH_2)_4CH_3$	4,163	995.0
Heptane	$CH_3(CH_2)_5CH_3$	4,817	1151.3
Octane	$CH_3(CH_2)_6CH_3$	5,471	1307.5
Nonane	$CH_3(CH_2)_7CH_3$	6,125	1463.9
Decane	$CH_3(CH_2)_8CH_3$	6,778	1620.1
Undecane	$CH_3(CH_2)_9CH_3$	7,431	1776.1
Dodecane	$CH_3(CH_2)_{10}CH_3$	8,086	1932.7
Hexadecane	$CH_3(CH_2)_{14}CH_3$	10,701	2557.6
2-Methyl-branched alkanes			
2-Methylpentane	$(CH_3)_2CHCH_2CH_2CH_3$	4,157	993.6
2-Methylhexane	$(CH_3)_2CH(CH_2)_3CH_3$	4,812	1150.0
2-Methylheptane	$(CH_3)_2CH(CH_2)_4CH_3$	5,466	1306.3

Heats of combustion can be used to measure the relative stability of isomeric hydrocarbons. They tell us not only which isomer is more stable than another, but by how much. Consider a group of C_8H_{18} alkanes:

$$CH_3(CH_2)_6CH_3 \qquad (CH_3)_2CHCH_2CH_2CH_2CH_2CH_3$$

Octane   2-Methylheptane

$$(CH_3)_3CCH_2CH_2CH_2CH_3 \qquad (CH_3)_3CC(CH_3)_3$$

2,2-Dimethylhexane   2,2,3,3-Tetramethylbutane

Figure 2.14 compares the heats of combustion of these C_8H_{18} isomers on a *potential energy diagram*. **Potential energy** is comparable with enthalpy; it is the energy a molecule has exclusive of its kinetic energy. A molecule with more potential energy is less stable than an isomer with less potential energy. These C_8H_{18} isomers all undergo combustion to the same final state according to the equation:

$$C_8H_{18} + \tfrac{25}{2}O_2 \longrightarrow 8CO_2 + 9H_2O$$

therefore, the differences in their heats of combustion translate directly to differences in their potential energies. *When comparing isomers, the one with the lowest potential energy (in this case, the lowest heat of combustion) is the most stable*. Among the C_8H_{18}

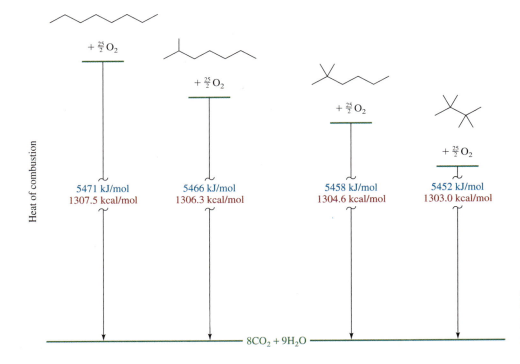

FIGURE 2.14 Energy diagram comparing heats of combustion of isomeric C_8H_{18} alkanes.

THERMOCHEMISTRY

Thermochemistry is the study of the heat changes that accompany chemical processes. It has a long history dating back to the work of the French chemist Antoine Laurent Lavoisier in the late eighteenth century. Thermochemistry provides quantitative information that complements the qualitative description of a chemical reaction and can help us understand why some reactions occur and others do not. It is of obvious importance when assessing the relative value of various materials as fuels, when comparing the stability of isomers, or when determining the practicality of a particular reaction. In the field of bioenergetics, thermochemical information is applied to the task of sorting out how living systems use chemical reactions to store and use the energy that originates in the sun.

By allowing compounds to react in a calorimeter, it is possible to measure the heat evolved in an exothermic reaction or the heat absorbed in an endothermic reaction. Thousands of reactions have been studied to produce a rich library of thermochemical data. These data take the form of **heats of reaction** and correspond to the value of the enthalpy change $\Delta H°$ for a particular reaction of a particular substance.

In this section you have seen how heats of combustion can be used to determine relative stabilities of isomeric alkanes. In later sections we shall expand our scope to include the experimentally determined heats of certain other reactions, such as *bond dissociation energies* (Section 4.16) and *heats of hydrogenation* (Section 6.2), to see how $\Delta H°$ values from various sources can aid our understanding of structure and reactivity.

Heat of formation ($\Delta H_f°$), the enthalpy change for formation of a compound directly from the elements, is one type of heat of reaction. In cases such as the formation of CO_2 or H_2O from the combustion of carbon or hydrogen, respectively, the heat of formation of a substance can be measured directly. In most other cases, heats of formation are not measured experimentally but are calculated from the measured heats of other reactions. Consider, for example, the heat of formation of methane. The reaction that defines the formation of methane from the elements,

$$C \text{ (graphite)} + 2H_2(g) \longrightarrow CH_4(g)$$

  Carbon   Hydrogen   Methane

can be expressed as the sum of three reactions:

(1) $C \text{ (graphite)} + O_2(g) \longrightarrow CO_2(g)$ $\Delta H° = -393 \text{ kJ}$
(2) $2H_2(g) + O_2(g) \longrightarrow 2H_2O(l)$ $\Delta H° = -572 \text{ kJ}$
(3) $CO_2(g) + 2H_2O(l) \longrightarrow CH_4(g) + 2O_2(g)$
                                              $\Delta H° = +890 \text{ kJ}$
_____
$C \text{ (graphite)} + 2H_2 \longrightarrow CH_4$ $\Delta H° = -75 \text{ kJ}$

Equations (1) and (2) are the heats of formation of carbon dioxide and water, respectively. Equation (3) is the reverse of the combustion of methane, and so the heat of reaction is equal to the heat of combustion but opposite in sign. The **molar heat of formation** of a substance is the enthalpy change for formation of one mole of the substance from the elements. For methane $\Delta H_f° = -75 \text{ kJ/mol}$.

The heats of formation of most organic compounds are derived from heats of reaction by arithmetic manipulations similar to that shown. Chemists find a table of $\Delta H_f°$ values to be convenient because it replaces many separate tables of $\Delta H°$ values for individual reaction types and permits $\Delta H°$ to be calculated for any reaction, real or imaginary, for which the heats of formation of reactants and products are available. It is more appropriate for our purposes, however, to connect thermochemical data to chemical processes as directly as possible, and therefore we will cite heats of particular reactions, such as heats of combustion and heats of hydrogenation, rather than heats of formation.

alkanes, the most highly branched isomer, 2,2,3,3-tetramethylbutane, is the most stable, and the unbranched isomer octane is the least stable. It is generally true for alkanes that a more branched isomer is more stable than a less branched one.

The small differences in stability between branched and unbranched alkanes result from an interplay between attractive and repulsive forces *within* a molecule (**intramolecular forces**). These forces are nucleus–nucleus repulsions, electron–electron repulsions, and nucleus–electron attractions, the same set of fundamental forces we met when talking about chemical bonding (Section 2.2) and van der Waals forces between molecules (Section 2.17). When the energy associated with these interactions is calculated for all of the nuclei and electrons within a molecule, it is found that the attractive forces increase more than the repulsive forces as the structure becomes more compact. Sometimes, though, two atoms in a molecule are held too closely together. We'll explore the consequences of that in Chapter 3.

PROBLEM 2.15 Without consulting Table 2.3, arrange the following compounds in order of decreasing heat of combustion: pentane, isopentane, neopentane, hexane.

2.19   OXIDATION–REDUCTION IN ORGANIC CHEMISTRY

As we have just seen, the reaction of alkanes with oxygen to give carbon dioxide and water is called *combustion*. A more fundamental classification of reaction types places it in the *oxidation–reduction* category. To understand why, let's review some principles of oxidation–reduction, beginning with the **oxidation number** (also known as **oxidation state**).

There are a variety of methods for calculating oxidation numbers. In compounds that contain a single carbon, such as methane (CH_4) and carbon dioxide (CO_2), the oxidation number of carbon can be calculated from the molecular formula. Both molecules are neutral, and so the algebraic sum of all the oxidation numbers must equal zero. Assuming, as is customary, that the oxidation state of hydrogen is $+1$, the oxidation state of carbon in CH_4 is calculated to be -4. Similarly, assuming an oxidation state of -2 for oxygen, carbon is $+4$ in CO_2. This kind of calculation provides an easy way to develop a list of one-carbon compounds in order of increasing oxidation state, as shown in Table 2.4.

The carbon in methane has the lowest oxidation number (-4) of any of the compounds in Table 2.4. Methane contains carbon in its most *reduced* form. Carbon dioxide and carbonic acid have the highest oxidation numbers ($+4$) for carbon, corresponding to its most *oxidized* state. When methane or any alkane undergoes combustion to form carbon dioxide, carbon is oxidized and oxygen is reduced.

A useful generalization from Table 2.4 is the following:

Oxidation of carbon corresponds to an increase in the number of bonds between carbon and oxygen or to a decrease in the number of carbon–hydrogen bonds. Conversely, *reduction corresponds to an increase in the number of carbon–hydrogen bonds or to a decrease in the number of carbon–oxygen bonds.* From Table 2.4 it can be seen that each successive increase in oxidation state increases the number of bonds between carbon and oxygen and decreases the number of carbon–hydrogen bonds. Methane has four C—H bonds and no C—O bonds; carbon dioxide has four C—O bonds and no C—H bonds.

Among the various classes of hydrocarbons, alkanes contain carbon in its most reduced state, and alkynes contain carbon in its most oxidized state.

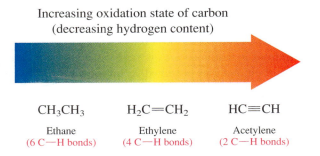

Increasing oxidation state of carbon
(decreasing hydrogen content)

CH_3CH_3	$H_2C{=}CH_2$	$HC{\equiv}CH$
Ethane	Ethylene	Acetylene
(6 C—H bonds)	(4 C—H bonds)	(2 C—H bonds)

TABLE 2.4	Oxidation Number of Carbon in One-Carbon Compounds		
Compound	Structural formula	Molecular formula	Oxidation number
Methane	CH_4	CH_4	-4
Methanol	CH_3OH	CH_4O	-2
Formaldehyde	$H_2C{=}O$	CH_2O	0
Formic acid	$\overset{\displaystyle O}{\overset{\displaystyle \|}{HCOH}}$	CH_2O_2	$+2$
Carbonic acid	$\overset{\displaystyle O}{\overset{\displaystyle \|}{HOCOH}}$	H_2CO_3	$+4$
Carbon dioxide	$O{=}C{=}O$	CO_2	$+4$

Many, indeed most, organic compounds contain carbon in more than one oxidation state. Consider ethanol (CH_3CH_2OH), for example. One carbon is connected to three hydrogens, the other to two hydrogens and one oxygen. Although it is an easy matter to associate more C—O bonds with higher oxidation state and more C—H bonds with lower oxidation state to correctly conclude that the oxidation state of carbon in CH_2OH is higher than in CH_3, how do we calculate the actual oxidation numbers? Table 2.5 describes a method.

> **PROBLEM 2.16** Can you calculate the oxidation number of carbon solely from the molecular formula of ethanol (C_2H_6O)? Explain. How does the oxidation number calculated from the molecular formula compare to the values obtained in Table 2.5?

Most of the time we are concerned only with whether a particular reaction is an oxidation or reduction rather than with determining the precise change in oxidation number. In general: *Oxidation of carbon occurs when a bond between carbon and an atom that is less electronegative than carbon is replaced by a bond to an atom that is more electronegative than carbon. The reverse process is reduction.*

$$-\overset{|}{\underset{|}{C}}-X \quad \underset{\xleftarrow{\text{reduction}}}{\xrightarrow{\text{oxidation}}} \quad -\overset{|}{\underset{|}{C}}-Y$$

X is less electronegative than carbon   Y is more electronegative than carbon

A similar approach is discussed in the January 1997 edition of the *Journal of Chemical Education*, pp. 69–72.

Notice that this generalization follows naturally from the method of calculating oxidation numbers outlined in Table 2.5. In a C—C bond one electron is assigned to one carbon, the second electron to the other. In a bond between carbon and some other element, none of the electrons in that bond are assigned to carbon when the element is more electronegative than carbon; both are assigned to carbon when the element is less electronegative than carbon.

TABLE 2.5	Oxidation Numbers in Compounds with More Than One Carbon

Step	Illustration of step for case of ethanol
1. Write the Lewis structure and include unshared electron pairs.	$\begin{array}{cc} H & H \\ | & | \\ H-C-C-\ddot{O}-H \\ | & | \\ H & H \end{array}$
2. Assign the electrons in a covalent bond between two atoms to the more electronegative partner.	Oxygen is the most electronegative atom in ethanol; hydrogen is the least electronegative. 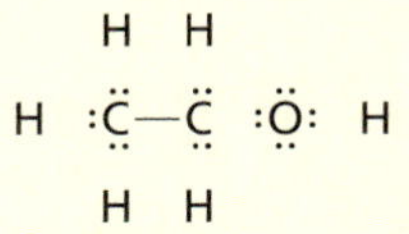 $\begin{array}{cc} H & H \\ H \; :\ddot{C}-\ddot{C} \; :\ddot{O}: \; H \\ H & H \end{array}$
3. For a bond between two atoms of the same element, assign the electrons in the bond equally.	Each carbon in the C—C bond is assigned one electron. 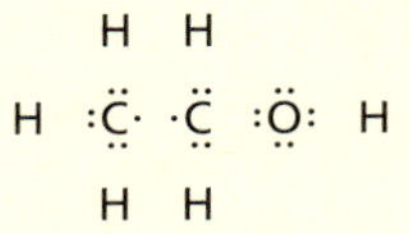 $\begin{array}{cc} H & H \\ H \; :\ddot{C}\cdot \; \cdot\ddot{C} \; :\ddot{O}: \; H \\ H & H \end{array}$
4. Count the number of electrons assigned to each atom and subtract that number from the number of electrons in the neutral atom; the result is the oxidation number.	A neutral carbon atom has four valence electrons. Five electrons are assigned to the CH_2OH carbon; therefore, it has an oxidation number of -1. Seven electrons are assigned to the CH_3 carbon; therefore, it has an oxidation number of -3. As expected, this method gives an oxidation number of -2 for oxygen and $+1$ for each hydrogen.

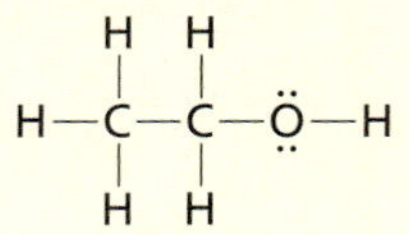

PROBLEM 2.17   Both of the following reactions will be encountered in Chapter 4. One is oxidation–reduction, the other is not. Which is which?

$(CH_3)_3COH + HCl \longrightarrow (CH_3)_3CCl + H_2O$

$(CH_3)_3CH \;\; + Br_2 \longrightarrow (CH_3)_3CBr + HBr$

The ability to recognize when oxidation or reduction occurs is of value when deciding on the kind of reactant with which an organic molecule must be treated to convert it into some desired product. Many of the reactions to be discussed in subsequent chapters involve oxidation–reduction.

2.20   *sp²* HYBRIDIZATION AND BONDING IN ETHYLENE

We conclude this introduction to hydrocarbons by describing the orbital hybridization model of bonding in ethylene and acetylene, parents of the alkene and alkyne families, respectively.

Ethylene is planar with bond angles close to 120° (Figure 2.15); therefore, some hybridization state other than sp^3 is required. The hybridization scheme is determined by the number of atoms to which carbon is directly attached. In sp^3 hybridization, four atoms are attached to carbon by σ bonds, and so four equivalent sp^3 hybrid orbitals are required. In ethylene, three atoms are attached to each carbon, so three equivalent hybrid orbitals

FIGURE 2.15 (*a*) All
the atoms of ethylene lie in
the same plane, the bond
angles are close to 120°, and
the carbon–carbon bond
distance is significantly
shorter than that of ethane.
(*b*) A space-filling model of
ethylene.

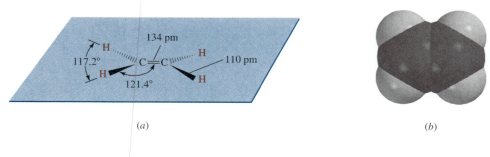

(*a*)

(*b*)

are needed. As shown in Figure 2.16, these three orbitals are generated by mixing the
carbon 2*s* orbital with *two* of the 2*p* orbitals and are called *sp²* **hybrid orbitals.** One of
the 2*p* orbitals is left unhybridized. The three *sp²* orbitals are of equal energy; each has
one-third *s* character and two-thirds *p* character. Their axes are coplanar, and each has a
shape much like that of an *sp³* orbital. The three *sp²* orbitals and the unhybridized *p*
orbital each contain one electron.

Each carbon of ethylene uses two of its *sp²* hybrid orbitals to form σ bonds to two
hydrogen atoms, as illustrated in the first part of Figure 2.17. The remaining *sp²* orbitals,
one on each carbon, overlap along the internuclear axis to give a σ bond connecting the
two carbons.

Each carbon atom still has, at this point, an unhybridized 2*p* orbital available for
bonding. These two half-filled 2*p* orbitals have their axes perpendicular to the framework of σ bonds of the molecule and overlap in a side-by-side manner to give what is

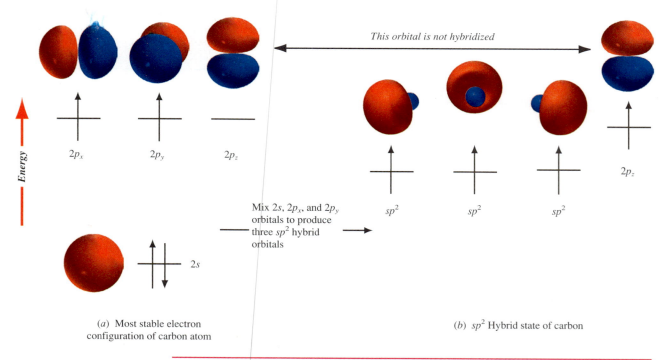

(*a*) Most stable electron
configuration of carbon atom

(*b*) *sp²* Hybrid state of carbon

FIGURE 2.16 *sp²* Hybridization. (*a*) Electron configuration of carbon in its most stable state.
(*b*) Mixing the *s* orbital with two of the three *p* orbitals generates three *sp²* hybrid orbitals and
leaves one of the 2*p* orbitals untouched. The axes of the three *sp²* orbitals lie in the same plane
and make angles of 120° with one another.

Begin with two *sp²* hybridized carbon atoms and four hydrogen atoms:

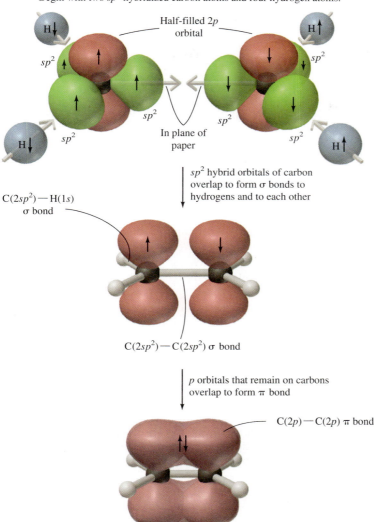

FIGURE 2.17 The carbon–carbon double bond in ethylene has a σ component and a π component. The σ component arises from overlap of *sp²*-hybridized orbitals along the internuclear axis. The π component results from a side-by-side overlap of 2*p* orbitals.

called a **pi (π) bond.** According to this analysis, the carbon–carbon double bond of ethylene is viewed as a combination of a σ bond plus a π bond. The additional increment of bonding makes a carbon–carbon double bond both stronger and shorter than a carbon–carbon single bond.

Electrons in a π bond are called **π electrons.** The probability of finding a π electron is highest in the region above and below the plane of the molecule. The plane of the molecule corresponds to a nodal plane, where the probability of finding a π electron is zero.

In general, you can expect that carbon will be sp²-hybridized when it is directly bonded to three atoms in a neutral molecule.

One measure of the strength of a bond is its *bond dissociation energy*. This topic will be introduced in Section 4.16 and applied to ethylene in Section 5.2.

PROBLEM 2.18   Identify the orbital overlaps involved in the indicated bond in the compound shown (propene). Is this a π bond or a σ bond?

$$H_2C=CH-CH_3$$
↑

2.21   *sp* HYBRIDIZATION AND BONDING IN ACETYLENE

One more hybridization scheme is important in organic chemistry. It is called *sp* **hybridization** and applies when carbon is directly bonded to two atoms, as in acetylene. The structure of acetylene is shown in Figure 2.18 along with its bond distances and bond angles. Its most prominent feature is its linear geometry.

Because each carbon in acetylene is bonded to two other atoms, the orbital hybridization model requires each carbon to have two equivalent orbitals available for σ bonds as outlined in Figure 2.19. According to this model the carbon 2*s* orbital and one of its 2*p* orbitals combine to generate two *sp* hybrid orbitals, each of which has 50% *s* character and 50% *p* character. These two *sp* orbitals share a common axis, but their major lobes are oriented at an angle of 180° to each other. Two of the original 2*p* orbitals remain unhybridized.

As portrayed in Figure 2.20, the two carbons of acetylene are connected to each other by a 2*sp*–2*sp* σ bond, and each is attached to a hydrogen substituent by a 2*sp*–1*s* σ bond. The unhybridized 2*p* orbitals on one carbon overlap with their counterparts on the other to form two π bonds. The carbon–carbon triple bond in acetylene is viewed as a multiple bond of the $\sigma + \pi + \pi$ type.

In general, you can expect that carbon will be sp-hybridized when it is directly bonded to two atoms in a neutral molecule.

PROBLEM 2.19   The hydrocarbon shown, called *vinylacetylene*, is used in the synthesis of neoprene, a synthetic rubber. Identify the orbital overlaps involved in the indicated bond. How many σ bonds are there in vinylacetylene? How many π bonds?

$$H_2C=CH-C\equiv CH$$
↑

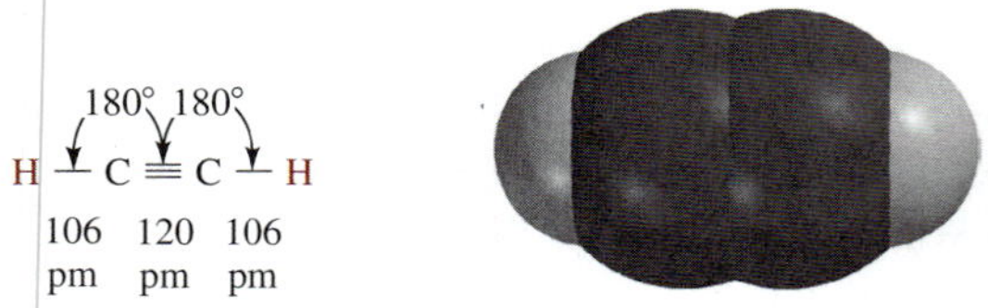

$$\overset{180°\ \ 180°}{H-C\equiv C-H}$$
106   120   106
pm   pm   pm

(*a*)

(*b*)

FIGURE 2.18 Acetylene is a linear molecule as indicated in (*a*) the structural formula and (*b*) a space-filling model.

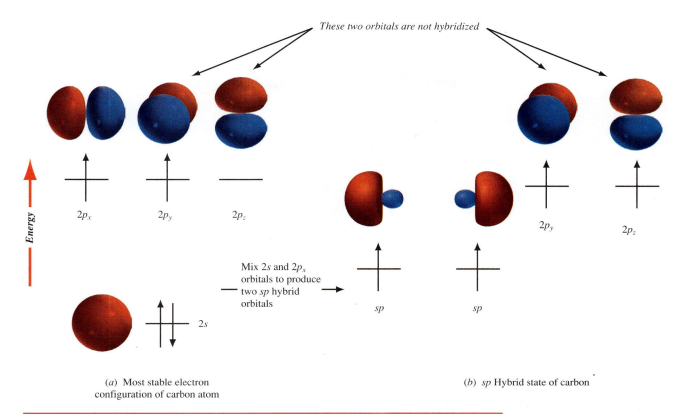

These two orbitals are not hybridized

Energy

$2p_x$ $2p_y$ $2p_z$

Mix 2s and $2p_x$ orbitals to produce two *sp* hybrid orbitals

sp sp

$2p_y$ $2p_z$

2s

(*a*) Most stable electron configuration of carbon atom

(*b*) *sp* Hybrid state of carbon

FIGURE 2.19 *sp* Hybridization. (*a*) Electron configuration of carbon in its most stable state. (*b*) Mixing the s orbital with one of the three *p* orbitals generates two *sp* hybrid orbitals and leaves two of the 2p orbitals untouched. The axes of the two *sp* orbitals make an angle of 180° with each other.

2.22   WHICH THEORY OF CHEMICAL BONDING IS BEST?

We have introduced three approaches to chemical bonding in this chapter:

1. The Lewis model
2. The orbital hybridization model (which is a type of valence bond model)
3. The molecular orbital model

Which one should you learn?

Generally speaking, the three models offer complementary information. Organic chemists use all three, emphasizing whichever one best suits a particular feature of structure or reactivity. Until recently, the Lewis and orbital hybridization models were used far more than the molecular orbital model. But that is changing.

The Lewis rules are relatively straightforward, easiest to master, and the most familiar. You will find that your ability to write Lewis formulas increases rapidly with experience. *Get as much practice as you can early in the course. Success in organic chemistry depends on writing correct Lewis structures.*

Orbital hybridization descriptions, because they too are based on the shared electron-pair bond, enhance the information content of Lewis formulas by distinguishing

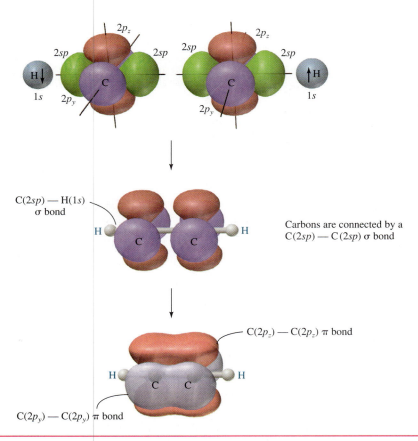

FIGURE 2.20 Bonding in acetylene based on *sp* hybridization of carbon. The carbon–carbon triple bond is viewed as consisting of one σ bond and two π bonds.

among various types of atoms, electrons, and bonds. As you become more familiar with a variety of structural types, you will find that the term *sp³-hybridized carbon* triggers associations in your mind that are different from those of some other term, such as *sp²-hybridized carbon,* for example.

Molecular orbital theory can provide insights into structure and reactivity that the Lewis and orbital hybridization models can't. It is the least intuitive of the three methods, however, and requires the most training, background, and chemical knowledge to apply. We have *discussed* molecular orbital theory so far only in the context of the bonding in H_2. We have *used* the results of molecular orbital theory, however, several times without acknowledging it until now. The electrostatic potential maps that open each chapter were obtained by molecular orbital calculations. Four molecular orbital calculations provided the drawings that we used in Figure 2.4 to illustrate how electron density builds up between the atoms in the valence bond (!) treatment of H_2. Molecular orbital theory is well suited to quantitative applications and is becoming increasingly available for routine use via software such as Spartan that runs on personal computers. You will see the results of molecular orbital theory often in this text, but the theory itself will be developed only at an introductory level.

2.23 SUMMARY

Section 2.1
The classes of hydrocarbons are **alkanes, alkenes, alkynes,** and **arenes.** Alkanes are hydrocarbons in which all of the bonds are *single* bonds and are characterized by the molecular formula C_nH_{2n+2}.

Section 2.2
Two theories of bonding, **valence bond** and **molecular orbital theory,** are based on the wave nature of an electron. Constructive interference between the electron wave of one atom and that of another gives a region between the two atoms in which the probability of sharing an electron is high—a bond.

Section 2.3
In valence bond theory a covalent bond is described in terms of in-phase overlap of a half-filled orbital of one atom with a half-filled orbital of another. When applied to bonding in H_2, the orbitals involved are the $1s$ orbitals of two hydrogen atoms and the bond is a σ bond.

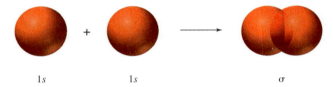

$1s$ $1s$ σ

Section 2.4
In molecular orbital theory, the molecular orbitals (MOs) are approximated by combining the atomic orbitals (AOs) of all of the atoms in a molecule. The number of MOs must equal the number of AOs that are combined.

Section 2.5
The first three alkanes are methane (CH_4), ethane (CH_3CH_3), and propane ($CH_3CH_2CH_3$).

Section 2.6
Bonding in methane is most often described by an **orbital hybridization** model, which is a modified form of valence bond theory. Four equivalent sp^3 hybrid orbitals of carbon are generated by mixing the $2s$, $2p_x$, $2p_y$, and $2p_z$ orbitals. Overlap of each half-filled sp^3 hybrid orbital with a half-filled hydrogen $1s$ orbital gives a σ bond.

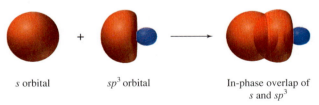

s orbital sp^3 orbital   In-phase overlap of *s* and sp^3

Section 2.7
The carbon–carbon bond in ethane is a σ bond in which an sp^3 hybrid orbital one carbon overlaps with an sp^3 hybrid orbital of the other.

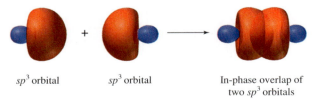

sp^3 orbital sp^3 orbital   In-phase overlap of two sp^3 orbitals

Section 2.8	Two constitutionally isomeric alkanes have the molecular formula C_4H_{10}. One has an unbranched chain ($CH_3CH_2CH_2CH_3$) and is called **n-butane;** the other has a branched chain [$(CH_3)_3CH$] and is called **isobutane.** Both *n*-butane and isobutane are **common names.**
Section 2.9	Unbranched alkanes of the type $CH_3(CH_2)_xCH_3$ are often referred to as *n*-alkanes, and are said to belong to a **homologous series.**
Section 2.10	There are three constitutional isomers of C_5H_{12}: **n-pentane** ($CH_3CH_2CH_2CH_2CH_3$), **isopentane** [$(CH_3)_2CHCH_2CH_3$], and **neopentane** [$(CH_3)_4C$].
Sections 2.11–2.15	A single alkane may have several different names; a name may be a common name, or it may be a *systematic name* developed by a well-defined set of rules. The most widely used system is **IUPAC nomenclature.** Table 2.6 summarizes the rules for alkanes and cycloalkanes. Table 2.7 gives the rules for naming alkyl groups.

TABLE 2.6   Summary of IUPAC Nomenclature of Alkanes and Cycloalkanes

Rule	Example
A. Alkanes	
1. Find the longest continuous chain of carbon atoms, and assign a basis name to the compound corresponding to the IUPAC name of the unbranched alkane having the same number of carbons.	The longest continuous chain in the alkane shown is six carbons. 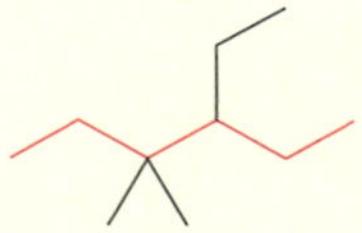 This alkane is named as a derivative of *hexane*.
2. List the substituents attached to the longest continuous chain in alphabetical order. Use the prefixes *di-, tri-, tetra-*, and so on, when the same substituent appears more than once. Ignore these prefixes when alphabetizing.	The alkane bears two methyl groups and an ethyl group. It is an *ethyldimethylhexane*. 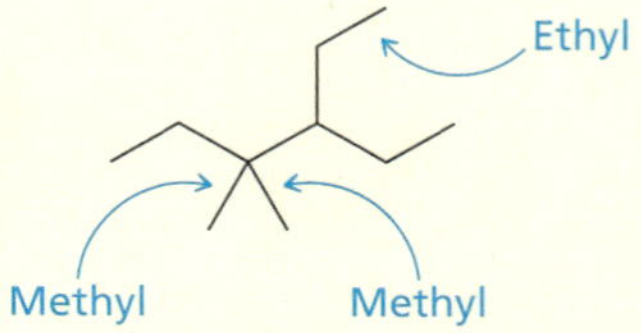
3. Number the chain in the direction that gives the lower locant to a substituent at the first point of difference.	When numbering from left to right, the substituents appear at carbons 3, 3, and 4. When numbering from right to left the locants are 3, 4, and 4; therefore, number from left to right. 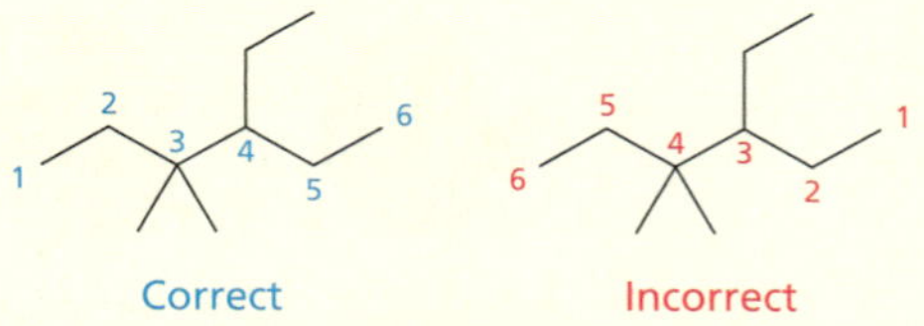 Correct Incorrect The correct name is *4-ethyl-3,3-dimethylhexane*.

(Continued)

Rule	Example
4. When two different numbering schemes give equivalent sets of locants, choose the direction that gives the lower locant to the group that appears first in the name.	In the following example, the substituents are located at carbons 3 and 4 regardless of the direction in which the chain is numbered. 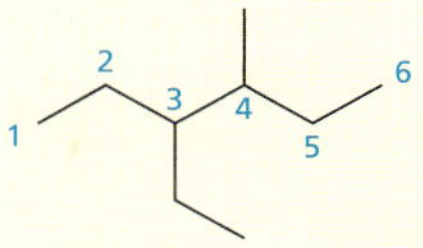

Correct   Incorrect

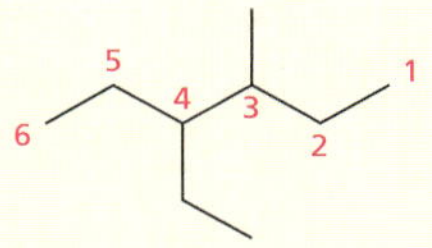

Ethyl precedes methyl in the name; therefore *3-ethyl-4-methylhexane* is correct.

Rule	Example
5. When two chains are of equal length, choose the one with the greater number of substituents as the parent. (Although this requires naming more substituents, the substituents have simpler names.)	Two different chains contain five carbons in the alkane: 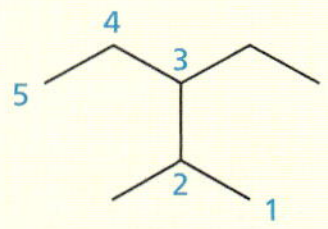

Correct   Incorrect

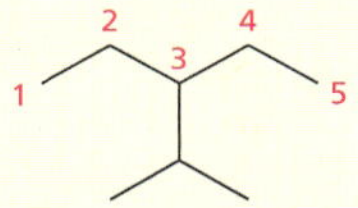

The correct name is *3-ethyl-2-methylpentane* (disubstituted chain), rather than 3-isopropylpentane (monosubstituted chain).

B. Cycloalkanes

Rule	Example
1. Count the number of carbons in the ring, and assign a basis name to the cycloalkane corresponding to the IUPAC name of the unbranched cycloalkane having the same number of carbons.	The compound shown contains five carbons in its ring. 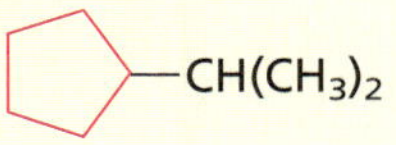

It is named as a derivative of *cyclopentane.*

Rule	Example
2. Name the alkyl group, and append it as a prefix to the cycloalkane. No locant is needed if the compound is a monosubstituted cycloalkane. It is understood that the alkyl group is attached to C-1.	The previous compound is *isopropylcyclopentane.* Alternatively, the alkyl group can be named according to the rules summarized in Table 2.7, whereupon the name becomes (*1-methylethyl*)*cyclopentane.* Parentheses are used to set off the name of the alkyl group as needed to avoid ambiguity.
3. When two or more different substituents are present, list them in alphabetical order, and number the ring in the direction that gives the lower number at the first point of difference.	The compound shown is *1,1-diethyl-4-hexylcyclooctane.* 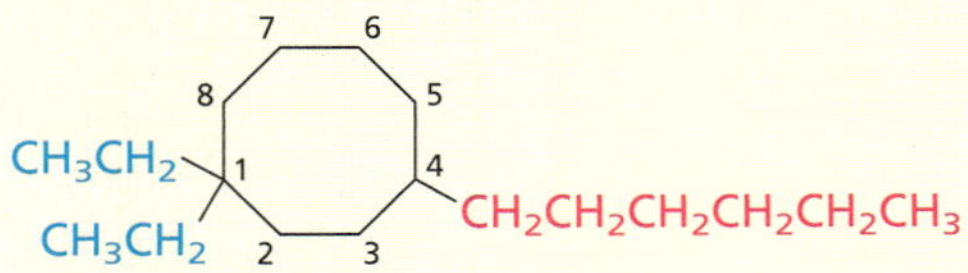

Rule	Example
4. Name the compound as a cycloalkyl-substituted alkane if the substituent has more carbons than the ring.	

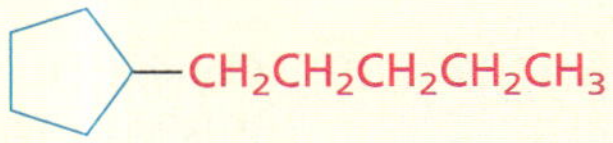

$CH_2CH_2CH_2CH_2CH_3$ is *pentylcyclopentane*

but

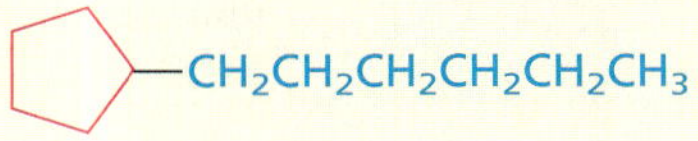

$CH_2CH_2CH_2CH_2CH_2CH_3$ is *1-cyclopentylhexane*

TABLE 2.7   Summary of IUPAC Nomenclature of Alkyl Groups

Rule	Example
1. Number the carbon atoms beginning at the point of attachment, proceeding in the direction that follows the longest continuous chain.	The longest continuous chain that begins at the point of attachment in the group shown contains six carbons.

$$CH_3CH_2CH_2\overset{1}{C}\overset{2}{C}H_2\overset{3}{C}H\overset{4}{C}H_2\overset{5}{C}H_2\overset{6}{C}H_3$$
$$\underset{CH_3}{|}\quad\underset{CH_3}{|}$$

Rule	Example
2. Assign a basis name according to the number of carbons in the corresponding unbranched alkane. Drop the ending *-ane* and replace it by *-yl*.	The alkyl group shown in step 1 is named as a substituted *hexyl* group.
3. List the substituents attached to the basis group in alphabetical order using replicating prefixes when necessary.	The alkyl group in step 1 is a *dimethylpropylhexyl* group.
4. Locate the substituents according to the numbering of the main chain described in step 1.	The alkyl group is a *1,3-dimethyl-1-propylhexyl* group.

Section 2.16   Natural gas is an abundant source of methane, ethane, and propane. Petroleum is a liquid mixture of many hydrocarbons, including alkanes. Alkanes also occur naturally in the waxy coating of leaves and fruits.

Section 2.17   Alkanes and cycloalkanes are nonpolar and insoluble in water. The forces of attraction between alkane molecules are **induced-dipole/induced-dipole** attractive forces. The boiling points of alkanes increase as the number of carbon atoms increases. Branched alkanes have lower boiling points than their unbranched isomers. There is a limit to how closely two molecules can approach each other, which is given by the sum of their **van der Waals radii.**

Section 2.18   Alkanes and cycloalkanes burn in air to give carbon dioxide, water, and heat. This process is called **combustion.**

$$(CH_3)_2CHCH_2CH_3 \;+\; 8O_2 \longrightarrow 5CO_2 \;+\; 6H_2O$$

| 2-Methylbutane | Oxygen | Carbon dioxide | Water |

$$\Delta H° = -3529 \text{ kJ} \,(-843.4 \text{ kcal})$$

The heat evolved on burning an alkane increases with the number of carbon atoms. The relative stability of isomers may be determined by comparing their respective **heats of combustion.** The more stable of two isomers has the lower heat of combustion.

Section 2.19   Combustion of alkanes is an example of **oxidation–reduction.** Although it is possible to calculate oxidation numbers of carbon in organic molecules, it is more convenient to regard oxidation of an organic substance as an increase in its oxygen content or a decrease in its hydrogen content.

Section 2.20   Carbon is **sp²-hybridized** in ethylene, and the double bond has a σ component and a π component. The sp^2 hybridization state is derived by mixing the 2s and two of the three 2p orbitals. Three equivalent sp^2 orbitals result, and their axes are coplanar. Overlap of an sp^2 orbital of one carbon with an sp^2 orbital of another produces a σ bond between them. Each carbon still has one unhybridized p orbital available for bonding, and "side-by-side" overlap of the p orbitals of adjacent carbons gives a π bond between them.

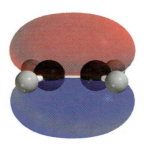

The π bond in ethylene generated by
overlap of p orbitals of adjacent carbons

Section 2.21   Carbon is **sp-hybridized** in acetylene, and the triple bond is of the σ + π + π type. The 2s orbital and one of the 2p orbitals combine to give two equivalent sp orbitals that have their axes in a straight line. A σ bond between the two carbons is supplemented by two π bonds formed by overlap of the remaining half-filled p orbitals.

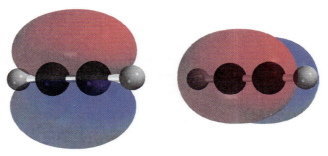

The triple bond of acetylene has a σ bond component and two π bonds;
the two π bonds are shown here and are perpendicular to each other.

Section 2.22   Lewis structures, orbital hybridization, and molecular orbital descriptions of bonding are all used in organic chemistry. Lewis structures are used the most, MO descriptions the least. All will be used in this text.

PROBLEMS

2.20 The general molecular formula for alkanes is C_nH_{2n+2}. What is the general molecular formula for:

(a) Cycloalkanes   (c) Alkynes

(b) Alkenes   (d) Cyclic hydrocarbons that contain one
                                        double bond

2.21   A certain hydrocarbon has a molecular formula of C_5H_8. Which of the following is **not** a structural possibility for this hydrocarbon?

(a) It is a cycloalkane.

(b) It contains one ring and one double bond.

(c) It contains two double bonds and no rings.

(d) It is an alkyne.

2.22   Which of the hydrocarbons in each of the following groups are isomers?

(a) $\diagup\diagdown\diagup\diagdown$ $\diamond$ $(CH_3)_3CH$

(b) $\underset{\underset{CH_3}{|}}{\overset{\overset{CH_3}{|}}{CH_3CCH_3}}$

(c)

2.23   Write structural formulas and give the IUPAC names for the nine alkanes that have the molecular formula C_7H_{16}.

2.24   From among the 18 constitutional isomers of C_8H_{18}, write structural formulas, and give the IUPAC names for those that are named as derivatives of

(a) Heptane   (c) Pentane

(b) Hexane   (d) Butane

2.25   *Pristane* is an alkane that is present to the extent of about 14% in shark liver oil. Its IUPAC name is 2,6,10,14-tetramethylpentadecane. Write its structural formula.

2.26   All the parts of this problem refer to the alkane having the carbon skeleton shown.

(a) What is the molecular formula of this alkane?

(b) What is its IUPAC name?

(c) How many methyl groups are present in this alkane? Methylene groups? Methine groups?

(d) How many carbon atoms are primary? Secondary? Tertiary? Quaternary?

2.27   Give the IUPAC name for each of the following compounds:

(a) $CH_3(CH_2)_{25}CH_3$ (c) $(CH_3CH_2)_3CCH(CH_2CH_3)_2$

(b) $(CH_3)_2CHCH_2(CH_2)_{14}CH_3$ (d)

(e) 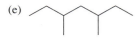   (f)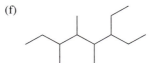

2.28 Write a structural formula for each of the following compounds:

(a) 6-Isopropyl-2,3-dimethylnonane

(b) 4-*tert*-Butyl-3-methylheptane

(c) 4-Isobutyl-1,1-dimethylcyclohexane

(d) *sec*-Butylcycloheptane

(e) Cyclobutylcyclopentane

2.29 Give the IUPAC name for each of the following alkyl groups, and classify each one as primary, secondary, or tertiary:

(a) $CH_3(CH_2)_{10}CH_2$—

(b) —$CH_2CH_2CHCH_2CH_2CH_3$
                     |
                  CH_2CH_3

(c) —$C(CH_2CH_3)_3$

(d) —$CHCH_2CH_2CH_3$
    (cyclopropyl substituent)

(e) (cyclohexyl)—CH_2CH_2—

(f) (cyclohexyl)—CH—
                |
             CH_3

2.30 Write the structural formula of a compound of molecular formula $C_4H_8Cl_2$ in which

(a) All the carbons belong to methylene groups

(b) None of the carbons belong to methylene groups

2.31 Female tiger moths signify their presence to male moths by giving off a sex attractant (pheromone). The sex attractant has been isolated and found to be a 2-methyl-branched alkane having a molecular weight of 254. What is this material?

2.32 Write a balanced chemical equation for the combustion of each of the following compounds:

(a) Decane

(b) Cyclodecane

(c) Methylcyclononane

(d) Cyclopentylcyclopentane

2.33 The heats of combustion of methane and butane are 890 kJ/mol (212.8 kcal/mol) and 2876 kJ/mol (687.4 kcal/mol), respectively. When used as a fuel, would methane or butane generate more heat for the same mass of gas? Which would generate more heat for the same volume of gas?

2.34 In each of the following groups of compounds, identify the one with the largest heat of combustion and the one with the smallest. (Try to do this problem without consulting Table 2.3.)

(a) Hexane, heptane, octane

(b) Isobutane, pentane, isopentane

(c) Isopentane, 2-methylpentane, neopentane

(d) Pentane, 3-methylpentane, 3,3-dimethylpentane

(e) Ethylcyclopentane, ethylcyclohexane, ethylcycloheptane

2.35 (a) Given $\Delta H°$ for the reaction

$$H_2(g) + \tfrac{1}{2}O_2(g) \longrightarrow H_2O(l) \qquad \Delta H° = -286 \text{ kJ}$$

along with the information that the heat of combustion of ethane is 1560 kJ/mol and that of ethylene is 1410 kJ/mol, calculate $\Delta H°$ for the hydrogenation of ethylene:

$$H_2C{=}CH_2(g) + H_2(g) \longrightarrow CH_3CH_3(g)$$

(b) If the heat of combustion of acetylene is 1300 kJ/mol, what is the value of $\Delta H°$ for its hydrogenation to ethylene? To ethane?

(c) What is the value of $\Delta H°$ for the hypothetical reaction

$$2H_2C{=}CH_2(g) \longrightarrow CH_3CH_3(g) + HC{\equiv}CH(g)$$

2.36 We have seen in this chapter that, among isomeric alkanes, the unbranched isomer is the least stable and has the highest boiling point; the most branched isomer is the most stable and has the lowest boiling point. Does this mean that one alkane boils lower than another *because* it is more stable? Explain.

2.37 Higher octane gasoline typically contains a greater proportion of branched alkanes relative to unbranched ones. Are branched alkanes better fuels because they give off more energy on combustion? Explain.

2.38 The reaction shown is important in the industrial preparation of dichlorodimethylsilane for eventual conversion to silicone polymers.

$$2CH_3Cl + Si \longrightarrow (CH_3)_2SiCl_2$$

(a) Is carbon oxidized, reduced, or neither in this reaction?

(b) On the basis of the molecular model of $(CH_3)_2SiCl_2$, deduce the hybridization state of silicon in this compound. What is the principal quantum number n of the silicon s and p orbitals that are hybridized? (You can view this model in more detail on *Learning By Modeling*.)

2.39 Alkanes spontaneously burst into flame in the presence of elemental fluorine. The reaction that takes place between pentane and F_2 gives CF_4 and HF as the only products.

(a) Write a balanced equation for this reaction.

(b) Is carbon oxidized, reduced, or does it undergo no change in oxidation state in this reaction?

2.40 Which atoms in the following reaction undergo changes in their oxidation state? Which atom is oxidized? Which one is reduced?

$$2CH_3CH_2OH + 2Na \longrightarrow 2CH_3CH_2ONa + H_2$$

2.41 Compound A undergoes the following reactions:

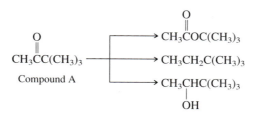

(a) Which of the reactions shown require(s) an oxidizing agent?

(b) Which of the reactions shown require(s) a reducing agent?

2.42 Each of the following reactions will be encountered at some point in this text. Classify each one according to whether the organic substrate is oxidized or reduced in the process.

(a) $CH_3C{\equiv}CH + 2Na + 2NH_3 \longrightarrow CH_3CH{=}CH_2 + 2NaNH_2$

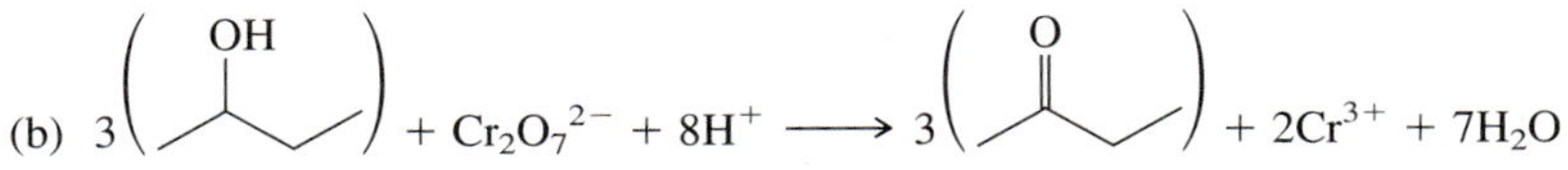

(b)

(c) $HOCH_2CH_2OH + HIO_4 \longrightarrow 2CH_2{=}O + HIO_3 + H_2O$

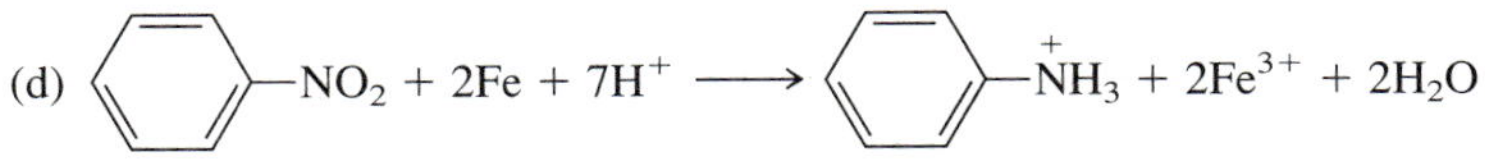

(d)

2.43 Of the overlaps between an *s* and a *p* orbital as shown in the illustration, one is bonding, one is antibonding, and the third is nonbonding (neither bonding nor antibonding). Which orbital overlap corresponds to which interaction? Why?

2.44 Does the overlap of two *p* orbitals in the fashion shown correspond to a σ bond or to a π bond? Explain.

2.45 The compound *cyanoacetylene* ($HC{\equiv}C{-}C{\equiv}N$) has been detected in interstellar space. Make a molecular model or sketch the approximate geometry expected for this compound. What is the hybridization of nitrogen and each carbon?

2.46 Molecular models such as the one shown often do not explicitly show double and triple bonds. Write a Lewis structure for this hydrocarbon showing the location of any multiple bonds. Specify the hybridization state of each carbon. (You can view this model in more detail on *Learning By Modeling*.)

2.47 The atoms in *methylketene* (C_3H_4O) are connected in the order and according to the geometry shown. (You can view this model in more detail on *Learning By Modeling*.) Determine the hybridization state of each carbon and write a Lewis structure for this neutral molecule.

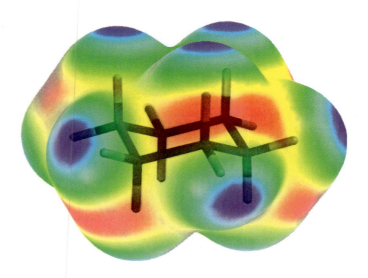

CHAPTER 3

CONFORMATIONS OF ALKANES AND CYCLOALKANES

Hydrogen peroxide is formed in the cells of plants and animals but is toxic to them. Consequently, living systems have developed mechanisms to rid themselves of hydrogen peroxide, usually by enzyme-catalyzed reduction to water. An understanding of how reactions take place, be they reactions in living systems or reactions in test tubes, begins with a thorough knowledge of the structure of the reactants, products, and catalysts. Even a simple molecule such as hydrogen peroxide may be structurally more complicated than you think. Suppose we wanted to write the structural formula for H_2O_2 in enough detail to show the positions of the atoms relative to one another. We could write two different planar geometries A and B that differ by a 180° rotation about the O—O bond. We could also write an infinite number of nonplanar structures, of which C is but one example, that differ from one another by tiny increments of rotation about the O—O bond.

Structures A, B, and C represent different **conformations** of hydrogen peroxide. *Conformations are different spatial arrangements of a molecule that are generated by rotation about single bonds.* Although we can't tell from simply looking at these structures, we now know from experimental studies that C is the most stable conformation.

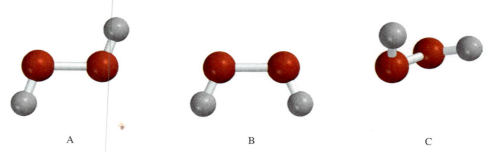

A   B   C

In this chapter we'll examine the conformations of various alkanes and cyclo-alkanes, focusing most of our attention on three of them: *ethane, butane,* and *cyclohexane.* A detailed study of even these three will take us a long way toward understanding the main ideas of *conformational analysis.*

Conformational analysis is the study of how conformational factors affect the structure of a molecule and its physical, chemical, and biological properties.

3.1 CONFORMATIONAL ANALYSIS OF ETHANE

Ethane is the simplest hydrocarbon that can have distinct conformations. Two, the **staggered conformation** and the **eclipsed conformation,** deserve special mention and are illustrated with molecular models in Figure 3.1.

In the staggered conformation, each C—H bond of one carbon bisects an H—C—H angle of the other carbon.

In the eclipsed conformation, each C—H bond of one carbon is aligned with a C—H bond of the other carbon.

The staggered and eclipsed conformations interconvert by rotation around the C—C bond, and do so very rapidly. We'll see just how rapidly later in this section.

Among the various ways in which the staggered and eclipsed forms are portrayed, wedge-and-dash, sawhorse, and Newman projection drawings are especially useful. These are shown for the staggered conformation of ethane in Figure 3.2 and for the eclipsed conformation in Figure 3.3.

We used *wedge-and-dash* drawings in earlier chapters, and so Figures 3.2*a* and 3.3*a* are familiar to us. A *sawhorse* drawing (Figures 3.2*b* and 3.3*b*) shows the conformation of a molecule without having to resort to different styles of bonds. In a *Newman projection* (Figures 3.2*c* and 3.3*c*), we sight down the C—C bond, and represent the front carbon by a point and the back carbon by a circle. Each carbon has three other bonds that are placed symmetrically around it.

The structural feature that Figures 3.2 and 3.3 illustrate is the spatial relationship between atoms on adjacent carbons. Each H—C—C—H unit in ethane is characterized by a *torsion angle* or *dihedral angle,* which is the angle between the H—C—C plane

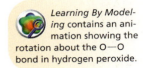
Learning By Modeling contains an animation showing the rotation about the O—O bond in hydrogen peroxide.

Different conformations of the same compound are sometimes called *conformers* or *rotamers.*

Newman projections were devised by Professor Melvin S. Newman of Ohio State University.

Staggered conformation of ethane

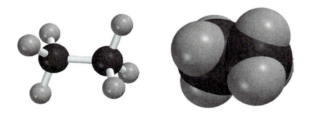

Eclipsed conformation of ethane

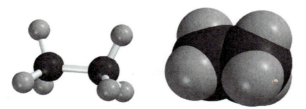

FIGURE 3.1 The staggered and eclipsed conformations of ethane shown as ball-and-spoke models (*left*) and as space-filling models (*right*).

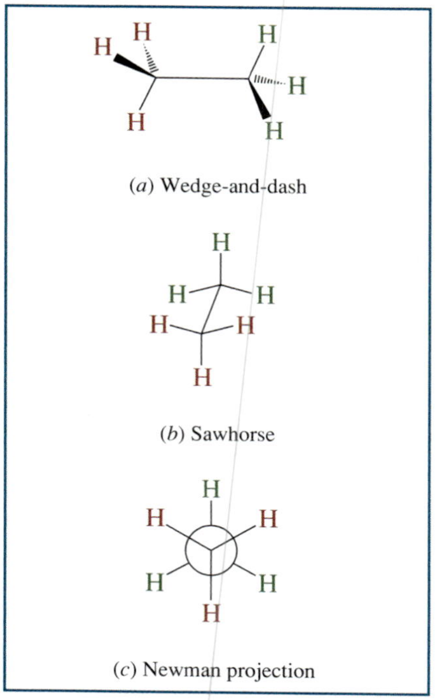

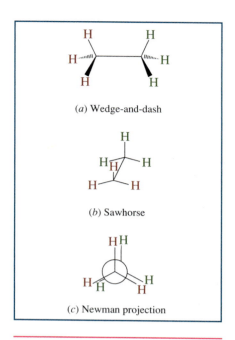

(a) Wedge-and-dash

(b) Sawhorse

(c) Newman projection

FIGURE 3.2 Some commonly used drawings of the staggered conformation of ethane.

(a) Wedge-and-dash

(b) Sawhorse

(c) Newman projection

FIGURE 3.3 Some commonly used drawings of the eclipsed conformation of ethane.

and the C—C—H plane. The torsion angle is easily seen in a Newman projection of ethane as the angle between C—H bonds of adjacent carbons.

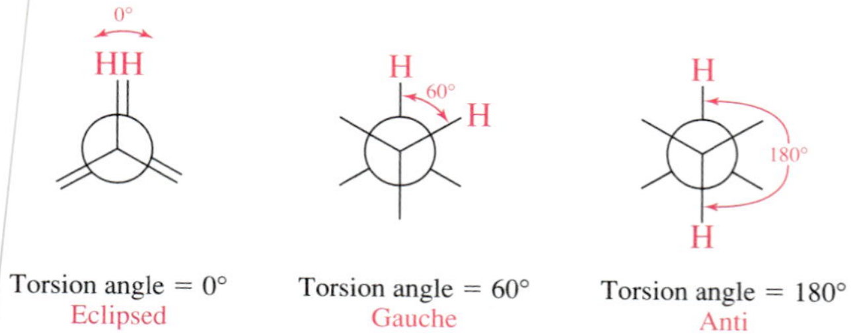

Torsion angle = 0°
Eclipsed

Torsion angle = 60°
Gauche

Torsion angle = 180°
Anti

Eclipsed bonds are characterized by a torsion angle of 0°. When the torsion angle is approximately 60°, we say that the spatial relationship is **gauche**; and when it is 180° we say that it is **anti**. Staggered conformations have only gauche or anti relationships between bonds on adjacent atoms.

PROBLEM 3.1 Use *Learning By Modeling* to make a molecular model of ethane. The staggered conformation will appear by default. Verify that the torsion (dihedral) angles are 60°. Convert the staggered to the eclipsed conformation by adjusting the dihedral angles. Rotate the models on the screen so that they correspond to the orientations shown for the wedge-and-dash, sawhorse, and Newman projection drawings shown in Figures 3.2 and 3.3.

Of the two conformations of ethane, the staggered is 12 kJ/mol (2.9 kcal/mol) more stable than the eclipsed. The staggered conformation is the most stable conformation; the eclipsed is the least stable conformation. Two main explanations have been offered for the difference in stability between the two conformations. One explanation holds that repulsions between bonds on adjacent atoms *destabilize* the eclipsed conformation. The other suggests that better electron delocalization *stabilizes* the staggered conformation. The latter of these two explanations is now believed to be the correct one.

Conformations in which the torsion angles between adjacent bonds are other than 60° are said to have **torsional strain.** Eclipsed bonds produce the most torsional strain; staggered bonds none. Because three pairs of eclipsed bonds are responsible for 12 kJ/mol (2.9 kcal/mol) of torsional strain in ethane, it is reasonable to assign an "energy cost" of 4 kJ/mol (1 kcal/mol) to each pair. In this chapter we'll learn of additional sources of strain in molecules, which together with torsional strain comprise **steric strain.**

In principle ethane has an infinite number of conformations that differ by only tiny increments in their torsion angles. Not only is the staggered conformation more stable than the eclipsed, it is the most stable of all of the conformations; the eclipsed is the least stable. Figure 3.4 shows how the potential energy of ethane changes for a 360° rotation about the carbon–carbon bond. Three equivalent eclipsed conformations and three equivalent staggered conformations occur during the 360° rotation; the eclipsed conformations appear at the highest points on the curve (*potential energy maxima*), the staggered ones at the lowest (*potential energy minima*).

At any instant, almost all of the molecules are in staggered conformations; hardly any are in eclipsed conformations.

Steric is derived from the Greek word *stereos* for "solid" and refers to the three-dimensional or spatial aspects of chemistry.

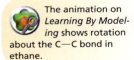 The animation on *Learning By Modeling* shows rotation about the C—C bond in ethane.

> **PROBLEM 3.2** Find the conformations in Figure 3.4 in which the red circles are (a) gauche and (b) anti.

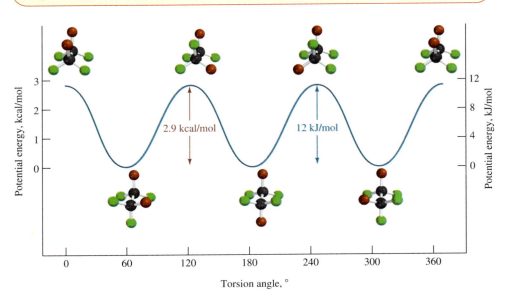

FIGURE 3.4 Potential energy diagram for rotation about the carbon–carbon bond in ethane. Two of the hydrogens are shown in red and four in green so as to indicate more clearly the bond rotation.

Diagrams such as Figure 3.4 help us understand how the potential energy of a system changes during a process. The process can be a simple one such as the one described here—rotation around a carbon–carbon bond. Or it might be more complicated—a chemical reaction, for example. We will see applications of potential energy diagrams to a variety of processes throughout the text.

Let's focus our attention on a portion of Figure 3.4. The region that lies between a torsion angle of 60° and 180° tracks the conversion of one staggered conformation of ethane to the next one. Both staggered conformations are equivalent and equal in energy, but for one staggered conformation to get to the next, it must first pass through an eclipsed conformation and needs to gain 12 kJ/mol (2.9 kcal/mol) of energy to reach it. This amount of energy is the **activation energy (E_{act})** for the process. Molecules must become energized in order to undergo a chemical reaction or, as in this case, to undergo rotation around a carbon–carbon bond. Kinetic (thermal) energy is absorbed by a molecule from collisions with other molecules and is transformed into potential energy. When the potential energy exceeds E_{act}, the unstable arrangement of atoms that exists at that instant can relax to a more stable structure, giving off its excess potential energy in collisions with other molecules or with the walls of a container. The point of maximum potential energy encountered by the reactants as they proceed to products is called the **transition state.** The eclipsed conformation is the transition state for the conversion of one staggered conformation of ethane to another.

> The structure that exists at the transition state is sometimes referred to as the *transition structure* or the *activated complex.*

Rotation around carbon–carbon bonds is one of the fastest processes in chemistry. Among the ways that we can describe the rate of a process is by its *half-life,* which is the length of time it takes for one half of the molecules to react. It takes less than 10^{-6} s for half of the molecules in a sample of ethane to go from one staggered conformation to another at 25°C.

As with all chemical processes, the rate of rotation about the carbon–carbon bond increases with temperature. The reason for this is apparent from Figure 3.5, where it can be seen that most of the molecules in a sample have energies that are clustered around some average value; some have less energy, a few have more. Only molecules with a potential energy greater than E_{act}, however, are able to go over the transition state and proceed on. The number of these molecules is given by the shaded areas under the curve in Figure 3.5. The energy distribution curve flattens out at higher temperatures, and a greater proportion of molecules have energies in excess of E_{act} at T_2 (higher) than at T_1

FIGURE 3.5 Distribution of energies. (*a*) The number of molecules with energy greater than E_{act} at temperature T_1 is shown as the darker green-shaded area. (*b*) At some higher temperature T_2, the curve is flatter, and more molecules have energies in excess of E_{act}.

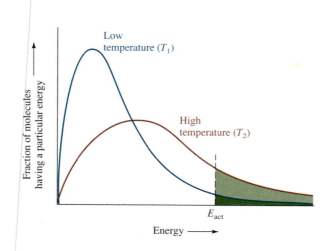

(lower). The effect of temperature is quite pronounced; an increase of only 10°C produces a two- to threefold increase in the rate of a typical chemical process.

3.2 CONFORMATIONAL ANALYSIS OF BUTANE

The next alkane that we will examine is butane. In particular, we consider conformations related by rotation about the bond between the middle two carbons (CH_3CH_2—CH_2CH_3). Unlike ethane, in which the staggered conformations are equivalent, two different staggered conformations occur in butane, shown in Figure 3.6. The methyl groups are gauche to each other in one, anti in the other. Both conformations are staggered, so are free of torsional strain, but two of the methyl hydrogens of the gauche conformation lie within 210 pm of each other. This distance is less than the sum of their van der Waals radii (240 pm), and there is a repulsive force between them. The destabilization of a molecule that results when two of its atoms are too close to each other is called **van der Waals strain,** or **steric hindrance** and contributes to the total steric strain. In the case of butane, van der Waals strain makes the gauche conformation approximately 3.3 kJ/mol (0.8 kcal/mol) less stable than the anti.

Figure 3.7 illustrates the potential energy relationships among the various conformations of butane. The staggered conformations are more stable than the eclipsed. At any instant, almost all the molecules exist in staggered conformations, and more are present in the anti conformation than in the gauche. The point of maximum potential

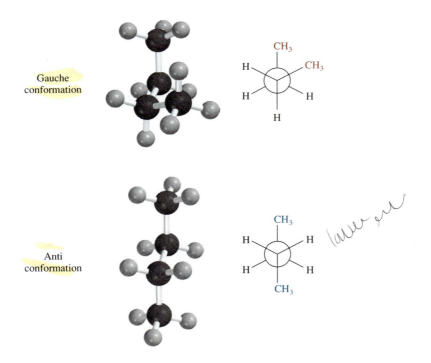

FIGURE 3.6 The gauche and anti conformations of butane shown as ball-and-spoke models (*left*) and as Newman projections (*right*). The gauche conformation is less stable than the anti because of the van der Waals strain between the methyl groups.

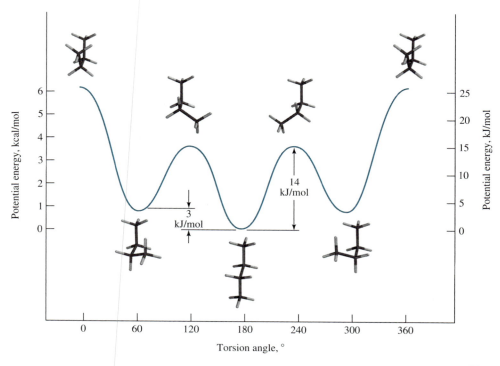

FIGURE 3.7 Potential energy diagram for rotation around the central carbon–carbon bond in butane.

energy lies some 25 kJ/mol (6.1 kcal/mol) above the anti conformation. The total strain in this structure is approximately equally divided between the torsional strain associated with three pairs of eclipsed bonds (12 kJ/mol; 2.9 kcal/mol) and the van der Waals strain between the methyl groups.

> **PROBLEM 3.3** Sketch a potential energy diagram for rotation around a carbon–carbon bond in propane. Clearly identify each potential energy maximum and minimum with a structural formula that shows the conformation of propane at that point. Does your diagram more closely resemble that of ethane or of butane? Would you expect the activation energy for bond rotation in propane to be more than or less than that of ethane? Of butane?

3.3   CONFORMATIONS OF HIGHER ALKANES

Higher alkanes having unbranched carbon chains are, like butane, most stable in their all-anti conformations. The energy difference between gauche and anti conformations is similar to that of butane, and appreciable quantities of the gauche conformation are present in liquid alkanes at 25°C. In depicting the conformations of higher alkanes it is often more helpful to look at them from the side rather than end-on as in a Newman projection. Viewed from this perspective, the most stable conformations of pentane and hexane

MOLECULAR MECHANICS APPLIED TO ALKANES AND CYCLOALKANES

Of the numerous applications of computer technology to chemistry, one that has been enthusiastically embraced by organic chemists examines molecular structure from a perspective similar to that gained by manipulating molecular models but with an additional quantitative dimension. *Molecular mechanics* is a computational method that allows us to assess the stability of a molecule by comparing selected features of its structure with those of ideal "unstrained" standards. Molecular mechanics makes no attempt to explain why the van der Waals radius of hydrogen is 120 pm, why the bond angles in methane are 109.5°, why the C—C bond distance in ethane is 153 pm, or why the staggered conformation of ethane is 12 kJ/mol more stable than the eclipsed, but instead uses these and other experimental observations as benchmarks to which the corresponding features of other substances are compared.

If we assume that there are certain "ideal" values for bond angles, bond distances, and so on, it follows that deviations from these ideal values will destabilize a particular structure and increase its potential energy. This increase in potential energy is referred to as the **strain energy** of the structure. Other terms for this increase include **steric energy** and **steric strain.** Arithmetically, the total strain energy (E_s) of an alkane or cycloalkane can be considered as

$$E_s = E_{bond\ stretching} + E_{angle\ bending} + E_{torsional} + E_{van\ der\ Waals}$$

where

$E_{bond\ stretching}$ is the strain that results when C—C and C—H bond distances are distorted from their ideal values of 153 pm and 111 pm, respectively.

$E_{angle\ bending}$ is the strain that results from the expansion or contraction of bond angles from the normal values of 109.5° for sp^3-hybridized carbon.

$E_{torsional}$ is the strain that results from deviation of torsion angles from their stable staggered relationship.

$E_{van\ der\ Waals}$ is the strain that results from "nonbonded interactions."

Nonbonded interactions are the forces between atoms that aren't bonded to one another; they may be either attractive or repulsive. It often happens that the shape of a molecule may cause two atoms to be close in space even though they are separated from each other by many bonds. Induced-dipole/induced-dipole interactions make van der Waals forces in alkanes weakly attractive at most distances, but when two atoms are closer to each other than the sum of their van der Waals radii, nuclear–nuclear and electron–electron repulsive forces between them dominate the $E_{van\ der\ Waals}$ term. The resulting destabilization is called van der Waals strain.

At its most basic level, separating the total strain of a structure into its components is a qualitative exercise. For example, a computer-drawn model of the eclipsed conformation of butane using ideal bond angles and bond distances (Figure 3.8) reveals that two pairs of hydrogens are separated by a distance of only 175 pm, a value considerably smaller than the sum of their van der Waals radii (2 × 120 pm = 240 pm). Thus, this conformation is destabilized not only by the torsional strain associated with its eclipsed bonds, but also by van der Waals strain.

At a higher level, molecular mechanics is applied quantitatively to strain energy calculations. Each component of strain is separately described by a mathematical expression developed and refined so that it gives solutions that match experimental observations for reference molecules. These empirically derived and tested expressions are then used to calculate the most stable structure of a substance. The various structural features are interdependent; van der Waals strain, for example, might be decreased at the expense of introducing some angle strain, torsional strain, or both. The computer program searches for the combination of bond angles, distances, torsion angles, and nonbonded interactions that gives the molecule the lowest total strain. This procedure is called *strain energy minimization* and is based on the common-sense notion that the most stable structure is the one that has the least strain.

—Cont.

The first widely used molecular mechanics program was developed by Professor N. L. Allinger of the University of Georgia and was known in its various versions as *MM2, MM3,* and so on. They have been refined to the extent that many structural features can be calculated more easily and more accurately than they can be measured experimentally.

Once requiring minicomputers and workstations, many molecular mechanics programs are available for personal computers. The information that strain energy calculations can provide is so helpful that molecular mechanics is no longer considered a novelty but rather as one more tool to be used by the practicing organic chemist. They have been joined by programs that calculate the energies of conformations by molecular orbital methods. The *Learning By Modeling* CD that accompanies this text contains molecular mechanics software that lets you seek out the most stable conformation of the structures you assemble. It also contains the most stable conformations of some molecules as determined by molecular orbital calculations.

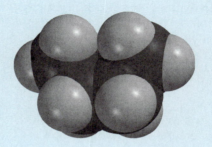

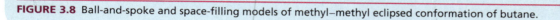

FIGURE 3.8 Ball-and-spoke and space-filling models of methyl–methyl eclipsed conformation of butane.

have their carbon "backbones" arranged in a zigzag fashion, as shown in Figure 3.9. All the bonds are staggered, and the chains are characterized by anti arrangements of C—C—C—C units.

3.4   THE SHAPES OF CYCLOALKANES: PLANAR OR NONPLANAR?

During the nineteenth century it was widely believed—incorrectly, as we'll soon see—that cycloalkane rings are planar. A leading advocate of this view was the German chemist Adolf von Baeyer. He noted that compounds containing rings other than those

Pentane

Hexane

FIGURE 3.9 Ball-and-spoke models of pentane and hexane in their all-anti (zigzag) conformations.

based on cyclopentane and cyclohexane were rarely encountered naturally and were difficult to synthesize. Baeyer connected both observations with cycloalkane stability, which he suggested was related to how closely the angles of planar rings match the tetrahedral value of 109.5°. For example, the 60° bond angle of cyclopropane and the 90° bond angles of a planar cyclobutane ring are much smaller than the tetrahedral angle of 109.5°. Baeyer suggested that three- and four-membered rings suffer from what we now call angle strain. **Angle strain** is the strain a molecule has because one or more of its bond angles deviate from the ideal value; in the case of alkanes the ideal value is 109.5°.

According to Baeyer, cyclopentane should be the most stable of all the cycloalkanes because the ring angles of a planar pentagon, 108°, are closer to the tetrahedral angle than those of any other cycloalkane. A prediction of the *Baeyer strain theory* is that the cycloalkanes beyond cyclopentane should become increasingly strained and correspondingly less stable. The angles of a regular hexagon are 120°, and the angles of larger polygons deviate more and more from the ideal tetrahedral angle.

Some of the inconsistencies in the Baeyer strain theory become evident when we use heats of combustion (Table 3.1) to probe the relative energies of cycloalkanes. The most important column in the table is the heat of combustion per methylene (CH_2) group. Because all of the cycloalkanes have molecular formulas of the type C_nH_{2n}, dividing the heat of combustion by n allows direct comparison of ring size and potential energy. Cyclopropane has the highest heat of combustion per methylene group, which is consistent with the idea that its potential energy is raised by angle strain. Cyclobutane has less angle strain at each of its carbon atoms and a lower heat of combustion per methylene group. Cyclopentane, as expected, has a lower value still. Notice, however, that contrary to the prediction of the Baeyer strain theory, cyclohexane has a smaller heat of combustion per methylene group than cyclopentane. If angle strain were greater in cyclohexane than in cyclopentane, the opposite would have been observed.

Furthermore, the heats of combustion per methylene group of the very large rings are all about the same and similar to that of cyclohexane. Rather than rising because of increasing angle strain in large rings, the heat of combustion per methylene group

Although better known now for his incorrect theory that cycloalkanes were planar, Baeyer was responsible for notable advances in the chemistry of organic dyes such as indigo and was awarded the 1905 Nobel Prize in chemistry for his work in that area.

TABLE 3.1	Heats of Combustion ($-\Delta H°$) of Cycloalkanes				
		Heat of combustion		Heat of combustion per CH_2 group	
Cycloalkane	**Number of CH_2 groups**	**kJ/mol**	**(kcal/mol)**	**kJ/mol**	**(kcal/mol)**
Cyclopropane	3	2,091	(499.8)	697	(166.6)
Cyclobutane	4	2,721	(650.3)	681	(162.7)
Cyclopentane	5	3,291	(786.6)	658	(157.3)
Cyclohexane	6	3,920	(936.8)	653	(156.0)
Cycloheptane	7	4,599	(1099.2)	657	(157.0)
Cyclooctane	8	5,267	(1258.8)	658	(157.3)
Cyclononane	9	5,933	(1418.0)	659	(157.5)
Cyclodecane	10	6,587	(1574.3)	659	(157.5)
Cycloundecane	11	7,237	(1729.8)	658	(157.3)
Cyclododecane	12	7,845	(1875.1)	654	(156.3)
Cyclotetradecane	14	9,139	(2184.2)	653	(156.0)
Cyclohexadecane	16	10,466	(2501.4)	654	(156.3)

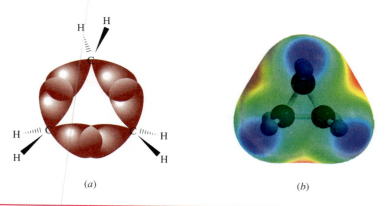

(a)   *(b)*

FIGURE 3.10 "Bent bonds" in cyclopropane. (*a*) The orbitals involved in carbon–carbon bond formation overlap in a region that is displaced from the internuclear axis. (*b*) The three areas of greatest negative electrostatic potential (red) correspond to those predicted by the bent-bond description.

remains constant at approximately 653 kJ/mol (156 kcal/mol), the value cited in Section 2.18 as the difference between successive members of a homologous series of alkanes. We conclude, therefore, that the bond angles of large cycloalkanes are not much different from the bond angles of alkanes themselves. The prediction of the Baeyer strain theory that angle strain increases steadily with ring size is contradicted by experimental fact.

The Baeyer strain theory is useful to us in identifying angle strain as a destabilizing effect. Its fundamental flaw is its assumption that the rings of cycloalkanes are planar. *With the exception of cyclopropane, cycloalkanes are nonplanar.* Sections 3.5–3.13 describe the shapes of cycloalkanes. We'll begin with cyclopropane.

3.5   SMALL RINGS: CYCLOPROPANE AND CYCLOBUTANE

Conformational analysis is far simpler in cyclopropane than in any other cycloalkane. Cyclopropane's three carbon atoms are, of geometric necessity, coplanar, and rotation about its carbon–carbon bonds is impossible. You saw in Section 3.4 how angle strain in cyclopropane leads to an abnormally large heat of combustion. Let's now look at cyclopropane in more detail to see how our orbital hybridization bonding model may be adapted to molecules of unusual geometry.

Strong sp^3–sp^3 σ bonds are not possible for cyclopropane, because the 60° bond angles of the ring do not permit the orbitals to be properly aligned for effective overlap (Figure 3.10). The less effective overlap that does occur leads to what chemists refer to as "bent" bonds. The electron density in the carbon–carbon bonds of cyclopropane does not lie along the internuclear axis but is distributed along an arc between the two carbon atoms. The ring bonds of cyclopropane are weaker than other carbon–carbon σ bonds.

In addition to angle strain, cyclopropane is destabilized by torsional strain. Each C—H bond of cyclopropane is eclipsed with two others.

In keeping with the "bent-bond" description of Figure 3.10, the carbon–carbon bond distance in cyclopropane (151 pm) is slightly shorter than that of ethane (153 pm) and cyclohexane (154 pm). The calculated values from molecular models (see *Learning By Modeling*) reproduce these experimental values.

All adjacent pairs
of bonds are eclipsed

FIGURE 3.11 Nonplanar ("puckered") conformation of cyclobutane. The nonplanar conformation avoids the eclipsing of bonds on adjacent carbons that characterizes the planar conformation.

Cyclobutane has less angle strain than cyclopropane and can reduce the torsional strain that goes with a planar geometry by adopting the nonplanar "puckered" conformation shown in Figure 3.11.

> **PROBLEM 3.4** The heats of combustion of ethylcyclopropane and methylcyclobutane have been measured as 3352 and 3384 kJ/mol (801.2 and 808.8 kcal/mol), respectively. Assign the correct heat of combustion to each isomer.

3.6 CYCLOPENTANE

Angle strain in the planar conformation of cyclopentane is relatively small because the 108° angles of a regular pentagon are not much different from the normal 109.5° bond angles of sp^3-hybridized carbon. The torsional strain, however, is substantial, because five bonds are eclipsed on the top face of the ring, and another set of five are eclipsed on the bottom face (Figure 3.12a). Some, but not all, of this torsional strain is relieved in nonplanar conformations. Two nonplanar conformations of cyclopentane, the **envelope** (Figure 3.12b) and the **half-chair** (Figure 3.12c) are of similar energy.

In the envelope conformation four of the carbon atoms are coplanar. The fifth carbon is out of the plane of the other four. There are three coplanar carbons in the half-chair conformation, with one carbon atom displaced above that plane and another below it. In both the envelope and the half-chair conformations, in-plane and out-of-plane carbons exchange positions rapidly. Equilibration between conformations of cyclopentane is very fast and occurs at rates similar to that of rotation about the carbon–carbon bond of ethane.

Neighboring C—H bonds are eclipsed in any planar cycloalkane. Thus all planar conformations are destabilized by torsional strain.

(a) Planar   (b) Envelope   (c) Half-Chair

FIGURE 3.12 The (a) planar, (b) envelope, and (c) half-chair conformations of cyclopentane.

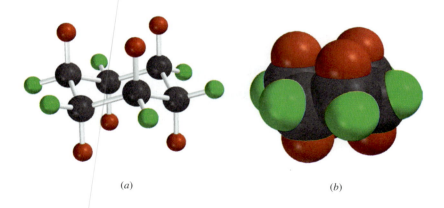

(*a*)   (*b*)

3.7   CONFORMATIONS OF CYCLOHEXANE

Experimental evidence indicating that six-membered rings are nonplanar began to accumulate in the 1920s. Eventually, Odd Hassel of the University of Oslo established that the most stable conformation of cyclohexane has the shape shown in Figure 3.13. This is called the **chair** conformation. With C—C—C bond angles of 111°, the chair conformation is nearly free of angle strain. All its bonds are staggered, making it free of torsional strain as well. The staggered arrangement of bonds in the chair conformation of cyclohexane is apparent in a Newman-style projection.

Hassel shared the 1969 Nobel Prize in chemistry with Sir Derek Barton of Imperial College (London). Barton demonstrated how Hassel's structural results could be extended to an analysis of conformational effects on chemical reactivity.

Staggered arrangement of bonds in chair conformation of cyclohexane

Make a molecular model of the chair conformation of cyclohexane, and turn it so that you can look down one of the C—C bonds.

A second, but much less stable, nonplanar conformation called the **boat** is shown in Figure 3.14. Like the chair, the boat conformation has bond angles that are approximately tetrahedral and is relatively free of angle strain. It is, however, destabilized by the torsional strain associated with eclipsed bonds on four of its carbons. The

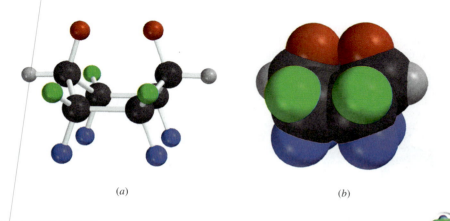

(*a*)   (*b*)

FIGURE 3.14 (*a*) A ball-and-spoke model and (*b*) a space-filling model of the boat conformation of cyclohexane. Torsional strain from eclipsed bonds and van der Waals strain involving the "flagpole" hydrogens (red) make the boat less stable than the chair.

FIGURE 3.15 (*a*) The boat and (*b*) skew boat conformations of cyclohexane. Some of the torsional strain in the boat is relieved by rotation about C—C bonds in going to the skew boat. This motion also causes the flagpole hydrogens to move away from one another, reducing the van der Waals strain between them.

(*a*)   (*b*)

close approach of the two "flagpole" hydrogens shown in Figure 3.14 contributes a small amount of van der Waals strain as well. Both sources of strain are reduced by rotation about the carbon–carbon bond to give the slightly more stable **twist boat, or skew boat,** conformation (Figure 3.15).

Sources of strain in the boat conformation are discussed in detail in a paper in the March 2000 issue of the *Journal of Chemical Education,* p. 332.

The various conformations of cyclohexane are in rapid equilibrium with one another, but at any moment almost all of the molecules exist in the chair conformation. Not more than one or two molecules per thousand are present in the skew boat conformation. Thus, the discussion of cyclohexane conformational analysis that follows focuses exclusively on the chair conformation.

3.8   AXIAL AND EQUATORIAL BONDS IN CYCLOHEXANE

One of the most important findings to come from conformational studies of cyclohexane is that its 12 hydrogen atoms can be divided into two groups, as shown in Figure 3.16. Six of the hydrogens, called **axial hydrogens,** have their bonds parallel to a vertical axis that passes through the ring's center. These axial bonds alternately are directed up and down on adjacent carbons. The second set of six hydrogens, called **equatorial** hydrogens, are located approximately along the equator of the molecule. Notice that the four bonds to each carbon are arranged tetrahedrally, consistent with an sp^3 hybridization of carbon.

The conformational features of six-membered rings are fundamental to organic chemistry, so it is essential that you have a clear understanding of the directional properties of axial and equatorial bonds and be able to represent them accurately. Figure 3.17 offers some guidance on the drawing of chair cyclohexane rings.

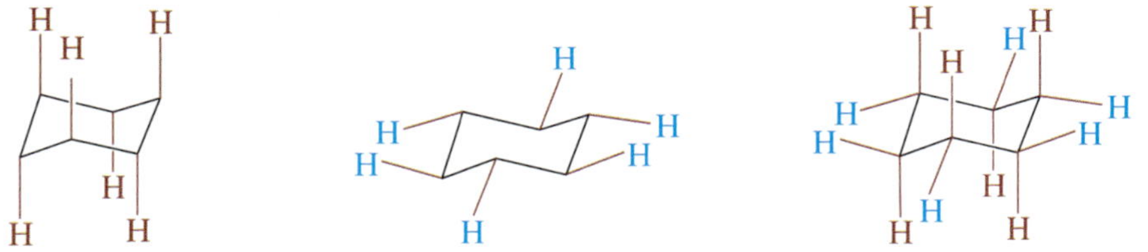

Axial C—H bonds   Equatorial C—H bonds   Axial and equatorial
bonds together

FIGURE 3.16 Axial and equatorial bonds in cyclohexane.

FIGURE 3.17 A guide to
representing the orienta-
tions of the bonds in the
chair conformation of cyclohexane.

(1) Begin with the chair conformation of cyclohexane.

(2) Draw the axial bonds before the equatorial ones, alternating their direction
on adjacent atoms. Always start by placing an axial bond "up" on the
uppermost carbon or "down" on the lowest carbon.

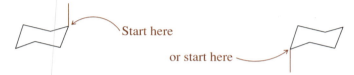

Start here

or start here

Then alternate to give

in which all the axial
bonds are parallel to
one another

(3) Place the equatorial bonds so as to approximate a tetrahedral arrangement of
the bonds to each carbon. The equatorial bond of each carbon should be
parallel to the ring bonds of its two nearest neighbor carbons.

Place equatorial bond
at C-1 so that it is
parallel to the bonds
between C-2 and C-3
and between C-5 and
C-6.

Following this pattern gives the complete set of equatorial bonds.

(4) Practice drawing cyclohexane chairs oriented in either direction.

and

It is no accident that sections of our chair cyclohexane drawings resemble saw-horse projections of staggered conformations of alkanes. The same spatial relationships seen in alkanes carry over to substituents on a six-membered ring. In the structure

(The substituted carbons have the spatial arrangement shown)

substituents A and B are anti to each other, and the other relationships—A and Y, X and Y, and X and B—are gauche.

PROBLEM 3.5 Given the following partial structure, add a substituent X to C-1 so that it satisfies the indicated stereochemical requirement. You may find it helpful to build a molecular model for reference.

(a) Anti to A   (c) Anti to C-3
(b) Gauche to A   (d) Gauche to C-3

SAMPLE SOLUTION (a) In order to be anti to A, substituent X must be axial. The blue lines in the drawing show the A—C—C—X torsion angle to be 180°.

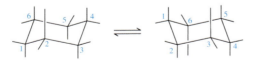

3.9   CONFORMATIONAL INVERSION (RING FLIPPING) IN CYCLOHEXANE

We have seen that alkanes are not locked into a single conformation. Rotation around the central carbon–carbon bond in butane occurs rapidly, interconverting anti and gauche conformations. Cyclohexane, too, is conformationally mobile. Through a process known as **ring inversion, chair–chair interconversion,** or, more simply, **ring flipping,** one chair conformation is converted to another chair.

A more detailed discussion of cyclohexane ring inversion can be found in the July 1997 issue of the *Journal of Chemical Education*, pp. 813–814.

The activation energy for cyclohexane ring inversion is 45 kJ/mol (10.8 kcal/mol). It is a very rapid process with a half-life of about 10^{-5} s at 25°C.

A potential energy diagram for ring inversion in cyclohexane is shown in Figure 3.18. In the first step the chair conformation is converted to a skew boat, which then proceeds to the inverted chair in the second step. The skew boat conformation is an *intermediate* in the process of ring inversion. Unlike a transition state, an **intermediate** is not a potential energy maximum but is a local minimum on the potential energy profile.

The best way to understand ring flipping in cyclohexane is to view the animation of Figure 3.18 in *Learning By Modeling*.

FIGURE 3.18 Energy diagram showing the interconversion of various conformations of cyclohexane.

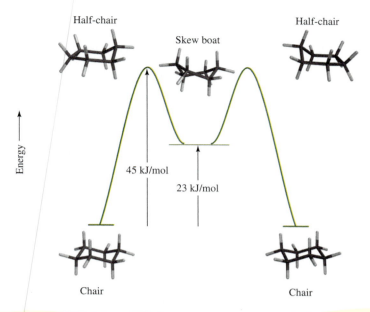

The most important result of ring inversion is that any substituent that is axial in the original chair conformation becomes equatorial in the ring-flipped form and vice versa.

X axial; Y equatorial   X equatorial; Y axial

The consequences of this point are developed for a number of monosubstituted cyclohexane derivatives in the following section, beginning with methylcyclohexane.

3.10   CONFORMATIONAL ANALYSIS OF MONOSUBSTITUTED CYCLOHEXANES

Ring inversion in methylcyclohexane differs from that of cyclohexane in that the two chair conformations are not equivalent. In one chair the methyl group is axial; in the other it is equatorial. At room temperature approximately 95% of the molecules of methylcyclohexane are in the chair conformation that has an equatorial methyl group, whereas only 5% of the molecules have an axial methyl group.

See *Learning By Modeling* for an animation of this process.

See the box entitled *Enthalpy, Free Energy, and Equilibrium Constant* accompanying this section for a discussion of these relationships.

5%   95%

When two conformations of a molecule are in equilibrium with each other, the one with the lower free energy predominates. Why is equatorial methylcyclohexane more stable than axial methylcyclohexane?

A methyl group is less crowded when it is equatorial than when it is axial. One of the hydrogens of an axial methyl group is within 190–200 pm of the axial hydrogens at C-3 and C-5. This distance is less than the sum of the van der Waals radii of two hydrogens (240 pm) and causes van der Waals strain in the axial conformation. When the methyl group is equatorial, it experiences no significant crowding.

Van der Waals strain
between hydrogen of axial
CH$_3$ and axial hydrogens
at C-3 and C-5

Smaller van der Waals
strain between hydrogen
at C-1 and axial hydrogens
at C-3 and C-5

Make a molecular model of each chair conformation of methylcyclohexane, and compare their energies.

The greater stability of an equatorial methyl group, compared with an axial one, is another example of a *steric effect* (Section 3.2). An axial substituent is said to be crowded because of **1,3-diaxial repulsions** between itself and the other two axial substituents located on the same side of the ring.

PROBLEM 3.6   The following questions relate to a cyclohexane ring depicted in the chair conformation shown.

(a) Is a methyl group at C-6 that is "down" axial or equatorial?

(b) Is a methyl group that is "up" at C-1 more or less stable than a methyl group that is up at C-4?

(c) Place a methyl group at C-3 in its most stable orientation. Is it up or down?

SAMPLE SOLUTION   (a) First indicate the directional properties of the bonds to the ring carbons. A substituent is down if it is below the other substituent on the same carbon atom. A methyl group that is down at C-6 is therefore axial.

We can relate the conformational preference for an equatorial methyl group in methylcyclohexane to the conformation of a noncyclic hydrocarbon we discussed earlier, butane. The red bonds in the following structural formulas trace paths through four carbons, beginning at an equatorial methyl group. The zigzag arrangement described by each path mimics the anti conformation of butane.

ENTHALPY, FREE ENERGY, AND EQUILIBRIUM CONSTANT

One of the fundamental equations of thermodynamics concerns systems at equilibrium and relates the equilibrium constant K to the difference in **standard free energy** ($\Delta G°$) between the products and the reactants.

$$\Delta G° = G°_{product} - G°_{reactants} = -RT \ln K$$

where T is the absolute temperature in kelvins and the constant R equals 8.314 J/mol · K (1.99 cal/mol · K).

For the equilibrium between the axial and equatorial conformations of a monosubstituted cyclohexane,

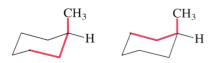

the equilibrium constant is given by the expression

$$K = \frac{[products]}{[reactants]}$$

Inserting the appropriate values for R, T (298 K), and K gives the values of $\Delta G°$ listed in the table (page 123) for the various substituents discussed in Section 3.10.

The relationship between $\Delta G°$ and K is plotted in Figure 3.19. A larger value of K is associated with a more negative $\Delta G°$.

Free energy and enthalpy are related by the expression

$$\Delta G° = \Delta H° - T\Delta S°$$

where $\Delta S°$ is the difference in *entropy* between the products and reactants. A positive $\Delta S°$ is accompanied by an increase in the disorder of a system. A positive $T\Delta S°$ term leads to a $\Delta G°$ that is more negative than $\Delta H°$ and a larger K than expected on the basis of enthalpy considerations alone. Conversely, a negative $\Delta S°$ gives a smaller K than expected. In the case of conformational equilibration between the chair forms of a substituted cyclohexane, $\Delta S°$ is close to zero and $\Delta G°$ and $\Delta H°$ are approximately equal.

—Cont.

When the methyl group is axial, each path mimics the gauche conformation of butane.

Qualitatively, the preference for an equatorial methyl group in methylcyclohexane is therefore analogous to the preference for the anti conformation in butane. Quantitatively, two gauche butane-like structural units are present in axial methylcyclohexane that are absent in equatorial methylcyclohexane. As we saw earlier in Figure 3.7, the anti conformation of butane is 3.3 kJ/mol (0.8 kcal/mol) lower in energy than the gauche. Therefore, the calculated energy difference between the equatorial and axial conformations of methylcyclohexane should be twice that, or 6.6 kJ/mol (1.6 kcal/mol). The experimentally measured difference of 7.1 kJ/mol (1.7 kcal/mol) is close to this estimate. This gives us confidence that the same factors that govern the conformations of noncyclic compounds also apply to cyclic ones. What we call 1,3-diaxial repulsions in substituted cyclohexanes are really the same as van der Waals strain in the gauche conformations of alkanes.

Other substituted cyclohexanes are similar to methylcyclohexane. Two chair conformations exist in rapid equilibrium, and the one in which the substituent is equatorial is more stable. The relative amounts of the two conformations depend on the effective size of the substituent. The size of a substituent, in the context of cyclohexane conformations, is related to the degree of branching at the atom connected to the ring. A single

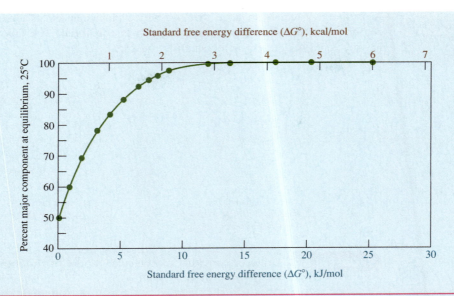

FIGURE 3.19 Distribution of two products at equilibrium at 25°C as a function of the standard free energy difference ($\Delta G°$) between them.

Substituent X	Percent axial	Percent equatorial	K	$\Delta G°_{298\,K}$ kJ/mol	(kcal/mol)
—F	40	60	1.5	−1.0	(−0.24)
—CH$_3$	5	95	19	−7.3	(−1.7)
—CH(CH$_3$)$_2$	3	97	32.3	−8.6	(−2.1)
—C(CH$_3$)$_3$	<0.01	>99.99	>9999	−22.8	(−5.5)

atom, such as a halogen, does not take up much space, and its preference for an equatorial orientation is less than that of a methyl group.

The halogens F, Cl, Br, and I do not differ much in their preference for the equatorial position. As the atomic radius increases in the order F < Cl < Br < I, so does the carbon–halogen bond distance, and the two effects tend to cancel.

40%   60%

A branched alkyl group such as isopropyl exhibits a greater preference for the equatorial orientation than does methyl.

3%   97%

Highly branched groups such as *tert*-butyl are commonly described as "bulky."

A *tert*-butyl group is so large that *tert*-butylcyclohexane exists almost entirely in the conformation in which the *tert*-butyl group is equatorial. The amount of axial *tert*-butylcyclohexane present is too small to measure.

Less than 0.01%
(Serious 1,3-diaxial repulsions involving *tert*-butyl group)

Greater than 99.99%
(Decreased van der Waals strain)

PROBLEM 3.7 Draw or construct a molecular model of the most stable conformation of 1-*tert*-butyl-1-methylcyclohexane.

3.11 DISUBSTITUTED CYCLOALKANES: STEREOISOMERS

When a cycloalkane bears two substituents on different carbons—methyl groups, for example—these substituents may be on the same or on opposite sides of the ring. When substituents are on the same side, we say they are ***cis*** to each other; if they are on opposite sides, they are ***trans*** to each other. Both terms come from the Latin, in which *cis* means "on this side" and *trans* means "across."

cis-1,2-Dimethylcyclopropane   *trans*-1,2-Dimethylcyclopropane

PROBLEM 3.8 Exclusive of compounds with double bonds, four hydrocarbons are *constitutional* isomers of *cis*- and *trans*-1,2-dimethylcyclopropane. Identify these compounds.

The prefix *stereo-* is derived from the Greek word *stereos,* meaning "solid." *Stereochemistry* is the term applied to the three-dimensional aspects of molecular structure and reactivity.

The cis and trans forms of 1,2-dimethylcyclopropane are stereoisomers. **Stereoisomers** are isomers that have their atoms bonded in the same order—that is, they have the same constitution, but they differ in the arrangement of atoms in space. Stereoisomers of the cis–trans type are sometimes referred to as *geometric isomers*. You learned in Section 2.18 that constitutional isomers could differ in stability. What about stereoisomers?

We can measure the energy difference between *cis*- and *trans*-1,2-dimethylcyclopropane by comparing their heats of combustion. As illustrated in Figure 3.20, the two compounds are isomers, and so the difference in their heats of combustion is a direct measure of the difference in their energies. Because the heat of combustion of *trans*-1,2-dimethylcyclopropane is 5 kJ/mol (1.2 kcal/mol) less than that of its cis stereoisomer, it follows that *trans*-1,2-dimethylcyclopropane is 5 kJ/mol (1.2 kcal/mol) more stable than *cis*-1,2-dimethylcyclopropane.

cis-1,2-Dimethylcyclopropane   *trans*-1,2-Dimethylcyclopropane

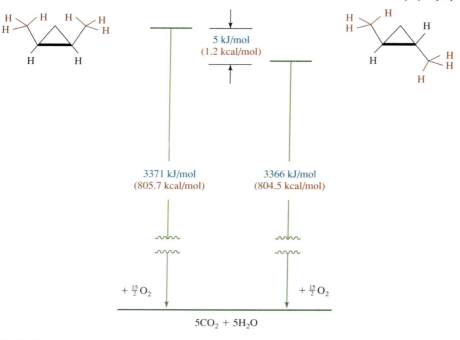

FIGURE 3.20 The enthalpy difference between *cis–* and *trans*-1,2-dimethylcyclopropane can be determined from their heats of combustion. Van der Waals strain between methyl groups on the same side of the ring make the cis isomer less stable than the trans.

In this case, the relationship between stability and stereochemistry is easily explained on the basis of van der Waals strain. The methyl groups on the same side of the ring in *cis*-1,2-dimethylcyclopropane crowd each other and increase the potential energy of this stereoisomer. Steric hindrance between methyl groups is absent in *trans*-1,2-dimethylcyclopropane.

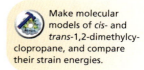

Make molecular models of *cis*- and *trans*-1,2-dimethylcyclopropane, and compare their strain energies.

Disubstituted cyclopropanes exemplify one of the simplest cases involving stability differences between stereoisomers. A three-membered ring has no conformational mobility, so the ring cannot therefore reduce the van der Waals strain between cis substituents on adjacent carbons without introducing other strain. The situation is different in disubstituted derivatives of cyclohexane.

3.12 CONFORMATIONAL ANALYSIS OF DISUBSTITUTED CYCLOHEXANES

We'll begin with *cis*- and *trans*-1,4-dimethylcyclohexane. A conventional method uses wedge-and-dash descriptions to represent cis and trans stereoisomers in cyclic systems.

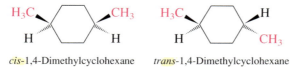

        cis-1,4-Dimethylcyclohexane   *trans*-1,4-Dimethylcyclohexane

Wedge-and-dash drawings fail to show conformation, and it's important to remember that the rings of *cis*- and *trans*-1,2-dimethylcyclohexane exist in a chair conformation. This fact must be taken into consideration when evaluating the relative stabilities of the stereoisomers.

Their heats of combustion (Table 3.2) reveal that *trans*-1,4-dimethylcyclohexane is 7 kJ/mol (1.7 kcal/mol) more stable than the cis stereoisomer. It is unrealistic to believe that van der Waals strain between cis substituents is responsible, because the methyl groups are too far away from each other. To understand why *trans*-1,4-dimethylcyclo-hexane is more stable than *cis*-1,4-dimethylcyclohexane, we need to examine each stereoisomer in its most stable conformation.

cis-1,4-Dimethylcyclohexane can adopt either of two equivalent chair conforma-tions, *each having one axial methyl group and one equatorial methyl group*. The two are in rapid equilibrium with each other by ring flipping. The equatorial methyl group becomes axial, and the axial methyl group becomes equatorial.

(One methyl group is
axial, the other
equatorial)

(One methyl group is
axial, the other
equatorial)

(Both methyl groups are up)
cis-1,4-Dimethylcyclohexane

The methyl groups are described as cis because both are up relative to the hydrogen present at each carbon. If both methyl groups were down, they would still be cis to each other. Notice that ring flipping does not alter the cis relationship between the methyl groups. Nor does it alter their up-versus-down quality; substituents that are up in one conformation remain up in the ring-flipped form.

The most stable conformation of trans-1,4-dimethylcyclohexane has both methyl groups in equatorial orientations. The two chair conformations of *trans*-1,4-dimethyl-cyclohexane are not equivalent to each other. One has two equatorial methyl groups; the other, two axial methyl groups.

TABLE 3.2	Heats of Combustion of Isomeric Dimethylcyclohexanes						
Compound	Orientation of methyl groups in most stable conformation	Heat of combustion		Difference in heat of combustion		More stable stereoisomer
		kJ/mol	(kcal/mol)	kJ/mol	(kcal/mol)	
cis-1,2-Dimethylcyclohexane	Axial–equatorial	5223	(1248.3)	6	(1.5)	trans
trans-1,2-Dimethylcyclohexane	Diequatorial	5217	(1246.8)			
cis-1,3-Dimethylcyclohexane	Diequatorial	5212	(1245.7)	7	(1.7)	cis
trans-1,3-Dimethylcyclohexane	Axial–equatorial	5219	(1247.4)			
cis-1,4-Dimethylcyclohexane	Axial–equatorial	5219	(1247.4)	7	(1.7)	trans
trans-1,4-Dimethylcyclohexane	Diequatorial	5212	(1245.7)			

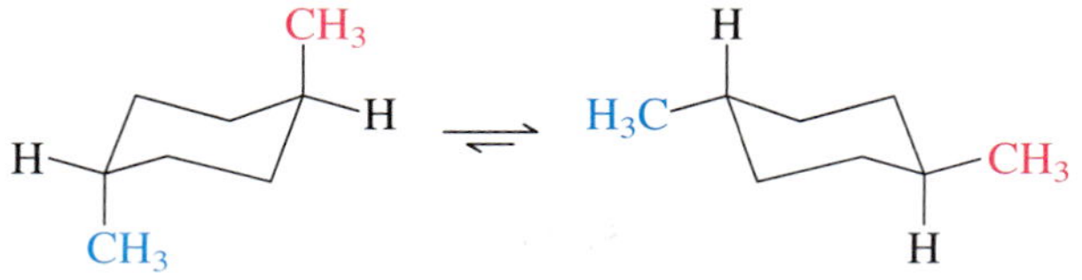

(Both methyl groups
are axial: less stable
chair conformation)

(Both methyl groups are
equatorial: more stable
chair conformation)

(One methyl group is up, the other down)

trans-1,4-Dimethylcyclohexane

The more stable chair—the one with both methyl groups equatorial—is the conformation adopted by most of the *trans*-1,4-dimethylcyclohexane molecules.

   trans-1,4-Dimethylcyclohexane is more stable than *cis*-1,4-dimethylcyclohexane because both methyl groups are equatorial in its most stable conformation. One methyl group must be axial in the cis stereoisomer. Remember, it is a general rule that any substituent is more stable in an equatorial orientation than in an axial one. It is worth pointing out that the 7 kJ/mol (1.7 kcal/mol) energy difference between *cis*- and *trans*-1,4-dimethylcyclohexane is the same as the energy difference between the axial and equatorial conformations of methylcyclohexane. There is a simple reason for this: in both instances the less stable structure has one axial methyl group, and the 7 kJ/mol (1.7 kcal/mol) energy difference can be considered the "energy cost" of having a methyl group in an axial rather than an equatorial orientation.

   Like the 1,4-dimethyl derivatives, *trans*-1,2-dimethylcyclohexane has a lower heat of combustion (see Table 3.2) and is more stable than *cis*-1,2-dimethylcyclohexane. The cis stereoisomer has two chair conformations of equal energy, each containing one axial and one equatorial methyl group.

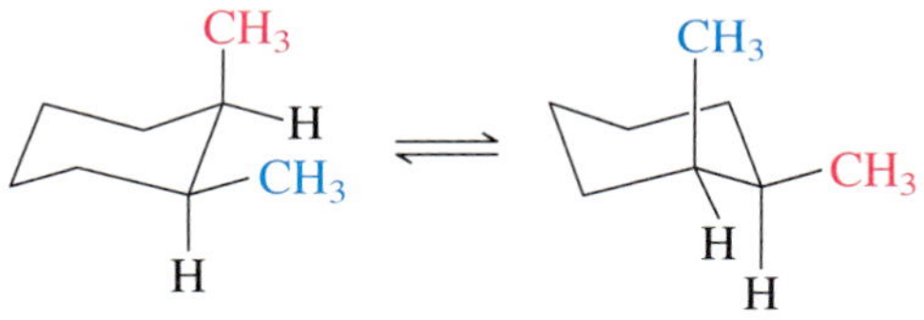

> Recall from Section 3.10 that the equatorial conformation of methylcyclohexane is 7 kJ/mol (1.7 kcal/mol) lower in energy than the conformation with an axial methyl group.

cis-1,2-Dimethylcyclohexane

Both methyl groups are equatorial in the most stable conformation of *trans*-1,2-dimethylcyclohexane.

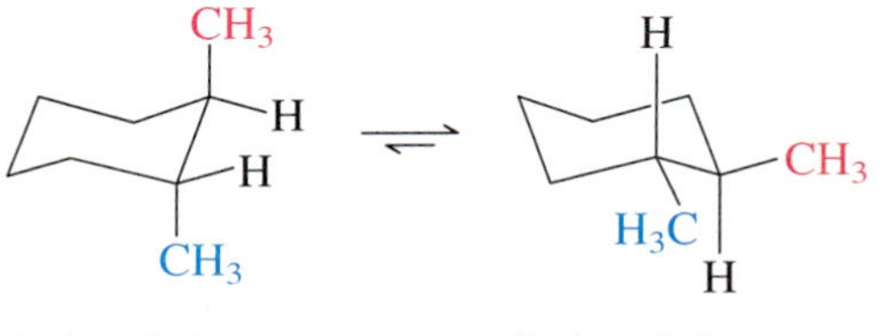

(Both methyl groups
are axial: less stable
chair conformation)

(Both methyl groups are
equatorial: more stable
chair conformation)

trans-1,2-Dimethylcyclohexane

   As in the 1,4-dimethylcyclohexanes, the 6 kJ/mol (1.5 kcal/mol) energy difference between the more stable (trans) and the less stable (cis) stereoisomer is attributed to the strain associated with the presence of an axial methyl group in the cis isomer.

Probably the most interesting observation in Table 3.2 concerns the 1,3-dimethylcyclohexanes. Unlike the 1,2- and 1,4-dimethylcyclohexanes, in which the trans stereoisomer is more stable than the cis, we find that *cis*-1,3-dimethylcyclohexane is 7 kJ/mol (1.7 kcal/mol) more stable than *trans*-1,3-dimethylcyclohexane. Why?

The most stable conformation of *cis*-1,3-dimethylcyclohexane has both methyl groups equatorial.

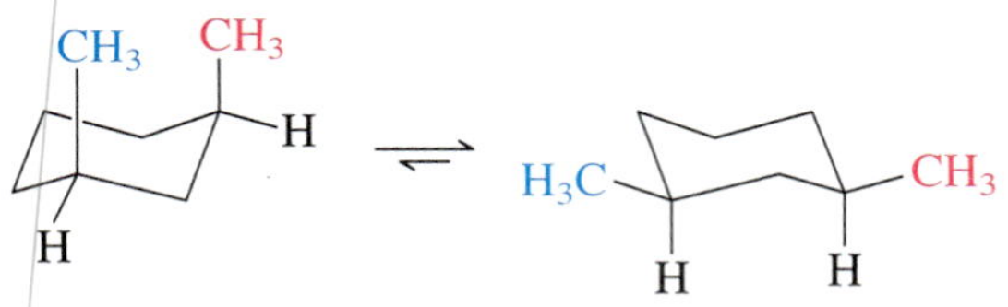

(Both methyl groups are axial: less stable chair conformation) (Both methyl groups are equatorial: more stable chair conformation)

cis-1,3-Dimethylcyclohexane

The two chair conformations of *trans*-1,3-dimethylcyclohexane are equivalent to each other. Both contain one axial and one equatorial methyl group.

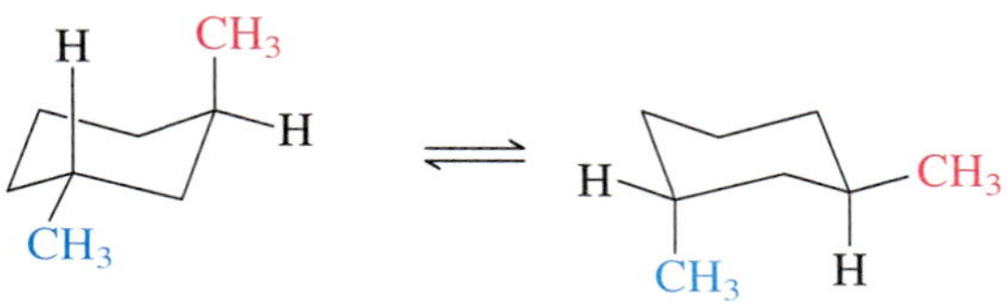

(One methyl group is axial, the other equatorial) (One methyl group is axial, the other equatorial)

trans-1,3-Dimethylcyclohexane

Thus the trans stereoisomer, with one axial methyl group, is less stable than *cis*-1,3-dimethylcyclohexane where both methyl groups are equatorial.

PROBLEM 3.9 Based on what you know about disubstituted cyclohexanes, which of the following two stereoisomeric 1,3,5-trimethylcyclohexanes would you expect to be more stable?

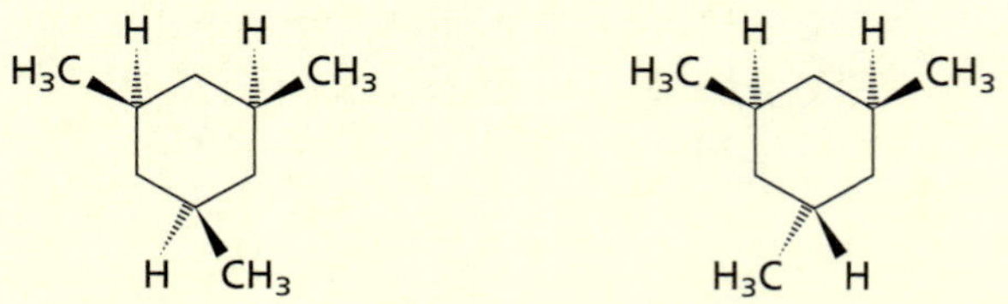

cis-1,3,5-Trimethylcyclohexane   *trans*-1,3,5-Trimethylcyclohexane

---

If a disubstituted cyclohexane has two different substituents, then the most stable conformation is the chair that has the larger substituent in an equatorial orientation. This is most apparent when one of the substituents is a bulky group such as *tert*-butyl. Thus, the most stable conformation of *cis*-1-*tert*-butyl-2-methylcyclohexane has an equatorial *tert*-butyl group and an axial methyl group.

C(CH₃)₃   H   CH₃   ⇌   CH₃   C(CH₃)₃   H   H

(Less stable conformation:
larger group is axial)

(More stable conformation:
larger group is equatorial)

cis-1-*tert*-Butyl-2-methylcyclohexane

PROBLEM 3.10 Write structural formulas or make molecular models for the most stable conformation of each of the following compounds:

(a) *trans*-1-*tert*-Butyl-3-methylcyclohexane

(b) *cis*-1-*tert*-Butyl-3-methylcyclohexane

(c) *trans*-1-*tert*-Butyl-4-methylcyclohexane

(d) *cis*-1-*tert*-Butyl-4-methylcyclohexane

SAMPLE SOLUTION (a) The most stable conformation is the one that has the larger substituent, the *tert*-butyl group, equatorial. Draw a chair conformation of cyclohexane, and place an equatorial *tert*-butyl group at one of its carbons. Add a methyl group at C-3 so that it is trans to the *tert*-butyl group.

H   C(CH₃)₃

Add methyl group
to axial position at
C-3 so that it is trans
to *tert*-butyl group

CH₃   H   C(CH₃)₃   H

tert-Butyl group
equatorial on
six-membered ring

trans-1-*tert*-Butyl-3-
methylcyclohexane

Cyclohexane rings that bear *tert*-butyl substituents are examples of conformationally biased molecules. A *tert*-butyl group has such a pronounced preference for the equatorial orientation that it will strongly bias the equilibrium to favor such conformations. This does not mean that ring inversion does not occur, however. Ring inversion does occur, but at any instant only a tiny fraction of the molecules exist in conformations having axial *tert*-butyl groups. It is not strictly correct to say that *tert*-butylcyclohexane and its derivatives are "locked" into a single conformation; conformations related by ring flipping are in rapid equilibrium with one another, but the distribution between them strongly favors those in which the *tert*-butyl group is equatorial.

3.13 MEDIUM AND LARGE RINGS

Beginning with cycloheptane, which has four conformations of similar energy, conformational analysis of cycloalkanes becomes more complicated. The same fundamental principles apply to medium and large rings as apply to smaller ones—but there are more atoms and more bonds to consider and more conformational possibilities.

In 1978, a German–Swiss team of organic chemists reported the synthesis of a cycloalkane with 96 carbons in its ring (cyclo-$C_{96}H_{192}$).

3.14 POLYCYCLIC RING SYSTEMS

Organic molecules in which *one* carbon atom is common to two rings are called **spirocyclic** compounds. The simplest spirocyclic hydrocarbon is *spiropentane*, a product of laboratory synthesis. More complicated spirocyclic hydrocarbons not only have been synthesized but also have been isolated from natural sources. α-*Alaskene,* for example, occurs in the fragrant oil given off by the needles of the Alaskan yellow cedar; one of its carbon atoms is common to both the six-membered ring and the five-membered ring.

> Make a molecular model of spiropentane. What feature of its geometry is more apparent from a model than from its structural formula?

Spiropentane 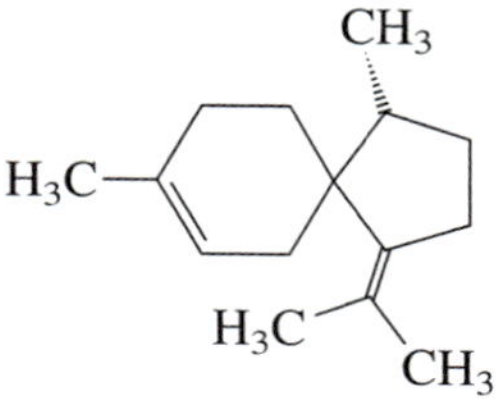   α-Alaskene

---

PROBLEM 3.11 Which of the following compounds are isomers of spiropentane?

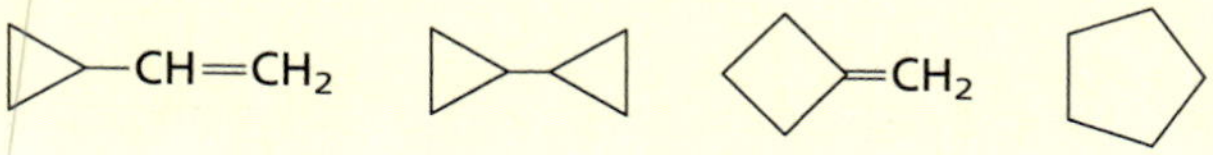

When *two* or more atoms are common to more than one ring, the compounds are called **polycyclic** ring systems. They are classified as *bicyclic, tricyclic, tetracyclic,* etc., according to the minimum number of bond cleavages required to generate a noncyclic structure. *Bicyclobutane* is the simplest bicyclic hydrocarbon; its four carbons form 2 three-membered rings that share a common side. *Camphene* is a naturally occurring bicyclic hydrocarbon obtained from pine oil. It is best regarded as a six-membered ring (indicated by blue bonds in the structure shown here) in which two of the carbons (designated by asterisks) are bridged by a CH₂ group.

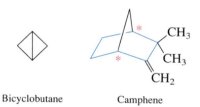

Bicyclobutane   Camphene

PROBLEM 3.12 Use the bond-cleavage criterion to verify that bicyclobutane and camphene are bicyclic.

Bicyclic compounds are named in the IUPAC system by counting the number of carbons in the ring system, assigning to the structure the base name of the unbranched alkane having the same number of carbon atoms, and attaching the prefix *bicyclo-*. The number of atoms in each of the bridges connecting the common atoms is then placed, in descending order, within brackets.

Bicyclo[3.2.0]heptane   Bicyclo[3.2.1]octane

Write structural formulas for each of the following bicyclic hydrocarbons:

(a) Bicyclo[2.2.1]heptane   (c) Bicyclo[3.1.1]heptane

(b) Bicyclo[5.2.0]nonane   (d) Bicyclo[3.3.0]octane

SAMPLE SOLUTION (a) The bicyclo[2.2.1]heptane ring system is one of the most frequently encountered bicyclic structural types. It contains seven carbon atoms, as indicated by the suffix -*heptane*. The bridging groups contain two, two, and one carbon, respectively.

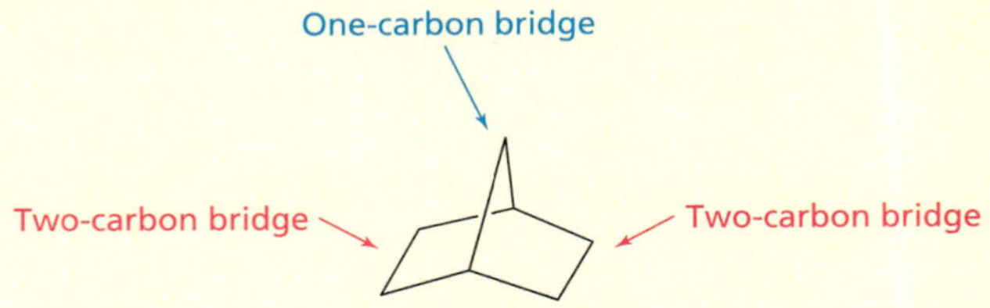

One-carbon bridge

Two-carbon bridge ← → Two-carbon bridge

Bicyclo[2.2.1]heptane

Among the most important of the bicyclic hydrocarbons are the two stereoisomeric bicyclo[4.4.0]decanes, called *cis-* and *trans*-decalin. The hydrogen atoms at the ring junctions are on the same side in *cis*-decalin and on opposite sides in *trans*-decalin. Both rings adopt the chair conformation in each stereoisomer.

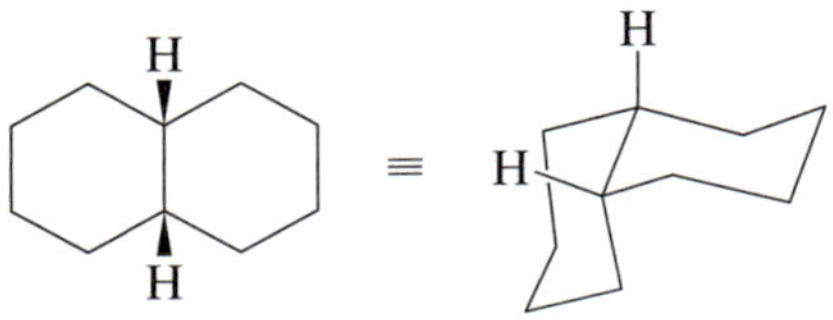

cis-Bicyclo[4.4.0]decane
(*cis*-decalin)

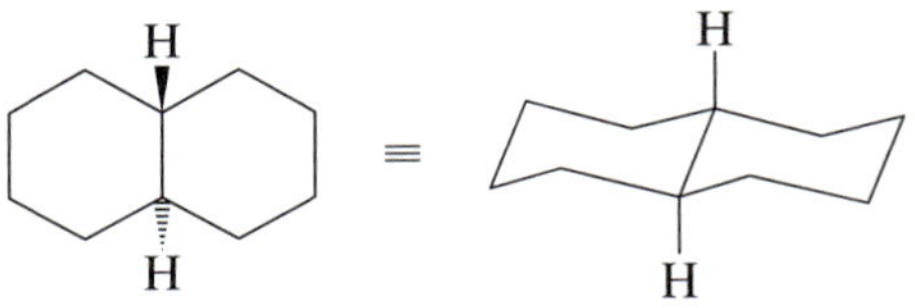

trans-Bicyclo[4.4.0]decane
(*trans*-decalin)

Make models of *cis*- and *trans*-decalin. Which is more stable? Why?

Decalin ring systems appear as structural units in a large number of naturally occurring substances, particularly the steroids. Cholic acid, for example, a steroid present in bile that promotes digestion, incorporates *cis*-decalin and *trans*-decalin units into a rather complex *tetracyclic* structure.

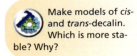

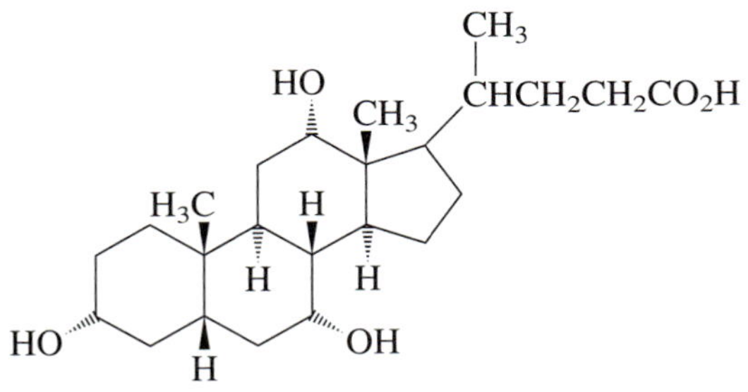

Cholic acid

3.15 HETEROCYCLIC COMPOUNDS

Not all cyclic compounds are hydrocarbons. Many substances include an atom other than carbon, called a *heteroatom* (Section 1.7), as part of a ring. A ring that contains at least one heteroatom is called a **heterocycle,** and a substance based on a heterocyclic ring is a **heterocyclic compound.** Each of the following heterocyclic ring systems will be encountered in this text:

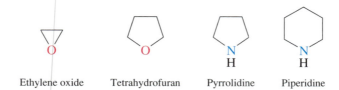

Ethylene oxide   Tetrahydrofuran   Pyrrolidine   Piperidine

The names cited are common names, which have been in widespread use for a long time and are acceptable in IUPAC nomenclature. We will introduce the systematic nomenclature of these ring systems as needed in later chapters.

The shapes of heterocyclic rings are very much like those of their all-carbon analogs. Thus, six-membered heterocycles such as piperidine exist in a chair conformation analogous to cyclohexane.

The hydrogen attached to nitrogen can be either axial or equatorial, and both chair conformations are approximately equal in stability.

PROBLEM 3.14 Draw or build a molecular model of what you would expect to be the most stable conformation of the piperidine derivative in which the hydrogen bonded to nitrogen has been replaced by methyl.

Sulfur-containing heterocycles are also common. Compounds in which sulfur is the heteroatom in three-, four-, five-, and six-membered rings, as well as larger rings, are all well known. Two interesting heterocyclic compounds that contain sulfur–sulfur bonds are *lipoic acid* and *lenthionine.*

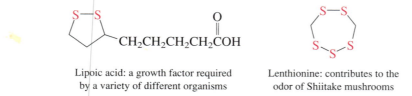

Lipoic acid: a growth factor required   Lenthionine: contributes to the
by a variety of different organisms   odor of Shiitake mushrooms

Many heterocyclic systems contain double bonds and are related to arenes. The most important representatives of this class are described in Sections 11.22 and 11.23.

3.16 SUMMARY

In this chapter we explored the three-dimensional shapes of alkanes and cycloalkanes. The most important point to be taken from the chapter is that a molecule adopts the shape that minimizes its total **strain.** The sources of strain in alkanes and cycloalkanes are:

1. *Bond length distortion:* destabilization of a molecule that results when one or more of its bond distances are different from the normal values

2. *Angle strain:* destabilization that results from distortion of bond angles from their normal values

3. *Torsional strain:* destabilization that results when adjacent atoms are not staggered

4. *Van der Waals strain:* destabilization that results when atoms or groups on non-adjacent atoms are too close to one another

The various spatial arrangements available to a molecule by rotation about single bonds are called **conformations,** and **conformational analysis** is the study of the differences in stability and properties of the individual conformations. Rotation around carbon–carbon single bonds is normally very fast, occurring hundreds of thousands of times per second at room temperature. Molecules are rarely frozen into a single conformation but engage in rapid equilibration among the conformations that are energetically accessible.

Section 3.1　The most stable conformation of ethane is the **staggered** conformation. It is approximately 12 kJ/mol (3 kcal/mol) more stable than the **eclipsed,** which is the least stable conformation.

Staggered conformation of ethane
(most stable conformation)

Eclipsed conformation of ethane
(least stable conformation)

The difference in energy between the staggered and eclipsed forms is due almost entirely to the torsional strain in the eclipsed conformation. At any instant, almost all the molecules of ethane reside in the staggered conformation.

Section 3.2　The two staggered conformations of butane are not equivalent. The **anti** conformation is more stable than the **gauche.**

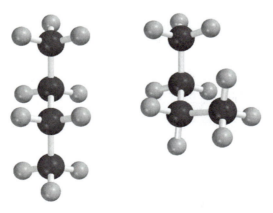

Anti conformation of butane　　　Gauche conformation of butane

Neither conformation suffers torsional strain, because each has a staggered arrangement of bonds. The gauche conformation is less stable because of van der Waals strain involving the methyl groups.

Section 3.3   Higher alkanes adopt a zigzag conformation of the carbon chain in which all the bonds are staggered.

Octane

Section 3.4   At one time all cycloalkanes were believed to be planar. It was expected that cyclopentane would be the least strained cycloalkane because the angles of a regular pentagon (108°) are closest to the tetrahedral angle of 109.5°. Heats of combustion established that this is not so. With the exception of cyclopropane, the rings of all cycloalkanes are nonplanar.

Section 3.5   Cyclopropane is planar and destabilized by angle strain and torsional strain. Cyclobutane is nonplanar and less strained than cyclopropane.

Cyclopropane   Cyclobutane

Section 3.6   Cyclopentane has two nonplanar conformations that are of similar stability: the **envelope** and the **half-chair.**

Envelope conformation of cyclopentane   Half-chair conformation of cyclopentane

Section 3.7   Three conformations of cyclohexane have approximately tetrahedral angles at carbon: the chair, the boat, and the skew boat. The chair is by far the most stable; it is free of torsional strain, but the boat and skew boat are not. When a cyclohexane ring is present in a compound, it almost always adopts a chair conformation.

Chair   Skew boat   Boat

Section 3.8 The C—H bonds in the chair conformation of cyclohexane are not all equivalent but are divided into two sets of six each, called **axial** and **equatorial.**

Axial bonds to H in cyclohexane   Equatorial bonds to H in cyclohexane

Section 3.9 Conformational inversion (ring flipping) is rapid in cyclohexane and causes all axial bonds to become equatorial and vice versa. As a result, a monosubstituted derivative of cyclohexane adopts the chair conformation in which the substituent is equatorial (see next section). *No bonds are made or broken in this process.*

Section 3.10 A substituent is less crowded and more stable when it is equatorial than when it is axial on a cyclohexane ring. Ring flipping of a monosubstituted cyclohexane allows the substituent to become equatorial.

Methyl group axial (less stable)   Methyl group equatorial (more stable)

Branched substituents, especially *tert*-butyl, have an increased preference for the equatorial position.

Sections 3.11–3.12 **Stereoisomers** are isomers that have the same constitution but differ in the arrangement of atoms in space. *Cis-* and *trans*-1,3-dimethylcyclohexane are stereoisomers. The cis isomer is more stable than the trans.

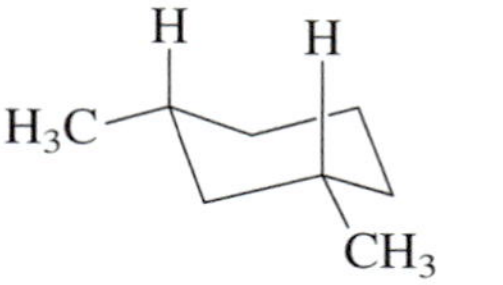

Most stable conformation of
cis-1,3-dimethylcyclohexane
(no axial methyl groups)

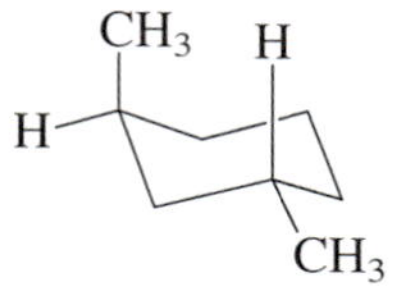

Most stable conformation of
trans-1,3-dimethylcyclohexane
(one axial methyl group)

Section 3.13   Higher cycloalkanes have angles at carbon that are close to tetrahedral and are sufficiently flexible to adopt conformations that reduce their torsional strain. They tend to be populated by several different conformations of similar stability.

Section 3.14   Cyclic hydrocarbons can contain more than one ring. **Spirocyclic** hydrocarbons are characterized by the presence of a single carbon that is common to two rings. **Bicyclic** alkanes contain two rings that share two or more atoms.

Section 3.15   Substances that contain one or more atoms other than carbon as part of a ring are called **heterocyclic** compounds. Rings in which the heteroatom is oxygen, nitrogen, or sulfur rank as both the most common and the most important.

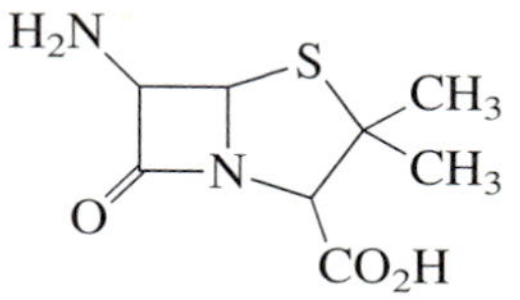

6-Aminopenicillanic acid
(bicyclic and heterocyclic)

PROBLEMS

3.15   Like hydrogen peroxide, the inorganic substances hydrazine (H_2NNH_2) and hydroxylamine (H_2NOH) possess conformational mobility. Write structural representations or build molecular models of two different staggered conformations of (a) hydrazine and (b) hydroxylamine.

3.16   Of the three conformations of propane shown, which one is the most stable? Which one is the least stable? Why?

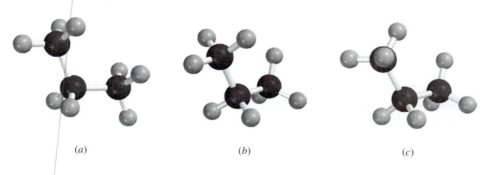

(a)   (b)   (c)

3.17 Sight down the C-2—C-3 bond, and draw Newman projection formulas for the

(a) Most stable conformation of 2,2-dimethylbutane

(b) Two most stable conformations of 2-methylbutane

(c) Two most stable conformations of 2,3-dimethylbutane

3.18 One of the staggered conformations of 2-methylbutane in Problem 3.17b is more stable than the other. Which one is more stable? Why?

3.19 Sketch an approximate potential energy diagram similar to that shown in Figures 3.4 and 3.7 for rotation about the carbon–carbon bond in 2,2-dimethylpropane. Does the form of the potential energy curve of 2,2-dimethylpropane more closely resemble that of ethane or that of butane?

3.20 Repeat Problem 3.19 for the case of 2-methylbutane.

3.21 The compound 2,2,4,4-tetramethylpentane $[(CH_3)_3CCH_2C(CH_3)_3]$ is distinctive because it has an unusually large C—C—C bond angle. What carbons are involved? How large is the angle? What steric factor is responsible for increasing the size of this angle? One of the other bond angles is unusually small. Which one?

3.22 Even though the methyl group occupies an equatorial site, the conformation shown is not the most stable one for methylcyclohexane. Explain.

3.23 Which of the structures shown for the axial conformation of methylcyclohexane do you think is more stable, A or B? Why?

A   B

3.24 Which do you expect to be the more stable conformation of *cis*-1,3-dimethylcyclobutane, A or B? Why?

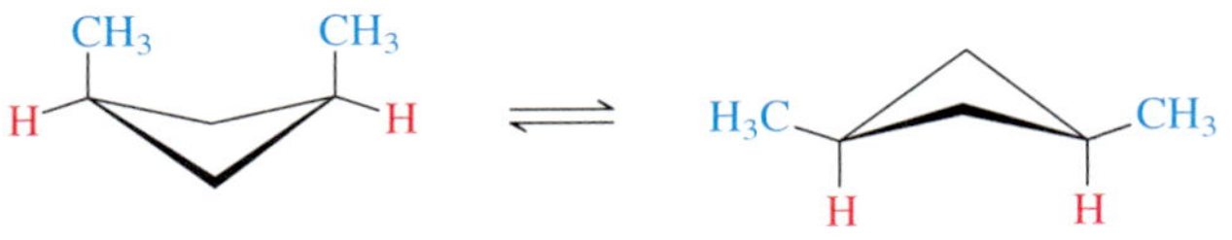

3.25 Determine whether the two structures in each of the following pairs represent *constitutional isomers*, different *conformations* of the same compound, or *stereoisomers* that cannot be interconverted by rotation about single bonds.

(a) 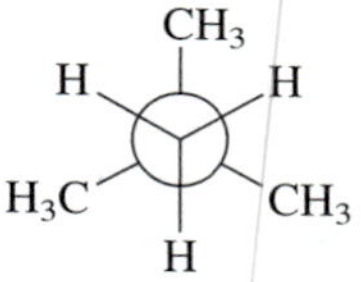 and

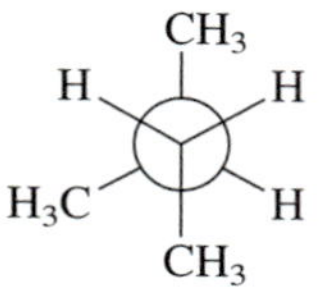

(b)  and

(c) 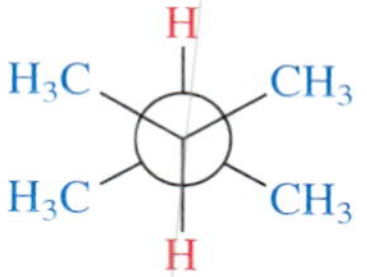 and

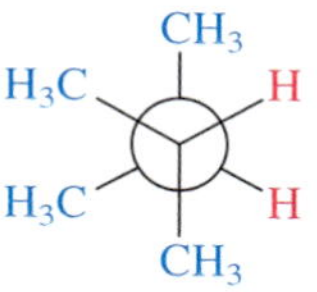

(d) *cis*-1,2-Dimethylcyclopentane and *trans*-1,3-dimethylcyclopentane

(e) 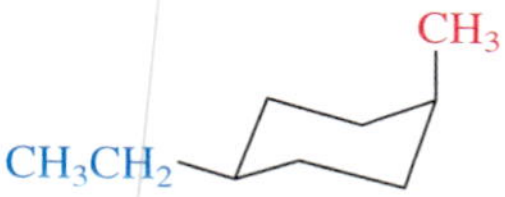 and

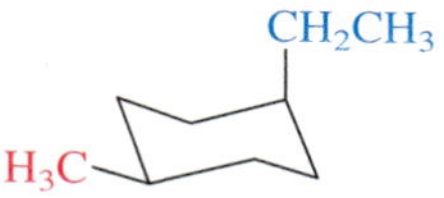

(f) 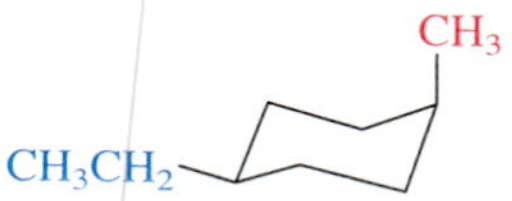 and

(g) 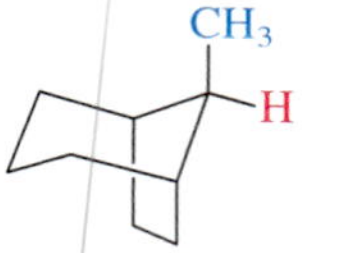 and

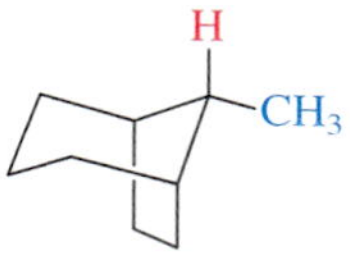

3.26 Excluding compounds that contain methyl or ethyl groups, write structural formulas for all the bicyclic isomers of (a) C_5H_8 and (b) C_6H_{10}.

3.27 In each of the following groups of compounds, identify the one with the largest heat of combustion and the one with the smallest. In which cases can a comparison of heats of combustion be used to assess relative stability?

(a) Cyclopropane, cyclobutane, cyclopentane

(b) *cis*-1,2-Dimethylcyclopentane, methylcyclohexane, 1,1,2,2-tetramethylcyclopropane

(c)

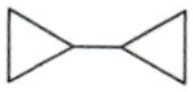

(d)

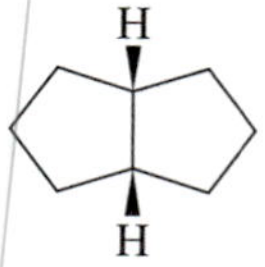

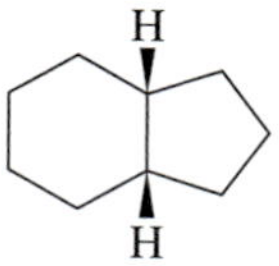

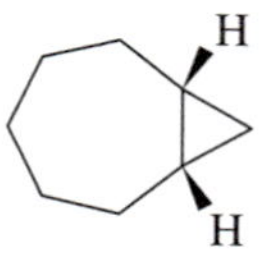

3.28 Write a structural formula for the most stable conformation of each of the following compounds:

(a) 2,2,5,5-Tetramethylhexane (Newman projection of conformation about C-3—C-4 bond)

(b) 2,2,5,5-Tetramethylhexane (zigzag conformation of entire molecule)

(c) *cis*-1-Isopropyl-3-methylcyclohexane

(d) *trans*-1-Isopropyl-3-methylcyclohexane

(e) *cis*-1-*tert*-Butyl-4-ethylcyclohexane

(f) *cis*-1,1,3,4-Tetramethylcyclohexane

(g)

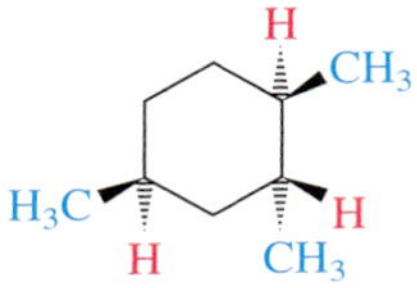

3.29 Identify the more stable stereoisomer in each of the following pairs, and give the reason for your choice:

(a) *cis*- or *trans*-1-Isopropyl-2-methylcyclohexane

(b) *cis*- or *trans*-1-Isopropyl-3-methylcyclohexane

(c) *cis*- or *trans*-1-Isopropyl-4-methylcyclohexane

(d)

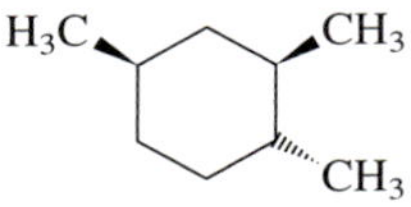

or

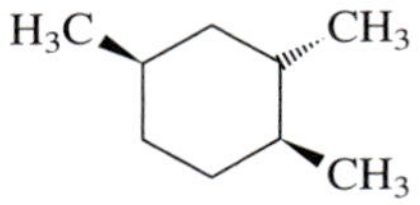

(e)

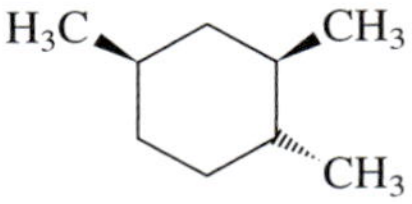

or

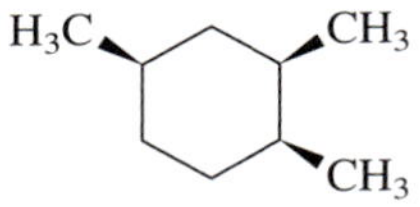

(f)

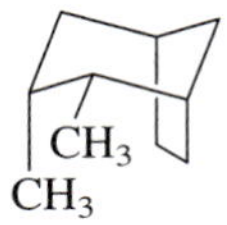

or

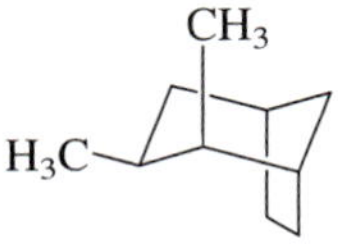

3.30 One stereoisomer of 1,1,3,5-tetramethylcyclohexane is 15 kJ/mol (3.7 kcal/mol) less stable than the other. Indicate which isomer is the less stable, and identify the reason for its decreased stability.

3.31 The heats of combustion of the more and less stable stereoisomers of the 1,2-, 1,3-, and 1,4-dimethylcyclohexanes are given here. The values are higher for the 1,2-dimethylcyclohexanes than for the 1,3- and 1,4-isomers. Suggest an explanation.

Dimethylcyclohexane	1,2	1,3	1,4
Heats of combustion (kJ/mol):			
More stable stereoisomer	5217	5212	5212
Less stable stereoisomer	5223	5219	5219

3.32   One of the following two stereoisomers is 20 kJ/mol (4.9 kcal/mol) less stable than the other. Indicate which isomer is the less stable, and identify the reason for its decreased stability.

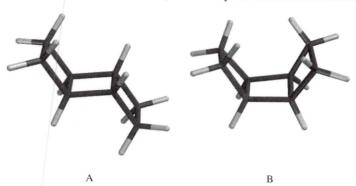

A   B

3.33   Cubane (C_8H_8) is the common name of a polycyclic hydrocarbon that was first synthesized in the early 1960s. As its name implies, its structure is that of a cube. How many rings are present in cubane?

Cubane

3.34   The following are representations of two forms of glucose. The six-membered ring is known to exist in a chair conformation in each form. Draw clear representations of the most stable conformation of each. Are they two different conformations of the same molecule, or are they stereoisomers? Which substituents (if any) occupy axial sites?

HOH$_2$C   HOH$_2$C

HO ·····| ||···· OH   HO ·····| ◄—OH

HO   OH   HO   OH

3.35   A typical steroid skeleton is shown along with the numbering scheme used for this class of compounds. Specify in each case whether the designated substituent is axial or equatorial.

(a) Substituent at C-1 cis to the methyl groups

(b) Substituent at C-4 cis to the methyl groups

(c) Substituent at C-7 trans to the methyl groups

(d) Substituent at C-11 trans to the methyl groups

(e) Substituent at C-12 cis to the methyl groups

3.36 Repeat Problem 3.35 for the stereoisomeric steroid skeleton having a cis ring fusion between the first two rings.

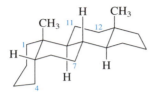

3.37 (a) Write Newman projections for the gauche and anti conformations of 1,2-dichloroethane ($ClCH_2CH_2Cl$).

(b) The measured dipole moment of $ClCH_2CH_2Cl$ is 1.12 D. Which one of the following statements about 1,2-dichloroethane is false?

(1) It may exist entirely in the anti conformation.

(2) It may exist entirely in the gauche conformation.

(3) It may exist as a mixture of anti and gauche conformations.

3.38 Compare the two staggered conformations of 1,1,2,2-tetrafluoroethane on *Learning By Modeling*. Do they differ in respect to their dipole moments? How?

3.39 Structural drawings (molecular models, too) can be deceiving. For example, the chlorine atoms in 1,2-dichlorocyclohexane seem much closer to each other in a drawing of the trans stereoisomer than in the cis. Make a molecular model of each, and measure the distance between the chlorines. What do you find?

trans-1,2-Dichlorocyclohexane   *cis*-1,2-Dichlorocyclohexane

3.40 Two stereoisomers of bicyclo[3.3.0]octane are possible. Make molecular models of both, and determine which is more stable.

Bicyclo[3.3.0]octane

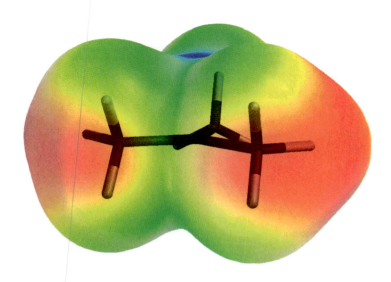

CHAPTER 4

ALCOHOLS AND ALKYL HALIDES

Our first three chapters established some fundamental principles concerning the *structure* of organic molecules and introduced the connection between structure and *reactivity* with a review of acid–base reactions. In this chapter we explore structure and reactivity in more detail by developing two concepts: *functional groups* and *reaction mechanisms*. A **functional group** is the atom or group in a molecule most responsible for the reaction the compound undergoes under a prescribed set of conditions. *How* the structure of the reactant is transformed to that of the product is what we mean by the **reaction mechanism.**

Organic compounds are grouped into families according to the functional groups they contain. Two of the most important families are *alcohols* and *alkyl halides*. Alcohols and alkyl halides are especially useful because they are versatile starting materials for preparing numerous other families. Indeed, alcohols or alkyl halides—often both—will appear in virtually all of the remaining chapters of this text.

The major portion of the present chapter concerns the conversion of alcohols to alkyl halides by reaction with hydrogen halides:

$$\text{R—OH} + \text{H—X} \longrightarrow \text{R—X} + \text{H—OH}$$

| Alcohol | Hydrogen halide | Alkyl halide | Water |

It is convenient in equations such as this to represent generic alcohols and alkyl halides as ROH and RX, respectively, where "R" stands for an alkyl group. In addition to convenience, this notation lets us focus more clearly on the functional group transformation that occurs; the OH functional group of an alcohol is replaced as a substituent on carbon by a halogen, usually chlorine (X = Cl) or bromine (X = Br).

While developing the connections between structure, reaction, and mechanism, we will also extend the fundamentals of IUPAC nomenclature to functional group families, beginning with alcohols and alkyl halides.

4.1   FUNCTIONAL GROUPS

The families of hydrocarbons—*alkanes, alkenes, alkynes,* and *arenes*—were introduced in Section 2.1. The double bond is a functional group in an alkene, the triple bond a functional group in an alkyne, and the benzene ring itself is a functional group in an arene. Alkanes (RH) are not considered to have a functional group, although as we'll see later in this chapter, reactions that replace a hydrogen atom can take place. In general though, hydrogen atoms of alkanes are relatively unreactive and any other group attached to the hydrocarbon framework (R) will be the functional group.

Table 4.1 lists the major families of organic compounds covered in this text and their functional groups.

TABLE 4.1	Functional Groups in Some Important Classes of Organic Compounds		
Class	**Generalized abbreviation***	**Representative example**	**Name of example[†]**
Alcohol	ROH	CH_3CH_2OH	Ethanol
Alkyl halide	RCl	CH_3CH_2Cl	Chloroethane
Amine[‡]	RNH_2	$CH_3CH_2NH_2$	Ethanamine
Epoxide	$R_2C\overset{\diagdown\diagup}{\underset{O}{\textstyle-\!-}}CR_2$	$H_2C\overset{\diagdown\diagup}{\underset{O}{\textstyle-\!-}}CH_2$	Oxirane
Ether	ROR	$CH_3CH_2OCH_2CH_3$	Diethyl ether
Nitrile	$RC\equiv N$	$CH_3CH_2C\equiv N$	Propanenitrile
Nitroalkane	RNO_2	$CH_3CH_2NO_2$	Nitroethane
Sulfide	RSR	CH_3SCH_3	Dimethyl sulfide
Thiol	RSH	CH_3CH_2SH	Ethanethiol
Aldehyde	$R\overset{O}{\overset{\|}{C}}H$	$CH_3\overset{O}{\overset{\|}{C}}H$	Ethanal
Ketone	$R\overset{O}{\overset{\|}{C}}R$	$CH_3\overset{O}{\overset{\|}{C}}CH_3$	2-Propanone
Carboxylic acid	$R\overset{O}{\overset{\|}{C}}OH$	$CH_3\overset{O}{\overset{\|}{C}}OH$	Ethanoic acid
Carboxylic acid derivatives			
Acyl halide	$R\overset{O}{\overset{\|}{C}}X$	$CH_3\overset{O}{\overset{\|}{C}}Cl$	Ethanoyl chloride
Acid anhydride	$R\overset{O}{\overset{\|}{C}}O\overset{O}{\overset{\|}{C}}R$	$CH_3\overset{O}{\overset{\|}{C}}O\overset{O}{\overset{\|}{C}}CH_3$	Ethanoic anhydride
Ester	$R\overset{O}{\overset{\|}{C}}OR$	$CH_3\overset{O}{\overset{\|}{C}}OCH_2CH_3$	Ethyl ethanoate
Amide	$R\overset{O}{\overset{\|}{C}}NR_2$	$CH_3\overset{O}{\overset{\|}{C}}NH_2$	Ethanamide

*When more than one R group is present, the groups may be the same or different.
[†]Most compounds have more than one acceptable name.
[‡]The example given is a *primary* amine (RNH_2). *Secondary* amines have the general structure R_2NH; *tertiary* amines are R_3N.

> **PROBLEM 4.1** (a) Write a structural formula for a sulfide having the molecular formula C_3H_8S. (b) What two thiols have the molecular formula C_3H_8S?
>
> **SAMPLE SOLUTION** (a) According to Table 4.1, sulfides have the general formula RSR and the Rs may be the same or different. The only possible connectivity for a sulfide with three carbons is C—S—C—C. Therefore, the sulfide is $CH_3SCH_2CH_3$.

We have already touched on some of these functional-group families in our discussion of acids and bases. We have seen that alcohols resemble water in their acidity and that carboxylic acids, although weak acids, are stronger acids than alcohols. Carboxylic acids belong to one of the most important groups of organic compounds—those that contain carbonyl groups (C=O). They and other carbonyl-containing compounds rank among the most abundant and biologically significant classes of naturally occurring substances.

Carbonyl group chemistry is discussed in a block of five chapters (Chapters 17–21).

> **PROBLEM 4.2** Many compounds contain more than one functional group. Prostaglandin E_1, a hormone that regulates the relaxation of smooth muscles, contains two different kinds of carbonyl groups. Classify each one (aldehyde, ketone, carboxylic acid, ester, amide, acyl chloride, or acid anhydride). Identify the most acidic proton in prostaglandin E_1 and use Table 1.7 to estimate its pK_a.
>
>
> Prostaglandin E_1

4.2 IUPAC NOMENCLATURE OF ALKYL HALIDES

The IUPAC rules permit alkyl halides to be named in two different ways, called *functional class* nomenclature and *substitutive* nomenclature. In **functional class nomenclature** the alkyl group and the halide (*fluoride, chloride, bromide,* or *iodide*) are designated as separate words. The alkyl group is named on the basis of its longest continuous chain beginning at the carbon to which the halogen is attached.

The IUPAC rules permit certain common alkyl group names to be used. These include n-propyl, isopropyl, n-butyl, sec-butyl, isobutyl, tert-butyl, and neopentyl (Section 2.13).

CH_3F $CH_3CH_2CH_2CH_2CH_2Cl$ $\overset{1}{C}H_3\overset{2}{C}H_2\overset{3}{C}HCH_2CH_2CH_3$
$\qquad\qquad\qquad\qquad\qquad\qquad\qquad\qquad\qquad\qquad\quad |$
$\qquad\qquad\qquad\qquad\qquad\qquad\qquad\qquad\qquad\qquad\ Br$

Methyl fluoride   Pentyl chloride   1-Ethylbutyl bromide   Cyclohexyl iodide

Substitutive nomenclature of alkyl halides treats the halogen as a *halo—(fluoro-, chloro-, bromo-,* or *iodo-) substituent* on an alkane chain. The carbon chain is numbered in the direction that gives the substituted carbon the lower number.

$\overset{5}{C}H_3\overset{4}{C}H_2\overset{3}{C}H_2\overset{2}{C}H_2\overset{1}{C}H_2F$ $\overset{1}{C}H_3\overset{2}{C}HCH_2CH_2CH_3$ $\overset{1}{C}H_3\overset{2}{C}H_2\overset{3}{C}HCH_2CH_3$
$\qquad\qquad\qquad\qquad\qquad\qquad\ |$
$\qquad\qquad\qquad\qquad\qquad\quad Br$

1-Fluoropentane   2-Bromopentane   3-Iodopentane

When the carbon chain bears both a halogen and an alkyl substituent, the two are considered of equal rank, and the chain is numbered so as to give the lower number to the substituent nearer the end of the chain.

$$\overset{1}{C}H_3\overset{2}{C}H\overset{3}{C}H_2\overset{4}{C}H_2\overset{5}{C}H\overset{6}{C}H_2\overset{7}{C}H_3 \qquad \overset{1}{C}H_3\overset{2}{C}H\overset{3}{C}H_2\overset{4}{C}H_2\overset{5}{C}H\overset{6}{C}H_2\overset{7}{C}H_3$$

$$\qquad\quad CH_3 \qquad\quad Cl \qquad\qquad\qquad Cl \qquad\qquad\quad CH_3$$

5-Chloro-2-methylheptane   2-Chloro-5-methylheptane

PROBLEM 4.3 Write structural formulas or build molecular models, and give the functional class and substitutive names of all the isomeric alkyl chlorides that have the molecular formula C_4H_9Cl.

Substitutive names are preferred, but functional class names are sometimes more convenient or more familiar and are frequently encountered in organic chemistry.

Functional class names are part of the IUPAC system; they are not common names.

4.3   IUPAC NOMENCLATURE OF ALCOHOLS

Functional class names of alcohols are derived by naming the alkyl group that bears the hydroxyl substituent (—OH) and then adding *alcohol* as a separate word. The chain is always numbered beginning at the carbon to which the hydroxyl group is attached.

Substitutive names of alcohols are developed by identifying the longest continuous chain that bears the hydroxyl group and replacing the -*e* ending of the corresponding alkane by the suffix -*ol*. The position of the hydroxyl group is indicated by number, choosing the sequence that assigns the lower locant to the carbon that bears the hydroxyl group.

Several alcohols are commonplace substances, well known by common names that reflect their origin (wood alcohol, grain alcohol) or use (rubbing alcohol). Wood alcohol is methanol *(methyl alcohol, CH_3OH), grain alcohol is* ethanol *(ethyl alcohol, CH_3CH_2OH), and rubbing alcohol is* 2-propanol *[isopropyl alcohol, $(CH_3)_2CHOH$].*

$$CH_3CH_2OH \qquad CH_3CHCH_2CH_2CH_2CH_3 \qquad \begin{matrix} CH_3 \\ | \\ CH_3CCH_2CH_2CH_3 \\ | \end{matrix}$$

$$\qquad\qquad\qquad\qquad\quad OH \qquad\qquad\qquad\qquad OH$$

Functional class name:	Ethyl alcohol	1-Methylpentyl alcohol	1,1-Dimethylbutyl alcohol
Substitutive name:	Ethanol	2-Hexanol	2-Methyl-2-pentanol

Hydroxyl groups take precedence over ("outrank") alkyl groups and halogen substituents in determining the direction in which a carbon chain is numbered. The OH group is assumed to be attached to C-1 of a cyclic alcohol, and is not numbered.

$$\overset{7}{C}H_3\overset{6}{C}H\overset{5}{C}H_2\overset{4}{C}H_2\overset{3}{C}H\overset{2}{C}H_2\overset{1}{C}H_3 \qquad\qquad\qquad FCH_2CH_2CH_2OH$$

$$\quad CH_3 \qquad\quad OH$$

6-Methyl-3-heptanol   *trans*-2-Methylcyclopentanol   3-Fluoro-1-propanol
(not 2-methyl-5-heptanol)

PROBLEM 4.4 Write structural formulas or build molecular models, and give the functional class and substitutive names of all the isomeric alcohols that have the molecular formula $C_4H_{10}O$.

4.4   CLASSES OF ALCOHOLS AND ALKYL HALIDES

Alcohols and alkyl halides are classified as primary, secondary, or tertiary according to the degree of substitution of the carbon that bears the functional group (Section 2.13). Thus, *primary alcohols* and *primary alkyl halides* are compounds of the type RCH_2G (where G is the functional group), *secondary alcohols* and *secondary alkyl halides* are compounds of the type R_2CHG, and *tertiary alcohols* and *tertiary alkyl halides* are compounds of the type R_3CG.

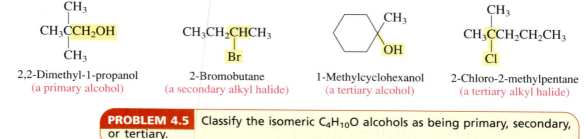

2,2-Dimethyl-1-propanol
(a primary alcohol)

2-Bromobutane
(a secondary alkyl halide)

1-Methylcyclohexanol
(a tertiary alcohol)

2-Chloro-2-methylpentane
(a tertiary alkyl halide)

PROBLEM 4.5   Classify the isomeric $C_4H_{10}O$ alcohols as being primary, secondary, or tertiary.

Many of the properties of alcohols and alkyl halides are affected by whether their functional groups are attached to primary, secondary, or tertiary carbons. We will see a number of cases in which a functional group attached to a primary carbon is more reactive than one attached to a secondary or tertiary carbon, as well as other cases in which the reverse is true.

4.5   BONDING IN ALCOHOLS AND ALKYL HALIDES

The carbon that bears the functional group is sp^3-hybridized in alcohols and alkyl halides. Figure 4.1 illustrates bonding in methanol. The bond angles at carbon are approximately tetrahedral, as is the C—O—H angle. A similar orbital hybridization model applies to alkyl halides, with the halogen connected to sp^3-hybridized carbon by a σ bond. Carbon–halogen bond distances in alkyl halides increase in the order C—F (140 pm) < C—Cl (179 pm) < C—Br (197 pm) < C—I (216 pm).

FIGURE 4.1 Orbital hybridization model of bonding in methanol. (*a*) The orbitals used in bonding are the 1*s* orbital of hydrogen and sp^3-hybridized orbitals of carbon and oxygen. (*b*) The bond angles at carbon and oxygen are close to tetrahedral, and the carbon–oxygen σ bond is about 10 pm shorter than a carbon–carbon single bond.

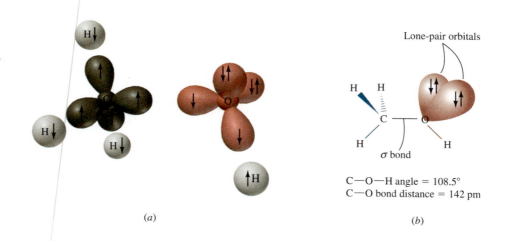

Lone-pair orbitals

C—O—H angle = 108.5°
C—O bond distance = 142 pm

(*a*)

(*b*)

Methanol (CH$_3$OH)   Chloromethane (CH$_3$Cl)

FIGURE 4.2 Electrostatic potential maps of methanol and chloromethane. The electrostatic potential is most negative near oxygen in methanol and near chlorine in chloromethane. The most positive region is near the O—H proton in methanol and near the methyl group in chloromethane.

Carbon–oxygen and carbon–halogen bonds are polar covalent bonds, and carbon bears a partial positive charge in alcohols ($^{\delta+}$C—O$^{\delta-}$) and in alkyl halides ($^{\delta+}$C—X$^{\delta-}$). Alcohols and alkyl halides are polar molecules. The dipole moments of methanol and chloromethane are very similar to each other and to water.

Water
(μ = 1.8 D)

Methanol
(μ = 1.7 D)

Chloromethane
(μ = 1.9 D)

PROBLEM 4.6 Bromine is less electronegative than chlorine, yet methyl bromide and methyl chloride have very similar dipole moments. Why?

Figure 4.2 maps the electrostatic potential in methanol and chloromethane. Both are similar in that the sites of highest negative potential (red) are near the electronegative atoms: oxygen and chlorine. The polarization of the bonds to oxygen and chlorine, as well as their unshared electron pairs, contribute to the concentration of negative charge on these atoms.

Relatively simple notions of attractive forces between opposite charges are sufficient to account for many of the properties of chemical substances. You will find it helpful to keep the polarity of carbon–oxygen and carbon–halogen bonds in mind as we develop the properties of alcohols and alkyl halides in later sections.

4.6 PHYSICAL PROPERTIES OF ALCOHOLS AND ALKYL HALIDES: INTERMOLECULAR FORCES

Boiling Point. When describing the effect of alkane structure on boiling point in Section 2.17, we pointed out that van der Waals attractive forces between neutral molecules are of three types. The first two involve induced dipoles and are often referred to as *dispersion forces,* or *London forces.*

1. Induced-dipole/induced-dipole forces
2. Dipole/induced-dipole forces
3. Dipole–dipole forces

FIGURE 4.3 A dipole–dipole attractive force. Two molecules of a polar substance associate so that the positively polarized region of one and the negatively polarized region of the other attract each other.

Induced-dipole/induced-dipole forces are the only intermolecular attractive forces available to nonpolar molecules such as alkanes. In addition to these forces, polar molecules engage in dipole–dipole and dipole/induced-dipole attractions. The **dipole–dipole attractive force** is easiest to visualize and is illustrated in Figure 4.3. Two molecules of a polar substance experience a mutual attraction between the positively polarized region of one molecule and the negatively polarized region of the other. As its name implies, the **dipole/induced-dipole force** combines features of both the induced-dipole/induced-dipole and dipole–dipole attractive forces. A polar region of one molecule alters the electron distribution in a nonpolar region of another in a direction that produces an attractive force between them.

Because so many factors contribute to the net intermolecular attractive force, it is not always possible to predict which of two compounds will have the higher boiling point. We can, however, use the boiling point behavior of selected molecules to inform us of the relative importance of various intermolecular forces and the structural features that influence them.

Consider three compounds similar in size and shape: the alkane propane, the alkyl halide fluoroethane, and the alcohol ethanol.

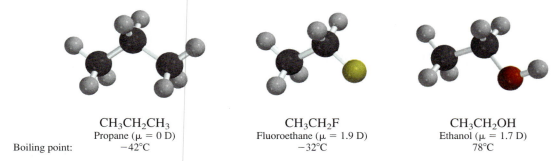

$CH_3CH_2CH_3$	CH_3CH_2F	CH_3CH_2OH
Propane ($\mu = 0$ D)	Fluoroethane ($\mu = 1.9$ D)	Ethanol ($\mu = 1.7$ D)
Boiling point: $-42°C$	$-32°C$	$78°C$

Both polar compounds, ethanol and fluoroethane, have higher boiling points than the nonpolar propane. We attribute this to a combination of dipole/induced-dipole and dipole–dipole attractive forces that are present in the liquid states of ethanol and fluoroethane, but absent in propane.

The most striking aspect of the data, however, is the much higher boiling point of ethanol compared with propane and fluoroethane. This suggests that the attractive forces in ethanol must be unusually strong. Figure 4.4 shows that this force results from a dipole–dipole attraction between the positively polarized proton of the —OH group of one ethanol molecule and the negatively polarized oxygen of another. The term **hydrogen bonding** is used to describe dipole–dipole attractive forces of this type. The proton involved must be bonded to an electronegative element, usually oxygen or nitrogen. Protons in C—H bonds do not participate in hydrogen bonding. Thus fluoroethane, even though it is a polar molecule and engages in dipole–dipole attractions, does *not* form hydrogen bonds and, therefore, has a lower boiling point than ethanol.

Hydrogen bonds between —OH groups are stronger than those between —NH groups, as a comparison of the boiling points of water (H_2O, 100°C) and ammonia (NH_3, −33°C) demonstrates.

Hydrogen bonding can be expected in molecules that have —OH or —NH groups. Individual hydrogen bonds are about 10–50 times weaker than typical covalent bonds, but their effects can be significant. More than other dipole–dipole attractive forces, intermolecular hydrogen bonds are strong enough to impose a relatively high degree of structural order on systems in which they are possible. As will be seen in Chapters 27 and 28, the three-dimensional structures adopted by proteins and nucleic acids, the organic molecules of life, are strongly influenced by hydrogen bonding.

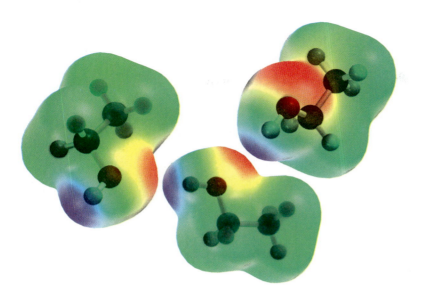

FIGURE 4.4 Hydrogen bonding in ethanol involves the oxygen of one molecule and the proton of the —OH group of another. Hydrogen bonding is much stronger than most other types of dipole–dipole attractive forces, but is not nearly as strong as a covalent bond.

PROBLEM 4.7 The constitutional isomer of ethanol, dimethyl ether (CH_3OCH_3), is a gas at room temperature. Suggest an explanation for this observation.

Table 4.2 lists the boiling points of some representative alkyl halides and alcohols. When comparing the boiling points of related compounds as a function of the *alkyl group*, we find that the boiling point increases with the number of carbon atoms, as it does with alkanes.

With respect to the *halogen* in a group of alkyl halides, the boiling point increases as one descends the periodic table; alkyl fluorides have the lowest boiling points, alkyl iodides the highest. This trend matches the order of increasing polarizability of the halogens. **Polarizability** is the ease with which the electron distribution around an atom is distorted by a nearby electric field and is a significant factor in determining the strength of induced-dipole/induced-dipole and dipole/induced-dipole attractions. Forces that depend on induced dipoles are strongest when the halogen is a highly polarizable iodine, and weakest when the halogen is a nonpolarizable fluorine.

For a discussion concerning the boiling point behavior of alkyl halides, see the January 1988 issue of the *Journal of Chemical Education*, pp. 62–64.

TABLE 4.2   Boiling Points of Some Alkyl Halides and Alcohols

Name of alkyl group	Formula	Substituent X and boiling point, °C (1 atm)				
		X = F	X = Cl	X = Br	X = I	X = OH
Methyl	CH_3X	−78	−24	3	42	65
Ethyl	CH_3CH_2X	−32	12	38	72	78
Propyl	$CH_3CH_2CH_2X$	−3	47	71	103	97
Pentyl	$CH_3(CH_2)_3CH_2X$	65	108	129	157	138
Hexyl	$CH_3(CH_2)_4CH_2X$	92	134	155	180	157

The boiling points of the chlorinated derivatives of methane increase with the number of chlorine atoms because of an increase in the induced-dipole/induced-dipole attractive forces.

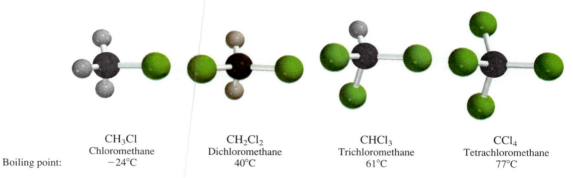

CH_3Cl	CH_2Cl_2	$CHCl_3$	CCl_4
Chloromethane	Dichloromethane	Trichloromethane	Tetrachloromethane
Boiling point: −24°C	40°C	61°C	77°C

Fluorine is unique among the halogens in that increasing the number of fluorines does not lead to higher and higher boiling points.

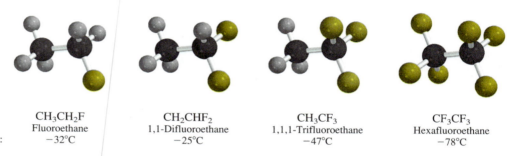

CH_3CH_2F	CH_2CHF_2	CH_3CF_3	CF_3CF_3
Fluoroethane	1,1-Difluoroethane	1,1,1-Trifluoroethane	Hexafluoroethane
Boiling point: −32°C	−25°C	−47°C	−78°C

These boiling points illustrate why we should do away with the notion that boiling points always increase with increasing molecular weight.

Thus, although the difluoride CH_3CHF_2 boils at a higher temperature than CH_3CH_2F, the trifluoride CH_3CF_3 boils at a lower temperature than either of them. Even more striking is the observation that the hexafluoride CF_3CF_3 is the lowest boiling of any of the fluorinated derivatives of ethane. The boiling point of CF_3CF_3 is, in fact, only 11°C higher than that of ethane itself. The reason for this behavior has to do with the very low polarizability of fluorine and a decrease in induced-dipole/induced-dipole forces that accompanies the incorporation of fluorine substituents into a molecule. Their weak intermolecular attractive forces give fluorinated hydrocarbons (**fluorocarbons**) certain desirable physical properties such as that found in the "no stick" *Teflon* coating of frying pans. Teflon is a *polymer* (Section 6.21) made up of long chains of —CF_2CF_2—units.

Solubility in Water.   Alkyl halides and alcohols differ markedly from one another in their solubility in water. All alkyl halides are insoluble in water, but low-molecular-weight alcohols (methyl, ethyl, *n*-propyl, and isopropyl) are soluble in water in all proportions. Their ability to participate in intermolecular hydrogen bonding not only affects the boiling points of alcohols, but also enhances their water solubility. Hydrogen-bonded networks of the type shown in Figure 4.5, in which alcohol and water molecules associate with one another, replace the alcohol–alcohol and water–water hydrogen-bonded networks present in the pure substances.

Higher alcohols become more "hydrocarbon-like" and less water-soluble. 1-Octanol, for example, dissolves to the extent of only 1 mL in 2000 mL of water. As the alkyl chain gets longer, the hydrophobic effect (Section 2.17) becomes more important, to the point that it, more than hydrogen bonding, governs the solubility of alcohols.

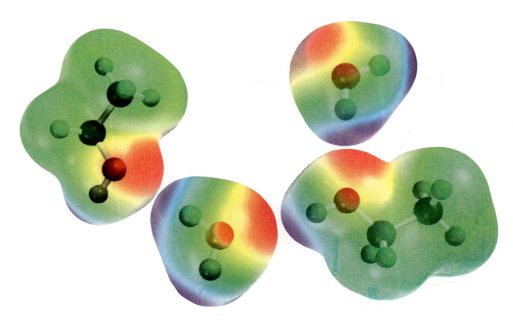

FIGURE 4.5 Hydrogen bonding between molecules of ethanol and water.

Density.   Alkyl fluorides and chlorides are less dense, and alkyl bromides and iodides more dense, than water.

	$CH_3(CH_2)_6CH_2F$	$CH_3(CH_2)_6CH_2Cl$	$CH_3(CH_2)_6CH_2Br$	$CH_3(CH_2)_6CH_2I$
Density (20°C):	0.80 g/mL	0.89 g/mL	1.12 g/mL	1.34 g/mL

Because alkyl halides are insoluble in water, a mixture of an alkyl halide and water separates into two layers. When the alkyl halide is a fluoride or chloride, it is the upper layer and water is the lower. The situation is reversed when the alkyl halide is a bromide or an iodide. In these cases the alkyl halide is the lower layer. Polyhalogenation increases the density. The compounds CH_2Cl_2, $CHCl_3$, and CCl_4, for example, are all more dense than water.

All liquid alcohols have densities of approximately 0.8 g/mL and are, therefore, less dense than water.

4.7   PREPARATION OF ALKYL HALIDES FROM ALCOHOLS AND HYDROGEN HALIDES

Much of what organic chemists do is directed toward practical goals. Chemists in the pharmaceutical industry synthesize new compounds as potential drugs for the treatment of disease. Agricultural chemicals designed to increase crop yields include organic compounds used for weed control, insecticides, and fungicides. Among the "building block" molecules used as starting materials to prepare new substances, alcohols and alkyl halides are especially valuable.

The procedures to be described in the remainder of this chapter use either an alkane or an alcohol as the starting material for preparing an alkyl halide. By knowing how to prepare alkyl halides, we can better appreciate the material in later chapters, where alkyl halides figure prominently in key chemical transformations. The preparation of alkyl halides also serves as a focal point to develop the principles of reaction mechanisms.

We'll begin with the preparation of alkyl halides from alcohols by reaction with hydrogen halides.

$$R-OH + H-X \longrightarrow R-X + H-OH$$

Alcohol   Hydrogen halide   Alkyl halide   Water

The order of reactivity of the hydrogen halides parallels their acidity: HI > HBr > HCl >> HF. Hydrogen iodide is used infrequently, however, and the reaction of alcohols with hydrogen fluoride is not a useful method for the preparation of alkyl fluorides.

Among the various classes of alcohols, tertiary alcohols are observed to be the most reactive and primary alcohols the least reactive.

Increasing reactivity of alcohols
toward hydrogen halides

$$CH_3OH \quad < \quad RCH_2OH \quad < \quad R_2CHOH \quad < \quad R_3COH$$

Methyl   Primary   Secondary   Tertiary
Least reactive   Most reactive

Tertiary alcohols are converted to alkyl chlorides in high yield within minutes on reaction with hydrogen chloride at room temperature and below.

$$(CH_3)_3COH \quad + \quad HCl \quad \xrightarrow{25°C} \quad (CH_3)_3CCl \quad + \quad H_2O$$

2-Methyl-2-propanol   Hydrogen chloride   2-Chloro-2-methylpropane   Water
(*tert*-butyl alcohol)   (*tert*-butyl chloride) (78–88%)

The efficiency of a synthetic transformation is normally expressed as a **percent yield,** or percentage of the theoretical yield. **Theoretical yield** is the amount of product that could be formed if the reaction proceeded to completion and did not lead to any products other than those given in the equation.

Secondary and primary alcohols do not react with HCl at rates fast enough to make the preparation of the corresponding alkyl chlorides a method of practical value. Therefore, the more reactive hydrogen halide HBr is used; even then, elevated temperatures are required to increase the rate of reaction.

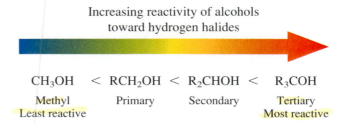

Cyclohexanol   Hydrogen bromide   Bromocyclohexane (73%)   Water

$$CH_3(CH_2)_5CH_2OH + HBr \xrightarrow{120°C} CH_3(CH_2)_5CH_2Br + H_2O$$

1-Heptanol   Hydrogen bromide   1-Bromoheptane   Water
(87–90%)

The same kind of transformation may be carried out by heating an alcohol with sodium bromide and sulfuric acid.

$$CH_3CH_2CH_2CH_2OH \xrightarrow[\text{heat}]{NaBr, H_2SO_4} CH_3CH_2CH_2CH_2Br$$

1-Butanol   1-Bromobutane (70–83%)
(*n*-butyl alcohol)   (*n*-butyl bromide)

We'll often write chemical equations in the abbreviated form shown here, in which reagents, especially inorganic ones, are not included in the body of the equation but instead are indicated over the arrow. Inorganic products—in this case, water—are usually omitted.

PROBLEM 4.8 Write chemical equations for the reaction that takes place between each of the following pairs of reactants:

(a) 2-Butanol and hydrogen bromide

(b) 3-Ethyl-3-pentanol and hydrogen chloride

(c) 1-Tetradecanol and hydrogen bromide

SAMPLE SOLUTION (a) An alcohol and a hydrogen halide react to form an alkyl halide and water. In this case 2-bromobutane was isolated in 73% yield.

$$CH_3CHCH_2CH_3 \ + \quad HBr \quad \longrightarrow \ CH_3CHCH_2CH_3 \ + \ H_2O$$
$$\qquad | \qquad\qquad\qquad\qquad\qquad\qquad\qquad\qquad | $$
$$\qquad OH \qquad\qquad\qquad\qquad\qquad\qquad\qquad\quad Br$$

| 2-Butanol | Hydrogen bromide | 2-Bromobutane | Water |

4.8 MECHANISM OF THE REACTION OF ALCOHOLS WITH HYDROGEN HALIDES

The reaction of an alcohol with a hydrogen halide is a **substitution.** A halogen, usually chlorine or bromine, replaces a hydroxyl group as a substituent on carbon. Calling the reaction a substitution tells us the relationship between the organic reactant and its product but does not reveal the mechanism. In developing a mechanistic picture for a particular reaction, we combine some basic principles of chemical reactivity with experimental observations to deduce the most likely sequence of steps.

Consider the reaction of *tert*-butyl alcohol with hydrogen chloride:

$$(CH_3)_3COH \ + \quad HCl \quad \longrightarrow \ (CH_3)_3CCl \ + \ H_2O$$

| *tert*-Butyl alcohol | Hydrogen chloride | *tert*-Butyl chloride | Water |

The generally accepted mechanism for this reaction is presented as a series of three equations in Figure 4.6. We say "generally accepted" because a reaction mechanism can never be proved to be correct. A mechanism is our best present assessment of how a reaction proceeds and must account for all experimental observations. If new experimental data appear that conflict with the mechanism, the mechanism must be modified to accommodate them. If the new data are consistent with the proposed mechanism, our confidence grows that the mechanism is likely to be correct.

Each equation in Figure 4.6 represents a single **elementary step.** An elementary step is one that involves only one transition state. A particular reaction might proceed by way of a single elementary step, in which case it is described as a **concerted reaction,** or by a series of elementary steps as in Figure 4.6. To be valid a proposed mechanism must meet a number of criteria, one of which is that the sum of the equations for the elementary steps must correspond to the equation for the overall reaction. Before we examine each step in detail, you should verify that the mechanism in Figure 4.6 satisfies this requirement.

Step 1: Proton transfer

We saw in Chapter 1, especially in Table 1.7, that alcohols resemble water in respect to their Brønsted acidity (ability to donate a proton *from oxygen*). They also resemble water in their Brønsted basicity (ability to accept a proton *on oxygen*). Just as proton transfer

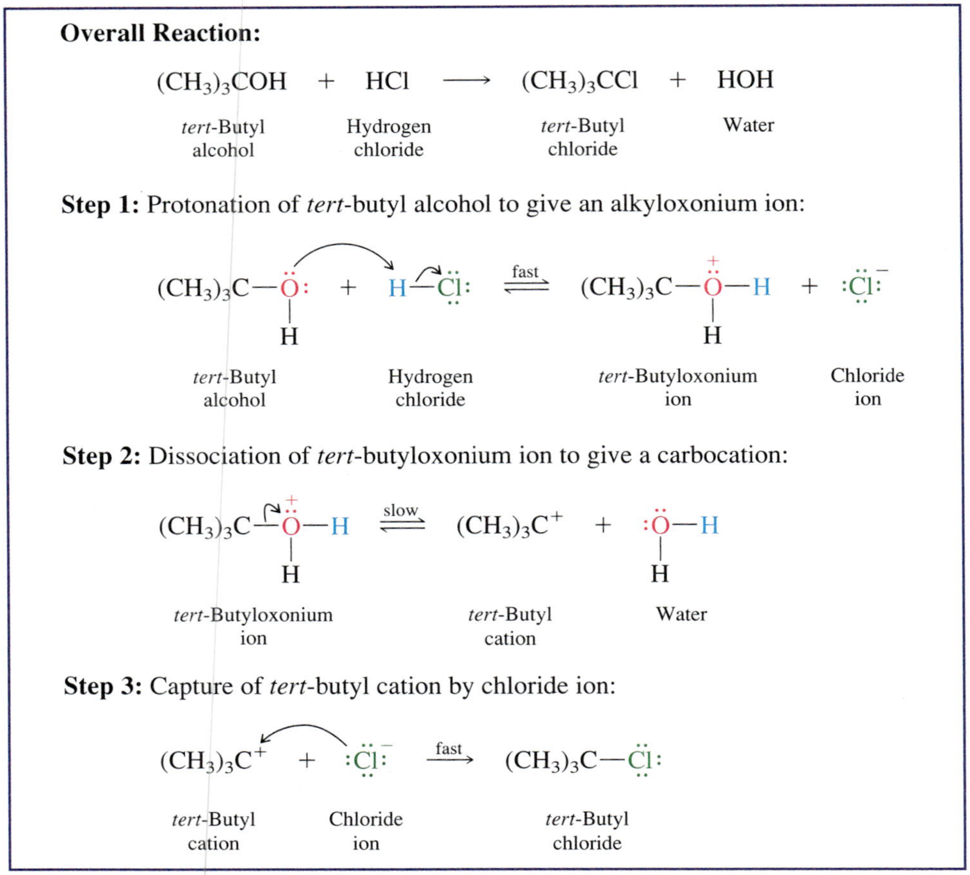

FIGURE 4.6 The mechanism
of formation of *tert*-butyl
chloride from *tert*-butyl alcohol and hydrogen chlo-
ride.

Overall Reaction:

$$(CH_3)_3COH \quad + \quad HCl \quad \longrightarrow \quad (CH_3)_3CCl \quad + \quad HOH$$

tert-Butyl alcohol	Hydrogen chloride	*tert*-Butyl chloride	Water

Step 1: Protonation of *tert*-butyl alcohol to give an alkyloxonium ion:

$$(CH_3)_3C-\overset{..}{\underset{|}{\overset{|}{O}}}: \quad + \quad H-\overset{..}{\underset{..}{Cl}}: \quad \underset{\text{fast}}{\rightleftharpoons} \quad (CH_3)_3C-\overset{+}{\underset{|}{\overset{|}{O}}}-H \quad + \quad :\overset{..}{\underset{..}{Cl}}:^{-}$$

tert-Butyl alcohol	Hydrogen chloride	*tert*-Butyloxonium ion	Chloride ion

Step 2: Dissociation of *tert*-butyloxonium ion to give a carbocation:

$$(CH_3)_3C-\overset{+}{\underset{|}{\overset{..}{O}}}-H \quad \underset{\text{slow}}{\rightleftharpoons} \quad (CH_3)_3C^{+} \quad + \quad :\overset{..}{O}-H$$

tert-Butyloxonium ion	*tert*-Butyl cation	Water

Step 3: Capture of *tert*-butyl cation by chloride ion:

$$(CH_3)_3C^{+} \quad + \quad :\overset{..}{\underset{..}{Cl}}:^{-} \quad \underset{\text{fast}}{\longrightarrow} \quad (CH_3)_3C-\overset{..}{\underset{..}{Cl}}:$$

tert-Butyl cation	Chloride ion	*tert*-Butyl chloride

to a water molecule gives oxonium ion (hydronium ion, H_3O^{+}), proton transfer to an alcohol gives an **alkyloxonium ion** (ROH_2^{+}).

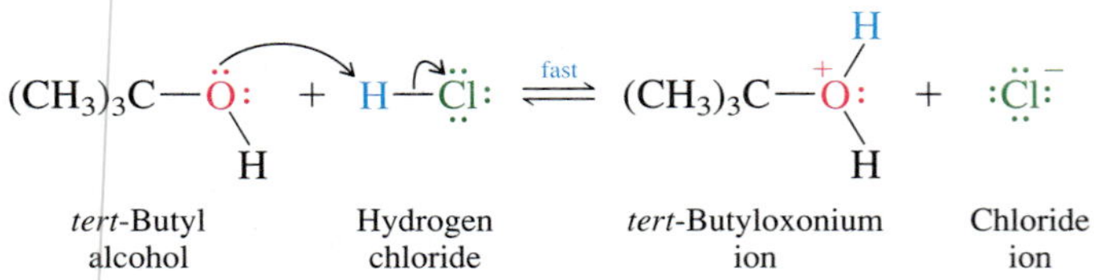

$$(CH_3)_3C-\overset{..}{\underset{|}{\overset{|}{O}}}: \quad + \quad H-\overset{..}{\underset{..}{Cl}}: \quad \underset{\text{fast}}{\rightleftharpoons} \quad (CH_3)_3C-\overset{+}{\overset{H}{\underset{|}{\overset{|}{O}}}}: \quad + \quad :\overset{..}{\underset{..}{Cl}}:^{-}$$

tert-Butyl alcohol	Hydrogen chloride	*tert*-Butyloxonium ion	Chloride ion

Furthermore, a substance such as HCl that dissociates completely when dissolved in water, also dissociates completely when dissolved in an alcohol. Many important reactions of alcohols involve strong acids either as reactants or as catalysts. In all these reactions the first step is formation of an alkyloxonium ion by proton transfer from the acid to the alcohol.

The **molecularity** of an elementary step is given by the number of species that undergo a chemical change in that step. Transfer of a proton from hydrogen chloride to *tert*-butyl alcohol is **bimolecular** because two molecules [HCl and $(CH_3)_3COH$] undergo chemical change.

The *tert*-butyloxonium ion formed in this step is an **intermediate.** It was not one of the initial reactants, nor is it formed as one of the final products. Rather it is formed

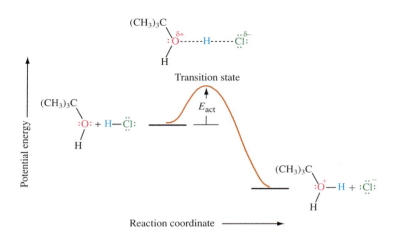

FIGURE 4.7 Potential energy diagram for proton transfer from hydrogen chloride to *tert*-butyl alcohol.

in one elementary step, consumed in another, and lies on the pathway from reactants to products.

Potential energy diagrams are especially useful when applied to reaction mechanisms. A potential energy diagram for proton transfer from hydrogen chloride to *tert*-butyl alcohol is shown in Figure 4.7. The potential energy of the system is plotted against the "reaction coordinate" which is a measure of the degree to which the reacting molecules have progressed on their way to products. Three aspects of the diagram are worth noting:

- Because this is an elementary step, it involves a single transition state.
- The step is known to be exothermic, so the products are placed lower in energy than the reactants.
- Proton transfers from strong acids to water and alcohols rank among the most rapid chemical processes and occur almost as fast as the molecules collide with one another. Thus the height of the energy barrier, the *activation energy* for proton transfer, must be quite low.

The 1967 Nobel Prize in chemistry was shared by Manfred Eigen, a German chemist who developed novel methods for measuring the rates of very fast reactions such as proton transfers.

The concerted nature of proton transfer contributes to its rapid rate. The energy cost of breaking the H—Cl bond is partially offset by the energy released in forming the new bond between the transferred proton and the oxygen of the alcohol. Thus, the activation energy is far less than it would be for a hypothetical two-step process in which the H—Cl bond breaks first, followed by bond formation between H^+ and the alcohol.

The species present at the transition state is not a stable structure and cannot be isolated or examined directly. Its structure is assumed to be one in which the proton being transferred is partially bonded to both chlorine and oxygen simultaneously, although not necessarily to the same extent.

$$(CH_3)_3C$$
$$\overset{\delta+}{:\!O}\text{----}H\text{----}\overset{\delta-}{\ddot{C}l:}$$
$$H$$

Dashed lines in transition-state structures represent *partial* bonds, that is, bonds in the process of being made or broken.

Inferring the structure of a transition state on the basis of the reactants and products of the elementary step in which it is involved is a time-honored practice in organic chemistry. Speaking specifically of transition states, George S. Hammond suggested that

Hammond made his proposal in 1955 while at Iowa State University. He later did pioneering work in organic photochemistry at Caltech.

if two states are similar in energy, they are similar in structure. This rationale is known as **Hammond's postulate.** One of its corollaries is that the structure of a transition state more closely resembles the immediately preceding or following state to which it is closer in energy. In the case of the exothermic proton transfer in Figure 4.7, the transition state is closer in energy to the reactants and so resembles them more closely than it does the products of this step. We often call this an "early" transition state. The next step of this mechanism will provide us with an example of a "late" transition state.

Step 2: Carbocation formation

In the second step of the mechanism described in Figure 4.6, the alkyloxonium ion dissociates to a molecule of water and a **carbocation,** an ion that contains a positively charged carbon.

One way to name carbocations in the IUPAC system is to add the word "cation" to the name of the alkyl group.

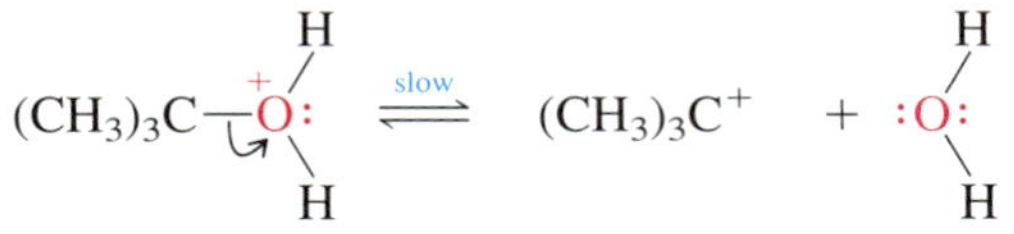

$$(CH_3)_3C\overset{+}{-}\overset{H}{\underset{H}{O}}: \quad \overset{slow}{\rightleftharpoons} \quad (CH_3)_3C^+ \quad + \quad \overset{H}{\underset{H}{:O}}:$$

$$\text{\textit{tert}-Butyloxonium ion} \qquad \text{\textit{tert}-Butyl cation} \qquad \text{Water}$$

Only one species, *tert*-butyloxonium ion, undergoes a chemical change in this step. Therefore, the step is **unimolecular.**

Like *tert*-butyloxonium ion, *tert*-butyl cation is an intermediate along the reaction pathway. It is, however, a relatively unstable species and its formation by dissociation of the alkyloxonium ion is endothermic. Step 2 is the slowest step in the mechanism and has the highest activation energy. Figure 4.8 shows a potential energy diagram for this step.

- Because this step is endothermic, the products of it are placed higher in energy than the reactants.

- The transition state is closer in energy to the carbocation (*tert*-butyl cation), so its structure more closely resembles the carbocation than it resembles *tert*-butyloxonium ion. The transition state has considerable "carbocation character" meaning that a significant degree of positive charge has developed at carbon.

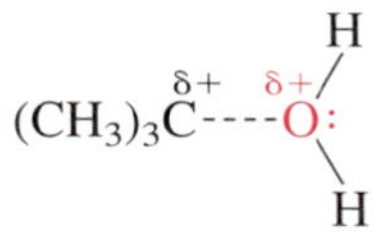

$$(CH_3)_3\overset{\delta+}{C}----\overset{\delta+}{\underset{H}{O}}\overset{H}{:}$$

There is ample evidence from a variety of sources that carbocations are intermediates in some chemical reactions, but they are almost always too unstable to isolate. The simplest reason for the instability of carbocations is that the positively charged carbon has only six electrons in its valence shell—the octet rule is not satisfied for the positively charged carbon.

The properties of *tert*-butyl cation can be understood by focusing on its structure, which is shown in Figure 4.9. With only six valence electrons, which are distributed among three coplanar σ bonds, the positively charged carbon is sp^2-hybridized. The unhybridized $2p$ orbital that remains on the positively charged carbon contains no electrons; its axis is perpendicular to the plane of the bonds connecting that carbon to the three methyl groups.

The positive charge on carbon and the vacant p orbital combine to make carbocations strongly **electrophilic** ("electron-loving" or "electron-seeking"). Electrophiles are Lewis acids (Section 1.17). They are electron-pair acceptors and react with Lewis bases (electron-pair donors). Step 3, which follows and completes the mechanism, is a Lewis

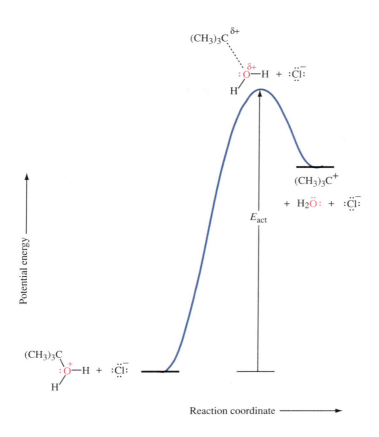

acid-Lewis base reaction. We'll return to carbocations and describe them in more detail in Section 4.10.

Step 3: Reaction of **tert-butyl cation with chloride ion**

The Lewis bases that react with electrophiles are called **nucleophiles** ("nucleus seekers"). They have an unshared electron pair that they can use in covalent bond formation. The nucleophile in Step 3 of Figure 4.6 is chloride ion.

$$(CH_3)_3C^+ \quad + \quad :\ddot{\underset{..}{Cl}:^- \quad \xrightarrow{\text{fast}} \quad (CH_3)_3C - \ddot{\underset{..}{Cl}}:$$

| *tert*-Butyl cation (electrophile) | Chloride ion (nucleophile) | *tert*-Butyl chloride |

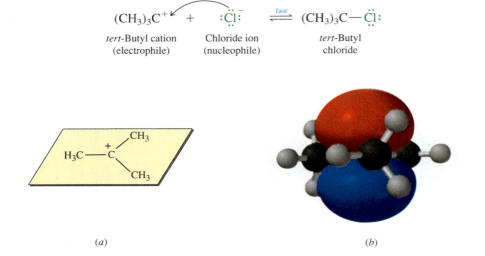

(a)   (b)

FIGURE 4.9 *tert*-Butyl cation. (*a*) The positively charged carbon is sp^2-hybridized. Each methyl group is attached to the positively charged carbon by a σ bond, and these three bonds lie in the same plane. (*b*) The sp^2-hybridized carbon has an empty $2p$ orbital, the axis of which is perpendicular to the plane of the carbon atoms.

FIGURE 4.10 Potential energy diagram for reaction of *tert*-butyl cation with chloride anion.

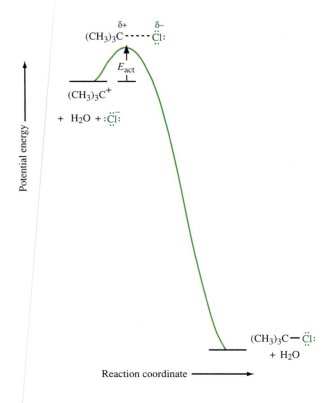

Step 3 is bimolecular because two species, the carbocation and chloride ion, react together. Figure 4.10 shows a potential energy diagram for this step.

- The step is exothermic; it leads from the carbocation intermediate to the isolated products of the reaction.
- The activation energy for this step is small, and bond formation between a positive ion and a negative ion occurs rapidly.
- The transition state for this step involves partial bond formation between *tert*-butyl cation and chloride ion.

$$(CH_3)_3\overset{\delta+}{C}\text{----}\overset{\delta-}{Cl}:$$

As shown in Figure 4.11, the crucial electronic interaction is between an unshared electron pair of Cl^- and the vacant $2p$ orbital of the positively charged carbon of $(CH_3)_3C^+$.

FIGURE 4.11 Combination of *tert*-butyl cation and chloride anion to give *tert*-butyl chloride. In-phase overlap between a vacant p orbital of $(CH_3)_3C^+$ and a filled p orbital of Cl^- gives a C—Cl σ bond.

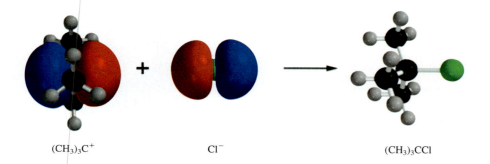

4.9 POTENTIAL ENERGY DIAGRAMS FOR MULTISTEP REACTIONS: THE S$_N$1 MECHANISM

We've just seen how the mechanism for the reaction of *tert*-butyl alcohol with hydrogen chloride, written as a series of elementary steps in Figure 4.6, can be supplemented with potential energy diagrams (Figures 4.7, 4.8, and 4.10). We'll complete the energy picture by combining the three separate diagrams into one that covers the entire process. This composite diagram (Figure 4.12) has three peaks and two valleys. The peaks correspond to transition states, one for each of the three elementary steps. The valleys correspond to the reactive intermediates—*tert*-butyloxonium ion and *tert*-butyl cation—species formed in one step and consumed in another.

With the potential energies shown on a common scale, we see that the transition state for formation of $(CH_3)_3C^+$ is the highest energy point on the diagram. A reaction can proceed no faster than its slowest step, which is referred to as the **rate-determining step.** In the reaction of *tert*-butyl alcohol with hydrogen chloride, formation of the

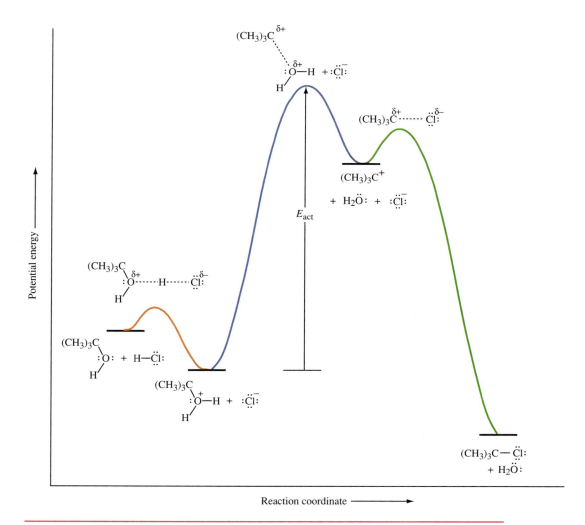

FIGURE 4.12 Potential energy diagram for the reaction of *tert*-butyl alcohol and hydrogen chloride according to the S$_N$1 mechanism.

carbocation by dissociation of the alkyloxonium ion has the highest activation energy and is rate determining.

Substitution reactions, of which the reaction of alcohols with hydrogen halides is but one example, will be discussed in more detail in Chapter 8. There, we will make extensive use of a notation originally introduced by Sir Christopher Ingold. Ingold proposed the symbol, S_N, to stand for *substitution nucleophilic*, to be followed by the number *1* or *2* according to whether the rate-determining step is unimolecular or bimolecular. The reaction of *tert*-butyl alcohol with hydrogen chloride, for example, is said to follow an S_N1 mechanism because its slow step (dissociation of *tert*-butyloxonium ion) is unimolecular. Only the oxonium ion undergoes a chemical change in this step.

4.10   STRUCTURE, BONDING, AND STABILITY OF CARBOCATIONS

As we have just seen, the rate-determining intermediate in the reaction of *tert*-butyl alcohol with hydrogen chloride is the carbocation $(CH_3)_3C^+$. Convincing evidence from a variety of sources tells us that carbocations can exist, but are relatively unstable. When carbocations are involved in chemical reactions, it is as reactive intermediates, formed slowly in one step and consumed rapidly in the next one.

Numerous other studies have shown that *alkyl groups directly attached to the positively charged carbon stabilize a carbocation*. Figure 4.13 illustrates this generalization for CH_3^+, $CH_3CH_2^+$, $(CH_3)_2CH^+$, and $(CH_3)_3C^+$. Among this group, CH_3^+ is the least stable and $(CH_3)_3C^+$ the most stable.

Carbocations are classified according to their degree of substitution at the positively charged carbon. The positive charge is on a primary carbon in $CH_3CH_2^+$, a secondary carbon in $(CH_3)_2CH^+$, and a tertiary carbon in $(CH_3)_3C^+$. Ethyl cation is a primary carbocation, isopropyl cation a secondary carbocation, and *tert*-butyl cation a tertiary carbocation.

As carbocations go, CH_3^+ is particularly unstable, and its existence as an intermediate in chemical reactions has never been demonstrated. Primary carbocations, although more stable than CH_3^+, are still too unstable to be involved as intermediates in chemical reactions. The threshold of stability is reached with secondary carbocations. Many reactions, including the reaction of secondary alcohols with hydrogen halides, are believed to involve secondary carbocations. The evidence in support of tertiary carbocation intermediates is stronger yet.

Increasing stability of carbocation

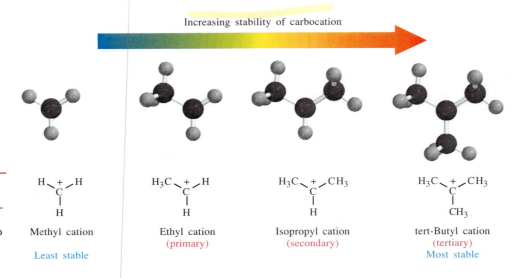

FIGURE 4.13 The order of carbocation stability is methyl < primary < secondary < tertiary. Alkyl groups that are directly attached to the positively charged carbon stabilize carbocations.

Methyl cation

Least stable

Ethyl cation
(primary)

Isopropyl cation
(secondary)

tert-Butyl cation
(tertiary)

Most stable

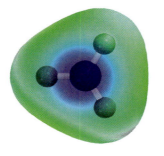

Methyl cation (CH_3^+)

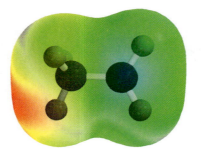

Ethyl cation ($CH_3CH_2^+$)

FIGURE 4.14 Electrostatic potential maps of methyl cation and ethyl cation. The region of highest positive charge is more concentrated in CH_3^+ and more spread out in $CH_3CH_2^+$. (The electrostatic potentials were mapped on the same scale in order to allow direct comparison.)

> **PROBLEM 4.9** Of the isomeric $C_5H_{11}^+$ carbocations, which one is the most stable?

Alkyl groups stabilize carbocations by releasing electron density to the positively charged carbon, thereby dispersing the positive charge. Figure 4.14 illustrates this charge dispersal by comparing the electrostatic potential maps of CH_3^+ and $CH_3CH_2^+$ where it can be seen that the blue region (positively charged) is more spread out and less intense in ethyl cation than in methyl cation.

There are two main ways in which alkyl groups disperse positive charge. One is the *inductive effect.* Recall from Section 1.15 that an inductive effect is the donation or withdrawal of electron density at some site transmitted by the polarization of σ bonds. As illustrated for $CH_3CH_2^+$ in Figure 4.15, the positively charged carbon draws the electrons in its σ bonds toward itself and away from the atoms attached to it. Electrons in a C—C bond are more polarizable than those in a C—H bond, so replacing hydrogens by alkyl groups reduces the net charge on the sp^2-hybridized carbon.

A second effect, called **hyperconjugation,** is also important. Again consider $CH_3CH_2^+$, this time directing your attention to the electrons in the C—H bonds of the CH_3 group. Figure 4.16 illustrates how an orbital associated with the CH_3 group of $CH_3CH_2^+$ can overlap with the vacant p orbital of the positively charged carbon to give an extended orbital that encompasses them both. This disperses the positive charge and allows the electrons of the CH_3 group to be shared by both carbons.

When applying hyperconjugation to carbocations more complicated than $CH_3CH_2^+$, it is helpful to keep track of the various bonds. Begin with the positively charged carbon and label the three bonds originating from it with the Greek letter α. Proceed down the chain, labeling the bonds extending from the next carbon β, those from the next carbon γ, and so on.

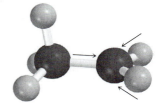

FIGURE 4.15 The charge in ethyl cation is stabilized by polarization of the electron distribution in the σ bonds to the positively charged carbon atom. Alkyl groups release electrons better than hydrogen.

$$\overset{\gamma|}{\underset{\gamma|}{\text{C}}}\overset{\beta|}{\underset{\beta|}{\text{C}}}\overset{\alpha}{\underset{\alpha}{\text{C}}}+$$

Only electrons in bonds that are β to the positively charged carbon can stabilize a carbocation by hyperconjugation. Moreover, it doesn't matter whether H or another carbon is at the far end of the β bond; stabilization by hyperconjugation will still operate. The key point is that electrons in bonds that are β to the positively charged carbon are more stabilizing than electrons in an α^+C—H bond. Thus, successive replacement of first one,

then two, then three hydrogens of CH_3^+ by alkyl groups increases the opportunities for hyperconjugation, which is consistent with the observed order of carbocation stability: $CH_3^+ < CH_3CH_2^+ < (CH_3)_2CH^+ < (CH_3)_3C^+$.

PROBLEM 4.10 For the general case of R = any alkyl group, how many bonded pairs of electrons are involved in stabilizing R_3C^+ by hyperconjugation? How many in R_2CH^+? In RCH_2^+?

To summarize, the most important factor to consider in assessing carbocation stability is the degree of substitution at the positively charged carbon.

$$CH_3^+ \quad < \quad RCH_2^+ \quad < \quad R_2CH^+ \quad < \quad R_3C^+$$

Methyl	Primary	Secondary	Tertiary
Least stable			Most stable

We will see numerous reactions that involve carbocation intermediates as we proceed through the text, so it is important to understand how their structure determines their stability.

4.11 EFFECT OF ALCOHOL STRUCTURE ON REACTION RATE

For a proposed reaction mechanism to be valid, the sum of its elementary steps must equal the equation for the overall reaction and the mechanism must be consistent with all experimental observations. The S_N1 mechanism set forth in Figure 4.6 satisfies the first criterion. What about the second?

One important experimental fact is that the rate of reaction of alcohols with hydrogen halides increases in the order methyl < primary < secondary < tertiary. This reactivity order parallels the carbocation stability order and is readily accommodated by the mechanism we have outlined.

The rate-determining step in the S_N1 mechanism is dissociation of the alkyloxonium ion to the carbocation.

Alkyloxonium ion	Transition state	Carbocation	Water

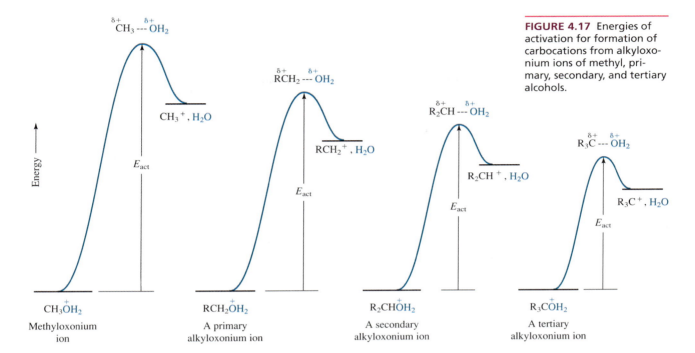

FIGURE 4.17 Energies of activation for formation of carbocations from alkyloxonium ions of methyl, primary, secondary, and tertiary alcohols.

The rate of this step is proportional to the concentration of the alkyloxonium ion:

$$\text{Rate} = k[\text{alkyloxonium ion}]$$

where k is a constant of proportionality called the *rate constant*. The value of k is related to the activation energy for alkyloxonium ion dissociation and is different for different alkyloxonium ions. A low activation energy implies a large value of k and a rapid rate of alkyloxonium ion dissociation. Conversely, a large activation energy is characterized by a small k for dissociation and a slow rate.

> The rate of any chemical reaction increases with increasing temperature. Thus the value of k for a reaction is not constant, but increases as the temperature increases.

The transition state is closer in energy to the carbocation and more closely resembles it than the alkyloxonium ion. Thus, structural features that stabilize carbocations stabilize transition states leading to them. It follows, therefore, that alkyloxonium ions derived from tertiary alcohols have a lower energy of activation for dissociation and are converted to their corresponding carbocations faster than those derived from secondary and primary alcohols. Simply put: *more stable carbocations are formed faster than less stable ones.* Figure 4.17 expresses this principle via a potential energy diagram.

The S$_N$1 mechanism is generally accepted to be correct for the reaction of tertiary and secondary alcohols with hydrogen halides. It is almost certainly *not* correct for methyl alcohol and primary alcohols because methyl and primary carbocations are believed to be much too unstable and the activation energies for their formation much too high for them to be reasonably involved. The next section describes how methyl and primary alcohols are converted to their corresponding halides by a mechanism related to, but different from S$_N$1.

4.12 REACTION OF PRIMARY ALCOHOLS WITH HYDROGEN HALIDES: THE S$_N$2 MECHANISM

Unlike tertiary and secondary carbocations, methyl and primary carbocations are too high in energy to be intermediates in chemical reactions. However, methyl and primary alcohols are converted, albeit rather slowly, to alkyl halides on treatment with hydrogen halides. Therefore, they must follow some other mechanism that avoids carbocation intermediates. This alternative mechanism is outlined in Figure 4.18 for the reaction of 1-heptanol with hydrogen bromide.

The first step of this new mechanism is exactly the same as that seen earlier for the reaction of *tert*-butyl alcohol with hydrogen chloride—formation of an alkyloxonium ion by proton transfer from the hydrogen halide to the alcohol. Like the earlier example, this is a rapid, reversible Brønsted acid–base reaction.

The major difference between the two mechanisms is the second step. The second step in the reaction of *tert*-butyl alcohol with hydrogen chloride is the unimolecular dissociation of *tert*-butyloxonium ion to *tert*-butyl cation and water. Heptyloxonium ion, however, instead of dissociating to an unstable primary carbocation, reacts differently. It is attacked by bromide ion, which acts as a nucleophile. We can represent the transition state for this step as:

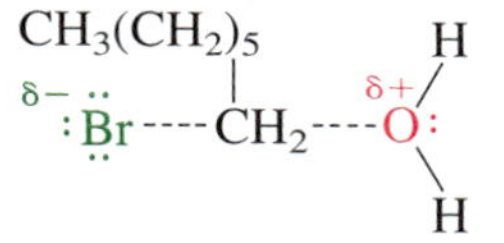

$$CH_3(CH_2)_5 \qquad H$$
$$\overset{\delta-}{:}\overset{..}{\underset{..}{Br}} ---- CH_2 ---- \overset{\delta+}{O}:$$
$$\qquad\qquad\qquad\qquad H$$

Bromide ion forms a bond to the primary carbon by "pushing off" a water molecule. This step is bimolecular because it involves both bromide and heptyloxonium ion. Step 2 is slower than the proton transfer in step 1, so it is rate-determining. Using Ingold's terminology, we classify nucleophilic substitutions that have a bimolecular rate-determining step by the mechanistic symbol S_N2.

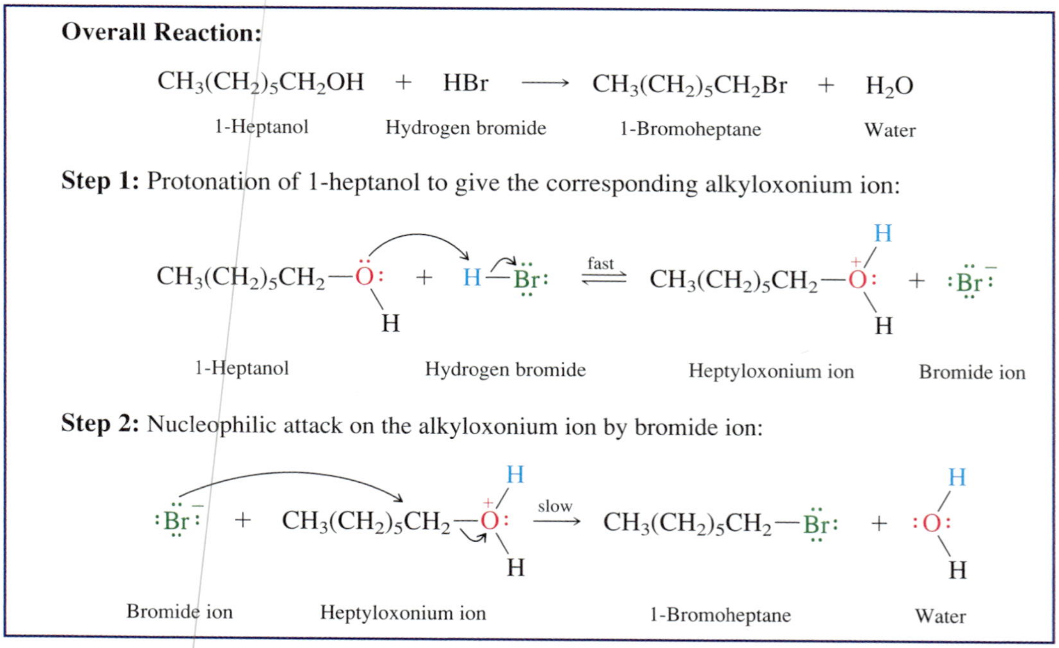

Overall Reaction:

$$CH_3(CH_2)_5CH_2OH \quad + \quad HBr \quad \longrightarrow \quad CH_3(CH_2)_5CH_2Br \quad + \quad H_2O$$

| 1-Heptanol | Hydrogen bromide | 1-Bromoheptane | Water |

Step 1: Protonation of 1-heptanol to give the corresponding alkyloxonium ion:

$$CH_3(CH_2)_5CH_2 \overset{..}{\underset{|}{O}}: \;+\; H \overset{..}{\underset{..}{Br}}: \;\overset{fast}{\rightleftharpoons}\; CH_3(CH_2)_5CH_2 \overset{+}{\overset{..}{O}}: \;+\; :\overset{..}{\underset{..}{Br}}:^-$$

| 1-Heptanol | Hydrogen bromide | Heptyloxonium ion | Bromide ion |

Step 2: Nucleophilic attack on the alkyloxonium ion by bromide ion:

$$:\overset{..}{\underset{..}{Br}}:^- \;+\; CH_3(CH_2)_5CH_2 \overset{+}{\overset{..}{O}}: \;\overset{slow}{\longrightarrow}\; CH_3(CH_2)_5CH_2 \overset{..}{\underset{..}{Br}}: \;+\; :\overset{..}{O}:$$

| Bromide ion | Heptyloxonium ion | 1-Bromoheptane | Water |

FIGURE 4.18 The mechanism of formation of 1-bromoheptane from 1-heptanol and hydrogen bromide.

PROBLEM 4.11 Sketch a potential energy diagram for the reaction of 1-heptanol with hydrogen bromide, paying careful attention to the positioning and structures of the intermediates and transition states.

PROBLEM 4.12 1-Butanol and 2-butanol are converted to their corresponding bromides on being heated with hydrogen bromide. Write a suitable mechanism for each reaction, and assign each the appropriate symbol (S_N1 or S_N2).

It is important to note that although methyl and primary alcohols react with hydrogen halides by a mechanism that involves fewer steps than the corresponding reactions of secondary and tertiary alcohols, fewer steps do not translate to faster reaction rates. Remember, the order of reactivity of alcohols with hydrogen halides is tertiary > secondary > primary > methyl. Reaction rate is governed by the activation energy of the slowest step, regardless of how many steps there are.

S_N1 and S_N2 are among the most fundamental and important mechanisms in organic chemistry. We will have much more to say about them in Chapter 8.

4.13 OTHER METHODS FOR CONVERTING ALCOHOLS TO ALKYL HALIDES

Alkyl halides are such useful starting materials for preparing other functional group types that chemists have developed several different methods for converting alcohols to alkyl halides. Two methods, based on the inorganic reagents *thionyl chloride* and *phosphorus tribromide*, bear special mention.

Thionyl chloride reacts with alcohols to give alkyl chlorides. The inorganic byproducts in the reaction, sulfur dioxide and hydrogen chloride, are both gases at room temperature and are easily removed, making it an easy matter to isolate the alkyl chloride.

$$ROH \ + \ SOCl_2 \ \longrightarrow \ RCl \ + \ SO_2 \ + \ HCl$$

| Alcohol | Thionyl chloride | Alkyl chloride | Sulfur dioxide | Hydrogen chloride |

Because tertiary alcohols are so readily converted to chlorides with hydrogen chloride, thionyl chloride is used mainly to prepare primary and secondary alkyl chlorides. Reactions with thionyl chloride are normally carried out in the presence of potassium carbonate or the weak organic base pyridine.

$$CH_3CH(CH_2)_5CH_3 \xrightarrow[K_2CO_3]{SOCl_2} CH_3CH(CH_2)_5CH_3$$

with OH below the first structure and Cl below the second

2-Octanol · 2-Chlorooctane (81%)

$$(CH_3CH_2)_2CHCH_2OH \xrightarrow[pyridine]{SOCl_2} (CH_3CH_2)_2CHCH_2Cl$$

2-Ethyl-1-butanol · 1-Chloro-2-ethylbutane (82%)

Phosphorus tribromide reacts with alcohols to give alkyl bromides and phosphorous acid.

$$3ROH + PBr_3 \longrightarrow 3RBr + H_3PO_3$$

| Alcohol | Phosphorus tribromide | Alkyl bromide | Phosphorous acid |

Phosphorous acid is water-soluble and may be removed by washing the alkyl halide with water or with dilute aqueous base.

$$(CH_3)_2CHCH_2OH \xrightarrow{PBr_3} (CH_3)_2CHCH_2Br$$

Isobutyl alcohol   Isobutyl bromide
                              (55–60%)

Cyclopentanol   Cyclopentyl bromide
                          (78–84%)

Thionyl chloride and phosphorus tribromide are specialized reagents used to bring about particular functional group transformations. For this reason, we won't present the mechanisms by which they convert alcohols to alkyl halides, but instead will limit ourselves to those mechanisms that have broad applicability and enhance our knowledge of fundamental principles. In those instances you will find that a mechanistic understanding is of great help in organizing the reaction types of organic chemistry.

4.14   HALOGENATION OF ALKANES

The rest of this chapter describes a completely different method for preparing alkyl halides, one that uses alkanes as reactants. It involves substitution of a halogen atom for one of the alkane's hydrogens.

$$R\text{—}H + X_2 \longrightarrow R\text{—}X + H\text{—}X$$

Alkane   Halogen   Alkyl halide   Hydrogen halide

The alkane is said to undergo *fluorination, chlorination, bromination,* or *iodination* according to whether X_2 is F_2, Cl_2, Br_2, or I_2, respectively. The general term is **halogenation. Chlorination** and **bromination** are the most widely used.

The reactivity of the halogens decreases in the order $F_2 > Cl_2 > Br_2 > I_2$. Fluorine is an extremely aggressive oxidizing agent, and its reaction with alkanes is strongly exothermic and difficult to control. Direct fluorination of alkanes requires special equipment and techniques, is not a reaction of general applicability, and will not be discussed further.

Chlorination of alkanes is less exothermic than fluorination, and bromination less exothermic than chlorination. Iodine is unique among the halogens in that its reaction with alkanes is endothermic and alkyl iodides are never prepared by iodination of alkanes.

Volume II of *Organic Reactions,* an annual series that reviews reactions of interest to organic chemists, contains the statement "Most organic compounds burn or explode when brought in contact with fluorine."

4.15   CHLORINATION OF METHANE

The gas-phase chlorination of methane is a reaction of industrial importance and leads to a mixture of chloromethane (CH_3Cl), dichloromethane (CH_2Cl_2), trichloromethane ($CHCl_3$), and tetrachloromethane (CCl_4) by sequential substitution of hydrogens.

$$CH_4 + Cl_2 \xrightarrow{400-440°C} CH_3Cl + HCl$$

Methane   Chlorine   Chloromethane   Hydrogen
                                 (bp −24°C)   chloride

$$CH_3Cl + Cl_2 \xrightarrow{400-440°C} CH_2Cl_2 + HCl$$

Chloromethane   Chlorine   Dichloromethane   Hydrogen
                                 (bp 40°C)   chloride

$$CH_2Cl_2 + Cl_2 \xrightarrow{400-440°C} CHCl_3 + HCl$$

Dichloromethane   Chlorine   Trichloromethane   Hydrogen
                                 (bp 61°C)   chloride

$$CHCl_3 + Cl_2 \xrightarrow{400-440°C} CCl_4 + HCl$$

Trichloromethane   Chlorine   Tetrachloromethane   Hydrogen
                                 (bp 77°C)   chloride

> Chlorination of methane provides approximately one third of the annual U.S. production of chloromethane. The reaction of methanol with hydrogen chloride is the major synthetic method for the preparation of chloromethane.

One of the chief uses of chloromethane is as a starting material from which silicone polymers are made. Dichloromethane is widely used as a paint stripper. Trichloromethane was once used as an inhalation anesthetic, but its toxicity caused it to be replaced by safer materials many years ago. Tetrachloromethane is the starting material for the preparation of several chlorofluorocarbons (CFCs), at one time widely used as refrigerant gases. Most of the world's industrialized nations have agreed to phase out all uses of CFCs because these compounds have been implicated in atmospheric processes that degrade the Earth's ozone layer.

> Dichloromethane, trichloromethane, and tetrachloromethane are widely known by their common names methylene chloride, chloroform, and carbon tetrachloride, respectively.

The chlorination of methane is carried out at rather high temperatures (400–440°C), even though each substitution in the series is exothermic. The high temperature provides the energy to initiate the reaction. The term *initiation step* has a specific meaning in organic chemistry, one that is related to the mechanism of the reaction. This mechanism, to be presented in Section 4.17, is fundamentally different from the mechanism by which alcohols react with hydrogen halides. Alcohols are converted to alkyl halides in reactions involving ionic (or "polar") intermediates—alkyloxonium ions and carbocations. The intermediates in the chlorination of methane and other alkanes are quite different; they are neutral ("nonpolar") species called *free radicals*.

4.16 STRUCTURE AND STABILITY OF FREE RADICALS

Free radicals are species that contain unpaired electrons. The octet rule notwithstanding, not all compounds have all of their electrons paired. Oxygen (O_2) is the most familiar example of a compound with unpaired electrons; it has two of them. Compounds that have an odd number of electrons, such as nitrogen dioxide (NO_2), must have at least one unpaired electron.

$$:\overset{..}{O}-\overset{..}{O}: \qquad :\overset{..}{O}=\overset{..}{N}-\overset{..}{O}: \qquad \cdot\overset{..}{N}=\overset{..}{O}:$$

Oxygen   Nitrogen dioxide   Nitrogen monoxide

Nitrogen monoxide ("nitric oxide") is another stable free radical. Although known for hundreds of years, NO has only recently been discovered to be an extremely important biochemical messenger and moderator of so many biological processes that it might be better to ask "Which ones is it not involved in?"

> The journal *Science* selected nitric oxide as its "Molecule of the Year" for 1992.

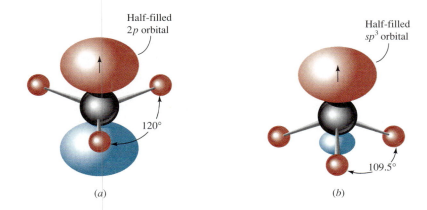

FIGURE 4.19 Bonding in methyl radical. (*a*) If the structure of the CH₃ radical is planar, then carbon is *sp²*-hybridized with an unpaired electron in 2*p* orbital. (*b*) If CH₃ is pyramidal, then carbon is *sp³*-hybridized with an electron in *sp³* orbital. Model (*a*) is more consistent with experimental observations.

The free radicals that we usually see in carbon chemistry are much less stable than these. Simple alkyl radicals, for example, require special procedures for their isolation and study. We will encounter them here only as reactive intermediates, formed in one step of a reaction mechanism and consumed in the next. Alkyl radicals are classified as primary, secondary, or tertiary according to the number of carbon atoms directly attached to the carbon that bears the unpaired electron.

Methyl radical   Primary radical   Secondary radical   Tertiary radical

An alkyl radical is neutral and has one more electron than the corresponding carbocation. Thus, bonding in methyl radical may be approximated by simply adding an electron to the vacant 2*p* orbital of *sp²*-hybridized carbon in methyl cation (Figure 4.19*a*). Alternatively, we could assume that carbon is *sp³*-hybridized and place the unpaired electron in an *sp³* orbital (Figure 4.19*b*).

Of the two extremes, experimental studies indicate that the planar *sp²* model describes the bonding in alkyl radicals better than the pyramidal *sp³* model. Methyl radical is planar, and more highly substituted radicals such as *tert*-butyl radical are flattened pyramids closer in shape to that expected for *sp²*-hybridized carbon than for *sp³*.

Free radicals, like carbocations, have an unfilled 2*p* orbital and are stabilized by substituents, such as alkyl groups, that release electrons. Consequently, the order of free-radical stability parallels that of carbocations.

Increasing stability of free radical

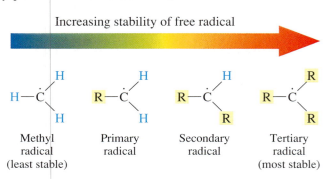

Methyl radical (least stable)   Primary radical   Secondary radical   Tertiary radical (most stable)

PROBLEM 4.13 Write a structural formula for the most stable of the free radicals that have the formula C_5H_{11}.

Some of the evidence indicating that alkyl substituents stabilize free radicals comes from bond energies. The strength of a bond is measured by the energy required to break it. A covalent bond can be broken in two ways. In a **homolytic cleavage** a bond between two atoms is broken so that each of them retains one of the electrons in the bond.

$$X{:}Y \longrightarrow X{\cdot} + {\cdot}Y$$

Homolytic bond cleavage

In contrast, in a **heterolytic cleavage** one fragment retains both electrons.

$$X{:}Y \longrightarrow X^+ + {:}Y^-$$

Heterolytic bond cleavage

We assess the relative stability of alkyl radicals by measuring the enthalpy change ($\Delta H°$) for the homolytic cleavage of a C—H bond in an alkane:

$$R{-}H \longrightarrow R{\cdot} + {\cdot}H$$

The more stable the radical, the lower the energy required to generate it by C—H bond homolysis.

The energy required for homolytic bond cleavage is called the **bond dissociation energy (BDE).** A list of some bond dissociation energies is given in Table 4.3.

As the table indicates, C—H bond dissociation energies in alkanes are approximately 375 to 435 kJ/mol (90–105 kcal/mol). Homolysis of the H—CH_3 bond in methane gives methyl radical and requires 435 kJ/mol (104 kcal/mol). The dissociation energy of the H—CH_2CH_3 bond in ethane, which gives a primary radical, is somewhat less (410 kJ/mol, or 98 kcal/mol) and is consistent with the notion that ethyl radical (primary) is more stable than methyl.

The dissociation energy of the terminal C—H bond in propane is exactly the same as that of ethane. The resulting free radical is primary ($R\overset{\cdot}{C}H_2$) in both cases.

$$CH_3CH_2CH_2{-}H \longrightarrow CH_3CH_2\overset{\cdot}{C}H_2 + \quad H{\cdot} \qquad \Delta H° = +410 \text{ kJ}$$
$$\text{(98 kcal)}$$

Propane	n-Propyl radical (primary)	Hydrogen atom

Note, however, that Table 4.3 includes two entries for propane. The second entry corresponds to the cleavage of a bond to one of the hydrogens of the methylene (CH_2) group. It requires slightly less energy to break a C—H bond in the methylene group than in the methyl group.

$$\underset{\underset{\text{H}}{|}}{CH_3CHCH_3} \longrightarrow CH_3\overset{\cdot}{C}HCH_3 + \quad H{\cdot} \qquad \Delta H° = +397 \text{ kJ}$$
$$\text{(95 kcal)}$$

Propane	Isopropyl radical (secondary)	Hydrogen atom

A curved arrow shown as a single-barbed fishhook ⤻ signifies the movement of *one* electron. "Normal" curved arrows ⤵ track the movement of a *pair* of electrons.

TABLE 4.3	Bond Dissociation Energies of Some Representative Compounds*				
Bond	**Bond dissociation energy**		**Bond**	**Bond dissociation energy**	
	kJ/mol	**(kcal/mol)**		**kJ/mol**	**(kcal/mol)**
Diatomic molecules					
H—H	435	(104)			
F—F	159	(38)	H—F	568	(136)
Cl—Cl	242	(58)	H—Cl	431	(103)
Br—Br	192	(46)	H—Br	366	(87.5)
I—I	150	(36)	H—I	297	(71)
Alkanes					
CH_3—H	435	(104)	CH_3—CH_3	368	(88)
CH_3CH_2—H	410	(98)	CH_3CH_2—CH_3	355	(85)
$CH_3CH_2CH_2$—H	410	(98)			
$(CH_3)_2CH$—H	397	(95)			
$(CH_3)_2CHCH_2$—H	410	(98)	$(CH_3)_2CH$—CH_3	351	(84)
$(CH_3)_3C$—H	380	(91)	$(CH_3)_3C$—CH_3	334	(80)
Alkyl halides					
CH_3—F	451	(108)	$(CH_3)_2CH$—F	439	(105)
CH_3—Cl	349	(83.5)	$(CH_3)_2CH$—Cl	339	(81)
CH_3—Br	293	(70)	$(CH_3)_2CH$—Br	284	(68)
CH_3—I	234	(56)	$(CH_3)_3C$—Cl	330	(79)
CH_3CH_2—Cl	338	(81)	$(CH_3)_3C$—Br	263	(63)
$CH_3CH_2CH_2$—Cl	343	(82)			
Water and alcohols					
HO—H	497	(119)	CH_3CH_2—OH	380	(91)
CH_3O—H	426	(102)	$(CH_3)_2CH$—OH	385	(92)
CH_3—OH	380	(91)	$(CH_3)_3C$—OH	380	(91)

*Bond dissociation energies refer to bond indicated in structural formula for each substance.

Because the starting material (propane) and one of the products (H·) are the same in both processes, the difference in bond dissociation energies is equal to the energy difference between an *n*-propyl radical (primary) and an isopropyl radical (secondary). As depicted in Figure 4.20, the secondary radical is 13 kJ/mol (3 kcal/mol) more stable than the primary radical.

Similarly, by comparing the bond dissociation energies of the two different types of C—H bonds in 2-methylpropane, we see that a tertiary radical is 30 kJ/mol (7 kcal/mol) more stable than a primary radical.

$$CH_3CHCH_2—H \longrightarrow CH_3CHCH_2 \; + \quad H· \qquad \Delta H° = +410 \text{ kJ}$$
$$| \qquad\qquad\qquad\qquad | \qquad\qquad\qquad\qquad\qquad (98 \text{ kcal})$$
$$CH_3 \qquad\qquad\qquad\quad CH_3$$

| 2-Methylpropane | Isobutyl radical (primary) | Hydrogen atom |

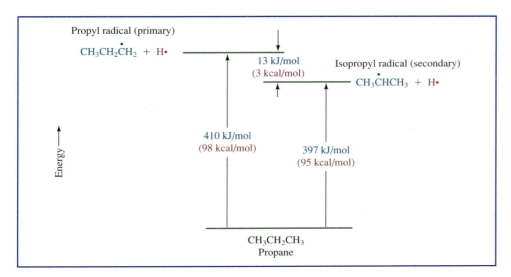

FIGURE 4.20 The bond dissociation energies of methylene and methyl C—H bonds in propane reveal difference in stabilities between two isomeric free radicals. The secondary radical is more stable than the primary.

$$
\underset{\substack{\text{2-Methylpropane}}}{\text{CH}_3\overset{\displaystyle \text{H}}{\underset{\displaystyle \text{CH}_3}{\text{CCH}_3}}} \longrightarrow \underset{\substack{\textit{tert}\text{-Butyl} \\ \text{radical} \\ \text{(tertiary)}}}{\text{CH}_3\overset{\displaystyle \cdot}{\underset{\displaystyle \text{CH}_3}{\text{CCH}_3}}} + \underset{\substack{\text{Hydrogen} \\ \text{atom}}}{\text{H}\cdot} \qquad \Delta H° = +380 \text{ kJ} \\ (91 \text{ kcal})
$$

PROBLEM 4.14 Carbon–carbon bond dissociation energies have been measured for alkanes. Without referring to Table 4.3, identify the alkane in each of the following pairs that has the lower carbon–carbon bond dissociation energy, and explain the reason for your choice.

(a) Ethane or propane

(b) Propane or 2-methylpropane

(c) 2-Methylpropane or 2,2-dimethylpropane

SAMPLE SOLUTION (a) First write the equations that describe homolytic carbon–carbon bond cleavage in each alkane.

$$
\underset{\text{Ethane}}{\text{CH}_3\text{—CH}_3} \longrightarrow \underset{\text{Two methyl radicals}}{\cdot\text{CH}_3 + \cdot\text{CH}_3}
$$

$$
\underset{\text{Propane}}{\text{CH}_3\text{CH}_2\text{—CH}_3} \longrightarrow \underset{\text{Ethyl radical}}{\text{CH}_3\overset{\displaystyle\cdot}{\text{CH}}_2} + \underset{\text{Methyl radical}}{\cdot\text{CH}_3}
$$

Cleavage of the carbon–carbon bond in ethane yields two methyl radicals, whereas propane yields an ethyl radical and one methyl radical. Ethyl radical is more stable than methyl, and so less energy is required to break the carbon–carbon bond in propane than in ethane. The measured carbon–carbon bond dissociation energy in ethane is 368 kJ/mol (88 kcal/mol), and that in propane is 355 kJ/mol (85 kcal/mol).

Like carbocations, most free radicals are exceedingly reactive species—too reactive to be isolated but capable of being formed as transient intermediates in chemical reactions. Methyl radical, as we shall see in the following section, is an intermediate in the chlorination of methane.

4.17   MECHANISM OF METHANE CHLORINATION

The generally accepted mechanism for the chlorination of methane is presented in Figure 4.21. As we noted earlier (Section 4.15), the reaction is normally carried out in the gas phase at high temperature. The reaction itself is strongly exothermic, but energy must be put into the system to get it going. This energy goes into breaking the weakest bond in the system, which, as we see from the bond dissociation energy data in Table 4.3, is the Cl—Cl bond with a bond dissociation energy of 242 kJ/mol (58 kcal/mol). The step in which Cl—Cl bond homolysis occurs is called the **initiation step.**

The bond dissociation energy of the other reactant, methane, is much higher. It is 435 kJ/mol (104 kcal/mol).

Each chlorine atom formed in the initiation step has seven valence electrons and is very reactive. Once formed, a chlorine atom abstracts a hydrogen atom from methane as shown in step 2 in Figure 4.21. Hydrogen chloride, one of the isolated products from

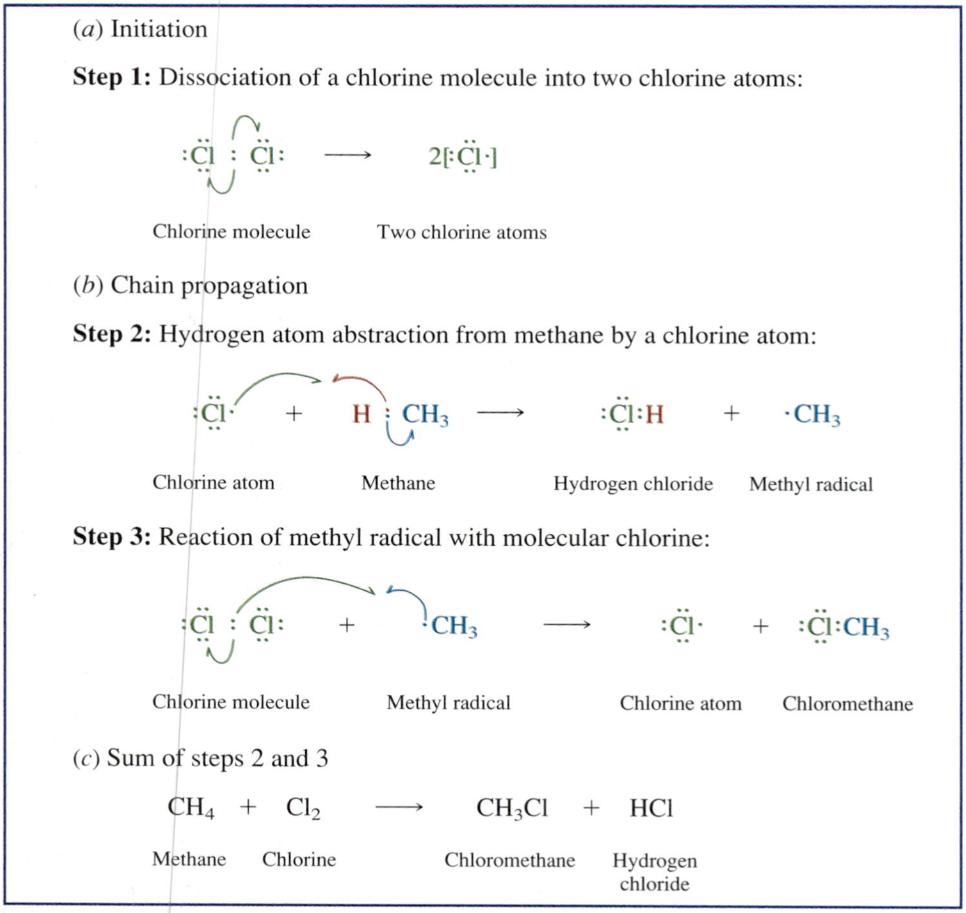

(*a*) Initiation

Step 1: Dissociation of a chlorine molecule into two chlorine atoms:

$$:\ddot{Cl} : \ddot{Cl}: \longrightarrow 2[:\ddot{Cl}\cdot]$$

Chlorine molecule   Two chlorine atoms

(*b*) Chain propagation

Step 2: Hydrogen atom abstraction from methane by a chlorine atom:

$$:\ddot{Cl}\cdot \quad + \quad H:CH_3 \longrightarrow :\ddot{Cl}:H \quad + \quad \cdot CH_3$$

Chlorine atom   Methane   Hydrogen chloride   Methyl radical

Step 3: Reaction of methyl radical with molecular chlorine:

$$:\ddot{Cl} : \ddot{Cl}: \quad + \quad \cdot CH_3 \longrightarrow :\ddot{Cl}\cdot \quad + \quad :\ddot{Cl}:CH_3$$

Chlorine molecule   Methyl radical   Chlorine atom   Chloromethane

(*c*) Sum of steps 2 and 3

$$CH_4 \quad + \quad Cl_2 \longrightarrow CH_3Cl \quad + \quad HCl$$

Methane   Chlorine   Chloromethane   Hydrogen chloride

FIGURE 4.21 The initiation and propagation steps in the free-radical mechanism for the chlorination of methane. Together the two propagation steps give the overall equation for the reaction.

the overall reaction, is formed in this step. A methyl radical is also formed, which then attacks a molecule of Cl_2 in step 3. Attack of methyl radical on Cl_2 gives chloromethane, the other product of the overall reaction, along with a chlorine atom which then cycles back to step 2, repeating the process. Steps 2 and 3 are called the **propagation steps** of the reaction and, when added together, give the overall equation for the reaction. Because one initiation step can result in a great many propagation cycles, the overall process is called a **free-radical chain reaction.**

PROBLEM 4.15 Write equations for the initiation and propagation steps for the formation of dichloromethane by free-radical chlorination of chloromethane.

In practice, side reactions intervene to reduce the efficiency of the propagation steps. The chain sequence is interrupted whenever two odd-electron species combine to give an even-electron product. Reactions of this type are called **chain-terminating steps.** Some commonly observed chain-terminating steps in the chlorination of methane are shown in the following equations.

Combination of a methyl radical with a chlorine atom:

$$\overset{\cdot}{C}H_3 \quad \cdot\overset{\cdot\cdot}{\underset{\cdot\cdot}{C}}l\!: \quad \longrightarrow \quad CH_3\overset{\cdot\cdot}{\underset{\cdot\cdot}{C}}l\!:$$

Methyl radical Chlorine atom   Chloromethane

Combination of two methyl radicals:

$$\overset{\cdot}{C}H_3 \quad \overset{\cdot}{C}H_3 \longrightarrow CH_3CH_3$$

Two methyl radicals   Ethane

Combination of two chlorine atoms:

$$:\overset{\cdot\cdot}{\underset{\cdot\cdot}{C}}l\cdot \quad \cdot\overset{\cdot\cdot}{\underset{\cdot\cdot}{C}}l\!: \longrightarrow \quad Cl_2$$

Two chlorine atoms   Chlorine molecule

Termination steps are, in general, less likely to occur than the propagation steps. Each of the termination steps requires two free radicals to encounter each other in a medium that contains far greater quantities of other materials (methane and chlorine molecules) with which they can react. Although some chloromethane undoubtedly arises via direct combination of methyl radicals with chlorine atoms, most of it is formed by the propagation sequence shown in Figure 4.21.

4.18 HALOGENATION OF HIGHER ALKANES

Like the chlorination of methane, chlorination of ethane is carried out on an industrial scale as a high-temperature gas-phase reaction.

$$CH_3CH_3 + Cl_2 \xrightarrow{420°C} CH_3CH_2Cl + HCl$$

Ethane   Chlorine   Chloroethane (78%)   Hydrogen chloride
                           (ethyl chloride)

FROM BOND ENERGIES TO HEATS OF REACTION

You have seen that measurements of heats of reaction, such as heats of combustion, can provide quantitative information concerning the relative stability of constitutional isomers (Section 2.18) and stereoisomers (Section 3.11). The box in Section 2.18 described how heats of reaction can be manipulated arithmetically to generate heats of formation (ΔH_f°) for many molecules. The following material shows how two different sources of thermochemical information, heats of formation and bond dissociation energies (see Table 4.3), can reveal whether a particular reaction is exothermic or endothermic and by how much.

Consider the chlorination of methane to chloromethane. The heats of formation of the reactants and products appear beneath the equation. These heats of formation for the chemical compounds are taken from published tabulations; the heat of formation of chlorine, as it is for all elements, is zero.

$$CH_4 \;+\; Cl_2 \longrightarrow CH_3Cl \;+\; HCl$$

$$\Delta H_f^\circ: \quad -74.8 \qquad 0 \qquad\quad -81.9 \qquad -92.3$$
(kJ/mol)

The overall heat of reaction is given by

$$\Delta H^\circ = \sum(\text{heats of formation of products}) -$$
$$\sum(\text{heats of formation of reactants})$$

$$\Delta H^\circ = (-81.9\ \text{kJ} - 92.3\ \text{kJ}) - (-74.8\ \text{kJ}) = -99.4\ \text{kJ}$$

Thus, the chlorination of methane is calculated to be an exothermic reaction on the basis of heat of formation data.

The same conclusion is reached using bond dissociation energies. The following equation shows the bond dissociation energies of the reactants and products taken from Table 4.3:

$$CH_4 \;+\; Cl_2 \longrightarrow CH_3Cl \;+\; HCl$$

$$\text{BDE:} \quad 435 \qquad 242 \qquad\quad 349 \qquad 431$$
(kJ/mol)

Because stronger bonds are formed at the expense of weaker ones, the reaction is exothermic and

$$\Delta H^\circ = \sum(\text{BDE of bonds broken}) -$$
$$\sum(\text{BDE of bonds formed})$$

$$\Delta H^\circ = (435\ \text{kJ} + 242\ \text{kJ}) -$$
$$(349\ \text{kJ} + 431\ \text{kJ}) = -103\ \text{kJ}$$

This value is in good agreement with that obtained from heat of formation data.

Compare chlorination of methane with iodination. The relevant bond dissociation energies are given in the equation.

$$CH_4 \;+\; I_2 \longrightarrow CH_3I \;+\; HI$$

$$\text{BDE:} \quad 435 \qquad 150 \qquad\quad 234 \qquad 297$$
(kJ/mol)

$$\Delta H^\circ = \sum(\text{BDE of bonds broken}) -$$
$$\sum(\text{BDE of bonds formed})$$

$$\Delta H^\circ = (435\ \text{kJ} + 150\ \text{kJ}) -$$
$$(234\ \text{kJ} + 297\ \text{kJ}) = +54\ \text{kJ}$$

A positive value for ΔH° signifies an **endothermic** reaction. The reactants are more stable than the products, and so iodination of alkanes is not a feasible reaction. You would not want to attempt the preparation of iodomethane by iodination of methane.

A similar analysis for fluorination of methane gives $\Delta H^\circ = -426\ \text{kJ}$ for its heat of reaction. Fluorination of methane is about four times as exothermic as chlorination. A reaction this exothermic, if it also occurs at a rapid rate, can proceed with explosive violence.

Bromination of methane is exothermic, but less exothermic than chlorination. The value calculated from bond dissociation energies is $\Delta H^\circ = -30\ \text{kJ}$. Although bromination of methane is energetically favorable, economic considerations cause most of the methyl bromide prepared commercially to be made from methanol by reaction with hydrogen bromide.

As in the chlorination of methane, it is often difficult to limit the reaction to monochlorination, and derivatives having more than one chlorine atom are also formed.

> **PROBLEM 4.16**   Chlorination of ethane yields, in addition to ethyl chloride, a mixture of two isomeric dichlorides. What are the structures of these two dichlorides?

In the laboratory it is more convenient to use light, either visible or ultraviolet, as the source of energy to initiate the reaction. Reactions that occur when light energy is absorbed by a molecule are called **photochemical reactions.** Photochemical techniques permit the reaction of alkanes with chlorine to be performed at room temperature.

$$\text{Cyclobutane} \quad + \quad Cl_2 \quad \xrightarrow{h\nu} \quad \text{Cl} \quad + \quad HCl$$

Cyclobutane	Chlorine	Chlorocyclobutane (73%) (cyclobutyl chloride)	Hydrogen chloride

> Photochemical energy is indicated by writing "light" or "*hν*" above or below the arrow. The symbol *hν* is equal to the energy of a light photon and will be discussed in more detail in Section 13.1.

Methane, ethane, and cyclobutane share the common feature that each one can give only a *single* monochloro derivative. All the hydrogens of cyclobutane, for example, are equivalent, and substitution of any one gives the same product as substitution of any other. Chlorination of alkanes in which the hydrogens are not all equivalent is more complicated in that a mixture of every possible monochloro derivative is formed, as the chlorination of butane illustrates:

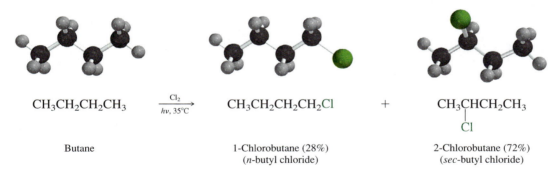

$$CH_3CH_2CH_2CH_3 \quad \xrightarrow[h\nu,\ 35°C]{Cl_2} \quad CH_3CH_2CH_2CH_2Cl \quad + \quad CH_3\underset{|}{\overset{}{C}}HCH_2CH_3 \\ \qquad\qquad\qquad\qquad\qquad\qquad\qquad\qquad\qquad\qquad\qquad\qquad\qquad\qquad\ Cl$$

Butane	1-Chlorobutane (28%) (*n*-butyl chloride)	2-Chlorobutane (72%) (*sec*-butyl chloride)

These two products arise because in one of the propagation steps a chlorine atom may abstract a hydrogen atom from either a methyl or a methylene group of butane.

> The percentages cited in the preceding equation reflect the composition of the monochloride fraction of the product mixture rather than the isolated yield of each component.

$$CH_3CH_2CH_2CH_2\text{—}H \ + \ \cdot\ddot{\underset{..}{Cl}}: \ \longrightarrow \ CH_3CH_2CH_2\dot{C}H_2 \ + \ H\ddot{\underset{..}{Cl}}:$$

Butane   *n*-Butyl radical

$$CH_3\underset{\underset{H}{|}}{\overset{}{C}}HCH_2CH_3 \ + \ \cdot\ddot{\underset{..}{Cl}}: \ \longrightarrow \ CH_3\dot{C}HCH_2CH_3 \ + \ H\ddot{\underset{..}{Cl}}:$$

Butane   *sec*-Butyl radical

The resulting free radicals react with chlorine to give the corresponding alkyl chlorides. Butyl radical gives only 1-chlorobutane; *sec*-butyl radical gives only 2-chlorobutane.

$$CH_3CH_2CH_2\overset{\bullet}{C}H_2 + Cl_2 \longrightarrow CH_3CH_2CH_2CH_2Cl + \cdot\overset{\bullet\bullet}{\underset{\bullet\bullet}{Cl}}:$$

n-Butyl radical   1-Chlorobutane
(*n*-butyl chloride)

$$CH_3\overset{\bullet}{C}HCH_2CH_3 + Cl_2 \longrightarrow CH_3\underset{\underset{Cl}{|}}{C}HCH_2CH_3 + \cdot\overset{\bullet\bullet}{\underset{\bullet\bullet}{Cl}}:$$

sec-Butyl radical   2-Chlorobutane
(*sec*-butyl chloride)

If every collision of a chlorine atom with a butane molecule resulted in hydrogen abstraction, the *n*-butyl/*sec*-butyl radical ratio and, therefore, the 1-chloro/2-chlorobutane ratio, would be given by the relative numbers of hydrogens in the two equivalent methyl groups of $CH_3CH_2CH_2CH_3$ (six) compared with those in the two equivalent methylene groups (four). The product distribution expected on a *statistical* basis would be 60% 1-chlorobutane and 40% 2-chlorobutane. The *experimentally observed* product distribution, however, is 28% 1-chlorobutane and 72% 2-chlorobutane. *sec*-Butyl radical is therefore formed in greater amounts, and *n*-butyl radical in lesser amounts, than expected statistically.

This behavior stems from the greater stability of secondary compared with primary free radicals. The transition state for the step in which a chlorine atom abstracts a hydrogen from carbon has free-radical character at carbon.

$$\overset{\delta\cdot}{Cl}\text{---}\overset{\delta\cdot}{H}\text{---}CH_2CH_2CH_2CH_3 \qquad \overset{\delta\cdot}{Cl}\text{---}\overset{\overset{\displaystyle CH_3}{\displaystyle |}}{\underset{}{\overset{\delta\cdot}{H}}}\text{---}CHCH_2CH_3$$

Transition state for abstraction   Transition state for abstraction
of a primary hydrogen   of a secondary hydrogen

A secondary hydrogen is abstracted faster than a primary hydrogen because the transition state with secondary radical character is more stable than the one with primary radical character. The same factors that stabilize a secondary radical stabilize a transition state with secondary radical character more than one with primary radical character. Hydrogen atom abstraction from a CH_2 group occurs faster than from a CH_3 group. We can calculate how much faster a *single* secondary hydrogen is abstracted compared with a *single* primary hydrogen from the experimentally observed product distribution.

$$\frac{72\% \text{ 2-chlorobutane}}{28\% \text{ 1-chlorobutane}} = \frac{\text{rate of secondary H abstraction} \times 4 \text{ secondary hydrogens}}{\text{rate of primary H abstraction} \times 6 \text{ primary hydrogens}}$$

$$\frac{\text{Rate of secondary H abstraction}}{\text{Rate of primary H abstraction}} = \frac{72}{28} \times \frac{6}{4} = \frac{3.9}{1}$$

A single secondary hydrogen in butane is abstracted by a chlorine atom 3.9 times faster than a single primary hydrogen.

PROBLEM 4.17 Assuming the relative rate of secondary to primary hydrogen atom abstraction to be the same in the chlorination of propane as it is in that of butane, calculate the relative amounts of propyl chloride and isopropyl chloride obtained in the free-radical chlorination of propane.

A similar study of the chlorination of 2-methylpropane established that a tertiary hydrogen is removed 5.2 times faster than each primary hydrogen.

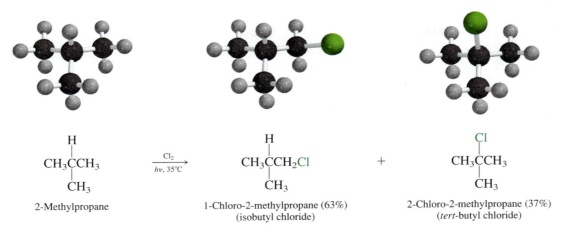

$$CH_3\overset{\displaystyle H}{\underset{\displaystyle CH_3}{C}}CH_3 \xrightarrow[hv,\ 35°C]{Cl_2} CH_3\overset{\displaystyle H}{\underset{\displaystyle CH_3}{C}}CH_2Cl \quad + \quad CH_3\overset{\displaystyle Cl}{\underset{\displaystyle CH_3}{C}}CH_3$$

2-Methylpropane   1-Chloro-2-methylpropane (63%)   2-Chloro-2-methylpropane (37%)
(isobutyl chloride)   (*tert*-butyl chloride)

In summary then, the chlorination of alkanes is not very selective. The various kinds of hydrogens present in a molecule (tertiary, secondary, and primary) differ by only a factor of 5 in the relative rate at which each reacts with a chlorine atom.

$$R_3CH \ > \ R_2CH_2 \ > \ RCH_3$$

(tertiary)   (secondary)   (primary)

Relative rate (chlorination)   5.2   3.9   1

Bromine reacts with alkanes by a free-radical chain mechanism analogous to that of chlorine. There is an important difference between chlorination and bromination, however. Bromination is highly selective for substitution of *tertiary hydrogens*. The spread in reactivity among primary, secondary, and tertiary hydrogens is greater than 10^3.

$$R_3CH \ > \ R_2CH_2 \ > \ RCH_3$$

(tertiary)   (secondary)   (primary)

Relative rate (bromination)   1640   82   1

In practice, this means that when an alkane contains primary, secondary, and tertiary hydrogens, it is usually only the tertiary hydrogen that is replaced by bromine.

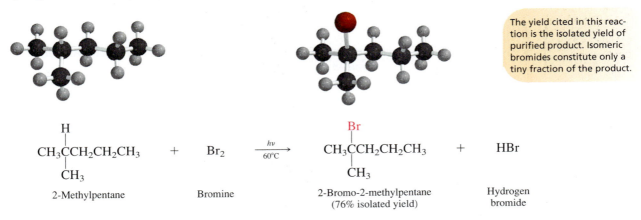

$$CH_3\overset{\displaystyle H}{\underset{\displaystyle CH_3}{C}}CH_2CH_2CH_3 \ + \ Br_2 \xrightarrow[60°C]{hv} CH_3\overset{\displaystyle Br}{\underset{\displaystyle CH_3}{C}}CH_2CH_2CH_3 \ + \ HBr$$

2-Methylpentane   Bromine   2-Bromo-2-methylpentane   Hydrogen
(76% isolated yield)   bromide

The yield cited in this reaction is the isolated yield of purified product. Isomeric bromides constitute only a tiny fraction of the product.

PROBLEM 4.18   Give the structure of the principal organic product formed by free-radical bromination of each of the following:

(a) Methylcyclopentane   (c) 2,2,4-Trimethylpentane

(b) 1-Isopropyl-1-methylcyclopentane

SAMPLE SOLUTION   (a) Write the structure of the starting hydrocarbon, and identify any tertiary hydrogens that are present. The only tertiary hydrogen in methylcyclopentane is the one attached to C-1. This is the one replaced by bromine.

Methylcyclopentane   1-Bromo-1-methylcyclopentane

This difference in selectivity between chlorination and bromination of alkanes needs to be kept in mind when one wishes to prepare an alkyl halide from an alkane:

1. Because chlorination of an alkane yields every possible monochloride, it is used only when all the hydrogens in an alkane are equivalent.

2. Bromination is normally used only to prepare tertiary alkyl bromides from alkanes.

Selectivity is not an issue in the conversion of alcohols to alkyl halides. Except for certain limitations to be discussed in Section 8.15, the location of the halogen substituent in the product corresponds to that of the hydroxyl group in the starting alcohol.

4.19   SUMMARY

Chemical reactivity and functional group transformations involving the preparation of alkyl halides from alcohols and from alkanes are the main themes of this chapter. Although the conversions of an alcohol or an alkane to an alkyl halide are both classified as substitutions, they proceed by very different mechanisms.

Section 4.1   **Functional groups** are the structural units responsible for the characteristic reactions of a molecule. The hydrocarbon chain to which a functional group is attached can often be considered as simply a supporting framework. The most common functional groups characterize the families of organic compounds listed on the inside front cover of the text.

Section 4.2   Alcohols and alkyl halides may be named using either **substitutive** or **functional class** IUPAC nomenclature. In substitutive nomenclature alkyl halides are named as halogen derivatives of alkanes. The parent is the longest continuous chain that bears the halogen substituent, and in the absence of other substituents the chain is numbered from the direction that gives the lowest number to the carbon that bears the halogen. The functional class names of alkyl halides begin with the name of the alkyl group and end with the halide as a separate word.

$$CH_3CHCH_2CH_2CH_2CH_3$$
|
Br

Substitutive name: 2-Bromohexane
Functional class name: 1-Methylpentyl bromide

Section 4.3

The substitutive names of alcohols are derived by replacing the *-e* ending of an alkane with *-ol*. The longest chain containing the OH group becomes the basis for the name. Functional class names of alcohols begin with the name of the alkyl group and end in the word *alcohol*.

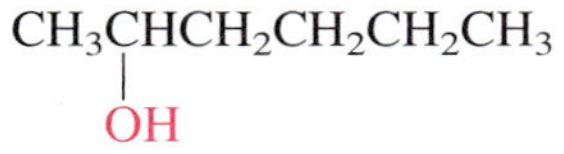

$$CH_3CHCH_2CH_2CH_2CH_3$$
|
OH

Substitutive name: 2-Hexanol
Functional class name: 1-Methylpentyl alcohol

Section 4.4

Alcohols (X = OH) and alkyl halides (X = F, Cl, Br, or I) are classified as primary, secondary, or tertiary according to the degree of substitution at the carbon that bears the functional group.

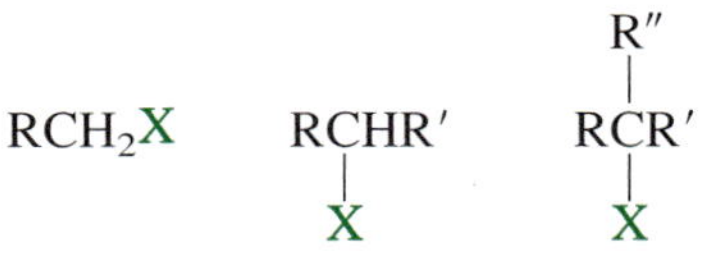

$$RCH_2X \qquad RCHR' \qquad RCR'$$

with R″ above the tertiary carbon, and X below the secondary and tertiary carbons.

Primary   Secondary   Tertiary

Section 4.5

The halogens (especially fluorine and chlorine) and oxygen are more electronegative than carbon, and the carbon–halogen bond in alkyl halides and the carbon–oxygen bond in alcohols are polar. Carbon is the positive end of the dipole and halogen or oxygen the negative end.

Section 4.6

Dipole/induced-dipole and dipole–dipole attractive forces make alcohols higher boiling than alkanes of similar molecular weight. The attractive force between —OH groups is called **hydrogen bonding.**

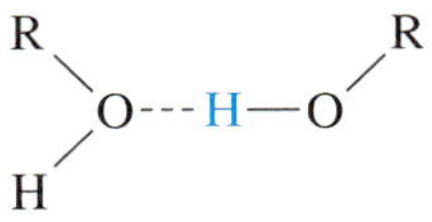

$$\underset{H}{\overset{R}{\diagdown}}O \text{---} H \overset{R}{—} O$$

Hydrogen bonding between the hydroxyl group of an alcohol and water makes the water-solubility of alcohols greater than that of hydrocarbons. Low-molecular-weight alcohols [CH_3OH, CH_3CH_2OH, $CH_3CH_2CH_2OH$, and $(CH_3)_2CHOH$] are soluble in water in all proportions. Alkyl halides are insoluble in water.

Section 4.7

See Table 4.4

Section 4.8

Secondary and tertiary alcohols react with hydrogen halides by a mechanism that involves formation of a carbocation intermediate in the rate-determining step.

$$(1) \quad ROH \ + \ HX \ \underset{\text{fast}}{\rightleftarrows} \ R\overset{+}{O}H_2 \ + \ X^-$$

Alcohol   Hydrogen   Alkyloxonium   Halide
    halide   ion   anion

$$(2) \quad R\overset{+}{O}H_2 \ \xrightarrow{\text{slow}} \ R^+ \ + \ H_2O$$

Alkyloxonium ion   Carbocation   Water

$$(3) \quad R^+ \ + \ X^- \ \xrightarrow{\text{fast}} \ RX$$

Carbocation   Halide ion   Alkyl halide

TABLE 4.4	Conversions of Alcohols and Alkanes to Alkyl Halides

Reaction (section) and comments	General equation and specific example(s)
Reactions of alcohols with hydrogen halides (Section 4.7) Alcohols react with hydrogen halides to yield alkyl halides. The reaction is useful as a synthesis of alkyl halides. The reactivity of hydrogen halides decreases in the order HI > HBr > HCl > HF. Alcohol reactivity decreases in the order tertiary > secondary > primary > methyl.	ROH + HX $\longrightarrow$ RX + H_2O Alcohol Hydrogen halide Alkyl halide Water 1-Methylcyclopentanol $\xrightarrow{HCl}$ 1-Chloro-1-methylcyclopentane (96%)
Reaction of alcohols with thionyl chloride (Section 4.13) Thionyl chloride is a synthetic reagent used to convert alcohols to alkyl chlorides.	ROH + $SOCl_2$ $\longrightarrow$ RCl + SO_2 + HCl Alcohol Thionyl chloride Alkyl chloride Sulfur dioxide Hydrogen chloride $CH_3CH_2CH_2CH_2CH_2OH \xrightarrow[\text{pyridine}]{SOCl_2} CH_3CH_2CH_2CH_2CH_2Cl$ 1-Pentanol 1-Chloropentane (80%)
Reaction of alcohols with phosphorus tribromide (Section 4.13) As an alternative to converting alcohols to alkyl bromides with hydrogen bromide, the inorganic reagent phosphorus tribromide is sometimes used.	$3ROH$ + PBr_3 $\longrightarrow$ $3RBr$ + H_3PO_3 Alcohol Phosphorus tribromide Alkyl bromide Phosphorous acid $CH_3CHCH_2CH_2CH_3 \xrightarrow{PBr_3} CH_3CHCH_2CH_2CH_3$ OH Br 2-Pentanol 2-Bromopentane (67%)
Free-radical halogenation of alkanes (Sections 4.14 through 4.18) Alkanes react with halogens by substitution of a halogen for a hydrogen on the alkane. The reactivity of the halogens decreases in the order $F_2 > Cl_2 > Br_2 > I_2$. The ease of replacing a hydrogen decreases in the order tertiary > secondary > primary > methyl. Chlorination is not very selective and so is used only when all the hydrogens of the alkane are equivalent. Bromination is highly selective, replacing tertiary hydrogens much more readily than secondary or primary ones.	RH + X_2 $\longrightarrow$ RX + HX Alkane Halogen Alkyl halide Hydrogen halide Cyclodecane $\xrightarrow[h\nu]{Cl_2}$ Cyclodecyl chloride (64%) $(CH_3)_2CHC(CH_3)_3 \xrightarrow[h\nu]{Br_2} (CH_3)_2CC(CH_3)_3$ Br 2,2,3-Trimethylbutane 2-Bromo-2,3,3-trimethylbutane (80%)

Section 4.9   The potential energy diagrams for separate elementary steps can be merged into a diagram for the overall process. The diagram for the reaction of a secondary or tertiary alcohol with a hydrogen halide is characterized by two intermediates and three transition states. The reaction is classified as a unimolecular **nucleophilic substitution,** abbreviated as S_N1.

Section 4.10   Carbocations contain a positively charged carbon with only three atoms or groups attached to it. This carbon is sp^2-hybridized and has a vacant $2p$ orbital.

Carbocations are stabilized by alkyl substituents attached directly to the positively charged carbon. Alkyl groups are *electron-releasing* substituents. Stability increases in the order:

(least stable) $CH_3{}^+ < RCH_2{}^+ < R_2CH^+ < R_3C^+$ (most stable)

Carbocations are strongly **electrophilic** (Lewis acids) and react with **nucleophiles** (Lewis bases).

Section 4.11   The rate at which alcohols are converted to alkyl halides depends on the rate of carbocation formation: tertiary alcohols are most reactive; primary alcohols and methanol are least reactive.

Section 4.12   Primary alcohols do not react with hydrogen halides by way of carbocation intermediates. The nucleophilic species (Br^- for example) attacks the alkyloxonium ion and "pushes off" a water molecule from carbon in a bimolecular step. This step is rate-determining, and the mechanism is S_N2.

Section 4.13   See Table 4.4

Section 4.14   See Table 4.4

Section 4.15   Methane reacts with Cl_2 to give chloromethane, dichloromethane, trichloromethane, and tetrachloromethane.

Section 4.16   Chlorination of methane, and halogenation of alkanes generally, proceed by way of **free-radical** intermediates. Alkyl radicals are neutral and have an unpaired electron on carbon.

Like carbocations, free radicals are stabilized by alkyl substituents. The order of free-radical stability parallels that of carbocation stability.

Section 4.17   The elementary steps (1) through (3) describe a free-radical chain mechanism for the reaction of an alkane with a halogen.

(1)   (initiation step) $X_2 \longrightarrow 2X\cdot$

           Halogen molecule   Two halogen atoms

(2) (propagation step)   RH +   X· ⟶ R· + HX

                        Alkane   Halogen   Alkyl   Hydrogen
                               atom   radical   halide

(3) (propagation step)   R· + X_2 ⟶ RX + ·X

                        Alkyl   Halogen   Alkyl   Halogen
                        radical   molecule   halide   atom

Section 4.18   See Table 4.4

PROBLEMS

4.19 Write structural formulas for each of the following alcohols and alkyl halides:

(a) Cyclobutanol

(b) *sec*-Butyl alcohol

(c) 3-Heptanol

(d) *trans*-2-Chlorocyclopentanol

(e) 2,6-Dichloro-4-methyl-4-octanol

(f) *trans*-4-*tert*-Butylcyclohexanol

(g) 1-Cyclopropylethanol

(h) 2-Cyclopropylethanol

4.20 Name each of the following compounds according to substitutive IUPAC nomenclature:

(a) $(CH_3)_2CHCH_2CH_2CH_2Br$

(b) $(CH_3)_2CHCH_2CH_2CH_2OH$

(c) Cl_3CCH_2Br

(d) $Cl_2CHCHBr$
        |
        Cl

(e) CF_3CH_2OH

(f)

(g)

(h)

(i)

4.21 Handbooks are notorious for listing compounds according to their common names. One gives the name "*sec*-isoamyl alcohol" for a compound which could be called 1,2-dimethylpropyl alcohol according to the IUPAC functional class rules. The best name for this compound is the substitutive IUPAC name. What is it?

4.22 Write structural formulas, or build molecular models for all the constitutionally isomeric alcohols of molecular formula $C_5H_{12}O$. Assign a substitutive and a functional class name to each one, and specify whether it is a primary, secondary, or tertiary alcohol.

4.23 A hydroxyl group is a somewhat "smaller" substituent on a six-membered ring than is a methyl group. That is, the preference of a hydroxyl group for the equatorial orientation is less pronounced than that of a methyl group. Given this information, write structural formulas or build molecular models for all the isomeric methylcyclohexanols, showing each one in its most stable conformation. Give the substitutive IUPAC name for each isomer.

4.24 By assuming that the heat of combustion of the cis isomer was larger than the trans, structural assignments were made many years ago for the stereoisomeric 2-, 3-, and 4-methylcyclohexanols. This assumption is valid for two of the stereoisomeric pairs but is incorrect for the other. For which pair of stereoisomers is the assumption incorrect? Why?

4.25 (a) *Menthol,* used to flavor various foods and tobacco, is the most stable stereoisomer of 2-isopropyl-5-methylcyclohexanol. Draw or make a molecular model of its most stable conformation. Is the hydroxyl group cis or trans to the isopropyl group? To the methyl group?

(b) *Neomenthol* is a stereoisomer of menthol. That is, it has the same constitution but differs in the arrangement of its atoms in space. Neomenthol is the second most stable stereoisomer of 2-isopropyl-5-methylcyclohexanol; it is less stable than menthol but more stable than any other stereoisomer. Write the structure, or make a molecular model of neomenthol in its most stable conformation.

4.26 *Epichlorohydrin* is the common name of an industrial chemical used as a component in epoxy cement. The molecular formula of epichlorohydrin is C_3H_5ClO. Epichlorohydrin has an epoxide functional group; it does not have a methyl group. Write a structural formula for epichlorohydrin.

4.27 (a) Complete the structure of the pain-relieving drug *ibuprofen* on the basis of the fact that ibuprofen is a carboxylic acid that has the molecular formula $C_{13}H_{18}O_2$, X is an isobutyl group, and Y is a methyl group.

$$X-\!\!\left\langle \!\!\bigcirc\!\! \right\rangle\!\!-\overset{\overset{\displaystyle Y}{|}}{CH}\!\!-\!\!Z$$

(b) *Mandelonitrile* may be obtained from peach flowers. Derive its structure from the template in part (a) given that X is hydrogen, Y is the functional group that characterizes alcohols, and Z characterizes nitriles.

4.28 *Isoamyl acetate* is the common name of the substance most responsible for the characteristic odor of bananas. Write a structural formula for isoamyl acetate, given the information that it is an ester in which the carbonyl group bears a methyl substituent and there is a 3-methylbutyl group attached to one of the oxygens.

4.29 *n-Butyl mercaptan* is the common name of a foul-smelling substance obtained from skunk fluid. It is a thiol of the type RX, where R is an *n*-butyl group and X is the functional group that characterizes a thiol. Write a structural formula for this substance.

4.30 Some of the most important organic compounds in biochemistry are the α-*amino acids,* represented by the general formula shown.

$$\overset{\overset{\displaystyle O}{\|}}{\underset{\underset{\displaystyle {}^+NH_3}{|}}{RCHCO^-}}$$

Write structural formulas for the following α-amino acids.

(a) Alanine (R = methyl)

(b) Valine (R = isopropyl)

(c) Leucine (R = isobutyl)

(d) Isoleucine (R = *sec*-butyl)

(e) Serine (R = XCH_2, where X is the functional group that characterizes alcohols)

(f) Cysteine (R = XCH_2, where X is the functional group that characterizes thiols)

(g) Aspartic acid (R = XCH_2, where X is the functional group that characterizes carboxylic acids)

4.31 Uscharidin is the common name of a poisonous natural product having the structure shown. Locate all of the following in uscharidin:

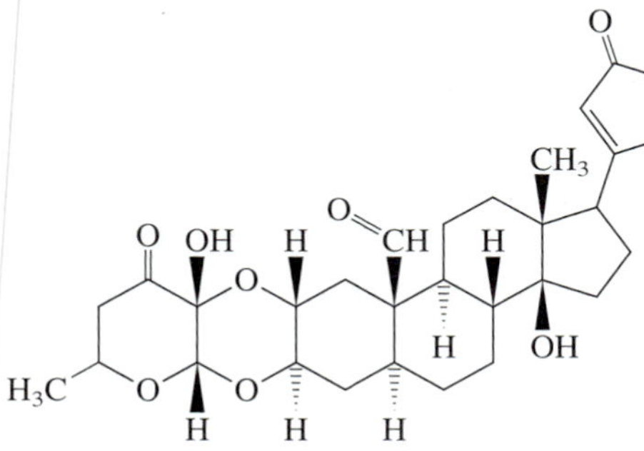

(a) Alcohol, aldehyde, ketone, and ester functional groups

(b) Methylene groups

(c) Primary carbons

4.32 Write a chemical equation for the reaction of 1-butanol with each of the following:

(a) Sodium amide ($NaNH_2$)   (d) Phosphorus tribromide

(b) Hydrogen bromide, heat   (e) Thionyl chloride

(c) Sodium bromide, sulfuric acid, heat

4.33 Each of the following reactions has been described in the chemical literature and involves an organic starting material somewhat more complex than those we have encountered so far. Nevertheless, on the basis of the topics covered in this chapter, you should be able to write the structure of the principal organic product of each reaction.

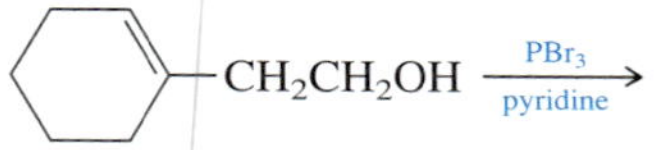

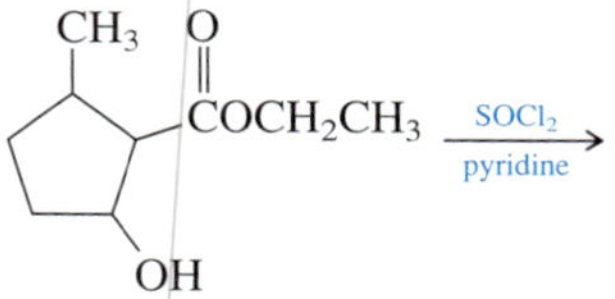

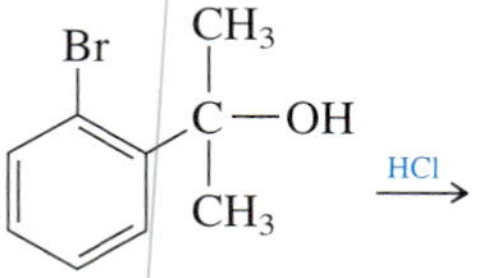

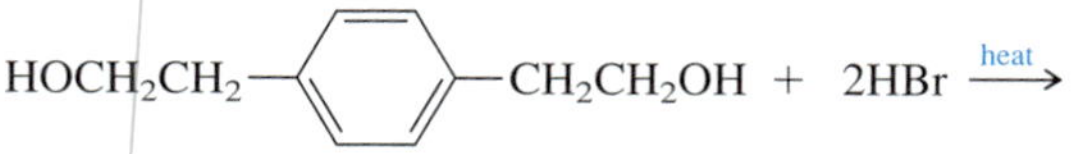

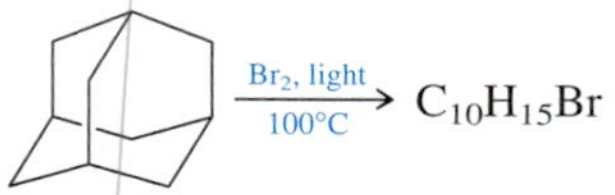

4.34 Select the compound in each of the following pairs that will be converted to the corresponding alkyl bromide more rapidly on being treated with hydrogen bromide. Explain the reason for your choice.

(a) 1-Butanol or 2-butanol

(b) 2-Methyl-1-butanol or 2-butanol

(c) 2-Methyl-2-butanol or 2-butanol

(d) 2-Methylbutane or 2-butanol

(e) 1-Methylcyclopentanol or cyclohexanol

(f) 1-Methylcyclopentanol or *trans*-2-methylcyclopentanol

(g) 1-Cyclopentylethanol or 1-ethylcyclopentanol

4.35 Assuming that the rate-determining step in the reaction of cyclohexanol with hydrogen bromide to give cyclohexyl bromide is unimolecular, write an equation for this step. Use curved arrows to show the flow of electrons.

4.36 Assuming that the rate-determining step in the reaction of 1-hexanol with hydrogen bromide to give 1-bromohexane is an attack by a nucleophile on an alkyloxonium ion, write an equation for this step. Use curved arrows to show the flow of electrons.

4.37 Two stereoisomers of 1-bromo-4-methylcyclohexane are formed when *trans*-4-methylcyclohexanol reacts with hydrogen bromide. Write structural formulas or make molecular models of:

(a) *trans*-4-Methylcylohexanol

(b) The carbocation intermediate in this reaction

(c) The two stereoisomers of 1-bromo-4-methylcyclohexane

4.38 Basing your answers on the bond dissociation energies in Table 4.3, calculate which of the following reactions are endothermic and which are exothermic:

(a) $(CH_3)_2CHOH + HF \rightarrow (CH_3)_2CHF + H_2O$

(b) $(CH_3)_2CHOH + HCl \rightarrow (CH_3)_2CHCl + H_2O$

(c) $CH_3CH_2CH_3 + HCl \rightarrow (CH_3)_2CHCl + H_2$

4.39 By carrying out the reaction at $-78°C$ it is possible to fluorinate 2,2-dimethylpropane to yield $(CF_3)_4C$. Write a balanced chemical equation for this reaction.

4.40 In a search for fluorocarbons having anesthetic properties, 1,2-dichloro-1,1-difluoropropane was subjected to photochemical chlorination. Two isomeric products were obtained, one of which was identified as 1,2,3-trichloro-1,1-difluoropropane. What is the structure of the second compound?

4.41 Among the isomeric alkanes of molecular formula C_5H_{12}, identify the one that on photochemical chlorination yields

(a) A single monochloride   (c) Four isomeric monochlorides

(b) Three isomeric monochlorides   (d) Two isomeric dichlorides

4.42 In both the following exercises, assume that all the methylene groups in the alkane are equally reactive as sites of free-radical chlorination.

(a) Photochemical chlorination of heptane gave a mixture of monochlorides containing 15% 1-chloroheptane. What other monochlorides are present? Estimate the percentage of each of these additional $C_7H_{15}Cl$ isomers in the monochloride fraction.

(b) Photochemical chlorination of dodecane gave a monochloride fraction containing 19% 2-chlorododecane. Estimate the percentage of 1-chlorododecane present in that fraction.

4.43   Photochemical chlorination of 2,2,4-trimethylpentane gives four isomeric monochlorides.

(a) Write structural formulas for these four isomers.

(b) The two primary chlorides make up 65% of the monochloride fraction. Assuming that all the primary hydrogens in 2,2,4-trimethylpentane are equally reactive, estimate the percentage of each of the two primary chlorides in the product mixture.

4.44   Photochemical chlorination of pentane gave a mixture of three isomeric monochlorides. The principal monochloride constituted 46% of the total, and the remaining 54% was approximately a 1:1 mixture of the other two isomers. Write structural formulas for the three monochloride isomers and specify which one was formed in greatest amount. (Recall that a secondary hydrogen is abstracted three times faster by a chlorine atom than a primary hydrogen.)

4.45   Cyclopropyl chloride has been prepared by the free-radical chlorination of cyclopropane. Write a stepwise mechanism for this reaction.

4.46   Deuterium oxide (D_2O) is water in which the protons (^1H) have been replaced by their heavier isotope deuterium (^2H). It is readily available and is used in a variety of mechanistic studies in organic chemistry and biochemistry. When D_2O is added to an alcohol (ROH), deuterium replaces the proton of the hydroxyl group.

$$ROH + D_2O \rightleftharpoons ROD + DOH$$

The reaction takes place extremely rapidly, and if D_2O is present in excess, all the alcohol is converted to ROD. This hydrogen–deuterium exchange can be catalyzed by either acids or bases. If D_3O^+ is the catalyst in acid solution and DO^- the catalyst in base, write reasonable reaction mechanisms for the conversion of ROH to ROD under conditions of (a) acid catalysis and (b) base catalysis.

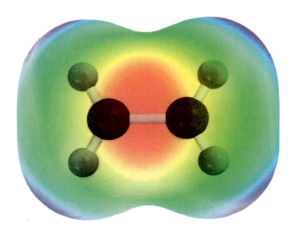

CHAPTER 5

STRUCTURE AND PREPARATION OF ALKENES: ELIMINATION REACTIONS

Alkenes are hydrocarbons that contain a carbon–carbon double bond. A carbon–carbon double bond is both an important structural unit and an important functional group in organic chemistry. The shape of an organic molecule is influenced by the presence of this bond, and the double bond is the site of most of the chemical reactions that alkenes undergo. Some representative alkenes include *isobutylene* (an industrial chemical), *α-pinene* (a fragrant liquid obtained from pine trees), and *farnesene* (a naturally occurring alkene with three double bonds).

$(CH_3)_2C{=}CH_2$

Isobutylene
(used in the production
of synthetic rubber)

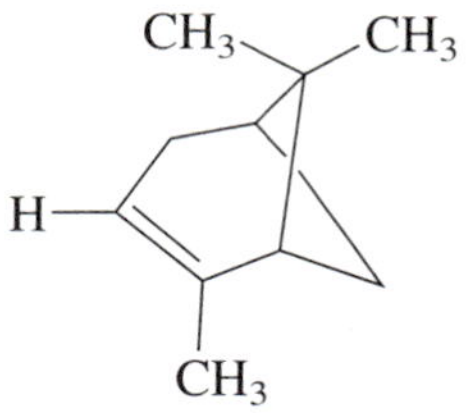

α-Pinene
(a major constituent
of turpentine)

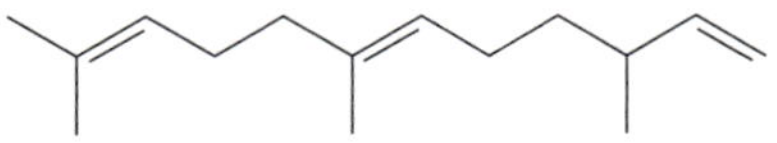

Farnesene
(present in the waxy coating
found on apple skins)

This chapter is the first of two dealing with alkenes; it describes their structure, bonding, and preparation. Chapter 6 discusses their chemical reactions.

5.1 ALKENE NOMENCLATURE

We give alkenes IUPAC names by replacing the *-ane* ending of the corresponding alkane with *-ene*. The two simplest alkenes are **ethene** and **propene.** Both are also well known by their common names *ethylene* and *propylene.*

$$H_2C=CH_2 \qquad\qquad CH_3CH=CH_2$$

<table>
<tr><td>IUPAC name: ethene</td><td>IUPAC name: propene</td></tr>
<tr><td>Common name: ethylene</td><td>Common name: propylene</td></tr>
</table>

Ethylene is an acceptable synonym for *ethene* in the IUPAC system. *Propylene, isobutylene,* and other common names ending in *-ylene* are not acceptable IUPAC names.

The longest continuous chain that includes the double bond forms the base name of the alkene, and the chain is numbered in the direction that gives the doubly bonded carbons their lower numbers. The locant (or numerical position) of only one of the doubly bonded carbons is specified in the name; it is understood that the other doubly bonded carbon must follow in sequence.

$$\overset{1}{H_2}C=\overset{2}{C}H\overset{3}{C}H_2\overset{4}{C}H_3 \qquad\qquad \overset{6}{C}H_3\overset{5}{C}H_2\overset{4}{C}H_2\overset{3}{C}H=\overset{2}{C}H\overset{1}{C}H_3$$

<table>
<tr><td>1-Butene</td><td>2-Hexene</td></tr>
<tr><td>(not 1,2-butene)</td><td>(not 4-hexene)</td></tr>
</table>

Carbon–carbon double bonds take precedence over alkyl groups and halogens in determining the main carbon chain and the direction in which it is numbered.

$$\overset{4}{C}H_3\overset{3}{C}H\overset{2}{C}H=\overset{1}{C}H_2 \qquad\qquad \overset{6}{B}rCH_2\overset{5}{C}H_2\overset{4}{C}H_2\overset{3}{C}HCH_2CH_2CH_3$$
$$\underset{CH_3}{|} \qquad\qquad\qquad\qquad\qquad \overset{2}{|}\underset{CH=CH_2}{}\overset{1}{}$$

<table>
<tr><td>3-Methyl-1-butene</td><td>6-Bromo-3-propyl-1-hexene</td></tr>
<tr><td>(not 2-methyl-3-butene)</td><td>(longest chain that contains double bond is six carbons)</td></tr>
</table>

Hydroxyl groups, however, outrank the double bond. Compounds that contain both a double bond and a hydroxyl group use the combined suffix *-en* + *-ol* to signify that both functional groups are present.

$$\overset{H}{\underset{\overset{1}{HOCH_2}\overset{2}{CH_2}\overset{3}{CH_2}}{\diagdown}}\overset{4}{C}=\overset{5}{C}\overset{\overset{6}{CH_3}}{\diagup}_{CH_3}$$

5-Methyl-4-hexen-1-ol
(not 2-methyl-2-hexen-6-ol)

PROBLEM 5.1 Name each of the following using IUPAC nomenclature:

(a) $(CH_3)_2C=C(CH_3)_2$

(b) $(CH_3)_3CCH=CH_2$

(c) $(CH_3)_2C=CHCH_2CH_2CH_3$

(d) $H_2C=CHCH_2CHCH_3$
　　　　　　　　　$\underset{Cl}{|}$

(e) $H_2C=CHCH_2CHCH_3$
　　　　　　　　　$\underset{OH}{|}$

SAMPLE SOLUTION (a) The longest continuous chain in this alkene contains four carbon atoms. The double bond is between C-2 and C-3, and so it is named as a derivative of 2-butene.

$$\overset{\overset{1}{H_3}C}{\underset{H_3C}{\diagdown}}\overset{2}{C}=\overset{3}{C}\overset{\overset{4}{CH_3}}{\diagup}_{CH_3}$$

2,3-Dimethyl-2-butene

Identifying the alkene as a derivative of 2-butene leaves two methyl groups to be accounted for as substituents attached to the main chain. This alkene is 2,3-dimethyl-2-butene.

ETHYLENE

Ethylene was known to chemists in the eighteenth century and isolated in pure form in 1795. An early name for ethylene was *gaz oléfiant* (French for "oil-forming gas"), to describe the fact that an oily liquid product is formed when two gases—ethylene and chlorine—react with each other.

$$H_2C{=}CH_2 \; + \quad Cl_2 \quad \longrightarrow \quad ClCH_2CH_2Cl$$

Ethylene	Chlorine	1,2-Dichloroethane
(bp: −104°C)	(bp: −34°C)	(bp: 83°C)

The term *gaz oléfiant* was the forerunner of the general term *olefin,* formerly used as the name of the class of compounds we now call *alkenes.*

Ethylene occurs naturally in small amounts as a plant hormone. Hormones are substances that act as messengers to regulate biological processes. Ethylene is involved in the ripening of many fruits, in which it is formed in a complex series of steps from a compound containing a cyclopropane ring:

1-Amino-cyclopropane-carboxylic acid	Ethylene

Even minute amounts of ethylene can stimulate ripening, and the rate of ripening increases with the concentration of ethylene. This property is used to advantage, for example, in the marketing of bananas. Bananas are picked green in the tropics, kept green by being stored with adequate ventilation to limit the amount of ethylene present, and then induced to ripen at their destination by passing ethylene over the fruit.*

Ethylene is the cornerstone of the world's mammoth petrochemical industry and is produced in vast quantities. In a typical year the amount of ethylene produced in the United States (5×10^{10} lb) exceeds the combined weight of all of its people. In one process, ethane from natural gas is heated to bring about its dissociation into ethylene and hydrogen:

$$CH_3CH_3 \xrightarrow{750°C} H_2C{=}CH_2 \; + \quad H_2$$

Ethane	Ethylene	Hydrogen

This reaction is known as **dehydrogenation** and is simultaneously both a source of ethylene and one of the methods by which hydrogen is prepared on an industrial scale. Most of the hydrogen so generated is subsequently used to reduce nitrogen to ammonia for the preparation of fertilizer.

Similarly, dehydrogenation of propane gives propene:

$$CH_3CH_2CH_3 \xrightarrow{750°C} CH_3CH{=}CH_2 \; + \quad H_2$$

Propane	Propene	Hydrogen

Propene is the second most important petrochemical and is produced on a scale about half that of ethylene.

Almost any hydrocarbon can serve as a starting material for production of ethylene and propene. Cracking of petroleum (Section 2.16) gives ethylene and propene by processes involving cleavage of carbon–carbon bonds of higher molecular weight hydrocarbons.

The major uses of ethylene and propene are as starting materials for the preparation of polyethylene and polypropylene plastics, fibers, and films. These and other applications will be described in Chapter 6.

*For a review, see "Ethylene—An Unusual Plant Hormone" in the April 1992 issue of the *Journal of Chemical Education* (pp. 315–318).

We noted in Section 2.13 that the common names of certain frequently encountered *alkyl* groups, such as isopropyl and *tert*-butyl, are acceptable in the IUPAC system. Three *alkenyl* groups—**vinyl, allyl,** and **isopropenyl**—are treated the same way.

Vinyl chloride is an industrial chemical produced in large amounts (10^{10} lb/year in the United States) and is used in the preparation of poly(vinyl chloride). Poly(vinyl chloride), often called simply *vinyl*, has many applications, including siding for houses, wall coverings, and PVC piping.

$$H_2C=CH-$$
Vinyl

as in

$$H_2C=CHCl$$
Vinyl chloride

$$H_2C=CHCH_2-$$
Allyl

as in

$$H_2C=CHCH_2OH$$
Allyl alcohol

$$H_2C=\underset{\underset{CH_3}{|}}{C}-$$
Isopropenyl

as in

$$H_2C=\underset{\underset{CH_3}{|}}{C}Cl$$
Isopropenyl chloride

When a CH_2 group is doubly bonded to a ring, the prefix *methylene* is added to the name of the ring.

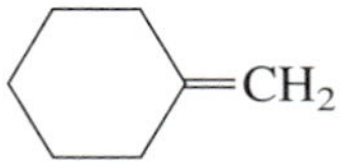

$$\bigcirc\!\!=CH_2$$

Methylenecyclohexane

Cycloalkenes and their derivatives are named by adapting cycloalkane terminology to the principles of alkene nomenclature.

Cyclopentene 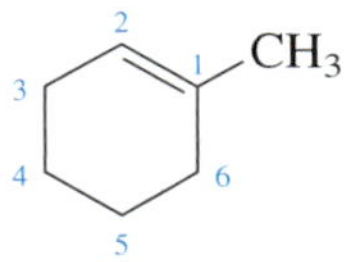   1-Methylcyclohexene 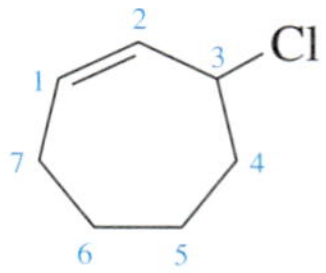   3-Chlorocycloheptene
(not 1-chloro-2-cycloheptene)

No locants are needed in the absence of substituents; it is understood that the double bond connects C-1 and C-2. Substituted cycloalkenes are numbered beginning with the double bond, proceeding through it, and continuing in sequence around the ring. The direction is chosen so as to give the lower of two possible numbers to the substituent.

PROBLEM 5.2   Write structural formulas or build molecular models and give the IUPAC names of all the monochloro-substituted derivatives of cyclopentene.

5.2   STRUCTURE AND BONDING IN ALKENES

The structure of ethylene and the orbital hybridization model for its double bond were presented in Section 2.20 and are briefly reviewed in Figure 5.1. Ethylene is planar, each carbon is sp^2-hybridized, and the double bond is considered to have a σ component and a π component. The σ component arises from overlap of sp^2 hybrid orbitals along a line connecting the two carbons, the π component via a "side-by-side" overlap of two p orbitals. Regions of high electron density, attributed to the π electrons, appear above and below the plane of the molecule and are clearly evident in the electrostatic potential map. Most of the reactions of ethylene and other alkenes involve these electrons.

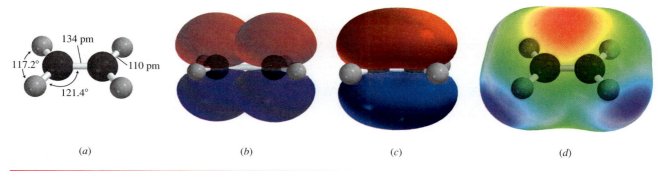

(a)   (b)   (c)   (d)

FIGURE 5.1 (a) The planar framework of σ bonds in ethylene showing bond distances and angles. (b) and (c) The p orbitals of two sp^2-hybridized carbons overlap to produce a π bond. (d) The electrostatic potential map shows a region of high negative potential due to the π electrons above and below the plane of the atoms.

The double bond in ethylene is stronger than the C—C single bond in ethane, but it is not twice as strong. Chemists do not agree on exactly how to apportion the total C=C bond energy between its σ and π components, but all agree that the π bond is weaker than the σ bond.

There are two different types of carbon–carbon bonds in propene, $CH_3CH=CH_2$. The double bond is of the σ + π type, and the bond to the methyl group is a σ bond formed by sp^3–sp^2 overlap.

> The simplest arithmetic approach subtracts the C—C σ bond energy of ethane (368 kJ/mol; 88 kcal/mol) from the C=C bond energy of ethylene (605 kJ/mol; 144.5 kcal/mol). This gives a value of 237 kJ/mol (56.5 kcal/mol) for the π bond energy.

sp^3-hybridized carbon

sp^2-hybridized carbon

C—C bond length = 150 pm
C=C bond length = 134 pm

PROBLEM 5.3 We can use bond-line formulas to represent alkenes in much the same way that we use them to represent alkanes. Consider the following alkene:

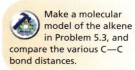

3-ethyl-3-hexene

(a) What is the molecular formula of this alkene?

(b) What is its IUPAC name?

(c) How many carbon atoms are sp^2-hybridized in this alkene? How many are sp^3-hybridized?

(d) How many σ bonds are of the sp^2–sp^3 type? How many are of the sp^3–sp^3 type?

> Make a molecular model of the alkene in Problem 5.3, and compare the various C—C bond distances.

SAMPLE SOLUTION (a) Recall when writing bond-line formulas for hydrocarbons that a carbon occurs at each end and at each bend in a carbon chain. The appropriate number of hydrogens are attached so that each carbon has four bonds. Thus the compound shown is

$$CH_3CH_2CH=C(CH_2CH_3)_2$$

The general molecular formula for an alkene is C_nH_{2n}. Ethylene is C_2H_4; propene is C_3H_6. Counting the carbons and hydrogens of the compound shown (C_8H_{16}) reveals that it, too, corresponds to C_nH_{2n}.

5.3   ISOMERISM IN ALKENES

Although ethylene is the only two-carbon alkene, and propene the only three-carbon alkene, there are *four* isomeric alkenes of molecular formula C_4H_8:

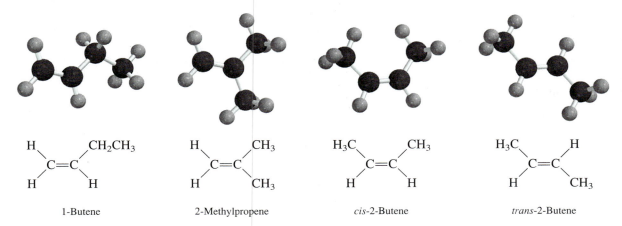

| 1-Butene | 2-Methylpropene | *cis*-2-Butene | *trans*-2-Butene |

1-Butene has an unbranched carbon chain with a double bond between C-1 and C-2. It is a constitutional isomer of the other three. Similarly, 2-methylpropene, with a branched carbon chain, is a constitutional isomer of the other three.

The pair of isomers designated *cis*- and *trans*-2-butene have the same constitution; both have an unbranched carbon chain with a double bond connecting C-2 and C-3. They differ from each other, however, in that the cis isomer has both of its methyl groups on the same side of the double bond, but the methyl groups in the trans isomer are on opposite sides of the double bond. Recall from Section 3.11 that isomers that have the same constitution but differ in the arrangement of their atoms in space are classified as *stereoisomers*. *cis*-2-Butene and *trans*-2-butene are stereoisomers, and the terms *cis* and *trans* specify the *configuration* of the double bond.

Cis–trans stereoisomerism in alkenes is not possible when one of the doubly bonded carbons bears two identical substituents. Thus, neither 1-butene nor 2-methylpropene can have stereoisomers.

> Stereoisomeric alkenes are sometimes referred to as *geometric isomers.*

Identical $\begin{Bmatrix} H \\ H \end{Bmatrix}$ C=C $\begin{matrix} CH_2CH_3 \\ H \end{matrix}$ Identical $\begin{Bmatrix} H \\ H \end{Bmatrix}$ C=C $\begin{Bmatrix} CH_3 \\ CH_3 \end{Bmatrix}$ Identical

1-Butene
(no stereoisomers possible)

2-Methylpropene
(no stereoisomers possible)

PROBLEM 5.4   How many alkenes have the molecular formula C_5H_{10}? Write their structures and give their IUPAC names. Specify the configuration of stereoisomers as cis or trans as appropriate.

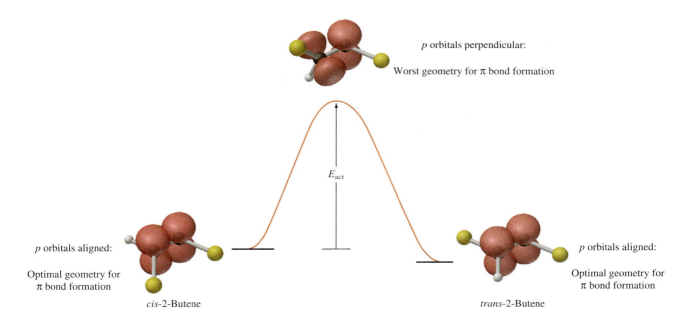

p orbitals perpendicular:

Worst geometry for π bond formation

E_{act}

p orbitals aligned:

Optimal geometry for
π bond formation

p orbitals aligned:

Optimal geometry for
π bond formation

cis-2-Butene

trans-2-Butene

FIGURE 5.2 Interconversion of *cis*- and *trans*-2-butene proceeds by cleavage of the π component of the double bond. The yellow balls represent methyl groups.

In principle, *cis*-2-butene and *trans*-2-butene may be interconverted by rotation about the C-2=C-3 *double* bond. However, unlike rotation about the C-2—C-3 *single* bond in butane, which is quite fast, interconversion of the stereoisomeric 2-butenes does not occur under normal circumstances. It is sometimes said that rotation about a carbon–carbon double bond is *restricted,* but this is an understatement. Conventional laboratory sources of heat do not provide enough energy for rotation about the double bond in alkenes. As shown in Figure 5.2, rotation about a double bond requires the *p* orbitals of C-2 and C-3 to be twisted from their stable parallel alignment—in effect, the π component of the double bond must be broken at the transition state.

> The activation energy for rotation about a typical carbon–carbon double bond is very high—on the order of 250 kJ/mol (about 60 kcal/mol). This quantity may be taken as a measure of the π bond contribution to the total C=C bond strength of 605 kJ/mol (144.5 kcal/mol) in ethylene and compares closely with the value estimated by manipulation of thermochemical data on page 191.

5.4 NAMING STEREOISOMERIC ALKENES BY THE *E–Z* NOTATIONAL SYSTEM

When the groups on either end of a double bond are the same or are structurally similar to each other, it is a simple matter to describe the configuration of the double bond as cis or trans. Oleic acid, for example, a compound that can be obtained from olive oil, has a cis double bond. Cinnamaldehyde, responsible for the characteristic odor of cinnamon, has a trans double bond.

CH₃(CH₂)₆CH₂   CH₂(CH₂)₆CO₂H   C₆H₅   H

C=C   C=C

H   H   H   CH
‖
O

Oleic acid   Cinnamaldehyde

PROBLEM 5.5 Female houseflies attract males by sending a chemical signal known as a *pheromone*. The substance emitted by the female housefly that attracts the male has been identified as *cis*-9-tricosene, $C_{23}H_{46}$. Write a structural formula, including stereochemistry, for this compound.

The terms *cis* and *trans* are ambiguous, however, when it is not obvious which substituent on one carbon is "similar" or "analogous" to a reference substituent on the other. Fortunately, a completely unambiguous system for specifying double bond stereochemistry has been developed based on an *atomic number* criterion for ranking substituents on the doubly bonded carbons. When atoms of higher atomic number are on the *same* side of the double bond, we say that the double bond has the **Z** configuration, where Z stands for the German word ***zusammen,*** meaning "together." When atoms of higher atomic number are on *opposite* sides of the double bond, we say that the configuration is *E*. The symbol *E* stands for the German word ***entgegen,*** meaning "opposite."

Higher ⟶ Cl, Br ⟵ Higher   Higher ⟶ Cl, F ⟵ Lower

$C=C$ $C=C$

Lower ⟶ H, F ⟵ Lower   Lower ⟶ H, Br ⟵ Higher

     Z configuration   *E* configuration

Higher ranked substituents (Cl and Br)   Higher ranked substituents (Cl and Br)
are on same side of double bond   are on opposite sides of double bond

The substituent groups on the double bonds of most alkenes are, of course, more complicated than in this example. The rules for ranking substituents, especially alkyl groups, are described in Table 5.1.

The priority rules in Table 5.1 were developed by R. S. Cahn and Sir Christopher Ingold (England) and Vladimir Prelog (Switzerland) in the context of a different aspect of organic stereochemistry; they will appear again in Chapter 7.

PROBLEM 5.6 Determine the configuration of each of the following alkenes as *Z* or *E* as appropriate:

(a) H_3C, CH_2OH (c) H_3C, CH_2CH_2OH
      $C=C$ $C=C$
      H, CH_3 H, $C(CH_3)_3$

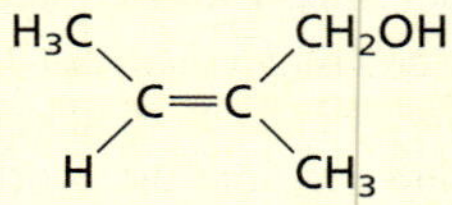

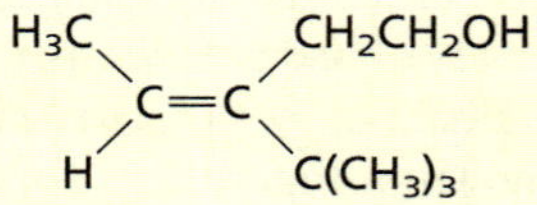

(b) H_3C, CH_2CH_2F (d)
      $C=C$
      H, $CH_2CH_2CH_2CH_3$ △ , H
                         $C=C$
                   CH_3CH_2, CH_3

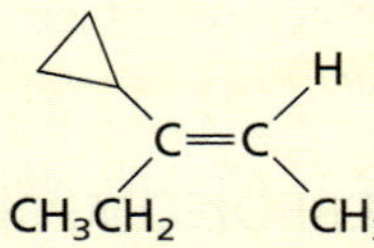

SAMPLE SOLUTION (a) One of the doubly bonded carbons bears a methyl group and a hydrogen. According to the rules of Table 5.1, methyl outranks hydrogen. The other carbon atom of the double bond bears a methyl and a —CH_2OH group. The —CH_2OH group is of higher priority than methyl.

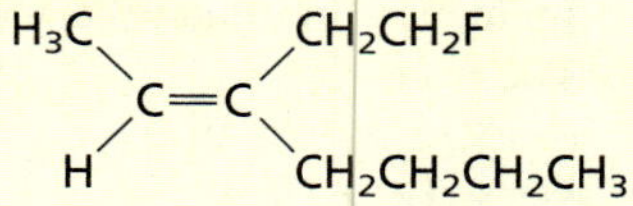

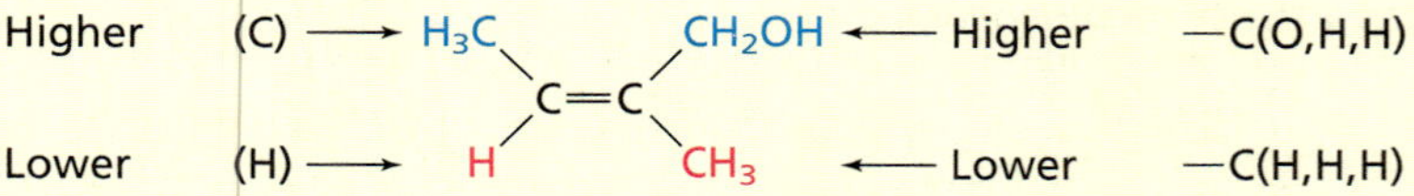

Higher   (C) ⟶ H_3C, CH_2OH ⟵ Higher   —C(O,H,H)

                 $C=C$

Lower   (H) ⟶ H, CH_3 ⟵ Lower   —C(H,H,H)

Higher ranked substituents are on the same side of the double bond; the configuration is *Z*.

TABLE 5.1	Cahn–Ingold–Prelog Priority Rules

Rule	Example
1. Higher atomic number takes precedence over lower. Bromine (atomic number 35) outranks chlorine (atomic number 17). Methyl (C, atomic number 6) outranks hydrogen (atomic number 1).	The compound 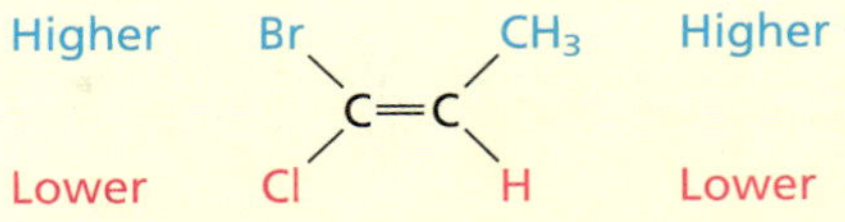 Higher Br＼ ／CH₃ Higher C＝C Lower Cl／ ＼H Lower has the *Z* configuration. Higher ranked atoms (Br and C of CH₃) are on the same side of the double bond.
2. When two atoms directly attached to the double bond are identical, compare the atoms attached with these two on the basis of their atomic numbers. Precedence is determined at the first point of difference: Ethyl [—C(**C**,H,H)] outranks methyl [—C(**H**,H,H)] Similarly, *tert*-butyl outranks isopropyl, and isopropyl outranks ethyl: 　—C(CH₃)₃ > —CH(CH₃)₂ > —CH₂CH₃ 　—C(C,C,C) > —C(C,C,H) > —C(C,H,H)	The compound 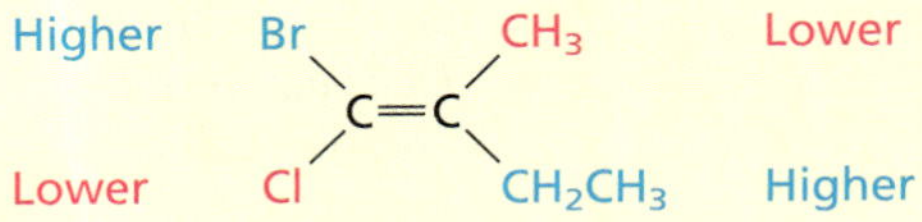 Higher Br＼ ／CH₃ Lower C＝C Lower Cl／ ＼CH₂CH₃ Higher has the *E* configuration.
3. Work outward from the point of attachment, comparing all the atoms attached to a particular atom before proceeding further along the chain: —CH(CH₃)₂ [—C(C,**C**,H)] outranks 　　　　　—CH₂CH₂OH [—C(C,**H**,H)]	The compound 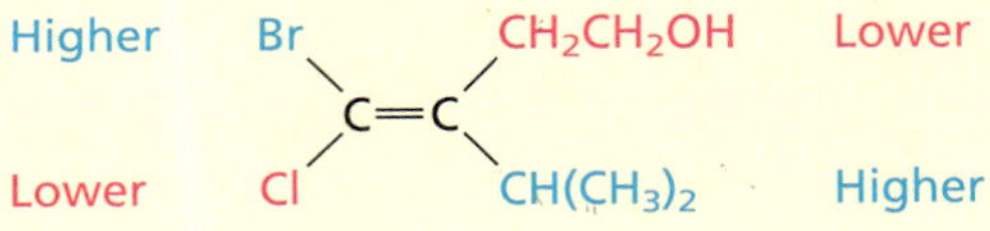 Higher Br＼ ／CH₂CH₂OH Lower C＝C Lower Cl／ ＼CH(CH₃)₂ Higher has the *E* configuration.
4. When working outward from the point of attachment, always evaluate substituent atoms one by one, never as a group. Because oxygen has a higher atomic number than carbon, —CH₂OH [—C(**O**,H,H)] outranks 　　　　　—C(CH₃)₃ [—C(**C**,C,C)]	The compound 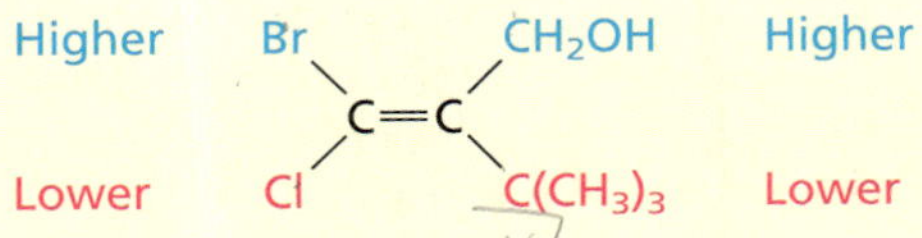 Higher Br＼ ／CH₂OH Higher C＝C Lower Cl／ ＼C(CH₃)₃ Lower has the *Z* configuration.

5. An atom that is multiply bonded to another atom is considered to be replicated as a substituent on that atom:
 	
<pre>   O
    ‖
  —CH</pre> is treated as if it were —C(O,O,H)
 	
The group —CH＝O [—C(O,**O**,H)] outranks —CH₂OH [—C(O,**H**,H)]
	The compound 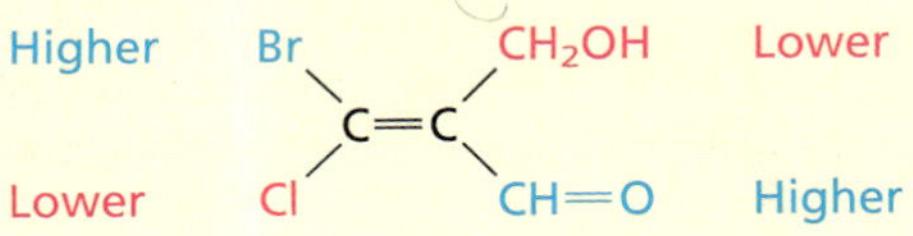
	
	Higher Br＼ ／CH₂OH Lower
	         C＝C
	Lower Cl／ ＼CH＝O Higher
	
	has the *E* configuration.

A table on the inside back cover (right page) lists some of the more frequently encountered atoms and groups in order of increasing precedence. You should not attempt to memorize this table, but should be able to derive the relative placement of one group versus another.

5.5   PHYSICAL PROPERTIES OF ALKENES

The physical properties of selected alkenes are collected in Appendix 1.

Alkenes resemble alkanes in most of their physical properties. The lower molecular weight alkenes through C_4H_8 are gases at room temperature and atmospheric pressure.

The dipole moments of most alkenes are quite small. Among the C_4H_8 isomers, 1-butene, *cis*-2-butene, and 2-methylpropene have dipole moments in the 0.3–0.5 D range; *trans*-2-butene has no dipole moment. Nevertheless, we can learn some things about alkenes by looking at the effect of substituents on dipole moments.

Experimental measurements of dipole moments give size, but not direction. We normally deduce the overall direction by examining the directions of individual bond dipoles. With alkenes the basic question concerns the alkyl groups attached to C=C. *Does an alkyl group donate electrons to or withdraw electrons from a double bond?* This question can be approached by comparing the effect of an alkyl group, methyl for example, with other substituents.

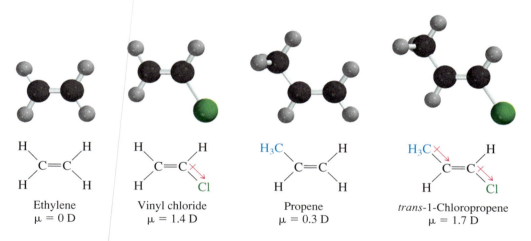

Ethylene
$\mu = 0$ D

Vinyl chloride
$\mu = 1.4$ D

Propene
$\mu = 0.3$ D

trans-1-Chloropropene
$\mu = 1.7$ D

Ethylene, of course, has no dipole moment. Replacing one of its hydrogens by chlorine gives vinyl chloride, which has a dipole moment of 1.4 D. The effect is much smaller when one of the hydrogens is replaced by methyl; propene has a dipole moment of only 0.3 D. Now place CH_3 and Cl trans to each other on the double bond. If methyl releases electrons better than H, then the dipole moment of *trans*-CH_3CH=$CHCl$ should be larger than that of H_2C=$CHCl$, because the effects of CH_3 and Cl reinforce each other. If methyl is electron attracting, the opposite should occur, and the dipole moment of *trans*-CH_3CH=$CHCl$ will be smaller than 1.4 D. In fact, the dipole moment of *trans*-CH_3CH=$CHCl$ is larger than that of H_2C=$CHCl$, indicating that a methyl group is an electron-donating substituent on the double bond.

A methyl group releases electrons to a double bond in much the same way that it releases electrons to the positively charged carbon of a carbocation—by an inductive effect and by hyperconjugation (Figure 5.3). Other alkyl groups behave similarly and, as we go along, we'll see several ways in which the electron-releasing effects of alkyl substituents influence the properties of alkenes. The first is described in the following section.

sp²-hybridized carbons of an alkene are more electronegative than sp³-hybridized carbon and are stabilized by electron-donating substituents.

Methyl group is a better electron-donating substituent than hydrogen.

$$H_3C \underset{H}{\overset{}{\diagdown}} C = C \diagup$$

FIGURE 5.3 Alkyl groups donate electrons to sp²-hybridized carbons of an alkene.

PROBLEM 5.7 Use *Learning By Modeling* to compare the calculated dipole moments of ethylene, propene, vinyl chloride, and *trans*-1-chloropropene. Unlike measured dipole moments, the calculated ones do show the direction of the dipole moment. How do the directions of the calculated dipole moments compare with those deduced by experiment?

5.6 RELATIVE STABILITIES OF ALKENES

Earlier (Sections 2.18, 3.11) we saw how to use heats of combustion to compare the stabilities of isomeric alkanes. We can do the same thing with isomeric alkenes. Consider the heats of combustion of the four isomeric alkenes of molecular formula C_4H_8. All undergo combustion according to the equation

$$C_4H_8 + 6O_2 \longrightarrow 4CO_2 + 4H_2O$$

When the heats of combustion of the isomers are plotted on a common scale as in Figure 5.4, we see that the isomer of highest energy (the least stable one) is 1-butene, $H_2C=CHCH_2CH_3$. The isomer of lowest energy (most stable) is 2-methylpropene $(CH_3)_2C=CH_2$.

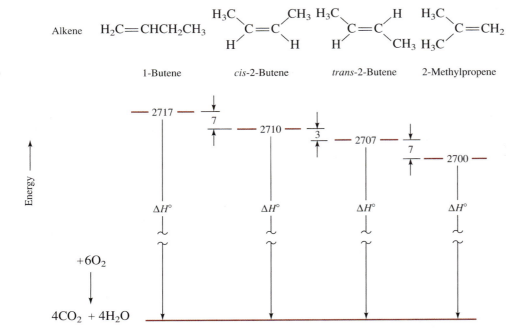

FIGURE 5.4 Heats of combustion of C_4H_8 alkene isomers. All energies are in kilojoules per mole. (An energy difference of 3 kJ/mol is equivalent to 0.7 kcal/mol; 7 kJ/mol is equivalent to 1.7 kcal/mol.)

Analogous data for a host of alkenes tell us that the most important factors governing alkene stability are:

1. *Degree of substitution* (alkyl substituents stabilize a double bond)
2. *Van der Waals strain* (destabilizing when alkyl groups are cis to each other)

Degree of substitution. We classify double bonds as **monosubstituted, disubstituted, trisubstituted,** or **tetrasubstituted** according to the number of carbon atoms *directly* attached to the $C=C$ structural unit.

Monosubstituted alkenes:

$$RCH=CH_2 \qquad \text{as in} \qquad CH_3CH_2CH=CH_2 \qquad \text{(1-butene)}$$

Disubstituted alkenes:
(R and R′ may be the same or different)

$$RCH=CHR' \qquad \text{as in} \qquad CH_3CH=CHCH_3 \qquad \text{(cis- or trans-2-butene)}$$

$$\begin{array}{c} R \\ \diagdown \\ \diagup \quad C=C \\ R' \end{array} \begin{array}{c} H \\ \diagup \\ \diagdown \\ H \end{array} \qquad \text{as in} \qquad (CH_3)_2C=CH_2 \qquad \text{(2-methylpropene)}$$

Trisubstituted alkenes:
(R, R′, and R″ may be the same or different)

$$\begin{array}{c} R \\ \diagdown \\ \diagup \quad C=C \\ R' \end{array} \begin{array}{c} R'' \\ \diagup \\ \diagdown \\ H \end{array} \qquad \text{as in} \qquad (CH_3)_2C=CHCH_2CH_3 \qquad \text{(2-methyl-2-pentene)}$$

Tetrasubstituted alkenes:
(R, R′, R″, and R‴ may be the same or different)

$$\begin{array}{c} R \\ \diagdown \\ \diagup \quad C=C \\ R' \end{array} \begin{array}{c} R'' \\ \diagup \\ \diagdown \\ R''' \end{array} \qquad \text{as in} \qquad \text{(1,2-dimethylcyclohexene)}$$

In the example shown, carbons 3 and 6 both count as substituents on the double bond.

PROBLEM 5.8 Write structural formulas or build molecular models and give the IUPAC names for all the alkenes of molecular formula C_6H_{12} that contain a trisubstituted double bond. (Don't forget to include stereoisomers.)

From the heats of combustion of the C_4H_8 alkenes in Figure 5.4 we see that each of the disubstituted alkenes

$$\begin{array}{c} H_3C \\ \diagdown \\ \diagup \quad C=C \\ H_3C \end{array} \begin{array}{c} H \\ \diagup \\ \diagdown \\ H \end{array} \qquad \begin{array}{c} H_3C \\ \diagdown \\ \diagup \quad C=C \\ H \end{array} \begin{array}{c} H \\ \diagup \\ \diagdown \\ CH_3 \end{array} \qquad \begin{array}{c} H_3C \\ \diagdown \\ \diagup \quad C=C \\ H \end{array} \begin{array}{c} CH_3 \\ \diagup \\ \diagdown \\ H \end{array}$$

2-Methylpropene   *trans*-2-Butene   *cis*-2-Butene is more stable than the monosubstituted alkene

$$
\underset{\text{1-Butene}}{
\begin{array}{c}
\text{H} \qquad\qquad \text{CH}_2\text{CH}_3 \\
\diagdown \qquad\qquad \diagup \\
\text{C}=\text{C} \\
\diagup \qquad\qquad \diagdown \\
\text{H} \qquad\qquad \text{H}
\end{array}
}
$$

In general, alkenes with more highly substituted double bonds are more stable than isomers with less substituted double bonds.

> **PROBLEM 5.9**   Give the structure or make a molecular model of the most stable C_6H_{12} alkene.

Like the sp^2-hybridized carbons of carbocations and free radicals, the sp^2-hybridized carbons of double bonds are electron attracting, and alkenes are stabilized by substituents that release electrons to these carbons. As we saw in the preceding section, alkyl groups are better electron-releasing substituents than hydrogen and are, therefore, better able to stabilize an alkene.

An effect that results when two or more atoms or groups interact so as to alter the electron distribution in a system is called an **electronic effect.** The greater stability of more highly substituted alkenes is an example of an electronic effect.

van der Waals strain.   *Alkenes are more stable when large substituents are trans to each other than when they are cis.* As we saw in Figure 5.4, *trans*-2-butene has a lower heat of combustion and is more stable than *cis*-2-butene. The energy difference between the two is 3 kJ/mol (0.7 kcal/mol). The source of this energy difference is illustrated in Figure 5.5, where, especially in the space-filling models, you can see that the methyl

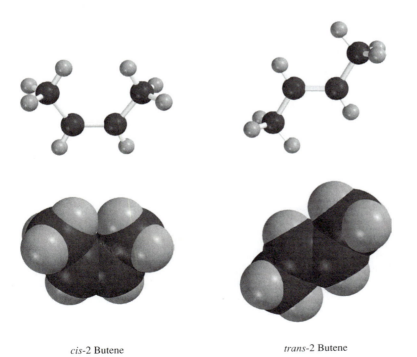

cis-2 Butene   *trans*-2 Butene

FIGURE 5.5 Ball-and-spoke and space-filling models of *cis*- and *trans*-2-butene. The space-filling model shows the serious van der Waals strain between two of the hydrogens in *cis*-2-butene. The molecule adjusts by expanding those bond angles that increase the separation between the crowded atoms. The combination of angle strain and van der Waals strain makes *cis*-2-butene less stable than *trans*-2-butene.

groups approach each other very closely in *cis*-2-butene, but the trans isomer is free of strain. An effect that results when two or more atoms are close enough in space that a repulsion occurs between them is one type of **steric effect.** The greater stability of trans alkenes compared with their cis counterparts is an example of a steric effect.

A similar steric effect was seen in Section 3.11, where van der Waals strain between methyl groups on the same side of the ring made *cis*-1,2-dimethylcyclopropane less stable than its trans stereoisomer.

PROBLEM 5.10   Arrange the following alkenes in order of decreasing stability: 1-pentene; (*E*)-2-pentene; (*Z*)-2-pentene; 2-methyl-2-butene.

The difference in stability between stereoisomeric alkenes is even more pronounced with larger alkyl groups on the double bond. A particularly striking example compares *cis*- and *trans*-2,2,5,5-tetramethyl-3-hexene, in which the heat of combustion of the cis stereoisomer is 44 kJ/mol (10.5 kcal/mol) higher than that of the trans. The cis isomer is destabilized by the large van der Waals strain between the bulky *tert*-butyl groups on the same side of the double bond.

The common names of these alkenes are *cis*- and *trans*-di-*tert*-butylethylene. In cases such as this the common names are somewhat more convenient than the IUPAC names because they are more readily associated with molecular structure.

Energy difference = 44 kJ/mol (10.5 kcal/mol)

cis-2,2,5,5-Tetramethyl-3-hexene (less stable)

trans-2,2,5,5-Tetramethyl-3-hexene (more stable)

PROBLEM 5.11   Despite numerous attempts, the alkene 3,4-di-*tert*-butyl-2,2,5,5-tetramethyl-3-hexene has never been synthesized. Can you explain why? Try making a space-filling model of this compound.

5.7   CYCLOALKENES

Double bonds are accommodated by rings of all sizes. The smallest cycloalkene, cyclopropene, was first synthesized in 1922. A cyclopropene ring is present in sterculic acid, a substance derived from one of the components of the oil present in the seeds of a tree (*Sterculia foelida*) that grows in the Philippines and Indonesia.

Sterculic acid and related substances are the subject of an article in the July 1982 issue of *Journal of Chemical Education* (pp. 539–543).

Cyclopropene

$CH_3(CH_2)_7$ $(CH_2)_7CO_2H$

Sterculic acid

As we saw in Section 3.5, cyclopropane is destabilized by angle strain because its 60° bond angles are much smaller than the normal 109.5° angles associated with sp^3-hybridized carbon. Cyclopropene is even more strained because of the distortion of the bond angles at its doubly bonded carbons from their normal sp^2-hybridization value of 120°. Cyclobutene has, of course, less angle strain than cyclopropene, and the angle strain in cyclopentene, cyclohexene, and higher cycloalkenes is negligible.

So far we have represented cycloalkenes by structural formulas in which the double bonds are of the cis configuration. If the ring is large enough, however, a trans stereoisomer is also possible. The smallest trans cycloalkene that is stable enough to be isolated and stored in a normal way is *trans*-cyclooctene.

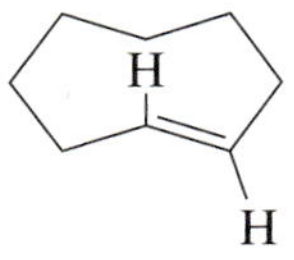

Energy difference = 39 kJ/mol (9.2 kcal/mol)

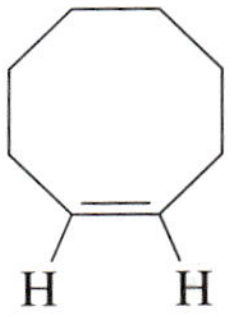

(*E*)-Cyclooctene
(*trans*-cyclooctene)
Less stable

(*Z*)-Cyclooctene
(*cis*-cyclooctene)
More stable

Make molecular models of (*E*)- and (*Z*)-cyclooctene, and compare their H—C=C—H dihedral angles.

trans-Cycloheptene has been prepared and studied at low temperature ($-90°C$) but is too reactive to be isolated and stored at room temperature. Evidence has also been presented for the fleeting existence of the even more strained *trans*-cyclohexene as a reactive intermediate in certain reactions.

PROBLEM 5.12 Place a double bond in the carbon skeleton shown so as to represent

(a) (*Z*)-1-Methylcyclodecene
(b) (*E*)-1-Methylcyclodecene
(c) (*Z*)-3-Methylcyclodecene
(d) (*E*)-3-Methylcyclodecene
(e) (*Z*)-5-Methylcyclodecene
(f) (*E*)-5-Methylcyclodecene

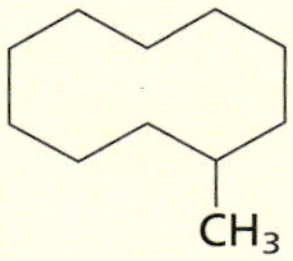

CH₃

SAMPLE SOLUTION (a) and (b) Because the methyl group must be at C-1, there are only two possible places to put the double bond:

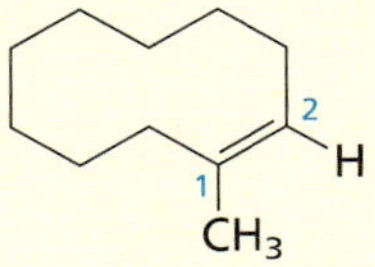

(*Z*)-1-Methylcyclodecene 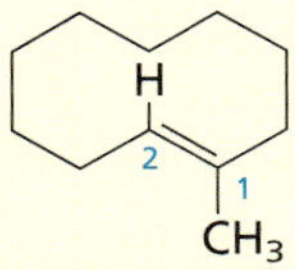   (*E*)-1-Methylcyclodecene

In the Z stereoisomer the two lower priority substituents—the methyl group and the hydrogen—are on the same side of the double bond. In the E stereoisomer these substituents are on opposite sides of the double bond. The ring carbons are the higher ranking substituents at each end of the double bond.

Because larger rings have more carbons with which to span the ends of a double bond, the strain associated with a trans cycloalkene decreases with increasing ring size. The strain eventually disappears when a 12-membered ring is reached and *cis*- and *trans*-cyclododecene are of approximately equal stability. When the rings are larger than 12 membered, trans cycloalkenes are more stable than cis. In these cases, the ring is large enough and flexible enough that it is energetically similar to a noncyclic alkene. As in noncyclic cis alkenes, van der Waals strain between carbons on the same side of the double bond destabilizes a cis cycloalkene.

5.8   PREPARATION OF ALKENES: ELIMINATION REACTIONS

The rest of this chapter describes how alkenes are prepared by reactions of the type:

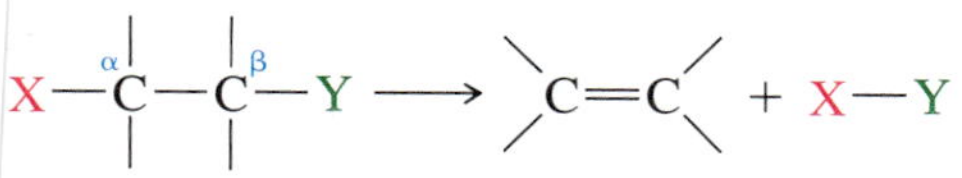

$$X-\underset{\alpha}{\overset{|}{C}}-\underset{\beta}{\overset{|}{C}}-Y \longrightarrow ^{\diagdown}C=C^{\diagup} + X-Y$$

Alkene formation requires that X and Y be substituents on adjacent carbon atoms. By making X the reference atom and identifying the carbon attached to it as the α carbon, we see that atom Y is a substituent on the β carbon. Carbons succeedingly more remote from the reference atom are designated γ, δ, and so on. Only β elimination reactions will be discussed in this chapter. [Beta (β) elimination reactions are also known as *1,2 eliminations*.]

You are already familiar with one type of β elimination, having seen in Section 5.1 that ethylene and propene are prepared on an industrial scale by the high-temperature *dehydrogenation* of ethane and propane. Both reactions involve β elimination of H_2.

$$CH_3CH_3 \xrightarrow{750°C} H_2C=CH_2 + H_2$$
$$\text{Ethane} \qquad\qquad \text{Ethylene} \qquad \text{Hydrogen}$$

$$CH_3CH_2CH_3 \xrightarrow{750°C} CH_3CH=CH_2 + H_2$$
$$\text{Propane} \qquad\qquad \text{Propene} \qquad \text{Hydrogen}$$

Many reactions classified as dehydrogenations occur within the cells of living systems at 25°C. H_2 is not one of the products, however. Instead, the hydrogens are lost in separate steps of an enzyme-catalyzed process. The enzyme indicated in the reaction:

A quote from a biochemistry text is instructive here. "This is not an easy reaction in organic chemistry. It is, however, a very important type of reaction in metabolic chemistry and is an integral step in the oxidation of carbohydrates, fats, and several amino acids." G. L. Zubay, *Biochemistry*, 4th ed., William C. Brown Publishers, 1996, p. 333.

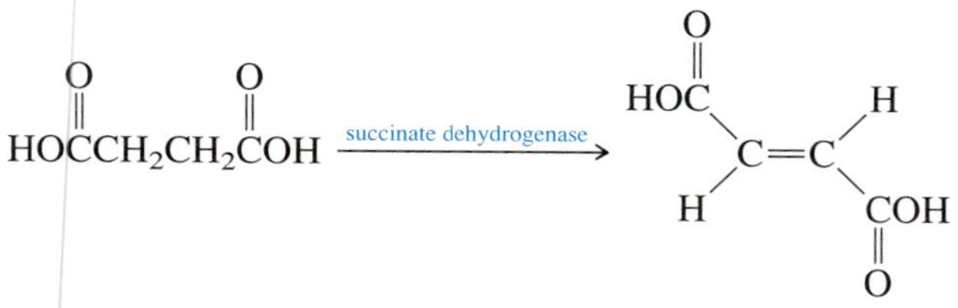

$$\underset{\text{Succinic acid}}{\overset{OO}{\overset{||||}{HOCCH_2CH_2COH}}} \xrightarrow{\text{succinate dehydrogenase}} \underset{\text{Fumaric acid}}{C=C}$$

is a special kind, known as a *flavoprotein*.

Dehydrogenation of alkanes is not a practical *laboratory* synthesis for the vast majority of alkenes. The principal methods by which alkenes are prepared in the laboratory are two other β eliminations: the **dehydration of alcohols** and the **dehydrohalogenation of alkyl halides.** A discussion of these two methods makes up the remainder of this chapter.

5.9   DEHYDRATION OF ALCOHOLS

In the dehydration of alcohols, the H and OH are lost from adjacent carbons. An acid catalyst is necessary.

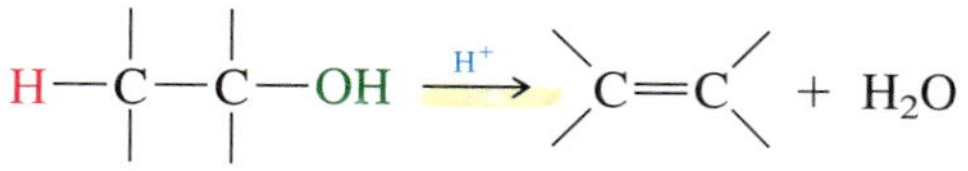

$$H-\overset{|}{C}-\overset{|}{C}-OH \xrightarrow{H^+} ^{\diagdown}C=C^{\diagup} + H_2O$$
$$\text{Alcohol} \qquad\qquad \text{Alkene} \quad \text{Water}$$

Before dehydrogenation of ethane became the dominant method, ethylene was prepared by heating ethyl alcohol with sulfuric acid.

$$CH_3CH_2OH \xrightarrow[160°C]{H_2SO_4} H_2C{=}CH_2 + H_2O$$

$$\text{Ethyl alcohol} \qquad \text{Ethylene} \quad \text{Water}$$

Other alcohols behave similarly. Secondary alcohols undergo elimination at lower temperatures than primary alcohols.

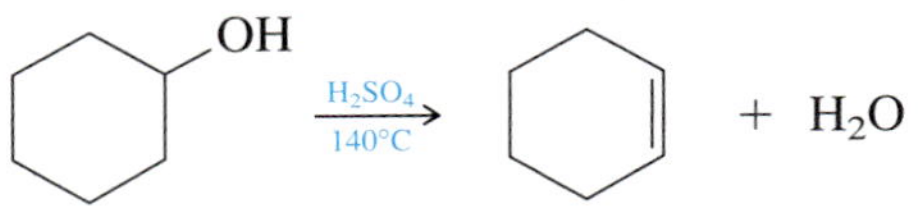

Cyclohexanol   Cyclohexene   Water
                      (79–87%)

and tertiary alcohols dehydrate at lower temperatures than secondary alcohols.

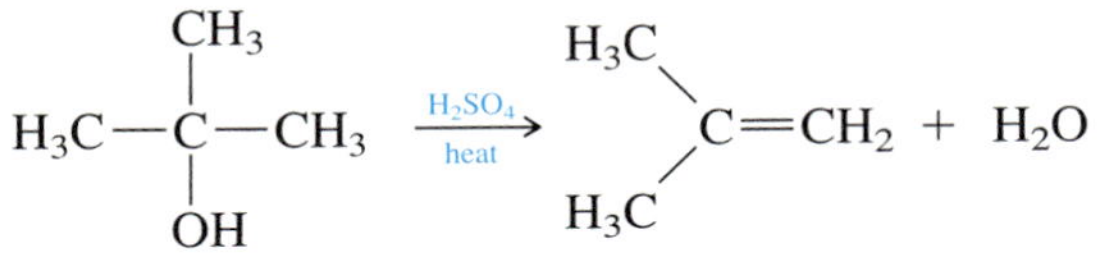

2-Methyl-2-propanol   2-Methylpropene   Water
                                (82%)

Sulfuric acid (H_2SO_4) and phosphoric acid (H_3PO_4) are the acids most frequently used in alcohol dehydrations. Potassium hydrogen sulfate ($KHSO_4$) is also often used.

HSO_4^- and H_3PO_4 are very similar in acid strength. Both are much weaker than H_2SO_4, which is a strong acid.

PROBLEM 5.13 Identify the alkene obtained on dehydration of each of the following alcohols:

(a) 3-Ethyl-3-pentanol

(b) 1-Propanol

(c) 2-Propanol

(d) 2,3,3-Trimethyl-2-butanol

SAMPLE SOLUTION (a) The hydrogen and the hydroxyl are lost from adjacent carbons in the dehydration of 3-ethyl-3-pentanol.

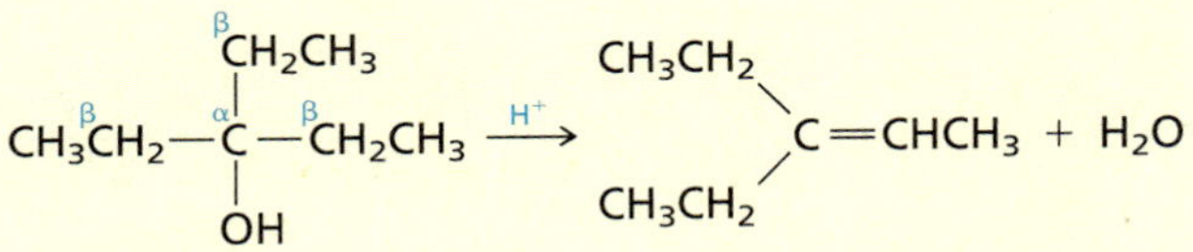

3-Ethyl-3-pentanol   3-Ethyl-2-pentene   Water

The hydroxyl group is lost from a carbon that bears three equivalent ethyl substituents. Beta elimination can occur in any one of three equivalent directions to give the same alkene, 3-ethyl-2-pentene.

Some biochemical processes involve alcohol dehydration as a key step. An example is the conversion of a compound called 3-dehydroquinic acid to 3-dehydroshikimic acid.

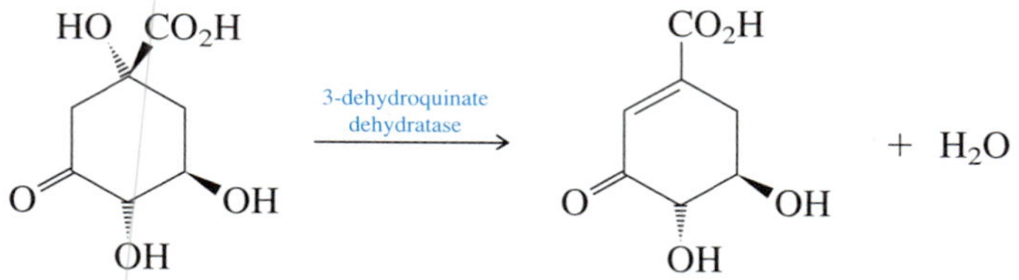

3-Dehydroquinic acid   →   3-Dehydroshikimic acid   Water

This reaction is catalyzed by an enzyme called a *dehydratase* and is one step along the pathway by which plants convert glucose to certain amino acids.

5.10   REGIOSELECTIVITY IN ALCOHOL DEHYDRATION: THE ZAITSEV RULE

Except for the biochemical example just cited, the structures of all of the alcohols in Section 5.9 (including those in Problem 5.13) were such that each one could give only a single alkene by β elimination. What about elimination in alcohols such as 2-methyl-2-butanol, in which dehydration can occur in two different directions to give alkenes that are constitutional isomers? Here, a double bond can be generated between C-1 and C-2 or between C-2 and C-3. Both processes occur but not nearly to the same extent. Under the usual reaction conditions 2-methyl-2-butene is the major product, and 2-methyl-1-butene the minor one.

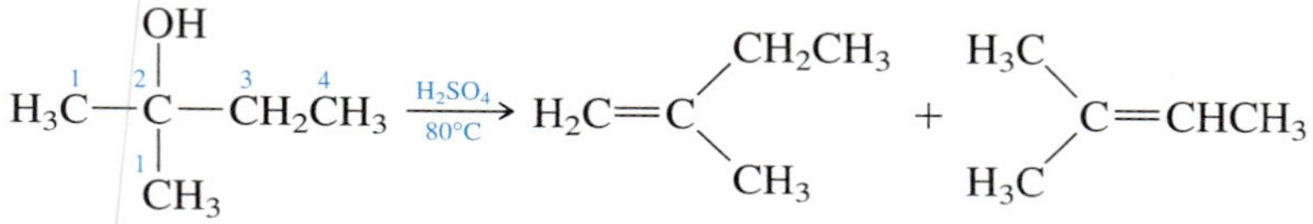

2-Methyl-2-butanol   2-Methyl-1-butene (10%)   2-Methyl-2-butene (90%)

Dehydration of this alcohol is selective in respect to its *direction*. Elimination occurs in the direction that leads to the double bond between C-2 and C-3 more than between C-2 and C-1. Reactions that can proceed in more than one direction, but in which one direction is preferred, are said to be **regioselective.**

As a second example, consider the regioselective dehydration of 2-methylcyclohexanol to yield a mixture of 1-methylcyclohexene (major) and 3-methylcyclohexene (minor).

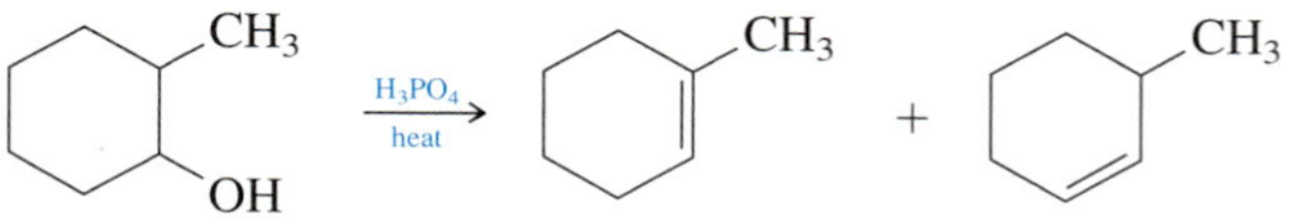

The term *regioselective* was coined by Alfred Hassner, then at the University of Colorado, in a paper published in the *Journal of Organic Chemistry* in 1968.

2-Methylcyclohexanol   1-Methylcyclohexene (84%)   3-Methylcyclohexene (16%)

Although Russian, Zaitsev published most of his work in German scientific journals, where his name was transliterated as *Saytzeff*. The spelling used here (*Zaitsev*) corresponds to the currently preferred style.

In 1875, Alexander M. Zaitsev of the University of Kazan (Russia) set forth a generalization describing the regioselectivity of β eliminations. **Zaitsev's rule** summarizes the results of numerous experiments in which alkene mixtures were produced by β elimination. In its original form, Zaitsev's rule stated that *the alkene formed in greatest amount is the one that corresponds to removal of the hydrogen from the β carbon having the fewest hydrogens.*

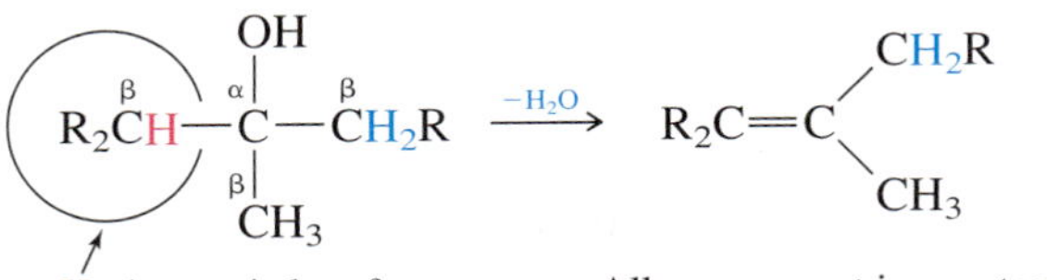

Hydrogen is lost from
β carbon having the fewest
attached hydrogens

Alkene present in greatest
amount in product

Zaitsev's rule as applied to the acid-catalyzed dehydration of alcohols is now more often expressed in a different way: *β elimination reactions of alcohols yield the most highly substituted alkene as the major product.* Because, as was discussed in Section 5.6, the most highly substituted alkene is also normally the most stable one, Zaitsev's rule is sometimes expressed as a preference for *predominant formation of the most stable alkene that could arise by β elimination.*

PROBLEM 5.14 Each of the following alcohols has been subjected to acid-catalyzed dehydration and yields a mixture of two isomeric alkenes. Identify the two alkenes in each case, and predict which one is the major product on the basis of the Zaitsev rule.

(a) $(CH_3)_2CCH(CH_3)_2$
    |
    OH

(b) H_3C OH

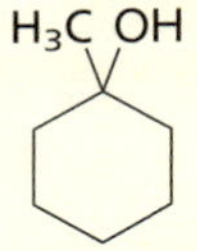

(c)

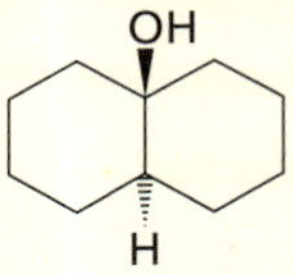

SAMPLE SOLUTION (a) Dehydration of 2,3-dimethyl-2-butanol can lead to either 2,3-dimethyl-1-butene by removal of a C-1 hydrogen or to 2,3-dimethyl-2-butene by removal of a C-3 hydrogen.

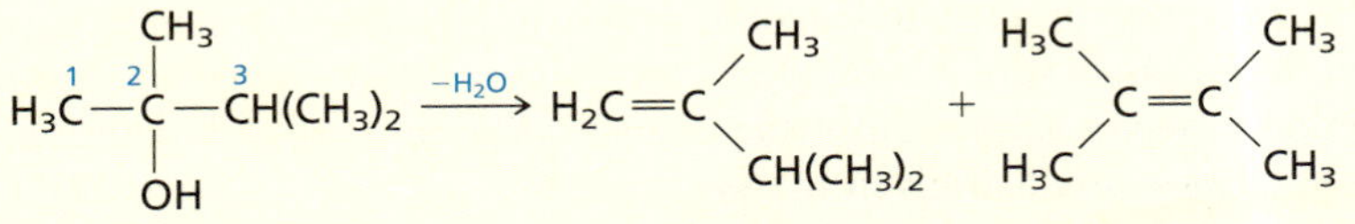

2,3-Dimethyl-2-butanol

2,3-Dimethyl-1-butene
(minor product)

2,3-Dimethyl-2-butene
(major product)

The major product is 2,3-dimethyl-2-butene. It has a tetrasubstituted double bond and is more stable than 2,3-dimethyl-1-butene, which has a disubstituted double bond. The major alkene arises by loss of a hydrogen from the β carbon that has fewer attached hydrogens (C-3) rather than from the β carbon that has the greater number of hydrogens (C-1).

5.11 STEREOSELECTIVITY IN ALCOHOL DEHYDRATION

In addition to being regioselective, alcohol dehydrations are stereoselective. A **stereoselective** reaction is one in which a single starting material can yield two or more stereoisomeric products, but gives one of them in greater amounts than any other. Alcohol dehydrations tend to produce the more stable stereoisomer of an alkene. Dehydration of 3-pentanol, for example, yields a mixture of *trans*-2-pentene and *cis*-2-pentene in which the more stable trans stereoisomer predominates.

$$CH_3CH_2CHCH_2CH_3 \xrightarrow[\text{heat}]{H_2SO_4}$$

(with OH on the carbon)

3-Pentanol

cis-2-Pentene (25%)
(minor product)

trans-2-Pentene (75%)
(major product)

PROBLEM 5.15 What three alkenes are formed in the acid-catalyzed dehydration of 2-pentanol?

The biological dehydrogenation of succinic acid described in Section 5.8 is 100% stereoselective. Only fumaric acid, which has a trans double bond, is formed. High levels of stereoselectivity are characteristic of enzyme-catalyzed reactions.

5.12   THE E1 AND E2 MECHANISMS OF ALCOHOL DEHYDRATION

The dehydration of alcohols resembles the reaction of alcohols with hydrogen halides (Section 4.7) in two important ways.

1. Both reactions are promoted by acids.
2. The relative reactivity of alcohols increases in the order primary < secondary < tertiary.

These common features suggest that carbocations are key intermediates in alcohol dehydrations, just as they are in the reaction of alcohols with hydrogen halides. Figure 5.6 portrays a three-step mechanism for the acid-catalyzed dehydration of *tert*-butyl alcohol. Steps 1 and 2 describe the generation of *tert*-butyl cation by a process similar to that which led to its formation as an intermediate in the reaction of *tert*-butyl alcohol with hydrogen chloride.

Like the reaction of *tert*-butyl alcohol with hydrogen chloride step 2, in which *tert*-butyloxonium ion dissociates to $(CH_3)_3C^+$ and water, is rate-determining. Because the rate-determining step is unimolecular, the overall dehydration process is referred to as a *unimolecular elimination* and given the symbol **E1.**

Step 3 is new to us. It is an acid–base reaction in which the carbocation acts as a Brønsted acid, transferring a proton to a Brønsted base (water). This is the property of carbocations that is of the most significance to elimination reactions. Carbocations are strong acids; they are the conjugate acids of alkenes and readily lose a proton to form alkenes. Even weak bases such as water are sufficiently basic to abstract a proton from a carbocation.

> Step 3 in Figure 5.6 shows water as the base which abstracts a proton from the carbocation. Other Brønsted bases present in the reaction mixture that can function in the same way include *tert*-butyl alcohol and hydrogen sulfate ion.

PROBLEM 5.16 Write a structural formula for the carbocation intermediate formed in the dehydration of each of the alcohols in Problem 5.14 (Section 5.10). Using curved arrows, show how each carbocation is deprotonated by water to give a mixture of alkenes.

SAMPLE SOLUTION (a) The carbon that bears the hydroxyl group in the starting alcohol is the one that becomes positively charged in the carbocation.

$$(CH_3)_2CCH(CH_3)_2 \xrightarrow[-H_2O]{H^+} (CH_3)_2\overset{+}{C}CH(CH_3)_2$$

(with OH below the first carbon)

Water may remove a proton from either C-1 or C-3 of this carbocation. Loss of a proton from C-1 yields the minor product 2,3-dimethyl-1-butene. (This alkene has a disubstituted double bond.)

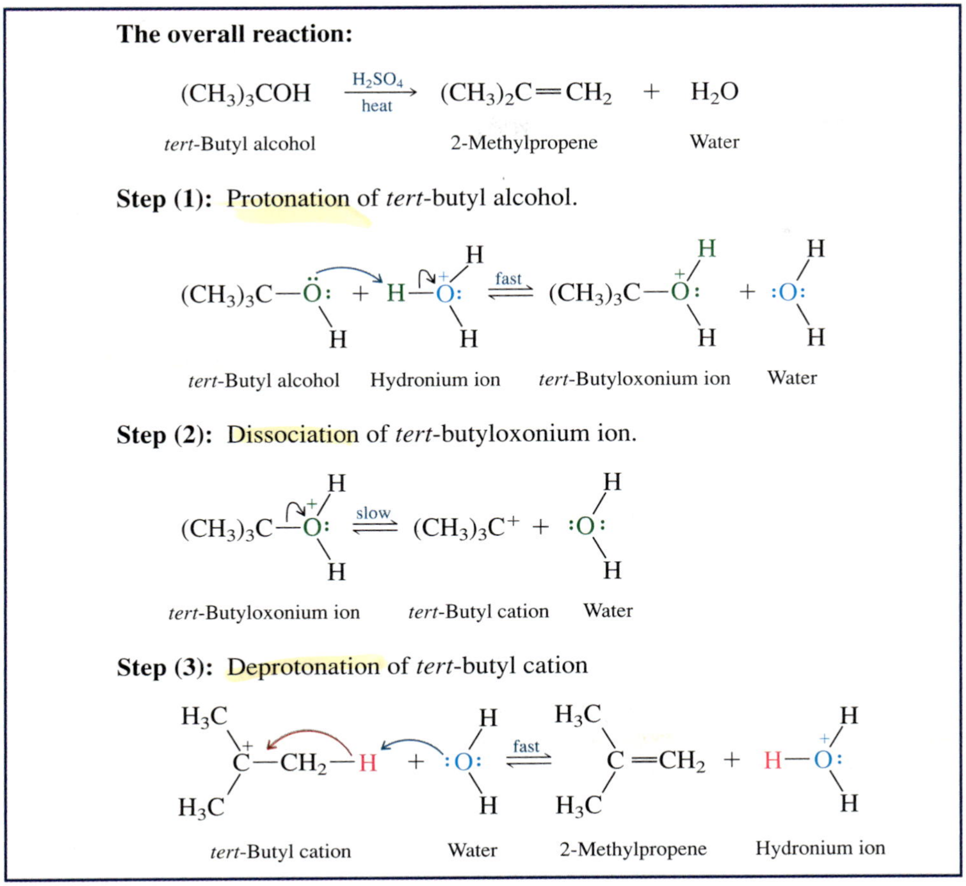

The overall reaction:

$$(CH_3)_3COH \xrightarrow[\text{heat}]{H_2SO_4} (CH_3)_2C\!=\!CH_2 \ + \ H_2O$$

tert-Butyl alcohol   2-Methylpropene   Water

Step (1): Protonation of *tert*-butyl alcohol.

$$(CH_3)_3C\!-\!\ddot{O}\!: \ + \ H\!-\!\overset{+}{\underset{H}{\ddot{O}}}\!: \ \overset{\text{fast}}{\rightleftharpoons} \ (CH_3)_3C\!-\!\overset{H}{\underset{H}{\overset{+}{O}}}\!: \ + \ :\!\overset{H}{\underset{H}{\ddot{O}}}$$

tert-Butyl alcohol   Hydronium ion   *tert*-Butyloxonium ion   Water

Step (2): Dissociation of *tert*-butyloxonium ion.

$$(CH_3)_3C\!-\!\overset{H}{\underset{H}{\overset{+}{O}}}\!: \ \overset{\text{slow}}{\rightleftharpoons} \ (CH_3)_3C^+ \ + \ :\!\overset{H}{\underset{H}{\ddot{O}}}$$

tert-Butyloxonium ion   *tert*-Butyl cation   Water

Step (3): Deprotonation of *tert*-butyl cation

tert-Butyl cation   Water   2-Methylpropene   Hydronium ion

FIGURE 5.6 The E1 mechanism for the acid-catalyzed dehydration of *tert*-butyl alcohol.

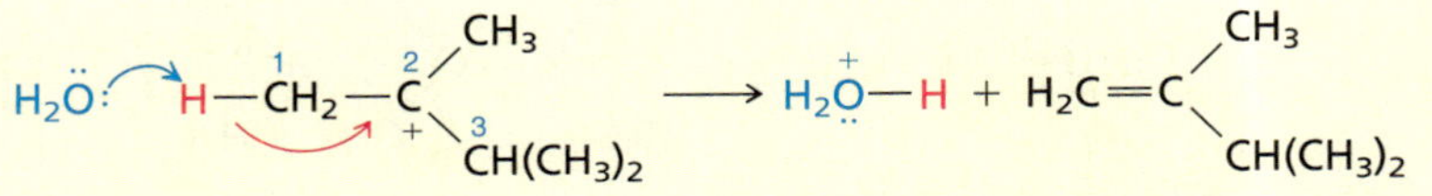

2,3-Dimethyl-1-butene

Loss of a proton from C-3 yields the major product 2,3-dimethyl-2-butene. (This alkene has a tetrasubstituted double bond.)

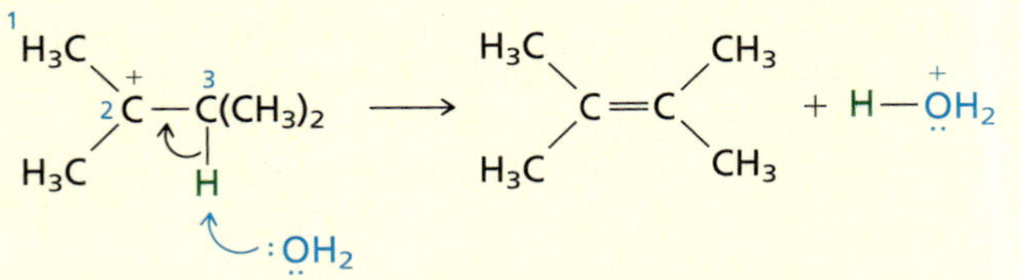

2,3-Dimethyl-2-butene

As noted earlier (Section 4.10) primary carbocations are too high in energy to be intermediates in most chemical reactions. If primary alcohols don't form primary carbocations, then how do they undergo elimination? A modification of our general mechanism for alcohol dehydration offers a reasonable explanation. For primary alcohols it is believed that a proton is lost from the alkyloxonium ion in the same step in which carbon–oxygen bond cleavage takes place. For example, the rate-determining step in the sulfuric acid-catalyzed dehydration of ethanol may be represented as:

$$H_2\ddot{O}: \; + \; H\!-\!CH_2\!-\!CH_2\!-\!\overset{+}{\underset{H}{O}}\!\!\diagup^{H} \quad \xrightarrow{\;slow\;} \quad H_2\overset{+}{O}\!-\!H \; + \; H_2C\!\!=\!\!CH_2 \; + \; :\!\underset{H}{\overset{H}{O}}\!:$$

| Water | Ethyloxonium ion | Hydronium ion | Ethylene | Water |

Because the rate-determining step involves two molecules—the alkyloxonium ion and water—the overall reaction is classified as a *bimolecular elimination* and given the symbol **E2.**

Like tertiary alcohols, secondary alcohols normally undergo dehydration by way of carbocation intermediates.

In Chapter 4 you learned that carbocations could be captured by halide anions to give alkyl halides. In the present chapter, a second type of carbocation reaction has been introduced—a carbocation can lose a proton to form an alkene. In the next section a third aspect of carbocation behavior will be described, the *rearrangement* of one carbocation to another.

5.13   REARRANGEMENTS IN ALCOHOL DEHYDRATION

Some alcohols undergo dehydration to yield alkenes having carbon skeletons different from the starting alcohols. Not only has elimination taken place, but the arrangement of atoms in the alkene is different from that in the alcohol. A **rearrangement** is said to have occurred. An example of an alcohol dehydration that is accompanied by rearrangement is the case of 3,3-dimethyl-2-butanol. This is one of many such experiments carried out by F. C. Whitmore and his students at Pennsylvania State University in the 1930s as part of a general study of rearrangement reactions.

| 3,3-Dimethyl-2-butanol | 3,3-Dimethyl-1-butene (3%) | 2,3-Dimethyl-2-butene (64%) | 2,3-Dimethyl-1-butene (33%) |

A mixture of three alkenes was obtained in 80% yield, having the composition shown. The alkene having the same carbon skeleton as the starting alcohol, 3,3-dimethyl-1-butene, constituted only 3% of the alkene mixture. The two alkenes present in greatest amount, 2,3-dimethyl-2-butene and 2,3-dimethyl-1-butene, both have carbon skeletons different from that of the starting alcohol.

Whitmore proposed that the carbon skeleton rearrangement occurred in a separate step following carbocation formation. Once the alcohol was converted to the corresponding carbocation, that carbocation could either lose a proton to give an alkene having the same carbon skeleton or rearrange to a different carbocation, as shown in Figure 5.7. The rearranged alkenes arise by loss of a proton from the rearranged carbocation.

Why do carbocations rearrange? The answer is straightforward once we recall that tertiary carbocations are more stable than secondary carbocations (Section 4.10). Thus, rearrangement of a secondary to a tertiary carbocation is energetically favorable. As

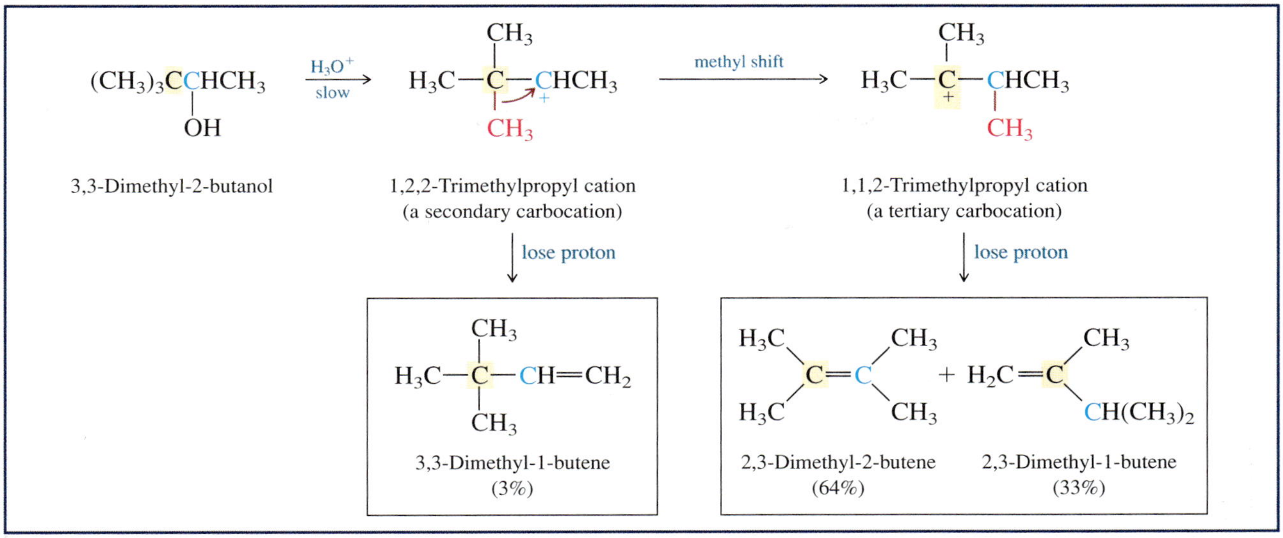

$$(CH_3)_3CCHCH_3 \xrightarrow[\text{slow}]{H_3O^+} H_3C-\underset{\underset{CH_3}{|}}{\overset{\overset{CH_3}{|}}{C}}\overset{+}{\underset{}{}}CHCH_3 \xrightarrow{\text{methyl shift}} H_3C-\underset{\underset{CH_3}{|}}{\overset{\overset{CH_3}{|}}{C}}\overset{+}{}CHCH_3$$

3,3-Dimethyl-2-butanol   1,2,2-Trimethylpropyl cation (a secondary carbocation)   1,1,2-Trimethylpropyl cation (a tertiary carbocation)

lose proton   lose proton

3,3-Dimethyl-1-butene (3%)   2,3-Dimethyl-2-butene (64%)   2,3-Dimethyl-1-butene (33%)

FIGURE 5.7 The first formed carbocation from 3,3-dimethyl-2-butanol is secondary and rearranges to a more stable tertiary carbocation by a methyl migration. The major portion of the alkene products is formed by way of the tertiary carbocation.

shown in Figure 5.7, the carbocation that is formed first in the dehydration of 3,3-dimethyl-2-butanol is secondary; the rearranged carbocation is tertiary. Rearrangement occurs, and almost all of the alkene products come from the tertiary carbocation.

How do carbocations rearrange? To understand this we need to examine the structural change that takes place at the transition state. Referring to the initial (secondary) carbocation intermediate in Figure 5.7, rearrangement occurs when a methyl group shifts from C-2 of the carbocation to the positively charged carbon. The methyl group migrates with the pair of electrons that made up its original σ bond to C-2. In the curved arrow notation for this methyl migration, the arrow shows the movement of both the methyl group and the electrons in the σ bond.

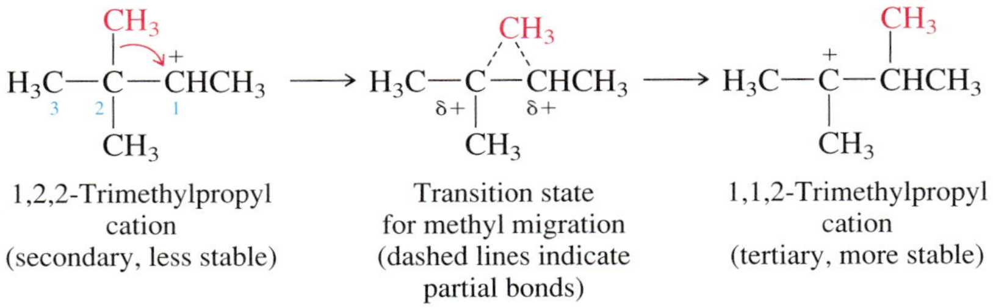

1,2,2-Trimethylpropyl cation (secondary, less stable)   Transition state for methyl migration (dashed lines indicate partial bonds)   1,1,2-Trimethylpropyl cation (tertiary, more stable)

At the transition state for rearrangement, the methyl group is partially bonded both to its point of origin and to the carbon that will be its destination.

This rearrangement is shown in orbital terms in Figure 5.8. The relevant orbitals of the secondary carbocation are shown in structure (a), those of the transition state for rearrangement in (b), and those of the tertiary carbocation in (c). Delocalization of the *electrons* of the C—CH_3 σ bond into the vacant p orbital of the positively charged carbon by hyperconjugation is present in both (a) and (c), requires no activation energy, and

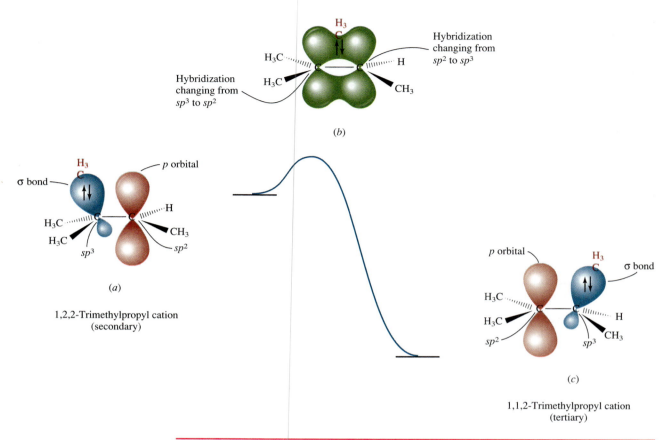

FIGURE 5.8 Methyl migration in 1,2,2-trimethylpropyl cation. Structure (a) is the initial secondary carbocation; structure (b) is the transition state for methyl migration, and structure (c) is the final tertiary carbocation.

Once a carbocation is formed, anything that happens afterward happens rapidly.

stabilizes each carbocation. Migration of the *atoms* of the methyl group, however, occurs only when sufficient energy is absorbed by (a) to achieve the transition state (b). The activation energy is modest, and carbocation rearrangements are normally quite fast.

PROBLEM 5.17 The alkene mixture obtained on dehydration of 2,2-dimethyl-cyclohexanol contains appreciable amounts of 1,2-dimethylcyclohexene. Give a mechanistic explanation for the formation of this product.

Alkyl groups other than methyl can also migrate to a positively charged carbon.

Many carbocation rearrangements involve migration of a hydrogen. These are called **hydride shifts.** The same requirements apply to hydride shifts as to alkyl group migrations; they proceed in the direction that leads to a more stable carbocation; the origin and destination of the migrating hydrogen are adjacent carbons, one of which must be positively charged; and the hydrogen migrates with a pair of electrons.

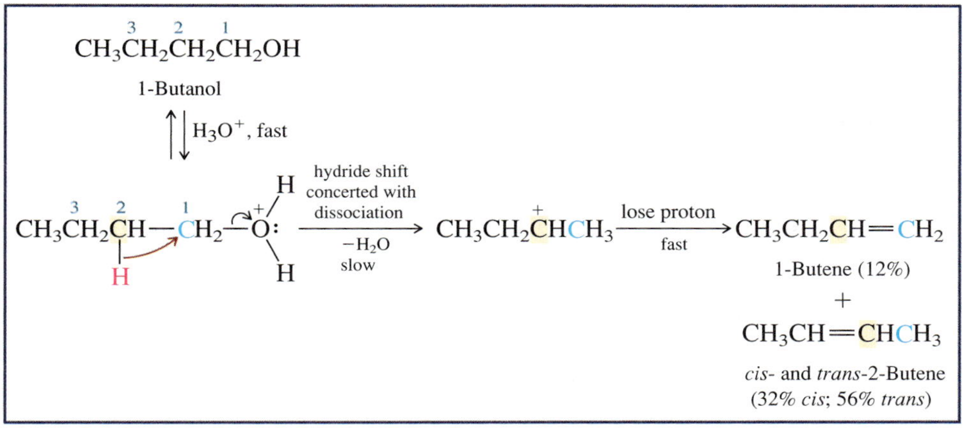

$$\overset{3}{\text{CH}_3}\overset{2}{\text{CH}_2}\overset{1}{\text{CH}_2}\text{CH}_2\text{OH}$$

1-Butanol

$\updownarrow \text{H}_3\text{O}^+, \text{fast}$

$$\overset{3}{\text{CH}_3}\overset{2}{\text{CH}_2}\overset{1}{\text{CH}}\text{—CH}_2\text{—}\overset{+}{\overset{\text{H}}{\underset{\text{H}}{\text{O}}}}: \quad\xrightarrow[\substack{-\text{H}_2\text{O} \\ \text{slow}}]{\substack{\text{hydride shift} \\ \text{concerted with} \\ \text{dissociation}}}\quad \text{CH}_3\text{CH}_2\overset{+}{\text{CH}}\text{CH}_3 \quad\xrightarrow[\text{fast}]{\text{lose proton}}\quad \text{CH}_3\text{CH}_2\text{CH}\!=\!\text{CH}_2$$

1-Butene (12%)

+

$$\text{CH}_3\text{CH}\!=\!\text{CH}\text{CH}_3$$

cis- and *trans*-2-Butene
(32% *cis*; 56% *trans*)

FIGURE 5.9 Dehydration of 1-butanol is accompanied by a hydride shift from C-2 to C-1.

Hydride shifts often occur during the dehydration of primary alcohols. Thus, although 1-butene would be expected to be the only alkene formed on dehydration of 1-butanol, it is in fact only a minor product. The major product is a mixture of *cis-* and *trans*-2-butene.

$$\text{CH}_3\text{CH}_2\text{CH}_2\text{CH}_2\text{OH} \xrightarrow[\text{140–170°C}]{\text{H}_2\text{SO}_4} \text{CH}_3\text{CH}_2\text{CH}\!=\!\text{CH}_2 \; + \quad \text{CH}_3\text{CH}\!=\!\text{CHCH}_3$$

$$\begin{array}{cc} \text{1-Butene} & \text{Mixture of } cis\text{-2-butene (32\%)} \\ (12\%) & \text{and } trans\text{-2-butene (56\%)} \end{array}$$

A mechanism for the formation of these three alkenes is shown in Figure 5.9. Dissociation of the primary alkyloxonium ion is accompanied by a shift of hydride from C-2 to C-1. This avoids the formation of a primary carbocation, leading instead to a secondary carbocation in which the positive charge is at C-2. Deprotonation of this carbocation yields the observed products. (Some 1-butene may also arise directly from the primary alkyloxonium ion.)

This concludes discussion of our second functional group transformation involving *alcohols:* the first was the conversion of alcohols to alkyl halides (Chapter 4), and the second the conversion of alcohols to alkenes. In the remaining sections of the chapter the conversion of *alkyl halides* to alkenes by dehydrohalogenation is described.

5.14   DEHYDROHALOGENATION OF ALKYL HALIDES

Dehydrohalogenation is the loss of a hydrogen and a halogen from an alkyl halide. It is one of the most useful methods for preparing alkenes by β elimination.

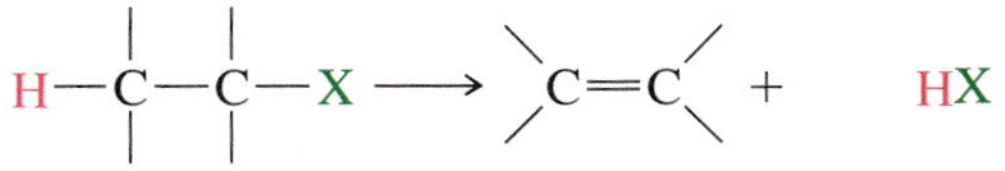

$$\text{H}\!-\!\overset{|}{\underset{|}{\text{C}}}\!-\!\overset{|}{\underset{|}{\text{C}}}\!-\!\text{X} \longrightarrow \;\; \overset{\diagdown}{\diagup}\text{C}\!=\!\text{C}\overset{\diagup}{\diagdown} \; + \quad \text{HX}$$

$$\begin{array}{ccc} \text{Alkyl halide} & \text{Alkene} & \text{Hydrogen halide} \end{array}$$

When applied to the preparation of alkenes, the reaction is carried out in the presence of a strong base, such as sodium ethoxide ($\text{NaOCH}_2\text{CH}_3$) in ethyl alcohol as solvent.

$$H-\overset{|}{\underset{|}{C}}-\overset{|}{\underset{|}{C}}-X \; + \; NaOCH_2CH_3 \longrightarrow \; \underset{/}{\overset{\backslash}{C}}=\overset{/}{\underset{\backslash}{C}} \; + \; CH_3CH_2OH \; + \; NaX$$

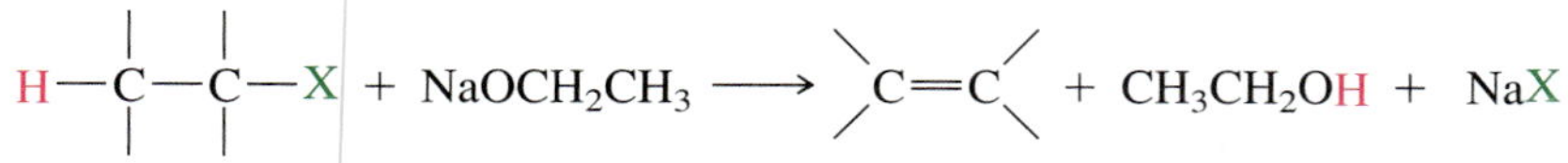

| Alkyl halide | Sodium ethoxide | Alkene | Ethyl alcohol | Sodium halide |

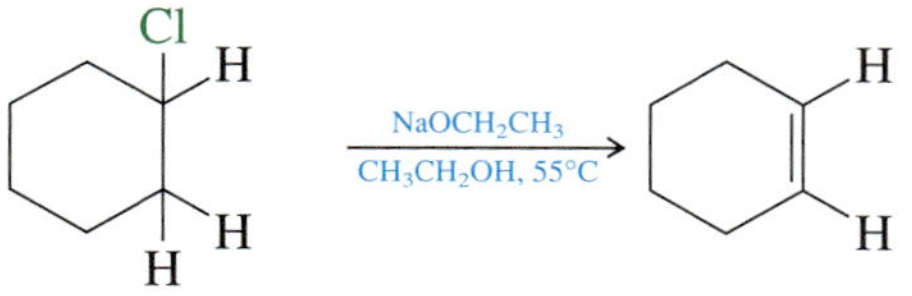

Cyclohexyl chloride   Cyclohexene (100%)

Similarly, sodium methoxide ($NaOCH_3$) is a suitable base and is used in methyl alcohol. Potassium hydroxide in ethyl alcohol is another base–solvent combination often employed in the dehydrohalogenation of alkyl halides. Potassium *tert*-butoxide [$KOC(CH_3)_3$] is the preferred base when the alkyl halide is primary; it is used in either *tert*-butyl alcohol or dimethyl sulfoxide as solvent.

$$CH_3(CH_2)_{15}CH_2CH_2Cl \xrightarrow[\text{DMSO, 25°C}]{KOC(CH_3)_3} CH_3(CH_2)_{15}CH=CH_2$$

1-Chlorooctadecane   1-Octadecene (86%)

The regioselectivity of dehydrohalogenation of alkyl halides follows the Zaitsev rule; β elimination predominates in the direction that leads to the more highly substituted alkene.

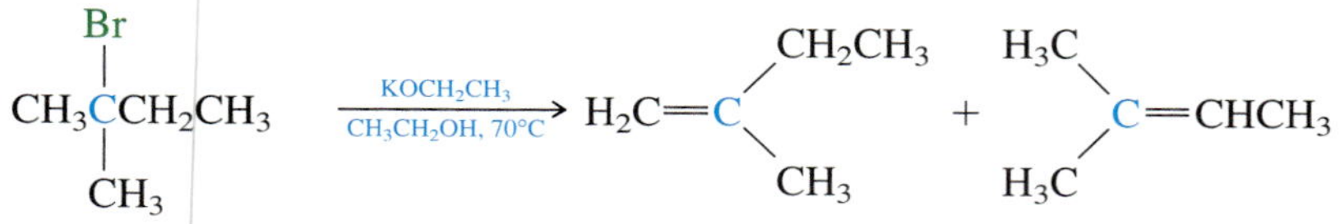

$$CH_3\overset{\overset{\displaystyle Br}{|}}{\underset{\underset{\displaystyle CH_3}{|}}{C}}CH_2CH_3 \xrightarrow[\text{CH}_3\text{CH}_2\text{OH, 70°C}]{KOCH_2CH_3} H_2C=\overset{CH_2CH_3}{\underset{CH_3}{C}} \; + \; \underset{H_3C}{\overset{H_3C}{C}}=CHCH_3$$

2-Bromo-2-methylbutane   2-Methyl-1-butene (29%)   2-Methyl-2-butene (71%)

PROBLEM 5.18 Write the structures of all the alkenes that can be formed by dehydrohalogenation of each of the following alkyl halides. Apply the Zaitsev rule to predict the alkene formed in greatest amount in each case.

(a) 2-Bromo-2,3-dimethylbutane   (d) 2-Bromo-3-methylbutane

(b) *tert*-Butyl chloride   (e) 1-Bromo-3-methylbutane

(c) 3-Bromo-3-ethylpentane   (f) 1-Iodo-1-methylcyclohexane

SAMPLE SOLUTION (a) First analyze the structure of 2-bromo-2,3-dimethyl-butane with respect to the number of possible β elimination pathways.

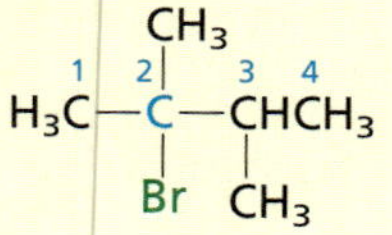

$$\overset{\overset{\displaystyle CH_3}{|}}{\underset{\underset{\displaystyle Br}{|}}{H_3C-\overset{2}{C}}}-\overset{\overset{\displaystyle}{}}{\underset{\underset{\displaystyle CH_3}{|}}{\overset{3}{C}HCH_3}}$$

$\overset{1}{}\overset{2}{}\overset{3}{}\overset{4}{}$

Bromine must be lost from C-2; hydrogen may be lost from C-1 or from C-3

The two possible alkenes are

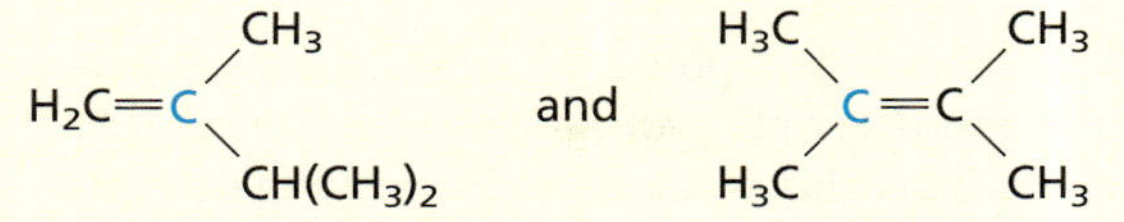

2,3-Dimethyl-1-butene
(minor product)

and

2,3-Dimethyl-2-butene
(major product)

The major product, predicted on the basis of Zaitsev's rule, is 2,3-dimethyl-2-butene. It has a tetrasubstituted double bond. The minor alkene has a disubstituted double bond.

In addition to being regioselective, dehydrohalogenation of alkyl halides is stereoselective and favors formation of the more stable stereoisomer. Usually, as in the case of 5-bromononane, the trans (or E) alkene is formed in greater amounts than its cis (or Z) stereoisomer.

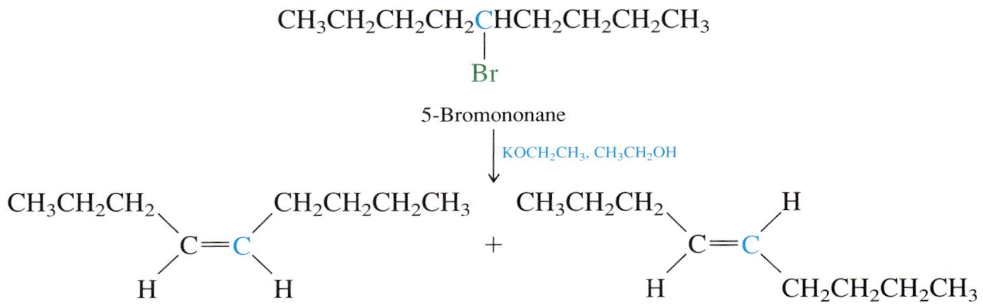

$$CH_3CH_2CH_2CH_2CHCH_2CH_2CH_2CH_3$$

Br

5-Bromononane

KOCH$_2$CH$_3$, CH$_3$CH$_2$OH

cis-4-Nonene
(23%)

+

trans-4-Nonene
(77%)

PROBLEM 5.19 Write structural formulas for all the alkenes that can be formed in the reaction of 2-bromobutane with potassium ethoxide.

Dehydrohalogenation of cycloalkyl halides lead exclusively to cis cycloalkenes when the ring has fewer than ten carbons. As the ring becomes larger, it can accommodate either a cis or a trans double bond, and large-ring cycloalkyl halides give mixtures of cis and trans cycloalkenes.

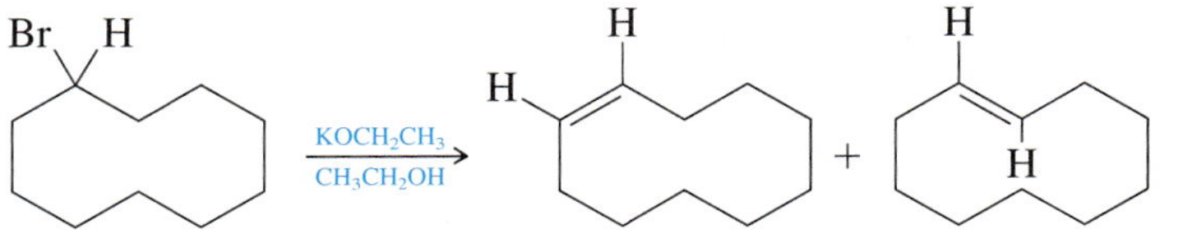

Bromocyclodecane

KOCH$_2$CH$_3$
CH$_3$CH$_2$OH

cis-Cyclodecene
[(Z)-cyclodecene]
(85%)

+

trans-Cyclodecene
[(E)-cyclodecene]
(15%)

5.15   THE E2 MECHANISM OF DEHYDROHALOGENATION OF ALKYL HALIDES

In the 1920s, Sir Christopher Ingold proposed a mechanism for dehydrohalogenation that is still accepted as the best description of how these reactions occur. Some of the information on which Ingold based his mechanism included these facts:

1. *The reaction exhibits second-order kinetics; it is first-order in alkyl halide and first-order in base.*

$$\text{Rate} = k[\text{alkyl halide}][\text{base}]$$

Doubling the concentration of either the alkyl halide or the base doubles the reaction rate. Doubling the concentration of both reactants increases the rate by a factor of 4.

2. *The rate of elimination depends on the halogen, the reactivity of alkyl halides increasing with decreasing strength of the carbon–halogen bond.*

Increasing rate of dehydrohalogenation

RF   <   RCl   <   RBr   <   RI

Alkyl fluoride   Alkyl iodide
(slowest rate of elimination;   (fastest rate of elimination;
strongest carbon–halogen bond)   weakest carbon–halogen bond)

Cyclohexyl bromide, for example, is converted to cyclohexene by sodium ethoxide in ethanol over 60 times faster than cyclohexyl chloride. Iodide is the best **leaving group** in a dehydrohalogenation reaction, fluoride the poorest. Fluoride is such a poor leaving group that alkyl fluorides are rarely used as starting materials in the preparation of alkenes.

What are the implications of second-order kinetics? Ingold reasoned that second-order kinetics suggest a bimolecular rate-determining step involving both a molecule of the alkyl halide and a molecule of base. He concluded that proton removal from the β carbon by the base occurs during the rate-determining step rather than in a separate step following the rate-determining step.

What are the implications of the effects of the various halide leaving groups? Because the halogen with the weakest bond to carbon reacts fastest, Ingold concluded that the carbon–halogen bond breaks in the rate-determining step. The weaker the carbon–halogen bond, the easier it breaks.

On the basis of these observations, Ingold proposed a one-step bimolecular E2 mechanism for dehydrohalogenation.

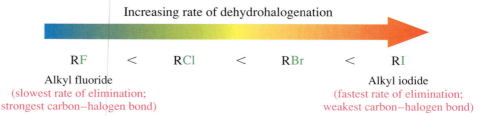

Transition state for bimolecular elimination

In the E2 mechanism the three key elements

1. C—H bond breaking
2. C=C π bond formation
3. C—X bond breaking

are all taking place at the same transition state. The carbon–hydrogen and carbon–halogen bonds are in the process of being broken, the base is becoming bonded to the hydrogen, a π bond is being formed, and the hybridization of carbon is changing from sp^3 to sp^2. An energy diagram for the E2 mechanism is shown in Figure 5.10.

The E2 mechanism is followed whenever an alkyl halide—be it primary, secondary, or tertiary—undergoes elimination in the presence of a strong base.

> **PROBLEM 5.20**   Use curved arrows to track electron movement in the dehydrohalogenation of *tert*-butyl chloride by sodium methoxide by the E2 mechanism.

The regioselectivity of elimination is accommodated in the E2 mechanism by noting that a partial double bond develops at the transition state. Because alkyl groups

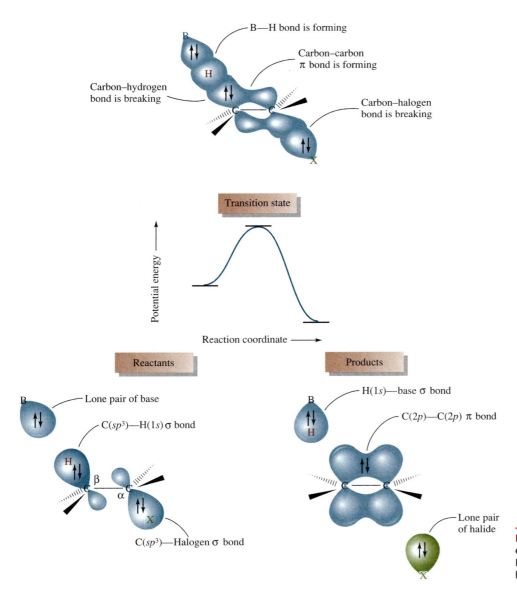

FIGURE 5.10 Potential energy diagram for concerted E2 elimination of an alkyl halide.

stabilize double bonds, they also stabilize a partially formed π bond in the transition state. The more stable alkene therefore requires a lower energy of activation for its formation and predominates in the product mixture because it is formed faster than a less stable one.

Ingold was a pioneer in applying quantitative measurements of reaction rates to the understanding of organic reaction mechanisms. Many of the reactions to be described in this text were studied by him and his students during the period of about 1920 to 1950. The facts disclosed by Ingold's experiments have been verified many times. His interpretations, although considerably refined during the decades that followed his original reports, still serve us well as a starting point for understanding how the fundamental processes of organic chemistry take place. Beta elimination of alkyl halides by the E2 mechanism is one of those fundamental processes.

5.16 ANTI ELIMINATION IN E2 REACTIONS: STEREOELECTRONIC EFFECTS

Further insight into the E2 mechanism comes from stereochemical studies. One such experiment compares the rates of elimination of the cis and trans isomers of 4-*tert*-butylcyclohexyl bromide.

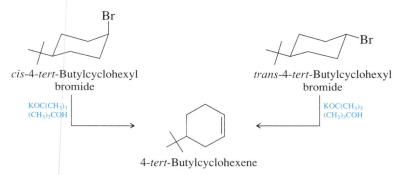

cis-4-*tert*-Butylcyclohexyl
bromide

trans-4-*tert*-Butylcyclohexyl
bromide

4-*tert*-Butylcyclohexene

Although both stereoisomers yield 4-*tert*-butylcyclohexene as the only alkene, they do so at quite different rates. The cis isomer reacts over 500 times faster than the trans.

The difference in reaction rate results from different degrees of π bond development in the E2 transition state. Since π overlap of p orbitals requires their axes to be parallel, π bond formation is best achieved when the four atoms of the H—C—C—X unit lie in the same plane at the transition state. The two conformations that permit this are termed *syn coplanar* and *anti coplanar*.

Syn coplanar;
orbitals aligned but
bonds are eclipsed

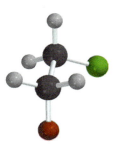

Gauche;
orbitals not aligned for
double bond formation

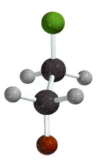

Anti coplanar;
orbitals aligned and
bonds are staggered

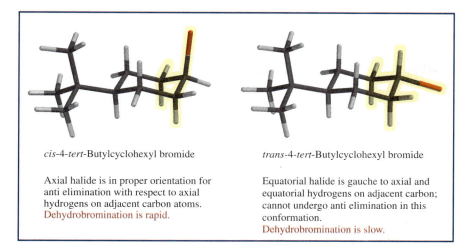

FIGURE 5.11 Confor-
mations of *cis*- and *trans*-4-
tert-butylcyclohexyl bromide
and their relationship to the
preference for an anti coplanar arrangement of proton
and leaving group.

cis-4-*tert*-Butylcyclohexyl bromide

Axial halide is in proper orientation for
anti elimination with respect to axial
hydrogens on adjacent carbon atoms.
Dehydrobromination is rapid.

trans-4-*tert*-Butylcyclohexyl bromide

Equatorial halide is gauche to axial and
equatorial hydrogens on adjacent carbon;
cannot undergo anti elimination in this
conformation.
Dehydrobromination is slow.

Because adjacent bonds are eclipsed when the H—C—C—X unit is syn coplanar, a
transition state with this geometry is less stable than one that has an anti coplanar relationship between the proton and the leaving group.

As Figure 5.11 shows, bromine is axial in the most stable conformation of *cis*-4-
tert-butylcyclohexyl bromide, but it is equatorial in the trans stereoisomer. An axial
bromine is anti coplanar with respect to the axial hydrogens at C-2 and C-6, and so the
proper geometry between the proton and the leaving group is already present in the cis
bromide, which undergoes E2 elimination rapidly. The less reactive stereoisomer, the
trans bromide, has an equatorial bromine in its most stable conformation. An equatorial
bromine is not anti coplanar with respect to any of the hydrogens that are β to it. The
relationship between an equatorial leaving group and all the C-2 and C-6 hydrogens is
gauche. To undergo E2 elimination, the trans bromide must adopt a geometry in which
the ring is strained. The transition state for its elimination is therefore higher in energy,
and reaction is slower.

> **PROBLEM 5.21**   Use curved arrows to show the bonding changes in the reaction
> of *cis*-4-*tert*-butylcyclohexyl bromide with potassium *tert*-butoxide. Be sure your
> drawing correctly represents the spatial relationship between the leaving group
> and the proton that is lost.

Effects that arise because one spatial arrangement of electrons (or orbitals or bonds)
is more stable than another are called **stereoelectronic effects.** *There is a stereoelectronic preference for the anti coplanar arrangement of proton and leaving group in E2
reactions.* Although coplanarity of the *p* orbitals is the best geometry for the E2 process,
modest deviations from this ideal can be tolerated. In such cases, the terms used are
syn periplanar and *anti periplanar.*

The *peri-* in *periplanar*
means "almost" or "nearly."
The coplanar/periplanar distinction is discussed in the
October, 2000 issue of the
*Journal of Chemical Educa-
tion*, p. 1366.

5.17   THE E1 MECHANISM OF DEHYDROHALOGENATION
OF ALKYL HALIDES

The E2 mechanism is a concerted process in which the carbon–hydrogen and
carbon–halogen bonds both break in the same elementary step. What if these bonds break
in separate steps?

The reaction:

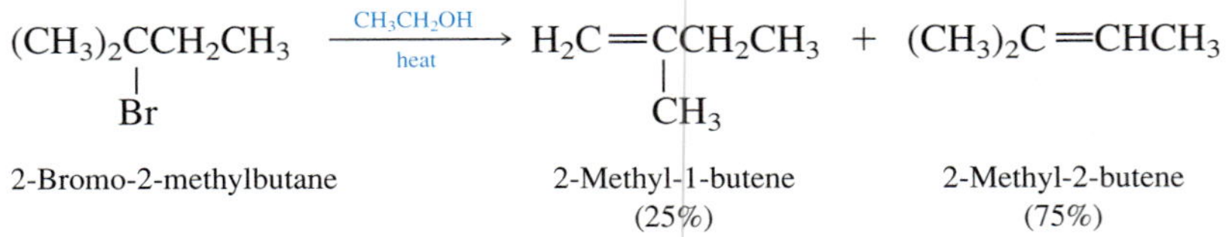

$(CH_3)_2CCH_2CH_3$ $\xrightarrow[\text{heat}]{\text{CH}_3\text{CH}_2\text{OH}}$ $H_2C{=}CCH_2CH_3$ + $(CH_3)_2C{=}CHCH_3$
| |
Br CH_3

2-Bromo-2-methylbutane   2-Methyl-1-butene   2-Methyl-2-butene
                                      (25%)   (75%)

The mechanism:

Step (1): Alkyl halide dissociates by heterolytic cleavage of carbon–halogen bond. (Ionization step)

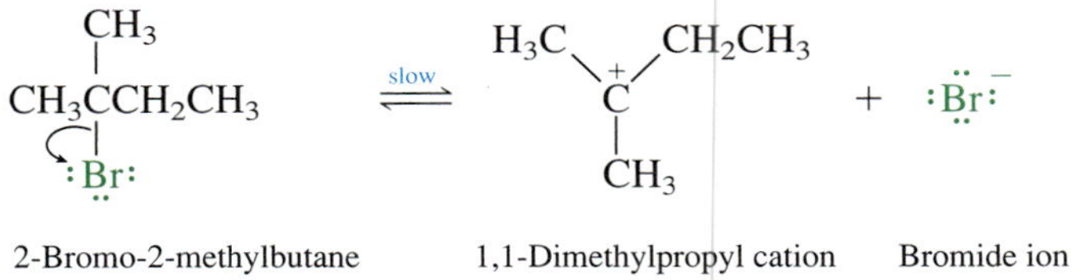

2-Bromo-2-methylbutane   1,1-Dimethylpropyl cation   Bromide ion

Step (2): Ethanol acts as a base to remove a proton from the carbocation to give the alkene products. (Deprotonation step)

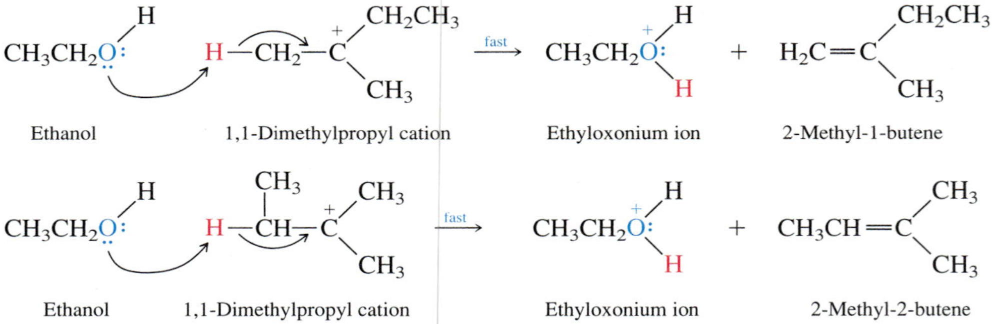

Ethanol   1,1-Dimethylpropyl cation   Ethyloxonium ion   2-Methyl-1-butene

Ethanol   1,1-Dimethylpropyl cation   Ethyloxonium ion   2-Methyl-2-butene

FIGURE 5.12 The E1 mechanism for the dehydrohalogenation of 2-bromo-2-methylbutane in ethanol.

One possibility is the two-step mechanism of Figure 5.12, in which the carbon–halogen bond breaks first to give a carbocation intermediate, followed by deprotonation of the carbocation in a second step.

The alkyl halide, in this case 2-bromo-2-methylbutane, ionizes to a carbocation and a halide anion by a heterolytic cleavage of the carbon–halogen bond. Like the dissociation of an alkyloxonium ion to a carbocation, this step is rate-determining. Because the rate-determining step is unimolecular—it involves only the alkyl halide and not the base—it is a type of E1 mechanism.

Typically, elimination by the E1 mechanism is observed only for tertiary and some secondary alkyl halides, and then only when the base is weak or in low concentration. Unlike eliminations that follow an E2 pathway and exhibit second-order kinetic behavior:

$$\text{Rate} = k[\text{alkyl halide}][\text{base}]$$

those that follow an E1 mechanism obey a first-order rate law.

$$\text{Rate} = k[\text{alkyl halide}]$$

The reactivity order parallels the ease of carbocation formation.

Increasing rate of elimination by the E1 mechanism

RCH₂X	<	R₂CHX	<	R₃CX

RCH_2X < R_2CHX < R_3CX

Primary alkyl halide
slowest rate of
E1 elimination

Tertiary alkyl halide
fastest rate of
E1 elimination

Because the carbon–halogen bond breaks in the slow step, the rate of the reaction depends on the leaving group. Alkyl iodides have the weakest carbon–halogen bond and are the most reactive; alkyl fluorides have the strongest carbon–halogen bond and are the least reactive.

The best examples of E1 eliminations are those carried out in the absence of added base. In the example cited in Figure 5.12, the base that abstracts the proton from the carbocation intermediate is a very weak one; it is a molecule of the solvent, ethyl alcohol. At even modest concentrations of strong base, elimination by the E2 mechanism is much faster than E1 elimination.

There is a strong similarity between the mechanism shown in Figure 5.12 and the one shown for alcohol dehydration in Figure 5.6. The main difference between the dehydration of 2-methyl-2-butanol and the dehydrohalogenation of 2-bromo-2-methylbutane is the source of the carbocation. When the alcohol is the substrate, it is the corresponding alkyloxonium ion that dissociates to form the carbocation. The alkyl halide ionizes directly to the carbocation.

Alkyloxonium ion		Carbocation		Alkyl halide

Like alcohol dehydrations, E1 reactions of alkyl halides can be accompanied by carbocation rearrangements. Eliminations by the E2 mechanism, on the other hand, normally proceed without rearrangement. Consequently, if one wishes to prepare an alkene from an alkyl halide, conditions favorable to E2 elimination should be chosen. In practice this simply means carrying out the reaction in the presence of a strong base.

5.18   SUMMARY

Section 5.1   Alkenes and cycloalkenes contain carbon–carbon double bonds. According to **IUPAC nomenclature,** alkenes are named by substituting -*ene* for the -*ane* suffix of the alkane that has the same number of carbon atoms as the longest continuous chain that includes the double bond. The chain is numbered in the direction that gives the lower number to the first-appearing carbon of the double bond. The double bond takes precedence over alkyl groups and halogens in dictating the direction of numbering, but is outranked by a hydroxyl group.

3-Ethyl-2-pentene   3-Bromocyclopentene   3-Buten-1-ol

Section 5.2   Bonding in alkenes is described according to an sp^2 orbital hybridization model. The double bond unites two sp^2-hybridized carbon atoms and is made of a σ component and a π component. The σ bond arises by overlap of an sp^2 hybrid orbital on each carbon. The π bond is weaker than the σ bond and results from a side-by-side overlap of *p* orbitals.

Sections 5.3–5.4   Isomeric alkenes may be either **constitutional isomers** or **stereoisomers.** There is a sizable barrier to rotation about a carbon–carbon double bond, which corresponds to the energy required to break the π component of the double bond. Stereoisomeric alkenes are configurationally stable under normal conditions. The **configurations** of stereoisomeric alkenes are described according to two notational systems. One system adds the prefix *cis-* to the name of the alkene when similar substituents are on the same side of the double bond and the prefix *trans-* when they are on opposite sides. The other ranks substituents according to a system of rules based on atomic number. The prefix *Z* is used for alkenes that have higher ranked substituents on the same side of the double bond; the prefix *E* is used when higher ranked substituents are on opposite sides.

cis-2-Pentene   *trans*-2-Pentene
[(*Z*)-2-pentene]   [(*E*)-2-pentene]

Section 5.5   Alkenes are relatively nonpolar. Alkyl substituents donate electrons to an sp^2-hybridized carbon to which they are attached slightly better than hydrogen does.

Section 5.6   Electron release from alkyl substituents stabilizes a double bond. In general, the order of alkene stability is:

1. Tetrasubstituted alkenes ($R_2C{=}CR_2$) are the most stable.

2. Trisubstituted alkenes ($R_2C{=}CHR$) are next.

3. Among disubstituted alkenes, *trans*-RCH$=$CHR is normally more stable than *cis*-RCH$=$CHR. Exceptions are cycloalkenes, cis cycloalkenes being more stable than trans when the ring contains fewer than 11 carbons.

4. Monosubstituted alkenes ($RCH{=}CH_2$) have a more stabilized double bond than ethylene (unsubstituted) but are less stable than disubstituted alkenes.

The greater stability of more highly substituted double bonds is an example of an **electronic effect.** The decreased stability that results from van der Waals strain between cis substituents is an example of a **steric effect.**

Section 5.7   Cycloalkenes that have trans double bonds in rings smaller than 12 members are less stable than their cis stereoisomers. *trans*-Cyclooctene can be isolated and stored at room temperature, but *trans*-cycloheptene is not stable above $-30°C$.

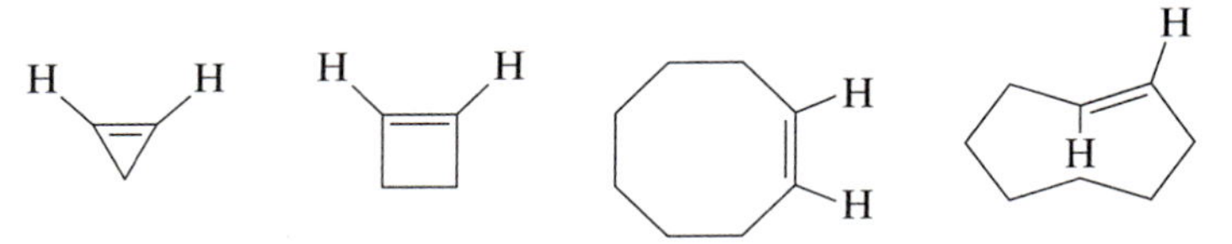

Cyclopropene   Cyclobutene   *cis*-Cyclooctene   *trans*-Cyclooctene

Section 5.8   Alkenes are prepared by **β elimination** of alcohols and alkyl halides. These reactions are summarized with examples in Table 5.2. In both cases, β elimination proceeds in the direction that yields the more highly substituted double bond (**Zaitsev's rule**).

Sections   See Table 5.2.
5.9–5.11

Section 5.12   Secondary and tertiary alcohols undergo **dehydration** by an E1 mechanism involving carbocation intermediates.

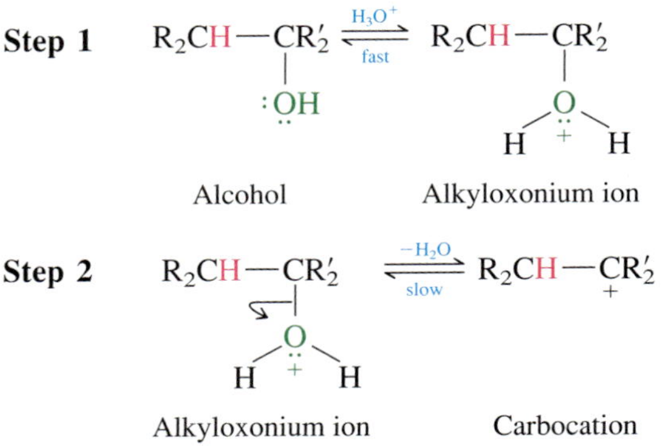

Step 1

Step 2

TABLE 5.2	Preparation of Alkenes by Elimination Reactions of Alcohols and Alkyl Halides

Reaction (section) and comments	General equation and specific example

Dehydration of alcohols (Sections 5.9–5.13) Dehydration requires an acid catalyst; the order of reactivity of alcohols is tertiary > secondary > primary. Elimination is regioselective and proceeds in the direction that produces the most highly substituted double bond. When stereoisomeric alkenes are possible, the more stable one is formed in greater amounts. An E1 (elimination unimolecular) mechanism via a carbocation intermediate is followed with secondary and tertiary alcohols. Primary alcohols react by an E2 (elimination bimolecular) mechanism. Sometimes elimination is accompanied by rearrangement.	$$R_2CHCR'_2 \xrightarrow{H^+} R_2C{=}CR'_2 + H_2O$$
	  $\quad\;\;\mid$
	  $\quad$ OH
	
	Alcohol $\qquad$ Alkene $\qquad$ Water
	
	2-Methyl-2-hexanol
	$\quad\downarrow$ H$_2$SO$_4$, 80°C
	
	2-Methyl-1-hexene $\qquad$ 2-Methyl-2-hexene
	(19%) $\qquad\qquad$ (81%)

Dehydrohalogenation of alkyl halides (Sections 5.14–5.16) Strong bases cause a proton and a halide to be lost from adjacent carbons of an alkyl halide to yield an alkene. Regioselectivity is in accord with the Zaitsev rule. The order of halide reactivity is I > Br > Cl > F. A concerted E2 reaction pathway is followed, carbocations are not involved, and rearrangements do not occur. An anti coplanar arrangement of the proton being removed and the halide being lost characterizes the transition state.	$$R_2CHCR'_2 + :B^- \longrightarrow R_2C{=}CR'_2 + \ H{-}B \ + \ X^-$$
	  $\quad\;\;\mid$
	  $\quad$ X
	
	Alkyl $\quad$ Base $\qquad$ Alkene $\qquad$ Conjugate $\quad$ Halide
	halide $\qquad\qquad\qquad\qquad$ acid of base
	
	1-Chloro-1-methylcyclohexane
	$\quad\downarrow$ KOCH$_2$CH$_3$, CH$_3$CH$_2$OH, 100°C
	
	Methylenecyclohexane $\qquad$ 1-Methylcyclohexene
	(6%) $\qquad\qquad\qquad$ (94%)

Step 3 $\quad$ $R_2C{-}CR'_2 \xrightarrow[\text{fast}]{-H^+} R_2C{=}CR'_2$
$\qquad\qquad\quad$ $\mid\!\nearrow\! +$
$\qquad\qquad\quad$ H

$\qquad\qquad$ Carbocation $\qquad\qquad$ Alkene

Primary alcohols do not dehydrate as readily as secondary or tertiary alcohols, and their dehydration does not involve a primary carbocation. A proton is lost from the β carbon in the same step in which carbon–oxygen bond cleavage occurs. The mechanism is E2.

Section 5.13 $\quad$ Alkene synthesis via alcohol dehydration is complicated by **carbocation rearrangements.** A less stable carbocation can rearrange to a more stable one by an alkyl group migration or by a hydride shift, opening the possibility for alkene formation from two different carbocations.

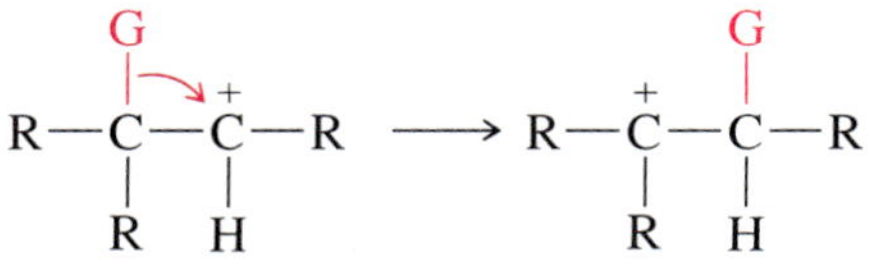

Secondary carbocation   Tertiary carbocation
(G is a migrating group; it may be either a hydrogen or an alkyl group)

Section 5.14   See Table 5.2.

Section 5.15   **Dehydrohalogenation** of alkyl halides by alkoxide bases is not complicated by rearrangements, because carbocations are not intermediates. The mechanism is E2. It is a concerted process in which the base abstracts a proton from the β carbon while the bond between the halogen and the α carbon undergoes heterolytic cleavage.

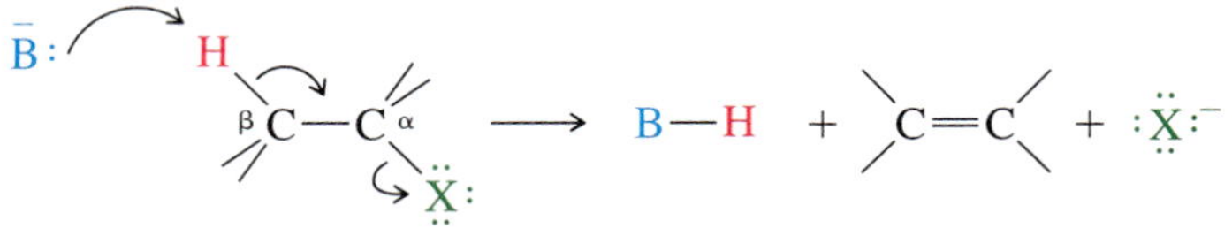

| Base | Alkyl halide | Conjugate acid of base | Alkene | Halide ion |

Section 5.16   The preceding equation shows the proton H and the halogen X in the **anti coplanar** relationship that is required for elimination by the E2 mechanism.

Section 5.17   In the absence of a strong base, alkyl halides eliminate by an E1 mechanism. Rate-determining ionization of the alkyl halide to a carbocation is followed by deprotonation of the carbocation.

Step 1

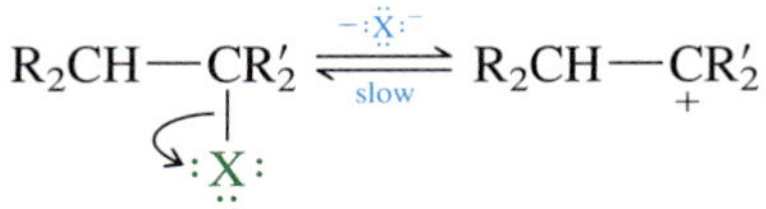

Alkyl halide   Carbocation

Step 2

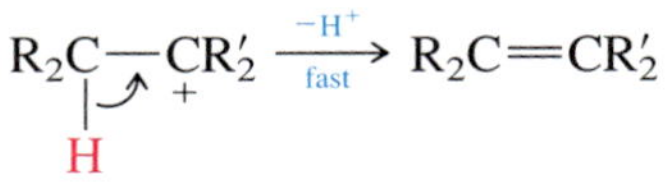

Carbocation   Alkene

PROBLEMS

5.22   Write structural formulas for each of the following:

(a) 1-Heptene

(b) 3-Ethyl-2-pentene

(c) *cis*-3-Octene

(d) *trans*-1,4-Dichloro-2-butene

(e) (*Z*)-3-Methyl-2-hexene

(f) (*E*)-3-Chloro-2-hexene

(g) 1-Bromo-3-methylcyclohexene

(h) 1-Bromo-6-methylcyclohexene

(i) 4-Methyl-4-penten-2-ol

(j) Vinylcycloheptane

(k) 1,1-Diallylcyclopropane

(l) *trans*-1-Isopropenyl-3-methylcyclohexane

5.23 Write a structural formula or build a molecular model and give a correct IUPAC name for each alkene of molecular formula C_7H_{14} that has a *tetrasubstituted* double bond.

5.24 Give the IUPAC names for each of the following compounds:

(a) $(CH_3CH_2)_2C{=}CHCH_3$

(b) $(CH_3CH_2)_2C{=}C(CH_2CH_3)_2$

(c) $(CH_3)_3CCH{=}CCl_2$

(d)

(e)

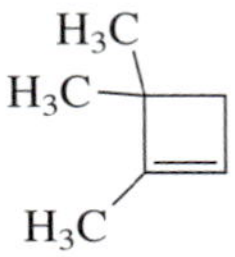

(f)

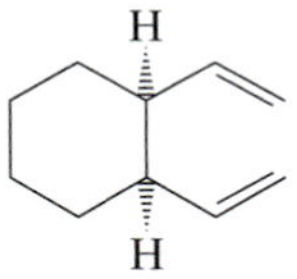

H

H

(g)

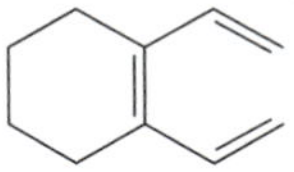

5.25 (a) A hydrocarbon isolated from fish oil and from plankton was identified as 2,6,10,14-tetramethyl-2-pentadecene. Write its structure.

(b) Alkyl isothiocyanates are compounds of the type $RN{=}C{=}S$. Write a structural formula for *allyl isothiocyanate*, a pungent-smelling compound isolated from mustard.

5.26 (a) The sex attractant of the Mediterranean fruit fly is (*E*)-6-nonen-1-ol. Write a structural formula or build a molecular model for this compound, showing the stereochemistry of the double bond.

(b) Geraniol is a naturally occurring substance present in the fragrant oil of many plants. It has a pleasing, roselike odor. Geraniol is the *E* isomer of

$$(CH_3)_2C{=}CHCH_2CH_2\underset{\underset{\displaystyle CH_3}{|}}{C}{=}CHCH_2OH$$

Write a structural formula or build a molecular model for geraniol, showing its stereochemistry.

(c) Nerol is a naturally occurring substance that is a stereoisomer of geraniol. Write its structure or build a molecular model.

(d) The sex attractant of the codling moth is the 2*Z*,6*E* stereoisomer of

$$CH_3CH_2CH_2\underset{\underset{\displaystyle CH_3}{|}}{C}{=}CHCH_2CH_2\underset{\underset{\displaystyle CH_2CH_3}{|}}{C}{=}CHCH_2OH$$

Write the structure of this substance or build a molecular model in a way that clearly shows its stereochemistry.

(e) The sex pheromone of the honeybee is the *E* stereoisomer of the compound shown. Write a structural formula or build a molecular model for this compound.

$$CH_3\overset{\overset{\displaystyle O}{\|}}{C}(CH_2)_4CH_2CH{=}CHCO_2H$$

(f) A growth hormone from the cecropia moth has the structure shown. Express the stereochemistry of the double bonds according to the E–Z system.

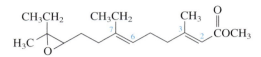

5.27 Which one of the following has the largest dipole moment (is the most polar)? Compare your answer with the calculated dipole moments on *Learning By Modeling*.

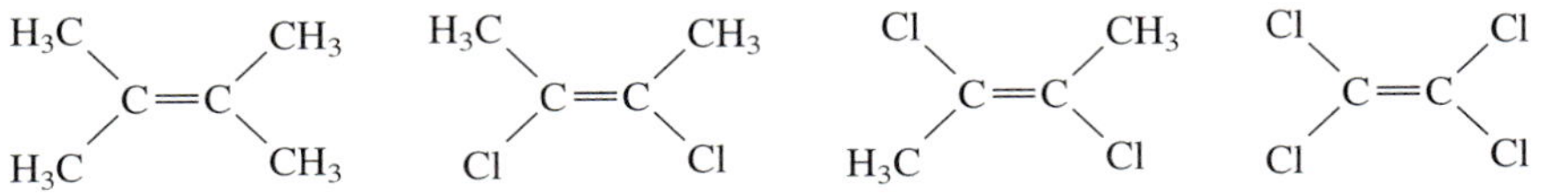

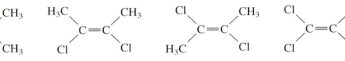

A   B   C   D

5.28 Match each alkene with the appropriate heat of combustion:

Heats of combustion (kJ/mol): 5293; 4658; 4650; 4638; 4632

Heats of combustion (kcal/mol): 1264.9; 1113.4; 1111.4; 1108.6; 1107.1

(a) 1-Heptene

(b) 2,4-Dimethyl-1-pentene

(c) 2,4-Dimethyl-2-pentene

(d) (Z)-4,4-Dimethyl-2-pentene

(e) 2,4,4-Trimethyl-2-pentene

5.29 Choose the more stable alkene in each of the following pairs. Explain your reasoning.

(a) 1-Methylcyclohexene or 3-methylcyclohexene

(b) Isopropenylcyclopentane or allylcyclopentane

(c) 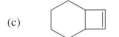   or

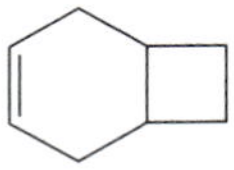

Bicyclo[4.2.0]oct-7-ene   Bicyclo[4.2.0]oct-3-ene

(d) (Z)-Cyclononene or (E)-cyclononene

(e) (Z)-Cyclooctadecene or (E)-cyclooctadecene

5.30 (a) Suggest an explanation for the fact that 1-methylcyclopropene is some 42 kJ/mol (10 kcal/mol) less stable than methylenecyclopropane.

—CH$_3$ is less stable than   =CH$_2$

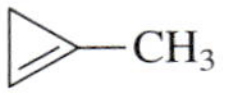

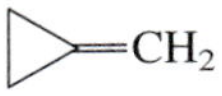

1-Methylcyclopropene   Methylenecyclopropane

(b) On the basis of your answer to part (a), compare the expected stability of 3-methyl-cyclopropene with that of 1-methylcyclopropene and that of methylenecyclopropane.

5.31 How many alkenes would you expect to be formed from each of the following alkyl bromides under conditions of E2 elimination? Identify the alkenes in each case.

(a) 1-Bromohexane

(b) 2-Bromohexane

(c) 3-Bromohexane

(d) 2-Bromo-2-methylpentane

(e) 2-Bromo-3-methylpentane

(f) 3-Bromo-2-methylpentane

(g) 3-Bromo-3-methylpentane

(h) 3-Bromo-2,2-dimethylbutane

5.32 Write structural formulas for all the alkene products that could reasonably be formed from each of the following compounds under the indicated reaction conditions. Where more than one alkene is produced, specify the one that is the major product.

(a) 1-Bromo-3,3-dimethylbutane (potassium *tert*-butoxide, *tert*-butyl alcohol, 100°C)

(b) 1-Methylcyclopentyl chloride (sodium ethoxide, ethanol, 70°C)

(c) 3-Methyl-3-pentanol (sulfuric acid, 80°C)

(d) 2,3-Dimethyl-2-butanol (phosphoric acid, 120°C)

(e) 3-Iodo-2,4-dimethylpentane (sodium ethoxide, ethanol, 70°C)

(f) 2,4-Dimethyl-3-pentanol (sulfuric acid, 120°C)

5.33 Choose the compound of molecular formula $C_7H_{13}Br$ that gives each alkene shown as the *exclusive* product of E2 elimination.

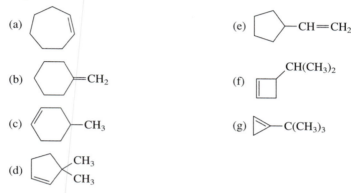

5.34 Give the structures of two different alkyl bromides both of which yield the indicated alkene as the *exclusive* product of E2 elimination.

(a) $CH_3CH=CH_2$

(b) $(CH_3)_2C=CH_2$

(c) $BrCH=CBr_2$

(d) with CH₃ groups

5.35 (a) Write the structures or build molecular models of all the isomeric alkyl bromides having the molecular formula $C_5H_{11}Br$.

(b) Which one undergoes E1 elimination at the fastest rate?

(c) Which one is incapable of reacting by the E2 mechanism?

(d) Which ones can yield only a single alkene on E2 elimination?

(e) For which isomer does E2 elimination give two alkenes that are not constitutional isomers?

(f) Which one yields the most complex mixture of alkenes on E2 elimination?

5.36 (a) Write the structures or build molecular models of all the isomeric alcohols having the molecular formula $C_5H_{12}O$.

(b) Which one will undergo acid-catalyzed dehydration most readily?

(c) Write the structure of the most stable C_5H_{11} carbocation.

(d) Which alkenes may be derived from the carbocation in part (c)?

(e) Which alcohols can yield the carbocation in part (c) by a process involving a hydride shift?

(f) Which alcohols can yield the carbocation in part (c) by a process involving a methyl shift?

5.37 Predict the major organic product of each of the following reactions. In spite of the structural complexity of some of the starting materials, the functional group transformations are all of the type described in this chapter.

(a)

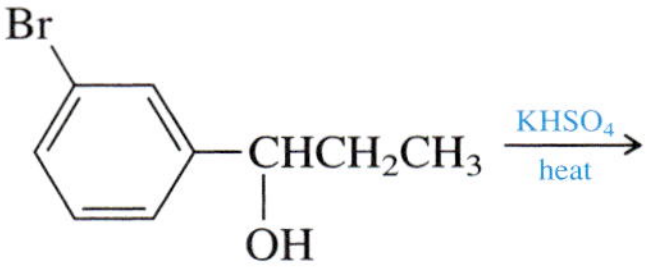

$$\text{—CHCH}_2\text{CH}_3 \xrightarrow[\text{heat}]{\text{KHSO}_4}$$

(b) $\text{ICH}_2\text{CH(OCH}_2\text{CH}_3)_2 \xrightarrow[\text{(CH}_3)_3\text{COH, heat}]{\text{KOC(CH}_3)_3}$

(c)

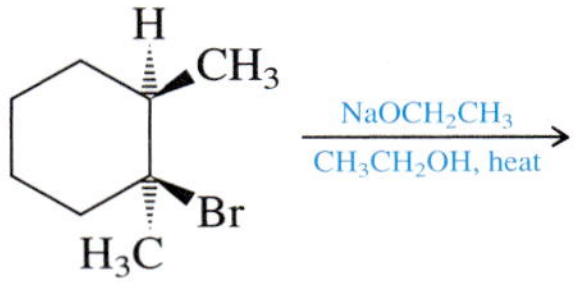

$$\xrightarrow[\text{CH}_3\text{CH}_2\text{OH, heat}]{\text{NaOCH}_2\text{CH}_3}$$

(d)

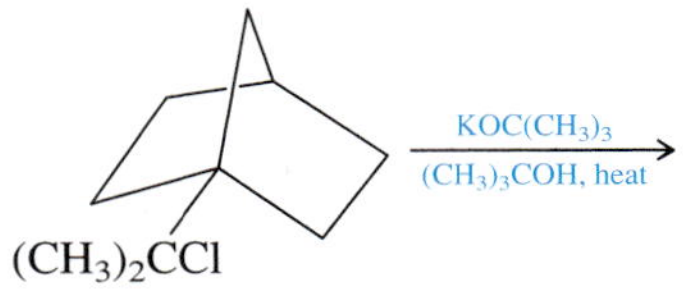

$$\xrightarrow[\text{(CH}_3)_3\text{COH, heat}]{\text{KOC(CH}_3)_3}$$

$(\text{CH}_3)_2\text{CCl}$

(e)

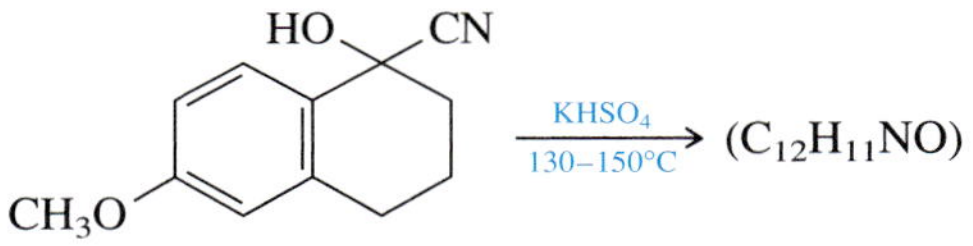

$$\xrightarrow[\text{130–150°C}]{\text{KHSO}_4} (\text{C}_{12}\text{H}_{11}\text{NO})$$

(f) 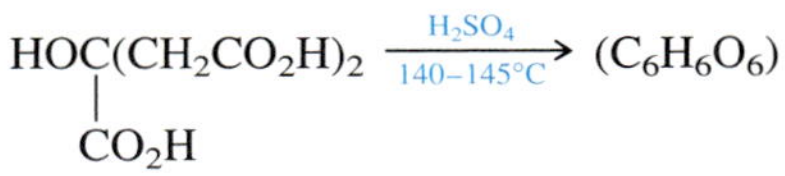 $\text{HOC(CH}_2\text{CO}_2\text{H})_2 \xrightarrow[\text{140–145°C}]{\text{H}_2\text{SO}_4} (\text{C}_6\text{H}_6\text{O}_6)$

   $\overset{|}{\text{CO}_2\text{H}}$

   Citric acid

(g)

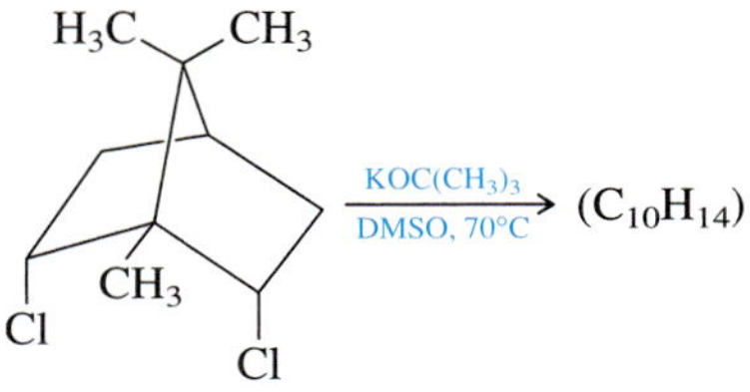

$$\xrightarrow[\text{DMSO, 70°C}]{\text{KOC(CH}_3)_3} (\text{C}_{10}\text{H}_{14})$$

(h) Br

$$\text{Br} \xrightarrow[\text{DMSO}]{\text{KOC(CH}_3)_3} (\text{C}_{14}\text{H}_{16}\text{O}_4)$$

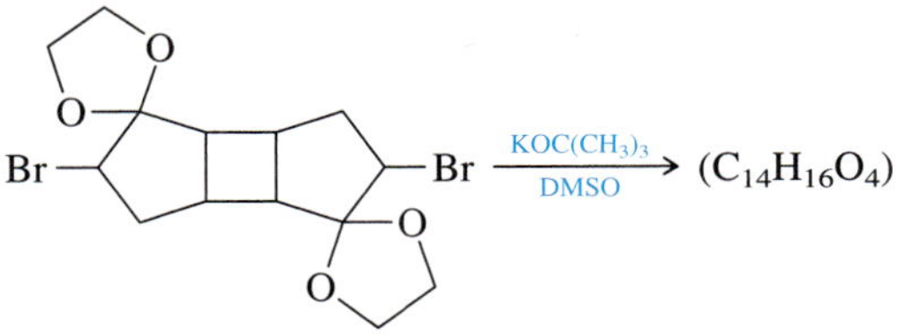

(i)

$$\xrightarrow[\text{heat}]{\text{KOH}} (\text{C}_{10}\text{H}_{18}\text{O}_5)$$

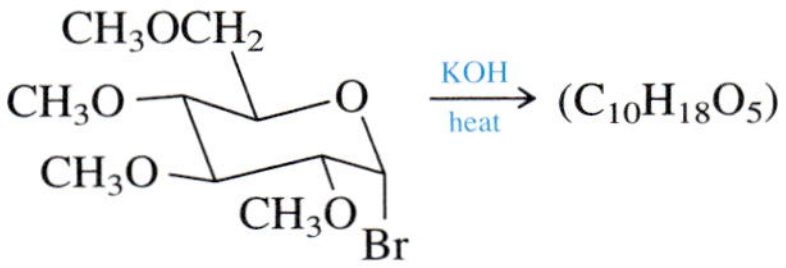

(j)

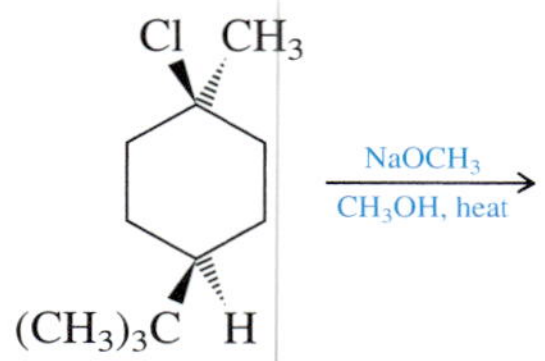

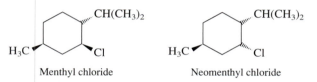

5.38 Evidence has been reported in the chemical literature that the reaction

$$(CH_3CH_2)_2CHCH_2Br + KNH_2 \rightarrow (CH_3CH_2)_2C{=}CH_2 + NH_3 + KBr$$

proceeds by the E2 mechanism. Use curved arrow notation to represent the flow of electrons for this process.

5.39 The rate of the reaction

$$(CH_3)_3CCl + NaSCH_2CH_3 \rightarrow (CH_3)_2C{=}CH_2 + CH_3CH_2SH + NaCl$$

is first-order in $(CH_3)_3CCl$ and first-order in $NaSCH_2CH_3$. Give the symbol (E1 or E2) for the most reasonable mechanism, and use curved arrow notation to represent the flow of electrons.

5.40 Menthyl chloride and neomenthyl chloride have the structures shown. One of these stereoisomers undergoes elimination on treatment with sodium ethoxide in ethanol much more readily than the other. Which reacts faster, menthyl chloride or neomenthyl chloride? Why? (Molecular models will help here.)

Menthyl chloride   Neomenthyl chloride

5.41 The stereoselectivity of elimination of 5-bromononane on treatment with potassium ethoxide was described in Section 5.14. Draw Newman projections or make molecular models of 5-bromononane showing the conformations that lead to *cis*-4-nonene and *trans*-4-nonene, respectively. Identify the proton that is lost in each case, and suggest a mechanistic explanation for the observed stereoselectivity.

5.42 In the acid-catalyzed dehydration of 2-methyl-1-propanol, what carbocation would be formed if a hydride shift accompanied cleavage of the carbon–oxygen bond in the alkyloxonium ion? What ion would be formed as a result of a methyl shift? Which pathway do you think will predominate, a hydride shift or a methyl shift?

5.43 Each of the following carbocations has the potential to rearrange to a more stable one. Write the structure of the rearranged carbocation.

(a) $CH_3CH_2CH_2{}^+$

(b) $(CH_3)_2CH\overset{+}{C}HCH_3$

(c) $(CH_3)_3C\overset{+}{C}HCH_3$

(d) $(CH_3CH_2)_3CCH_2{}^+$

(e) 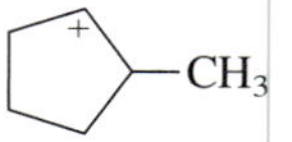 (cyclopentyl cation)$-CH_3$

5.44 Write a sequence of steps depicting the mechanisms of each of the following reactions:

(a)

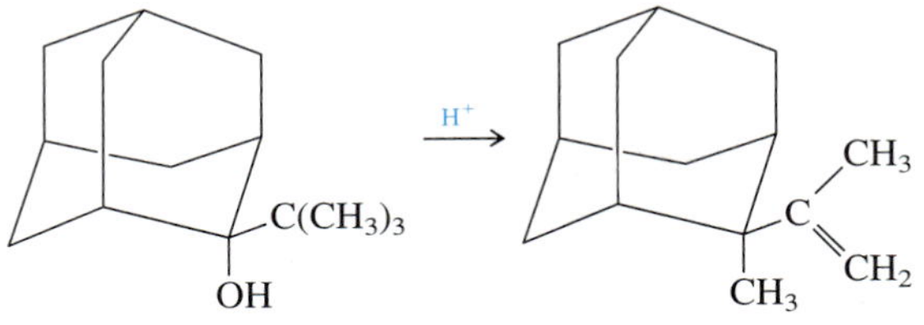

(b)

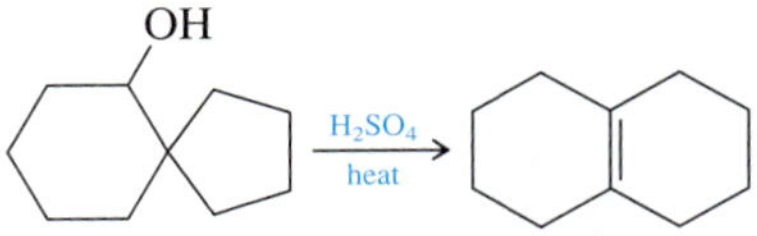

(c)

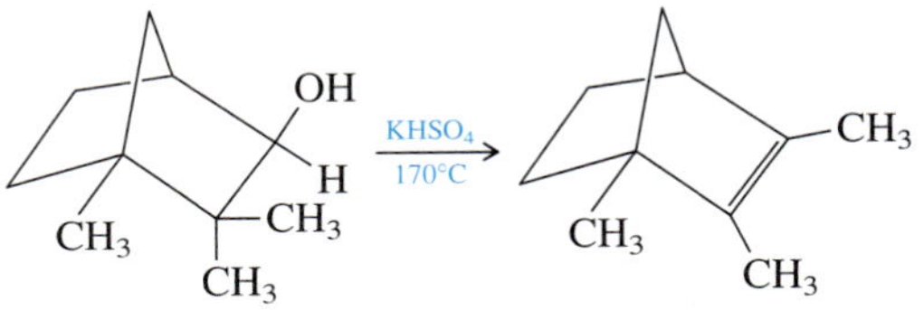

5.45 In Problem 5.17 (Section 5.13) we saw that acid-catalyzed dehydration of 2,2-dimethyl-cyclohexanol afforded 1,2-dimethylcyclohexene. To explain this product we must write a mechanism for the reaction in which a methyl shift transforms a secondary carbocation to a tertiary one. Another product of the dehydration of 2,2-dimethylcyclohexanol is isopropylidenecyclopentane. Write a mechanism to rationalize its formation.

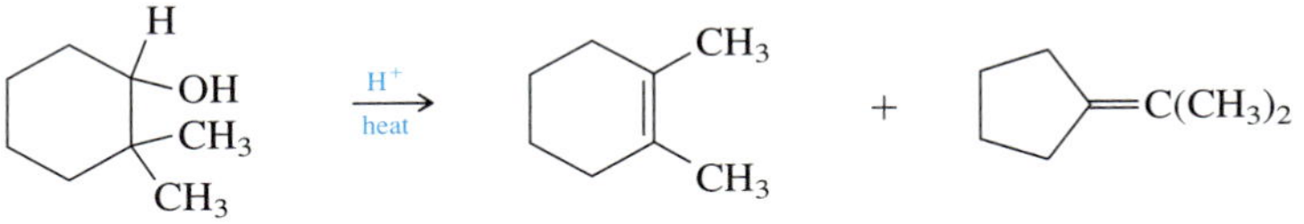

2,2-Dimethylcyclohexanol   1,2-Dimethylcyclohexene   Isopropylidenecyclopentane

5.46 Acid-catalyzed dehydration of 2,2-dimethyl-1-hexanol gave a number of isomeric alkenes including 2-methyl-2-heptene as shown in the following formula.

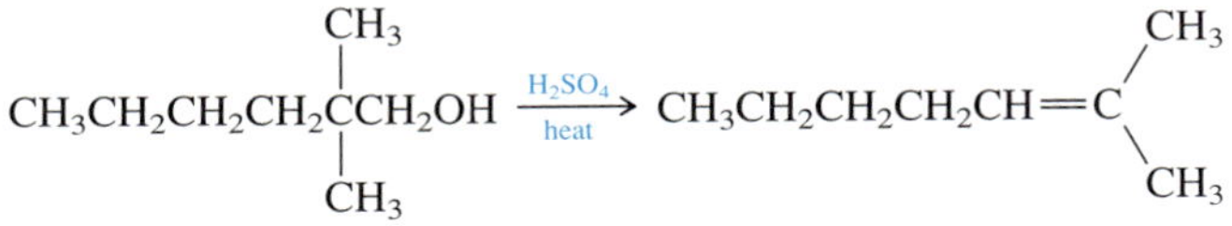

(a) Write a stepwise mechanism for the formation of 2-methyl-2-heptene.

(b) What other alkenes do you think are formed in this reaction?

5.47 Compound A (C_4H_{10}) gives two different monochlorides on photochemical chlorination. Treatment of either of these monochlorides with potassium *tert*-butoxide in dimethyl sulfoxide gives the same alkene B (C_4H_8) as the only product. What are the structures of compound A, the two monochlorides, and alkene B?

5.48 Compound A (C_6H_{14}) gives three different monochlorides on photochemical chlorination. One of these monochlorides is inert to E2 elimination. The other two monochlorides yield the same alkene B (C_6H_{12}) on being heated with potassium *tert*-butoxide in *tert*-butyl alcohol. Identify compound A, the three monochlorides, and alkene B.

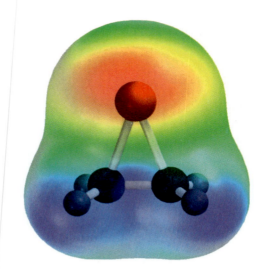

CHAPTER 6

REACTIONS OF ALKENES: ADDITION REACTIONS

N ow that we're familiar with the structure and preparation of alkenes, let's look at their chemical reactions. The characteristic reaction of alkenes is **addition** to the double bond according to the general equation:

$$A-B + \ \underset{\diagup}{\overset{\diagdown}{C}}=\underset{\diagdown}{\overset{\diagup}{C}} \longrightarrow A-\overset{|}{\underset{|}{C}}-\overset{|}{\underset{|}{C}}-B$$

The range of compounds represented as A—B in this equation is quite large, and their variety offers a wealth of opportunity for converting alkenes to a number of other structural types.

Alkenes are commonly described as **unsaturated hydrocarbons** because they have the capacity to react with substances which add to them. Alkanes, on the other hand, are said to be **saturated** hydrocarbons and are incapable of undergoing addition reactions.

6.1 HYDROGENATION OF ALKENES

The relationship between reactants and products in addition reactions can be illustrated by the *hydrogenation* of alkenes to yield alkanes. **Hydrogenation** is the addition of H_2 to a multiple bond. An example is the reaction of hydrogen with ethylene to form ethane.

$$\underset{H}{\overset{H}{\diagdown}}C\underset{\pi}{=}C\underset{\diagdown H}{\overset{\diagup H}{}} + \ H\underset{\sigma}{-}H \ \xrightarrow{\text{Pt, Pd, Ni, or Rh}} \ H-\underset{\underset{H}{\overset{|}{\underset{\sigma}{C}}}}{\overset{H}{\overset{|}{C}}}-\underset{\underset{H}{\overset{|}{\underset{\sigma}{C}}}}{\overset{H}{\overset{|}{C}}}-H \qquad \begin{matrix}\Delta H^\circ = -136 \text{ kJ} \\ (-32.6 \text{ kcal})\end{matrix}$$

Ethylene   Hydrogen   Ethane

The bonds in the product are stronger than the bonds in the reactants; two C—H σ bonds of an alkane are formed at the expense of the H—H σ bond and the π component of the alkene's double bond. The reaction is *exothermic,* and hydrogenation is characterized by a negative sign for $\Delta H°$ for all alkenes. The heat given off on hydrogenation of an alkene is called its **heat of hydrogenation.** Defining them in terms of heat evolved allows heats of hydrogenation (like heats of combustion) to be cited without a sign.

The uncatalyzed addition of hydrogen to an alkene, although exothermic, is very slow. The rate of hydrogenation increases dramatically, however, in the presence of certain finely divided metal catalysts. *Platinum* is the hydrogenation catalyst most often used, although *palladium, nickel,* and *rhodium* are also effective. Metal-catalyzed addition of hydrogen is normally rapid at room temperature, and the alkane is produced in high yield, usually as the only product.

The French chemist Paul Sabatier received the 1912 Nobel Prize in chemistry for his discovery that finely divided nickel is an effective hydrogenation catalyst.

$$(CH_3)_2C{=}CHCH_3 \ + \ H_2 \ \xrightarrow{\text{Pt}} \ (CH_3)_2CHCH_2CH_3$$

2-Methyl-2-butene   Hydrogen   2-Methylbutane (100%)

5,5-Dimethyl(methylene)cyclononane   Hydrogen   1,1,5-Trimethylcyclononane (73%)

PROBLEM 6.1 What three alkenes yield 2-methylbutane on catalytic hydrogenation?

The solvent used in catalytic hydrogenation is chosen for its ability to dissolve the alkene and is typically ethanol, hexane, or acetic acid. The metal catalysts are insoluble in these solvents (or, indeed, in any solvent). Two phases, the solution and the metal, are present, and the reaction takes place at the interface between them. Reactions involving a substance in one phase with a different substance in a second phase are called **heterogeneous reactions.**

Catalytic hydrogenation of an alkene is believed to proceed by the series of steps shown in Figure 6.1. As already noted, addition of hydrogen to the alkene is very slow in the absence of a metal catalyst, meaning that any uncatalyzed mechanism must have a very high activation energy. The metal catalyst accelerates the rate of hydrogenation by providing an alternative pathway that involves a sequence of several low activation energy steps.

6.2 HEATS OF HYDROGENATION

In much the same way as heats of combustion, heats of hydrogenation are used to compare the relative stabilities of alkenes. Both methods measure the differences in the energy of *isomers* by converting them to a product or products common to all. Catalytic hydrogenation of 1-butene, *cis*-2-butene, or *trans*-2-butene yields the same product— butane. As Figure 6.2 shows, the measured heats of hydrogenation reveal that *trans*-2-butene is 4 kJ/mol (1.0 kcal/mol) lower in energy than *cis*-2-butene and that *cis*-2-butene is 7 kJ/mol (1.7 kcal/mol) lower in energy than 1-butene.

Heats of hydrogenation can be used to *estimate* the stability of double bonds as structural units, even in alkenes that are not isomers. Table 6.1 lists the heats of hydrogenation for a representative collection of alkenes.

Remember that a catalyst affects the rate of a reaction but not the energy relationships between reactants and products. Thus, the heat of hydrogenation of a particular alkene is the same irrespective of what catalyst is used.

FIGURE 6.1 A mechanism for heterogeneous catalysis in the hydrogenation of alkenes.

Step 1: Hydrogen molecules react with metal atoms at the catalyst surface. The relatively strong hydrogen–hydrogen σ bond is broken and replaced by two weak metal–hydrogen bonds.

Step 2: The alkene reacts with the metal catalyst. The π component of the double bond between the two carbons is replaced by two relatively weak carbon–metal σ bonds.

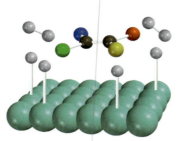

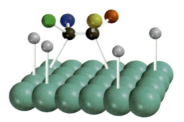

Step 3: A hydrogen atom is transferred from the catalyst surface to one of the carbons of the double bond.

Step 4: The second hydrogen atom is transferred, forming the alkane. The sites on the catalyst surface at which the reaction occurred are free to accept additional hydrogen and alkene molecules.

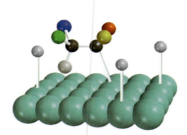

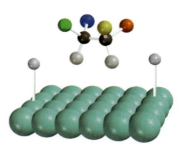

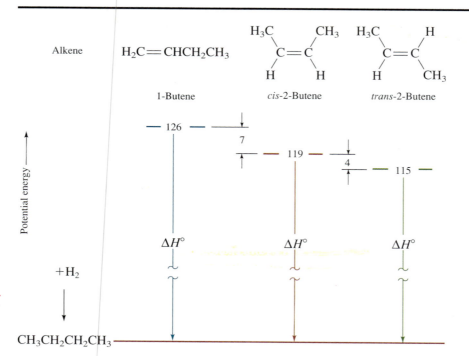

FIGURE 6.2 Heats of hydrogenation of butene isomers. All energies are in kilojoules per mole.

TABLE 6.1	Heats of Hydrogenation of Some Alkenes		
		Heat of hydrogenation	
Alkene	**Structure**	**kJ/mol**	**kcal/mol**
Ethylene	$H_2C{=}CH_2$	136	32.6
Monosubstituted alkenes			
Propene	$H_2C{=}CHCH_3$	125	29.9
1-Butene	$H_2C{=}CHCH_2CH_3$	126	30.1
1-Hexene	$H_2C{=}CHCH_2CH_2CH_2CH_3$	126	30.2
Cis-disubstituted alkenes			
cis-2-Butene	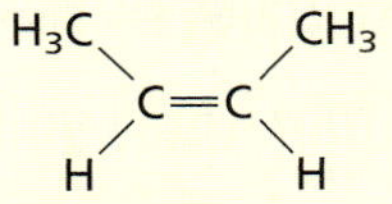 $\begin{array}{c}H_3C \quad\quad CH_3 \\ \diagdown C{=}C \diagup \\ \diagup \quad\quad\quad \diagdown \\ H \quad\quad\quad H\end{array}$	119	28.4
cis-2-Pentene	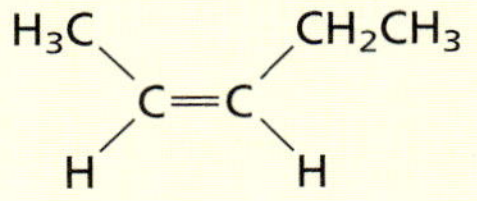 $\begin{array}{c}H_3C \quad\quad CH_2CH_3 \\ \diagdown C{=}C \diagup \\ \diagup \quad\quad\quad \diagdown \\ H \quad\quad\quad H\end{array}$	117	28.1
Trans-disubstituted alkenes			
trans-2-Butene	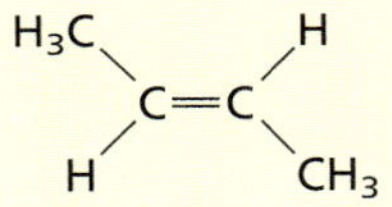 $\begin{array}{c}H_3C \quad\quad H \\ \diagdown C{=}C \diagup \\ \diagup \quad\quad\quad \diagdown \\ H \quad\quad\quad CH_3\end{array}$	115	27.4
trans-2-Pentene	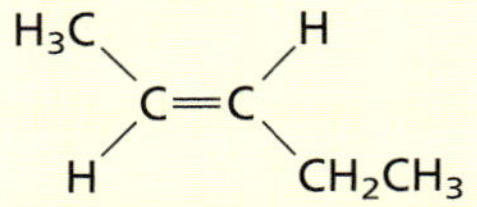 $\begin{array}{c}H_3C \quad\quad H \\ \diagdown C{=}C \diagup \\ \diagup \quad\quad\quad \diagdown \\ H \quad\quad\quad CH_2CH_3\end{array}$	114	27.2
Trisubstituted alkenes			
2-Methyl-2-pentene	$(CH_3)_2C{=}CHCH_2CH_3$	112	26.7
Tetrasubstituted alkenes			
2,3-Dimethyl-2-butene	$(CH_3)_2C{=}C(CH_3)_2$	110	26.4

The pattern of alkene stability determined from heats of hydrogenation parallels exactly the pattern deduced from heats of combustion.

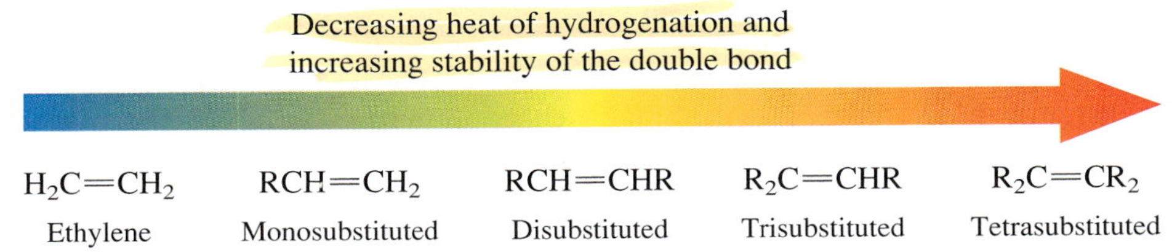

Decreasing heat of hydrogenation and
increasing stability of the double bond

$H_2C{=}CH_2$	$RCH{=}CH_2$	$RCH{=}CHR$	$R_2C{=}CHR$	$R_2C{=}CR_2$
Ethylene	Monosubstituted	Disubstituted	Trisubstituted	Tetrasubstituted

Ethylene, which has no alkyl substituents to stabilize its double bond, has the highest heat of hydrogenation. Alkenes that are similar in structure to one another have similar heats of hydrogenation. For example, the heats of hydrogenation of the monosubstituted (terminal) alkenes propene, 1-butene, and 1-hexene are almost identical. Cis-disubstituted alkenes have lower heats of hydrogenation than monosubstituted alkenes but higher heats of hydrogenation than their more stable trans stereoisomers. Alkenes with trisubstituted double bonds have lower heats of hydrogenation than disubstituted alkenes, and tetra-substituted alkenes have the lowest heats of hydrogenation.

> **PROBLEM 6.2** Match each alkene of Problem 6.1 with its correct heat of hydro-genation.
> *Heats of hydrogenation in kJ/mol (kcal/mol):* 112 (26.7); 118 (28.2); 126 (30.2)

6.3   STEREOCHEMISTRY OF ALKENE HYDROGENATION

In the mechanism for alkene hydrogenation shown in Figure 6.1, hydrogen atoms are transferred from the catalyst's surface to the alkene. Although the two hydrogens are not transferred simultaneously, they both add to the same face of the double bond.

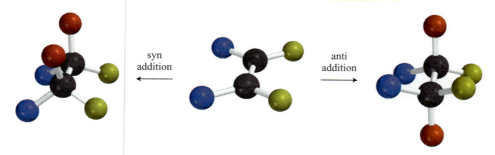

Dimethyl cyclohexene-1,2-dicarboxylate

Dimethyl cyclohexane-*cis*-1,2-dicarboxylate (100%)

The term **syn addition** describes the stereochemistry of reactions such as this in which two atoms or groups add to the *same face* of a double bond. When atoms or groups add to *opposite faces* of the double bond, the process is called **anti addition.**

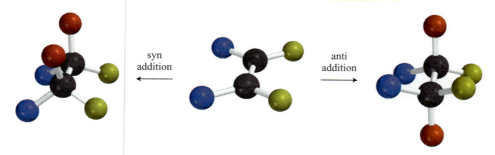

syn addition

anti addition

A second stereochemical aspect of alkene hydrogenation concerns its **stereoselectivity.** A reaction in which a single starting material can give two or more stereoisomeric products but yields one of them in greater amounts than the other (or even to the exclusion of the other) is said to be **stereoselective.** The catalytic hydrogenation of α-pinene (a constituent of turpentine) is an example of a stereoselective reaction. Syn addition of hydrogen can in principle lead to either *cis*-pinane or *trans*-pinane, depending on which face of the double bond accepts the hydrogen atoms (shown in red in the equation).

Stereoselectivity was defined and introduced in connec-tion with the formation of stereoisomeric alkenes in elimination reactions (Sec-tion 5.11).

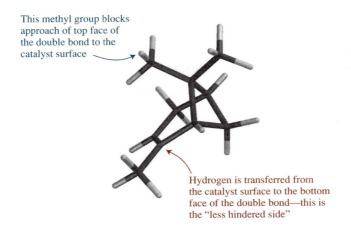

This methyl group blocks approach of top face of the double bond to the catalyst surface

Hydrogen is transferred from the catalyst surface to the bottom face of the double bond—this is the "less hindered side"

FIGURE 6.3 The methyl group that lies over the double bond of α-pinene shields one face of it, preventing a close approach to the surface of the catalyst. Hydrogenation of α-pinene occurs preferentially from the bottom face of the double bond.

α-Pinene

$\xrightarrow[\text{Ni}]{H_2}$

cis-Pinane
(only product)

trans-Pinane
(not formed)

cis-Pinane and trans-pinane are common names that denote the relationship between the pair of methyl groups on the bridge and the third methyl group.

In practice, hydrogenation of α-pinene is observed to be 100% stereoselective. The only product obtained is cis-pinane. No trans-pinane is formed.

The stereoselectivity of this reaction depends on how the alkene approaches the catalyst surface. As the molecular model in Figure 6.3 shows, one of the methyl groups on the bridge carbon lies directly over the double bond and blocks that face from easy access to the catalyst. The bottom face of the double bond is more exposed, and both hydrogens are transferred from the catalyst surface to that face.

Reactions such as catalytic hydrogenation that take place at the "less hindered" side of a reactant are common in organic chemistry and are examples of steric effects on *reactivity*. Previously we saw steric effects on *structure* and *stability* in the case of cis and trans stereoisomers and in the preference for equatorial substituents on cyclohexane rings.

6.4 ELECTROPHILIC ADDITION OF HYDROGEN HALIDES TO ALKENES

In many addition reactions the attacking reagent, unlike H_2, is a polar molecule. Hydrogen halides are among the simplest examples of polar substances that add to alkenes.

$$\overset{\diagup}{\underset{\diagdown}{C}}=\overset{\diagup}{\underset{\diagdown}{C}} + {}^{\delta+}H—X^{\delta-} \longrightarrow H—\overset{|}{\underset{|}{C}}—\overset{|}{\underset{|}{C}}—X$$

Alkene   Hydrogen halide   Alkyl halide

Addition occurs rapidly in a variety of solvents, including pentane, benzene, dichloromethane, chloroform, and acetic acid.

$$\underset{\substack{\text{cis-3-Hexene}}}{\underset{\text{H}}{\overset{\text{CH}_3\text{CH}_2}{\diagdown}}\text{C}=\text{C}\underset{\text{H}}{\overset{\text{CH}_2\text{CH}_3}{\diagup}}} + \underset{\text{Hydrogen bromide}}{\text{HBr}} \xrightarrow[\text{CHCl}_3]{-30°\text{C}} \underset{\text{3-Bromohexane (76\%)}}{\text{CH}_3\text{CH}_2\text{CH}_2\underset{\underset{\text{Br}}{|}}{\text{CH}}\text{CH}_2\text{CH}_3}$$

The reactivity of the hydrogen halides reflects their ability to donate a proton. Hydrogen iodide is the strongest acid of the hydrogen halides and reacts with alkenes at the fastest rate.

Increasing reactivity of hydrogen halides
in addition to alkenes

$$\text{HF} \ll \text{HCl} < \text{HBr} < \text{HI}$$

Slowest rate of addition;   Fastest rate of addition;
      least acidic   most acidic

We can gain a general understanding of the mechanism of hydrogen halide addition to alkenes by extending some of the principles of reaction mechanisms introduced earlier. In Section 5.12 we pointed out that carbocations are the conjugate acids of alkenes. Therefore, strong acids such as HCl, HBr, and HI can protonate the double bond of an alkene to form a carbocation.

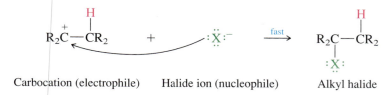

$$\underset{\substack{\text{Alkene}\\\text{(base)}}}{\text{R}_2\text{C}=\text{CR}_2} + \underset{\substack{\text{Hydrogen halide}\\\text{(acid)}}}{\text{H}-\ddot{\text{X}}:} \xrightleftharpoons{\text{slow}} \underset{\substack{\text{Carbocation}\\\text{(conjugate acid)}}}{\overset{\overset{\text{H}}{|}}{\text{R}_2\overset{+}{\text{C}}-\text{CR}_2}} + \underset{\substack{\text{Anion}\\\text{(conjugate base)}}}{:\ddot{\text{X}}:^-}$$

Figure 6.4 shows the complementary nature of the electrostatic potentials of an alkene and a hydrogen halide. We also know (from Section 4.8) that carbocations, when generated in the presence of halide anions, react with them to form alkyl halides.

$$\underset{\substack{\text{Carbocation (electrophile)}}}{\overset{\overset{\text{H}}{|}}{\text{R}_2\overset{+}{\text{C}}-\text{CR}_2}} + \underset{\substack{\text{Halide ion (nucleophile)}}}{:\ddot{\text{X}}:^-} \xrightarrow{\text{fast}} \underset{\substack{\text{Alkyl halide}}}{\overset{\overset{\text{H}}{|}}{\text{R}_2\text{C}-\underset{\underset{:\ddot{\text{X}}:}{|}}{\text{CR}_2}}}$$

Both steps in this general mechanism are based on precedent. It is called **electrophilic addition** because the reaction is triggered by the attack of an acid acting as an electrophile on the π electrons of the double bond. Using the two π electrons to form a bond to an electrophile generates a carbocation as a reactive intermediate; normally this is the rate-determining step.

6.5   REGIOSELECTIVITY OF HYDROGEN HALIDE ADDITION: MARKOVNIKOV'S RULE

In principle a hydrogen halide can add to an unsymmetrical alkene (an alkene in which the two carbons of the double bond are not equivalently substituted) in either of two directions. In practice, addition is so highly regioselective as to be considered regiospecific.

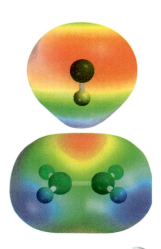

FIGURE 6.4 Electrostatic potential maps of HCl and ethylene. When the two react, the interaction is between the electron-rich site (red) of ethylene and electron-poor region (blue) of HCl. The electron-rich region of ethylene is associated with the π electrons of the double bond, and H is the electron-poor atom of HCl.

$$RCH{=}CH_2 + H{-}X \longrightarrow \underset{\underset{X}{|}\;\;\underset{H}{|}}{RCH{-}CH_2} \quad \text{rather than} \quad \underset{\underset{H}{|}\;\;\underset{X}{|}}{RCH{-}CH_2}$$

$$R_2C{=}CH_2 + H{-}X \longrightarrow \underset{\underset{X}{|}\;\;\underset{H}{|}}{R_2C{-}CH_2} \quad \text{rather than} \quad \underset{\underset{H}{|}\;\;\underset{X}{|}}{R_2C{-}CH_2}$$

$$R_2C{=}CHR + H{-}X \longrightarrow \underset{\underset{X}{|}\;\;\underset{H}{|}}{R_2C{-}CHR} \quad \text{rather than} \quad \underset{\underset{H}{|}\;\;\underset{X}{|}}{R_2C{-}CHR}$$

In 1870, Vladimir Markovnikov, a colleague of Alexander Zaitsev at the University of Kazan, noticed a pattern in the hydrogen halide addition to alkenes and organized his observations into a simple statement. **Markovnikov's rule** states that *when an unsymmetrically substituted alkene reacts with a hydrogen halide, the hydrogen adds to the carbon that has the greater number of hydrogens, and the halogen adds to the carbon having fewer hydrogens.* The preceding general equations illustrate regioselective addition according to Markovnikov's rule, and the equations that follow provide some examples.

An article in the December 1988 issue of the *Journal of Chemical Education* traces the historical development of Markovnikov's rule. In that article Markovnikov's name is spelled *Markownikoff,* which is the way it appeared in his original paper written in German.

$$\underset{\text{1-Butene}}{CH_3CH_2CH{=}CH_2} + \underset{\text{Hydrogen bromide}}{HBr} \xrightarrow{\text{acetic acid}} \underset{\text{2-Bromobutane (80\%)}}{\underset{\underset{Br}{|}}{CH_3CH_2CHCH_3}}$$

$$\underset{\text{2-Methylpropene}}{\overset{H_3C}{\underset{H_3C}{>}}C{=}CH_2} + \underset{\text{Hydrogen bromide}}{HBr} \xrightarrow{\text{acetic acid}} \underset{\text{2-Bromo-2-methylpropane (90\%)}}{\overset{CH_3}{\underset{CH_3}{|}}{H_3C{-}C{-}Br}}$$

$$\underset{\text{1-Methylcyclopentene}}{\text{(cyclopentene)}{-}CH_3} + \underset{\text{Hydrogen chloride}}{HCl} \xrightarrow{0°C} \underset{\text{1-Chloro-1-methylcyclopentane (100\%)}}{\text{(cyclopentane)}\overset{CH_3}{\underset{Cl}{<}}}$$

PROBLEM 6.3 Write the structure of the major organic product formed in the reaction of hydrogen chloride with each of the following:

(a) 2-Methyl-2-butene

(b) 2-Methyl-1-butene

(c) *cis*-2-Butene

(d) $CH_3CH{=}$(cyclohexane)

SAMPLE SOLUTION (a) Hydrogen chloride adds to the double bond of 2-methyl-2-butene in accordance with Markovnikov's rule. The proton adds to the carbon that has one attached hydrogen, chlorine to the carbon that has none.

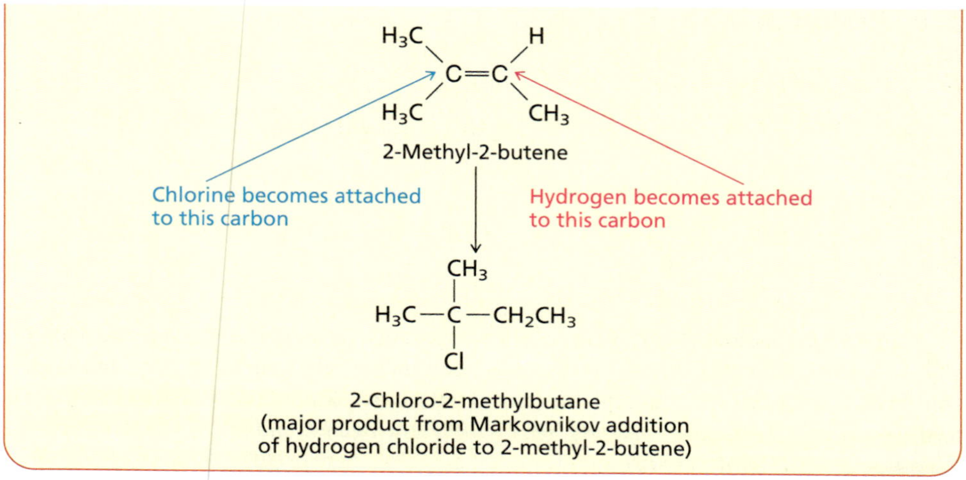

$$H_3C\diagdown\diagup H$$
$$C\!=\!C$$
$$H_3C\diagup\diagdown CH_3$$

2-Methyl-2-butene

Chlorine becomes attached
to this carbon

Hydrogen becomes attached
to this carbon

$$\begin{array}{c} CH_3 \\ | \\ H_3C\!-\!C\!-\!CH_2CH_3 \\ | \\ Cl \end{array}$$

2-Chloro-2-methylbutane
(major product from Markovnikov addition
of hydrogen chloride to 2-methyl-2-butene)

Markovnikov's rule, like Zaitsev's, organizes experimental observations in a form suitable for predicting the major product of a reaction. The reasons why it works will appear when we examine the mechanism of electrophilic addition in more detail.

6.6   MECHANISTIC BASIS FOR MARKOVNIKOV'S RULE

Let's compare the carbocation intermediates for addition of a hydrogen halide (HX) to an unsymmetrical alkene of the type $RCH\!=\!CH_2$ (a) according to Markovnikov's rule and (b) opposite to Markovnikov's rule.

(a) *Addition according to Markovnikov's rule:*

$$RCH\!=\!CH_2 \longrightarrow \overset{+}{RCH}\!-\!CH_2 + :\!\overset{..}{\underset{..}{X}}\!:^- \longrightarrow RCHCH_3$$
$$H\!-\!\overset{..}{\underset{..}{X}}\!: \qquad\qquad\quad | \qquad\qquad\qquad\quad | $$
$$H \qquad\qquad\qquad\quad :\!\overset{..}{\underset{..}{X}}\!:$$

Secondary   Halide   Observed product
carbocation   ion

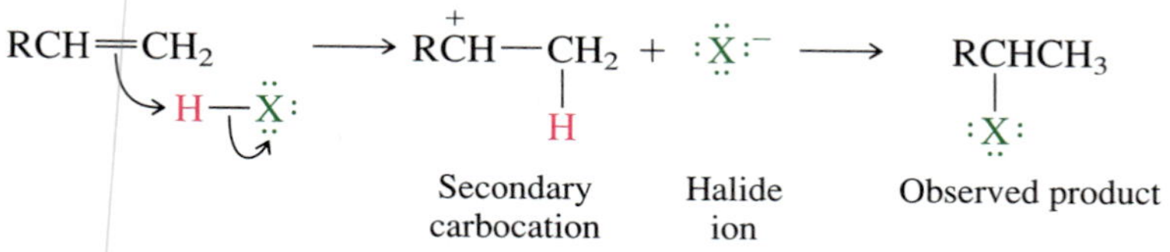

(b) *Addition opposite to Markovnikov's rule:*

$$RCH\!=\!CH_2 \longrightarrow RCH\!-\!\overset{+}{CH_2} + :\!\overset{..}{\underset{..}{X}}\!:^- \longrightarrow RCH_2CH_2\!-\!\overset{..}{\underset{..}{X}}\!:$$
$$:\!\overset{..}{\underset{..}{X}}\!-\!H \qquad\qquad\quad | $$
$$H$$

Primary   Halide   Not formed
carbocation   ion

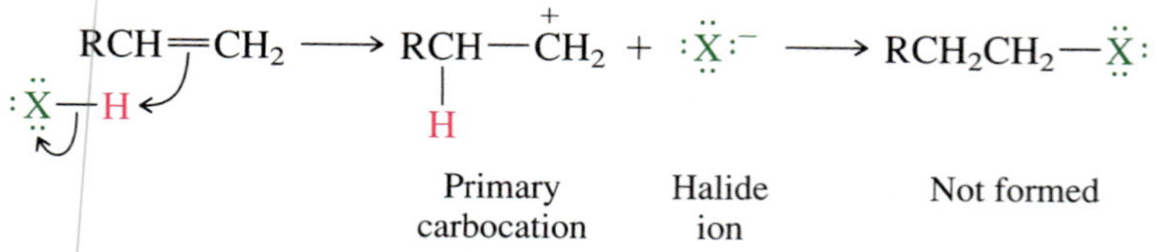

The transition state for protonation of the double bond has much of the character of a carbocation, and the activation energy for formation of the more stable carbocation (secondary) is less than that for formation of the less stable (primary) one. Figure 6.5 uses a potential energy diagram to illustrate these two competing modes of addition. Both carbocations are rapidly captured by X^- to give an alkyl halide, with the major product derived from the carbocation that is formed faster. The energy difference between a primary carbocation and a secondary carbocation is so great and their rates of formation are so different that essentially all the product is derived from the secondary carbocation.

RULES, LAWS, THEORIES, AND THE SCIENTIFIC METHOD

As we have just seen, Markovnikov's rule can be expressed in two ways:

1. When a hydrogen halide adds to an alkene, hydrogen adds to the carbon of the alkene that has the greater number of hydrogens attached to it, and the halogen to the carbon that has the fewer hydrogens.

2. When a hydrogen halide adds to an alkene, protonation of the double bond occurs in the direction that gives the more stable carbocation.

The first of these statements is close to the way Vladimir Markovnikov expressed it in 1870; the second is the way we usually phrase it now. These two statements differ in an important way—a way that is related to the *scientific method.*

Adherence to the scientific method is what defines science. The scientific method has four major elements: observation, law, theory, and hypothesis.

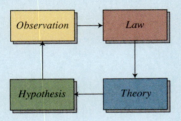

Most *observations* in chemistry come from experiments. If we do enough experiments we may see a pattern running through our observations. A *law* is a mathematical (the law of gravity) or verbal (the law of diminishing returns) description of that pattern. Establishing a law can lead to the framing of a *rule* that lets us predict the results of future experiments. This is what the 1870 version of Markovnikov's rule is: a statement based on experimental observations that has predictive value.

A *theory* is our best present interpretation of why things happen the way they do. The modern version of Markovnikov's rule, which is based on mechanistic reasoning and carbocation stability, recasts the rule in terms of theoretical ideas. Mechanisms, and explanations grounded in them, belong to the theory part of the scientific method.

It is worth remembering that a theory can never be proven correct. It can only be proven incorrect, incomplete, or inadequate. Thus, theories are always being tested and refined. As important as anything else in the scientific method is the *testable hypothesis.* Once a theory is proposed, experiments are designed to test its validity. If the results are consistent with the theory, our belief in its soundness is strengthened. If the results conflict with it, the theory is flawed and must be modified. Section 6.7 describes some observations that support the theory that carbocations are intermediates in the addition of hydrogen halides to alkenes.

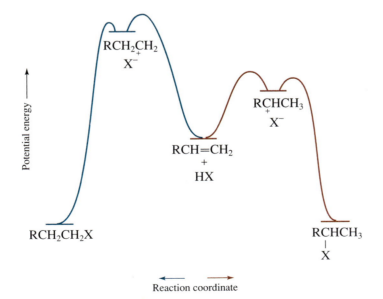

FIGURE 6.5 Energy diagrams comparing addition of a hydrogen halide to an alkene according to Markovnikov's rule with addition in the direction opposite to Markovnikov's rule. The alkene and hydrogen halide are shown in the center of the diagram. The lower energy pathway that corresponds to Markovnikov's rule proceeds to the right and is shown in red; the higher energy pathway proceeds to the left and is shown in blue.

FIGURE 6.6 Electron flow and orbital interactions in the transfer of a proton from a hydrogen halide to an alkene of the type $H_2C=CHR$.

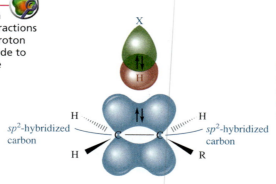

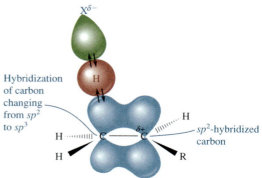

(*a*) The hydrogen halide (HX) and the alkene ($H_2C=CHR$) approach each other. The electrophile is the hydrogen halide, and the site of electrophilic attack is the orbital containing the π electrons of the double bond.

(*b*) Electrons flow from the π orbital of the alkene to the hydrogen halide. The π electrons flow in the direction that generates a partial positive charge on the carbon atom that bears the electron-releasing alkyl group (R). The hydrogen–halogen bond is partially broken and a C—H σ bond is partially formed at the transition state.

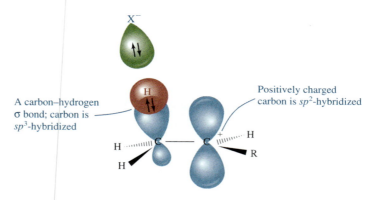

(*c*) Loss of the halide ion (X^-) from the hydrogen halide and C—H σ bond formation complete the formation of the more stable carbocation intermediate $CH_3\overset{+}{C}HR$.

Figure 6.6 focuses on the orbitals involved and shows how the π electrons of the double bond flow in the direction that generates the more stable of the two possible carbocations.

PROBLEM 6.4 Give a structural formula for the carbocation intermediate that leads to the major product in each of the reactions of Problem 6.3.

SAMPLE SOLUTION (a) Protonation of the double bond of 2-methyl-2-butene can give a tertiary carbocation or a secondary carbocation.

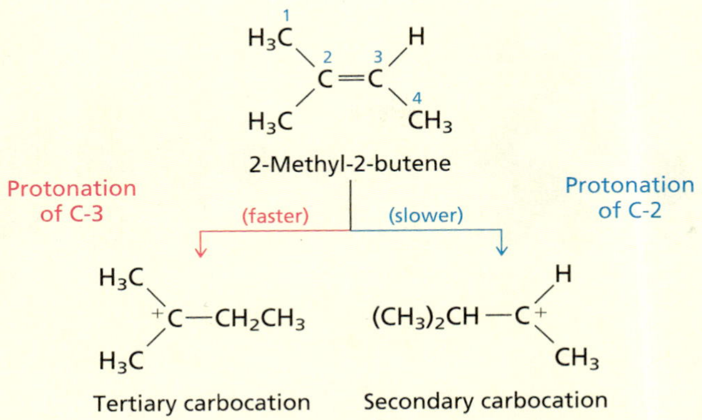

$$H_2C = \overset{1}{C}\overset{2}{H_3C}$$

2-Methyl-2-butene

Protonation of C-3   (faster) | (slower)   Protonation of C-2

$$\overset{H_3C}{\underset{H_3C}{>}}\overset{+}{C}-CH_2CH_3 \qquad (CH_3)_2CH-\overset{H}{\underset{CH_3}{\overset{+}{C}}}$$

Tertiary carbocation   Secondary carbocation

The product of the reaction is derived from the more stable carbocation—in this case, it is a tertiary carbocation that is formed more rapidly than a secondary one.

In general, alkyl substituents increase the reactivity of a double bond toward electrophilic addition. Alkyl groups are electron-releasing, and the more *electron-rich* a double bond, the better it can share its π electrons with an electrophile. Along with the observed regioselectivity of addition, this supports the idea that carbocation formation, rather than carbocation capture, is rate-determining.

6.7 CARBOCATION REARRANGEMENTS IN HYDROGEN HALIDE ADDITION TO ALKENES

Our belief that carbocations are intermediates in the addition of hydrogen halides to alkenes is strengthened by the fact that rearrangements sometimes occur. For example, the reaction of hydrogen chloride with 3-methyl-1-butene is expected to produce 2-chloro-3-methylbutane. Instead, a mixture of 2-chloro-3-methylbutane and 2-chloro-2-methylbutane results.

$$H_2C = CHCH(CH_3)_2 \xrightarrow[0°C]{HCl} \underset{\underset{Cl}{|}}{CH_3CHCH(CH_3)_2} + \underset{\underset{Cl}{|}}{CH_3CH_2C(CH_3)_2}$$

3-Methyl-1-butene   2-Chloro-3-methylbutane (40%)   2-Chloro-2-methylbutane (60%)

Addition begins in the usual way, by protonation of the double bond to give, in this case, a secondary carbocation. This carbocation can be captured by chloride to give 2-chloro-3-methylbutane (40%) or it can rearrange by way of a hydride shift to give a tertiary carbocation. The tertiary carbocation reacts with chloride ion to give 2-chloro-2-methylbutane (60%).

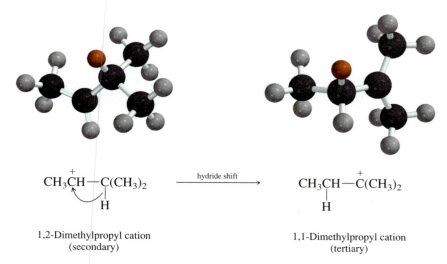

$$\underset{\substack{\text{1,2-Dimethylpropyl cation}\\(\text{secondary})}}{\overset{+}{CH_3CH} \!-\! \underset{H}{\overset{|}{C}}(CH_3)_2} \quad \xrightarrow{\text{hydride shift}} \quad \underset{\substack{\text{1,1-Dimethylpropyl cation}\\(\text{tertiary})}}{CH_3CH \!-\! \underset{H}{\overset{|}{\overset{+}{C}}}(CH_3)_2}$$

The similar yields of the two alkyl chloride products indicate that the rate of attack by chloride on the secondary carbocation and the rate of rearrangement must be very similar.

> **PROBLEM 6.5**   Addition of hydrogen chloride to 3,3-dimethyl-1-butene gives a mixture of two isomeric chlorides in approximately equal amounts. Suggest reasonable structures for these two compounds, and offer a mechanistic explanation for their formation.

6.8   FREE-RADICAL ADDITION OF HYDROGEN BROMIDE TO ALKENES

Addition opposite to Markovnikov's rule is sometimes termed "anti-Markovnikov addition."

For a long time the regioselectivity of addition of hydrogen bromide to alkenes was unpredictable. Sometimes addition occurred according to Markovnikov's rule, but at other times, seemingly under the same conditions, it occurred opposite to Markovnikov's rule. In 1929, Morris S. Kharasch and his students at the University of Chicago began a systematic investigation of this puzzle. After hundreds of experiments, Kharasch concluded that addition occurred opposite to Markovnikov's rule when peroxides, that is, organic compounds of the type ROOR, were present in the reaction mixture. He and his colleagues found, for example, that carefully purified 1-butene reacted with hydrogen bromide to give only 2-bromobutane—the product expected on the basis of Markovnikov's rule.

$$\underset{\text{1-Butene}}{H_2C\!=\!CHCH_2CH_3} \; + \; \underset{\text{Hydrogen bromide}}{HBr} \quad \xrightarrow[\text{peroxides}]{\text{no}} \quad \underset{\substack{\text{2-Bromobutane}\\(\text{only product; 90\% yield})}}{CH_3\underset{\underset{Br}{|}}{CH}CH_2CH_3}$$

On the other hand, when the same reaction was performed in the presence of an added peroxide, only 1-bromobutane was formed.

$$\underset{\text{1-Butene}}{H_2C\!=\!CHCH_2CH_3} \; + \; \underset{\text{Hydrogen bromide}}{HBr} \quad \xrightarrow{\text{peroxides}} \quad \underset{\substack{\text{1-Bromobutane}\\(\text{only product; 95\% yield})}}{BrCH_2CH_2CH_2CH_3}$$

Kharasch called this **the peroxide effect** and demonstrated that it could occur even if peroxides were not deliberately added to the reaction mixture. Unless alkenes are protected from atmospheric oxygen, they become contaminated with small amounts of alkyl hydroperoxides, compounds of the type ROOH. These alkyl hydroperoxides act in the same way as deliberately added peroxides, promoting addition in the direction opposite to that predicted by Markovnikov's rule.

PROBLEM 6.6 Kharasch's earliest studies in this area were carried out in collaboration with graduate student Frank R. Mayo. Mayo performed over 400 experiments in which allyl bromide (3-bromo-1-propene) was treated with hydrogen bromide under a variety of conditions, and determined the distribution of the "normal" and "abnormal" products formed during the reaction. What two products were formed? Which is the product of addition in accordance with Markovnikov's rule? Which one corresponds to addition opposite to the rule?

Kharasch proposed that hydrogen bromide can add to alkenes by two different mechanisms, both of which are regiospecific. The first mechanism is electrophilic addition and follows Markovnikov's rule.

The second mechanism is the one followed when addition occurs opposite to Markovnikov's rule. Unlike electrophilic addition via a carbocation intermediate, this alternative mechanism is a chain reaction involving free-radical intermediates. It is presented in Figure 6.7.

Peroxides are *initiators;* they are not incorporated into the product but act as a source of radicals necessary to get the chain reaction started. The oxygen–oxygen bond of a peroxide is relatively weak, and the free-radical addition of hydrogen bromide to alkenes begins when a peroxide molecule breaks apart, giving two alkoxy radicals. This is depicted in step 1 of Figure 6.7. A bromine atom is generated in step 2 when one of these alkoxy radicals abstracts a hydrogen atom from hydrogen bromide. Once a bromine atom becomes available, the propagation phase of the chain reaction begins. In the propagation phase as shown in step 3, a bromine atom adds to the alkene in the direction that produces the more stable alkyl radical.

Addition of a bromine atom to C-1 gives a secondary alkyl radical.

$$\overset{4}{C}H_3\overset{3}{C}H_2\overset{2}{C}H{=}\overset{1}{C}H_2 \longrightarrow CH_3CH_2\overset{\cdot}{C}H{-}CH_2$$
$$:\!\overset{\cdot\cdot}{Br}\!:$$
$$:\!\overset{\cdot\cdot}{Br}\!:$$

Secondary alkyl radical

Addition of a bromine atom to C-2 gives a primary alkyl radical.

$$\overset{4}{C}H_3\overset{3}{C}H_2\overset{2}{C}H{=}\overset{1}{C}H_2 \longrightarrow CH_3CH_2\overset{\cdot}{C}H{-}\overset{\cdot}{C}H_2$$
$$:\!\overset{\cdot\cdot}{Br}\!:$$
$$:\!\overset{\cdot\cdot}{Br}\!:$$

Primary alkyl radical

A secondary alkyl radical is more stable than a primary radical. Bromine therefore adds to C-1 of 1-butene faster than it adds to C-2. Once the bromine atom has added to the double bond, the regioselectivity of addition is set. The alkyl radical then abstracts a hydrogen atom from hydrogen bromide to give the alkyl bromide product as shown in

The overall reaction:

$$CH_3CH_2CH=CH_2 \quad + \quad HBr \quad \xrightarrow[\text{light or heat}]{ROOR} \quad CH_3CH_2CH_2CH_2Br$$

1-Butene   Hydrogen bromide   1-Bromobutane

The mechanism:

(*a*) Initiation

Step 1: Dissociation of a peroxide into two alkoxy radicals:

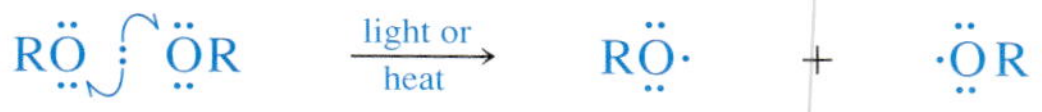

$$R\ddot{O} \overset{\frown}{} \ddot{O}R \quad \xrightarrow[\text{heat}]{\text{light or}} \quad R\ddot{O}\cdot \quad + \quad \cdot\ddot{O}R$$

Peroxide   Two alkoxy radicals

Step 2: Hydrogen atom abstraction from hydrogen bromide by an alkoxy radical:

$$R\ddot{O}\cdot \qquad H \;:\; \ddot{B}r\!: \quad \longrightarrow \quad R\ddot{O}-H \quad + \quad \cdot\ddot{B}r\!:$$

Alkoxy   Hydrogen   Alcohol   Bromine
radical   bromide   atom

(*b*) Chain propagation

Step 3: Addition of a bromine atom to the alkene:

$$CH_3CH_2CH=CH_2 \qquad \cdot\ddot{B}r\!: \quad \longrightarrow \quad CH_3CH_2\overset{\cdot}{C}H-CH_2-\ddot{B}r\!:$$

1-Butene   Bromine atom   1-(Bromomethyl)propyl radical

Step 4: Abstraction of a hydrogen atom from hydrogen bromide by the free radical formed in step 3:

$$CH_3CH_2\overset{\cdot}{C}H-CH_2Br \qquad H \;:\; \ddot{B}r\!: \quad \longrightarrow \quad CH_3CH_2CH_2CH_2Br \quad + \quad \cdot\ddot{B}r\!:$$

1-(Bromomethyl)propyl   Hydrogen   1-Bromobutane   Bromine
radical   bromide   atom

FIGURE 6.7 Initiation and propagation steps in the free-radical addition of hydrogen bromide to 1-butene.

step 4 of Figure 6.7. Steps 3 and 4 propagate the chain, making 1-bromobutane the major product.

The regioselectivity of addition of HBr to alkenes under normal (electrophilic addition) conditions is controlled by the tendency of a *proton* to add to the double bond so as to produce the more stable *carbocation*. Under free-radical conditions the regioselectivity is governed by addition of a *bromine atom* to give the more stable *alkyl radical*.

Free-radical addition of hydrogen bromide to the double bond can also be initiated photochemically, either with or without added peroxides.

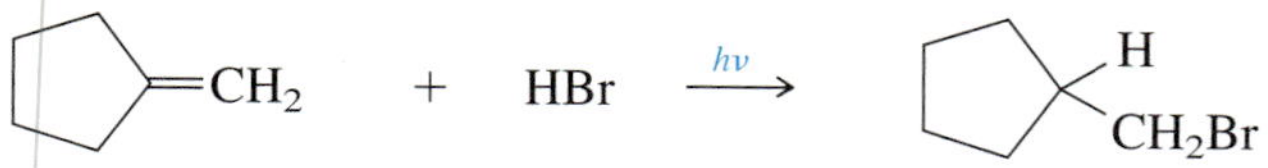

Using an *sp²*-hybridized carbon for the carbon that has the unpaired electron, make a molecular model of the free-radical intermediate in this reaction.

$$\text{(methylenecyclopentane)} =CH_2 \quad + \quad HBr \quad \xrightarrow{h\nu} \quad \text{(cyclopentane)}\overset{H}{\underset{CH_2Br}{\big<}}$$

Methylenecyclopentane   Hydrogen   (Bromomethyl)cyclopentane
                                       bromide   (60%)

Among the hydrogen halides, only hydrogen bromide reacts with alkenes by both electrophilic and free-radical addition mechanisms. Hydrogen iodide and hydrogen chloride always add to alkenes by electrophilic addition and follow Markovnikov's rule. Hydrogen bromide normally reacts by electrophilic addition, but if peroxides are present or if the reaction is initiated photochemically, the free-radical mechanism is followed.

PROBLEM 6.7   Give the major organic product formed when hydrogen bromide reacts with each of the alkenes in Problem 6.3 in the absence of peroxides and in their presence.

SAMPLE SOLUTION   (a) The addition of hydrogen bromide in the absence of peroxides exhibits a regioselectivity just like that of hydrogen chloride addition; Markovnikov's rule is followed.

$$H_3C, H, C=C, H_3C, CH_3 \quad + \quad HBr \quad \xrightarrow{\text{no peroxides}} \quad H_3C-\underset{\underset{Br}{|}}{\overset{\overset{CH_3}{|}}{C}}-CH_2CH_3$$

2-Methyl-2-butene   Hydrogen bromide   2-Bromo-2-methylbutane

Under free-radical conditions in the presence of peroxides, addition takes place with a regioselectivity opposite to that of Markovnikov's rule.

$$H_3C, H, C=C, H_3C, CH_3 \quad + \quad HBr \quad \xrightarrow{\text{peroxides}} \quad H_3C-\underset{\underset{H}{|}}{\overset{\overset{CH_3}{|}}{C}}-\underset{\underset{Br}{|}}{C}HCH_3$$

2-Methyl-2-butene   Hydrogen bromide   2-Bromo-3-methylbutane

Although the possibility of having two different reaction paths available to an alkene and hydrogen bromide may seem like a complication, it can be an advantage in organic synthesis. From a single alkene one may prepare either of two different alkyl bromides, with control of regioselectivity, simply by choosing reaction conditions that favor ionic addition or free-radical addition of hydrogen bromide.

6.9   ADDITION OF SULFURIC ACID TO ALKENES

Acids other than hydrogen halides also add to the carbon–carbon bond of alkenes. Concentrated sulfuric acid, for example, reacts with certain alkenes to form alkyl hydrogen sulfates.

$$C=C \quad + \quad H-OSO_2OH \longrightarrow H-\underset{|}{\overset{|}{C}}-\underset{|}{\overset{|}{C}}-OSO_2OH$$

Alkene   Sulfuric acid   Alkyl hydrogen sulfate

Notice in the following example that a proton adds to the carbon that has the greater number of hydrogens, and the hydrogen sulfate anion ($^-OSO_2OH$) adds to the carbon that has the fewer hydrogens.

$$CH_3CH=CH_2 + HOSO_2OH \longrightarrow CH_3\underset{\underset{OSO_2OH}{|}}{C}HCH_3$$

Propene   Sulfuric acid   Isopropyl hydrogen sulfate

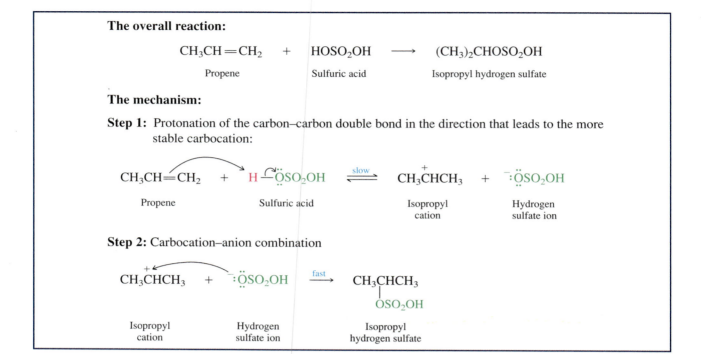

FIGURE 6.8 Mechanism of addition of sulfuric acid to propene.

Markovnikov's rule is obeyed because the mechanism of sulfuric acid addition to alkenes, illustrated for the case of propene in Figure 6.8, is analogous to that described earlier for the electrophilic addition of hydrogen halides.

Alkyl hydrogen sulfates can be converted to alcohols by heating them with water. This is called **hydrolysis,** because a bond is cleaved by reaction with water. It is the oxygen–sulfur bond that is broken when an alkyl hydrogen sulfate undergoes hydrolysis.

Cleavage occurs
here during hydrolysis

$$\underset{\text{Alkyl hydrogen sulfate}}{H-\overset{|}{\underset{|}{C}}-\overset{|}{\underset{|}{C}}-O\!+\!SO_2OH} \;+\; \underset{\text{Water}}{H_2O} \;\xrightarrow{\text{heat}}\; \underset{\text{Alcohol}}{H-\overset{|}{\underset{|}{C}}-\overset{|}{\underset{|}{C}}-OH} \;+\; \underset{\text{Sulfuric acid}}{HOSO_2OH}$$

The combination of sulfuric acid addition to propene, followed by hydrolysis of the resulting isopropyl hydrogen sulfate, is the major method by which over 10^9 lb of isopropyl alcohol is prepared each year in the United States.

$$\underset{\text{Propene}}{CH_3CH\!=\!CH_2} \;\xrightarrow{H_2SO_4}\; \underset{\substack{\text{Isopropyl}\\\text{hydrogen sulfate}}}{\underset{OSO_2OH}{CH_3\overset{|}{\underset{|}{C}}HCH_3}} \;\xrightarrow[\text{heat}]{H_2O}\; \underset{\substack{\text{Isopropyl}\\\text{alcohol}}}{\underset{OH}{CH_3\overset{|}{\underset{|}{C}}HCH_3}}$$

We say that propene has undergone **hydration.** Overall, H and OH have added across the carbon–carbon double bond. In the same manner, cyclohexanol has been prepared by hydration of cyclohexene:

Cyclohexene   Cyclohexanol
                                (75%)

It is convenient in synthetic transformations involving more than one step simply to list all the reagents with a single arrow. Individual synthetic steps are indicated by number. Numbering the individual steps is essential so as to avoid the implication that everything is added to the reaction mixture at the same time.

PROBLEM 6.8 Write a structural formula for the compound formed on electrophilic addition of sulfuric acid to cyclohexene (step 1 in the two-step transformation shown in the preceding equation).

Hydration of alkenes by this method, however, is limited to monosubstituted alkenes and disubstituted alkenes of the type RCH=CHR. Disubstituted alkenes of the type R_2C=CH_2, along with trisubstituted and tetrasubstituted alkenes, do not form alkyl hydrogen sulfates under these conditions but instead react in a more complicated way with concentrated sulfuric acid (to be discussed in Section 6.21).

6.10 ACID-CATALYZED HYDRATION OF ALKENES

Another method for the hydration of alkenes is by reaction with water under conditions of acid catalysis.

Alkene   Water   Alcohol

Unlike the addition of concentrated sulfuric acid to form alkyl hydrogen sulfates, this reaction is carried out in a *dilute acid* medium. A 50% water/sulfuric acid solution is often used, yielding the alcohol directly without the necessity of a separate hydrolysis step. Markovnikov's rule is followed:

2-Methyl-2-butene   2-Methyl-2-butanol (90%)

Page 396 of the March 2000 issue of the *Journal of Chemical Education* outlines some molecular modeling exercises concerning the regioselectivity of alkene hydration.

Methylenecyclobutane   1-Methylcyclobutanol (80%)

We can extend the general principles of electrophilic addition to acid-catalyzed hydration. In the first step of the mechanism shown in Figure 6.9, proton transfer to 2-methylpropene forms *tert*-butyl cation. This is followed in step 2 by reaction of the carbocation with a molecule of water acting as a nucleophile. The alkyloxonium ion formed in this step is simply the conjugate acid of *tert*-butyl alcohol. Deprotonation of the alkyloxonium ion in step 3 yields the alcohol and regenerates the acid catalyst.

The overall reaction:

$$(CH_3)_2C=CH_2 \quad + \quad H_2O \quad \xrightarrow{H_3O^+} \quad (CH_3)_3COH$$

2-Methylpropene   Water   *tert*-Butyl alcohol

The mechanism:

Step 1: Protonation of the carbon–carbon double bond in the direction that leads to the more stable carbocation:

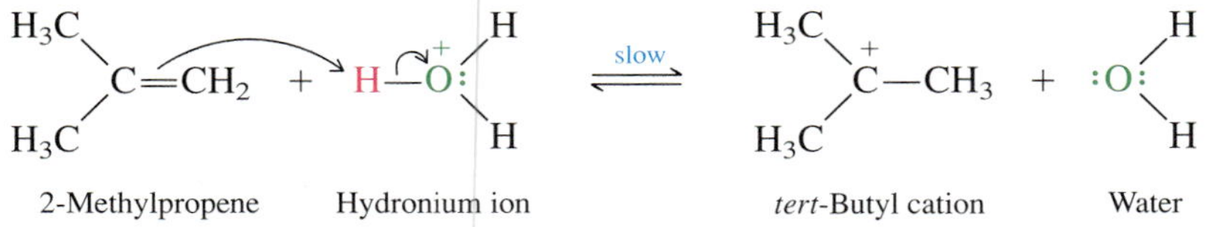

2-Methylpropene   Hydronium ion   *tert*-Butyl cation   Water

Step 2: Water acts as a nucleophile to capture *tert*-butyl cation:

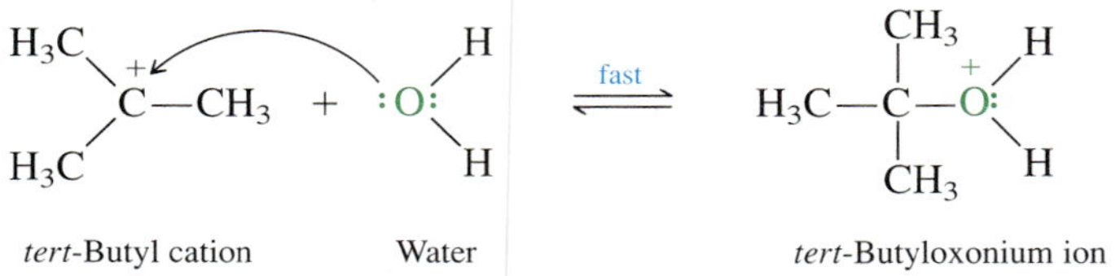

tert-Butyl cation   Water   *tert*-Butyloxonium ion

Step 3: Deprotonation of *tert*-butyloxonium ion. Water acts as a Brønsted base:

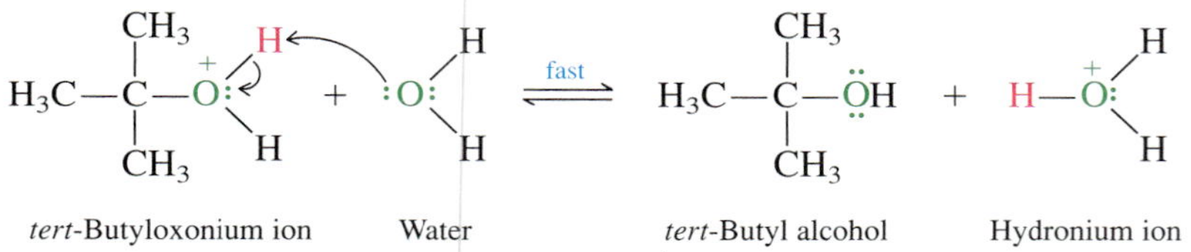

tert-Butyloxonium ion   Water   *tert*-Butyl alcohol   Hydronium ion

FIGURE 6.9 Mechanism of acid-catalyzed hydration of 2-methylpropene.

PROBLEM 6.9 Instead of the three-step mechanism of Figure 6.9, the following two-step mechanism might be considered:

1. $(CH_3)_2C=CH_2 + H_3O^+ \xrightarrow{slow} (CH_3)_3C^+ + H_2O$

2. $(CH_3)_3C^+ + HO^- \xrightarrow{fast} (CH_3)_3COH$

This mechanism cannot be correct! What is its fundamental flaw?

The notion that carbocation formation is rate-determining follows from our previous experience and by observing how the reaction rate is affected by the structure of the alkene. Table 6.2 gives some data showing that alkenes that yield relatively stable carbocations react faster than those that yield less stable carbocations. Protonation of ethylene, the least reactive alkene in the table, yields a primary carbocation; protonation of 2-methylpropene, the most reactive in the table, yields a tertiary carbocation. As we have seen on other occasions, the more stable the carbocation, the faster is its rate of formation.

TABLE 6.2	Relative Rates of Acid-Catalyzed Hydration of Some Representative Alkenes	

Alkene	Structural formula	Relative rate of acid-catalyzed hydration*
Ethylene	$H_2C{=}CH_2$	1.0
Propene	$CH_3CH{=}CH_2$	1.6×10^6
2-Methylpropene	$(CH_3)_2C{=}CH_2$	2.5×10^{11}

*In water, 25°C.

PROBLEM 6.10   The rates of hydration of the two alkenes shown differ by a factor of over 7000 at 25°C. Which isomer is the more reactive? Why?

$$\text{trans-}\triangleright\!\!-CH{=}CHCH_3 \quad \text{and} \quad \triangleright\!\!-\underset{\displaystyle \underset{CH_3}{|}}{C}{=}CH_2$$

   You may have noticed that the acid-catalyzed hydration of an alkene and the acid-catalyzed dehydration of an alcohol are the reverse of each other.

$$\underset{\text{Alkene}}{\overset{\displaystyle \diagup}{\underset{\displaystyle \diagup}{C}}{=}\overset{\displaystyle \diagdown}{\underset{\displaystyle \diagdown}{C}} \; + \; \underset{\text{Water}}{H_2O} \;\; \overset{H^+}{\rightleftharpoons} \;\; \underset{\text{Alcohol}}{H{-}\overset{|}{\underset{|}{C}}{-}\overset{|}{\underset{|}{C}}{-}OH}$$

According to **Le Châtelier's principle,** *a system at equilibrium adjusts so as to minimize any stress applied to it.* When the concentration of water is increased, the system responds by consuming water. This means that proportionally more alkene is converted to alcohol; the position of equilibrium shifts to the right. Thus, when we wish to prepare an alcohol from an alkene, we employ a reaction medium in which the molar concentration of water is high—dilute sulfuric acid, for example.

   On the other hand, alkene formation is favored when the concentration of water is kept low. The system responds to the absence of water by causing more alcohol molecules to suffer dehydration, and when alcohol molecules dehydrate, they form more alkene. The amount of water in the reaction mixture is kept low by using concentrated strong acids as catalysts. Distilling the reaction mixture is an effective way of removing water as it is formed, causing the equilibrium to shift toward products. If the alkene is low-boiling, it too can be removed by distillation. This offers the additional benefit of protecting the alkene from acid-catalyzed isomerization after it is formed.

   In any equilibrium process, the sequence of intermediates and transition states encountered as reactants proceed to products in one direction must also be encountered, and in precisely the reverse order, in the opposite direction. This is called the **principle of microscopic reversibility.** Just as the reaction

$$\underset{\text{2-Methylpropene}}{(CH_3)_2C{=}CH_2} \; + \; \underset{\text{Water}}{H_2O} \;\; \overset{H^+}{\rightleftharpoons} \;\; \underset{\text{2-Methyl-2-propanol}}{(CH_3)_3COH}$$

is reversible with respect to reactants and products, so each tiny increment of progress along the reaction coordinate is reversible. Once we know the mechanism for the forward phase of a particular reaction, we also know what the intermediates and transition states must be for the reverse. In particular, the three-step mechanism for the acid-catalyzed hydration of 2-methylpropene in Figure 6.9 is the reverse of that for the acid-catalyzed dehydration of *tert*-butyl alcohol in Figure 5.6.

> **PROBLEM 6.11** Is the electrophilic addition of hydrogen chloride to 2-methylpropene the reverse of the E1 or the E2 elimination reaction of *tert*-butyl chloride?

6.11 HYDROBORATION–OXIDATION OF ALKENES

Acid-catalyzed hydration converts alkenes to alcohols with regioselectivity according to Markovnikov's rule. Frequently, however, one needs an alcohol having a structure that corresponds to hydration of an alkene with a regioselectivity opposite to that of Markovnikov's rule. The conversion of 1-decene to 1-decanol is an example of such a transformation.

$$CH_3(CH_2)_7CH{=}CH_2 \longrightarrow CH_3(CH_2)_7CH_2CH_2OH$$

| 1-Decene | 1-Decanol |

The synthetic method used to accomplish this is an indirect one known as **hydroboration–oxidation.** It was developed by Professor Herbert C. Brown and his coworkers at Purdue University in the 1950s as part of a broad program designed to apply boron-containing reagents to organic chemical synthesis. The number of applications is so large (hydroboration–oxidation is just one of them) and the work so novel that Brown was a corecipient of the 1979 Nobel Prize in chemistry.

Hydroboration is a reaction in which a boron hydride, a compound of the type R_2BH, adds to a carbon–carbon bond. A carbon–hydrogen bond and a carbon–boron bond result.

$$\overset{}{\underset{}{C}}{=}\overset{}{\underset{}{C} \quad + \quad R_2B{-}H \quad \longrightarrow \quad H{-}\overset{|}{\underset{|}{C}}{-}\overset{|}{\underset{|}{C}}{-}BR_2}$$

| Alkene | Boron hydride | Organoborane |

Following hydroboration, the organoborane is oxidized by treatment with hydrogen peroxide in aqueous base. This is the **oxidation** stage of the sequence; hydrogen peroxide is the oxidizing agent, and the organoborane is converted to an alcohol.

With sodium hydroxide as the base, boron of the alkylborane is converted to the water-soluble and easily removed sodium salt of boric acid.

$$H{-}\overset{|}{\underset{|}{C}}{-}\overset{|}{\underset{|}{C}}{-}BR_2 \quad + \quad 3H_2O_2 \quad + \quad HO^- \quad \longrightarrow$$

| Organoborane | Hydrogen peroxide | Hydroxide ion |

$$H{-}\overset{|}{\underset{|}{C}}{-}\overset{|}{\underset{|}{C}}{-}OH \quad + \quad 2ROH \quad + \quad B(OH)_4^-$$

| Alcohol | Alcohol | Borate ion |

The combination of hydroboration and oxidation leads to the overall hydration of an alkene. Notice, however, that water is not a reactant. The hydrogen that becomes bonded to carbon comes from the organoborane, and the hydroxyl group from hydrogen peroxide.

   With this as introduction, let us now look at the individual steps in more detail for the case of hydroboration–oxidation of 1-decene. A boron hydride that is often used is *diborane* (B_2H_6). Diborane adds to 1-decene to give tridecylborane according to the balanced equation:

$$6CH_3(CH_2)_7CH=CH_2 + B_2H_6 \xrightarrow{\text{diglyme}} 2[CH_3(CH_2)_7CH_2CH_2]_3B$$

1-Decene   Diborane   Tridecylborane

Diglyme, shown above the arrow in the equation is the solvent in this example. Diglyme is an acronym for *diethylene glycol dimethyl ether,* and its structure is $CH_3OCH_2CH_2OCH_2CH_2OCH_3$.

There is a pronounced tendency for boron to become bonded to the less substituted carbon of the double bond. Thus, the hydrogen atoms of diborane add to C-2 of 1-decene, and boron to C-1. This is believed to be mainly a steric effect, but the regioselectivity of addition does correspond to Markovnikov's rule in the sense that hydrogen is the negatively polarized atom in a B—H bond and boron the positively polarized one.

   Oxidation of tridecylborane gives 1-decanol. The net result is the conversion of an alkene to an alcohol with a regioselectivity opposite to that of acid-catalyzed hydration.

$$[CH_3(CH_2)_7CH_2CH_2]_3B \xrightarrow[\text{NaOH}]{H_2O_2} CH_3(CH_2)_7CH_2CH_2OH$$

Tridecylborane   1-Decanol

It is customary to combine the two stages, hydroboration and oxidation, in a single equation with the operations numbered sequentially above and below the arrow.

$$CH_3(CH_2)_7CH=CH_2 \xrightarrow[\text{2. } H_2O_2,\ HO^-]{\text{1. } B_2H_6,\ \text{diglyme}} CH_3(CH_2)_7CH_2CH_2OH$$

1-Decene   1-Decanol (93%)

A more convenient hydroborating agent is the borane–tetrahydrofuran complex ($H_3B{\cdot}THF$). It is very reactive, adding to alkenes within minutes at 0°C, and is used in tetrahydrofuran as the solvent.

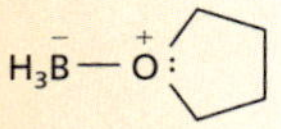

$$(CH_3)_2C=CHCH_3 \xrightarrow[\text{2. } H_2O_2,\ HO^-]{\text{1. } H_3B{\cdot}THF} (CH_3)_2CHCHCH_3$$
$$|$$
$$OH$$

2-Methyl-2-butene   3-Methyl-2-butanol (98%)

$H_3\overset{-}{B}—\overset{+}{O}$⟨⟩

Borane–tetrahydrofuran complex

Carbocation intermediates are not involved in hydroboration–oxidation. Hydration of double bonds takes place without rearrangement, even in alkenes as highly branched as the following:

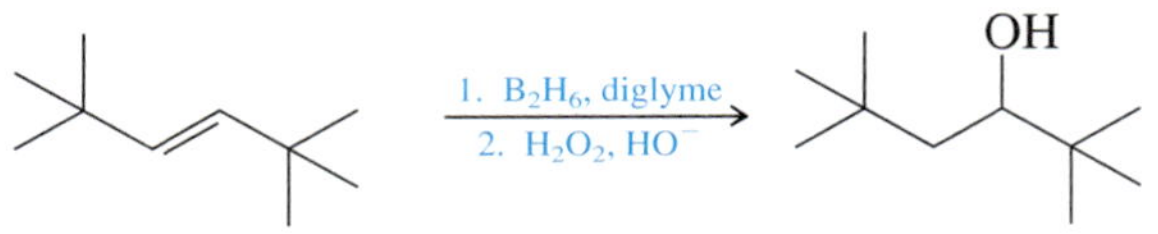

(E)-2,2,5,5-Tetramethyl-3-hexene → [1. B_2H_6, diglyme; 2. H_2O_2, HO⁻] → 2,2,5,5-Tetramethyl-3-hexanol (82%)

PROBLEM 6.12 Write the structure of the major organic product obtained by hydroboration–oxidation of each of the following alkenes:

(a) 2-Methylpropene   (b) *cis*-2-Butene

(c) ◁=CH₂   (e) 3-Ethyl-2-pentene

(d) Cyclopentene   (f) 3-Ethyl-1-pentene

SAMPLE SOLUTION   (a) In hydroboration–oxidation H and OH are introduced with a regioselectivity opposite to that of Markovnikov's rule. In the case of 2–methylpropene, this leads to 2-methyl-1-propanol as the product.

$$(CH_3)_2C=CH_2 \xrightarrow[\text{2. oxidation}]{\text{1. hydroboration}} (CH_3)_2CH-CH_2OH$$

2-Methylpropene   2-Methyl-1-propanol

Hydrogen becomes bonded to the carbon that has the fewer hydrogens, hydroxyl to the carbon that has the greater number of hydrogens.

6.12   STEREOCHEMISTRY OF HYDROBORATION–OXIDATION

A second aspect of hydroboration–oxidation concerns its stereochemistry. As illustrated for the case of 1-methylcyclopentene, H and OH add to the same face of the double bond.

1-Methylcyclopentene   → (1. B₂H₆, diglyme 2. H₂O₂, HO⁻)   trans-2-Methylcyclopentanol
                                                                       (only product, 86% yield)

Overall, the reaction leads to syn addition of H and OH to the double bond. This fact has an important bearing on the mechanism of the process.

PROBLEM 6.13   Hydroboration–oxidation of α-pinene (page 235), like catalytic hydrogenation, is stereoselective. Addition takes place at the less hindered face of the double bond, and a single alcohol is produced in high yield (89%). Suggest a reasonable structure for this alcohol.

6.13   MECHANISM OF HYDROBORATION–OXIDATION

The regioselectivity and syn stereochemistry of hydroboration–oxidation, coupled with a knowledge of the chemical properties of alkenes and boranes, contribute to our understanding of the reaction mechanism.

   We can consider the hydroboration step as though it involved borane (BH_3). It simplifies our mechanistic analysis and is at variance with reality only in matters of detail. Borane is electrophilic; it has a vacant $2p$ orbital and can accept a pair of electrons into that orbital. The source of this electron pair is the π bond of an alkene. It is believed, as shown in Figure 6.10 for the example of the hydroboration of 1-methylcyclopentene, that the first step produces an unstable intermediate called a *π complex*. In this π complex boron and the two carbon atoms of the double bond are joined by a *three-center two-electron bond*, by which we mean that three atoms share two electrons. Three-center two-electron bonds are frequently encountered in boron chemistry. The π complex is formed by a transfer of electron density from the π orbital of the alkene to the $2p$ orbital

Borane (BH_3) does not exist as such under normal conditions of temperature and atmospheric pressure. Two molecules of BH_3 combine to give diborane (B_2H_6), which is the more stable form.

Step 1: A molecule of borane (BH_3) attacks the alkene. Electrons flow from the π orbital of the alkene to the $2p$ orbital of boron. A π complex is formed.

FIGURE 6.10 Orbital interactions and electron flow in the hydroboration of 1-methylcyclopentene.

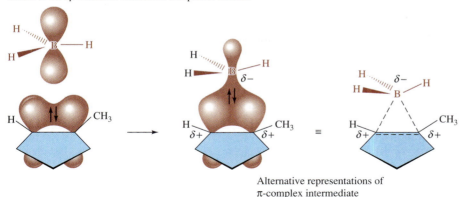

Alternative representations of π-complex intermediate

Step 2: The π complex rearranges to an organoborane. Hydrogen migrates from boron to carbon, carrying with it the two electrons in its bond to boron. Development of the transition state for this process is shown in 2(*a*), and its transformation to the organoborane is shown in 2(*b*).

2(*a*)

Representations of transition state for hydride migration in π-complex intermediate

2(*b*)

Product of addition of borane (BH_3) to 1-methylcyclopentene

of boron. This leaves each carbon of the complex with a small positive charge, while boron is slightly negative. The negative character of boron in this intermediate makes it easy for one of its hydrogens to migrate with a pair of electrons (a hydride shift) from boron to carbon. The transition state for this process is shown in step 2(*a*) of Figure 6.10; completion of the migration in step 2(*b*) yields the alkylborane. According to this mechanism, the carbon–boron bond and the carbon–hydrogen bond are formed on the same side of the alkene. Hydroboration is a syn addition.

The regioselectivity of addition is consistent with the electron distribution in the complex. Hydrogen is transferred with a pair of electrons to the carbon atom that can best support a positive charge, namely, the one that bears the methyl group.

Steric effects may be an even more important factor in controlling the regioselectivity of addition. Boron, with its attached substituents, is much larger than a hydrogen atom and becomes bonded to the less crowded carbon of the double bond, whereas hydrogen becomes bonded to the more crowded carbon.

The electrophilic character of boron is again evident when we consider the oxidation of organoboranes. In the oxidation phase of the hydroboration–oxidation sequence, as presented in Figure 6.11, the conjugate base of hydrogen peroxide attacks boron. Hydroperoxide ion is formed in an acid–base reaction in step 1 and attacks boron in step 2. The empty $2p$ orbital of boron makes it electrophilic and permits nucleophilic reagents such as HOO^- to add to it.

The combination of a negative charge on boron and the weak oxygen–oxygen bond causes an alkyl group to migrate from boron to oxygen in step 3. This alkyl group migration occurs with loss of hydroxide ion and is the step in which the critical carbon–oxygen bond is formed. What is especially significant about this alkyl group migration is that the stereochemical orientation of the new carbon–oxygen bond is the same as that of the original carbon–boron bond. This is crucial to the overall syn stereochemistry of the hydroboration–oxidation sequence. Migration of the alkyl group from boron to oxygen is said to have occurred with *retention of configuration* at carbon. The alkoxyborane intermediate formed in step 3 undergoes subsequent base-promoted oxygen-boron bond cleavage in step 4 to give the alcohol product.

The mechanistic complexity of hydroboration–oxidation stands in contrast to the simplicity with which these reactions are carried out experimentally. Both the hydroboration and oxidation steps are extremely rapid reactions and are performed at room temperature with conventional laboratory equipment. Ease of operation, along with the fact that hydroboration–oxidation leads to syn hydration of alkenes and occurs with a regioselectivity opposite to Markovnikov's rule, makes this procedure one of great value to the synthetic chemist.

6.14   ADDITION OF HALOGENS TO ALKENES

In contrast to the free-radical substitution observed when halogens react with *alkanes*, halogens normally react with *alkenes* by electrophilic addition.

$$\overset{\diagup}{\underset{\diagdown}{C}}=\overset{\diagup}{\underset{\diagdown}{C}} \;+\; X_2 \;\longrightarrow\; X-\overset{|}{\underset{|}{C}}-\overset{|}{\underset{|}{C}}-X$$

Alkene   Halogen   Vicinal dihalide

The products of these reactions are called **vicinal** dihalides. Two substituents, in this case the halogens, are vicinal if they are attached to adjacent carbons. The word is derived from the Latin *vicinalis,* which means "neighboring." The halogen is either chlorine (Cl_2) or bromine (Br_2), and addition takes place rapidly at room temperature and below in a variety of solvents, including acetic acid, carbon tetrachloride, chloroform, and dichloromethane.

$$CH_3CH{=}CHCH(CH_3)_2 \;+\; Br_2 \;\xrightarrow[0°C]{CHCl_3}\; CH_3CH{-}CHCH(CH_3)_2$$
$$\hspace{7cm}\underset{Br}{|}\;\;\underset{Br}{|}$$

4-Methyl-2-pentene   Bromine   2,3-Dibromo-4-methylpentane (100%)

Step 1: Hydrogen peroxide is converted to its anion in basic solution:

H—O—O—H   +   ⁻OH   ⇌   H—O—O⁻   +   H—O—H

Hydrogen   Hydroxide   Hydroperoxide   Water
peroxide   ion   ion

FIGURE 6.11 The oxidation
phase in the hydroboration–
oxidation of 1-methylcyclo-
pentene.

Step 2: Anion of hydrogen peroxide acts as a nucleophile, attacking boron and forming
an oxygen–boron bond:

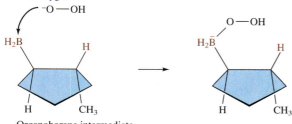

Organoborane intermediate
from hydroboration of
1-methylcyclopentene

Step 3: Carbon migrates from boron to oxygen, displacing hydroxide ion. Carbon migrates
with the pair of electrons in the carbon–boron bond; these become the electrons in
the carbon–oxygen bond:

Representation of transition
state for migration of carbon
from boron to oxygen

Alkoxyborane

Step 4: Hydrolysis cleaves the boron–oxygen bond, yielding the alcohol:

+   H₂B—OH

Alkoxyborane   *trans*-2-Methylcyclopentanol

Rearrangements do not normally occur, which can mean either of two things. Either carbocations are not intermediates, or if they are, they are captured by a nucleophile faster than they rearrange. We shall see in Section 6.16 that the first of these is believed to be the case.

Fluorine addition to alkenes is a violent reaction, difficult to control, and accompanied by substitution of hydrogens by fluorine. Vicinal diiodides, on the other hand, tend to lose I_2 and revert to alkenes, making them an infrequently encountered class of compounds.

6.15   STEREOCHEMISTRY OF HALOGEN ADDITION

The reaction of chlorine and bromine with cycloalkenes illustrates an important stereochemical feature of halogen addition. *Anti addition is observed;* the two bromine atoms of Br_2 or the two chlorines of Cl_2 add to opposite faces of the double bond.

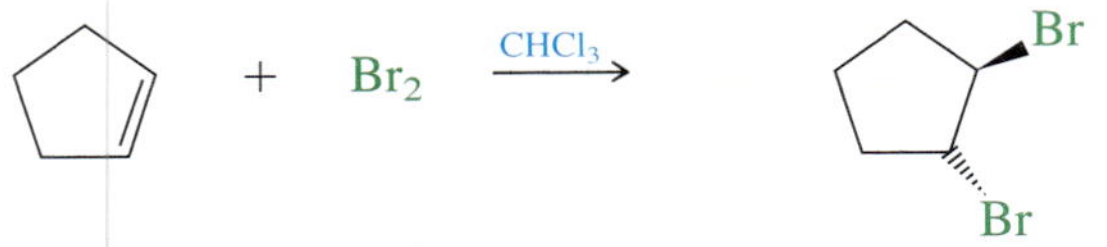

Cyclopentene   Bromine   *trans*-1,2-Dibromocyclopentane
(80% yield; none of the cis isomer is formed)

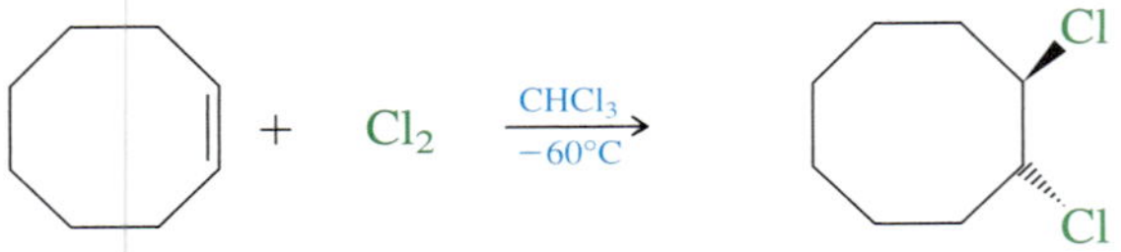

Cyclooctene   Chlorine   *trans*-1,2-Dichlorocyclooctane
(73% yield; none of the cis isomer is formed)

These observations must be taken into account when considering the mechanism of halogen addition. They force the conclusion that a simple one-step "bond-switching" process of the following type cannot be correct. A process of this type requires syn addition; it is *not* consistent with the anti addition that we actually see.

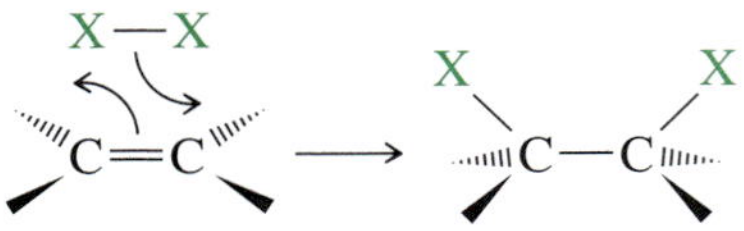

PROBLEM 6.14 The mass 82 isotope of bromine (^{82}Br) is radioactive and is used as a tracer to identify the origin and destination of individual atoms in chemical reactions and biological transformations. A sample of 1,1,2-tribromocyclohexane was prepared by adding $^{82}Br—^{82}Br$ to ordinary (nonradioactive) 1-bromocyclohexene. How many of the bromine atoms in the 1,1,2-tribromocyclohexane produced are radioactive? Which ones are they?

6.16   MECHANISM OF HALOGEN ADDITION TO ALKENES: HALONIUM IONS

Many of the features of the generally accepted mechanism for the addition of halogens to alkenes can be introduced by referring to the reaction of ethylene with bromine:

$$H_2C=CH_2 \ + \ Br_2 \ \longrightarrow \ BrCH_2CH_2Br$$

| Ethylene | Bromine | 1,2-Dibromoethane |

Neither bromine nor ethylene is a polar molecule, but both are *polarizable*, and an induced-dipole/induced-dipole force causes them to be mutually attracted to each other. This induced-dipole/induced-dipole attraction sets the stage for Br_2 to act as an electrophile. Electrons flow from the π system of ethylene to Br_2, causing the weak bromine–bromine bond to break. By analogy to the customary mechanisms for electrophilic addition, we might represent this as the formation of a carbocation in a bimolecular elementary step.

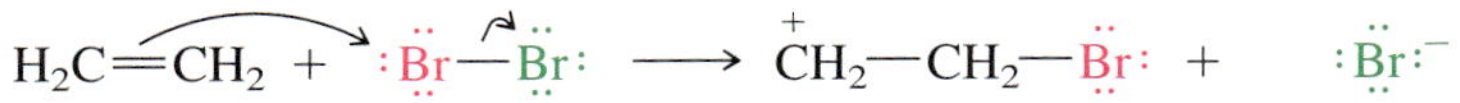

$$H_2C=CH_2 \ + \ :\!\ddot{B}r\!-\!\ddot{B}r\!: \ \longrightarrow \ \overset{+}{C}H_2\!-\!CH_2\!-\!\ddot{B}r\!: \ + \ :\!\ddot{B}r\!:^-$$

| Ethylene (nucleophile) | Bromine (electrophile) | 2-Bromoethyl cation | Bromide ion (leaving group) |

Such a carbocation, however, has been demonstrated to be less stable than an alternative structure called a **cyclic bromonium ion,** in which the positive charge resides on bromine, not carbon.

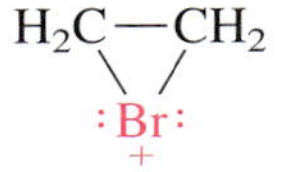

$$\begin{array}{c} H_2C-CH_2 \\ \diagdown \ \diagup \\ :\!\ddot{B}r\!: \\ + \end{array}$$

Ethylenebromonium ion

The chief reason why ethylenebromonium ion, in spite of its strained three-membered ring, is more stable than 2-bromoethyl cation is that both carbons and bromine have octets of electrons, whereas one carbon has only six electrons in the carbocation.

Thus, the mechanism for electrophilic addition of Br_2 to ethylene as presented in Figure 6.12 is characterized by the direct formation of a cyclic bromonium ion as its

Until it was banned in the United States in 1984, 1,2-dibromoethane (ethylene dibromide, or EDB) was produced on a large scale for use as a pesticide and soil fumigant.

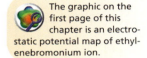

The graphic on the first page of this chapter is an electrostatic potential map of ethylenebromonium ion.

The overall reaction:

$$H_2C=CH_2 \ + \ Br_2 \ \longrightarrow \ BrCH_2CH_2Br$$

| Ethylene | Bromine | 1,2-Dibromoethane |

The mechanism:

Step 1: Reaction of ethylene and bromine to form a bromonium ion intermediate:

$$H_2C=CH_2 \ + \ :\!\ddot{B}r\!-\!\ddot{B}r\!: \ \longrightarrow \ \begin{array}{c} H_2C-CH_2 \\ \diagdown \ \diagup \\ :\!\ddot{B}r\!: \\ + \end{array} \ + \ :\!\ddot{B}r\!:^-$$

| Ethylene | Bromine | Ethylenebromonium ion | Bromide ion |

Step 2: Nucleophilic attack of bromide anion on the bromonium ion:

$$:\!\ddot{B}r\!:^- \qquad \begin{array}{c} H_2C-CH_2 \\ \diagdown \ \diagup \\ :\!\ddot{B}r\!: \\ + \end{array} \ \longrightarrow \ :\!\ddot{B}r\!-\!CH_2\!-\!CH_2\!-\!\ddot{B}r\!:$$

| Bromide ion | Ethylenebromonium ion | 1,2-Dibromoethane |

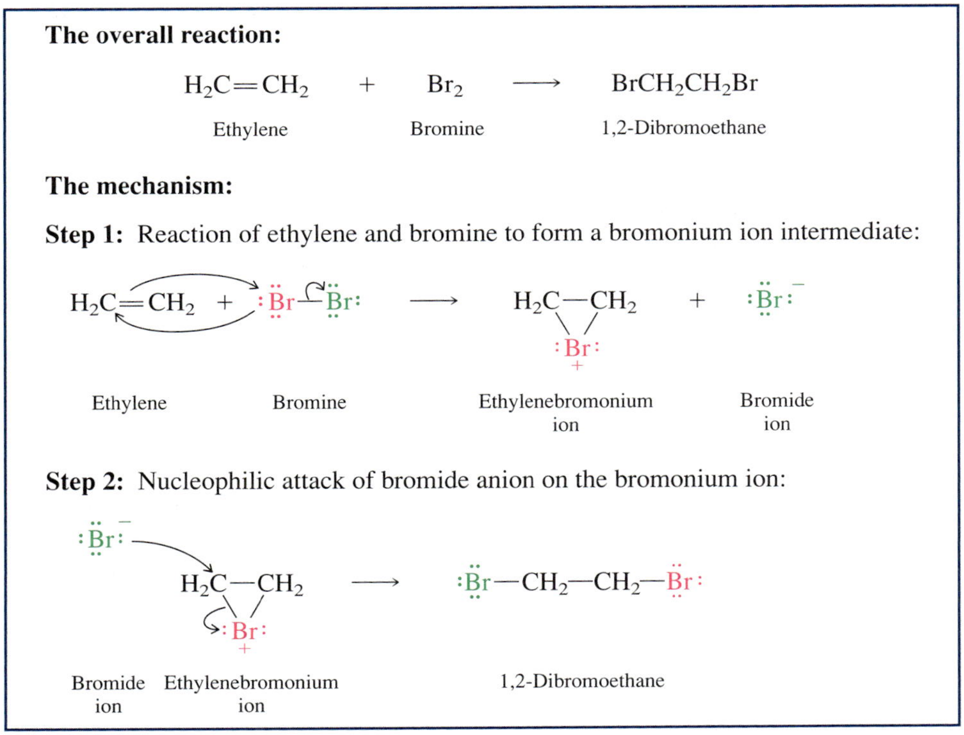

FIGURE 6.12 Mechanism of electrophilic addition of bromine to ethylene.

TABLE 6.3	Relative Rates of Reaction of Some Representative Alkenes with Bromine	
Alkene	**Structural formula**	**Relative rate of reaction with bromine***
Ethylene	$H_2C{=}CH_2$	1.0
Propene	$CH_3CH{=}CH_2$	61
2-Methylpropene	$(CH_3)_2C{=}CH_2$	5,400
2,3-Dimethyl-2-butene	$(CH_3)_2C{=}C(CH_3)_2$	920,000

*In methanol, 25°C.

first elementary step via the transition state:

Transition state for bromonium
ion formation from an alkene
and bromine

$$\delta+ C \cdots \quad \delta- \\ \qquad\qquad Br \cdots Br$$
$$\delta+ C$$

Step 2 is the conversion of the bromonium ion to 1,2-dibromoethane by reaction with bromide ion (Br^-).

Table 6.3 shows that the effect of substituents on the rate of addition of bromine to alkenes is substantial and consistent with a rate-determining step in which electrons flow from the alkene to the halogen. Alkyl groups on the carbon–carbon double bond release electrons, stabilize the transition state for bromonium ion formation, and increase the reaction rate.

PROBLEM 6.15   Arrange the compounds 2-methyl-1-butene, 2-methyl-2-butene, and 3-methyl-1-butene in order of decreasing reactivity toward bromine.

Step 2 of the mechanism in Figure 6.12 is a nucleophilic attack by Br^- at one of the carbons of the cyclic bromonium ion. For reasons that will be explained in Chapter 8, reactions of this type normally take place via a transition state in which the nucleophile approaches carbon from the side opposite the bond that is to be broken. Recalling that the vicinal dibromide formed from cyclopentene is exclusively the trans stereoisomer, we see that attack by Br^- from the side opposite the C—Br bond of the bromonium ion intermediate can give only *trans*-1,2-dibromocyclopentane in accordance with the experimental observations.

Bromonium ion intermediate   *trans*-1,2-Dibromocyclopentane

Some supporting evidence is described in the article "The Bromonium Ion," in the August 1963 issue of the *Journal of Chemical Education* (pp. 392–395).

The idea that a cyclic bromonium ion was an intermediate was a novel concept when it was first proposed. Much additional evidence, including the isolation of a stable cyclic bromonium ion, has been obtained since then to support it. Similarly, **cyclic**

chloronium ions are believed to be involved in the addition of chlorine to alkenes. In the next section we shall see how cyclic chloronium and bromonium ions **(halonium ions)** are intermediates in a second reaction involving alkenes and halogens.

6.17 CONVERSION OF ALKENES TO VICINAL HALOHYDRINS

In *aqueous* solution chlorine and bromine react with alkenes to form **vicinal halohydrins,** compounds that have a halogen and a hydroxyl group on adjacent carbons.

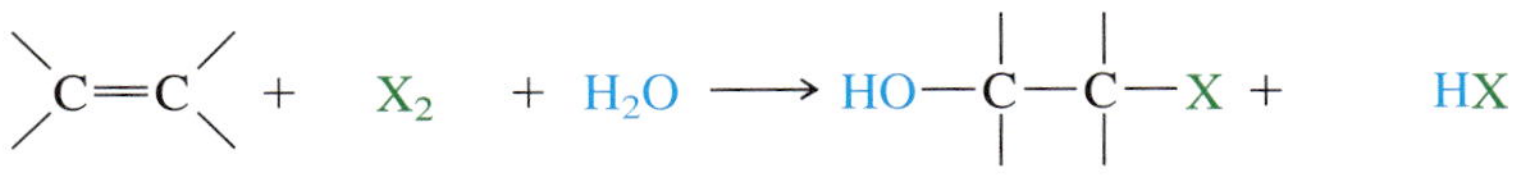

$$\underset{\text{Alkene}}{\underset{\diagup}{\overset{\diagdown}{C}}=\underset{\diagdown}{\overset{\diagup}{C}}} + \underset{\text{Halogen}}{X_2} + \underset{\text{Water}}{H_2O} \longrightarrow \underset{\text{Halohydrin}}{HO-\overset{|}{\underset{|}{C}}-\overset{|}{\underset{|}{C}}-X} + \underset{\text{Hydrogen halide}}{HX}$$

$$\underset{\text{Ethylene}}{H_2C=CH_2} + \underset{\text{Bromine}}{Br_2} \overset{H_2O}{\longrightarrow} \underset{\text{2-Bromoethanol (70\%)}}{HOCH_2CH_2Br}$$

Anti addition occurs. The halogen and the hydroxyl group add to opposite faces of the double bond.

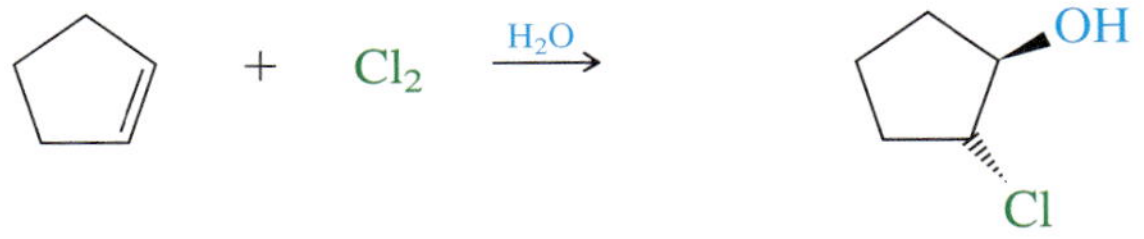

Cyclopentene   Chlorine   *trans*-2-Chlorocyclopentanol
(52–56% yield; cis isomer not formed)

Halohydrin formation, as depicted in Figure 6.13, is mechanistically related to halogen addition to alkenes. A halonium ion intermediate is formed, which is attacked by water in aqueous solution.

The regioselectivity of addition is established when water attacks one of the carbons of the halonium ion. In the following example, the structure of the product tells us that water attacks the more highly substituted carbon.

$$\underset{\text{2-Methylpropene}}{(CH_3)_2C=CH_2} \overset{Br_2}{\underset{H_2O}{\longrightarrow}} \underset{\substack{\text{1-Bromo-2-methyl-}\\\text{2-propanol (77\%)}}}{(CH_3)_2\overset{|}{\underset{|}{\underset{OH}{C}}}-CH_2Br}$$

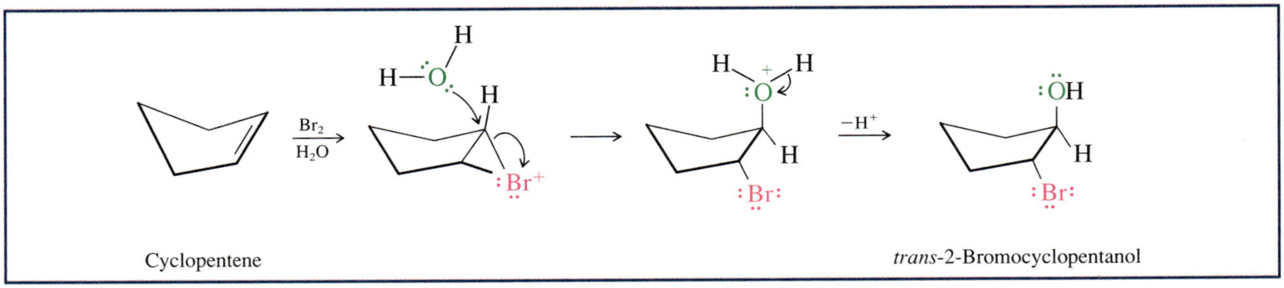

FIGURE 6.13 Mechanism of bromohydrin formation from cyclopentene. A bridged bromonium ion is formed and is attacked by a water molecule from the side opposite the carbon–bromine bond. The bromine and the hydroxyl group are trans to each other in the product.

Cyclopentene   *trans*-2-Bromocyclopentanol

This suggests that, as water attacks the bromonium ion, positive charge develops on the carbon from which the bromine departs. The transition state has some of the character of a carbocation. We know that more substituted carbocations are more stable than less substituted ones; therefore, when the bromonium ion ring opens, it does so by breaking the bond between bromine and the more substituted carbon.

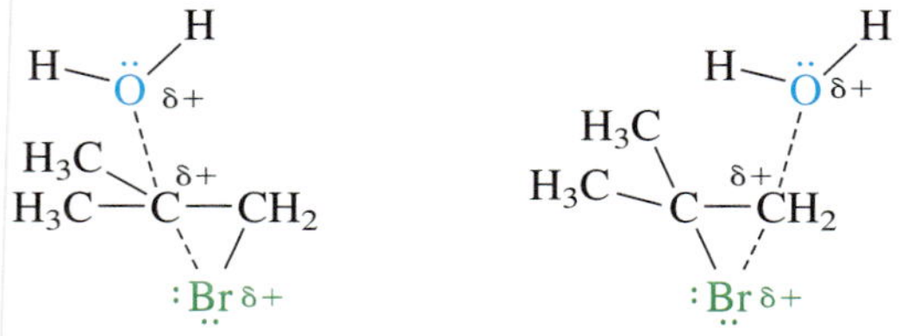

More stable transition state; has some of the character of a tertiary carbocation

Less stable transition state; has some of the character of a primary carbocation

---

PROBLEM 6.16   Give the structure of the product formed when each of the following alkenes reacts with bromine in water:

(a) 2-Methyl-1-butene

(c) 3-Methyl-1-butene

(b) 2-Methyl-2-butene

(d) 1-Methylcyclopentene

SAMPLE SOLUTION   (a) The hydroxyl group becomes bonded to the more substituted carbon of the double bond, and bromine bonds to the less substituted one.

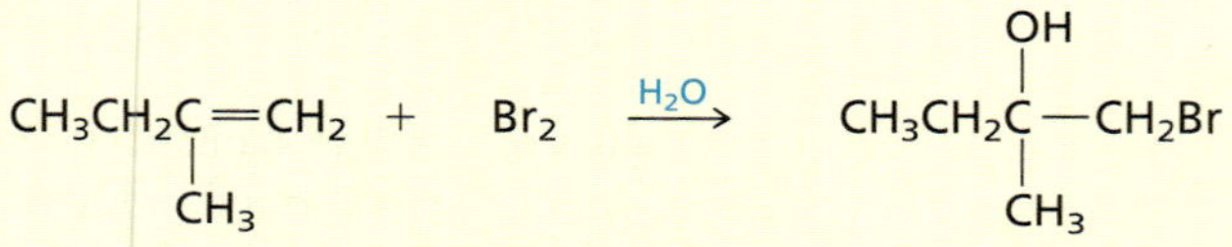

2-Methyl-1-butene   Bromine   1-Bromo-2-methyl-2-butanol

6.18   EPOXIDATION OF ALKENES

You have just seen that cyclic halonium ion intermediates are formed when sources of electrophilic *halogen* attack a double bond. Likewise, three-membered oxygen-containing rings are formed by the reaction of alkenes with sources of electrophilic *oxygen*.

Three-membered rings that contain oxygen are called *epoxides*. At one time, epoxides were named as oxides of alkenes. Ethylene oxide and propylene oxide, for example, are the common names of two industrially important epoxides.

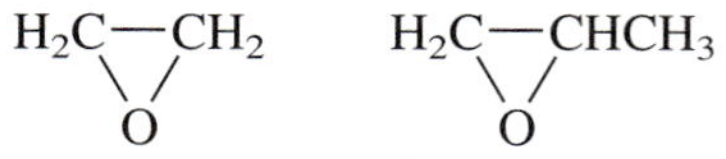

Ethylene oxide   Propylene oxide

A second method for naming epoxides in the IUPAC system is described in Section 16.1.

Substitutive IUPAC nomenclature names epoxides as *epoxy* derivatives of alkanes. According to this system, ethylene oxide becomes epoxyethane, and propylene oxide becomes 1,2-epoxypropane. The prefix *epoxy-* always immediately precedes the alkane ending; it is not listed in alphabetical order like other substituents.

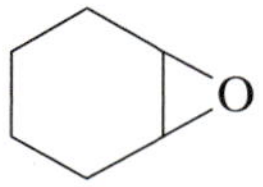

1,2-Epoxycyclohexane 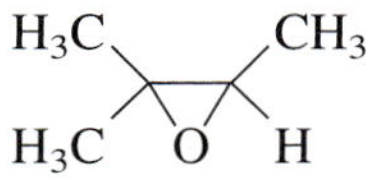   2-Methyl-2,3-epoxybutane

Functional group transformations of epoxides rank among the fundamental reactions of organic chemistry, and epoxides are commonplace natural products. The female gypsy moth, for example, attracts the male by emitting an epoxide known as *disparlure*. On detecting the presence of this pheromone, the male follows the scent to its origin and mates with the female.

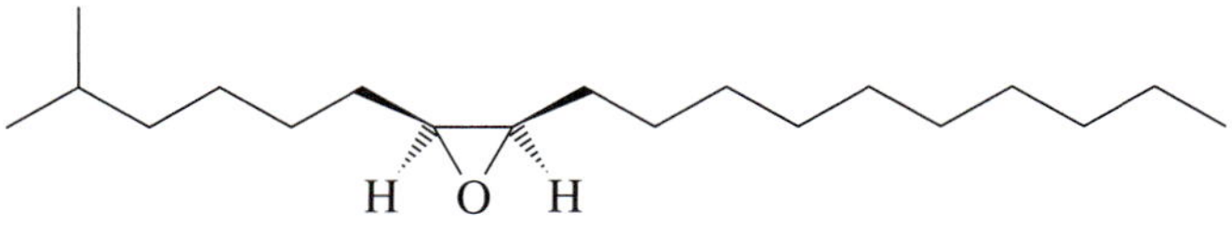

Disparlure

In one strategy designed to control the spread of the gypsy moth, infested areas are sprayed with synthetic disparlure. With the sex attractant everywhere, male gypsy moths become hopelessly confused as to the actual location of individual females. Many otherwise fertile female gypsy moths then live out their lives without producing hungry gypsy moth caterpillars.

PROBLEM 6.17 Give the substitutive IUPAC name, including stereochemistry, for disparlure.

Epoxides are very easy to prepare via the reaction of an alkene with a peroxy acid. This process is known as **epoxidation.**

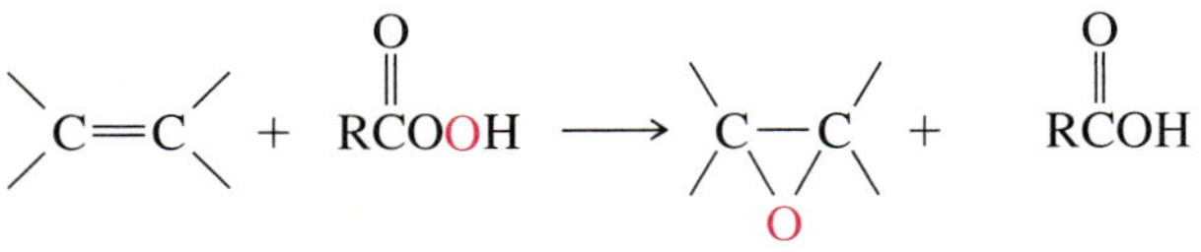

Alkene   Peroxy acid   Epoxide   Carboxylic acid

A commonly used peroxy acid is peroxyacetic acid (CH_3CO_2OH). Peroxyacetic acid is normally used in acetic acid as the solvent, but epoxidation reactions tolerate a variety of solvents and are often carried out in dichloromethane or chloroform.

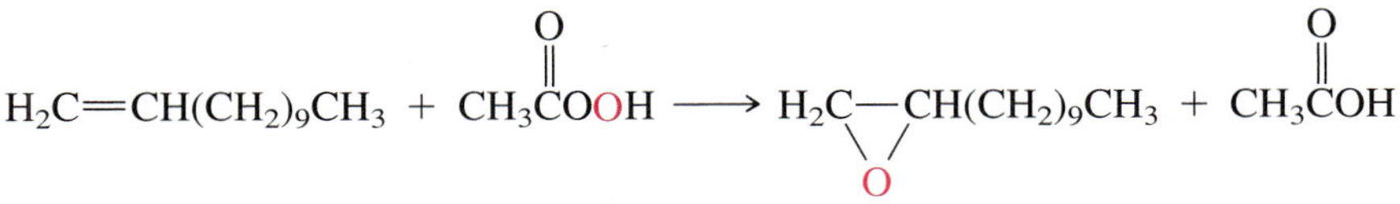

1-Dodecene   Peroxyacetic   1,2-Epoxydodecane   Acetic
                         acid   (52%)   acid

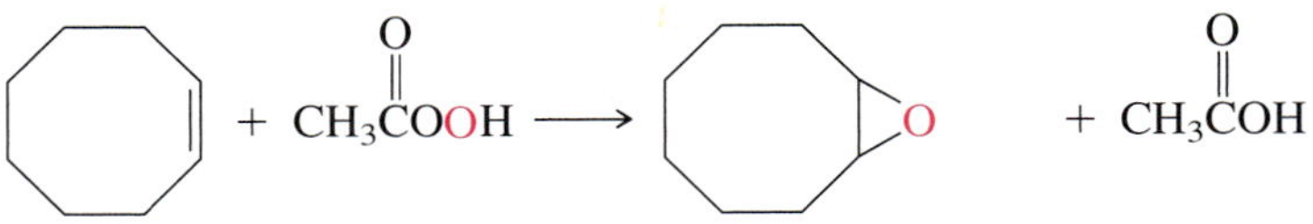

Cyclooctene   Peroxyacetic   1,2-Epoxycyclooctane   Acetic
                         acid   (86%)   acid

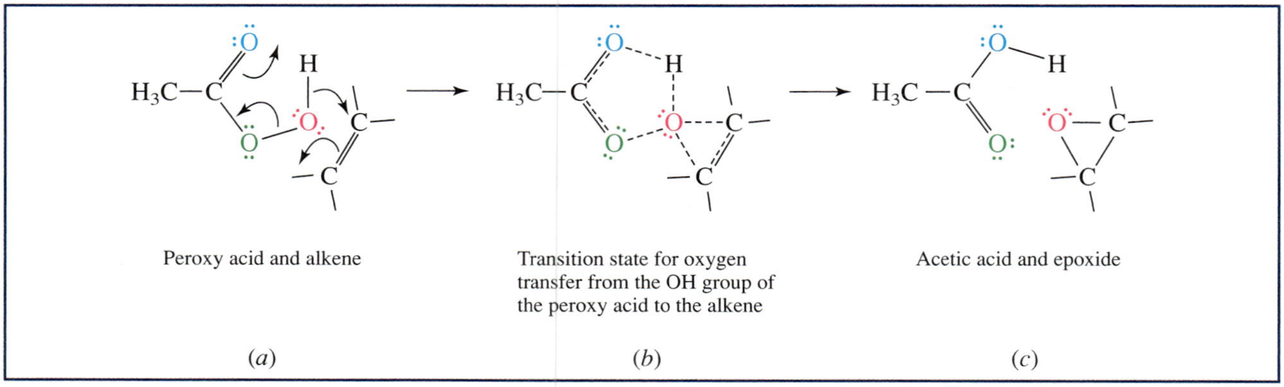

Peroxy acid and alkene

(a)

Transition state for oxygen
transfer from the OH group of
the peroxy acid to the alkene

(b)

Acetic acid and epoxide

(c)

FIGURE 6.14 A one-step mechanism for epoxidation of alkenes by peroxyacetic acid. In (a) the starting peroxy acid is shown in a conformation in which the proton of the OH group is hydrogen bonded to the oxygen of the C=O group. (b) The weak O—O bond of the peroxy acid breaks and both C—O bonds of the epoxide form in the same transition state leading to products (c).

TABLE 6.4   Relative Rates of Epoxidation of Some Representative Alkenes with Peroxyacetic Acid

Alkene	Structural formula	Relative rate of epoxidation*
Ethylene	$H_2C=CH_2$	1.0
Propene	$CH_3CH=CH_2$	22
2-Methylpropene	$(CH_3)_2C=CH_2$	484
2-Methyl-2-butene	$(CH_3)_2C=CHCH_3$	6526

*In acetic acid, 26°C.

Epoxidation of alkenes with peroxy acids is a syn addition to the double bond. Substituents that are cis to each other in the alkene remain cis in the epoxide; substituents that are trans in the alkene remain trans in the epoxide.

PROBLEM 6.18 Give the structure of the alkene, including stereochemistry, that you would choose as the starting material in a preparation of synthetic disparlure.

The structure of disparlure is shown on page 261.

As shown in Table 6.4, electron-releasing alkyl groups on the double bond increase the rate of epoxidation. This suggests that the peroxy acid acts as an electrophilic reagent toward the alkene.

The mechanism of alkene epoxidation is believed to be a concerted process involving a single bimolecular elementary step, as shown in Figure 6.14.

6.19   OZONOLYSIS OF ALKENES

Ozone (O_3) is the triatomic form of oxygen. It is a neutral but polar molecule that can be represented as a hybrid of its two most stable Lewis structures.

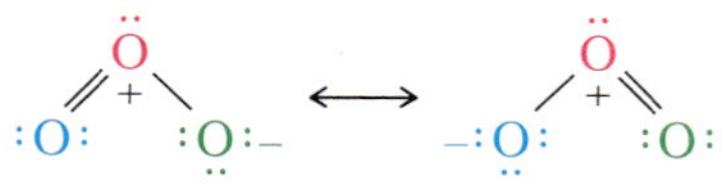

Ozone is a powerful electrophile and undergoes a remarkable reaction with alkenes in

which both the σ and π components of the carbon–carbon double bond are cleaved to give a product referred to as an **ozonide.**

$$\text{C=C} \;+\; O_3 \;\longrightarrow\; \text{Ozonide}$$

Alkene   Ozone   Ozonide

Ozonides undergo hydrolysis in water, giving carbonyl compounds.

$$\text{Ozonide} \;+\; H_2O \;\longrightarrow\; \text{C=O} + \text{O=C} \;+\; H_2O_2$$

Ozonide   Water   Two carbonyl compounds   Hydrogen peroxide

Two aldehydes, two ketones, or one aldehyde and one ketone may be formed. Let's recall the classes of carbonyl compounds from Table 4.1. Aldehydes have at least one hydrogen on the carbonyl group; ketones have two carbon substituents—alkyl groups, for example—on the carbonyl. Carboxylic acids have a hydroxyl substituent attached to the carbonyl group.

$$\underset{\text{Formaldehyde}}{\overset{O}{\underset{H}{\parallel}{\overset{C}{}}H}} \qquad \underset{\text{Aldehyde}}{\overset{O}{\underset{R}{\parallel}{\overset{C}{}}H}} \qquad \underset{\text{Ketone}}{\overset{O}{\underset{R}{\parallel}{\overset{C}{}}R'}} \qquad \underset{\text{Carboxylic acid}}{\overset{O}{\underset{R}{\parallel}{\overset{C}{}}OH}}$$

Aldehydes are easily oxidized to carboxylic acids under conditions of ozonide hydrolysis. When one wishes to isolate the aldehyde itself, a reducing agent such as zinc is included during the hydrolysis step. Zinc reduces the ozonide and reacts with any oxidants present (excess ozone and hydrogen peroxide) to prevent them from oxidizing any aldehyde formed. An alternative, more modern technique follows ozone treatment of the alkene in methanol with reduction by dimethyl sulfide (CH_3SCH_3).

The two-stage reaction sequence is called **ozonolysis** and is represented by the general equation

$$\underset{\text{Alkene}}{\overset{R}{\underset{H}{}}\text{C=C}\overset{R'}{\underset{R''}{}}} \xrightarrow[\substack{\text{or} \\ \text{1. } O_3, CH_3OH; \text{ 2. } (CH_3)_2S}]{\text{1. } O_3; \text{ 2. } H_2O, Zn} \underset{\text{Aldehyde}}{\overset{R}{\underset{H}{}}\text{C=O}} + \underset{\text{Ketone}}{\text{O=C}\overset{R'}{\underset{R''}{}}}$$

Each carbon of the double bond becomes the carbon of a carbonyl group.

Ozonolysis has both synthetic and analytical applications in organic chemistry. In synthesis, ozonolysis of alkenes provides a method for the preparation of aldehydes and ketones.

$$\underset{\text{1-Octene}}{CH_3(CH_2)_5CH\text{=}CH_2} \xrightarrow[\text{2. } (CH_3)_2S]{\text{1. } O_3, CH_3OH} \underset{\text{Heptanal (75\%)}}{CH_3(CH_2)_5\overset{O}{\overset{\parallel}{C}}H} + \underset{\text{Formaldehyde}}{H\overset{O}{\overset{\parallel}{C}}H}$$

$$CH_3CH_2CH_2CH_2C{=}CH_2 \xrightarrow[\text{2. H}_2\text{O, Zn}]{\text{1. O}_3} CH_3CH_2CH_2CH_2\overset{\overset{\displaystyle O}{\|}}{C}CH_3 + \overset{\overset{\displaystyle O}{\|}}{H}CH$$

$$\underset{\text{CH}_3}{|}$$

2-Methyl-1-hexene   2-Hexanone (60%)   Formaldehyde

When the objective is analytical, the products of ozonolysis are isolated and identified, thereby allowing the structure of the alkene to be deduced. In one such example, an alkene having the molecular formula C_8H_{16} was obtained from a chemical reaction and was then subjected to ozonolysis, giving acetone and 2,2-dimethylpropanal as the products.

$$\overset{\overset{\displaystyle O}{\|}}{CH_3CCH_3} \qquad\qquad (CH_3)_3\overset{\overset{\displaystyle O}{\|}}{C}CH$$

Acetone   2,2-Dimethylpropanal

Together, these two products contain all eight carbons of the starting alkene. The two carbonyl carbons correspond to those that were doubly bonded in the original alkene. One of the doubly bonded carbons therefore bears two methyl substituents; the other bears a hydrogen and a *tert*-butyl group. The alkene is identified as 2,4,4-trimethyl-2-pentene, $(CH_3)_2C{=}CHC(CH_3)_3$, as shown in Figure 6.15.

PROBLEM 6.19 The same reaction that gave 2,4,4-trimethyl-2-pentene also yielded an isomeric alkene. This second alkene produced formaldehyde and 4,4-dimethyl-2-pentanone on ozonolysis. Identify this alkene.

$$\overset{\overset{\displaystyle O}{\|}}{CH_3CCH_2C(CH_3)_3}$$

4,4-Dimethyl-2-pentanone

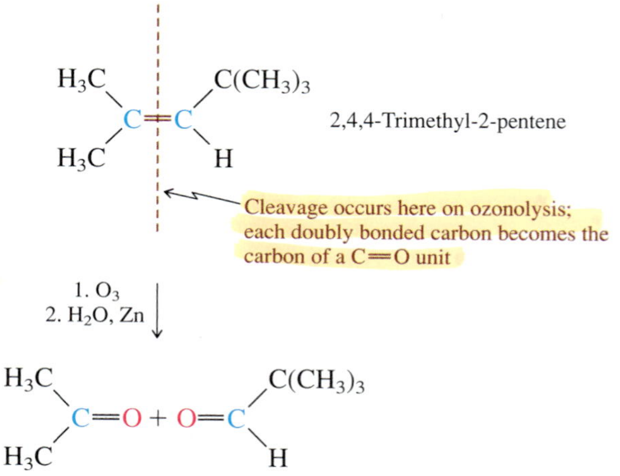

FIGURE 6.15 Ozonolysis of 2,4,4-trimethyl-2-pentene. On cleavage, each of the doubly bonded carbons becomes the carbon of a carbonyl (C=O) group.

6.20   INTRODUCTION TO ORGANIC CHEMICAL SYNTHESIS

An important concern to chemists is *synthesis,* the challenge of preparing a particular compound in an economical way with confidence that the method chosen will lead to the desired structure. In this section we will introduce the topic of synthesis, emphasizing the need for systematic planning to decide what is the best sequence of steps to convert a specified starting material to a desired product (the **target molecule**).

A critical feature of synthetic planning is *to reason backward from the target to the starting material.* A second is to *always use reactions that you know will work.*

Let's begin with a simple example. Suppose you wanted to prepare cyclohexane, given cyclohexanol as the starting material. We haven't encountered any reactions so far that permit us to carry out this conversion in a single step.

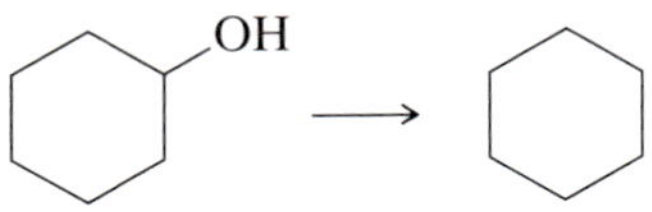

        Cyclohexanol   Cyclohexane

Reasoning backward, however, we know that we can prepare cyclohexane by hydrogenation of cyclohexene. We'll therefore use this reaction as the last step in our proposed synthesis.

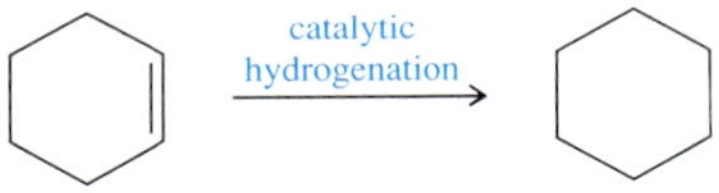

        Cyclohexene   Cyclohexane

Recognizing that cyclohexene may be prepared by dehydration of cyclohexanol, a practical synthesis of cyclohexane from cyclohexanol becomes apparent.

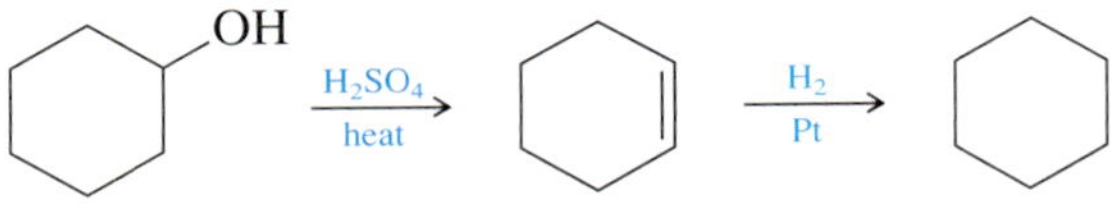

    Cyclohexanol   Cyclohexene   Cyclohexane

As a second example, consider the preparation of 1-bromo-2-methyl-2-propanol from *tert*-butyl alcohol.

$$(CH_3)_3COH \longrightarrow (CH_3)_2CCH_2Br$$
$$\underset{\textstyle OH}{|}$$

   tert-Butyl alcohol   1-Bromo-2-methyl-2-propanol

Begin by asking the question, "What kind of compound is the target molecule, and what methods can I use to prepare that kind of compound?" The desired product has a bromine and a hydroxyl on adjacent carbons; it is a *vicinal bromohydrin.* The only method we have learned so far for the preparation of vicinal bromohydrins involves the reaction of alkenes with Br_2 in water. Thus, a reasonable last step is:

$$(CH_3)_2C{=}CH_2 \xrightarrow[H_2O]{Br_2} (CH_3)_2CCH_2Br$$
$$\underset{\textstyle OH}{|}$$

    2-Methylpropene   1-Bromo-2-methyl-2-propanol

We now have a new problem: Where does the necessary alkene come from? Alkenes are prepared from alcohols by acid-catalyzed dehydration (Section 5.9) or from alkyl halides by dehydrohalogenation (Section 5.14). Because our designated starting material is *tert*-butyl alcohol, we can combine its dehydration with bromohydrin formation to give the correct sequence of steps:

$$(CH_3)_3COH \xrightarrow[\text{heat}]{H_2SO_4} (CH_3)_2C{=}CH_2 \xrightarrow[H_2O]{Br_2} (CH_3)_2\underset{\underset{OH}{|}}{C}CH_2Br$$

tert-Butyl alcohol   2-Methylpropene   1-Bromo-2-methyl-2-propanol

PROBLEM 6.20 Write a series of equations describing a synthesis of 1-bromo-2-methyl-2-propanol from *tert*-butyl bromide.

Often more than one synthetic route may be available to prepare a particular compound. Indeed, it is normal to find in the chemical literature that the same compound has been synthesized in a number of different ways. As we proceed through the text and develop a larger inventory of functional group transformations, our ability to evaluate alternative synthetic plans will increase. In most cases the best synthetic plan is the one with the fewest steps.

6.21 REACTIONS OF ALKENES WITH ALKENES: POLYMERIZATION

Although 2-methylpropene undergoes acid-catalyzed hydration in *dilute* sulfuric acid to form *tert*-butyl alcohol (Section 6.10), a different reaction occurs in more concentrated solutions of sulfuric acid. Rather than form the expected alkyl hydrogen sulfate (see Section 6.9), 2-methylpropene is converted to a mixture of two isomeric C_8H_{16} alkenes.

The structures of these two C_8H_{16} alkenes were determined by ozonolysis as described in Section 6.19.

$$2(CH_3)_2C{=}CH_2 \xrightarrow{65\% \ H_2SO_4} H_2C{=}\underset{\underset{CH_3}{|}}{C}CH_2C(CH_3)_3 \ + \ (CH_3)_2C{=}CHC(CH_3)_3$$

2-Methylpropene   2,4,4-Trimethyl-1-pentene   2,4,4-Trimethyl-2-pentene

With molecular formulas corresponding to twice that of the starting alkene, the products of this reaction are referred to as **dimers** of 2-methylpropene, which is, in turn, called the **monomer.** The suffix *-mer* is derived from the Greek *meros,* meaning "part." Three monomeric units produce a **trimer,** four a **tetramer,** and so on. A high-molecular-weight material comprising a large number of monomer subunits is called a **polymer.**

PROBLEM 6.21 The two dimers of 2-methylpropene shown in the equation can be converted to 2,2,4-trimethylpentane (known by its common name *isooctane*) for use as a gasoline additive. Can you suggest a method for this conversion?

The two dimers of $(CH_3)_2C{=}CH_2$ are formed by the mechanism shown in Figure 6.16. In step 1 protonation of the double bond generates a small amount of *tert*-butyl cation in equilibrium with the alkene. The carbocation is an electrophile and attacks a second molecule of 2-methylpropene in step 2, forming a new carbon–carbon bond and generating a C_8 carbocation. This new carbocation loses a proton in step 3 to form a mixture of 2,4,4-trimethyl-1-pentene and 2,4,4-trimethyl-2-pentene.

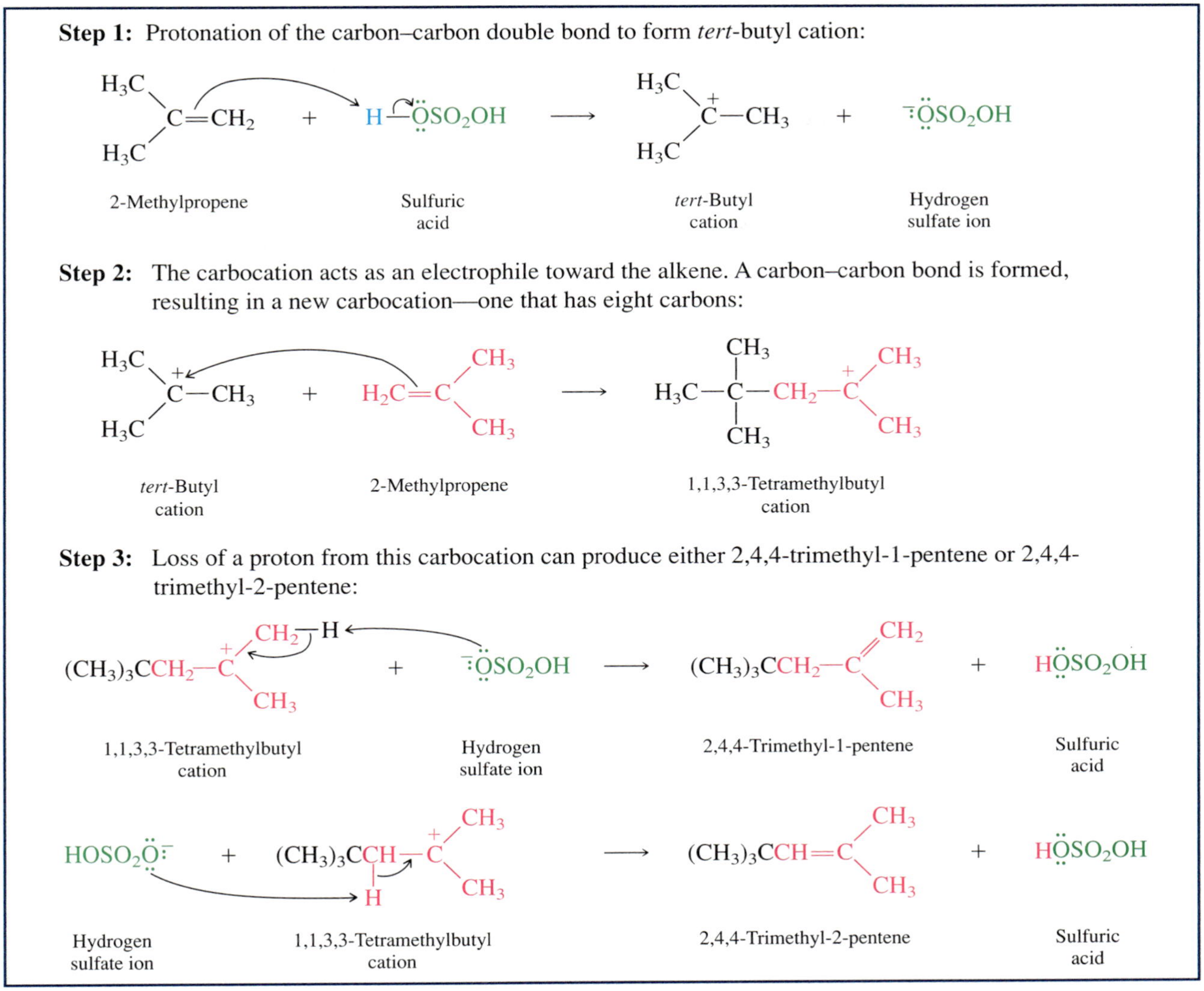

Step 1: Protonation of the carbon–carbon double bond to form *tert*-butyl cation:

| 2-Methylpropene | Sulfuric acid | *tert*-Butyl cation | Hydrogen sulfate ion |

Step 2: The carbocation acts as an electrophile toward the alkene. A carbon–carbon bond is formed, resulting in a new carbocation—one that has eight carbons:

| *tert*-Butyl cation | 2-Methylpropene | 1,1,3,3-Tetramethylbutyl cation |

Step 3: Loss of a proton from this carbocation can produce either 2,4,4-trimethyl-1-pentene or 2,4,4-trimethyl-2-pentene:

| 1,1,3,3-Tetramethylbutyl cation | Hydrogen sulfate ion | 2,4,4-Trimethyl-1-pentene | Sulfuric acid |

| Hydrogen sulfate ion | 1,1,3,3-Tetramethylbutyl cation | 2,4,4-Trimethyl-2-pentene | Sulfuric acid |

FIGURE 6.16 Mechanism of acid-catalyzed dimerization of 2-methylpropene.

Dimerization in concentrated sulfuric acid occurs mainly with those alkenes that form tertiary carbocations. In some cases reaction conditions can be developed that favor the formation of higher molecular-weight polymers. Because these reactions proceed by way of carbocation intermediates, the process is referred to as **cationic polymerization.**

We made special mention in Section 5.1 of the enormous volume of ethylene and propene production in the petrochemical industry. The accompanying box summarizes the principal uses of these alkenes. Most of the ethylene is converted to **polyethylene,** a high-molecular-weight polymer of ethylene. Polyethylene cannot be prepared by cationic polymerization, but is the simplest example of a polymer that is produced on a large scale by **free-radical polymerization.**

In the free-radical polymerization of ethylene, ethylene is heated at high pressure in the presence of oxygen or a peroxide.

The uses to which ethylene and its relatives are put are summarized in an article entitled "Alkenes and Their Derivatives: The Alchemists' Dream Come True," in the August 1989 issue of the *Journal of Chemical Education* (pp. 670–672).

Step 1: Homolytic dissociation of a peroxide produces alkoxy radicals that serve as free-radical initiators:

$$R\ddot{O} : \ddot{O}R \longrightarrow R\ddot{O}\cdot \; + \; \cdot\ddot{O}R$$

Peroxide   Two alkoxy radicals

Step 2: An alkoxy radical adds to the carbon–carbon double bond:

$$R\ddot{O}\cdot \; + \; H_2C\!=\!CH_2 \longrightarrow R\ddot{O}\!-\!CH_2\!-\!\dot{C}H_2$$

Alkoxy   Ethylene   2-Alkoxyethyl
radical   radical

Step 3: The radical produced in step 2 adds to a second molecule of ethylene:

$$R\ddot{O}\!-\!CH_2\!-\!\dot{C}H_2 \; + \; H_2C\!=\!CH_2 \longrightarrow R\ddot{O}\!-\!CH_2\!-\!CH_2\!-\!CH_2\!-\!\dot{C}H_2$$

2-Alkoxyethyl   Ethylene   4-Alkoxybutyl radical
radical

The radical formed in step 3 then adds to a third molecule of ethylene, and the process continues, forming a long chain of methylene groups.

FIGURE 6.17 Mechanism of peroxide-initiated free-radical polymerization of ethylene.

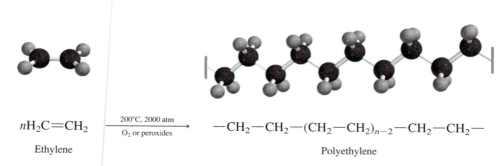

$$n\,H_2C\!=\!CH_2 \xrightarrow[\text{O}_2 \text{ or peroxides}]{200°C, 2000 \text{ atm}} -CH_2-CH_2-(CH_2-CH_2)_{n-2}-CH_2-CH_2-$$

Ethylene   Polyethylene

In this reaction n can have a value of thousands.

The mechanism of free-radical polymerization of ethylene is outlined in Figure 6.17. Dissociation of a peroxide initiates the process in step 1. The resulting peroxy radical adds to the carbon–carbon double bond in step 2, giving a new radical, which then adds to a second molecule of ethylene in step 3. The carbon–carbon bond-forming process in step 3 can be repeated thousands of times to give long carbon chains.

In spite of the *-ene* ending to its name, polyethylene is much more closely related to alk*anes* than to alk*enes*. It is simply a long chain of CH_2 groups bearing at its ends an alkoxy group (from the initiator) or a carbon–carbon double bond.

The properties that make polyethylene so useful come from its alkane-like structure. Except for the ends of the chain, which make up only a tiny portion of the molecule, polyethylene has no functional groups so is almost completely inert to most substances with which it comes in contact.

ETHYLENE AND PROPENE: THE MOST IMPORTANT INDUSTRIAL ORGANIC CHEMICALS

Having examined the properties of alkenes and introduced the elements of polymers and polymerization, let's now look at some commercial applications of ethylene and propene.

ETHYLENE We discussed ethylene production in an earlier boxed essay (Section 5.1), where it was pointed out that the output of the U.S. petrochemical industry exceeds 5×10^{10} lb/year. Approximately 90% of this material is used for the preparation of four compounds (polyethylene, ethylene oxide, vinyl chloride, and styrene), with polymerization to polyethylene accounting for half the total. Both vinyl chloride and styrene are polymerized to give poly(vinyl chloride) and polystyrene, respectively (see Table 6.5). Ethylene oxide is a starting material for the preparation of ethylene glycol for use as an antifreeze in automobile radiators and in the production of polyester fibers (see the boxed essay "Condensation Polymers: Polyamides and Polyesters" in Chapter 20).

PROPENE The major use of propene is in the production of polypropylene. Two other propene-derived organic chemicals, acrylonitrile and propylene oxide, are also starting materials for polymer synthesis. Acrylonitrile is used to make acrylic fibers (see Table 6.5), and propylene oxide is one component in the preparation of *polyurethane* polymers. Cumene itself has no direct uses but rather serves as the starting material in a process that yields two valuable industrial chemicals: acetone and phenol.

We have not indicated the reagents employed in the reactions by which ethylene and propene are converted to the compounds shown. Because of patent requirements, different companies often use different processes. Although the processes may be different, they share the common characteristic of being extremely efficient. The industrial chemist faces the challenge of producing valuable materials, at low cost. Success in the industrial environment requires both an understanding of chemistry and an

$(—CH_2CH_2—)_n$	Polyethylene	(50%)
$H_2C—CH_2$ (epoxide, O)	Ethylene oxide	(20%)
$H_2C=CHCl$	Vinyl chloride	(15%)
⬡—CH=CH₂	Styrene	(5%)
Other chemicals		(10%)

$H_2C=CH_2$ (Ethylene) →

Among the "other chemicals" prepared from ethylene are ethanol and acetaldehyde:

CH_3CH_2OH

Ethanol (industrial solvent; used in preparation of ethyl acetate; unleaded gasoline additive)

$$CH_3\overset{\displaystyle O}{\overset{\displaystyle \|}{C}}H$$

Acetaldehyde (used in preparation of acetic acid)

appreciation of the economics associated with alternative procedures. One measure of how successfully these challenges have been met can be seen in the fact that the United States maintains a positive trade balance in chemicals each year. In 2000 that surplus amounted to $8.9 billion in chemicals versus an overall trade deficit of $436.5 billion.

$(—CH_2—\overset{\displaystyle CH_3}{\underset{\displaystyle |}{CH}}—)_n$	Polypropylene	(35%)
$H_2C=CH—C\equiv N$	Acrylonitrile	(20%)
$H_2C—CHCH_3$ (epoxide, O)	Propylene oxide	(10%)
⬡—CH(CH₃)₂	Cumene	(10%)
Other chemicals		(25%)

$CH_3CH=CH_2$ (Propene) →

TABLE 6.5   Some Compounds with Carbon–Carbon Double Bonds Used to Prepare Polymers

A. Alkenes of the type $H_2C=CH-X$ used to form polymers of the type $(-CH_2-CH-)_n$
$$\underset{X}{|}$$

Compound	Structure	—X in polymer	Application
Ethylene	$H_2C=CH_2$	—H	Polyethylene films as packaging material; "plastic" squeeze bottles are molded from high-density polyethylene.
Propene	$H_2C=CH-CH_3$	—CH_3	Polypropylene fibers for use in carpets and automobile tires; consumer items (luggage, appliances, etc.); packaging material.
Styrene	$H_2C=CH-$⬡ 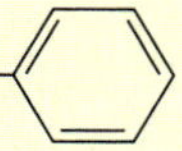	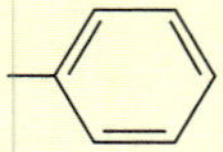 ⬡	Polystyrene packaging, housewares, luggage, radio and television cabinets.
Vinyl chloride	$H_2C=CH-Cl$	—Cl	Poly(vinyl chloride) (PVC) has replaced leather in many of its applications; PVC tubes and pipes are often used in place of copper.
Acrylonitrile	$H_2C=CH-C\equiv N$	—$C\equiv N$	Wool substitute in sweaters, blankets, etc.

B. Alkenes of the type $H_2C=CX_2$ used to form polymers of the type $(-CH_2-CX_2-)_n$

Compound	Structure	X in polymer	Application
1,1-Dichloroethene (vinylidene chloride)	$H_2C=CCl_2$	Cl	Saran used as air- and water-tight packaging film.
2-Methylpropene	$H_2C=C(CH_3)_2$	CH_3	Polyisobutene is component of "butyl rubber," one of earliest synthetic rubber substitutes.

C. Others

Compound	Structure	Polymer	Application
Tetrafluoroethene	$F_2C=CF_2$	$(-CF_2-CF_2-)_n$ (Teflon)	Nonstick coating for cooking utensils; bearings, gaskets, and fittings.
Methyl methacrylate	$H_2C=CCO_2CH_3$ $\|$ CH_3	$(-CH_2-\overset{\overset{\textstyle CO_2CH_3}{\|}}{\underset{\underset{\textstyle CH_3}{\|}}{C}}-)_n$	When cast in sheets, is transparent; used as glass substitute (Lucite, Plexiglas).
2-Methyl-1,3-butadiene	$H_2C=CCH=CH_2$ $\|$ CH_3	$(-CH_2C=CH-CH_2-)_n$ $\|$ CH_3 (Polyisoprene)	Synthetic rubber.

Source: R. C. Atkins and F. A. Carey, *Organic Chemistry: A Brief Course,* 3rd ed. McGraw-Hill, New York, 2002, p. 237.

Teflon is made in a similar way by free-radical polymerization of tetrafluoroethene.

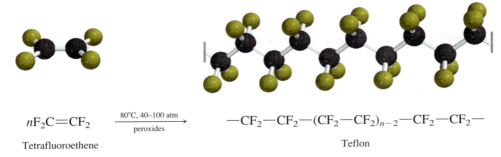

$$n\text{F}_2\text{C}{=}\text{CF}_2 \xrightarrow[\text{peroxides}]{80°\text{C, }40\text{–}100\text{ atm}} {-}\text{CF}_2{-}\text{CF}_2{-}(\text{CF}_2{-}\text{CF}_2)_{n-2}{-}\text{CF}_2{-}\text{CF}_2{-}$$

Tetrafluoroethene   Teflon

Carbon–fluorine bonds are quite strong (slightly stronger than C—H bonds), and like polyethylene, Teflon is a very stable, inert material. We are all familiar with the most characteristic property of Teflon, its "nonstick" surface. This can be understood by comparing Teflon and polyethylene. The high electronegativity of fluorine makes C—F bonds less polarizable than C—H bonds, causing the dispersion forces in Teflon to be less than those in polyethylene. Thus, the surface of Teflon is even less "sticky" than the already slick surface of polyethylene.

A large number of compounds with carbon–carbon double bonds have been polymerized to yield materials having useful properties. Some of the more important or familiar of these are listed in Table 6.5. Not all these monomers are effectively polymerized under free-radical conditions, and much research has been carried out to develop alternative polymerization techniques. One of these, **coordination polymerization,** employs novel transition-metal catalysts. Polyethylene produced by coordination polymerization has a higher density than that produced by free-radical polymerization and somewhat different—in many applications, more desirable—properties. Coordination polymerization was developed independently by Karl Ziegler in Germany and Giulio Natta in Italy in the early 1950s. They shared the Nobel Prize in chemistry in 1963 for this work. Coordination polymerization gives a form of **polypropylene** suitable for plastics and fibers. When propene is polymerized under free-radical conditions, the polypropylene has physical properties (such as a low melting point) that make it useless for most applications.

> Coordination polymerization is described in more detail in Sections 7.15 and 14.15.

6.22 SUMMARY

Alkenes are **unsaturated hydrocarbons** and react with substances that add to the double bond.

Section 6.1   See Table 6.6.

Section 6.2   Hydrogenation of alkenes is exothermic. Heats of hydrogenation can be measured and used to assess the stability of various types of double bonds. The information parallels that obtained from heats of combustion.

Section 6.3   Hydrogenation of alkenes is a syn addition.

Sections 6.4–6.7   See Table 6.6. Hydrogen halide addition to alkenes proceeds by electrophilic attack of the reagent on the π electrons of the double bond. Carbocations are intermediates.

$$\underset{\text{Alkene}}{\diagdown\!\!\!\diagup\!\!C=C\!\!\diagup\!\!\!\diagdown} + \underset{\substack{\text{Hydrogen} \\ \text{halide}}}{H-X} \longrightarrow \underset{\text{Carbocation}}{\overset{+}{\diagup\!\!C-\diagup\!\!C-H}} + \underset{\substack{\text{Halide} \\ \text{ion}}}{X^-} \longrightarrow \underset{\text{Alkyl halide}}{X-\overset{|}{\underset{|}{C}}-\overset{|}{\underset{|}{C}}-H}$$

Protonation of the double bond occurs in the direction that gives the more stable of two possible carbocations.

TABLE 6.6   Addition Reactions of Alkenes

Reaction (section) and comments	General equation and specific example
Catalytic hydrogenation (Sections 6.1–6.3) Alkenes react with hydrogen in the presence of a platinum, palladium, rhodium, or nickel catalyst to form the corresponding alkane.	$\underset{\text{Alkene}}{R_2C=CR_2} + \underset{\text{Hydrogen}}{H_2} \xrightarrow{\text{Pt, Pd, Rh, or Ni}} \underset{\text{Alkane}}{R_2CHCHR_2}$ cis-Cyclododecene $\xrightarrow[\text{Pt}]{H_2}$ Cyclododecane (100%)
Addition of hydrogen halides (Sections 6.4–6.7) A proton and a halogen add to the double bond of an alkene to yield an alkyl halide. Addition proceeds in accordance with Markovnikov's rule; hydrogen adds to the carbon that has the greater number of hydrogens, halide to the carbon that has the fewer hydrogens.	$\underset{\text{Alkene}}{RCH=CR_2'} + \underset{\substack{\text{Hydrogen} \\ \text{halide}}}{HX} \longrightarrow \underset{\substack{\text{Alkyl} \\ \text{halide}}}{RCH_2-\overset{|}{\underset{X}{CR_2'}}}$ Methylenecyclohexane $=CH_2$ + Hydrogen chloride HCl $\longrightarrow$ 1-Chloro-1-methylcyclohexane (75–80%)
Addition of sulfuric acid (Section 6.9) Alkenes react with sulfuric acid to form alkyl hydrogen sulfates. A proton and a hydrogen sulfate ion add to the double bond in accordance with Markovnikov's rule. Alkenes that yield tertiary carbocations on protonation tend to polymerize in concentrated sulfuric acid (Section 6.21).	$\underset{\text{Alkene}}{RCH=CR_2'} + \underset{\text{Sulfuric acid}}{HOSO_2OH} \longrightarrow \underset{\text{Alkyl hydrogen sulfate}}{RCH_2-\overset{|}{\underset{OSO_2OH}{CR_2'}}}$ $\underset{\text{1-Butene}}{H_2C=CHCH_2CH_3} + \underset{\text{Sulfuric acid}}{HOSO_2OH} \longrightarrow \underset{\substack{\textit{sec}\text{-Butyl hydrogen} \\ \text{sulfate}}}{CH_3-\overset{|}{\underset{OSO_2OH}{CHCH_2CH_3}}}$
Acid-catalyzed hydration (Section 6.10) Addition of water to the double bond of an alkene takes place in aqueous acid. Addition occurs according to Markovnikov's rule. A carbocation is an intermediate and is captured by a molecule of water acting as a nucleophile.	$\underset{\text{Alkene}}{RCH=CR_2'} + \underset{\text{Water}}{H_2O} \xrightarrow{H^+} \underset{\text{Alcohol}}{RCH_2\overset{|}{\underset{OH}{CR_2'}}}$ $\underset{\text{2-Methylpropene}}{H_2C=C(CH_3)_2} \xrightarrow{50\%\ H_2SO_4/H_2O} \underset{\substack{\textit{tert}\text{-Butyl alcohol} \\ (55\text{–}58\%)}}{(CH_3)_3COH}$

(Continued)

TABLE 6.6	Addition Reactions of Alkenes *(Continued)*

Reaction (section) and comments	General equation and specific example
Hydroboration–oxidation (Sections 6.11–6.13) This two-step sequence achieves hydration of alkenes in a stereospecific syn manner, with a regioselectivity opposite to Markovnikov's rule. An organoborane is formed by electrophilic addition of diborane to an alkene. Oxidation of the organoborane intermediate with hydrogen peroxide completes the process. Rearrangements do not occur.	$RCH{=}CR_2' \xrightarrow[\text{2. } H_2O_2,\ HO^-]{\text{1. } B_2H_6,\ \text{diglyme}} RCHCHR_2'$ 　　　　　　　　　　　　　　　　　$\mid$ 　　　　　　　　　　　　　　　　　OH 　Alkene　　　　　　　　　　　Alcohol $(CH_3)_2CHCH_2CH{=}CH_2 \xrightarrow[\text{2. } H_2O_2,\ HO^-]{\text{1. } H_3B{\cdot}THF} (CH_3)_2CHCH_2CH_2CH_2OH$ 　4-Methyl-1-pentene　　　　　　　　4-Methyl-1-pentanol 　　　　　　　　　　　　　　　　　　　　(80%)
Addition of halogens (Sections 6.14–6.16) Bromine and chlorine add to alkenes to form vicinal dihalides. A cyclic halonium ion is an intermediate. Stereospecific anti addition is observed.	$R_2C{=}CR_2 + X_2 \longrightarrow$　　$\overset{\displaystyle R\ \ \ R}{\underset{\displaystyle R\ \ \ R}{X{-}C{-}C{-}X}}$ Alkene　　Halogen　　Vicinal dihalide $H_2C{=}CHCH_2CH_2CH_2CH_3 + Br_2 \longrightarrow BrCH_2{-}CHCH_2CH_2CH_2CH_3$ 　　　　　　　　　　　　　　　　　　　　　　　　　$\mid$ 　　　　　　　　　　　　　　　　　　　　　　　　Br 　1-Hexene　　　　　　　Bromine　　1,2-Dibromohexane (100%)
Halohydrin formation (Section 6.17) When treated with bromine or chlorine in aqueous solution, alkenes are converted to vicinal halohydrins. A halonium ion is an intermediate. The halogen adds to the carbon that has the greater number of hydrogens. Addition is anti.	$RCH{=}CR_2' + X_2 + H_2O \longrightarrow \overset{\displaystyle R'}{\underset{\displaystyle R\ \ \ R'}{X{-}CH{-}C{-}OH}} + HX$ Alkene　Halogen　Water　　Vicinal　　Hydrogen 　　　　　　　　　　　　　halohydrin　halide Methylenecyclohexane $\xrightarrow[H_2O]{Br_2}$ (1-Bromomethyl)cyclohexanol (89%)
Epoxidation (Section 6.18) Peroxy acids transfer oxygen to the double bond of alkenes to yield epoxides. The reaction is a stereospecific syn addition.	$R_2C{=}CR_2 + R'\overset{\displaystyle O}{\overset{\displaystyle \|}{C}}OOH \longrightarrow R_2C{-}CR_2 + R'\overset{\displaystyle O}{\overset{\displaystyle \|}{C}}OH$ 　　　　　　　　　　　　　　　　$\underset{O}{\diagdown\diagup}$ Alkene　　Peroxy　　　　Epoxide　Carboxylic 　　　　　acid　　　　　　　　　　acid 1-Methylcycloheptene + $CH_3\overset{\displaystyle O}{\overset{\displaystyle \|}{C}}OOH \longrightarrow$ 1-Methyl-1,2-epoxycycloheptane (65%) + $CH_3\overset{\displaystyle O}{\overset{\displaystyle \|}{C}}OH$ 　　　　　　　Peroxyacetic　　　　　　　　　　　　Acetic 　　　　　　　acid　　　　　　　　　　　　　　　acid

Section 6.8   Hydrogen bromide is unique among the hydrogen halides in that it can add to alkenes either by electrophilic or free-radical addition. Under photochemical conditions or in the presence of peroxides, free-radical addition is observed, and HBr adds to the double bond with a regio-selectivity opposite to that of Markovnikov's rule.

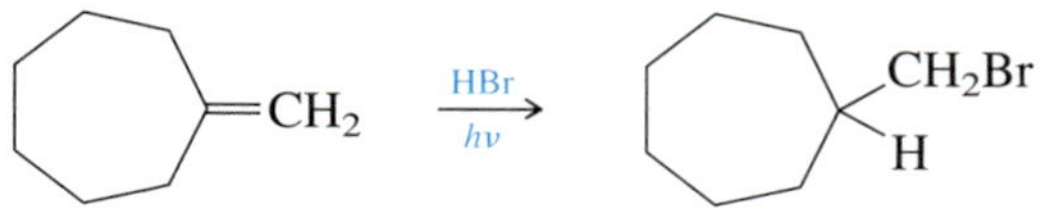

Methylenecycloheptane   (Bromomethyl)cycloheptane
                                            (61%)

Sections   See Table 6.6.
6.9–6.18

Section 6.19   Alkenes are cleaved to carbonyl compounds by **ozonolysis.** This reaction is useful both for synthesis (preparation of aldehydes, ketones, or car-boxylic acids) and analysis. When applied to analysis, the carbonyl com-pounds are isolated and identified, allowing the substituents attached to the double bond to be deduced.

$$CH_3CH{=}C(CH_2CH_3)_2 \xrightarrow[\text{2. Zn, H}_2\text{O}]{\text{1. O}_3} CH_3\overset{\overset{\text{O}}{\|}}{C}H \;+\; CH_3CH_2\overset{\overset{\text{O}}{\|}}{C}CH_2CH_3$$

3-Ethyl-2-pentene   Acetaldehyde   3-Pentanone

Section 6.20   The reactions described so far can be carried out sequentially to prepare compounds of prescribed structure from some given starting material. The best way to approach a synthesis is to reason backward from the desired target molecule and to always use reactions that you are sure will work. The 11 exercises that make up Problem 6.32 at the end of this chapter provide some opportunities for practice.

Section 6.21   In their **polymerization,** many individual alkene molecules combine to give a high-molecular-weight product. Among the methods for alkene polymerization, *cationic polymerization, coordination polymerization,* and *free-radical polymerization* are the most important. An example of cationic polymerization is:

$$n(CH_3)_2C{=}CH_2 \xrightarrow{H^+} \left(\!\begin{array}{c}CH_3\\|\\C\\|\\CH_3\end{array}\!\!-CH_2-\!\!\begin{array}{c}CH_3\\|\\C\\|\\CH_3\end{array}\!\!-CH_2\!\right)_{\!n/2}$$

2-Methylpropene   Polyisobutylene

PROBLEMS

6.22 Write the structure of the major organic product formed in the reaction of 1-pentene with each of the following:

   (a) Hydrogen chloride

   (b) Hydrogen bromide

   (c) Hydrogen bromide in the presence of peroxides

(d) Hydrogen iodide

(e) Dilute sulfuric acid

(f) Diborane in diglyme, followed by basic hydrogen peroxide

(g) Bromine in carbon tetrachloride

(h) Bromine in water

(i) Peroxyacetic acid

(j) Ozone

(k) Product of part (j) treated with zinc and water

6.23 Repeat Problem 6.22 for 2-methyl-2-butene.

6.24 Repeat Problem 6.22 for 1-methylcyclohexene.

6.25 Match the following alkenes with the appropriate heats of hydrogenation:

(a) 1-Pentene

(b) (*E*)-4,4-Dimethyl-2-pentene

(c) (*Z*)-4-Methyl-2-pentene

(d) (*Z*)-2,2,5,5-Tetramethyl-3-hexene

(e) 2,4-Dimethyl-2-pentene

Heats of hydrogenation in kJ/mol (kcal/mol): 151(36.2); 122(29.3); 114(27.3); 111(26.5); 105(25.1).

6.26 (a) How many alkenes yield 2,2,3,4,4-pentamethylpentane on catalytic hydrogenation?

(b) How many yield 2,3-dimethylbutane?

(c) How many yield methylcyclobutane?

6.27 Two alkenes undergo hydrogenation to yield a mixture of *cis*- and *trans*-1,4-dimethyl-cyclohexane. A third, however, gives only *cis*-1,4-dimethylcyclohexane. What compound is this?

6.28 Specify reagents suitable for converting 3-ethyl-2-pentene to each of the following:

(a) 2,3-Dibromo-3-ethylpentane

(b) 3-Chloro-3-ethylpentane

(c) 2-Bromo-3-ethylpentane

(d) 3-Ethyl-3-pentanol

(e) 3-Ethyl-2-pentanol

(f) 3-Ethyl-2,3-epoxypentane

(g) 3-Ethylpentane

6.29 (a) Which primary alcohol of molecular formula $C_5H_{12}O$ cannot be prepared from an alkene? Why?

(b) Write equations describing the preparation of three isomeric primary alcohols of molecular formula $C_5H_{12}O$ from alkenes.

(c) Write equations describing the preparation of the tertiary alcohol of molecular formula $C_5H_{12}O$ from two different alkenes.

6.30 All the following reactions have been reported in the chemical literature. Give the structure of the principal organic product in each case.

(a) $CH_3CH_2CH\!=\!CHCH_2CH_3 \ + \ HBr \ \xrightarrow{\text{no peroxides}}$

(b) $(CH_3)_2CHCH_2CH_2CH_2CH\!=\!CH_2 \ \xrightarrow[\text{peroxides}]{\text{HBr}}$

(c) 2-*tert*-Butyl-3,3-dimethyl-1-butene $\xrightarrow[\text{2. H}_2\text{O}_2,\ \text{HO}^-]{\text{1. B}_2\text{H}_6}$

(d) 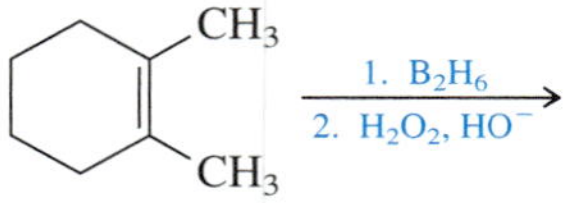 $\xrightarrow[\text{2. H}_2\text{O}_2,\ \text{HO}^-]{\text{1. B}_2\text{H}_6}$

(e) $H_2C{=}CCH_2CH_2CH_3$ + Br_2 $\xrightarrow{\text{CHCl}_3}$
       with CH$_3$ substituent

(f) $(CH_3)_2C{=}CHCH_3$ + Br_2 $\xrightarrow{\text{H}_2\text{O}}$

(g) 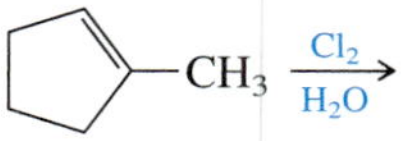 —CH_3 $\xrightarrow[\text{H}_2\text{O}]{\text{Cl}_2}$

(h) $(CH_3)_2C{=}C(CH_3)_2$ + $CH_3\overset{\overset{\text{O}}{\|}}{C}OOH$ $\longrightarrow$

(i) 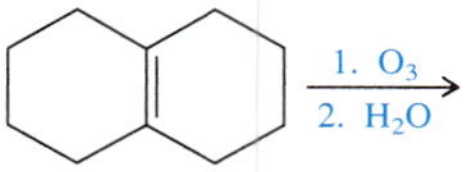 $\xrightarrow[\text{2. H}_2\text{O}]{\text{1. O}_3}$

6.31 A single epoxide was isolated in 79–84% yield in the following reaction. Was this epoxide A or B? Explain your reasoning.

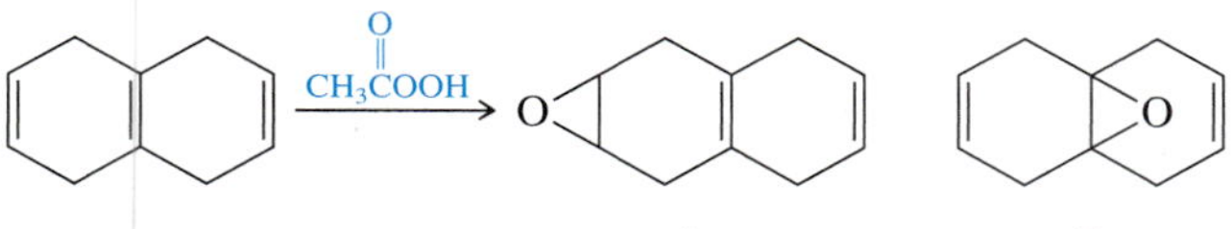

A   B

6.32 Suggest a sequence of reactions suitable for preparing each of the following compounds from the indicated starting material. You may use any necessary organic or inorganic reagents.

(a) 1-Propanol from 2-propanol

(b) 1-Bromopropane from 2-bromopropane

(c) 1,2-Dibromopropane from 2-bromopropane

(d) 1-Bromo-2-propanol from 2-propanol

(e) 1,2-Epoxypropane from 2-propanol

(f) *tert*-Butyl alcohol from isobutyl alcohol

(g) *tert*-Butyl iodide from isobutyl iodide

(h) *trans*-2-Chlorocyclohexanol from cyclohexyl chloride

(i) Cyclopentyl iodide from cyclopentane

(j) *trans*-1,2-Dichlorocyclopentane from cyclopentane

(k) $H\overset{\overset{\text{O}}{\|}}{C}CH_2CH_2CH_2\overset{\overset{\text{O}}{\|}}{C}H$ from cyclopentanol

6.33 Two different compounds having the molecular formula $C_8H_{15}Br$ are formed when 1,6-dimethylcyclohexene reacts with hydrogen bromide in the dark and in the absence of peroxides. The same two compounds are formed from 1,2-dimethylcyclohexene. What are these two compounds?

6.34 On catalytic hydrogenation over a rhodium catalyst, the compound shown gave a mixture containing *cis*-1-*tert*-butyl-4-methylcyclohexane (88%) and *trans*-1-*tert*-butyl-4-methylcyclo-hexane (12%). With this stereochemical result in mind, consider the reactions in (a) and (b).

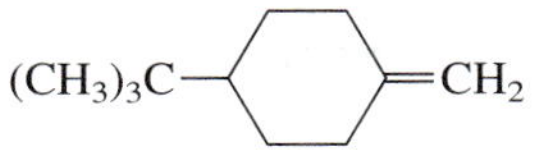

$$(CH_3)_3C-\!\!\bigcirc\!\!=CH_2$$

4-*tert*-Butyl(methylene)cyclohexane

(a) What two products are formed in the epoxidation of 4-*tert*-butyl(methylene)cyclo-hexane? Which one do you think will predominate?

(b) What two products are formed in the hydroboration–oxidation of 4-*tert*-butyl(methyl-ene)cyclohexane? Which one do you think will predominate?

6.35 Compound A undergoes catalytic hydrogenation much faster than does compound B. Why? Making molecular models will help.

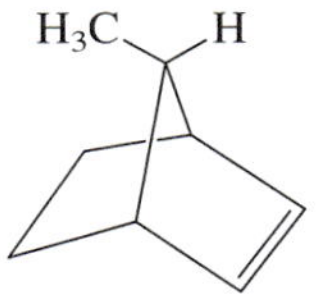

    A 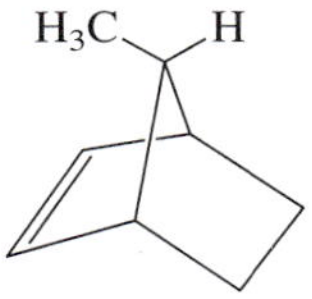   B

6.36 Catalytic hydrogenation of 1,4-dimethylcyclopentene yields a mixture of two products. Iden-tify them. One of them is formed in much greater amounts than the other (observed ratio =10:1). Which one is the major product?

6.37 There are two products that can be formed by syn addition of hydrogen to 2,3-dimethylbi-cyclo[2.2.1]-2-heptene. Write or make molecular models of their structures.

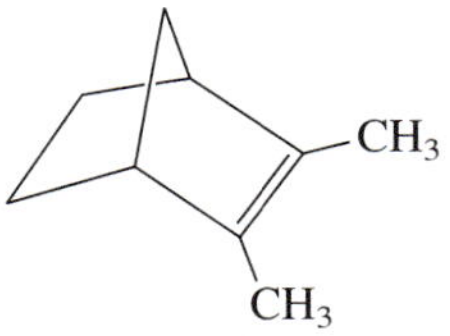

2,3-Dimethylbicyclo[2.2.1]-2-heptene

6.38 Hydrogenation of 3-carene is, in principle, capable of yielding two stereoisomeric products. Write their structures. Only one of them was actually obtained on catalytic hydrogenation over platinum. Which one do you think is formed? Explain your reasoning with the aid of a drawing or a molecular model.

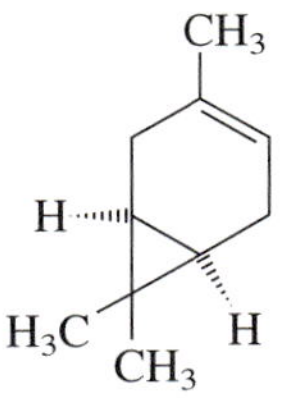

3-Carene

6.39 In a widely used industrial process, the mixture of ethylene and propene that is obtained by dehydrogenation of natural gas is passed into concentrated sulfuric acid. Water is added, and the solution is heated to hydrolyze the alkyl hydrogen sulfate. The product is almost exclusively a sin-gle alcohol. Is this alcohol ethanol, 1-propanol, or 2-propanol? Why is this particular one formed almost exclusively?

6.40 On the basis of the mechanism of acid-catalyzed hydration, can you suggest a reason why the reaction

$$H_2C{=}CHCH(CH_3)_2 \xrightarrow[\text{H}_2\text{O}]{\text{H}_2\text{SO}_4} CH_3\underset{\underset{\text{OH}}{|}}{C}HCH(CH_3)_2$$

would probably *not* be a good method for the synthesis of 3-methyl-2-butanol?

6.41 As a method for the preparation of alkenes, a weakness in the acid-catalyzed dehydration of alcohols is that the initially formed alkene (or mixture of alkenes) sometimes isomerizes under the conditions of its formation. Write a stepwise mechanism showing how 2-methyl-1-butene might isomerize to 2-methyl-2-butene in the presence of sulfuric acid.

6.42 When bromine is added to a solution of 1-hexene in methanol, the major products of the reaction are as shown:

$$H_2C{=}CHCH_2CH_2CH_2CH_3 \xrightarrow[\text{CH}_3\text{OH}]{\text{Br}_2} BrCH_2\underset{\underset{\text{Br}}{|}}{C}HCH_2CH_2CH_2CH_3 + BrCH_2\underset{\underset{\text{OCH}_3}{|}}{C}HCH_2CH_2CH_2CH_3$$

1-Hexene	1,2-Dibromohexane	1-Bromo-2-methoxyhexane

1,2-Dibromohexane is not converted to 1-bromo-2-methoxyhexane under the reaction conditions. Suggest a reasonable mechanism for the formation of 1-bromo-2-methoxyhexane.

6.43 The reaction of thiocyanogen (N≡CS—SC≡N) with *cis*-cyclooctene proceeds by anti addition.

A bridged *sulfonium ion* is presumed to be an intermediate. Write a stepwise mechanism for this reaction.

6.44 On the basis of the mechanism of cationic polymerization, predict the alkenes of molecular formula $C_{12}H_{24}$ that can most reasonably be formed when 2-methylpropene [$(CH_3)_2C{=}CH_2$] is treated with sulfuric acid.

6.45 On being heated with a solution of sodium ethoxide in ethanol, compound A ($C_7H_{15}Br$) yielded a mixture of two alkenes B and C, each having the molecular formula C_7H_{14}. Catalytic hydrogenation of the major isomer B or the minor isomer C gave only 3-ethylpentane. Suggest structures for compounds A, B, and C consistent with these observations.

6.46 Compound A ($C_7H_{15}Br$) is not a primary alkyl bromide. It yields a single alkene (compound B) on being heated with sodium ethoxide in ethanol. Hydrogenation of compound B yields 2,4-dimethylpentane. Identify compounds A and B.

6.47 Compounds A and B are isomers of molecular formula $C_9H_{19}Br$. Both yield the same alkene C as the exclusive product of elimination on being treated with potassium *tert*-butoxide in dimethyl sulfoxide. Hydrogenation of alkene C gives 2,3,3,4-tetramethylpentane. What are the structures of compounds A and B and alkene C?

6.48 Alcohol A ($C_{10}H_{18}O$) is converted to a mixture of alkenes B and C on being heated with potassium hydrogen sulfate ($KHSO_4$). Catalytic hydrogenation of B and C yields the same product. Assuming that dehydration of alcohol A proceeds without rearrangement, deduce the structures of alcohol A and alkene C.

Compound B

6.49 Reaction of 3,3-dimethyl-1-butene with hydrogen iodide yields two compounds A and B, each having the molecular formula $C_6H_{13}I$, in the ratio A:B $=$ 90:10. Compound A, on being heated with potassium hydroxide in *n*-propyl alcohol, gives only 3,3-dimethyl-1-butene. Compound B undergoes elimination under these conditions to give 2,3-dimethyl-2-butene as the major product. Suggest structures for compounds A and B, and write a reasonable mechanism for the formation of each.

6.50 Dehydration of 2,2,3,4,4-pentamethyl-3-pentanol gave two alkenes A and B. Ozonolysis of the lower boiling alkene A gave formaldehyde ($H_2C\!=\!O$) and 2,2,4,4-tetramethyl-3-pentanone. Ozonolysis of B gave formaldehyde and 3,3,4,4-tetramethyl-2-pentanone. Identify A and B, and suggest an explanation for the formation of B in the dehydration reaction.

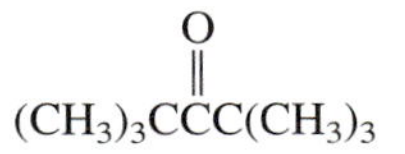

$$ \underset{\text{2,2,4,4-Tetramethyl-3-pentanone}}{(CH_3)_3CCC(CH_3)_3} \qquad \underset{\text{3,3,4,4-Tetramethyl-2-pentanone}}{CH_3CCC(CH_3)_3} $$

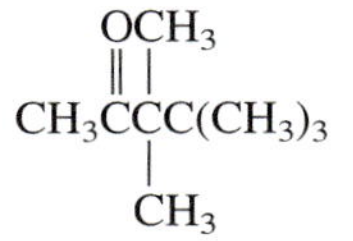

6.51 Compound A ($C_7H_{13}Br$) is a tertiary bromide. On treatment with sodium ethoxide in ethanol, A is converted into B (C_7H_{12}). Ozonolysis of B gives C as the only product. Deduce the structures of A and B. What is the symbol for the reaction mechanism by which A is converted to B under the reaction conditions?

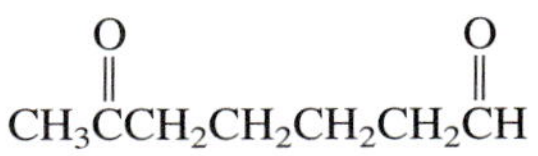

$$ CH_3CCH_2CH_2CH_2CH_2CH $$

Compound C

6.52 East Indian sandalwood oil contains a hydrocarbon given the name *santene* (C_9H_{14}). Ozonation of santene followed by hydrolysis gives compound A. What is the structure of santene?

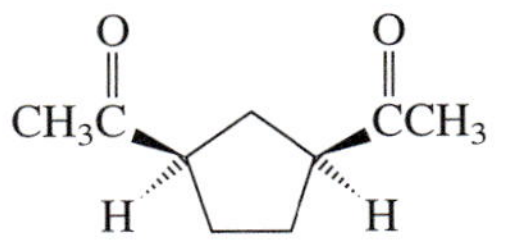

Compound A

6.53 *Sabinene* and Δ^3-*carene* are isomeric natural products with the molecular formula $C_{10}H_{16}$. (a) Ozonolysis of sabinene followed by hydrolysis in the presence of zinc gives compound A. What is the structure of sabinene? What other compound is formed on ozonolysis? (b) Ozonolysis of Δ^3-carene followed by hydrolysis in the presence of zinc gives compound B. What is the structure of Δ^3-carene?

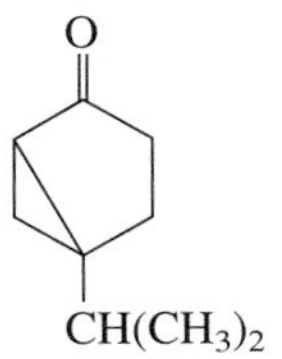

Compound A 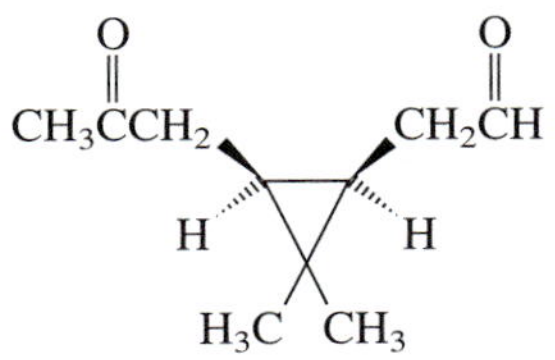   Compound B

6.54 The sex attractant by which the female housefly attracts the male has the molecular formula $C_{23}H_{46}$. Catalytic hydrogenation yields an alkane of molecular formula $C_{23}H_{48}$. Ozonolysis yields

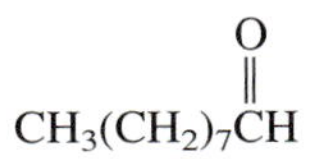

$$ CH_3(CH_2)_7CH \qquad \text{and} \qquad CH_3(CH_2)_{12}CH $$

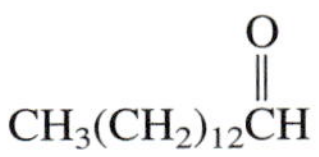

What is the structure of the housefly sex attractant?

6.55 A certain compound of molecular formula $C_{19}H_{38}$ was isolated from fish oil and from plankton. On hydrogenation it gave 2,6,10,14-tetramethylpentadecane. Ozonolysis gave $(CH_3)_2C{=}O$ and a 16-carbon aldehyde. What is the structure of the natural product? What is the structure of the aldehyde?

6.56 The sex attractant of the female arctiid moth contains, among other components, a compound of molecular formula $C_{21}H_{40}$ that yields

$$CH_3(CH_2)_{10}\overset{\displaystyle O}{\overset{\displaystyle \|}{C}}H \qquad CH_3(CH_2)_4\overset{\displaystyle O}{\overset{\displaystyle \|}{C}}H \qquad \text{and} \qquad H\overset{\displaystyle O}{\overset{\displaystyle \|}{C}}CH_2\overset{\displaystyle O}{\overset{\displaystyle \|}{C}}H$$

on ozonolysis. What is the constitution of this material?

6.57 Construct a molecular model of the product formed by catalytic hydrogenation of 1,2-dimethylcyclohexene. Assume syn addition occurs.

6.58 Construct a molecular model of the product formed by anti addition of Br_2 to 1,2-dimethylcyclohexene.

6.59 Examine the electrostatic potential map of $H_3B \cdot THF$ (borane–tetrahydrofuran complex) on *Learning By Modeling*. How does the electrostatic potential of the hydrogens bonded to boron differ from the potential of the hydrogens of the tetrahydrofuran ring?

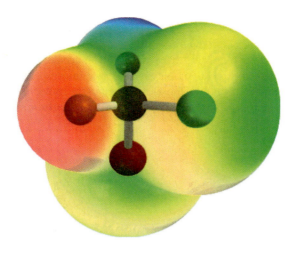

CHAPTER 7

STEREOCHEMISTRY

*S*tereochemistry refers to chemistry in three dimensions. Its foundations were laid by Jacobus van't Hoff[*] and Joseph Achille Le Bel in 1874. Van't Hoff and Le Bel independently proposed that the four bonds to carbon were directed toward the corners of a tetrahedron. One consequence of a tetrahedral arrangement of bonds to carbon is that two compounds may be different because the arrangement of their atoms in space is different. Isomers that have the same constitution but differ in the spatial arrangement of their atoms are called **stereoisomers.** We have already had considerable experience with certain types of stereoisomers—those involving cis and trans substitution patterns in alkenes and in cycloalkanes.

Our major objectives in this chapter are to develop a feeling for molecules as three-dimensional objects and to become familiar with stereochemical principles, terms, and notation. A full understanding of organic and biological chemistry requires an awareness of the spatial requirements for interactions between molecules; this chapter provides the basis for that understanding.

7.1 MOLECULAR CHIRALITY: ENANTIOMERS

Everything has a mirror image, but not all things are superimposable on their mirror images. Mirror-image superimposability characterizes many objects we use every day. Cups and saucers, forks and spoons, chairs and beds are all identical with their mirror images. Many other objects though—and this is the more interesting case—are not. Your left hand and your right hand, for example, are mirror images of each other but can't be made to coincide point for point, palm to palm, knuckle to knuckle, in three dimensions. In 1894, William Thomson (Lord Kelvin) coined a word for this property. He defined an object as **chiral** if it is not superimposable on its mirror image. Applying Thomson's term to chemistry, we

---

[*]Van't Hoff was the recipient of the first Nobel Prize in chemistry in 1901 for his work in chemical dynamics and osmotic pressure—two topics far removed from stereochemistry.

say that a *molecule is chiral if its two mirror-image forms are not superimposable in three dimensions*. The word *chiral* is derived from the Greek word *cheir*, meaning "hand," and it is entirely appropriate to speak of the "handedness" of molecules. The opposite of chiral is **achiral.** A molecule that *is* superimposable on its mirror image is achiral.

In organic chemistry, chirality most often occurs in molecules that contain a carbon that is attached to four different groups. An example is bromochlorofluoromethane (BrClFCH).

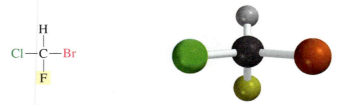

Bromochlorofluoromethane is a known compound, and samples selectively enriched in each enantiomer have been described in the chemical literature. In 1989 two chemists at Polytechnic University (Brooklyn, New York) described a method for the preparation of BrClFCH that is predominantly one enantiomer.

Bromochlorofluoromethane

As shown in Figure 7.1, the two mirror images of bromochlorofluoromethane cannot be superimposed on each other. *Because the two mirror images of bromochlorofluoromethane are not superimposable, BrClFCH is chiral.*

The mirror images of bromochlorofluoromethane have the same constitution. That is, the atoms are connected in the same order. But they differ in the arrangement of their atoms in space; they are **stereoisomers.** Stereoisomers that are related as an object and its nonsuperimposable mirror image are classified as **enantiomers.** The word *enantiomer* describes a particular relationship between two objects. One cannot look at a single molecule in isolation and ask if it is an enantiomer any more than one can look at an individual human being and ask, "Is that person a cousin?" Furthermore, just as an object has one, and only one, mirror image, a chiral molecule can have one, and only one, enantiomer.

Notice in Figure 7.1*c*, where the two enantiomers of bromochlorofluoromethane are similarly oriented, that the difference between them corresponds to an interchange of the positions of bromine and chlorine. It will generally be true for species of the type C(*w, x, y, z*), where *w, x, y,* and *z* are different atoms or groups, that an exchange of two of them converts a structure to its enantiomer, but an exchange of three returns the original structure, albeit in a different orientation.

Consider next a molecule such as chlorodifluoromethane (ClF_2CH), in which two of the atoms attached to carbon are the same. Figure 7.2 shows two molecular models of ClF_2CH drawn so as to be mirror images. As is evident from these drawings, it is a simple matter to merge the two models so that all the atoms match. *Because mirror-image representations of chlorodifluoromethane are superimposable on each other, ClF_2CH is achiral.*

The surest test for chirality is a careful examination of mirror-image forms for superimposability. Working with models provides the best practice in dealing with molecules as three-dimensional objects and is strongly recommended.

7.2 THE CHIRALITY CENTER

As we've just seen, molecules of the general type

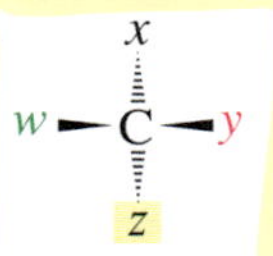

(*a*) Structures A and B are mirror-image representations of bromochlorofluoromethane (BrClFCH).

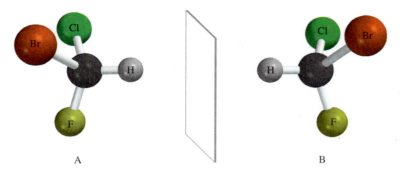

A   B

(*b*) To test for superimposability, reorient B by turning it 180°.

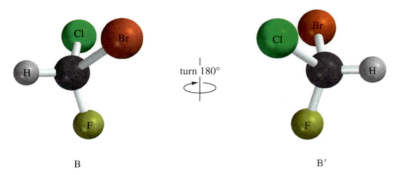

turn 180°

B   B′

(*c*) Compare A and B′. The two do not match. A and B′ cannot be superimposed on each other. Bromochlorofluoromethane is therefore a chiral molecule. The two mirror-image forms are enantiomers of each other.

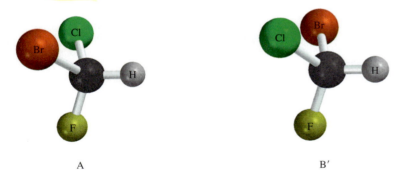

A   B′

FIGURE 7.1 A molecule with four different groups attached to a single carbon is chiral. Its two mirror-image forms are not superimposable.

are chiral when *w*, *x*, *y*, and *z* are different. In 1996, the IUPAC recommended that a tetrahedral carbon atom that bears four different atoms or groups be called a **chirality center,** which is the term that we will use. Several earlier terms, including *asymmetric center, asymmetric carbon, chiral center, stereogenic center,* and *stereocenter,* are still widely used.

Noting the presence of one (but not more than one) chirality center is a simple, rapid way to determine if a molecule is chiral. For example, C-2 is a chirality center in

The 1996 IUPAC recommendations for stereochemical terms can be viewed at http://www.chem.qmw.ac.uk/iupac/stereo

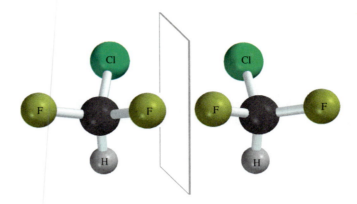

2-butanol; it bears a hydrogen atom and methyl, ethyl, and hydroxyl groups as its four different substituents. By way of contrast, none of the carbon atoms bear four different groups in the achiral alcohol 2-propanol.

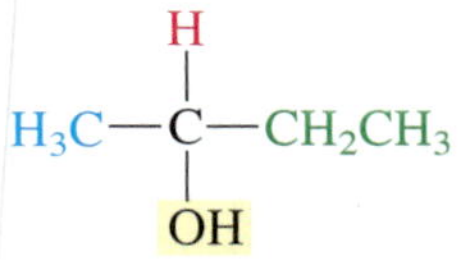

$$H_3C-\underset{\underset{OH}{|}}{\overset{\overset{H}{|}}{C}}-CH_2CH_3 \qquad H_3C-\underset{\underset{OH}{|}}{\overset{\overset{H}{|}}{C}}-CH_3$$

2-Butanol
Chiral; four different
substituents at C-2

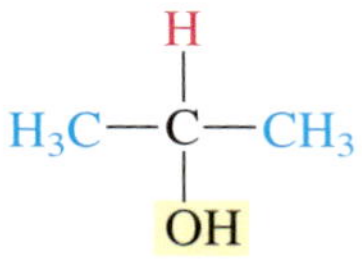

2-Propanol
Achiral; two of the substituents
at C-2 are the same

PROBLEM 7.1   Examine the following for chirality centers:

(a) 2-Bromopentane   (c) 1-Bromo-2-methylbutane
(b) 3-Bromopentane   (d) 2-Bromo-2-methylbutane

SAMPLE SOLUTION   A carbon with four different groups attached to it is a chirality center. (a) In 2-bromopentane, C-2 satisfies this requirement. (b) None of the carbons in 3-bromopentane has four different substituents, and so none of its atoms is a chirality center.

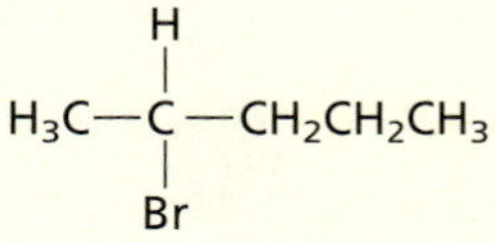

$$H_3C-\underset{\underset{Br}{|}}{\overset{\overset{H}{|}}{C}}-CH_2CH_2CH_3 \qquad CH_3CH_2-\underset{\underset{Br}{|}}{\overset{\overset{H}{|}}{C}}-CH_2CH_3$$

2-Bromopentane 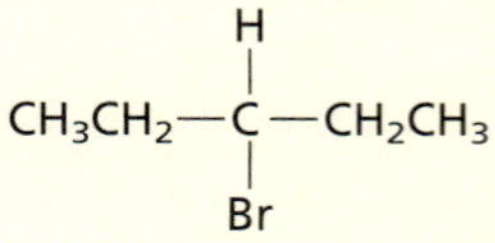   3-Bromopentane

Molecules with chirality centers are very common, both as naturally occurring substances and as the products of chemical synthesis. (Carbons that are part of a double bond or a triple bond can't be chirality centers.)

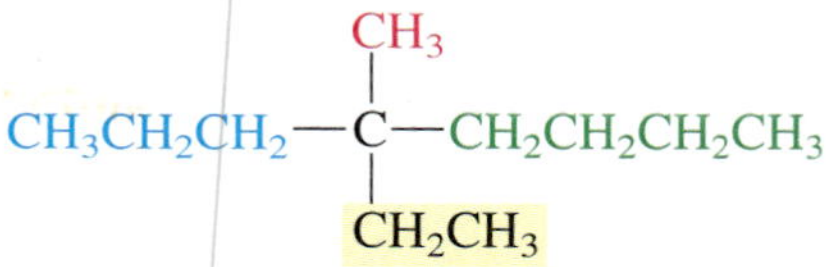

$$CH_3CH_2CH_2-\underset{\underset{CH_2CH_3}{|}}{\overset{\overset{CH_3}{|}}{C}}-CH_2CH_2CH_2CH_3 \qquad (CH_3)_2C=CHCH_2CH_2-\underset{\underset{OH}{|}}{\overset{\overset{CH_3}{|}}{C}}-CH=CH_2$$

4-Ethyl-4-methyloctane
(a chiral alkane)

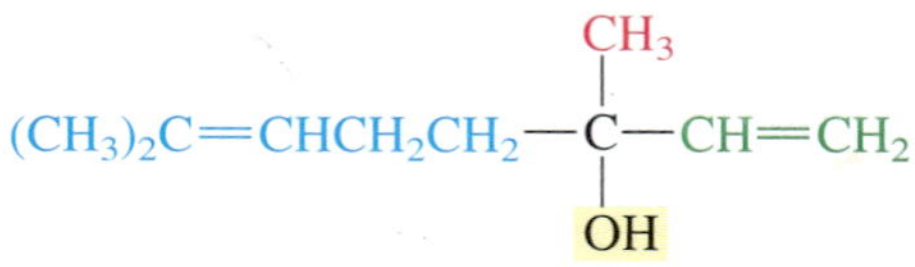

Linalool
(a pleasant-smelling oil
obtained from orange flowers)

A carbon atom in a ring can be a chirality center if it bears two different substituents and the path traced around the ring from that carbon in one direction is different from that traced in the other. The carbon atom that bears the methyl group in 1,2-epoxypropane, for example, is a chirality center. The sequence of groups is $O—CH_2$ as one proceeds clockwise around the ring from that atom, but is $CH_2—O$ in the counterclockwise direction. Similarly, C-4 is a chirality center in limonene.

$$H_2C—CHCH_3$$
$$O$$

1-2-Epoxypropane
(product of epoxidation of propene)

Limonene
(a constituent of lemon oil)

> Examine the molecular models of the two enantiomers of 1,2-epoxypropane on *Learning By Modeling* and test them for superimposability.

PROBLEM 7.2   Identify the chirality centers, if any, in

(a) 2-Cyclopentenol and 3-cyclopentenol

(b) 1,1,2-Trimethylcyclobutane and 1,1,3-trimethylcyclobutane

SAMPLE SOLUTION   (a) The hydroxyl-bearing carbon in 2-cyclopentenol is a chirality center. There is no chirality center in 3-cyclopentenol, because the sequence of atoms $1 \rightarrow 2 \rightarrow 3 \rightarrow 4 \rightarrow 5$ is equivalent regardless of whether one proceeds clockwise or counterclockwise.

2-Cyclopentenol

3-Cyclopentenol
(does not have a chirality center)

Even isotopes qualify as different substituents at a chirality center. The stereochemistry of biological oxidation of a derivative of ethane that is chiral because of deuterium ($D = {}^2H$) and tritium ($T = {}^3H$) atoms at carbon, has been studied and shown to proceed as follows:

$$\overset{T}{\underset{H}{D}}C—CH_3 \xrightarrow{\text{biological oxidation}} \overset{T}{\underset{HO}{D}}C—CH_3$$

The stereochemical relationship between the reactant and the product, revealed by the isotopic labeling, shows that oxygen becomes bonded to carbon on the same side from which H is lost. As you will see in this and the chapters to come, determining the three-dimensional aspects of a chemical or biochemical transformation can be a subtle, yet powerful, tool for increasing our understanding of how these reactions occur.

One final, very important point: *Everything we have said in this section concerns molecules that have one and only one chirality center; molecules with more than one chirality center may or may not be chiral.* Molecules that have more than one chirality center will be discussed in Sections 7.10 through 7.13.

7.3   SYMMETRY IN ACHIRAL STRUCTURES

Certain structural features can sometimes help us determine by inspection whether a molecule is chiral or achiral. For example, a molecule that has a *plane of symmetry* or a *center of symmetry* is superimposable on its mirror image and is achiral.

A **plane of symmetry** bisects a molecule so that one half of the molecule is the mirror image of the other half. The achiral molecule chlorodifluoromethane, for example, has the plane of symmetry shown in Figure 7.3.

A point in a molecule is a **center of symmetry** if any line drawn from it to some element of the structure will, when extended an equal distance in the opposite direction, encounter an identical element. The cyclobutane derivative in Figure 7.4 lacks a plane of symmetry, yet is achiral because it possesses a center of symmetry.

PROBLEM 7.3   Locate any planes of symmetry or centers of symmetry in each of the following compounds. Which of the compounds are chiral? Which are achiral?

(a) (*E*)-1,2-Dichloroethene

(b) (*Z*)-1,2,Dichloroethene

(c) *cis*-1,2-Dichlorocyclopropane

(d) *trans*-1,2-Dichlorocyclopropane

SAMPLE SOLUTION   (a) (*E*)-1,2-Dichloroethene is planar. The molecular plane is a plane of symmetry. Identifying a plane of symmetry tells us the molecule is achiral.

Furthermore, (*E*)-1,2-dichloroethene has a center of symmetry located at the midpoint of the carbon–carbon double bond. This too tells us the molecule is achiral.

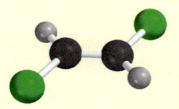

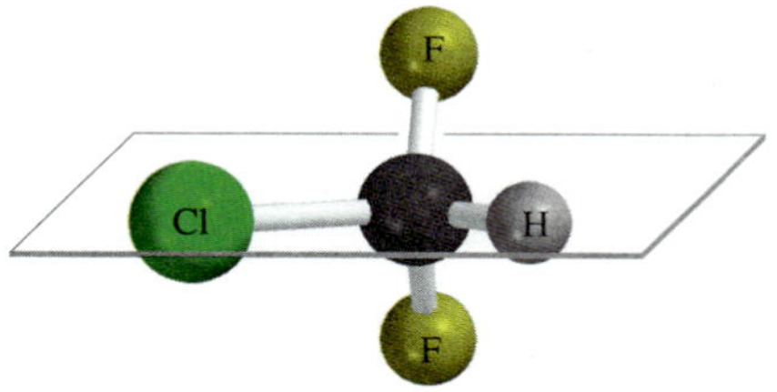

FIGURE 7.3 A plane of symmetry defined by the atoms H—C—Cl divides chlorodifluoromethane into two mirror-image halves.

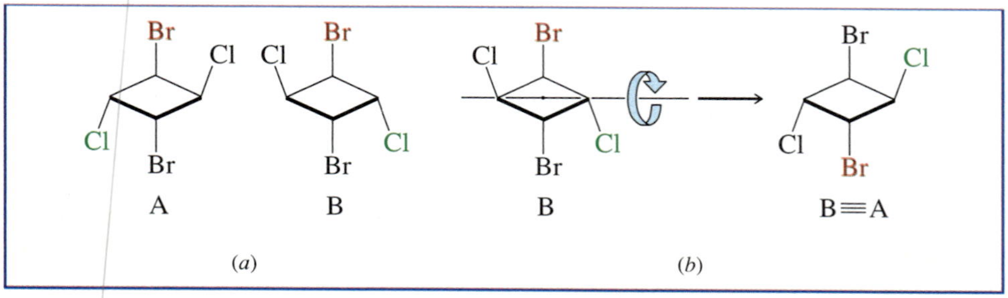

FIGURE 7.4 (*a*) Structural formulas A and B are drawn as mirror images. (*b*) The two mirror images are superimposable by rotating form B 180° about an axis passing through the center of the molecule. The center of the molecule is a center of symmetry.

Any molecule with a plane of symmetry or a center of symmetry is achiral, but their absence is not sufficient for a molecule to be chiral. A molecule lacking a center of symmetry or a plane of symmetry is *likely* to be chiral, but the superimposability test should be applied to be certain.

7.4   OPTICAL ACTIVITY

The experimental facts that led van't Hoff and Le Bel to propose that molecules having the same constitution could differ in the arrangement of their atoms in space concerned the physical property of optical activity. **Optical activity** is the ability of a chiral substance to rotate the plane of **plane-polarized light** and is measured using an instrument called a **polarimeter.** (Figure 7.5).

The light used to measure optical activity has two properties: it consists of a single wavelength and it is plane-polarized. The wavelength used most often is 589 nm (called the *D line*), which corresponds to the yellow light produced by a sodium lamp. Except for giving off light of a single wavelength, a sodium lamp is like any other lamp in that its light is unpolarized, meaning that the plane of its electric field vector can have any orientation along the line of travel. A beam of unpolarized light is transformed to plane-polarized light by passing it through a polarizing filter, which removes all the waves except those that have their electric field vector in the same plane. This plane-polarized light now passes through the sample tube containing the substance to be examined, either in the liquid phase or as a solution in a suitable solvent (usually water, ethanol, or chloroform). The sample is "optically active" if it rotates the plane of polarized light. The direction and magnitude of rotation are measured using a second polarizing filter (the "analyzer") and cited as α, the observed rotation. *To be optically active, the sample must contain a chiral substance and one enantiomer must be present in excess of the other.* A substance that does not rotate the plane of polarized light is said to be optically inactive. *All achiral substances are optically inactive.*

What causes optical rotation? The plane of polarization of a light wave undergoes a minute rotation when it encounters a chiral molecule. Enantiomeric forms of a chiral molecule cause a rotation of the plane of polarization in exactly equal amounts but in

> The phenomenon of optical activity was discovered by the French physicist Jean-Baptiste Biot in 1815.

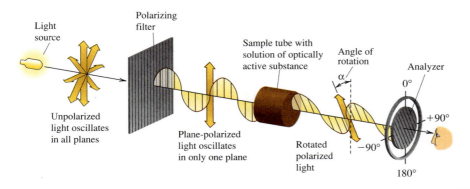

FIGURE 7.5 The sodium lamp emits light moving in all planes. When the light passes through the first polarizing filter, only one plane emerges. The plane-polarized beam enters the sample compartment, which contains a solution enriched in one of the enantiomers of a chiral substance. The plane rotates as it passes through the solution. A second polarizing filter (called the analyzer) is attached to a movable ring calibrated in degrees that is used to measure the angle of rotation α. (Adapted from M. Silberberg, *Chemistry,* 2nd ed., McGraw-Hill Higher Education, New York, 2000, p. 616.)

opposite directions. A solution containing equal quantities of enantiomers therefore exhibits no net rotation because all the tiny increments of clockwise rotation produced by molecules of one "handedness" are canceled by an equal number of increments of counterclockwise rotation produced by molecules of the opposite handedness.

Mixtures containing equal quantities of enantiomers are called **racemic mixtures.** Racemic mixtures are optically inactive. Conversely, when one enantiomer is present in excess, a net rotation of the plane of polarization is observed. At the limit, where all the molecules are of the same handedness, we say the substance is **optically pure.** Optical purity, or *percent enantiomeric excess,* is defined as:

Optical purity = percent enantiomeric excess

= percent of one enantiomer − percent of other enantiomer

Thus, a material that is 50% optically pure contains 75% of one enantiomer and 25% of the other.

Rotation of the plane of polarized light in the clockwise sense is taken as positive (+), and rotation in the counterclockwise sense is taken as a negative (−) rotation. Older terms for positive and negative rotations were *dextrorotatory* and *levorotatory,* from the Latin prefixes *dextro-* ("to the right") and *levo-* ("to the left"), respectively. At one time, the symbols *d* and *l* were used to distinguish between enantiomeric forms of a substance. Thus the dextrorotatory enantiomer of 2-butanol was called *d*-2-butanol, and the levorotatory form *l*-2-butanol; a racemic mixture of the two was referred to as *dl*-2-butanol. Current custom favors using algebraic signs instead, as in (+)-2-butanol, (−)-2-butanol, and (±)-2-butanol, respectively.

The observed rotation α of an optically pure substance depends on how many molecules the light beam encounters. A filled polarimeter tube twice the length of another produces twice the observed rotation, as does a solution twice as concentrated. To account for the effects of path length and concentration, chemists have defined the term **specific rotation,** given the symbol $[\alpha]$. Specific rotation is calculated from the observed rotation according to the expression

$$[\alpha] = \frac{100\alpha}{cl}$$

> If concentration is expressed as grams per milliliter of solution instead of grams per 100 mL, an equivalent expression is
>
> $$[\alpha] = \frac{\alpha}{cl}$$

where c is the concentration of the sample in grams per 100 mL of solution, and l is the length of the polarimeter tube in decimeters. (One decimeter is 10 cm.)

Specific rotation is a physical property of a substance, just as melting point, boiling point, density, and solubility are. For example, the lactic acid obtained from milk is exclusively a single enantiomer. We cite its specific rotation in the form $[\alpha]_D^{25} = +3.8°$. The temperature in degrees Celsius and the wavelength of light at which the measurement was made are indicated as superscripts and subscripts, respectively.

> **PROBLEM 7.4** Cholesterol, when isolated from natural sources, is obtained as a single enantiomer. The observed rotation α of a 0.3-g sample of cholesterol in 15 mL of chloroform solution contained in a 10-cm polarimeter tube is −0.78°. Calculate the specific rotation of cholesterol.

> **PROBLEM 7.5** A sample of synthetic cholesterol was prepared consisting entirely of (+)-cholesterol. This synthetic (+)-cholesterol was mixed with some natural (−)-cholesterol. The mixture had a specific rotation $[\alpha]_D^{20}$ of −13°. What fraction of the mixture was (+)-cholesterol? (*Note:* You need to use the solution to Problem 7.4 for the specific rotation of (−)-cholesterol.)

It is convenient to distinguish between enantiomers by prefixing the sign of rotation to the name of the substance. For example, we refer to one of the enantiomers of 2-butanol as (+)-2-butanol and the other as (−)-2-butanol. Optically pure (+)-2-butanol has a specific rotation $[\alpha]_D^{27}$ of +13.5°; optically pure (−)-2-butanol has an exactly opposite specific rotation $[\alpha]_D^{27}$ of −13.5°.

7.5 ABSOLUTE AND RELATIVE CONFIGURATION

The spatial arrangement of substituents at a chirality center is its **absolute configuration.** Neither the sign nor the magnitude of rotation by itself can tell us the absolute configuration of a substance. Thus, one of the following structures is (+)-2-butanol and the other is (−)-2-butanol, but without additional information we can't tell which is which.

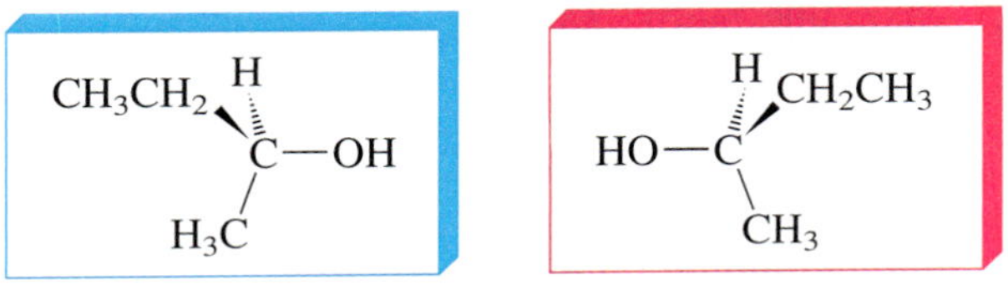

> In several places throughout the chapter we will use red and blue frames to call attention to structures that are enantiomeric.

Although no absolute configuration was known for any substance until the mid-twentieth century, organic chemists had experimentally determined the configurations of thousands of compounds relative to one another (their **relative configurations**) through chemical interconversion. To illustrate, consider (+)-3-buten-2-ol. Hydrogenation of this compound yields (+)-2-butanol.

$$CH_3CHCH{=}CH_2 \ + \ H_2 \ \xrightarrow{Pd} \ CH_3CHCH_2CH_3$$
$$\underset{OH}{|} \qquad\qquad\qquad\qquad \underset{OH}{|}$$

3-Buten-2-ol   Hydrogen   2-Butanol
$[\alpha]_D^{27}$ +33.2° $[\alpha]_D^{27}$ +13.5°

> Make a molecular model of one of the enantiomers of 3-buten-2-ol and the 2-butanol formed from it.

Because hydrogenation of the double bond does not involve any of the bonds to the chirality center, the spatial arrangement of substituents in (+)-3-buten-2-ol must be the same as that of the substituents in (+)-2-butanol. The fact that these two compounds have the same sign of rotation when they have the same relative configuration is established by the hydrogenation experiment; it could not have been predicted in advance of the experiment.

Compounds that have the same relative configuration often have optical rotations of opposite sign. For example, treatment of (−)-2-methyl-1-butanol with hydrogen bromide converts it to (+)-1-bromo-2-methylbutane.

$$CH_3CH_2CHCH_2OH \ + \ HBr \ \longrightarrow \ CH_3CH_2CHCH_2Br \ + \ H_2O$$
$$\qquad \underset{CH_3}{|} \qquad\qquad\qquad\qquad\qquad \underset{CH_3}{|}$$

2-Methyl-1-butanol   Hydrogen   1-Bromo-2-methylbutane   Water
$[\alpha]_D^{25}$ −5.8°   bromide $[\alpha]_D^{25}$ +4.0°

> Make a molecular model of one of the enantiomers of 2-methyl-1-butanol and the 1-bromo-2-methylbutane formed from it.

This reaction does not involve any of the bonds to the chirality center, and so both the starting alcohol (−) and the product bromide (+) have the same relative configuration.

An elaborate network connecting signs of rotation and relative configurations was developed that included the most important compounds of organic and biological chemistry. When, in 1951, the absolute configuration of a salt of (+)-tartaric acid was determined, the absolute configurations of all the compounds whose configurations had been related to (+)-tartaric acid stood revealed as well. Thus, returning to the pair of 2-butanol enantiomers that began this section, their absolute configurations are now known to be as shown.

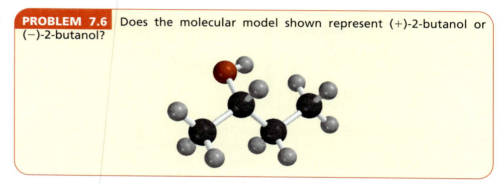

(+)-2-Butanol   (−)-2-Butanol

PROBLEM 7.6 Does the molecular model shown represent (+)-2-butanol or (−)-2-butanol?

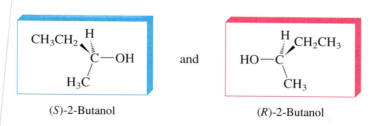

7.6   THE CAHN–INGOLD–PRELOG *R–S* NOTATIONAL SYSTEM

Just as it makes sense to have a nomenclature system by which we can specify the constitution of a molecule in words rather than pictures, so too is it helpful to have one that lets us describe stereochemistry. We have already had some experience with this idea when we distinguished between *E* and *Z* stereoisomers of alkenes.

In the *E–Z* system, substituents are ranked by atomic number according to a set of rules devised by R. S. Cahn, Sir Christopher Ingold, and Vladimir Prelog (Section 5.4). Actually, Cahn, Ingold, and Prelog first developed their ranking system to deal with the problem of the absolute configuration at a chirality center, and this is the system's major application. Table 7.1 shows how the Cahn–Ingold–Prelog system, called the **sequence rules,** is used to specify the absolute configuration at the chirality center in (+)-2-butanol.

As outlined in Table 7.1, (+)-2-butanol has the *S* configuration. Its mirror image is (−)-2-butanol, which has the *R* configuration.

The January 1994 issue of the *Journal of Chemical Education* contains an article that describes how to use your hands to assign *R* and *S* configurations.

(S)-2-Butanol   and   (R)-2-Butanol

Often, the *R* or *S* configuration and the sign of rotation are incorporated into the name of the compound, as in (R)-(−)-2-butanol and (S)-(+)-2-butanol.

TABLE 7.1	Absolute Configuration According to the Cahn–Ingold–Prelog Notational System

Step number	Example

Given that the absolute configuration of (+)-2-butanol is

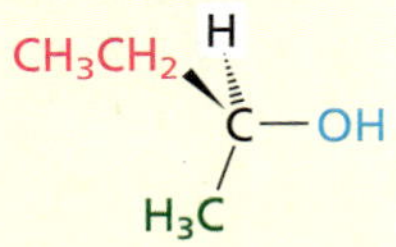

$$CH_3CH_2\underset{H_3C}{\overset{H}{\diagdown}}C-OH$$

(+)-2-Butanol

1. Identify the substituents at the chirality center, and rank them in order of decreasing precedence according to the system described in Section 5.4. Precedence is determined by atomic number, working outward from the point of attachment at the chirality center.

In order of decreasing precedence, the four substituents attached to the chirality center of 2-butanol are

$$HO- \; > \; CH_3CH_2- \; > \; CH_3- \; > \; H-$$
(highest)   (lowest)

2. Orient the molecule so that the lowest ranked substituent points away from you.

As represented in the wedge-and-dash drawing at the top of this table, the molecule is already appropriately oriented. Hydrogen is the lowest ranked atom attached to the chirality center and points away from us.

3. Draw the three highest ranked substituents as they appear to you when the molecule is oriented so that the lowest ranked group points away from you.

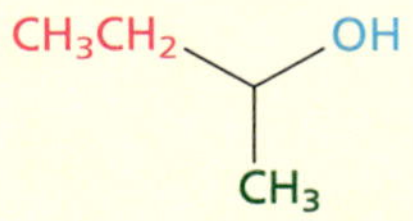

$$CH_3CH_2\diagdown\quad\diagup OH$$
$$CH_3$$

4. If the order of decreasing precedence of the three highest ranked substituents appears in a clockwise sense, the absolute configuration is *R* (Latin *rectus*, "right," "correct"). If the order of decreasing precedence is counterclockwise, the absolute configuration is *S* (Latin *sinister*, "left").

The order of decreasing precedence is *counterclockwise*. The configuration at the chirality center is *S*.

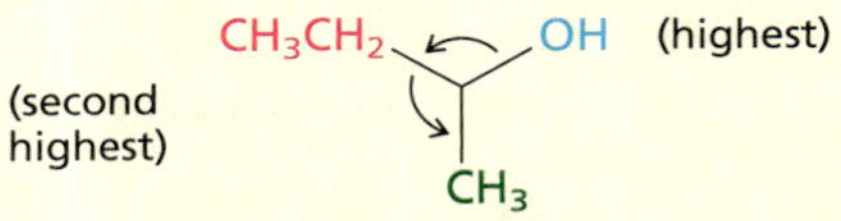

$$CH_3CH_2\diagdown\quad\diagup OH \quad \text{(highest)}$$
(second highest)
$$CH_3$$
(third highest)

PROBLEM 7.7   Assign absolute configurations as *R* or *S* to each of the following compounds:

(a) 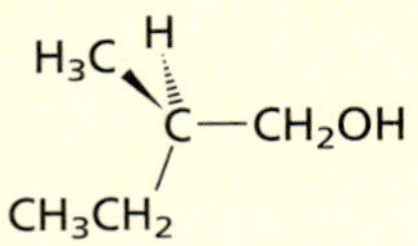
$$H_3C\underset{CH_3CH_2}{\overset{H}{\diagdown}}C-CH_2OH$$
(+)-2-Methyl-1-butanol

(b) 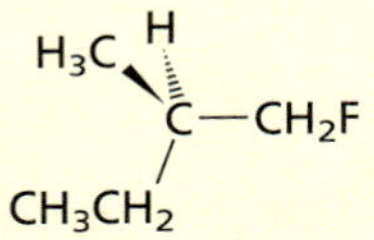
$$H_3C\underset{CH_3CH_2}{\overset{H}{\diagdown}}C-CH_2F$$
(+)-1-Fluoro-2-methylbutane

(c) 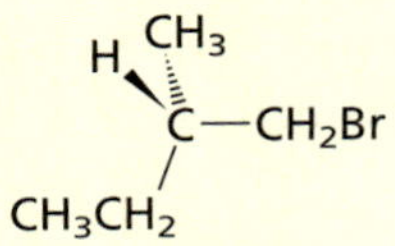
$$H\underset{CH_3CH_2}{\overset{CH_3}{\diagdown}}C-CH_2Br$$
(+)-1-Bromo-2-methylbutane

(d) 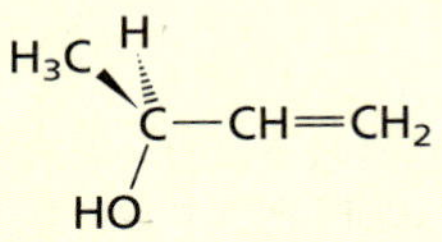
$$H_3C\underset{HO}{\overset{H}{\diagdown}}C-CH=CH_2$$
(+)-3-Buten-2-ol

SAMPLE SOLUTION (a) The highest ranking substituent at the chirality center of 2-methyl-1-butanol is CH_2OH; the lowest is H. Of the remaining two, ethyl outranks methyl.

Order of precedence: $CH_2OH > CH_3CH_2 > CH_3 > H$

The lowest ranking substituent (hydrogen) points away from us in the drawing. The three highest ranking groups trace a clockwise path from $CH_2OH \rightarrow CH_3CH_2 \rightarrow CH_3$.

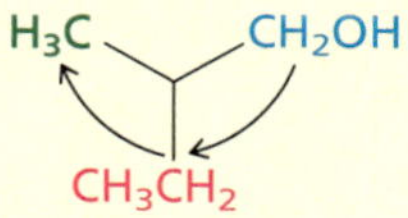

$$H_3C \qquad CH_2OH$$
$$CH_3CH_2$$

This compound therefore has the *R* configuration. It is (*R*)-(+)-2-methyl-1-butanol.

Compounds in which a chirality center is part of a ring are handled in an analogous fashion. To determine, for example, whether the configuration of (+)-4-methylcyclohexene is *R* or *S*, treat the right- and left-hand paths around the ring as if they were independent substituents.

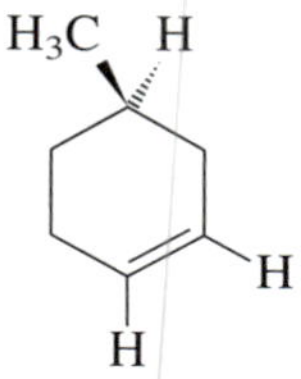

$$H_3C \quad H$$

is treated as

Lower priority path $\left(\begin{array}{cc} H_2C & CH_2 \\ H_2C & C-H \\ C & C \end{array} \right)$ Higher priority path

$$H_3C \quad H$$

H

H

(+)-4-Methylcyclohexene

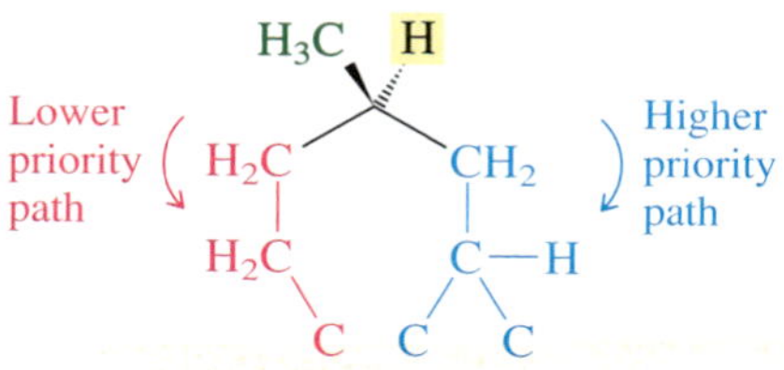

With the lowest ranked group (hydrogen) directed away from us, we see that the order of decreasing sequence rule precedence is *clockwise*. The absolute configuration is *R*.

PROBLEM 7.8 Draw three-dimensional representations or make molecular models of

(a) The *R* enantiomer of

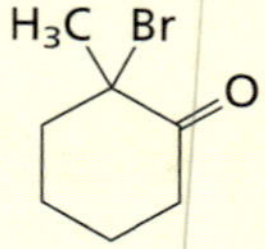

$$H_3C \quad Br$$
$$O$$

(b) The *S* enantiomer of

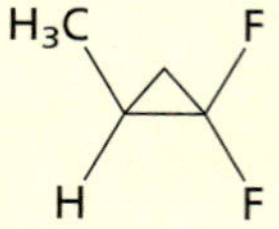

$$H_3C \qquad F$$
$$H \qquad F$$

SAMPLE SOLUTION (a) The chirality center is the one that bears the bromine. In order of decreasing precedence, the substituents attached to the chirality center are

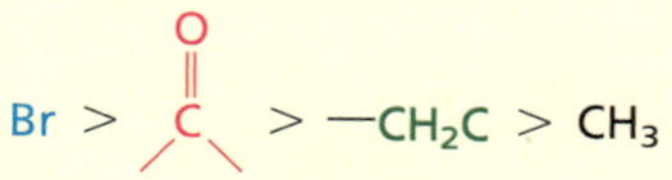

$$Br > \overset{O}{\underset{\|}{C}} > -CH_2C > CH_3$$

When the lowest ranked substituent (the methyl group) is away from us, the order of decreasing precedence of the remaining groups must appear in a clockwise sense in the *R* enantiomer.

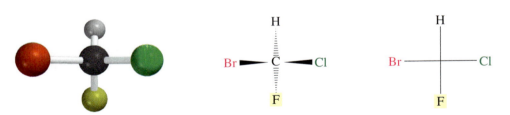

which leads to the structure

(R)-2-Bromo-2-methylcyclohexanone

Since its introduction in 1956, the Cahn–Ingold–Prelog system has become the standard method of stereochemical notation. Prior to that a system based on analogies to specified reference compounds using the prefixes D and L was used. As long as the substance in question is structurally similar to the reference, the D,L system is adequate. It is ambiguous, however, when applied to structures that are much different from the references, and as a *general* method for specifying configuration, the D,L system is largely obsolete. When *limited* to carbohydrates and amino acids though, D,L notation is firmly entrenched and works very well. We will use it when we get to Chapters 25–28, but won't need it until then.

7.7 FISCHER PROJECTIONS

Stereochemistry deals with the three-dimensional arrangement of a molecule's atoms, and we have attempted to show stereochemistry with wedge-and-dash drawings and computer-generated models. It is possible, however, to convey stereochemical information in an abbreviated form using a method devised by the German chemist Emil Fischer.

Let's return to bromochlorofluoromethane as a simple example of a chiral molecule. The two enantiomers of BrClFCH are shown as ball-and-stick models, as wedge-and-dash drawings, and as **Fischer projections** in Figure 7.6. Fischer projections are always generated the same way: the molecule is oriented so that the vertical bonds at the chirality center are directed away from you and the horizontal bonds point toward you. A projection of the bonds onto the page is a cross. The chirality center lies at the center of the cross but is not explicitly shown.

Fischer was the foremost organic chemist of the late nineteenth century. He won the 1902 Nobel Prize in chemistry for his pioneering work in carbohydrate and protein chemistry.

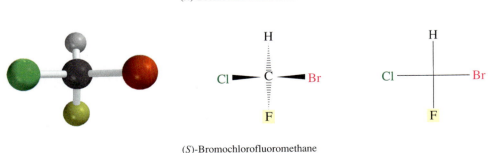

(R)-Bromochlorofluoromethane

(S)-Bromochlorofluoromethane

FIGURE 7.6 Ball-and-spoke models (*left*), wedge-and-dash drawings (*center*), and Fischer projections (*right*) of the R and S enantiomers of bromochloro-fluoromethane.

It is customary to orient the molecule so that the carbon chain is vertical with the lowest numbered carbon at the top as shown for the Fischer projection of (R)-2-butanol.

The Fischer projection

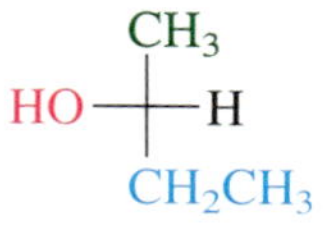

$$CH_3$$
$$HO \overline{|} H$$
$$CH_2CH_3$$

corresponds to

$$CH_3$$
$$HO - C - H$$
$$CH_2CH_3$$

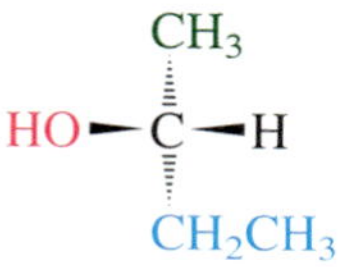

(R)-2-Butanol

To verify that the Fischer projection has the R configuration at its chirality center, rotate the three-dimensional representation so that the lowest-ranked atom (H) points away from you. Be careful to maintain the proper stereochemical relationships during the operation.

$$CH_3$$
$$HO - C - H$$
$$CH_2CH_3$$

→ *rotate 180° around vertical axis* →

$$CH_3$$
$$H \cdots C \cdots OH$$
$$CH_2CH_3$$

With H pointing away from us, we can see that the order of decreasing precedence OH > CH_2CH_3 > CH_3 traces a clockwise path, verifying the configuration as R.

$$CH_3$$
$$H \cdots C \cdots OH$$
$$CH_2CH_3$$

$$H_3C \qquad OH$$
$$CH_2CH_3$$

You may find it helpful to use a molecular model of (R)-2-butanol to guide you through this procedure.

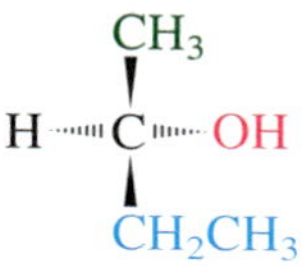

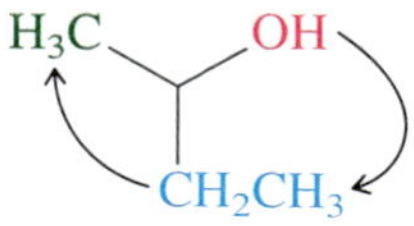

PROBLEM 7.9 What is the absolute configuration (R or S) of the compound represented by the Fischer projection shown here?

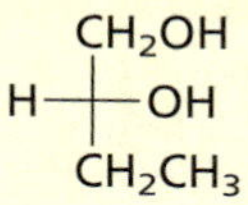

$$CH_2OH$$
$$H \overline{|} OH$$
$$CH_2CH_3$$

H ⁞ C

As you work with Fischer projections, you may notice that some routine structural changes lead to predictable outcomes—outcomes that may reduce the number of manipulations you need to do to solve stereochemistry problems. Instead of listing these shortcuts, Problem 7.10 invites you to discover some of them for yourself.

PROBLEM 7.10 Using the Fischer projection of (R)-2-butanol that was given at the top of this page, explain how each of the following affects the configuration of the chirality center.

(a) Switching the positions of H and OH.

(b) Switching the positions of CH_3 and CH_2CH_3.

(c) Switching the positions of three groups.

(d) Switching H with OH, and CH_3 with CH_2CH_3.

(e) Rotating the Fischer projection 180° around an axis perpendicular to the page.

SAMPLE SOLUTION (a) Exchanging the positions of H and OH in the Fischer projection of (R)-2-butanol converts it to the mirror-image Fischer projection. The configuration of the chirality center goes from R to S.

$$CH_3$$
$$HO\text{---}H \xrightarrow{\text{exchange the positions of H and OH}} H\text{---}OH$$
$$CH_2CH_3 \qquad\qquad CH_2CH_3$$

(R)-2-butanol   (S)-2-butanol

Switching the positions of two groups in a Fischer projection reverses the configuration of the chirality center.

We mentioned in Section 7.6 that the D,L system of stereochemical notation, while outdated for most purposes, is still widely used for carbohydrates and amino acids. Likewise, Fischer projections find their major application in these same two families of compounds.

7.8   PROPERTIES OF ENANTIOMERS

The usual physical properties such as density, melting point, and boiling point are identical for both enantiomers of a chiral compound.

Enantiomers can have striking differences, however, in properties that depend on the arrangement of atoms in space. Take, for example, the enantiomeric forms of carvone. (R)-(−)-Carvone is the principal component of spearmint oil. Its enantiomer, (S)-(+)-carvone, is the principal component of caraway seed oil. The two enantiomers do not smell the same; each has its own characteristic odor.

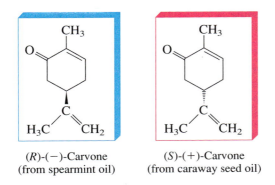

(R)-(−)-Carvone   (S)-(+)-Carvone
(from spearmint oil)   (from caraway seed oil)

The difference in odor between (R)- and (S)-carvone results from their different behavior toward receptor sites in the nose. It is believed that volatile molecules occupy only those odor receptors that have the proper shape to accommodate them. Because the receptor sites are themselves chiral, one enantiomer may fit one kind of receptor while the other enantiomer fits a different kind. An analogy that can be drawn is to hands and gloves. Your left hand and your right hand are enantiomers. You can place your left hand into a left glove but not into a right one. The receptor (the glove) can accommodate one enantiomer of a chiral object (your hand) but not the other.

The term *chiral recognition* refers to a process in which some chiral receptor or reagent interacts selectively with one of the enantiomers of a chiral molecule. Very high levels of chiral recognition are common in biological processes. (−)-Nicotine, for example, is much more toxic than (+)-nicotine, and (+)-adrenaline is more active than (−)-adrenaline in constricting blood vessels. (−)-Thyroxine, an amino acid of the thyroid gland that speeds up metabolism, is one of the most widely used of all prescription

An article entitled "When Drug Molecules Look in the Mirror" in the June 1996 issue of the *Journal of Chemical Education* (pp. 481–484) describes numerous examples of common drugs in which the two enantiomers have different biological properties.

CHIRAL DRUGS

A recent estimate places the number of prescription and over-the-counter drugs marketed throughout the world at about 2000. Approximately one third of these are either naturally occurring substances themselves or are prepared by chemical modification of natural products. Most of the drugs derived from natural sources are chiral and are almost always obtained as a single enantiomer rather than as a racemic mixture. Not so with the over 500 chiral substances represented among the more than 1300 drugs that are the products of synthetic organic chemistry. Until recently, such substances were, with few exceptions, prepared, sold, and administered as racemic mixtures even though the desired therapeutic activity resided in only one of the enantiomers. Spurred by a number of factors ranging from safety and efficacy to synthetic methodology and economics, this practice is undergoing rapid change as more and more chiral synthetic drugs become available in enantiomerically pure form.

Because of the high degree of chiral recognition inherent in most biological processes (Section 7.8), it is unlikely that both enantiomers of a chiral drug will exhibit the same level, or even the same kind, of effect. At one extreme, one enantiomer has the desired effect, and the other exhibits no biological activity at all. In this case, which is relatively rare, the racemic form is simply a drug that is 50% pure and contains 50% "inert ingredients." Real cases are more complicated. For example, the S enantiomer is responsible for the pain-relieving properties of ibuprofen, normally sold as a racemic mixture. The 50% of racemic ibuprofen that is the R enantiomer is not completely wasted, however, because enzyme-catalyzed reactions in our body convert much of it to active (S)-ibuprofen.

$(CH_3)_2CHCH_2$-⟨benzene ring⟩-$CHCOH$ (with O double-bonded to C and CH_3 below)

Ibuprofen

A much more serious drawback to using chiral drugs as racemic mixtures is illustrated by thalidomide, briefly employed as a sedative and antinausea drug in Europe during the period 1959–1962. The desired properties are those of (R)-thalidomide. (S)-Thalidomide, however, has a very different spectrum of biological activity and was shown to be responsible for over 2000 cases of serious birth defects in children born to women who took it while pregnant.

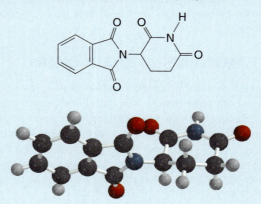

Thalidomide

Basic research aimed at controlling the stereochemistry of chemical reactions has led to novel methods for the synthesis of chiral molecules in enantiomerically pure form. Aspects of this work were recognized with the award of the 2001 Nobel Prize in chemistry to William S. Knowles (Monsanto), Ryoji Noyori (Nagoya University), and K. Barry Sharpless (Scripps Research Institute). Most major pharmaceutical companies are examining their existing drugs to see which are the best candidates for synthesis as single enantiomers and, when preparing a new drug, design its synthesis so as to provide only the desired enantiomer. One incentive to developing enantiomerically pure versions of existing drugs is that the novel production methods they require may make them eligible for patent protection separate from that of the original drugs. Thus the temporary monopoly position that patent law views as essential to fostering innovation can be extended by transforming a successful chiral, but racemic, drug into an enantiomerically pure version.

drugs—about 10 million people in the United States take $(-)$-thyroxine on a daily basis. Its enantiomer, $(+)$-thyroxine has none of the metabolism-regulating effects, but was formerly given to heart patients to lower their cholesterol levels before being replaced by better drugs.

HOCHCH$_2$NHCH$_3$

HO— —O— —CH$_2$CHCO$_2^-$

| Nicotine | Adrenaline | Thyroxine |

(Can you find the chirality center in each of these?)

7.9   REACTIONS THAT CREATE A CHIRALITY CENTER

Many of the reactions we've already encountered can yield a chiral product from an achiral starting material. Epoxidation of propene, for example, creates a chirality center by adding oxygen to the double bond.

$$H_2C=CHCH_3 \xrightarrow{CH_3CO_2OH} H_2C\!\!-\!\!CHCH_3$$

| Propene | 1,2-Epoxypropane |
| (achiral) | (chiral) |

In this, as in other reactions in which achiral reactants yield chiral products, the product is formed as a *racemic mixture* and is *optically inactive*. Remember, for a substance to be optically active, not only must it be chiral but one enantiomer must be present in excess of the other.

It is a general principle that *optically active products cannot be formed when optically inactive substrates react with optically inactive reagents.* This principle holds irrespective of whether the addition is syn or anti, concerted or stepwise. No matter how many steps are involved in a reaction, if the reactants are achiral, formation of one enantiomer is just as likely as the other, and a racemic mixture results.

Figure 7.7 shows why equal amounts of (R)- and (S)-1,2-epoxypropane are formed in the epoxidation of propene. There is no difference between the top face of the double bond and the bottom face. Peroxyacetic acid can transfer oxygen to either face with equal facility, the rates of formation of the R and S enantiomers of the product are the same, and the product is racemic.

In this example, addition to the double bond of an alkene converted an achiral molecule to a chiral one. The general term for a structural feature, the alteration of which introduces a chirality center in a molecule, is **prochiral.** A chirality center is introduced when the double bond of propene reacts with a peroxy acid. The double bond is a prochiral structural unit, and we speak of the top and bottom faces of the double bond as *prochiral faces*. Because attack at one prochiral face gives the enantiomer of the compound formed by attack at the other face, we classify the relationship between the two faces as **enantiotopic.**

In a second example, addition of hydrogen bromide converts 2-butene, which is achiral, to 2-bromobutane, which is chiral. But, as before, the product is racemic because

The *pro-* in *prochiral* means "before" or "in advance of" in roughly the same way as in "proactive."

FIGURE 7.7 Epoxidation of propene produces equal amounts of (*R*)- and (*S*)-1,2-epoxypropane.

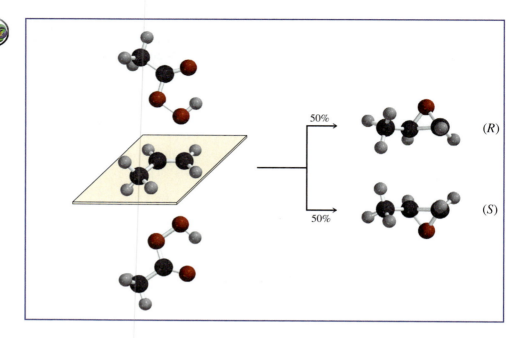

both enantiomers are formed at equal rates. This is true regardless of whether the starting alkene is *cis-* or *trans-*2-butene or whether the mechanism is electrophilic addition or free-radical addition of HBr.

$$CH_3CH{=}CHCH_3 \;+\; HBr \longrightarrow CH_3CHCH_2CH_3$$
$$\underset{Br}{|}$$

cis- or *trans-*2-Butene	Hydrogen bromide	2-Bromobutane
(achiral)	(achiral)	(chiral, but racemic)

Whatever happens at one enantiotopic face of the double bond of *cis-* or *trans-*2-butene happens at the same rate at the other, resulting in a 1:1 mixture of (*R*)- and (*S*)-2-bromobutane.

PROBLEM 7.11 What two stereoisomeric alkanes are formed in the catalytic hydrogenation of (*E*)-3-methyl-2-hexene? What are the relative amounts of each?

Addition to double bonds is not the only kind of reaction that converts an achiral molecule to a chiral one. Other possibilities include substitution reactions such as the formation of 2-chlorobutane by free radical chlorination of butane. Here again, the product is chiral, but racemic.

1-Chlorobutane is also formed in this reaction.

$$CH_3CH_2CH_2CH_3 \;+\; Cl_2 \xrightarrow{\text{heat}} CH_3CHCH_2CH_3 \;+\; HCl$$
$$\underset{Cl}{|}$$

Butane	Chlorine	2-Chlorobutane	Hydrogen chloride
(achiral)	(achiral)	(chiral, but racemic)	

We can view this reaction as the replacement of one or the other of the two methylene protons at C-2 of butane. These protons are *prochiral atoms* and, as the red and blue protons in the Newman projection indicate, occupy mirror-image environments.

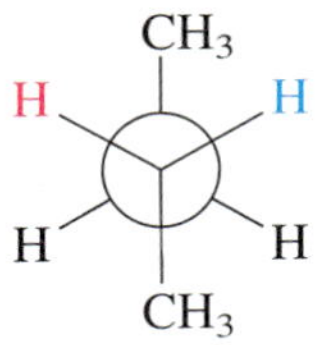

$$CH_3$$

H   H

H   H

$$CH_3$$

Replacing one of them by some different atom or group gives the enantiomer of the structure obtained by replacing the other; therefore, the methylene hydrogens at C-2 of butane are enantiotopic. The same is true for the hydrogens at C-3.

PROBLEM 7.12 Citric acid has three CO_2H groups. Which, if any, of them are enantiotopic?

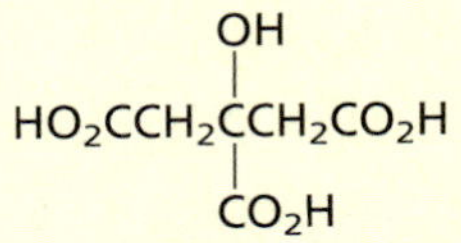

OH
|
$HO_2CCH_2CCH_2CO_2H$
|
CO_2H

Citric acid

When a reactant is chiral but optically inactive because it is *racemic,* any products derived from its reactions with optically inactive reagents will be *optically inactive.* For example, 2-butanol is chiral and may be converted with hydrogen bromide to 2-bromobutane, which is also chiral. If racemic 2-butanol is used, each enantiomer will react at the same rate with the achiral reagent. Whatever happens to (R)-$(-)$-2-butanol is mirrored in a corresponding reaction of (S)-$(+)$-2-butanol, and a racemic, optically inactive product results.

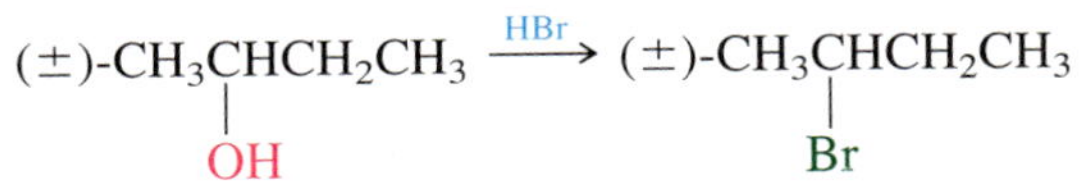

$$(\pm)\text{-}CH_3CHCH_2CH_3 \xrightarrow{HBr} (\pm)\text{-}CH_3CHCH_2CH_3$$

OH   Br

2-Butanol   2-Bromobutane
(chiral but racemic)   (chiral but racemic)

Optically inactive starting materials can give optically active products *only* if they are treated with an optically active reagent or if the reaction is catalyzed by an optically active substance. The best examples are found in biochemical processes. Most biochemical reactions are catalyzed by enzymes. Enzymes are chiral and enantiomerically homogeneous; they provide an asymmetric environment in which chemical reaction can take place. Ordinarily, enzyme-catalyzed reactions occur with such a high level of stereoselectivity that one enantiomer of a substance is formed exclusively even when the substrate is achiral. The enzyme *fumarase,* for example, catalyzes hydration of the double bond of fumaric acid to malic acid in apples and other fruits. Only the *S* enantiomer of malic acid is formed in this reaction.

$$HO_2C \diagdown C = C \diagup H \quad + \; H_2O \; \underset{\text{fumarase}}{\rightleftharpoons} \quad HO_2C \diagdown \underset{HO_2CCH_2}{\overset{H}{C}} -OH$$

Fumaric acid   (S)-(−)-Malic acid

The reaction is reversible, and its stereochemical requirements are so pronounced that neither the cis isomer of fumaric acid (maleic acid) nor the *R* enantiomer of malic acid can serve as a substrate for the fumarase-catalyzed hydration–dehydration equilibrium.

PROBLEM 7.13 Biological reduction of pyruvic acid, catalyzed by the enzyme lactate dehydrogenase, gives (+)-lactic acid, represented by the Fischer projection shown. What is the configuration of (+)-lactic acid according to the Cahn–Ingold–Prelog *R–S* notational system? Making a molecular model of the Fischer projection will help.

$$CH_3\overset{\overset{\textstyle O}{\|}}{C}CO_2H \xrightarrow{\text{biological reduction}} HO\!-\!\!\!\overset{\textstyle CO_2H}{\underset{\textstyle CH_3}{\vline}}\!\!\!-\!H$$

Pyruvic acid   (+)-Lactic acid

We'll continue with the three-dimensional details of chemical reactions later in this chapter. First though, we need to develop some additional stereochemical principles concerning structures with more than one chirality center.

7.10 CHIRAL MOLECULES WITH TWO CHIRALITY CENTERS

When a molecule contains two chirality centers, as does 2,3-dihydroxybutanoic acid, how many stereoisomers are possible?

$$\underset{HO \; \; OH}{\overset{4 \quad 3 \quad 2 \quad 1}{CH_3CHCHC}} \diagup \overset{\textstyle O}{\underset{\textstyle OH}{\diagdown}}$$

2,3-Dihydroxybutanoic acid

We can use straightforward reasoning to come up with the answer. The absolute configuration at C-2 may be *R* or *S*. Likewise, C-3 may have either the *R* or the *S* configuration. The four possible combinations of these two chirality centers are

(2R,3R) (stereoisomer I)   (2S,3S) (stereoisomer II)
(2R,3S) (stereoisomer III)   (2S,3R) (stereoisomer IV)

Figure 7.8 presents structural formulas for these four stereoisomers. Stereoisomers I and II are enantiomers of each other; the enantiomer of (*R,R*) is (*S,S*). Likewise stereoisomers III and IV are enantiomers of each other, the enantiomer of (*R,S*) being (*S,R*).

Stereoisomer I is not a mirror image of III or IV, so it is not an enantiomer of either one. Stereoisomers that are not related as an object and its mirror image are called diastereomers; *diastereomers are stereoisomers that are not enantiomers*. Thus, stereoisomer I is a diastereomer of III and a diastereomer of IV. Similarly, II is a diastereomer of III and IV.

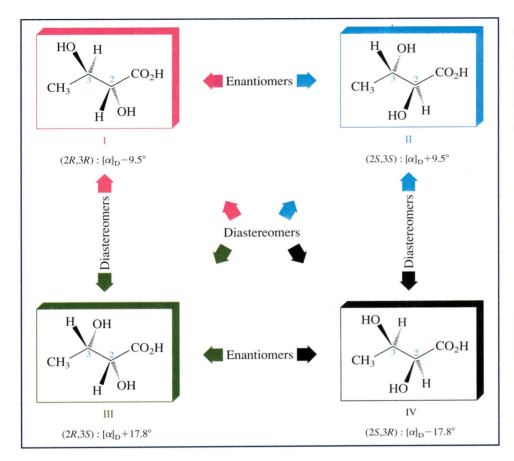

FIGURE 7.8 Stereo-isomeric 2,3-dihydroxybu-tanoic acids. Stereoisomers I and II are enantiomers. Stereoisomers III and IV are enantiomers. All other relationships are diastereomeric (see text).

A molecule framed in black is an enantiomer of a green-framed one. Both are diastereomers of their red or blue-framed stereoisomers.

To convert a molecule with two chirality centers to its enantiomer, the configuration at *both* centers must be changed. Reversing the configuration at only one chirality center converts it to a diastereomeric structure.

Enantiomers must have equal and opposite specific rotations. Diastereomers can have different rotations, with respect to both sign and magnitude. Thus, as Figure 7.8 shows, the (2R,3R) and (2S,3S) enantiomers (I and II) have specific rotations that are equal in magnitude but opposite in sign. The (2R,3S) and (2S,3R) enantiomers (III and IV) likewise have specific rotations that are equal to each other but opposite in sign. The magnitudes of rotation of I and II are different, however, from those of their diastereomers III and IV.

In writing Fischer projections of molecules with two chirality centers, the molecule is arranged in an *eclipsed* conformation for projection onto the page, as shown in Figure 7.9. Again, horizontal lines in the projection represent bonds coming toward you; vertical bonds point away.

Organic chemists use an informal nomenclature system based on Fischer projections to distinguish between diastereomers. When the carbon chain is vertical and like substituents are on the same side of the Fischer projection, the molecule is described as the **erythro** diastereomer. When like substituents are on opposite sides of the Fischer projection, the molecule is described as the **threo** diastereomer. Thus, as seen in the

FIGURE 7.9 Repre-
sentations of (2R,3R)-dihy-
droxybutanoic acid. (a) The
staggered conformation is
the most stable, but is not
properly arranged to show
stereochemistry as a Fischer
projection. (b) Rotation
about the C-2–C-3 bond
gives the eclipsed conformation, and projection of the
eclipsed conformation onto
the page gives (c) a correct
Fischer projection.

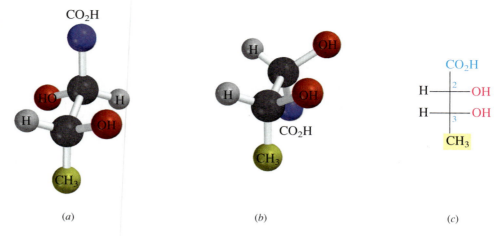

(a)   (b)   (c)

Fischer projections of the stereoisomeric 2,3-dihydroxybutanoic acids, compounds I and II are erythro stereoisomers and III and IV are threo.

> Erythro and threo describe
> the *relative configuration*
> (Section 7.5) of two chirality
> centers within a single
> molecule.

I	II	III	IV
erythro	erythro	threo	threo

Because diastereomers are not mirror images of each other, they can have quite different physical and chemical properties. For example, the (2R,3R) stereoisomer of 3-amino-2-butanol is a liquid, but the (2R,3S) diastereomer is a crystalline solid.

(2R,3R)-3-Amino-2-butanol   (2R,3S)-3-Amino-2-butanol
(liquid)   (solid, mp 49°C)

PROBLEM 7.14   Draw Fischer projections or make molecular models of the four stereoisomeric 3-amino-2-butanols, and label each erythro or threo as appropriate.

PROBLEM 7.15   One other stereoisomer of 3-amino-2-butanol is a crystalline solid. Which one?

The situation is the same when the two chirality centers are present in a ring. There are four stereoisomeric 1-bromo-2-chlorocyclopropanes: a pair of enantiomers in which the halogens are trans and a pair in which they are cis. The cis compounds are diaste-reomers of the trans.

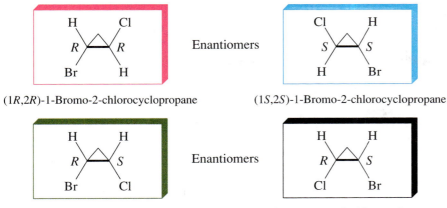

(1R,2R)-1-Bromo-2-chlorocyclopropane   (1S,2S)-1-Bromo-2-chlorocyclopropane

(1R,2S)-1-Bromo-2-chlorocyclopropane   (1S,2R)-1-Bromo-2-chlorocyclopropane

A good thing to remember is that the cis and trans isomers of a particular compound are diastereomers of each other.

7.11 ACHIRAL MOLECULES WITH TWO CHIRALITY CENTERS

Now think about a molecule, such as 2,3-butanediol, which has two chirality centers that are equivalently substituted.

$$CH_3CHCHCH_3$$
$$HO \quad OH$$

2,3-Butanediol

Only *three,* not four, stereoisomeric 2,3-butanediols are possible. These three are shown in Figure 7.10. The (2R,3R) and (2S,3S) forms are enantiomers of each other and have equal and opposite optical rotations. A third combination of chirality centers, (2R,3S), however, gives an *achiral* structure that is superimposable on its (2S,3R) mirror image. Because it is achiral, this third stereoisomer is *optically inactive.* We call achiral molecules that have chirality centers **meso forms.** The meso form in Figure 7.10 is known as *meso*-2,3-butanediol.

One way to demonstrate that *meso*-2,3-butanediol is achiral is to recognize that its eclipsed conformation has a plane of symmetry that passes through and is perpendicular to the C-2–C-3 bond, as illustrated in Figure 7.11a. The anti conformation is achiral as

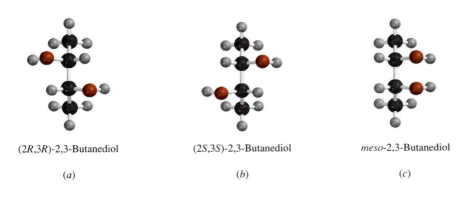

(2R,3R)-2,3-Butanediol   (2S,3S)-2,3-Butanediol   *meso*-2,3-Butanediol

(a)   (b)   (c)

FIGURE 7.10 Stereoisomeric 2,3-butanediols shown in their eclipsed conformations for convenience. Stereoisomers (a) and (b) are enantiomers of each other. Structure (c) is a diastereomer of (a) and (b), and is achiral. It is called *meso*-2,3-butanediol.

FIGURE 7.11 (*a*) The eclipsed conformation of *meso*-2,3-butanediol has a plane of symmetry. (*b*) The anti conformation of *meso*-2,3-butanediol has a center of symmetry.

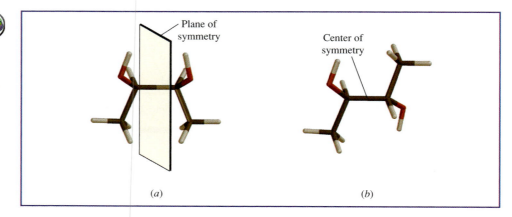

well. As Figure 7.11*b* shows, this conformation is characterized by a center of symmetry at the midpoint of the C-2–C-3 bond.

Fischer projection formulas can help us identify meso forms. Of the three stereoisomeric 2,3-butanediols, notice that only in the meso stereoisomer does a dashed line through the center of the Fischer projection divide the molecule into two mirror-image halves.

In the same way that a Fischer formula is a projection of the eclipsed conformation onto the page, the line drawn through its center is a projection of the plane of symmetry that is present in the eclipsed conformation of *meso*-2,3-butanediol.

$$
\begin{array}{ccc}
\text{CH}_3 & \text{CH}_3 & \text{CH}_3 \\
\text{HO}\!\!-\!\!\!-\!\!\text{H} & \text{H}\!\!-\!\!\!-\!\!\text{OH} & \text{H}\!\!-\!\!\!-\!\!\text{OH} \\
\text{H}\!\!-\!\!\!-\!\!\text{OH} & \text{HO}\!\!-\!\!\!-\!\!\text{H} & \text{H}\!\!-\!\!\!-\!\!\text{OH} \\
\text{CH}_3 & \text{CH}_3 & \text{CH}_3 \\
(2R,3R)\text{-2,3-Butanediol} & (2S,3S)\text{-2,3-Butanediol} & meso\text{-2,3-Butanediol}
\end{array}
$$

When using Fischer projections for this purpose, however, be sure to remember what three-dimensional objects they stand for. One should not, for example, test for superimposition of the two chiral stereoisomers by a procedure that involves moving any part of a Fischer projection out of the plane of the paper in any step.

PROBLEM 7.16 A meso stereoisomer is possible for one of the following compounds. Which one?

2,3-Dibromopentane; 2,4-dibromopentane; 3-bromo-2-pentanol; 4-bromo-2-pentanol

Turning to cyclic compounds, we see that there are three, not four, stereoisomeric 1,2-dibromocyclopropanes. Of these, two are enantiomeric *trans*-1,2-dibromocyclopropanes. The cis diastereomer is a meso form; it has a plane of symmetry.

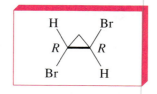

(1*R*,2*R*)-1,2-Dibromocyclopropane

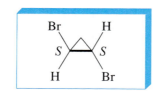

(1*S*,2*S*)-Dibromocyclopropane

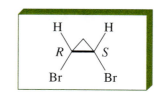

meso-1,2-Dibromocyclopropane

PROBLEM 7.17 One of the stereoisomers of 1,3-dimethylcyclohexane is a meso form. Which one?

CHIRALITY OF DISUBSTITUTED CYCLOHEXANES

Disubstituted cyclohexanes present us with a challenging exercise in stereochemistry. Consider the seven possible dichlorocyclohexanes: 1,1-; cis- and trans-1,2-; cis- and trans-1,3-; and cis- and trans-1,4-. Which are chiral? Which are achiral?

Four isomers—the ones that are achiral because they have a plane of symmetry—are relatively easy to identify:

ACHIRAL DICHLOROCYCLOHEXANES

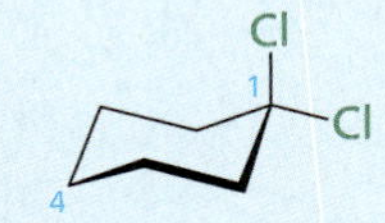

1,1
(plane of symmetry through C-1 and C-4)

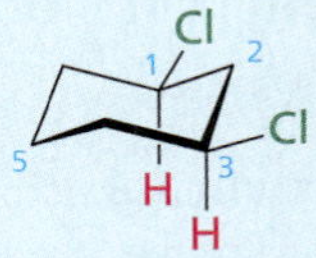

cis-1,3
(plane of symmetry through C-2 and C-5)

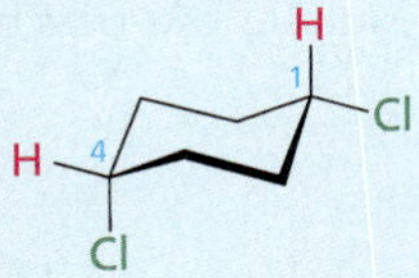

cis-1,4
(plane of symmetry through C-1 and C-4)

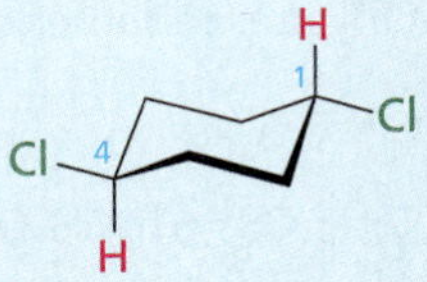

trans-1,4
(plane of symmetry through C-1 and C-4)

The remaining three isomers are chiral:

CHIRAL DICHLOROCYCLOHEXANES

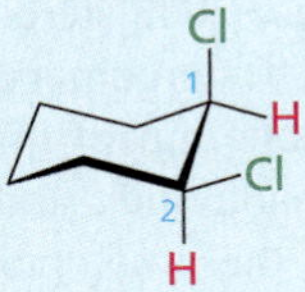

cis-1,2

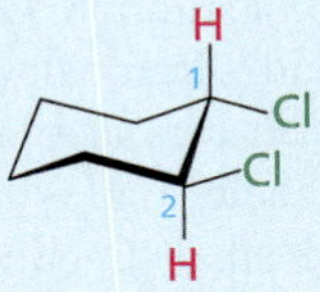

trans-1,2

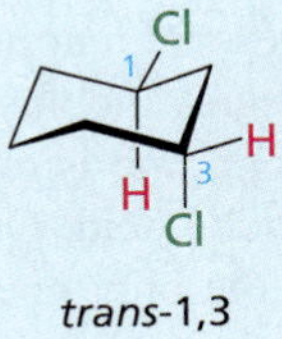

trans-1,3

Among all the isomers, cis-1,2-dichlorocyclohexane is unique in that the ring-flipping process typical of cyclohexane derivatives (Section 3.9) converts it to its enantiomer.

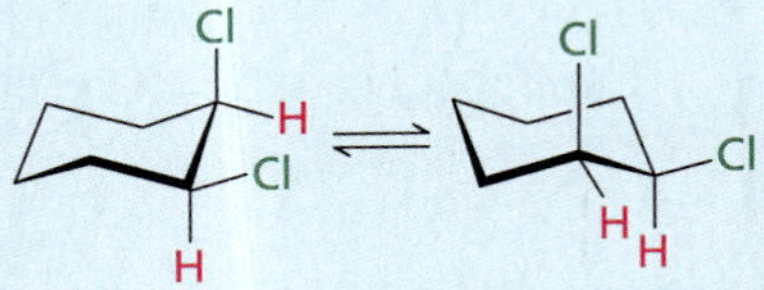

which is equivalent to

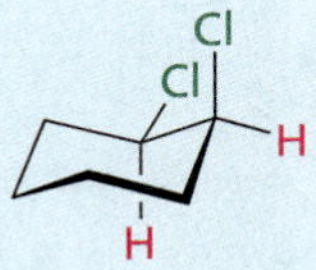

Structures A and A' are nonsuperimposable mirror images of each other. Thus although cis-1,2-dichlorocyclohexane is chiral, it is optically inactive when chair–chair interconversion occurs. Such interconversion is rapid at room temperature and converts optically active A to a racemic mixture of A and A'. Because A and A' are enantiomers interconvertible by a conformational change, they are sometimes referred to as **conformational enantiomers**.

The same kind of spontaneous racemization occurs for any cis-1,2 disubstituted cyclohexane in which both substituents are the same. Because such compounds are chiral, it is incorrect to speak of them as meso compounds, which are achiral by definition. Rapid chair–chair interconversion, however, converts them to a 1:1 mixture of enantiomers, and this mixture is optically inactive.

7.12   MOLECULES WITH MULTIPLE CHIRALITY CENTERS

Many naturally occurring compounds contain several chirality centers. By an analysis similar to that described for the case of two chirality centers, it can be shown that the maximum number of stereoisomers for a particular constitution is 2^n, where n is equal to the number of chirality centers.

> **PROBLEM 7.18** Using *R* and *S* descriptors, write all the possible combinations for a molecule with three chirality centers.

When two or more of a molecule's chirality centers are equivalently substituted, meso forms are possible, and the number of stereoisomers is then less than 2^n. Thus, 2^n represents the *maximum* number of stereoisomers for a molecule containing n chirality centers.

The best examples of substances with multiple chirality centers are the *carbohydrates* (Chapter 25). One class of carbohydrates, called *hexoses,* has the constitution

$$HOCH_2CH-CH-CH-CH-C\overset{\displaystyle O}{\underset{\displaystyle H}{<}}$$
$$\underset{OH}{|}\ \underset{OH}{|}\ \underset{OH}{|}\ \underset{OH}{|}$$

A hexose

Because there are four chirality centers and no possibility of meso forms, there are 2^4, or 16, stereoisomeric hexoses. All 16 are known, having been isolated either as natural products or as the products of chemical synthesis.

> **PROBLEM 7.19** A second category of six-carbon carbohydrates, called *2-hexuloses,* has the constitution shown. How many stereoisomeric 2-hexuloses are possible?
>
> $$HOCH_2\overset{\displaystyle O}{\overset{\|}{C}}CH-CH-CHCH_2OH$$
> $$\underset{OH}{|}\ \underset{OH}{|}\ \underset{OH}{|}$$
>
> A 2-hexulose

Steroids are another class of natural products with multiple chirality centers. One such compound is *cholic acid,* which can be obtained from bile. Its structural formula is given in Figure 7.12. Cholic acid has 11 chirality centers, and so a total (including cholic acid) of 2^{11}, or 2048, stereoisomers have this constitution. Of these 2048 stereoisomers, how many are diastereomers of cholic acid? Remember! Diastereomers are stereoisomers that are not enantiomers, and any object can have only one mirror image. Therefore, of the 2048 stereoisomers, one is cholic acid, one is its enantiomer, and the other 2046 are diastereomers of cholic acid. Only a small fraction of these compounds are known, and (+)-cholic acid is the only one ever isolated from natural sources.

Eleven chirality centers may seem like a lot, but it is nowhere close to a world record. It is a modest number when compared with the more than 100 chirality centers typical for most small proteins and the thousands of chirality centers present in nucleic acids.

A molecule that contains both chirality centers and double bonds has additional opportunities for stereoisomerism. For example, the configuration of the chirality center in 3-penten-2-ol may be either *R* or *S,* and the double bond may be either *E* or *Z.* Therefore 3-penten-2-ol has four stereoisomers even though it has only one chirality center.

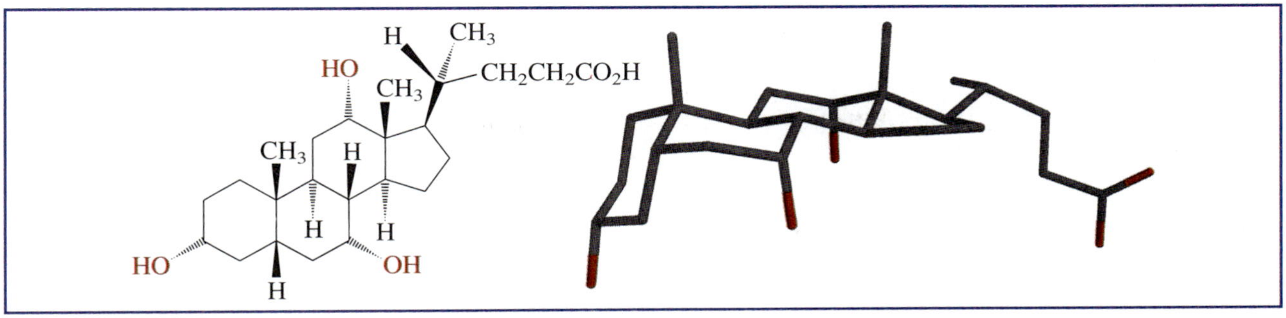

FIGURE 7.12 Cholic acid. Its 11 chirality centers are those carbons at which stereochemistry is indicated in the structural drawing at the left. The molecular model at the right more clearly shows the overall shape of the molecule.

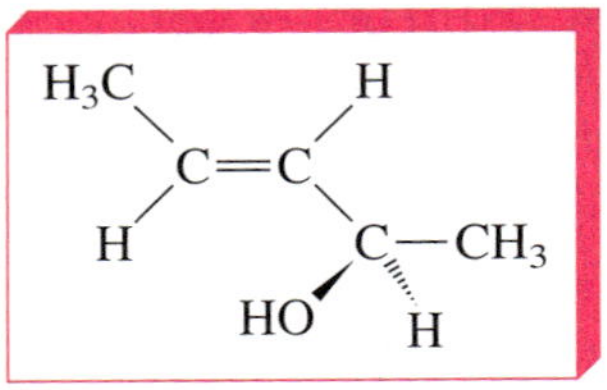

n in 2^n includes double bonds capable of stereochemical variation (*E*, *Z*) as well as chirality centers.

(2*R*,3*E*)-3-Penten-2-ol

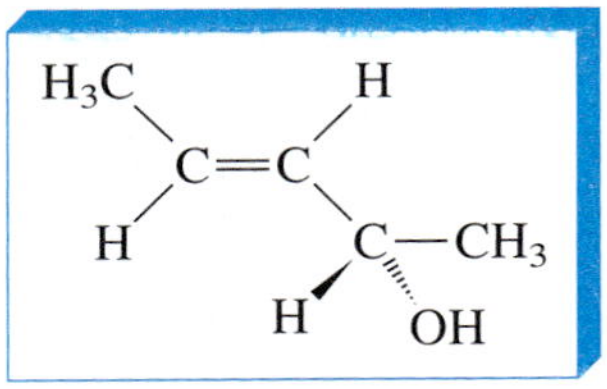

(2*S*,3*E*)-3-Penten-2-ol

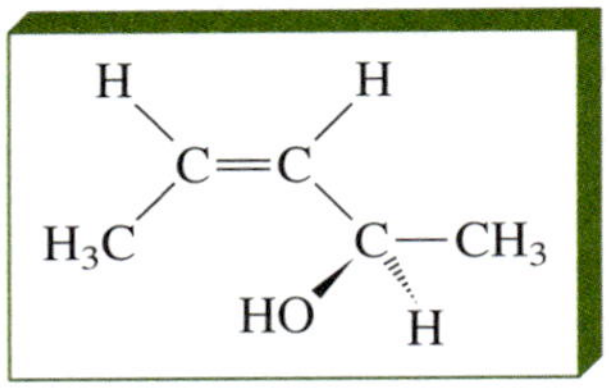

(2*R*,3*Z*)-3-Penten-2-ol

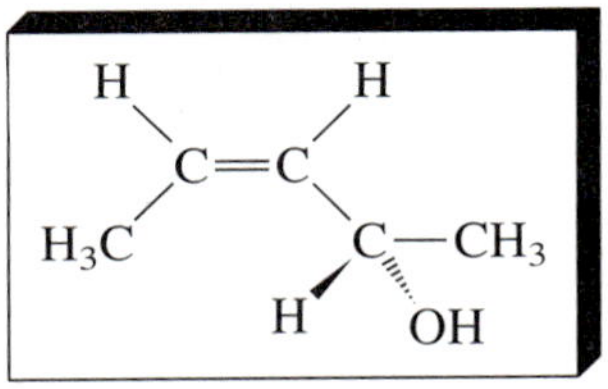

(2*S*,3*Z*)-3-Penten-2-ol

The relationship of the (2*R*,3*E*) stereoisomer to the others is that it is the enantiomer of (2*S*,3*E*)-3-penten-2-ol and is a diastereomer of the (2*R*,3*Z*) and (2*S*,3*Z*) isomers.

7.13 REACTIONS THAT PRODUCE DIASTEREOMERS

Once we grasp the idea of stereoisomerism in molecules with two or more chirality centers, we can explore further details of addition reactions of alkenes.

When bromine adds to (*Z*)- or (*E*)-2-butene, the product 2,3-dibromobutane contains two equivalently substituted chirality centers:

$$CH_3CH=CHCH_3 \xrightarrow{Br_2} \underset{\substack{| \ | \\ Br \ Br}}{CH_3CHCHCH_3}$$

(*Z*)- or (*E*)-2-butene   2,3-Dibromobutane

Three stereoisomers are possible: a pair of enantiomers and a meso form.

Two factors combine to determine which stereoisomers are actually formed in the reaction.

1. The (*E*)- or (*Z*)-configuration of the starting alkene
2. The anti stereochemistry of addition

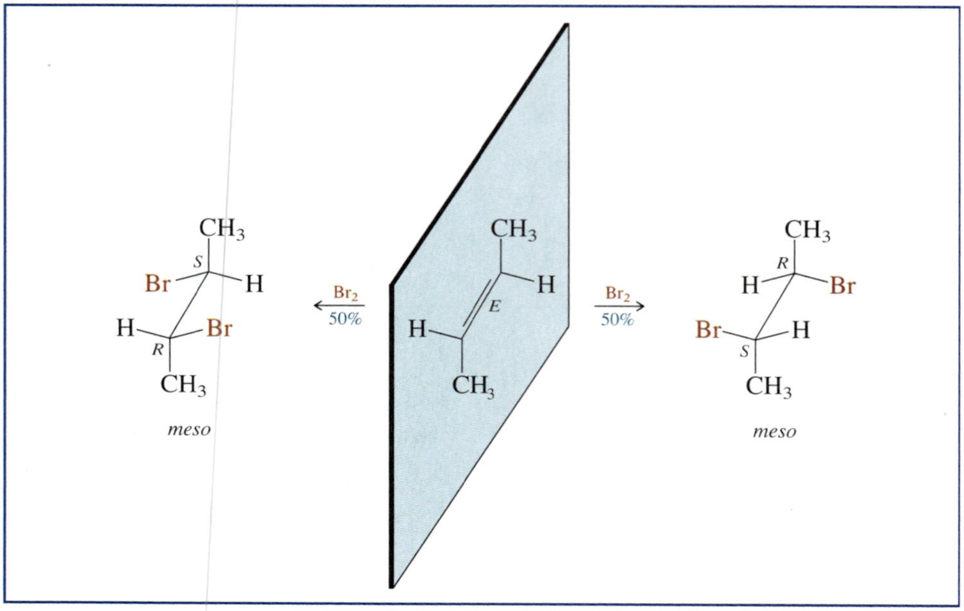

FIGURE 7.13 Anti addition of Br$_2$ to (*E*)-2-butene gives *meso*-2,3-dibromobutane.

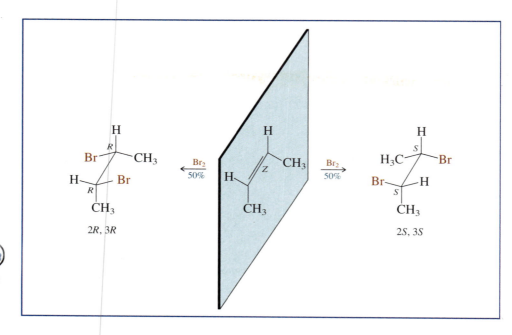

FIGURE 7.13 Anti addition of Br$_2$ to (*E*)-2-butene gives *meso*-2,3-dibromobutane.

Figures 7.13 and 7.14 depict the stereochemical relationships associated with anti addition of bromine to (*E*)- and (*Z*)-2-butene, respectively. The trans alkene (*E*)-2-butene yields only *meso*-2,3-dibromobutane, but the cis alkene (*Z*)-2-butene gives a racemic mixture of (2*R*,3*R*)- and (2*S*,3*S*)-2,3-dibromobutane.

Bromine addition to alkenes is an example of a stereospecific reaction. A **stereospecific reaction** is one in which stereoisomeric starting materials yield products

FIGURE 7.14 Anti addition of Br$_2$ to (*Z*)-2-butene gives a racemic mixture of (2*R*,3*R*)- and (2*S*,3*S*)-2,3-dibromobutane.

that are stereoisomers of each other. In this case the starting materials, in separate reactions, are the *E* and *Z* stereoisomers of 2-butene. The chiral dibromides from (*Z*)-2-butene are stereoisomers (diastereomers) of the meso dibromide formed from (*E*)-2-butene.

Notice further that, consistent with the principle developed in Section 7.9, optically inactive starting materials (achiral alkenes and bromine) yield optically inactive products (a racemic mixture or a meso structure) in these reactions.

PROBLEM 7.20 Epoxidation of alkenes is a stereospecific syn addition. Which stereoisomer of 2-butene reacts with peroxyacetic acid to give *meso*-2,3-epoxybutane? Which one gives a racemic mixture of (2*R*,3*R*)- and (2*S*,3*S*)-2,3-epoxybutane?

A reaction that introduces a second chirality center into a starting material that already has one need not produce equal quantities of two possible diastereomers. Consider catalytic hydrogenation of 2-methyl(methylene)cyclohexane. As you might expect, both *cis*- and *trans*-1,2-dimethylcyclohexane are formed.

| 2-Methyl(methylene)cyclo-hexane | *cis*-1,2-Dimethylcyclo-hexane (68%) | *trans*-1,2-Dimethylcyclo-hexane (32%) |

Make molecular models of the reactant and both products shown in the equation.

The relative amounts of the two products, however, are not equal; more *cis*-1,2-dimethylcyclohexane is formed than *trans*-. The reason for this is that it is the less hindered face of the double bond that approaches the catalyst surface and is the face to which hydrogen is transferred. Hydrogenation of 2-methyl(methylene)cyclohexane occurs preferentially at the side of the double bond opposite that of the methyl group and leads to a faster rate of formation of the cis stereoisomer of the product.

The double bond in 2-methyl(methylene)cyclohexane is prochiral. The two faces, however, are not enantiotopic as they were for the alkenes we discussed in Section 7.9. In those earlier examples, when addition to the double bond created a new chirality center, attack at one face gave one enantiomer; attack at the other gave the other enantiomer. In the case of 2-methyl(methylene)cyclohexane, which already has one chirality center, attack at opposite faces of the double bond gives two products that are diastereomers of each other. Prochiral faces of this type are called **diastereotopic.**

PROBLEM 7.21 Could the fact that hydrogenation of 2-methyl(methylene)cyclohexane gives more *cis*-1,2-dimethylcyclohexane than *trans*- be explained on the basis of the relative stabilities of the two stereoisomeric products?

The hydrogenation of 2-methyl(methylene)cyclohexane is an example of a *stereoselective reaction*, meaning one in which stereoisomeric products are formed in unequal amounts from a single starting material (Section 5.11).

A common misconception is that a stereospecific reaction is simply one that is 100% stereoselective. The two terms are not synonymous, however. A stereospecific reaction is one which, when carried out with stereoisomeric starting materials, gives a product from one reactant that is a stereoisomer of the product from the other. A stereoselective reaction is one in which a single starting material gives a predominance of a

Note that the terms *regioselective* and *regiospecific,* however, are defined in terms of each other. A regiospecific reaction is one that is 100% regioselective.

single stereoisomer when two or more are possible. *Stereospecific* is more closely connected with features of the reaction than with the reactant. Thus terms such as syn *addition* and anti *elimination* describe the stereospecificity of reactions. *Stereoselective* is more closely connected with structural effects in the reactant as expressed in terms such as *addition to the less hindered side*. A stereospecific reaction can also be stereoselective. For example, syn addition describes stereospecificity in the catalytic hydrogenation of alkenes, whereas the preference for addition to the less hindered face of the double bond describes stereoselectivity.

7.14 RESOLUTION OF ENANTIOMERS

The separation of a racemic mixture into its enantiomeric components is termed **resolution.** The first resolution, that of tartaric acid, was carried out by Louis Pasteur in 1848. Tartaric acid is a byproduct of wine making and is almost always found as its dextrorotatory 2*R*,3*R* stereoisomer, shown here in a perspective drawing and in a Fischer projection.

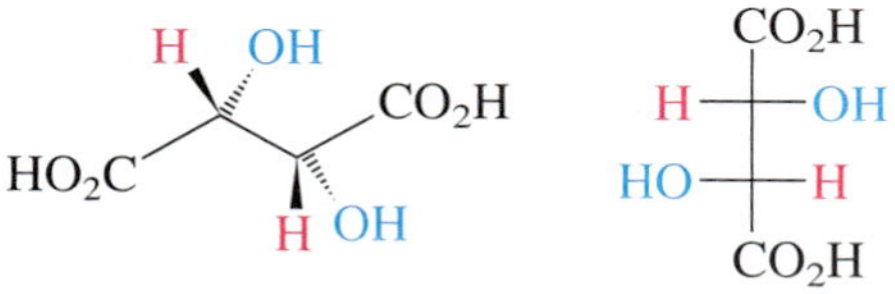

(2*R*,3*R*)-Tartaric acid (mp 170°C, $[\alpha]_D$ +12°)

PROBLEM 7.22 There are two other stereoisomeric tartaric acids. Write their Fischer projections, and specify the configuration at their chirality centers.

A description of Pasteur's work, as part of a broader discussion concerning crystal structure, can be found in the article "Molecules, Crystals, and Chirality" in the July 1997 issue of the *Journal of Chemical Education*, pp. 800–806.

Occasionally, an optically inactive sample of tartaric acid was obtained. Pasteur noticed that the sodium ammonium salt of optically inactive tartaric acid was a mixture of two mirror-image crystal forms. With microscope and tweezers, Pasteur carefully separated the two. He found that one kind of crystal (in aqueous solution) was dextrorotatory, whereas the mirror-image crystals rotated the plane of polarized light an equal amount but were levorotatory.

Although Pasteur was unable to provide a structural explanation—that had to wait for van't Hoff and Le Bel a quarter of a century later—he correctly deduced that the enantiomeric quality of the crystals was the result of enantiomeric molecules. The rare form of tartaric acid was optically inactive because it contained equal amounts of (+)-tartaric acid and (−)-tartaric acid. It had earlier been called *racemic acid* (from Latin *racemus,* meaning "a bunch of grapes"), a name that subsequently gave rise to our present term for an equal mixture of enantiomers.

PROBLEM 7.23 Could the unusual, optically inactive form of tartaric acid studied by Pasteur have been *meso*-tartaric acid?

Pasteur's technique of separating enantiomers not only is laborious but requires that the crystals of the enantiomers be distinguishable. This happens very rarely. Consequently, alternative and more general approaches for resolving enantiomers have been developed. Most are based on a strategy of temporarily converting the enantiomers of a racemic mixture to diastereomeric derivatives, separating these diastereomers, then regenerating the enantiomeric starting materials.

Figure 7.15 illustrates this strategy. Say we have a mixture of enantiomers, which, for simplicity, we label as C(+) and C(−). Assume that C(+) and C(−) bear some

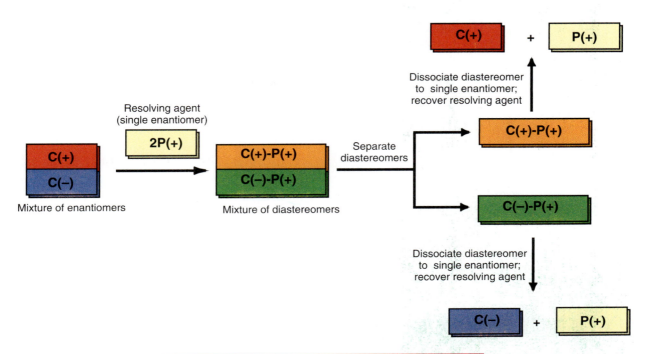

FIGURE 7.15 The general procedure for resolving a chiral substance into its enantiomers. Reaction with a single enantiomer of a chiral resolving agent P(+) converts the racemic mixture of enantiomers C(+) and C(−) to a mixture of diastereomers C(+)-P(+) and C(−)-P(+). The mixture of diastereomers is separated—by fractional crystallization, for example. A chemical reaction is then carried out to convert diastereomer C(+)-P(+) to C(+) and the resolving agent P(+). Likewise, diastereomer C(−)-P(+) is converted to C(−) and P(+). C(+) has been separated from C(−), and the resolving agent P(+) can be recovered for further use.

functional group that can combine with a reagent P to yield adducts C(+)-P and C(−)-P. Now, if reagent P is chiral, and if only a single enantiomer of P, say, P(+), is added to a racemic mixture of C(+) and C(−), as shown in the first step of Figure 7.15, then the products of the reaction are C(+)-P(+) and C(−)-P(+). These products are not mirror images; they are diastereomers. Diastereomers can have different physical properties, which can serve as a means of separating them. The mixture of diastereomers is separated, usually by recrystallization from a suitable solvent. In the last step, an appropriate chemical transformation liberates the enantiomers and restores the resolving agent.

Whenever possible, the chemical reactions involved in the formation of diastereomers and their conversion to separate enantiomers are simple acid–base reactions. For example, naturally occurring (S)-(−)-malic acid is often used to resolve amines. One such amine that has been resolved in this way is 1-phenylethylamine. Amines are bases, and malic acid is an acid. Proton transfer from (S)-(−)-malic acid to a racemic mixture of (R)- and (S)-1-phenylethylamine gives a mixture of diastereomeric salts.

$$C_6H_5\underset{\underset{CH_3}{|}}{C}HNH_2 \ + \ HO_2CCH_2\underset{\underset{OH}{|}}{C}HCO_2H \ \longrightarrow \ C_6H_5\underset{\underset{CH_3}{|}}{C}H\overset{+}{N}H_3 \quad HO_2CCH_2\underset{\underset{OH}{|}}{C}HCO_2^-$$

1-Phenylethylamine   (S)-(−)-Malic acid   1-Phenylethylammonium (S)-malate
(racemic mixture)   (resolving agent)   (mixture of diastereomeric salts)

The diastereomeric salts are separated and the individual enantiomers of the amine liberated by treatment with a base:

$$\overset{+}{\underset{\underset{\text{CH}_3}{|}}{\text{C}_6\text{H}_5\text{CHNH}_3}} \quad \underset{\underset{\text{OH}}{|}}{\text{HO}_2\text{CCH}_2\text{CHCO}_2}^- \;+\; 2\text{OH}^- \longrightarrow$$

1-Phenylethylammonium (S)-malate
(mixture of diastereomeric salts)   Hydroxide ion

$$\underset{\underset{\text{CH}_3}{|}}{\text{C}_6\text{H}_5\text{CHNH}_2} \quad + \quad ^-\text{O}_2\text{CCH}_2\underset{\underset{\text{OH}}{|}}{\text{CHCO}_2}^- \quad + \; 2\text{H}_2\text{O}$$

1-Phenylethylamine   (S)-(−)-Malic acid   Water
(a single enantiomer)   (recovered resolving agent)

PROBLEM 7.24 In the resolution of 1-phenylethylamine using (−)-malic acid, the compound obtained by recrystallization of the mixture of diastereomeric salts is (R)-1-phenylethylammonium (S)-malate. The other component of the mixture is more soluble and remains in solution. What is the configuration of the more soluble salt?

This method is widely used for the resolution of chiral amines and carboxylic acids. Analogous methods based on the formation and separation of diastereomers have been developed for other functional groups; the precise approach depends on the kind of chemical reactivity associated with the functional groups present in the molecule.

As the experimental tools for biochemical transformations have become more powerful and procedures for carrying out these transformations in the laboratory more routine, the application of biochemical processes to mainstream organic chemical tasks including the production of enantiomerically pure chiral molecules has grown.

One approach, called **enzymatic resolution,** involves treating a racemic mixture with an enzyme that catalyzes the reaction of only one of the enantiomers. Some of the most commonly used ones are *lipases* and *esterases,* enzymes that catalyze the hydrolysis of esters. In a typical procedure, one enantiomer of the acetate ester of a racemic alcohol undergoes hydrolysis and the other is left unchanged when hydrolyzed in the presence of an esterase from hog liver.

Acetate ester of racemic alcohol   (S)-Alcohol   Acetate ester of (R)-alcohol

High yields of the enantiomerically pure alcohol and enantiomerically pure ester are regularly achieved. The growing interest in chiral drugs (see the boxed essay on this topic, p. 296) has stimulated the development of large-scale enzymatic resolution as a commercial process.

7.15 STEREOREGULAR POLYMERS

Before the development of the Ziegler–Natta catalyst systems (Section 6.21), polymerization of propene was not a reaction of much value. The reason for this has a stereochemical basis. Consider a section of *polypropylene:*

$$\left(\begin{matrix} \underset{|}{CH_3} & \underset{|}{CH_3} & \underset{|}{CH_3} & \underset{|}{CH_3} & \underset{|}{CH_3} & \underset{|}{CH_3} \\ -CH_2CHCH_2CHCH_2CHCH_2CHCH_2CHCH_2CH- \end{matrix}\right)$$

Representation of the polymer chain in an extended zigzag conformation, as shown in Figure 7.16, reveals several distinct structural possibilities differing with respect to the relative configurations of the carbons that bear the methyl groups.

One structure, represented in Figure 7.16a, has all the methyl groups oriented in the same direction with respect to the polymer chain. This stereochemical arrangement is said to be **isotactic.** Another form, shown in Figure 7.16b, has its methyl groups alternating front and back along the chain. This arrangement is described as **syndiotactic.** Both the isotactic and the syndiotactic forms of polypropylene are known as **stereoregular polymers** because each is characterized by a precise stereochemistry at the carbon atom that bears the methyl group. A third possibility, shown in Figure 7.16c, is described as **atactic.** Atactic polypropylene has a random orientation of its methyl groups; it is not a stereoregular polymer.

Polypropylene chains associate with one another because of attractive van der Waals forces. The extent of this association is relatively large for isotactic and

(*a*) Isotactic polypropylene

(*b*) Syndiotactic polypropylene

(*c*) Atactic polypropylene

FIGURE 7.16 Polymers of propene. The main chain is shown in a zigzag conformation. Every other carbon bears a methyl substituent and is a chirality center. (*a*) All the methyl groups are on the same side of the carbon chain in isotactic polypropylene. (*b*) Methyl groups alternate from one side to the other in syndiotactic polypropylene. (*c*) The spatial orientation of the methyl groups is random in atactic polypropylene.

syndiotactic polymers, because the stereoregularity of the polymer chains permits efficient packing. Atactic polypropylene, on the other hand, does not associate as strongly. It has a lower density and lower melting point than the stereoregular forms. The physical properties of stereoregular polypropylene are more useful for most purposes than those of atactic polypropylene.

When propene is polymerized under free-radical conditions, the polypropylene that results is atactic. Catalysts of the Ziegler–Natta type, however, permit the preparation of either isotactic or syndiotactic polypropylene. We see here an example of how proper choice of experimental conditions can affect the stereochemical course of a chemical reaction to the extent that entirely new materials with unique properties result.

7.16 CHIRALITY CENTERS OTHER THAN CARBON

Our discussion to this point has been limited to molecules in which the chirality center is carbon. Atoms other than carbon may also be chirality centers. Silicon, like carbon, has a tetrahedral arrangement of bonds when it bears four substituents. A large number of organosilicon compounds in which silicon bears four different groups have been resolved into their enantiomers.

Trigonal pyramidal molecules are chiral if the central atom bears three different groups. If one is to resolve substances of this type, however, the pyramidal inversion that interconverts enantiomers must be slow at room temperature. Pyramidal inversion at nitrogen is so fast that attempts to resolve chiral amines fail because of their rapid racemization.

> Verify that $CH_3NHCH_2CH_3$ is chiral by trying to superimpose models of both enantiomers.

Phosphorus is in the same group of the periodic table as nitrogen, and tricoordinate phosphorus compounds (phosphines), like amines, are trigonal pyramidal. Phosphines, however, undergo pyramidal inversion much more slowly than amines, and a number of optically active phosphines have been prepared.

Tricoordinate sulfur compounds are chiral when sulfur bears three different substituents. The rate of pyramidal inversion at sulfur is rather slow. The most common compounds in which sulfur is a chirality center are sulfoxides such as:

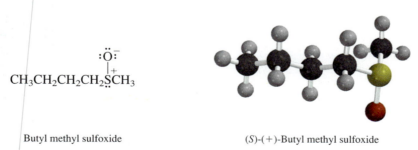

$$\overset{\displaystyle \ddot{\text{:}\overset{\_}{\text{O}}\text{:}}}{\underset{}{|}}$$
$$CH_3CH_2CH_2CH_2\overset{+}{\underset{\displaystyle \ddot{}}{S}}CH_3$$

Butyl methyl sulfoxide

(S)-(+)-Butyl methyl sulfoxide

The absolute configuration at sulfur is specified by the Cahn–Ingold–Prelog method with the provision that the unshared electron pair is considered to be the lowest ranking substituent.

TABLE 7.2	Classification of Isomers*

Definition	Example
1. *Constitutional isomers* are isomers that differ in the order in which their atoms are connected.	Three constitutionally isomeric compounds have the molecular formula C_3H_8O:
	$CH_3CH_2CH_2OH$ CH_3CHCH_3 $CH_3CH_2OCH_3$
	$\overset{\textstyle |}{OH}$
	1-Propanol 2-Propanol Ethyl methyl ether
2. *Stereoisomers* are isomers that have the same constitution but differ in the arrangement of their atoms in space.	
(a) *Enantiomers* are stereoisomers that are related as an object and its nonsuperimposable mirror image.	The two enantiomeric forms of 2-chlorobutane are
	(*R*)-(−)-2-Chlorobutane (*S*)-(+)-2-Chlorobutane
(b) *Diastereomers* are stereoisomers that are not enantiomers.	The cis and trans isomers of 4-methylcyclohexanol are stereoisomers, but they are not related as an object and its mirror image; they are diastereomers.
	cis-4-Methylcyclohexanol *trans*-4-Methylcyclohexanol

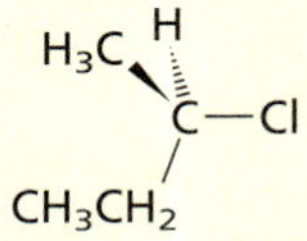

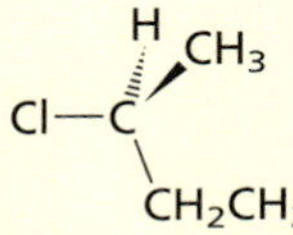

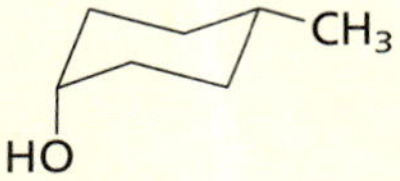

*Isomers are different compounds that have the same molecular formula. They may be either constitutional isomers or stereoisomers.

7.17 SUMMARY

Chemistry in three dimensions is known as **stereochemistry.** At its most fundamental level, stereochemistry deals with molecular structure; at another level, it is concerned with chemical reactivity. Table 7.2 summarizes some basic definitions relating to molecular structure and stereochemistry.

A detailed flowchart describing a more finely divided set of subcategories of isomers appears in the February 1990 issue of the *Journal of Chemical Education.*

Section 7.1   A molecule is **chiral** if it cannot be superimposed on its mirror image. *Nonsuperimposable mirror images* are **enantiomers** of one another. Molecules in which mirror images are superimposable are achiral.

$$CH_3CHCH_2CH_3 \qquad CH_3CHCH_3$$
$$\overset{\textstyle |}{Cl} \qquad\qquad \overset{\textstyle |}{Cl}$$

2-Chlorobutane   2-Chloropropane
(chiral)   (achiral)

Section 7.2   The most common kind of chiral molecule contains a carbon atom that bears four different atoms or groups. Such an atom is called a **chirality center.** Table 7.2 shows the enantiomers of 2-chlorobutane. C-2 is a chirality center in 2-chlorobutane.

Section 7.3   A molecule that has a plane of symmetry or a center of symmetry is achiral. *cis*-4-Methylcyclohexanol (Table 7.2) has a plane of symmetry that bisects the molecule into two mirror-image halves and is achiral. The same can be said for *trans*-4-methylcyclohexanol.

Section 7.4   **Optical activity,** or the degree to which a substance rotates the plane of polarized light, is a physical property used to characterize chiral substances. Enantiomers have equal and opposite **optical rotations.** To be optically active a substance must be chiral, and one enantiomer must be present in excess of the other. A **racemic mixture** is optically inactive and contains equal quantities of enantiomers.

Section 7.5   **Relative configuration** compares the arrangement of atoms in space to some reference. The prefix *cis* in *cis*-4-methylcyclohexanol, for example, describes relative configuration by referencing the orientation of the CH_3 group to the OH. **Absolute configuration** is an exact description of the arrangement of atoms in space.

Section 7.6   Absolute configuration in chiral molecules is best specified using the prefixes *R* and *S* of the Cahn–Ingold–Prelog notational system. Substituents at a chirality center are ranked in order of decreasing precedence. If the three highest ranked substituents trace a clockwise path (highest→second highest→third highest) when the lowest ranked substituent is held away from you, the configuration is *R*. If the path is counterclockwise, the configuration is *S*. Table 7.2 shows the *R* and *S* enantiomers of 2-chlorobutane.

Section 7.7   A **Fischer projection** shows how a molecule would look if its bonds were projected onto a flat surface. Horizontal lines represent bonds coming toward you; vertical bonds point away from you. The projection is normally drawn so that the carbon chain is vertical, with the lowest numbered carbon at the top.

(*R*)-2-Chlorobutane   (*S*)-2-Chlorobutane

Section 7.8   Both enantiomers of the same substance are identical in most of their physical properties. The most prominent differences are biological ones, such as taste and odor, in which the substance interacts with a chiral receptor site in a living system. Enantiomers also have important consequences in medicine, in which the two enantiomeric forms of a drug can have much different effects on a patient.

Section 7.9   A chemical reaction can convert an achiral substance to a chiral one. If the product contains a single chirality center, it is formed as a racemic mixture. Optically active products can be formed from optically inactive starting materials only if some optically active agent is present. The best examples are biological processes in which enzymes catalyze the formation of only a single enantiomer.

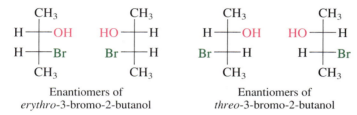

Section 7.10   When a molecule has two chirality centers and these two chirality centers are not equivalent, four stereoisomers are possible.

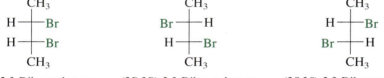

Stereoisomers that are not enantiomers are classified as **diastereomers.** Each enantiomer of *erythro*-3-bromo-2-butanol is a diastereomer of each enantiomer of *threo*-3-bromo-2-butanol.

Section 7.11   Achiral molecules that contain chirality centers are called **meso forms.** Meso forms typically contain (but are not limited to) two equivalently substituted chirality centers. They are optically inactive.

Section 7.12   For a particular constitution, the maximum number of stereoisomers is 2^n, where n is the number of structural units capable of stereochemical variation—usually this is the number of chirality centers, but can include E and Z double bonds as well. The number of stereoisomers is reduced to less than 2^n when there are meso forms.

Section 7.13   Addition reactions of alkenes may generate one (Section 7.9) or two (Section 7.13) chirality centers. When two chirality centers are produced, their relative stereochemistry depends on the configuration (E or Z) of the alkene and whether the addition is syn or anti.

Section 7.14   **Resolution** is the separation of a racemic mixture into its enantiomers. It is normally carried out by converting the mixture of enantiomers to a mixture of diastereomers, separating the diastereomers, then regenerating the enantiomers.

Section 7.15   Certain polymers such as polypropylene contain chirality centers, and the relative configurations of these centers affect the physical properties of the polymers. Like substituents appear on the same side of a zigzag carbon chain in an **isotactic** polymer, alternate along the chain in a **syndiotactic** polymer, and appear in a random manner in an **atactic** polymer. Isotactic and syndiotactic polymers are referred to as **stereoregular** polymers.

Section 7.16   Atoms other than carbon can be chirality centers. Examples include those based on tetracoordinate silicon and tricoordinate sulfur as the chirality center. In principle, tricoordinate nitrogen can be a chirality center in compounds of the type N(x, y, z), where x, y, and z are different, but inversion of the nitrogen pyramid is so fast that racemization occurs virtually instantly at room temperature.

PROBLEMS

7.25 Which of the isomeric alcohols having the molecular formula $C_5H_{12}O$ are chiral? Which are achiral?

7.26 Write structural formulas or make molecular models for all the compounds that are trichloro derivatives of cyclopropane. (Don't forget to include stereoisomers.) Which are chiral? Which are achiral?

7.27 In each of the following pairs of compounds one is chiral and the other is achiral. Identify each compound as chiral or achiral, as appropriate.

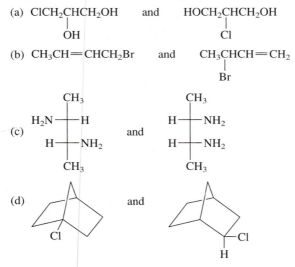

7.28 Compare 2,3-pentanediol and 2,4-pentanediol with respect to the number of stereoisomers possible for each constitution. Which stereoisomers are chiral? Which are achiral?

7.29 In 1996, it was determined that the absolute configuration of ($-$)-bromochlorofluoromethane is R. Which of the following is (are) ($-$)-BrClFCH?

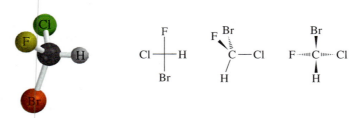

7.30 Specify the configuration of the chirality center as *R* or *S* in each of the following.

(a) (−)-2-Octanol

(b) Monosodium L-glutamate (only this stereoisomer is of any value as a flavor-enhancing agent)

$$CO_2^-$$
$$H_3\overset{+}{N} \rule{1cm}{0.4pt} H$$
$$CH_2CH_2CO_2^- \ Na^+$$

7.31 A subrule of the Cahn–Ingold–Prelog system specifies that higher mass number takes precedence over lower when distinguishing between isotopes.

(a) Determine the absolute configurations of the reactant and product in the biological oxidation of isotopically labeled ethane described in Section 7.2.

$$\underset{H}{\overset{D \quad T}{C}} - CH_3 \xrightarrow{\text{biological oxidation}} \underset{HO}{\overset{D \quad T}{C}} - CH_3$$

(b) Because OH becomes bonded to carbon at the same side from which H is lost, the oxidation proceeds with retention of configuration (Section 6.13). Compare this fact with the *R* and *S* configurations you determined in part (a) and reconcile any *apparent* conflicts.

7.32 Identify the relationship in each of the following pairs. Do the drawings represent constitutional isomers or stereoisomers, or are they just different ways of drawing the same compound? If they are stereoisomers, are they enantiomers or diastereomers? (Molecular models may prove useful in this problem.)

(a)
$$\underset{HO}{\overset{H \quad CH_3}{C}} - CH_2Br \quad \text{and} \quad \underset{Br}{\overset{H_3C \quad H}{C}} - CH_2OH$$

(b)
$$\underset{CH_3CH_2}{\overset{H \quad CH_3}{C}} - Br \quad \text{and} \quad \underset{CH_3CH_2}{\overset{H \quad Br}{C}} - CH_3$$

(c)
$$\underset{CH_3CH_2}{\overset{H \quad CH_3}{C}} - Br \quad \text{and} \quad \underset{Br}{\overset{H_3C \quad H}{C}} - CH_2CH_3$$

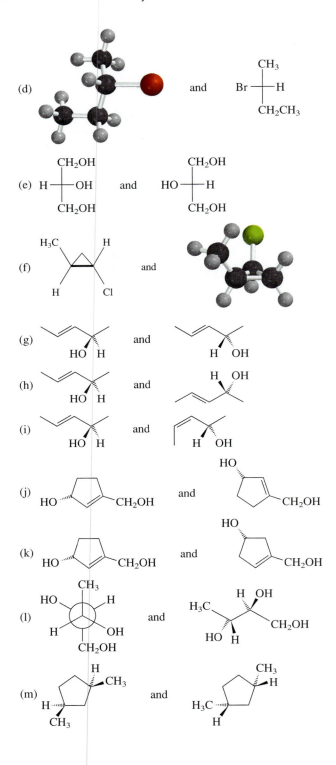

(d) and

$$CH_3$$
$$Br \text{---} H$$
$$CH_2CH_3$$

(e)
$$CH_2OH$$
$$H \text{---} OH$$
$$CH_2OH$$
and
$$CH_2OH$$
$$HO \text{---} H$$
$$CH_2OH$$

(f) and

(g) and

(h) and

(i) and

(j) and

(k) and

(l) and

(m) and

(n)

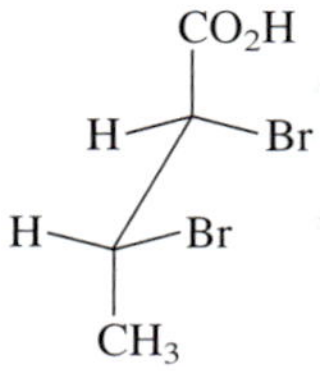

CO$_2$H
H—Br
H—Br
CH$_3$

and

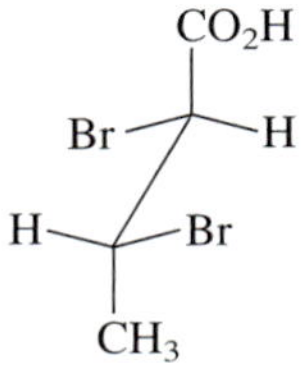

CO$_2$H
Br—H
H—Br
CH$_3$

(o)

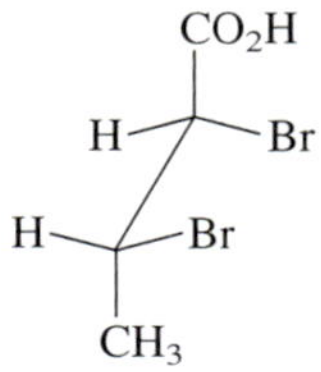

CO$_2$H
H—Br
H—Br
CH$_3$

and

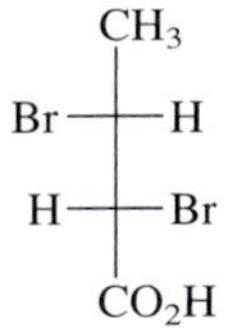

CH$_3$
Br—H
H—Br
CO$_2$H

(p)

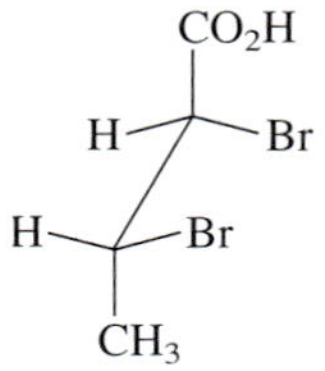

CO$_2$H
H—Br
H—Br
CH$_3$

and

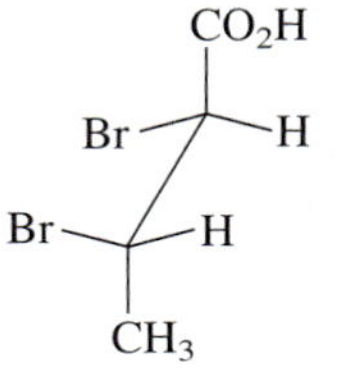

CO$_2$H
Br—H
Br—H
CH$_3$

(q) 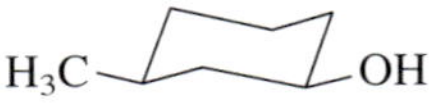 H$_3$C⟍⟍OH   and   H$_3$C⟍⟍OH

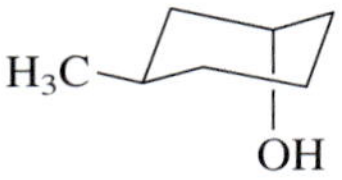

(r)  (CH$_3$)$_3$C⟍⟍I   and   (CH$_3$)$_3$C⟍⟍I

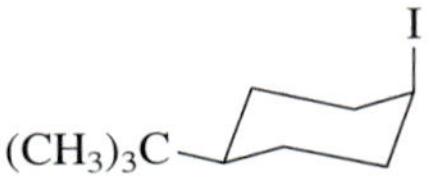

(s)

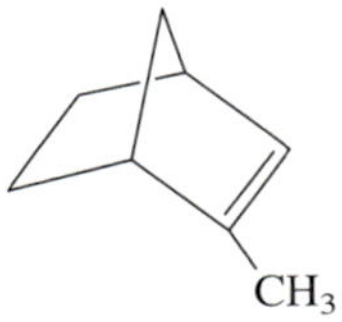

CH$_3$

and

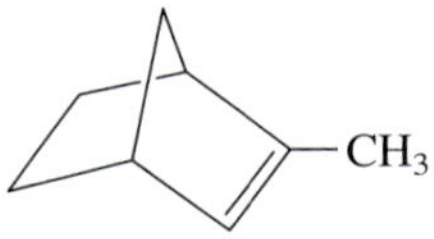

CH$_3$

(t)

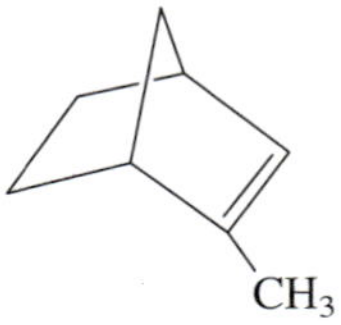

CH$_3$

and

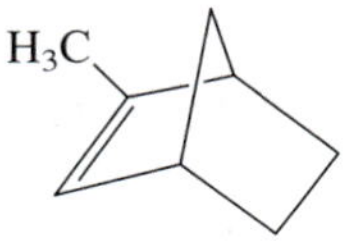

H$_3$C

(u)

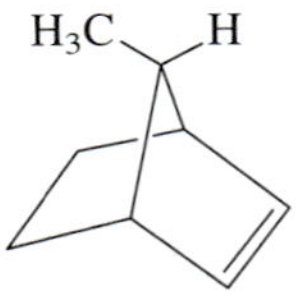

H$_3$C   H

and

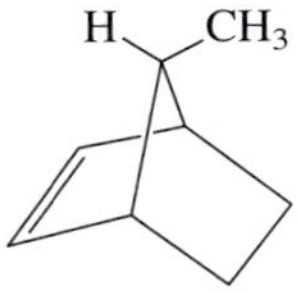

H   CH$_3$

(v)

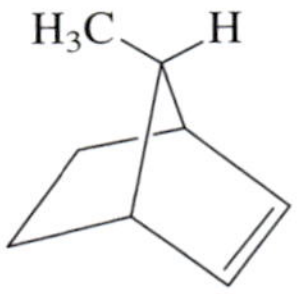

H$_3$C   H

and

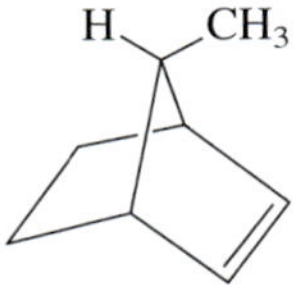

H   CH$_3$

7.33 Chemical degradation of chlorophyll gives a number of substances including *phytol*. The constitution of phytol is given by the name 3,7,11,15-tetramethyl-2-hexadecen-1-ol. How many stereoisomers have this constitution?

7.34 *Muscarine* is a poisonous substance present in the mushroom *Amanita muscaria*. Its structure is represented by the constitution shown here.

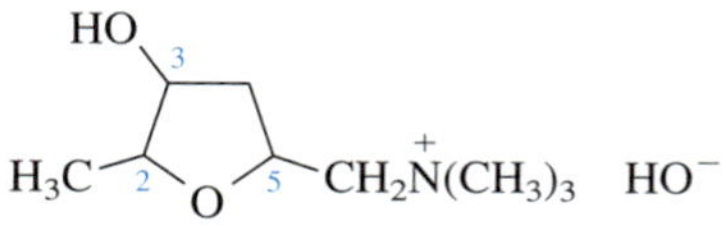

$$HO \overset{3}{\diagdown} $$

$$H_3C \overset{2}{\diagup} \underset{O}{\diagup} \overset{5}{\diagdown} CH_2\overset{+}{N}(CH_3)_3 \quad HO^-$$

(a) Including muscarine, how many stereoisomers have this constitution?

(b) One of the substituents on the ring of muscarine is trans to the other two. How many of the stereoisomers satisfy this requirement?

(c) Muscarine has the configuration 2*S*,3*R*,5*S*. Write a structural formula or build a molecular model of muscarine showing its correct stereochemistry.

7.35 *Ectocarpene* is a volatile, sperm cell-attracting material released by the eggs of the seaweed *Ectocarpus siliculosus*. Its constitution is

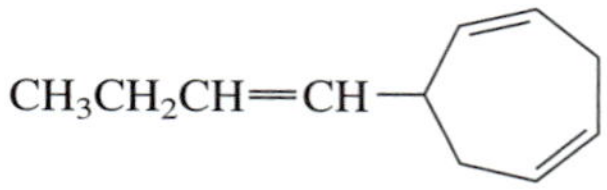

$$CH_3CH_2CH=CH-$$

All the double bonds are cis, and the absolute configuration of the chirality center is *S*. Write a stereochemically accurate representation of ectocarpene.

7.36 *Multifidene* is a sperm cell-attracting substance released by the female of a species of brown algae (*Cutleria multifida*). The constitution of multifidene is

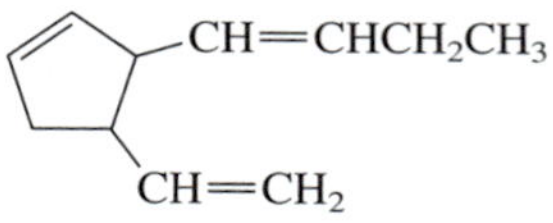

$$-CH=CHCH_2CH_3$$

$$CH=CH_2$$

(a) How many stereoisomers are represented by this constitution?

(b) Multifidene has a cis relationship between its alkenyl substituents. Given this information, how many stereoisomers are possible?

(c) The butenyl side chain has the *Z* configuration of its double bond. On the basis of all the data, how many stereoisomers are possible?

(d) Draw stereochemically accurate representations of all the stereoisomers that satisfy the structural requirements of multifidene.

(e) How are these stereoisomeric multifidenes related (enantiomers or diastereomers)?

7.37 *Streptimidone* is an antibiotic and has the structure shown. How many diastereomers of streptimidone are possible? How many enantiomers? Using the *E*,*Z* and *R*,*S* descriptors, specify all essential elements of stereochemistry of streptimidone.

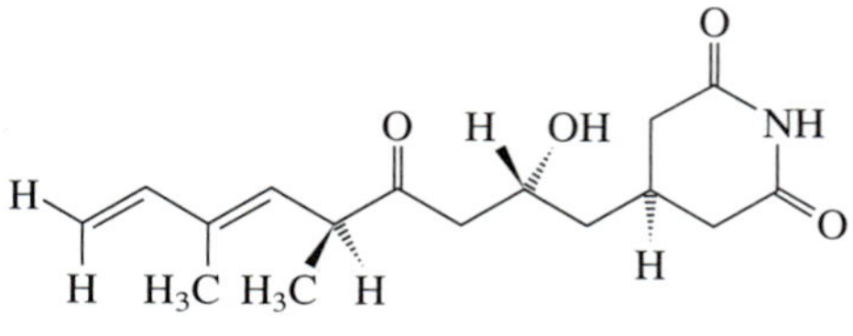

7.38 In Problem 4.25 you were asked to draw the preferred conformation of menthol on the basis of the information that menthol is the most stable stereoisomer of 2-isopropyl-5-methylcyclo-hexanol. We can now completely describe (−)-menthol structurally by noting that it has the *R* configuration at the hydroxyl-substituted carbon.

    (a) Draw or construct a molecular model of the preferred conformation of (−)-menthol.

    (b) (+)-Isomenthol has the same constitution as (−)-menthol. The configurations at C-1 and C-2 of (+)-isomenthol are the opposite of the corresponding chirality centers of (−)-menthol. Write the preferred conformation of (+)-isomenthol.

7.39 A certain natural product having [α]$_D$ + 40.3° was isolated. Two structures have been independently proposed for this compound. Which one do you think is more likely to be correct? Why?

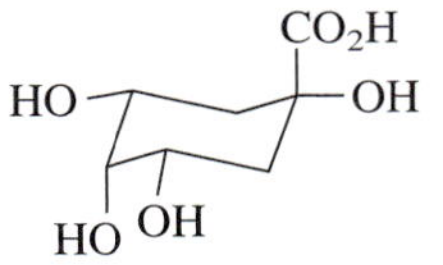

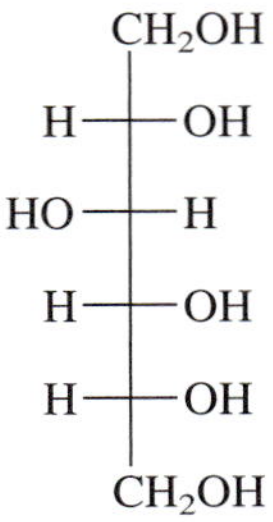

7.40 One of the principal substances obtained from archaea (one of the oldest forms of life on earth) is derived from a 40-carbon diol. Given the fact that this diol is optically active, is it compound A or is it compound B?

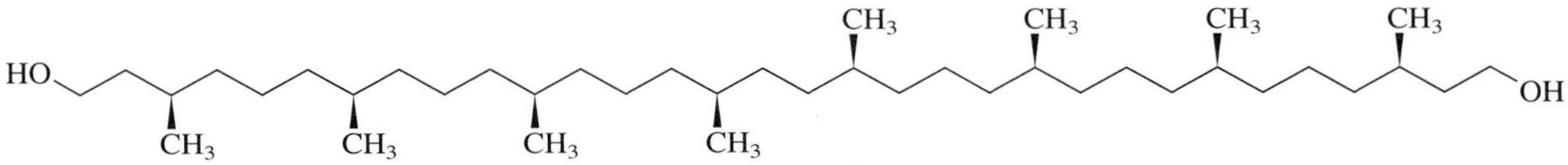

Compound A

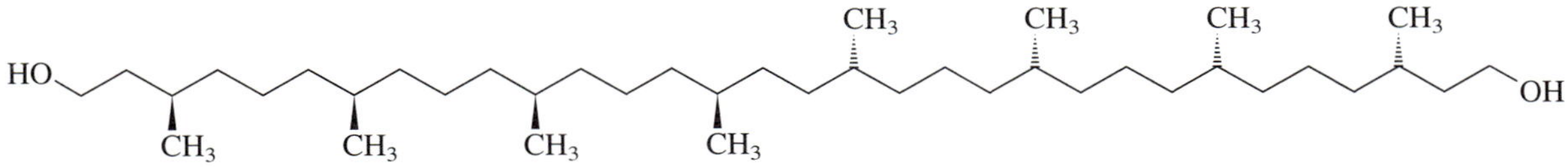

Compound B

7.41 (a) An aqueous solution containing 10 g of optically pure fructose was diluted to 500 mL with water and placed in a polarimeter tube 20 cm long. The measured rotation was −5.20°. Calculate the specific rotation of fructose.

    (b) If this solution were mixed with 500 mL of a solution containing 5 g of racemic fructose, what would be the specific rotation of the resulting fructose mixture? What would be its optical purity?

7.42 Write the organic products of each of the following reactions. If two stereoisomers are formed, show both. Label all chirality centers *R* or *S* as appropriate.

    (a) 1-Butene and hydrogen iodide

    (b) (*E*)-2-Pentene and bromine in carbon tetrachloride

    (c) (*Z*)-2-Pentene and bromine in carbon tetrachloride

    (d) 1-Butene and peroxyacetic acid in dichloromethane

    (e) (*Z*)-2-Pentene and peroxyacetic acid in dichloromethane

(f) 1,5,5-Trimethylcyclopentene and hydrogen in the presence of platinum

(g) 1,5,5-Trimethylcyclopentene and diborane in tetrahydrofuran followed by oxidation with hydrogen peroxide

7.43 The enzyme *aconitase* catalyzes the hydration of aconitic acid to two products: citric acid and isocitric acid. Isocitric acid is optically active; citric acid is not. What are the respective constitutions of citric acid and isocitric acid?

$$HO_2CCH_2 \diagdown \hspace{2em} \diagup CO_2H$$
$$C=C$$
$$HO_2C \diagup \hspace{2em} \diagdown H$$

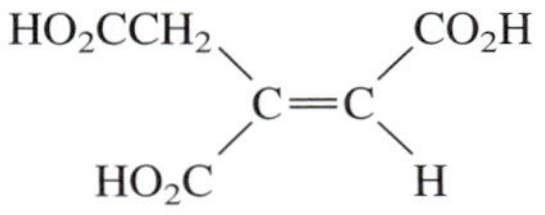

Aconitic acid

7.44 Consider the ozonolysis of *trans*-4,5-dimethylcyclohexene having the configuration shown.

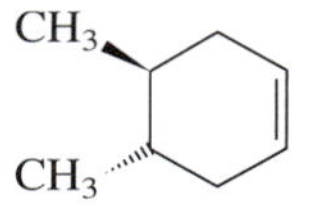

CH$_3$

CH$_3$

Structures A, B, and C are three stereoisomeric forms of the reaction product.

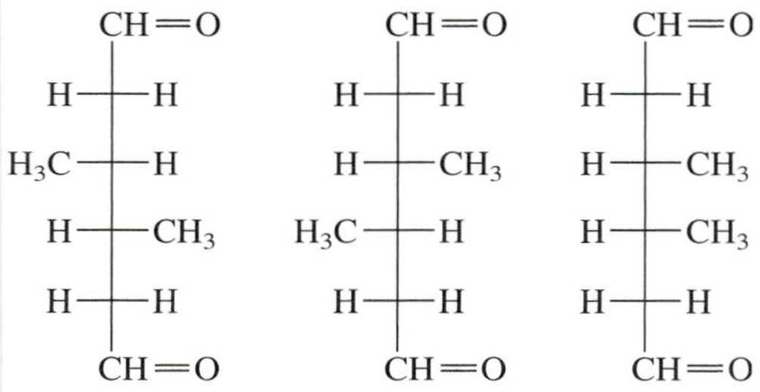

```
    CH=O            CH=O            CH=O
  H─┼─H           H─┼─H           H─┼─H
 H₃C─┼─H           H─┼─CH₃         H─┼─CH₃
  H─┼─CH₃        H₃C─┼─H           H─┼─CH₃
  H─┼─H           H─┼─H           H─┼─H
    CH=O            CH=O            CH=O

     A               B               C
```

(a) Which, if any, of the compounds A, B, and C are chiral?

(b) What product is formed in the reaction?

(c) What product would be formed if the methyl groups were cis to each other in the starting alkene?

7.45 (a) On being heated with potassium ethoxide in ethanol (70°C), the deuterium-labeled alkyl bromide shown gave a mixture of 1-butene, *cis*-2-butene, and *trans*-2-butene. On the basis of your knowledge of the E2 mechanism, predict which alkene(s), if any, contained deuterium.

(b) The bromide shown in part (a) is the erythro diastereomer. How would the deuterium content of the alkenes formed by dehydrohalogenation of the threo diastereomer differ from those produced in part (a)?

7.46 A compound (C_6H_{10}) contains a five-membered ring. When Br_2 adds to it, two diastereomeric dibromides are formed. Suggest reasonable structures for the compound and the two dibromides.

7.47 When optically pure 2,3-dimethyl-2-pentanol was subjected to dehydration, a mixture of two alkenes was obtained. Hydrogenation of this alkene mixture gave 2,3-dimethylpentane, which was 50% optically pure. What were the two alkenes formed in the elimination reaction, and what were the relative amounts of each?

7.48 When (R)-3-buten-2-ol is treated with a peroxy acid, two stereoisomeric epoxides are formed in a 60:40 ratio. The minor stereoisomer has the structure shown.

(a) Write the structure of the major stereoisomer.

(b) What is the relationship between the two epoxides? Are they enantiomers or diastereomers?

(c) What four stereoisomeric products are formed when racemic 3-buten-2-ol is epoxidized under the same conditions? How much of each stereoisomer is formed?

7.49 Verify that dibromochloromethane is achiral by superimposing models of its two mirror image forms. In the same way, verify that bromochlorofluoromethane is chiral.

7.50 Construct a molecular model of (S)-3-chlorocyclopentene.

7.51 Construct a molecular model corresponding to the Fischer projection of *meso*-2,3-dibromobutane. Convert this molecular model to a staggered conformation in which the bromines are anti to one another. Are the methyl groups anti or gauche to one another in this staggered conformation?

7.52 What alkene gives a racemic mixture of (2R,3S) and (2S,3R)-3-bromo-2-butanol on treatment with Br_2 in aqueous solution? (*Hint:* Make a molecular model of one of the enantiomeric 3-bromo-2-butanols, arrange it in a conformation in which the Br and OH groups are anti to one another, then disconnect them.)

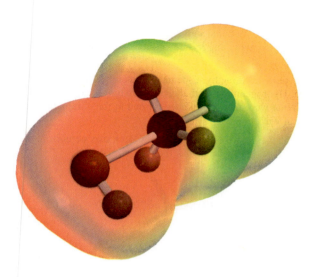

CHAPTER 8

NUCLEOPHILIC SUBSTITUTION

When we discussed elimination reactions in Chapter 5, we learned that a Lewis base can react with an alkyl halide to form an alkene. In the present chapter, you will find that the same kinds of reactants can also undergo a different reaction, one in which the Lewis base acts as a **nucleophile** to substitute for the halogen substituent on carbon.

$$R—\overset{..}{\underset{..}{X}}: \ + \quad Y:^{-} \quad \longrightarrow \quad R—Y \quad + \ :\overset{..}{\underset{..}{X}}:^{-}$$

| Alkyl halide | Lewis base | Product of nucleophilic substitution | Halide anion |

We first encountered nucleophilic substitution in Chapter 4, in the reaction of alcohols with hydrogen halides to form alkyl halides. Now we'll see how alkyl halides can themselves be converted to other classes of organic compounds by nucleophilic substitution.

This chapter has a mechanistic emphasis designed to achieve a practical result. By understanding the mechanisms by which alkyl halides undergo nucleophilic substitution, we can choose experimental conditions best suited to carrying out a particular functional group transformation. The difference between a successful reaction that leads cleanly to a desired product and one that fails is often a subtle one. Mechanistic analysis helps us to appreciate these subtleties and use them to our advantage.

8.1 FUNCTIONAL GROUP TRANSFORMATION BY NUCLEOPHILIC SUBSTITUTION

Nucleophilic substitution reactions of alkyl halides are related to elimination reactions in that the halogen acts as a leaving group on carbon and is lost as an anion. The carbon–halogen bond of the alkyl halide is broken **heterolytically:** the two electrons in that bond are lost with the leaving group.

The carbon–halogen bond in an alkyl halide is polar

$$\overset{\delta+\quad\delta-}{R-X} \qquad X = I, Br, Cl, F$$

and is cleaved on attack by a nucleophile so that the two electrons in the bond are retained by the halogen

$$^-Y\!: \quad R-\!\overset{..}{\underset{..}{X}}\!: \longrightarrow R-Y + :\overset{..}{\underset{..}{X}}\!:^-$$

The most frequently encountered nucleophiles in functional group transformations are anions, which are used as their lithium, sodium, or potassium salts. If we use M to represent lithium, sodium, or potassium, some representative nucleophilic reagents are

MOR (a metal *alkoxide,* a source of the nucleophilic anion $R\overset{..}{\underset{..}{O}}\!:^-$)

$$\overset{\displaystyle O}{\underset{\displaystyle \parallel}{MOCR}}$$ (a metal *carboxylate,* a source of the nucleophilic anion $R\overset{\displaystyle :\overset{..}{O}:}{\overset{\parallel}{C}}-\overset{..}{\underset{..}{O}}\!:^-$)

MSH (a metal *hydrogen sulfide,* a source of the nucleophilic anion $H\overset{..}{\underset{..}{S}}\!:^-$)

MCN (a metal *cyanide,* a source of the nucleophilic anion $^-\!:C\equiv N\!:$)

MN$_3$ (a metal *azide,* a source of the nucleophilic anion $^-\!:\overset{..}{N}=\overset{+}{N}=\overset{..}{\underset{..}{N}}\!:$)

Table 8.1 illustrates an application of each of these to a functional group transformation. The anionic portion of the salt substitutes for the halogen of an alkyl halide. The metal cation portion becomes a lithium, sodium, or potassium halide.

$$M^+ \;\; ^-Y\!: + R-\!\overset{..}{\underset{..}{X}}\!: \longrightarrow \;\; R-Y \;\; + \; M^+ \; :\overset{..}{\underset{..}{X}}\!:^-$$

| Nucleophilic reagent | Alkyl halide | Product of nucleophilic substitution | Metal halide |

Notice that all the examples in Table 8.1 involve **alkyl halides,** that is, compounds in which the halogen is attached to an sp^3-hybridized carbon. **Alkenyl halides** and **aryl halides,** compounds in which the halogen is attached to sp^2-hybridized carbons, are essentially unreactive under these conditions, and the principles to be developed in this chapter do not apply to them.

Alkenyl halides are also referred to as *vinylic halides.*

sp^3-hybridized carbon sp^2-hybridized carbon

Alkyl halide   Alkenyl halide   Aryl halide

To ensure that reaction occurs in homogeneous solution, solvents are chosen that dissolve both the alkyl halide and the ionic salt. The alkyl halide substrates are soluble in organic solvents, but the salts often are not. Inorganic salts are soluble in water, but alkyl

| TABLE 8.1 | Representative Functional Group Transformations by Nucleophilic Substitution Reactions of Alkyl Halides |

Nucleophile and comments	General equation and specific example

Alkoxide ion ($R\ddot{O}:^-$) The oxygen atom of a metal alkoxide acts as a nucleophile to replace the halogen of an alkyl halide. The product is an *ether*.

$$R'\ddot{O}:^- \quad + \quad R-\ddot{X}: \longrightarrow R'\ddot{O}R \quad + \quad :\ddot{X}:^-$$

Alkoxide ion   Alkyl halide   Ether   Halide ion

$$(CH_3)_2CHCH_2ONa + CH_3CH_2Br \xrightarrow{\text{isobutyl alcohol}} (CH_3)_2CHCH_2OCH_2CH_3 + NaBr$$

Sodium isobutoxide   Ethyl bromide   Ethyl isobutyl ether (66%)   Sodium bromide

Carboxylate ion ($RC{-}\overset{:O:}{\overset{\|}{}}\ddot{O}:^-$) An *ester* is formed when the negatively charged oxygen of a carboxylate replaces the halogen of an alkyl halide.

$$R'\overset{:O:}{\overset{\|}{C}}\ddot{O}:^- \quad + \quad R-\ddot{X}: \longrightarrow R'\overset{:O:}{\overset{\|}{C}}\ddot{O}R \quad + \quad :\ddot{X}:^-$$

Carboxylate ion   Alkyl halide   Ester   Halide ion

$$KO\overset{O}{\overset{\|}{C}}(CH_2)_{16}CH_3 + CH_3CH_2I \xrightarrow{\text{acetone water}} CH_3CH_2O\overset{O}{\overset{\|}{C}}(CH_2)_{16}CH_3 + KI$$

Potassium octadecanoate   Ethyl iodide   Ethyl octadecanoate (95%)   Potassium iodide

Hydrogen sulfide ion ($H\ddot{S}:^-$) Use of hydrogen sulfide as a nucleophile permits the conversion of alkyl halides to compounds of the type RSH. These compounds are the sulfur analogs of alcohols and are known as *thiols*.

$$H\ddot{S}:^- \quad + \quad R-\ddot{X}: \longrightarrow R\ddot{S}H \quad + \quad :\ddot{X}:^-$$

Hydrogen sulfide ion   Alkyl halide   Thiol   Halide ion

$$KSH + CH_3CH(CH_2)_6CH_3 \xrightarrow{\text{ethanol water}} CH_3CH(CH_2)_6CH_3 + KBr$$
$$\qquad\qquad\quad | \qquad\qquad\qquad\qquad\qquad |$$
$$\qquad\qquad\quad Br \qquad\qquad\qquad\qquad\qquad SH$$

Potassium hydrogen sulfide   2-Bromononane   2-Nonanethiol (74%)   Potassium bromide

Cyanide ion ($:\overset{-}{C}\equiv N:$) The negatively charged carbon atom of cyanide ion is usually the site of its nucleophilic character. Use of cyanide ion as a nucleophile permits the extension of a carbon chain by carbon–carbon bond formation. The product is an *alkyl cyanide, or nitrile*.

$$:N\equiv \overset{-}{C}: \quad + \quad R-\ddot{X}: \longrightarrow RC\equiv N: \quad + \quad :\ddot{X}:^-$$

Cyanide ion   Alkyl halide   Alkyl cyanide   Halide ion

$$NaCN + \langle\text{cyclopentyl}\rangle{-}Cl \xrightarrow{\text{DMSO}} \langle\text{cyclopentyl}\rangle{-}CN + NaCl$$

Sodium cyanide   Cyclopentyl chloride   Cyclopentyl cyanide (70%)   Sodium chloride

Azide ion ($:\overset{-}{N}=\overset{+}{N}=\overset{-}{N}:$) Sodium azide is a reagent used for carbon–nitrogen bond formation. The product is an *alkyl azide*.

$$:\overset{-}{N}=\overset{+}{N}=\overset{-}{N}: + R-\ddot{X}: \longrightarrow R\overset{-}{N}=\overset{+}{N}=\overset{-}{N}: + :\ddot{X}:^-$$

Azide ion   Alkyl halide   Alkyl azide   Halide ion

$$NaN_3 + CH_3(CH_2)_4I \xrightarrow{\text{1-propanol– water}} CH_3(CH_2)_4N_3 + NaI$$

Sodium azide   Pentyl iodide   Pentyl azide (52%)   Sodium iodide

(Continued)

TABLE 8.1	Representative Functional Group Transformations by Nucleophilic Substitution Reactions of Alkyl Halides *(Continued)*

Nucleophile and comments	General equation and specific example
Iodide ion (:Ï:⁻) Alkyl chlorides and bromides are converted to *alkyl iodides* by treatment with sodium iodide in acetone. NaI is soluble in acetone, but NaCl and NaBr are insoluble and crystallize from the reaction mixture, making the reaction irreversible.	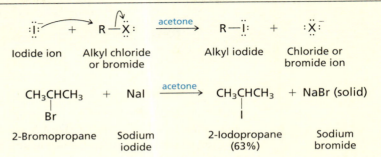

halides are not. Mixed solvents such as ethanol–water mixtures that can dissolve enough of both the substrate and the nucleophile to give fairly concentrated solutions are frequently used. Many salts, as well as most alkyl halides, possess significant solubility in dimethyl sulfoxide (DMSO), which makes this a good medium for carrying out nucleophilic substitution reactions.

The use of DMSO as a solvent in *elimination* reactions was mentioned earlier, in Section 5.14.

PROBLEM 8.1 Write a structural formula for the principal organic product formed in the reaction of methyl bromide with each of the following compounds:

(a) NaOH (sodium hydroxide)

(b) KOCH₂CH₃ (potassium ethoxide)

(c) NaOC(=O)—C₆H₅ (sodium benzoate)

(d) LiN₃ (lithium azide)

(e) KCN (potassium cyanide)

(f) NaSH (sodium hydrogen sulfide)

(g) NaI (sodium iodide)

SAMPLE SOLUTION (a) The nucleophile in sodium hydroxide is the negatively charged hydroxide ion. The reaction that occurs is nucleophilic substitution of bromide by hydroxide. The product is methyl alcohol.

$$HO:^- \quad + \quad CH_3-Br: \quad \longrightarrow \quad CH_3-OH \quad + \quad :Br:^-$$

Hydroxide ion (nucleophile)   Methyl bromide (substrate)   Methyl alcohol (product)   Bromide ion (leaving group)

With Table 8.1 as background, you can begin to see how useful alkyl halides are in synthetic organic chemistry. Alkyl halides may be prepared from alcohols by nucleophilic substitution, from alkanes by free-radical halogenation, and from alkenes by addition of hydrogen halides. They then become available as starting materials for the preparation of other functionally substituted organic compounds by replacement of the halide leaving group with a nucleophile. The range of compounds that can be prepared by nucleophilic substitution reactions of alkyl halides is quite large; the examples shown in Table 8.1 illustrate only a few of them. Numerous other examples will be added to the list in this and subsequent chapters.

8.2   RELATIVE REACTIVITY OF HALIDE LEAVING GROUPS

Among alkyl halides, alkyl iodides undergo nucleophilic substitution at the fastest rate, alkyl fluorides the slowest.

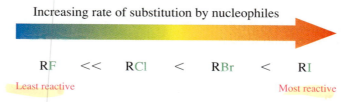

Increasing rate of substitution by nucleophiles

$$RF \quad << \quad RCl \quad < \quad RBr \quad < \quad RI$$

Least reactive   Most reactive

The order of alkyl halide reactivity in nucleophilic substitutions is the same as their order in eliminations. Iodine has the weakest bond to carbon, and iodide is the best leaving group. Alkyl iodides are several times more reactive than alkyl bromides and from 50 to 100 times more reactive than alkyl chlorides. Fluorine has the strongest bond to carbon, and fluoride is the poorest leaving group. Alkyl fluorides are rarely used as substrates in nucleophilic substitution because they are several thousand times less reactive than alkyl chlorides.

> **PROBLEM 8.2** A single organic product was obtained when 1-bromo-3-chloropropane was allowed to react with one molar equivalent of sodium cyanide in aqueous ethanol. What was this product?

The relationship between leaving group ability and basicity is explored in more detail in Section 8.14.

Leaving-group ability is also related to basicity. A strongly basic anion is usually a poorer leaving group than a weakly basic one. Fluoride is the most basic and the poorest leaving group among the halide anions, iodide the least basic and the best leaving group.

8.3   THE S$_N$2 MECHANISM OF NUCLEOPHILIC SUBSTITUTION

The mechanisms by which nucleophilic substitution takes place have been the subject of much study. Extensive research by Sir Christopher Ingold and Edward D. Hughes and their associates at University College, London, during the 1930s emphasized kinetic and stereochemical measurements to probe the mechanisms of these reactions.

Recall that the term *kinetics* refers to how the rate of a reaction varies with changes in concentration. Consider the nucleophilic substitution in which sodium hydroxide reacts with methyl bromide to form methyl alcohol and sodium bromide:

$$CH_3Br \quad + \quad HO^- \quad \longrightarrow \quad CH_3OH \quad + \quad Br^-$$

Methyl bromide   Hydroxide ion   Methyl alcohol   Bromide ion

The rate of this reaction is observed to be directly proportional to the concentration of both methyl bromide and sodium hydroxide. It is first-order in each reactant, or *second-order* overall.

$$\text{Rate} = k[CH_3Br][HO^-]$$

The S$_N$2 mechanism was introduced earlier in Section 4.12.

Hughes and Ingold interpreted second-order kinetic behavior to mean that the rate-determining step is *bimolecular*, that is, that both hydroxide ion and methyl bromide are involved at the transition state. The symbol given to the detailed description of the mechanism that they developed is S$_N$2, standing for **substitution nucleophilic bimolecular.**

The Hughes and Ingold S$_N$2 mechanism is a single-step process in which both the alkyl halide and the nucleophile are involved at the transition state. Cleavage of the bond between carbon and the leaving group is assisted by formation of a bond between carbon and the nucleophile. In effect, the nucleophile "pushes off" the leaving group from its point of attachment to carbon. The S$_N$2 mechanism for the hydrolysis of methyl bromide may be represented by a single elementary step:

$$HO{:}^- \ + \ CH_3\ddot{B}r{:} \ \longrightarrow \ \overset{\delta-}{HO}{-}{-}{-}CH_3{-}{-}{-}\overset{\delta-}{Br}{:} \ \longrightarrow \ HOCH_3 \ + \ {:}\ddot{B}r{:}^-$$

| Hydroxide ion | Methyl bromide | Transition state | Methyl alcohol | Bromide ion |

Carbon is partially bonded to both the incoming nucleophile and the departing halide at the transition state. Progress is made toward the transition state as the nucleophile begins to share a pair of its electrons with carbon and the halide ion leaves, taking with it the pair of electrons in its bond to carbon.

PROBLEM 8.3 Is the two-step sequence depicted in the following equations consistent with the second-order kinetic behavior observed for the hydrolysis of methyl bromide?

$$CH_3Br \xrightarrow{\text{slow}} CH_3^+ + Br^-$$

$$CH_3^+ + HO^- \xrightarrow{\text{fast}} CH_3OH$$

The S$_N$2 mechanism is believed to describe most substitutions in which simple primary and secondary alkyl halides react with anionic nucleophiles. All the examples cited in Table 8.1 proceed by the S$_N$2 mechanism (or a mechanism very much like S$_N$2—remember, mechanisms can never be established with certainty but represent only our best present explanations of experimental observations). We'll examine the S$_N$2 mechanism, particularly the structure of the transition state, in more detail in Section 8.5 after first looking at some stereochemical studies carried out by Hughes and Ingold.

8.4 STEREOCHEMISTRY OF S$_N$2 REACTIONS

What is the structure of the transition state in an S$_N$2 reaction? In particular, what is the spatial arrangement of the nucleophile in relation to the leaving group as reactants pass through the transition state on their way to products?

Two stereochemical possibilities present themselves. In the pathway shown in Figure 8.1a, the nucleophile simply assumes the position occupied by the leaving group. It attacks the substrate at the same face from which the leaving group departs. This is called *front-side displacement,* or substitution with **retention of configuration.**

In a second possibility, illustrated in Figure 8.1b, the nucleophile attacks the substrate from the side opposite the bond to the leaving group. This is called *back-side displacement,* or substitution with **inversion of configuration.**

Which of these two opposite stereochemical possibilities operates was determined in experiments with optically active alkyl halides. In one such experiment, Hughes and Ingold determined that the reaction of 2-bromooctane with hydroxide ion gave 2-octanol having a configuration opposite that of the starting alkyl halide.

FIGURE 8.1 Two contrasting stereochemical pathways for substitution of a leaving group (red) by a nucleophile (blue). In (*a*) the nucleophile attacks carbon at the same side from which the leaving group departs. In (*b*) nucleophilic attack occurs at the side opposite the bond to the leaving group.

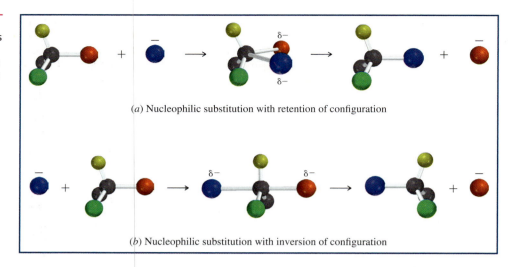

(*a*) Nucleophilic substitution with retention of configuration

(*b*) Nucleophilic substitution with inversion of configuration

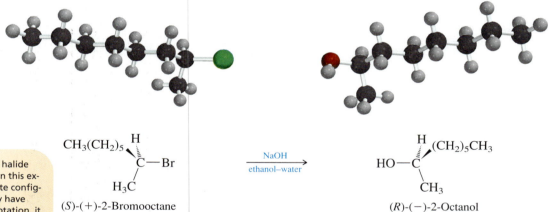

$$CH_3(CH_2)_5 \overset{H}{\underset{H_3C}{\underset{|}{\overset{|}{C}}}}{-}Br$$

(*S*)-(+)-2-Bromooctane

$$\xrightarrow[\text{ethanol–water}]{\text{NaOH}}$$

$$HO{-}\overset{H}{\underset{CH_3}{\underset{|}{\overset{|}{C}}}}(CH_2)_5CH_3$$

(*R*)-(−)-2-Octanol

Although the alkyl halide and alcohol given in this example have opposite configurations when they have opposite signs of rotation, it cannot be assumed that this will be true for all alkyl halide/alcohol pairs.

Nucleophilic substitution had occurred with inversion of configuration, consistent with the following transition state:

$$\underset{CH_3}{\overset{CH_3(CH_2)_5}{\underset{|}{\overset{H}{\underset{\delta^-}{HO{-}{-}{-}}}}}}\overset{H}{\underset{|}{C}}\overset{\delta^-}{{-}{-}{-}Br}$$

For a change of pace, try doing Problem 8.4 with molecular models instead of making structural drawings.

PROBLEM 8.4 The Fischer projection formula for (+)-2-bromooctane is shown. Write the Fischer projection of the (−)-2-octanol formed from it by nucleophilic substitution with inversion of configuration.

$$\begin{array}{c} CH_3 \\ H{-}\!\!\!\!-\!\!\!\!-\!\!\!\!-\!\!\!\!\!|\!\!\!\!\!-\!\!\!\!-\!\!\!\!-\!\!\!\!-Br \\ CH_2(CH_2)_4CH_3 \end{array}$$

PROBLEM 8.5 Would you expect the 2-octanol formed by S$_N$2 hydrolysis of (−)-2-bromooctane to be optically active? If so, what will be its absolute configuration and sign of rotation? What about the 2-octanol formed by hydrolysis of racemic 2-bromooctane?

Numerous similar experiments have demonstrated the generality of this observation. Substitution by the S$_N$2 mechanism is stereospecific and proceeds with inversion of configuration at the carbon that bears the leaving group. *There is a stereoelectronic requirement for the nucleophile to approach carbon from the side opposite the bond to the leaving group.* Organic chemists often speak of this as a **Walden inversion,** after the Latvian chemist Paul Walden, who described the earliest experiments in this area in the 1890s.

The first example of a stereoelectronic effect in this text concerned anti elimination in E2 reactions of alkyl halides (Section 5.16).

8.5 HOW S$_N$2 REACTIONS OCCUR

When we consider the overall reaction stereochemistry along with the kinetic data, a fairly complete picture of the bonding changes that take place during S$_N$2 reactions emerges. The potential energy diagram of Figure 8.2 for the hydrolysis of (S)-(+)-2-bromooctane is one that is consistent with the experimental observations.

Hydroxide ion acts as a nucleophile, using an unshared electron pair to attack carbon from the side opposite the bond to the leaving group. The hybridization of the carbon at which substitution occurs changes from sp^3 in the alkyl halide to sp^2 in the transition state. Both the nucleophile (hydroxide) and the leaving group (bromide) are partially bonded to this carbon in the transition state. Carbon is fully bonded to three substituents and partially bonded to both the leaving group and the incoming nucleophile. The bonds to the nucleophile and the leaving group are relatively long and weak at the transition state.

Once past the transition state, the leaving group is expelled and carbon becomes tetracoordinate, its hybridization returning to sp^3.

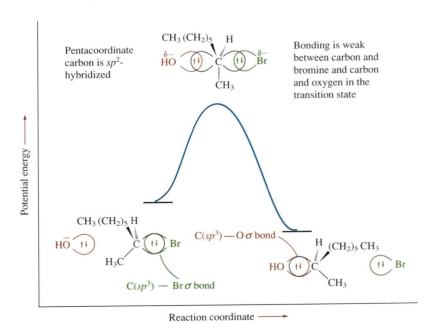

FIGURE 8.2 Hybrid orbital description of the bonding changes that take place at carbon during nucleophilic substitution by the S$_N$2 mechanism.

During the passage of starting materials to products, three structural changes take place:

1. Stretching, then breaking, of the bond to the leaving group
2. Formation of a bond to the nucleophile from the opposite side of the bond that is broken
3. Stereochemical inversion of the tetrahedral arrangement of bonds to the carbon at which substitution occurs

Although this mechanistic picture developed from experiments involving optically active alkyl halides, chemists speak even of methyl halides as undergoing nucleophilic substitution with *inversion*. By this they mean that tetrahedral inversion of the bonds to carbon occurs as the reactant proceeds to the product.

> The graphic that opened this chapter is an electrostatic potential map of the S_N2 transition state for the reaction of hydroxide ion with methyl chloride.

| Hydroxide ion | Methyl chloride | Transition state | Methyl alcohol | Chloride ion |

We saw in Section 8.2 that the rate of nucleophilic substitution depends strongly on the leaving group—alkyl iodides are the most reactive, alkyl fluorides the least. In the next section, we'll see that the structure of the alkyl group can have an even greater effect.

8.6   STERIC EFFECTS IN S_N2 REACTIONS

There are very large differences in the rates at which the various kinds of alkyl halides—methyl, primary, secondary, or tertiary—undergo nucleophilic substitution. As Table 8.2 shows for the reaction of a series of alkyl bromides:

$$RBr \quad + \quad LiI \quad \xrightarrow{\text{acetone}} \quad RI \quad + \quad LiBr$$

| Alkyl bromide | Lithium iodide | | Alkyl iodide | Lithium bromide |

the rates of nucleophilic substitution of a series of alkyl bromides differ by a factor of over 10^6 when comparing the most reactive member of the group (methyl bromide) and the least reactive member (*tert*-butyl bromide).

TABLE 8.2   Reactivity of Some Alkyl Bromides Toward Substitution by the S_N2 Mechanism*

Alkyl bromide	Structure	Class	Relative rate[†]
Methyl bromide	CH_3Br	Unsubstituted	221,000
Ethyl bromide	CH_3CH_2Br	Primary	1,350
Isopropyl bromide	$(CH_3)_2CHBr$	Secondary	1
tert-Butyl bromide	$(CH_3)_3CBr$	Tertiary	Too small to measure

*Substitution of bromide by lithium iodide in acetone.
[†]Ratio of second-order rate constant k for indicated alkyl bromide to k for isopropyl bromide at 25°C.

Least crowded–
most reactive

Most crowded–
least reactive

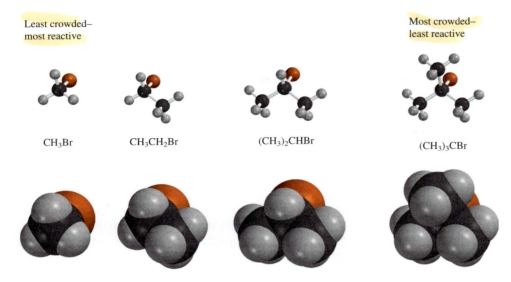

CH$_3$Br   CH$_3$CH$_2$Br   (CH$_3$)$_2$CHBr   (CH$_3$)$_3$CBr

FIGURE 8.3 Ball-and-spoke (top) and space-filling (bottom) models of alkyl bromides, showing how substituents shield the carbon atom that bears the leaving group from attack by a nucleophile. The nucleophile must attack from the side opposite the bond to the leaving group.

The large rate difference between methyl, ethyl, isopropyl, and *tert*-butyl bromides reflects the **steric hindrance** each offers to nucleophilic attack. The nucleophile must approach the alkyl halide from the side opposite the bond to the leaving group, and, as illustrated in Figure 8.3, this approach is hindered by alkyl substituents on the carbon that is being attacked. The three hydrogens of methyl bromide offer little resistance to approach of the nucleophile, and a rapid reaction occurs. Replacing one of the hydrogens by a methyl group somewhat shields the carbon from attack by the nucleophile and causes ethyl bromide to be less reactive than methyl bromide. Replacing all three hydrogens by methyl groups almost completely blocks back-side approach to the tertiary carbon of (CH$_3$)$_3$CBr and shuts down bimolecular nucleophilic substitution.

In general, S$_N$2 reactions of alkyl halides show the following dependence of rate on structure:

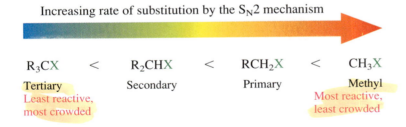

Increasing rate of substitution by the S$_N$2 mechanism

R$_3$CX   <   R$_2$CHX   <   RCH$_2$X   <   CH$_3$X

Tertiary   Secondary   Primary   Methyl
Least reactive,   Most reactive,
most crowded   least crowded

PROBLEM 8.6   Identify the compound in each of the following pairs that reacts with sodium iodide in acetone at the faster rate:

(a) 1-Chlorohexane or cyclohexyl chloride
(b) 1-Bromopentane or 3-bromopentane
(c) 2-Chloropentane or 2-fluoropentane
(d) 2-Bromo-2-methylhexane or 2-bromo-5-methylhexane
(e) 2-Bromopropane or 1-bromodecane

SAMPLE SOLUTION   (a) Compare the structures of the two chlorides. 1-Chlorohexane is a primary alkyl chloride; cyclohexyl chloride is secondary. Primary alkyl halides are less crowded at the site of substitution than secondary ones and react faster in substitution by the S_N2 mechanism. 1-Chlorohexane is more reactive.

$$CH_3CH_2CH_2CH_2CH_2CH_2Cl$$

1-Chlorohexane
(primary, more reactive)

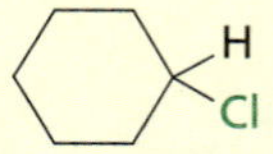

Cyclohexyl chloride
(secondary, less reactive)

Alkyl groups at the carbon atom *adjacent* to the point of nucleophilic attack also decrease the rate of the S_N2 reaction. Compare the rates of nucleophilic substitution in the series of primary alkyl bromides shown in Table 8.3. Taking ethyl bromide as the standard and successively replacing its C-2 hydrogens by methyl groups, we see that each additional methyl group decreases the rate of displacement of bromide by iodide. The effect is slightly smaller than for alkyl groups that are attached directly to the carbon that bears the leaving group, but it is still substantial. When C-2 is completely substituted by methyl groups, as it is in neopentyl bromide [(CH_3)_3CCH_2Br], we see the unusual case of a primary alkyl halide that is practically inert to substitution by the S_N2 mechanism because of steric hindrance.

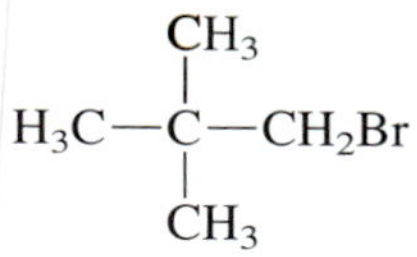

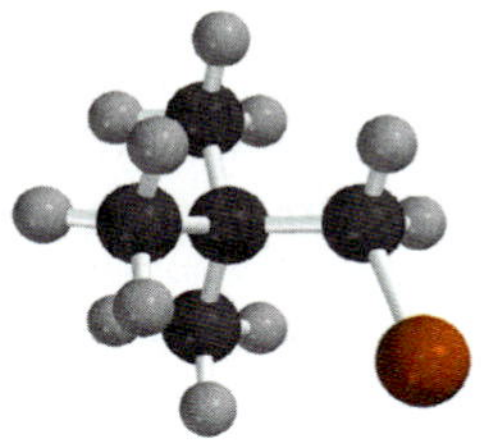

$$H_3C-\overset{\overset{\displaystyle CH_3}{|}}{\underset{\underset{\displaystyle CH_3}{|}}{C}}-CH_2Br$$

Neopentyl bromide
(1-Bromo-2,2-dimethylpropane)

TABLE 8.3	Effect of Chain Branching on Reactivity of Primary Alkyl Bromides Toward Substitution Under S_N2 Conditions*	
Alkyl bromide	**Structure**	**Relative rate[†]**
Ethyl bromide	CH_3CH_2Br	1.0
Propyl bromide	$CH_3CH_2CH_2Br$	0.8
Isobutyl bromide	$(CH_3)_2CHCH_2Br$	0.036
Neopentyl bromide	$(CH_3)_3CCH_2Br$	0.00002

*Substitution of bromide by lithium iodide in acetone.
[†]Ratio of second-order rate constant k for indicated alkyl bromide to k for ethyl bromide at 25°C.

8.7   NUCLEOPHILES AND NUCLEOPHILICITY

The Lewis base that acts as the nucleophile often is, but need not always be, an anion. Neutral Lewis bases can also serve as nucleophiles. Common examples of substitutions involving neutral nucleophiles include solvolysis reactions. **Solvolysis** reactions are substitutions in which the nucleophile is the solvent in which the reaction is carried out. Solvolysis in water (*hydrolysis*) converts an alkyl halide to an *alcohol*.

$$RX + 2H_2O \longrightarrow ROH + H_3O^+ + X^-$$

| Alkyl halide | Water | | Alcohol | Hydronium ion | Halide ion |

The reaction occurs in two stages. Only the first stage involves nucleophilic substitution. It is the rate-determining step.

$$\underset{\substack{\text{Water}}}{\overset{\substack{H\\ \diagdown \\ :O: \\ \diagup \\ H}}{}} + \underset{\substack{\text{Alkyl}\\ \text{halide}}}{R-\overset{..}{\underset{..}{X}}:} \xrightarrow{\text{slow}} \underset{\substack{\text{Alkyloxonium}\\ \text{ion}}}{\overset{\substack{H\\ \diagdown + \\ :O-R \\ \diagup \\ H}}{}} + \underset{\substack{\text{Halide}\\ \text{ion}}}{:\overset{..}{\underset{..}{X}}:^-}$$

The second stage is a Brønsted acid–base reaction and is fast.

$$\underset{\substack{\text{Water}}}{\overset{\substack{H\\ \diagdown \\ :O: \\ \diagup \\ H}}{}} + \underset{\substack{\text{Alkyloxonium}\\ \text{ion}}}{\overset{\substack{H\\ \diagdown + \\ :O-R \\ \diagup \\ H}}{}} \xrightleftharpoons{\text{fast}} \underset{\substack{\text{Hydronium}\\ \text{ion}}}{\overset{\substack{H\\ \diagdown + \\ :O-H \\ \diagup \\ H}}{}} + \underset{\substack{\text{Alcohol}}}{\overset{\substack{H\\ \diagdown \\ :\overset{..}{O}-R \\ \diagup}}{}}$$

Analogous reactions take place in other solvents which, like water, contain an —OH group. Solvolysis in methanol (*methanolysis*) gives a methyl ether.

$$RX + 2CH_3OH \longrightarrow ROCH_3 + CH_3\overset{+}{O}H_2 + X^-$$

| Alkyl halide | Methanol | | Alkyl methyl ether | Methyloxonium ion | Halide ion |

> **PROBLEM 8.7** Adapt the preceding mechanism for the hydrolysis of RX so that it describes the methanolysis of ethyl bromide.

As we have seen, the nucleophile attacks the substrate in the rate-determining step of the S_N2 mechanism, it therefore follows that the rate of substitution may vary from nucleophile to nucleophile. Just as some alkyl halides are more reactive than others, some nucleophiles are more reactive than others. Nucleophilic strength, or **nucleophilicity,** is a measure of how fast a Lewis base displaces a leaving group from a suitable substrate. By measuring the rate at which various Lewis bases react with methyl iodide in methanol, a list of their nucleophilicities relative to methanol as the standard nucleophile has been compiled. It is presented in Table 8.4.

Neutral Lewis bases such as water, alcohols, and carboxylic acids are much weaker nucleophiles than their conjugate bases. When comparing species that have the same nucleophilic atom, a negatively charged nucleophile is more reactive than a neutral one.

$$R-\overset{..}{\underset{..}{O}}:^- \quad \text{is more nucleophilic than} \quad R-\overset{..}{\underset{..}{O}}-H$$

| Alkoxide ion | | Alcohol |

$$\underset{\substack{\text{Carboxylate ion}}}{R\overset{\overset{\displaystyle :O:}{\|}}{C}-\overset{..}{\underset{..}{O}}:^-} \quad \text{is more nucleophilic than} \quad \underset{\substack{\text{Carboxylic acid}}}{R-\overset{\overset{\displaystyle :O:}{\|}}{C}-\overset{..}{\underset{..}{O}}-H}$$

TABLE 8.4	Nucleophilicity of Some Common Nucleophiles	
Reactivity class	**Nucleophile**	**Relative reactivity***
Very good nucleophiles	I^-, HS^-, RS^-	$>10^5$
Good nucleophiles	Br^-, HO^-, RO^-, CN^-, N_3^-	10^4
Fair nucleophiles	NH_3, Cl^-, F^-, RCO_2^-	10^3
Weak nucleophiles	H_2O, ROH	1
Very weak nucleophiles	RCO_2H	10^{-2}

*Relative reactivity is k(nucleophile)/k(methanol) for typical S_N2 reactions and is approximate. Data pertain to methanol as the solvent.

As long as the nucleophilic atom is the same, the more basic the nucleophile, the more reactive it is. An alkoxide ion (RO^-) is more basic and more nucleophilic than a carboxylate ion (RCO_2^-).

$$R-\overset{..}{\underset{..}{O}}{:}^-$$

Stronger base
Conjugate acid is ROH:
$pK_a = 16$

is more nucleophilic than

$$\overset{\displaystyle :O:}{\overset{\|}{RC}}-\overset{..}{\underset{..}{O}}{:}^-$$

Weaker base
Conjugate acid is RCO_2H:
$pK_a = 5$

The connection between basicity and nucleophilicity holds when comparing atoms in the *same row* of the periodic table. Thus, HO^- is more basic and more nucleophilic than F^-, and H_3N is more basic and more nucleophilic than H_2O. *It does not hold when proceeding down a column in the periodic table.* For example, I^- is the least basic of the halide ions but is the most nucleophilic. F^- is the most basic halide ion but the least nucleophilic.

The factor that seems most responsible for the inverse relationship between basicity and nucleophilicity among the halide ions is the degree to which they are *solvated* by hydrogen bonds of the type illustrated in Figure 8.4. Smaller anions, because of their high charge-to-size ratio, are more strongly solvated than larger ones. In order to act as a nucleophile, the halide must shed some of the solvent molecules that surround it. Among the halide anions, F^- forms the strongest hydrogen bonds to water and alcohols, and I^- the weakest. Thus, the nucleophilicity of F^- is suppressed more than that of Cl^-, Cl^- more than Br^-, and Br^- more than I^-. Similarly, HO^- is smaller, more solvated, and less nucleophilic than HS^-. The importance of solvation in reducing the nucleophilicity of small anions more than larger ones can be seen in the fact that, when measured in the gas phase where solvation forces don't exist, the order of halide nucleophilicity reverses and tracks basicity: $F^- > Cl^- > Br^- > I^-$.

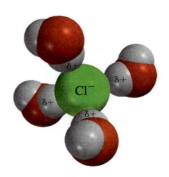

FIGURE 8.4 Solvation of a chloride ion by hydrogen bonding with water.

AN ENZYME-CATALYZED NUCLEOPHILIC SUBSTITUTION OF AN ALKYL HALIDE

Nucleophilic substitution is one of a variety of mechanisms by which living systems detoxify halogenated organic compounds introduced into the environment. Enzymes that catalyze these reactions are known as *haloalkane dehalogenases*. The hydrolysis of 1,2-dichloroethane to 2-chloroethanol, for example, is a biological nucleophilic substitution catalyzed by a dehalogenase.

ClCH$_2$CH$_2$Cl + 2H$_2$O $\xrightarrow{\text{dehalogenase enzyme}}$

1,2-Dichloroethane Water

ClCH$_2$CH$_2$OH + H$_3$O$^+$ + Cl$^-$

2-Chloroethanol Hydronium ion Chloride ion

The haloalkane dehydrogenase is believed to act by using one of its side-chain carboxylates to displace chloride by an S$_N$2 mechanism. (Recall the reaction of carboxylate ions with alkyl halides from Table 8.1.)

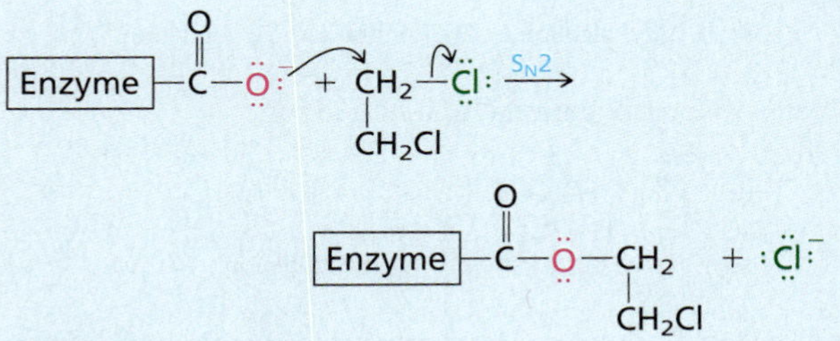

The product of this nucleophilic substitution then reacts with water, restoring the enzyme to its original state and giving the observed products of the reaction.

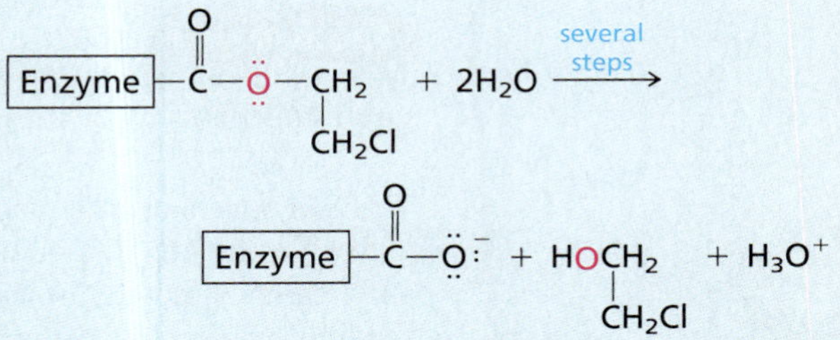

This stage of the reaction proceeds by a mechanism that will be discussed in Chapter 20. Both stages are faster than the reaction of 1,2-dichloroethane with water in the absence of the enzyme.

Some of the most common biological S$_N$2 reactions involve attack at methyl groups, especially a methyl group of *S-adenosylmethionine*. Examples of these will be given in Chapter 16.

8.8 THE S$_N$1 MECHANISM OF NUCLEOPHILIC SUBSTITUTION

Having just learned that tertiary alkyl halides are practically inert to substitution by the S$_N$2 mechanism because of steric hindrance, we might wonder whether they undergo nucleophilic substitution at all. We'll see in this section that they do, but by a mechanism different from S$_N$2.

Hughes and Ingold observed that the hydrolysis of *tert*-butyl bromide, which occurs readily, is characterized by a *first-order* rate law:

(CH$_3$)$_3$CBr + 2H$_2$O $\longrightarrow$ (CH$_3$)$_3$COH + H$_3$O$^+$ + Br$^-$

tert-Butyl bromide Water *tert*-Butyl alcohol Hydronium ion Bromide ion

$$\text{Rate} = k[(\text{CH}_3)_3\text{CBr}]$$

They found that the rate of hydrolysis depends only on the concentration of *tert*-butyl bromide. Adding the stronger nucleophile hydroxide ion, moreover, causes no change in the rate of substitution, nor does this rate depend on the concentration of hydroxide. Just as second-order kinetics was interpreted as indicating a bimolecular rate-determining step, first-order kinetics was interpreted as evidence for a *unimolecular* rate-determining step—a step that involves only the alkyl halide.

The proposed mechanism is outlined in Figure 8.5 and is called S_N1, standing for **substitution nucleophilic unimolecular.** The first step, a unimolecular dissociation of the alkyl halide to form a carbocation as the key intermediate, is rate-determining. An energy diagram for the process is shown in Figure 8.6.

The S_N1 mechanism was earlier introduced in Section 4.9.

> **PROBLEM 8.8** Suggest a structure for the product of nucleophilic substitution obtained on solvolysis of *tert*-butyl bromide in methanol, and outline a reasonable mechanism for its formation.

The S_N1 mechanism is an *ionization* mechanism. The nucleophile does not participate until after the rate-determining step has taken place. Thus, the effects of nucleophile and alkyl halide structure are expected to be different from those observed for reactions proceeding by the S_N2 pathway. How the structure of the alkyl halide affects the rate of S_N1 reactions is the topic of the next section.

The Overall Reaction:

$$(CH_3)_3CBr + 2H_2O \longrightarrow (CH_3)_3COH + H_3O^+ + Br^-$$

| *tert*-Butyl bromide | Water | *tert*-Butyl alcohol | Hydronium ion | Bromide ion |

Step 1: The alkyl halide dissociates to a carbocation and a halide ion.

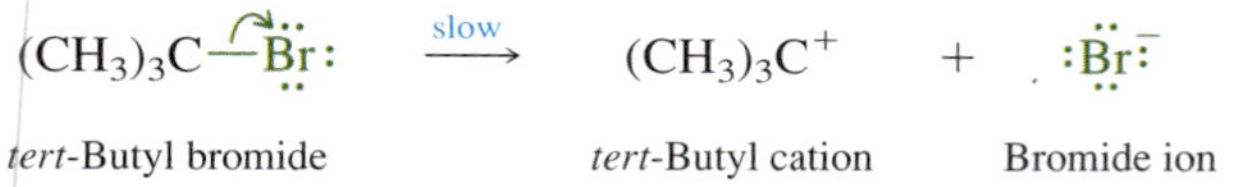

$$(CH_3)_3C\!-\!\ddot{B}r\!: \xrightarrow{\text{slow}} (CH_3)_3C^+ + :\!\ddot{B}r\!:^-$$

| *tert*-Butyl bromide | *tert*-Butyl cation | Bromide ion |

Step 2: The carbocation formed in step 1 reacts rapidly with a water molecule. Water is a nucleophile. This step completes the nucleophilic substitution stage of the mechanism and yields an alkyloxonium ion.

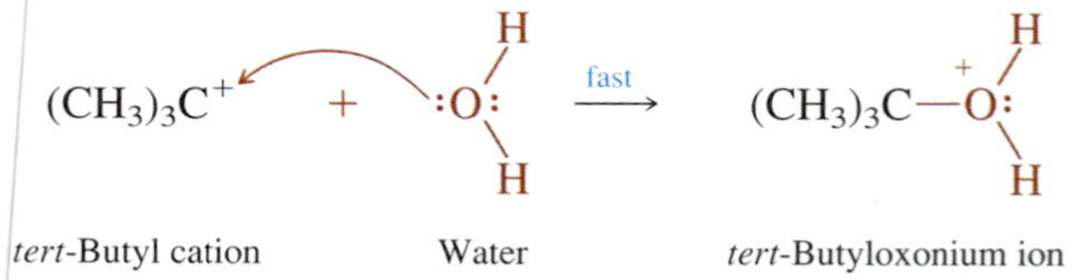

$$(CH_3)_3C^+ + :\!\overset{\displaystyle H}{\underset{\displaystyle H}{\ddot{O}}} \xrightarrow{\text{fast}} (CH_3)_3C\!-\!\overset{+}{\underset{\displaystyle H}{\overset{\displaystyle H}{O}}}$$

| *tert*-Butyl cation | Water | *tert*-Butyloxonium ion |

Step 3: This step is a fast acid-base reaction that follows the nucleophilic substitution. Water acts as a base to remove a proton from the alkyloxonium ion to give the observed product of the reaction, *tert*-butyl alcohol.

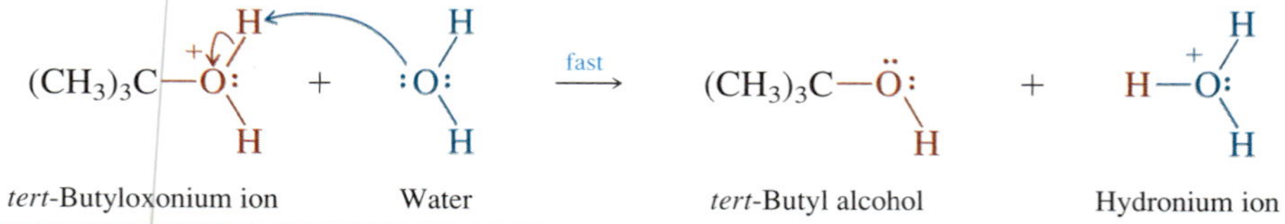

$$(CH_3)_3C\!-\!\overset{+}{\underset{\displaystyle H}{\overset{\displaystyle H}{O}}} + :\!\overset{\displaystyle H}{\underset{\displaystyle H}{\ddot{O}}} \xrightarrow{\text{fast}} (CH_3)_3C\!-\!\overset{\displaystyle H}{\underset{\displaystyle H}{\ddot{O}}} + H\!-\!\overset{+}{\underset{\displaystyle H}{\overset{\displaystyle H}{O}}}$$

| *tert*-Butyloxonium ion | Water | *tert*-Butyl alcohol | Hydronium ion |

FIGURE 8.5 The S_N1 mechanism for hydrolysis of *tert*-butyl bromide.

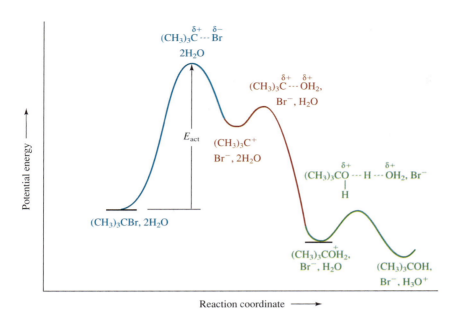

FIGURE 8.6 Energy diagram illustrating the S_N1 mechanism for hydrolysis of *tert*-butyl bromide.

8.9   CARBOCATION STABILITY AND S_N1 REACTION RATES

In order to compare S_N1 substitution rates in a range of alkyl halides, experimental conditions are chosen in which competing substitution by the S_N2 route is very slow. One such set of conditions is solvolysis in aqueous formic acid (HCO_2H):

$$RX \quad + \quad 2H_2O \xrightarrow{\text{formic acid}} ROH \quad + \quad H_3O^+ \quad + \quad X^-$$

Alkyl halide   Water   Alcohol   Hydronium ion   Halide ion

Neither formic acid nor water is very nucleophilic, and so S_N2 substitution is suppressed. The relative rates of hydrolysis of a group of alkyl bromides under these conditions are presented in Table 8.5.

The relative reactivity of alkyl halides in S_N1 reactions is exactly the opposite of S_N2:

S_N1 reactivity: methyl < primary < secondary < tertiary

S_N2 reactivity: tertiary < secondary < primary < methyl

TABLE 8.5	Reactivity of Some Alkyl Bromides Toward Substitution by the S_N1 Mechanism*		
Alkyl bromide	**Structure**	**Class**	**Relative rate[†]**
Methyl bromide	CH_3Br	Unsubstituted	1
Ethyl bromide	CH_3CH_2Br	Primary	2
Isopropyl bromide	$(CH_3)_2CHBr$	Secondary	43
tert-Butyl bromide	$(CH_3)_3CBr$	Tertiary	100,000,000

*Solvolysis in aqueous formic acid.
[†]Ratio of rate constant k for indicated alkyl bromide to k for methyl bromide at 25°C.

Clearly, the steric crowding that influences reaction rates in S_N2 processes plays no role in S_N1 reactions. The order of alkyl halide reactivity in S_N1 reactions is the same as the order of carbocation stability: the more stable the carbocation, the more reactive the alkyl halide.

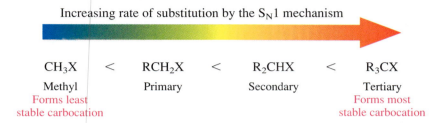

Increasing rate of substitution by the S_N1 mechanism

CH_3X	$<$	RCH_2X	$<$	R_2CHX	$<$	R_3CX
Methyl		Primary		Secondary		Tertiary
Forms least stable carbocation						Forms most stable carbocation

We have seen this situation before in the reaction of alcohols with hydrogen halides (Section 4.11), in the acid-catalyzed dehydration of alcohols (Section 5.12), and in the conversion of alkyl halides to alkenes by the E1 mechanism (Section 5.17). As in these other reactions, an electronic effect, specifically, the stabilization of the carbocation intermediate by alkyl substituents, is the decisive factor. The more stable the carbocation, the faster it is formed.

PROBLEM 8.9 Identify the compound in each of the following pairs that reacts at the faster rate in an S_N1 reaction:

(a) Isopropyl bromide or isobutyl bromide

(b) Cyclopentyl iodide or 1-methylcyclopentyl iodide

(c) Cyclopentyl bromide or 1-bromo-2,2-dimethylpropane

(d) *tert*-Butyl chloride or *tert*-butyl iodide

SAMPLE SOLUTION (a) Isopropyl bromide, $(CH_3)_2CHBr$, is a secondary alkyl halide, whereas isobutyl bromide, $(CH_3)_2CHCH_2Br$, is primary. Because the rate-determining step in an S_N1 reaction is carbocation formation and secondary carbocations are more stable than primary ones, isopropyl bromide is more reactive than isobutyl bromide in nucleophilic substitution by the S_N1 mechanism.

Primary carbocations are so high in energy that their intermediacy in nucleophilic substitution reactions is unlikely. When ethyl bromide undergoes hydrolysis in aqueous formic acid, substitution probably takes place by an S_N2-like process, in which water is the nucleophile.

8.10 STEREOCHEMISTRY OF S_N1 REACTIONS

Although S_N2 reactions are stereospecific and proceed with inversion of configuration at carbon, the situation is not as clear-cut for S_N1 reactions. When the leaving group is attached to the chirality center of an optically active halide, ionization gives a carbocation intermediate that is achiral. It is achiral because the three bonds to the positively charged carbon lie in the same plane, and this plane is a plane of symmetry for the carbocation. As shown in Figure 8.7, such a carbocation should react with a nucleophile at the same rate at either of its two faces. We expect the product of substitution by the S_N1 mechanism to be racemic and optically inactive. This outcome is rarely observed in practice, however. Normally, the product is formed with predominant, but not complete, inversion of configuration.

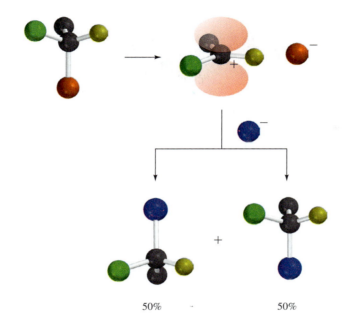

50%   50%

FIGURE 8.7 Formation of a racemic product by nucleophilic substitution via a carbocation intermediate.

For example, the hydrolysis of optically active 2-bromooctane in the absence of added base follows a first-order rate law, but the resulting 2-octanol is formed with 66% inversion of configuration.

$$
\begin{array}{c}
\underset{\substack{\text{H}_3\text{C} \quad \text{H} \\ \text{C}-\text{Br} \\ \text{CH}_3(\text{CH}_2)_5}}{} \xrightarrow[\text{ethanol}]{\text{H}_2\text{O}} \underset{\substack{\text{H} \quad \text{CH}_3 \\ \text{HO}-\text{C} \\ (\text{CH}_2)_5\text{CH}_3}}{} + \underset{\substack{\text{H}_3\text{C} \quad \text{H} \\ \text{C}-\text{OH} \\ \text{CH}_3(\text{CH}_2)_5}}{}
\end{array}
$$

(R)-(−)-2-Bromooctane   (S)-(+)-2-Octanol   (R)-(−)-2-Octanol
66% net inversion corresponds
to 83% S, 17% R

Partial but not complete loss of optical activity in S$_N$1 reactions probably results from the carbocation not being completely "free" when it is attacked by the nucleophile. Ionization of the alkyl halide gives a carbocation–halide ion pair, as depicted in Figure 8.8. The halide ion shields one side of the carbocation, and the nucleophile captures the carbocation faster from the opposite side. More product of inverted configuration is formed than product of retained configuration. In spite of the observation that the products of S$_N$1 reactions are only partially racemic, the fact that these reactions are not stereospecific is more consistent with a carbocation intermediate than a concerted bimolecular mechanism.

PROBLEM 8.10 What two stereoisomeric substitution products would you expect to isolate from the hydrolysis of *cis*-1,4-dimethylcyclohexyl bromide? From hydrolysis of *trans*-1,4-dimethylcyclohexyl bromide?

FIGURE 8.8 Inversion
of configuration predomi-
nates in S_N1 reactions
because one face of the carbocation is shielded by the
leaving group (red).

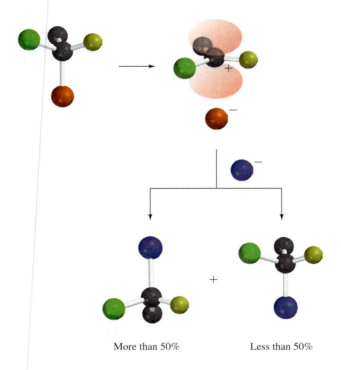

More than 50%   Less than 50%

8.11   CARBOCATION REARRANGEMENTS IN S_N1 REACTIONS

Additional evidence for carbocation intermediates in certain nucleophilic substitutions comes from observing rearrangements of the kind normally associated with such species. For example, hydrolysis of the secondary alkyl bromide 2-bromo-3-methylbutane yields the rearranged tertiary alcohol 2-methyl-2-butanol as the only substitution product.

$$\underset{\substack{| \\ Br \\ \text{2-Bromo-3-methylbutane}}}{\overset{\substack{CH_3 \\ |}}{CH_3CHCHCH_3}} \quad \xrightarrow{H_2O} \quad \underset{\substack{| \\ OH \\ \text{2-Methyl-2-butanol (93\%)}}}{\overset{\substack{CH_3 \\ |}}{CH_3CCH_2CH_3}}$$

A reasonable mechanism for this observation assumes rate-determining ionization of the substrate as the first step followed by a hydride shift that converts the secondary carbocation to a more stable tertiary one.

$$\underset{\substack{| \quad \quad | \\ H \;\; :Br: \\ \text{2-Bromo-3-methylbutane}}}{\overset{\substack{CH_3 \\ |}}{CH_3C-CHCH_3}} \xrightarrow[-:Br:^-]{slow} \underset{\substack{| \\ H \\ \text{1,2-Dimethylpropyl cation} \\ \text{(secondary carbocation)}}}{\overset{\substack{CH_3 \\ |}}{CH_3C-\overset{+}{C}HCH_3}} \xrightarrow{fast} \underset{\substack{| \\ H \\ \text{1,1-Dimethylpropyl cation} \\ \text{(tertiary carbocation)}}}{\overset{\substack{CH_3 \\ |}}{CH_3\overset{+}{C}-CHCH_3}}$$

The tertiary carbocation then reacts with water to yield the observed product.

CH₃   H   CH₃   CH₃
| | | |
CH₃C—CH₂CH₃   + :O: ──fast──▶ CH₃C—CH₂CH₃   ──fast──▶   CH₃C—CH₂CH₃
 +| | | :ÖH₂ |
                             H   +Ö   O:
                                                  H   H   H

1,1-Dimethylpropyl cation   Water   2-Methyl-2-butanol

PROBLEM 8.11 Why does the carbocation intermediate in the hydrolysis of 2-bromo-3-methylbutane rearrange by way of a hydride shift rather than a methyl shift?

Rearrangements, when they do occur, are taken as evidence for carbocation intermediates and point to the S_N1 mechanism as the reaction pathway. Rearrangements are never observed in S_N2 reactions.

8.12   EFFECT OF SOLVENT ON THE RATE OF NUCLEOPHILIC SUBSTITUTION

The major effect of the solvent is on the *rate* of nucleophilic substitution, not on what the products are. Thus we need to consider two related questions:

1. What properties of the *solvent* influence the rate most?
2. How does the rate-determining step of the *mechanism* respond to this property of the solvent?

Because the S_N1 and S_N2 mechanisms are so different from each other, let's examine each one separately.

Solvent Effects on the Rate of Substitution by the S_N1 Mechanism. Table 8.6 lists the relative rate of solvolysis of *tert*-butyl chloride in several media in order of increasing **dielectric constant** (ε). Dielectric constant is a measure of the ability of a material, in this case the solvent, to moderate the force of attraction between oppositely charged particles compared with that of a standard. The standard dielectric is a vacuum, which is assigned a value ε of exactly 1. The higher the dielectric constant ε, the better the medium is able to support separated positively and negatively charged species. Solvents

TABLE 8.6	Relative Rate of S_N1 Solvolysis of *tert*-Butyl Chloride as a Function of Solvent Polarity*	
Solvent	**Dielectric constant ε**	**Relative rate**
Acetic acid	6	1
Methanol	33	4
Formic acid	58	5,000
Water	78	150,000

*Ratio of first-order rate constant for solvolysis in indicated solvent to that for solvolysis in acetic acid at 25°C.

with high dielectric constants are classified as *polar solvents*. As Table 8.6 illustrates, the rate of solvolysis of *tert*-butyl chloride (which is equal to its rate of ionization) increases dramatically as the dielectric constant of the solvent increases.

According to the S_N1 mechanism, a molecule of an alkyl halide ionizes to a positively charged carbocation and a negatively charged halide ion in the rate-determining step. As the alkyl halide approaches the transition state for this step, a partial positive charge develops on carbon and a partial negative charge on the halogen. Figure 8.9 contrasts the behavior of a nonpolar and a polar solvent on the energy of the transition state. Polar and nonpolar solvents are similar in their interaction with the starting alkyl halide, but differ markedly in how they affect the transition state. A solvent with a low dielectric constant has little effect on the energy of the transition state, whereas one with a high dielectric constant stabilizes the charge-separated transition state, lowers the activation energy, and increases the rate of reaction.

Solvent Effects on the Rate of Substitution by the S_N2 Mechanism.   Polar solvents are required in typical bimolecular substitutions because ionic substances, such as the sodium and potassium salts cited earlier in Table 8.1, are not sufficiently soluble in nonpolar solvents to give a high enough concentration of the nucleophile to allow the reaction to occur at a rapid rate. Other than the requirement that the solvent be polar enough to dissolve ionic compounds, however, the effect of solvent polarity on the rate of S_N2 reactions is small. What is most important is whether or not the polar solvent is **protic** or **aprotic.**

Water (HOH), alcohols (ROH), and carboxylic acids (RCO_2H) are classified as *polar protic solvents;* they all have OH groups that allow them to form hydrogen bonds

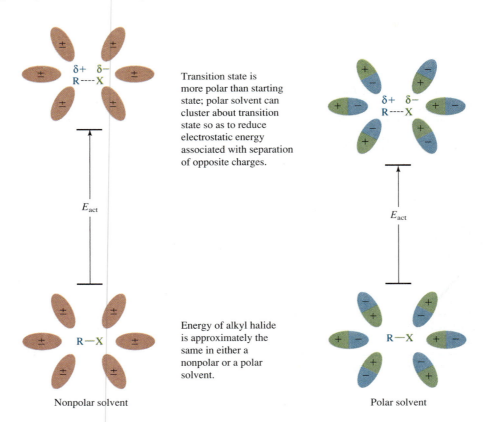

Transition state is more polar than starting state; polar solvent can cluster about transition state so as to reduce electrostatic energy associated with separation of opposite charges.

Energy of alkyl halide is approximately the same in either a nonpolar or a polar solvent.

E_{act}

E_{act}

$\delta+ \quad \delta-$
R----X

$\delta+ \quad \delta-$
R----X

R—X

R—X

Nonpolar solvent

Polar solvent

FIGURE 8.9 A polar solvent stabilizes the transition state of an S_N1 reaction and increases its rate.

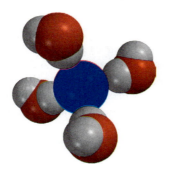

FIGURE 8.10 Hydrogen bonding of the solvent to the nucleophile stabilizes the nucleophile and makes it less reactive.

to anionic nucleophiles as shown in Figure 8.10. Solvation forces such as these stabilize the anion and suppress its nucleophilicity. *Aprotic solvents,* on the other hand, lack OH groups and do not solvate anions very strongly, leaving them much more able to express their nucleophilic character. Table 8.7 compares the second-order rate constants k for S_N2 substitution of 1-bromobutane by azide ion (a good nucleophile) in some common polar aprotic solvents with the corresponding k's for the much slower reactions observed in the polar protic solvents methanol and water.

$$CH_3CH_2CH_2CH_2Br \ + \ \ N_3^- \ \longrightarrow \ CH_3CH_2CH_2CH_2N_3 \ + \ \ Br^-$$

| 1-Bromobutane | Azide ion | 1-Azidobutane | Bromide ion |

The large rate enhancements observed for bimolecular nucleophilic substitutions in polar aprotic solvents are used to advantage in synthetic applications. An example can be seen in the preparation of alkyl cyanides (nitriles) by the reaction of sodium cyanide with alkyl halides:

$$CH_3(CH_2)_4CH_2X \ + \ \ \ NaCN \ \ \ \longrightarrow \ CH_3(CH_2)_4CH_2CN \ + \ \ \ NaX$$

| Hexyl halide | Sodium cyanide | Hexyl cyanide | Sodium halide |

When the reaction was carried out in aqueous methanol as the solvent, hexyl bromide was converted to hexyl cyanide in 71% yield by heating with sodium cyanide. Although this is a perfectly acceptable synthetic reaction, a period of over *20 hours* was required. Changing the solvent to dimethyl sulfoxide brought about an increase in the reaction rate

TABLE 8.7	Relative Rate of S_N2 Displacement of 1-Bromobutane by Azide in Various Solvents*			
Solvent	**Structural formula**	**Dielectric constant ϵ**	**Type of solvent**	**Relative rate**
Methanol	CH_3OH	32.6	Polar protic	1
Water	H_2O	78.5	Polar protic	7
Dimethyl sulfoxide	$(CH_3)_2S{=}O$	48.9	Polar aprotic	1300
N,N-Dimethylformamide	$(CH_3)_2NCH{=}O$	36.7	Polar aprotic	2800
Acetonitrile	$CH_3C{\equiv}N$	37.5	Polar aprotic	5000

*Ratio of second-order rate constant for substitution in indicated solvent to that for substitution in methanol at 25°C.

sufficient to allow the less reactive substrate hexyl chloride to be used instead, and the reaction was complete (91% yield) in only *20 minutes.*

The *rate* at which reactions occur can be important in the laboratory, and understanding how solvents affect rate is of practical value. As we proceed through the text, however, and see how nucleophilic substitution is applied to a variety of functional group transformations, be aware that it is the nature of the substrate and the nucleophile that, more than anything else, determines what *product* is formed.

8.13   SUBSTITUTION AND ELIMINATION AS COMPETING REACTIONS

We have seen that an alkyl halide and a Lewis base can react together in either a substitution or an elimination reaction.

$$-\underset{|}{\overset{H}{\underset{|}{C}}}-\underset{|}{\overset{|}{\underset{X}{C}}}- + Y^- \longrightarrow \begin{cases} \xrightarrow{\beta\text{ elimination}} \quad \underset{\diagup}{\overset{\diagdown}{C}}{=}\underset{\diagdown}{\overset{\diagup}{C}} + H{-}Y + X^- \\[3em] \xrightarrow[\substack{\text{nucleophilic}\\\text{substitution}}]{} \quad -\underset{|}{\overset{H}{\underset{|}{C}}}-\underset{|}{\overset{|}{\underset{Y}{C}}}- + X^- \end{cases}$$

Substitution can take place by the S_N1 or the S_N2 mechanism, elimination by E1 or E2.

How can we predict whether substitution or elimination will be the principal reaction observed with a particular combination of reactants? The two most important factors are the *structure of the alkyl halide* and the *basicity of the anion.* It is useful to approach the question from the premise that the characteristic reaction of alkyl halides with Lewis bases is *elimination,* and that substitution predominates only under certain special circumstances. In a typical reaction, a typical secondary alkyl halide such as isopropyl bromide reacts with a typical Lewis base such as sodium ethoxide mainly by elimination:

$$\underset{\substack{| \\ Br \\ \text{Isopropyl bromide}}}{CH_3CHCH_3} \xrightarrow[\text{CH}_3\text{CH}_2\text{OH, 55°C}]{\text{NaOCH}_2\text{CH}_3} \underset{\text{Propene (87\%)}}{CH_3CH{=}CH_2} + \underset{\substack{| \\ OCH_2CH_3 \\ \text{Ethyl isopropyl ether (13\%)}}}{CH_3CHCH_3}$$

Figure 8.11 illustrates the close relationship between the E2 and S_N2 pathways for this case, and the results cited in the preceding equation clearly show that E2 is faster than S_N2 when a secondary alkyl halide reacts with a strong base.

As crowding at the carbon that bears the leaving group decreases, the rate of nucleophilic attack by the Lewis base increases. A low level of steric hindrance to approach of the nucleophile is one of the special circumstances that permit substitution to predominate, and primary alkyl halides react with alkoxide bases by an S_N2 mechanism in preference to E2:

$$\underset{\text{Propyl bromide}}{CH_3CH_2CH_2Br} \xrightarrow[\text{CH}_3\text{CH}_2\text{OH, 55°C}]{\text{NaOCH}_2\text{CH}_3} \underset{\text{Propene (9\%)}}{CH_3CH{=}CH_2} + \underset{\text{Ethyl propyl ether (91\%)}}{CH_3CH_2CH_2OCH_2CH_3}$$

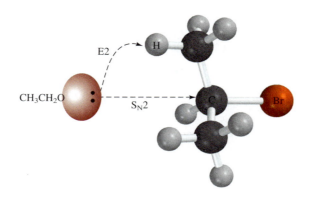

FIGURE 8.11 When a Lewis base reacts with an alkyl halide, either substitution or elimination can occur. Substitution (S_N2) occurs when the Lewis base acts as a nucleophile and attacks carbon to displace bromide. Elimination (E2) occurs when the Lewis base abstracts a proton from the β carbon. The alkyl halide shown is isopropyl bromide, and elimination (E2) predominates over substitution with alkoxide bases.

If, however, the base itself is a crowded one, such as potassium *tert*-butoxide, even primary alkyl halides undergo elimination rather than substitution:

$$CH_3(CH_2)_{15}CH_2CH_2Br \xrightarrow[\text{(CH}_3)_3\text{COH, 40°C}]{\text{KOC(CH}_3)_3} CH_3(CH_2)_{15}CH=CH_2 \ + \ CH_3(CH_2)_{15}CH_2CH_2OC(CH_3)_3$$

1-Bromooctadecane   1-Octadecene (87%)   *tert*-Butyl octadecyl ether (13%)

    A second factor that can tip the balance in favor of substitution is weak basicity of the nucleophile. Nucleophiles that are less basic than hydroxide react with both primary and secondary alkyl halides to give the product of nucleophilic substitution in high yield. To illustrate, cyanide ion is much less basic than hydroxide and reacts with 2-chlorooctane to give the corresponding alkyl cyanide as the major product.

> Cyanide is a weaker base than hydroxide because its conjugate acid HCN (pK_a 9.1) is a stronger acid than water (pK_a 15.7).

$$\underset{\underset{\text{Cl}}{|}}{CH_3CH(CH_2)_5CH_3} \xrightarrow[\text{DMSO}]{\text{KCN}} \underset{\underset{\text{CN}}{|}}{CH_3CH(CH_2)_5CH_3}$$

2-Chlorooctane   2-Cyanooctane (70%)

    Azide ion ($:\overset{-}{N}=\overset{+}{N}=\overset{-}{N}:$) is a good nucleophile and an even weaker base than cyanide. It reacts with secondary alkyl halides mainly by substitution:

> The conjugate acid of azide ion is called *hydrazoic acid* (HN$_3$). It has a pK_a of 4.6, and so is similar to acetic acid in its acidity.

Cyclohexyl iodide   Cyclohexyl azide (75%)

    Hydrogen sulfide ion HS$^-$, and anions of the type RS$^-$, are substantially less basic than hydroxide ion and react with both primary and secondary alkyl halides to give mainly substitution products.

> Hydrogen sulfide (pK_a 7.0) is a stronger acid than water (pK_a 15.7). Therefore HS$^-$ is a much weaker base than HO$^-$.

    Tertiary alkyl halides are so sterically hindered to nucleophilic attack that the presence of any anionic Lewis base favors elimination. Usually substitution predominates over elimination in tertiary alkyl halides only when anionic Lewis bases are absent. In the solvolysis of the tertiary bromide 2-bromo-2-methylbutane, for example, the ratio of substitution to elimination is 64:36 in pure ethanol but falls to 1:99 in the presence of 2 M sodium ethoxide.

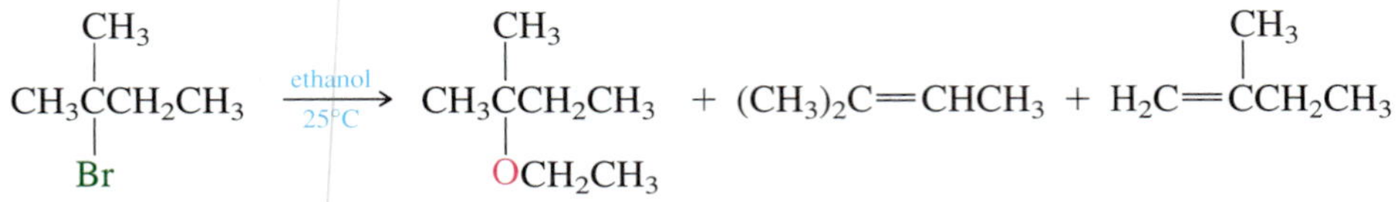

$$\underset{\substack{\text{2-Bromo-2-methyl-}\\\text{butane}}}{\overset{\displaystyle \underset{\substack{|\\ \text{Br}}}{\overset{\substack{\text{CH}_3\\|}}{\text{CH}_3\text{CCH}_2\text{CH}_3}}}{}} \xrightarrow[\substack{25^\circ\text{C}}]{\text{ethanol}} \underset{\substack{\text{2-Ethoxy-2-}\\\text{methylbutane}\\\text{(Major product in}\\\text{absence of sodium}\\\text{ethoxide)}}}{\overset{\displaystyle \underset{\substack{|\\ \text{OCH}_2\text{CH}_3}}{\overset{\substack{\text{CH}_3\\|}}{\text{CH}_3\text{CCH}_2\text{CH}_3}}}{}} + \underset{\substack{\text{2-Methyl-2-butene}}}{(\text{CH}_3)_2\text{C}=\text{CHCH}_3} + \underset{\substack{\text{2-Methyl-1-butene}}}{\overset{\displaystyle \overset{\substack{\text{CH}_3\\|}}{\text{H}_2\text{C}=\text{CCH}_2\text{CH}_3}}{}}$$

(Alkene mixture is major product
in presence of sodium ethoxide)

PROBLEM 8.12   Predict the major organic product of each of the following reactions:

(a) Cyclohexyl bromide and potassium ethoxide

(b) Ethyl bromide and potassium cyclohexanolate

(c) *sec*-Butyl bromide solvolysis in methanol

(d) *sec*-Butyl bromide solvolysis in methanol containing 2 M sodium methoxide

SAMPLE SOLUTION   (a) Cyclohexyl bromide is a secondary halide and reacts with alkoxide bases by elimination rather than substitution. The major organic products are cyclohexene and ethanol.

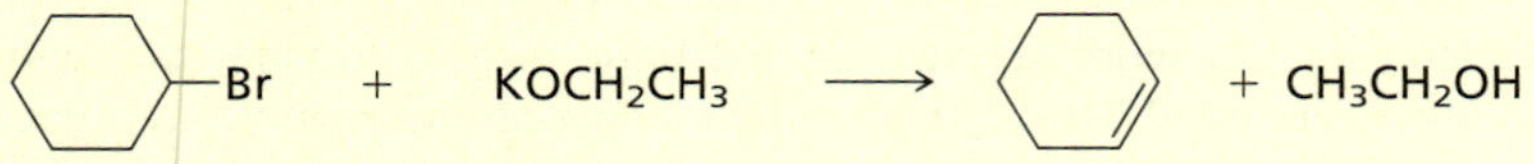

$$\text{Cyclohexyl—Br} \;+\; \text{KOCH}_2\text{CH}_3 \;\longrightarrow\; \text{Cyclohexene} \;+\; \text{CH}_3\text{CH}_2\text{OH}$$

Cyclohexyl bromide   Potassium ethoxide   Cyclohexene   Ethanol

---

Regardless of the alkyl halide, raising the temperature increases both the rate of substitution and the rate of elimination. The rate of elimination, however, usually increases faster than substitution, so that at higher temperatures the proportion of elimination products increases at the expense of substitution products.

As a practical matter, elimination can always be made to occur quantitatively. Strong bases, especially bulky ones such as *tert*-butoxide ion, react even with primary alkyl halides by an E2 process at elevated temperatures. The more difficult task is to find conditions that promote substitution. In general, the best approach is to choose conditions that favor the S_N2 mechanism—an unhindered substrate, a good nucleophile that is not strongly basic, and the lowest practical temperature consistent with reasonable reaction rates.

Functional group transformations that rely on substitution by the S_N1 mechanism are not as generally applicable as those of the S_N2 type. Hindered substrates are prone to elimination, and rearrangement is possible when carbocation intermediates are involved. Only in cases in which elimination is impossible are S_N1 reactions used for functional group transformations.

8.14   SULFONATE ESTERS AS SUBSTRATES IN NUCLEOPHILIC SUBSTITUTION

Two kinds of starting materials have been examined in nucleophilic substitution reactions to this point. In Chapter 4 we saw that alcohols can be converted to alkyl halides by reaction with hydrogen halides and pointed out that this process is a nucleophilic substitution taking place on the protonated form of the alcohol, with water serving as the leaving group. In the present chapter the substrates have been alkyl halides, and halide ions have been the leaving groups. A few other classes of organic compounds undergo nucleophilic substitution reactions analogous to those of alkyl halides, the most important of these being esters of sulfonic acids.

Sulfonic acids such as methanesulfonic acid and *p*-toluenesulfonic acid are strong acids, comparable in acidity with sulfuric acid.

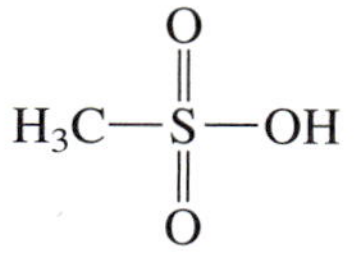

Methanesulfonic acid 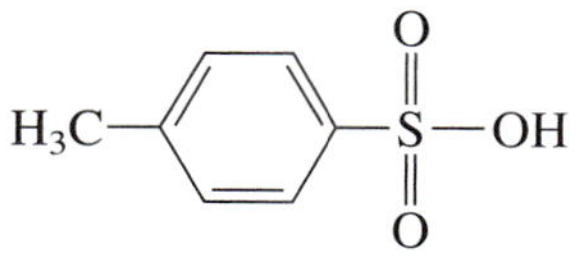   *p*-Toluenesulfonic acid

Alkyl sulfonates are derivatives of sulfonic acids in which the proton of the hydroxyl group is replaced by an alkyl group. They are prepared by treating an alcohol with the appropriate sulfonyl chloride, usually in the presence of pyridine.

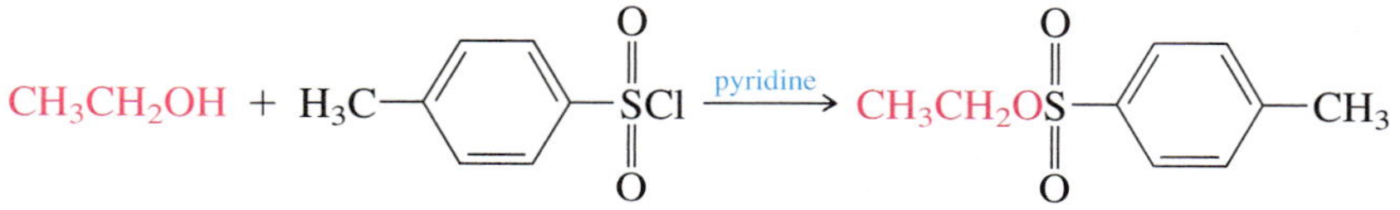

Ethanol   *p*-Toluenesulfonyl chloride   Ethyl *p*-toluenesulfonate (72%)

Alkyl sulfonate esters resemble alkyl halides in their ability to undergo elimination and nucleophilic substitution.

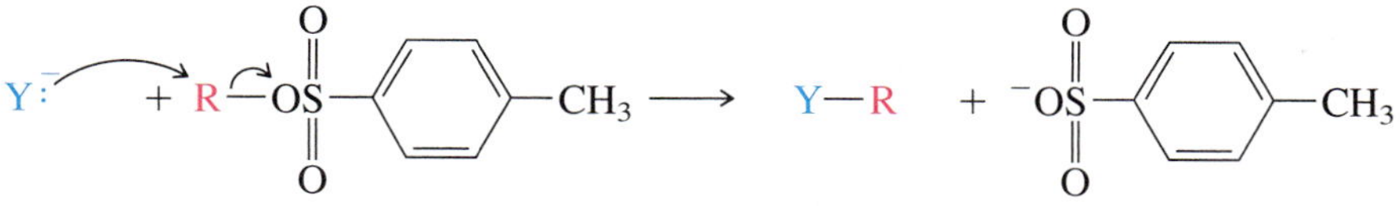

Nucleophile   *p*-Toluenesulfonate ester   Product of nucleophilic substitution   *p*-Toluenesulfonate anion

The sulfonate esters used most frequently are the *p*-toluenesulfonates. They are commonly known as *tosylates* abbreviated as ROTs.

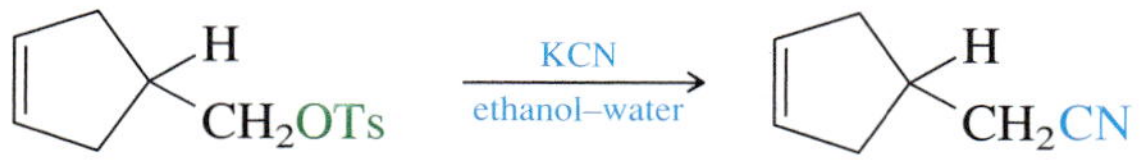

(3-Cyclopentenyl)methyl *p*-toluenesulfonate   4-(Cyanomethyl)cyclopentene (86%)

p-Toluenesulfonate (TsO⁻) is a very good leaving group. As Table 8.8 reveals, alkyl *p*-toluenesulfonates undergo nucleophilic substitution at rates that are even faster than those of alkyl iodides. A correlation of leaving-group abilities with carbon–halogen bond strengths was noted earlier, in Section 8.2. Note also the correlation with the basicity of the leaving group. Iodide is the weakest base among the halide anions and is the best leaving group, fluoride the strongest base and the poorest leaving group. A similar correlation with basicity is seen among oxygen-containing leaving groups. The weaker

TABLE 8.8	Approximate Relative Leaving-Group Abilities*		
Leaving group	Relative rate	Conjugate acid of leaving group	pK$_a$ of conjugate acid
F$^-$	10^{-5}	HF	3.5
Cl$^-$	10^0	HCl	-7
Br$^-$	10^1	HBr	-9
I$^-$	10^2	HI	-10
H$_2$O	10^1	H$_3$O$^+$	-1.7
TsO$^-$	10^5	TsOH	-2.8
CF$_3$SO$_2$O$^-$	10^8	CF$_3$SO$_2$OH	-6

*Values are approximate and vary according to substrate.

Trifluoromethanesulfonate
esters are called *triflates*.

the base, the better the leaving group. Trifluoromethanesulfonic acid (CF$_3$SO$_2$OH) is a much stronger acid than *p*-toluenesulfonic acid, and therefore trifluoromethanesulfonate is a much weaker base than *p*-toluenesulfonate and a much better leaving group.

Notice too that strongly basic leaving groups are absent from Table 8.8. In general, any species that has pK$_a$ greater than about 2 for its conjugate acid cannot be a leaving group in a nucleophilic substitution. Thus, hydroxide (HO$^-$) is far too strong a base to be displaced from an alcohol (ROH), and alcohols do not undergo nucleophilic substitution. In strongly acidic media, alcohols are protonated to give alkyloxonium ions, and these do undergo nucleophilic substitution, because the leaving group is a weakly basic water molecule.

Because halides are poorer leaving groups than *p*-toluenesulfonate, alkyl *p*-toluenesulfonates can be converted to alkyl halides by S$_N$2 reactions involving chloride, bromide, or iodide as the nucleophile.

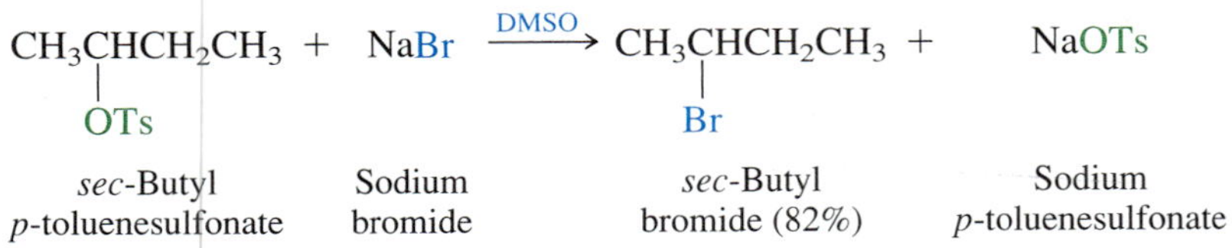

$$CH_3CHCH_2CH_3 \; + \; NaBr \; \xrightarrow{DMSO} \; CH_3CHCH_2CH_3 \; + \; NaOTs$$
$$\qquad | \qquad\qquad\qquad\qquad\qquad\qquad\qquad | $$
$$\;\;OTs \qquad\qquad\qquad\qquad\qquad\qquad\qquad Br$$

| *sec*-Butyl *p*-toluenesulfonate | Sodium bromide | *sec*-Butyl bromide (82%) | Sodium *p*-toluenesulfonate |

PROBLEM 8.13 Write a chemical equation showing the preparation of octadecyl *p*-toluenesulfonate.

PROBLEM 8.14 Write equations showing the reaction of octadecyl *p*-toluenesulfonate with each of the following reagents:

(a) Potassium acetate (KOCCH$_3$, with =O above the C)

(b) Potassium iodide (KI)

(c) Potassium cyanide (KCN)

(d) Potassium hydrogen sulfide (KSH)

(e) Sodium butanethiolate (NaSCH$_2$CH$_2$CH$_2$CH$_3$)

SAMPLE SOLUTION   All these reactions of octadecyl *p*-toluenesulfonate have been reported in the chemical literature, and all proceed in synthetically useful yield. You should begin by identifying the nucleophile in each of the parts to this problem. The nucleophile replaces the *p*-toluenesulfonate leaving group in an S_N2 reaction. In part (a) the nucleophile is acetate ion, and the product of nucleophilic substitution is octadecyl acetate.

$$\underset{\substack{\\ \text{Acetate ion}}}{CH_3\overset{\displaystyle :O:}{\overset{\|}{C}}\overset{..}{\underset{..}{O}}:^-} \; + \; \underset{\substack{| \\ (CH_2)_{16}CH_3 \\ \text{Octadecyl tosylate}}}{CH_2{-}\overset{..}{\underset{..}{O}}Ts} \longrightarrow \underset{\text{Octadecyl acetate}}{CH_3\overset{\displaystyle :O:}{\overset{\|}{C}}OCH_2(CH_2)_{16}CH_3} \; + \; \underset{\text{Tosylate ion}}{^-:\overset{..}{\underset{..}{O}}Ts}$$

Sulfonate esters are subject to the same limitations as alkyl halides. Competition from elimination needs to be considered when planning a functional group transformation that requires an anionic nucleophile, because tosylates undergo elimination reactions, just as alkyl halides do.

An advantage that sulfonate esters have over alkyl halides is that their preparation from alcohols does not involve any of the bonds to carbon. The alcohol oxygen becomes the oxygen that connects the alkyl group to the sulfonyl group. Thus, the configuration of a sulfonate ester is exactly the same as that of the alcohol from which it was prepared. If we wish to study the stereochemistry of nucleophilic substitution in an optically active substrate, for example, we know that a tosylate ester will have the same configuration and the same optical purity as the alcohol from which it was prepared.

(S)-(+)-2-Octanol
[α]²⁵_D +9.9°
(optically pure)

(S)-(+)-1-Methylheptyl p-toluenesulfonate
[α]²⁵_D +7.9°
(optically pure)

The same cannot be said about reactions with alkyl halides as substrates. The conversion of optically active 2-octanol to the corresponding halide *does* involve a bond to the chirality center, and so the optical purity and absolute configuration of the alkyl halide need to be independently established.

The mechanisms by which sulfonate esters undergo nucleophilic substitution are the same as those of alkyl halides. Inversion of configuration is observed in S_N2 reactions of alkyl sulfonates and predominant inversion accompanied by racemization in S_N1 processes.

PROBLEM 8.15   The hydrolysis of sulfonate esters of 2-octanol is stereospecific and proceeds with complete inversion of configuration. Write a structural formula that shows the stereochemistry of the 2-octanol formed by hydrolysis of an optically pure sample of (S)-(+)-1-methylheptyl *p*-toluenesulfonate, identify the product as *R* or *S*, and deduce its specific rotation.

8.15   LOOKING BACK: REACTIONS OF ALCOHOLS WITH HYDROGEN HALIDES

The principles developed in this chapter can be applied to a more detailed examination of the reaction of alcohols with hydrogen halides than was possible when this reaction was first introduced in Chapter 4.

$$\text{ROH} \; + \quad \text{HX} \quad \longrightarrow \quad \text{RX} \quad + \; \text{H}_2\text{O}$$

Alcohol   Hydrogen halide   Alkyl halide   Water

As pointed out in Chapter 4, the first step in the reaction is proton transfer to the alcohol from the hydrogen halide to yield an alkyloxonium ion. This is an acid–base reaction.

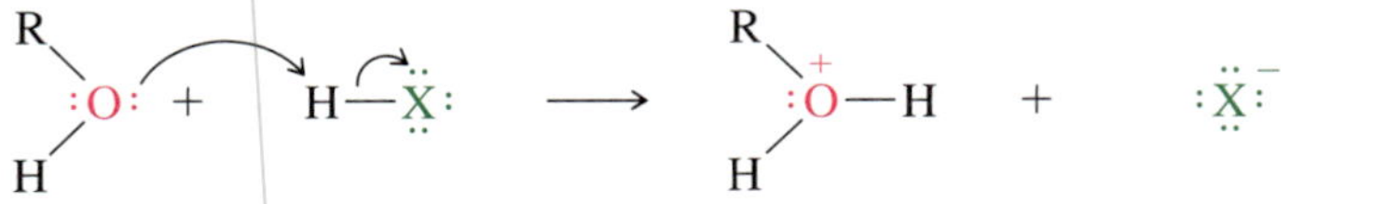

Alcohol   Hydrogen halide   Alkyloxonium ion   Halide ion
(base)   (acid)   (conjugate acid)   (conjugate base)

With primary alcohols, the next stage is an S_N2 reaction in which the halide ion, bromide, for example, displaces a molecule of water from the alkyloxonium ion.

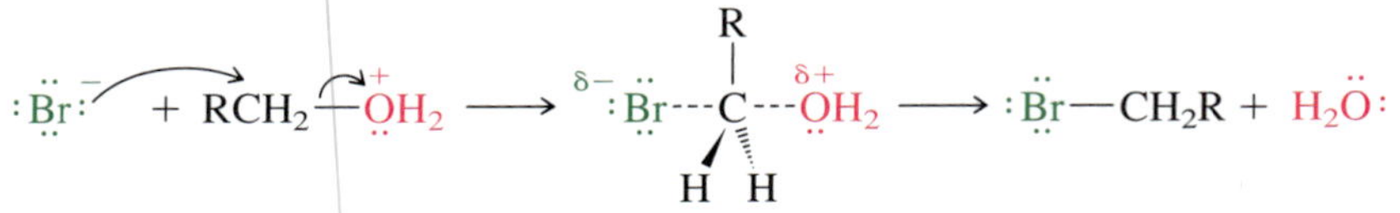

Bromide   Primary alkyl- S_N2 transition state   Primary   Water
ion   oxonium ion   alkyl bromide

With secondary and tertiary alcohols, this stage is an S_N1 reaction in which the alkyloxonium ion dissociates to a carbocation and water.

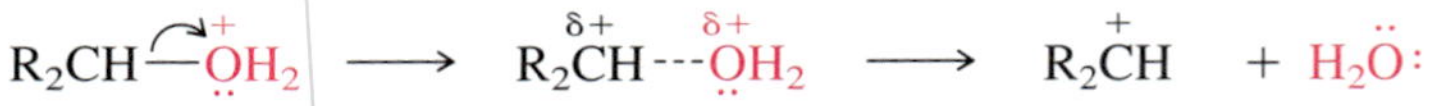

Secondary S_N1 transition state   Secondary   Water
alkyloxonium ion   carbocation

Following its formation, the carbocation is captured by halide.

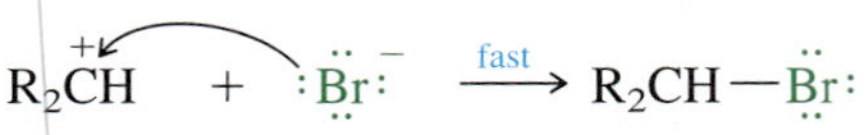

Secondary   Bromide   Secondary
carbocation   ion   alkyl bromide

With optically active secondary alcohols the reaction proceeds with predominant, but incomplete, inversion of configuration.

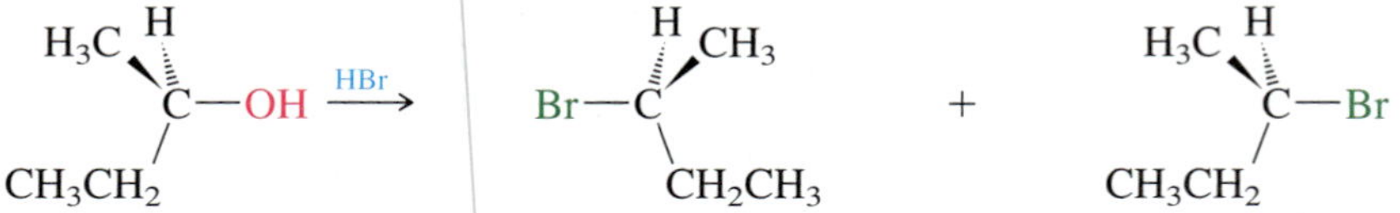

(*R*)-(−)-2-Butanol   (*S*)-(+)-2-Bromobutane (87%)   (*R*)-(−)-2-Bromobutane (13%)

The few studies that have been carried out with optically active tertiary alcohols indicate that almost complete racemization accompanies the preparation of tertiary alkyl halides by this method.

Rearrangement can occur, and the desired alkyl halide is sometimes accompanied by an isomeric halide. An example is seen in the case of the secondary alcohol 2-octanol, which yields a mixture of 2- and 3-bromooctane:

$$\underset{\underset{\text{OH}}{|}}{CH_3CHCH_2(CH_2)_4CH_3} \xrightarrow{HBr} \overset{+}{CH_3CHCH_2(CH_2)_4CH_3} \longrightarrow CH_3CH_2\overset{+}{CH}(CH_2)_4CH_3$$

2-Octanol	1-Methylheptyl cation	1-Ethylhexyl cation

$$\underset{\underset{\text{Br}}{|}}{CH_3CHCH_2(CH_2)_4CH_3} \qquad \underset{\underset{\text{Br}}{|}}{CH_3CH_2CH(CH_2)_4CH_3}$$

	2-Bromooctane (93%)	3-Bromooctane (7%)

PROBLEM 8.16   Treatment of 3-methyl-2-butanol with hydrogen chloride yielded only a trace of 2-chloro-3-methylbutane. An isomeric chloride was isolated in 97% yield. Suggest a reasonable structure for this product.

Unbranched primary alcohols and tertiary alcohols tend to react with hydrogen halides without rearrangement. The alkyloxonium ions from primary alcohols react rapidly with bromide ion, for example, in an S_N2 process. Tertiary alcohols give tertiary alkyl halides because tertiary carbocations are stable and show little tendency to rearrange.

When it is necessary to prepare secondary alkyl halides with assurance that no trace of rearrangement accompanies their formation, the corresponding alcohol is first converted to its *p*-toluenesulfonate ester and this ester is then allowed to react with sodium chloride, bromide, or iodide, as described in Section 8.14.

8.16   SUMMARY

Section 8.1   Nucleophilic substitution is an important reaction type in synthetic organic chemistry because it is one of the main methods for functional group transformations. Examples of synthetically useful nucleophilic substitutions were given in Table 8.1. It is a good idea to return to that table and review its entries now that the details of nucleophilic substitution have been covered.

Sections 8.2–8.12   These sections show how a variety of experimental observations led to the proposal of the S_N1 and the S_N2 mechanisms for nucleophilic substitution. Summary Table 8.9 integrates the material in these sections.

Section 8.13   When nucleophilic substitution is used for synthesis, the competition between substitution and elimination must be favorable. However, *the normal reaction of a secondary alkyl halide with a base as strong or stronger than hydroxide is elimination (E2)*. Substitution by the S_N2 mechanism predominates only when the base is weaker than hydroxide or the alkyl halide is primary. Elimination predominates when tertiary alkyl halides react with any anion.

TABLE 8.9	Comparison of S_N1 and S_N2 Mechanisms of Nucleophilic Substitution in Alkyl Halides	
	S_N1	**S_N2**
Characteristics of mechanism	Two elementary steps:	Single step:
	Step 1: $R—\ddot{X}: \rightleftharpoons R^+ + :\ddot{X}:^-$	$^-Nu:\frown R—\ddot{X}: \longrightarrow Nu—R + :\ddot{X}:^-$
	Step 2: $R^+ + :Nu^- \longrightarrow R—Nu$	Nucleophile displaces leaving group; bonding to the incoming nucleophile accompanies cleavage of the bond to the leaving group. (Sections 8.3 and 8.5)
	Ionization of alkyl halide (step 1) is rate-determining. (Section 8.8)	
Rate-determining transition state	$^{\delta+}R\text{---}\ddot{X}:^{\delta-}$	$^{\delta-}Nu\text{---}R\text{---}\ddot{X}:^{\delta-}$
	(Section 8.8)	(Sections 8.3 and 8.5)
Molecularity	Unimolecular (Section 8.8)	Bimolecular (Section 8.3)
Kinetics and rate law	First order: Rate = k[alkyl halide] (Section 8.8)	Second order: Rate = k[alkyl halide][nucleophile] (Section 8.3)
Relative reactivity of halide leaving groups	RI > RBr > RCl >> RF (Section 8.2)	RI > RBr > RCl >> RF (Section 8.2)
Effect of structure on rate	$R_3CX > R_2CHX > RCH_2X > CH_3X$	$CH_3X > RCH_2X > R_2CHX > R_3CX$
	Rate is governed by stability of carbocation that is formed in ionization step. Tertiary alkyl halides can react only by the S_N1 mechanism; they never react by the S_N2 mechanism. (Section 8.9)	Rate is governed by steric effects (crowding in transition state). Methyl and primary alkyl halides can react only by the S_N2 mechanism; they never react by the S_N1 mechanism. (Section 8.6)
Effect of nucleophile on rate	Rate of substitution is independent of both concentration and nature of nucleophile. Nucleophile does not participate until after rate-determining step. (Section 8.8)	Rate depends on both nature of nucleophile and its concentration. (Sections 8.3 and 8.7)
Effect of solvent on rate	Rate increases with increasing polarity of solvent as measured by its dielectric constant ϵ. (Section 8.12)	Polar aprotic solvents give fastest rates of substitution; solvation of $Nu:^-$ is minimal and nucleophilicity is greatest. (Section 8.12)
Stereochemistry	Not stereospecific: racemization accompanies inversion when leaving group is located at a chirality center. (Section 8.10)	Stereospecific: 100% inversion of configuration at reaction site. Nucleophile attacks carbon from side opposite bond to leaving group. (Section 8.4)
Potential for rearrangements	Carbocation intermediate capable of rearrangement. (Section 8.11)	No carbocation intermediate; no rearrangement.

Section 8.14 Nucleophilic substitution can occur with leaving groups other than halide. Alkyl p-toluenesulfonates (*tosylates*), which are prepared from alcohols by reaction with p-toulenesulfonyl chloride, are often used.

$$ROH + H_3C\!-\!\!\left\langle\bigcirc\right\rangle\!\!-\!SO_2Cl \xrightarrow{\text{pyridine}} ROS\!-\!\!\left\langle\bigcirc\right\rangle\!\!-\!CH_3 \text{ (ROTs)}$$

Alcohol   p-Toluenesulfonyl chloride   Alkyl p-toluenesulfonate (alkyl tosylate)

In its ability to act as a leaving group, p-toluenesulfonate is even more reactive than iodide.

$$\overline{Nu:}\;\;\;\;R\!-\!OTs \longrightarrow Nu\!-\!R + \;\;\;\;^-OTs$$

Nucleophile   Alkyl p-toluenesulfonate   Substitution product   p-Toluenesulfonate ion

Section 8.15 The reactions of alcohols with hydrogen halides to give alkyl halides (Chapter 4) are nucleophilic substitution reactions of alkyloxonium ions in which water is the leaving group. Primary alcohols react by an S_N2-like displacement of water from the alkyloxonium ion by halide. Secondary and tertiary alcohols give alkyloxonium ions which form carbocations in an S_N1-like process. Rearrangements are possible with secondary alcohols, and substitution takes place with predominant, but not complete, inversion of configuration.

PROBLEMS

8.17 Write the structure of the principal organic product to be expected from the reaction of 1-bromopropane with each of the following:

(a) Sodium iodide in acetone

(b) Sodium acetate ($CH_3\overset{\displaystyle O}{\overset{\|}{C}}ONa$) in acetic acid

(c) Sodium ethoxide in ethanol

(d) Sodium cyanide in dimethyl sulfoxide

(e) Sodium azide in aqueous ethanol

(f) Sodium hydrogen sulfide in ethanol

(g) Sodium methanethiolate ($NaSCH_3$) in ethanol

8.18 All the reactions of 1-bromopropane in the preceding problem give the product of nucleophilic substitution in high yield. High yields of substitution products are also obtained in all but one of the analogous reactions using 2-bromopropane as the substrate. In one case, however, 2-bromopropane is converted to propene, especially when the reaction is carried out at elevated temperature (about 55°C). Which reactant is most effective in converting 2-bromopropane to propene?

8.19 Each of the following nucleophilic substitution reactions has been reported in the chemical literature. Many of them involve reactants that are somewhat more complex than those we have dealt with to this point. Nevertheless, you should be able to predict the product by analogy to what you know about nucleophilic substitution in simple systems.

(a) $BrCH_2COCH_2CH_3$ $\xrightarrow[\text{acetone}]{\text{NaI}}$

(b) 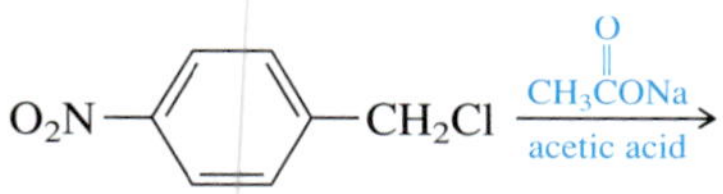 O_2N-⟨benzene ring⟩$-CH_2Cl$ $\xrightarrow[\text{acetic acid}]{CH_3CONa}$

(c) $CH_3CH_2OCH_2CH_2Br$ $\xrightarrow[\text{ethanol–water}]{\text{NaCN}}$

(d) 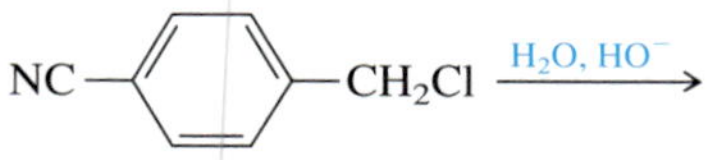 $NC-$⟨benzene ring⟩$-CH_2Cl$ $\xrightarrow{H_2O,\ HO^-}$

(e) $ClCH_2COC(CH_3)_3$ $\xrightarrow[\text{acetone–water}]{\text{NaN}_3}$

(f) 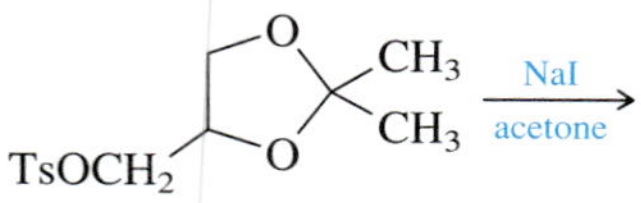 $TsOCH_2$⟨dioxolane ring with CH_3, CH_3⟩ $\xrightarrow[\text{acetone}]{\text{NaI}}$

(g) 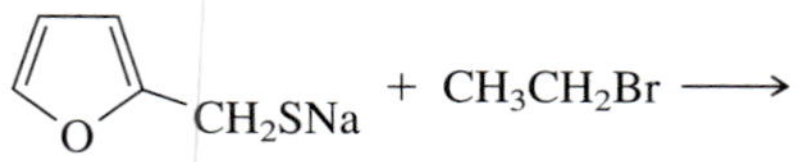 ⟨furan⟩CH_2SNa $+ CH_3CH_2Br \longrightarrow$

(h) 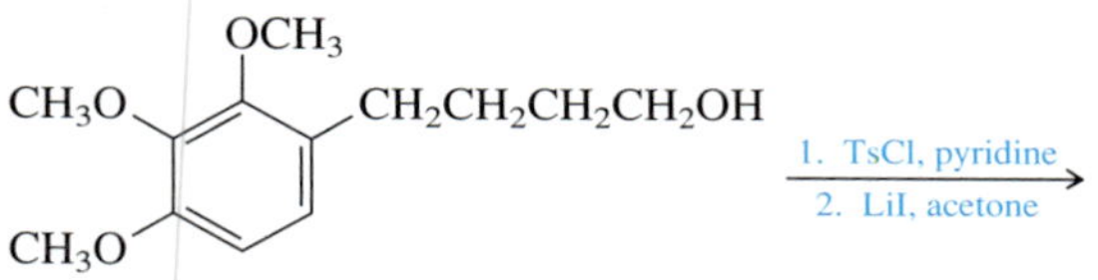 ⟨trimethoxy aromatic ring⟩$CH_2CH_2CH_2CH_2OH$ $\xrightarrow[\text{2.\ LiI, acetone}]{\text{1.\ TsCl, pyridine}}$

8.20 Each of the reactions shown involves nucleophilic substitution. The product of reaction (a) is an isomer of the product of reaction (b). What kind of isomer? By what mechanism does nucleophilic substitution occur? Write the structural formula of the product of each reaction.

(a) 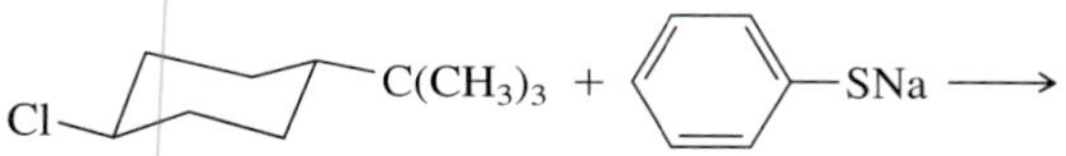 ⟨cyclohexane ring with Cl and $C(CH_3)_3$⟩ $+$ ⟨benzene ring⟩$-SNa \longrightarrow$

(b) 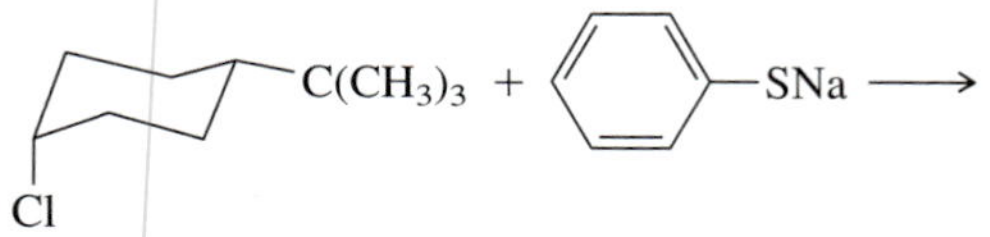 ⟨cyclohexane ring with Cl and $C(CH_3)_3$⟩ $+$ ⟨benzene ring⟩$-SNa \longrightarrow$

8.21 Arrange the isomers of molecular formula C_4H_9Cl in order of decreasing rate of reaction with sodium iodide in acetone.

8.22 There is an overall 29-fold difference in reactivity of 1-chlorohexane, 2-chlorohexane, and 3-chlorohexane toward potassium iodide in acetone.

   (a) Which one is the most reactive? Why?

   (b) Two of the isomers differ by only a factor of 2 in reactivity. Which two are these? Which one is the more reactive? Why?

8.23 In each of the following indicate which reaction will occur faster. Explain your reasoning.

   (a) $CH_3CH_2CH_2CH_2Br$ or $CH_3CH_2CH_2CH_2I$ with sodium cyanide in dimethyl sulfoxide

   (b) 1-Chloro-2-methylbutane or 1-chloropentane with sodium iodide in acetone

   (c) Hexyl chloride or cyclohexyl chloride with sodium azide in aqueous ethanol

   (d) Solvolysis of 1-bromo-2,2-dimethylpropane or tert-butyl bromide in ethanol

(e) Solvolysis of isobutyl bromide or *sec*-butyl bromide in aqueous formic acid

(f) Reaction of 1-chlorobutane with sodium acetate in acetic acid or with sodium methoxide in methanol

(g) Reaction of 1-chlorobutane with sodium azide or sodium *p*-toluenesulfonate in aqueous ethanol

8.24 Under conditions of photochemical chlorination, $(CH_3)_3CCH_2C(CH_3)_3$ gave a mixture of two monochlorides in a 4:1 ratio. The structures of these two products were assigned on the basis of their S_N1 hydrolysis rates in aqueous ethanol. The major product (compound A) underwent hydrolysis much more slowly than the minor one (compound B). Deduce the structures of compounds A and B.

8.25 The compound KSCN is a source of *thiocyanate* ion.

(a) Write the two most stable Lewis structures for thiocyanate ion and identify the atom in each that bears a formal charge of -1.

(b) Two constitutionally isomeric products of molecular formula C_5H_9NS were isolated in a combined yield of 87% in the reaction shown. (DMF stands for *N,N*-dimethylformamide, a polar aprotic solvent.) Suggest reasonable structures for these two compounds.

$$CH_3CH_2CH_2CH_2Br \xrightarrow[\text{DMF}]{\text{KSCN}}$$

8.26 Sodium nitrite ($NaNO_2$) reacted with 2-iodooctane to give a mixture of two constitutionally isomeric compounds of molecular formula $C_8H_{17}NO_2$ in a combined yield of 88%. Suggest reasonable structures for these two isomers.

8.27 Reaction of ethyl iodide with triethylamine $[(CH_3CH_2)_3N:]$ yields a crystalline compound $C_8H_{20}NI$ in high yield. This compound is soluble in polar solvents such as water but insoluble in nonpolar ones such as diethyl ether. It does not melt below about 200°C. Suggest a reasonable structure for this product.

8.28 Write an equation, clearly showing the stereochemistry of the starting material and the product, for the reaction of (*S*)-1-bromo-2-methylbutane with sodium iodide in acetone. What is the configuration (*R* or *S*) of the product?

8.29 Identify the product in each of the following reactions:

(a) $ClCH_2CH_2CHCH_2CH_3 \xrightarrow[\text{acetone}]{\text{NaI (1 mole)}} C_5H_{10}ClI$
       $\quad\quad\quad\quad\;\; |$
       $\quad\quad\quad\quad\;\; Cl$

(b) $BrCH_2CH_2Br + NaSCH_2CH_2SNa \longrightarrow C_4H_8S_2$

(c) $ClCH_2CH_2CH_2CH_2Cl + Na_2S \longrightarrow C_4H_8S$

8.30 Give the mechanistic symbols (S_N1, S_N2, E1, E2) that are most consistent with each of the following statements:

(a) Methyl halides react with sodium ethoxide in ethanol only by this mechanism.

(b) Unhindered primary halides react with sodium ethoxide in ethanol mainly by this mechanism.

(c) When cyclohexyl bromide is treated with sodium ethoxide in ethanol, the major product is formed by this mechanism.

(d) The substitution product obtained by solvolysis of *tert*-butyl bromide in ethanol arises by this mechanism.

(e) In ethanol that contains sodium ethoxide, *tert*-butyl bromide reacts mainly by this mechanism.

(f) These reaction mechanisms represent concerted processes.

(g) Reactions proceeding by these mechanisms are stereospecific.

(h) These reaction mechanisms involve carbocation intermediates.

(i) These reaction mechanisms are the ones most likely to have been involved when the products are found to have a different carbon skeleton from the substrate.

(j) Alkyl iodides react faster than alkyl bromides in reactions that proceed by these mechanisms.

8.31 Outline an efficient synthesis of each of the following compounds from the indicated starting material and any necessary organic or inorganic reagents:

(a) Cyclopentyl cyanide from cyclopentane

(b) Cyclopentyl cyanide from cyclopentene

(c) Cyclopentyl cyanide from cyclopentanol

(d) $NCCH_2CH_2CN$ from ethyl alcohol

(e) Isobutyl iodide from isobutyl chloride

(f) Isobutyl iodide from *tert*-butyl chloride

(g) Isopropyl azide from isopropyl alcohol

(h) Isopropyl azide from 1-propanol

(i) (S)-*sec*-Butyl azide from (R)-*sec*-butyl alcohol

(j) (S)-$CH_3CH_2CHCH_3$ from (R)-*sec*-butyl alcohol
      |
      SH

8.32 Select the combination of alkyl bromide and potassium alkoxide that would be the most effective in the syntheses of the following ethers:

(a) $CH_3OC(CH_3)_3$

(b) 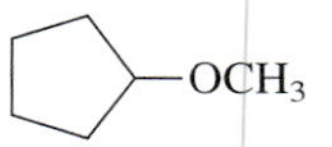 ⬠—OCH_3

(c) $(CH_3)_3CCH_2OCH_2CH_3$

8.33 (*Note to the student:* This problem previews an important aspect of Chapter 9 and is well worth attempting in order to get a head start on the material presented there.)

Alkynes of the type $RC{\equiv}CH$ may be prepared by nucleophilic substitution reactions in which one of the starting materials is sodium acetylide $(Na^+ : C{\equiv}CH)$.

(a) Devise a method for the preparation of $CH_3CH_2C{\equiv}CH$ from sodium acetylide and any necessary organic or inorganic reagents.

(b) Given the information that the pK_a for acetylene $(HC{\equiv}CH)$ is 26, comment on the scope of this preparative procedure with respect to R in $RC{\equiv}CH$. Could you prepare $(CH_3)_2CHC{\equiv}CH$ or $(CH_3)_3CC{\equiv}CH$ in good yield by this method?

8.34 Give the structures, including stereochemistry, of compounds A and B in the following sequence of reactions:

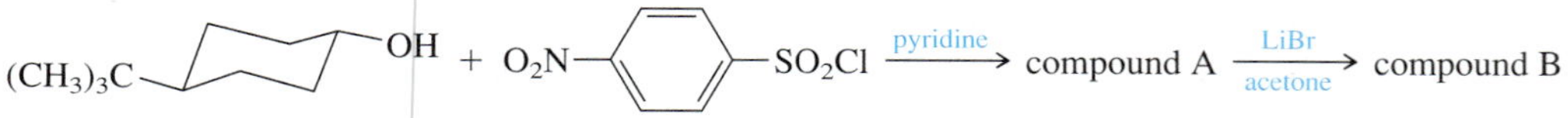

$(CH_3)_3C$—◯—OH + O_2N—◯—SO_2Cl $\xrightarrow[\text{}]{\text{pyridine}}$ compound A $\xrightarrow[\text{acetone}]{\text{LiBr}}$ compound B

8.35 (a) Suggest a reasonable series of synthetic transformations for converting *trans*-2-methyl-cyclopentanol to *cis*-2-methylcyclopentyl acetate.

cis-2-Methylcyclopentyl acetate

(b) How could you prepare *cis*-2-methylcyclopentyl acetate from 1-methylcyclopentanol?

8.36 Optically pure (*S*)-(+)-2-butanol was converted to its methanesulfonate ester according to the reaction shown.

(a) Write the Fischer projection of the *sec*-butyl methanesulfonate formed in this reaction.

(b) The *sec*-butyl methanesulfonate in part (a) was treated with $NaSCH_2CH_3$ to give a product having an optical rotation α_D of $-25°$. Write the Fischer projection of this product. By what mechanism is it formed? What is its absolute configuration (*R* or *S*)?

(c) When treated with PBr_3, optically pure (*S*)-(+)-2-butanol gave 2-bromobutane having an optical rotation $\alpha_D = -38°$. This bromide was then allowed to react with $NaSCH_2CH_3$ to give a product having an optical rotation α_D of $+23°$. Write the Fischer projection for (−)-2-bromobutane and specify its configuration as *R* or *S*. Does the reaction of 2-butanol with PBr_3 proceed with predominant inversion or retention of configuration?

(d) What is the optical rotation of optically pure 2-bromobutane?

8.37 In a classic experiment, Edward Hughes (a colleague of Ingold's at University College, London) studied the rate of racemization of 2-iodooctane by sodium iodide in acetone and compared it with the rate of incorporation of radioactive iodine into 2-iodooctane.

$$RI + [I^*]^- \longrightarrow RI^* + I^-$$

(I^* = radioactive iodine)

How will the rate of racemization compare with the rate of incorporation of radioactivity if

(a) Each act of exchange proceeds stereospecifically with retention of configuration?

(b) Each act of exchange proceeds stereospecifically with inversion of configuration?

(c) Each act of exchange proceeds in a stereorandom manner, in which retention and inversion of configuration are equally likely?

8.38 The ratio of elimination to substitution is exactly the same (26% elimination) for 2-bromo-2-methylbutane and 2-iodo-2-methylbutane in 80% ethanol/20% water at 25°C.

(a) By what mechanism does substitution most likely occur in these compounds under these conditions?

(b) By what mechanism does elimination most likely occur in these compounds under these conditions?

(c) Which substrate undergoes substitution faster?

(d) Which substrate undergoes elimination faster?

(e) What two substitution products are formed from each substrate?

(f) What two elimination products are formed from each substrate?

(g) Why do you suppose the ratio of elimination to substitution is the same for the two substrates?

8.39 The reaction of 2,2-dimethyl-1-propanol with HBr is very slow and gives 2-bromo-2-methyl-propane as the major product.

$$
\underset{\substack{\text{CH}_3 \\ | \\ \text{CH}_3\text{CCH}_2\text{OH} \\ | \\ \text{CH}_3}}{} \xrightarrow[\text{65°C}]{\text{HBr}} \underset{\substack{\text{CH}_3 \\ | \\ \text{CH}_3\text{CCH}_2\text{CH}_3 \\ | \\ \text{Br}}}{}
$$

Give a mechanistic explanation for these observations.

8.40 Solvolysis of 2-bromo-2-methylbutane in acetic acid containing potassium acetate gave three products. Identify them.

8.41 Solvolysis of 1,2-dimethylpropyl *p*-toluenesulfonate in acetic acid (75°C) yields five different products: three are alkenes and two are substitution products. Suggest reasonable structures for these five products.

8.42 Solution A was prepared by dissolving potassium acetate in methanol. Solution B was prepared by adding potassium methoxide to acetic acid. Reaction of methyl iodide either with solution A or with solution B gave the same major product. Why? What was this product?

8.43 If the temperature is not kept below 25°C during the reaction of primary alcohols with *p*-toluenesulfonyl chloride in pyridine, it is sometimes observed that the isolated product is not the desired alkyl *p*-toluenesulfonate but is instead the corresponding alkyl chloride. Suggest a mechanistic explanation for this observation.

8.44 The reaction of cyclopentyl bromide with sodium cyanide to give cyclopentyl cyanide

$$
\underset{\text{Cyclopentyl bromide}}{\overset{\text{H}}{\underset{\text{Br}}{\bigcirc}}} \xrightarrow[\text{ethanol–water}]{\text{NaCN}} \underset{\text{Cyclopentyl cyanide}}{\overset{\text{H}}{\underset{\text{CN}}{\bigcirc}}}
$$

proceeds faster if a small amount of sodium iodide is added to the reaction mixture. Can you suggest a reasonable mechanism to explain the catalytic function of sodium iodide?

8.45 Illustrate the stereochemistry associated with unimolecular nucleophilic substitution by constructing molecular models of *cis*-4-*tert*-butylcyclohexyl bromide, its derived carbocation, and the alcohols formed from it by hydrolysis under S$_N$1 conditions.

8.46 Given the molecular formula C$_6$H$_{11}$Br, construct a molecular model of the isomer that is a primary alkyl bromide yet relatively unreactive toward bimolecular nucleophilic substitution.

8.47 Cyclohexyl bromide is less reactive than noncyclic secondary alkyl halides toward S$_N$2 substitution. Construct a molecular model of cyclohexyl bromide and suggest a reason for its low reactivity.

8.48 1-Bromobicyclo[2.2.1]heptane (the structure of which is shown) is exceedingly unreactive toward nucleophilic substitution by either the S$_N$1 or S$_N$2 mechanism. Use molecular models to help you understand why.

1-Bromobicyclo[2.2.1]heptane

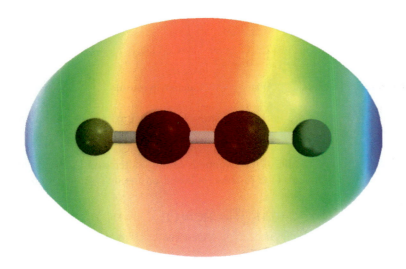

CHAPTER 9

ALKYNES

ydrocarbons that contain a carbon–carbon triple bond are called **alkynes.** Non-cyclic alkynes have the molecular formula C_nH_{2n-2}. *Acetylene* (HC≡CH) is the simplest alkyne. We call compounds that have their triple bond at the end of a carbon chain (RC≡CH) *monosubstituted,* or *terminal, alkynes.* Disubstituted alkynes (RC≡CR′) have *internal* triple bonds. You will see in this chapter that a carbon–carbon triple bond is a functional group, reacting with many of the same reagents that react with the double bonds of alkenes.

The most distinctive aspect of the chemistry of acetylene and terminal alkynes is their acidity. As a class, compounds of the type RC≡CH are the most acidic of all hydrocarbons. The structural reasons for this property, as well as the ways in which it is used to advantage in chemical synthesis, are important elements of this chapter.

9.1 SOURCES OF ALKYNES

Acetylene was discovered in 1836 by Edmund Davy and characterized by the French chemist P. E. M. Berthelot in 1862. It did not command much attention until its large-scale preparation from calcium carbide in the last decade of the nineteenth century stimulated interest in industrial applications. In the first stage of that synthesis, limestone and coke, a material rich in elemental carbon obtained from coal, are heated in an electric furnace to form calcium carbide.

$$CaO \quad + \quad 3C \quad \xrightarrow{1800-2100°C} \quad CaC_2 \quad + \quad CO$$

Calcium oxide	Carbon		Calcium carbide	Carbon monoxide
(from limestone)	(from coke)			

Calcium carbide is the calcium salt of the doubly negative carbide ion ($:\bar{C}≡\bar{C}:$). Carbide ion is strongly basic and reacts with water to form acetylene:

$$\mathrm{Ca^{2+}} \begin{bmatrix} \ddot{\mathrm{C}} \\ \| \\ \underset{\cdot\cdot}{\mathrm{C}} \end{bmatrix}^{2-} + 2\mathrm{H_2O} \longrightarrow \mathrm{Ca(OH)_2} + \mathrm{HC\equiv CH}$$

Calcium carbide   Water   Calcium hydroxide   Acetylene

> **PROBLEM 9.1** Use curved arrows to show how calcium carbide reacts with water to give acetylene.

Beginning in the middle of the twentieth century, alternative methods of acetylene production became practical. One of these is based on the dehydrogenation of ethylene.

$$\mathrm{H_2C=CH_2} \overset{heat}{\rightleftharpoons} \mathrm{HC\equiv CH} + \mathrm{H_2}$$

Ethylene   Acetylene   Hydrogen

The reaction is endothermic, and the equilibrium favors ethylene at low temperatures but shifts to favor acetylene above 1150°C. Indeed, at very high temperatures most hydrocarbons, even methane, are converted to acetylene. Acetylene has value not only by itself but is also the starting material from which higher alkynes are prepared.

Natural products that contain carbon–carbon triple bonds are numerous. Two examples are *tariric acid,* from the seed fat of a Guatemalan plant, and *cicutoxin,* a poisonous substance isolated from water hemlock.

> Its linear geometry makes it easy to pick out a triple bond in a molecular model.

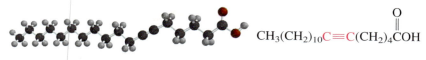

$$\mathrm{CH_3(CH_2)_{10}C\equiv C(CH_2)_4\overset{\overset{\displaystyle O}{\|}}{C}OH}$$

Tariric acid

$$\mathrm{HOCH_2CH_2CH_2C\equiv C-C\equiv CCH=CHCH=CHCH=CHCHCH_2CH_2CH_3}$$
$$\underset{\mathrm{OH}}{|}$$

Cicutoxin

Diacetylene ($\mathrm{HC\equiv C-C\equiv CH}$) has been identified as a component of the hydrocarbon-rich atmospheres of Uranus, Neptune, and Pluto. It is also present in the atmospheres of Titan and Triton, satellites of Saturn and Neptune, respectively.

9.2   NOMENCLATURE

In naming alkynes the usual IUPAC rules for hydrocarbons are followed, and the suffix *-ane* is replaced by *-yne.* Both acetylene and ethyne are acceptable IUPAC names for $\mathrm{HC\equiv CH}$. The position of the triple bond along the chain is specified by number in a manner analogous to alkene nomenclature.

$$\mathrm{HC\equiv CCH_3} \qquad \mathrm{HC\equiv CCH_2CH_3} \qquad \mathrm{CH_3C\equiv CCH_3} \qquad \mathrm{(CH_3)_3CC\equiv CCH_3}$$

Propyne   1-Butyne   2-Butyne   4,4-Dimethyl-2-pentyne

> **PROBLEM 9.2** Write structural formulas and give the IUPAC names for all the alkynes of molecular formula $\mathrm{C_5H_8}$.

When the —C≡CH group is named as a substituent, it is designated as an *ethynyl* group.

9.3 PHYSICAL PROPERTIES OF ALKYNES

Alkynes resemble alkanes and alkenes in their physical properties. They share with these other hydrocarbons the properties of low density and low water-solubility. They are slightly more polar and generally have slightly higher boiling points than the corresponding alkanes and alkenes.

> Examples of physical properties of alkynes are given in Appendix 1.

9.4 STRUCTURE AND BONDING IN ALKYNES: *sp* HYBRIDIZATION

Acetylene is linear, with a carbon–carbon bond distance of 120 pm and carbon–hydrogen bond distances of 106 pm.

Acetylene

Linear geometries characterize the H—C≡C—C and C—C≡C—C units of terminal and internal triple bonds, respectively, as well. This linear geometry is responsible for the relatively small number of known *cycloalkynes*. Figure 9.1 shows a molecular model for cyclononyne in which the bending of the C—C≡C—C unit is clearly evident. Angle strain destabilizes cycloalkynes to the extent that cyclononyne is the smallest one that is stable enough to be stored for long periods. The next smaller one, cyclooctyne, has been isolated, but is relatively reactive and polymerizes on standing.

In spite of the fact that few cycloalkynes occur naturally, they gained recent attention when it was discovered that some of them hold promise as anticancer drugs. (See the boxed essay *Natural and "Designed" Enediyne Antibiotics* following this section.)

FIGURE 9.1 Molecular model of cyclononyne showing bending of bond angles associated with triply bonded carbons. This model represents the structure obtained when the strain energy is minimized according to molecular mechanics, and closely matches the structure determined experimentally. Notice too the degree to which the staggering of bonds on adjacent atoms governs the overall shape of the ring.

FIGURE 9.2 The carbon atoms of acetylene are connected by a σ + π + π triple bond. (*a*) Both carbon atoms are *sp*-hybridized, and each is bonded to a hydrogen by a σ bond. The two π bonds are perpendicular to each other and are shown separately in (*b*) and (*c*).

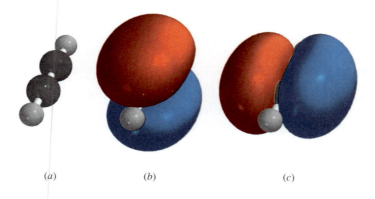

(*a*)   (*b*)   (*c*)

An *sp* hybridization model for the carbon–carbon triple bond was developed in Section 2.21 and is reviewed for acetylene in Figure 9.2. Figure 9.3 compares the electrostatic potential maps of ethylene and acetylene and shows how the second π bond in acetylene causes a band of high electron density to encircle the molecule.

At this point, it's useful to compare some structural features of alkanes, alkenes, and alkynes. Table 9.1 gives some of the most fundamental ones. To summarize, as we progress through the series in the order ethane → ethylene → acetylene:

1. The geometry at carbon changes from tetrahedral → trigonal planar → linear.
2. The C—C and C—H bonds become shorter and stronger.
3. The acidity of the C—H bonds increases.

All of these trends can be accommodated by the orbital hybridization model. The bond angles are characteristic for the sp^3, sp^2, and *sp* hybridization states of carbon and don't require additional comment. The bond distances, bond strengths, and acidities are related to the *s* character in the orbitals used for bonding. *s* Character is a simple concept, being nothing more than the percentage of the hybrid orbital contributed by an *s* orbital. Thus, an sp^3 orbital has one quarter *s* character and three quarters *p*, an sp^2 orbital has one third *s* and two thirds *p*, and an *sp* orbital one half *s* and one half *p*. We then use this information to analyze how various qualities of the hybrid orbital reflect those of its *s* and *p* contributors.

Take C—H bond distance and bond strength, for example. Recalling that an electron in a 2*s* orbital is, on average, closer to the nucleus and more strongly held than an

FIGURE 9.3 Electrostatic potential maps of ethylene and acetylene. The region of highest negative charge (*red*) is associated with the π bonds and lies between the two carbons in both. This electron-rich region is above and below the plane of the molecule in ethylene. Because acetylene has two π bonds, a band of high electron density encircles the molecule.

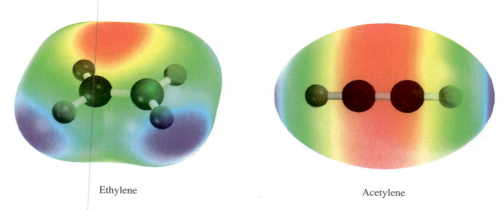

Ethylene   Acetylene

TABLE 9.1	Structural Features of Ethane, Ethylene, and Acetylene		
Feature	**Ethane**	**Ethylene**	**Acetylene**
Systematic name	Ethane	Ethene	Ethyne
Molecular formula	C_2H_6	C_2H_4	C_2H_2
Structural formula			$H\!-\!C\!\equiv\!C\!-\!H$
C—C bond distance, pm	153	134	120
C—H bond distance, pm	111	110	106
H—C—C bond angles	111.0°	121.4°	180°
C—C bond dissociation energy, kJ/mol (kcal/mol)	368 (88)	611 (146)	820 (196)
C—H bond dissociation energy, kJ/mol (kcal/mol)	410 (98)	452 (108)	536 (128)
Hybridization of carbon	sp^3	sp^2	sp
s character in C—H bonds	25%	33%	50%
Approximate pK_a	62	45	26

electron in a $2p$ orbital, it follows that an electron in an orbital with more *s* character will be closer to the nucleus and more strongly held than an electron in an orbital with less *s* character. Thus, when an *sp* orbital of carbon overlaps with a hydrogen $1s$ orbital to give a C—H σ bond, the electrons are held more strongly and the bond is stronger and shorter than electrons in a bond between hydrogen and sp^2-hybridized carbon. Similar reasoning holds for the shorter C—C bond distance of acetylene compared with ethylene, although here the additional π bond in acetylene is also a factor.

The pattern is repeated in higher alkynes as shown when comparing propyne and propene. The bonds to the *sp*-hybridized carbons of propyne are shorter than the corresponding bonds to the sp^2-hybridized carbons of propene.

An easy way to keep track of the effect of the *s* character of carbon is to associate it with electronegativity. As the *s* character of carbon increases, so does that carbon's electronegativity (the electrons in the bond involving that orbital are closer to carbon). The hydrogens in C—H bonds behave as if they are attached to an increasingly more electronegative carbon in the series ethane → ethylene → acetylene.

> How do the bond distances of molecular models of propene and propyne compare with the experimental values?

PROBLEM 9.3 How do bond distances and bond strengths change with electronegativity in the series NH_3, H_2O, and HF?

The property that most separates acetylene from ethane and ethylene is its acidity. It, too, can be explained on the basis of the greater electronegativity of *sp*-hybridized carbon compared with sp^3 and sp^2.

Beginning in the 1980s, research directed toward the isolation of new drugs derived from natural sources identified a family of tumor-inhibitory antibiotic substances characterized by novel structures containing a C≡C—C=C—C≡C unit as part of a nine- or ten-membered ring. With one double bond and two triple bonds (-ene + di- + -yne), these compounds soon became known as *enediyne* antibiotics. The simplest member of the class is *dynemicin A**; most of the other enediynes have even more complicated structures.

Enediynes hold substantial promise as anticancer drugs because of their potency and selectivity. Not only do they inhibit cell growth, they have a greater tendency to kill cancer cells than they do normal cells. The mechanism by which enediynes act involves novel chemistry unique to the C≡C—C=C—C≡C unit, which leads to a species that cleaves DNA and halts tumor growth.

The history of drug development has long been based on naturally occurring substances. Often, however, compounds that might be effective drugs are produced by plants and microorganisms in such small amounts that their isolation from natural sources is impractical. If the structure is relatively simple, chemical synthesis provides an alternative source of the drug, making it more available at a lower price. Equally important, chemical synthesis, modification, or both can improve the effectiveness of a drug. Building on the enediyne core of dynemicin A, for example, Professor Kyriacos C. Nicolaou and his associates at the Scripps Research Institute and the University of California at San Diego have prepared a simpler analog that is both more potent and more selective than dynemicin A. It is a "designed enediyne" in that its structure was conceived on the basis of chemical reasoning so as to carry out its biochemical task. The designed enediyne offers the additional advantage of being more amenable to large-scale synthesis.

Dynemicin A

"Designed" enediyne

*Learning By Modeling contains a model of dynemicin A, which shows that the C≡C—C=C—C≡C unit can be incorporated into the molecule without much angle strain.

9.5 ACIDITY OF ACETYLENE AND TERMINAL ALKYNES

The C—H bonds of hydrocarbons show little tendency to ionize, and alkanes, alkenes, and alkynes are all very weak acids. The acid-dissociation constant K_a for methane, for example, is too small to be measured directly but is estimated to be about 10^{-60} (pK_a 60).

$$H-\underset{\underset{H}{|}}{\overset{\overset{H}{|}}{C}}-H \rightleftharpoons H^+ + H-\underset{\underset{H}{|}}{\overset{\overset{H}{|}}{C}}{:}^-$$

Methane   Proton   Methide anion (a *carbanion*)

The conjugate base of a hydrocarbon is called a **carbanion.** It is an anion in which the negative charge is borne by carbon. Because it is derived from a very weak acid, a carbanion such as $^-$:CH_3 is an exceptionally strong base.

As we saw in Section 1.15, the ability of an atom to bear a negative charge is related to its electronegativity. Both the electronegativity of an atom X and the acidity of H—X increase across a row in the periodic table.

$$CH_4 \quad < \quad NH_3 \quad < \quad H_2O \quad < \quad HF$$

Methane	Ammonia	Water	Hydrogen fluoride
$pK_a \approx 60$	36	15.7	3.1
(weakest acid)			(strongest acid)

Using the relationship from the preceding section that the effective electronegativity of carbon in a C—H bond increases with its s character ($sp^3 < sp^2 < sp$), the order of hydrocarbon acidity behaves much like the preceding methane, ammonia, water, hydrogen fluoride series.

Increasing acidity

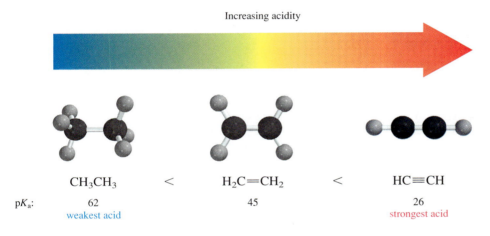

$$CH_3CH_3 \quad < \quad H_2C{=}CH_2 \quad < \quad HC{\equiv}CH$$

pK_a:	62	45	26
	weakest acid		strongest acid

The acidity increases as carbon becomes more electronegative. Ionization of acetylene gives an anion in which the unshared electron pair occupies an orbital with 50% s character.

$$H{-}C{\equiv}C{-}H \rightleftharpoons H^+ + H{-}C{\equiv}C{:}^- \; sp$$

Acetylene	Proton	Acetylide ion

In the corresponding ionizations of ethylene and ethane, the unshared pair occupies an orbital with 33% (sp^2) and 25% (sp^3) s character, respectively.

Terminal alkynes (RC≡CH) resemble acetylene in acidity.

$$(CH_3)_3CC{\equiv}CH \qquad pK_a = 25.5$$

3,3-Dimethyl-1-butyne

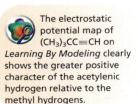

The electrostatic potential map of $(CH_3)_3CC{\equiv}CH$ on *Learning By Modeling* clearly shows the greater positive character of the acetylenic hydrogen relative to the methyl hydrogens.

Although acetylene and terminal alkynes are far stronger acids than other hydrocarbons, we must remember that they are, nevertheless, very weak acids—much weaker than water and alcohols, for example. Hydroxide ion is too weak a base to convert acetylene to its anion in meaningful amounts. The position of the equilibrium described by the following equation lies overwhelmingly to the left:

$$H—C\equiv C—H + \quad :\ddot{O}H \quad \rightleftharpoons \quad H—C\equiv C:^- + \quad H—\ddot{O}H$$

Acetylene	Hydroxide ion	Acetylide ion	Water
(weaker acid)	(weaker base)	(stronger base)	(stronger acid)
$pK_a = 26$			$pK_a = 15.7$

Because acetylene is a far weaker acid than water and alcohols, these substances are not suitable solvents for reactions involving acetylide ions. Acetylide is instantly converted to acetylene by proton transfer from compounds that contain —OH groups.

Amide ion is a much stronger base than acetylide ion and converts acetylene to its conjugate base quantitatively.

$$H—C\equiv C—H + \quad :\ddot{N}H_2 \quad \rightleftharpoons \quad H—C\equiv C:^- + \quad H—\ddot{N}H_2$$

Acetylene	Amide ion	Acetylide ion	Ammonia
(stronger acid)	(stronger base)	(weaker base)	(weaker acid)
$pK_a = 26$			$pK_a = 36$

Solutions of sodium acetylide ($HC\equiv CNa$) may be prepared by adding *sodium amide* ($NaNH_2$) to acetylene in liquid ammonia as the solvent. Terminal alkynes react similarly to give species of the type $RC\equiv CNa$.

PROBLEM 9.4 Complete each of the following equations to show the conjugate acid and the conjugate base formed by proton transfer between the indicated species. Use curved arrows to show the flow of electrons, and specify whether the position of equilibrium lies to the side of reactants or products.

(a) $CH_3C\equiv CH + \;^-:\ddot{O}CH_3 \rightleftharpoons$

(b) $HC\equiv CH + H_2\ddot{C}CH_3 \rightleftharpoons$

(c) $H_2C=CH_2 + \;:\ddot{N}H_2 \rightleftharpoons$

(d) $CH_3C\equiv CCH_2OH + \;:\ddot{N}H_2 \rightleftharpoons$

SAMPLE SOLUTION (a) The equation representing the acid–base reaction between propyne and methoxide ion is:

$$CH_3C\equiv C—H + \quad :\ddot{O}CH_3 \quad \rightleftharpoons \quad CH_3C\equiv C:^- + \quad H—\ddot{O}CH_3$$

Propyne	Methoxide ion	Propynide ion	Methanol
(weaker acid)	(weaker base)	(stronger base)	(stronger acid)

Alcohols are stronger acids than acetylene, and so the position of equilibrium lies to the left. Methoxide ion is not a strong enough base to remove a proton from acetylene.

Anions of acetylene and terminal alkynes are nucleophilic and react with methyl and primary alkyl halides to form carbon–carbon bonds by nucleophilic substitution. Some useful applications of this reaction will be discussed in the following section.

9.6   PREPARATION OF ALKYNES BY ALKYLATION OF ACETYLENE AND TERMINAL ALKYNES

Organic synthesis makes use of two major reaction types:

1. Functional group transformations
2. Carbon–carbon bond-forming reactions

Both strategies are applied to the preparation of alkynes. In this section we shall see how to prepare alkynes while building longer carbon chains. By attaching alkyl groups to acetylene, more complex alkynes can be prepared.

$$H-C\equiv C-H \longrightarrow R-C\equiv C-H \longrightarrow R-C\equiv C-R'$$

| Acetylene | Monosubstituted or terminal alkyne | Disubstituted derivative of acetylene |

Reactions that attach alkyl groups to molecular fragments are called **alkylation** reactions. One way in which alkynes are prepared is by alkylation of acetylene.

Alkylation of acetylene involves a sequence of two separate operations. In the first one, acetylene is converted to its conjugate base by treatment with sodium amide.

$$HC\equiv CH + NaNH_2 \longrightarrow HC\equiv CNa + NH_3$$

| Acetylene | Sodium amide | Sodium acetylide | Ammonia |

Next, an alkyl halide (the *alkylating agent*) is added to the solution of sodium acetylide. Acetylide ion acts as a nucleophile, displacing halide from carbon and forming a new carbon–carbon bond. Substitution occurs by an S_N2 mechanism.

$$HC\equiv CNa + RX \longrightarrow HC\equiv CR + NaX \quad \text{via} \quad HC\equiv C:\frown R\frown X$$

| Sodium acetylide | Alkyl halide | Alkyne | Sodium halide |

The synthetic sequence is usually carried out in liquid ammonia as the solvent. Alternatively, diethyl ether or tetrahydrofuran may be used.

$$HC\equiv CNa + CH_3CH_2CH_2CH_2Br \xrightarrow{NH_3} CH_3CH_2CH_2CH_2C\equiv CH$$

| Sodium acetylide | 1-Bromobutane | 1-Hexyne (70–77%) |

An analogous sequence using terminal alkynes as starting materials yields alkynes of the type $RC\equiv CR'$.

$$(CH_3)_2CHCH_2C\equiv CH \xrightarrow[NH_3]{NaNH_2} (CH_3)_2CHCH_2C\equiv CNa \xrightarrow{CH_3Br} (CH_3)_2CHCH_2C\equiv CCH_3$$

| 4-Methyl-1-pentyne | | 5-Methyl-2-hexyne (81%) |

Dialkylation of acetylene can be achieved by carrying out the sequence twice.

$$HC\equiv CH \xrightarrow[2.\ CH_3CH_2Br]{1.\ NaNH_2,\ NH_3} HC\equiv CCH_2CH_3 \xrightarrow[2.\ CH_3Br]{1.\ NaNH_2,\ NH_3} CH_3C\equiv CCH_2CH_3$$

| Acetylene | 1-Butyne | 2-Pentyne (81%) |

As in other nucleophilic substitution reactions, alkyl *p*-toluenesulfonates may be used in place of alkyl halides.

PROBLEM 9.5 Outline efficient syntheses of each of the following alkynes from acetylene and any necessary organic or inorganic reagents:

(a) 1-Heptyne

(b) 2-Heptyne

(c) 3-Heptyne

SAMPLE SOLUTION   (a) An examination of the structural formula of 1-heptyne reveals it to have a pentyl group attached to an acetylene unit. Alkylation of acetylene, by way of its anion, with a pentyl halide is a suitable synthetic route to 1-heptyne.

$$HC{\equiv}CH \xrightarrow[\text{NH}_3]{\text{NaNH}_2} HC{\equiv}CNa \xrightarrow{\text{CH}_3\text{CH}_2\text{CH}_2\text{CH}_2\text{CH}_2\text{Br}} HC{\equiv}CCH_2CH_2CH_2CH_2CH_3$$

Acetylene   Sodium acetylide   1-Heptyne

The major limitation to this reaction is that synthetically acceptable yields are obtained only with methyl halides and primary alkyl halides. Acetylide anions are very basic, much more basic than hydroxide, for example, and react with secondary and tertiary alkyl halides by elimination.

$$HC{\equiv}C{:}^{-} \quad H{-}CH_2{-}\underset{\underset{\text{CH}_3}{|}}{\overset{\overset{\text{CH}_3}{|}}{C}}{-}Br \xrightarrow{\text{E2}} HC{\equiv}CH + H_2C{=}C\begin{matrix}\text{CH}_3\\\text{CH}_3\end{matrix} \quad + \quad Br^{-}$$

Acetylide   *tert*-Butyl bromide   Acetylene   2-Methylpropene   Bromide

The desired S_N2 substitution pathway is observed only with methyl and primary alkyl halides.

PROBLEM 9.6   Which of the alkynes of molecular formula C_5H_8 can be prepared in good yield by alkylation or dialkylation of acetylene? Explain why the preparation of the other C_5H_8 isomers would not be practical.

A second strategy for alkyne synthesis, involving functional group transformation reactions, is described in the following section.

9.7   PREPARATION OF ALKYNES BY ELIMINATION REACTIONS

Just as it is possible to prepare alkenes by dehydrohalogenation of alkyl halides, so may alkynes be prepared by a *double dehydrohalogenation* of dihaloalkanes. The dihalide may be a **geminal dihalide,** one in which both halogens are on the same carbon, or it may be a **vicinal dihalide,** one in which the halogens are on adjacent carbons.

Double dehydrohalogenation of a geminal dihalide

$$R{-}\underset{\underset{\text{H}}{|}}{\overset{\overset{\text{H}}{|}}{C}}{-}\underset{\underset{\text{X}}{|}}{\overset{\overset{\text{X}}{|}}{C}}{-}R' + 2NaNH_2 \longrightarrow R{-}C{\equiv}C{-}R' + 2NH_3 + 2NaX$$

Geminal dihalide   Sodium amide   Alkyne   Ammonia   Sodium halide

Double dehydrohalogenation of a vicinal dihalide

$$R{-}\underset{\underset{\text{X}}{|}}{\overset{\overset{\text{H}}{|}}{C}}{-}\underset{\underset{\text{X}}{|}}{\overset{\overset{\text{H}}{|}}{C}}{-}R' + 2NaNH_2 \longrightarrow R{-}C{\equiv}C{-}R' + 2NH_3 + 2NaX$$

Vicinal dihalide   Sodium amide   Alkyne   Ammonia   Sodium halide

The most frequent applications of these procedures lie in the preparation of terminal alkynes. Because the terminal alkyne product is acidic enough to transfer a proton to amide anion, one equivalent of base in addition to the two equivalents required for double dehydrohalogenation is needed. Adding water or acid after the reaction is complete converts the sodium salt to the corresponding alkyne.

Double dehydrohalogenation of a geminal dihalide

$$(CH_3)_3CCH_2CHCl_2 \xrightarrow[NH_3]{3NaNH_2} (CH_3)_3CC{\equiv}CNa \xrightarrow{H_2O} (CH_3)_3CC{\equiv}CH$$

1,1-Dichloro-3,3-dimethylbutane	Sodium salt of alkyne product (not isolated)	3,3-Dimethyl-1-butyne (56–60%)

Double dehydrohalogenation of a vicinal dihalide

$$CH_3(CH_2)_7CHCH_2Br \xrightarrow[NH_3]{3NaNH_2} CH_3(CH_2)_7C{\equiv}CNa \xrightarrow{H_2O} CH_3(CH_2)_7C{\equiv}CH$$
$$\underset{Br}{|}$$

1,2-Dibromodecane	Sodium salt of alkyne product (not isolated)	1-Decyne (54%)

Double dehydrohalogenation to form terminal alkynes may also be carried out by heating geminal and vicinal dihalides with potassium *tert*-butoxide in dimethyl sulfoxide.

> **PROBLEM 9.7** Give the structures of three isomeric dibromides that could be used as starting materials for the preparation of 3,3-dimethyl-1-butyne.

Because vicinal dihalides are prepared by addition of chlorine or bromine to alkenes (Section 6.14), alkenes, especially terminal alkenes, can serve as starting materials for the preparation of alkynes as shown in the following example:

$$(CH_3)_2CHCH{=}CH_2 \xrightarrow{Br_2} (CH_3)_2CHCHCH_2Br \xrightarrow[2.\ H_2O]{1.\ NaNH_2,\ NH_3} (CH_3)_2CHC{\equiv}CH$$
$$\underset{Br}{|}$$

3-Methyl-1-butene	1,2-Dibromo-3-methylbutane	3-Methyl-1-butyne (52%)

> **PROBLEM 9.8** Show, by writing an appropriate series of equations, how you could prepare propyne from each of the following compounds as starting materials. You may use any necessary organic or inorganic reagents.
>
> (a) 2-Propanol   (d) 1,1-Dichloroethane
>
> (b) 1-Propanol   (e) Ethyl alcohol
>
> (c) Isopropyl bromide
>
> **SAMPLE SOLUTION** (a) Because we know that we can convert propene to propyne by the sequence of reactions
>
> $$CH_3CH{=}CH_2 \xrightarrow{Br_2} CH_3CHCH_2Br \xrightarrow[2.\ H_2O]{1.\ NaNH_2,\ NH_3} CH_3C{\equiv}CH$$
> $$\underset{Br}{|}$$
>
> | Propene | 1,2-Dibromopropane | Propyne |
> |---|---|---|
>
> all that remains to completely describe the synthesis is to show the preparation of propene from 2-propanol. Acid-catalyzed dehydration is suitable.

$$(CH_3)_2CHOH \xrightarrow[\text{heat}]{H^+} CH_3CH{=}CH_2$$

2-Propanol   Propene

9.8   REACTIONS OF ALKYNES

We have already discussed one important chemical property of alkynes, the acidity of acetylene and terminal alkynes. In the remaining sections of this chapter several other reactions of alkynes will be explored. Most of them will be similar to reactions of alkenes. Like alkenes, alkynes undergo addition reactions. We'll begin with a reaction familiar to us from our study of alkenes, namely, catalytic hydrogenation.

9.9   HYDROGENATION OF ALKYNES

The conditions for hydrogenation of alkynes are similar to those employed for alkenes. In the presence of finely divided platinum, palladium, nickel, or rhodium, two molar equivalents of hydrogen add to the triple bond of an alkyne to yield an alkane.

$$RC{\equiv}CR' + 2H_2 \xrightarrow{\text{Pt, Pd, Ni, or Rh}} RCH_2CH_2R'$$

Alkyne   Hydrogen   Alkane

$$CH_3CH_2\underset{\underset{CH_3}{|}}{CH}CH_2C{\equiv}CH + 2H_2 \xrightarrow{\text{Ni}} CH_3CH_2\underset{\underset{CH_3}{|}}{CH}CH_2CH_2CH_3$$

4-Methyl-1-hexyne   Hydrogen   3-Methylhexane (77%)

> **PROBLEM 9.9** Write a series of equations showing how you could prepare octane from acetylene and any necessary organic and inorganic reagents.

The heat of hydrogenation of an alkyne is greater than twice the heat of hydrogenation of an alkene. When two moles of hydrogen add to an alkyne, addition of the first mole (triple bond → double bond) is more exothermic than the second (double bond → single bond).

Substituents affect the heats of hydrogenation of alkynes in the same way they affect alkenes. Compare the heats of hydrogenation of 1-butyne and 2-butyne, both of which give butane on taking up two moles of H_2.

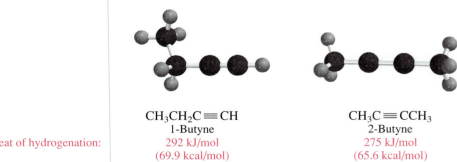

	$CH_3CH_2C{\equiv}CH$	$CH_3C{\equiv}CCH_3$
	1-Butyne	2-Butyne
Heat of hydrogenation:	292 kJ/mol	275 kJ/mol
	(69.9 kcal/mol)	(65.6 kcal/mol)

The internal triple bond of 2-butyne is stabilized relative to the terminal triple bond of 1-butyne. Alkyl groups release electrons to sp-hybridized carbon, stabilizing the alkyne and decreasing the heat of hydrogenation.

Like the hydrogenation of alkenes, hydrogenation of alkynes is a syn addition; cis alkenes are intermediates in the hydrogenation of alkynes to alkanes.

$$RC \equiv CR' \xrightarrow[\text{catalyst}]{H_2} \underset{\substack{| \quad | \\ H \quad H}}{\overset{\substack{R \quad R' \\ \diagdown \quad \diagup}}{C=C}} \xrightarrow[\text{catalyst}]{H_2} RCH_2CH_2R'$$

Alkyne   cis Alkene   Alkane

Noting that cis alkenes are intermediates in the hydrogenation of alkynes leads us to consider the possibility of halting hydrogenation at the cis alkene stage. If partial hydrogenation of an alkyne could be achieved, it would provide us with methods for preparing:

1. Alkenes from alkynes, and

2. cis Alkenes free of their trans stereoisomers

Both objectives have been met by designing special hydrogenation catalysts. The most frequently used one is the **Lindlar catalyst,** a palladium on calcium carbonate combination to which lead acetate and quinoline have been added. Lead acetate and quinoline partially deactivate ("poison") the catalyst, making it a poor catalyst for alkene hydrogenation while retaining its ability to catalyze the addition of H_2 to the triple bond.

The structure of quinoline is shown on page 460. In subsequent equations, we will simply use the term *Lindlar Pd* to stand for all of the components of the Lindlar catalyst.

1-Ethynylcyclohexanol   Hydrogen   1-Vinylcyclohexanol (90–95%)

Hydrogenation of alkynes with internal triple bonds gives cis alkenes.

$$CH_3(CH_2)_3C \equiv C(CH_2)_3CH_3 \xrightarrow[\text{Lindlar Pd}]{H_2} \underset{\substack{| \quad | \\ H \quad H}}{\overset{\substack{CH_3(CH_2)_3 \quad (CH_2)_3CH_3 \\ \diagdown \quad \diagup}}{C=C}}$$

5-Decyne   *cis*-5-Decene (87%)

PROBLEM 9.10 Write a series of equations showing how to prepare *cis*-5-decene from acetylene and 1-bromobutane as the source of all its carbons, using any necessary organic or inorganic reagents. (*Hint:* You may find it helpful to review Section 9.6.)

Hydrogenation of alkynes to alkenes using the Lindlar catalyst is attractive because it sidesteps the regioselectivity and stereoselectivity issues that accompany the dehydration of alcohols and dehydrohalogenation of alkyl halides. In terms of regioselectivity, the position of the double bond is never in doubt—it appears in the carbon chain at exactly the same place where the triple bond was. In terms of stereoselectivity, only the cis alkene forms. Recall that dehydration and dehydrohalogenation normally give a cis–trans mixture in which the cis isomer is the minor product.

In the following section, we'll see another method for converting alkynes to alkenes. The reaction conditions are very different from those of Lindlar hydrogenation. So is the stereochemistry.

9.10 METAL–AMMONIA REDUCTION OF ALKYNES

A useful alternative to catalytic partial hydrogenation for converting alkynes to alkenes is reduction by a Group I metal (lithium, sodium, or potassium) in liquid ammonia. The unique feature of metal–ammonia reduction is that it converts alkynes to trans alkenes, whereas catalytic hydrogenation yields cis alkenes. Thus, from the same alkyne one can prepare either a cis or a trans alkene by choosing the appropriate reaction conditions.

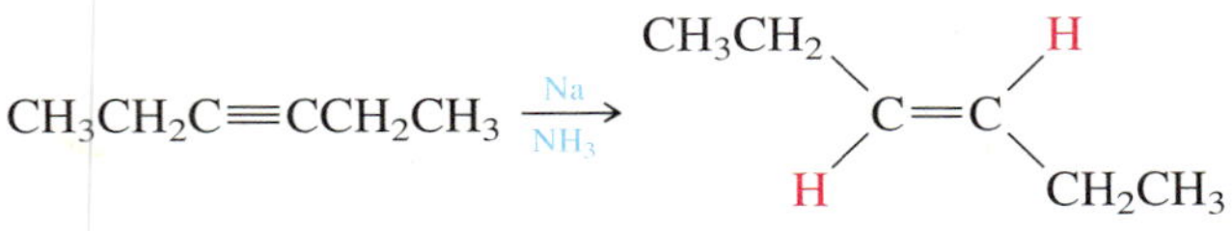

$$CH_3CH_2C\equiv CCH_2CH_3 \xrightarrow[NH_3]{Na} \underset{\substack{CH_3CH_2 \\ H}}{} C=C \underset{\substack{H \\ CH_2CH_3}}{}$$

3-Hexyne   *trans*-3-Hexene (82%)

> **PROBLEM 9.11** Suggest an efficient synthesis of *trans*-2-heptene from propyne and any necessary organic or inorganic reagents.

The stereochemistry of metal–ammonia reduction of alkynes differs from that of catalytic hydrogenation because the mechanisms of the two reactions are different. The mechanism of hydrogenation of alkynes is similar to that of catalytic hydrogenation of alkenes (Sections 6.1–6.3). A mechanism for metal–ammonia reduction of alkynes is outlined in Figure 9.4.

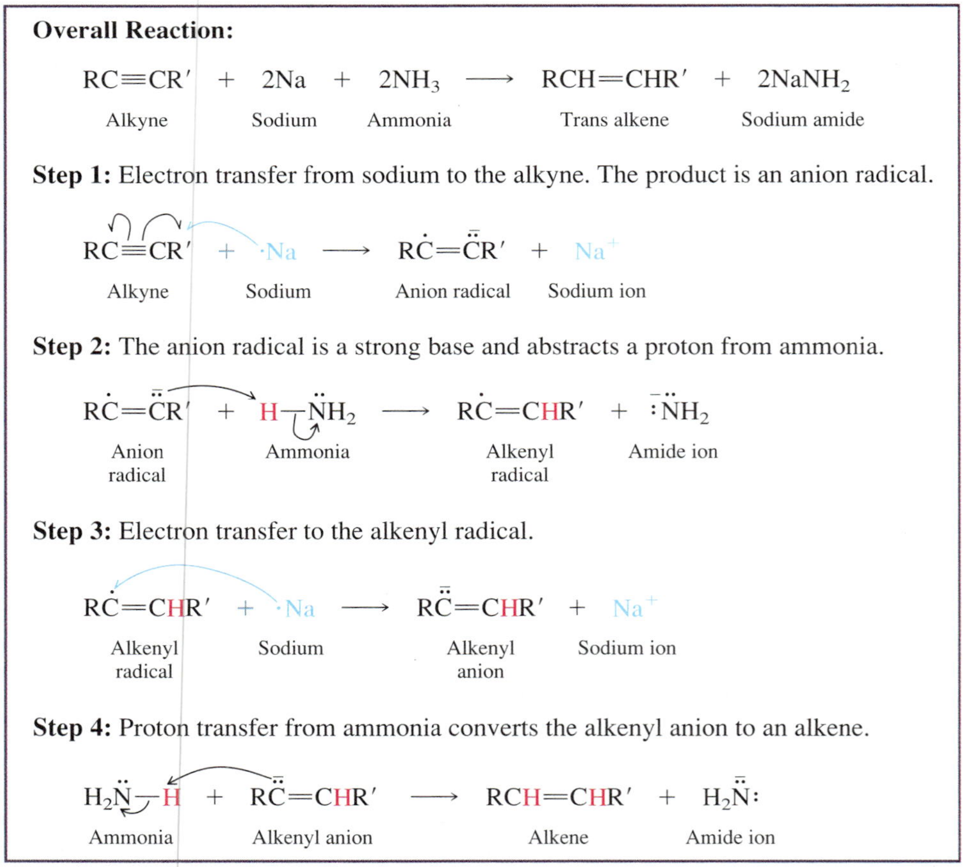

Overall Reaction:

$$RC\equiv CR' \ + \ 2Na \ + \ 2NH_3 \longrightarrow RCH=CHR' \ + \ 2NaNH_2$$

Alkyne   Sodium   Ammonia   Trans alkene   Sodium amide

Step 1: Electron transfer from sodium to the alkyne. The product is an anion radical.

$$RC\equiv CR' \ + \ \cdot Na \longrightarrow R\overset{\cdot}{C}=\overset{\cdot\cdot}{C}R' \ + \ Na^+$$

Alkyne   Sodium   Anion radical   Sodium ion

Step 2: The anion radical is a strong base and abstracts a proton from ammonia.

$$R\overset{\cdot}{C}=\overset{\cdot\cdot}{C}R' \ + \ H-\overset{\cdot\cdot}{N}H_2 \longrightarrow R\overset{\cdot}{C}=CHR' \ + \ \overset{-\ \cdot\cdot}{:}NH_2$$

Anion   Ammonia   Alkenyl   Amide ion
radical   radical

Step 3: Electron transfer to the alkenyl radical.

$$R\overset{\cdot}{C}=CHR' \ + \ \cdot Na \longrightarrow R\overset{\cdot\cdot}{C}=CHR' \ + \ Na^+$$

Alkenyl   Sodium   Alkenyl   Sodium ion
radical   anion

Step 4: Proton transfer from ammonia converts the alkenyl anion to an alkene.

$$H_2\overset{\cdot\cdot}{N}-H \ + \ R\overset{\cdot\cdot}{C}=CHR' \longrightarrow RCH=CHR' \ + \ H_2\overset{\cdot\cdot}{N}:$$

Ammonia   Alkenyl anion   Alkene   Amide ion

FIGURE 9.4 Mechanism of the sodium–ammonia reduction of an alkyne.

The mechanism includes two single-electron transfers (steps 1 and 3) and two proton transfers (steps 2 and 4). Experimental evidence indicates that step 2 is rate-determining, and it is believed that the observed trans stereochemistry reflects the distribution of the two stereoisomeric alkenyl radical intermediates formed in this step.

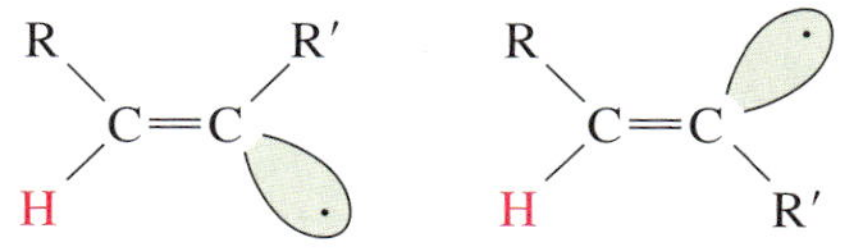

(Z)-Alkenyl radical   (E)-Alkenyl radical
(less stable)   (more stable)

The more stable (E)-alkenyl radical, in which the alkyl groups R and R′ are trans to each other, is formed faster than its Z stereoisomer. Steps 3 and 4, which follow, are fast, and the product distribution is determined by the E–Z ratio of radicals produced in step 2.

9.11   ADDITION OF HYDROGEN HALIDES TO ALKYNES

Alkynes react with many of the same electrophilic reagents that add to the carbon–carbon double bond of alkenes. Hydrogen halides, for example, add to alkynes to form alkenyl halides.

$$RC{\equiv}CR' \;+\; HX \;\longrightarrow\; \underset{\underset{X}{|}}{RCH{=}CR'}$$

Alkyne   Hydrogen halide   Alkenyl halide

The regioselectivity of addition follows Markovnikov's rule. A proton adds to the carbon that has the greater number of hydrogens, and halide adds to the carbon with the fewer hydrogens.

$$CH_3CH_2CH_2CH_2C{\equiv}CH \;+\; HBr \;\longrightarrow\; \underset{\underset{Br}{|}}{CH_3CH_2CH_2CH_2C{=}CH_2}$$

1-Hexyne   Hydrogen bromide   2-Bromo-1-hexene (60%)

When formulating a mechanism for the reaction of alkynes with hydrogen halides, we could propose a process analogous to that of electrophilic addition to alkenes in which the first step is formation of a carbocation and is rate-determining. The second step according to such a mechanism would be nucleophilic capture of the carbocation by a halide ion.

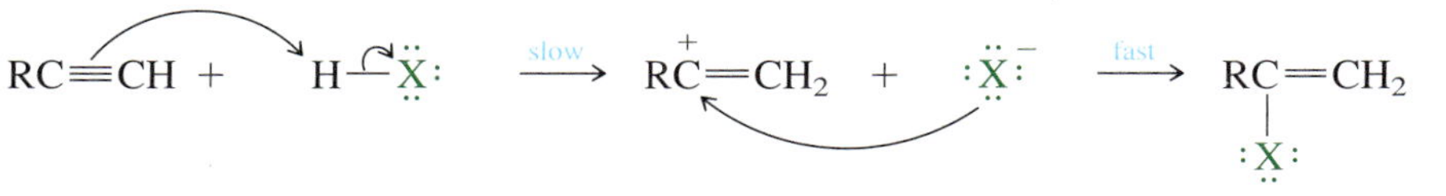

Alkyne   Hydrogen halide   Alkenyl cation   Halide ion   Alkenyl halide

Evidence from a variety of sources, however, indicates that alkenyl cations (also called *vinylic cations*) are much less stable than simple alkyl cations, and their involvement in these additions has been questioned. For example, although electrophilic addition of hydrogen halides to alkynes occurs more slowly than the corresponding additions

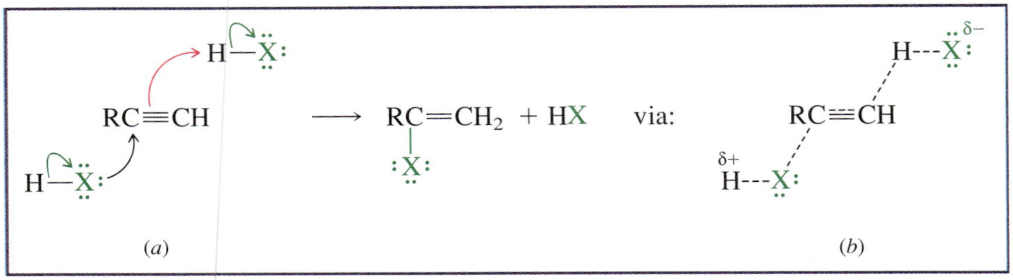

FIGURE 9.5 (*a*) Curved arrow notation, and (*b*) transition-state for electrophilic addition of a hydrogen halide HX to an alkyne.

to alkenes, the difference is not nearly as great as the difference in carbocation stabilities would suggest.

Furthermore, kinetic studies reveal that electrophilic addition of hydrogen halides to alkynes follows a rate law that is third-order overall and second-order in hydrogen halide.

$$\text{Rate} = k[\text{alkyne}][\text{HX}]^2$$

This third-order rate dependence suggests a termolecular transition state, one that involves two molecules of the hydrogen halide. Figure 9.5 depicts such a termolecular process using curved arrows to show the flow of electrons, and dashed-lines to indicate the bonds being made and broken at the transition state. This mechanism, called Ad_E3 for *addition-electrophilic-termolecular,* avoids the formation of a very unstable alkenyl cation intermediate by invoking nucleophilic participation by the halogen at an early stage. Nevertheless, because Markovnikov's rule is observed, it seems likely that some degree of positive character develops at carbon and controls the regioselectivity of addition.

In the presence of excess hydrogen halide, geminal dihalides are formed by sequential addition of two molecules of hydrogen halide to the carbon–carbon triple bond.

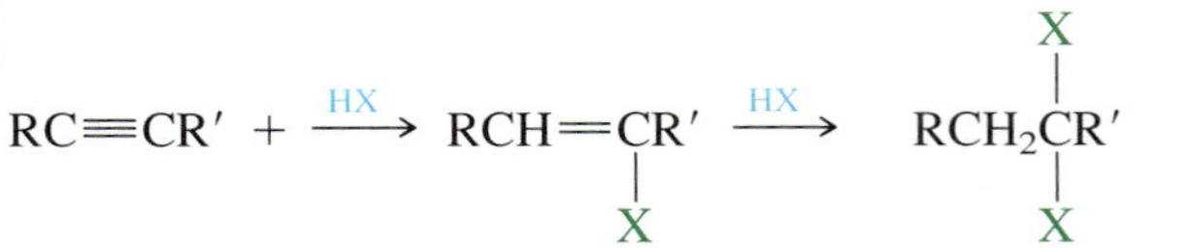

For further discussion of this topic, see the article "The Electrophilic Addition to Alkynes" in the November 1993 edition of the *Journal of Chemical Education* (p. 873). Additional commentary appeared in the November 1996 issue.

$$RC{\equiv}CR' + \xrightarrow{\text{HX}} RCH{=}CR' \xrightarrow{\text{HX}} RCH_2CR'$$

| Alkyne | Alkenyl halide | Geminal dihalide |

The hydrogen halide adds to the initially formed alkenyl halide in accordance with Markovnikov's rule. Overall, both protons become bonded to the same carbon and both halogens to the adjacent carbon.

$$CH_3CH_2C{\equiv}CCH_2CH_3 +\quad 2HF \quad\longrightarrow\quad CH_3CH_2CH_2CCH_2CH_3$$

| 3-Hexyne | Hydrogen fluoride | 3,3-Difluorohexane (76%) |

PROBLEM 9.12 Write a series of equations showing how to prepare 1,1-dichloroethane from

(a) Ethylene

(b) Vinyl chloride ($H_2C{=}CHCl$)

(c) 1,1-Dibromoethane

SAMPLE SOLUTION (a) Reasoning backward, we recognize 1,1-dichloroethane as the product of addition of two molecules of hydrogen chloride to acetylene. Thus, the synthesis requires converting ethylene to acetylene as a key feature. As described in Section 9.7, this may be accomplished by conversion of ethylene to a vicinal dihalide, followed by double dehydrohalogenation. A suitable synthesis based on this analysis is as shown:

$$H_2C{=}CH_2 \xrightarrow{Br_2} BrCH_2CH_2Br \xrightarrow[2.\ H_2O]{1.\ NaNH_2} HC{\equiv}CH \xrightarrow{2HCl} CH_3CHCl_2$$

Ethylene   1,2-Dibromoethane   Acetylene   1,1-Dichloroethane

Hydrogen bromide (but not hydrogen chloride or hydrogen iodide) adds to alkynes by a free-radical mechanism when peroxides are present in the reaction mixture. As in the free-radical addition of hydrogen bromide to alkenes (Section 6.8), a regioselectivity opposite to Markovnikov's rule is observed.

$$CH_3CH_2CH_2CH_2C{\equiv}CH\ +\qquad HBr \xrightarrow{peroxides} CH_3CH_2CH_2CH_2CH{=}CHBr$$

1-Hexyne   Hydrogen bromide   1-Bromo-1-hexene (79%)

9.12 HYDRATION OF ALKYNES

By analogy to the hydration of alkenes, hydration of an alkyne is expected to yield an alcohol. The kind of alcohol, however, would be of a special kind, one in which the hydroxyl group is a substituent on a carbon–carbon double bond. This type of alcohol is called an **enol** (the double bond suffix -*ene* plus the alcohol suffix -*ol*). An important property of enols is their rapid isomerization to aldehydes or ketones under the conditions of their formation.

$$RC{\equiv}CR'\ +\ H_2O \xrightarrow{slow} \underset{\substack{\text{Enol}\\(\text{not isolated})}}{RCH{=}\overset{\overset{\textstyle OH}{|}}{C}R'} \xrightarrow{fast} \underset{\substack{R'=H;\ \text{aldehyde}\\R'=\text{alkyl};\ \text{ketone}}}{RCH_2\overset{\overset{\textstyle O}{\|}}{C}R'}$$

Alkyne   Water

The aldehyde or ketone is called the **keto** form, and the keto ⇌ enol equilibration referred to as *keto–enol isomerism* or *keto–enol tautomerism.* **Tautomers** are constitutional isomers that equilibrate by migration of an atom or group, and their equilibration is called **tautomerism.** The mechanism of keto–enol isomerism involves the sequence of proton transfers shown in Figure 9.6.

The first step, protonation of the double bond of the enol, is analogous to the protonation of the double bond of an alkene. It takes place more readily, however, because the carbocation formed in this step is stabilized by resonance involving delocalization of a lone pair of oxygen.

Of the two resonance forms A and B, A has only six electrons around its positively charged carbon. B satisfies the octet rule for both carbon and oxygen. It is more stable than A and more stable than a carbocation formed by protonation of a typical alkene.

FIGURE 9.6 Conversion of an enol to a ketone takes place by way of two solvent-mediated proton transfers. A proton is transferred to carbon in the first step, then removed from oxygen in the second.

Overall Reaction:

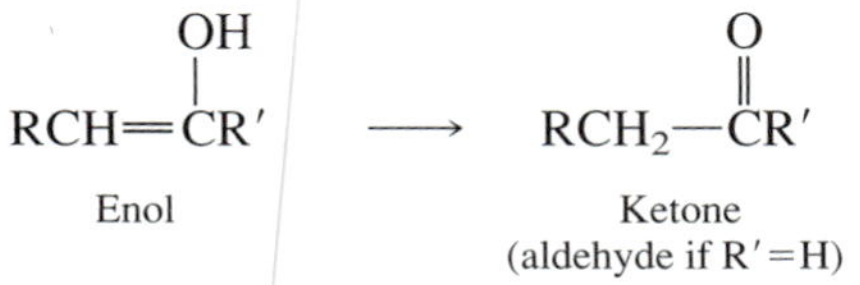

$$\underset{\substack{\text{Enol}}}{RCH\!=\!\overset{\displaystyle OH}{\overset{|}{CR'}}} \longrightarrow \underset{\substack{\text{Ketone}\\ \text{(aldehyde if R'=H)}}}{RCH_2\!-\!\overset{\displaystyle O}{\overset{\|}{CR'}}}$$

Step 1: The enol is formed in aqueous acidic solution. The first step of its transformation to a ketone is proton transfer to the carbon–carbon double bond.

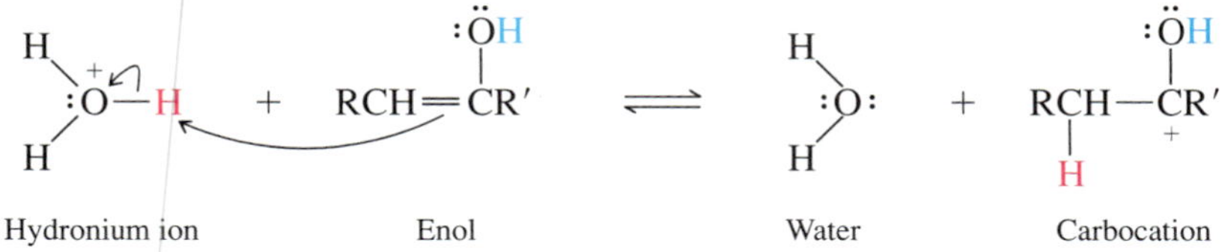

$$\underset{\text{Hydronium ion}}{\overset{H}{\underset{H}{\overset{|}{\,\,\overset{+}{O}\!-\!H}}}} + \underset{\text{Enol}}{RCH\!=\!\overset{:\ddot{O}H}{\overset{|}{CR'}}} \rightleftharpoons \underset{\text{Water}}{\overset{H}{\underset{H}{\overset{|}{\,\,\ddot{O}:}}}} + \underset{\text{Carbocation}}{RCH\!-\!\overset{:\ddot{O}H}{\overset{|}{\underset{H}{\overset{+}{CR'}}}}}$$

Step 2: The carbocation transfers a proton from oxygen to a water molecule, yielding a ketone.

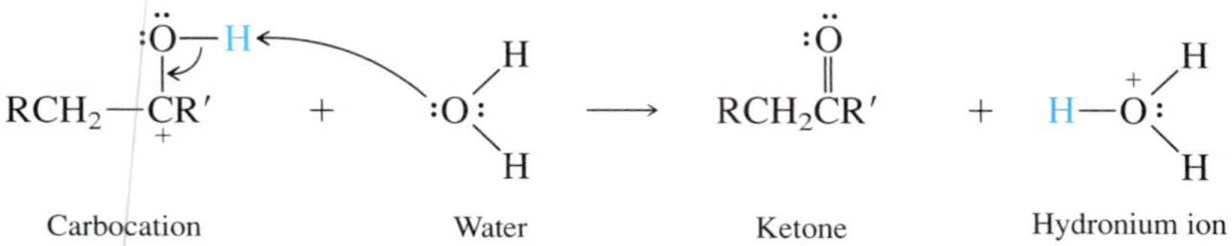

$$\underset{\text{Carbocation}}{RCH_2\!-\!\overset{:\ddot{O}\!-\!H}{\underset{+}{\overset{|}{CR'}}}} + \underset{\text{Water}}{\overset{H}{\underset{H}{\overset{}{:\ddot{O}:}}}} \longrightarrow \underset{\text{Ketone}}{RCH_2\overset{:\ddot{O}}{\overset{\|}{CR'}}} + \underset{\text{Hydronium ion}}{\overset{H}{\underset{H}{H\!-\!\overset{+}{\overset{}{\ddot{O}:}}}}}$$

PROBLEM 9.13 Give the structure of the enol formed by hydration of 2-butyne, and write a series of equations showing its conversion to its corresponding ketone isomer.

In general, ketones are more stable than their enol precursors and are the products actually isolated when alkynes undergo acid-catalyzed hydration. The standard method for alkyne hydration employs aqueous sulfuric acid as the reaction medium and mercury(II) sulfate or mercury(II) oxide as a catalyst.

Mercury(II) sulfate and mercury(II) oxide are also known as *mercuric* sulfate and oxide, respectively.

$$\underset{\text{4-Octyne}}{CH_3CH_2CH_2C\!\equiv\!CCH_2CH_2CH_3} + H_2O \xrightarrow{H^+,\ Hg^{2+}} \underset{\text{4-Octanone (89\%)}}{CH_3CH_2CH_2CH_2\overset{\displaystyle O}{\overset{\|}{C}}CH_2CH_2CH_3}$$

Hydration of alkynes follows Markovnikov's rule; terminal alkynes yield methyl-substituted ketones.

$$\underset{\text{1-Octyne}}{HC\!\equiv\!CCH_2CH_2CH_2CH_2CH_2CH_3} + H_2O \xrightarrow[\text{HgSO}_4]{\text{H}_2\text{SO}_4} \underset{\text{2-Octanone (91\%)}}{CH_3\overset{\displaystyle O}{\overset{\|}{C}}CH_2CH_2CH_2CH_2CH_2CH_3}$$

PROBLEM 9.14 Show by a series of equations how you could prepare 2-octanone from acetylene and any necessary organic or inorganic reagents. How could you prepare 4-octanone?

Because of the regioselectivity of alkyne hydration, acetylene is the only alkyne structurally capable of yielding an aldehyde under these conditions.

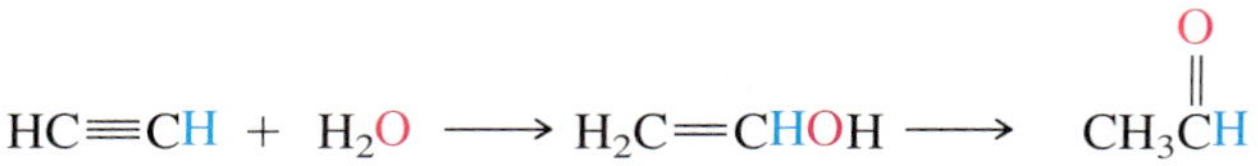

$$HC{\equiv}CH + H_2O \longrightarrow H_2C{=}CHOH \longrightarrow CH_3\overset{\displaystyle O}{\overset{\|}{C}}H$$

Acetylene   Water   Vinyl alcohol   Acetaldehyde
(not isolated)

At one time acetaldehyde was prepared on an industrial scale by this method. Modern methods involve direct oxidation of ethylene and are more economical.

The industrial synthesis of acetaldehyde from ethylene is shown on page 644.

9.13 ADDITION OF HALOGENS TO ALKYNES

Alkynes react with chlorine and bromine to yield tetrahaloalkanes. Two molecules of the halogen add to the triple bond.

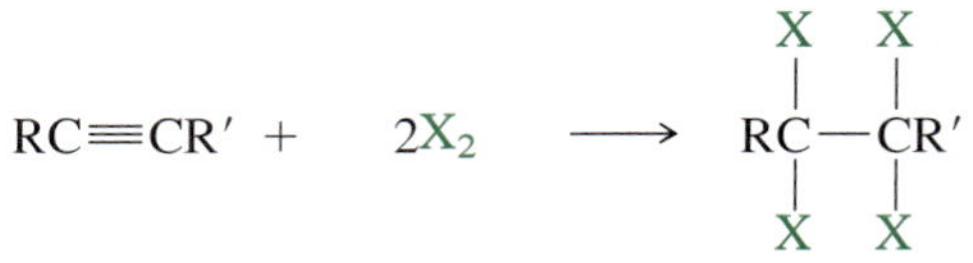

$$RC{\equiv}CR' + 2X_2 \longrightarrow R\overset{\displaystyle X}{\underset{\displaystyle X}{\overset{|}{\underset{|}{C}}}}{-}\overset{\displaystyle X}{\underset{\displaystyle X}{\overset{|}{\underset{|}{C}}}}R'$$

Alkyne   Halogen   Tetrahaloalkane
(chlorine or
bromine)

$$CH_3C{\equiv}CH + 2Cl_2 \longrightarrow CH_3\overset{\displaystyle Cl}{\underset{\displaystyle Cl}{\overset{|}{\underset{|}{C}}}}CHCl_2$$

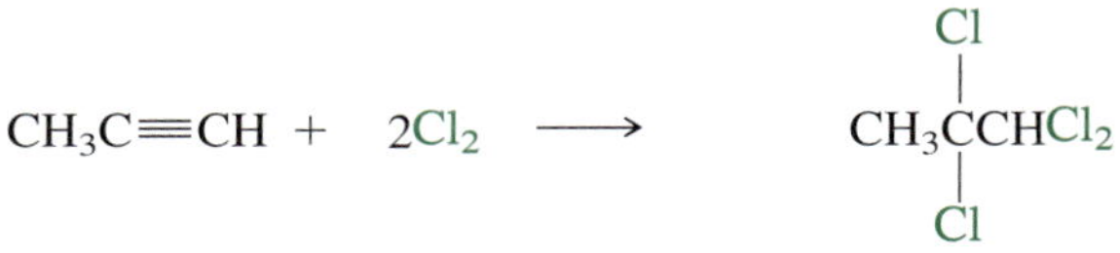

Propyne   Chlorine   1,1,2,2-Tetrachloropropane (63%)

A dihaloalkene is an intermediate and is the isolated product when the alkyne and the halogen are present in equimolar amounts. The stereochemistry of addition is anti.

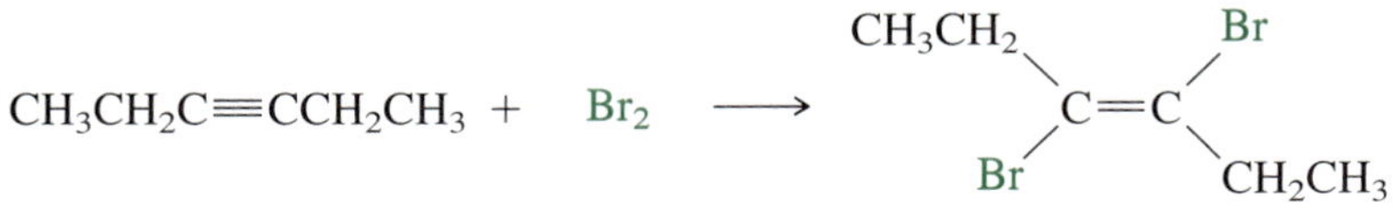

$$CH_3CH_2C{\equiv}CCH_2CH_3 + Br_2 \longrightarrow$$

$$\underset{Br}{\overset{CH_3CH_2}{\diagdown}}C{=}C\underset{CH_2CH_3}{\overset{Br}{\diagup}}$$

3-Hexyne   Bromine   (E)-3,4-Dibromo-3-hexene (90%)

9.14 OZONOLYSIS OF ALKYNES

Carboxylic acids are produced when alkynes are subjected to ozonolysis.

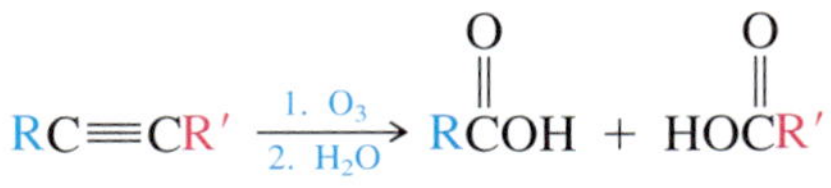

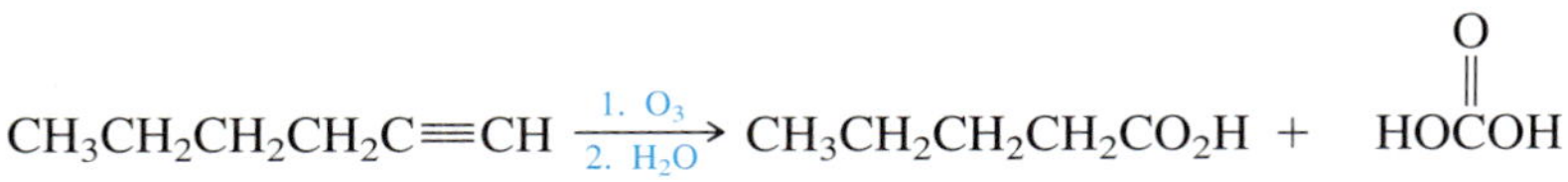

$$RC{\equiv}CR' \xrightarrow[\text{2. } H_2O]{\text{1. } O_3} R\overset{\displaystyle O}{\overset{\|}{C}}OH + HO\overset{\displaystyle O}{\overset{\|}{C}}R'$$

$$CH_3CH_2CH_2CH_2C{\equiv}CH \xrightarrow[\text{2. } H_2O]{\text{1. } O_3} CH_3CH_2CH_2CH_2CO_2H + HO\overset{\displaystyle O}{\overset{\|}{C}}OH$$

1-Hexyne   Pentanoic acid (51%)   Carbonic acid

Recall that when carbonic acid is formed as a reaction product, it dissociates to carbon dioxide and water.

Ozonolysis is sometimes used as a tool in structure determination. By identifying the carboxylic acids produced, we can deduce the structure of the alkyne. As with many other chemical methods of structure determination, however, it has been superseded by spectroscopic methods.

PROBLEM 9.15 A certain hydrocarbon had the molecular formula $C_{16}H_{26}$ and contained two triple bonds. Ozonolysis gave $CH_3(CH_2)_4CO_2H$ and $HO_2CCH_2CH_2CO_2H$ as the only products. Suggest a reasonable structure for this hydrocarbon.

9.15   SUMMARY

Section 9.1   **Alkynes** are hydrocarbons that contain a carbon–carbon *triple bond*. Simple alkynes having no other functional groups or rings have the general formula C_nH_{2n-2}. Acetylene is the simplest alkyne.

Section 9.2   Alkynes are named in much the same way as alkenes, using the suffix *-yne* instead of *-ene*.

4,4-Dimethyl-2-pentyne

Section 9.3   The physical properties (boiling point, solubility in water, dipole moment) of alkynes resemble those of alkanes and alkenes.

Section 9.4   Acetylene is linear and alkynes have a linear geometry of their X—C≡C—Y units. The carbon–carbon triple bond in alkynes is composed of a σ and two π components. The triply bonded carbons are *sp*-hybridized. The σ component of the triple bond contains two electrons in an orbital generated by the overlap of *sp*-hybridized orbitals on adjacent carbons. Each of these carbons also has two $2p$ orbitals, which overlap in pairs so as to give two π orbitals, each of which contains two electrons.

Section 9.5   Acetylene and terminal alkynes are more *acidic* than other hydrocarbons. They have pK_a's of approximately 26, compared with about 45 for alkenes and about 60 for alkanes. Sodium amide is a strong enough base to remove a proton from acetylene or a terminal alkyne, but sodium hydroxide is not.

$$CH_3CH_2C\!\equiv\!CH \;+\; NaNH_2 \;\longrightarrow\; CH_3CH_2C\!\equiv\!CNa \;+\; NH_3$$

       1-Butyne   Sodium amide   Sodium 1-butynide   Ammonia

Sections 9.6–9.7   Table 9.2 summarizes the methods for preparing alkynes.

Section 9.8   Like alkenes, alkynes undergo addition reactions.

TABLE 9.2   Preparation of Alkynes

Reaction (section) and comments	General equation and specific example
Alkylation of acetylene and terminal alkynes (Section 9.6) The acidity of acetylene and terminal alkynes permits them to be converted to their conjugate bases on treatment with sodium amide. These anions are good nucleophiles and react with methyl and primary alkyl halides to form carbon–carbon bonds. Secondary and tertiary alkyl halides cannot be used, because they yield only elimination products under these conditions.	$RC{\equiv}CH$ + $NaNH_2$ $\longrightarrow$ $RC{\equiv}CNa$ + NH_3 Alkyne — Sodium amide — Sodium alkynide — Ammonia $RC{\equiv}CNa$ + $R'CH_2X$ $\longrightarrow$ $RC{\equiv}CCH_2R'$ + NaX Sodium alkynide — Primary alkyl halide — Alkyne — Sodium halide $(CH_3)_3CC{\equiv}CH$ $\xrightarrow[\text{2. CH}_3\text{I}]{\text{1. NaNH}_2,\ \text{NH}_3}$ $(CH_3)_3CC{\equiv}CCH_3$ 3,3-Dimethyl-1-butyne — 4,4-Dimethyl-2-pentyne (96%)
Double dehydrohalogenation of geminal dihalides (Section 9.7) An E2 elimination reaction of a geminal dihalide yields an alkenyl halide. If a strong enough base is used, sodium amide, for example, a second elimination step follows the first and the alkenyl halide is converted to an alkyne.	$RC(H)(X){-}CR'(H)(X)$ + $2NaNH_2$ $\longrightarrow$ $RC{\equiv}CR'$ + $2NaX$ Geminal dihalide — Sodium amide — Alkyne — Sodium halide $(CH_3)_3CCH_2CHCl_2$ $\xrightarrow[\text{2. H}_2\text{O}]{\text{1. 3NaNH}_2,\ \text{NH}_3}$ $(CH_3)_3CC{\equiv}CH$ 1,1-Dichloro-3,3-dimethylbutane — 3,3-Dimethyl-1-butyne (56–60%)
Double dehydrohalogenation of vicinal dihalides (Section 9.7) Dihalides in which the halogens are on adjacent carbons undergo two elimination processes analogous to those of geminal dihalides.	$RC(H)(X){-}CR'(H)(X)$ + $2NaNH_2$ $\longrightarrow$ $RC{\equiv}CR'$ + $2NaX$ Vicinal dihalide — Sodium amide — Alkyne — Sodium halide $CH_3CH_2CHBrCH_2Br$ $\xrightarrow[\text{2. H}_2\text{O}]{\text{1. 3NaNH}_2,\ \text{NH}_3}$ $CH_3CH_2C{\equiv}CH$ 1,2-Dibromobutane — 1-Butyne (78–85%)

Sections 9.9–9.10   Table 9.3 summarizes reactions that reduce alkynes to alkenes and alkanes.

Sections 9.11–9.13   Table 9.4 summarizes electrophilic addition to alkynes.

Section 9.14   Carbon–carbon triple bonds can be cleaved by ozonolysis. The cleavage products are carboxylic acids.

$$CH_3CH_2CH_2C{\equiv}CCH_3 \xrightarrow[\text{2. H}_2\text{O}]{\text{1. O}_3} CH_3CH_2CH_2\overset{\overset{\displaystyle O}{\|}}{C}OH + HO\overset{\overset{\displaystyle O}{\|}}{C}CH_3$$

2-Hexyne   Butanoic acid   Acetic acid

TABLE 9.3	Conversion of Alkynes to Alkenes and Alkanes

Reaction (section) and comments	General equation and specific example
Hydrogenation of alkynes to alkanes (Section 9.9) Alkynes are completely hydrogenated, yielding alkanes, in the presence of the customary metal hydrogenation catalysts.	$RC{\equiv}CR' + 2H_2 \xrightarrow[\text{catalyst}]{\text{metal}} RCH_2CH_2R'$ Alkyne Hydrogen Alkane 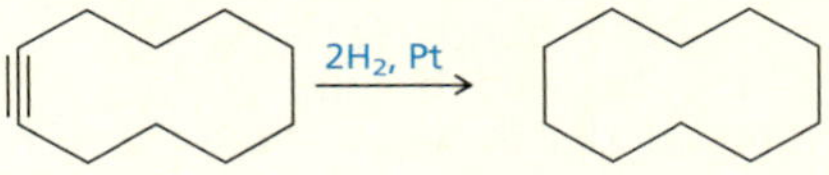 Cyclodecyne $\xrightarrow{2H_2,\ Pt}$ Cyclodecane (71%)
Hydrogenation of alkynes to alkenes (Section 9.9) Hydrogenation of alkynes may be halted at the alkene stage by using special catalysts. Lindlar palladium is the metal catalyst employed most often. Hydrogenation occurs with syn stereochemistry and yields a cis alkene.	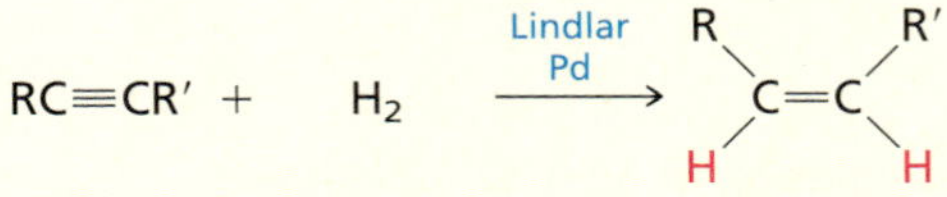 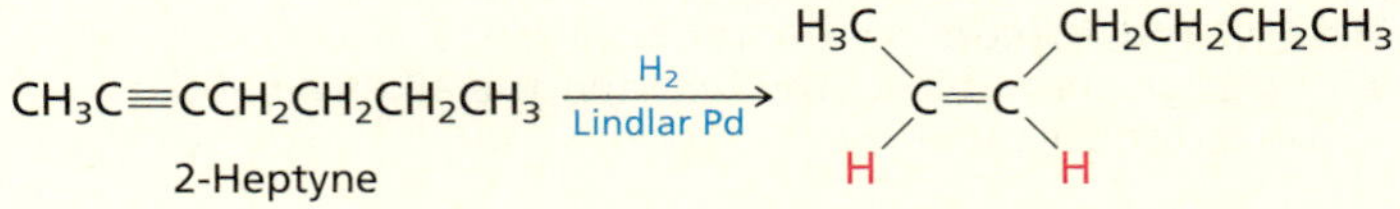 $RC{\equiv}CR' + H_2 \xrightarrow[\text{Pd}]{\text{Lindlar}}$ Cis alkene Alkyne Hydrogen $CH_3C{\equiv}CCH_2CH_2CH_3 \xrightarrow[\text{Lindlar Pd}]{H_2}$ 2-Heptyne cis-2-Heptene (59%)
Metal–ammonia reduction (Section 9.10) Group I metals—sodium is the one usually employed—in liquid ammonia as the solvent convert alkynes to trans alkenes. The reaction proceeds by a four-step sequence in which electron-transfer and proton-transfer steps alternate.	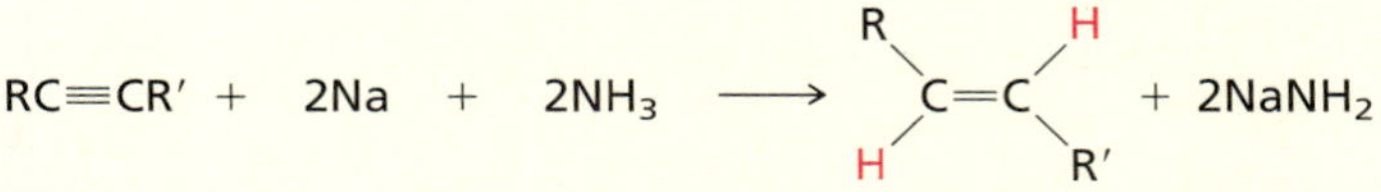 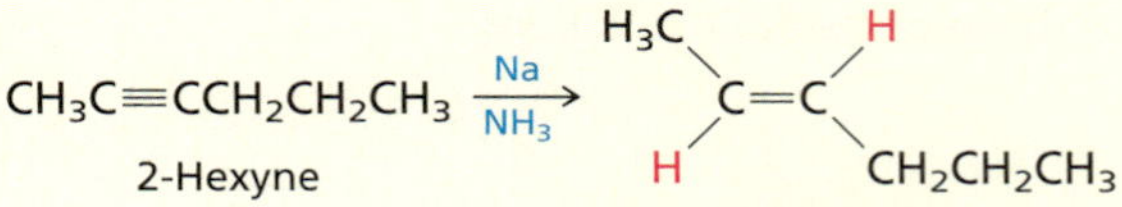 $RC{\equiv}CR' + 2Na + 2NH_3 \longrightarrow$ Trans alkene $+ 2NaNH_2$ Alkyne Sodium Ammonia Sodium amide $CH_3C{\equiv}CCH_2CH_2CH_3 \xrightarrow[\text{NH}_3]{\text{Na}}$ 2-Hexyne trans-2-Hexene (69%)

PROBLEMS

9.16 Write structural formulas and give the IUPAC names for all the alkynes of molecular formula C_6H_{10}.

9.17 Provide the IUPAC name for each of the following alkynes:

(a) $CH_3CH_2CH_2C{\equiv}CH$

(b) $CH_3CH_2C{\equiv}CCH_3$

(c) $CH_3C{\equiv}CCHCH(CH_3)_2$
                         |
                       CH_3

(d)  $\triangleright{-}CH_2CH_2CH_2C{\equiv}CH$

(e) 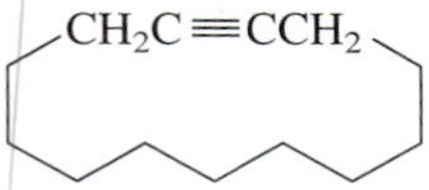 $CH_2C{\equiv}CCH_2$

TABLE 9.4   Electrophilic Addition to Alkynes

Reaction (section) and comments	General equation and specific example
Addition of hydrogen halides (Section 9.11) Hydrogen halides add to alkynes in accordance with Markovnikov's rule to give alkenyl halides. In the presence of 2 moles of hydrogen halide, a second addition occurs to give a geminal dihalide.	$RC{\equiv}CR' \xrightarrow{HX} RCH{=}CR' \xrightarrow{HX} RCH_2\overset{\displaystyle X}{\underset{\displaystyle X}{C}}R'$ Alkyne Alkenyl halide Geminal dihalide $CH_3C{\equiv}CH + 2HBr \longrightarrow CH_3\overset{\displaystyle Br}{\underset{\displaystyle Br}{C}}CH_3$ Propyne Hydrogen bromide 2,2-Dibromopropane (100%)
Acid-catalyzed hydration (Section 9.12) Water adds to the triple bond of alkynes to yield ketones by way of an unstable enol intermediate. The enol arises by Markovnikov hydration of the alkyne. Enol formation is followed by rapid isomerization of the enol to a ketone.	$RC{\equiv}CR' + H_2O \xrightarrow[Hg^{2+}]{H_2SO_4} RCH_2\overset{\displaystyle O}{C}R'$ Alkyne Water Ketone $HC{\equiv}CCH_2CH_2CH_2CH_3 + H_2O \xrightarrow[HgSO_4]{H_2SO_4} CH_3\overset{\displaystyle O}{C}CH_2CH_2CH_2CH_3$ 1-Hexyne Water 2-Hexanone (80%)
Halogenation (Section 9.13) Addition of 1 mole of chlorine or bromine to an alkyne yields a trans dihaloalkene. A tetrahalide is formed on addition of a second equivalent of the halogen.	$RC{\equiv}CR' \xrightarrow{X_2} \underset{X}{\overset{R}{C}}{=}\underset{R'}{\overset{X}{C}} \xrightarrow{X_2} R\overset{X\ X}{\underset{X\ X}{C-C}}R'$ Alkyne Dihaloalkene Tetrahaloalkane $CH_3C{\equiv}CH + 2Cl_2 \longrightarrow CH_3\overset{\displaystyle Cl}{\underset{\displaystyle Cl}{C}}CHCl_2$ Propyne Chlorine 1,1,2,2-Tetrachloropropane (63%)

(f) $CH_3CH_2CH_2CH_2\underset{\displaystyle C{\equiv}CCH_3}{CH}CH_2CH_2CH_2CH_3$

(g) $(CH_3)_3CC{\equiv}CC(CH_3)_3$

9.18 Write a structural formula or build a molecular model of each of the following:

(a) 1-Octyne

(b) 2-Octyne

(c) 3-Octyne

(d) 4-Octyne

(e) 2,5-Dimethyl-3-hexyne

(f) 4-Ethyl-1-hexyne

(g) Ethynylcyclohexane

(h) 3-Ethyl-3-methyl-1-pentyne

9.19 All the compounds in Problem 9.18 are isomers except one. Which one?

9.20 Write structural formulas for all the alkynes of molecular formula C_8H_{14} that yield 3-ethylhexane on catalytic hydrogenation.

9.21 An unknown acetylenic amino acid obtained from the seed of a tropical fruit has the molecular formula $C_7H_{11}NO_2$. On catalytic hydrogenation over platinum this amino acid yielded homoleucine (an amino acid of known structure shown here) as the only product. What is the structure of the unknown amino acid?

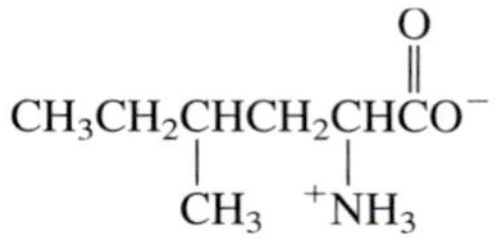

$$CH_3CH_2\underset{\underset{CH_3}{|}}{C}HCH_2\underset{\underset{+NH_3}{|}}{C}H\overset{\overset{O}{\|}}{C}O^-$$

Homoleucine

9.22 Show by writing appropriate chemical equations how each of the following compounds could be converted to 1-hexyne:

   (a) 1,1-Dichlorohexane   (c) Acetylene

   (b) 1-Hexene   (d) 1-Iodohexane

9.23 Show by writing appropriate chemical equations how each of the following compounds could be converted to 3-hexyne:

   (a) 1-Butene

   (b) 1,1-Dichlorobutane

   (c) Acetylene

9.24 When 1,2-dibromodecane was treated with potassium hydroxide in aqueous ethanol, it yielded a mixture of three isomeric compounds of molecular formula $C_{10}H_{19}Br$. Each of these compounds was converted to 1-decyne on reaction with sodium amide in dimethyl sulfoxide. Identify these three compounds.

9.25 Write the structure of the major organic product isolated from the reaction of 1-hexyne with

   (a) Hydrogen (2 mol), platinum

   (b) Hydrogen (1 mol), Lindlar palladium

   (c) Lithium in liquid ammonia

   (d) Sodium amide in liquid ammonia

   (e) Product in part (d) treated with 1-bromobutane

   (f) Product in part (d) treated with *tert*-butyl bromide

   (g) Hydrogen chloride (1 mol)

   (h) Hydrogen chloride (2 mol)

   (i) Chlorine (1 mol)

   (j) Chlorine (2 mol)

   (k) Aqueous sulfuric acid, mercury(II) sulfate

   (l) Ozone followed by hydrolysis

9.26 Write the structure of the major organic product isolated from the reaction of 3-hexyne with

(a) Hydrogen (2 mol), platinum

(b) Hydrogen (1 mol), Lindlar palladium

(c) Lithium in liquid ammonia

(d) Hydrogen chloride (1 mol)

(e) Hydrogen chloride (2 mol)

(f) Chlorine (1 mol)

(g) Chlorine (2 mol)

(h) Aqueous sulfuric acid, mercury(II) sulfate

(i) Ozone followed by hydrolysis

9.27 When 2-heptyne was treated with aqueous sulfuric acid containing mercury(II) sulfate, two products, each having the molecular formula $C_7H_{14}O$, were obtained in approximately equal amounts. What are these two compounds?

9.28 The alkane formed by hydrogenation of (S)-4-methyl-1-hexyne is optically active, but the one formed by hydrogenation of (S)-3-methyl-1-pentyne is not. Explain. Would you expect the products of hydrogenation of these two compounds in the presence of Lindlar palladium to be optically active?

9.29 All the following reactions have been described in the chemical literature and proceed in good yield. In some cases the reactants are more complicated than those we have so far encountered. Nevertheless, on the basis of what you have already learned, you should be able to predict the principal product in each case.

(a) $NaC\equiv CH + ClCH_2CH_2CH_2CH_2CH_2CH_2I \longrightarrow$

(b) 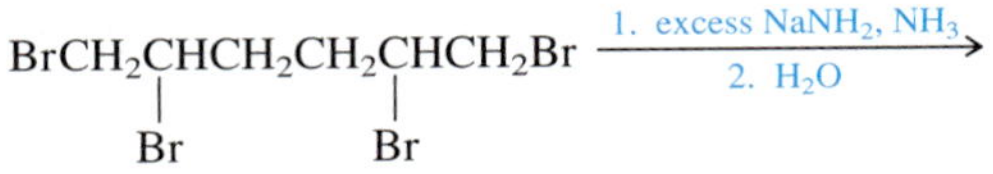 $BrCH_2CHCH_2CH_2CHCH_2Br$ $\xrightarrow[\text{2. } H_2O]{\text{1. excess NaNH}_2, \text{ NH}_3}$

with Br and Br substituents

(c) 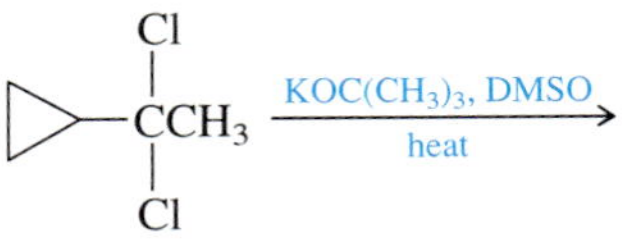 cyclopropyl–$\overset{Cl}{\underset{Cl}{\overset{|}{\underset{|}{C}}}}CH_3$ $\xrightarrow[\text{heat}]{\text{KOC(CH}_3)_3, \text{ DMSO}}$

(d) 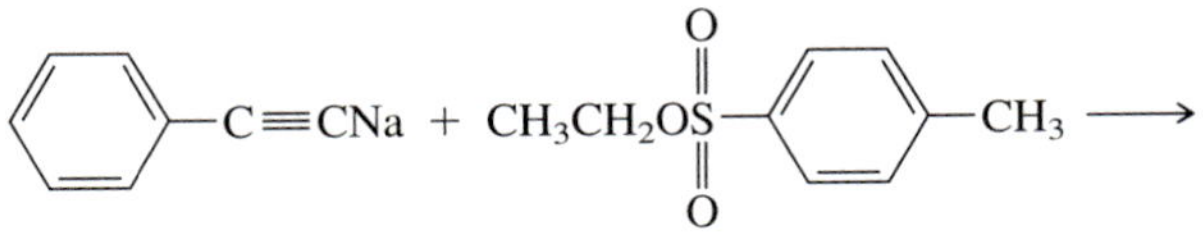 phenyl–$C\equiv CNa + CH_3CH_2O\overset{O}{\underset{O}{\overset{\|}{\underset{\|}{S}}}}$–phenyl–$CH_3 \longrightarrow$

(e) Cyclodecyne $\xrightarrow[\text{2. } H_2O]{\text{1. } O_3}$

(f) 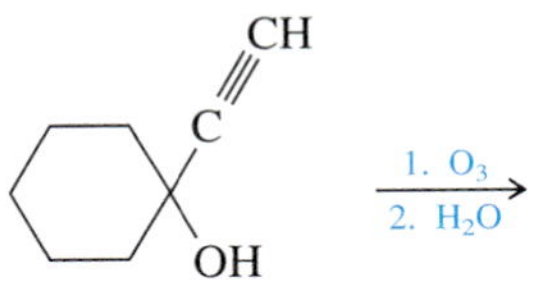 cyclohexane ring with $\overset{CH}{\underset{\|}{\overset{\|}{C}}}$ and OH substituents $\xrightarrow[\text{2. } H_2O]{\text{1. } O_3}$

(g) 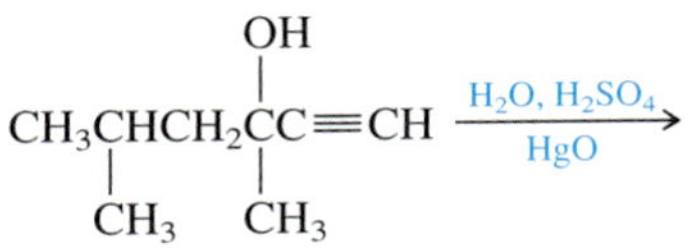 $CH_3CHCH_2\overset{OH}{\underset{CH_3}{\overset{|}{\underset{|}{C}}}}C\equiv CH$ $\xrightarrow[\text{HgO}]{H_2O, \text{ } H_2SO_4}$

with CH_3 substituent

(h) (Z)-$CH_3CH_2CH_2CH_2CH$=$CHCH_2(CH_2)_7C$≡CCH_2CH_2OH $\xrightarrow[\text{2. H}_2\text{O}]{\text{1. Na, NH}_3}$

(i) 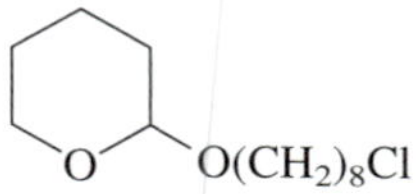 + NaC≡$CCH_2CH_2CH_2CH_3$ $\longrightarrow$

(j) Product of part (i) $\xrightarrow[\text{Lindlar Pd}]{\text{H}_2}$

9.30 (a) Oleic acid and stearic acid are naturally occurring compounds, which can be isolated from various fats and oils. In the laboratory, each can be prepared by hydrogenation of a compound known as *stearolic acid,* which has the formula $CH_3(CH_2)_7C$≡$C(CH_2)_7CO_2H$. Oleic acid is obtained by hydrogenation of stearolic acid over Lindlar palladium; stearic acid is obtained by hydrogenation over platinum. What are the structures of oleic acid and stearic acid?

(b) Sodium–ammonia reduction of stearolic acid yields a compound known as *elaidic acid.* What is the structure of elaidic acid?

9.31 The ketone 2-heptanone has been identified as contributing to the odor of a number of dairy products, including condensed milk and cheddar cheese. Describe a synthesis of 2-heptanone from acetylene and any necessary organic or inorganic reagents.

$$CH_3\overset{\overset{\displaystyle O}{\|}}{C}CH_2CH_2CH_2CH_2CH_3$$

2-Heptanone

9.32 (Z)-9-Tricosene $[(Z)$-$CH_3(CH_2)_7CH$=$CH(CH_2)_{12}CH_3]$ is the sex pheromone of the female housefly. Synthetic (Z)-9-tricosene is used as bait to lure male flies to traps that contain insecticide. Using acetylene and alcohols of your choice as starting materials, along with any necessary inorganic reagents, show how you could prepare (Z)-9-tricosene.

9.33 Show by writing a suitable series of equations how you could prepare each of the following compounds from the designated starting materials and any necessary organic or inorganic reagents:

(a) 2,2-Dibromopropane from 1,1-dibromopropane

(b) 2,2-Dibromopropane from 1,2-dibromopropane

(c) 1,1,2,2-Tetrachloropropane from 1,2-dichloropropane

(d) 2,2-Diiodobutane from acetylene and ethyl bromide

(e) 1-Hexene from 1-butene and acetylene

(f) Decane from 1-butene and acetylene

(g) Cyclopentadecyne from cyclopentadecene

(h) 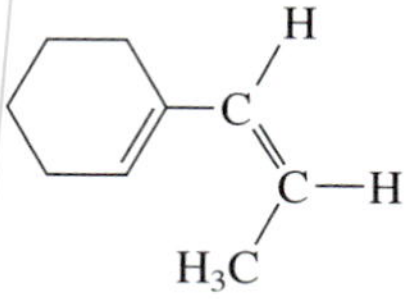 from —C≡CH and methyl bromide

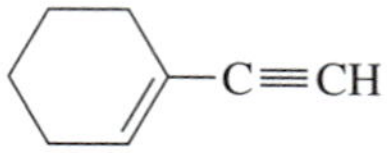

(i) *meso*-2,3-Dibromobutane from 2-butyne

9.34 Assume that you need to prepare 4-methyl-2-pentyne and discover that the only alkynes on hand are acetylene and propyne. You also have available methyl iodide, isopropyl bromide, and 1,1-dichloro-3-methylbutane. Which of these compounds would you choose in order to perform your synthesis, and how would you carry it out?

9.35 Compound A has the molecular formula $C_{14}H_{25}Br$ and was obtained by reaction of sodium acetylide with 1,12-dibromododecane. On treatment of compound A with sodium amide, it was converted to compound B ($C_{14}H_{24}$). Ozonolysis of compound B gave the diacid $HO_2C(CH_2)_{12}CO_2H$. Catalytic hydrogenation of compound B over Lindlar palladium gave compound C ($C_{14}H_{26}$), and hydrogenation over platinum gave compound D ($C_{14}H_{28}$). Sodium–ammonia reduction of compound B gave compound E ($C_{14}H_{26}$). Both C and E yielded $O=CH(CH_2)_{12}CH=O$ on ozonolysis. Assign structures to compounds A through E so as to be consistent with the observed transformations.

9.36 Use molecular models to compare $-C\equiv CH$, $-CH=CH_2$, and $-CH_2CH_3$ with respect to their preference for an equatorial orientation when attached to a cyclohexane ring. One of these groups is very much different from the other two. Which one? Why?

9.37 Try making a model of a hydrocarbon that contains three carbons, only one of which is *sp*-hybridized. What is its molecular formula? Is it an alkyne? What must be the hybridization state of the other two carbons? (You will learn more about compounds of this type in Chapter 10.)

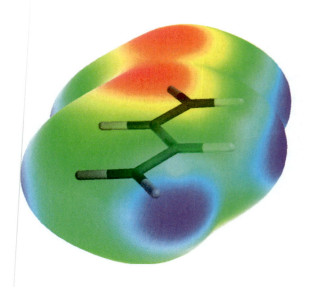

CHAPTER 10

CONJUGATION IN ALKADIENES AND ALLYLIC SYSTEMS

ot all the properties of alkenes are revealed by focusing exclusively on the functional group behavior of the double bond. A double bond can affect the properties of a second functional unit to which it is directly attached. It can be a substituent, for example, on a positively charged carbon in an **allylic carbocation,** or on a carbon that bears an unpaired electron in an **allylic free radical,** or it can be a substituent on a second double bond in a **conjugated diene.**

Allylic carbocation   Allylic free radical   Conjugated diene

Conjugare is a Latin verb meaning "to link or yoke together," and allylic carbocations, allylic free radicals, and conjugated dienes are all examples of **conjugated systems.** In this chapter we'll see how conjugation permits two functional units within a molecule to display a kind of reactivity that is qualitatively different from that of either unit alone.

10.1 THE ALLYL GROUP

The group $H_2C=CHCH_2-$ is known as **allyl**[*], which is both a common name and a permissible IUPAC name. It is most often encountered in functionally substituted derivatives, and compounds containing this group are much better known by their functional class IUPAC names than by their substitutive ones:

[*]*Allyl* is derived from the botanical name for garlic (*Allium sativum*). It was found in 1892 that the major component obtained by distilling garlic oil is $H_2C=CHCH_2SSCH_2CH=CH_2$, and the word *allyl* was coined for the $H_2C=CHCH_2-$ group on the basis of this origin.

$$H_2C=CHCH_2OH \qquad H_2C=CHCH_2Cl \qquad H_2C=CHCH_2Br$$

Allyl alcohol   Allyl chloride   Allyl bromide
(2-propen-1-ol)   (3-chloro-1-propene)   (3-bromo-1-propene)

The term *allylic* refers to a C=C—C unit. The singly bonded carbon is called the **allylic carbon,** and an **allylic substituent** is one that is attached to an allylic carbon. Conversely, doubly bonded carbons are called **vinylic carbons,** and substituents attached to either one of them are referred to as **vinylic substituents.**

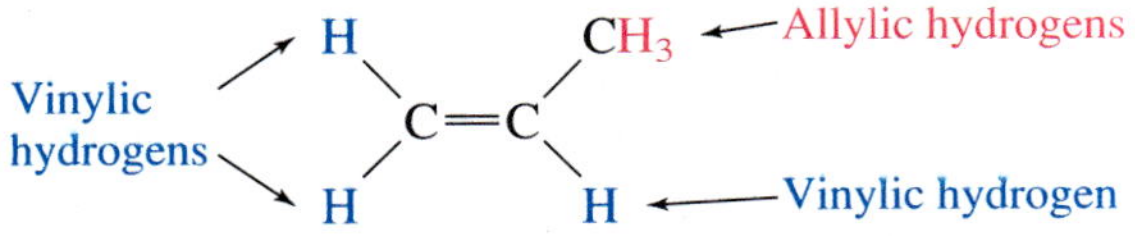

Allylic is often used as a general term for molecules that have a functional group at an allylic position. Thus, the following compounds represent an *allylic alcohol* and an *allylic chloride,* respectively.

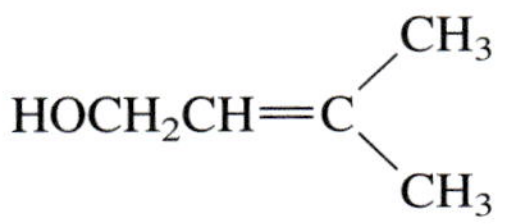

3-Methyl-2-buten-1-ol   3-Chloro-3-methyl-1-butene
(an allylic alcohol)   (an allylic chloride)

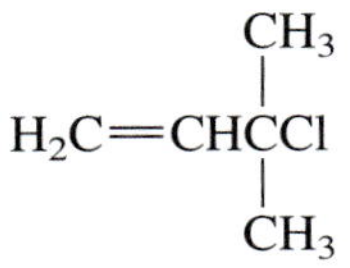

10.2 ALLYLIC CARBOCATIONS

Allylic carbocations are carbocations in which the positive charge is on an allylic carbon. Allyl cation is the simplest allylic carbocation.

Representative allylic carbocations

$$H_2C=CH-\overset{+}{C}H_2 \qquad CH_3CH=CH-\overset{+}{C}HCH_3$$

Allyl cation   1-Methyl-2-butenyl   2-Cyclopentenyl
               cation   cation

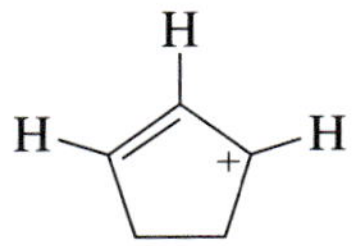

A substantial body of evidence indicates that allylic carbocations are more stable than simple alkyl cations. For example, the rate of solvolysis of a chloride that is both tertiary and allylic is much faster than that of a typical tertiary alkyl chloride.

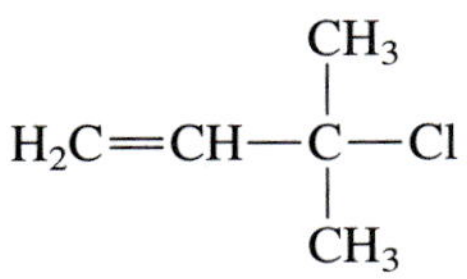

3-Chloro-3-methyl-1-butene   *tert*-Butyl chloride
More reactive: *k*(rel) 123   Less reactive: *k*(rel) 1.0

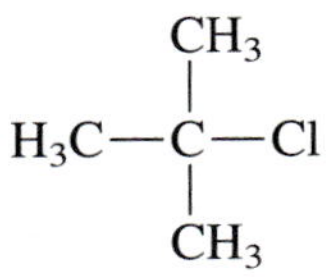

The first-order rate constant for ethanolysis of the allylic chloride 3-chloro-3-methyl-1-butene is over 100 times greater than that of *tert*-butyl chloride at the same temperature.

Both compounds react by an S_N1 mechanism, and their relative rates reflect their activation energies for carbocation formation. Because the allylic chloride is more reactive, we reason that it ionizes more rapidly because it forms a more stable carbocation. Structurally, the two carbocations differ in that the allylic carbocation has a vinyl substituent on its positively charged carbon in place of one of the methyl groups of *tert*-butyl cation.

$$H_2C=CH-\overset{+}{C}\underset{CH_3}{\overset{CH_3}{\Big<}} \qquad H_3C-\overset{+}{C}\underset{CH_3}{\overset{CH_3}{\Big<}}$$

1,1-Dimethylallyl cation   *tert*-Butyl cation
(more stable)   (less stable)

A vinyl group stabilizes a carbocation more than does a methyl group. Why?

A vinyl group is an extremely effective electron-releasing substituent. Resonance of the type shown delocalizes the π electrons of the double bond and disperses the positive charge.

> A rule of thumb is that a C=C substituent stabilizes a carbocation about as well as two methyl groups. Although allyl cation ($H_2C=CHCH_2^+$) is a primary carbocation, it is about as stable as a typical secondary carbocation such as isopropyl cation, $(CH_3)_2CH^+$.

$$H_2C=CH-\overset{+}{C}\underset{CH_3}{\overset{CH_3}{\Big<}} \quad\longleftrightarrow\quad H_2\overset{+}{C}-CH=C\underset{CH_3}{\overset{CH_3}{\Big<}}$$

It's important to recognize that the positive charge is shared by the two end carbons of the $C=C-C^+$ unit; the center carbon does not bear a positive charge in either of the resonance structures that we just wrote. Keep that fact in mind as you answer Problem 10.1.

PROBLEM 10.1 Write a second resonance structure for each of the following carbocations:

(a) $CH_3CH=\overset{+}{CH}\overset{+}{CH_2}$ (b) $H_2C=\overset{+}{C}\underset{CH_3}{\overset{|}{C}}H_2$ (c) ⬡=$C(CH_3)_2$

SAMPLE SOLUTION   (a) When writing resonance forms of allylic carbocations, electrons are moved in pairs from the double bond toward the positively charged carbon.

$$CH_3CH=CH-\overset{+}{CH_2} \quad\longleftrightarrow\quad CH_3\overset{+}{CH}-CH=CH_2$$

Electron delocalization in allylic carbocations can be indicated using a dashed line to show the sharing of a pair of π electrons by the three carbons. The structural formula is completed by placing a positive charge above the dashed line or by adding partial positive charges to the carbons at the end of the allylic system.

$$\underset{H}{\overset{H}{\underset{|}{\big|}}}\!C\cdots\overset{+}{C}\cdots C\underset{CH_3}{\overset{CH_3}{\Big<}} \qquad \text{or} \qquad \underset{H}{\overset{H}{\underset{|}{\big|}}}\!\overset{\delta+}{C}\cdots C\cdots \overset{+\delta}{C}\underset{CH_3}{\overset{CH_3}{\Big<}}$$

Dashed-line representations of 1,1-dimethylallyl cation

In the case of $H_2C=CH-CH_2^+$ both the terminal carbons are equivalently substituted, and so each bears exactly half of a unit positive charge.

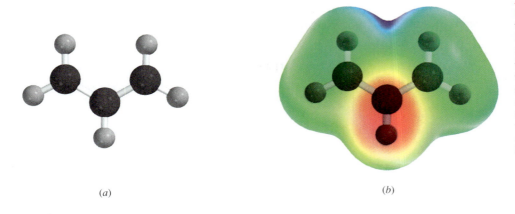

(*a*)   (*b*)

Allyl cation

Figure 10.1*a* shows the geometry of $H_2C{=}CH{-}\overset{+}{C}H_2$; the corresponding electrostatic potential map [Figure 10.1*b*] illustrates the sharing of the positive charge between the first and third carbons. This sharing of positive charge is responsible for the characteristic properties of allyl cation.

An orbital overlap description of electron delocalization in 1,1-dimethylallyl cation $H_2C{=}CH{-}\overset{+}{C}(CH_3)_2$ is given in Figure 10.2. Figure 10.2*a* shows the π bond and the vacant *p* orbital as independent units. Figure 10.2*b* shows how the units can overlap to give an extended π orbital that encompasses all three carbons. This permits the two π electrons to be delocalized over three carbons and disperses the positive charge.

Because the positive charge in an allylic carbocation is shared by two carbons, there are two potential sites for attack by a nucleophile. Thus, hydrolysis of 3-chloro-3-methyl-1-butene gives a mixture of two allylic alcohols:

$$(CH_3)_2CCH{=}CH_2 \xrightarrow[Na_2CO_3]{H_2O} (CH_3)_2CCH{=}CH_2 + (CH_3)_2C{=}CHCH_2OH$$
$$\quad\quad\;\; | \quad\quad\quad\quad\quad\quad\quad\quad\quad\quad\quad | $$
$$\quad\quad\;\; Cl \quad\quad\quad\quad\quad\quad\quad\quad\quad\quad OH$$

3-Chloro-3-methyl-   2-Methyl-3-buten-2-ol   3-Methyl-2-buten-1-ol
1-butene   (85%)   (15%)

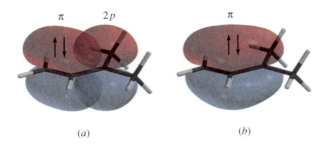

π   2*p*   π

(*a*)   (*b*)

FIGURE 10.2 Electron delocalization in an allylic carbocation. (*a*) The π orbital of the double bond, and the vacant 2*p* orbital of the positively charged carbon. (*b*) Overlap of the π orbital and the 2*p* orbital gives an extended π orbital that encompasses all three carbons. The two electrons in the π bond are delocalized over two carbons in part (*a*) and over three carbons in part (*b*).

Both alcohols are formed from the same carbocation. Water may react with the carbocation to give either a primary alcohol or a tertiary alcohol.

Use *Learning By Modeling* to view the carbocation represented by resonance structures A and B. How is the positive charge distributed among its carbons?

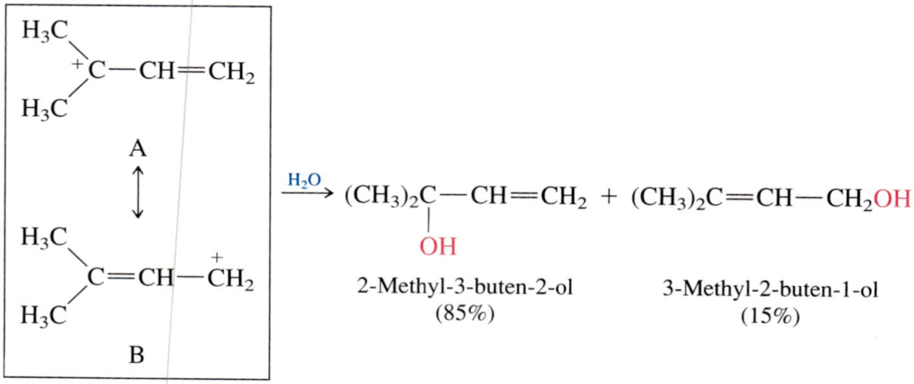

$$
\begin{array}{c}
H_3C \\
{}^{+}\!C\!-\!CH\!=\!CH_2 \\
H_3C
\end{array}
\quad A
$$

$$
\updownarrow
$$

$$
\begin{array}{c}
H_3C \\
C\!=\!CH\!-\!\overset{+}{C}H_2 \\
H_3C
\end{array}
\quad B
$$

$$
\xrightarrow{H_2O} \quad (CH_3)_2C\!-\!CH\!=\!CH_2 \;+\; (CH_3)_2C\!=\!CH\!-\!CH_2OH
$$

$$
\underset{OH}{|}
$$

2-Methyl-3-buten-2-ol (85%)   3-Methyl-2-buten-1-ol (15%)

It must be emphasized that we are not dealing with an equilibrium between two isomeric carbocations. *There is only one carbocation.* Its structure is not adequately represented by either of the individual resonance forms but is a hybrid having qualities of both of them. The carbocation has more of the character of A than B because resonance structure A is more stable than B. Water attacks faster at the tertiary carbon because it bears a greater share of the positive charge.

The same two alcohols are formed in the hydrolysis of 1-chloro-3-methyl-2-butene:

$$
(CH_3)_2C\!=\!CHCH_2Cl \xrightarrow[Na_2CO_3]{H_2O} (CH_3)_2CCH\!=\!CH_2 \;+\; (CH_3)_2C\!=\!CHCH_2OH
$$

$$
\underset{OH}{|}
$$

1-Chloro-3-methyl-2-butene   2-Methyl-3-buten-2-ol (85%)   3-Methyl-2-buten-1-ol (15%)

The carbocation formed on ionization of 1-chloro-3-methyl-2-butene is the same allylic carbocation as the one formed on ionization of 3-chloro-3-methyl-1-butene and gives the same mixture of products.

Reactions of allylic systems that yield products in which double-bond migration has occurred are said to have proceeded with **allylic rearrangement,** or by way of an **allylic shift.**

PROBLEM 10.2   From among the following compounds, choose the two that yield the same carbocation on ionization.

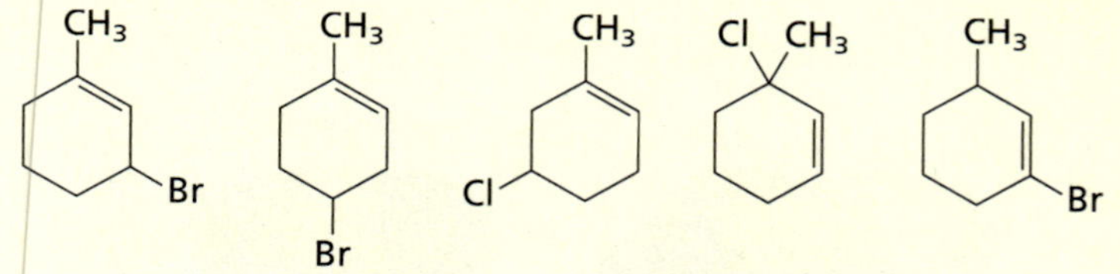

Later in this chapter we'll see how allylic carbocations are involved in electrophilic addition to dienes and how the principles developed in this section apply there as well.

10.3   ALLYLIC FREE RADICALS

Just as allyl cation is stabilized by electron delocalization, so is allyl radical:

$$H_2C\!\!=\!\!CH\!-\!\dot{C}H_2 \longleftrightarrow H_2\dot{C}\!-\!CH\!\!=\!\!CH_2 \quad \text{or}$$

Allyl radical

Allyl radical is a conjugated system in which three electrons are delocalized over three carbons. The resonance structures indicate that the unpaired electron has an equal probability of being found at C-1 or C-3. C-2 shares none of the unpaired electron.

Figure 10.3 shows another way of representing electron delocalization in free radicals. **Spin density** is a measure of the unpaired electron distribution in a molecule. In Figure 10.3a we see that, except for a small amount of spin density on the hydrogen at C-2, C-1 and C-3 share equally all of the spin density of allyl radical. Looking at allyl radical from a different direction in Figure 10.3b shows that the region of space where the spin density is greatest corresponds to the 2p orbitals of C-1 and C-3 that are part of the allylic π electron system.

The degree to which allylic radicals are stabilized by delocalization of the unpaired electron causes reactions that generate them to proceed more readily than those that give simple alkyl radicals. Compare, for example, the bond dissociation energies of the primary C—H bonds of propane and propene:

$$CH_3CH_2CH_2\!:\!H \longrightarrow CH_3CH_2\dot{C}H_2 + \quad \cdot H \qquad \Delta H° = +410 \text{ kJ } (+98 \text{ kcal})$$

| Propane | Propyl radical | Hydrogen atom |

$$H_2C\!\!=\!\!CHCH_2\!:\!H \longrightarrow H_2C\!\!=\!\!CH\dot{C}H_2 + \quad \cdot H \qquad \Delta H° = +368 \text{ kJ } (+88 \text{ kcal})$$

| Propene | Allyl radical | Hydrogen atom |

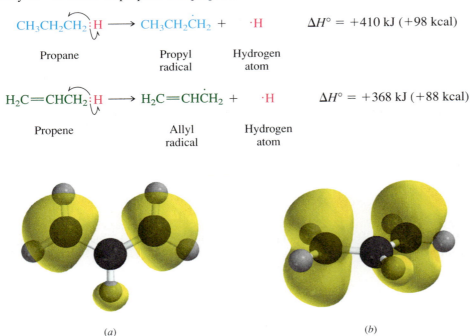

(a)   (b)

FIGURE 10.3 (a) The spin density (yellow) in allyl radical is equally divided between the two allylic carbons. There is a much smaller spin density at C-2 hydrogen. (b) The odd electron is in an orbital that is part of the allylic π system.

Breaking a bond to a primary hydrogen atom in propene requires less energy, by 42 kJ/mol (10 kcal/mol), than in propane. The free radical produced from propene is allylic and stabilized by electron delocalization; the one from propane is not.

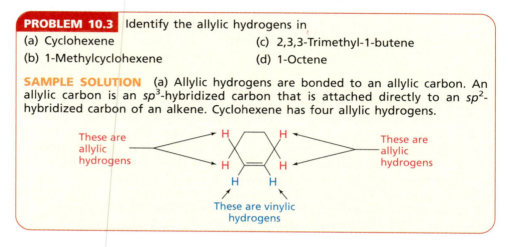

PROBLEM 10.3 Identify the allylic hydrogens in
(a) Cyclohexene   (c) 2,3,3-Trimethyl-1-butene
(b) 1-Methylcyclohexene   (d) 1-Octene

SAMPLE SOLUTION (a) Allylic hydrogens are bonded to an allylic carbon. An allylic carbon is an sp^3-hybridized carbon that is attached directly to an sp^2-hybridized carbon of an alkene. Cyclohexene has four allylic hydrogens.

These are allylic hydrogens

These are allylic hydrogens

These are vinylic hydrogens

10.4   ALLYLIC HALOGENATION

Of the reactions that involve carbon radicals, the most familiar are the chlorination and bromination of alkanes (Sections 4.14 through 4.18):

$$RH \; + \; X_2 \; \xrightarrow{\text{heat or light}} \; RX \; + \; HX$$

Alkane   Halogen   Alkyl   Hydrogen
                            halide   halide

Although alkenes typically react with chlorine and bromine by *addition* at room temperature and below (Section 6.14), *substitution* becomes competitive at higher temperatures, especially when the concentration of the halogen is low. When substitution does occur, it is highly selective for the allylic position. This forms the basis of an industrial preparation of allyl chloride:

$$H_2C{=}CHCH_3 \; + \; Cl_2 \; \xrightarrow{500°C} \; H_2C{=}CHCH_2Cl \; + \; HCl$$

Propene   Chlorine   Allyl chloride   Hydrogen chloride
                                   (80–85%)

The reaction proceeds by a free-radical chain mechanism, involving the following propagation steps:

$$H_2C{=}CHCH_2{:}H \; + \; \overset{\cdot\cdot}{\underset{\cdot\cdot}{Cl}}{:} \; \longrightarrow \; H_2C{=}CH\overset{\cdot}{C}H_2 \; + \; H{:}\overset{\cdot\cdot}{\underset{\cdot\cdot}{Cl}}{:}$$

Propene   Chlorine atom   Allyl radical   Hydrogen chloride

$$H_2C{=}CH\overset{\cdot}{C}H_2 \; + \; {:}\overset{\cdot\cdot}{\underset{\cdot\cdot}{Cl}}{:}\overset{\cdot\cdot}{\underset{\cdot\cdot}{Cl}}{:} \; \longrightarrow \; H_2C{=}CHCH_2\overset{\cdot\cdot}{\underset{\cdot\cdot}{Cl}}{:} \; + \; \cdot\overset{\cdot\cdot}{\underset{\cdot\cdot}{Cl}}{:}$$

Allyl radical   Chlorine   Allyl chloride   Chlorine atom

Can you write an equation for the initiation step that precedes these propagation steps?

Allyl chloride is quite reactive toward nucleophilic substitutions, especially those that proceed by the S_N2 mechanism, and is used as a starting material in the synthesis of a variety of drugs and agricultural and industrial chemicals.

Allylic brominations are normally carried out using one of a number of specialized reagents developed for that purpose. *N*-Bromosuccinimide (NBS) is the most frequently used of these reagents. An alkene is dissolved in carbon tetrachloride, *N*-bromosuccinimide is added, and the reaction mixture is heated, illuminated with a sunlamp, or both. The products are an allylic halide and succinimide.

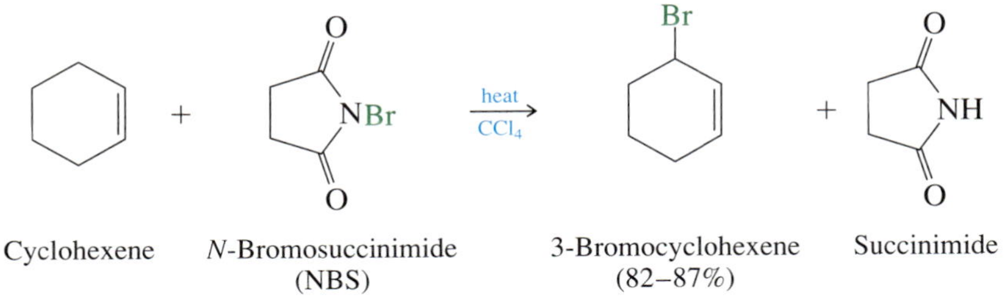

| Cyclohexene | *N*-Bromosuccinimide (NBS) | | 3-Bromocyclohexene (82–87%) | Succinimide |

N-Bromosuccinimide will be seen again as a reagent for selective bromination in Section 11.12.

N-Bromosuccinimide provides a low concentration of molecular bromine, which reacts with alkenes by a mechanism analogous to that of other free-radical halogenations.

PROBLEM 10.4 Assume that *N*-bromosuccinimide serves as a source of Br_2, and write equations for the propagation steps in the formation of 3-bromocyclohexene by allylic bromination of cyclohexene.

Although allylic brominations and chlorinations offer a method for attaching a reactive functional group to a hydrocarbon framework, we need to be aware of two important limitations. For allylic halogenation to be effective in a particular synthesis:

1. All the allylic hydrogens in the starting alkene must be equivalent.
2. Both resonance forms of the allylic radical must be equivalent.

In the two examples cited so far, the chlorination of propene and the bromination of cyclohexene, both criteria are met.

All the allylic hydrogens of propene are equivalent.

$$H_2C\!=\!CH\!-\!CH_3$$

The two resonance forms of allyl radical are equivalent.

$$H_2C\!=\!CH\!-\!\overset{\cdot}{C}H_2 \longleftrightarrow H_2\overset{\cdot}{C}\!-\!CH\!=\!CH_2$$

All the allylic hydrogens of cyclohexene are equivalent.

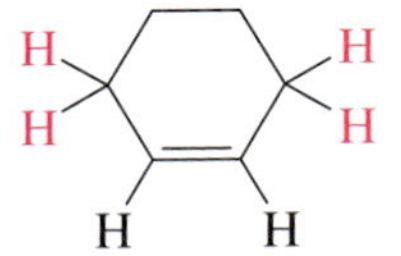

The two resonance forms of 2-cyclohexenyl radical are equivalent.

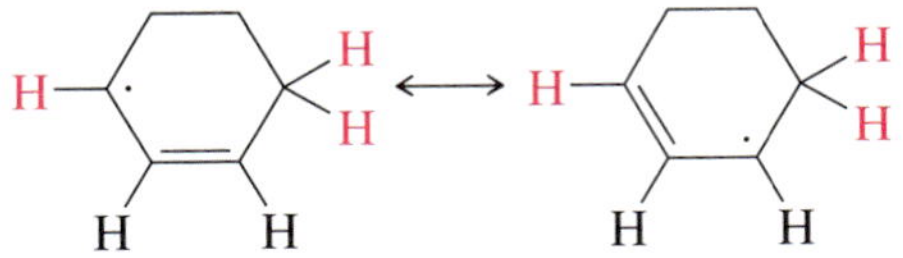

Unless both criteria are met, mixtures of constitutionally isomeric allylic halides result.

PROBLEM 10.5 The two alkenes 2,3,3-trimethyl-1-butene and 1-octene were each subjected to allylic halogenation with *N*-bromosuccinimide. One of these alkenes yielded a single allylic bromide, whereas the other gave a mixture of two constitutionally isomeric allylic bromides. Match the chemical behavior to the correct alkene and give the structure of the allylic bromide(s) formed from each.

10.5   CLASSES OF DIENES

Allylic carbocations and allylic radicals are conjugated systems involved as reactive intermediates in chemical reactions. The third type of conjugated system that we will examine, **conjugated dienes,** consists of stable molecules.

A hydrocarbon that contains two double bonds is called an **alkadiene,** and the relationship between the double bonds may be described as *isolated, conjugated,* or *cumulated.* **Isolated diene** units are those in which two carbon–carbon double bond units are separated from each other by one or more sp^3-hybridized carbon atoms. 1,4-Pentadiene and 1,5-cyclooctadiene have isolated double bonds:

$$H_2C=CHCH_2CH=CH_2$$

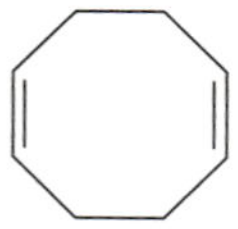

1,4-Pentadiene   1,5-Cyclooctadiene

Conjugated dienes are those in which two carbon–carbon double bond units are connected to each other by a single bond. 1,3-Pentadiene and 1,3-cyclooctadiene contain conjugated double bonds:

$$H_2C=CH-CH=CHCH_3$$

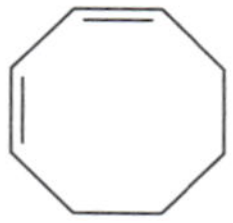

1,3-Pentadiene   1,3-Cyclooctadiene

PROBLEM 10.6 Are the double bonds in 1,4-cyclooctadiene isolated or conjugated?

Cumulated dienes are those in which one carbon atom is common to two carbon–carbon double bonds. The simplest cumulated diene is 1,2-propadiene, also called *allene,* and compounds of this class are generally referred to as *allenes.*

Allene is an acceptable IUPAC name for 1,2-propadiene.

$$H_2C=C=CH_2$$

1,2-Propadiene

PROBLEM 10.7 Many naturally occurring substances contain several carbon–carbon double bonds: some isolated, some conjugated, and some cumulated. Identify the types of carbon–carbon double bonds found in each of the following substances:

(a) β-Springene (a scent substance from the dorsal gland of springboks)

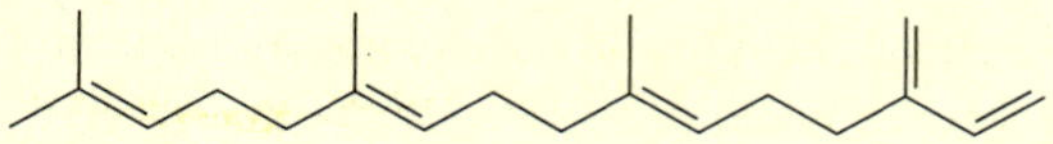

(b) Cembrene (occurs in pine resin)

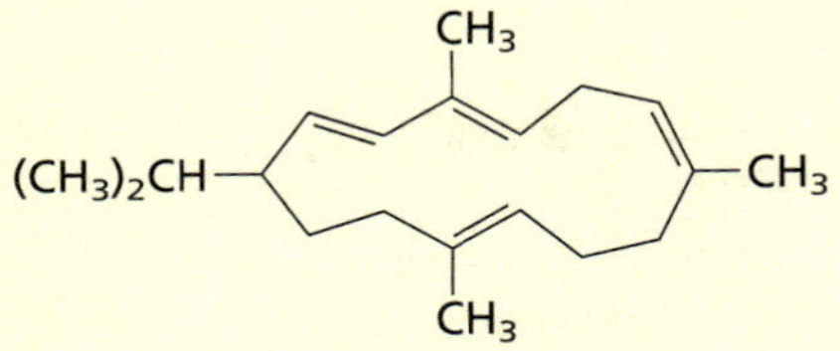

(c) The sex attractant of the male dried-bean beetle

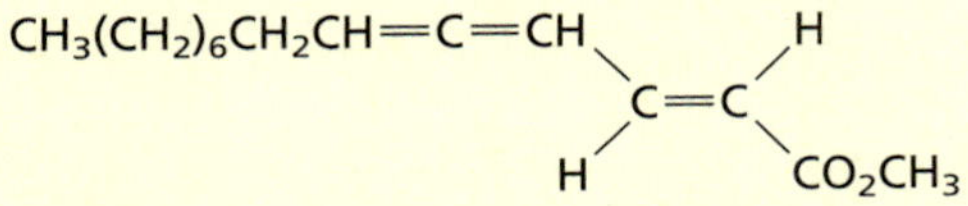

SAMPLE SOLUTION   (a) β-Springene has three isolated double bonds and a pair of conjugated double bonds:

Conjugated double bonds

Isolated double bonds

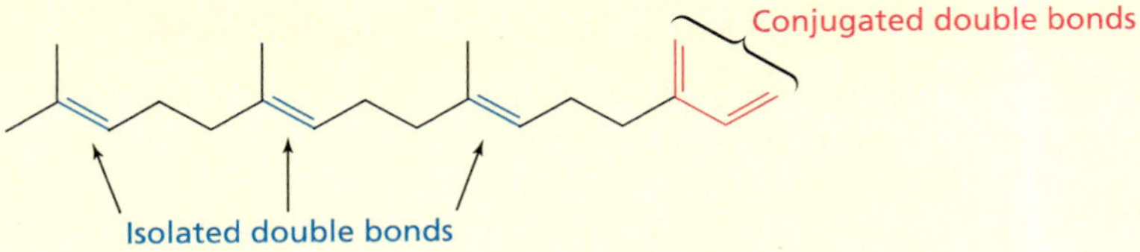

Isolated double bonds are separated from other double bonds by at least one sp^3-hybridized carbon. Conjugated double bonds are joined by a single bond.

Alkadienes are named according to the IUPAC rules by replacing the -*ane* ending of an alkane with -*adiene* and locating the position of each double bond by number. Compounds with three carbon–carbon double bonds are called *alkatrienes* and named accordingly, those with four double bonds are *alkatetraenes,* and so on.

10.6   RELATIVE STABILITIES OF DIENES

Which is the most stable arrangement of double bonds in an alkadiene—isolated, conjugated, or cumulated?

As we saw in Chapter 6, the stabilities of alkenes may be assessed by comparing their heats of hydrogenation. Figure 10.4 depicts the heats of hydrogenation of an isolated diene (1,4-pentadiene) and a conjugated diene (1,3-pentadiene), along with the alkenes 1-pentene and (*E*)-2-pentene. The figure shows that an isolated pair of double bonds behaves much like two independent alkene units. The measured heat of hydrogenation of the two double bonds in 1,4-pentadiene is 252 kJ/mol (60.2 kcal/mol), exactly twice the heat of hydrogenation of 1-pentene. Furthermore, the heat evolved on hydrogenation of each double bond must be 126 kJ/mol (30.1 kcal/mol) because 1-pentene is an intermediate in the hydrogenation of 1,4-pentadiene to pentane.

By the same reasoning, hydrogenation of the terminal double bond in the conjugated diene (*E*)-1,3-pentadiene releases only 111 kJ/mol (26.5 kcal/mol) when it is hydrogenated to (*E*)-2-pentene. Hydrogenation of the terminal double bond in the conjugated diene evolves 15 kJ/mol (3.6 kcal/mol) less heat than hydrogenation of a terminal double bond in the diene with isolated double bonds. *A conjugated double bond is 15 kJ/mol (3.6 kcal/mol) more stable than an isolated double bond.* We call this increased stability due to conjugation the **delocalization energy, resonance energy,** or **conjugation energy.**

FIGURE 10.4 Heats of hydrogenation are used to assess the stabilities of isolated versus conjugated double bonds. Comparing the measured heats of hydrogenation (solid lines) of the four compounds shown gives the values shown by the dashed lines for the heats of hydrogenation of the terminal double bond of 1,4-pentadiene and (E)-1,3-pentadiene. A conjugated double bond is approximately 15 kJ/mol more stable than an isolated double bond.

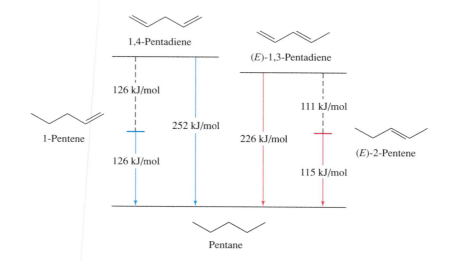

The cumulated double bonds of an allenic system are of relatively high energy. The heat of hydrogenation of allene is more than twice that of propene.

$$H_2C=C=CH_2 + 2H_2 \longrightarrow CH_3CH_2CH_3 \qquad \Delta H° = -295 \text{ kJ } (-70.5 \text{ kcal})$$

Allene $\qquad$ Hydrogen $\qquad$ Propane

$$CH_3CH=CH_2 + H_2 \longrightarrow CH_3CH_2CH_3 \qquad \Delta H° = -125 \text{ kJ } (-29.9 \text{ kcal})$$

Propene $\qquad$ Hydrogen $\qquad$ Propane

PROBLEM 10.8 Another way in which energies of isomers may be compared is by their heats of combustion. Match the heat of combustion with the appropriate diene.

Dienes: 1,2-Pentadiene, (E)-1,3-pentadiene, 1,4-pentadiene
Heats of combustion: 3186 kJ/mol, 3217 kJ/mol, 3251 kJ/mol

Thus, the order of alkadiene stability decreases in the order: conjugated diene (most stable) → isolated diene → cumulated diene (least stable). To understand this ranking, we need to look at structure and bonding in alkadienes in more detail.

10.7 BONDING IN CONJUGATED DIENES

At 146 pm the C-2—C-3 distance in 1,3-butadiene is relatively short for a carbon–carbon single bond. This is most reasonably seen as a hybridization effect. In ethane both carbons are sp^3-hybridized and are separated by a distance of 153 pm. The carbon–carbon single bond in propene unites sp^3- and sp^2-hybridized carbons and is shorter than that of ethane. Both C-2 and C-3 are sp^2-hybridized in 1,3-butadiene, and a decrease in bond distance between them reflects the tendency of carbon to attract electrons more strongly as its s character increases.

$$sp^3 \quad sp^3 \qquad\qquad sp^3 \quad sp^2 \qquad\qquad sp^2 \qquad sp^2$$
$$H_3C{-}CH_3 \qquad\qquad H_3C{-}CH{=}CH_2 \qquad\qquad H_2C{=}CH{-}CH{=}CH_2$$

153 pm $\qquad\qquad\qquad$ 151 pm $\qquad\qquad\qquad$ 146 pm

The factor most responsible for the increased stability of conjugated double bonds is the greater delocalization of their π electrons compared with the π electrons of isolated double bonds. As shown in Figure 10.5a, the π electrons of an isolated diene system occupy, in pairs, two noninteracting π orbitals. Each of these π orbitals encompasses two carbon atoms. An sp^3-hybridized carbon isolates the two π orbitals from each other, preventing the exchange of electrons between them. In a conjugated diene, however, mutual overlap of the two π orbitals, represented in Figure 10.5b, gives an orbital system in which each π electron is delocalized over four carbon atoms. Delocalization of electrons lowers their energy and gives a more stable molecule.

Additional evidence for electron delocalization in 1,3-butadiene can be obtained by considering its conformations. Overlap of the two π electron systems is optimal when the four carbon atoms are coplanar. Two conformations allow this coplanarity: they are called the *s*-cis and *s*-trans conformations.

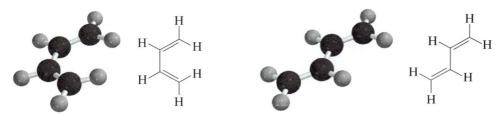

s-Cis conformation of 1,3-butadiene   s-Trans conformation of 1,3-butadiene

The letter *s* in *s*-cis and *s*-trans refers to conformations around the C—C single bond in the diene. The *s*-trans conformation of 1,3-butadiene is 12 kJ/mol (2.8 kcal/mol) more stable than the *s*-cis, which is destabilized by van der Waals strain between the hydrogens at C-1 and C-4.

The *s*-cis and *s*-trans conformations of 1,3-butadiene interconvert by rotation around the C-2—C-3 bond, as illustrated in Figure 10.6. The conformation at the midpoint of this rotation, the *perpendicular conformation*, has its 2p orbitals in a geometry that prevents extended conjugation. It has localized double bonds. This is an example of a stereoelectronic effect (Section 5.16); electron delocalization in a conjugated system is most effective when the interacting orbitals are suitably aligned. In the case of conjugated dienes, the most favorable alignment occurs when the axes of the π orbitals are parallel. The main contributor to the energy of activation for rotation about the single bond in 1,3-butadiene is the decrease in electron delocalization that accompanies conversion of the *s*-cis or *s*-trans conformation to the perpendicular conformation.

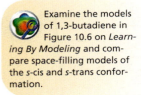

Examine the models of 1,3-butadiene in Figure 10.6 on *Learning By Modeling* and compare space-filling models of the *s*-cis and *s*-trans conformation.

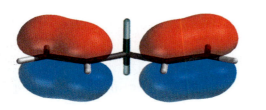

(*a*) Isolated double bonds   (*b*) Conjugated double bonds

FIGURE 10.5 (*a*) Isolated double bonds are separated from one another by one or more sp^3-hybridized carbons and cannot overlap to give an extended π orbital. (*b*) In a conjugated diene, overlap of two π orbitals gives an extended π system encompassing four carbon atoms.

FIGURE 10.6 Conformations and electron delocalization in 1,3-butadiene. The *s*-cis and the *s*-trans conformations permit the 2*p* orbitals to be aligned parallel to one another for maximum π electron delocalization. The *s*-trans conformation is more stable than the *s*-cis. Stabilization resulting from π electron delocalization is least in the perpendicular conformation, which is a transition state for rotation about the C-2—C-3 single bond. The green and yellow colors are meant to differentiate the orbitals and do not indicate their phases.

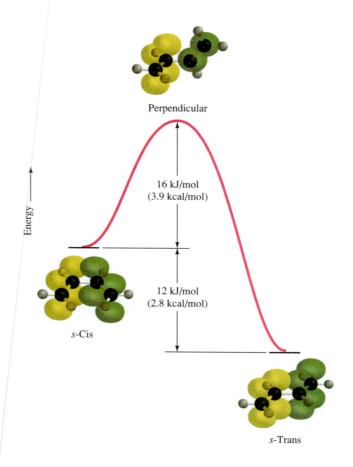

Perpendicular

16 kJ/mol
(3.9 kcal/mol)

12 kJ/mol
(2.8 kcal/mol)

s-Cis

s-Trans

Energy

10.8   BONDING IN ALLENES

The three carbons of allene lie in a straight line, with relatively short carbon–carbon bond distances of 131 pm. The central carbon, because it bears only two substituents, is *sp*-hybridized. The terminal carbons of allene are *sp²*-hybridized.

$$\underset{\text{Allene}}{\overset{\displaystyle \underset{H}{\overset{H}{\diagdown}}\,C{=}C{=}CH_2}{}}$$

118.4°

sp   *sp²*

108 pm   131 pm

Allene

Structural studies show allene to be nonplanar. As Figure 10.7 illustrates, the plane of one HCH unit is perpendicular to the plane of the other. Figure 10.7 also portrays the reason for the molecular geometry of allene. The 2*p* orbital of each of the terminal carbons overlaps with a different 2*p* orbital of the central carbon. Because the 2*p* orbitals of the central carbon are perpendicular to each other, the perpendicular nature of the two HCH units follows naturally.

(a) Planes defined by H(C-1)H and H(C-3)H are mutually perpendicular.

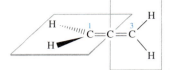

FIGURE 10.7 Bonding and geometry in 1,2-propa-diene (allene). The green and yellow colors are meant to differentiate the orbitals and do not indicate their phases.

(b) The p orbital of C-1 and one of the p orbitals of C-2 can overlap so as to participate in π bonding.

(c) The p orbital of C-3 and one of the p orbitals of C-2 can overlap so as to participate in a second π orbital perpendicular to the one in (b).

(d) Allene is a nonplanar molecule characterized by a linear carbon chain and two mutually perpendicular π bonds.

The nonplanarity of allenes has an interesting stereochemical consequence. 1,3-Disubstituted allenes are chiral; they are not superimposable on their mirror images. Even an allene as simple as 2,3-pentadiene ($CH_3CH=C=CHCH_3$) has been obtained as separate enantiomers.

Examine models of both enantiomers of 2,3-pentadiene to verify that they are non-superimposable.

(+)-2,3-Pentadiene   (−)-2,3-Pentadiene

The enantiomers shown are related as a right-hand and left-hand screw, respectively.

Chiral allenes are examples of a small group of molecules that are chiral, but don't have a chirality center. What they do have is a **chirality axis,** which in the case of 2,3-pentadiene is a line passing through the three carbons of the allene unit (carbons 2, 3, and 4).

The Cahn–Ingold–Prelog *R–S* notation has been extended to chiral allenes and other molecules that have a chirality axis. Such compounds are so infrequently encountered, however, we will not cover the rules for specifying their stereochemistry in this text.

PROBLEM 10.9 Is 2-methyl-2,3-pentadiene chiral? What about 2-chloro-2,3-pentadiene?

Because of the linear geometry required of cumulated dienes, cyclic allenes, like cycloalkynes, are strained unless the rings are fairly large. 1,2-Cyclononadiene is the smallest cyclic allene that is sufficiently stable to be isolated and stored conveniently.

10.9 PREPARATION OF DIENES

The use of 1,3-butadiene in the preparation of synthetic rubber is discussed in the boxed essay *Diene Polymers* that appears later in this chapter.

The conjugated diene 1,3-butadiene is used in the manufacture of synthetic rubber and is prepared on an industrial scale in vast quantities. Production in the United States is currently 4×10^9 lb/year. One industrial process is similar to that used for the preparation of ethylene: In the presence of a suitable catalyst, butane undergoes thermal dehydrogenation to yield 1,3-butadiene.

$$CH_3CH_2CH_2CH_3 \xrightarrow[\text{chromia–alumina}]{590-675°C} H_2C{=}CHCH{=}CH_2 + 2H_2$$

Laboratory syntheses of conjugated dienes can be achieved by elimination reactions of unsaturated alcohols and alkyl halides. In the two examples that follow, the conjugated diene is produced in high yield even though an isolated diene is also possible.

$$\underset{\substack{| \\ OH}}{H_2C{=}CHCH_2\overset{\overset{\displaystyle CH_3}{|}}{C}CH_2CH_3} \xrightarrow{\text{KHSO}_4,\text{ heat}} H_2C{=}CHCH{=}\overset{\overset{\displaystyle CH_3}{|}}{C}CH_2CH_3$$

3-Methyl-5-hexen-3-ol   4-Methyl-1,3-hexadiene (88%)

$$\underset{\substack{| \\ Br}}{H_2C{=}CHCH_2\overset{\overset{\displaystyle CH_3}{|}}{C}CH_2CH_3} \xrightarrow{\text{KOH, heat}} H_2C{=}CHCH{=}\overset{\overset{\displaystyle CH_3}{|}}{C}CH_2CH_3$$

4-Bromo-4-methyl-1-hexene   4-Methyl-1,3-hexadiene (78%)

As we saw in Chapter 5, dehydrations and dehydrohalogenations are typically regioselective in the direction that leads to the most stable double bond. Conjugated dienes are more stable than isolated dienes and are formed faster via a lower energy transition state.

PROBLEM 10.10 What dienes containing isolated double bonds are capable of being formed, but are not observed, in the two preceding equations describing elimination in 3-methyl-5-hexen-3-ol and 4-bromo-4-methyl-1-hexene?

Dienes with isolated double bonds can be formed when the structure of the alkyl halide doesn't permit the formation of a conjugated diene.

2,6-Dichlorocamphane   Bornadiene (83%)

We will not discuss the preparation of cumulated dienes. They are prepared less readily than isolated or conjugated dienes and require special methods.

10.10 ADDITION OF HYDROGEN HALIDES TO CONJUGATED DIENES

Our discussion of chemical reactions of alkadienes will be limited to those of conjugated dienes. The reactions of isolated dienes are essentially the same as those of individual alkenes. The reactions of cumulated dienes are—like their preparation—so specialized that their treatment is better suited to an advanced course in organic chemistry.

Electrophilic addition is the characteristic chemical reaction of alkenes, and conjugated dienes undergo addition reactions with the same electrophiles that react with alkenes, and by similar mechanisms. As we saw in the reaction of hydrogen halides with alkenes (Section 6.6), the regioselectivity of electrophilic addition is governed by protonation of the double bond in the direction that gives the more stable of two possible carbocations. With conjugated dienes it is one of the terminal carbons that is protonated because the species that results is an allylic carbocation which is stabilized by electron delocalization. Thus, when 1,3-cyclopentadiene reacts with hydrogen chloride, the product is 3-chlorocyclopentene.

| 1,3-Cyclopentadiene | 3-Chlorocyclopentene (70–90%) | | 4-Chlorocyclopentene |

The carbocation that leads to the observed product is secondary and allylic; the other is secondary but not allylic.

Protonation at end of diene unit gives a carbocation that is both secondary and allylic; most stable carbocation; product is formed from this carbocation.

Protonation at C-2 gives a carbocation that is secondary but not allylic; less stable carbocation; not formed as rapidly

Both resonance forms of the allylic carbocation from 1,3-cyclopentadiene are equivalent, and so attack at either of the carbons that share the positive charge gives the same product, 3-chlorocyclopentene. This is not the case with 1,3-butadiene, and so hydrogen halides add to 1,3-butadiene to give a mixture of two regioisomeric allylic halides. For the case of electrophilic addition of hydrogen bromide at −80°C,

$$H_2C{=}CH{-}CH{=}CH_2 \xrightarrow[-80°C]{HBr} CH_3CH{-}CH{=}CH_2 \;+\; CH_3CH{=}CH{-}CH_2Br$$

$$\overset{\displaystyle |}{\underset{\displaystyle Br}{}}$$

| 1,3-Butadiene | 3-Bromo-1-butene (81%) | 1-Bromo-2-butene (19%) |

The major product corresponds to addition of a proton at C-1 and bromide at C-2. This mode of addition is called **1,2 addition,** or **direct addition.** The minor product has its proton and bromide at C-1 and C-4, respectively, of the original diene system. This mode of addition is called **1,4 addition,** or **conjugate addition.** The double bond that was between C-3 and C-4 in the starting material remains there in the product from 1,2 addition but migrates to a position between C-2 and C-3 in the product from 1,4 addition.

Both the 1,2-addition product and the 1,4-addition product are derived from the same allylic carbocation.

Use *Learning By Modeling* to view the charge distribution in the allylic carbocation shown in the equation.

$$
\begin{array}{c}
\overset{+}{CH_3CH}-CH=CH_2 \\
\updownarrow \\
CH_3CH=CH-\overset{+}{CH_2}
\end{array}
\xrightarrow{Br^-}
CH_3CH-CH=CH_2 + CH_3CH=CH-CH_2Br
$$

$$
\underset{\substack{\text{3-Bromo-1-butene} \\ \text{(major)}}}{\overset{|}{\underset{Br}{}}}
\qquad
\underset{\substack{\text{1-Bromo-2-butene} \\ \text{(minor)}}}{}
$$

The secondary carbon bears more of the positive charge than does the primary carbon, and attack by the nucleophilic bromide ion is faster there. Hence, the major product is the secondary bromide.

When the major product of a reaction is the one that is formed at the fastest rate, we say that the reaction is governed by **kinetic control.** Most organic reactions fall into this category, and the electrophilic addition of hydrogen bromide to 1,3-butadiene at low temperature is a kinetically controlled reaction.

When, however, the ionic addition of hydrogen bromide to 1,3-butadiene is carried out at room temperature, the ratio of isomeric allylic bromides observed is different from that which is formed at $-80°C$. At room temperature, the 1,4-addition product predominates.

$$
H_2C=CH-CH=CH_2 \xrightarrow[\text{room temperature}]{HBr} CH_3CH-CH=CH_2 + CH_3CH=CH-CH_2Br
$$
$$
\underset{\substack{\text{1,3-Butadiene}}}{} \qquad \underset{\substack{\text{3-Bromo-1-butene (44\%)}}}{\overset{|}{\underset{Br}{}}} \qquad \underset{\substack{\text{1-Bromo-2-butene (56\%)}}}{}
$$

Clearly, the temperature at which the reaction occurs exerts a major influence on the product composition. To understand why, an important fact must be added. The 1,2- and 1,4-addition products *interconvert rapidly* by allylic rearrangement at elevated temperature in the presence of hydrogen bromide. Heating the product mixture to 45°C in the presence of hydrogen bromide leads to a mixture in which the ratio of 3-bromo-1-butene to 1-bromo-2-butene is 15:85.

$$
CH_3CH-CH=CH_2 \underset{\substack{\text{cation–anion} \\ \text{combination}}}{\overset{\text{ionization}}{\rightleftharpoons}} CH_3CH\overset{H}{\underset{:\overset{..}{Br}:}{\overset{|}{\underset{}{C}}}}CH_2 \underset{\substack{\text{ionization}}}{\overset{\substack{\text{cation–anion} \\ \text{combination}}}{\rightleftharpoons}} CH_3CH=CH-CH_2\overset{..}{\underset{..}{Br}}:
$$

$$
\underset{\substack{\text{3-Bromo-1-butene} \\ \text{(less stable isomer)}}}{:\overset{..}{Br}:}
\qquad
\underset{\substack{\text{Carbocation} \\ \text{+ bromide anion}}}{}
\qquad
\underset{\substack{\text{1-Bromo-2-butene} \\ \text{(more stable isomer)}}}{}
$$

The product of 1,4 addition, 1-bromo-2-butene, contains an internal double bond and so is *more stable* than the product of 1,2 addition, 3-bromo-1-butene, which has a terminal double bond.

When addition occurs under conditions in which the products can equilibrate, the composition of the reaction mixture no longer reflects the relative rates of formation of

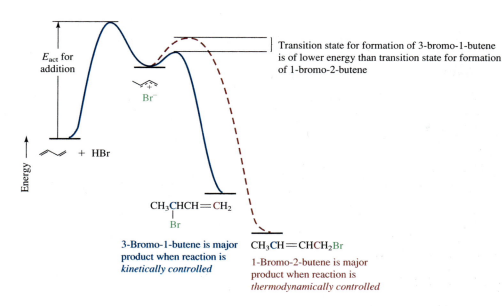

FIGURE 10.8 Energy diagram showing relationship of kinetic control to thermodynamic control in addition of hydrogen bromide to 1,3-butadiene.

the products but tends to reflect their *relative stabilities.* Reactions of this type are said to be governed by **thermodynamic control.** One way to illustrate kinetic and thermodynamic control in the addition of hydrogen bromide to 1,3-butadiene is by way of the energy diagram of Figure 10.8. At low temperature, addition takes place irreversibly. Isomerization is slow because insufficient thermal energy is available to permit the products to surmount the energy barrier for ionization. At higher temperatures isomerization is possible, and the more stable product predominates.

PROBLEM 10.11   Addition of hydrogen chloride to 2-methyl-1,3-butadiene is a kinetically controlled reaction and gives one product in much greater amounts than any isomers. What is this product?

10.11   HALOGEN ADDITION TO DIENES

Mixtures of 1,2- and 1,4-addition products are obtained when 1,3-butadiene reacts with chlorine or bromine.

$$H_2C=CHCH=CH_2 \ + \ Br_2 \ \xrightarrow{CHCl_3} \ BrCH_2CHCH=CH_2 \ + \ \underset{H}{\overset{BrCH_2}{}}C=C\underset{CH_2Br}{\overset{H}{}}$$

| 1,3-Butadiene | Bromine | 3,4-Dibromo-1-butene (37%) | (E)-1,4-Dibromo-2-butene (63%) |

The tendency for conjugate addition is pronounced, and E double bonds are generated almost exclusively.

PROBLEM 10.12   Exclusive of stereoisomers, how many products are possible in the electrophilic addition of 1 mole of bromine to 2-methyl-1,3-butadiene?

DIENE POLYMERS

Some 500 years ago during Columbus's second voyage to what are now the Americas, he and his crew saw children playing with balls made from the latex of trees that grew there. Later, Joseph Priestley called this material "rubber" to describe its ability to erase pencil marks by rubbing, and in 1823 Charles Macintosh demonstrated how rubber could be used to make waterproof coats and shoes. Shortly thereafter Michael Faraday determined an empirical formula of C_5H_8 for rubber. It was eventually determined that rubber is a polymer of 2-methyl-1,3-butadiene.

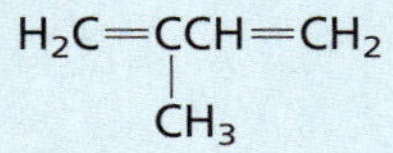

$$H_2C=CCH=CH_2$$
$$\overset{|}{CH_3}$$

2-Methyl-1,3-butadiene (common name: *isoprene*)

The structure of rubber corresponds to 1,4 addition of several thousand isoprene units to one another:

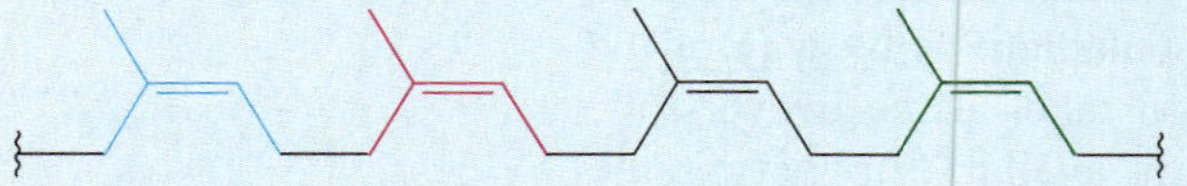

All the double bonds in rubber have the *Z* (or cis) configuration. A different polymer of isoprene, called *gutta-percha,* has shorter polymer chains and *E* (or trans) double bonds. Gutta-percha is a tough, horn-like substance once used as a material for golf ball covers.*

In natural rubber the attractive forces between neighboring polymer chains are relatively weak, and there is little overall structural order. The chains slide easily past one another when stretched and return, in time, to their disordered state when the distorting force is removed. The ability of a substance to recover its original shape after distortion is its *elasticity.* The elasticity of natural rubber is satisfactory only within a limited temperature range; it is too rigid when cold and too sticky when warm to be very useful. Rubber's elasticity is improved by *vulcanization,* a process discovered by Charles Goodyear in 1839. When natural rubber is heated with sulfur, a chemical reaction occurs in which neighboring polyisoprene chains become connected through covalent bonds to sulfur. Although these sulfur "bridges" permit only limited movement of one chain with respect to another, their presence ensures that the rubber will snap back to its original shape once the distorting force is removed.

As the demand for rubber increased, so did the chemical industry's efforts to prepare a synthetic substitute. One of the first **elastomers** (a synthetic polymer that possesses elasticity) to find a commercial niche was *neoprene,* discovered by chemists at Du Pont in 1931. Neoprene is produced by free-radical polymerization of 2-chloro-1,3-butadiene and has the greatest variety of applications of any elastomer. Some uses include electrical insulation, conveyer belts, hoses, and weather balloons.

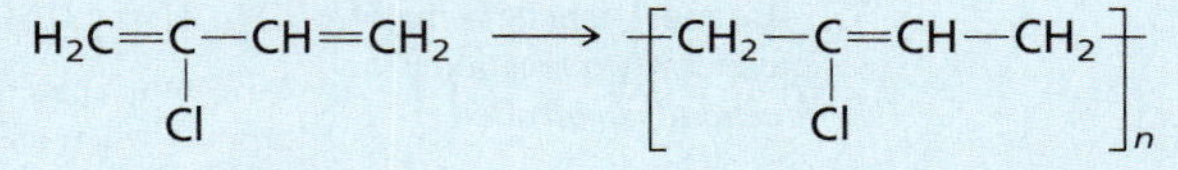

$$H_2C=\overset{|}{\underset{Cl}{C}}-CH=CH_2 \longrightarrow \left[CH_2-\overset{|}{\underset{Cl}{C}}=CH-CH_2\right]_n$$

2-Chloro-1,3-butadiene   Neoprene

The elastomer produced in greatest amount is *styrene-butadiene rubber* (SBR). Annually, just under 10^9 lb of SBR is produced in the United States, and almost all of it is used in automobile tires. As its name suggests, SBR is prepared from styrene and 1,3-butadiene. It is an example of a **copolymer,** a polymer assembled from two or more different monomers. Free-radical polymerization of a mixture of styrene and 1,3-butadiene gives SBR.

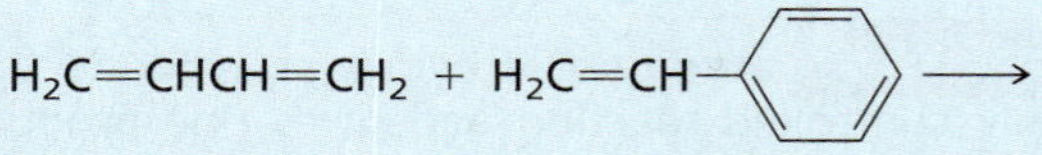

$$H_2C=CHCH=CH_2 + H_2C=CH- \bigcirc \longrightarrow$$

1,3-Butadiene   Styrene

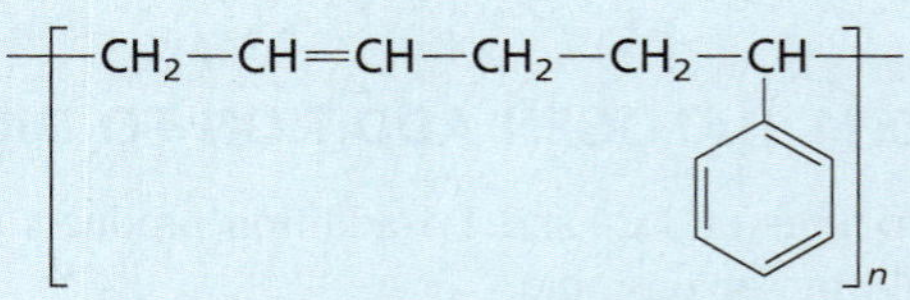

$$\left[CH_2-CH=CH-CH_2-CH_2-CH\right]_n$$

Styrene-butadiene rubber

Coordination polymerization of isoprene using Ziegler–Natta catalyst systems (Section 6.21) gives a material similar in properties to natural rubber, as does polymerization of 1,3-butadiene. Poly(1,3-butadiene) is produced in about two thirds the quantity of SBR each year. It, too, finds its principal use in tires.

---

* A detailed discussion of the history, structure, and applications of natural rubber appears in the May 1990 issue of the *Journal of Chemical Education.*

10.12 THE DIELS–ALDER REACTION

A particular kind of conjugate addition reaction earned the Nobel Prize in chemistry for Otto Diels and Kurt Alder of the University of Kiel (Germany) in 1950. The Diels–Alder reaction is the *conjugate addition of an alkene to a diene.* Using 1,3-butadiene as a typical diene, the Diels–Alder reaction may be represented by the general equation:

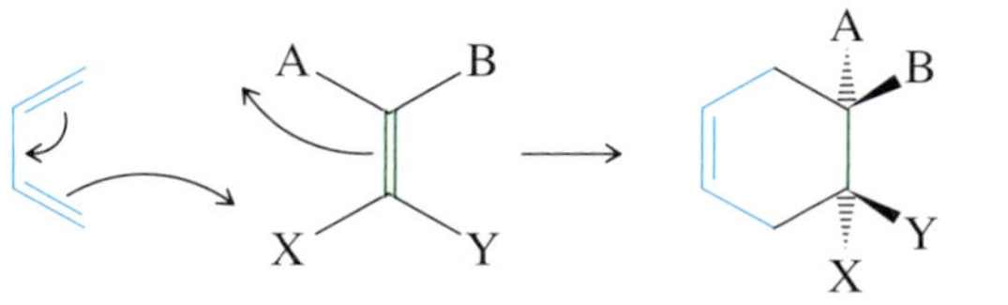

1,3-Butadiene   Dienophile   Diels–Alder adduct

For an animation of this reaction, see *Learning By Modeling.*

The alkene that adds to the diene is called the **dienophile.** Because the Diels–Alder reaction leads to the formation of a ring, it is termed a **cycloaddition** reaction. The product contains a cyclohexene ring as a structural unit.

The Diels–Alder cycloaddition is one example of a **pericyclic reaction,** which is a one-step reaction that proceeds through a cyclic transition state. Bond formation occurs at both ends of the diene system, and the Diels–Alder transition state involves a cyclic array of six carbons and six π electrons. The diene must adopt the *s-cis* conformation in the transition state.

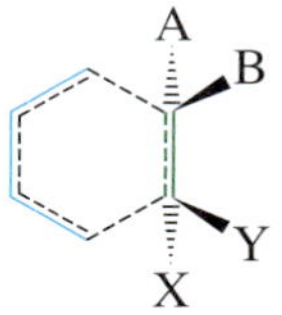

Epoxidation of alkenes (Section 6.18) is another example of a cycloaddition.

Transition state for Diels–Alder cycloaddition

The simplest of all Diels–Alder reactions, cycloaddition of ethylene to 1,3-butadiene, does not proceed readily. It has a high activation energy and a low reaction rate. Substituents such as C=O or C≡N, however, when *directly* attached to the double bond of the dienophile, increase its reactivity, and compounds of this type give high yields of Diels–Alder adducts at modest temperatures.

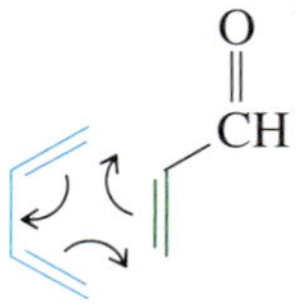

1,3-Butadiene   Acrolein   Cyclohexene-4-carboxaldehyde (100%)

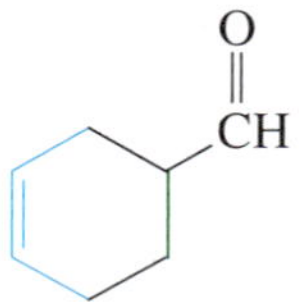

The product of a Diels–Alder cycloaddition always contains one more ring than was present in the reactants. The dienophile *maleic anhydride* contains one ring, so the product of its addition to a diene contains two.

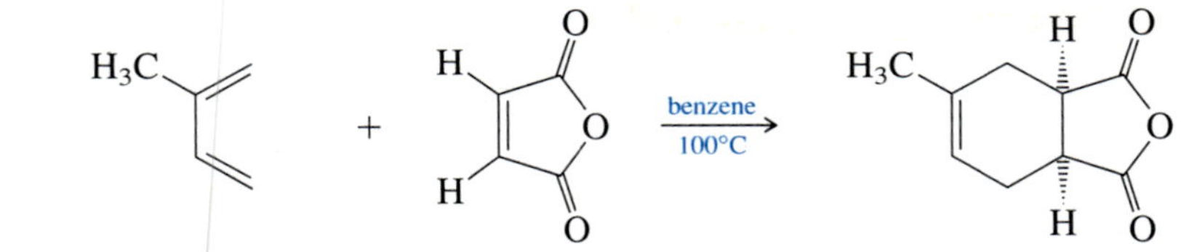

2-Methyl-1,3-butadiene   Maleic anhydride   1-Methylcyclohexene-4,5-
dicarboxylic anhydride (100%)

PROBLEM 10.13 Benzoquinone is a very reactive dienophile. It reacts with 2-chloro-1,3-butadiene to give a single product, $C_{10}H_9ClO_2$, in 95% yield. Write a structural formula for this product.

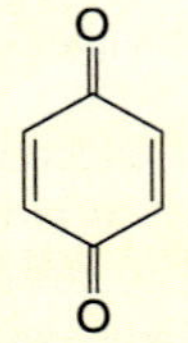

Benzoquinone

Acetylene, like ethylene, is a poor dienophile, but alkynes that bear C=O or C≡N substituents react readily with dienes. A cyclohexadiene derivative is the product.

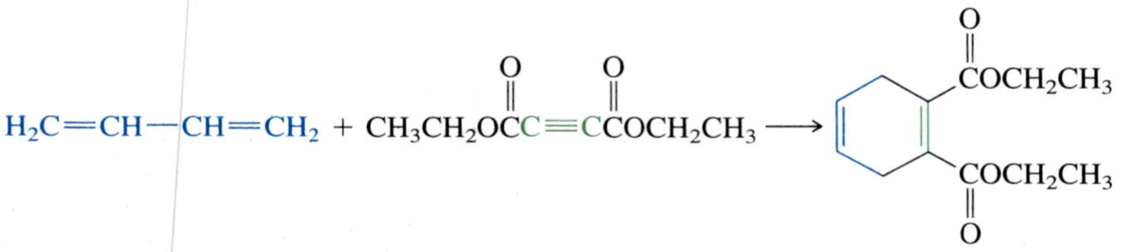

1,3-Butadiene   Diethyl acetylenedicarboxylate   Diethyl 1,4-cyclohexadiene-
1,2-dicarboxylate (98%)

The Diels–Alder reaction is stereospecific. Substituents that are cis in the dienophile remain cis in the product; substituents that are trans in the dienophile remain trans in the product.

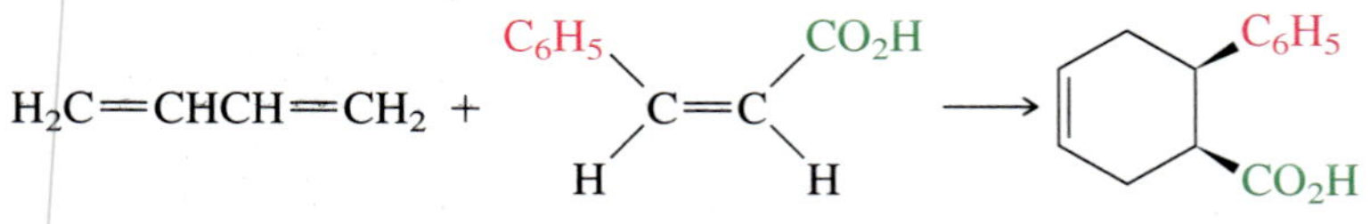

1,3-Butadiene   *cis*-Cinnamic acid   Only product

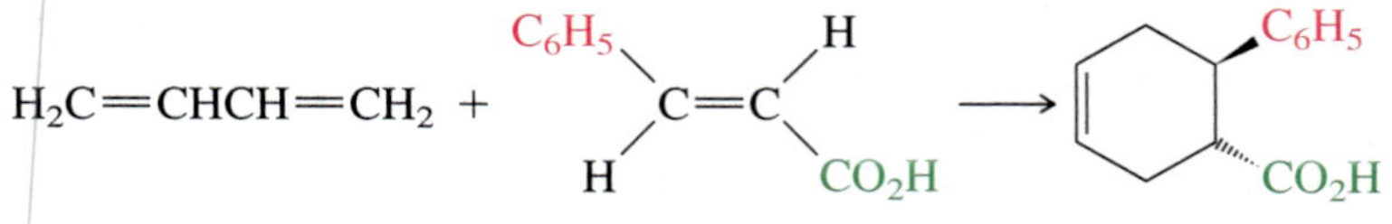

1,3-Butadiene   *trans*-Cinnamic acid   Only product

Recall from Section 7.13 that a stereospecific reaction is one in which each stereoisomer of a particular starting material yields a different stereoisomeric form of the reaction product. In the examples shown, the product from Diels–Alder cycloaddition of 1,3-butadiene to *cis*-cinnamic acid is a stereoisomer of the product from *trans*-cinnamic acid. Each product, although chiral, is formed as a racemic mixture.

PROBLEM 10.14 What combination of diene and dienophile would you choose in order to prepare each of the following compounds?

(a) (b) 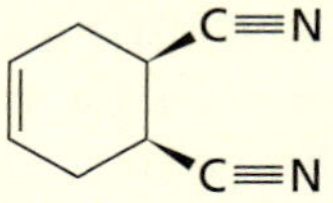 (c)

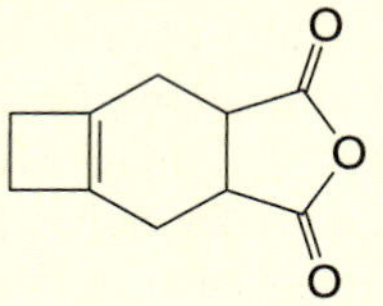

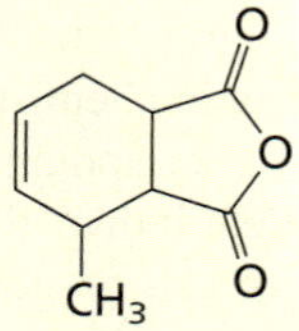

SAMPLE SOLUTION (a) Using curved arrows, we represent a Diels–Alder reaction as

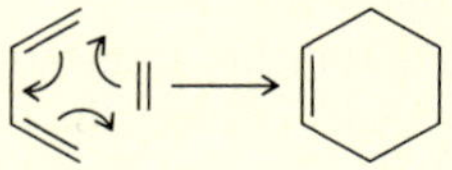

To deduce the identity of the diene and dienophile that lead to a particular Diels–Alder adduct, we use curved arrows in the reverse fashion to "undo" the cyclohexene derivative. Start with the π component of the double bond in the six-membered ring, and move electrons in pairs.

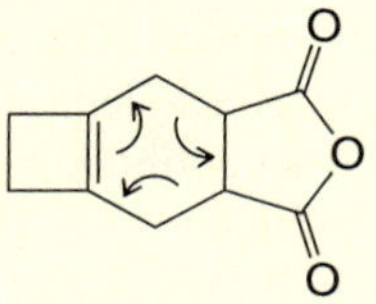

is derived from

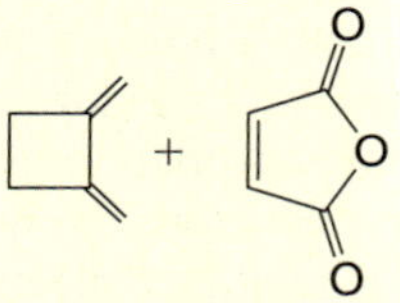

Diels–Alder adduct   Diene   Dienophile

Cyclic dienes yield bridged bicyclic Diels–Alder adducts.

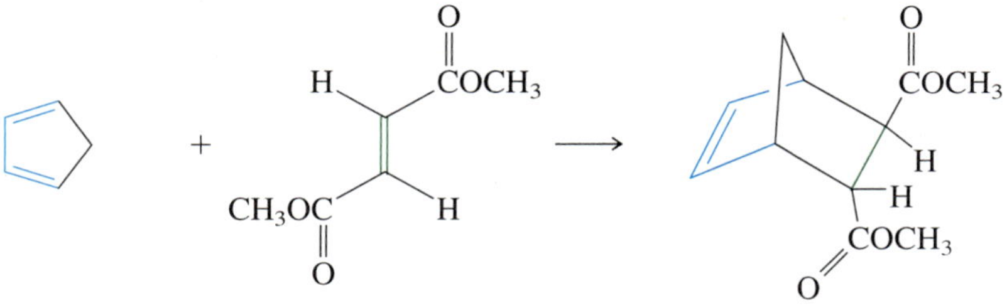

1,3-Cyclopentadiene   Dimethyl fumarate   Dimethyl bicyclo[2.2.1]hept-2-ene-
                                                        trans-5,6-dicarboxylate

PROBLEM 10.15 The Diels–Alder reaction of 1,3-cyclopentadiene with methyl

acrylate ($H_2C=CHCOCH_3$) gives a mixture of two diastereomers. Write their structural formulas.

   The importance of the Diels–Alder reaction is in synthesis. It gives us a method to form *two* new carbon–carbon bonds in a single operation and requires no reagents, such as acids or bases, that might affect other functional groups in the molecule.

   The mechanism of the Diels–Alder reaction is best understood on the basis of a molecular orbital approach. To understand this approach we need to take a more detailed look at the π orbitals of alkenes and dienes.

10.13 THE π MOLECULAR ORBITALS OF ETHYLENE AND 1,3-BUTADIENE

The valence bond approach has served us well to this point as a tool to probe structure and reactivity in organic chemistry. An appreciation for the delocalization of π electrons through a system of overlapping p orbitals has given us insights into conjugated systems that are richer in detail than those obtained by examining Lewis formulas. An even deeper understanding can be gained by applying qualitative molecular orbital theory to these π electron systems. We shall see that useful information can be gained by directing attention to what are called the **frontier orbitals** of molecules. The frontier orbitals are the *highest occupied molecular orbital* (the *HOMO*) and the *lowest unoccupied molecular orbital* (the *LUMO*). When electrons are transferred *from* a molecule, it is the electrons in the HOMO that are involved, because they are the most weakly held. When electrons are transferred *to* a molecule, they go into the LUMO, because that is the lowest energy vacant orbital.

Ethylene. Let's begin by examining the π molecular orbitals of ethylene. Recall from Section 2.4 that the number of molecular orbitals is equal to the number of atomic orbitals that combine to form them. We saw that the $1s$ orbitals of two hydrogen atoms overlap to give both a bonding (σ) and an antibonding (σ^*) orbital. The same principle applies to π orbitals. As Figure 10.9 illustrates for the case of ethylene, the $2p$ orbitals of adjacent carbons overlap to give both a bonding (π) and an antibonding (π^*) orbital. Notice that the σ electrons are not explicitly considered in Figure 10.9. These electrons are strongly held, and the collection of σ bonds can be thought of as an inert framework that supports the valence electrons of the π orbital.

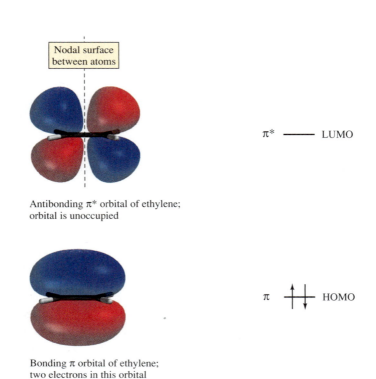

Nodal surface between atoms

Energy →

π^* ——— LUMO

Antibonding π^* orbital of ethylene; orbital is unoccupied

π ⥮ HOMO

Bonding π orbital of ethylene; two electrons in this orbital

FIGURE 10.9 The bonding (π) and antibonding (π^*) molecular orbitals of ethylene. The wave function changes sign (red to blue) on passing through a nodal surface. The plane of the molecule is a nodal surface in both orbitals; the antibonding orbital has an additional nodal surface perpendicular to the plane of the molecule.

Both the π and π* molecular orbitals of ethylene are *antisymmetric* with respect to the plane of the molecule. By this we mean that the wave function changes sign on passing through the molecular plane. It's convenient to designate the signs of *p* orbital wave functions by shading one lobe of a *p* orbital in red and the other in blue instead of using plus (+) and minus (−) signs that might be confused with electronic charges. The plane of the molecule corresponds to a nodal plane where the probability of finding the π electrons is zero. The bonding π orbital has no nodes other than this plane, whereas the antibonding π* orbital has a nodal plane between the two carbons. The more nodes an orbital has, the higher is its energy.

As is true for all orbitals, a π orbital may contain a maximum of two electrons. Ethylene has two π electrons, and these occupy the bonding π molecular orbital, which is the HOMO. The antibonding π* molecular orbital is vacant, and is the LUMO.

PROBLEM 10.16   Which molecular orbital of ethylene (π or π*) is the most important one to look at in a reaction in which ethylene is attacked by an electrophile?

1,3-Butadiene.   The π molecular orbitals of 1,3-butadiene are shown in Figure 10.10. The four sp^2-hybridized carbons contribute four $2p$ atomic orbitals, and their overlap

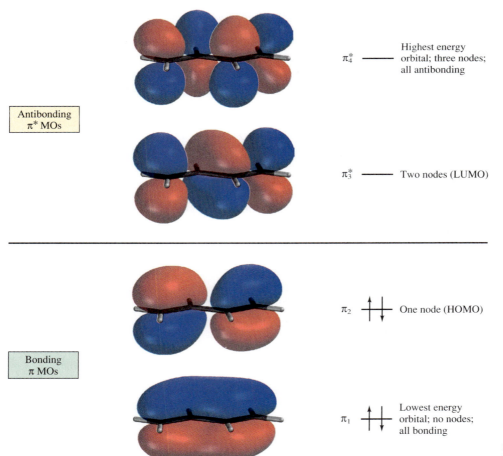

Antibonding π* MOs	π_4^* ——— Highest energy orbital; three nodes; all antibonding
	π_3^* ——— Two nodes (LUMO)
Bonding π MOs	π_2 ⥮ One node (HOMO)
	π_1 ⥮ Lowest energy orbital; no nodes; all bonding

FIGURE 10.10 The π molecular orbitals of 1,3-butadiene.

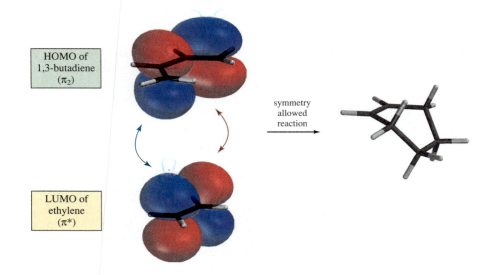

FIGURE 10.11 The HOMO of 1,3-butadiene and the LUMO of ethylene have the proper symmetry to allow σ bond formation to occur at both ends of the diene chain in the same transition state.

leads to four π molecular orbitals. Two are bonding (π_1 and π_2) and two are antibonding (π_3^* and π_4^*). Each π molecular orbital encompasses all four carbons of the diene. There are four π electrons, and these are distributed in pairs between the two orbitals of lowest energy (π_1 and π_2). Both bonding orbitals are occupied; π_2 is the HOMO. Both antibonding orbitals are vacant; π_3^* is the LUMO.

10.14 A π MOLECULAR ORBITAL ANALYSIS OF THE DIELS–ALDER REACTION

Let us now examine the Diels–Alder cycloaddition from a molecular orbital perspective. Chemical experience, such as the observation that the substituents that increase the reactivity of a dienophile tend to be those that attract electrons, suggests that electrons flow from the diene to the dienophile during the reaction. Thus, the orbitals to be considered are the HOMO of the diene and the LUMO of the dienophile. As shown in Figure 10.11 for the case of ethylene and 1,3-butadiene, the symmetry properties of the HOMO of the diene and the LUMO of the dienophile permit bond formation between the ends of the diene system and the two carbons of the dienophile double bond because the necessary orbitals overlap "in phase" with each other. Cycloaddition of a diene and an alkene is said to be a **symmetry-allowed** reaction.

Contrast the Diels–Alder reaction with a cycloaddition reaction that looks superficially similar, the combination of two ethylene molecules to give cyclobutane.

Ethylene   Ethylene   Cyclobutane

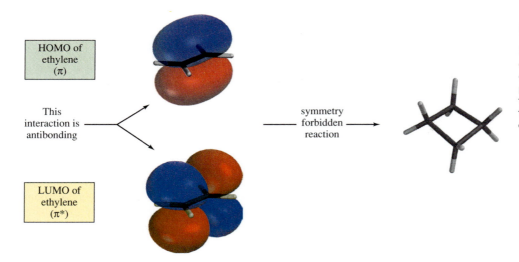

HOMO of ethylene (π)

This interaction is antibonding

LUMO of ethylene (π*)

symmetry forbidden reaction

FIGURE 10.12 The HOMO of one ethylene molecule and the LUMO of another do not have the proper symmetry to permit two σ bonds to be formed in the same transition state for concerted cycloaddition.

Reactions of this type are rather rare and seem to proceed in a stepwise fashion rather than by way of a concerted mechanism involving a single transition state.

Figure 10.12 shows the interaction between the HOMO of one ethylene molecule and the LUMO of another. In particular, notice that two of the carbons that are to become σ-bonded to each other in the product experience an antibonding interaction during the cycloaddition process. This raises the activation energy for cycloaddition and leads the reaction to be classified as a **symmetry-forbidden** reaction. Reaction, were it to occur, would take place slowly and by a mechanism in which the two new σ bonds are formed in separate steps rather than by way of a concerted process involving a single transition state.

PROBLEM 10.17 Use frontier orbital analysis to decide whether the dimerization of 1,3-butadiene shown here is symmetry allowed or forbidden.

$$2H_2C=CH-CH=CH_2 \xrightarrow{heat}$$

Frontier orbital analysis is a powerful theory that aids our understanding of a great number of organic reactions. Its early development is attributed to Professor Kenichi Fukui of Kyoto University, Japan. The application of frontier orbital methods to Diels–Alder reactions represents one part of what organic chemists refer to as the *Woodward–Hoffmann rules,* a beautifully simple analysis of organic reactions by Professor R. B. Woodward of Harvard University and Professor Roald Hoffmann of Cornell University. Professors Fukui and Hoffmann were corecipients of the 1981 Nobel Prize in chemistry for their work.

Woodward's death in 1979 prevented his being considered for a share of the 1981 prize with Fukui and Hoffmann. Woodward had earlier won a Nobel Prize (1965) for his achievements in organic synthesis.

10.15 SUMMARY

This chapter focused on the effect of a carbon–carbon double bond as a stabilizing substituent on a positively charged carbon in an **allylic carbocation,** on a carbon bearing an odd electron in an **allylic free radical,** and on a second double bond as in a **conjugated diene.**

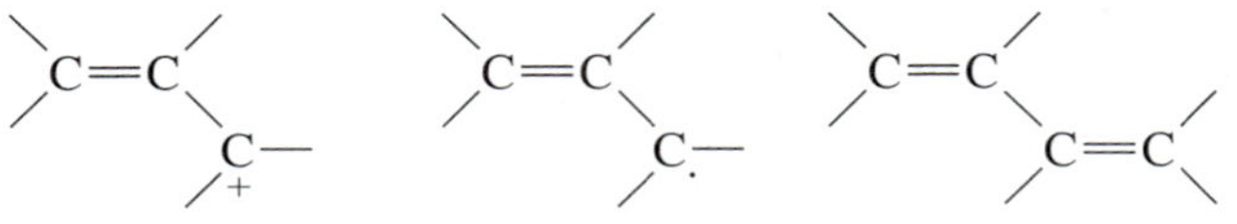

Allylic carbocation   Allylic radical   Conjugated diene

Section 10.1   **Allyl** is the common name of the parent group $H_2C\!=\!CHCH_2\!-$ and is an acceptable name in IUPAC nomenclature.

Section 10.2   The carbocations formed as intermediates when allylic halides undergo S_N1 reactions have their positive charge shared by the two end carbons of the allylic system and may be attacked by nucleophiles at either site. Products may be formed with the same pattern of bonds as the starting allylic halide or with *allylic rearrangement.*

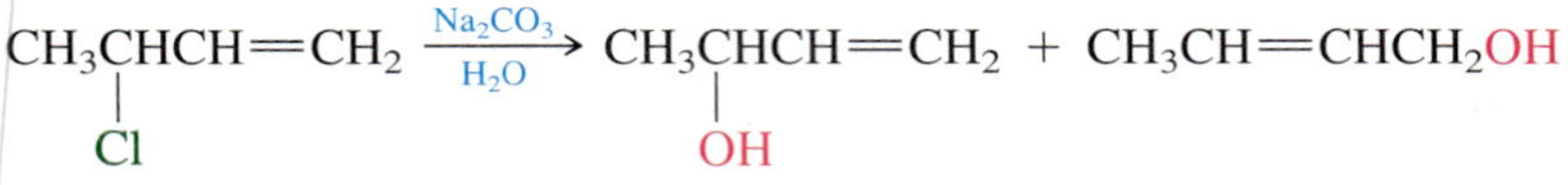

$$CH_3CHCH\!=\!CH_2 \xrightarrow[\text{H}_2\text{O}]{\text{Na}_2\text{CO}_3} CH_3CHCH\!=\!CH_2 + CH_3CH\!=\!CHCH_2OH$$
$$\underset{\text{Cl}}{\mid} \qquad\qquad \underset{\text{OH}}{\mid}$$

3-Chloro-1-butene   3-Buten-2-ol (65%)   2-Buten-1-ol (35%)

via: $CH_3\overset{+}{C}H\!-\!CH\!=\!CH_2 \longleftrightarrow CH_3CH\!=\!CH\!-\!\overset{+}{C}H_2$

Sections 10.3–10.4   Alkenes react with *N*-bromosuccinimide (NBS) to give allylic bromides. NBS serves as a source of Br_2, and substitution occurs by a free-radical mechanism. The reaction is used for synthetic purposes only when the two resonance forms of the allylic radical are equivalent. Otherwise a mixture of isomeric allylic bromides is produced.

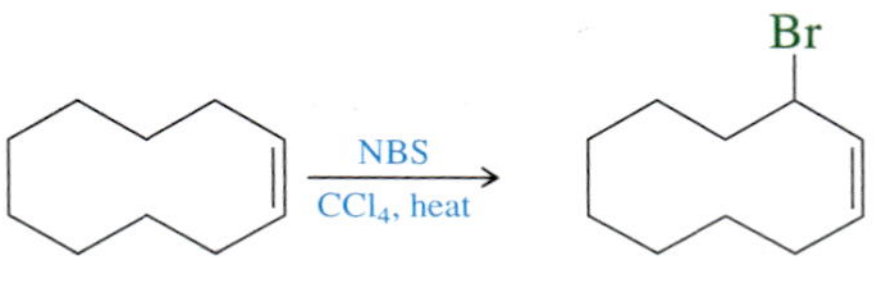

Cyclodecene   3-Bromocyclodecene (56%)

via:

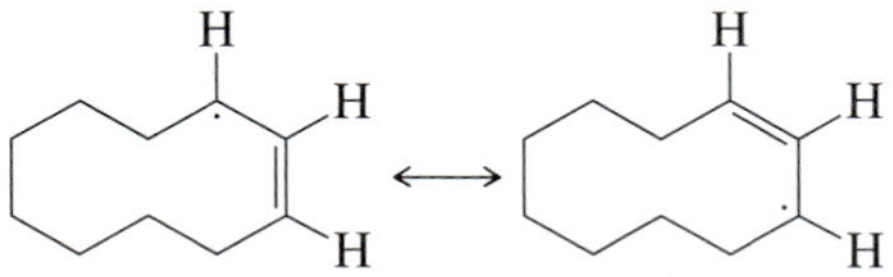

Section 10.5   Dienes are classified as having **isolated, conjugated,** or **cumulated** double bonds.

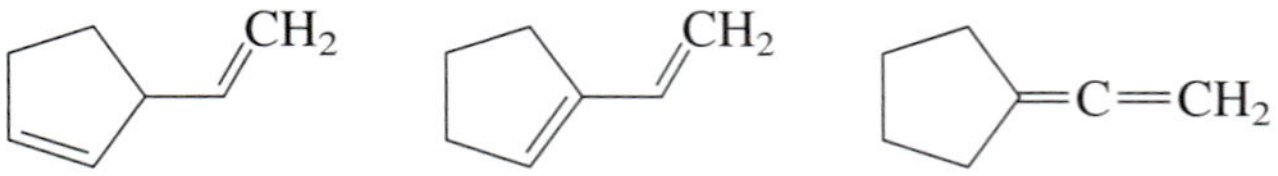

Isolated   Conjugated   Cumulated

Section 10.6   Conjugated dienes are more stable than isolated dienes, and cumulated dienes are the least stable of all.

Section 10.7   Conjugated dienes are stabilized by electron delocalization to the extent of 12–16 kJ/mol (3–4 kcal/mol). Overlap of the *p* orbitals of four adjacent *sp²*-hybridized carbons in a conjugated diene gives an extended π system through which the electrons are delocalized.

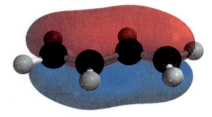

The two most stable conformations of conjugated dienes are the *s*-cis and *s*-trans. The *s*-trans conformation is normally more stable than the *s*-cis. Both conformations are planar, which allows the *p* orbitals to overlap to give an extended π system.

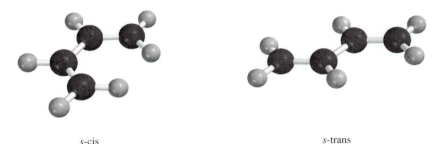

s-cis   s-trans

Section 10.8   1,2-Propadiene ($H_2C{=}C{=}CH_2$), also called **allene,** is the simplest cumulated diene. The two π bonds in an allene share an *sp*-hybridized carbon and are at right angles to each other. Certain allenes such as 2,3-pentadiene ($CH_3CH{=}C{=}CHCH_3$) possess a *chirality axis* and are chiral.

Section 10.9   1,3-Butadiene is an industrial chemical and is prepared by dehydrogenation of butane. Elimination reactions such as dehydration and dehydrohalogenation are common routes to alkadienes.

$$H_2C{=}CHCH_2\underset{\underset{\displaystyle OH}{|}}{\overset{\overset{\displaystyle CH_3}{|}}{C}}CH_2CH_3 \xrightarrow[\text{heat}]{KHSO_4} H_2C{=}CHCH{=}\overset{\overset{\displaystyle CH_3}{|}}{C}CH_2CH_3$$

3-Methyl-5-hexen-3-ol   4-Methyl-1,3-hexadiene (88%)

Elimination is typically regioselective and gives a conjugated diene rather than an isolated or cumulated diene system of double bonds.

Section 10.10   Protonation at the terminal carbon of a conjugated diene system gives an allylic carbocation that can be captured by the halide nucleophile at either of the two sites that share the positive charge. Nucleophilic attack at the carbon adjacent to the one that is protonated gives the product of *direct addition* (*1,2 addition*). Capture at the other site gives the product of *conjugate addition* (*1,4 addition*).

$$\underset{\text{1,3-Butadiene}}{H_2C{=}CHCH{=}CH_2} \xrightarrow{HCl} \underset{\substack{\text{3-Chloro-1-butene}\\(78\%)}}{CH_3\underset{\displaystyle Cl}{\underset{|}{C}}HCH{=}CH_2} + \underset{\substack{\text{1-Chloro-2-butene}\\(22\%)}}{CH_3CH{=}CHCH_2Cl}$$

$$\textit{via:}\quad CH_3\overset{+}{C}H{-}CH{=}CH_2 \longleftrightarrow CH_3CH{=}CH{-}\overset{+}{C}H_2$$

Section 10.11 1,4-Addition predominates when Cl_2 and Br_2 add to conjugated dienes.

Section 10.12 Conjugate addition of an alkene (the *dienophile*) to a conjugated diene gives a cyclohexene derivative in a process called the *Diels–Alder reaction*. It is concerted and stereospecific; substituents that are cis to each other on the dienophile remain cis in the product.

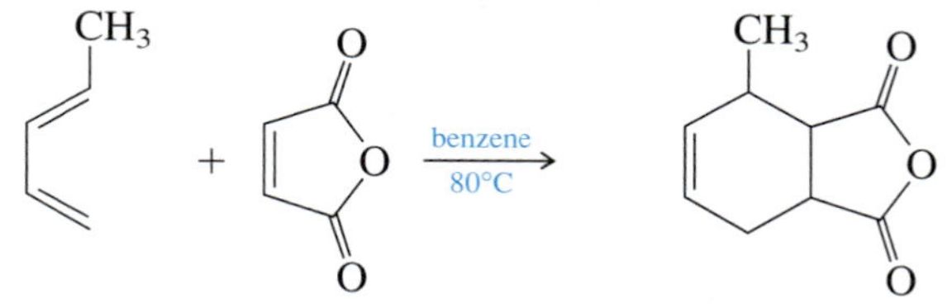

trans-1,3-Pentadiene — Maleic anhydride — 3-Methylcyclohexene-4,5-dicarboxylic anhydride (81%)

Sections 10.13–10.14 The Diels–Alder reaction is believed to proceed in a single step. A deeper level of understanding of the bonding changes in the transition state can be obtained by examining the nodal properties of the highest occupied molecular orbital (HOMO) of the diene and the lowest unoccupied molecular orbital (LUMO) of the dienophile.

PROBLEMS

10.18 Write structural formulas for each of the following:

(a) 3,4-Octadiene
(b) (*E*,*E*)-3,5-Octadiene
(c) (*Z*,*Z*)-1,3-Cyclooctadiene
(d) (*Z*,*Z*)-1,4-Cyclooctadiene
(e) (*E*,*E*)-1,5-Cyclooctadiene
(f) (2*E*,4*Z*,6*E*)-2,4,6-Octatriene
(g) 5-Allyl-1,3-cyclopentadiene
(h) *trans*-1,2-Divinylcyclopropane
(i) 2,4-Dimethyl-1,3-pentadiene

10.19 Give the IUPAC names for each of the following compounds:

(a) $H_2C{=}CH(CH_2)_5CH{=}CH_2$

(b) $(CH_3)_2C{=}\overset{\displaystyle CH_3}{\overset{|}{C}}\underset{\displaystyle CH_3}{\underset{|}{C}}{=}C(CH_3)_2$

(c) $(H_2C{=}CH)_3CH$

(d) 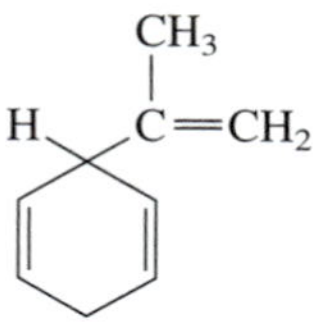

(e) 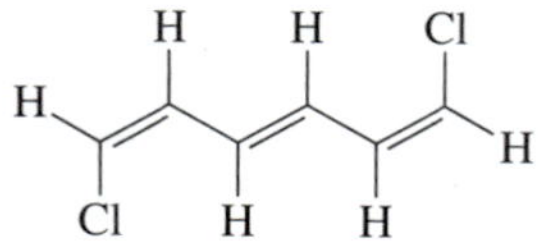

(f) $H_2C{=}C{=}CHCH{=}CHCH_3$

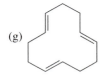

(g)

(h)

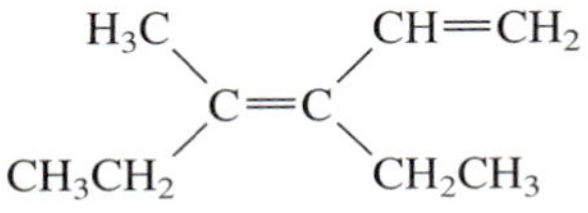

H_3C $CH=CH_2$

$C=C$

CH_3CH_2 CH_2CH_3

10.20 (a) What compound of molecular formula C_6H_{10} gives 2,3-dimethylbutane on catalytic hydrogenation over platinum?

    (b) What two compounds of molecular formula $C_{11}H_{20}$ give 2,2,6,6-tetramethylheptane on catalytic hydrogenation over platinum?

10.21 Write structural formulas for all the

    (a) Conjugated dienes   (b) Isolated dienes   (c) Cumulated dienes

that give 2,4-dimethylpentane on catalytic hydrogenation.

10.22 A certain species of grasshopper secretes an allenic substance of molecular formula $C_{13}H_{20}O_3$ that acts as an ant repellent. The carbon skeleton and location of various substituents in this substance are indicated in the partial structure shown. Complete the structure, adding double bonds where appropriate.

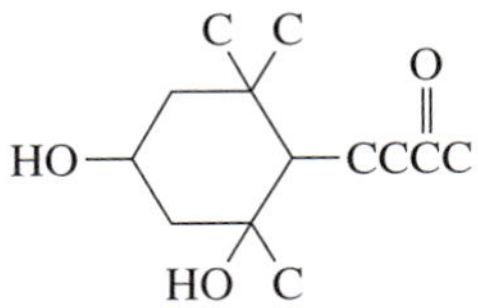

$$\begin{array}{c} C \quad C \\ HO \text{—} \overset{\displaystyle\text{—}}{\bigcirc} \text{—} CC\overset{\displaystyle O}{\overset{\|}{C}}C \\ HO \quad C \end{array}$$

10.23 Show how to prepare each of the following compounds from propene and any necessary organic or inorganic reagents:

    (a) Allyl bromide   (e) 1,2,3,-Tribromopropane

    (b) 1,2-Dibromopropane   (f) Allyl alcohol

    (c) 1,3-Dibromopropane   (g) 1-Penten-4-yne ($H_2C=CHCH_2C\equiv CH$)

    (d) 1-Bromo-2-chloropropane   (h) 1,4-Pentadiene

10.24 Show, by writing a suitable sequence of chemical equations, how to prepare each of the following compounds from cyclopentene and any necessary organic or inorganic reagents:

    (a) 2-Cyclopentenol   (d) 1,3-Cyclopentadiene

    (b) 3-Iodocyclopentene

    (c) 3-Cyanocyclopentene   (e)

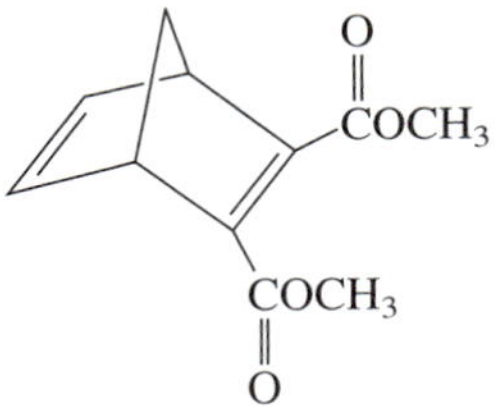

10.25 Give the structure, exclusive of stereochemistry, of the principal organic product formed on reaction of 2,3-dimethyl-1,3-butadiene with each of the following:

    (a) 2 mol H_2, platinum catalyst   (f) 2 mol Br_2

    (b) 1 mol HCl (product of direct addition)

    (c) 1 mol HCl (product of conjugate addition)

    (d) 1 mol Br_2 (product of direct addition)   (g)

    (e) 1 mol Br_2 (product of conjugate addition)

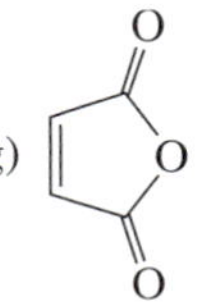

10.26 Repeat the previous problem for the reactions of 1,3-cyclohexadiene.

10.27 Give the structure of the Diels–Alder adduct of 1,3-cyclohexadiene and dimethyl acetylenedicarboxylate ($CH_3OCC\equiv CCOCH_3$).

10.28 Two constitutional isomers of molecular formula $C_8H_{12}O$ are formed in the following reaction. Ignoring stereochemistry suggest reasonable structures for these Diels–Alder adducts.

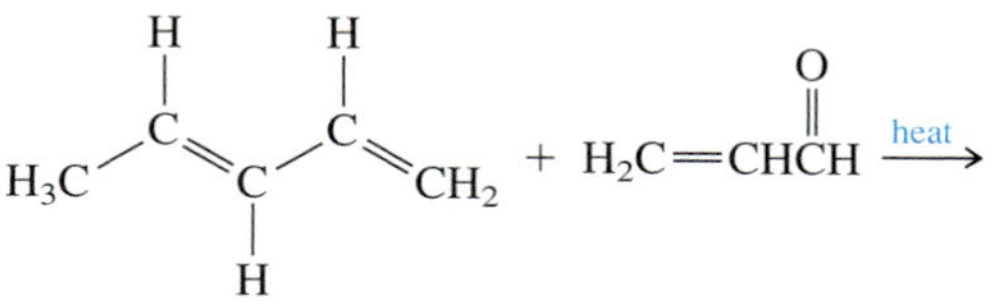

10.29 Allene can be converted to a trimer (compound A) of molecular formula C_9H_{12}. Compound A reacts with dimethyl acetylenedicarboxylate to give compound B. Deduce the structure of compound A.

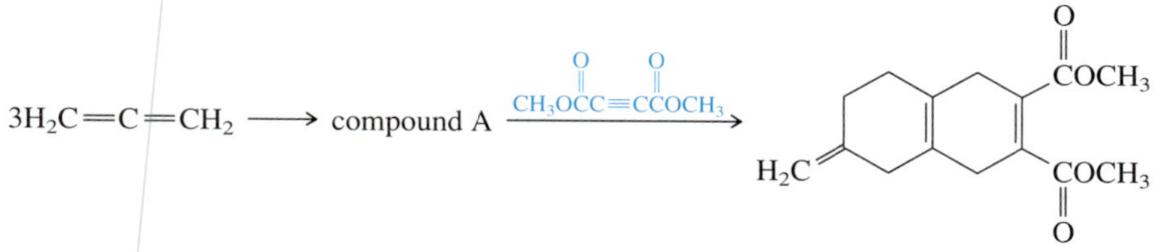

Compound B

10.30 The following reaction gives only the product indicated. By what mechanism does this reaction most likely occur?

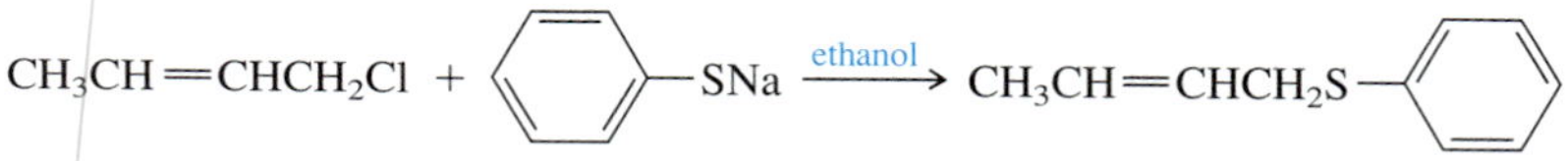

10.31 Suggest reasonable explanations for each of the following observations:

(a) The first-order rate constant for the solvolysis of $(CH_3)_2C\!\!=\!\!CHCH_2Cl$ in ethanol is over 6000 times greater than that of allyl chloride (25°C).

(b) After a solution of 3-buten-2-ol in aqueous sulfuric acid had been allowed to stand for 1 week, it was found to contain both 3-buten-2-ol and 2-buten-1-ol.

(c) Treatment of $CH_3CH\!\!=\!\!CHCH_2OH$ with hydrogen bromide gave a mixture of 1-bromo-2-butene and 3-bromo-1-butene.

(d) Treatment of 3-buten-2-ol with hydrogen bromide gave the same mixture of bromides as in part (c).

(e) The major product in parts (c) and (d) was 1-bromo-2-butene.

10.32 2-Chloro-1,3-butadiene (chloroprene) is the monomer from which the elastomer *neoprene* is prepared. 2-Chloro-1,3-butadiene is the thermodynamically controlled product formed by addition of hydrogen chloride to vinylacetylene ($H_2C\!\!=\!\!CHC\equiv CH$). The principal product under conditions of kinetic control is the allenic chloride 4-chloro-1,2-butadiene. Suggest a mechanism to account for the formation of each product.

10.33 (a) Write equations expressing the *s*-trans ⇌ *s*-cis conformational equilibrium for (*E*)-1,3-pentadiene and for (*Z*)-1,3-pentadiene.

(b) For which stereoisomer will the equilibrium favor the *s*-trans conformation more strongly? Why? Support your prediction by making molecular models.

10.34 Which of the following are chiral?

(a) 2-Methyl-2,3-hexadiene   (c) 2,4-Dimethyl-2,3-pentadiene

(b) 4-Methyl-2,3-hexadiene

10.35 (a) Describe the molecular geometry expected for 1,2,3-butatriene (H_2C=C=C=CH_2).

(b) Two stereoisomers are expected for 2,3,4-hexatriene (CH_3CH=C=C=$CHCH_3$). What should be the relationship between these two stereoisomers?

10.36 Suggest reagents suitable for carrying out each step in the following synthetic sequence:

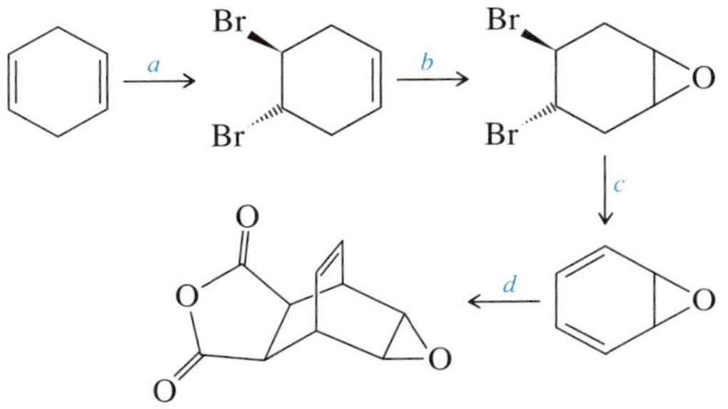

10.37 A very large number of Diels–Alder reactions are recorded in the chemical literature, many of which involve relatively complicated dienes, dienophiles, or both. On the basis of your knowledge of Diels–Alder reactions, predict the constitution of the Diels–Alder adduct that you would expect to be formed from the following combinations of dienes and dienophiles:

(a) 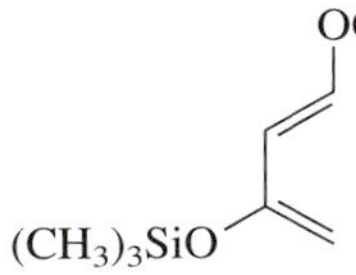 + CH_3O_2CC≡CCO_2CH_3

(b) 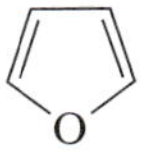 + CH_3O_2CC≡CCO_2CH_3

(c) 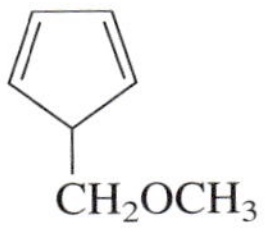 + H_2C=$CHNO_2$

10.38 On standing, 1,3-cyclopentadiene is transformed into a new compound called *dicyclopentadiene*, having the molecular formula $C_{10}H_{12}$. Hydrogenation of dicyclopentadiene gives the compound shown. Suggest a structure for dicyclopentadiene. What kind of reaction is occurring in its formation?

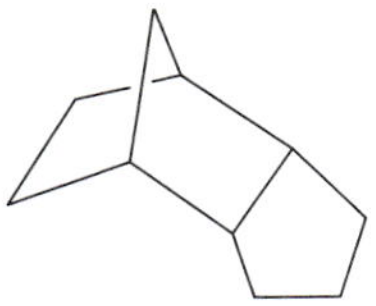

1,3-Cyclopentadiene $C_{10}H_{12}$ $C_{10}H_{16}$

dicyclopentadiene $\xrightarrow[Pt]{H_2}$

10.39 Refer to the molecular orbital diagrams of allyl cation (Figure 10.13) and those presented earlier in this chapter for ethylene and 1,3-butadiene (Figures 10.9 and 10.10) to decide which of the following cycloaddition reactions are allowed and which are forbidden according to the Woodward–Hoffmann rules.

(a) 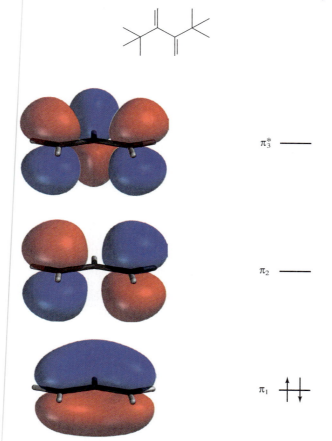   (b)

10.40 Alkenes slowly undergo a reaction in air called *autoxidation* in which allylic hydroperoxides are formed.

Cyclohexene   Oxygen   3-Hydroperoxycyclohexene

Keeping in mind that oxygen has two unpaired electrons ($\cdot \ddot{O} \colon \ddot{O} \cdot$), suggest a reasonable mechanism for this reaction.

 10.41 Make molecular models of:

(a) 1,2-Pentadiene   (b) (*E*)-1,3-Pentadiene   (c) 1,4-Pentadiene

Examine the C—C bond distances in these substances. Is there a correlation with the hybridization states of the bonded carbons?

 10.42 The compound shown is quite unreactive in Diels–Alder reactions. Make a space-filling model of it in the conformation required for the Diels–Alder reaction to see why.

π_3^* ——

π_2 ——

π_1

FIGURE 10.13 The π molecular orbitals of allyl cation. The allyl cation has two π electrons and they are in the orbital marked π_1.

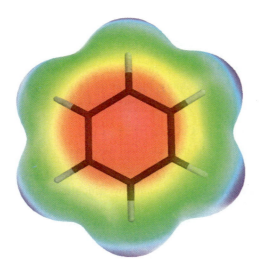

ARENES AND AROMATICITY

In this chapter and the next we extend our coverage of conjugated systems to include **arenes.** Arenes are hydrocarbons based on the benzene ring as a structural unit. Benzene, toluene, and naphthalene, for example, are arenes.

Benzene   Toluene   Naphthalene

One factor that makes conjugation in arenes special is its cyclic nature. A conjugated system that closes on itself can have properties that are much different from those of open-chain polyenes. Arenes are also referred to as **aromatic hydrocarbons.** Used in this sense, the word *aromatic* has nothing to do with odor but means instead that arenes are much more stable than we expect them to be based on their formulation as conjugated trienes. Our goal in this chapter is to develop an appreciation for the concept of **aromaticity**—to see what properties of benzene and its derivatives reflect its special stability and to explore the reasons for it. This chapter develops the idea of the benzene ring as a fundamental structural unit and examines the effect of a benzene ring as a substituent. The chapter following this one describes reactions that involve the ring itself.

Let's begin by tracing the history of benzene, its origin, and its structure. Many of the terms we use, including *aromaticity* itself, are of historical origin. We'll begin with the discovery of benzene.

11.1   BENZENE

In 1825, Michael Faraday isolated a new hydrocarbon from illuminating gas, which he called "bicarburet of hydrogen." Nine years later Eilhardt Mitscherlich of the University of Berlin prepared the same substance by heating benzoic acid with lime and found it to be a hydrocarbon having the empirical formula C_nH_n.

$$C_6H_5CO_2H \quad + \quad CaO \quad \xrightarrow{\text{heat}} \quad C_6H_6 \quad + \quad CaCO_3$$

Benzoic acid   Calcium oxide   Benzene   Calcium carbonate

Eventually, because of its relationship to benzoic acid, this hydrocarbon came to be named *benzin,* then later *benzene,* the name by which it is known today.

Benzoic acid had been known for several hundred years by the time of Mitscherlich's experiment. Many trees exude resinous materials called *balsams* when cuts are made in their bark. Some of these balsams are very fragrant, which once made them highly prized articles of commerce, especially when the trees that produced them could be found only in exotic, faraway lands. *Gum benzoin* is a balsam obtained from a tree that grows in Java and Sumatra. *Benzoin* is a word derived from the French equivalent, *benjoin,* which in turn comes from the Arabic *luban jawi,* meaning "incense from Java." Benzoic acid is itself odorless but can easily be isolated from gum benzoin.

Compounds related to benzene were obtained from similar plant extracts. For example, a pleasant-smelling resin known as *tolu balsam* was obtained from the South American tolu tree. In the 1840s it was discovered that distillation of tolu balsam gave a methyl derivative of benzene, which, not surprisingly, came to be named *toluene.*

Although benzene and toluene are not particularly fragrant compounds themselves, their origins in aromatic plant extracts led them and compounds related to them to be classified as *aromatic hydrocarbons.* Alkanes, alkenes, and alkynes belong to another class, the **aliphatic hydrocarbons.** The word *aliphatic* comes from the Greek *aleiphar* (meaning "oil" or "unguent") and was given to hydrocarbons that were obtained by the chemical degradation of fats.

Benzene was prepared from coal tar by August W. von Hofmann in 1845. Coal tar remained the primary source for the industrial production of benzene for many years, until petroleum-based technologies became competitive about 1950. Current production is about 6 million tons per year in the United States. A substantial portion of this benzene is converted to styrene for use in the preparation of polystyrene plastics and films.

Toluene is also an important organic chemical. Like benzene, its early industrial production was from coal tar, but most of it now comes from petroleum.

11.2   KEKULÉ AND THE STRUCTURE OF BENZENE

The classification of hydrocarbons as aliphatic or aromatic took place in the 1860s when it was already apparent that there was something special about benzene, toluene, and their derivatives. Their molecular formulas (benzene is C_6H_6, toluene is C_7H_8) indicate that, like alkenes and alkynes, they are unsaturated and should undergo addition reactions. Under conditions in which bromine, for example, reacts rapidly with alkenes and alkynes, however, benzene proved to be inert. Benzene does react with Br_2 in the presence of iron(III) bromide as a catalyst, but even then addition isn't observed. Substitution occurs instead!

$$C_6H_6 \; + \; Br_2 \xrightarrow{\;CCl_4\;} \text{no observable reaction}$$

Benzene   Bromine

$$\xrightarrow{\;FeBr_3\;} C_6H_5Br \; + \; HBr$$

Bromobenzene   Hydrogen bromide

Furthermore, only one monobromination product of benzene was ever obtained, which suggests that all the hydrogen atoms of benzene are equivalent. Substitution of one hydrogen by bromine gives the same product as substitution of any of the other hydrogens.

Chemists came to regard the six carbon atoms of benzene as a fundamental structural unit. Reactions could be carried out that altered its substituents, but the integrity of the benzene unit remained undisturbed. There must be something "special" about benzene that makes it inert to many of the reagents that add to alkenes and alkynes.

In 1866, only a few years after publishing his ideas concerning what we now recognize as the structural theory of organic chemistry, August Kekulé applied it to the structure of benzene. He based his reasoning on three premises:

1. Benzene is C_6H_6.

2. All the hydrogens of benzene are equivalent.

3. The structural theory requires that there be four bonds to each carbon.

Kekulé advanced the venturesome notion that the six carbon atoms of benzene were joined together in a ring. Four bonds to each carbon could be accommodated by a system of alternating single and double bonds with one hydrogen on each carbon.

In 1861, Johann Josef Loschmidt, who was later to become a professor at the University of Vienna, privately published a book containing a structural formula for benzene similar to the one Kekulé would propose five years later. Loschmidt's book reached few readers, and his ideas were not well known.

A flaw in Kekulé's structure for benzene was soon discovered. Kekulé's structure requires that 1,2- and 1,6-disubstitution patterns create different compounds (isomers).

How many isomers of C_6H_6 can you write? An article in the March 1994 issue of the *Journal of Chemical Education* (pp. 222–224) claims that there are several hundred and draws structural formulas for 25 of them.

1,2-Disubstituted
derivative of benzene

1,6-Disubstituted
derivative of benzene

The two substituted carbons are connected by a double bond in one structure but by a single bond in the other. Because no such cases of isomerism in benzene derivatives were known, and none could be found, Kekulé suggested that two isomeric structures could exist but interconverted too rapidly to be separated.

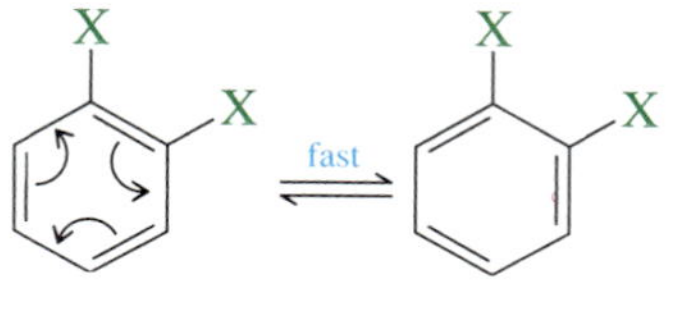

Kekulé's ideas about the structure of benzene left an important question unanswered. What is it about benzene that makes it behave so much differently from other unsaturated compounds? We'll see in this chapter that the answer is a simple one—the low reactivity of benzene and its derivatives reflects their special stability. Kekulé was wrong. *Benzene is not cyclohexatriene, nor is it a pair of rapidly equilibrating cyclohexatriene isomers.* But there was no way that Kekulé could have gotten it right given the state of chemical knowledge at the time. After all, the electron hadn't even been discovered yet. It remained for twentieth-century electronic theories of bonding to provide insight into why benzene is so stable. We'll outline these theories shortly. First, however, let's look at the structure of benzene in more detail.

Benzene is planar and its carbon skeleton has the shape of a regular hexagon. There is no evidence that it has alternating single and double bonds. As shown in Figure 11.1, all the carbon–carbon bonds are the same length (140 pm) and the 120° bond angles

BENZENE, DREAMS, AND CREATIVE THINKING

At ceremonies in Berlin in 1890 celebrating the twenty-fifth anniversary of his proposed structure of benzene, August Kekulé recalled the origins of his view of the benzene structure.

There I sat and wrote for my textbook, but things did not go well; my mind was occupied with other matters. I turned the chair towards the fireplace and began to doze. Once again the atoms danced before my eyes. This time smaller groups modestly remained in the background. My mental eye, sharpened by repeated apparitions of similar kind, now distinguished larger units of various shapes. Long rows, frequently joined more densely; everything in motion, twisting and turning like snakes. And behold, what was that? One of the snakes caught hold of its own tail and mockingly whirled round before my eyes. I awoke, as if by lightning; this time, too, I spent the rest of the night working out the consequences of this hypothesis.*

Concluding his remarks, Kekulé merged his advocacy of creative imagination with the rigorous standards of science by reminding his audience:

Let us learn to dream, then perhaps we shall find the truth. But let us beware of publishing our dreams before they have been put to the proof by the waking understanding.

The imagery of a whirling circle of snakes evokes a vivid picture that engages one's attention when first exposed to Kekulé's model of the benzene structure. Recently, however, the opinion has been expressed that Kekulé might have engaged in some hyperbole during his speech. Professor John Wotiz of Southern Illinois University suggests that discoveries in science are the result of a disciplined analysis of a sufficient body of experimental observations to progress to a higher level of understanding. Wotiz' view that Kekulé's account is more fanciful than accurate has sparked a controversy with ramifications that go beyond the history of organic chemistry. How does creative thought originate? What can we do to become more creative? Because these questions have concerned psychologists for decades, the idea of a sleepy Kekulé being more creative than an alert Kekulé becomes more than simply a charming story he once told about himself.

*The Kekulé quotes are taken from the biographical article of K. Hafner published in *Angew. Chem. Internat. ed. Engl.* **18,** 641–651 (1979).

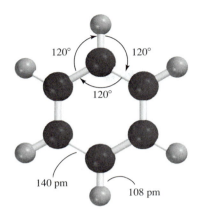

FIGURE 11.1 Bond distances and bond angles of benzene.

correspond to perfect sp^2 hybridization. Interestingly, the 140-pm bond distances in benzene are exactly midway between the typical sp^2–sp^2 single-bond distance of 146 pm and the sp^2–sp^2 double-bond distance of 134 pm. If bond distances are related to bond type, what kind of carbon–carbon bond is it that lies halfway between a single bond and a double bond in length?

11.3   A RESONANCE PICTURE OF BONDING IN BENZENE

Twentieth-century theories of bonding in benzene gave us a clearer picture of aromaticity. We'll start with a resonance description of benzene.

The two Kekulé structures for benzene have the same arrangement of atoms, but differ in the placement of electrons. Thus they are resonance forms, and neither one by itself correctly describes the bonding in the actual molecule. As a hybrid of the two Kekulé structures, benzene is often represented by a hexagon containing an inscribed circle.

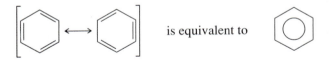

The circle-in-a-hexagon symbol was first suggested by the British chemist Sir Robert Robinson to represent what he called the "aromatic sextet"—the six delocalized π electrons of the three double bonds. Robinson's symbol is a convenient time-saving shorthand device, but Kekulé-type formulas are better for counting and keeping track of electrons, especially in chemical reactions.

Robinson won the 1947 Nobel Prize in chemistry for his studies of natural products. He may also have been the first to use curved arrows to track electron movement.

PROBLEM 11.1 Write structural formulas for toluene ($C_6H_5CH_3$) and for benzoic acid ($C_6H_5CO_2H$) (a) as resonance hybrids of two Kekulé forms and (b) with the Robinson symbol.

Because the carbons that are singly bonded in one resonance form are doubly bonded in the other, the resonance description is consistent with the observed carbon–carbon bond distances in benzene. These distances not only are all identical but also are intermediate between typical single-bond and double-bond lengths.

We have come to associate electron delocalization with increased stability. On that basis alone, benzene ought to be stabilized. It differs from other conjugated systems that we have seen, however, in that its π electrons are delocalized over a *cyclic conjugated* system. Both Kekulé structures of benzene are of equal energy, and one of the principles of resonance theory is that stabilization is greatest when the contributing structures are of similar energy. Cyclic conjugation in benzene, then, leads to a greater stabilization than is observed in noncyclic conjugated trienes. How much greater can be estimated from heats of hydrogenation.

11.4 THE STABILITY OF BENZENE

Hydrogenation of benzene and other arenes is more difficult than hydrogenation of alkenes and alkynes. Two of the more active catalysts are rhodium and platinum, and it is possible to hydrogenate arenes in the presence of these catalysts at room temperature and modest pressure. Benzene consumes three molar equivalents of hydrogen to give cyclohexane.

| Benzene | Hydrogen (2–3 atm pressure) | Cyclohexane (100%) |

Nickel catalysts, although less expensive than rhodium and platinum, are also less active. Hydrogenation of arenes in the presence of nickel requires high temperatures (100–200°C) and pressures (100 atm).

The measured heat of hydrogenation of benzene to cyclohexane is, of course, the same regardless of the catalyst and is 208 kJ/mol (49.8 kcal/mol). To put this value into perspective, compare it with the heats of hydrogenation of cyclohexene and 1,3-cyclohexadiene, as shown in Figure 11.2. The most striking feature of Figure 11.2 is that the heat of hydrogenation of benzene, with three "double bonds," is less than the heat of hydrogenation of the two double bonds of 1,3-cyclohexadiene.

Our experience has been that some 125 kJ/mol (30 kcal/mol) is given off whenever a double bond is hydrogenated. When benzene combines with three molecules of hydrogen, the reaction is far less exothermic than we would expect it to be on the basis of a 1,3,5-cyclohexatriene structure for benzene.

How much less? Because 1,3,5-cyclohexatriene does not exist (if it did, it would instantly relax to benzene), we cannot measure its heat of hydrogenation in order to compare it with benzene. We can approximate the heat of hydrogenation of 1,3,5-cyclohexatriene as being equal to three times the heat of hydrogenation of cyclohexene, or a total of 360 kJ/mol (85.8 kcal/mol). The heat of hydrogenation of benzene is 152 kJ/mol (36 kcal/mol) *less* than expected for a hypothetical 1,3,5-cyclohexatriene with noninteracting double bonds. This is the **resonance energy** of benzene. It is a measure of how much more stable benzene is than would be predicted on the basis of its formulation as a pair of rapidly interconverting 1,3,5-cyclohexatrienes.

We reach a similar conclusion when comparing benzene with the open-chain conjugated triene (Z)-1,3,5-hexatriene. Here we compare two real molecules, both conjugated trienes, but one is cyclic and the other is not. The heat of hydrogenation of

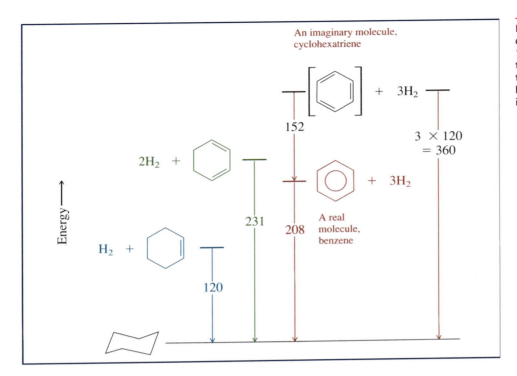

FIGURE 11.2 Heats of hydrogenation of cyclohexene, 1,3-cyclohexadiene, a hypothetical 1,3,5-cyclohexatriene, and benzene. All heats of hydrogenation are in kilojoules per mole.

(Z)-1,3,5-hexatriene is 337 kJ/mol (80.5 kcal/mol), a value which is 129 kJ/mol (30.7 kcal/mol) greater than that of benzene.

$$\text{(Z)-1,3,5-Hexatriene} + 3H_2 \longrightarrow CH_3(CH_2)_4CH_3 \qquad \Delta H° = -337 \text{ kJ} \\ (-80.5 \text{ kcal})$$

(Z)-1,3,5-Hexatriene   Hydrogen   Hexane

    The precise value of the resonance energy of benzene depends, as comparisons with 1,3,5-cyclohexatriene and (Z)-1,3,5-hexatriene illustrate, on the compound chosen as the reference. What is important is that the resonance energy of benzene is quite large, six to ten times that of a conjugated triene. It is this very large increment of resonance energy that places benzene and related compounds in a separate category that we call *aromatic*.

> **PROBLEM 11.2** The heats of hydrogenation of cycloheptene and 1,3,5-cyclo-heptatriene are 110 kJ/mol (26.3 kcal/mol) and 305 kJ/mol (73.0 kcal/mol), respectively. In both cases cycloheptane is the product. What is the resonance energy of 1,3,5-cycloheptatriene? How does it compare with the resonance energy of benzene?

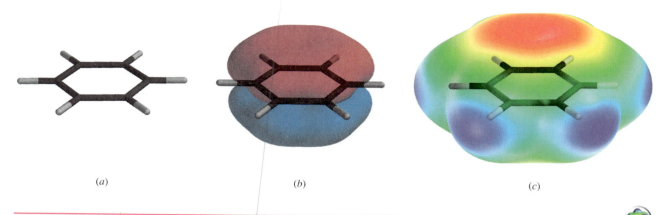

(a) (b) (c)

FIGURE 11.3 (*a*) The framework of bonds shown in the tube model of benzene are σ bonds. (*b*) Each carbon is sp^2-hybridized and has a 2*p* orbital perpendicular to the σ framework. Overlap of the 2*p* orbitals generates a π system encompassing the entire ring. (*c*) Electrostatic potential map of benzene. The red area in the center corresponds to the region above and below the plane of the ring where the π electrons are concentrated.

11.5   AN ORBITAL HYBRIDIZATION VIEW OF BONDING IN BENZENE

The structural facts that benzene is planar, all of the bond angles are 120°, and each carbon is bonded to three other atoms, suggest sp^2 hybridization for carbon and the framework of σ bonds shown in Figure 11.3*a*.

In addition to its three sp^2 hybrid orbitals, each carbon has a half-filled 2*p* orbital that can participate in π bonding. Figure 11.3*b* shows the continuous π system that encompasses all of the carbons that result from overlap of these 2*p* orbitals. The six π electrons of benzene are delocalized over all six carbons.

The electrostatic potential map of benzene (Figure 11.3*c*) shows regions of high electron density above and below the plane of the ring, which is where we expect the most loosely held electrons (the π electrons) to be. In Chapter 12 we will see how this region of high electron density is responsible for the characteristic chemical reactivity of benzene and its relatives.

11.6   THE π MOLECULAR ORBITALS OF BENZENE

The picture of benzene as a planar framework of σ bonds with six electrons in a delocalized π orbital is a useful, but superficial, one. Six electrons cannot simultaneously occupy any one orbital, be it an atomic orbital or a molecular orbital. We can fix this with the more accurate molecular orbital picture shown in Figure 11.4. We learned in Section 2.4 that when atomic orbitals (AOs) combine to give molecular orbitals (MOs), the final number of MOs must equal the original number of AOs. Thus, the six 2*p* AOs of six sp^2-hybridized carbons combine to give six π MOs of benzene.

The orbitals in Figure 11.4 are arranged in order of increasing energy. Three orbitals are bonding; three are antibonding. Each of the three bonding MOs contains two electrons, accounting for the six π electrons of benzene. There are no electrons in the antibonding MOs. Benzene is said to have a **closed-shell** π-electron configuration.

Figure 11.4 also shows the orbital overlaps and nodal properties of the benzene MOs. Recall that a wave function changes sign on passing through a nodal plane and is

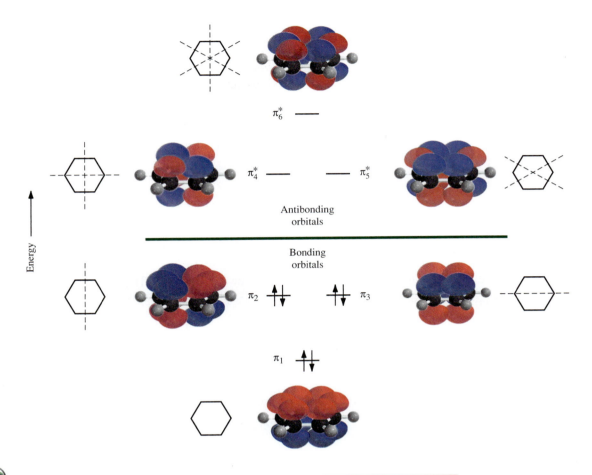

🌐 **FIGURE 11.4** The π molecular orbitals of benzene arranged in order of increasing energy and showing nodal surfaces. The six π electrons of benzene occupy the three lowest energy orbitals, all of which are bonding.

zero at a node (Section 1.1). All of the orbital interactions in the lowest energy orbital π_1 are bonding; therefore, π_1 has no nodes. The other two bonding orbitals π_2 and π_3 each have one nodal plane. The first two antibonding orbitals π_4^* and π_5^* each have two nodal planes. The highest energy orbital π_6^* has three nodal planes. All adjacent p orbitals are out of phase with one another in π_6^*—all of the interactions are antibonding.

The pattern of orbital energies is different for benzene than it would be if the six π electrons were confined to three noninteracting double bonds. The delocalization provided by cyclic conjugation in benzene causes its π electrons to be held more strongly than they would be in the absence of cyclic conjugation. Stronger binding of its π electrons is the factor most responsible for the special stability—the aromaticity—of benzene.

Later in this chapter we'll explore the criteria for aromaticity in more detail to see how they apply to cyclic polyenes of different ring sizes. The next several sections introduce us to the chemistry of compounds that contain a benzene ring as a structural unit. We'll start with how we name them.

11.7   SUBSTITUTED DERIVATIVES OF BENZENE AND THEIR NOMENCLATURE

All compounds that contain a benzene ring are aromatic, and substituted derivatives of benzene make up the largest class of aromatic compounds. Many such compounds are named by attaching the name of the substituent as a prefix to *benzene*.

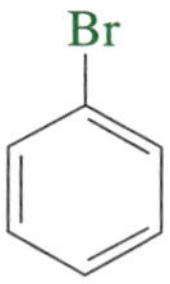

Bromobenzene 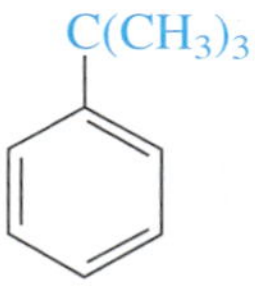   *tert*-Butylbenzene 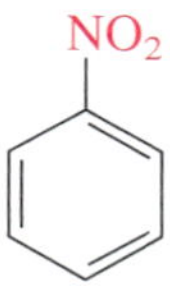   Nitrobenzene

Many simple monosubstituted derivatives of benzene have common names of long standing that have been retained in the IUPAC system. Table 11.1 lists some of the most important ones.

Dimethyl derivatives of benzene are called *xylenes*. There are three xylene isomers, the *ortho* (*o*)-, *meta* (*m*)-, and *para* (*p*)- substituted derivatives.

TABLE 11.1	Names of Some Frequently Encountered Derivatives of Benzene	
Structure	**Systematic Name**	**Common Name***
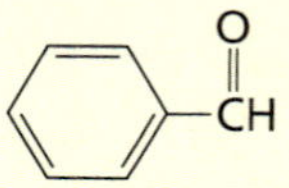	Benzenecarbaldehyde	Benzaldehyde
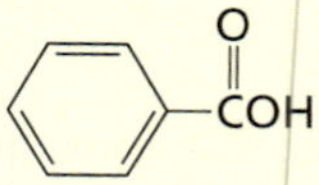	Benzenecarboxylic acid	Benzoic acid
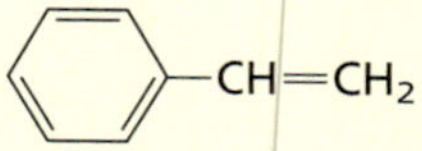	Vinylbenzene	Styrene
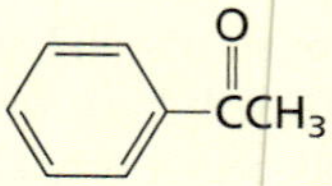	Methyl phenyl ketone	Acetophenone
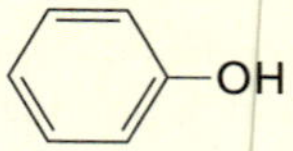	Benzenol	Phenol
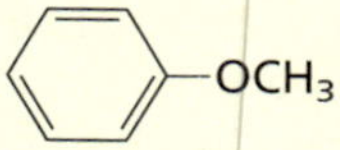	Methoxybenzene	Anisole
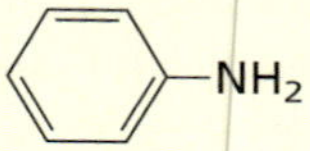	Benzenamine	Aniline

*These common names are acceptable in IUPAC nomenclature and are the names that will be used in this text.

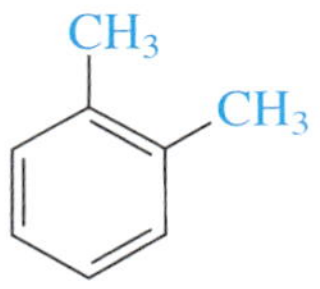

o-Xylene
(1,2-dimethylbenzene)

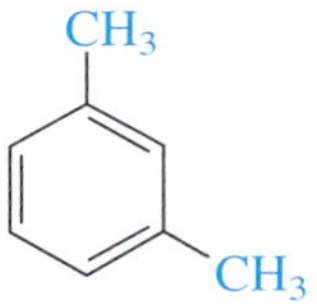

m-Xylene
(1,3-dimethylbenzene)

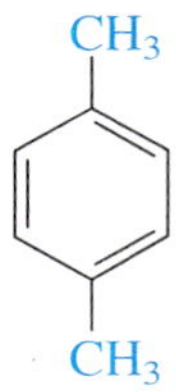

p-Xylene
(1,4-dimethylbenzene)

The prefix *ortho* signifies a 1,2-disubstituted benzene ring, *meta* signifies 1,3-disubstitution, and *para* signifies 1,4-disubstitution. The prefixes *o, m,* and *p* can be used when a substance is named as a benzene derivative or when a specific base name (such as acetophenone) is used. For example,

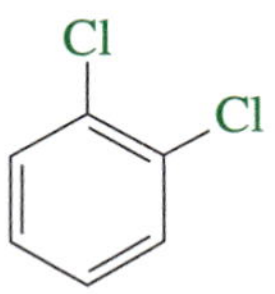

o-Dichlorobenzene
(1,2-dichlorobenzene)

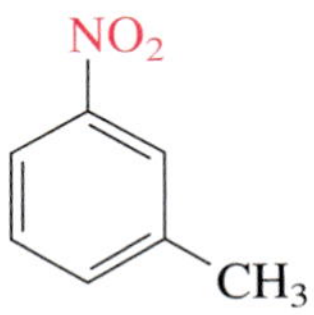

m-Nitrotoluene
(3-nitrotoluene)

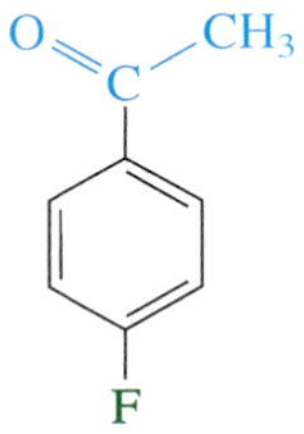

p-Fluoroacetophenone
(4-fluoroacetophenone)

PROBLEM 11.3 Write a structural formula for each of the following compounds:

(a) *o*-Ethylanisole   (b) *m*-Chlorostyrene   (c) *p*-Nitroaniline

SAMPLE SOLUTION (a) The parent compound in *o*-ethylanisole is anisole. Anisole, as shown in Table 11.1, has a methoxy (CH_3O-) substituent on the benzene ring. The ethyl group in *o*-ethylanisole is attached to the carbon adjacent to the one that bears the methoxy substituent.

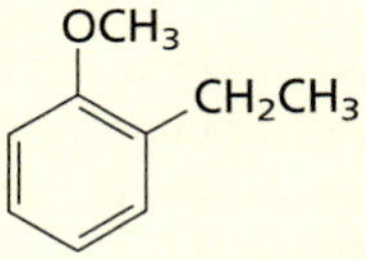

o-Ethylanisole

The *o, m,* and *p* prefixes are *not* used when three or more substituents are present on benzene; numerical locants must be used instead.

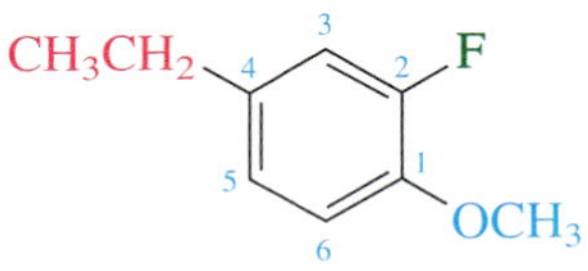

4-Ethyl-2-fluoroanisole

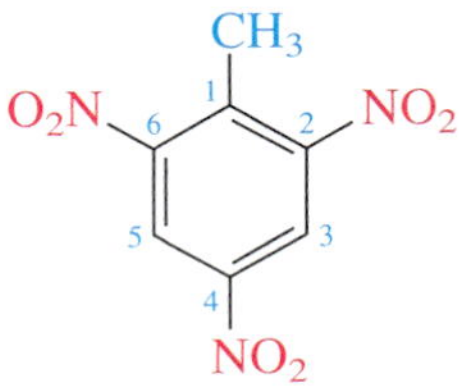

2,4,6-Trinitrotoluene

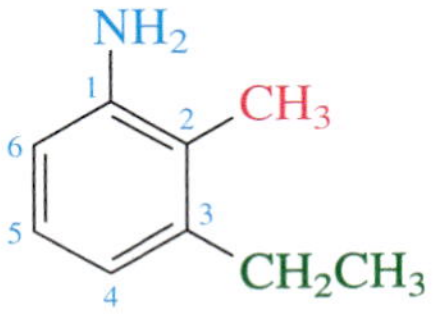

3-Ethyl-2-methylaniline

In these examples the base name of the benzene derivative determines the carbon at which numbering begins: anisole has its methoxy group at C-1, toluene its methyl group

The "first point of difference" rule was introduced in Section 2.14.

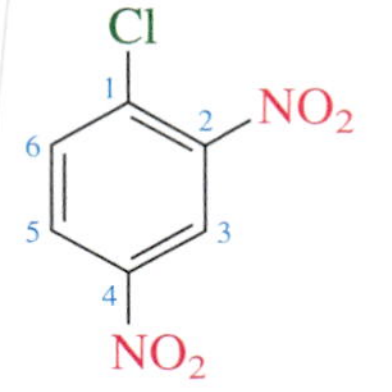

at C-1, and aniline its amino group at C-1. The direction of numbering is chosen to give the next substituted position the lowest number irrespective of what substituent it bears. *The order of appearance of substituents in the name is alphabetical.* When no simple base name other than benzene is appropriate, positions are numbered so as to give the lowest locant at the first point of difference. Thus, each of the following examples is named as a 1,2,4-trisubstituted derivative of benzene rather than as a 1,3,4-derivative:

1-Chloro-2,4-dinitrobenzene 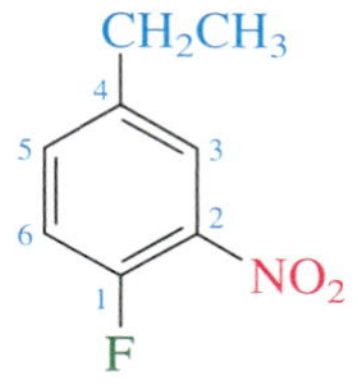   4-Ethyl-1-fluoro-2-nitrobenzene

When the benzene ring is named as a substituent, the word *phenyl* stands for C_6H_5—. Similarly, an arene named as a substituent is called an *aryl* group. A *benzyl* group is $C_6H_5CH_2$—.

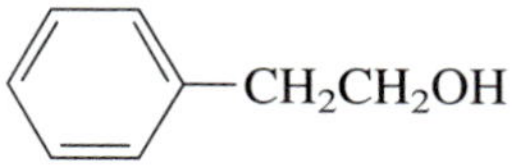

2-Phenylethanol 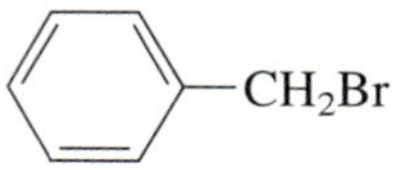   Benzyl bromide

Biphenyl is the accepted IUPAC name for the compound in which two benzene rings are connected by a single bond.

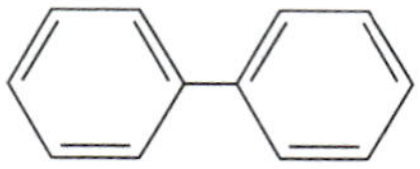

Biphenyl 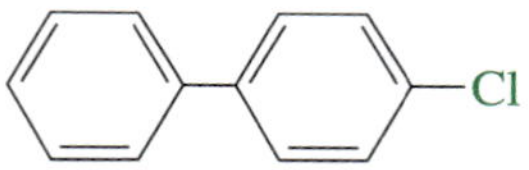   *p*-Chlorobiphenyl

11.8 POLYCYCLIC AROMATIC HYDROCARBONS

Members of a class of arenes called **polycyclic aromatic hydrocarbons** possess substantial resonance energies because each is a collection of benzene rings fused together.

Naphthalene, anthracene, and phenanthrene are the three simplest members of this class. They are all present in **coal tar,** a mixture of organic substances formed when coal is converted to coke by heating at high temperatures (about 1000°C) in the absence of air. Naphthalene is **bicyclic** (has two rings), and its two benzene rings share a common side. Anthracene and phenanthrene are both **tricyclic** aromatic hydrocarbons. Anthracene has three rings fused in a "linear" fashion; an "angular" fusion characterizes phenanthrene. The structural formulas of naphthalene, anthracene, and phenanthrene are shown along with the numbering system used to name their substituted derivatives:

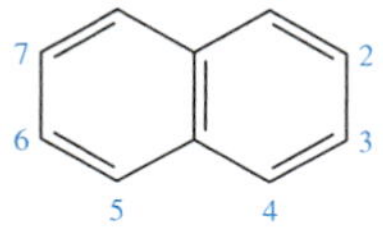

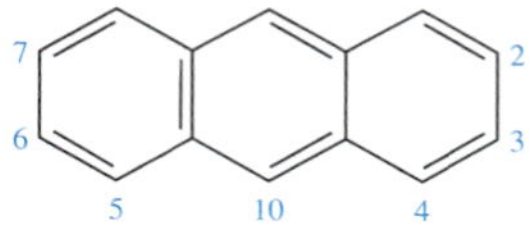

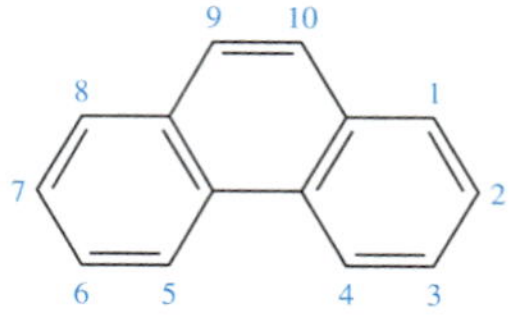

Naphthalene is a white crystalline solid melting at 80°C that sublimes readily. It has a characteristic odor and was formerly used as a moth repellent.

Arene:	Naphthalene	Anthracene	Phenanthrene
Resonance energy:	255 kJ/mol (61 kcal/mol)	347 kJ/mol (83 kcal/mol)	381 kJ/mol (91 kcal/mol)

In general, the most stable resonance structure for a polycyclic aromatic hydrocarbon is the one with the greatest number of rings that correspond to Kekulé formulations of benzene. Naphthalene provides a fairly typical example:

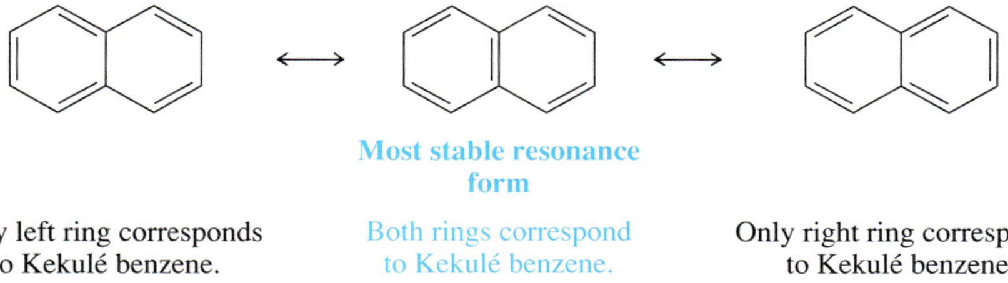

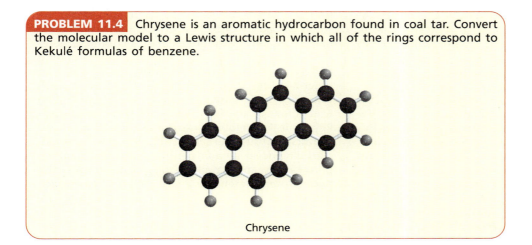

Only left ring corresponds to Kekulé benzene.

Most stable resonance form
Both rings correspond to Kekulé benzene.

Only right ring corresponds to Kekulé benzene.

Notice that anthracene cannot be represented by any single Lewis structure in which all three rings correspond to Kekulé formulations of benzene, but phenanthrene can.

> **PROBLEM 11.4** Chrysene is an aromatic hydrocarbon found in coal tar. Convert the molecular model to a Lewis structure in which all of the rings correspond to Kekulé formulas of benzene.
>
> Chrysene

A large number of polycyclic aromatic hydrocarbons are known. Many have been synthesized in the laboratory, and several of the others are products of combustion. Benzo[a]pyrene, for example, is present in tobacco smoke, contaminates food cooked on barbecue grills, and collects in the soot of chimneys. Benzo[a]pyrene is a **carcinogen** (a cancer-causing substance). It is converted in the liver to an epoxy diol that can induce mutations leading to the uncontrolled growth of certain cells.

In 1775, the British surgeon Sir Percivall Pott suggested that scrotal cancer in chimney sweeps was caused by soot. This was the first proposal that cancer could be caused by chemicals present in the workplace.

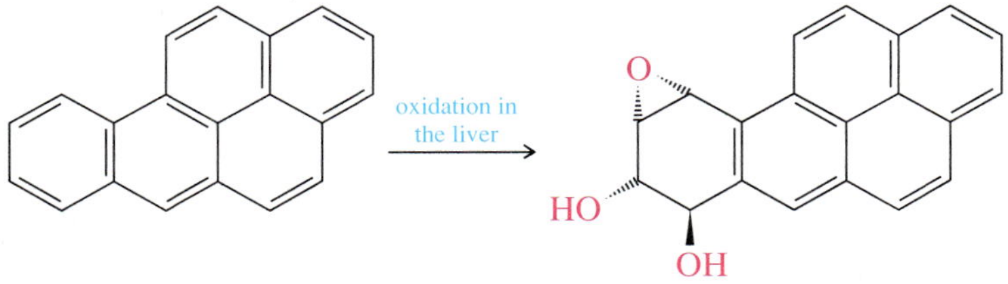

Benzo[a]pyrene

oxidation in the liver

7,8-Dihydroxy-9,10-epoxy-7,8,9,10-tetrahydrobenzo[a]pyrene

CARBON CLUSTERS, FULLERENES, AND NANOTUBES

The 1996 Nobel Prize in chemistry was awarded to Professors Harold W. Kroto (University of Sussex), Robert F. Curl, and Richard E. Smalley (both of Rice University) for groundbreaking work involving elemental carbon that opened up a whole new area of chemistry. The work began when Kroto wondered whether polyacetylenes of the type $HC{\equiv}C{-}(C{\equiv}C)_n{-}C{\equiv}CH$ might be present in interstellar space and discussed experiments to test this idea while visiting Curl and Smalley at Rice in the spring of 1984. Smalley had developed a method for the laser-induced evaporation of metals at very low pressure and was able to measure the molecular weights of the various clusters of atoms produced. Kroto, Curl, and Smalley felt that by applying this technique to graphite (Figure 11.5) the vaporized carbon produced might be similar to that produced by a carbon-rich star.

When the experiment was carried out in the fall of 1985, Kroto, Curl, and Smalley found that under certain conditions a species with a molecular formula of C_{60} was present in amounts much greater than any other. On speculating about what C_{60} might be, they concluded that its most likely structure is the spherical cluster of carbon atoms shown in Figure 11.6 and suggested it be called *buckminsterfullerene* because of its similarity to the geodesic domes popularized by the American architect and inventor R. Buckminster Fuller. (It is also often referred to as a "buckyball.") Other carbon clusters, some larger than C_{60} and some smaller, were also formed in the experiment, and the general term *fullerene* refers to such carbon clusters.

All of the carbon atoms in buckminsterfullerene are equivalent and are sp^2-hybridized; each one simultaneously belongs to one five-membered ring and two benzene-like six-membered rings. The strain caused by distortion of the rings from coplanarity is equally distributed among all of the carbons.

Confirming its structure required isolating enough C_{60} to apply modern techniques of structure determination. A quantum leap in fullerene research came in 1990 when a team led by Wolfgang Krätschmer of the Max Planck Institute for Nuclear Physics in Heidelberg and Donald Huffman of the University of Arizona successfully prepared buckminsterfullerene in amounts sufficient for its isolation, purification, and detailed study. Not only was the buckminsterfullerene structure shown to be correct,

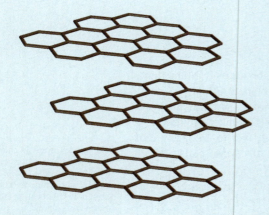
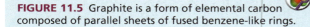

FIGURE 11.5 Graphite is a form of elemental carbon composed of parallel sheets of fused benzene-like rings.

FIGURE 11.6 Buckminsterfullerene (C_{60}). All carbons are equivalent, and no five-membered rings are adjacent to one another.

—*Cont.*

but academic and industrial scientists around the world seized the opportunity afforded by the availability of C_{60} in quantity to study its properties.

Speculation about the stability of C_{60} centered on the extent to which the aromaticity associated with its 20 benzene rings is degraded by their nonplanarity and the accompanying angle strain. It is now clear that C_{60} is a relatively reactive substance, reacting with many substances toward which benzene itself is inert. Many of these reactions are characterized by addition to buckminsterfullerene, converting sp^2-hybridized carbons to sp^3-hybridized ones and reducing the overall strain.

The field of fullerene chemistry expanded in an unexpected direction in 1991 when Sumio Iijima of the NEC Fundamental Research Laboratories in Japan discovered fibrous carbon clusters in one of his fullerene preparations. This led, within a short time, to substances of the type portrayed in Figure 11.7 called *single-walled nanotubes*. The best way to think about this material is as a "stretched" fullerene. Take a molecule of C_{60}, cut it in half, and place a cylindrical tube of fused six-membered carbon rings between the two halves.

Thus far, the importance of carbon cluster chemistry has been in the discovery of new knowledge. Many scientists feel that the earliest industrial applications of the fullerenes will be based on their novel electrical properties. Buckminsterfullerene is an insulator, but has a high electron affinity and is a superconductor in its reduced form. Nanotubes have aroused a great deal of interest for their electrical properties and as potential sources of carbon fibers of great strength.

The question that began the fullerene story, the possibility that carbon clusters are formed in stars, moved closer to an answer in 2000 when a team of geochemists led by Luann Becker (University of Hawaii) and Robert J. Poreda (University of Rochester) reported finding fullerenes in ancient sediments. These particular sediments were deposited 251 million years ago, a time that coincides with the Permian–Triassic (P-T) boundary, marking the greatest mass extinction in Earth's history. The P-T extinction was far more devastating than the better known Cretaceous–Tertiary (K-T) extinction in which the dinosaurs disappeared 65 million years ago. It is widely believed that the K-T extinction was triggered when a comet smashed into the Earth, sending up debris clouds that darkened the sky and led to the loss of vegetation on which the dinosaurs and other large animals fed.

The possibility that an even larger impact caused the P-T extinction received support when Becker and Poreda found that helium and argon atoms were present in the inner cores of some of the fullerenes from the P-T boundary sediments. (The cover of this book shows a helium atom inside a molecule of C_{60}.) What is special about the fullerene-trapped atoms is that the mixtures of both helium and argon isotopes resemble extraterrestrial isotopic mixtures more than earthly ones. The $^3He/^4He$ ratio in the P-T boundary fullerenes, for example, is 50 times larger than the "natural abundance" ratio.

What a story! Fullerenes formed during the explosion of a star travel through interstellar space as passengers on a comet or asteroid that eventually smashes into Earth. Some of the fullerenes carry passengers themselves—atoms of helium and argon from the dying star. The fullerenes and the noble gas atoms silently wait for 251 million years to tell us where they came from and what happened when they got here.

FIGURE 11.7 A portion of a nanotube. The closed end is one half of a buckyball. The main length cannot close as long as all of the rings are hexagons.

11.9   PHYSICAL PROPERTIES OF ARENES

Selected physical properties for a number of arenes are listed in Appendix 1.

In general, arenes resemble other hydrocarbons in their physical properties. They are nonpolar, insoluble in water, and less dense than water. In the absence of polar substituents, intermolecular forces are weak and limited to van der Waals attractions of the induced-dipole/induced-dipole type.

At one time, benzene was widely used as a solvent. This use virtually disappeared when statistical studies revealed an increased incidence of leukemia among workers exposed to atmospheric levels of benzene as low as 1 ppm. Toluene has replaced benzene as an inexpensive organic solvent because it has similar solvent properties but has not been determined to be carcinogenic in the cell systems and at the dose levels that benzene is.

11.10   REACTIONS OF ARENES: A PREVIEW

We'll examine the chemical properties of aromatic compounds from two different perspectives:

1. *One mode of chemical reactivity involves the ring itself as a functional group and includes*

   (a) Reduction

   (b) Electrophilic aromatic substitution

Reduction of arenes by catalytic hydrogenation was described in Section 11.4. A different method using Group I metals as reducing agents, which gives 1,4-cyclohexadiene derivatives, will be presented in Section 11.11. **Electrophilic aromatic substitution** is the most important reaction type exhibited by benzene and its derivatives and constitutes the entire subject matter of Chapter 12.

2. *The second family of reactions are those in which the aryl group acts as a substituent and affects the reactivity of a functional unit to which it is attached.*

A carbon atom that is directly attached to a benzene ring is called a **benzylic** carbon (analogous to the allylic carbon of C=C—C). A phenyl group (C_6H_5—) is an even better conjugating substituent than a vinyl group (H_2C=CH—), and benzylic carbocations and radicals are more highly stabilized than their allylic counterparts. The double bond of an alkenylbenzene is stabilized to about the same extent as that of a conjugated diene.

Benzylic carbocation   Benzylic radical   Alkenylbenzene

Reactions involving benzylic cations, benzylic radicals, and alkenylbenzenes will be discussed in Sections 11.12 through 11.17.

11.11   THE BIRCH REDUCTION

We saw in Section 9.10 that the combination of a Group I metal and liquid ammonia is a powerful reducing system capable of reducing alkynes to *trans*-alkenes. In the presence of an alcohol, this same combination reduces arenes to *nonconjugated dienes*. Thus, treatment of benzene with sodium and methanol or ethanol in liquid ammonia converts it to 1,4-cyclohexadiene.

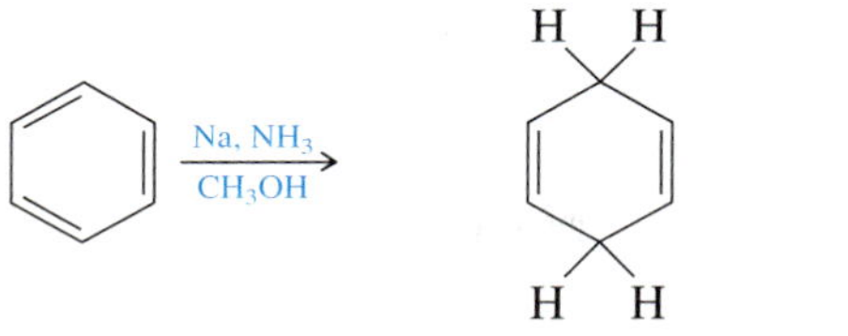

Benzene   1,4-Cyclohexadiene (80%)

Metal–ammonia–alcohol reductions of aromatic rings are known as **Birch reductions,** after the Australian chemist Arthur J. Birch, who demonstrated their usefulness beginning in the 1940s.

The mechanism by which the Birch reduction of benzene takes place (Figure 11.8) is analogous to the mechanism for the metal–ammonia reduction of alkynes. It involves a sequence of four steps in which steps 1 and 3 are single-electron transfers from the metal and steps 2 and 4 are proton transfers from the alcohol.

The Birch reduction not only provides a method to prepare dienes from arenes, which cannot be accomplished by catalytic hydrogenation, but also gives a nonconjugated diene system rather than the more stable conjugated one.

Alkyl-substituted arenes give 1,4-cyclohexadienes in which the alkyl group is a substituent on the double bond.

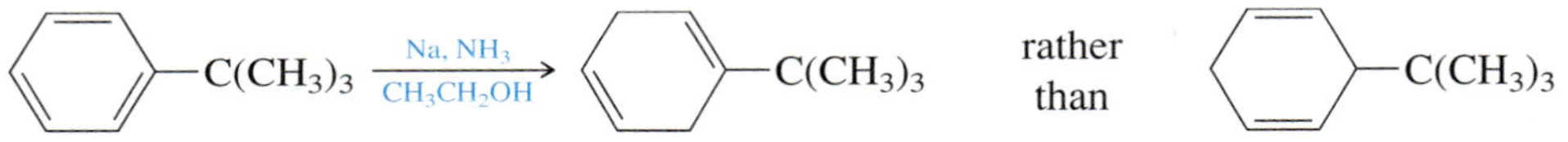

tert-Butylbenzene	1-tert-Butyl-1,4-cyclohexadiene (86%)	3-tert-Butyl-1,4-cyclohexadiene

PROBLEM 11.5   A single organic product was isolated after Birch reduction of *p*-xylene. Suggest a reasonable structure for this substance.

Substituents other than alkyl groups may also be present on the aromatic ring, but their reduction is beyond the scope of the present discussion.

11.12   FREE-RADICAL HALOGENATION OF ALKYLBENZENES

The benzylic position in alkylbenzenes is analogous to the allylic position in alkenes. Thus a benzylic C—H bond, like an allylic one, is weaker than a C—H bond of an alkane, as the bond dissociation energies of toluene, propene, and 2-methylpropane attest:

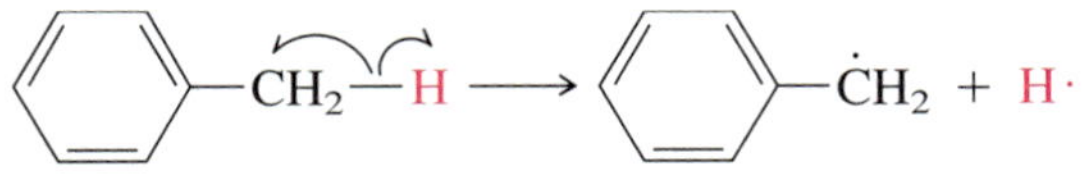

Toluene   Benzyl radical

$$H_2C=CHCH_2 \overset{\frown}{} H \longrightarrow H_2C=CH\dot{C}H_2 + H\cdot \qquad \Delta H° = 368 \text{ kJ (88 kcal)}$$

Propene   Allyl radical

$$(CH_3)_3C \overset{\frown}{} H \longrightarrow (CH_3)_3C\cdot \;\; + H\cdot \qquad \Delta H° = 380 \text{ kJ (91 kcal)}$$

2-Methylpropane   *tert*-Butyl radical

The overall reaction:

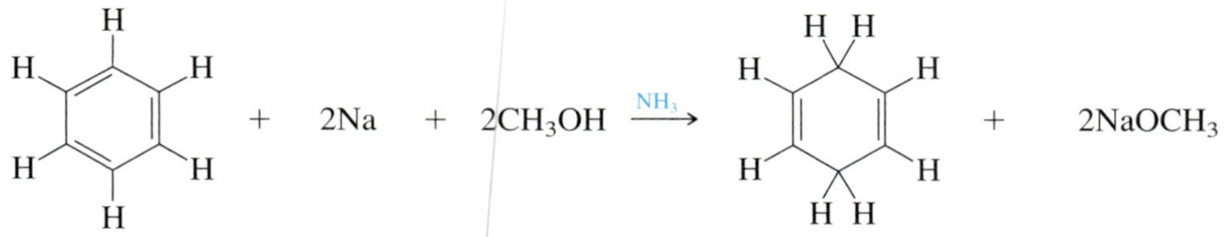

Benzene   Sodium   Methanol   1,4-Cyclohexadiene   Sodium methoxide

The mechanism:

Step 1: An electron is transferred from sodium (the reducing agent) to the π system of the aromatic ring. The product is an anion radical.

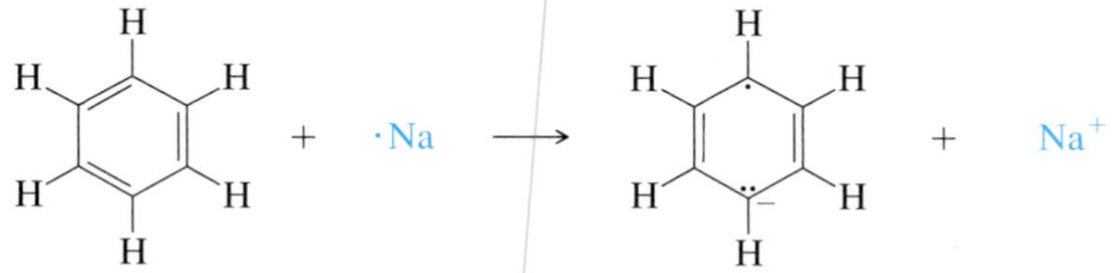

Benzene   Sodium   Benzene anion radical   Sodium ion

Step 2: The anion radical is a strong base and abstracts a proton from methanol.

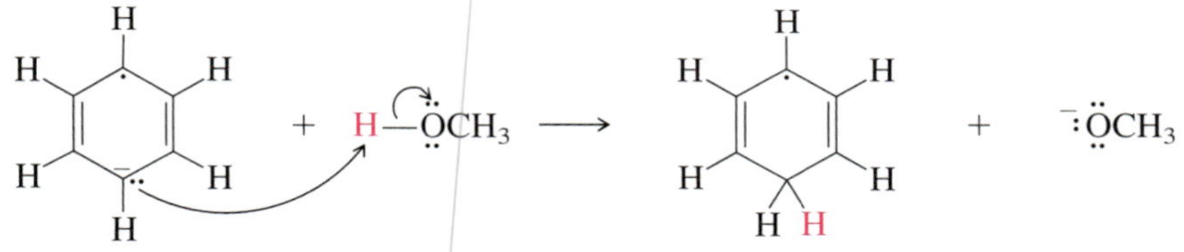

Benzene anion radical   Methanol   Cyclohexadienyl radical   Methoxide ion

Step 3: The cyclohexadienyl radical produced in step 2 is converted to an anion by electron transfer from sodium.

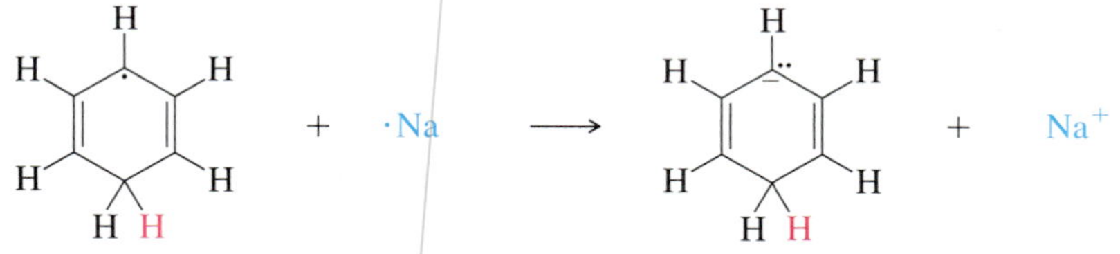

Cyclohexadienyl radical   Sodium   Cyclohexadienyl anion   Sodium ion

Step 4: Proton transfer from methanol to the anion gives 1,4-cyclohexadiene.

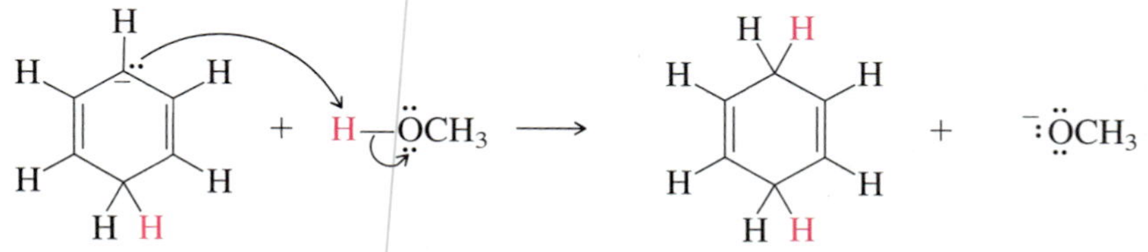

Cyclohexadienyl anion   1,4-Cyclohexadiene   Methoxide ion

FIGURE 11.8 Mechanism of the Birch reduction.

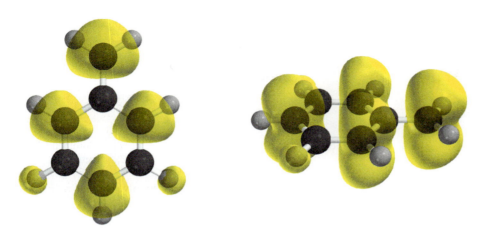

FIGURE 11.9 Two veiws of the spin density in benzyl radical. The unpaired electron is shared mainly by the benzylic carbon and the ortho and para carbons of the ring.

We attributed the decreased bond dissociation energy in propene to stabilization of allyl radical by electron delocalization. Similarly, electron delocalization stabilizes benzyl radical and weakens the benzylic C—H bond.

The unpaired electron in benzyl radical is shared by the benzylic carbon and by the ring carbons that are ortho and para to it as shown by the spin density surface in Figure 11.9. Delocalization of the unpaired electron from the benzylic carbon to the ortho and para positions can be explained on the basis of resonance contributions from the following structures:

Most stable Lewis structure of benzyl radical

Notice that, in converting one resonance form to the next, electrons are moved in exactly the same way as was done with allyl radical.

In orbital terms, as represented in Figure 11.10, benzyl radical is stabilized by delocalization of electrons throughout the extended π system formed by overlap of the p orbital of the benzylic carbon with the π system of the ring.

The comparative ease with which a benzylic hydrogen is abstracted leads to high selectivity in free-radical halogenations of alkylbenzenes. Thus, chlorination of toluene

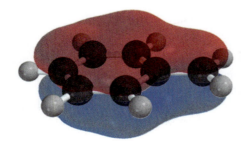

FIGURE 11.10 The lowest energy π molecular orbital of benzyl radical shows the interaction of the 2p orbital of the benzylic carbon with the π system of the aromatic ring.

takes place exclusively at the benzylic carbon and is an industrial process for the preparation of the compounds shown.

The common names of (dichloromethyl)benzene and (trichloromethyl)benzene are benzal chloride and benzotrichloride, respectively.

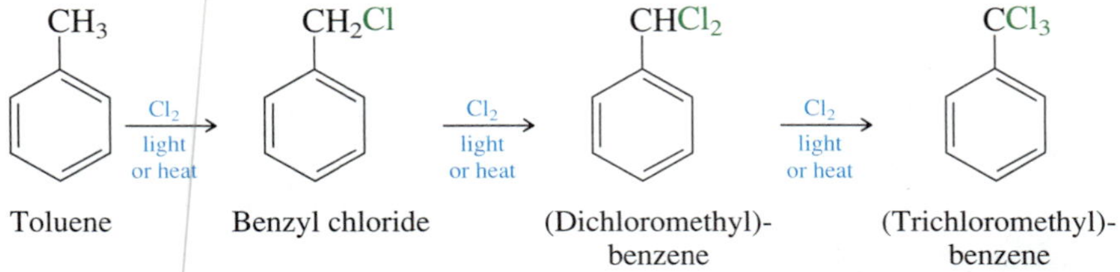

Toluene Benzyl chloride (Dichloromethyl)-benzene (Trichloromethyl)-benzene

The propagation steps in the formation of benzyl chloride involve benzyl radical as an intermediate.

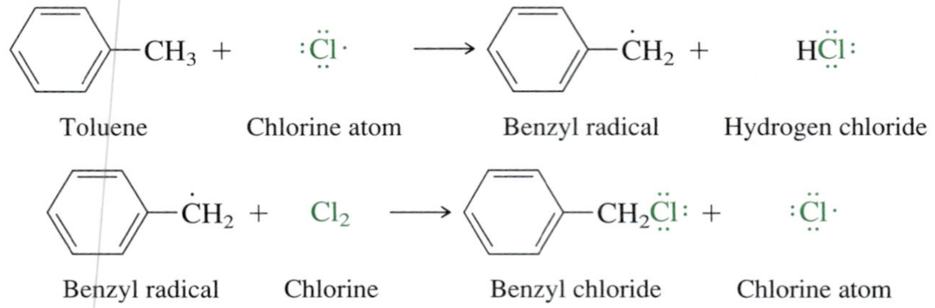

Toluene Chlorine atom Benzyl radical Hydrogen chloride

Benzyl radical Chlorine Benzyl chloride Chlorine atom

(Dichloromethyl)benzene and (trichloromethyl)benzene arise by further side-chain chlorination of benzyl chloride.

PROBLEM 11.6 The unpaired electron in benzyl radical is shared by the benzylic carbon plus the ortho and para carbons of the ring. Yet, chlorine becomes attached only to the benzylic carbon. Can you think of a reason why? (*Hint:* Write a structural formula for the compound formed by attachment of chlorine to one of the ring carbons.)

Benzylic bromination is a more commonly used laboratory procedure than chlorination and is typically carried out under conditions of photochemical initiation.

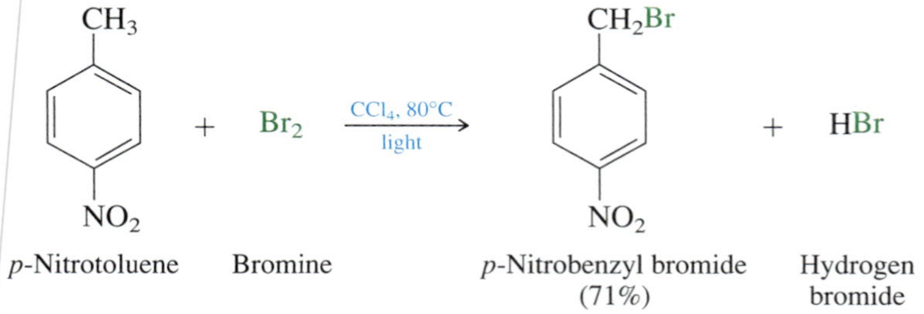

p-Nitrotoluene Bromine *p*-Nitrobenzyl bromide (71%) Hydrogen bromide

Benzoyl peroxide is a commonly used free-radical initiator. It has the formula

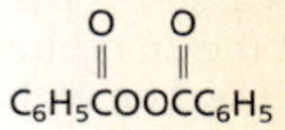

$$C_6H_5COOCC_6H_5$$

As we saw when discussing allylic bromination in Section 10.4, *N*-bromosuccinimide (NBS) is a convenient free-radical brominating agent. Benzylic brominations with NBS are normally performed in carbon tetrachloride as the solvent in the presence of peroxides, which are added as initiators. As the example illustrates, free-radical bromination is selective for substitution of benzylic hydrogens.

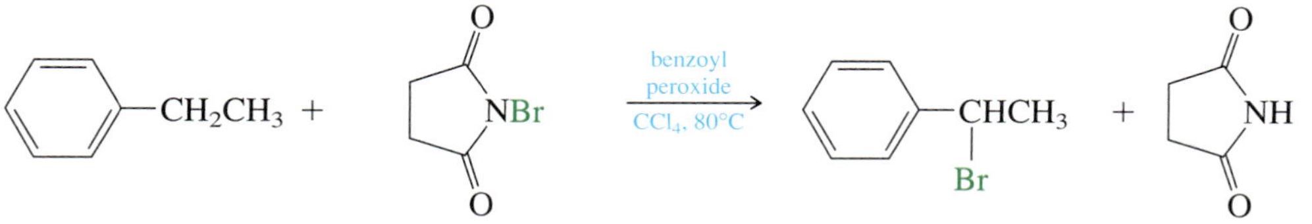

Ethylbenzene + N-Bromosuccinimide (NBS) reacting with benzoyl peroxide, CCl₄, 80°C to give 1-Bromo-1-phenylethane (87%) + Succinimide

Ethylbenzene   N-Bromosuccinimide   1-Bromo-1-phenylethane   Succinimide
                              (NBS)   (87%)

PROBLEM 11.7 The reaction of N-bromosuccinimide with the following compounds has been reported in the chemical literature. Each compound yields a single product in 95% yield. Identify the product formed from each starting material.

(a) p-tert-Butyltoluene   (b) 4-Methyl-3-nitroanisole

SAMPLE SOLUTION (a) The only benzylic hydrogens in p-tert-butyltoluene are those of the methyl group that is attached directly to the ring. Substitution occurs there to give p-tert-butylbenzyl bromide.

$(CH_3)_3C$—⟨ring⟩—CH_3 →[NBS, CCl₄, 80°C, free-radical initiator]→ $(CH_3)_3C$—⟨ring⟩—CH_2Br

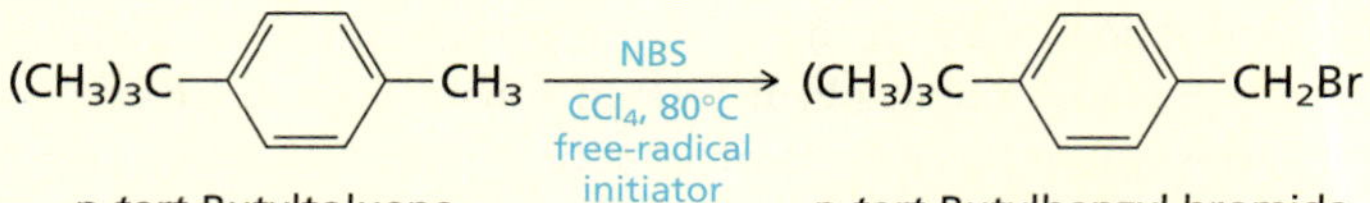

      p-tert-Butyltoluene   p-tert-Butylbenzyl bromide

11.13 OXIDATION OF ALKYLBENZENES

A striking example of the activating effect that a benzene ring has on reactions that take place at benzylic positions may be found in the reactions of alkylbenzenes with oxidizing agents. Chromic acid, for example, prepared by adding sulfuric acid to aqueous sodium dichromate, is a strong oxidizing agent but does not react either with benzene or with alkanes.

$$RCH_2CH_2R' \xrightarrow[\text{H}_2\text{O, H}_2\text{SO}_4, \text{ heat}]{\text{Na}_2\text{Cr}_2\text{O}_7} \text{no reaction}$$

$$\text{⟨benzene⟩} \xrightarrow[\text{H}_2\text{O, H}_2\text{SO}_4, \text{ heat}]{\text{Na}_2\text{Cr}_2\text{O}_7} \text{no reaction}$$

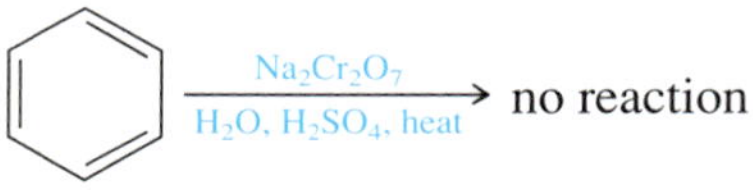

An alternative oxidizing agent, similar to chromic acid in its reactions with organic compounds, is potassium permanganate (KMnO₄).

On the other hand, an alkyl side chain on a benzene ring is oxidized on being heated with chromic acid. The product is benzoic acid or a substituted derivative of benzoic acid.

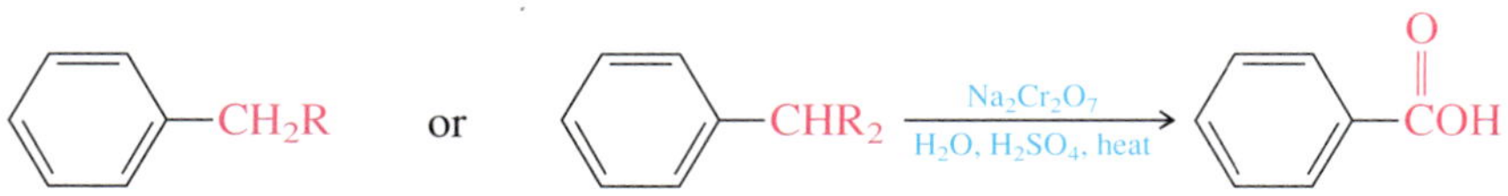

⟨ring⟩—CH_2R or   ⟨ring⟩—CHR_2 →[Na₂Cr₂O₇, H₂O, H₂SO₄, heat]→ ⟨ring⟩—$\overset{O}{\overset{\|}{C}}OH$

          Alkylbenzene   Benzoic acid

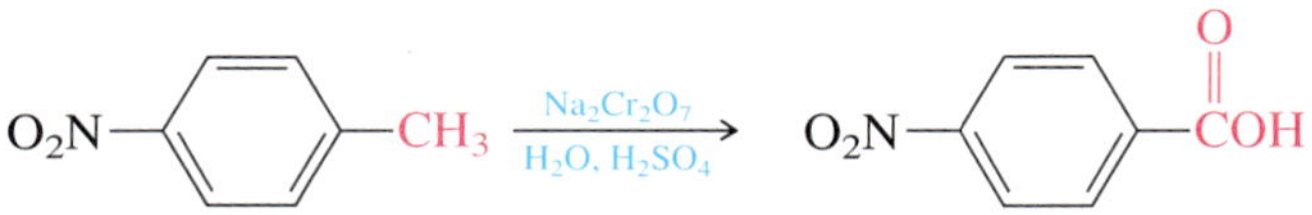

O_2N—⟨ring⟩—CH_3 →[Na₂Cr₂O₇, H₂O, H₂SO₄]→ O_2N—⟨ring⟩—$\overset{O}{\overset{\|}{C}}OH$

      p-Nitrotoluene   p-Nitrobenzoic acid (82–86%)

When two alkyl groups are present on the ring, both are oxidized.

$$CH_3 \overset{}{\longleftarrow} CH(CH_3)_2 \xrightarrow[\text{H}_2\text{O, H}_2\text{SO}_4, \text{ heat}]{\text{Na}_2\text{Cr}_2\text{O}_7} HOC \overset{}{\longleftarrow} COH$$

p-Isopropyltoluene p-Benzenedicarboxylic acid (45%)

Note that alkyl groups, regardless of their chain length, are converted to carboxyl groups ($-CO_2H$) attached directly to the ring. An exception is a substituent of the type $-CR_3$. Because it lacks benzylic hydrogens, such a group is not susceptible to oxidation under these conditions.

> **PROBLEM 11.8** Chromic acid oxidation of 4-*tert*-butyl-1,2-dimethylbenzene yielded a single compound having the molecular formula $C_{12}H_{14}O_4$. What was this compound?

Side-chain oxidation of alkylbenzenes is important in certain metabolic processes. One way in which the body rids itself of foreign substances is by oxidation in the liver to compounds that are more polar and hence more easily excreted in the urine. Toluene, for example, is oxidized to benzoic acid by this process and is eliminated rather readily.

$$\overset{}{\longleftarrow} CH_3 \xrightarrow[\substack{\text{cytochrome P-450} \\ \text{(an enzyme in} \\ \text{the liver)}}]{\text{O}_2} \overset{}{\longleftarrow} COH$$

Toluene   Benzoic acid

Benzene, with no alkyl side chain and no benzylic hydrogens, undergoes a different reaction under these conditions. Oxidation of the ring occurs to convert benzene to its epoxide.

$$\text{Benzene} \xrightarrow[\text{cytochrome P-450}]{\text{O}_2} \text{Benzene oxide}$$

Benzene   Benzene oxide

Benzene oxide and compounds derived from it are carcinogenic and can react with DNA to induce mutations. This difference in the site of biological oxidation—ring versus side-chain—seems to be responsible for the fact that benzene is carcinogenic but toluene is not.

11.14 NUCLEOPHILIC SUBSTITUTION IN BENZYLIC HALIDES

Primary benzylic halides are ideal substrates for S_N2 reactions because they are very reactive toward good nucleophiles and cannot undergo competing elimination.

$$O_2N \overset{}{\longleftarrow} CH_2Cl \xrightarrow[\text{acetic acid}]{\text{CH}_3\text{CO}_2^- \text{ Na}^+} O_2N \overset{}{\longleftarrow} CH_2OCCH_3$$

p-Nitrobenzyl chloride p-Nitrobenzyl acetate (78–82%)

Benzylic halides that are secondary resemble secondary alkyl halides in that they undergo substitution only when the nucleophile is weakly basic. If the nucleophile is a strong base such as sodium ethoxide, elimination by the E2 mechanism is faster than substitution.

PROBLEM 11.9 Give the structure of the principal organic product formed on reaction of benzyl bromide with each of the following reagents:

(a) Sodium ethoxide

(b) Potassium *tert*-butoxide

(c) Sodium azide

(d) Sodium hydrogen sulfide

(e) Sodium iodide (in acetone)

SAMPLE SOLUTION (a) Benzyl bromide is a primary bromide and undergoes S_N2 reactions readily. It has no hydrogens β to the leaving group and so cannot undergo elimination. Ethoxide ion acts as a nucleophile, displacing bromide and forming benzyl ethyl ether.

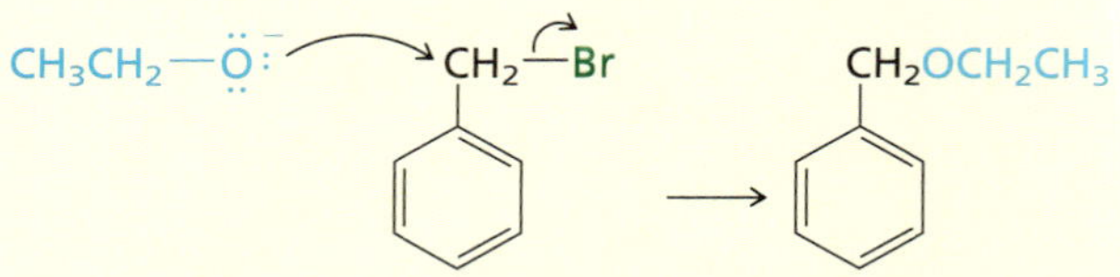

Ethoxide ion   Benzyl bromide   Benzyl ethyl ether

Benzylic halides resemble allylic halides in the readiness with which they form carbocations. On comparing the rate of S_N1 hydrolysis in aqueous acetone of the following two tertiary chlorides, we find that the benzylic chloride reacts over 600 times faster than does *tert*-butyl chloride.

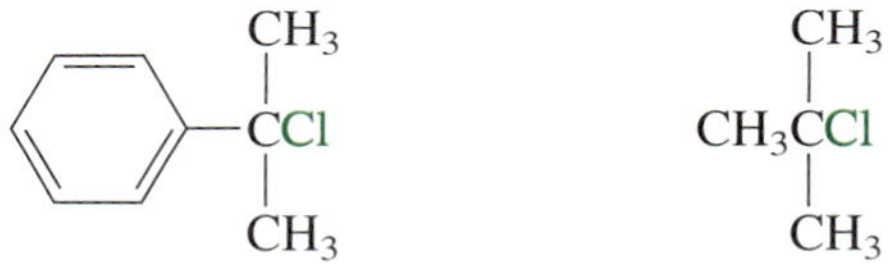

2-Chloro-2-phenylpropane   2-Chloro-2-methylpropane

The positive charge in benzyl cation is shared by the carbons ortho and para to the benzylic carbon.

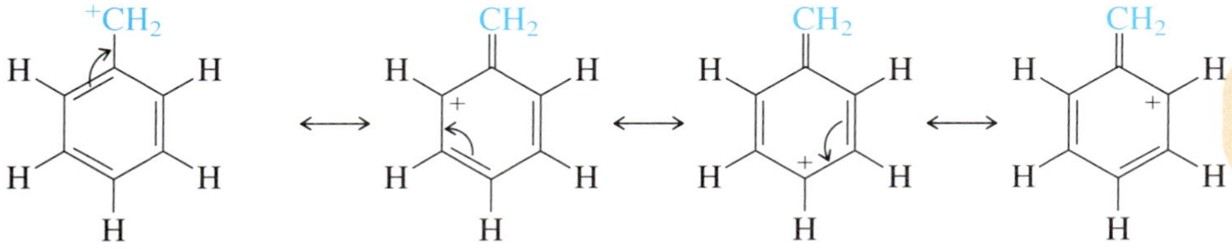

Most stable Lewis structure
of benzyl cation

See *Learning By Modeling* for an electrostatic potential map of benzyl cation.

Unlike the case with allylic carbocations, however, dispersal of the positive charge does not result in nucleophilic attack at more than one carbon. There is no "benzylic rearrangement" analogous to allylic rearrangement (Section 10.2), because the aromatic stabilization would be lost if the nucleophile became bonded to one of the ring carbons. Thus, when conditions are chosen that favor S_N1 substitution over E2 elimination (solvolysis, weakly basic nucleophile), benzylic halides give a single substitution product in high yield.

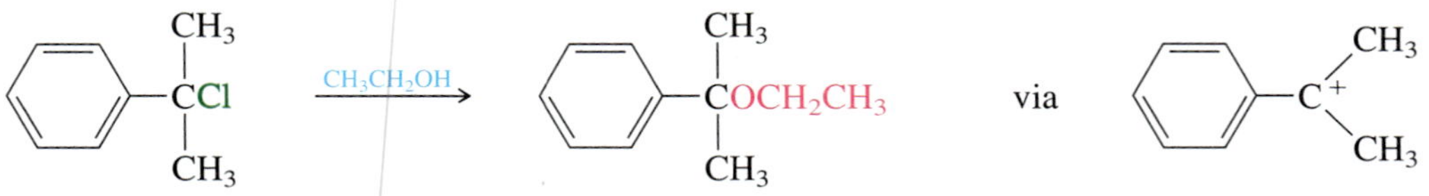

2-Chloro-2-phenylpropane   2-Ethoxy-2-phenylpropane (87%)   via

The triphenylmethyl group is often referred to as a *trityl* group.

Additional phenyl substituents stabilize carbocations even more. Triphenylmethyl cation is particularly stable. Its perchlorate salt is ionic and stable enough to be isolated and stored indefinitely.

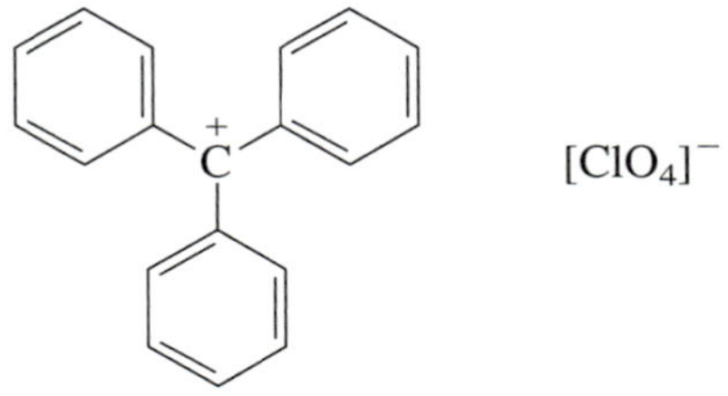

$[ClO_4]^-$

Triphenylmethyl perchlorate

11.15 PREPARATION OF ALKENYLBENZENES

Alkenylbenzenes are prepared by the various methods described in Chapter 5 for the preparation of alkenes: *dehydrogenation, dehydration,* and *dehydrohalogenation.*

Dehydrogenation of alkylbenzenes is not a convenient laboratory method but is used industrially to convert ethylbenzene to styrene.

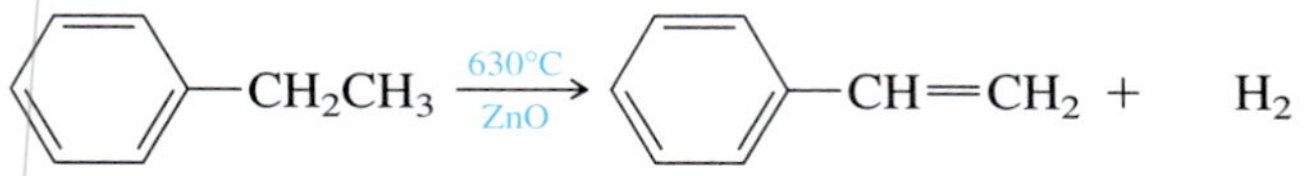

Ethylbenzene   Styrene   Hydrogen

Practically all of the 1.3×10^{10} lb of ethylbenzene produced annually in the United States is converted to styrene.

Acid-catalyzed dehydration of benzylic alcohols is a useful route to alkenylbenzenes, as is dehydrohalogenation under E2 conditions.

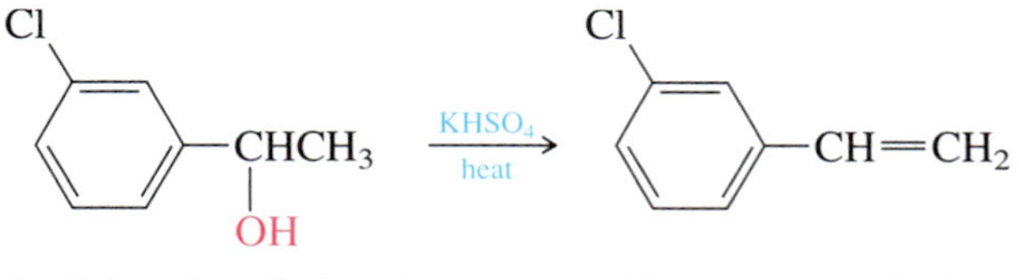

1-(*m*-Chlorophenyl)ethanol   *m*-Chlorostyrene (80–82%)

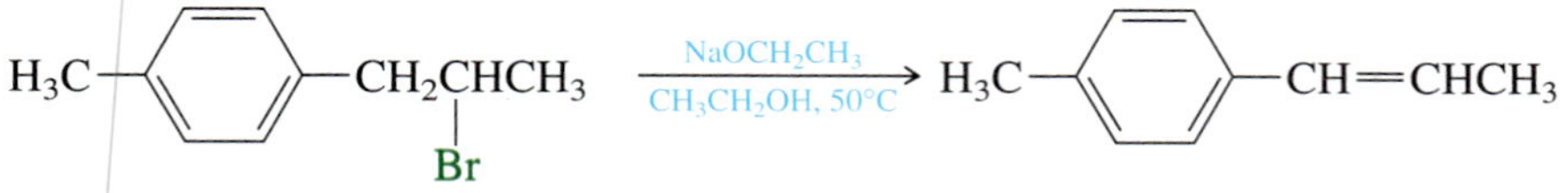

2-Bromo-1-(*p*-methylphenyl)propane   1-(*p*-Methylphenyl)propene (99%)

11.16 ADDITION REACTIONS OF ALKENYLBENZENES

Most of the reactions of alkenes that were discussed in Chapter 6 find a parallel in the reactions of alkenylbenzenes.

Hydrogenation of the side-chain double bond of an alkenylbenzene is much easier than hydrogenation of the aromatic ring and can be achieved with high selectivity, leaving the ring unaffected.

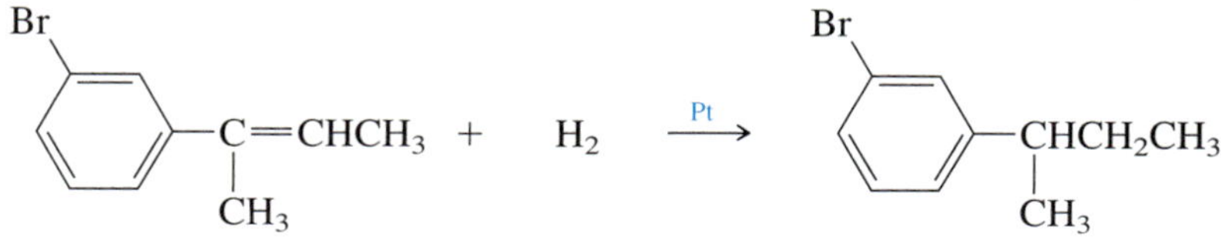

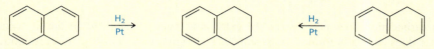

2-(*m*-Bromophenyl)-2-butene   Hydrogen   2-(*m*-Bromophenyl)butane (92%)

PROBLEM 11.10   Both 1,2-dihydronaphthalene and 1,4-dihydronaphthalene may be selectively hydrogenated to 1,2,3,4-tetrahydronaphthalene.

1,2-Dihydronaphthalene   1,2,3,4-Tetrahydronaphthalene   1,4-Dihydronaphthalene

One of these isomers has a heat of hydrogenation of 101 kJ/mol (24.1 kcal/mol), and the heat of hydrogenation of the other is 113 kJ/mol (27.1 kcal/mol). Match the heat of hydrogenation with the appropriate dihydronaphthalene.

The double bond in the alkenyl side chain undergoes addition reactions that are typical of alkenes when treated with electrophilic reagents.

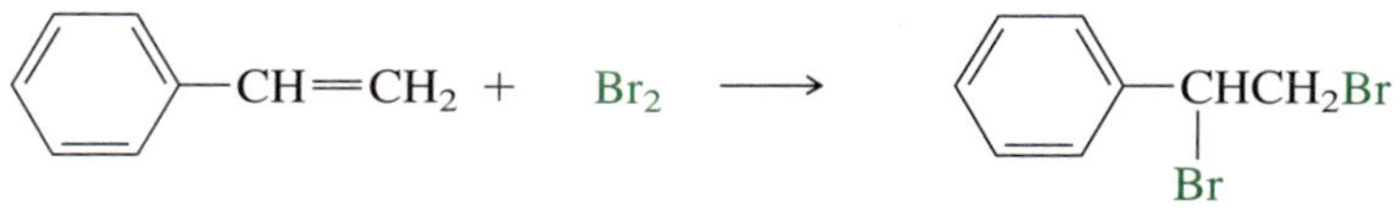

Styrene   Bromine   1,2-Dibromo-1-phenylethane (82%)

The regioselectivity of electrophilic addition is governed by the ability of an aromatic ring to stabilize an adjacent carbocation. This is clearly seen in the addition of hydrogen chloride to indene. Only a single chloride is formed.

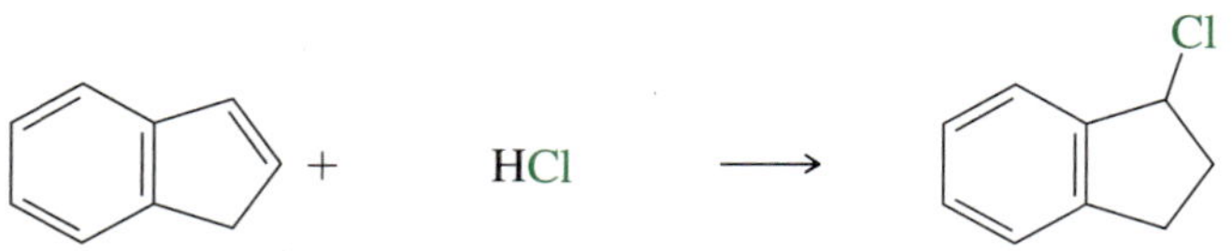

Indene   Hydrogen chloride   1-Chloroindane (75–84%)

Only the benzylic chloride is formed because protonation of the double bond occurs in the direction that gives a carbocation that is both secondary and benzylic.

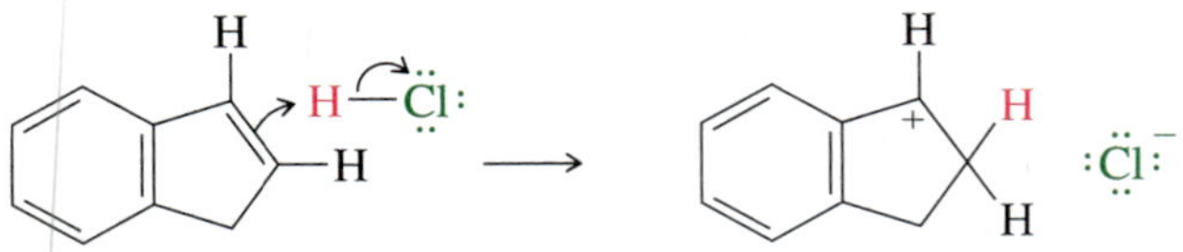

Carbocation is secondary and benzylic
and gives the observed product

Protonation in the opposite direction also gives a secondary carbocation, but it is not benzylic.

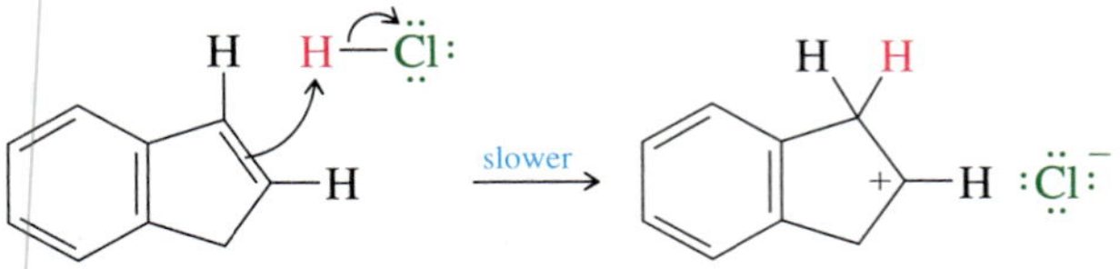

slower

Less stable carbocation
is secondary but not benzylic

This carbocation does not receive the extra increment of stabilization that its benzylic isomer does and so is formed more slowly. The regioselectivity of addition is controlled by the rate of carbocation formation; the more stable benzylic carbocation is formed faster and is the one that determines the reaction product.

PROBLEM 11.11 Each of the following reactions has been reported in the chemical literature and gives a single organic product in high yield. Write the structure of the product for each reaction.

(a) 2-Phenylpropene + hydrogen chloride

(b) 2-Phenylpropene treated with diborane in tetrahydrofuran followed by oxidation with basic hydrogen peroxide

(c) Styrene + bromine in aqueous solution

(d) Styrene + peroxybenzoic acid (two organic products in this reaction; identify both by writing a balanced equation.)

SAMPLE SOLUTION (a) Addition of hydrogen chloride to the double bond takes place by way of a tertiary benzylic carbocation.

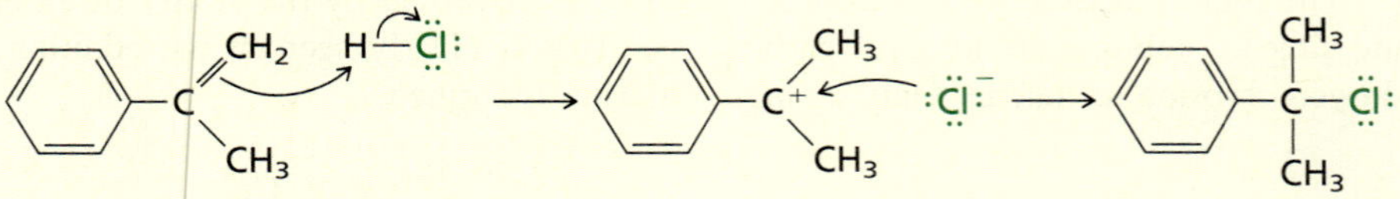

2-Phenylpropene	Hydrogen chloride		2-Chloro-2-phenylpropane

In the presence of peroxides, hydrogen bromide adds to the double bond of styrene with a regioselectivity opposite to Markovnikov's rule. The reaction is a free-radical addition, and the regiochemistry is governed by preferential formation of the more stable radical.

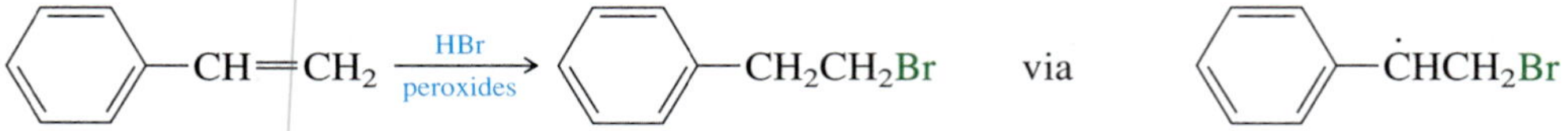

Styrene $\xrightarrow[\text{peroxides}]{\text{HBr}}$ —CH$_2$CH$_2$Br   via   —ĊHCH$_2$Br

Styrene   1-Bromo-2-phenylethane   2-Bromo-1-phenylethyl
                 (major product)   radical (secondary; benzylic)

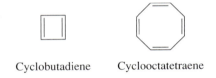

FIGURE 11.11 Chain propagation in polymerization of styrene. The growing polymer chain has a free-radical site at the benzylic carbon. It adds to a molecule of styrene to extend the chain by one styrene unit. The new polymer chain is also a benzylic radical; it attacks another molecule of styrene and the process repeats over and over again.

11.17 POLYMERIZATION OF STYRENE

The annual production of styrene in the United States is approximately 1.2×10^{10} lb, with about 65% of this output used to prepare polystyrene plastics and films. Styrofoam coffee cups are made from polystyrene. Polystyrene can also be produced in a form that is very strong and impact-resistant and is used widely in luggage, television and radio cabinets, and furniture.

Polymerization of styrene is carried out under free-radical conditions, often with benzoyl peroxide as the initiator. Figure 11.11 illustrates a step in the growth of a polystyrene chain by a mechanism analogous to that of the polymerization of ethylene (Section 6.21).

As described in the box "Diene Polymers" in Chapter 10, most synthetic rubber is a copolymer of styrene and 1,3-butadiene.

11.18 CYCLOBUTADIENE AND CYCLOOCTATETRAENE

During our discussion of benzene and its derivatives, it may have occurred to you that cyclobutadiene and cyclooctatetraene might be stabilized by cyclic π electron delocalization in a manner analogous to that of benzene.

Cyclobutadiene   Cyclooctatetraene

The same thought occurred to early chemists. However, the complete absence of naturally occurring compounds based on cyclobutadiene and cyclooctatetraene contrasted starkly with the abundance of compounds containing a benzene unit. Attempts to synthesize cyclobutadiene and cyclooctatetraene met with failure and reinforced the growing conviction that these compounds would prove to be quite unlike benzene if, in fact, they could be isolated at all.

The first breakthrough came in 1911 when Richard Willstätter prepared cyclooctatetraene by a lengthy degradation of *pseudopelletierine*, a natural product obtained from the bark of the pomegranate tree. Today, cyclooctatetraene is prepared from acetylene in a reaction catalyzed by nickel cyanide.

Willstätter's most important work, for which he won the 1915 Nobel Prize in chemistry, was directed toward determining the structure of chlorophyll.

$$4HC\equiv CH \xrightarrow[\text{heat, pressure}]{Ni(CN)_2}$$

Acetylene   Cyclooctatetraene (70%)

Cyclooctatetraene is relatively stable, but lacks the "special stability" of benzene. Unlike benzene, which we saw has a heat of hydrogenation that is 152 kJ/mol (36 kcal/mol) less than three times the heat of hydrogenation of cyclohexene, cyclooctatetraene's heat of hydrogenation is only 26 kJ/mol (6 kJ/mol) less than four times that of *cis*-cyclooctene.

cis-Cyclooctene   Hydrogen   Cyclooctane $\Delta H° = -96$ kJ (-23 kcal)

Cyclooctatetraene   Hydrogen   Cyclooctane $\Delta H° = -410$ kJ (-98 kcal)

> **PROBLEM 11.12** Both cyclooctatetraene and styrene have the molecular formula C_8H_8 and undergo combustion according to the equation
>
> $$C_8H_8 + 10O_2 \rightarrow 8CO_2 + 4H_2O$$
>
> The measured heats of combustion are 4393 and 4543 kJ/mol (1050 and 1086 kcal/mol). Which heat of combustion belongs to which compound?

Thermodynamically, cyclooctatetraene does not qualify as aromatic. Nor does its structure offer any possibility of the π electron delocalization responsible for aromaticity. As shown in Figure 11.12, cyclooctatetraene is *nonplanar* with four short and four long carbon–carbon bond distances. Cyclooctatetraene is satisfactorily represented by a single Lewis structure having alternating single and double bonds in a tub-shaped eight-membered ring.

All of the evidence indicates that cyclooctatetraene is not aromatic and is better considered as a conjugated polyene than as an aromatic hydrocarbon.

The April 1993 issue of the *Journal of Chemical Education* (pp. 291–293) contained an article about cyclooctatetraene entitled "Don't Stop with Benzene!" "Keep Going with Cyclooctatetraene" appeared in the January 2000 issue (pp. 55–57).

133 pm

146 pm

FIGURE 11.12 Molecular geometry of cyclooctatetraene. The ring is not planar, and the bond distances alternate between short double bonds and long single bonds.

What about cyclobutadiene?

Cyclobutadiene escaped chemical characterization for more than 100 years. Despite numerous attempts, all synthetic efforts met with failure. It became apparent not only that cyclobutadiene was not aromatic but that it was exceedingly unstable. Beginning in the 1950s, a variety of novel techniques succeeded in generating cyclobutadiene as a transient, reactive intermediate.

PROBLEM 11.13   One of the chemical properties that makes cyclobutadiene difficult to isolate is that it reacts readily with itself to give a dimer:

What reaction of dienes does this resemble?

High-level molecular orbital calculations of cyclobutadiene itself and experimentally measured bond distances of a stable, highly substituted derivative both reveal a pattern of alternating short and long bonds characteristic of a rectangular, rather than square, geometry.

Cyclobutadiene   Sterically hindered cyclobutadiene derivative

Thus cyclobutadiene, like cyclooctatetraene, is not aromatic. More than this, cyclobutadiene is even less stable than its Lewis structure would suggest. It belongs to a class of compounds called antiaromatic. An **antiaromatic** compound is one that is *destabilized* by cyclic conjugation.

Cyclic conjugation, although necessary for aromaticity, is not sufficient for it. Some other factor or factors must contribute to the special stability of benzene and compounds based on the benzene ring. To understand these factors, let's return to the molecular orbital description of benzene.

11.19   HÜCKEL'S RULE

One of molecular orbital theories' early successes came in 1931 when Erich Hückel discovered an interesting pattern in the π orbital energy levels of benzene, cyclobutadiene, and cyclooctatetraene. By limiting his analysis to monocyclic conjugated polyenes and restricting the structures to planar geometries, Hückel found that whether a hydrocarbon of this type was aromatic depended on its number of π electrons. He set forth what we now call **Hückel's rule:**

> *Among planar, monocyclic, fully conjugated polyenes, only those possessing (4n + 2) π electrons, where n is a whole number, will have special stability; that is, be aromatic.*

Hückel was a German physical chemist. Before his theoretical studies of aromaticity, Hückel collaborated with Peter Debye in developing what remains the most widely accepted theory of electrolyte solutions.

Thus for this group of hydrocarbons, those with $(4n + 2) = 6, 10, 14, \ldots$ π electrons will be aromatic. These values correspond to $(4n + 2)$ when $n = 1, 2, 3. \ldots$

Hückel proposed his theory before ideas of antiaromaticity emerged. We can amplify his generalization by noting that among the hydrocarbons covered by Hückel's rule, those with $(4n)$ π electrons not only are not aromatic, they are antiaromatic.

Benzene, cyclobutadiene, and cyclooctatetraene provide clear examples of Hückel's rule. Benzene, with six π electrons is a $(4n + 2)$ system and is predicted to be aromatic by the rule. Square cyclobutadiene and planar cyclooctatetraene are $4n$ systems with four and eight π electrons, respectively, and are antiaromatic.

The $(4n + 2)$ π electron standard follows from the pattern of orbital energies in monocyclic, completely conjugated polyenes. The π energy levels were shown for benzene earlier in Figure 11.4 and are repeated in Figure 11.13b. Figure 11.13a and 11.13c show the π energy levels for square cyclobutadiene and planar cyclooctatetraene, respectively.

The energy diagrams in Figure 11.13 illustrate a simple method, called the **Frost circle** for setting out the Hückel MOs of "planar, monocyclic, completely conjugated polyenes." By inscribing a polygon having the appropriate number of sides within a circle so that one of its vertices lies at the bottom, the location of each of the polygon's corners defines a π electron energy level. Their vertical separation is proportional to the energy difference between the MOs. A horizontal line drawn through the center of the circle separates the bonding and antibonding MOs; an orbital that lies directly on the line is nonbonding.

For qualitative purposes, the circle itself isn't even necessary. We could locate the Hückel MOs by simply working with the polygons themselves. The circle is needed only when Frost's method is used quantitatively. In those cases the radius of the circle has a prescribed value, allowing each MO to be assigned a specific energy.

The pattern of orbital energies in Figure 11.13 provides a convincing explanation for why benzene is aromatic while square cyclobutadiene and planar cyclooctatetraene are not. We start by counting π electrons; cyclobutadiene has four, benzene six, and cyclooctatetraene has eight. These π electrons are assigned to MOs in accordance with the usual rules—lowest energy orbitals first, a maximum of two electrons per orbital,

> The circle mnemonic was devised by Arthur A. Frost, a theoretical chemist at Northwestern University.

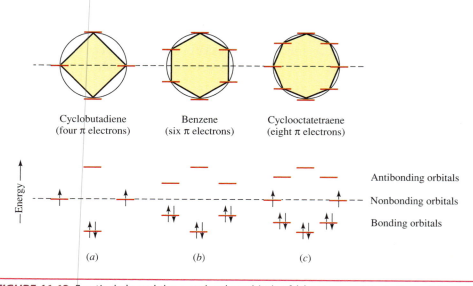

FIGURE 11.13 Frost's circle and the π molecular orbitals of (a) square cyclobutadiene, (b) benzene, and (c) planar cyclooctatetraene.

and when two orbitals are of equal energy, each gets one electron before either orbital gets two (Hund's rule).

Benzene   As seen earlier in Figure 11.4 (Section 11.6), the six π electrons of benzene are distributed in pairs among its three bonding π MOs, giving a closed-shell electron configuration. All the bonding orbitals are filled, and all the electron spins are paired.

Cyclobutadiene   Square cyclobutadiene has one bonding π MO, two equal-energy nonbonding π MOs and one antibonding π* MO. After the bonding MO is filled, the remaining two electrons are assigned to different nonbonding MOs in accordance with Hund's rule. This results in a species with two unpaired electrons—a **diradical.** In a square geometry, cyclobutadiene lacks a closed-shell electron configuration. It is not stabilized and, with two unpaired electrons, should be very reactive.

Cyclooctatetraene   Six of the eight π electrons of planar cyclooctatetraene occupy three bonding orbitals. The remaining two π electrons occupy, one each, the two equal-energy nonbonding orbitals. Planar cyclooctatetraene should, like square cyclobutadiene, be a diradical.

An important conclusion we draw from the qualitative MO diagrams is that the geometry required for maximum π electron delocalization, a planar ring with p orbitals aligned and equal C—C bond distances, gives relatively unstable electron configurations for square cyclobutadiene and planar cyclooctatetraene. Both escape to alternative geometries that have electron configurations which, although not aromatic, at least have all their electron spins paired. For cyclobutadiene the stable geometry is rectangular; for cyclooctatetraene it is tub-shaped.

Benzene's structure allows effective π electron conjugation and gives a closed-shell electron configuration. To understand why it also conveys special stability, we need to go one step further and compare the Hückel π MOs of benzene to those of a hypothetical "cyclohexatriene" with alternating single and double bonds. Without going into quantitative detail, we'll simply note that the occupied orbitals of a structure in which the π electrons are restricted to three noninteracting double bonds are of higher energy (less stable) than the occupied Hückel MOs of benzene.

Before looking at other applications of Hückel's rule, it is worth pointing out that its opening phrase: "Among planar, monocyclic, fully conjugated polyenes" does *not* mean that *only* "planar, monocyclic, fully conjugated polyenes" can be aromatic. It merely limits the rule to compounds of this type. There are thousands of aromatic compounds that are not monocyclic—naphthalene and related polycyclic aromatic hydrocarbons (Section 11.8), for example. All compounds based on benzene rings are aromatic. Cyclic conjugation *is* a requirement for aromaticity, however, and in those cases the conjugated system must contain $(4n + 2)$ π electrons.

PROBLEM 11.14 Give an explanation for each of the following observations:
(a) Compound A has six π electrons but is not aromatic.
(b) Compound B has six π electrons but is not aromatic.
(c) Compound C has 12 π electrons and is aromatic.

Compound A   Compound B   Compound C

SAMPLE SOLUTION   (a) Cycloheptatriene (compound A) is not aromatic because, although it does contain six π electrons, its conjugated system of three double bonds does not close on itself—it lacks cyclic conjugation. The CH_2 group prevents cyclic delocalization of the π electrons.

In the next section we'll explore Hückel's rule for values of n greater than 1 to see how it can be extended beyond cyclobutadiene, benzene, and cyclooctatetraene.

11.20 ANNULENES

The general term **annulene** has been coined to apply to completely conjugated mono-cyclic hydrocarbons with more than six carbons. Cyclobutadiene and benzene retain their names, but higher members of the group are named **[x]annulene** where x is the number of carbons in the ring. Thus, cyclooctatetraene becomes [8]annulene, cyclodecapentaene becomes [10]annulene and so on.

PROBLEM 11.15   Use Frost's circle to construct orbital energy diagrams for (a) [10]annulene and (b) [12]annulene. Is either aromatic according to Hückel's rule?

SAMPLE SOLUTION   (a) [10]Annulene is a ten-membered ring with five conjugated double bonds. Drawing a polygon with ten sides with its vertex pointing downward within a circle gives the orbital template. Place the orbitals at the positions where each vertex contacts the circle. The ten π electrons of [10]annulene satisfy the (4n + 2) rule for n = 2 and occupy the five bonding orbitals in pairs. [10]Annulene is aromatic according to Hückel's rule.

[10]Annulene   Frost's circle π-Electron configuration

The prospect of observing aromatic character in conjugated polyenes having 10, 14, 18, and so on π electrons spurred efforts toward the synthesis of higher annulenes. A problem immediately arises in the case of the all-cis isomer of [10]annulene, the structure of which is shown in the preceding problem. Geometry requires a ten-sided regular polygon to have 144° bond angles; sp^2 hybridization at carbon requires 120° bond angles. Therefore, aromatic stabilization due to conjugation in all-*cis*-[10]annulene is opposed by the destabilizing effect of 24° of angle strain at each of its carbon atoms. All-*cis*-[10]annulene has been prepared. It is not very stable and is highly reactive.

The size of each angle of a regular polygon is given by the expression

$$180° \times \frac{\text{(number of sides)} - 2}{\text{(number of sides)}}$$

A second isomer of [10]annulene (the cis, trans, cis, cis, trans stereoisomer) can have bond angles close to 120° but is destabilized by a close contact between two hydrogens directed toward the interior of the ring. To minimize the van der Waals strain between these hydrogens, the ring adopts a nonplanar geometry, which limits its ability to be stabilized by π electron delocalization. It, too, has been prepared and is not very stable. Similarly, the next higher ($4n + 2$) system, [14]annulene, is also somewhat destabilized by van der Waals strain and is nonplanar.

Planar geometry required for aromaticity destabilized by van der Waals repulsions between indicated hydrogens

cis,trans,cis,cis,trans-
[10]Annulene

[14]Annulene

When the ring contains 18 carbon atoms, it is large enough to be planar while still allowing its interior hydrogens to be far enough apart that they do not interfere with one another. The [18]annulene shown is planar, or nearly so, and has all its carbon–carbon bond distances in the range 137–143 pm, very much like those of benzene. Its resonance energy is estimated to be about 418 kJ/mol (100 kcal/mol). Although its structure and resonance energy attest to the validity of Hückel's rule, which predicts "special stability" for [18]annulene, its chemical reactivity does not. [18]Annulene behaves more like a polyene than like benzene in that it is hydrogenated readily, undergoes addition rather than substitution with bromine, and forms a Diels–Alder adduct with maleic anhydride.

No serious repulsions among six interior hydrogens; molecule is planar and aromatic.

[18]Annulene

As noted earlier, planar annulenes with $4n$ π electrons are antiaromatic. A member of this group, [16]annulene, has been prepared. It is nonplanar and shows a pattern of alternating short (average 134 pm) and long (average 146 pm) bonds typical of a nonaromatic cyclic polyene.

[16]Annulene

Molecular models of [10]-, [14]-, [16]-, and [18]annulene can be viewed on *Learning By Modeling.*

Most of the synthetic work directed toward the higher annulenes was carried out by Franz Sondheimer and his students, first at Israel's Weizmann Institute and later at the University of London. Sondheimer's research systematically explored the chemistry of these hydrocarbons and provided experimental verification of Hückel's rule.

11.21 AROMATIC IONS

Hückel realized that his molecular orbital analysis of conjugated systems could be extended beyond neutral hydrocarbons. He pointed out that cycloheptatrienyl cation, also called *tropylium ion,* contained a completely conjugated closed-shell six-π electron system analogous to that of benzene.

Benzene:
completely conjugated,
six π electrons delocalized over six carbons

Cycloheptatrienyl cation:
completely conjugated,
six π electrons delocalized over seven carbons

Figure 11.14 shows a molecular orbital diagram for cycloheptatrienyl cation. There are seven π MOs, three of which are bonding and contain the six π electrons of the cation. Cycloheptatrienyl cation is a Hückel ($4n + 2$) system and is an aromatic ion.

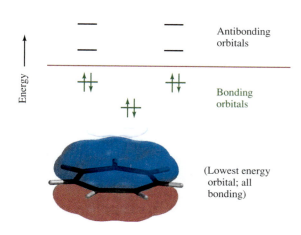

FIGURE 11.14 The π molecular orbitals of cycloheptatrienyl cation.

It is important to recognize the difference between the hydrocarbon cyclohepta-triene and cycloheptatrienyl cation.

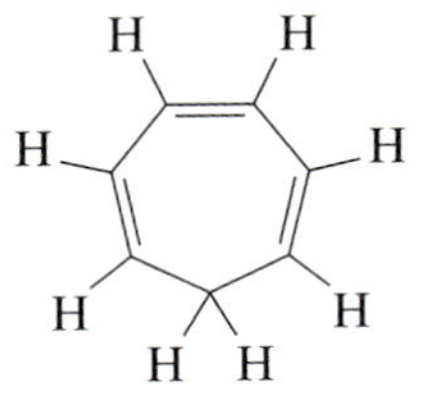

Cycloheptatriene:
lacks cyclic conjugation,
interrupted by CH$_2$ group

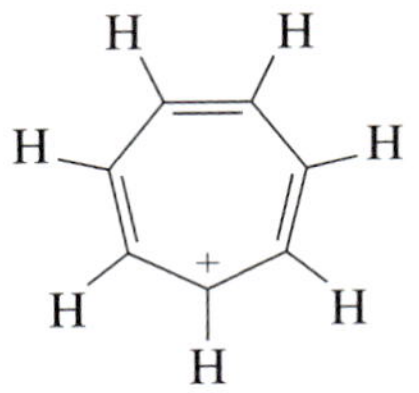

Cycloheptatrienyl cation:
completely conjugated,
six π electrons delocalized over seven carbons

The carbocation is aromatic; the hydrocarbon is not. Although cycloheptatriene has six π electrons in a conjugated system, the ends of the triene system are separated by an sp^3-hybridized carbon, which prevents continuous π electron delocalization.

PROBLEM 11.18 Cycloheptatrienyl radical (C$_7$H$_7$·) contains a cyclic, completely conjugated system of π electrons. Is it aromatic? Explain.

When we say cycloheptatriene is not aromatic but cycloheptatrienyl cation is, we are not comparing the stability of the two to each other. Cycloheptatriene is a stable hydrocarbon but does not possess the *special stability* required to be called *aromatic*. Cycloheptatrienyl cation, although aromatic, is still a carbocation and reasonably reactive toward nucleophiles. Its special stability does not imply a rock-like passivity, but rather a much greater ease of formation than expected on the basis of the Lewis structure drawn for it. A number of observations indicate that cycloheptatrienyl cation is far more stable than most other carbocations. To emphasize its aromatic nature, chemists often write the structure of cycloheptatrienyl cation in the Robinson circle-in-a-ring style.

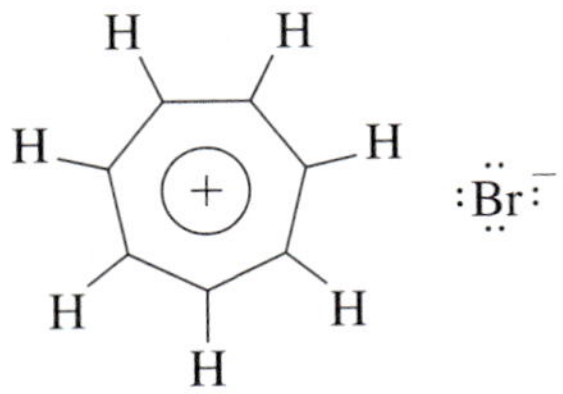

Tropylium bromide

Tropylium bromide was first prepared, but not recognized as such, in 1891. The work was repeated in 1954, and the ionic properties of tropylium bromide were demonstrated. The ionic properties of tropylium bromide are apparent in its unusually high melting point (203°C), its solubility in water, and its complete lack of solubility in diethyl ether.

PROBLEM 11.19 Write resonance structures for tropylium cation sufficient to show the delocalization of the positive charge over all seven carbons.

The five-membered cyclopentadienyl system contrasts with cycloheptatrienyl. Here, the cation has four π electrons, is antiaromatic, very unstable, and very difficult to generate. Cyclopentadienyl anion, however, has six π electrons delocalized over five carbons and is aromatic.

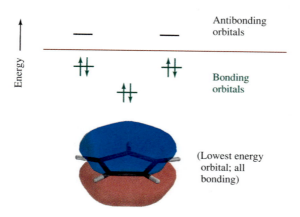

Cyclopentadienyl cation   Cyclopentadienyl anion

Figure 11.15 shows the Hückel MOs of cyclopentadienyl anion. Like benzene and cyclo-heptatrienyl cation, cyclopentadienyl anion has six π electrons and a closed-shell electron configuration.

> **PROBLEM 11.20** Show how you could adapt Frost's circle to generate the orbital energy level diagram shown in Figure 11.15 for cyclopentadienyl anion.

The acidity of cyclopentadiene provides convincing evidence for the special stability of cyclopentadienyl anion.

Cyclopentadiene   Water   Cyclopentadienyl anion   Hydronium ion
$pK_a = 16$

With a pK_a of 16, cyclopentadiene is only a slightly weaker acid than water ($pK_a = 15.7$). It is much more acidic than other hydrocarbons—its K_a for ionization is 10^{10} times greater than acetylene, for example—because its conjugate base is aromatic and stabilized by electron delocalization.

Energy →

Antibonding orbitals

Bonding orbitals

(Lowest energy orbital; all bonding)

FIGURE 11.15 The π molecular orbitals of cyclopentadienyl anion.

PROBLEM 11.21   Write resonance structures for cyclopentadienyl anion sufficient to show the delocalization of the negative charge over all five carbons.

There is a striking difference in the acidity of cyclopentadiene compared with cycloheptatriene. Cycloheptatriene has a pK_a of 36, which makes it 10^{20} times weaker in acid strength than cyclopentadiene.

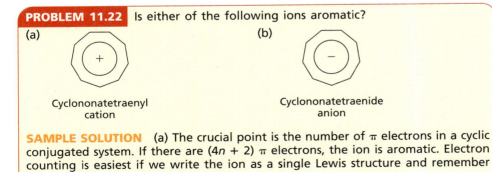

Cycloheptatriene   Water   Cycloheptatrienyl anion   Hydronium ion
$pK_a = 36$

Even though resonance tells us that the negative charge in cycloheptatrienyl anion can be shared by all seven of its carbons, this delocalization offers little in the way of stabilization. Indeed with eight π electrons, cycloheptatrienyl anion is antiaromatic and relatively unstable.

Hückel's rule is now taken to apply to planar, monocyclic, completely conjugated systems generally, not just to neutral hydrocarbons.

A planar, monocyclic, continuous system of p orbitals possesses aromatic stability when it contains (4n + 2) π electrons.

Other aromatic ions include cyclopropenyl cation (two π electrons) and cyclooctatetraene dianion (ten π electrons).

Cyclopropenyl   Cyclooctatetraene
cation   dianion

Here, liberties have been taken with the Robinson symbol. Instead of restricting its use to a sextet of electrons, organic chemists have come to adopt it as an all-purpose symbol for cyclic electron delocalization.

PROBLEM 11.22   Is either of the following ions aromatic?

(a)   (b)

Cyclononatetraenyl   Cyclononatetraenide
cation   anion

SAMPLE SOLUTION   (a) The crucial point is the number of π electrons in a cyclic conjugated system. If there are (4n + 2) π electrons, the ion is aromatic. Electron counting is easiest if we write the ion as a single Lewis structure and remember that each double bond contributes two π electrons, a negatively charged carbon contributes two, and a positively charged carbon contributes none.

Cyclononatetraenyl cation has eight π electrons; it is *not aromatic.*

11.22 HETEROCYCLIC AROMATIC COMPOUNDS

Cyclic compounds that contain at least one atom other than carbon within their ring are called **heterocyclic compounds,** and those that possess aromatic stability are called **heterocyclic aromatic compounds.** Some representative heterocyclic aromatic compounds are *pyridine, pyrrole, furan,* and *thiophene.* The structures and the IUPAC numbering system used in naming their derivatives are shown. In their stability and chemical behavior, all these compounds resemble benzene more than they resemble alkenes.

Pyridine   Pyrrole   Furan   Thiophene

Pyridine, pyrrole, and thiophene, like benzene, are present in coal tar. Furan is prepared from a substance called *furfural* obtained from corncobs.

Heterocyclic aromatic compounds can be polycyclic as well. A benzene ring and a pyridine ring, for example, can share a common side in two different ways. One way gives a compound called *quinoline;* the other gives *isoquinoline.*

Quinoline   Isoquinoline

Analogous compounds derived by fusion of a benzene ring to a pyrrole, furan, or thiophene nucleus are called *indole, benzofuran,* and *benzothiophene.*

Indole   Benzofuran   Benzothiophene

PROBLEM 11.23 Unlike quinoline and isoquinoline, which are of comparable stability, the compounds indole and isoindole are quite different from each other. Which one is more stable? Explain the reason for your choice.

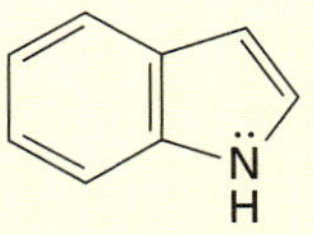

Indole 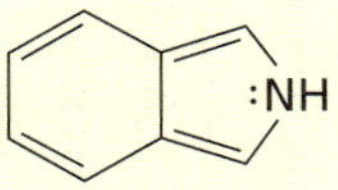   Isoindole

A large group of heterocyclic aromatic compounds are related to pyrrole by replacement of one of the ring carbons β to nitrogen by a second heteroatom. Compounds of this type are called *azoles*.

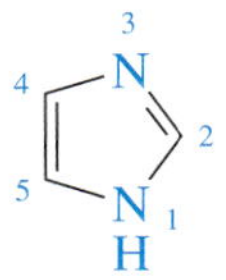

Imidazole 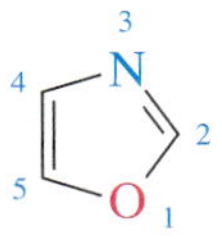   Oxazole 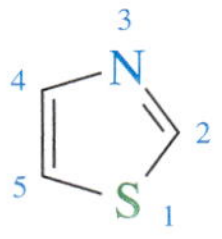   Thiazole

A widely prescribed drug for the treatment of gastric ulcers with the generic name *cimetidine* is a synthetic imidazole derivative. *Firefly luciferin* is a thiazole derivative that is the naturally occurring light-emitting substance present in fireflies.

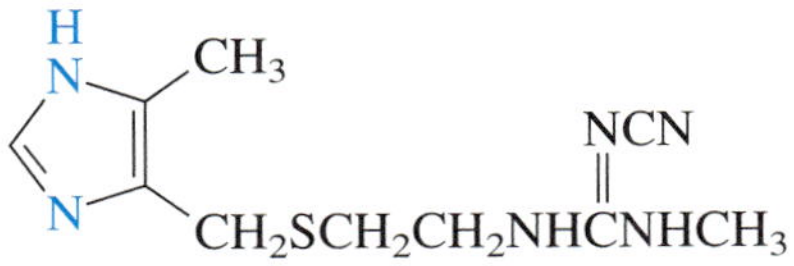

Cimetidine 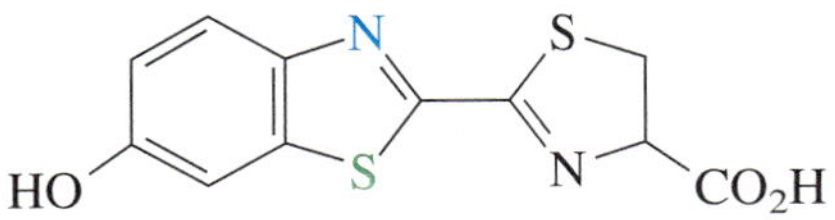   Firefly luciferin

Firefly luciferin is an example of an azole that contains a benzene ring fused to the five-membered ring. Such structures are fairly common. Another example is *benzimidazole*, present as a structural unit in vitamin B_{12}. Some compounds related to benzimidazole include *purine* and its amino-substituted derivative *adenine*, one of the so-called heterocyclic bases found in DNA and RNA (Chapter 28).

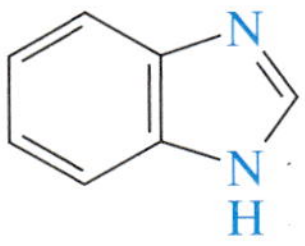

Benzimidazole 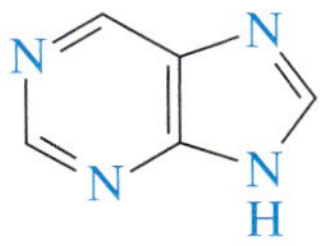   Purine 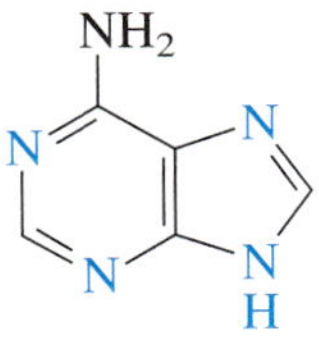   Adenine

PROBLEM 11.24 Can you deduce the structural formulas of *benzoxazole* and *benzothiazole*?

The structural types described in this section are but a tiny fraction of those possible. The chemistry of heterocyclic aromatic compounds is a rich and varied field with numerous applications.

11.23 HETEROCYCLIC AROMATIC COMPOUNDS AND HÜCKEL'S RULE

Hückel's rule can be extended to heterocyclic aromatic compounds. A single heteroatom can contribute either 0 or 2 of its lone-pair electrons as needed to the π system so as to satisfy the $(4n + 2)$ π electron requirement. The lone pair in pyridine, for example, is associated entirely with nitrogen and is not delocalized into the aromatic π system. As shown in Figure 11.16a, pyridine is simply a benzene ring in which a nitrogen atom has replaced a CH group. The nitrogen is sp^2-hybridized, and the three double bonds of the ring contribute the necessary six π electrons to make pyridine a heterocyclic aromatic compound. The unshared electron pair of nitrogen occupies an sp^2 orbital in the plane of the ring, not a p orbital aligned with the π system.

In pyrrole, on the other hand, the unshared pair belonging to nitrogen must be added to the four π electrons of the two double bonds in order to meet the six-π elec-tron requirement. As shown in Figure 11.16b, the nitrogen of pyrrole is sp^2-hybridized and the pair of electrons occupies a p orbital where both electrons can participate in the aromatic π system.

Pyridine and pyrrole are both weak bases, but pyridine is much more basic than pyrrole. When pyridine is protonated, its unshared pair is used to bond to a proton and,

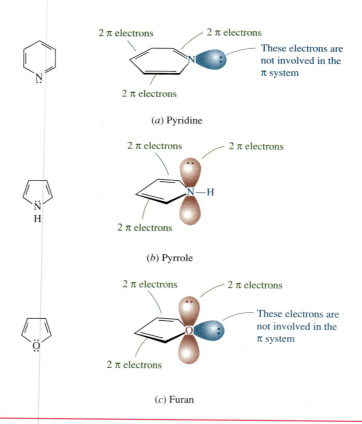

(a) Pyridine

(b) Pyrrole

(c) Furan

FIGURE 11.16 (a) Pyridine has six π electrons plus an unshared pair in a nitrogen sp^2 orbital. (b) Pyrrole has six π electrons. (c) Furan has six π electrons plus an unshared pair in an oxygen sp^2 orbital, which is perpendicular to the π system and does not interact with it.

because the unshared pair is not involved in the π system, the aromatic character of the ring is little affected. When pyrrole acts as a base, the two electrons used to form a bond to hydrogen must come from the π system, and the aromaticity of the molecule is sacrificed on protonation.

PROBLEM 11.25 Imidazole is a much stronger base than pyrrole. Predict which nitrogen is protonated when imidazole reacts with an acid, and write a structural formula for the species formed.

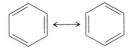

Imidazole

The oxygen in furan has two unshared electron pairs (Figure 11.16c). One pair is like the pair in pyrrole, occupying a *p* orbital and contributing two electrons to complete the six-π electron requirement for aromatic stabilization. The other electron pair in furan is an "extra" pair, not needed to satisfy the $4n + 2$ rule for aromaticity, and occupies an sp^2-hybridized orbital like the unshared pair in pyridine.

The bonding in thiophene is similar to that of furan.

11.24 SUMMARY

Section 11.1 Benzene is the parent of a class of hydrocarbons called **arenes,** or **aromatic hydrocarbons.**

Section 11.2 An important property of aromatic hydrocarbons is that they are much more stable and less reactive than other unsaturated compounds. Benzene, for example, does not react with many of the reagents that react rapidly with alkenes. When reaction does take place, substitution rather than addition is observed. The Kekulé formulas for benzene seem inconsistent with its low reactivity and with the fact that all of the C—C bonds in benzene are the same length (140 pm).

Section 11.3 One explanation for the structure and stability of benzene and other arenes is based on resonance, according to which benzene is regarded as a hybrid of the two Kekulé structures.

The article "A History of the Structural Theory of Benzene—The Aromatic Sextet and Hückel's Rule" in the February 1997 issue of the *Journal of Chemical Education* (pp. 194–201) is a rich source of additional information about this topic.

Section 11.4 The extent to which benzene is more stable than either of the Kekulé structures is its **resonance energy,** which is estimated to be 125–150 kJ/mol (30–36 kcal/mol) from heats of hydrogenation data.

Section 11.5 According to the orbital hybridization model, benzene has six π electrons, which are shared by all six sp^2-hybridized carbons. Regions of high π electron density are located above and below the plane of the ring.

Section 11.6   A molecular orbital description of benzene has three π orbitals that are bonding and three that are antibonding. Each of the bonding orbitals is fully occupied (two electrons each), and the antibonding orbitals are vacant.

Section 11.7   Many aromatic compounds are simply substituted derivatives of benzene and are named accordingly. Many others have names based on some other parent aromatic compound.

tert-Butylbenzene   *m*-Chlorotoluene   2,6-Dimethylphenol

Section 11.8   **Polycyclic aromatic hydrocarbons,** of which anthracene is an example, contain two or more benzene rings fused together.

Anthracene

Section 11.9   The physical properties of arenes resemble those of other hydrocarbons.

Section 11.10 Chemical reactions of arenes can take place on the ring itself, or on a side chain. Reactions that take place on the side chain are strongly influenced by the stability of **benzylic radicals** and **benzylic carbocations.**

Benzylic free radical   Benzylic carbocation

Section 11.11 An example of a reaction in which the ring itself reacts is the **Birch reduction.** The ring of an arene is reduced to a nonconjugated diene by treatment with a Group I metal (usually sodium) in liquid ammonia in the presence of an alcohol.

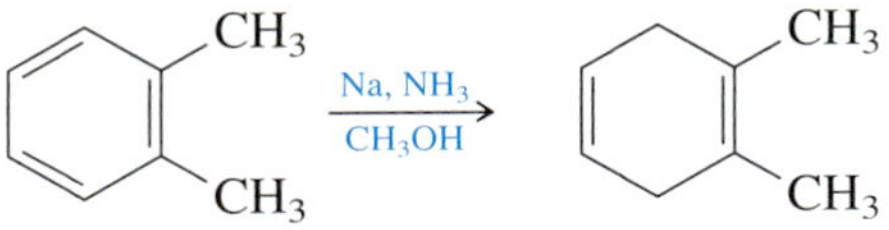

o-Xylene $\xrightarrow[\text{CH}_3\text{OH}]{\text{Na, NH}_3}$ 1,2-Dimethyl-1,4-cyclohexadiene (92%)

Sections 11.12–11.13 Free-radical halogenation and oxidation involve reactions at the benzylic carbon. See Table 11.2.

Section 11.14 Benzylic carbocations are intermediates in S_N1 reactions of benzylic halides and are stabilized by electron delocalization.

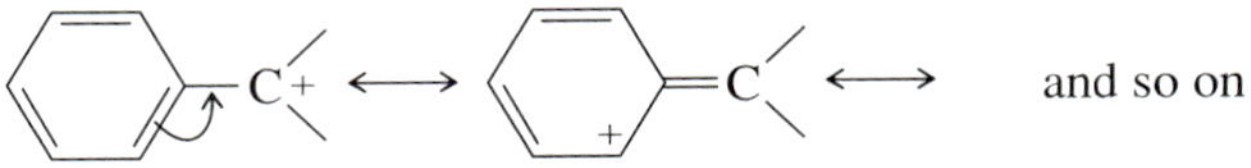

and so on

Section 11.15 The simplest alkenylbenzene is styrene ($C_6H_5CH{=}CH_2$). An aryl group stabilizes a double bond to which it is attached. Alkenylbenzenes are usually prepared by dehydration of benzylic alcohols or dehydrohalogenation of benzylic halides.

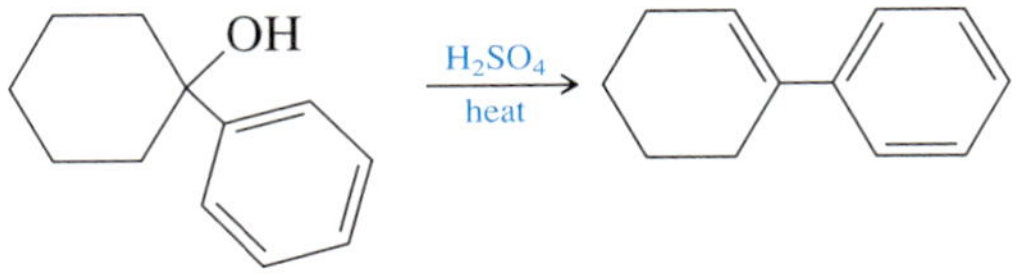

1-Phenylcyclohexanol $\xrightarrow[\text{heat}]{\text{H}_2\text{SO}_4}$ 1-Phenylcyclohexene

Section 11.16 Addition reactions to alkenylbenzenes occur at the double bond of the alkenyl substituent, and the regioselectivity of electrophilic addition is governed by carbocation formation at the benzylic carbon. See Table 11.2.

Section 11.17 Polystyrene is a widely used vinyl polymer prepared by the free-radical polymerization of styrene.

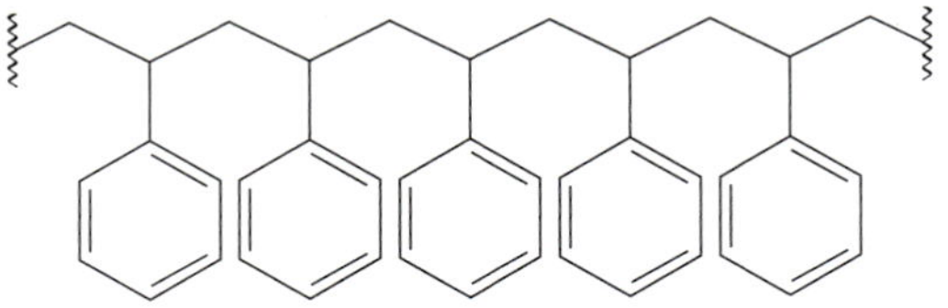

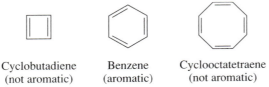

Polystyrene

Section 11.18 Although cyclic conjugation is a necessary requirement for aromaticity, this alone is not sufficient. If it were, cyclobutadiene and cyclooctatetraene would be aromatic. They are not.

Cyclobutadiene (not aromatic)   Benzene (aromatic)   Cyclooctatetraene (not aromatic)

TABLE 11.2   Reactions Involving Alkyl and Alkenyl Side Chains in Arenes and Arene Derivatives

Reaction (section) and comments	General equation and specific example

Halogenation (Section 11.12) Free-radical halogenation of alkylbenzenes is highly selective for substitution at the benzylic position. In the example shown, elemental bromine was used. Alternatively, *N*-bromosuccinimide is a convenient reagent for benzylic bromination.

$$ArCHR_2 \xrightarrow[\substack{benzoyl\ peroxide \\ CCl_4,\ 80°C}]{NBS} ArCR_2$$

Arene → 1-Arylalkyl bromide (with Br substituent)

$$O_2N-\langle\ \rangle-CH_2CH_3 \xrightarrow[\substack{CCl_4 \\ light}]{Br_2} O_2N-\langle\ \rangle-CHCH_3$$

p-Ethylnitrobenzene → 1-(*p*-Nitrophenyl)ethyl bromide (77%)

Oxidation (Section 11.13) Oxidation of alkylbenzenes occurs at the benzylic position of the alkyl group and gives a benzoic acid derivative. Oxidizing agents include sodium or potassium dichromate in aqueous sulfuric acid. Potassium permanganate ($KMnO_4$) is also an effective oxidant.

$$ArCHR_2 \xrightarrow{oxidize} ArCO_2H$$

Arene → Arenecarboxylic acid

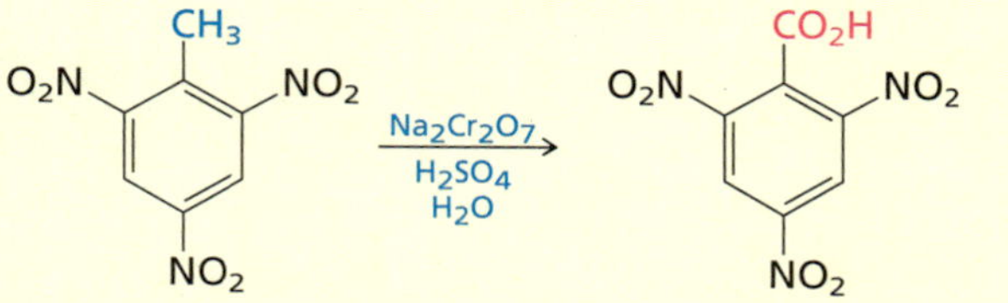

2,4,6-Trinitrotoluene $\xrightarrow[\substack{H_2SO_4 \\ H_2O}]{Na_2Cr_2O_7}$ 2,4,6-Trinitrobenzoic acid (57–69%)

Hydrogenation (Section 11.16) Hydrogenation of aromatic rings is somewhat slower than hydrogenation of alkenes, and it is a simple matter to reduce the double bond of an unsaturated side chain in an arene while leaving the ring intact.

$$ArCH{=}CR_2 + H_2 \xrightarrow{Pt} ArCH_2CHR_2$$

Alkenylarene   Hydrogen → Alkylarene

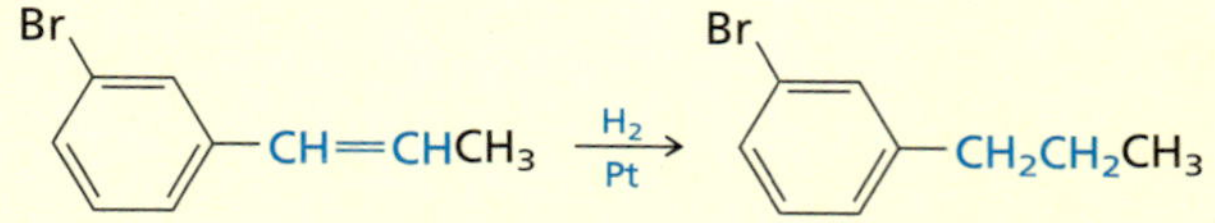

1-(*m*-Bromophenyl)propene $\xrightarrow[Pt]{H_2}$ *m*-Bromopropylbenzene (85%)

Electrophilic addition (Section 11.16) An aryl group stabilizes a benzylic carbocation and controls the regioselectivity of addition to a double bond involving the benzylic carbon. Markovnikov's rule is obeyed.

$$ArCH{=}CH_2 \xrightarrow{\overset{\delta+}{E}{-}\overset{\delta-}{Y}} ArCH{-}CH_2E$$

Alkenylarene → Product of electrophilic addition (with Y substituent)

Styrene $\xrightarrow{HBr}$ 1-Phenylethyl bromide (85%)

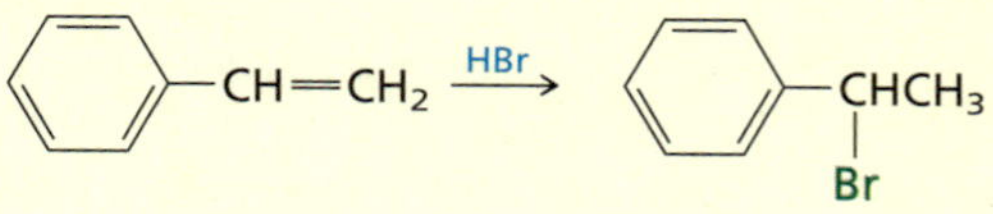

Section 11.19 An additional requirement for aromaticity is that the number of π electrons in conjugated, planar, monocyclic species must be equal to $4n + 2$, where n is an integer. This is called **Hückel's rule.** Benzene, with six π electrons, satisfies Hückel's rule for $n = 1$. Square cyclobutadiene (four π electrons) and planar cyclooctatetraene (eight π electrons) do not. Both are examples of systems with $4n$ π electrons and are antiaromatic.

Section 11.20 **Annulenes** are monocyclic, completely conjugated polyenes synthesized for the purpose of testing Hückel's rule. They are named by using a bracketed numerical prefix to indicate the number of carbons, followed by the word *annulene*. [4n]Annulenes are characterized by rings with alternating short (double) and long (single) bonds and are *antiaromatic*. The expected aromaticity of [4n + 2]annulenes is diminished by angle and van der Waals strain unless the ring contains approximately 18 carbons.

Section 11.21 Species with six π electrons that possess "special stability" include certain ions, such as *cyclopentadienide* anion and *cycloheptatrienyl* cation.

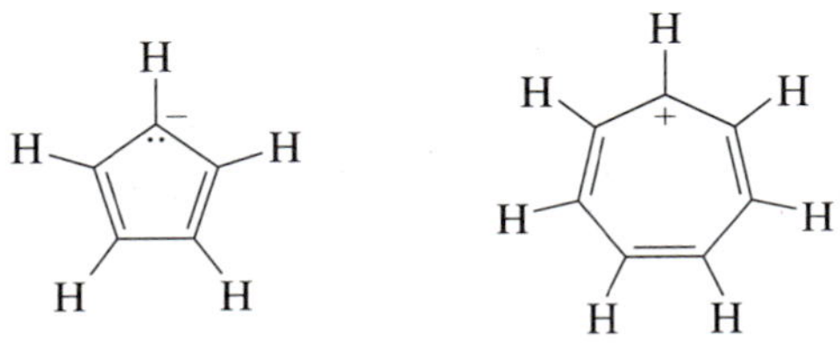

Cyclopentadienide anion   Cycloheptatrienyl cation
(six π electrons)   (six π electrons)

Section 11.22 **Heterocyclic aromatic compounds** are compounds that contain at least one atom other than carbon within an aromatic ring.

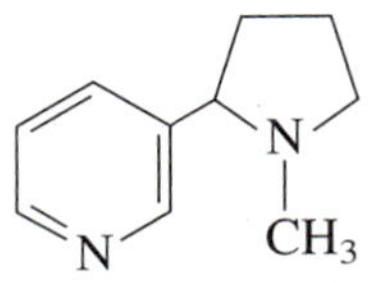

Nicotine

Section 11.23 Hückel's rule can be extended to heterocyclic aromatic compounds. Unshared electron pairs of the heteroatom may be used as π electrons as necessary to satisfy the $4n + 2$ rule.

PROBLEMS

11.26 Write structural formulas and give the IUPAC names for all the isomers of $C_6H_5C_4H_9$ that contain a monosubstituted benzene ring.

11.27 Write a structural formula corresponding to each of the following:

(a) Allylbenzene

(b) (*E*)-1-Phenyl-1-butene

(c) (*Z*)-2-Phenyl-2-butene

(d) (*R*)-1-Phenylethanol

(e) *o*-Chlorobenzyl alcohol

(f) *p*-Chlorophenol

(g) 2-Nitrobenzenecarboxylic acid

(h) *p*-Diisopropylbenzene

(i) 2,4,6-Tribromoaniline

(j) *m*-Nitroacetophenone

(k) 4-Bromo-3-ethylstyrene

11.28 Using numerical locants and the names in Table 11.1 as a guide, give an acceptable IUPAC name for each of the following compounds:

(a) Estragole (principal component of wormwood oil)

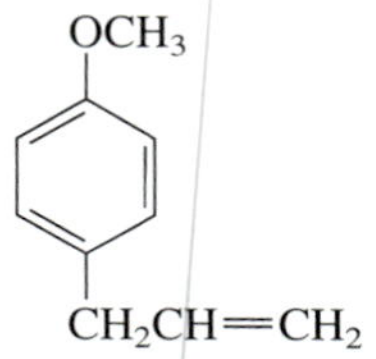

(b) Diosphenol (used in veterinary medicine to control parasites in animals)

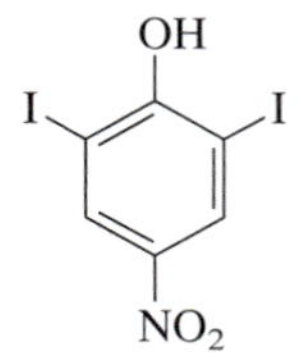

(c) *m*-Xylidine (used in synthesis of lidocaine, a local anesthetic)

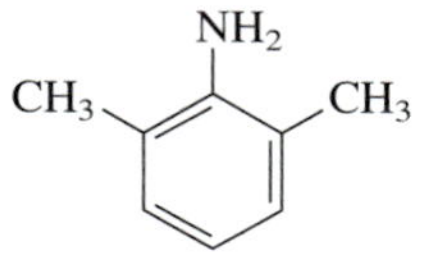

11.29 Write structural formulas and give acceptable names for all the isomeric

(a) Nitrotoluenes

(b) Dichlorobenzoic acids

(c) Tribromophenols

(d) Tetrafluorobenzenes

(e) Naphthalenecarboxylic acids

(f) Bromoanthracenes

11.30 Each of the following may be represented by at least one alternative resonance structure in which all the six-membered rings correspond to Kekulé forms of benzene. Write such a resonance form for each.

(a)　　　　(c)　　　　(e)

(b)　　　　(d)

The common name of isopropylbenzene is *cumene*.

11.31 Give the structure of the expected product from the reaction of isopropylbenzene with

(a) Hydrogen (3 mol), Pt

(b) Sodium and ethanol in liquid ammonia

(c) Sodium dichromate, water, sulfuric acid, heat

(d) *N*-Bromosuccinimide in CCl_4, heat, benzoyl peroxide

(e) The product of part (d) treated with sodium ethoxide in ethanol

11.32 Each of the following reactions has been described in the chemical literature and gives a single organic product in good yield. Identify the product of each reaction.

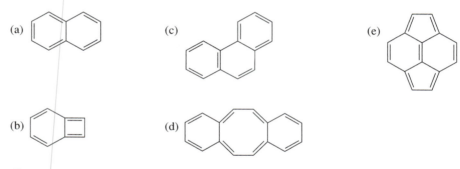

(a) 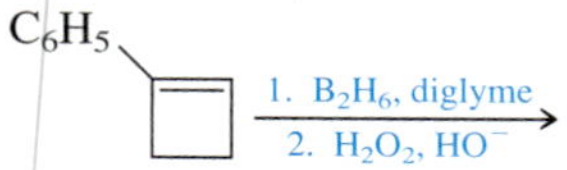 $\xrightarrow[\text{2. } H_2O_2,\ HO^-]{\text{1. } B_2H_6,\ \text{diglyme}}$

(b) $+\ H_2$ (1 mol) $\xrightarrow{\text{Pt}}$

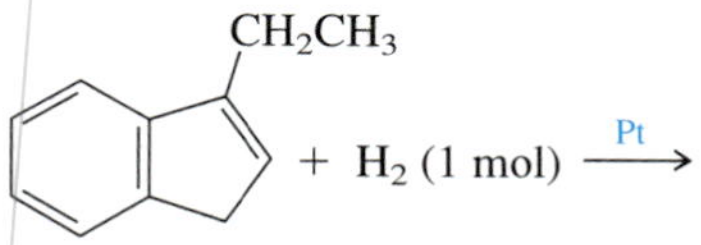

(c) 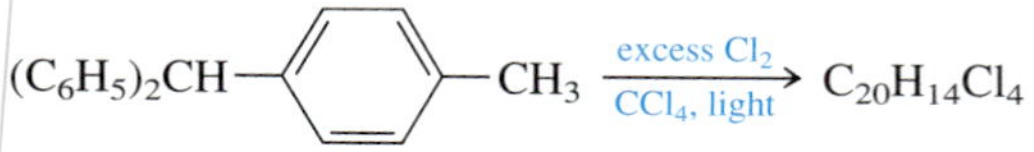 $(C_6H_5)_2CH{-}\!\!\!\!\!\bigcirc\!\!\!\!\!{-}CH_3 \xrightarrow[CCl_4,\ \text{light}]{\text{excess } Cl_2} C_{20}H_{14}Cl_4$

(d) (E)-C_6H_5CH=CHC_6H_5 $\xrightarrow[\text{acetic acid}]{CH_3CO_2OH}$

(e) 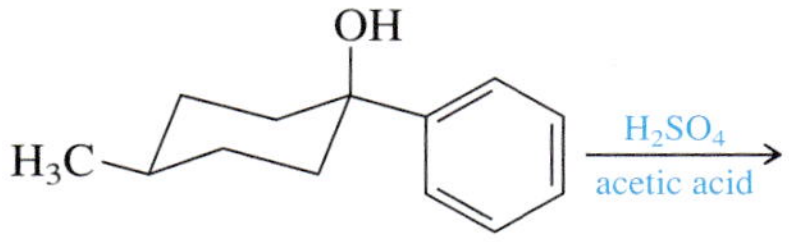 H_3C—⟨cyclohexane with OH and phenyl⟩ $\xrightarrow[\text{acetic acid}]{H_2SO_4}$

(f) 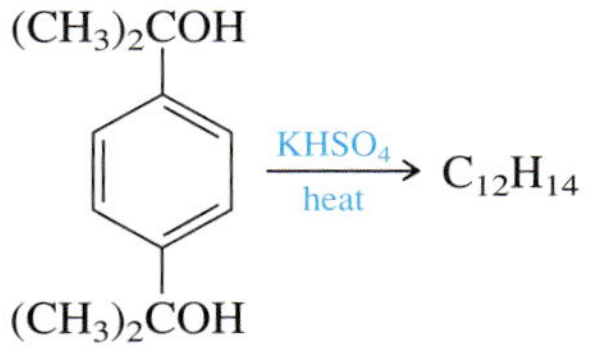 $(CH_3)_2\overset{OH}{\underset{}{C}}$—⟨benzene ring⟩—$(CH_3)_2COH$ $\xrightarrow[\text{heat}]{KHSO_4}$ $C_{12}H_{14}$

(g)  $(Cl$—⟨benzene ring⟩—$)_2CHCCl_3$ $\xrightarrow[\text{CH}_3\text{OH}]{NaOCH_3}$ $C_{14}H_8Cl_4$

(DDT)

(h) 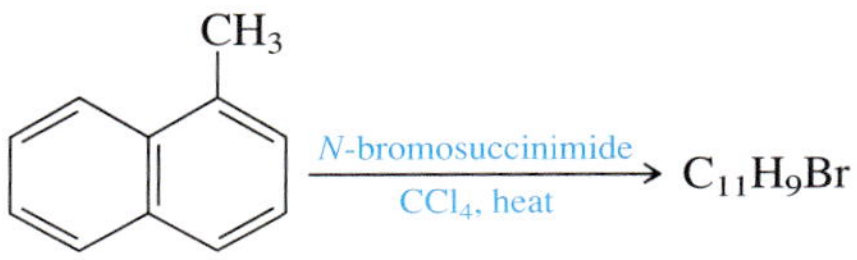 ⟨naphthalene with CH$_3$⟩ $\xrightarrow[\text{CCl}_4,\text{ heat}]{N\text{-bromosuccinimide}}$ $C_{11}H_9Br$

(i) 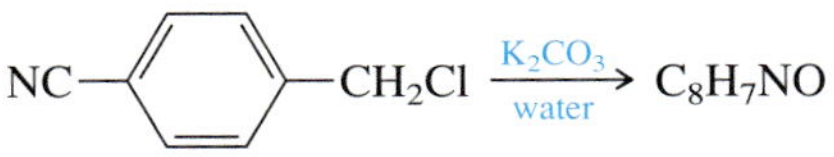 NC—⟨benzene ring⟩—CH_2Cl $\xrightarrow[\text{water}]{K_2CO_3}$ C_8H_7NO

11.33 A certain compound A, when treated with N-bromosuccinimide and benzoyl peroxide under photochemical conditions in refluxing carbon tetrachloride, gave 3,4,5-tribromobenzyl bromide in excellent yield. Deduce the structure of compound A.

11.34 A compound was obtained from a natural product and had the molecular formula $C_{14}H_{20}O_3$. It contained three methoxy (—OCH_3) groups and a —CH_2CH=$C(CH_3)_2$ substituent. Oxidation with either chromic acid or potassium permanganate gave 2,3,5-trimethoxybenzoic acid. What is the structure of the compound?

11.35 Hydroboration–oxidation of (E)-2-(p-anisyl)-2-butene yielded an alcohol A, mp 60°C, in 72% yield. When the same reaction was performed on the Z alkene, an isomeric liquid alcohol B was obtained in 77% yield. Suggest reasonable structures for A and B, and describe the relationship between them.

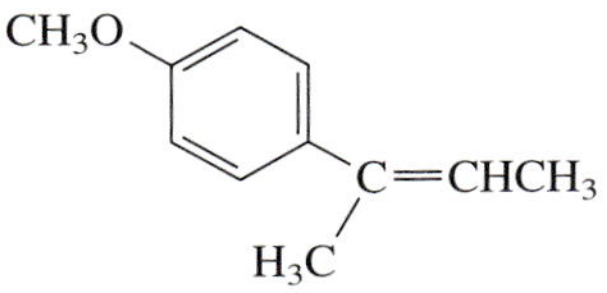

CH_3O—⟨benzene ring⟩—$\underset{H_3C}{\overset{}{C}}$=$CHCH_3$

2-(p-Anisyl)-2-butene

11.36 Dehydrohalogenation of the diastereomeric forms of 1-chloro-1,2-diphenylpropane is stereospecific. One diastereomer yields (E)-1,2-diphenylpropene, and the other yields the Z isomer. Which diastereomer yields which alkene? Why?

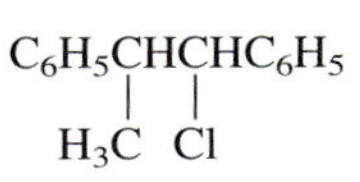

$C_6H_5\underset{H_3C}{\overset{}{C}}H\underset{Cl}{\overset{}{C}}HC_6H_5$ $\underset{H_3C}{\overset{C_6H_5}{}}C$=$CHC_6H_5$

1-Chloro-1,2-diphenylpropane 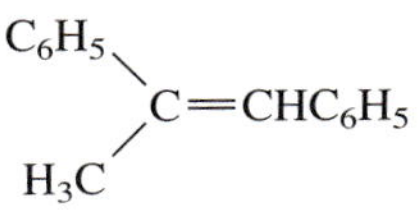   1,2-Diphenylpropene

11.37 Suggest reagents suitable for carrying out each of the following conversions. In most cases more than one synthetic operation will be necessary.

(a) $C_6H_5CH_2CH_3 \longrightarrow C_6H_5\underset{\underset{Br}{|}}{C}HCH_3$

(b) $C_6H_5\underset{\underset{Br}{|}}{C}HCH_3 \longrightarrow C_6H_5\underset{\underset{Br}{|}}{C}HCH_2Br$

(c) $C_6H_5CH{=}CH_2 \longrightarrow C_6H_5C{\equiv}CH$

(d) $C_6H_5C{\equiv}CH \longrightarrow C_6H_5CH_2CH_2CH_2CH_3$

(e) $C_6H_5CH_2CH_2OH \longrightarrow C_6H_5CH_2CH_2C{\equiv}CH$

(f) $C_6H_5CH_2CH_2Br \longrightarrow C_6H_5\underset{\underset{OH}{|}}{C}HCH_2Br$

11.38 The relative rates of reaction of ethane, toluene, and ethylbenzene with bromine atoms have been measured. The most reactive hydrocarbon undergoes hydrogen atom abstraction a million times faster than does the least reactive one. Arrange these hydrocarbons in order of decreasing reactivity.

11.39 Write the principal resonance structures of *o*-methylbenzyl cation and *m*-methylbenzyl cation. Which one has a tertiary carbocation as a contributing resonance form?

11.40 Suggest an explanation for the observed order of S_N1 reactivity of the following compounds.

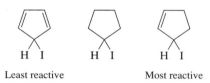

Least reactive   Most reactive

11.41 A standard method for preparing sodium cyclopentadienide (C_5H_5Na) is by the reaction of cyclopentadiene with a solution of $NaNH_2$ in liquid ammonia. Write a net ionic equation for this reaction, identify the acid and the base, and use curved arrows to track the flow of electrons.

11.42 The same anion is formed by loss of the most acidic proton from 1-methyl-1,3-cyclopentadiene as from 5-methyl-1,3-cyclopentadiene. Explain.

11.43 Cyclooctatetraene has two different tetramethyl derivatives with methyl groups on four adjacent carbon atoms. They are both completely conjugated and are not stereoisomers. Write their structures.

11.44 Evaluate each of the following processes applied to cyclooctatetraene, and decide whether the species formed is aromatic or not.

(a) Addition of one more π electron, to give $C_8H_8^-$

(b) Addition of two more π electrons, to give $C_8H_8^{2-}$

(c) Removal of one π electron, to give $C_8H_8^+$

(d) Removal of two π electrons, to give $C_8H_8^{2+}$

11.45 Evaluate each of the following processes applied to cyclononatetraene, and decide whether the species formed is aromatic or not:

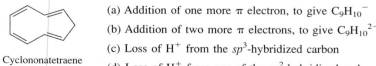

Cyclononatetraene

(a) Addition of one more π electron, to give $C_9H_{10}^-$

(b) Addition of two more π electrons, to give $C_9H_{10}^{2-}$

(c) Loss of H^+ from the sp^3-hybridized carbon

(d) Loss of H^+ from one of the sp^2-hybridized carbons

11.46 From among the molecules and ions shown, all of which are based on cycloundecapentaene, identify those which satisfy the criteria for aromaticity as prescribed by Hückel's rule.

(a) Cycloundecapentaene

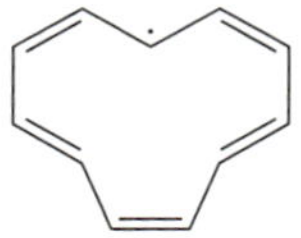

(c) Cycloundecapentaenyl cation

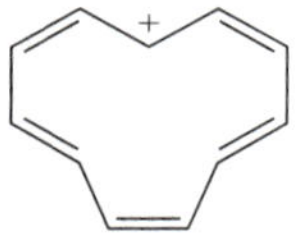

(b) Cycloundecapentaenyl radical

(d) Cycloundecapentaenyl anion

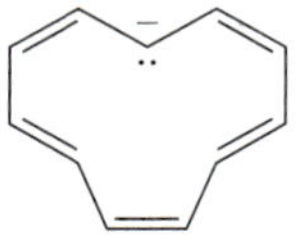

11.47 (a) Figure 11.17 is an electrostatic potential map of *calicene,* so named because its shape resembles a chalice (*calix* is the Latin word for "cup"). Both the electrostatic potential map and its calculated dipole moment (μ = 4.3 D) indicate that calicene is an unusually polar hydrocarbon. Which of the dipolar resonance forms, A or B, better corresponds to the electron distribution in the molecule? Why is this resonance form more important than the other?

Calicene   A   B

(b) Which one of the following should be stabilized by resonance to a greater extent? (*Hint:* Consider the reasonableness of dipolar resonance forms.)

   or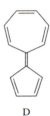

C   D

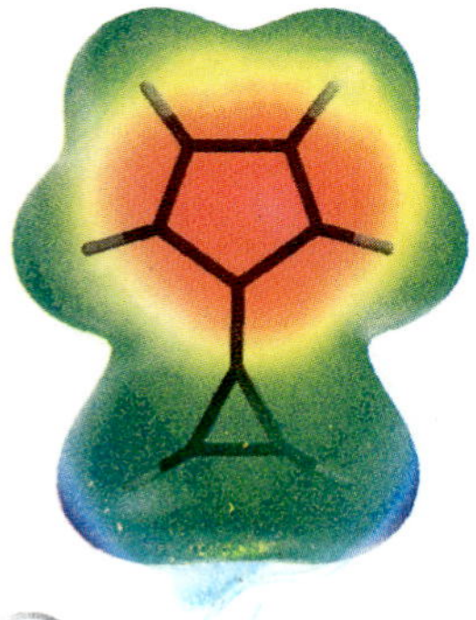

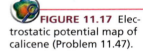

FIGURE 11.17 Electrostatic potential map of calicene (Problem 11.47).

11.48 Classify each of the following heterocyclic molecules as aromatic or not, according to Hückel's rule:

(a)

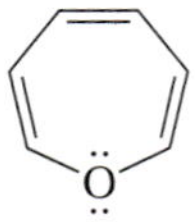

(c)

(b)

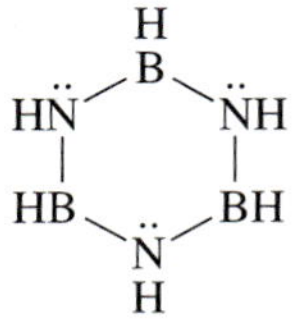

(d)

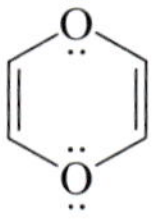

11.49 Pellagra is a disease caused by a deficiency of *niacin* ($C_6H_5NO_2$) in the diet. Niacin can be synthesized in the laboratory by the side-chain oxidation of 3-methylpyridine with chromic acid or potassium permanganate. Suggest a reasonable structure for niacin.

11.50 *Nitroxoline* is the generic name by which 5-nitro-8-hydroxyquinoline is sold as an antibacterial drug. Write its structural formula.

11.51 *Acridine* is a heterocyclic aromatic compound obtained from coal tar that is used in the synthesis of dyes. The molecular formula of acridine is $C_{13}H_9N$, and its ring system is analogous to that of anthracene except that one CH group has been replaced by N. The two most stable resonance structures of acridine are equivalent to each other, and both contain a pyridine-like structural unit. Write a structural formula for acridine.

11.52 Identify the longest and the shortest carbon–carbon bond in styrene, and check your answer by making a minimized molecular model using *Learning By Modeling*.

11.53 Mesitylene (1,3,5-trimethylbenzene) is the most stable of the trimethylbenzene isomers. Why? Which isomer do you think is the least stable? Make a molecular model of each isomer and compare their calculated strain energies with your predictions. Do space-filling models support your explanation?

11.54 Which one of the dichlorobenzene isomers does not have a dipole moment? Which one has the largest dipole moment? Compare your answers with the dipole moments calculated using the molecular-modeling software in *Learning By Modeling*.

11.55 Make molecular models of the two chair conformations of *cis*-1-*tert*-butyl-4-phenylcyclohexane. What is the strain energy calculated for each conformation by molecular mechanics? Which has a greater preference for the equatorial orientation, phenyl or *tert*-butyl?

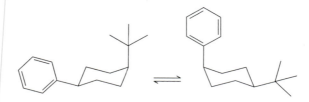

11.56 Use *Learning By Modeling* to carry out strain energy minimization routines for three conformations of biphenyl that differ in respect to the dihedral angle defined by the bonds shown in red. Start with dihedral angles of 0°, 45°, and 90°. What are the respective dihedral angles after minimization? Which conformation is the most stable? What factors are important in determining the most stable conformation?

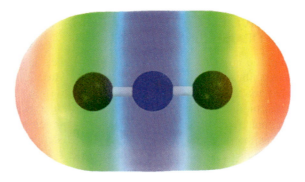

CHAPTER 12

REACTIONS OF ARENES:
ELECTROPHILIC AROMATIC SUBSTITUTION

In the preceding chapter the *special stability* of benzene was described, along with reactions in which an aromatic ring was present as a substituent. Now we'll examine the aromatic ring as a functional group. What kind of reactions are available to benzene and its derivatives? What sort of reagents react with arenes, and what products are formed in those reactions?

Characteristically, the reagents that react with the aromatic ring of benzene and its derivatives are *electrophiles*. We already have some experience with electrophilic reagents, particularly with respect to how they react with alkenes. Electrophilic reagents *add* to alkenes.

$$\underset{\text{Alkene}}{\diagdown \!\! C \!\! = \!\! C \diagup} \;\; + \;\; \underset{\text{Electrophilic}}{\overset{\delta+ \;\;\; \delta-}{E \!\! - \!\! Y}} \;\; \longrightarrow \;\; \underset{\substack{\text{Product of} \\ \text{electrophilic addition}}}{E \!\! - \!\! C \!\! - \!\! C \!\! - \!\! Y}$$

A different reaction takes place when electrophiles react with arenes. *Substitution is observed instead of addition.* If we represent an arene by the general formula ArH, where Ar stands for an aryl group, the electrophilic portion of the reagent replaces one of the hydrogens on the ring:

$$\underset{\text{Arene}}{Ar \!\! - \!\! H} \;\; + \;\; \underset{\substack{\text{Electrophilic} \\ \text{reagent}}}{\overset{\delta+ \;\;\; \delta-}{E \!\! - \!\! Y}} \;\; \longrightarrow \;\; \underset{\substack{\text{Product of} \\ \text{electrophilic aromatic} \\ \text{substitution}}}{Ar \!\! - \!\! E \; + \; H \!\! - \!\! Y}$$

We call this reaction **electrophilic aromatic substitution;** it is one of the fundamental processes of organic chemistry.

12.1   REPRESENTATIVE ELECTROPHILIC AROMATIC SUBSTITUTION REACTIONS OF BENZENE

The scope of electrophilic aromatic substitution is quite large; both the aromatic compound and the electrophilic reagent are capable of wide variation. Indeed, it is this breadth of scope that makes electrophilic aromatic substitution so important. Electrophilic aromatic substitution is the method by which substituted derivatives of benzene are prepared. We can gain a feeling for these reactions by examining a few typical examples in which benzene is the substrate. These examples are listed in Table 12.1, and each will be discussed in more detail in Sections 12.3 through 12.7. First, however, let us look at the general mechanism of electrophilic aromatic substitution.

12.2   MECHANISTIC PRINCIPLES OF ELECTROPHILIC AROMATIC SUBSTITUTION

Recall from Chapter 6 the general mechanism for electrophilic addition to alkenes:

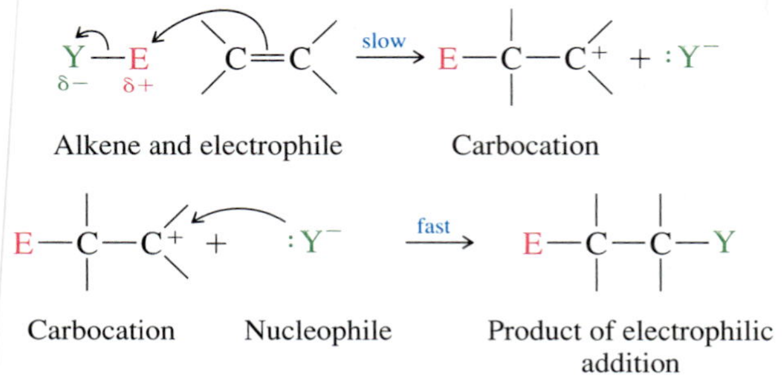

Alkene and electrophile   Carbocation

Carbocation   Nucleophile   Product of electrophilic addition

The first step is rate-determining. In it a carbocation forms when the pair of π electrons of the alkene is used to form a bond with the electrophile. Following its formation, the carbocation undergoes rapid capture by some Lewis base present in the medium.

The first step in the reaction of electrophilic reagents with benzene is similar. An electrophile accepts an electron pair from the π system of benzene to form a carbocation:

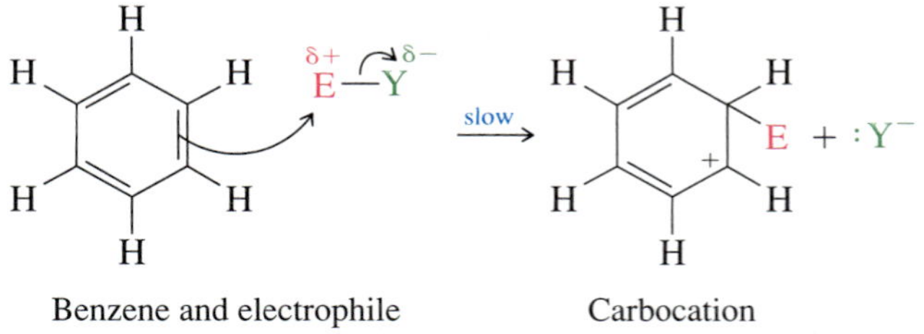

Recall that an electron-pair acceptor is a Lewis acid. Electrophiles are Lewis acids.

Benzene and electrophile   Carbocation

The carbocation formed in this step is a **cyclohexadienyl cation.** Other commonly used terms include **arenium ion** and **σ-complex.** It is an allylic carbocation and is stabilized by electron delocalization which can be represented by resonance.

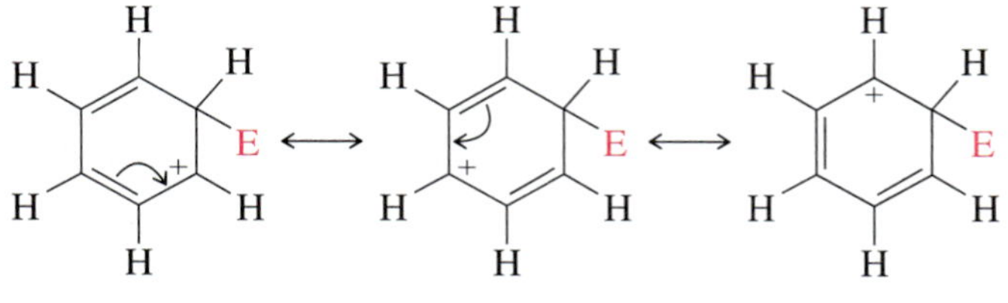

Resonance structures of a cyclohexadienyl cation

| TABLE 12.1 | Representative Electrophilic Aromatic Substitution Reactions of Benzene |

Reaction and comments	Equation
1. Nitration Warming benzene with a mixture of nitric acid and sulfuric acid gives nitrobenzene. A nitro group (—NO_2) replaces one of the ring hydrogens.	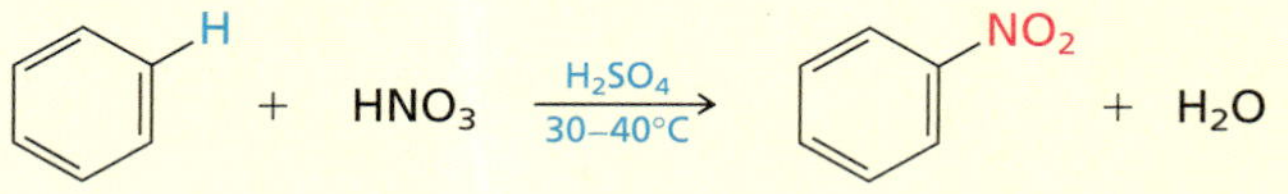 Benzene + HNO_3 $\xrightarrow[30-40°C]{H_2SO_4}$ Nitrobenzene (95%) + H_2O Water
2. Sulfonation Treatment of benzene with hot concentrated sulfuric acid gives benzenesulfonic acid. A sulfonic acid group (—SO_2OH) replaces one of the ring hydrogens.	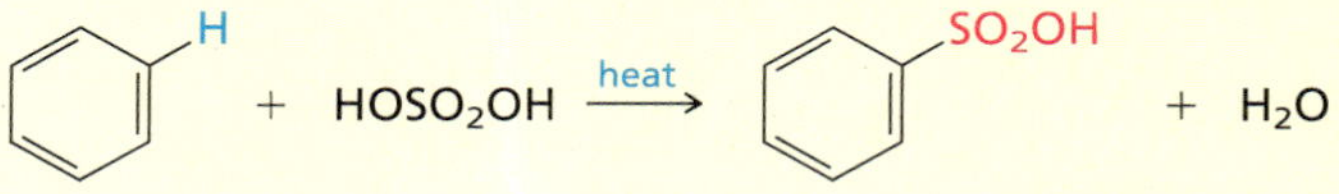 Benzene + $HOSO_2OH$ $\xrightarrow{heat}$ Benzenesulfonic acid (100%) + H_2O Water
3. Halogenation Bromine reacts with benzene in the presence of iron(III) bromide as a catalyst to give bromobenzene. Chlorine reacts similarly in the presence of iron(III) chloride to give chlorobenzene.	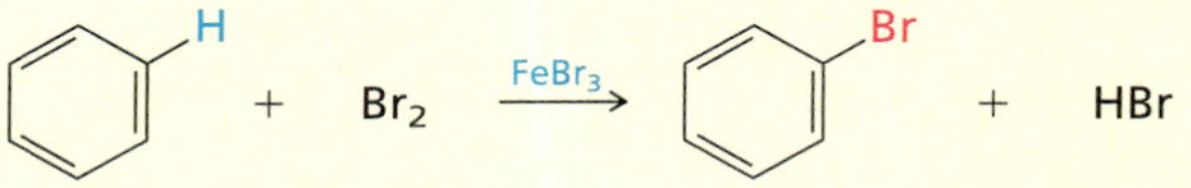 Benzene + Br_2 $\xrightarrow{FeBr_3}$ Bromobenzene (65–75%) + HBr Hydrogen bromide
4. Friedel–Crafts alkylation Alkyl halides react with benzene in the presence of aluminum chloride to yield alkylbenzenes.	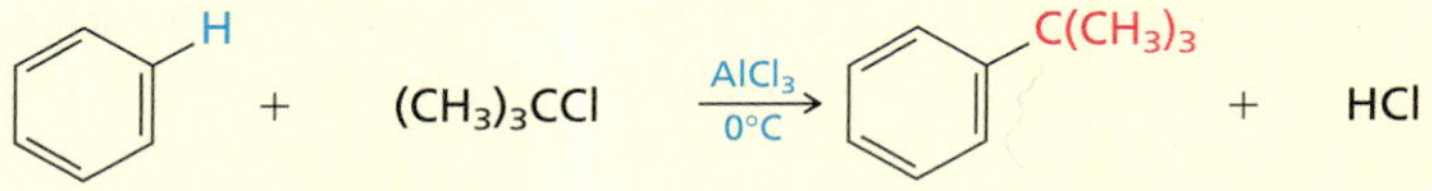 Benzene + $(CH_3)_3CCl$ $\xrightarrow[0°C]{AlCl_3}$ tert-Butylbenzene (60%) + HCl Hydrogen chloride
5. Friedel–Crafts acylation An analogous reaction occurs when acyl halides react with benzene in the presence of aluminum chloride. The products are acylbenzenes.	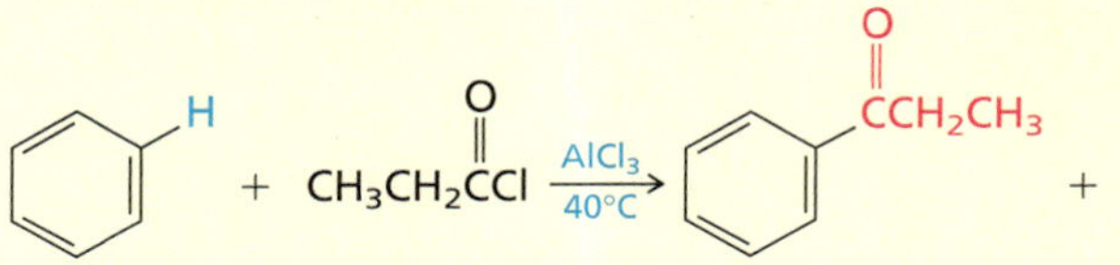 Benzene + CH_3CH_2CCl (O) $\xrightarrow[40°C]{AlCl_3}$ 1-Phenyl-1-propanone (88%) + HCl Hydrogen chloride

PROBLEM 12.1 When a molecular orbital method was used to calculate the charge distribution in cyclohexadienyl cation, it gave the results indicated. How does the charge at each carbon compare with that deduced by examining the resonance structures for cyclohexadienyl cation?

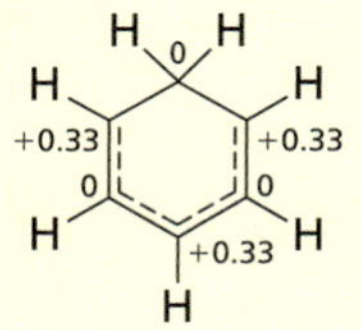

A model showing the electrostatic potential of this carbocation can be viewed on *Learning By Modeling.*

Most of the resonance stabilization of benzene is lost when it is converted to the cyclohexadienyl cation intermediate. In spite of being allylic, a cyclohexadienyl cation is *not* aromatic and possesses only a fraction of the resonance stabilization of benzene.

Once formed, it rapidly loses a proton, restoring the aromaticity of the ring and giving the product of electrophilic aromatic substitution.

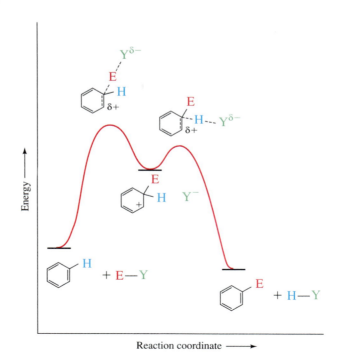

If the Lewis base ($:Y^-$) had acted as a nucleophile and bonded to carbon, the product would have been a nonaromatic cyclohexadiene derivative. Addition and substitution products arise by alternative reaction paths of a cyclohexadienyl cation. Substitution occurs preferentially because there is a substantial driving force favoring rearomatization.

Figure 12.1 is a potential energy diagram describing the general mechanism of electrophilic aromatic substitution. For electrophilic aromatic substitution reactions to

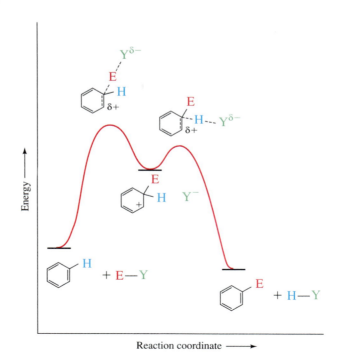

FIGURE 12.1 Potential energy diagram for electrophilic aromatic substitution.

overcome the high activation energy that characterizes the first step, the electrophile must be a fairly reactive one. Many of the electrophilic reagents that react rapidly with alkenes do not react at all with benzene. Peroxy acids and diborane, for example, fall into this category. Others, such as bromine, react with benzene only in the presence of catalysts that increase their electrophilicity. The low level of reactivity of benzene toward electrophiles stems from the substantial loss of resonance stabilization that accompanies transfer of a pair of its six π electrons to an electrophile.

With this as background, let us now examine each of the electrophilic aromatic substitution reactions presented in Table 12.1 in more detail, especially with respect to the electrophile that attacks benzene.

12.3 NITRATION OF BENZENE

Now that we've outlined the general mechanism for electrophilic aromatic substitution, we need only identify the specific electrophile in the nitration of benzene to have a fairly clear idea of how the reaction occurs.

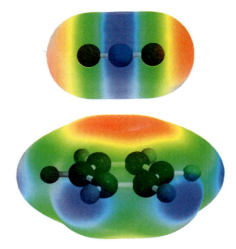

| Benzene | Nitric acid | Nitrobenzene (95%) | Water |

The electrophile (E^+) in this reaction is *nitronium ion* ($:\overset{..}{O}=\overset{+}{N}=\overset{..}{O}:$). The charge distribution in nitronium ion is evident both in its Lewis structure and in the electrostatic potential map of Figure 12.2. There we see the complementary relationship between the electron-poor region near nitrogen of NO_2^+ and the electron-rich region associated with the π electrons of benzene.

Figure 12.3 adapts the general mechanism of electrophilic aromatic substitution to the nitration of benzene. The first step is rate-determining; in it benzene reacts with nitronium ion to give the cyclohexadienyl cation intermediate. In the second step, the aromaticity of the ring is restored by loss of a proton from the cyclohexadienyl cation.

The purpose of sulfuric acid in the reaction is to increase the concentration of nitronium ion. Nitric acid alone does not furnish a high enough concentration of nitronium

The role of nitronium ion in the nitration of benzene was demonstrated by Sir Christopher Ingold—the same person who suggested the S_N1 and S_N2 mechanisms of nucleophilic substitution and who collaborated with Cahn and Prelog on the *R* and *S* notational system.

FIGURE 12.2 Electrostatic potential maps of NO_2^+ (top) and benzene (bottom). The region of greatest positive potential in NO_2^+ is associated with nitrogen. The region of greatest negative potential in benzene is associated with the π electrons above and below the ring.

Step 1: Reaction of nitronium cation with the π system of the aromatic ring

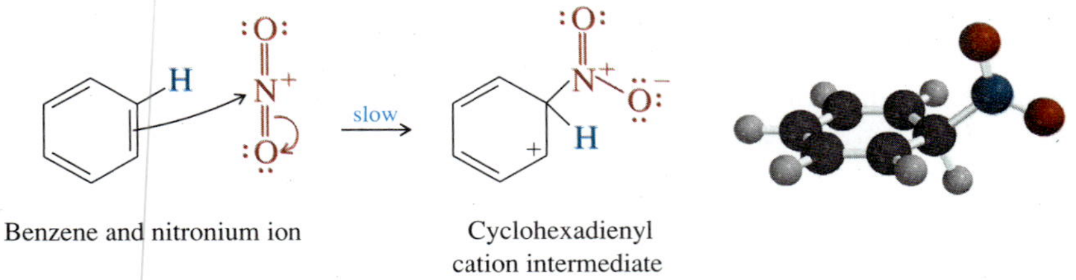

Benzene and nitronium ion $\xrightarrow{slow}$ Cyclohexadienyl cation intermediate

Step 2: Loss of a proton from the cyclohexadienyl cation

Cyclohexadienyl cation intermediate   Water $\xrightarrow{fast}$ Nitrobenzene   Hydronium ion

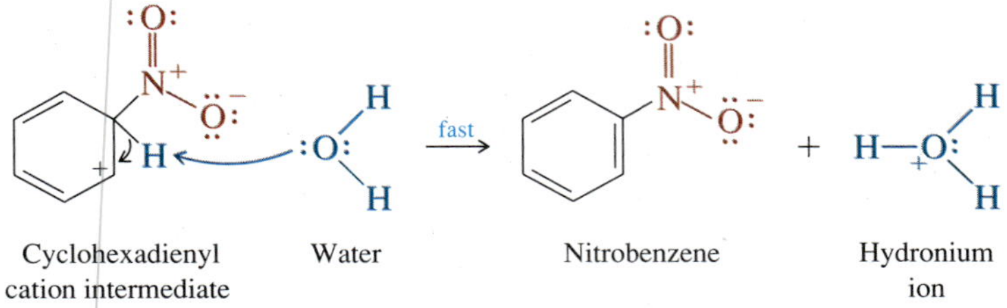

ion for the reaction to proceed at a convenient rate. Nitric acid reacts with sulfuric acid to give nitronium ion according to the equation:

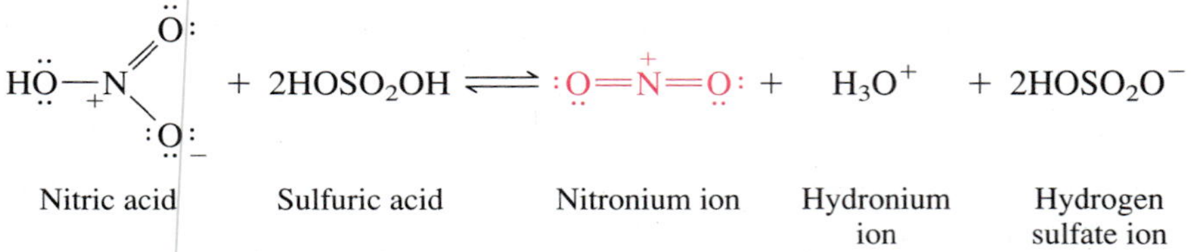

Nitric acid   Sulfuric acid   Nitronium ion   Hydronium ion   Hydrogen sulfate ion

Nitration by electrophilic aromatic substitution is not limited to benzene alone, but is a general reaction of compounds that contain a benzene ring. It would be a good idea to write out the answer to the following problem to ensure that you understand the relationship of starting materials to products in aromatic nitration before continuing to the next section.

PROBLEM 12.2 Nitration of 1,4-dimethylbenzene (*p*-xylene) gives a single product having the molecular formula $C_8H_9NO_2$ in high yield. What is this product?

12.4 SULFONATION OF BENZENE

The reaction of benzene with sulfuric acid to produce benzenesulfonic acid:

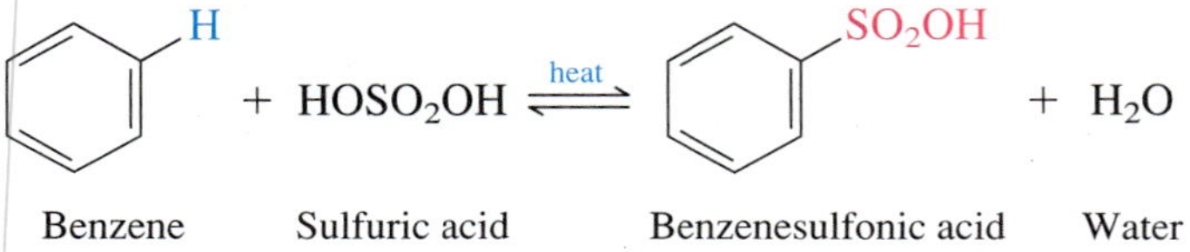

Benzene   Sulfuric acid   Benzenesulfonic acid   Water

is reversible but can be driven to completion by several techniques. Removing the water formed in the reaction, for example, allows benzenesulfonic acid to be obtained in virtually quantitative yield. When a solution of sulfur trioxide in sulfuric acid is used as the sulfonating agent, the rate of sulfonation is much faster and the equilibrium is displaced entirely to the side of products, according to the equation

Benzene   Sulfur   Benzenesulfonic acid
                 trioxide

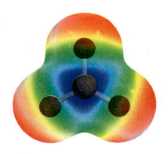

Among the variety of electrophilic species present in concentrated sulfuric acid, sulfur trioxide (Figure 12.4) is probably the actual electrophile in aromatic sulfonation. We can represent the mechanism of sulfonation of benzene by sulfur trioxide by the sequence of steps shown in Figure 12.5.

Step 1: Sulfur trioxide attacks benzene in the rate-determining step

Benzene and sulfur trioxide   Cyclohexadienyl cation intermediate

Step 2: A proton is lost from the sp^3-hybridized carbon of the intermediate to restore the aromaticity of the ring. The species shown that abstracts the proton is a hydrogen sulfate ion formed by ionization of sulfuric acid.

Cyclohexadienyl   Hydrogen   Benzenesulfonate ion   Sulfuric acid
cation intermediate   sulfate ion

Step 3: A rapid proton transfer from the oxygen of sulfuric acid to the oxygen of benzenesulfonate completes the process.

Benzenesulfonate   Sulfuric acid   Benzenesulfonic acid   Hydrogen
ion   sulfate ion

FIGURE 12.5 The mechanism of sulfonation of benzene. The molecular model depicts the cyclohexadienyl cation intermediate formed in step 1.

PROBLEM 12.3 On being heated with sulfur trioxide in sulfuric acid, 1,2,4,5-tetramethylbenzene was converted to a product of molecular formula $C_{10}H_{14}O_3S$ in 94% yield. Suggest a reasonable structure for this product.

12.5 HALOGENATION OF BENZENE

According to the usual procedure for preparing bromobenzene, bromine is added to benzene in the presence of metallic iron (customarily a few carpet tacks) and the reaction mixture is heated.

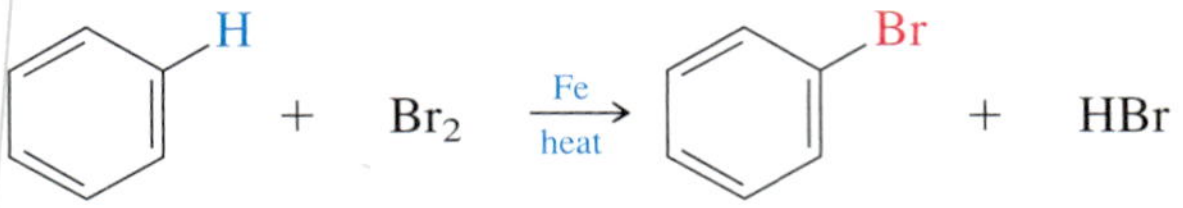

Benzene   Bromine   Bromobenzene   Hydrogen
                               (65–75%)   bromide

Bromine, although it adds rapidly to alkenes, is too weak an electrophile to react at an appreciable rate with benzene. A catalyst that increases the electrophilic properties of bromine must be present. Somehow carpet tacks can do this. How?

The active catalyst is not iron itself but iron(III) bromide, formed by reaction of iron and bromine.

$$2Fe + 3Br_2 \longrightarrow 2FeBr_3$$

Iron   Bromine   Iron(III) bromide

Iron(III) bromide is a weak Lewis acid. It combines with bromine to form a Lewis acid-Lewis base complex.

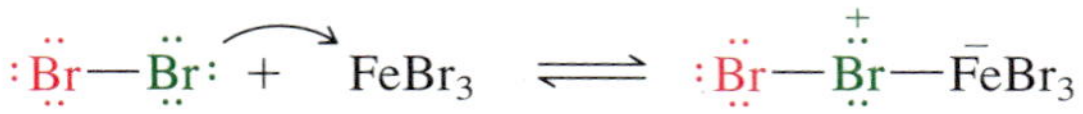

Lewis base   Lewis acid   Lewis acid-Lewis base
                                    complex

Complexation of bromine with iron(III) bromide makes bromine more electrophilic, and it attacks benzene to give a cyclohexadienyl intermediate as shown in step 1 of the mechanism (Figure 12.6). In step 2, as in nitration and sulfonation, loss of a proton from the cyclohexadienyl cation is rapid and gives the product of electrophilic aromatic substitution.

Only small quantities of iron(III) bromide are required. It is a catalyst for the bromination and, as Figure 12.6 indicates, is regenerated in the course of the reaction. We'll see later in this chapter that some aromatic substrates are much more reactive than benzene and react rapidly with bromine even in the absence of a catalyst.

Chlorination is carried out in a manner similar to bromination and provides a ready route to chlorobenzene and related aryl chlorides. Fluorination and iodination of benzene and other arenes are rarely performed. Fluorine is so reactive that its reaction with benzene is difficult to control. Iodination is very slow and has an unfavorable equilibrium constant. Syntheses of aryl fluorides and aryl iodides are normally carried out by way of functional group transformations of arylamines; these reactions will be described in Chapter 22.

Iron(III) bromide (FeBr₃) is also called *ferric bromide*.

Step 1: The bromine–iron(III) bromide complex is the active electrophile that attacks benzene. Two of the π electrons of benzene are used to form a bond to bromine and give a cyclohexadienyl cation intermediate.

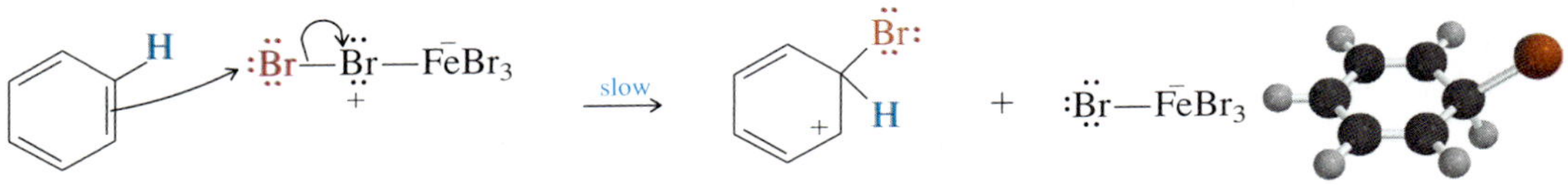

Benzene and bromine–iron(III) bromide complex → (slow) → Cyclohexadienyl cation intermediate + Tetrabromoferrate ion

Step 2: Loss of a proton from the cyclohexadienyl cation yields bromobenzene.

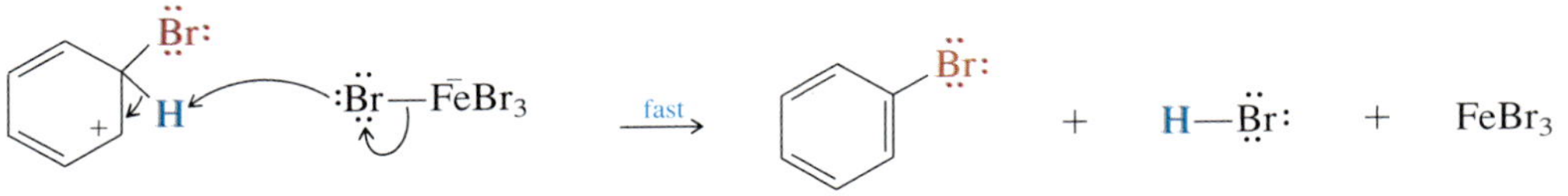

Cyclohexadienyl cation intermediate + Tetrabromoferrate ion → (fast) → Bromobenzene + Hydrogen bromide + Iron(III) bromide

12.6 FRIEDEL–CRAFTS ALKYLATION OF BENZENE

Alkyl halides react with benzene in the presence of aluminum chloride to yield alkylbenzenes.

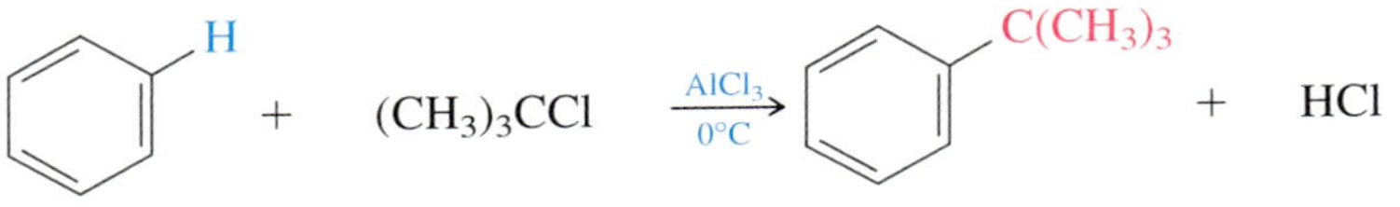

Benzene + $(CH_3)_3CCl$ $\xrightarrow[0°C]{AlCl_3}$ tert-Butylbenzene (60%) + HCl

Benzene + tert-Butyl chloride → tert-Butylbenzene (60%) + Hydrogen chloride

Alkylation of benzene with alkyl halides in the presence of aluminum chloride was discovered by Charles Friedel and James M. Crafts in 1877. Crafts, who later became president of the Massachusetts Institute of Technology, collaborated with Friedel at the Sorbonne in Paris, and together they developed what we now call the **Friedel–Crafts reaction** into one of the most useful synthetic methods in organic chemistry.

Alkyl halides by themselves are insufficiently electrophilic to react with benzene. Aluminum chloride serves as a Lewis acid catalyst to enhance the electrophilicity of the alkylating agent. With tertiary and secondary alkyl halides, the addition of aluminum chloride leads to the formation of carbocations, which then attack the aromatic ring.

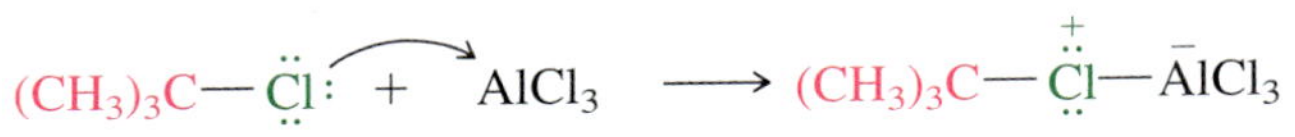

$(CH_3)_3C-\overset{..}{\underset{..}{Cl}}: + AlCl_3 \longrightarrow (CH_3)_3C-\overset{+}{\underset{..}{Cl}}-\overset{-}{AlCl_3}$

tert-Butyl chloride / Aluminum chloride / Lewis acid-Lewis base complex

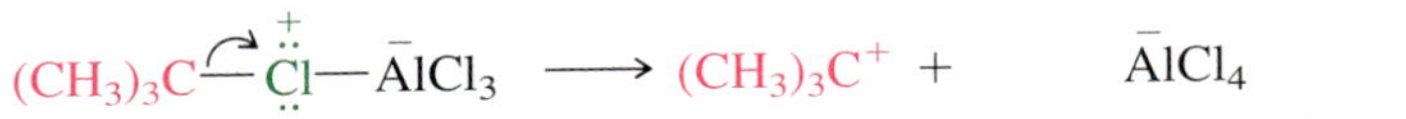

$(CH_3)_3C-\overset{+}{\underset{..}{Cl}}-\overset{-}{AlCl_3} \longrightarrow (CH_3)_3C^+ + \overset{-}{AlCl_4}$

tert-Butyl chloride–aluminum chloride complex / tert-Butyl cation / Tetrachloroaluminate anion

Figure 12.7 illustrates attack on the benzene ring by *tert*-butyl cation (step 1) and subsequent formation of *tert*-butylbenzene by loss of a proton from the cyclohexadienyl cation intermediate (step 2).

Secondary alkyl halides react by a similar mechanism involving attack on benzene by a secondary carbocation. Methyl and ethyl halides do not form carbocations when treated with aluminum chloride, but do alkylate benzene under Friedel–Crafts conditions. The aluminum chloride complexes of methyl and ethyl halides contain highly polarized carbon–halogen bonds, and these complexes are the electrophilic species that react with benzene.

$$CH_3 - \overset{+}{\underset{..}{\ddot{X}}} - \bar{A}lX_3 \qquad CH_3CH_2 - \overset{+}{\underset{..}{\ddot{X}}} - \bar{A}lX_3$$

Methyl halide–aluminum   Ethyl halide–aluminum
halide complex   halide complex

Other limitations to Friedel–Crafts reactions will be encountered in this chapter and are summarized in Table 12.4 (page 511).

One drawback to Friedel–Crafts alkylation is that rearrangements can occur, especially when primary alkyl halides are used. For example, Friedel–Crafts alkylation of benzene with isobutyl chloride (a primary alkyl halide) yields only *tert*-butylbenzene.

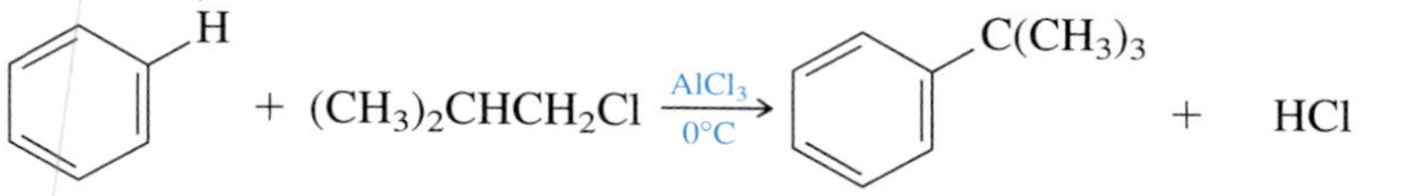

Benzene   Isobutyl chloride   *tert*-Butylbenzene   Hydrogen
                           (66%)   chloride

Step 1: Once generated by the reaction of *tert*-butyl chloride and aluminum chloride, *tert*-butyl cation attacks the π electrons of benzene, and a carbon–carbon bond is formed.

Benzene and *tert*-butyl cation   Cyclohexadienyl
                             cation intermediate

Step 2: Loss of a proton from the cyclohexadienyl cation intermediate yields *tert*-butylbenzene.

Cyclohexadienyl   Tetrachloroaluminate   *tert*-Butylbenzene   Hydrogen   Aluminum
cation intermediate   ion   chloride   chloride

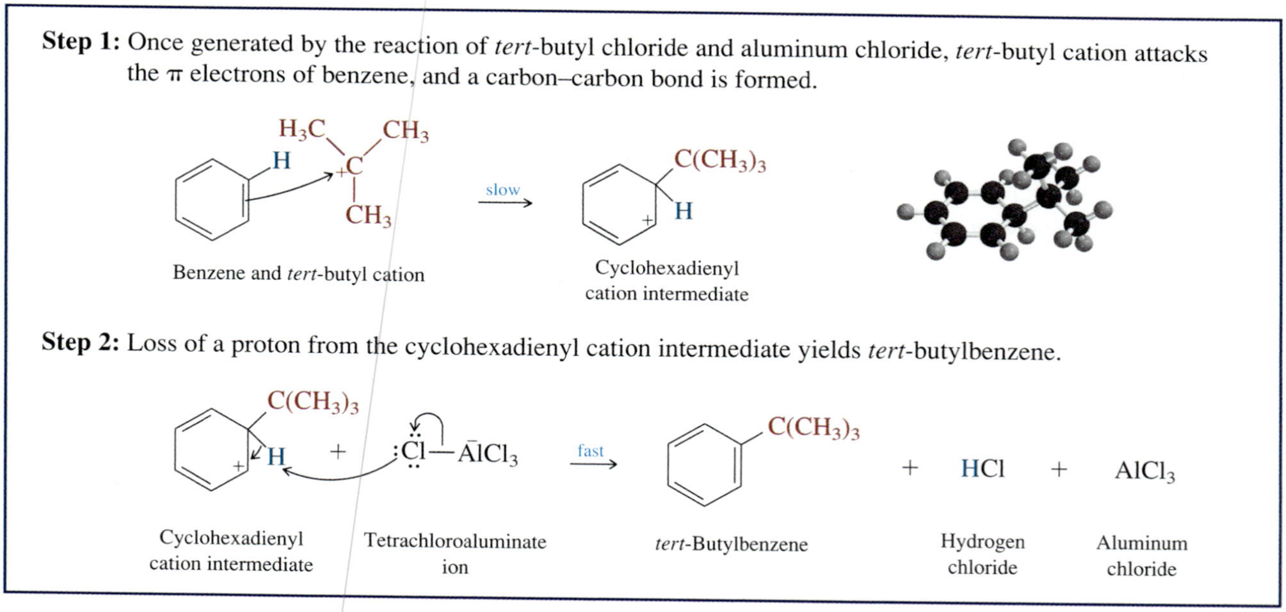

FIGURE 12.7 The mechanism of Friedel–Crafts alkylation. The molecular model depicts the cyclohexadienyl cation intermediate formed in step 1.

Here, the electrophile is *tert*-butyl cation formed by a hydride migration that accompanies ionization of the carbon–chlorine bond.

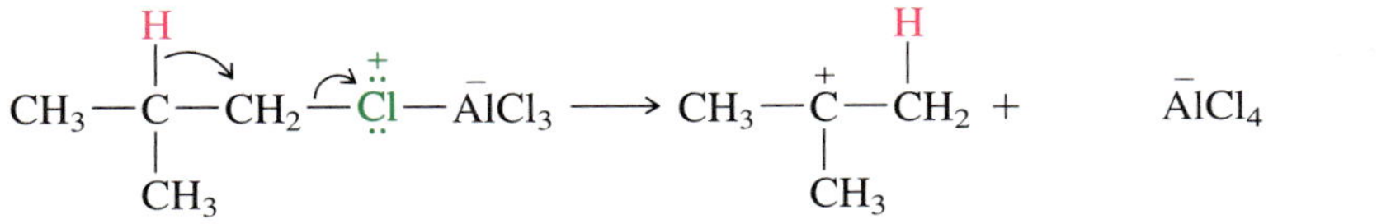

$$CH_3-\underset{\underset{CH_3}{|}}{\overset{\overset{H}{|}}{C}}-CH_2-\overset{+}{\ddot{\underset{..}{Cl}}}-\bar{A}lCl_3 \longrightarrow CH_3-\underset{\underset{CH_3}{|}}{\overset{+}{C}}-CH_2 + \quad \bar{A}lCl_4$$

| Isobutyl chloride–aluminum chloride complex | *tert*-Butyl cation | Tetrachloroaluminate ion |

> We saw rearrangements involving hydride shifts earlier in Sections 5.13 and 6.7.

PROBLEM 12.4 In an attempt to prepare propylbenzene, a chemist alkylated benzene with 1-chloropropane and aluminum chloride. However, two isomeric hydrocarbons were obtained in a ratio of 2:1, the desired propylbenzene being the minor component. What do you think was the major product? How did it arise?

Because electrophilic attack on benzene is simply another reaction available to a carbocation, other carbocation precursors can be used in place of alkyl halides. For example, alkenes, which are converted to carbocations by protonation, can be used to alkylate benzene.

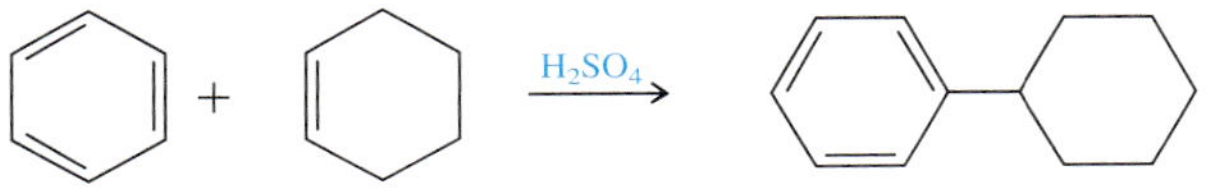

| Benzene | Cyclohexene | | Cyclohexylbenzene (65–68%) |

PROBLEM 12.5 Write a reasonable mechanism for the formation of cyclohexylbenzene from the reaction of benzene, cyclohexene, and sulfuric acid.

Alkenyl halides such as vinyl chloride ($H_2C{=}CHCl$) do *not* form carbocations on treatment with aluminum chloride and so cannot be used in Friedel–Crafts reactions. Thus, the industrial preparation of styrene from benzene and ethylene does not involve vinyl chloride but proceeds by way of ethylbenzene.

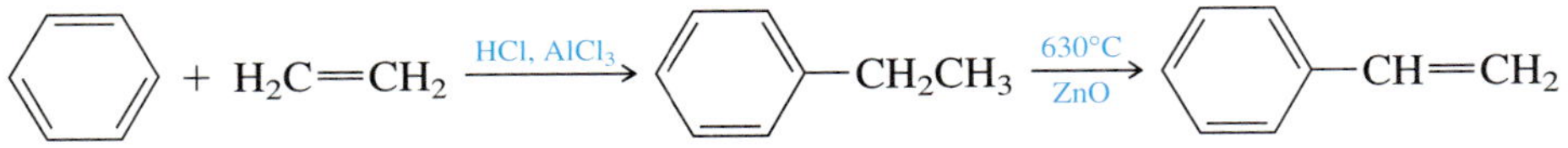

| Benzene | Ethylene | | Ethylbenzene | | Styrene |

Dehydrogenation of alkylbenzenes, although useful in the industrial preparation of styrene, is not a general procedure and is not well suited to the laboratory preparation of alkenylbenzenes. In such cases an alkylbenzene is subjected to benzylic bromination (Section 11.12), and the resulting benzylic bromide is treated with base to effect dehydrohalogenation.

PROBLEM 12.6 Outline a synthesis of 1-phenylcyclohexene from benzene and cyclohexene.

12.7 FRIEDEL–CRAFTS ACYLATION OF BENZENE

An acyl group has the general formula

$$\overset{O}{\underset{\parallel}{RC}}—$$

Another version of the Friedel–Crafts reaction uses **acyl halides** instead of alkyl halides and yields aryl ketones.

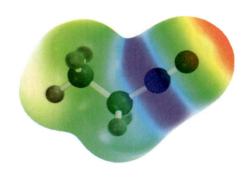

| Benzene | Propanoyl chloride | | 1-Phenyl-1-propanone (88%) | Hydrogen chloride |

The electrophile in a Friedel–Crafts acylation reaction is an **acyl cation** (also referred to as an **acylium ion**). Acyl cations are stabilized by resonance. The acyl cation derived from propanoyl chloride is represented by the two resonance forms

$$CH_3CH_2\overset{+}{C}=\ddot{O}: \longleftrightarrow CH_3CH_2C\equiv\overset{+}{O}:$$

Most stable resonance form;
oxygen and carbon have octets of electrons

Acyl cations form by coordination of an acyl chloride with aluminum chloride, followed by cleavage of the carbon–chlorine bond.

$$CH_3CH_2\overset{:O:}{\underset{\parallel}{C}}—\overset{..}{\underset{..}{Cl}}: + \quad AlCl_3 \longrightarrow CH_3CH_2\overset{:O:}{\underset{\parallel}{C}}\overset{+}{\overset{..}{Cl}}—\bar{A}lCl_3 \longrightarrow CH_3CH_2C\equiv\overset{+}{O}: + \quad \bar{A}lCl_4$$

| Propanoyl chloride | Aluminum chloride | Lewis acid–Lewis base complex | Propanoyl cation | Tetrachloroaluminate ion |

The electrophilic site of an acyl cation is its acyl carbon. An electrostatic potential map of the acyl cation from propanoyl chloride (Figure 12.8) illustrates nicely the concentration of positive charge at the acyl carbon, as shown by the blue color. The mechanism of the reaction between this cation and benzene is analogous to that of other electrophilic reagents (Figure 12.9).

FIGURE 12.8 Electrostatic potential map of propanoyl cation [(CH_3CH_2C=O)^+]. The region of greatest positive charge is associated with the carbon of the C=O group.

Step 1: The acyl cation attacks benzene. A pair of π electrons of benzene is used to form a covalent bond to the carbon of the acyl cation.

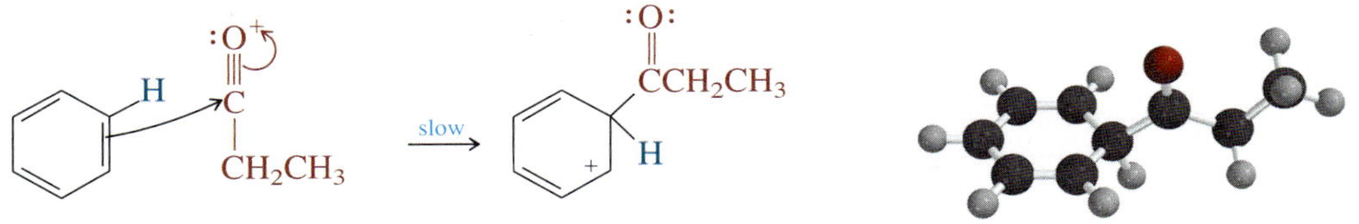

Benzene and propanoyl cation

Cyclohexadienyl
cation intermediate

Step 2: Aromaticity of the ring is restored when it loses a proton to give the aryl ketone.

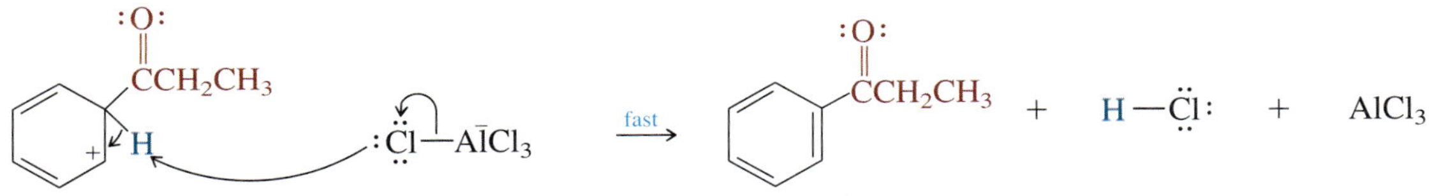

Cyclohexadienyl   Tetrachloroaluminate   1-Phenyl-1-propanone   Hydrogen   Aluminum
cation intermediate   ion   chloride   chloride

FIGURE 12.9 The mechanism of Friedel–Crafts acylation. The molecular model depicts the cyclohexadienyl cation intermediate formed in step 1.

PROBLEM 12.7 The reaction shown gives a single product in 88% yield. What is that product?

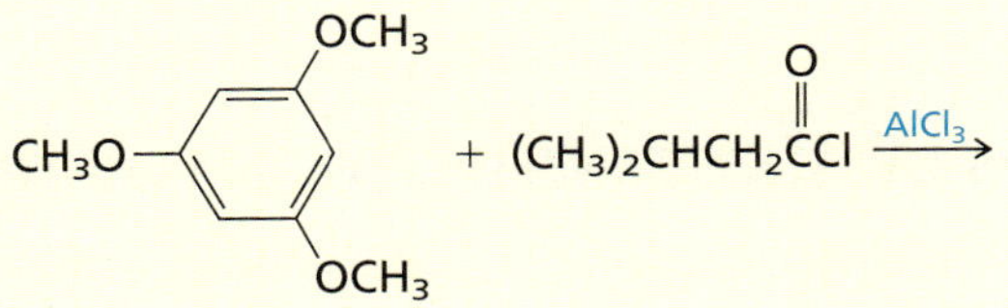

Acyl chlorides are readily available. They are prepared from carboxylic acids by reaction with thionyl chloride.

$$\underset{\substack{\text{Carboxylic acid}}}{\overset{\overset{\displaystyle O}{\|}}{\text{RCOH}}} \; + \; \underset{\substack{\text{Thionyl}\\\text{chloride}}}{\text{SOCl}_2} \longrightarrow \underset{\substack{\text{Acyl chloride}}}{\overset{\overset{\displaystyle O}{\|}}{\text{RCCl}}} \; + \; \underset{\substack{\text{Sulfur}\\\text{dioxide}}}{\text{SO}_2} \; + \; \underset{\substack{\text{Hydrogen}\\\text{chloride}}}{\text{HCl}}$$

Carboxylic acid anhydrides, compounds of the type $\overset{\overset{\displaystyle O \;\; O}{\| \;\; \|}}{\text{RCOCR}}$, can also serve as sources of acyl cations and, in the presence of aluminum chloride, acylate benzene. One acyl unit of an acid anhydride becomes attached to the benzene ring, and the other becomes part of a carboxylic acid.

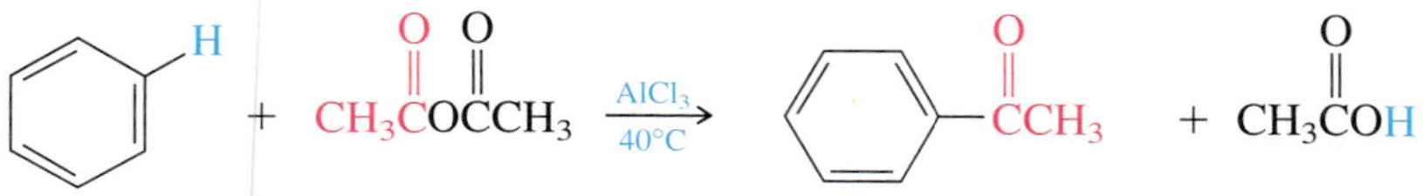

Benzene   Acetic anhydride   Acetophenone (76–83%)   Acetic acid

Acetophenone is one of the commonly encountered benzene derivatives listed in Table 11.1.

PROBLEM 12.8 *Succinic anhydride,* the structure of which is shown, is a cyclic anhydride often used in Friedel–Crafts acylations. Give the structure of the product obtained when benzene is acylated with succinic anhydride in the presence of aluminum chloride.

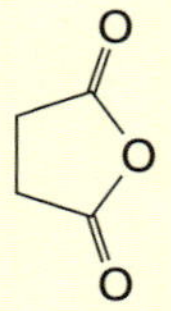

An important difference between Friedel–Crafts alkylations and acylations is that acyl cations *do not rearrange.* The acyl group of the acyl chloride or acid anhydride is transferred to the benzene ring unchanged. The reason for this is that an acyl cation is so strongly stabilized by resonance that it is more stable than any ion that could conceivably arise from it by a hydride or alkyl group shift.

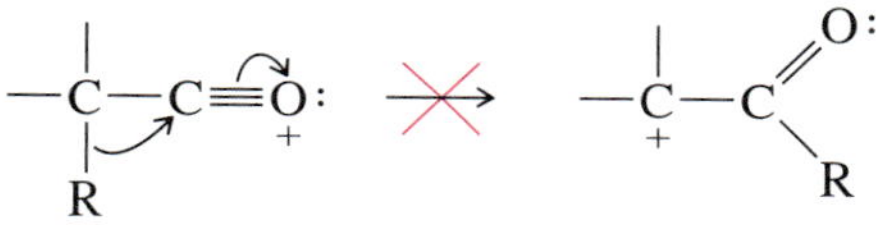

More stable cation; all atoms have octets of electrons

Less stable cation; six electrons at carbon

12.8   SYNTHESIS OF ALKYLBENZENES BY ACYLATION–REDUCTION

Because acylation of an aromatic ring can be accomplished without rearrangement, it is frequently used as the first step in a procedure for the *alkylation* of aromatic compounds by *acylation–reduction.* As we saw in Section 12.6, Friedel–Crafts alkylation of benzene with primary alkyl halides normally yields products having rearranged alkyl groups as substituents. When a compound of the type $ArCH_2R$ is desired, a two-step sequence is used in which the first step is a Friedel–Crafts acylation.

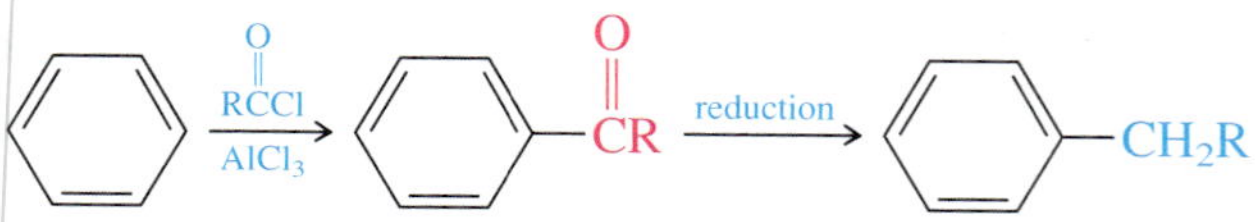

Benzene   Aryl ketone   Alkylbenzene

The second step is a reduction of the carbonyl group (C=O) to a methylene group (CH_2).

The most commonly used method for reducing an aryl ketone to an alkylbenzene employs a zinc–mercury amalgam in concentrated hydrochloric acid and is called the **Clemmensen reduction.** Zinc is the reducing agent.

The synthesis of butylbenzene illustrates the acylation–reduction sequence.

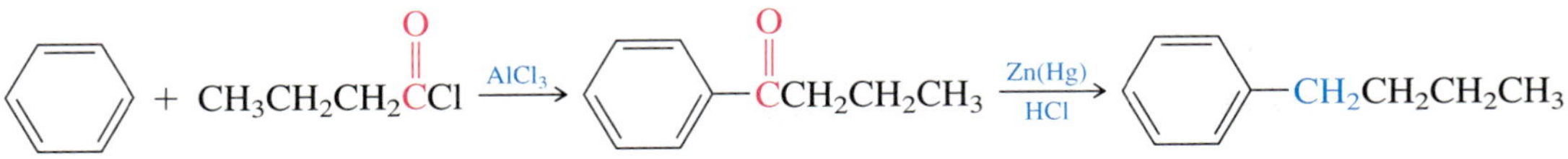

Benzene   Butanoyl chloride   1-Phenyl-1-butanone (86%)   Butylbenzene (73%)

Direct alkylation of benzene using 1-chlorobutane and aluminum chloride would yield *sec*-butylbenzene by rearrangement and so could not be used.

> **PROBLEM 12.9** Using benzene and any necessary organic or inorganic reagents, suggest efficient syntheses of
> (a) Isobutylbenzene, $C_6H_5CH_2CH(CH_3)_2$
> (b) Neopentylbenzene, $C_6H_5CH_2C(CH_3)_3$
>
> **SAMPLE SOLUTION** (a) Friedel–Crafts alkylation of benzene with isobutyl chloride is not suitable, because it yields *tert*-butylbenzene by rearrangement. 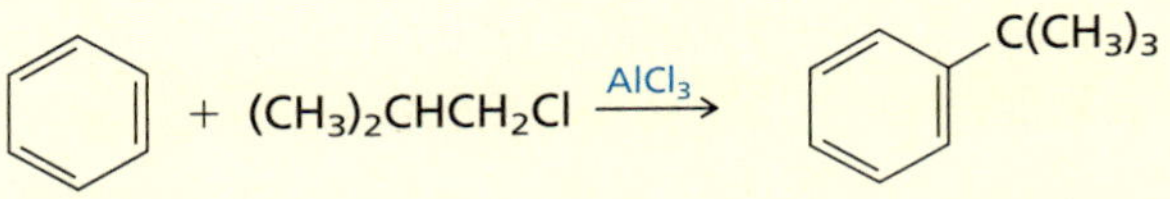
>
>
> Benzene   Isobutyl chloride   *tert*-Butylbenzene (66%)
>
> The two-step acylation–reduction sequence is required. Acylation of benzene puts the side chain on the ring with the correct carbon skeleton. Clemmensen reduction converts the carbonyl group to a methylene group. 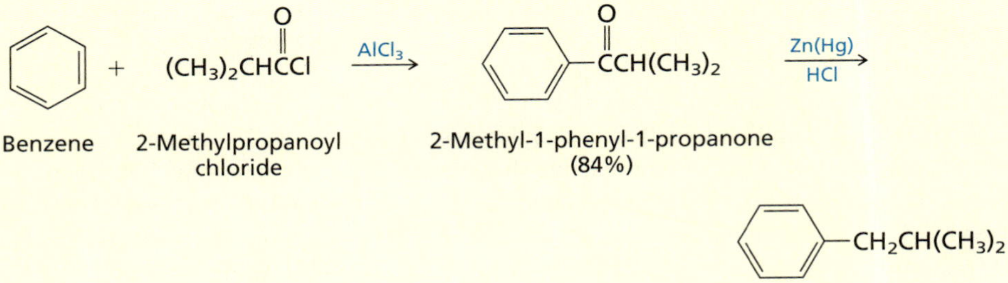
>
>
> Benzene   2-Methylpropanoyl chloride   2-Methyl-1-phenyl-1-propanone (84%)
>
> Isobutylbenzene (80%)

Another way to reduce aldehyde and ketone carbonyl groups is by **Wolff–Kishner reduction.** Heating an aldehyde or a ketone with hydrazine (H_2NNH_2) and sodium or potassium hydroxide in a high-boiling alcohol such as triethylene glycol ($HOCH_2CH_2OCH_2CH_2OCH_2CH_2OH$, bp 287°C) converts the carbonyl to a CH_2 group.

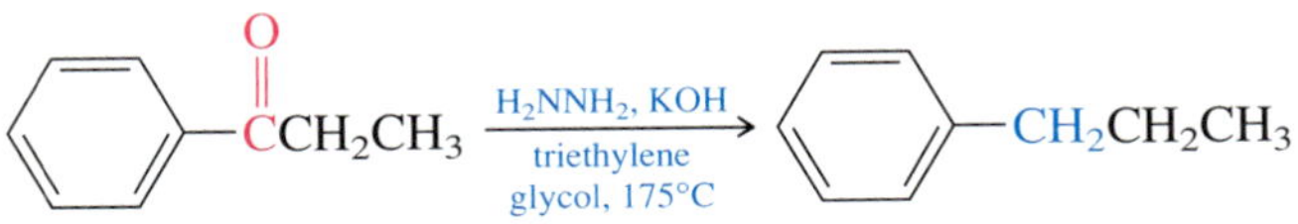

1-Phenyl-1-propanone   Propylbenzene (82%)

Both the Clemmensen and the Wolff–Kishner reductions are designed to carry out a specific functional group transformation, the reduction of an aldehyde or ketone carbonyl to a methylene group. Neither one will reduce the carbonyl group of a carboxylic acid, nor are carbon–carbon double or triple bonds affected by these methods. We will not discuss the mechanism of either the Clemmensen reduction or the Wolff–Kishner reduction; both involve chemistry that is beyond the scope of what we have covered to this point.

12.9   RATE AND REGIOSELECTIVITY IN ELECTROPHILIC AROMATIC SUBSTITUTION

So far we've been concerned only with electrophilic substitution of benzene. Two important questions arise when we turn to substitution on rings that already bear at least one substituent:

1. What is the effect of a substituent on the *rate* of electrophilic aromatic substitution?

2. What is the effect of a substituent on the *regioselectivity* of electrophilic aromatic substitution?

To illustrate substituent effects on rate, consider the nitration of benzene, toluene, and (trifluoromethyl)benzene.

Increasing rate of nitration

(Trifluoromethyl)benzene   Benzene   Toluene
(least reactive)   (most reactive)

> Examine the molecular models of toluene and (trifluoromethyl)benzene on *Learning By Modeling.* In which molecule is the electrostatic potential of the ring most negative? How should this affect the rate of nitration?

The range of rates of nitration among these three compounds is quite large; it covers a spread of approximately 1-millionfold. Toluene undergoes nitration some 20–25 times faster than benzene. Because toluene is more reactive than benzene, we say that a methyl group *activates* the ring toward electrophilic aromatic substitution. (Trifluoromethyl)benzene, on the other hand, undergoes nitration about 40,000 times more slowly than benzene. We say that a trifluoromethyl group *deactivates* the ring toward electrophilic aromatic substitution.

Just as there is a marked difference in how methyl and trifluoromethyl substituents affect the rate of electrophilic aromatic substitution, so too there is a marked difference in how they affect its regioselectivity.

Three products are possible from nitration of toluene: *o*-nitrotoluene, *m*-nitrotoluene, and *p*-nitrotoluene. All are formed, but not in equal amounts. Together, the ortho- and para-substituted isomers make up 97% of the product mixture; the meta only 3%.

> How do the charges on the ring carbons of toluene and (trifluoromethyl)benzene relate to the regioselectivity of nitration?

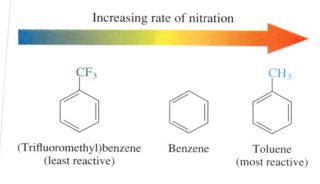

Toluene

o-Nitrotoluene
(63%)

m-Nitrotoluene
(3%)

p-Nitrotoluene
(34%)

Because substitution in toluene occurs primarily at positions ortho and para to methyl, we say that *a methyl substituent is an* **ortho, para director.**

Nitration of (trifluoromethyl)benzene, on the other hand, yields almost exclusively *m*-nitro(trifluoromethyl)benzene (91%). The ortho- and para-substituted isomers are minor components of the reaction mixture.

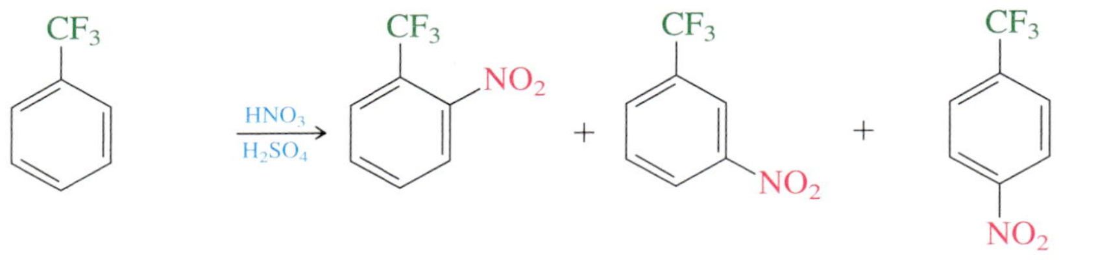

| (Trifluoromethyl)benzene | *o*-Nitro(trifluoro-methyl)benzene (6%) | *m*-Nitro(trifluoro-methyl)benzene (91%) | *p*-Nitro(trifluoro-methyl)benzene (3%) |

Because substitution in (trifluoromethyl)benzene occurs primarily at positions meta to the substituent, we say that *a trifluoromethyl group is a* **meta director.**

The regioselectivity of substitution, like the rate, is strongly affected by the substituent. In the following several sections we will examine the relationship between the structure of the substituent and its effect on rate and regioselectivity of electrophilic aromatic substitution.

12.10 RATE AND REGIOSELECTIVITY IN THE NITRATION OF TOLUENE

Why is there such a marked difference between methyl and trifluoromethyl substituents in their influence on electrophilic aromatic substitution? Methyl is activating and ortho, para-directing; trifluoromethyl is deactivating and meta-directing. The first point to remember is that the regioselectivity of substitution is set once the cyclohexadienyl cation intermediate is formed. If we can explain why

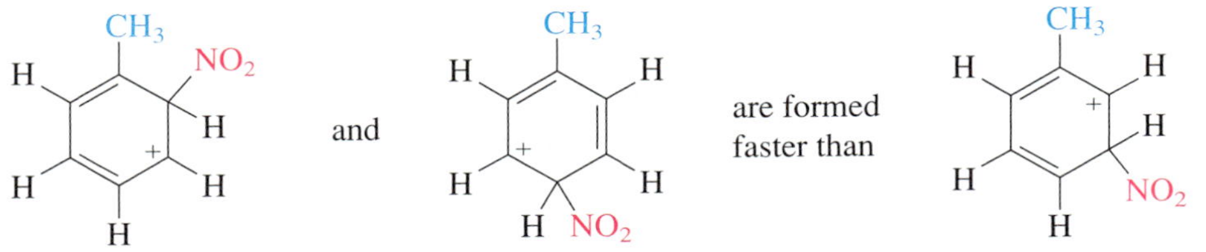

in the rate-determining step, we will understand the reasons for the regioselectivity. A principle we have used before serves us well here: *a more stable carbocation is formed faster than a less stable one.* The most likely reason for the directing effect of a CH_3 group must be that the cyclohexadienyl cation precursors to *o*- and *p*-nitrotoluene are more stable than the one leading to *m*-nitrotoluene.

One way to assess the relative stabilities of these various intermediates is to examine electron delocalization in them using a resonance description. The cyclohexadienyl cations leading to *o*- and *p*-nitrotoluene have tertiary carbocation character. Each has a resonance form in which the positive charge resides on the carbon that bears the methyl group.

Ortho attack

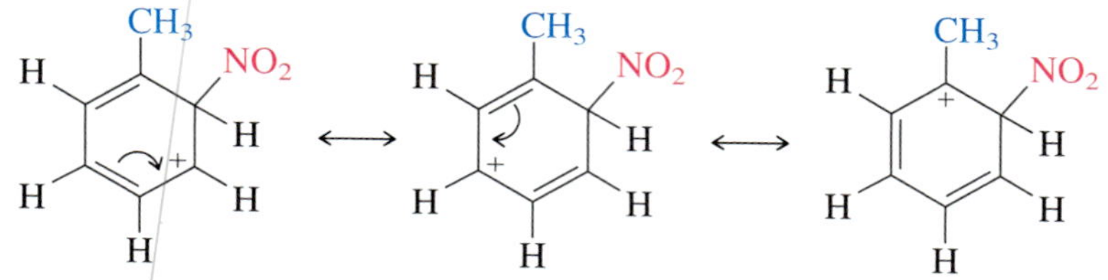

This resonance form
is a tertiary carbocation

Para attack

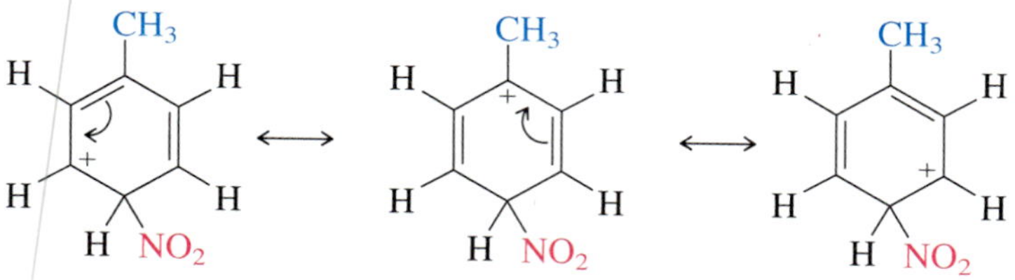

This resonance form
is a tertiary carbocation

The three resonance forms of the intermediate leading to meta substitution are all secondary carbocations.

Meta attack

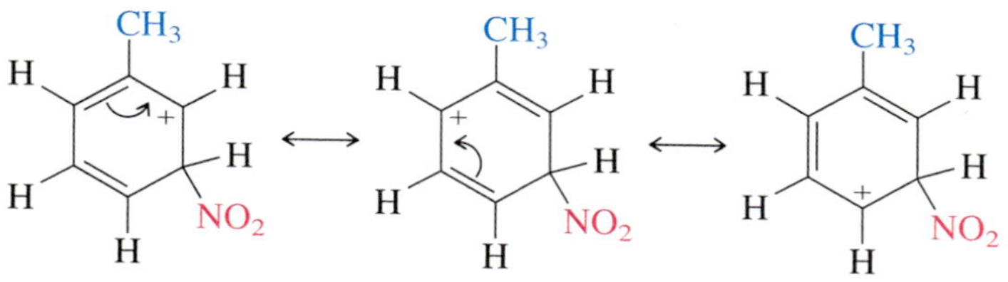

Because of their tertiary carbocation character the intermediates leading to ortho and to para substitution are more stable and are formed faster than the one leading to meta substitution. They are also more stable than the secondary cyclohexadienyl cation intermediate formed during nitration of benzene. A methyl group is an activating substituent because it stabilizes the carbocation intermediate formed in the rate-determining step more than a hydrogen does. It is ortho, para-directing because it stabilizes the carbocation formed by electrophilic attack at these positions more than it stabilizes the intermediate formed by attack at the meta position. Figure 12.10 compares the energies of activation for attack at the various positions of toluene.

A methyl group is an *electron-releasing* substituent and activates *all* of the ring carbons of toluene toward electrophilic attack. The ortho and para positions are activated more than the meta positions. The relative rates of attack at the various positions in toluene compared with a single position in benzene are as follows (for nitration at 25°C):

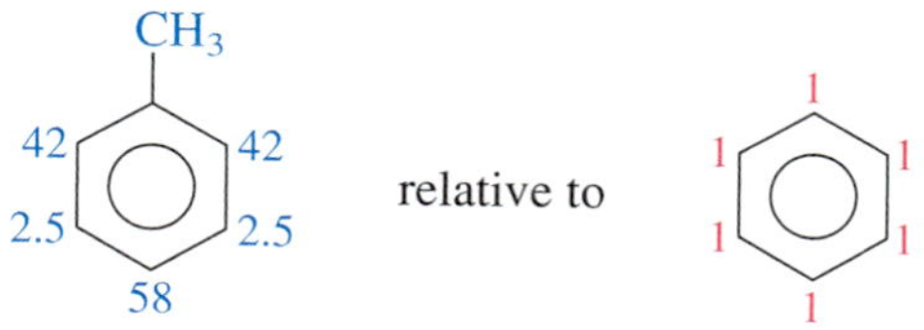

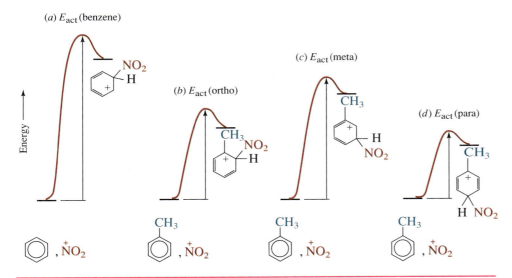

FIGURE 12.10 Comparative energy diagrams for nitronium ion attack on (a) benzene and at the (b) ortho, (c) meta, and (d) para positions of toluene. E_{act} (benzene) > E_{act} (meta) > E_{act} (ortho) > E_{act} (para).

These relative rate data per position are experimentally determined and are known as *partial rate factors*. They offer a convenient way to express substituent effects in electrophilic aromatic substitution reactions.

   The major influence of the methyl group is *electronic*. The most important factor is relative carbocation stability. To a small extent, the methyl group sterically hinders the ortho positions, making attack slightly more likely at the para carbon than at a single ortho carbon. However, para substitution is at a statistical disadvantage because there are two equivalent ortho positions but only one para position.

PROBLEM 12.10   The partial rate factors for nitration of *tert*-butylbenzene are as shown.

$$C(CH_3)_3$$

4.5   4.5
3   3
75

(a) How reactive is *tert*-butylbenzene toward nitration compared with benzene?

(b) How reactive is *tert*-butylbenzene toward nitration compared with toluene?

(c) Predict the distribution among the various mononitration products of *tert*-butylbenzene.

SAMPLE SOLUTION   (a) Benzene has six equivalent sites at which nitration can occur. Summing the individual relative rates of attack at each position in *tert*-butylbenzene compared to benzene, we obtain

$$\frac{tert\text{-Butylbenzene}}{Benzene} = \frac{2(4.5) + 2(3) + 75}{6(1)} = \frac{90}{6} = 15$$

tert-Butylbenzene undergoes nitration 15 times faster than benzene.

All alkyl groups, not just methyl, are activating substituents and ortho, para directors. This is because any alkyl group, be it methyl, ethyl, isopropyl, *tert*-butyl, or any other, stabilizes a carbocation site to which it is directly attached. When R = alkyl,

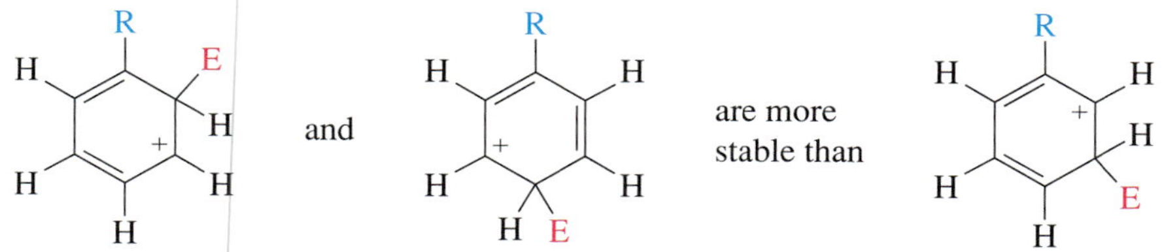

where E^+ is any electrophile. All three structures are more stable for R = alkyl than for R = H and are formed more quickly.

12.11 RATE AND REGIOSELECTIVITY IN THE NITRATION OF (TRIFLUOROMETHYL)BENZENE

Turning now to electrophilic aromatic substitution in (trifluoromethyl)benzene, we consider the electronic properties of a trifluoromethyl group. Because of their high electronegativity the three fluorine atoms polarize the electron distribution in their σ bonds to carbon, so that carbon bears a partial positive charge.

Recall from Section 1.15 that effects that are transmitted by the polarization of σ bonds are called *inductive effects*.

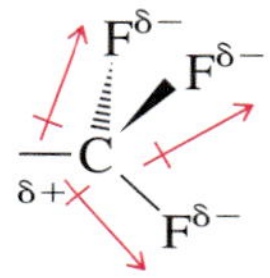

Unlike a methyl group, which is slightly electron-releasing, a trifluoromethyl group is a powerful electron-withdrawing substituent. Consequently, a CF_3 group *destabilizes* a carbocation site to which it is attached.

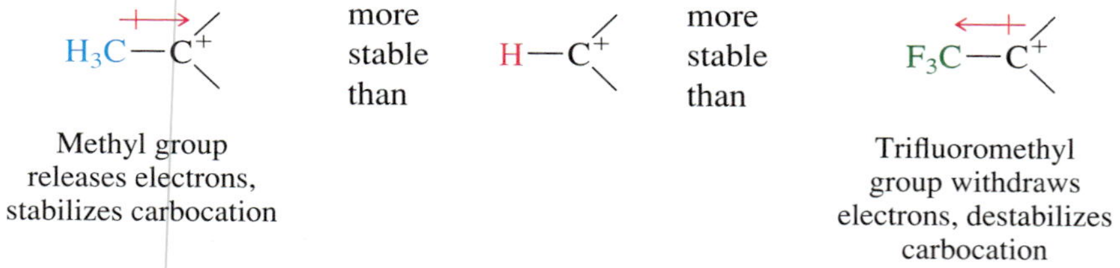

Methyl group releases electrons, stabilizes carbocation		Trifluoromethyl group withdraws electrons, destabilizes carbocation

When we examine the cyclohexadienyl cation intermediates involved in the nitration of (trifluoromethyl)benzene, we find that those leading to ortho and para substitution are strongly *destabilized*.

Ortho attack

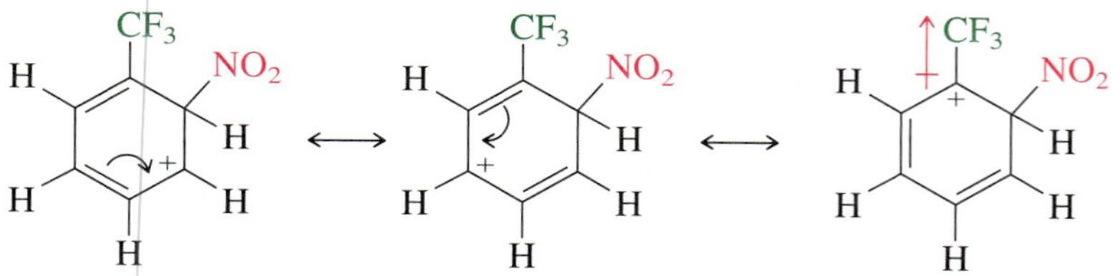

Positive charge on carbon bearing CF_3 group (very unstable)

Para attack

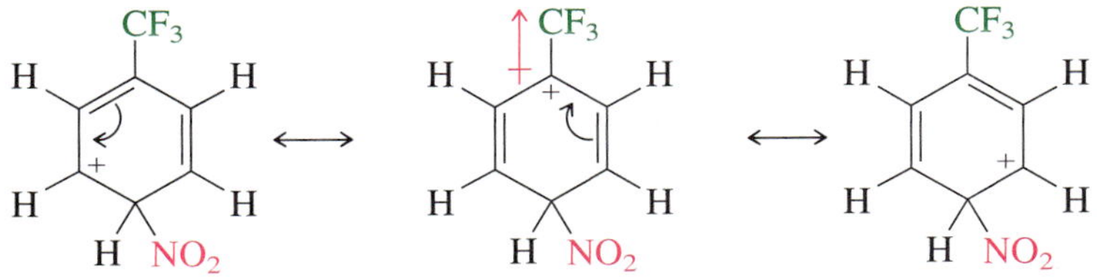

Positive charge on carbon
bearing CF₃ group
(very unstable)

None of the three major resonance forms of the intermediate formed by attack at the meta position has a positive charge on the carbon bearing the —CF₃ group.

Meta attack

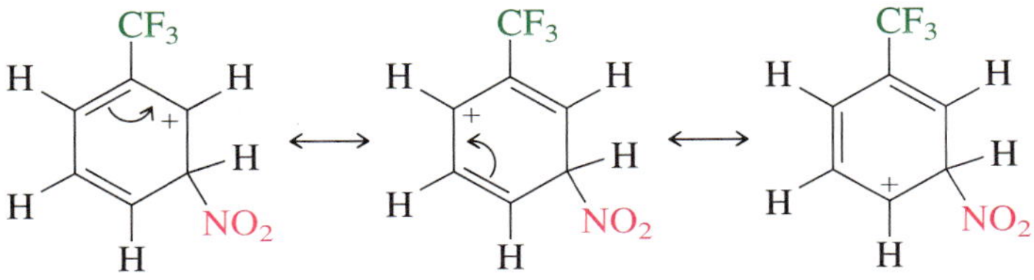

Attack at the meta position leads to a more stable intermediate than attack at either the ortho or the para position, and so meta substitution predominates. Even the intermediate corresponding to meta attack, however, is very unstable and is formed with difficulty. The trifluoromethyl group is only one bond farther removed from the positive charge here than it is in the ortho and para intermediates and so still exerts a significant, although somewhat diminished, destabilizing inductive effect.

All the ring positions of (trifluoromethyl)benzene are deactivated compared with benzene. The meta position is simply deactivated *less* than the ortho and para positions. The partial rate factors for nitration of (trifluoromethyl)benzene are

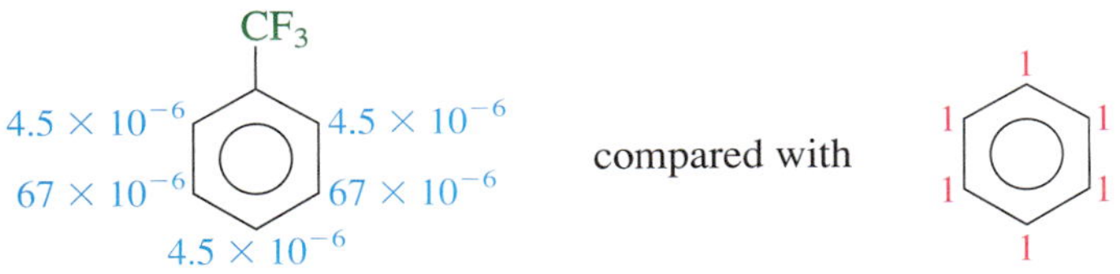

Figure 12.11 compares the energy profile for nitration of benzene with those for attack at the ortho, meta, and para positions of (trifluoromethyl)benzene. The presence of the electron-withdrawing trifluoromethyl group raises the activation energy for attack at all the ring positions, but the increase is least for attack at the meta position.

PROBLEM 12.11   The compounds benzyl chloride ($C_6H_5CH_2Cl$), (dichloromethyl)benzene ($C_6H_5CHCl_2$), and (trichloromethyl)benzene ($C_6H_5CCl_3$) all undergo nitration more slowly than benzene. The proportion of *m*-nitro-substituted product is 4% in one, 34% in another, and 64% in another. Classify the substituents —CH_2Cl, —$CHCl_2$, and —CCl_3 according to each one's effect on rate and regioselectivity in electrophilic aromatic substitution.

FIGURE 12.11 Comparative energy diagrams for nitronium ion attack on (a) benzene and at the (b) ortho, (c) meta, and (d) para positions of (trifluoromethyl)-benzene. E_{act} (ortho) > E_{act} (para) > E_{act} (meta) > E_{act} (benzene).

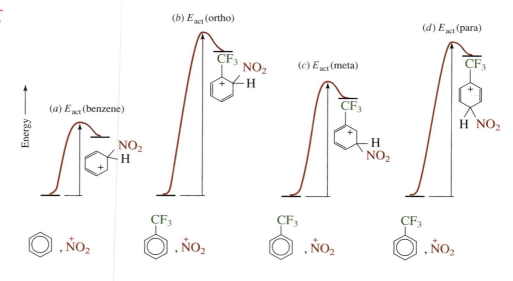

12.12 SUBSTITUENT EFFECTS IN ELECTROPHILIC AROMATIC SUBSTITUTION: ACTIVATING SUBSTITUENTS

Our analysis of substituent effects has so far centered on two groups: methyl and trifluoromethyl. We have seen that a methyl substituent is activating and ortho, para-directing. A trifluoromethyl group is strongly deactivating and meta-directing. What about other substituents?

Table 12.2 summarizes orientation and rate effects in electrophilic aromatic substitution reactions for a variety of frequently encountered substituents. It is arranged in order of decreasing activating power: the most strongly *activating* substituents are at the top, the most strongly *deactivating* substituents are at the bottom. The main features of the table can be summarized as follows:

1. All activating substituents are ortho, para directors.

2. Halogen substituents are slightly deactivating but are ortho, para-directing.

3. Strongly deactivating substituents are meta directors.

Some of the most powerful *activating* substituents are those in which an oxygen atom is attached directly to the ring. These substituents include the hydroxyl group as well as alkoxy and acyloxy groups. All are ortho, para directors.

$$\text{H}\ddot{\text{O}}— \qquad \text{R}\ddot{\text{O}}— \qquad \overset{\overset{\text{O}}{\|}}{\text{R}\ddot{\text{C}}\ddot{\text{O}}}—$$
Hydroxyl   Alkoxy   Acyloxy

Phenol and *anisole* are among the commonly encountered benzene derivatives listed in Table 11.1. Electrophilic aromatic substitution in phenol is discussed in more detail in Section 24.8.

Phenol $+ HNO_3 \xrightarrow{\text{acetic acid}}$ *o*-Nitrophenol (44%) $+$ *p*-Nitrophenol (56%)

TABLE 12.2	Classification of Substituents in Electrophilic Aromatic Substitution Reactions	
Effect on rate	**Substituent**	**Effect on orientation**
Very strongly activating	—$\ddot{N}H_2$ (amino)	Ortho, para-directing
	—$\ddot{N}HR$ (alkylamino)	
	—$\ddot{N}R_2$ (dialkylamino)	
	—$\ddot{O}H$ (hydroxyl)	
Strongly activating	$-\overset{\overset{\displaystyle O}{\|}}{\ddot{N}HCR}$ (acylamino)	Ortho, para-directing
	—$\ddot{O}R$ (alkoxy)	
	$-\overset{\overset{\displaystyle O}{\|}}{\ddot{O}CR}$ (acyloxy)	
Activating	—R (alkyl)	Ortho, para-directing
	—Ar (aryl)	
	—CH=CR$_2$ (alkenyl)	
Standard of comparison	—H (hydrogen)	
Deactivating	—X (halogen) (X = F, Cl, Br, I)	Ortho, para-directing
	—CH$_2$X (halomethyl)	
Strongly deactivating	$-\overset{\overset{\displaystyle O}{\|}}{CH}$ (formyl)	Meta-directing
	$-\overset{\overset{\displaystyle O}{\|}}{CR}$ (acyl)	
	$-\overset{\overset{\displaystyle O}{\|}}{COH}$ (carboxylic acid)	
	$-\overset{\overset{\displaystyle O}{\|}}{COR}$ (ester)	
	$-\overset{\overset{\displaystyle O}{\|}}{CCl}$ (acyl chloride)	
	—C≡N (cyano)	
	—SO$_3$H (sulfonic acid)	
Very strongly deactivating	—CF$_3$ (trifluoromethyl)	Meta-directing
	—NO$_2$ (nitro)	

Hydroxyl, alkoxy, and acyloxy groups activate the ring to such an extent that bromination occurs rapidly even in the absence of a catalyst.

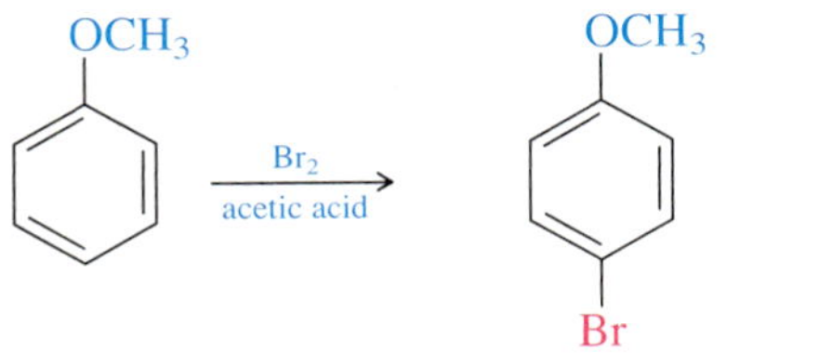

Anisole → p-Bromoanisole (90%)

The inductive effect of hydroxyl and alkoxy groups, because of the electronegativity of oxygen, is to withdraw electrons and would seem to require that such substituents be deactivating. The electron-withdrawing inductive effect, however, is overcome by a much larger electron-releasing effect involving the unshared electron pairs of oxygen. Attack at positions ortho and para to a carbon that bears a substituent of the type —ÖR gives a cation stabilized by delocalization of an unshared electron pair of oxygen into the π system of the ring (a *resonance* or *conjugation* effect).

Ortho attack

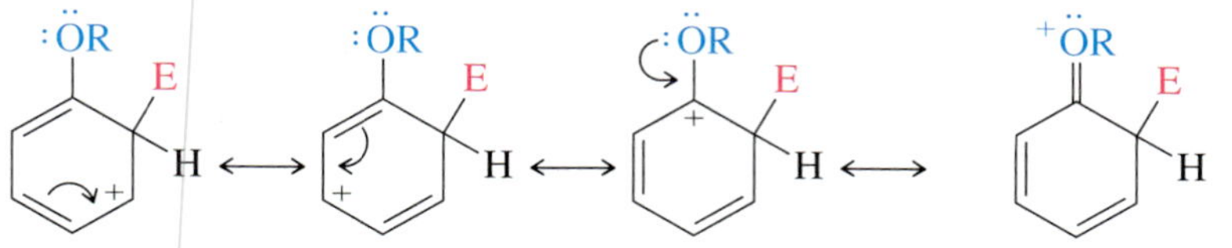

Most stable resonance form; oxygen and all carbons have octets of electrons

Para attack

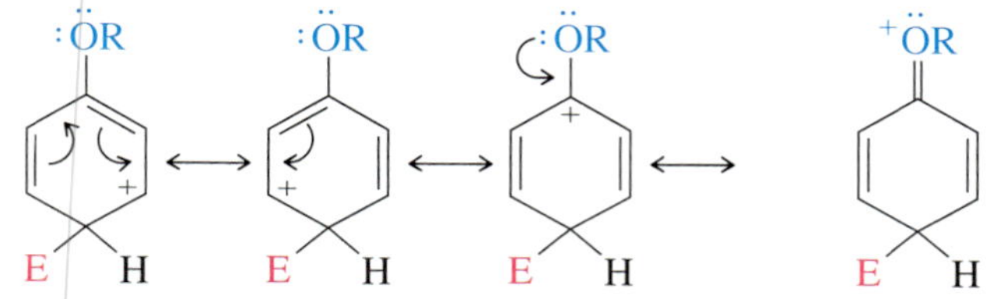

Most stable resonance form; oxygen and all carbons have octets of electrons

Oxygen-stabilized carbocations of this type are far more stable than tertiary carbocations. They are best represented by structures in which the positive charge is on oxygen because all the atoms have octets of electrons in such a structure. Their stability permits them to be formed rapidly, resulting in rates of electrophilic aromatic substitution that are much faster than that of benzene.

The lone pair on oxygen cannot be directly involved in carbocation stabilization when attack is meta to the substituent.

Meta attack

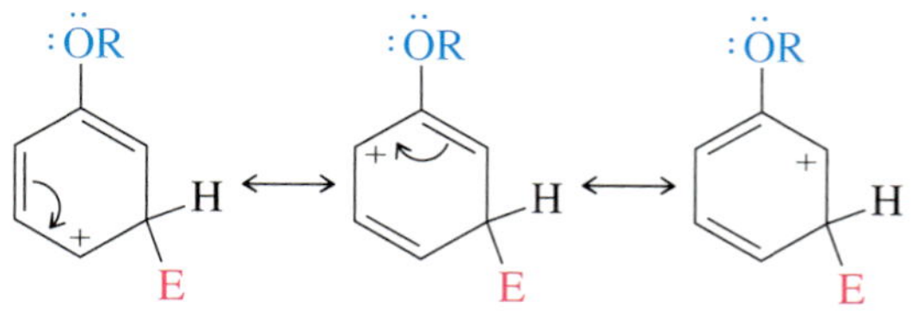

Oxygen lone pair cannot be used to stabilize positive charge in any of these structures; all have six electrons around positively charged carbon.

The greater stability of the intermediates arising from attack at the ortho and para positions compared with those formed by attack at the position meta to the oxygen substituent explains the ortho, para-directing property of hydroxyl, alkoxy, and acyloxy groups.

Nitrogen-containing substituents related to the amino group are even more strongly activating than the corresponding oxygen-containing substituents.

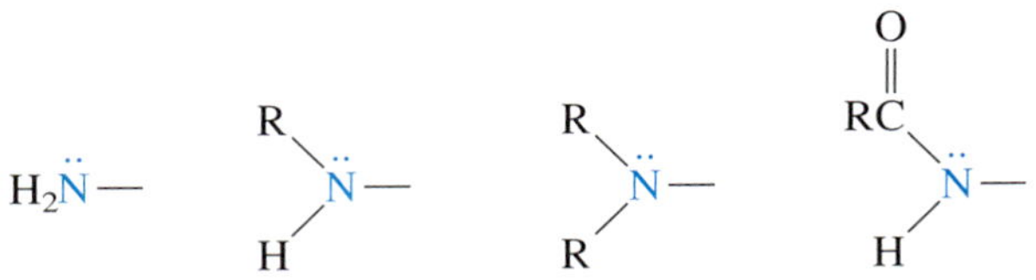

The nitrogen atom in each of these groups bears an electron pair that, like the unshared pairs of an oxygen substituent, stabilizes a carbocation site to which it is attached. Because nitrogen is less electronegative than oxygen, it is a better electron pair donor and stabilizes the cyclohexadienyl cation intermediates in electrophilic aromatic substitution to an even greater degree.

PROBLEM 12.12   Write structural formulas for the cyclohexadienyl cations formed from aniline ($C_6H_5NH_2$) during

(a) Ortho bromination (four resonance structures)

(b) Meta bromination (three resonance structures)

(c) Para bromination (four resonance structures)

SAMPLE SOLUTION   (a) There are the customary three resonance structures for the cyclohexadienyl cation plus a resonance structure (the most stable one) derived by delocalization of the nitrogen lone pair into the ring.

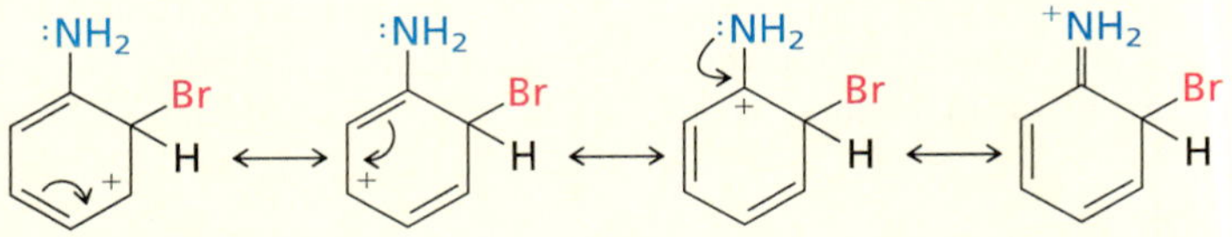

Most stable
resonance
structure

Aniline and its derivatives are so reactive in electrophilic aromatic substitution that special strategies are usually necessary to carry out these reactions effectively. This topic is discussed in Section 22.14.

Alkyl groups are, as we saw when we discussed the nitration of toluene in Section 12.10, activating and ortho, para-directing substituents. Aryl and alkenyl substituents resemble alkyl groups in this respect; they too are activating and ortho, para-directing.

PROBLEM 12.13   Treatment of biphenyl (see Section 11.7 to remind yourself of its structure) with a mixture of nitric acid and sulfuric acid gave two principal products both having the molecular formula $C_{12}H_9NO_2$. What are these two products?

The next group of substituents in Table 12.2 that we'll discuss are the ones near the bottom of the table, those that are meta-directing and strongly deactivating.

12.13 SUBSTITUENT EFFECTS IN ELECTROPHILIC AROMATIC SUBSTITUTION: STRONGLY DEACTIVATING SUBSTITUENTS

As Table 12.2 indicates, a variety of substituent types are *meta-directing and strongly deactivating*. We have already discussed one of these, the trifluoromethyl group. Several of the others have a carbonyl group attached directly to the aromatic ring.

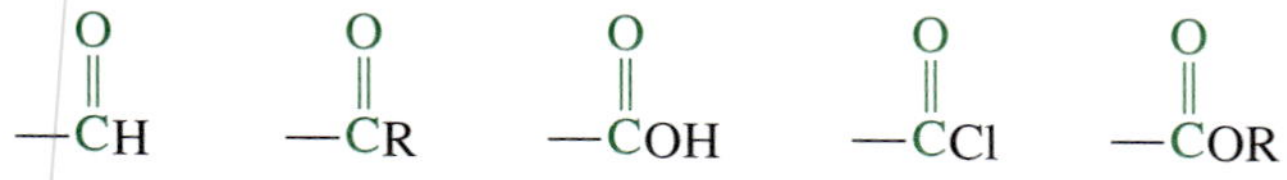

| Aldehyde | Ketone | Carboxylic acid | Acyl chloride | Ester |

The behavior of aromatic aldehydes is typical. Nitration of benzaldehyde takes place several thousand times more slowly than that of benzene and yields *m*-nitrobenzaldehyde as the major product.

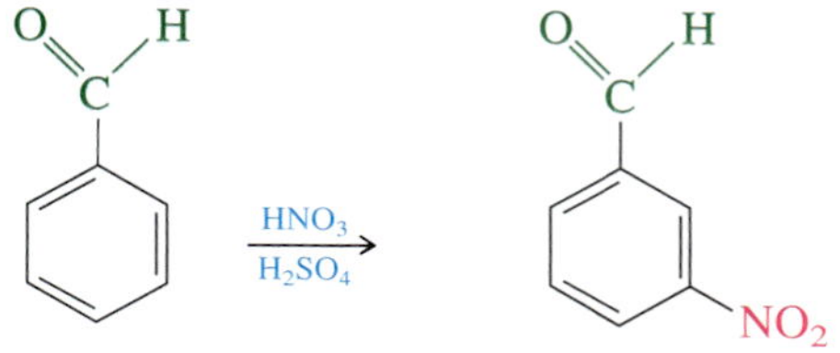

Benzaldehyde   *m*-Nitrobenzaldehyde (75–84%)

To understand the effect of a carbonyl group attached directly to the ring, consider its polarization. The electrons in the carbon–oxygen double bond are drawn toward oxygen and away from carbon, leaving the carbon attached to the ring with a partial positive charge. Using benzaldehyde as an example,

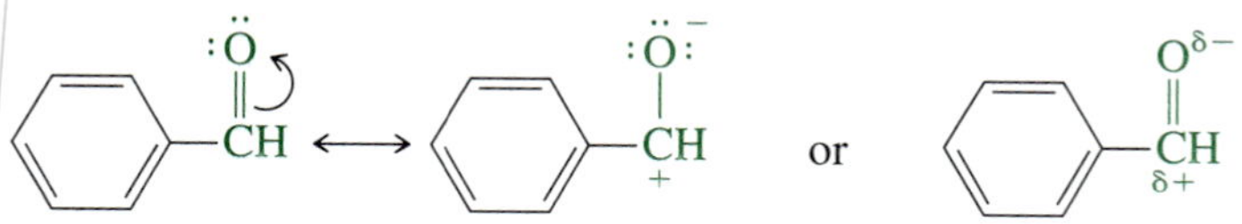

Because the carbon atom attached to the ring is positively polarized, a carbonyl group behaves in much the same way as a trifluoromethyl group and *destabilizes* all the cyclohexadienyl cation intermediates in electrophilic aromatic substitution reactions. Attack at any ring position in benzaldehyde is slower than attack in benzene. The intermediates for ortho and para substitution are particularly unstable because each has a resonance structure in which there is a positive charge on the carbon that bears the electron-withdrawing substituent. The intermediate for meta substitution avoids this unfavorable juxtaposition of positive charges, is not as unstable, and gives rise to most of the product.

For the nitration of benzaldehyde:

Ortho attack	*Meta attack*	*Para attack*

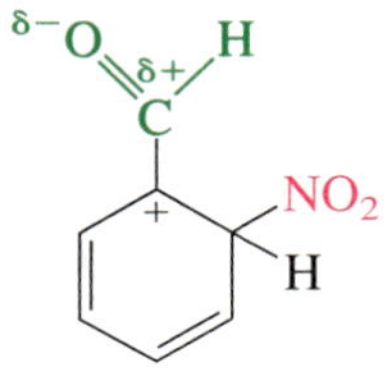

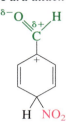

Unstable because of adjacent positively polarized atoms	Positively polarized atoms not adjacent; most stable intermediate	Unstable because of adjacent positively polarized atoms

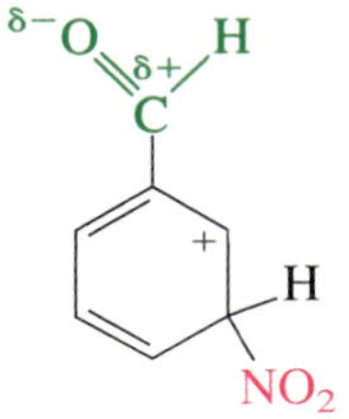

PROBLEM 12.14 Each of the following reactions has been reported in the chemical literature, and the major organic product has been isolated in good yield. Write a structural formula for the product of each reaction.

(a) Treatment of benzoyl chloride ($C_6H_5\overset{O}{\overset{\|}{C}}Cl$) with chlorine and iron(III) chloride

(b) Treatment of methyl benzoate ($C_6H_5\overset{O}{\overset{\|}{C}}OCH_3$) with nitric acid and sulfuric acid

(c) Nitration of 1-phenyl-1-propanone ($C_6H_5\overset{O}{\overset{\|}{C}}CH_2CH_3$)

SAMPLE SOLUTION (a) Benzoyl chloride has a carbonyl group attached directly to the ring. A $-\overset{O}{\overset{\|}{C}}Cl$ substituent is meta-directing. The combination of chlorine and iron(III) chloride introduces a chlorine onto the ring. The product is *m*-chlorobenzoyl chloride.

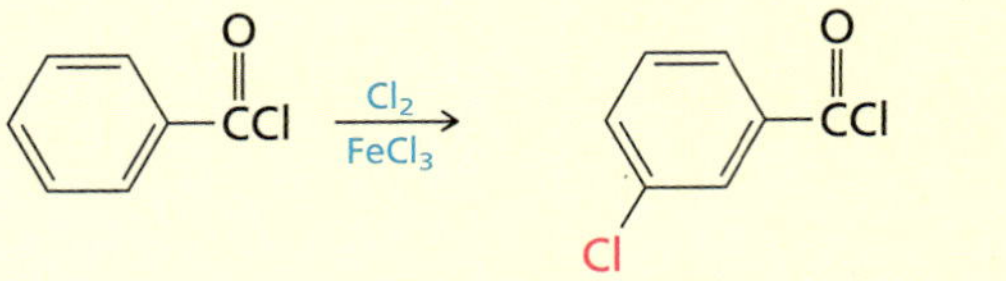

Benzoyl chloride $\xrightarrow[\text{FeCl}_3]{\text{Cl}_2}$ *m*-Chlorobenzoyl chloride (isolated in 62% yield)

A cyano group is similar to a carbonyl for analogous reasons involving resonance of the type shown for benzonitrile.

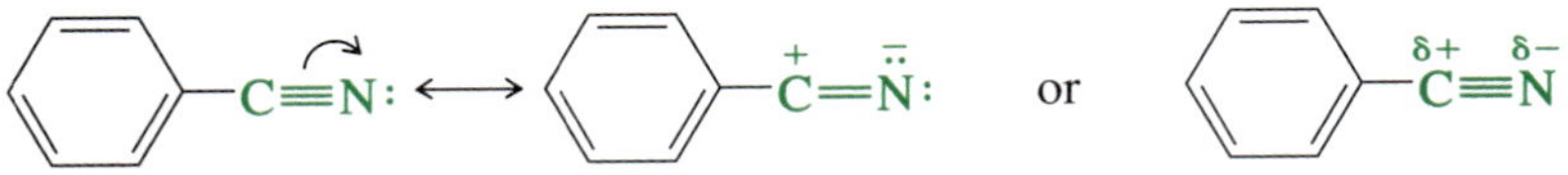

Cyano groups are electron-withdrawing, deactivating, and meta-directing.

Sulfonic acid groups are electron-withdrawing because sulfur has a formal positive charge in several of the resonance forms of benzenesulfonic acid.

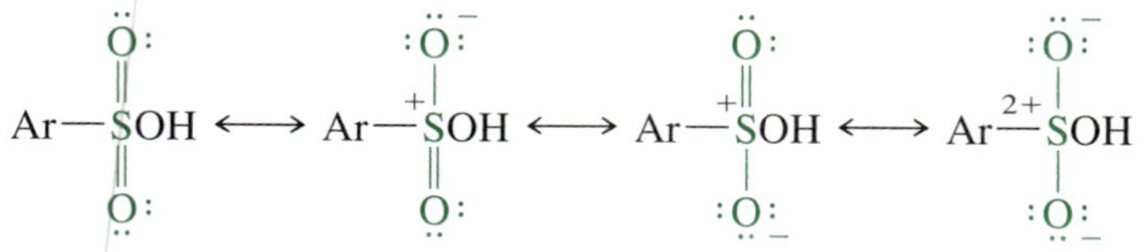

$$Ar-\overset{\overset{\ddot{O}:}{\|}}{\underset{\underset{\ddot{O}:}{\|}}{S}}OH \longleftrightarrow Ar-\overset{\overset{:\ddot{O}:^-}{\|}}{\underset{\underset{\ddot{O}:}{\|}}{\overset{+}{S}}}OH \longleftrightarrow Ar-\overset{\overset{\ddot{O}:}{\|}}{\underset{\underset{:\underset{..}{O}:^-}{\|}}{\overset{+}{S}}}OH \longleftrightarrow Ar-\overset{\overset{:\ddot{O}:^-}{\|}}{\underset{\underset{:\underset{..}{O}:^-}{\|}}{\overset{2+}{S}}}OH$$

When benzene undergoes disulfonation, *m*-benzenedisulfonic acid is formed. The first sulfonic acid group to go on directs the second one meta to itself.

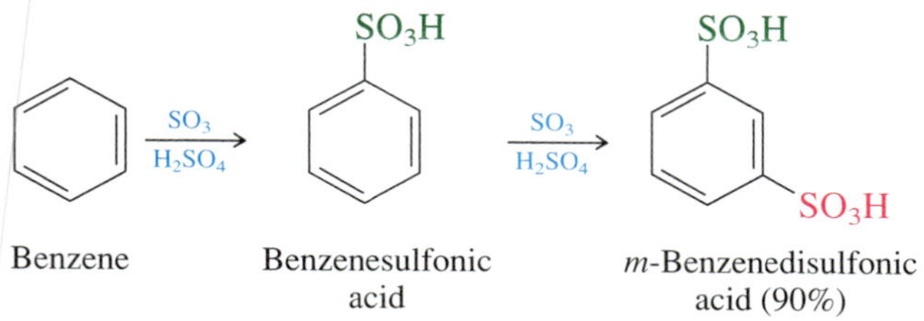

Benzene   Benzenesulfonic   *m*-Benzenedisulfonic
                    acid   acid (90%)

The nitrogen atom of a nitro group bears a full positive charge in its two most stable Lewis structures.

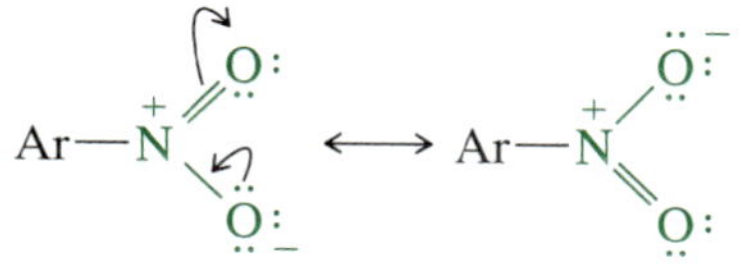

This makes the nitro group a powerful electron-withdrawing deactivating substituent and a meta director.

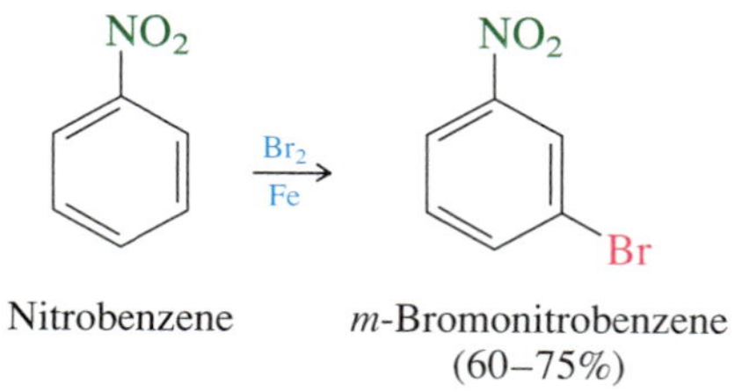

Nitrobenzene   *m*-Bromonitrobenzene
                        (60–75%)

PROBLEM 12.15 Would you expect the substituent $-\overset{+}{N}(CH_3)_3$ to more closely resemble $-\overset{..}{N}(CH_3)_2$ or $-NO_2$ in its effect on rate and regioselectivity in electrophilic aromatic substitution? Why?

12.14 SUBSTITUENT EFFECTS IN ELECTROPHILIC AROMATIC SUBSTITUTION: HALOGENS

Returning to Table 12.2, notice that *halogen substituents direct an incoming electrophile to the ortho and para positions but deactivate the ring toward substitution.* Nitration of chlorobenzene is a typical example of electrophilic aromatic substitution in a halobenzene;

its rate is some 30 times slower than the corresponding nitration of benzene. The major products are *o*-chloronitrobenzene and *p*-chloronitrobenzene.

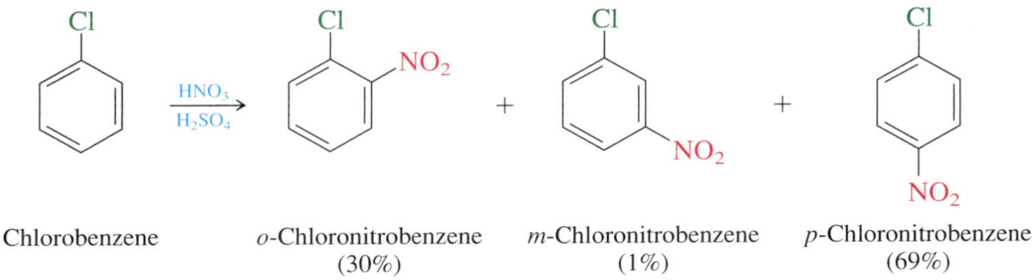

Chlorobenzene   *o*-Chloronitrobenzene   *m*-Chloronitrobenzene   *p*-Chloronitrobenzene
                              (30%)   (1%)   (69%)

PROBLEM 12.16 Reaction of chlorobenzene with *p*-chlorobenzyl chloride and aluminum chloride gave a mixture of two products in good yield (76%). What were these two products?

Because we have come to associate activating substituents with ortho, para-directing effects and deactivating substituents with meta, the properties of the halogen substituents appear on initial inspection to be unusual.

This seeming inconsistency between regioselectivity and rate can be understood by analyzing the two ways that a halogen substituent can affect the stability of a cyclohexadienyl cation. First, halogens are electronegative, and their inductive effect is to draw electrons away from the carbon to which they are bonded. Thus, all the intermediates formed by electrophilic attack on a halobenzene are less stable than the corresponding cyclohexadienyl cation for benzene, causing halobenzenes to react more slowly than benzene.

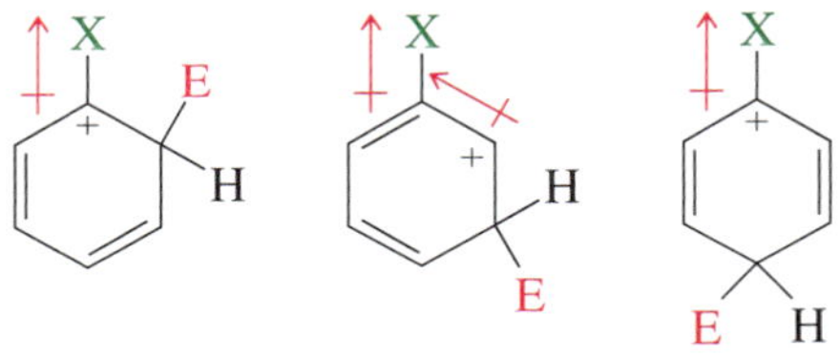

All these ions are less stable when X = F, Cl, Br, or I than when X = H

Like hydroxyl groups and amino groups, however, halogen substituents possess unshared electron pairs that can be donated to a positively charged carbon. This electron donation into the π system stabilizes the intermediates derived from ortho and from para attack.

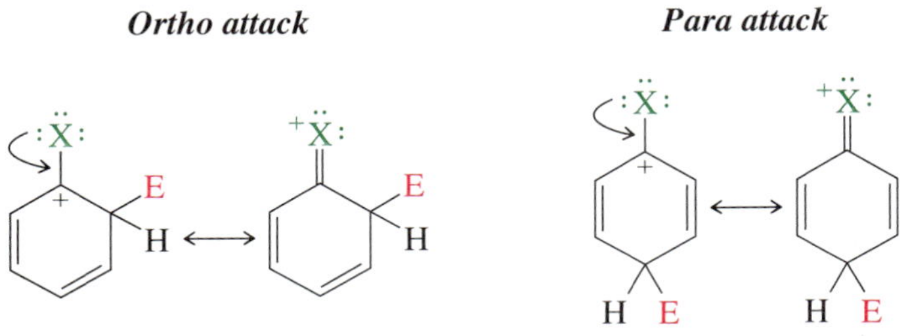

Ortho attack   *Para attack*

Comparable stabilization of the intermediate leading to meta substitution is not possible. Thus, resonance involving halogen lone pairs causes electrophilic attack to be favored at the ortho and para positions but is weak and insufficient to overcome the electron-withdrawing inductive effect of the halogen, which deactivates all the ring positions. The experimentally observed partial rate factors for nitration of chlorobenzene result from this blend of inductive and resonance effects.

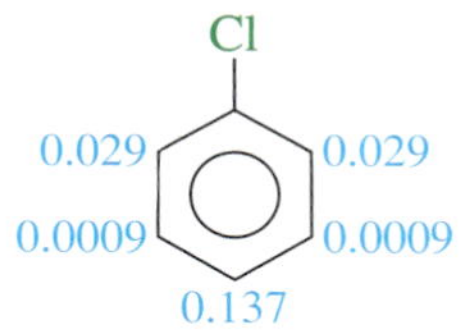

The mix of inductive and resonance effects varies from one halogen to another, but the net result is that fluorine, chlorine, bromine, and iodine are weakly deactivating, ortho, para-directing substituents.

12.15 MULTIPLE SUBSTITUENT EFFECTS

When a benzene ring bears two or more substituents, both its reactivity and the site of further substitution can usually be predicted from the cumulative effects of its substituents.

In the simplest cases all the available sites are equivalent, and substitution at any one of them gives the same product.

Problems 12.2, 12.3, and 12.7 offer additional examples of reactions in which only a single product of electrophilic aromatic substitution is possible.

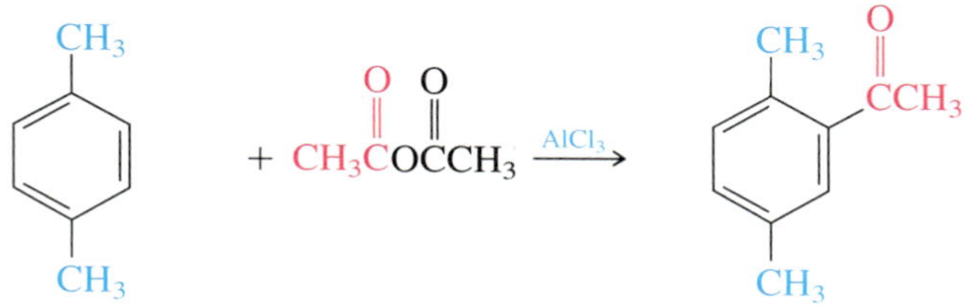

1,4-Dimethylbenzene (*p*-xylene)

2,5-Dimethylacetophenone (99%)

Often the directing effects of substituents reinforce each other. Bromination of *p*-nitrotoluene, for example, takes place at the position that is ortho to the ortho, para-directing methyl group and meta to the meta-directing nitro group.

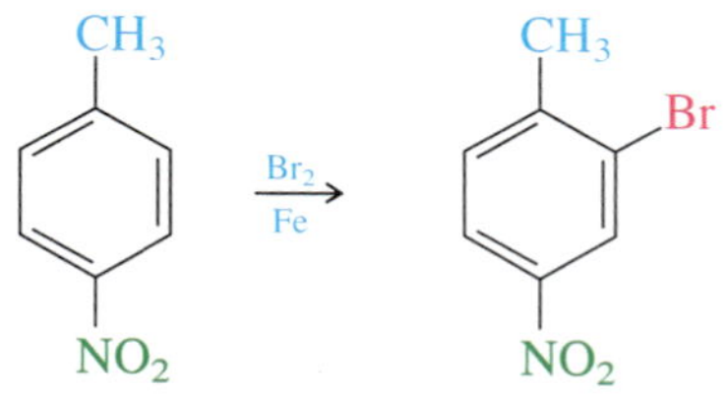

p-Nitrotoluene

2-Bromo-4-nitrotoluene (86–90%)

In almost all cases, including most of those in which the directing effects of individual substituents oppose each other, *it is the more activating substituent that controls the regioselectivity of electrophilic aromatic substitution.* Thus, bromination occurs ortho to the *N*-methylamino group in 4-chloro-*N*-methylaniline because this group is a very powerful activating substituent while the chlorine is weakly deactivating.

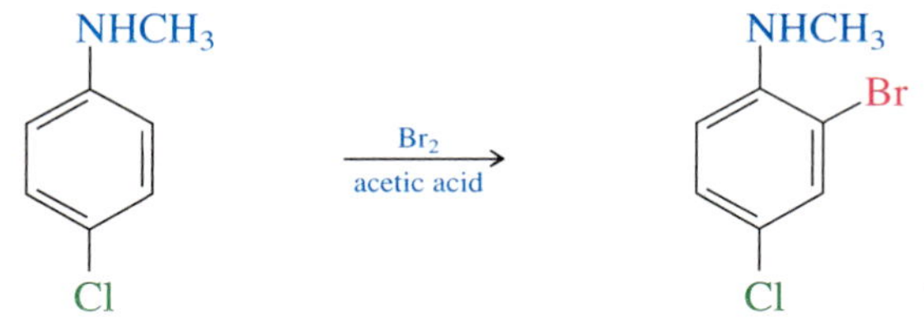

4-Chloro-*N*-methylaniline

2-Bromo-4-chloro-*N*-methylaniline
(87%)

When two positions are comparably activated by alkyl groups, substitution usually occurs at the less hindered site. Nitration of *p-tert*-butyltoluene takes place at positions ortho to the methyl group in preference to those ortho to the larger *tert*-butyl group. This is an example of a *steric effect*.

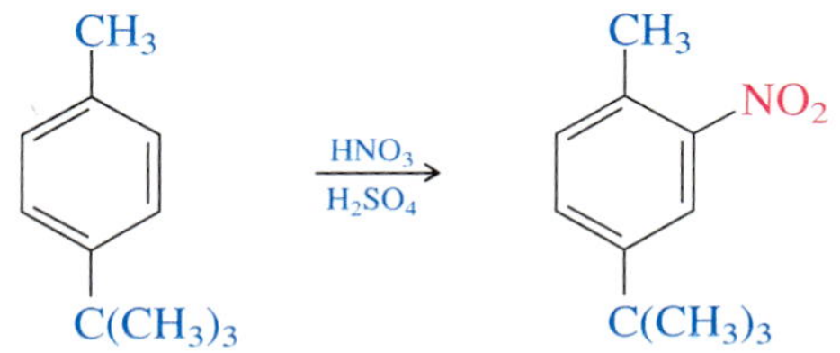

p-tert-Butyltoluene

4-*tert*-Butyl-2-nitrotoluene
(88%)

Nitration of *m*-xylene is directed ortho to one methyl group and para to the other.

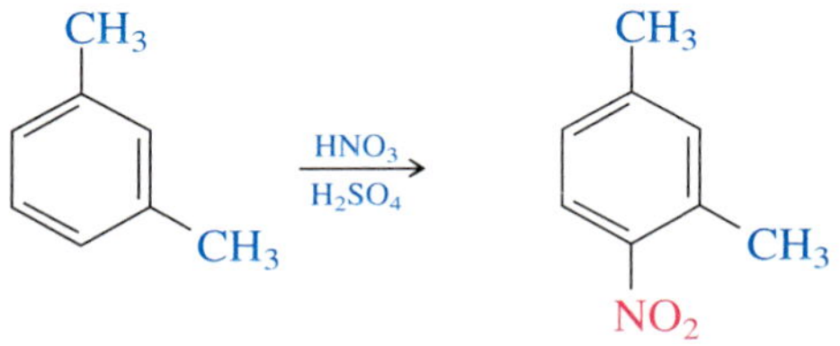

m-Xylene

2,4-Dimethyl-1-nitrobenzene
(98%)

The ortho position between the two methyl groups is less reactive because it is more sterically hindered.

PROBLEM 12.17 Write the structure of the principal organic product obtained on nitration of each of the following:

(a) *p*-Methylbenzoic acid
(b) *m*-Dichlorobenzene
(c) *m*-Dinitrobenzene
(d) *p*-Methoxyacetophenone
(e) *p*-Methylanisole
(f) 2,6-Dibromoanisole

SAMPLE SOLUTION (a) Of the two substituents in *p*-methylbenzoic acid, the methyl group is more activating and so controls the regioselectivity of electrophilic aromatic substitution. The position para to the ortho, para-directing methyl group already bears a substituent (the carboxyl group), and so substitution occurs ortho to the methyl group. This position is meta to the *m*-directing carboxyl group, and the orienting properties of the two substituents reinforce each other. The product is 4-methyl-3-nitrobenzoic acid.

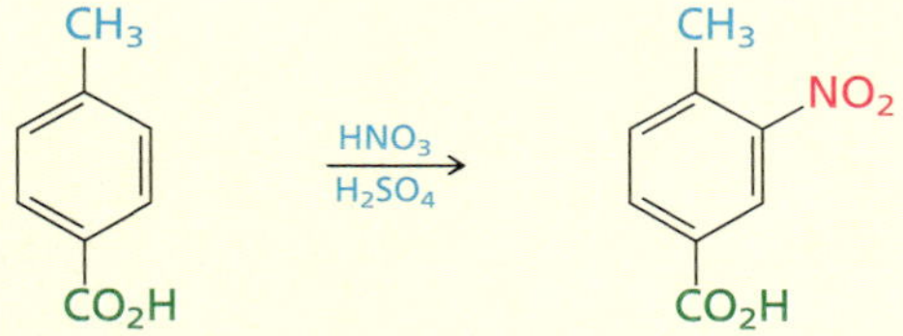

p-Methylbenzoic acid   4-Methyl-3-nitrobenzoic acid

Problem 12.39 illustrates how partial rate factor data may be applied to such cases.

An exception to the rule that regioselectivity is controlled by the most activating substituent occurs when the directing effects of alkyl groups and halogen substituents oppose each other. Alkyl groups and halogen substituents are weakly activating and weakly deactivating, respectively, and the difference between them is too small to allow a simple generalization.

12.16 REGIOSELECTIVE SYNTHESIS OF DISUBSTITUTED AROMATIC COMPOUNDS

Because the position of electrophilic attack on an aromatic ring is controlled by the directing effects of substituents already present, the preparation of disubstituted aromatic compounds requires that careful thought be given to the order of introduction of the two groups.

Compare the independent preparations of *m*-bromoacetophenone and *p*-bromoacetophenone from benzene. Both syntheses require a Friedel–Crafts acylation step and a bromination step, but the major product is determined by the *order* in which the two steps are carried out. When the meta-directing acetyl group is introduced first, the final product is *m*-bromoacetophenone.

Aluminum chloride is a stronger Lewis acid than iron(III) bromide and has been used as a catalyst in electrophilic bromination when, as in the example shown, the aromatic ring bears a strongly deactivating substituent.

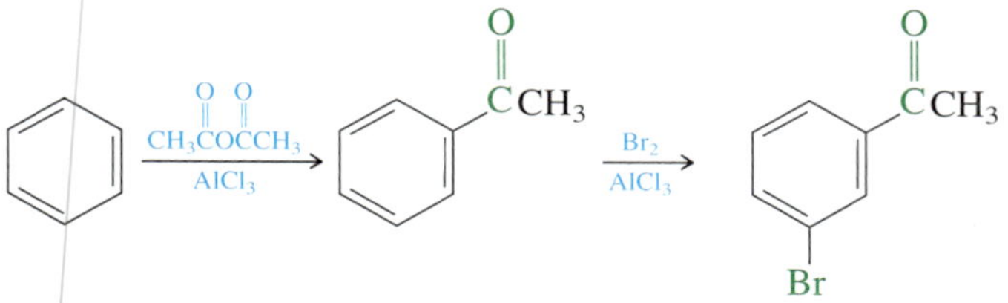

Benzene   Acetophenone   *m*-Bromoacetophenone
                      (76–83%)   (59%)

When the ortho, para-directing bromine is introduced first, the major product is *p*-bromoacetophenone (along with some of its ortho isomer, from which it is separated by distillation).

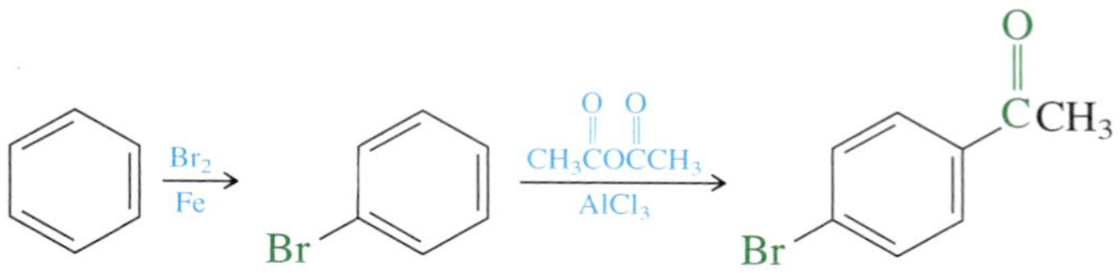

Benzene   Bromobenzene   p-Bromoacetophenone
              (65–75%)   (69–79%)

PROBLEM 12.18 Write chemical equations showing how you could prepare *m*-bromonitrobenzene as the principal organic product, starting with benzene and using any necessary organic or inorganic reagents. How could you prepare *p*-bromonitrobenzene?

A less obvious example of a situation in which the success of a synthesis depends on the order of introduction of substituents is illustrated by the preparation of *m*-nitroacetophenone. Here, even though both substituents are meta-directing, the only practical synthesis is the one in which Friedel–Crafts acylation is carried out first.

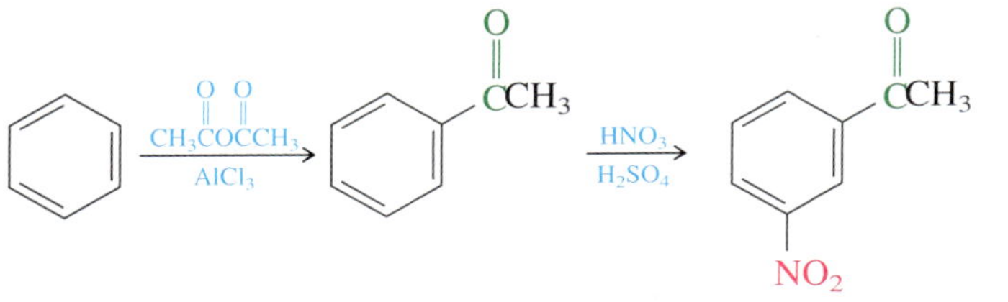

Benzene	Acetophenone (76–83%)	*m*-Nitroacetophenone (55%)

When the reverse order of steps is attempted, it is observed that the Friedel–Crafts acylation of nitrobenzene fails.

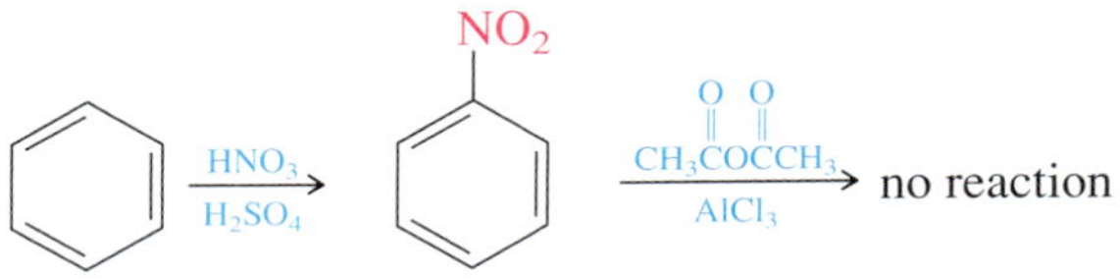

Benzene	Nitrobenzene (95%)

Neither Friedel–Crafts acylation nor alkylation reactions can be carried out on nitrobenzene. The presence of a strongly deactivating substituent such as a nitro group on an aromatic ring so depresses its reactivity that Friedel–Crafts reactions do not take place. Nitrobenzene is so unreactive that it is sometimes used as a solvent in Friedel–Crafts reactions. The practical limit for Friedel–Crafts alkylation and acylation reactions is effectively a monohalobenzene. *An aromatic ring more deactivated than a monohalobenzene cannot be alkylated or acylated under Friedel–Crafts conditions.*

Sometimes the orientation of two substituents in an aromatic compound precludes its straightforward synthesis. *m*-Chloroethylbenzene, for example, has two ortho, para-directing groups in a meta relationship and so can't be prepared either from chlorobenzene or ethylbenzene. In cases such as this we couple electrophilic aromatic substitution with functional group manipulation to produce the desired compound.

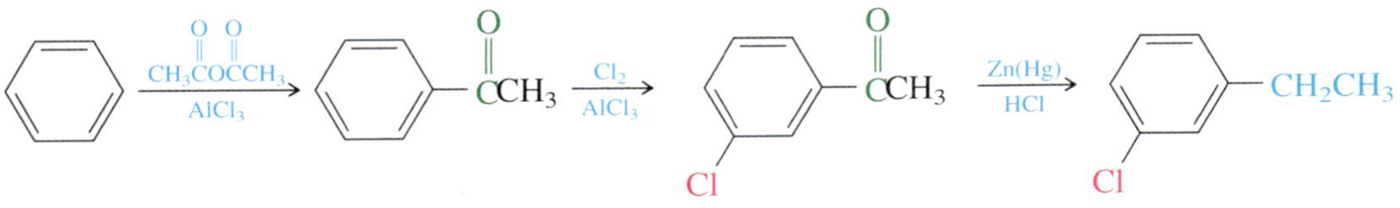

Benzene	Acetophenone	*m*-Chloroacetophenone	*m*-Chloroethylbenzene

The key here is to recognize that an ethyl substituent can be introduced by Friedel–Crafts acylation followed by a Clemmensen or Wolff–Kishner reduction step later in the synthesis. If the chlorine is introduced prior to reduction, it will be directed meta to the acetyl group, giving the correct substitution pattern.

A related problem concerns the synthesis of *p*-nitrobenzoic acid. Here, two meta-directing substituents are para to each other. This compound has been prepared from toluene according to the procedure shown:

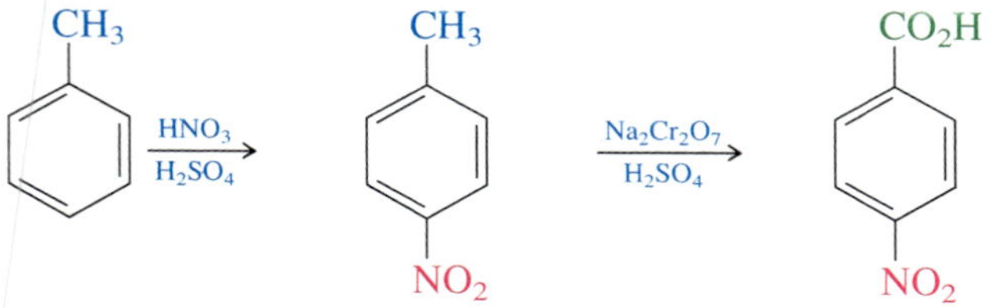

p-Nitrotoluene
(separate from ortho
isomer)

p-Nitrobenzoic acid
(82–86%)

Because it may be oxidized to a carboxyl group (Section 11.13), a methyl group can be used to introduce the nitro substituent in the proper position.

PROBLEM 12.19 Suggest an efficient synthesis of *m*-nitrobenzoic acid from toluene.

12.17 SUBSTITUTION IN NAPHTHALENE

Polycyclic aromatic hydrocarbons undergo electrophilic aromatic substitution when treated with the same reagents that react with benzene. In general, polycyclic aromatic hydrocarbons are more reactive than benzene. Most lack the symmetry of benzene, however, and mixtures of products may be formed even on monosubstitution. Among polycyclic aromatic hydrocarbons, we will discuss only naphthalene, and that only briefly.

Two sites are available for substitution in naphthalene, C-1 and C-2, C-1 being normally the preferred site of electrophilic attack.

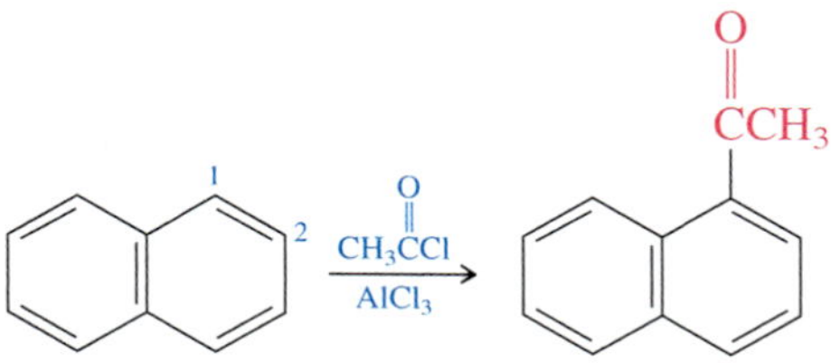

Naphthalene

1-Acetylnaphthalene
(90%)

C-1 is more reactive because the intermediate formed by electrophilic attack there is a relatively stable carbocation. A benzene-type pattern of bonds is retained in one ring, and the positive charge is delocalized by allylic resonance.

Attack at C-1

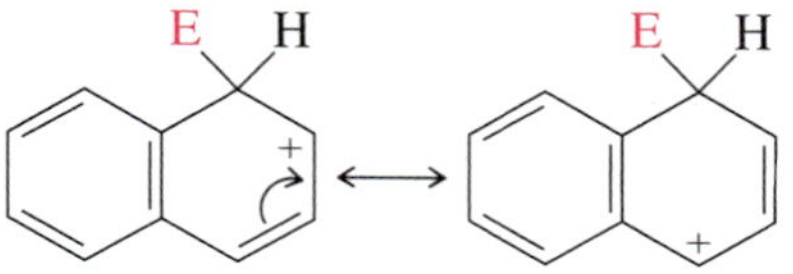

Attack at C-2

To involve allylic resonance in stabilizing the arenium ion formed during attack at C-2, the benzenoid character of the other ring is sacrificed.

> **PROBLEM 12.20**   Sulfonation of naphthalene is reversible at elevated temperature. A different isomer of naphthalenesulfonic acid is the major product at 160°C than is the case at 0°C. Which isomer is the product of kinetic control? Which one is formed under conditions of thermodynamic control? Can you think of a reason why one isomer is more stable than the other? (*Hint:* Build space-filling models of both isomers.)

12.18 SUBSTITUTION IN HETEROCYCLIC AROMATIC COMPOUNDS

The great variety of available structural types causes heterocyclic aromatic compounds to range from exceedingly reactive to practically inert toward electrophilic aromatic substitution.

Pyridine lies near one extreme in being far less reactive than benzene toward substitution by electrophilic reagents. In this respect it resembles strongly deactivated aromatic compounds such as nitrobenzene. It is incapable of being acylated or alkylated under Friedel–Crafts conditions, but can be sulfonated at high temperature. Electrophilic substitution in pyridine, when it does occur, takes place at C-3.

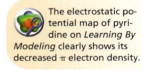

Pyridine   →   Pyridine-3-sulfonic acid
(71%)

The electrostatic potential map of pyridine on *Learning By Modeling* clearly shows its decreased π electron density.

One reason for the low reactivity of pyridine is that its nitrogen atom, because it is more electronegative than a CH in benzene, causes the π electrons to be held more tightly and raises the activation energy for attack by an electrophile. Another is that the nitrogen of pyridine is protonated in sulfuric acid and the resulting pyridinium ion is even more deactivated than pyridine itself.

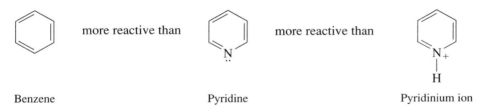

Benzene   Pyridine   Pyridinium ion

Lewis acid catalysts such as aluminum chloride and iron(III) halides also bond to nitrogen to strongly deactivate the ring toward Friedel–Crafts reactions and halogenation.

Pyrrole, furan, and thiophene, on the other hand, have electron-rich aromatic rings and are extremely reactive toward electrophilic aromatic substitution—more like phenol and aniline than benzene. Like benzene they have six π electrons, but these π electrons are delocalized over *five* atoms, not six, and are not held as strongly as those of benzene. Even when the ring atom is as electronegative as oxygen, substitution takes place readily.

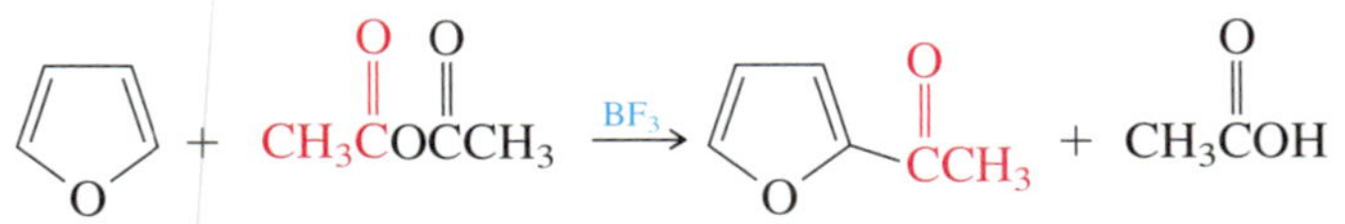

| Furan | Acetic anhydride | 2-Acetylfuran (75–92%) | Acetic acid |

The regioselectivity of substitution in furan is explained using a resonance description. When the electrophile attacks C-2, the positive charge is shared by three atoms: C-3, C-5, and O.

Attack at C-2

Carbocation *more stable;* positive charge shared by C-3, C-5, and O.

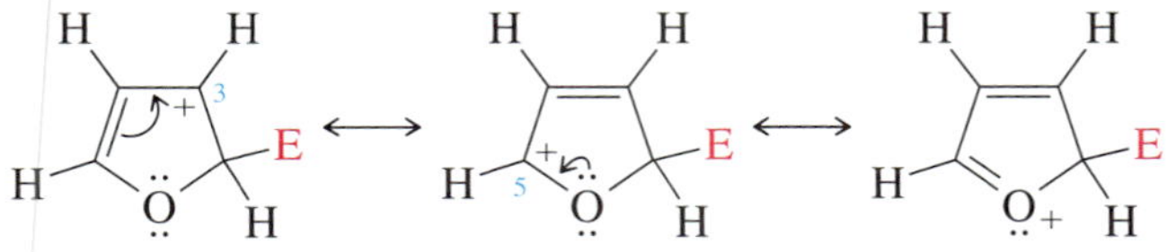

When the electrophile attacks at C-3, the positive charge is shared by only two atoms, C-2 and O, and the carbocation intermediate is less stable and formed more slowly.

Attack at C-3

Carbocation *less stable;* positive charge shared by C-2 and O.

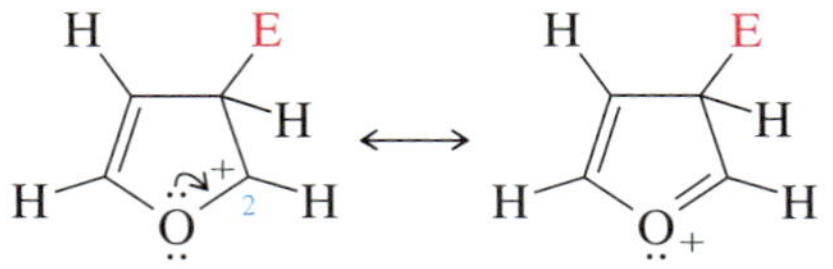

The regioselectivity of substitution in pyrrole and thiophene is like that of furan and for similar reasons.

PROBLEM 12.21 When benzene is prepared from coal tar, it is contaminated with thiophene, from which it cannot be separated by distillation because of very similar boiling points. Shaking a mixture of benzene and thiophene with sulfuric acid causes sulfonation of the thiophene ring but leaves benzene untouched. The sulfonation product of thiophene dissolves in the sulfuric acid layer, from which the benzene layer is separated; the benzene layer is then washed with water and distilled. Give the structure of the sulfonation product of thiophene.

12.19 SUMMARY

Section 12.1   On reaction with electrophilic reagents, compounds that contain a benzene ring undergo **electrophilic aromatic substitution.** Table 12.1 in Section 12.1 and Table 12.3 in this summary give examples.

Section 12.2   The mechanism of electrophilic aromatic substitution involves two stages: attack of the electrophile on the π electrons of the ring (slow, rate-determining), followed by loss of a proton to restore the aromaticity of the ring.

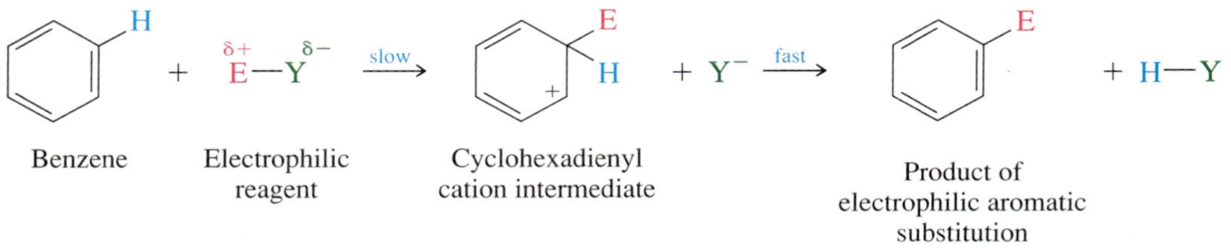

| Benzene | Electrophilic reagent | Cyclohexadienyl cation intermediate | Product of electrophilic aromatic substitution |

Sections 12.3–12.5	See Table 12.3
Sections 12.6–12.7	See Tables 12.3 and 12.4
Section 12.8	Friedel–Crafts acylation, followed by Clemmensen or Wolff–Kishner reduction is a standard sequence used to introduce a primary alkyl group onto an aromatic ring.

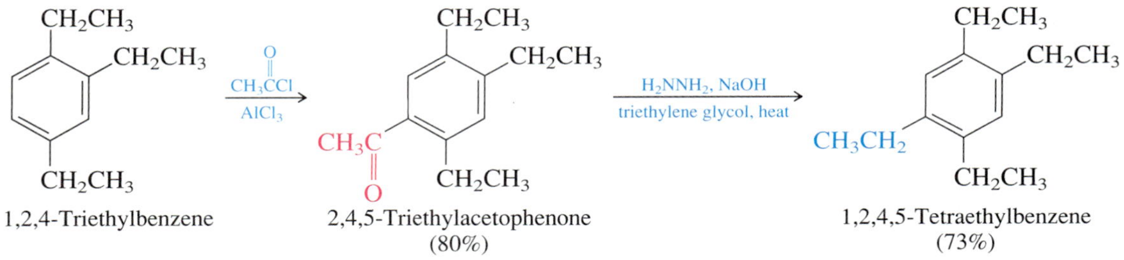

1,2,4-Triethylbenzene

2,4,5-Triethylacetophenone
(80%)

1,2,4,5-Tetraethylbenzene
(73%)

Section 12.9 Substituents on an aromatic ring can influence both the *rate* and *regioselectivity* of electrophilic aromatic substitution. Substituents are classified as *activating* or *deactivating* according to whether they cause the ring to react more rapidly or less rapidly than benzene. With respect to regioselectivity, substituents are either *ortho, para-directing* or *meta-directing*. A methyl group is activating and ortho, para-directing. A trifluoromethyl group is deactivating and meta-directing.

Sections 12.10–12.14 How substituents control rate and regioselectivity in electrophilic aromatic substitution results from their effect on carbocation stability. An electron-releasing substituent stabilizes the cyclohexadienyl cation intermediates corresponding to ortho and para attack more than meta.

| Stabilized when G is electron-releasing | Less stabilized when G is electron-releasing | Stabilized when G is electron-releasing |

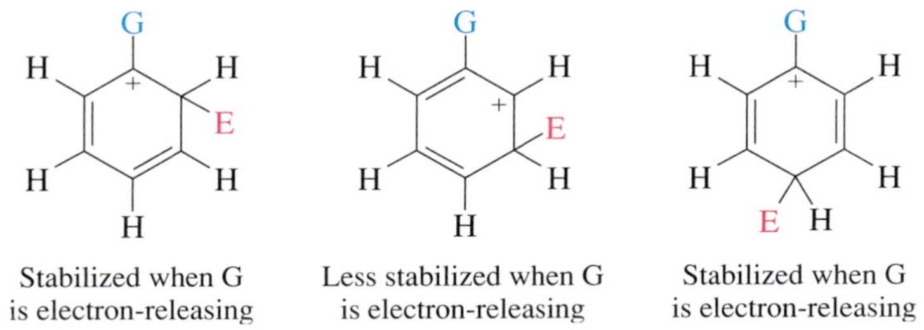

Conversely, an electron-withdrawing substituent destabilizes the cyclohexadienyl cations corresponding to ortho and para attack more than meta. Thus, meta substitution predominates.

TABLE 12.3   Representative Electrophilic Aromatic Substitution Reactions

Reaction (section) and comments	General equation and specific example
Nitration (Section 12.3) The active electrophile in the nitration of benzene and its derivatives is nitronium cation ($:\ddot{O}{=}\overset{+}{N}{=}\ddot{O}:$). It is generated by reaction of nitric acid and sulfuric acid. Very reactive arenes—those that bear strongly activating substituents—undergo nitration in nitric acid alone.	$ArH + HNO_3 \xrightarrow{H_2SO_4} ArNO_2 + H_2O$ Arene Nitric acid Nitroarene Water F—⟨benzene⟩ $\xrightarrow[H_2SO_4]{HNO_3}$ F—⟨benzene⟩—NO$_2$ Fluorobenzene *p*-Fluoronitrobenzene (80%)
Sulfonation (Section 12.4) Sulfonic acids are formed when aromatic compounds are treated with sources of sulfur trioxide. These sources can be concentrated sulfuric acid (for very reactive arenes) or solutions of sulfur trioxide in sulfuric acid (for benzene and arenes less reactive than benzene).	$ArH + SO_3 \longrightarrow ArSO_3H$ Arene Sulfur trioxide Arenesulfonic acid 1,2,4,5-Tetramethylbenzene $\xrightarrow[H_2SO_4]{SO_3}$ 2,3,5,6-Tetramethylbenzenesulfonic acid (94%)
Halogenation (Section 12.5) Chlorination and bromination of arenes are carried out by treatment with the appropriate halogen in the presence of a Lewis acid catalyst. Very reactive arenes undergo halogenation in the absence of a catalyst.	$ArH + X_2 \xrightarrow{FeX_3} ArX + HX$ Arene Halogen Aryl halide Hydrogen halide HO—⟨benzene⟩ $\xrightarrow[CS_2]{Br_2}$ HO—⟨benzene⟩—Br Phenol *p*-Bromophenol (80–84%)
Friedel–Crafts alkylation (Section 12.6) Carbocations, usually generated from an alkyl halide and aluminum chloride, attack the aromatic ring to yield alkylbenzenes. The arene must be at least as reactive as a halobenzene. Carbocation rearrangements can occur, especially with primary alkyl halides.	$ArH + RX \xrightarrow{AlCl_3} ArR + HX$ Arene Alkyl halide Alkylarene Hydrogen halide Benzene + Cyclopentyl bromide —Br $\xrightarrow{AlCl_3}$ Cyclopentylbenzene (54%)
Friedel–Crafts acylation (Section 12.7) Acyl cations (acylium ions) generated by treating an acyl chloride or acid anhydride with aluminum chloride attack aromatic rings to yield ketones. The arene must be at least as reactive as a halobenzene. Acyl cations are relatively stable, and do not rearrange.	$ArH + RCCl \xrightarrow{AlCl_3} ArCR + HCl$ (with C=O) Arene Acyl chloride Ketone Hydrogen chloride $ArH + RCOCR \xrightarrow{AlCl_3} ArCR + RCOH$ (anhydride, ketone, carboxylic acid) Arene Acid anhydride Ketone Carboxylic acid CH$_3$O—⟨benzene⟩ $\xrightarrow[AlCl_3]{CH_3COCCH_3}$ CH$_3$O—⟨benzene⟩—CCH$_3$ Anisole *p*-Methoxyacetophenone (90–94%)

TABLE 12.4	Limitations on Friedel–Crafts Reactions

1. The organic halide that reacts with the arene must be an alkyl halide (Section 12.6) or an acyl halide (Section 12.7).

These will react with benzene under Friedel–Crafts conditions:

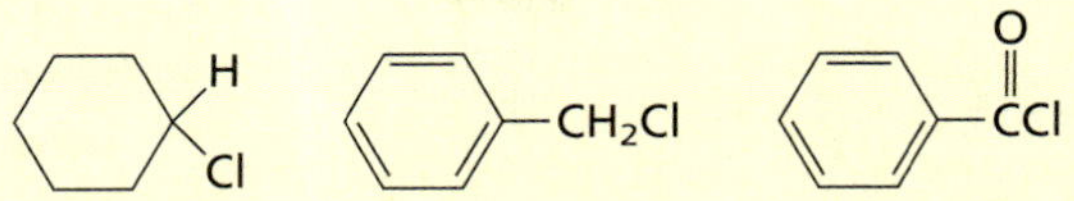

Alkyl halide   Benzylic halide   Acyl halide

Vinylic halides and aryl halides do not form carbocations under conditions of the Friedel–Crafts reaction and so cannot be used in place of an alkyl halide or an acyl halide.

These will not react with benzene under Friedel–Crafts conditions:

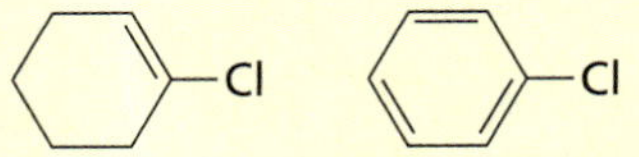

Vinylic halide   Aryl halide

2. Rearrangement of alkyl groups can occur (Section 12.6).

Rearrangement is especially prevalent with primary alkyl halides of the type RCH_2CH_2X and R_2CHCH_2X. Aluminum chloride induces ionization with rearrangement to give a more stable carbocation. Benzylic halides and acyl halides do not rearrange.

3. Strongly deactivated aromatic rings do not undergo Friedel–Crafts alkylation or acylation (Section 12.16). Friedel–Crafts alkylations and acylations fail when applied to compounds of the following type, where EWG is a strongly electron-withdrawing group:

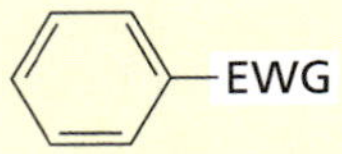

EWG:

$$-CF_3, \quad -NO_2, \quad -SO_3H, \quad -C\equiv N,$$

4. It is sometimes difficult to limit Friedel–Crafts alkylation to monoalkylation.

The first alkyl group that goes on makes the ring more reactive toward further substitution because alkyl groups are activating substituents. Monoacylation is possible because the first acyl group to go on is strongly electron-withdrawing and deactivates the ring toward further substitution.

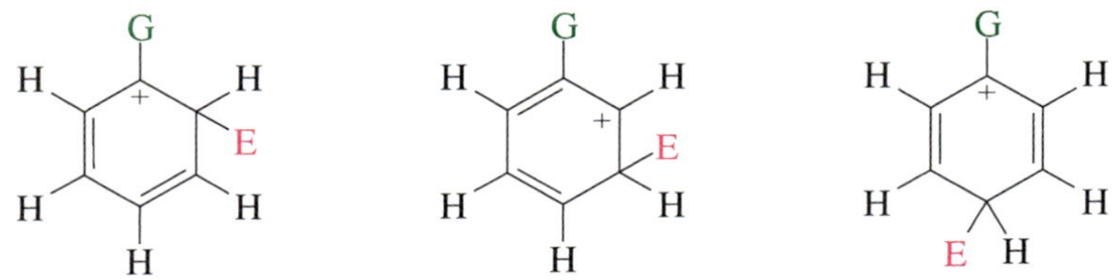

Destabilized when G is electron-withdrawing   Less destabilized when G is electron-withdrawing   Destabilized when G is electron-withdrawing

Substituents can be arranged into three major categories:

1. **Activating and ortho, para-directing:** These substituents stabilize the cyclohexadienyl cation formed in the rate-determining step. They include $-\ddot{N}R_2$, $-\ddot{O}R$, $-R$, $-Ar$, and related species. The most strongly activating members of this group are bonded to the ring by a nitrogen or oxygen atom that bears an unshared pair of electrons.

2. **Deactivating and ortho, para-directing:** The halogens are the most prominent members of this class. They withdraw electron density from all the ring positions by an inductive effect, making halobenzenes less reactive than benzene. Lone-pair electron donation stabilizes the cyclohexadienyl cations corresponding to attack at the ortho and para positions more than those formed by attack at the meta positions, giving rise to the observed regioselectivity.

3. **Deactivating and meta-directing:** These substituents are strongly electron-withdrawing and destabilize carbocations. They include

$$-CF_3, \quad \overset{\overset{\displaystyle O}{\|}}{-CR}, \quad -C{\equiv}N, \quad -NO_2$$

and related species. All the ring positions are deactivated, but because the *meta* positions are deactivated less than the ortho and para, meta substitution is favored.

Section 12.15   When two or more substituents are present on a ring, the regioselectivity of electrophilic aromatic substitution is generally controlled by the directing effect of the more powerful *activating* substituent.

Section 12.16   The order in which substituents are introduced onto a benzene ring needs to be considered in order to prepare the desired isomer in a multistep synthesis.

Section 12.17   Polycyclic aromatic hydrocarbons undergo the same kind of electrophilic aromatic substitution reactions as benzene.

Section 12.18   Heterocyclic aromatic compounds may be more reactive or less reactive than benzene. Pyridine is much less reactive than benzene, but pyrrole, furan, and thiophene are more reactive.

PROBLEMS

12.22 Give reagents suitable for carrying out each of the following reactions, and write the major organic products. If an ortho, para mixture is expected, show both. If the meta isomer is the expected major product, write only that isomer.

  (a) Nitration of benzene

  (b) Nitration of the product of part (a)

  (c) Bromination of toluene

  (d) Bromination of (trifluoromethyl)benzene

  (e) Sulfonation of anisole

  (f) Sulfonation of acetanilide ($C_6H_5NH\overset{\overset{\displaystyle O}{\|}}{C}CH_3$)

  (g) Chlorination of bromobenzene

  (h) Friedel–Crafts alkylation of anisole with benzyl chloride

  (i) Friedel–Crafts acylation of benzene with benzoyl chloride ($C_6H_5\overset{\overset{\displaystyle O}{\|}}{C}Cl$)

  (j) Nitration of the product from part (i)

(k) Clemmensen reduction of the product from part (i)

(l) Wolff–Kishner reduction of the product from part (i)

12.23 Write a structural formula for the most stable cyclohexadienyl cation intermediate formed in each of the following reactions. Is this intermediate more or less stable than the one formed by electrophilic attack on benzene?

(a) Bromination of *p*-xylene

(b) Chlorination of *m*-xylene

(c) Nitration of acetophenone

(d) Friedel–Crafts acylation of anisole with $CH_3\overset{\overset{\displaystyle O}{\|}}{C}Cl$

(e) Nitration of isopropylbenzene

(f) Bromination of nitrobenzene

(g) Sulfonation of furan

(h) Bromination of pyridine

12.24 In each of the following pairs of compounds choose which one will react faster with the indicated reagent, and write a chemical equation for the faster reaction:

(a) Toluene or chlorobenzene with a mixture of nitric acid and sulfuric acid

(b) Fluorobenzene or (trifluoromethyl)benzene with benzyl chloride and aluminum chloride

(c) Methyl benzoate ($C_6H_5\overset{\overset{\displaystyle O}{\|}}{C}OCH_3$) or phenyl acetate ($C_6H_5O\overset{\overset{\displaystyle O}{\|}}{C}CH_3$) with bromine in acetic acid

(d) Acetanilide ($C_6H_5NH\overset{\overset{\displaystyle O}{\|}}{C}CH_3$) or nitrobenzene with sulfur trioxide in sulfuric acid

(e) *p*-Dimethylbenzene (*p*-xylene) or *p*-di-*tert*-butylbenzene with acetyl chloride and aluminum chloride

(f) Benzophenone ($C_6H_5\overset{\overset{\displaystyle O}{\|}}{C}C_6H_5$) or biphenyl ($C_6H_5—C_6H_5$) with chlorine and iron(III) chloride

12.25 Arrange the following five compounds in order of decreasing rate of bromination: benzene, toluene, *o*-xylene, *m*-xylene, 1,3,5-trimethylbenzene (the relative rates are 2×10^7, 5×10^4, 5×10^2, 60, and 1).

12.26 Each of the following reactions has been carried out under conditions such that disubstitution or trisubstitution occurred. Identify the principal organic product in each case.

(a) Nitration of *p*-chlorobenzoic acid (dinitration)

(b) Bromination of aniline (tribromination)

(c) Bromination of *o*-aminoacetophenone (dibromination)

(d) Nitration of benzoic acid (dinitration)

(e) Bromination of *p*-nitrophenol (dibromination)

(f) Reaction of biphenyl with *tert*-butyl chloride and iron(III) chloride (dialkylation)

(g) Sulfonation of phenol (disulfonation)

12.27 Write equations showing how to prepare each of the following from benzene or toluene and any necessary organic or inorganic reagents. If an ortho, para mixture is formed in any step of your synthesis, assume that you can separate the two isomers.

(a) Isopropylbenzene

(b) *p*-Isopropylbenzenesulfonic acid

(c) 2-Bromo-2-phenylpropane

(d) 4-*tert*-Butyl-2-nitrotoluene

(e) *m*-Chloroacetophenone

(f) *p*-Chloroacetophenone

(g) 3-Bromo-4-methylacetophenone

(h) 2-Bromo-4-ethyltoluene

(i) 1-Bromo-3-nitrobenzene

(j) 1-Bromo-2,4-dinitrobenzene

(k) 3-Bromo-5-nitrobenzoic acid

(l) 2-Bromo-4-nitrobenzoic acid

(m) Diphenylmethane

(n) 1-Phenyloctane

(o) 1-Phenyl-1-octene

(p) 1-Phenyl-1-octyne

(q) 1,4-Di-*tert*-butyl-1,4-cyclohexadiene

12.28 Write equations showing how you could prepare each of the following from anisole and any necessary organic or inorganic reagents. If an ortho, para mixture is formed in any step of your synthesis, assume that you can separate the two isomers.

(a) *p*-Methoxybenzenesulfonic acid

(b) 2-Bromo-4-nitroanisole

(c) 4-Bromo-2-nitroanisole

(d) *p*-Methoxystyrene

12.29 How many products are capable of being formed from toluene in each of the following reactions?

(a) Mononitration (HNO_3, H_2SO_4, 40°C).

(b) Dinitration (HNO_3, H_2SO_4, 80°C).

(c) Trinitration (HNO_3, H_2SO_4, 110°C). The explosive TNT (trinitrotoluene) is the major product obtained on trinitration of toluene. Which trinitrotoluene isomer is TNT?

12.30 Friedel–Crafts acylation of the individual isomers of xylene with acetyl chloride and aluminum chloride yields a single product, different for each xylene isomer, in high yield in each case. Write the structures of the products of acetylation of *o*-, *m*-, and *p*-xylene.

12.31 Reaction of benzanilide ($C_6H_5NHCC_6H_5$) with chlorine in acetic acid yields a mixture of two monochloro derivatives formed by electrophilic aromatic substitution. Suggest reasonable structures for these two isomers.

12.32 Each of the following reactions has been reported in the chemical literature and gives a predominance of a single product in synthetically acceptable yield. Write the structure of the product. Only monosubstitution is involved in each case, unless otherwise indicated.

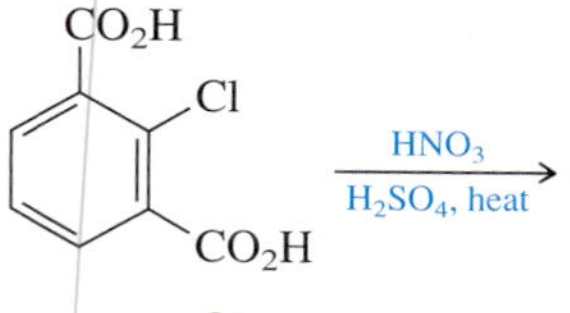

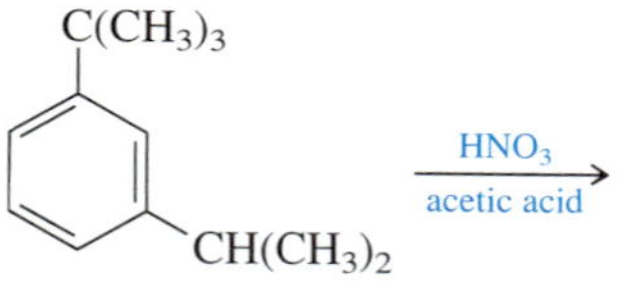

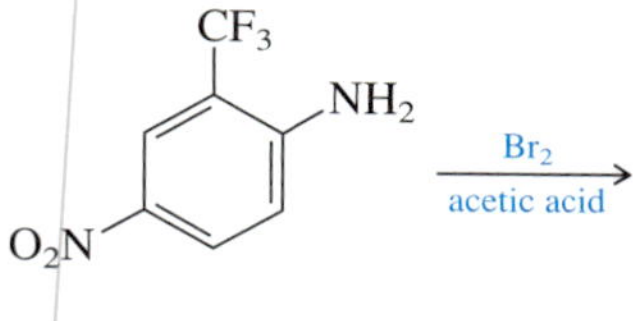

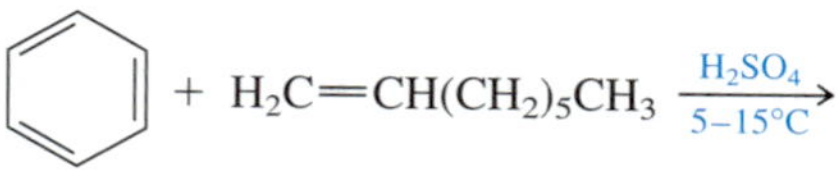

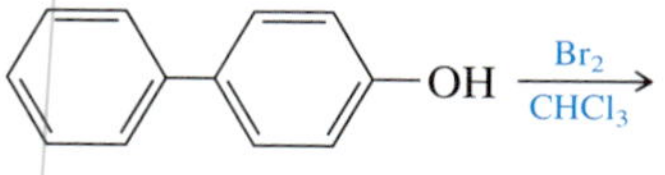

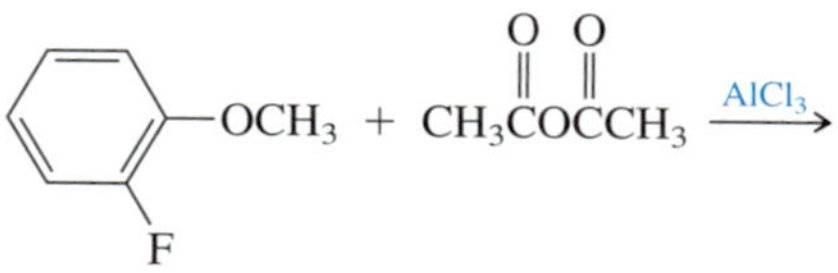

(g)

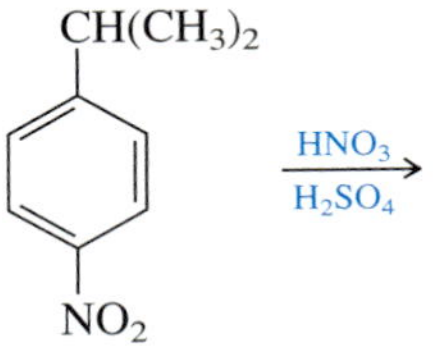

CH(CH$_3$)$_2$

$\xrightarrow[\text{H}_2\text{SO}_4]{\text{HNO}_3}$

NO$_2$

(k) ⬡—F + ⬡—CH$_2$Cl $\xrightarrow{\text{AlCl}_3}$

(h)

OCH$_3$

CH$_3$

+ (CH$_3$)$_2$C=CH$_2$ $\xrightarrow{\text{H}_2\text{SO}_4}$

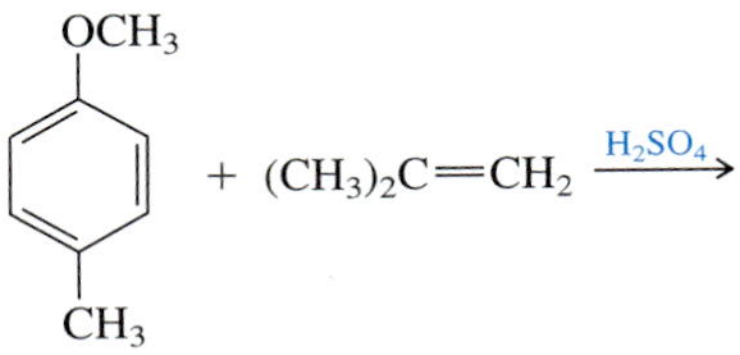

(l)

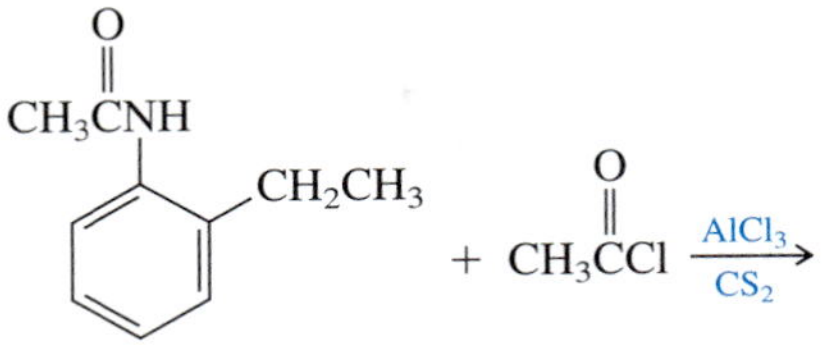

O
‖
CH$_3$CNH

CH$_2$CH$_3$

+ CH$_3$CCl (O) $\xrightarrow[\text{CS}_2]{\text{AlCl}_3}$

(i) 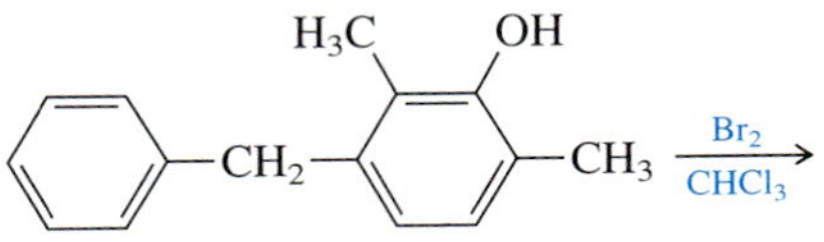 ⬡—CH$_2$—

H$_3$C   OH

—CH$_3$ $\xrightarrow[\text{CHCl}_3]{\text{Br}_2}$

(m)

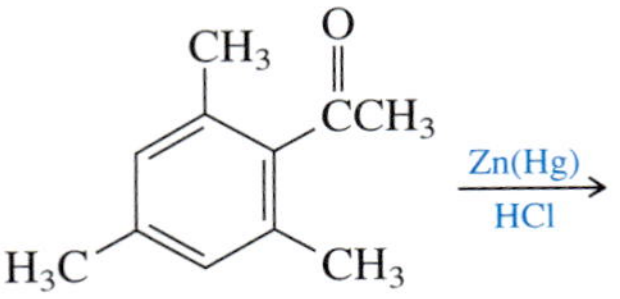

CH$_3$ O
         ‖
         CCH$_3$

H$_3$C   CH$_3$

$\xrightarrow[\text{HCl}]{\text{Zn(Hg)}}$

(j)

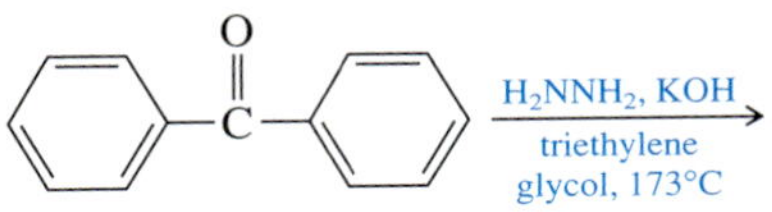

O
‖
⬡—C—⬡

$\xrightarrow[\substack{\text{triethylene}\\\text{glycol, 173°C}}]{\text{H}_2\text{NNH}_2,\ \text{KOH}}$

(n)

CO$_2$H

S

$\xrightarrow[\text{acetic acid}]{\text{Br}_2}$

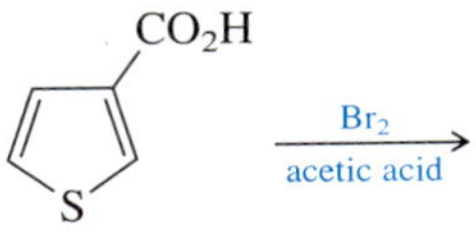

12.33 What combination of acyl chloride or acid anhydride and arene would you choose to prepare each of the following compounds by Friedel–Crafts acylation?

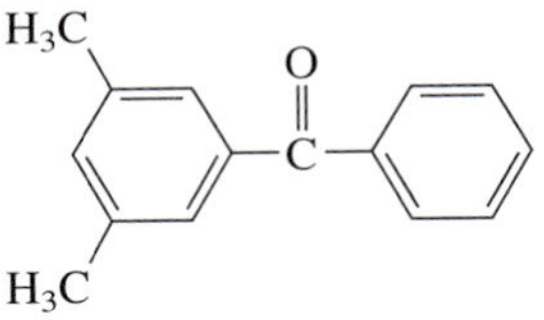

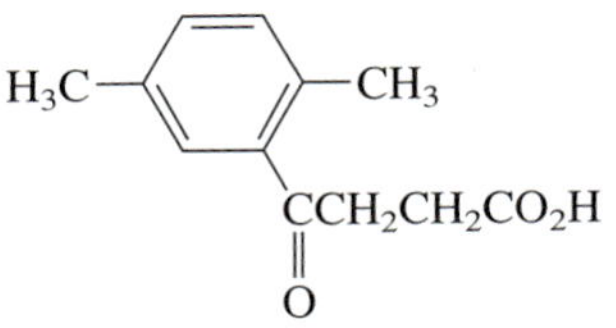

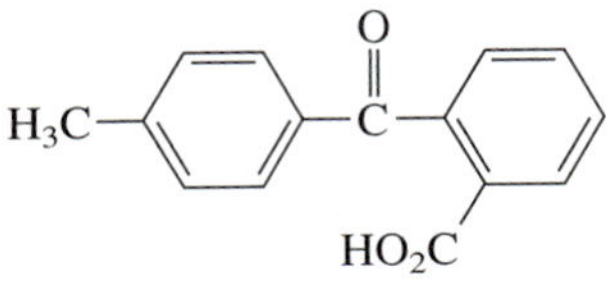

(a) 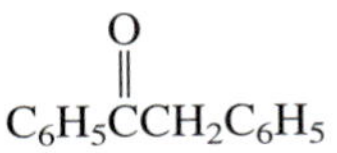 C$_6$H$_5$CCH$_2$C$_6$H$_5$ (with O double bond)

(d)

H$_3$C

O
‖
C—⬡

H$_3$C

(b) H$_3$C—⬡—CH$_3$ with CCH$_2$CH$_2$CO$_2$H (O below)

(e) H$_3$C—⬡—C(=O)—⬡ with HO$_2$C

(c) O$_2$N—⬡—C(=O)—⬡

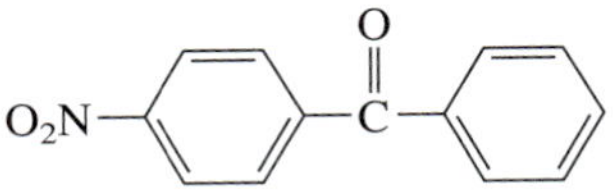

12.34 Suggest a suitable series of reactions for carrying out each of the following synthetic transformations:

(a)

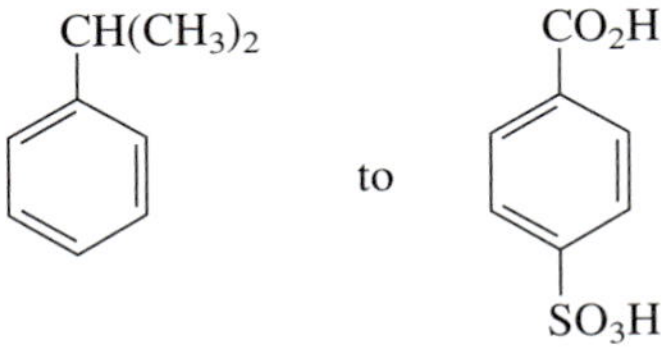

CH(CH$_3$)$_2$
⬡

to

CO$_2$H
⬡
SO$_3$H

(b) 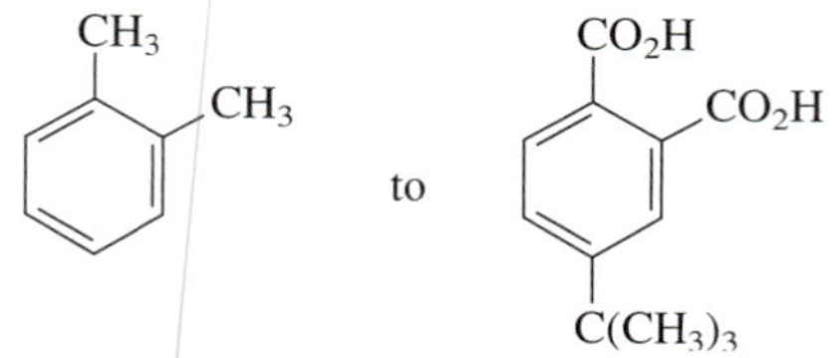   to

(c) 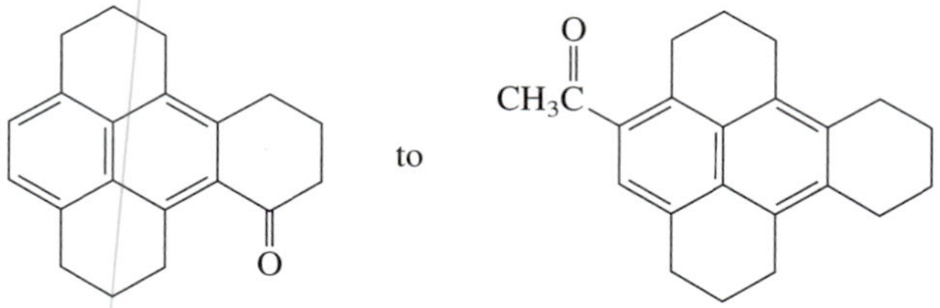   to

(d) 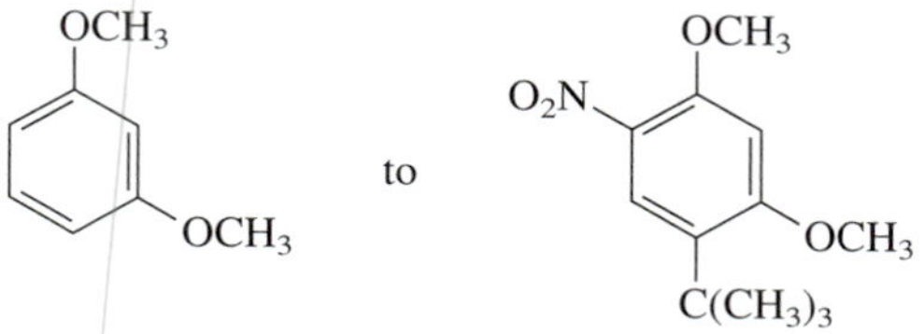   to

12.35 A standard synthetic sequence for building a six-membered cyclic ketone onto an existing aromatic ring is shown in outline as follows. Specify the reagents necessary for each step.

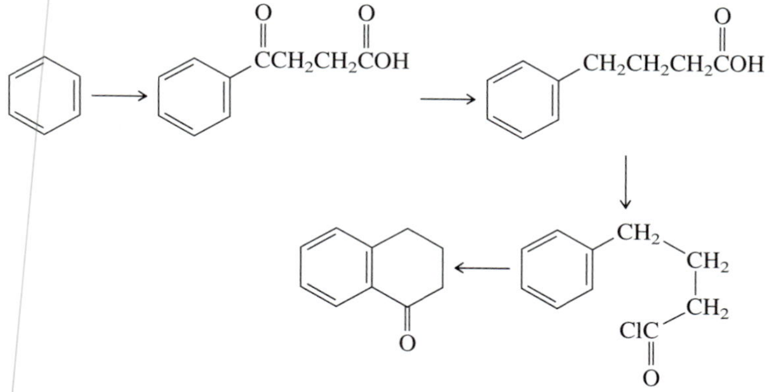

12.36 Each of the compounds indicated undergoes an intramolecular Friedel–Crafts acylation reaction to yield a cyclic ketone. Write the structure of the expected product in each case.

(a)

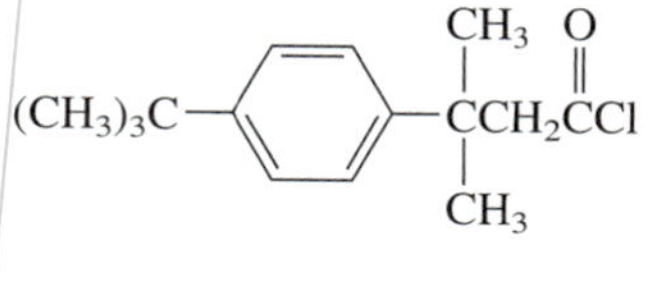

(c)

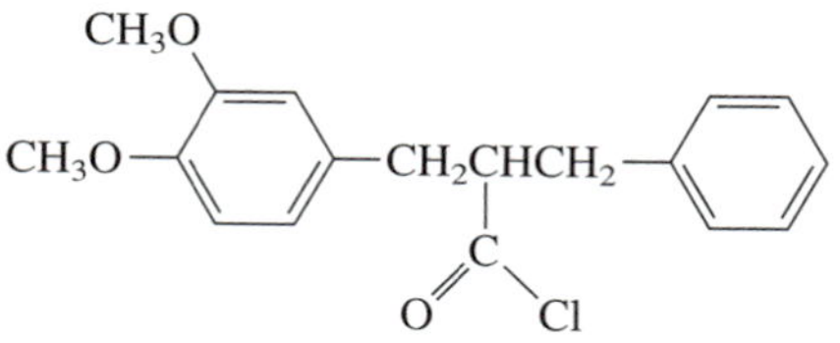

(b)

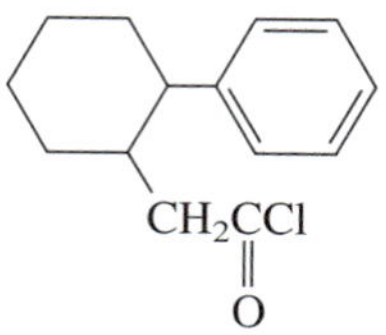

12.37 Of the groups shown, which is the most likely candidate for substituent X based on the partial rate factors for chlorination?

$—CF_3$ $—C(CH_3)_3$ $—Br$ $—SO_3H$ $—CH{=}O$

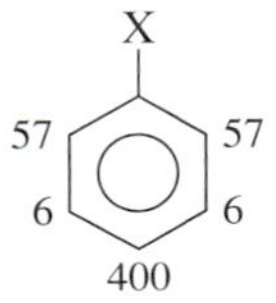

12.38 The partial rate factors for chlorination of biphenyl are as shown.

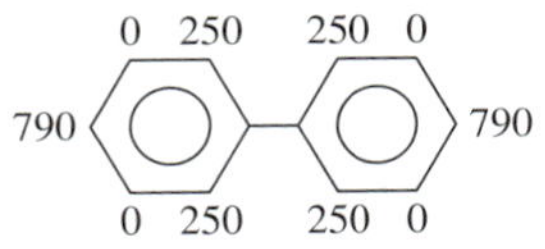

(a) What is the relative rate of chlorination of biphenyl compared with benzene?

(b) If, in a particular chlorination reaction, 10 g of *o*-chlorobiphenyl was formed, how much *p*-chlorobiphenyl would you expect to find?

12.39 Partial rate factors may be used to estimate product distributions in disubstituted benzene derivatives. The reactivity of a particular position in *o*-bromotoluene, for example, is given by the product of the partial rate factors for the corresponding position in toluene and bromobenzene. On the basis of the partial rate factor data given here for Friedel–Crafts acylation, predict the major product of the reaction of *o*-bromotoluene with acetyl chloride and aluminum chloride.

Partial rate factors for reaction of toluene and bromobenzene with $CH_3\overset{\mathrm{O}}{\overset{\|}{\mathrm{C}}}Cl$, $AlCl_3$

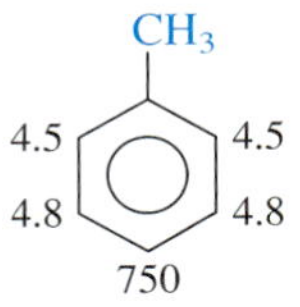

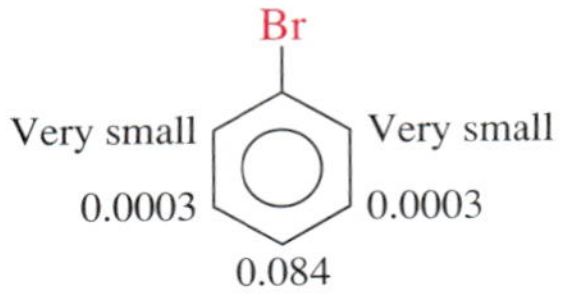

12.40 When 2-isopropyl-1,3,5-trimethylbenzene is heated with aluminum chloride (trace of HCl present) at 50°C, the major material present after 4 h is 1-isopropyl-2,4,5-trimethylbenzene. Suggest a reasonable mechanism for this isomerization.

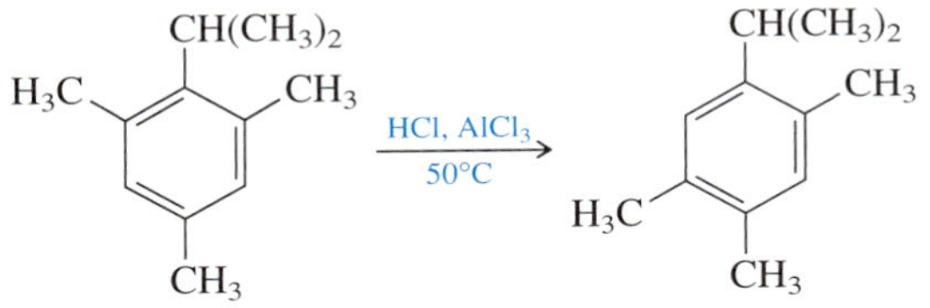

12.41 When a dilute solution of 6-phenylhexanoyl chloride in carbon disulfide was slowly added (over a period of eight days!) to a suspension of aluminum chloride in the same solvent, it yielded a product A ($C_{12}H_{14}O$) in 67% yield. Oxidation of A gave benzene-1,2-dicarboxylic acid.

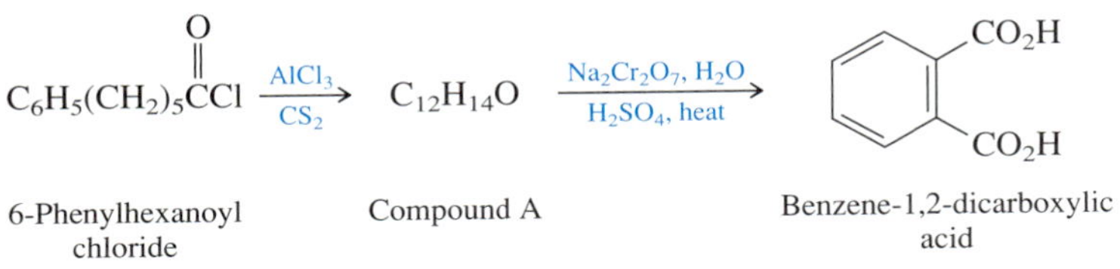

Formulate a reasonable structure for compound A.

12.42 Reaction of hexamethylbenzene with methyl chloride and aluminum chloride gave a salt A, which, on being treated with aqueous sodium bicarbonate solution, yielded compound B. Suggest a mechanism for the conversion of hexamethylbenzene to B by correctly inferring the structure of A.

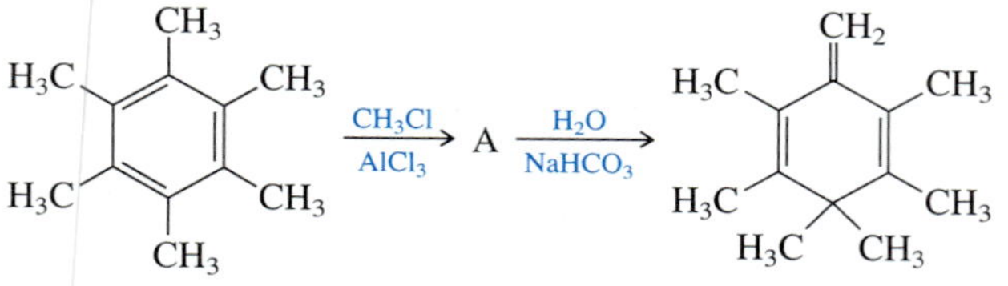

Hexamethylbenzene   Compound B

12.43 The synthesis of compound C was achieved by using compounds A and B as the source of all carbon atoms. Suggest a synthetic sequence involving no more than three steps by which A and B may be converted to C.

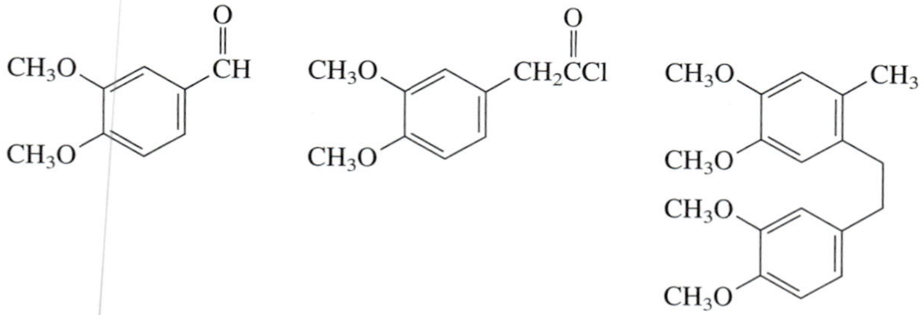

Compound A   Compound B   Compound C

12.44 When styrene is refluxed with aqueous sulfuric acid, two "styrene dimers" are formed as the major products. One of these styrene dimers is 1,3-diphenyl-1-butene; the other is 1-methyl-3-phenylindan. Suggest a reasonable mechanism for the formation of each of these compounds.

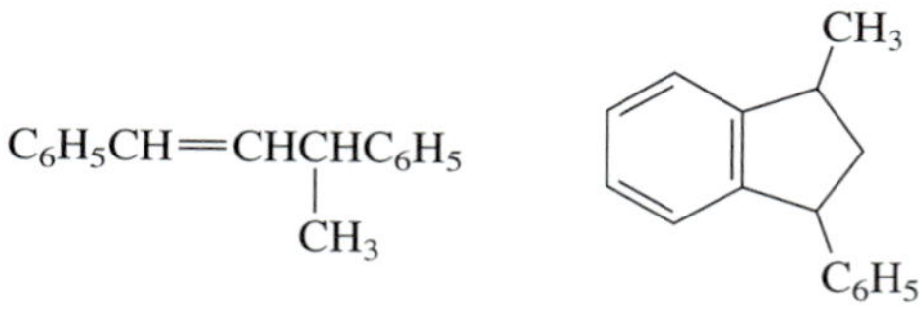

$C_6H_5CH{=}CHCHC_6H_5$
             |
            CH_3

1,3-Diphenyl-1-butene   1-Methyl-3-phenylindan

12.45 Treatment of the alcohol whose structure is shown here with sulfuric acid gave as the major organic product a tricyclic hydrocarbon of molecular formula $C_{16}H_{16}$. Suggest a reasonable structure for this hydrocarbon.

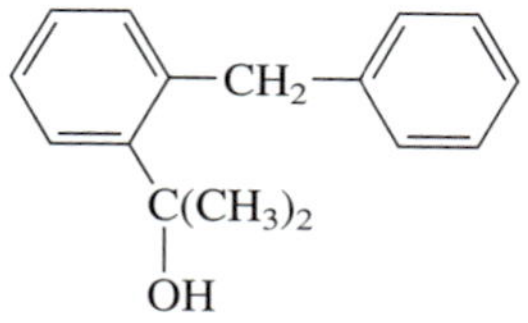

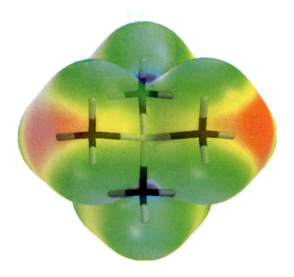

CHAPTER 13

SPECTROSCOPY

Until the second half of the twentieth century, the structure of a substance—a newly discovered natural product, for example—was determined using information obtained from chemical reactions. This information included the identification of functional groups by chemical tests, along with the results of experiments in which the substance was broken down into smaller, more readily identifiable fragments. Typical of this approach is the demonstration of the presence of a double bond in an alkene by catalytic hydrogenation and subsequent determination of its location by ozonolysis. After considering all the available chemical evidence, the chemist proposed a candidate structure (or structures) consistent with the observations. Proof of structure was provided either by converting the substance to some already known compound or by an independent synthesis.

Qualitative tests and chemical degradation have been to a large degree replaced by instrumental methods of structure determination. The most prominent methods and the structural clues they provide are:

- **Nuclear magnetic resonance (NMR) spectroscopy,** which tells us about the carbon skeleton and the environments of the hydrogens attached to it.
- **Infrared (IR) spectroscopy,** which reveals the presence or absence of key functional groups.
- **Ultraviolet-visible (UV-VIS) spectroscopy,** which probes the electron distribution, especially in molecules that have conjugated π electron systems.
- **Mass spectrometry (MS),** which gives the molecular weight and formula, both of the molecule itself and various structural units within it.

As diverse as these techniques are, all of them are based on the absorption of energy by a molecule, and all measure how a molecule responds to that absorption. In describing these techniques our emphasis will be on their application to structure determination. We'll start with a brief discussion of electromagnetic radiation, which is the source of the energy that a molecule absorbs in NMR, IR, and UV-VIS spectroscopy.

13.1   PRINCIPLES OF MOLECULAR SPECTROSCOPY: ELECTROMAGNETIC RADIATION

Electromagnetic radiation, of which visible light is but one example, has the properties of both particles and waves. The particles are called **photons,** and each possesses an amount of energy referred to as a **quantum.** In 1900, the German physicist Max Planck proposed that the energy of a photon (E) is directly proportional to its frequency (ν).

$$E = h\nu$$

"Modern" physics dates from Planck's proposal that energy is quantized, which set the stage for the development of quantum mechanics. Planck received the 1918 Nobel Prize in physics.

The SI units of frequency are reciprocal seconds (s^{-1}), given the name *hertz* and the symbol Hz in honor of the nineteenth-century physicist Heinrich R. Hertz. The constant of proportionality h is called **Planck's constant** and has the value

$$h = 6.63 \times 10^{-34} \text{ J} \cdot \text{s}$$

Electromagnetic radiation travels at the speed of light ($c = 3.0 \times 10^8$ m/s), which is equal to the product of its frequency ν and its wavelength λ:

$$c = \nu\lambda$$

The range of photon energies is called the *electromagnetic spectrum* and is shown in Figure 13.1. Visible light occupies a very small region of the electromagnetic spectrum. It is characterized by wavelengths of 4×10^{-7} m (violet) to 8×10^{-7} m (red). When examining Figure 13.1 be sure to keep the following two relationships in mind:

1. *Frequency is inversely proportional to wavelength;* the greater the frequency, the shorter the wavelength.

2. *Energy is directly proportional to frequency;* electromagnetic radiation of higher frequency possesses more energy than radiation of lower frequency.

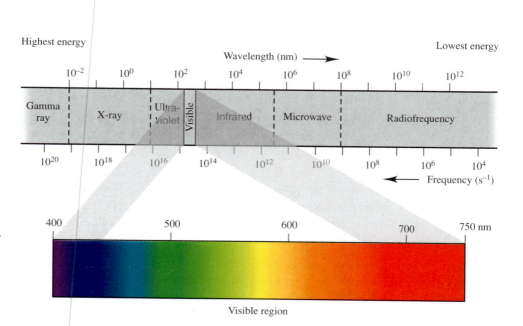

FIGURE 13.1 The electromagnetic spectrum. (Reprinted, with permission, from M. Silberberg, *Chemistry,* 2nd ed., WCB/McGraw-Hill, 2000, p. 260.)

Depending on its source, a photon can have a vast amount of energy; gamma rays and X-rays are streams of very high energy photons. Radio waves are of relatively low energy. Ultraviolet radiation is of higher energy than the violet end of visible light. Infrared radiation is of lower energy than the red end of visible light. When a molecule is exposed to electromagnetic radiation, it may absorb a photon, increasing its energy by an amount equal to the energy of the photon. Molecules are highly selective with respect to the frequencies they absorb. Only photons of certain specific frequencies are absorbed by a molecule. The particular photon energies absorbed by a molecule depend on molecular structure and can be measured with instruments called **spectrometers.** The data obtained are very sensitive indicators of molecular structure and have revolutionized the practice of chemical analysis.

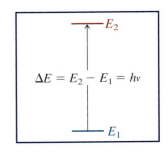

FIGURE 13.2 Two energy states of a molecule. Absorption of energy equal to $E_2 - E_1$ excites a molecule from its lower energy state to the next higher state.

13.2 PRINCIPLES OF MOLECULAR SPECTROSCOPY: QUANTIZED ENERGY STATES

What determines whether a photon is absorbed by a molecule? The most important requirement is that the energy of the photon must equal the energy difference between two states, such as two nuclear spin states, two vibrational states, or two electronic states. In physics, the term for this is *resonance*—the transfer of energy between two objects that occurs when their frequencies are matched. In molecular spectroscopy, we are concerned with the transfer of energy from a photon to a molecule, but the idea is the same. Consider, for example, two energy states of a molecule designated E_1 and E_2 in Figure 13.2. The energy difference between them is $E_2 - E_1$, or ΔE. In nuclear magnetic resonance (NMR) spectroscopy these are two different spin states of an atomic nucleus; in infrared (IR) spectroscopy, they are two different vibrational energy states; in ultraviolet-visible (UV-VIS) spectroscopy, they are two different electronic energy states. Unlike kinetic energy, which is continuous, meaning that all values of kinetic energy are available to a molecule, only certain energies are possible for electronic, vibrational, and nuclear spin states. These energy states are said to be **quantized.** More of the molecules exist in the lower energy state E_1 than in the higher energy state E_2. Excitation of a molecule from a lower state to a higher one requires the addition of an increment of energy equal to ΔE. Thus, when electromagnetic radiation is incident upon a molecule, only the frequency whose corresponding energy equals ΔE is absorbed. All other frequencies are transmitted.

Spectrometers are designed to measure the absorption of electromagnetic radiation by a sample. Basically, a spectrometer consists of a source of radiation, a compartment containing the sample through which the radiation passes, and a detector. The frequency of radiation is continuously varied, and its intensity at the detector is compared with that at the source. When the frequency is reached at which the sample absorbs radiation, the detector senses a decrease in intensity. The relation between frequency and absorption is plotted as a **spectrum,** which consists of a series of peaks at characteristic frequencies. Its interpretation can furnish structural information. Each type of spectroscopy developed independently of the others, and so the data format is different for each one. An NMR spectrum looks different from an IR spectrum, and both look different from a UV-VIS spectrum.

With this as background, we will now discuss spectroscopic techniques individually. NMR, IR, and UV-VIS spectroscopy provide complementary information, and all are useful. Among them, NMR provides the information that is most directly related to molecular structure and is the one we'll examine first.

13.3   INTRODUCTION TO ^{1}H NMR SPECTROSCOPY

Nuclear magnetic resonance spectroscopy depends on the absorption of energy when the nucleus of an atom is excited from its lowest energy spin state to the next higher one. We should first point out that many elements are difficult to study by NMR, and some can't be studied at all. Fortunately though, the two elements that are the most common in organic molecules (carbon and hydrogen) have isotopes (^{1}H and ^{13}C) capable of giving NMR spectra that are rich in structural information. A proton nuclear magnetic resonance (^{1}H NMR) spectrum tells us about the environments of the various hydrogens in a molecule; a carbon-13 nuclear magnetic resonance (^{13}C NMR) spectrum does the same for the carbon atoms. Separately and together ^{1}H and ^{13}C NMR take us a long way toward determining a substance's molecular structure. We'll develop most of the general principles of NMR by discussing ^{1}H NMR, then extend them to ^{13}C NMR. The ^{13}C NMR discussion is shorter, not because it is less important than ^{1}H NMR, but because many of the same principles apply to both techniques.

Like an electron, a proton has two spin states with quantum numbers of $+\frac{1}{2}$ and $-\frac{1}{2}$. There is no difference in energy between these two nuclear spin states; a proton is just as likely to have a spin of $+\frac{1}{2}$ as $-\frac{1}{2}$. Absorption of electromagnetic radiation can only occur when the two spin states have different energies. A way to make them different is to place the sample in a magnetic field. A spinning proton behaves like a tiny bar magnet and has a magnetic moment associated with it (Figure 13.3). In the presence of an external magnetic field $\mathcal{H}_0$, the spin state in which the magnetic moment of the nucleus is aligned with $\mathcal{H}_0$ is lower in energy than the one in which it opposes $\mathcal{H}_0$.

As shown in Figure 13.4, the energy difference between the two states is directly proportional to the strength of the applied field. Net absorption of electromagnetic radiation requires that the lower state be more highly populated than the higher one, and quite strong magnetic fields are required to achieve the separation necessary to give a detectable signal. A magnetic field of 4.7 T, which is about 100,000 times stronger than

Nuclear magnetic resonance of protons was first detected in 1946 by Edward Purcell (Harvard) and by Felix Bloch (Stanford). Purcell and Bloch shared the 1952 Nobel Prize in physics.

The SI unit for magnetic field strength is the tesla (T), named after Nikola Tesla, a contemporary of Thomas Edison and who, like Edison, was an inventor of electrical devices.

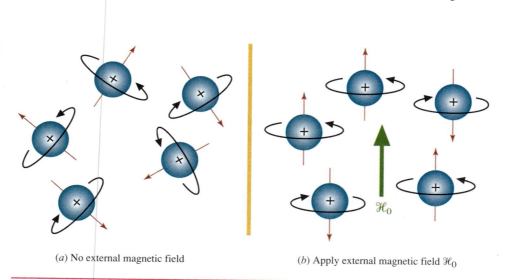

(a) No external magnetic field   (b) Apply external magnetic field $\mathcal{H}_0$

FIGURE 13.3 (a) In the absence of an external magnetic field, the nuclear spins of the protons are randomly oriented. (b) In the presence of an external magnetic field $\mathcal{H}_0$, the nuclear spins are oriented so that the resulting nuclear magnetic moments are aligned either parallel or antiparallel to $\mathcal{H}_0$. The lower energy orientation is the one parallel to $\mathcal{H}_0$, and more nuclei have this orientation.

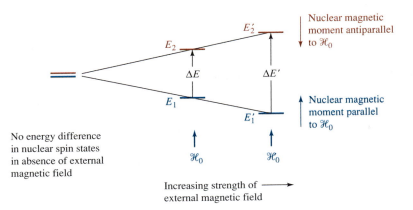

FIGURE 13.4 An external magnetic field causes the two nuclear spin states to have different energies. The difference in energy ΔE is proportional to the strength of the applied field.

earth's magnetic field, for example, separates the two spin states of ^1H by only 8×10^{-5} kJ/mol (1.9×10^{-5} kcal/mol). From Planck's equation $\Delta E = h\nu$, this energy gap corresponds to radiation having a frequency of 2×10^8 Hz (200 MHz), which lies in the radiofrequency (rf) region of the electromagnetic spectrum (see Figure 13.1).

Frequency of electromagnetic radiation (s^{-1} or Hz)	is proportional to	Energy difference between nuclear spin states (kJ/mol or kcal/mol)	is proportional to	Magnetic field (T)

PROBLEM 13.1 Most of the NMR spectra in this text were recorded on a spectrometer having a field strength of 4.7 T (200 MHz for ^1H). The first generation of widely used NMR spectrometers were 60-MHz instruments. What was the magnetic field strength of these earlier spectrometers?

The response of an atom to the strength of the external magnetic field is different for different elements, and for different isotopes of the same element. The resonance frequencies of most nuclei are sufficiently different that an NMR experiment is sensitive only to a particular isotope of a single element. The frequency for ^1H is 200 MHz at 4.7 T, but that of ^{13}C is 50.4 MHz. Thus, when recording the NMR spectrum of an organic compound, we see signals only for ^1H or ^{13}C, but not both; ^1H and ^{13}C NMR spectra are recorded in separate experiments with different instrument settings.

PROBLEM 13.2 What will be the ^{13}C frequency of an NMR spectrometer that operates at 100 MHz for protons?

The essential features of an NMR spectrometer, shown in Figure 13.5, are not hard to understand. They consist of a magnet to align the nuclear spins, a radiofrequency (rf) transmitter as a source of energy to excite a nucleus from its lowest energy state to the next higher one, a receiver to detect the absorption of rf radiation, and a recorder to print out the spectrum.

It turns out though that there are several possible variations on this general theme. We could, for example, keep the magnetic field constant and continuously vary the radiofrequency until it matched the energy difference between the nuclear spin states. Or, we could keep the rf constant and adjust the energy levels by varying the magnetic

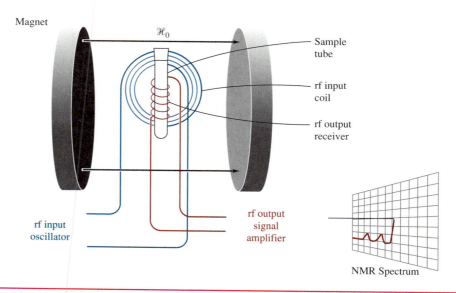

Magnet

$\mathcal{H}_0$

Sample tube

rf input coil

rf output receiver

rf input oscillator

rf output signal amplifier

NMR Spectrum

FIGURE 13.5 Diagram of a nuclear magnetic resonance spectrometer. (Reprinted, with permission, from S. H. Pine, J. B. Hendrickson, D. J. Cram, and G. S. Hammond, *Organic Chemistry*, 4th ed., McGraw-Hill, New York, 1980, p. 136.)

field strength. Both methods work, and the instruments based on them are called *continuous wave* (CW) spectrometers. Many of the terms we use in NMR spectroscopy have their origin in the way CW instruments operate, but CW instruments are rarely used anymore.

CW-NMR spectrometers have been replaced by *pulsed Fourier-transform* nuclear magnetic resonance (FT-NMR) spectrometers. FT-NMR spectrometers are far more versatile than CW instruments and are more complicated. Most of the visible differences between them lie in computerized data acquisition and analysis components that are fundamental to FT-NMR spectroscopy. But there is an important difference in how a pulsed FT-NMR experiment is carried out as well. Rather than sweeping through a range of frequencies (or magnetic field strengths), the sample is placed in a magnetic field and irradiated with a short, intense burst of radiofrequency radiation (the *pulse*) that excites all of the protons in the molecule. The magnetic field associated with the new orientation of nuclear spins induces an electrical signal in the receiver that decreases with time as the nuclei return to their original orientation. The resulting *free-induction decay* (FID) is a composite of the decay patterns of all of the protons in the molecule. The FID pattern is stored in a computer and converted into a spectrum by a mathematical process known as a *Fourier transform*. The pulse-relaxation sequence takes only about a second, but usually gives signals too weak to distinguish from background noise. The signal-to-noise ratio is enhanced by repeating the sequence many times, then averaging the data. Noise is random and averaging causes it to vanish; signals always appear at the same place and accumulate. All of the operations—the interval between pulses, collecting, storing, and averaging the data and converting it to a spectrum by a Fourier transform—are under computer control, which makes the actual taking of an FT-NMR spectrum a fairly routine operation.

Not only is pulsed FT-NMR the best method for obtaining proton spectra, it is the only practical method for many other nuclei, including ^{13}C. It also makes possible a large number of sophisticated techniques that have revolutionized NMR spectroscopy.

Richard R. Ernst of the Swiss Federal Institute of Technology won the 1991 Nobel Prize in chemistry for devising pulse-relaxation NMR techniques.

13.4   NUCLEAR SHIELDING AND ^{1}H CHEMICAL SHIFTS

Our discussion so far has concerned ^{1}H nuclei in general without regard for the environments of individual protons in a molecule. Protons in a molecule are connected to other atoms—carbon, oxygen, nitrogen, and so on—by covalent bonds. The electrons in these bonds, indeed all the electrons in a molecule, affect the magnetic environment of the protons. Alone, a proton would feel the full strength of the external field, but a proton in an organic molecule responds to both the external field plus any local fields within the molecule. An external magnetic field affects the motion of the electrons in a molecule, inducing local fields characterized by lines of force that circulate in the *opposite* direction from the applied field (Figure 13.6). Thus, the net field felt by a proton in a molecule will always be less than the applied field, and the proton is said to be **shielded.** All of the protons of a molecule are shielded from the applied field by the electrons, but some are less shielded than others. Sometimes the term *deshielded* is used to describe this decreased shielding of one proton relative to another.

The more shielded a proton is, the greater must be the strength of the applied field in order to achieve resonance and produce a signal. A more shielded proton absorbs rf radiation at higher field strength (**upfield**) compared with one at lower field strength (**downfield**). Different protons give signals at different field strengths. *The dependence of the resonance position of a nucleus that results from its molecular environment is called its* **chemical shift.** This is where the real power of NMR lies. The chemical shifts of various protons in a molecule can be different and are characteristic of particular structural features.

Figure 13.7 shows the ^{1}H NMR spectrum of chloroform ($CHCl_3$) to illustrate how the terminology just developed applies to a real spectrum.

Instead of measuring chemical shifts in absolute terms, we measure them with respect to a standard—*tetramethylsilane* $(CH_3)_4Si$, abbreviated *TMS*. The protons of TMS are more shielded than those of most organic compounds, so all of the signals in

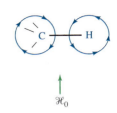

FIGURE 13.6 The induced magnetic field of the electrons in the carbon–hydrogen bond opposes the external magnetic field. The resulting magnetic field experienced by the proton and the carbon is slightly less than $\mathscr{H}_0$.

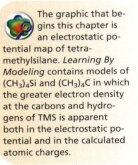

The graphic that begins this chapter is an electrostatic potential map of tetramethylsilane. *Learning By Modeling* contains models of $(CH_3)_4Si$ and $(CH_3)_4C$ in which the greater electron density at the carbons and hydrogens of TMS is apparent both in the electrostatic potential and in the calculated atomic charges.

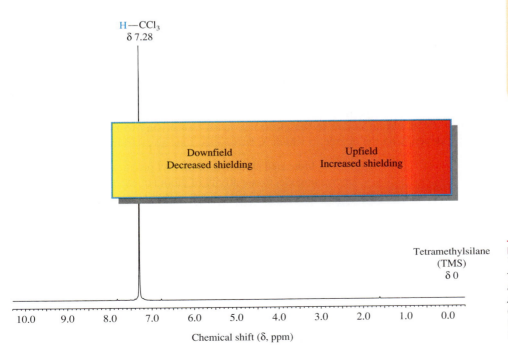

H—CCl$_3$
δ 7.28

Downfield
Decreased shielding

Upfield
Increased shielding

Tetramethylsilane
(TMS)
δ 0

| | | | | | | | | | | | |
|10.0|9.0|8.0|7.0|6.0|5.0|4.0|3.0|2.0|1.0|0.0|

Chemical shift (δ, ppm)

FIGURE 13.7 The 200-MHz ^{1}H NMR spectrum of chloroform (HCCl$_3$). Chemical shifts are measured along the x-axis in parts per million (ppm) from tetramethylsilane as the reference, which is assigned a value of zero.

a sample ordinarily appear at lower field than those of the TMS reference. When measured using a 100-MHz instrument, the signal for the proton in chloroform (CHCl$_3$), for example, appears 728 Hz downfield from the TMS signal. But because frequency is proportional to magnetic field strength, the same signal would appear 1456 Hz downfield from TMS on a 200-MHz instrument. We simplify the reporting of chemical shifts by converting them to parts per million (ppm) downfield from TMS, which is assigned a value of 0. The TMS need not actually be present in the sample, nor even appear in the spectrum in order to serve as a reference. When chemical shifts are reported this way, they are identified by the symbol δ and are independent of the field strength.

$$\text{Chemical shift } (\delta) = \frac{\text{position of signal} - \text{position of TMS peak}}{\text{spectrometer frequency}} \times 10^6$$

Thus, the chemical shift for the proton in chloroform is:

$$\delta = \frac{1456 \text{ Hz} - 0 \text{ Hz}}{200 \times 10^6 \text{ Hz}} \times 10^6 = 7.28$$

> **PROBLEM 13.3**   The ^{1}H NMR signal for bromoform (CHBr$_3$) appears at 2065 Hz when recorded on a 300-MHz NMR spectrometer. (a) What is the chemical shift of this proton? (b) Is the proton in CHBr$_3$ more shielded or less shielded than the proton in CHCl$_3$?

NMR spectra are usually run in solution and, although chloroform is a good solvent for most organic compounds, it's rarely used because its own signal at δ 7.28 would be so intense that it would obscure signals in the sample. Because the magnetic properties of deuterium (D = ^{2}H) are different from those of ^{1}H, CDCl$_3$ gives no signals at all in an ^{1}H NMR spectrum and is used instead. Indeed, CDCl$_3$ is the most commonly used solvent in ^{1}H NMR spectroscopy. Likewise, D$_2$O is used instead of H$_2$O for water-soluble substances such as carbohydrates.

13.5   EFFECTS OF MOLECULAR STRUCTURE ON ^{1}H CHEMICAL SHIFTS

Problem 13.3 in the preceding section was based on the chemical shift difference between the proton in CHCl$_3$ and the proton in CHBr$_3$ and its relation to shielding.

Nuclear magnetic resonance spectroscopy is such a powerful tool for structure determination because *protons in different environments experience different degrees of shielding and have different chemical shifts.* In compounds of the type CH$_3$X, for example, the shielding of the methyl protons increases as X becomes less electronegative. Inasmuch as the shielding is due to the electrons, it isn't surprising to find that the chemical shift depends on the degree to which X draws electrons away from the methyl group.

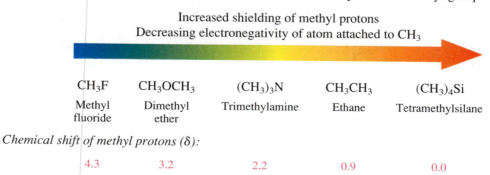

Increased shielding of methyl protons
Decreasing electronegativity of atom attached to CH$_3$

CH$_3$F	CH$_3$OCH$_3$	(CH$_3$)$_3$N	CH$_3$CH$_3$	(CH$_3$)$_4$Si
Methyl fluoride	Dimethyl ether	Trimethylamine	Ethane	Tetramethylsilane

Chemical shift of methyl protons (δ):

4.3	3.2	2.2	0.9	0.0

A similar trend is seen in the methyl halides, in which the protons in CH_3F are the least shielded (δ 4.3) and those of CH_3I (δ 2.2) are the most.

The decreased shielding caused by electronegative substituents is primarily an inductive effect, and like other inductive effects falls off rapidly as the number of bonds between the substituent and the proton increases. Compare the chemical shifts of the protons in propane and 1-nitropropane.

$$\delta\ 0.9 \quad \delta\ 1.3 \quad \delta\ 0.9 \qquad\qquad \delta\ 4.3 \quad \delta\ 2.0 \quad \delta\ 1.0$$

$$H_3C\!-\!CH_2\!-\!CH_3 \qquad\qquad O_2N\!-\!CH_2\!-\!CH_2\!-\!CH_3$$

$$\text{Propane} \qquad\qquad\qquad \text{1-Nitropropane}$$

The strongly electron-withdrawing nitro group deshields the protons on C-1 by 3.4 ppm (δ 4.3–0.9). The effect is smaller on the protons at C-2 (0.7 ppm), and almost completely absent at C-3.

The deshielding effects of electronegative substituents are cumulative, as the chemical shifts for various chlorinated derivatives of methane indicate.

$$CH_3Cl \qquad\qquad CH_2Cl_2 \qquad\qquad CHCl_3$$

$$\text{Methyl chloride} \quad\ \text{Methylene chloride} \quad\ \text{Chloroform}$$

Chemical shift (δ):   3.1   5.3   7.3

PROBLEM 13.4 Identify the most shielded and least shielded protons in

(a) 2-Bromobutane

(b) 1,1,2-Trichloropropane

(c) Tetrahydrofuran:

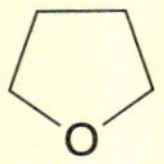

SAMPLE SOLUTION (a) Bromine is electronegative and will have its greatest electron-withdrawing effect on protons that are separated from it by the fewest bonds. Therefore, the proton at C-2 will be the least shielded, and those at C-4 the most shielded.

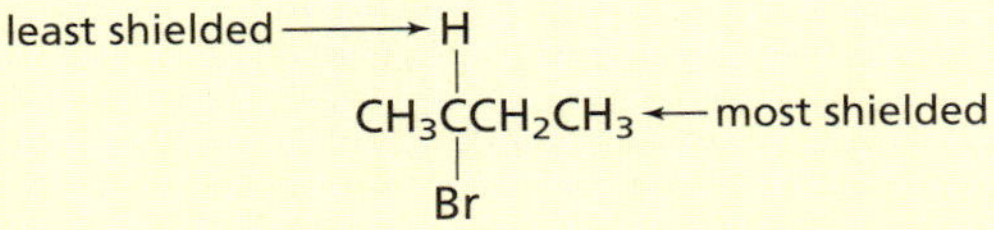

$$\text{least shielded} \longrightarrow \underset{\underset{Br}{|}}{\overset{\overset{H}{|}}{CH_3CCH_2CH_3}} \longleftarrow \text{most shielded}$$

The observed chemical shifts are δ 4.1 for the proton at C-2 and δ 1.1 for the protons at C-4. The protons at C-1 and C-3 appear in the range δ 1.7–2.0.

Table 13.1 collects chemical-shift information for protons of various types. The beginning and major portion of the table concerns protons bonded to carbon. Within each type, methyl (CH_3) protons are more shielded than methylene (CH_2) protons, and methylene protons are more shielded than methine (CH) protons. These differences are small as the following two examples illustrate.

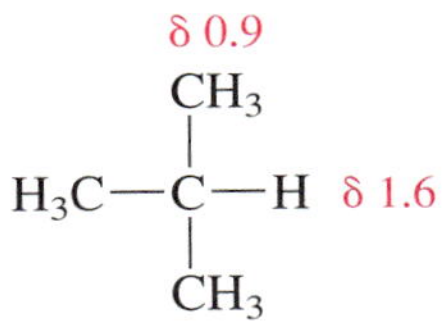

$$\delta\ 0.9 \qquad\qquad\qquad \delta\ 0.9$$

$$\underset{\underset{CH_3}{|}}{\overset{\overset{CH_3}{|}}{H_3C\!-\!C\!-\!H}}\ \delta\ 1.6 \qquad \underset{\underset{CH_3}{|}}{\overset{\overset{CH_3}{|}}{H_3C\!-\!C\!-\!CH_2\!-\!CH_3}}\ \ \delta\ 1.2\quad\delta\ 0.8$$

$$\text{2-Methylpropane} \qquad\qquad \text{2,2-Dimethylbutane}$$

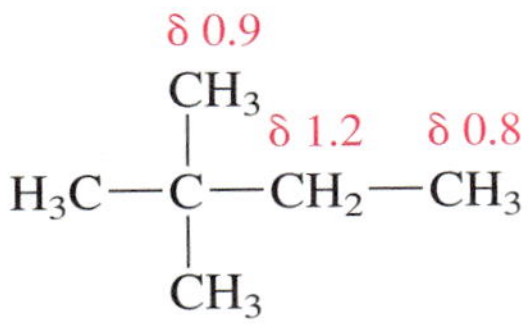

TABLE 13.1 Approximate Chemical Shifts of Representative Protons

Compound class or type of proton		Chemical shift (δ), ppm*
Protons bonded to carbon		
Alkane	RCH_3, R_2CH_2, R_3CH	0.9–1.8
Allylic	$H-C$ $C=C$	1.5–2.6
C—H adjacent to C=O	$H-C$ $C=O$	2.0–2.5
C—H adjacent to C≡N	$H-C-C{\equiv}N$	2.1–2.3
Alkyne	$H-C{\equiv}C-R$	2.5
Benzylic	$H-C-Ar$	2.3–2.8
Amine	$H-C-NR_2$	2.2–2.9
Alkyl chloride	$H-C-Cl$	3.1–4.1
Alkyl bromide	$H-C-Br$	2.7–4.1
Alcohol or ether	$H-C-O$	3.3–3.7
Vinylic	$\overset{H}{\underset{}{C}}=C$	4.5–6.5
Aryl	$H-Ar$	6.5–8.5
Aldehyde	$\overset{R}{\underset{H}{C}}=O$	9–10
Protons bonded to nitrogen or oxygen		
Amine	$H-NR_2$	1–3[†]
Alcohol	$H-OR$	0.5–5[†]
Phenol	$H-OAr$	6–8[†]
Carboxylic acid	$H-O\overset{O}{\overset{\|}{C}}R$	10–13[†]

*Approximate values relative to tetramethylsilane; other groups within the molecule can cause a proton signal to appear outside of the range cited.
[†]The chemical shifts of O—H and N—H protons are temperature- and concentration-dependent.

With the additional information that the chemical shift of methane is δ 0.2, we can attribute the decreased shielding of the protons of RCH_3, R_2CH_2, and R_3CH to the number of carbons attached to primary, secondary, and tertiary carbons. Carbon is more electronegative than hydrogen, so replacing the hydrogens of CH_4 by one, then two, then three carbons decreases the shielding of the remaining protons.

Likewise, the generalization that sp^2-hybridized carbon is more electronegative than sp^3-hybridized carbon is consistent with the decreased shielding of allylic and benzylic protons.

δ 1.5 — H_3C δ 0.8 — CH_3

1,5-Dimethylcyclohexene

δ 1.2   δ 2.6
H_3C—CH_2—

Ethylbenzene

Hydrogens that are directly attached to double bonds (vinylic protons) or to aromatic rings (aryl protons) are especially deshielded.

δ 5.3

Ethylene

δ 7.3

Benzene

The main reason for the deshielded nature of vinylic and aryl protons is related to the induced magnetic fields associated with π electrons. We saw earlier in Section 13.4 that the local field resulting from electrons in a C—H σ bond opposes the applied field and shields a molecule's protons. The hydrogens of ethylene and benzene, however, lie in a region of the molecule where the induced magnetic field of the π electrons reinforces the applied field, *deshielding the protons* (Figure 13.8). In the case of benzene, this is described as a **ring current** effect that originates in the circulating π electrons. It has interesting consequences, some of which are described in the boxed essay *Ring Currents—Aromatic and Antiaromatic.*

> **PROBLEM 13.5** Assign the chemical shifts δ 1.6, δ 2.2, and δ 4.8 to the appropriate protons of methylenecyclopentane
>
> ⬡=CH_2

Acetylenic hydrogens are unusual in that they are more shielded than we would expect for protons bonded to *sp*-hybridized carbon. This is because the π electrons circulate around the triple bond, not along it (Figure 13.9a). Therefore, the induced magnetic field is parallel to the long axis of the triple bond and shields the acetylenic proton (Figure 13.9b). Acetylenic protons typically have chemical shifts near δ 2.5.

δ 2.4   δ 4.0   δ 3.3
H—$C≡C$—CH_2OCH_3

Methyl 2-propynyl ether

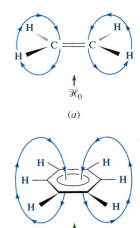

FIGURE 13.8 The induced magnetic field of the π electrons of (a) an alkene and (b) an arene reinforces the applied field in the regions where vinyl and aryl protons are located.

RING CURRENTS—AROMATIC AND ANTIAROMATIC

We saw in Chapter 12 that aromaticity reveals itself in various ways. Qualitatively, aromatic compounds are more stable and less reactive than alkenes. Quantitatively, their heats of hydrogenation are smaller than expected. Theory, especially Hückel's rule, furnishes a structural basis for aromaticity. Now let's examine some novel features of their NMR spectra.

We mentioned in Section 13.5 that the protons in benzene appear at relatively low field because of the deshielding effect of the magnetic field associated with the circulating π electrons. The amount of deshielding is sufficiently large—on the order of 2 ppm more than the corresponding effect in alkenes—that its presence is generally accepted as evidence for aromaticity. We speak of the deshielding as resulting from an *aromatic ring current*.

Something interesting happens when we go beyond benzene to apply the aromatic ring current test to annulenes.

[18]Annulene satisfies the Hückel $(4n+2)$ π electron rule for aromaticity, and many of its properties indicate aromaticity (Section 11.20). As shown in Figure 13.10a, [18]annulene contains two different kinds of protons; 12 lie on the ring's periphery ("outside"), and 6 reside near the middle of the molecule ("inside"). The 2:1 ratio of outside/inside protons makes it easy to assign the signals in the ^1H NMR spectrum. The outside protons have a chemical shift δ of 9.3 ppm, which makes them even less shielded than those of benzene. The six inside protons, on the other hand, have a *negative* chemical shift (δ -3.0), meaning that the signal for these protons appears at *higher field* (to the right) of the TMS peak. *The inside protons of [18]annulene are more than 12 ppm more shielded than the outside protons.*

As shown in Figure 13.10a, both the shielding of the inside protons and the deshielding of the outside ones result from the same aromatic ring current. When the molecule is placed in an external magnetic field $\mathcal{H}_0$, its circulating π electrons produce their own magnetic field. This induced field opposes the applied field $\mathcal{H}_0$ in the center of the molecule, shielding the inside protons. Because the induced magnetic field closes on itself, the outside protons lie in a region where the induced field reinforces $\mathcal{H}_0$. The aromatic ring current in [18]annulene shields the 6 inside protons and deshields the 12 outside ones.

Exactly the opposite happens in [16]annulene (Figure 13.10b). Now it is the outside protons (δ 5.3) that are more shielded. The inside protons (δ 10.6) are less shielded than the outside ones and less shielded than the protons of both benzene and [18]annulene. This reversal of the shielding and deshielding regions in going from [18] to [16]annulene can only mean that the directions of their induced magnetic fields are reversed. Thus [16]annulene, which is antiaromatic, not only lacks an aromatic ring current, its π electrons produce exactly the opposite effect when placed in a magnetic field.

Score one for Hückel.

—*Cont.*

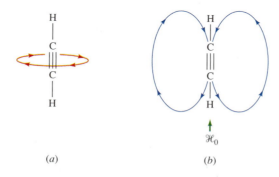

FIGURE 13.9 (a) The π electrons of acetylene circulate in a region surrounding the long axis of the molecule. (b) The induced magnetic field associated with the π electrons opposes the applied field and shields the protons.

The induced field of a carbonyl group (C=O) deshields protons in much the same way that a carbon–carbon double bond does, and the presence of oxygen makes it even more electron withdrawing. Thus, protons attached to C=O in aldehydes are the least shielded of any protons bonded to carbon. They have chemical shifts in the range δ 9–10.

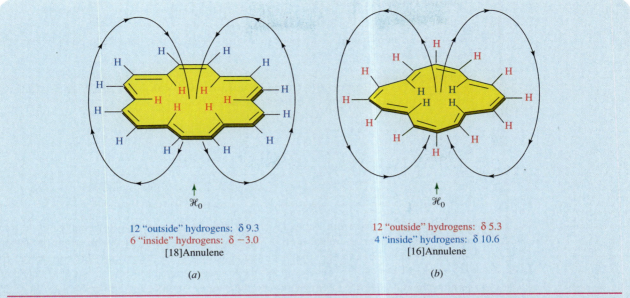

12 "outside" hydrogens: δ 9.3
6 "inside" hydrogens: δ −3.0
[18]Annulene

(a)

12 "outside" hydrogens: δ 5.3
4 "inside" hydrogens: δ 10.6
[16]Annulene

(b)

FIGURE 13.10 More shielded (red) and less shielded (blue) protons in (a) [18]annulene and (b) [16]annulene. The induced magnetic field associated with the aromatic ring current in [18]annulene shields the inside protons and deshields the outside protons. The opposite occurs in [16]annulene, which is antiaromatic.

δ 2.4

δ 1.1   H   O   δ 9.7

H₃C—C—C—H

CH₃

2-Methylpropanal

δ 6.9   δ 7.8

δ 1.4   H   H   O   δ 9.9

CH₃CH₂O—⟨benzene ring⟩—C—H

δ 4.1   H   H

p-Ethoxybenzaldehyde

Protons on carbons adjacent to a carbonyl group are deshielded slightly more than allylic hydrogens.

> **PROBLEM 13.6** Assign the chemical shifts δ 1.1, δ 1.7, δ 2.0, and δ 2.3 to the appropriate protons of 2-pentanone.
>
> O
> ‖
> CH₃CCH₂CH₂CH₃

The second portion of Table 13.1 deals with O—H and N—H protons. As the table indicates, the chemical shifts of these vary much more than for protons bonded to carbon. This is because O—H and N—H groups can be involved in intermolecular hydrogen bonding, the extent of which depends on molecular structure, temperature, concentration, and solvent. Generally, an increase in hydrogen bonding decreases the shielding. This is especially evident in carboxylic acids. With δ values in the 10–12 ppm range, O—H protons of carboxylic acids are the least shielded of all of the protons in Table 13.1. We'll discuss hydrogen bonding in carboxylic acids in more detail in Chapter 19, but point out here that it is stronger than in most other classes of compounds that contain O—H groups.

PROBLEM 13.7 Assign the chemical shifts δ 1.6, δ 4.0, δ 7.5, δ 8.2, and δ 12.0 to the appropriate protons of 2-(*p*-nitrophenyl)propanoic acid.

As you can see from Table 13.1, it is common for several different kinds of protons to have similar chemical shifts. The range covered for ^1H chemical shifts is only 12 ppm, which is relatively small compared with (as we'll see) the 200-ppm range for ^{13}C chemical shifts. The ability of an NMR spectrometer to separate signals that have similar chemical shifts is termed its *resolving power* and is directly related to the magnetic field strength of the instrument. Even though the δ values of their chemical shifts don't change, two signals that are closely spaced at 60 MHz become well separated at 300 MHz.

13.6 INTERPRETING ^1H NMR SPECTRA

Analyzing an NMR spectrum in terms of a unique molecular structure begins with the information contained in Table 13.1. By knowing the chemical shifts characteristic of various proton environments, the presence of a particular structural unit in an unknown compound may be inferred. An NMR spectrum also provides other useful information, including:

1. *The number of signals,* which tells us how many different kinds of protons there are.

2. *The intensity of the signals* as measured by the area under each peak, which tells us the relative ratios of the different kinds of protons.

3. *The multiplicity, or splitting, of each signal,* which tells us how many protons are vicinal to the one giving the signal.

Protons that have different chemical shifts are said to be **chemical-shift-nonequivalent** (or **chemically nonequivalent**). A separate NMR signal is given for each chemical-shift-nonequivalent proton in a substance. Figure 13.11 shows the 200-MHz ^1H NMR spectrum of methoxyacetonitrile (CH_3OCH_2CN), a molecule with protons in two different environments. The three protons in the CH_3O group constitute one set, the two protons in the OCH_2CN group the other. These two sets of protons give rise to the two peaks that we see in the NMR spectrum and can be assigned on the basis of their chemical shifts. The protons in the OCH_2CN group are connected to a carbon that bears two electronegative substituents (O and C≡N) and are less shielded than those of the CH_3O group, which are attached to a carbon that bears only one electronegative atom (O). The signal for the protons in the OCH_2CN group appears at δ 4.1; the signal corresponding to the CH_3O protons is at δ 3.3.

Another way to assign the peaks is by comparing their intensities. The three equivalent protons of the CH_3O group give rise to a more intense peak than the two equivalent protons of the OCH_2CN group. This is clear by simply comparing the heights of the peaks in the spectrum. It is better, though, to compare peak areas by a process called

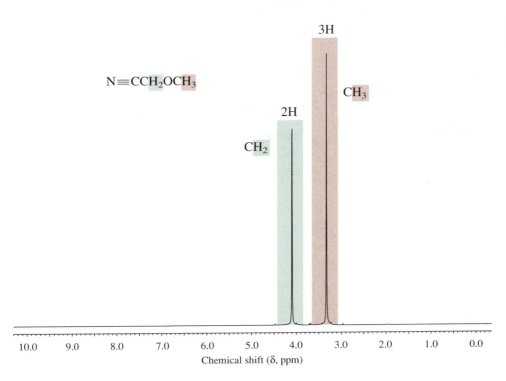

FIGURE 13.11 The 200-MHz ^{1}H NMR spectrum of methoxyacetonitrile (CH_3OCH_2CN).

integration. This is done electronically at the time the NMR spectrum is recorded, and the integrated areas are displayed on the computer screen or printed out. Peak areas are proportional to the number of equivalent protons responsible for that signal.

It is important to remember that integration of peak areas gives relative, not absolute, proton counts. Thus, a 3:2 ratio of areas can, as in the case of CH_3OCH_2CN, correspond to a 3:2 ratio of protons. But in some other compound a 3:2 ratio of areas might correspond to a 6:4 or 9:6 ratio of protons.

PROBLEM 13.8 The 200-MHz ^{1}H NMR spectrum of 1,4-dimethylbenzene looks exactly like that of CH_3OCH_2CN except the chemical shifts of the two peaks are δ 2.2 and δ 7.0. Assign the peaks to the appropriate protons of 1,4-dimethylbenzene.

Protons are equivalent to one another and have the same chemical shift when they are in equivalent environments. Often it is an easy matter to decide, simply by inspection, when protons are equivalent or not. In more difficult cases, mentally replacing a proton in a molecule by a "test group" can help. We'll illustrate the procedure for a simple case—the protons of propane. To see if they have the same chemical shift, replace one of the methyl protons at C-1 by chlorine, then do the same thing for a proton at C-3. Both replacements give the same molecule, 1-chloropropane. Therefore the methyl protons at C-1 are equivalent to those at C-3.

$$CH_3CH_2CH_3 \qquad ClCH_2CH_2CH_3 \qquad CH_3CH_2CH_2Cl$$

Propane   1-Chloropropane   1-Chloropropane

If the two structures produced by mental replacement of two different hydrogens in a molecule by a test group are the same, the hydrogens are chemically equivalent. Thus, the six methyl protons of propane are all chemically equivalent to one another and have the same chemical shift.

   Replacement of either one of the methylene protons of propane generates 2-chloro-propane. Both methylene protons are equivalent. Neither of them is equivalent to any of the methyl protons.

   The ^{1}H NMR spectrum of propane contains two signals: one for the six equivalent methyl protons, the other for the pair of equivalent methylene protons.

PROBLEM 13.9   How many signals would you expect to find in the ^{1}H NMR spectrum of each of the following compounds?

(a) 1-Bromobutane

(b) 1-Butanol

(c) Butane

(d) 1,4-Dibromobutane

(e) 2,2-Dibromobutane

(f) 2,2,3,3-Tetrabromobutane

(g) 1,1,4-Tribromobutane

(h) 1,1,1-Tribromobutane

SAMPLE SOLUTION   (a) To test for chemical-shift equivalence, replace the protons at C-1, C-2, C-3, and C-4 of 1-bromobutane by some test group such as chlorine. Four constitutional isomers result:

CH₃CH₂CH₂CHBr	CH₃CH₂CHCH₂Br	CH₃CHCH₂CH₂Br	ClCH₂CH₂CH₂CH₂Br
Cl	Cl	Cl	
1-Bromo-1-chlorobutane	1-Bromo-2-chlorobutane	1-Bromo-3-chlorobutane	1-Bromo-4-chlorobutane

Thus, separate signals will be seen for the protons at C-1, C-2, C-3, and C-4. Barring any accidental overlap, we expect to find four signals in the NMR spectrum of 1-bromobutane.

   Chemical-shift nonequivalence can occur when two environments are stereochemically different. The two vinyl protons of 2-bromopropene have different chemical shifts.

2-Bromopropene

   One of the vinyl protons is cis to bromine; the other trans. Replacing one of the vinyl protons by some test group, say, chlorine, gives the *Z* isomer of 2-bromo-1-chloro-propene; replacing the other gives the *E* stereoisomer. The *E* and *Z* forms of 2-bromo-1-chloropropene are diastereomers. Protons that yield diastereomers on being replaced by some test group are described as *diastereotopic* (Section 7.13) and can have different chemical shifts. Because their environments are similar, however, the chemical shift difference is usually small, and it sometimes happens that two diastereotopic protons accidentally have the same chemical shift. Recording the spectrum on a higher field NMR spectrometer is often helpful in resolving signals with similar chemical shifts.

PROBLEM 13.10   How many signals would you expect to find in the ^{1}H NMR spectrum of each of the following compounds?

(a) Vinyl bromide

(b) 1,1-Dibromoethene

(c) *cis*-1,2-Dibromoethene

(e) Allyl bromide

(d) *trans*-1,2-Dibromoethene

(f) 2-Methyl-2-butene

SAMPLE SOLUTION   (a) Each proton of vinyl bromide is unique and has a chemical shift different from the other two. The least shielded proton is attached to the carbon that bears the bromine. The pair of protons at C-2 are diastereotopic with respect to each other; one is cis to bromine and the other is trans to bromine. There are three proton signals in the NMR spectrum of vinyl bromide. Their observed chemical shifts are as indicated.

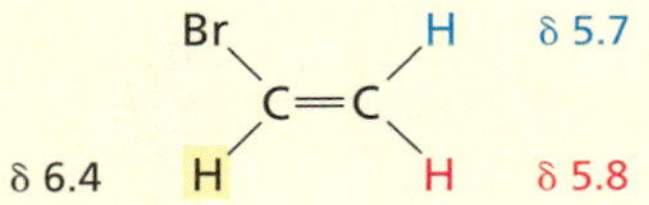

When enantiomers are generated by replacing first one proton and then another by a test group, the pair of protons are *enantiotopic* (Section 7.9). The methylene protons at C-2 of 1-propanol, for example, are enantiotopic.

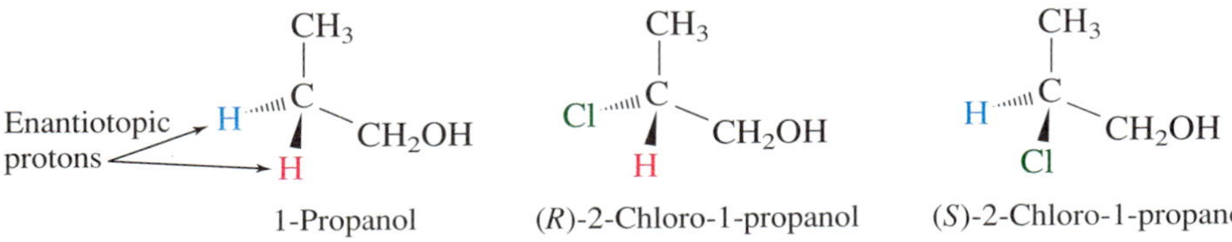

Replacing one of these protons by chlorine as a test group gives (*R*)-2-chloro-1-propanol; replacing the other gives (*S*)-2-chloro-1-propanol. Enantiotopic protons have the same chemical shift, regardless of the field strength of the NMR spectrometer.

At the beginning of this section we noted that an NMR spectrum provides structural information based on chemical shift, the number of peaks, their relative areas, and the multiplicity, or splitting, of the peaks. We have discussed the first three of these features of ^{1}H NMR spectroscopy. Let's now turn our attention to peak splitting to see what kind of information it offers.

Enantiotopic protons can have different chemical shifts in a chiral solvent. Because the customary solvent ($CDCl_3$) used in NMR measurements is achiral, this phenomenon is not observed in routine work.

13.7   SPIN–SPIN SPLITTING IN ^{1}H NMR SPECTROSCOPY

The ^{1}H NMR spectrum of CH_3OCH_2CN (see Figure 13.11) discussed in the preceding section is relatively simple because both signals are **singlets;** that is, each one consists of a single peak. It is quite common though to see a signal for a particular proton appear not as a singlet, but as a collection of peaks. The signal may be split into two peaks (a **doublet**), three peaks (a **triplet**), four peaks (a **quartet**), or even more. Figure 13.12 shows the ^{1}H NMR spectrum of 1,1-dichloroethane (CH_3CHCl_2), which is characterized by a doublet centered at δ 2.1 for the methyl protons and a quartet at δ 5.9 for the methine proton.

The number of peaks into which the signal for a particular proton is split is called its **multiplicity.** For simple cases the rule that allows us to predict splitting in ^{1}H NMR spectroscopy is

$$\text{Multiplicity of signal for } H_a = n + 1$$

where n is equal to the number of equivalent protons that are vicinal to H_a. Two protons are vicinal to each other when they are bonded to adjacent atoms. Protons vicinal

More complicated splitting patterns conform to an extension of the "$n + 1$" rule and will be discussed in Section 13.11.

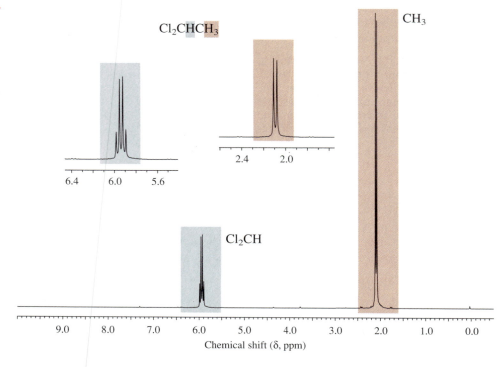

to H_a are separated from H_a by three bonds. The three methyl protons of 1,1-dichloroethane are vicinal to the methine proton and split its signal into a quartet. The single methine proton, in turn, splits the methyl protons' signal into a doublet.

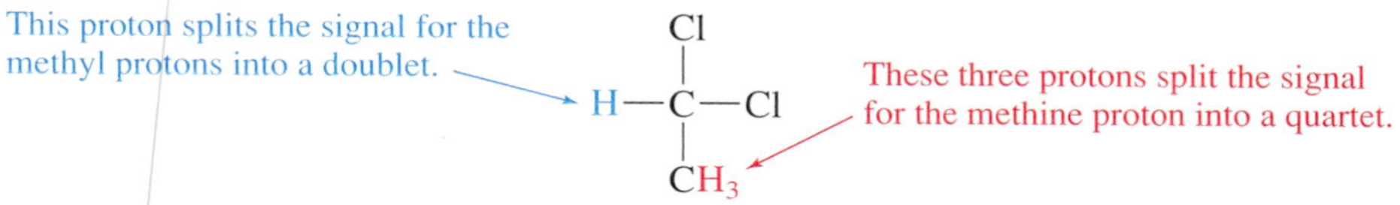

This proton splits the signal for the methyl protons into a doublet.

These three protons split the signal for the methine proton into a quartet.

$$H-\underset{\underset{CH_3}{|}}{\overset{\overset{Cl}{|}}{C}}-Cl$$

The physical basis for peak splitting in 1,1-dichloroethane can be explained with the aid of Figure 13.13, which examines how the chemical shift of the methyl protons is affected by the spin of the methine proton. There are two magnetic environments for

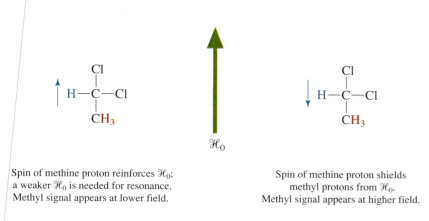

the methyl protons: one in which the magnetic moment of the methine proton is parallel to the applied field, and the other in which it is antiparallel to it. When the magnetic moment of the methine proton is parallel to the applied field, it reinforces it. This decreases the shielding of the methyl protons and causes their signal to appear at slightly lower field strength. Conversely, when the magnetic moment of the methine proton is antiparallel to the applied field, it opposes it and increases the shielding of the methyl protons. Instead of a single peak for the methyl protons, there are two of approximately equal intensity: one at slightly higher field than the "true" chemical shift, the other at slightly lower field.

Turning now to the methine proton, its signal is split by the methyl protons into a quartet. The same kind of analysis applies here and is outlined in Figure 13.14. The methine proton "sees" eight different combinations of nuclear spins for the methyl protons. In one combination, the magnetic moments of all three methyl protons reinforce the applied field. At the other extreme, the magnetic moments of all three methyl protons oppose the applied field. There are three combinations in which the magnetic moments of two methyl protons reinforce the applied field, whereas one opposes it. Finally, there are three combinations in which the magnetic moments of two methyl protons oppose the applied field and one reinforces it. These eight possible combinations give rise to four distinct peaks for the methine proton, with a ratio of intensities of 1:3:3:1.

We describe the observed splitting of NMR signals as **spin–spin splitting** and the physical basis for it as **spin–spin coupling.** It has its origin in the communication of nuclear spin information via the electrons in the bonds that intervene between the nuclei. Its effect is greatest when the number of bonds is small. Vicinal protons are separated by three bonds, and coupling between vicinal protons, as in 1,1-dichloroethane, is called **three-bond coupling** or **vicinal coupling.** Four-bond couplings are weaker and not normally observable.

A very important characteristic of spin–spin splitting is that protons that have the same chemical shift do not split each other's signal. Ethane, for example, shows only a single sharp peak in its NMR spectrum. Even though there is a vicinal relationship between the protons of one methyl group and those of the other, they do not split each other's signal because they are equivalent.

There are eight possible combinations of the nuclear spins of the three methyl protons in CH_3CHCl_2.

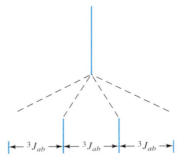

These eight combinations cause the signal of the $CHCl_2$ proton to be split into a quartet, in which the intensities of the peaks are in the ratio 1:3:3:1.

FIGURE 13.14 The methyl protons of 1,1-dichloroethane split the signal of the methine proton into a quartet.

PROBLEM 13.11 Describe the appearance of the ^{1}H NMR spectrum of each of the following compounds. How many signals would you expect to find, and into how many peaks will each signal be split?

(a) 1,2-Dichloroethane

(b) 1,1,1-Trichloroethane

(c) 1,1,2-Trichloroethane

(d) 1,2,2-Trichloropropane

(e) 1,1,1,2-Tetrachloropropane

SAMPLE SOLUTION (a) All the protons of 1,2-dichloroethane ($ClCH_2CH_2Cl$) are chemically equivalent and have the same chemical shift. Protons that have the same chemical shift do not split each other's signal, and so the NMR spectrum of 1,2-dichloroethane consists of a single sharp peak.

Coupling of nuclear spins requires that the nuclei split each other's signal equally. The separation between the two halves of the methyl doublet in 1,1-dichloroethane is equal to the separation between any two adjacent peaks of the methine quartet. The extent to which two nuclei are coupled is given by the **coupling constant J** and in simple cases is equal to the separation between adjacent lines of the signal of a particular proton. The three-bond coupling constant $^3J_{ab}$ in 1,1-dichloroethane has a value of 7 Hz. *The size of the coupling constant is independent of the field strength;* the separation between adjacent peaks in 1,1-dichloroethane is 7 Hz, irrespective of whether the spectrum is recorded at 200 MHz or 500 MHz.

13.8 SPLITTING PATTERNS: THE ETHYL GROUP

At first glance, splitting may seem to complicate the interpretation of NMR spectra. In fact, it makes structure determination easier because it provides additional information. It tells us how many protons are vicinal to a proton responsible for a particular signal. With practice, we learn to pick out characteristic patterns of peaks, associating them with particular structural types. One of the most common of these patterns is that of the ethyl group, represented in the NMR spectrum of ethyl bromide in Figure 13.15.

In compounds of the type CH_3CH_2X, especially where X is an electronegative atom or group, such as bromine in ethyl bromide, the ethyl group appears as a *triplet–quartet pattern*. The signal for the methylene protons is split into a quartet by coupling with the methyl protons. The signal for the methyl protons is a triplet because of vicinal coupling to the two protons of the adjacent methylene group.

$$Br-CH_2-CH_3$$

These two protons split the methyl signal into a triplet.

These three protons split the methylene signal into a quartet.

We have discussed in the preceding section why methyl groups split the signals due to vicinal protons into a quartet. Splitting by a methylene group gives a triplet corresponding to the spin combinations shown in Figure 13.16 for ethyl bromide. The relative intensities of the peaks of this triplet are 1:2:1.

PROBLEM 13.12 Describe the appearance of the ^{1}H NMR spectrum of each of the following compounds. How many signals would you expect to find, and into how many peaks will each signal be split?

(a) $ClCH_2OCH_2CH_3$

(b) $CH_3CH_2OCH_3$

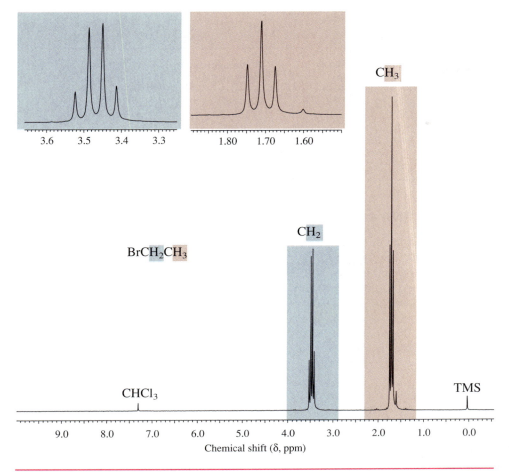

FIGURE 13.15 The 200-MHz ^{1}H NMR spectrum of ethyl bromide ($BrCH_2CH_3$), showing the characteristic triplet–quartet pattern of an ethyl group. The small peak at δ 1.6 is an impurity.

There are four possible combinations of the nuclear spins of the two methylene protons in CH_3CH_2Br.

These four combinations cause the signal of the CH_3 protons to be split into a triplet, in which the intensities of the peaks are in the ratio 1:2:1.

FIGURE 13.16 The methylene protons of ethyl bromide split the signal of the methyl protons into a triplet.

(c) $CH_3CH_2OCH_2CH_3$

(d) *p*-Diethylbenzene

(e) $ClCH_2CH_2OCH_2CH_3$

SAMPLE SOLUTION (a) Along with the triplet–quartet pattern of the ethyl group, the NMR spectrum of this compound will contain a singlet for the two protons of the chloromethyl group.

$$ClCH_2—O—CH_2—CH_3$$

Singlet; no protons vicinal to these; therefore, no splitting

Split into quartet by three protons of methyl group

Split into triplet by two protons of adjacent methylene group

Table 13.2 summarizes the splitting patterns and peak intensities expected for coupling to various numbers of protons.

TABLE 13.2   Splitting Patterns of Common Multiplets

The intensities correspond to the coefficients of a binomial expansion (Pascal's triangle).

Number of equivalent protons to which nucleus is coupled	Appearance of multiplet	Intensities of lines in multiplet
1	Doublet	1:1
2	Triplet	1:2:1
3	Quartet	1:3:3:1
4	Pentet	1:4:6:4:1
5	Sextet	1:5:10:10:5:1
6	Septet	1:6:15:20:15:6:1

13.9   SPLITTING PATTERNS: THE ISOPROPYL GROUP

The NMR spectrum of isopropyl chloride (Figure 13.17) illustrates the appearance of an isopropyl group. The signal for the six equivalent methyl protons at δ 1.5 is split into a doublet by the proton of the H—C—Cl unit. In turn, the H—C—Cl proton signal at δ 4.2 is split into a septet by the six methyl protons. A *doublet–septet* pattern is characteristic of an isopropyl group.

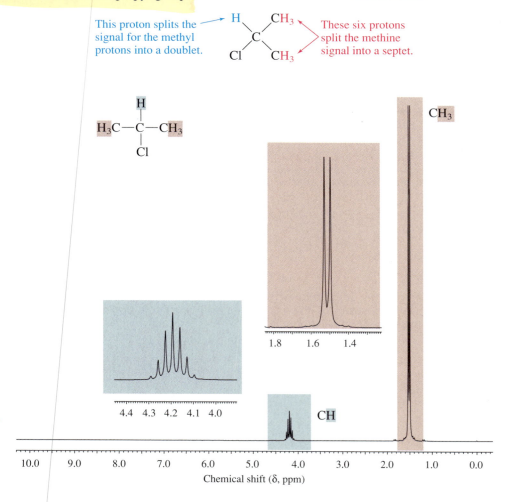

FIGURE 13.17 The 200-MHz ^{1}H NMR spectrum of isopropyl chloride, showing the doublet–septet pattern of an isopropyl group.

13.10 SPLITTING PATTERNS: PAIRS OF DOUBLETS

We often see splitting patterns in which the intensities of the individual peaks do not match those given in Table 13.2, but are distorted in that the signals for coupled protons "lean" toward each other. This leaning is a general phenomenon, but is most easily illustrated for the case of two nonequivalent vicinal protons as shown in Figure 13.18.

$$
\begin{array}{ccc}
 & X & Y \\
 & | & | \\
W- & C-C & -Z \\
 & | & | \\
 & H_1 & H_2
\end{array}
$$

The appearance of the splitting pattern of protons 1 and 2 depends on their coupling constant J and the chemical shift difference $\Delta\nu$ between them. When the ratio $\Delta\nu/J$ is large, two symmetrical 1:1 doublets are observed. We refer to this as the "AX" case, using two letters that are remote in the alphabet to stand for signals well removed from each other on the spectrum. Keeping the coupling constant the same while reducing $\Delta\nu$ leads to a steady decrease in the intensity of the outer two peaks with a simultaneous increase in the inner two as we progress from AX through AM to AB. At the extreme (A_2), the two protons have the same chemical shift, the outermost lines have disappeared, and no

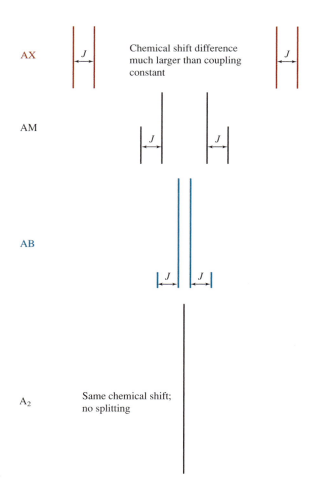

FIGURE 13.18 The appearance of the splitting pattern of two coupled protons depends on their coupling constant J and the chemical shift difference $\Delta\nu$ between them. As the ratio $\Delta\nu/J$ decreases, the doublets become increasingly distorted. When the two protons have the same chemical shift, no splitting is observed.

FIGURE 13.19 The 200-MHz ^{1}H NMR spectrum of 2,3,4-trichloroanisole, showing the splitting of the ring protons into a pair of doublets that "lean" toward each other.

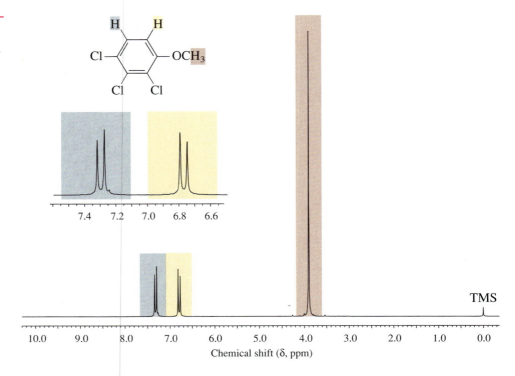

splitting is observed. Because of its appearance, it is easy to misinterpret an AB or AM pattern as a quartet, rather than the pair of skewed doublets it really is.

A skewed doublet of doublets is clearly visible in the ^{1}H NMR spectrum of 2,3,4-trichloroanisole (Figure 13.19). In addition to the singlet at δ 3.9 for the protons of the —OCH$_3$ group, we see doublets at δ 6.8 and δ 7.3 for the two protons of the aromatic ring.

Doublet
δ 7.3

Doublet
δ 6.8

Singlet
δ 3.9

2,3,4-Trichloroanisole

A similar pattern can occur with *geminal* protons (protons bonded to the same carbon). Geminal protons are separated by two bonds, and geminal coupling is referred to as *two-bond coupling* (^2J) in the same way that vicinal coupling is referred to as *three-bond coupling* (^3J). An example of geminal coupling is provided by the compound 1-chloro-1-cyanoethene, in which the two hydrogens appear as a pair of doublets. The splitting in each doublet is 2 Hz.

The protons in 1-chloro-1-cyanoethene are *diastereotopic* (Section 13.6). They are nonequivalent and have different chemical shifts. Remember, splitting can only occur between protons that have different chemical shifts.

Doublet

$^2J = 2$ Hz

Doublet

1-Chloro-1-cyanoethene

Splitting due to geminal coupling is seen only in CH_2 groups and only when the two protons have different chemical shifts. All three protons of a methyl (CH_3) group are equivalent and cannot split one another's signal, and, of course, there are no protons geminal to a single methine (CH) proton.

13.11 COMPLEX SPLITTING PATTERNS

All the cases we've discussed so far have involved splitting of a proton signal by coupling to other protons that were equivalent to one another. Indeed, we have stated the splitting rule in terms of the multiplicity of a signal as being equal to $n + 1$, where n is equal to the number of equivalent protons to which the proton that gives the signal is coupled. What if all the vicinal protons are *not* equivalent?

Figure 13.20a shows the signal for the proton marked $ArCH_a{=}CH_2$ in *m*-nitrostyrene, which appears as a set of four peaks in the range δ 6.7–6.9. These four peaks are in fact a "doublet of doublets." The proton in question is *unequally coupled* to the two protons at the end of the vinyl side chain. The size of the vicinal coupling constant between protons trans to each other on a double bond is normally larger than that between cis protons. In this case the trans coupling constant is 16 Hz and the cis coupling constant is 12 Hz. Thus, as shown in Figure 13.20b, the signal is split into a doublet with a spacing of 16 Hz by one vicinal proton, and each line of this doublet is then split into another doublet with a spacing of 12 Hz.

You will find it revealing to construct a splitting diagram similar to that of Figure 13.20 for the case in which the cis and trans **H—C=C—H** coupling constants are equal. Under those circumstances the four-line pattern simplifies to a triplet, as it should for a proton equally coupled to two vicinal protons.

> **PROBLEM 13.13** In addition to the proton marked H_a in *m*-nitrostyrene in Figure 13.20, there are two other vinylic protons. Assuming that the coupling constant between the two geminal protons in ArCH=CH$_2$ is 2 Hz and the vicinal coupling constants are 12 Hz (cis) and 16 Hz (trans), describe the splitting pattern for each of these other two vinylic hydrogens.

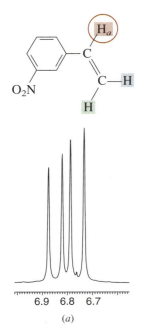

(a)

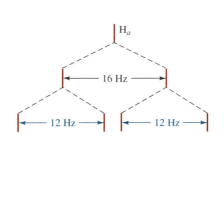

(b)

FIGURE 13.20 Splitting of a signal into a doublet of doublets by unequal coupling to two vicinal protons. (a) Appearance of the signal for the proton marked H_a in *m*-nitrostyrene as a set of four peaks. (b) Origin of these four peaks through successive splitting of the signal for H_a.

The "$n + 1$ rule" should be amended to read: *When a proton* H_a *is coupled to* H_b, H_c, H_d, *etc., and* $J_{ab} \neq J_{ac}$, $\neq J_{ad}$, *etc., the original signal for* H_a *is split into* $n + 1$ *peaks by* n H_b *protons, each of these lines is further split into* $n + 1$ *peaks by* n H_c *protons, and each of these into* $n + 1$ *lines by* n H_d *protons, and so on.* Bear in mind that because of overlapping peaks, the number of lines actually observed can be less than that expected on the basis of the splitting rule.

PROBLEM 13.14 Describe the splitting pattern expected for the proton at

(a) C-2 in (*Z*)-1,3-dichloropropene

(b) C-2 in CH_3CHCH (with O double bonded to the carbon and Br below the CH)

$$CH_3\overset{\displaystyle O}{\overset{\displaystyle \|}{C}}H\overset{}{\underset{\displaystyle Br}{|}}CH$$

SAMPLE SOLUTION (a) The signal of the proton at C-2 is split into a doublet by coupling to the proton cis to it on the double bond, and each line of this doublet is split into a triplet by the two protons of the CH_2Cl group.

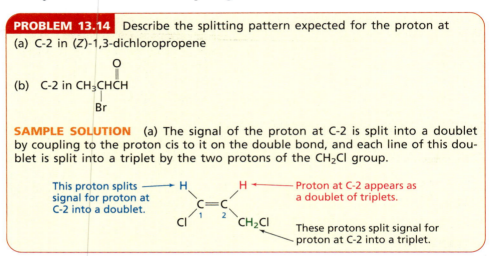

This proton splits signal for proton at C-2 into a doublet.

Proton at C-2 appears as a doublet of triplets.

These protons split signal for proton at C-2 into a triplet.

13.12 ^{1}H NMR SPECTRA OF ALCOHOLS

The —OH proton of a primary alcohol RCH_2OH is vicinal to two protons, and its signal would be expected to be split into a triplet. Under certain conditions signal splitting of alcohol protons is observed, but usually it is not. Figure 13.21 presents the NMR spectrum of benzyl alcohol, showing the methylene and hydroxyl protons as singlets at δ 4.7 and 2.5, respectively. (The aromatic protons also appear as a singlet, but that is because they all accidentally have the same chemical shift and so cannot split each other.) The reason that splitting of the hydroxyl proton of an alcohol is not observed is that it is involved in rapid exchange reactions with other alcohol molecules. Transfer of a proton from an oxygen of one alcohol molecule to the oxygen of another is quite fast and effectively *decouples* it from other protons in the molecule. Factors that slow down this exchange of OH protons, such as diluting the solution, lowering the temperature, or increasing the crowding around the OH group, can cause splitting of hydroxyl resonances.

The chemical shift of the hydroxyl proton is variable, with a range of δ 0.5–5, depending on the solvent, the temperature at which the spectrum is recorded, and the concentration of the solution. The alcohol proton shifts to lower field in more concentrated solutions.

An easy way to verify that a particular signal belongs to a hydroxyl proton is to add D_2O. The hydroxyl proton is replaced by deuterium according to the equation:

$$RCH_2OH + D_2O \rightleftharpoons RCH_2OD + DOH$$

Deuterium does not give a signal under the conditions of ^{1}H NMR spectroscopy. Thus, replacement of a hydroxyl proton by deuterium leads to the disappearance of the OH peak of the alcohol. Protons bonded to nitrogen and sulfur also undergo exchange with

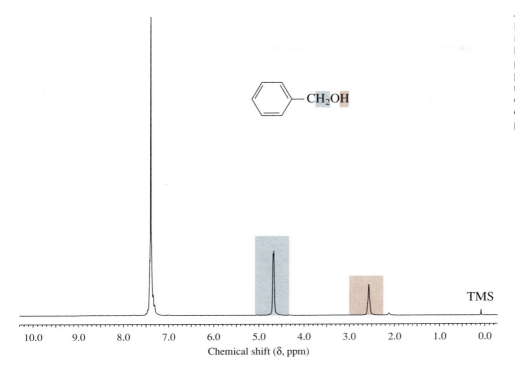

FIGURE 13.21 The 200-MHz ^{1}H NMR spectrum of benzyl alcohol. The hydroxyl proton and the methylene protons are vicinal but do not split each other because of the rapid intermolecular exchange of hydroxyl protons.

D_2O. Those bound to carbon normally do not, and so this technique is useful for assigning the proton resonances of OH, NH, and SH groups.

13.13 NMR AND CONFORMATIONS

We know from Chapter 3 that the protons in cyclohexane exist in two different environments: axial and equatorial. The NMR spectrum of cyclohexane, however, shows only a single sharp peak at δ 1.4. All the protons of cyclohexane appear to be equivalent in the NMR spectrum. Why?

The answer is related to the very rapid rate of ring flipping in cyclohexane.

One property of NMR spectroscopy is that it is too slow a technique to "see" the individual conformations of cyclohexane. What NMR sees is the *average* environment of the protons. Because chair–chair interconversion in cyclohexane converts each axial proton to an equatorial one and vice versa, the average environments of all the protons are the same. A single peak is observed that has a chemical shift midway between the true chemical shifts of the axial and the equatorial protons.

The rate of ring flipping can be slowed down by lowering the temperature. At temperatures on the order of $-100°C$, separate signals are seen for the axial and equatorial protons of cyclohexane.

MAGNETIC RESONANCE IMAGING

Like all photographs, a chest X-ray is a two-dimensional projection of a three-dimensional object. It is literally a collection of shadows produced by all the organs that lie between the source of the X-rays and the photographic plate. The clearest images in a chest X-ray are not the lungs (the customary reason for taking the X-ray in the first place) but rather the ribs and backbone. It would be desirable if we could limit X-ray absorption to two dimensions at a time rather than three. This is, in fact, what is accomplished by a technique known as *computerized axial tomography,* which yields its information in a form called a CT (or CAT) scan. With the aid of a computer, a CT scanner controls the movement of an X-ray source and detector with respect to the patient and to each other, stores the X-ray absorption pattern, and converts it to an image that is equivalent to an X-ray photograph of a thin section of tissue. It is a *noninvasive* diagnostic method, meaning that surgery is not involved nor are probes inserted into the patient's body.

As useful as the CT scan is, it has some drawbacks. Prolonged exposure to X-rays is harmful, and CT scans often require contrast agents to make certain organs more opaque to X-rays. Some patients are allergic to these contrast agents. An alternative technique that is not only safer but more versatile than X-ray tomography is *magnetic resonance imaging,* or MRI. MRI is an application of nuclear magnetic resonance spectroscopy that makes it possible to examine the inside of the human body using radiofrequency radiation, which is lower in energy (see Figure 13.1) and less damaging than X-rays and requires no imaging or contrast agents. By all rights MRI should be called NMRI, but the word *nuclear* was dropped from the name so as to avoid confusion with nuclear medicine, which involves radioactive isotopes.

Although the technology of an MRI scanner is rather sophisticated, it does what we have seen other NMR spectrometers do; it detects protons. Thus, MRI is especially sensitive to biological materials such as water and lipids that are rich in hydrogen. Figure 13.22 shows an example of the use of MRI to detect a brain tumor. Regions of the image are lighter or darker according to the relative concentration of protons and to their environments.

Using MRI as a substitute for X-ray tomography is only the first of what are many medical applications. More lie on the horizon. If, for example, the rate of data acquisition could be increased, then it would become possible to make the leap from the equivalent of still photographs to motion pictures. One could watch the inside of the body as it works—see the heart beat, see the lungs expand and contract—rather than merely examine the structure of an organ.

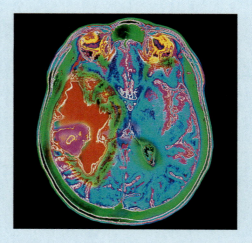

FIGURE 13.22 A magnetic resonance image of a section of a patient's brain showing a tumor in the left hemisphere. The image has been computer enhanced to show the tumor and the surrounding liquid in different shades of red, fatty tissues in green, the normal part of the brain in blue, and the eyeballs in yellow. (*Photograph courtesy of Simon Fraser SPL/Photo Researchers, Inc.*)

13.14 ^{13}C NMR SPECTROSCOPY

We pointed out in Section 13.3 that both ^{1}H and ^{13}C are nuclei that can provide useful structural information when studied by NMR. Although a ^{1}H NMR spectrum helps us infer much about the carbon skeleton of a molecule, a ^{13}C NMR spectrum has the obvious advantage of probing the carbon skeleton directly. ^{13}C NMR spectroscopy is analogous to ^{1}H NMR in that the number of signals informs us about the number of different kinds of carbons, and their chemical shifts are related to particular chemical environments.

However, unlike ^{1}H, which is the most abundant of the hydrogen isotopes (99.985%), only 1.1% of the carbon atoms in a sample are ^{13}C. Moreover, the intensity of the signal produced by ^{13}C nuclei is far weaker than the signal produced by the same number of ^{1}H nuclei. In order for ^{13}C NMR to be a useful technique in structure determination, a vast increase in the signal-to-noise ratio is required. Pulsed FT-NMR provides for this, and its development was the critical breakthrough that led to ^{13}C NMR becoming the routine tool that it is today.

To orient ourselves in the information that ^{13}C NMR provides, let's compare the ^{1}H and ^{13}C NMR spectra of 1-chloropentane (Figures 13.23a and 13.23b, respectively). The ^{1}H NMR spectrum shows reasonably well defined triplets for the protons of the CH$_3$ and CH$_2$Cl groups (δ 0.9 and 3.55, respectively). The signals for the six CH$_2$ protons at C-2, C-3, and C-4 of CH$_3$CH$_2$CH$_2$CH$_2$CH$_2$Cl, however, appear as two unresolved multiplets at δ 1.4 and 1.8.

The ^{13}C NMR spectrum, on the other hand, is very simple: *a separate, distinct peak is observed for each carbon.*

Notice, too, how well-separated these ^{13}C signals are: they cover a range of over 30 ppm, compared with less than 3 ppm for the proton signals of the same compound. In general, the window for proton signals in organic molecules is about 12 ppm; ^{13}C chemical shifts span a range of over 200 ppm. The greater spread of ^{13}C chemical shifts makes it easier to interpret the spectra.

PROBLEM 13.15   How many signals would you expect to see in the ^{13}C NMR spectrum of each of the following compounds?

(a) Propylbenzene

(d) 1,2,4-Trimethylbenzene

(b) Isopropylbenzene

(e) 1,3,5-Trimethylbenzene

(c) 1,2,3-Trimethylbenzene

SAMPLE SOLUTION   (a) The two ring carbons that are ortho to the propyl substituent are equivalent and so must have the same chemical shift. Similarly, the two ring carbons that are meta to the propyl group are equivalent to each other. The carbon atom para to the substituent is unique, as is the carbon that bears the substituent. Thus, there will be four signals for the ring carbons, designated w, x, y, and z in the structural formula. These four signals for the ring carbons added to those for the three nonequivalent carbons of the propyl group yield a total of seven signals.

Propylbenzene

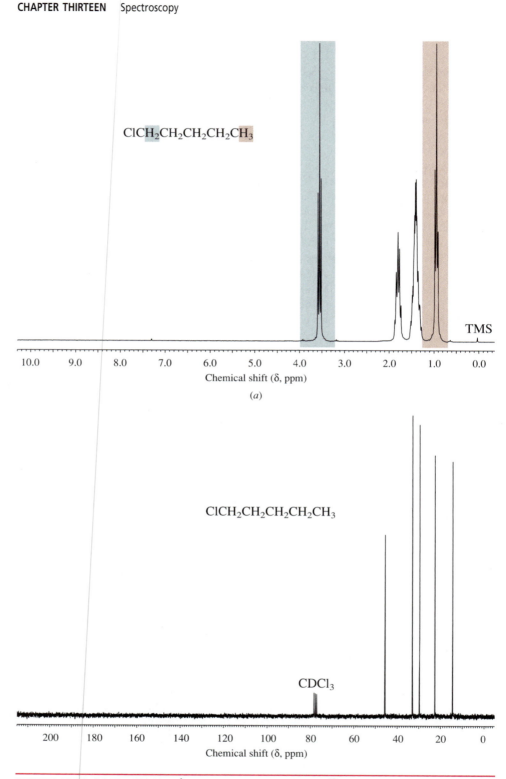

FIGURE 13.23 (*a*) The 200-MHz ^{1}H NMR spectrum and (*b*) the ^{13}C NMR spectrum of 1-chloropentane.

13.15 ^{13}C CHEMICAL SHIFTS

Just as chemical shifts in ^{1}H NMR are measured relative to the *protons* of tetramethylsilane, chemical shifts in ^{13}C NMR are measured relative to the *carbons* of tetramethylsilane. Table 13.3 lists typical chemical-shift ranges for some representative types of carbon atoms.
    In general, the factors that most affect ^{13}C chemical shifts are

1. The electronegativity of the groups attached to carbon
2. The hybridization of carbon

Electronegativity Effects.   Electronegative substituents affect ^{13}C chemical shifts in the same way as they affect ^{1}H chemical shifts, by withdrawing electrons. For ^{1}H NMR, recall that because carbon is more electronegative than hydrogen, the protons in methane (CH_4) are more shielded than primary hydrogens (RCH_3), primary hydrogens are more shielded than secondary (R_2CH_2), and secondary more shielded than tertiary (R_3CH). The same holds true for carbons in ^{13}C NMR, but the effects can be 10–20 times greater.

	CH_4	CH_3CH_3	$CH_3CH_2CH_3$	$(CH_3)_3CH$	$(CH_3)_4C$
Classification:		Primary	Secondary	Tertiary	Quaternary

Chemical shift (δ), ppm:

H	0.2	0.9	1.3	1.7	
C	−2	8	16	25	28

Likewise, for functionally substituted methyl groups:

	CH_4	CH_3NH_2	CH_3OH	CH_3F

Chemical shift (δ), ppm:

H	0.2	2.5	3.4	4.3
C	−2	27	50	75

TABLE 13.3	**Chemical Shifts of Representative Carbons**		
Type of carbon	**Chemical shift (δ) ppm***	**Type of carbon**	**Chemical shift (δ) ppm***
Hydrocarbons		**Functionally substituted carbons**	
RCH_3	0–35	RCH_2Br	20–40
R_2CH_2	15–40	RCH_2Cl	25–50
R_3CH	25–50	RCH_2NH_2	35–50
R_4C	30–40	RCH_2OH and RCH_2OR	50–65
$RC{\equiv}CR$	65–90	$RC{\equiv}N$	110–125
$R_2C{=}CR_2$	100–150	$\overset{O}{\overset{\|}{R}}COH$ and $\overset{O}{\overset{\|}{R}}COR$	160–185
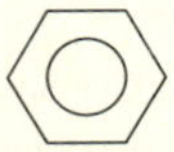	110–175	$\overset{O}{\overset{\|}{R}}CH$ and $\overset{O}{\overset{\|}{R}}CR$	190–220

*Approximate values relative to tetramethylsilane.

Figure 13.23 compared the appearance of the ^{1}H and ^{13}C NMR spectra of 1-chloropentane and drew attention to the fact each carbon gave a separate peak, well separated from the others. Let's now take a closer look at the ^{13}C NMR spectrum of 1-chloropentane with respect to assigning these peaks to individual carbons.

$$Cl-CH_2-CH_2-CH_2-CH_2-CH_3$$

^{13}C chemical shift (δ), ppm: 45   33   29   22   14

The most obvious feature of these ^{13}C chemical shifts is that the closer the carbon is to the electronegative chlorine, the more deshielded it is. Peak assignments will not always be this easy, but the correspondence with electronegativity is so pronounced that *spectrum simulators* are available that allow reliable prediction of ^{13}C chemical shifts from structural formulas. These simulators are based on arithmetic formulas that combine experimentally derived chemical shift increments for the various structural units within a molecule.

> **PROBLEM 13.16**   The ^{13}C NMR spectrum of 1-bromo-3-chloropropane contains peaks at δ 30, δ 35, and δ 43. Assign these signals to the appropriate carbons.

Hybridization Effects.   Here again, the effects are similar to those seen in ^{1}H NMR. As illustrated by 4-phenyl-1-butene, sp^3-hybridized carbons are more shielded than sp^2-hybridized ones.

$$H_2C=CH-CH_2-CH_2-\text{(phenyl ring)}$$

^{13}C chemical shift (δ), ppm: 114   138   36   36   126–142

Of the sp^2-hybridized carbons, C-1 is the most shielded because it is bonded to only one other carbon. The least shielded carbon is the ring carbon to which the side chain is attached. It is the only sp^2-hybridized carbon connected to three others.

> **PROBLEM 13.17**   Consider carbons x, y, and z in p-methylanisole. One has a chemical shift of δ 20, another has δ 55, and the third δ 157. Match the chemical shifts with the appropriate carbons.
>
> $$H_3C-\overset{x}{\text{(ring)}}-\overset{y}{}\overset{z}{OCH_3}$$

Acetylenes are anomalous in ^{13}C, as in ^{1}H NMR. sp-Hybridized carbons are less shielded than sp^3-hybridized ones, but more shielded than sp^2-hybridized ones.

$$H-C\equiv C-CH_2-CH_2-CH_3$$

^{13}C chemical shift (δ), ppm: 68   84   22   20   13

Electronegativity and hybridization effects combine to make the carbon of a carbonyl group especially deshielded. Normally, the carbon of C=O is the least shielded one in a ^{13}C NMR spectrum.

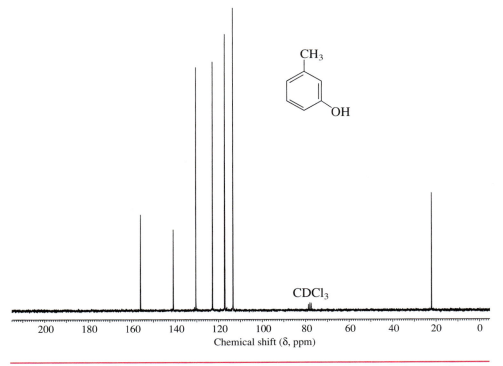

^{13}C chemical shift (δ), ppm: 127–134 ¦ 41   171   61   14

PROBLEM 13.18 Which would you expect to be more shielded, the carbonyl carbon of an aldehyde or ketone? Why?

We will have more to say about ^{13}C chemical shifts in later chapters when various families of compounds, especially those that contain carbonyl groups, are discussed in more detail.

13.16 ^{13}C NMR AND PEAK INTENSITIES

Two features that are fundamental to ^{1}H NMR spectroscopy—integrated areas and splitting patterns—are not very important in ^{13}C NMR.

Although it is a simple matter to integrate ^{13}C signals, it is rarely done because the observed ratios can be more misleading than helpful. The pulsed FT technique that is standard for ^{13}C NMR has the side effect of distorting the signal intensities, especially for carbons that lack attached hydrogens. Examine Figure 13.24, which shows the ^{13}C

FIGURE 13.24 The ^{13}C NMR spectrum of *m*-cresol. Each of the seven carbons of *m*-cresol gives a separate peak. Integrating the spectrum would not provide useful information because the intensities of the peaks are so different, even though each one corresponds to a single carbon.

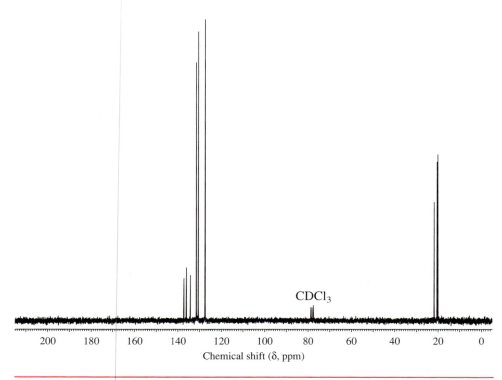

CDCl$_3$

200   180   160   140   120   100   80   60   40   20   0

Chemical shift (δ, ppm)

FIGURE 13.25 The ^{13}C NMR spectrum of the unknown compound of Problem 13.19.

NMR spectrum of 3-methylphenol (*m*-cresol). Notice that, contrary to what we might expect for a compound with seven peaks for seven different carbons, the intensities of these peaks are not nearly the same. The two least intense signals, those at δ 140 and δ 157, correspond to carbons that lack attached hydrogens.

PROBLEM 13.19 To which of the compounds of Problem 13.15 does the ^{13}C NMR spectrum of Figure 13.25 belong?

13.17 ^{13}C—^{1}H COUPLING

You may have noticed another characteristic of ^{13}C NMR spectra—all of the peaks are singlets. With a spin of $\pm\frac{1}{2}$, a ^{13}C nucleus is subject to the same splitting rules that apply to ^{1}H, and we might expect to see splittings due to ^{13}C—^{13}C and ^{13}C—^{1}H couplings. We don't. Why?

The lack of splitting due to ^{13}C—^{13}C coupling is easy to understand. ^{13}C NMR spectra are measured on samples that contain ^{13}C at the "natural abundance" level. Only 1% of all the carbons in the sample are ^{13}C, and the probability that any molecule contains more than one ^{13}C atom is quite small.

Splitting due to ^{13}C—^{1}H coupling is absent for a different reason, one that has to do with the way the spectrum is run. Because a ^{13}C signal can be split not only by the protons to which it is directly attached, but also by protons separated from it by two, three, or even more bonds, the number of splittings might be so large as to make the spectrum too complicated to interpret. Thus, the spectrum is measured under conditions, called **broadband decoupling,** that suppress such splitting. In addition to pulsing the sample by a radiofrequency tuned for ^{13}C, the sample is continuously irradiated by a second rf transmitter that covers the entire frequency range for all the ^1H nuclei. The effect of this second rf is to decouple the ^1H spins from the ^{13}C spins, which causes all the ^{13}C signals to collapse to singlets.

What we gain from broadband decoupling in terms of a simple-looking spectrum comes at the expense of some useful information. For example, being able to see splitting corresponding to one-bond ^{13}C—^1H coupling would immediately tell us the number of hydrogens directly attached to each carbon. The signal for a carbon with no attached hydrogens (a *quaternary* carbon) would be a singlet, the hydrogen of a CH group would split the carbon signal into a doublet, and the signals for the carbons of a CH_2 and a CH_3 group would appear as a triplet and a quartet, respectively. Although it is possible, with a technique called *off-resonance decoupling,* to observe such one-bond couplings, identifying a signal as belonging to a quaternary carbon or to the carbon of a CH, CH_2, or CH_3 group is normally done by a method called DEPT, which is described in the next section.

13.18 USING DEPT TO COUNT HYDROGENS ATTACHED TO ^{13}C

In general, a simple pulse FT-NMR experiment involves the following stages:

1. Equilibration of the nuclei between the lower and higher spin states under the influence of a magnetic field

2. Application of a radiofrequency pulse to give an excess of nuclei in the higher spin state

3. Acquisition of free-induction decay data during the time interval in which the equilibrium distribution of nuclear spins is restored

4. Mathematical manipulation (Fourier transform) of the data to plot a spectrum

The pulse sequence (stages 2–3) can be repeated hundreds of times to enhance the signal-to-noise ratio. The duration of time for stage 2 is on the order of milliseconds, and that for stage 3 is about 1 second.

Major advances in NMR have been made by using a second rf transmitter to irradiate the sample at some point during the sequence. There are several such techniques, of which we'll describe just one, called **distortionless enhancement of polarization transfer,** abbreviated as **DEPT.**

In the DEPT routine, a second transmitter excites ^1H, which affects the appearance of the ^{13}C spectrum. A typical DEPT experiment is illustrated for the case of 1-phenyl-1-pentanone in Figure 13.26. In addition to the normal spectrum shown in Figure 13.26*a,* four more spectra are run using prescribed pulse sequences. In one (Figure 13.26*b*), the signals for carbons of CH_3 and CH groups appear normally, whereas those for CH_2 groups are inverted and those for C without any attached hydrogens are nulled. In the others (not shown) different pulse sequences produce combinations of normal, nulled, and inverted peaks that allow assignments to be made to the various types of carbons with confidence.

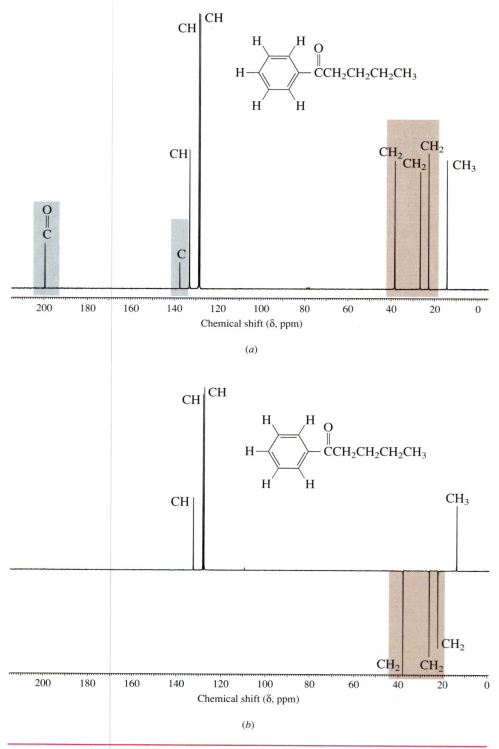

(a)

(b)

FIGURE 13.26 ^{13}C NMR spectra of 1-phenyl-1-pentanone. (a) Normal spectrum. (b) DEPT spectrum recorded using a pulse sequence in which CH$_3$ and CH carbons appear as positive peaks, CH$_2$ carbons as negative peaks, and carbons without any attached hydrogens are nulled.

SPECTRA BY THE THOUSANDS

The best way to get good at interpreting spectra is by experience. Look at as many spectra and do as many spectroscopy problems as you can.

Among Web sites that offer spectroscopic problems, two stand out (Figure 13.27). One, called *WebSpectra,* was developed by Professor Craig A. Merlic (UCLA):*

   http://www.chem.ucla.edu/~webspectra

The other is the *Organic Structure Elucidation* workbook, created by Professor Bradley D. Smith (Notre Dame):

   http://www.nd.edu/~smithgrp/
   structure/workbook.html

WebSpectra includes 75 problems. All the problems display the ^{1}H and ^{13}C spectra, several with DEPT or COSY enhancements. A number include IR spectra. *Organic Structure Elucidation* contains 64 problems, all with ^{1}H and ^{13}C NMR, IR, and mass spectra. The exercises in both *WebSpectra* and *Organic Structure Elucidation* are graded according to difficulty. Give them a try.

Several excellent print collections of spectra are available, but are beyond the budgets of most college libraries. Fortunately, vast numbers of NMR, IR, and mass spectra are freely accessible via the *Spectral Data Base System* (SDBS) maintained by the Japanese National Institute of Advanced Industrial Science and Technology at:

   http://www.aist.go.jp/RIODBS/SDBS/menu-e.html

The SDBS contains 13,000 ^{1}H NMR, 11,800 ^{13}C NMR,

47,300 IR, and 20,500 mass spectra.[†] Not only does the SDBS contain more spectra than anyone could possibly browse through, it incorporates some very useful search features. If you want spectra for a particular compound, entering the name of the compound calls up links to its spectra, which can then be displayed. If you don't know what the compound is, but know one or more of the following:

- Molecular formula
- ^{1}H or ^{13}C chemical shift of one or more peaks
- Mass number of mass spectra fragments

entering the values singly or in combination returns the names of the best matches in the database. You can then compare the spectra of these candidate compounds with the spectra of the sample to identify it.

As extensive as the SDBS is, don't be disappointed if the exact compound you are looking for is not there. There are, after all, millions of organic compounds. However, much of structure determination (and organic chemistry in general) is based on analogy. Finding the spectrum of a related compound can be almost as helpful as finding the one you really want.

These Web resources, in conjunction with the figures and problems in your text afford a wealth of opportunities to gain practice and experience in modern techniques of structure determination.

*For a complete description of *WebSpectra* see pp. 118–120 of the January 2001 issue of the *Journal of Chemical Education.*
[†]Using the SDBS as the basis for student exercises in organic spectroscopy is described in the September 2001 issue of the *Journal of Chemical Education,* pp. 1208–1209.

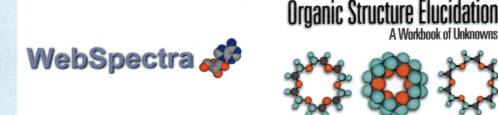

FIGURE 13.27 These two welcome screens open the door to almost 150 spectroscopy problems. The screens are used with permission of Professors Craig A. Merlic (*WebSpectra*) and Bradley D. Smith (*Organic Structure Elucidation*). See the text for the respective URL's.

13.19 2D NMR: COSY AND HETCOR

The theoretical aspects, including pulse sequences, that underlie 2D NMR are discussed in the May 1990 issue of the *Journal of Chemical Education*, pp. A125–A137.

The more information you can extract from an NMR spectrum, the better your chances at arriving at a unique structure. Like spin–spin splitting, which complicates the appearance of a ^{1}H NMR spectrum but provides additional information, 2D NMR looks more complicated than it is while making structure determination easier.

The key dimension in NMR is the frequency axis. All of the spectra we have seen so far are 1D spectra because they have only one frequency axis. In 2D NMR a standard pulse sequence adds a second frequency axis. Only pulsed FT-NMR spectrometers are capable of carrying out 2D experiments.

One kind of 2D NMR is called **COSY**, which stands for **correlated spectroscopy.** With a COSY spectrum you can determine by inspection which signals correspond to spin-coupled protons. Identifying coupling relationships is a valuable aid to establishing a molecule's *connectivity.*

Figure 13.28 is the COSY spectrum of 2-hexanone. Both the *x*- and *y*-axes are frequency axes expressed as chemical shifts. Displaying the 1D ^{1}H NMR spectrum of 2-hexanone along the *x*- and *y*-axes makes it easier to interpret the 2D information, which is the collection of contoured objects contained within the axes. To orient ourselves, first note that many of the contours lie along the diagonal that runs from the lower left to the upper right. This diagonal bisects the 2D NMR into two mirror-image halves. The off-diagonal contours are called *cross peaks* and contain the connectivity information we need.

Each cross peak has *x* and *y* coordinates. One coordinate corresponds to the chemical shift of a proton, the other to the chemical shift to a proton to which it is coupled. Because the diagonal splits the 2D spectrum in half, each cross peak is duplicated on the other side of the other diagonal with the same coordinates, except in reverse order. This redundancy means that we really need to examine only half of the cross peaks.

To illustrate, start with the lowest field signal (δ 2.4) of 2-hexanone. We assign this signal, a triplet, to the protons at C-3 on the basis of its chemical shift and the splitting evident in the 1D spectrum.

$$
\begin{array}{c}
\overset{\displaystyle O}{\underset{\displaystyle \|}{}} \\
CH_3CCH_2CH_2CH_2CH_3 \\
\uparrow \\
\delta\ 2.4
\end{array}
$$

We look for cross peaks with the same *x* coordinate by drawing a vertical line from δ 2.4, finding a cross peak with a *y* coordinate of δ 1.6. *This means that the protons responsible for the signal at δ 2.4 are coupled to the ones at δ 1.6.* Therefore, the chemical shift of the C-4 protons is δ 1.6.

Now work from these C-4 protons. Drawing a vertical line from δ 1.6 on the *x*-axis finds two cross peaks. One cross peak simply confirms the coupling to the protons at C-3. The other has a *y* coordinate of δ 1.3 and, therefore, must correspond to the protons at C-5.

A vertical line drawn from δ 1.3 intersects the cross peaks at both δ 1.6 and δ 0.9. The former confirms the coupling of C-5 to C-4; the latter corresponds to the C-5 to C-6 coupling and identifies the signal at δ 0.9 as belonging to the protons at C-6.

Finally, a vertical line drawn from δ 2.1 intersects no cross peaks. The singlet at δ 2.1, as expected, is due to the protons at C-1, which are not coupled to any of the other protons in the molecule.

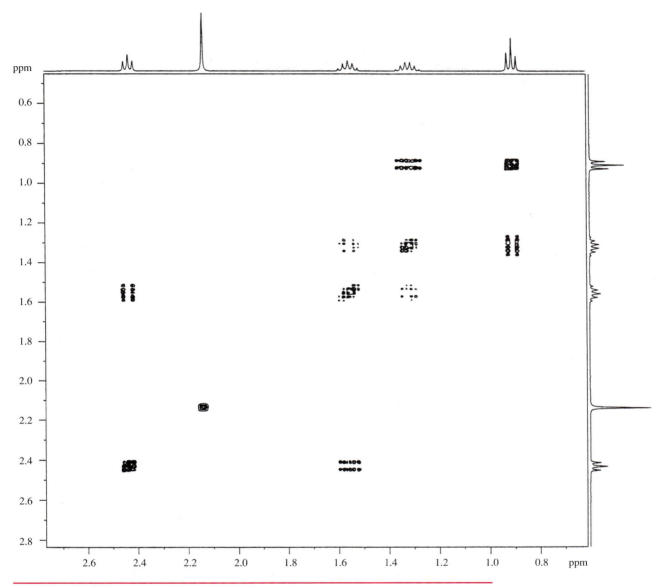

FIGURE 13.28 ^{1}H-^{1}H COSY NMR spectrum of 2-hexanone.

The complete connectivity and assignment of ^{1}H chemical shifts is

$$H_3C-\underset{\substack{\| \\ O}}{C}-CH_2-CH_2-CH_2-CH_3$$

2.1   2.4   1.6   1.3   0.9

The July 1995 issue of the *Journal of Chemical Education* (pp. 659–661) contains an undergraduate laboratory experiment in which COSY is used to analyze the products of a chemical reaction.

Although the 1D ^{1}H spectrum of 2-hexanone is simple enough to be interpreted directly, you can see that COSY offers one more tool we can call on in more complicated cases.

A second 2D NMR method called **HETCOR (heteronuclear chemical shift correlation)** is a type of COSY in which the two frequency axes are the chemical shifts for different nuclei, usually ^1H and ^{13}C. With HETCOR it is possible to relate a peak in a ^{13}C spectrum to the ^1H signal of the protons attached to that carbon. As we did with COSY, we'll use 2-hexanone to illustrate the technique.

The HETCOR spectrum of 2-hexanone is shown in Figure 13.29. It is considerably simpler than a COSY spectrum, lacking diagonal peaks and contoured cross peaks. Instead, we see objects that are approximately as tall as a ^1H signal is wide, and as wide as a ^{13}C signal. As with the COSY cross peaks, however, it is their coordinates that matter, not their size or shape. Interpreting the spectrum is straightforward. The ^{13}C peak at δ 30 correlates with the ^1H singlet at δ 2.1, which because of its multiplicity and chemical shift corresponds to the protons at C-1. Therefore, this ^{13}C peak can be assigned to C-1 of 2-hexanone. Repeating this procedure for the other carbons gives:

FIGURE 13.29 ^1H-^{13}C HETCOR NMR spectrum of 2-hexanone.

$$H_3C-\underset{\underset{O}{\|}}{C}-CH_2-CH_2-CH_2-CH_3$$

^1H chemical shift (δ), ppm:	2.1		2.4	1.6	1.3	0.7
^{13}C chemical shift (δ), ppm:	30		43	26	22	14

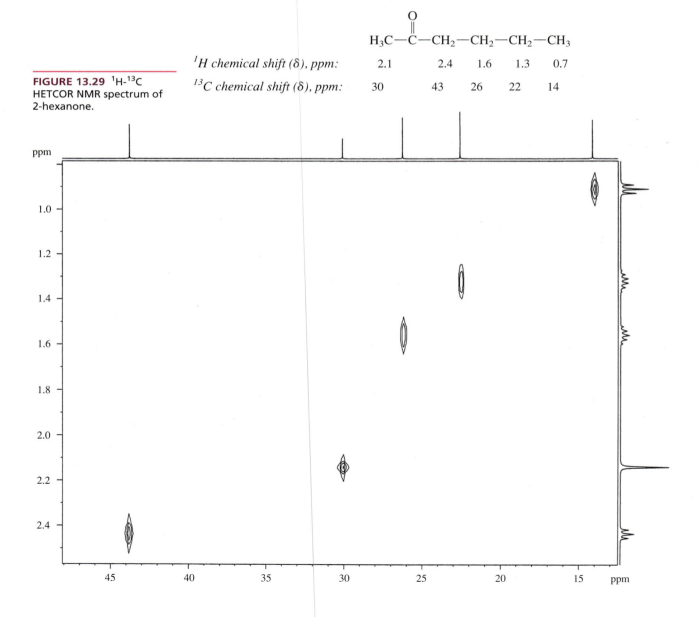

The chemical shift of the carbonyl carbon (δ 209) is not included because it has no attached hydrogens.

Because the digitized areas of the ^{1}H spectrum give the relative number of protons responsible for each signal, HETCOR serves as an alternative to DEPT for counting the number of protons bonded to each carbon.

A number of 2D NMR techniques are available for a variety of purposes. They are especially valuable when attempting to determine the structure of complicated natural products and the conformations of biomolecules.

13.20 INFRARED SPECTROSCOPY

Before the advent of NMR spectroscopy, infrared (IR) spectroscopy was the instrumental method most often applied to determine the structure of organic compounds. Although NMR spectroscopy, in general, tells us more about the structure of an unknown compound, IR still retains an important place in the chemist's inventory of spectroscopic methods because of its usefulness in identifying the presence of certain *functional groups* within a molecule.

Infrared radiation is the portion of the electromagnetic spectrum (see Figure 13.1) between microwaves and visible light. The fraction of the infrared region of most use for structure determination lies between 2.5×10^{-6} m and 16×10^{-6} m in wavelength. Two units commonly employed in IR spectroscopy are the *micrometer* and the *wavenumber*. One micrometer (μm) is 10^{-6} m, and IR spectra record the region from 2.5 μm to 16 μm. Wavenumbers are reciprocal centimeters (cm^{-1}), so that the region 2.5–16 μm corresponds to 4000–625 cm^{-1}. An advantage to using wavenumbers is that they are directly proportional to energy. Thus, 4000 cm^{-1} is the high-energy end of the scale, and 625 cm^{-1} is the low-energy end.

Electromagnetic radiation in the 4000–625 cm^{-1} region corresponds to the separation between adjacent **vibrational energy states** in organic molecules. Absorption of a photon of infrared radiation excites a molecule from its lowest, or *ground,* vibrational state to a higher one. These vibrations include stretching and bending modes of the type illustrated for a methylene group in Figure 13.30. A single molecule can have a large number of distinct vibrations available to it, and IR spectra of different molecules, like fingerprints, are different. Superimposability of their IR spectra is commonly offered as proof that two compounds are the same.

A typical IR spectrum, such as that of hexane in Figure 13.31, appears as a series of absorption peaks of varying shape and intensity. Almost all organic compounds exhibit a peak or group of peaks near 3000 cm^{-1} due to carbon–hydrogen stretching. The peaks at 1460, 1380, and 725 cm^{-1} are due to various bending vibrations.

IR spectra can be recorded on a sample regardless of its physical state—solid, liquid, gas, or dissolved in some solvent. The spectrum in Figure 13.31 was taken on the neat sample, meaning the pure liquid. A drop or two of hexane was placed between two sodium chloride disks, through which the IR beam is passed. Solids may be dissolved in a suitable solvent such as carbon tetrachloride or chloroform. More commonly, though, a solid sample is mixed with potassium bromide and the mixture pressed into a thin wafer, which is placed in the path of the IR beam.

In using IR spectroscopy for structure determination, peaks in the range 1600–4000 cm^{-1} are usually emphasized because this is the region in which the vibrations characteristic of particular functional groups are found. The region 1300–625 cm^{-1} is known as the **fingerprint region**; it is here that the pattern of peaks varies most from compound to compound. Table 13.4 lists the frequencies (in wavenumbers) associated with a variety of groups commonly found in organic compounds.

Like NMR spectrometers, some IR spectrometers operate in a continuous-sweep mode, whereas others employ pulse Fourier-transform (FT-IR) technology. All the IR spectra in this text were obtained on an FT-IR instrument.

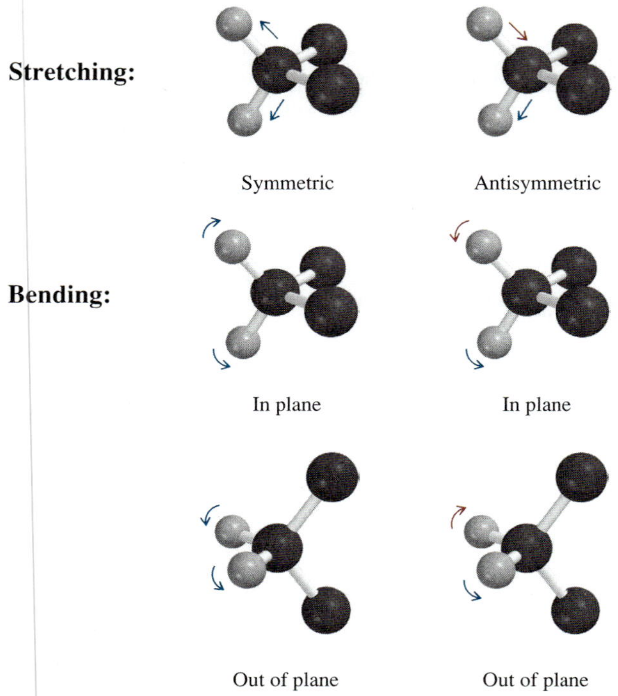

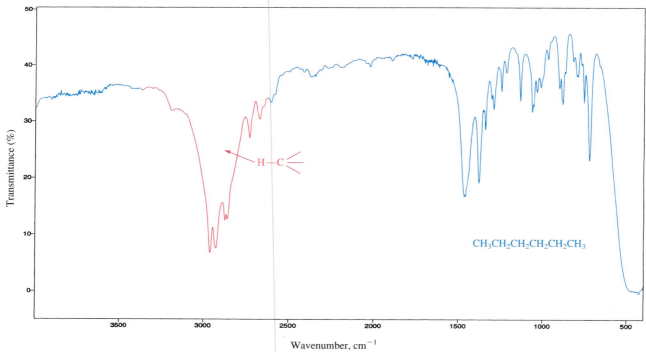

Stretching:
Symmetric
Antisymmetric
Bending:
In plane
In plane
Out of plane
Out of plane

FIGURE 13.30 Stretching and bending vibrations of a methylene unit.

50
40
30
20
10
0
Transmittance (%)
3500
3000
2500
2000
1500
1000
500
Wavenumber, cm^{-1}
H—C
$CH_3CH_2CH_2CH_2CH_2CH_3$

FIGURE 13.31 The IR spectrum of hexane.

TABLE 13.4 Infrared Absorption Frequencies of Some Common Structural Units

Structural unit	Frequency, cm^{-1}	Structural unit	Frequency, cm^{-1}
Stretching vibrations			
Single bonds		**Double bonds**	
—O—H (alcohols)	3200–3600	$\diagdown$C$=$C$\diagup$	1620–1680
—O—H (carboxylic acids)	2500–3600		
$\diagdown$N—H	3350–3500	$\diagdown$C$=$O	
sp C—H	3310–3320	Aldehydes and ketones	1710–1750
sp^2 C—H	3000–3100	Carboxylic acids	1700–1725
sp^3 C—H	2850–2950	Acid anhydrides	1800–1850 and 1740–1790
		Acyl halides	1770–1815
sp^2 C—O	1200	Esters	1730–1750
sp^3 C—O	1025–1200	Amides	1680–1700
		Triple bonds	
		—C$\equiv$C—	2100–2200
		—C$\equiv$N	2240–2280
Bending vibrations of diagnostic value			
Alkenes:		*Substituted derivatives of benzene:*	
RCH$=$CH$_2$	910, 990	Monosubstituted	730–770 and 690–710
R$_2$C$=$CH$_2$	890	Ortho-disubstituted	735–770
cis-RCH$=$CHR′	665–730	Meta-disubstituted	750–810 and 680–730
trans-RCH$=$CHR′	960–980	Para-disubstituted	790–840
R$_2$C$=$CHR′	790–840		

To illustrate how structural features affect IR spectra, compare the spectrum of hexane (Figure 13.31) with that of 1-hexene (Figure 13.32). The two are quite different. In the C—H stretching region of 1-hexene, there is a peak at 3095 cm^{-1}, whereas all the C—H stretching vibrations of hexane appear below 3000 cm^{-1}. A peak or peaks above 3000 cm^{-1} is characteristic of a hydrogen bonded to sp^2-hybridized carbon. The IR spectrum of 1-hexene also displays a peak at 1640 cm^{-1} corresponding to its C$=$C stretching vibration. The peaks near 1000 and 900 cm^{-1} in the spectrum of 1-hexene, absent in the spectrum of hexane, are bending vibrations involving the hydrogens of the doubly bonded carbons.

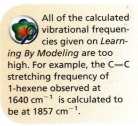

All of the calculated vibrational frequencies given on *Learning By Modeling* are too high. For example, the C$=$C stretching frequency of 1-hexene observed at 1640 cm^{-1} is calculated to be at 1857 cm^{-1}.

Carbon–hydrogen stretching vibrations with frequencies above 3000 cm^{-1} are also found in arenes such as *tert*-butylbenzene, as shown in Figure 13.33. This spectrum also contains two intense bands at 760 and 700 cm^{-1}, which are characteristic of monosubstituted benzene rings. Other substitution patterns, some of which are listed in Table 13.4, give different combinations of peaks.

In addition to sp^2 C—H stretching modes, there are other stretching vibrations that appear at frequencies above 3000 cm^{-1}. The most important of these is the O—H stretch of alcohols. Figure 13.34 shows the IR spectrum of 2-hexanol. It contains a broad peak at 3300 cm^{-1} ascribable to O—H stretching of hydrogen-bonded alcohol groups. In

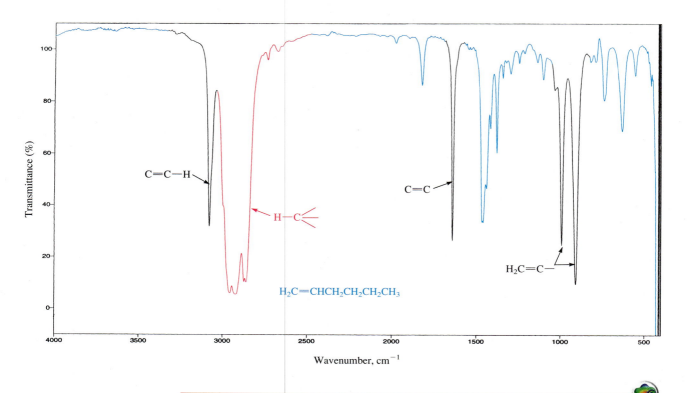

FIGURE 13.32 The IR spectrum of 1-hexene.

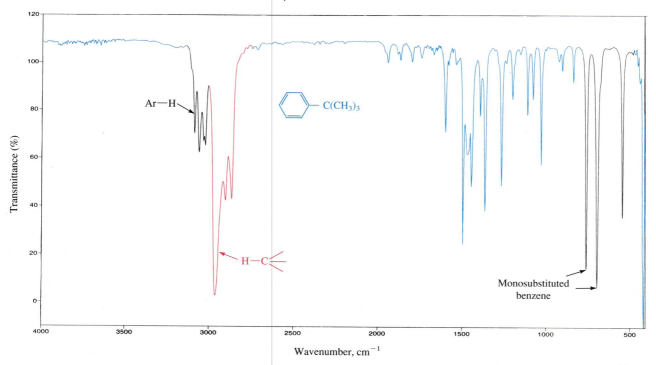

FIGURE 13.33 The IR spectrum of *tert*-butylbenzene.

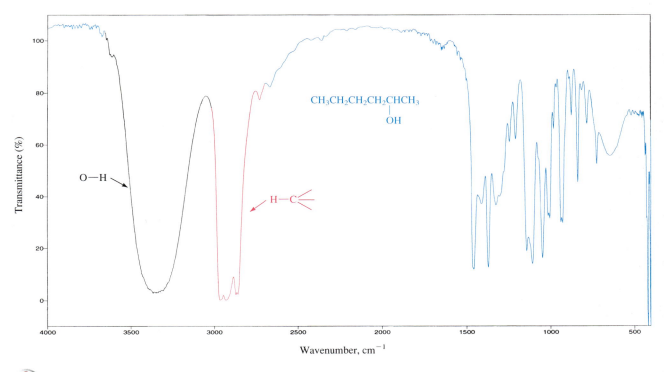

FIGURE 13.34 The IR spectrum of 2-hexanol.

dilute solution, where hydrogen bonding is less and individual alcohol molecules are present as well as hydrogen-bonded aggregates, an additional peak appears at approximately 3600 cm^{-1}.

Carbonyl groups rank among the structural units most readily revealed by IR spectroscopy. The carbon–oxygen double bond stretching mode gives rise to a very strong peak in the 1650–1800 cm^{-1} region. This peak is clearly evident in the spectrum of 2-hexanone, shown in Figure 13.35. The position of the carbonyl peak varies with the nature of the substituents on the carbonyl group. Thus, characteristic frequencies are associated with aldehydes and ketones, amides, esters, and so forth, as summarized in Table 13.4.

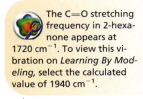

The C=O stretching frequency in 2-hexanone appears at 1720 cm^{-1}. To view this vibration on *Learning By Modeling*, select the calculated value of 1940 cm^{-1}.

PROBLEM 13.20 Which one of the following compounds is most consistent with the IR spectrum given in Figure 13.36? Explain your reasoning.

Phenol	Acetophenone	Benzoic acid	Benzyl alcohol

In later chapters, when families of compounds are discussed in detail, the IR frequencies associated with each type of functional group will be described.

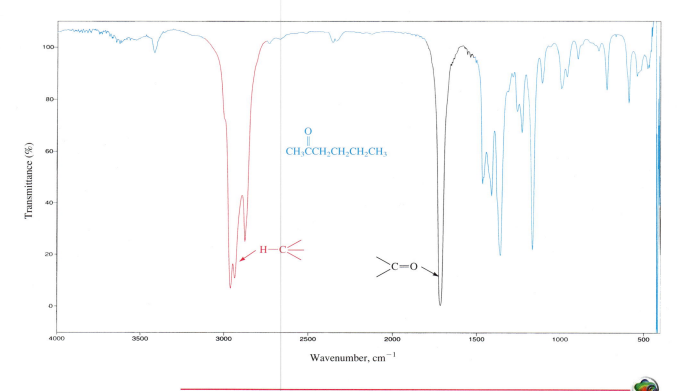

FIGURE 13.35 The IR spectrum of 2-hexanone.

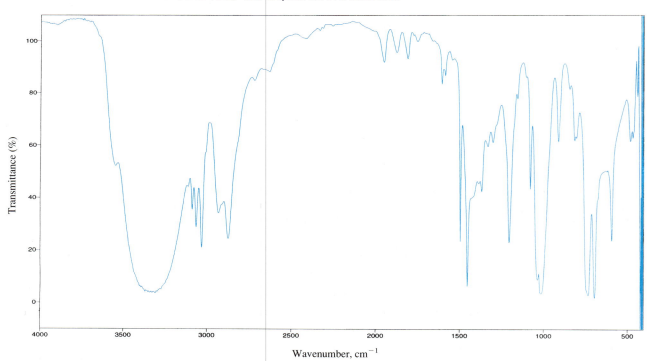

FIGURE 13.36 The IR spectrum of the unknown compound in Problem 13.20.

13.21   ULTRAVIOLET-VISIBLE (UV-VIS) SPECTROSCOPY

The main application of UV-VIS spectroscopy, which depends on transitions between electronic energy levels, is in identifying conjugated π electron systems.

Much greater energies separate vibrational states than nuclear spin states, and the energy differences between electronic states are greater yet. The energy required to promote an electron from one electronic state to the next lies in the visible and ultraviolet range of the electromagnetic spectrum (see Figure 13.1). We usually identify radiation in the UV-VIS range by its wavelength in nanometers (1 nm $= 10^{-9}$ m). Thus, the visible region corresponds to 400–800 nm. Red light is the low-energy (long wavelength) end of the visible spectrum, violet light the high-energy (short wavelength) end. Ultraviolet light lies beyond the visible spectrum with wavelengths in the 200–400-nm range.

Figure 13.37 shows the UV spectrum of the conjugated diene *cis,trans*-1,3-cyclooctadiene, measured in ethanol as the solvent. As is typical of most UV spectra, the absorption is rather broad and is often spoken of as a "band" rather than a "peak." The wavelength at an absorption maximum is referred to as the λ_{max} of the band. There is only one band in the UV spectrum of 1,3-cyclooctadiene; its λ_{max} is 230 nm. In addition to λ_{max}, UV-VIS bands are characterized by their **absorbance** (*A*), which is a measure of how much of the radiation that passes through the sample is absorbed. To correct for concentration and path length effects, absorbance is converted to **molar absorptivity** (ϵ) by dividing it by the concentration *c* in moles per liter and the path length *l* in centimeters.

> An important enzyme in biological electron transport called *cytochrome P450* gets its name from its UV absorption. The "P" stands for "pigment" because it is colored, and the "450" corresponds to the 450-nm absorption of one of its derivatives.

> Molar absorptivity used to be called the *molar extinction coefficient*.

$$\epsilon = \frac{A}{c \cdot l}$$

Molar absorptivity, when measured at λ_{max}, is cited as ϵ_{max}. It is normally expressed without units. Both λ_{max} and ϵ_{max} are affected by the solvent, which is therefore included when reporting UV-VIS spectroscopic data. Thus, you might find a literature reference expressed in the form

cis,trans-1,3-Cyclooctadiene

$\lambda_{max}^{ethanol}$ 230 nm

$\epsilon_{max}^{ethanol}$ 2630

Figure 13.38 illustrates the transition between electronic energy states responsible for the 230-nm UV band of *cis,trans*-1,3-cyclooctadiene. Absorption of UV radiation excites an electron from the highest occupied molecular orbital (HOMO) to the lowest unoccupied molecular orbital (LUMO). In alkenes and polyenes, both the HOMO and LUMO are π type orbitals (rather than σ); the HOMO is the highest energy π orbital and the LUMO is the lowest energy π^* orbital. Exciting one of the π electrons from a bonding π orbital to an antibonding π^* orbital is referred to as a $\pi \rightarrow \pi^*$ transition.

> **PROBLEM 13.21** λ_{max} for the $\pi \rightarrow \pi^*$ transition in ethylene is 170 nm. Is the HOMO–LUMO energy difference in ethylene greater than or less than that of *cis,trans*-1,3-cyclooctadiene?

The HOMO–LUMO energy gap and, consequently, λ_{max} for the $\pi \rightarrow \pi^*$ transition varies with the substituents on the double bonds. The data in Table 13.5 illustrate two substituent effects: adding methyl substituents to the double bond, and extending conjugation. Both cause λ_{max} to shift to longer wavelengths, but the effect of conjugation is the larger of the two. Based on data collected for many dienes it has been found that

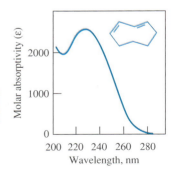

FIGURE 13.37   The UV spectrum of *cis,trans*-1,3-cyclooctadiene.

FIGURE 13.38 The $\pi \rightarrow \pi^*$ transition in *cis,trans*-1,3-cyclooctadiene involves excitation of an electron from the highest occupied molecular orbital (HOMO) to the lowest unoccupied molecular orbital (LUMO).

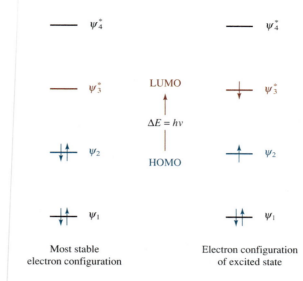

each methyl substituent on the double bonds causes a shift to longer wavelengths of about 5 nm, whereas extending the conjugation causes a shift of about 36 nm for each additional double bond.

> **PROBLEM 13.22** Which one of the C_5H_8 isomers shown has its λ_{max} at the longest wavelength?

A striking example of the effect of conjugation on light absorption occurs in *lycopene*, which is one of the pigments in ripe tomatoes. Lycopene has a conjugated system of 11 double bonds and absorbs *visible light*. It has several UV-VIS bands, each characterized by a separate λ_{max}. Its longest wavelength absorption is at 505 nm.

TABLE 13.5   Absorption Maxima of Some Representative Alkenes and Polyenes*

Compound	Structure	λ_{max} (nm)
Ethylene	$H_2C{=}CH_2$	170
2-Methylpropene	$H_2C{=}C(CH_3)_2$	188
1,3-Butadiene	$H_2C{=}CHCH{=}CH_2$	217
4-Methyl-1,3-pentadiene	$H_2C{=}CHCH{=}C(CH_3)_2$	234
2,5-Dimethyl-2,4-hexadiene	$(CH_3)_2C{=}CHCH{=}C(CH_3)_2$	241
(2E,4E,6E)-2,4,6-Octatriene	$CH_3CH{=}CHCH{=}CHCH{=}CHCH_3$	263
(2E,4E,6E,8E)-2,4,6,8-Decatetraene	$CH_3CH{=}CH(CH{=}CH)_2CH{=}CHCH_3$	299
(2E,4E,6E,8E,10E)-2,4,6,8,10-Dodecapentaene	$CH_3CH{=}CH(CH{=}CH)_3CH{=}CHCH_3$	326

*The value of λ_{max} refers to the longest wavelength $\pi \rightarrow \pi^*$ transition.

Lycopene

Many organic compounds such as lycopene are colored because their HOMO–LUMO energy gap is small enough that λ_{max} appears in the visible range of the spectrum. All that is required for a compound to be colored, however, is that it possess some absorption in the visible range. It often happens that a compound will have its λ_{max} in the UV region but that the peak is broad and extends into the visible. Absorption of the blue-to-violet components of visible light occurs, and the compound appears yellow.

A second type of absorption that is important in UV-VIS examination of organic compounds is the $n \rightarrow \pi^*$ transition of the carbonyl (C=O) group. One of the electrons in a lone-pair orbital of oxygen is excited to an antibonding orbital of the carbonyl group. The n in $n \rightarrow \pi^*$ identifies the electron as one of the nonbonded electrons of oxygen. This transition gives rise to relatively weak absorption peaks ($\epsilon_{max} < 100$) in the region 270–300 nm.

> Don't confuse the n in $n \rightarrow \pi^*$ with the n of Hückel's rule.

The structural unit associated with an electronic transition in UV-VIS spectroscopy is called a **chromophore.** Chemists often refer to *model compounds* to help interpret UV-VIS spectra. An appropriate model is a simple compound of known structure that incorporates the chromophore suspected of being present in the sample. Because remote substituents do not affect λ_{max} of the chromophore, a strong similarity between the spectrum of the model compound and that of the unknown can serve to identify the kind of π electron system present in the sample. There is a substantial body of data concerning the UV-VIS spectra of a great many chromophores, as well as empirical correlations of substituent effects on λ_{max}. Such data are helpful when using UV-VIS spectroscopy as a tool for structure determination.

13.22 MASS SPECTROMETRY

Mass spectrometry differs from the other instrumental methods discussed in this chapter in a fundamental way. It does not depend on the absorption of electromagnetic radiation but rather examines what happens when a molecule is bombarded with high-energy electrons. If an electron having an energy of about 10 electronvolts (10 eV = 230.5 kcal/mol) collides with an organic molecule, the energy transferred as a result of that collision is sufficient to dislodge one of the molecule's electrons.

$$\text{A:B} \quad + \quad e^- \quad \longrightarrow \quad \text{A} \cdot \overset{+}{\text{B}} \quad + \quad 2e^-$$

Molecule   Electron   Cation radical   Two electrons

We say the molecule AB has been ionized by **electron impact.** The species that results, called the **molecular ion,** is positively charged and has an odd number of electrons—it is a **cation radical.** The molecular ion has the same mass (less the negligible mass of a single electron) as the molecule from which it is formed.

Although energies of about 10 eV are required, energies of about 70 eV are used. Electrons this energetic not only cause ionization of a molecule but impart a large amount of energy to the molecular ion, enough energy to break chemical bonds. The molecular ion dissipates this excess energy by dissociating into smaller fragments. Dissociation of a cation radical produces a neutral fragment and a positively charged fragment.

$$\overset{+}{A \cdot B} \overset{\curvearrowright}{\longrightarrow} A^+ + B\cdot$$

Cation radical   Cation   Radical

Ionization and fragmentation produce a mixture of particles, some neutral and some positively charged. To understand what follows, we need to examine the design of an electron-impact mass spectrometer, shown in a schematic diagram in Figure 13.39. The sample is bombarded with 70-eV electrons, and the resulting positively charged ions (the molecular ion as well as fragment ions) are directed into an analyzer tube surrounded by a magnet. This magnet deflects the ions from their original trajectory, causing them to adopt a circular path, the radius of which depends on their mass-to-charge ratio (m/z). Ions of small m/z are deflected more than those of larger m/z. By varying either the magnetic field strength or the degree to which the ions are accelerated on entering the analyzer, ions of a particular m/z can be selectively focused through a narrow slit onto a detector, where they are counted. Scanning all m/z values gives the distribution of positive ions, called a **mass spectrum,** characteristic of a particular compound.

Modern mass spectrometers are interfaced with computerized data-handling systems capable of displaying the mass spectrum according to a number of different formats. Bar graphs on which relative intensity is plotted versus m/z are the most common. Figure 13.40 shows the mass spectrum of benzene in bar graph form.

FIGURE 13.39 Diagram of a mass spectrometer. Only positive ions are detected. The cation X^+ has the lowest mass-to-charge ratio and its path is deflected most by the magnet. The cation Z^+ has the highest mass-to-charge ratio and its path is deflected least. (Adapted, with permission, from M. Silberberg, *Chemistry,* WCB/McGraw-Hill, 2000, p. 56.)

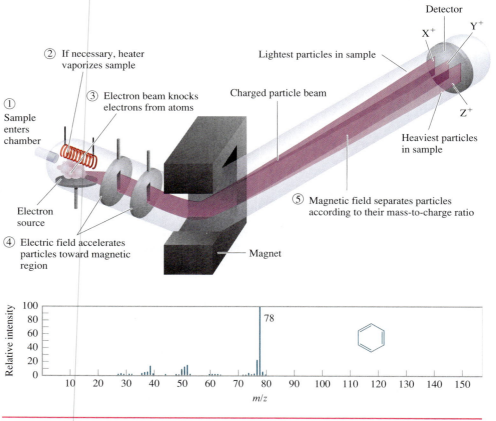

FIGURE 13.40 The mass spectrum of benzene. The peak at $m/z = 78$ corresponds to the C_6H_6 molecular ion.

The mass spectrum of benzene is relatively simple and illustrates some of the information that mass spectrometry provides. The most intense peak in the mass spectrum is called the **base peak** and is assigned a relative intensity of 100. Ion abundances are proportional to peak intensities and are reported as intensities relative to the base peak. The base peak in the mass spectrum of benzene corresponds to the molecular ion (M^+) at $m/z = 78$.

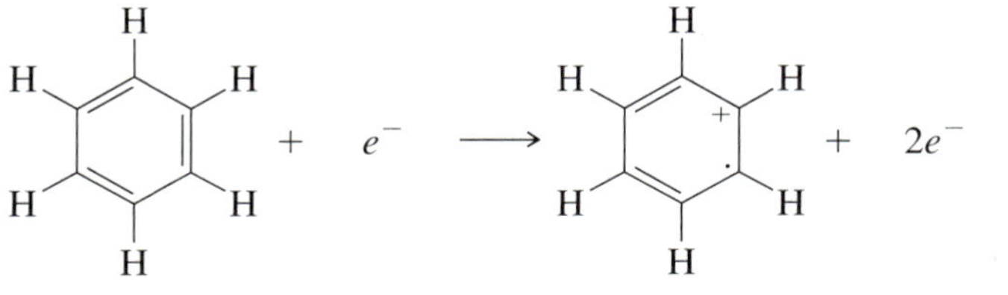

Benzene   Electron   Molecular ion   Two
                             of benzene   electrons

Benzene does not undergo extensive fragmentation; none of the fragment ions in its mass spectrum are as abundant as the molecular ion.

There is a small peak one mass unit higher than M^+ in the mass spectrum of benzene. What is the origin of this peak? What we see in Figure 13.40 as a single mass spectrum is actually a superposition of the spectra of three isotopically distinct benzenes. Most of the benzene molecules contain only ^{12}C and ^{1}H and have a molecular mass of 78. Smaller proportions of benzene molecules contain ^{13}C in place of one of the ^{12}C atoms or ^{2}H in place of one of the protons. Both these species have a molecular mass of 79.

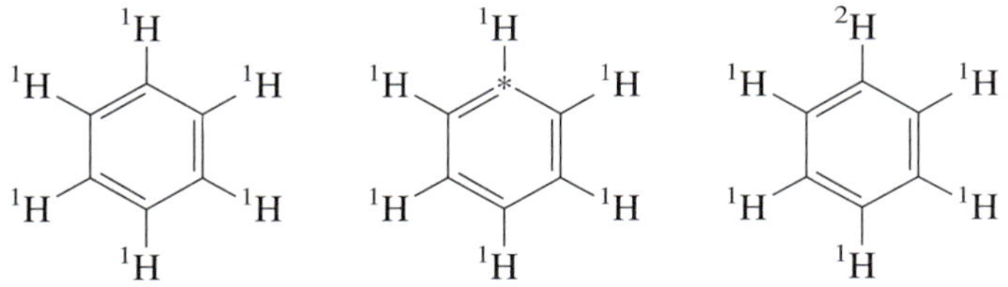

93.4%   6.5%   0.1%
(all carbons are ^{12}C)   (* = ^{13}C)   (all carbons are ^{12}C)
Gives M^+ 78   Gives M^+ 79   Gives M^+ 79

Not only the molecular ion peak but all the peaks in the mass spectrum of benzene are accompanied by a smaller peak one mass unit higher. Indeed, because all organic compounds contain carbon and most contain hydrogen, similar **isotopic clusters** will appear in the mass spectra of all organic compounds.

Isotopic clusters are especially apparent when atoms such as bromine and chlorine are present in an organic compound. The natural ratios of isotopes in these elements are

$$\frac{^{35}Cl}{^{37}Cl} = \frac{100}{32.7} \qquad \frac{^{79}Br}{^{81}Br} = \frac{100}{97.5}$$

Figure 13.41 presents the mass spectrum of chlorobenzene. There are two prominent molecular ion peaks, one at m/z 112 for $C_6H_5{}^{35}Cl$ and the other at m/z 114 for $C_6H_5{}^{37}Cl$. The peak at m/z 112 is three times as intense as the one at m/z 114.

FIGURE 13.41 The mass spectrum of chlorobenzene.

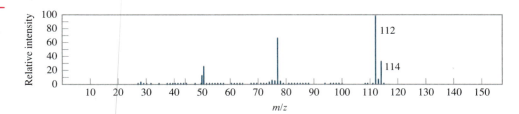

PROBLEM 13.23 Knowing what to look for with respect to isotopic clusters can aid in interpreting mass spectra. How many peaks would you expect to see for the molecular ion in each of the following compounds? At what *m/z* values would these peaks appear? (Disregard the small peaks due to ^{13}C and ^{2}H.)

(a) *p*-Dichlorobenzene   (c) *p*-Dibromobenzene

(b) *o*-Dichlorobenzene   (d) *p*-Bromochlorobenzene

SAMPLE SOLUTION (a) The two isotopes of chlorine are ^{35}Cl and ^{37}Cl. There will be three isotopically different forms of *p*-dichlorobenzene present. They have the structures shown as follows. Each one will give an M$^+$ peak at a different value of *m/z*.

^{35}Cl ^{35}Cl ^{37}Cl

^{35}Cl ^{37}Cl ^{37}Cl

m/z 146   *m/z* 148   *m/z* 150

Unlike the case of benzene, in which ionization involves loss of a π electron from the ring, electron-impact-induced ionization of chlorobenzene involves loss of an electron from an unshared pair of chlorine. The molecular ion then fragments by carbon–chlorine bond cleavage.

| Chlorobenzene | Molecular ion of chlorobenzene | Chlorine atom | Phenyl cation *m/z* 77 |

The peak at *m/z* 77 in the mass spectrum of chlorobenzene in Figure 13.41 is attributed to this fragmentation. Because there is no peak of significant intensity two atomic mass units higher, we know that the cation responsible for the peak at *m/z* 77 cannot contain chlorine.

Some classes of compounds are so prone to fragmentation that the molecular ion peak is very weak. The base peak in most unbranched alkanes, for example, is *m/z* 43, which is followed by peaks of decreasing intensity at *m/z* values of 57, 71, 85, and so on. These peaks correspond to cleavage of each possible carbon–carbon bond in the molecule. This pattern is evident in the mass spectrum of decane, depicted in Figure 13.42. The points of cleavage are indicated in the following diagram:

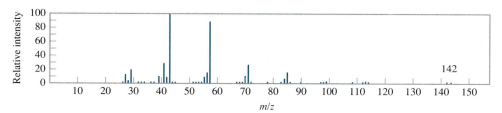

FIGURE 13.42 The mass
spectrum of decane. The
peak for the molecular ion is
extremely small. The most
prominent peaks arise by
fragmentation.

$H_3C—CH_2—CH_2—CH_2—CH_2—CH_2—CH_2—CH_2—CH_2—CH_3$ M^+ 142

127

113

99

85

71

57

43

Many fragmentations in mass spectrometry proceed so as to form a stable carbocation, and the principles that we have developed regarding carbocation stability apply. Alkylbenzenes of the type $C_6H_5CH_2R$ undergo cleavage of the bond to the benzylic carbon to give m/z 91 as the base peak. The mass spectrum in Figure 13.43 and the following fragmentation diagram illustrate this for propylbenzene.

$—CH_2—CH_2—CH_3$ M^+ 120

91

Although this cleavage is probably driven by the stability of benzyl cation, evidence has been obtained suggesting that tropylium cation, formed by rearrangement of benzyl cation, is actually the species responsible for the peak.

The structure of tropylium cation is given in Section 11.21.

PROBLEM 13.24 The base peak appears at m/z 105 for one of the following compounds and at m/z 119 for the other two. Match the compounds with the appropriate m/z values for their base peaks.

CH_2CH_3 $CH_2CH_2CH_3$ CH_3CHCH_3

H_3C CH_3 CH_3 CH_3

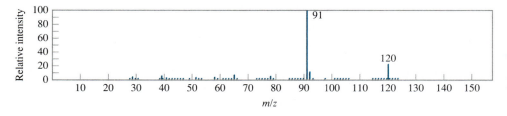

FIGURE 13.43 The mass
spectrum of propylbenzene.
The most intense peak is
$C_7H_7^+$.

GAS CHROMATOGRAPHY, GC/MS, AND MS/MS

All of the spectra in this chapter (^{1}H NMR, ^{13}C NMR, IR, UV-VIS, and MS) were obtained using pure substances. It is much more common, however, to encounter an organic substance, either formed as the product of a chemical reaction or isolated from natural sources, as but one component of a mixture. Just as the last half of the twentieth century saw a revolution in the methods available for the *identification* of organic compounds, so too has it seen remarkable advances in methods for their *separation* and *purification.*

Classical methods for separation and purification include fractional distillation of liquids and recrystallization of solids, and these two methods are routinely included in the early portions of laboratory courses in organic chemistry. Because they are capable of being adapted to work on a large scale, fractional distillation and recrystallization are the preferred methods for purifying organic substances in the pharmaceutical and chemical industries.

Some other methods are more appropriate when separating small amounts of material in laboratory-scale work and are most often encountered there. Indeed, it is their capacity to deal with exceedingly small quantities that is the strength of a number of methods that together encompass the various forms of **chromatography.** The first step in all types of chromatography involves absorbing the sample onto some material called the *stationary phase.* Next, a second phase (the *mobile phase*) is allowed to move across the stationary phase. Depending on the properties of the two phases and the components of the mixture, the mixture is separated into its components according to the rate at which each is removed from the stationary phase by the mobile phase.

In **gas chromatography** (GC), the stationary phase consists of beads of an inert solid support coated with a high-boiling liquid, and the mobile phase is a gas, usually helium. Figure 13.44 shows a typical gas chromatograph. The sample is injected by syringe onto a heated block where a stream of helium carries it onto a coiled column packed with the stationary phase. The components of the mixture move through the column at different rates. They are said to have different *retention times.* Gas chromatography is also referred to as *gas–liquid partition chromatography,* because the technique depends on how different substances partition themselves between the gas phase (dispersed in the helium carrier gas) and the liquid phase (dissolved in the coating on the beads of solid support).

Typically the effluent from a gas chromatograph is passed through a detector, which feeds a signal to a recorder whenever a substance different from pure carrier gas leaves the column. Thus, one

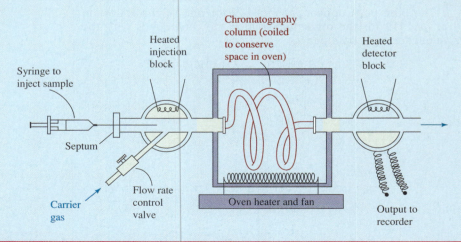

FIGURE 13.44 Diagram of a gas chromatograph. When connected to a mass spectrometer as in GC/MS, the effluent is split into two streams as it leaves the column. One stream goes to the detector, the other to the mass spectrometer. (Adapted, with permission, from H. D. Durst and G. W. Gokel, *Experimental Organic Chemistry,* 2nd ed., McGraw-Hill, New York, 1987.)

determines the number of components in a mixture by counting the number of peaks on a strip chart. It is good practice to carry out the analysis under different conditions by varying the liquid phase, the temperature, and the flow rate of the carrier gas so as to ensure that two substances have not eluted together and given a single peak under the original conditions. Gas chromatography can also be used to identify the components of a mixture by comparing their retention times with those of authentic samples.

In **gas chromatography/mass spectrometry** (GC/MS), the effluent from a gas chromatograph is passed into a mass spectrometer and a mass spectrum is taken every few milliseconds. Thus gas chromatography is used to separate a mixture, and mass spectrometry used to analyze it. GC/MS is a very powerful analytical technique. One of its more visible applications involves the testing of athletes for steroids, stimulants, and other performance-enhancing drugs. These drugs are converted in the body to derivatives called *metabolites,* which are then excreted in the urine. When the urine is subjected to GC/MS analysis, the mass spectra of its organic components are identified by comparison with the mass spectra of known metabolites stored in the instrument's computer. Using a similar procedure, the urine of newborn infants is monitored by GC/MS for metabolite markers of genetic disorders that can be treated if detected early in life. GC/MS is also used to detect and measure the concentration of halogenated hydrocarbons in drinking water.

Although GC/MS is the most widely used analytical method that combines a chromatographic separation with the identification power of mass spectrometry, it is not the only one. Chemists have coupled mass spectrometers to most of the instruments that are used to separate mixtures. Perhaps the ultimate is **mass spectrometry/mass spectrometry** (MS/MS), in which one mass spectrometer generates and separates the molecular ions of the components of a mixture and a second mass spectrometer examines their fragmentation patterns!

Understanding how molecules fragment upon electron impact permits a mass spectrum to be analyzed in sufficient detail to deduce the structure of an unknown compound. Thousands of compounds of known structure have been examined by mass spectrometry, and the fragmentation patterns that characterize different classes are well documented. As various groups are covered in subsequent chapters, aspects of their fragmentation behavior under conditions of electron impact will be described.

13.23 MOLECULAR FORMULA AS A CLUE TO STRUCTURE

As we have just seen, interpreting the fragmentation patterns in a mass spectrum in terms of a molecule's structural units makes mass spectrometry much more than just a tool for determining molecular weights. Nevertheless, even the molecular weight can provide more information than you might think.

A relatively simple example is the **nitrogen rule.** A molecule with an odd number of nitrogens has an odd molecular weight; a molecule with only C, H, and O or with an even number of nitrogens has an even molecular weight.

Aniline (C_6H_7N)   *p*-Nitroaniline ($C_6H_6N_2O_2$)   2,4-Dinitroaniline ($C_6H_5N_3O_4$)

Molecular weight: 93   138   183

A second example concerns different compounds that have the same molecular weight, but different molecular formulas, such as heptane and cyclopropyl acetate.

$$CH_3(CH_2)_5CH_3 \qquad \overset{\displaystyle O}{\underset{\displaystyle \parallel}{CH_3C}}O-\triangleleft$$

Heptane (C_7H_{16})   Cyclopropyl acetate ($C_5H_8O_2$)

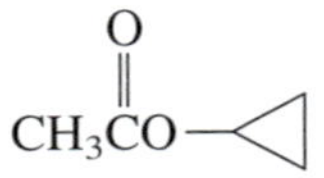

Because we normally round off molecular weights to whole numbers, both have a molecular weight of 100 and both have a peak for their molecular ion at m/z 100 in a typical mass spectrum. Recall, however, that mass spectra contain isotopic clusters that differ according to the isotopes present in each ion. Using the exact values for the major isotopes of C, H, and O, we calculate *exact masses* of m/z of 100.1253 and 100.0524 for the molecular ions of heptane (C_7H_{16}) and cyclopropyl acetate ($C_5H_8O_2$), respectively. As similar as these values are, it is possible to distinguish between them using a *high-resolution mass spectrometer*. This means that the exact mass of a molecular ion can usually be translated into a unique molecular formula.

You can't duplicate these molecular weights for C_7H_{16} and $C_5H_8O_2$ by using the atomic weights given in the periodic table. Those values are for the natural-abundance mixture of isotopes. The exact values are 12.00000 for ^{12}C, 1.00783 for ^{1}H, and 15.9949 for ^{16}O.

Once we have the molecular formula, it can provide information that limits the amount of trial-and-error structure writing we have to do. Consider, for example, heptane and its molecular formula of C_7H_{16}. We know immediately that the molecular formula belongs to an alkane because it corresponds to C_nH_{2n+2}.

What about a substance with the molecular formula C_7H_{14}? This compound cannot be an alkane but may be either a cycloalkane or an alkene, because both these classes of hydrocarbons correspond to the general molecular formula C_nH_{2n}. *Any time a ring or a double bond is present in an organic molecule, its molecular formula has two fewer hydrogen atoms than that of an alkane with the same number of carbons.*

The relationship between molecular formulas, multiple bonds, and rings is referred to as the *index of hydrogen deficiency* and can be expressed by the equation:

Other terms that mean the same thing as the index of hydrogen deficiency include *elements of unsaturation, sites of unsaturation,* and *the sum of double bonds and rings.*

$$\text{Index of hydrogen deficiency} = \tfrac{1}{2}(C_nH_{2n+2} - C_nH_x)$$

where C_nH_x is the molecular formula of the compound.

A molecule that has a molecular formula of C_7H_{14} has an index of hydrogen deficiency of 1:

$$\text{Index of hydrogen deficiency} = \tfrac{1}{2}(C_7H_{16} - C_7H_{14})$$

$$\text{Index of hydrogen deficiency} = \tfrac{1}{2}(2) = 1$$

Thus, the compound has one ring or one double bond. It can't have a triple bond.

A molecule of molecular formula C_7H_{12} has four fewer hydrogens than the corresponding alkane. It has an index of hydrogen deficiency of 2 and can have two rings, two double bonds, one ring and one double bond, or one triple bond.

What about substances other than hydrocarbons, 1-heptanol [$CH_3(CH_2)_5CH_2OH$], for example? Its molecular formula ($C_7H_{16}O$) contains the same carbon-to-hydrogen ratio as heptane and, like heptane, it has no double bonds or rings. Cyclopropyl acetate ($C_5H_8O_2$), the structure of which was given at the beginning of this section, has one ring and one double bond and an index of hydrogen deficiency of 2. Oxygen atoms have no effect on the index of hydrogen deficiency.

A halogen substituent, like hydrogen is monovalent, and when present in a molecular formula is treated as if it were hydrogen for counting purposes.

A more detailed discussion can be found in the May 1995 issue of the *Journal of Chemical Education*, pp. 245–248.

How does one distinguish between rings and double bonds? This additional piece of information comes from catalytic hydrogenation experiments in which the amount of hydrogen consumed is measured exactly. Each of a molecule's double bonds consumes one molar equivalent of hydrogen, but rings are unaffected. For example, a substance with a hydrogen deficiency of 5 that takes up 3 moles of hydrogen must have two rings.

PROBLEM 13.25 How many rings are present in each of the following compounds? Each consumes 2 moles of hydrogen on catalytic hydrogenation.

(a) $C_{10}H_{18}$

(b) C_8H_8

(c) $C_8H_8Cl_2$

(d) C_8H_8O

(e) $C_8H_{10}O_2$

(f) C_8H_9ClO

SAMPLE SOLUTION (a) The molecular formula $C_{10}H_{18}$ contains four fewer hydrogens than the alkane having the same number of carbon atoms ($C_{10}H_{22}$). Therefore, the index of hydrogen deficiency of this compound is 2. Because it consumes two molar equivalents of hydrogen on catalytic hydrogenation, it must have either two double bonds and no rings or a triple bond.

13.24 SUMMARY

Section 13.1   Structure determination in modern-day organic chemistry relies heavily on instrumental methods. Several of the most widely used ones depend on the absorption of electromagnetic radiation.

Section 13.2   Absorption of electromagnetic radiation causes a molecule to be excited from its most stable state (the *ground* state) to a higher energy state (an *excited* state).

Spectroscopic method	*Transitions between*
Nuclear magnetic resonance	Spin states of an atom's nucleus
Infrared	Vibrational states
Ultraviolet-visible	Electronic states

Mass spectrometry is not based on absorption of electromagnetic radiation, but monitors what happens when a substance is ionized by collision with a high-energy electron.

¹H Nuclear Magnetic Resonance Spectroscopy

Section 13.3   In the presence of an external magnetic field, the $+\frac{1}{2}$ and $-\frac{1}{2}$ nuclear spin states of a proton have slightly different energies.

Section 13.4   The energy required to "flip" the spin of a proton from the lower energy spin state to the higher state depends on the extent to which a nucleus is shielded from the external magnetic field by the molecule's electrons.

Section 13.5   Protons in different environments within a molecule have different **chemical shifts;** that is, they experience different degrees of shielding. Chemical shifts (δ) are reported in parts per million (ppm) from tetramethylsilane (TMS). Table 13.1 lists characteristic chemical shifts for various types of protons.

Section 13.6   In addition to *chemical shift,* an ¹H NMR spectrum provides structural information based on:

Number of signals, which tells how many different kinds of protons there are

Integrated areas, which tells the ratios of the various kinds of protons

Splitting pattern, which gives information about the number of protons that are within two or three bonds of the one giving the signal

Section 13.7   **Spin–spin splitting** of NMR signals results from coupling of the nuclear spins that are separated by two bonds (*geminal coupling*) or three bonds (*vicinal coupling*).

Geminal hydrogens are separated by two bonds	Vicinal hydrogens are separated by three bonds

In the simplest cases, the number of peaks into which a signal is split is equal to $n + 1$, where n is the number of protons to which the proton in question is coupled. *Protons that have the same chemical shift do not split each other's signal.*

Section 13.8   The methyl protons of an ethyl group appear as a *triplet* and the methylene protons as a *quartet* in compounds of the type CH_3CH_2X.

Section 13.9   The methyl protons of an isopropyl group appear as a *doublet* and the methine proton as a *septet* in compounds of the type $(CH_3)_2CHX$.

Section 13.10   A *doublet of doublets* characterizes the signals for the protons of the type shown (where W, X, Y, and Z are not H or atoms that split H themselves).

Section 13.11   Complicated splitting patterns can result when a proton is unequally coupled to two or more protons that are different from one another.

Section 13.12   Splitting resulting from coupling to the O—H proton of alcohols is not normally observed, because the hydroxyl proton undergoes rapid intermolecular exchange with other alcohol molecules, which "decouples" it from other protons in the molecule.

Section 13.13   Many processes such as conformational changes take place faster than they can be detected by NMR. Consequently, NMR provides information about the *average* environment of a proton. For example, cyclohexane gives a single peak for its 12 protons even though, at any instant, 6 are axial and 6 are equatorial.

13C Nuclear Magnetic Resonance Spectroscopy

Section 13.14 ^{13}C has a nuclear spin of $\pm\frac{1}{2}$ but only about 1% of all the carbons in a sample are ^{13}C. Nevertheless, high-quality ^{13}C NMR spectra can be obtained by pulse FT techniques and are a useful complement to ^1H NMR spectra.

Section 13.15 ^{13}C signals are more widely separated from one another than proton signals, and ^{13}C NMR spectra are relatively easy to interpret. Table 13.3 gives chemical shift values for carbon in various environments.

Section 13.16 ^{13}C NMR spectra are rarely integrated because the pulse FT technique distorts the signal intensities.

Section 13.17   Carbon signals normally appear as singlets, but several techniques are available that allow one to distinguish among the various kinds of carbons shown.

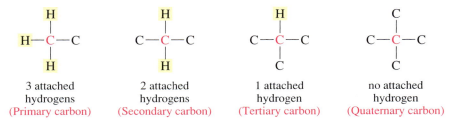

| 3 attached hydrogens (Primary carbon) | 2 attached hydrogens (Secondary carbon) | 1 attached hydrogen (Tertiary carbon) | no attached hydrogen (Quaternary carbon) |

Section 13.18   One of the special techniques for distinguishing carbons according to the number of their attached hydrogens is called **DEPT.** A series of NMR measurements using different pulse sequences gives normal, nulled, and inverted peaks that allow assignment of primary, secondary, tertiary, and quaternary carbons.

Section 13.19   2D NMR techniques are enhancements that are sometimes useful in gaining additional structural information. A ^{1}H-^{1}H COSY spectrum reveals which protons are spin-coupled to other protons, which helps in determining connectivity. A HETCOR spectrum shows the C—H connections by correlating ^{13}C and ^{1}H chemical shifts.

Infrared Spectroscopy

Section 13.20   IR spectroscopy probes molecular structure by examining transitions between vibrational energy levels using electromagnetic radiation in the 625–4000-cm^{-1} range. The presence or absence of a peak at a characteristic frequency tells us whether a certain *functional group* is present. Table 13.4 lists IR absorption frequencies for common structural units.

Ultraviolet-Visible Spectroscopy

Section 13.21   Transitions between electronic energy levels involving electromagnetic radiation in the 200–800-nm range form the basis of UV-VIS spectroscopy. The absorption peaks tend to be broad but are often useful in indicating the presence of particular π *electron* systems within a molecule.

Mass Spectrometry

Section 13.22   Mass spectrometry exploits the information obtained when a molecule is ionized by electron impact and then dissociates to smaller fragments. Positive ions are separated and detected according to their mass-to-charge (m/z) ratio. By examining the fragments and by knowing how classes of molecules dissociate on electron impact, one can deduce the structure of a compound. Mass spectrometry is quite sensitive; as little as 10^{-9} g of compound is sufficient for analysis.

Section 13.23   A compound's molecular formula gives information about the number of double bonds and rings it contains and is a useful complement to spectroscopic methods of structure determination.

PROBLEMS

13.26 Each of the following compounds is characterized by a ^{1}H NMR spectrum that consists of only a single peak having the chemical shift indicated. Identify each compound.

(a) C_8H_{18}; δ 0.9

(b) C_5H_{10}; δ 1.5

(c) C_8H_8; δ 5.8

(d) C_4H_9Br; δ 1.8

(e) $C_2H_4Cl_2$; δ 3.7

(f) $C_2H_3Cl_3$; δ 2.7

(g) $C_5H_8Cl_4$; δ 3.7

(h) $C_{12}H_{18}$; δ 2.2

(i) $C_3H_6Br_2$; δ 2.6

13.27 Each of the following compounds is characterized by a ^{1}H NMR spectrum that consists of two peaks, both singlets, having the chemical shifts indicated. Identify each compound.

(a) C_6H_8; δ 2.7 (4H) and 5.6 (4H)

(b) $C_5H_{11}Br$; δ 1.1 (9H) and 3.3 (2H)

(c) $C_6H_{12}O$; δ 1.1 (9H) and 2.1 (3H)

(d) $C_6H_{10}O_2$; δ 2.2 (6H) and 2.7 (4H)

13.28 Deduce the structure of each of the following compounds on the basis of their ^{1}H NMR spectra and molecular formulas:

(a) C_8H_{10}; δ 1.2 (triplet, 3H)

δ 2.6 (quartet, 2H)

δ 7.1 (broad singlet, 5H)

(b) $C_{10}H_{14}$; δ 1.3 (singlet, 9H)

δ 7.0 to 7.5 (multiplet, 5H)

(c) C_6H_{14}; δ 0.8 (doublet, 12H)

δ 1.4 (heptet, 2H)

(d) C_6H_{12}; δ 0.9 (triplet, 3H)

δ 1.6 (singlet, 3H)

δ 1.7 (singlet, 3H)

δ 2.0 (pentet, 2H)

δ 5.1 (triplet, 1H)

(e) $C_4H_6Cl_4$; δ 3.9 (doublet, 4H)

δ 4.6 (triplet, 2H)

(f) $C_4H_6Cl_2$; δ 2.2 (singlet, 3H)

δ 4.1 (doublet, 2H)

δ 5.7 (triplet, 1H)

(g) C_3H_7ClO; δ 2.0 (pentet, 2H)

δ 2.8 (singlet, 1H)

δ 3.7 (triplet, 2H)

δ 3.8 (triplet, 2H)

(h) $C_{14}H_{14}$; δ 2.9 (singlet, 4H)

δ 7.1 (broad singlet, 10H)

13.29 From among the isomeric compounds of molecular formula C_4H_9Cl, choose the one having a ^{1}H NMR spectrum that

(a) Contains only a single peak

(b) Has several peaks including a doublet at δ 3.4

(c) Has several peaks including a triplet at δ 3.5

(d) Has several peaks including two distinct three-proton signals, one of them a triplet at δ 1.0 and the other a doublet at δ 1.5

13.30 Identify the C_3H_5Br isomers on the basis of the following information:

(a) Isomer A has the ^{1}H NMR spectrum shown in Figure 13.45.

(b) Isomer B has three peaks in its ^{13}C NMR spectrum: δ 32.6 (CH_2); 118.8 (CH_2); and 134.2 (CH).

(c) Isomer C has two peaks in its ^{13}C NMR spectrum: δ 12.0 (CH_2) and 16.8 (CH). The peak at lower field is only half as intense as the one at higher field.

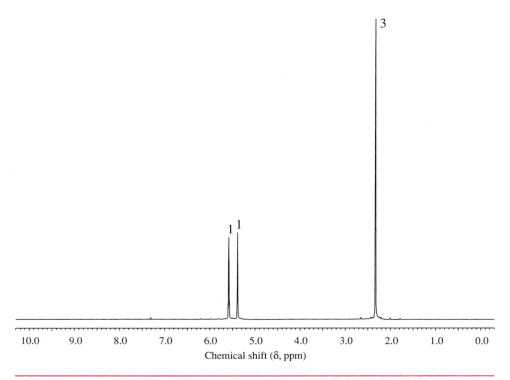

FIGURE 13.45 The 200-MHz ^{1}H NMR spectrum of isomer A of C_3H_5Br (Problem 13.30a).

13.31 Identify each of the $C_4H_{10}O$ isomers on the basis of their ^{13}C NMR spectra:

(a) δ 18.9 (CH$_3$) (two carbons)
    δ 30.8 (CH) (one carbon)
    δ 69.4 (CH$_2$) (one carbon)

(b) δ 10.0 (CH$_3$)
    δ 22.7 (CH$_3$)
    δ 32.0 (CH$_2$)
    δ 69.2 (CH)

(c) δ 31.2 (CH$_3$) (three carbons)
    δ 68.9 (C) (one carbon)

13.32 Identify the C_6H_{14} isomers on the basis of their ^{13}C NMR spectra:

(a) δ 19.1 (CH$_3$)
    δ 33.9 (CH)

(b) δ 13.7 (CH$_3$)
    δ 22.8 (CH$_2$)
    δ 31.9 (CH$_2$)

(c) δ 11.1 (CH$_3$)
    δ 18.4 (CH$_3$)
    δ 29.1 (CH$_2$)
    δ 36.4 (CH)

(d) δ 8.5 (CH$_3$)
    δ 28.7 (CH$_3$)
    δ 30.2 (C)
    δ 36.5 (CH$_2$)

(e) δ 14.0 (CH$_3$)
    δ 20.5 (CH$_2$)
    δ 22.4 (CH$_3$)
    δ 27.6 (CH)
    δ 41.6 (CH$_2$)

13.33 A compound (C_4H_6) has two signals of approximately equal intensity in its ^{13}C NMR spectrum; one is a CH_2 carbon at δ 30.2, the other a CH at δ 136. Identify the compound.

13.34 A compound ($C_3H_7ClO_2$) exhibited three peaks in its ^{13}C NMR spectrum at δ 46.8 (CH_2), δ 63.5 (CH_2), and δ 72.0 (CH). Excluding compounds that have Cl and OH on the same carbon, which are unstable, what is the most reasonable structure for this compound?

13.35 From among the compounds chlorobenzene, *o*-dichlorobenzene, and *p*-dichlorobenzene, choose the one that

(a) Gives the simplest ^1H NMR spectrum

(b) Gives the simplest ^{13}C NMR spectrum

(c) Has three peaks in its ^{13}C NMR spectrum

(d) Has four peaks in its ^{13}C NMR spectrum

13.36 Compounds A and B are isomers of molecular formula $C_{10}H_{14}$. Identify each one on the basis of the ^{13}C NMR spectra presented in Figure 13.46.

13.37 A compound ($C_8H_{10}O$) has the IR and ^1H NMR spectra presented in Figure 13.47. What is its structure?

13.38 Deduce the structure of a compound having the mass spectrum and ^1H NMR spectrum presented in Figure 13.48.

13.39 Figure 13.49 presents several types of spectroscopic data (IR, ^1H NMR, ^{13}C NMR, and mass spectra) for a particular compound. What is it?

13.40 Which would you predict to be more shielded, the inner or outer protons of [24]annulene?

13.41 ^{19}F is the only isotope of fluorine that occurs naturally, and it has a nuclear spin of $\pm\frac{1}{2}$.

(a) Into how many peaks will the proton signal in the ^1H NMR spectrum of methyl fluoride be split?

(b) Into how many peaks will the fluorine signal in the ^{19}F NMR spectrum of methyl fluoride be split?

(c) The chemical shift of the protons in methyl fluoride is δ 4.3. Given that the geminal ^1H—^{19}F coupling constant is 45 Hz, specify the δ values at which peaks are observed in the proton spectrum of this compound at 200 MHz.

The dependence of ^3J on dihedral angle is referred to as the **Karplus relationship** after Martin Karplus (Harvard University) who offered the presently accepted theoretical treatment of it.

13.42 In general, the vicinal coupling constant between two protons varies with the angle between the C—H bonds of the H—C—C—H unit. The coupling constant is greatest when the protons are coplanar (dihedral angle = 0° or 180°) and smallest when the angle is approximately 90°. Describe, with the aid of molecular models, how you could distinguish between *cis*-1-bromo-2-chlorocyclopropane and its trans stereoisomer on the basis of their ^1H NMR spectra.

13.43 The $\pi\rightarrow\pi^*$ transition in the UV spectrum of *trans*-stilbene (*trans*-C_6H_5CH=CHC_6H_5) appears at 295 nm compared with 283 nm for the cis stereoisomer. The extinction coefficient ϵ_{max} is approximately twice as great for *trans*-stilbene as for *cis*-stilbene. Both facts are normally interpreted in terms of more effective conjugation of the π electron system in *trans*-stilbene. Construct a molecular model of each stereoisomer, and identify the reason for the decreased effectiveness of conjugation in *cis*-stilbene.

13.44 ^{31}P is the only phosphorus isotope present at natural abundance and has a nuclear spin of $\pm\frac{1}{2}$. The ^1H NMR spectrum of trimethyl phosphite, $(CH_3O)_3P$, exhibits a doublet for the methyl protons with a splitting of 12 Hz.

(a) Into how many peaks is the ^{31}P signal split?

(b) What is the difference in chemical shift (in hertz) between the lowest and highest field peaks of the ^{31}P multiplet?

See page 586 for problems 13.45–13.50.

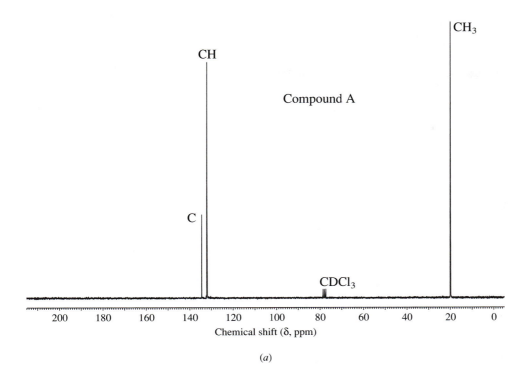

(*a*)

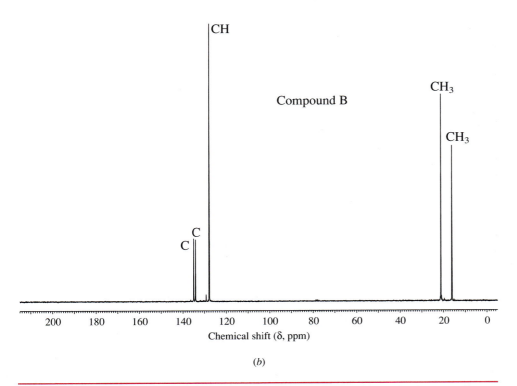

(*b*)

FIGURE 13.46 The ^{13}C NMR spectrum of (*a*) compound A and (*b*) compound B, isomers of $C_{10}H_{14}$ (Problem 13.36).

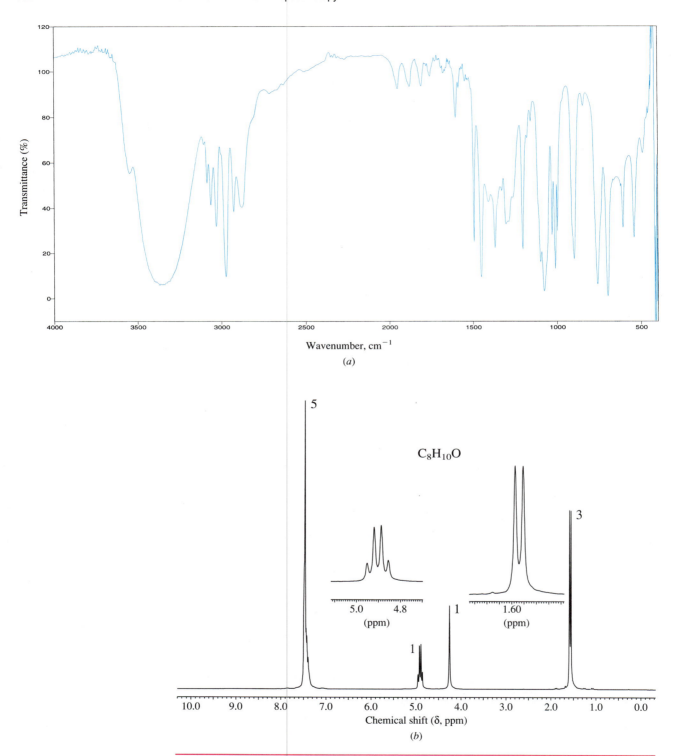

FIGURE 13.47 (a) IR and (b) 200-MHz ^{1}H NMR spectra of compound A $C_8H_{10}O$ (Problem 13.37).

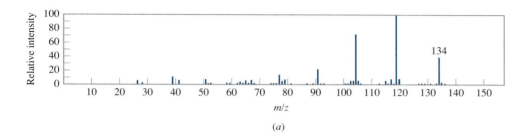

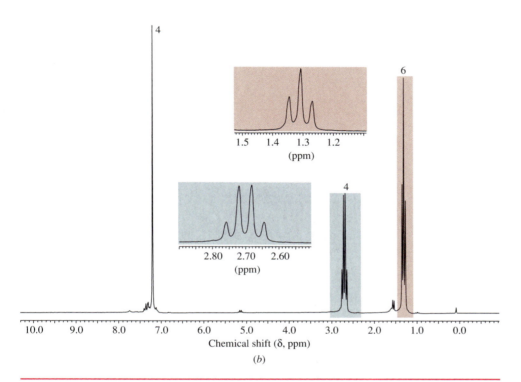

FIGURE 13.48 (a) Mass spectrum and (b) 200-MHz ^{1}H NMR spectrum of compound A (Problem 13.38).

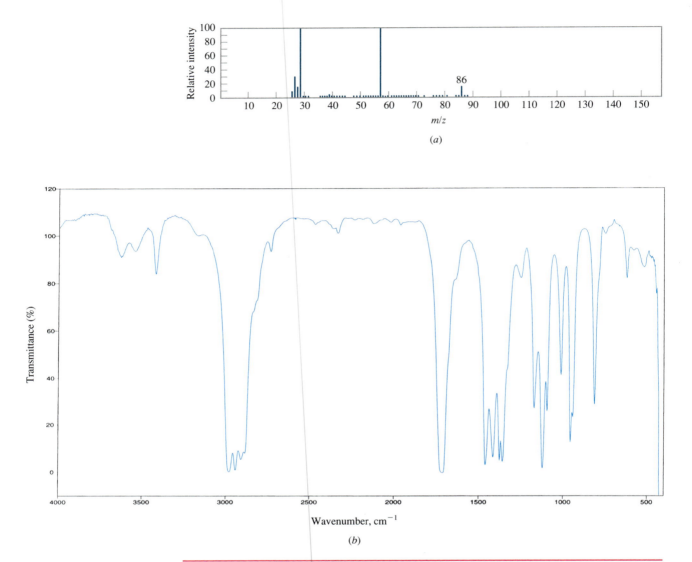

FIGURE 13.49 (*a*) Mass, (*b*) IR, (*c*) 200-MHz ^{1}H NMR, and (*d*) ^{13}C NMR spectra for the compound of Problem 13.39.

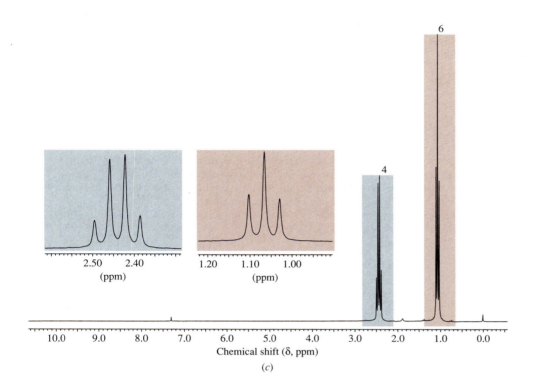

2.50   2.40
(ppm)

1.20   1.10   1.00
(ppm)

10.0   9.0   8.0   7.0   6.0   5.0   4.0   3.0   2.0   1.0   0.0

Chemical shift (δ, ppm)

(c)

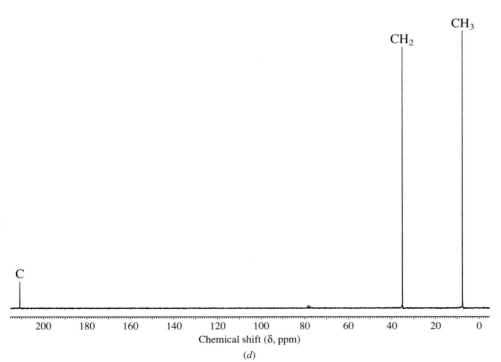

CH₃

CH₂

C

200   180   160   140   120   100   80   60   40   20   0

Chemical shift (δ, ppm)

(d)

13.45 We noted in Section 13.13 that an NMR spectrum is an average spectrum of the conformations populated by a molecule. From the following data, estimate the percentages of axial and equatorial bromine present in bromocyclohexane.

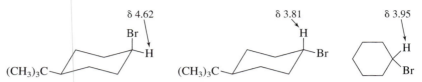

13.46 IR spectroscopy is an inherently "faster" method than NMR, and an IR spectrum is a superposition of the spectra of the various conformations, rather than an average of them. When 1,2-dichloroethane is cooled below its freezing point, the crystalline material gives an IR spectrum consistent with a single species that has a center of symmetry. At room temperature, the IR spectrum of liquid 1,2-dichloroethane retains the peaks present in the solid, but includes new peaks as well. Explain these observations.

13.47 *Microwave spectroscopy* is used to probe transitions between rotational energy levels in molecules.

(a) A typical wavelength for microwaves is 10^{-2} m, compared with 10^{-5} m for IR radiation. Is the energy separation between rotational energy levels in a molecule greater or less than the separation between vibrational energy levels?

(b) Microwave ovens cook food by heating the water in the food. Absorption of microwave radiation by the water excites it to a higher rotational energy state, and it gives off this excess energy as heat when it relaxes to its ground state. Why are vibrational and electronic energy states not involved in this process?

13.48 The peak in the UV-VIS spectrum of acetone $[(CH_3)_2C{=}O]$ corresponding to the $n{\rightarrow}\pi^*$ transition appears at 279 nm when hexane is the solvent, but shifts to 262 nm in water. Which is more polar, the ground electronic state or the excited state?

13.49 A particular vibration will give an absorption peak in the IR spectrum only if the dipole moment of the molecule changes during the vibration. Which vibration of carbon dioxide, the symmetric stretch or the antisymmetric stretch, is "infrared-active"?

$$\overset{\leftarrow}{O}{=}\vec{C}{=}O \qquad \vec{O}{=}\vec{C}{=}O$$

Symmetric stretch   Antisymmetric stretch

13.50 The protons in the methyl group shown in blue in the following structure are highly shielded and give a signal 0.38 *upfield* from TMS. The other methyl group on the same carbon has a more normal chemical shift of 0.86 downfield from TMS. Why is the blue methyl group so highly shielded? (Building a molecular model can help.)

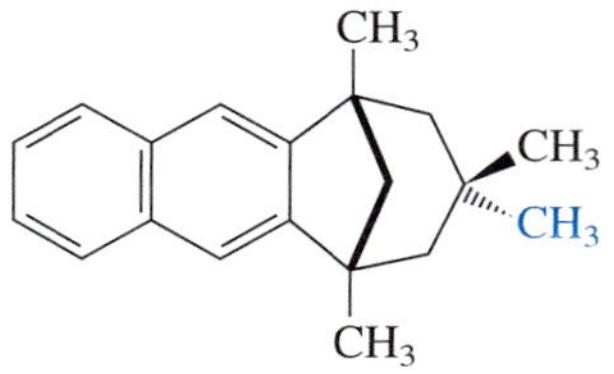

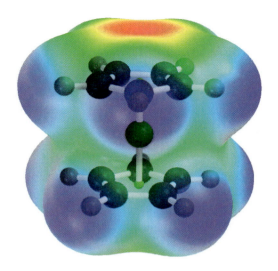

CHAPTER 14

ORGANOMETALLIC COMPOUNDS

O *rganometallic compounds are compounds that have a carbon–metal bond;* they lie at the place where organic and inorganic chemistry meet. You are already familiar with at least one organometallic compound, sodium acetylide (Na$C\equiv$CH), which has an ionic bond between carbon and sodium. But just because a compound contains both a metal and carbon isn't enough to classify it as organometallic. Like sodium acetylide, sodium methoxide (NaOCH_3) is an ionic compound. Unlike sodium acetylide, however, the negative charge in sodium methoxide resides on oxygen, not carbon.

$$Na^+ \ :\bar{C}\equiv CH \qquad\qquad Na^+ \ :\bar{\ddot{O}}CH_3$$

Sodium acetylide
(has a carbon-to-metal bond)

Sodium methoxide
(does not have a carbon-to-metal bond)

The properties of organometallic compounds are much different from those of the other classes we have studied to this point. Most important, many organometallic compounds are powerful sources of nucleophilic carbon, something that makes them especially valuable to the synthetic organic chemist. For example, the preparation of alkynes by the reaction of sodium acetylide with alkyl halides (Section 9.6) depends on the presence of a negatively charged, nucleophilic carbon in acetylide ion.

Synthetic procedures that use organometallic reagents are among the most important methods for carbon–carbon bond formation in organic chemistry. In this chapter you will learn how to prepare organic derivatives of lithium, magnesium, copper, and zinc and see how their novel properties can be used in organic synthesis. We will also finish the story of polyethylene and polypropylene begun in Chapter 6 and continued in Chapter 7 to see the unique mechanism by which organometallic compounds catalyze alkene polymerization.

14.1 ORGANOMETALLIC NOMENCLATURE

Organometallic compounds are named as substituted derivatives of metals. The metal is the parent, and the attached alkyl groups are identified by the appropriate prefix.

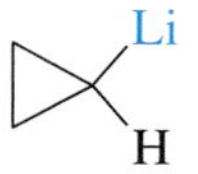

$H_2C=CHNa$ $(CH_3CH_2)_2Mg$

Cyclopropyllithium   Vinylsodium   Diethylmagnesium

When the metal bears a substituent other than carbon, the substituent is treated as if it were an anion and named separately.

CH_3MgI $(CH_3CH_2)_2AlCl$

Methylmagnesium iodide   Diethylaluminum chloride

PROBLEM 14.1 Both of the following organometallic reagents will be encountered later in this chapter. Suggest a suitable name for each.

(a) $(CH_3)_3CLi$ (b)

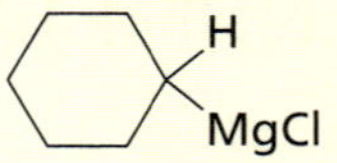

SAMPLE SOLUTION (a) The metal lithium provides the base name for $(CH_3)_3CLi$. The alkyl group to which lithium is bonded is *tert*-butyl, and so the name of this organometallic compound is *tert*-butyllithium. An alternative, equally correct name is 1,1-dimethylethyllithium.

An exception to this type of nomenclature is $NaC\equiv CH$, which is normally referred to as *sodium acetylide*. Both sodium acetylide and ethynylsodium are acceptable IUPAC names.

14.2 CARBON–METAL BONDS IN ORGANOMETALLIC COMPOUNDS

With an electronegativity of 2.5 (Table 14.1), carbon is neither strongly electropositive nor strongly electronegative. When carbon is bonded to an element more electronegative

---

TABLE 14.1	**Electronegativities of Some Elements**

Element	Electronegativity
F	4.0
O	3.5
Cl	3.0
N	3.0
C	2.5
H	2.1
Cu	1.9
Zn	1.6
Al	1.5
Mg	1.2
Li	1.0
Na	0.9
K	0.8

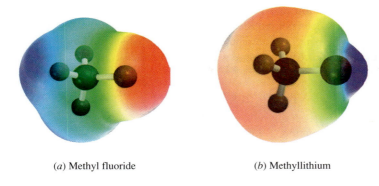

(a) Methyl fluoride   (b) Methyllithium

than itself, such as oxygen or chlorine, the electron distribution in the bond is polarized so that carbon is slightly positive and the more electronegative atom is slightly negative. Conversely, when carbon is bonded to a less electronegative element, such as a metal, the electrons in the bond are more strongly attracted toward carbon.

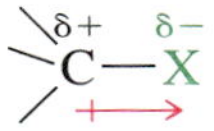

$$\overset{\delta+}{C}-\overset{\delta-}{X} \qquad \overset{\delta-}{C}-\overset{\delta+}{M}$$

X is more electronegative   M is less electronegative
than carbon   than carbon

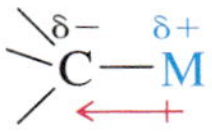

Figure 14.1 uses electrostatic potential maps to show how different the electron distribution is between methyl fluoride (CH_3F) and methyllithium (CH_3Li).

An anion that contains a negatively charged carbon is referred to as a **carbanion.** Covalently bonded organometallic compounds are said to have *carbanionic character.* As the metal becomes more electropositive, the ionic character of the carbon–metal bond becomes more pronounced. Organosodium and organopotassium compounds have ionic carbon–metal bonds; organolithium and organomagnesium compounds tend to have covalent, but rather polar, carbon–metal bonds with significant carbanionic character. *It is the carbanionic character of such compounds that is responsible for their usefulness as synthetic reagents.*

14.3 PREPARATION OF ORGANOLITHIUM COMPOUNDS

Before we describe the applications of organometallic reagents to organic synthesis, let us examine their preparation. Organolithium compounds and other Group I organometallic compounds are prepared by the reaction of an alkyl halide with the appropriate metal.

The reaction of an alkyl halide with lithium is an *oxidation–reduction* reaction. Group I metals are powerful reducing agents.

$$RX \ + \ 2M \ \longrightarrow \ RM \ + \ M^+X^-$$

Alkyl   Group I   Group I   Metal
halide   metal   organometallic   halide
                          compound

$$(CH_3)_3CCl \ + \ 2Li \ \xrightarrow[-30°C]{\text{diethyl ether}} \ (CH_3)_3CLi \ + \ LiCl$$

tert-Butyl chloride   Lithium   *tert*-Butyllithium   Lithium
                                      (75%)   chloride

The alkyl halide can be primary, secondary, or tertiary. Alkyl iodides are the most reactive, followed by bromides, then chlorides. Fluorides are relatively unreactive.

Unlike elimination and nucleophilic substitution reactions, formation of organo-lithium compounds does not require that the halogen be bonded to sp^3-hybridized carbon. Compounds such as vinyl halides and aryl halides, in which the halogen is bonded to sp^2-hybridized carbon, react in the same way as alkyl halides, but at somewhat slower rates.

$$\text{C}_6\text{H}_5\text{—Br} + 2\text{Li} \xrightarrow[\text{35°C}]{\text{diethyl ether}} \text{C}_6\text{H}_5\text{—Li} + \text{LiBr}$$

| Bromobenzene | Lithium | Phenyllithium (95–99%) | Lithium bromide |

Organolithium compounds are sometimes prepared in hydrocarbon solvents such as pentane and hexane, but normally diethyl ether is used. *It is especially important that the solvent be anhydrous.* Even trace amounts of water or alcohols react with lithium to form insoluble lithium hydroxide or lithium alkoxides that coat the surface of the metal and prevent it from reacting with the alkyl halide. Furthermore, organolithium reagents are strong bases and react rapidly with even weak proton sources to form hydrocarbons. We shall discuss this property of organolithium reagents in Section 14.5.

PROBLEM 14.2 Write an equation showing the formation of each of the following from the appropriate bromide:

(a) Isopropenyllithium   (b) *sec*-Butyllithium

SAMPLE SOLUTION (a) In the preparation of organolithium compounds from organic halides, lithium becomes bonded to the carbon that bore the halogen. Therefore, isopropenyllithium must arise from isopropenyl bromide.

$$\text{H}_2\text{C}=\overset{\displaystyle |}{\underset{\displaystyle \text{Br}}{\text{C}}}\text{CH}_3 + 2\text{Li} \xrightarrow{\text{diethyl ether}} \text{H}_2\text{C}=\overset{\displaystyle |}{\underset{\displaystyle \text{Li}}{\text{C}}}\text{CH}_3 + \text{LiBr}$$

| Isopropenyl bromide | Lithium | Isopropenyllithium | Lithium bromide |

Reaction with an alkyl halide takes place at the metal surface. In the first step, an electron is transferred from lithium to the alkyl halide.

$$\text{R:}\ddot{\text{X}}\text{:} + \text{Li·} \longrightarrow [\text{R:}\ddot{\text{X}}\text{:}]^{\bar{}} + \text{Li}^+$$

| Alkyl halide | Lithium | Anion radical | Lithium cation |

Having gained one electron, the alkyl halide is now negatively charged and has an odd number of electrons. It is an *anion radical*. The extra electron occupies an antibonding orbital. This anion radical fragments to an alkyl radical and a halide anion.

$$[\text{R:}\ddot{\text{X}}\text{:}]^{\bar{}} \longrightarrow \text{R·} + \text{:}\ddot{\text{X}}\text{:}^{\bar{}}$$

| Anion radical | Alkyl radical | Halide anion |

Following fragmentation, the alkyl radical rapidly combines with a lithium atom to form the organometallic compound.

$$\text{R·} + \text{Li·} \longrightarrow \text{R:Li}$$

| Alkyl radical | Lithium | Alkyllithium |

14.4 PREPARATION OF ORGANOMAGNESIUM COMPOUNDS: GRIGNARD REAGENTS

The most important organometallic reagents in organic chemistry are organomagnesium compounds. They are called **Grignard reagents** after the French chemist Victor Grignard. Grignard developed efficient methods for the preparation of organic derivatives of magnesium and demonstrated their application in the synthesis of alcohols. For these achievements he was a corecipient of the 1912 Nobel Prize in chemistry.

Grignard reagents are prepared from organic halides by reaction with magnesium, a Group II metal.

$$RX \quad + \quad Mg \quad \longrightarrow \quad RMgX$$

Organic halide   Magnesium   Organomagnesium halide

(R may be methyl or primary, secondary, or tertiary alkyl; it may also be a cycloalkyl, alkenyl, or aryl group.)

Cyclohexyl chloride   Magnesium   Cyclohexylmagnesium chloride
(96%)

diethyl ether
35°C

Bromobenzene   Magnesium   Phenylmagnesium bromide
(95%)

diethyl ether
35°C

Anhydrous diethyl ether is the customary solvent used when preparing organomagnesium compounds. Sometimes the reaction does not begin readily, but once started, it is exothermic and maintains the temperature of the reaction mixture at the boiling point of diethyl ether (35°C).

The order of halide reactivity is I > Br > Cl > F, and alkyl halides are more reactive than aryl and vinyl halides. Indeed, aryl and vinyl chlorides do not form Grignard reagents in diethyl ether. When more vigorous reaction conditions are required, tetrahydrofuran (THF) is used as the solvent.

$$H_2C{=}CHCl \xrightarrow[\text{THF, 60°C}]{Mg} H_2C{=}CHMgCl$$

Vinyl chloride   Vinylmagnesium chloride
(92%)

THF forms a more stable complex with the Grignard reagent and, with a boiling point of 60°C, allows the reaction to be carried out at a higher temperature.

> Grignard shared the prize with Paul Sabatier, who, as was mentioned in Chapter 6, showed that finely divided nickel could be used to catalyze the hydrogenation of alkenes.

> Recall the structure of tetrahydrofuran from Section 3.15:
>
>

PROBLEM 14.3 Write the structure of the Grignard reagent formed from each of the following compounds on reaction with magnesium in diethyl ether:

(a) *p*-Bromofluorobenzene   (c) Iodocyclobutane
(b) Allyl chloride   (d) 1-Bromocyclohexene

SAMPLE SOLUTION (a) Of the two halogen substituents on the aromatic ring, bromine reacts much faster than fluorine with magnesium. Therefore, fluorine is left intact on the ring, but the carbon–bromine bond is converted to a carbon–magnesium bond.

$$F-\text{⟨benzene ring⟩}-Br \ + \ Mg \xrightarrow{\text{diethyl ether}} F-\text{⟨benzene ring⟩}-MgBr$$

p-Bromofluorobenzene   Magnesium   *p*-Fluorophenylmagnesium bromide

The formation of a Grignard reagent is analogous to that of organolithium reagents except that each magnesium atom can participate in two separate one-electron transfer steps:

$$R\!:\!\ddot{X}\!: \ + \ Mg\!: \ \longrightarrow \ [R\!:\!\ddot{X}\!:]^{\overline{}} \ + \ \overset{+}{Mg}\!\cdot$$

Alkyl halide   Magnesium   Anion radical

$$[R\!:\!\ddot{X}\!:]^{\overline{}} \ \longrightarrow \ R\!\cdot \ + \ :\!\ddot{X}\!:^{\overline{}} \ \xrightarrow{\overset{+}{Mg}\cdot} \ R\!:\!Mg^{+}\!:\!\ddot{X}\!:^{\overline{}}$$

Anion radical   Alkyl   Halide   Alkylmagnesium halide
                    radical   ion

Organolithium and organomagnesium compounds find their chief use in the preparation of alcohols by reaction with aldehydes and ketones. Before discussing these reactions, let us first examine the reactions of these organometallic compounds with proton donors.

14.5   ORGANOLITHIUM AND ORGANOMAGNESIUM COMPOUNDS AS BRØNSTED BASES

Organolithium and organomagnesium compounds are stable species when prepared in suitable solvents such as diethyl ether. They are strongly basic, however, and react instantly with proton donors even as weakly acidic as water and alcohols. A proton is transferred from the hydroxyl group to the negatively polarized carbon of the organometallic compound to form a hydrocarbon.

$$R\!-\!M \ + \ R'\!-\!\ddot{O}\!-\!H \ \longrightarrow \ R\!-\!H \ + \ R'\!-\!\ddot{O}\!:^{\overline{}} \ M^{+}$$

stronger   stronger   weaker   weaker
base   acid   acid   base

$$via \qquad \overset{\delta-}{R} \quad \overset{\delta+}{H}\!\!-\!\!\overset{\cdot\cdot}{\underset{\cdot\cdot}{O}}R'$$
$$\underset{\delta+}{\overset{|}{M}}$$

$$CH_3CH_2CH_2CH_2Li \ + \ H_2O \ \longrightarrow \ CH_3CH_2CH_2CH_3 \ + \qquad LiOH$$

Butyllithium   Water   Butane (100%)   Lithium hydroxide

$$\text{⟨benzene ring⟩}-MgBr \ + \ CH_3OH \ \longrightarrow \ \text{⟨benzene ring⟩} \ + \ CH_3OMgBr$$

Phenylmagnesium   Methanol   Benzene   Methoxymagnesium
bromide   (100%)   bromide

PROBLEM 14.4   Use curved arrows to show the flow of electrons in the reaction of butyllithium with water, and that of phenylmagnesium bromide with methanol.

Because of their basicity organolithium compounds and Grignard reagents cannot be prepared or used in the presence of any material that bears an —OH group. Nor are these reagents compatible with —NH or —SH groups, which can also convert an organolithium or organomagnesium compound to a hydrocarbon by proton transfer.

The carbon–metal bonds of organolithium and organomagnesium compounds have appreciable carbanionic character. Carbanions rank among the strongest bases that we'll see in this text. Their conjugate acids are hydrocarbons—very weak acids indeed. The equilibrium constants K_a for ionization of hydrocarbons are much smaller than the K_a's for water and alcohols, thus hydrocarbons have much larger pK_a's.

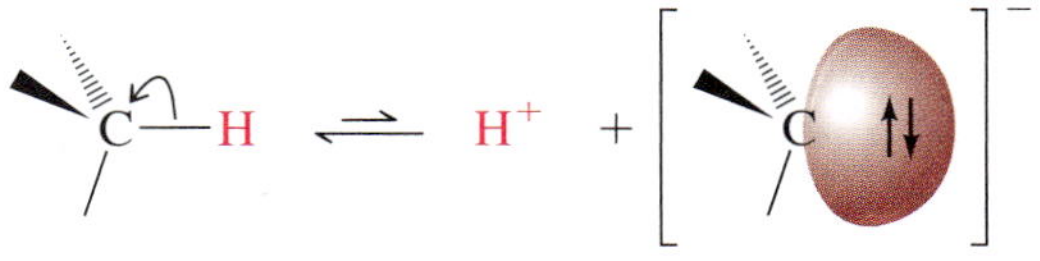

Hydrocarbon (very weak acid)   Proton   Carbanion (very strong base)

Table 14.2 repeats some approximate data presented earlier in Table 1.7 for the acid strengths of representative hydrocarbons and reference compounds.

TABLE 14.2	**Approximate Acidities of Some Hydrocarbons and Reference Materials**		
Compound	**pK_a**	**Formula***	**Conjugate Base**
Methanol	15.2	$CH_3O{-}H$	$CH_3\ddot{O}{:}^-$
Water	15.7	$HO{-}H$	$H\ddot{O}{:}^-$
Ethanol	16	$CH_3CH_2O{-}H$	$CH_3CH_2\ddot{O}{:}^-$
Acetylene	26	$HC{\equiv}C{-}H$	$HC{\equiv}C{:}^-$
Ammonia	36	$H_2N{-}H$	$H_2\ddot{N}{:}^-$
Diethylamine	36	$(CH_3CH_2)_2N{-}H$	$(CH_3CH_2)_2\ddot{N}{:}^-$
Benzene	43	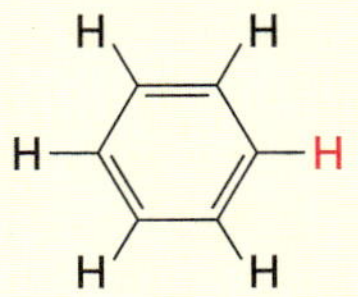 (see structure)	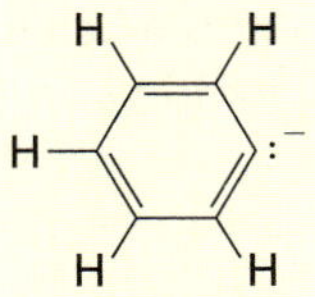 (see structure)
Ethylene	45	$H_2C{=}CH{-}H$	$H_2C{=}\ddot{C}H^-$
Methane	60	$H_3C{-}H$	$H_3C{:}^-$
Ethane	62	$CH_3CH_2{-}H$	$CH_3\ddot{C}H_2{}^-$
2-Methylpropane	71	$(CH_3)_3C{-}H$	$(CH_3)_3C{:}^-$

*The acidic proton in each compound is marked in red.

Acidity decreases from the top of Table 14.2 to the bottom. An acid will transfer a proton to the conjugate base of any acid below it in the table. Organolithium compounds and Grignard reagents act like carbanions and will abstract a proton from any substance more acidic than a hydrocarbon. Thus, N—H groups and terminal alkynes (RC≡C—H) are converted to their conjugate bases by proton transfer to organolithium and organomagnesium compounds.

$$CH_3Li \quad + \quad NH_3 \quad \longrightarrow \quad CH_4 \quad + \quad LiNH_2$$

Methyllithium	Ammonia	Methane	Lithium amide
(stronger base)	(stronger acid: $pK_a = 36$)	(weaker acid: $pK_a = 60$)	(weaker base)

$$CH_3CH_2MgBr \; + \quad HC≡CH \quad \longrightarrow \quad CH_3CH_3 \quad + \quad HC≡CMgBr$$

Ethylmagnesium bromide	Acetylene	Ethane	Ethynylmagnesium bromide
(stronger base)	(stronger acid: $pK_a = 26$)	(weaker acid: $pK_a = 62$)	(weaker base)

PROBLEM 14.5 Butyllithium is commercially available and is frequently used by organic chemists as a strong base. Show how you could use butyllithium to prepare solutions containing

(a) Lithium diethylamide, $(CH_3CH_2)_2NLi$

(b) Lithium 1-hexanolate, $CH_3(CH_2)_4CH_2OLi$

(c) Lithium benzenethiolate, C_6H_5SLi

SAMPLE SOLUTION When butyllithium is used as a base, it abstracts a proton, in this case a proton attached to nitrogen. The source of lithium diethylamide must be diethylamine.

$$(CH_3CH_2)_2NH \; + \; CH_3CH_2CH_2CH_2Li \; \longrightarrow \; (CH_3CH_2)_2NLi \; + \; CH_3CH_2CH_2CH_3$$

Diethylamine	Butyllithium	Lithium diethylamide	Butane
(stronger acid) $pK_a = 36$	(stronger base)	(weaker base)	(weaker acid) $pK_a = 62$

Although butane is not specifically listed in either Table 1.7 or 14.2, we would expect its pK_a to be similar to that of ethane.

It is sometimes necessary in a synthesis to reduce an alkyl halide to a hydrocarbon. In such cases converting the halide to a Grignard reagent and then adding water or an alcohol as a proton source is a satisfactory procedure. Adding D_2O to a Grignard reagent is a commonly used method for introducing deuterium into a molecule at a specific location.

> Deuterium is the mass 2 isotope of hydrogen.

$$CH_3CH=CHBr \xrightarrow[THF]{Mg} CH_3CH=CHMgBr \xrightarrow{D_2O} CH_3CH=CHD$$

1-Bromopropene	Propenylmagnesium bromide	1-Deuteriopropene (70%)

14.6 SYNTHESIS OF ALCOHOLS USING GRIGNARD REAGENTS

The main synthetic application of Grignard reagents is their reaction with certain carbonyl-containing compounds to produce alcohols. Carbon–carbon bond formation is rapid and exothermic when a Grignard reagent reacts with an aldehyde or ketone.

$$\overset{\delta+}{\underset{R-MgX}{\overset{}{C}}}\!\!=\!\!\overset{..\,\delta-}{\overset{}{O}}: \quad\longrightarrow\quad -\overset{|}{\underset{|}{C}}-\overset{..}{\underset{}{O}}:^{-} \qquad \text{normally} \qquad -\overset{|}{\underset{R}{C}}OMgX$$

A carbonyl group is quite polar, and its carbon atom is electrophilic. Grignard reagents are nucleophilic and add to carbonyl groups, forming a new carbon–carbon bond. This addition step leads to an alkoxymagnesium halide, which in the second stage of the synthesis is converted to an alcohol by adding aqueous acid.

$$R-\overset{|}{\underset{|}{C}}-OMgX \; + \; H_3O^+ \;\longrightarrow\; R-\overset{|}{\underset{|}{C}}-OH \; + \; Mg^{2+} \; + \; X^- \; + \; H_2O$$

| Alkoxymagnesium halide | Hydronium ion | Alcohol | Magnesium ion | Halide ion | Water |

The type of alcohol produced depends on the carbonyl compound. Substituents present on the carbonyl group of an aldehyde or ketone stay there—they become substituents on the carbon that bears the hydroxyl group in the product. Thus as shown in Table 14.3 (following page), formaldehyde reacts with Grignard reagents to yield primary alcohols, aldehydes yield secondary alcohols, and ketones yield tertiary alcohols.

PROBLEM 14.6 Write the structure of the product of the reaction of propylmagnesium bromide with each of the following. Assume that the reactions are worked up by the addition of dilute aqueous acid.

(a) Formaldehyde, HCH (with O double bonded to C)

(c) Cyclohexanone,

(b) Benzaldehyde, C$_6$H$_5$CH (with O double bonded to C)

(d) 2-Butanone, CH$_3$CCH$_2$CH$_3$ (with O double bonded to C)

SAMPLE SOLUTION (a) Grignard reagents react with formaldehyde to give primary alcohols having one more carbon atom than the alkyl halide from which the Grignard reagent was prepared. The product is 1-butanol.

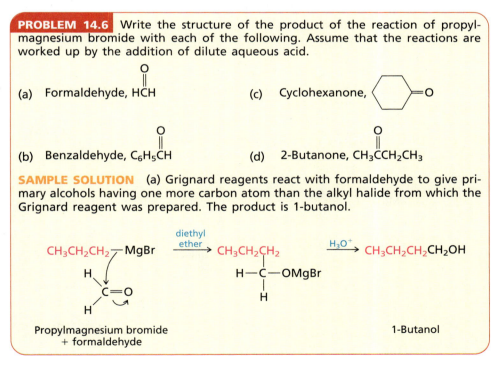

Propylmagnesium bromide + formaldehyde   1-Butanol

An ability to form carbon–carbon bonds is fundamental to organic synthesis. The addition of Grignard reagents to aldehydes and ketones is one of the most frequently used reactions in synthetic organic chemistry. Not only does it permit the extension of carbon chains, but because the product is an alcohol, a wide variety of subsequent functional group transformations is possible.

TABLE 14.3	Reactions of Grignard Reagents with Aldehydes and Ketones

Reaction	General equation and specific example

Reaction with formaldehyde
Grignard reagents react with formaldehyde (H_2C=O) to give *primary alcohols* having one more carbon than the Grignard reagent.

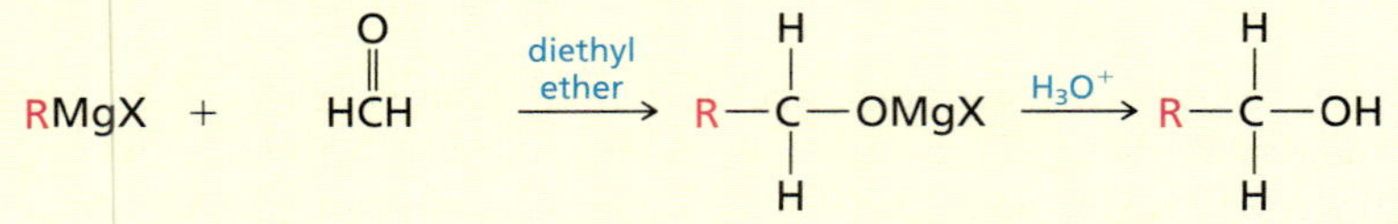

$$RMgX \ + \ \underset{\text{HCH}}{\overset{O}{\parallel}} \ \xrightarrow[\text{ether}]{\text{diethyl}} \ R\!-\!\overset{\displaystyle H}{\underset{\displaystyle H}{\overset{|}{\underset{|}{C}}}}\!-\!OMgX \ \xrightarrow{H_3O^+} \ R\!-\!\overset{\displaystyle H}{\underset{\displaystyle H}{\overset{|}{\underset{|}{C}}}}\!-\!OH$$

Grignard reagent   Formaldehyde   Primary alkoxymagnesium halide   Primary alcohol

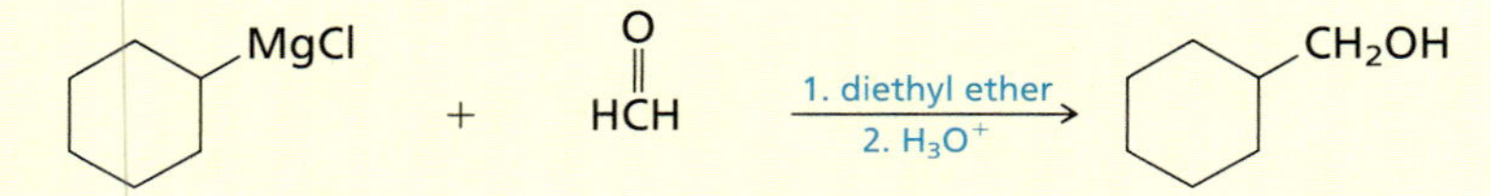

Cyclohexylmagnesium chloride   Formaldehyde   Cyclohexylmethanol (64–69%)

Reaction with aldehydes Grignard reagents react with aldehydes ($R'CH$=O) to give *secondary alcohols*.

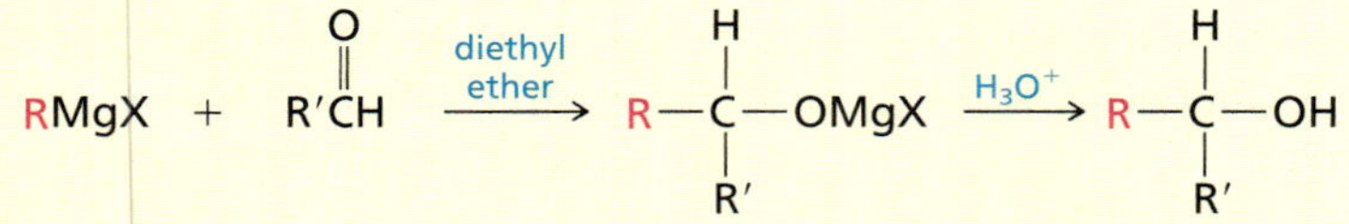

$$RMgX \ + \ \underset{\text{R'CH}}{\overset{O}{\parallel}} \ \xrightarrow[\text{ether}]{\text{diethyl}} \ R\!-\!\overset{\displaystyle H}{\underset{\displaystyle R'}{\overset{|}{\underset{|}{C}}}}\!-\!OMgX \ \xrightarrow{H_3O^+} \ R\!-\!\overset{\displaystyle H}{\underset{\displaystyle R'}{\overset{|}{\underset{|}{C}}}}\!-\!OH$$

Grignard reagent   Aldehyde   Secondary alkoxymagnesium halide   Secondary alcohol

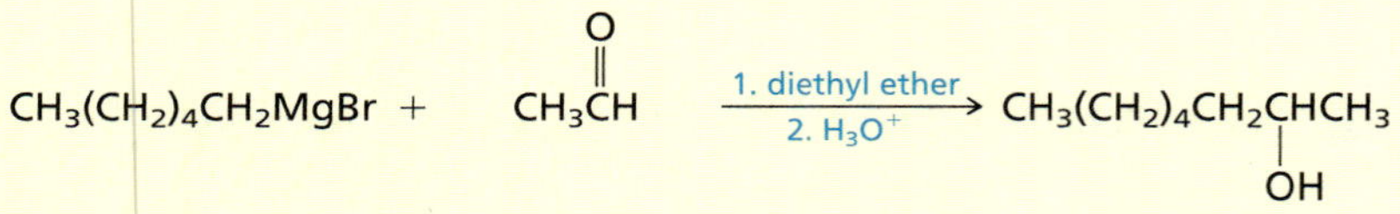

$$CH_3(CH_2)_4CH_2MgBr \ + \ \underset{\text{CH}_3\text{CH}}{\overset{O}{\parallel}} \ \xrightarrow[\text{2. H}_3\text{O}^+]{\text{1. diethyl ether}} \ CH_3(CH_2)_4CH_2\underset{\displaystyle \underset{OH}{|}}{C}HCH_3$$

Hexylmagnesium bromide   Ethanal (acetaldehyde)   2-Octanol (84%)

Reaction with ketones Grignard reagents react with ketones ($R'\overset{O}{\overset{\parallel}{C}}R''$) to give *tertiary alcohols*.

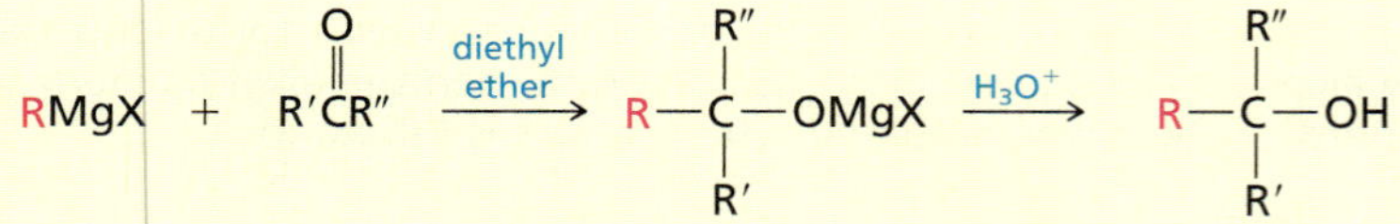

$$RMgX \ + \ \underset{\text{R'CR''}}{\overset{O}{\parallel}} \ \xrightarrow[\text{ether}]{\text{diethyl}} \ R\!-\!\overset{\displaystyle R''}{\underset{\displaystyle R'}{\overset{|}{\underset{|}{C}}}}\!-\!OMgX \ \xrightarrow{H_3O^+} \ R\!-\!\overset{\displaystyle R''}{\underset{\displaystyle R'}{\overset{|}{\underset{|}{C}}}}\!-\!OH$$

Grignard reagent   Ketone   Tertiary alkoxymagnesium halide   Tertiary alcohol

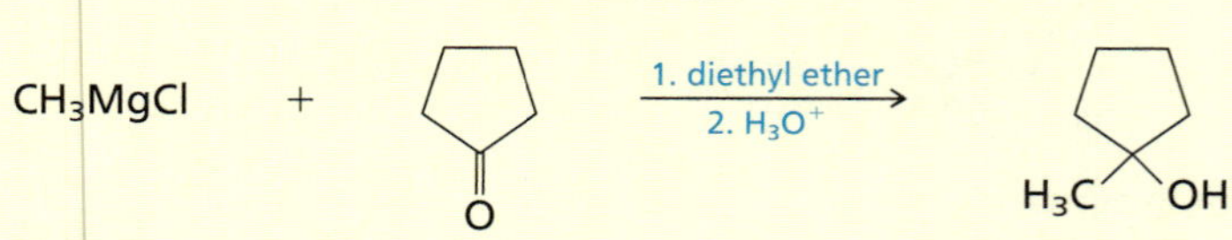

Methylmagnesium chloride   Cyclopentanone   1-Methylcyclopentanol (62%)

14.7 SYNTHESIS OF ALCOHOLS USING ORGANOLITHIUM REAGENTS

Organolithium reagents react with carbonyl groups in the same way that Grignard reagents do. In their reactions with aldehydes and ketones, organolithium reagents are somewhat more reactive than Grignard reagents.

$$RLi \ + \ \underset{\text{Aldehyde}}{\underset{\text{or ketone}}{C=O}} \longrightarrow R-\overset{|}{\underset{|}{C}}-OLi \xrightarrow{H_3O^+} R-\overset{|}{\underset{|}{C}}-OH$$

| Alkyllithium compound | Aldehyde or ketone | Lithium alkoxide | Alcohol |

$$H_2C=CHLi \ + \ \underset{\text{Benzaldehyde}}{C_6H_5-\overset{O}{\overset{\|}{CH}}} \xrightarrow[\text{2. } H_3O^+]{\text{1. diethyl ether}} C_6H_5-\underset{\underset{OH}{|}}{CH}CH=CH_2$$

| Vinyllithium | Benzaldehyde | 1-Phenyl-2-propen-1-ol (76%) |

In this example, the product can be variously described as a *secondary* alcohol, a *benzylic* alcohol, and an *allylic* alcohol. Can you identify the structural reason for each classification?

14.8 SYNTHESIS OF ACETYLENIC ALCOHOLS

The first organometallic compounds we encountered were compounds of the type $RC{\equiv}CNa$ obtained by treatment of terminal alkynes with sodium amide in liquid ammonia (Section 9.6):

$$\underset{\substack{\text{Terminal} \\ \text{alkyne}}}{RC{\equiv}CH} + \underset{\substack{\text{Sodium} \\ \text{amide}}}{NaNH_2} \xrightarrow[-33°C]{NH_3} \underset{\substack{\text{Sodium} \\ \text{alkynide}}}{RC{\equiv}CNa} + \underset{\text{Ammonia}}{NH_3}$$

These compounds are sources of the nucleophilic anion $RC{\equiv}C:^-$, and their reaction with primary alkyl halides provides an effective synthesis of alkynes (Section 9.6). The nucleophilicity of acetylide anions is also evident in their reactions with aldehydes and ketones, which are entirely analogous to those of Grignard and organolithium reagents.

$$\underset{\substack{\text{Sodium} \\ \text{alkynide}}}{RC{\equiv}CNa} + \underset{\substack{\text{Aldehyde} \\ \text{or ketone}}}{R'\overset{O}{\overset{\|}{C}}R''} \xrightarrow{NH_3} \underset{\substack{\text{Sodium salt of an} \\ \text{alkynyl alcohol}}}{RC{\equiv}C-\underset{\underset{R'}{|}}{\overset{\overset{R''}{|}}{C}}-ONa} \xrightarrow{H_3O^+} \underset{\substack{\text{Alkynyl} \\ \text{alcohol}}}{RC{\equiv}C\underset{\underset{R'}{|}}{\overset{\overset{R''}{|}}{C}}OH}$$

These reactions are normally carried out in liquid ammonia because that is the solvent in which the sodium salt of the alkyne is prepared.

$$\underset{\substack{\text{Sodium acetylide}}}{HC{\equiv}CNa} + \underset{\text{Cyclohexanone}}{\text{(cyclohexanone)}} \xrightarrow[\text{2. } H_3O^+]{\text{1. } NH_3} \text{1-Ethynylcyclohexanol}$$

1-Ethynylcyclohexanol
(65–75%)

Acetylenic Grignard reagents of the type $RC\equiv CMgBr$ are prepared, not from an acetylenic halide, but by an acid–base reaction in which a Grignard reagent abstracts a proton from a terminal alkyne.

Which is the stronger acid in this reaction? The weaker acid?

$$CH_3(CH_2)_3C\equiv CH + CH_3CH_2MgBr \xrightarrow{\text{diethyl ether}} CH_3(CH_2)_3C\equiv CMgBr + CH_3CH_3$$

| 1-Hexyne | Ethylmagnesium bromide | 1-Hexynylmagnesium bromide | Ethane |

$$CH_3(CH_2)_3C\equiv CMgBr + \overset{\overset{\displaystyle O}{\parallel}}{HCH} \xrightarrow[\text{2. } H_3O^+]{\text{1. diethyl ether}} CH_3(CH_2)_3C\equiv CCH_2OH$$

| 1-Hexynylmagnesium bromide | Formaldehyde | 2-Heptyn-1-ol (82%) |

PROBLEM 14.7 Write the equation for the reaction of 1-hexyne with ethylmagnesium bromide as if it involved ethyl anion ($CH_3\ddot{C}H_2{}^-$) instead of CH_3CH_2MgBr and use curved arrows to represent the flow of electrons.

14.9 RETROSYNTHETIC ANALYSIS

In our earlier discussions of synthesis, we stressed the value of reasoning backward from the target molecule to suitable starting materials. A name for this process is *retrosynthetic analysis*. Organic chemists have employed this approach for many years, but the term was invented and a formal statement of its principles was set forth only relatively recently by E. J. Corey at Harvard University. Beginning in the 1960s, Corey began studies aimed at making the strategy of organic synthesis sufficiently systematic so that the power of electronic computers could be applied to assist synthetic planning.

Corey was honored with the 1990 Nobel Prize for his achievements in synthetic organic chemistry.

A symbol used to indicate a retrosynthetic step is an open arrow written from product to suitable precursors or fragments of those precursors.

Target molecule ⟹ precursors

Problem 14.7 at the end of the preceding section introduced this idea with the suggestion that ethylmagnesium bromide be represented as ethyl anion.

Often the precursor is not defined completely, but rather its chemical nature is emphasized by writing it as a species to which it is equivalent for synthetic purposes. Thus, a Grignard reagent or an organolithium reagent might be considered synthetically equivalent to a carbanion:

RMgX   or   RLi   is synthetically equivalent to   R:⁻

Figure 14.2 illustrates how retrosynthetic analysis can guide you in planning the synthesis of alcohols by identifying suitable Grignard reagent and carbonyl-containing precursors. In the first step, locate the carbon of the target alcohol that bears the hydroxyl group, remembering that this carbon originated in the $C={O}$ group. Next, as shown in step 2, mentally disconnect a bond between that carbon and one of its attached groups (other than hydrogen). The attached group is the one that is to be transferred from the Grignard reagent. Once you recognize these two structural fragments, the carbonyl partner and the carbanion that attacks it (step 3), you can readily determine the synthetic mode wherein a Grignard reagent is used as the synthetic equivalent of a carbanion (step 4).

Step 1: Locate the hydroxyl-bearing carbon.

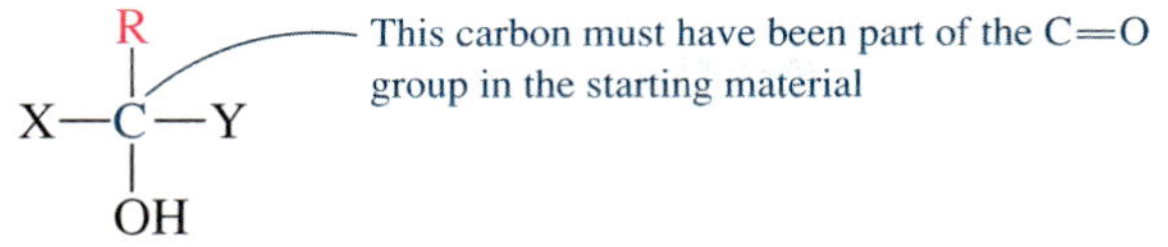

This carbon must have been part of the C=O group in the starting material

Step 2: Disconnect one of the organic substituents attached to the carbon that bears the hydroxyl group.

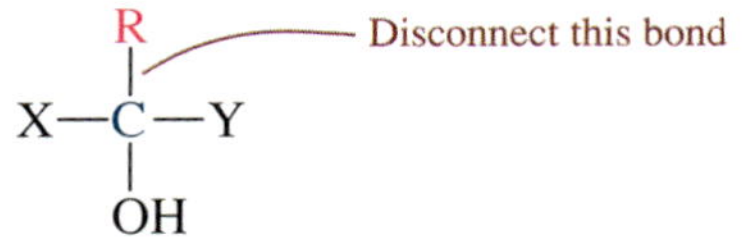

Disconnect this bond

Step 3: Steps 1 and 2 reveal the carbonyl-containing substrate and the carbanionic fragment.

Step 4: Since a Grignard reagent may be considered as synthetically equivalent to a carbanion, this suggests the synthesis shown.

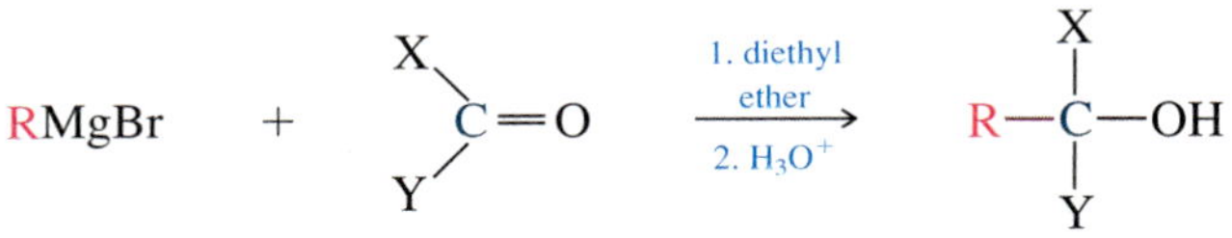

$$RMgBr \quad + \quad \underset{Y}{\overset{X}{>}}C=O \quad \xrightarrow[\text{2. } H_3O^+]{\text{1. diethyl ether}} \quad R-\underset{Y}{\overset{X}{\underset{|}{C}}}-OH$$

FIGURE 14.2 A retrosynthetic analysis of alcohol preparation by way of the addition of a Grignard reagent to an aldehyde or ketone.

Primary alcohols, by this analysis, are seen to be the products of Grignard addition to formaldehyde:

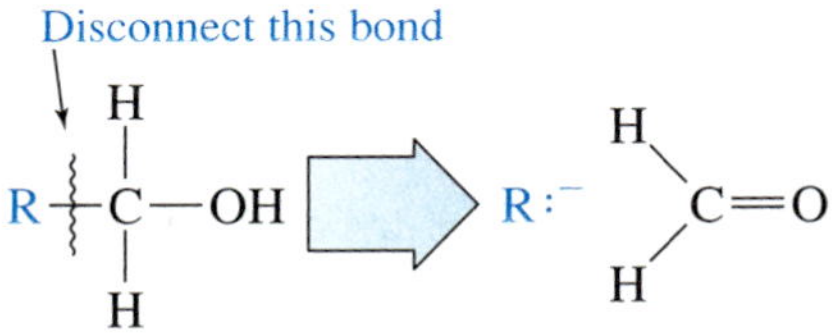

Disconnect this bond

Secondary alcohols may be prepared by *two* different combinations of Grignard reagent and aldehyde:

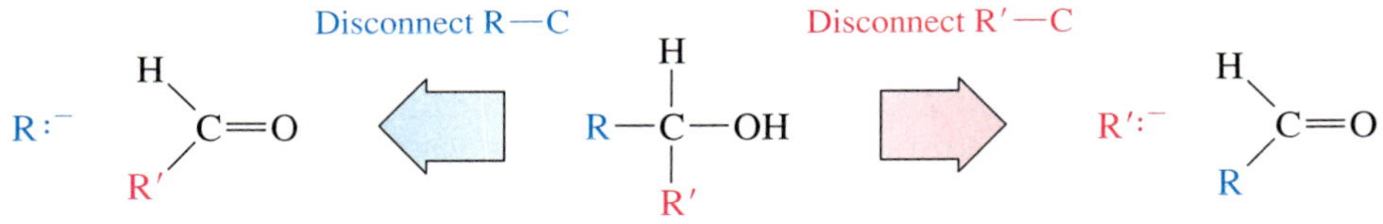

Disconnect R—C   Disconnect R′—C

Three combinations of Grignard reagent and ketone give rise to tertiary alcohols:

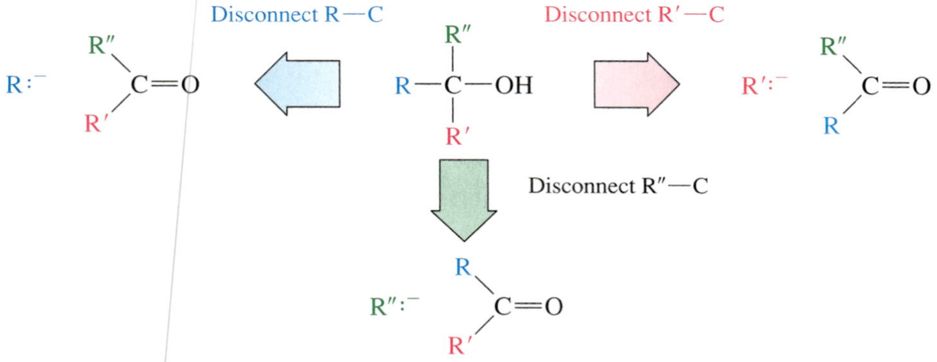

Usually, there is little advantage in choosing one route over another to prepare a particular target alcohol. For example, all three of the following combinations have been used to prepare the tertiary alcohol 2-phenyl-2-butanol:

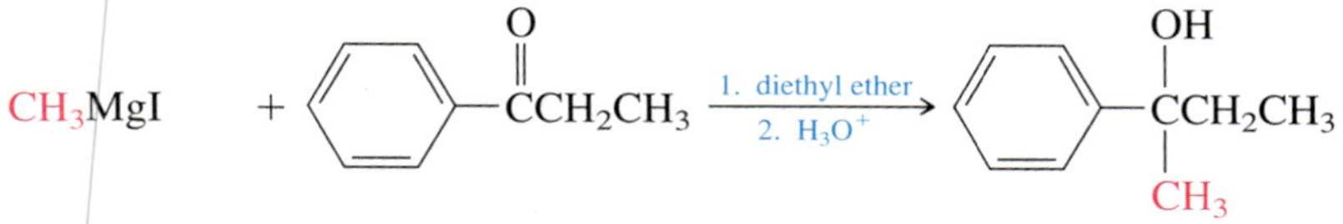

Methylmagnesium iodide   1-Phenyl-1-propanone   2-Phenyl-2-butanol

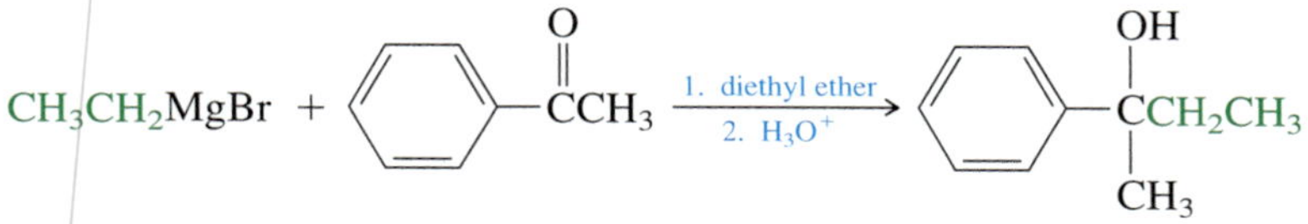

Ethylmagnesium bromide   Acetophenone   2-Phenyl-2-butanol

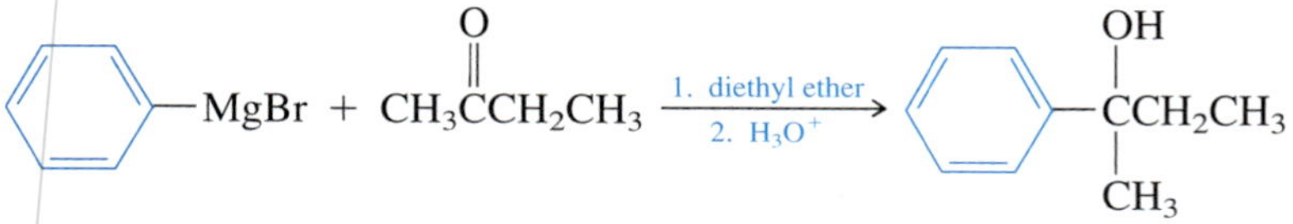

Phenylmagnesium bromide   2-Butanone   2-Phenyl-2-butanol

PROBLEM 14.8 Suggest two ways in which each of the following alcohols might be prepared by using a Grignard reagent:

(a) 2-Hexanol, $CH_3CHCH_2CH_2CH_2CH_3$
                                |
                               OH

(b) 2-Phenyl-2-propanol, $C_6H_5C(CH_3)_2$
                                      |
                                      OH

SAMPLE SOLUTION (a) Because 2-hexanol is a secondary alcohol, we consider the reaction of a Grignard reagent with an aldehyde. Disconnection of bonds to the hydroxyl-bearing carbon generates two pairs of structural fragments:

$$CH_3\overset{\displaystyle |}{\underset{\displaystyle OH}{C}}HCH_2CH_2CH_2CH_3 \quad \Longrightarrow \quad :\overset{-}{C}H_3 \qquad H\overset{\displaystyle O}{\underset{\displaystyle \|}{C}}CH_2CH_2CH_2CH_3$$

$$\text{and} \quad CH_3\overset{\displaystyle |}{\underset{\displaystyle OH}{C}}HCH_2CH_2CH_2CH_3 \quad \Longrightarrow \quad CH_3\overset{\displaystyle O}{\underset{\displaystyle \|}{C}}H \quad :\overset{-}{C}H_2CH_2CH_2CH_3$$

Therefore, one route involves the addition of a methyl Grignard reagent to a five-carbon aldehyde:

$$CH_3MgI \quad + \quad CH_3CH_2CH_2CH_2\overset{\displaystyle O}{\overset{\displaystyle \|}{C}}H \quad \xrightarrow[\text{2. } H_3O^+]{\text{1. diethyl ether}} \quad CH_3CH_2CH_2CH_2\overset{\displaystyle |}{\underset{\displaystyle OH}{C}}HCH_3$$

| Methylmagnesium iodide | Pentanal | 2-Hexanol |

The other requires addition of a butylmagnesium halide to a two-carbon aldehyde:

$$CH_3CH_2CH_2CH_2MgBr \quad + \quad CH_3\overset{\displaystyle O}{\overset{\displaystyle \|}{C}}H \quad \xrightarrow[\text{2. } H_3O^+]{\text{1. diethyl ether}} \quad CH_3CH_2CH_2CH_2\overset{\displaystyle |}{\underset{\displaystyle OH}{C}}HCH_3$$

| Butylmagnesium bromide | Acetaldehyde | 2-Hexanol |

All that has been said in this section applies with equal force to the use of organolithium reagents in the synthesis of alcohols. Grignard reagents are one source of nucleophilic carbon; organolithium reagents are another. Both have substantial carbanionic character in their carbon–metal bonds and undergo the same kind of reaction with aldehydes and ketones.

14.10 PREPARATION OF TERTIARY ALCOHOLS FROM ESTERS AND GRIGNARD REAGENTS

Tertiary alcohols can be prepared by a variation of the Grignard synthesis that uses esters as the source of the carbonyl group. Methyl and ethyl esters are readily available and are the types most often used. Two moles of a Grignard reagent are required per mole of ester; the first mole reacts with the ester, converting it to a ketone.

$$RMgX \; + \; R'\overset{\displaystyle O}{\overset{\displaystyle \|}{C}}OCH_3 \; \xrightarrow{\text{diethyl ether}} \; R'\overset{\displaystyle O-MgX}{\underset{\displaystyle R}{\overset{\displaystyle |}{C}}}OCH_3 \; \longrightarrow \; R'\overset{\displaystyle O}{\overset{\displaystyle \|}{C}}R \; + \; CH_3OMgX$$

| Grignard reagent | Methyl ester | | Ketone | Methoxymagnesium halide |

The ketone is not isolated, but reacts rapidly with the Grignard reagent to give, after adding aqueous acid, a tertiary alcohol. Ketones are more reactive than esters toward

Grignard reagents, and so it is not normally possible to interrupt the reaction at the ketone stage even if only one equivalent of the Grignard reagent is used.

$$\underset{\substack{\text{Ketone}}}{R'\overset{\displaystyle O}{\overset{\|}{C}}R} \;+\; \underset{\substack{\text{Grignard}\\\text{reagent}}}{RMgX} \xrightarrow[\text{2. } H_3O^+]{\text{1. diethyl ether}} \underset{\substack{\text{Tertiary}\\\text{alcohol}}}{R'\underset{\displaystyle R}{\overset{\displaystyle OH}{\overset{|}{\underset{|}{C}}}}R}$$

Two of the groups bonded to the hydroxyl-bearing carbon of the alcohol are the same because they are derived from the Grignard reagent. For example,

$$2CH_3MgBr \;+\; (CH_3)_2CH\overset{\displaystyle O}{\overset{\|}{C}}OCH_3 \xrightarrow[\text{2. } H_3O^+]{\text{1. diethyl ether}} (CH_3)_2CH\underset{\displaystyle CH_3}{\overset{\displaystyle OH}{\overset{|}{\underset{|}{C}}}}CH_3 \;+\; CH_3OH$$

| Methylmagnesium bromide | Methyl 2-methylpropanoate | 2,3-Dimethyl-2-butanol (73%) | Methanol |

PROBLEM 14.9 What combination of ester and Grignard reagent could you use to prepare each of the following tertiary alcohols?

(a) $C_6H_5\underset{\displaystyle OH}{\overset{|}{\underset{|}{C}}}(CH_2CH_3)_2$ (b) $(C_6H_5)_2\underset{\displaystyle OH}{\overset{|}{\underset{|}{C}}}\!\!-\!\!\triangleleft$

SAMPLE SOLUTION (a) To apply the principles of retrosynthetic analysis to this case, we disconnect both ethyl groups from the tertiary carbon and identify them as arising from the Grignard reagent. The phenyl group originates in an ester of the type $C_6H_5CO_2R$ (a benzoate ester).

$$C_6H_5\underset{\displaystyle OH}{\overset{|}{\underset{|}{C}}}(CH_2CH_3)_2 \;\Longrightarrow\; C_6H_5\overset{\displaystyle O}{\overset{\|}{C}}OR \;+\; 2CH_3CH_2MgX$$

An appropriate synthesis would be

$$2CH_3CH_2MgBr \;+\; C_6H_5\overset{\displaystyle O}{\overset{\|}{C}}OCH_3 \xrightarrow[\text{2. } H_3O^+]{\text{1. diethyl ether}} C_6H_5\underset{\displaystyle OH}{\overset{|}{\underset{|}{C}}}(CH_2CH_3)_2$$

| Ethylmagnesium bromide | Methyl benzoate | 3-Phenyl-3-pentanol |

14.11 ALKANE SYNTHESIS USING ORGANOCOPPER REAGENTS

Organometallic compounds of copper were known for a long time before their versatility in synthetic organic chemistry was fully appreciated. The most useful ones are the lithium dialkylcuprates, which result when a copper(I) halide reacts with two equivalents of an alkyllithium in diethyl ether or tetrahydrofuran.

$$2RLi \quad + \quad CuX \quad \xrightarrow[\text{or THF}]{\text{diethyl ether}} \quad R_2CuLi \quad + \quad LiX$$

Alkyllithium	Cu(I) halide	Lithium	Lithium
	(X = Cl, Br, I)	dialkylcuprate	halide

Organocopper compounds used for carbon–carbon bond formation are called *Gilman reagents* in honor of Henry Gilman who first studied them. Gilman's career in teaching and research at Iowa State spanned more than half a century (1919–1975).

In the first stage of the preparation, one molar equivalent of alkyllithium displaces halide from copper to give an alkylcopper(I) species:

$$R\!-\!Li \longrightarrow RCu \quad + \quad LiI$$
$$\overset{\displaystyle }{\underset{\displaystyle Cu\!-\!I}{}}$$

Alkylcopper   Lithium iodide

The second molar equivalent of the alkyllithium adds to the alkylcopper to give a negatively charged dialkyl-substituted derivative of copper(I) called a *dialkylcuprate*. It is formed as its lithium salt, a lithium dialkylcuprate.

$$Li\!-\!R \quad + \quad Cu\!-\!R \longrightarrow [R\!-\!\bar{Cu}\!-\!R]Li^+$$

Alkyllithium	Alkylcopper	Lithium dialkylcuprate
		(soluble in diethyl ether and in THF)

Lithium dialkylcuprates react with alkyl halides to produce alkanes by carbon–carbon bond formation between the alkyl group of the alkyl halide and the alkyl group of the dialkylcuprate:

$$R_2CuLi \quad + \quad R'X \longrightarrow R\!-\!R' \quad + \quad RCu \quad + \quad LiX$$

Lithium dialkylcuprate	Alkyl halide	Alkane	Alkylcopper	Lithium halide

Methyl and primary alkyl halides, especially iodides, work best. Elimination becomes a problem with secondary and tertiary alkyl halides:

$$(CH_3)_2CuLi \quad + \quad CH_3(CH_2)_8CH_2I \quad \xrightarrow[\text{0°C}]{\text{diethyl ether}} \quad CH_3(CH_2)_8CH_2CH_3$$

Lithium dimethylcuprate	1-Iododecane	Undecane (90%)

Lithium diarylcuprates are prepared in the same way as lithium dialkylcuprates and undergo comparable reactions with primary alkyl halides:

$$(C_6H_5)_2CuLi \quad + \quad ICH_2(CH_2)_6CH_3 \quad \xrightarrow{\text{diethyl ether}} \quad C_6H_5CH_2(CH_2)_6CH_3$$

Lithium diphenylcuprate	1-Iodooctane	1-Phenyloctane (99%)

   The most frequently used organocuprates are those in which the alkyl group is primary. Steric hindrance makes secondary and tertiary dialkylcuprates less reactive, and they tend to decompose before they react with the alkyl halide. The reaction of cuprate reagents with alkyl halides follows the usual S_N2 order: $CH_3 >$ primary $>$ secondary $>$ tertiary, and $I > Br > Cl > F$. *p*-Toluenesulfonates are somewhat more reactive than halides. Because the alkyl halide and dialkylcuprate reagent should both be primary in order to produce satisfactory yields of coupled products, the reaction is limited to the formation of $RCH_2\!-\!CH_2R'$ and $RCH_2\!-\!CH_3$ bonds in alkanes.

A key step in the reaction mechanism appears to be nucleophilic attack on the alkyl halide by the negatively charged copper atom, but the details of the mechanism are not well understood. Indeed, there is probably more than one mechanism by which cuprates react with organic halogen compounds. Vinyl halides and aryl halides are known to be very unreactive toward nucleophilic attack, yet react with lithium dialkylcuprates:

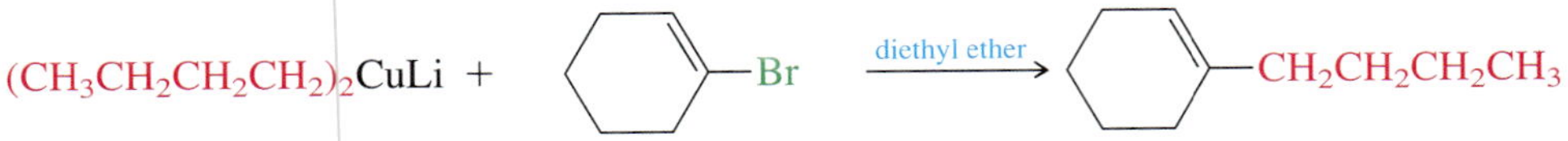

$(CH_3CH_2CH_2CH_2)_2CuLi$ + [1-Bromocyclohexene] $\xrightarrow{\text{diethyl ether}}$ [1-Butylcyclohexene] $CH_2CH_2CH_2CH_3$

Lithium dibutylcuprate · · · 1-Bromocyclohexene · · · 1-Butylcyclohexene (80%)

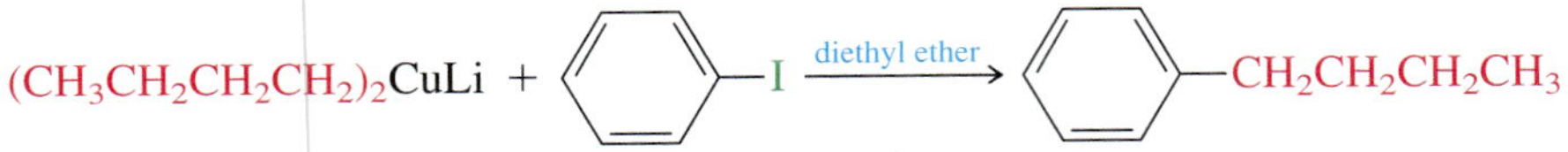

$(CH_3CH_2CH_2CH_2)_2CuLi$ + [Iodobenzene]—I $\xrightarrow{\text{diethyl ether}}$ [Butylbenzene]—$CH_2CH_2CH_2CH_3$

Lithium dibutylcuprate · · · Iodobenzene · · · Butylbenzene (75%)

PROBLEM 14.10 Suggest a combination of organic halide and cuprate reagent appropriate for the preparation of each of the following compounds:

(a) 2-Methylbutane

(b) 1,3,3-Trimethylcyclopentene

SAMPLE SOLUTION (a) First inspect the target molecule to see which bonds are capable of being formed by reaction of an alkyl halide and a cuprate, bearing in mind that neither the alkyl halide nor the alkyl group of the lithium dialkylcuprate should be secondary or tertiary.

A bond between a methyl group and a methylene group can be formed.

$$H_3C-CH_2-\underset{\underset{CH_3}{|}}{CH}-CH_3$$

None of the bonds to the methine group can be formed efficiently.

There are two combinations, both acceptable, that give the H_3C-CH_2 bond:

$(CH_3)_2CuLi$ + $BrCH_2CH(CH_3)_2$ $\longrightarrow$ $CH_3CH_2CH(CH_3)_2$

Lithium dimethylcuprate · · · 1-Bromo-2-methylpropane · · · 2-Methylbutane

CH_3I + $LiCu[CH_2CH(CH_3)_2]_2$ $\longrightarrow$ $CH_3CH_2CH(CH_3)_2$

Iodomethane · · · Lithium diisobutylcuprate · · · 2-Methylbutane

---

14.12 AN ORGANOZINC REAGENT FOR CYCLOPROPANE SYNTHESIS

Zinc reacts with alkyl halides in a manner similar to that of magnesium.

$$RX + Zn \xrightarrow{\text{diethyl ether}} RZnX$$

Alkyl halide · · · Zinc · · · Alkylzinc halide

Victor Grignard was led to study organomagnesium compounds because of earlier work he performed with organic derivatives of zinc.

Organozinc reagents are not nearly as reactive toward aldehydes and ketones as Grignard reagents and organolithium compounds but are intermediates in certain reactions of alkyl halides.

An organozinc compound that occupies a special niche in organic synthesis is *iodomethylzinc iodide* (ICH_2ZnI). It is prepared by the reaction of zinc–copper couple [Zn(Cu), zinc that has had its surface activated with a little copper] with diiodomethane in diethyl ether.

$$CH_2I_2 \quad + \quad Zn \quad \xrightarrow[\text{Cu}]{\text{diethyl ether}} \quad ICH_2ZnI$$

Diiodomethane   Zinc   Iodomethylzinc iodide

What makes iodomethylzinc iodide such a useful reagent is that it reacts with alkenes to give cyclopropanes.

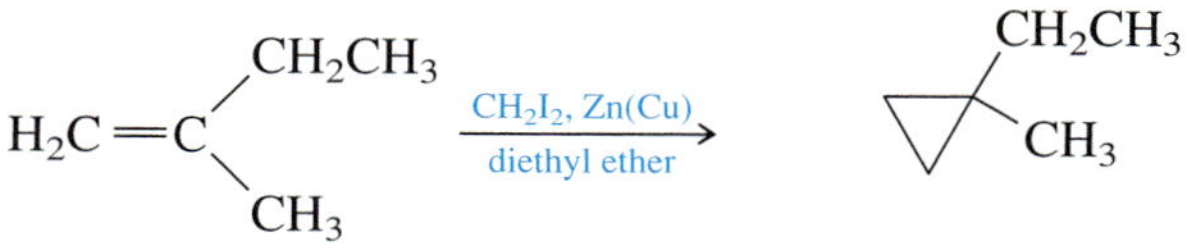

$$H_2C=C\Big\langle\begin{smallmatrix}CH_2CH_3\\CH_3\end{smallmatrix} \quad \xrightarrow[\text{diethyl ether}]{CH_2I_2,\ Zn(Cu)} \quad \triangle\begin{smallmatrix}CH_2CH_3\\CH_3\end{smallmatrix}$$

2-Methyl-1-butene   1-Ethyl-1-methylcyclopropane
                                (79%)

This reaction is called the *Simmons–Smith reaction* and is one of the few methods available for the synthesis of cyclopropanes. Mechanistically, the Simmons–Smith reaction seems to proceed by a single-step cycloaddition of a methylene (CH_2) unit from iodomethylzinc iodide to the alkene:

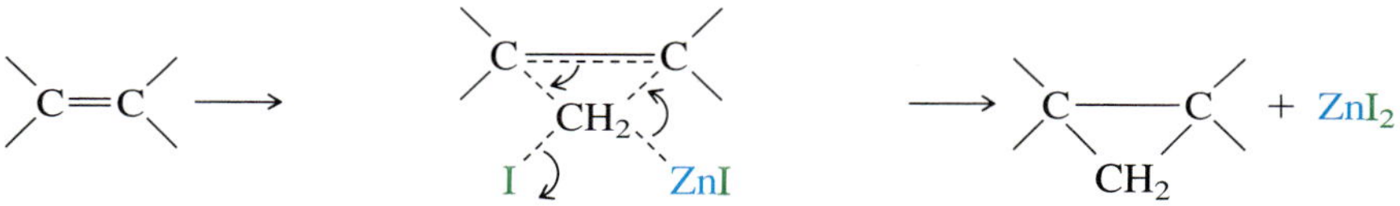

$$\sideset{}{}{\mathop{C=C}} \longrightarrow \underset{\underset{I\qquad ZnI}{CH_2}}{\overset{}{C-----C}} \longrightarrow \underset{CH_2}{C---C} + ZnI_2$$

ICH_2ZnI Transition state for methylene transfer

PROBLEM 14.11   What alkenes would you choose as starting materials in order to prepare each of the following cyclopropane derivatives by reaction with iodomethylzinc iodide?

(a) 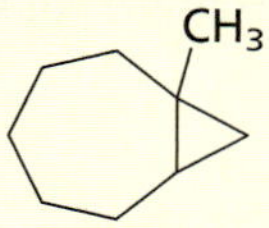   CH$_3$ (b)

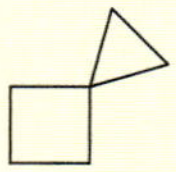

SAMPLE SOLUTION   (a) In a cyclopropane synthesis using the Simmons–Smith reagent, you should remember that a CH_2 unit is transferred. Therefore, retrosynthetically disconnect the bonds to a CH_2 group of a three-membered ring to identify the starting alkene.

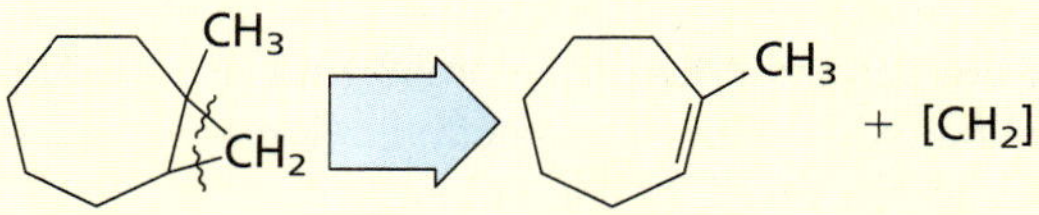

The complete synthesis is

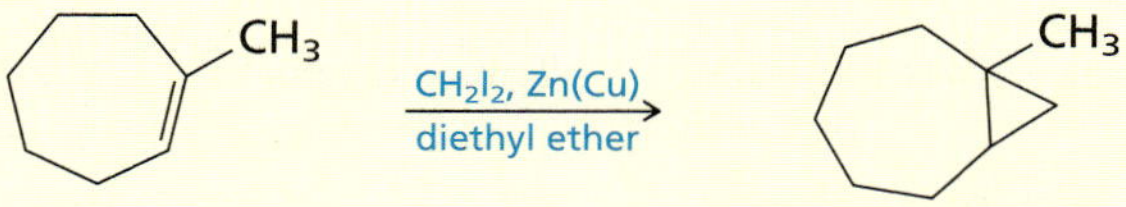

1-Methylcycloheptene $\xrightarrow[\text{diethyl ether}]{CH_2I_2,\ Zn(Cu)}$ 1-Methylbicyclo[5.1.0]octane
                                    (55%)

Methylene transfer from iodomethylzinc iodide is *stereospecific*. Substituents that were cis in the alkene remain cis in the cyclopropane.

$$CH_3CH_2 \quad CH_2CH_3$$
(Z)-3-Hexene $\xrightarrow[\text{ether}]{\substack{CH_2I_2 \\ Zn(Cu)}}$ cis-1,2-Diethylcyclopropane (34%)

(E)-3-Hexene $\xrightarrow[\text{ether}]{\substack{CH_2I_2 \\ Zn(Cu)}}$ trans-1,2-Diethylcyclopropane (15%)

Yields in Simmons–Smith reactions are sometimes low. Nevertheless, because it often provides the only feasible route to a particular cyclopropane derivative, it is a valuable addition to the organic chemist's store of synthetic methods.

14.13 CARBENES AND CARBENOIDS

Divalent carbon species first received attention with the work of the Swiss-American chemist J. U. Nef in the late nineteenth century; they were then largely ignored until the 1950s.

Iodomethylzinc iodide is often referred to as a **carbenoid,** meaning that it resembles a **carbene** in its chemical reactions. Carbenes are neutral molecules in which one of the carbon atoms has six valence electrons. Such carbons are *divalent;* they are directly bonded to only two other atoms and have no multiple bonds. Iodomethylzinc iodide reacts as if it were a source of the carbene $H—\ddot{C}—H$.

It is clear that free :CH_2 is not involved in the Simmons–Smith reaction, but there is substantial evidence to indicate that carbenes are formed as intermediates in certain other reactions that convert alkenes to cyclopropanes. The most studied examples of these reactions involve dichlorocarbene and dibromocarbene.

Dichlorocarbene   Dibromocarbene

Carbenes are too reactive to be isolated and stored, but have been trapped in frozen argon for spectroscopic study at very low temperatures.

Dihalocarbenes are formed when trihalomethanes are treated with a strong base, such as potassium *tert*-butoxide. The trihalomethyl anion produced on proton abstraction dissociates to a dihalocarbene and a halide anion:

$$Br_3C—H \quad + \quad \bar{:}\ddot{O}C(CH_3)_3 \longrightarrow Br_3\bar{C}: \quad + \quad H—\ddot{O}C(CH_3)_3$$

| Tribromomethane | *tert*-Butoxide ion | Tribromomethide ion | *tert*-Butyl alcohol |

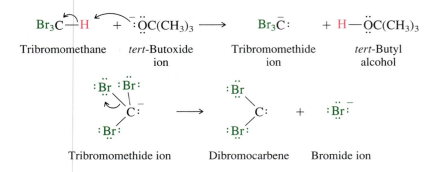

Tribromomethide ion   Dibromocarbene   Bromide ion

The process in which a dihalocarbene is formed from a trihalomethane is an elimination in which a proton and a halide are lost from the same carbon. It is an *α-elimination*.

When generated in the presence of an alkene, dihalocarbenes undergo cycloaddition to the double bond to give dihalocyclopropanes.

Cyclohexene   Tribromomethane   7,7-Dibromobicyclo[4.1.0]heptane
                                                                    (75%)

$$\text{Cyclohexene} + CHBr_3 \xrightarrow[\text{(CH}_3)_3\text{COH}]{\text{KOC(CH}_3)_3}$$

The reaction of dihalocarbenes with alkenes is stereospecific, and syn addition is observed.

> **PROBLEM 14.12**   The syn stereochemistry of dibromocarbene cycloaddition was demonstrated in experiments using *cis*- and *trans*-2-butene. Give the structure of the product obtained from addition of dibromocarbene to each alkene.

Bonding in dihalocarbenes is based on sp^2 hybridization of carbon, shown for CCl_2 in Figure 14.3a. Two of carbon's sp^2 hybrid orbitals are involved in σ bonds to the halogen. The third sp^2 orbital contains the unshared electron pair, and the unhybridized $2p$ orbital is vacant. The electrostatic potential map in Figure 14.3b illustrates this nicely with the highest negative character (red) concentrated in the region of the lone pair orbital, and the region of highest positive charge (blue) situated above and below the plane of the molecule.

Thus, with dihalocarbenes, we have the interesting case of a species that resembles both a carbanion (unshared pair of electrons on carbon) and a carbocation (empty p orbital). Which structural feature controls its reactivity? Does its empty p orbital cause it to react as an electrophile? Does its unshared pair make it nucleophilic? By comparing the rate of reaction of CBr_2 toward a series of alkenes with that of typical electrophiles toward the same alkenes (Table 14.4), we see that the reactivity of CBr_2

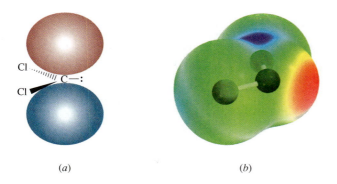

(a)   (b)

FIGURE 14.3 (a) The unshared electron pair occupies an sp^2-hybridized orbital in dichlorocarbene. There are no electrons in the unhybridized p orbital. (b) An electrostatic potential map of dichlorocarbene shows negative charge is concentrated in the region of the unshared pair, and positive charge above and below the carbon.

TABLE 14.4   Relative Reactivity Toward Alkenes

Alkene	CBr_2	Br_2	Epoxidation
$(CH_3)_2C{=}C(CH_3)_2$	3.5	2.5	very fast
$(CH_3)_2C{=}CHCH_3$	3.2	1.9	13.5
$(CH_3)_2C{=}CH_2$	1.00	1.00	1.00
$CH_3CH_2CH_2CH_2CH{=}CH_2$	0.07	0.36	0.05

parallels that of typical electrophilic reagents such as Br_2 and peroxy acids. Therefore, dibromocarbene is electrophilic, and it is reasonable to conclude that electrons flow from the π system of the alkene to the empty p orbital of the carbene in the rate-determining step of cyclopropane formation.

14.14 TRANSITION-METAL ORGANOMETALLIC COMPOUNDS

A large number of organometallic compounds are based on transition metals. Examples include organic derivatives of iron, nickel, chromium, platinum, and rhodium. Many important industrial processes are catalyzed by transition metals or their complexes. Before we look at these processes, a few words about the structures of transition-metal complexes are in order.

A transition-metal complex consists of a transition-metal to which are attached groups called **ligands.** Essentially, anything attached to a metal is a ligand. A ligand can be an element (O_2, N_2), a compound (NO), or an ion (CN^-); it can be inorganic as in the examples just cited or it can be an organic ligand. Ligands differ in the number of electrons that they share with the transition metal to which they are attached. Carbon monoxide is a frequently encountered ligand in transition-metal complexes and contributes two electrons; it is best thought of in terms of the Lewis structure $:C{\equiv}\overset{+}{O}:$ in which carbon is the reactive site. An example of a carbonyl complex of a transition metal is nickel carbonyl, a very toxic substance, which was first prepared over a hundred years ago and is an intermediate in the purification of nickel. It forms spontaneously when carbon monoxide is passed over elemental nickel.

$$Ni \quad + \quad 4CO \quad \longrightarrow \quad Ni(CO)_4$$

Nickel   Carbon monoxide   Nickel carbonyl

Many transition-metal complexes, including $Ni(CO)_4$, obey the **18-electron rule,** which is to transition-metal complexes as the octet rule is to main-group elements like carbon and oxygen. It states that

> For transition-metal complexes, *the number of ligands that can be attached to a metal will be such that the sum of the electrons brought by the ligands plus the valence electrons of the metal equals 18.*

With an atomic number of 28, nickel has the electron configuration $[Ar]4s^23d^8$ (ten valence electrons). The 18-electron rule is satisfied by adding to these ten the eight electrons from four carbon monoxide ligands. A useful point to remember about the 18-electron rule when we discuss some reactions of transition-metal complexes is that if the number is less than 18, the metal is considered *coordinatively unsaturated* and can accept additional ligands.

PROBLEM 14.13   Like nickel, iron reacts with carbon monoxide to form a compound having the formula M(CO)$_n$ that obeys the 18-electron rule. What is the value of n in the formula Fe(CO)$_n$?

Not all ligands use just two electrons to bond to transition metals. Chromium has the electron configuration [Ar]$4s^2 3d^4$ (6 valence electrons) and needs 12 more to satisfy the 18-electron rule. In the compound (benzene)tricarbonylchromium, 6 of these 12 are the π electrons of the benzene ring; the remaining 6 are from the three carbonyl ligands.

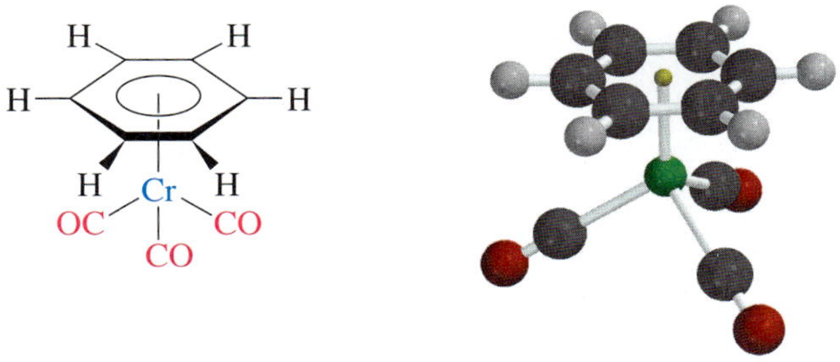

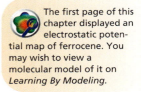

(Benzene)tricarbonylchromium

Ferrocene has an even more interesting structure. A central iron is π-bonded to two cyclopentadienyl ligands in what is aptly described as a *sandwich*. It, too, obeys the 18-electron rule. Each cyclopentadienyl ligand contributes five electrons for a total of ten and iron, with an electron configuration of [Ar]$4s^2 3d^6$ contributes eight. Alternatively, ferrocene can be viewed as being derived from Fe^{2+} (six valence electrons) and two aromatic cyclopentadienide rings (six electrons each).

The first page of this chapter displayed an electrostatic potential map of ferrocene. You may wish to view a molecular model of it on *Learning By Modeling*.

Ferrocene

Indeed, ferrocene was first prepared by adding iron(II) chloride to cyclopentadienylsodium. Instead of the expected σ-bonded species shown in the equation, ferrocene was formed.

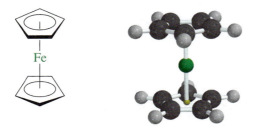

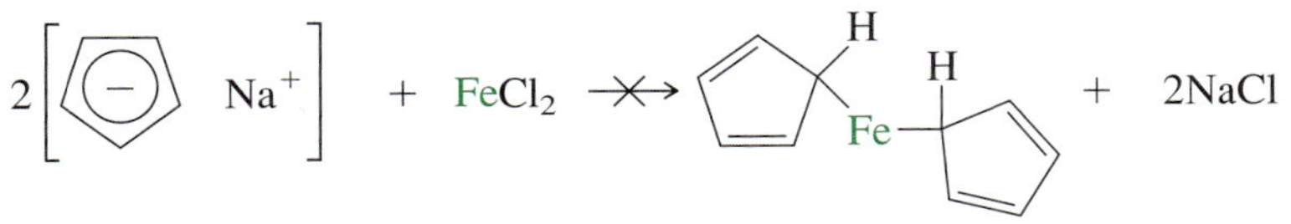

Cyclopentadienylsodium   Iron(II)   (Not formed)
                          chloride

Cyclopentadienylsodium is ionic. Its anion is the aromatic cyclopentadienide ion, which contains six π electrons.

The preparation and structure determination of ferrocene marked the beginning of metallocene chemistry. **Metallocenes** are organometallic compounds that bear cyclopentadienide ligands. A large number are known, even some in which uranium is the metal. Metallocenes are not only stucturally interesting, but many of them have useful applications as catalysts for industrial processes. Zirconium-based metallocenes, for example, are the most widely used catalysts for Ziegler–Natta polymerization of alkenes. We'll have more to say about them in Section 14.15.

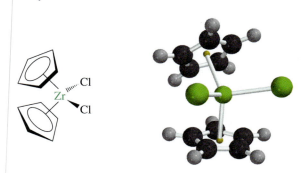

(Bis)-cyclopentadienylzirconium dichloride

Naturally occurring compounds with carbon–metal bonds are very rare. The best example of such an organometallic compound is coenzyme B_{12}, which has a carbon–cobalt σ bond (Figure 14.4). Pernicious anemia results from a coenzyme B_{12} deficiency and can be treated by adding sources of cobalt to the diet. One source of cobalt is vitamin B_{12}, a compound structurally related to, but not identical with, coenzyme B_{12}.

14.15 ZIEGLER–NATTA CATALYSIS OF ALKENE POLYMERIZATION

In Section 6.21 we listed three main methods for polymerizing alkenes: cationic, free-radical, and coordination polymerization. In Section 7.15 we extended our knowledge of polymers to their stereochemical aspects by noting that although free-radical polymerization of propene gives atactic polypropylene, coordination polymerization produces a stereoregular polymer with superior physical properties. Because the catalysts responsible for coordination polymerization are organometallic compounds, we are now in a position to examine coordination polymerization in more detail, especially with respect to how the catalyst works.

In the early 1950s, Karl Ziegler, then at the Max Planck Institute for Coal Research in Germany, was studying the use of aluminum compounds as catalysts for the oligomerization of ethylene.

$$n\text{H}_2\text{C}{=}\text{CH}_2 \xrightarrow{\text{Al(CH}_2\text{CH}_3)_3} \text{CH}_3\text{CH}_2(\text{CH}_2\text{CH}_2)_{n-2}\text{CH}{=}\text{CH}_2$$

Ethylene   Ethylene oligomers

Ziegler found that adding certain metals or their compounds to the reaction mixture led to the formation of ethylene oligomers with 6–18 carbons, but others promoted the formation of very long carbon chains giving polyethylene. Both were major discoveries. The 6–18 carbon ethylene oligomers constitute a class of industrial organic chemicals known as *linear α olefins* that are produced at a rate of 3×10^9 pounds/year in the

AN ORGANOMETALLIC COMPOUND THAT OCCURS NATURALLY: COENZYME B₁₂

Pernicious anemia is a disease characterized, as are all anemias, by a deficiency of red blood cells. Unlike ordinary anemia, pernicious anemia does not respond to treatment with sources of iron, and before effective treatments were developed, was often fatal. Injection of liver extracts was one such treatment, and in 1948 chemists succeeded in isolating the "antipernicious anemia factor" from beef liver as a red crystalline compound, which they called **vitamin B₁₂**. This compound had the formula $C_{63}H_{88}CoN_{14}O_{14}P$. Its complexity precluded structure determination by classical degradation techniques, and spectroscopic methods were too primitive to be of much help. The structure was solved by Dorothy Crowfoot Hodgkin of Oxford University in 1955 using X-ray diffraction techniques and is shown in Figure 14.4a. Structure determination by X-ray crystallography can be superficially considered as taking a photograph of a molecule with X-rays. It is a demanding task and earned Hodgkin the 1964 Nobel Prize in chemistry. Modern structural studies by X-ray crystallography use computers to collect and analyze the diffraction data and take only a fraction of the time required years ago to solve the vitamin B₁₂ structure.

The structure of vitamin B₁₂ is interesting in that it contains a central cobalt atom that is surrounded by six atoms in an octahedral geometry. One substituent, the cyano (—CN) group, is what is known as an "artifact." It appears to be introduced into the molecule during the isolation process and leads to the synonym **cyanocobalamin** for vitamin B₁₂. This is the material used to treat pernicious anemia, but is not the form in which it exerts its activity. The biologically active substance is called **coenzyme B₁₂** and differs from vitamin B₁₂ in the ligand attached to cobalt (Figure 14.4b). Coenzyme B₁₂ is the only known naturally occurring substance that has a carbon–metal bond. Moreover, coenzyme B₁₂ was discovered before any compound containing an alkyl group σ-bonded to cobalt had ever been isolated in the laboratory!

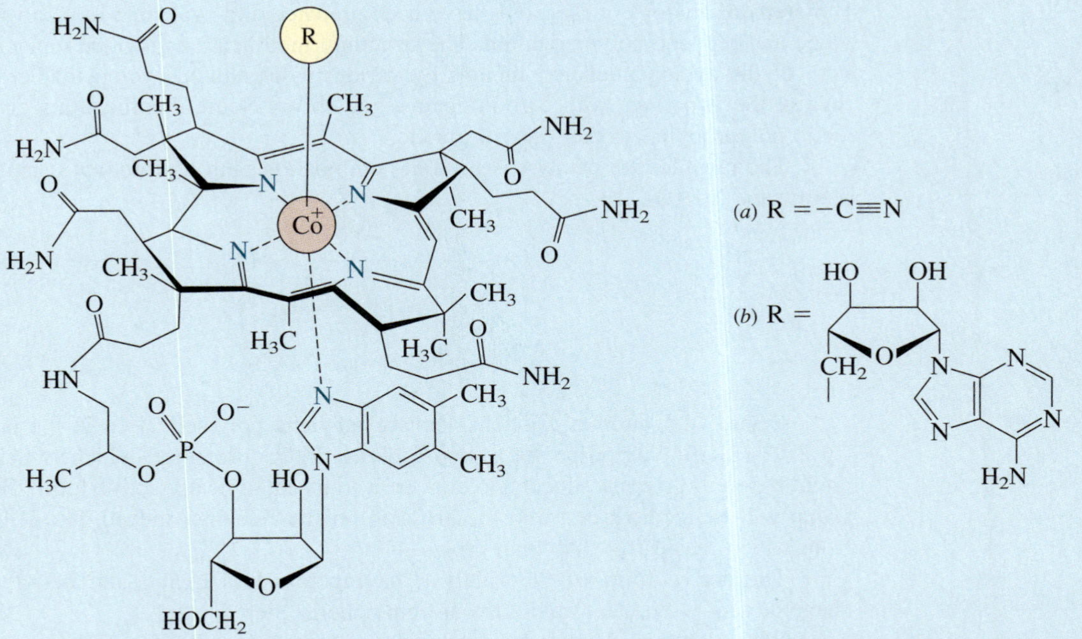

FIGURE 14.4 The structures of (a) vitamin B₁₂ and (b) coenzyme B₁₂.

611

United States. The Ziegler route to polyethylene is even more important because it occurs at modest temperatures and pressures and gives *high-density polyethylene,* which has properties superior to the low-density material formed by the free-radical polymerization described in Section 6.21.

Ziegler had a working relationship with the Italian chemical company Montecatini, for which Giulio Natta of the Milan Polytechnic Institute was a consultant. When Natta used Ziegler's catalyst to polymerize propene, he discovered that the catalyst was not only effective but that it gave mainly isotactic polypropylene. (Recall from Section 7.15 that free-radical polymerization of propene gives atactic polypropylene.) Isotactic polypropylene has a higher melting point than the atactic form and can be drawn into fibers or molded into hard, durable materials.

The earliest Ziegler–Natta catalysts were combinations of titanium tetrachloride $(TiCl_4)$ and diethylaluminum chloride $[(CH_3CH_2)_2AlCl]$, but these have given way to more effective zirconium-based metallocenes, the simplest of which is bis(cyclopentadienyl)zirconium dichloride (Section 14.14).

<aside>
Zirconium lies below titanium in the periodic table, so was an obvious choice in the search for other Ziegler–Natta catalysts.
</aside>

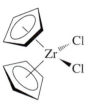

Bis(cyclopentadienyl)zirconium dichloride (Cp_2ZrCl_2)

Hundreds of analogs of Cp_2ZrCl_2 have been prepared and evaluated as catalysts for ethylene and propene polymerization. The structural modifications include replacing one or both of the cyclopentadienyl ligands by variously substituted cyclopentadienyl groups, linking the two rings with carbon chains, and so on. Some modifications give syndiotactic polypropylene, others give isotactic.

The metallocene catalyst is used in combination with a promoter, usually methylalumoxane (MAO).

$$\left[O-Al-O-Al \atop \underset{CH_3}{|} \quad \underset{CH_3}{|} \right]_n$$

Methylalumoxane (MAO)

Figure 14.5 outlines a mechanism for ethylene polymerization in the presence of Cp_2ZrCl_2. Step 1 describes the purpose of the MAO promoter, which is to transfer a methyl group to the metallocene to convert it to its catalytically active form. This methyl group will be incorporated into the growing polymer chain—indeed, it will be the end from which the rest of the chain grows.

The active form of the catalyst, having one less ligand and being positively charged, acts as an electrophile toward ethylene in Step 2.

With electrons flowing from ethylene to zirconium, the $Zr-CH_3$ bond weakens, the carbons of ethylene become positively polarized, and the methyl group migrates from zirconium to one of the carbons of ethylene. Cleavage of the $Zr-CH_3$ bond is accompanied by formation of a σ bond between zirconium and one of the carbons of ethylene in Step 3. The product of this step is a chain-extended form of the active catalyst, ready to accept another ethylene ligand and repeat the chain extending steps.

Step 1: Cp_2ZrCl_2 is converted to the active catalyst by reaction with the promoter methylalumoxane (MAO). A methyl group from MAO displaces one of the chlorine ligands of Cp_2ZrCl_2. The second chlorine is lost as chloride by ionization, giving a positively charged metallocene.

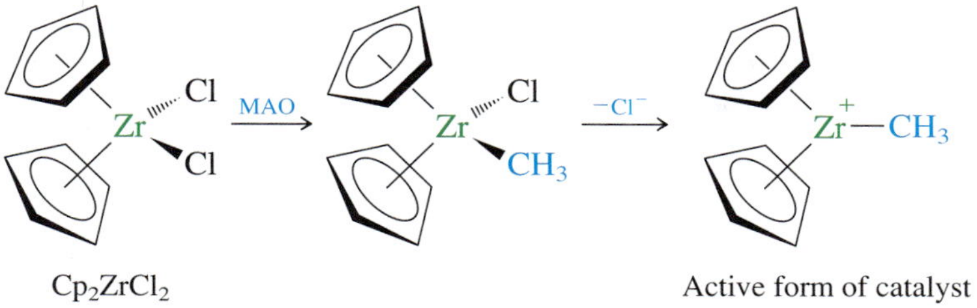

Cp_2ZrCl_2 Active form of catalyst

Step 2: Ethylene reacts with the active form of the catalyst. The two π electrons of ethylene are used to bind it as a ligand to zirconium.

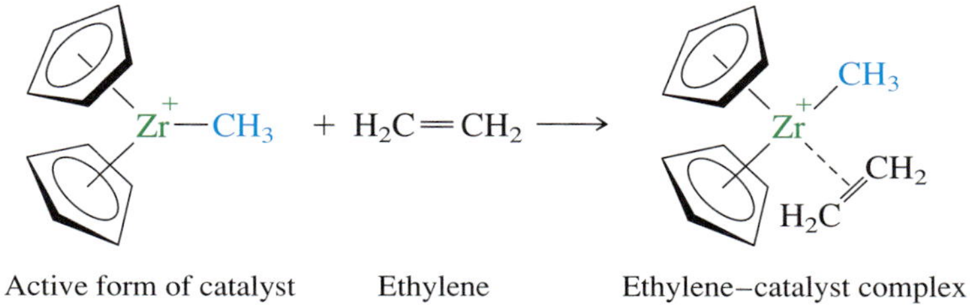

Active form of catalyst   Ethylene   Ethylene–catalyst complex

Step 3: The methyl group migrates from zirconium to one of the carbons of the ethylene ligand. At the same time, the π electrons of the ethylene ligand are used to form a σ bond between the other carbon and zirconium.

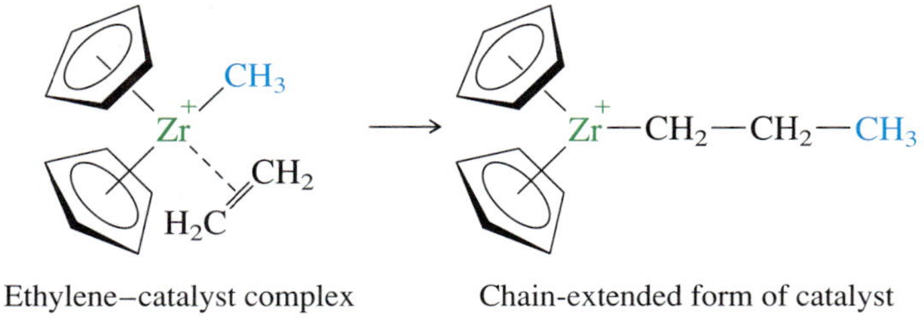

Ethylene–catalyst complex   Chain-extended form of catalyst

Step 4: The catalyst now has a propyl group on zirconium instead of a methyl group. Repeating Steps 2 and 3 converts the propyl group to a pentyl group, then a heptyl group, and so on. After thousands of repetitions, polyethylene is formed.

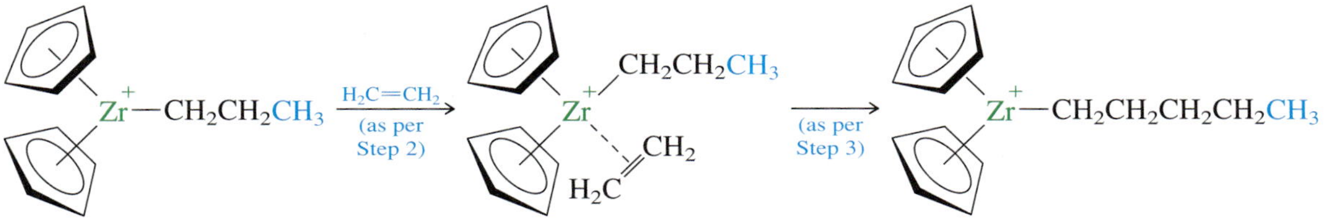

FIGURE 14.5 Mechanism for the polymerization of ethylene in the presence of a Ziegler–Natta catalyst.

Before coordination polymerization was discovered by Ziegler and applied to propene by Natta, there was no polypropylene industry. Now, more than 10^{10} pounds of it are prepared each year in the United States. Ziegler and Natta shared the 1963 Nobel Prize in chemistry: Ziegler for discovering novel catalytic systems for alkene polymerization and Natta for stereoregular polymerization.

14.16 SUMMARY

Section 14.1   Organometallic compounds contain a carbon–metal bond. They are named as alkyl (or aryl) derivatives of metals.

$$CH_3CH_2CH_2CH_2Li \qquad C_6H_5MgBr$$

Butyllithium   Phenylmagnesium bromide

Section 14.2   Carbon is more electronegative than metals and carbon–metal bonds are polarized so that carbon bears a partial to complete negative charge and the metal bears a partial to complete positive charge.

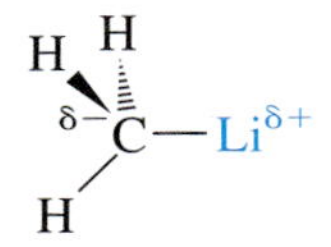

Methyllithium has a polar covalent carbon–lithium bond.

Sodium acetylide has an ionic bond between carbon and sodium.

Section 14.3   See Table 14.5

Section 14.4   See Table 14.5

Section 14.5   Organolithium compounds and Grignard reagents are strong bases and react instantly with compounds that have —OH groups.

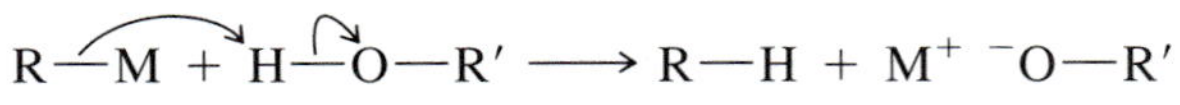

$$R \frown M + H \frown O—R' \longrightarrow R—H + M^+ \ ^-O—R'$$

These organometallic compounds cannot therefore be formed or used in solvents such as water and ethanol. The most commonly employed solvents are diethyl ether and tetrahydrofuran.

Section 14.6   See Tables 14.3 and 14.6

Section 14.7   See Table 14.6

Section 14.8   See Table 14.6

Section 14.9   When planning the synthesis of a compound using an organometallic reagent, or indeed any synthesis, the best approach is to reason backward from the product. This method is called **retrosynthetic analysis.** Retrosynthetic analysis of 1-methylcyclohexanol suggests it can be prepared by the reaction of methylmagnesium bromide and cyclohexanone.

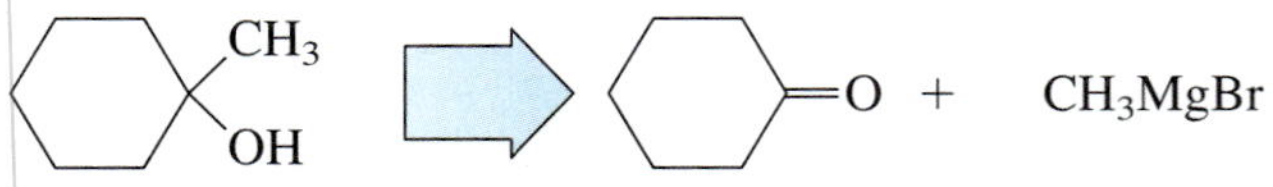

1-Methylcyclohexanol   Cyclohexanone   Methylmagnesium bromide

TABLE 14.5 Preparation of Organometallic Reagents Used in Synthesis

Type of organometallic reagent (section) and comments	General equation for preparation and specific example
Organolithium reagents (Section 14.3) Lithium metal reacts with organic halides to produce organolithium compounds. The organic halide may be alkyl, alkenyl, or aryl. Iodides react most and fluorides least readily; bromides are used most often. Suitable solvents include hexane, diethyl ether, and tetrahydrofuran.	$RX \ + \ 2Li \ \longrightarrow \ RLi \ + \ LiX$ Alkyl halide Lithium Alkyllithium Lithium halide $CH_3CH_2CH_2Br \xrightarrow[\text{diethyl ether}]{\text{Li}} CH_3CH_2CH_2Li$ Propyl bromide Propyllithium (78%)
Grignard reagents (Section 14.4) Grignard reagents are prepared in a manner similar to that used for organolithium compounds. Diethyl ether and tetrahydrofuran are appropriate solvents.	$RX \ + \ Mg \ \longrightarrow \ RMgX$ Alkyl halide Magnesium Alkylmagnesium halide (Grignard reagent) $C_6H_5CH_2Cl \xrightarrow[\text{diethyl ether}]{\text{Mg}} C_6H_5CH_2MgCl$ Benzyl chloride Benzylmagnesium chloride (93%)
Lithium dialkylcuprates (Section 14.11) These reagents contain a negatively charged copper atom and are formed by the reaction of a copper(I) salt with two equivalents of an organolithium reagent.	$2RLi \ + \ CuX \longrightarrow R_2CuLi \ + \ LiX$ Alkyllithium Copper(I) halide Lithium dialkylcuprate Lithium halide $2CH_3Li \ + \ CuI \xrightarrow{\text{diethyl ether}} (CH_3)_2CuLi \ + \ LiI$ Methyllithium Copper(I) iodide Lithium dimethylcuprate Lithium iodide
Iodomethylzinc iodide (Section 14.12) This is the Simmons–Smith reagent. It is prepared by the reaction of zinc (usually in the presence of copper) with diiodomethane.	$CH_2I_2 \ + \ Zn \xrightarrow[\text{Cu}]{\text{diethyl ether}} ICH_2ZnI$ Diiodomethane Zinc Iodomethylzinc iodide

Section 14.10 See Table 14.6

Section 14.11 See Tables 14.5 and 14.6

Section 14.12 See Tables 14.5 and 14.6

Section 14.13 Carbenes are species that contain a *divalent carbon;* that is, a carbon with only two bonds. One of the characteristic reactions of carbenes is with alkenes to give cyclopropane derivatives.

$$H_2C\!=\!C\!\!\begin{array}{c}CH_3\\\\CH_3\end{array} + \ CHCl_3 \xrightarrow[\text{(CH}_3)_3\text{COH}]{\text{KOC(CH}_3)_3}$$

2-Methylpropene 1,1-Dichloro-2,2-dimethylcyclopropane (65%)

Certain organometallic compounds resemble carbenes in their reactions and are referred to as **carbenoids.** Iodomethylzinc iodide (Section 14.12) is an example.

TABLE 14.6   Carbon–Carbon Bond-Forming Reactions of Organometallic Reagents

Reaction (section) and comments	General equation and specific example
Alcohol synthesis via the reaction of Grignard reagents with carbonyl compounds (Section 14.6) This is one of the most useful reactions in synthetic organic chemistry. Grignard reagents react with formaldehyde to yield primary alcohols, with aldehydes to give secondary alcohols, and with ketones to form tertiary alcohols.	$$RMgX \ + \ R'\overset{\displaystyle O}{\overset{\|}{C}}R'' \ \xrightarrow[\text{2. } H_3O^+]{\text{1. diethyl ether}} \ R\overset{\displaystyle R'}{\underset{\displaystyle R''}{C}}OH$$ Grignard Aldehyde Alcohol reagent or ketone $$CH_3MgI \ + \ CH_3CH_2CH_2\overset{\displaystyle O}{\overset{\|}{C}}H \ \xrightarrow[\text{2. } H_3O^+]{\text{1. diethyl ether}} \ CH_3CH_2CH_2\underset{\displaystyle OH}{CH}CH_3$$ Methylmagnesium Butanal 2-Pentanol (82%) iodide
Reaction of Grignard reagents with esters (Section 14.10) Tertiary alcohols in which two of the substituents on the hydroxyl carbon are the same may be prepared by the reaction of an ester with two equivalents of a Grignard reagent.	$$2RMgX \ + \ R'\overset{\displaystyle O}{\overset{\|}{C}}OR'' \ \xrightarrow[\text{2. } H_3O^+]{\text{1. diethyl ether}} \ R\overset{\displaystyle R'}{\underset{\displaystyle R}{C}}OH$$ Grignard Ester Tertiary reagent alcohol $$2C_6H_5MgBr \ + \ C_6H_5\overset{\displaystyle O}{\overset{\|}{C}}OCH_2CH_3 \ \xrightarrow[\text{2. } H_3O^+]{\text{1. diethyl ether}} \ (C_6H_5)_3COH$$ Phenylmagnesium Ethyl benzoate Triphenylmethanol bromide (89–93%)

Synthesis of alcohols using organolithium reagents (Section 14.7) Organolithium reagents react with aldehydes and ketones in a manner similar to that of Grignard reagents to produce alcohols.	$$RLi \ + \ R'\overset{\displaystyle O}{\overset{\|}{C}}R'' \ \xrightarrow[\text{2. } H_3O^+]{\text{1. diethyl ether}} \ R\overset{\displaystyle R'}{\underset{\displaystyle R''}{C}}OH$$ Alkyllithium Aldehyde Alcohol
	             or ketone
	
	 $$\triangleright\!\!-Li \ + \ CH_3\overset{\displaystyle O}{\overset{\|}{C}}C(CH_3)_3 \ \xrightarrow[\text{2. } H_3O^+]{\text{1. diethyl ether}} \ \triangleright\!\!-\underset{\displaystyle CH_3}{\overset{\displaystyle OH}{C}}C(CH_3)_3$$ Cyclopropyllithium 3,3-Dimethyl- 2-Cyclopropyl-
	              2-butanone 3,3-dimethyl-
	                            2-butanol (71%)

(Continued)

Section 14.14   Transition-metal complexes that contain one or more organic ligands offer a rich variety of structural types and reactivity. Organic ligands can be bonded to a metal by a σ bond or through its π system. **Metallocenes** are transition-metal complexes in which one or more of the ligands is a cyclopentadienyl ring. Ferrocene was the first metallocene synthesized; its electrostatic potential map opens this chapter.

TABLE 14.6   Carbon–Carbon Bond-Forming Reactions of Organometallic Reagents (Continued)

Reaction (section) and comments	General equation and specific example
Synthesis of acetylenic alcohols (Section 14.8) Sodium acetylide and acetylenic Grignard reagents react with aldehydes and ketones to give alcohols of the type $C{\equiv}C-\overset{\|}{\underset{\|}{C}}-OH$.	$$NaC{\equiv}CH + RCR' \xrightarrow[\text{2. }H_3O^+]{\text{1. }NH_3,\ -33°C} HC{\equiv}CCR'$$ with OH and R on product carbon. Sodium acetylide Aldehyde or ketone Alcohol $$NaC{\equiv}CH + CH_3CCH_2CH_3 \xrightarrow[\text{2. }H_3O^+]{\text{1. }NH_3,\ -33°C} HC{\equiv}CCCH_2CH_3$$ with OH and CH_3 on product carbon. Sodium acetylide 2-Butanone 3-Methyl-1-pentyn-3-ol (72%)
Preparation of alkanes using lithium dialkylcuprates (Section 14.11) Two alkyl groups may be coupled together to form an alkane by the reaction of an alkyl halide with a lithium dialkylcuprate. Both alkyl groups must be primary (or methyl). Aryl and vinyl halides may be used in place of alkyl halides.	$$R_2CuLi + R'CH_2X \longrightarrow RCH_2R'$$ Lithium dialkylcuprate Primary alkyl halide Alkane $$(CH_3)_2CuLi + C_6H_5CH_2Cl \xrightarrow{\text{diethyl ether}} C_6H_5CH_2CH_3$$ Lithium dimethylcuprate Benzyl chloride Ethylbenzene (80%)
The Simmons–Smith reaction (Section 14.12) Methylene transfer from iodomethylzinc iodide converts alkenes to cyclopropanes. The reaction is a stereospecific syn addition of a CH_2 group to the double bond.	$$R_2C{=}CR_2 + ICH_2ZnI \xrightarrow{\text{diethyl ether}} \text{cyclopropane} + ZnI_2$$ Alkene Iodomethylzinc iodide Cyclopropane derivative Zinc iodide Cyclopentene $\xrightarrow[\text{diethyl ether}]{CH_2I_2,\ Zn(Cu)}$ Bicyclo[3.1.0]hexane (53%)

Section 14.15 Coordination polymerization of ethylene and propene has the biggest economic impact of any organic chemical process. Ziegler–Natta polymerization is carried out using catalysts derived from transition metals such as titanium and zirconium. π-Bonded and σ-bonded organometallic compounds are intermediates in coordination polymerization.

PROBLEMS

14.14 Write structural formulas for each of the following compounds. Specify which compounds qualify as organometallic compounds.

  (a) Cyclopentyllithium

  (b) Ethoxymagnesium chloride

  (c) 2-Phenylethylmagnesium iodide

  (d) Lithium divinylcuprate

  (e) Sodium carbonate

  (f) Benzylpotassium

14.15 *Dibal* is an informal name given to the organometallic compound [(CH$_3$)$_2$CHCH$_2$]$_2$AlH, used as a reducing agent in certain reactions. Can you figure out the systematic name from which "dibal" is derived?

14.16 Suggest appropriate methods for preparing each of the following compounds from the starting material of your choice.

(a) CH$_3$CH$_2$CH$_2$CH$_2$CH$_2$MgI

(b) CH$_3$CH$_2$C≡CMgI

(c) CH$_3$CH$_2$CH$_2$CH$_2$CH$_2$Li

(d) (CH$_3$CH$_2$CH$_2$CH$_2$CH$_2$)$_2$CuLi

14.17 Which compound in each of the following pairs would you expect to have the more polar carbon–metal bond? Compare the models on *Learning By Modeling* with respect to the charge on the carbon bonded to the metal.

(a) CH$_3$CH$_2$Li or (CH$_3$CH$_2$)$_3$Al

(b) (CH$_3$)$_2$Zn or (CH$_3$)$_2$Mg

(c) CH$_3$CH$_2$MgBr or HC≡CMgBr

14.18 Write the structure of the principal organic product of each of the following reactions:

(a) 1-Bromopropane with lithium in diethyl ether

(b) 1-Bromopropane with magnesium in diethyl ether

(c) 2-Iodopropane with lithium in diethyl ether

(d) 2-Iodopropane with magnesium in diethyl ether

(e) Product of part (a) with copper(I) iodide

(f) Product of part (e) with 1-bromobutane

(g) Product of part (e) with iodobenzene

(h) Product of part (b) with D$_2$O and DCl

(i) Product of part (c) with D$_2$O and DCl

(j) Product of part (a) with formaldehyde in diethyl ether, followed by dilute acid

(k) Product of part (b) with benzaldehyde in diethyl ether, followed by dilute acid

(l) Product of part (c) with cycloheptanone in diethyl ether, followed by dilute acid

(m) Product of part (d) with $\overset{\overset{\textstyle O}{\|}}{CH_3CCH_2CH_3}$ in diethyl ether, followed by dilute acid

(n) Product of part (b) (2 mol) with $\overset{\overset{\textstyle O}{\|}}{C_6H_5COCH_3}$ in diethyl ether, followed by dilute acid

(o) 1-Octene with diiodomethane and zinc–copper couple in diethyl ether

(p) (*E*)-2-Decene with diiodomethane and zinc–copper couple in diethyl ether

(q) (*Z*)-3-Decene with diiodomethane and zinc–copper couple in diethyl ether

(r) 1-Pentene with tribromomethane and potassium *tert*-butoxide in *tert*-butyl alcohol

14.19 Using 1-bromobutane and any necessary organic or inorganic reagents, suggest efficient syntheses of each of the following alcohols:

(a) 1-Pentanol

(b) 2-Hexanol

(c) 1-Phenyl-1-pentanol

(d) 3-Methyl-3-heptanol

(e) 1-Butylcyclobutanol

14.20 Using bromobenzene and any necessary organic or inorganic reagents, suggest efficient syntheses of each of the following:

(a) Benzyl alcohol
(b) 1-Phenyl-1-hexanol
(c) Bromodiphenylmethane
(d) 4-Phenyl-4-heptanol
(e) 1-Phenylcyclooctanol
(f) *trans*-2-Phenylcyclooctanol

14.21 Analyze the following structures so as to determine all the practical combinations of Grignard reagent and carbonyl compound that will give rise to each:

(a) $CH_3CH_2\underset{\displaystyle OH}{\underset{|}{C}H}CH_2CH(CH_3)_2$

(b) 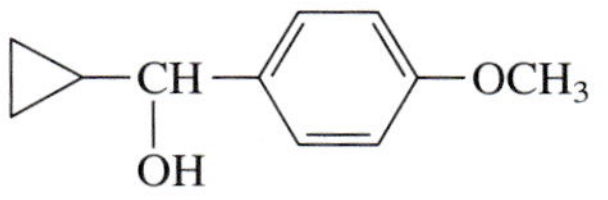
CH
OH
OCH3

(c) $(CH_3)_3CCH_2OH$

(d) 6-Methyl-5-hepten-2-ol

(e) 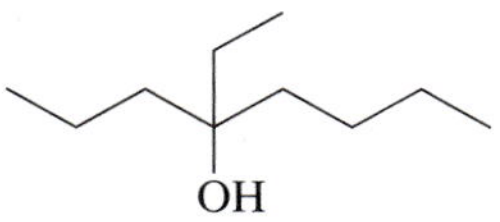
OH

14.22 A number of drugs are prepared by reactions of the type described in this chapter. Indicate what you believe would be a reasonable last step in the synthesis of each of the following:

(a) $CH_3CH_2\overset{\displaystyle OH}{\overset{|}{\underset{\displaystyle CH_3}{\underset{|}{C}}}}C\equiv CH$ *Meparfynol,* a mild hypnotic or sleep-inducing agent

(b) 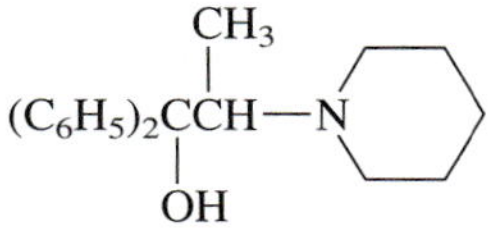
CH3
(C6H5)2CCH—N
OH
Diphepanol, an antitussive (cough suppressant)

(c) 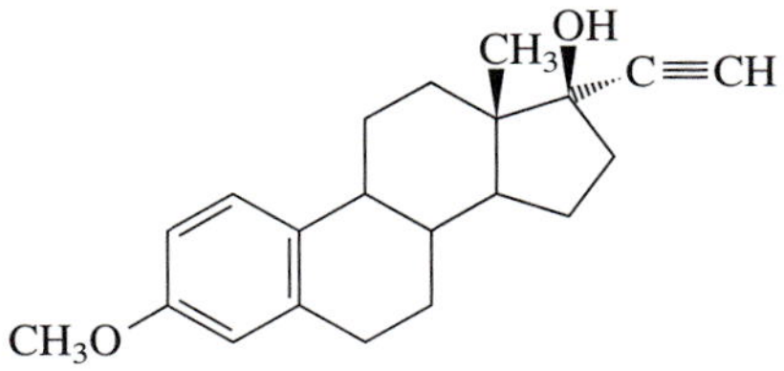
CH3
OH
C≡CH
CH3O
Mestranol, an estrogenic component of oral contraceptive drugs

14.23 Predict the principal organic product of each of the following reactions:

(a) 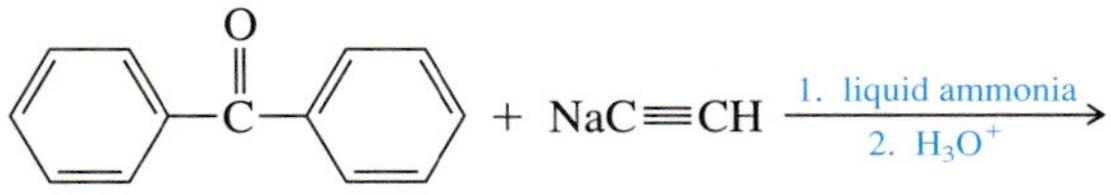
O
C
+ NaC≡CH
1. liquid ammonia
2. H3O+

(b) 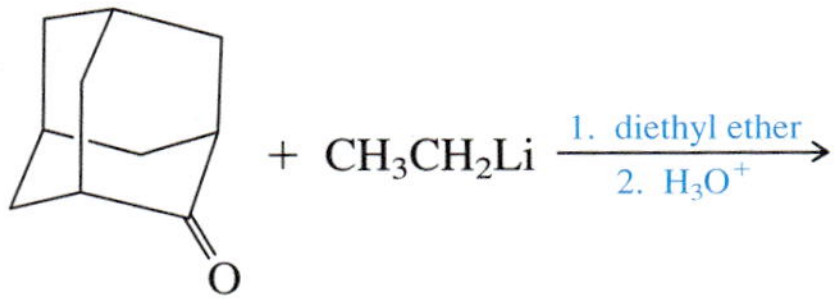
O
+ CH3CH2Li
1. diethyl ether
2. H3O+

(c)

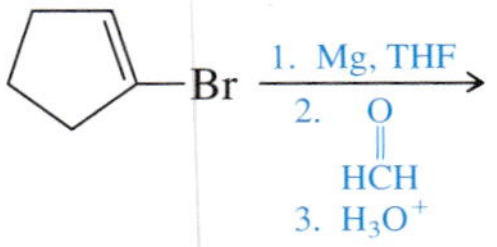

$$\xrightarrow[\substack{2. \quad \overset{O}{\underset{\|}{HCH}} \\ 3. \ H_3O^+}]{1. \ Mg, \ THF}$$

(d)

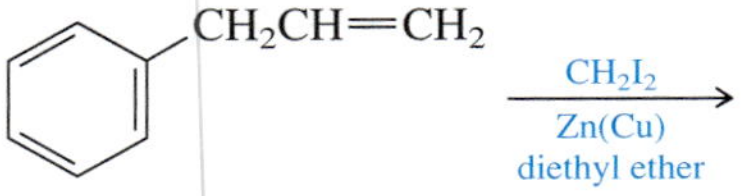

$$\xrightarrow[\substack{Zn(Cu) \\ diethyl \ ether}]{CH_2I_2}$$

(e)

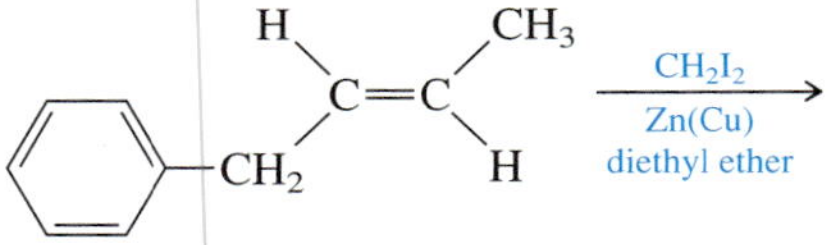

$$\xrightarrow[\substack{Zn(Cu) \\ diethyl \ ether}]{CH_2I_2}$$

(f)

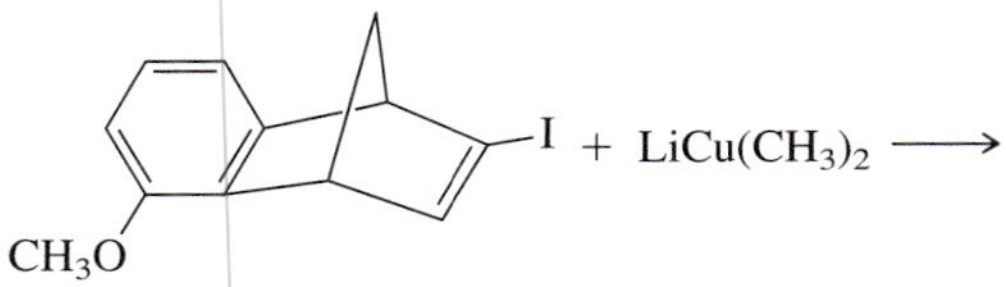

$$-I \ + \ LiCu(CH_3)_2 \longrightarrow$$

(g)

$$-CH_3 \ + \ LiCu(CH_2CH_2CH_2CH_3)_2 \longrightarrow$$

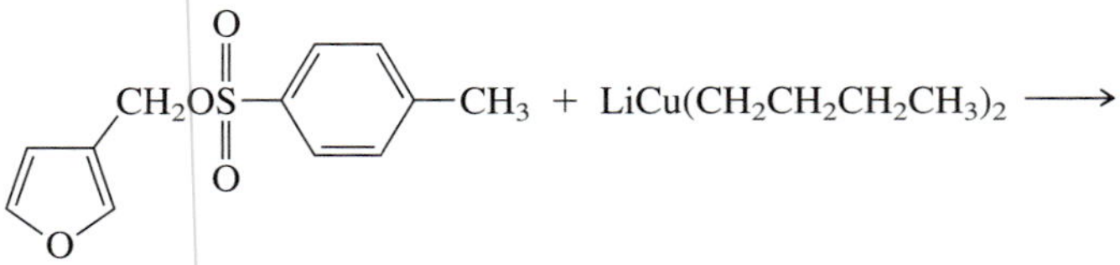

14.24 Addition of phenylmagnesium bromide to 4-*tert*-butylcyclohexanone gives two isomeric tertiary alcohols as products. Both alcohols yield the same alkene when subjected to acid-catalyzed dehydration. Suggest reasonable structures for these two alcohols.

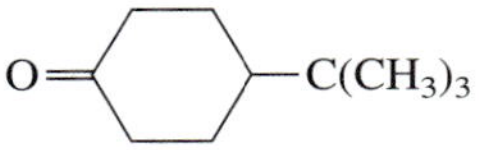

4-*tert*-Butylcyclohexanone

14.25 (a) Unlike other esters which react with Grignard reagents to give tertiary alcohols, ethyl

formate ($H\overset{O}{\overset{\|}{C}}OCH_2CH_3$) yields a different class of alcohols on treatment with Grignard reagents. What kind of alcohol is formed in this case and why?

(b) Diethyl carbonate ($CH_3CH_2O\overset{O}{\overset{\|}{C}}OCH_2CH_3$) reacts with excess Grignard reagent to yield alcohols of a particular type. What is the structural feature that characterizes alcohols prepared in this way?

14.26 Reaction of lithium diphenylcuprate with optically active 2-bromobutane yields 2-phenylbutane, with high net inversion of configuration. When the 2-bromobutane used has the stereostructure shown, will the 2-phenylbutane formed have the *R* or the *S* configuration?

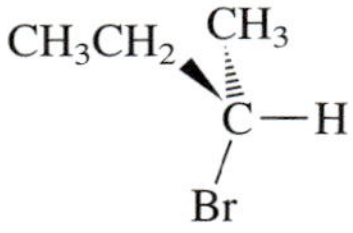

14.27 Suggest reasonable structures for compounds A, B, and C in the following reactions:

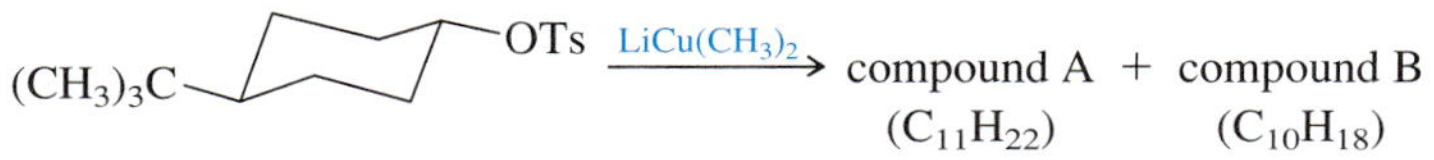

$(CH_3)_3C$—⟨ring⟩—OTs $\xrightarrow{\text{LiCu(CH}_3)_2}$ compound A + compound B
                                                            $(C_{11}H_{22})$ $(C_{10}H_{18})$

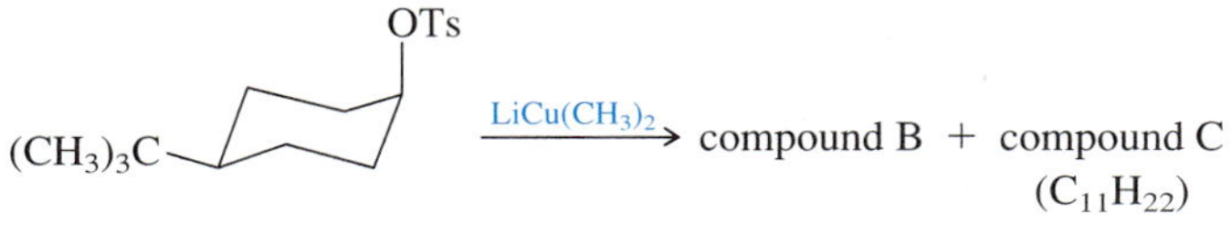

$(CH_3)_3C$—⟨ring with OTs⟩ $\xrightarrow{\text{LiCu(CH}_3)_2}$ compound B + compound C
                                                            $(C_{11}H_{22})$

Compound C is more stable than compound A. OTs stands for toluenesulfonate.

14.28 Isonitriles are stable, often naturally occurring, compounds that contain a divalent carbon. An example is axisonitrile-3, which can be isolated from a species of sponge and possesses anti-malarial activity. Write a resonance form for axisonitrile-3 that satisfies the octet rule. Don't forget to include formal charges.

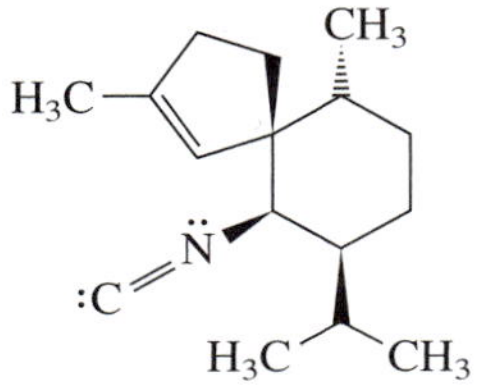

Axisonitrile-3

14.29 The following conversion has been reported in the chemical literature. It was carried out in two steps, the first of which involved formation of a *p*-toluenesulfonate ester. Indicate the reagents for this step, and show how you could convert the *p*-toluenesulfonate to the desired product.

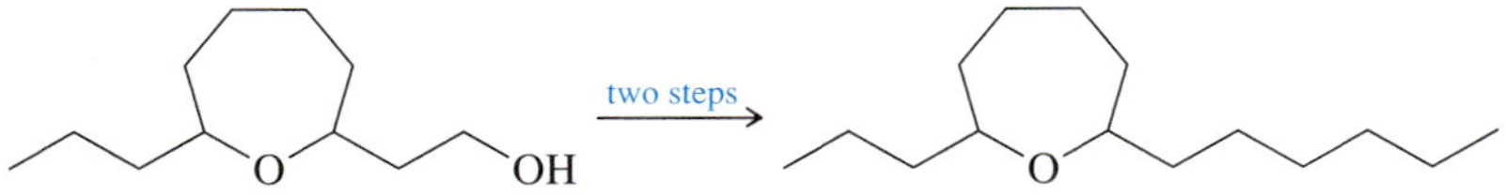

14.30 Sometimes the strongly basic properties of Grignard reagents can be turned to synthetic advantage. A chemist needed samples of butane specifically labeled with deuterium, the mass 2 isotope of hydrogen, as shown:

(a) $CH_3CH_2CH_2CH_2D$ (b) $CH_3CHDCH_2CH_3$

Suggest methods for the preparation of each of these using D_2O as the source of deuterium, butanols of your choice, and any necessary organic or inorganic reagents.

14.31 Diphenylmethane is significantly more acidic than benzene, and triphenylmethane is more acidic than either. Identify the most acidic proton in each compound, and suggest a reason for the trend in acidity.

C_6H_6	$(C_6H_5)_2CH_2$	$(C_6H_5)_3CH$
Benzene	Diphenylmethane	Triphenylmethane
$pK_a \approx 43$	$pK_a \approx 34$	$pK_a \approx 32$

14.32 The 18-electron rule is a general, but not universal, guide for assessing whether a certain transition-metal complex is stable or not. Both of the following are stable compounds, but only one obeys the 18-electron rule. Which one?

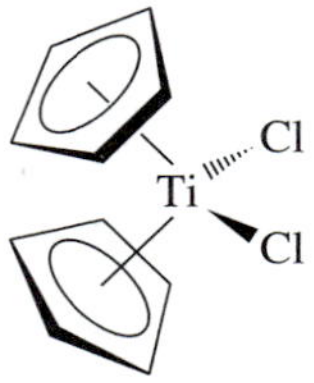

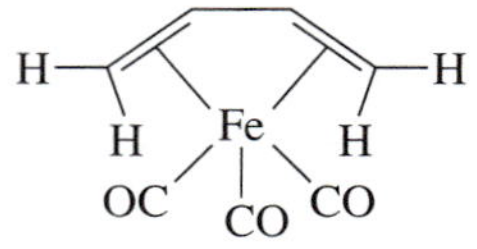

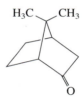

14.33 One of the main uses of the "linear α-olefins" prepared by oligomerization of ethylene is in the preparation of *linear low-density polyethylene*. Linear low-density polyethylene is a copolymer produced when ethylene is polymerized in the presence of a "linear α-olefin" such as 1-decene [$H_2C=CH(CH_2)_7CH_3$]. 1-Decene replaces ethylene at random points in the growing polymer chain. Can you deduce how the structure of linear low-density polyethylene differs from a linear chain of CH_2 units?

14.34 Make a molecular model of 7,7-dimethylbicyclo[2.2.1]heptan-2-one. Two diastereomeric alcohols may be formed when it reacts with methylmagnesium bromide. Which one is formed in greater amounts?

7,7-Dimethylbicyclo[2.2.1]heptan-2-one

14.35 Make molecular models of the product of addition of dichlorocarbene to:

    (a) *trans*-2-Butene

    (b) *cis*-2-Butene

Which product is achiral? Which one is formed as a racemic mixture?

14.36 Examine the molecular model of ferrocene on *Learning By Modeling*. Does ferrocene have a dipole moment? Would you expect the cyclopentadienyl rings of ferrocene to be more reactive toward nucleophiles or electrophiles? Where is the region of highest electrostatic potential?

14.37 Inspect the electrostatic potential map of the benzyl anion structure given on *Learning By Modeling*. What is the hybridization state of the benzylic carbon? Does the region of most negative electrostatic potential lie in the plane of the molecule or perpendicular to it? Which ring carbons bear the greatest share of negative charge?

CHAPTER 15

ALCOHOLS, DIOLS, AND THIOLS

The next several chapters deal with the chemistry of various oxygen-containing functional groups. The interplay of these important classes of compounds—alcohols, ethers, aldehydes, ketones, carboxylic acids, and derivatives of carboxylic acids—is fundamental to organic chemistry and biochemistry.

$$
\underset{\text{Alcohol}}{\text{ROH}} \qquad
\underset{\text{Ether}}{\text{ROR}'} \qquad
\underset{\text{Aldehyde}}{\overset{\displaystyle O}{\overset{\|}{\text{RCH}}}} \qquad
\underset{\text{Ketone}}{\overset{\displaystyle O}{\overset{\|}{\text{RCR}'}}} \qquad
\underset{\text{Carboxylic acid}}{\overset{\displaystyle O}{\overset{\|}{\text{RCOH}}}}
$$

We'll start by discussing in more detail a class of compounds already familiar to us, *alcohols*. Alcohols were introduced in Chapter 4 and have appeared regularly since then. With this chapter we extend our knowledge of alcohols, particularly with respect to their relationship to carbonyl-containing compounds. In the course of studying alcohols, we shall also look at some relatives. **Diols** are alcohols in which two hydroxyl groups (—OH) are present; **thiols** are compounds that contain an —SH group. **Phenols,** compounds of the type ArOH, share many properties in common with alcohols but are sufficiently different from them to warrant separate discussion in Chapter 24.

This chapter is a transitional one. It ties together much of the material encountered earlier and sets the stage for our study of other oxygen-containing functional groups in the chapters that follow.

15.1 SOURCES OF ALCOHOLS

Until the 1920s, the major source of *methanol* was as a byproduct in the production of charcoal from wood—hence, the name *wood alcohol*. Now, most of the more than 10 billion lb of methanol used annually in the United States is synthetic, prepared by reduction of carbon monoxide with hydrogen.

$$CO \quad + \quad 2H_2 \quad \xrightarrow[400°C]{ZnO/Cr_2O_3} \quad CH_3OH$$

Carbon monoxide   Hydrogen   Methanol

A mixture of the two reactants, carbon monoxide and hydrogen, is called *synthesis gas* and is prepared by several processes. The most widely used route to synthesis gas employs methane (from natural gas) and gives a 3:1 hydrogen-to-carbon monoxide ratio.

$$CH_4 \quad + \quad H_2O \quad \xrightarrow[800-1000°C]{Ni} \quad CO \quad + \quad 3H_2$$

Methane   Water   Carbon monoxide   Hydrogen

Some of the H_2 formed in the reaction goes to make methanol, the rest is used to reduce nitrogen to ammonia.

The major uses of methanol are in the preparation of formaldehyde and *tert*-butyl methyl ether (known commercially as MTBE). Formaldehyde is a starting material for various resins and plastics, including the first completely synthetic plastic bakelite. MTBE is an effective gasoline additive, but problems with it leaking from underground tanks and contaminating groundwater make it unsuitable for continued use. Minor amounts of methanol are used as a solvent and as a convenient clean-burning liquid fuel. This last property makes it a candidate as a fuel for automobiles—methanol is already used to power Indianapolis-class race cars.

Methanol is a colorless liquid, boiling at 65°C, and is miscible with water in all proportions. It is poisonous; drinking as little as 30 mL has been fatal. Smaller amounts can produce blindness.

When vegetable matter ferments, its carbohydrates are converted to *ethanol* and carbon dioxide by enzymes present in yeast. Fermentation of barley produces beer; grapes give wine. The maximum ethanol content is on the order of 15%, because higher concentrations inactivate the enzymes, halting fermentation. Because ethanol boils at 78°C and water at 100°C, distillation of the fermentation broth gives "distilled spirits" of increased ethanol content. Whiskey is the aged distillate of fermented grain and contains slightly less than 50% ethanol. Brandy and cognac are made by aging the distilled spirits from fermented grapes and other fruits. The characteristic flavors, odors, and colors of the various alcoholic beverages depend on both their origin and the way they are aged.

Synthetic ethanol is derived from petroleum by hydration of ethylene. In the United States, some 700 million lb of synthetic ethanol is produced annually. It is relatively inexpensive and useful for industrial applications. To make it unfit for drinking, it is *denatured* by adding any of a number of noxious materials, exempting it from the high taxes most governments impose on ethanol used in beverages.

Some of the substances used to denature ethanol include methanol, benzene, pyridine, castor oil, and gasoline.

Our bodies are reasonably well equipped to metabolize ethanol, making it less dangerous than methanol. Alcohol abuse and alcoholism, however, have been and remain persistent problems.

Isopropyl alcohol is prepared from petroleum by hydration of propene. With a boiling point of 82°C, isopropyl alcohol evaporates quickly from the skin, producing a cooling effect. Often containing dissolved oils and fragrances, it is the major component of rubbing alcohol. Isopropyl alcohol possesses weak antibacterial properties and is used to maintain medical instruments in a sterile condition and to clean the skin before minor surgery.

Methanol, ethanol, and isopropyl alcohol are included among the readily available starting materials commonly found in laboratories where organic synthesis is carried out. So, too, are many other alcohols. All alcohols of four carbons or fewer, as well as most of the five- and six-carbon alcohols and many higher alcohols, are commercially available at low cost. Some occur naturally; others are the products of efficient synthe- ses. Figure 15.1 presents the structures of a few naturally occurring alcohols. Table 15.1 summarizes the reactions encountered in earlier chapters that give alcohols and illustrates a thread that runs through the fabric of organic chemistry: *a reaction that is character- istic of one functional group often serves as a synthetic method for preparing another.*

As Table 15.1 indicates, reactions leading to alcohols are not in short supply. Nevertheless, several more will be added to the list in the present chapter—testimony to the importance of alcohols in synthetic organic chemistry. Some of these methods involve reduction of carbonyl groups:

> Recall from Section 2.19 that reduction corresponds to a decrease in the number of bonds between carbon and oxygen or an increase in the number of bonds between carbon and hydrogen (or both).

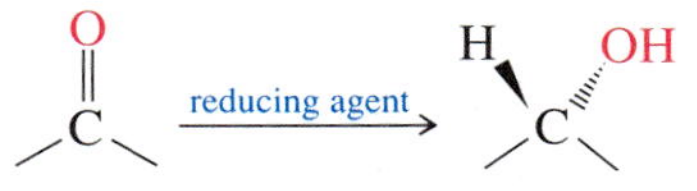

$$
\underset{\text{C}}{\overset{\text{O}}{\parallel}} \quad \xrightarrow{\text{reducing agent}} \quad \underset{\text{C}}{\overset{\text{H} \quad \text{OH}}{}}
$$

We will begin with the reduction of aldehydes and ketones.

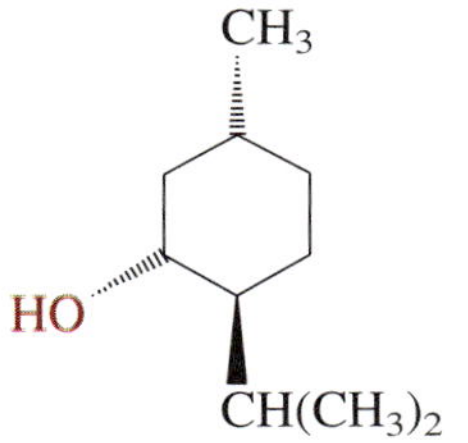

Menthol (obtained from oil of peppermint and used to flavor tobacco and food)

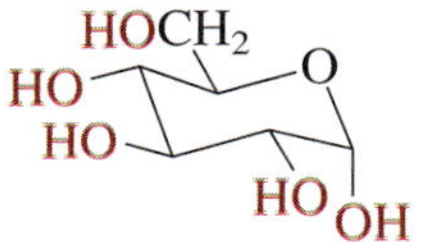

Glucose (a carbohydrate)

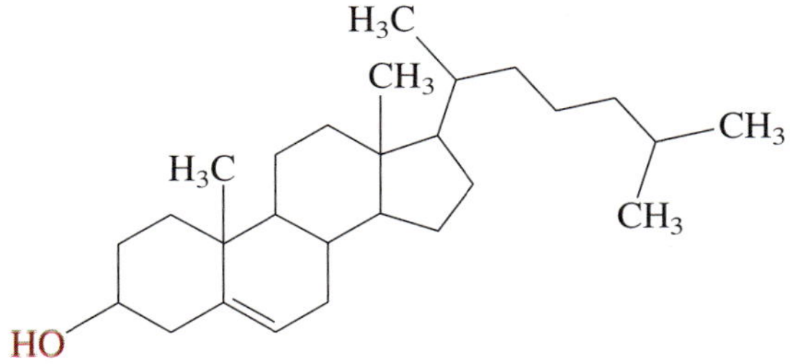

Cholesterol (principal constituent of gallstones and biosynthetic precursor of the steroid hormones)

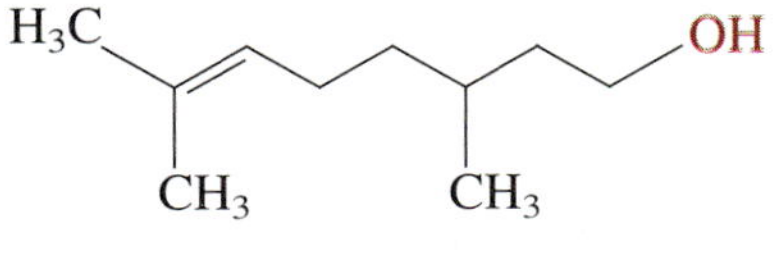

Citronellol (found in rose and geranium oil and used in perfumery)

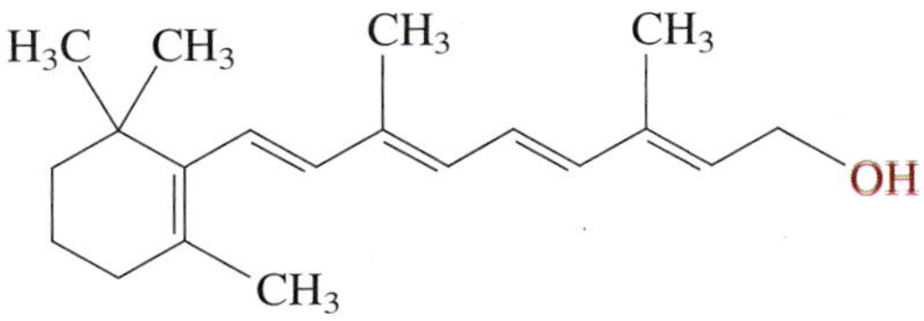

Retinol (vitamin A, an important substance in vision)

FIGURE 15.1 Some naturally occurring alcohols.

TABLE 15.1   Summary of Reactions Discussed in Earlier Chapters That Yield Alcohols

Reaction (section) and comments	General equation and specific example

Acid-catalyzed hydration of alkenes (Section 6.10) Water adds to the double bond in accordance with Markovnikov's rule.

$$R_2C{=}CR_2 + H_2O \xrightarrow{H^+} R_2CHCR_2$$
$$\qquad\qquad\qquad\qquad\quad |$$
$$\qquad\qquad\qquad\qquad\quad OH$$

Alkene   Water   Alcohol

$$(CH_3)_2C{=}CHCH_3 \xrightarrow[H_2SO_4]{H_2O} CH_3\overset{\overset{\displaystyle CH_3}{|}}{C}CH_2CH_3$$
$$\qquad\qquad\qquad\qquad\qquad\qquad |$$
$$\qquad\qquad\qquad\qquad\qquad\qquad OH$$

2-Methyl-2-butene   2-Methyl-2-butanol (90%)

Hydroboration–oxidation of alkenes (Section 6.11) H and OH add to the double bond with a regioselectivity opposite to that of Markovnikov's rule. This is a very good synthetic method; addition is syn, and no rearrangements are observed.

$$R_2C{=}CR_2 \xrightarrow[\text{2. }H_2O_2,\text{ }HO^-]{\text{1. }B_2H_6} R_2CHCR_2$$
$$\qquad\qquad\qquad\qquad\qquad\qquad |$$
$$\qquad\qquad\qquad\qquad\qquad\qquad OH$$

Alkene   Alcohol

$$CH_3(CH_2)_7CH{=}CH_2 \xrightarrow[\text{2. }H_2O_2,\text{ }HO^-]{\text{1. }B_2H_6,\text{ diglyme}} CH_3(CH_2)_7CH_2CH_2OH$$

1-Decene   1-Decanol (93%)

Hydrolysis of alkyl halides (Section 8.1) A reaction useful only with substrates that do not undergo E2 elimination readily. It is rarely used for the synthesis of alcohols, since alkyl halides are normally prepared from alcohols.

$$RX + HO^- \longrightarrow ROH + X^-$$

Alkyl   Hydroxide   Alcohol   Halide
halide   ion   ion

2,4,6-Trimethylbenzyl chloride $\xrightarrow[\text{heat}]{H_2O,\text{ }Ca(OH)_2}$ 2,4,6-Trimethylbenzyl alcohol (78%)

Reaction of Grignard reagents with aldehydes and ketones (Section 14.6) A method that allows for alcohol preparation with formation of new carbon–carbon bonds. Primary, secondary, and tertiary alcohols can all be prepared.

$$RMgX + R'\overset{\overset{\displaystyle O}{\|}}{C}R'' \xrightarrow[\text{2. }H_3O^+]{\text{1. diethyl ether}} R\overset{\overset{\displaystyle R'}{|}}{\underset{\underset{\displaystyle R''}{|}}{C}}OH$$

Grignard   Aldehyde   Alcohol
reagent   or ketone

Cyclopentylmagnesium bromide + Formaldehyde $\xrightarrow[\text{2. }H_3O^+]{\text{1. diethyl ether}}$ Cyclopentylmethanol (62–64%)

(Continued)

TABLE 15.1	**Summary of Reactions Discussed in Earlier Chapters That Yield Alcohols (Continued)**
Reaction (section) and comments	**General equation and specific example**
Reaction of organolithium reagents with aldehydes and ketones (Section 14.7) Organolithium reagents react with aldehydes and ketones in a manner similar to that of Grignard reagents to form alcohols.	$$RLi \quad + \quad R'\overset{\overset{O}{\|}}{C}R'' \xrightarrow[\text{2. } H_3O^+]{\text{1. diethyl ether}} R\overset{\overset{R'}{\|}}{\underset{\overset{\|}{R''}}{C}}OH$$ Organolithium reagent ・ Aldehyde or ketone ・ Alcohol $$CH_3CH_2CH_2CH_2Li + \text{(phenyl)}\overset{\overset{O}{\|}}{C}CH_3 \xrightarrow[\text{2. } H_3O^+]{\text{1. diethyl ether}} CH_3CH_2CH_2CH_2{-}\overset{\overset{\text{(phenyl)}}{\|}}{\underset{\overset{\|}{CH_3}}{C}}{-}OH$$ Butyllithium ・ Acetophenone ・ 2-Phenyl-2-hexanol (67%)
Reaction of Grignard reagents with esters (Section 14.10) Produces tertiary alcohols in which two of the substituents on the hydroxyl-bearing carbon are derived from the Grignard reagent.	$$2RMgX + R'\overset{\overset{O}{\|}}{C}OR'' \xrightarrow[\text{2. } H_3O^+]{\text{1. diethyl ether}} R\overset{\overset{R'}{\|}}{\underset{\overset{\|}{R}}{C}}OH + R''OH$$ $$2CH_3CH_2CH_2CH_2CH_2MgBr + CH_3\overset{\overset{O}{\|}}{C}OCH_2CH_3 \xrightarrow[\text{2. } H_3O^+]{\text{1. diethyl ether}}$$ Pentylmagnesium bromide ・ Ethyl acetate $$CH_3\overset{\overset{OH}{\|}}{\underset{\overset{\|}{CH_2CH_2CH_2CH_2CH_3}}{C}}CH_2CH_2CH_2CH_2CH_3$$ 6-Methyl-6-undecanol (75%)

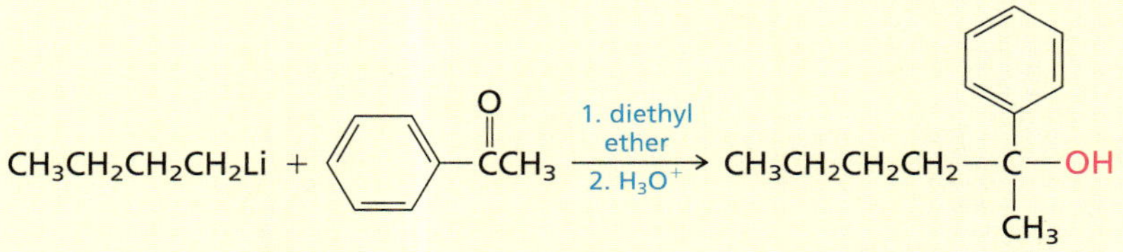

15.2 PREPARATION OF ALCOHOLS BY REDUCTION OF ALDEHYDES AND KETONES

The most obvious way to reduce an aldehyde or a ketone to an alcohol is by hydrogenation of the carbon–oxygen double bond. Like the hydrogenation of alkenes, the reaction is exothermic but exceedingly slow in the absence of a catalyst. Finely divided metals such as platinum, palladium, nickel, and ruthenium are effective catalysts for the hydrogenation of aldehydes and ketones. Aldehydes yield primary alcohols:

$$R\overset{\overset{O}{\|}}{C}H \quad + \quad H_2 \xrightarrow{\text{Pt, Pd, Ni, or Ru}} RCH_2OH$$

Aldehyde ・ Hydrogen ・ Primary alcohol

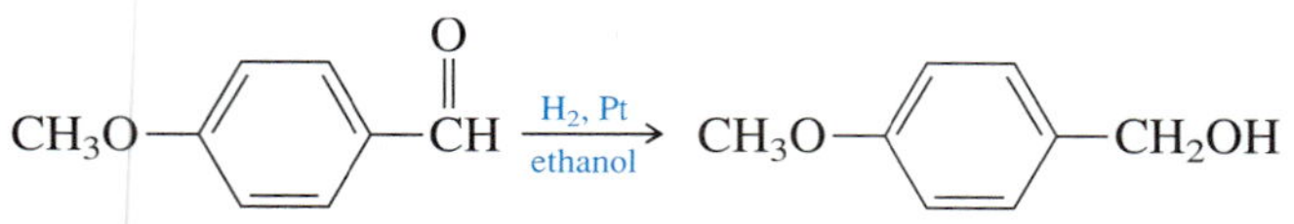

$$CH_3O\text{—}\underset{}{\bigcirc}\text{—}\overset{\overset{\displaystyle O}{\|}}{CH} \xrightarrow[\text{ethanol}]{H_2, Pt} CH_3O\text{—}\underset{}{\bigcirc}\text{—}CH_2OH$$

p-Methoxybenzaldehyde   *p*-Methoxybenzyl alcohol (92%)

Ketones yield secondary alcohols:

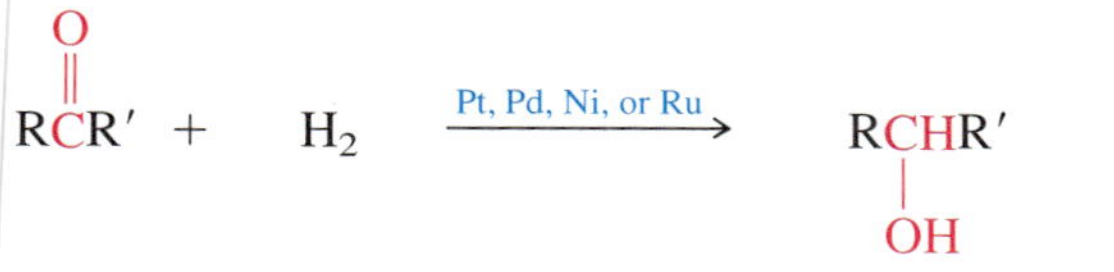

$$\underset{\text{Ketone}}{\overset{\overset{\displaystyle O}{\|}}{RCR'}} + \underset{\text{Hydrogen}}{H_2} \xrightarrow{\text{Pt, Pd, Ni, or Ru}} \underset{\text{Secondary alcohol}}{\overset{\displaystyle RCHR'}{\underset{\displaystyle OH}{|}}}$$

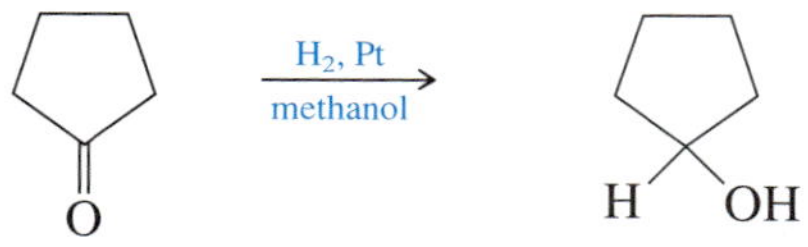

Cyclopentanone   Cyclopentanol (93–95%)

> **PROBLEM 15.1** Which of the isomeric $C_4H_{10}O$ alcohols can be prepared by hydrogenation of aldehydes? Which can be prepared by hydrogenation of ketones? Which cannot be prepared by hydrogenation of a carbonyl compound?

For most laboratory-scale reductions of aldehydes and ketones, catalytic hydrogenation has been replaced by methods based on metal hydride reducing agents. The two most common reagents are sodium borohydride and lithium aluminum hydride.

$$Na^+ \begin{bmatrix} & H & \\ H-&B&-H \\ & | & \\ & H & \end{bmatrix} \qquad Li^+ \begin{bmatrix} & H & \\ H-&Al&-H \\ & | & \\ & H & \end{bmatrix}$$

Sodium borohydride ($NaBH_4$)   Lithium aluminum hydride ($LiAlH_4$)

Sodium borohydride is especially easy to use, needing only to be added to an aqueous or alcoholic solution of an aldehyde or a ketone:

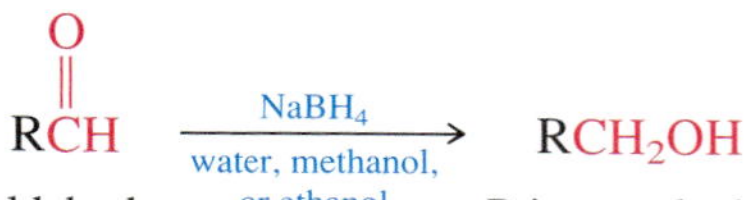

$$\underset{\text{Aldehyde}}{\overset{\overset{\displaystyle O}{\|}}{RCH}} \xrightarrow[\substack{\text{water, methanol,} \\ \text{or ethanol}}]{NaBH_4} \underset{\text{Primary alcohol}}{RCH_2OH}$$

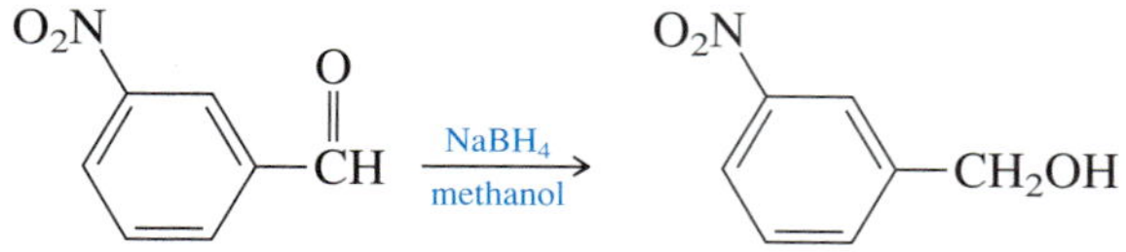

m-Nitrobenzaldehyde   *m*-Nitrobenzyl alcohol (82%)

Compare the distribution of charge among the atoms in CH_4, BH_4^-, and AlH_4^- on *Learning By Modeling*. Notice how different they are.

$$\underset{\text{Ketone}}{\overset{\displaystyle\overset{O}{\|}}{\text{RCR}'}} \xrightarrow[\substack{\text{water, methanol,} \\ \text{or ethanol}}]{\text{NaBH}_4} \underset{\text{Secondary alcohol}}{\overset{\displaystyle\overset{OH}{|}}{\text{RCHR}'}}$$

$$\underset{\text{4,4-Dimethyl-2-pentanone}}{\overset{\displaystyle\overset{O}{\|}}{\text{CH}_3\text{CCH}_2\text{C(CH}_3)_3}} \xrightarrow[\text{ethanol}]{\text{NaBH}_4} \underset{\substack{\text{4,4-Dimethyl-2-pentanol} \\ (85\%)}}{\overset{\displaystyle\overset{OH}{|}}{\text{CH}_3\text{CHCH}_2\text{C(CH}_3)_3}}$$

Lithium aluminum hydride reacts violently with water and alcohols, so it must be used in solvents such as anhydrous diethyl ether or tetrahydrofuran. Following reduction, a separate hydrolysis step is required to liberate the alcohol product:

The same kinds of solvents are used for LiAlH$_4$ as for Grignard reagents.

$$\underset{\text{Aldehyde}}{\overset{\displaystyle\overset{O}{\|}}{\text{RCH}}} \xrightarrow[\text{2. H}_2\text{O}]{\text{1. LiAlH}_4, \text{ diethyl ether}} \underset{\text{Primary alcohol}}{\text{RCH}_2\text{OH}}$$

$$\underset{\text{Heptanal}}{\overset{\displaystyle\overset{O}{\|}}{\text{CH}_3(\text{CH}_2)_5\text{CH}}} \xrightarrow[\text{2. H}_2\text{O}]{\text{1. LiAlH}_4, \text{ diethyl ether}} \underset{\text{1-Heptanol (86\%)}}{\text{CH}_3(\text{CH}_2)_5\text{CH}_2\text{OH}}$$

$$\underset{\text{Ketone}}{\overset{\displaystyle\overset{O}{\|}}{\text{RCR}'}} \xrightarrow[\text{2. H}_2\text{O}]{\text{1. LiAlH}_4, \text{ diethyl ether}} \underset{\text{Secondary alcohol}}{\overset{\displaystyle\underset{OH}{|}}{\text{RCHR}'}}$$

$$\underset{\text{1,1-Diphenyl-2-propanone}}{\overset{\displaystyle\overset{O}{\|}}{(\text{C}_6\text{H}_5)_2\text{CHCCH}_3}} \xrightarrow[\text{2. H}_2\text{O}]{\text{1. LiAlH}_4, \text{ diethyl ether}} \underset{\text{1,1-Diphenyl-2-propanol (84\%)}}{\overset{\displaystyle\underset{OH}{|}}{(\text{C}_6\text{H}_5)_2\text{CHCHCH}_3}}$$

Sodium borohydride and lithium aluminum hydride react with carbonyl compounds in much the same way that Grignard reagents do, except that they function as *hydride donors* rather than as carbanion sources. Figure 15.2 outlines the general mechanism for the sodium borohydride reduction of an aldehyde or ketone ($R_2C\!=\!O$). Two points are especially important about this process.

1. At no point is H_2 involved. The reducing agent is borohydride ion ($BH_4{}^-$).

2. In the reduction $R_2C\!=\!O \rightarrow R_2CHOH$, the hydrogen bonded to carbon comes from $BH_4{}^-$; the hydrogen on oxygen comes from an OH group of the solvent (water, methanol, or ethanol).

The mechanism of lithium aluminum hydride reduction of aldehydes and ketones is analogous to that of sodium borohydride except that the reduction and hydrolysis

Overall Reaction:

$$4R_2C{=}O + BH_4^- + 4H_2O \longrightarrow 4R_2CHOH + B(OH)_4^-$$

Aldehyde or ketone	Borohydride ion	Water	Alcohol	Borate ion

Mechanism:

Step 1: Hydride (hydrogen + two electrons) is transferred from boron to the positively polarized carbon of the carbonyl group. The carbonyl oxygen bonds to boron.

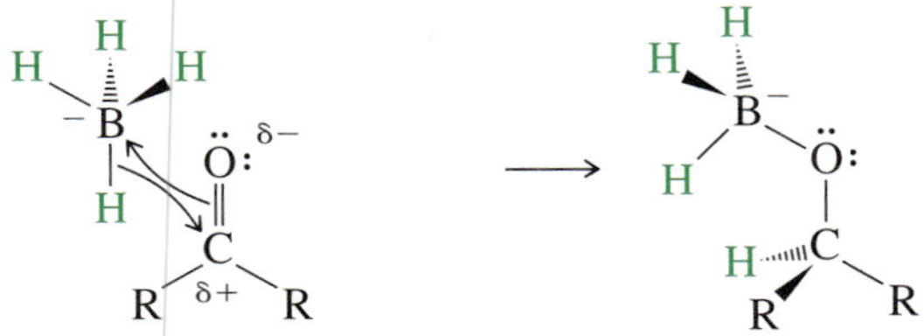

Borohydride ion + aldehyde or ketone   Alkoxyborohydride

Steps 2–4: The alkoxyborohydride formed in the first step contains three more hydrogens that can be donated to carbonyl groups. It reacts with three more molecules of the starting aldehyde or ketone.

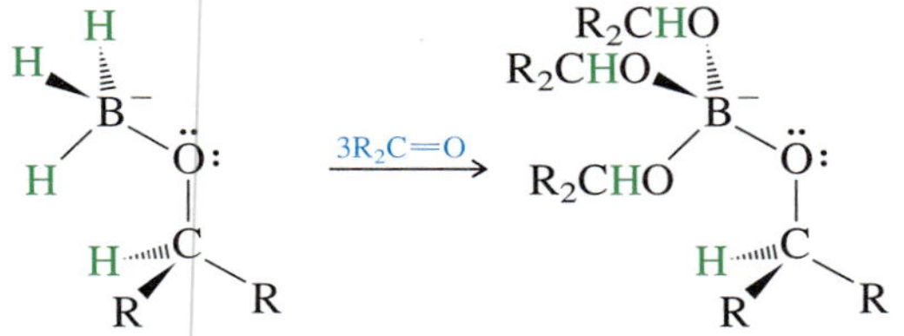

Alkoxyborohydride   Tetraalkoxyborate

Step 5: When the reaction is carried out in water as the solvent, the tetraalkoxyborate undergoes spontaneous hydrolysis.

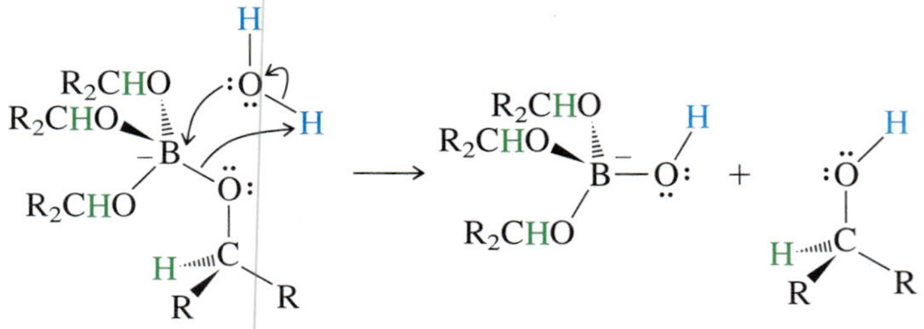

Tetraalkoxyborate + water   Trialkoxyborate   Alcohol

Steps 6–8: Three more hydrolysis steps convert the trialkoxyborate to three more molecules of R_2CHOH and $(HO)_4B^-$.

FIGURE 15.2 Mechanism of sodium borohydride reduction of an aldehyde or ketone.

stages are independent operations. The reduction is carried out in diethyl ether, followed by a separate hydrolysis step when aqueous acid is added to the reaction mixture.

$$4R_2C=O \xrightarrow[\text{diethyl ether}]{\text{LiAlH}_4} (R_2CHO)_4Al^- \xrightarrow{4H_2O} 4R_2CHOH + Al(OH)_4^-$$

Aldehyde or ketone   Tetraalkoxyaluminate   Alcohol

PROBLEM 15.2 Sodium borodeuteride ($NaBD_4$) and lithium aluminum deuteride ($LiAlD_4$) are convenient reagents for introducing deuterium, the mass-2 isotope of hydrogen, into organic compounds. Write the structure of the organic product of the following reactions, clearly showing the position of all the deuterium atoms in each:

(a) Reduction of $CH_3\overset{\displaystyle O}{\overset{\|}{C}}H$ (acetaldehyde) with $NaBD_4$ in H_2O

(b) Reduction of $CH_3\overset{\displaystyle O}{\overset{\|}{C}}CH_3$ (acetone) with $NaBD_4$ in CH_3OD

(c) Reduction of $C_6H_5\overset{\displaystyle O}{\overset{\|}{C}}H$ (benzaldehyde) with $NaBD_4$ in CD_3OH

(d) Reduction of $H\overset{\displaystyle O}{\overset{\|}{C}}H$ (formaldehyde) with $LiAlD_4$ in diethyl ether, followed by addition of D_2O

SAMPLE SOLUTION (a) Sodium borodeuteride transfers deuterium to the carbonyl group of acetaldehyde, forming a C—D bond.

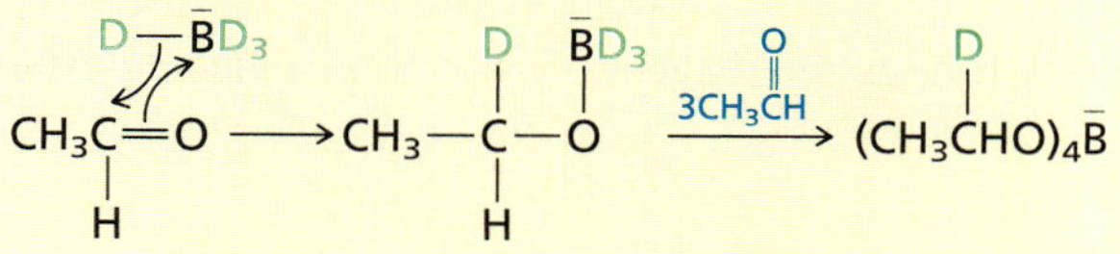

Hydrolysis of $(CH_3CHDO)_4B^-$ in H_2O leads to the formation of ethanol, retaining the C—D bond formed in the preceding step while forming an O—H bond.

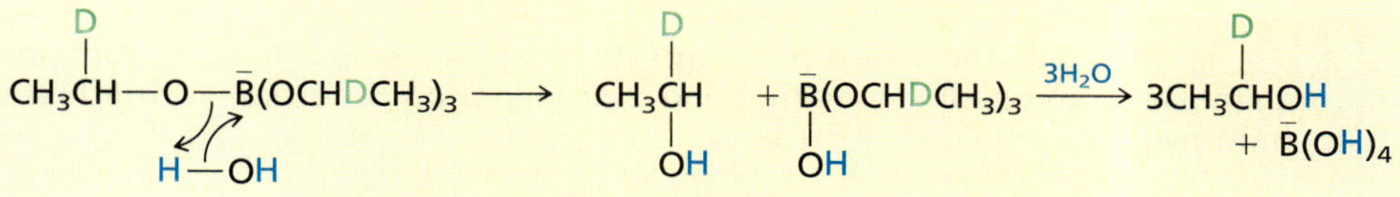

Ethanol-1-d

An undergraduate laboratory experiment related to Problem 15.2 appears in the March 1996 issue of the *Journal of Chemical Education,* pp. 264–266.

Neither sodium borohydride nor lithium aluminum hydride reduces isolated carbon–carbon double bonds. This makes possible the selective reduction of a carbonyl group in a molecule that contains both carbon–carbon and carbon–oxygen double bonds.

$$(CH_3)_2C=CHCH_2CH_2\overset{\displaystyle O}{\overset{\|}{C}}CH_3 \xrightarrow[\text{2. H}_2O]{\text{1. LiAlH}_4\text{, diethyl ether}} (CH_3)_2C=CHCH_2CH_2\overset{\displaystyle OH}{\overset{|}{C}}HCH_3$$

6-Methyl-5-hepten-2-one   6-Methyl-5-hepten-2-ol (90%)

Catalytic hydrogenation would not be suitable for this transformation, because H_2 adds to carbon–carbon double bonds faster than it reduces carbonyl groups.

15.3 PREPARATION OF ALCOHOLS BY REDUCTION OF CARBOXYLIC ACIDS AND ESTERS

Carboxylic acids are exceedingly difficult to reduce. Acetic acid, for example, is often used as a solvent in catalytic hydrogenations because it is inert under the reaction conditions. A very powerful reducing agent is required to convert a carboxylic acid to a primary alcohol. Lithium aluminum hydride is that reducing agent.

$$
\underset{\text{Carboxylic acid}}{\text{R}\overset{\overset{\displaystyle O}{\|}}{\text{C}}\text{OH}} \quad \xrightarrow[\text{2. H}_2\text{O}]{\text{1. LiAlH}_4,\ \text{diethyl ether}} \quad \underset{\text{Primary alcohol}}{\text{RCH}_2\text{OH}}
$$

$$
\underset{\substack{\text{Cyclopropanecarboxylic}\\\text{acid}}}{\triangleright\!\!-\!\text{CO}_2\text{H}} \quad \xrightarrow[\text{2. H}_2\text{O}]{\text{1. LiAlH}_4,\ \text{diethyl ether}} \quad \underset{\substack{\text{Cyclopropylmethanol}\\(78\%)}}{\triangleright\!\!-\!\text{CH}_2\text{OH}}
$$

Sodium borohydride is not nearly as potent a hydride donor as lithium aluminum hydride and does not reduce carboxylic acids.

Esters are more easily reduced than carboxylic acids. Two alcohols are formed from each ester molecule. The acyl group of the ester is cleaved, giving a primary alcohol.

$$
\underset{\text{Ester}}{\text{R}\overset{\overset{\displaystyle O}{\|}}{\text{C}}\text{OR}'} \quad \longrightarrow \quad \underset{\text{Primary alcohol}}{\text{RCH}_2\text{OH}} \quad + \quad \underset{\text{Alcohol}}{\text{R}'\text{OH}}
$$

Lithium aluminum hydride is the reagent of choice for reducing esters to alcohols.

$$
\underset{\text{Ethyl benzoate}}{\text{C}_6\text{H}_5\overset{\overset{\displaystyle O}{\|}}{\text{C}}\text{OCH}_2\text{CH}_3} \quad \xrightarrow[\text{2. H}_2\text{O}]{\text{1. LiAlH}_4,\ \text{diethyl ether}} \quad \underset{\text{Benzyl alcohol (90\%)}}{\text{C}_6\text{H}_5\text{CH}_2\text{OH}} \quad + \quad \underset{\text{Ethanol}}{\text{CH}_3\text{CH}_2\text{OH}}
$$

PROBLEM 15.3 Give the structure of an ester that will yield a mixture containing equimolar amounts of 1-propanol and 2-propanol on reduction with lithium aluminum hydride.

Sodium borohydride reduces esters, but the reaction is too slow to be useful. Hydrogenation of esters requires a special catalyst and extremely high pressures and temperatures; it is used in industrial settings but rarely in the laboratory.

15.4 PREPARATION OF ALCOHOLS FROM EPOXIDES

Although the chemical reactions of epoxides will not be covered in detail until the following chapter, we shall introduce their use in the synthesis of alcohols here.

Grignard reagents react with ethylene oxide to yield primary alcohols containing two more carbon atoms than the alkyl halide from which the organometallic compound was prepared.

$$RMgX + H_2C\overset{}{\underset{O}{\diagup\!\!\diagdown}}CH_2 \xrightarrow[\text{2. } H_3O^+]{\text{1. diethyl ether}} RCH_2CH_2OH$$

Grignard   Ethylene oxide   Primary alcohol
reagent

$$CH_3(CH_2)_4CH_2MgBr + H_2C\overset{}{\underset{O}{\diagup\!\!\diagdown}}CH_2 \xrightarrow[\text{2. } H_3O^+]{\text{1. diethyl ether}} CH_3(CH_2)_4CH_2CH_2CH_2OH$$

Hexylmagnesium   Ethylene oxide   1-Octanol (71%)
bromide

Organolithium reagents react with epoxides in a similar manner.

PROBLEM 15.4 Each of the following alcohols has been prepared by reaction of a Grignard reagent with ethylene oxide. Select the appropriate Grignard reagent in each case.

(a) [structure: o-methylphenyl group]—CH$_2$CH$_2$OH

(b) [structure: cyclohexylmethyl group]CH$_2$CH$_2$OH

SAMPLE SOLUTION (a) Reaction with ethylene oxide results in the addition of a —CH$_2$CH$_2$OH unit to the Grignard reagent. The Grignard reagent derived from o-bromotoluene (or o-chlorotoluene or o-iodotoluene) is appropriate here.

[structure] —MgBr   + $H_2C\overset{}{\underset{O}{\diagup\!\!\diagdown}}CH_2$ $\xrightarrow[\text{2. } H_3O^+]{\text{1. diethyl ether}}$ [structure] —CH$_2$CH$_2$OH

o-Methylphenylmagnesium   Ethylene oxide   2-(o-Methylphenyl)ethanol
bromide   (66%)

Epoxide rings are readily opened with cleavage of the carbon–oxygen bond when attacked by nucleophiles. Grignard reagents and organolithium reagents react with ethylene oxide by serving as sources of nucleophilic carbon.

$$\overset{\delta-}{R}\!-\!\overset{\delta+}{MgX} \longrightarrow R\!-\!CH_2\!-\!CH_2\!-\!\overset{..}{\underset{..}{O}}\!:^- \overset{+}{MgX} \xrightarrow{H_3O^+} RCH_2CH_2OH$$

$$H_2C\overset{}{\underset{\overset{..}{\underset{..}{O}}}{\diagup\!\!\diagdown}}CH_2$$

(may be written as
RCH$_2$CH$_2$OMgX)

This kind of chemical reactivity of epoxides is rather general. Nucleophiles other than Grignard reagents react with epoxides, and epoxides more elaborate than ethylene oxide may be used. All these features of epoxide chemistry will be discussed in Sections 16.11–16.13.

15.5 PREPARATION OF DIOLS

Much of the chemistry of diols—compounds that bear two hydroxyl groups—is analogous to that of alcohols. Diols may be prepared, for example, from compounds that contain two carbonyl groups, using the same reducing agents employed in the preparation of alcohols. The following example shows the conversion of a dialdehyde to a diol by catalytic hydrogenation. Alternatively, the same transformation can be achieved by reduction with sodium borohydride or lithium aluminum hydride.

$$\underset{\substack{| \\ CH_3 \\ \text{3-Methylpentanedial}}}{\overset{\substack{O \\ \| \quad\quad\quad \| \\ HCCH_2CHCH_2CH}}{}} \xrightarrow[\text{Ni, 125°C}]{H_2\ (100\ atm)} \underset{\substack{| \\ CH_3 \\ \text{3-Methyl-1,5-pentanediol (81–83\%)}}}{HOCH_2CH_2CHCH_2CH_2OH}$$

Diols are almost always given substitutive IUPAC names. As the name of the product in the example indicates, the substitutive nomenclature of diols is similar to that of alcohols. The suffix *-diol* replaces *-ol*, and two locants, one for each hydroxyl group, are required. Note that the final *-e* of the parent alkane name is retained when the suffix begins with a consonant *(-diol)*, but dropped when the suffix begins with a vowel *(-ol)*.

> **PROBLEM 15.5** Write equations showing how 3-methyl-1,5-pentanediol could be prepared from a dicarboxylic acid or a diester.

Vicinal diols are diols that have their hydroxyl groups on adjacent carbons. Two commonly encountered vicinal diols are 1,2-ethanediol and 1,2-propanediol.

$$\underset{\substack{\text{1,2-Ethanediol} \\ \text{(ethylene glycol)}}}{HOCH_2CH_2OH} \qquad \underset{\substack{| \\ OH \\ \text{1,2-Propanediol} \\ \text{(propylene glycol)}}}{CH_3CHCH_2OH}$$

Ethylene glycol and *propylene glycol* are common names for these two diols and are acceptable IUPAC names. Aside from these two compounds, the IUPAC system does not use the word *glycol* for naming diols.

In the laboratory, vicinal diols are normally prepared from alkenes using the reagent *osmium tetraoxide* (OsO_4). Osmium tetraoxide reacts rapidly with alkenes to give cyclic osmate esters.

Ethylene glycol and propylene glycol are prepared industrially from the corresponding alkenes by way of their epoxides. Some applications were given in the box in Section 6.21.

$$R_2C{=}CR_2 \ + \ OsO_4 \ \longrightarrow \ \underset{\substack{\text{Cyclic osmate ester}}}{R_2C{-}CR_2}$$

Alkene   Osmium   Cyclic osmate ester
tetraoxide

Osmate esters are fairly stable but are readily cleaved in the presence of an oxidizing agent such as *tert*-butyl hydroperoxide.

$$R_2C{-}CR_2 \ + \ 2(CH_3)_3COOH \xrightarrow[\substack{\text{tert-butyl} \\ \text{alcohol}}]{HO^-} \underset{\substack{HO \quad OH}}{R_2C{-}CR_2} \ + \ OsO_4 \ + \ 2(CH_3)_3COH$$

<table>
<tr><td>tert-Butyl
hydroperoxide</td><td>Vicinal
diol</td><td>Osmium
tetraoxide</td><td>tert-Butyl
alcohol</td></tr>
</table>

Because osmium tetraoxide is regenerated in this step, alkenes can be converted to vicinal diols using only catalytic amounts of osmium tetraoxide, which is both toxic and expensive. The entire process is performed in a single operation by simply allowing a solution of the alkene and *tert*-butyl hydroperoxide in *tert*-butyl alcohol containing a small amount of osmium tetraoxide and base to stand for several hours.

$$CH_3(CH_2)_7CH{=}CH_2 \xrightarrow[\textit{tert-butyl alcohol, HO}^-]{(CH_3)_3COOH,\ OsO_4(cat)} CH_3(CH_2)_7\underset{\underset{\displaystyle OH}{|}}{C}HCH_2OH$$

1-Decene   1,2-Decanediol (73%)

For developing osmium-catalyzed oxidation methods for preparing chiral compounds of high optical purity, Professor K. Barry Sharpless (Scripps Research Institute) shared the 2001 Nobel Prize in chemistry.

Overall, the reaction leads to addition of two hydroxyl groups to the double bond and is referred to as **hydroxylation.** Both oxygens of the diol come from osmium tetraoxide via the cyclic osmate ester. The reaction of OsO_4 with the alkene is a syn addition, and the conversion of the cyclic osmate to the diol involves cleavage of the bonds between oxygen and osmium. Thus, both hydroxyl groups of the diol become attached to the same face of the double bond; *syn hydroxylation of the alkene is observed.*

Cyclohexene   *cis*-1,2-Cyclohexanediol (62%)

Construct a molecular model of *cis*-1,2-cyclohexanediol. What is the orientation of the OH groups, axial or equatorial?

PROBLEM 15.6 Give the structures, including stereochemistry, for the diols obtained by hydroxylation of *cis*-2-butene and *trans*-2-butene.

A complementary method, one that gives anti hydroxylation of alkenes by way of the hydrolysis of epoxides, will be described in Section 16.13.

15.6   REACTIONS OF ALCOHOLS: A REVIEW AND A PREVIEW

Alcohols are versatile starting materials for the preparation of a variety of organic functional groups. Several reactions of alcohols have already been seen in earlier chapters and are summarized in Table 15.2. The remaining sections of this chapter add to the list.

15.7   CONVERSION OF ALCOHOLS TO ETHERS

Primary alcohols are converted to ethers on heating in the presence of an acid catalyst, usually sulfuric acid.

$$2RCH_2OH \xrightarrow{H^+,\ heat} RCH_2OCH_2R\ +\ H_2O$$

Primary alcohol   Dialkyl ether   Water

This kind of reaction is called a **condensation.** A condensation is a reaction in which two molecules combine to form a larger one while liberating a small molecule. In this case two alcohol molecules combine to give an ether and water.

$$2CH_3CH_2CH_2CH_2OH \xrightarrow[130°C]{H_2SO_4} CH_3CH_2CH_2CH_2OCH_2CH_2CH_2CH_3\ +\ H_2O$$

1-Butanol   Dibutyl ether (60%)   Water

TABLE 15.2 — Summary of Reactions of Alcohols Discussed in Earlier Chapters

Reaction (section) and comments	General equation and specific example
Reaction with hydrogen halides (Section 4.7) The order of alcohol reactivity parallels the order of carbocation stability: $R_3C^+ > R_2CH^+ > RCH_2^+ > CH_3^+$. Benzylic alcohols react readily.	$ROH + HX \longrightarrow RX + H_2O$ Alcohol Hydrogen halide Alkyl halide Water *m*-Methoxybenzyl alcohol $\xrightarrow{HBr}$ *m*-Methoxybenzyl bromide (98%)
Reaction with thionyl chloride (Section 4.13) Thionyl chloride converts alcohols to alkyl chlorides.	$ROH + SOCl_2 \longrightarrow RCl + SO_2 + HCl$ Alcohol Thionyl chloride Alkyl chloride Sulfur dioxide Hydrogen chloride $(CH_3)_2C{=}CHCH_2CH_2CHCH_3$ (OH) $\xrightarrow[\text{diethyl ether}]{\text{SOCl}_2,\ \text{pyridine}}$ $(CH_3)_2C{=}CHCH_2CH_2CHCH_3$ (Cl) 6-Methyl-5-hepten-2-ol → 6-Chloro-2-methyl-2-heptene (67%)
Reaction with phosphorus trihalides (Section 4.13) Phosphorus trichloride and phosphorus tribromide convert alcohols to alkyl halides.	$3ROH + PX_3 \longrightarrow 3RX + H_3PO_3$ Alcohol Phosphorus trihalide Alkyl halide Phosphorous acid Cyclopentylmethanol—$CH_2OH \xrightarrow{PBr_3}$ —CH_2Br (Bromomethyl)cyclopentane (50%)
Acid-catalyzed dehydration (Section 5.9) This is a frequently used procedure for the preparation of alkenes. The order of alcohol reactivity parallels the order of carbocation stability: $R_3C^+ > R_2CH^+ > RCH_2^+$. Benzylic alcohols react readily. Rearrangements are sometimes observed.	R_2CCHR_2 (OH) $\xrightarrow[\text{heat}]{H^+}$ $R_2C{=}CR_2 + H_2O$ Alcohol Alkene Water 1-(*m*-Bromophenyl)-1-propanol —CHCH_2CH_3 (OH) $\xrightarrow[\text{heat}]{KHSO_4}$ —CH=CHCH_3 1-(*m*-Bromophenyl)propene (71%)
Conversion to *p*-toluenesulfonate esters (Section 8.14) Alcohols react with *p*-toluenesulfonyl chloride to give *p*-toluenesulfonate esters. Sulfonate esters are reactive substrates for nucleophilic substitution and elimination reactions. The *p*-toluenesulfonate group is often abbreviated —OTs.	$ROH + H_3C{-}\langle\rangle{-}SO_2Cl \longrightarrow ROS(=O)(=O){-}\langle\rangle{-}CH_3 + HCl$ Alcohol *p*-Toluenesulfonyl chloride Alkyl *p*-toluenesulfonate Hydrogen chloride Cycloheptanol —OH $\xrightarrow[\text{pyridine}]{p\text{-toluenesulfonyl chloride}}$ —OTs Cycloheptyl *p*-toluenesulfonate (83%)

When applied to the synthesis of ethers, the reaction is effective only with primary alcohols. Elimination to form alkenes predominates with secondary and tertiary alcohols.

Diethyl ether is prepared on an industrial scale by heating ethanol with sulfuric acid at 140°C. At higher temperatures elimination predominates, and ethylene is the major product. A mechanism for the formation of diethyl ether is outlined in Figure 15.3. The individual steps of this mechanism are analogous to those seen earlier. Nucleophilic attack on a protonated alcohol was encountered in the reaction of primary alcohols with hydrogen halides (Section 4.12), and the nucleophilic properties of alcohols were discussed in the context of solvolysis reactions (Section 8.7). Both the first and the last steps are proton-transfer reactions between oxygens.

Diols react intramolecularly to form cyclic ethers when a five-membered or six-membered ring can result.

$$HOCH_2CH_2CH_2CH_2CH_2OH \xrightarrow[\text{heat}]{H_2SO_4} \qquad + \quad H_2O$$

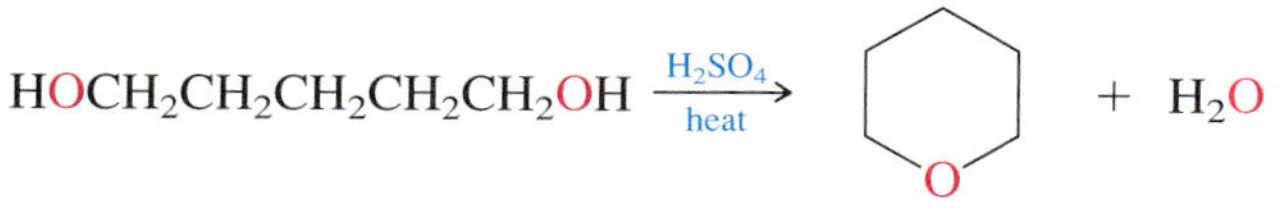

1,5-Pentanediol   Oxane (76%)   Water

Oxane is also called tetrahydropyran.

FIGURE 15.3 The mechanism of acid-catalyzed formation of diethyl ether from ethyl alcohol.

Overall Reaction:

$$2CH_3CH_2OH \xrightarrow[140°C]{H_2SO_4} CH_3CH_2OCH_2CH_3 + H_2O$$

Ethanol   Diethyl ether   Water

Step 1: Proton transfer from the acid catalyst (sulfuric acid) to the oxygen of the alcohol to produce an alkyloxonium ion.

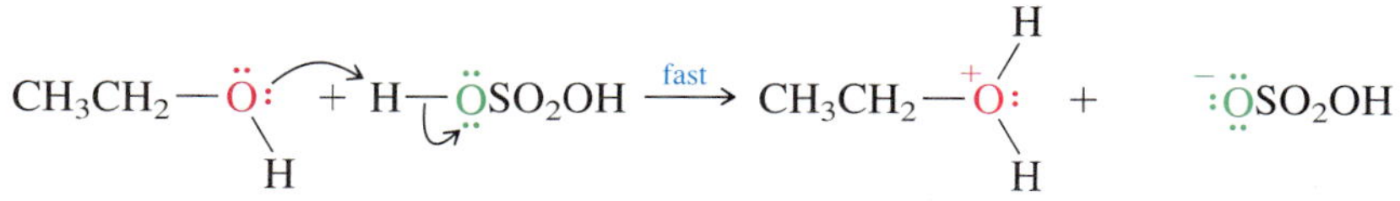

$$CH_3CH_2\ddot{O}\!:\ +\ H\!-\!\ddot{O}SO_2OH \xrightarrow{\text{fast}} CH_3CH_2\overset{+}{\ddot{O}}\!:\ +\ \ ^-\!:\ddot{O}SO_2OH$$

Ethanol   Sulfuric acid   Ethyloxonium ion   Hydrogen sulfate ion

Step 2: Nucleophilic attack by a molecule of alcohol on the alkyloxonium ion formed in step 1.

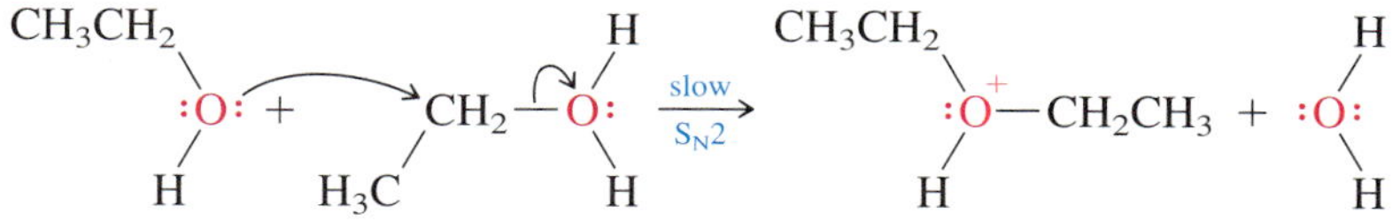

$$\underset{\text{Ethanol}}{CH_3CH_2\ \ddot{O}\!:\ +} \quad \underset{\text{Ethyloxonium ion}}{CH_2\!-\!\overset{+}{\ddot{O}}\!:} \xrightarrow[\text{S}_N2]{\text{slow}} \underset{\text{Diethyloxonium ion}}{CH_3CH_2\ \overset{+}{O}\!-\!CH_2CH_3} \ + \ \underset{\text{Water}}{:\ddot{O}}$$

Step 3: The product of step 2 is the conjugate acid of the dialkyl ether. It is deprotonated in the final step of the process to give the ether.

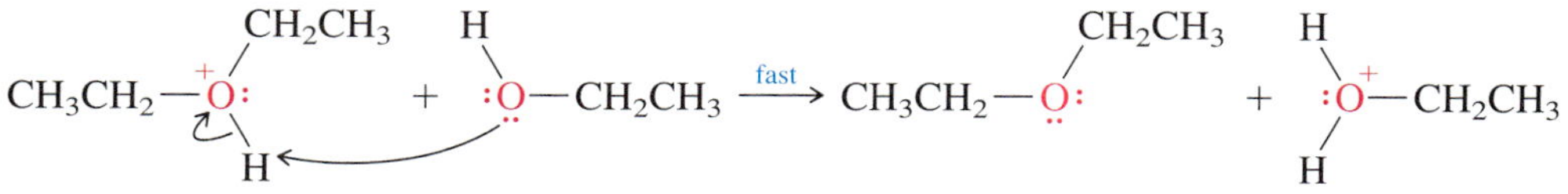

$$\underset{\text{Diethyloxonium ion}}{CH_3CH_2\overset{+}{\ddot{O}}} \ + \ \underset{\text{Ethanol}}{:\ddot{O}\!-\!CH_2CH_3} \xrightarrow{\text{fast}} \underset{\text{Diethyl ether}}{CH_3CH_2\ddot{O}\!:} \ + \ \underset{\text{Ethyloxonium ion}}{:\overset{+}{\ddot{O}}\!-\!CH_2CH_3}$$

In these intramolecular ether-forming reactions, the alcohol may be primary, secondary, or tertiary.

> **PROBLEM 15.7** On the basis of the mechanism for the acid-catalyzed formation of diethyl ether from ethanol in Figure 15.3, write a stepwise mechanism for the formation of oxane from 1,5-pentanediol (see the equation on page 637).

15.8 ESTERIFICATION

Acid-catalyzed condensation of an alcohol and a carboxylic acid yields an ester and water and is known as the **Fischer esterification.**

$$ \underset{\text{Alcohol}}{ROH} + \underset{\text{Carboxylic acid}}{R'\overset{\overset{\textstyle O}{\|}}{C}OH} \underset{H^+}{\rightleftharpoons} \underset{\text{Ester}}{R'\overset{\overset{\textstyle O}{\|}}{C}OR} + \underset{\text{Water}}{H_2O} $$

Fischer esterification is reversible, and the position of equilibrium lies slightly to the side of products when the reactants are simple alcohols and carboxylic acids. When the Fischer esterification is used for preparative purposes, the position of equilibrium can be made more favorable by using either the alcohol or the carboxylic acid in excess. In the following example, in which an excess of the alcohol was employed, the yield indicated is based on the carboxylic acid as the limiting reactant.

$$ \underset{\substack{\text{Methanol} \\ \text{(0.6 mol)}}}{CH_3OH} + \underset{\substack{\text{Benzoic acid} \\ \text{(0.1 mol)}}}{C_6H_5\overset{\overset{\textstyle O}{\|}}{C}OH} \xrightarrow[\text{heat}]{H_2SO_4} \underset{\substack{\text{Methyl benzoate} \\ \text{(isolated in 70\%} \\ \text{yield based on} \\ \text{benzoic acid)}}}{C_6H_5\overset{\overset{\textstyle O}{\|}}{C}OCH_3} + \underset{\text{Water}}{H_2O} $$

Another way to shift the position of equilibrium to favor the formation of ester is by removing water from the reaction mixture. This can be accomplished by adding benzene as a cosolvent and distilling the azeotropic mixture of benzene and water.

An *azeotropic mixture* contains two or more substances that distill together at a constant boiling point. The benzene–water azeotrope contains 9% water and boils at 69°C.

$$ \underset{\substack{\textit{sec}\text{-Butyl alcohol} \\ \text{(0.20 mol)}}}{\underset{\overset{|}{OH}}{CH_3CHCH_2CH_3}} + \underset{\substack{\text{Acetic acid} \\ \text{(0.25 mol)}}}{CH_3\overset{\overset{\textstyle O}{\|}}{C}OH} \xrightarrow[\text{benzene, heat}]{H^+} \underset{\substack{\textit{sec}\text{-Butyl acetate} \\ \text{(isolated in 71\%} \\ \text{yield based on} \\ \textit{sec}\text{-butyl alcohol)}}}{\underset{\overset{|}{CH_3}}{CH_3\overset{\overset{\textstyle O}{\|}}{C}OCHCH_2CH_3}} + \underset{\substack{\text{Water} \\ \text{(codistills} \\ \text{with benzene)}}}{H_2O} $$

For steric reasons, the order of alcohol reactivity in the Fischer esterification is $CH_3OH >$ primary $>$ secondary $>$ tertiary.

PROBLEM 15.8 Write the structure of the ester formed in each of the following reactions:

(a) $CH_3CH_2CH_2CH_2OH$ + $CH_3CH_2\overset{O}{\overset{||}{C}}OH$ $\xrightarrow[\text{heat}]{H_2SO_4}$

(b) 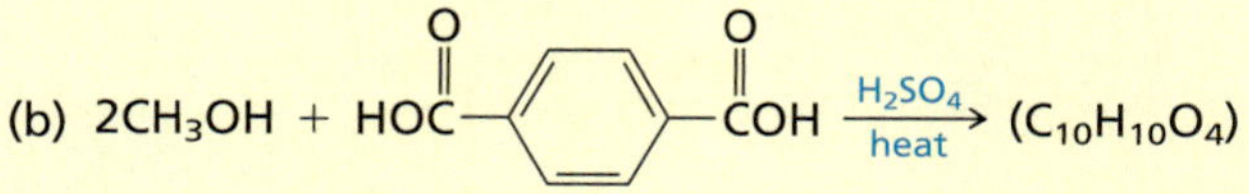 $2CH_3OH$ + $HO\overset{O}{\overset{||}{C}}$—⟨benzene⟩—$\overset{O}{\overset{||}{C}}OH$ $\xrightarrow[\text{heat}]{H_2SO_4}$ $(C_{10}H_{10}O_4)$

SAMPLE SOLUTION (a) By analogy to the general equation and to the examples cited in this section, we can write the equation

$CH_3CH_2CH_2CH_2OH$ + $CH_3CH_2\overset{O}{\overset{||}{C}}OH$ $\xrightarrow[\text{heat}]{H_2SO_4}$ $CH_3CH_2\overset{O}{\overset{||}{C}}OCH_2CH_2CH_2CH_3$ + H_2O

1-Butanol   Propanoic acid   Butyl propanoate   Water

As actually carried out in the laboratory, 3 moles of propanoic acid was used per mole of 1-butanol, and the desired ester was obtained in 78% yield.

Esters are also formed by the reaction of alcohols with acyl chlorides:

ROH + $R'\overset{O}{\overset{||}{C}}Cl$ ⟶ $R'\overset{O}{\overset{||}{C}}OR$ + HCl

Alcohol   Acyl chloride   Ester   Hydrogen chloride

This reaction is normally carried out in the presence of a weak base such as pyridine, which reacts with the hydrogen chloride that is formed.

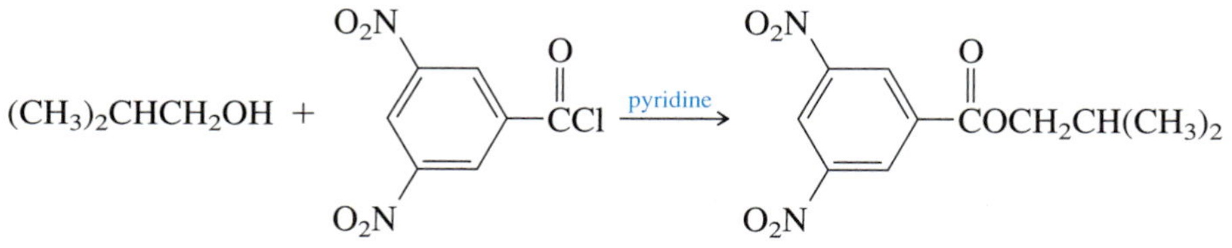

$(CH_3)_2CHCH_2OH$ + [3,5-dinitrobenzoyl chloride] $\xrightarrow{\text{pyridine}}$ [isobutyl 3,5-dinitrobenzoate]

Isobutyl alcohol   3,5-Dinitrobenzoyl chloride   Isobutyl 3,5-dinitrobenzoate (86%)

Carboxylic acid anhydrides react similarly to acyl chlorides.

ROH + $R'\overset{O}{\overset{||}{C}}O\overset{O}{\overset{||}{C}}R'$ ⟶ $R'\overset{O}{\overset{||}{C}}OR$ + $R'\overset{O}{\overset{||}{C}}OH$

Alcohol   Carboxylic acid anhydride   Ester   Carboxylic acid

$C_6H_5CH_2CH_2OH$ + $CF_3\overset{O}{\overset{||}{C}}O\overset{O}{\overset{||}{C}}CF_3$ $\xrightarrow{\text{pyridine}}$ $C_6H_5CH_2CH_2O\overset{O}{\overset{||}{C}}CF_3$ + $CF_3\overset{O}{\overset{||}{C}}OH$

2-Phenylethanol   Trifluoroacetic anhydride   2-Phenylethyl trifluoroacetate (83%)   Trifluoroacetic acid

The mechanisms of the Fischer esterification and the reactions of alcohols with acyl chlorides and acid anhydrides will be discussed in detail in Chapters 19 and 20 after some fundamental principles of carbonyl group reactivity have been developed. For the present, it is sufficient to point out that most of the reactions that convert alcohols to esters leave the C—O bond of the alcohol intact.

$$H\!-\!O\!-\!R \longrightarrow R'\overset{\displaystyle O}{\underset{\displaystyle \|}{C}}\!-\!O\!-\!R$$

This is the same oxygen that was attached to the group R in the starting alcohol.

The acyl group of the carboxylic acid, acyl chloride, or acid anhydride is transferred to the oxygen of the alcohol. This fact is most clearly evident in the esterification of chiral alcohols, where, because none of the bonds to the chirality center is broken in the process, *retention of configuration is observed.*

Make a molecular model corresponding to the stereochemistry of the Fischer projection of 2-phenyl-2-butanol shown in the equation and verify that it has the *R* configuration.

(R)-(+)-2-Phenyl-
2-butanol   +   p-Nitrobenzoyl chloride $\xrightarrow{\text{pyridine}}$ (R)-(−)-1-Methyl-1-phenylpropyl
p-nitrobenzoate (63% yield)

PROBLEM 15.9 A similar conclusion may be drawn by considering the reactions of the cis and trans isomers of 4-*tert*-butylcyclohexanol with acetic anhydride. On the basis of the information just presented, predict the product formed from each stereoisomer.

The reaction of alcohols with acyl chlorides is analogous to their reaction with *p*-toluenesulfonyl chloride described earlier (Section 8.14 and Table 15.2). In those reactions, a *p*-toluenesulfonate ester was formed by displacement of chloride from the sulfonyl group by the oxygen of the alcohol. Carboxylic esters arise by displacement of chloride from a carbonyl group by the alcohol oxygen.

15.9 ESTERS OF INORGANIC ACIDS

Although the term *ester,* used without a modifier, is normally taken to mean an ester of a carboxylic acid, alcohols can react with inorganic acids in a process similar to the Fischer esterification. The products are esters of inorganic acids. For example, *alkyl nitrates* are esters formed by the reaction of alcohols with *nitric acid.*

$$\underset{\text{Alcohol}}{ROH} \;+\; \underset{\text{Nitric acid}}{HONO_2} \;\xrightarrow{H^+}\; \underset{\text{Alkyl nitrate}}{RONO_2} \;+\; \underset{\text{Water}}{H_2O}$$

$$\underset{\text{Methanol}}{CH_3OH} \;+\; \underset{\text{Nitric acid}}{HONO_2} \;\xrightarrow{H_2SO_4}\; \underset{\substack{\text{Methyl nitrate}\\(66-80\%)}}{CH_3ONO_2} \;+\; \underset{\text{Water}}{H_2O}$$

PROBLEM 15.10 Alfred Nobel's fortune was based on his 1866 discovery that nitroglycerin, which is far too shock-sensitive to be transported or used safely, can be stabilized by adsorption onto a substance called *kieselguhr* to give what is familiar to us as *dynamite.* Nitroglycerin is the trinitrate of glycerol

(1,2,3-propanetriol). Write a structural formula or construct a molecular model of nitroglycerin.

Dialkyl sulfates are esters of *sulfuric acid,* **trialkyl phosphites** are esters of *phosphorous acid* (H_3PO_3), and **trialkyl phosphates** are esters of *phosphoric acid* (H_3PO_4).

$$\underset{\text{Dimethyl sulfate}}{CH_3O\overset{\displaystyle O}{\underset{\displaystyle O}{\overset{\|}{\underset{\|}{S}}}}OCH_3} \qquad \underset{\text{Trimethyl phosphite}}{(CH_3O)_3P\colon} \qquad \underset{\text{Trimethyl phosphate}}{(CH_3O)_3\overset{+}{P}-\overset{\cdot\cdot}{\underset{\cdot\cdot}{O}}\colon^-}$$

Some esters of inorganic acids, such as dimethyl sulfate, are used as reagents in synthetic organic chemistry. Certain naturally occurring alkyl phosphates play an important role in biological processes.

15.10 OXIDATION OF ALCOHOLS

Oxidation of an alcohol yields a carbonyl compound. Whether the resulting carbonyl compound is an aldehyde, a ketone, or a carboxylic acid depends on the alcohol and on the oxidizing agent.

Primary alcohols may be oxidized either to an aldehyde or to a carboxylic acid:

$$\underset{\text{Primary alcohol}}{RCH_2OH} \xrightarrow{\text{oxidize}} \underset{\text{Aldehyde}}{R\overset{\displaystyle O}{\overset{\|}{C}}H} \xrightarrow{\text{oxidize}} \underset{\text{Carboxylic acid}}{R\overset{\displaystyle O}{\overset{\|}{C}}OH}$$

Vigorous oxidation leads to the formation of a carboxylic acid, but a number of methods permit us to stop the oxidation at the intermediate aldehyde stage. The reagents most commonly used for oxidizing alcohols are based on high-oxidation-state transition metals, particularly chromium(VI).

Chromic acid (H_2CrO_4) is a good oxidizing agent and is formed when solutions containing chromate (CrO_4^{2-}) or dichromate ($Cr_2O_7^{2-}$) are acidified. Sometimes it is possible to obtain aldehydes in satisfactory yield before they are further oxidized, but in most cases carboxylic acids are the major products isolated on treatment of primary alcohols with chromic acid.

Potassium permanganate ($KMnO_4$) will also oxidize primary alcohols to carboxylic acids. What is the oxidation state of manganese in $KMnO_4$?

$$\underset{\text{3-Fluoro-1-propanol}}{FCH_2CH_2CH_2OH} \xrightarrow[\text{H_2SO_4, H_2O}]{K_2Cr_2O_7} \underset{\substack{\text{3-Fluoropropanoic acid}\\(74\%)}}{FCH_2CH_2\overset{\displaystyle O}{\overset{\|}{C}}OH}$$

Conditions that do permit the easy isolation of aldehydes in good yield by oxidation of primary alcohols employ various Cr(VI) species as the oxidant in *anhydrous* media. Two such reagents are **pyridinium chlorochromate (PCC),** $C_5H_5NH^+ ClCrO_3^-$, and **pyridinium dichromate (PDC),** $(C_5H_5NH)_2^{2+} Cr_2O_7^{2-}$; both are used in dichloromethane.

$$CH_3(CH_2)_5CH_2OH \xrightarrow[CH_2Cl_2]{PCC} CH_3(CH_2)_5\overset{\displaystyle O}{\overset{\|}{C}}H$$

1-Heptanol   Heptanal (78%)

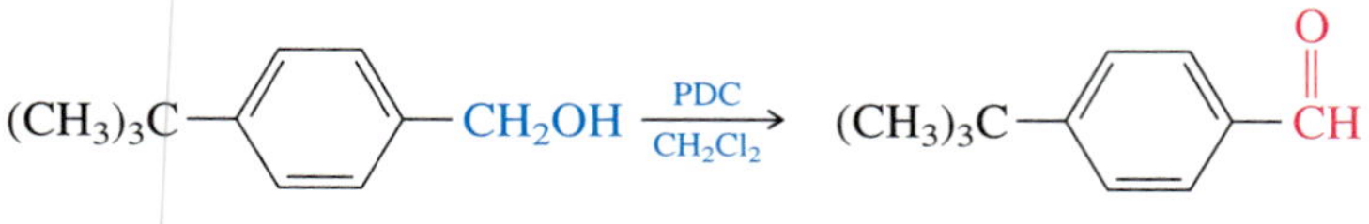

$$(CH_3)_3C-\!\!\!\left\langle\;\right\rangle\!\!\!-CH_2OH \xrightarrow[CH_2Cl_2]{PDC} (CH_3)_3C-\!\!\!\left\langle\;\right\rangle\!\!\!-\overset{\displaystyle O}{\overset{\|}{C}}H$$

p-tert-Butylbenzyl alcohol   *p-tert*-Butylbenzaldehyde (94%)

Secondary alcohols are oxidized to ketones by the same reagents that oxidize primary alcohols:

$$\underset{\substack{\text{Secondary alcohol}}}{\overset{\displaystyle OH}{\overset{|}{R\text{CH}R'}}} \xrightarrow{\text{oxidize}} \underset{\substack{\text{Ketone}}}{\overset{\displaystyle O}{\overset{\|}{R\text{C}R'}}}$$

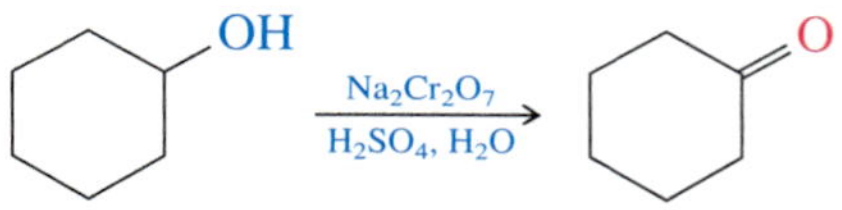

Cyclohexanol   Cyclohexanone (85%)

$$\underset{\substack{\text{1-Octen-3-ol}}}{\overset{\displaystyle OH}{\overset{|}{H_2C=CHCHCH_2CH_2CH_2CH_2CH_3}}} \xrightarrow[CH_2Cl_2]{PDC} \underset{\substack{\text{1-Octen-3-one (80%)}}}{\overset{\displaystyle O}{\overset{\|}{H_2C=CHCCH_2CH_2CH_2CH_2CH_3}}}$$

Tertiary alcohols have no hydrogen on their hydroxyl-bearing carbon and do not undergo oxidation readily:

$$\underset{\substack{}}{\overset{\displaystyle R'}{\overset{|}{R-\underset{\displaystyle R''}{\underset{|}{C}}-OH}}} \xrightarrow{\text{oxidize}} \text{no reaction except under forcing conditions}$$

In the presence of strong oxidizing agents at elevated temperatures, oxidation of tertiary alcohols leads to cleavage of the various carbon–carbon bonds at the hydroxyl-bearing carbon atom, and a complex mixture of products results.

PROBLEM 15.11 Predict the principal organic product of each of the following reactions:

(a) $ClCH_2CH_2CH_2CH_2OH \xrightarrow[H_2SO_4, H_2O]{K_2Cr_2O_7}$

(b) $CH_3CHCH_2CH_2CH_2CH_2CH_2CH_3 \xrightarrow[H_2SO_4, H_2O]{Na_2Cr_2O_7}$
       |
       OH

(c) $CH_3CH_2CH_2CH_2CH_2CH_2CH_2OH \xrightarrow[CH_2Cl_2]{PCC}$

SAMPLE SOLUTION (a) The reactant is a primary alcohol and so can be oxidized either to an aldehyde or to a carboxylic acid. Aldehydes are the major products only when the oxidation is carried out in anhydrous media. Carboxylic acids are formed when water is present. The reaction shown produced 4-chlorobutanoic acid in 56% yield.

$$ClCH_2CH_2CH_2CH_2OH \xrightarrow[H_2SO_4, H_2O]{K_2Cr_2O_7} ClCH_2CH_2CH_2\overset{\displaystyle O}{\overset{\displaystyle \|}{C}}OH$$

4-Chloro-1-butanol   4-Chlorobutanoic acid

The mechanisms by which transition-metal oxidizing agents convert alcohols to aldehydes and ketones are complicated with respect to their *inorganic* chemistry. The organic chemistry is clearer and one possible mechanism is outlined in Figure 15.4. The key intermediate is an alkyl chromate, an ester of an alcohol and chromic acid.

FIGURE 15.4 A mechanism for chromic acid oxidation of an alcohol.

Step 1: Reaction of the alcohol with chromic acid gives an alkyl chromate.

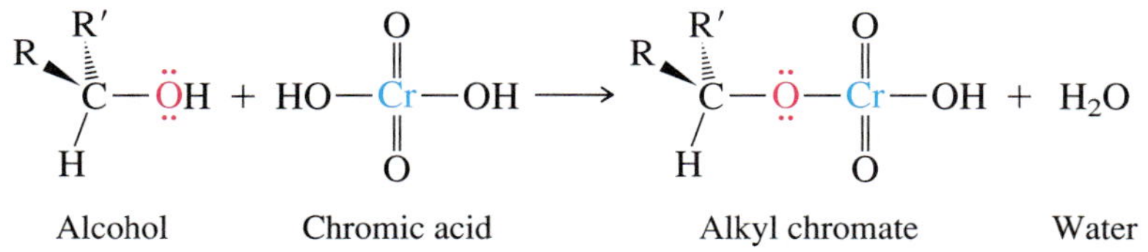

Alcohol   Chromic acid   Alkyl chromate   Water

Step 2: The oxidation step can be viewed as a β elimination. Water acts as a base to remove a proton from carbon while the O—Cr bond breaks.

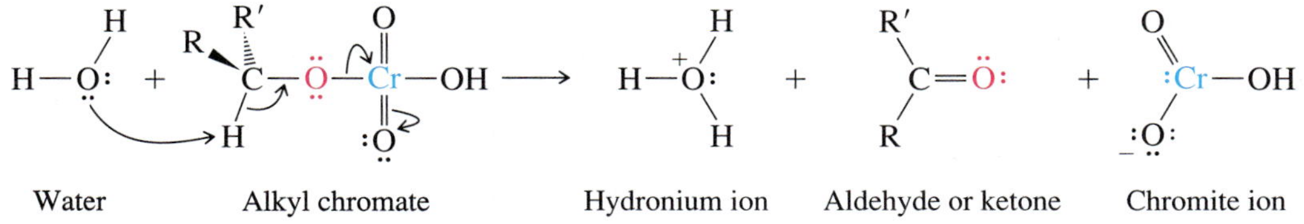

Water   Alkyl chromate   Hydronium ion   Aldehyde or ketone   Chromite ion

Step 3: A series of redox reactions converts chromium from the 4+ oxidation state in $HCrO_3^-$ to the 3+ oxidation state.

ECONOMIC AND ENVIRONMENTAL FACTORS IN ORGANIC SYNTHESIS

Beyond the obvious difference in scale that is evident when one compares preparing tons of a compound versus preparing just a few grams of it, there are sharp distinctions between "industrial" and "laboratory" syntheses. On a laboratory scale, a chemist is normally concerned only with obtaining a modest amount of a substance. Sometimes making the compound is an end in itself, but on other occasions the compound is needed for some further study of its physical, chemical, or biological properties. Considerations such as the cost of reagents and solvents tend to play only a minor role when planning most laboratory syntheses. Faced with a choice between two synthetic routes to a particular compound, one based on the cost of chemicals and the other on the efficient use of a chemist's time, the decision is almost always made in favor of the latter.

Not so for synthesis in the chemical industry, where a compound must be prepared not only on a large scale, but at low cost. There is a pronounced bias toward reactants and reagents that are both abundant and inexpensive. The oxidizing agent of choice, for example, in the chemical industry is O_2, and extensive research has been devoted to developing catalysts for preparing various compounds by air oxidation of readily available starting materials. To illustrate, air and ethylene are the reactants for the industrial preparation of both acetaldehyde and ethylene oxide. Which of the two products is obtained depends on the catalyst employed.

$$\text{H}_2\text{C}=\text{CH}_2 + \tfrac{1}{2}\text{O}_2 \quad\begin{array}{c}\xrightarrow[\text{H}_2\text{O}]{\text{PdCl}_2,\ \text{CuCl}_2}\ \ \text{CH}_3\overset{\displaystyle O}{\overset{\|}{\text{CH}}}\ \ \text{Acetaldehyde}\\[2em]\xrightarrow[\text{300°C}]{\text{Ag}}\ \ \text{H}_2\text{C}{-}\text{CH}_2\!\diagdown\!{}_{\text{O}}\ \ \text{Ethylene oxide}\end{array}$$

Ethylene   Oxygen

Dating approximately from the creation of the U.S. Environmental Protection Agency (EPA) in 1970,

dealing with the byproducts of synthetic procedures has become an increasingly important consideration in designing a chemical synthesis. In terms of changing the strategy of synthetic planning, the chemical industry actually had a shorter road to travel than the pharmaceutical industry, academic laboratories, and research institutes. Simple business principles had long dictated that waste chemicals represented wasted opportunities. It made better sense for a chemical company to recover the solvent from a reaction and use it again than to throw it away and buy more. Similarly, it was far better to find a "value-added" use for a byproduct from a reaction than to throw it away. By raising the cost of generating chemical waste, environmental regulations increased the economic incentive to design processes that produced less of it.

The terms *green chemistry* and *environmentally benign synthesis* have been coined to refer to procedures explicitly designed to minimize the formation of byproducts that present disposal problems. Both the National Science Foundation and the Environmental Protection Agency have allocated a portion of their grant budgets to encourage efforts in this vein.

The application of environmentally benign principles to laboratory-scale synthesis can be illustrated by revisiting the oxidation of alcohols. As noted in Section 15.10, the most widely used methods involve Cr(VI)-based oxidizing agents. Cr(VI) compounds are carcinogenic, however, and appear on the EPA list of compounds requiring special disposal methods. The best way to replace Cr(VI)-based oxidants would be to develop catalytic methods analogous to those used in industry. Another approach would be to use oxidizing agents that are less hazardous, such as sodium hypochlorite. Aqueous solutions of sodium hypochlorite are available as "swimming-pool chlorine," and procedures for their use in oxidizing secondary alcohols to ketones have been developed. One is described on page 71 of the January 1991 edition of the *Journal of Chemical Education*.

—*Cont.*

$$(CH_3)_2CHCH_2CHCH_2CH_2CH_3 \xrightarrow[\text{acetic acid-water}]{\text{NaOCl}} (CH_3)_2CHCH_2\overset{\overset{\displaystyle O}{\|}}{C}CH_2CH_2CH_3$$

| | OH | |

2-Methyl-4-heptanol   2-Methyl-4-heptanone (77%)

There is a curious irony in the nomination of hypochlorite as an environmentally benign oxidizing agent. It comes at a time of increasing pressure to eliminate chlorine and chlorine-containing compounds from the environment to as great a degree as possible. Any all-inclusive assault on chlorine needs to be carefully scrutinized, especially when one remembers that chlorination of the water supply has probably done more to extend human life than any other public health measure ever undertaken. (The role of chlorine in the formation of chlorinated hydrocarbons in water is discussed in Section 18.7.)

15.11 BIOLOGICAL OXIDATION OF ALCOHOLS

Many biological processes involve oxidation of alcohols to carbonyl compounds or the reverse process, reduction of carbonyl compounds to alcohols. Ethanol, for example, is metabolized in the liver to acetaldehyde. Such processes are catalyzed by enzymes; the enzyme that catalyzes the oxidation of ethanol is called *alcohol dehydrogenase.*

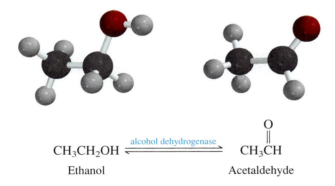

$$CH_3CH_2OH \underset{\text{alcohol dehydrogenase}}{\rightleftharpoons} CH_3\overset{\overset{\displaystyle O}{\|}}{C}H$$

Ethanol   Acetaldehyde

In addition to enzymes, biological oxidations require substances known as *coenzymes.* Coenzymes are organic molecules that, in concert with an enzyme, act on a substrate to bring about chemical change. Most of the substances that we call vitamins are coenzymes. The coenzyme contains a functional group that is complementary to a functional group of the substrate; the enzyme catalyzes the interaction of these mutually complementary functional groups. If ethanol is oxidized, some other substance must be reduced. This other substance is the oxidized form of the coenzyme *nicotinamide adenine dinucleotide* (NAD). Chemists and biochemists abbreviate the oxidized form of this coenzyme as NAD^+ and its reduced form as NADH. More completely, the chemical equation for the biological oxidation of ethanol may be written:

$$CH_3CH_2OH + NAD^+ \underset{\text{alcohol dehydrogenase}}{\rightleftharpoons} CH_3\overset{\overset{\displaystyle O}{\|}}{C}H + NADH + H^+$$

Ethanol   Oxidized form   Acetaldehyde   Reduced
         of NAD coenzyme   form of NAD
                                                         coenzyme

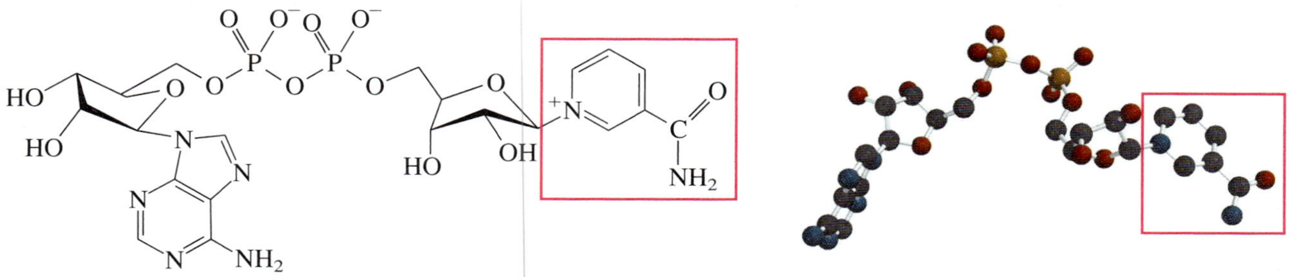

FIGURE 15.5 Structure of NAD^+, the oxidized form of the coenzyme nicotinamide adenine dinucleotide. The functional part of the coenzyme is framed in red.

The structure of the oxidized form of nicotinamide adenine dinucleotide is shown in Figure 15.5. The only portion of the coenzyme that undergoes chemical change in the reaction is the substituted pyridine ring of the nicotinamide unit (framed in red in Figure 15.5). If the remainder of the coenzyme molecule is represented by R, its role as an oxidizing agent is shown in the equation in Figure 15.6, which tracks the flow of electrons in the oxidation of ethanol. The key feature of this mechanism is that hydrogen is transferred from C-1 of ethanol not as a proton (H^+), but as hydride ($H:^-$). The ability of ethanol to transfer hydride is enhanced by removal of the O—H proton by a basic site of the enzyme. Hydride is never free, but is transferred directly from ethanol to the positively charged pyridinium ring of NAD^+ to give the dihydropyridine ring of NADH.

PROBLEM 15.12 The mechanism of enzymatic oxidation has been studied by isotopic labeling with the aid of deuterated derivatives of ethanol. Specify the number of deuterium atoms that you would expect to find attached to the dihydropyridine ring of the reduced form of the nicotinamide adenine dinucleotide coenzyme following enzymatic oxidation of each of the alcohols given:

(a) CD_3CH_2OH (b) CH_3CD_2OH (c) CH_3CH_2OD

SAMPLE SOLUTION According to the proposed mechanism for biological oxidation of ethanol, the hydrogen that is transferred to the coenzyme comes from C-1 of ethanol. Therefore, the dihydropyridine ring will bear no deuterium atoms when CD_3CH_2OH is oxidized, because all the deuterium atoms of the alcohol are attached to C-2.

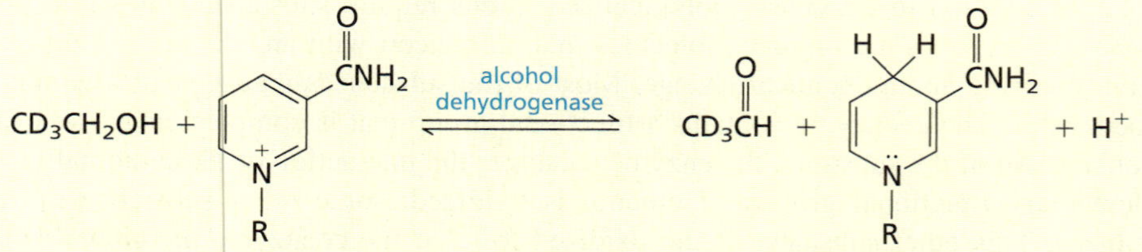

2,2,2-Trideuterioethanol — NAD^+ — 2,2,2-Trideuterioethanal — NADH

The reverse reaction also occurs in living systems; NADH reduces acetaldehyde to ethanol in the presence of alcohol dehydrogenase. In this process, NADH serves as a hydride donor and is oxidized to NAD^+ while acetaldehyde is reduced.

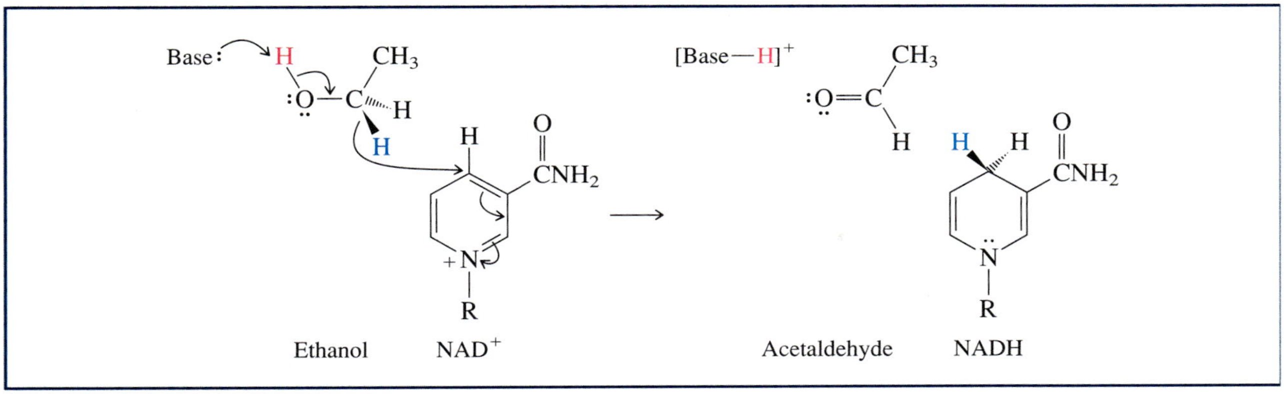

FIGURE 15.6 Bonding changes during the enzyme catalyzed oxidation of ethanol in the presence of the coenzyme NAD$^+$.

The NAD$^+$–NADH coenzyme system is involved in a large number of biological oxidation–reductions. Another reaction similar to the ethanol–acetaldehyde conversion is the oxidation of lactic acid to pyruvic acid by NAD$^+$ and the enzyme *lactic acid dehydrogenase*:

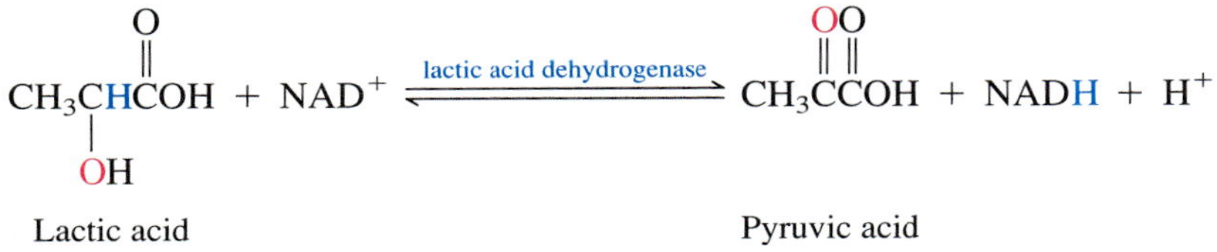

We shall encounter other biological processes in which the NAD$^+$ ⇌ NADH interconversion plays a prominent role in biological oxidation–reduction.

15.12 OXIDATIVE CLEAVAGE OF VICINAL DIOLS

A reaction characteristic of vicinal diols is their oxidative cleavage on treatment with periodic acid (HIO$_4$). The carbon–carbon bond of the vicinal diol unit is broken and two carbonyl groups result. Periodic acid is reduced to iodic acid (HIO$_3$).

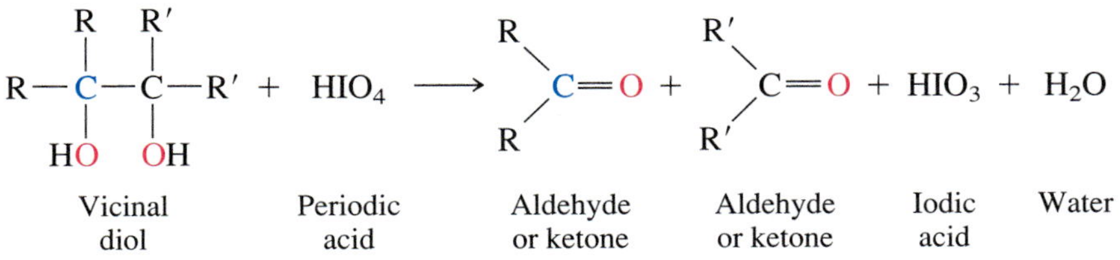

What is the oxidation state of iodine in HIO$_4$? In HIO$_3$?

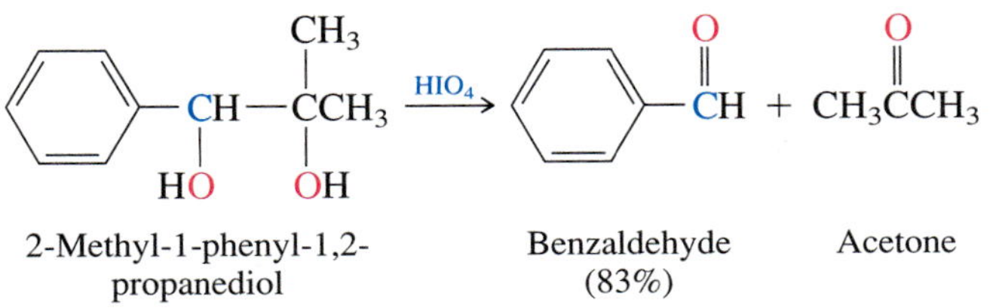

Can you remember what reaction of an alkene would give the same products as the periodic acid cleavage shown here?

This reaction occurs only when the hydroxyl groups are on adjacent carbons.

PROBLEM 15.13 Predict the products formed on oxidation of each of the following with periodic acid:

(a) HOCH$_2$CH$_2$OH

(b) (CH$_3$)$_2$CHCH$_2$CHCHCH$_2$C$_6$H$_5$
            | |
           HO OH

(c)

SAMPLE SOLUTION (a) The carbon–carbon bond of 1,2-ethanediol is cleaved by periodic acid to give two molecules of formaldehyde:

$$\text{HOCH}_2\text{CH}_2\text{OH} \xrightarrow{\text{HIO}_4} \quad 2\text{HCH}$$

1,2-Ethanediol   Formaldehyde

Cyclic diols give dicarbonyl compounds. The reactions are faster when the hydroxyl groups are cis than when they are trans, but both stereoisomers are oxidized by periodic acid.

1,2-Cyclopentanediol   Pentanedial
(either stereoisomer)

Periodic acid cleavage of vicinal diols is often used for analytical purposes as an aid in structure determination. By identifying the carbonyl compounds produced, the constitution of the starting diol may be deduced. This technique finds its widest application with carbohydrates and will be discussed more fully in Chapter 25.

15.13 THIOLS

Sulfur lies just below oxygen in the periodic table, and many oxygen-containing organic compounds have sulfur analogs. The sulfur analogs of alcohols (ROH) are **thiols (RSH).** Thiols are given substitutive IUPAC names by appending the suffix -*thiol* to the name of the corresponding alkane, numbering the chain in the direction that gives the lower locant to the carbon that bears the —SH group. As with diols (Section 15.5), the final -*e* of the alkane name is retained. When the —SH group is named as a substituent, it is called a *mercapto* group. It is also often referred to as a *sulfhydryl* group, but this is a generic term, not used in systematic nomenclature.

(CH$_3$)$_2$CHCH$_2$CH$_2$SH   HSCH$_2$CH$_2$OH   HSCH$_2$CH$_2$CH$_2$SH

3-Methyl-1-butanethiol   2-Mercaptoethanol   1,3-Propanedithiol

Thiols have a marked tendency to bond to mercury, and the word *mercaptan* comes from the Latin *mercurium captans,* which means "seizing mercury." The drug *dimercaprol* is used to treat mercury and lead poisoning; it is 2,3-dimercapto-1-propanol.

At one time thiols were named *mercaptans.* Thus, CH$_3$CH$_2$SH was called "ethyl mercaptan" according to this system. This nomenclature was abandoned beginning with the 1965 revision of the IUPAC rules but is still sometimes encountered. When one encounters a thiol for the first time, especially a low-molecular-weight thiol, its most obvious property is its foul odor. Ethanethiol is added to natural gas so that leaks can be detected without special equipment—your nose is so sensitive that it can detect less than one part of ethanethiol in 10,000,000,000 parts of air! The odor of thiols weakens with the number of carbons, because both the volatility and the sulfur content decrease. 1-Dodecanethiol, for example, has only a faint odor. On the positive side, of the hundreds of substances that contribute to the aroma of freshly brewed coffee, the one most responsible for its characteristic odor is the thiol 2-mercaptomethylfuran.

2-Mercaptomethylfuran

> **PROBLEM 15.14** The main components of a skunk's scent fluid are 3-methyl-1-butanethiol and *cis*- and *trans*-2-butene-1-thiol. Write structural formulas for each of these compounds.

A historical account of the analysis of skunk scent and a modern determination of its composition appear in the March 1978 issue of the *Journal of Chemical Education.*

The S—H bond is less polar than the O—H bond, as is clearly seen in the electrostatic potential maps of Figure 15.7. The decreased polarity of the S—H bond, especially the decreased positive character of the proton, causes hydrogen bonding to be absent in thiols. Thus, methanethiol (CH_3SH) is a gas at room temperature (bp 6°C), whereas methanol (CH_3OH) is a liquid (bp 65°C).

Compare the boiling point of H_2S (−60°C) and H_2O (100°C).

In spite of S—H bonds being less polar than O—H bonds, thiols are stronger acids than alcohols. This is largely because S—H bonds are weaker than O—H bonds. We have seen that most alcohols have pK_a's of 16–18. The corresponding value for a thiol is about 10. The significance of this difference is that a thiol can be quantitatively converted to its conjugate base (RS^-), called an **alkanethiolate,** by hydroxide. Consequently, thiols dissolve in aqueous base.

Alkanethiol (stronger acid) ($pK_a = 11$)	Hydroxide ion (stronger base)	Alkanethiolate ion (weaker base)	Water (weaker acid) ($pK_a = 15.7$)

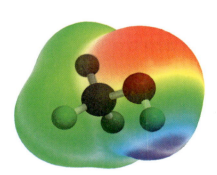 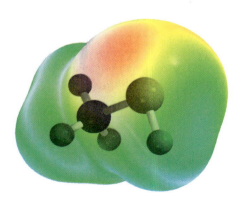

(*a*) Methanol (CH_3OH)   (*b*) Methanethiol (CH_3SH)

FIGURE 15.7 Electrostatic potential maps of (*a*) methanol, and (*b*) methanethiol. The color scales were adjusted to be the same for both molecules to allow for direct comparison. The development of charge is more pronounced in the region surrounding the —OH group in methanol than it is for the —SH group in methanethiol.

Alkanethiolate ions (RS^-) are weaker bases than alkoxide ions (RO^-) and undergo synthetically useful S_N2 reactions even with secondary alkyl halides.

> Recall from Section 8.13 that the major pathway for reaction of *alkoxide* ions with secondary alkyl halides is E2, not S_N2.

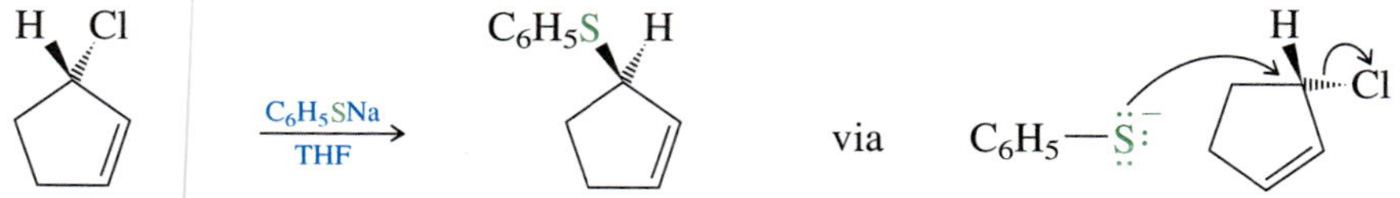

3-Chlorocyclopentene → 2-Cyclopentenyl phenyl sulfide (75%)   via $C_6H_5-\ddot{S}:^-$

Thiols themselves are sometimes prepared by nucleophilic substitution using the conjugate base of H_2S.

$$CH_3(CH_2)_4CH_2Br \xrightarrow[\text{ethanol}]{KSH} CH_3(CH_2)_4CH_2SH$$

1-Bromohexane   1-Hexanethiol (67%)

PROBLEM 15.15   Outline a synthesis of 1-hexanethiol from 1-hexanol.

A major difference between alcohols and thiols concerns their oxidation. We have seen earlier in this chapter that oxidation of alcohols gives compounds having carbonyl groups. Analogous oxidation of thiols to compounds with C=S functions does *not* occur. Only sulfur is oxidized, not carbon, and compounds containing sulfur in various oxidation states are possible. These include a series of acids classified as *sulfenic, sulfinic,* and *sulfonic* according to the number of oxygens attached to sulfur.

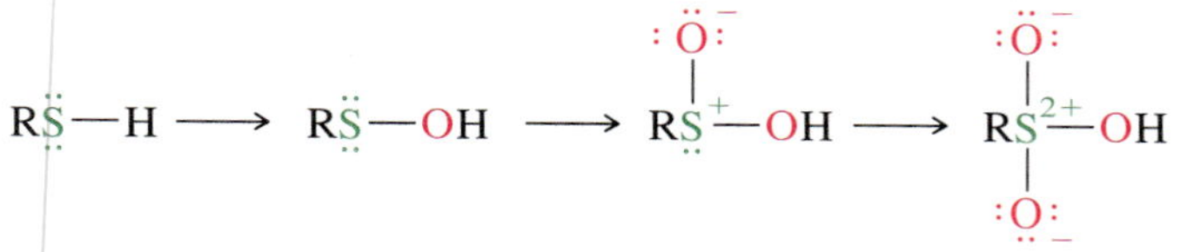

Thiol   Sulfenic acid   Sulfinic acid   Sulfonic acid

Of these the most important are the sulfonic acids. In general, however, sulfonic acids are not prepared by oxidation of thiols. Arenesulfonic acids ($ArSO_3H$), for example, are prepared by sulfonation of arenes (Section 12.4).

The most important oxidation, from a biochemical perspective, is the conversion of thiols to **disulfides.**

$$2RSH \underset{\text{reduction}}{\overset{\text{oxidation}}{\rightleftharpoons}} RSSR$$

Thiol   Disulfide

Although a variety of oxidizing agents are available for this transformation, it occurs so readily that thiols are slowly converted to disulfides by the oxygen in the air. Dithiols give cyclic disulfides by intramolecular sulfur–sulfur bond formation. An example of a cyclic disulfide is the coenzyme α-*lipoic acid.* The last step in the laboratory synthesis of α-lipoic acid is an iron(III)-catalyzed oxidation of the dithiol shown:

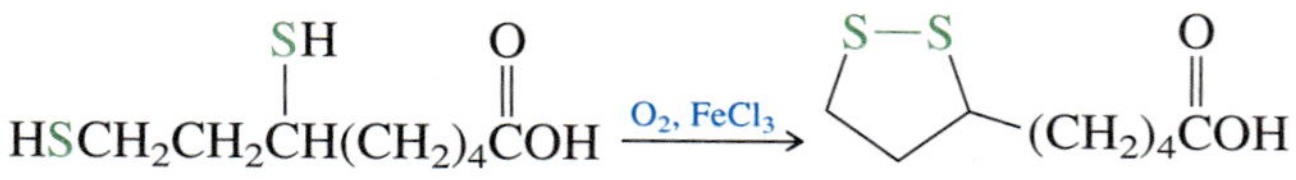

6,8-Dimercaptooctanoic acid   α-Lipoic acid (78%)

Rapid and reversible making and breaking of the sulfur–sulfur bond is essential to the biological function of α-lipoic acid.

An important aspect of S—S bonds in disulfides is that they are intermediate in strength between typical covalent bonds and weaker interactions such as hydrogen bonds. Covalent bonds involving C, H, N, and O have bond strengths on the order of 330–420 kJ/mol. The S—S bond energy in a disulfide is about 220 kJ/mol, and hydrogen bond strengths are usually less than 30 kJ/mol. Thus S—S bonds provide more structural stability than a hydrogen bond, but can be broken while leaving the covalent framework intact.

All mammalian cells contain a thiol called *glutathione*. Glutathione protects the cell by scavenging harmful oxidants. It reacts with these oxidants by forming a disulfide, which is eventually converted back to glutathione.

$$2\left(\begin{array}{c} \overset{+}{\text{H}_3}\text{NCHCH}_2\text{CH}_2\overset{\text{O}}{\overset{\|}{\text{C}}}\text{NHCHCH}\overset{\text{O}}{\overset{\|}{\text{C}}}\text{NHCH}_2\text{CO}_2^- \\ | \qquad\qquad\qquad | \\ \text{CO}_2^- \qquad\qquad \text{CH}_2\text{SH} \end{array} \right) \underset{\text{reduction}}{\overset{\text{oxidation}}{\rightleftarrows}}$$

Glutathione (reduced form)   Glutathione (oxidized form)

The three-dimensional shapes of many proteins are governed and stabilized by S—S bonds connecting what would ordinarily be remote segments of the molecule. We'll have more to say about these *disulfide bridges* in Chapter 27.

15.14 SPECTROSCOPIC ANALYSIS OF ALCOHOLS AND THIOLS

Infrared: We discussed the most characteristic features of the infrared spectra of *alcohols* earlier (Section 13.20). The O—H stretching vibration is especially easy to identify, appearing in the 3200–3650 cm^{-1} region. As the infrared spectrum of cyclohexanol, presented in Figure 15.8 demonstrates, this peak is seen as a broad absorption of moderate intensity. The C—O bond stretching of alcohols gives rise to a moderate to strong absorbance between 1025 and 1200 cm^{-1}. It appears at 1070 cm^{-1} in cyclohexanol, a typical secondary alcohol, but is shifted to slightly higher energy in tertiary alcohols and slightly lower energy in primary alcohols.

The S—H stretching frequency of *thiols* gives rise to a weak band in the range 2550–2700 cm^{-1}.

¹H NMR: The most helpful signals in the ¹H NMR spectrum of *alcohols* result from the O—H proton and the proton in the **H—C—O** unit of primary and secondary alcohols.

H—C—O—H

δ 3.3–4.0   δ 0.5–5

The chemical shift of the hydroxyl proton signal is variable, depending on solvent, temperature, and concentration. Its precise position is not particularly significant in structure determination. Because the signals due to hydroxyl protons are not usually split by other protons in the molecule and are often rather broad, they are often fairly easy to

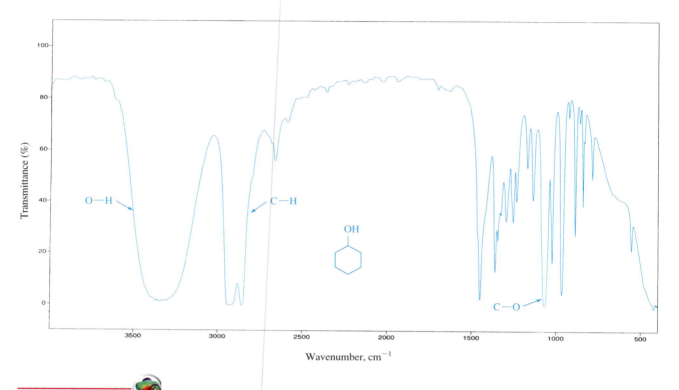

FIGURE 15.8 The infrared spectrum of cyclohexanol.

identify. To illustrate, Figure 15.9 shows the ^{1}H NMR spectrum of 2-phenylethanol, in which the hydroxyl proton signal appears as a singlet at δ 4.5. Of the two triplets in this spectrum, the one at lower field strength (δ 4.0) corresponds to the protons of the CH_2O unit. The higher-field strength triplet at δ 3.1 arises from the benzylic CH_2 group. The assignment of a particular signal to the hydroxyl proton can be confirmed by adding D_2O. The hydroxyl proton is replaced by deuterium, and its ^{1}H NMR signal disappears.

Because of its lower electronegativity, sulfur shields neighboring protons more than oxygen does. Thus, the protons of a CH_2S group appear at higher field than those of a CH_2OH group.

$$CH_3CH_2CH_2{-}CH_2{-}OH \qquad CH_3CH_2CH_2{-}CH_2{-}SH$$

^{1}H Chemical shift: δ 3.6 δ 2.5

^{13}C NMR: The electronegative oxygen of an *alcohol* decreases the shielding of the carbon to which it is attached. The chemical shift for the carbon of the C—OH is 60–75 ppm for most alcohols. Carbon of a C—S group is more shielded than carbon of C—O.

$$CH_3{-}CH_2{-}CH_2{-}CH_2{-}OH \qquad CH_3{-}CH_2{-}CH_2{-}CH_2{-}SH$$

^{13}C Chemical shift: δ 14 δ 19 δ 35 δ 62 δ 13 δ 21 δ 36 δ 24

UV-VIS: Unless the molecule has other chromophores, alcohols are transparent above about 200 nm; λ_{max} for methanol, for example, is 177 nm.

Mass Spectrometry: The molecular ion peak is usually quite small in the mass spectrum of an alcohol. A peak corresponding to loss of water is often evident. Alcohols also fragment readily by a pathway in which the molecular ion loses an alkyl group from the

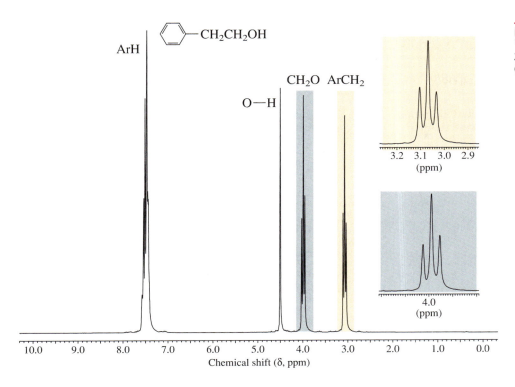

FIGURE 15.9 The 200-MHz ^{1}H NMR spectrum of 2-phenylethanol $(C_6H_5CH_2CH_2OH)$.

hydroxyl-bearing carbon to form a stable cation. Thus, the mass spectra of most primary alcohols exhibit a prominent peak at m/z 31.

$$RCH_2\ddot{O}H \longrightarrow R-CH_2-\overset{+}{\underset{..}{O}}H \longrightarrow R\cdot \; + \; H_2C=\overset{+}{\underset{..}{O}}H$$

| Primary alcohol | Molecular ion | Alkyl radical | Conjugate acid of formaldehyde, m/z 31 |

> **PROBLEM 15.16** Three of the most intense peaks in the mass spectrum of 2-methyl-2-butanol appear at m/z 59, 70, and 73. Explain the origin of these peaks.

Interpreting the mass spectra of sulfur compounds is aided by the observation of an M+2 peak because of the presence of the mass-34 isotope of sulfur. The major cleavage pathway of *thiols* is analogous to that of alcohols.

15.15 SUMMARY

Section 15.1 Functional group interconversions involving alcohols either as reactants or as products are the focus of this chapter. Alcohols are commonplace natural products. Table 15.1 summarizes reactions discussed in earlier sections that can be used to prepare alcohols.

Section 15.2 Alcohols can be prepared from carbonyl compounds by reduction of aldehydes and ketones. See Table 15.3.

Section 15.3 Alcohols can be prepared from carbonyl compounds by reduction of carboxylic acids and esters. See Table 15.3.

TABLE 15.3   Preparation of Alcohols by Reduction of Carbonyl Functional Groups

Carbonyl compound	Product of reduction of carbonyl compound by specified reducing agent		
	Lithium aluminum hydride (LiAlH$_4$)	Sodium borohydride (NaBH$_4$)	Hydrogen (in the presence of a catalyst)
Aldehyde $\overset{\overset{O}{\|}}{R}CH$ (Section 15.2)	Primary alcohol RCH$_2$OH	Primary alcohol RCH$_2$OH	Primary alcohol RCH$_2$OH
Ketone $\overset{\overset{O}{\|}}{R}CR'$ (Section 15.2)	Secondary alcohol RCHR' | OH	Secondary alcohol RCHR' | OH	Secondary alcohol RCHR' | OH
Carboxylic acid $\overset{\overset{O}{\|}}{R}COH$ (Section 15.3)	Primary alcohol RCH$_2$OH	Not reduced	Not reduced
Carboxylic ester $\overset{\overset{O}{\|}}{R}COR'$ (Section 15.3)	Primary alcohol RCH$_2$OH plus R'OH	Reduced too slowly to be of practical value	Requires special catalyst, high pressures and temperatures

Section 15.4   Grignard and organolithium reagents react with ethylene oxide to give primary alcohols.

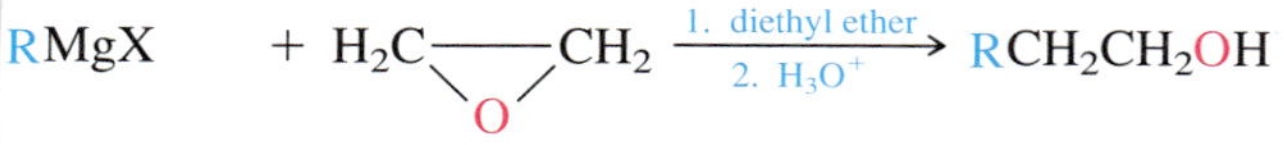

$$RMgX \quad + \quad H_2C\!\!-\!\!CH_2 \xrightarrow[\text{2. } H_3O^+]{\text{1. diethyl ether}} RCH_2CH_2OH$$

Grignard reagent   Ethylene oxide   Primary alcohol

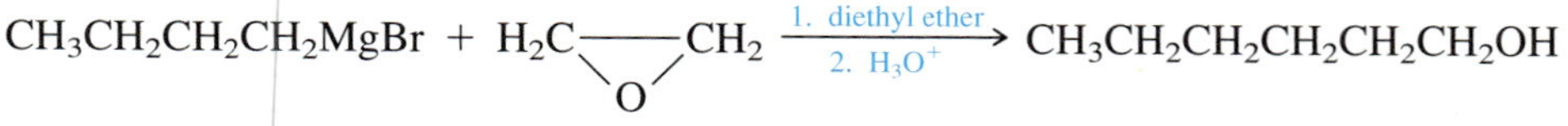

$$CH_3CH_2CH_2CH_2MgBr + H_2C\!\!-\!\!CH_2 \xrightarrow[\text{2. } H_3O^+]{\text{1. diethyl ether}} CH_3CH_2CH_2CH_2CH_2CH_2OH$$

Butylmagnesium bromide   Ethylene oxide   1-Hexanol (60–62%)

Section 15.5   Osmium tetraoxide is a key reactant in the conversion of alkenes to vicinal diols.

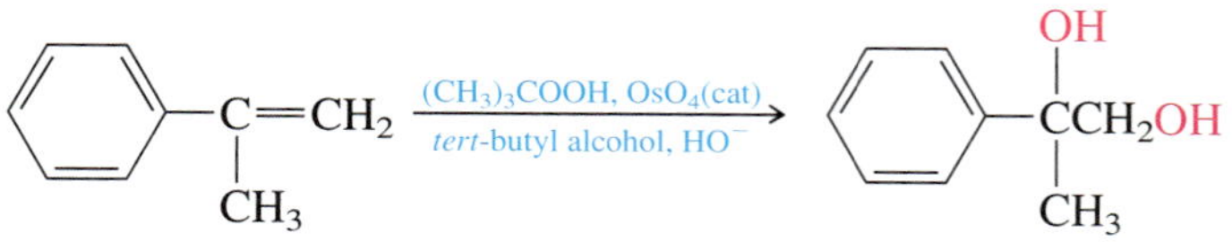

2-Phenylpropene   2-Phenyl-1,2-propanediol (71%)

The reaction is called **hydroxylation** and proceeds by syn addition to the double bond.

Section 15.6   Table 15.2 summarizes reactions of alcohols that were introduced in earlier chapters.

Section 15.7   See Table 15.4

Section 15.8   See Table 15.4

Section 15.9   See Table 15.4

Section 15.10   See Table 15.5

Section 15.11   Oxidation of alcohols to aldehydes and ketones is a common biological reaction. Most require a coenzyme such as the oxidized form of nicotinamide adenine dinucleotide (NAD^+).

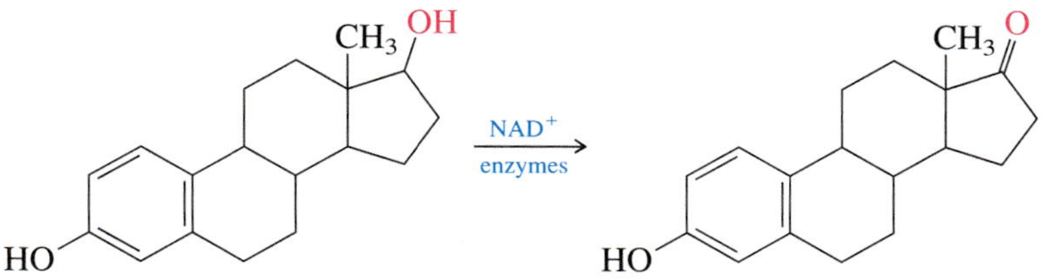

Estradiol   Estrone

Section 15.12   Periodic acid cleaves vicinal diols; two aldehydes, two ketones, or an aldehyde and a ketone are formed.

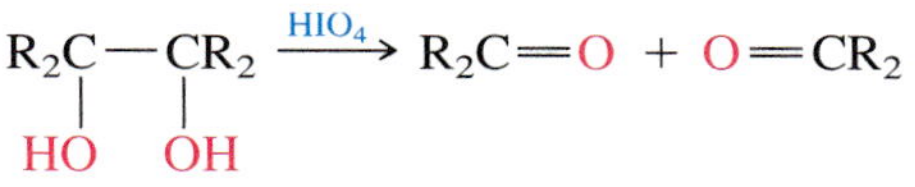

$$R_2C\!-\!CR_2 \xrightarrow{HIO_4} R_2C\!=\!O \ + \ O\!=\!CR_2$$
$$\quad\ \ |\quad\ |$$
$$\quad\ HO\quad OH$$

Diol   Two carbonyl-containing
                             compounds

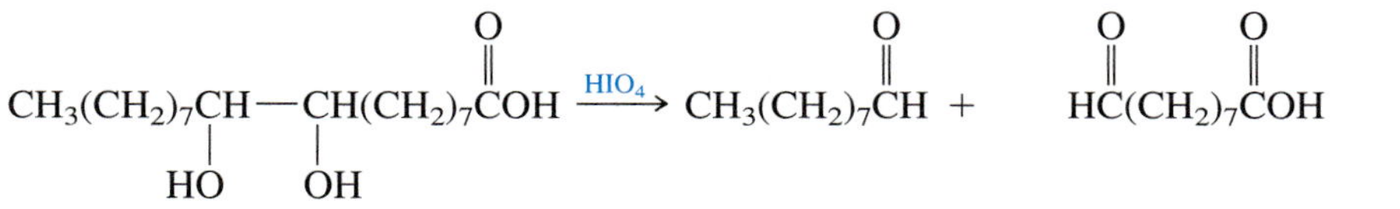

9,10-Dihydroxyoctadecanoic acid   Nonanal (89%)   9-Oxononanoic acid (76%)

Section 15.13   **Thiols** are compounds of the type RSH. They are more acidic than alcohols and are readily deprotonated by reaction with aqueous base. Thiols can be oxidized to sulfenic acids (RSOH), sulfinic acids (RSO_2H), and sulfonic acids (RSO_3H). The redox relationship between thiols and disulfides is important in certain biochemical processes.

$$2RSH \underset{\text{reduction}}{\overset{\text{oxidation}}{\rightleftarrows}} RSSR$$

Thiol   Disulfide

Section 15.14   The hydroxyl group of an alcohol has its O—H and C—O stretching vibrations at 3200–3650 and 1025–1200 cm^{-1}, respectively.
The chemical shift of the proton of an O—H group is variable (δ 1–5) and depends on concentration, temperature, and solvent. Oxygen deshields both the proton and the carbon of an H—C—O unit. Typical NMR chemical shifts are δ 3.3–4.0 for ^1H and δ 60–75 for ^{13}C of H—C—O.

TABLE 15.4   Summary of Reactions of Alcohols Presented in This Chapter

Reaction (section) and comments	General equation and specific example
Conversion to dialkyl ethers (Section 15.7) On being heated in the presence of an acid catalyst, two molecules of a primary alcohol combine to form an ether and water. Diols can undergo an intramolecular condensation if a five-membered or six-membered cyclic ether results.	$2RCH_2OH \xrightarrow[\text{heat}]{H^+} RCH_2OCH_2R + H_2O$ Alcohol ⟶ Dialkyl ether + Water $2(CH_3)_2CHCH_2CH_2OH \xrightarrow[150°C]{H_2SO_4} (CH_3)_2CHCH_2CH_2OCH_2CH_2CH(CH_3)_2$ 3-Methyl-1-butanol ⟶ Di-(3-methylbutyl) ether (27%)
Fischer esterification (Section 15.8) Alcohols and carboxylic acids yield an ester and water in the presence of an acid catalyst. The reaction is an equilibrium process that can be driven to completion by using either the alcohol or the acid in excess or by removing the water as it is formed.	$ROH + R'\overset{O}{\overset{\|}{C}}OH \xrightarrow{H^+} R'\overset{O}{\overset{\|}{C}}OR + H_2O$ Alcohol + Carboxylic acid ⟶ Ester + Water $CH_3CH_2CH_2CH_2CH_2OH + CH_3\overset{O}{\overset{\|}{C}}OH \xrightarrow{H^+} CH_3\overset{O}{\overset{\|}{C}}OCH_2CH_2CH_2CH_2CH_3$ 1-Pentanol + Acetic acid ⟶ Pentyl acetate (71%)
Esterification with acyl chlorides (Section 15.8) Acyl chlorides react with alcohols to give esters. The reaction is usually carried out in the presence of pyridine.	$ROH + R'\overset{O}{\overset{\|}{C}}Cl \longrightarrow R'\overset{O}{\overset{\|}{C}}OR + HCl$ Alcohol + Acyl chloride ⟶ Ester + Hydrogen chloride $(CH_3)_3COH + CH_3\overset{O}{\overset{\|}{C}}Cl \xrightarrow{\text{pyridine}} CH_3\overset{O}{\overset{\|}{C}}OC(CH_3)_3$ tert-Butyl alcohol + Acetyl chloride ⟶ tert-Butyl acetate (62%)
Esterification with carboxylic acid anhydrides (Section 15.8) Carboxylic acid anhydrides react with alcohols to form esters in the same way that acyl chlorides do.	$ROH + R'\overset{O}{\overset{\|}{C}}O\overset{O}{\overset{\|}{C}}R' \longrightarrow R'\overset{O}{\overset{\|}{C}}OR + R'\overset{O}{\overset{\|}{C}}OH$ Alcohol + Carboxylic acid anhydride ⟶ Ester + Carboxylic acid m-Methoxybenzyl alcohol + Acetic anhydride $\xrightarrow{\text{pyridine}}$ m-Methoxybenzyl acetate (99%)
Formation of esters of inorganic acids (Section 15.9) Alkyl nitrates, dialkyl sulfates, trialkyl phosphites, and trialkyl phosphates are examples of alkyl esters of inorganic acids. In some cases, these compounds are prepared by the direct reaction of an alcohol and the inorganic acid.	$ROH + HONO_2 \xrightarrow{H^+} RONO_2 + H_2O$ Alcohol + Nitric acid ⟶ Alkyl nitrate + Water Cyclopentanol $\xrightarrow[H_2SO_4]{HNO_3}$ Cyclopentyl nitrate (69%)

TABLE 15.5 Oxidation of Alcohols

Class of alcohol	Desired product	Suitable oxidizing agent(s)
Primary, RCH_2OH	Aldehyde $\;\;RCH{=}O$	PCC* PDC*
Primary, RCH_2OH	Carboxylic acid $\;\;RCOH{=}O$	$Na_2Cr_2O_7$, H_2SO_4, H_2O H_2CrO_4
Secondary, $RCHR'$ $\quad\quad\;\;OH$	Ketone $\;\;RCR'{=}O$	PCC* PDC* $Na_2Cr_2O_7$, H_2SO_4, H_2O H_2CrO_4

*PCC is pyridinium chlorochromate; PDC is pyridinium dichromate. Both are used in dichloromethane.

The most intense peaks in the mass spectrum of an alcohol correspond to the ion formed according to carbon–carbon cleavage of the type shown:

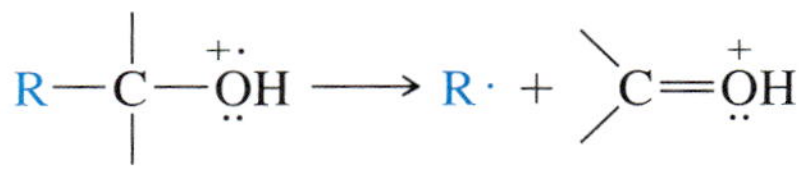

$$R\!-\!\overset{|}{\underset{|}{C}}\!-\!\overset{+\cdot}{\underset{\cdot\cdot}{O}}H \longrightarrow R\cdot \;+\; {\underset{\diagup}{\overset{\diagdown}{}}}C{=}\overset{+}{\underset{\cdot\cdot}{O}}H$$

PROBLEMS

15.17 Write chemical equations, showing all necessary reagents, for the preparation of 1-butanol by each of the following methods:

    (a) Hydroboration–oxidation of an alkene

    (b) Use of a Grignard reagent

    (c) Use of a Grignard reagent in a way different from part (b)

    (d) Reduction of a carboxylic acid

    (e) Reduction of a methyl ester

    (f) Reduction of a butyl ester

    (g) Hydrogenation of an aldehyde

    (h) Reduction with sodium borohydride

15.18 Write chemical equations, showing all necessary reagents, for the preparation of 2-butanol by each of the following methods:

    (a) Hydroboration–oxidation of an alkene

    (b) Use of a Grignard reagent

    (c) Use of a Grignard reagent different from that used in part (b)

    (d–f) Three different methods for reducing a ketone

15.19 Write chemical equations, showing all necessary reagents, for the preparation of *tert*-butyl alcohol by:

(a) Reaction of a Grignard reagent with a ketone

(b) Reaction of a Grignard reagent with an ester of the type $RCOCH_3$ (with a carbonyl O above the C)

15.20 Which of the isomeric $C_5H_{12}O$ alcohols can be prepared by lithium aluminum hydride reduction of:

(a) An aldehyde

(c) A carboxylic acid

(b) A ketone

(d) An ester of the type $RCOCH_3$ (with a carbonyl O above the C)

15.21 Evaluate the feasibility of the route

$$RH \xrightarrow[\text{light or heat}]{Br_2} RBr \xrightarrow{KOH} ROH$$

as a method for preparing

(a) 1-Butanol from butane

(b) 2-Methyl-2-propanol from 2-methylpropane

(c) Benzyl alcohol from toluene

(d) (*R*)-1-Phenylethanol from ethylbenzene

15.22 Sorbitol is a sweetener often substituted for cane sugar, because it is better tolerated by diabetics. It is also an intermediate in the commercial synthesis of vitamin C. Sorbitol is prepared by high-pressure hydrogenation of glucose over a nickel catalyst. What is the structure (including stereochemistry) of sorbitol?

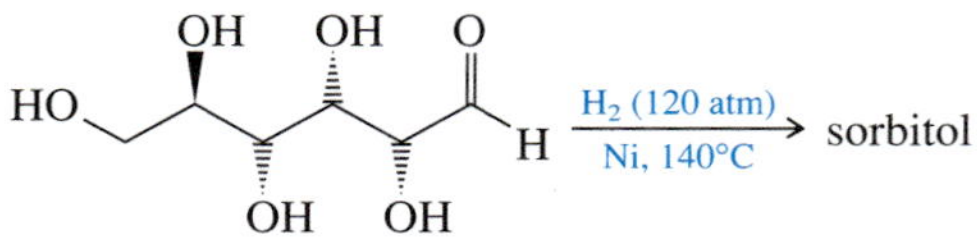

Glucose

15.23 Write equations showing how 1-phenylethanol ($C_6H_5CHCH_3$, with OH on the CH) could be prepared from each of the following starting materials:

(a) Bromobenzene

(d) Acetophenone

(b) Benzaldehyde

(e) Benzene

(c) Benzyl alcohol

15.24 Write equations showing how 2-phenylethanol ($C_6H_5CH_2CH_2OH$) could be prepared from each of the following starting materials:

(a) Bromobenzene

(b) Styrene

(c) 2-Phenylethanal ($C_6H_5CH_2CHO$)

(d) Ethyl 2-phenylethanoate ($C_6H_5CH_2CO_2CH_2CH_3$)

(e) 2-Phenylethanoic acid ($C_6H_5CH_2CO_2H$)

15.25 Outline practical syntheses of each of the following compounds from alcohols containing no more than four carbon atoms and any necessary organic or inorganic reagents. In many cases the desired compound can be made from one prepared in an earlier part of the problem.

(a) 1-Butanethiol

(b) 1-Hexanol

(c) 2-Hexanol

(d) Hexanal, $CH_3CH_2CH_2CH_2CH_2CH{=}O$

(e) 2-Hexanone, $CH_3\overset{\overset{\displaystyle O}{\|}}{C}CH_2CH_2CH_2CH_3$

(f) Hexanoic acid, $CH_3(CH_2)_4CO_2H$

(g) Ethyl hexanoate, $CH_3(CH_2)_4\overset{\overset{\displaystyle O}{\|}}{C}OCH_2CH_3$

(h) 2-Methyl-1,2-propanediol

(i) 2,2-Dimethylpropanal, $(CH_3)_3C\overset{\overset{\displaystyle O}{\|}}{C}H$

15.26 Outline practical syntheses of each of the following compounds from benzene, alcohols, and any necessary organic or inorganic reagents:

(a) 1-Chloro-2-phenylethane

(b) 2-Methyl-1-phenyl-1-propane, $C_6H_5\overset{\overset{\displaystyle O}{\|}}{C}CH(CH_3)_2$

(c) Isobutylbenzene, $C_6H_5CH_2CH(CH_3)_2$

15.27 Show how each of the following compounds can be synthesized from cyclopentanol and any necessary organic or inorganic reagents. In many cases the desired compound can be made from one prepared in an earlier part of the problem.

(a) 1-Phenylcyclopentanol

(b) 1-Phenylcyclopentene

(c) *trans*-2-Phenylcyclopentanol

(d)

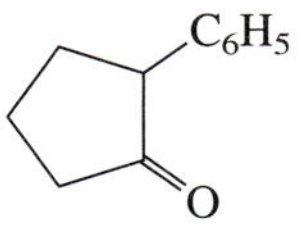

(e)

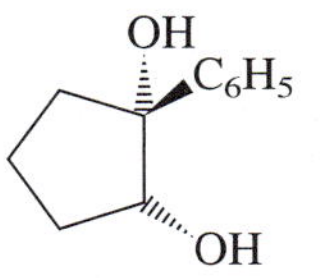

(f) $C_6H_5\overset{\overset{\displaystyle O}{\|}}{C}CH_2CH_2CH_2\overset{\overset{\displaystyle O}{\|}}{C}H$

(g) 1-Phenyl-1,5-pentanediol

15.28 Write the structure of the principal organic product formed in the reaction of 1-propanol with each of the following reagents:

(a) Sulfuric acid (catalytic amount), heat at 140°C

(b) Sulfuric acid (catalytic amount), heat at 200°C

(c) Nitric acid (H_2SO_4 catalyst)

(d) Pyridinium chlorochromate (PCC) in dichloromethane

(e) Potassium dichromate ($K_2Cr_2O_7$) in aqueous sulfuric acid, heat

(f) Sodium amide ($NaNH_2$)

(g) Acetic acid ($CH_3\overset{\overset{\displaystyle O}{\|}}{C}OH$) in the presence of dissolved hydrogen chloride

(h) 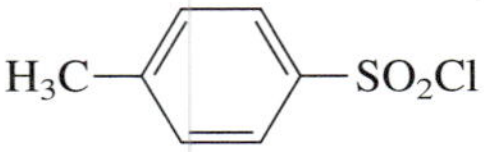 H_3C—⬡—SO_2Cl in the presence of pyridine

(i) 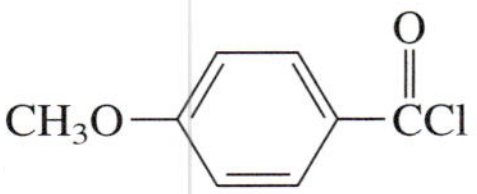 CH_3O—⬡—$\overset{\overset{O}{\|}}{C}Cl$ in the presence of pyridine

(j) $C_6H_5\overset{\overset{O}{\|}}{C}O\overset{\overset{O}{\|}}{C}C_6H_5$ in the presence of pyridine

(k) 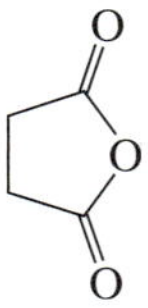 [cyclic anhydride structure] in the presence of pyridine

15.29 Each of the following reactions has been reported in the chemical literature. Predict the product in each case, showing stereochemistry where appropriate.

(a) 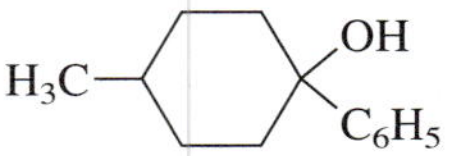 H_3C—[cyclohexane ring with OH and C_6H_5] $\xrightarrow[\text{heat}]{H_2SO_4}$

(b) $(CH_3)_2C{=}C(CH_3)_2$ $\xrightarrow[\text{(CH}_3\text{)}_3\text{COH, HO}^-]{\text{(CH}_3\text{)}_3\text{COOH, OsO}_4\text{(cat)}}$

(c) 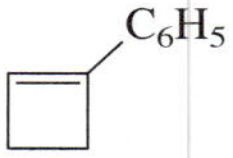 [cyclobutene ring with C_6H_5] $\xrightarrow[\text{2. H}_2\text{O}_2, \text{ HO}^-]{\text{1. B}_2\text{H}_6, \text{ diglyme}}$

(d)  [cyclopentene ring]—CO_2H $\xrightarrow[\text{2. H}_2\text{O}]{\text{1. LiAlH}_4, \text{ diethyl ether}}$

(e) $CH_3\overset{\overset{}{\underset{\underset{OH}{|}}{C}}}{H}C{\equiv}C(CH_2)_3CH_3$ $\xrightarrow[\text{H}_2\text{SO}_4, \text{ H}_2\text{O, acetone}]{\text{H}_2\text{CrO}_4}$

(f) $CH_3\overset{\overset{O}{\|}}{C}CH_2CH{=}CHCH_2\overset{\overset{O}{\|}}{C}CH_3$ $\xrightarrow[\text{2. H}_2\text{O}]{\text{1. LiAlH}_4, \text{ diethyl ether}}$

(g) 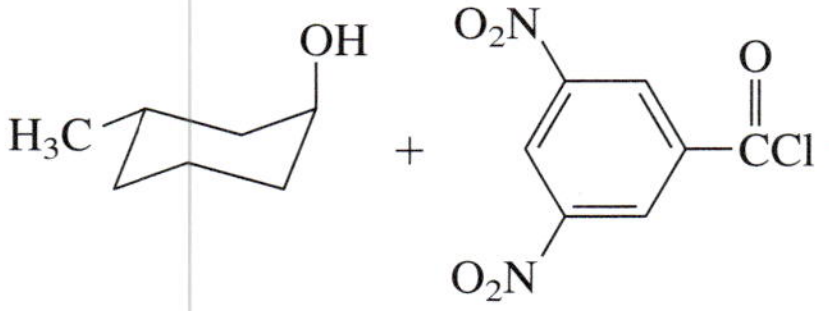 H_3C—[cyclohexane ring with OH] $+$ [aromatic ring with two O_2N groups and $\overset{\overset{O}{\|}}{C}Cl$] $\xrightarrow{\text{pyridine}}$

(h) 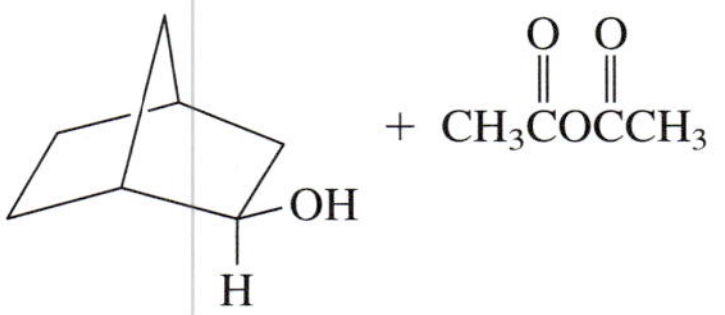 [bicyclic structure with OH and H] $+ CH_3\overset{\overset{O}{\|}}{C}O\overset{\overset{O}{\|}}{C}CH_3$ $\longrightarrow$

(i) Cl—[aromatic ring with two O_2N groups and $\overset{\overset{O}{\|}}{C}OH$] $\xrightarrow[\text{H}_2\text{SO}_4]{\text{CH}_3\text{OH}}$

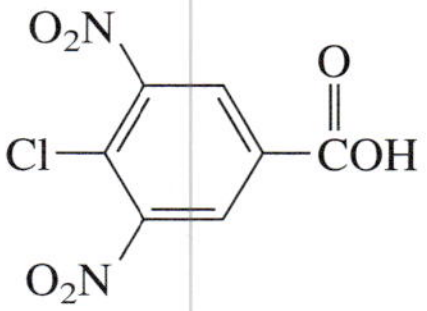

(j)

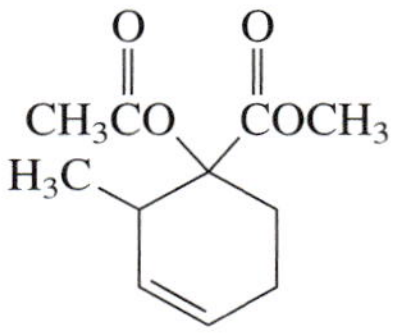

O   O
‖   ‖
CH₃CO   COCH₃
H₃C

1. LiAlH₄
2. H₂O
→

(k) Product of part (j) $\xrightarrow[\text{CH}_3\text{OH, H}_2\text{O}]{\text{HIO}_4}$

15.30 On heating 1,2,4-butanetriol in the presence of an acid catalyst, a cyclic ether of molecular formula $C_4H_8O_2$ was obtained in 81–88% yield. Suggest a reasonable structure for this product.

15.31 *myo*-Inositol is a member of a group of stereoisomeric 1,2,3,4,5,6-cyclohexanehexols.

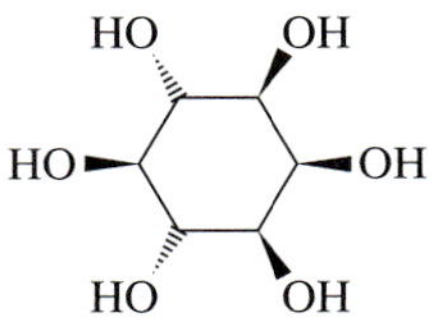

HO   OH

HO   OH

HO   OH

myo-Inositol

(a) Draw a structural formula or make a molecular model of *myo*-inositol in its most stable conformation.

(b) Is *myo*-inositol chiral?

15.32 Suggest reaction sequences and reagents suitable for carrying out each of the following conversions. Two synthetic operations are required in each case.

(a) 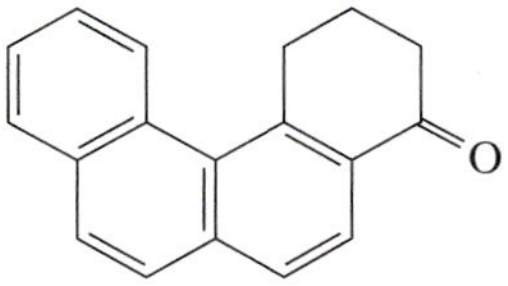 [structure] to [structure]

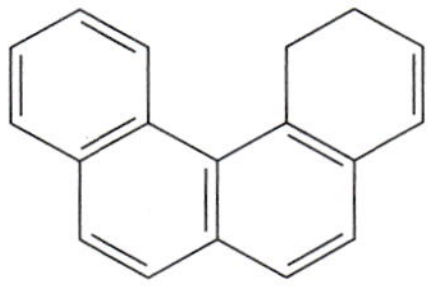

(b) 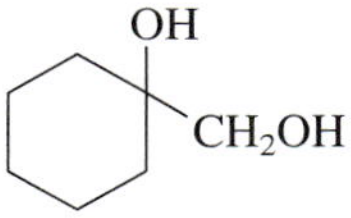 [structure with OH and CH₂OH] to [structure with OH]

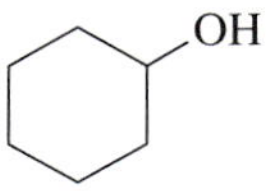

(c) 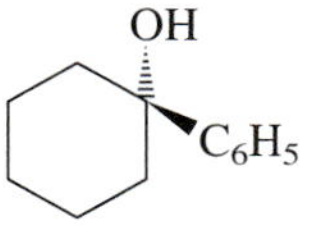 [structure with OH and C₆H₅] to [structure with OH, C₆H₅ and OH]

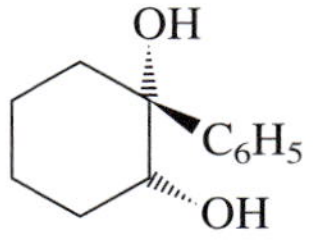

15.33 The fungus responsible for Dutch elm disease is spread by European bark beetles when they burrow into the tree. Other beetles congregate at the site, attracted by the scent of a mixture of chemicals, some emitted by other beetles and some coming from the tree. One of the compounds given off by female bark beetles is 4-methyl-3-heptanol. Suggest an efficient synthesis of this pheromone from alcohols of five carbon atoms or fewer.

15.34 Show by a series of equations how you could prepare 3-methylpentane from ethanol and any necessary inorganic reagents.

15.35 (a) The cis isomer of 3-hexen-1-ol ($CH_3CH_2CH{=}CHCH_2CH_2OH$) has the characteristic odor of green leaves and grass. Suggest a synthesis for this compound from acetylene and any necessary organic or inorganic reagents.

(b) One of the compounds responsible for the characteristic odor of ripe tomatoes is the cis isomer of $CH_3CH_2CH{=}CHCH_2CH{=}O$. How could you prepare this compound?

15.36 R. B. Woodward was one of the leading organic chemists of the middle part of the twentieth century. Known primarily for his achievements in the synthesis of complex natural products, he was awarded the Nobel Prize in chemistry in 1965. He entered Massachusetts Institute of Technology as a 16-year-old freshman in 1933 and four years later was awarded the Ph.D. While a student there he carried out a synthesis of *estrone,* a female sex hormone. The early stages of Woodward's estrone synthesis required the conversion of *m*-methoxybenzaldehyde to *m*-methoxybenzyl cyanide, which was accomplished in three steps:

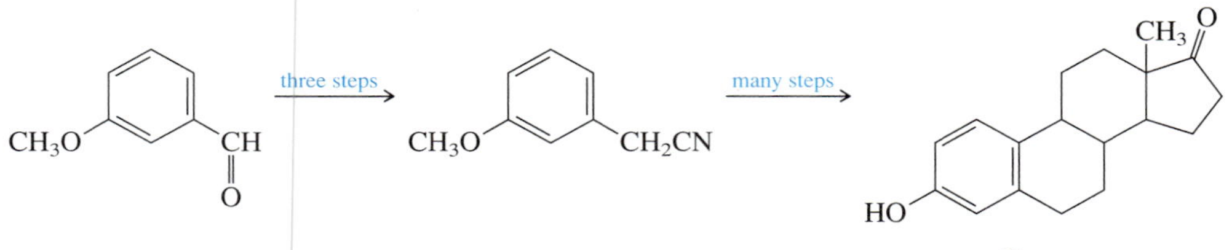

Estrone

Suggest a reasonable three-step sequence, showing all necessary reagents, for the preparation of *m*-methoxybenzyl cyanide from *m*-methoxybenzaldehyde.

15.37 Complete the following series of equations by writing structural formulas for compounds A through I:

(a)

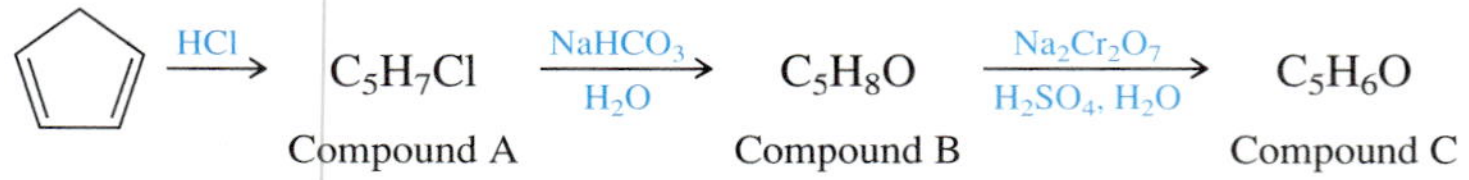

(b)

$$H_2C{=}CHCH_2CH_2\underset{\displaystyle OH}{\underset{|}{C}}HCH_3 \xrightarrow[\text{pyridine}]{SOCl_2} \underset{\text{Compound D}}{C_6H_{11}Cl} \xrightarrow[\text{2. Zn, }H_2O]{\text{1. }O_3} \underset{\text{Compound E}}{C_5H_9ClO} \xrightarrow{NaBH_4} \underset{\text{Compound F}}{C_5H_{11}ClO}$$

(c)

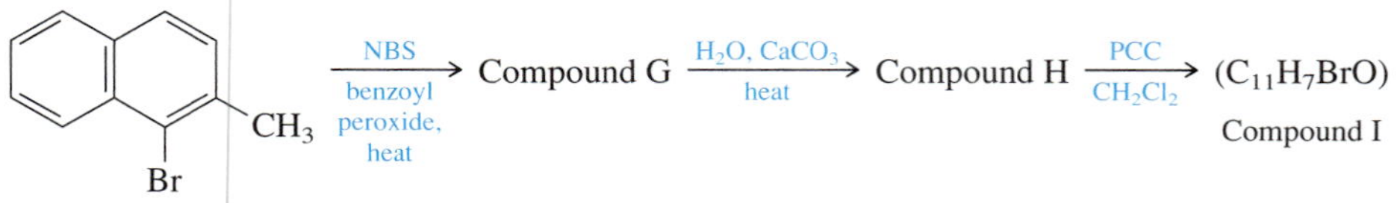

15.38 When 2-phenyl-2-butanol is allowed to stand in ethanol containing a few drops of sulfuric acid, the following ether is formed:

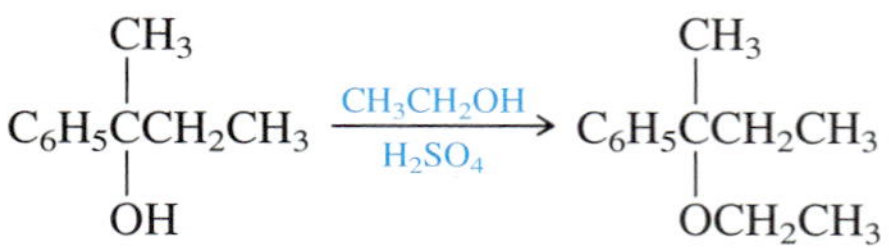

$$C_6H_5\underset{\displaystyle OH}{\underset{|}{\overset{\displaystyle CH_3}{\overset{|}{C}}}}CH_2CH_3 \xrightarrow[H_2SO_4]{CH_3CH_2OH} C_6H_5\underset{\displaystyle OCH_2CH_3}{\underset{|}{\overset{\displaystyle CH_3}{\overset{|}{C}}}}CH_2CH_3$$

Suggest a reasonable mechanism for this reaction based on the observation that the ether produced from optically active alcohol is racemic, and that alkenes can be shown not to be intermediates in the reaction.

15.39 Suggest a chemical test that would permit you to distinguish between the two glycerol monobenzyl ethers shown.

$$C_6H_5CH_2OCH_2CHCH_2OH \qquad HOCH_2CHCH_2OH$$

$$\overset{|}{OH} \qquad\qquad \overset{|}{OCH_2C_6H_5}$$

1-*O*-Benzylglycerol   2-*O*-Benzylglycerol

15.40 Choose the correct enantiomer of 2-butanol that would permit you to prepare (*R*)-2-butanethiol by way of a *p*-toluenesulfonate ester.

15.41 The amino acid *cysteine* has the structure shown:

$$\overset{\displaystyle O}{\underset{\underset{\textstyle {}^{+}NH_3}{|}}{\overset{\|}{HSCH_2CHCO^-}}}$$

Cysteine

(a) A second sulfur-containing amino acid called *cystine* ($C_6H_{12}N_2O_4S_2$) is formed when cysteine undergoes biological oxidation. Suggest a reasonable structure for cystine.

(b) Another metabolic pathway converts cysteine to *cysteine sulfinic acid* ($C_3H_7NO_4S$), then to *cysteic acid* ($C_3H_7NO_5S$). What are the structures of these two compounds?

15.42 A diol ($C_8H_{18}O_2$) does not react with periodic acid. Its ^1H NMR spectrum contains three singlets at δ 1.2 (12 protons), 1.6 (4 protons), and 2.0 (2 protons). What is the structure of this diol?

15.43 Identify compound A ($C_8H_{10}O$) on the basis of its ^1H NMR spectrum (Figure 15.10). The broad peak at δ 2.1 disappears when D_2O is added.

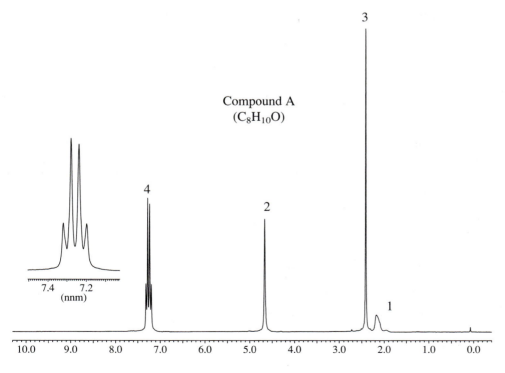

FIGURE 15.10 The 200-MHz ^1H NMR spectrum of compound A ($C_8H_{10}O$) (Problem 15.43).

15.44 Identify each of the following ($C_4H_{10}O$) isomers on the basis of their ^{13}C NMR spectra:

(a) δ 31.2: CH_3

    δ 68.9: C

(b) δ 10.0: CH_3

    δ 22.7: CH_3

    δ 32.0: CH_2

    δ 69.2: CH

(c) δ 18.9: CH_3, area 2

    δ 30.8: CH, area 1

    δ 69.4: CH_2, area 1

15.45 A compound $C_3H_7ClO_2$ exhibited three peaks in its ^{13}C NMR spectrum at δ 46.8 (CH_2), δ 63.5 (CH_2), and δ 72.0 (CH). What is the structure of this compound?

15.46 A compound $C_6H_{14}O$ has the ^{13}C NMR spectrum shown in Figure 15.11. Its mass spectrum has a prominent peak at m/z 31. Suggest a reasonable structure for this compound.

 15.47 Refer to *Learning By Modeling* and compare the properties calculated for CH_3CH_2OH and CH_3CH_2SH. Which has the greater dipole moment? Compare the charges at carbon and hydrogen in C—O—H versus C—S—H. Why does ethanol have a higher boiling point than ethanethiol?

 15.48 Construct molecular models of the gauche and anti conformations of 1,2-ethanediol and explore the possibility of intramolecular hydrogen bond formation in each one.

 15.49 Intramolecular hydrogen bonding is present in the chiral diastereomer of 2,2,5,5-tetra-methylhexane-3,4-diol, but absent in the meso diastereomer. Construct molecular models of each, and suggest a reason for the difference between the two.

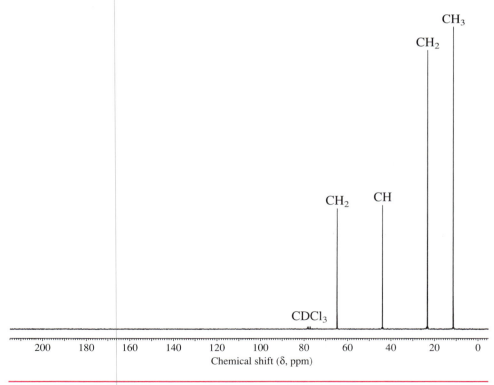

FIGURE 15.11 The ^{13}C NMR spectrum of the compound $C_6H_{14}O$ (Problem 15.46).

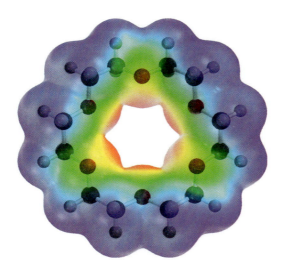

CHAPTER 16

ETHERS, EPOXIDES, AND SULFIDES

I n contrast to alcohols with their rich chemical reactivity, **ethers** (compounds containing a C—O—C unit) undergo relatively few chemical reactions. As you saw when we discussed Grignard reagents in Chapter 14 and lithium aluminum hydride reductions in Chapter 15, this lack of reactivity of ethers makes them valuable as solvents in a number of synthetically important transformations. In the present chapter you will learn of the conditions in which an ether linkage acts as a functional group, as well as the methods by which ethers are prepared.

Unlike most ethers, **epoxides** (compounds in which the C—O—C unit forms a three-membered ring) are very reactive substances. The principles of nucleophilic substitution are important in understanding the preparation and properties of epoxides.

Sulfides (RSR′) are the sulfur analogs of ethers. Just as in the preceding chapter, where we saw that the properties of thiols (RSH) are different from those of alcohols, we will explore differences between sulfides and ethers in this chapter.

16.1 NOMENCLATURE OF ETHERS, EPOXIDES, AND SULFIDES

Ethers are named, in substitutive IUPAC nomenclature, as *alkoxy* derivatives of alkanes. Functional class IUPAC names of ethers are derived by listing the two alkyl groups in the general structure ROR′ in alphabetical order as separate words, and then adding the word *ether* at the end. When both alkyl groups are the same, the prefix *di-* precedes the name of the alkyl group.

	$CH_3CH_2OCH_2CH_3$	$CH_3CH_2OCH_3$	$CH_3CH_2OCH_2CH_2CH_2Cl$
Substitutive IUPAC name:	Ethoxyethane	Methoxyethane	1-Chloro-3-ethoxypropane
Functional class IUPAC name:	Diethyl ether	Ethyl methyl ether	3-Chloropropyl ethyl ether

Ethers are described as **symmetrical** or **unsymmetrical** depending on whether the two groups bonded to oxygen are the same or different. Unsymmetrical ethers are also called

mixed ethers. Diethyl ether is a symmetrical ether; ethyl methyl ether is an unsymmetrical ether.

Cyclic ethers have their oxygen as part of a ring—they are *heterocyclic compounds* (Section 3.15). Several have specific IUPAC names.

Oxirane (Ethylene oxide)	Oxetane	Oxolane (Tetrahydrofuran)	Oxane (Tetrahydropyran)

In each case the ring is numbered starting at the oxygen. The IUPAC rules also permit oxirane (without substituents) to be called *ethylene oxide*. Tetrahydrofuran and *tetrahydropyran* are acceptable synonyms for oxolane and oxane, respectively.

PROBLEM 16.1 Each of the following ethers has been shown to be or is suspected to be a *mutagen*, which means it can induce mutations in test cells. Write the structure of each of these ethers.

(a) Chloromethyl methyl ether

(b) 2-(Chloromethyl)oxirane (also known as epichlorohydrin)

(c) 3,4-Epoxy-1-butene (2-vinyloxirane)

SAMPLE SOLUTION (a) Chloromethyl methyl ether has a chloromethyl group ($ClCH_2$—) and a methyl group (CH_3—) attached to oxygen. Its structure is $ClCH_2OCH_3$.

Recall from Section 6.18 that epoxides may be named as -epoxy derivatives of alkanes in substitutive IUPAC nomenclature.

Many substances have more than one ether linkage. Two such compounds, often used as solvents, are the *diethers* 1,2-dimethoxyethane and 1,4-dioxane. Diglyme, also a commonly used solvent, is a *triether.*

$$CH_3OCH_2CH_2OCH_3 \qquad\qquad\qquad CH_3OCH_2CH_2OCH_2CH_2OCH_3$$

1,2-Dimethoxyethane	1,4-Dioxane	Diethylene glycol dimethyl ether (diglyme)

Molecules that contain several ether functions are referred to as *polyethers*. Polyethers have some novel properties and will appear in Section 16.4.

The sulfur analogs (RS—) of alkoxy groups are called *alkylthio* groups. The first two of the following examples illustrate the use of alkylthio prefixes in substitutive nomenclature of sulfides. Functional class IUPAC names of sulfides are derived in exactly the same way as those of ethers but end in the word *sulfide*. Sulfur heterocycles have names analogous to their oxygen relatives, except that *ox-* is replaced by *thi-*. Thus the sulfur heterocycles containing three-, four-, five-, and six-membered rings are named *thiirane*, *thietane*, *thiolane*, and *thiane*, respectively.

Sulfides are sometimes informally referred to as thioethers, but this term is not part of systematic IUPAC nomenclature.

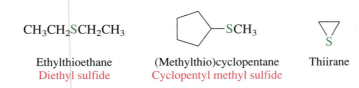

$CH_3CH_2SCH_2CH_3$	(Methylthio)cyclopentane	Thiirane
Ethylthioethane Diethyl sulfide	Cyclopentyl methyl sulfide	

16.2 STRUCTURE AND BONDING IN ETHERS AND EPOXIDES

Bonding in ethers is readily understood by comparing ethers with water and alcohols. Van der Waals strain involving alkyl groups causes the bond angle at oxygen to be larger in ethers than in alcohols, and larger in alcohols than in water. An extreme example is di-*tert*-butyl ether, where steric hindrance between the *tert*-butyl groups is responsible for a dramatic increase in the C—O—C bond angle.

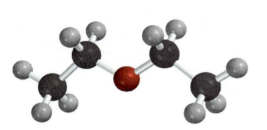

| Water | Methanol | Dimethyl ether | Di-*tert*-butyl ether |

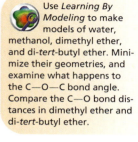

Use *Learning By Modeling* to make models of water, methanol, dimethyl ether, and di-*tert*-butyl ether. Minimize their geometries, and examine what happens to the C—O—C bond angle. Compare the C—O bond distances in dimethyl ether and di-*tert*-butyl ether.

Typical carbon–oxygen bond distances in ethers are similar to those of alcohols ($\approx$142 pm) and are shorter than carbon–carbon bond distances in alkanes ($\approx$153 pm).

An ether oxygen affects the conformation of a molecule in much the same way that a CH_2 unit does. The most stable conformation of diethyl ether is the all-staggered anti conformation. Tetrahydropyran is most stable in the chair conformation—a fact that has an important bearing on the structures of many carbohydrates.

Anti conformation of diethyl ether   Chair conformation of tetrahydropyran

Incorporating an oxygen atom into a three-membered ring requires its bond angle to be seriously distorted from the normal tetrahedral value. In ethylene oxide, for example, the bond angle at oxygen is 61.5°.

Thus epoxides, like cyclopropanes, have significant angle strain. They tend to undergo reactions that open the three-membered ring by cleaving one of the carbon–oxygen bonds.

PROBLEM 16.2 The heats of combustion of 1,2-epoxybutane (2-ethyloxirane) and tetrahydrofuran have been measured: one is 2499 kJ/mol; the other is 2546 kJ/mol. Match the heats of combustion with the respective compounds.

Ethers, like water and alcohols, are polar molecules. Diethyl ether, for example, has a dipole moment of 1.2 D. Cyclic ethers have larger dipole moments; ethylene oxide and tetrahydrofuran have dipole moments in the 1.7- to 1.8-D range—about the same as that of water (1.8D).

16.3 PHYSICAL PROPERTIES OF ETHERS

It is instructive to compare the physical properties of ethers with alkanes and alcohols. With respect to boiling point, ethers resemble alkanes more than alcohols. With respect to solubility in water the reverse is true; ethers resemble alcohols more than alkanes. Why?

	$CH_3CH_2OCH_2CH_3$	$CH_3CH_2CH_2CH_2CH_3$	$CH_3CH_2CH_2CH_2OH$
	Diethyl ether	Pentane	1-Butanol
Boiling point:	35°C	36°C	117°C
Solubility in water:	7.5 g/100 mL	Insoluble	9 g/100 mL

In general, the boiling points of alcohols are unusually high because of hydrogen bonding (Section 4.6). Attractive forces in the liquid phases of ethers and alkanes, which lack —OH groups and cannot form intermolecular hydrogen bonds, are much weaker, and their boiling points lower.

As shown in Figure 16.1, however, the presence of an oxygen atom permits ethers to participate in hydrogen bonds to water molecules. These attractive forces cause ethers to dissolve in water to approximately the same extent as comparably constituted alcohols. Alkanes cannot engage in hydrogen bonding to water.

> **PROBLEM 16.3** Ethers tend to dissolve in alcohols and vice versa. Represent the hydrogen-bonding interaction between an alcohol molecule and an ether molecule.

16.4 CROWN ETHERS

Their polar carbon–oxygen bonds and the presence of unshared electron pairs at oxygen contribute to the ability of ethers to form Lewis acid-Lewis base complexes with metal ions.

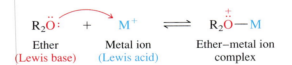

$$R_2\ddot{O}: \quad + \quad M^+ \quad \rightleftharpoons \quad R_2\overset{+}{\ddot{O}}-M$$

Ether	Metal ion	Ether–metal ion
(Lewis base)	(Lewis acid)	complex

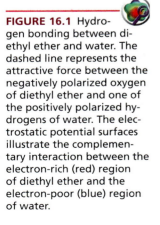

FIGURE 16.1 Hydrogen bonding between diethyl ether and water. The dashed line represents the attractive force between the negatively polarized oxygen of diethyl ether and one of the positively polarized hydrogens of water. The electrostatic potential surfaces illustrate the complementary interaction between the electron-rich (red) region of diethyl ether and the electron-poor (blue) region of water.

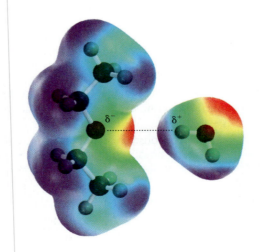

The strength of this bonding depends on the kind of ether. Simple ethers form relatively weak complexes with metal ions, but Charles J. Pedersen of Du Pont discovered that certain *polyethers* form much more stable complexes with metal ions than do simple ethers.

Pedersen prepared a series of *macrocyclic polyethers,* cyclic compounds containing four or more oxygens in a ring of 12 or more atoms. He called these compounds **crown ethers,** because their molecular models resemble crowns. Systematic nomenclature of crown ethers is somewhat cumbersome, and so Pedersen devised a shorthand description whereby the word *crown* is preceded by the total number of atoms in the ring and is followed by the number of oxygen atoms.

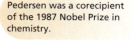

Pedersen was a corecipient of the 1987 Nobel Prize in chemistry.

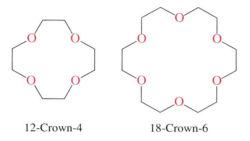

12-Crown-4   18-Crown-6

12-Crown-4 and 18-crown-6 are a cyclic tetramer and hexamer, respectively, of repeating —OCH$_2$CH$_2$— units; they are polyethers based on ethylene glycol (HOCH$_2$CH$_2$OH) as the parent alcohol.

> **PROBLEM 16.4** What organic compound mentioned earlier in this chapter is a cyclic dimer of —OCH$_2$CH$_2$— units?

The metal–ion complexing properties of crown ethers are clearly evident in their effects on the solubility and reactivity of ionic compounds in nonpolar media. Potassium fluoride (KF) is ionic and practically insoluble in benzene alone, but dissolves in it when 18-crown-6 is present. This happens because of the electron distribution of 18-crown-6 as shown in Figure 16.2a. The electrostatic potential surface consists of essentially two regions: an electron-rich interior associated with the oxygens and a hydrocarbon-like exterior associated with the CH$_2$ groups. When KF is added to a solution of 18-crown-6 in benzene, potassium ion (K$^+$) interacts with the oxygens of the crown ether to form a Lewis acid-Lewis base complex. As can be seen in the space-filling model of this

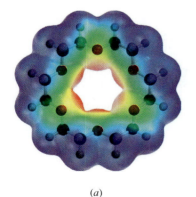

(a)

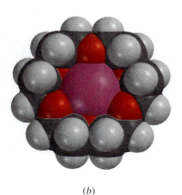

(b)

FIGURE 16.2 (a) An electrostatic potential map of 18-crown-6. The region of highest electron density (red) is associated with the negatively polarized oxygens and their lone pairs. The outer periphery of the crown ether (blue) is relatively nonpolar (hydrocarbon-like) and causes the molecule to be soluble in nonpolar solvents such as benzene. (b) A space-filling model of the complex formed between 18-crown-6 and potassium ion (K$^+$). K$^+$ fits into the cavity of the crown ether where it is bound by Lewis acid-Lewis base interaction with the oxygens.

POLYETHER ANTIBIOTICS

One way in which pharmaceutical companies search for new drugs is by growing colonies of microorganisms in nutrient broths and assaying the substances produced for their biological activity. This method has yielded thousands of antibiotic substances, of which hundreds have been developed into effective drugs. Antibiotics are, by definition, toxic (*anti* = "against"; *bios* = "life"), and the goal is to find substances that are more toxic to infectious organisms than to their human hosts.

Since 1950, a number of **polyether antibiotics** have been discovered using fermentation technology. They are characterized by the presence of several cyclic ether structural units, as illustrated for the case of *monensin* in Figure 16.3*a*. Monensin and other naturally occurring polyethers are similar to crown ethers in their ability to form stable complexes with metal ions. The structure of the sodium salt of monensin is depicted in Figure 16.3*b*, where it can be seen that four ether oxygens and two hydroxyl groups surround a sodium ion. The alkyl groups are oriented toward the outside of the complex, and the polar oxygens and the metal ion are on the inside. The hydrocarbon-like surface of the complex permits it to carry its sodium ion through the hydrocarbon-like interior of a cell membrane. This disrupts the normal balance of sodium ions within the cell and interferes with important processes of cellular respiration. Small amounts of monensin are added to poultry feed to kill parasites that live in the intestines of chickens. Compounds such as monensin and the crown ethers that affect metal ion transport are referred to as **ionophores** ("ion carriers").

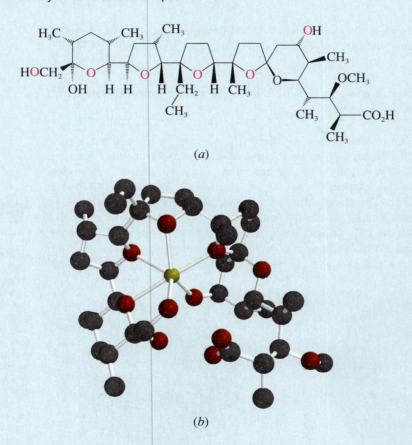

FIGURE 16.3 (*a*) The structure of monensin; (*b*) The structure of the sodium salt of monensin showing coordination of Na$^+$ (*yellow*) to the six oxygens shown in red in the structural formula. Hydrogen atoms have been omitted from the model for clarity.

complex (Figure 16.2b), K^+, with an ionic radius of 266 pm, fits comfortably within the 260–320 pm internal cavity of 18-crown-6. Nonpolar CH_2 groups dominate the outer surface of the complex, mask its polar interior, and permit the complex to dissolve in nonpolar solvents. Every K^+ that is carried into benzene brings a fluoride ion with it, resulting in a solution containing strongly complexed potassium ions and relatively unsolvated fluoride ions.

| 18-Crown-6 | Potassium fluoride (solid) | 18-Crown-6-potassium fluoride complex (in solution) |

In media such as water and alcohols, fluoride ion is strongly solvated by hydrogen bonding and is neither very basic nor very nucleophilic. On the other hand, the poorly solvated, or "naked," fluoride ions that are present when potassium fluoride dissolves in benzene in the presence of a crown ether are better able to express their anionic reactivity. Thus, alkyl halides react with potassium fluoride in benzene containing 18-crown-6, thereby providing a method for the preparation of otherwise difficultly accessible alkyl fluorides.

$$CH_3(CH_2)_6CH_2Br \xrightarrow[\text{18-crown-6}]{\text{KF, benzene, 90°C}} CH_3(CH_2)_6CH_2F$$

1-Bromooctane   1-Fluorooctane (92%)

> The reaction proceeds in the direction indicated because a C—F bond is much stronger than a C—Br bond.

No reaction is observed when the process is carried out under comparable conditions but with the crown ether omitted.

Catalysis by crown ethers has been used to advantage to increase the rate of many organic reactions that involve anions as reactants. Just as important, though, is the increased understanding that studies of crown ether catalysis have brought to our knowledge of biological processes in which metal ions, including Na^+ and K^+, are transported through the nonpolar interiors of cell membranes.

16.5 PREPARATION OF ETHERS

Because they are widely used as solvents, many simple dialkyl ethers are commercially available. Diethyl ether and dibutyl ether, for example, are prepared by acid-catalyzed condensation of the corresponding alcohols, as described earlier in Section 15.7.

> The mechanism for the formation of diethyl ether from ethanol under conditions of acid catalysis was shown in Figure 15.3.

$$2CH_3CH_2CH_2CH_2OH \xrightarrow[\text{130°C}]{\text{H}_2\text{SO}_4} CH_3CH_2CH_2CH_2OCH_2CH_2CH_2CH_3 + H_2O$$

1-Butanol   Dibutyl ether (60%)   Water

In general, this method is limited to the preparation of symmetrical ethers in which both alkyl groups are primary. Isopropyl alcohol, however, is readily available at low cost and gives high enough yields of diisopropyl ether to justify making $(CH_3)_2CHOCH(CH_3)_2$ by this method on an industrial scale.

tert-Butyl methyl ether is often referred to as MTBE, standing for the incorrect name "methyl *tert*-butyl ether." Remember, italicized prefixes are ignored when alphabetizing, and *tert*-butyl precedes methyl.

Acid-catalyzed addition of alcohols to alkenes is sometimes used. Indeed, before its use as a gasoline additive was curtailed, billions of pounds of *tert*-butyl methyl ether (MTBE) was prepared by the reaction:

$$(CH_3)_2C{=\!\!=}CH_2 + CH_3OH \xrightarrow{\ H^+\ } (CH_3)_3COCH_3$$

2-Methylpropene   Methanol   *tert*-Butyl methyl ether

Small amounts of *tert*-butyl methyl ether increase the octane rating of gasoline. Before environmental concerns placed limits on its use, the demand for MTBE exceeded the supply.

> **PROBLEM 16.5** Outline a reasonable mechanism for the formation of *tert*-butyl methyl ether according to the preceding equation.

The following section describes a versatile method for preparing either symmetrical or unsymmetrical ethers that is based on the principles of bimolecular nucleophilic substitution.

16.6 THE WILLIAMSON ETHER SYNTHESIS

The reaction is named for Alexander Williamson, a British chemist who used it to prepare diethyl ether in 1850.

A long-standing method for the preparation of ethers is the **Williamson ether synthesis.** Nucleophilic substitution of an alkyl halide by an alkoxide gives the carbon–oxygen bond of an ether:

$$R\ddot{O}\!:^{-} \ \ R'{-}\ddot{X}\!: \xrightarrow{\ S_N2\ } R\ddot{O}R' + \ :\ddot{X}\!:^{-}$$

Alkoxide   Alkyl   Ether   Halide ion
ion   halide

Preparation of ethers by the Williamson ether synthesis is most successful with methyl and primary alkyl halides.

$$CH_3CH_2CH_2CH_2ONa + CH_3CH_2I \longrightarrow CH_3CH_2CH_2CH_2OCH_2CH_3 + NaI$$

Sodium butoxide   Iodoethane   Butyl ethyl ether (71%)   Sodium
iodide

> **PROBLEM 16.6** Write equations describing two different ways in which benzyl ethyl ether could be prepared by a Williamson ether synthesis.

Secondary and tertiary *alkyl halides* are not suitable, because they react with alkoxide bases by E2 elimination rather than by S_N2 substitution. Whether the *alkoxide base* is primary, secondary, or tertiary is much less important than the nature of the alkyl halide. Thus benzyl isopropyl ether is prepared in high yield from benzyl chloride, a primary chloride that is incapable of undergoing elimination, and sodium isopropoxide:

$$(CH_3)_2CHONa + \text{[benzyl]}{-}CH_2Cl \longrightarrow (CH_3)_2CHOCH_2{-}\text{[phenyl]} + NaCl$$

Sodium   Benzyl chloride   Benzyl isopropyl ether   Sodium
isopropoxide   (84%)   chloride

The alternative synthetic route using the sodium salt of benzyl alcohol and an isopropyl halide would be much less effective, because of increased competition from elimination as the alkyl halide becomes more sterically hindered.

PROBLEM 16.7 Only one combination of alkyl halide and alkoxide is appropriate for the preparation of each of the following ethers by the Williamson ether synthesis. What is the correct combination in each case?

(a) CH_3CH_2O—⬠

(c) $(CH_3)_3COCH_2C_6H_5$

(b) $H_2C=CHCH_2OCH(CH_3)_2$

SAMPLE SOLUTION (a) The ether linkage of cyclopentyl ethyl ether involves a primary carbon and a secondary one. Choose the alkyl halide corresponding to the primary alkyl group, leaving the secondary alkyl group to arise from the alkoxide nucleophile.

⬠—ONa   + CH_3CH_2Br $\xrightarrow{S_N2}$ ⬠—OCH_2CH_3

Sodium cyclopentanolate   Ethyl bromide   Cyclopentyl ethyl ether

The alternative combination, cyclopentyl bromide and sodium ethoxide, is not appropriate because elimination will be the major reaction:

CH_3CH_2ONa +   ⬠—Br $\xrightarrow{E2}$ CH_3CH_2OH +   ⬠

Sodium ethoxide   Bromocyclopentane   Ethanol   Cyclopentene
(major products)

Both reactants in the Williamson ether synthesis usually originate in alcohol precursors. Sodium and potassium alkoxides are prepared by reaction of an alcohol with the appropriate metal, and alkyl halides are most commonly made from alcohols by reaction with a hydrogen halide (Section 4.7), thionyl chloride (Section 4.13), or phosphorus tribromide (Section 4.13). Alternatively, alkyl *p*-toluenesulfonates may be used in place of alkyl halides; alkyl *p*-toluenesulfonates are also prepared from alcohols as their immediate precursors (Section 8.14).

16.7   REACTIONS OF ETHERS: A REVIEW AND A PREVIEW

Up to this point, we haven't seen any reactions of dialkyl ethers. Indeed, ethers are one of the least reactive of the functional groups we shall study. It is this low level of reactivity, along with an ability to dissolve nonpolar substances, that makes ethers so often used as solvents when carrying out organic reactions. Nevertheless, most ethers are hazardous materials, and precautions must be taken when using them. Diethyl ether is extremely flammable and because of its high volatility can form explosive mixtures in air relatively quickly. Open flames must never be present in laboratories where diethyl ether is being used. Other low-molecular-weight ethers must also be treated as fire hazards.

PROBLEM 16.8 Combustion in air is, of course, a chemical property of ethers that is shared by many other organic compounds. Write a balanced chemical equation for the complete combustion (in air) of diethyl ether.

The hazards of working with diisopropyl ether are described in the *Journal of Chemical Education*, p. 469 (1963).

A second dangerous property of ethers is the ease with which they undergo oxidation in air to form explosive peroxides. Air oxidation of diisopropyl ether proceeds according to the equation

$$(CH_3)_2CHOCH(CH_3)_2 \ + \ O_2 \ \longrightarrow \ \underset{\underset{HOO}{|}}{(CH_3)_2COCH(CH_3)_2}$$

Diisopropyl ether   Oxygen   Diisopropyl ether hydroperoxide

The reaction follows a free-radical mechanism and gives a hydroperoxide, a compound of the type ROOH. Hydroperoxides tend to be unstable and shock-sensitive. On standing, they form related peroxidic derivatives, which are also prone to violent decomposition. Air oxidation leads to peroxides within a few days if ethers are even briefly exposed to atmospheric oxygen. For this reason, one should never use old bottles of dialkyl ethers, and extreme care must be exercised in their disposal.

16.8   ACID-CATALYZED CLEAVAGE OF ETHERS

Just as the carbon–oxygen bond of alcohols is cleaved on reaction with hydrogen halides (Section 4.8), so too is an ether linkage broken:

$$ROH \ + \ HX \ \longrightarrow \ RX \ + \ H_2O$$

Alcohol   Hydrogen   Alkyl   Water
         halide   halide

$$ROR' \ + \ HX \ \longrightarrow \ RX \ + \ R'OH$$

Ether   Hydrogen   Alkyl   Alcohol
    halide   halide

The cleavage of ethers is normally carried out under conditions (excess hydrogen halide, heat) that convert the alcohol formed as one of the original products to an alkyl halide. Thus, the reaction typically leads to two alkyl halide molecules:

$$ROR' \ + \ 2HX \ \xrightarrow{\text{heat}} \ RX + R'X \ + \ H_2O$$

Ether   Hydrogen   Two alkyl halides   Water
    halide

$$\underset{\underset{OCH_3}{|}}{CH_3CHCH_2CH_3} \ \xrightarrow[\text{heat}]{HBr} \ \underset{\underset{Br}{|}}{CH_3CHCH_2CH_3} \ + \ CH_3Br$$

sec-Butyl methyl ether   2-Bromobutane (81%)   Bromomethane

The order of hydrogen halide reactivity is HI > HBr >> HCl. Hydrogen fluoride is not effective.

> **PROBLEM 16.9** A series of dialkyl ethers was allowed to react with excess hydrogen bromide, with the following results. Identify the ether in each case.
>
> (a) One ether gave a mixture of bromocyclopentane and 1-bromobutane.
> (b) Another ether gave only benzyl bromide.
> (c) A third ether gave one mole of 1,5-dibromopentane per mole of ether.

SAMPLE SOLUTION (a) In the reaction of dialkyl ethers with excess hydrogen bromide, each alkyl group of the ether function is cleaved and forms an alkyl bromide. Because bromocyclopentane and 1-bromobutane are the products, the starting ether must be butyl cyclopentyl ether.

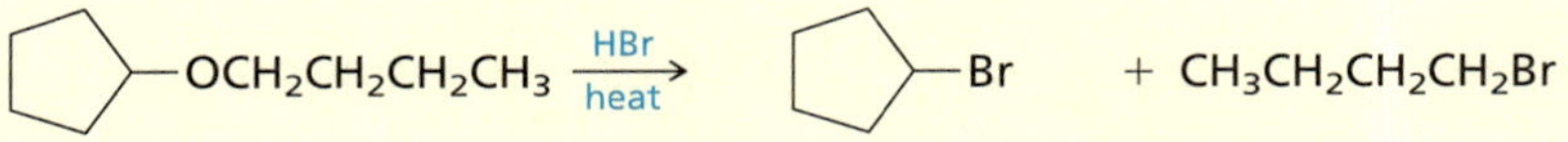

| Butyl cyclopentyl ether | Bromocyclopentane | 1-Bromobutane |

A mechanism for the cleavage of diethyl ether by hydrogen bromide is outlined in Figure 16.4. The key step is an S_N2-like attack on a dialkyloxonium ion by bromide (step 2).

Overall Reaction:

$$CH_3CH_2OCH_2CH_3 \quad + \quad HBr \quad \xrightarrow{heat} \quad 2CH_3CH_2Br \quad + \quad H_2O$$

| Diethyl ether | Hydrogen bromide | Ethyl bromide | Water |

Mechanism:

Step 1: Proton transfer to the oxygen of the ether to give a dialkyloxonium ion.

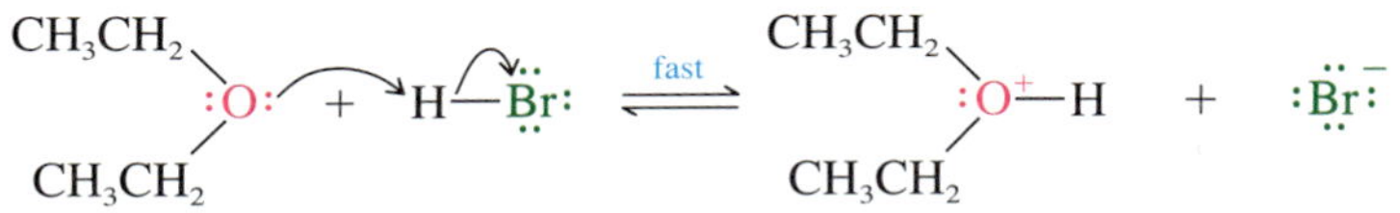

| Diethyl ether | Hydrogen bromide | Diethyloxonium ion | Bromide ion |

Step 2: Nucleophilic attack of the halide anion on carbon of the dialkyloxonium ion. This step gives one molecule of an alkyl halide and one molecule of an alcohol.

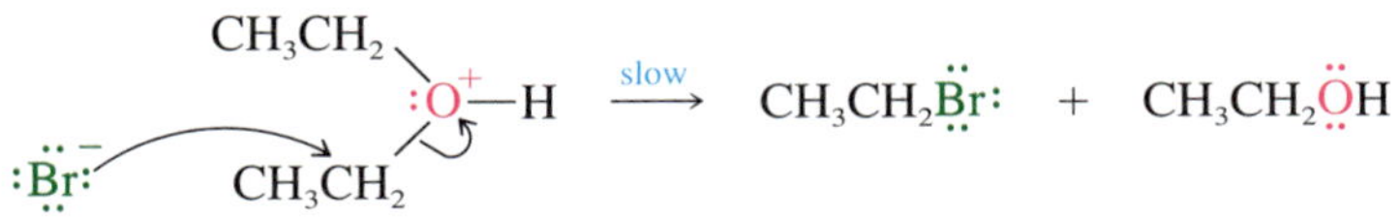

| Bromide ion | Diethyloxonium ion | Ethyl bromide | Ethanol |

Step 3 and Step 4: These two steps do not involve an ether at all. They correspond to those in which an alcohol is converted to an alkyl halide (Sections 4.8–4.13).

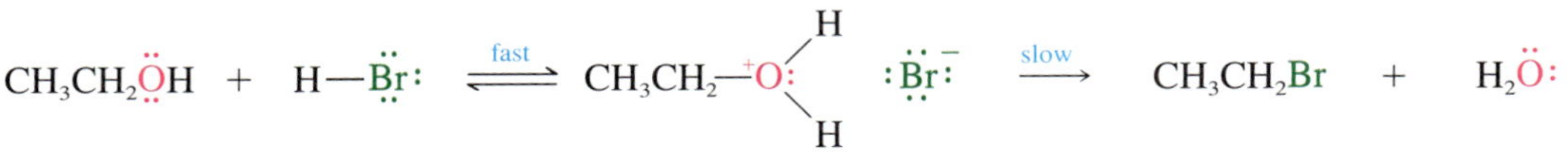

| Ethanol | Hydrogen bromide | | Ethyl bromide | Water |

FIGURE 16.4 The mechanism for the cleavage of ethers by hydrogen halides, using the reaction of diethyl ether with hydrogen bromide as an example.

PROBLEM 16.10 Adapt the mechanism shown in Figure 16.4 to the reaction:

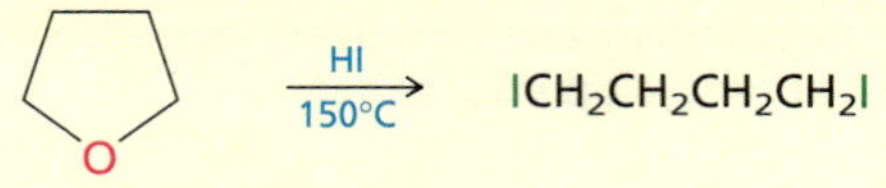

$$\text{(tetrahydrofuran)} \xrightarrow[150°C]{HI} ICH_2CH_2CH_2CH_2I$$

Tetrahydrofuran   1,4-Diiodobutane (65%)

With mixed ethers of the type ROR′, the question of which carbon–oxygen bond is broken first is not one that we need examine at our level of study.

16.9 PREPARATION OF EPOXIDES: A REVIEW AND A PREVIEW

There are two main methods for the preparation of epoxides:

1. Epoxidation of alkenes by reaction with peroxy acids
2. Base-promoted ring closure of vicinal halohydrins

Epoxidation of alkenes was discussed in Section 6.18 and is represented by the general equation

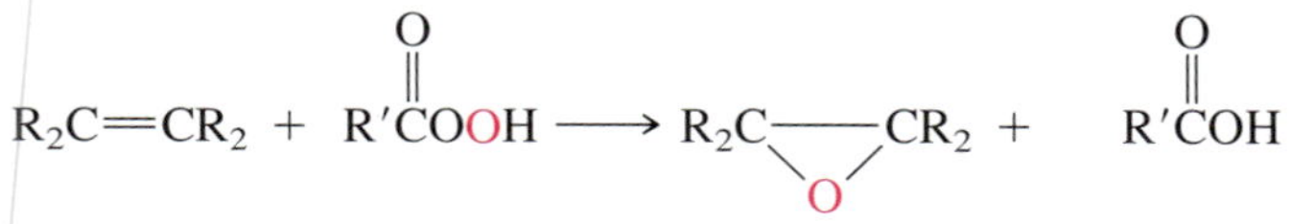

$$R_2C{=}CR_2 + R'COOH \longrightarrow R_2C{-}CR_2 + R'COH$$

Alkene   Peroxy acid   Epoxide   Carboxylic acid

The reaction is easy to carry out, and yields are usually high. Epoxidation is a stereospecific *syn* addition.

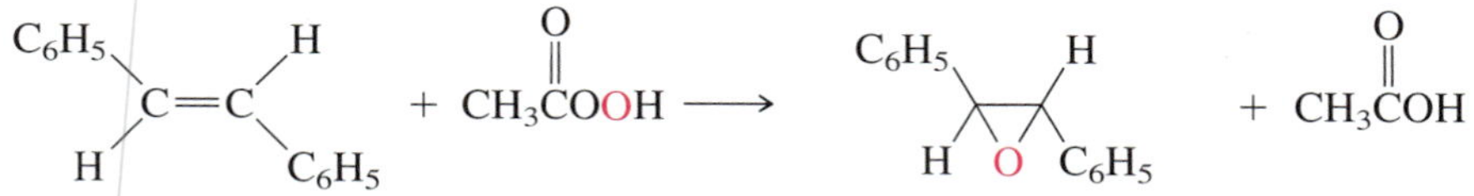

$$\begin{array}{c}C_6H_5\\ \end{array}C{=}C\begin{array}{c}H\\C_6H_5\end{array} + CH_3COOH \longrightarrow \begin{array}{c}C_6H_5\\ \end{array}\triangle\begin{array}{c}H\\C_6H_5\end{array} + CH_3COH$$

(*E*)-1,2-Diphenylethene   Peroxyacetic   *trans*-2,3-Diphenyloxirane   Acetic acid
                               acid   (78–83%)

The following section describes the preparation of epoxides by the base-promoted ring closure of vicinal halohydrins. Because vicinal halohydrins are customarily prepared from alkenes (Section 6.17), both methods—epoxidation using peroxy acids and ring closure of halohydrins—are based on alkenes as the starting materials for preparing epoxides.

16.10 CONVERSION OF VICINAL HALOHYDRINS TO EPOXIDES

The formation of vicinal halohydrins from alkenes was described in Section 6.17. Halohydrins are readily converted to epoxides on treatment with base:

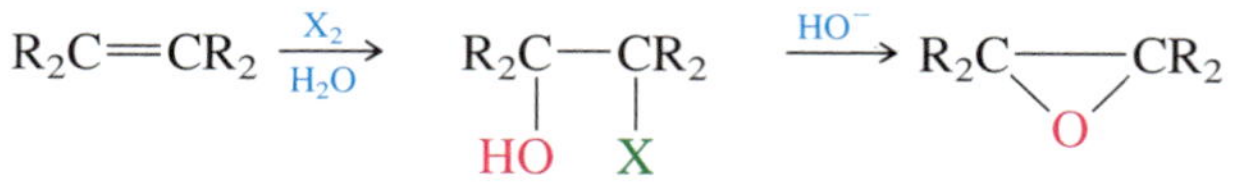

$$R_2C{=}CR_2 \xrightarrow[H_2O]{X_2} R_2C{-}CR_2 \xrightarrow{HO^-} R_2C{-}CR_2$$
$$ \underset{HO \;\; X}{|\quad|} \triangle$$

Alkene   Vicinal halohydrin   Epoxide

Reaction with base brings the alcohol function of the halohydrin into equilibrium with its corresponding alkoxide:

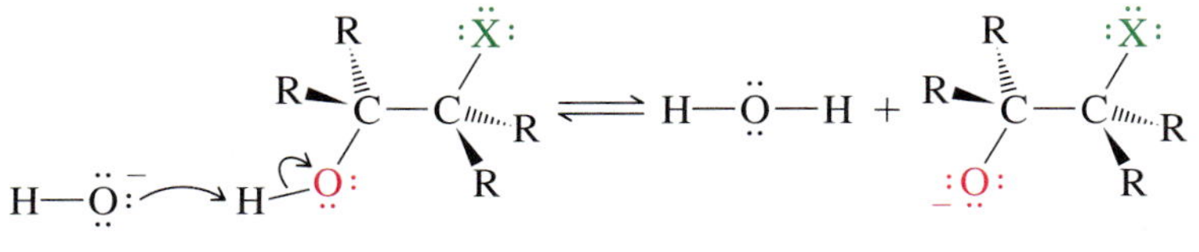

Vicinal halohydrin

Next, in what amounts to an *intramolecular* Williamson ether synthesis, the alkoxide oxygen attacks the carbon that bears the halide leaving group, giving an epoxide. As in other nucleophilic substitutions, the nucleophile approaches carbon from the side opposite the bond to the leaving group:

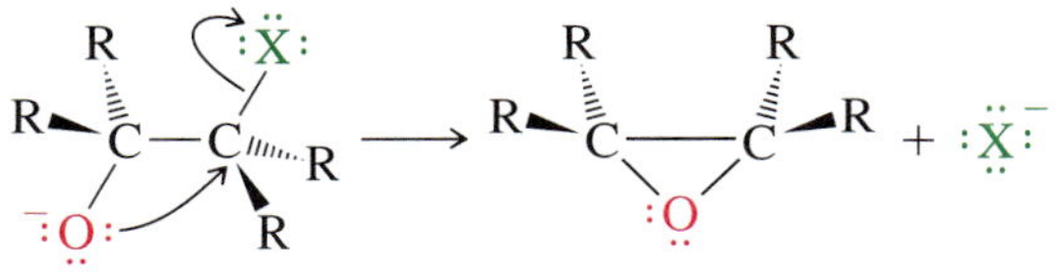

Epoxide

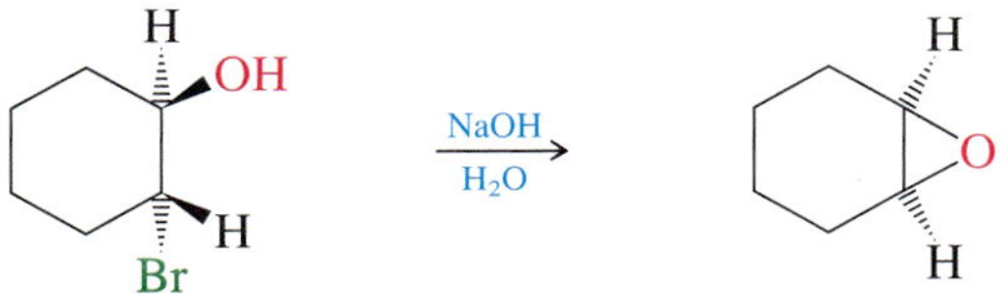

trans-2-Bromocyclohexanol   1,2-Epoxycyclohexane (81%)

Overall, the stereospecificity of this method is the same as that observed in peroxy acid oxidation of alkenes. Substituents that are cis to each other in the alkene remain cis in the epoxide. This is because formation of the bromohydrin involves anti addition, and the ensuing intramolecular nucleophilic substitution reaction takes place with inversion of configuration at the carbon that bears the halide leaving group.

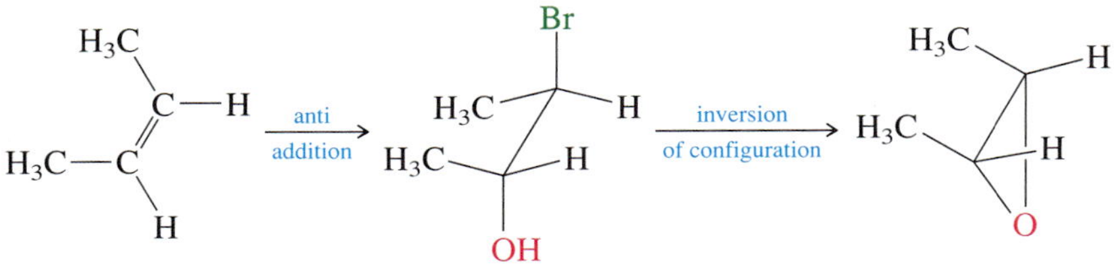

cis-2-Butene   *cis*-2,3-Epoxybutane

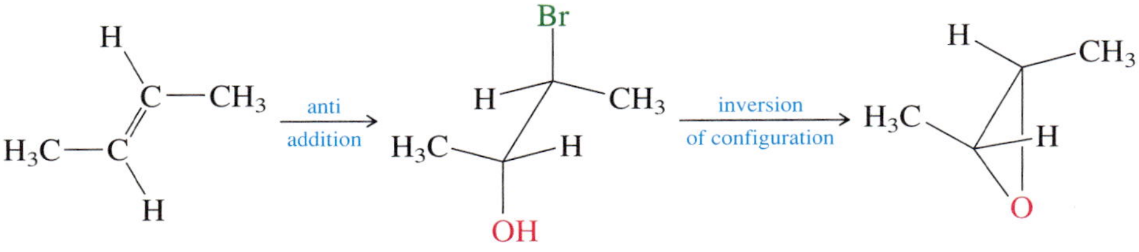

trans-2-Butene   *trans*-2,3-Epoxybutane

> **PROBLEM 16.11** Is either of the epoxides formed in the preceding reactions chiral? Is either epoxide optically active when prepared from the alkene by this method?

About 2×10^9 lb/year of 1,2-epoxypropane is produced in the United States as an intermediate in the preparation of various polymeric materials, including polyurethane plastics and foams and polyester resins. A large fraction of the 1,2-epoxypropane is made from propene by way of its chlorohydrin.

16.11 REACTIONS OF EPOXIDES: A REVIEW AND A PREVIEW

The most striking chemical property of epoxides is their far greater reactivity toward nucleophilic reagents compared with that of simple ethers. Epoxides react rapidly with nucleophiles under conditions in which other ethers are inert. This enhanced reactivity results from the angle strain of epoxides. Reactions that open the ring relieve this strain.

Angle strain is the main source of strain in epoxides, but torsional strain that results from the eclipsing of bonds on adjacent carbons is also present. Both kinds of strain are relieved when a ring-opening reaction occurs.

We saw an example of nucleophilic ring opening of epoxides in Section 15.4, where the reaction of Grignard reagents with ethylene oxide was described as a synthetic route to primary alcohols:

$$\text{RMgX} \quad + \quad \text{H}_2\text{C}\underset{\text{O}}{\text{———}}\text{CH}_2 \quad \xrightarrow[\text{2. H}_3\text{O}^+]{\text{1. diethyl ether}} \quad \text{RCH}_2\text{CH}_2\text{OH}$$

Grignard reagent   Ethylene oxide   Primary alcohol

$$\bigcirc\!\!-\text{CH}_2\text{MgCl} \ + \ \text{H}_2\text{C}\underset{\text{O}}{\text{———}}\text{CH}_2 \ \xrightarrow[\text{2. H}_3\text{O}^+]{\text{1. diethyl ether}} \ \bigcirc\!\!-\text{CH}_2\text{CH}_2\text{CH}_2\text{OH}$$

Benzylmagnesium chloride   Ethylene oxide   3-Phenyl-1-propanol (71%)

Nucleophiles other than Grignard reagents also open epoxide rings. These reactions are carried out in two different ways. The first (Section 16.12) involves anionic nucleophiles in neutral or basic solution.

$$\text{Y}\!:\!^- \ + \ \text{R}_2\text{C}\underset{\text{Ö}}{\text{———}}\text{CR}_2 \ \longrightarrow \ \text{R}_2\overset{\text{Y}}{\text{C}}\text{——}\underset{:\ddot{\text{O}}:^-}{\text{CR}_2} \ \xrightarrow{\text{H}_2\text{O}} \ \text{R}_2\overset{\text{Y}}{\text{C}}\text{——}\underset{:\ddot{\text{O}}\text{H}}{\text{CR}_2}$$

Nucleophile   Epoxide   Alkoxide ion   Alcohol

These reactions are usually performed in water or alcohols as solvents, and the alkoxide ion intermediate is rapidly transformed to an alcohol by proton transfer.

The other involves acid catalysis. Here the nucleophile is often

$$\text{HY}\!:\! \ + \ \text{R}_2\text{C}\underset{\ddot{\text{O}}}{\text{———}}\text{CR}_2 \ \xrightarrow{\text{H}^+} \ \text{R}_2\overset{\text{Y}:}{\text{C}}\text{——}\underset{:\ddot{\text{O}}\text{H}}{\text{CR}_2}$$

Epoxide   Alcohol

Acid-catalyzed ring opening of epoxides is discussed in Section 16.13.

There is an important difference in the regiochemistry of ring-opening reactions of epoxides depending on the reaction conditions. Unsymmetrically substituted epoxides tend to react with anionic nucleophiles at the less hindered carbon of the ring. Under conditions of acid catalysis, however, the more highly substituted carbon is attacked.

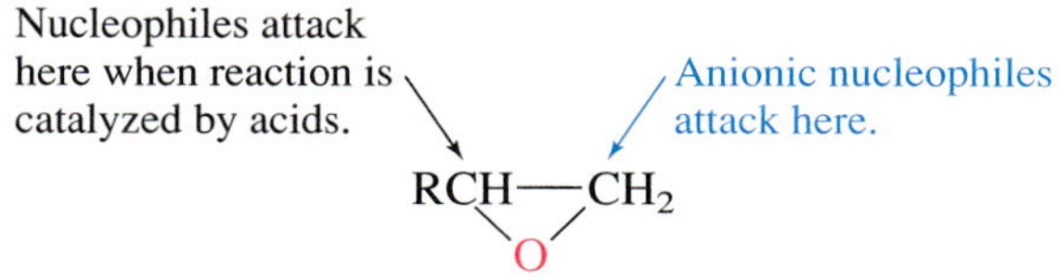

Nucleophiles attack here when reaction is catalyzed by acids.

Anionic nucleophiles attack here.

$$RCH\!\!-\!\!CH_2$$
$$O$$

The underlying reasons for this difference in regioselectivity will be explained in Section 16.13.

16.12   NUCLEOPHILIC RING-OPENING OF EPOXIDES

Ethylene oxide is a very reactive substance. It reacts rapidly and exothermically with anionic nucleophiles to yield 2-substituted derivatives of ethanol by cleaving the carbon–oxygen bond of the ring:

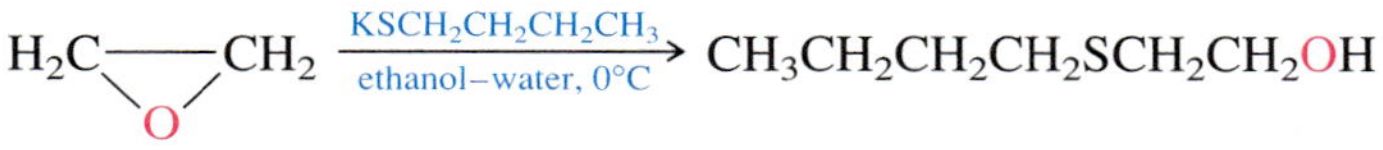

$$H_2C\!\!-\!\!CH_2 \xrightarrow[\text{ethanol–water, 0°C}]{KSCH_2CH_2CH_2CH_3} CH_3CH_2CH_2CH_2SCH_2CH_2OH$$
$$O$$

Ethylene oxide   2-(Butylthio)ethanol (99%)
(oxirane)

PROBLEM 16.12   What is the principal organic product formed in the reaction of ethylene oxide with each of the following?

(a) Sodium cyanide (NaCN) in aqueous ethanol
(b) Sodium azide (NaN$_3$) in aqueous ethanol
(c) Sodium hydroxide (NaOH) in water
(d) Phenyllithium (C$_6$H$_5$Li) in ether, followed by addition of dilute sulfuric acid
(e) 1-Butynylsodium (CH$_3$CH$_2$C≡CNa) in liquid ammonia

SAMPLE SOLUTION   (a) Sodium cyanide is a source of the nucleophilic cyanide anion. Cyanide ion attacks ethylene oxide, opening the ring and forming 2-cyanoethanol:

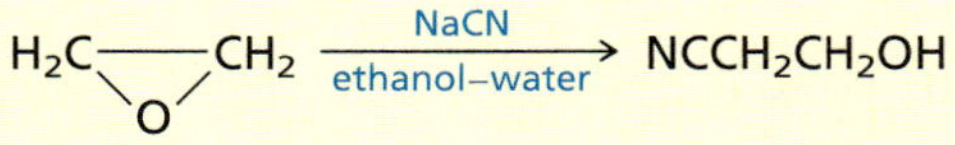

$$H_2C\!\!-\!\!CH_2 \xrightarrow[\text{ethanol–water}]{NaCN} NCCH_2CH_2OH$$
$$O$$

Ethylene oxide   2-Cyanoethanol

Nucleophilic ring opening of epoxides has many of the features of an S$_N$2 reaction. Inversion of configuration is observed at the carbon at which substitution occurs.

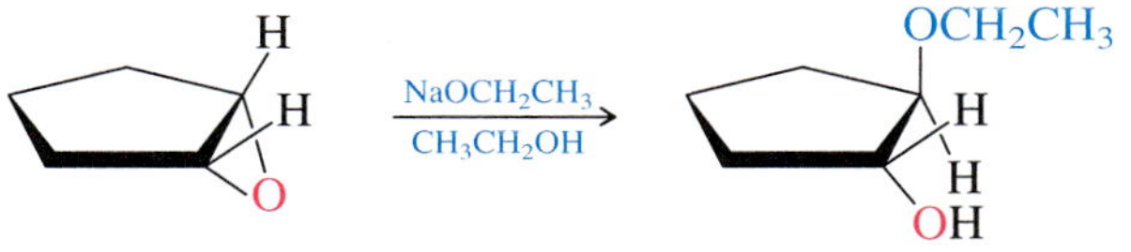

1,2-Epoxycyclopentane   *trans*-2-Ethoxycyclopentanol
                                        (67%)

Manipulating models of these compounds can make it easier to follow the stereochemistry.

(2R,3R)-2,3-Epoxybutane →[NH₃, H₂O] (2R,3S)-3-Amino-2-butanol (70%)

Unsymmetrical epoxides are attacked at the less substituted, less sterically hindered carbon of the ring:

2,2,3-Trimethyloxirane →[NaOCH₃ / CH₃OH] CH_3CHCCH_3 with CH_3O, CH_3, and OH — 3-Methoxy-2-methyl-2-butanol (53%)

PROBLEM 16.13 Given the starting material 1-methyl-1,2-epoxycyclopentane, of absolute configuration as shown, decide which one of the compounds A through C correctly represents the product of its reaction with sodium methoxide in methanol.

1-Methyl-1,2-epoxycyclopentane

Compound A   Compound B   Compound C

The experimental observations combine with the principles of nucleophilic substitution to give the picture of epoxide ring opening shown in Figure 16.5. The nucleophile attacks the less crowded carbon from the side opposite the carbon–oxygen bond. Bond

FIGURE 16.5 Nucleophilic ring opening of an epoxide.

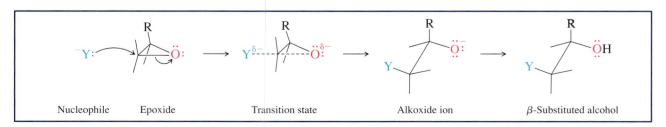

Nucleophile   Epoxide   Transition state   Alkoxide ion   β-Substituted alcohol

formation with the nucleophile accompanies carbon–oxygen bond breaking, and a substantial portion of the strain in the three-membered ring is relieved as it begins to open at the transition state. The initial product of nucleophilic substitution is an alkoxide anion, which rapidly abstracts a proton from the solvent to give a β-substituted alcohol as the isolated product.

The reaction of Grignard reagents with epoxides is regioselective in the same sense. Attack occurs at the less substituted carbon of the ring.

$$C_6H_5MgBr \quad + \quad H_2C\overset{}{\underset{O}{\diagdown\diagup}}CHCH_3 \xrightarrow[\text{2. } H_3O^+]{\text{1. diethyl ether}} C_6H_5CH_2\underset{\underset{OH}{|}}{C}HCH_3$$

Phenylmagnesium   1,2-Epoxypropane   1-Phenyl-2-propanol
bromide   (60%)

Epoxides are reduced to alcohols on treatment with lithium aluminum hydride. Hydride is transferred to the less substituted carbon.

$$H_2C\overset{}{\underset{O}{\diagdown\diagup}}CH(CH_2)_7CH_3 \xrightarrow[\text{2. } H_2O]{\text{1. LiAlH}_4} CH_3\underset{\underset{OH}{|}}{C}H(CH_2)_7CH_3$$

1,2-Epoxydecane   2-Decanol (90%)

Epoxidation of an alkene, followed by lithium aluminum hydride reduction of the resulting epoxide, gives the same alcohol that would be obtained by acid-catalyzed hydration (Section 6.10) of the alkene.

16.13 ACID-CATALYZED RING-OPENING OF EPOXIDES

As we've just seen, nucleophilic ring opening of ethylene oxide yields 2-substituted derivatives of ethanol. Those reactions involved nucleophilic attack on the carbon of the ring under neutral or basic conditions. Other nucleophilic ring-openings of epoxides likewise give 2-substituted derivatives of ethanol but either involve an acid as a reactant or occur under conditions of acid catalysis:

$$H_2C\overset{}{\underset{O}{\diagdown\diagup}}CH_2 \xrightarrow[\text{10°C}]{\text{HBr}} BrCH_2CH_2OH$$

Ethylene oxide   2-Bromoethanol
                                  (87–92%)

$$H_2C\overset{}{\underset{O}{\diagdown\diagup}}CH_2 \xrightarrow[\text{H}_2\text{SO}_4,\ 25°C]{\text{CH}_3\text{CH}_2\text{OH}} CH_3CH_2OCH_2CH_2OH$$

Ethylene oxide   2-Ethoxyethanol (85%)

A third example is the industrial preparation of ethylene glycol ($HOCH_2CH_2OH$) by hydrolysis of ethylene oxide in dilute sulfuric acid. This reaction and its mechanism (Figure 16.6) illustrate the difference between the ring openings of epoxides discussed in the preceding section and the acid-catalyzed ones described here. Under conditions of acid catalysis, the species that is attacked by the nucleophile is not the epoxide itself, but rather its conjugate acid. The transition state for ring opening has a fair measure of carbocation character. Breaking of the ring carbon–oxygen bond is more advanced than formation of the bond to the nucleophile.

FIGURE 16.6 The mechanism for the acid-catalyzed nucleophilic ring opening of ethylene oxide by water.

Overall Reaction:

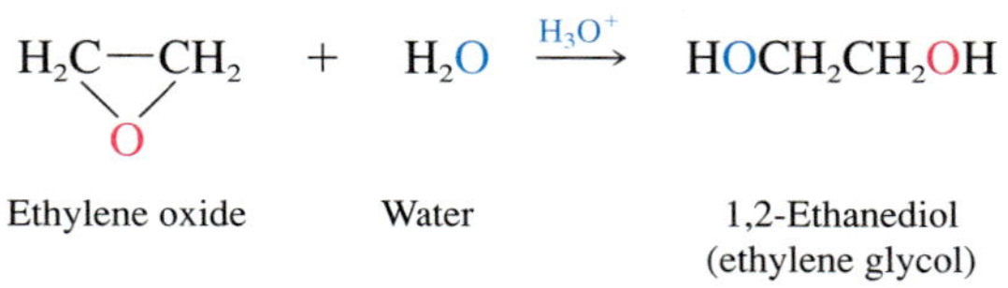

$$H_2C\text{—}CH_2 \ (O) \ + \ H_2O \ \xrightarrow{H_3O^+} \ HOCH_2CH_2OH$$

Ethylene oxide   Water   1,2-Ethanediol
(ethylene glycol)

Mechanism:

Step 1: Proton transfer to the oxygen of the epoxide to give an oxonium ion.

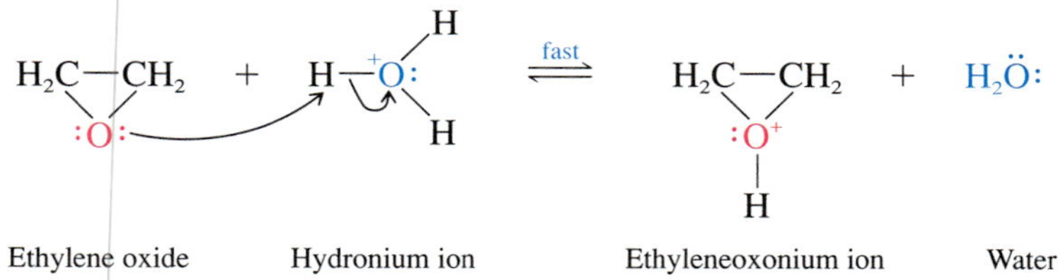

Ethylene oxide   Hydronium ion   Ethyleneoxonium ion   Water

Step 2: Nucleophilic attack by water on carbon of the oxonium ion. The carbon–oxygen bond of the ring is broken in this step and the ring opens.

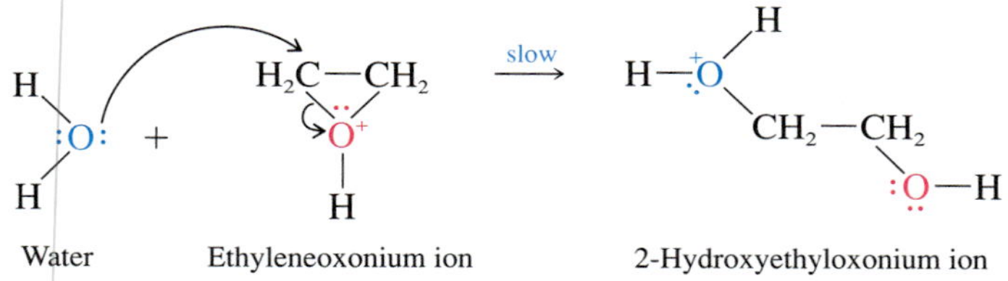

Water   Ethyleneoxonium ion   2-Hydroxyethyloxonium ion

Step 3: Proton transfer to water completes the reaction and regenerates the acid catalyst.

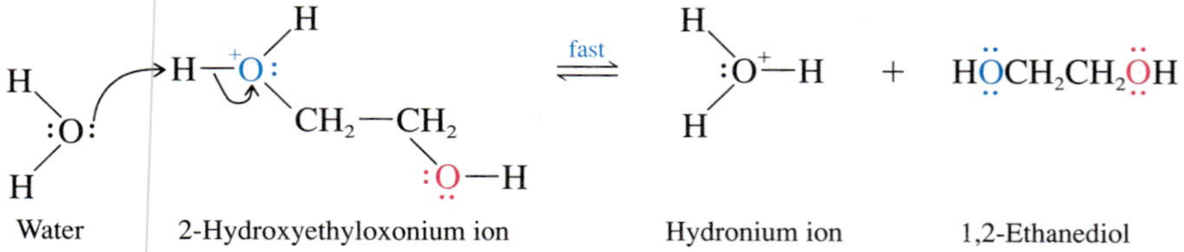

Water   2-Hydroxyethyloxonium ion   Hydronium ion   1,2-Ethanediol

Transition state for attack by water on conjugate acid of ethylene oxide

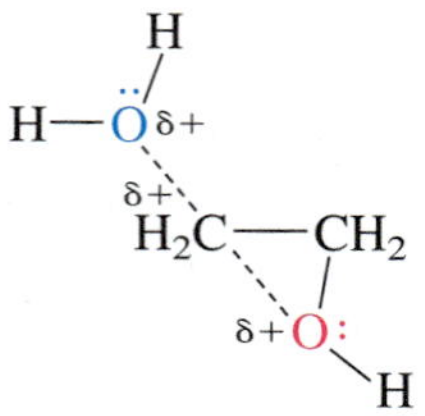

Because *carbocation* character develops at the transition state, substitution is favored at the carbon that can better support a developing positive charge. Thus, in

contrast to the reaction of epoxides with relatively basic nucleophiles, in which S_N2-like attack is faster at the less crowded carbon of the three-membered ring, acid catalysis promotes substitution at the position that bears the greater number of alkyl groups:

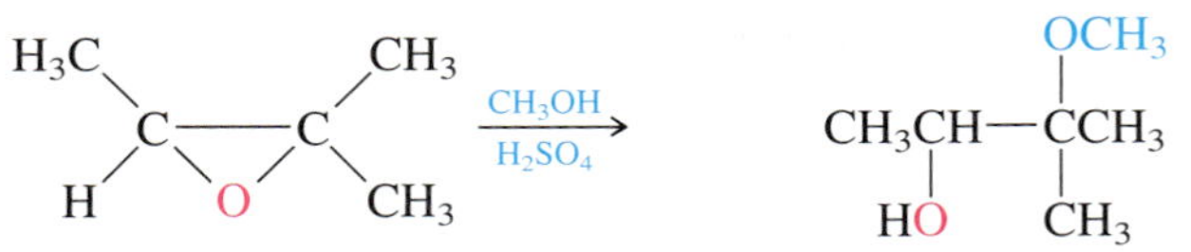

2,2,3-Trimethyloxirane   3-Methoxy-3-methyl-2-butanol (76%)

Although nucleophilic participation at the transition state is slight, it is enough to ensure that substitution proceeds with inversion of configuration.

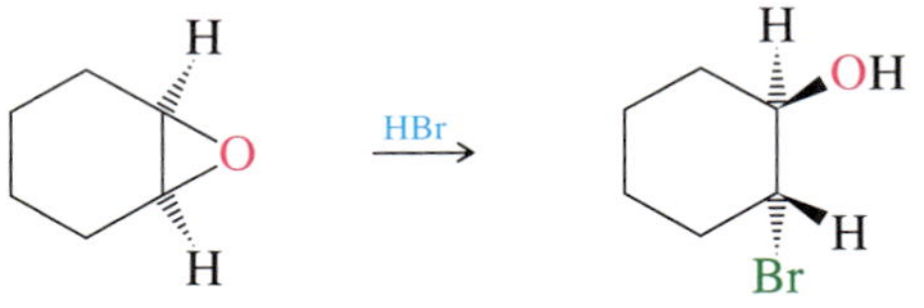

1,2-Epoxycyclohexane   *trans*-2-Bromocyclohexanol
(73%)

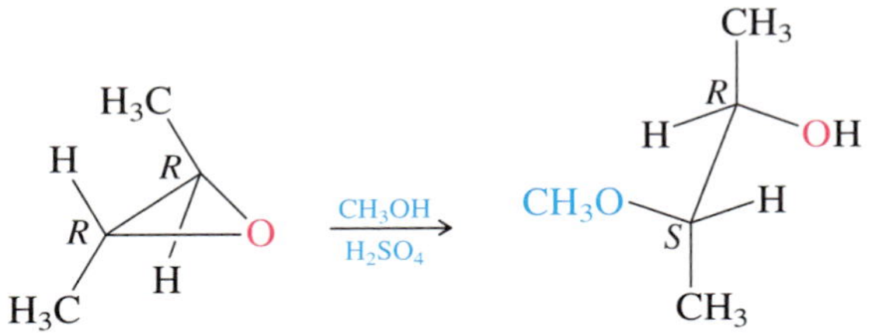

(2R,3R)-2,3-Epoxybutane   (2R,3S)-3-Methoxy-2-butanol
(57%)

PROBLEM 16.14 Which product, compound A, B, or C, would you expect to be formed when 1-methyl-1,2-epoxycyclopentane of the absolute configuration shown is allowed to stand in methanol containing a few drops of sulfuric acid? Compare your answer with that given for Problem 16.13.

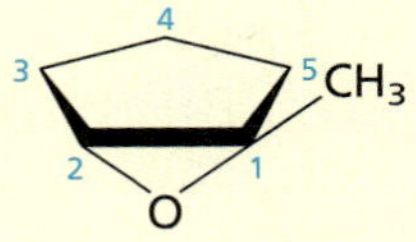

1-Methyl-1,2-
epoxycyclopentane

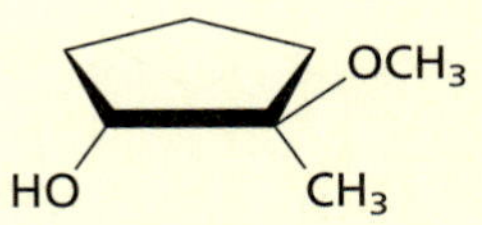

Compound A 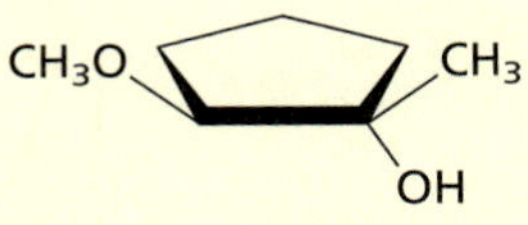   Compound B 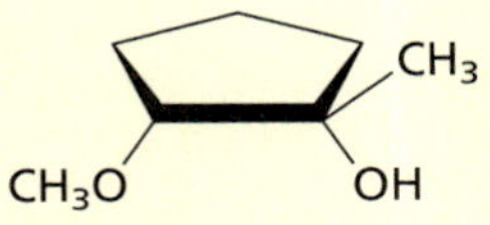   Compound C

A method for achieving net anti hydroxylation of alkenes combines two stereo-specific processes: epoxidation of the double bond and hydrolysis of the derived epoxide.

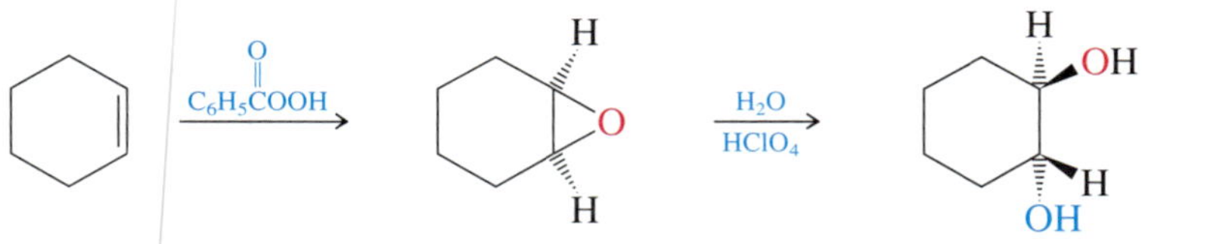

Cyclohexene   1,2-Epoxycyclohexane   *trans*-1,2-Cyclohexanediol
                                                                        (80%)

PROBLEM 16.15 Which alkene, *cis*-2-butene or *trans*-2-butene, would you choose in order to prepare *meso*-2,3-butanediol by epoxidation followed by acid-catalyzed hydrolysis? Which alkene would yield *meso*-2,3,-butanediol by osmium tetraoxide hydroxylation?

16.14 EPOXIDES IN BIOLOGICAL PROCESSES

Many naturally occurring substances are epoxides. You have seen two examples of such compounds already in disparlure, the sex attractant of the gypsy moth (Section 6.18), and in the carcinogenic epoxydiol formed from benzo[a]pyrene (Section 11.8). In most cases, epoxides are biosynthesized by the enzyme-catalyzed transfer of one of the oxygen atoms of an O_2 molecule to an alkene. Because only one of the atoms of O_2 is transferred to the substrate, the enzymes that catalyze such transfers are classified as *monooxygenases*. A biological reducing agent, usually the coenzyme NADH (Section 15.11), is required as well.

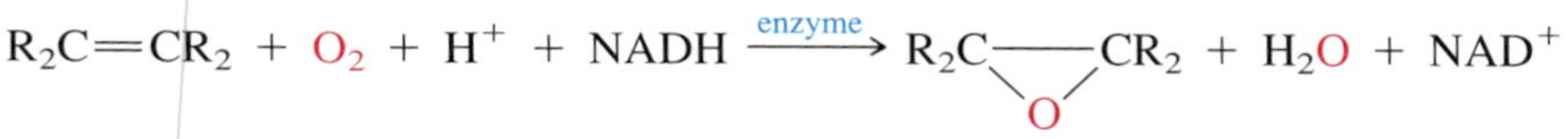

$$R_2C{=}CR_2 + O_2 + H^+ + NADH \xrightarrow{\text{enzyme}} R_2C{-}CR_2 + H_2O + NAD^+$$

A prominent example of such a reaction is the biological epoxidation of the polyene squalene.

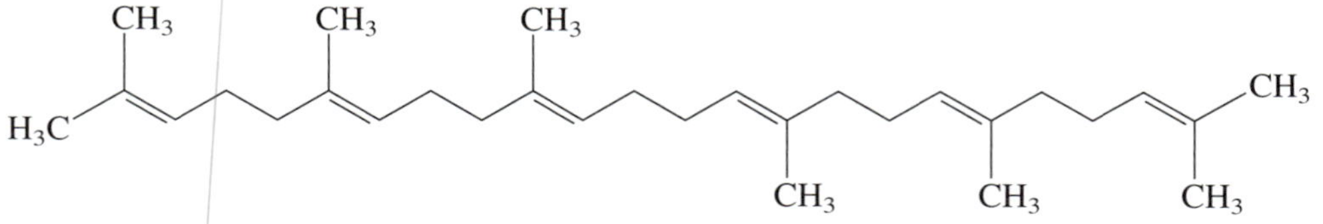

Squalene

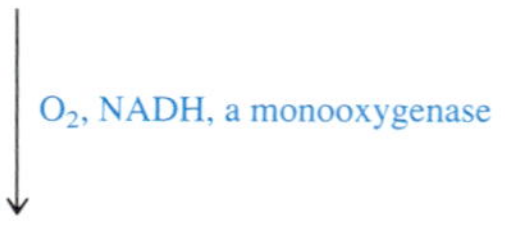

O_2, NADH, a monooxygenase

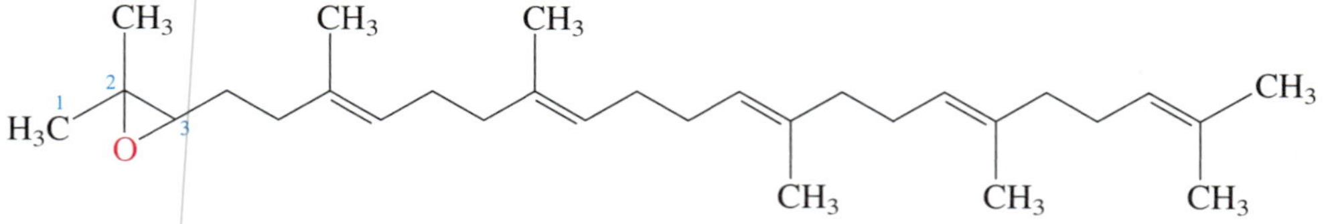

Squalene 2,3-epoxide

The reactivity of epoxides toward nucleophilic ring opening is responsible for one of the biological roles they play. Squalene 2,3-epoxide, for example, is the biological precursor to cholesterol and the steroid hormones, including testosterone, progesterone, estrone, and cortisone. The pathway from squalene 2,3-epoxide to these compounds is triggered by epoxide ring opening and will be described in Chapter 26.

16.15 PREPARATION OF SULFIDES

Sulfides, compounds of the type RSR′, are prepared by nucleophilic substitution reactions. Treatment of a primary or secondary alkyl halide with an alkanethiolate ion (RS⁻) gives a sulfide:

$$RS:^- \ Na^+ \ + \ R'\!-\!X: \ \xrightarrow{S_N2} \ R\ddot{S}R' \ + \ Na^+ \ :\ddot{X}:^-$$

Sodium alkanethiolate   Alkyl halide   Sulfide   Sodium halide

$$CH_3CHCH{=}CH_2 \ \xrightarrow[\text{methanol}]{\text{NaSCH}_3} \ CH_3CHCH{=}CH_2$$
$$\overset{|}{Cl} \qquad\qquad\qquad\qquad \overset{|}{SCH_3}$$

3-Chloro-1-butene   Methyl 1-methylallyl sulfide
                                    (62%)

It is not necessary to prepare and isolate the sodium alkanethiolate in a separate operation. Because thiols are more acidic than water, they are quantitatively converted to their alkanethiolate anions by sodium hydroxide. Thus, all that is normally done is to add a thiol to sodium hydroxide in a suitable solvent (water or an alcohol) followed by the alkyl halide.

The pK_a for CH_3SH is 10.7.

> **PROBLEM 16.16** The p-toluenesulfonate derived from (R)-2-octanol and p-toluenesulfonyl chloride was allowed to react with sodium benzenethiolate (C_6H_5SNa). Give the structure, including stereochemistry and the appropriate R or S descriptor, of the product.

16.16 OXIDATION OF SULFIDES: SULFOXIDES AND SULFONES

We saw in Section 15.13 that thiols differ from alcohols with respect to their behavior toward oxidation. Similarly, sulfides differ from ethers in their behavior toward oxidizing agents. Whereas ethers tend to undergo oxidation at carbon to give hydroperoxides (Section 16.7), sulfides are oxidized at sulfur to give **sulfoxides.** If the oxidizing agent is strong enough and present in excess, oxidation can proceed further to give **sulfones.**

$$R\!-\!\ddot{S}\!-\!R' \ \xrightarrow{\text{oxidize}} \ R\!-\!\overset{\overset{:\ddot{O}:^-}{|}}{\underset{\cdot\cdot}{\overset{+}{S}}}\!-\!R' \ \xrightarrow{\text{oxidize}} \ R\!-\!\overset{\overset{:\ddot{O}:^-}{|}}{\underset{\underset{:\ddot{O}:^-}{|}}{\overset{2+}{S}}}\!-\!R'$$

Sulfide   Sulfoxide   Sulfone

Third-row elements such as sulfur can expand their valence shell beyond eight electrons, and so sulfur–oxygen bonds in sulfoxides and sulfones are sometimes represented as double bonds.

When the desired product is a sulfoxide, sodium metaperiodate ($NaIO_4$) is an ideal reagent. It oxidizes sulfides to sulfoxides in high yield but shows no tendency to oxidize sulfoxides to sulfones.

Methyl phenyl sulfide + Sodium metaperiodate → Methyl phenyl sulfoxide (91%) + Sodium iodate

Peroxy acids, usually in dichloromethane as the solvent, are also reliable reagents for converting sulfides to sulfoxides.

One equivalent of a peroxy acid or of hydrogen peroxide converts sulfides to sulfoxides; two equivalents gives the corresponding sulfone.

Phenyl vinyl sulfide + Hydrogen peroxide → Phenyl vinyl sulfone (74–78%) + Water

> **PROBLEM 16.17** Verify, by making molecular models, that the bonds to sulfur are arranged in a trigonal pyramidal geometry in sulfoxides and in a tetrahedral geometry in sulfones. Is phenyl vinyl sulfoxide chiral? What about phenyl vinyl sulfone?

Oxidation of sulfides occurs in living systems as well. Among naturally occurring sulfoxides, one that has received recent attention is *sulforaphane*, which is present in broccoli and other vegetables. Sulforaphane holds promise as a potential anticancer agent because, unlike most anticancer drugs, which act by killing rapidly dividing tumor cells faster than they kill normal cells, sulforaphane is nontoxic and may simply inhibit the formation of tumors.

$$CH_3\overset{+}{\underset{}{S}}CH_2CH_2CH_2CH_2N=C=S$$

Sulforaphane

16.17 ALKYLATION OF SULFIDES: SULFONIUM SALTS

Sulfur is more nucleophilic than oxygen (Section 8.7), and sulfides react with alkyl halides much faster than do ethers. The products of these reactions, called **sulfonium salts,** are also more stable than the corresponding oxygen analogs.

Use Learning By Modeling to view the geometry of sulfur in trimethylsulfonium ion.

Sulfide + Alkyl halide $\xrightarrow{S_N2}$ Sulfonium salt

$$CH_3(CH_2)_{10}CH_2SCH_3 + CH_3I \longrightarrow CH_3(CH_2)_{10}CH_2\overset{+}{\underset{CH_3}{S}}CH_3 \ I^-$$

Dodecyl methyl sulfide   Methyl iodide   Dodecyldimethylsulfonium iodide

PROBLEM 16.18 What other combination of alkyl halide and sulfide will yield the same sulfonium salt shown in the preceding example? Predict which combination will yield the sulfonium salt at the faster rate.

A naturally occurring sulfonium salt, *S-adenosylmethionine (SAM),* is a key substance in certain biological processes. It is formed by a nucleophilic substitution in which the sulfur atom of methionine attacks the primary carbon of adenosine triphosphate, displacing the triphosphate leaving group as shown in Figure 16.7.

S-Adenosylmethionine acts as a biological methyl-transfer agent. Nucleophiles, particularly nitrogen atoms of amines, attack the methyl carbon of SAM, breaking the carbon–sulfur bond. The equation shown on the next page represents the biological formation of *epinephrine* by methylation of *norepinephrine.* Only the methyl group and the sulfur of SAM are shown explicitly in the equation to draw attention to the similarity of this reaction, which occurs in living systems, to the more familiar S_N2 reactions we have studied.

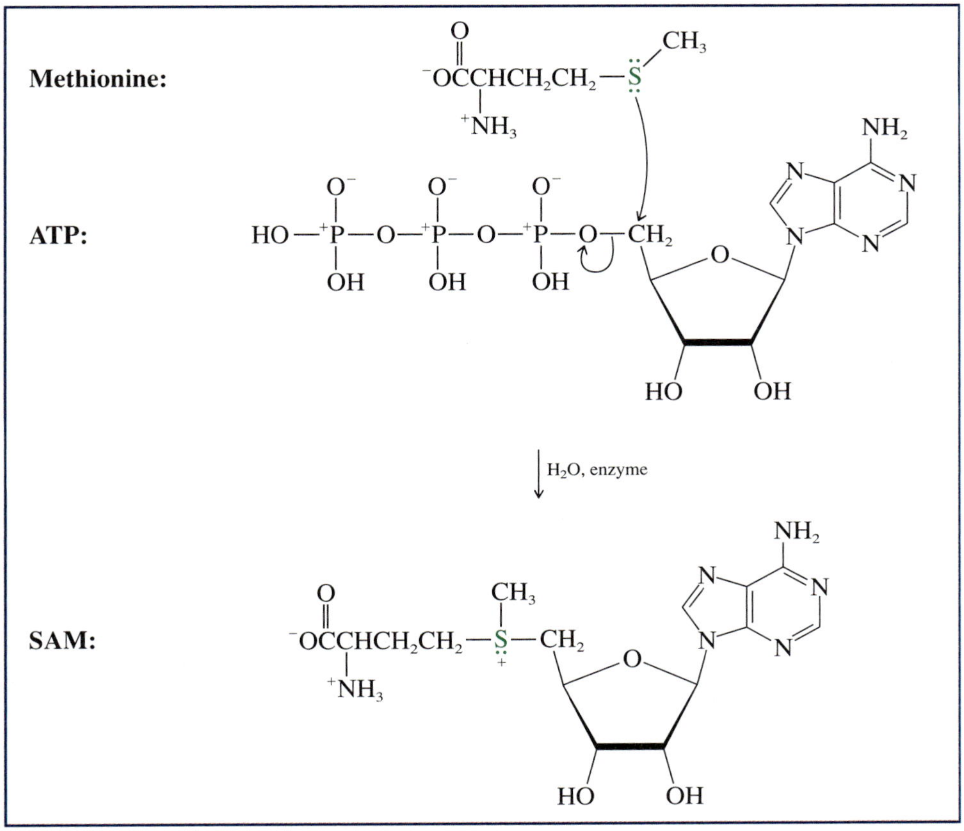

The *S* in *S*-adenosylmethionine indicates that the adenosyl group is bonded to sulfur. It does *not* stand for the Cahn–Ingold–Prelog stereochemical descriptor.

FIGURE 16.7 Nucleophilic substitution at the primary carbon of adenosine triphosphate (ATP) by the sulfur atom of methionine yields *S*-adenosylmethionine (SAM). The reaction is catalyzed by an enzyme.

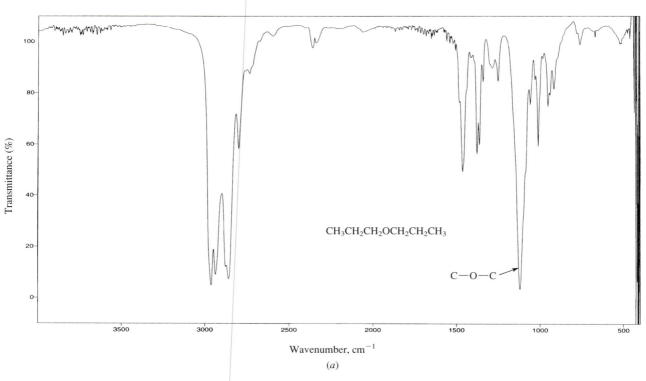

Norepinephrine   SAM

Epinephrine is also known as *adrenaline* and is a hormone with profound physiological effects designed to prepare the body for "fight or flight."

Epinephrine

16.18 SPECTROSCOPIC ANALYSIS OF ETHERS, EPOXIDES, AND SULFIDES

The IR, ^{1}H NMR, and ^{13}C NMR spectra of dipropyl ether, which appear in parts *a*, *b*, and *c*, respectively of Figure 16.8, illustrate some of the spectroscopic features of ethers.

$CH_3CH_2CH_2OCH_2CH_2CH_3$

C—O—C

Wavenumber, cm^{-1}

(*a*)

FIGURE 16.8 The (*a*) infrared, (*b*) 200-MHz ^{1}H NMR, and (*c*) ^{13}C NMR spectra of dipropyl ether ($CH_3CH_2CH_2OCH_2CH_2CH_3$).

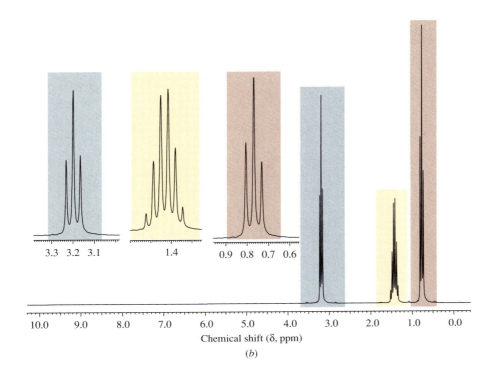

3.3 3.2 3.1
1.4
0.9 0.8 0.7 0.6
10.0 9.0 8.0 7.0 6.0 5.0 4.0 3.0 2.0 1.0 0.0
Chemical shift (δ, ppm)
(*b*)

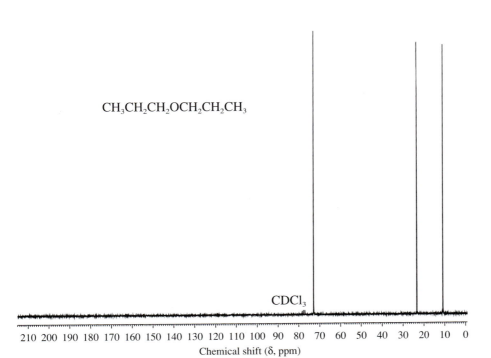

$CH_3CH_2CH_2OCH_2CH_2CH_3$
$CDCl_3$
210 200 190 180 170 160 150 140 130 120 110 100 90 80 70 60 50 40 30 20 10 0
Chemical shift (δ, ppm)
(*c*)

Infrared: The infrared spectra of *ethers* are characterized by a strong, rather broad band due to antisymmetric C—O—C stretching between 1070 and 1150 cm^{-1}. Dialkyl ethers exhibit this band consistently at 1120 cm^{-1}, as shown in the IR spectrum of dipropyl ether.

$$CH_3CH_2CH_2OCH_2CH_2CH_3$$

Dipropyl ether

C—O—C: $\nu = 1121$ cm^{-1}

The analogous band in alkyl aryl ethers (ROAr) appears at 1200–1275 cm^{-1} (Section 24.15).

Epoxides typically exhibit three bands. Two bands, one at 810–950 cm^{-1} and the other near 1250 cm^{-1}, correspond to asymmetric and symmetric stretching of the ring, respectively. The third band appears in the range 750–840 cm^{-1}.

H$_2$C—CH(CH$_2$)$_9$CH$_3$
\ /
O

1,2-Epoxydodecane

Epoxide vibrations: $\nu = 837, 917,$ and 1265 cm^{-1}

The C—S—C stretching vibration of *sulfides* gives a weak peak in the 600–700 cm^{-1} range. *Sulfoxides* show a strong peak due to S—O stretching at 1030–1070 cm^{-1}. With two oxygens attached to sulfur, *sulfones* exhibit strong bands due to symmetric (1120–1160 cm^{-1}) and asymmetric (1290–1350 cm^{-1}) S—O stretching.

O$^-$
|+
H$_3$C—S—CH$_3$

O$^-$
|2+
H$_3$C—S—CH$_3$
|
O$_-$

Dimethyl sulfoxide
S—O: $\nu = 1050$ cm^{-1}

Dimethyl sulfone
S—O: $\nu = 1139$ and 1298 cm^{-1}

^{1}H NMR: The chemical shift of the proton in the **H**—C—O—C unit of an *ether* is very similar to that of the proton in the **H**—C—OH unit of an alcohol. A range of δ 3.2–4.0 is typical. The proton in the **H**—C—S—C unit of a *sulfide* appears at higher field than the corresponding proton of an ether because sulfur is less electronegative than oxygen.

^{1}H Chemical shift (δ):

CH$_3$CH$_2$CH$_2$—O—CH$_2$CH$_2$CH$_3$
└── 3.2 ──┘

CH$_3$CH$_2$CH$_2$—S—CH$_2$CH$_2$CH$_3$
└── 2.5 ──┘

Oxidation of a sulfide to a *sulfoxide* or *sulfone* is accompanied by a decrease in shielding of the **H**—C—S—C proton by about 0.3–0.5 ppm for each oxidation.

Epoxides are unusual in that the protons on the ring are more shielded than expected. The protons in ethylene oxide, for example, appear at δ 2.5 instead of the δ 3.2–4.0 range just cited for dialkyl ethers.

^{13}C NMR: The carbons of the C—O—C group in an *ether* are about 10 ppm less shielded than those of an alcohol and appear in the range δ 57–87. The carbons of the C—S—C group in a *sulfide* are significantly more shielded than those of an ether.

^{13}C Chemical shift (δ):

CH$_3$CH$_2$CH$_2$—O—CH$_2$CH$_2$CH$_3$
└── 73 ──┘

CH$_3$CH$_2$CH$_2$—S—CH$_2$CH$_2$CH$_3$
└── 34 ──┘

The ring carbons of an *epoxide* are somewhat more shielded than the carbons of a C—O—C unit of larger rings or dialkyl ethers.

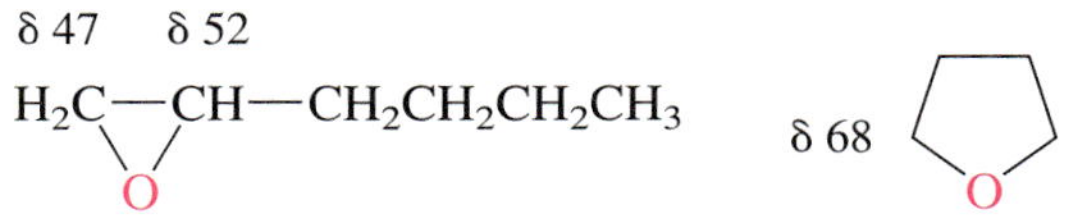

δ 47   δ 52

H₂C—CH—CH₂CH₂CH₂CH₃   δ 68
    \ /
     O

UV-VIS: Simple ethers have their absorption maximum at about 185 nm and are transparent to ultraviolet radiation above about 220 nm.

Mass Spectrometry: Ethers, like alcohols, lose an alkyl radical from their molecular ion to give an oxygen-stabilized cation. Thus, m/z 73 and m/z 87 are both more abundant than the molecular ion in the mass spectrum of *sec*-butyl ethyl ether.

$$CH_3CH_2\overset{\cdot\cdot+}{O}—CHCH_2CH_3$$
$$|$$
$$CH_3$$

m/z 102

$CH_3CH_2\overset{+}{O}{=}CHCH_3 \;+\; \cdot CH_2CH_3$ $CH_3CH_2\overset{+}{O}{=}CHCH_2CH_3 \;+\; \cdot CH_3$

m/z 73 m/z 87

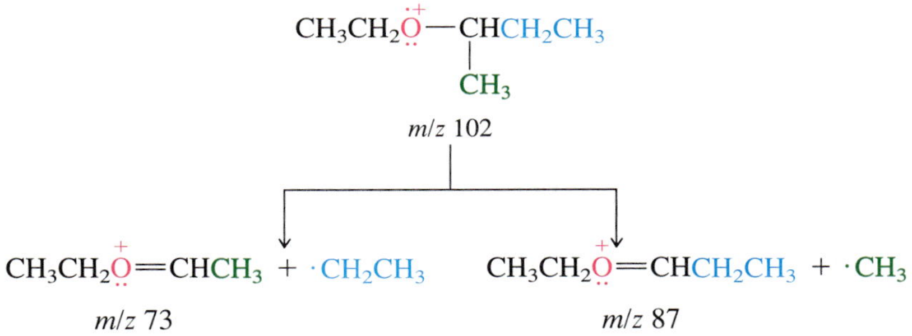

PROBLEM 16.19   There is another oxygen-stabilized cation of m/z 87 capable of being formed by fragmentation of the molecular ion in the mass spectrum of *sec*-butyl ethyl ether. Suggest a reasonable structure for this ion.

An analogous fragmentation process occurs in the mass spectra of sulfides. As with other sulfur-containing compounds, the presence of sulfur can be inferred by a peak at m/z of M+2.

16.19 SUMMARY

Section 16.1   **Ethers** are compounds that contain a C—O—C linkage. In substitutive IUPAC nomenclature, they are named as *alkoxy* derivatives of alkanes. In functional class IUPAC nomenclature, we name each alkyl group as a separate word (in alphabetical order) followed by the word *ether.*

$CH_3OCH_2CH_2CH_2CH_2CH_2CH_3$ **Substitutive IUPAC name:** 1-Methoxyhexane
Functional class name: Hexyl methyl ether

Epoxides are normally named as *epoxy* derivatives of alkanes or as substituted *oxiranes.*

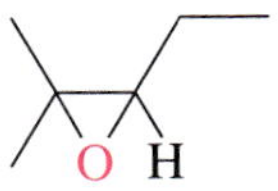

2-Methyl-2,3-epoxypentane
3-Ethyl-2,2-dimethyloxirane

Sulfides are sulfur analogs of ethers: they contain the C—S—C functional group. They are named as *alkylthio* derivatives of alkanes in substitutive IUPAC nomenclature. The functional class IUPAC names of sulfides are derived in the same manner as those of ethers, but the concluding word is *sulfide.*

$$CH_3SCH_2CH_2CH_2CH_2CH_2CH_3$$

Substitutive IUPAC name: 1-(Methylthio)hexane
Functional class name: Hexyl methyl sulfide

Section 16.2   The oxygen atom in an ether or epoxide affects the shape of the molecule in much the same way as an sp^3-hybridized carbon of an alkane or cycloalkane.

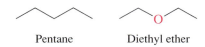

Pentane   Diethyl ether

Section 16.3   The carbon–oxygen bond of ethers is polar, and ethers can act as proton *acceptors* in hydrogen bonds with water and alcohols.

$$\begin{array}{c} R \\ \diagdown \; {}^{\delta-} \quad {}^{\delta+} \\ \quad :\ddot{O}:\text{---}H\text{---}\ddot{O}R' \\ \diagup \\ R \end{array}$$

But ethers lack OH groups and cannot act as proton *donors* in forming hydrogen bonds.

Section 16.4   Ethers form Lewis acid-Lewis base complexes with metal ions. Certain cyclic polyethers, called **crown ethers,** are particularly effective in coordinating with Na^+ and K^+, and salts of these cations can be dissolved in nonpolar solvents when crown ethers are present. Under these conditions the rates of many reactions that involve anions are accelerated.

$$CH_3(CH_2)_4CH_2Br \xrightarrow[\text{acetonitrile, heat}]{KOCCH_3, \text{ 18-crown-6}} CH_3(CH_2)_4CH_2OCCH_3$$

1-Bromohexane   Hexyl acetate (96%)

Sections 16.5 and 16.6   The two major methods for preparing ethers are summarized in Table 16.1.

Section 16.7   Dialkyl ethers are useful solvents for organic reactions, but must be used cautiously due to their tendency to form explosive hydroperoxides by air oxidation in opened bottles.

Section 16.8   The only important reaction of ethers is their cleavage by hydrogen halides.

$$\underset{\text{Ether}}{ROR'} + \underset{\substack{\text{Hydrogen} \\ \text{halide}}}{2HX} \longrightarrow \underset{\substack{\text{Alkyl} \\ \text{halide}}}{RX} + \underset{\substack{\text{Alkyl} \\ \text{halide}}}{R'X} + \underset{\text{Water}}{H_2O}$$

The order of hydrogen halide reactivity is HI > HBr > HCl.

$$\text{benzyl ethyl ether} \xrightarrow[\text{heat}]{HBr} \text{benzyl bromide} + CH_3CH_2Br$$

Benzyl ethyl ether   Benzyl bromide   Ethyl bromide

Sections 16.9 and 16.10   Epoxides are prepared by the methods listed in Table 16.2.

TABLE 16.1   Preparation of Ethers

Reaction (section) and comments	General equation and specific example
Acid-catalyzed condensation of alcohols (Sections 15.7 and 16.5) Two molecules of an alcohol condense in the presence of an acid catalyst to yield a dialkyl ether and water. The reaction is limited to the synthesis of symmetrical ethers from primary alcohols.	$2RCH_2OH \xrightarrow{\text{H}^+} RCH_2OCH_2R + H_2O$ Alcohol — Ether — Water $CH_3CH_2CH_2OH \xrightarrow[\text{heat}]{\text{H}_2\text{SO}_4} CH_3CH_2CH_2OCH_2CH_2CH_3$ Propyl alcohol — Dipropyl ether
The Williamson ether synthesis (Section 16.6) An alkoxide ion displaces a halide or similar leaving group in an S_N2 reaction. The alkyl halide cannot be one that is prone to elimination, and so this reaction is limited to methyl and primary alkyl halides. There is no limitation on the alkoxide ion that can be used.	$RO^- + R'CH_2X \longrightarrow ROCH_2R' + X^-$ Alkoxide ion — Primary alkyl halide — Ether — Halide ion $(CH_3)_2CHCH_2ONa + CH_3CH_2Br \longrightarrow (CH_3)_2CHCH_2OCH_2CH_3 + NaBr$ Sodium isobutoxide — Ethyl bromide — Ethyl isobutyl ether (66%) — Sodium bromide

TABLE 16.2   Preparation of Epoxides

Reaction (section) and comments	General equation and specific example
Peroxy acid oxidation of alkenes (Sections 6.18 and 16.9) Peroxy acids transfer oxygen to alkenes to yield epoxides. Stereospecific syn addition is observed.	$R_2C{=}CR_2 + R'\overset{O}{\overset{\|}{C}}OOH \longrightarrow R_2\overset{O}{\underset{\diagup\diagdown}{C{-}C}}R_2 + R'\overset{O}{\overset{\|}{C}}OH$ Alkene — Peroxy acid — Epoxide — Carboxylic acid $(CH_3)_2C{=}C(CH_3)_2 \xrightarrow{CH_3COOH}$ 2,3-Dimethyl-2-butene → 2,2,3,3-Tetramethyloxirane (70–80%)
Base-promoted cyclization of vicinal halohydrins (Section 16.10) This reaction is an intramolecular version of the Williamson ether synthesis. The alcohol function of a vicinal halohydrin is converted to its conjugate base, which then displaces halide from the adjacent carbon to give an epoxide.	Vicinal halohydrin $\xrightarrow{HO^-}$ → Epoxide $(CH_3)_2C{-}CHCH_3$ with HO, Br $\xrightarrow[\text{H}_2\text{O}]{\text{NaOH}}$ $(CH_3)_2C{-}CHCH_3$ epoxide 3-Bromo-2-methyl-2-butanol — 2,2,3-Trimethyloxirane (78%)

Section 16.11 Epoxides are much more reactive than ethers, especially in reactions that lead to cleavage of their three-membered ring.

Section 16.12 Anionic nucleophiles usually attack the less substituted carbon of the epoxide in an S_N2-like fashion.

$$\text{RCH}-\text{CR}_2 + \text{Y}^- \longrightarrow \overset{\text{Y}}{\underset{\text{O}^-}{\text{RCH}-\text{CR}_2}} \longrightarrow \overset{\text{Y}}{\underset{\text{OH}}{\text{RCH}-\text{CR}_2}}$$

Epoxide   Nucleophile   β-substituted alcohol

Nucleophile attacks this carbon.

$$\overset{\text{H}_3\text{C}}{\underset{\text{H}}{}}\text{C}-\text{C}\overset{\text{CH}_3}{\underset{\text{CH}_3}{}} \xrightarrow[\text{CH}_3\text{OH}]{\text{NaOCH}_3} \overset{\text{CH}_3\text{O}\quad\text{CH}_3}{\underset{\text{OH}}{\text{CH}_3\text{CH}-\text{CCH}_3}}$$

2,2,3-Trimethyloxirane

3-Methoxy-2-methyl-2-butanol (53%)

Section 16.13 Under conditions of acid catalysis, nucleophiles attack the carbon that can better support a positive charge. Carbocation character is developed in the transition state

$$\text{RCH}-\text{CR}_2 + \text{H}^+ \rightleftharpoons \text{RCH}-\text{CR}_2 \xrightarrow{\text{HY}} \overset{\overset{+}{\text{YH}}}{\underset{\text{OH}}{\text{RCH}-\text{CR}_2}} \xrightarrow{-\text{H}^+} \overset{\text{Y}}{\underset{\text{OH}}{\text{RCHCR}_2}}$$

Epoxide   β-substituted alcohol

Nucleophile attacks this carbon.

$$\overset{\text{H}_3\text{C}}{\underset{\text{H}}{}}\text{C}-\text{C}\overset{\text{CH}_3}{\underset{\text{CH}_3}{}} \xrightarrow[\text{H}_2\text{SO}_4]{\text{CH}_3\text{OH}} \overset{\text{OCH}_3}{\underset{\text{HO}\quad\text{CH}_3}{\text{CH}_3\text{CH}-\text{CCH}_3}}$$

2,2,3-Trimethyloxirane

3-Methoxy-3-methyl-2-butanol (76%)

Inversion of configuration is observed at the carbon that is attacked by the nucleophile, irrespective of whether the reaction takes place in acidic or basic solution.

Section 16.14 Epoxide functions are present in a great many natural products, and epoxide ring opening is sometimes a key step in the biosynthesis of other substances.

Section 16.15 Sulfides are prepared by nucleophilic substitution (S_N2) in which an alkanethiolate ion attacks an alkyl halide.

$$\text{R}\ddot{\text{S}}\text{:}^- + \text{R}-\overset{..}{\underset{..}{\text{X}}}\text{:} \longrightarrow \text{R}\ddot{\text{S}}-\text{R}' + \text{:}\overset{..}{\underset{..}{\text{X}}}\text{:}^-$$

Alkanethiolate   Alkyl halide   Sulfide   Halide

$$C_6H_5SH \xrightarrow{\text{NaOCH}_2\text{CH}_3} C_6H_5SNa \xrightarrow{\text{C}_6\text{H}_5\text{CH}_2\text{Cl}} C_6H_5SCH_2C_6H_5$$

| Benzenethiol | Sodium benzenethiolate | Benzyl phenyl sulfide (60%) |

Section 16.16 Oxidation of sulfides yields sulfoxides, then sulfones. Sodium metaperiodate is specific for the oxidation of sulfides to sulfoxides, and no further. Hydrogen peroxide or peroxy acids can yield sulfoxides (1 mole of oxidant per mole of sulfide) or sulfone (2 moles of oxidant per mole of sulfide).

R—S—R′ $\xrightarrow{\text{oxidize}}$ R—S—R′ $\xrightarrow{\text{oxidize}}$ R—S—R′

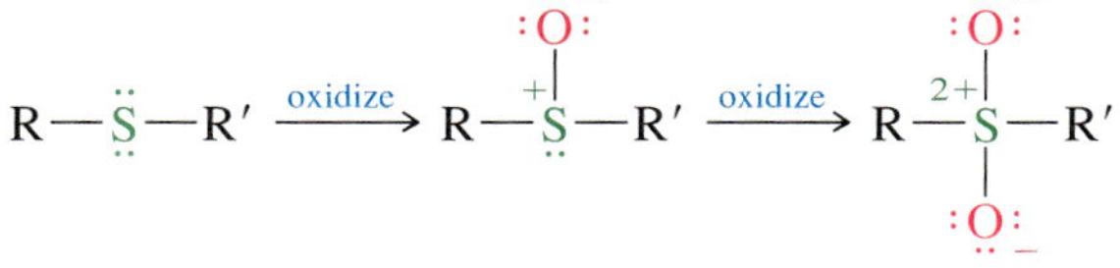

| Sulfide | Sulfoxide | Sulfone |

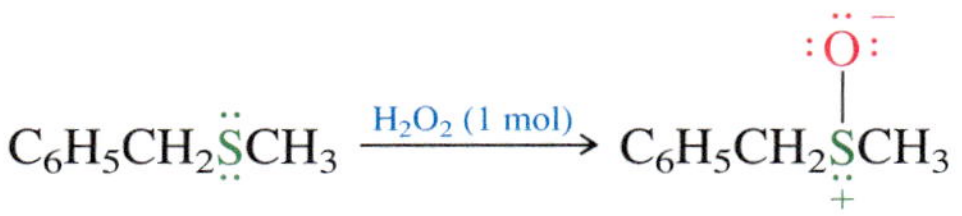

$$C_6H_5CH_2\overset{..}{\underset{..}{S}}CH_3 \xrightarrow{\text{H}_2\text{O}_2 \ (1 \ \text{mol})} C_6H_5CH_2\overset{:\overset{..}{O}:^-}{\underset{+}{S}}CH_3$$

| Benzyl methyl sulfide | Benzyl methyl sulfoxide (94%) |

Section 16.17 Sulfides react with alkyl halides to give sulfonium salts.

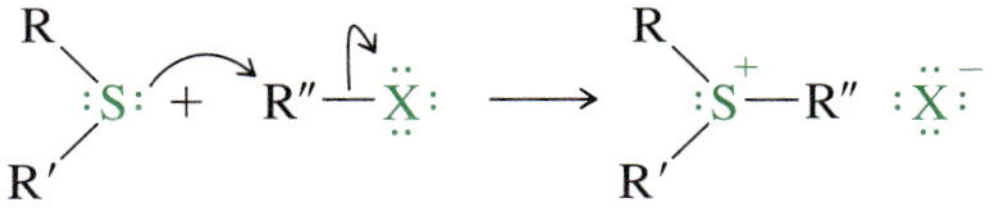

R\
 :S: + R″—X: ⟶ :S⁺—R″ :X:⁻\
R′   R′

| Sulfide | Alkyl halide | Sulfonium salt |

$$CH_3—\overset{..}{\underset{..}{S}}—CH_3 + CH_3I \longrightarrow CH_3—\overset{\overset{CH_3}{|}}{\underset{+}{S}}—CH_3 \ I^-$$

| Dimethyl sulfide | Methyl iodide | Trimethylsulfonium iodide (100%) |

Section 16.18 An H—C—O—C structural unit in an ether resembles an H—C—O—H unit of an alcohol with respect to the C—O stretching frequency in its infrared spectrum and the H—C chemical shift in its ^{1}H NMR spectrum. Because sulfur is less electronegative than oxygen, the ^{1}H and ^{13}C chemical shifts of H—C—S—C units appear at higher field than those of H—C—O—C.

PROBLEMS

16.20 Write the structures of all the constitutionally isomeric ethers of molecular formula $C_5H_{12}O$, and give an acceptable name for each.

16.21 Many ethers, including diethyl ether, are effective as general anesthetics. Because simple ethers are quite flammable, their place in medical practice has been taken by highly halogenated nonflammable ethers. Two such general anesthetic agents are *isoflurane* and *enflurane*. These compounds are isomeric; isoflurane is 1-chloro-2,2,2-trifluoroethyl difluoromethyl ether; enflurane is 2-chloro-1,1,2-trifluoroethyl difluoromethyl ether. Write the structural formulas of isoflurane and enflurane.

16.22 Although epoxides are always considered to have their oxygen atom as part of a three-membered ring, the prefix *epoxy* in the IUPAC system of nomenclature can be used to denote a cyclic ether of various sizes. Thus

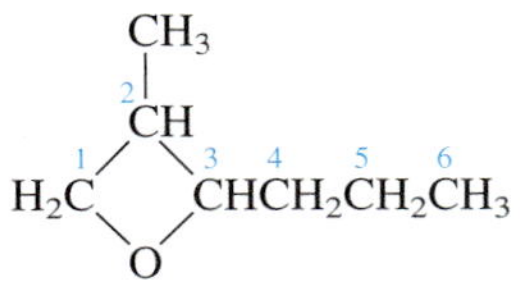

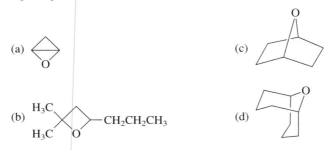

may be named 2-methyl-1,3-epoxyhexane. Using the epoxy prefix in this way, name each of the following compounds:

(a)

(c)

(b)   H₃C— , —CH₂CH₂CH₃, H₃C   O

(d)

16.23 The name of the parent six-membered sulfur-containing heterocycle is *thiane*. It is numbered beginning at sulfur. Multiple incorporation of sulfur in the ring is indicated by the prefixes *di-*, *tri-*, and so on.

    (a) How many methyl-substituted thianes are there? Which ones are chiral?

    (b) Write structural formulas for 1,4-dithiane and 1,3,5-trithiane.

    (c) Which dithiane isomer (1,2-, 1,3-, or 1,4-) is a disulfide?

    (d) Draw the two most stable conformations of the sulfoxide derived from thiane.

 16.24 The most stable conformation of 1,3-dioxan-5-ol is the chair form that has its hydroxyl group in an axial orientation. Suggest a reasonable explanation for this fact. Building a molecular model is helpful.

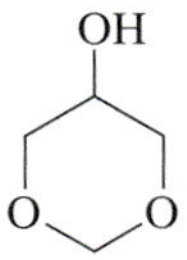

OH

O   O

1,3-Dioxan-5-ol

16.25 Outline the steps in the preparation of each of the constitutionally isomeric ethers of molecular formula $C_4H_{10}O$, starting with the appropriate alcohols. Use the Williamson ether synthesis as your key reaction.

16.26 Predict the principal organic product of each of the following reactions. Specify stereochemistry where appropriate.

(a) 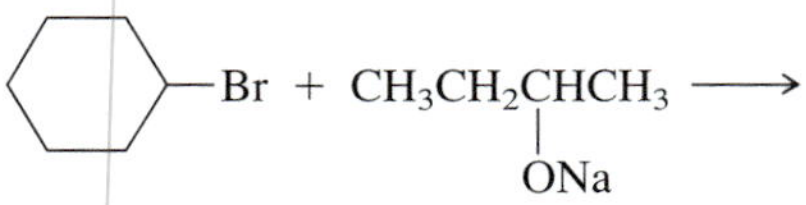   ⬡—Br + CH₃CH₂CHCH₃ ⟶
                   |
                  ONa

(b) 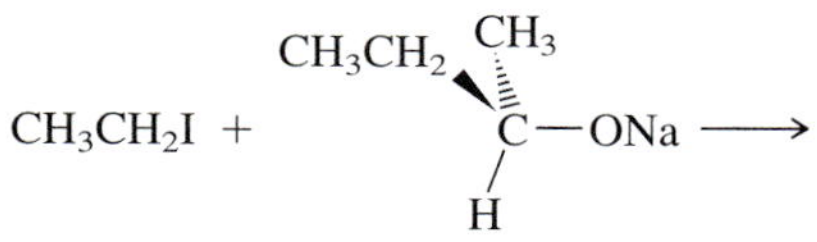 CH_3CH_2I +

(c) 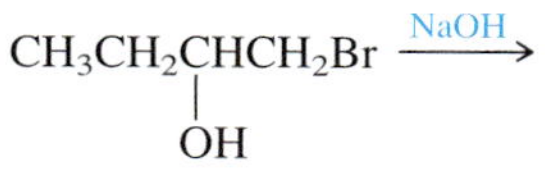 $CH_3CH_2CHCH_2Br$ $\xrightarrow{\text{NaOH}}$
     |
     OH

(d) 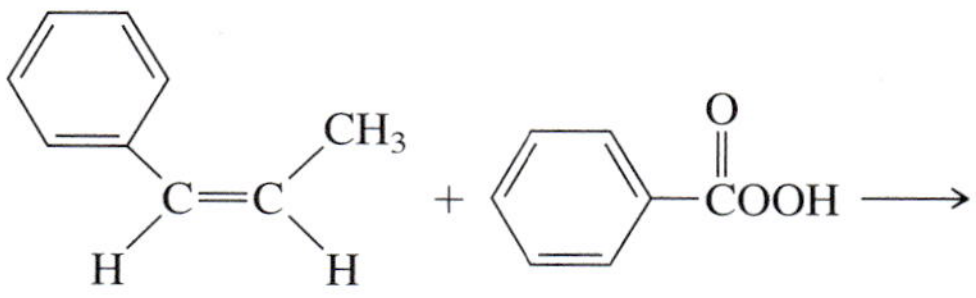 + $\longrightarrow$

(e) 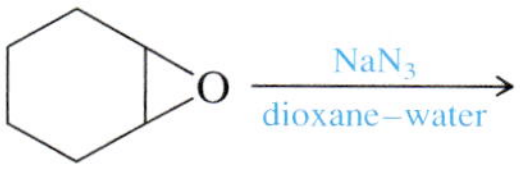 $\xrightarrow[\text{dioxane–water}]{\text{NaN}_3}$

(f) 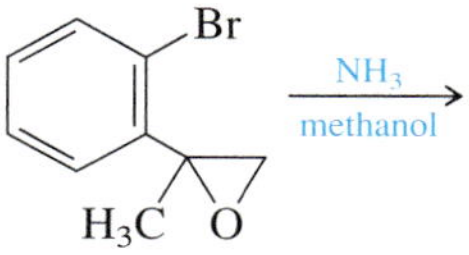 $\xrightarrow[\text{methanol}]{\text{NH}_3}$

(g) 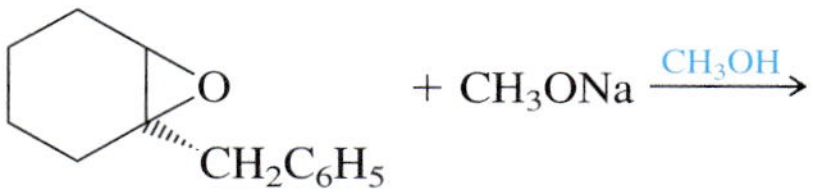 + CH_3ONa $\xrightarrow{\text{CH}_3\text{OH}}$

(h) 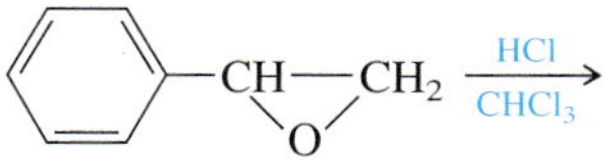 $\xrightarrow[\text{CHCl}_3]{\text{HCl}}$

(i) $CH_3(CH_2)_{16}CH_2OTs$ + $CH_3CH_2CH_2CH_2SNa$ $\longrightarrow$

(j) 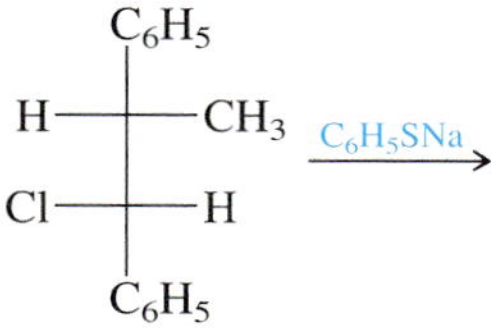 $\xrightarrow{\text{C}_6\text{H}_5\text{SNa}}$

16.27 Oxidation of 4-*tert*-butylthiane (see Problem 16.23 for the structure of thiane) with sodium metaperiodate gives a mixture of two compounds of molecular formula $C_9H_{18}OS$. Both products give the same sulfone on further oxidation with hydrogen peroxide. What is the relationship between the two compounds?

16.28 When (*R*)-(+)-2-phenyl-2-butanol is allowed to stand in methanol containing a few drops of sulfuric acid, racemic 2-methoxy-2-phenylbutane is formed. Suggest a reasonable mechanism for this reaction.

16.29 Select reaction conditions that would allow you to carry out each of the following stereospecific transformations:

(a) 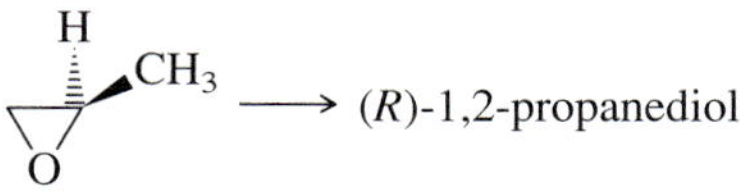 $\longrightarrow$ (*R*)-1,2-propanediol   (b) 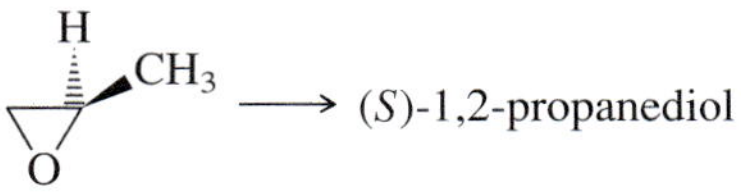 $\longrightarrow$ (*S*)-1,2-propanediol

16.30 The last step in the synthesis of divinyl ether (used as an anesthetic under the name *Vinethene*) involves heating $ClCH_2CH_2OCH_2CH_2Cl$ with potassium hydroxide. Show how you could prepare the necessary starting material $ClCH_2CH_2OCH_2CH_2Cl$ from ethylene.

16.31 Suggest short, efficient reaction sequences suitable for preparing each of the following compounds from the given starting materials and any necessary organic or inorganic reagents:

(a) 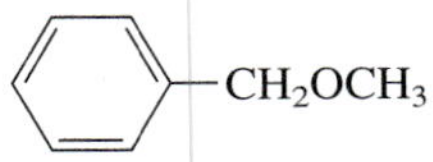 $C_6H_5\text{—}CH_2OCH_3$ from $C_6H_5\text{—}COCH_3$ (with C=O)

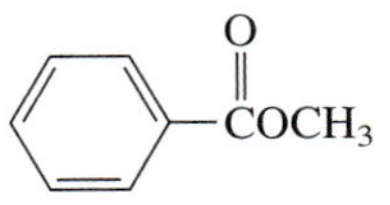

(b) 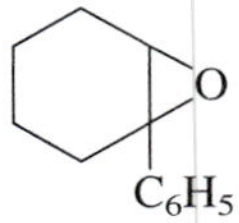 (cyclohexane epoxide, C_6H_5)   from bromobenzene and cyclohexanol

(c) $C_6H_5CH_2CHCH_3$ (OH)   from bromobenzene and isopropyl alcohol

(d) $C_6H_5CH_2CH_2CH_2OCH_2CH_3$ from benzyl alcohol and ethanol

(e) 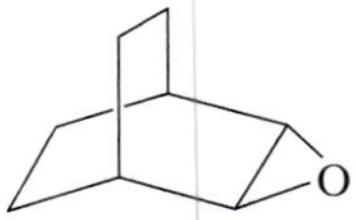 (bicyclic ether structure)   from 1,3-cyclohexadiene and ethanol

(f) $C_6H_5CHCH_2SCH_2CH_3$ (OH)   from styrene and ethanol

16.32 Among the ways in which 1,4-dioxane may be prepared are the methods expressed in the equations shown:

(a) $2HOCH_2CH_2OH \xrightarrow[\text{heat}]{H_2SO_4}$ 1,4-Dioxane $+ 2H_2O$

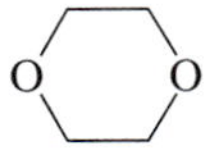

Ethylene glycol   1,4-Dioxane   Water

(b) $ClCH_2CH_2OCH_2CH_2Cl \xrightarrow{NaOH}$ 1,4-Dioxane

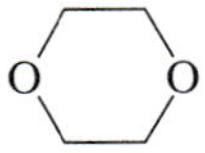

Bis(2-chloroethyl) ether   1,4-Dioxane

Suggest reasonable mechanisms for each of these reactions.

16.33 Deduce the identity of the missing compounds in the following reaction sequences. Show stereochemistry in parts (b) through (d).

(a) $H_2C{=}CHCH_2Br \xrightarrow[\text{2. }H_2C=O]{\text{1. Mg}}{}_{\text{3. }H_3O^+}$ Compound A (C_4H_8O) $\xrightarrow{Br_2}$ Compound B $(C_4H_8Br_2O)$ $\xrightarrow{\text{KOH, 25°C}}$ Compound C (C_4H_7BrO) $\xrightarrow[\text{heat}]{KOH}$ 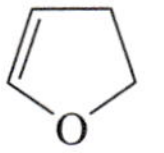 Compound D

(b)
$$\text{Cl}\overset{\text{CO}_2\text{H}}{\underset{\text{CH}_3}{\rule{0pt}{0pt}}}\text{H}$$
$\xrightarrow[\text{2. H}_2\text{O}]{\text{1. LiAlH}_4}$ Compound E $\xrightarrow{\text{KOH, H}_2\text{O}}$ Compound F
(C$_3$H$_7$ClO)   (C$_3$H$_6$O)

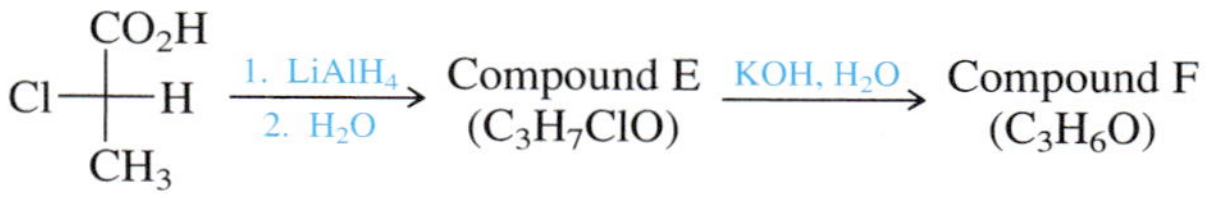

(c)
$$\overset{\text{CH}_3}{\underset{\text{CH}_3}{\overset{\text{H}---\text{Cl}}{\text{H}---\text{OH}}}}$$
$\xrightarrow{\text{NaOH}}$ Compound G $\xrightarrow{\text{NaSCH}_3}$ Compound H
(C$_4$H$_8$O)   (C$_5$H$_{12}$OS)

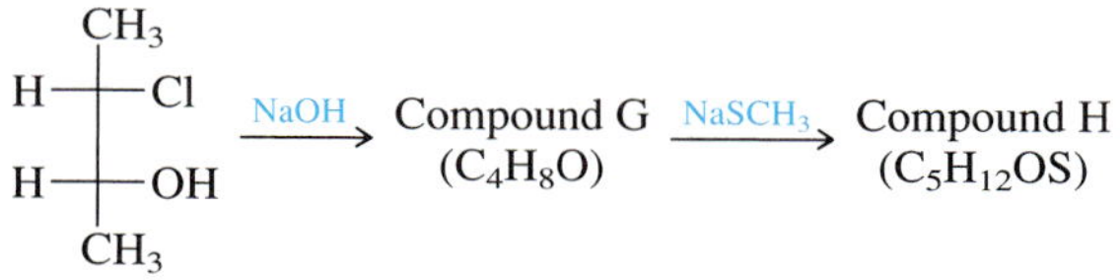

(d) Compound I (C$_7$H$_{12}$) $\xrightarrow[\text{(CH}_3)_3\text{COH, HO}]{\text{OsO}_4\text{, (CH}_3)_3\text{COOH}}$ Compound J (C$_7$H$_{14}$O$_2$)

↓ C$_6$H$_5$CO$_2$OH   (a liquid)

$$\text{H}_3\text{C}\diagdown\underset{\text{O}}{\bigtriangleup}\diagup\text{CH}_3$$ $\xrightarrow[\text{H}_2\text{SO}_4]{\text{H}_2\text{O}}$ Compound L (C$_7$H$_{14}$O$_2$)

(mp 99.5–101°C)

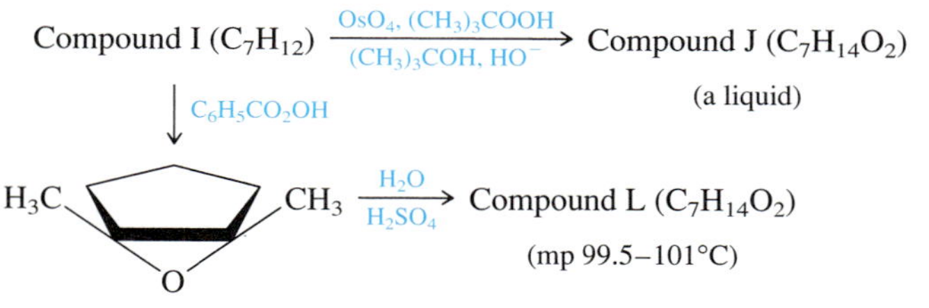

Compound K

16.34 Cineole is the chief component of eucalyptus oil; it has the molecular formula C$_{10}$H$_{18}$O and contains no double or triple bonds. It reacts with hydrochloric acid to give the dichloride shown:

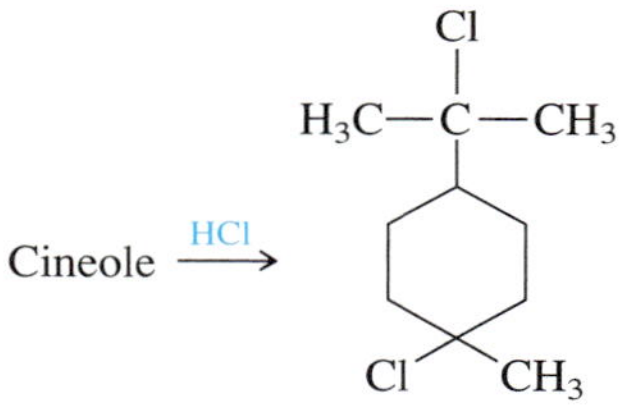

Cineole $\xrightarrow{\text{HCl}}$

$$\text{H}_3\text{C}-\overset{\overset{\text{Cl}}{|}}{\text{C}}-\text{CH}_3$$

Deduce the structure of cineole.

16.35 The *p*-toluenesulfonate shown undergoes an intramolecular Williamson reaction on treatment with base to give a spirocyclic ether. Demonstrate your understanding of the terminology used in the preceding sentence by writing the structure, including stereochemistry, of the product.

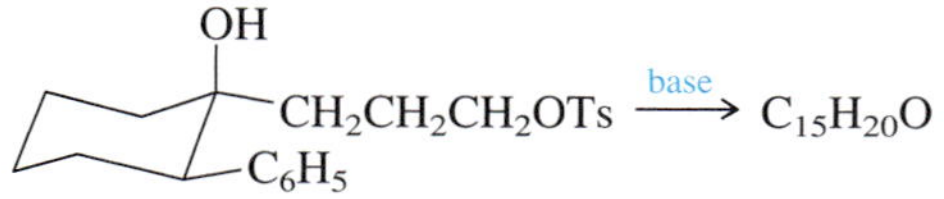

$\xrightarrow{\text{base}}$ C$_{15}$H$_{20}$O

—CH$_2$CH$_2$CH$_2$OTs
—C$_6$H$_5$

16.36 This problem is adapted from an experiment designed for undergraduate organic chemistry laboratories published in the January 2001 issue of the *Journal of Chemical Education*, pp. 77–78,

(a) Reaction of (*E*)-1-(*p*-methoxyphenyl)propene with *m*-chloroperoxybenzoic acid converted the alkene to its corresponding epoxide. Give the structure, including stereochemistry, of this epoxide.

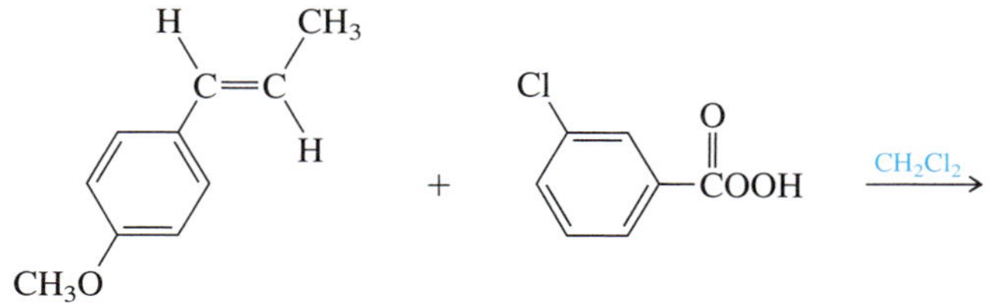

$$\underset{\text{CH}_3\text{O}}{\rule{0pt}{0pt}}\overset{\text{H}\quad\text{CH}_3}{\underset{\text{H}}{\text{C}=\text{C}}}$$

+
$$\overset{\text{Cl}}{\underset{}{\rule{0pt}{0pt}}}\overset{\text{O}}{\underset{}{-\overset{||}{\text{C}}\text{OOH}}}$$
$\xrightarrow{\text{CH}_2\text{Cl}_2}$

(*E*)-1-(*p*-Methoxyphenyl)propene   *m*-Chloroperoxybenzoic acid

(b) Assign the signals in the ^{1}H NMR spectrum of the epoxide to the appropriate hydrogens.

δ 1.4 (doublet, 3H)   δ 3.8 (singlet, 3H)

δ 3.0 (quartet of doublets, 1H)   δ 6.9 (doublet, 2H)

δ 3.5 (doublet, 1H)   δ 7.2 (doublet, 2H)

(c) Three signals appear in the range δ 55–60 in the ^{13}C NMR spectrum of the epoxide. To which carbons of the epoxide do these signals correspond?

(d) The epoxide is isolated only when the reaction is carried out under conditions (added Na_2CO_3) that ensure that the reaction mixture does not become acidic. Unless this precaution is taken, the isolated product has the molecular formula $C_{17}H_{17}O_4Cl$. Suggest a reasonable structure for this product and write a reasonable mechanism for its formation.

16.37 All the following questions pertain to ^{1}H NMR spectra of isomeric ethers having the molecular formula $C_5H_{12}O$.

(a) Which one has only singlets in its ^{1}H NMR spectrum?

(b) Along with other signals, this ether has a coupled doublet–septet pattern. None of the protons responsible for this pattern are coupled to protons anywhere else in the molecule. Identify this ether.

(c) In addition to other signals in its ^{1}H NMR spectrum, this ether exhibits two signals at relatively low field. One is a singlet; the other is a doublet. What is the structure of this ether?

(d) In addition to other signals in its ^{1}H NMR spectrum, this ether exhibits two signals at relatively low field. One is a triplet; the other is a quartet. Which ether is this?

16.38 The ^{1}H NMR spectrum of compound A (C_8H_8O) consists of two singlets of equal area at δ 5.1 (sharp) and 7.2 ppm (broad). On treatment with excess hydrogen bromide, compound A is converted to a single dibromide ($C_8H_8Br_2$). The ^{1}H NMR spectrum of the dibromide is similar to that of A in that it exhibits two singlets of equal area at δ 4.7 (sharp) and 7.3 ppm (broad). Suggest reasonable structures for compound A and the dibromide derived from it.

16.39 The ^{1}H NMR spectrum of a compound ($C_{10}H_{13}BrO$) is shown in Figure 16.9. The compound gives benzyl bromide, along with a second compound $C_3H_6Br_2$, when heated with HBr. What is the first compound?

16.40 A compound is a cyclic ether of molecular formula $C_9H_{10}O$. Its ^{13}C NMR spectrum is shown in Figure 16.10. Oxidation of the compound with sodium dichromate and sulfuric acid gave 1,2-benzenedicarboxylic acid. What is the compound?

16.41 Make a molecular model of dimethyl sulfide. How does its bond angle at sulfur compare with the C—O—C bond angle in dimethyl ether?

16.42 View molecular models of dimethyl ether and ethylene oxide on *Learning By Modeling*. Which one has the greater dipole moment? Do the calculated dipole moments bear any relationship to the observed boiling points (ethylene oxide: +10°C; dimethyl ether: −25°C)?

16.43 Find the molecular model of 18-crown-6 (see Figure 16.2) on *Learning By Modeling*, and examine its electrostatic potential map. View the map in various modes (dots, contours, and as a transparent surface). Does 18-crown-6 have a dipole moment? Are vicinal oxygens anti or gauche to one another?

16.44 Find the model of dimethyl sulfoxide [$(CH_3)_2S$=O] on *Learning By Modeling*, and examine its electrostatic potential map. To which atom (S or O) would you expect a proton to bond?

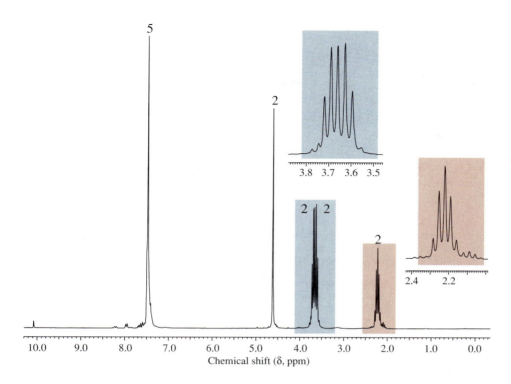

FIGURE 16.9 The 200-MHz
^{1}H NMR spectrum of a compound, $C_{10}H_{13}BrO$ (Problem 16.39). The integral ratios of the signals reading from left to right (low field to high field) are 5:2:2:2:2. The signals centered at δ 3.6 and δ 3.7 are two overlapping triplets.

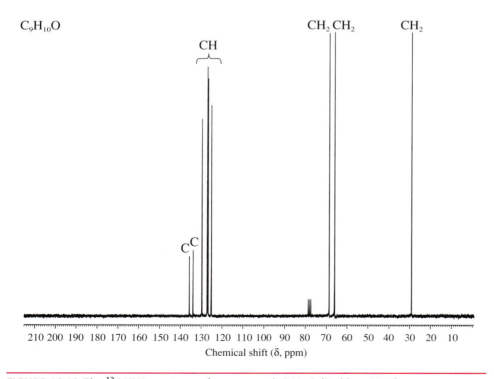

FIGURE 16.10 The ^{13}C NMR spectrum of a compound, $C_9H_{10}O$ (Problem 16.40).

16.45 Construct a molecular model of *trans*-2-bromocyclohexanol in its most stable conformation. This conformation is ill-suited to undergo epoxide formation on treatment with base. Why? What must happen in order to produce 1,2-epoxycyclohexane from *trans*-2-bromocyclohexanol?

16.46 Construct a molecular model of *threo*-3-bromo-2-butanol. What is the stereochemistry (cis or trans) of the 2,3-epoxybutane formed on treatment of *threo*-3-bromo-2-butanol with base? Repeat the exercise for *erythro*-3-bromo-2-butanol.

16.47 Use *Learning By Modeling* to compare the C—O bond distances in 1,2-epoxypropane and its protonated form.

$$\text{CH}_3$$

1,2-Epoxypropane

How do the bond distances of 1,2-epoxypropane change on protonation of the ring oxygen? Assuming that the longer C—O bond is the weaker of the two, do the bond distances in the protonated form correlate with the regioselectivity of acid-catalyzed ring opening?

16.48 Predict which carbon undergoes nucleophilic attack on acid-catalyzed ring opening of *cis*-3,3,3-trifluoro-2,3-epoxybutane. Examine the C—O bond distances of the protonated form of the epoxide on *Learning By Modeling*. How do these bond distances compare with your prediction?

$$\text{F}_3\text{C} \qquad \text{CH}_3$$

cis-1,1,1-Trifluoro-2,3-epoxybutane

CHAPTER 17

ALDEHYDES AND KETONES: NUCLEOPHILIC ADDITION TO THE CARBONYL GROUP

$$\text{O}$$
$$\|$$

Aldehydes and ketones contain an acyl group RC— bonded either to hydrogen or to another carbon.

$$\text{O} \qquad \text{O} \qquad \text{O}$$
$$\| \qquad\quad \| \qquad\quad \|$$
$$\text{HCH} \qquad \text{RCH} \qquad \text{RCR}'$$

Formaldehyde · · · · · Aldehyde · · · · · Ketone

Although the present chapter includes the usual collection of topics designed to acquaint us with a particular class of compounds, its central theme is a fundamental reaction type, *nucleophilic addition to carbonyl groups*. The principles of nucleophilic addition to aldehydes and ketones developed here will be seen to have broad applicability in later chapters when transformations of various derivatives of carboxylic acids are discussed.

17.1 NOMENCLATURE

$$\text{O}$$
$$\|$$

The longest continuous chain that contains the —CH group provides the base name for aldehydes. The -*e* ending of the corresponding alkane name is replaced by -*al,* and substituents are specified in the usual way. It is not necessary to specify the location of

$$\text{O}$$
$$\|$$

the —CH group in the name, because the chain must be numbered by starting with this group as C-1. The suffix -*dial* is added to the appropriate alkane name when the compound contains two aldehyde functions.*

---

*The -*e* ending of an alkane name is dropped before a suffix beginning with a vowel (-*al*) and retained before one beginning with a consonant (-*dial*).

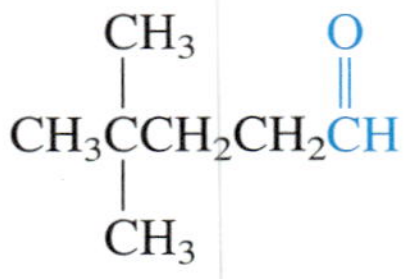

$$\underset{\overset{\displaystyle |}{CH_3}}{\overset{\overset{\displaystyle CH_3}{|}}{CH_3CCH_2CH_2CH}} \overset{\overset{\displaystyle O}{\|}}{}$$

$$\underset{}{H_2C{=}CHCH_2CH_2CH_2CH} \overset{\overset{\displaystyle O}{\|}}{}$$

$$\underset{}{HCCHCH_2CH} \overset{\overset{\displaystyle O}{\|}}{} \overset{\overset{\displaystyle O}{\|}}{}$$

4,4-Dimethylpentanal    5-Hexenal 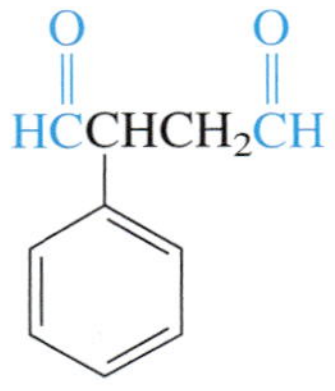   2-Phenylbutanedial

Notice that, because they define the ends of the carbon chain in 2-phenylbutanedial, the aldehyde positions are not designated by numerical locants in the name.

When a formyl group (—CH=O) is attached to a ring, the ring name is followed by the suffix *-carbaldehyde*.

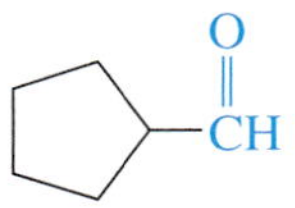

Cyclopentanecarbaldehyde 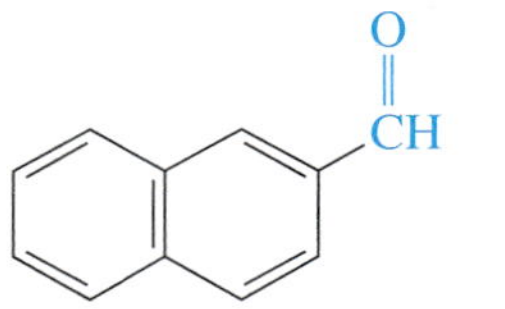   2-Naphthalenecarbaldehyde

Certain common names of familiar aldehydes are acceptable as IUPAC names. A few examples include

$$\underset{}{HCH} \overset{\overset{\displaystyle O}{\|}}{}$$

$$\underset{}{CH_3CH} \overset{\overset{\displaystyle O}{\|}}{}$$

Formaldehyde   Acetaldehyde   Benzaldehyde
(methanal)   (ethanal)   (benzenecarbaldehyde)

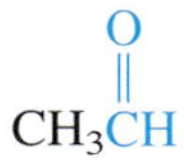

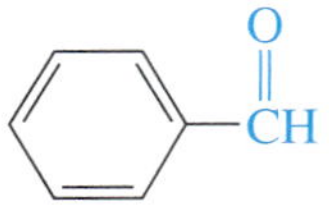

Among oxygen containing groups, a higher oxidation state takes precedence over a lower one in determining the suffix of the substitutive name. Thus, a compound that contains both an alcohol and an aldehyde function is named as an aldehyde.

$$\underset{}{HOCH_2CH_2CH_2CH_2CH} \overset{\overset{\displaystyle O}{\|}}{}$$

5-Hydroxypentanal   *trans*-4-Hydroxycyclohexanecarbaldehyde   *p*-Hydroxybenzaldehyde

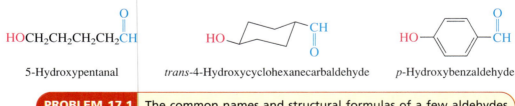

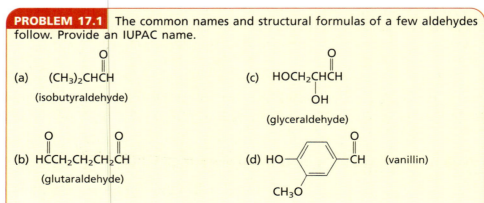

PROBLEM 17.1 The common names and structural formulas of a few aldehydes follow. Provide an IUPAC name.

(a) $(CH_3)_2CHCH$ $\overset{\overset{\displaystyle O}{\|}}{}$

(isobutyraldehyde)

(b) $HCCH_2CH_2CH_2CH$ $\overset{\overset{\displaystyle O}{\|}}{}\quad\overset{\overset{\displaystyle O}{\|}}{}$

(glutaraldehyde)

(c) $HOCH_2CHCH$ $\overset{\overset{\displaystyle O}{\|}}{}$ with OH below

(glyceraldehyde)

(d) HO— (ring) —CH with CH_3O (vanillin)

SAMPLE SOLUTION (a) Don't be fooled by the fact that the common name is isobutyraldehyde. The longest continuous chain has three carbons, and so the base name is *propanal.* There is a methyl group at C-2; thus the compound is 2-methylpropanal.

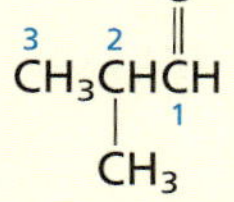

$$\underset{\substack{|\\ CH_3 \\ 1}}{\overset{\overset{O}{\|}}{\underset{3}{CH_3}\underset{2}{CH}CH}}$$

2-Methylpropanal
(isobutyraldehyde)

With ketones, the *-e* ending of an alkane is replaced by *-one* in the longest continuous chain containing the carbonyl group. The chain is numbered in the direction that provides the lower number for this group. The carbonyl carbon of a cyclic ketone is C-1 and the number does not appear in the name.

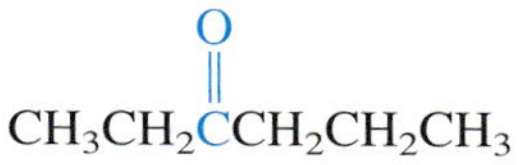

$CH_3CH_2\overset{O}{\overset{\|}{C}}CH_2CH_2CH_3$ $CH_3\underset{CH_3}{CH}CH_2\overset{O}{\overset{\|}{C}}CH_3$ (4-Methylcyclohexanone structure)

3-Hexanone 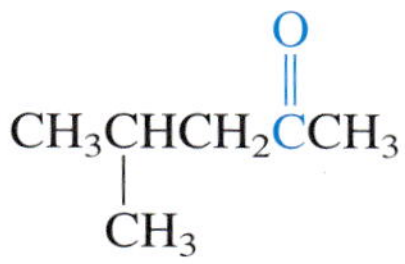   4-Methyl-2-pentanone 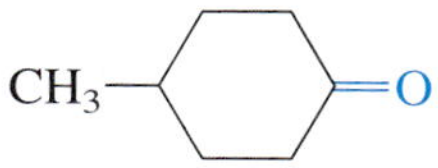   4-Methylcyclohexanone

Like aldehydes, ketone functions take precedence over alcohol functions, double bonds, halogens, and alkyl groups in determining the parent name and direction of numbering. Aldehydes outrank ketones, however, and a compound that contains both an aldehyde and a ketone carbonyl group is named as an aldehyde. In such cases, the carbonyl oxygen of the ketone is considered an *oxo*-substituent on the main chain.

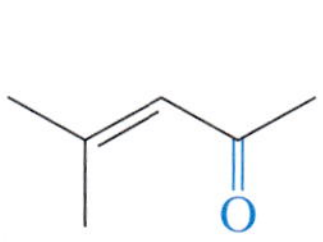

4-Methyl-3-penten-2-one 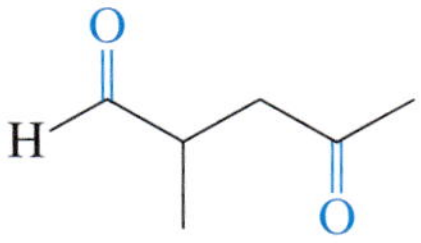   2-Methyl-4-oxopentanal

Although substitutive names of the type just described are preferred, the IUPAC rules also permit ketones to be named by functional class nomenclature. The groups attached to the carbonyl group are named as separate words followed by the word *ketone.* The groups are listed alphabetically.

There are no functional class names for aldehydes in the IUPAC system.

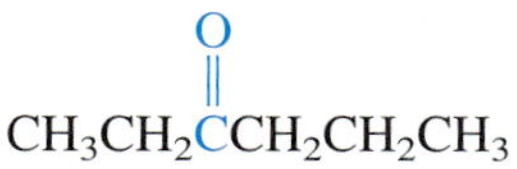

$CH_3CH_2\overset{O}{\overset{\|}{C}}CH_2CH_2CH_3$ (benzene ring)—$CH_2\overset{O}{\overset{\|}{C}}CH_2CH_3$ $H_2C{=}CH\overset{O}{\overset{\|}{C}}CH{=}CH_2$

Ethyl propyl ketone 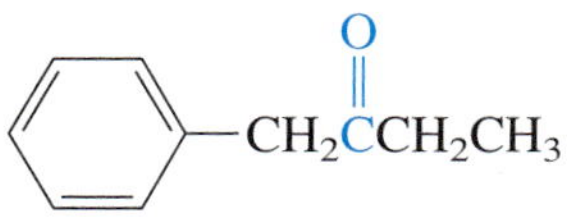   Benzyl ethyl ketone    Divinyl ketone

PROBLEM 17.2 Convert each of the following functional class IUPAC names to a substitutive name.
(a) Dibenzyl ketone   (b) Ethyl isopropyl ketone

(c) Methyl 2,2-dimethylpropyl ketone   (d) Allyl methyl ketone

SAMPLE SOLUTION   (a) First write the structure corresponding to the name. Dibenzyl ketone has two benzyl groups attached to a carbonyl.

$$\text{—CH}_2\overset{\displaystyle O}{\overset{\displaystyle \|}{\underset{2}{C}}}\text{CH}_2\text{—}$$

Dibenzyl ketone

The longest continuous chain contains three carbons, and C-2 is the carbon of the carbonyl group. The substitutive IUPAC name for this ketone is *1,3-diphenyl-2-propanone*.

A few of the common names acceptable for ketones in the IUPAC system are

$$\underset{\text{Acetone}}{CH_3\overset{\displaystyle O}{\overset{\displaystyle \|}{C}}CH_3}\qquad\qquad \underset{\text{Acetophenone}}{\overset{\displaystyle O}{\overset{\displaystyle \|}{C}}CH_3}\qquad\qquad \underset{\text{Benzophenone}}{\overset{\displaystyle O}{\overset{\displaystyle \|}{C}}}$$

(The suffix *-phenone* indicates that the acyl group is attached to a benzene ring.)

17.2   STRUCTURE AND BONDING: THE CARBONYL GROUP

Two notable aspects of the carbonyl group are its *geometry* and *polarity*. The coplanar geometry of the bonds to the carbonyl group is seen in the molecular models of formaldehyde, acetaldehyde, and acetone in Figure 17.1. The bond angles involving the carbonyl group are approximately 120°, but vary somewhat from compound to compound as shown by the examples in Figure 17.1. The C=O bond distance in aldehydes and ketones is 122 pm, which is significantly shorter than the typical C—O bond distance of 141 pm seen in alcohols and ethers.

Bonding in formaldehyde can be described according to an sp^2-hybridization model analogous to that of ethylene (Figure 17.2). According to this model, the carbon–

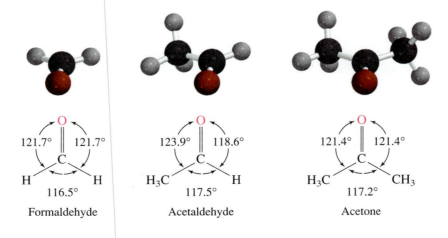

FIGURE 17.1 The bonds to the carbon of the carbonyl group lie in the same plane and at angles of approximately 120° with respect to each other.

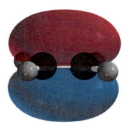

(a) Ethylene   (b) Formaldehyde

FIGURE 17.2 Both
(a) ethylene and (b) formaldehyde have the same num-
ber of electrons, and carbon
is sp^2-hybridized in both. In
formaldehyde, one of the
carbons is replaced by an
sp^2-hybridized oxygen. Like
the carbon–carbon double
bond of ethylene, the
carbon–oxygen double bond
of formaldehyde is com-
posed of a σ component and
a π component.

oxygen double bond is viewed as one of the $\sigma + \pi$ type. Overlap of half-filled sp^2 hybrid orbitals of carbon and oxygen gives the σ component, whereas side-by-side overlap of half-filled $2p$ orbitals gives the π bond. The oxygen lone pairs of formaldehyde occupy sp^2 hybrid orbitals, the axes of which lie in the plane of the molecule.

The carbonyl group makes aldehydes and ketones rather polar molecules, with dipole moments that are substantially higher than alkenes.

$$CH_3CH_2CH{=}CH_2 \qquad CH_3CH_2CH{=}O$$

1-Butene   Propanal
Dipole moment: 0.3 D   Dipole moment: 2.5 D

How much a carbonyl group affects the charge distribution in a molecule is apparent in the electrostatic potential maps of 1-butene and propanal (Figure 17.3). When the color scale is adjusted to be the same for both molecules, the much greater separation of positive and negative charge in propanal relative to 1-butene is readily apparent. The region of greatest negative potential in 1-butene is associated with the π electrons of the double bond, and MO calculations indicate a buildup of negative charge on the doubly bonded carbons. On the other hand, both the electrostatic potential map and MO calculations show the carbonyl carbon of propanal as positively polarized and the oxygen as negatively polarized.

The various ways of representing the polarization in a carbonyl group include

$$\overset{\delta+}{\underset{}{C}}{=}\overset{\delta-}{O} \qquad or \qquad \overset{+\longrightarrow}{C}{=}O$$

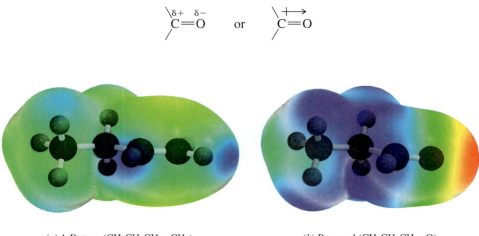

(a) 1-Butene ($CH_3CH_2CH{=}CH_2$)   (b) Propanal ($CH_3CH_2CH{=}O$)

FIGURE 17.3 Elec-
trostatic potential maps
of (a) 1-butene and
(b) propanal. The color
ranges are adjusted to a
common scale so that the
charge distributions in the
two compounds can be com-
pared directly. The region of
highest negative potential
in 1-butene is associated
with the π electrons of the
double bond. The charge
separation is greater in
propanal. The carbon of the
carbonyl group is a site of
positive potential. The
region of highest negative
potential is near oxygen.

and the resonance description.

$$\overset{\diagup}{\underset{\diagup}{C}}=\overset{..}{\underset{..}{O}}: \longleftrightarrow \overset{\diagup}{\underset{\diagup}{^+C}}-\overset{..}{\underset{..}{O}}:^-$$

The structural features, especially the very polar nature of the carbonyl group, point clearly to the kind of chemistry we will see for aldehydes and ketones in this chapter. The partially positive carbon of C=O has carbocation character and is electrophilic. The planar arrangement of its bonds make this carbon relatively uncrowded and susceptible to attack by nucleophiles. Oxygen is partially negative and weakly basic.

nucleophiles
attack carbon

$$\overset{..}{\underset{\diagdown}{C}}=\overset{..}{\underset{..}{O}}: \longrightarrow$$ electrophiles, especially
protons, bond to oxygen

Alkyl substituents stabilize a carbonyl group in much the same way that they stabilize carbon–carbon double bonds and carbocations—by releasing electrons to sp^2-hybridized carbon. Thus, as their heats of combustion reveal, the ketone 2-butanone is more stable than its aldehyde isomer butanal.

$$\overset{O}{\overset{\|}{CH_3CH_2CH_2CH}}$$

Butanal

$$\overset{O}{\overset{\|}{CH_3CH_2CCH_3}}$$

2-Butanone

Heat of combustion: 2475 kJ/mol (592 kcal/mol)   2442 kJ/mol (584 kcal/mol)

The carbonyl carbon of a ketone bears two electron-releasing alkyl groups; an aldehyde carbonyl group has only one. Just as a disubstituted double bond in an alkene is more stable than a monosubstituted double bond, a ketone carbonyl is more stable than an aldehyde carbonyl. We'll see later in this chapter that structural effects on the relative *stability* of carbonyl groups in aldehydes and ketones are an important factor in their relative *reactivity*.

17.3 PHYSICAL PROPERTIES

Physical constants such as melting point, boiling point, and solubility in water are collected for a variety of aldehydes and ketones in Appendix 1.

In general, aldehydes and ketones have higher boiling points than alkenes because they are more polar and the dipole–dipole attractive forces between molecules are stronger. But they have lower boiling points than alcohols because, unlike alcohols, two carbonyl groups can't form hydrogen bonds to each other.

	$CH_3CH_2CH{=}CH_2$	$CH_3CH_2CH{=}O$	$CH_3CH_2CH_2OH$
	1-Butene	Propanal	1-Propanol
bp (1 atm)	−6°C	49°C	97°C
Solubility in water (g/100 mL)	Negligible	20	Miscible in all proportions

The carbonyl oxygen of aldehydes and ketones can form hydrogen bonds with the protons of OH groups. This makes them more soluble in water than alkenes, but less soluble than alcohols.

PROBLEM 17.3 Sketch the hydrogen bonding between benzaldehyde and water.

17.4 SOURCES OF ALDEHYDES AND KETONES

As we'll see later in this chapter and the next, aldehydes and ketones are involved in many of the most used reactions in synthetic organic chemistry. Where do aldehydes and ketones come from?

Many occur naturally. In terms of both variety and quantity, aldehydes and ketones rank among the most common and familiar natural products. Several are shown in Figure 17.4.

Many aldehydes and ketones are made in the laboratory from alkenes, alkynes, arenes, and alcohols by reactions that you already know about and are summarized in Table 17.1.

To the synthetic chemist, the most important of the reactions in Table 17.1 are the last two: the oxidation of primary alcohols to aldehydes and secondary alcohols to ketones. *Indeed, when combined with reactions that yield alcohols, the oxidation methods are so versatile that it will not be necessary to introduce any new methods for preparing aldehydes and ketones in this chapter.* A few examples will illustrate this point.

Let's first consider how to prepare an aldehyde from a carboxylic acid. There are no good methods for going from RCO_2H to $RCHO$ directly. Instead, we do it indirectly

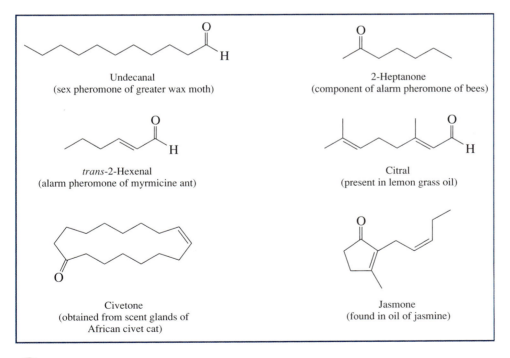

Undecanal
(sex pheromone of greater wax moth)

2-Heptanone
(component of alarm pheromone of bees)

trans-2-Hexenal
(alarm pheromone of myrmicine ant)

Citral
(present in lemon grass oil)

Civetone
(obtained from scent glands of African civet cat)

Jasmone
(found in oil of jasmine)

 FIGURE 17.4 Some naturally occurring aldehydes and ketones.

Reaction (section) and comments	General equation and specific example
Ozonolysis of alkenes (Section 6.19) This cleavage reaction is more often seen in structural analysis than in synthesis. The substitution pattern around a double bond is revealed by identifying the carbonyl-containing compounds that make up the product. Hydrolysis of the ozonide intermediate in the presence of zinc (reductive workup) permits aldehyde products to be isolated without further oxidation.	$$\underset{\text{Alkene}}{\overset{R}{\underset{R'}{>}}C=C\overset{H}{\underset{R''}{<}}} \xrightarrow[\text{2. H}_2\text{O, Zn}]{\text{1. O}_3} \underset{\text{Two carbonyl compounds}}{RCR' + R''CH}$$ $$\underset{\text{2,6-Dimethyl-2-octene}}{} \xrightarrow[\text{2. H}_2\text{O, Zn}]{\text{1. O}_3} \underset{\text{Acetone}}{CH_3CCH_3} + \underset{\text{4-Methylhexanal (91\%)}}{HCCH_2CH_2CHCH_2CH_3}$$
Hydration of alkynes (Section 9.12) Reaction occurs by way of an enol intermediate formed by Markovnikov addition of water to the triple bond.	$$\underset{\text{Alkyne}}{RC\equiv CR'} + H_2O \xrightarrow[\text{HgSO}_4]{\text{H}_2\text{SO}_4} \underset{\text{Ketone}}{RCCH_2R'}$$ $$\underset{\text{1-Octyne}}{HC\equiv C(CH_2)_5CH_3} + H_2O \xrightarrow[\text{HgSO}_4]{\text{H}_2\text{SO}_4} \underset{\text{2-Octanone (91\%)}}{CH_3C(CH_2)_5CH_3}$$
Friedel–Crafts acylation of aromatic compounds (Section 12.7) Acyl chlorides and carboxylic acid anhydrides acylate aromatic rings in the presence of aluminum chloride. The reaction is electrophilic aromatic substitution in which acylium ions are generated and attack the ring.	$$ArH + RCCl \xrightarrow{\text{AlCl}_3} ArCR + HCl \quad \text{or}$$ $$ArH + RCOCR \xrightarrow{\text{AlCl}_3} ArCR + RCO_2H$$ $$\underset{\text{Anisole}}{CH_3O-\bigcirc} + \underset{\text{Acetic anhydride}}{CH_3COCCH_3} \xrightarrow{\text{AlCl}_3} \underset{\substack{p\text{-Methoxyacetophenone}\\(90\text{--}94\%)}}{CH_3O-\bigcirc-CCH_3}$$
Oxidation of primary alcohols to aldehydes (Section 15.10) Pyridinium dichromate (PDC) or pyridinium chlorochromate (PCC) in anhydrous media such as dichloromethane oxidizes primary alcohols to aldehydes while avoiding overoxidation to carboxylic acids.	$$\underset{\text{Primary alcohol}}{RCH_2OH} \xrightarrow[\text{CH}_2\text{Cl}_2]{\text{PDC or PCC}} \underset{\text{Aldehyde}}{RCH}$$ $$\underset{\text{1-Decanol}}{CH_3(CH_2)_8CH_2OH} \xrightarrow[\text{CH}_2\text{Cl}_2]{\text{PDC}} \underset{\text{Decanal (98\%)}}{CH_3(CH_2)_8CH}$$
Oxidation of secondary alcohols to ketones (Section 15.10) Many oxidizing agents are available for converting secondary alcohols to ketones. PDC or PCC may be used, as well as other Cr(VI)-based agents such as chromic acid or potassium dichromate and sulfuric acid.	$$\underset{\text{Secondary alcohol}}{\underset{\overset{|}{OH}}{RCHR'}} \xrightarrow{\text{Cr(VI)}} \underset{\text{Ketone}}{RCR'}$$ $$\underset{\text{1-Phenyl-1-pentanol}}{\underset{\overset{|}{OH}}{C_6H_5CHCH_2CH_2CH_2CH_3}} \xrightarrow[\substack{\text{acetic acid/}\\\text{water}}]{\text{CrO}_3} \underset{\text{1-Phenyl-1-pentanone (93\%)}}{C_6H_5CCH_2CH_2CH_2CH_3}$$

by first reducing the carboxylic acid to the corresponding primary alcohol, then oxidizing the primary alcohol to the aldehyde.

$$RCO_2H \xrightarrow{\text{reduce}} RCH_2OH \xrightarrow{\text{oxidize}} \overset{O}{\overset{\|}{R\overset{}{C}H}}$$

Carboxylic acid   Primary alcohol   Aldehyde

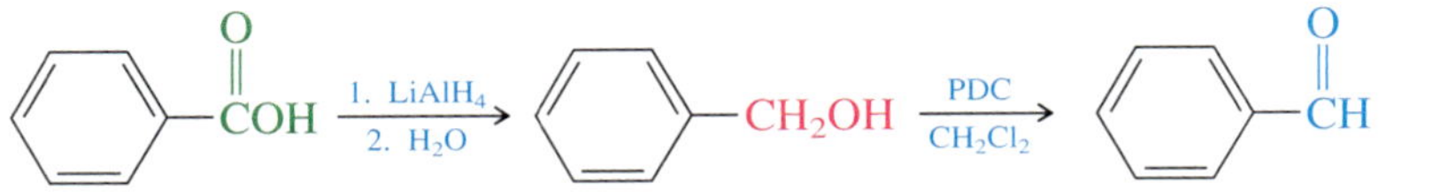

Benzoic acid   Benzyl alcohol (81%)   Benzaldehyde (83%)

> **PROBLEM 17.4** Can catalytic hydrogenation be used to reduce a carboxylic acid to a primary alcohol in the first step of this sequence?

It is often necessary to prepare ketones by processes involving carbon–carbon bond formation. In such cases the standard method combines addition of a Grignard reagent to an aldehyde with oxidation of the resulting secondary alcohol:

$$\overset{O}{\overset{\|}{R\overset{}{C}H}} \xrightarrow[\text{2. } H_3O^+]{\text{1. } R'MgX, \text{ diethyl ether}} \overset{OH}{\overset{|}{R\overset{}{C}HR'}} \xrightarrow{\text{oxidize}} \overset{O}{\overset{\|}{R\overset{}{C}R'}}$$

Aldehyde   Secondary alcohol   Ketone

$$\overset{O}{\overset{\|}{CH_3CH_2\overset{}{C}H}} \xrightarrow[\text{2. } H_3O^+]{\substack{\text{1. } CH_3(CH_2)_3MgBr \\ \text{diethyl ether}}} \overset{OH}{\overset{|}{CH_3CH_2\overset{}{C}H(CH_2)_3CH_3}} \xrightarrow{H_2CrO_4} \overset{O}{\overset{\|}{CH_3CH_2\overset{}{C}(CH_2)_3CH_3}}$$

Propanal   3-Heptanol   3-Heptanone (57% from propanal)

> **PROBLEM 17.5** Show how 2-butanone could be prepared by a procedure in which all of the carbons originate in acetic acid (CH_3CO_2H).

Many low-molecular-weight aldehydes and ketones are important industrial chemicals. Formaldehyde, a starting material for a number of plastics, is prepared by oxidation of methanol over a silver or iron oxide/molybdenum oxide catalyst at elevated temperature.

The name *aldehyde* was invented to stand for *al*cohol *dehyd*rogenatum, indicating that aldehydes are related to alcohols by loss of hydrogen.

$$CH_3OH + \tfrac{1}{2}O_2 \xrightarrow[\text{500°C}]{\text{catalyst}} \overset{O}{\overset{\|}{HCH}} + H_2O$$

Methanol   Oxygen   Formaldehyde   Water

Similar processes are used to convert ethanol to acetaldehyde and isopropyl alcohol to acetone.

The "linear α-olefins" described in Section 14.15 are starting materials for the preparation of a variety of aldehydes by reaction with carbon monoxide. The process is called **hydroformylation.**

$$RCH{=}CH_2 + \underset{\substack{\text{Carbon} \\ \text{monoxide}}}{CO} + \underset{\text{Hydrogen}}{H_2} \xrightarrow{Co_2(CO)_8} \underset{\text{Aldehyde}}{RCH_2CH_2\overset{\displaystyle O}{\overset{\|}{C}}H}$$

$$\underset{\text{Alkene}}{}$$

Excess hydrogen brings about the hydrogenation of the aldehyde and allows the process to be adapted to the preparation of primary alcohols. Over 2×10^9 lb/year of a variety of aldehydes and alcohols is prepared in the United States by hydroformylation.

A number of aldehydes and ketones are prepared both in industry and in the laboratory by a reaction known as the *aldol condensation,* which will be discussed in detail in Chapter 18.

17.5 REACTIONS OF ALDEHYDES AND KETONES: A REVIEW AND A PREVIEW

Table 17.2 summarizes the reactions of aldehydes and ketones that you've seen in earlier chapters. All are valuable tools to the synthetic chemist. Carbonyl groups provide access to hydrocarbons by Clemmensen or Wolff–Kishner reduction (Section 12.8), to alcohols by reduction (Section 15.2) or by reaction with Grignard or organolithium reagents (Sections 14.6 and 14.7).

The most important chemical property of the carbonyl group is its tendency to undergo *nucleophilic addition* reactions of the type represented in the general equation:

$$\underset{\substack{\text{Aldehyde} \\ \text{or ketone}}}{\overset{\delta+}{C}{=}\overset{\delta-}{O}} + \overset{\delta+}{X}{-}\overset{\delta-}{Y} \longrightarrow \underset{\substack{\text{Product of} \\ \text{nucleophilic addition}}}{\overset{\displaystyle O{-}X}{\underset{\displaystyle Y}{C}}}$$

A negatively polarized atom or group attacks the positively polarized carbon of the carbonyl group in the rate-determining step of these reactions. Grignard reagents, organolithium reagents, lithium aluminum hydride, and sodium borohydride, for example, all react with carbonyl compounds by nucleophilic addition.

The next section explores the mechanism of nucleophilic addition to aldehydes and ketones. There we'll discuss their *hydration,* a reaction in which water adds to the C=O group. After we use this reaction to develop some general principles, we'll then survey a number of related reactions of synthetic, mechanistic, or biological interest.

17.6 PRINCIPLES OF NUCLEOPHILIC ADDITION: HYDRATION OF ALDEHYDES AND KETONES

Effects of Structure on Equilibrium: Aldehydes and ketones react with water in a rapid equilibrium. The product is a geminal diol.

$$\underset{\substack{\text{Aldehyde} \\ \text{or ketone}}}{\overset{\displaystyle O}{\overset{\|}{R}CR'}} + \underset{\text{Water}}{H_2O} \overset{\text{fast}}{\rightleftharpoons} \underset{\substack{\text{Geminal diol} \\ \text{(hydrate)}}}{\overset{\displaystyle OH}{\underset{\displaystyle OH}{R\overset{\|}{C}R'}}} \qquad K_{\text{hydr}} = \frac{[\text{hydrate}]}{[\text{carbonyl compound}]}$$

TABLE 17.2 Summary of Reactions of Aldehydes and Ketones Discussed in Earlier Chapters

Reaction (section) and comments	General equation and specific example
Reduction to hydrocarbons (Section 12.8) Two methods for converting carbonyl groups to methylene units are the Clemmensen reduction (zinc amalgam and concentrated hydrochloric acid) and the Wolff–Kishner reduction (heat with hydrazine and potassium hydroxide in a high-boiling alcohol).	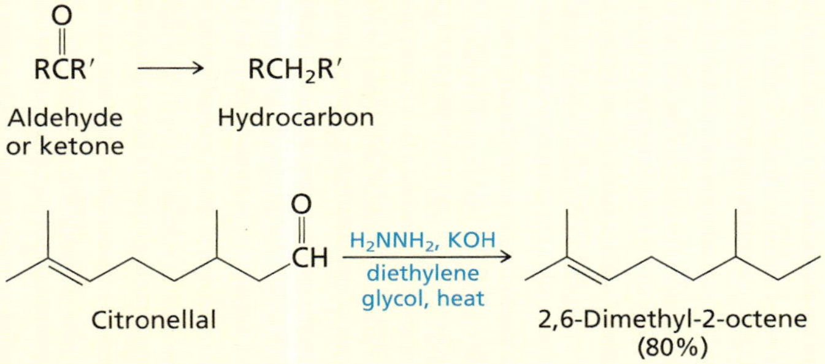 $$RCR' \longrightarrow RCH_2R'$$ Aldehyde or ketone → Hydrocarbon Citronellal $\xrightarrow[\text{diethylene glycol, heat}]{H_2NNH_2,\ KOH}$ 2,6-Dimethyl-2-octene (80%)
Reduction to alcohols (Section 15.2) Aldehydes are reduced to primary alcohols, and ketones are reduced to secondary alcohols by a variety of reducing agents. Catalytic hydrogenation over a metal catalyst and reduction with sodium borohydride or lithium aluminum hydride are general methods.	$$RCR' \longrightarrow RCHR'$$ $$|$$ $$OH$$ Aldehyde or ketone → Alcohol CH_3O–⟨⟩–CH $\xrightarrow[CH_3OH]{NaBH_4}$ CH_3O–⟨⟩–CH_2OH p-Methoxybenzaldehyde → p-Methoxybenzyl alcohol (96%)
Addition of Grignard reagents and organolithium compounds (Sections 14.6–14.7) Aldehydes are converted to secondary alcohols and ketones to tertiary alcohols.	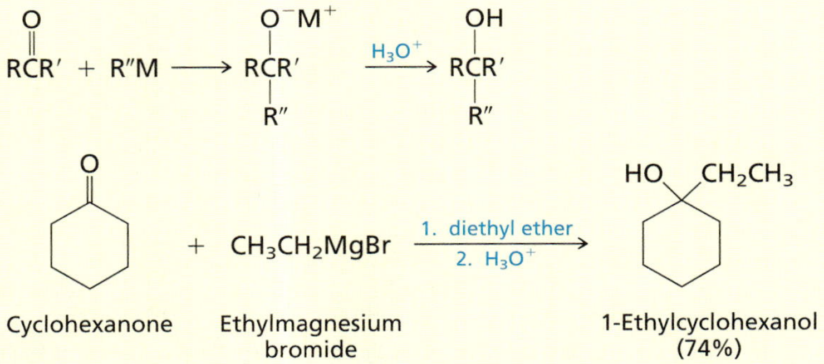 $$RCR' + R''M \longrightarrow RCR' \xrightarrow{H_3O^+} RCR'$$ with $O^- M^+$, then OH, and R'' substituents Cyclohexanone + CH_3CH_2MgBr $\xrightarrow[2.\ H_3O^+]{1.\ \text{diethyl ether}}$ 1-Ethylcyclohexanol (74%) Ethylmagnesium bromide

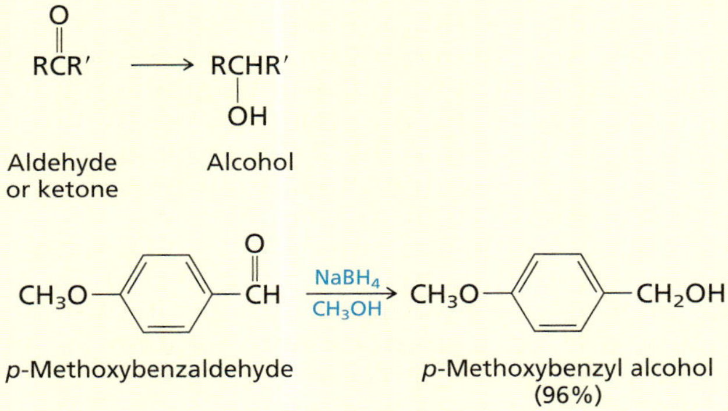

Overall, the reaction is classified as an *addition*. Water adds to the carbonyl group. Hydrogen becomes bonded to the negatively polarized carbonyl oxygen, hydroxyl to the positively polarized carbon.

Table 17.3 compares the equilibrium constants K_{hydr} for hydration of some simple aldehydes and ketones. The position of equilibrium depends on what groups are attached to C=O and how they affect its *steric* and *electronic* environment. Both effects contribute, but the electronic effect controls K_{hydr} more than the steric effect.

Consider first the electronic effect of alkyl groups versus hydrogen atoms attached to C=O. Recall from Section 17.2 that alkyl substituents stabilize C=O, making a ketone carbonyl more stable than an aldehyde carbonyl. As with all equilibria, factors

TABLE 17.3	Equilibrium Constants (K_{hydr}) and Relative Rates of Hydration of Some Aldehydes and Ketones			
Carbonyl compound	**Hydrate**	K_{hydr}*	**Percent conversion to hydrate**	**Relative rate**[†]
$\overset{\displaystyle O}{\overset{\|}{HCH}}$	$CH_2(OH)_2$	2300	> 99.9	2200
$\overset{\displaystyle O}{\overset{\|}{CH_3CH}}$	$CH_3CH(OH)_2$	1.0	50	1.0
$\overset{\displaystyle O}{\overset{\|}{(CH_3)_3CCH}}$	$(CH_3)_3CCH(OH)_2$	0.2	17	0.09
$\overset{\displaystyle O}{\overset{\|}{CH_3CCH_3}}$	$(CH_3)_2C(OH)_2$	0.0014	0.14	0.0018

*$K_{hydr} = \dfrac{[\text{hydrate}]}{[\text{carbonyl compound}]}$.

[†]Neutral solution, 25°C.

that stabilize the reactants decrease the equilibrium constant. Thus, the extent of hydration decreases as the number of alkyl groups on the carbonyl increase.

Increasing stabilization of carbonyl group; decreasing K for hydration

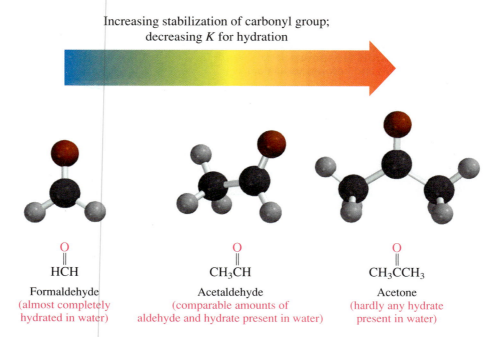

$\overset{\displaystyle O}{\overset{\|}{HCH}}$	$\overset{\displaystyle O}{\overset{\|}{CH_3CH}}$	$\overset{\displaystyle O}{\overset{\|}{CH_3CCH_3}}$
Formaldehyde (almost completely hydrated in water)	Acetaldehyde (comparable amounts of aldehyde and hydrate present in water)	Acetone (hardly any hydrate present in water)

A striking example of an electronic effect on carbonyl group stability and its relation to the equilibrium constant for hydration is seen in the case of hexafluoroacetone. In contrast to the almost negligible hydration of acetone, hexafluoroacetone is completely hydrated.

$$CF_3\overset{\overset{\textstyle O}{\|}}{C}CF_3 \quad + \quad H_2O \;\rightleftharpoons\; CF_3\overset{\overset{\textstyle OH}{|}}{\underset{\underset{\textstyle OH}{|}}{C}}CF_3 \qquad K_{hydr} = 22{,}000$$

Hexafluoroacetone   Water   1,1,1,3,3,3-Hexafluoro-
2,2-propanediol

Instead of stabilizing the carbonyl group by electron donation as alkyl substituents do, trifluoromethyl groups destabilize it by withdrawing electrons. A less stabilized carbonyl group is associated with a greater equilibrium constant for addition.

> **PROBLEM 17.6** *Chloral* is one of the common names for *trichloroethanal.* A solution of chloral in water is called *chloral hydrate;* this material has featured prominently in countless detective stories as the notorious "Mickey Finn" knock-out drops. Write a structural formula for chloral hydrate.

Now let's turn our attention to steric effects by looking at how the size of the groups that were attached to C=O affect K_{hydr}. The bond angles at carbon shrink from $\approx120°$ to $\approx109.5°$ as the hybridization changes from sp^2 in the reactant (aldehyde or ketone) to sp^3 in the product (hydrate). The increased crowding this produces in the hydrate is better tolerated, and K_{hydr} is greater when the groups are small (hydrogen) than when they are large (alkyl).

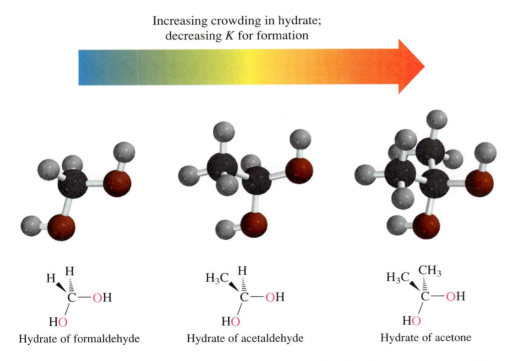

Increasing crowding in hydrate;
decreasing *K* for formation

Hydrate of formaldehyde   Hydrate of acetaldehyde   Hydrate of acetone

Electronic and steric effects operate in the same direction. Both cause the equilibrium constants for hydration of aldehydes to be greater than those of ketones.

Effects of Structure on Rate: Electronic and steric effects influence the rate of hydration in the same way that they affect equilibrium. Indeed, the rate and equilibrium data of Table 17.3 parallel each other almost exactly.

Hydration of aldehydes and ketones is a rapid reaction, quickly reaching equilibrium, but faster in acid or base than in neutral solution. Thus, instead of a single mechanism for hydration, we'll look at two mechanisms, one for basic and the other for acidic solution.

Mechanism of Base-Catalyzed Hydration: The base-catalyzed mechanism (Figure 17.5) is a two-step process in which the first step is rate-determining. In step 1, the nucleophilic hydroxide ion attacks the carbonyl group, forming a bond to carbon. An alkoxide ion is the product of step 1. This alkoxide ion abstracts a proton from water in step 2, yielding the geminal diol. The second step, like all other proton transfers between oxygen that we have seen, is fast.

The role of the basic catalyst (HO^-) is to increase the rate of the nucleophilic addition step. Hydroxide ion, the nucleophile in the base-catalyzed reaction, is much more reactive than a water molecule, the nucleophile in neutral solutions.

Aldehydes react faster than ketones for almost the same reasons that their equilibrium constants for hydration are more favorable. The $sp^2 \rightarrow sp^3$ hybridization change that the carbonyl carbon undergoes on hydration is partially developed in the transition state for the rate-determining nucleophilic addition step (Figure 17.6). Alkyl groups at the reaction site increase the activation energy by simultaneously lowering the energy of the starting state (ketones have a more stabilized carbonyl group than aldehydes) and raising the energy of the transition state (a steric crowding effect).

Mechanism of Acid-Catalyzed Hydration: Three steps are involved in acid-catalyzed hydration (Figure 17.7 on page 718). The first and last are rapid proton transfers between

---

Overall Reaction:

Aldehyde   Water   Geminal
or ketone   diol

Step 1: Nucleophilic addition of hydroxide ion to the carbonyl group.

Hydroxide   Aldehyde   Alkoxide ion
              or ketone   intermediate

Step 2: Proton transfer from water to the intermediate formed in step 1.

Alkoxide ion   Water   Geminal diol   Hydroxide
intermediate

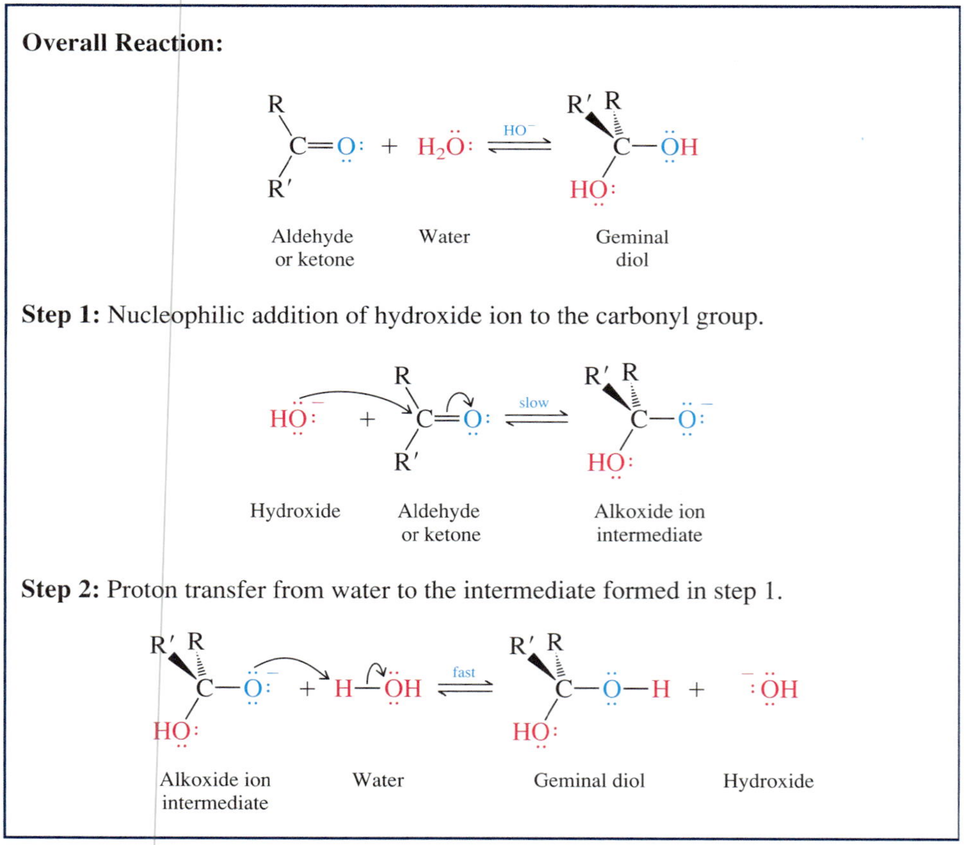

FIGURE 17.5 The mechanism of hydration of an aldehyde or ketone in basic solution. Hydroxide ion is a catalyst; it is consumed in the first step, and regenerated in the second.

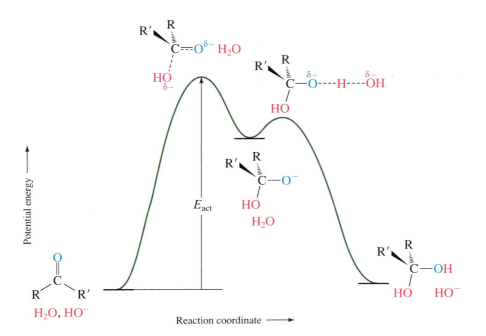

FIGURE 17.6 Potential energy diagram for base-catalyzed hydration of an aldehyde or ketone.

oxygens. The second is a nucleophilic addition. The acid catalyst activates the carbonyl group toward attack by a weakly nucleophilic water molecule. Protonation of oxygen makes the carbonyl carbon of an aldehyde or a ketone much more electrophilic. Expressed in resonance terms, the protonated carbonyl has a greater degree of carbocation character than an unprotonated carbonyl.

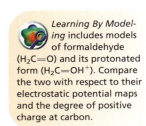

Learning By Modeling includes models of formaldehyde ($H_2C{=}O$) and its protonated form ($H_2C{=}OH^+$). Compare the two with respect to their electrostatic potential maps and the degree of positive charge at carbon.

Steric and electronic effects influence the rate of nucleophilic addition to a protonated carbonyl group in much the same way as they do for the case of a neutral one, and protonated aldehydes react faster than protonated ketones.

With this as background, let us now examine how the principles of nucleophilic addition apply to the characteristic reactions of aldehydes and ketones. We'll begin with the addition of hydrogen cyanide.

17.7 CYANOHYDRIN FORMATION

The product of addition of hydrogen cyanide to an aldehyde or a ketone contains both a hydroxyl group and a cyano group bonded to the same carbon. Compounds of this type are called **cyanohydrins.**

$$\underset{\substack{\text{Aldehyde} \\ \text{or ketone}}}{\overset{\overset{\textstyle O}{\|}}{RCR'}} \quad + \quad \underset{\substack{\text{Hydrogen} \\ \text{cyanide}}}{HC{\equiv}N} \quad \longrightarrow \quad \underset{\text{Cyanohydrin}}{\overset{\textstyle OH}{\underset{\textstyle C{\equiv}N}{RCR'}}}$$

Overall Reaction:

Aldehyde or ketone   Water   Geminal diol

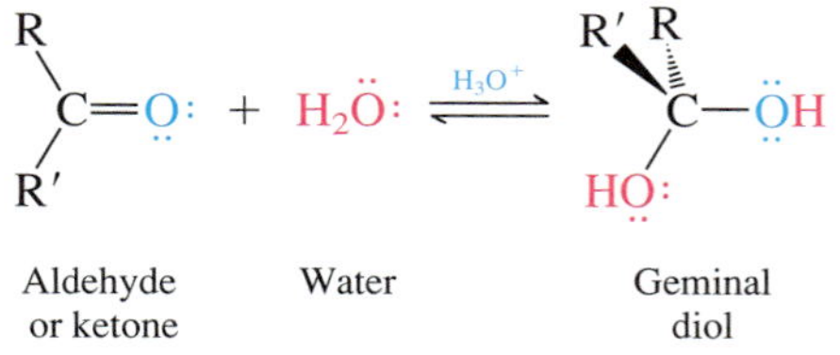

Step 1: Protonation of the carbonyl oxgyen

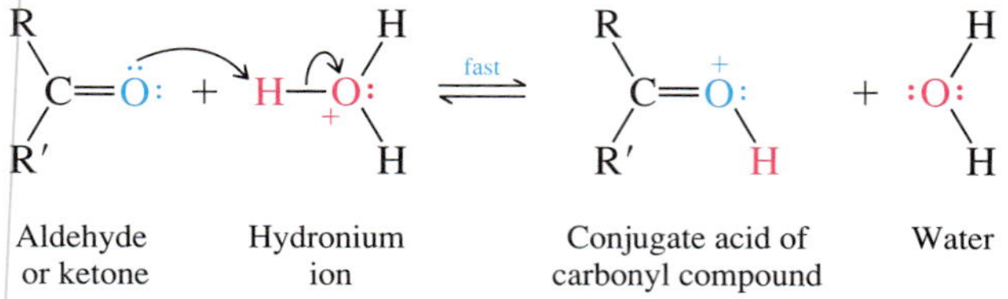

Aldehyde or ketone   Hydronium ion   Conjugate acid of carbonyl compound   Water

Step 2: Nucleophilic addition to the protonated aldehyde or ketone

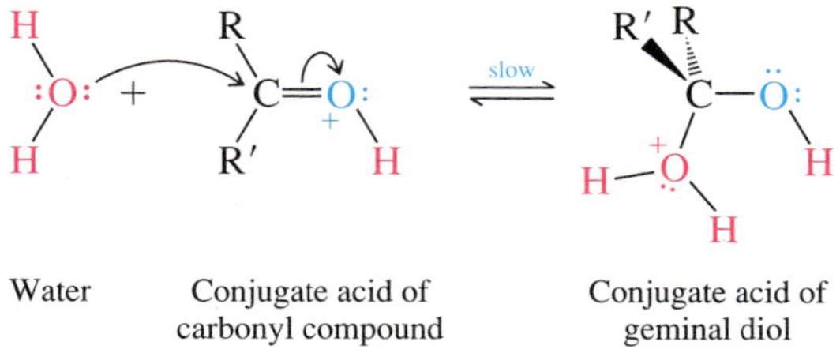

Water   Conjugate acid of carbonyl compound   Conjugate acid of geminal diol

Step 3: Proton transfer from the conjugate acid of the geminal diol to a water molecule

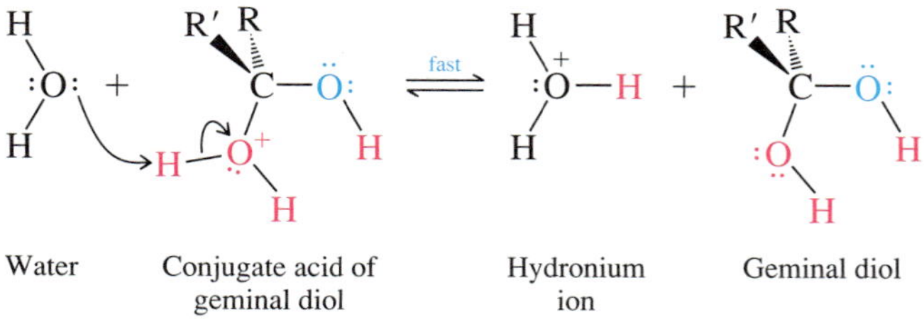

Water   Conjugate acid of geminal diol   Hydronium ion   Geminal diol

The mechanism of this reaction is outlined in Figure 17.8. It is analogous to the mechanism of base-catalyzed hydration in that the nucleophile (cyanide ion) attacks the carbonyl carbon in the first step of the reaction, followed by proton transfer to the carbonyl oxygen in the second step.

The addition of hydrogen cyanide is catalyzed by cyanide ion, but HCN is too weak an acid to provide enough $:\overline{C}\equiv N:$ for the reaction to proceed at a reasonable rate. Cyanohydrins are therefore normally prepared by adding an acid to a solution containing the carbonyl compound and sodium or potassium cyanide. This procedure ensures that free cyanide ion is always present in amounts sufficient to increase the rate of the reaction.

FIGURE 17.8 The mechanism of cyanohydrin formation from an aldehyde or ketone. Cyanide ion is a catalyst; it is consumed in the first step and regenerated in the second.

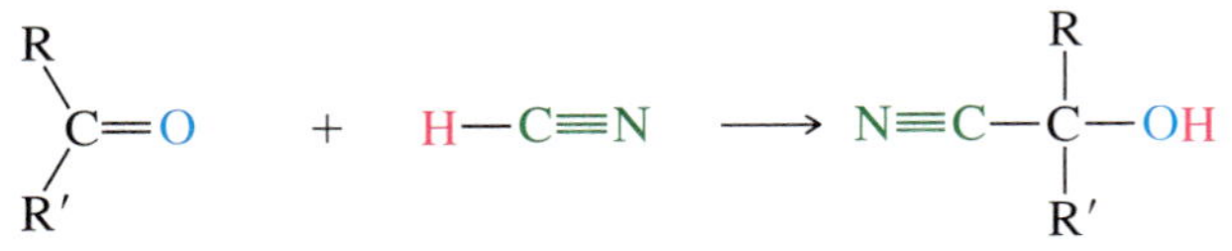

Overall reaction:

Aldehyde or ketone	Hydrogen cyanide	Cyanohydrin

Step 1: Nucleophilic attack by the negatively charged carbon of cyanide ion at the carbonyl carbon of the aldehyde or ketone. Hydrogen cyanide itself is not very nucleophilic and does not ionize to form cyanide ion to a significant extent. Thus, a source of cyanide ion such as NaCN or KCN is used.

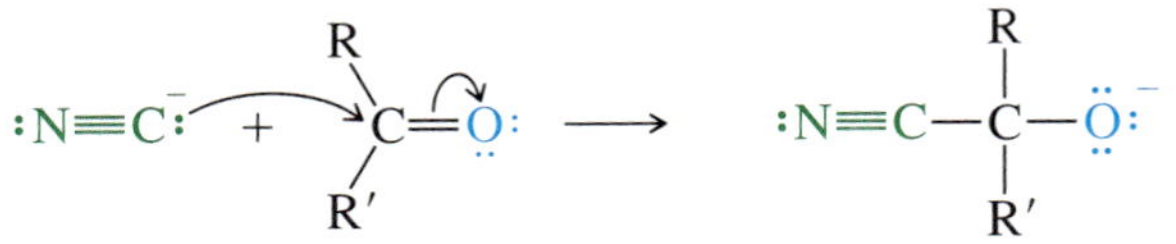

Cyanide ion	Aldehyde or ketone	Conjugate base of cyanohydrin

Step 2: The alkoxide ion formed in the first step abstracts a proton from hydrogen cyanide. This step yields the cyanohydrin product and regenerates cyanide ion.

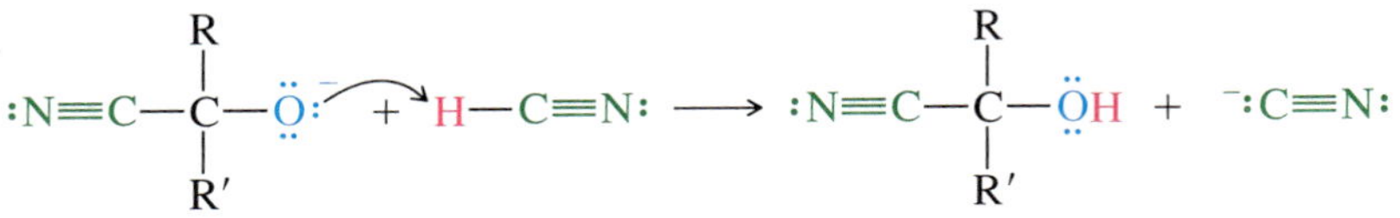

Conjugate base of cyanohydrin	Hydrogen cyanide	Cyanohydrin	Cyanide ion

Cyanohydrin formation is reversible, and the position of equilibrium depends on the steric and electronic factors governing nucleophilic addition to carbonyl groups described in the preceding section. Aldehydes and unhindered ketones give good yields of cyanohydrins.

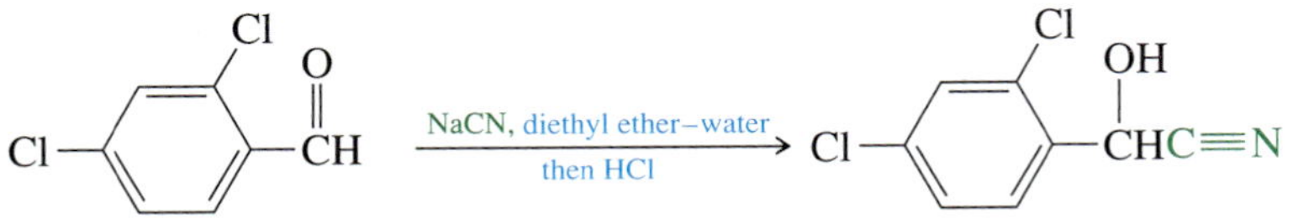

2,4-Dichlorobenzaldehyde	2,4-Dichlorobenzaldehyde cyanohydrin (100%)

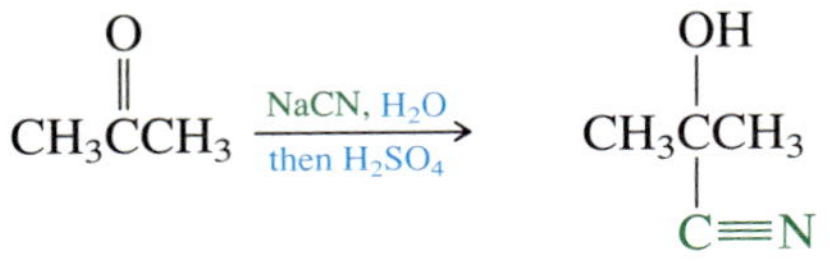

Acetone	Acetone cyanohydrin (77–78%)

In substitutive IUPAC nomenclature, cyanohydrins are named as hydroxy derivatives of nitriles. Because nitrile nomenclature will not be discussed until Section 20.1, we will refer to cyanohydrins as derivatives of the parent aldehyde or ketone as shown in the examples. This conforms to the practice of most chemists.

Converting aldehydes and ketones to cyanohydrins is of synthetic value for two reasons: (1) a new carbon–carbon bond is formed, and (2) the cyano group in the product can be converted to a carboxylic acid function (CO_2H) by hydrolysis (to be discussed in Section 19.12) or to an amine of the type CH_2NH_2 by reduction (to be discussed in Section 22.9).

> **PROBLEM 17.7** The hydroxyl group of a cyanohydrin is also a potentially reactive site. *Methacrylonitrile* is an industrial chemical used in the production of plastics and fibers. One method for its preparation is the acid-catalyzed dehydration of acetone cyanohydrin. Deduce the structure of *methacrylonitrile*.

A few cyanohydrins and ethers of cyanohydrins occur naturally. One species of millipede stores benzaldehyde cyanohydrin, along with an enzyme that catalyzes its cleavage to benzaldehyde and hydrogen cyanide, in separate compartments above its legs. When attacked, the insect ejects a mixture of the cyanohydrin and the enzyme, repelling the invader by spraying it with hydrogen cyanide.

17.8 ACETAL FORMATION

Many of the most interesting and useful reactions of aldehydes and ketones involve transformation of the initial product of nucleophilic addition to some other substance under the reaction conditions. An example is the reaction of aldehydes with alcohols under conditions of acid catalysis. The expected product of nucleophilic addition of the alcohol to the carbonyl group is called a **hemiacetal.** The product actually isolated, however, corresponds to reaction of one mole of the aldehyde with *two* moles of alcohol to give *geminal diethers* known as **acetals:**

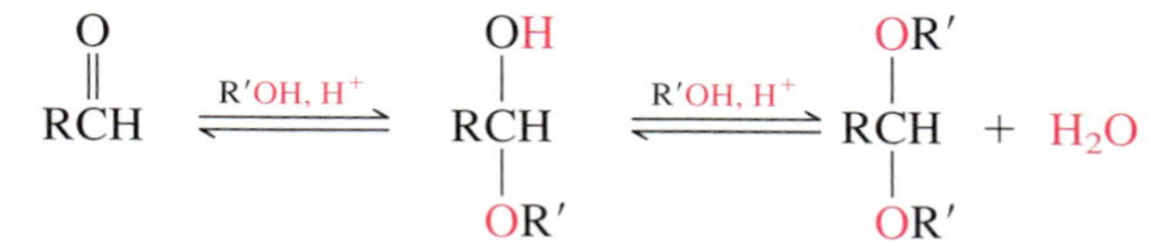

| Aldehyde | Hemiacetal | Acetal | Water |

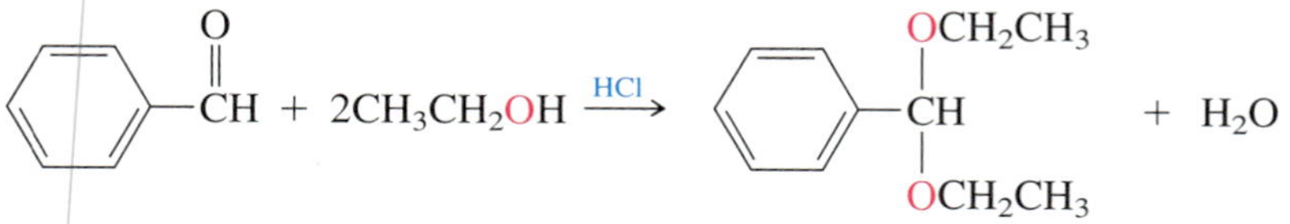

Benzaldehyde   Ethanol   Benzaldehyde diethyl acetal   Water
(66%)

The mechanism for formation of benzaldehyde diethyl acetal, which proceeds in two stages, is presented in Figure 17.9. The first stage (steps 1–3) involves formation of a hemiacetal; in the second stage (steps 4–7) the hemiacetal is converted to the acetal. Nucleophilic addition to the carbonyl group characterizes the first stage, carbocation chemistry the second. The key carbocation intermediate is stabilized by electron release from oxygen.

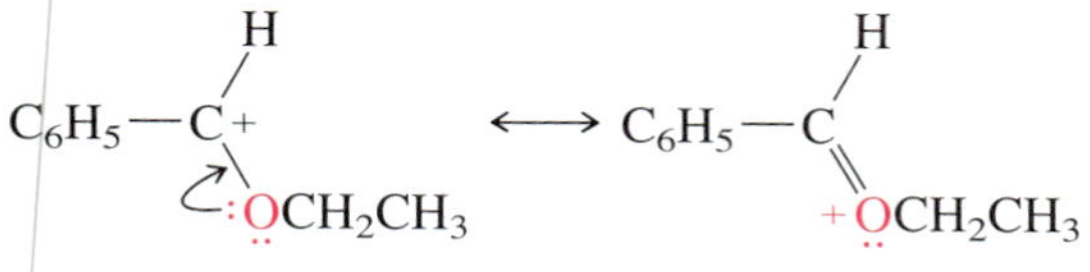

A particularly stable resonance form; satisfies the octet rule for carbon and oxygen.

Overall Reaction:

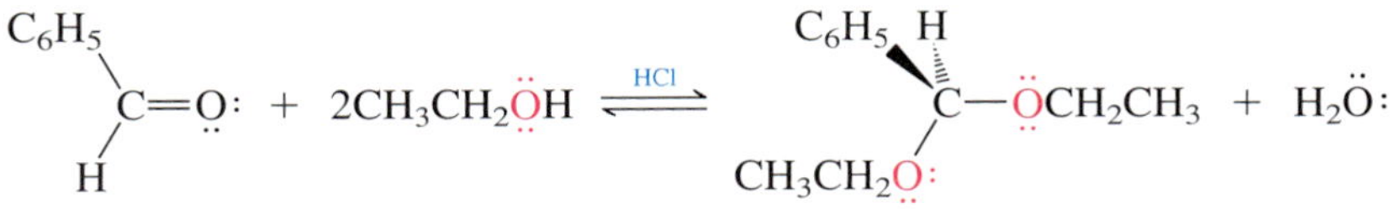

Steps 1–3: Acid-catalyzed nucleophilic addition of 1 mole of ethanol to the carbonyl group. The details of these steps are analogous to the three steps of acid-catalyzed hydration in Figure 17.7. The product of these three steps is a hemiacetal.

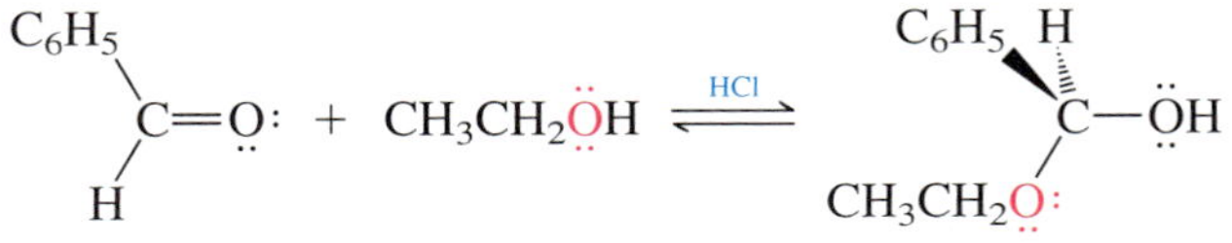

Steps 4–5: Conversion of hemiacetal to carbocation. These steps are analogous to the formation of carbocations in acid-catalyzed reactions of alcohols.

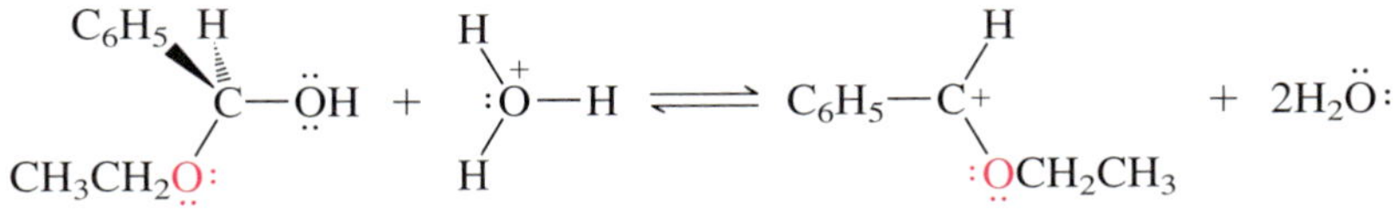

Step 6: Nucleophilic attack by ethanol on the carbocation.

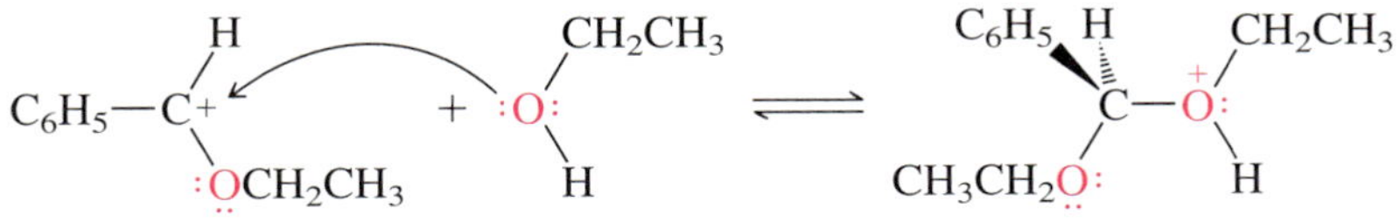

Step 7: Proton transfer from the conjugate acid of the product to ethanol.

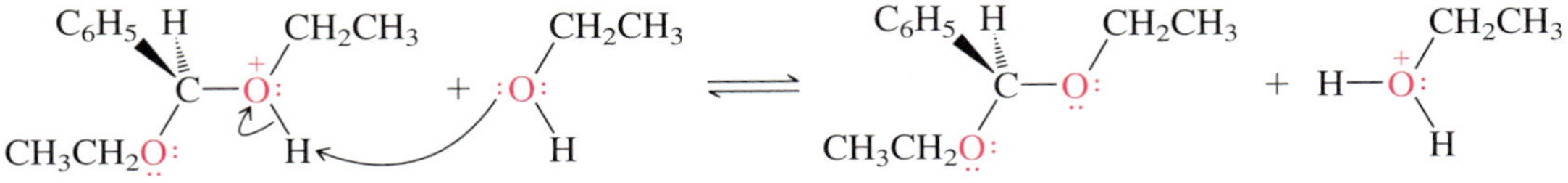

FIGURE 17.9 The mechanism of acetal formation from benzaldehyde and ethanol.

PROBLEM 17.8 Be sure you fully understand the mechanism outlined in Figure 17.9 by writing equations for steps 1–3 and 4–5. Use curved arrows to show electron flow.

The position of equilibrium is favorable for acetal formation from most aldehydes, especially when excess alcohol is present as the reaction solvent. For most ketones the position of equilibrium is unfavorable, and other methods must be used for the preparation of acetals from ketones.

Diols that bear two hydroxyl groups in a 1,2 or 1,3 relationship to each other yield *cyclic acetals* on reaction with either aldehydes or ketones. The five-membered cyclic acetals derived from ethylene glycol (1,2-ethanediol) are the most commonly encountered examples. Often the position of equilibrium is made more favorable by removing the water formed in the reaction by azeotropic distillation with benzene or toluene:

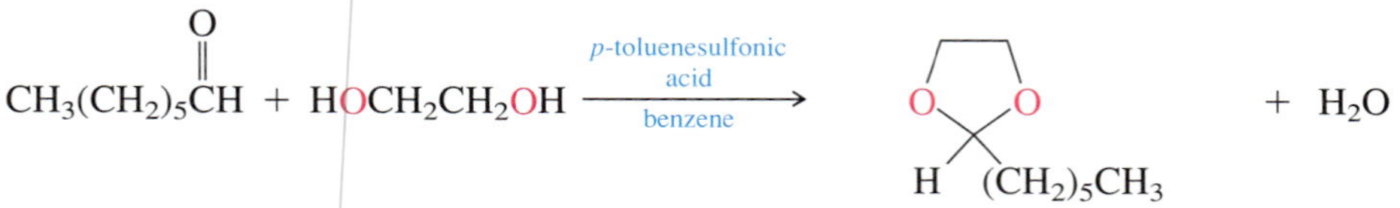

Ketal is an acceptable subcategory term for acetals formed from ketones. It was dropped from IUPAC nomenclature, but continues to be so widely used that it has been reinstated.

$$CH_3(CH_2)_5CH + HOCH_2CH_2OH \xrightarrow[\text{benzene}]{\substack{\text{p-toluenesulfonic}\\\text{acid}}} + H_2O$$

Heptanal Ethylene glycol 2-Hexyl-1,3-dioxolane (81%) Water

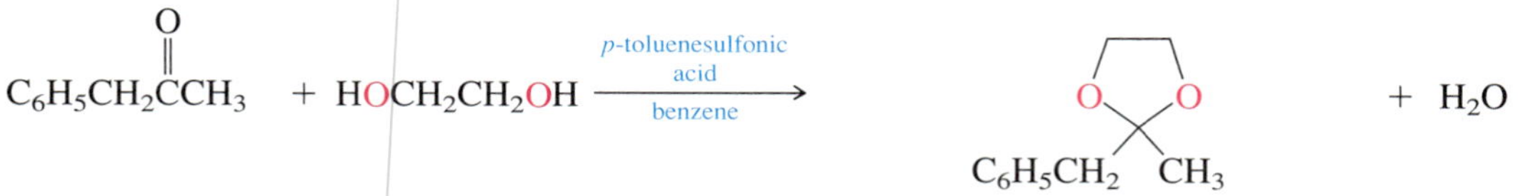

$$C_6H_5CH_2CCH_3 + HOCH_2CH_2OH \xrightarrow[\text{benzene}]{\substack{\text{p-toluenesulfonic}\\\text{acid}}} + H_2O$$

Benzyl methyl ketone Ethylene glycol 2-Benzyl-2-methyl-1,3-dioxolane (78%) Water

PROBLEM 17.9 Write the structures of the cyclic acetals derived from each of the following.
(a) Cyclohexanone and ethylene glycol
(b) Benzaldehyde and 1,3-propanediol
(c) Isobutyl methyl ketone and ethylene glycol
(d) Isobutyl methyl ketone and 2,2-dimethyl-1,3-propanediol

SAMPLE SOLUTION (a) The cyclic acetals derived from ethylene glycol contain a five-membered 1,3-dioxolane ring.

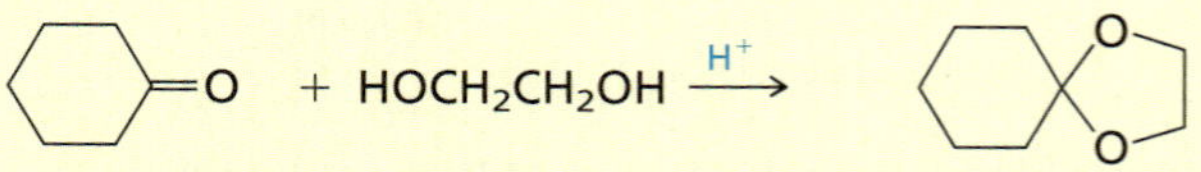

Cyclohexanone Ethylene glycol Acetal of cyclohexanone and ethylene glycol

Acetals are susceptible to hydrolysis in aqueous acid:

$$\underset{\substack{|\\OR''}}{\overset{\substack{OR''\\|}}{RCR'}} + H_2O \underset{}{\overset{H^+}{\rightleftharpoons}} RCR' + 2R''OH$$

Acetal Water Aldehyde or ketone Alcohol

This reaction is simply the reverse of the reaction by which acetals are formed—acetal formation is favored by excess alcohol, acetal hydrolysis by excess water. Acetal formation and acetal hydrolysis share the same mechanistic pathway but travel along that pathway in opposite directions. In the following section you'll see a clever way in which acetal formation and hydrolysis have been applied to synthetic organic chemistry.

> **PROBLEM 17.10** Problem 17.8 asked you to write details of the mechanism describing formation of benzaldehyde diethyl acetal from benzaldehyde and ethanol. Write a stepwise mechanism for the acid hydrolysis of this acetal.

17.9   ACETALS AS PROTECTING GROUPS

In an organic synthesis, it sometimes happens that one of the reactants contains a functional group that is incompatible with the reaction conditions. Consider, for example, the conversion

$$\underset{\text{5-Hexyn-2-one}}{CH_3CCH_2CH_2C{\equiv}CH} \longrightarrow \underset{\text{5-Heptyn-2-one}}{CH_3CCH_2CH_2C{\equiv}CCH_3}$$

It looks as though all that is needed is to prepare the acetylenic anion, then alkylate it with methyl iodide (Section 9.6). There is a complication, however. The carbonyl group in the starting alkyne will neither tolerate the strongly basic conditions required for anion formation nor survive in a solution containing carbanions. Acetylide ions add to carbonyl groups (Section 14.8). Thus, the necessary anion $CH_3CCH_2CH_2C{\equiv}\overset{-}{C}\colon$ is inaccessible.

The strategy that is routinely followed is to *protect* the carbonyl group during the reactions with which it is incompatible and then to *remove* the protecting group in a subsequent step. Acetals, especially those derived from ethylene glycol, are among the most useful groups for carbonyl protection, because they can be introduced and removed readily. A key fact is that acetals resemble ethers in being inert to many of the reagents, such as hydride reducing agents and organometallic compounds, that react readily with carbonyl groups. The following sequence is the one that was actually used to bring about the desired transformation.

(a) Protection of carbonyl group

$$\underset{\text{5-Hexyn-2-one}}{CH_3CCH_2CH_2C{\equiv}CH} \xrightarrow[\substack{p\text{-toluenesulfonic}\\ \text{acid, benzene}}]{HOCH_2CH_2OH} \underset{\text{Acetal of reactant}}{H_3C \quad CH_2CH_2C{\equiv}CH}$$

(b) Alkylation of alkyne

$$\underset{H_3C \quad CH_2CH_2C{\equiv}CH}{} \xrightarrow[\text{2. } CH_3I]{\text{1. } NaNH_2, NH_3} \underset{\substack{H_3C \quad CH_2CH_2C{\equiv}CCH_3 \\ \text{Acetal of product}}}{}$$

(c) Unmasking of the carbonyl group by hydrolysis

$$H_3C \quad CH_2CH_2C{\equiv}CCH_3 \xrightarrow[\text{HCl}]{H_2O} CH_3CCH_2CH_2C{\equiv}CCH_3$$

5-Heptyn-2-one (96%)

Although protecting and unmasking the carbonyl group adds two steps to the synthetic procedure, both steps are essential to its success. The tactic of functional group protection is frequently encountered in preparative organic chemistry, and considerable attention has been paid to the design of effective protecting groups for a variety of functionalities.

PROBLEM 17.11 Acetal formation is a characteristic reaction of aldehydes and ketones, but not of carboxylic acids. Show how you could advantageously use a cyclic acetal protecting group in the following synthesis:

$$\text{Convert} \quad CH_3\overset{O}{\overset{\|}{C}}{-}\!\!\!\!\raisebox{-3pt}{}\!\!\!\!{-}\overset{O}{\overset{\|}{C}}OH \quad \text{to} \quad CH_3\overset{O}{\overset{\|}{C}}{-}\!\!\!\!\raisebox{-3pt}{}\!\!\!\!{-}CH_2OH$$

17.10 REACTION WITH PRIMARY AMINES: IMINES

A second two-stage reaction that begins with nucleophilic addition to aldehydes and ketones is their reaction with primary amines, compounds of the type RNH_2 or $ArNH_2$. In the first stage of the reaction the amine adds to the carbonyl group to give a species known as a **carbinolamine.** Once formed, the carbinolamine undergoes dehydration to yield the product of the reaction, an *N*-alkyl- or *N*-aryl-substituted **imine:**

N-substituted imines are sometimes called **Schiff's bases,** after Hugo Schiff, a German chemist who described their formation in 1864.

$$\overset{O}{\overset{\|}{R{-}C{-}R'}} + R''\ddot{N}H_2 \underset{\text{addition}}{\rightleftharpoons} \overset{OH}{\underset{HN{-}R''}{R{-}C{-}R'}} \underset{\text{elimination}}{\rightleftharpoons} \overset{:NR''}{\overset{\|}{R{-}C{-}R'}} + H_2O$$

Aldehyde or ketone	Primary amine	Carbinolamine	*N*-substituted imine	Water

$$\text{(benzaldehyde)}\overset{O}{\overset{\|}{{-}CH}} + CH_3NH_2 \longrightarrow {-}CH{=}NCH_3$$

Benzaldehyde   Methylamine   *N*-Benzylidenemethylamine (70%)

The December, 2000 issue of the *Journal of Chemical Education* (pp. 1644–1648) contains an article entitled "Carbinolamines and Geminal Diols in Aqueous Environmental Organic Chemistry."

$${=}O + (CH_3)_2CHCH_2NH_2 \longrightarrow {=}NCH_2CH(CH_3)_2$$

Cyclohexanone   Isobutylamine   *N*-Cyclohexylideneisobutylamine (79%)

Figure 17.10 presents the mechanism for the reaction between benzaldehyde and methylamine given in the first example. The first two steps lead to the carbinolamine, the last three show the dehydration of the carbinolamine to the imine. Step 4, the key

Overall Reaction:

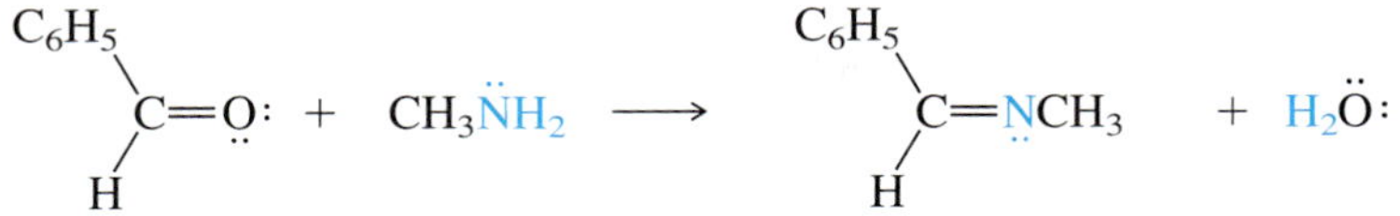

Benzaldehyde　Methylamine　*N*-Benzylidenemethylamine　Water

Step 1: The amine acts as a nucleophile, attacking the carbonyl group and forming a C—N bond.

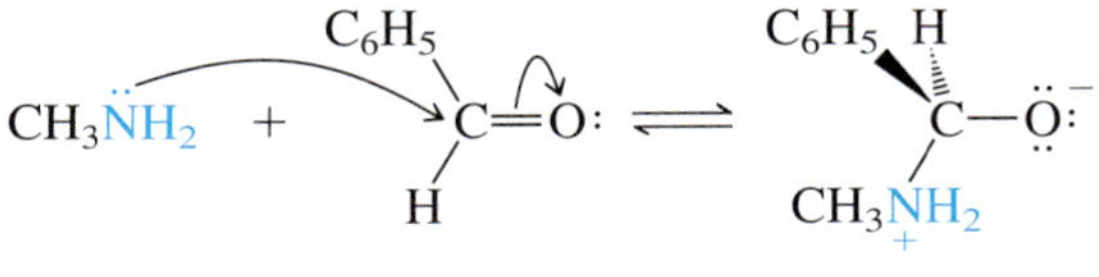

Methylamine　Benzaldehyde　First intermediate

Step 2: In a solvent such as water, proton transfers convert the dipolar intermediate to the carbinolamine.

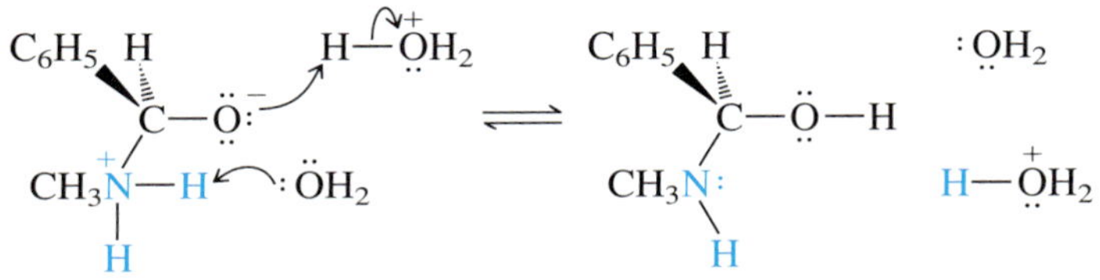

First intermediate　Carbinolamine

Step 3: The dehydration stage begins with protonation of the carbinolamine on oxygen.

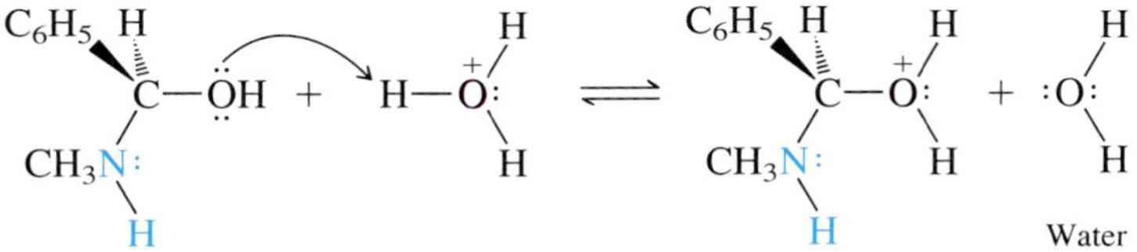

Carbinolamine intermediate　Hydronium ion　*O*-Protonated carbinolamine　Water

Step 4: The oxygen-protonated carbinolamine loses water to give a nitrogen-stabilized carbocation.

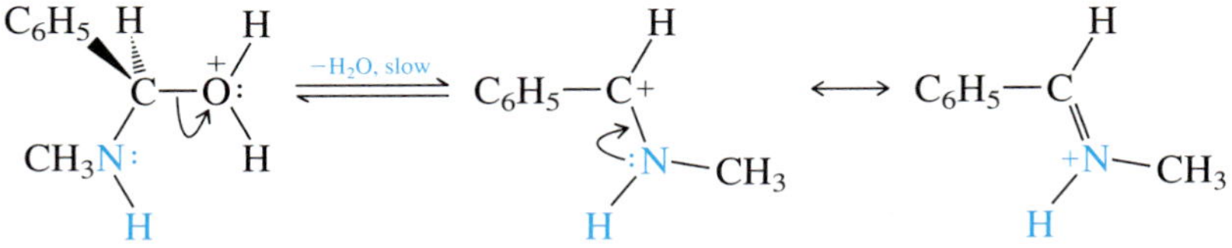

O-Protonated carbinolamine　Nitrogen-stabilized carbocation

—Cont.

FIGURE 17.10 The mechanism of imine formation from benzaldehyde and methylamine.

(Continued)

Step 5: The nitrogen-stabilized carbocation is the conjugate acid of the imine. Proton transfer to water gives the imine.

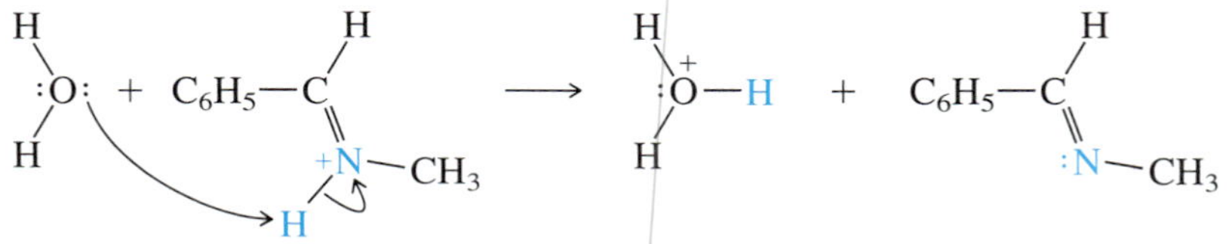

| Water | Nitrogen-stabilized carbocation | Hydronium ion | N-Benzylidenemethylamine |

FIGURE 17.10 *(Continued)*

step in the dehydration phase, is rate-determining when the reaction is carried out in acid solution. If the solution is too acidic, however, protonation of the amine blocks step 1. Therefore there is some optimum pH, usually about 5, at which the reaction rate is a maximum. Too basic a solution reduces the rate of step 4; too acidic a solution reduces the rate of step 1.

PROBLEM 17.12 Write the structure of the carbinolamine intermediate and the imine product formed in the reaction of each of the following:

(a) Acetaldehyde and benzylamine, $C_6H_5CH_2NH_2$

(b) Benzaldehyde and butylamine, $CH_3CH_2CH_2CH_2NH_2$

(c) Cyclohexanone and *tert*-butylamine, $(CH_3)_3CNH_2$

(d) Acetophenone and cyclohexylamine, ⬡—NH_2

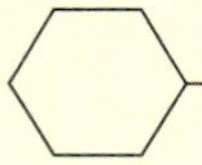

SAMPLE SOLUTION The carbinolamine is formed by nucleophilic addition of the amine to the carbonyl group. Its dehydration gives the imine product.

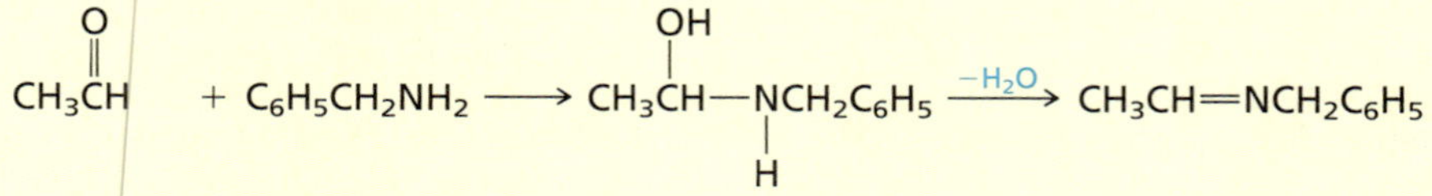

| Acetaldehyde | Benzylamine | Carbinolamine intermediate | Imine product (N-ethylidene-benzylamine) |

A number of compounds of the general type H_2NZ react with aldehydes and ketones in a manner analogous to that of primary amines. The carbonyl group (C=O) is converted to C=NZ, and a molecule of water is formed. Table 17.4 presents examples of some of these reactions. The mechanism by which each proceeds is similar to the nucleophilic addition–elimination mechanism described for the reaction of primary amines with aldehydes and ketones.

The reactions listed in Table 17.4 are reversible and have been extensively studied from a mechanistic perspective because of their relevance to biological processes.

TABLE 17.4	Reaction of Aldehydes and Ketones with Derivatives of Ammonia: $\underset{\text{RCR}'}{\overset{\text{O}}{\parallel}} + H_2NZ \longrightarrow \underset{\text{RCR}'}{\overset{\text{NZ}}{\parallel}} + H_2O$			
Reagent (H$_2$NZ)	**Name of reagent**	**Type of product**	**Example**	
H$_2$NOH	Hydroxylamine	Oxime	$\underset{\text{Heptanal}}{CH_3(CH_2)_5\overset{\overset{\displaystyle O}{\parallel}}{C}H} \xrightarrow{\text{H}_2\text{NOH}} \underset{\substack{\text{Heptanal oxime (81–93\%)}}}{CH_3(CH_2)_5\overset{\overset{\displaystyle NOH}{\parallel}}{C}H}$	
H$_2$NNHC$_6$H$_5$*	Phenylhydrazine	Phenylhydrazone	Acetophenone $\xrightarrow{\text{H}_2\text{NNHC}_6\text{H}_5}$ Acetophenone phenylhydrazone (87–91%)	
$\underset{H_2N\overset{\overset{\displaystyle O}{\parallel}}{N}H\overset{\displaystyle}{C}NH_2}{}$	Semicarbazide	Semicarbazone	$\underset{\text{2-Dodecanone}}{CH_3\overset{\overset{\displaystyle O}{\parallel}}{C}(CH_2)_9CH_3} \xrightarrow{\text{H}_2\text{NNHCNH}_2} \underset{\substack{\text{2-Dodecanone}\\\text{semicarbazone (93\%)}}}{CH_3\overset{\overset{\displaystyle O}{\parallel}}{C}(CH_2)_9CH_3}$	

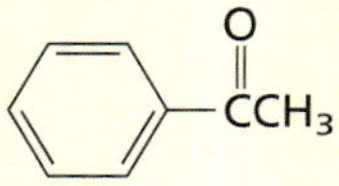

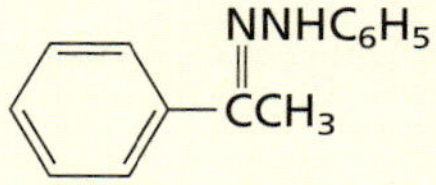

*Compounds related to phenylhydrazine react in an analogous way. *p*-Nitrophenylhydrazine yields *p*-nitrophenylhydrazones; 2,4-dinitrophenylhydrazine yields 2,4-dinitrophenylhydrazones.

Many biological reactions involve initial binding of a carbonyl compound to an enzyme or coenzyme via imine formation. The boxed essay *Imines in Biological Chemistry* gives some important examples.

17.11   REACTION WITH SECONDARY AMINES: ENAMINES

Secondary amines are compounds of the type R$_2$NH. They add to aldehydes and ketones to form carbinolamines, but their carbinolamine intermediates can dehydrate to a stable product only in the direction that leads to a carbon–carbon double bond:

$$\underset{\substack{\text{Aldehyde}\\\text{or ketone}}}{\overset{\overset{\displaystyle O}{\parallel}}{RCH_2CR'}} + \underset{\substack{\text{Secondary}\\\text{amine}}}{R''_2\ddot{N}H} \rightleftharpoons \underset{\text{Carbinolamine}}{RCH_2\underset{:NR''_2}{\overset{\overset{\displaystyle OH}{|}}{C}}{-}R'} \underset{}{\overset{-H_2O}{\rightleftharpoons}} \underset{\text{Enamine}}{RCH{=}\underset{:NR''_2}{C}R'}$$

via:

(structure showing the iminium intermediate)

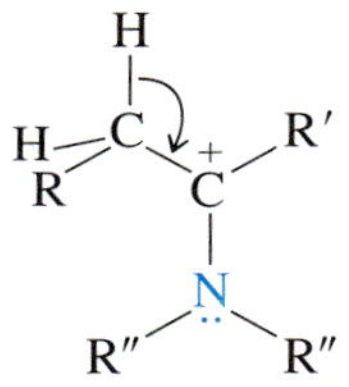

The product of this dehydration is an alkenyl-substituted amine, or **enamine.**

| Cyclopentanone | Pyrrolidine | *N*-(1-Cyclopentenyl)-pyrrolidine (80–90%) | Water |

IMINES IN BIOLOGICAL CHEMISTRY

Many biological processes involve an "association" between two species in a step prior to some subsequent transformation. This association can take many forms. It can be a weak association of the attractive van der Waals type, or a stronger interaction such as a hydrogen bond. It can be an electrostatic attraction between a positively charged atom of one molecule and a negatively charged atom of another. Covalent bond formation between two species of complementary chemical reactivity represents an extreme kind of association. It often occurs in biological processes in which aldehydes or ketones react with amines via imine intermediates.

An example of a biologically important aldehyde is *pyridoxal phosphate,* which is the active form of *vitamin B₆* and a coenzyme for many of the reactions of α-amino acids. In these reactions the amino acid binds to the coenzyme by reacting with it to form an imine of the kind shown in the equation. Reactions then take place at the amino acid portion of the imine, modifying the amino acid. In the last step, enzyme-catalyzed hydrolysis cleaves the imine to pyridoxal and the modified amino acid.

A key step in the chemistry of vision is binding of an aldehyde to an enzyme via an imine. An outline of the steps involved is presented in Figure 17.11. It starts with *β-carotene,* a pigment that occurs naturally in several fruits and vegetables, including carrots. β-Carotene undergoes oxidative cleavage in the liver to give an alcohol known as *retinol,* or *vitamin A.* Oxidation of vitamin A, followed by isomerization of one of its double bonds, gives the aldehyde 11-*cis*-retinal. In the eye, the aldehyde function of 11-*cis*-retinal combines with an amino group of the protein *opsin* to form an imine called *rhodopsin.* When rhodopsin absorbs a photon of visible light, the cis double bond of the retinal unit undergoes a photochemical cis-to-trans isomerization, which is attended by a dramatic change in its shape and a change in the conformation of rhodopsin. This conformational change is translated into a nerve impulse perceived by the brain as a visual image. Enzyme-promoted hydrolysis of the photochemically isomerized rhodopsin regenerates opsin and a molecule of all-*trans*-retinal. Once all-*trans*-retinal has been enzymatically converted to its 11-cis isomer, it and opsin reenter the cycle.

| Pyridoxal phosphate | α-Amino acid | Imine |

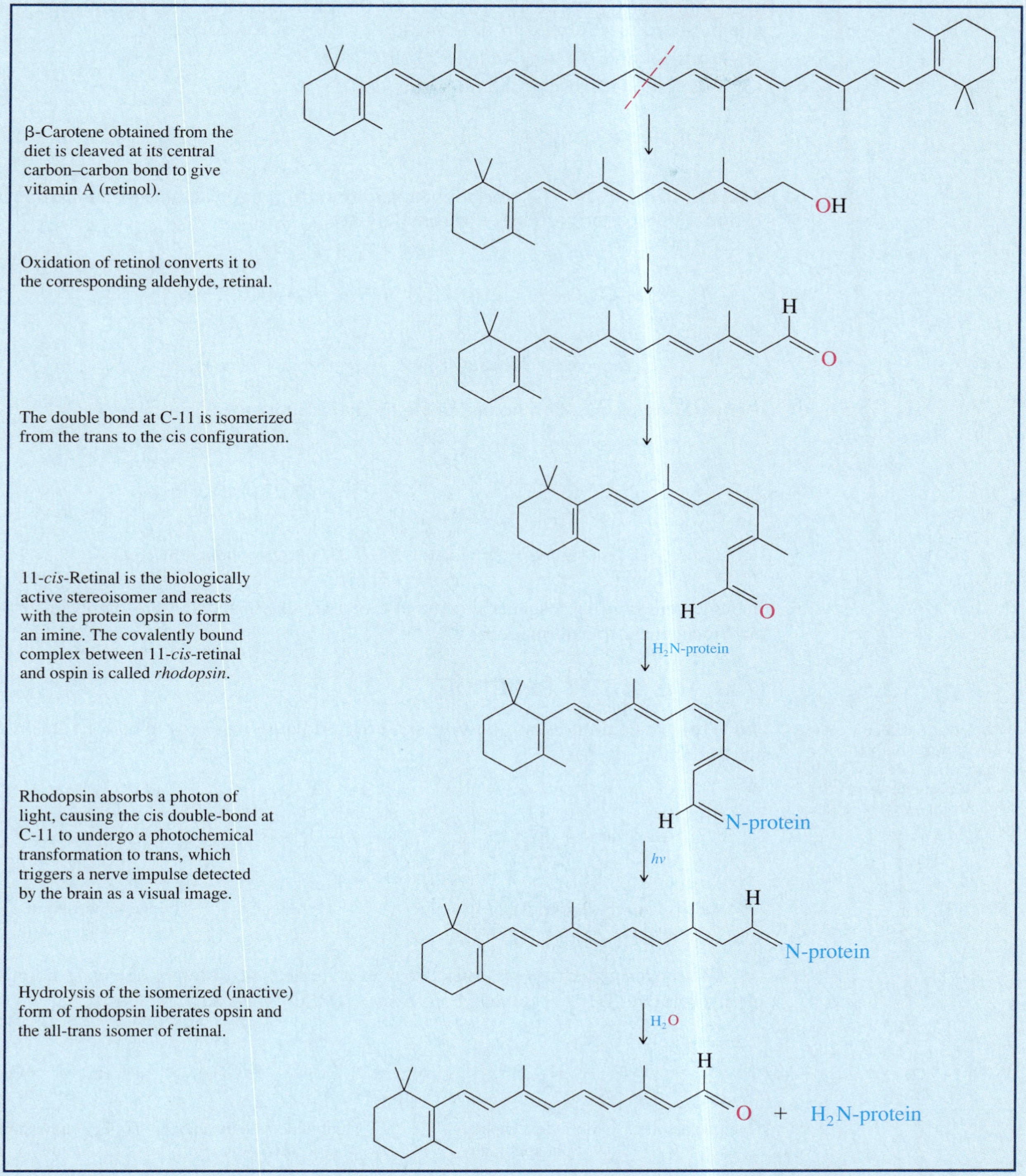
β-Carotene obtained from the diet is cleaved at its central carbon–carbon bond to give vitamin A (retinol).
OH
Oxidation of retinol converts it to the corresponding aldehyde, retinal.
H
O
The double bond at C-11 is isomerized from the trans to the cis configuration.
H
O
11-*cis*-Retinal is the biologically active stereoisomer and reacts with the protein opsin to form an imine. The covalently bound complex between 11-*cis*-retinal and ospin is called *rhodopsin*.
H₂N-protein
H
N-protein
Rhodopsin absorbs a photon of light, causing the cis double-bond at C-11 to undergo a photochemical transformation to trans, which triggers a nerve impulse detected by the brain as a visual image.
hν
H
N-protein
Hydrolysis of the isomerized (inactive) form of rhodopsin liberates opsin and the all-trans isomer of retinal.
H₂O
H
O
+
H₂N-protein

FIGURE 17.11 Imine formation between the aldehyde function of 11-*cis*-retinal and an amino group of a protein (opsin) is involved in the chemistry of vision. The numbering scheme in retinal is specifically developed for carotenes and related compounds.

PROBLEM 17.13 Write the structure of the carbinolamine intermediate and the enamine product formed in the reaction of each of the following:

(a) Propanal and dimethylamine, CH_3NHCH_3

(b) 3-Pentanone and pyrrolidine (page 728)

(c) Acetophenone and HN⬡

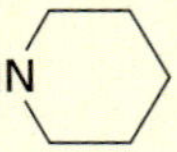

SAMPLE SOLUTION (a) Nucleophilic addition of dimethylamine to the carbonyl group of propanal produces a carbinolamine:

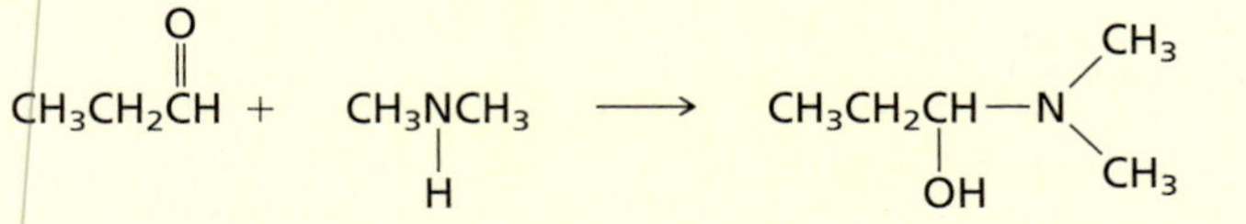

$$CH_3CH_2\overset{\overset{\displaystyle O}{\|}}{C}H + CH_3\underset{\underset{\displaystyle H}{|}}{N}CH_3 \longrightarrow CH_3CH_2\underset{\underset{\displaystyle OH}{|}}{C}H-N\overset{\diagup CH_3}{\diagdown CH_3}$$

Propanal	Dimethylamine	Carbinolamine intermediate

Dehydration of this carbinolamine yields the enamine:

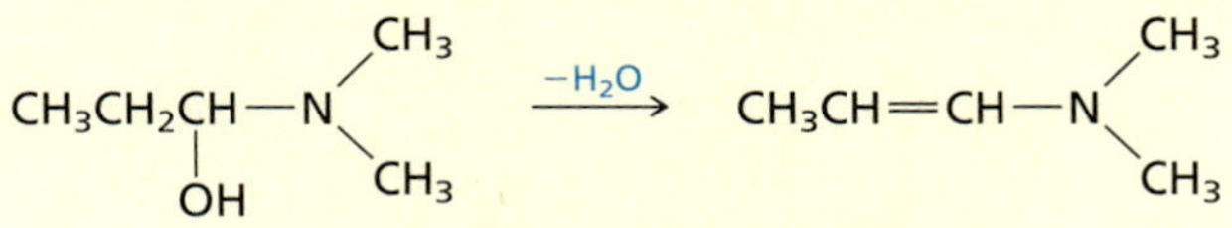

$$CH_3CH_2\underset{\underset{\displaystyle OH}{|}}{C}H-N\overset{\diagup CH_3}{\diagdown CH_3} \xrightarrow{-H_2O} CH_3CH=CH-N\overset{\diagup CH_3}{\diagdown CH_3}$$

Carbinolamine intermediate	*N*-(1-Propenyl)dimethylamine

Enamines are used as reagents in synthetic organic chemistry and are involved in certain biochemical transformations.

17.12 THE WITTIG REACTION

The **Wittig reaction** uses *phosphorus ylides* (called *Wittig reagents*) to convert aldehydes and ketones to alkenes.

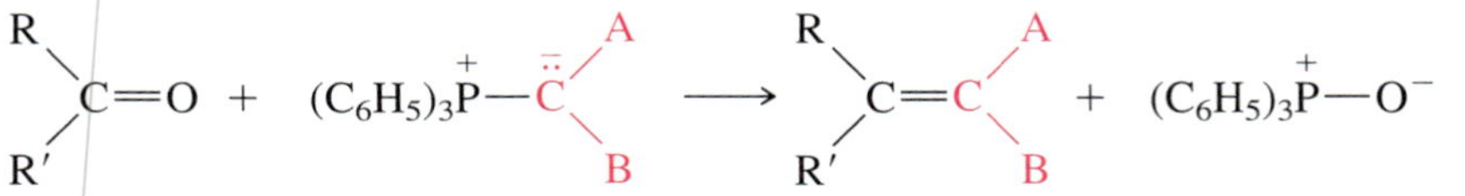

The reaction is named after Georg Wittig, a German chemist who shared the 1979 Nobel Prize in chemistry for demonstrating its synthetic potential.

$$\underset{R'}{\overset{R}{>}}C=O + (C_6H_5)_3\overset{+}{P}-\overset{..}{\underset{B}{\overset{A}{C}}} \longrightarrow \underset{R'}{\overset{R}{>}}C=C\overset{A}{\underset{B}{<}} + (C_6H_5)_3\overset{+}{P}-O^-$$

Aldehyde or ketone	Triphenylphosphonium ylide	Alkene	Triphenylphosphine oxide

Wittig reactions may be carried out in a number of different solvents; normally tetrahydrofuran (THF) or dimethyl sulfoxide (DMSO) is used.

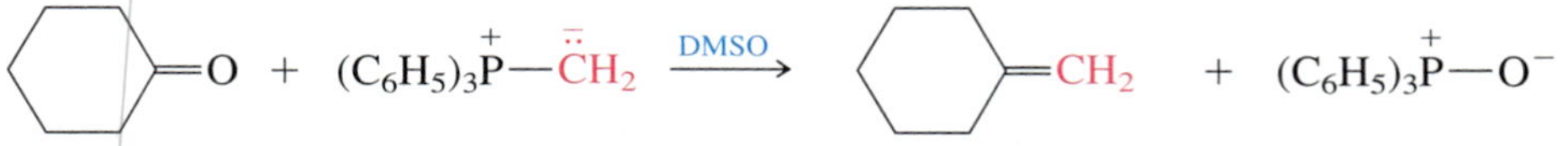

$$⬡{=}O + (C_6H_5)_3\overset{+}{P}-\overset{..}{C}H_2 \xrightarrow{DMSO} ⬡{=}CH_2 + (C_6H_5)_3\overset{+}{P}-O^-$$

Cyclohexanone	Methylenetriphenyl-phosphorane	Methylenecyclohexane (86%)	Triphenylphosphine oxide

The most attractive feature of the Wittig reaction is its regiospecificity. The location of the double bond is never in doubt. The double bond connects the carbon of the original C=O group of the aldehyde or ketone and the negatively charged carbon of the ylide.

PROBLEM 17.14 Identify the alkene product in each of the following Wittig reactions:

(a) Benzaldehyde + $(C_6H_5)_3\overset{+}{P}$—⟨⟩:

(b) Butanal + $(C_6H_5)_3\overset{+}{P}$—$\overset{..}{C}HCH$=CH_2

(c) Cyclohexyl methyl ketone + $(C_6H_5)_3\overset{+}{P}$—$\overset{..}{C}H_2$

SAMPLE SOLUTION (a) In a Wittig reaction the negatively charged substituent attached to phosphorus is transferred to the aldehyde or ketone, replacing the carbonyl oxygen. The reaction shown has been used to prepare the indicated alkene in 65% yield.

Benzaldehyde	Cyclopentylidenetriphenyl-phosphorane		Benzylidenecyclopentane (65%)

To understand the mechanism of the Wittig reaction, we need to examine the structure and properties of ylides. **Ylides** are neutral molecules that have two oppositely charged atoms, each with an octet of electrons, directly bonded to each other. In an ylide such as $(C_6H_5)_3\overset{+}{P}$—$\overset{..}{C}H_2$, phosphorus has eight electrons and is positively charged; its attached carbon has eight electrons and is negatively charged.

PROBLEM 17.15 Can you write a resonance structure for $(C_6H_5)_3\overset{+}{P}$—$\overset{-}{C}H_2$ in which neither phosphorus nor carbon has a formal charge? (*Hint:* Remember phosphorus can have more than eight electrons in its valence shell.)

We can focus on the charge distribution in an ylide by replacing the phenyl groups in $(C_6H_5)_3\overset{+}{P}$—$\overset{-}{C}H_2$ by hydrogens. Figure 17.12 shows the electrostatic potential map of $H_3\overset{+}{P}$—$\overset{..}{C}H_2$, where it can be seen that the electron distribution is highly polarized in the direction that makes carbon electron-rich. The carbon has much of the character of a carbanion and can act as a nucleophile toward C=O.

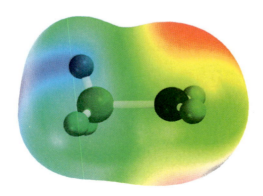

FIGURE 17.12 An electrostatic potential map of the ylide $H_3\overset{+}{P}$—$\overset{..}{C}H_2$. The region of greatest negative charge is concentrated at carbon.

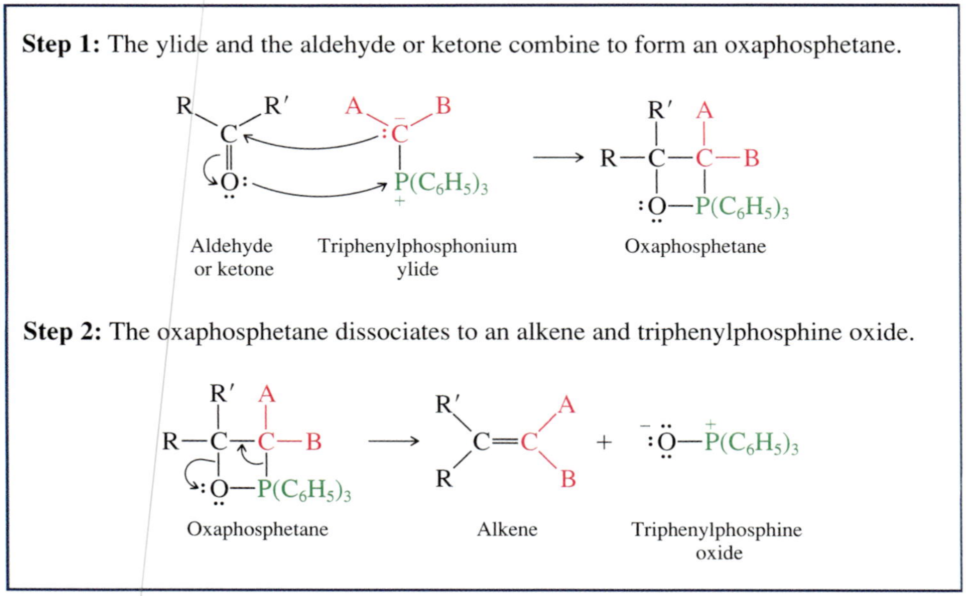

FIGURE 17.13 The mechanism of the Wittig reaction.

Step 1: The ylide and the aldehyde or ketone combine to form an oxaphosphetane.

Aldehyde or ketone   Triphenylphosphonium ylide   Oxaphosphetane

Step 2: The oxaphosphetane dissociates to an alkene and triphenylphosphine oxide.

Oxaphosphetane   Alkene   Triphenylphosphine oxide

The Wittig reaction is one that is still undergoing mechanistic investigation. Another possibility is that the oxaphosphetane intermediate is formed by a two-step process, rather than the one-step process shown in Figure 17.13.

Figure 17.13 outlines a mechanism for the Wittig reaction. The first stage is a cycloaddition in which the ylide reacts with the carbonyl group to give an intermediate containing a four-membered ring called an **oxaphosphetane.** This oxaphosphetane then dissociates to give an alkene and triphenylphosphine oxide. Presumably the direction of dissociation of the oxaphosphetane is dictated by the strong phosphorus–oxygen bond that results. The P—O bond strength in triphenylphosphine oxide has been estimated to be greater than 540 kJ/mol (130 kcal/mol).

17.13 PLANNING AN ALKENE SYNTHESIS VIA THE WITTIG REACTION

To identify the carbonyl compound and the ylide required to produce a given alkene, mentally disconnect the double bond so that one of its carbons is derived from a carbonyl group and the other is derived from an ylide. Taking styrene as a representative example, we see that two such disconnections are possible; either benzaldehyde or formaldehyde is an appropriate precursor.

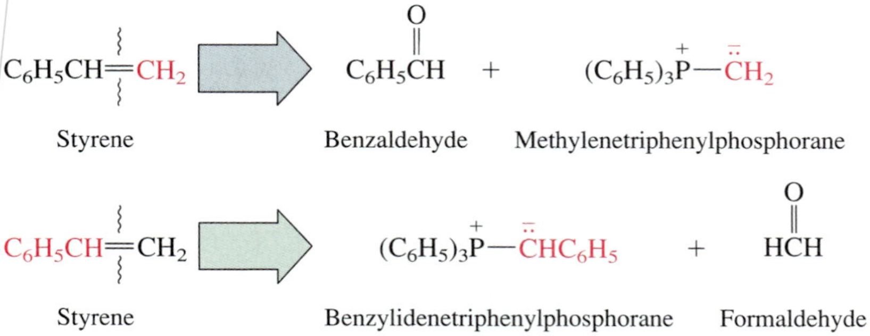

Styrene   Benzaldehyde   Methylenetriphenylphosphorane

Styrene   Benzylidenetriphenylphosphorane   Formaldehyde

Either route is feasible, and indeed styrene has been prepared from both combinations of reactants. Typically there will be two Wittig routes to an alkene, and any choice between them is made on the basis of availability of the particular starting materials.

PROBLEM 17.16 What combinations of carbonyl compound and ylide could you use to prepare each of the following alkenes?

(a) $CH_3CH_2CH_2CH{=}CCH_2CH_3$
         $\underset{\displaystyle CH_3}{|}$

(b) $CH_3CH_2CH_2CH{=}CH_2$

SAMPLE SOLUTION (a) Two Wittig reaction routes lead to the target molecule.

$$CH_3CH_2CH_2CH{=}\underset{\underset{\textstyle CH_3}{|}}{C}CH_2CH_3 \implies CH_3CH_2CH_2\overset{\displaystyle O}{\overset{\|}{C}}H \ + \ (C_6H_5)_3\overset{+}{P}{-}\underset{\underset{\textstyle CH_3}{|}}{\overset{..}{C}}CH_2CH_3$$

3-Methyl-3-heptene   Butanal   1-Methylpropylidenetriphenyl-
                                                            phosphorane

and

$$CH_3CH_2CH_2CH{=}\underset{\underset{\textstyle CH_3}{|}}{C}CH_2CH_3 \implies CH_3CH_2CH_2\overset{..}{C}H{-}\overset{+}{P}(C_6H_5)_3 \ + \ CH_3\overset{\displaystyle O}{\overset{\|}{C}}CH_2CH_3$$

3-Methyl-3-heptene   Butylidenetriphenylphosphorane   2-Butanone

Phosphorus ylides are prepared from alkyl halides by a two-step sequence. The first step is a nucleophilic substitution of the S_N2 type by triphenylphosphine on an alkyl halide to give an alkyltriphenylphosphonium salt:

$$(C_6H_5)_3P\!: \overset{\displaystyle A}{\underset{\displaystyle B}{\underset{\frown}{CH}}}{-}X \xrightarrow{\ S_N2\ } (C_6H_5)_3\overset{+}{P}{-}\overset{\displaystyle A}{\underset{\displaystyle |}{CH}}{-}B \ \ :X^-$$

Triphenylphosphine   Alkyl halide   Alkyltriphenylphosphonium
                                                    halide

Triphenylphosphine is a very powerful nucleophile, yet is not strongly basic. Methyl, primary, and secondary alkyl halides are all suitable substrates.

$$(C_6H_5)_3P\!: \ + \ CH_3Br \xrightarrow{\ benzene\ } (C_6H_5)_3\overset{+}{P}{-}CH_3 \ \ Br^-$$

Triphenylphosphine   Bromomethane   Methyltriphenylphosphonium
                                                bromide (99%)

The alkyltriphenylphosphonium salt products are ionic and crystallize in high yield from the nonpolar solvents in which they are prepared. After isolation, the alkyltriphenylphosphonium halide is converted to the desired ylide by deprotonation with a strong base:

$$(C_6H_5)_3\overset{+}{P}{-}\overset{\displaystyle A}{\underset{\displaystyle H}{\overset{|}{\underset{|}{C}}}}{-}B \ + \ Y^- \longrightarrow (C_6H_5)_3\overset{+}{P}{-}\overset{..}{\underset{\displaystyle B}{\overset{\displaystyle A}{C}}} \ + \ HY$$

Alkyltriphenylphosphonium salt   Base   Triphenylphosphonium ylide   Conjugate acid
                                                                          of base used

Suitable strong bases include the sodium salt of dimethyl sulfoxide (in dimethyl sulfoxide as the solvent) and organolithium reagents (in diethyl ether or tetrahydrofuran).

$$(C_6H_5)_3\overset{+}{P}\!-\!CH_3 \; Br^- \; + \; NaCH_2SCH_3 \; \xrightarrow{\text{DMSO}} \; (C_6H_5)_3\overset{+}{P}\!-\!\overset{..}{\overset{..}{C}}H_2 \; + \; CH_3SCH_3 \; + \; NaBr$$

Methyltriphenylphos-phonium bromide	Sodiomethyl methyl sulfoxide	Methylenetri-phenylphos-phorane	Dimethyl sulfoxide	Sodium bromide

PROBLEM 17.17 The sample solution to Problem 17.16(a) showed the preparation of 3-methyl-3-heptene by a Wittig reaction involving the ylide shown. Write equations showing the formation of this ylide beginning with 2-bromobutane.

$$(C_6H_5)_3\overset{+}{P}\!-\!\overset{..}{\overset{..}{C}}CH_2CH_3$$
$$\underset{\displaystyle CH_3}{|}$$

Normally the ylides are not isolated. Instead, the appropriate aldehyde or ketone is added directly to the solution in which the ylide was generated.

17.14 STEREOSELECTIVE ADDITION TO CARBONYL GROUPS

Nucleophilic addition to carbonyl groups sometimes leads to a mixture of stereoisomeric products. The direction of attack is often controlled by steric factors, with the nucleophile approaching the carbonyl group at its less hindered face. Sodium borohydride reduction of 7,7-dimethylbicyclo[2.2.1]heptan-2-one illustrates this point:

You may find it helpful to make molecular models of the reactant and products in the reaction shown.

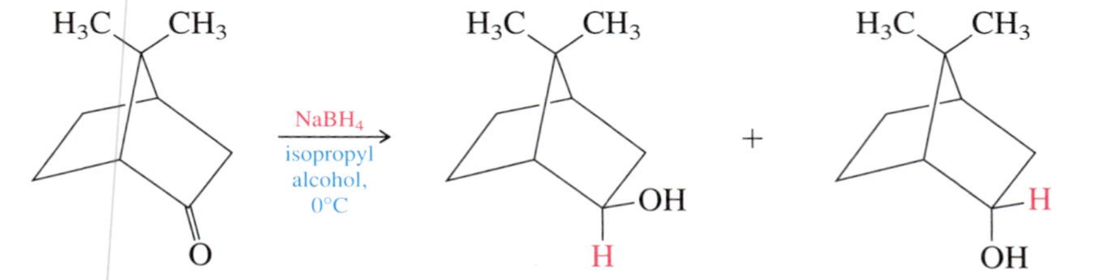

7,7-Dimethylbicyclo-[2.2.1]heptan-2-one → (NaBH₄, isopropyl alcohol, 0°C) → *exo*-7,7-Dimethylbicyclo-[2.2.1]heptan-2-ol (80%) + *endo*-7,7-Dimethylbicyclo-[2.2.1]heptan-2-ol (20%)

Approach of borohydride to the top face of the carbonyl group is sterically hindered by one of the methyl groups. The bottom face of the carbonyl group is less congested, and the major product is formed by hydride transfer from this direction.

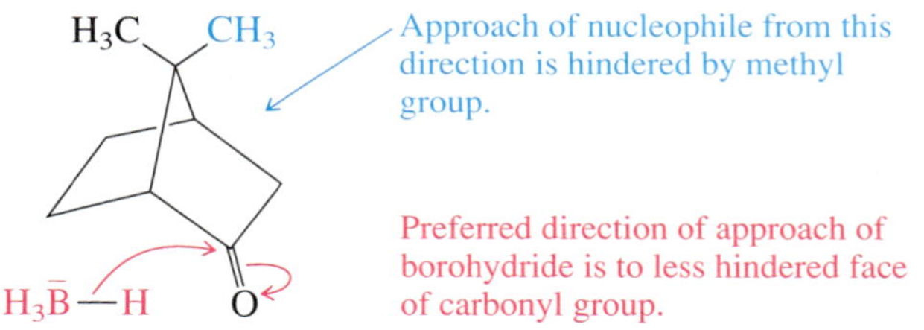

Approach of nucleophile from this direction is hindered by methyl group.

Preferred direction of approach of borohydride is to less hindered face of carbonyl group.

The reduction is *stereoselective*. A single starting material can form two stereoisomers of the product but yields one of them in greater amounts than the other or even to the exclusion of the other.

> **PROBLEM 17.18**   What is the relationship between the products of the reaction just described? Are they enantiomers or diastereomers?

Enzyme-catalyzed reductions of carbonyl groups are, more often than not, completely stereoselective. Pyruvic acid, for example, is converted exclusively to (S)-$(+)$-lactic acid by the lactate dehydrogenase-NADH system (Section 15.11). The enantiomer (R)-$(-)$-lactic acid is not formed.

$$CH_3\overset{\displaystyle O}{\overset{\|}{C}}-\overset{\displaystyle O}{\overset{\|}{C}}OH \; + \; NADH \; + \; H^+ \; \xrightarrow[\text{dehydrogenase}]{\text{lactate}} \; H_3C\underset{HO}{\overset{H}{\diagdown}}\!C-\overset{\displaystyle O}{\overset{\|}{C}}OH \; + \; NAD^+$$

Pyruvic acid	Reduced form of coenzyme		(S)-$(+)$-lactic acid	Oxidized form of coenzyme

The enzyme is a single enantiomer of a chiral molecule and binds the coenzyme and substrate in such a way that hydride is transferred exclusively to the face of the carbonyl group that leads to (S)-$(+)$-lactic acid. Reduction of pyruvic acid in the *absence* of an enzyme however, say with sodium borohydride, also gives lactic acid but as a racemic mixture containing equal quantities of the R and S enantiomers.

The stereoselectivity of enzyme-catalyzed reactions can be understood on the basis of a relatively simple model. Consider the case of an sp^2-hybridized carbon with prochiral faces as in Figure 17.14a. If structural features on the enzyme are complementary in some respect to the groups attached to this carbon, one prochiral face can bind to the enzyme better than the other—there will be a preferred geometry of the enzyme–substrate complex. The binding forces are the usual ones: electrostatic, van der Waals, and so on. If a reaction occurs that converts the sp^2-hybridized carbon to sp^3, there will be a bias toward adding the fourth group from a particular direction as shown in Figure 17.14b. As a result, an achiral molecule is converted to a single enantiomer of a chiral one. The reaction is stereoselective because it occurs preferentially at one prochiral face.

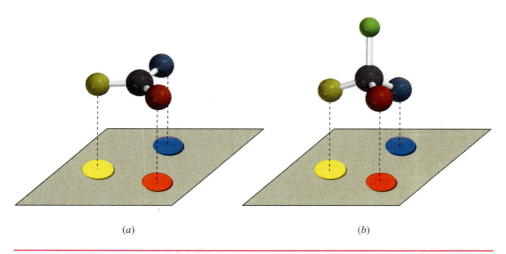

(a)   (b)

FIGURE 17.14 (a) Binding sites of enzyme discriminate between prochiral faces of substrate. One prochiral face can bind to the enzyme better than the other. (b) Reaction attaches fourth group to substrate producing only one enantiomer of chiral product.

17.15 OXIDATION OF ALDEHYDES

Aldehydes are readily oxidized to carboxylic acids by a number of reagents, including those based on Cr(VI) in aqueous media.

$$\underset{\text{Aldehyde}}{\overset{\overset{\displaystyle O}{\|}}{\text{RCH}}} \xrightarrow{\text{oxidize}} \underset{\text{Carboxylic acid}}{\overset{\overset{\displaystyle O}{\|}}{\text{RCOH}}}$$

Furfural $\xrightarrow[\text{H}_2\text{SO}_4,\ \text{H}_2\text{O}]{\text{K}_2\text{Cr}_2\text{O}_7}$ Furoic acid (75%)

Mechanistically, these reactions probably proceed through the hydrate of the aldehyde and follow a course similar to that of alcohol oxidation.

$$\underset{\text{Aldehyde}}{\overset{\overset{\displaystyle O}{\|}}{\text{RCH}}} + \text{H}_2\text{O} \rightleftharpoons \underset{\substack{\text{Geminal diol} \\ \text{(hydrate)}}}{\overset{\overset{\displaystyle OH}{|}}{\underset{\underset{\displaystyle OH}{|}}{\text{RCH}}}} \xrightarrow{\text{oxidize}} \underset{\substack{\text{Carboxylic} \\ \text{acid}}}{\overset{\overset{\displaystyle O}{\|}}{\text{RCOH}}}$$

Aldehydes are more easily oxidized than alcohols, which is why special reagents such as PCC and PDC (Section 15.10) have been developed for oxidizing primary alcohols to aldehydes and no further. PCC and PDC are effective because they are sources of Cr(VI), but are used in nonaqueous media (dichloromethane). By keeping water out of the reaction mixture, the aldehyde is not converted to its hydrate, which is the necessary intermediate that leads to the carboxylic acid.

17.16 BAEYER–VILLIGER OXIDATION OF KETONES

Peroxy acids have been seen before as reagents for the epoxidation of alkenes (Section 6.18).

The reaction of ketones with peroxy acids is both novel and synthetically useful. An oxygen from the peroxy acid is inserted between the carbonyl group and one of the attached carbons of the ketone to give an *ester*. Reactions of this type were first described by Adolf von Baeyer and Victor Villiger in 1899 and are known as **Baeyer–Villiger oxidations.**

$$\underset{\text{Ketone}}{\overset{\overset{\displaystyle O}{\|}}{\text{RCR}'}} + \underset{\text{Peroxy acid}}{\overset{\overset{\displaystyle O}{\|}}{\text{R}''\text{COOH}}} \longrightarrow \underset{\text{Ester}}{\overset{\overset{\displaystyle O}{\|}}{\text{RCOR}'}} + \underset{\text{Carboxylic acid}}{\overset{\overset{\displaystyle O}{\|}}{\text{R}''\text{COH}}}$$

Methyl ketones give esters of acetic acid; that is, oxygen insertion occurs between the carbonyl carbon and the larger of the two groups attached to it.

Cyclohexyl methyl ketone $\xrightarrow[\text{CHCl}_3]{\text{C}_6\text{H}_5\text{CO}_2\text{OH}}$ Cyclohexyl acetate (67%)

The mechanism of the Baeyer–Villiger oxidation is shown in Figure 17.15. It begins with nucleophilic addition of the peroxy acid to the carbonyl group of the ketone, followed by migration of an alkyl group from the carbonyl group to oxygen. In general, it is the more substituted group that migrates. The *migratory aptitude* of the various alkyl groups is:

Tertiary alkyl > secondary alkyl > primary alkyl > methyl

PROBLEM 17.19 Using Figure 17.15 as a guide, write a mechanism for the Baeyer–Villiger oxidation of cyclohexyl methyl ketone by peroxybenzoic acid.

The reaction is stereospecific; the alkyl group migrates with retention of configuration.

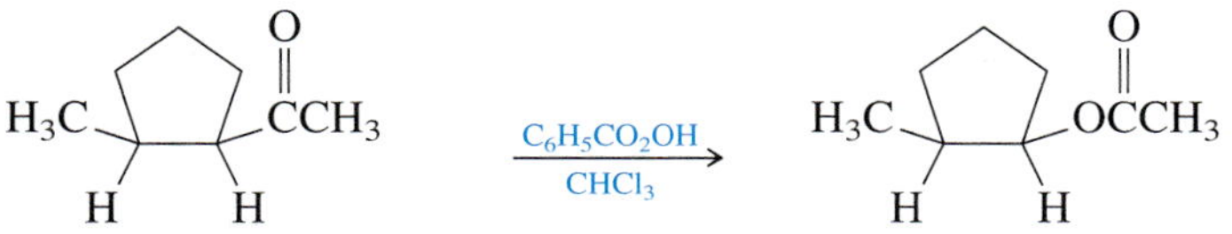

cis-1-Acetyl-2-methylcyclopentane

cis-2-Methylcyclopentyl acetate (only product; 66% yield)

Overall reaction:

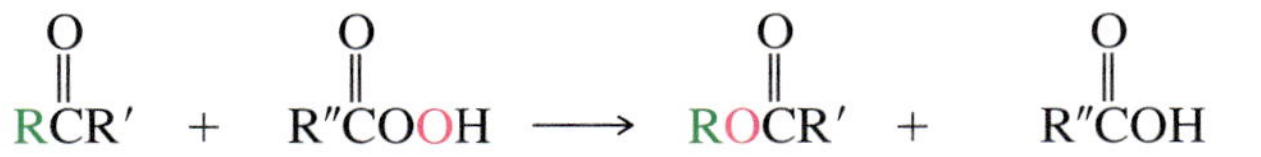

Ketone | Peroxy acid | Ester | Carboxylic acid

Step 1: The peroxy acid adds to the carbonyl group of the ketone. This step is a nucleophilic addition analogous to *gem*-diol and hemiacetal formation.

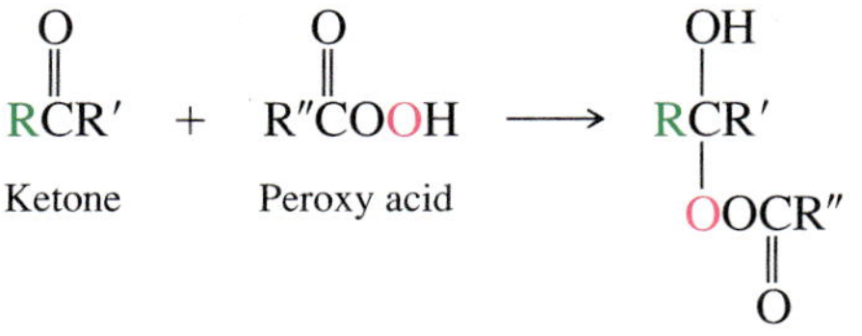

(Peroxy monoester of *gem*-diol)

Ketone | Peroxy acid

Step 2: The intermediate from step 1 undergoes rearrangement. Cleavage of the weak O—O bond of the peroxy ester is assisted by migration of one of the substituents from the carbonyl group to oxygen. The group R migrates with its pair of electrons in much the same way as alkyl groups migrate in carbocation rearrangements.

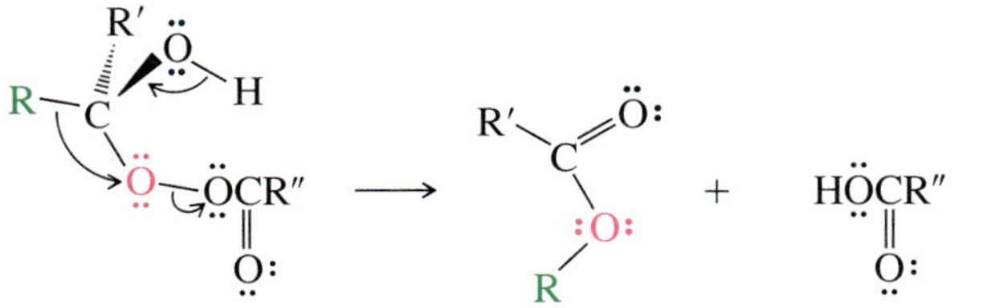

Ester | Carboxylic acid

FIGURE 17.15 Mechanism of the Baeyer–Villiger oxidation of a ketone.

In the companion experiment carried out on the trans stereoisomer of the ketone, only the trans acetate was formed.

As unusual as the Baeyer–Villiger reaction may seem, it is even more remarkable that an analogous reaction occurs in living systems. Certain bacteria, including those of the *Pseudomonas* and *Acinetobacter* type, can use a variety of organic compounds, even hydrocarbons, as a carbon source. With cyclohexane, for example, the early stages of this process proceed by oxidation to cyclohexanone, which then undergoes the "biological Baeyer–Villiger reaction."

The November, 2001 issue of the *Journal of Chemical Education* (pp. 1533–1534) describes an introductory biochemistry laboratory experiment involving cyclohexanone monooxygenase oxidation of cyclic ketones.

Cyclohexane   Cyclohexanone   6-Hexanolide

The product (6-hexanolide) is a cyclic ester, or *lactone* (Section 19.15). Like the Baeyer–Villiger oxidation, an oxygen atom is inserted between the carbonyl group and a carbon attached to it. But peroxy acids are not involved in any way; the oxidation of cyclohexanone is catalyzed by an enzyme called *cyclohexanone monooxygenase* with the aid of certain coenzymes.

17.17 SPECTROSCOPIC ANALYSIS OF ALDEHYDES AND KETONES

Infrared: Carbonyl groups are among the easiest functional groups to detect by IR spectroscopy. The $C=O$ stretching vibration of aldehydes and ketones gives rise to strong absorption in the region 1710–1750 cm^{-1} as illustrated for butanal in Figure 17.16. In addition to a peak for $C=O$ stretching, the $CH=O$ group of an aldehyde exhibits two weak bands for $C-H$ stretching near 2720 and 2820 cm^{-1}.

^{1}H NMR: Aldehydes are readily identified by the presence of a signal for the hydrogen of $CH=O$ at δ 9–10. This is a region where very few other protons ever appear. Figure 17.17 shows the ^{1}H NMR spectrum of 2-methylpropanal [$(CH_3)_2CHCH=O)$], where the large chemical shift difference between the aldehyde proton and the other protons in the molecule is clearly evident. As seen in the expanded-scale inset, the aldehyde proton is a doublet, split by the proton as C-2. Coupling between the protons in $HC-CH=O$ is much smaller than typical vicinal couplings, making the multiplicity of the aldehyde peak difficult to see without expanding the scale.

Methyl ketones, such as 2-butanone in Figure 17.18, are characterized by sharp singlets near δ 2 for the protons of $CH_3C=O$. Similarly, the deshielding effect of the carbonyl causes the protons of $CH_2C=O$ to appear at lower field (δ 2.4) than in a CH_2 group of an alkane.

^{13}C NMR: The signal for the carbon of $C=O$ in aldehydes and ketones appears at very low field, some 190–220 ppm downfield from tetramethylsilane. Figure 17.19 illustrates this for 3-heptanone, in which separate signals appear for each of the seven carbons. The six sp^3-hybridized carbons appear in the range δ 8–42, and the carbon of the $C=O$ group is at δ 210. Note, too, that the intensity of the peak for the $C=O$ carbon is much less than all the others, even though each peak corresponds to a single carbon. This decreased intensity is a characteristic of pulsed Fourier transform (FT) spectra for carbons that don't have attached hydrogens.

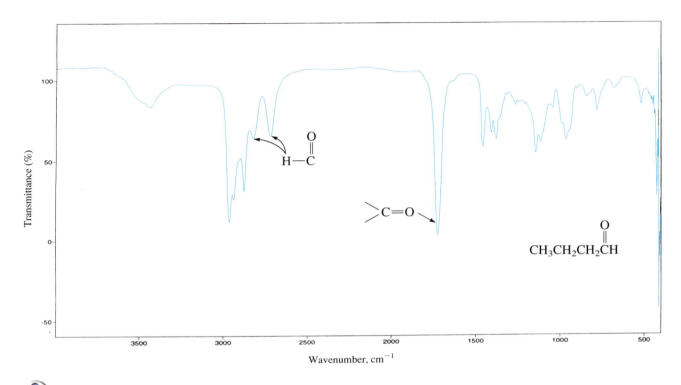

FIGURE 17.16 IR spectrum of butanal showing peaks characteristic of the CH=O unit at 2700 and 2800 cm^{-1} (C—H) and at 1720 cm^{-1} (C=O).

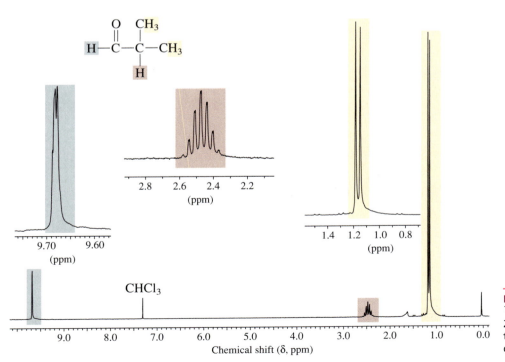

FIGURE 17.17 The 200-MHz ^{1}H NMR spectrum of 2-methylpropanal, showing the aldehyde proton as a doublet at low field (δ 9.7).

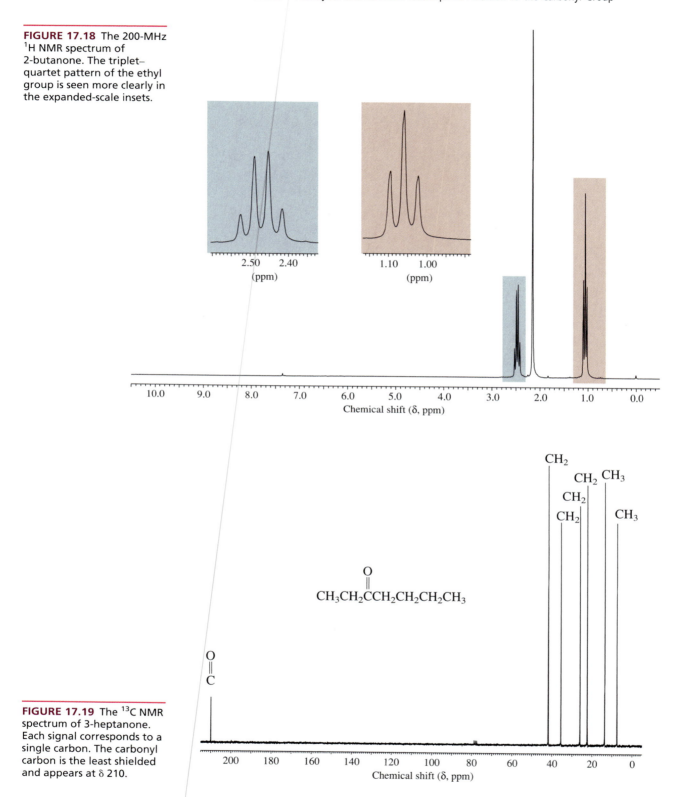

FIGURE 17.18 The 200-MHz ¹H NMR spectrum of 2-butanone. The triplet–quartet pattern of the ethyl group is seen more clearly in the expanded-scale insets.

FIGURE 17.19 The ¹³C NMR spectrum of 3-heptanone. Each signal corresponds to a single carbon. The carbonyl carbon is the least shielded and appears at δ 210.

UV-VIS: Aldehydes and ketones have two absorption bands in the ultraviolet region. Both involve excitation of an electron to an antibonding π^* orbital. In one, called a $\pi \rightarrow \pi^*$ transition, the electron is one of the π electrons of the C=O group. In the other, called an $n \rightarrow \pi^*$ transition, it is one of the oxygen lone-pair electrons. Because the π electrons are more strongly held than the lone-pair electrons, the $\pi \rightarrow \pi^*$ transition is of higher energy and shorter wavelength than the $n \rightarrow \pi^*$ transition. For simple aldehydes and ketones, the $\pi \rightarrow \pi^*$ transition is below 200 nm and of little use in structure determination. The $n \rightarrow \pi^*$ transition, although weak, is of more diagnostic value.

$$\text{H}_3\text{C} \diagdown \text{C}=\ddot{\text{O}}: \quad \begin{array}{l} \pi \rightarrow \pi^* \ \lambda_{max} \ 187 \text{ nm} \\ n \rightarrow \pi^* \ \lambda_{max} \ 270 \text{ nm} \end{array}$$
$$\text{H}_3\text{C} \diagup$$

Acetone

Mass Spectrometry: Aldehydes and ketones typically give a prominent molecular ion peak in their mass spectra. Aldehydes also exhibit an M−1 peak. A major fragmentation pathway for both aldehydes and ketones leads to formation of acyl cations (acylium ions) by cleavage of an alkyl group from the carbonyl. The most intense peak in the mass spectrum of diethyl ketone, for example, is m/z 57, corresponding to loss of ethyl radical from the molecular ion.

$$\overset{:\overset{+}{\text{O}}\cdot}{\overset{\|}{\text{CH}_3\text{CH}_2\text{CCH}_2\text{CH}_3}} \longrightarrow \text{CH}_3\text{CH}_2\text{C}{=}\ddot{\text{O}}^+ + \cdot\text{CH}_2\text{CH}_3$$

m/z 86 m/z 57

17.18 SUMMARY

The chemistry of the carbonyl group is probably the single most important aspect of organic chemical reactivity. Classes of compounds that contain the carbonyl group include many derived from carboxylic acids (acyl chlorides, acid anhydrides, esters, and amides) as well as the two related classes discussed in this chapter: *aldehydes* and *ketones.*

Section 17.1   The substitutive IUPAC names of aldehydes and ketones are developed by identifying the longest continuous chain that contains the carbonyl group and replacing the final *-e* of the corresponding alkane by *-al* for aldehydes and *-one* for ketones. The chain is numbered in the direction that gives the lowest locant to the carbon of the carbonyl group.

3-Methylbutanal   3-Methyl-2-butanone

Ketones may also be named using functional class IUPAC nomenclature by citing the two groups attached to the carbonyl in alphabetical order followed by the word *ketone.* Thus, 3-methyl-2-butanone (substitutive) becomes isopropyl methyl ketone (functional class).

Section 17.2   The carbonyl carbon is sp^2-hybridized, and it and the atoms attached to it are coplanar. Aldehydes and ketones are polar molecules. Nucleophiles attack C=O at carbon (positively polarized) and electrophiles, especially protons, attack oxygen (negatively polarized).

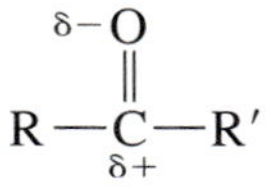

$$
\begin{array}{c}
\overset{\delta-}{O} \\
\parallel \\
R-\underset{\delta+}{C}-R'
\end{array}
$$

Section 17.3   Aldehydes and ketones have higher boiling points than hydrocarbons, but have lower boiling points than alcohols.

Section 17.4   The numerous reactions that yield aldehydes and ketones discussed in earlier chapters and reviewed in Table 17.1 are sufficient for most syntheses.

Sections 17.5–17.13   The characteristic reactions of aldehydes and ketones involve *nucleophilic addition* to the carbonyl group and are summarized in Table 17.5. Reagents of the type HY react according to the general equation

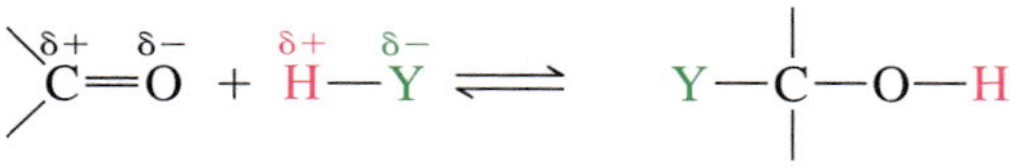

$$
\overset{\delta+}{\underset{}{C}}{=}\overset{\delta-}{O} + \overset{\delta+}{H}{-}\overset{\delta-}{Y} \rightleftharpoons Y-\overset{|}{\underset{|}{C}}-O-H
$$

Aldehyde   Product of nucleophilic
or ketone   addition to carbonyl group

Aldehydes undergo nucleophilic addition more readily and have more favorable equilibrium constants for addition than do ketones.

The step in which the nucleophile attacks the carbonyl carbon is rate-determining in both base-catalyzed and acid-catalyzed nucleophilic addition. In the base-catalyzed mechanism this is the first step.

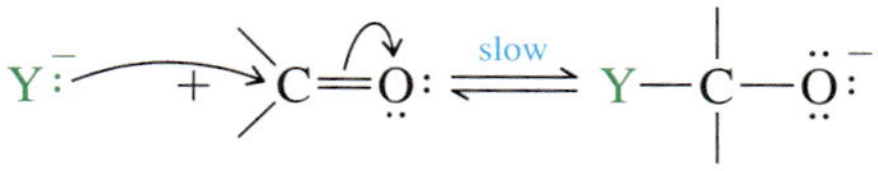

$$
Y:^- + \quad C{=}\ddot{O}: \xrightarrow{\text{slow}} Y-\overset{|}{\underset{|}{C}}-\ddot{O}:^-
$$

Nucleophile   Aldehyde
                   or ketone

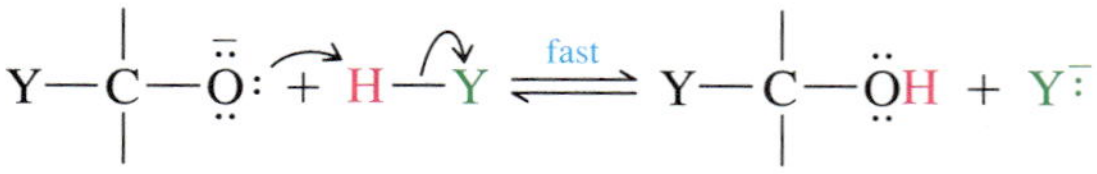

$$
Y-\overset{|}{\underset{|}{C}}-\ddot{O}:^- \; + \; H{-}Y \xrightarrow{\text{fast}} Y-\overset{|}{\underset{|}{C}}-\ddot{O}H \; + \; Y:^-
$$

Product of
nucleophilic
addition

Under conditions of acid catalysis, the nucleophilic addition step follows protonation of the carbonyl oxygen. Protonation increases the carbocation character of a carbonyl group and makes it more electrophilic.

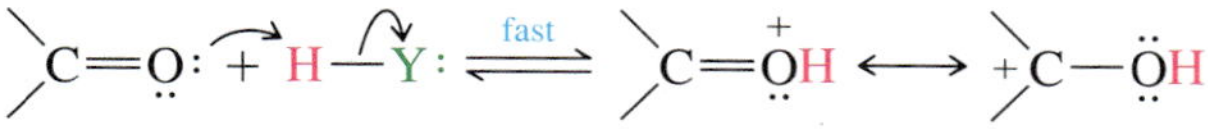

$$
C{=}\ddot{O}: \; + \; H{-}Y: \xrightarrow{\text{fast}} C{=}\overset{+}{O}H \longleftrightarrow \overset{+}{C}-\ddot{O}H
$$

Aldehyde   Resonance forms of protonated
or ketone   aldehyde or ketone

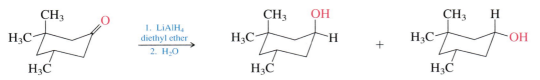

Often the product of nucleophilic addition is not isolated but is an intermediate leading to the ultimate product. Most of the reactions in Table 17.5 are of this type.

Section 17.14   Nucleophilic addition to the carbonyl group is *stereoselective*. When one direction of approach to the carbonyl group is less hindered than the other, the nucleophile normally attacks at the less hindered face.

3,3,5-Trimethylcyclohexanone

trans-3,3,5-Trimethylcyclohexanol
(83%)

cis-3,3,5-Trimethylcyclohexanol
(17%)

TABLE 17.5   Nucleophilic Addition to Aldehydes and Ketones

Reaction (section) and comments	General equation and typical example
Hydration (Section 17.6) Can be either acid- or base-catalyzed. Equilibrium constant is normally unfavorable for hydration of ketones unless R, R′, or both are strongly electron-withdrawing.	$\underset{\text{Aldehyde or ketone}}{\overset{O}{\underset{\|}{RCR'}}}$ + $\underset{\text{Water}}{H_2O}$ ⇌ $\underset{\text{Geminal diol}}{\overset{OH}{\underset{\|}{RCR'}}}\overset{}{\underset{OH}{}}$ $\underset{\substack{\text{Chloroacetone}\\(90\%\text{ at equilibrium})}}{\overset{O}{\underset{\|}{ClCH_2CCH_3}}}$ $\overset{H_2O}{\rightleftharpoons}$ $\underset{\substack{\text{Chloroacetone hydrate}\\(10\%\text{ at equilibrium})}}{\overset{OH}{\underset{\|}{ClCH_2CCH_3}}\overset{}{\underset{OH}{}}}$
Cyanohydrin formation (Section 17.7) Reaction is catalyzed by cyanide ion. Cyanohydrins are useful synthetic intermediates; cyano group can be hydrolyzed to —CO_2H or reduced to —CH_2NH_2.	$\underset{\substack{\text{Aldehyde}\\\text{or ketone}}}{\overset{O}{\underset{\|}{RCR'}}}$ + $\underset{\substack{\text{Hydrogen}\\\text{cyanide}}}{HCN}$ ⇌ $\underset{\text{Cyanohydrin}}{\overset{OH}{\underset{\|}{RCR'}}\overset{}{\underset{CN}{}}}$ $\underset{\text{3-Pentanone}}{\overset{O}{\underset{\|}{CH_3CH_2CCH_2CH_3}}}$ $\xrightarrow[H^+]{KCN}$ $\underset{\text{3-Pentanone cyanohydrin (75\%)}}{\overset{OH}{\underset{\|}{CH_3CH_2CCH_2CH_3}}\overset{}{\underset{CN}{}}}$

(Continued)

TABLE 17.5   Nucleophilic Addition to Aldehydes and Ketones *(Continued)*

Reaction (section) and comments	General equation and typical example

Acetal formation (Sections 17.8–17.9) Reaction is acid-catalyzed. Equilibrium constant normally favorable for aldehydes, unfavorable for ketones. Cyclic acetals from vicinal diols form readily.

$$\underset{\substack{\text{Aldehyde} \\ \text{or ketone}}}{\overset{\displaystyle O}{\underset{\displaystyle \|}{RCR'}}} + \underset{\text{Alcohol}}{2R''OH} \underset{}{\overset{H^+}{\rightleftharpoons}} \underset{\text{Acetal}}{\overset{\displaystyle OR''}{\underset{\displaystyle OR''}{RCR'}}} + \underset{\text{Water}}{H_2O}$$

m-Nitrobenzaldehyde + CH_3OH →(HCl)→ m-Nitrobenzaldehyde dimethyl acetal (76–85%)

[Structure: benzaldehyde with NO_2 substituent, CHO group] + CH_3OH →(HCl)→ [benzene ring with NO_2 and CH(OCH_3)_2]

m-Nitrobenzaldehyde   Methanol   m-Nitrobenzaldehyde dimethyl acetal (76–85%)

Reaction with primary amines (Section 17.10) Isolated product is an imine (Schiff's base). A carbinolamine intermediate is formed, which undergoes dehydration to an imine.

$$\underset{\substack{\text{Aldehyde or ketone}}}{\overset{\displaystyle O}{\underset{\displaystyle \|}{RCR'}}} + \underset{\text{Primary amine}}{R''NH_2} \rightleftharpoons \underset{\text{Imine}}{\overset{\displaystyle NR''}{\underset{\displaystyle \|}{RCR'}}} + \underset{\text{Water}}{H_2O}$$

$$\underset{\text{2-Methylpropanal}}{\overset{\displaystyle O}{\underset{\displaystyle \|}{(CH_3)_2CHCH}}} + \underset{\text{\textit{tert}-Butylamine}}{(CH_3)_3CNH_2} \longrightarrow \underset{\substack{\text{\textit{N}-(2-Methyl-1-propylidene)-} \\ \text{\textit{tert}-butylamine (50\%)}}}{(CH_3)_2CHCH\!=\!NC(CH_3)_3}$$

Reaction with secondary amines (Section 17.11) Isolated product is an enamine. Carbinolamine intermediate cannot dehydrate to a stable imine.

$$\underset{\substack{\text{Aldehyde} \\ \text{or ketone}}}{\overset{\displaystyle O}{\underset{\displaystyle \|}{RCCH_2R'}}} + \underset{\substack{\text{Secondary} \\ \text{amine}}}{R''_2NH} \rightleftharpoons \underset{\text{Enamine}}{\overset{\displaystyle R''NR''}{\underset{\displaystyle |}{RC\!=\!CHR'}}} + \underset{\text{Water}}{H_2O}$$

Cyclohexanone + Morpholine →(benzene heat)→ 1-Morpholinocyclohexene (85%)

The Wittig reaction (Sections 17.12–17.13) Reaction of a phosphorus ylide with aldehydes and ketones leads to the formation of an alkene. A versatile method for the regiospecific preparation of alkenes.

$$\underset{\substack{\text{Aldehyde} \\ \text{or ketone}}}{\overset{\displaystyle O}{\underset{\displaystyle \|}{RCR'}}} + \underset{\substack{\text{Wittig} \\ \text{reagent (an ylide)}}}{(C_6H_5)_3\overset{+}{P}\!-\!\overset{..}{\underset{\displaystyle B}{\overset{\displaystyle A}{C}}}} \longrightarrow \underset{\text{Alkene}}{\overset{\displaystyle R\quad A}{\underset{\displaystyle R'\quad B}{C\!=\!C}}} + \underset{\substack{\text{Triphenylphosphine} \\ \text{oxide}}}{(C_6H_5)_3\overset{+}{P}\!-\!O^-}$$

$$\underset{\text{Acetone}}{\overset{\displaystyle O}{\underset{\displaystyle \|}{CH_3CCH_3}}} + \underset{\text{1-Pentylidenetriphenylphosphorane}}{(C_6H_5)_3\overset{+}{P}\!-\!\overset{..}{C}HCH_2CH_2CH_2CH_3} \overset{DMSO}{\longrightarrow}$$

$$\underset{\substack{\text{2-Methyl-2-heptene} \\ \text{(56\%)}}}{(CH_3)_2C\!=\!CHCH_2CH_2CH_2CH_3} + \underset{\substack{\text{Triphenylphosphine} \\ \text{oxide}}}{(C_6H_5)_3\overset{+}{P}\!-\!O^-}$$

Section 17.15   Aldehydes are easily oxidized to carboxylic acids.

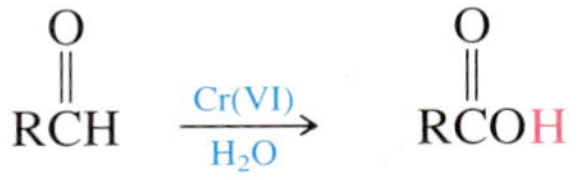

$$\underset{\text{Aldehyde}}{\overset{\overset{\displaystyle O}{\|}}{\text{RCH}}} \quad \xrightarrow[\text{H}_2\text{O}]{\text{Cr(VI)}} \quad \underset{\text{Carboxylic acid}}{\overset{\overset{\displaystyle O}{\|}}{\text{RCOH}}}$$

Section 17.16   The oxidation of ketones with peroxy acids is called the *Baeyer–Villiger oxidation* and is a useful method for preparing esters.

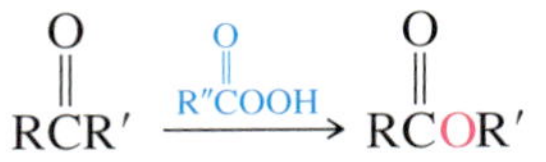

$$\underset{\text{Ketone}}{\overset{\overset{\displaystyle O}{\|}}{\text{RCR}'}} \quad \xrightarrow{\text{R}''\text{COOH}} \quad \underset{\text{Ester}}{\overset{\overset{\displaystyle O}{\|}}{\text{RCOR}'}}$$

Section 17.17   A strong peak near 1700 cm^{-1} in the IR spectrum is characteristic of compounds that contain a C=O group. The ^{1}H and ^{13}C NMR spectra of aldehydes and ketones are affected by the deshielding of a C=O group. The proton of an H—C=O group appears in the δ 8–10 range. The carbon of a C=O group is at δ 190–210.

PROBLEMS

17.20 (a) Write structural formulas and provide IUPAC names for all the isomeric aldehydes and ketones that have the molecular formula $C_5H_{10}O$. Include stereoisomers.

    (b) Which of the isomers in part (a) yield chiral alcohols on reaction with sodium borohydride?

    (c) Which of the isomers in part (a) yield chiral alcohols on reaction with methylmagnesium iodide?

17.21 Each of the following aldehydes or ketones is known by a common name. Its substitutive IUPAC name is provided in parentheses. Write a structural formula for each one.

    (a) Chloral (2,2,2-trichloroethanal)

    (b) Pivaldehyde (2,2-dimethylpropanal)

    (c) Acrolein (2-propenal)

    (d) Crotonaldehyde [(*E*)-2-butenal]

    (e) Citral [(*E*)-3,7-dimethyl-2,6-octadienal]

    (f) Diacetone alcohol (4-hydroxy-4-methyl-2-pentanone)

    (g) Carvone (5-isopropenyl-2-methyl-2-cyclohexenone)

    (h) Biacetyl (2,3-butanedione)

17.22 The African dwarf crocodile secretes a volatile substance believed to be a sex pheromone. It is a mixture of two stereoisomers, one of which is shown:

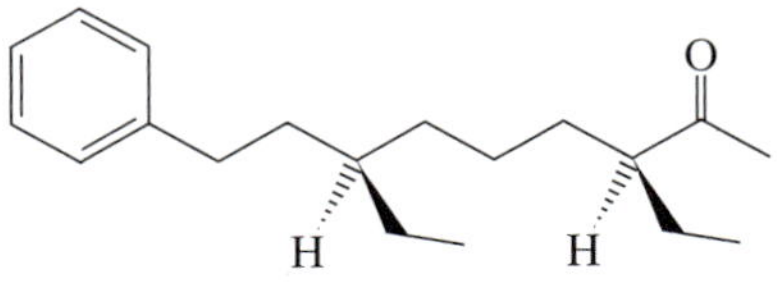

    (a) Give the IUPAC name for this compound, including *R* and *S* descriptors for its chirality centers.

(b) One component of the scent substance has the *S* configuration at both chirality centers. How is this compound related to the one shown? Is the compound shown its enantiomer, or a diastereomer?

17.23 Predict the product of the reaction of propanal with each of the following:

(a) Lithium aluminum hydride

(b) Sodium borohydride

(c) Hydrogen (nickel catalyst)

(d) Methylmagnesium iodide, followed by dilute acid

(e) Sodium acetylide, followed by dilute acid

(f) Phenyllithium, followed by dilute acid

(g) Methanol containing dissolved hydrogen chloride

(h) Ethylene glycol, *p*-toluenesulfonic acid, benzene

(i) Aniline ($C_6H_5NH_2$)

(j) Dimethylamine, *p*-toluenesulfonic acid, benzene

(k) Hydroxylamine

(l) Hydrazine

(m) Product of part (l) heated in triethylene glycol with sodium hydroxide

(n) *p*-Nitrophenylhydrazine

(o) Semicarbazide

(p) Ethylidenetriphenylphosphorane $[(C_6H_5)_3\overset{+}{P}—\overset{..}{C}HCH_3]$

(q) Sodium cyanide with addition of sulfuric acid

(r) Chromic acid

17.24 Repeat the preceding problem for cyclopentanone instead of propanal.

17.25 Hydride reduction (with $LiAlH_4$ or $NaBH_4$) of each of the following ketones has been reported in the chemical literature and gives a mixture of two diastereomeric alcohols in each case. Give the structures or build molecular models of both alcohol products for each ketone.

(a) (*S*)-3-Phenyl-2-butanone

(b) 4-*tert*-Butylcyclohexanone

(c)

(d)

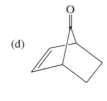

17.26 Choose which member in each of the following pairs reacts faster or has the more favorable equilibrium constant for reaction with the indicated reagent. Explain your reasoning.

(a) $C_6H_5\overset{O}{\overset{||}{C}}H$ or $C_6H_5\overset{O}{\overset{||}{C}}CH_3$ (rate of reduction with sodium borohydride)

(b) $Cl_3C\overset{O}{\overset{||}{C}}H$ or $CH_3\overset{O}{\overset{||}{C}}H$ (equilibrium constant for hydration)

(c) Acetone or 3,3-dimethyl-2-butanone (equilibrium constant for cyanohydrin formation)

(d) Acetone or 3,3-dimethyl-2-butanone (rate of reduction with sodium borohydride)

(e) $CH_2(OCH_2CH_3)_2$ or $(CH_3)_2C(OCH_2CH_3)_2$ (rate of acid-catalyzed hydrolysis)

17.27 Equilibrium constants for the dissociation (K_{diss}) of cyanohydrins according to the equation

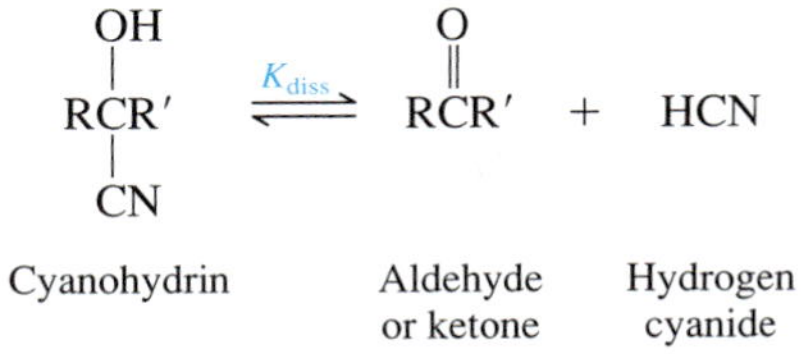

$$\underset{\substack{| \\ CN}}{\overset{\substack{OH \\ |}}{RCR'}} \quad \xrightarrow{\; K_{diss}\;} \quad \overset{\substack{O \\ \parallel}}{RCR'} \;+\; HCN$$

Cyanohydrin   Aldehyde   Hydrogen
                or ketone   cyanide

have been measured for a number of cyanohydrins. Which cyanohydrin in each of the following pairs has the greater dissociation constant?

(a) 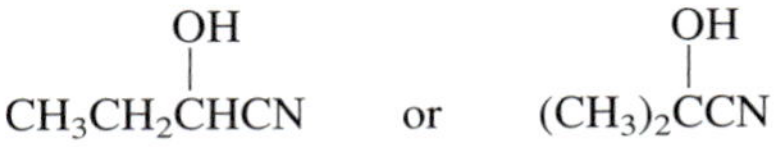 $CH_3CH_2\overset{\substack{OH \\ |}}{C}HCN$ or $(CH_3)_2\overset{\substack{OH \\ |}}{C}CN$

(b) 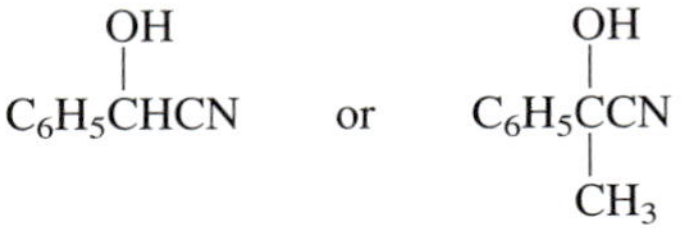 $C_6H_5\overset{\substack{OH \\ |}}{C}HCN$ or $C_6H_5\underset{\substack{| \\ CH_3}}{\overset{\substack{OH \\ |}}{C}}CN$

17.28 Each of the following reactions has been reported in the chemical literature and gives a single organic product in good yield. What is the principal product in each reaction?

(a)

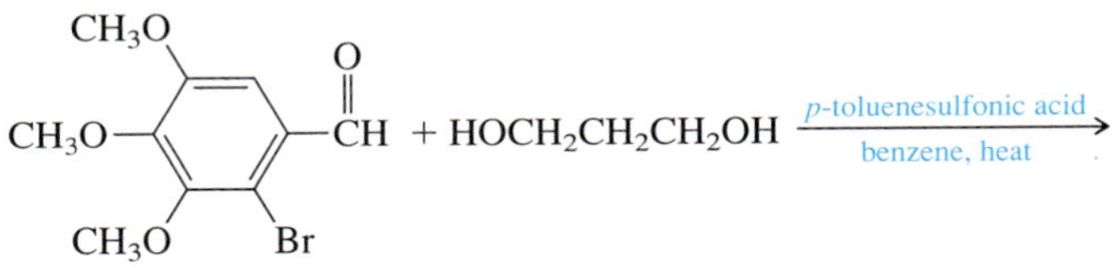

$+\; HOCH_2CH_2CH_2OH \xrightarrow[\text{benzene, heat}]{\textit{p}-\text{toluenesulfonic acid}}$

(b)

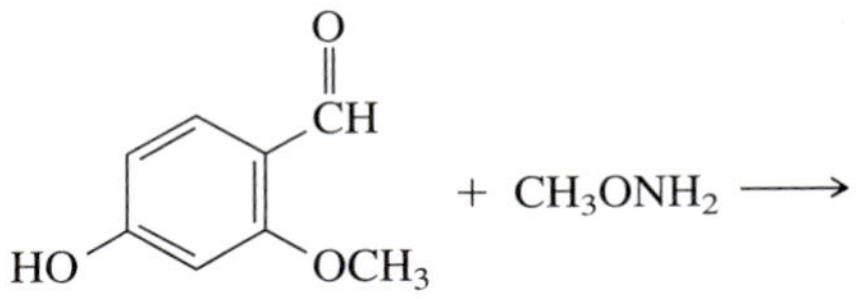

$+\; CH_3ONH_2 \longrightarrow$

(c) 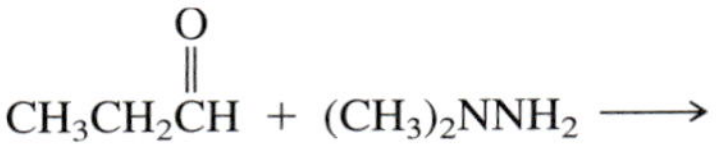 $CH_3CH_2\overset{\substack{O \\ \parallel}}{C}H \;+\; (CH_3)_2NNH_2 \longrightarrow$

(d)

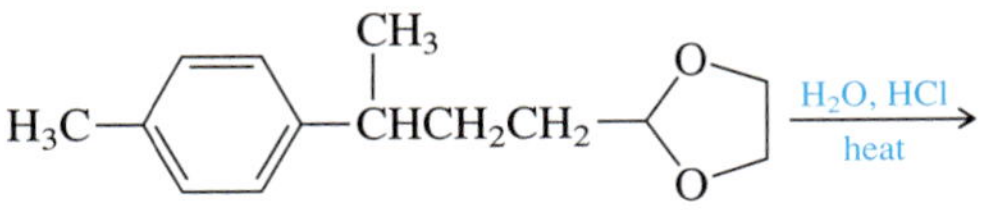

$\xrightarrow[\text{heat}]{H_2O,\, HCl}$

(e) 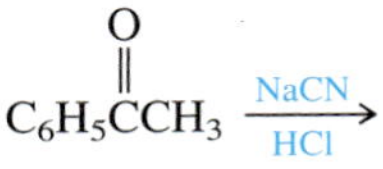 $C_6H_5\overset{\substack{O \\ \parallel}}{C}CH_3 \xrightarrow[\text{HCl}]{\text{NaCN}}$

(f) 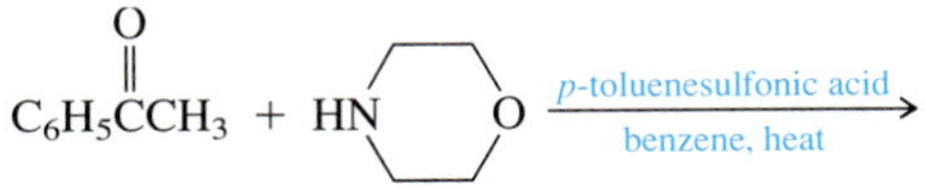 $C_6H_5\overset{\substack{O \\ \parallel}}{C}CH_3 \;+\; HN\!\!\left(\text{morpholine}\right)\!O \xrightarrow[\text{benzene, heat}]{\textit{p}-\text{toluenesulfonic acid}}$

(g)

$+\; C_6H_5\overset{\substack{O \\ \parallel}}{C}OOH \xrightarrow{CHCl_3}$

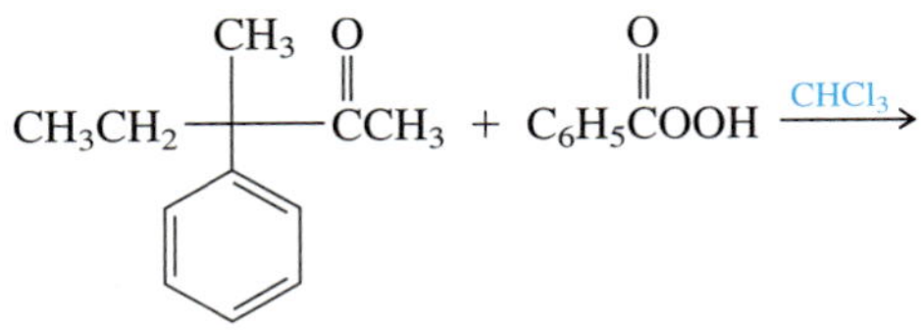

(h) $(CH_3)_2CHCCH(CH_3)_2$ + $HOCH_2CH_2SH$ $\xrightarrow[\text{benzene, heat}]{p\text{-toluenesulfonic acid}}$ $C_9H_{18}OS$

17.29 Wolff–Kishner reduction (hydrazine, KOH, ethylene glycol, 130°C) of the compound shown gave compound A. Treatment of compound A with *m*-chloroperoxybenzoic acid gave compound B, which on reduction with lithium aluminum hydride gave compound C. Oxidation of compound C with chromic acid gave compound D ($C_9H_{14}O$). Identify compounds A through D in this sequence.

17.30 On standing in ^{17}O-labeled water, both formaldehyde and its hydrate are found to have incorporated the ^{17}O isotope of oxygen. Suggest a reasonable explanation for this observation.

17.31 Reaction of benzaldehyde with 1,2-octanediol in benzene containing a small amount of *p*-toluenesulfonic acid yields almost equal quantities of two products in a combined yield of 94%. Both products have the molecular formula $C_{15}H_{22}O_2$. Suggest reasonable structures for these products.

17.32 Compounds that contain both carbonyl and alcohol functional groups are often more stable as cyclic hemiacetals or cyclic acetals than as open-chain compounds. Examples of several of these are shown. Deduce the structure of the open-chain form of each.

(a)

(c)

Brevicomin (sex attractant of
Western pine beetle)

(b)

(d)

HOCH₂

Talaromycin A (a toxic
substance produced by a
fungus that grows on
poultry house litter)

17.33 Compounds that contain a carbon–nitrogen double bond are capable of stereoisomerism much like that seen in alkenes. The structures

$$\underset{R'}{\overset{R}{\diagdown}}C=N\overset{\displaystyle{\cdot\cdot}}{\diagdown}X \quad \text{and} \quad \underset{R'}{\overset{R}{\diagdown}}C=N\overset{X}{\diagup}$$

are stereoisomeric. Specifying stereochemistry in these systems is best done by using *E–Z* descriptors and considering the nitrogen lone pair to be the lowest priority group. Write the structures or build molecular models, clearly showing stereochemistry, of the following:

(a) (*Z*)-CH₃CH=NCH₃

(b) (*E*)-Acetaldehyde oxime

(c) (*Z*)-2-Butanone hydrazone

(d) (*E*)-Acetophenone semicarbazone

17.34 Compounds known as *lactones,* which are cyclic esters, are formed on Baeyer–Villiger oxidation of cyclic ketones. Suggest a mechanism for the Baeyer–Villiger oxidation shown.

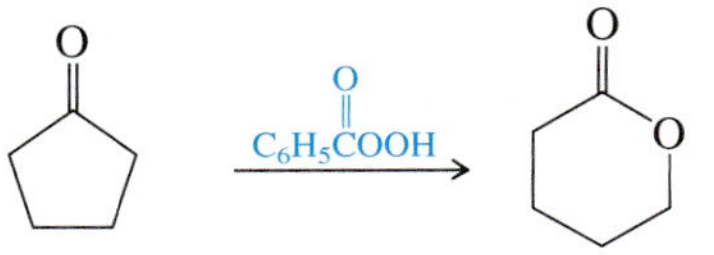

Cyclopentanone   5-Pentanolide
                        (78%)

17.35 Organic chemists often use enantiomerically homogeneous starting materials for the synthesis of complex molecules (see *Chiral Drugs,* p. 296). A novel preparation of the *S* enantiomer of compound B has been described using a bacterial cyclohexanone monooxygenase enzyme system.

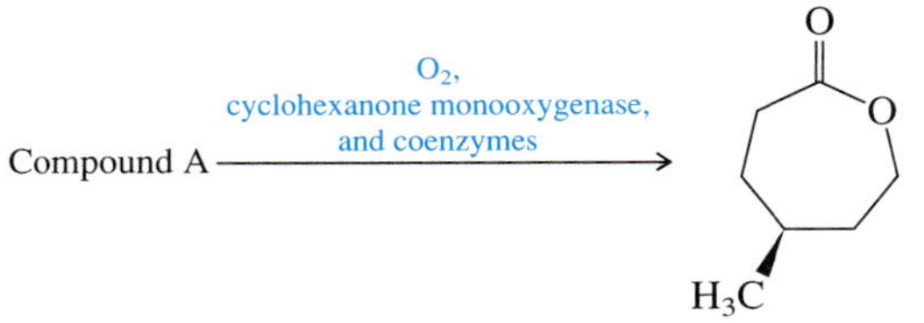

Compound A

Compound B

(a) What is compound A?

(b) How would the product obtained by treatment of compound A with peroxyacetic acid differ from that shown in the equation?

17.36 Suggest a reasonable mechanism for each of the following reactions:

(a)

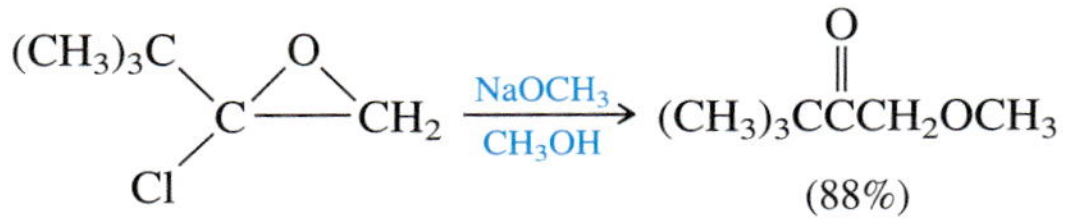

$\xrightarrow[\text{CH}_3\text{OH}]{\text{NaOCH}_3}$ $(CH_3)_3CCCH_2OCH_3$

(88%)

(b) 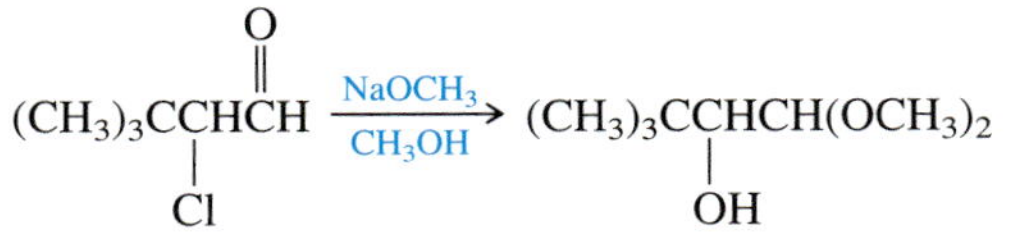 $(CH_3)_3CCHCH$ $\xrightarrow[\text{CH}_3\text{OH}]{\text{NaOCH}_3}$ $(CH_3)_3CCHCH(OCH_3)_2$
                Cl   OH

(72%)

17.37 *Amygdalin,* a substance present in peach, plum, and almond pits, is a derivative of the *R* enantiomer of benzaldehyde cyanohydrin. Give the structure of (*R*)-benzaldehyde cyanohydrin.

17.38 Using ethanol as the source of all the carbon atoms, describe efficient syntheses of each of the following, using any necessary organic or inorganic reagents:

(a) $CH_3CH(OCH_2CH_3)_2$

(b)

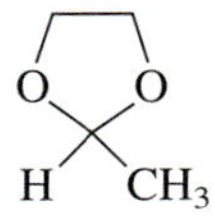

(c)

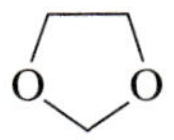

(d) 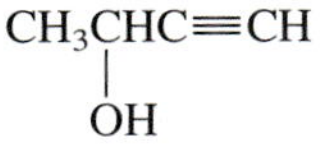 $CH_3CHC\equiv CH$
          OH

(e) 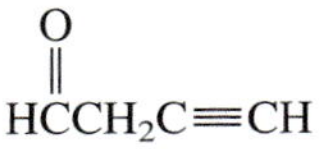 $HCCH_2C\equiv CH$

(f) $CH_3CH_2CH_2CH_2OH$

17.39 Describe reasonable syntheses of benzophenone, $C_6H_5CC_6H_5$, from each of the following starting materials and any necessary inorganic reagents.

(a) Benzoyl chloride and benzene

(b) Benzyl alcohol and bromobenzene

(c) Bromodiphenylmethane, $(C_6H_5)_2CHBr$

(d) Dimethoxydiphenylmethane, $(C_6H_5)_2C(OCH_3)_2$

(e) 1,1,2,2-Tetraphenylethene, $(C_6H_5)_2C{=}C(C_6H_5)_2$

17.40 The sex attractant of the female winter moth has been identified as the tetraene $CH_3(CH_2)_8CH{=}CHCH_2CH{=}CHCH_2CH{=}CHCH{=}CH_2$. Devise a synthesis of this material from 3,6-hexadecadien-1-ol and allyl alcohol.

17.41 Hydrolysis of a compound A in dilute aqueous hydrochloric acid gave (along with methanol) a compound B, mp 164–165°C. Compound B had the molecular formula $C_{16}H_{16}O_4$; it exhibited hydroxyl absorption in its IR spectrum at 3550 cm^{-1} but had no peaks in the carbonyl region. What is a reasonable structure for compound B?

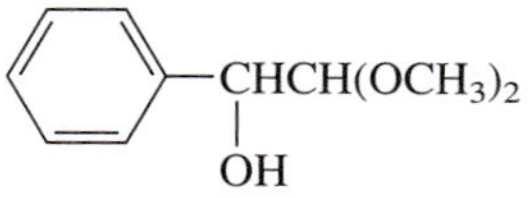

Compound A

17.42 Syntheses of each of the following compounds have been reported in the chemical literature. Using the indicated starting material and any necessary organic or inorganic reagents, describe short sequences of reactions that would be appropriate for each transformation.

(a) 1,1,5-Trimethylcyclononane from 5,5-dimethylcyclononanone

(b) 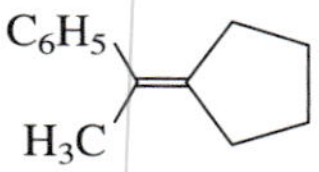   from

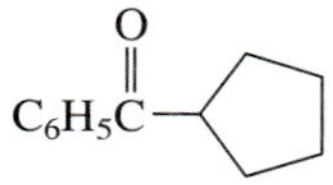

(c) 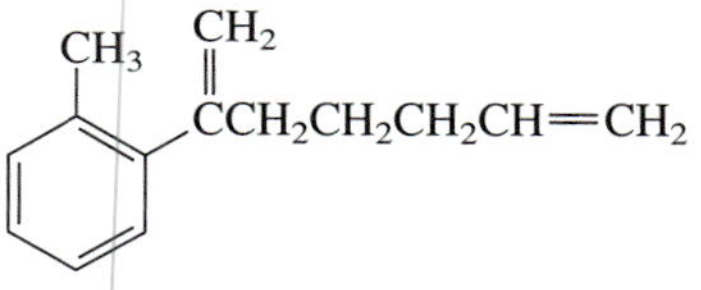   from   *o*-bromotoluene and 5-hexenal

(d) $CH_3CCH_2CH_2C(CH_2)_5CH_3$ from $HC{\equiv}CCH_2CH_2CH_2OH$

(e) 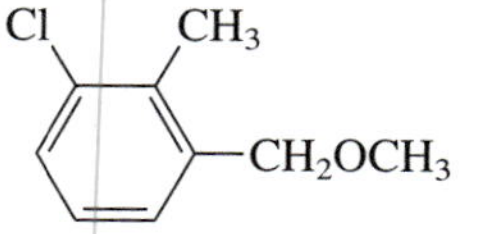   from   3-chloro-2-methylbenzaldehyde

17.43 The following five-step synthesis has been reported in the chemical literature. Suggest reagents appropriate for each step.

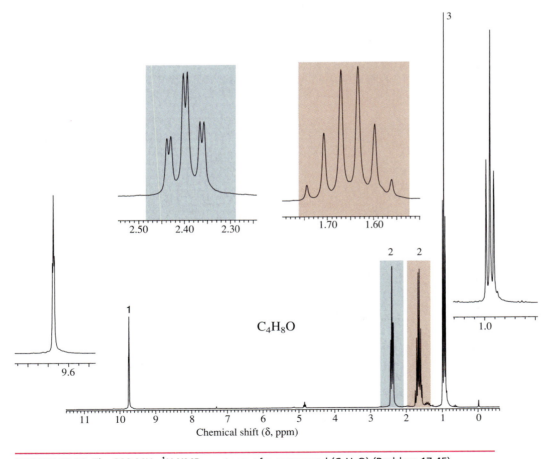

17.44 Increased "single-bond character" in a carbonyl group is associated with a decreased carbon–oxygen stretching frequency. Among the three compounds benzaldehyde, 2,4,6-trimethoxybenzaldehyde, and 2,4,6-trinitrobenzaldehyde, which one will have the lowest frequency carbonyl absorption? Which one will have the highest?

17.45 A compound has the molecular formula C_4H_8O and contains a carbonyl group. Identify the compound on the basis of its ^1H NMR spectrum shown in Figure 17.20.

FIGURE 17.20 The 200-MHz ^1H NMR spectrum of a compound (C_4H_8O) (Problem 17.45).

17.46 A compound ($C_7H_{14}O$) has a strong peak in its IR spectrum at 1710 cm^{-1}. Its ^{1}H NMR spectrum consists of three singlets in the ratio 9:3:2 at δ 1.0, 2.1, and 2.3, respectively. Identify the compound.

17.47 Compounds A and B are isomeric diketones of molecular formula $C_6H_{10}O_2$. The ^{1}H NMR spectrum of compound A contains two signals, both singlets, at δ 2.2 (six protons) and 2.8 (four protons). The ^{1}H NMR spectrum of compound B contains two signals, one at δ 1.3 (triplet, six protons) and the other at δ 2.8 (quartet, four protons). What are the structures of compounds A and B?

17.48 A compound ($C_{11}H_{14}O$) has a strong peak in its IR spectrum near 1700 cm^{-1}. Its 200-MHz ^{1}H NMR spectrum is shown in Figure 17.21. What is the structure of the compound?

17.49 A compound is a ketone of molecular formula $C_7H_{14}O$. Its ^{13}C NMR spectrum is shown in Figure 17.22. What is the structure of the compound?

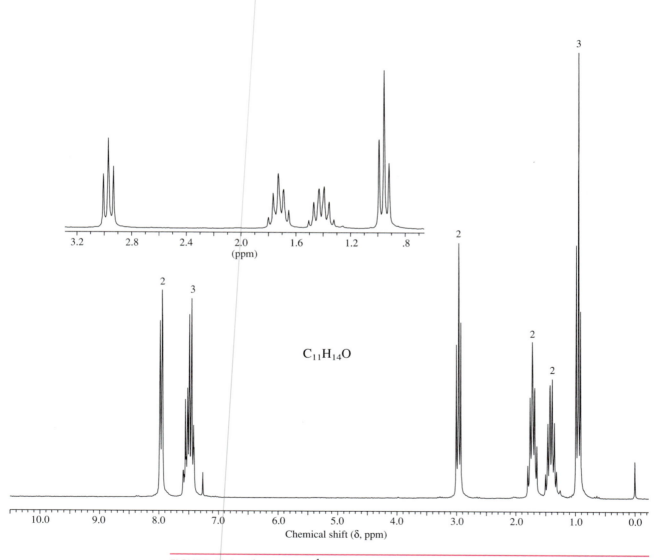

FIGURE 17.21 The 200-MHz ^{1}H NMR spectrum of a compound ($C_{11}H_{14}O$) (Problem 17.48).

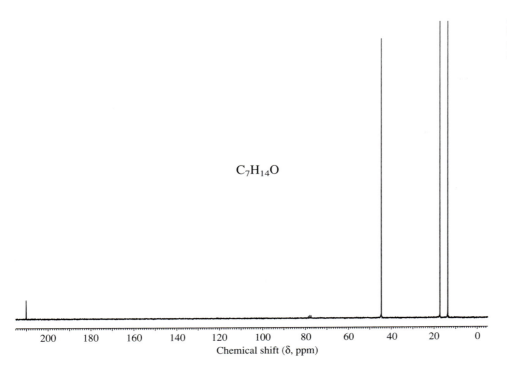

$C_7H_{14}O$

Chemical shift (δ, ppm)

FIGURE 17.22 The ^{13}C NMR spectrum of a compound ($C_7H_{14}O$) (Problem 17.49).

17.50 Compound A and compound B are isomers having the molecular formula $C_{10}H_{12}O$. The mass spectrum of each compound contains an abundant peak at m/z 105. The ^{13}C NMR spectra of compound A (Figure 17.23) and compound B (Figure 17.24) are shown. Identify these two isomers.

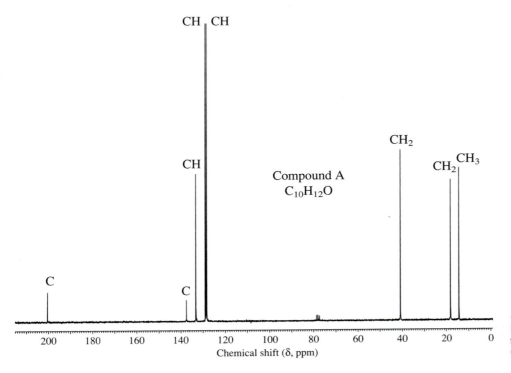

CH CH

CH

Compound A
$C_{10}H_{12}O$

CH$_2$

CH$_2$ CH$_3$

C

C

Chemical shift (δ, ppm)

FIGURE 17.23 The ^{13}C NMR spectrum of compound A ($C_{10}H_{12}O$) (Problem 17.50).

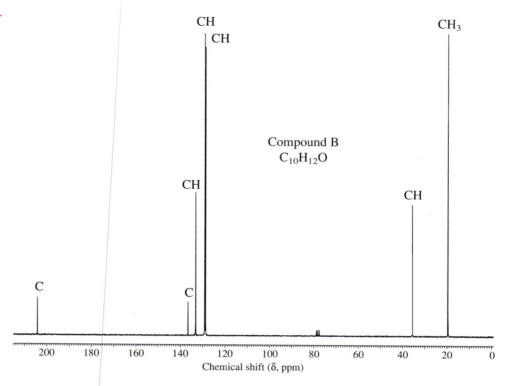

FIGURE 17.24 The ^{13}C NMR spectrum of compound B ($C_{10}H_{12}O$) (Problem 17.50).

17.51 The most stable conformation of acetone has one of the hydrogens of each methyl group eclipsed with the carbonyl oxygen. Construct a model of this conformation.

17.52 Construct a molecular model of cyclohexanone. Do either of the hydrogens of C-2 eclipse the carbonyl oxygen?

CHAPTER 18

ENOLS AND ENOLATES

In the preceding chapter you learned that nucleophilic addition to the carbonyl group is one of the fundamental reaction types of organic chemistry. In addition to its own reactivity, a carbonyl group can affect the chemical properties of aldehydes and ketones in other ways. Aldehydes and ketones having at least one hydrogen on a carbon next to the carbonyl are in equilibrium with their **enol** isomers.

$$\underset{\substack{\text{Aldehyde or}\\\text{ketone}}}{R_2CHCR'} \ \overset{O}{\underset{}{\|}} \ \rightleftharpoons \ \underset{\text{Enol}}{R_2C=CR'} \ \overset{OH}{\underset{}{|}}$$

In this chapter you'll see a number of processes in which the enol, rather than the aldehyde or a ketone, is the reactive species.

There is also an important group of reactions in which the carbonyl group acts as a powerful electron-withdrawing substituent, increasing the acidity of protons on the adjacent carbons.

$$\underset{\overset{|}{H}}{\overset{O}{\overset{\|}{R_2CCR'}}}$$

This proton is far more acidic than a hydrogen in an alkane.

As an electron-withdrawing group on a carbon–carbon double bond, a carbonyl group renders the double bond susceptible to nucleophilic attack:

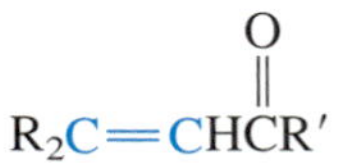

$$\overset{\displaystyle O}{\underset{\displaystyle R_2C=CHCR'}{\|}}$$

Normally, carbon–carbon double bonds are attacked by electrophiles; a carbon–carbon double bond that is conjugated to a carbonyl group is attacked by nucleophiles.

The presence of a carbonyl group in a molecule makes possible a number of chemical reactions that are of great synthetic, biochemical, and mechanistic importance. This chapter is complementary to the preceding one; the two chapters taken together demonstrate the extraordinary range of chemical reactions available to aldehydes and ketones.

18.1 THE α-CARBON ATOM AND ITS HYDROGENS

It is convenient to use the Greek letters α, β, γ, and so forth, to locate the carbons in a molecule in relation to the carbonyl group. The carbon atom adjacent to the carbonyl carbon is the α-carbon atom, the next one down the chain is the β carbon, and so on. Butanal, for example, has an α carbon, a β carbon, and a γ carbon.

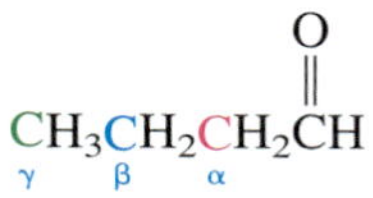

$$\overset{\displaystyle O}{\underset{\displaystyle \underset{\gamma\quad\beta\quad\alpha}{CH_3CH_2CH_2CH}}{\|}}$$ Carbonyl group is reference point; no Greek letter assigned to it.

Hydrogens take the same Greek letter as the carbon atom to which they are attached. A hydrogen connected to the α-carbon atom is an α hydrogen. Butanal has two α hydrogens, two β hydrogens, and three γ hydrogens. No Greek letter is assigned to the hydrogen attached directly to the carbonyl group of an aldehyde.

PROBLEM 18.1 How many α hydrogens are in each of the following?

(a) 3,3-Dimethyl-2-butanone (c) Benzyl methyl ketone
(b) 2,2-Dimethylpropanal (d) Cyclohexanone

SAMPLE SOLUTION (a) This ketone has two different α carbons, but only one of them has attached hydrogens. There are three equivalent α hydrogens. The other nine hydrogens are attached to β-carbon atoms.

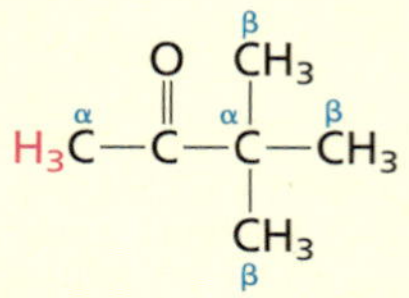

$$H_3C-\overset{O}{\overset{\|}{C}}-\overset{\overset{\beta}{CH_3}}{\underset{\underset{\beta}{CH_3}}{C}}-CH_3$$

3,3-Dimethyl-2-butanone

Other than nucleophilic addition to the carbonyl group, the most important reactions of aldehydes and ketones involve replacing an α hydrogen. A particularly well studied example is halogenation of aldehydes and ketones.

18.2 α HALOGENATION OF ALDEHYDES AND KETONES

Aldehydes and ketones react with halogens by *substitution* of one of the α hydrogens:

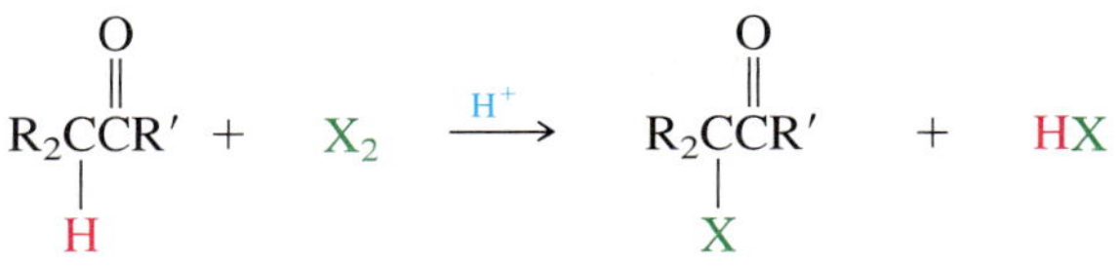

| Aldehyde or ketone | Halogen | α-Halo aldehyde or ketone | Hydrogen halide |

The reaction is *regiospecific* for substitution of an α hydrogen. None of the hydrogens farther removed from the carbonyl group are affected.

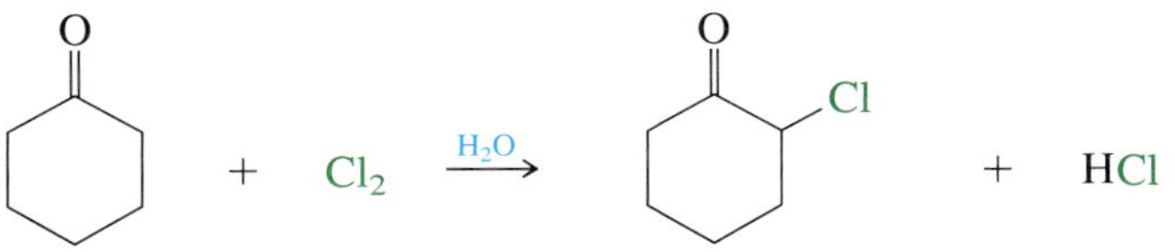

| Cyclohexanone | Chlorine | 2-Chlorocyclohexanone (61–66%) | Hydrogen chloride |

Nor is the hydrogen directly attached to the carbonyl group in aldehydes affected. Only the α hydrogen is replaced.

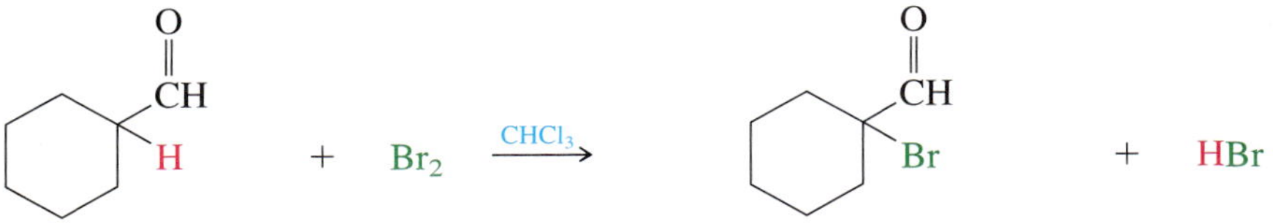

| Cyclohexanecarbaldehyde | Bromine | 1-Bromocyclohexanecarbaldehyde (80%) | Hydrogen bromide |

> **PROBLEM 18.2** Chlorination of 2-butanone yields two isomeric products, each having the molecular formula C_4H_7ClO. Identify these two compounds.

α Halogenation of aldehydes and ketones can be carried out in a variety of solvents (water and chloroform are shown in the examples, but acetic acid and diethyl ether are also often used). The reaction is catalyzed by acids. Because one of the reaction products, the hydrogen halide, is an acid and therefore a catalyst for the reaction, the process is said to be **autocatalytic.** Free radicals are *not* involved, and the reactions occur at room temperature in the absence of initiators. Mechanistically, acid-catalyzed halogenation of aldehydes and ketones is much different from free-radical halogenation of alkanes. Although both processes lead to the replacement of a hydrogen by a halogen, they do so by completely different pathways.

18.3 MECHANISM OF α HALOGENATION OF ALDEHYDES AND KETONES

In one of the earliest mechanistic investigations in organic chemistry, Arthur Lapworth discovered in 1904 that the rates of chlorination and bromination of acetone were the same. Later he found that iodination of acetone proceeded at the same rate as chlorination and bromination. Moreover, the rates of all three halogenation reactions, although first-order in acetone, are independent of the halogen concentration. *Thus, the halogen does not participate in the reaction until after the rate-determining step.* These kinetic observations, coupled with the fact that substitution occurs exclusively at the α-carbon atom, led Lapworth to propose that the rate-determining step is the conversion of acetone to a more reactive form, its enol isomer:

The graphic that opened this chapter is an electrostatic potential map of the enol of acetone.

$$\underset{\substack{\text{Acetone}}}{\text{CH}_3\overset{\displaystyle O}{\overset{\|}{\text{C}}}\text{CH}_3} \xrightarrow[\text{slow}]{\rightleftharpoons} \underset{\substack{\text{Propen-2-ol (enol} \\ \text{form of acetone)}}}{\text{CH}_3\overset{\displaystyle OH}{\overset{|}{\text{C}}}\!\!=\!\!\text{CH}_2}$$

Once formed, this enol reacts rapidly with the halogen to form an α-halo ketone:

$$\underset{\substack{\text{Propen-2-ol (enol} \\ \text{form of acetone)}}}{\text{CH}_3\overset{\displaystyle OH}{\overset{|}{\text{C}}}\!\!=\!\!\text{CH}_2} + \underset{\substack{\text{Halogen}}}{\text{X}_2} \xrightarrow{\text{fast}} \underset{\substack{\text{α-Halo derivative} \\ \text{of acetone}}}{\text{CH}_3\overset{\displaystyle O}{\overset{\|}{\text{C}}}\text{CH}_2\text{X}} + \underset{\substack{\text{Hydrogen} \\ \text{halide}}}{\text{HX}}$$

> **PROBLEM 18.3**   Write the structures of the enol forms of 2-butanone that react with chlorine to give 1-chloro-2-butanone and 3-chloro-2-butanone.

Lapworth was far ahead of his time in understanding how organic reactions occur. For an account of Lapworth's contributions to mechanistic organic chemistry, see the November 1972 issue of the *Journal of Chemical Education*, pp. 750–752.

Both parts of the Lapworth mechanism, enol formation and enol halogenation, are new to us. Let's examine them in reverse order. We can understand enol halogenation by analogy to halogen addition to alkenes. An enol is a very reactive kind of alkene. Its carbon–carbon double bond bears an electron-releasing hydroxyl group, which makes it "electron-rich" and activates it toward attack by electrophiles.

$$\underset{\substack{\text{Propen-2-ol} \\ \text{(enol form} \\ \text{of acetone)}}}{\text{CH}_3\overset{\displaystyle :\!\ddot{O}H}{\overset{|}{\text{C}}}\!\!=\!\!\text{CH}_2} + \underset{\substack{\text{Bromine}}}{:\!\ddot{\text{Br}}\!-\!\ddot{\text{Br}}:} \xrightarrow[\text{fast}]{\text{very}} \underset{\substack{\text{Stabilized carbocation}}}{\text{H}_3\text{C}\!-\!\overset{\displaystyle :\!\ddot{O}H}{\underset{+}{\text{C}}}\!-\!\text{CH}_2\ddot{\text{Br}}:} + \underset{\substack{\text{Bromide} \\ \text{ion}}}{:\!\ddot{\text{Br}}:^-}$$

The hydroxyl group stabilizes the carbocation by delocalization of one of the unshared electron pairs of oxygen:

$$\underset{\substack{\text{Less stable resonance} \\ \text{form; 6 electrons on} \\ \text{positively charged} \\ \text{carbon.}}}{\text{H}_3\text{C}\!-\!\overset{\displaystyle :\!\ddot{O}\!-\!\text{H}}{\underset{+}{\text{C}}}\!-\!\text{CH}_2\text{Br}} \longleftrightarrow \underset{\substack{\text{More stable resonance} \\ \text{form; all atoms (except} \\ \text{hydrogen) have octets} \\ \text{of electrons.}}}{\text{H}_3\text{C}\!-\!\overset{\displaystyle \overset{+}{\ddot{O}}\!-\!\text{H}}{\overset{\|}{\text{C}}}\!-\!\text{CH}_2\text{Br}}$$

Participation by the oxygen lone pairs is responsible for the rapid attack on the carbon–carbon double bond of an enol by bromine. We can represent this participation explicitly:

$$CH_3C=CH_2 \quad \overset{:\ddot{O}H}{|} \quad \longrightarrow \quad H_3C-\overset{+\ddot{O}H}{\underset{\|}{C}}-CH_2\ddot{Br}: \; + \; :\ddot{Br}:^-$$
$$:\ddot{Br}-\ddot{Br}:$$

Writing the bromine addition step in this way emphasizes the increased nucleophilicity of the enol double bond and identifies the source of that increased nucleophilicity as the enolic oxygen.

> **PROBLEM 18.4** Represent the reaction of chlorine with each of the enol forms of 2-butanone (see Problem 18.3) according to the curved arrow notation just described.

The cationic intermediate is simply the protonated form (conjugate acid) of the α-halo ketone. Deprotonation of the cationic intermediate gives the products.

$$\underset{\text{Cationic intermediate}}{CH_3\overset{\overset{+}{:}\ddot{O}-H}{\underset{\|}{C}}CH_2Br} \quad :\ddot{Br}:^- \longrightarrow \underset{\text{Bromoacetone}}{CH_3\overset{:O:}{\underset{\|}{C}}CH_2Br} \; + \; \underset{\text{Hydrogen bromide}}{H-\ddot{Br}:}$$

Having now seen how an enol, once formed, reacts with a halogen, let us consider the process of enolization itself.

18.4   ENOLIZATION AND ENOL CONTENT

Enols are related to an aldehyde or a ketone by a proton-transfer equilibrium known as **keto–enol tautomerism.** (*Tautomerism* refers to an interconversion between two structures that differ by the placement of an atom or a group.)

> The keto and enol forms are constitutional isomers. Using older terminology they are referred to as *tautomers* of each other.

$$\underset{\text{Keto form}}{RCH_2\overset{O}{\underset{\|}{C}}R'} \; \underset{\text{tautomerism}}{\rightleftharpoons} \; \underset{\text{Enol form}}{RCH=\overset{OH}{\underset{|}{C}}R'}$$

The mechanism of enolization involves two separate proton-transfer steps rather than a one-step process in which a proton jumps from carbon to oxygen. It is relatively slow in neutral media. The rate of enolization is catalyzed by acids as shown by the mechanism in Figure 18.1. In aqueous acid, a hydronium ion transfers a proton to the carbonyl oxygen in step 1, and a water molecule acts as a Brønsted base to remove a proton from the α-carbon atom in step 2. The second step is slower than the first. The first step involves proton transfer between oxygens, and the second is a proton transfer from carbon to oxygen.

You have had earlier experience with enols in their role as intermediates in the hydration of alkynes (Section 9.12). The mechanism of enolization of aldehydes and ketones is precisely the reverse of the mechanism by which an enol is converted to a carbonyl compound.

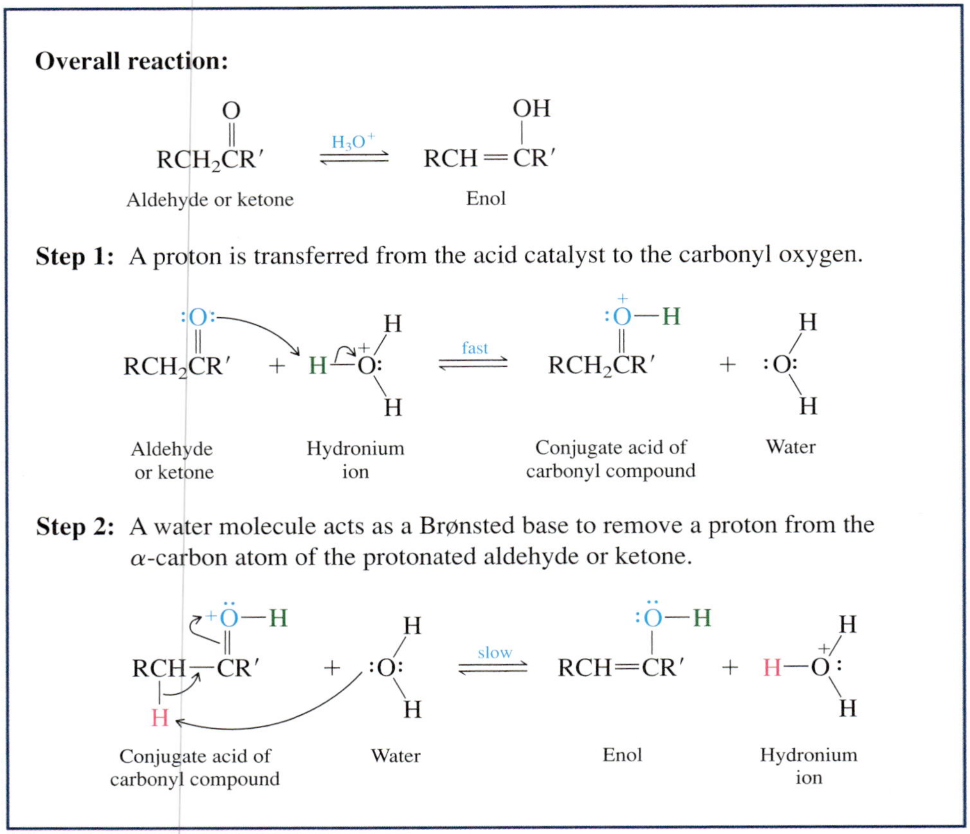

Overall reaction:

$$RCH_2CR' \xrightarrow{\text{H}_3\text{O}^+} RCH=CR'$$

Aldehyde or ketone   Enol

Step 1: A proton is transferred from the acid catalyst to the carbonyl oxygen.

$$RCH_2CR' + H-O: \xrightleftharpoons[\text{fast}] RCH_2CR' + :O$$

Aldehyde   Hydronium   Conjugate acid of   Water
or ketone   ion   carbonyl compound

Step 2: A water molecule acts as a Brønsted base to remove a proton from the α-carbon atom of the protonated aldehyde or ketone.

$$RCH-CR' + :O: \xrightleftharpoons[\text{slow}] RCH=CR' + H-O:$$

Conjugate acid of   Water   Enol   Hydronium
carbonyl compound   ion

FIGURE 18.1 Mechanism of acid-catalyzed enolization of an aldehyde or ketone in aqueous solution.

The amount of enol present at equilibrium, the *enol content,* is quite small for simple aldehydes and ketones. The equilibrium constants for enolization, as shown by the following examples, are much less than 1.

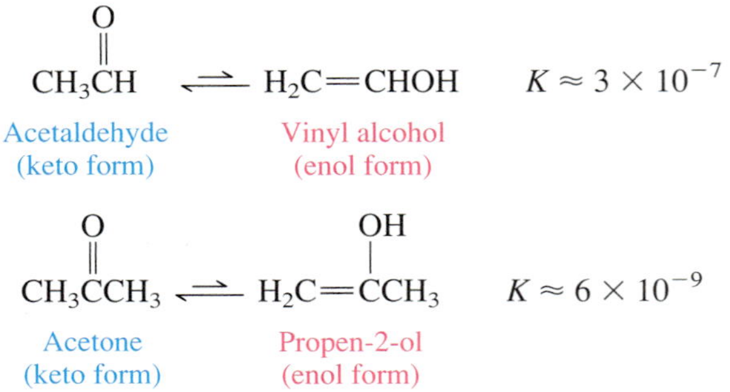

$$CH_3CH \rightleftharpoons H_2C=CHOH \qquad K \approx 3 \times 10^{-7}$$

Acetaldehyde   Vinyl alcohol
(keto form)   (enol form)

$$CH_3CCH_3 \rightleftharpoons H_2C=CCH_3 \qquad K \approx 6 \times 10^{-9}$$

Acetone   Propen-2-ol
(keto form)   (enol form)

In these and numerous other simple cases, the keto form is more stable than the enol by some 45–60 kJ/mol (11–14 kcal/mol). The chief reason for this difference is that a carbon–oxygen double bond is stronger than a carbon–carbon double bond.

With unsymmetrical ketones, enolization may occur in either of two directions:

$$\underset{\substack{\text{1-Buten-2-ol} \\ \text{(enol form)}}}{\overset{\overset{\displaystyle OH}{|}}{H_2C{=}CCH_2CH_3}} \;\;\rightleftharpoons\;\; \underset{\substack{\text{2-Butanone} \\ \text{(keto form)}}}{\overset{\overset{\displaystyle O}{\|}}{CH_3CCH_2CH_3}} \;\;\rightleftharpoons\;\; \underset{\substack{\text{2-Buten-2-ol} \\ \text{(enol form)}}}{\overset{\overset{\displaystyle OH}{|}}{CH_3C{=}CHCH_3}}$$

The ketone is by far the most abundant species present at equilibrium. Both enols are also present, but in very small concentrations.

PROBLEM 18.5   Write structural formulas corresponding to

(a) The enol form of 2,4-dimethyl-3-pentanone

(b) The enol form of acetophenone

(c) The two enol forms of 2-methylcyclohexanone

SAMPLE SOLUTION   (a) Remember that enolization involves the α-carbon atom. The ketone 2,4-dimethyl-3-pentanone gives a single enol because the two α carbons are equivalent.

$$\underset{\substack{\text{2,4-Dimethyl-3-pentanone} \\ \text{(keto form)}}}{\overset{\overset{\displaystyle O}{\|}}{(CH_3)_2CHCCH(CH_3)_2}} \;\;\rightleftharpoons\;\; \underset{\substack{\text{2,4-Dimethyl-2-penten-3-ol} \\ \text{(enol form)}}}{\overset{\overset{\displaystyle OH}{|}}{(CH_3)_2C{=}CCH(CH_3)_2}}$$

It is important to recognize that an enol is a real substance, capable of independent existence. An enol is *not* a resonance form of a carbonyl compound; the two are constitutional isomers of each other.

18.5   STABILIZED ENOLS

Certain structural features can make the keto–enol equilibrium more favorable by stabilizing the enol form. Enolization of 2,4-cyclohexadienone is one such example:

K is too large to measure.

2,4-Cyclohexadienone (keto form; not aromatic)   Phenol (enol form; aromatic)

The enol form is *phenol,* and the stabilization gained by forming an aromatic ring is more than enough to overcome the normal preference for the keto form.

A 1,3 arrangement of two carbonyl groups (compounds called **β-diketones**) leads to a situation in which the keto and enol forms are of comparable stability.

$$\underset{\substack{\text{2,4-Pentanedione (20\%)} \\ \text{(keto form)}}}{\overset{\overset{\displaystyle O \quad\; O}{\| \quad\; \|}}{CH_3CCH_2CCH_3}} \;\;\rightleftharpoons\;\; \underset{\substack{\text{4-Hydroxy-3-penten-2-one (80\%)} \\ \text{(enol form)}}}{\overset{\overset{\displaystyle OH \quad\; O}{| \qquad \|}}{CH_3C{=}CHCCH_3}} \qquad K = 4$$

The two most important structural features that stabilize the enol of a β-dicarbonyl compound are

1. Conjugation of its double bond with the remaining carbonyl group, and
2. Intramolecular hydrogen bond between the —OH group and the carbonyl oxygen

Both of these features are apparent in the structure of the enol of 2,4-pentanedione shown in Figure 18.2.

> **PROBLEM 18.6** Discuss the relative importance of intramolecular hydrogen bonding versus conjugation in light of the fact that 1,3-cyclohexanedione exists mainly in its enol form.

In β-diketones it is the methylene group flanked by two carbonyls that is involved in enolization. The alternative enol

$$\underset{\underset{H_2C=CCH_2CCH_3}{}}{\overset{\overset{OH\quad O}{|\quad\;\;\|}}{}} \quad \text{is less stable than} \quad \underset{\underset{CH_3C=CHCCH_3}{}}{\overset{\overset{OH\quad\;\; O}{|\quad\;\;\;\|}}{}}$$

because its double bond is not conjugated with the carbonyl group. It is present in negligible amounts at equilibrium.

> **PROBLEM 18.7** Write structural formulas corresponding to
>
> (a) The two most stable enol forms of $CH_3\overset{\overset{O}{\|}}{C}CH_2\overset{\overset{O}{\|}}{C}H$
> (b) The two most stable enol forms of 1-phenyl-1,3-butanedione
>
> **SAMPLE SOLUTION** (a) Enolization of this 1,3-dicarbonyl compound can involve either of the two carbonyl groups:
>
>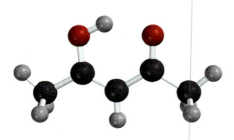
>
> Both enols have their carbon–carbon double bonds conjugated to a carbonyl group and can form an intramolecular hydrogen bond. They are of comparable stability.

FIGURE 18.2 (a) A molecular model and (b) bond distances in the enol of 2,4-pentanedione.

O---H separation in intramolecular hydrogen bond is 166 pm

103 pm
133 pm
124 pm
134 pm
141 pm

(a)

(b)

18.6   BASE-CATALYZED ENOLIZATION. ENOLATE ANIONS

The proton-transfer equilibrium that interconverts a carbonyl compound and its enol can be catalyzed by bases as well as by acids. Figure 18.3 illustrates the roles of hydroxide ion and water in a base-catalyzed enolization. As in acid-catalyzed enolization, protons are transferred sequentially rather than in a single step. First (step 1), the base abstracts a proton from the α-carbon atom to yield an anion. This anion is a resonance-stabilized species. Its negative charge is shared by the α-carbon atom and the carbonyl oxygen.

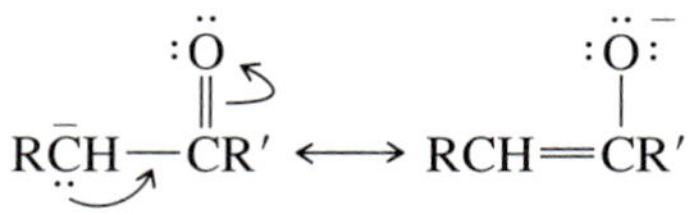

Electron delocalization
in conjugate base of ketone

Protonation of this anion can occur either at the α carbon or at oxygen. Protonation of the α carbon simply returns the anion to the starting aldehyde or ketone. Protonation of oxygen, as shown in step 2 of Figure 18.3, produces the enol.

The key intermediate in this process, the conjugate base of the carbonyl compound, is referred to as an **enolate ion** because it is the conjugate base of an enol. The term

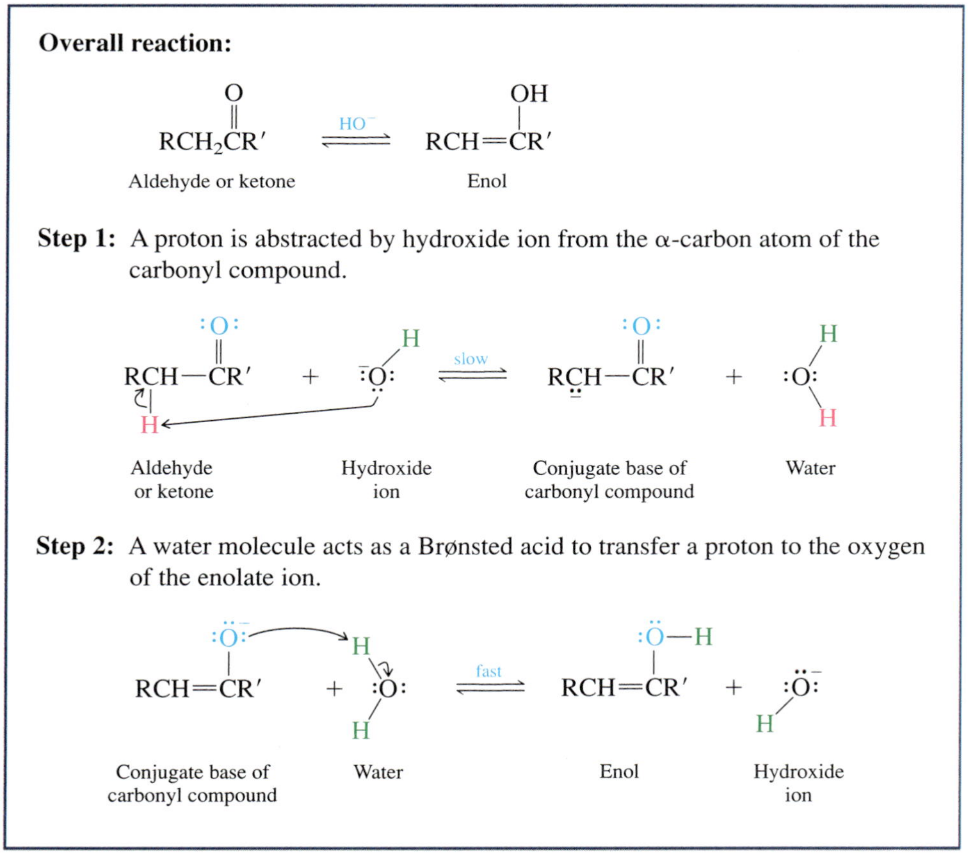

Overall reaction:

Step 1: A proton is abstracted by hydroxide ion from the α-carbon atom of the carbonyl compound.

Step 2: A water molecule acts as a Brønsted acid to transfer a proton to the oxygen of the enolate ion.

FIGURE 18.3 Mechanism of the base-catalyzed enolization of an aldehyde or ketone in aqueous solution.

Examine the enolate of acetone on *Learning By Modeling.* How is the negative charge distributed between oxygen and the α carbon?

enolate is more descriptive of the electron distribution in this intermediate in that oxygen bears a greater share of the negative charge than does the α-carbon atom.

The slow step in base-catalyzed enolization is formation of the enolate ion. The second step, proton transfer from water to the enolate oxygen, is very fast, as are almost all proton transfers from one oxygen atom to another.

Our experience to this point has been that C—H bonds are not very acidic. Compared with most hydrocarbons, however, aldehydes and ketones have relatively acidic protons on their α-carbon atoms. pK_a's for enolate formation from simple aldehydes and ketones are in the 16 to 20 range.

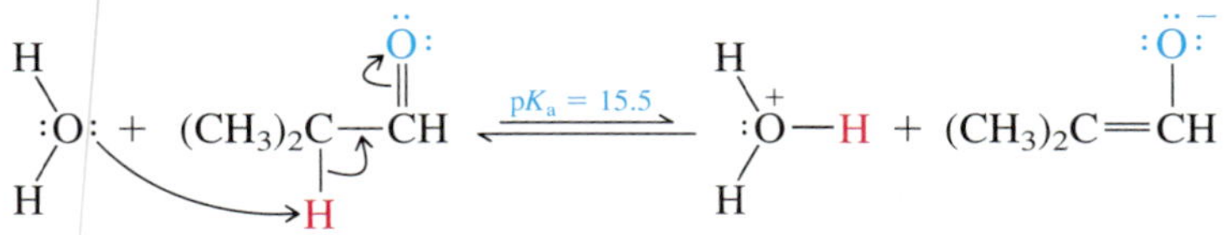

2-Methylpropanal

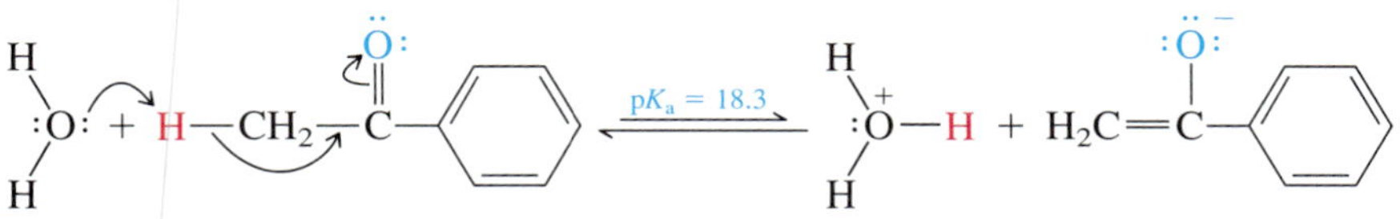

Acetophenone

Delocalization of the negative charge onto the electronegative oxygen is responsible for the enhanced acidity of aldehydes and ketones. With pK_a's in the 16 to 20 range, aldehydes and ketones are about as acidic as water and alcohols. Thus, hydroxide ion and alkoxide ions are sufficiently strong bases to produce solutions containing significant concentrations of enolate ions at equilibrium.

β-Diketones are even more acidic:

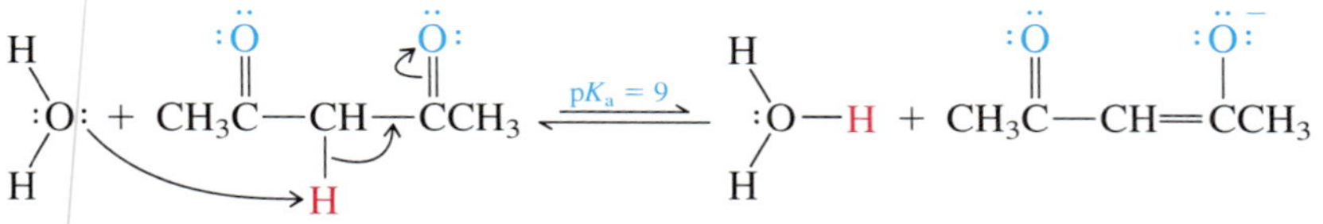

2,4-Pentanedione

In the presence of bases such as hydroxide, methoxide, and ethoxide, these β-diketones are converted completely to their enolate ions. Notice that it is the methylene group flanked by the two carbonyl groups that is deprotonated. Both carbonyl groups participate in stabilizing the enolate by delocalizing its negative charge.

Learning By Modeling contains molecular models of the enolates of acetone and 2,4-pentanedione. Compare the two with respect to the distribution of negative charge.

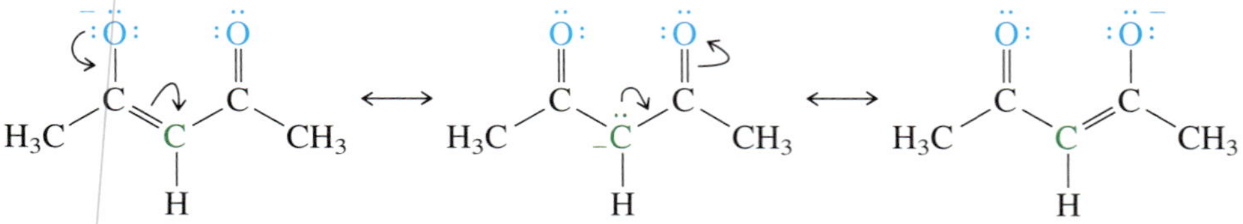

PROBLEM 18.8 Write the structure of the enolate ion derived from each of the following β-dicarbonyl compounds. Give the three most stable resonance forms of each enolate.

(a) 2-Methyl-1,3-cyclopentanedione

(b) 1-Phenyl-1,3-butanedione

(c) 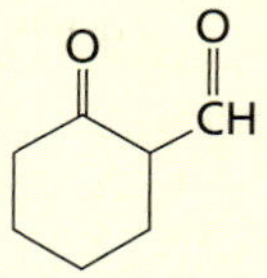

SAMPLE SOLUTION (a) First identify the proton that is removed by the base. It is on the carbon between the two carbonyl groups.

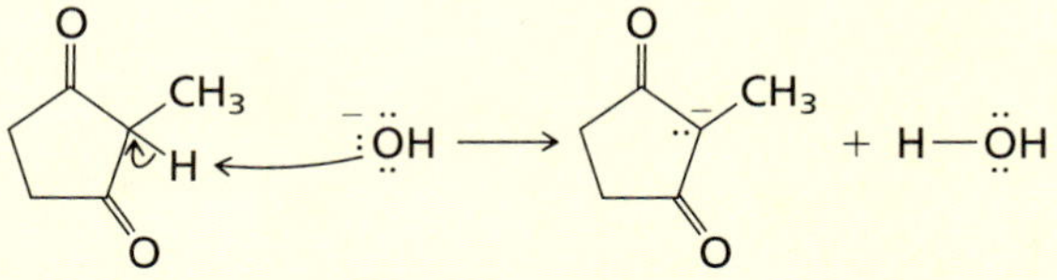

The three most stable resonance forms of this anion are

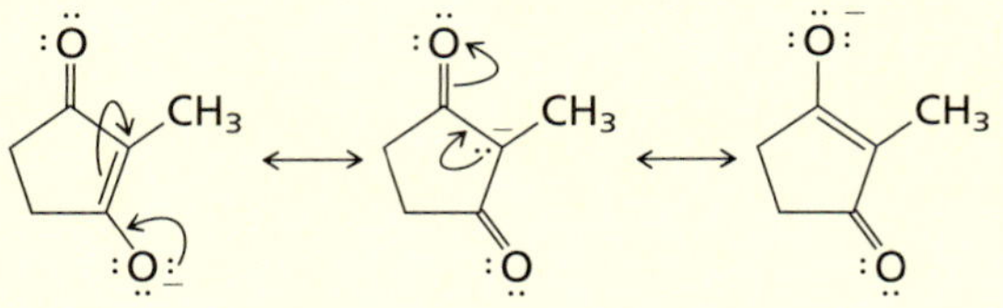

Enolate ions of β-dicarbonyl compounds are useful intermediates in organic synthesis. We shall see some examples of how they are employed in this way later in the chapter.

18.7 THE HALOFORM REACTION

Rapid halogenation of the α-carbon atom takes place when an enolate ion is generated in the presence of chlorine, bromine, or iodine.

$$\underset{\text{Aldehyde or ketone}}{R_2CH\overset{O}{\overset{\|}{C}}R'} \xrightleftharpoons[]{HO^-,\ \text{slow}} \underset{\text{Enolate}}{R_2C{=}\overset{O^-}{\overset{|}{C}}R'} \xrightarrow{X_2,\ \text{fast}} \underset{\alpha\text{-Halo aldehyde or ketone}}{R_2\underset{X}{\underset{|}{C}}\overset{O}{\overset{\|}{C}}R'}$$

As in the acid-catalyzed halogenation of aldehydes and ketones, the reaction rate is independent of the concentration of the halogen; chlorination, bromination, and iodination all occur at the same rate. Formation of the enolate is rate-determining, and, once formed, the enolate ion reacts rapidly with the halogen.

Unlike its acid-catalyzed counterpart, α halogenation in base cannot normally be limited to monohalogenation. Methyl ketones, for example, undergo a novel polyhalogenation and cleavage on treatment with a halogen in aqueous base.

$$\underset{\text{Methyl ketone}}{R\overset{O}{\overset{\|}{C}}CH_3} + \underset{\text{Halogen}}{3X_2} + \underset{\text{Hydroxide ion}}{4HO^-} \longrightarrow \underset{\text{Carboxylate ion}}{R\overset{O}{\overset{\|}{C}}O^-} + \underset{\text{Trihalomethane}}{CHX_3} + \underset{\text{Halide ion}}{3X^-} + \underset{\text{Water}}{3H_2O}$$

This is called the *haloform reaction* because the trihalomethane produced is chloroform ($CHCl_3$), bromoform ($CHBr_3$), or iodoform (CHI_3), depending on the halogen used.

The mechanism of the haloform reaction begins with α halogenation via the enolate. The electron-attracting effect of an α halogen increases the acidity of the protons on the carbon to which it is bonded, making each subsequent halogenation *at that carbon* faster than the preceding one.

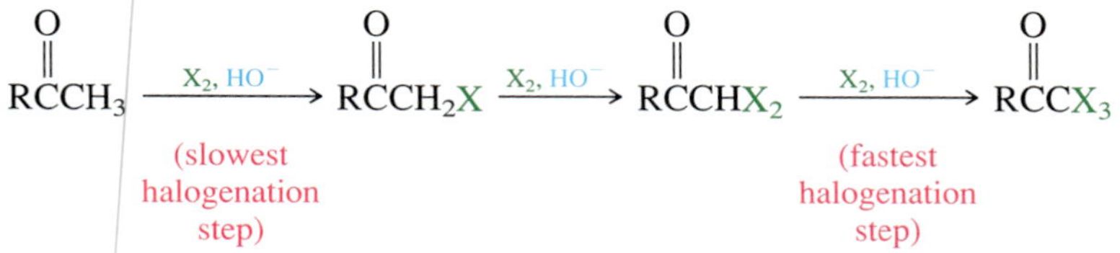

$$
\underset{\text{(slowest halogenation step)}}{RCCH_3} \xrightarrow{X_2,\ HO^-} RCCH_2X \xrightarrow{X_2,\ HO^-} RCCHX_2 \xrightarrow{X_2,\ HO^-} \underset{\text{(fastest halogenation step)}}{RCCX_3}
$$

The trihalomethyl ketone ($RCCX_3$) so formed then undergoes nucleophilic addition of hydroxide ion to its carbonyl group, triggering its dissociation.

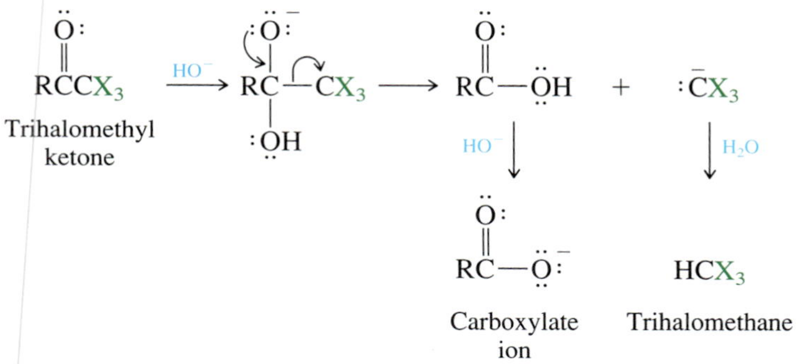

$$
\underset{\substack{\text{Trihalomethyl} \\ \text{ketone}}}{RCCX_3} \xrightarrow{HO^-} RC(\overset{:O:^-}{\underset{:OH}{})CX_3 \longrightarrow RC\text{—}\ddot{O}H + :\bar{C}X_3
$$

$$
\underset{\substack{\text{Carboxylate} \\ \text{ion}}}{RC\text{—}\ddot{O}:^-} \qquad \underset{\text{Trihalomethane}}{HCX_3}
$$

The three electron-withdrawing halogen substituents stabilize the negative charge of the trihalomethide ion ($^-:CX_3$), permitting it to act as a leaving group in the carbon–carbon bond cleavage step.

The haloform reaction is sometimes used for the preparation of carboxylic acids from methyl ketones.

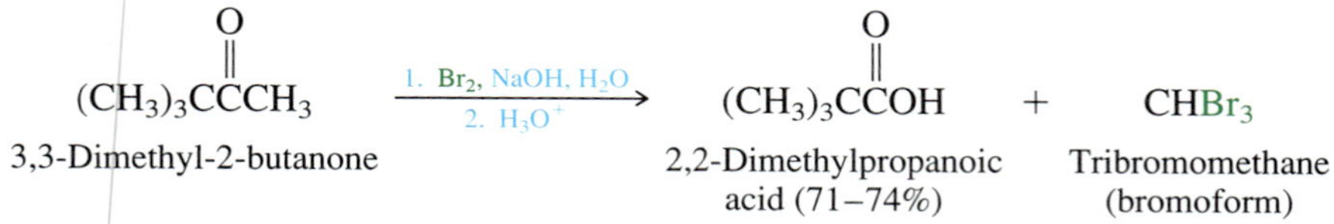

$$
\underset{\text{3,3-Dimethyl-2-butanone}}{(CH_3)_3CCCH_3} \xrightarrow[\text{2. } H_3O^+]{\text{1. } Br_2,\ NaOH,\ H_2O} \underset{\substack{\text{2,2-Dimethylpropanoic} \\ \text{acid (71–74\%)}}}{(CH_3)_3CCOH} + \underset{\substack{\text{Tribromomethane} \\ \text{(bromoform)}}}{CHBr_3}
$$

The methyl ketone shown in the example can enolize in only one direction and typifies the kind of reactant that can be converted to a carboxylic acid in synthetically acceptable yield by the haloform reaction. When C-3 of a methyl ketone bears enolizable hydrogens, as in $CH_3CH_2CCH_3$, the first halogenation step is not very regioselective and the isolated yield of $CH_3CH_2CO_2H$ is only about 50%.

The haloform reaction, using iodine, was once used as an analytical test in which the formation of a yellow precipitate of iodoform was taken as evidence that a substance was a methyl ketone. This application has been superseded by spectroscopic methods of structure determination. Interest in the haloform reaction has returned with the realization that chloroform and bromoform occur naturally and are biosynthesized by an analogous process. (See the boxed essay *The Haloform Reaction and the Biosynthesis of Trihalomethanes*.)

THE HALOFORM REACTION AND THE BIOSYNTHESIS OF TRIHALOMETHANES

Until scientists started looking specifically for them, it was widely believed that naturally occurring organohalogen compounds were rare. We now know that more than 2000 such compounds occur naturally, with the oceans being a particularly rich source.[*] Over 50 organohalogen compounds, including $CHBr_3$, $CHBrClI$, $BrCH_2CH_2I$, CH_2I_2, $Br_2CHCH=O$, I_2CHCO_2H, and $(Cl_3C)_2C=O$, have been found in a single species of Hawaiian red seaweed, for example. It is not surprising that organisms living in the oceans have adapted to their halide-rich environment by incorporating chlorine, bromine, and iodine into their metabolic processes. Chloromethane (CH_3Cl), bromomethane (CH_3Br), and iodomethane (CH_3I) are all produced by marine algae and kelp, but land-based plants and fungi also contribute their share to the more than 5 million tons of the methyl halides formed each year by living systems. The ice plant, which grows in arid regions throughout the world and is cultivated as a ground cover along coastal highways in California, biosynthesizes CH_3Cl by a process in which nucleophilic attack by chloride ion (Cl^-) on the methyl group of S-adenosylmethionine is the key step (Section 16.17).

Interestingly, the trihalomethanes chloroform ($CHCl_3$), bromoform ($CHBr_3$), and iodoform (CHI_3) are biosynthesized by an entirely different process, one that is equivalent to the haloform reaction (Section 18.7) and begins with the formation of an α-halo ketone. Unlike the biosynthesis of methyl halides, which requires attack by a halide nucleophile (X^-), α halogenation of a ketone requires attack by an electrophilic form of the halogen. For chlorination, the electrophilic form of the halogen is generated by oxidation of Cl^- in the presence of the enzyme *chloroperoxidase*. Thus, the overall equation for the enzyme-catalyzed chlorination of a methyl ketone may be written as

$$CH_3CR + Cl^- + \tfrac{1}{2}O_2 \xrightarrow{\text{chloroperoxidase}}$$

Methyl ketone   Chloride   Oxygen

$$ClCH_2CR + HO^-$$

Chloromethyl ketone   Hydroxide

Further chlorination of the chloromethyl ketone gives the corresponding trichloromethyl ketone, which then undergoes hydrolysis to form chloroform.

$$ClCH_2CR \xrightarrow[Cl^-, O_2]{\text{chloro-peroxidase}} Cl_2CHCR \xrightarrow[Cl^-, O_2]{\text{chloro-peroxidase}}$$

Chloromethyl ketone   Dichloromethyl ketone

$$Cl_3CCR \xrightarrow[HO^-]{H_2O} Cl_3CH + RCO_2^-$$

Trichloromethyl ketone   Chloroform   Carboxylate

Purification of drinking water, by adding Cl_2 to kill bacteria, is a source of electrophilic chlorine and contributes a nonenzymatic pathway for α chlorination and subsequent chloroform formation. Although some of the odor associated with tap water may be due to chloroform, more of it probably results from chlorination of algae-produced organic compounds.

[*]The November 1994 edition of the *Journal of Chemical Education* contains as its cover story the article "Natural Organohalogens. Many More Than You Think!"

18.8 SOME CHEMICAL AND STEREOCHEMICAL CONSEQUENCES OF ENOLIZATION

A number of novel reactions involving the α-carbon atom of aldehydes and ketones involve enol and enolate anion intermediates.

Substitution of deuterium for hydrogen at the α-carbon atom of an aldehyde or a ketone is a convenient way to introduce an isotopic label into a molecule and is readily carried out by treating the carbonyl compound with deuterium oxide (D_2O) and base.

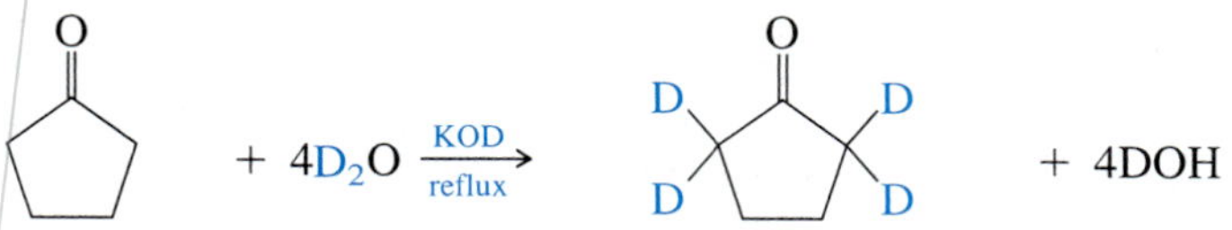

Cyclopentanone   Cyclopentanone-2,2,5,5-d_4

Only the α hydrogens are replaced by deuterium in this reaction. The key intermediate is the enolate ion formed by proton abstraction from the α-carbon atom of cyclopentanone. Transfer of deuterium from the solvent D_2O to the enolate gives cyclopentanone containing a deuterium atom in place of one of the hydrogens at the α carbon.

Formation of the enolate

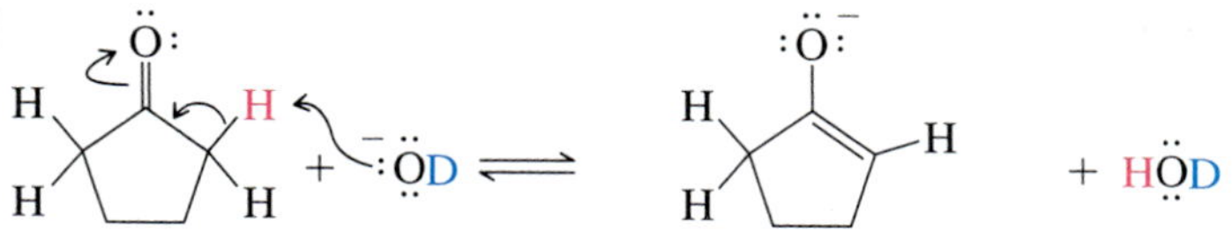

Cyclopentanone   Enolate of cyclopentanone

Deuterium transfer to the enolate

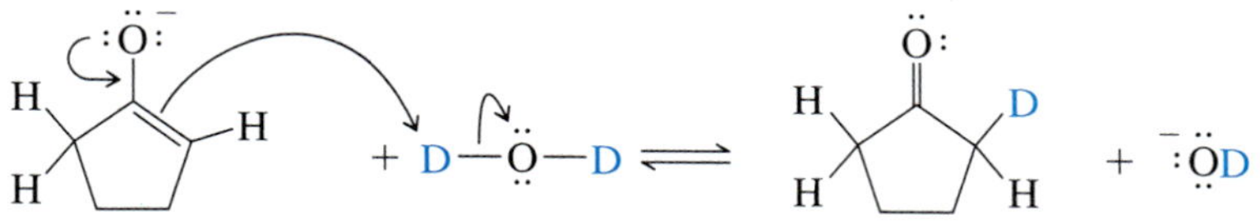

Enolate of cyclopentanone   Cyclopentanone-2-d_1

In excess D_2O the process continues until all four α protons are eventually replaced by deuterium.

PROBLEM 18.9 After the compound shown was heated in D_2O containing K_2CO_3 at 70°C the only signals that could be found in its ^1H NMR spectrum were at δ 3.9 (6H) and δ 6.7–6.9 (3H). What happened?

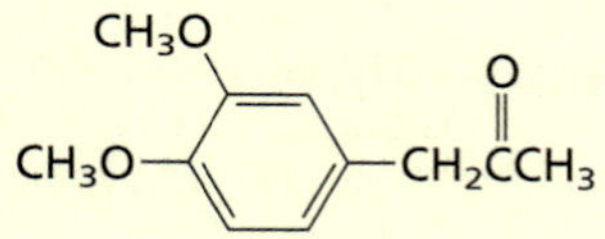

If the α-carbon atom of an aldehyde or a ketone is a chirality center, its stereochemical integrity is lost on enolization. Enolization of optically active *sec*-butyl phenyl ketone leads to its racemization by way of the achiral enol form.

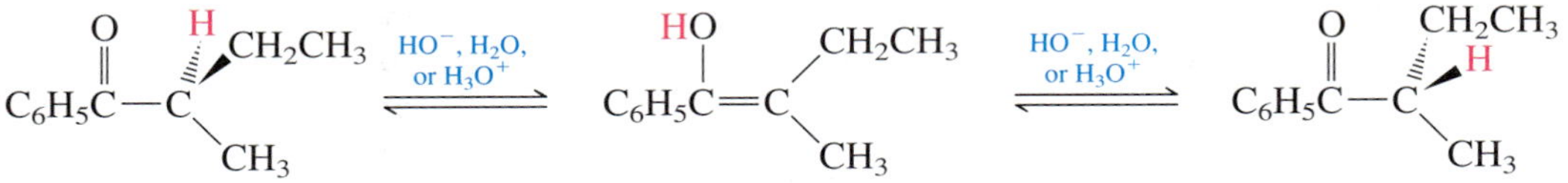

(R)-sec-Butyl phenyl ketone

Enol form [achiral; may be converted to either (R)- or (S)- sec-butyl phenyl ketone]

(S)-sec-Butyl phenyl ketone

Each act of proton abstraction from the α carbon converts a chiral molecule to an achiral enol or enolate ion. The sp^3-hybridized carbon that is the chirality center in the starting ketone becomes sp^2-hybridized in the enol or enolate. Careful kinetic studies have established that the rate of loss of optical activity of *sec*-butyl phenyl ketone is equal to its rate of hydrogen–deuterium exchange, its rate of bromination, and its rate of iodination. In each case, the rate-determining step is conversion of the starting ketone to the enol or enolate anion.

PROBLEM 18.10 Is the product from the α chlorination of (R)-sec-butyl phenyl ketone with Cl_2 in acetic acid chiral? Is it optically active?

18.9   THE ALDOL CONDENSATION

As noted earlier, an aldehyde possessing at least one α hydrogen is partially converted to its enolate anion by bases such as hydroxide ion and alkoxide ions.

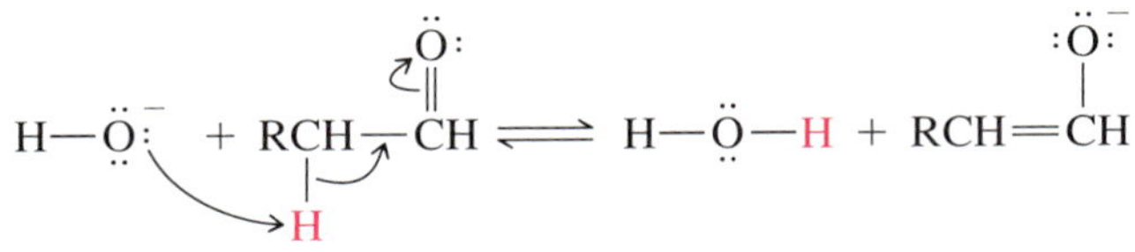

Hydroxide

Aldehyde
$pK_a = 16$

Water
$pK_a = 15.7$

Enolate

Because the pK_a's of the aldehyde and water are similar, the solution contains significant quantities of both the aldehyde and its enolate. Moreover, their reactivities are complementary. The aldehyde is capable of undergoing nucleophilic addition to its carbonyl group, and the enolate is a nucleophile capable of adding to a carbonyl group. And as shown in Figure 18.4, this is exactly what happens. The product of this step is an alkoxide, which abstracts a proton from the solvent (usually water or ethanol) to yield a β-hydroxy aldehyde. A compound of this type is known as an *aldol* because it contains both an aldehyde function and a hydroxyl group *(ald + ol = aldol)*. The reaction is called **aldol addition.**

Examples include:

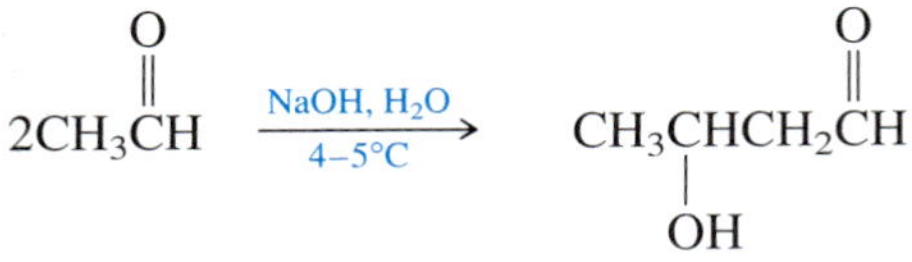

Acetaldehyde

3-Hydroxybutanal (50%) (acetaldol)

Some of the earliest studies of the aldol reaction were carried out by Aleksander Borodin. Though a physician by training and a chemist by profession, Borodin is remembered as the composer of some familiar works in Russian music. See pp. 326–327 in the April 1987 issue of the *Journal of Chemical Education* for a biographical sketch of Borodin.

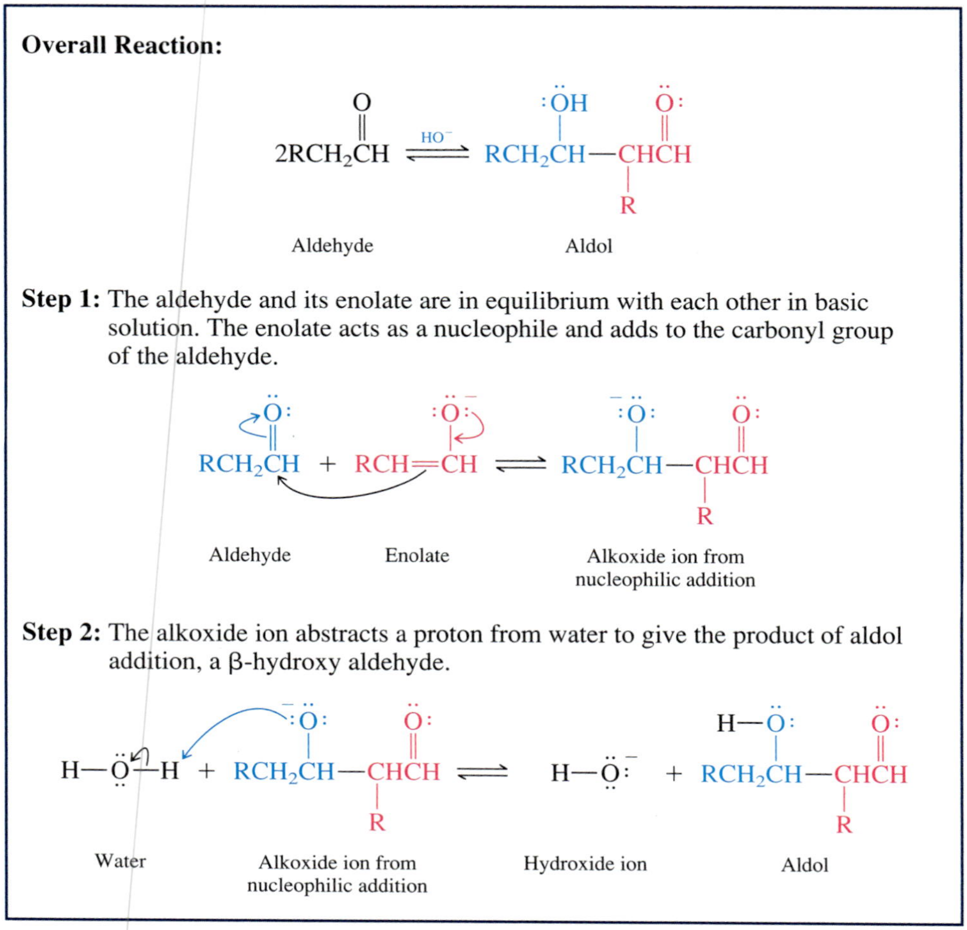
Overall Reaction:
Aldehyde
Aldol
Step 1: The aldehyde and its enolate are in equilibrium with each other in basic solution. The enolate acts as a nucleophile and adds to the carbonyl group of the aldehyde.
Aldehyde
Enolate
Alkoxide ion from nucleophilic addition
Step 2: The alkoxide ion abstracts a proton from water to give the product of aldol addition, a β-hydroxy aldehyde.
Water
Alkoxide ion from nucleophilic addition
Hydroxide ion
Aldol

FIGURE 18.4 Mechanism of aldol addition.

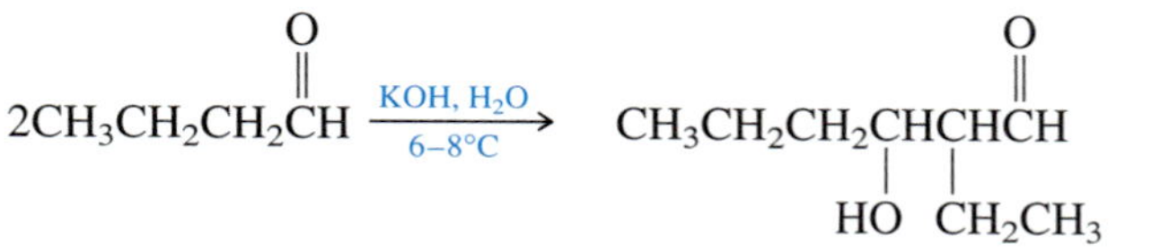

Butanal 2-Ethyl-3-hydroxyhexanal (75%)

An important feature of aldol addition is that carbon–carbon bond formation occurs between the α-carbon atom of one aldehyde and the carbonyl group of another. This is because carbanion (enolate) generation can involve proton abstraction *only* from the α-carbon atom. The overall transformation can be represented schematically, as shown in Figure 18.5.

One of these protons is removed by base to form an enolate

$$RCH_2CH + CH_2CH \xrightarrow{base} RCH_2CH-CHCH$$
$$\qquad\qquad | \qquad\qquad\qquad |$$
$$\qquad\qquad R \qquad\qquad\qquad R$$

Carbonyl group to which enolate adds

This is the carbon–carbon bond that is formed in the reaction

FIGURE 18.5 The reactive sites in aldol addition are the carbonyl group of one aldehyde molecule and the α-carbon atom of another.

PROBLEM 18.11 Write the structure of the aldol addition product of

(a) Pentanal, $CH_3CH_2CH_2CH_2CH$ with O double bond

(c) 3-Methylbutanal, $(CH_3)_2CHCH_2CH$ with O double bond

(b) 2-Methylbutanal, CH_3CH_2CHCH with O double bond and CH_3

SAMPLE SOLUTION (a) A good way to correctly identify the aldol addition product of any aldehyde is to work through the process mechanistically. Remember that the first step is enolate formation and that this *must* involve proton abstraction from the α carbon.

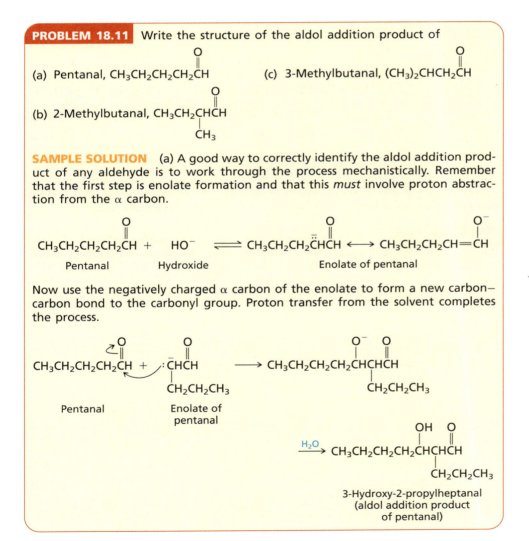

$CH_3CH_2CH_2CH_2CH$ + HO^- $\rightleftharpoons$ $CH_3CH_2CH_2\overset{..}{C}HCH$ $\longleftrightarrow$ $CH_3CH_2CH_2CH=CH$

Pentanal   Hydroxide   Enolate of pentanal

Now use the negatively charged α carbon of the enolate to form a new carbon–carbon bond to the carbonyl group. Proton transfer from the solvent completes the process.

$CH_3CH_2CH_2CH_2CH$ + $:CHCH$ $\longrightarrow$ $CH_3CH_2CH_2CH_2CHCHCH$
$\qquad\qquad\qquad\qquad CH_2CH_2CH_3 \qquad\qquad\qquad CH_2CH_2CH_3$

Pentanal   Enolate of pentanal

$\xrightarrow{H_2O}$ $CH_3CH_2CH_2CH_2CHCHCH$
$\qquad\qquad\qquad\qquad\qquad CH_2CH_2CH_3$

3-Hydroxy-2-propylheptanal (aldol addition product of pentanal)

The β-hydroxy aldehyde products of aldol addition undergo dehydration on heating, to yield *α,β-unsaturated aldehydes:*

$$\underset{\substack{\text{β-Hydroxy aldehyde}}}{\text{RCH}_2\text{CHCHCH}} \xrightarrow{\text{heat}} \underset{\substack{\text{α,β-Unsaturated} \\ \text{aldehyde}}}{\text{RCH}_2\text{CH}{=}\text{CCH}} + \underset{\text{Water}}{\text{H}_2\text{O}}$$

Conjugation of the newly formed double bond with the carbonyl group stabilizes the α,β-unsaturated aldehyde, provides the driving force for the dehydration, and controls its regioselectivity. Dehydration can be effected by heating the aldol with acid or base. Normally, if the α,β-unsaturated aldehyde is the desired product, all that is done is to carry out the base-catalyzed aldol addition reaction at elevated temperature. Under these conditions, once the aldol addition product is formed, it rapidly loses water to form the α,β-unsaturated aldehyde.

$$\underset{\text{Butanal}}{2\text{CH}_3\text{CH}_2\text{CH}_2\text{CH}} \xrightarrow[\text{80–100°C}]{\text{NaOH, H}_2\text{O}} \underset{\substack{\text{2-Ethyl-2-hexenal (86\%)}}}{\text{CH}_3\text{CH}_2\text{CH}_2\text{CH}{=}\text{CCH}} \quad \textit{via} \quad \underset{\substack{\text{2-Ethyl-3-hydroxyhexanal} \\ \text{(not isolated; dehydrates} \\ \text{under reaction conditions)}}}{\text{CH}_3\text{CH}_2\text{CH}_2\text{CHCHCH}}$$

Reactions in which two molecules of an aldehyde combine to form an α,β-unsaturated aldehyde and a molecule of water are called **aldol condensations.**

> Recall from Section 15.7 that a condensation is a reaction in which two molecules combine to give a product along with some small (usually inorganic) molecule such as water.

PROBLEM 18.12   Write the structure of the aldol condensation product of each of the aldehydes in Problem 18.11. One of these aldehydes can undergo aldol addition, but not aldol condensation. Which one? Why?

SAMPLE SOLUTION   (a) Dehydration of the product of aldol addition of pentanal introduces the double bond between C-2 and C-3 to give an α,β-unsaturated aldehyde.

$$\underset{\substack{\text{Product of aldol addition of} \\ \text{pentanal (3-hydroxy-2-} \\ \text{propylheptanal)}}}{\text{CH}_3\text{CH}_2\text{CH}_2\text{CH}_2\text{CHCHCH}} \xrightarrow{-\text{H}_2\text{O}} \underset{\substack{\text{Product of aldol condensation} \\ \text{of pentanal (2-propyl-2-} \\ \text{heptenal)}}}{\text{CH}_3\text{CH}_2\text{CH}_2\text{CH}_2\text{CH}{=}\text{CCH}}$$

The point was made earlier (Section 5.9) that alcohols require acid catalysis in order to undergo dehydration to alkenes. Thus, it may seem strange that aldol addition products can be dehydrated in base. This is another example of the way in which the enhanced acidity of protons at the α-carbon atom affects the reactions of carbonyl compounds. Elimination may take place in a concerted E2 fashion or it may be stepwise and proceed through an enolate ion.

$$\underset{\substack{\text{OH} \quad \text{O} \\ | \quad \quad \| \\ \text{R}}}{\text{RCH}_2\text{CHCHCH}} + \text{HO}^- \underset{\text{fast}}{\rightleftharpoons} \underset{\substack{\text{OH} \quad \quad \text{O} \\ | \quad \quad \quad \| \\ \text{R}}}{\text{RCH}_2\text{CHC}\overset{..}{-}\text{CH}} + \text{HOH}$$

β-Hydroxy aldehyde   Enolate ion of
                               β-hydroxy aldehyde

$$\underset{\substack{\text{OH} \quad \quad \text{O} \\ | \quad \quad \quad \| \\ \text{R}}}{\text{RCH}_2\text{CH}\overset{..}{-}\overset{..}{\text{C}}\text{—CH}} \xrightarrow{\text{slow}} \underset{\substack{\quad \quad \text{O} \\ \quad \quad \| \\ \text{R}}}{\text{RCH}_2\text{CH}=\text{CCH}} + \text{HO}^-$$

Enolate ion of   α,β-Unsaturated
β-hydroxy aldehyde   aldehyde

As with other reversible nucleophilic addition reactions, the equilibria for aldol additions are less favorable for ketones than for aldehydes. For example, only 2% of the aldol addition product of acetone is present at equilibrium.

$$2\text{CH}_3\overset{\overset{\text{O}}{\|}}{\text{C}}\text{CH}_3 \underset{98\%}{\overset{2\%}{\rightleftharpoons}} \underset{\substack{\quad | \\ \quad \text{CH}_3}}{\text{CH}_3\overset{\overset{\text{OH}}{|}}{\text{C}}\text{CH}_2\overset{\overset{\text{O}}{\|}}{\text{C}}\text{CH}_3}$$

Acetone   4-Hydroxy-4-methyl-2-pentanone

The situation is similar for other ketones. Special procedures for aldol addition and self-condensation of ketones have been developed, but are rarely used.

Aldol condensations of dicarbonyl compounds—even diketones—occur intramolecularly when five- or six-membered rings are possible.

1,6-Cyclodecanedione $\xrightarrow[\text{reflux}]{\text{Na}_2\text{CO}_3, \text{H}_2\text{O}}$ Not isolated; dehydrates under reaction conditions   Bicyclo[5.3.0]dec-1(7)-en-2-one (96%)

Aldol condensations are one of the fundamental carbon–carbon bond-forming processes of synthetic organic chemistry. Furthermore, because the products of these aldol condensations contain functional groups capable of subsequent modification, access to a host of useful materials is gained.

To illustrate how aldol condensation may be coupled to functional group modification, consider the synthesis of 2-ethyl-1,3-hexanediol, a compound used as an insect repellent. This 1,3-diol is prepared by reduction of the aldol addition product of butanal:

$$\underset{\text{Butanal}}{\text{CH}_3\text{CH}_2\text{CH}_2\overset{\overset{\text{O}}{\|}}{\text{C}}\text{H}} \xrightarrow[\text{addition}]{\text{aldol}} \underset{\substack{\quad \quad \quad \quad | \\ \quad \quad \quad \quad \text{CH}_2\text{CH}_3 \\ \text{2-Ethyl-3-hydroxyhexanal}}}{\text{CH}_3\text{CH}_2\text{CH}_2\overset{\overset{\text{OH}}{|}}{\text{CH}}\overset{\overset{\text{O}}{\|}}{\text{CHCH}}} \xrightarrow[\text{Ni}]{\text{H}_2} \underset{\substack{\quad \quad \quad \quad | \\ \quad \quad \quad \quad \text{CH}_2\text{CH}_3 \\ \text{2-Ethyl-1,3-hexanediol}}}{\text{CH}_3\text{CH}_2\text{CH}_2\overset{\overset{\text{OH}}{|}}{\text{CH}}\text{CHCH}_2\text{OH}}$$

PROBLEM 18.13 Outline a synthesis of 2-ethyl-1-hexanol from butanal.

The carbon–carbon bond-forming potential of the aldol condensation has been extended beyond the self-condensations described in this section to cases in which two different carbonyl compounds react in what are called *mixed aldol condensations*.

18.10 MIXED ALDOL CONDENSATIONS

Mixed aldol condensations can be effective only if we limit the number of reaction possibilities. It would not be useful, for example, to treat a solution of acetaldehyde and propanal with base. A mixture of four aldol addition products forms under these conditions. Two of the products are those of self-addition:

$$\underset{\substack{| \\ \text{OH}}}{\text{CH}_3\text{CHCH}_2\overset{\displaystyle\overset{\text{O}}{\|}}{\text{CH}}}$$

3-Hydroxybutanal
(from addition of enolate
of acetaldehyde to acetaldehyde)

$$\text{CH}_3\text{CH}_2\underset{\substack{| \\ \text{CH}_3}}{\overset{\text{HO}}{\text{CH}}}\text{CH}\overset{\displaystyle\overset{\text{O}}{\|}}{\text{CH}}$$

3-Hydroxy-2-methylpentanal
(from addition of enolate
of propanal to propanal)

Two are the products of mixed addition:

$$\text{CH}_3\underset{\substack{| \\ \text{CH}_3}}{\overset{\text{OH}}{\text{CH}}}\text{CH}\overset{\displaystyle\overset{\text{O}}{\|}}{\text{CH}}$$

3-Hydroxy-2-methylbutanal
(from addition of enolate
of propanal to acetaldehyde)

$$\text{CH}_3\text{CH}_2\underset{\substack{| \\ \text{OH}}}{\text{CH}}\text{CH}_2\overset{\displaystyle\overset{\text{O}}{\|}}{\text{CH}}$$

3-Hydroxypentanal
(from addition of enolate
of acetaldehyde to propanal)

PROBLEM 18.14 Use curved arrows to show the carbon–carbon bond-forming processes that lead to the four aldol addition products just shown.

The mixed aldol condensations that are the most synthetically useful are those in which:

1. Only one of the reactants can form an enolate; or
2. One of the reactants is more reactive toward nucleophilic addition than the other.

Formaldehyde, for example, cannot form an enolate but can react with the enolate of an aldehyde or ketone that can.

$$\overset{\displaystyle\overset{\text{O}}{\|}}{\text{HCH}} \;+\; (\text{CH}_3)_2\text{CHCH}_2\overset{\displaystyle\overset{\text{O}}{\|}}{\text{CH}} \;\xrightarrow[\text{water–ether}]{\text{K}_2\text{CO}_3}\; (\text{CH}_3)_2\underset{\substack{| \\ \text{CH}_2\text{OH}}}{\text{CH}}\text{CH}\overset{\displaystyle\overset{\text{O}}{\|}}{\text{CH}}$$

Formaldehyde   3-Methylbutanal   2-Hydroxymethyl-3-
                                                            methylbutanal (52%)

Indeed, formaldehyde is so reactive toward nucleophilic addition that it suppresses the self-condensation of the other component by reacting rapidly with any enolate present.

Aromatic aldehydes cannot form enolates, and a large number of mixed aldol condensations have been carried out in which an aromatic aldehyde reacts with an enolate.

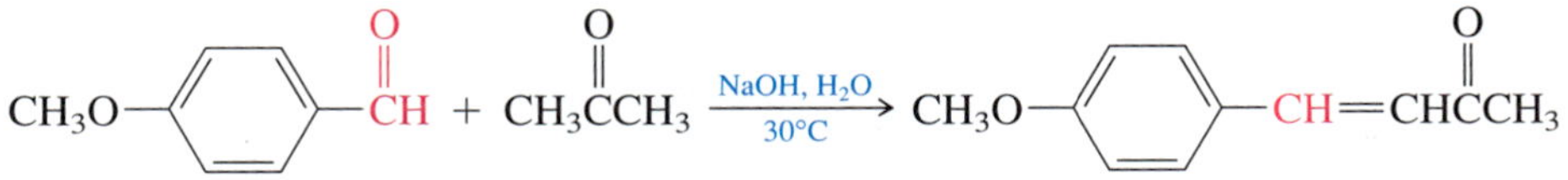

$$CH_3O-\!\!\!\bigcirc\!\!\!-\overset{\overset{O}{\parallel}}{CH} + CH_3\overset{\overset{O}{\parallel}}{C}CH_3 \xrightarrow[30°C]{NaOH,\ H_2O} CH_3O-\!\!\!\bigcirc\!\!\!-CH\!=\!CH\overset{\overset{O}{\parallel}}{C}CH_3$$

p-Methoxybenzaldehyde	Acetone	4-*p*-Methoxyphenyl-3-buten-2-one (83%)

Recall that ketones do not readily undergo self-condensation. Thus, in the preceding example, the enolate of acetone reacts preferentially with the aromatic aldehyde and gives the mixed aldol condensation product in good yield. Mixed aldol condensations using aromatic aldehydes always involve dehydration of the product of mixed addition and yield a product in which the double bond is conjugated to both the aromatic ring and the carbonyl group.

> Mixed aldol condensations in which a ketone reacts with an aromatic aldehyde are known as *Claisen–Schmidt condensations*.

PROBLEM 18.15   Give the structure of the mixed aldol condensation product of benzaldehyde with

(a)   Acetophenone, $C_6H_5\overset{\overset{O}{\parallel}}{C}CH_3$

(b)   *tert*-Butyl methyl ketone, $(CH_3)_3C\overset{\overset{O}{\parallel}}{C}CH_3$

(c)   Cyclohexanone

SAMPLE SOLUTION   (a) The enolate of acetophenone reacts with benzaldehyde to yield the product of mixed addition. Dehydration of the intermediate occurs, giving the α,β-unsaturated ketone.

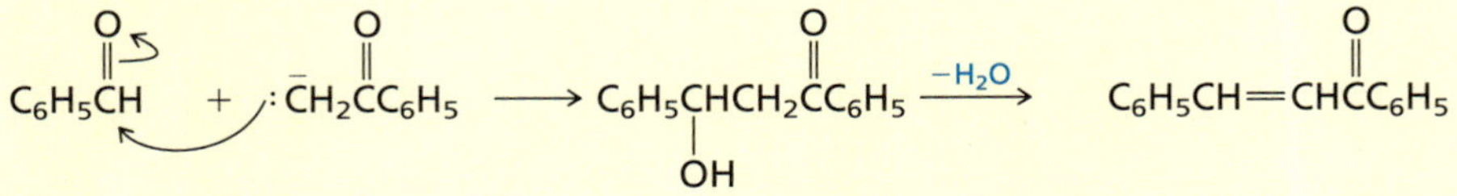

$$C_6H_5\overset{\overset{O}{\parallel}}{CH} + \ddot{:}CH_2\overset{\overset{O}{\parallel}}{C}C_6H_5 \longrightarrow C_6H_5\underset{\underset{OH}{|}}{CH}CH_2\overset{\overset{O}{\parallel}}{C}C_6H_5 \xrightarrow{-H_2O} C_6H_5CH\!=\!CH\overset{\overset{O}{\parallel}}{C}C_6H_5$$

Benzaldehyde	Enolate of acetophenone	1,3-Diphenyl-2-propen-1-one

As actually carried out, the mixed aldol condensation product, 1,3-diphenyl-2-propen-1-one, has been isolated in 85% yield on treating benzaldehyde with acetophenone in an aqueous ethanol solution of sodium hydroxide at 15–30°C.

18.11 EFFECTS OF CONJUGATION IN α,β-UNSATURATED ALDEHYDES AND KETONES

Aldol condensation offers an effective route to α,β-unsaturated aldehydes and ketones. These compounds have some interesting properties that result from conjugation of the carbon–carbon double bond with the carbonyl group. As shown in Figure 18.6, the π systems of the carbon–carbon and carbon–oxygen double bonds overlap to form an extended π system that permits increased electron delocalization.

FIGURE 18.6 Acrolein (H_2C=CHCH=O) is a planar molecule. Oxygen and each carbon is sp^2-hybridized, and each contributes one electron to a conjugated π electron system analogous to that of 1,3-butadiene.

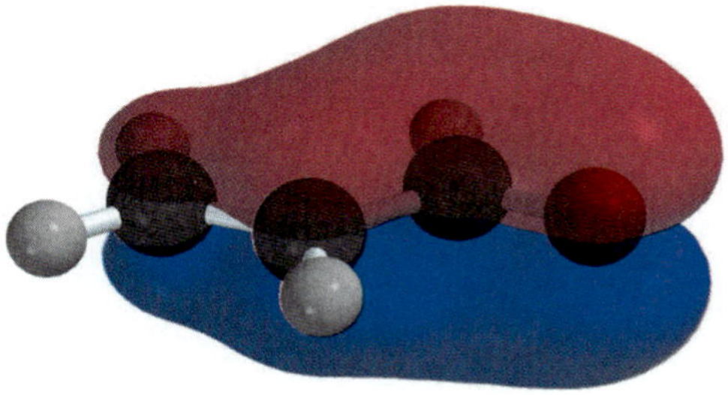

This electron delocalization stabilizes a conjugated system. Under conditions chosen to bring about their interconversion, the equilibrium between a β,γ-unsaturated ketone and an α,β-unsaturated analog favors the conjugated isomer.

Figure 3.19 (page 123) shows how the composition of an equilibrium mixture of two components varies according to the free-energy difference between them. For the equilibrium shown in the accompanying equation, $\Delta G° = -4$ kJ/mol (-1 kcal/mol).

$$CH_3CH=CHCH_2CCH_3 \underset{25°C}{\overset{K = 4.8}{\rightleftharpoons}} CH_3CH_2CH=CHCCH_3$$

4-Hexen-2-one (17%)   3-Hexen-2-one (83%)
(β,γ-unsaturated ketone)   (α,β-unsaturated ketone)

PROBLEM 18.16 Commercial mesityl oxide, $(CH_3)_2C$=CHCCH$_3$, is often contaminated with about 10% of an isomer having the same carbon skeleton. What is a likely structure for this compound?

In resonance terms, electron delocalization in α,β-unsaturated carbonyl compounds is represented by contributions from three principal resonance structures:

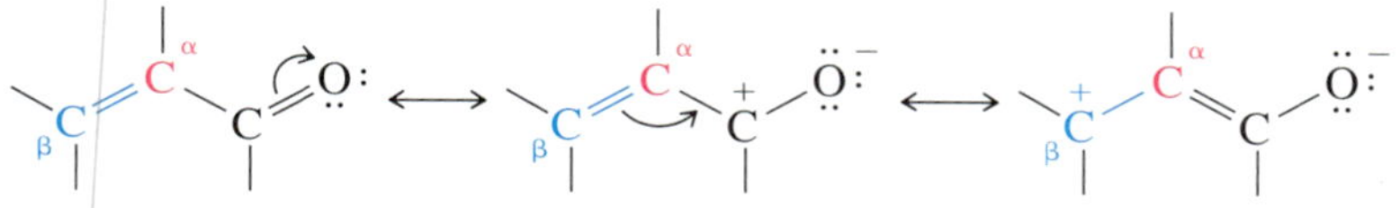

Most stable structure

The carbonyl group withdraws π electron density from the double bond, and both the carbonyl carbon and the β carbon are positively polarized. Their greater degree of charge separation makes the dipole moments of α,β-unsaturated carbonyl compounds significantly larger than those of comparable aldehydes and ketones.

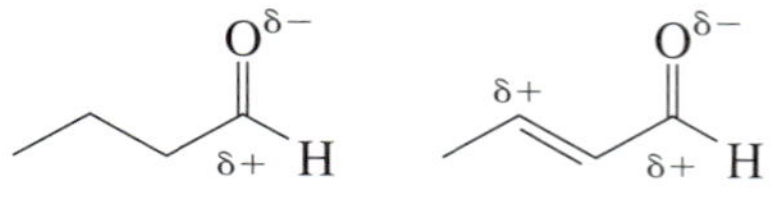

Butanal   *trans*-2-Butenal
$\mu = 2.7$ D $\mu = 3.7$ D

The diminished π electron density in the double bond makes α,β-unsaturated aldehydes and ketones less reactive than alkenes toward electrophilic addition. Electrophilic reagents—bromine and peroxy acids, for example—react more slowly with the carbon–carbon double bond of α,β-unsaturated carbonyl compounds than with simple alkenes.

On the other hand, the polarization of electron density in α,β-unsaturated carbonyl compounds makes their β-carbon atoms rather electrophilic. Some chemical consequences of this enhanced electrophilicity are described in the following section.

18.12 CONJUGATE ADDITION TO α,β-UNSATURATED CARBONYL COMPOUNDS

α,β-Unsaturated carbonyl compounds contain two electrophilic sites: the carbonyl carbon and the carbon atom that is β to it. Nucleophiles such as organolithium and Grignard reagents and lithium aluminum hydride tend to react by nucleophilic addition to the carbonyl group, as shown in the following example:

$$
\underset{\substack{\text{2-Butenal}}}{CH_3CH{=}CHCH} + \underset{\substack{\text{Ethynylmagnesium}\\\text{bromide}}}{HC{\equiv}CMgBr} \xrightarrow[\text{2. } H_3O^+]{\text{1. THF}} \underset{\substack{\text{4-Hexen-1-yn-3-ol}\\(84\%)}}{CH_3CH{=}CHCHC{\equiv}CH}
$$

This is called *direct addition*, or *1,2 addition*. (The "1" and "2" do not refer to IUPAC locants but are used in a manner analogous to that employed in Section 10.10 to distinguish between direct and conjugate addition to conjugated dienes.)

With certain other nucleophiles, addition takes place at the carbon–carbon double bond rather than at the carbonyl group. Such reactions proceed via enol intermediates and are described as *conjugate addition*, or *1,4-addition*, reactions.

α,β-Unsaturated aldehyde or ketone → Enol formed by 1,4 addition → Isolated product of 1,4-addition pathway

The nucleophilic portion of the reagent (Y in HY) becomes bonded to the β carbon. For reactions carried out under conditions in which the attacking species is the anion :Y⁻, an enolate ion precedes the enol.

Enolate ion formed by nucleophilic addition of :Y⁻
to β carbon

Ordinarily, nucleophilic addition to the carbon–carbon double bond of an alkene is very rare. It occurs with α,β-unsaturated carbonyl compounds because the carbanion that results is an enolate, which is more stable than a simple alkyl anion.

Conjugate addition is most often observed when the nucleophile (Y:⁻) is weakly basic. The nucleophiles in the two examples that follow are ⁻:C≡N: and $C_6H_5CH_2\ddot{S}$:⁻, respectively. Both are much weaker bases than acetylide ion, which was the nucleophile used in the example illustrating direct addition.

Hydrogen cyanide and alkanethiols have pK_a values in the 9 to 10 range, and the pK_a for acetylene is 26.

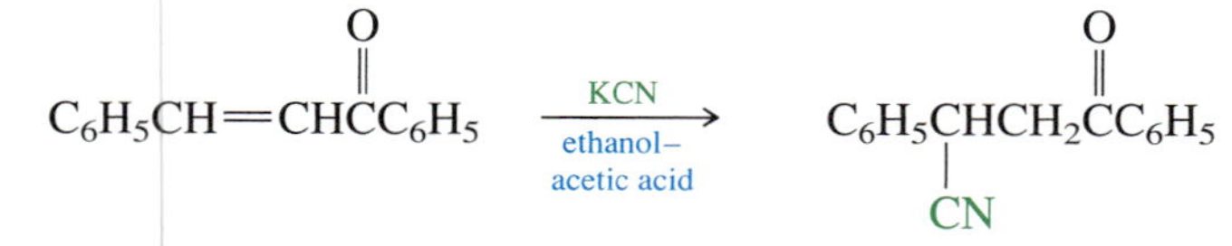

$$C_6H_5CH{=}CHCC_6H_5 \xrightarrow[\substack{\text{ethanol--}\\\text{acetic acid}}]{\text{KCN}} C_6H_5CHCH_2CC_6H_5$$

1,3-Diphenyl-2-propen-1-one

4-Oxo-2,4-diphenylbutanenitrile
(93–96%)

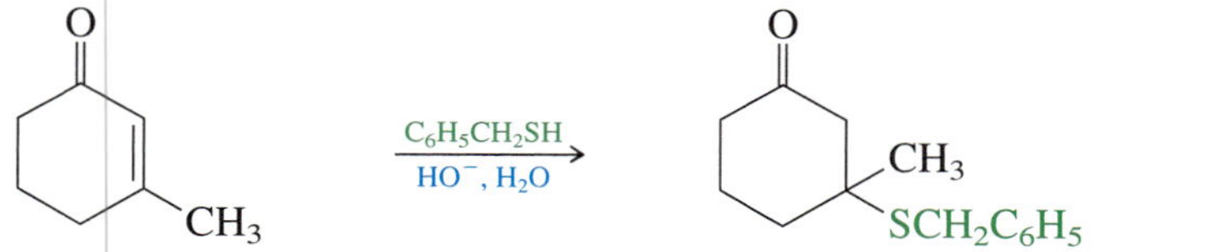

3-Methyl-2-cyclohexenone

3-Benzylthio-3-methylcyclohexanone
(58%)

One explanation for these observations is presented in Figure 18.7. Nucleophilic addition to α,β-unsaturated aldehydes and ketones may be governed either by *kinetic control* or by *thermodynamic control* (Section 10.10). Under conditions in which the 1,2- and 1,4-addition products do not equilibrate, 1,2 addition predominates because it is faster than 1,4 addition. Kinetic control operates with strongly basic nucleophiles to give the 1,2-addition product. A weakly basic nucleophile, however, goes on and off the carbonyl carbon readily and permits the 1,2-addition product to equilibrate with the more slowly formed, but more stable, 1,4-addition product. Thermodynamic control is observed with weakly basic nucleophiles. The product of 1,4 addition, which retains the carbon–oxygen double bond, is more stable than the product of 1,2 addition, which retains the carbon–carbon double bond. In general, carbon–oxygen double bonds are stronger than carbon–carbon double bonds because the greater electronegativity of oxygen permits the π electrons to be bound more strongly.

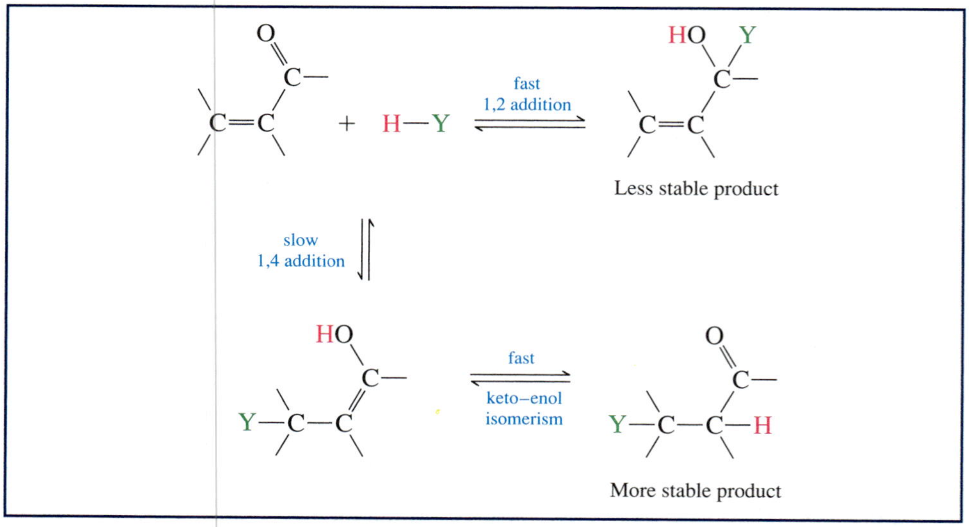

FIGURE 18.7 Nucleophilic addition to α,β-unsaturated aldehydes and ketones may take place either in a 1,2 or 1,4 manner. Direct addition (1,2) occurs faster than conjugate addition (1,4) but gives a less stable product. The product of 1,4 addition retains the carbon–oxygen double bond, which is, in general, stronger than a carbon–carbon double bond.

PROBLEM 18.17 *Acrolein* (H_2C=CHCH=O) reacts with sodium azide (NaN_3) in aqueous acetic acid to form a compound, $C_3H_5N_3O$ in 71% yield. Propanal (CH_3CH_2CH=O), when subjected to the same reaction conditions, is recovered unchanged. Suggest a structure for the product formed from acrolein, and offer an explanation for the difference in reactivity between acrolein and propanal.

18.13 ADDITION OF CARBANIONS TO α,β-UNSATURATED KETONES: THE MICHAEL REACTION

A synthetically useful reaction known as the **Michael reaction,** or **Michael addition,** involves nucleophilic addition of carbanions to α,β-unsaturated ketones. The most common types of carbanions used are enolate ions derived from β-diketones. These enolates are weak bases (Section 18.6) and react with α,β-unsaturated ketones by *conjugate addition.*

Arthur Michael, for whom the reaction is named, was an American chemist whose career spanned the period between the 1870s and the 1930s. He was independently wealthy and did much of his research in his own private laboratory.

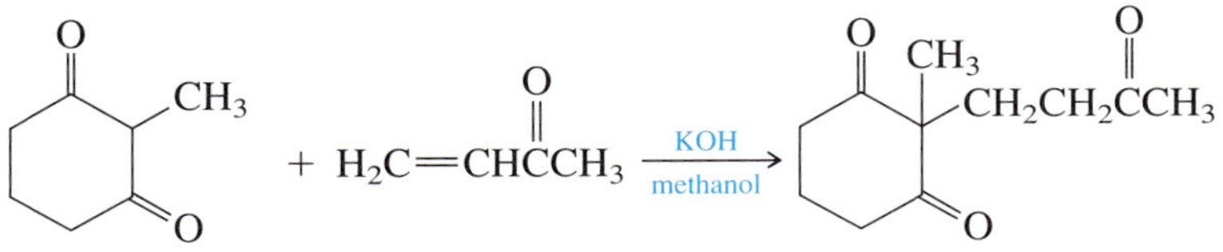

2-Methyl-1,3-cyclohexanedione

Methyl vinyl ketone

2-Methyl-2-(3′-oxobutyl)-1,3-cyclohexanedione (85%)

The product of this Michael addition has the necessary functionality to undergo an intramolecular aldol condensation:

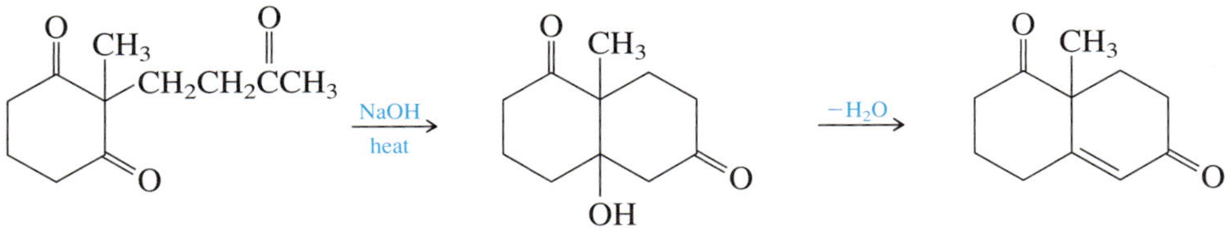

2-Methyl-2-(3′-oxobutyl)-1,3-cyclohexanedione

Intramolecular aldol addition product; not isolated

Δ^4-9-Methyloctalin-3,8-dione

The synthesis of cyclohexenone derivatives by Michael addition followed by intramolecular aldol condensation is called the **Robinson annulation,** after Sir Robert Robinson, who popularized its use. By *annulation* we mean the building of a ring onto some starting molecule. (The alternative spelling "annelation" is also often used.)

PROBLEM 18.18 Both the conjugate addition step and the intramolecular aldol condensation step can be carried out in one synthetic operation without isolating any of the intermediates along the way. For example, consider the reaction

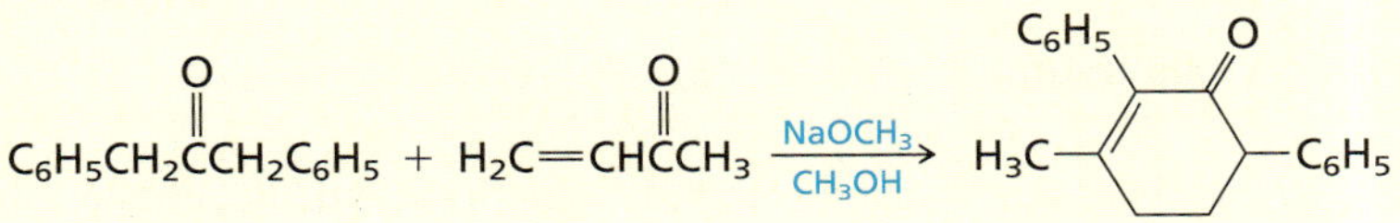

Dibenzyl ketone

Methyl vinyl ketone

3-Methyl-2,6-diphenyl-2-cyclohexenone (55%)

Write structural formulas corresponding to the intermediates formed in the conjugate addition step and in the aldol addition step.

18.14 CONJUGATE ADDITION OF ORGANOCOPPER REAGENTS TO α,β-UNSATURATED CARBONYL COMPOUNDS

The preparation and some synthetic applications of lithium dialkylcuprates were described earlier (Section 14.11). The most prominent feature of these reagents is their capacity to undergo conjugate addition to α,β-unsaturated aldehydes and ketones.

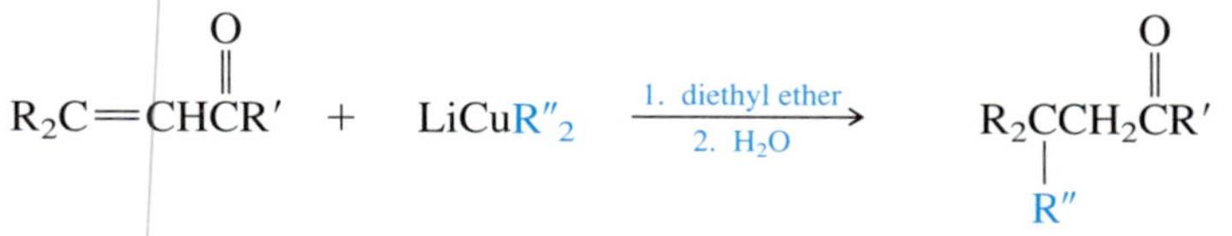

$$R_2C{=}CHCR' \quad + \quad LiCuR''_2 \quad \xrightarrow[\text{2. } H_2O]{\text{1. diethyl ether}} \quad R_2CCH_2CR'$$

α,β-Unsaturated aldehyde or ketone	Lithium dialkylcuprate	Aldehyde or ketone alkylated at the β position

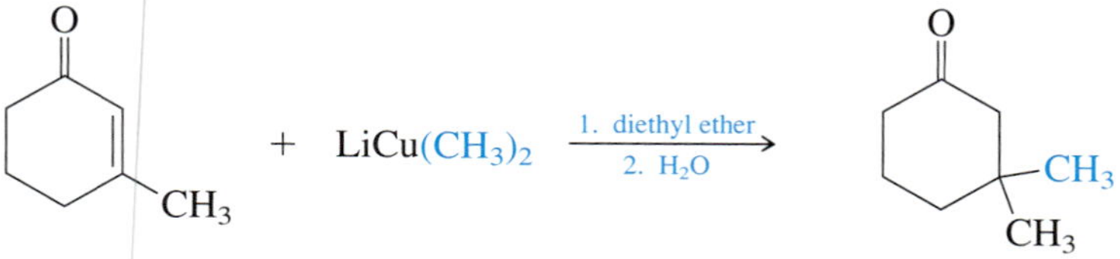

3-Methyl-2-cyclohexenone   +   Lithium dimethylcuprate $LiCu(CH_3)_2$ $\xrightarrow[\text{2. } H_2O]{\text{1. diethyl ether}}$ 3,3-Dimethylcyclohexanone (98%)

PROBLEM 18.19   Outline two ways in which 4-methyl-2-octanone can be prepared by conjugate addition of an organocuprate to an α,β-unsaturated ketone.

SAMPLE SOLUTION   Mentally disconnect one of the bonds to the β carbon so as to identify the group that comes from the lithium dialkylcuprate.

Disconnect this bond

$$CH_3CH_2CH_2CH_2{-}CHCH_2CCH_3 \implies CH_3CH_2CH_2\overset{..}{C}H_2 \;+\; CH_3CH{=}CHCCH_3$$
$$\mid$$
$$CH_3$$

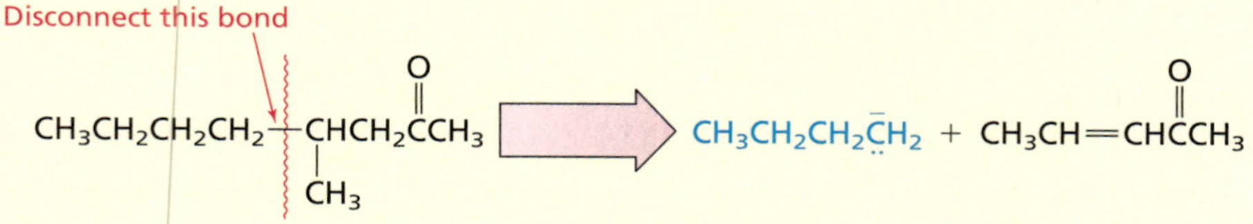

4-Methyl-2-octanone

According to this disconnection, the butyl group is derived from lithium dibutylcuprate. A suitable preparation is

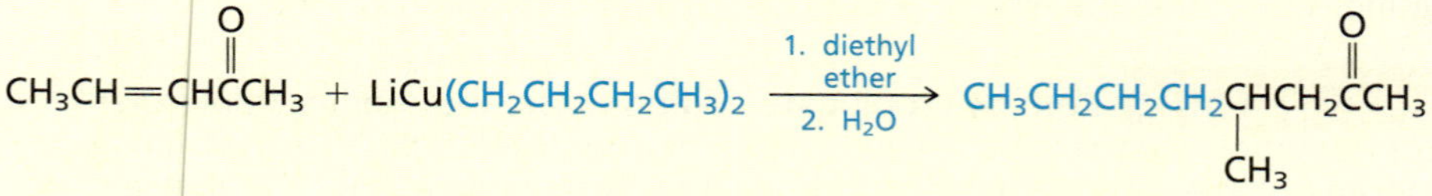

$$CH_3CH{=}CHCCH_3 + LiCu(CH_2CH_2CH_2CH_3)_2 \xrightarrow[\text{2. } H_2O]{\text{1. diethyl ether}} CH_3CH_2CH_2CH_2CHCH_2CCH_3$$
$$\mid$$
$$CH_3$$

3-Penten-2-one   Lithium dibutylcuprate   4-Methyl-2-octanone

Now see if you can identify the second possibility.

Like other carbon–carbon bond-forming reactions, organocuprate addition to enones is a powerful tool in organic synthesis.

18.15 ALKYLATION OF ENOLATE ANIONS

Because enolate anions are sources of nucleophilic carbon, one potential use in organic synthesis is their reaction with alkyl halides to give α-alkyl derivatives of aldehydes and ketones:

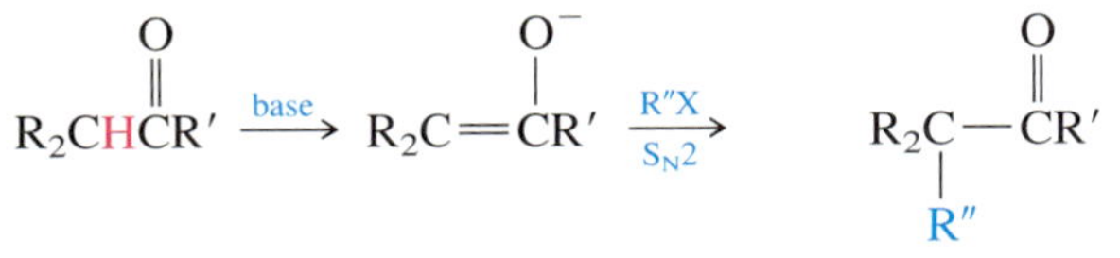

$$\underset{\substack{\text{Aldehyde}\\\text{or ketone}}}{R_2CHCR'} \xrightarrow{\text{base}} \underset{\text{Enolate anion}}{R_2C=CR'} \xrightarrow[S_N2]{R''X} \underset{\substack{\alpha\text{-Alkyl derivative}\\\text{of an aldehyde or a ketone}}}{R_2C\overset{\displaystyle O}{-}CR'}$$

Alkylation occurs by an S_N2 mechanism in which the enolate ion acts as a nucleophile toward the alkyl halide.

In practice, this reaction is difficult to carry out with simple aldehydes and ketones because aldol condensation competes with alkylation. Furthermore, it is not always possible to limit the reaction to the introduction of a single alkyl group. The most successful alkylation procedures use β-diketones as starting materials. Because they are relatively acidic, β-diketones can be converted quantitatively to their enolate ions by weak bases and do not self-condense. Ideally, the alkyl halide should be a methyl or primary alkyl halide.

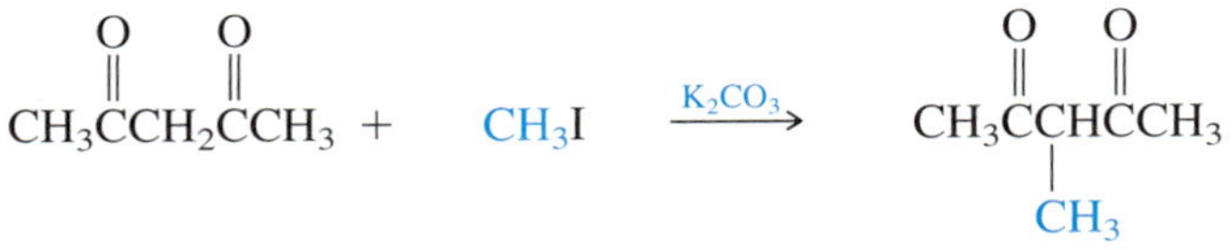

$$\underset{\text{2,4-Pentanedione}}{CH_3CCH_2CCH_3} + \underset{\text{Iodomethane}}{CH_3I} \xrightarrow{K_2CO_3} \underset{\substack{\text{3-Methyl-2,4-pentanedione}\\(75-77\%)}}{CH_3CCHCCH_3}$$

18.16 SUMMARY

Section 18.1 Greek letters are commonly used to identify various carbons in aldehydes and ketones. Using the carbonyl group as a reference, the adjacent carbon is designated α, the next one β, and so on as one moves down the chain. Attached groups take the same Greek letter as the carbon to which they are connected.

Sections 18.2–18.15 Because aldehydes and ketones exist in equilibrium with their corresponding enol isomers, they can express a variety of different kinds of chemical reactivity.

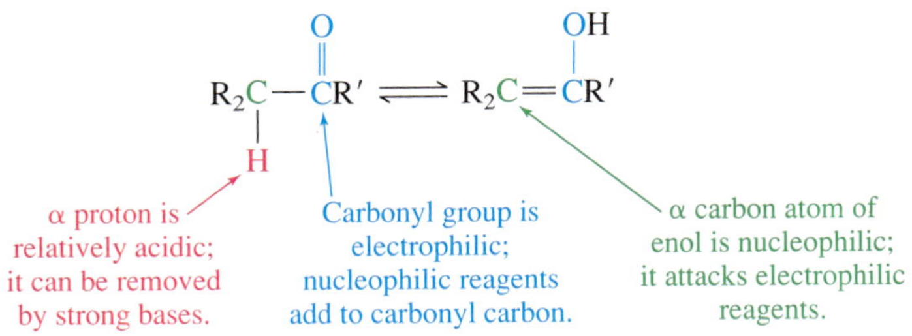

$$R_2C\overset{\displaystyle O}{\underset{\displaystyle H}{-}}CR' \rightleftharpoons R_2C=CR' \;\;\overset{\displaystyle OH}{}$$

α proton is relatively acidic; it can be removed by strong bases.

Carbonyl group is electrophilic; nucleophilic reagents add to carbonyl carbon.

α carbon atom of enol is nucleophilic; it attacks electrophilic reagents.

Reactions that proceed via enol or enolate intermediates are summarized in Table 18.1.

TABLE 18.1   Reactions of Aldehydes and Ketones That Involve Enol or Enolate Ion Intermediates

Reaction (section) and comments	General equation and typical example

α Halogenation (Sections 18.2 and 18.3) Halogens react with aldehydes and ketones by substitution; an α hydrogen is replaced by a halogen. Reaction occurs by electrophilic attack of the halogen on the carbon–carbon double bond of the enol form of the aldehyde or ketone. An acid catalyst increases the rate of enolization, which is the rate-determining step.

$$R_2CHCR' + X_2 \longrightarrow R_2CCR' + HX$$
$$\qquad\qquad\qquad\qquad\quad |$$
$$\qquad\qquad\qquad\qquad\quad X$$

Aldehyde Halogen   α-Halo aldehyde Hydrogen
or ketone   or ketone   halide

$$Br\!-\!\!\bigcirc\!\!-\!CCH_3 + Br_2 \xrightarrow[\text{acid}]{\text{acetic}} Br\!-\!\!\bigcirc\!\!-\!CCH_2Br + HBr$$

p-Bromoacetophenone   Bromine   p-Bromophenacyl   Hydrogen
                                   bromide (69–72%)   bromide

Enolization (Sections 18.4 through 18.6) Aldehydes and ketones having at least one α hydrogen exist in equilibrium with their enol forms. The rate at which equilibrium is achieved is increased by acidic or basic catalysts. The enol content of simple aldehydes and ketones is quite small; β-diketones, however, are extensively enolized.

$$R_2CH\!-\!CR' \rightleftharpoons R_2C\!=\!CR'$$

Aldehyde   Enol
or ketone

Cyclopentanone   Cyclopenten-1-ol $K = 1 \times 10^{-8}$

Enolate ion formation (Section 18.6) An α hydrogen of an aldehyde or a ketone is more acidic than most other protons bound to carbon. Aldehydes and ketones are weak acids, with pK_a's in the 16 to 20 range. Their enhanced acidity is due to the electron-withdrawing effect of the carbonyl group and the resonance stabilization of the enolate anion.

$$R_2CHCR' + HO^- \rightleftharpoons R_2C\!=\!CR' + H_2O$$

Aldehyde Hydroxide   Enolate   Water
or ketone   ion   anion

$$CH_3CH_2CCH_2CH_3 + HO^- \rightleftharpoons CH_3CH\!=\!CCH_2CH_3 + H_2O$$

3-Pentanone   Hydroxide ion   Enolate anion Water
                                 of 3-pentanone

Haloform reaction (Section 18.7) Methyl ketones are cleaved on reaction with excess halogen in the presence of base. The products are a trihalomethane (haloform) and a carboxylate salt.

$$RCCH_3 + 3X_2 \xrightarrow{HO^-} RCO^- + HCX_3$$

Methyl   Halogen   Carboxylate Trihalomethane
ketone   ion   (haloform)

$$(CH_3)_3CCH_2CCH_3 \xrightarrow[\text{2. } H^+]{\text{1. } Br_2, \text{ NaOH}} (CH_3)_3CCH_2CO_2H + CHBr_3$$

4,4-Dimethyl-2-pentanone   3,3-Dimethylbutanoic   Bromoform
                                      acid (89%)

(Continued)

| TABLE 18.1 | Reactions of Aldehydes and Ketones That Involve Enol or Enolate Ion Intermediates *(Continued)* |

Reaction (section) and comments	General equation and typical example
Aldol condensation (Section 18.9) A reaction of great synthetic value for carbon–carbon bond formation. Nucleophilic addition of an enolate ion to a carbonyl group, followed by dehydration of the β-hydroxy aldehyde, yields an α,β-unsaturated aldehyde.	$2RCH_2CR' \xrightarrow{HO^-} RCH_2C{=}CCR' + H_2O$ with substituents R', R Aldehyde α,β-Unsaturated aldehyde Water $CH_3(CH_2)_6CH \xrightarrow[CH_3CH_2OH]{NaOCH_2CH_3} CH_3(CH_2)_6CH{=}C(CH_2)_5CH_3$ with $HC{=}O$ Octanal 2-Hexyl-2-decenal (79%)
Claisen–Schmidt reaction (Section 18.10) A mixed aldol condensation in which an aromatic aldehyde reacts with an enolizable aldehyde or ketone.	$ArCH + RCH_2CR' \xrightarrow{HO^-} ArCH{=}CCR' + H_2O$ with R Aromatic aldehyde; Aldehyde or ketone; α,β-Unsaturated carbonyl compound; Water $C_6H_5CH + (CH_3)_3CCCH_3 \xrightarrow[ethanol–water]{NaOH} C_6H_5CH{=}CHCC(CH_3)_3$ Benzaldehyde; 3,3-Dimethyl-2-butanone; 4,4-Dimethyl-1-phenyl-1-penten-3-one (88–93%)
Conjugate addition to α,β-unsaturated carbonyl compounds (Sections 18.11 through 18.14) The β-carbon atom of an α,β-unsaturated carbonyl compound is electrophilic; nucleophiles, especially weakly basic ones, yield the products of conjugate addition to α,β-unsaturated aldehydes and ketones.	$R_2C{=}CHCR' + HY{:} \longrightarrow R_2CCH_2CR'$ with Y α,β-Unsaturated aldehyde or ketone; Nucleophile; Product of conjugate addition $(CH_3)_2C{=}CHCCH_3 \xrightarrow[H_2O]{NH_3} (CH_3)_2CCH_2CCH_3$ with NH_2 4-Methyl-3-penten-2-one (mesityl oxide); 4-Amino-4-methyl-2-pentanone (63–70%)
Robinson annulation (Section 18.13) A combination of conjugate addition of an enolate anion to an α,β-unsaturated ketone with subsequent intramolecular aldol condensation.	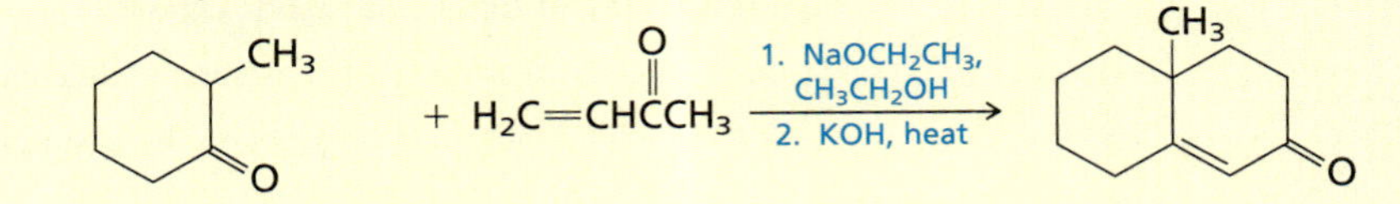 2-Methylcyclohexanone $+ H_2C{=}CHCCH_3 \xrightarrow[\text{2. KOH, heat}]{\text{1. NaOCH}_2\text{CH}_3,\ \text{CH}_3\text{CH}_2\text{OH}}$ 6-Methylbicyclo[4.4.0]-1-decen-3-one (46%) Methyl vinyl ketone

(Continued)

TABLE 18.1	Reactions of Aldehydes and Ketones That Involve Enol or Enolate Ion Intermediates *(Continued)*

Reaction (section) and comments	General equation and typical example
Conjugate addition of organocopper compounds (Section 18.14) The principal synthetic application of lithium dialkylcuprate reagents is their reaction with α,β-unsaturated carbonyl compounds. Alkylation of the β carbon occurs.	$R_2C{=}CHCR'$ (α,β-Unsaturated aldehyde or ketone) $+$ R''_2CuLi (Lithium dialkylcuprate) $\xrightarrow[\text{2. } H_2O]{\text{1. diethyl ether}}$ $R_2C{-}CH_2CR'$ with R'' (β-Alkyl aldehyde or ketone) 6-Methylcyclohept-2-enone $\xrightarrow[\text{2. } H_2O]{\text{1. LiCu(CH}_3)_2}$ 3,6-Dimethylcycloheptanone (85%)
α Alkylation of aldehydes and ketones (Section 18.15) Alkylation of simple aldehydes and ketones via their enolates is difficult. β-Diketones can be converted quantitatively to their enolate anions, which react efficiently with primary alkyl halides.	$RCCH_2CR$ (β-Diketone) $\xrightarrow{R'CH_2X,\ HO^-}$ $RCCHCR$ with CH_2R' (α-Alkyl-β-diketone) 2-Benzyl-1,3-cyclohexanedione $+\ C_6H_5CH_2Cl$ (Benzyl chloride) $\xrightarrow[\text{ethanol}]{KOCH_2CH_3}$ 2,2-Dibenzyl-1,3-cyclohexanedione (69%)

PROBLEMS

18.20 (a) Write structural formulas or build molecular models for all the noncyclic aldehydes and ketones of molecular formula C_4H_6O.

   (b) Are any of these compounds stereoisomeric?

   (c) Are any of these compounds chiral?

   (d) Which of these are α,β-unsaturated aldehydes or α,β-unsaturated ketones?

   (e) Which of these can be prepared by a simple (i.e., not mixed) aldol condensation?

18.21 The main flavor component of the hazelnut is (2*E*,5*S*)-5-methyl-2-hepten-4-one. Write a structural formula or build a molecular model showing its stereochemistry.

18.22 The simplest α,β-unsaturated aldehyde *acrolein* is prepared by heating glycerol with an acid catalyst. Suggest a mechanism for this reaction.

$$HOCH_2CHCH_2OH \xrightarrow[\text{heat}]{KHSO_4} H_2C{=}CHCH + H_2O$$
$$\text{with } OH$$

18.23 In each of the following pairs of compounds, choose the one that has the greater enol content, and write the structure of its enol form:

(a) 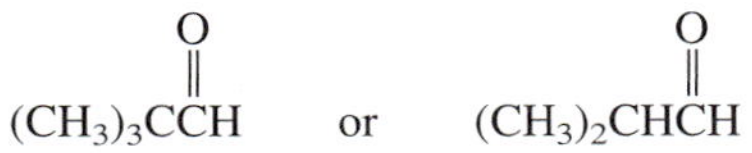 $(CH_3)_3CCH{=}O$ or $(CH_3)_2CHCH{=}O$

(b) $C_6H_5CC_6H_5$ (C=O)   or $C_6H_5CH_2CCH_2C_6H_5$ (C=O)

(c) $C_6H_5CCH_2CC_6H_5$ (two C=O)   or $C_6H_5CH_2CCH_2C_6H_5$ (C=O)

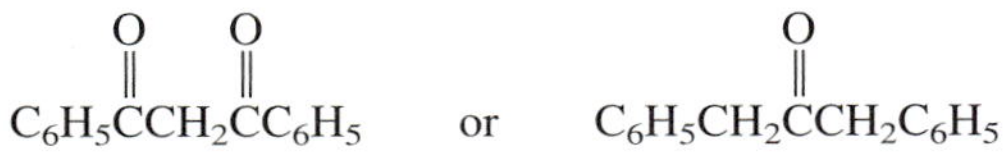

(d) 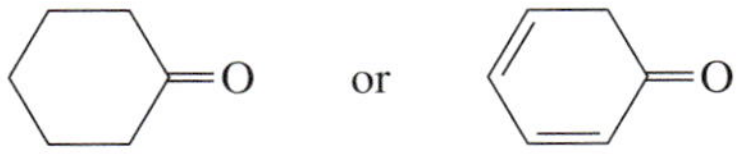 cyclohexanone (=O)   or   cyclohex-3-enone (=O)

(e) 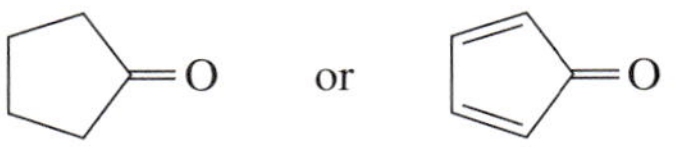 cyclopentanone (=O)   or   cyclopentadienone (=O)

(f) 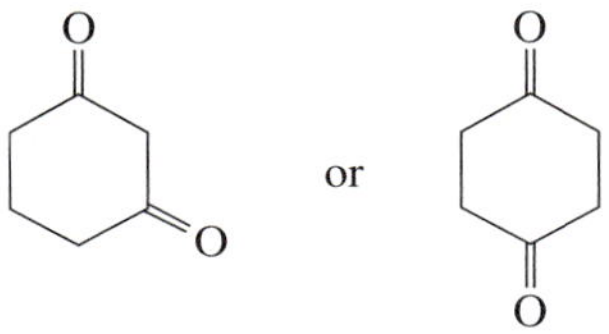 1,3-cyclohexanedione   or   1,4-cyclohexanedione

18.24 Give the structure of the expected organic product in the reaction of 3-phenylpropanal with each of the following:

(a) Chlorine in acetic acid

(b) Sodium hydroxide in ethanol, 10°C

(c) Sodium hydroxide in ethanol, 70°C

(d) Product of part (c) with lithium aluminum hydride; then H_2O

(e) Product of part (c) with sodium cyanide in acidic ethanol

18.25 Each of the following reactions has been reported in the chemical literature. Write the structure of the product(s) formed in each case.

(a) 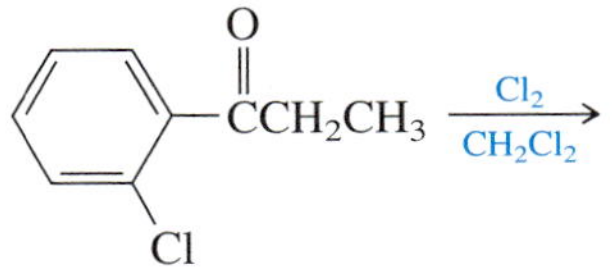 (2-chlorophenyl)–CCH_2CH_3 (C=O) $\xrightarrow[CH_2Cl_2]{Cl_2}$

(b) H_3C–(cyclohexanone ring)=$C(CH_3)_2$ $\xrightarrow[NaOH,\ H_2O]{C_6H_5CH_2SH}$

(c) (cyclopentanone with C_6H_5, C_6H_5) $\xrightarrow[\text{diethyl ether}]{Br_2}$

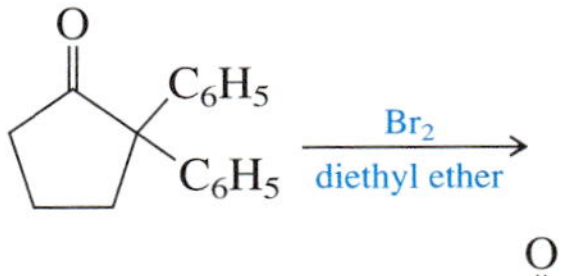

(d) 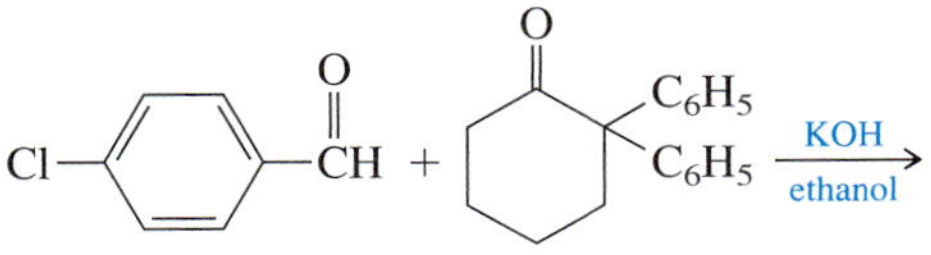 Cl–(phenyl)–$CH{=}O$ (O) + (cyclohexanone with C_6H_5, C_6H_5) $\xrightarrow[\text{ethanol}]{KOH}$

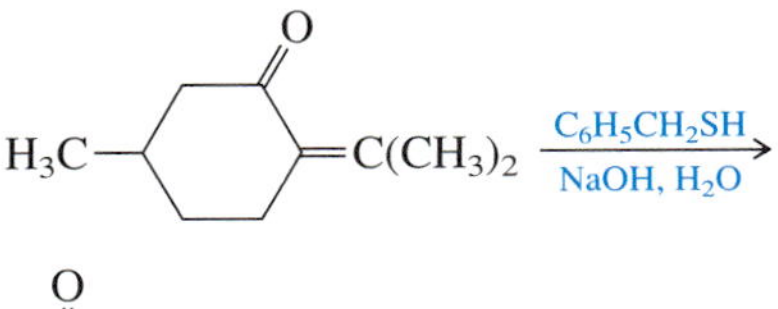

(e) 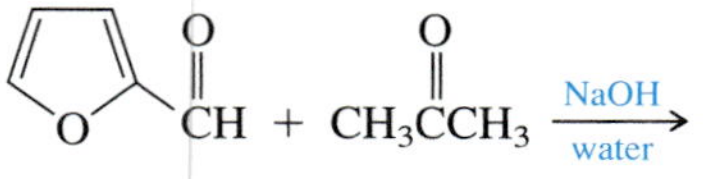 $+ CH_3CCH_3 \xrightarrow[\text{water}]{\text{NaOH}}$

(f) 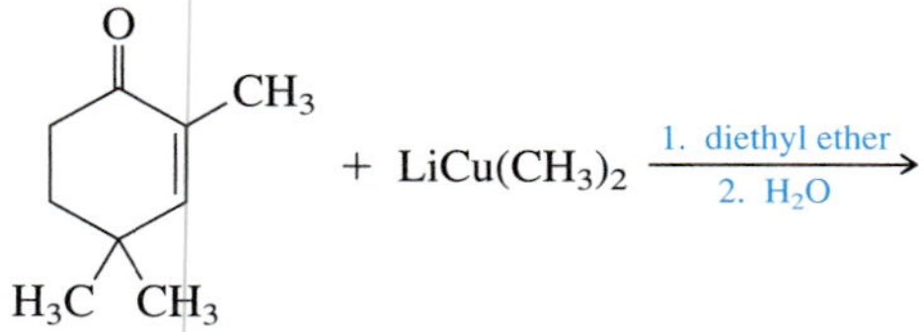 $+ LiCu(CH_3)_2 \xrightarrow[\text{2. H}_2\text{O}]{\text{1. diethyl ether}}$

(g) $+ C_6H_5CH \xrightarrow[\text{ethanol–water}]{\text{NaOH}}$

(h) $+ H_2C={=}CHCH_2Br \xrightarrow{\text{KOH}}$

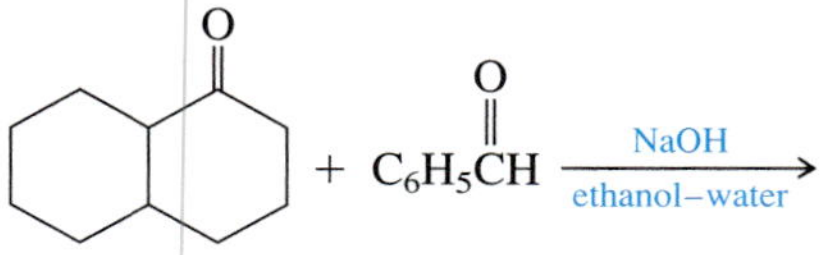

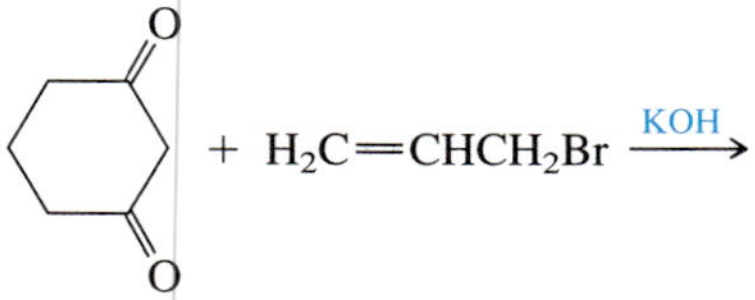

18.26 Show how each of the following compounds could be prepared from 3-pentanone. In most cases more than one synthetic transformation will be necessary.

(a) 2-Bromo-3-pentanone

(b) 1-Penten-3-one

(c) 1-Penten-3-ol

(d) 3-Hexanone

(e) 2-Methyl-1-phenyl-1-penten-3-one

18.27 (a) A synthesis that begins with 3,3-dimethyl-2-butanone gives the epoxide shown. Suggest reagents appropriate for each step in the synthesis.

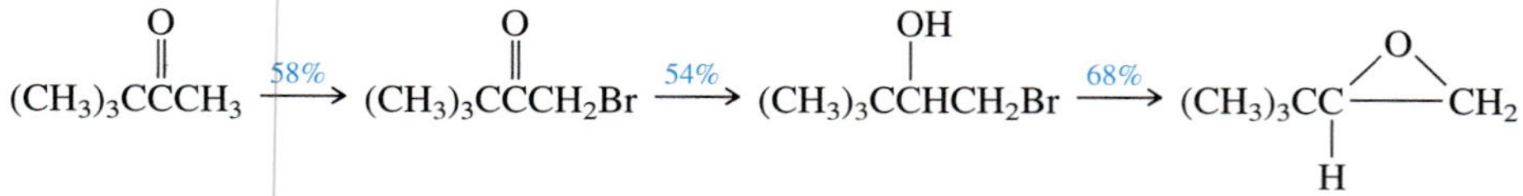

$$(CH_3)_3CCCH_3 \xrightarrow{58\%} (CH_3)_3CCCH_2Br \xrightarrow{54\%} (CH_3)_3CCHCH_2Br \xrightarrow{68\%} (CH_3)_3CC\overset{O}{\underset{H}{-}}CH_2$$

(b) The yield for each step as actually carried out in the laboratory is given above each arrow. What is the overall yield for the three-step sequence?

18.28 Using benzene, acetic anhydride, and 1-propanethiol as the source of all the carbon atoms, along with any necessary inorganic reagents, outline a synthesis of the compound shown.

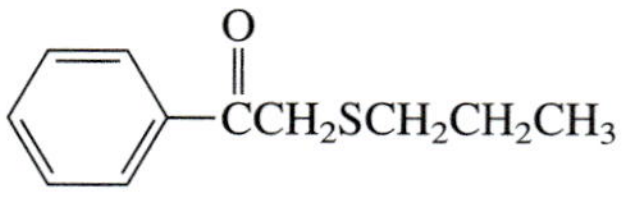

18.29 Show how you could prepare each of the following compounds from cyclopentanone, D_2O, and any necessary organic or inorganic reagents.

(a)

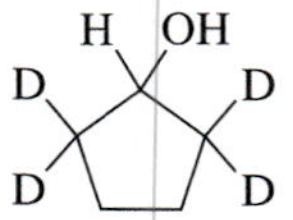

(b)

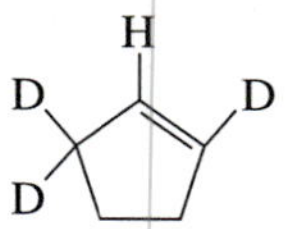

(c)

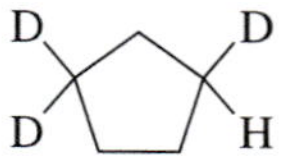

(d)

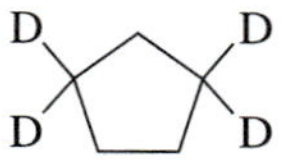

18.30 (a) At present, butanal is prepared industrially by hydroformylation of propene (Section 17.4). Write a chemical equation for this industrial synthesis.

(b) Before about 1970, the principal industrial preparation of butanal was from acetaldehyde. Outline a practical synthesis of butanal from acetaldehyde.

18.31 Identify the reagents appropriate for each step in the following syntheses:

(a)

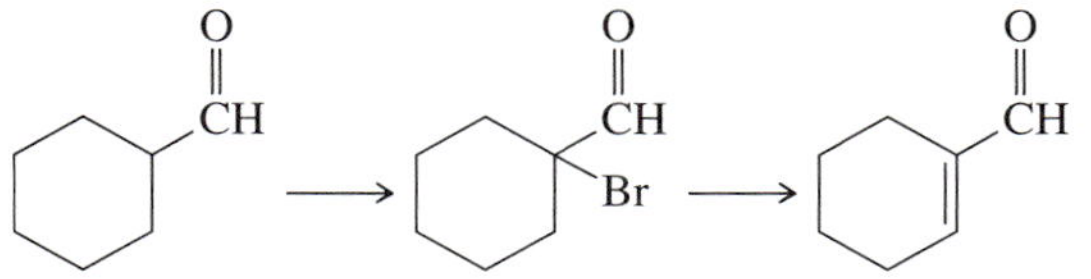

(b)

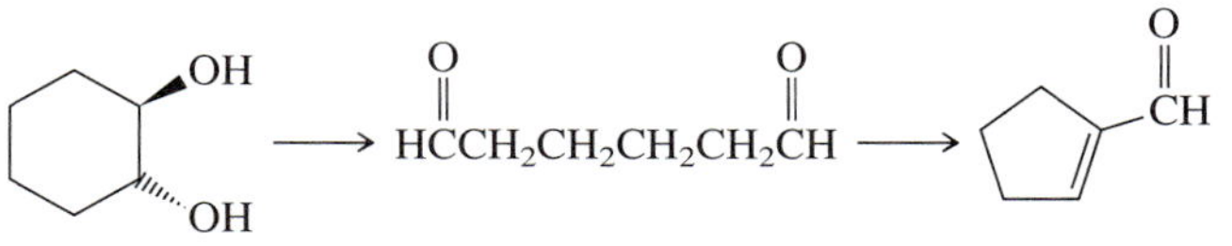

(c)

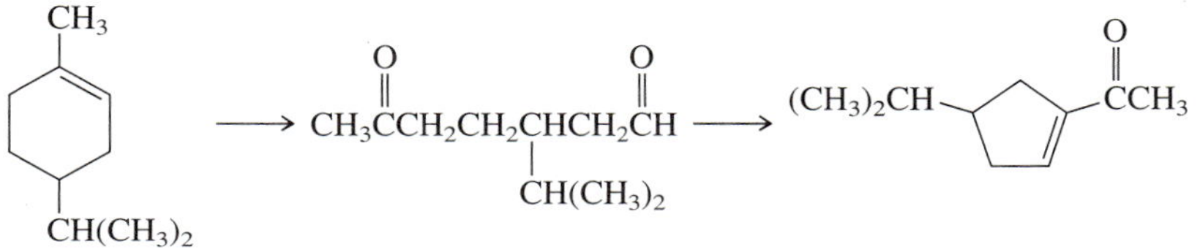

(d)

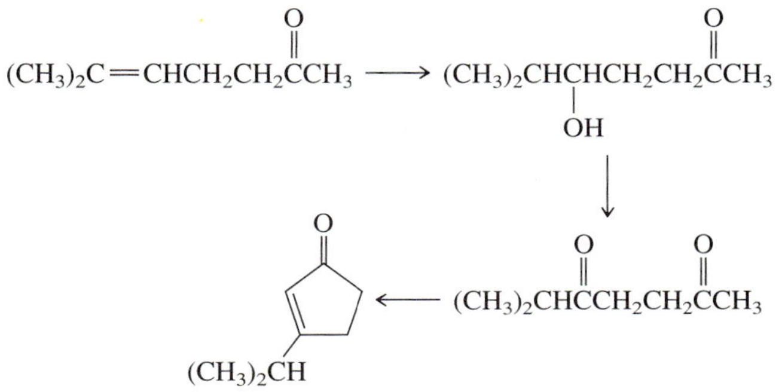

18.32 Give the structure of the product derived by intramolecular aldol condensation of the keto aldehyde shown:

18.33 Prepare each of the following compounds from the starting materials given and any necessary organic or inorganic reagents:

(a) $(CH_3)_2CHCHCHCH_2OH$ from $(CH_3)_2CHCH_2OH$
with CH_3 and HO CH_3 substituents

(b) $C_6H_5CH{=}CCH_2OH$ from benzyl alcohol and 1-propanol
with CH_3 substituent

(c)

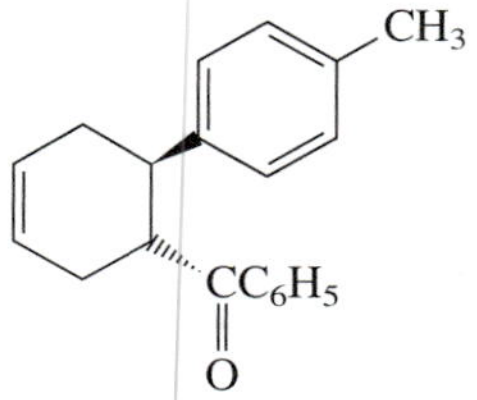

from acetophenone,
4-methylbenzyl alcohol,
and 1,3-butadiene

18.34 *Terreic acid* is a naturally occurring antibiotic substance. Its actual structure is an enol isomer of the structure shown. Write the two most stable enol forms of terreic acid, and choose which of those two is more stable.

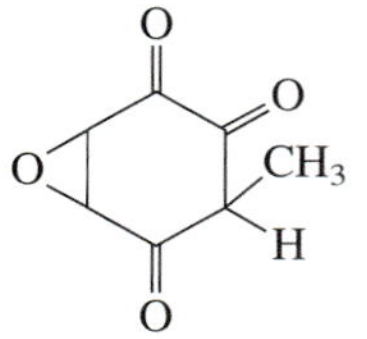

18.35 In each of the following, the indicated observations were made before any of the starting material was transformed to aldol addition or condensation products:

(a) In aqueous acid, only 17% of $(C_6H_5)_2CHCH=O$ is present as the aldehyde; 2% of the enol is present. Some other species accounts for 81% of the material. What is it?

(b) In aqueous base, 97% of $(C_6H_5)_2CHCH=O$ is present as a species different from any of those in part (a). What is this species?

18.36 (a) For a long time attempts to prepare compound A were thwarted by its ready isomerization to compound B. The isomerization is efficiently catalyzed by traces of base. Write a reasonable mechanism for this isomerization.

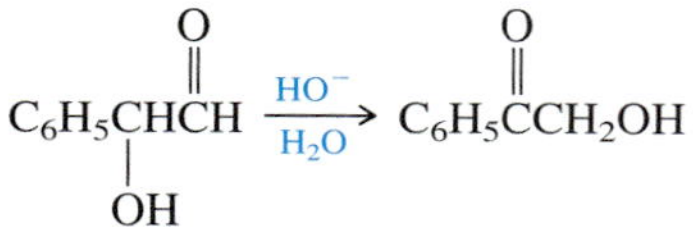

Compound A   Compound B

(b) Another attempt to prepare compound A by hydrolysis of its diethyl acetal gave only the 1,4-dioxane derivative C. How was compound C formed?

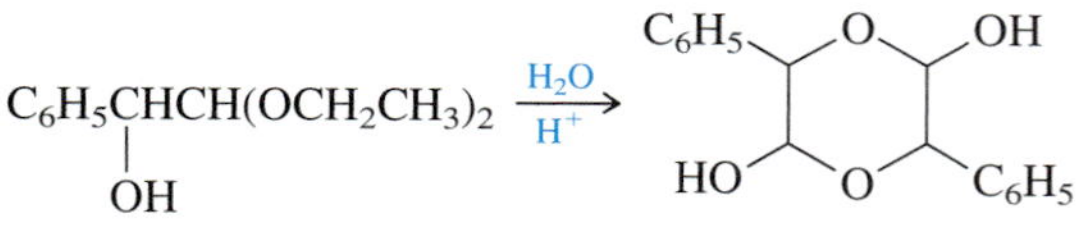

Compound C

18.37 Consider the ketones piperitone, menthone, and isomenthone.

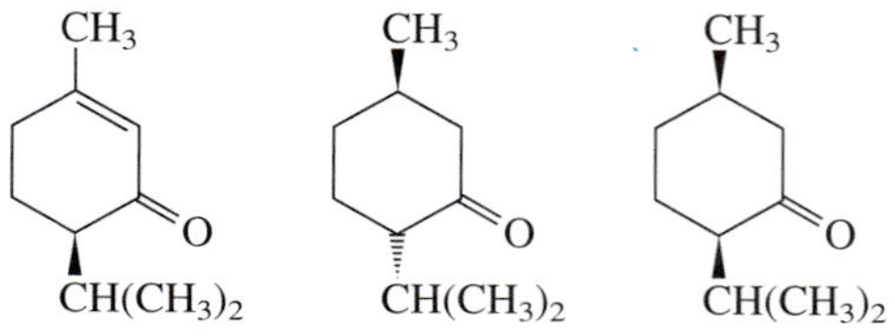

(−)-Piperitone   Menthone   Isomenthone

Suggest reasonable explanations for each of the following observations:

(a) Optically active piperitone ($\alpha_D = -32°$) is converted to racemic piperitone on standing in a solution of sodium ethoxide in ethanol.

(b) Menthone is converted to a mixture of menthone and isomenthone on treatment with 90% sulfuric acid.

18.38 Many nitrogen-containing compounds engage in a proton-transfer equilibrium that is analogous to keto–enol tautomerism:

$$HX-N=Z \rightleftharpoons X=N-ZH$$

Each of the following compounds is the less stable partner of such a tautomeric pair. Write the structure of the more stable partner for each one.

(a) $CH_3CH_2N=O$

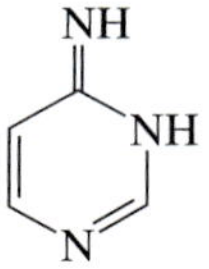

(b) $(CH_3)_2C=CHNHCH_3$

(c) 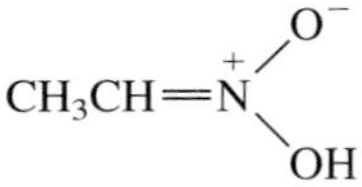 $CH_3CH=\overset{+}{N}\overset{O^-}{\underset{OH}{\diagdown}}$

(d)

(e) 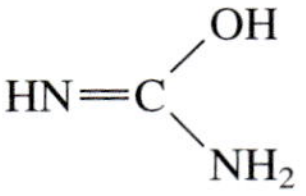 $HN=C\overset{OH}{\underset{NH_2}{\diagdown}}$

18.39 Outline reasonable mechanisms for each of the following reactions:

(a)

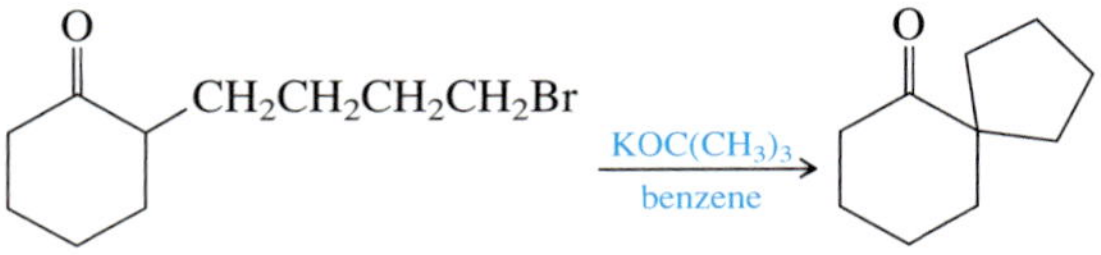

(76%)

(b) 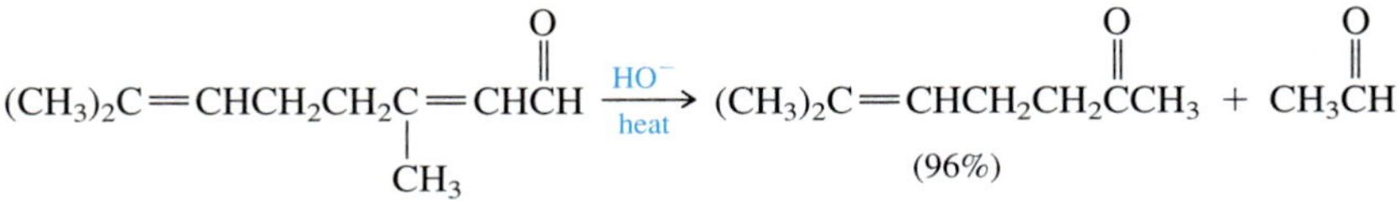 $(CH_3)_2C=CHCH_2CH_2\overset{O}{\overset{\|}{C}}=CHCH \xrightarrow[\text{heat}]{HO^-} (CH_3)_2C=CHCH_2CH_2\overset{O}{\overset{\|}{C}}CH_3 + CH_3\overset{O}{\overset{\|}{C}}H$

with CH_3 substituent on the starting material

(96%)

(c) 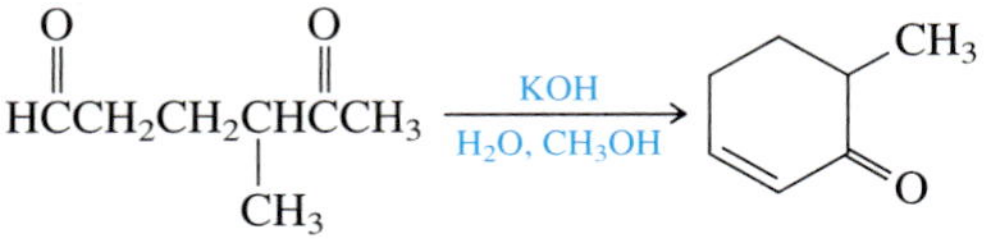 $H\overset{O}{\overset{\|}{C}}CH_2CH_2\overset{O}{\underset{CH_3}{C}}H\overset{\|}{C}CH_3 \xrightarrow[H_2O, CH_3OH]{KOH}$

(40%)

(d)

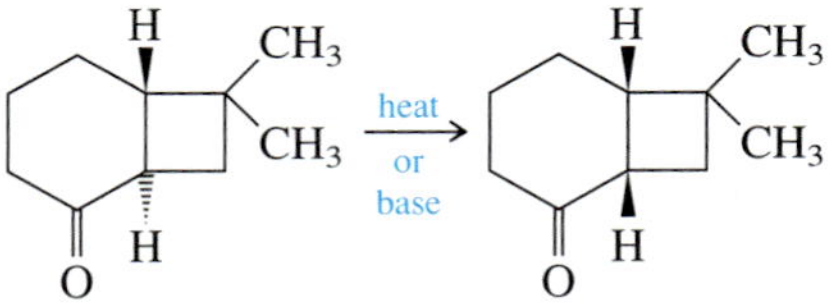

(e) 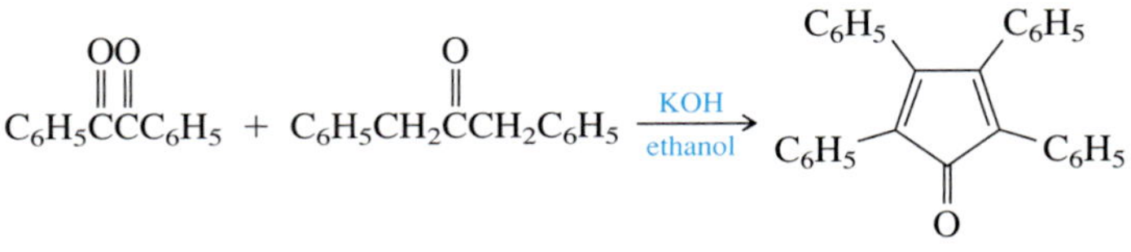 $C_6H_5\overset{O\ O}{\overset{\|\ \|}{C}CC}_6H_5 + C_6H_5CH_2\overset{O}{\overset{\|}{C}}CH_2C_6H_5 \xrightarrow[\text{ethanol}]{KOH}$

(91–96%)

$$\underset{(f)}{C_6H_5CH_2\overset{\displaystyle O}{\overset{\|}{C}}CH_2CH_3} + H_2C=\overset{\displaystyle O}{\overset{\|}{C}}\underset{\displaystyle C_6H_5}{\overset{\displaystyle}{C}}C_6H_5 \xrightarrow[\text{CH}_3\text{OH}]{\text{NaOCH}_3}$$

(51%)

18.40 Suggest reasonable explanations for each of the following observations:

(a) The C=O stretching frequency of α,β-unsaturated ketones (about 1675 cm^{-1}) is less than that of typical dialkyl ketones (1710–1750 cm^{-1}).

(b) The C=O stretching frequency of cyclopropenone (1640 cm^{-1}) is lower than that of typical α,β-unsaturated ketones (1675 cm^{-1}).

(c) The dipole moment of diphenylcyclopropenone ($\mu = 5.1$ D) is substantially larger than that of benzophenone ($\mu = 3.0$ D)

(d) The β carbon of an α,β-unsaturated ketone is less shielded than the corresponding carbon of an alkene. Typical ^{13}C NMR chemical shift values are

$$\underset{(\delta \approx 129)}{H_2C=CH\overset{\displaystyle O}{\overset{\|}{C}}R} \qquad \underset{(\delta \approx 114)}{H_2C=CHCH_2R}$$

18.41 Bromination of 3-methyl-2-butanone yielded two compounds, each having the molecular formula C$_5$H$_9$BrO, in a 95:5 ratio. The ^{1}H NMR spectrum of the major isomer A was characterized by a doublet at δ 1.2 (six protons), a septet at δ 3.0 (one proton), and a singlet at δ 4.1 (two protons). The ^{1}H NMR spectrum of the minor isomer B exhibited two singlets, one at δ 1.9 and the other at δ 2.5. The lower field singlet had half the area of the higher field one. Suggest reasonable structures for these two compounds.

18.42 Treatment of 2-butanone (1 mole) with Br$_2$ (2 moles) in aqueous HBr gave C$_4$H$_6$Br$_2$O. The ^{1}H NMR spectrum of the product was characterized by signals at δ 1.9 (doublet, three protons), 4.6 (singlet, two protons), and 5.2 (quartet, one proton). Identify this compound.

18.43 2-Phenylpropanedial [C$_6$H$_5$CH(CHO)$_2$] exists in the solid state as an enol in which the configuration of the double bond is *E*. In solution (CDCl$_3$), an enol form again predominates but this time the configuration is *Z*. Make molecular models of these two enols, and suggest an explanation for the predominance of the *Z* enol in solution. (*Hint:* Think about intermolecular versus intramolecular hydrogen bonding.)

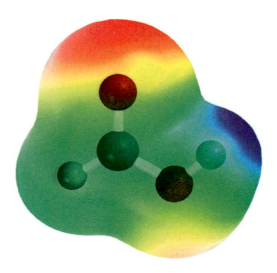

CARBOXYLIC ACIDS

$$\overset{\displaystyle O}{\underset{\displaystyle \|}{}}$$

Carboxylic acids, compounds of the type $RCOH$, constitute one of the most frequently encountered classes of organic compounds. Countless natural products are carboxylic acids or are derived from them. Some carboxylic acids, such as acetic acid, have been known for centuries. Others, such as the prostaglandins, which are powerful regulators of numerous biological processes, remained unknown until relatively recently. Still others, aspirin for example, are the products of chemical synthesis. The therapeutic effects of aspirin, welcomed long before the discovery of prostaglandins, are now understood to result from aspirin's ability to inhibit the biosynthesis of prostaglandins.

$$CH_3\overset{O}{\overset{\|}{C}}OH$$

Acetic acid
(present in
vinegar)

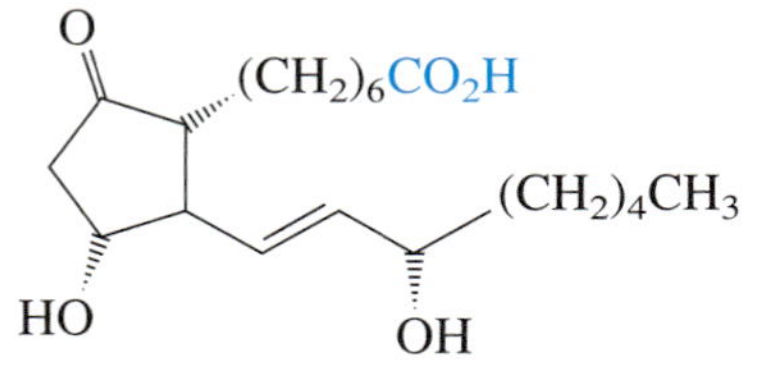

PGE$_1$ (a prostaglandin; a small amount
of PGE$_1$ lowers blood pressure
significantly)

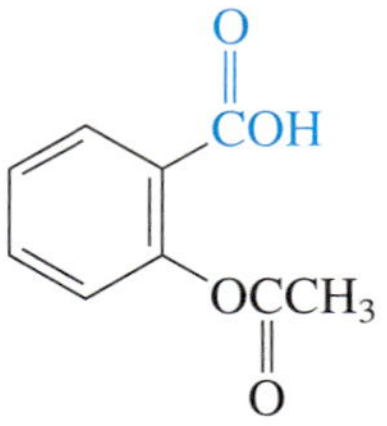

Aspirin

The chemistry of carboxylic acids is the central theme of this chapter. The importance of carboxylic acids is magnified when we realize that they are the parent compounds of a large group of derivatives that includes acyl chlorides, acid anhydrides, esters, and amides. Those classes of compounds will be discussed in Chapter 20. Together, this chapter and the next tell the story of some of the most fundamental structural types and functional group transformations in organic and biological chemistry.

19.1 CARBOXYLIC ACID NOMENCLATURE

It is hard to find a class of compounds in which the common names of its members have influenced organic nomenclature more than carboxylic acids. Not only are the common names of carboxylic acids themselves abundant and widely used, but the names of many other compounds are derived from them. Benzene took its name from benzoic acid and propane from propionic acid, not the other way around. The name butane comes from butyric acid, present in rancid butter. The common names of most aldehydes are derived from the common names of carboxylic acids—valeraldehyde from valeric acid, for example. Many carboxylic acids are better known by common names than by their systematic ones, and the framers of the IUPAC rules have taken a liberal view toward accepting these common names as permissible alternatives to the systematic ones. Table 19.1 lists both common and systematic names for a number of important carboxylic acids.

Systematic names for carboxylic acids are derived by counting the number of carbons in the longest continuous chain that includes the carboxyl group and replacing the -e ending of the corresponding alkane by -oic acid. The first three acids in Table 19.1, methanoic (1 carbon), ethanoic (2 carbons), and octadecanoic acid (18 carbons), illustrate this point. When substituents are present, their locations are identified by number;

TABLE 19.1   Systematic and Common Names of Some Carboxylic Acids

	Structural formula	Systematic name	Common name
1.	HCO_2H	Methanoic acid	Formic acid
2.	CH_3CO_2H	Ethanoic acid	Acetic acid
3.	$CH_3(CH_2)_{16}CO_2H$	Octadecanoic acid	Stearic acid
4.	CH_3CHCO_2H OH	2-Hydroxypropanoic acid	Lactic acid
5.	(phenyl)—$CHCO_2H$ OH	2-Hydroxy-2-phenylethanoic acid	Mandelic acid
6.	$H_2C=CHCO_2H$	Propenoic acid	Acrylic acid

7.	$CH_3(CH_2)_7$ $(CH_2)_7CO_2H$
	     $C=C$
	     H H
		(Z)-9-Octadecenoic acid	Oleic acid

8.	(phenyl)—CO_2H	Benzenecarboxylic acid	Benzoic acid
9.	(phenyl with OH and CO_2H)	o-Hydroxybenzenecarboxylic acid	Salicylic acid
10.	$HO_2CCH_2CO_2H$	Propanedioic acid	Malonic acid
11.	$HO_2CCH_2CH_2CO_2H$	Butanedioic acid	Succinic acid
12.	(phenyl with two CO_2H)	1,2-Benzenedicarboxylic acid	Phthalic acid

numbering of the carbon chain always begins at the carboxyl group. This is illustrated in entries 4 and 5 in the table.

Notice that compounds 4 and 5 are named as hydroxy derivatives of carboxylic acids, rather than as carboxyl derivatives of alcohols. This parallels what we saw earlier in Section 17.1 where an aldehyde or ketone function took precedence over a hydroxyl group in defining the main chain. In keeping with the principle that precedence is determined by oxidation state, carboxylic acid groups not only outrank hydroxyl groups, they take precedence over aldehyde and ketone functions as well. Carboxylic acids outrank all the common groups we have encountered to this point in respect to defining the main chain.

Double bonds in the main chain are signaled by the ending *-enoic acid,* and their position is designated by a numerical prefix. Entries 6 and 7 are representative carboxylic acids that contain double bonds. Double-bond stereochemistry is specified by using either the cis–trans or the *E–Z* notation.

When a carboxyl group is attached to a ring, the parent ring is named (retaining the final *-e*) and the suffix *-carboxylic acid* is added, as shown in entries 8 and 9.

Compounds with two carboxyl groups, as illustrated by entries 10 through 12, are distinguished by the suffix *-dioic acid* or *-dicarboxylic acid* as appropriate. The final *-e* in the base name of the alkane is retained.

PROBLEM 19.1 The list of carboxylic acids in Table 19.1 is by no means exhaustive insofar as common names are concerned. Many others are known by their common names, a few of which follow. Give a systematic IUPAC name for each.

(a) $H_2C=CCO_2H$
          |
          CH_3

(Methacrylic acid)

(b) H_3C H
          $C=C$
      H CO_2H

(Crotonic acid)

(c) HO_2CCO_2H

(Oxalic acid)

(d) H_3C-⟨⟩$-CO_2H$

(*p*-Toluic acid)

SAMPLE SOLUTION (a) Methacrylic acid is an industrial chemical used in the preparation of transparent plastics such as *Lucite* and *Plexiglas.* The carbon chain that includes both the carboxylic acid and the double bond is three carbon atoms in length. The compound is named as a derivative of *propenoic acid.* It is not necessary to locate the position of the double bond by number, as in "2-propenoic acid," because no other positions are structurally possible for it. The methyl group is at C-2, and so the correct systematic name for methacrylic acid is *2-methylpropenoic acid.*

19.2   STRUCTURE AND BONDING

The structural features of the carboxyl group are most apparent in formic acid. Formic acid is planar, with one of its carbon–oxygen bonds shorter than the other, and with bond angles at carbon close to 120°.

Bond Distances			**Bond Angles**	
C=O	120 pm		H—C=O	124°
C—O	134 pm		H—C—O	111°
			O—C=O	125°

FIGURE 19.1 Carbon and both oxygens are sp^2-hybridized in formic acid. The π component of the C=O group and the p orbital of the OH oxygen overlap to form an extended π system that includes carbon and both oxygens.

Examine the electrostatic potential map of butanoic acid on *Learning By Modeling* and notice how much more intense the blue color (positive charge) is on the OH hydrogen than on the hydrogens bonded to carbon.

This suggests sp^2 hybridization at carbon, and a $\sigma + \pi$ carbon–oxygen double bond analogous to that of aldehydes and ketones.

Additionally, sp^2 hybridization of the hydroxyl oxygen allows one of its unshared electron pairs to be delocalized by orbital overlap with the π system of the carbonyl group (Figure 19.1). In resonance terms, this electron delocalization is represented as:

$$
\text{H}-\text{C}\!\!\overset{\displaystyle \text{O:}}{\underset{\displaystyle \text{OH}}{\diagdown}} \longleftrightarrow \text{H}-\overset{+}{\text{C}}\!\!\overset{\displaystyle \text{O:}^{-}}{\underset{\displaystyle \text{OH}}{\diagdown}} \longleftrightarrow \text{H}-\text{C}\!\!\overset{\displaystyle \text{O:}^{-}}{\underset{\displaystyle \overset{+}{\text{OH}}}{\diagdown}}
$$

Lone-pair donation from the hydroxyl oxygen makes the carbonyl group less electrophilic than that of an aldehyde or ketone. The graphic that opened this chapter is an electrostatic potential map of formic acid that shows the most electron-rich site to be the oxygen of the carbonyl group and the most electron-poor one to be, as expected, the OH hydrogen.

Carboxylic acids are fairly polar, and simple ones such as acetic acid, propanoic acid, and benzoic acid have dipole moments in the range 1.7–1.9 D.

19.3 PHYSICAL PROPERTIES

The melting points and boiling points of carboxylic acids are higher than those of hydrocarbons and oxygen-containing organic compounds of comparable size and shape and indicate strong intermolecular attractive forces.

	2-Methyl-1-butene	2-Butanone	2-Butanol	Propanoic acid
bp (1 atm):	31°C	80°C	99°C	141°C

A summary of physical properties of some representative carboxylic acids is presented in Appendix 1.

A unique hydrogen-bonding arrangement, shown in Figure 19.2, contributes to these attractive forces. The hydroxyl group of one carboxylic acid molecule acts as a proton donor toward the carbonyl oxygen of a second. In a reciprocal fashion, the hydroxyl proton of the second carboxyl function interacts with the carbonyl oxygen of the first. The result is that the two carboxylic acid molecules are held together by *two* hydrogen bonds. So efficient is this hydrogen bonding that some carboxylic acids exist as hydrogen-bonded dimers even in the gas phase. In the pure liquid a mixture of hydrogen-bonded dimers and higher aggregates is present.

FIGURE 19.2 Attractions between regions of positive (blue) and negative (red) electrostatic potential are responsible for intermolecular hydrogen bonding between two molecules of acetic acid.

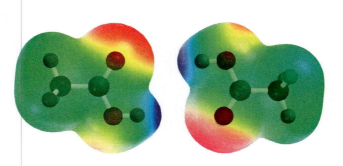

In aqueous solution intermolecular association between carboxylic acid molecules is replaced by hydrogen bonding to water. The solubility properties of carboxylic acids are similar to those of alcohols. Carboxylic acids of four carbon atoms or fewer are miscible with water in all proportions.

19.4   ACIDITY OF CARBOXYLIC ACIDS

Carboxylic acids are the most acidic class of compounds that contain only carbon, hydrogen, and oxygen. With pK_a's of about 5, they are much stronger acids than water and alcohols. The case should not be overstated, however. Carboxylic acids are weak acids; a 0.1 M solution of acetic acid in water, for example, is only 1.3% ionized.

To understand the greater acidity of carboxylic acids compared with water and alcohols, compare the structural changes that accompany the ionization of a representative alcohol (ethanol) and a representative carboxylic acid (acetic acid).

Ionization of ethanol

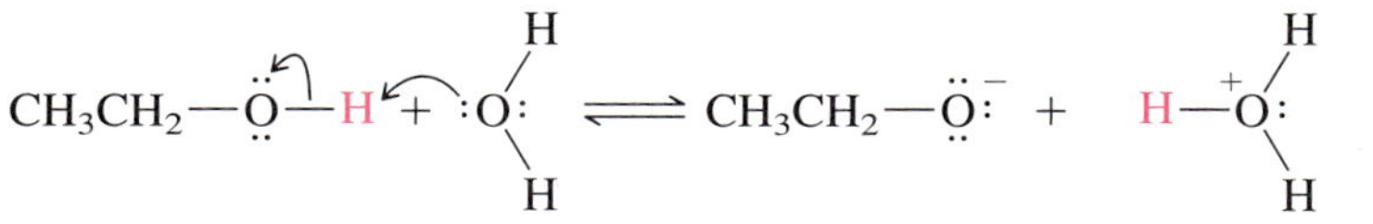

$$CH_3CH_2{-}\ddot{O}{-}H + :\ddot{O}\!\diagdown^{H}_{H} \rightleftharpoons CH_3CH_2{-}\ddot{O}:^- + H{-}\overset{+}{O}\!\diagdown^{H}_{H} \qquad pK_a = 16$$

Ethanol   Water   Ethoxide ion   Hydronium ion

Ionization of acetic acid

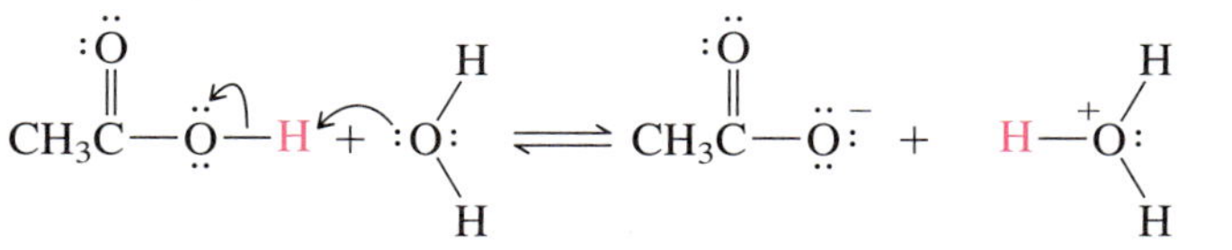

$$CH_3\overset{:\ddot{O}:}{\overset{\|}{C}}{-}\ddot{O}{-}H + :\ddot{O}\!\diagdown^{H}_{H} \rightleftharpoons CH_3\overset{:\ddot{O}:}{\overset{\|}{C}}{-}\ddot{O}:^- + H{-}\overset{+}{O}\!\diagdown^{H}_{H} \qquad pK_a = 4.7$$

Acetic acid   Water   Acetate ion   Hydronium ion

From these pK_a values, the calculated free energies of ionization ($\Delta G°$) are 91 kJ/mol (21.7 kcal/mol) for ethanol versus 27 kJ/mol (6.5 kcal/mol) for acetic acid. An energy diagram portraying these relationships is presented in Figure 19.3. Because the *equilibria,* and not the *rates* of ionization are being compared, the diagram shows only the initial and final states. It is not necessary to be concerned about the energy of activation because that affects only the rate of ionization, not the extent of ionization.

The large difference in the free energies of ionization of ethanol and acetic acid reflects a greater stabilization of acetate ion relative to ethoxide ion. Ionization of ethanol yields an alkoxide ion in which the negative charge is localized on oxygen. Solvation forces are the chief means by which ethoxide ion is stabilized. Acetate ion is also stabilized by solvation, but has two additional mechanisms for dispersing its negative charge that are not available to ethoxide ion:

1. *The inductive effect of the carbonyl group.* The carbonyl group of acetate ion is electron-withdrawing, and by attracting electrons away from the negatively charged oxygen, acetate anion is stabilized. This is an inductive effect, arising in the polarization of the electron distribution in the σ bond between the carbonyl carbon and the negatively charged oxygen.

Free energies of ionization are calculated from equilibrium constants according to the relationship

$$\Delta G° = -RT \ln K_a$$

Positively polarized carbon attracts electrons from negatively charged oxygen.

CH$_2$ group has negligible effect on electron density at negatively charged oxygen.

2. *The resonance effect of the carbonyl group.* Electron delocalization, expressed by resonance between the following Lewis structures, causes the negative charge in acetate to be shared equally by both oxygens. Electron delocalization of this type is not available to ethoxide ion.

PROBLEM 19.2 Peroxyacetic acid (CH$_3$COOH) is a weaker acid than acetic acid; its pK_a is 8.2 versus 4.7 for acetic acid. Why are peroxy acids weaker than carboxylic acids?

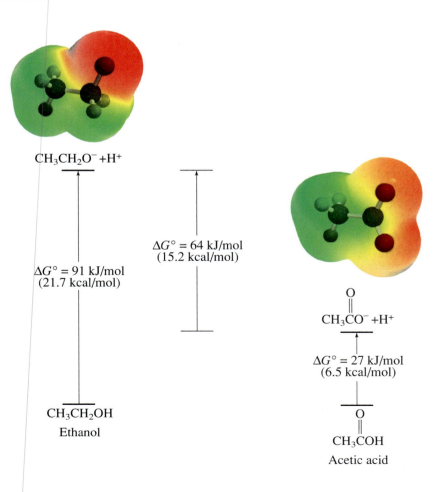

$$CH_3CH_2O^- + H^+$$

$$\Delta G^\circ = 64 \text{ kJ/mol}$$
$$(15.2 \text{ kcal/mol})$$

$$\Delta G^\circ = 91 \text{ kJ/mol}$$
$$(21.7 \text{ kcal/mol})$$

$$CH_3CO^- + H^+$$

$$\Delta G^\circ = 27 \text{ kJ/mol}$$
$$(6.5 \text{ kcal/mol})$$

CH$_3$CH$_2$OH
Ethanol

CH$_3$COH
Acetic acid

FIGURE 19.3 The free energies of ionization of ethanol and acetic acid in water. The electrostatic potential maps of ethoxide and acetate ion show the concentration of negative charge in ethoxide versus dispersal of charge in acetate. The color ranges are equal in both models to allow direct comparison.

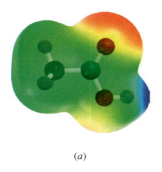

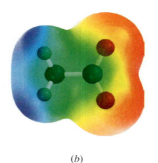

(*a*)   (*b*)

Electron delocalization in carboxylate ions is nicely illustrated with the aid of electrostatic potential maps. As Figure 19.4 shows, the electrostatic potential is different for the two different oxygens of acetic acid, but is the same for the two equivalent oxygens of acetate ion.

Likewise, the experimentally measured pattern of carbon–oxygen bond lengths in acetic acid is different from that of acetate ion. Acetic acid has a short C=O and a long C—O distance. In ammonium acetate, though, both carbon–oxygen distances are equal.

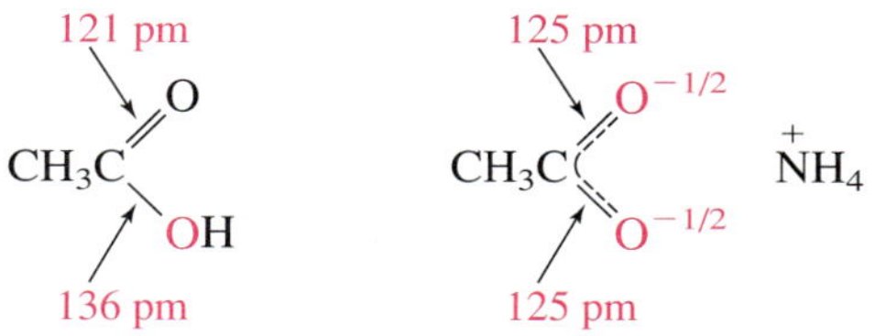

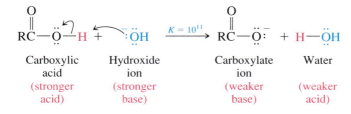

For many years, resonance in carboxylate ions was emphasized when explaining the acidity of carboxylic acids. Recently, however, it has been suggested that the inductive effect of the carbonyl group may be more important. It seems clear that, even though their relative contributions may be a matter of debate, both play major roles.

19.5   SALTS OF CARBOXYLIC ACIDS

In the presence of strong bases such as sodium hydroxide, carboxylic acids are neutralized rapidly and quantitatively:

$$RC\overset{O}{\underset{\|}{}}-\ddot{O}-H \; + \; :\ddot{O}H \; \xrightarrow{K\,=\,10^{11}} \; RC\overset{O}{\underset{\|}{}}-\ddot{O}:^- \; + \; H-\ddot{O}H$$

Carboxylic acid	Hydroxide ion	Carboxylate ion	Water
(stronger acid)	(stronger base)	(weaker base)	(weaker acid)

PROBLEM 19.3   Write an ionic equation for the reaction of acetic acid with each of the following, and specify whether the equilibrium favors starting materials or products:

(a) Sodium ethoxide   (d) Sodium acetylide

(b) Potassium *tert*-butoxide   (e) Potassium nitrate

(c) Sodium bromide   (f) Lithium amide

QUANTITATIVE RELATIONSHIPS INVOLVING CARBOXYLIC ACIDS

Suppose you take two flasks, one containing pure water and the other a buffer solution maintained at a pH of 7.0. If you add 0.1 mole of acetic acid to each one and the final volume in each flask is 1 L, how much acetic acid is present at equilibrium? How much acetate ion? In other words, what is the extent of ionization of acetic acid in an unbuffered medium and in a buffered one?

The first case simply involves the ionization of a weak acid and is governed by the expression that defines K_a for acetic acid:

$$K_a = \frac{[H^+][CH_3CO_2^-]}{[CH_3CO_2H]} = 1.8 \times 10^{-5}$$

Because ionization of acetic acid gives one H^+ for each $CH_3CO_2^-$, the concentrations of the two ions are equal, and setting each one equal to x gives:

$$K_a = \frac{x^2}{0.1 - x} = 1.8 \times 10^{-5}$$

Solving for x gives the acetate ion concentration as:

$$x = 1.3 \times 10^{-3}$$

Thus when acetic acid is added to pure water, the ratio of acetate ion to acetic acid is

$$\frac{[CH_3CO_2^-]}{[CH_3CO_2H]} = \frac{1.3 \times 10^{-3}}{0.1} = 0.013$$

Only 1.3% of the acetic acid has ionized. Most of it (98.7%) remains unchanged.

Now think about what happens when the same amount of acetic acid is added to water that is buffered at pH = 7.0. Before doing the calculation, let us recognize that it is the $[CH_3CO_2^-]/[CH_3CO_2H]$ ratio in which we are interested and do a little algebraic manipulation. Because

$$K_a = \frac{[H^+][CH_3CO_2^-]}{[CH_3CO_2H]}$$

then

$$\frac{[CH_3CO_2^-]}{[CH_3CO_2H]} = \frac{K_a}{[H^+]}$$

This relationship is one form of the **Henderson–Hasselbalch equation.** It is a useful relationship in chemistry and biochemistry. One rarely needs to calculate the pH of a solution—pH is more often measured than calculated. It is much more common that one needs to know the degree of ionization of an acid at a particular pH, and the Henderson–Hasselbalch equation gives that ratio.

For the case at hand, the solution is buffered at pH = 7.0. Therefore,

$$\frac{[CH_3CO_2^-]}{[CH_3CO_2H]} = \frac{1.8 \times 10^{-5}}{10^{-7}} = 180$$

A very different situation exists in an aqueous solution maintained at pH = 7.0 from the situation in pure water. We saw earlier that almost all the acetic acid in a 0.1 M solution in pure water was nonionized. At pH 7.0, however, hardly any nonionized acetic acid remains; it is almost completely converted to its carboxylate ion.

This difference in behavior for acetic acid in pure water versus water buffered at pH = 7.0 has some important practical consequences. Biochemists usually do not talk about acetic acid (or lactic acid, or salicylic acid, etc.). They talk about acetate (and lactate, and salicylate). Why? It's because biochemists are concerned with carboxylic acids as they exist in dilute aqueous solution at what is called *biological pH*. Biological fluids are naturally buffered. The pH of blood, for example, is maintained at 7.2, and at this pH carboxylic acids are almost entirely converted to their carboxylate anions.

An alternative form of the Henderson–Hasselbalch equation for acetic acid is

$$pH = pK_a + \log \frac{[CH_3CO_2^-]}{[CH_3CO_2H]}$$

From this equation it can be seen that when $[CH_3CO_2^-] = [CH_3CO_2H]$, then the second term is log 1 = 0, and pH = pK_a. This means that when the pH of a solution is equal to the pK_a of a weak acid, the concentration of the acid and its conjugate base are equal. This is a relationship worth remembering.

SAMPLE SOLUTION   (a) This is an acid–base reaction; ethoxide ion is the base.

$$CH_3CO_2H \ + \ CH_3CH_2O^- \longrightarrow CH_3CO_2^- \ + \ CH_3CH_2OH$$

Acetic acid	Ethoxide ion	Acetate ion	Ethanol
(stronger acid)	(stronger base)	(weaker base)	(weaker acid)

The position of equilibrium lies well to the right. Ethanol, with $pK_a = 16$, is a much weaker acid than acetic acid ($pK_a = 4.7$).

The metal carboxylate salts formed on neutralization of carboxylic acids are named by first specifying the metal ion and then adding the name of the acid modified by replacing -ic acid by -ate. Monocarboxylate salts of diacids are designated by naming both the cation and the hydrogen of the CO_2H group.

$$\underset{\text{Lithium}\atop\text{acetate}}{CH_3COLi} \qquad \underset{\text{Sodium }p\text{-chlorobenzoate}}{Cl-\!\!\langle \ \rangle\!\!-CONa} \qquad \underset{\text{Potassium hydrogen}\atop\text{hexanedioate}}{HOC(CH_2)_4COK}$$

Metal carboxylates are ionic, and when the molecular weight isn't too high, the sodium and potassium salts of carboxylic acids are soluble in water. Carboxylic acids therefore may be extracted from ether solutions into aqueous sodium or potassium hydroxide.

The solubility behavior of salts of carboxylic acids having 12–18 carbons is unusual and can be illustrated by considering sodium stearate (sodium octadecanoate). As seen by the structural formula of its sodium salt

Sodium stearate
(sodium octadecanoate)

stearate ion contains two very different structural units—a long nonpolar hydrocarbon chain and a polar carboxylate group. The electrostatic potential map of sodium stearate in Figure 19.5 illustrates how different most of the molecule is from its polar carboxylate end.

FIGURE 19.5 Electrostatic potential map of sodium stearate. Most of the molecule is comprised of a nonpolar hydrocarbon chain (green). One end is very polar as indicated by the red and blue associated with the carboxylate and sodium ions, respectively.

Carboxylate groups are **hydrophilic** ("water-loving") and tend to confer water solubility on species that contain them. Long hydrocarbon chains are **lipophilic** ("fatloving") and tend to associate with other hydrocarbon chains. Sodium stearate is an example of an **amphiphilic** substance; both hydrophilic and lipophilic groups occur within the same molecule.

"Hydrophobic" is often used instead of "lipophilic."

When sodium stearate is placed in water, the hydrophilic carboxylate group encourages the formation of a solution; the lipophilic alkyl chain discourages it. The compromise achieved is to form a colloidal dispersion of aggregates called **micelles** (Figure 19.6). Micelles form spontaneously when the carboxylate concentration exceeds a certain minimum value called the **critical micelle concentration.** Each micelle is composed of 50–100 individual molecules, with the polar carboxylate groups directed toward its outside where they experience attractive forces with water and sodium ions. The nonpolar hydrocarbon chains are directed toward the interior of the micelle, where individually weak but cumulatively significant induced-dipole/induced-dipole forces bind them together. Micelles are approximately spherical because a sphere exposes the minimum surface for a given volume of material and disrupts the water structure least. Because their surfaces are negatively charged, two micelles repel each other rather than clustering to form higher aggregates.

The formation of micelles and their properties are responsible for the cleansing action of soaps. Water that contains sodium stearate removes grease by enclosing it in the hydrocarbon-like interior of the micelles. The grease is washed away with the water, not because it dissolves in the water but because it dissolves in the micelles that are dispersed in the water. Sodium stearate is an example of a soap; sodium and potassium salts of other C_{12}–C_{18} unbranched carboxylic acids possess similar properties.

Detergents are substances, including soaps, that cleanse by micellar action. A large number of synthetic detergents are known. One example is sodium lauryl sulfate. Sodium lauryl sulfate has a long hydrocarbon chain terminating in a polar sulfate ion and forms soap-like micelles in water.

Compare the electrostatic potential maps of sodium lauryl sulfate and sodium stearate on *Learning By Modeling.*

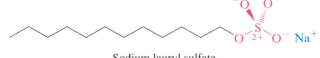

Sodium lauryl sulfate
(sodium dodecyl sulfate)

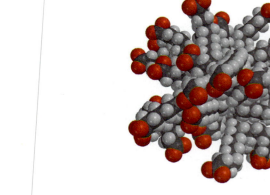

FIGURE 19.6 Spacefilling model of a micelle formed by association of carboxylate ions derived from a long-chain carboxylic acid. The hydrocarbon chains tend to be on the inside and the carboxylate ions on the surface where they are in contact with water molecules and metal cations.

Detergents are designed to be effective in hard water, meaning water containing calcium salts that form insoluble calcium carboxylates with soaps. These precipitates rob the soap of its cleansing power and form an unpleasant scum. The calcium salts of synthetic detergents such as sodium lauryl sulfate, however, are soluble and retain their micelle-forming ability even in hard water.

19.6   SUBSTITUENTS AND ACID STRENGTH

The effect of structure on acidity was introduced in Section 1.15 where we developed the generalization that electronegative substituents near an ionizable hydrogen increase its acidity. Substituent effects on the acidity of carboxylic acids have been extensively studied.

Alkyl groups have little effect. The ionization constants of all acids that have the general formula $C_nH_{2n+1}CO_2H$ are very similar to one another and equal approximately 10^{-5} ($pK_a = 5$). Table 19.2 gives a few examples.

An electronegative substituent, particularly if it is attached to the α carbon, increases the acidity of a carboxylic acid. As the data in Table 19.2 show, all the mono-haloacetic acids are about 100 times more acidic than acetic acid. Multiple halogen substitution increases the acidity even more; trichloroacetic acid is 7000 times more acidic than acetic acid!

TABLE 19.2   Effect of Substituents on Acidity of Carboxylic Acids*

Name of acid	Structure	pK_a
Standard of comparison.		
Acetic acid	CH_3CO_2H	4.7
Alkyl substituents have a negligible effect on acidity.		
Propanoic acid	$CH_3CH_2CO_2H$	4.9
2-Methylpropanoic acid	$(CH_3)_2CHCO_2H$	4.8
2,2-Dimethylpropanoic acid	$(CH_3)_3CCO_2H$	5.1
Heptanoic acid	$CH_3(CH_2)_5CO_2H$	4.9
α-Halogen substituents increase acidity.		
Fluoroacetic acid	FCH_2CO_2H	2.6
Chloroacetic acid	$ClCH_2CO_2H$	2.9
Bromoacetic acid	$BrCH_2CO_2H$	2.9
Dichloroacetic acid	Cl_2CHCO_2H	1.3
Trichloroacetic acid	Cl_3CCO_2H	0.9
Electron-attracting groups increase acidity.		
Methoxyacetic acid	$CH_3OCH_2CO_2H$	3.6
Cyanoacetic acid	$N{\equiv}CCH_2CO_2H$	2.5
Nitroacetic acid	$O_2NCH_2CO_2H$	1.7

*In water at 25°C.

The acid-strengthening effect of electronegative atoms or groups was introduced in Section 1.15 and is easily seen as an inductive effect of the substituent transmitted through the σ bonds of the molecule. According to this model, the σ electrons in the carbon–chlorine bond of chloroacetate ion are drawn toward chlorine, leaving the α-carbon atom with a slight positive charge. The α carbon, because of this positive character, attracts electrons from the negatively charged carboxylate, thus dispersing the charge and stabilizing the anion. The more stable the anion, the greater the equilibrium constant for its formation.

Learning By Modeling contains molecular models of $CH_3CO_2^-$ (acetate) and $Cl_3CCO_2^-$ (trichloroacetate). Compare these two ions with respect to the amount of negative charge on their oxygens.

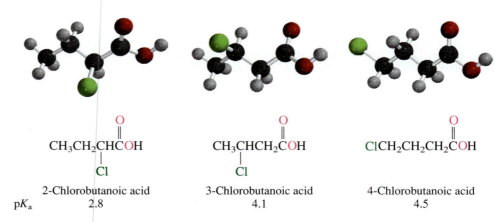

Chloroacetate anion is
stabilized by electron-withdrawing
effect of chlorine.

Inductive effects depend on the electronegativity of the substituent and the number of σ bonds between it and the affected site. As the number of bonds increases, the inductive effect decreases.

$$CH_3CH_2\overset{O}{\overset{\|}{C}}HCOH$$
                Cl

2-Chlorobutanoic acid
pK_a 2.8

$$CH_3CHCH_2\overset{O}{\overset{\|}{C}}OH$$
        Cl

3-Chlorobutanoic acid
4.1

$$ClCH_2CH_2CH_2\overset{O}{\overset{\|}{C}}OH$$

4-Chlorobutanoic acid
4.5

PROBLEM 19.4   Which is the stronger acid in each of the following pairs?

(a) $(CH_3)_3CCH_2CO_2H$ or $(CH_3)_3\overset{+}{N}CH_2CO_2H$

(b) $CH_3CH_2CO_2H$ or $CH_3\underset{OH}{CH}CO_2H$

(c) $CH_3\overset{O}{\overset{\|}{C}}CO_2H$ or $H_2C=CHCO_2H$

(d) $CH_3CH_2CH_2CO_2H$ or $CH_3\overset{O}{\underset{O}{\overset{\|}{\underset{\|}{S}}}}CH_2CO_2H$

SAMPLE SOLUTION   (a) Think of the two compounds as substituted derivatives of acetic acid. A *tert*-butyl group is slightly electron-releasing and has only a modest effect on acidity. The compound $(CH_3)_3CCH_2CO_2H$ is expected to have an acid strength similar to that of acetic acid. A trimethylammonium substituent, on the other hand, is positively charged and is a powerful electron-withdrawing substituent. The compound $(CH_3)_3\overset{+}{N}CH_2CO_2H$ is expected to be a much stronger acid than $(CH_3)_3CCH_2CO_2H$. The measured ionization constants, shown as follows, confirm this prediction.

$$(CH_3)_3CCH_2CO_2H \qquad (CH_3)_3\overset{+}{N}CH_2CO_2H$$

Weaker acid	Stronger acid
$pK_a = 5.3$	$pK_a = 1.8$

Closely related to the inductive effect, and operating in the same direction, is the **field effect.** In the field effect the electronegativity of a substituent is communicated, not by successive polarization of bonds but via the medium, usually the solvent. A substituent in a molecule polarizes surrounding solvent molecules and this polarization is transmitted through other solvent molecules to the remote site.

It is a curious fact that substituents affect the entropy of ionization more than they do the enthalpy term in the expression

$$\Delta G^\circ = \Delta H^\circ - T\Delta S^\circ$$

The enthalpy term ΔH° is close to zero for the ionization of most carboxylic acids, regardless of their strength. The free energy of ionization ΔG° is dominated by the $-T\Delta S^\circ$ term. Ionization is accompanied by an increase in solvation forces, leading to a decrease in the entropy of the system; ΔS° is negative, and $-T\Delta S^\circ$ is positive. Anions that incorporate substituents capable of dispersing negative charge impose less order on the solvent (water), and less entropy is lost in their production.

19.7 IONIZATION OF SUBSTITUTED BENZOIC ACIDS

A considerable body of data is available on the acidity of substituted benzoic acids. Benzoic acid itself is a somewhat stronger acid than acetic acid. Its carboxyl group is attached to an sp^2-hybridized carbon and ionizes to a greater extent than one that is attached to an sp^3-hybridized carbon. Remember, carbon becomes more electron-withdrawing as its s character increases.

$$CH_3CO_2H \qquad H_2C{=}CHCO_2H \qquad \text{(benzene ring)}{-}CO_2H$$

Acetic acid	Acrylic acid	Benzoic acid
$pK_a = 4.8$	$pK_a = 4.3$	$pK_a = 4.2$

PROBLEM 19.5 What is the most acidic neutral molecule characterized by the formula $C_3H_xO_2$?

Table 19.3 lists the ionization constants of some substituted benzoic acids. The largest effects are observed when strongly electron-withdrawing substituents are ortho to the carboxyl group. An *o*-nitro substituent, for example, increases the acidity of benzoic acid 100-fold. Substituent effects are small at positions meta and para to the carboxyl group. In those cases the pK_a values are clustered in the range 3.5–4.5.

TABLE 19.3	Acidity of Some Substituted Benzoic Acids*		
Substituent in	**pK$_a$ for different positions of substituent X**		
XC$_6$H$_4$CO$_2$H	**Ortho**	**Meta**	**Para**
H	4.2	4.2	4.2
CH$_3$	3.9	4.3	4.4
F	3.3	3.9	4.1
Cl	2.9	3.8	4.0
Br	2.8	3.8	4.0
I	2.9	3.9	4.0
CH$_3$O	4.1	4.1	4.5
O$_2$N	2.2	3.5	3.4

*In water at 25°C.

19.8 DICARBOXYLIC ACIDS

Separate ionization constants, designated K_1 and K_2, respectively, characterize the two successive ionization steps of a dicarboxylic acid.

$$\underset{\text{Oxalic acid}}{\text{HOC}-\text{COH}} \overset{K_1}{\rightleftharpoons} \text{H}^+ + \underset{\substack{\text{Hydrogen oxalate} \\ \text{(monoanion)}}}{\text{HOC}-\text{CO}^-} \qquad pK_1 = 1.2$$

> Oxalic acid is poisonous and occurs naturally in a number of plants including sorrel and begonia. It is a good idea to keep houseplants out of the reach of small children, who might be tempted to eat the leaves or berries.

$$\underset{\substack{\text{Hydrogen oxalate} \\ \text{(monoanion)}}}{\text{HOC}-\text{CO}^-} \overset{K_2}{\rightleftharpoons} \text{H}^+ + \underset{\substack{\text{Oxalate} \\ \text{(Dianion)}}}{{}^-\text{OC}-\text{CO}^-} \qquad pK_2 = 4.3$$

The first ionization constant of dicarboxylic acids is larger than K_a for monocarboxylic analogs. One reason is statistical. There are two potential sites for ionization rather than one, making the effective concentration of carboxyl groups twice as large. Furthermore, one carboxyl group acts as an electron-withdrawing group to facilitate dissociation of the other. This is particularly noticeable when the two carboxyl groups are separated by only a few bonds. Oxalic and malonic acid, for example, are several orders of magnitude stronger than simple alkyl derivatives of acetic acid. Heptanedioic acid, in which the carboxyl groups are well separated from each other, is only slightly stronger than acetic acid.

HO$_2$CCO$_2$H	HO$_2$CCH$_2$CO$_2$H	HO$_2$C(CH$_2$)$_5$CO$_2$H
Oxalic acid	Malonic acid	Heptanedioic acid
pK$_1$ = 1.2	pK$_1$ = 2.8	pK$_1$ = 4.3

19.9 CARBONIC ACID

Through an accident of history, the simplest dicarboxylic acid, carbonic acid, $\overset{\text{O}}{\overset{\|}{\text{HOCOH}}}$, is not even classified as an organic compound. Because many minerals are carbonate salts, nineteenth-century chemists placed carbonates, bicarbonates, and carbon dioxide in the inorganic realm. Nevertheless, the essential features of carbonic acid and its salts are easily understood on the basis of our knowledge of carboxylic acids.

Carbonic acid is formed when carbon dioxide reacts with water. Hydration of carbon dioxide is far from complete, however. Almost all the carbon dioxide that is dissolved in water exists as carbon dioxide; only 0.3% of it is converted to carbonic acid. Carbonic acid is a weak acid and ionizes to a small extent to bicarbonate ion.

> The systematic name for bicarbonate ion is *hydrogen carbonate*. Thus, the systematic name for sodium bicarbonate (NaHCO$_3$) is *sodium hydrogen carbonate*.

$$CO_2 \ + \ H_2O \ \rightleftharpoons \ \underset{\substack{\text{Carbonic} \\ \text{acid}}}{HO\overset{\overset{\displaystyle O}{\|}}{C}OH} \ \rightleftharpoons \ H^+ \ + \ \underset{\substack{\text{Bicarbonate} \\ \text{ion}}}{HO\overset{\overset{\displaystyle O}{\|}}{C}O^-}$$

$$\underset{\substack{\text{Carbon} \\ \text{dioxide}}}{} \quad \underset{\text{Water}}{}$$

The equilibrium constant for the overall reaction is related to an apparent equilibrium constant K_1 for carbonic acid ionization by the expression

$$K_1 = \frac{[H^+][HCO_3^-]}{[CO_2]} = 4.3 \times 10^{-7} \qquad pK_1 = 6.4$$

These equations tell us that the reverse process, proton transfer from acids to bicarbonate to form carbon dioxide, will be favorable when K_a of the acid exceeds 4.3×10^{-7} ($pK_a < 6.4$). Among compounds containing carbon, hydrogen, and oxygen, only carboxylic acids are acidic enough to meet this requirement. They dissolve in aqueous sodium bicarbonate with the evolution of carbon dioxide. This behavior is the basis of a qualitative test for carboxylic acids.

PROBLEM 19.6 The value cited for the "apparent K_1" of carbonic acid, 4.3×10^{-7}, is the one normally given in reference books. It is determined by measuring the pH of water to which a known amount of carbon dioxide has been added. When we recall that only 0.3% of carbon dioxide is converted to carbonic acid in water, what is the "true K_1" of carbonic acid?

Carbonic anhydrase is an enzyme that catalyzes the hydration of carbon dioxide to bicarbonate. The uncatalyzed hydration of carbon dioxide is too slow to be effective in transporting carbon dioxide from the tissues to the lungs, and so animals have developed catalysts to speed this process. The activity of carbonic anhydrase is remarkable; it has been estimated that one molecule of this enzyme can catalyze the hydration of 3.6×10^7 molecules of carbon dioxide per minute.

As with other dicarboxylic acids, the second ionization constant of carbonic acid is far smaller than the first.

$$\underset{\substack{\text{Bicarbonate ion}}}{HO\overset{\overset{\displaystyle O}{\|}}{C}O^-} \ \underset{K_2}{\rightleftharpoons} \ H^+ \ + \ \underset{\substack{\text{Carbonate ion}}}{^-O\overset{\overset{\displaystyle O}{\|}}{C}O^-}$$

The value of pK_2 is 10.2. Bicarbonate is a weaker acid than carboxylic acids but a stronger acid than water and alcohols.

19.10 SOURCES OF CARBOXYLIC ACIDS

Many carboxylic acids were first isolated from natural sources and were given names based on their origin. Formic acid (Latin *formica,* meaning "ant") was obtained by distilling ants. Since ancient times acetic acid (Latin *acetum,* for "vinegar") has been known to be present in wine that has turned sour. Butyric acid (Latin *butyrum,* meaning "butter") contributes to the odor of both rancid butter and ginkgo berries, and lactic acid (Latin *lac,* for "milk") has been isolated from sour milk.

Although these humble origins make interesting historical notes, in most cases the large-scale preparation of carboxylic acids relies on chemical synthesis. Virtually none of the 3×10^9 lb of acetic acid produced in the United States each year is obtained from vinegar. Instead, most industrial acetic acid comes from the reaction of methanol with carbon monoxide.

$$\underset{\text{Methanol}}{CH_3OH} \;+\; \underset{\substack{\text{Carbon}\\\text{monoxide}}}{CO} \;\xrightarrow[\text{heat, pressure}]{\substack{\text{cobalt or}\\\text{rhodium catalyst}}}\; \underset{\text{Acetic acid}}{CH_3CO_2H}$$

The principal end use of acetic acid is in the production of vinyl acetate for paints and adhesives.

The carboxylic acid produced in the greatest amounts is 1,4-benzenedicarboxylic acid (terephthalic acid). About 5×10^9 lb/year is produced in the United States as a starting material for the preparation of polyester fibers. One important process converts *p*-xylene to terephthalic acid by oxidation with nitric acid:

$$\underset{\textit{p}\text{-Xylene}}{H_3C-\!\!\!\bigcirc\!\!\!-CH_3} \;\xrightarrow{HNO_3}\; \underset{\substack{\text{1,4-Benzenedicarboxylic acid}\\\text{(terephthalic acid)}}}{HO_2C-\!\!\!\bigcirc\!\!\!-CO_2H}$$

You will recognize the side-chain oxidation of *p*-xylene to terephthalic acid as a reaction type discussed previously (Section 11.13). Examples of other reactions encountered earlier that can be applied to the synthesis of carboxylic acids are collected in Table 19.4.

The examples in the table give carboxylic acids that have the same number of carbon atoms as the starting material. The reactions to be described in the next two sections permit carboxylic acids to be prepared by extending a chain by one carbon atom and are of great value in laboratory syntheses of carboxylic acids.

19.11 SYNTHESIS OF CARBOXYLIC ACIDS BY THE CARBOXYLATION OF GRIGNARD REAGENTS

We've seen how Grignard reagents add to the carbonyl group of aldehydes, ketones, and esters. Grignard reagents react in much the same way with *carbon dioxide* to yield magnesium salts of carboxylic acids. Acidification converts these magnesium salts to the desired carboxylic acids.

TABLE 19.4 Summary of Reactions Discussed in Earlier Chapters That Yield Carboxylic Acids

Reaction (section) and comments	General equation and specific example
Side-chain oxidation of alkylbenzenes (Section 11.13) A primary or secondary alkyl side chain on an aromatic ring is converted to a carboxyl group by reaction with a strong oxidizing agent such as potassium permanganate or chromic acid.	$ArCHR_2 \xrightarrow[K_2Cr_2O_7,\ H_2SO_4]{KMnO_4\ or} ArCO_2H$ Alkylbenzene → Arenecarboxylic acid 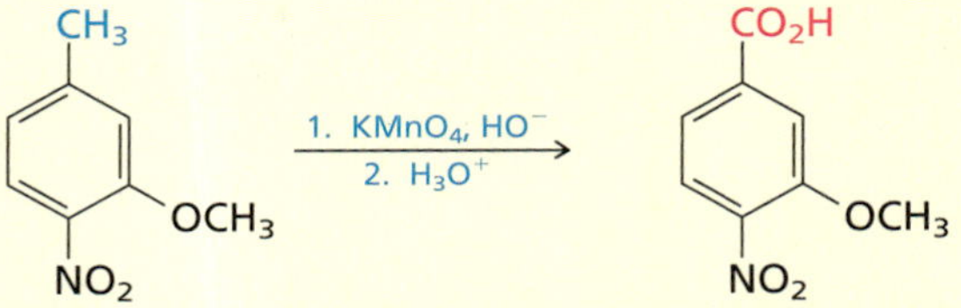 3-Methoxy-4-nitrotoluene $\xrightarrow[2.\ H_3O^+]{1.\ KMnO_4,\ HO^-}$ 3-Methoxy-4-nitrobenzoic acid (100%)
Oxidation of primary alcohols (Section 15.10) Potassium permanganate and chromic acid convert primary alcohols to carboxylic acids by way of the corresponding aldehyde.	$RCH_2OH \xrightarrow[K_2Cr_2O_7,\ H_2SO_4]{KMnO_4\ or} RCO_2H$ Primary alcohol → Carboxylic acid $(CH_3)_3CCHC(CH_3)_3 \xrightarrow[H_2O,\ H_2SO_4]{H_2CrO_4} (CH_3)_3CCHC(CH_3)_3$ 2-*tert*-Butyl-3,3-dimethyl-1-butanol → 2-*tert*-Butyl-3,3-dimethylbutanoic acid (82%)
Oxidation of aldehydes (Section 17.15) Aldehydes are particularly sensitive to oxidation and are converted to carboxylic acids by a number of oxidizing agents, including potassium permanganate and chromic acid.	$RCH=O \xrightarrow{\text{oxidizing agent}} RCO_2H$ Aldehyde → Carboxylic acid 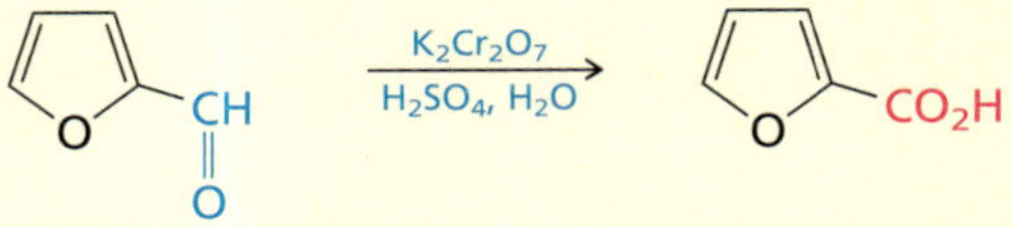 Furan-2-carbaldehyde (furfural) $\xrightarrow[H_2SO_4,\ H_2O]{K_2Cr_2O_7}$ Furan-2-carboxylic acid (furoic acid) (75%)

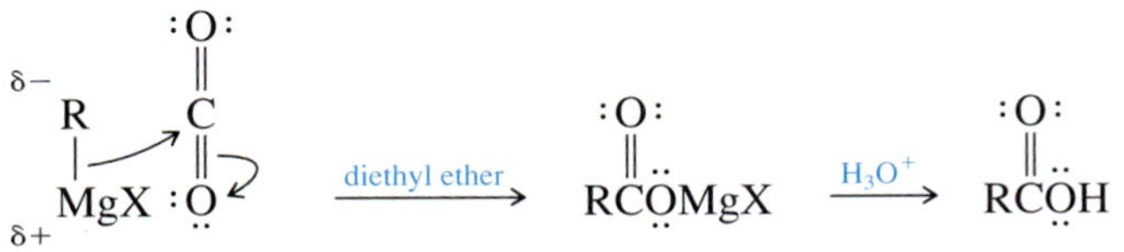

Grignard reagent acts as a nucleophile toward carbon dioxide → Halomagnesium carboxylate → Carboxylic acid

Overall, the carboxylation of Grignard reagents transforms an alkyl or aryl halide to a carboxylic acid in which the carbon skeleton has been extended by one carbon atom.

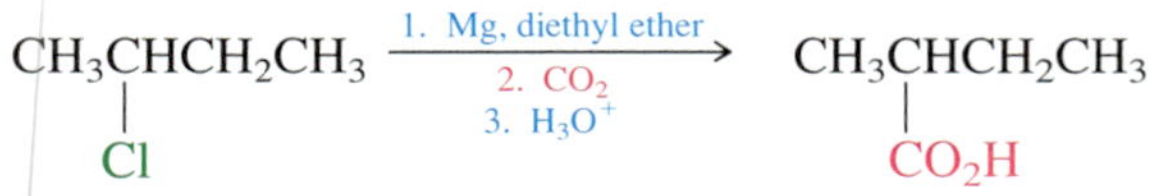

$$CH_3CHCH_2CH_3 \xrightarrow[\substack{2.\ CO_2 \\ 3.\ H_3O^+}]{1.\ Mg,\ diethyl\ ether} CH_3CHCH_2CH_3$$

Cl	CO_2H
2-Chlorobutane	2-Methylbutanoic acid (76–86%)

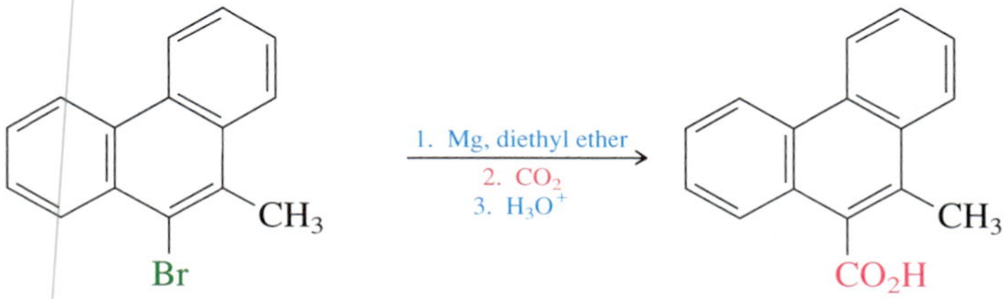

$$\xrightarrow[\substack{2.\ CO_2 \\ 3.\ H_3O^+}]{1.\ Mg,\ diethyl\ ether}$$

9-Bromo-10-methylphenanthrene

10-Methylphenanthrene-9-carboxylic acid (82%)

The major limitation to this procedure is that the alkyl or aryl halide must not bear substituents that are incompatible with Grignard reagents, such as OH, NH, SH, or C=O.

19.12 SYNTHESIS OF CARBOXYLIC ACIDS BY THE PREPARATION AND HYDROLYSIS OF NITRILES

Primary and secondary alkyl halides may be converted to the next higher carboxylic acid by a two-step synthetic sequence involving the preparation and hydrolysis of *nitriles.* Nitriles, also known as *alkyl cyanides,* are prepared by nucleophilic substitution.

$$:\ddot{X}-R \ + \ :\bar{C}\equiv N: \ \xrightarrow{S_N2} \ RC\equiv N \ + \ :\ddot{X}:^-$$

Primary or secondary alkyl halide	Cyanide ion	Nitrile (alkyl cyanide)	Halide ion

The reaction is of the S_N2 type and works best with primary and secondary alkyl halides. Elimination is the only reaction observed with tertiary alkyl halides. Aryl and vinyl halides do not react. Dimethyl sulfoxide is the preferred solvent for this reaction, but alcohols and water–alcohol mixtures have also been used.

Once the cyano group has been introduced, the nitrile is subjected to hydrolysis. Usually this is carried out in aqueous acid at reflux.

The mechanism of nitrile hydrolysis will be described in Section 20.19.

$$RC\equiv N \ + \ 2H_2O \ + \ H^+ \ \xrightarrow{heat} \ \overset{\displaystyle O}{\overset{\displaystyle \|}{RCOH}} \ + \ NH_4^+$$

Nitrile	Water		Carboxylic acid	Ammonium ion

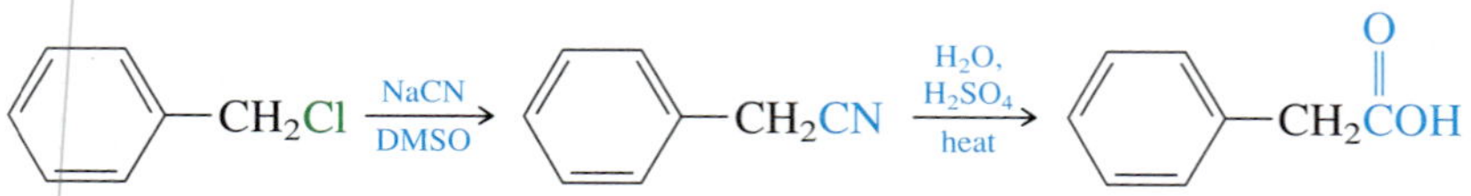

$$\text{—CH}_2\text{Cl} \xrightarrow[DMSO]{NaCN} \text{—CH}_2\text{CN} \xrightarrow[heat]{H_2O,\ H_2SO_4} \text{—CH}_2\overset{\displaystyle O}{\overset{\displaystyle \|}{\text{COH}}}$$

Benzyl chloride	Benzyl cyanide (92%)	Phenylacetic acid (77%)

Dicarboxylic acids have been prepared from dihalides by this method:

$$BrCH_2CH_2CH_2Br \xrightarrow[\text{H}_2\text{O}]{\text{NaCN}} NCCH_2CH_2CH_2CN \xrightarrow[\text{heat}]{\text{H}_2\text{O, HCl}} \overset{\displaystyle O}{\overset{\displaystyle \|}{HOCCH}}_2CH_2CH_2\overset{\displaystyle O}{\overset{\displaystyle \|}{C}}OH$$

1,3-Dibromopropane	1,5-Pentanedinitrile	1,5-Pentanedioic acid
	(77–86%)	(83–85%)

PROBLEM 19.7   Of the two procedures just described, preparation and carboxy-lation of a Grignard reagent or formation and hydrolysis of a nitrile, only one is appropriate to each of the following RX → RCO₂H conversions. Identify the correct procedure in each case, and specify why the other will fail.

(a) Bromobenzene → benzoic acid

(b) 2-Chloroethanol → 3-hydroxypropanoic acid

(c) *tert*-Butyl chloride → 2,2-dimethylpropanoic acid

SAMPLE SOLUTION   (a) Bromobenzene is an aryl halide and is unreactive toward nucleophilic substitution by cyanide ion. The route $C_6H_5Br \rightarrow C_6H_5CN \rightarrow C_6H_5CO_2H$ fails because the first step fails. The route proceeding through the Grignard reagent is perfectly satisfactory and appears as an experiment in a number of introductory organic chemistry laboratory texts.

Bromobenzene	Phenylmagnesium bromide	Benzoic acid

Nitrile groups in cyanohydrins are hydrolyzed under conditions similar to those of alkyl cyanides. Cyanohydrin formation followed by hydrolysis provides a route to the preparation of α-hydroxy carboxylic acids.

Recall the preparation of cyanohydrins in Section 17.7.

$$\overset{\displaystyle O}{\overset{\displaystyle \|}{CH_3C}}CH_2CH_2CH_3 \xrightarrow[\text{2. H}_3\text{O}^+]{\text{1. NaCN}} \underset{\displaystyle CN}{\overset{\displaystyle OH}{CH_3\overset{\displaystyle |}{\underset{\displaystyle |}{C}}CH_2CH_2CH_3}} \xrightarrow[\text{heat}]{\text{H}_2\text{O, HCl}} \underset{\displaystyle CO_2H}{\overset{\displaystyle OH}{CH_3\overset{\displaystyle |}{\underset{\displaystyle |}{C}}CH_2CH_2CH_3}}$$

2-Pentanone	2-Pentanone cyanohydrin	2-Hydroxy-2-methyl-pentanoic acid (60% from 2-pentanone)

19.13   REACTIONS OF CARBOXYLIC ACIDS: A REVIEW AND A PREVIEW

The most apparent chemical property of carboxylic acids, their acidity, has already been examined in earlier sections of this chapter. Three reactions of carboxylic acids—conversion to acyl chlorides, reduction, and esterification—have been encountered in previous chapters and are reviewed in Table 19.5. Acid-catalyzed esterification of carboxylic acids is one of the fundamental reactions of organic chemistry, and this portion of the chapter begins with an examination of the mechanism by which it occurs. Later, in Sections 19.16 and 19.17, two new reactions of carboxylic acids that are of synthetic value will be described.

TABLE 19.5	Summary of Reactions of Carboxylic Acids Discussed in Earlier Chapters

Reaction (section) and comments	**General equation and specific example**
Formation of acyl chlorides (Section 12.7) Thionyl chloride reacts with carboxylic acids to yield acyl chlorides.	RCO_2H + $SOCl_2$ ⟶ $RCCl$ + SO_2 + HCl Carboxylic acid Thionyl chloride Acyl chloride Sulfur dioxide Hydrogen chloride 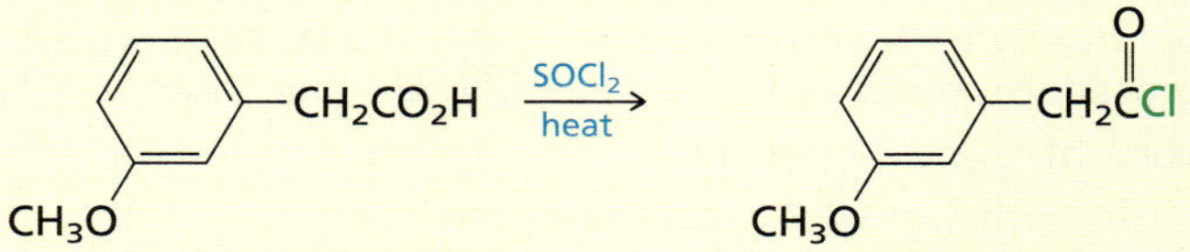 *m*-Methoxyphenylacetic acid $\xrightarrow[\text{heat}]{SOCl_2}$ *m*-Methoxyphenylacetyl chloride (85%)
Lithium aluminum hydride reduction (Section 15.3) Carboxylic acids are reduced to primary alcohols by the powerful reducing agent lithium aluminum hydride.	RCO_2H $\xrightarrow[\text{2. }H_2O]{\text{1. LiAlH}_4\text{, diethyl ether}}$ RCH_2OH Carboxylic acid Primary alcohol 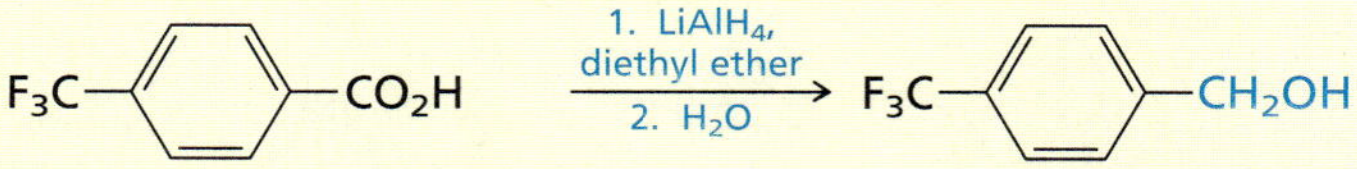 F_3C—CO₂H $\xrightarrow[\text{2. }H_2O]{\text{1. LiAlH}_4\text{, diethyl ether}}$ F_3C—CH_2OH *p*-(Trifluoromethyl)benzoic acid *p*-(Trifluoromethyl)benzyl alcohol (96%)
Esterification (Section 15.8) In the presence of an acid catalyst, carboxylic acids and alcohols react to form esters. The reaction is an equilibrium process but can be driven to favor the ester by removing the water that is formed.	RCO_2H + $R'OH$ $\underset{}{\overset{H^+}{\rightleftharpoons}}$ $RCOR'$ + H_2O Carboxylic acid Alcohol Ester Water 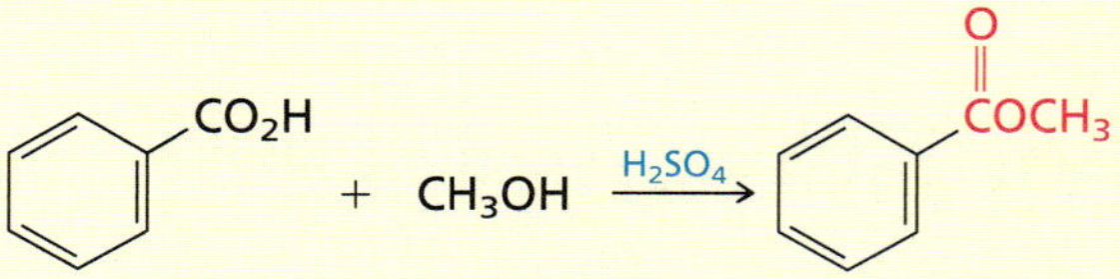 Benzoic acid Methanol $\xrightarrow{H_2SO_4}$ Methyl benzoate (70%)

19.14 MECHANISM OF ACID-CATALYZED ESTERIFICATION

An important question about the mechanism of acid-catalyzed esterification concerns the origin of the alkoxy oxygen. For example, does the methoxy oxygen in methyl benzoate come from methanol, or is it derived from benzoic acid?

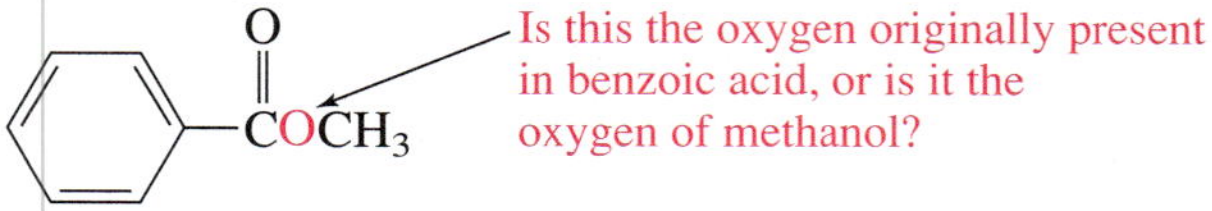

Is this the oxygen originally present in benzoic acid, or is it the oxygen of methanol?

The answer to this question is critical because it tells us whether the carbon–oxygen bond of the alcohol or a carbon–oxygen of the carboxylic acid is broken during esterification.

A clear-cut answer was provided by Irving Roberts and Harold C. Urey of Columbia University in 1938. They prepared methanol that had been enriched in the mass-18 isotope of oxygen. When this sample of methanol was esterified with benzoic acid, the methyl benzoate product contained all the ^{18}O label that was originally present in the methanol.

$$
\underset{\text{Benzoic acid}}{C_6H_5\overset{\displaystyle O}{\overset{\|}{C}}OH} + \underset{\substack{^{18}\text{O-enriched}\\\text{methanol}}}{CH_3OH} \xrightarrow{H^+} \underset{\substack{^{18}\text{O-enriched}\\\text{methyl benzoate}}}{C_6H_5\overset{\displaystyle O}{\overset{\|}{C}}OCH_3} + \underset{\text{Water}}{H_2O}
$$

In this equation, the red O signifies oxygen enriched in its mass -18 isotope; analysis of isotopic enrichment was performed by mass spectrometry.

The results of the Roberts–Urey experiment tell us that the C—O bond of the alcohol is preserved during esterification. The oxygen that is lost as a water molecule must come from the carboxylic acid.

A mechanism consistent with these facts is presented in Figure 19.7. The six steps are best viewed as a combination of two distinct stages. *Formation* of a **tetrahedral intermediate** characterizes the first stage (steps 1–3), and *dissociation* of this tetrahedral intermediate characterizes the second (steps 4–6).

$$
\underset{\substack{\\\text{Benzoic acid}}}{C_6H_5C\overset{\displaystyle O}{\underset{\displaystyle OH}{}}} + \underset{\substack{\\\text{Methanol}}}{CH_3OH} \underset{H^+}{\overset{\text{steps 1–3}}{\rightleftharpoons}} \underset{\substack{\\\text{Tetrahedral}\\\text{intermediate}}}{C_6H_5\overset{\displaystyle OH}{\underset{\displaystyle OH}{C}}-OCH_3} \underset{H^+}{\overset{\text{steps 4–6}}{\rightleftharpoons}} \underset{\substack{\\\text{Methyl}\\\text{benzoate}}}{C_6H_5C\overset{\displaystyle O}{\underset{\displaystyle OCH_3}{}}} + \underset{\substack{\\\text{Water}}}{H_2O}
$$

The species connecting the two stages is called a *tetrahedral intermediate* because the hybridization at carbon has changed from sp^2 in the carboxylic acid to sp^3 in the intermediate before returning to sp^2 in the ester product. *The tetrahedral intermediate is formed by nucleophilic addition of an alcohol to a carboxylic acid and is analogous to a hemiacetal formed by nucleophilic addition of an alcohol to an aldehyde or a ketone.* The three steps that lead to the tetrahedral intermediate in the first stage of esterification are analogous to those in the mechanism for acid-catalyzed nucleophilic addition of an alcohol to an aldehyde or a ketone (Section 17.8). The tetrahedral intermediate cannot be isolated. It is unstable under the conditions of its formation and undergoes acid-catalyzed dehydration to form the ester.

Notice that the oxygen of methanol becomes incorporated into the methyl benzoate product according to the mechanism outlined in Figure 19.7, as the observations of the Roberts–Urey experiment require it to be.

Notice, too, that the carbonyl oxygen of the carboxylic acid is protonated in the first step and not the hydroxyl oxygen. The species formed by protonation of the carbonyl oxygen is more stable because it is stabilized by electron delocalization. The positive charge is shared equally by both oxygens.

$$
C_6H_5C\overset{\displaystyle \overset{+}{\ddot{O}}H}{\underset{\displaystyle \ddot{\ddot{O}}H}{}} \longleftrightarrow C_6H_5C\overset{\displaystyle \ddot{O}H}{\underset{\displaystyle \overset{}{\underset{+}{\ddot{O}}H}}{}}
$$

Electron delocalization in carbonyl-protonated benzoic acid

FIGURE 19.7 The mechanism of acid-catalyzed esterification of benzoic acid with methanol.

Overall reaction:

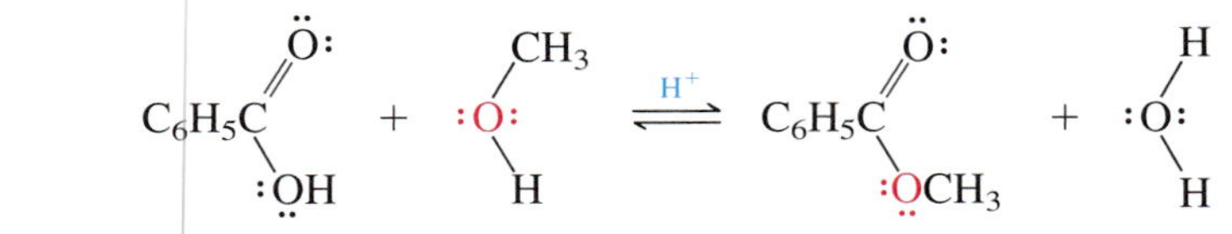

Benzoic acid   Methanol   Methyl benzoate   Water

Step 1: The carboxylic acid is protonated on its carbonyl oxygen. The proton donor shown in the equation for this step is an alkyloxonium ion formed by proton transfer from the acid catalyst to the alcohol.

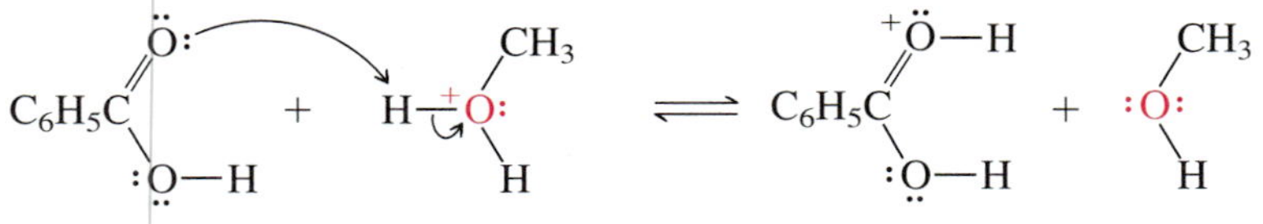

Benzoic acid   Methyloxonium ion   Conjugate acid of benzoic acid   Methanol

Step 2: Protonation of the carboxylic acid increases the positive character of its carbonyl group. A molecule of the alcohol acts as a nucleophile and attacks the carbonyl carbon.

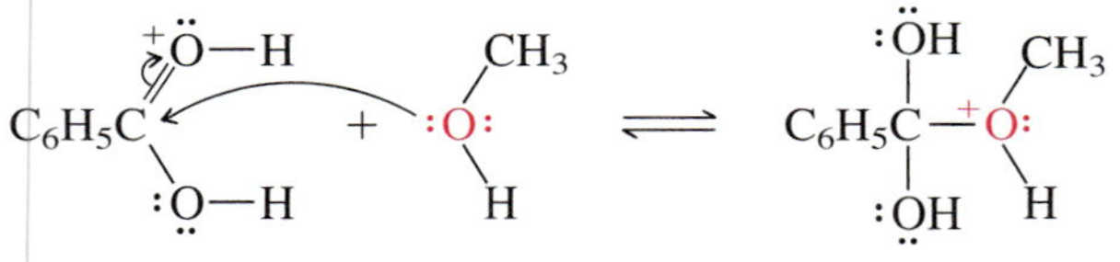

Conjugate acid of benzoic acid   Methanol   Protonated form of tetrahedral intermediate

Step 3: The oxonium ion formed in step 2 loses a proton to give the tetrahedral intermediate in its neutral form. This step concludes the first stage in the mechanism.

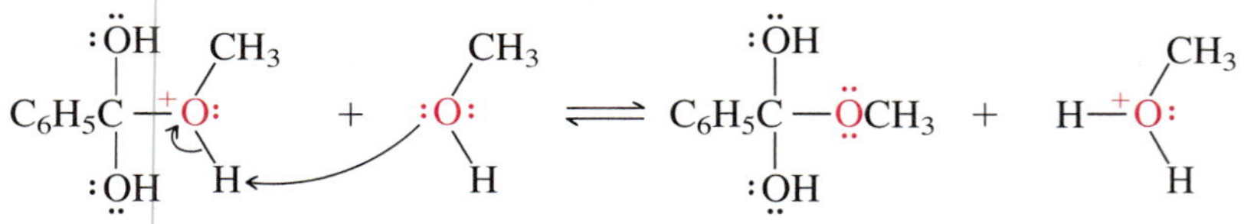

Protonated form of tetrahedral intermediate   Methanol   Tetrahedral intermediate   Methyloxonium ion

—*Cont.*

Protonation of the hydroxyl oxygen, on the other hand, yields a less stable cation:

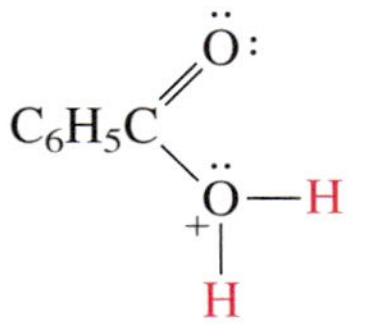

Localized positive charge in hydroxyl-protonated benzoic acid

(Continued)

Step 4: The second stage begins with protonation of the tetrahedral intermediate on one of its hydroxyl oxygens.

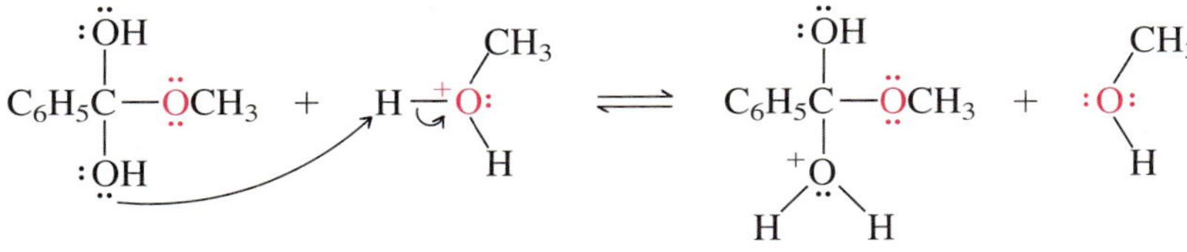

| Tetrahedral intermediate | Methyloxonium ion | Hydroxyl-protonated tetrahedral intermediate | Methanol |

Step 5: This intermediate loses a molecule of water to give the protonated form of the ester.

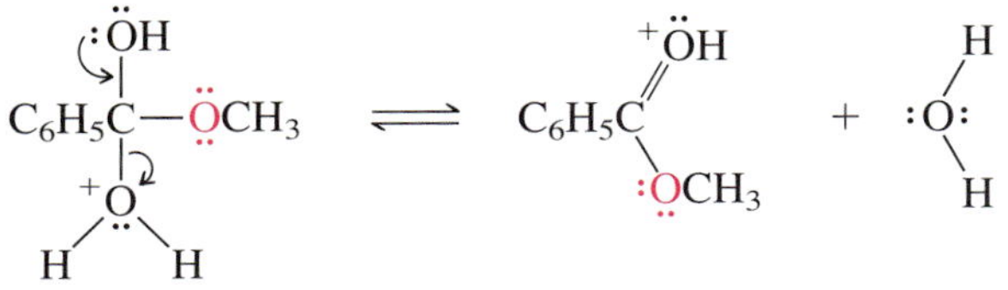

| Hydroxyl-protonated tetrahedral intermediate | Conjugate acid of methyl benzoate | Water |

Step 6: Deprotonation of the species formed in step 5 gives the neutral form of the ester product.

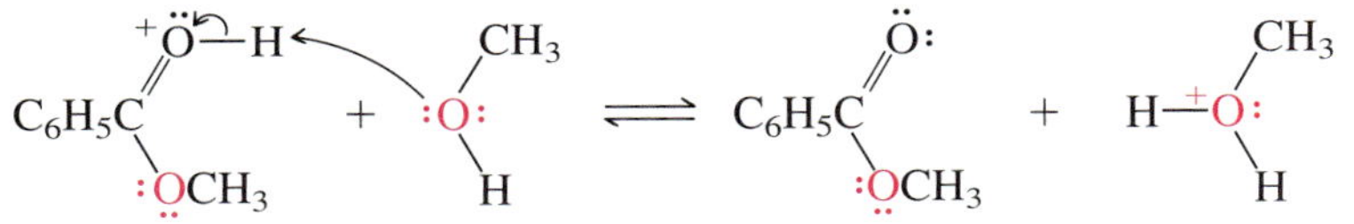

| Conjugate acid of methyl benzoate | Methanol | Methyl benzoate | Methyloxonium ion |

The positive charge in this cation cannot be shared by the two oxygens; it is localized on one of them. Because protonation of the carbonyl oxygen gives a more stable cation, that cation is formed preferentially.

> **PROBLEM 19.8** When benzoic acid is allowed to stand in water enriched in ^{18}O, the isotopic label becomes incorporated into the benzoic acid. The reaction is catalyzed by acids. Suggest an explanation for this observation.

In the next chapter the three elements of the mechanism just described will be seen again as part of the general theme that unites the chemistry of carboxylic acid derivatives. These elements are

1. Activation of the carbonyl group by protonation of the carbonyl oxygen
2. Nucleophilic addition to the protonated carbonyl to form a tetrahedral intermediate
3. Elimination from the tetrahedral intermediate to restore the carbonyl group

This sequence is one of the fundamental mechanistic patterns of organic chemistry.

19.15 INTRAMOLECULAR ESTER FORMATION: LACTONES

Hydroxy acids, compounds that contain both a hydroxyl and a carboxylic acid function, have the capacity to form cyclic esters called *lactones*. This intramolecular esterification takes place spontaneously when the ring that is formed is five- or six-membered. Lactones that contain a five-membered cyclic ester are referred to as **γ-lactones**; their six-membered analogs are known as **δ-lactones.**

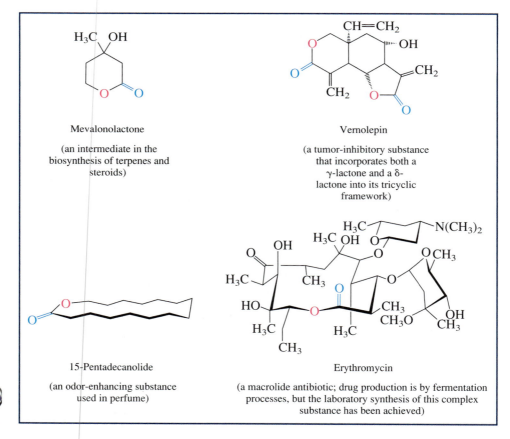

4-Hydroxybutanoic acid   4-Butanolide   Water

5-Hydroxypentanoic acid   5-Pentanolide   Water

A lactone is named by replacing the *-oic acid* ending of the parent carboxylic acid by *-olide* and identifying its oxygenated carbon by number. This system is illustrated in the lactones shown in the preceding equations. Both 4-butanolide and 5-pentanolide are

Mevalonolactone

(an intermediate in the
biosynthesis of terpenes and
steroids)

Vernolepin

(a tumor-inhibitory substance
that incorporates both a
γ-lactone and a δ-
lactone into its tricyclic
framework)

15-Pentadecanolide

(an odor-enhancing substance
used in perfume)

Erythromycin

(a macrolide antibiotic; drug production is by fermentation
processes, but the laboratory synthesis of this complex
substance has been achieved)

FIGURE 19.8 Some
naturally occurring lactones.

better known by their common names, γ-butyrolactone and δ-valerolactone, respectively, and these two common names are permitted by the IUPAC rules.

Reactions that are expected to produce hydroxy acids often yield the derived lactones instead if a five- or six-membered ring can be formed.

$$CH_3CCH_2CH_2CH_2COH \xrightarrow[\text{2. } H_3O^+]{\text{1. NaBH}_4}$$
        via $$CH_3CHCH_2CH_2CH_2COH$$

5-Oxohexanoic acid   5-Hexanolide (78%)   5-Hydroxyhexanoic acid

Many natural products are lactones, and it is not unusual to find examples in which the ring size is rather large. A few naturally occurring lactones are shown in Figure 19.8. The *macrolide antibiotics,* of which erythromycin is one example, are macrocyclic (large-ring) lactones. The lactone ring of erythromycin is 14-membered.

PROBLEM 19.9 Write the structure of the hydroxy acid corresponding to each of the following lactones. The structure of each lactone is given in Figure 19.8.

(a) Mevalonolactone

(b) Pentadecanolide

(c) Vernolepin

SAMPLE SOLUTION (a) The ring oxygen of the lactone is derived from the hydroxyl group of the hydroxy acid, whereas the carbonyl group corresponds to that of the carboxyl function. To identify the hydroxy acid, disconnect the O—C(O) bond of the ester.

$$HOCH_2CH_2CCH_2C$$

Mevalonolactone   Mevalonic acid
(disconnect bond indicated)

Lactones whose rings are three- or four-membered (α-lactones and β-lactones) are very reactive, making their isolation difficult. Special methods are normally required for the laboratory synthesis of small-ring lactones as well as those that contain rings larger than six-membered.

The compound *anisatin* is an example of a naturally occurring β-lactone. Its isolation and structure determination were described in the journal *Tetrahedron Letters* (1982), p. 5111.

19.16 α HALOGENATION OF CARBOXYLIC ACIDS: THE HELL–VOLHARD–ZELINSKY REACTION

Esterification of carboxylic acids involves nucleophilic addition to the carbonyl group as a key step. In this respect the carbonyl group of a carboxylic acid resembles that of an aldehyde or a ketone. Do carboxylic acids resemble aldehydes and ketones in other ways? Do they, for example, form *enols,* and can they be halogenated at their α-carbon atom via an enol in the way that aldehydes and ketones can?

The enol content of a carboxylic acid is far less than that of an aldehyde or ketone, and introduction of a halogen substituent at the α-carbon atom requires a different set of reaction conditions. Bromination is the reaction that is normally carried out, and the usual procedure involves treatment of the carboxylic acid with bromine in the presence of a small amount of phosphorus trichloride as a catalyst.

$$R_2\underset{\underset{H}{|}}{C}CO_2H \quad + \quad Br_2 \quad \xrightarrow{PCl_3} \quad R_2\underset{\underset{Br}{|}}{C}CO_2H \quad + \quad HBr$$

| Carboxylic acid | Bromine | α-Bromo carboxylic acid | Hydrogen bromide |

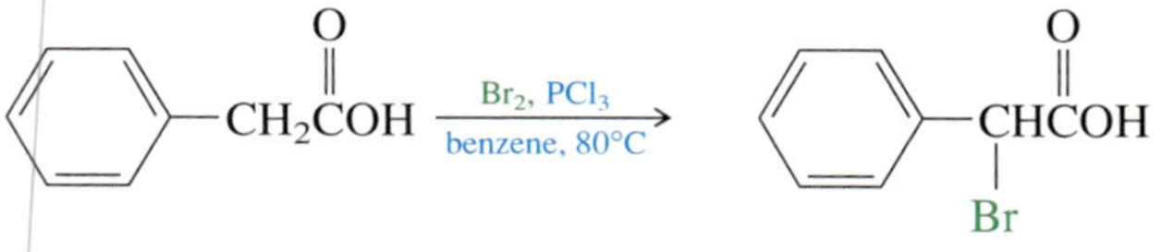

Phenylacetic acid → 2-Bromo-2-phenylacetic acid (60–62%)

This method of α bromination of carboxylic acids is called the **Hell–Volhard–Zelinsky reaction.** This reaction is sometimes carried out by using a small amount of phosphorus instead of phosphorus trichloride. Phosphorus reacts with bromine to yield phosphorus tribromide as the active catalyst under these conditions.

The Hell–Volhard–Zelinsky reaction is of synthetic value in that the α halogen can be displaced by nucleophilic substitution:

$$CH_3CH_2CH_2CO_2H \xrightarrow[P]{Br_2} CH_3CH_2\underset{\underset{Br}{|}}{C}HCO_2H \xrightarrow[H_2O,\ heat]{K_2CO_3} CH_3CH_2\underset{\underset{OH}{|}}{C}HCO_2H$$

| Butanoic acid | 2-Bromobutanoic acid (77%) | 2-Hydroxybutanoic acid (69%) |

A standard method for the preparation of an α-amino acid uses α-bromo carboxylic acids as the substrate and aqueous ammonia as the nucleophile:

$$(CH_3)_2CHCH_2CO_2H \xrightarrow[PCl_3]{Br_2} (CH_3)_2CH\underset{\underset{Br}{|}}{C}HCO_2H \xrightarrow[H_2O]{NH_3} (CH_3)_2CH\underset{\underset{NH_2}{|}}{C}HCO_2H$$

| 3-Methylbutanoic acid | 2-Bromo-3-methylbutanoic acid (88%) | 2-Amino-3-methylbutanoic acid (48%) |

PROBLEM 19.10 α-Iodo acids are not normally prepared by direct iodination of carboxylic acids under conditions of the Hell–Volhard–Zelinsky reaction. Show how you could convert octadecanoic acid to its 2-iodo derivative by an efficient sequence of reactions.

19.17 DECARBOXYLATION OF MALONIC ACID AND RELATED COMPOUNDS

The loss of a molecule of carbon dioxide from a carboxylic acid is known as **decarboxylation.**

$$RCO_2H \longrightarrow RH + CO_2$$

| Carboxylic acid | Alkane | Carbon dioxide |

Decarboxylation of simple carboxylic acids takes place with great difficulty and is rarely encountered.

Compounds that readily undergo thermal decarboxylation include those related to malonic acid. On being heated above its melting point, malonic acid is converted to acetic acid and carbon dioxide.

$$HO_2CCH_2CO_2H \xrightarrow{150°C} CH_3CO_2H + CO_2$$

Malonic acid   Acetic acid   Carbon dioxide
(propanedioic acid)   (ethanoic acid)

It is important to recognize that only one carboxyl group is lost in this process. The second carboxyl group is retained. A mechanism recognizing the assistance that one carboxyl group gives to the departure of the other is represented by the equation

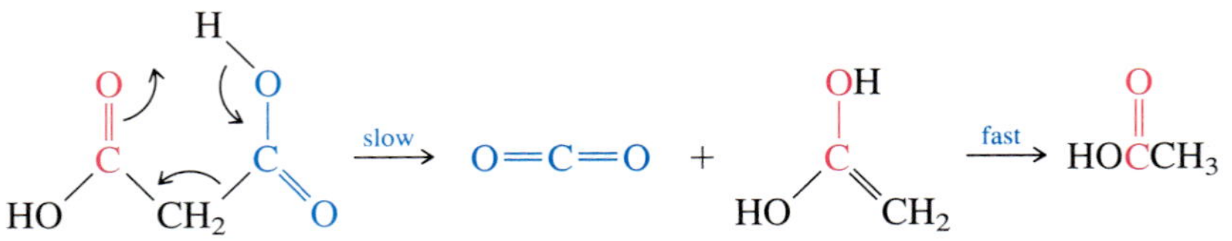

Carbon dioxide   Enol form of   Acetic acid
acetic acid

The transition state involves the carbonyl oxygen of one carboxyl group—the one that stays behind—acting as a proton acceptor toward the hydroxyl group of the carboxyl that is lost. Carbon–carbon bond cleavage leads to the enol form of acetic acid, along with a molecule of carbon dioxide.

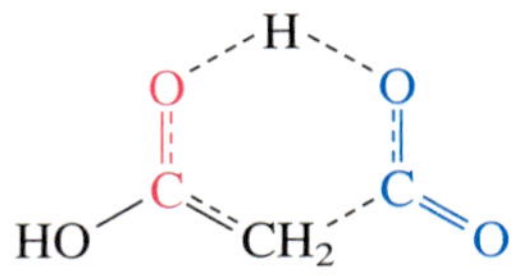

Transition state in
thermal decarboxylation of
malonic acid

The enol intermediate subsequently tautomerizes to acetic acid.

The protons attached to C-2 of malonic acid are not directly involved in the process and so may be replaced by other substituents without much effect on the ease of decarboxylation. Analogs of malonic acid substituted at C-2 undergo efficient thermal decarboxylation.

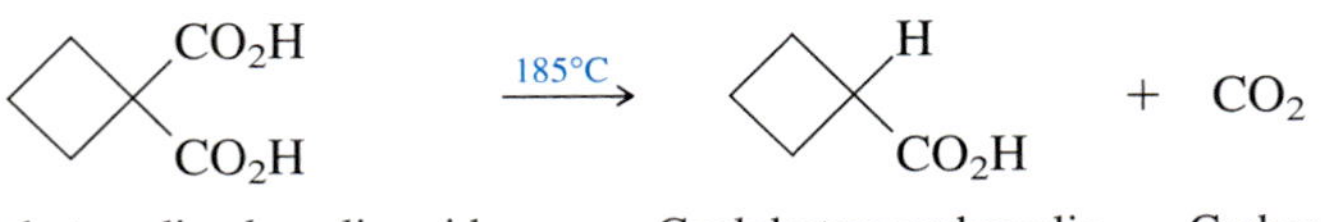

1,1-Cyclobutanedicarboxylic acid   Cyclobutanecarboxylic   Carbon
acid (74%)   dioxide

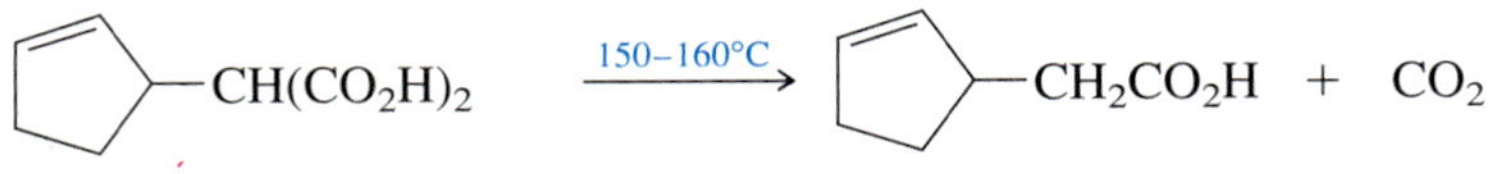

2-(2-Cyclopentenyl)malonic acid   (2-Cyclopentenyl)acetic   Carbon
acid (96–99%)   dioxide

PROBLEM 19.11 What will be the product isolated after thermal decarboxylation of each of the following? Using curved arrows, represent the bond changes that take place at the transition state.

(a) $(CH_3)_2C(CO_2H)_2$

(b) $CH_3(CH_2)_6CHCO_2H$
       |
       CO_2H

(c)

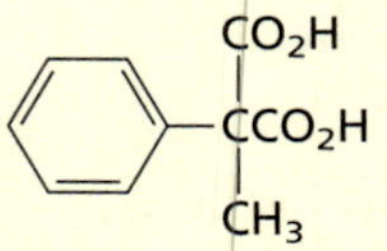

SAMPLE SOLUTION (a) Thermal decarboxylation of malonic acid derivatives leads to the replacement of one of the carboxyl groups by a hydrogen.

$$(CH_3)_2C(CO_2H)_2 \xrightarrow{\text{heat}} (CH_3)_2CHCO_2H \; + \; CO_2$$

| 2,2-Dimethylmalonic acid | 2-Methylpropanoic acid | Carbon dioxide |

The transition state incorporates a cyclic array of six atoms:

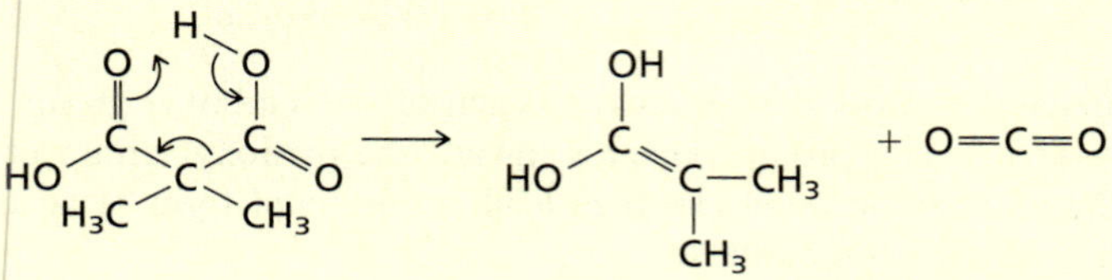

| 2,2-Dimethylmalonic acid | Enol form of 2-methylpropanoic acid | Carbon dioxide |

Tautomerization of the enol form to 2-methylpropanoic acid completes the process.

The thermal decarboxylation of malonic acid derivatives is the last step in a multistep synthesis of carboxylic acids known as the *malonic ester synthesis*. This synthetic method will be described in Section 21.7.

Notice that the carboxyl group that stays behind during the decarboxylation of malonic acid has a hydroxyl function that is not directly involved in the process. Compounds that have substituents other than hydroxyl groups at this position undergo an analogous decarboxylation.

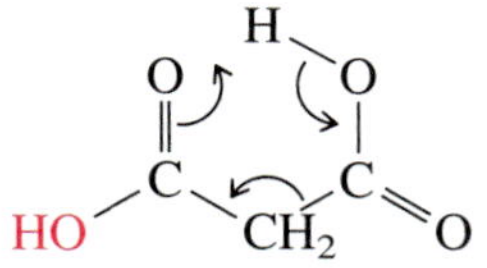

| Bonding changes during decarboxylation of malonic acid | Bonding changes during decarboxylation of a β-keto acid |

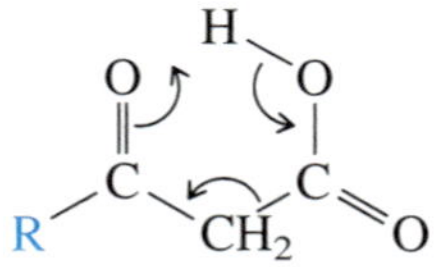

The compounds most frequently encountered in this reaction are β-keto acids, that is, carboxylic acids in which the β carbon is a carbonyl function. Decarboxylation of β-keto acids leads to ketones.

$$\underset{\substack{\text{β-Keto acid}}}{\text{RCCH}_2\text{CO}_2\text{H}} \xrightarrow{\text{heat}} \underset{\substack{\text{Carbon dioxide}}}{\text{CO}_2} + \underset{\substack{\text{Enol form of ketone}}}{\text{RC}{=}\text{CH}_2} \xrightarrow{\text{fast}} \underset{\substack{\text{Ketone}}}{\text{RCCH}_3}$$

$$\underset{\substack{\text{2,2-Dimethylacetoacetic}\\\text{acid}}}{\text{CH}_3\text{CCCO}_2\text{H}} \xrightarrow{\text{heat}} \underset{\substack{\text{3-Methyl-2-butanone}}}{\text{CH}_3\text{CCH(CH}_3)_2} + \underset{\substack{\text{Carbon dioxide}}}{\text{CO}_2}$$

PROBLEM 19.12   Show the bonding changes that occur, and write the structure of the intermediate formed in the thermal decarboxylation of

(a) Benzoylacetic acid

(b) 2,2-Dimethylacetoacetic acid

SAMPLE SOLUTION   (a) By analogy to the thermal decarboxylation of malonic acid, we represent the corresponding reaction of benzoylacetic acid as

| Benzoylacetic acid | Enol form of acetophenone | Carbon dioxide |

Acetophenone is the isolated product; it is formed from its enol by proton transfers.

The thermal decarboxylation of β-keto acids is the last step in a ketone synthesis known as the *acetoacetic ester synthesis*. The acetoacetic ester synthesis is discussed in Section 21.6.

19.18 SPECTROSCOPIC ANALYSIS OF CARBOXYLIC ACIDS

Infrared: The most characteristic peaks in the IR spectra of carboxylic acids are those of the hydroxyl and carbonyl groups. As shown in the IR spectrum of 4-phenylbutanoic acid (Figure 19.9) the O—H and C—H stretching frequencies overlap to produce a broad absorption in the 3500–2500 cm^{-1} region. The carbonyl group gives a strong band for C=O stretching at 1700 cm^{-1}.

^{1}H NMR: The hydroxyl proton of a CO$_2$H group is normally the least shielded of all the protons in an NMR spectrum, appearing 10–12 ppm downfield from tetramethylsilane, often as a broad peak. Figure 19.10 illustrates this for 4-phenylbutanoic acid. As with other hydroxyl protons, the proton of a carboxyl group can be identified by adding D$_2$O to the sample. Hydrogen–deuterium exchange converts —CO$_2$H to —CO$_2$D, and the signal corresponding to the carboxyl group disappears.

^{13}C NMR: Like other carbonyl groups, the carbon of the —CO$_2$H group of a carboxylic acid is strongly deshielded (δ 160–185), but not as much as that of an aldehyde or ketone (δ 190–215).

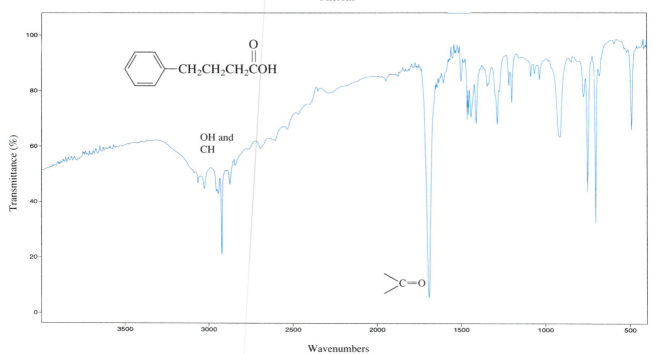

FIGURE 19.9 The IR spectrum of 4-phenylbutanoic acid.

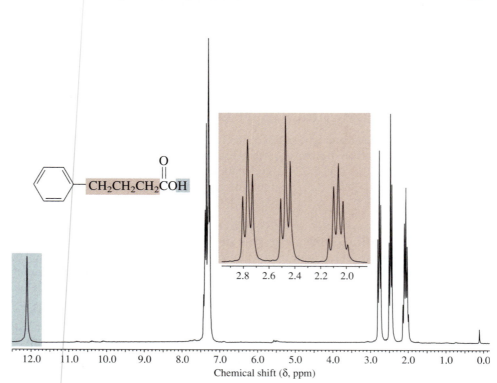

FIGURE 19.10 The 200-MHz ¹H NMR spectrum of 4-phenylbutanoic acid. The peak for the proton of the CO₂H group is at δ 12.

UV-VIS: In the absence of any additional chromophores, carboxylic acids absorb at a wavelength (210 nm) that is not very useful for diagnostic purposes.

Mass Spectrometry: Aside from a peak for the molecular ion, which is normally easy to pick out, aliphatic carboxylic acids undergo a variety of fragmentation processes. The dominant fragmentation in aromatic acids corresponds to loss of OH, then loss of CO.

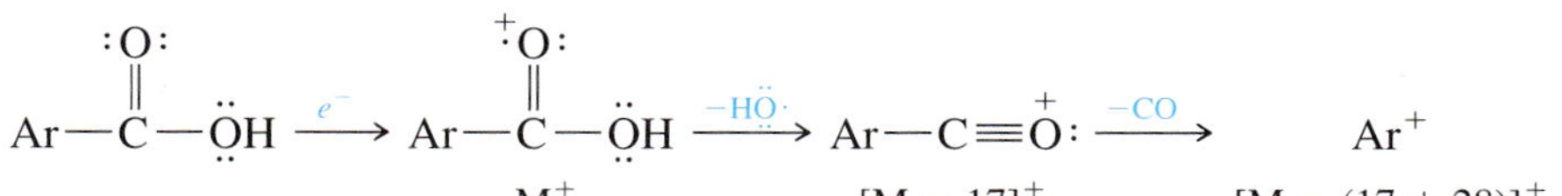

$$Ar-\overset{\overset{:O:}{\|}}{C}-\overset{..}{\underset{..}{O}}H \xrightarrow{e^-} Ar-\overset{\overset{+O:}{\|}}{C}-\overset{..}{\underset{..}{O}}H \xrightarrow{-H\overset{..}{O}\cdot} Ar-C\overset{+}{\equiv}\overset{..}{O}: \xrightarrow{-CO} Ar^+$$

M⁺   [M − 17]⁺   [M − (17 + 28)]⁺

19.19 SUMMARY

Section 19.1   Carboxylic acids take their names from the alkane that contains the same number of carbons as the longest continuous chain that contains the —CO₂H group. The *-e* ending is replaced by *-oic acid*. Numbering begins at the carbon of the —CO₂H group.

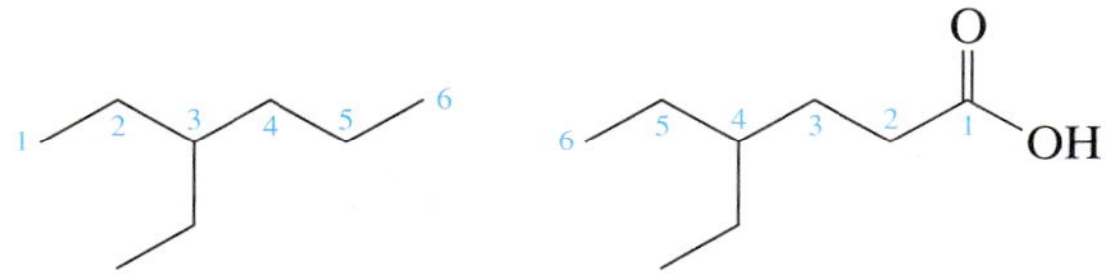

3-Ethylhexane   4-Ethylhexanoic acid

Section 19.2   Like the carbonyl group of aldehydes and ketones, the carbon of a C=O unit in a carboxylic acid is sp^2-hybridized. Compared with the carbonyl group of an aldehyde or ketone, the C=O unit of a carboxylic acid receives an extra degree of stabilization from its attached OH group.

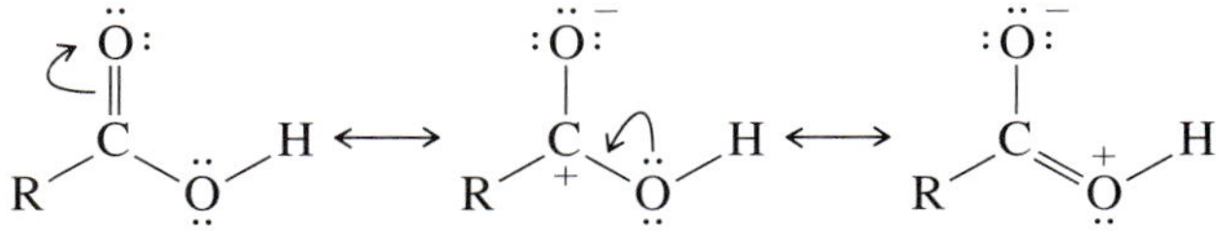

Section 19.3   Hydrogen bonding in carboxylic acids raises their melting points and boiling points above those of comparably constituted alkanes, alcohols, aldehydes, and ketones.

Section 19.4   Carboxylic acids are weak acids and, in the absence of electron-attracting substituents, have pK_a's of approximately 5. Carboxylic acids are much stronger acids than alcohols because of the electron-withdrawing power of the carbonyl group (inductive effect) and its ability to delocalize negative charge in the carboxylate anion (resonance effect).

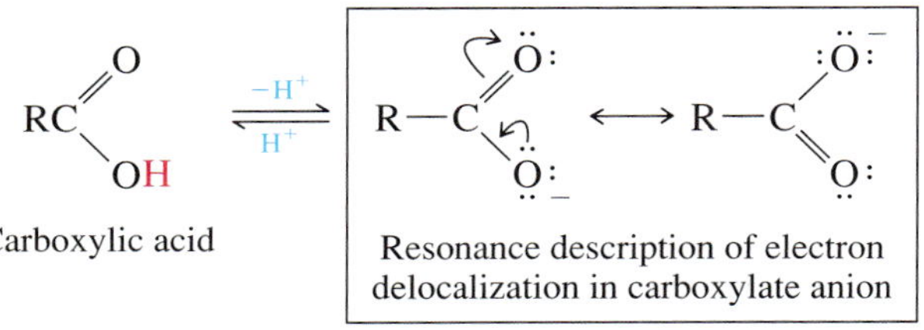

Carboxylic acid   Resonance description of electron delocalization in carboxylate anion

Section 19.5   Although carboxylic acids dissociate to only a small extent in water, they are deprotonated almost completely in basic solution.

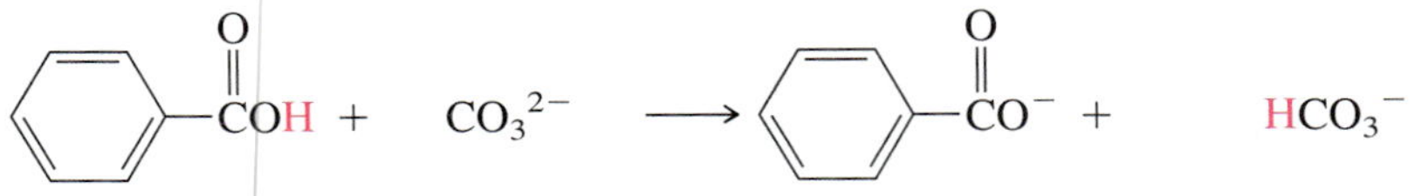

Benzoic acid
$pK_a = 4.2$
(stronger acid)

Carbonate ion

Benzoate ion

Hydrogen carbonate ion
$pK_a = 10.2$
(weaker acid)

Sections 19.6–19.7   Electronegative substituents, especially those within a few bonds of the carboxyl group, increase the acidity of carboxylic acids.

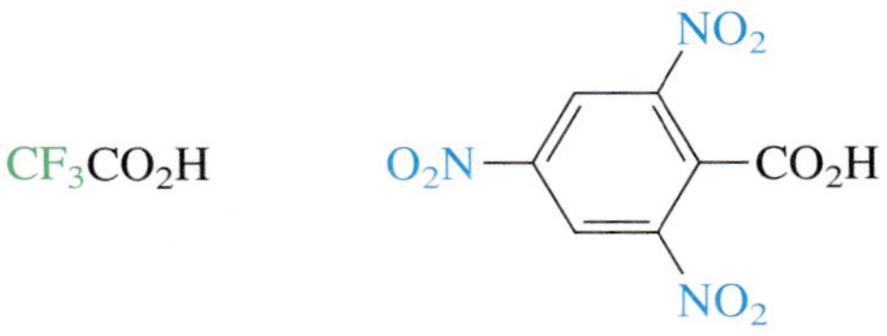

CF_3CO_2H

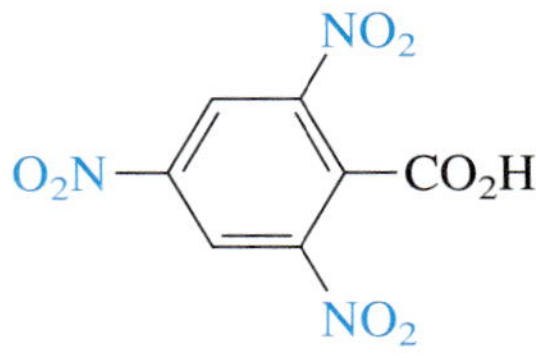

Trifluoroacetic acid
$pK_a = 0.2$

2,4,6-Trinitrobenzoic acid
$pK_a = 0.6$

Section 19.8   Dicarboxylic acids have separate pK_a values for their first and second ionizations.

Section 19.9   Carbon dioxide and carbonic acid are in equilibrium in water. Carbon dioxide is the major component.

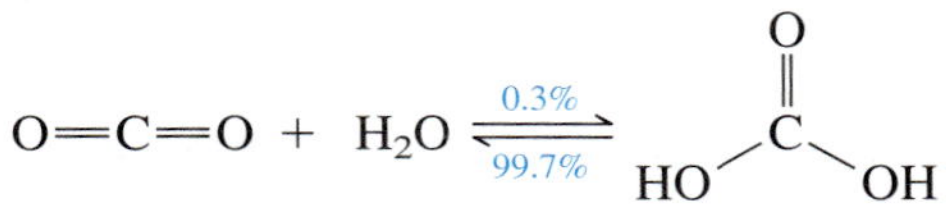

$$O{=}C{=}O \ + \ H_2O \underset{99.7\%}{\overset{0.3\%}{\rightleftharpoons}}$$

Section 19.10   Several of the reactions introduced in earlier chapters can be used to prepare carboxylic acids (See Table 19.4).

Section 19.11   Carboxylic acids can be prepared by the reaction of Grignard reagents with carbon dioxide.

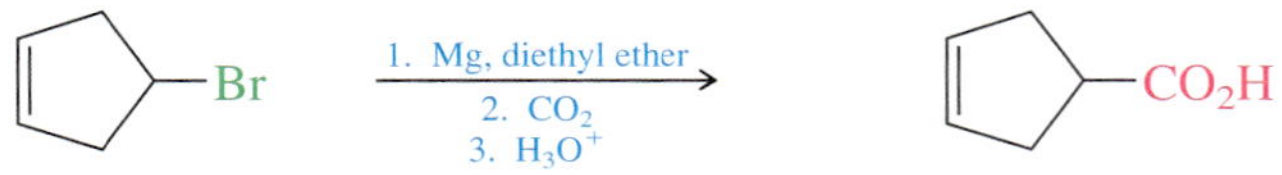

4-Bromocyclopentene

1. Mg, diethyl ether
2. CO_2
3. H_3O^+

Cyclopentene-4-carboxylic acid (66%)

Section 19.12   Nitriles, which can be prepared from primary and secondary alkyl halides by nucleophilic substitution with cyanide ion, can be converted to carboxylic acids by hydrolysis.

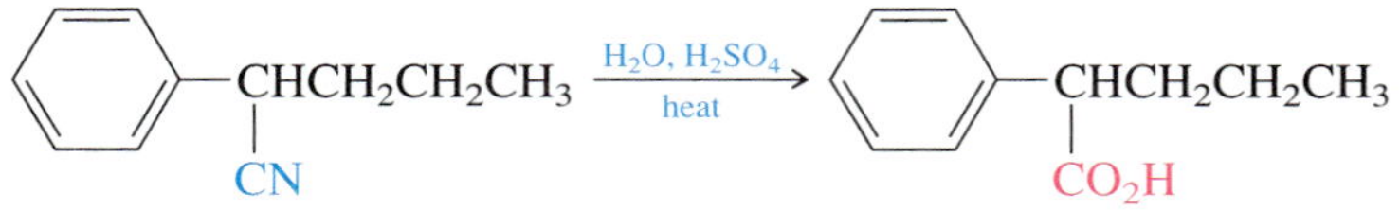

2-Phenylpentanenitrile

H_2O, H_2SO_4
heat

2-Phenylpentanoic acid (52%)

Likewise, the cyano group of a cyanohydrin can be hydrolyzed to $-CO_2H$.

Section 19.13 Among the reactions of carboxylic acids, their conversions to acyl chlorides, primary alcohols, and esters were introduced in earlier chapters and were reviewed in Table 19.5.

Section 19.14 The mechanism of acid-catalyzed esterification involves some key features that are fundamental to the chemistry of carboxylic acids and their derivatives.

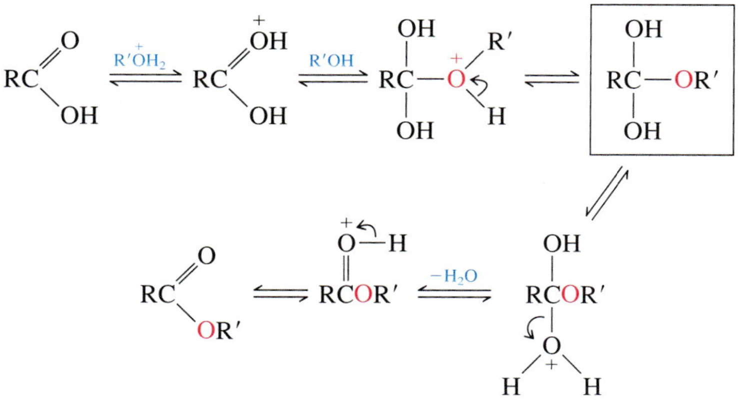

Protonation of the carbonyl oxygen activates the carbonyl group toward nucleophilic addition. Addition of an alcohol gives a tetrahedral intermediate (shown in the box in the preceding equation), which has the capacity to revert to starting materials or to undergo dehydration to yield an ester.

Section 19.15 An intramolecular esterification can occur when a molecule contains both a hydroxyl and a carboxyl group. Cyclic esters are called *lactones* and are most stable when the ring is five- or six-membered.

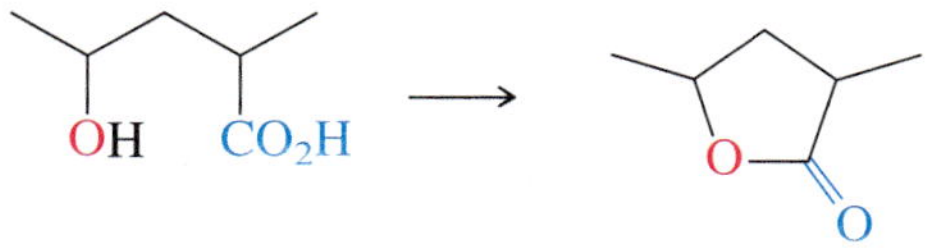

4-Hydroxy-2-   2-Methyl-4-pentanolide
methylpentanoic acid

Section 19.16 Halogenation at the α-carbon atom of carboxylic acids can be accomplished by the *Hell–Volhard–Zelinsky reaction*. An acid is treated with chlorine or bromine in the presence of a catalytic quantity of phosphorus or a phosphorus trihalide:

$$R_2CHCO_2H \ + \ X_2 \xrightarrow{\text{P or } PX_3} R_2CCO_2H \ + \ H{-}X$$
$$\underset{\text{X}}{|}$$

Carboxylic   Halogen α-Halo acid
acid

This reaction is of synthetic value in that α-halo acids are reactive substrates in nucleophilic substitution reactions.

Section 19.17 1,1-Dicarboxylic acids (malonic acids) and β-keto acids undergo thermal decarboxylation by a mechanism in which a β-carbonyl group assists the departure of carbon dioxide.

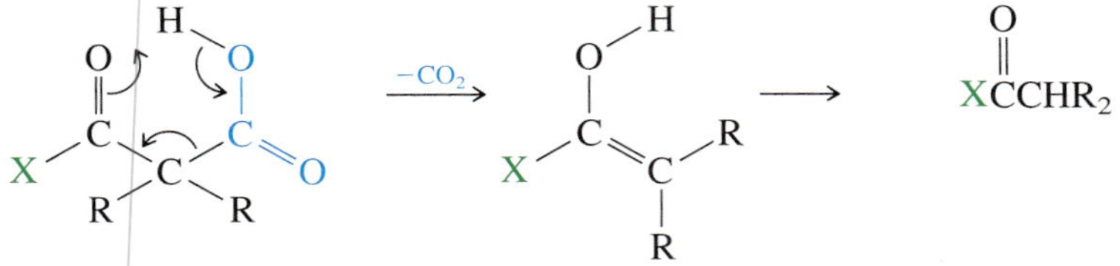

X = OH: malonic acid derivative	Enol form of product	X = OH: carboxylic acid
X = alkyl or aryl: β-keto acid		X = alkyl or aryl: ketone

Section 19.18 Carboxylic acids are readily identified by the presence of strong IR absorptions at 1700 cm^{-1} (C=O) and between 2500 and 3500 cm^{-1} (OH), an ^{1}H NMR signal for the hydroxyl proton at δ 10–12, and a ^{13}C signal for the carbonyl carbon near δ 180.

PROBLEMS

19.13 Many carboxylic acids are much better known by their common names than by their systematic names. Some of these follow. Provide a structural formula for each one on the basis of its systematic name.

(a) 2-Hydroxypropanoic acid (better known as *lactic acid,* it is found in sour milk and is formed in the muscles during exercise)

(b) 2-Hydroxy-2-phenylethanoic acid (also known as *mandelic acid,* it is obtained from plums, peaches, and other fruits)

(c) Tetradecanoic acid (also known as *myristic acid,* it can be obtained from a variety of fats)

(d) 10-Undecenoic acid (also called *undecylenic acid,* it is used, in combination with its zinc salt, to treat fungal infections such as athlete's foot)

(e) 3,5-Dihydroxy-3-methylpentanoic acid (also called *mevalonic acid,* it is an important intermediate in the biosynthesis of terpenes and steroids)

(f) (*E*)-2-Methyl-2-butenoic acid (also known as *tiglic acid,* it is a constituent of various natural oils)

(g) 2-Hydroxybutanedioic acid (also known as *malic acid,* it is found in apples and other fruits)

(h) 2-Hydroxy-1,2,3-propanetricarboxylic acid (better known as *citric acid,* it contributes to the tart taste of citrus fruits)

(i) 2-(*p*-Isobutylphenyl)propanoic acid (an antiinflammatory drug better known as *ibuprofen*)

(j) *o*-Hydroxybenzenecarboxylic acid (better known as *salicylic acid,* it is obtained from willow bark)

19.14 Give an acceptable IUPAC name for each of the following:

(a) $CH_3(CH_2)_6CO_2H$

(b) $CH_3(CH_2)_6CO_2K$

(c) $H_2C{=}CH(CH_2)_5CO_2H$

(d) 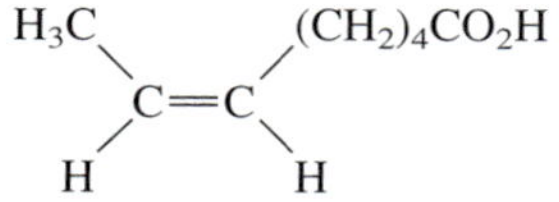
$$\underset{H}{\overset{H_3C}{}}C=C\underset{H}{\overset{(CH_2)_4CO_2H}{}}$$

(g) [cycloheptane ring with CO_2H]

(e) $HO_2C(CH_2)_6CO_2H$

(f) $CH_3(CH_2)_4CH(CO_2H)_2$

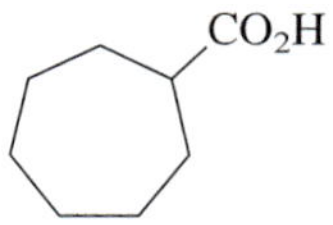

(h) 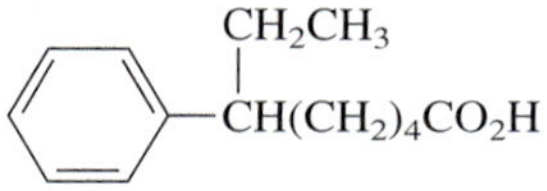 [benzene ring with $-CH(CH_2)_4CO_2H$ bearing a CH_2CH_3 substituent]

19.15 Rank the compounds in each of the following groups in order of decreasing acidity:

(a) Acetic acid, ethane, ethanol

(b) Benzene, benzoic acid, benzyl alcohol

(c) Propanedial, 1,3-propanediol, propanedioic acid, propanoic acid

(d) Acetic acid, ethanol, trifluoroacetic acid, 2,2,2-trifluoroethanol, trifluoromethane-sulfonic acid (CF_3SO_2OH)

(e) Cyclopentanecarboxylic acid, 2,4-pentanedione, cyclopentanone, cyclopentene

19.16 Identify the more acidic compound in each of the following pairs:

(a) $CF_3CH_2CO_2H$ or $CF_3CH_2CH_2CO_2H$

(b) $CH_3CH_2CH_2CO_2H$ or $CH_3C{\equiv}CCO_2H$

(c) 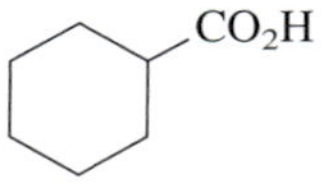 [cyclohexane ring with CO_2H]   or   [benzene ring with CO_2H]

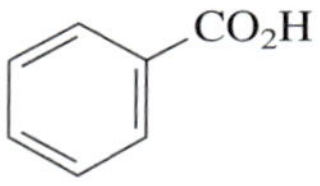

(d) 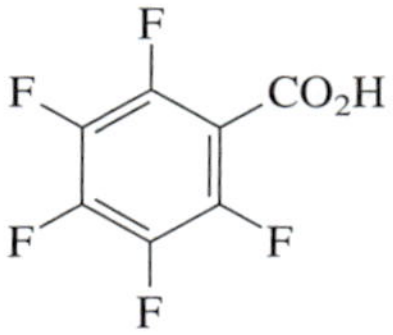 [tetrafluoro-substituted benzene ring with CO_2H]   or   [benzene ring with CO_2H]

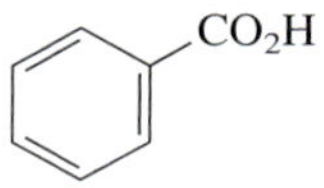

(e) 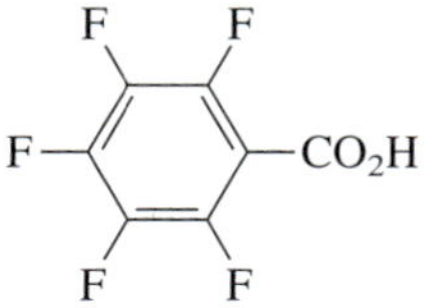 [pentafluorophenyl $-CO_2H$]   or   [tetrafluorophenyl-phenyl $-CO_2H$]

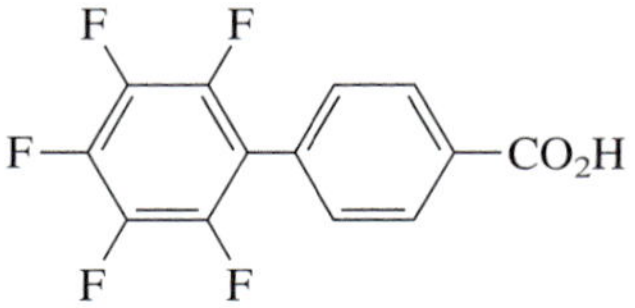

(f) 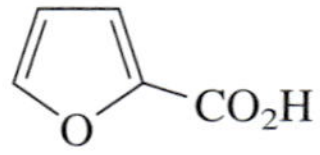 [furan-2-carboxylic acid, CO_2H]   or   [furan-3-carboxylic acid, CO_2H]

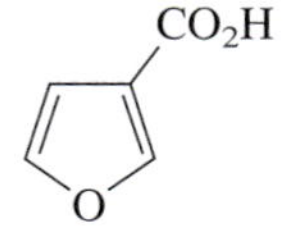

(g) 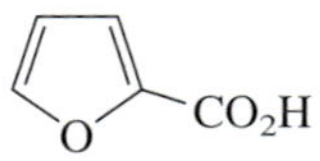 [furan-2-carboxylic acid, CO_2H]   or   [pyrrole-2-carboxylic acid, N–H, CO_2H]

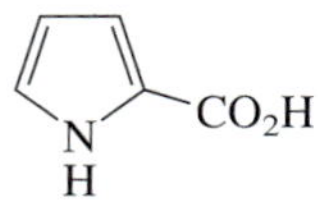

19.17 Propose methods for preparing butanoic acid from each of the following:

(a) 1-Butanol

(b) Butanal

(c) 1-Butene

(d) 1-Propanol

(e) 2-Propanol

(f) Acetaldehyde

(g) $CH_3CH_2CH(CO_2H)_2$

19.18 It is sometimes necessary to prepare isotopically labeled samples of organic substances for probing biological transformations and reaction mechanisms. Various sources of the radioactive mass-14 carbon isotope are available. Describe synthetic procedures by which benzoic acid, labeled with ^{14}C at its carbonyl carbon, could be prepared from benzene and the following ^{14}C-labeled precursors. You may use any necessary organic or inorganic reagents. (In the formulas shown, an asterisk (*) indicates ^{14}C.)

(a) $\overset{*}{C}H_3Cl$

(b) $H\overset{*}{C}H$, with $=O$ (HCH with carbonyl, asterisk on the carbon)

(c) $\overset{*}{C}O_2$

19.19 Give the product of the reaction of pentanoic acid with each of the following reagents:

(a) Sodium hydroxide

(b) Sodium bicarbonate

(c) Thionyl chloride

(d) Phosphorus tribromide

(e) Benzyl alcohol, sulfuric acid (catalytic amount)

(f) Chlorine, phosphorus tribromide (catalytic amount)

(g) Bromine, phosphorus trichloride (catalytic amount)

(h) Product of part (g) treated with sodium iodide in acetone

(i) Product of part (g) treated with aqueous ammonia

(j) Lithium aluminum hydride, then hydrolysis

(k) Phenylmagnesium bromide

19.20 Show how butanoic acid may be converted to each of the following compounds:

(a) 1-Butanol

(b) Butanal

(c) 1-Chlorobutane

(d) Butanoyl chloride

(e) Phenyl propyl ketone

(f) 4-Octanone

(g) 2-Bromobutanoic acid

(h) 2-Butenoic acid

19.21 Show by a series of equations, using any necessary organic or inorganic reagents, how acetic acid can be converted to each of the following compounds:

(a) $H_2NCH_2CO_2H$

(b) $C_6H_5OCH_2CO_2H$

(c) $NCCH_2CO_2H$

(d) $HO_2CCH_2CO_2H$

(e) ICH_2CO_2H

(f) $BrCH_2CO_2CH_2CH_3$

(g) $(C_6H_5)_3\overset{+}{P}—\overset{-}{C}HCO_2CH_2CH_3$

(h) $C_6H_5CH=CHCO_2CH_2CH_3$

19.22 Each of the following reactions has been reported in the chemical literature and gives a single product in good yield. What is the product in each reaction?

(a)
$$\begin{array}{c} H_3C \\ \\ H \end{array} C=C \begin{array}{c} CH_3 \\ \\ CO_2H \end{array} \xrightarrow{\text{ethanol, } H_2SO_4}$$

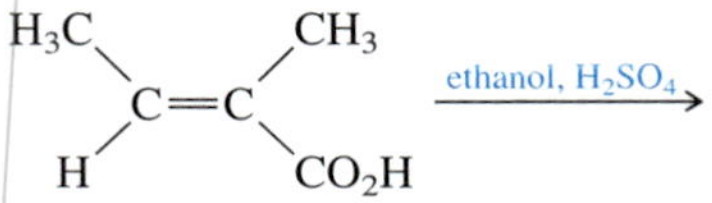

(b) (cyclopropyl)$—CO_2H \xrightarrow[\text{2. } H_2O]{\text{1. LiAlD}_4}$

(c) (cyclohexyl)$CO_2H \xrightarrow[P]{Br_2}$

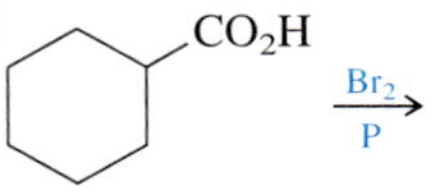

(d) (aryl with CF_3 and Br) $\xrightarrow[\text{3. } H_3O^+]{\begin{array}{c}\text{1. Mg, diethyl ether}\\\text{2. } CO_2\end{array}}$

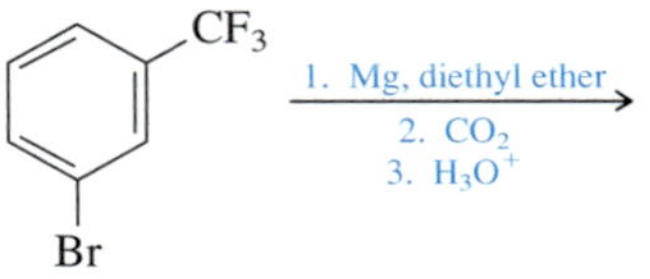

(e) (aryl with CH_2CN and Cl) $\xrightarrow[\text{H}_2SO_4,\text{ heat}]{H_2O,\text{ acetic acid}}$

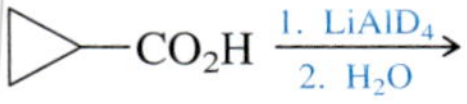

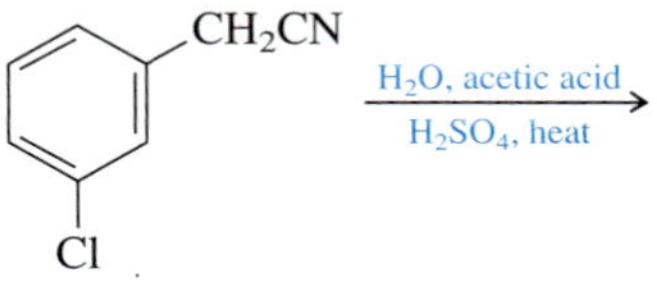

(f) $H_2C=CH(CH_2)_8CO_2H \xrightarrow[\text{benzoyl peroxide}]{HBr}$

19.23 Show by a series of equations how you could synthesize each of the following compounds from the indicated starting material and any necessary organic or inorganic reagents:

(a) 2-Methylpropanoic acid from *tert*-butyl alcohol

(b) 3-Methylbutanoic acid from *tert*-butyl alcohol

(c) 3,3-Dimethylbutanoic acid from *tert*-butyl alcohol

(d) $HO_2C(CH_2)_5CO_2H$ from $HO_2C(CH_2)_3CO_2H$

(e) 3-Phenyl-1-butanol from CH_3CHCH_2CN
$\qquad\qquad\qquad\qquad\qquad\qquad\ \ \ |$
$\qquad\qquad\qquad\qquad\qquad\qquad\ C_6H_5$

(f)  [structure: cyclopentane with Br and CO₂H] from cyclopentyl bromide

(g) 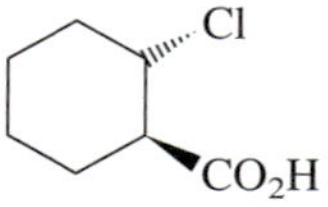 [structure: cyclohexane with Cl and CO₂H] from (E)-ClCH=CHCO_2H

(h) 2,4-Dimethylbenzoic acid from *m*-xylene

(i) 4-Chloro-3-nitrobenzoic acid from *p*-chlorotoluene

(j) (Z)-CH_3CH=$CHCO_2H$ from propyne

19.24 (a) Which stereoisomer of 4-hydroxycyclohexanecarboxylic acid (cis or trans) can form a lactone? Make a molecular model of this lactone. What is the conformation of the cyclohexane ring in the starting hydroxy acid? In the lactone?

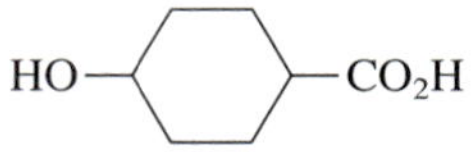

HO—[cyclohexane ring]—CO_2H

(b) Repeat part (a) for the case of 3-hydroxycyclohexanecarboxylic acid.

19.25 Suggest reasonable explanations for each of the following observations.

(a) Both hydrogens are anti to each other in the most stable conformation of formic acid.

(b) Oxalic acid has a dipole moment of zero in the gas phase.

(c) The dissociation constant of *o*-hydroxybenzoic acid is greater (by a factor of 12) than that of *o*-methoxybenzoic acid.

(d) Ascorbic acid (vitamin C), although not a carboxylic acid, is sufficiently acidic to cause carbon dioxide liberation on being dissolved in aqueous sodium bicarbonate.

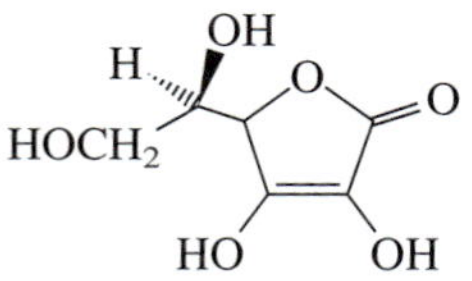

[structure of ascorbic acid with OH, H, HOCH₂, O, HO, OH groups]

Ascorbic acid

19.26 When compound A is heated, two isomeric products are formed. What are these two products?

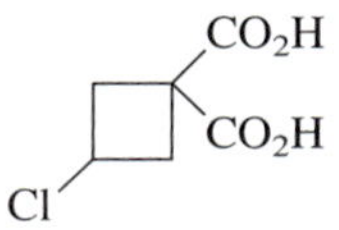

[structure: cyclobutane ring with two CO₂H groups and Cl]

Compound A

19.27 A certain carboxylic acid ($C_{14}H_{26}O_2$), which can be isolated from whale blubber or sardine oil, yields nonanal and $O{=}CH(CH_2)_3CO_2H$ on ozonolysis. What is the structure of this acid?

19.28 When levulinic acid ($CH_3\overset{\overset{\displaystyle O}{\|}}{C}CH_2CH_2CO_2H$) was hydrogenated at high pressure over a nickel catalyst at 220°C, a single product, $C_5H_8O_2$, was isolated in 94% yield. This compound lacks hydroxyl absorption in its IR spectrum and does not immediately liberate carbon dioxide on being shaken with sodium bicarbonate. What is a reasonable structure for the compound?

19.29 On standing in dilute aqueous acid, compound A is smoothly converted to mevalonolactone.

Compound A   Mevalonolactone

Suggest a reasonable mechanism for this reaction. What other organic product is also formed?

19.30 Suggest reaction conditions suitable for the preparation of compound A from 5-hydroxy-2-hexynoic acid.

$CH_3CHCH_2C{\equiv}CCO_2H \longrightarrow$

OH

5-Hydroxy-2-hexynoic acid   Compound A

19.31 In the presence of the enzyme *aconitase*, the double bond of aconitic acid undergoes hydration. The reaction is reversible, and the following equilibrium is established:

Isocitric acid $\underset{H_2O}{\rightleftharpoons}$ $\underset{H_2O}{\rightleftharpoons}$ Citric acid

$(C_6H_8O_7)$ Aconitic acid $(C_6H_8O_7)$
(6% at equilibrium)   (4% at equilibrium)   (90% at equilibrium)

(a) The major tricarboxylic acid present is *citric acid,* the substance responsible for the tart taste of citrus fruits. Citric acid is achiral. What is its structure?

(b) What must be the constitution of isocitric acid? (Assume that no rearrangements accompany hydration.) How many stereoisomers are possible for isocitric acid?

19.32 The ^1H NMR spectra of formic acid (HCO_2H), maleic acid (*cis*-$HO_2CCH{=}CHCO_2H$), and malonic acid ($HO_2CCH_2CO_2H$) are similar in that each is characterized by two singlets of equal intensity. Match these compounds with the designations A, B, and C on the basis of the appropriate ^1H NMR chemical shift data.

Compound A: signals at δ 3.2 and 12.1

Compound B: signals at δ 6.3 and 12.4

Compound C: signals at δ 8.0 and 11.4

19.33 Compounds A and B are isomers having the molecular formula $C_4H_8O_3$. Identify A and B on the basis of their ^1H NMR spectra.

Compound A: δ 1.3 (3H, triplet); 3.6 (2H, quartet); 4.1 (2H, singlet); 11.1 (1H, broad singlet)

Compound B: δ 2.6 (2H, triplet); 3.4 (3H, singlet); 3.7 (2H triplet); 11.3 (1H, broad singlet)

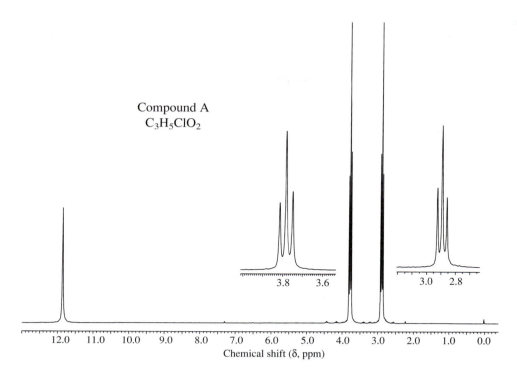

FIGURE 19.11 The 200-MHz ^{1}H NMR spectrum of compound A ($C_3H_5ClO_2$) (Problem 19.34a).

19.34 Compounds A and B are carboxylic acids. Identify each one on the basis of its ^{1}H NMR spectrum.

    (a) Compound A ($C_3H_5ClO_2$) (Figure 19.11).

    (b) Compound B ($C_9H_9NO_4$) has a nitro group attached to an aromatic ring (Figure 19.12).

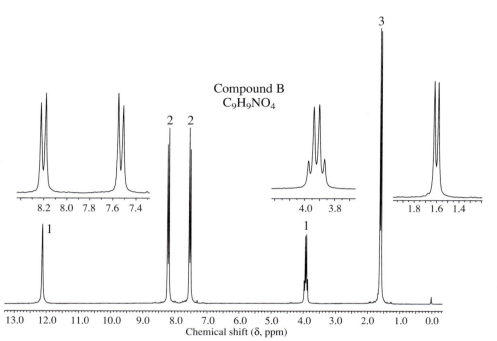

FIGURE 19.12 The 200-MHz ^{1}H NMR spectrum of compound B ($C_9H_9NO_4$) (Problem 19.34b).

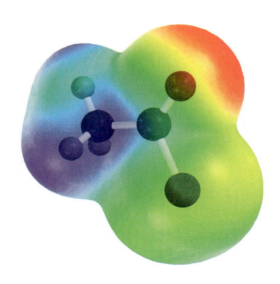

CHAPTER 20

CARBOXYLIC ACID DERIVATIVES: NUCLEOPHILIC ACYL SUBSTITUTION

This chapter differs from preceding ones in that it deals with several related classes of compounds rather than just one. Although the compounds may encompass several functional group types, they share the common feature of yielding carboxylic acids on hydrolysis and, for this reason, are called **carboxylic acid derivatives.**

$$\textit{Acyl chloride:} \quad \underset{\|}{\overset{O}{\text{RCCl}}} \; + \; H_2O \longrightarrow \underset{\|}{\overset{O}{\text{RCOH}}} \; + \; HCl$$

$$\textit{Carboxylic acid} \atop \textit{anhydride:} \quad \underset{\|\;\;\|}{\overset{O\;\;O}{\text{RCOCR}}} \; + \; H_2O \longrightarrow 2\underset{\|}{\overset{O}{\text{RCOH}}}$$

$$\textit{Thioester:} \quad \underset{\|}{\overset{O}{\text{RCSR}'}} \; + \; H_2O \longrightarrow \underset{\|}{\overset{O}{\text{RCOH}}} \; + \; R'SH$$

$$\textit{Ester:} \quad \underset{\|}{\overset{O}{\text{RCOR}'}} \; + \; H_2O \longrightarrow \underset{\|}{\overset{O}{\text{RCOH}}} \; + \; R'OH$$

$$\textit{Amide:} \quad \underset{\|}{\overset{O}{\text{RCNR}'_2}} \; + \; H_2O \longrightarrow \underset{\|}{\overset{O}{\text{RCOH}}} \; + \; R'_2NH \longrightarrow \underset{\|}{\overset{O}{\text{RCO}^-}} \; R'_2NH_2{}^+$$

$$\textit{Nitrile:} \quad \text{RC}\equiv\text{N} \; + \; H_2O \longrightarrow \underset{\|}{\overset{O}{\text{RCOH}}} \; + \; NH_3 \longrightarrow \underset{\|}{\overset{O}{\text{RCO}^-}} \; NH_4{}^+$$

The hydrolysis of a carboxylic acid derivative is but one example of a **nucleophilic acyl substitution.** The mechanism of nucleophilic acyl substitution is one of the major themes of this chapter. All of the carboxylic acid derivatives share the common feature of a two-stage mechanism. The first stage in the hydrolysis of a carboxylic acid derivative is nucleophilic addition to the carbonyl group to give what is called a **tetrahedral interme-diate.** The second stage is the dissociation of the tetrahedral intermediate to the products.

$$\underset{\text{Reactants}}{\overset{\overset{\displaystyle O}{\|}}{RCX}} + H_2O \xrightarrow{\text{slow}} \underset{\substack{\text{Tetrahedral}\\\text{intermediate}}}{\overset{\displaystyle OH}{\underset{\displaystyle OH}{RC-X}}} \xrightarrow{\text{fast}} \underset{\text{Products}}{\overset{\overset{\displaystyle O}{\|}}{RCOH}} + HX$$

Both stages involve more than one step, and these steps differ in detail among the various carboxylic acid derivatives and for different reaction conditions. This chapter is organized to place the various nucleophilic acyl substitutions into a common mechanistic framework and to point out the ways in which individual classes differ from the rest.

20.1 NOMENCLATURE OF CARBOXYLIC ACID DERIVATIVES

With the exception of nitriles ($RC{\equiv}N$), all carboxylic acid derivatives consist of an acyl group ($\overset{\overset{\displaystyle O}{\|}}{RC}-$) attached to an electronegative atom. *Acyl groups* are named by replacing the *-ic acid* ending of the corresponding carboxylic acid by *-yl*.

Acyl halides are named by placing the name of the appropriate halide after that of the acyl group.

$$\overset{\overset{\displaystyle O}{\|}}{CH_3CCl} \qquad\qquad \overset{\overset{\displaystyle O}{\|}}{H_2C{=}CHCH_2CCl} \qquad\qquad F{-}\!\!\left\langle\!\!\bigcirc\!\!\right\rangle\!\!{-}\overset{\overset{\displaystyle O}{\|}}{CBr}$$

Acetyl chloride   3-Butenoyl   *p*-Fluorobenzoyl bromide
                                chloride

Although acyl fluorides, bromides, and iodides are all known classes of organic compounds, they are encountered far less frequently than are acyl chlorides. Acyl chlorides will be the only acyl halides discussed in this chapter.

In naming *carboxylic acid anhydrides* in which both acyl groups are the same, we simply specify the acid and replace *acid* by *anhydride*. When the acyl groups are different, they are cited in alphabetical order.

$$\overset{\overset{\displaystyle O}{\|}\overset{\displaystyle O}{\|}}{CH_3COCCH_3} \qquad\quad \overset{\overset{\displaystyle O}{\|}\overset{\displaystyle O}{\|}}{C_6H_5COCC_6H_5} \qquad\quad \overset{\overset{\displaystyle O}{\|}\overset{\displaystyle O}{\|}}{C_6H_5COC(CH_2)_5CH_3}$$

Acetic anhydride   Benzoic anhydride   Benzoic heptanoic anhydride

The alkyl group and the acyl group of an *ester* are specified independently. Esters are named as *alkyl alkanoates*. The alkyl group R′ of $RC\overset{\overset{\displaystyle O}{\|}}{O}R'$ is cited first, followed by the acyl portion $R\overset{\overset{\displaystyle O}{\|}}{C}-$. The acyl portion is named by substituting the suffix *-ate* for the *-ic acid* ending of the corresponding acid.

$$CH_3COCH_2CH_3 \qquad CH_3CH_2COCH_3 \qquad \text{(phenyl)}-COCH_2CH_2Cl$$

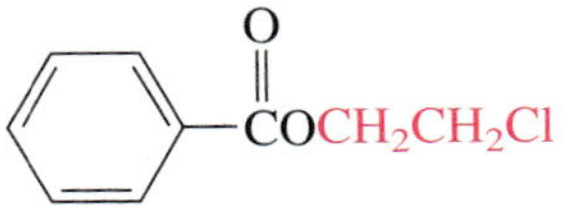

Ethyl acetate   Methyl propanoate   2-Chloroethyl benzoate

Aryl esters, that is, compounds of the type $RCOAr$, are named in an analogous way.

The names of *amides* of the type $RCNH_2$ are derived from carboxylic acids by replacing the suffix *-oic acid* or *-ic acid* by *-amide*.

$$CH_3CNH_2 \qquad C_6H_5CNH_2 \qquad (CH_3)_2CHCH_2CNH_2$$

Acetamide   Benzamide   3-Methylbutanamide

We name compounds of the type $RCNHR'$ and $RCNR'_2$ as *N*-alkyl- and *N,N*-dialkyl-substituted derivatives of a parent amide.

$$CH_3CNHCH_3 \qquad C_6H_5CN(CH_2CH_3)_2 \qquad CH_3CH_2CH_2CNCH(CH_3)_2$$
$$\qquad\qquad\qquad\qquad\qquad\qquad\qquad\qquad CH_3$$

N-Methylacetamide   *N,N*-Diethylbenzamide   *N*-Isopropyl-*N*-methyl-
                                                    butanamide

Substitutive IUPAC names for *nitriles* add the suffix *-nitrile* to the name of the parent hydrocarbon chain that includes the carbon of the cyano group. Nitriles may also be named by replacing the *-ic acid* or *-oic acid* ending of the corresponding carboxylic acid with *-onitrile*. Alternatively, they are sometimes given functional class IUPAC names as alkyl cyanides.

$$CH_3C\equiv N \qquad C_6H_5C\equiv N \qquad CH_3CHCH_3$$
$$\qquad\qquad\qquad\qquad\qquad\qquad\qquad C\equiv N$$

Ethanenitrile   Benzonitrile   2-Methylpropanenitrile
(acetonitrile)   (isopropyl cyanide)

PROBLEM 20.1 Write a structural formula for each of the following compounds:

(a) 2-Phenylbutanoyl chloride
(b) 2-Phenylbutanoic anhydride
(c) Butyl 2-phenylbutanoate
(d) 2-Phenylbutyl butanoate
(e) 2-Phenylbutanamide
(f) *N*-Ethyl-2-phenylbutanamide
(g) 2-Phenylbutanenitrile

SAMPLE SOLUTION (a) A 2-phenylbutanoyl group is a four-carbon acyl unit that bears a phenyl substituent at C-2. When the name of an acyl group is followed by the name of a halide, it designates an *acyl halide*.

$$CH_3CH_2\overset{\displaystyle O}{\overset{\displaystyle \|}{C}}HCCl$$
$$| \atop C_6H_5$$

2-Phenylbutanoyl chloride

20.2 STRUCTURE AND REACTIVITY OF CARBOXYLIC ACID DERIVATIVES

The number of reactions in this chapter is quite large and keeping track of all of them can be difficult—or it can be manageable. The key to making it manageable is the same as always; *structure determines properties.*

Figure 20.1 shows the structures of various derivatives of acetic acid (acetyl chloride, acetic anhydride, ethyl thioacetate, ethyl acetate and acetamide) arranged in order

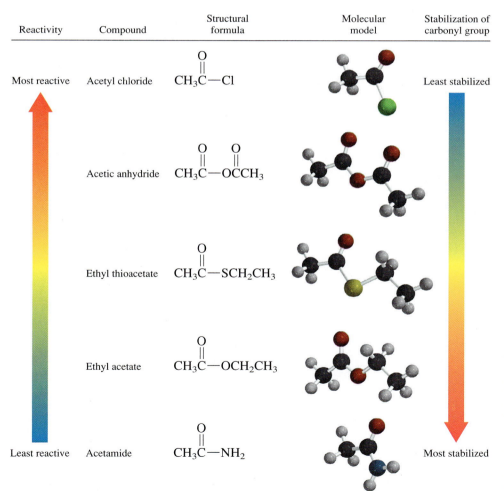

Reactivity	Compound	Structural formula	Molecular model	Stabilization of carbonyl group
Most reactive	Acetyl chloride	$CH_3\overset{\displaystyle O}{\overset{\displaystyle \|}{C}}—Cl$		Least stabilized
	Acetic anhydride	$CH_3\overset{\displaystyle O}{\overset{\displaystyle \|}{C}}—O\overset{\displaystyle O}{\overset{\displaystyle \|}{C}}CH_3$		
	Ethyl thioacetate	$CH_3\overset{\displaystyle O}{\overset{\displaystyle \|}{C}}—SCH_2CH_3$		
	Ethyl acetate	$CH_3\overset{\displaystyle O}{\overset{\displaystyle \|}{C}}—OCH_2CH_3$		
Least reactive	Acetamide	$CH_3\overset{\displaystyle O}{\overset{\displaystyle \|}{C}}—NH_2$		Most stabilized

FIGURE 20.1 Structure, reactivity, and carbonyl-group stabilization in carboxylic acid derivatives. Acyl chlorides are the most reactive, amides the least reactive. Acyl chlorides have the least stabilized carbonyl group, amides the most. Conversion of one class of compounds to another is feasible only in the direction that leads to a more stabilized carbonyl group; that is, from more reactive to less reactive.

of decreasing reactivity toward nucleophilic acyl substitution. Acyl chlorides are the most reactive, amides the least reactive. The reactivity order:

$$\text{acyl chloride} > \text{anhydride} > \text{thioester} > \text{ester} > \text{amide}$$

is general for nucleophilic acyl substitution and well worth remembering. The range of reactivities is quite large; a factor of about 10^{13} in relative rate separates acyl chlorides from amides.

This difference in reactivity, especially toward hydrolysis, has an important result. We'll see in Chapter 27 that the structure and function of proteins are critical to life itself. The bonds mainly responsible for the structure of proteins are amide bonds, which are about 100 times more stable to hydrolysis than ester bonds. These amide bonds are stable enough to maintain the structural integrity of proteins in an aqueous environment, but susceptible enough to hydrolysis to be broken when the occasion demands.

What structural features are responsible for the reactivity order of carboxylic acid derivatives? Like the other carbonyl-containing compounds that we've studied, they all have a planar arrangement of bonds to the carbonyl group. Thus, all are about the same in offering relatively unhindered access to the approach of a nucleophile. They differ in the degree to which the atom attached to the carbonyl group can stabilize the carbonyl group by electron donation.

Electron release from the substituent X not only stabilizes the carbonyl group, it decreases the positive character of the carbonyl carbon and makes the carbonyl group less electrophilic.

The order of reactivity of carboxylic acid derivatives toward nucleophilic acyl substitution can be explained on the basis of the electron-donating properties of substituent X. The greater the electron donating powers of X, the slower the rate.

1. **Acyl chlorides:** Although chlorine has unshared electron pairs, it is a poor electron-pair donor in resonance of the type:

Weak resonance
stabilization

The C—Cl bond is so long that the lone-pair orbital ($3p$) of chlorine is too far away to overlap effectively with the π orbital of the carbonyl group. The carbonyl group of an acyl chloride feels the normal electron-withdrawing effect of an electronegative substituent without much compensating electron donation by resonance. This destabilizes the carbonyl group and makes it more reactive.

2. **Acid anhydrides:** The carbonyl group of an acid anhydride is better stabilized by electron donation than the carbonyl group of an acyl chloride. Even though oxygen

is more electronegative than chlorine, it is a far better electron-pair donor toward sp^2-hybridized carbon.

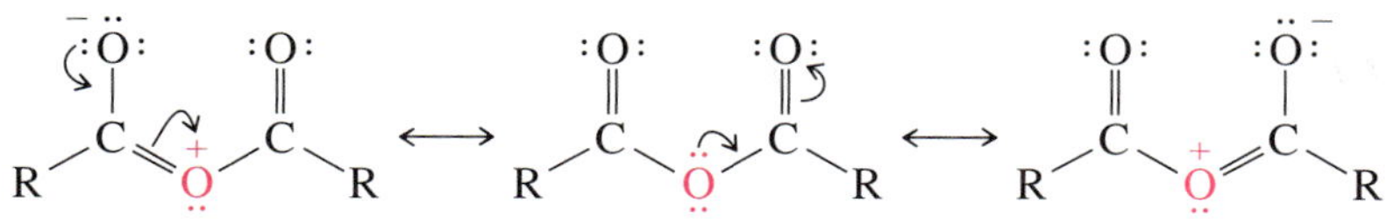

Working against this electron-delocalization is the fact that both carbonyl groups are competing for the same electron pair. Thus, the extent to which each one is stabilized is reduced.

3. **Thioesters:** Like chlorine, sulfur is a third-row element with limited ability to donate a pair of $3p$ electrons into the carbonyl π system. With an electronegativity that is much less than Cl or O, however, its destabilizing effect on the carbonyl group is slight, and thioesters lie in the middle of the group of carboxylic acid derivatives in respect to reactivity.

4. **Esters:** Like acid anhydrides, the carbonyl group of an ester is stabilized by electron release from oxygen. Because there is only one carbonyl group, versus two in anhydrides, esters are stabilized more and are less reactive than anhydrides.

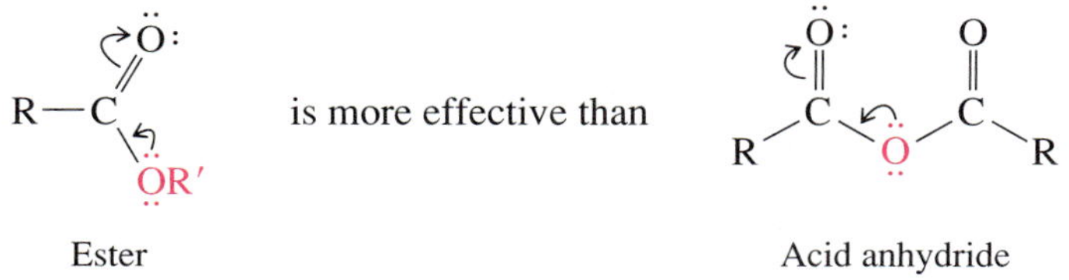

Ester   is more effective than   Acid anhydride

5. **Amides:** Nitrogen is less electronegative than oxygen; therefore, the carbonyl group of an amide is stabilized more than that of an ester.

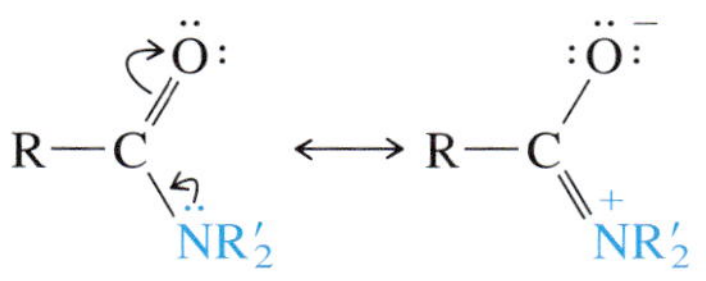

Very effective
resonance stabilization

Amide resonance is a powerful stabilizing force and gives rise to a number of structural effects. Unlike the pyramidal arrangement of bonds in ammonia and amines, the bonds to nitrogen in amides lie in the same plane. The carbon–nitrogen bond has considerable double-bond character and, at 135 pm, is substantially shorter than the normal 147-pm carbon–nitrogen single-bond distance observed in amines.

The barrier to rotation about the carbon–nitrogen bond in amides is 75 to 85 kJ/mol (18–20 kcal/mol).

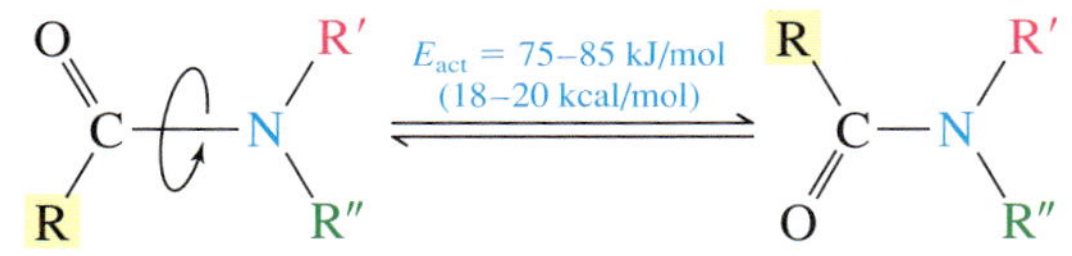

$E_{act} = 75\text{–}85$ kJ/mol
(18–20 kcal/mol)

Recall (Section 3.1) that the rotational barrier in ethane is only 12 kJ/mol (3 kcal/mol).

This is an unusually high rotational energy barrier for a single bond and indicates that the carbon–nitrogen bond has significant double-bond character, as the resonance picture suggests.

> **PROBLEM 20.2** The ^{1}H NMR spectrum of *N,N*-dimethylformamide shows a separate signal for each of the two methyl groups. Can you explain why?

Electron release from nitrogen stabilizes the carbonyl group of amides and decreases the rate at which nucleophiles attack the carbonyl carbon.

An extreme example of carbonyl group stabilization is seen in carboxylate anions:

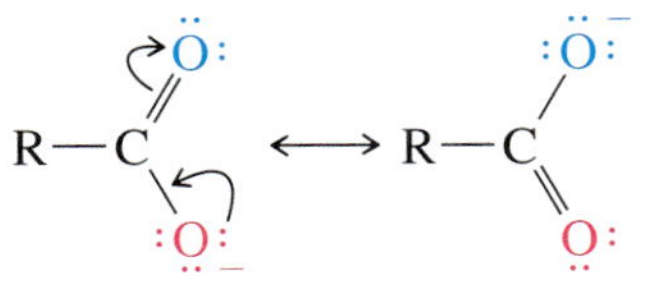

The negatively charged oxygen substituent is a powerful electron donor to the carbonyl group. Resonance in carboxylate anions is more effective than resonance in carboxylic acids, acyl chlorides, anhydrides, thioesters, esters, and amides.

Most methods for their preparation convert one class of carboxylic acid derivative to another, and the order of carbonyl group stabilization given in Figure 20.1 bears directly on the means by which these transformations may be achieved. A reaction that converts one carboxylic acid derivative to another that lies below it in the figure is practical; a reaction that converts it to one that lies above it is not. This is another way of saying that *one carboxylic acid derivative can be converted to another if the reaction leads to a more stabilized carbonyl group*. Numerous examples of reactions of this type will be presented in the sections that follow.

20.3   GENERAL MECHANISM FOR NUCLEOPHILIC ACYL SUBSTITUTION

Nucleophilic acyl substitutions follow a two-stage mechanism and proceed by way of a tetrahedral intermediate.

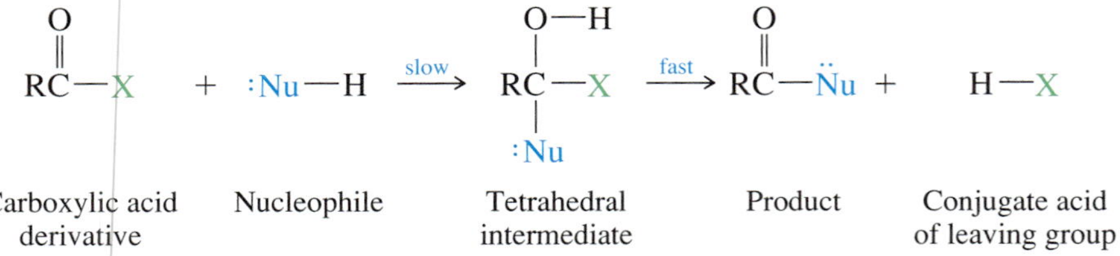

We have seen this theme before in Section 19.14 when we presented the mechanism of the Fischer esterification. As was the case then, formation of the tetrahedral intermediate is rate-determining.

It is important to remember that each stage consists of more than one elementary step. Therefore, a complete mechanism can have many steps and look complicated if we try to absorb all of it at once. If we keep the two stages separate in our minds and build on what we already know, our job becomes easier. Two points are helpful:

1. The first stage of the mechanism is exactly the same as for nucleophilic *addition* to the carbonyl group of an aldehyde or ketone. Many of the same nucleophiles that add to aldehydes and ketones—water (Section 17.6), alcohols (Section 17.8), amines (Sections 17.10–17.11)—add to the carbonyl groups of carboxylic acid derivatives.
2. The features that complicate the mechanism of nucleophilic acyl substitution are almost entirely related to acid–base chemistry. We try to keep track, as best we can, of the form in which the various species—reactants, intermediates, and products—exist under the reaction conditions.

With regard to the second point, we already know a good bit about the acid–base chemistry of the reactants and products; that of the tetrahedral intermediate is less familiar. We can, for example, imagine the following species in equilibrium with the tetrahedral intermediate (TI).

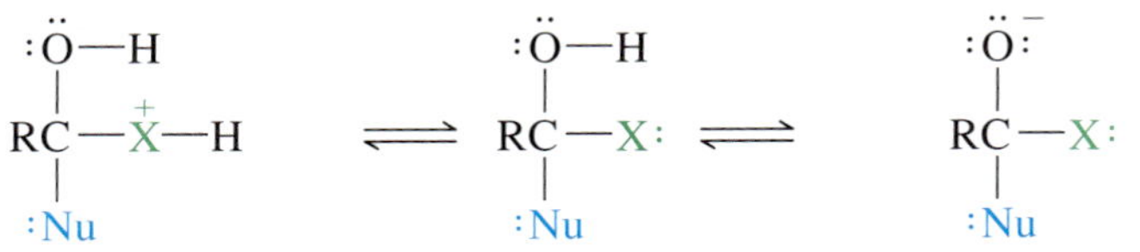

TI—H⁺ Conjugate acid of tetrahedral intermediate | TI Tetrahedral intermediate | TI⁻ Conjugate base of tetrahedral intermediate

Each one of these can proceed to the product of nucleophilic acyl substitution.

Dissociation of TI—H⁺:

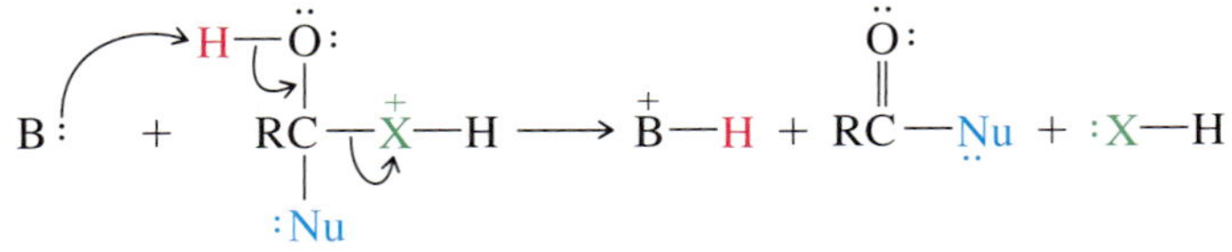

Brønsted base | TI—H⁺ | Product

Dissociation of TI:

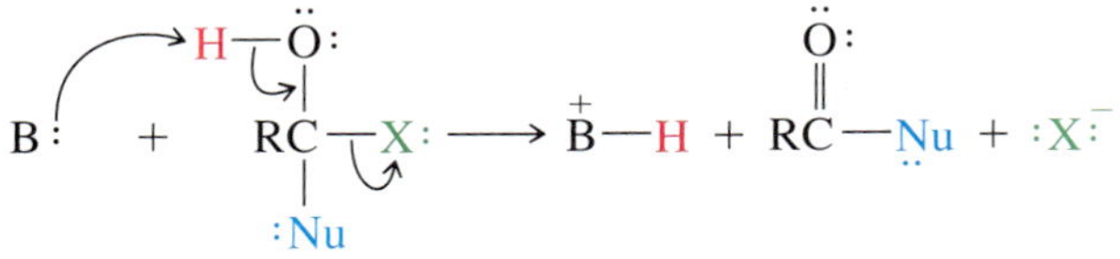

Brønsted base | TI | Product

Dissociation of TI⁻:

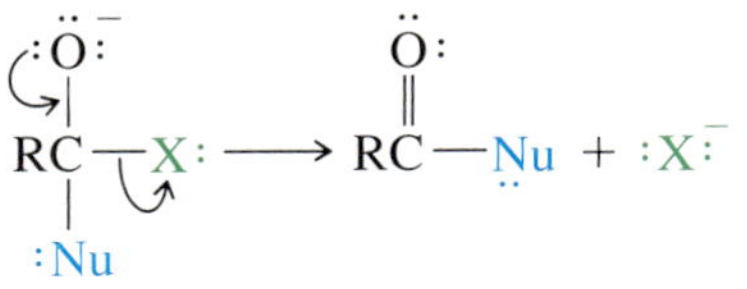

TI⁻ | Product

More than one form of the tetrahedral intermediate can be present at a particular pH, and the most abundant form need not be the one that gives most of the product. A less abundant form may react at a faster rate than a more abundant one.

Mechanisms for a number of nucleophilic acyl substitutions will appear in the sections that follow. It is better to look for the important ways in which they are similar than to search for details in which they differ.

20.4 NUCLEOPHILIC SUBSTITUTION IN ACYL CHLORIDES

> One of the most useful reactions of acyl chlorides was presented in Section 12.7. Friedel–Crafts acylation of aromatic rings takes place when arenes are treated with acyl chlorides in the presence of aluminum chloride.

Acyl chlorides are readily prepared from carboxylic acids by reaction with thionyl chloride (Section 12.7).

$$\underset{\substack{\text{Carboxylic}\\\text{acid}}}{\text{RCOH}} + \underset{\substack{\text{Thionyl}\\\text{chloride}}}{\text{SOCl}_2} \longrightarrow \underset{\substack{\text{Acyl}\\\text{chloride}}}{\text{RCCl}} + \underset{\substack{\text{Sulfur}\\\text{dioxide}}}{\text{SO}_2} + \underset{\substack{\text{Hydrogen}\\\text{chloride}}}{\text{HCl}}$$

$$\underset{\text{2-Methylpropanoic acid}}{(CH_3)_2CHCOH} \xrightarrow[\text{heat}]{SOCl_2} \underset{\text{2-Methylpropanoyl chloride (90\%)}}{(CH_3)_2CHCCl}$$

On treatment with the appropriate nucleophile, an acyl chloride may be converted to an acid anhydride, an ester, an amide, or a carboxylic acid. Examples are presented in Table 20.1.

> **PROBLEM 20.3** Apply the knowledge gained by studying Table 20.1 to help you predict the major organic product obtained by reaction of benzoyl chloride with each of the following:
>
> (a) Acetic acid
> (b) Benzoic acid
> (c) Ethanol
> (d) Methylamine, CH_3NH_2
> (e) Dimethylamine, $(CH_3)_2NH$
> (f) Water
>
> **SAMPLE SOLUTION** (a) As noted in Table 20.1, the reaction of an acyl chloride with a carboxylic acid yields an acid anhydride.
>
> $$\underset{\text{Benzoyl chloride}}{C_6H_5CCl} + \underset{\text{Acetic acid}}{CH_3COH} \longrightarrow \underset{\text{Acetic benzoic anhydride}}{C_6H_5COCCH_3}$$
>
> The product is a mixed anhydride. Acetic acid acts as a nucleophile and substitutes for chloride on the benzoyl group.

The mechanisms of all the reactions cited in Table 20.1 are similar to the mechanism of hydrolysis of an acyl chloride outlined in Figure 20.2. They differ with respect to the nucleophile that attacks the carbonyl group.

In the first stage of the hydrolysis mechanism, water undergoes nucleophilic addition to the carbonyl group to form a tetrahedral intermediate. This stage of the process is analogous to the hydration of aldehydes and ketones discussed in Section 17.6.

TABLE 20.1   Conversion of Acyl Chlorides to Other Carboxylic Acid Derivatives

Reaction (section) and comments	General equation and specific example
Reaction with carboxylic acids (Section 20.5) Acyl chlorides react with carboxylic acids to yield acid anhydrides. When this reaction is used for preparative purposes, a weak organic base such as pyridine is normally added. Pyridine is a catalyst for the reaction and also acts as a base to neutralize the hydrogen chloride that is formed.	$$\underset{\text{Acyl chloride}}{RCCl} + \underset{\text{Carboxylic acid}}{R'COH} \longrightarrow \underset{\text{Acid anhydride}}{RCOCR'} + \underset{\text{Hydrogen chloride}}{HCl}$$ $$\underset{\substack{\text{Heptanoyl}\\\text{chloride}}}{CH_3(CH_2)_5CCl} + \underset{\substack{\text{Heptanoic}\\\text{acid}}}{CH_3(CH_2)_5COH} \xrightarrow{\text{pyridine}} \underset{\substack{\text{Heptanoic anhydride}\\(78-83\%)}}{CH_3(CH_2)_5COC(CH_2)_5CH_3}$$
Reaction with alcohols (Section 15.8) Acyl chlorides react with alcohols to form esters. The reaction is typically carried out in the presence of pyridine.	$$\underset{\substack{\text{Acyl}\\\text{chloride}}}{RCCl} + \underset{\text{Alcohol}}{R'OH} \longrightarrow \underset{\text{Ester}}{RCOR'} + \underset{\substack{\text{Hydrogen}\\\text{chloride}}}{HCl}$$ $$\underset{\substack{\text{Benzoyl}\\\text{chloride}}}{C_6H_5CCl} + \underset{\substack{tert\text{-Butyl}\\\text{alcohol}}}{(CH_3)_3COH} \xrightarrow{\text{pyridine}} \underset{\substack{tert\text{-Butyl}\\\text{benzoate (80\%)}}}{C_6H_5COC(CH_3)_3}$$
Reaction with ammonia and amines (Section 20.14) Acyl chlorides react with ammonia and amines to form amides. A base such as sodium hydroxide is normally added to react with the hydrogen chloride produced.	$$\underset{\substack{\text{Acyl}\\\text{chloride}}}{RCCl} + \underset{\substack{\text{Ammonia}\\\text{or amine}}}{R_2'NH} + \underset{\text{Hydroxide}}{HO^-} \longrightarrow \underset{\text{Amide}}{RCNR_2'} + \underset{\text{Water}}{H_2O} + \underset{\substack{\text{Chloride}\\\text{ion}}}{Cl^-}$$ $$\underset{\substack{\text{Benzoyl}\\\text{chloride}}}{C_6H_5CCl} + \underset{\text{Piperidine}}{HN\bigcirc} \xrightarrow[\text{H}_2\text{O}]{\text{NaOH}} \underset{\substack{N\text{-Benzoylpiperidine}\\(87-91\%)}}{C_6H_5C-N\bigcirc}$$
Hydrolysis (Section 20.4) Acyl chlorides react with water to yield carboxylic acids. In base, the acid is converted to its carboxylate salt. The reaction has little preparative value because the acyl chloride is nearly always prepared from the carboxylic acid rather than vice versa.	$$\underset{\substack{\text{Acyl}\\\text{chloride}}}{RCCl} + \underset{\text{Water}}{H_2O} \longrightarrow \underset{\substack{\text{Carboxylic}\\\text{acid}}}{RCOH} + \underset{\substack{\text{Hydrogen}\\\text{chloride}}}{HCl}$$ $$\underset{\substack{\text{Phenylacetyl}\\\text{chloride}}}{C_6H_5CH_2CCl} + \underset{\text{Water}}{H_2O} \longrightarrow \underset{\substack{\text{Phenylacetic}\\\text{acid}}}{C_6H_5CH_2COH} + \underset{\substack{\text{Hydrogen}\\\text{chloride}}}{HCl}$$

The tetrahedral intermediate has three potential leaving groups on carbon: two hydroxyl groups and a chlorine. In the second stage of the reaction, the tetrahedral intermediate dissociates, restoring the resonance-stabilized carbonyl group. Loss of chloride from the tetrahedral intermediate is faster than loss of hydroxide; chloride is less basic than hydroxide and is a better leaving group.

First stage: Formation of the tetrahedral intermediate by nucleophilic addition of water to the carbonyl group

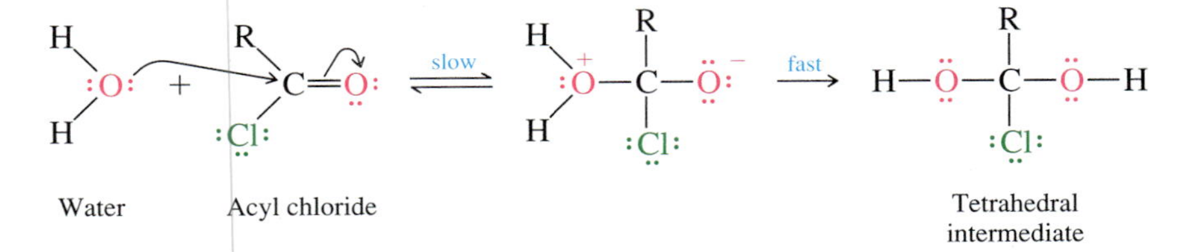

Water   Acyl chloride   Tetrahedral intermediate

Second stage: Dissociation of the tetrahedral intermediate by dehydrohalogenation

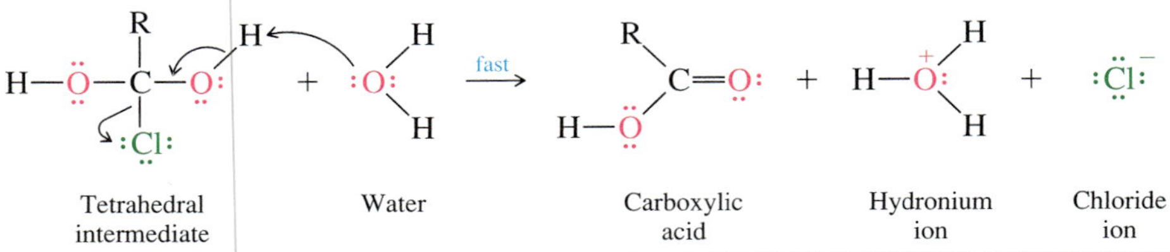

Tetrahedral   Water   Carboxylic   Hydronium   Chloride
intermediate   acid   ion   ion

FIGURE 20.2 Hydrolysis of an acyl chloride proceeds by way of a tetrahedral intermediate. Formation of the tetrahedral intermediate is rate-determining.

PROBLEM 20.4 Write the structure of the tetrahedral intermediate formed in each of the reactions given in Problem 20.3. Using curved arrows, show how each tetrahedral intermediate dissociates to the appropriate products. Use the symbol B: to stand for a Brønsted base.

SAMPLE SOLUTION (a) The tetrahedral intermediate arises by nucleophilic addition of acetic acid to benzoyl chloride.

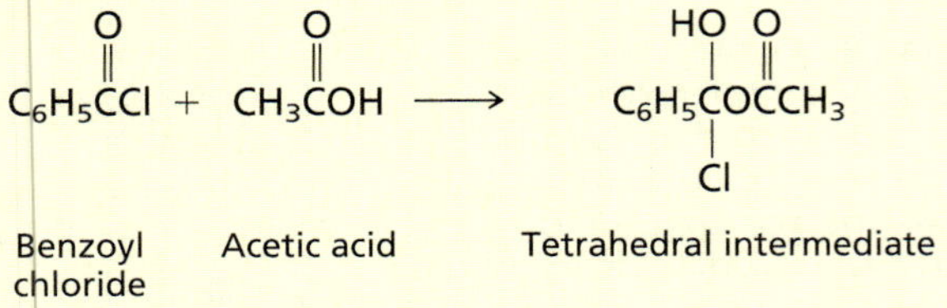

Benzoyl   Acetic acid   Tetrahedral intermediate
chloride

Loss of a proton and of chloride ion from the tetrahedral intermediate yields the mixed anhydride.

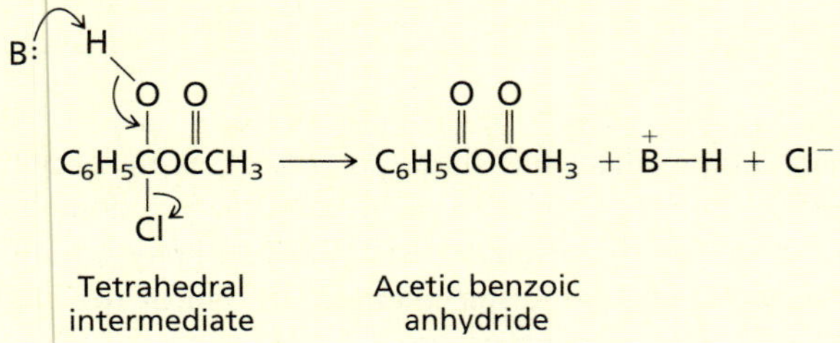

Tetrahedral   Acetic benzoic
intermediate   anhydride

Nucleophilic substitution in *acyl* chlorides is much faster than in *alkyl* chlorides.

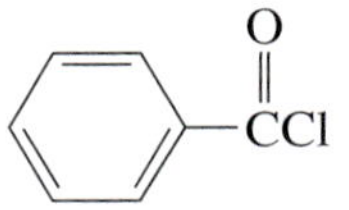

	Benzoyl chloride	Benzyl chloride
Relative rate of hydrolysis (80% ethanol–20% water; 25°C)	1000	1

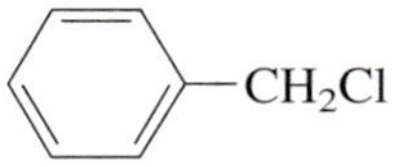

The sp^2-hybridized carbon of an acyl chloride is less sterically hindered than the sp^3-hybridized carbon of an alkyl chloride, making an acyl chloride more open toward nucleophilic attack. Also, unlike the S_N2 transition state or a carbocation intermediate in an S_N1 reaction, the tetrahedral intermediate in nucleophilic acyl substitution has a stable arrangement of bonds and can be formed via a lower energy transition state.

20.5 PREPARATION OF CARBOXYLIC ACID ANHYDRIDES

After acyl halides, acid anhydrides are the most reactive carboxylic acid derivatives. Three of them, acetic anhydride, phthalic anhydride, and maleic anhydride, are industrial chemicals and are encountered far more often than others. Phthalic anhydride and maleic anhydride have their anhydride function incorporated into a ring and are referred to as *cyclic anhydrides.*

Acetic anhydride	Phthalic anhydride	Maleic anhydride

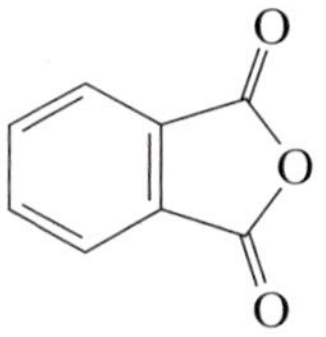

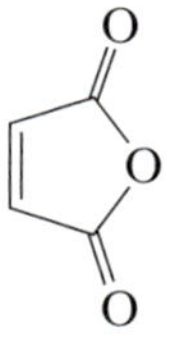

The customary method for the laboratory synthesis of acid anhydrides is the reaction of acyl chlorides with carboxylic acids (Table 20.1).

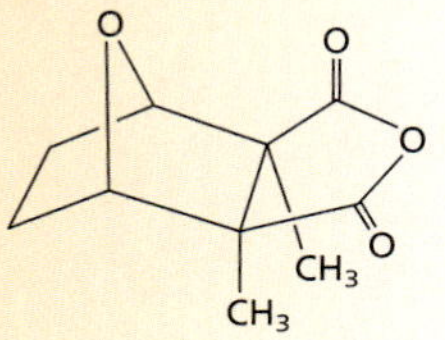

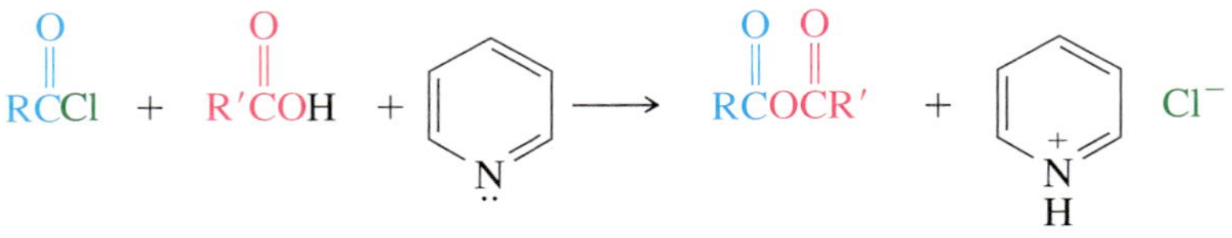

Acyl chloride	Carboxylic acid	Pyridine	Carboxylic acid anhydride	Pyridinium chloride

Acid anhydrides rarely occur naturally. One example is the putative aphrodisiac *cantharidin*, obtained from a species of beetle.

This procedure is applicable to the preparation of both symmetrical anhydrides (R and R′ the same) and mixed anhydrides (R and R′ different).

PROBLEM 20.5 Benzoic anhydride has been prepared in excellent yield by adding one molar equivalent of water to two molar equivalents of benzoyl chloride. How do you suppose this reaction takes place?

Cyclic anhydrides in which the ring is five- or six-membered are sometimes prepared by heating the corresponding dicarboxylic acids in an inert solvent:

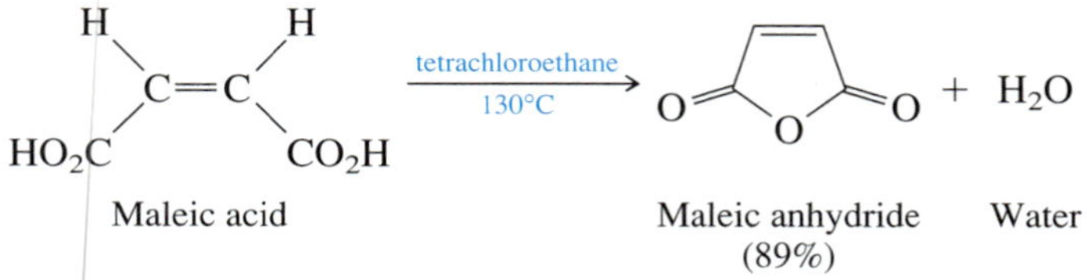

Maleic acid   Maleic anhydride   Water
                        (89%)

20.6   REACTIONS OF CARBOXYLIC ACID ANHYDRIDES

Nucleophilic acyl substitution in acid anhydrides involves cleavage of a bond between oxygen and one of the carbonyl groups. One acyl group is transferred to an attacking nucleophile; the other retains its single bond to oxygen and becomes the acyl group of a carboxylic acid.

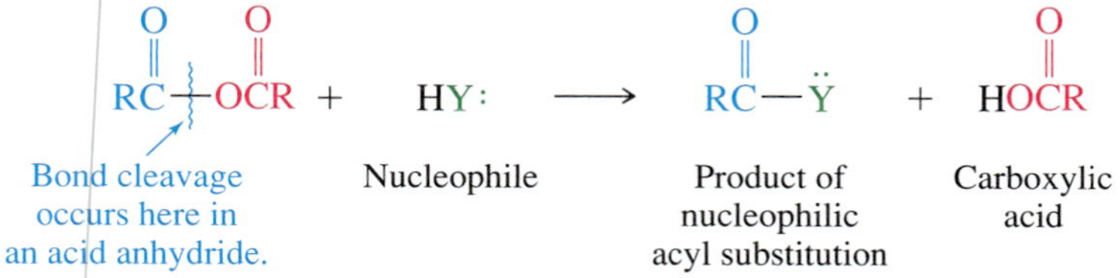

Bond cleavage   Nucleophile   Product of   Carboxylic
occurs here in   nucleophilic   acid
an acid anhydride.   acyl substitution

One reaction of this type, Friedel–Crafts acylation (Section 12.7), is already familiar to us.

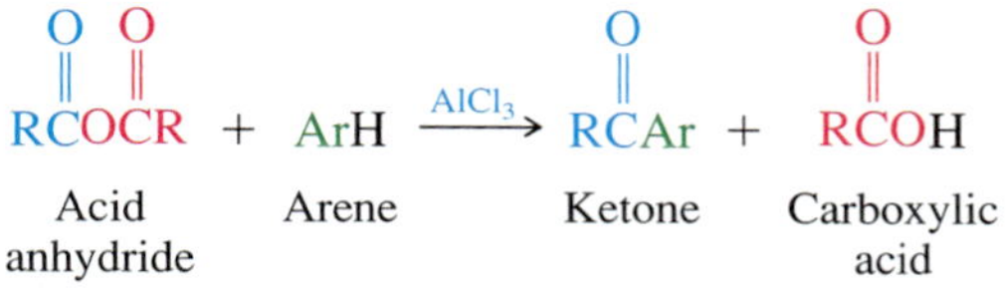

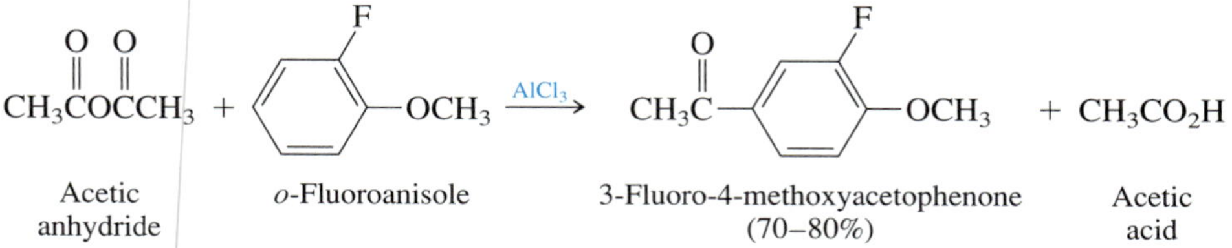

Acid   Arene   Ketone   Carboxylic
anhydride   acid

Acetic   o-Fluoroanisole   3-Fluoro-4-methoxyacetophenone   Acetic
anhydride   (70–80%)   acid

An acyl cation is an intermediate in Friedel–Crafts acylation reactions.

> **PROBLEM 20.6** Write a structural formula for the acyl cation intermediate in the preceding reaction.

Conversions of acid anhydrides to other carboxylic acid derivatives are illustrated in Table 20.2. Because a more highly stabilized carbonyl group must result in order for nucleophilic acyl substitution to be effective, acid anhydrides are readily converted to carboxylic acids, esters, and amides but not to acyl chlorides.

> **PROBLEM 20.7** Apply the knowledge gained by studying Table 20.2 to help you predict the major organic product of each of the following reactions:
>
> (a) Benzoic anhydride + methanol $\xrightarrow{\text{H}_2\text{SO}_4}$
>
> (b) Acetic anhydride + ammonia (2 mol) $\longrightarrow$
>
> (c) Phthalic anhydride + $(\text{CH}_3)_2\text{NH}$ (2 mol) $\longrightarrow$
>
> (d) Phthalic anhydride + sodium hydroxide (2 mol) $\longrightarrow$

SAMPLE SOLUTION (a) Nucleophilic acyl substitution by an alcohol on an acid anhydride yields an ester.

$$\underset{\substack{\text{Benzoic} \\ \text{anhydride}}}{C_6H_5\overset{O}{\overset{\|}{C}}O\overset{O}{\overset{\|}{C}}C_6H_5} + \underset{\text{Methanol}}{CH_3OH} \xrightarrow{H_2SO_4} \underset{\text{Methyl benzoate}}{C_6H_5\overset{O}{\overset{\|}{C}}OCH_3} + \underset{\text{Benzoic acid}}{C_6H_5\overset{O}{\overset{\|}{C}}OH}$$

TABLE 20.2   Conversion of Acid Anhydrides to Other Carboxylic Acid Derivatives

Reaction (section) and comments	General equation and specific example
Reaction with alcohols (Section 15.8) Acid anhydrides react with alcohols to form esters. The reaction may be carried out in the presence of pyridine or it may be catalyzed by acids. In the example shown, only one acyl group of acetic anhydride becomes incorporated into the ester; the other becomes the acyl group of an acetic acid molecule.	$\underset{\substack{\text{Acid} \\ \text{anhydride}}}{R\overset{O}{\overset{\|}{C}}O\overset{O}{\overset{\|}{C}}R} + \underset{\text{Alcohol}}{R'OH} \longrightarrow \underset{\text{Ester}}{R\overset{O}{\overset{\|}{C}}OR'} + \underset{\substack{\text{Carboxylic} \\ \text{acid}}}{R\overset{O}{\overset{\|}{C}}OH}$ $\underset{\substack{\text{Acetic} \\ \text{anhydride}}}{CH_3\overset{O}{\overset{\|}{C}}O\overset{O}{\overset{\|}{C}}CH_3} + \underset{\substack{\text{sec-Butyl} \\ \text{alcohol}}}{HOCHCH_2CH_3} \xrightarrow{H_2SO_4} \underset{\substack{\text{sec-Butyl} \\ \text{acetate (60\%)}}}{CH_3\overset{O}{\overset{\|}{C}}OCHCH_2CH_3}$ (with CH_3 substituents)
Reaction with ammonia and amines (Section 20.14) Acid anhydrides react with ammonia and amines to form amides. Two molar equivalents of amine are required. In the example shown, only one acyl group of acetic anhydride becomes incorporated into the amide; the other becomes the acyl group of the amine salt of acetic acid.	$\underset{\substack{\text{Acid} \\ \text{anhydride}}}{R\overset{O}{\overset{\|}{C}}O\overset{O}{\overset{\|}{C}}R} + \underset{\text{Amine}}{2R'_2NH} \longrightarrow \underset{\text{Amide}}{R\overset{O}{\overset{\|}{C}}NR'_2} + \underset{\substack{\text{Ammonium} \\ \text{carboxylate} \\ \text{salt}}}{R\overset{O}{\overset{\|}{C}}O^- \; H_2\overset{+}{N}R'_2}$ $\underset{\substack{\text{Acetic} \\ \text{anhydride}}}{CH_3\overset{O}{\overset{\|}{C}}O\overset{O}{\overset{\|}{C}}CH_3} + \underset{\substack{\text{p-Isopropylaniline}}}{H_2N{-}\langle\text{aryl}\rangle{-}CH(CH_3)_2} \longrightarrow \underset{\substack{\text{p-Isopropylacetanilide} \\ (98\%)}}{CH_3\overset{O}{\overset{\|}{C}}NH{-}\langle\text{aryl}\rangle{-}CH(CH_3)_2}$
Hydrolysis (Section 20.6) Acid anhydrides react with water to yield two carboxylic acid functions. Cyclic anhydrides yield dicarboxylic acids.	$\underset{\substack{\text{Acid} \\ \text{anhydride}}}{R\overset{O}{\overset{\|}{C}}O\overset{O}{\overset{\|}{C}}R'} + \underset{\text{Water}}{H_2O} \longrightarrow \underset{\substack{\text{Carboxylic} \\ \text{acid}}}{2R\overset{O}{\overset{\|}{C}}OH}$ Phthalic anhydride + Water $\longrightarrow$ Phthalic acid

The first example in Table 20.2 introduces a new aspect of nucleophilic acyl substitution that applies not only to acid anhydrides but also to acyl chlorides, thioesters, esters, and amides. Nucleophilic acyl substitutions can be catalyzed by acids.

We can see how an acid catalyst increases the rate of nucleophilic acyl substitution by considering the hydrolysis of an acid anhydride (Figure 20.3). Formation of the tetrahedral intermediate is rate-determining and is the stage that is accelerated by the catalyst. The acid anhydride is activated toward nucleophilic addition by protonation of one of its carbonyl groups in step 1 of Figure 20.3. The protonated form of the anhydride is present to only a very small extent, but it is quite electrophilic. Water and other nucleophiles add to a protonated carbonyl group (step 2) much faster than they add to a neutral one. Thus, the rate-determining nucleophilic addition of water to form a tetrahedral intermediate takes place more rapidly in the presence of an acid than in its absence.

This pattern of increased reactivity resulting from carbonyl group protonation has been seen before in nucleophilic additions to aldehydes and ketones (Section 17.6) and

---

Step 1: The acid catalyst activates the anhydride toward nucleophilic addition by protonation of the carbonyl oxygen.

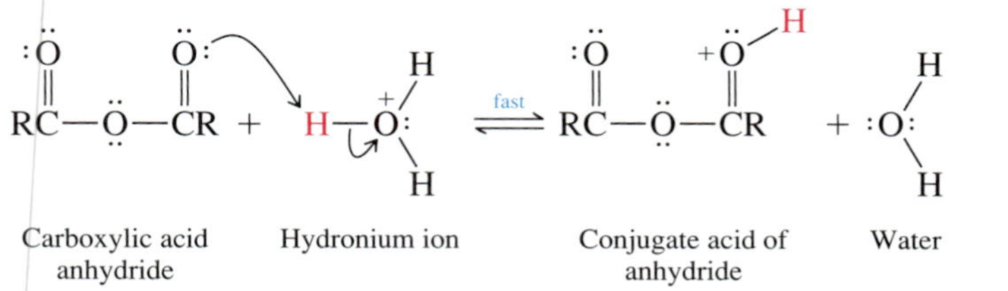

| Carboxylic acid anhydride | Hydronium ion | Conjugate acid of anhydride | Water |

Step 2: The nucleophile, in this case a water molecule, adds to the carbonyl group. This is the rate-determining step.

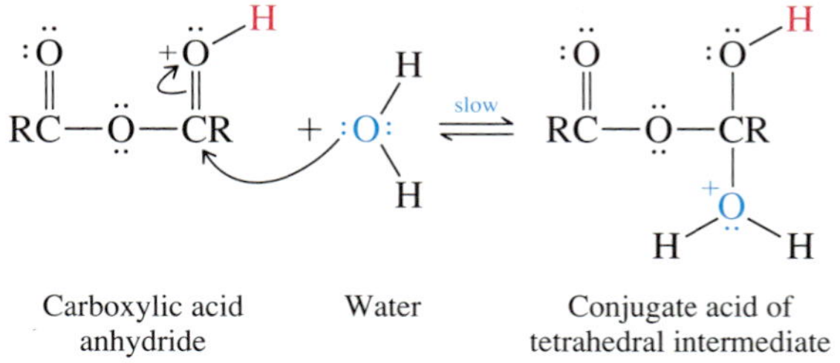

| Carboxylic acid anhydride | Water | Conjugate acid of tetrahedral intermediate |

Step 3: The product of step 2 is the conjugate acid of the tetrahedral intermediate. It transfers a proton to water, giving the neutral form of the tetrahedral intermediate and regenerating the acid catalyst.

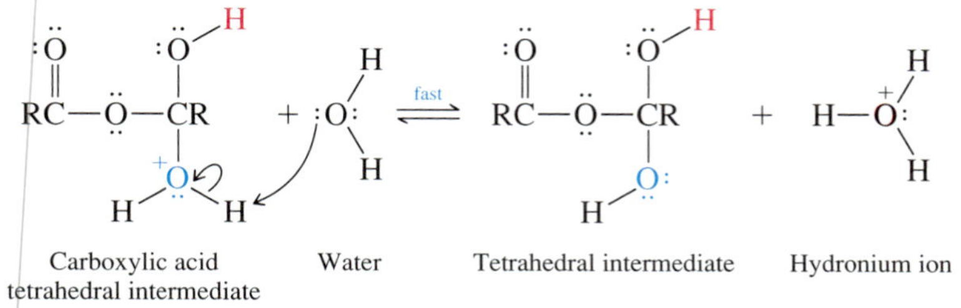

| Carboxylic acid tetrahedral intermediate | Water | Tetrahedral intermediate | Hydronium ion |

FIGURE 20.3 An acid catalyzes the hydrolysis of a carboxylic acid anhydride by increasing the rate of the first stage of the mechanism. The faster the tetrahedral intermediate is formed, the faster the rate of hydrolysis.

in the mechanism of the acid-catalyzed esterification of carboxylic acids (Section 19.14). Many biological reactions involve nucleophilic acyl substitution; some are catalyzed by enzymes that act by donating a proton to the carbonyl oxygen; others involve coordination of the carbonyl oxygen with a metal cation.

PROBLEM 20.8   Write the structure of the tetrahedral intermediate formed in each of the reactions given in Problem 20.7.

SAMPLE SOLUTION   (a) The reaction given is the acid-catalyzed esterification of methanol by benzoic anhydride. The tetrahedral intermediate is formed by addition of a molecule of methanol to one of the carbonyl groups of the anhydride. This reaction is analogous to the acid-catalyzed formation of a hemiacetal by reaction of methanol with an aldehyde or ketone.

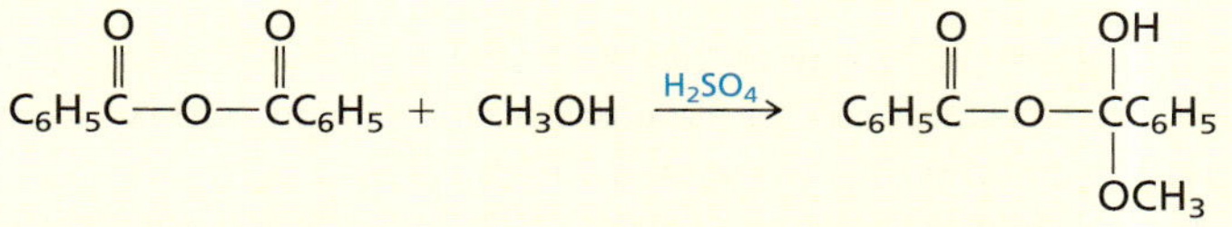

| Benzoic anhydride | Methanol | | Tetrahedral intermediate |

Acid anhydrides are more stable and less reactive than acyl chlorides. Acetyl chloride, for example, undergoes hydrolysis about 100,000 times more rapidly than acetic anhydride at 25°C.

20.7   SOURCES OF ESTERS

Many esters occur naturally. Those of low molecular weight are fairly volatile, and many have pleasing odors. Esters often form a significant fraction of the fragrant oil of fruits and flowers. The aroma of oranges, for example, contains 30 different esters along with 10 carboxylic acids, 34 alcohols, 34 aldehydes and ketones, and 36 hydrocarbons.

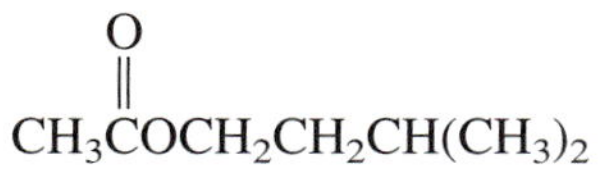

3-Methylbutyl acetate
(contributes to characteristic
odor of bananas)

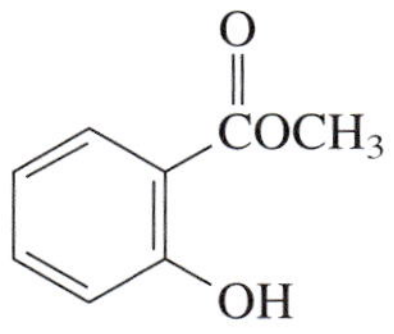

Methyl salicylate
(principal component of oil
of wintergreen)

3-Methylbutyl acetate is more commonly known as isoamyl acetate.

Among the chemicals used by insects to communicate with one another, esters occur frequently.

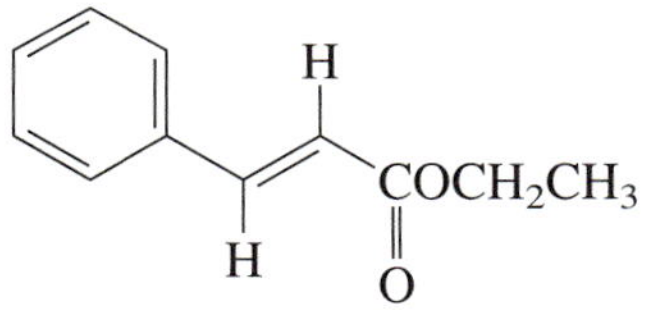

Ethyl cinnamate
(one of the constituents of
the sex pheromone of the
male oriental fruit moth)

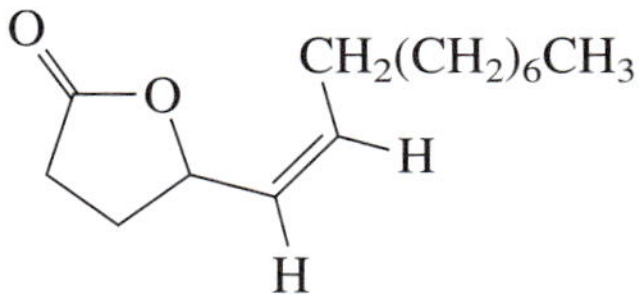

(Z)-5-Tetradecen-4-olide
(sex pheromone of female
Japanese beetle)

Notice that (Z)-5-tetradecen-4-olide is a cyclic ester. Recall from Section 19.15 that cyclic esters are called *lactones* and that the suffix -olide is characteristic of IUPAC names for lactones.

Esters of glycerol, called *glycerol triesters, triacylglycerols,* or *triglycerides,* are abundant natural products. The most important group of glycerol triesters includes those in which each acyl group is unbranched and has 14 or more carbon atoms.

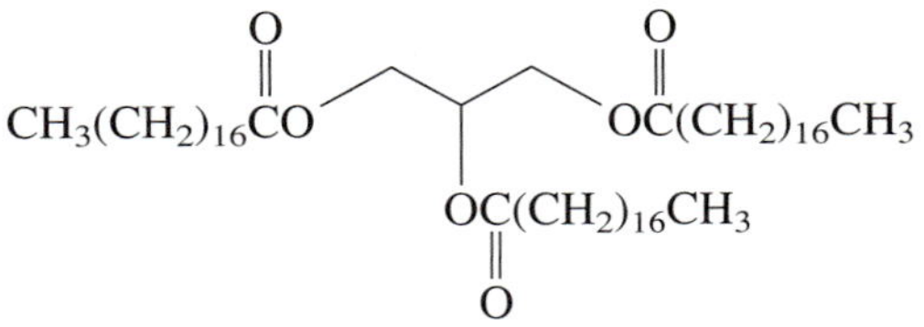

$$CH_3(CH_2)_{16}\overset{\overset{\displaystyle O}{\|}}{C}O \diagdown \diagup OC(CH_2)_{16}CH_3$$

$$OC(CH_2)_{16}CH_3$$

Tristearin, a trioctadecanoyl ester
of glycerol found in many animal and
vegetable fats

A molecular model of
tristearin is shown in
Figure 26.2.

Fats and **oils** are naturally occurring mixtures of glycerol triesters. Fats are mixtures that are solids at room temperature; oils are liquids. The long-chain carboxylic acids obtained from fats and oils by hydrolysis are known as **fatty acids.**

The chief methods used to prepare esters in the laboratory were all described earlier and are summarized in Table 20.3.

20.8 PHYSICAL PROPERTIES OF ESTERS

Esters are moderately polar, with dipole moments in the 1.5 to 2.0-D range. Dipole–dipole attractive forces give esters higher boiling points than hydrocarbons of similar shape and molecular weight. Because they lack hydroxyl groups, however, ester molecules cannot form hydrogen bonds to each other; consequently, esters have lower boiling points than alcohols of comparable molecular weight.

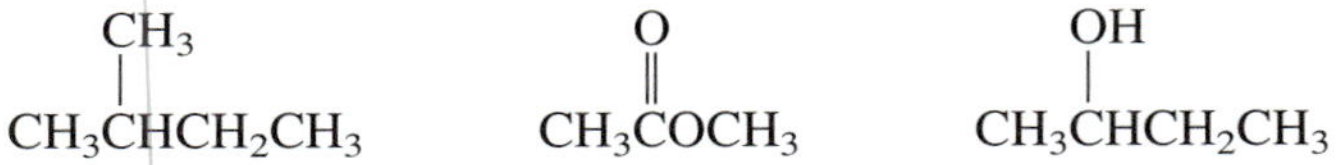

$$\underset{\underset{CH_3CHCH_2CH_3}{|}}{CH_3} \qquad \underset{CH_3COCH_3}{\overset{\overset{\displaystyle O}{\|}}{}} \qquad \underset{\underset{CH_3CHCH_2CH_3}{|}}{OH}$$

| 2-Methylbutane: | Methyl acetate: | 2-Butanol: |
| mol wt 72, bp 28°C | mol wt 74, bp 57°C | mol wt 74, bp 99°C |

Esters can participate in hydrogen bonds with substances that contain hydroxyl groups (water, alcohols, carboxylic acids). This confers some measure of water solubility on low-molecular-weight esters; methyl acetate, for example, dissolves in water to the extent of 33 g/100 mL. Water solubility decreases as the carbon content of the ester increases. Fats and oils, the glycerol esters of long-chain carboxylic acids, are practically insoluble in water.

20.9 REACTIONS OF ESTERS: A REVIEW AND A PREVIEW

The reaction of esters with Grignard reagents and with lithium aluminum hydride, both useful in the synthesis of alcohols, were described earlier. They are reviewed in Table 20.4 on page 848.

Nucleophilic acyl substitutions at the ester carbonyl group are summarized in Table 20.5 on page 849. Esters are less reactive than acyl chlorides and acid anhydrides. Nucleophilic acyl substitution in esters, especially ester hydrolysis, has been extensively investigated from a mechanistic perspective. Indeed, much of what we know concerning the general topic of nucleophilic acyl substitution comes from studies carried out on esters. The following sections describe those mechanistic studies.

TABLE 20.3 Preparation of Esters

Reaction (section) and comments	General equation and specific example
From carboxylic acids (Sections 15.8 and 19.14) In the presence of an acid catalyst, alcohols and carboxylic acids react to form an ester and water. This is the Fischer esterification.	
From acyl chlorides (Sections 15.8 and 20.4) Alcohols react with acyl chlorides by nucleophilic acyl substitution to yield esters. These reactions are typically performed in the presence of a weak base such as pyridine.	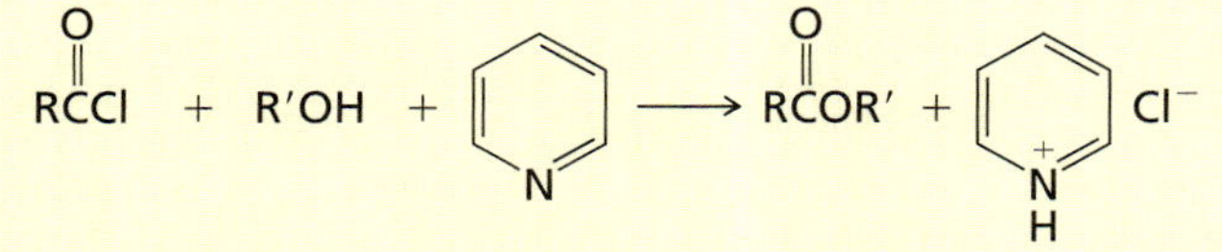 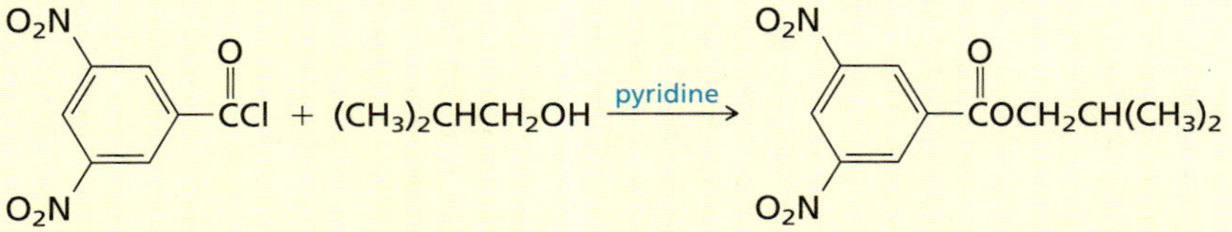
From carboxylic acid anhydrides (Sections 15.8 and 20.6) Acyl transfer from an acid anhydride to an alcohol is a standard method for the preparation of esters. The reaction is subject to catalysis by either acids (H_2SO_4) or bases (pyridine).	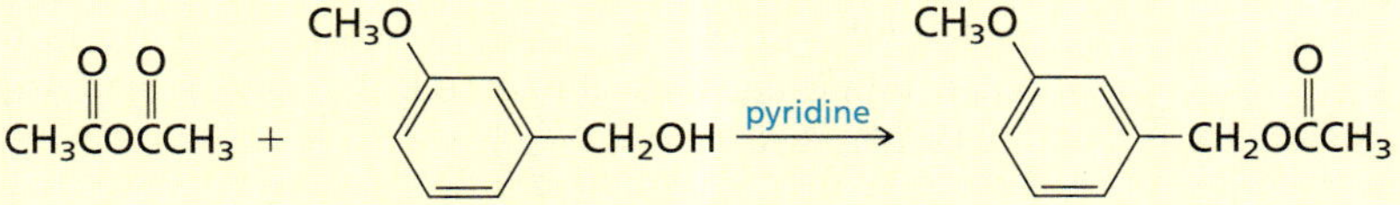
Baeyer–Villiger oxidation of ketones (Section 17.16) Ketones are converted to esters on treatment with peroxy acids. The reaction proceeds by migration of the group R' from carbon to oxygen. It is the more highly substituted group that migrates. Methyl ketones give acetate esters.	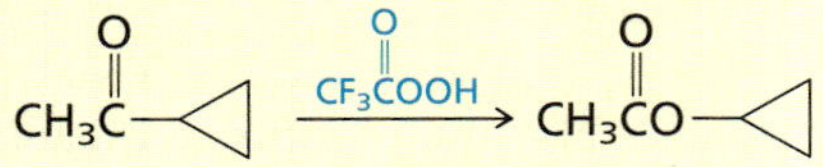

From carboxylic acids:

$$RCOH + R'OH \xrightleftharpoons{H_2SO_4} RCOR' + H_2O$$

Carboxylic acid · Alcohol · Ester · Water

$$CH_3CH_2COH + CH_3CH_2CH_2CH_2OH \xrightarrow{H_2SO_4} CH_3CH_2COCH_2CH_2CH_2CH_3 + H_2O$$

Propanoic acid · 1-Butanol · Butyl propanoate (85%) · Water

From acyl chlorides:

$$RCCl + R'OH + \text{Pyridine} \longrightarrow RCOR' + \text{Pyridinium chloride} + Cl^-$$

Acyl chloride · Alcohol · Pyridine · Ester · Pyridinium chloride

3,5-Dinitrobenzoyl chloride + Isobutyl alcohol $(CH_3)_2CHCH_2OH$ $\xrightarrow{pyridine}$ Isobutyl 3,5-dinitrobenzoate (85%)

From carboxylic acid anhydrides:

$$RCOCR + R'OH \longrightarrow RCOR' + RCOH$$

Acid anhydride · Alcohol · Ester · Carboxylic acid

$$CH_3COCCH_3 + \text{m-Methoxybenzyl alcohol} \xrightarrow{pyridine} \text{m-Methoxybenzyl acetate (99%)}$$

Acetic anhydride

Baeyer–Villiger oxidation of ketones:

$$RCR' + R''COOH \longrightarrow RCOR' + R''COH$$

Ketone · Peroxy acid · Ester · Carboxylic acid

$$CH_3C\text{—cyclopropyl} \xrightarrow{CF_3COOH} CH_3CO\text{—cyclopropyl}$$

Cyclopropyl methyl ketone · Cyclopropyl acetate (53%)

TABLE 20.4	Summary of Reactions of Esters Discussed in Earlier Chapters

Reaction (section) and comments	General equation and specific example
Reaction with Grignard reagents (Section 14.10) Esters react with two equivalents of a Grignard reagent to produce tertiary alcohols. Two of the groups bonded to the carbon that bears the hydroxyl group in the tertiary alcohol are derived from the Grignard reagent.	$$\underset{\text{Ester}}{\text{RCOR}'} + \underset{\substack{\text{Grignard} \\ \text{reagent}}}{2\text{R}''\text{MgX}} \xrightarrow[\text{2. H}_3\text{O}^+]{\text{1. diethyl ether}} \underset{\substack{\text{Tertiary} \\ \text{alcohol}}}{\text{RCR}''} + \underset{\text{Alcohol}}{\text{R}'\text{OH}}$$ where the central product bears OH and R'' substituents. $$\underset{\substack{\text{Ethyl} \\ \text{cyclopropanecarboxylate}}}{\triangleright\!-\!\text{COCH}_2\text{CH}_3} + \underset{\substack{\text{Methylmagnesium} \\ \text{iodide}}}{2\text{CH}_3\text{MgI}} \xrightarrow[\text{2. H}_3\text{O}^+]{\text{1. diethyl ether}} \underset{\substack{\text{2-Cyclopropyl-2-} \\ \text{propanol (93\%)}}}{\triangleright\!-\!\overset{\text{OH}}{\underset{\text{CH}_3}{\text{CCH}_3}}} + \underset{\text{Ethanol}}{\text{CH}_3\text{CH}_2\text{OH}}$$
Reduction with lithium aluminum hydride (Section 15.3) Lithium aluminum hydride cleaves esters to yield two alcohols.	$$\underset{\text{Ester}}{\text{RCOR}'} \xrightarrow[\text{2. H}_2\text{O}]{\text{1. LiAlH}_4} \underset{\substack{\text{Primary} \\ \text{alcohol}}}{\text{RCH}_2\text{OH}} + \underset{\text{Alcohol}}{\text{R}'\text{OH}}$$ $$\underset{\text{Ethyl benzoate}}{\text{C}_6\text{H}_5\text{COCH}_2\text{CH}_3} \xrightarrow[\text{2. H}_2\text{O}]{\text{1. LiAlH}_4} \underset{\substack{\text{Benzyl} \\ \text{alcohol (90\%)}}}{\text{C}_6\text{H}_5\text{CH}_2\text{OH}} + \underset{\substack{\text{Ethyl} \\ \text{alcohol}}}{\text{CH}_3\text{CH}_2\text{OH}}$$

20.10 ACID-CATALYZED ESTER HYDROLYSIS

Ester hydrolysis is the most studied and best understood of all nucleophilic acyl substitutions. Esters are fairly stable in neutral aqueous media but are cleaved when heated with water in the presence of strong acids or bases. The hydrolysis of esters in dilute aqueous acid is the reverse of the Fischer esterification (Sections 15.8 and 19.14):

$$\underset{\text{Ester}}{\text{RCOR}'} + \underset{\text{Water}}{\text{H}_2\text{O}} \underset{}{\overset{\text{acid}}{\rightleftharpoons}} \underset{\substack{\text{Carboxylic} \\ \text{acid}}}{\text{RCOH}} + \underset{\text{Alcohol}}{\text{R}'\text{OH}}$$

When esterification is the objective, water is removed from the reaction mixture to encourage ester formation. When ester hydrolysis is the objective, the reaction is carried out in the presence of a generous excess of water. Both reactions illustrate the application of Le Châtelier's principle (Section 6.10) to organic synthesis.

TABLE 20.5 Conversion of Esters to Other Carboxylic Acid Derivatives

Reaction (section) and comments	General equation and specific example
Reaction with ammonia and amines (Section 20.12) Esters react with ammonia and amines to form amides. Methyl and ethyl esters are the most reactive.	$$\underset{\text{Ester}}{RCOR'} + \underset{\text{Amine}}{R''_2NH} \longrightarrow \underset{\text{Amide}}{RCNR''_2} + \underset{\text{Alcohol}}{R'OH}$$ $$\underset{\substack{\text{Ethyl}\\\text{fluoroacetate}}}{FCH_2COCH_2CH_3} + \underset{\text{Ammonia}}{NH_3} \xrightarrow{H_2O} \underset{\substack{\text{Fluoroacetamide}\\(90\%)}}{FCH_2CNH_2} + \underset{\text{Ethanol}}{CH_3CH_2OH}$$
Hydrolysis (Sections 20.10 and 20.11) Ester hydrolysis may be catalyzed either by acids or by bases. Acid-catalyzed hydrolysis is an equilibrium-controlled process, the reverse of the Fischer esterification. Hydrolysis in base is irreversible and is the method usually chosen for preparative purposes.	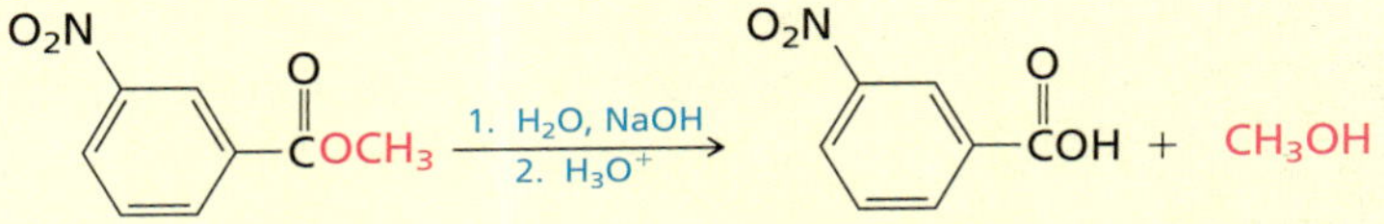 $$\underset{\text{Ester}}{RCOR'} + \underset{\text{Water}}{H_2O} \longrightarrow \underset{\substack{\text{Carboxylic}\\\text{acid}}}{RCOH} + \underset{\text{Alcohol}}{R'OH}$$ (structures) $\xrightarrow[\text{2. } H_3O^+]{\text{1. } H_2O, NaOH}$ (structures) $+ CH_3OH$ Methyl m-nitrobenzoate → m-Nitrobenzoic acid (90–96%) + Methanol

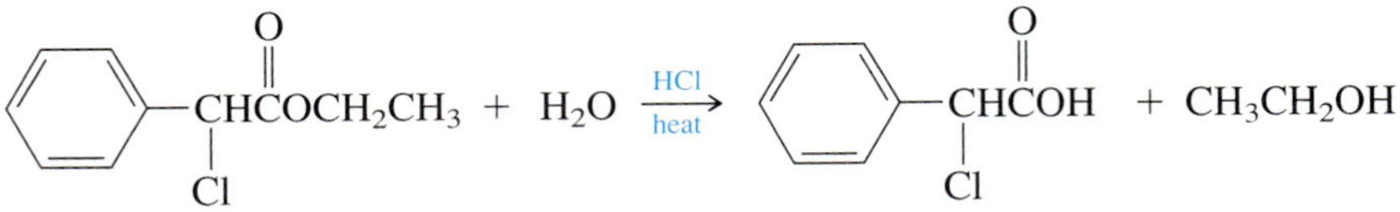

(structure) $-\underset{Cl}{CH}COCH_2CH_3 + H_2O \xrightarrow[\text{heat}]{HCl}$ (structure) $-\underset{Cl}{CH}COH + CH_3CH_2OH$

Ethyl   Water   2-Chloro-2-phenylacetic   Ethyl
2-chloro-2-phenylacetate   acid (80–82%)   alcohol

PROBLEM 20.9 The compound having the structure shown was heated with dilute sulfuric acid to give a product having the molecular formula $C_5H_{12}O_3$ in 63–71% yield. Propose a reasonable structure for this product. What other organic compound is formed in this reaction?

$$CH_3COCH_2\underset{\underset{O}{\overset{|}{\underset{\|}{OCCH_3}}}}{CH}CH_2CH_2CH_2OCCH_3 \xrightarrow[\text{heat}]{H_2O, H_2SO_4} \ ?$$

The mechanism of acid-catalyzed ester hydrolysis is presented in Figure 20.4. It is precisely the reverse of the mechanism given for acid-catalyzed ester formation in Section 19.14. Like other nucleophilic acyl substitutions, it proceeds in two stages. A

FIGURE 20.4 The mechanism of acid-catalyzed ester hydrolysis. Steps 1 through 3 show the formation of the tetrahedral intermediate. Dissociation of the tetrahedral intermediate is shown in steps 4 through 6.

Step 1: Protonation of the carbonyl oxygen of the ester

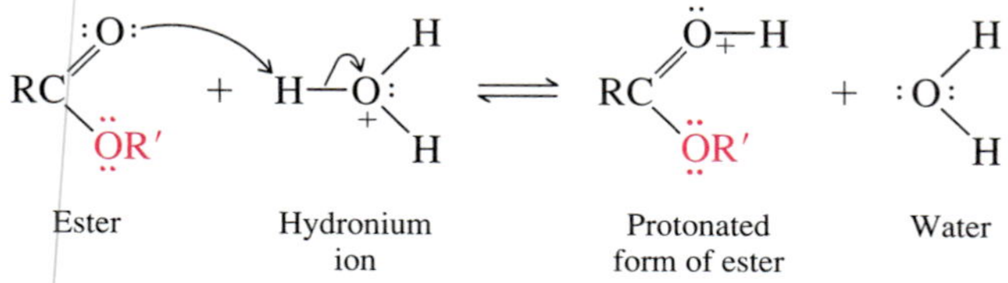

Ester   Hydronium ion   Protonated form of ester   Water

Step 2: Nucleophilic addition of water to protonated form of ester

Water   Protonated form of ester   Conjugate acid of tetrahedral intermediate

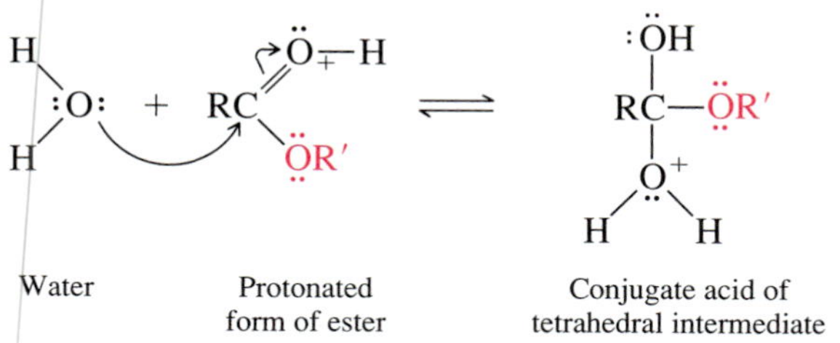

Step 3: Deprotonation of the oxonium ion to give the neutral form of the tetrahedral intermediate

Conjugate acid of tetrahedral intermediate   Water   Tetrahedral intermediate   Hydronium ion

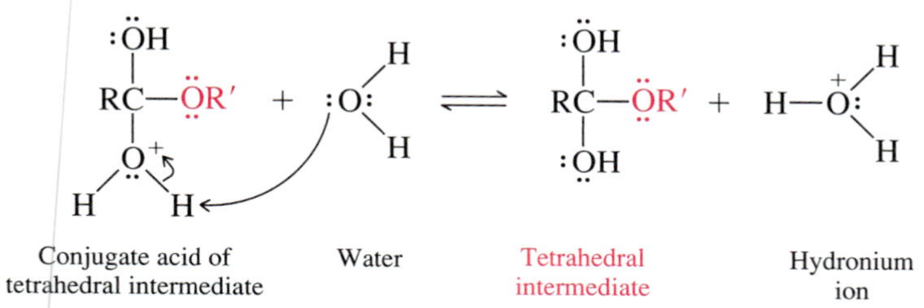

—*Cont.*

tetrahedral intermediate is formed in the first stage, and this tetrahedral intermediate dissociates to products in the second stage.

A key feature of the first stage is the site at which the starting ester is protonated. Protonation of the carbonyl oxygen, as shown in step 1 of Figure 20.4, gives a cation that is stabilized by electron delocalization. The alternative site of protonation, the alkoxy oxygen, gives rise to a much less stable cation.

Protonation of carbonyl oxygen   *Protonation of alkoxy oxygen*

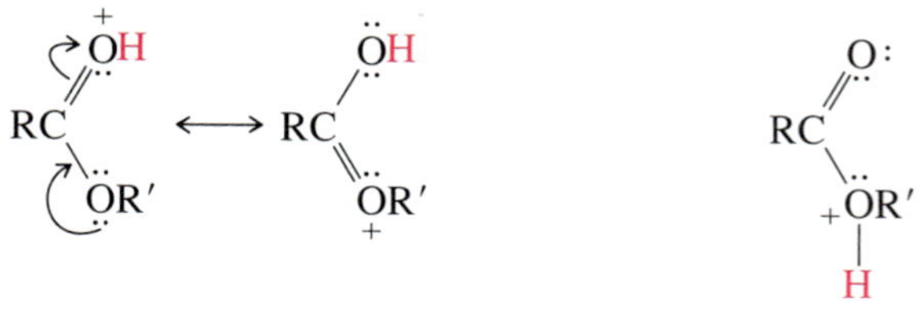

Positive charge is delocalized.   Positive charge is localized on a single oxygen.

FIGURE 20.4 (Continued)

Step 4: Protonation of the tetrahedral intermediate at its alkoxy oxygen

| Tetrahedral intermediate | Hydronium ion | Conjugate acid of tetrahedral intermediate | Water |

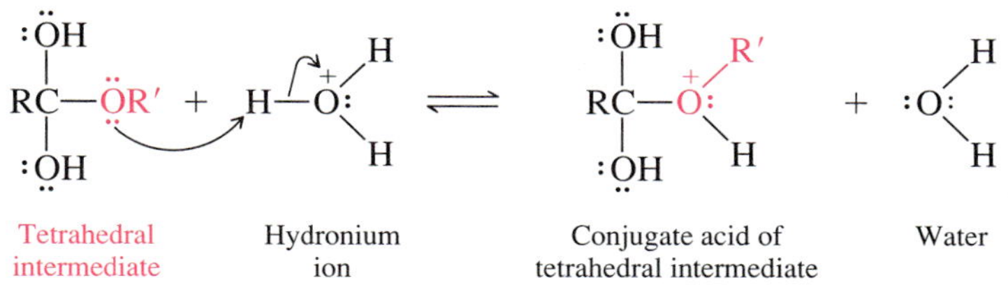

Step 5: Dissociation of the protonated form of the tetrahedral intermediate to an alcohol and the protonated form of the carboxylic acid

| Conjugate acid of tetrahedral intermediate | Protonated form of carboxylic acid | Alcohol |

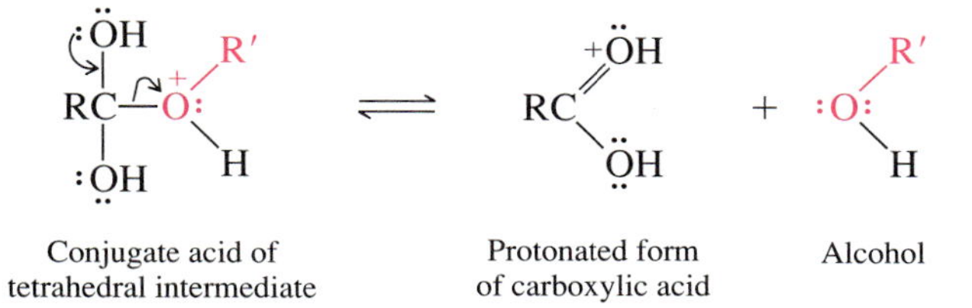

Step 6: Deprotonation of the protonated carboxylic acid

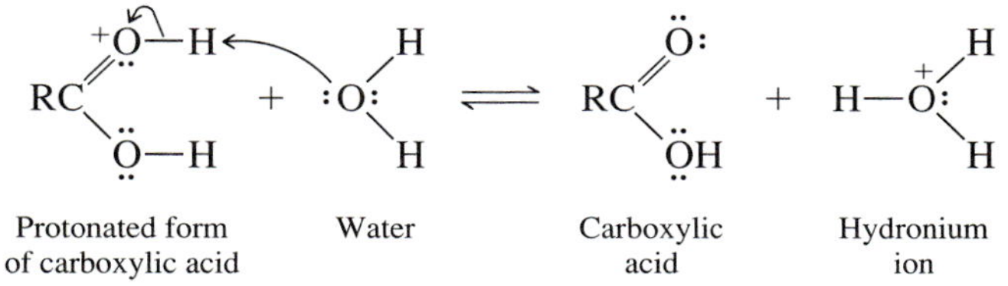

| Protonated form of carboxylic acid | Water | Carboxylic acid | Hydronium ion |

Protonation of the carbonyl oxygen, as emphasized earlier, makes the carbonyl group more susceptible to nucleophilic attack. A water molecule adds to the carbonyl group of the protonated ester in step 2. Loss of a proton from the resulting oxonium ion gives the neutral form of the tetrahedral intermediate in step 3 and completes the first stage of the mechanism.

Once formed, the tetrahedral intermediate can revert to starting materials by merely reversing the reactions that formed it, or it can continue onward to products. In the second stage of ester hydrolysis, the tetrahedral intermediate dissociates to an alcohol and a carboxylic acid. In step 4 of Figure 20.4, protonation of the tetrahedral intermediate at its alkoxy oxygen gives a new oxonium ion, which loses a molecule of alcohol in step 5. Along with the alcohol, the protonated form of the carboxylic acid arises by dissociation of the tetrahedral intermediate. Its deprotonation in step 6 completes the process.

PROBLEM 20.10 On the basis of the general mechanism for acid-catalyzed ester hydrolysis shown in Figure 20.4, write an analogous sequence of steps for the specific case of ethyl benzoate hydrolysis.

The most important species in the mechanism for ester hydrolysis is the tetrahedral intermediate. Evidence in support of the existence of the tetrahedral intermediate was developed by Professor Myron Bender on the basis of isotopic labeling experiments he carried out at the University of Chicago. Bender prepared ethyl benzoate, labeled with the mass-18 isotope of oxygen at the carbonyl oxygen, then subjected it to acid-catalyzed hydrolysis in ordinary (unlabeled) water. He found that ethyl benzoate, recovered from the reaction before hydrolysis was complete, had lost a portion of its isotopic label. This observation is consistent only with the reversible formation of a tetrahedral intermediate under the reaction conditions:

$$
\underset{\substack{\text{Ethyl benzoate}\\ \text{(labeled with }^{18}\text{O)}}}{\underset{C_6H_5}{\overset{O}{\overset{\|}{C}}}\!-\!OCH_2CH_3} + \underset{\text{Water}}{H_2O} \underset{\rightleftharpoons}{\overset{H_3O^+}{}} \underset{\substack{\text{Tetrahedral}\\ \text{intermediate}}}{\underset{C_6H_5}{\overset{HO\quad OH}{\overset{\backslash\ /}{C}}}\!-\!OCH_2CH_3} \underset{\rightleftharpoons}{\overset{H_3O^+}{}} \underset{\text{Ethyl benzoate}}{\underset{C_6H_5}{\overset{O}{\overset{\|}{C}}}\!-\!OCH_2CH_3} + \underset{\substack{\text{Water}\\ \text{(labeled with }^{18}\text{O)}}}{H_2O}
$$

The two OH groups in the tetrahedral intermediate are equivalent, and so either the labeled or the unlabeled one can be lost when the tetrahedral intermediate reverts to ethyl benzoate. Both are retained when the tetrahedral intermediate goes on to form benzoic acid.

PROBLEM 20.11 In a similar experiment, unlabeled 4-butanolide was allowed to stand in an acidic solution in which the water had been labeled with ^{18}O. When the lactone was extracted from the solution after four days, it was found to contain ^{18}O. Which oxygen of the lactone do you think became isotopically labeled?

4-Butanolide

20.11 ESTER HYDROLYSIS IN BASE: SAPONIFICATION

Unlike its acid-catalyzed counterpart, ester hydrolysis in aqueous base is *irreversible*.

$$
\underset{\text{Ester}}{\overset{O}{\overset{\|}{R\text{C}}}\text{OR}'} + \underset{\text{Hydroxide ion}}{HO^-} \longrightarrow \underset{\substack{\text{Carboxylate}\\ \text{ion}}}{\overset{O}{\overset{\|}{R\text{C}}}\text{O}^-} + \underset{\text{Alcohol}}{R'\text{OH}}
$$

> Because it is consumed, hydroxide ion is a reactant, not a catalyst.

This is because carboxylic acids are converted to their corresponding carboxylate anions, which are stable under the reaction conditions.

$$
\underset{\substack{o\text{-Methylbenzyl}\\ \text{acetate}}}{\text{C}_6H_4(\text{CH}_3)\!-\!\text{CH}_2\text{O}\overset{O}{\overset{\|}{\text{C}}}\text{CH}_3} + \underset{\substack{\text{Sodium}\\ \text{hydroxide}}}{\text{NaOH}} \xrightarrow[\text{heat}]{\text{water–}\\ \text{methanol}} \underset{\substack{\text{Sodium}\\ \text{acetate}}}{\text{NaO}\overset{O}{\overset{\|}{\text{C}}}\text{CH}_3} + \underset{\substack{o\text{-Methylbenzyl alcohol}\\ (95\text{–}97\%)}}{\text{C}_6H_4(\text{CH}_3)\!-\!\text{CH}_2\text{OH}}
$$

To isolate the carboxylic acid, a separate acidification step following hydrolysis is necessary. Acidification converts the carboxylate salt to the free acid.

$$H_2C=\overset{\underset{\displaystyle CH_3}{|}}{\underset{\displaystyle \|}{\overset{\displaystyle O}{C}}}COCH_3 \quad \xrightarrow[\text{2. } H_2SO_4]{\text{1. NaOH, } H_2O, \text{ heat}} \quad H_2C=\overset{\underset{\displaystyle CH_3}{|}}{\underset{\displaystyle \|}{\overset{\displaystyle O}{C}}}COH \quad + \quad CH_3OH$$

Methyl 2-methylpropenoate   2-Methylpropenoic   Methyl alcohol
(methyl methacrylate)   acid (87%)
                                 (methacrylic acid)

Ester hydrolysis in base is called **saponification,** which means "soap making." Over 2000 years ago, the Phoenicians made soap by heating animal fat with wood ashes. Animal fat is rich in glycerol triesters, and wood ashes are a source of potassium carbonate. Basic hydrolysis of the fats produced a mixture of long-chain carboxylic acids as their potassium salts.

Procedures for making a variety of soaps are given in the May 1998 issue of the *Journal of Chemical Education,* pp. 612–614.

$$CH_3(CH_2)_xCO\text{---}\cdots\text{---}OC(CH_2)_zCH_3 \quad \xrightarrow[\text{heat}]{K_2CO_3, H_2O}$$

$$\overset{|}{\underset{\displaystyle \|}{\underset{\displaystyle O}{O}}}C(CH_2)_yCH_3$$

$$HOCH_2\overset{\underset{\displaystyle OH}{|}}{C}HCH_2OH \; + \; KO\overset{\displaystyle O}{\overset{\displaystyle \|}{C}}(CH_2)_xCH_3 \; + \; KO\overset{\displaystyle O}{\overset{\displaystyle \|}{C}}(CH_2)_yCH_3 \; + \; KO\overset{\displaystyle O}{\overset{\displaystyle \|}{C}}(CH_2)_zCH_3$$

Glycerol   Potassium carboxylate salts

Potassium and sodium salts of long-chain carboxylic acids form micelles that dissolve grease (Section 19.5) and have cleansing properties. The carboxylic acids obtained by saponification of fats are called *fatty acids.*

PROBLEM 20.12 *Trimyristin* is obtained from coconut oil and has the molecular formula $C_{45}H_{86}O_6$. On being heated with aqueous sodium hydroxide followed by acidification, trimyristin was converted to glycerol and tetradecanoic acid as the only products. What is the structure of trimyristin?

In one of the earliest kinetic studies of an organic reaction, carried out in the nineteenth century, the rate of hydrolysis of ethyl acetate in aqueous sodium hydroxide was found to be first order in ester and first order in base.

$$CH_3\overset{\displaystyle O}{\overset{\displaystyle \|}{C}}OCH_2CH_3 \; + \; NaOH \; \longrightarrow \; CH_3\overset{\displaystyle O}{\overset{\displaystyle \|}{C}}ONa \; + \; CH_3CH_2OH$$

Ethyl acetate   Sodium   Sodium acetate   Ethanol
               hydroxide

$$\text{Rate} = k[CH_3\overset{\displaystyle O}{\overset{\displaystyle \|}{C}}OCH_2CH_3][NaOH]$$

Overall, the reaction exhibits second-order kinetics. Both the ester and the base are involved in the rate-determining step or in a rapid step that precedes it.

Two processes that are consistent with second-order kinetics both involve hydroxide ion as a nucleophile but differ in the site of nucleophilic attack. One of these processes is an S_N2 reaction in which hydroxide displaces carboxylate from the alkyl group of the ester.

S_N2

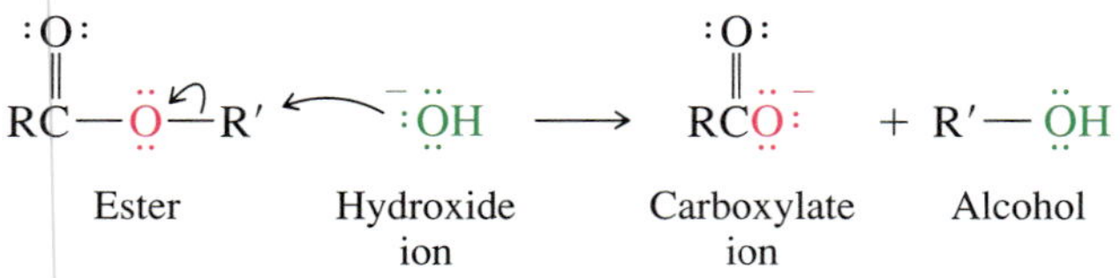

| Ester | Hydroxide ion | Carboxylate ion | Alcohol |

The other process is a nucleophilic acyl substitution triggered by hydroxide attack at the carbonyl group.

Nucleophilic acyl substitution

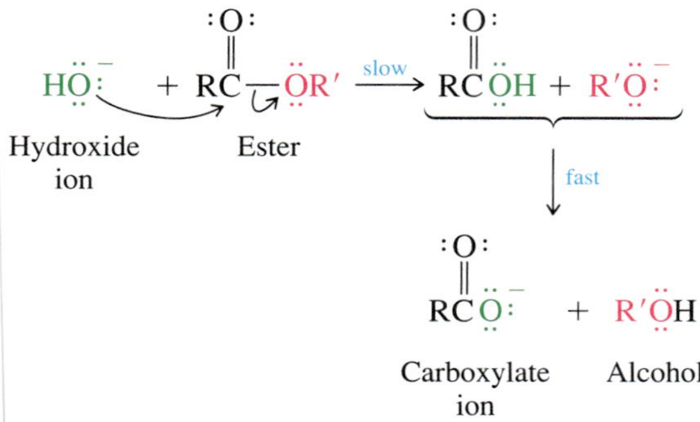

Convincing evidence that ester hydrolysis in base proceeds by the second of these two paths, namely, nucleophilic acyl substitution, has been obtained from several sources. In one experiment, ethyl propanoate labeled with ^{18}O in the ethoxy group was hydrolyzed. On isolating the products, all the ^{18}O was found in the ethyl alcohol; there was no ^{18}O enrichment in the sodium propanoate.

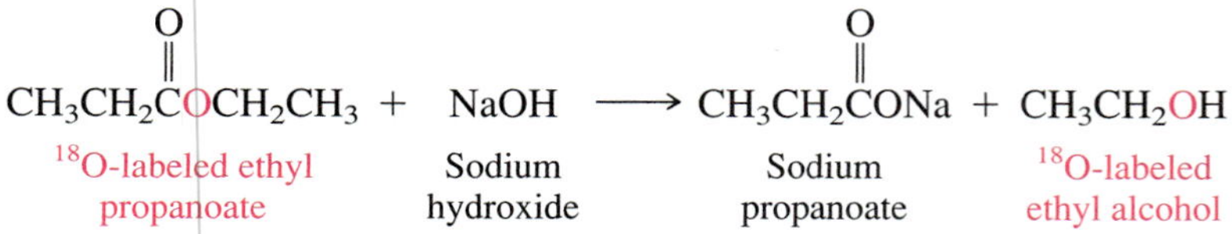

| ^{18}O-labeled ethyl propanoate | Sodium hydroxide | Sodium propanoate | ^{18}O-labeled ethyl alcohol |

The carbon–oxygen bond broken in the process is therefore the one between oxygen and the acyl group. The bond between oxygen and the ethyl group remains intact. An S_N2 reaction at the ethyl group would have broken this bond.

> **PROBLEM 20.13**   In a similar experiment, pentyl acetate was subjected to saponification with ^{18}O-labeled hydroxide in ^{18}O-labeled water. What product do you think became isotopically labeled here, acetate ion or 1-pentanol?

Identical conclusions come from stereochemical studies. Saponification of esters of optically active alcohols proceeds with *retention of configuration*.

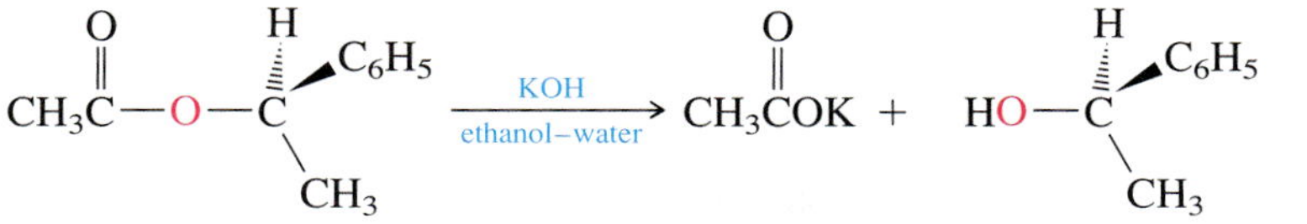

(R)-(+)-1-Phenylethyl
acetate

Potassium
acetate

(R)-(+)-1-Phenylethyl
alcohol
(80% yield; same
optical purity as ester)

None of the bonds to the chirality center is broken when hydroxide attacks the carbonyl group. Had an S_N2 reaction occurred instead, inversion of configuration at the chirality center would have taken place to give (S)-(−)-1-phenylethyl alcohol.

Once it was established that hydroxide ion attacks the carbonyl group in basic ester hydrolysis, the next question to be addressed concerned whether the reaction is concerted or involves a tetrahedral intermediate. In a concerted reaction the bond to the leaving group breaks at the same time that hydroxide ion attacks the carbonyl.

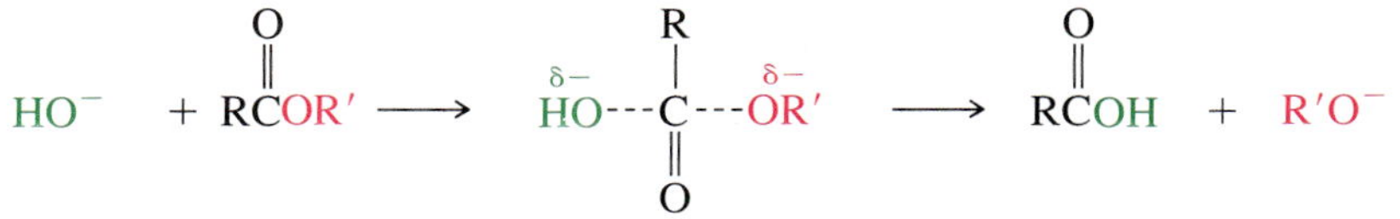

Hydroxide
ion

Ester

Transition state for
concerted displacement

Carboxylic
acid

Alkoxide
ion

In an extension of the work described in the preceding section, Bender showed that basic ester hydrolysis was *not* concerted and, like acid hydrolysis, took place by way of a tetrahedral intermediate. The nature of the experiment was the same, and the results were similar to those observed in the acid-catalyzed reaction. Ethyl benzoate enriched in ^{18}O at the carbonyl oxygen was subjected to hydrolysis in base, and samples were isolated before saponification was complete. The recovered ethyl benzoate was found to have lost a portion of its isotopic label, consistent with the formation of a tetrahedral intermediate:

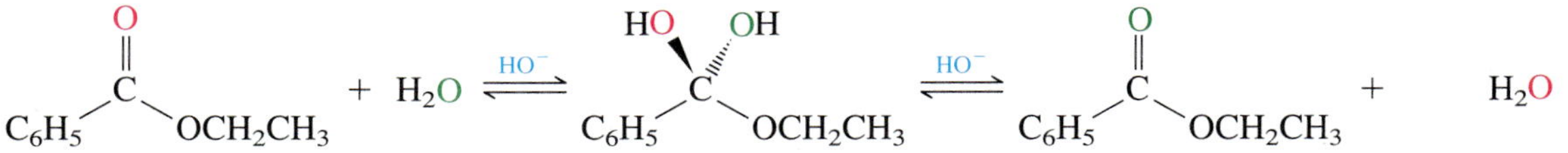

Ethyl benzoate
(labeled with ^{18}O)

Water

Tetrahedral
intermediate

Ethyl benzoate

Water
(labeled with ^{18}O)

All these facts—the observation of second-order kinetics, nucleophilic attack at the carbonyl group, and the involvement of a tetrahedral intermediate—are accommodated by the reaction mechanism shown in Figure 20.5. Like the acid-catalyzed mechanism, it has two distinct stages, namely, formation of the tetrahedral intermediate and its subsequent dissociation. All the steps are reversible except the last one. The equilibrium constant for proton abstraction from the carboxylic acid by hydroxide is so large that step 4 is, for all intents and purposes, irreversible, and this makes the overall reaction irreversible.

Steps 2 and 4 are proton-transfer reactions and are very fast. Nucleophilic addition to the carbonyl group has a higher activation energy than dissociation of the tetrahedral intermediate; step 1 is rate-determining.

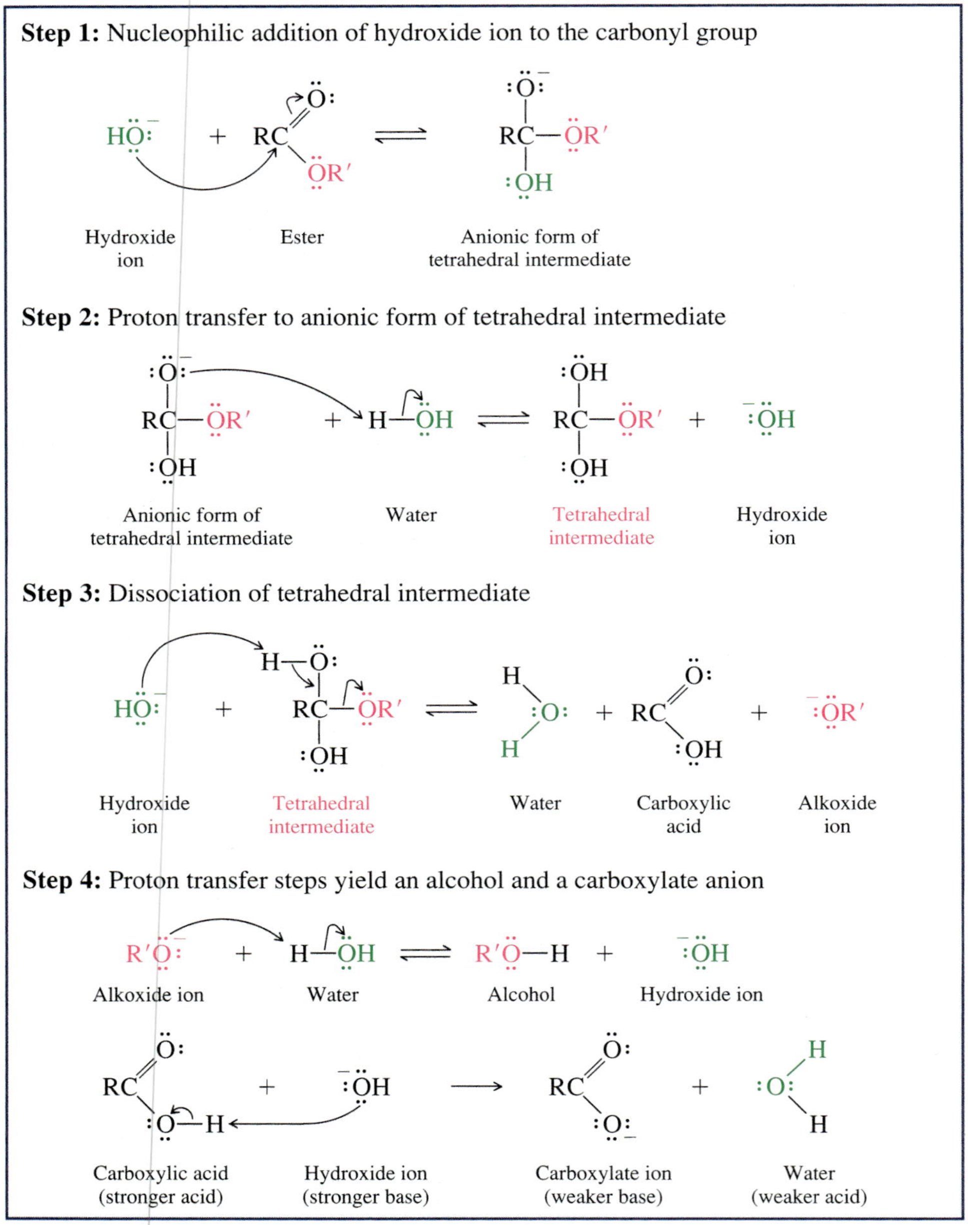

Step 1: Nucleophilic addition of hydroxide ion to the carbonyl group

| Hydroxide ion | Ester | Anionic form of tetrahedral intermediate |

Step 2: Proton transfer to anionic form of tetrahedral intermediate

| Anionic form of tetrahedral intermediate | Water | Tetrahedral intermediate | Hydroxide ion |

Step 3: Dissociation of tetrahedral intermediate

| Hydroxide ion | Tetrahedral intermediate | Water | Carboxylic acid | Alkoxide ion |

Step 4: Proton transfer steps yield an alcohol and a carboxylate anion

| Alkoxide ion | Water | Alcohol | Hydroxide ion |

| Carboxylic acid (stronger acid) | Hydroxide ion (stronger base) | Carboxylate ion (weaker base) | Water (weaker acid) |

FIGURE 20.5 The mechanism of ester hydrolysis in basic solution.

PROBLEM 20.14 On the basis of the general mechanism for basic ester hydrolysis shown in Figure 20.5, write an analogous sequence of steps for the saponification of ethyl benzoate.

20.12 REACTION OF ESTERS WITH AMMONIA AND AMINES

Esters react with ammonia to form amides.

$$\underset{\text{Ester}}{\text{RCOR}'} + \underset{\text{Ammonia}}{\text{NH}_3} \longrightarrow \underset{\text{Amide}}{\text{RCNH}_2} + \underset{\text{Alcohol}}{\text{R}'\text{OH}}$$

Ammonia is more nucleophilic than water, making it possible to carry out this reaction using aqueous ammonia.

$$\underset{\substack{\text{Methyl 2-methylpropenoate}}}{\text{H}_2\text{C}{=}\overset{\text{CH}_3}{\underset{|}{\text{C}}}\text{COCH}_3} + \underset{\text{Ammonia}}{\text{NH}_3} \xrightarrow{\text{H}_2\text{O}} \underset{\substack{\text{2-Methylpropenamide}\\(75\%)}}{\text{H}_2\text{C}{=}\overset{\text{CH}_3}{\underset{|}{\text{C}}}\text{CNH}_2} + \underset{\text{Methyl alcohol}}{\text{CH}_3\text{OH}}$$

Amines, which are substituted derivatives of ammonia, react similarly:

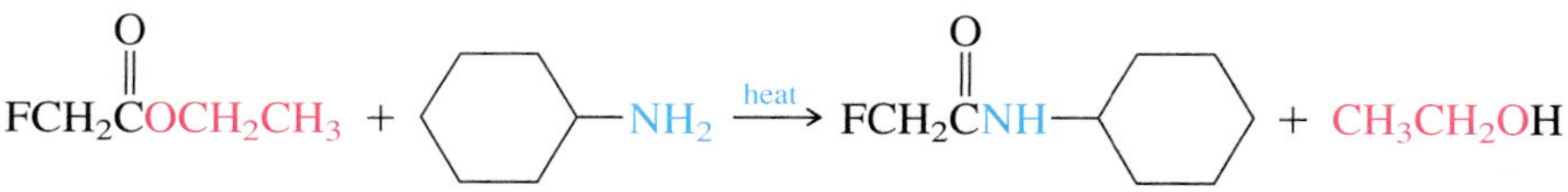

$$\underset{\text{Ethyl fluoroacetate}}{\text{FCH}_2\text{COCH}_2\text{CH}_3} + \underset{\text{Cyclohexylamine}}{\bigcirc\!\!-\text{NH}_2} \xrightarrow{\text{heat}} \underset{\substack{N\text{-Cyclohexyl-}\\\text{fluoroacetamide (61\%)}}}{\text{FCH}_2\text{CNH}\!-\!\bigcirc} + \underset{\text{Ethyl alcohol}}{\text{CH}_3\text{CH}_2\text{OH}}$$

The amine must be primary (RNH_2) or secondary (R_2NH). Tertiary amines (R_3N) cannot form amides because they have no proton on nitrogen that can be replaced by an acyl group.

PROBLEM 20.15 Give the structure of the expected product of the following reaction:

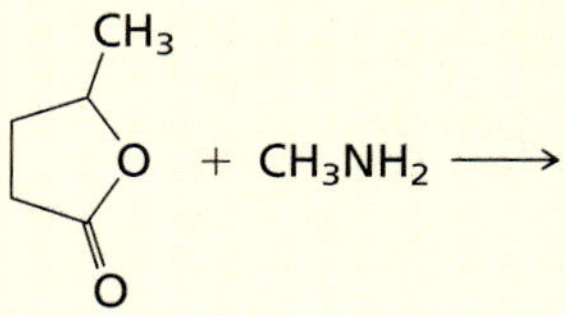

$$\text{(structure)} + \text{CH}_3\text{NH}_2 \longrightarrow$$

The reaction of ammonia and amines with esters follows the same general mechanistic course as other nucleophilic acyl substitution reactions (Figure 20.6). A tetrahedral intermediate is formed in the first stage of the process and dissociates in the second stage.

Overall Reaction:

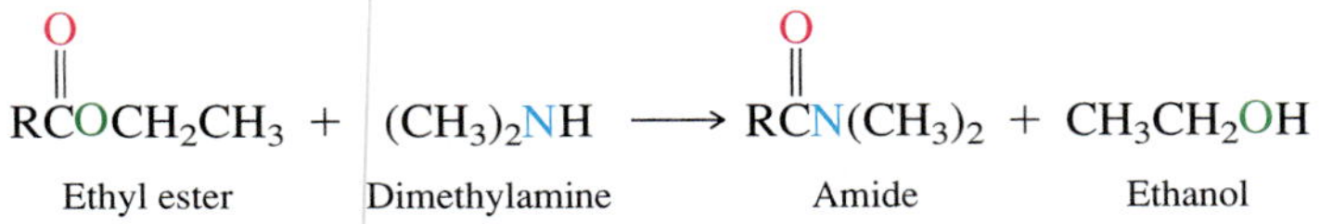

$$\underset{\text{Ethyl ester}}{RCOCH_2CH_3} + \underset{\text{Dimethylamine}}{(CH_3)_2NH} \longrightarrow \underset{\text{Amide}}{RCN(CH_3)_2} + \underset{\text{Ethanol}}{CH_3CH_2OH}$$

Stage 1: The amine adds to the carbonyl group of the ester to give a tetrahedral intermediate. The tetrahedral intermediate is analogous to the carbinolamine formed by addition of an amine to an aldehyde or ketone

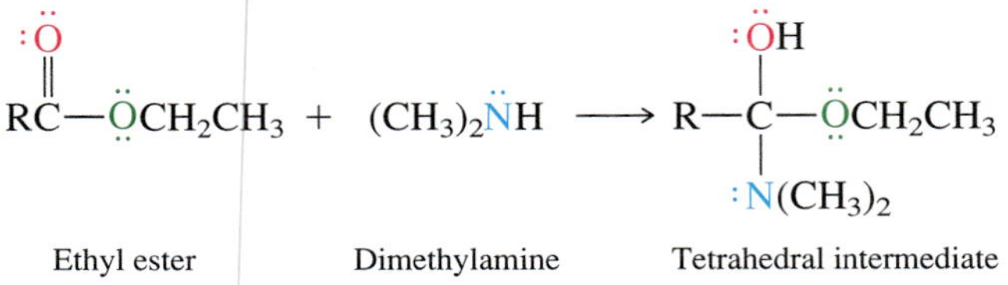

$$\underset{\text{Ethyl ester}}{RC-\ddot{O}CH_2CH_3} + \underset{\text{Dimethylamine}}{(CH_3)_2\ddot{N}H} \longrightarrow \underset{\text{Tetrahedral intermediate}}{R-\overset{\text{:}\ddot{O}H}{\underset{\text{:}N(CH_3)_2}{C}}-\ddot{O}CH_2CH_3}$$

Stage 2: The tetrahedral intermediate dissociates to give an amide. The O—H and C—O bonds may break in the same step, or in separate steps.

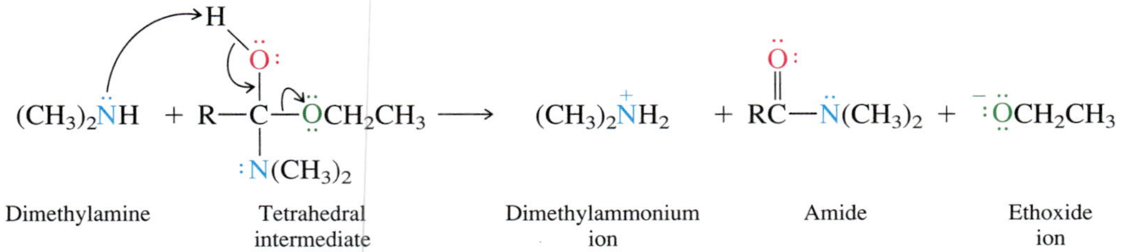

$$\underset{\text{Dimethylamine}}{(CH_3)_2\ddot{N}H} + \underset{\substack{\text{Tetrahedral} \\ \text{intermediate}}}{R-\overset{H}{\underset{:N(CH_3)_2}{\overset{\ddot{O}:}{C}}}-\ddot{O}CH_2CH_3} \longrightarrow \underset{\substack{\text{Dimethylammonium} \\ \text{ion}}}{(CH_3)_2\overset{+}{\ddot{N}}H_2} + \underset{\text{Amide}}{RC-\ddot{N}(CH_3)_2} + \underset{\substack{\text{Ethoxide} \\ \text{ion}}}{:\ddot{O}CH_2CH_3}$$

Proton transfers convert the ammonium ion and ethoxide ion to their stable forms under the reaction conditions.

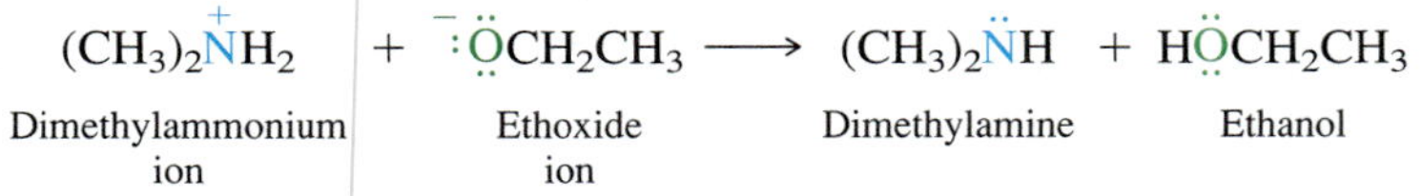

$$\underset{\substack{\text{Dimethylammonium} \\ \text{ion}}}{(CH_3)_2\overset{+}{\ddot{N}}H_2} + \underset{\substack{\text{Ethoxide} \\ \text{ion}}}{^-:\ddot{O}CH_2CH_3} \longrightarrow \underset{\text{Dimethylamine}}{(CH_3)_2\ddot{N}H} + \underset{\text{Ethanol}}{H\ddot{O}CH_2CH_3}$$

FIGURE 20.6 Mechanism of amide formation in the reaction of a secondary amine with an ethyl ester.

20.13 THIOESTERS

Thioesters undergo the same kinds of reactions as esters and by similar mechanisms. Nucleophilic acyl substitution of a thioester gives a *thiol* along with the product of acyl transfer. For example:

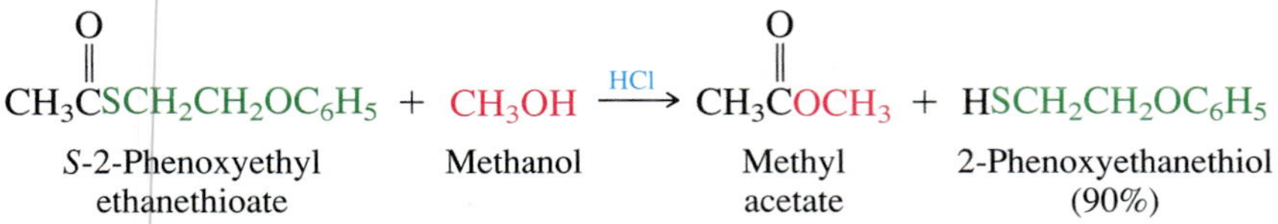

$$\underset{\substack{\text{S-2-Phenoxyethyl} \\ \text{ethanethioate}}}{CH_3CSCH_2CH_2OC_6H_5} + \underset{\text{Methanol}}{CH_3OH} \xrightarrow{\text{HCl}} \underset{\substack{\text{Methyl} \\ \text{acetate}}}{CH_3COCH_3} + \underset{\substack{\text{2-Phenoxyethanethiol} \\ (90\%)}}{HSCH_2CH_2OC_6H_5}$$

PROBLEM 20.16 Write the structure of the tetrahedral intermediate formed in the reaction just described.

A number of important biological processes involve thioesters; several of these are described in Chapter 26.

20.14 PREPARATION OF AMIDES

Amides are readily prepared by acylation of ammonia and amines with acyl chlorides, acid anhydrides, or esters.

Acylation of *ammonia* (NH$_3$) yields an amide (R'CNH$_2$).

$$\underset{}{\overset{O}{\overset{\|}{}}}$$

Acylation of *ammonia* (NH$_3$) yields an amide (R'$\overset{\overset{\textstyle O}{\|}}{C}NH_2$).

Primary amines (RNH$_2$) yield *N*-substituted amides (R'$\overset{\overset{\textstyle O}{\|}}{C}$NHR).

Secondary amines (R$_2$NH) yield *N,N*-disubstituted amides (R'$\overset{\overset{\textstyle O}{\|}}{C}NR_2$).

Examples illustrating these reactions may be found in Tables 20.1, 20.2, and 20.5.

Two molar equivalents of amine are required in the reaction with acyl chlorides and acid anhydrides; one molecule of amine acts as a nucleophile, the second as a Brønsted base.

$$2R_2NH \ + \ R'\overset{\overset{\textstyle O}{\|}}{C}Cl \ \longrightarrow \ R'\overset{\overset{\textstyle O}{\|}}{C}NR_2 \ + \ R_2\overset{+}{N}H_2 \ Cl^-$$

Amine	Acyl chloride	Amide	Hydrochloride salt of amine

$$2R_2NH \ + \ R'\overset{\overset{\textstyle O\ O}{\|\ \|}}{CO}CR' \ \longrightarrow \ R'\overset{\overset{\textstyle O}{\|}}{C}NR_2 \ + \ R_2\overset{+}{N}H_2 \ {}^-O\overset{\overset{\textstyle O}{\|}}{C}R'$$

Amine	Acid anhydride	Amide	Carboxylate salt of amine

It is possible to use only one molar equivalent of amine in these reactions if some other base, such as sodium hydroxide, is present in the reaction mixture to react with the hydrogen chloride or carboxylic acid that is formed. This is a useful procedure in those cases in which the amine is a valuable one or is available only in small quantities.

Esters and amines react in a 1:1 molar ratio to give amides. No acidic product is formed from the ester, and so no additional base is required.

$$R_2NH \ + \ R'\overset{\overset{\textstyle O}{\|}}{C}OCH_3 \ \longrightarrow \ R'\overset{\overset{\textstyle O}{\|}}{C}NR_2 \ + \ CH_3OH$$

Amine	Methyl ester	Amide	Methanol

PROBLEM 20.17 Write an equation showing the preparation of the following amides from the indicated carboxylic acid derivative:

(a) (CH$_3$)$_2$CH$\overset{\overset{\textstyle O}{\|}}{C}NH_2$ from an acyl chloride

(b) $CH_3\overset{\overset{\displaystyle O}{\|}}{C}NHCH_3$ from an acid anhydride

(c) $H\overset{\overset{\displaystyle O}{\|}}{C}N(CH_3)_2$ from a methyl ester

SAMPLE SOLUTION   (a) Amides of the type $R\overset{\overset{\displaystyle O}{\|}}{C}NH_2$ are derived by acylation of ammonia.

$$(CH_3)_2CH\overset{\overset{\displaystyle O}{\|}}{C}Cl \;+\; 2NH_3 \longrightarrow (CH_3)_2CH\overset{\overset{\displaystyle O}{\|}}{C}NH_2 \;+\; NH_4Cl$$

| 2-Methylpropanoyl chloride | Ammonia | 2-Methylpropanamide | Ammonium chloride |

Two molecules of ammonia are needed because its acylation produces, in addition to the desired amide, a molecule of hydrogen chloride. Hydrogen chloride (an acid) reacts with ammonia (a base) to give ammonium chloride.

All these reactions proceed by nucleophilic addition of the amine to the carbonyl group. Dissociation of the tetrahedral intermediate proceeds in the direction that leads to an amide.

$$R\overset{\overset{\displaystyle :O:}{\|}}{C}X \;+\; R'_2\ddot{N}H \;\rightleftharpoons\; R\overset{\overset{\displaystyle :\ddot{O}-H}{|}}{\underset{:NR'_2}{\overset{|}{C}}}X \;\rightleftharpoons\; R\overset{\overset{\displaystyle :O:}{\|}}{C}NR'_2 \;+\; HX$$

| Acylating agent | Amine | Tetrahedral intermediate | Amide | Conjugate acid of leaving group |

The carbonyl group of an amide is stabilized to a greater extent than that of an acyl chloride, acid anhydride, or ester; amides are formed rapidly and in high yield from each of these carboxylic acid derivatives.

Amides are sometimes prepared directly from carboxylic acids and amines by a two-step process. The first step is an acid–base reaction in which the acid and the amine combine to form an ammonium carboxylate salt. On heating, the ammonium carboxylate salt loses water to form an amide.

$$R\overset{\overset{\displaystyle O}{\|}}{C}OH \;+\; R'_2NH \longrightarrow R\overset{\overset{\displaystyle O}{\|}}{C}O^- \; R'_2\overset{+}{N}H_2 \;\overset{heat}{\longrightarrow}\; R\overset{\overset{\displaystyle O}{\|}}{C}NR'_2 \;+\; H_2O$$

| Carboxylic acid | Amine | Ammonium carboxylate salt | Amide | Water |

In practice, both steps may be combined in a single operation by simply heating a carboxylic acid and an amine together:

$$C_6H_5\overset{\overset{\displaystyle O}{\|}}{C}OH \;+\; C_6H_5NH_2 \;\overset{225°C}{\longrightarrow}\; C_6H_5\overset{\overset{\displaystyle O}{\|}}{C}NHC_6H_5 \;+\; H_2O$$

| Benzoic acid | Aniline | N-Phenylbenzamide (80–84%) | Water |

A similar reaction in which ammonia and carbon dioxide are heated under pressure is the basis of the industrial synthesis of *urea*. Here, the reactants first combine, yielding a salt called *ammonium carbamate:*

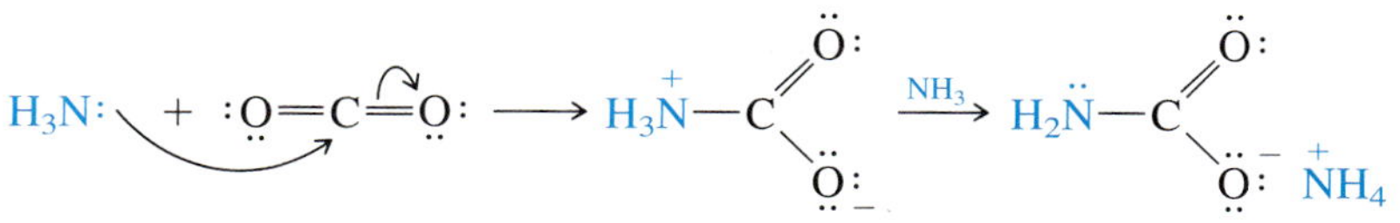

Ammonia   Carbon dioxide   Ammonium carbamate

On being heated, ammonium carbamate undergoes dehydration to form urea:

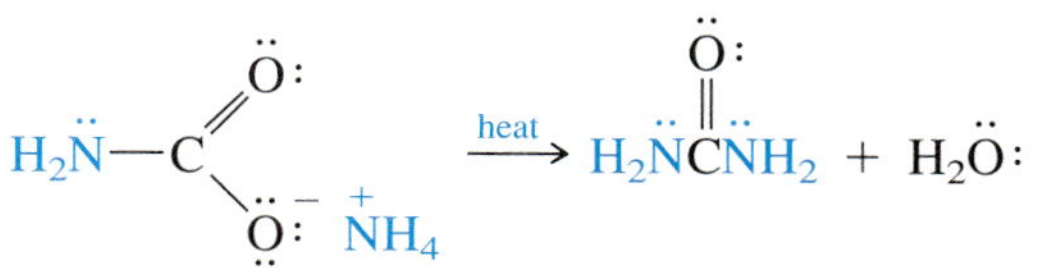

Ammonium carbamate   Urea   Water

Over 10^{10} lb of urea—most of it used as fertilizer—is produced annually in the United States by this method.

These thermal methods for preparing amides are limited in their generality. Most often amides are prepared in the laboratory from acyl chlorides, acid anhydrides, or esters, and these are the methods that you should apply to solving synthetic problems.

20.15 LACTAMS

Lactams are cyclic amides and are analogous to lactones, which are cyclic esters. Most lactams are known by their common names, as the examples shown illustrate.

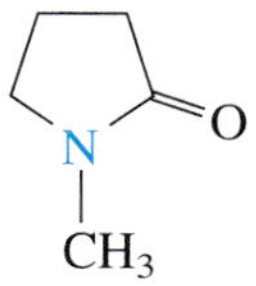

N-Methylpyrrolidone
(a polar aprotic
solvent)

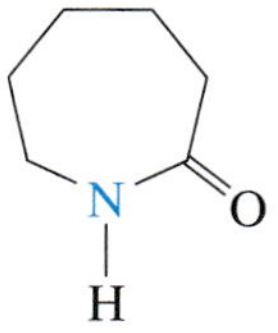

ε-Caprolactam
(industrial chemical
used to prepare a type of nylon)

Just as amides are more stable than esters, lactams are more stable than lactones. Thus, although β-lactones are rare (Section 19.15), β-lactams are among the best known products of the pharmaceutical industry. The penicillin and cephalosporin antibiotics, which are so useful in treating bacterial infections, are β-lactams and are customarily referred to as *β-lactam antibiotics.*

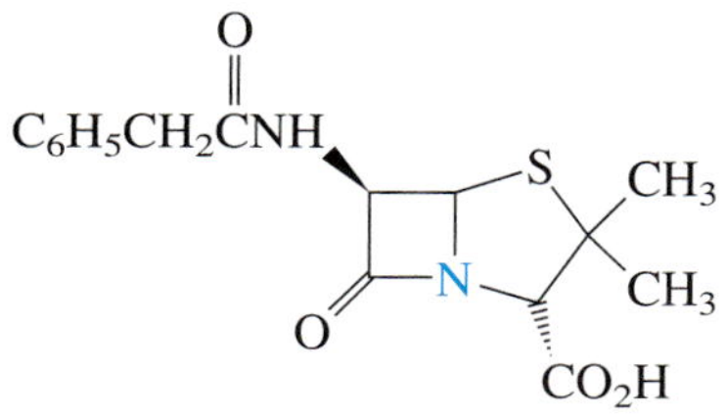

Penicillin G 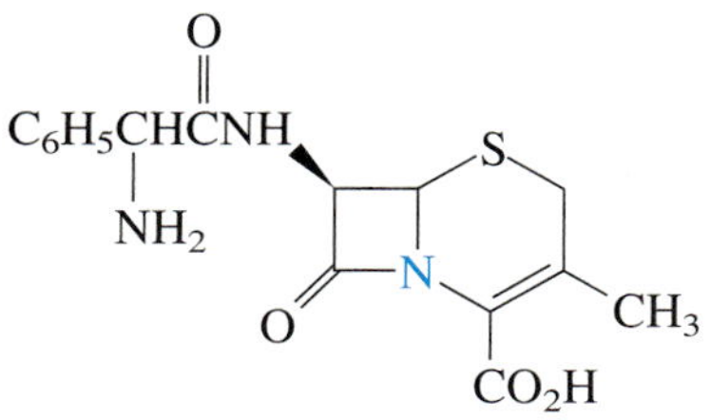   Cephalexin

These antibiotics inhibit a bacterial enzyme that is essential for cell wall formation. A nucleophilic site on the enzyme reacts with the carbonyl group in the four-membered ring, and the ring opens to acylate the enzyme. Once its nucleophilic site is acylated, the enzyme is no longer active and the bacteria die. The β-lactam rings of the penicillins and cephalosporins combine just the right level of stability in aqueous media with reactivity toward nucleophilic substitution to be effective acylating agents toward this critical bacterial enzyme.

20.16 IMIDES

Compounds that have two acyl groups bonded to a single nitrogen are known as **imides.** The most common imides are cyclic ones:

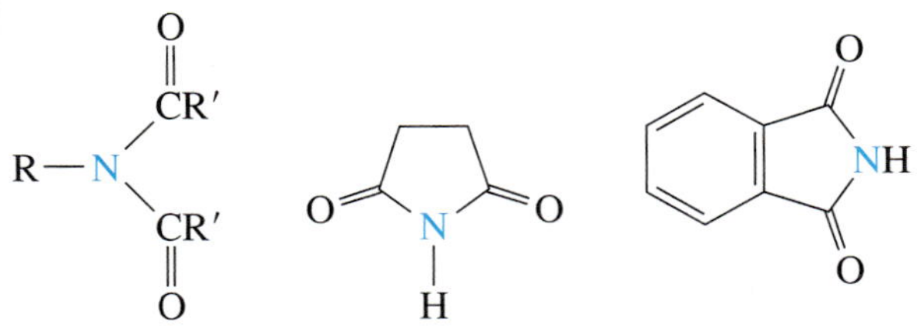

Imide   Succinimide   Phthalimide

Cyclic imides can be prepared by heating the ammonium salts of dicarboxylic acids:

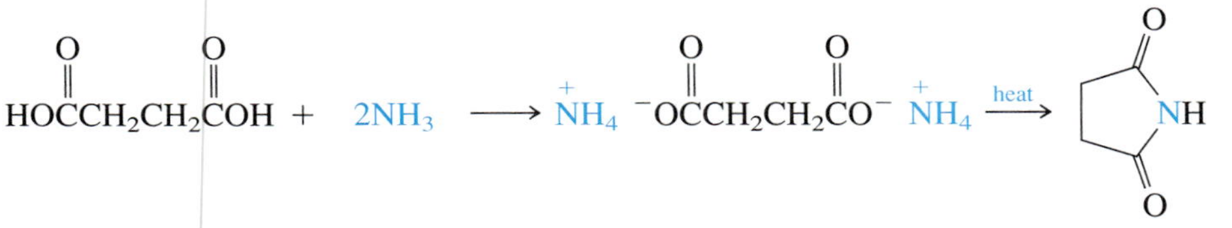

Replacement of the proton on nitrogen in succinimide by bromine gives *N*-bromo-succinimide, a reagent used for allylic and benzylic brominations (Sections 10.4 and 11.12).

Succinic acid   Ammonia   Ammonium succinate   Succinimide
                                                                     (82–83%)

PROBLEM 20.18   Phthalimide has been prepared in 95% yield by heating the compound formed on reaction of phthalic anhydride (Section 20.5) with excess ammonia. This compound has the molecular formula $C_8H_{10}N_2O_3$. What is its structure?

20.17 HYDROLYSIS OF AMIDES

Amides are the least reactive carboxylic acid derivative, and the only nucleophilic acyl substitution reaction they undergo is hydrolysis. Amides are fairly stable in water, but the amide bond is cleaved on heating in the presence of strong acids or bases. Nominally, this cleavage produces an amine and a carboxylic acid.

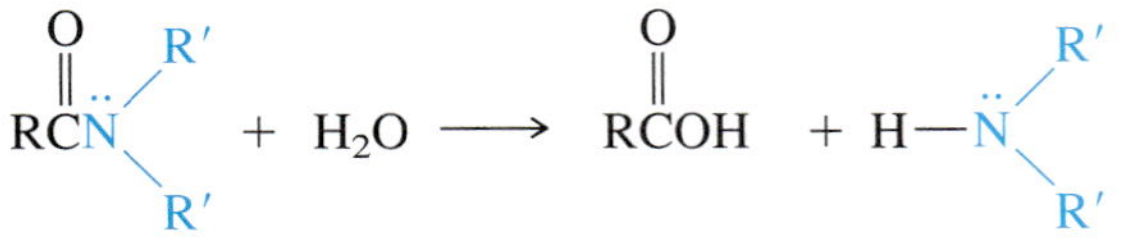

Amide   Water   Carboxylic   Amine
                       acid

In acid, however, the amine is protonated, giving an ammonium ion, $R_2'\overset{+}{N}H_2$:

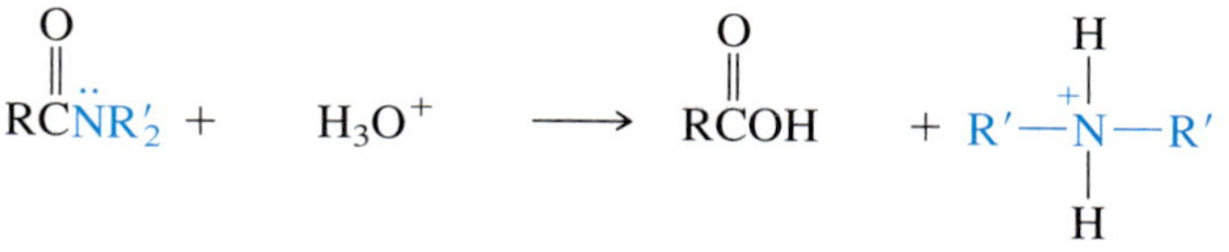

| Amide | Hydronium ion | Carboxylic acid | Ammonium ion |

In base the carboxylic acid is deprotonated, giving a carboxylate ion:

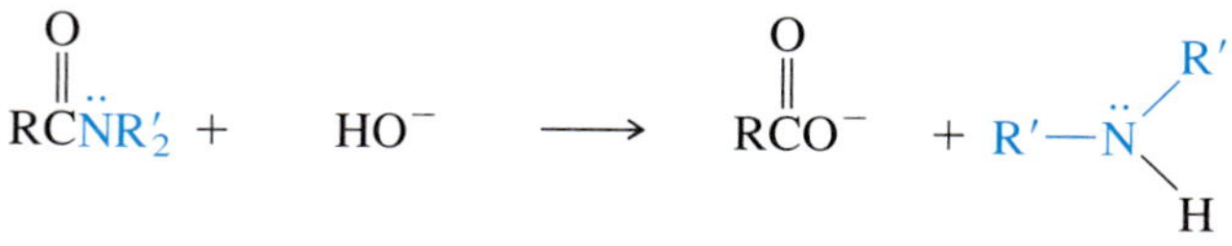

| Amide | Hydroxide ion | Carboxylate ion | Amine |

The acid–base reactions that occur after the amide bond is broken make the overall hydrolysis irreversible in both cases. The amine product is protonated in acid; the carboxylic acid is deprotonated in base.

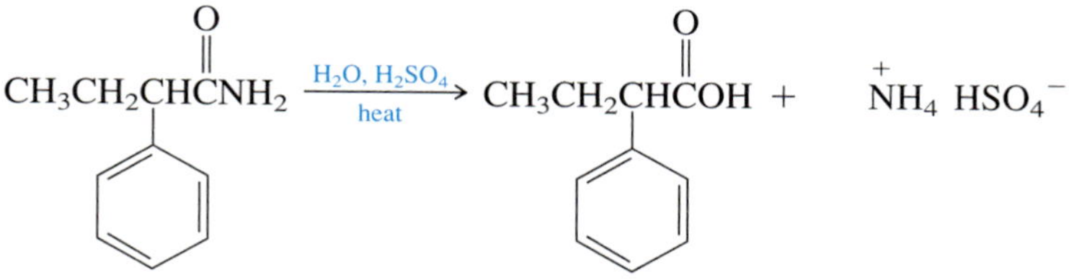

| 2-Phenylbutanamide | 2-Phenylbutanoic acid (88–90%) | Ammonium hydrogen sulfate |

| N-(4-Bromophenyl)acetamide (p-bromoacetanilide) | Potassium acetate | p-Bromoaniline (95%) |

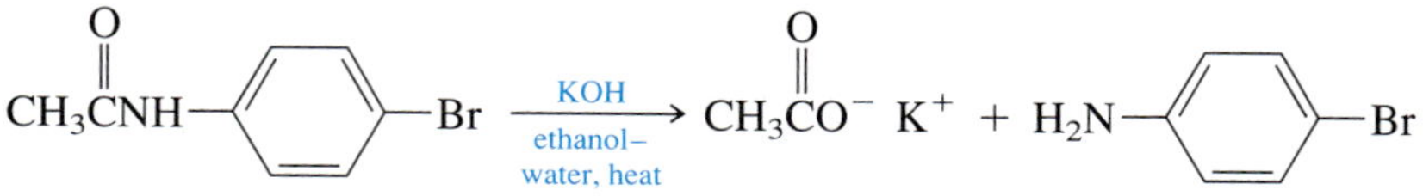

Mechanistically, amide hydrolysis is similar to the hydrolysis of other carboxylic acid derivatives. The mechanism of the hydrolysis in acid is presented in Figure 20.7. It proceeds in two stages; a tetrahedral intermediate is formed in the first stage and dissociates in the second.

The amide is activated toward nucleophilic attack by protonation of its carbonyl oxygen. The cation produced in this step is stabilized by resonance involving the nitrogen lone pair and is more stable than the intermediate in which the amide nitrogen is protonated.

FIGURE 20.7 The mecha-
nism of amide hydrolysis in
acid solution. Steps 1
through 3 show the for-
mation of the tetrahedral
intermediate. Dissociation
of the tetrahedral inter-
mediate is shown in steps
4 through 6.

Step 1: Protonation of the carbonyl oxygen of the amide

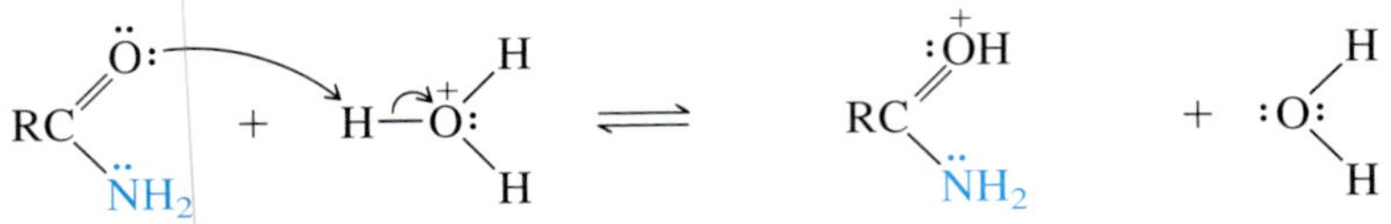

Amide   Hydronium ion   Protonated form of amide   Water

Step 2: Nucleophilic addition of water to the protonated form of the amide

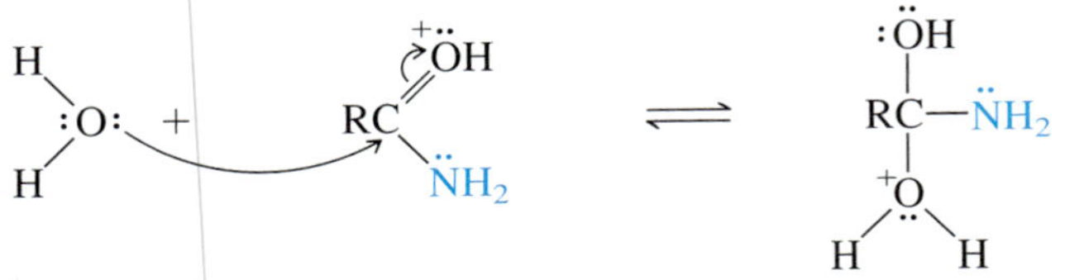

Water   Protonated form of amide   Conjugate acid of tetrahedral intermediate

Step 3: Deprotonation of the oxonium ion to give the neutral form of the tetrahedral intermediate

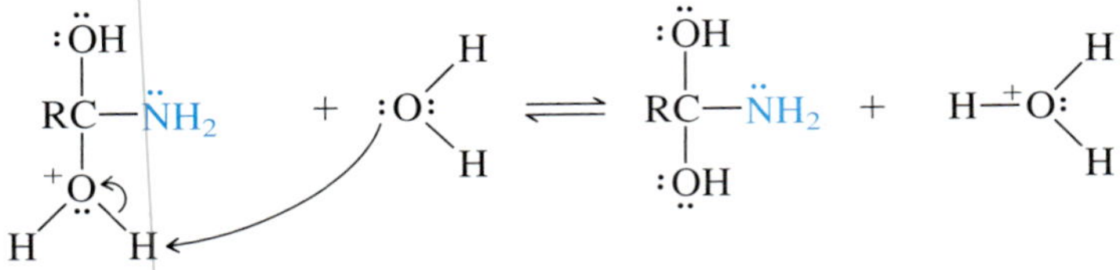

Conjugate acid of tetrahedral intermediate   Water   Tetrahedral intermediate   Hydronium ion

Step 4: Protonation of the tetrahedral intermediate at its amino nitrogen

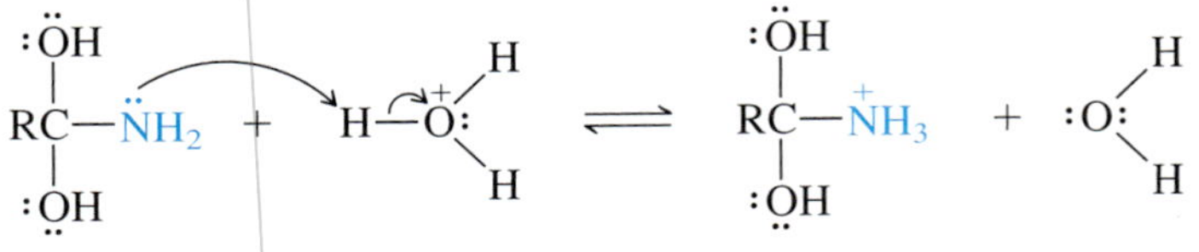

Tetrahedral intermediate   Hydronium ion   Ammonium ion   Water

Step 5: Dissociation of the N-protonated form of the tetrahedral intermediate to give ammonia and the protonated form of the carboxylic acid

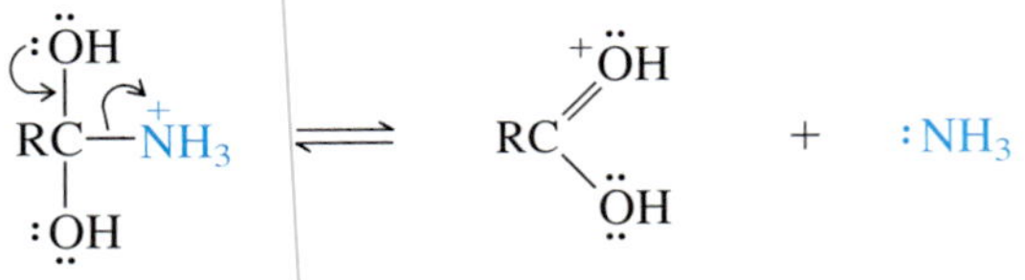

Ammonium ion   Protonated form of carboxylic acid   Ammonia

—*Cont.*

FIGURE 20.7 *(Continued)*

Step 6: Proton transfer processes yielding ammonium ion and the carboxylic acid

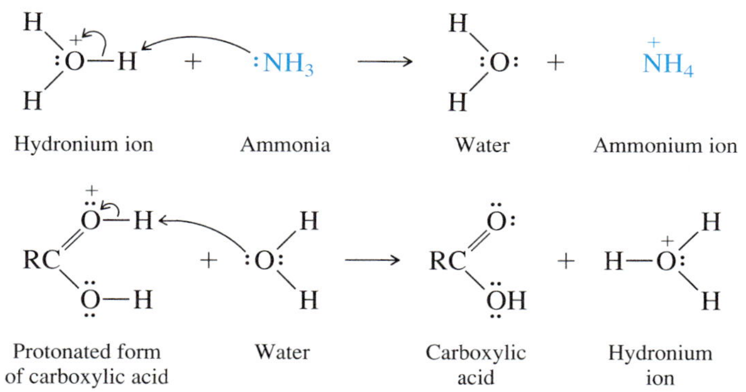

| Hydronium ion | Ammonia | Water | Ammonium ion |

| Protonated form of carboxylic acid | Water | Carboxylic acid | Hydronium ion |

Protonation of carbonyl oxygen   *Protonation of amide nitrogen*

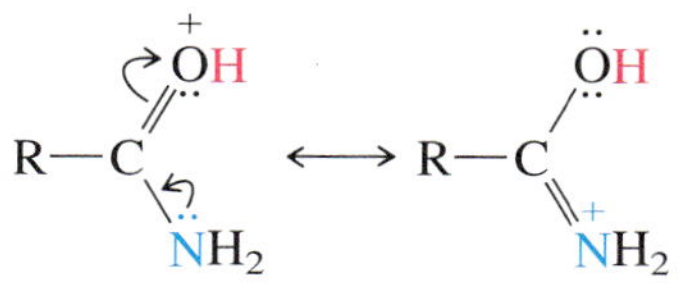

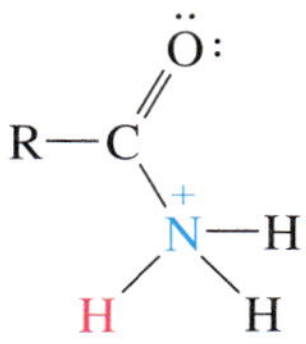

Most stable resonance forms of an *O*-protonated amide

An acylammonium ion; the positive charge is localized on nitrogen

Once formed, the *O*-protonated intermediate is attacked by a water molecule in step 2. The intermediate formed in this step loses a proton in step 3 to give the neutral form of the tetrahedral intermediate. The tetrahedral intermediate has its amino group (—NH$_2$) attached to sp^3-hybridized carbon, and this amino group is the site at which protonation occurs in step 4. Cleavage of the carbon–nitrogen bond in step 5 yields the protonated form of the carboxylic acid, along with a molecule of ammonia. In acid solution ammonia is immediately protonated to give ammonium ion, as shown in step 6. This protonation step has such a large equilibrium constant that it makes the overall reaction irreversible.

PROBLEM 20.19   On the basis of the general mechanism for amide hydrolysis in acidic solution shown in Figure 20.7, write an analogous sequence of steps for the hydrolysis of acetanilide, CH$_3$CNHC$_6$H$_5$.

In base the tetrahedral intermediate is formed in a manner analogous to that proposed for ester saponification. Steps 1 and 2 in Figure 20.8 show the formation of the tetrahedral intermediate in the basic hydrolysis of amides. In step 3 the basic amino group of the tetrahedral intermediate abstracts a proton from water, and in step 4 the derived ammonium ion dissociates. Conversion of the carboxylic acid to its corresponding carboxylate anion in step 5 completes the process and renders the overall reaction irreversible.

FIGURE 20.8 The mechanism of amide hydrolysis in basic solution.

Step 1: Nucleophilic addition of hydroxide ion to the carbonyl group

Hydroxide ion   Amide   Anionic form of tetrahedral intermediate

Step 2: Proton transfer to anionic form of tetrahedral intermediate

Anionic form of tetrahedral intermediate   Water   Tetrahedral intermediate   Hydroxide ion

Step 3: Protonation of amino nitrogen of tetrahedral intermediate

Tetrahedral intermediate   Water   Ammonium ion   Hydroxide ion

Step 4: Dissociation of *N*-protonated form of tetrahedral intermediate

Hydroxide ion   Ammonium ion   Water   Carboxylic acid   Ammonia

Step 5: Irreversible formation of carboxylate anion

Carboxylic acid (stronger acid)   Hydroxide ion (stronger base)   Carboxylate ion (weaker base)   Water (weaker acid)

FIGURE 20.8 The mechanism of amide hydrolysis in basic solution.

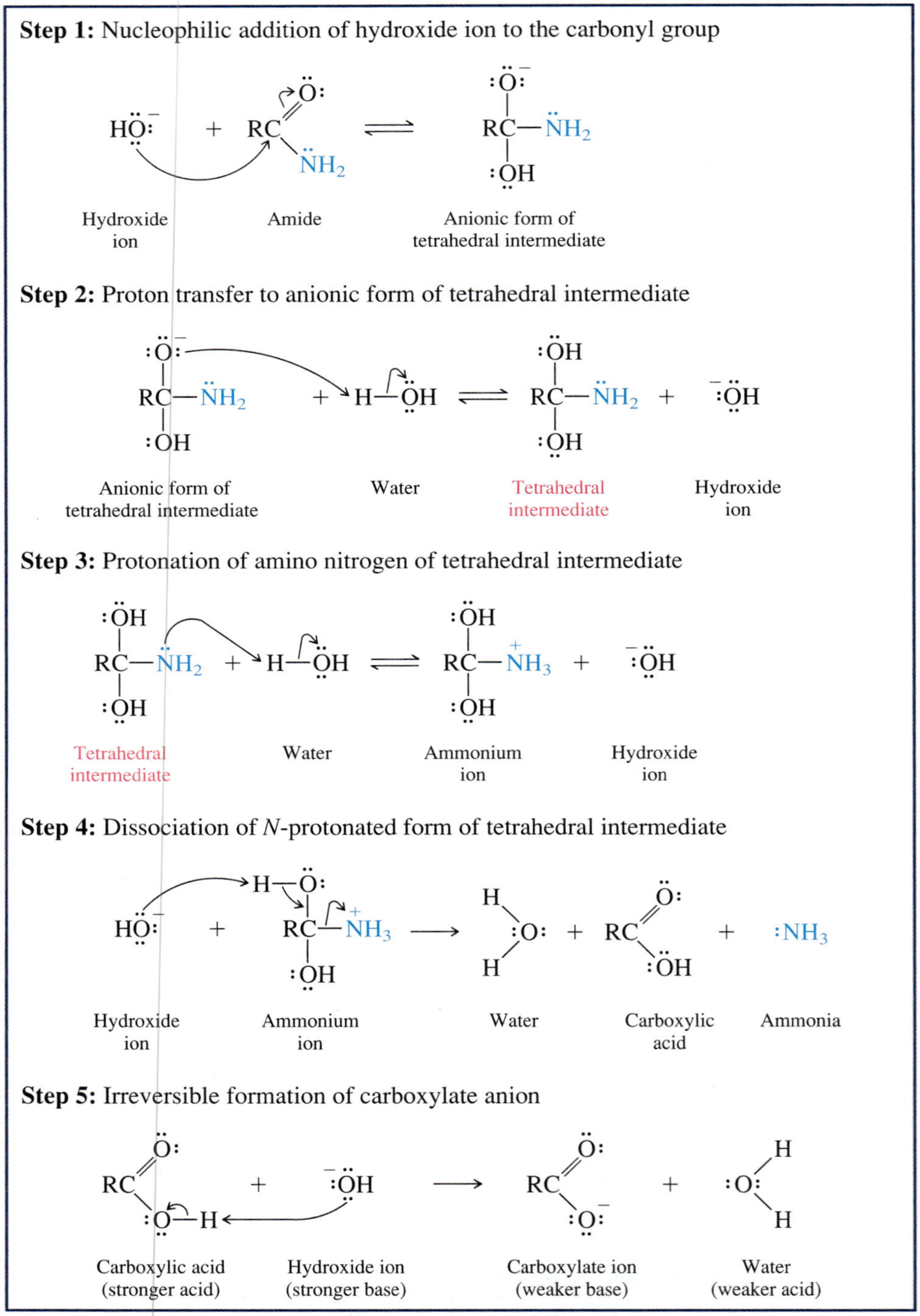

PROBLEM 20.20 On the basis of the general mechanism for basic hydrolysis shown in Figure 20.8, write an analogous sequence for the hydrolysis of *N,N*-

$$\overset{O}{\overset{\|}{\text{dimethylformamide, HCN(CH}_3)_2}}$$

20.18 PREPARATION OF NITRILES

Nitriles contain the $-C\equiv N$ functional group. We have already discussed the two main procedures by which they are prepared, namely, the nucleophilic substitution of alkyl halides by cyanide and the conversion of aldehydes and ketones to cyanohydrins. Table 20.6 reviews aspects of these reactions. Neither of the reactions in Table 20.6 is suitable for aryl nitriles (ArC$\equiv$N); these compounds are readily prepared by a reaction to be discussed in Chapter 22.

Both alkyl and aryl nitriles are accessible by dehydration of amides.

$$\underset{\substack{\text{Amide}\\\text{(R may be alkyl}\\\text{or aryl)}}}{\overset{O}{\overset{\|}{\text{RCNH}_2}}} \longrightarrow \underset{\substack{\text{Nitrile}\\\text{(R may be alkyl}\\\text{or aryl)}}}{\text{RC}\equiv\text{N}} + \underset{\text{Water}}{\text{H}_2\text{O}}$$

TABLE 20.6	Preparation of Nitriles
Reaction (section) and comments	**General equation and specific example**
Nucleophilic substitution by cyanide ion (Sections 8.1, 8.13) Cyanide ion is a good nucleophile and reacts with alkyl halides to give nitriles. The reaction is of the S_N2 type and is limited to primary and secondary alkyl halides. Tertiary alkyl halides undergo elimination; aryl and vinyl halides do not react.	$\underset{\substack{\text{Cyanide}\\\text{ion}}}{:\text{N}\equiv\bar{\text{C}}:} + \underset{\substack{\text{Alkyl}\\\text{halide}}}{\text{R}-\text{X}} \longrightarrow \underset{\substack{\text{Nitrile}}}{\text{RC}\equiv\text{N}} + \underset{\substack{\text{Halide}\\\text{ion}}}{\text{X}^-}$ $\underset{\text{1-Chlorodecane}}{\text{CH}_3(\text{CH}_2)_8\text{CH}_2\text{Cl}} \xrightarrow[\substack{\text{ethanol}-\\\text{water}}]{\text{KCN}} \underset{\text{Undecanenitrile (95\%)}}{\text{CH}_3(\text{CH}_2)_8\text{CH}_2\text{CN}}$
Cyanohydrin formation (Section 17.7) Hydrogen cyanide adds to the carbonyl group of aldehydes and ketones.	$\underset{\substack{\text{Aldehyde or}\\\text{ketone}}}{\overset{O}{\overset{\|}{\text{RCR}'}}} + \underset{\substack{\text{Hydrogen}\\\text{cyanide}}}{\text{HCN}} \longrightarrow \underset{\substack{\text{Cyanohydrin}}}{\overset{\text{OH}}{\underset{\overset{\|}{C\equiv N}}{\text{R}\overset{\|}{\text{C}}\text{R}'}}}$ $\underset{\text{3-Pentanone}}{\overset{O}{\overset{\|}{\text{CH}_3\text{CH}_2\text{CCH}_2\text{CH}_3}}} \xrightarrow[\text{H}_3\text{O}^+]{\text{KCN}} \underset{\substack{\text{3-Pentanone}\\\text{cyanohydrin (75\%)}}}{\overset{\text{OH}}{\underset{\text{CN}}{\text{CH}_3\text{CH}_2\overset{\|}{\text{C}}\text{CH}_2\text{CH}_3}}}$

CONDENSATION POLYMERS: POLYAMIDES AND POLYESTERS

All fibers are polymers of one kind or another. Cotton, for example, is cellulose, and cellulose is a naturally occurring polymer of glucose. Silk and wool are naturally occurring polymers of amino acids. An early goal of inventors and entrepreneurs was to produce fibers from other naturally occurring polymers. Their earliest efforts consisted of chemically modifying the short cellulose fibers obtained from wood so that they could be processed into longer fibers more like cotton and silk. These efforts were successful, and the resulting fibers of modified cellulose, known generically as *rayon*, have been produced by a variety of techniques since the late nineteenth century.

A second approach involved direct chemical synthesis of polymers by connecting appropriately chosen small molecules together into a long chain. In 1938, E. I. Du Pont de Nemours and Company announced the development of *nylon*, the first synthetic polymer fiber.

The leader of DuPont's effort was Wallace H. Carothers,[*] who reasoned that he could reproduce the properties of silk by constructing a polymer chain held together, as is silk, by amide bonds. The necessary amide bonds were formed by heating a dicarboxylic acid with a diamine. Hexanedioic acid (*adipic acid*) and 1,6-hexanediamine (*hexamethylenediamine*) react to give a salt that, when heated, gives a **polyamide** called *nylon 66*. The amide bonds form by a condensation reaction, and nylon 66 is an example of a **condensation polymer.**

$$\underset{\text{Adipic acid}}{\text{HOC(CH}_2)_4\text{COH}} + \underset{\text{Hexamethylenediamine}}{\text{H}_2\text{N(CH}_2)_6\text{NH}_2} \longrightarrow {}^-\text{OC(CH}_2)_4\text{CO}^- \quad {}^+\text{H}_3\text{N(CH}_2)_6\overset{+}{\text{N}}\text{H}_3$$

$${}^-\text{OC(CH}_2)_4\text{C} \left[\text{NH(CH}_2)_6\text{NHC(CH}_2)_4\text{C} \right]_n \text{NH(CH}_2)_6\overset{+}{\text{N}}\text{H}_3 \quad \xleftarrow{\text{heat, } -\text{H}_2\text{O}}$$

Nylon 66

The first "6" in nylon 66 stands for the number of carbons in the diamine, the second for the number of carbons in the dicarboxylic acid. Nylon 66 was an immediate success and fostered the development of a large number of related polyamides, many of which have also found their niche in the marketplace.

A slightly different class of polyamides is the *aramids* (*aromatic polyamides*). Like the nylons, the aramids are prepared from a dicarboxylic acid and a diamine, but the functional groups are anchored to benzene rings. An example of an aramid is *Kevlar*, which is a polyamide derived from 1,4-benzenedicarboxylic acid (*terephthalic acid*) and 1,4-benzenediamine (*p-phenylenediamine*):

Kevlar (a polyamide of the aramid class)

Kevlar fibers are very strong, which makes Kevlar a popular choice in applications where the ratio of strength to weight is important. For example, a cable made from Kevlar weighs only one fifth as much as a steel one but is just as strong. Kevlar is also used to make lightweight bulletproof vests.

Nomex is another aramid fiber. Kevlar and Nomex differ only in that the substitution pattern in the aromatic rings is para in Kevlar but meta in Nomex. Nomex is best known for its fire-resistant properties and is used in protective clothing for firefighters, astronauts, and race-car drivers.

[*]For an account of Carothers' role in the creation of nylon, see the September 1988 issue of the *Journal of Chemical Education* (pp. 803–808).

—*Cont.*

Polyesters are a second class of condensation polymers, and the principles behind their synthesis parallel those of polyamides. Ester formation between the functional groups of a dicarboxylic acid and a diol serve to connect small molecules together into a long polyester. The most familiar example of a polyester is *Dacron,* which is prepared from 1,4-benzenedicarboxylic acid and 1,2-ethanediol (*ethylene glycol*):

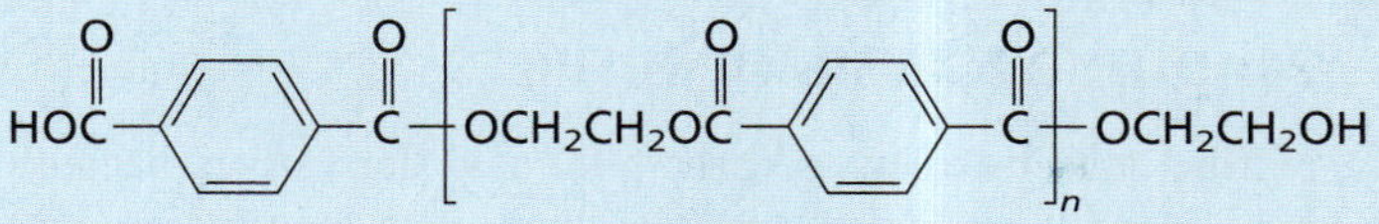

Dacron (a polyester)

The production of polyester fibers leads that of all other types. Annual United States production of polyester fibers is 1.6 million tons versus 1.4 million tons for cotton and 1.0 million tons for nylon. Wool and silk trail far behind at 0.04 and 0.01 million tons, respectively.

Not all synthetic polymers are used as fibers. *Mylar,* for example, is chemically the same as Dacron, but is prepared in the form of a thin film instead of a fiber. *Lexan* is a polyester which, because of its impact resistance, is used as a shatterproof substitute for glass. It is a **polycarbonate** having the structure shown:

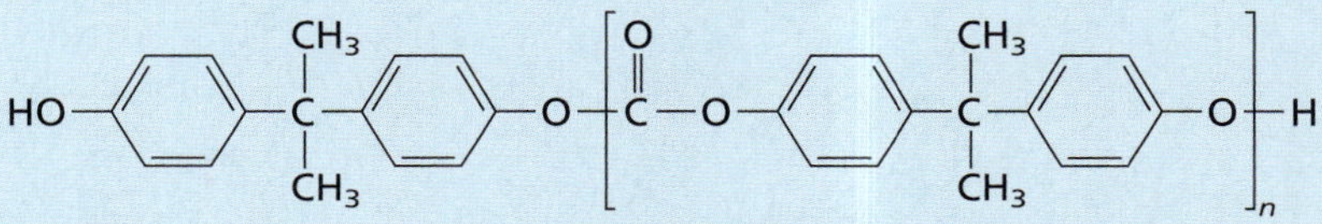

Lexan (a polycarbonate)

In terms of the number of scientists and engineers involved, research and development in polymer chemistry is the principal activity of the chemical industry. The initial goal of making synthetic materials that are the equal of natural fibers has been more than met; it has been far exceeded. What is also important is that all of this did not begin with a chance discovery. It began with a management decision to do basic research in a specific area, and to support it in the absence of any guarantee that success would be quickly achieved.[†]

[†]The April 1988 issue of the *Journal of Chemical Education* contains a number of articles on polymers, including a historical review entitled "Polymers Are Everywhere" (pp. 327–334) and a glossary of terms (pp. 314–319).

Among the reagents used to effect the dehydration of amides is the compound P_4O_{10}, known by the common name *phosphorus pentoxide* because it was once thought to have the molecular formula P_2O_5. Phosphorus pentoxide is the anhydride of phosphoric acid and is used in a number of reactions requiring dehydrating agents.

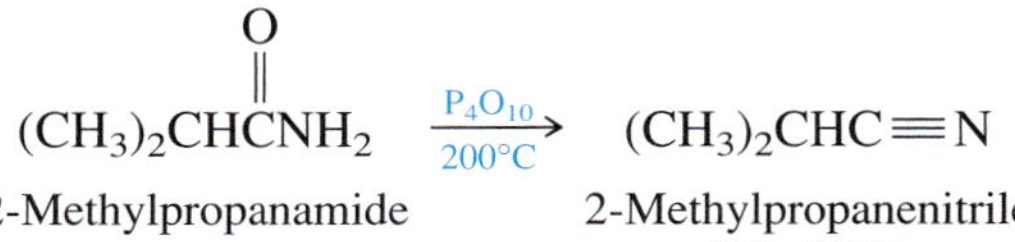

$$(CH_3)_2CHCNH_2 \xrightarrow[200°C]{P_4O_{10}} (CH_3)_2CHC{\equiv}N$$

2-Methylpropanamide → 2-Methylpropanenitrile (69–86%)

PROBLEM 20.21 Show how ethyl alcohol could be used to prepare (a) CH_3CN and (b) CH_3CH_2CN. Along with ethyl alcohol you may use any necessary inorganic reagents.

An important nitrile is *acrylonitrile*, $H_2C=CHCN$. It is prepared industrially from propene, ammonia, and oxygen in the presence of a special catalyst. Polymers of acrylonitrile have many applications, the most prominent being their use in the preparation of acrylic fibers.

20.19 HYDROLYSIS OF NITRILES

Nitriles are classified as carboxylic acid derivatives because they are converted to carboxylic acids on hydrolysis. The conditions required are similar to those for the hydrolysis of amides, namely, heating in aqueous acid or base for several hours. Like the hydrolysis of amides, nitrile hydrolysis is irreversible in the presence of acids or bases. Acid hydrolysis yields ammonium ion and a carboxylic acid.

$$RC\equiv N \; + \; H_2O \; + \; H_3O^+ \; \longrightarrow \; RCOH \; + \; \overset{+}{N}H_4$$

| Nitrile | Water | Hydronium ion | | Carboxylic acid | Ammonium ion |

$$O_2N-\!\!\!\left\langle\underset{}{\bigcirc}\right\rangle\!\!\!-CH_2CN \; \xrightarrow[\text{heat}]{H_2O,\ H_2SO_4} \; O_2N-\!\!\!\left\langle\underset{}{\bigcirc}\right\rangle\!\!\!-CH_2COH$$

p-Nitrobenzyl cyanide   *p*-Nitrophenylacetic acid (92–95%)

In aqueous base, hydroxide ion abstracts a proton from the carboxylic acid. Isolating the acid requires a subsequent acidification step.

$$RC\equiv N \; + \; H_2O \; + \; HO^- \; \longrightarrow \; RCO^- \; + \; NH_3$$

| Nitrile | Water | Hydroxide ion | | Carboxylate ion | Ammonia |

$$CH_3(CH_2)_9CN \; \xrightarrow[\text{2. } H_3O^+]{\text{1. KOH, } H_2O,\ \text{heat}} \; CH_3(CH_2)_9COH$$

Undecanenitrile   Undecanoic acid (80%)

Nitriles are susceptible to nucleophilic addition. In their hydrolysis, water adds to the carbon–nitrogen triple bond. In a series of proton-transfer steps, an amide is produced:

$$RC\equiv N\!: \; + \; H_2O \; \rightleftharpoons \; RC\overset{OH}{\underset{NH}{\diagdown\!\!\diagup}} \; \rightleftharpoons \; RC\overset{O}{\underset{NH_2}{\diagdown\!\!\diagup}}$$

Nitrile   Water   Imino acid   Amide

We already discussed both the acidic and basic hydrolysis of amides (see Section 20.17). All that remains to complete the mechanistic picture of nitrile hydrolysis is to examine the conversion of the nitrile to the corresponding amide.

Nucleophilic addition to the nitrile may be either acid- or base-catalyzed. In aqueous base, hydroxide adds to the carbon–nitrogen triple bond:

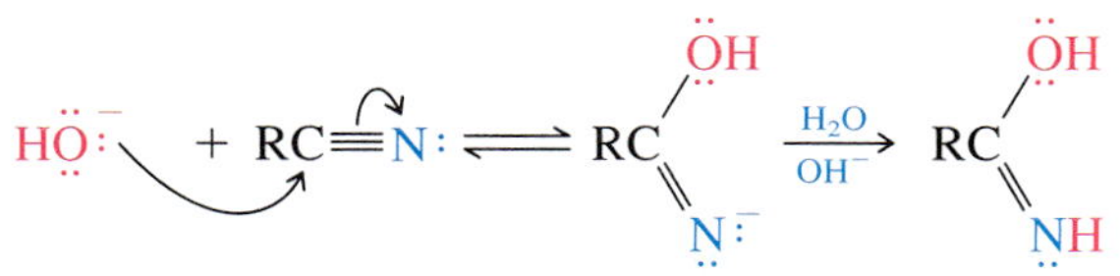

Hydroxide ion	Nitrile	Imino acid

The imino acid is transformed to the amide by the sequence

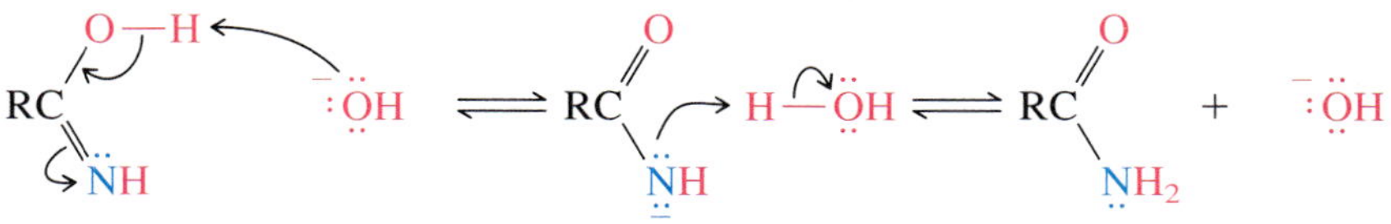

Imino acid	Hydroxide ion	Amide anion	Water	Amide	Hydroxide ion

> **PROBLEM 20.22** Suggest a reasonable mechanism for the conversion of a nitrile (RCN) to the corresponding amide in aqueous acid.

Nucleophiles other than water can also add to the carbon–nitrogen triple bond of nitriles. In the following section we will see a synthetic application of such a nucleophilic addition.

20.20 ADDITION OF GRIGNARD REAGENTS TO NITRILES

The carbon–nitrogen triple bond of nitriles is much less reactive toward nucleophilic addition than is the carbon–oxygen double bond of aldehydes and ketones. Strongly basic nucleophiles such as Grignard reagents, however, do react with nitriles in a reaction that is of synthetic value:

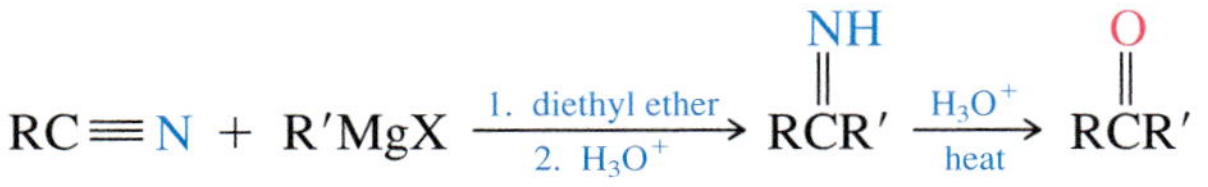

Nitrile	Grignard reagent		Imine	Ketone

The imine formed by nucleophilic addition of the Grignard reagent to the nitrile is normally not isolated but is hydrolyzed directly to a ketone. The overall sequence is used as a means of preparing ketones.

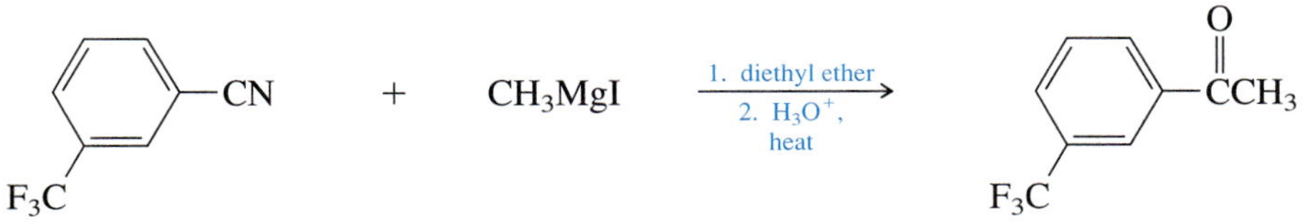

m-(Trifluoromethyl)benzonitrile	Methylmagnesium iodide	m-(Trifluoromethyl)acetophenone (79%)

> **PROBLEM 20.23** Write an equation showing how you could prepare ethyl phenyl ketone from propanenitrile and a Grignard reagent. What is the structure of the imine intermediate?

Organolithium reagents react in the same way and are often used instead of Grignard reagents.

20.21 SPECTROSCOPIC ANALYSIS OF CARBOXYLIC ACID DERIVATIVES

The C=O stretching vibrations of these compounds may be viewed on Learning By Modeling.

Infrared: IR spectroscopy is quite useful in identifying carboxylic acid derivatives. The carbonyl stretching vibration is very strong, and its position is sensitive to the nature of the carbonyl group. In general, electron donation from the substituent decreases the double-bond character of the bond between carbon and oxygen and decreases the stretching frequency. Two distinct absorptions are observed for the symmetric and antisymmetric stretching vibrations of the anhydride function.

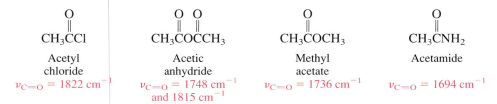

Acetyl chloride	Acetic anhydride	Methyl acetate	Acetamide
$\nu_{C=O} = 1822$ cm^{-1}	$\nu_{C=O} = 1748$ cm^{-1} and 1815 cm^{-1}	$\nu_{C=O} = 1736$ cm^{-1}	$\nu_{C=O} = 1694$ cm^{-1}

Nitriles are readily identified by absorption due to —C≡N stretching in the 2210–2260 cm^{-1} region.

^{1}H NMR: Chemical-shift differences in their ^{1}H NMR spectra aid the structure determination of esters. Consider the two isomeric esters: ethyl acetate and methyl propanoate. As Figure 20.9 shows, the number of signals and their multiplicities are the same for both esters. Both have a methyl singlet and a triplet–quartet pattern for their ethyl group.

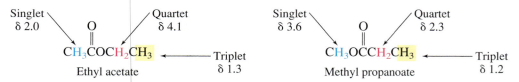

Notice, however, that there is a significant difference in the chemical shifts of the corresponding signals in the two spectra. The methyl singlet is more shielded (δ 2.0) when it is bonded to the carbonyl group of ethyl acetate than when it is bonded to the oxygen of methyl propanoate (δ 3.6). The methylene quartet is more shielded (δ 2.3) when it is bonded to the carbonyl group of methyl propanoate than when it is bonded to the oxygen of ethyl acetate (δ 4.1). Analysis of only the number of peaks and their splitting patterns will not provide an unambiguous answer to structure assignment in esters; chemical-shift data must also be considered.

The chemical shift of the N—H proton of amides appears in the range δ 5–8. It is often a very broad peak; sometimes it is so broad that it does not rise much over the baseline and can be lost in the background noise.

^{13}C NMR: The ^{13}C NMR spectra of carboxylic acid derivatives, like the spectra of carboxylic acids themselves, are characterized by a low-field resonance for the carbonyl

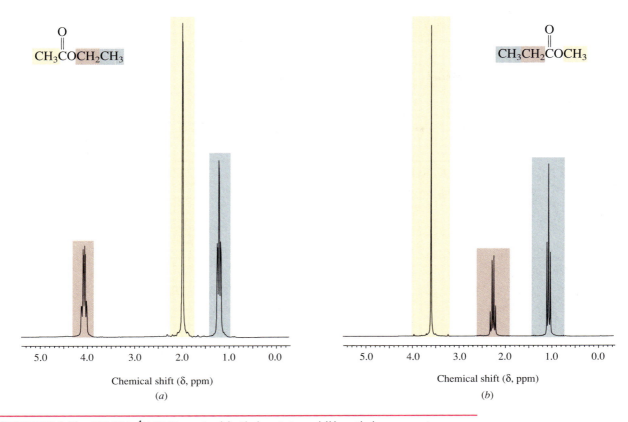

FIGURE 20.9 The 200-MHz ^{1}H NMR spectra (a) ethyl acetate and (b) methyl propanoate.

carbon in the range δ 160–180. The carbonyl carbons of carboxylic acid derivatives are more shielded than those of aldehydes and ketones, but less shielded than the sp^2-hybridized carbons of alkenes and arenes.

The carbon of a C≡N group appears near δ 120.

UV-VIS: The following values are typical for the $n{\rightarrow}\pi^*$ absorption associated with the C=O group of carboxylic acid derivatives.

	$\underset{\substack{\text{O}\\\|\|}}{\text{CH}_3\text{CCl}}$	$\underset{\substack{\text{O O}\\\|\|\ \|\|}}{\text{CH}_3\text{COCCH}_3}$	$\underset{\substack{\text{O}\\\|\|}}{\text{CH}_3\text{COCH}_3}$	$\underset{\substack{\text{O}\\\|\|}}{\text{CH}_3\text{CNH}_2}$
	Acetyl chloride	Acetic anhydride	Methyl acetate	Acetamide
λ_{max}	235 nm	225 nm	207 nm	214 nm

Mass Spectrometry: A prominent peak in the mass spectra of most carboxylic acid derivatives corresponds to an acylium ion derived by cleavage of the bond to the carbonyl group:

$$R{-}C\overset{\ddot{\text{O}}{:}^+}{\underset{\text{X}{:}}{}} \longrightarrow R{-}C{\equiv}\overset{+}{\text{O}}{:} \ + \ \cdot\text{X}{:}$$

Amides, however, tend to cleave in the opposite direction to produce a nitrogen-stabilized acylium ion:

$$R-\overset{+}{C}\overset{..\,..}{O:} \longrightarrow R\cdot + [:\overset{+}{O}\!\!\equiv\!\!C-\overset{..}{N}R'_2 \longleftrightarrow :\overset{..}{O}\!\!=\!\!C\!\!=\!\!\overset{+}{N}R'_2]$$

20.22 SUMMARY

Section 20.1   This chapter concerns the preparation and reactions of *acyl chlorides, acid anhydrides, thioesters, esters, amides,* and *nitriles*. These compounds are generally classified as carboxylic acid derivatives, and their nomenclature is based on that of carboxylic acids.

$\underset{\text{Acyl chloride}}{\overset{O}{\underset{\|}{RCCl}}}$	$\underset{\text{Carboxylic acid anhydride}}{\overset{O\ \ \ O}{\underset{\| \ \ \ \|}{RCOCR}}}$	$\underset{\text{Thioester}}{\overset{O}{\underset{\|}{RCSR'}}}$	$\underset{\text{Ester}}{\overset{O}{\underset{\|}{RCOR'}}}$	$\underset{\text{Amide}}{\overset{O}{\underset{\|}{RCNR'_2}}}$	$\underset{\text{Nitrile}}{RC\!\!\equiv\!\!N}$

Section 20.2   The structure and reactivity of carboxylic acid derivatives depend on how well the atom bonded to the carbonyl group donates electrons to it.

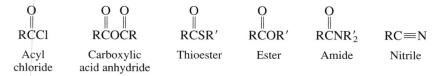

Electron-pair donation stabilizes the carbonyl group and makes it less reactive toward nucleophilic acyl substitution.

Most reactive   Least reactive

$$\overset{O}{\underset{\|}{RCCl}} > \overset{O\ \ \ O}{\underset{\| \ \ \ \|}{RCOCR}} > \overset{O}{\underset{\|}{RCSR'}} > \overset{O}{\underset{\|}{RCOR'}} > \overset{O}{\underset{\|}{RCNR'_2}}$$

Least stabilized carbonyl group   Most stabilized carbonyl group

Nitrogen is a better electron-pair donor than oxygen, and amides have a more stabilized carbonyl group than esters and anhydrides. Chlorine is the poorest electron-pair donor, and acyl chlorides have the least stabilized carbonyl group and are the most reactive.

Section 20.3   The characteristic reaction of acyl chlorides, acid anhydrides, esters, and amides is **nucleophilic acyl substitution.** Addition of a nucleophilic reagent :Nu—H to the carbonyl group leads to a tetrahedral intermediate that dissociates to give the product of substitution:

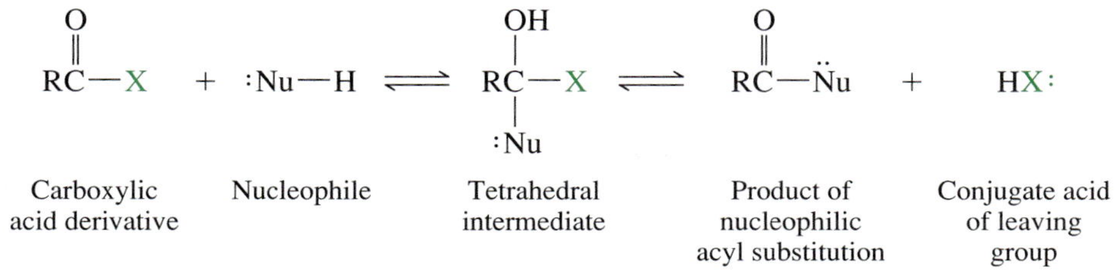

$$
\underset{\substack{\text{Carboxylic} \\ \text{acid derivative}}}{RC\overset{O}{\overset{\|}{-}}X} + \underset{\text{Nucleophile}}{:Nu-H} \rightleftharpoons \underset{\substack{\text{Tetrahedral} \\ \text{intermediate}}}{\overset{OH}{\underset{:Nu}{RC-X}}} \rightleftharpoons \underset{\substack{\text{Product of} \\ \text{nucleophilic} \\ \text{acyl substitution}}}{RC\overset{O}{\overset{\|}{-}}\ddot{N}u} + \underset{\substack{\text{Conjugate acid} \\ \text{of leaving} \\ \text{group}}}{HX:}
$$

Section 20.4   Acyl chlorides are converted to acid anhydrides, esters, and amides by nucleophilic acyl substitution.

$$
\underset{\substack{\text{Acyl} \\ \text{chloride}}}{RCCl} + \underset{\substack{\text{Carboxylic} \\ \text{acid}}}{R'COH} \longrightarrow \underset{\substack{\text{Acid} \\ \text{anhydride}}}{RCOCR'} + \underset{\substack{\text{Hydrogen} \\ \text{chloride}}}{HCl}
$$

$$
\underset{\text{Acyl chloride}}{RCCl} + \underset{\text{Alcohol}}{R'OH} \longrightarrow \underset{\text{Ester}}{RCOR'} + \underset{\substack{\text{Hydrogen} \\ \text{chloride}}}{HCl}
$$

$$
\underset{\substack{\text{Acyl} \\ \text{chloride}}}{RCCl} + \underset{\text{Amine}}{2R'_2NH} \longrightarrow \underset{\text{Amide}}{RCNR'_2} + \underset{\substack{\text{Ammonium} \\ \text{chloride salt}}}{R'_2\overset{+}{N}H_2\ Cl^-}
$$

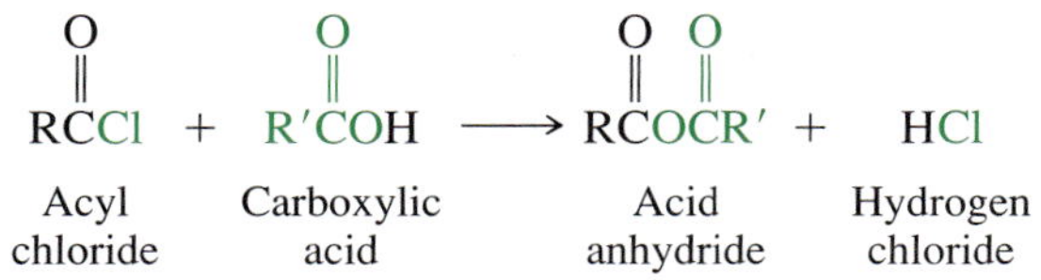

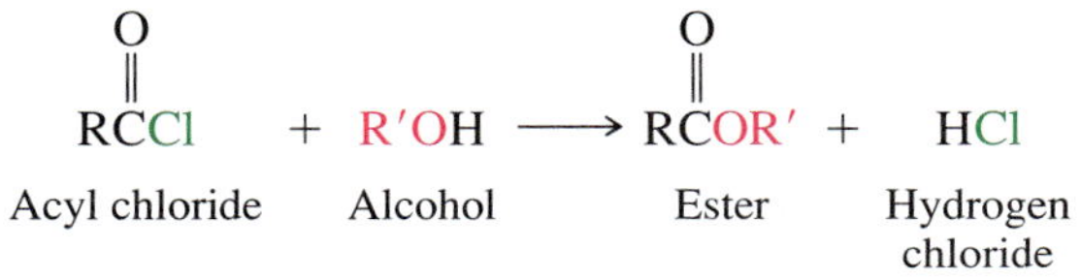

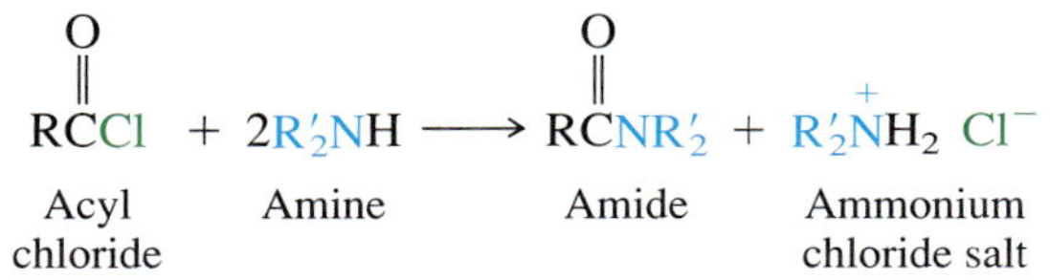

Examples of each of these reactions may be found in Table 20.1.

Section 20.5   Acid anhydrides may be prepared from acyl chlorides in the laboratory, but the most commonly encountered ones (acetic anhydride, phthalic anhydride, and maleic anhydride) are industrial chemicals prepared by specialized methods.

Section 20.6   Acid anhydrides are less reactive toward nucleophilic acyl substitution than acyl chlorides, but are useful reagents for preparing esters and amides.

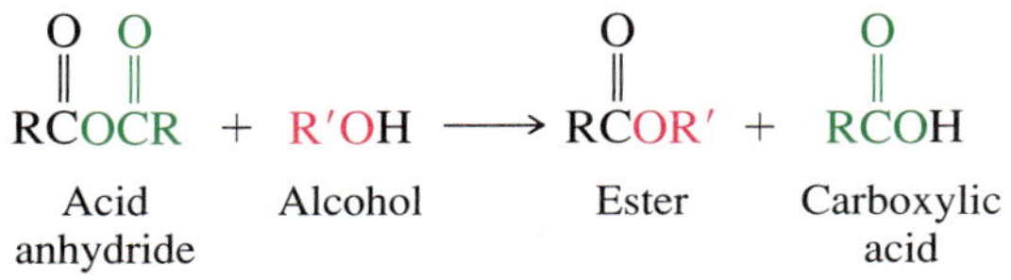

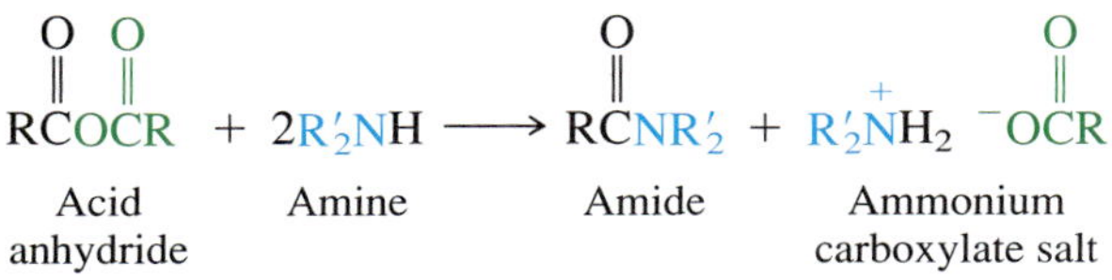

$$
\underset{\substack{\text{Acid} \\ \text{anhydride}}}{RCOCR} + \underset{\text{Alcohol}}{R'OH} \longrightarrow \underset{\text{Ester}}{RCOR'} + \underset{\substack{\text{Carboxylic} \\ \text{acid}}}{RCOH}
$$

$$
\underset{\substack{\text{Acid} \\ \text{anhydride}}}{RCOCR} + \underset{\text{Amine}}{2R'_2NH} \longrightarrow \underset{\text{Amide}}{RCNR'_2} + \underset{\substack{\text{Ammonium} \\ \text{carboxylate salt}}}{R'_2\overset{+}{N}H_2\ {}^-OCR}
$$

Table 20.2 presents examples of these reactions.

Section 20.7   Esters occur naturally or are prepared from alcohols by Fischer esterification or by acylation with acyl chlorides or acid anhydrides (see Table 20.3).

Section 20.8   Esters are polar and have higher boiling points than alkanes of comparable size and shape. Esters don't form hydrogen bonds to other ester molecules so have lower boiling points than analogous alcohols. They can form hydrogen bonds to water and so are comparable to alcohols in their solubility in water.

Section 20.9   Esters react with Grignard reagents and are reduced by lithium aluminum hydride (Table 20.4).

Section 20.10 Ester hydrolysis can be catalyzed by acids and its mechanism (Figure 20.4) is the reverse of the mechanism for Fischer esterification. The reaction proceeds via a tetrahedral intermediate.

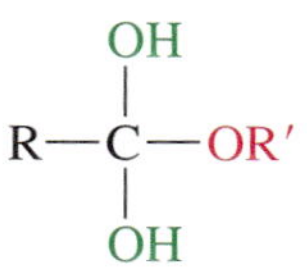

$$R-\underset{\underset{OH}{|}}{\overset{\overset{OH}{|}}{C}}-OR'$$

Tetrahedral intermediate
in ester hydrolysis

Section 20.11 Ester hydrolysis in basic solution is called *saponification* and proceeds through the same tetrahedral intermediate (Figure 20.5) as in acid-catalyzed hydrolysis. Unlike acid-catalyzed hydrolysis, saponification is irreversible because the carboxylic acid is deprotonated under the reaction conditions.

$$\underset{\text{Ester}}{\overset{\overset{O}{\|}}{RCOR'}} + \underset{\substack{\text{Hydroxide}\\\text{ion}}}{HO^-} \longrightarrow \underset{\substack{\text{Carboxylate}\\\text{ion}}}{\overset{\overset{O}{\|}}{RCO^-}} + \underset{\text{Alcohol}}{R'OH}$$

Section 20.12 Esters react with amines to give amides.

$$\underset{\text{Ester}}{\overset{\overset{O}{\|}}{RCOR'}} + \underset{\text{Amine}}{R''_2NH} \longrightarrow \underset{\text{Amide}}{\overset{\overset{O}{\|}}{RCNR''_2}} + \underset{\text{Alcohol}}{R'OH}$$

Section 20.13 Thioesters undergo reactions analogous to those of esters, but at faster rates. A sulfur atom stabilizes a carbonyl group less effectively than an oxygen.

$$\underset{\text{Ester}}{\overset{\overset{O}{\|}}{\underset{R}{\diagup}C\underset{OR'}{\diagdown}}} \qquad \underset{\text{Thioester}}{\overset{\overset{O}{\|}}{\underset{R}{\diagup}C\underset{SR'}{\diagdown}}}$$

Section 20.14 Amides are normally prepared by the reaction of amines with acyl chlorides, anhydrides, or esters.

Section 20.15 Lactams are cyclic amides.

Section 20.16   Imides are compounds that have two acyl groups attached to nitrogen.

Section 20.17   Like ester hydrolysis, amide hydrolysis can be achieved in either aqueous acid or aqueous base. The process is irreversible in both media. In base, the carboxylic acid is converted to the carboxylate anion; in acid, the amine is protonated to an ammonium ion:

$$
\underset{\substack{\text{Amide}}}{\text{RCNR}_2'} + \underset{\substack{\text{Water}}}{\text{H}_2\text{O}}
$$

$$
\xrightarrow{\text{H}_3\text{O}^+} \underset{\substack{\text{Carboxylic} \\ \text{acid}}}{\text{RCOH}} + \underset{\substack{\text{Ammonium} \\ \text{ion}}}{\overset{+}{\text{R}_2'\text{NH}_2}}
$$

$$
\xrightarrow{\text{HO}^-} \underset{\substack{\text{Carboxylate} \\ \text{ion}}}{\text{RCO}^-} + \underset{\substack{\text{Amine}}}{\text{R}_2'\text{NH}}
$$

Section 20.18   Nitriles are prepared by nucleophilic substitution (S_N2) of alkyl halides with cyanide ion, by converting aldehydes or ketones to cyanohydrins (Table 20.6), or by dehydration of amides.

Section 20.19   The hydrolysis of nitriles to carboxylic acids is irreversible in both acidic and basic solution.

$$
\underset{\substack{\text{Nitrile}}}{\text{RC}\equiv\text{N}} \xrightarrow[\substack{\text{or} \\ \text{1. H}_2\text{O, HO}^-,\ \text{heat} \\ \text{2. H}_3\text{O}^+}]{\text{H}_3\text{O}^+,\ \text{heat}} \underset{\substack{\text{Carboxylic acid}}}{\text{RCOH}}
$$

Section 20.20   Nitriles are useful starting materials for the preparation of ketones by reaction with Grignard reagents.

$$
\underset{\substack{\text{Nitrile}}}{\text{RC}\equiv\text{N}} + \underset{\substack{\text{Grignard reagent}}}{\text{R}'\text{MgX}} \xrightarrow[\substack{\text{2. H}_3\text{O}^+,\ \text{heat}}]{\text{1. diethyl ether}} \underset{\substack{\text{Ketone}}}{\text{RCR}'}
$$

Section 20.21   Acyl chlorides, anhydrides, esters, and amides all show a strong band for C=O stretching in the infrared. The range extends from about 1820 cm^{-1} (acyl chlorides) to 1690 cm^{-1} (amides). Their ^{13}C NMR spectra are characterized by a peak near δ 180 for the carbonyl carbon. ^{1}H NMR spectroscopy is useful for distinguishing between the groups R and R′ in esters (RCO$_2$R′). The protons on the carbon bonded to O in R′ appear at lower field (less shielded) than those on the carbon bonded to C=O.

PROBLEMS

20.24 Write a structural formula for each of the following compounds:

(a) *m*-Chlorobenzoyl chloride

(b) Trifluoroacetic anhydride

(c) *cis*-1,2-Cyclopropanedicarboxylic anhydride

(d) Ethyl cycloheptanecarboxylate

(e) 1-Phenylethyl acetate

(f) 2-Phenylethyl acetate

(g) p-Ethylbenzamide

(h) N-Ethylbenzamide

(i) 2-Methylhexanenitrile

20.25 Give an acceptable IUPAC name for each of the following compounds:

(a) $CH_3CHCH_2\overset{\overset{\displaystyle O}{\|}}{C}Cl$
$\quad\quad\;\; |$
$\quad\quad\; Cl$

(b) 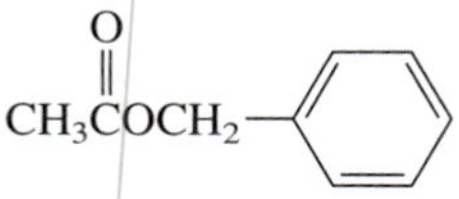 $CH_3\overset{\overset{\displaystyle O}{\|}}{C}OCH_2{-}\bigcirc$

(c) $CH_3O\overset{\overset{\displaystyle O}{\|}}{C}CH_2{-}\bigcirc$

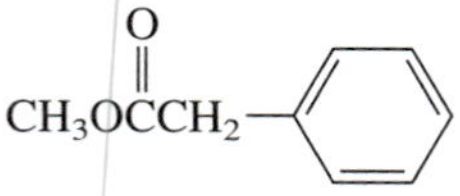

(d) $ClCH_2CH_2\overset{\overset{\displaystyle O}{\|}}{C}O\overset{\overset{\displaystyle O}{\|}}{C}CH_2CH_2Cl$

(e)

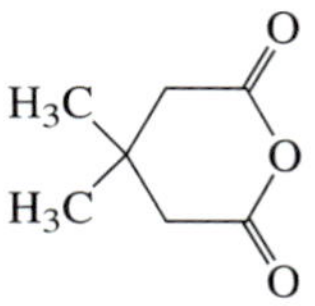

(f) $(CH_3)_2CHCH_2CH_2C{\equiv}N$

(g) $(CH_3)_2CHCH_2CH_2\overset{\overset{\displaystyle O}{\|}}{C}NH_2$

(h) $(CH_3)_2CHCH_2CH_2\overset{\overset{\displaystyle O}{\|}}{C}NHCH_3$

(i) $(CH_3)_2CHCH_2CH_2\overset{\overset{\displaystyle O}{\|}}{C}N(CH_3)_2$

20.26 Write a structural formula for the principal organic product or products of each of the following reactions:

(a) Acetyl chloride and bromobenzene, $AlCl_3$

(b) Acetyl chloride and 1-butanethiol

(c) Propanoyl chloride and sodium propanoate

(d) Butanoyl chloride and benzyl alcohol

(e) p-Chlorobenzoyl chloride and ammonia

(f) 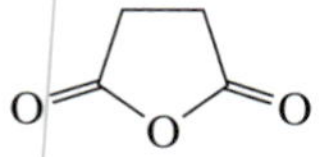   and water

(g) 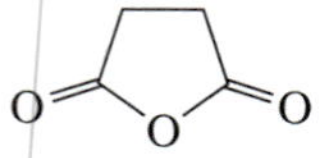   and aqueous sodium hydroxide

(h) 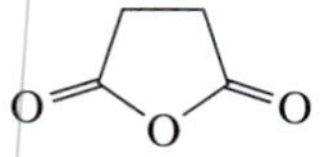   and aqueous ammonia

(i) 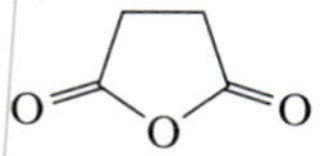   and benzene, $AlCl_3$

(j) 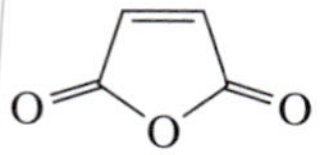   and 1,3-pentadiene

(k) Acetic anhydride and 3-pentanol

(l) and aqueous sodium hydroxide

(m) 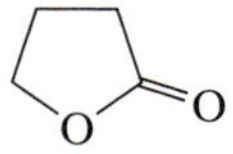 and aqueous ammonia

(n) 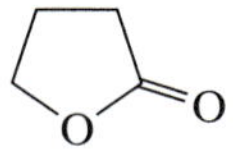 and lithium aluminum hydride, then H$_2$O

(o) 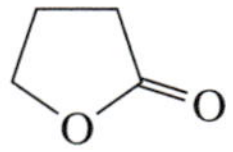 and excess methylmagnesium bromide, then H$_3$O$^+$

(p) Ethyl phenylacetate and methylamine (CH$_3$NH$_2$)

(q) 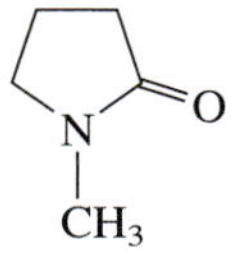 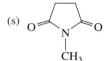 and aqueous sodium hydroxide

(r) 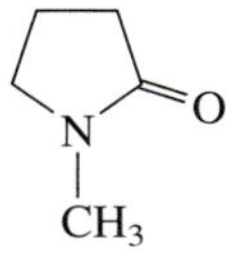 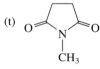 and aqueous hydrochloric acid, heat

(s) and aqueous sodium hydroxide

(t) 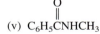 and aqueous hydrochloric acid, heat

(u) C$_6$H$_5$NHCCH$_3$ and aqueous hydrochloric acid, heat

(v) C$_6$H$_5$CNHCH$_3$ and aqueous sulfuric acid, heat

(w) 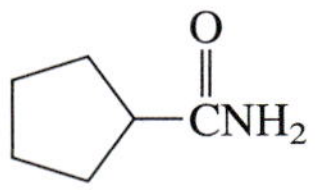 [cyclopentyl]—CNH$_2$ and P$_4$O$_{10}$

(x) (CH$_3$)$_2$CHCH$_2$C≡N and aqueous hydrochloric acid, heat

(y) *p*-Methoxybenzonitrile and aqueous sodium hydroxide, heat

(z) Propanenitrile and methylmagnesium bromide, then H$_3$O$^+$, heat

20.27 Using ethanol as the ultimate source of all the carbon atoms, along with any necessary inorganic reagents, show how you could prepare each of the following:

(a) Acetyl chloride

(b) Acetic anhydride

(c) Ethyl acetate

(d) Ethyl bromoacetate

(e) 2-Bromoethyl acetate

(f) Ethyl cyanoacetate

(g) Acetamide

(h) 2-Hydroxypropanoic acid

20.28 Using toluene as the ultimate source of all the carbon atoms, along with any necessary inorganic reagents, show how you could prepare each of the following:

(a) Benzoyl chloride

(b) Benzoic anhydride

(c) Benzyl benzoate

(d) Benzamide

(e) Benzonitrile

(f) Benzyl cyanide

(g) Phenylacetic acid

(h) *p*-Nitrobenzoyl chloride

(i) *m*-Nitrobenzoyl chloride

20.29 The saponification of ^{18}O-labeled ethyl propanoate was described in Section 20.11 as one of the significant experiments that demonstrated acyl–oxygen cleavage in ester hydrolysis. The ^{18}O-labeled ethyl propanoate used in this experiment was prepared from ^{18}O-labeled ethyl alcohol, which in turn was obtained from acetaldehyde and ^{18}O-enriched water. Write a series of equations showing the preparation of $CH_3CH_2\overset{\overset{\displaystyle O}{\|}}{C}OCH_2CH_3$ (where $O = {}^{18}O$) from these starting materials.

20.30 Suggest a reasonable explanation for each of the following observations:

(a) The second-order rate constant k for saponification of ethyl trifluoroacetate is over 1 million times greater than that for ethyl acetate (25°C).

(b) The second-order rate constant for saponification of ethyl 2,2-dimethylpropanoate, $(CH_3)_3CCO_2CH_2CH_3$, is almost 100 times smaller than that for ethyl acetate (30°C).

(c) The second-order rate constant k for saponification of methyl acetate is 100 times greater than that for *tert*-butyl acetate (25°C).

(d) The second-order rate constant k for saponification of methyl *m*-nitrobenzoate is 40 times greater than that for methyl benzoate (25°C).

(e) The second-order rate constant k for saponification of 5-pentanolide is over 20 times greater than that for 4-butanolide (25°C).

5-Pentanolide   4-Butanolide

(f) The second-order rate constant k for saponification of ethyl *trans*-4-*tert*-butylcyclohexanecarboxylate is 20 times greater than that for its cis diastereomer (25°C).

Ethyl *trans*-4-*tert*-
butylcyclohexanecarboxylate

Ethyl *cis*-4-*tert*-
butylcyclohexanecarboxylate

20.31 The preparation of *cis*-4-*tert*-butylcyclohexanol from its trans stereoisomer was carried out by the following sequence of steps. Write structural formulas, including stereochemistry, for compounds A and B.

Step 1: + H₃C— ⬡ —SO₂Cl $\xrightarrow{\text{pyridine}}$ Compound A

($C_{17}H_{26}O_3S$)

Step 2: Compound A + ⬡—$\overset{\overset{\displaystyle O}{\|}}{C}ONa$ $\xrightarrow[\text{heat}]{\textit{N,N}\text{-dimethylformamide}}$ Compound B

($C_{17}H_{24}O_2$)

Step 3: Compound B $\xrightarrow[\text{H}_2\text{O}]{\text{NaOH}}$

(structure showing cyclohexane ring with OH group and tert-butyl substituent)

20.32 The ketone shown was prepared in a three-step sequence from ethyl trifluoroacetate. The first step in the sequence involved treating ethyl trifluoroacetate with ammonia to give a compound A. Compound A was in turn converted to the desired ketone by way of a compound B. Fill in the missing reagents in the sequence shown, and give the structures of compounds A and B.

$$\underset{\text{CF}_3\text{COCH}_2\text{CH}_3}{\overset{O}{\|}} \xrightarrow{\text{NH}_3} \text{Compound A} \longrightarrow \text{Compound B} \longrightarrow \underset{\text{CF}_3\text{CC(CH}_3)_3}{\overset{O}{\|}}$$

20.33 *Ambrettolide* is obtained from hibiscus and has a musk-like odor. Its preparation from a compound A is outlined in the table that follows. Write structural formulas, ignoring stereochemistry, for compounds B through G in this synthesis. (*Hint:* Zinc, as used in step 4, converts vicinal dibromides to alkenes.)

$$\underset{\text{Compound A}}{\overset{O}{\|}} \text{HOC(CH}_2)_5\text{CH}-\text{CH(CH}_2)_7\text{CH}_2\text{OH} \longrightarrow$$

(structure of Ambrettolide, a macrocyclic lactone with a double bond)

Compound A   Ambrettolide

Step	Reactant	Reagents	Product
1.	Compound A	H_2O, H^+, heat	Compound B ($C_{16}H_{32}O_5$)
2.	Compound B	HBr	Compound C ($C_{16}H_{29}Br_3O_2$)
3.	Compound C	Ethanol, H_2SO_4	Compound D ($C_{18}H_{33}Br_3O_2$)
4.	Compound D	Zinc, ethanol	Compound E ($C_{18}H_{33}BrO_2$)
5.	Compound E	Sodium acetate, acetic acid	Compound F ($C_{20}H_{36}O_4$)
6.	Compound F	KOH, ethanol, then H^+	Compound G ($C_{16}H_{30}O_3$)
7.	Compound G	Heat	Ambrettolide ($C_{16}H_{28}O_2$)

20.34 The preparation of the sex pheromone of the bollworm moth, (*E*)-9,11-dodecadien-1-yl acetate, from compound A has been described. Suggest suitable reagents for each step in this sequence.

(a) $\text{HOCH}_2\text{CH}=\text{CH(CH}_2)_7\text{CO}_2\text{CH}_3 \longrightarrow \underset{}{\overset{O}{\|}}\text{HCCH}=\text{CH(CH}_2)_7\text{CO}_2\text{CH}_3$

    Compound A (*E* isomer)   Compound B

(b) Compound B $\longrightarrow$ $\text{H}_2\text{C}=\text{CHCH}=\text{CH(CH}_2)_7\text{CO}_2\text{CH}_3$

                    Compound C

(c) Compound C $\longrightarrow$ $\text{H}_2\text{C}=\text{CHCH}=\text{CH(CH}_2)_7\text{CH}_2\text{OH}$

                    Compound D

(d) Compound D $\longrightarrow$ H$_2$C=CHCH=CH(CH$_2$)$_7$CH$_2$OCCH$_3$

(E)-9,11-Dodecadien-1-yl acetate

20.35 Outline reasonable mechanisms for each of the following reactions:

(a)

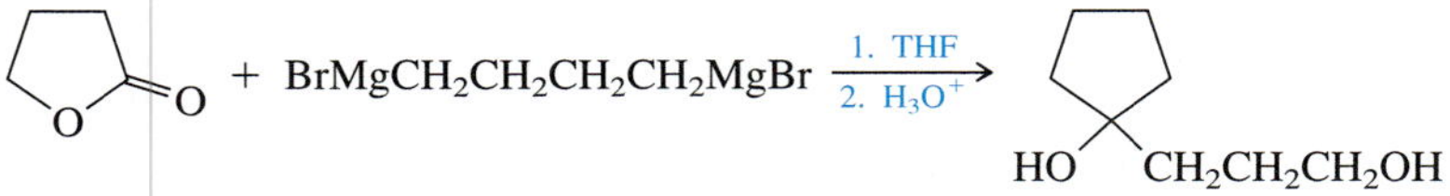

(b)

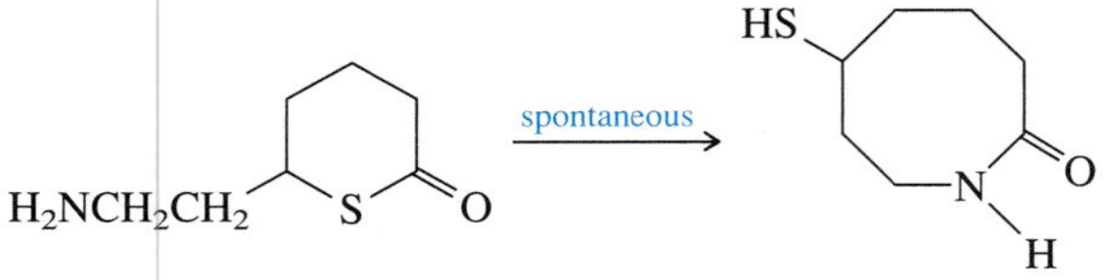

20.36 Identify compounds A through D in the following equations:

(a)

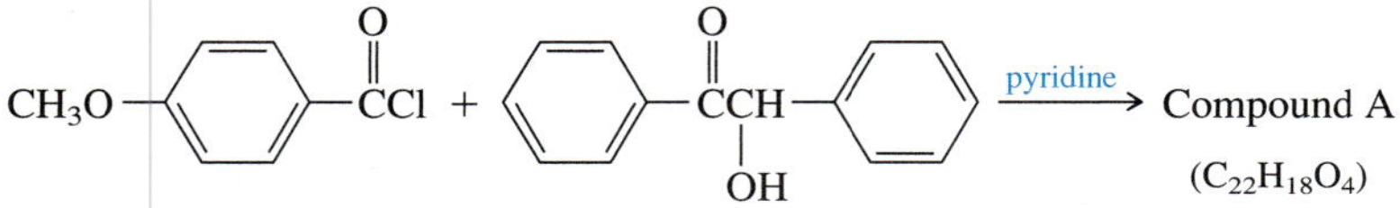

(b) CH$_3$CCH$_2$CH$_2$COCH$_2$CH$_3$ $\xrightarrow[\text{2. H}_3\text{O}^+]{\text{1. CH}_3\text{MgI (1 equiv), diethyl ether}}$ Compound B

(a lactone, C$_6$H$_{10}$O$_2$)

(c)

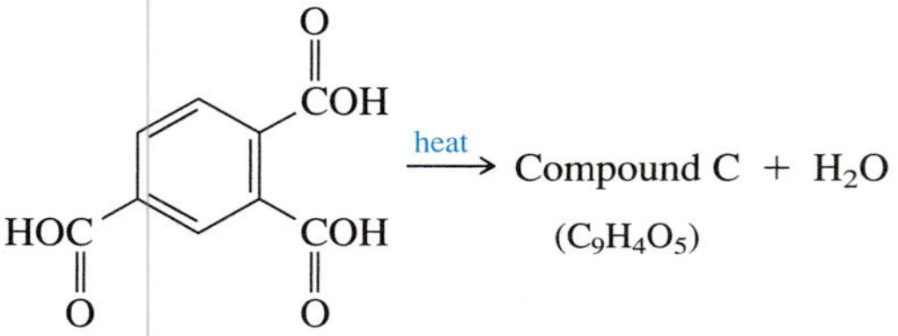

(d)

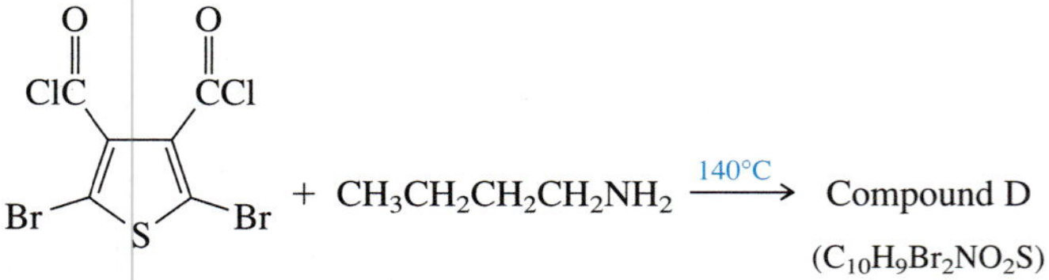

20.37 When compounds of the type represented by A are allowed to stand in pentane, they are converted to a constitutional isomer.

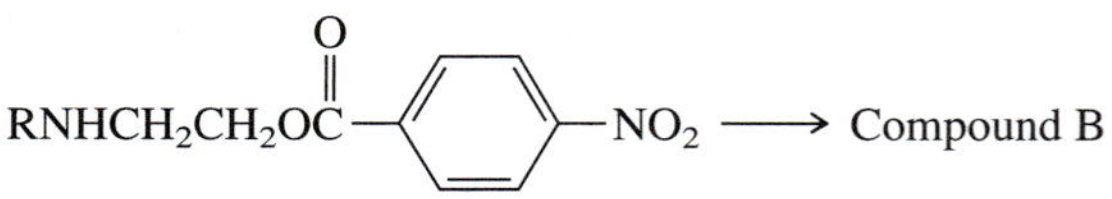

Compound A

Hydrolysis of either A or B yields RNHCH$_2$CH$_2$OH and p-nitrobenzoic acid. Suggest a reasonable structure for compound B, and demonstrate your understanding of the mechanism of this reaction by writing the structure of the key intermediate in the conversion of compound A to compound B.

20.38 (a) In the presence of dilute hydrochloric acid, compound A is converted to a constitutional isomer, compound B.

$$\text{HO} \diagdown \quad \overset{O}{\underset{\parallel}{\text{NHC}}} \diagdown \diagdown \text{—NO}_2 \quad \xrightarrow{\text{H}^+} \quad \text{Compound B}$$

Compound A

Suggest a reasonable structure for compound B.

(b) The trans stereoisomer of compound A is stable under the reaction conditions. Why does it not rearrange?

20.39 Poly(vinyl alcohol) is a useful water-soluble polymer. It cannot be prepared directly from vinyl alcohol because of the rapidity with which vinyl alcohol (H_2C=CHOH) isomerizes to acetaldehyde. Vinyl acetate, however, does not rearrange and can be polymerized to poly(vinyl acetate). How could you make use of this fact to prepare poly(vinyl alcohol)?

$$\left(\text{CH}_2\text{CHCH}_2\text{CH} \atop \underset{\text{OH}}{|} \quad \underset{\text{OH}}{|} \right)_n \qquad \left(\text{CH}_2\text{CHCH}_2\text{CH} \atop \underset{\underset{\text{O}}{\parallel}}{\overset{\text{CH}_3\text{CO}}{|}} \quad \underset{\underset{\text{O}}{\parallel}}{\overset{\text{OCCH}_3}{|}} \right)_n$$

Poly(vinyl alcohol)   Poly(vinyl acetate)

20.40 *Lucite* is a polymer of methyl methacrylate.

(a) Assuming the first step in the polymerization of methyl methacrylate is as shown,

$$\text{R—O}\cdot \; + \; \underset{\underset{\text{CH}_3}{|}}{\text{H}_2\text{C}=\overset{\overset{O}{\parallel}}{\text{C}}\text{COCH}_3} \; \longrightarrow \; \text{ROCH}_2-\underset{\underset{\text{CH}_3}{|}}{\overset{\cdot\;\overset{O}{\parallel}}{\text{C}}\text{COCH}_3}$$

Methyl methacrylate

write a structural formula for the free radical produced after the next two propagation steps.

(b) Outline a synthesis of methyl methacrylate from acetone, sodium cyanide, and any necessary organic or inorganic reagents.

20.41 A certain compound has a molecular weight of 83 and contains nitrogen. Its infrared spectrum contains a moderately strong peak at 2270 cm^{-1}. Its ^{1}H and ^{13}C NMR spectra are shown in Figure 20.10. What is the structure of this compound?

20.42 A compound has a molecular formula of $C_8H_{14}O_4$, and its IR spectrum contains an intense peak at 1730 cm^{-1}. The ^{1}H NMR spectrum of the compound is shown in Figure 20.11. What is its structure?

20.43 A compound ($C_4H_6O_2$) has a strong band in the infrared at 1760 cm^{-1}. Its ^{13}C NMR spectrum exhibits signals at δ 20.2 (CH$_3$), 96.8 (CH$_2$), 141.8 (CH), and 167.6 (C). The ^{1}H NMR spectrum of the compound has a three-proton singlet at δ 2.1 along with three other signals, each of which is a doublet of doublets, at δ 4.7, 4.9, and 7.3 . What is the structure of the compound?

20.44 Excluding enantiomers, there are three isomeric cyclopropanedicarboxylic acids. Two of them, A and B, are constitutional isomers of each other, and each forms a cyclic anhydride on being heated. The third diacid, C, does not form a cyclic anhydride. C is a constitutional isomer of A and a stereoisomer of B. Identify A, B, and C. Construct molecular models of the cyclic anhydrides formed on heating A and B. Why doesn't C form a cyclic anhydride?

FIGURE 20.10 The
(a) 200-MHz ^{1}H and
(b) ^{13}C NMR spectra of the
compound in Problem 20.41.

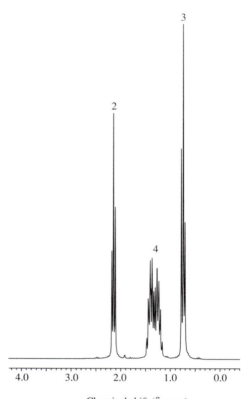

Chemical shift (δ, ppm)

(a)

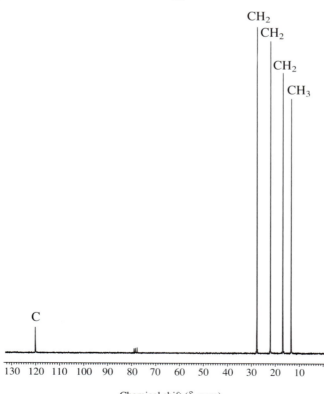

Chemical shift (δ, ppm)

(b)

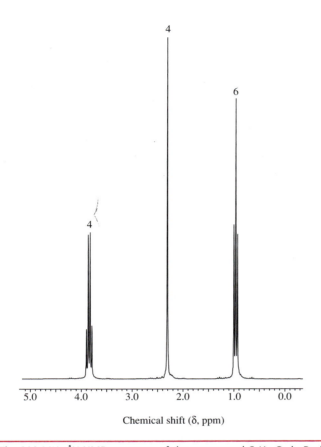

Chemical shift (δ, ppm)

FIGURE 20.11 The 200-MHz ^{1}H NMR spectrum of the compound $C_8H_{14}O_4$ in Problem 20.42.

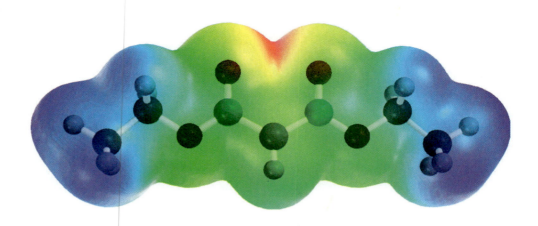

CHAPTER 21

ESTER ENOLATES

You have already had considerable experience with carbanionic compounds and their applications in synthetic organic chemistry. The first was acetylide ion in Chapter 9, followed in Chapter 14 by organometallic compounds—Grignard reagents, for example—that act as sources of negatively polarized carbon. In Chapter 18 you learned that enolate ions—reactive intermediates generated from aldehydes and ketones—are nucleophilic, and that this property can be used to advantage as a method for carbon–carbon bond formation.

The present chapter extends our study of carbanions to the enolate ions derived from esters. **Ester enolates** are important reagents in synthetic organic chemistry. The stabilized enolates derived from **β-keto esters** are particularly useful.

$$R \overset{\overset{\displaystyle O}{\parallel}}{\underset{\beta}{C}} \overset{\alpha}{CH_2} \overset{\overset{\displaystyle O}{\parallel}}{C} OR'$$

β-Keto ester: a ketone carbonyl is β to
the carbonyl group of the ester.

A hydrogen attached to the α-carbon atom of a β-keto ester is relatively acidic. Typical β-keto esters have pK_a values of about 11. Because the α-carbon atom is flanked by two electron-withdrawing carbonyl groups, a carbanion formed at this site is highly stabilized. The electron delocalization in the anion of a β-keto ester is represented by the resonance structures

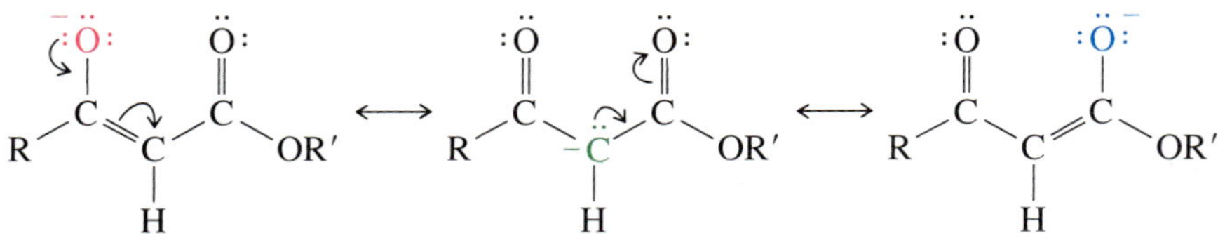

Principal resonance structures of the anion of a β-keto ester

We'll begin by describing the preparation and properties of β-keto esters, proceed to a discussion of their synthetic applications, continue to an examination of related species, and conclude by exploring some recent developments in the active field of synthetic carbanion chemistry.

21.1 THE CLAISEN CONDENSATION

Before describing how β-keto esters are used as reagents for organic synthesis, we need to see how these compounds themselves are prepared. The main method for the preparation of β-keto esters is the **Claisen condensation:**

On treatment with alkoxide bases, esters undergo self-condensation to give a β-keto ester and an alcohol. Ethyl acetate, for example, undergoes a Claisen condensation on treatment with sodium ethoxide to give a β-keto ester known by its common name *ethyl acetoacetate* (also called *acetoacetic ester*):

The systematic IUPAC name of ethyl acetoacetate is *ethyl 3-oxobutanoate*. The presence of a ketone carbonyl group is indicated by the designation "*oxo*" along with the appropriate locant. Thus, there are four carbon atoms in the acyl group of ethyl 3-oxobutanoate, C-3 being the carbonyl carbon of the ketone function.

The mechanism of the Claisen condensation of ethyl acetate is presented in Figure 21.1. The first two steps of the mechanism are analogous to those of aldol addition (Section 18.9). An enolate ion is generated in step 1, which undergoes nucleophilic addition to the carbonyl group of a second ester molecule in step 2. The species formed in this step is a tetrahedral intermediate of the same type that we encountered in our discussion of nucleophilic acyl substitution of esters. It dissociates by expelling an ethoxide ion, as shown in step 3, which restores the carbonyl group to give the β-keto ester. Steps 1 to 3 show two different types of ester reactivity: one molecule of the ester gives rise to an enolate; the second molecule acts as an acylating agent.

Claisen condensations involve two distinct experimental operations. The first stage concludes in step 4 of Figure 21.1, where the base removes a proton from C-2 of the β-keto ester. Because this hydrogen is relatively acidic, the position of equilibrium for step 4 lies far to the right.

Ludwig Claisen was a German chemist who worked during the last two decades of the nineteenth century and the first two decades of the twentieth. His name is associated with three reactions. The *Claisen–Schmidt reaction* was presented in Section 18.10, the *Claisen condensation* is discussed in this section, and the *Claisen rearrangement* will be introduced in Section 24.13.

Overall reaction:

$$2 \ CH_3COCH_2CH_3 \xrightarrow[\text{2. } H_3O^+]{\text{1. } NaOCH_2CH_3} CH_3CCH_2COCH_2CH_3 + CH_3CH_2OH$$

Ethyl acetate   Ethyl 3-oxobutanoate   Ethanol
(ethyl acetoacetate)

Step 1: Proton abstraction from the α carbon atom of ethyl acetate to give the corresponding enolate.

$$CH_3CH_2\ddot{O}: \ + \ {}^-CH_2C\overset{\ddot{O}:}{\underset{OCH_2CH_3}{}} \ \rightleftharpoons \ CH_3CH_2\ddot{O} \ + \ H_2\ddot{C}{-}C\overset{\ddot{O}:}{\underset{OCH_2CH_3}{}} \ \longleftrightarrow \ H_2C{=}C\overset{:\ddot{O}:^-}{\underset{OCH_2CH_3}{}}$$

Ethoxide   Ethyl acetate   Ethanol   Enolate of ethyl acetate

Step 2: Nucleophilic addition of the ester enolate to the carbonyl group of the neutral ester. The product is the anionic form of the tetrahedral intermediate.

$$CH_3\overset{\ddot{O}:}{C}OCH_2CH_3 \ + \ H_2C{=}C\overset{:\ddot{O}:^-}{\underset{OCH_2CH_3}{}} \ \rightleftharpoons \ CH_3\overset{:\ddot{O}:^-}{\underset{OCH_2CH_3}{C}}CH_2\overset{\ddot{O}:}{C}OCH_2CH_3$$

Ethyl acetate   Enolate of ethyl acetate   Anionic form of tetrahedral intermediate

Step 3: Dissociation of the tetrahedral intermediate.

$$CH_3\overset{:\ddot{O}:^-}{\underset{:\ddot{O}CH_2CH_3}{C}}CH_2\overset{\ddot{O}:}{C}OCH_2CH_3 \ \rightleftharpoons \ CH_3\overset{\ddot{O}:}{C}CH_2\overset{\ddot{O}:}{C}OCH_2CH_3 \ + \ {}^-:\ddot{O}CH_2CH_3$$

Anionic form of   Ethyl   Ethoxide
tetrahedral intermediate   3-oxobutanoate   ion

Step 4: Deprotonation of the β-keto ester product.

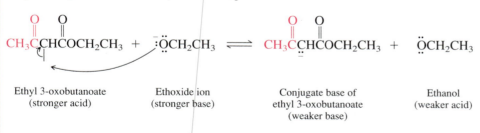

Ethyl 3-oxobutanoate   Ethoxide ion   Conjugate base of   Ethanol
(stronger acid)   (stronger base)   ethyl 3-oxobutanoate   (weaker acid)
                                                  (weaker base)

—*Cont.*

FIGURE 21.1 The mechanism of the Claisen condensation of ethyl acetate.

Step 5: Acidification of the reaction mixture. This is performed in a separate synthetic operation to give the product in its neutral form for eventual isolation.

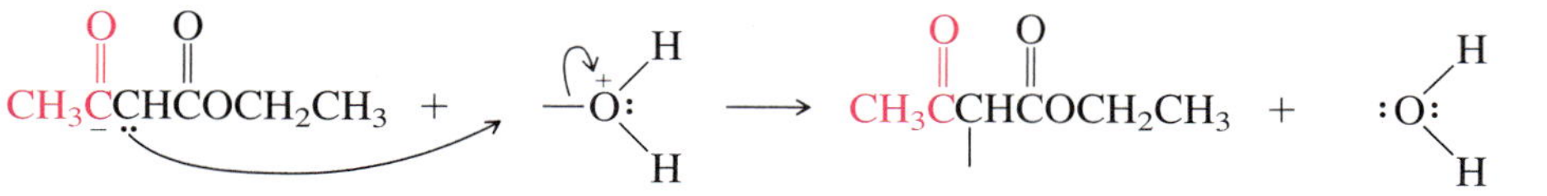

$$CH_3CCHCOCH_2CH_3 \; + \quad \overset{+}{O:}\overset{H}{\underset{H}{\diagup}} \quad \longrightarrow \quad CH_3CCHCOCH_2CH_3 \; + \quad :\overset{H}{\underset{H}{O}:}$$

| Conjugate base of ethyl 3-oxobutanoate (stronger base) | Hydronium ion (stronger acid) | Ethyl 3-oxobutanoate (weaker acid) | Water (weaker base) |

FIGURE 21.1 (Continued)

In general, the equilibrium represented by the sum of steps 1 to 3 is unfavorable. (Two ester carbonyl groups are more stable than one ester plus one ketone carbonyl.) However, because the β-keto ester is deprotonated under the reaction conditions, the equilibrium represented by the sum of steps 1 to 4 does lie to the side of products. On subsequent acidification (step 5), the anion of the β-keto ester is converted to its neutral form and isolated.

Organic chemists sometimes write equations for the Claisen condensation in a form that shows both stages explicitly:

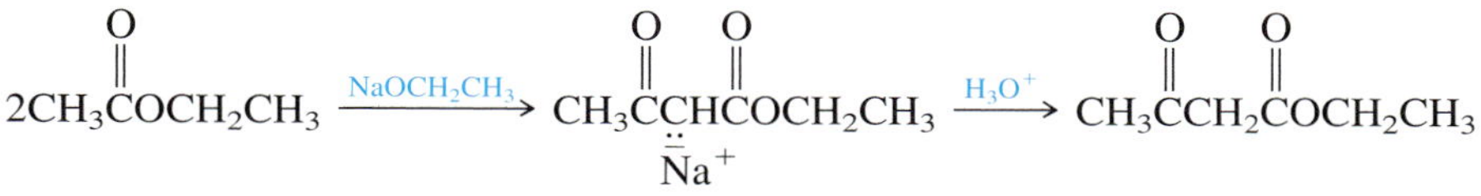

$$2CH_3COCH_2CH_3 \xrightarrow{NaOCH_2CH_3} CH_3CCHCOCH_2CH_3 \xrightarrow{H_3O^+} CH_3CCH_2COCH_2CH_3$$
$$\overset{..}{\underset{Na^+}{}}$$

| Ethyl acetate | Sodium salt of ethyl acetoacetate | Ethyl acetoacetate |

Like aldol condensations, Claisen condensations always involve bond formation between the α-carbon atom of one molecule and the carbonyl carbon of another:

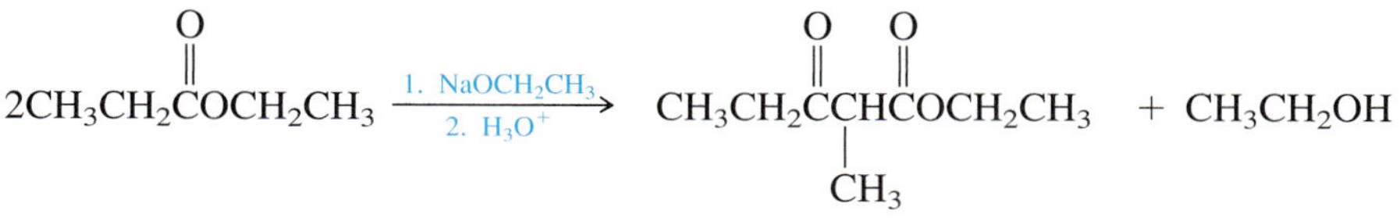

$$2CH_3CH_2COCH_2CH_3 \xrightarrow[\text{2. } H_3O^+]{\text{1. } NaOCH_2CH_3} CH_3CH_2CCHCOCH_2CH_3 \; + \; CH_3CH_2OH$$
$$\underset{CH_3}{|}$$

| Ethyl propanoate | Ethyl 2-methyl-3-oxopentanoate (81%) | Ethanol |

PROBLEM 21.1 One of the following esters cannot undergo the Claisen condensation. Which one? Write structural formulas for the Claisen condensation products of the other two.

$CH_3CH_2CH_2CH_2CO_2CH_2CH_3$ $C_6H_5CO_2CH_2CH_3$ $C_6H_5CH_2CO_2CH_2CH_3$

Ethyl pentanoate   Ethyl benzoate   Ethyl phenylacetate

Unless the β-keto ester can form a stable anion by deprotonation as in step 4 of Figure 21.1, the Claisen condensation product is present in only trace amounts at equilibrium. Ethyl 2-methylpropanoate, for example, does not give any of its condensation product under the customary conditions of the Claisen condensation.

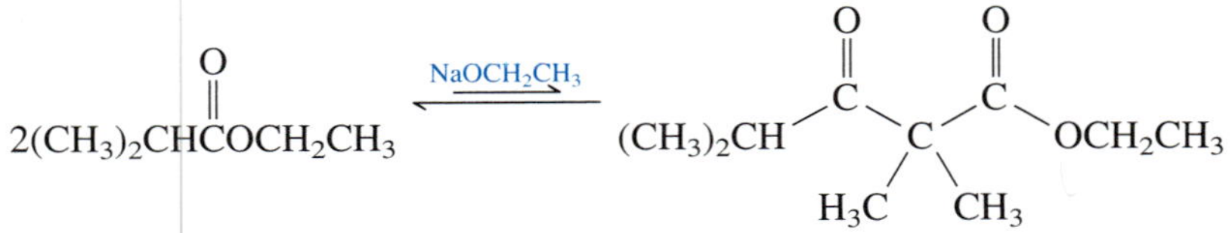

$$2(CH_3)_2CHCOCH_2CH_3 \xrightarrow{NaOCH_2CH_3} (CH_3)_2CH-C-C-OCH_2CH_3$$

Ethyl 2-methylpropanoate

Ethyl 2,2,4-trimethyl-3-oxopentanoate
(cannot form a stable anion; formed in
no more than trace amounts)

At least two protons must be present at the α carbon for the equilibrium to favor product formation. Claisen condensation is possible for esters of the type RCH_2CO_2R', but not for R_2CHCO_2R'.

21.2 INTRAMOLECULAR CLAISEN CONDENSATION: THE DIECKMANN REACTION

Esters of *dicarboxylic acids* undergo an intramolecular version of the Claisen condensation when a five- or six-membered ring can be formed.

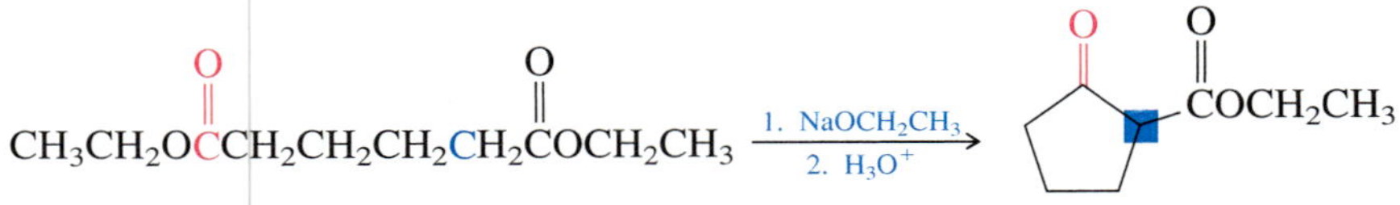

$$CH_3CH_2OCCH_2CH_2CH_2CH_2COCH_2CH_3 \xrightarrow[2.\ H_3O^+]{1.\ NaOCH_2CH_3} $$

Diethyl hexanedioate

Ethyl (2-oxocyclopentane)-
carboxylate (74–81%)

This reaction is an example of a **Dieckmann cyclization.** The anion formed by proton abstraction at the carbon α to one carbonyl group attacks the other carbonyl to form a five-membered ring.

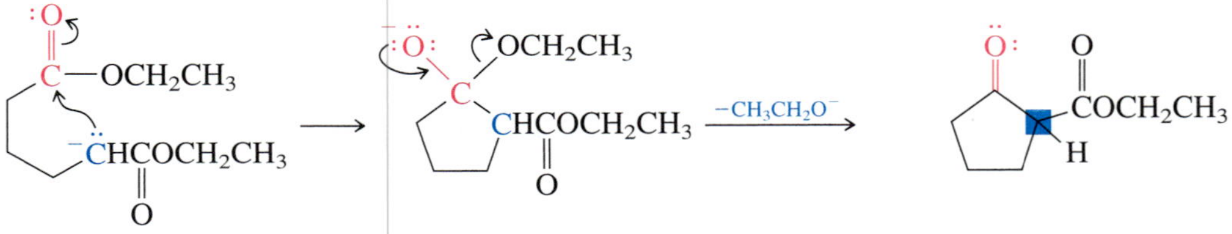

Enolate of diethyl
hexanedioate

Ethyl (2-oxocyclopentane)carboxylate

Walter Dieckmann was a German chemist and a contemporary of Claisen.

PROBLEM 21.2 Write the structure of the Dieckmann cyclization product formed on treatment of each of the following diesters with sodium ethoxide, followed by acidification.

(a) $CH_3CH_2OCCH_2CH_2CH_2CH_2CH_2COCH_2CH_3$

(b) $CH_3CH_2OCCH_2CH_2CHCH_2CH_2COCH_2CH_3$
$\qquad\qquad\qquad\qquad CH_3$

(c) 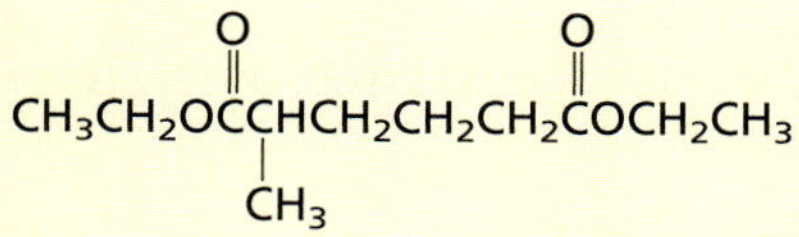
$$CH_3CH_2O\overset{O}{\overset{\|}{C}}CHCH_2CH_2CH_2C\overset{O}{\overset{\|}{O}}OCH_2CH_3$$
$\quad\quad\quad\quad\quad |$
$\quad\quad\quad\quad\; CH_3$

SAMPLE SOLUTION (a) Diethyl heptanedioate has one more methylene group in its chain than the diester cited in the example (diethyl hexanedioate). Its Dieckmann cyclization product contains a six-membered ring instead of the five-membered ring formed from diethyl hexanedioate.

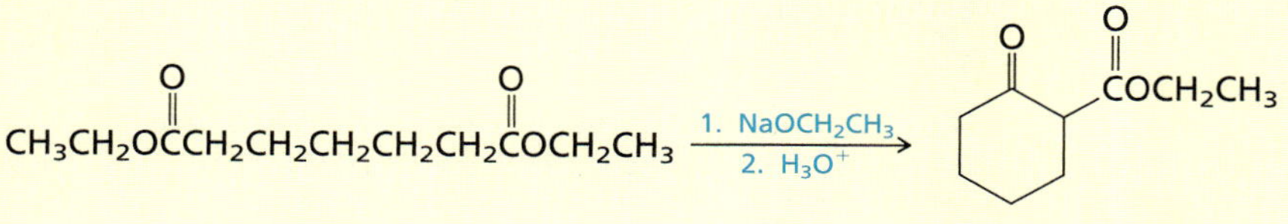

$$CH_3CH_2O\overset{O}{\overset{\|}{C}}CH_2CH_2CH_2CH_2CH_2C\overset{O}{\overset{\|}{O}}OCH_2CH_3 \xrightarrow[\text{2. }H_3O^+]{\text{1. }NaOCH_2CH_3}$$

Diethyl heptanedioate

Ethyl (2-oxocyclohexane)-carboxylate

21.3 MIXED CLAISEN CONDENSATIONS

Mixed Claisen condensations are analogous to mixed aldol condensations and involve carbon–carbon bond formation between the α-carbon atom of one ester and the carbonyl carbon of another.

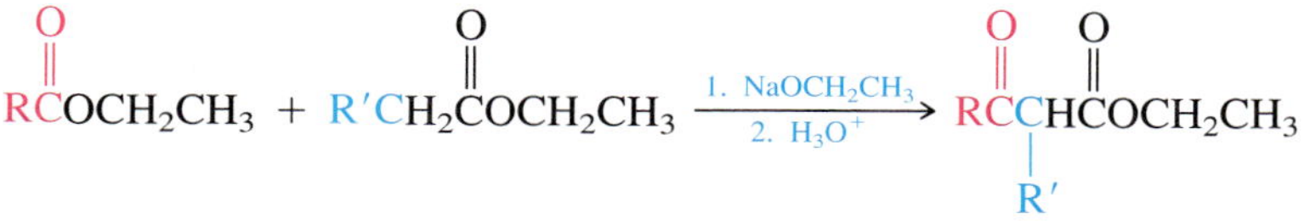

$$R\overset{O}{\overset{\|}{C}}OCH_2CH_3 \; + \; R'CH_2\overset{O}{\overset{\|}{C}}OCH_2CH_3 \xrightarrow[\text{2. }H_3O^+]{\text{1. }NaOCH_2CH_3} R\overset{O}{\overset{\|}{C}}CH\overset{O}{\overset{\|}{C}}OCH_2CH_3$$
$$\quad\quad\quad\quad\quad\quad\quad\quad\quad\quad\quad\quad\quad\quad\quad\quad\quad\quad\quad\quad\quad\quad |$$
$$\quad\quad\quad\quad\quad\quad\quad\quad\quad\quad\quad\quad\quad\quad\quad\quad\quad\quad\quad\quad\quad\; R'$$

Ester   Another ester   β-Keto ester

The best results are obtained when one of the ester components is incapable of forming an enolate. Esters of this type include the following:

$$H\overset{O}{\overset{\|}{C}}OR \quad\quad RO\overset{O}{\overset{\|}{C}}OR \quad\quad RO\overset{O\,O}{\overset{\|\;\|}{C\,C}}OR \quad\quad $$

HCOR   ROCOR   ROCCOR   COR

Formate esters   Carbonate esters   Oxalate esters   Benzoate esters

The following equation shows an example of a mixed Claisen condensation in which a benzoate ester is used as the nonenolizable component:

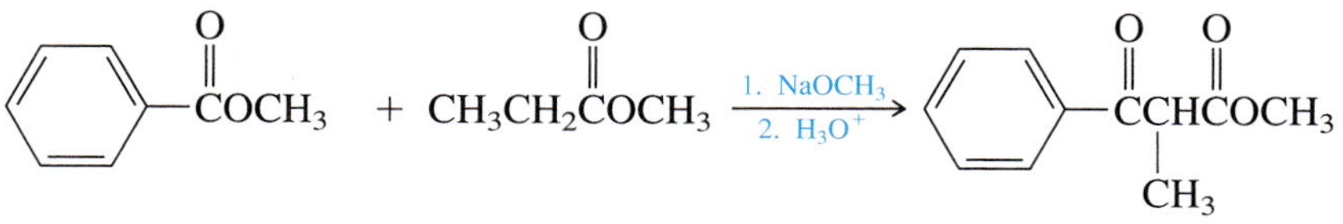

Methyl benzoate
(cannot form an enolate)

Methyl propanoate

Methyl 2-methyl-3-oxo-3-phenylpropanoate (60%)

PROBLEM 21.3 Give the structure of the product obtained when ethyl phenylacetate ($C_6H_5CH_2CO_2CH_2CH_3$) is treated with each of the following esters under conditions of the mixed Claisen condensation:
(a) Diethyl carbonate   (b) Diethyl oxalate   (c) Ethyl formate

SAMPLE SOLUTION (a) Diethyl carbonate cannot form an enolate, but ethyl phenylacetate can. Nucleophilic acyl substitution on diethyl carbonate by the enolate of ethyl phenylacetate yields a *diester.*

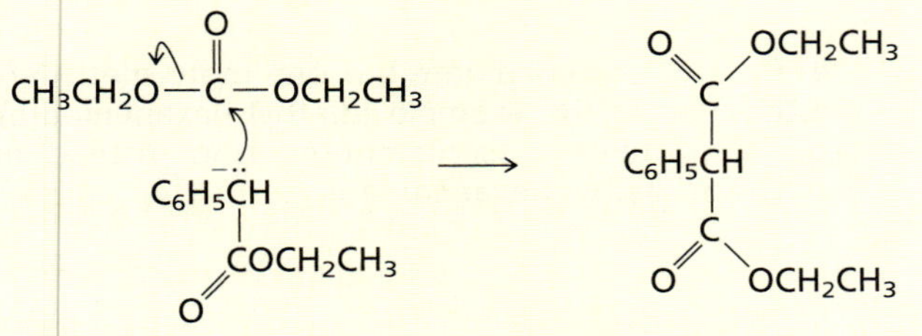

Diethyl 2-phenylpropanedioate
(diethyl phenylmalonate)

The reaction proceeds in good yield (86%), and the product is a useful one in further synthetic transformations of the type to be described in Section 21.7.

21.4 ACYLATION OF KETONES WITH ESTERS

In a reaction related to the mixed Claisen condensation, nonenolizable esters are used as acylating agents for ketone enolates. Ketones (via their enolates) are converted to β-keto esters by reaction with diethyl carbonate.

Sodium hydride was used as the base in this example. It is often used instead of sodium ethoxide in these reactions.

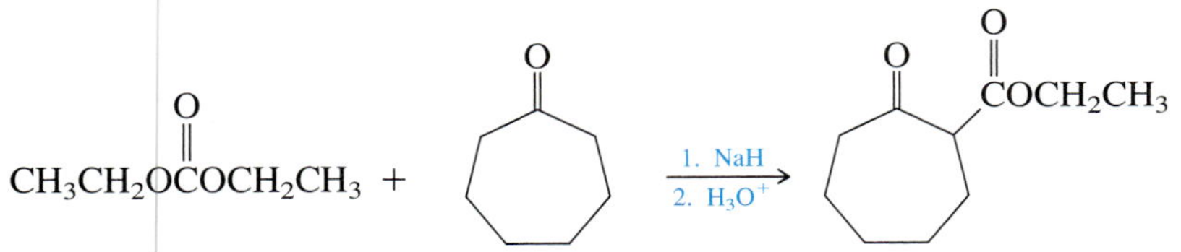

Diethyl carbonate   Cycloheptanone   Ethyl (2-oxocycloheptane)-
                                              carboxylate (91–94%)

Esters of nonenolizable monocarboxylic acids such as ethyl benzoate give β-diketones on reaction with ketone enolates:

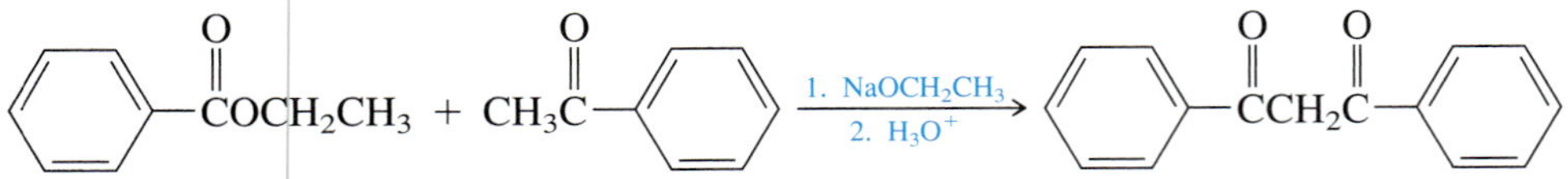

Ethyl benzoate   Acetophenone   1,3-Diphenyl-1,3-
                                                  propanedione (62–71%)

Intramolecular acylation of ketones yields cyclic β-diketones when the ring that is formed is five- or six-membered.

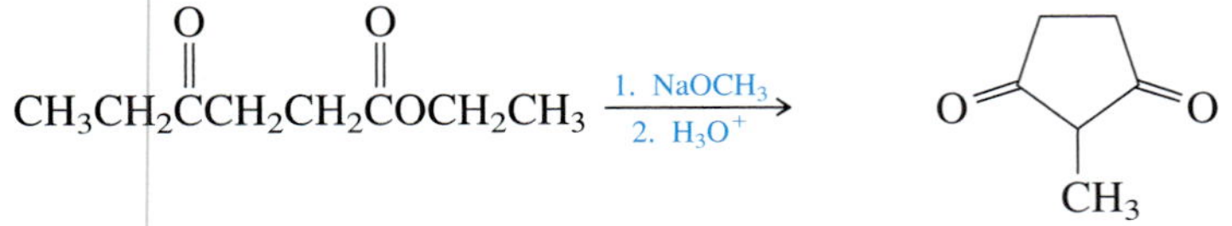

Ethyl 4-oxohexanoate   2-Methyl-1,3-cyclopentanedione
                                            (70–71%)

PROBLEM 21.4 Write an equation for the carbon–carbon bond-forming step in the cyclization reaction just cited. Show clearly the structure of the enolate ion, and use curved arrows to represent its nucleophilic addition to the appropriate carbonyl group. Write a second equation showing dissociation of the tetrahedral intermediate formed in the carbon–carbon bond-forming step.

Even though ketones have the potential to react with themselves by aldol addition, recall that the position of equilibrium for such reactions lies to the side of the starting materials (Section 18.9). On the other hand, acylation of ketone enolates gives products (β-keto esters or β-diketones) that are converted to stabilized anions under the reaction conditions. Consequently, ketone acylation is observed to the exclusion of aldol addition when ketones are treated with base in the presence of esters.

21.5 KETONE SYNTHESIS VIA β-KETO ESTERS

The carbon–carbon bond-forming potential inherent in the Claisen and Dieckmann reactions has been extensively exploited in organic synthesis. Subsequent transformations of the β-keto ester products permit the synthesis of other functional groups. One of these transformations converts β-keto esters to ketones; it is based on the fact that β-keto *acids* (not esters!) undergo decarboxylation readily (Section 19.17). Indeed, β-keto *acids,* and their corresponding carboxylate anions as well, lose carbon dioxide so easily that they tend to decarboxylate under the conditions of their formation.

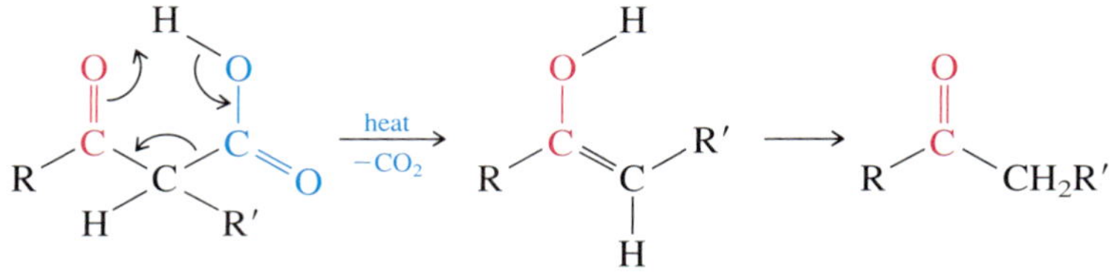

β-Keto acid Enol form of ketone Ketone

Thus, 5-nonanone has been prepared from ethyl pentanoate by the sequence

$$\underset{\text{Ethyl pentanoate}}{CH_3CH_2CH_2CH_2\overset{O}{\overset{\|}{C}}OCH_2CH_3} \xrightarrow[2.\ H_3O^+]{1.\ NaOCH_2CH_3} \underset{\text{Ethyl 3-oxo-2-propylheptanoate (80\%)}}{CH_3CH_2CH_2CH_2\overset{O}{\overset{\|}{C}}\underset{CH_2CH_2CH_3}{\underset{|}{C}H}\overset{O}{\overset{\|}{C}}OCH_2CH_3}$$

$$\downarrow \text{1. KOH, } H_2O\text{, 70–80°C; 2. } H_3O^+$$

$$\underset{\text{5-Nonanone (81\%)}}{CH_3CH_2CH_2CH_2\overset{O}{\overset{\|}{C}}CH_2CH_2CH_2CH_3} \xleftarrow[-CO_2]{70\text{–}80°C} \underset{\text{3-Oxo-2-propylheptanoic acid}}{CH_3CH_2CH_2CH_2\overset{O}{\overset{\|}{C}}\underset{CH_2CH_2CH_3}{\underset{|}{C}H}\overset{O}{\overset{\|}{C}}OH}$$

(not isolated; decarboxylates under conditions of its formation)

The sequence begins with a Claisen condensation of ethyl pentanoate to give a β-keto ester. The ester is hydrolyzed, and the resulting β-keto acid decarboxylates to yield the desired ketone.

> **PROBLEM 21.5** Write appropriate chemical equations showing how you could prepare cyclopentanone from diethyl hexanedioate.

The major application of β-keto esters to organic synthesis employs a similar pattern of ester saponification and decarboxylation as its final stage, as described in the following section.

21.6 THE ACETOACETIC ESTER SYNTHESIS

Ethyl acetoacetate (acetoacetic ester), available by the Claisen condensation of ethyl acetate, has properties that make it a useful starting material for the preparation of ketones. These properties are

1. The acidity of the α hydrogen
2. The ease with which acetoacetic acid undergoes thermal decarboxylation

Ethyl acetoacetate is a stronger acid than ethanol and is quantitatively converted to its anion on treatment with sodium ethoxide in ethanol.

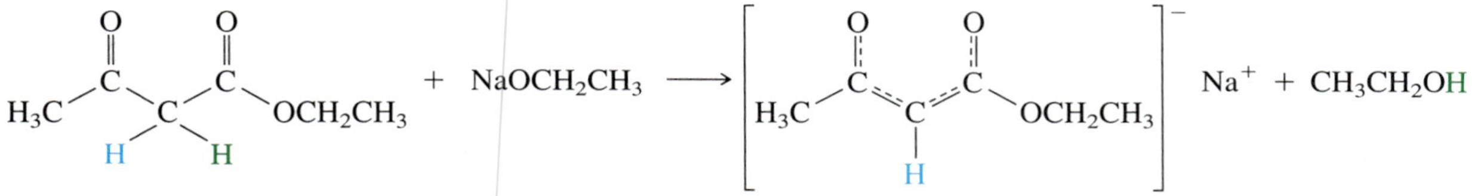

Ethyl acetoacetate	Sodium ethoxide	Sodium salt of ethyl	Ethanol
(stronger acid)	(stronger base)	acetoacetate	(weaker acid)
$pK_a = 11$		(weaker base)	$pK_a = 16$

The anion produced by proton abstraction from ethyl acetoacetate is nucleophilic. Adding an alkyl halide to a solution of the sodium salt of ethyl acetoacetate leads to alkylation of the α carbon.

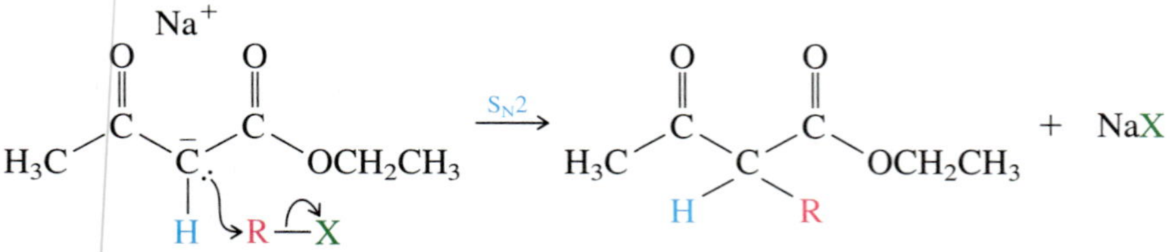

| Sodium salt of ethyl acetoacetate; | 2-Alkyl derivative of | Sodium |
| alkyl halide | ethyl acetoacetate | halide |

Because the new carbon–carbon bond is formed by an S_N2-type reaction the alkyl halide must not be sterically hindered. Methyl and primary alkyl halides work best; secondary alkyl halides give lower yields. Tertiary alkyl halides fail, reacting only by elimination, not substitution.

Saponification and decarboxylation of the alkylated derivative of ethyl acetoacetate yields a ketone.

$$\underset{\substack{\text{2-Alkyl derivative of}\\\text{ethyl acetoacetate}}}{H_3C-\overset{\displaystyle O}{\overset{\|}{C}}-\underset{\underset{R}{\overset{|}{\underset{|}{H}}}}{C}-\overset{\displaystyle O}{\overset{\|}{C}}-OCH_2CH_3} \xrightarrow[\text{2. } H_3O^+]{\text{1. } HO^-, H_2O} \underset{\substack{\text{2-Alkyl derivative of}\\\text{acetoacetic acid}}}{H_3C-\overset{\displaystyle O}{\overset{\|}{C}}-\underset{\underset{R}{\overset{|}{\underset{|}{H}}}}{C}-\overset{\displaystyle O}{\overset{\|}{C}}-OH} \xrightarrow[-CO_2]{\text{heat}} \underset{\text{Ketone}}{CH_3\overset{\displaystyle O}{\overset{\|}{C}}CH_2R}$$

This reaction sequence is called the **acetoacetic ester synthesis.** It is a standard procedure for the preparation of ketones from alkyl halides, as the conversion of 1-bromobutane to 2-heptanone illustrates.

$$\underset{\text{acetoacetate}}{\underset{\text{Ethyl}}{CH_3\overset{\displaystyle O}{\overset{\|}{C}}CH_2\overset{\displaystyle O}{\overset{\|}{C}}OCH_2CH_3}} \xrightarrow[\text{2. } CH_3CH_2CH_2CH_2Br]{\substack{\text{1. NaOCH}_2CH_3,\\ \text{ethanol}}} \underset{\substack{\text{Ethyl 2-butyl-3-}\\\text{oxobutanoate (70\%)}}}{CH_3\overset{\displaystyle O}{\overset{\|}{C}}\underset{\underset{CH_2CH_2CH_2CH_3}{\overset{|}{\underset{|}{H}}}}{C}\overset{\displaystyle O}{\overset{\|}{C}}OCH_2CH_3} \xrightarrow[\substack{\text{3. heat}\\ -CO_2}]{\substack{\text{1. NaOH, } H_2O\\ \text{2. } H_3O^+}} \underset{\substack{\text{2-Heptanone}\\(60\%)}}{CH_3\overset{\displaystyle O}{\overset{\|}{C}}CH_2CH_2CH_2CH_2CH_3}$$

The acetoacetic ester synthesis brings about the overall transformation of an alkyl halide to an alkyl derivative of acetone.

$$\underset{\substack{\text{Primary or secondary}\\\text{alkyl halide}}}{R-X} \longrightarrow \underset{\substack{\alpha\text{-Alkylated derivative}\\\text{of acetone}}}{R-CH_2\overset{\displaystyle O}{\overset{\|}{C}}CH_3}$$

We call a structural unit in a molecule that is related to a synthetic operation a **synthon.** The three-carbon unit $-CH_2\overset{\displaystyle O}{\overset{\|}{C}}CH_3$ is a synthon that alerts us to the possibility that a particular molecule may be accessible by the acetoacetic ester synthesis.

E. J. Corey (Section 14.9) invented the word *synthon* in connection with his efforts to formalize synthetic planning.

PROBLEM 21.6 Show how you could prepare each of the following ketones from ethyl acetoacetate and any necessary organic or inorganic reagents:

(a) 1-Phenyl-1,4-pentanedione   (c) 5-Hexen-2-one

(b) 4-Phenyl-2-butanone

SAMPLE SOLUTION (a) Approach these syntheses in a retrosynthetic way. Identify the synthon $-CH_2\overset{\displaystyle O}{\overset{\|}{C}}CH_3$ and mentally disconnect the bond to the α-carbon atom. The $-CH_2\overset{\displaystyle O}{\overset{\|}{C}}CH_3$ synthon is derived from ethyl acetoacetate; the remainder of the molecule originates in the alkyl halide.

Disconnect here

1-Phenyl-1,4-pentanedione   Required alkyl halide   Derived from ethyl acetoacetate

Can you think of how bromomethyl phenyl ketone might be prepared?

Analyzing the target molecule in this way reveals that the required alkyl halide is an α-halo ketone. Thus, a suitable starting material would be bromomethyl phenyl ketone.

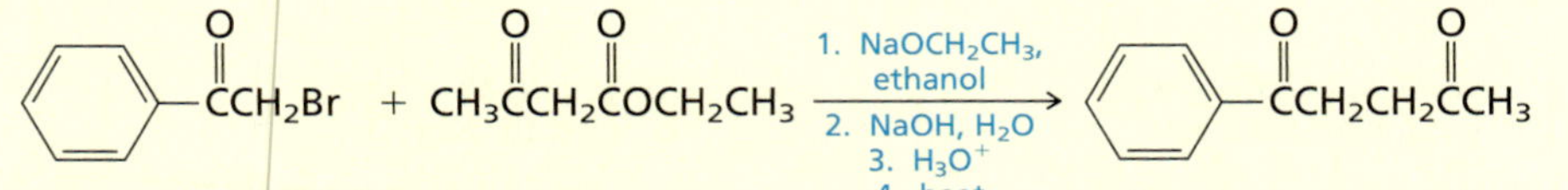

Bromomethyl phenyl ketone — Ethyl acetoacetate — 1-Phenyl-1,4-pentanedione

Dialkylation of ethyl acetoacetate can also be accomplished, opening the way to ketones with two alkyl substituents at the α carbon:

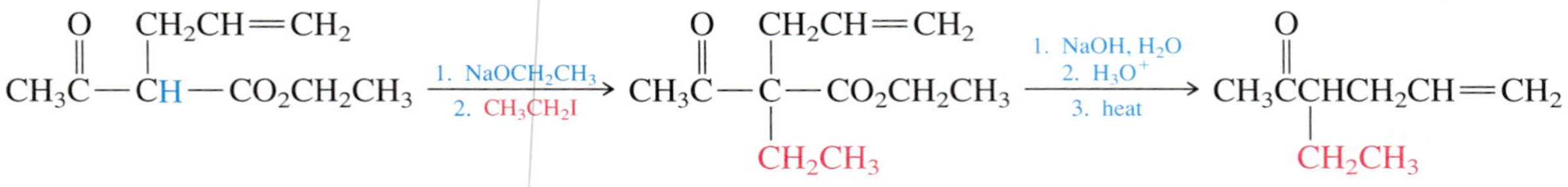

Ethyl 2-allylacetoacetate — Ethyl 2-allyl-2-ethylacetoacetate (75%) — 3-Ethyl-5-hexen-2-one (48%)

The starting material in the example is obtained by alkylation of ethyl acetoacetate with allyl bromide.

Recognize, too, that the reaction sequence is one that is characteristic of β-keto esters in general and not limited to just ethyl acetoacetate and its derivatives. Thus,

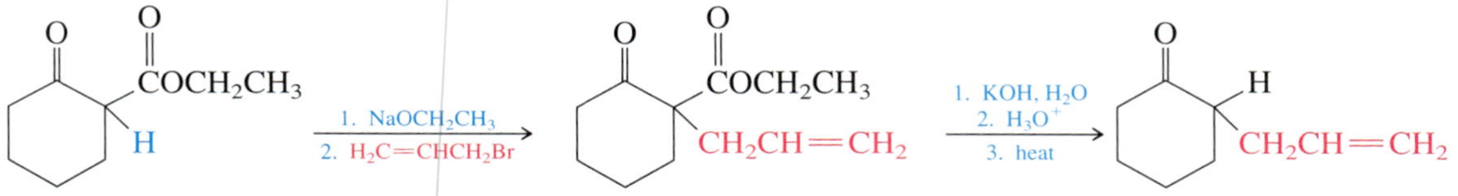

Ethyl 2-oxocyclohexanecarboxylate — Ethyl 1-allyl-2-oxocyclohexanecarboxylate (89%) — 2-Allylcyclohexanone (66%)

The starting material in this example is the Dieckmann cyclization product of diethyl heptanedioate (see Problem 21.2a).

It's reasonable to ask why one would prepare a ketone by way of a keto ester (ethyl acetoacetate, for example) rather than by direct alkylation of the enolate of a ketone. One reason is that the monoalkylation of ketones via their enolates is a difficult reaction to carry out in good yield. (Remember, however, that *acylation* of ketone enolates as described in Section 21.4 is achieved readily.) A second reason is that the delocalized enolates of β-keto esters, being far less basic than ketone enolates, give a higher substitution–elimination ratio when they react with alkyl halides. This can be quite important in those syntheses in which the alkyl halide is expensive or difficult to obtain.

Anions of β-keto esters are said to be *synthetically equivalent* to the enolates of ketones. The anion of ethyl acetoacetate is synthetically equivalent to the enolate of acetone, for example. The use of synthetically equivalent groups is a common tactic in synthetic organic chemistry. One of the skills that characterize the most creative practitioners of organic synthesis is an ability to recognize situations in which otherwise difficult transformations can be achieved through the use of synthetically equivalent reagents.

21.7 THE MALONIC ESTER SYNTHESIS

The **malonic ester synthesis** is a method for the preparation of carboxylic acids and is represented by the general equation

$$\underset{\substack{\text{Alkyl}\\\text{halide}}}{RX} + \underset{\substack{\text{Diethyl malonate}\\\text{(malonic ester)}}}{CH_2(COOCH_2CH_3)_2} \xrightarrow[\text{ethanol}]{NaOCH_2CH_3} \underset{\substack{\alpha\text{-Alkylated}\\\text{derivative of}\\\text{diethyl malonate}}}{RCH(COOCH_2CH_3)_2} \xrightarrow[\substack{\text{2. } H_3O^+\\\text{3. heat}}]{\text{1. } HO^-,\, H_2O} \underset{\text{Carboxylic acid}}{RCH_2COOH}$$

The malonic ester synthesis is conceptually analogous to the acetoacetic ester synthesis. The overall transformation is

$$\underset{\substack{\text{Primary or secondary}\\\text{alkyl halide}}}{R-X} \longrightarrow \underset{\substack{\alpha\text{-Alkylated derivative}\\\text{of acetic acid}}}{R-CH_2\overset{\displaystyle O}{\overset{\|}{C}}OH}$$

Diethyl malonate (also known as malonic ester) serves as a source of the synthon $-CH_2\overset{O}{\overset{\|}{C}}OH$ in the same way that the ethyl acetoacetate serves as a source of the synthon $-CH_2\overset{O}{\overset{\|}{C}}CH_3$.

The properties of diethyl malonate that make the malonic ester synthesis a useful procedure are the same as those responsible for the synthetic value of ethyl acetoacetate. The hydrogens at C-2 of diethyl malonate are relatively acidic, and one is readily removed on treatment with sodium ethoxide.

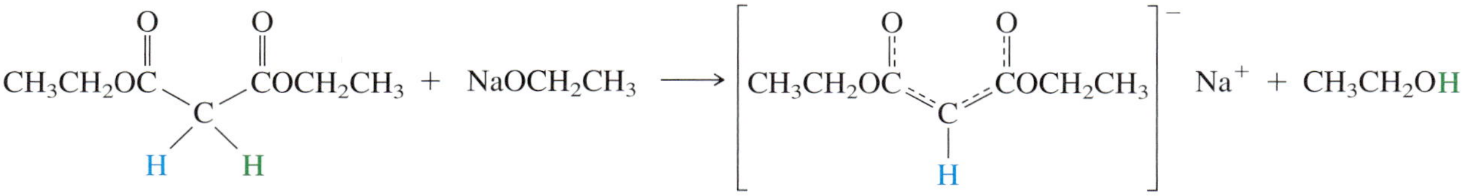

> Among the methods for preparing carboxylic acids, carboxylation of a Grignard reagent and preparation and hydrolysis of a nitrile convert RBr to RCO_2H. The malonic ester synthesis converts RBr to RCH_2CO_2H.

| Diethyl malonate (stronger acid) $pK_a = 13$ | Sodium ethoxide (stronger base) | Sodium salt of diethyl malonate (weaker base) | Ethanol (weaker acid) $pK_a = 16$ |

Treatment of the anion of diethyl malonate with alkyl halides leads to alkylation at C-2.

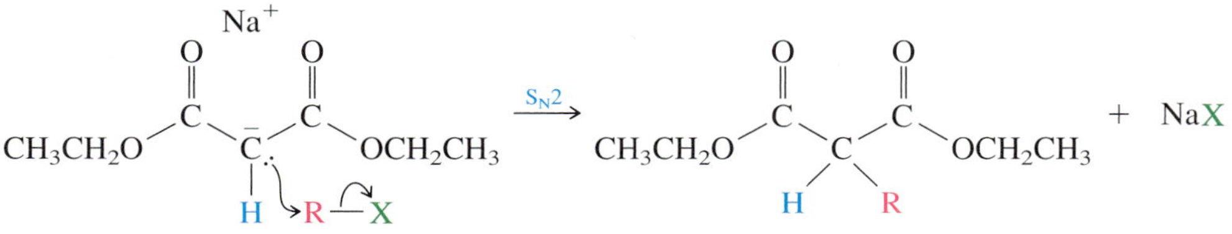

| Sodium salt of diethyl malonate; alkyl halide | 2-Alkyl derivative of diethyl malonate | Sodium halide |

Converting the C-2 alkylated derivative to the corresponding malonic acid derivative by ester hydrolysis gives a compound susceptible to thermal decarboxylation. Temperatures of approximately 180°C are normally required.

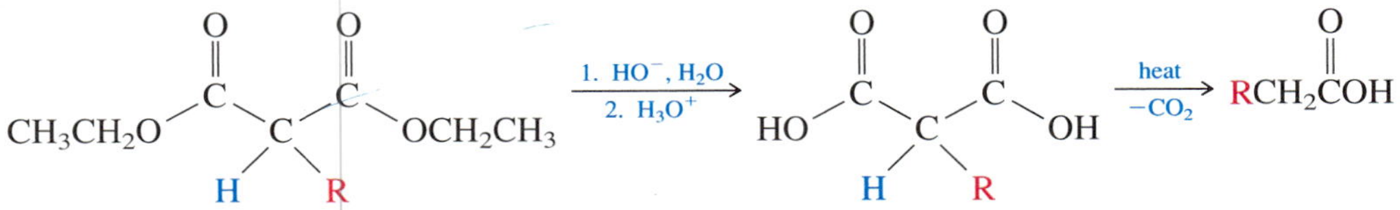

| 2-Alkyl derivative of diethyl malonate | 2-Alkyl derivative of malonic acid | Carboxylic acid |

In a typical example of the malonic ester synthesis, 6-heptenoic acid has been prepared from 5-bromo-1-pentene:

$$H_2C=CHCH_2CH_2CH_2Br + CH_2(COOCH_2CH_3)_2 \xrightarrow[\text{ethanol}]{NaOCH_2CH_3} H_2C=CHCH_2CH_2CH_2CH(COOCH_2CH_3)_2$$

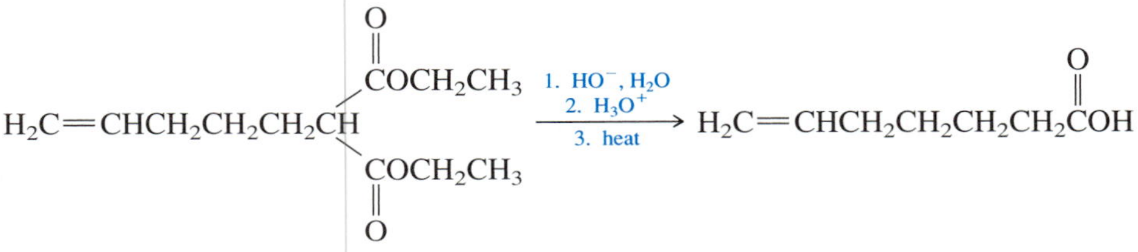

| 5-Bromo-1-pentene | Diethyl malonate | Diethyl 2-(4-pentenyl)malonate (85%) |

| Diethyl 2-(4-pentenyl)malonate | 6-Heptenoic acid (75%) |

PROBLEM 21.7 Show how you could prepare each of the following carboxylic acids from diethyl malonate and any necessary organic or inorganic reagents:

(a) 3-Methylpentanoic acid   (c) 4-Methylhexanoic acid

(b) Nonanoic acid   (d) 3-Phenylpropanoic acid

SAMPLE SOLUTION (a) Analyze the target molecule retrosynthetically by mentally disconnecting a bond to the α-carbon atom.

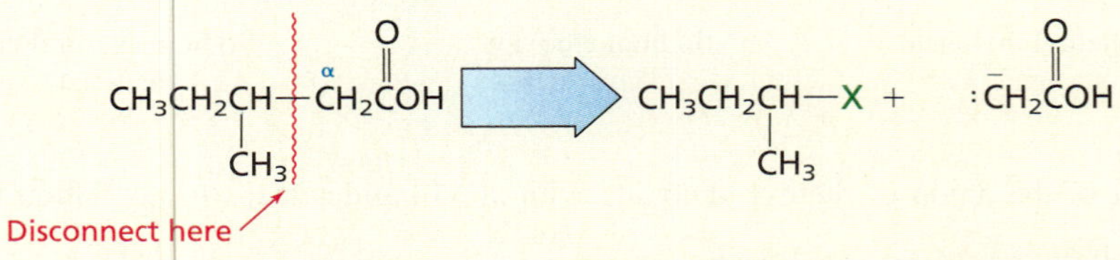

| 3-Methylpentanoic acid | Required alkyl halide | Derived from diethyl malonate |

We see that a secondary alkyl halide is needed as the alkylating agent. The anion of diethyl malonate is a weaker base than ethoxide ion and reacts with secondary alkyl halides by substitution rather than elimination. Thus, the synthesis of 3-methylpentanoic acid begins with the alkylation of the anion of diethyl malonate by 2-bromobutane.

$$CH_3CH_2\overset{\underset{\displaystyle CH_3}{|}}{CHBr} + CH_2(COOCH_2CH_3)_2 \xrightarrow[\substack{3.\ H_3O^+ \\ 4.\ heat}]{\substack{1.\ NaOCH_2CH_3,\ ethanol \\ 2.\ NaOH,\ H_2O}} CH_3CH_2\overset{\underset{\displaystyle CH_3}{|}}{CH}CH_2\overset{\displaystyle O}{\overset{\|}{C}}OH$$

2-Bromobutane   Diethyl malonate   3-Methylpentanoic acid

As actually carried out and reported in the chemical literature, diethyl malonate has been alkylated with 2-bromobutane in 83–84% yield and the product of that reaction converted to 3-methylpentanoic acid by saponification, acidification, and decarboxylation in 62–65% yield.

By performing two successive alkylation steps, the malonic ester synthesis can be applied to the synthesis of α,α-disubstituted derivatives of acetic acid:

$$CH_2(COOCH_2CH_3)_2 \xrightarrow[\substack{2.\ CH_3Br}]{\substack{1.\ NaOCH_2CH_3,\ ethanol}} CH_3CH(COOCH_2CH_3)_2$$

Diethyl malonate   Diethyl 2-methyl-1,3-propanedioate (79–83%)

$$\downarrow \substack{1.\ NaOCH_2CH_3,\ ethanol \\ 2.\ CH_3(CH_2)_8CH_2Br}$$

$$\underset{\substack{\text{2-Methyldodecanoic acid} \\ (61–74\%)}}{\overset{H_3C}{\underset{CH_3(CH_2)_8CH_2}{>}}C\overset{H}{\underset{COOH}{<}}} \xleftarrow[\substack{3.\ heat}]{\substack{1.\ KOH,\ ethanol–water \\ 2.\ H_3O^+}} \underset{\substack{\text{Diethyl} \\ \text{2-decyl-2-methyl-1,3-propanedioate}}}{\overset{H_3C}{\underset{CH_3(CH_2)_8CH_2}{>}}C\overset{COOCH_2CH_3}{\underset{COOCH_2CH_3}{<}}}$$

PROBLEM 21.8 Ethyl acetoacetate may also be subjected to double alkylation. Show how you could prepare 3-methyl-2-butanone by double alkylation of ethyl acetoacetate.

The malonic ester synthesis has been adapted to the preparation of cyclo-alkanecarboxylic acids from dihaloalkanes:

$$CH_2(COOCH_2CH_3)_2 \xrightarrow[\substack{2.\ BrCH_2CH_2CH_2Br}]{\substack{1.\ NaOCH_2CH_3,\ ethanol}} H_2C\overset{CH_2}{\underset{\underset{\underset{Br}{|}}{CH_2}}{<}}CH(COOCH_2CH_3)_2$$

Diethyl malonate   (Not isolated; cyclizes in the presence of sodium ethoxide)

$$\downarrow$$

$$\underset{\substack{\text{Cyclobutanecarboxylic} \\ \text{acid (80\% from diester)}}}{\text{（環）}\overset{H}{\underset{COOH}{}}} \xleftarrow[\substack{2.\ heat}]{\substack{1.\ H_3O^+}} \underset{\substack{\text{Diethyl} \\ \text{1,1-cyclobutanedicarboxylate (60–65\%)}}}{H_2C\overset{\overset{\displaystyle CH_2}{}}{\underset{\underset{\displaystyle CH_2}{}}{>}}C\overset{COOCH_2CH_3}{\underset{COOCH_2CH_3}{<}}}$$

The cyclization step is limited to the formation of rings of seven carbons or fewer.

> **PROBLEM 21.9** Cyclopentyl methyl ketone has been prepared from 1,4-dibromobutane and ethyl acetoacetate. Outline the steps in this synthesis by writing a series of equations showing starting materials, reagents, and isolated intermediates.

21.8 BARBITURATES

Barbituric acid was first prepared in 1864 by Adolf von Baeyer (page 112). A historical account of his work and the later development of barbiturates as sedative–hypnotics appeared in the October 1951 issue of the *Journal of Chemical Education* (pp. 524–526).

Diethyl malonate has uses other than in the synthesis of carboxylic acids. One particularly valuable application lies in the preparation of *barbituric acid* by nucleophilic acyl substitution with urea:

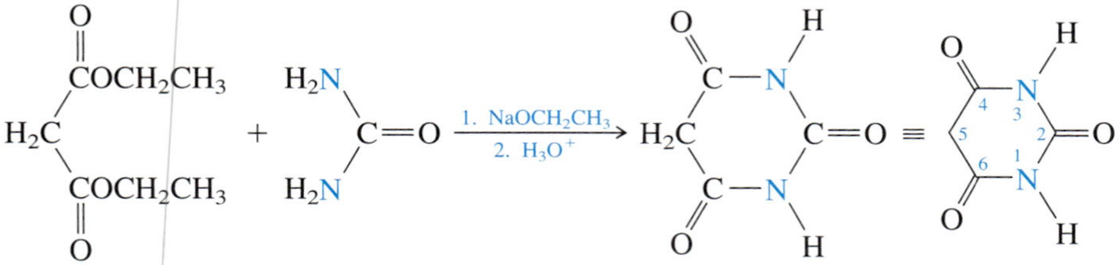

Diethyl malonate   Urea   Barbituric acid
                                    (72–78%)

Barbituric acid is the parent of a group of compounds known as **barbiturates.** The barbiturates are classified as *sedative–hypnotic agents,* meaning that they decrease the responsiveness of the central nervous system and promote sleep. Thousands of derivatives of the parent ring system of barbituric acid have been tested for sedative–hypnotic activity; the most useful are the 5,5-disubstituted derivatives.

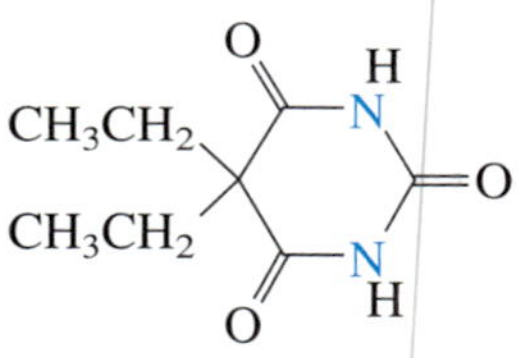

5,5-Diethylbarbituric acid
(barbital; Veronal)

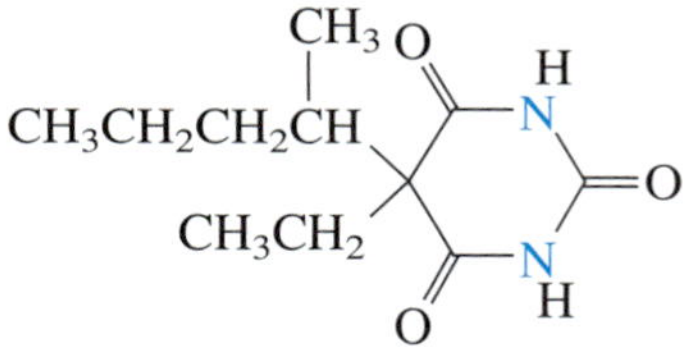

5-Ethyl-5-(1-methylbutyl)
barbituric acid
(pentobarbital; Nembutal)

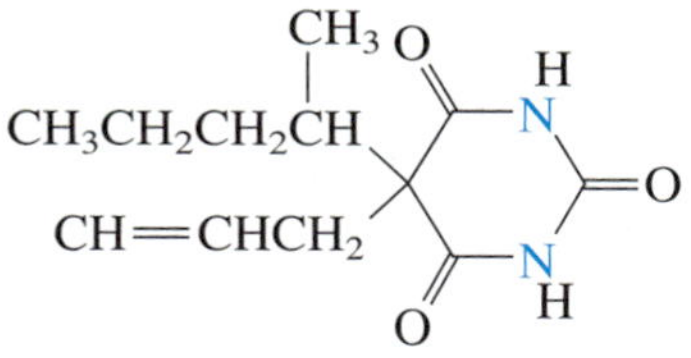

5-Allyl-5-(1-methylbutyl)
barbituric acid
(secobarbital; Seconal)

These compounds are prepared in a manner analogous to that of barbituric acid itself. Diethyl malonate is alkylated twice, then treated with urea.

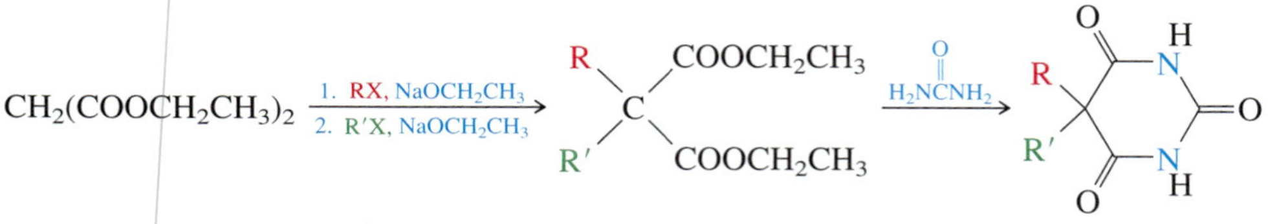

Diethyl malonate

Dialkylated
derivative of
diethyl malonate

5,5-Disubstituted
derivative of
barbituric acid

PROBLEM 21.10   Show, by writing a suitable sequence of reactions, how you could prepare pentobarbital from diethyl malonate. (The structure of pentobarbital was shown in this section.)

Barbituric acids, as their name implies, are weakly acidic and are converted to their sodium salts (sodium barbiturates) in aqueous sodium hydroxide. Sometimes the drug is dispensed in its neutral form; sometimes the sodium salt is used. The salt is designated by appending the word "sodium" to the name of the barbituric acid—*pentobarbital sodium,* for example.

PROBLEM 21.11   Thiourea (H_2NCNH_2) reacts with diethyl malonate and its alkyl derivatives in the same way that urea does. Give the structure of the product obtained when thiourea is used instead of urea in the synthesis of pentobarbital. The anesthetic *thiopental (Pentothal) sodium* is the sodium salt of this product. What is the structure of this compound?

PROBLEM 21.12   Aryl halides react too slowly to undergo substitution by the S_N2 mechanism with the sodium salt of diethyl malonate, and so the phenyl substituent of *phenobarbital* cannot be introduced in the way that alkyl substituents can.

5-Ethyl-5-phenylbarbituric acid
(phenobarbital)

One synthesis of phenobarbital begins with ethyl phenylacetate and diethyl carbonate. Using these starting materials and any necessary organic or inorganic reagents, devise a synthesis of phenobarbital. (*Hint:* See the sample solution to Problem 21.3a.)

The various barbiturates differ in the time required for the onset of sleep and in the duration of their effects. All the barbiturates must be used only in strict accordance with instructions to avoid potentially lethal overdoses. Drug dependence in some individuals is also a problem.

21.9   MICHAEL ADDITIONS OF STABILIZED ANIONS

Stabilized anions exhibit a pronounced tendency to undergo conjugate addition to α,β-unsaturated carbonyl compounds. This reaction, called the *Michael reaction,* has been described for anions derived from β-diketones in Section 18.13. The enolates of ethyl acetoacetate and diethyl malonate also undergo Michael addition to the β-carbon atom of α,β-unsaturated aldehydes, ketones, and esters. For example,

$$CH_3\overset{O}{\overset{\|}{C}}CH=CH_2 \ + \ CH_2(COOCH_2CH_3)_2 \ \xrightarrow[\text{ethanol}]{KOH} \ CH_3\overset{O}{\overset{\|}{C}}CH_2CH_2CH(COOCH_2CH_3)_2$$

Methyl vinyl ketone   Diethyl malonate   Ethyl 2-carboethoxy-5-oxohexanoate
(83%)

In this reaction the enolate of diethyl malonate adds to the β carbon of methyl vinyl ketone.

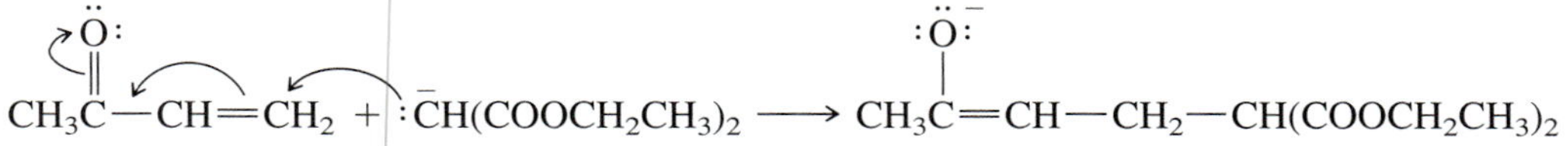

$$CH_3C\text{—}CH\text{=}CH_2 + \ ^{-}\!:CH(COOCH_2CH_3)_2 \longrightarrow CH_3C\text{=}CH\text{—}CH_2\text{—}CH(COOCH_2CH_3)_2$$

The intermediate formed in the nucleophilic addition step abstracts a proton from the solvent to give the observed product.

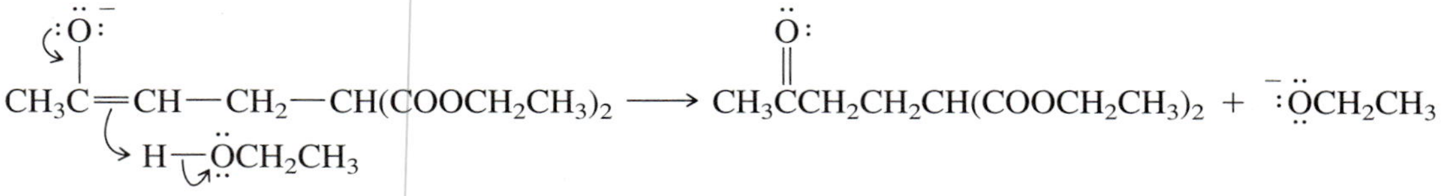

$$CH_3C\text{=}CH\text{—}CH_2\text{—}CH(COOCH_2CH_3)_2 \longrightarrow CH_3CCH_2CH_2CH(COOCH_2CH_3)_2 + \ ^{-}\!:OCH_2CH_3$$
$$H\text{—}OCH_2CH_3$$

After isolation, the Michael adduct may be subjected to ester hydrolysis and decarboxylation. When α,β-unsaturated ketones are carried through this sequence, the final products are 5-keto acids (δ-keto acids).

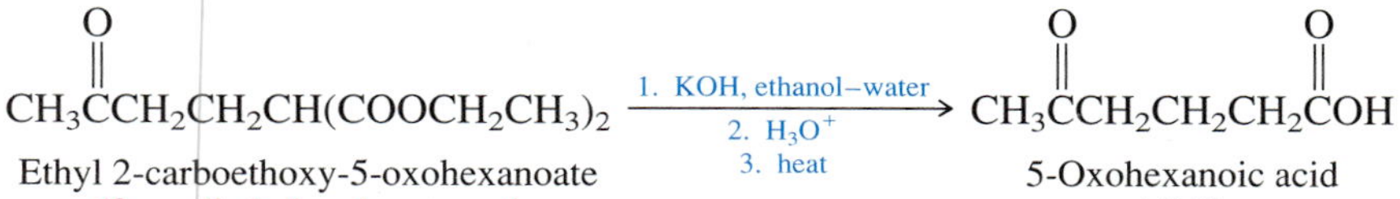

$$CH_3CCH_2CH_2CH(COOCH_2CH_3)_2 \xrightarrow[\substack{2.\ H_3O^+ \\ 3.\ heat}]{1.\ KOH,\ ethanol\text{–}water} CH_3CCH_2CH_2CH_2COH$$

Ethyl 2-carboethoxy-5-oxohexanoate
(from diethyl malonate and
methyl vinyl ketone)

5-Oxohexanoic acid
(42%)

PROBLEM 21.13 Ethyl acetoacetate behaves similarly to diethyl malonate in its reactivity toward α,β-unsaturated carbonyl compounds. Give the structure of the product of the following reaction sequence:

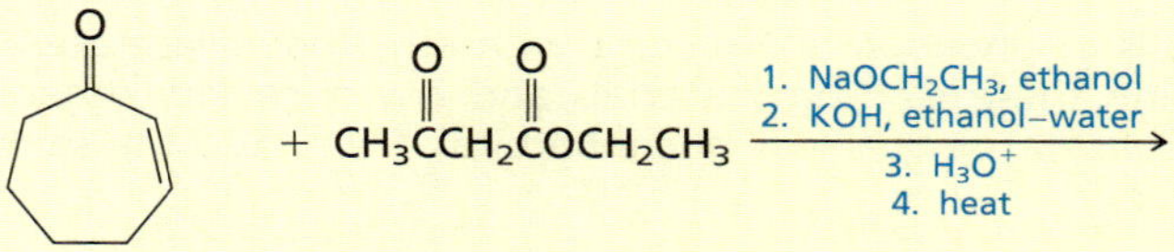

2-Cycloheptenone $+ CH_3CCH_2COCH_2CH_3$ (Ethyl acetoacetate) $\xrightarrow[\substack{3.\ H_3O^+ \\ 4.\ heat}]{\substack{1.\ NaOCH_2CH_3,\ ethanol \\ 2.\ KOH,\ ethanol\text{–}water}}$

21.10 α DEPROTONATION OF CARBONYL COMPOUNDS BY LITHIUM DIALKYLAMIDES

Most of the reactions of ester enolates described so far have centered on stabilized enolates derived from 1,3-dicarbonyl compounds such as diethyl malonate and ethyl acetoacetate. Although the synthetic value of these and related stabilized enolates is clear, chemists have long been interested in extending the usefulness of nonstabilized enolates derived from simple esters. Consider the deprotonation of an ester as represented by the acid—base reaction

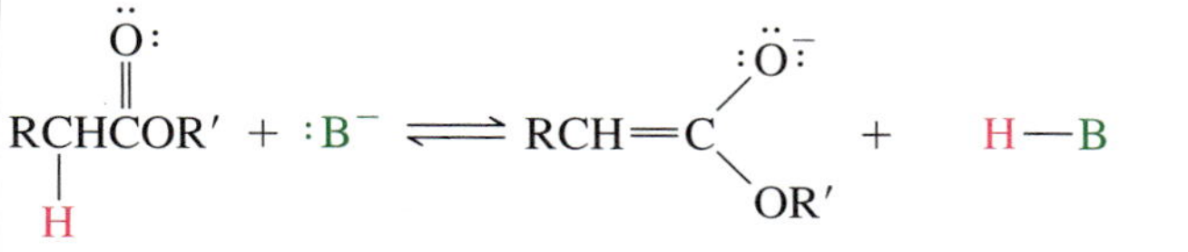

$$RCHCOR' + \ :B^- \rightleftharpoons RCH\text{=}C(OR') + H\text{—}B$$

Ester   Base   Ester enolate   Conjugate acid of base

We already know what happens when simple esters are treated with alkoxide bases—they undergo the Claisen condensation (Section 21.1). Simple esters have pK_a's of approximately 22 and give only a small amount of enolate when treated with alkoxide bases. The small amount of enolate that is formed reacts by nucleophilic addition to the carbonyl group of the ester.

What happens if the base is much stronger than an alkoxide ion?

If the base is strong enough, it will convert the ester completely to its enolate.

Under these conditions the Claisen condensation is suppressed because there is no neutral ester present for the enolate to add to. A very strong base is one that is derived from a very weak acid. Referring to Table 1.7, we see that ammonia is quite a weak acid; its pK_a is 36. Therefore, amide ion ($H_2\ddot{N}{:}^-$) is a very strong base—more than strong enough to deprotonate an ester quantitatively. Amide ion, however, also tends to add to the carbonyl group of esters; to avoid this complication, highly hindered analogs of $H_2\ddot{N}{:}^-$ are used instead. The most frequently used base for ester enolate formation is *lithium diisopropylamide* (LDA):

$$\text{Li}^+ \ (CH_3)_2CH\overset{\cdot\cdot}{-}\overset{\overline{}}{N}-CH(CH_3)_2$$

Lithium diisopropylamide

Lithium diisopropylamide is commercially available. Alternatively, it may be prepared by the reaction of butyllithium with $[(CH_3)_2CH]_2NH$ (see Problem 14.4a for a related reaction).

Lithium diisopropylamide is a strong enough base to abstract a proton from the α-carbon atom of an ester, but because it is so sterically hindered, it does not add readily to the carbonyl group. To illustrate,

$$\underset{\substack{\text{Methyl} \\ \text{butanoate} \\ \text{(stronger acid)} \\ pK_a = 22}}{CH_3CH_2\overset{O}{\overset{\|}{C}}OCH_3} + \underset{\substack{\text{Lithium} \\ \text{diisopropylamide} \\ \text{(stronger base)}}}{[(CH_3)_2CH]_2NLi} \longrightarrow \underset{\substack{\text{Lithium enolate of} \\ \text{methyl butanoate} \\ \text{(weaker base)}}}{CH_3CH_2CH{=}C\overset{OLi}{\underset{OCH_3}{\diagup}}} + \underset{\substack{\text{Diisopropylamine} \\ \text{(weaker acid)} \\ pK_a = 36}}{[(CH_3)_2CH]_2NH}$$

Direct alkylation of esters can be carried out by forming the enolate with LDA followed by addition of an alkyl halide. Tetrahydrofuran (THF) is the solvent most often used in these reactions.

$$\underset{\substack{\text{Methyl butanoate}}}{CH_3CH_2CH_2\overset{O}{\overset{\|}{C}}OCH_3} \xrightarrow[\text{THF}]{\text{LDA}} \underset{\substack{\text{Lithium enolate of} \\ \text{methyl butanoate}}}{CH_3CH_2CH{=}C\overset{OLi}{\underset{OCH_3}{\diagup}}} \xrightarrow{CH_3CH_2I} \underset{\substack{\text{Methyl 2-ethylbutanoate} \\ (92\%)}}{CH_3CH_2\underset{CH_3CH_2}{\overset{O}{\overset{\|}{C}H}C}OCH_3}$$

Ester enolates generated by proton abstraction with dialkylamide bases add to aldehydes and ketones to give β-hydroxy esters.

$$\underset{\substack{\text{Ethyl acetate}}}{CH_3COCH_2CH_3} \xrightarrow[\text{THF}]{LiNR_2} \underset{\substack{\text{Lithium enolate of} \\ \text{ethyl acetate}}}{H_2C=C\underset{OCH_2CH_3}{\overset{OLi}{\big<}}} \xrightarrow[\text{2. } H_3O^+]{\text{1. } (CH_3)_2C=O} \underset{\substack{\text{Ethyl 3-hydroxy-} \\ \text{3-methylbutanoate} \\ (90\%)}}{CH_3\underset{CH_3}{\overset{HO}{\underset{|}{C}}}CH_2COCH_2CH_3}$$

Lithium dialkylamides are excellent bases for making ketone enolates as well. Ketone enolates generated in this way can be alkylated with alkyl halides or, as illustrated in the following equation, treated with an aldehyde or a ketone.

$$\underset{\substack{\text{2,2-Dimethyl-} \\ \text{3-pentanone}}}{CH_3CH_2CC(CH_3)_3} \xrightarrow[\text{THF}]{LDA} \underset{\substack{\text{Lithium enolate of} \\ \text{2,2-dimethyl-3-pentanone}}}{CH_3CH=C\underset{C(CH_3)_3}{\overset{OLi}{\big<}}} \xrightarrow[\text{2. } H_3O^+]{\text{1. } CH_3CH_2CH} \underset{\substack{\text{5-Hydroxy-2,2,4-} \\ \text{trimethyl-3-heptanone} \\ (81\%)}}{CH_3\underset{HOCHCH_2CH_3}{\overset{O}{CHCC(CH_3)_3}}}$$

Thus, mixed aldol additions can be achieved by the tactic of quantitative enolate formation using LDA followed by addition of a different aldehyde or ketone.

PROBLEM 21.14 Outline efficient syntheses of each of the following compounds from readily available aldehydes, ketones, esters, and alkyl halides according to the methods described in this section:

(a) $(CH_3)_2CHCHCOCH_2CH_3$
            $\overset{|}{CH_3}$ (with $\overset{O}{\overset{\|}{}}$ above)

(b) $C_6H_5CHCOCH_3$
          $\overset{|}{CH_3}$ (with $\overset{O}{\overset{\|}{}}$ above)

(c) cyclohexanone with $\overset{OH}{\underset{}{\overset{|}{CHC_6H_5}}}$ substituent

(d) cyclohexane ring with $\overset{OH}{\underset{}{}}$ and $CH_2COC(CH_3)_3$ (with $\overset{O}{\overset{\|}{}}$)

SAMPLE SOLUTION (a) The α-carbon atom of the ester has two different alkyl groups attached to it.

Disconnect bond ⓐ → $(CH_3)_2CHX$ + $CH_3\overset{..}{C}HCOCH_2CH_3$ (with $\overset{O}{\overset{\|}{}}$)

$(CH_3)_2CH\overset{ⓐ}{-}\underset{\underset{CH_3}{\overset{|}{\overset{ⓑ}{}}}}{CHCOCH_2CH_3}$ (with $\overset{O}{\overset{\|}{}}$)

Disconnect bond ⓑ → CH_3X + $(CH_3)_2CH\overset{..}{C}HCOCH_2CH_3$ (with $\overset{O}{\overset{\|}{}}$)

The critical carbon–carbon bond-forming step requires nucleophilic substitution on an alkyl halide by an ester enolate. Methyl halides are more reactive than isopropyl halides in S_N2 reactions and cannot undergo elimination as a competing process; therefore, choose the synthesis in which bond ⓑ is formed by alkylation.

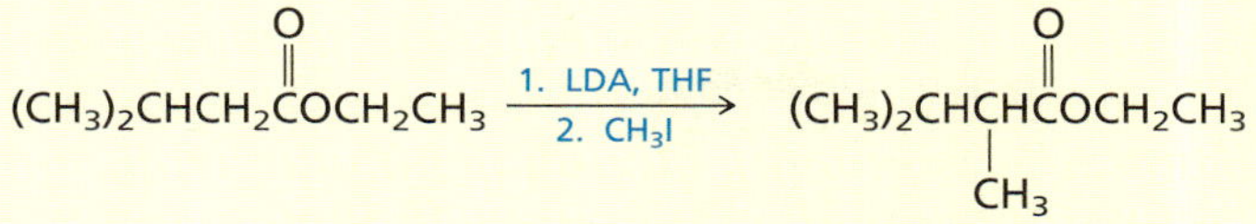

$$(CH_3)_2CHCH_2\overset{O}{\overset{\|}{C}}OCH_2CH_3 \xrightarrow[\text{2. CH}_3\text{I}]{\text{1. LDA, THF}} (CH_3)_2CHCHCOCH_2CH_3$$

Ethyl 3-methylbutanoate   Ethyl 2,3-dimethylbutanoate

(This synthesis has been reported in the chemical literature and gives the desired product in 95% yield.)

21.11 SUMMARY

Sections 21.1–21.4 β-Keto esters, which are useful reagents for a number of carbon–carbon bond-forming reactions, are prepared by the methods shown in Table 21.1.

Section 21.5 Hydrolysis of β-keto esters, such as those shown in Table 21.1, gives β-keto acids which undergo rapid decarboxylation, forming ketones.

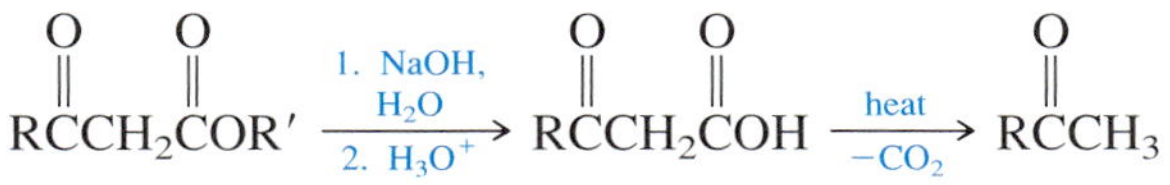

$$RCCH_2COR' \xrightarrow[\text{2. H}_3\text{O}^+]{\text{1. NaOH, H}_2\text{O}} RCCH_2COH \xrightarrow[-CO_2]{\text{heat}} RCCH_3$$

β-Keto ester   β-Keto acid   Ketone

β-Keto esters are characterized by pK_a's of about 11 and are quantitatively converted to their enolates on treatment with alkoxide bases.

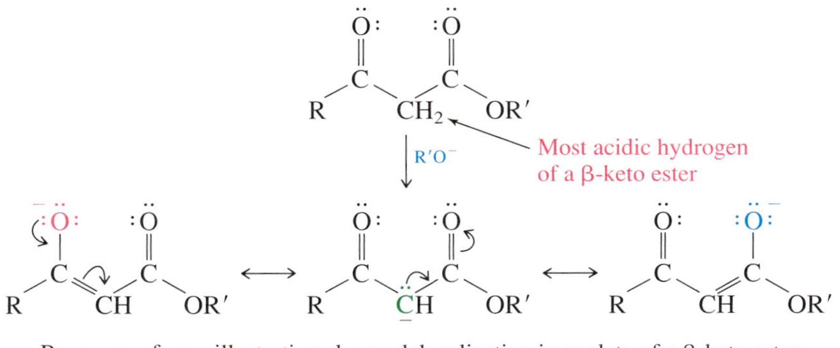

Most acidic hydrogen of a β-keto ester

Resonance forms illustrating charge delocalization in enolate of a β-keto ester

The anion of a β-keto ester may be alkylated at carbon with an alkyl halide and the product of this reaction subjected to ester hydrolysis and decarboxylation to give a ketone.

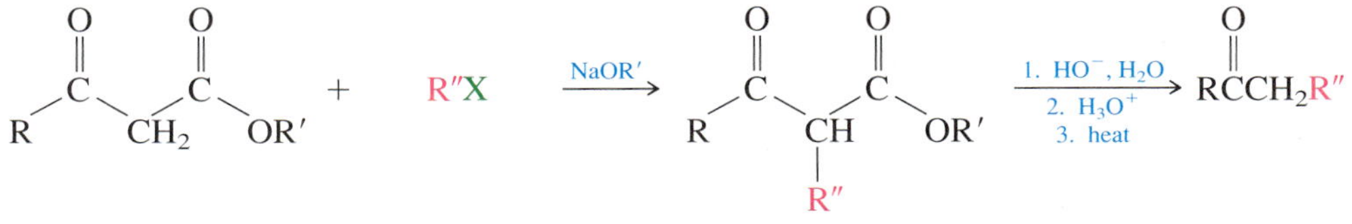

β-Keto ester   Alkyl halide   Alkylated β-keto ester   Ketone

TABLE 21.1   Preparation of β-Keto Esters

Reaction (section) and comments	General equation and specific example
Claisen condensation (Section 21.1) Esters of the type RCH_2COR' are converted to β-keto esters on treatment with alkoxide bases. One molecule of an ester is converted to its enolate; a second molecule of ester acts as an acylating agent toward the enolate.	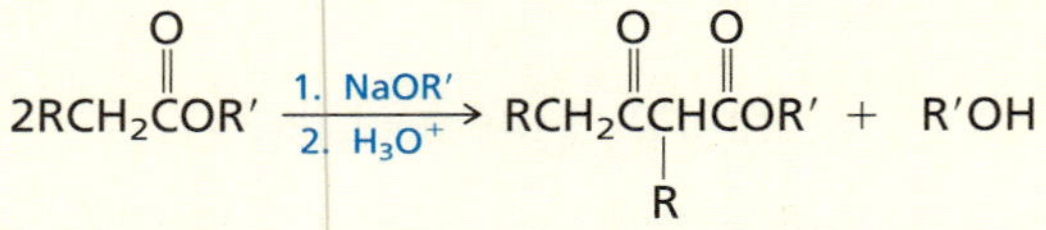 $2RCH_2COR' \xrightarrow[\text{2. }H_3O^+]{\text{1. NaOR}'} RCH_2CCHCOR' + R'OH$ Ester β-Keto ester Alcohol 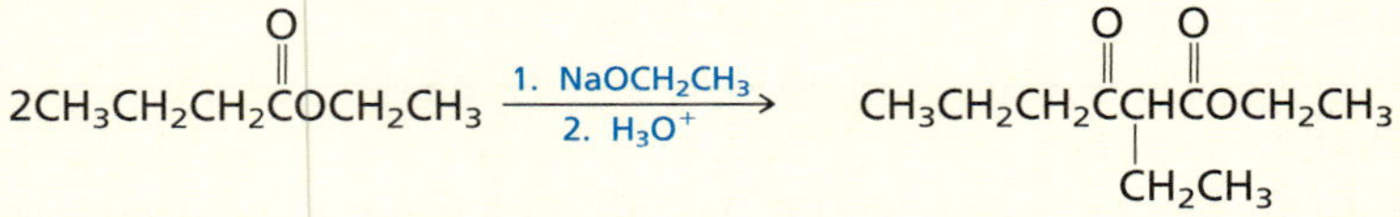 $2CH_3CH_2CH_2COCH_2CH_3 \xrightarrow[\text{2. }H_3O^+]{\text{1. NaOCH}_2CH_3} CH_3CH_2CH_2CCHCOCH_2CH_3$ with CH_2CH_3 branch Ethyl butanoate Ethyl 2-ethyl-3-oxohexanoate (76%)
Dieckmann cyclization (Section 21.2) An intramolecular analog of the Claisen condensation. Cyclic β-keto esters in which the ring is five- to seven-membered may be formed by using this reaction.	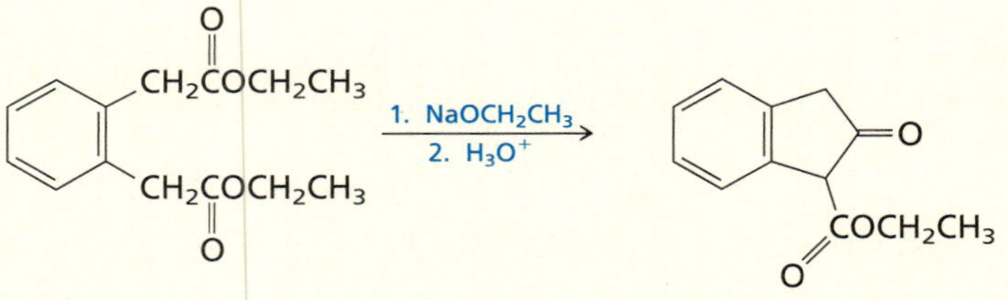 (Diethyl 1,2-benzenediacetate) $\xrightarrow[\text{2. }H_3O^+]{\text{1. NaOCH}_2CH_3}$ (Ethyl indan-2-one-1-carboxylate) (70%) Diethyl 1,2-benzenediacetate Ethyl indan-2-one-1-carboxylate (70%)
Mixed Claisen condensations (Section 21.3) Diethyl carbonate, diethyl oxalate, ethyl formate, and benzoate esters cannot form ester enolates but can act as acylating agents toward other ester enolates.	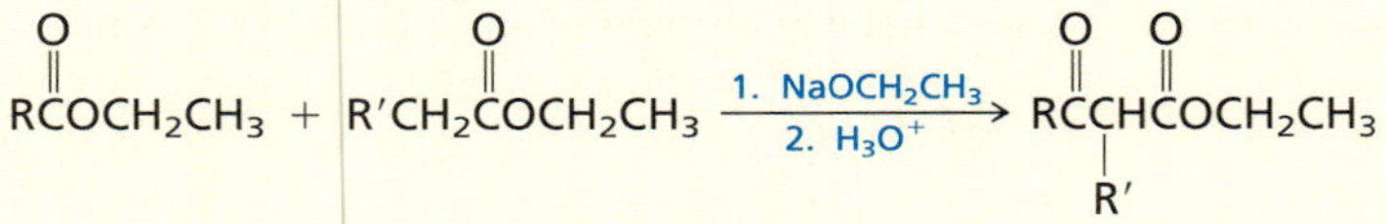 $RCOCH_2CH_3 + R'CH_2COCH_2CH_3 \xrightarrow[\text{2. }H_3O^+]{\text{1. NaOCH}_2CH_3} RCCHCOCH_2CH_3$ with R' branch Ester Another ester β-Keto ester 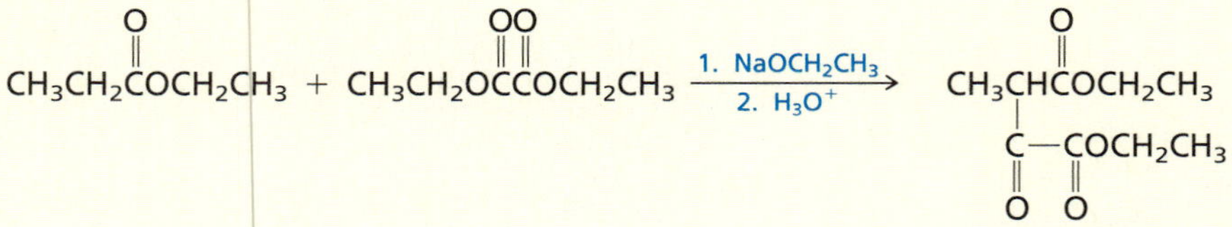 $CH_3CH_2COCH_2CH_3 + CH_3CH_2OCCOCH_2CH_3 \xrightarrow[\text{2. }H_3O^+]{\text{1. NaOCH}_2CH_3} CH_3CHCOCH_2CH_3$ with $C{-}COCH_2CH_3$ / $O\ O$ branch Ethyl propanoate Diethyl oxalate Diethyl 3-methyl-2-oxobutanedioate (60–70%)
Acylation of ketones (Section 21.4) Diethyl carbonate and diethyl oxalate can be used to acylate ketone enolates to give β-keto esters.	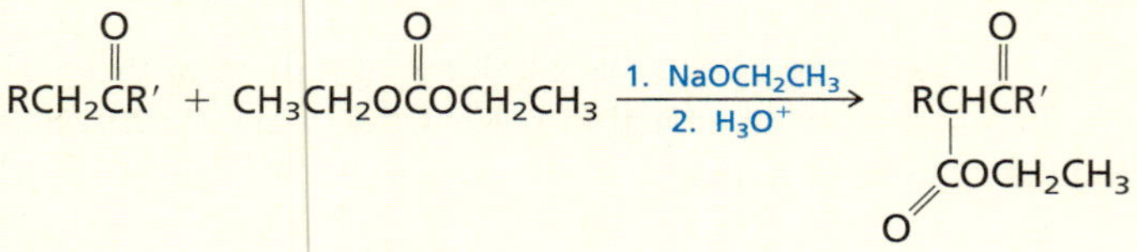 $RCH_2CR' + CH_3CH_2OCOCH_2CH_3 \xrightarrow[\text{2. }H_3O^+]{\text{1. NaOCH}_2CH_3} RCHCR'$ with $COCH_2CH_3$ branch Ketone Diethyl carbonate β-Keto ester 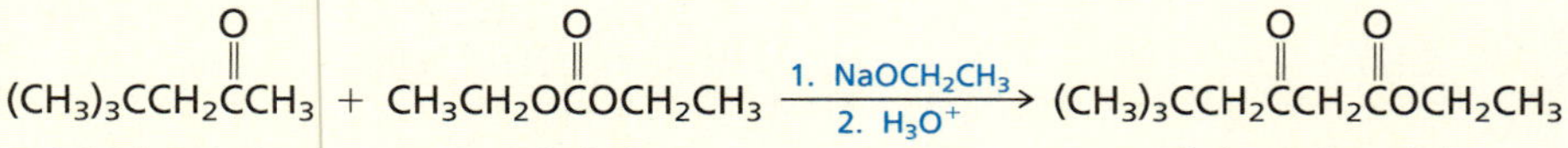 $(CH_3)_3CCH_2CCH_3 + CH_3CH_2OCOCH_2CH_3 \xrightarrow[\text{2. }H_3O^+]{\text{1. NaOCH}_2CH_3} (CH_3)_3CCH_2CCH_2COCH_2CH_3$ 4,4-Dimethyl-2-pentanone Diethyl carbonate Ethyl 5,5-dimethyl-3-oxohexanoate (66%)

Section 21.6   The **acetoacetic ester synthesis** is a procedure in which ethyl acetoacetate is alkylated with an alkyl halide as the first step in the preparation of ketones of the type $CH_3\overset{O}{\overset{\|}{C}}CH_2R$.

$$CH_3\overset{O}{\overset{\|}{C}}CH_2\overset{O}{\overset{\|}{C}}OCH_2CH_3 \xrightarrow[\text{CH}_3\text{CH}=\text{CHCH}_2\text{Br}]{\text{NaOCH}_2\text{CH}_3} CH_3\overset{O}{\overset{\|}{C}}\underset{\underset{CH_2CH=CHCH_3}{|}}{CH}\overset{O}{\overset{\|}{C}}OCH_2CH_3 \xrightarrow[\substack{2.\ H_3O^+ \\ 3.\ heat}]{1.\ HO^-,\ H_2O} CH_3\overset{O}{\overset{\|}{C}}CH_2CH_2CH=CHCH_3$$

Ethyl acetoacetate   5-Hepten-2-one (81%)

Section 21.7   The **malonic ester synthesis** is related to the acetoacetic ester synthesis. Alkyl halides (RX) are converted to carboxylic acids of the type RCH_2COOH by reaction with the enolate ion derived from diethyl malonate, followed by saponification and decarboxylation.

$$CH_2(COOCH_2CH_3)_2 \xrightarrow{\text{NaOCH}_2\text{CH}_3} \text{(cyclopentenyl)}-CH(COOCH_2CH_3)_2 \xrightarrow[\substack{2.\ H_3O^+ \\ 3.\ heat}]{1.\ HO^-,\ H_2O} \text{(cyclopentenyl)}-CH_2\overset{O}{\overset{\|}{C}}OH$$

Diethyl malonate   (2-Cyclopentenyl)acetic acid (66%)

Section 21.8   Alkylation of diethyl malonate, followed by reaction with urea, gives derivatives of barbituric acid, called **barbiturates,** which are useful sleep-promoting drugs.

$$CH_2(COOCH_2CH_3)_2 \xrightarrow{\text{RX, NaOCH}_2\text{CH}_3} RCH(COOCH_2CH_3)_2 \xrightarrow{H_2N\overset{O}{\overset{\|}{C}}NH_2} \text{(barbiturate ring)}$$

Diethyl malonate   Alkylated derivative of diethyl malonate   Alkylated derivative of barbituric acid

Section 21.9   **Michael addition** of the enolate ions derived from ethyl acetoacetate and diethyl malonate provides an alternative method for preparing their α-alkyl derivatives.

$$CH_2(COOCH_2CH_3)_2\ +\ CH_3CH=CH\overset{O}{\overset{\|}{C}}OCH_2CH_3 \xrightarrow[\text{CH}_3\text{CH}_2\text{OH}]{\text{NaOCH}_2\text{CH}_3} CH_3\underset{\underset{CH(COOCH_2CH_3)_2}{|}}{CH}CH_2\overset{O}{\overset{\|}{C}}OCH_2CH_3$$

Diethyl malonate   Ethyl 2-butenoate   Triethyl 2-methylpropane-1,1,3-tricarboxylate (95%)

Section 21.10 It is possible to generate ester enolates by deprotonation provided that the base used is very strong. Lithium diisopropylamide (LDA) is often used for this purpose. It also converts ketones quantitatively to their enolates.

$$\underset{\text{2,2-Dimethyl-3-pentanone}}{CH_3CH_2\overset{\overset{O}{\|}}{C}C(CH_3)_3} \xrightarrow[\text{THF}]{\text{LDA}} CH_3CH\overset{\overset{OLi}{|}}{=}CC(CH_3)_3 \xrightarrow[\text{2. } H_3O^+]{\text{1. } C_6H_5CH=O} \underset{\underset{CH_3}{|}}{C_6H_5\overset{\overset{OH}{|}}{C}H\overset{}{C}H\overset{\overset{O}{\|}}{C}C(CH_3)_3}$$

2,2-Dimethyl-3-pentanone

1-Hydroxy-2,4,4-trimethyl-1-phenyl-3-pentanone (78%)

PROBLEMS

21.15 The following questions pertain to the esters shown and their behavior under conditions of the Claisen condensation.

$$\underset{\text{Ethyl pentanoate}}{CH_3CH_2CH_2CH_2\overset{\overset{O}{\|}}{C}OCH_2CH_3} \qquad \underset{\underset{CH_3}{\underset{|}{\text{2-methylbutanoate}}}}{CH_3CH_2\overset{}{C}H\overset{\overset{O}{\|}}{C}OCH_2CH_3} \qquad \underset{\underset{CH_3}{\underset{|}{\text{3-methylbutanoate}}}}{CH_3\overset{}{C}HCH_2\overset{\overset{O}{\|}}{C}OCH_2CH_3} \qquad \underset{\text{2,2-dimethylpropanoate}}{(CH_3)_3C\overset{\overset{O}{\|}}{C}OCH_2CH_3}$$

Ethyl pentanoate   Ethyl 2-methylbutanoate   Ethyl 3-methylbutanoate   Ethyl 2,2-dimethylpropanoate

(a) Two of these esters are converted to β-keto esters in good yield on treatment with sodium ethoxide and subsequent acidification of the reaction mixture. Which two are these? Write the structure of the Claisen condensation product of each one.

(b) One ester is capable of being converted to a β-keto ester on treatment with sodium ethoxide, but the amount of β-keto ester that can be isolated after acidification of the reaction mixture is quite small. Which ester is this?

(c) One ester is incapable of reaction under conditions of the Claisen condensation. Which one? Why?

21.16 (a) Give the structure of the Claisen condensation product of ethyl phenylacetate ($C_6H_5CH_2COOCH_2CH_3$).

(b) What ketone would you isolate after saponification and decarboxylation of this Claisen condensation product?

(c) What ketone would you isolate after treatment of the Claisen condensation product of ethyl phenylacetate with sodium ethoxide and allyl bromide, followed by saponification and decarboxylation?

(d) Give the structure of the mixed Claisen condensation product of ethyl phenylacetate and ethyl benzoate.

(e) What ketone would you isolate after saponification and decarboxylation of the product in part (d)?

(f) What ketone would you isolate after treatment of the product in part (d) with sodium ethoxide and allyl bromide, followed by saponification and decarboxylation?

21.17 All the following questions concern ethyl (2-oxocyclohexane)carboxylate.

Ethyl (2-oxocyclohexane)carboxylate

(a) Write a chemical equation showing how you could prepare ethyl (2 oxocyclohexane)-carboxylate by a Dieckmann reaction.

(b) Write a chemical equation showing how you could prepare ethyl (2-oxocyclohexane)-carboxylate by acylation of a ketone.

(c) Write structural formulas for the two most stable enol forms of ethyl (2-oxocyclo-hexane)carboxylate.

(d) Write the three most stable resonance forms for the most stable enolate derived from ethyl (2-oxocyclohexane)carboxylate.

(e) Show how you could use ethyl (2-oxocyclohexane)carboxylate to prepare 2-methyl-cyclohexanone.

(f) Give the structure of the product formed on treatment of ethyl (2-oxocyclohexane)car-

boxylate with acrolein ($H_2C\!\!=\!\!CHCH$) in ethanol in the presence of sodium ethoxide.

$$\overset{\displaystyle O}{\overset{\|}{}}$$

21.18 Give the structure of the product formed on reaction of ethyl acetoacetate with each of the following:

(a) 1-Bromopentane and sodium ethoxide

(b) Saponification and decarboxylation of the product in part (a)

(c) Methyl iodide and the product in part (a) treated with sodium ethoxide

(d) Saponification and decarboxylation of the product in part (c)

(e) 1-Bromo-3-chloropropane and one equivalent of sodium ethoxide

(f) Product in part (e) treated with a second equivalent of sodium ethoxide

(g) Saponification and decarboxylation of the product in part (f)

(h) Phenyl vinyl ketone and sodium ethoxide

(i) Saponification and decarboxylation of the product in part (h)

21.19 Repeat the preceding problem for diethyl malonate.

21.20 (a) Only a small amount (less than 0.01%) of the enol form of diethyl malonate is present at equilibrium. Write a structural formula for this enol.

(b) Enol forms are present to the extent of about 8% in ethyl acetoacetate. There are three constitutionally isomeric enols possible. Write structural formulas for these three enols. Which one do you think is the most stable? The least stable? Why?

(c) Bromine reacts rapidly with both diethyl malonate and ethyl acetoacetate. The reaction is acid-catalyzed and liberates hydrogen bromide. What is the product formed in each reaction?

21.21 (a) On addition of one equivalent of methylmagnesium iodide to ethyl acetoacetate, the Grignard reagent is consumed, but the only organic product obtained after working up the reaction mixture is ethyl acetoacetate. Why? What happens to the Grignard reagent?

(b) On repeating the reaction but using D_2O and DCl to work up the reaction mixture, it is found that the recovered ethyl acetoacetate contains deuterium. Where is this deuterium located?

21.22 Give the structure of the principal organic product of each of the following reactions:

(a) Ethyl octanoate $\xrightarrow[\text{2. } H_3O^+]{\text{1. } NaOCH_2CH_3}$

(b) Product of part (a) $\xrightarrow[\substack{\text{2. } H_3O^+ \\ \text{3. heat}}]{\text{1. } NaOH, H_2O}$

(c) Ethyl acetoacetate + 1-bromobutane $\xrightarrow{NaOCH_2CH_3, \text{ ethanol}}$

(d) Product of part (c) $\xrightarrow[\substack{\text{2. } H_3O^+ \\ \text{3. heat}}]{\text{1. } NaOH, H_2O}$

(e) Product of part (c) + 1-iodobutane $\xrightarrow{NaOCH_2CH_3, \text{ ethanol}}$

(f) Product of part (e) $\xrightarrow[\substack{\text{2. } H_3O^+ \\ \text{3. heat}}]{\text{1. } NaOH, H_2O}$

(g) Acetophenone + diethyl carbonate $\xrightarrow[\text{2. } H_3O^+]{\text{1. } NaOCH_2CH_3}$

(h) Acetone + diethyl oxalate $\xrightarrow[\text{2. } H_3O^+]{\text{1. } NaOCH_2CH_3}$

(i) Diethyl malonate + 1-bromo-2-methylbutane $\xrightarrow{NaOCH_2CH_3, \text{ ethanol}}$

(j) Product of part (i) $\xrightarrow[\substack{\text{2. } H_3O^+ \\ \text{3. heat}}]{\text{1. } NaOH, H_2O}$

(k) Diethyl malonate + 6-methyl-2-cyclohexenone $\xrightarrow{NaOCH_2CH_3, \text{ ethanol}}$

(l) Product of part (k) $\xrightarrow{H_2O, HCl, \text{ heat}}$

(m) *tert*-Butyl acetate $\xrightarrow[\substack{\text{2. benzaldehyde} \\ \text{3. } H_3O^+}]{\text{1. } [(CH_3)_2CH]_2NLi, THF}$

21.23 Give the structure of the principal organic product of each of the following reactions:

(a)

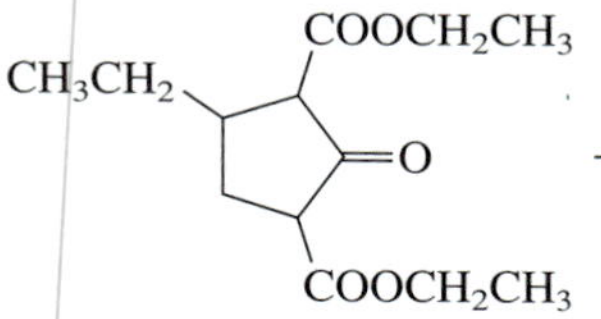

$\xrightarrow[\text{heat}]{H_2O, H_2SO_4} C_7H_{12}O$

(b) 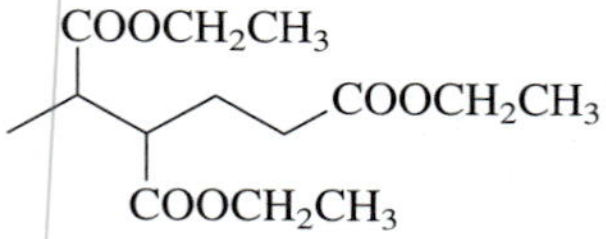 $\xrightarrow[\text{2. } H_3O^+]{\text{1. } NaOCH_2CH_3} C_{12}H_{18}O_5$

(c) Product of part (b) $\xrightarrow[\text{heat}]{H_3O^+} C_7H_{10}O_3$

(d) 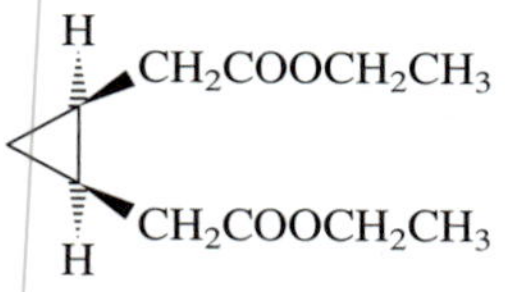 $\xrightarrow[\text{2. } H_3O^+]{\text{1. } NaOCH_2CH_3} C_9H_{12}O_3$

(e) Product of part (d) $\xrightarrow[\substack{\text{2. } H_3O^+ \\ \text{3. heat}}]{\text{1. } HO^-, H_2O} C_6H_8O$

21.24 The spicy flavor of cayenne pepper is due mainly to a substance called *capsaicin*. The following sequence of steps was used in a synthesis of capsaicin. See if you can deduce the structure of capsaicin on the basis of this synthesis.

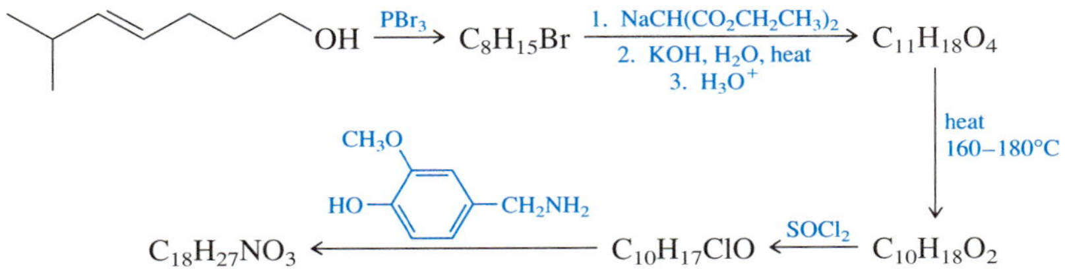

Capsaicin

21.25 Show how you could prepare each of the following compounds. Use the starting material indicated along with ethyl acetoacetate or diethyl malonate and any necessary inorganic reagents. Assume also that the customary organic solvents are freely available.

(a) 4-Phenyl-2-butanone from benzyl alcohol

(b) 3-Phenylpropanoic acid from benzyl alcohol

(c) 2-Allyl-1,3-propanediol from propene

(d) 4-Penten-1-ol from propene

(e) 5-Hexen-2-ol from propene

(f) Cyclopropanecarboxylic acid from 1,2-dibromoethane

(g) 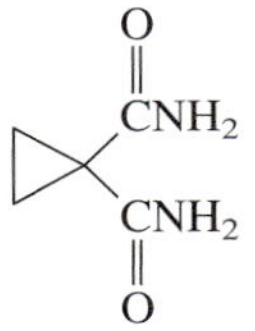   from 1,2-dibromoethane

(h) $HO_2C(CH_2)_{10}CO_2H$ from $HO_2C(CH_2)_6CO_2H$

21.26 *Diphenadione* inhibits the clotting of blood; that is, it is an *anticoagulant*. It is used to control vampire bat populations in South America by a "Trojan horse" strategy. A few bats are trapped, smeared with diphenadione, and then released back into their normal environment. Other bats, in the course of grooming these diphenadione-coated bats, ingest the anticoagulant and bleed to death, either internally or through accidental bites and scratches.

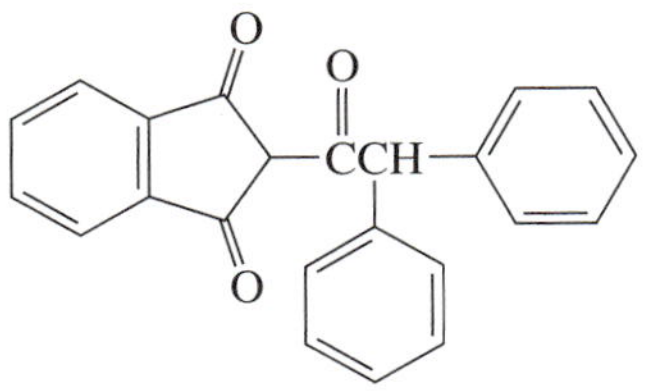

Diphenadione

Suggest a synthesis of diphenadione from 1,1-diphenylacetone and dimethyl 1,2-benzenedicarboxylate.

21.27 *Phenylbutazone* is a frequently prescribed antiinflammatory drug. It is prepared by the reaction shown.

$$CH_3CH_2CH_2CH_2CH(COOCH_2CH_3)_2 \ + \ C_6H_5NHNHC_6H_5 \ \longrightarrow \ C_{19}H_{20}N_2O_2$$

Diethyl butylmalonate   1,2-Diphenylhydrazine   Phenylbutazone

What is the structure of phenylbutazone?

21.28 The use of epoxides as alkylating agents for diethyl malonate provides a useful route to γ-lactones. Write equations illustrating such a sequence for styrene oxide as the starting epoxide. Is the lactone formed by this reaction 3-phenylbutanolide, or is it 4-phenylbutanolide?

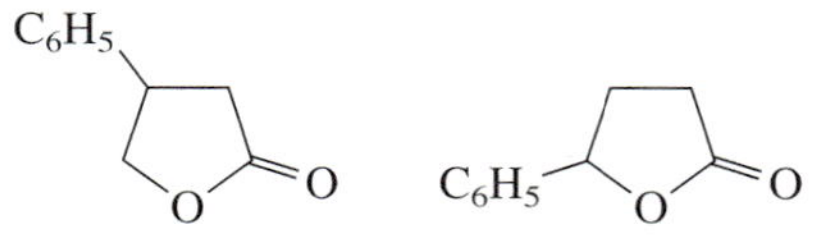

3-Phenylbutanolide   4-Phenylbutanolide

21.29 Diethyl malonate is prepared commercially by hydrolysis and esterification of ethyl cyanoacetate.

$$N{\equiv}CCH_2\overset{\displaystyle O}{\overset{\|}{C}}OCH_2CH_3$$

Ethyl cyanoacetate

The preparation of ethyl cyanoacetate proceeds via ethyl chloroacetate and begins with acetic acid. Write a sequence of reactions describing this synthesis.

21.30 The tranquilizing drug *meprobamate* has the structure shown.

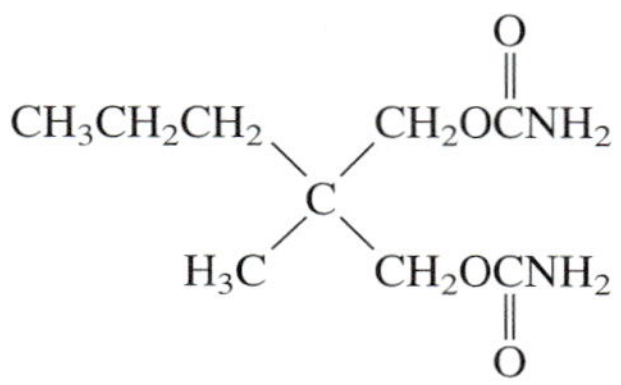

Meprobamate

Devise a synthesis of meprobamate from diethyl malonate and any necessary organic or inorganic reagents. *Hint: Carbamate esters,* that is, compounds of the type $RO\overset{\displaystyle O}{\overset{\|}{C}}NH_2$, are prepared from alcohols by the sequence of reactions

$$ROH \;+\; Cl\overset{\displaystyle O}{\overset{\|}{C}}Cl \longrightarrow RO\overset{\displaystyle O}{\overset{\|}{C}}Cl \xrightarrow{\;NH_3,\ H_2O\;} RO\overset{\displaystyle O}{\overset{\|}{C}}NH_2$$

Alcohol   Phosgene   Chlorocarbonate ester   Carbamate ester

21.31 When the compound shown was heated in refluxing aqueous hydrochloric acid for 60 hours, a product with the molecular formula $C_5H_6O_3$ was isolated in 97% yield. Identify this product. Along with this product, three other carbon-containing substances are formed. What are they?

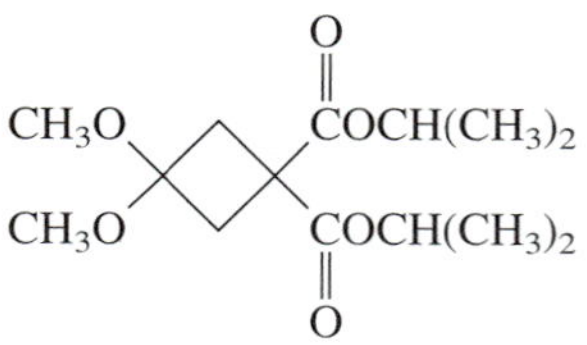

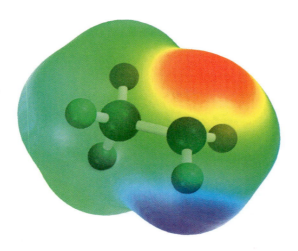

CHAPTER 22

AMINES

Nitrogen-containing compounds are essential to life. Their ultimate source is atmospheric nitrogen which, by a process known as *nitrogen fixation,* is reduced to ammonia, then converted to organic nitrogen compounds. This chapter describes the chemistry of **amines,** organic derivatives of ammonia. **Alkylamines** have their nitrogen attached to sp^3-hybridized carbon; **arylamines** have their nitrogen attached to an sp^2-hybridized carbon of a benzene or benzene-like ring.

$$R \overset{..}{-} \overset{..}{N}\diagdown \qquad Ar \overset{..}{-} \overset{..}{N}\diagdown$$

R = alkyl group: Ar = aryl group:
alkylamine   arylamine

Amines, like ammonia, are weak bases. They are, however, the strongest uncharged bases found in significant quantities under physiological conditions. Amines are usually the bases involved in biological acid–base reactions; they are often the nucleophiles in biological nucleophilic substitutions.

Our word *vitamin* was coined in 1912 in the belief that the substances present in the diet that prevented scurvy, pellagra, beriberi, rickets, and other diseases were "vital amines." In many cases, that belief was confirmed; certain vitamins did prove to be amines. In many other cases, however, vitamins were not amines. Nevertheless, the name *vitamin* entered our language and stands as a reminder that early chemists recognized the crucial place occupied by amines in biological processes.

22.1 AMINE NOMENCLATURE

Unlike alcohols and alkyl halides, which are classified as primary, secondary, or tertiary according to the degree of substitution at the carbon that bears the functional group, amines are classified according to their *degree of substitution at nitrogen.* An amine with one carbon attached to nitrogen is a *primary amine*, an amine with two is a *secondary amine*, and an amine with three is a *tertiary amine*.

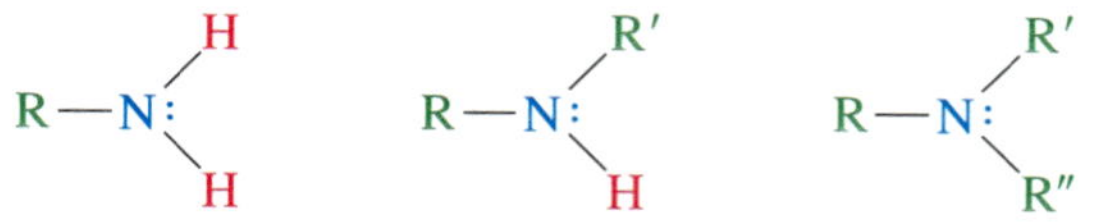

$$R-\overset{H}{\underset{H}{N:}} \qquad R-\overset{R'}{\underset{H}{N:}} \qquad R-\overset{R'}{\underset{R''}{N:}}$$

Primary amine   Secondary amine   Tertiary amine

The groups attached to nitrogen may be any combination of alkyl or aryl groups.

Amines are named in two main ways in the IUPAC system, either as *alkylamines* or as *alkanamines*. When primary amines are named as alkylamines, the ending *-amine* is added to the name of the alkyl group that bears the nitrogen. When named as alkanamines, the alkyl group is named as an alkane and the *-e* ending replaced by *-amine*.

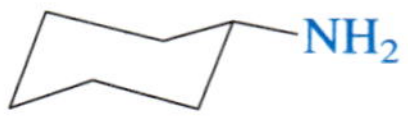

$$CH_3CH_2NH_2 \qquad \qquad \qquad CH_3CHCH_2CH_2CH_3$$
$$\qquad \qquad \qquad \qquad \qquad \qquad \qquad \overset{|}{NH_2}$$

Ethylamine	Cyclohexylamine	1-Methylbutylamine
(ethanamine)	(cyclohexanamine)	(2-pentanamine)

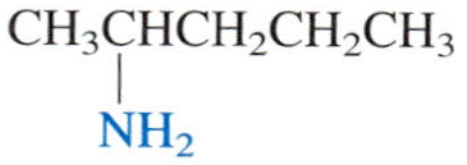

PROBLEM 22.1 Give an acceptable alkylamine or alkanamine name for each of the following amines:

(a) $C_6H_5CH_2CH_2NH_2$

(b) $C_6H_5\underset{\underset{CH_3}{|}}{C}HNH_2$

(c) $H_2C{=}CHCH_2NH_2$

SAMPLE SOLUTION   (a) The amino substituent is bonded to an ethyl group that bears a phenyl substituent at C-2. The compound $C_6H_5CH_2CH_2NH_2$ may be named as either 2-phenylethylamine or 2-phenylethanamine.

Aniline is the parent IUPAC name for amino-substituted derivatives of benzene. Substituted derivatives of aniline are numbered beginning at the carbon that bears the amino group. Substituents are listed in alphabetical order, and the direction of numbering is governed by the usual "first point of difference" rule.

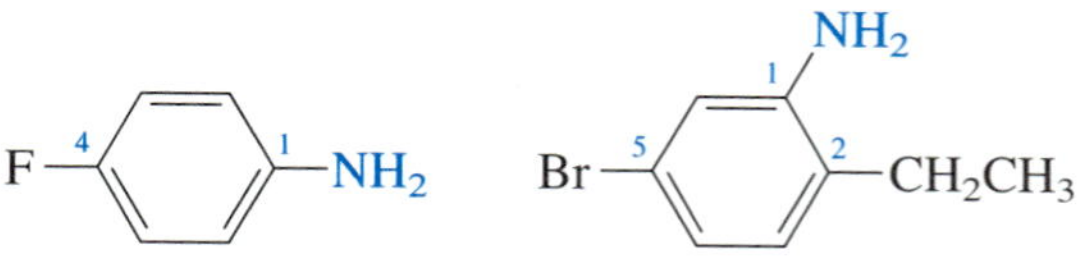

p-Fluoroaniline   5-Bromo-2-ethylaniline

Arylamines may also be named as *arenamines*. Thus, *benzenamine* is an alternative, but rarely used, name for aniline.

Compounds with two amino groups are named by adding the suffix *-diamine* to the name of the corresponding alkane or arene. The final *-e* of the parent hydrocarbon is retained.

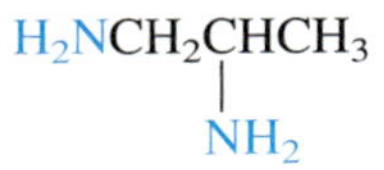

$H_2NCH_2CHCH_3$
       |
       NH_2

$H_2NCH_2CH_2CH_2CH_2CH_2CH_2NH_2$

H_2N—⟨benzene⟩—NH_2

1,2-Propanediamine   1,6-Hexanediamine   1,4-Benzenediamine

Amino groups rank rather low in seniority when the parent compound is identified for naming purposes. Hydroxyl groups and carbonyl groups outrank amino groups. In these cases, the amino group is named as a substituent.

$HOCH_2CH_2NH_2$

$$\underset{\text{HC}}{\overset{\overset{\displaystyle O}{\|}}{}}\!\!-\!\!\overset{1}{⟨benzene⟩}\overset{4}{}\!\!-\!\!NH_2$$

2-Aminoethanol   *p*-Aminobenzaldehyde
                    (4-Aminobenzenecarbaldehyde)

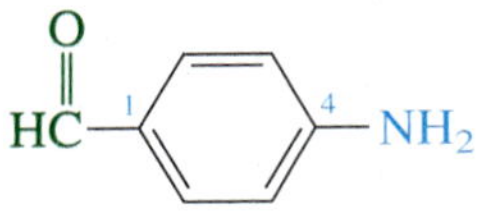

Secondary and tertiary amines are named as *N*-substituted derivatives of primary amines. The parent primary amine is taken to be the one with the longest carbon chain. The prefix *N*- is added as a locant to identify substituents on the amino nitrogen as needed.

$CH_3NHCH_2CH_3$

$NHCH_2CH_3$ (on benzene ring, positions 1,3,4 with NO_2 at 3 and Cl at 4)

$N(CH_3)_2$ (on cycloheptane ring)

N-Methylethylamine   4-Chloro-*N*-ethyl-3-nitroaniline   *N,N*-Dimethylcyclo-heptylamine

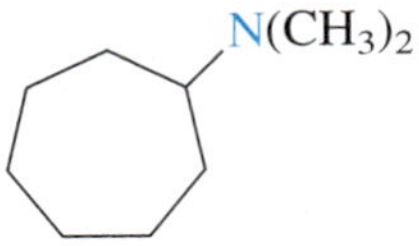

(a secondary amine)   (a secondary amine)   (a tertiary amine)

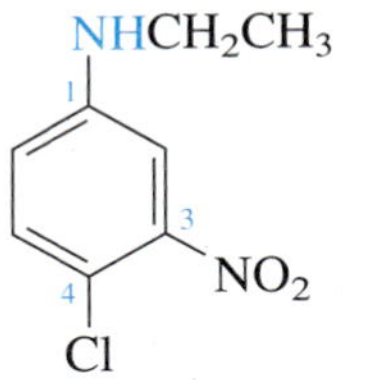

PROBLEM 22.2 Assign alkanamine names to *N*-methylethylamine and to *N,N*-dimethylcycloheptylamine.

SAMPLE SOLUTION *N*-Methylethylamine (given as $CH_3NHCH_2CH_3$ in the preceding example) is an *N*-substituted derivative of ethanamine; it is *N*-methylethanamine.

PROBLEM 22.3 Classify the following amine as primary, secondary, or tertiary, and give it an acceptable IUPAC name.

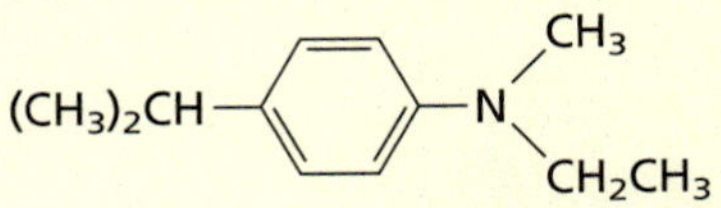

$$(CH_3)_2CH-⟨benzene⟩-N\!\!\begin{array}{l}\diagup CH_3\\\diagdown CH_2CH_3\end{array}$$

A nitrogen that bears four substituents is positively charged and is named as an *ammonium* ion. The anion that is associated with it is also identified in the name.

$$CH_3\overset{+}{N}H_3 \ \ Cl^-$$

Methylammonium
chloride

N-Ethyl-N-methylcyclopentyl-
ammonium trifluoroacetate

$$C_6H_5CH_2\overset{+}{N}(CH_3)_3 \ \ I^-$$

Benzyltrimethyl-
ammonium iodide
(a quaternary ammonium
salt)

Ammonium salts that have four alkyl groups bonded to nitrogen are called **quaternary ammonium salts.**

22.2 STRUCTURE AND BONDING

Alkylamines: As shown in Figure 22.1 methylamine, like ammonia, has a pyramidal arrangement of bonds to nitrogen. Its H—N—H angles (106°) are slightly smaller than the tetrahedral value of 109.5°, whereas the C—N—H angle (112°) is slightly larger. The C—N bond distance of 147 pm lies between typical C—C bond distances in alkanes (153 pm) and C—O bond distances in alcohols (143 pm).

An orbital hybridization description of bonding in methylamine is shown in Figure 22.2. Nitrogen and carbon are both sp^3-hybridized and are joined by a σ bond. The unshared electron pair on nitrogen occupies an sp^3-hybridized orbital. This lone pair is involved in reactions in which amines act as bases or nucleophiles. The graphic that opened this chapter is an electrostatic potential map that clearly shows the concentration of electron density at nitrogen in methylamine.

You can examine the structure of methylamine, including its electrostatic potential map, in more detail on *Learning By Modeling.*

Arylamines: Aniline, like alkylamines, has a pyramidal arrangement of bonds around nitrogen, but its pyramid is somewhat shallower. One measure of the extent of this flattening is given by the angle between the carbon–nitrogen bond and the bisector of the H—N—H angle.

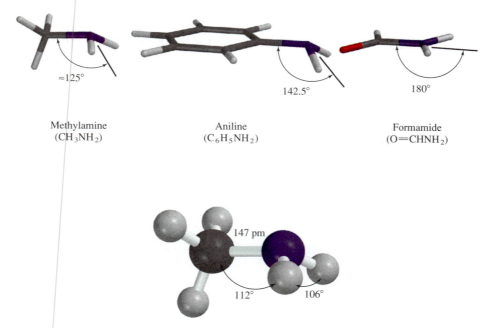

≈125°

Methylamine
(CH_3NH_2)

142.5°

Aniline
($C_6H_5NH_2$)

180°

Formamide
($O{=}CHNH_2$)

FIGURE 22.1 A ball-and-spoke model of methylamine showing the trigonal pyramidal arrangement of bonds to nitrogen. The most stable conformation has the staggered arrangement of bonds shown. Other alkylamines have similar geometries.

147 pm

112°   106°

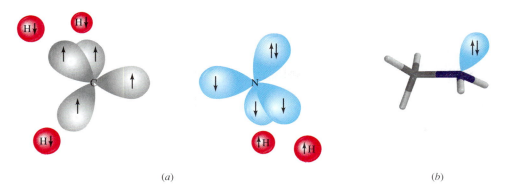

(a)   (b)

FIGURE 22.2 Orbital hybridization description of bonding in methylamine. (a) Carbon has four valence electrons; each of four equivalent sp^3-hybridized orbitals contains one electron. Nitrogen has five valence electrons. Three of its sp^3 hybrid orbitals contain one electron each; the fourth sp^3 hybrid orbital contains two electrons. (b) Nitrogen and carbon are connected by a σ bond in methylamine. This σ bond is formed by overlap of an sp^3 hybrid orbital on each atom. The five hydrogen atoms of methylamine are joined to carbon and nitrogen by σ bonds. The two remaining electrons of nitrogen occupy an sp^3-hybridized orbital.

For sp^3-hybridized nitrogen, this angle (not the same as the C—N—H bond angle) is 125°, and the measured angles in simple alkylamines are close to that. The corresponding angle for sp^2 hybridization at nitrogen with a planar arrangement of bonds, as in amides, for example, is 180°. The measured value for this angle in aniline is 142.5°, suggesting a hybridization somewhat closer to sp^3 than to sp^2.

The structure of aniline reflects a compromise between two modes of binding the nitrogen lone pair (Figure 22.3). The electrons are more strongly attracted to nitrogen when they are in an orbital with some s character—an sp^3-hybridized orbital, for example—than when they are in a p orbital. On the other hand, delocalization of these electrons into the aromatic π system is better achieved if they occupy a p orbital. A p orbital of nitrogen is better aligned for overlap with the p orbitals of the benzene ring to form an extended π system than is an sp^3-hybridized orbital. As a result of these two opposing forces, nitrogen adopts an orbital hybridization that is between sp^3 and sp^2.

The corresponding resonance description shows the delocalization of the nitrogen lone-pair electrons in terms of contributions from dipolar structures.

The geometry at nitrogen in amines is discussed in an article entitled "What Is the Geometry at Trigonal Nitrogen?" in the January 1998 issue of the *Journal of Chemical Education*, pp. 108–109.

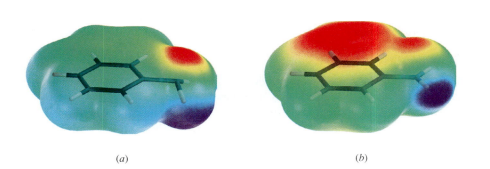

Most stable Lewis structure for aniline

Dipolar resonance forms of aniline

(a)   (b)

FIGURE 22.3 Electrostatic potential maps of aniline in which the geometry at nitrogen is (a) nonplanar and (b) planar. In the nonplanar geometry, the unshared pair occupies an sp³ hybrid orbital of nitrogen. The region of highest electron density in (a) is associated with nitrogen. In the planar geometry, nitrogen is sp²-hybridized, and the electron pair is delocalized between a p orbital of nitrogen and the π system of the ring. The region of highest electron density in (b) encompasses both the ring and nitrogen. The actual structure combines features of both; nitrogen adopts a hybridization state between sp³ and sp².

The orbital and resonance models for bonding in arylamines are simply alternative ways of describing the same phenomenon. Delocalization of the nitrogen lone pair decreases the electron density at nitrogen while increasing it in the π system of the aromatic ring. We've already seen one chemical consequence of this in the high level of reactivity of aniline in electrophilic aromatic substitution reactions (Section 12.12). Other ways in which electron delocalization affects the properties of arylamines are described in later sections of this chapter.

PROBLEM 22.4 As the extent of electron delocalization into the ring increases, the geometry at nitrogen flattens. *p*-Nitroaniline, for example, is planar. Write a resonance form for *p*-nitroaniline that shows how the nitro group increases electron delocalization. Examine the electrostatic potential of the *p*-nitroaniline model on *Learning By Modeling*. Where is the greatest concentration of negative charge?

22.3 PHYSICAL PROPERTIES

We have often seen that the polar nature of a substance can affect physical properties such as boiling point. This is true for amines, which are more polar than alkanes but less polar than alcohols. For similarly constituted compounds, alkylamines have boiling points higher than those of alkanes but lower than those of alcohols.

A collection of physical properties of some representative amines is given in Appendix 1. Most commonly encountered alkylamines are liquids with unpleasant, "fishy" odors.

$CH_3CH_2CH_3$	$CH_3CH_2NH_2$	CH_3CH_2OH
Propane	Ethylamine	Ethanol
$\mu = 0$ D	$\mu = 1.2$ D	$\mu = 1.7$ D
bp $-42°$C	bp $17°$C	bp $78°$C

Dipole–dipole interactions, especially hydrogen bonding, are present in amines but absent in alkanes. But because nitrogen is less electronegative than oxygen, an N—H bond is less polar than an O—H bond and hydrogen bonding is weaker in amines than in alcohols.

Among isomeric amines, primary amines have the highest boiling points, and tertiary amines the lowest.

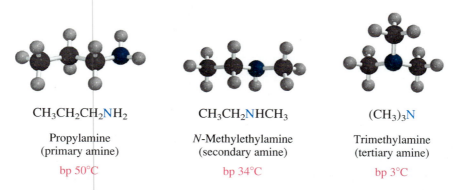

$CH_3CH_2CH_2NH_2$	$CH_3CH_2NHCH_3$	$(CH_3)_3N$
Propylamine (primary amine)	*N*-Methylethylamine (secondary amine)	Trimethylamine (tertiary amine)
bp $50°$C	bp $34°$C	bp $3°$C

Primary and secondary amines can participate in intermolecular hydrogen bonding, but tertiary amines lack N—H bonds and so cannot.

Amines that have fewer than six or seven carbon atoms are soluble in water. All amines, even tertiary amines, can act as proton acceptors in hydrogen bonding to water molecules.

The simplest arylamine, aniline, is a liquid at room temperature and has a boiling point of 184°C. Almost all other arylamines have higher boiling points. Aniline is only slightly soluble in water (3 g/100 mL). Substituted derivatives of aniline tend to be even less water-soluble.

22.4 BASICITY OF AMINES

As we discussed in Section 1.14, it is more useful to describe the basicity of amines in terms of the pK_a's of their conjugate acids than as basicity constants K_b. Always bear in mind that:

The more basic the amine, the weaker its conjugate acid.

The more basic the amine, the larger the pK_a of its conjugate acid.

Citing amine basicity according to the pK_a of the conjugate acid permits acid–base reactions involving amines to be analyzed according to the usual Brønsted relationships. For example, we see that amines are converted to ammonium ions by acids even as weak as acetic acid:

| Methylamine | Acetic acid (stronger acid; $pK_a = 4.7$) | | Methylammonium ion (weaker acid; $pK_a = 10.7$) | Acetate ion |

Recall that acid–base reactions are favorable when the stronger acid is on the left and the weaker acid on the right.

Conversely, adding sodium hydroxide to an ammonium salt converts it to the free amine:

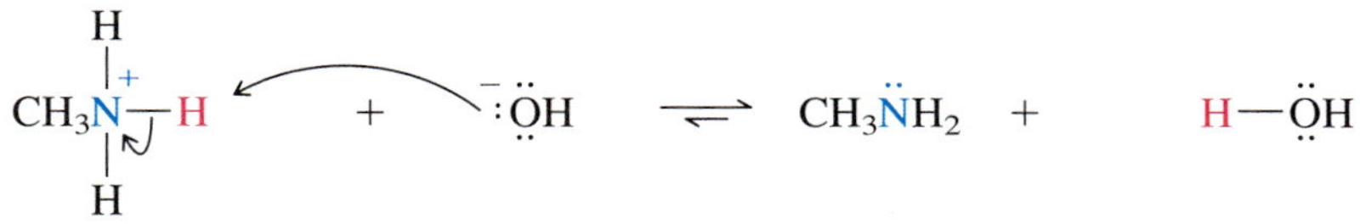

| Methylammonium ion (stronger acid; $pK_a = 10.7$) | Hydroxide ion | Methylamine | Water (weaker acid; $pK_a = 15.7$) |

PROBLEM 22.5 Apply the Henderson–Hasselbalch equation (see "Quantitative Relationships Involving Carboxylic Acids," the box accompanying Section 19.4) to calculate the $CH_3NH_3^+/CH_3NH_2$ ratio in water buffered at pH 7.

Their basicity provides a means by which amines may be separated from neutral organic compounds. A mixture containing an amine is dissolved in diethyl ether and shaken with dilute hydrochloric acid to convert the amine to an ammonium salt. The ammonium salt, being ionic, dissolves in the aqueous phase, which is separated from the ether layer. Adding sodium hydroxide to the aqueous layer converts the ammonium salt back to the free amine, which is then removed from the aqueous phase by extraction with a fresh portion of ether.

Amines are weak bases, but as a class, *amines are the strongest bases of all neutral molecules.* Table 22.1 lists basicity data for a number of amines. The most important relationships to be drawn from the data are

1. Alkylamines are slightly stronger bases than ammonia.

2. Alkylamines differ very little among themselves in basicity. Their basicities cover a range of less than 10 in equilibrium constant (1 pK unit).

	Basicity of Amines As Measured by the pK_a of Their Conjugate Acids*	
TABLE 22.1		

Compound	Structure	pK_a of conjugate acid
Ammonia	NH_3	9.3
Primary amines		
Methylamine	CH_3NH_2	10.6
Ethylamine	$CH_3CH_2NH_2$	10.8
Isopropylamine	$(CH_3)_2CHNH_2$	10.6
tert-Butylamine	$(CH_3)_3CNH_2$	10.4
Aniline	$C_6H_5NH_2$	4.6
Secondary amines		
Dimethylamine	$(CH_3)_2NH$	10.7
Diethylamine	$(CH_3CH_2)_2NH$	11.1
N-Methylaniline	$C_6H_5NHCH_3$	4.8
Tertiary amines		
Trimethylamine	$(CH_3)_3N$	9.7
Triethylamine	$(CH_3CH_2)_3N$	10.8
N,N-Dimethylaniline	$C_6H_5N(CH_3)_2$	5.1

*In water at 25°C.

3. Arylamines are about 1 million times (6 pK units) weaker bases than ammonia and alkylamines.

The small differences in basicity between ammonia and alkylamines, and among the various classes of alkylamines (primary, secondary, tertiary), come from a mix of effects. Replacing hydrogens of ammonia by alkyl groups affects both sides of the acid–base equilibrium in ways that largely cancel.

Replacing hydrogens by aryl groups is a different story, however. An aryl group affects the base much more than the conjugate acid, and the overall effect is large. One way to compare alkylamines and arylamines is by examining the Brønsted equilibrium for proton transfer *to* an alkylamine *from* the conjugate acid of an arylamine.

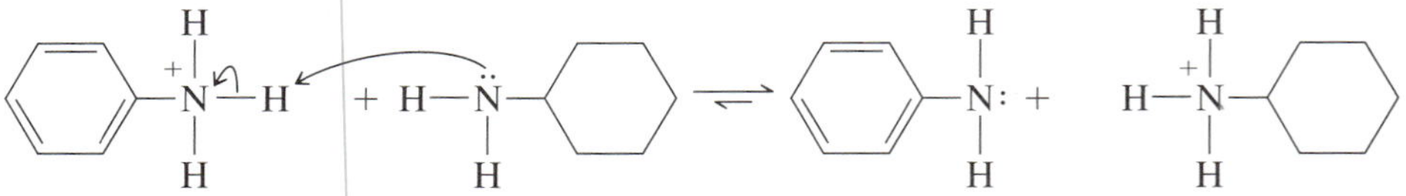

Anilinium ion
(stronger acid; $pK_a = 4.6$)   Cyclohexylamine   Aniline   Cyclohexylammonium ion
(weaker acid; $pK_a = 10.6$)

K_{eq} can be obtained by raising 10 to the power of the number obtained by subtracting the pK_a of the acid on the left from the acid on the right (10.6 − 4.6).

The equilibrium shown in the equation lies to the right. $K_{eq} = 10^6$ for proton transfer from the conjugate acid of aniline to cyclohexylamine, making cyclohexylamine 1,000,000 times more basic than aniline.

Reading the equation from left to right, we can say that anilinium ion is a stronger acid than cyclohexylammonium ion because loss of a proton from anilinium ion creates an unshared electron pair of aniline. Conjugation of this unshared pair with the aromatic ring stabilizes the right-hand side of the equation and biases the equilibrium in that direction.

Reading the equation from right to left, we can say that aniline is a weaker base than cyclohexylamine because the electron pair on nitrogen of aniline is strongly held by virtue of being delocalized into the π system of the aromatic ring. The unshared pair in cyclohexylamine is localized on nitrogen, less strongly held, and therefore "more available" in an acid–base reaction.

Even though they are weaker bases, arylamines, like alkylamines, can be completely protonated by strong acids. Aniline is extracted from an ether solution into 1 M hydrochloric acid by being completely converted to a water-soluble anilinium salt under these conditions.

Compare the calcu-
lated charge on
nitrogen in cyclo-
hexylamine and aniline on
Learning By Modeling.

PROBLEM 22.6 The pK_a's of the conjugate acids of the two amines shown differ by a factor of 40,000. Which amine is the stronger base? Why? Using *Learning By Modeling*, what are the calculated charges on the two nitrogens?

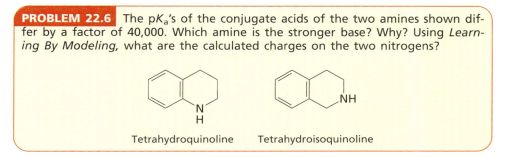

Tetrahydroquinoline   Tetrahydroisoquinoline

Conjugation of the amino group of an arylamine with a second aromatic ring, then a third, reduces its basicity even further. Diphenylamine is 6300 times less basic than aniline, whereas triphenylamine is scarcely a base at all, being estimated as 10^{10} times less basic than aniline and 10^{14} times less basic than ammonia.

	$C_6H_5NH_2$	$(C_6H_5)_2NH$	$(C_6H_5)_3N$
	Aniline	Diphenylamine	Triphenylamine
pK_a of conjugate acid:	4.6	0.8	≈ -5

In general, electron-donating substituents on the aromatic ring increase the basicity of arylamines only slightly. Thus, as shown in Table 22.2, an electron-donating methyl group in the para position *increases* the basicity of aniline by less than 1 pK unit. Electron-withdrawing groups are base-weakening and can exert large effects. A *p*-trifluoromethyl group *decreases* the basicity of aniline by a factor of 200 and a *p*-nitro group by a factor of 3800. In the case of *p*-nitroaniline a resonance interaction of the type shown provides for extensive delocalization of the unshared electron pair of the amine group.

TABLE 22.2   Effect of para Substituents on the Basicity of Aniline

X—⟨benzene⟩—NH_2	X	pK_a of conjugate acid
	H	4.6
	CH_3	5.3
	CF_3	3.5
	O_2N	1.0

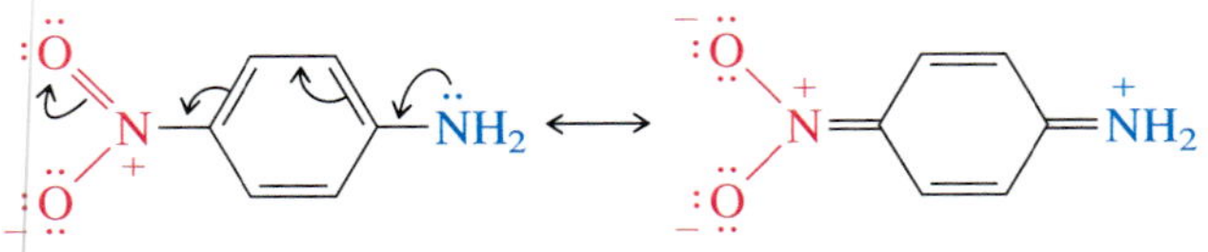

Electron delocalization in *p*-nitroaniline

Just as aniline is much less basic than alkylamines because the unshared electron pair of nitrogen is delocalized into the π system of the ring, *p*-nitroaniline is even less basic because the extent of this delocalization is greater and involves the oxygens of the nitro group.

PROBLEM 22.7   Each of the following is a much weaker base than aniline. Present a resonance argument to explain the effect of the substituent in each case.

(a) *o*-Cyanoaniline

(c) *p*-Aminoacetophenone

(b) $C_6H_5NHCCH_3$ (with O double-bonded above the C)

SAMPLE SOLUTION   (a) A cyano substituent is strongly electron-withdrawing. When present at a position ortho to an amino group on an aromatic ring, a cyano substituent increases the delocalization of the amine lone-pair electrons by a direct resonance interaction.

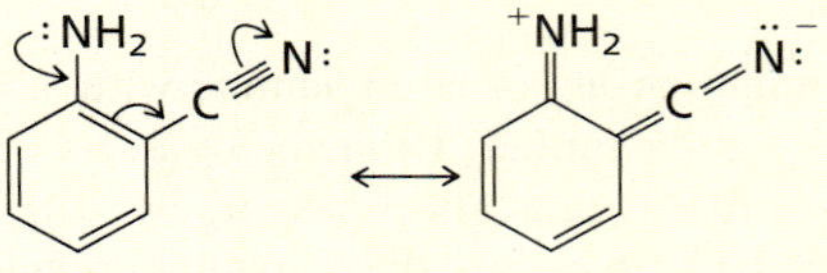

This resonance stabilization is lost when the amine group becomes protonated, and *o*-cyanoaniline is therefore a weaker base than aniline.

Multiple substitution by strongly electron-withdrawing groups diminishes the basicity of arylamines still more. As just noted, aniline is 3800 times as strong a base as *p*-nitroaniline; however, it is 10^9 times more basic than 2,4-dinitroaniline. A practical consequence of this is that arylamines that bear two or more strongly electron-withdrawing groups are often not capable of being extracted from a diethyl ether solution into dilute aqueous acid.

Nonaromatic heterocyclic compounds, piperidine, for example, are similar in basicity to alkylamines. When nitrogen is part of an aromatic ring, however, its basicity decreases markedly. Pyridine, for example, resembles arylamines in being almost 1 million times less basic than piperidine.

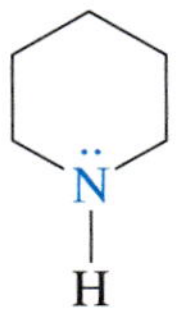

is more basic than

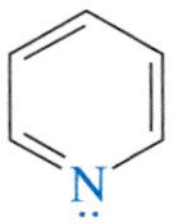

Piperidine
pK_a of conjugate acid = 11.2

Pyridine
pK_a of conjugate acid = 5.2

Pyridine and imidazole were two of the heterocyclic aromatic compounds described in Section 11.22.

Imidazole and its derivatives form an interesting and important class of heterocyclic aromatic amines. Imidazole is approximately 100 times more basic than pyridine.

Protonation of imidazole yields an ion that is stabilized by the electron delocalization represented in the resonance structures shown:

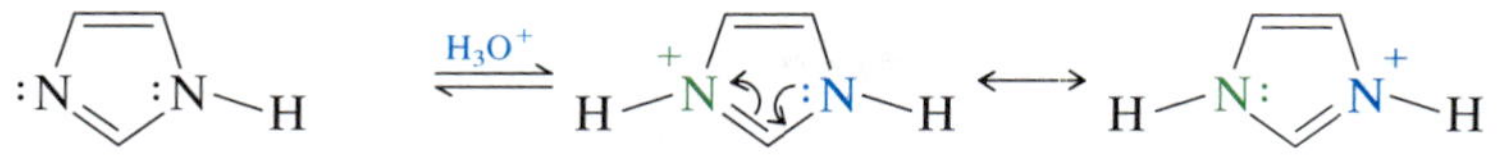

Imidazole

pK_a of conjugate acid $= 7$

Imidazolium ion

An imidazole ring is a structural unit in the amino acid *histidine* (Section 27.1) and is involved in a large number of biological processes as a base and as a nucleophile.

> **PROBLEM 22.8** Given that the pK_a of imidazolium ion is 7, is a 1 M aqueous solution of imidazolium chloride acidic, basic, or neutral? What about a 1 M solution of imidazole? A solution containing equal molar quantities of imidazole and imidazolium chloride?

22.5   TETRAALKYLAMMONIUM SALTS AS PHASE-TRANSFER CATALYSTS

In spite of being ionic, many quaternary ammonium salts dissolve in nonpolar media. The four alkyl groups attached to nitrogen shield its positive charge and impart *lipophilic* character to the tetraalkylammonium ion. The following two quaternary ammonium salts, for example, are soluble in solvents of low polarity such as benzene, decane, and halogenated hydrocarbons:

$$CH_3\overset{+}{N}(CH_2CH_2CH_2CH_2CH_2CH_2CH_2CH_3)_3 \ Cl^-$$

Methyltrioctylammonium chloride

Benzyltriethylammonium chloride

This property of quaternary ammonium salts is used to advantage in an experimental technique known as **phase-transfer catalysis.** Imagine that you wish to carry out the reaction

$$CH_3CH_2CH_2CH_2Br \ + \ NaCN \longrightarrow CH_3CH_2CH_2CH_2CN \ + \ NaBr$$

Butyl bromide   Sodium cyanide   Pentanenitrile   Sodium bromide

Sodium cyanide does not dissolve in butyl bromide. The two reactants contact each other only at the surface of the solid sodium cyanide, and the rate of reaction under these conditions is too slow to be of synthetic value. Dissolving the sodium cyanide in water is of little help because butyl bromide is not soluble in water and reaction can occur only at the interface between the two phases. Adding a small amount of benzyltrimethylammonium chloride, however, causes pentanenitrile to form rapidly even at room temperature. The quaternary ammonium salt is acting as a *catalyst;* it increases the reaction rate. How?

Quaternary ammonium salts catalyze the reaction between an anion and an organic substrate by transferring the anion from the aqueous phase, where it cannot contact the substrate, to the organic phase. In the example just cited, the first step occurs in the

AMINES AS NATURAL PRODUCTS

The ease with which amines are extracted into aqueous acid, combined with their regeneration on treatment with base, makes it a simple matter to separate amines from other plant materials, and nitrogen-containing natural products were among the earliest organic compounds to be studied.* Their basic properties led amines obtained from plants to be called **alkaloids**. The number of known alkaloids exceeds 5000. They are of special interest because most are characterized by a high level of biological activity. Some examples include *cocaine, coniine,* and *morphine.*

Cocaine
(A central nervous system stimulant obtained from the leaves of the coca plant.)

Coniine
(Present along with other alkaloids in the hemlock extract used to poison Socrates.)

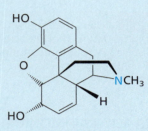

Morphine
(An opium alkaloid. Although it is an excellent analgesic, its use is restricted because of the potential for addiction. Heroin is the diacetate ester of morphine.)

Many alkaloids, such as *nicotine* and *quinine,* contain two (or more) nitrogen atoms. The nitrogens shown in red in quinine and nicotine are part of a substituted quinoline and pyridine ring, respectively.

Quinine
(Alkaloid of cinchona bark used to treat malaria)

Nicotine
(An alkaloid present in tobacco; a very toxic compound sometimes used as an insecticide)

Several naturally occurring amines mediate the transmission of nerve impulses and are referred to as **neurotransmitters**. Two examples are *epinephrine* and *serotonin.* (Strictly speaking, these compounds are not classified as alkaloids because they are not isolated from plants.)

*The isolation of alkaloids from plants is reviewed in the August 1991 issue of the *Journal of Chemical Education,* pp, 700–703.

—Cont.

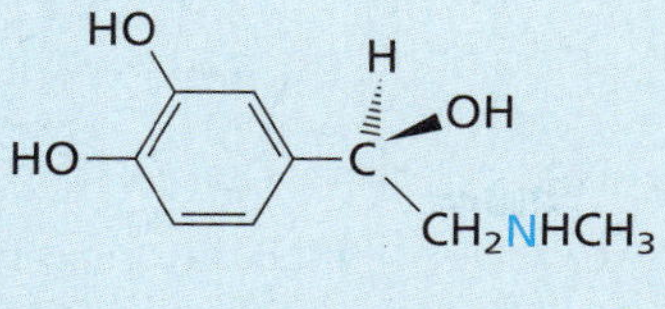

Epinephrine

(Also called *adrenaline;* a hormone secreted by the adrenal gland that prepares the organism for "flight or fight.")

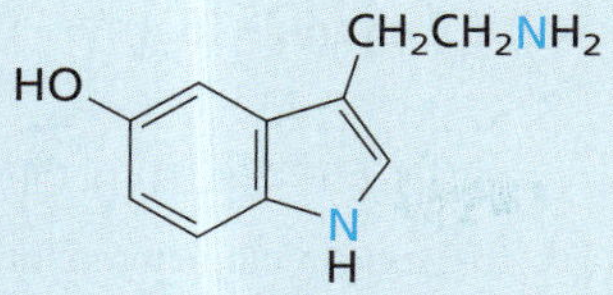

Serotonin

(A hormone synthesized in the pineal gland. Certain mental disorders are believed to be related to serotonin levels in the brain.)

Bioactive amines are also widespread in animals. A variety of structures and properties have been found in substances isolated from frogs, for example. One, called epibatidine, is a naturally occurring painkiller isolated from the skin of an Ecuadoran frog. Another family of frogs produces a toxic mixture of several stereoisomeric amines, called dendrobines, on their skin that protects them from attack.

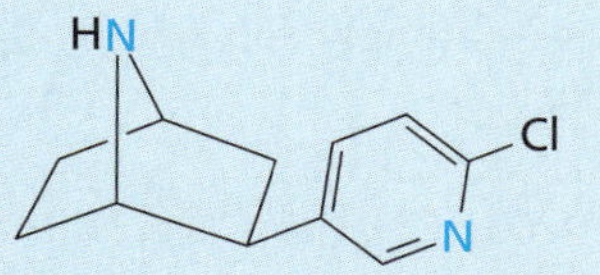

Epibatidine

(Once used as an arrow poison, it is hundreds of times more powerful than morphine in relieving pain. It is too toxic to be used as a drug, however.)

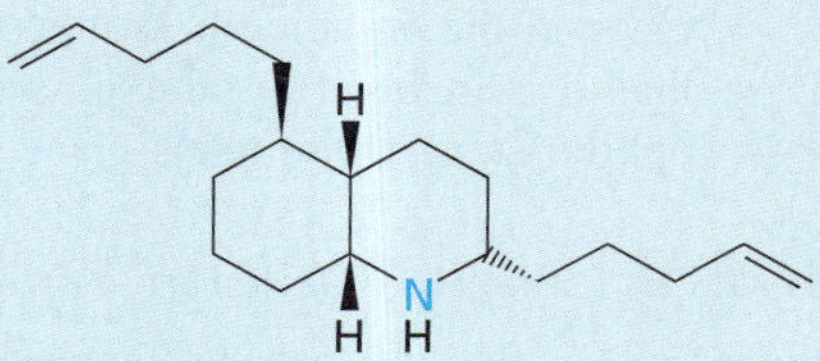

Dendrobine

(Isolated from frogs of the Dendrobatidae family. Related compounds have also been isolated from certain ants.)

Among the more important amine derivatives found in the body are a group of compounds known as **polyamines,** which contain two to four nitrogen atoms separated by several methylene units:

Putrescine

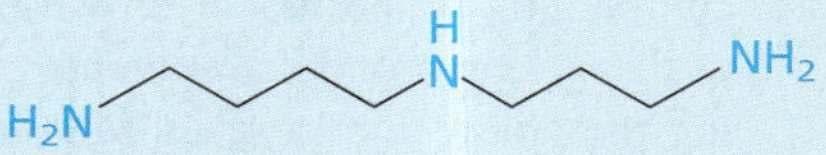

Spermidine

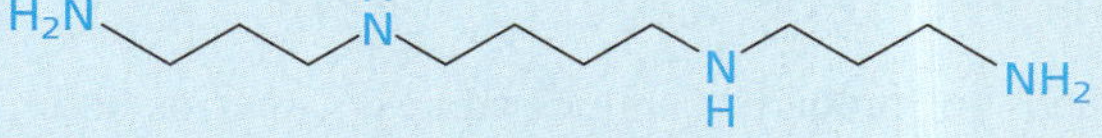

Spermine

These compounds are present in almost all mammalian cells, where they are believed to be involved in cell differentiation and proliferation. Because each nitrogen of a polyamine is protonated at physiological pH (7.4), putrescine, spermidine, and spermine exist as cations with a charge of $+2$, $+3$, and $+4$, respectively, in body fluids. Structural studies suggest that these polyammonium ions affect the conformation of biological macromolecules by electrostatic binding to specific anionic sites—the negatively charged phosphate groups of DNA, for example.

aqueous phase and is an exchange of the anionic partner of the quaternary ammonium salt for cyanide ion:

$$C_6H_5CH_2\overset{+}{N}(CH_3)_3 \ Cl^- \ + \ CN^- \ \underset{}{\overset{fast}{\rightleftharpoons}} \ C_6H_5CH_2\overset{+}{N}(CH_3)_3 \ CN^- \ + \ Cl^-$$

Benzyltrimethylammonium chloride	Cyanide ion	Benzyltrimethylammonium cyanide	Chloride ion
(aqueous)	(aqueous)	(aqueous)	(aqueous)

The benzyltrimethylammonium ion migrates to the butyl bromide phase, carrying a cyanide ion along with it.

$$C_6H_5CH_2\overset{+}{N}(CH_3)_3 \ CN^- \ \overset{fast}{\rightleftharpoons} \ C_6H_5CH_2\overset{+}{N}(CH_3)_3 \ CN^-$$

Benzyltrimethylammonium cyanide	Benzyltrimethylammonium cyanide
(aqueous)	(in butyl bromide)

Once in the organic phase, cyanide ion is only weakly solvated and is far more reactive than it is in water or ethanol, where it is strongly solvated by hydrogen bonding. Nucleophilic substitution takes place rapidly.

$$CH_3CH_2CH_2CH_2Br \ + \ C_6H_5CH_2\overset{+}{N}(CH_3)_3 \ CN^- \ \longrightarrow \ CH_3CH_2CH_2CH_2CN \ + \ C_6H_5CH_2\overset{+}{N}(CH_3)_3 \ Br^-$$

Butyl bromide	Benzyltrimethylammonium cyanide	Pentanenitrile	Benzyltrimethylammonium bromide
	(in butyl bromide)	(in butyl bromide)	(in butyl bromide)

The benzyltrimethylammonium bromide formed in this step returns to the aqueous phase, where it can repeat the cycle.

Phase-transfer catalysis succeeds for two reasons. First, it provides a mechanism for introducing an anion into the medium that contains the reactive substrate. More important, the anion is introduced in a weakly solvated, highly reactive state. You've already seen phase-transfer catalysis in another form in Section 16.4, where the metal-complexing properties of crown ethers were described. Crown ethers permit metal salts to dissolve in nonpolar solvents by surrounding the cation with a lipophilic cloak, leaving the anion free to react without the encumbrance of strong solvation forces.

> Phase-transfer catalysis is the subject of an article in the April 1978 issue of the *Journal of Chemical Education* (pp. 235–238). This article includes examples of a variety of reactions carried out under phase-transfer conditions.

22.6 REACTIONS THAT LEAD TO AMINES: A REVIEW AND A PREVIEW

Methods for preparing amines address either or both of the following questions:

1. How is the required carbon–nitrogen bond to be formed?

2. Given a nitrogen-containing organic compound such as an amide, a nitrile, or a nitro compound, how is the correct oxidation state of the desired amine to be achieved?

A number of reactions that lead to carbon–nitrogen bond formation were presented in earlier chapters and are summarized in Table 22.3. Among the reactions in the table, the nucleophilic ring opening of epoxides and the reaction of α-halo acids with ammonia give amines directly. The other reactions in Table 22.3 yield products that are converted to amines by some subsequent procedure. As these procedures are described in the following sections, you will see that they are largely applications of principles that you've already learned. You will encounter some new reagents and some new uses for familiar reagents, but very little in the way of new reaction types is involved.

TABLE 22.3   Methods for Carbon–Nitrogen Bond Formation Discussed in Earlier Chapters

Reaction (section) and comments	General equation and specific example
Nucleophilic substitution by azide ion on an alkyl halide (Sections 8.1, 8.13) Azide ion is a very good nucleophile and reacts with primary and secondary alkyl halides to give alkyl azides. Phase-transfer catalysts accelerate the rate of reaction.	$:\overset{..}{N}{=}\overset{+}{N}{=}\overset{..}{N}:$ + R—X $\longrightarrow$ $:\overset{..}{N}{=}\overset{+}{N}{=}\overset{..}{N}{-}R$ + X$^-$ Azide ion Alkyl halide Alkyl azide Halide ion $CH_3CH_2CH_2CH_2CH_2Br \xrightarrow[\text{catalyst}]{\substack{NaN_3 \\ \text{phase-transfer}}} CH_3CH_2CH_2CH_2CH_2N_3$ Pentyl bromide Pentyl azide (89%) (1-bromopentane) (1-azidopentane)
Nitration of arenes (Section 12.3) The standard method for introducing a nitrogen atom as a substituent on an aromatic ring is nitration with a mixture of nitric acid and sulfuric acid. The reaction proceeds by electrophilic aromatic substitution.	ArH + HNO$_3$ $\xrightarrow{H_2SO_4}$ ArNO$_2$ + H$_2$O Arene Nitric acid Nitroarene Water Benzaldehyde $\xrightarrow[\text{H}_2\text{SO}_4]{\text{HNO}_3}$ *m*-Nitrobenzaldehyde (75–84%)
Nucleophilic ring opening of epoxides by ammonia (Section 16.12) The strained ring of an epoxide is opened on nucleophilic attack by ammonia and amines to give β-amino alcohols. Azide ion also reacts with epoxides; the products are β-azido alcohols.	$H_3N:$ + $R_2C{-}CR_2$ $\longrightarrow$ $H_2\overset{..}{N}{-}\overset{R}{\underset{R}{C}}{-}\overset{R}{\underset{R}{C}}{-}OH$ \O/ Ammonia Epoxide β-Amino alcohol (2R,3R)-2,3-Epoxybutane $\xrightarrow[\text{H}_2\text{O}]{\text{NH}_3}$ (2R,3S)-3-Amino-2-butanol (70%)
Nucleophilic addition of amines to aldehydes and ketones (Sections 17.10, 17.11) Primary amines undergo nucleophilic addition to the carbonyl group of aldehydes and ketones to form carbinolamines. These carbinolamines dehydrate under the conditions of their formation to give N-substituted imines. Secondary amines yield enamines.	RNH_2 + $R'\overset{O}{\overset{\|}{C}}R''$ $\longrightarrow$ $R'\overset{:NR}{\overset{\|}{C}}R''$ + H$_2$O Primary Aldehyde Imine Water amine or ketone CH_3NH_2 + $C_6H_5\overset{O}{\overset{\|}{C}}H$ $\longrightarrow$ $C_6H_5CH{=}NCH_3$ Methylamine Benzaldehyde N-Benzylidenemethylamine (70%)

(Continued)

TABLE 22.3	Methods for Carbon–Nitrogen Bond Formation Discussed in Earlier Chapters *(Continued)*

Reaction (section) and comments	General equation and specific example
Nucleophilic substitution by ammonia on α-halo acids (Section 19.16) The α-halo acids obtained by halogenation of carboxylic acids under conditions of the Hell–Volhard–Zelinsky reaction are reactive substrates in nucleophilic substitution processes. A standard method for the preparation of α-amino acids is displacement of halide from α-halo acids by nucleophilic substitution using excess aqueous ammonia.	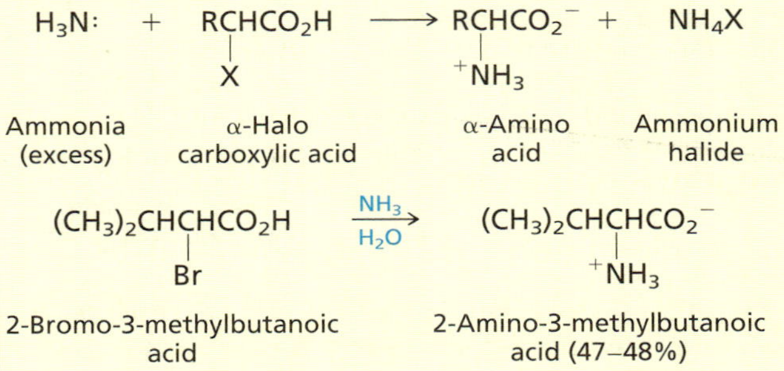 $H_3N: + RCHCO_2H \longrightarrow RCHCO_2^- + NH_4X$ (with X and $^+NH_3$ groups) Ammonia (excess) α-Halo carboxylic acid α-Amino acid Ammonium halide $(CH_3)_2CHCHCO_2H \xrightarrow[H_2O]{NH_3} (CH_3)_2CHCHCO_2^-$ (Br; $^+NH_3$) 2-Bromo-3-methylbutanoic acid 2-Amino-3-methylbutanoic acid (47–48%)
Nucleophilic acyl substitution (Sections 20.4, 20.6, and 20.12) Acylation of ammonia and amines by an acyl chloride, acid anhydride, or ester is an exceptionally effective method for the formation of carbon–nitrogen bonds.	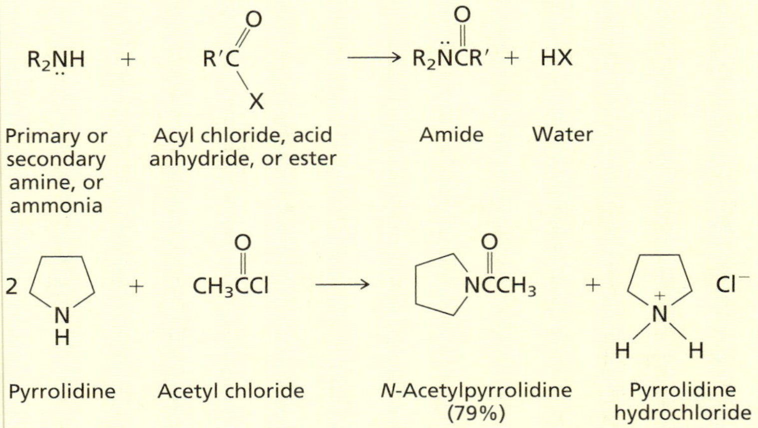 $R_2NH + R'CX(=O) \longrightarrow R_2NCR'(=O) + HX$ Primary or secondary amine, or ammonia Acyl chloride, acid anhydride, or ester Amide Water $2 \text{(pyrrolidine)} + CH_3CCl(=O) \longrightarrow \text{(N-acetylpyrrolidine)} + \text{(pyrrolidine hydrochloride) } Cl^-$ Pyrrolidine Acetyl chloride N-Acetylpyrrolidine (79%) Pyrrolidine hydrochloride

22.7   PREPARATION OF AMINES BY ALKYLATION OF AMMONIA

Alkylamines are, in principle, capable of being prepared by nucleophilic substitution reactions of alkyl halides with ammonia.

$$RX + 2NH_3 \longrightarrow RNH_2 + \overset{+}{N}H_4 \ X^-$$

Alkyl halide   Ammonia   Primary amine   Ammonium halide salt

Although this reaction is useful for preparing α-amino acids (Table 22.3, fifth entry), it is not a general method for the synthesis of amines. Its major limitation is that the expected primary amine product is itself a nucleophile and competes with ammonia for the alkyl halide.

$$RX + RNH_2 + NH_3 \longrightarrow RNHR + \overset{+}{N}H_4 \ X^-$$

Alkyl halide   Primary amine   Ammonia   Secondary amine   Ammonium halide salt

When 1-bromooctane, for example, is allowed to react with ammonia, both the primary amine and the secondary amine are isolated in comparable amounts.

$$CH_3(CH_2)_6CH_2Br \xrightarrow{\text{NH}_3\ (2\ \text{mol})} CH_3(CH_2)_6CH_2NH_2\ +\ [CH_3(CH_2)_6CH_2]_2NH$$

1-Bromooctane	Octylamine	*N,N*-Dioctylamine
(1 mol)	(45%)	(43%)

In a similar manner, competitive alkylation may continue, resulting in formation of a trialkylamine.

$$RX\ +\ R_2NH\ +\ NH_3\ \longrightarrow\ R_3N\ +\ \overset{+}{N}H_4\ X^-$$

Alkyl halide	Secondary amine	Ammonia	Tertiary amine	Ammonium halide salt

Even the tertiary amine competes with ammonia for the alkylating agent. The product is a quaternary ammonium salt.

$$RX\ +\ R_3N\ \longrightarrow\ \overset{+}{R}_4N\ X^-$$

Alkyl halide	Tertiary amine	Quaternary ammonium salt

Because alkylation of ammonia can lead to a complex mixture of products, it is used to prepare primary amines only when the starting alkyl halide is not particularly expensive and the desired amine can be easily separated from the other components of the reaction mixture.

> **PROBLEM 22.9**   Alkylation of ammonia is sometimes employed in industrial processes; the resulting mixture of amines is separated by distillation. The ultimate starting materials for the industrial preparation of allylamine are propene, chlorine, and ammonia. Write a series of equations showing the industrial preparation of allylamine from these starting materials. (Allylamine has a number of uses, including the preparation of the diuretic drugs *meralluride* and *mercaptomerin*.)

Aryl halides do not normally react with ammonia under these conditions. The few exceptions are special cases and will be described in Section 23.5.

22.8   THE GABRIEL SYNTHESIS OF PRIMARY ALKYLAMINES

A method that achieves the same end result as that desired by alkylation of ammonia but which avoids the formation of secondary and tertiary amines as byproducts is the **Gabriel synthesis.** Alkyl halides are converted to primary alkylamines without contamination by secondary or tertiary amines. The key reagent is the potassium salt of phthalimide, prepared by the reaction

The Gabriel synthesis is based on work carried out by Siegmund Gabriel at the University of Berlin in the 1880s. A detailed discussion of each step in the Gabriel synthesis of benzylamine can be found in the October 1975 *Journal of Chemical Education* (pp. 670–671).

Phthalimide	*N*-Potassiophthalimide	Water

Phthalimide, with a pK_a of 8.3, can be quantitatively converted to its potassium salt with potassium hydroxide. The potassium salt of phthalimide has a negatively charged nitrogen atom, which acts as a nucleophile toward primary alkyl halides in a bimolecular nucleophilic substitution (S_N2) process.

DMF is an abbreviation for N,N-dimethylformamide,

$HCN(CH_3)_2$. DMF is a polar aprotic solvent (Section 8.12) and an excellent medium for S_N2 reactions.

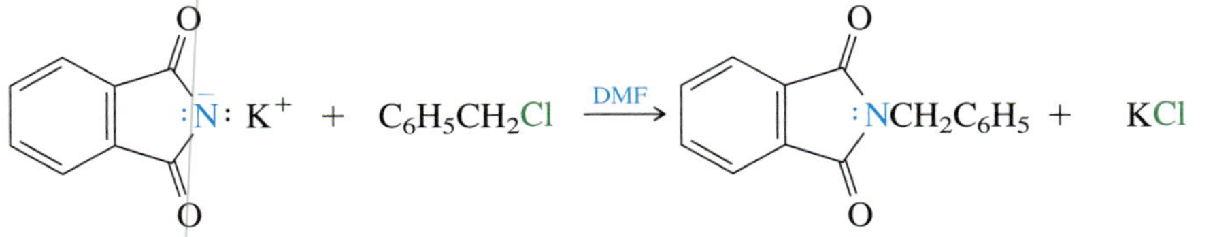

| N-Potassiophthalimide | Benzyl chloride | N-Benzylphthalimide (74%) | Potassium chloride |

The product of this reaction is an imide (Section 20.16), a diacyl derivative of an amine. Either aqueous acid or aqueous base can be used to hydrolyze its two amide bonds and liberate the desired primary amine. A more effective method of cleaving the two amide bonds is by acyl transfer to hydrazine:

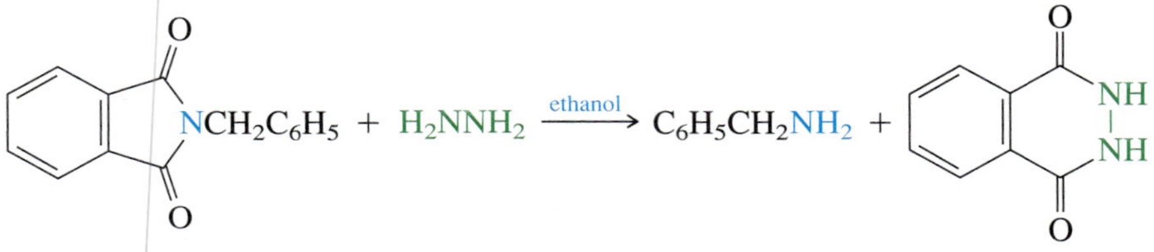

| N-Benzylphthalimide | Hydrazine | Benzylamine (97%) | Phthalhydrazide |

Aryl halides cannot be converted to arylamines by the Gabriel synthesis because they do not undergo nucleophilic substitution with N-potassiophthalimide in the first step of the procedure.

Among compounds other than simple alkyl halides, α-halo ketones and α-halo esters have been employed as substrates in the Gabriel synthesis. Alkyl p-toluenesulfonate esters have also been used. Because phthalimide can undergo only a single alkylation, the formation of secondary and tertiary amines does not occur, and the Gabriel synthesis is a valuable procedure for the laboratory preparation of primary amines.

PROBLEM 22.10　Which of the following amines can be prepared by the Gabriel synthesis? Which ones cannot? Write equations showing the successful applications of this method.

(a) Butylamine

(b) Isobutylamine

(c) *tert*-Butylamine

(d) 2-Phenylethylamine

(e) N-Methylbenzylamine

(f) Aniline

SAMPLE SOLUTION　(a) The Gabriel synthesis is limited to preparation of amines of the type RCH_2NH_2, that is, primary alkylamines in which the amino group is bonded to a primary carbon. Butylamine may be prepared from butyl bromide by this method.

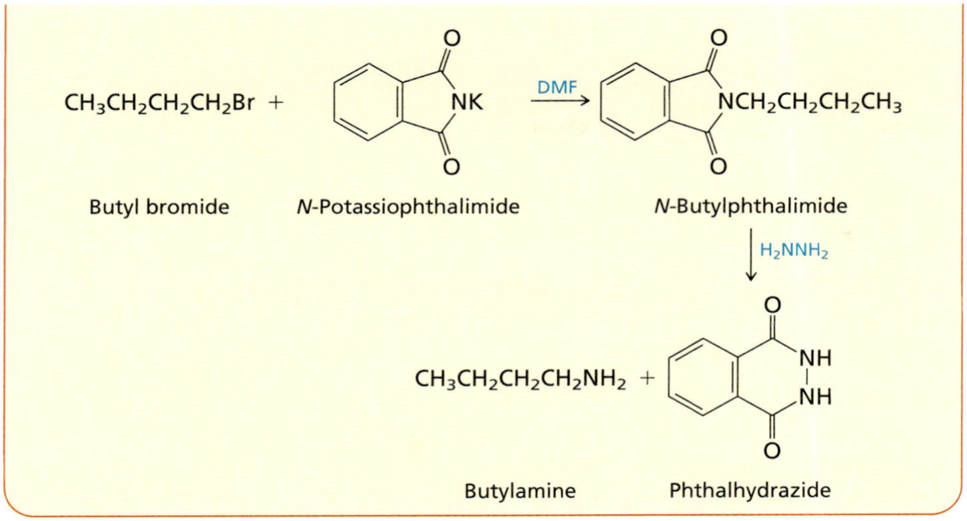

$CH_3CH_2CH_2CH_2Br$ +
NK
DMF
$NCH_2CH_2CH_2CH_3$
Butyl bromide
N-Potassiophthalimide
N-Butylphthalimide
H_2NNH_2
$CH_3CH_2CH_2CH_2NH_2$ +
NH
NH
Butylamine
Phthalhydrazide

22.9 PREPARATION OF AMINES BY REDUCTION

Almost any nitrogen-containing organic compound can be reduced to an amine. The synthesis of amines then becomes a question of the availability of suitable precursors and the choice of an appropriate reducing agent.

Alkyl *azides,* prepared by nucleophilic substitution of alkyl halides by sodium azide, as shown in the first entry of Table 22.3, are reduced to alkylamines by a variety of reagents, including lithium aluminum hydride.

$$R{-}\ddot{N}{=}\overset{+}{N}{=}\overset{-}{\ddot{N}}{:} \xrightarrow{\text{reduce}} R\ddot{N}H_2$$

Alkyl azide   Primary amine

$$C_6H_5CH_2CH_2N_3 \xrightarrow[\text{2. } H_2O]{\text{1. } LiAlH_4\text{, diethyl ether}} C_6H_5CH_2CH_2NH_2$$

2-Phenylethyl azide   2-Phenylethylamine (89%)

Catalytic hydrogenation is also effective:

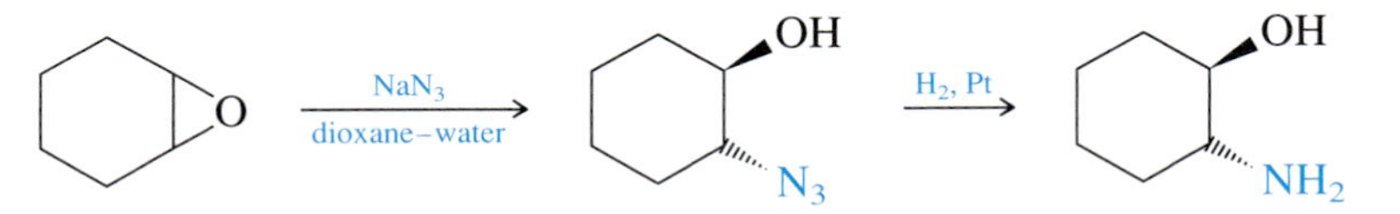

O
NaN_3
dioxane–water
OH
N_3
H_2, Pt
OH
NH_2

1,2-Epoxycyclohexane   *trans*-2-Azidocyclohexanol (61%)   *trans*-2-Aminocyclohexanol (81%)

In its overall design, this procedure is similar to the Gabriel synthesis; a nitrogen nucleophile is used in a carbon–nitrogen bond-forming operation and then converted to an amino group in a subsequent transformation.

The same reduction methods may be applied to the conversion of *nitriles* to primary amines.

$$RC\equiv N \xrightarrow[\text{H}_2,\text{ catalyst}]{\text{LiAlH}_4 \text{ or}} RCH_2NH_2$$

Nitrile   Primary amine

$$F_3C-\text{(benzene ring)}-CH_2CN \xrightarrow[\text{2. H}_2\text{O}]{\substack{\text{1. LiAlH}_4,\\ \text{diethyl ether}}} F_3C-\text{(benzene ring)}-CH_2CH_2NH_2$$

p-(Trifluoromethyl)benzyl   2-(p-Trifluoromethyl)phenylethyl-
cyanide   amine (53%)

The preparation of pen-
tanenitrile under phase-
transfer conditions was
described in Section 22.5.

$$CH_3CH_2CH_2CH_2CN \xrightarrow[\text{diethyl ether}]{\text{H}_2 \text{ (100 atm), Ni}} CH_3CH_2CH_2CH_2CH_2NH_2$$

Pentanenitrile   1-Pentanamine (56%)

Because nitriles can be prepared from alkyl halides by nucleophilic substitution with cyanide ion, the overall process $RX \rightarrow RC\equiv N \rightarrow RCH_2NH_2$ leads to primary amines that have one more carbon atom than the starting alkyl halide.

Cyano groups in *cyanohydrins* (Section 17.7) are reduced under the same reaction conditions.

Nitro groups are readily reduced to primary amines by a variety of methods. Catalytic hydrogenation over platinum, palladium, or nickel is often used, as is reduction by iron or tin in hydrochloric acid. The ease with which nitro groups are reduced is especially useful in the preparation of arylamines, where the sequence $ArH \rightarrow ArNO_2 \rightarrow ArNH_2$ is the standard route to these compounds.

CH(CH₃)₂ / NO₂
$$\xrightarrow[\text{methanol}]{\text{H}_2,\text{ Ni}}$$
CH(CH₃)₂ / NH₂

o-Isopropylnitrobenzene   o-Isopropylaniline (92%)

For reductions carried out in
acidic media, a pH adjustment with sodium hydroxide
is required in the last step in
order to convert ArNH$_3^+$ to
ArNH$_2$.

$$Cl-\text{(benzene ring)}-NO_2 \xrightarrow[\text{2. NaOH}]{\text{1. Fe, HCl}} Cl-\text{(benzene ring)}-NH_2$$

p-Chloronitrobenzene   p-Chloroaniline (95%)

m-Nitroacetophenone $\xrightarrow[\text{2. NaOH}]{\text{1. Sn, HCl}}$ m-Aminoacetophenone (82%)

O₂N—(benzene ring)—CCH₃ (with =O)   H₂N—(benzene ring)—CCH₃ (with =O)

m-Nitroacetophenone   m-Aminoacetophenone (82%)

PROBLEM 22.11 Outline syntheses of each of the following arylamines from benzene:

(a) o-Isopropylaniline   (d) p-Chloroaniline
(b) p-Isopropylaniline   (e) m-Aminoacetophenone
(c) 4-Isopropyl-1,3-benzenediamine

SAMPLE SOLUTION   (a) The last step in the synthesis of o-isopropylaniline, the reduction of the corresponding nitro compound by catalytic hydrogenation, is given as one of the three preceding examples. The necessary nitroarene is obtained by fractional distillation of the ortho–para mixture formed during nitration of isopropylbenzene.

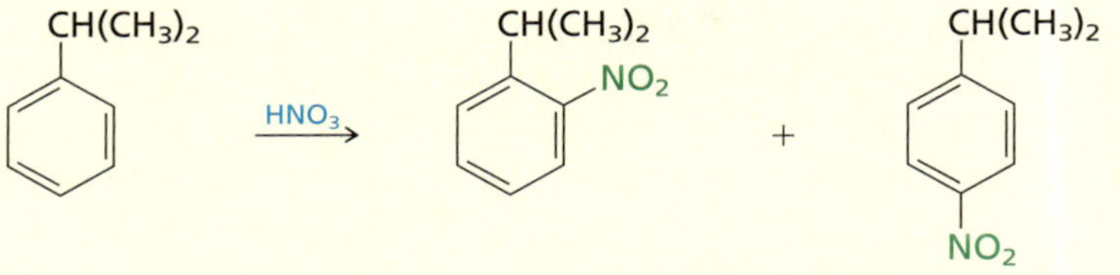

| Isopropylbenzene | o-Isopropylnitrobenzene (bp 110°C) | p-Isopropylnitrobenzene (bp 131°C) |

As actually performed, a 62% yield of a mixture of ortho and para nitration products has been obtained with an ortho–para ratio of about 1:3.

Isopropylbenzene is prepared by the Friedel–Crafts alkylation of benzene using isopropyl chloride and aluminum chloride (Section 12.6).

Reduction of an azide, a nitrile, or a nitro compound furnishes a primary amine. A method that provides access to primary, secondary, or tertiary amines is reduction of the carbonyl group of an amide by lithium aluminum hydride.

$$\underset{\text{Amide}}{\text{RCNR}'_2} \xrightarrow[\text{2. H}_2\text{O}]{\text{1. LiAlH}_4} \underset{\text{Amine}}{\text{RCH}_2\text{NR}'_2}$$

In this general equation, R and R′ may be either alkyl or aryl groups. When R′ = H, the product is a primary amine:

$$\underset{\substack{| \\ \text{CH}_3 \\ \text{3-Phenylbutanamide}}}{\text{C}_6\text{H}_5\text{CHCH}_2\text{CNH}_2} \xrightarrow[\text{2. H}_2\text{O}]{\substack{\text{1. LiAlH}_4, \\ \text{diethyl ether}}} \underset{\substack{| \\ \text{CH}_3 \\ \text{3-Phenyl-1-butanamine (59\%)}}}{\text{C}_6\text{H}_5\text{CHCH}_2\text{CH}_2\text{NH}_2}$$

N-Substituted amides yield secondary amines:

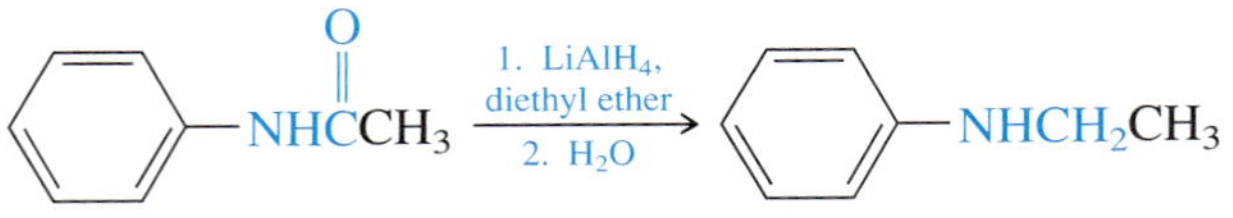

| Acetanilide | *N*-Ethylaniline (92%) |

N,N-Disubstituted amides yield tertiary amines:

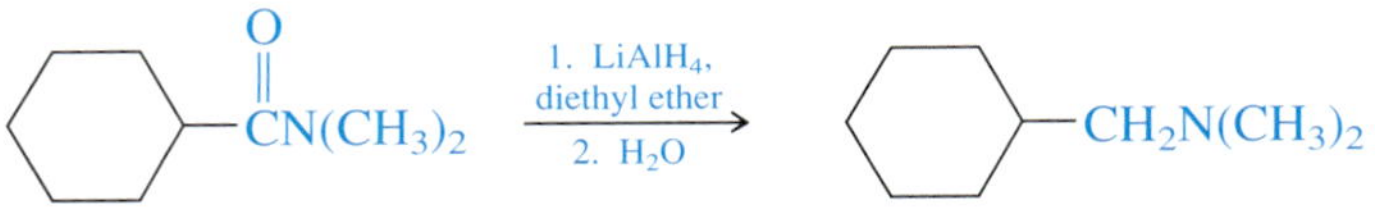

| *N,N*-Dimethylcyclohexane-carboxamide | *N,N*-Dimethyl(cyclohexylmethyl)-amine (88%) |

Acetanilide is an acceptable IUPAC synonym for *N*-phenylethanamide.

Because amides are so easy to prepare, this is a versatile method for the preparation of amines.

The preparation of amines by the methods described in this section involves the prior synthesis and isolation of some reducible material that has a carbon–nitrogen bond: an azide, a nitrile, a nitro-substituted arene, or an amide. The following section describes a method that combines the two steps of carbon–nitrogen bond formation and reduction into a single operation. Like the reduction of amides, it offers the possibility of preparing primary, secondary, or tertiary amines by proper choice of starting materials.

22.10 REDUCTIVE AMINATION

A class of nitrogen-containing compounds that was omitted from the section just discussed includes *imines* and their derivatives. Imines are formed by the reaction of aldehydes and ketones with ammonia. Imines can be reduced to primary amines by catalytic hydrogenation.

$$
\underset{\substack{\text{Aldehyde} \\ \text{or ketone}}}{\overset{\overset{\displaystyle O}{\|}}{R C R'}} + \underset{\text{Ammonia}}{NH_3} \longrightarrow \underset{\text{Imine}}{\overset{\overset{\displaystyle NH}{\|}}{R C R'}} \xrightarrow[\text{catalyst}]{H_2} \underset{\text{Primary amine}}{\overset{\overset{\displaystyle NH_2}{|}}{R C H R'}}
$$

The overall reaction converts a carbonyl compound to an amine by carbon–nitrogen bond formation and reduction; it is commonly known as **reductive amination.** What makes it a particularly valuable synthetic procedure is that it can be carried out in a single operation by hydrogenation of a solution containing both ammonia and the carbonyl compound along with a hydrogenation catalyst. The intermediate imine is not isolated but undergoes reduction under the conditions of its formation. Also, the reaction is broader in scope than implied by the preceding equation. All classes of amines—primary, secondary, and tertiary—may be prepared by reductive amination.

When primary amines are desired, the reaction is carried out as just described:

Cyclohexanone   Ammonia   Cyclohexylamine (80%)   via

Secondary amines are prepared by hydrogenation of a carbonyl compound in the presence of a primary amine. An *N*-substituted imine, or *Schiff's base,* is an intermediate:

Heptanal   Aniline   *N*-Heptylaniline (65%)   via

Reductive amination has been successfully applied to the preparation of tertiary amines from carbonyl compounds and secondary amines even though a neutral imine is not possible in this case.

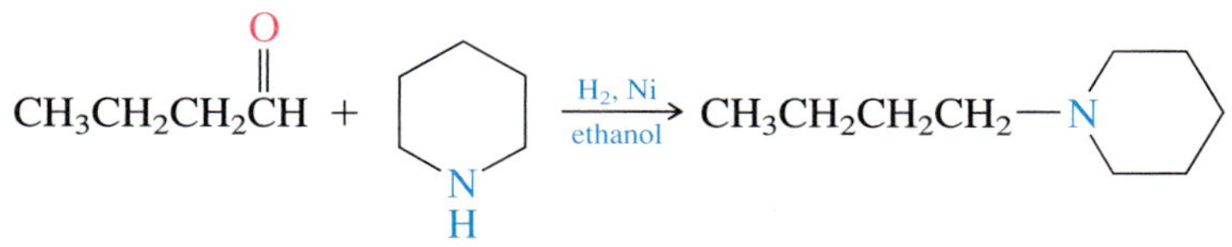

$$CH_3CH_2CH_2\overset{\displaystyle O}{\overset{\|}{C}}H \; + \; \underset{\underset{H}{N}}{\bigcirc} \quad \xrightarrow[\text{ethanol}]{H_2,\ Ni} \quad CH_3CH_2CH_2CH_2-N\bigcirc$$

Butanal	Piperidine	N-Butylpiperidine (93%)

Presumably, the species that undergoes reduction here is a carbinolamine, an iminium ion derived from it, or an enamine.

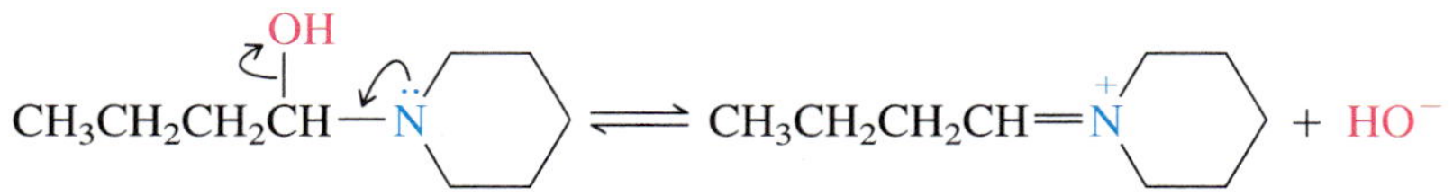

$$CH_3CH_2CH_2\overset{\overset{\displaystyle OH}{|}}{CH}-\ddot{N}\bigcirc \;\rightleftharpoons\; CH_3CH_2CH_2CH=\overset{+}{N}\bigcirc \; + \; HO^-$$

Carbinolamine	Iminium ion

PROBLEM 22.12 Show how you could prepare each of the following amines from benzaldehyde by reductive amination:

(a) Benzylamine
(b) Dibenzylamine
(c) N,N-Dimethylbenzylamine
(d) N-Benzylpiperidine

SAMPLE SOLUTION (a) Because benzylamine is a primary amine, it is derived from ammonia and benzaldehyde.

$$C_6H_5\overset{\displaystyle O}{\overset{\|}{C}}H \; + \; NH_3 \; + \; H_2 \quad \xrightarrow{Ni} \quad C_6H_5CH_2NH_2 \; + \; H_2O$$

Benzaldehyde	Ammonia	Hydrogen	Benzylamine (89%)	Water

The reaction proceeds by initial formation of the imine $C_6H_5CH=NH$, followed by its hydrogenation.

A variation of the classical reductive amination procedure uses sodium cyanoborohydride ($NaBH_3CN$) instead of hydrogen as the reducing agent and is better suited to amine syntheses in which only a few grams of material are needed. All that is required is to add sodium cyanoborohydride to an alcohol solution of the carbonyl compound and an amine.

$$C_6H_5\overset{\displaystyle O}{\overset{\|}{C}}H \; + \; CH_3CH_2NH_2 \quad \xrightarrow[\text{methanol}]{NaBH_3CN} \quad C_6H_5CH_2NHCH_2CH_3$$

Benzaldehyde	Ethylamine	N-Ethylbenzylamine (91%)

22.11 REACTIONS OF AMINES: A REVIEW AND A PREVIEW

The noteworthy properties of amines are their *basicity* and their *nucleophilicity*. The basicity of amines has been discussed in Section 22.4. Several reactions in which amines act as nucleophiles have already been encountered in earlier chapters. These are summarized in Table 22.4.

Both the basicity and the nucleophilicity of amines originate in the unshared electron pair of nitrogen. When an amine acts as a base, this electron pair abstracts a

TABLE 22.4   Reactions of Amines Discussed in Previous Chapters*

Reaction (section) and comments	General equation and specific example

Reaction of primary amines with aldehydes and ketones (Section 17.10) Imines are formed by nucleophilic addition of a primary amine to the carbonyl group of an aldehyde or a ketone. The key step is formation of a carbinolamine intermediate, which then dehydrates to the imine.

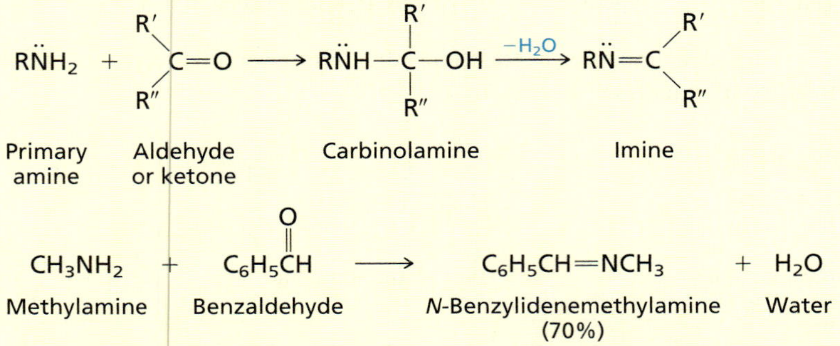

$$RNH_2 \;+\; \underset{R''}{\overset{R'}{C}}{=}O \longrightarrow R\ddot{N}H{-}\underset{R''}{\overset{R'}{C}}{-}OH \xrightarrow{-H_2O} R\ddot{N}{=}\underset{R''}{\overset{R'}{C}}$$

Primary amine · Aldehyde or ketone · Carbinolamine · Imine

$$CH_3NH_2 \;+\; C_6H_5CH{=}O \longrightarrow C_6H_5CH{=}NCH_3 \;+\; H_2O$$

Methylamine · Benzaldehyde · N-Benzylidenemethylamine (70%) · Water

Reaction of secondary amines with aldehydes and ketones (Section 17.11) Enamines are formed in the corresponding reaction of secondary amines with aldehydes and ketones.

$$R_2\ddot{N}H \;+\; \underset{R''}{\overset{R'CH_2}{C}}{=}O \longrightarrow R_2\ddot{N}{-}\underset{R''}{\overset{CH_2R'}{C}}{-}OH \xrightarrow{-H_2O} R_2\ddot{N}{-}\underset{R''}{\overset{CHR'}{C}}$$

Secondary amine · Aldehyde or ketone · Carbinolamine · Enamine

Pyrrolidine + Cyclohexanone $\xrightarrow[\text{heat}]{\text{benzene}}$ N-(1-Cyclohexenyl)pyrrolidine (85–90%) + H_2O

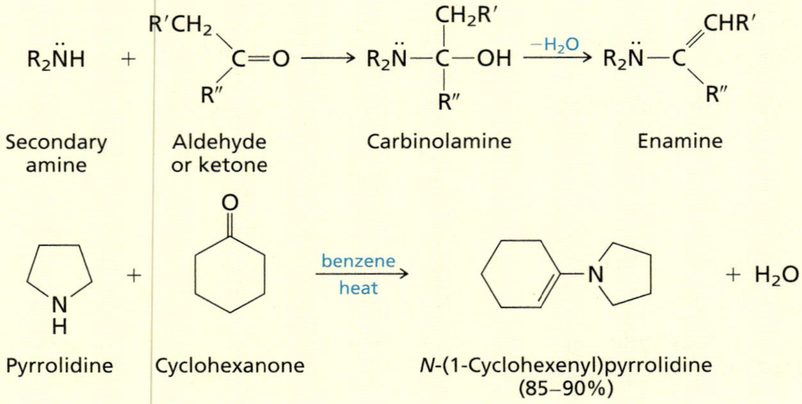

Reaction of amines with acyl chlorides (Section 20.4) Amines are converted to amides on reaction with acyl chlorides. Other acylating agents, such as carboxylic acid anhydrides and esters, may also be used but are less reactive.

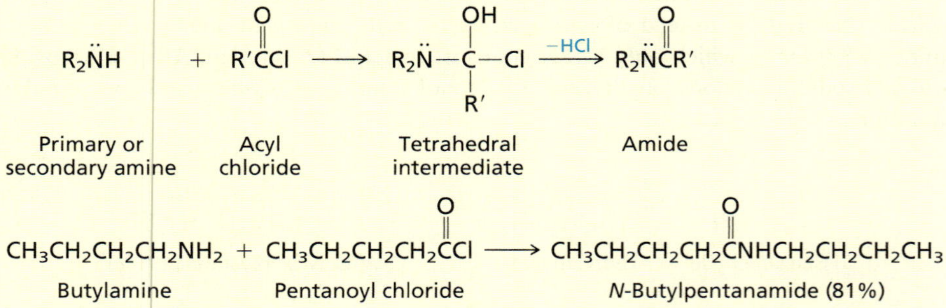

$$R_2\ddot{N}H \;+\; R'\overset{O}{C}Cl \longrightarrow R_2\ddot{N}{-}\underset{R'}{\overset{OH}{C}}{-}Cl \xrightarrow{-HCl} R_2\ddot{N}\overset{O}{C}R'$$

Primary or secondary amine · Acyl chloride · Tetrahedral intermediate · Amide

$$CH_3CH_2CH_2CH_2NH_2 + CH_3CH_2CH_2CH_2CCl \longrightarrow CH_3CH_2CH_2CH_2CNHCH_2CH_2CH_2CH_3$$

Butylamine · Pentanoyl chloride · N-Butylpentanamide (81%)

*Both alkylamines and arylamines undergo these reactions.

proton from a Brønsted acid. When an amine undergoes the reactions summarized in Table 22.4, the first step in each case is the attack of the unshared electron pair on the positively polarized carbon of a carbonyl group.

Amine acting as a base   Amine acting as a nucleophile

In addition to being more basic than arylamines, alkylamines are also more nucleophilic. All the reactions in Table 22.4 take place faster with alkylamines than with arylamines.

The sections that follow introduce some additional reactions of amines. In all cases our understanding of how these reactions take place starts with a consideration of the role of the unshared electron pair of nitrogen.

We will begin with an examination of the reactivity of amines as nucleophiles in S_N2 reactions.

22.12   REACTION OF AMINES WITH ALKYL HALIDES

Nucleophilic substitution results when primary alkyl halides are treated with amines.

The reaction of amines with alkyl halides was seen earlier (Section 22.7) as a complicating factor in the preparation of amines by alkylation of ammonia.

| Primary amine | Primary alkyl halide | Ammonium halide salt | Secondary amine | Hydrogen halide |

$$C_6H_5NH_2 \ + \ C_6H_5CH_2Cl \ \xrightarrow[90°C]{NaHCO_3} \ C_6H_5NHCH_2C_6H_5$$

Aniline (4 mol)   Benzyl chloride (1 mol)   N-Benzylaniline (85–87%)

A second alkylation may follow, converting the secondary amine to a tertiary amine. Alkylation need not stop there; the tertiary amine may itself be alkylated, giving a quaternary ammonium salt.

$$R\ddot{N}H_2 \ \xrightarrow[S_N2]{R'CH_2X} \ R\ddot{N}HCH_2R' \ \xrightarrow[S_N2]{R'CH_2X} \ R\ddot{N}(CH_2R')_2 \ \xrightarrow[S_N2]{R'CH_2X} \ \overset{+}{R}N(CH_2R')_3 \ X^-$$

Primary amine   Secondary amine   Tertiary amine   Quaternary ammonium salt

Because of its high reactivity toward nucleophilic substitution, methyl iodide is the alkyl halide most often used to prepare quaternary ammonium salts.

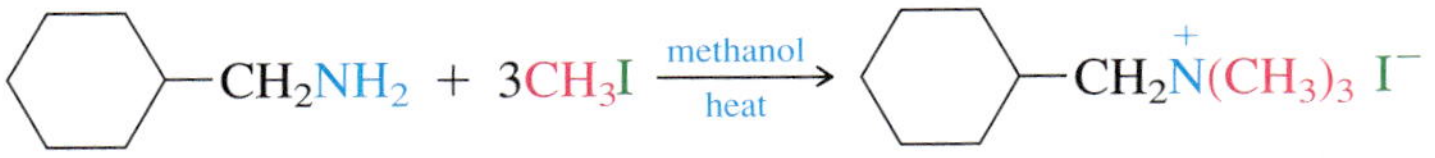

(Cyclohexylmethyl)- amine   Methyl iodide   (Cyclohexylmethyl)trimethyl- ammonium iodide (99%)

Quaternary ammonium salts, as we have seen, are useful in synthetic organic chemistry as phase-transfer catalysts. In another, more direct application, quaternary ammonium *hydroxides* are used as substrates in an elimination reaction to form alkenes.

22.13 THE HOFMANN ELIMINATION

The halide anion of quaternary ammonium iodides may be replaced by hydroxide by treatment with an aqueous slurry of silver oxide. Silver iodide precipitates, and a solution of the quaternary ammonium hydroxide is formed.

$$2(R_4\overset{+}{N}\ I^-)\quad +\ Ag_2O\ +\ H_2O\ \longrightarrow\quad 2(R_4\overset{+}{N}\ \overset{-}{OH})\quad +\ 2AgI$$

| Quaternary ammonium iodide | Silver oxide | Water | Quaternary ammonium hydroxide | Silver iodide |

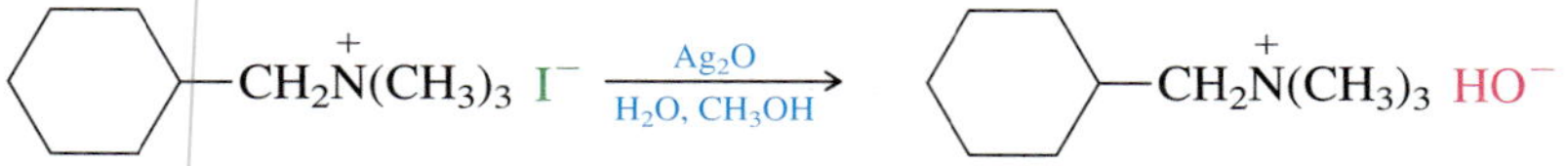

(Cyclohexylmethyl)trimethyl-ammonium iodide   (Cyclohexylmethyl)trimethylammonium hydroxide

When quaternary ammonium hydroxides are heated, they undergo β elimination to form an alkene and an amine.

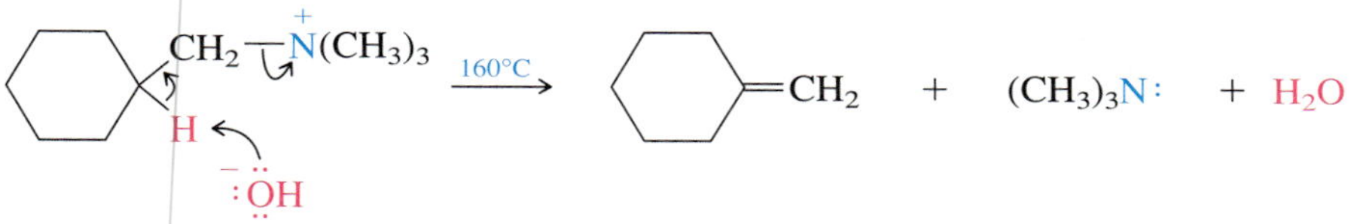

(Cyclohexylmethyl)trimethyl-ammonium hydroxide   Methylenecyclohexane (69%)   Trimethylamine   Water

This reaction is known as the **Hofmann elimination;** it was developed by August W. Hofmann in the middle of the nineteenth century and is both a synthetic method to prepare alkenes and an analytical tool for structure determination.

A novel aspect of the Hofmann elimination is its regioselectivity. Elimination in alkyltrimethylammonium hydroxides proceeds in the direction that gives the *less* substituted alkene.

$$CH_3\overset{|}{\underset{+N(CH_3)_3}{C}}HCH_2CH_3\ HO^- \xrightarrow[\substack{-H_2O \\ -(CH_3)_3N}]{heat} H_2C{=}CHCH_2CH_3\ +\ CH_3CH{=}CHCH_3$$

sec-Butyltrimethylammonium hydroxide   1-Butene (95%)   2-Butene (5%) (cis and trans)

The least sterically hindered β hydrogen is removed by the base in Hofmann elimination reactions. Methyl groups are deprotonated in preference to methylene groups, and methylene groups are deprotonated in preference to methines. The regioselectivity of Hofmann elimination is opposite to that predicted by the Zaitsev rule (Section 5.10). Elimination reactions of alkyltrimethylammonium hydroxides are said to obey the **Hofmann rule;** they yield the less substituted alkene.

PROBLEM 22.13 Give the structure of the major alkene formed when the hydroxide of each of the following quaternary ammonium ions is heated.

(a)

$$CH_3$$

$$\overset{+}{N}(CH_3)_3$$

(b) $(CH_3)_3CCH_2\underset{\overset{|}{+N(CH_3)_3}}{C}(CH_3)_2$

(c) $CH_3CH_2\overset{\overset{\displaystyle CH_3}{\overset{|}{+}}}{\underset{\underset{\displaystyle CH_3}{|}}{N}}CH_2CH_2CH_2CH_3$

SAMPLE SOLUTION (a) Two alkenes are capable of being formed by β elimination: methylenecyclopentane and 1-methylcyclopentene.

(1-Methylcyclopentyl)trimethyl-   Methylenecyclopentane   1-Methylcyclopentene
     ammonium hydroxide

Methylenecyclopentane has the less substituted double bond and is the major product. The reported isomer distribution is 91% methylenecyclopentane and 9% 1-methylcyclopentene.

We can understand the regioselectivity of the Hofmann elimination by comparing steric effects in the E2 transition states for formation of 1-butene and *trans*-2-butene from *sec*-butyltrimethylammonium hydroxide. In terms of its size, $(CH_3)_3\overset{+}{N}$— (trimethylammonio) is comparable to $(CH_3)_3C$— (*tert*-butyl). As Figure 22.4 illustrates, the E2 transition state requires an anti relationship between the proton that is removed and the trimethylammonio group. No serious van der Waals repulsions are evident in the transition state geometry for formation of 1-butene. The conformation leading to *trans*-2-butene, however, is destabilized by van der Waals strain between the trimethylammonio group and a methyl group gauche to it. Thus, the activation energy for formation of *trans*-2-butene exceeds that of 1-butene, which becomes the major product because it is formed faster.

With a regioselectivity opposite to that of the Zaitsev rule, the Hofmann elimination is sometimes used in synthesis to prepare alkenes not accessible by dehydrohalogenation of alkyl halides. This application decreased in importance once the Wittig reaction (Section 17.12) became established as a synthetic method. Similarly, most of the analytical applications of Hofmann elimination have been replaced by spectroscopic methods.

22.14 ELECTROPHILIC AROMATIC SUBSTITUTION IN ARYLAMINES

Arylamines contain two functional groups, the amine group and the aromatic ring; they are **difunctional compounds.** The reactivity of the amine group is affected by its aryl substituent, and the reactivity of the ring is affected by its amine substituent. The same electron delocalization that reduces the basicity and the nucleophilicity of an arylamine nitrogen increases the electron density in the aromatic ring and makes arylamines extremely reactive toward electrophilic aromatic substitution.

The reactivity of arylamines was noted in Section 12.12, where it was pointed out that —N̈H₂, —N̈HR, and —N̈R₂ are ortho, para-directing and exceedingly powerful

FIGURE 22.4 New-
man projections showing
the conformations leading
to (*a*) 1-butene, and (*b*)
trans-2-butene by Hofmann
elimination of *sec*-butyl-
trimethylammonium hy-
droxide. The major product
is 1-butene.

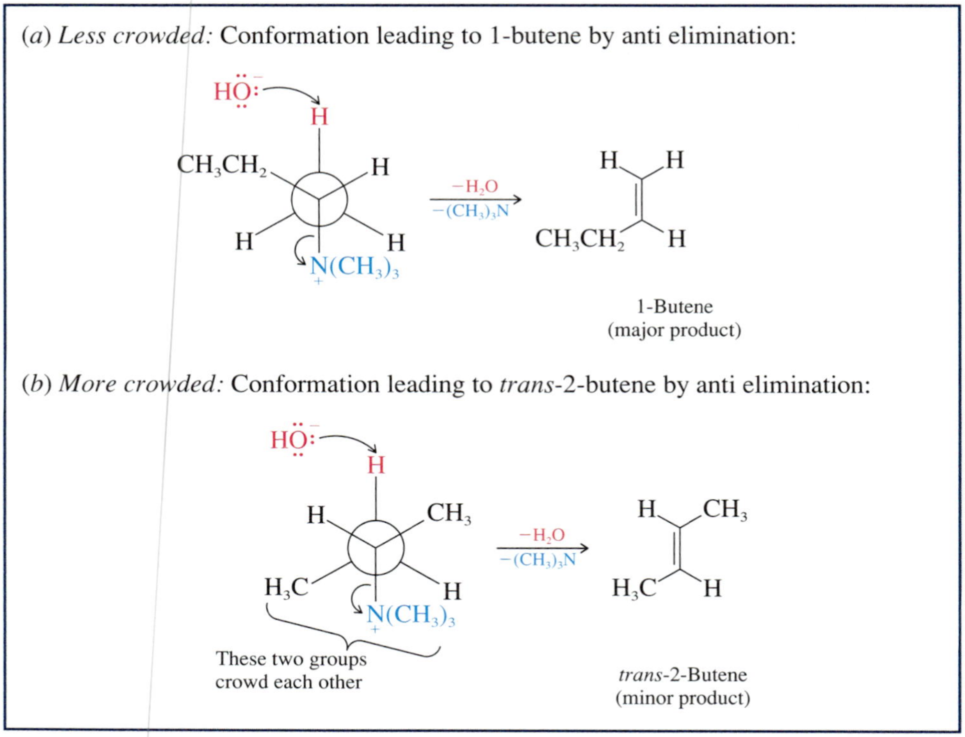

(*a*) *Less crowded:* Conformation leading to 1-butene by anti elimination:

1-Butene
(major product)

(*b*) *More crowded:* Conformation leading to *trans*-2-butene by anti elimination:

These two groups
crowd each other

trans-2-Butene
(minor product)

activating groups. These substituents are such powerful activators that electrophilic aromatic substitution is only rarely performed directly on arylamines.

Direct nitration of aniline and other arylamines fails because oxidation leads to the
formation of dark-colored "tars." As a solution to this problem it is standard practice to
first protect the amino group by acylation with either acetyl chloride or acetic anhydride.

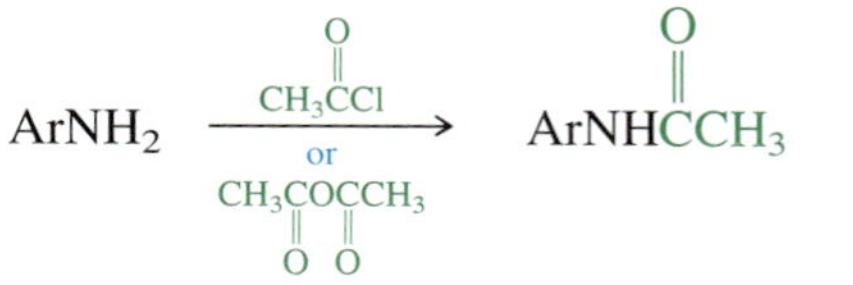

Arylamine   *N*-Acetylarylamine

Amide resonance within the *N*-acetyl group competes with delocalization of the nitrogen lone pair into the ring.

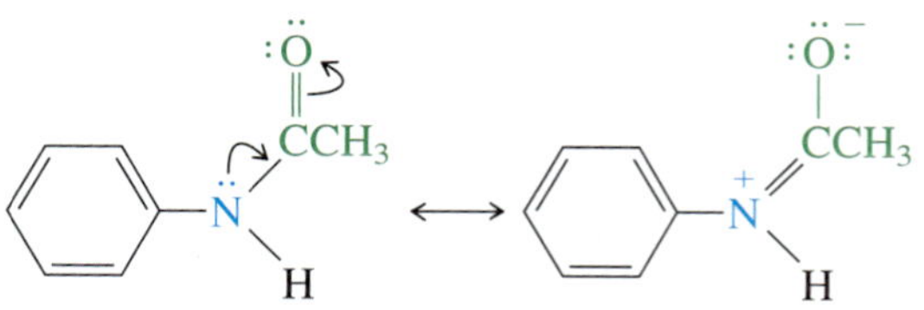

Amide resonance in acetanilide

Protecting the amino group of an arylamine in this way moderates its reactivity and permits nitration of the ring to be achieved. The acetamido group is activating toward electrophilic aromatic substitution and is ortho, para-directing.

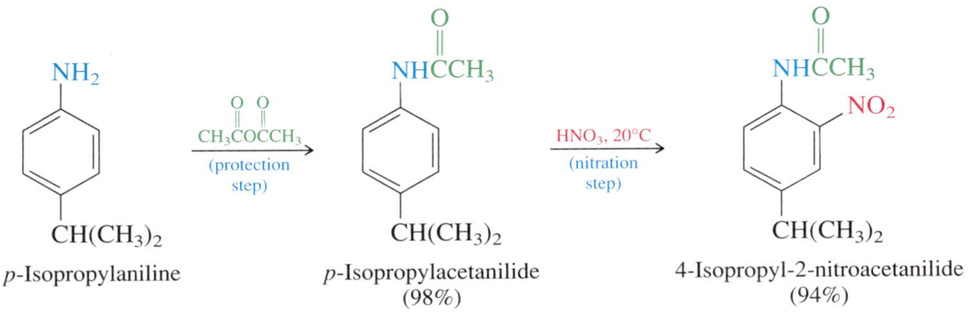

$$\text{p-Isopropylaniline} \xrightarrow[\text{(protection step)}]{\text{CH}_3\text{COCOCH}_3} \text{p-Isopropylacetanilide (98\%)} \xrightarrow[\text{(nitration step)}]{\text{HNO}_3,\ 20^\circ\text{C}} \text{4-Isopropyl-2-nitroacetanilide (94\%)}$$

p-Isopropylaniline

p-Isopropylacetanilide
(98%)

4-Isopropyl-2-nitroacetanilide
(94%)

After the N-acetyl-protecting group has served its purpose, it may be removed by hydrolysis, liberating the amino group:

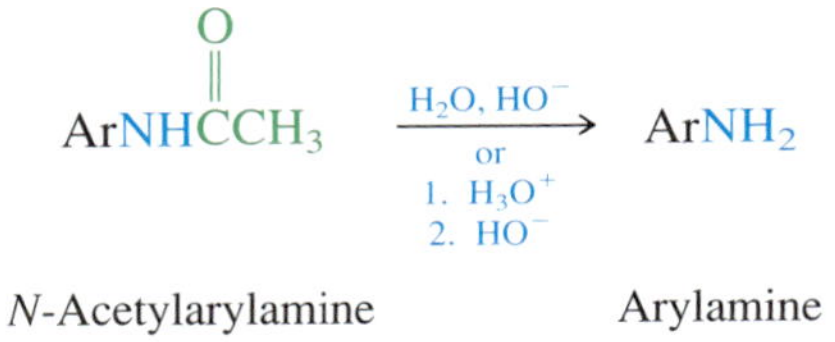

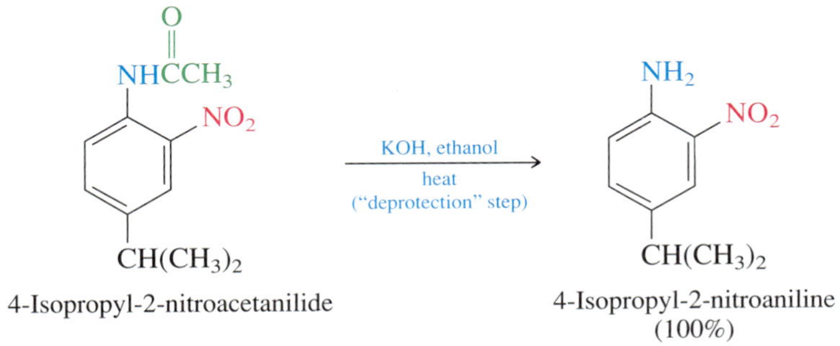

$$\text{ArNHCCH}_3 \xrightarrow[\substack{\text{or} \\ 1.\ \text{H}_3\text{O}^+ \\ 2.\ \text{HO}^-}]{\text{H}_2\text{O, HO}^-} \text{ArNH}_2$$

N-Acetylarylamine   Arylamine

$$\text{4-Isopropyl-2-nitroacetanilide} \xrightarrow[\substack{\text{heat} \\ \text{("deprotection" step)}}]{\text{KOH, ethanol}} \text{4-Isopropyl-2-nitroaniline (100\%)}$$

4-Isopropyl-2-nitroacetanilide

4-Isopropyl-2-nitroaniline
(100%)

The net effect of the sequence *protect–nitrate–deprotect* is the same as if the substrate had been nitrated directly. Because direct nitration is impossible, however, the indirect route is the only practical method.

PROBLEM 22.14   Outline syntheses of each of the following from aniline and any necessary organic or inorganic reagents:

(a) *p*-Nitroaniline   (c) *p*-Aminoacetanilide

(b) 2,4-Dinitroaniline

SAMPLE SOLUTION   (a) Because direct nitration of aniline is not a practical reaction, the amino group must first be protected as its *N*-acetyl derivative.

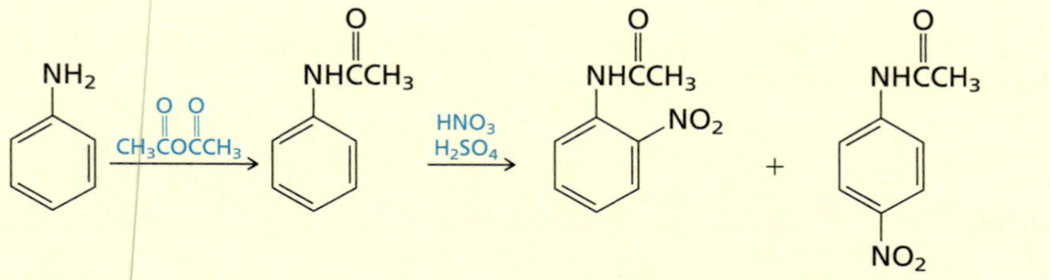

Aniline   Acetanilide   o-Nitroacetanilide   p-Nitroacetanilide

Nitration of acetanilide yields a mixture of ortho and para substitution products. The para isomer is separated, then subjected to hydrolysis to give p-nitroaniline.

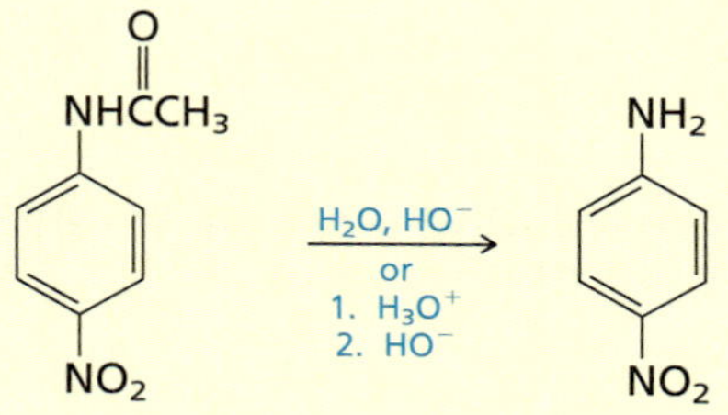

p-Nitroacetanilide   p-Nitroaniline

Unprotected arylamines are so reactive toward halogenation that it is difficult to limit the reaction to monosubstitution. Generally, halogenation proceeds rapidly to replace all the available hydrogens that are ortho or para to the amino group.

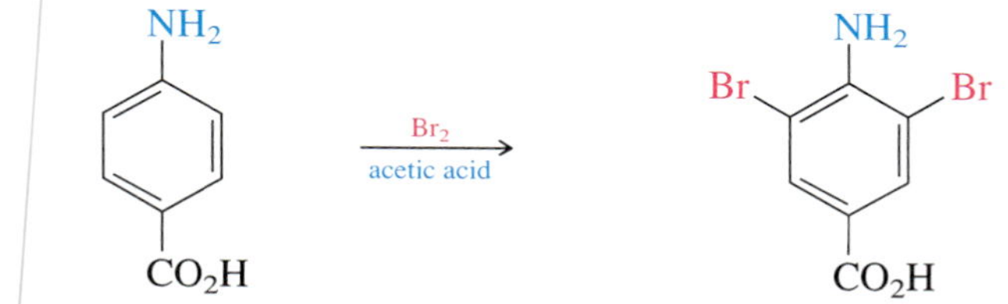

p-Aminobenzoic acid   4-Amino-3,5-dibromobenzoic acid (82%)

Decreasing the electron-donating ability of an amino group by acylation makes it possible to limit halogenation to monosubstitution.

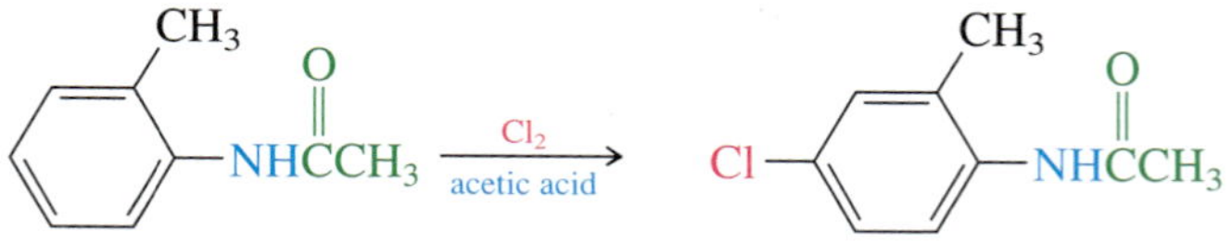

2-Methylacetanilide   4-Chloro-2-methylacetanilide (74%)

Friedel–Crafts reactions are normally not successful when attempted on an arylamine, but can be carried out readily once the amino group is protected.

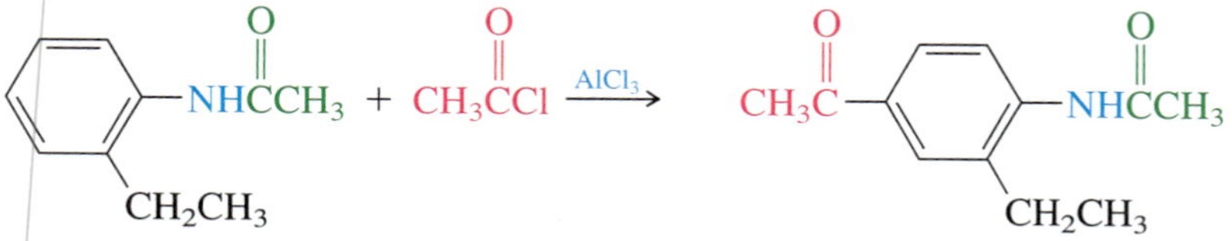

2-Ethylacetanilide   4-Acetamido-3-ethylacetophenone (57%)

22.15 NITROSATION OF ALKYLAMINES

When solutions of sodium nitrite ($NaNO_2$) are acidified, a number of species are formed that act as **nitrosating agents.** That is, they react as sources of nitrosyl cation, $:\overset{+}{N}{=}\ddot{O}:$. For simplicity, organic chemists group all these species together and speak of the chemistry of one of them, *nitrous acid,* as a generalized precursor to nitrosyl cation.

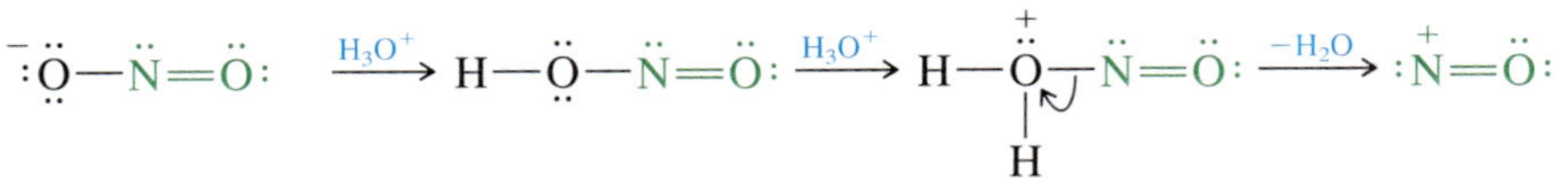

Nitrite ion (from sodium nitrite) | Nitrous acid | Nitrosyl cation

Nitrosyl cation is also called *nitrosonium* ion. It can be represented by the two resonance structures

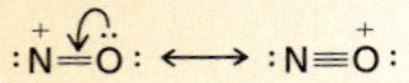

Nitrosation of amines is best illustrated by examining what happens when a secondary amine "reacts with nitrous acid." The amine acts as a nucleophile, attacking the nitrogen of nitrosyl cation. The intermediate that is formed in the first step loses a proton to give an *N*-nitroso amine as the isolated product.

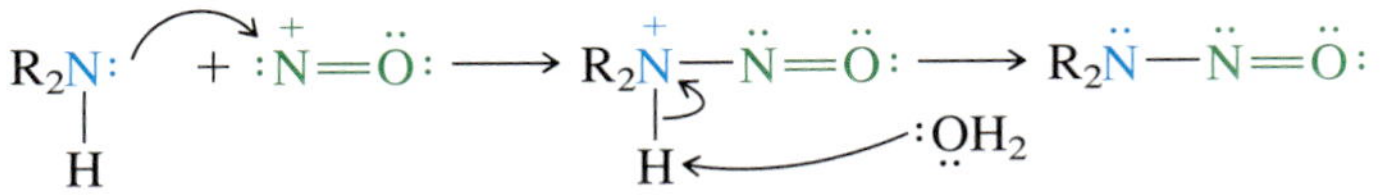

Secondary alkylamine | Nitrosyl cation | *N*-Nitroso amine

Refer to the molecular model of nitrosyl cation on *Learning By Modeling* to verify that the region of positive electrostatic potential is concentrated at nitrogen.

For example,

$$(CH_3)_2\ddot{N}H \xrightarrow[H_2O]{NaNO_2,\ HCl} (CH_3)_2\ddot{N}{-}\ddot{N}{=}\ddot{O}:$$

Dimethylamine → *N*-Nitrosodimethylamine (88–90%)

PROBLEM 22.15 *N*-Nitroso amines are stabilized by electron delocalization. Write the two most stable resonance forms of *N*-nitrosodimethylamine, $(CH_3)_2NNO$.

N-Nitroso amines are more often called *nitrosamines,* and because many of them are potent carcinogens, they have been the object of much investigation. We encounter nitrosamines in the environment on a daily basis. A few of these, all of which are known carcinogens, are:

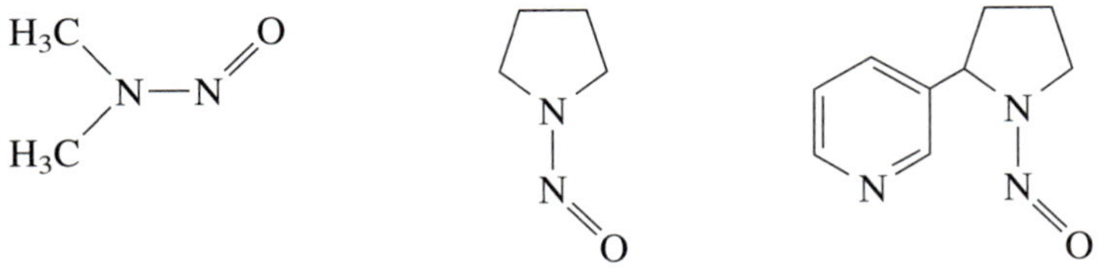

N-Nitrosodimethylamine (formed during tanning of leather; also found in beer and herbicides)

N-Nitrosopyrrolidine (formed when bacon that has been cured with sodium nitrite is fried)

N-Nitrosonornicotine (present in tobacco smoke)

Nitrosamines are formed whenever nitrosating agents come in contact with secondary amines. Indeed, more nitrosamines are probably synthesized within our body than enter it

The July 1977 issue of the *Journal of Chemical Education* contains an article entitled "Formation of Nitrosamines in Food and in the Digestive System."

by environmental contamination. Enzyme-catalyzed reduction of nitrate (NO_3^-) produces nitrite (NO_2^-), which combines with amines present in the body to form *N*-nitroso amines.

When primary amines are nitrosated, their *N*-nitroso compounds can't be isolated because they react further.

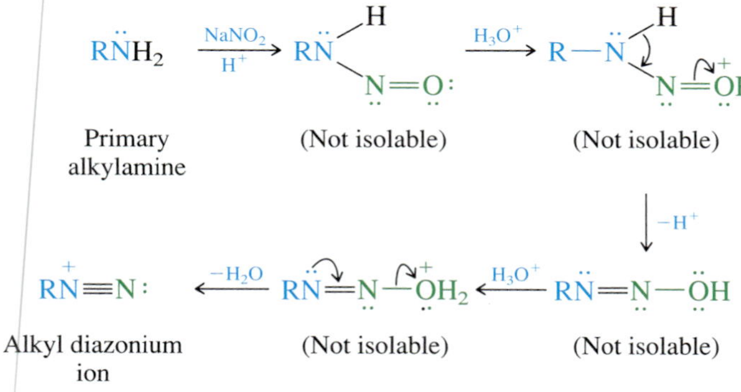

$$R\overset{\cdot\cdot}{N}H_2 \xrightarrow[H^+]{NaNO_2} R\overset{\cdot\cdot}{N}\diagdown\!\!\!\!\diagup^{H}_{N=\overset{\cdot\cdot}{O}:} \xrightarrow{H_3O^+} R-\overset{\cdot\cdot}{N}\diagdown\!\!\!\!\diagup^{H}_{N=\overset{\cdot\cdot}{O}H}$$

Primary   (Not isolable)   (Not isolable)
alkylamine

$$\downarrow -H^+$$

$$R\overset{+}{N}\equiv N: \quad\xleftarrow{-H_2O}\quad R\overset{\cdot\cdot}{N}=N-\overset{+}{O}H_2 \xleftarrow{H_3O^+} R\overset{\cdot\cdot}{N}=N-\overset{\cdot\cdot}{O}H$$

Alkyl diazonium   (Not isolable)   (Not isolable)
ion

The product of this series of steps is an alkyl **diazonium ion,** and the amine is said to have been **diazotized.** Alkyl diazonium ions are not very stable, decomposing rapidly under the conditions of their formation. Molecular nitrogen is a leaving group par excellence, and the reaction products arise by solvolysis of the diazonium ion. Usually, a carbocation intermediate is involved.

Recall from Section 8.14 that decreasing basicity is associated with increasing leaving-group ability. Molecular nitrogen is an exceedingly weak base and an excellent leaving group.

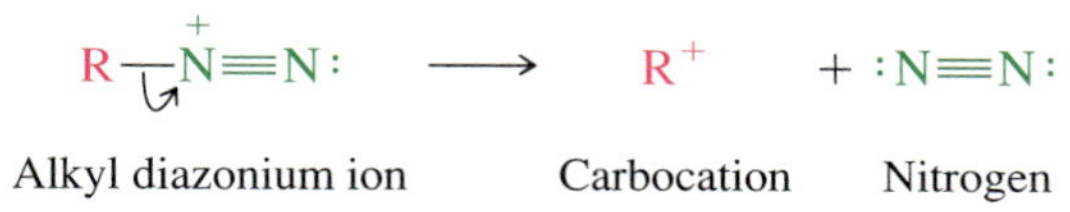

$$R-\overset{+}{N}\equiv N: \longrightarrow R^+ \quad + \quad :N\equiv N:$$

Alkyl diazonium ion   Carbocation   Nitrogen

Figure 22.5 shows what happens when a typical primary alkylamine reacts with nitrous acid.

Because nitrogen-free products result from the formation and decomposition of diazonium ions, these reactions are often referred to as **deamination reactions.** Alkyl

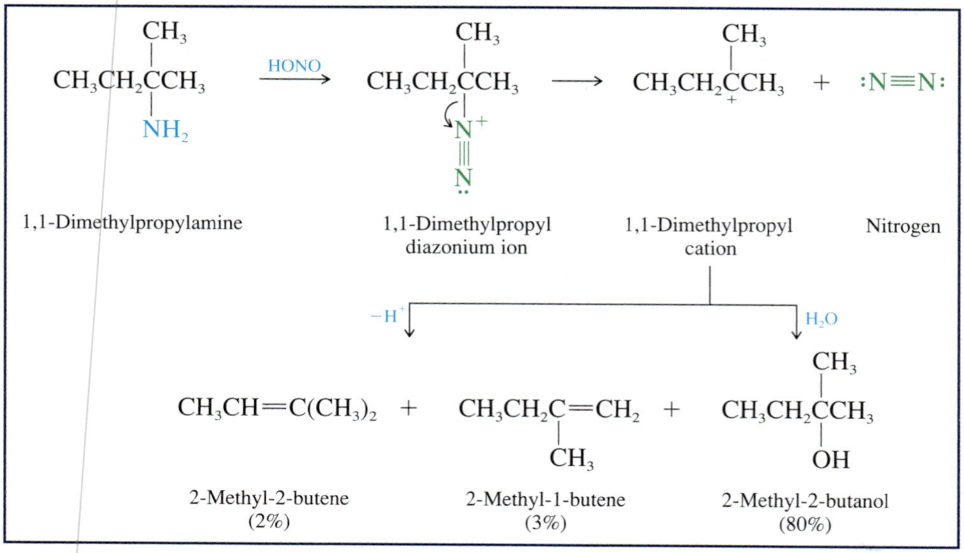

$$\underset{\substack{\text{1,1-Dimethylpropylamine}}}{\overset{\displaystyle CH_3}{\underset{\displaystyle NH_2}{CH_3CH_2C\!-\!CH_3}}} \xrightarrow{HONO} \underset{\substack{\text{1,1-Dimethylpropyl}\\\text{diazonium ion}}}{\overset{\displaystyle CH_3}{CH_3CH_2C\!-\!CH_3}} \longrightarrow \underset{\substack{\text{1,1-Dimethylpropyl}\\\text{cation}}}{\overset{\displaystyle CH_3}{\underset{+}{CH_3CH_2C\!-\!CH_3}}} + \underset{\text{Nitrogen}}{:N\equiv N:}$$

$$\underset{\substack{\text{2-Methyl-2-butene}\\(2\%)}}{CH_3CH=C(CH_3)_2} + \underset{\substack{\text{2-Methyl-1-butene}\\(3\%)}}{\underset{\displaystyle CH_3}{CH_3CH_2C=CH_2}} + \underset{\substack{\text{2-Methyl-2-butanol}\\(80\%)}}{\overset{\displaystyle CH_3}{\underset{\displaystyle OH}{CH_3CH_2C\!-\!CH_3}}}$$

FIGURE 22.5 The diazonium ion generated by treatment of a primary alkylamine with nitrous acid loses nitrogen to give a carbocation. The isolated products are derived from the carbocation and include, in this example, alkenes (by loss of a proton) and an alcohol (nucleophilic capture by water).

diazonium ions are rarely used in synthetic work but have been studied extensively to probe the behavior of carbocations generated under conditions in which the leaving group is lost rapidly and irreversibly.

PROBLEM 22.16 Nitrous acid deamination of 2,2-dimethylpropylamine, $(CH_3)_3CCH_2NH_2$, gives the same products as were indicated as being formed from 1,1-dimethylpropylamine in Figure 22.5. Suggest a mechanism for the formation of these compounds from 2,2-dimethylpropylamine.

Aryl diazonium ions, prepared by nitrous acid diazotization of primary arylamines, are substantially more stable than alkyl diazonium ions and are of enormous synthetic value. Their use in the synthesis of substituted aromatic compounds is described in the following two sections.

The nitrosation of tertiary alkylamines is rather complicated, and no generally useful chemistry is associated with reactions of this type.

22.16 NITROSATION OF ARYLAMINES

We learned in the preceding section that different reactions are observed when the various classes of alkylamines—primary, secondary, and tertiary—react with nitrosating agents. Although no useful chemistry attends the nitrosation of tertiary alkylamines, electrophilic aromatic substitution by nitrosyl cation ($: N \equiv \overset{+}{O} :$) takes place with *N,N*-dialkylarylamines.

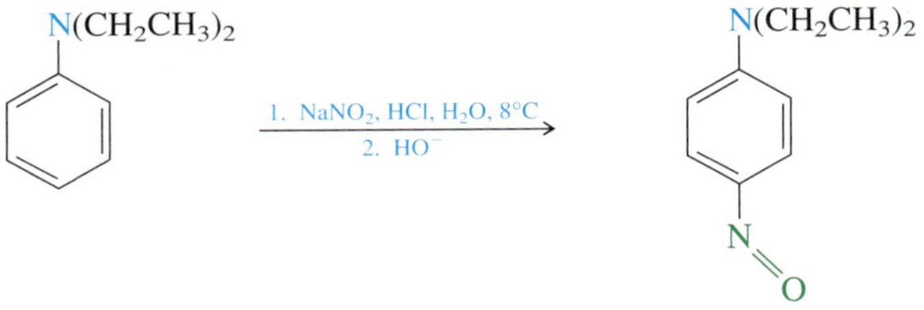

N,N-Diethylaniline   N,N-Diethyl-p-nitrosoaniline (95%)

Nitrosyl cation is a relatively weak electrophile and attacks only very strongly activated aromatic rings.

N-Alkylarylamines resemble secondary alkylamines in that they form *N*-nitroso compounds on reaction with nitrous acid.

$$C_6H_5NHCH_3 \xrightarrow[\text{H}_2\text{O, 10°C}]{\text{NaNO}_2,\ \text{HCl}} C_6H_5N-N=O$$
$$\underset{CH_3}{|}$$

N-Methylaniline   N-Methyl-N-nitrosoaniline
                              (87–93%)

Primary arylamines, like primary alkylamines, form diazonium ion salts on nitrosation. Aryl diazonium ions are considerably more stable than their alkyl counterparts. Whereas alkyl diazonium ions decompose under the conditions of their formation, aryl diazonium salts are stable enough to be stored in aqueous solution at 0–5°C for a reasonable time. Loss of nitrogen from an aryl diazonium ion generates an unstable aryl cation and is much slower than loss of nitrogen from an alkyl diazonium ion.

$$C_6H_5NH_2 \xrightarrow[\text{H}_2\text{O, 0–5°C}]{\text{NaNO}_2, \text{ HCl}} C_6H_5\overset{+}{N}\equiv N\colon Cl^-$$

Aniline   Benzenediazonium chloride

$$(CH_3)_2CH-\!\!\!\bigcirc\!\!\!-NH_2 \xrightarrow[\text{H}_2\text{O, 0–5°C}]{\text{NaNO}_2, \text{ H}_2\text{SO}_4} (CH_3)_2CH-\!\!\!\bigcirc\!\!\!-\overset{+}{N}\equiv N\colon HSO_4^-$$

p-Isopropylaniline   *p*-Isopropylbenzenediazonium
                                    hydrogen sulfate

Aryl diazonium ions undergo a variety of reactions that make them versatile intermediates for preparing a host of ring-substituted aromatic compounds. In these reactions, summarized in Figure 22.6 and discussed individually in the following section, molecular nitrogen acts as a leaving group and is replaced by another atom or group. All the reactions are regiospecific; the entering group becomes bonded to the same carbon from which nitrogen departs.

22.17 SYNTHETIC TRANSFORMATIONS OF ARYL DIAZONIUM SALTS

An important reaction of aryl diazonium ions is their conversion to *phenols* by hydrolysis:

$$Ar\overset{+}{N}\equiv N\colon \quad + \quad H_2O \longrightarrow ArOH \quad + H^+ + \colon\!N\equiv N\colon$$

Aryl diazonium ion   Water   A phenol   Nitrogen

This is the most general method for preparing phenols. It is easily performed; the aqueous acidic solution in which the diazonium salt is prepared is heated and gives the phenol directly. An aryl cation is probably generated, which is then captured by water acting as a nucleophile.

$$(CH_3)_2CH-\!\!\!\bigcirc\!\!\!-NH_2 \xrightarrow[\text{2. H}_2\text{O, heat}]{\text{1. NaNO}_2, \text{ H}_2\text{SO}_4, \text{ H}_2\text{O}} (CH_3)_2CH-\!\!\!\bigcirc\!\!\!-OH$$

p-Isopropylaniline   *p*-Isopropylphenol (73%)

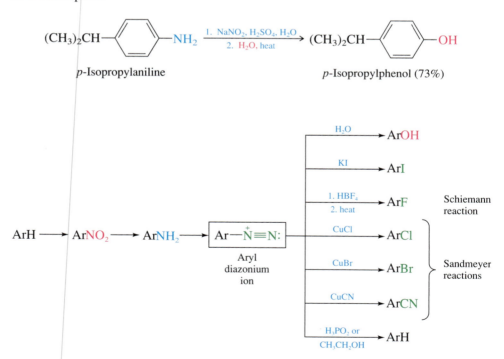

FIGURE 22.6 Flowchart showing the synthetic origin of aryl diazonium ions and their most useful transformations.

Sulfuric acid is normally used instead of hydrochloric acid in the diazotization step so as to minimize the competition with water for capture of the cationic intermediate. Hydrogen sulfate anion ($HSO_4{}^-$) is less nucleophilic than chloride.

PROBLEM 22.17 Design a synthesis of *m*-bromophenol from benzene.

The reaction of an aryl diazonium salt with potassium iodide is the standard method for the preparation of *aryl iodides*. The diazonium salt is prepared from a primary aromatic amine in the usual way, a solution of potassium iodide is then added, and the reaction mixture is brought to room temperature or heated to accelerate the reaction.

$$Ar\overset{+}{-}N\equiv N: \; + \; I^- \; \longrightarrow \; ArI \; + \; :N\equiv N:$$

| Aryl diazonium ion | Iodide ion | Aryl iodide | Nitrogen |

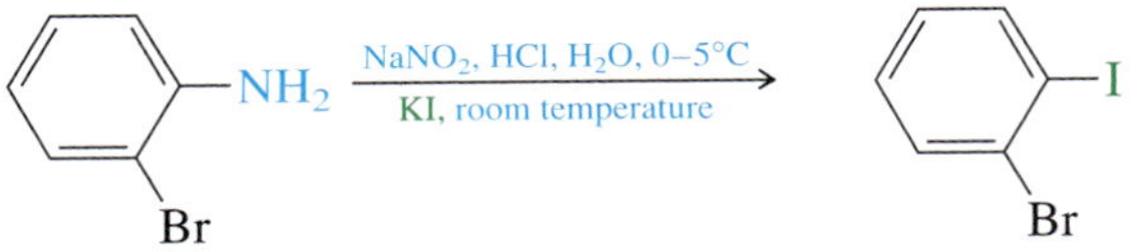

o-Bromoaniline → *o*-Bromoiodobenzene (72–83%)

Reagents: NaNO₂, HCl, H₂O, 0–5°C; KI, room temperature

PROBLEM 22.18 Show by a series of equations how you could prepare *m*-bromoiodobenzene from benzene.

Diazonium salt chemistry provides the principal synthetic method for the preparation of *aryl fluorides* through a process known as the **Schiemann reaction.** In this procedure the aryl diazonium ion is isolated as its fluoroborate salt, which then yields the desired aryl fluoride on being heated.

$$Ar\overset{+}{-}N\equiv N: \bar{B}F_4 \; \overset{heat}{\longrightarrow} \; ArF \; + \; BF_3 \; + \; :N\equiv N:$$

| Aryl diazonium fluoroborate | Aryl fluoride | Boron trifluoride | Nitrogen |

A standard way to form the aryl diazonium fluoroborate salt is to add fluoroboric acid (HBF_4) or a fluoroborate salt to the diazotization medium.

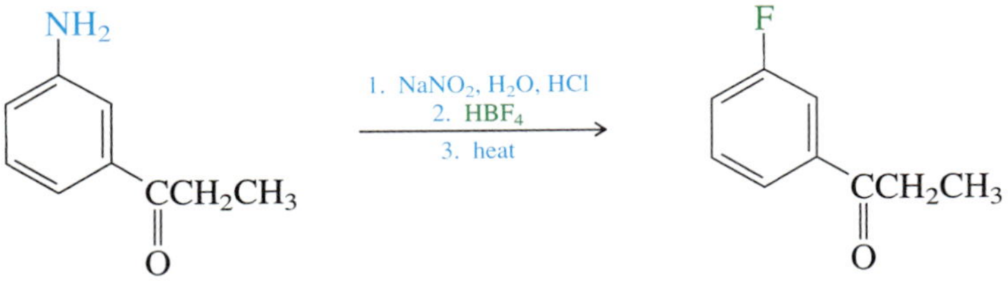

m-Aminophenyl ethyl ketone → Ethyl *m*-fluorophenyl ketone (68%)

Reagents: 1. NaNO₂, H₂O, HCl 2. HBF₄ 3. heat

PROBLEM 22.19 Show the proper sequence of synthetic transformations in the conversion of benzene to ethyl *m*-fluorophenyl ketone.

Although it is possible to prepare *aryl chlorides* and *aryl bromides* by electrophilic aromatic substitution, it is often necessary to prepare these compounds from an aromatic amine. The amine is converted to the corresponding diazonium salt and then treated with copper(I) chloride or copper(I) bromide as appropriate.

$$\overset{+}{Ar-N}\equiv N: \quad \xrightarrow{CuX} \quad ArX \quad + \quad :N\equiv N:$$

Aryl diazonium   Aryl chloride   Nitrogen
ion   or bromide

m-Nitroaniline

1. NaNO$_2$, HCl, H$_2$O, 0–5°C
2. CuCl, heat

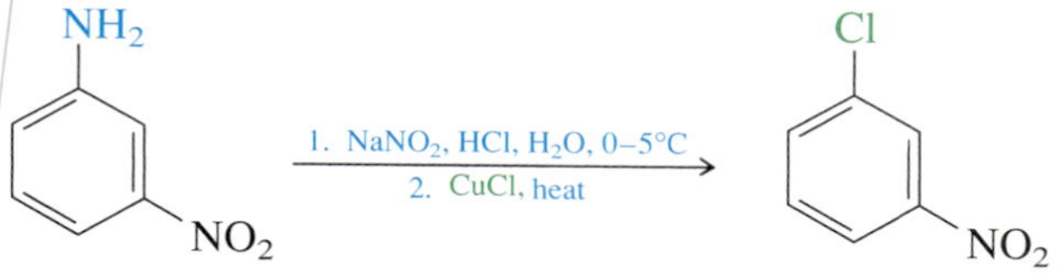

m-Chloronitrobenzene
(68–71%)

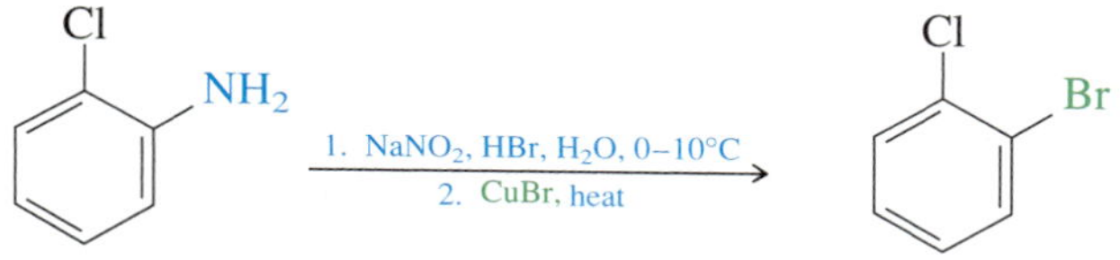

o-Chloroaniline

1. NaNO$_2$, HBr, H$_2$O, 0–10°C
2. CuBr, heat

o-Bromochlorobenzene
(89–95%)

Reactions that employ copper(I) salts as reagents for replacement of nitrogen in diazonium salts are called **Sandmeyer reactions.** The Sandmeyer reaction using copper(I) cyanide is a good method for the preparation of aromatic *nitriles:*

$$\overset{+}{Ar-N}\equiv N: \quad \xrightarrow{CuCN} \quad ArCN \quad + \quad :N\equiv N:$$

Aryl diazonium   Aryl   Nitrogen
ion   nitrile

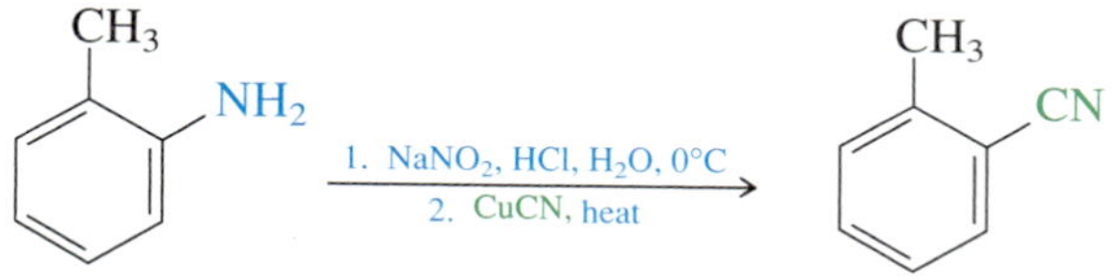

o-Toluidine

1. NaNO$_2$, HCl, H$_2$O, 0°C
2. CuCN, heat

o-Methylbenzonitrile
(64–70%)

Because cyano groups may be hydrolyzed to carboxylic acids (Section 20.19), the Sandmeyer preparation of aryl nitriles is a key step in the conversion of arylamines to substituted benzoic acids. In the example just cited, the *o*-methylbenzonitrile that was formed was subsequently subjected to acid-catalyzed hydrolysis and gave *o*-methylbenzoic acid in 80–89% yield.

The preparation of aryl chlorides, bromides, and cyanides by the Sandmeyer reaction is mechanistically complicated and may involve arylcopper intermediates.

It is possible to replace amino substituents on an aromatic nucleus by hydrogen by reducing a diazonium salt with hypophosphorous acid (H$_3$PO$_2$) or with ethanol. These reductions are free-radical reactions in which ethanol or hypophosphorous acid acts as a hydrogen atom donor:

$$\overset{+}{Ar-N} \equiv N: \xrightarrow[\text{CH}_3\text{CH}_2\text{OH}]{\text{H}_3\text{PO}_2 \text{ or}} ArH \ + \ :N \equiv N:$$

Aryl diazonium ion   Arene   Nitrogen

Reactions of this type are called **reductive deaminations.**

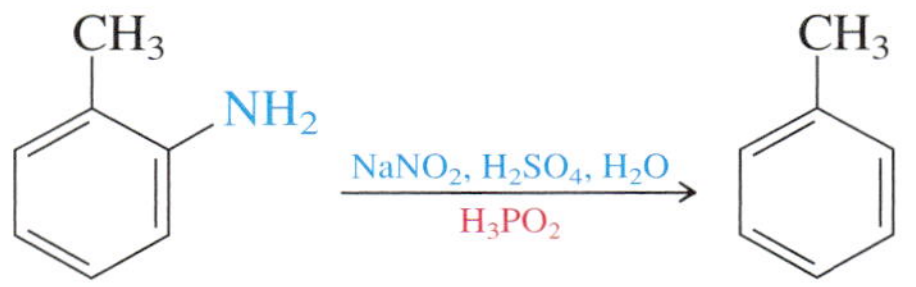

o-Toluidine

Toluene
(70–75%)

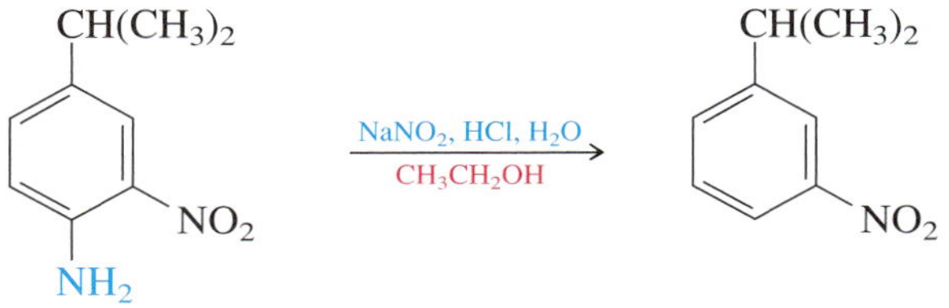

4-Isopropyl-2-nitroaniline

m-Isopropylnitrobenzene
(59%)

Sodium borohydride has also been used to reduce aryl diazonium salts in reductive deamination reactions.

PROBLEM 22.20 Cumene (isopropylbenzene) is a relatively inexpensive commercially available starting material. Show how you could prepare m-isopropylnitrobenzene from cumene.

The value of diazonium salts in synthetic organic chemistry rests on two main points. Through the use of diazonium salt chemistry:

1. Substituents that are otherwise accessible only with difficulty, such as fluoro, iodo, cyano, and hydroxyl, may be introduced onto a benzene ring.

2. Compounds that have substitution patterns not directly available by electrophilic aromatic substitution can be prepared.

The first of these two features is readily apparent and is illustrated by Problems 22.17 to 22.19. If you have not done these problems yet, you are strongly encouraged to attempt them now.

The second point is somewhat less obvious but is readily illustrated by the synthesis of 1,3,5-tribromobenzene. This particular substitution pattern cannot be obtained by direct bromination of benzene because bromine is an ortho, para director. Instead, advantage is taken of the powerful activating and ortho, para-directing effects of the amino group in aniline. Bromination of aniline yields 2,4,6-tribromoaniline in quantitative yield. Diazotization of the resulting 2,4,6-tribromoaniline and reduction of the diazonium salt gives the desired 1,3,5-tribromobenzene.

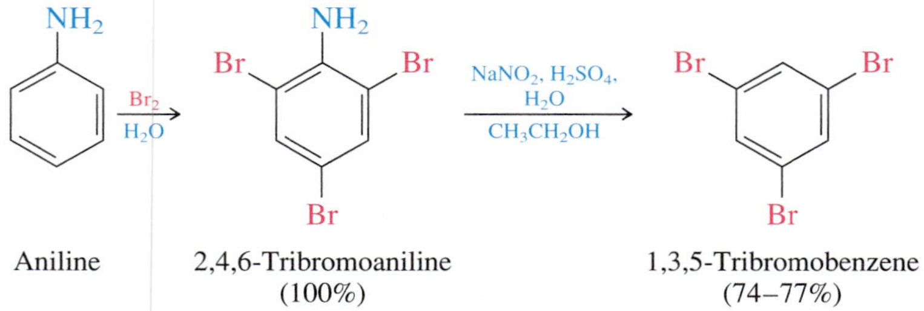

Aniline   2,4,6-Tribromoaniline   1,3,5-Tribromobenzene
(100%)   (74–77%)

To exploit the synthetic versatility of aryl diazonium salts, be prepared to reason backward. When you see a fluorine attached to a benzene ring, for example, realize that it probably will have to be introduced by a Schiemann reaction of an arylamine; realize that the required arylamine is derived from a nitroarene, and that the nitro group is introduced by nitration. Be aware that an unsubstituted position of a benzene ring need not have always been that way. It might once have borne an amino group that was used to control the orientation of electrophilic aromatic substitution reactions before being removed by reductive deamination. The strategy of synthesis is intellectually demanding, and a considerable sharpening of your reasoning power can be gained by attacking the synthesis problems at the end of each chapter. Remember, plan your sequence of accessible intermediates by reasoning backward from the target; then fill in the details on how each transformation is to be carried out.

22.18 AZO COUPLING

A reaction of aryl diazonium salts that does not involve loss of nitrogen takes place when they react with phenols and arylamines. Aryl diazonium ions are relatively weak electrophiles but have sufficient reactivity to attack strongly activated aromatic rings. The reaction is known as *azo coupling;* two aryl groups are joined together by an azo (—N=N—) function.

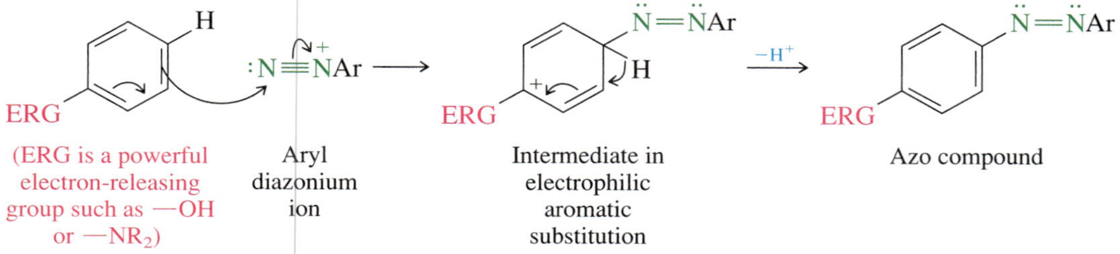

(ERG is a powerful   Aryl   Intermediate in   Azo compound
electron-releasing   diazonium   electrophilic
group such as —OH   ion   aromatic
or —NR$_2$)   substitution

Azo compounds are often highly colored, and many of them are used as dyes.

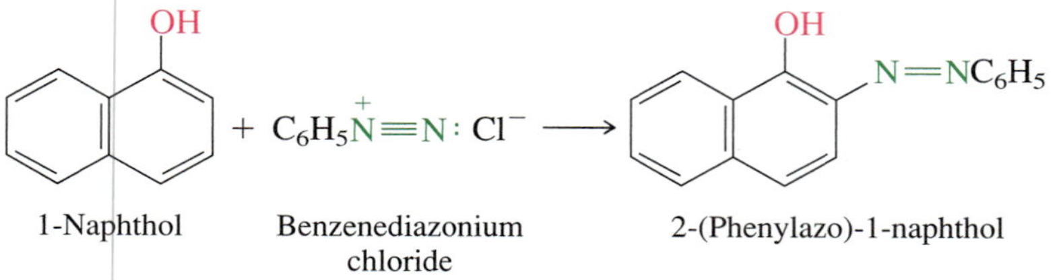

1-Naphthol   Benzenediazonium   2-(Phenylazo)-1-naphthol
chloride

A number of pH indicators— methyl red, for example— are azo compounds.

The colors of azo compounds vary with the nature of the aryl group, with its substituents, and with pH. Substituents also affect the water-solubility of azo dyes and how well they

FROM DYES TO SULFA DRUGS

The medicine cabinet was virtually bare of antibacterial agents until **sulfa drugs** burst on the scene in the 1930s. Before sulfa drugs became available, bacterial infection might transform a small cut or puncture wound to a life-threatening event. The story of how sulfa drugs were developed is an interesting example of being right for the wrong reasons. It was known that many bacteria absorbed dyes, and staining was a standard method for making bacteria more visible under the microscope. Might there not be some dye that is both absorbed by bacteria and toxic to them? Acting on this hypothesis, scientists at the German dyestuff manufacturer I. G. Farbenindustrie undertook a program to test the thousands of compounds in their collection for their antibacterial properties.

In general, in vitro testing of drugs precedes in vivo testing. The two terms mean, respectively, "in glass" and "in life." In vitro testing of antibiotics is carried out using bacterial cultures in test tubes or Petri dishes. Drugs that are found to be active in vitro progress to the stage of in vivo testing. In vivo testing is carried out in living organisms: laboratory animals or human volunteers. The I. G. Farben scientists found that some dyes did possess antibacterial properties, both in vitro and in vivo. Others were active in vitro but were converted to inactive substances in vivo and therefore of no use as drugs. Unexpectedly, an azo dye called *Prontosil* was inactive in vitro but active in vivo. In 1932, a member of the I. G. Farben research group, Gerhard Domagk used Prontosil to treat a young child suffering from a serious, potentially fatal staphylococcal infection. According to many accounts, the child was Domagk's own daughter; her infection was cured and her recovery was rapid and complete. Systematic testing followed and Domagk was awarded the 1939 Nobel Prize in medicine or physiology.

In spite of the rationale on which the testing of dyestuffs as antibiotics rested, subsequent research revealed that the antibacterial properties of Prontosil had nothing at all to do with its being a dye! In the body, Prontosil undergoes a reductive cleavage of its azo linkage to form *sulfanilamide*, which is the substance actually responsible for the observed biological activity. This is why Prontosil is active in vivo, but not in vitro.

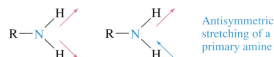

Prontosil   Sulfanilamide

—Cont.

bind to a particular fabric. Countless combinations of diazonium salts and aromatic substrates have been examined with a view toward obtaining azo dyes suitable for a particular application.

22.19 SPECTROSCOPIC ANALYSIS OF AMINES

Infrared: The absorptions of interest in the IR spectra of amines are those associated with N—H vibrations. Primary alkyl- and arylamines exhibit two peaks in the range $3000–3500 \text{ cm}^{-1}$, which are due to symmetric and antisymmetric N—H stretching modes.

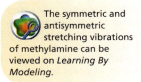

The symmetric and antisymmetric stretching vibrations of methylamine can be viewed on *Learning By Modeling.*

Symmetric N—H stretching of a primary amine   Antisymmetric N—H stretching of a primary amine

Bacteria require *p*-aminobenzoic acid to biosynthesize *folic acid,* a growth factor. Structurally, sulfanilamide resembles *p*-aminobenzoic acid and is mistaken for it by the bacteria. Folic acid biosynthesis is inhibited and bacterial growth is slowed sufficiently to allow the body's natural defenses to effect a cure. Because animals do not biosynthesize folic acid but obtain it in their food, sulfanilamide halts the growth of bacteria without harm to the host.

Identification of the mechanism by which Prontosil combats bacterial infections was an early triumph of **pharmacology,** a branch of science at the interface of physiology and biochemistry that studies the mechanism of drug action. By recognizing that sulfanilamide was the active agent, the task of preparing structurally modified analogs with potentially superior properties was considerably simplified. Instead of preparing Prontosil analogs, chemists synthesized sulfanilamide analogs. They did this with a vengeance; over 5000 compounds related to sulfanilamide were prepared during the period 1935–1946. Two of the most widely used sulfa drugs are *sulfathiazole* and *sulfadiazine.*

Sulfathiazole

Sulfadiazine

We tend to take the efficacy of modern drugs for granted. One comparison with the not-too-distant past might put this view into better perspective. Once sulfa drugs were introduced in the United States, the number of pneumonia deaths alone decreased by an estimated 25,000 per year. The sulfa drugs are used less now than they were in the mid-twentieth century. Not only are more-effective, less-toxic antibiotics available, such as the penicillins and tetracyclines, but many bacteria that were once susceptible to sulfa drugs have become resistant.

These two vibrations are clearly visible at 3270 and 3380 cm^{-1} in the IR spectrum of butylamine, shown in Figure 22.7*a*. Secondary amines such as diethylamine, shown in Figure 22.7*b*, exhibit only one peak, which is due to N—H stretching, at 3280 cm^{-1}. Tertiary amines, of course, are transparent in this region because they have no N—H bonds.

^{1}H NMR: Characteristics of the nuclear magnetic resonance spectra of amines may be illustrated by comparing 4-methylbenzylamine (Figure 22.8*a*) with 4-methylbenzyl

FIGURE 22.7 Portions of the IR spectra of (*a*) butylamine and (*b*) diethylamine. Primary amines exhibit two peaks due to N—H stretching in the 3300 to 3350 cm^{-1} region, whereas secondary amines show only one.

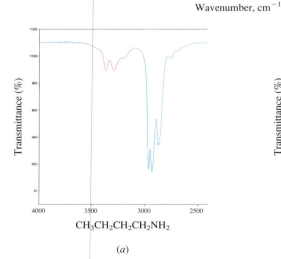

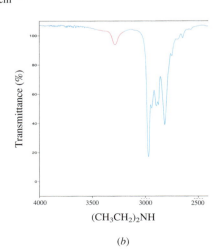

alcohol (Figure 22.8*b*). Nitrogen is less electronegative than oxygen and so shields neighboring nuclei to a greater extent. The benzylic methylene group attached to nitrogen in 4-methylbenzylamine appears at higher field (δ 3.8) than the benzylic methylene of 4-methylbenzyl alcohol (δ 4.6). The N—H protons are somewhat more shielded than the O—H protons of an alcohol. In 4-methylbenzylamine the protons of the amino group correspond to the signal at δ 1.5, whereas the hydroxyl proton signal of 4-methylbenzyl alcohol is found at δ 2.1. The chemical shifts of amino group protons, like those of hydroxyl protons, are variable and are sensitive to solvent, concentration, and temperature.

^{13}C *NMR:*   Similarly, carbons that are bonded to nitrogen are more shielded than those bonded to oxygen, as revealed by comparing the ^{13}C chemical shifts of methylamine and methanol.

δ 26.9 CH_3NH_2 δ 48.0 CH_3OH

Methylamine   Methanol

UV-VIS: In the absence of any other chromophore, the UV-VIS spectrum of an alkylamine is not very informative. The longest wavelength absorption involves promoting one of the unshared electrons of nitrogen to an antibonding σ^* orbital ($n \rightarrow \sigma^*$) with a λ_{max} in the relatively inaccessible region near 200 nm. In arylamines the interaction of the nitrogen lone pair with the π-electron system of the ring shifts the ring's absorptions to longer wavelength. Tying up the lone pair by protonation causes the UV-VIS spectrum of anilinium ion to resemble benzene.

	X	λ_{max} (nm)
Benzene	H	204, 256
Aniline	NH_2	230, 280
Anilinium ion	NH_3^+	203, 254

Mass Spectrometry: A number of features make amines easily identifiable by mass spectrometry.

First, the peak for the molecular ion M^+ for all compounds that contain only carbon, hydrogen, and oxygen has an *m/z* value that is an even number. The presence of a nitrogen atom in the molecule requires that the *m/z* value for the molecular ion be odd. An odd number of nitrogens corresponds to an odd value of the molecular weight; an even number of nitrogens corresponds to an even molecular weight.

Second, nitrogen is exceptionally good at stabilizing adjacent carbocation sites. The fragmentation pattern seen in the mass spectra of amines is dominated by cleavage of groups from the carbon atom attached to the nitrogen, as the data for the following pair of constitutionally isomeric amines illustrate:

Recall the "nitrogen rule" from Section 13.23.

$(CH_3)_2\ddot{N}CH_2CH_2CH_2CH_3 \xrightarrow{e^-} (CH_3)_2\overset{+}{\ddot{N}}{-}CH_2{-}CH_2CH_2CH_3 \longrightarrow (CH_3)_2\overset{+}{N}{=}CH_2 + \cdot CH_2CH_2CH_3$

N,N-Dimethyl-1-butanamine M^+ (*m/z* 101)   (*m/z* 58)
(most intense peak)

$CH_3\ddot{N}HCH_2CH_2CH(CH_3)_2 \xrightarrow{e^-} CH_3\overset{+}{\ddot{N}}H{-}CH_2{-}CH_2CH(CH_3)_2 \longrightarrow CH_3\overset{+}{N}H{=}CH_2 + \cdot CH_2CH(CH_3)_2$

N,3-Dimethyl-1-butanamine M^+ (*m/z* 101)   (*m/z* 44)
(most intense peak)

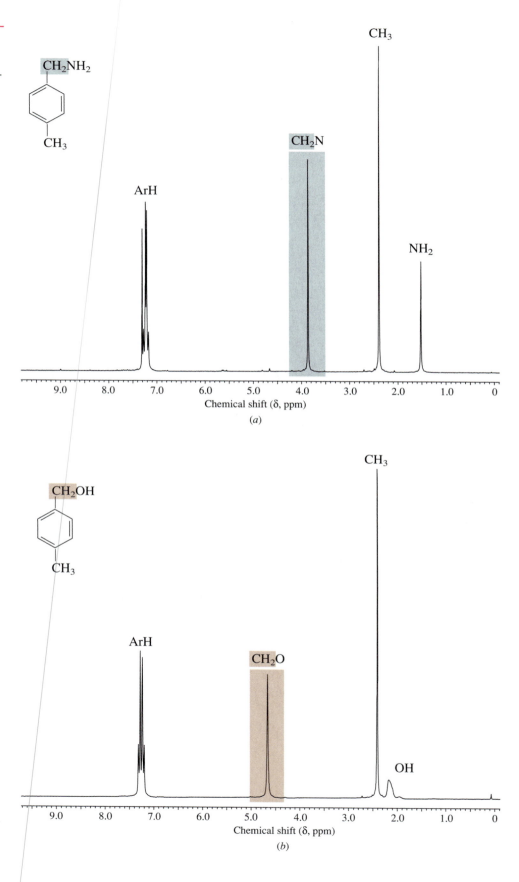

FIGURE 22.8 The 200-MHz ¹H NMR spectra of (a) 4-methylbenzylamine and (b) 4-methylbenzyl alcohol. The singlet corresponding to CH₂N in (a) is more shielded than that of CH₂O in (b).

22.20 SUMMARY

Section 22.1   Alkylamines are compounds of the type shown, where R, R', and R" are alkyl groups. One or more of these groups is an aryl group in arylamines.

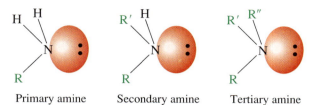

Primary amine   Secondary amine   Tertiary amine

Alkylamines are named in two ways. One method adds the ending -*amine* to the name of the alkyl group. The other applies the principles of substitutive nomenclature by replacing the -*e* ending of an alkane name by -*amine* and uses appropriate locants to identify the position of the amino group. Arylamines are named as derivatives of aniline.

Section 22.2   Nitrogen's unshared electron pair is of major importance in understanding the structure and properties of amines. Alkylamines have a pyramidal arrangement of bonds to nitrogen, and the unshared electron pair resides in an sp^3-hybridized orbital. The geometry at nitrogen in arylamines is somewhat flatter than in alkylamines, and the unshared electron pair is delocalized into the π system of the ring. Delocalization binds the electron pair more strongly in arylamines than in alkylamines. Arylamines are less basic and less nucleophilic than alkylamines.

Section 22.3   Amines are less polar than alcohols. Hydrogen bonding in amines is weaker than in alcohols because nitrogen is less electronegative than oxygen. Amines have lower boiling points than alcohols, but higher boiling points than alkanes. Primary amines have higher boiling points than isomeric secondary amines; tertiary amines, which cannot form intermolecular hydrogen bonds, have the lowest boiling points. Amines resemble alcohols in their solubility in water.

Section 22.4   The basicity of amines is conveniently expressed in terms of the pK_a of their conjugate acids.

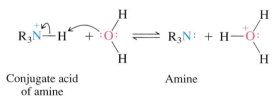

Conjugate acid   Amine
of amine

The stronger base is associated with the weaker conjugate acid. The greater the pK_a of the conjugate acid, the stronger the base. The pK_a's of the conjugate acids of alkylamines lie in the 9–11 range. Arylamines are much weaker bases than alkylamines. The pK_a's of the conjugate acids of arylamines are usually 3–5. Strong electron-withdrawing groups can weaken the basicity of arylamines even more.

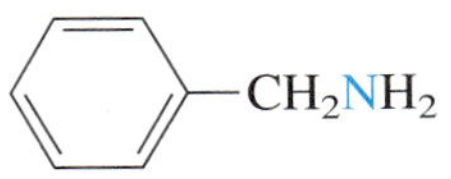

Benzylamine
(alkylamine: pK_a of conjugate acid = 9.3)

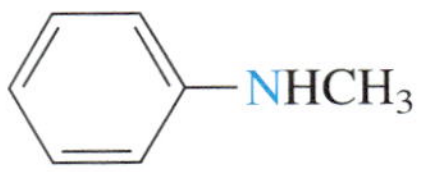

N-Methylaniline
(arylamine: pK_a of conjugate acid = 3.2)

Section 22.5 Quaternary ammonium salts, compounds of the type $R_4N^+ X^-$, find application in a technique called **phase-transfer catalysis.** A small amount of a quaternary ammonium salt promotes the transfer of an anion from aqueous solution, where it is highly solvated, to an organic solvent, where it is much less solvated and much more reactive.

Sections 22.6–22.10 Methods for the preparation of amines are summarized in Table 22.5.

TABLE 22.5   Preparation of Amines

Reaction (section) and comments	General equation and specific example
Alkylation methods	
Alkylation of ammonia (Section 22.7) Ammonia can act as a nucleophile toward primary and some secondary alkyl halides to give primary alkylamines. Yields tend to be modest because the primary amine is itself a nucleophile and undergoes alkylation. Alkylation of ammonia can lead to a mixture containing a primary amine, a secondary amine, a tertiary amine, and a quaternary ammonium salt.	$RX\ +\ 2NH_3\ \longrightarrow\ RNH_2\ +\ NH_4X$ Alkyl halide Ammonia Alkylamine Ammonium halide $C_6H_5CH_2Cl\ \xrightarrow{NH_3\ (8\ mol)}\ C_6H_5CH_2NH_2\ +\ (C_6H_5CH_2)_2NH$ Benzyl chloride (1 mol) Benzylamine (53%) Dibenzylamine (39%)
Alkylation of phthalimide. The Gabriel synthesis (Section 22.8) The potassium salt of phthalimide reacts with alkyl halides to give *N*-alkylphthalimide derivatives. Hydrolysis or hydrazinolysis of this derivative yields a primary alkylamine.	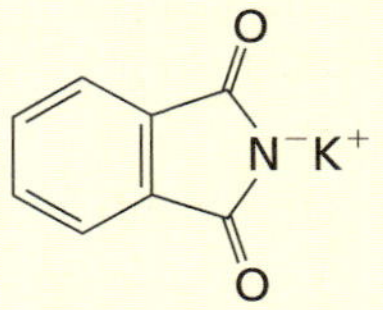 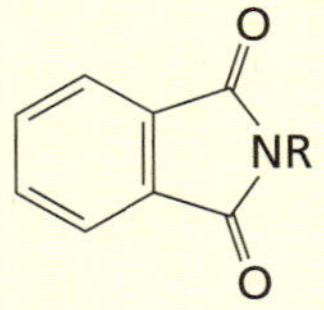 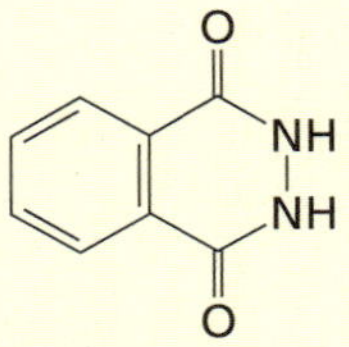 Alkyl halide *N*-Potassiophthalimide *N*-Alkylphthalimide 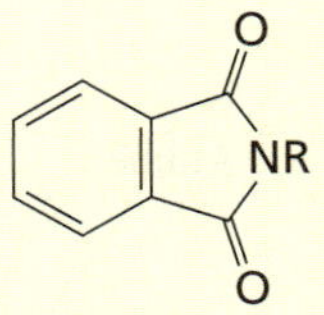 *N*-Alkylphthalimide Hydrazine Primary amine Phthalhydrazide $CH_3CH{=}CHCH_2Cl\ \xrightarrow[\text{2. }H_2NNH_2,\ \text{ethanol}]{\text{1. }N\text{-potassiophthalimide, DMF}}\ CH_3CH{=}CHCH_2NH_2$ 1-Chloro-2-butene 2-Buten-1-amine (95%)

(Continued)

TABLE 22.5 Preparation of Amines (Continued)

Reaction (section) and comments	General equation and specific example

Reduction methods

Reduction of alkyl azides (Section 22.9) Alkyl azides, prepared by nucleophilic substitution by azide ion in primary or secondary alkyl halides, are reduced to primary alkylamines by lithium aluminum hydride or by catalytic hydrogenation.

$$R\ddot{N}=\overset{+}{N}=\ddot{\underset{..}{N}}: \xrightarrow{\text{reduce}} R\ddot{N}H_2$$

Alkyl azide   Primary amine

$$CF_3CH_2\underset{N_3}{CH}CO_2CH_2CH_3 \xrightarrow{H_2,\ Pd} CF_3CH_2\underset{NH_2}{CH}CO_2CH_2CH_3$$

Ethyl 2-azido-4,4,4-trifluorobutanoate   Ethyl 2-amino-4,4,4-trifluorobutanoate (96%)

Reduction of nitriles (Section 22.9) Nitriles are reduced to primary amines by lithium aluminum hydride or by catalytic hydrogenation.

$$RC\equiv N \xrightarrow{\text{reduce}} RCH_2NH_2$$

Nitrile   Primary amine

$$\triangleright\!-CN \xrightarrow[2.\ H_2O]{1.\ LiAlH_4} \triangleright\!-CH_2NH_2$$

Cyclopropyl cyanide   Cyclopropylmethanamine (75%)

Reduction of aryl nitro compounds (Section 22.9) The standard method for the preparation of an arylamine is by nitration of an aromatic ring, followed by reduction of the nitro group. Typical reducing agents include iron or tin in hydrochloric acid or catalytic hydrogenation.

$$ArNO_2 \xrightarrow{\text{reduce}} ArNH_2$$

Nitroarene   Arylamine

$$C_6H_5NO_2 \xrightarrow[2.\ HO^-]{1.\ Fe,\ HCl} C_6H_5NH_2$$

Nitrobenzene   Aniline (97%)

Reduction of amides (Section 22.9) Lithium aluminum hydride reduces the carbonyl group of an amide to a methylene group. Primary, secondary, or tertiary amines may be prepared by proper choice of the starting amide. R and R′ may be either alkyl or aryl.

$$\underset{\substack{\parallel \\ O}}{R}CNR'_2 \xrightarrow{\text{reduce}} RCH_2NR'_2$$

Amide   Amine

$$CH_3\overset{O}{\overset{\parallel}{C}}NHC(CH_3)_3 \xrightarrow[2.\ H_2O]{1.\ LiAlH_4} CH_3CH_2NHC(CH_3)_3$$

N-tert-Butylacetamide   N-Ethyl-tert-butylamine (60%)

Reductive amination (Section 22.10) Reaction of ammonia or an amine with an aldehyde or a ketone in the presence of a reducing agent is an effective method for the preparation of primary, secondary, or tertiary amines. The reducing agent may be either hydrogen in the presence of a metal catalyst or sodium cyanoborohydride. R, R′, and R″ may be either alkyl or aryl.

$$\underset{\substack{\parallel \\ O}}{R}CR' + R''_2NH \xrightarrow{\substack{\text{reducing} \\ \text{agent}}} \underset{\substack{| \\ H}}{R}\overset{NR''_2}{\underset{}{C}}R'$$

Aldehyde or ketone   Ammonia or an amine   Amine

$$CH_3\overset{O}{\overset{\parallel}{C}}CH_3 + \text{(cyclohexylamine)} \xrightarrow{H_2,\ Pt} \text{(N-Isopropylcyclohexylamine)}$$

Acetone   Cyclohexylamine   N-Isopropylcyclohexylamine (79%)

Sections 22.11–22.18 The reactions of amines are summarized in Tables 22.6 and 22.7.

Section 22.19 The N—H stretching frequency of primary and secondary amines appears in the infrared in the 3000–3500 cm^{-1} region. In the NMR spectra of amines, protons and carbons of the type H—C—N are more shielded than H—C—O.

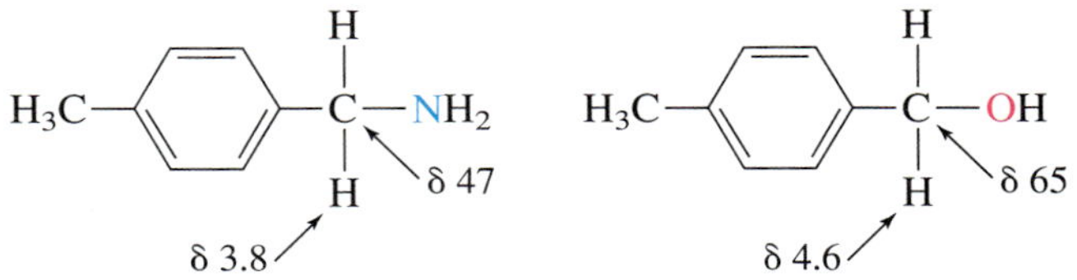

Amines have odd-numbered molecular weights, which helps identify them by mass spectrometry. Fragmentation tends to be controlled by the formation of a nitrogen-stabilized cation.

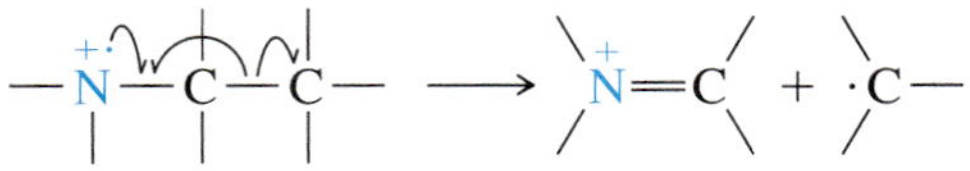

TABLE 22.6 Reactions of Amines Discussed in This Chapter

Reaction (section) and comments	General equation and specific example
Alkylation (Section 22.12) Amines act as nucleophiles toward alkyl halides. Primary amines yield secondary amines, secondary amines yield tertiary amines, and tertiary amines yield quaternary ammonium salts.	$R\ddot{N}H_2 \xrightarrow{R'CH_2X} R\ddot{N}HCH_2R'$ (Primary amine → Secondary amine) $R\ddot{N}HCH_2R' \xrightarrow{R'CH_2X} R\ddot{N}(CH_2R')_2$ (Secondary amine → Tertiary amine) $R\ddot{N}(CH_2R')_2 \xrightarrow{R'CH_2X} R\overset{+}{N}(CH_2R')_3\ X^-$ (Tertiary amine → Quaternary ammonium salt) 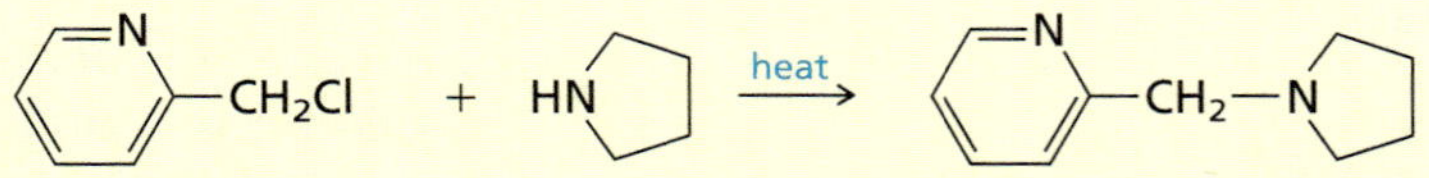 2-Chloromethylpyridine + Pyrrolidine → 2-(Pyrrolidinylmethyl)pyridine (93%)
Hofmann elimination (Section 22.13) Quaternary ammonium hydroxides undergo elimination on being heated. It is an anti elimination of the E2 type. The regioselectivity of the Hofmann elimination is opposite to that of the Zaitsev rule and leads to the less highly substituted alkene.	$RCH_2CH(\overset{+}{N}(CH_3)_3)R'\ HO^- \xrightarrow{heat} RCH{=}CHR' + :N(CH_3)_3 + H_2O$ Alkyltrimethylammonium hydroxide → Alkene + Trimethylamine + Water 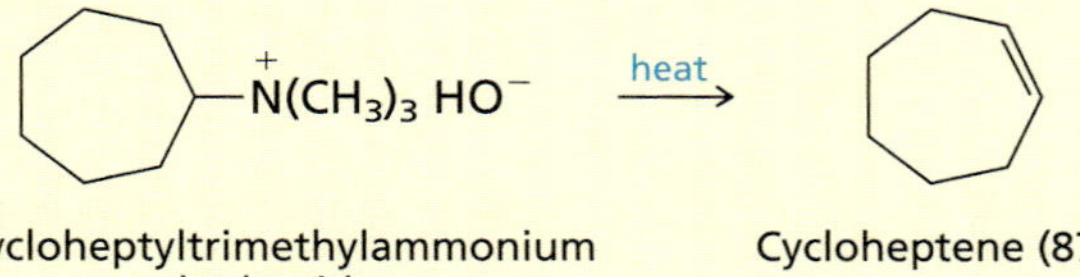 Cycloheptyltrimethylammonium hydroxide → Cycloheptene (87%)

(Continued)

TABLE 22.6 Reactions of Amines Discussed in This Chapter (Continued)

Reaction (section) and comments	General equation and specific example
Electrophilic aromatic substitution (Section 22.14) Arylamines are very reactive toward electrophilic aromatic substitution. It is customary to protect arylamines as their *N*-acyl derivatives before carrying out ring nitration, chlorination, bromination, sulfonation, or Friedel–Crafts reactions.	ArH + E^+ ⟶ ArE + H^+ Arylamine Electrophile Product of electrophilic aromatic substitution Proton 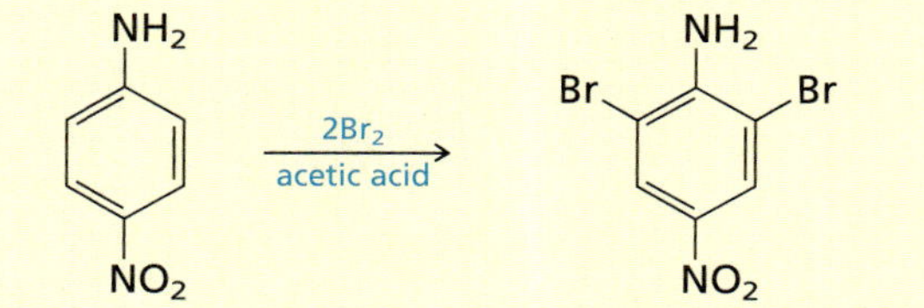 *p*-Nitroaniline $\xrightarrow[\text{acetic acid}]{2Br_2}$ 2,6-Dibromo-4-nitroaniline (95%)
Nitrosation (Section 22.15) Nitrosation of amines occurs when sodium nitrite is added to a solution containing an amine and an acid. *Primary amines* yield alkyl diazonium salts. Alkyl diazonium salts are very unstable and yield carbocation-derived products. Aryl diazonium salts are exceedingly useful synthetic intermediates. Their reactions are described in Table 22.7.	$RNH_2 \xrightarrow[H_3O^+]{NaNO_2} R\overset{+}{N}\equiv N:$ Primary amine Diazonium ion 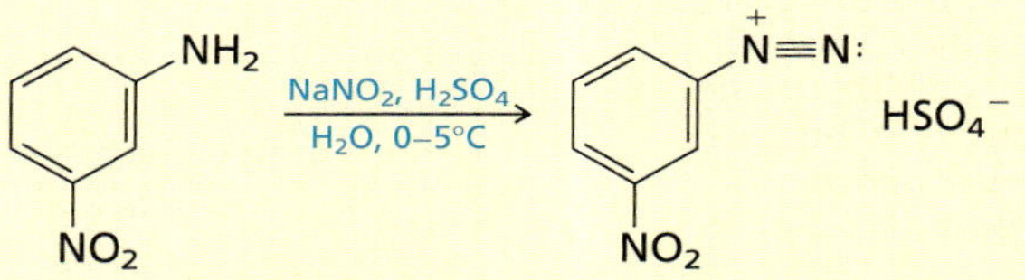 *m*-Nitroaniline $\xrightarrow[\text{H}_2\text{O, 0–5°C}]{NaNO_2,\ H_2SO_4}$ $\overset{+}{N}\equiv N:$ HSO_4^- *m*-Nitrobenzenediazonium hydrogen sulfate
Secondary alkylamines and *secondary arylamines* yield *N*-nitroso amines.	$R_2NH \xrightarrow[H_3O^+]{NaNO_2} R_2N-N{=}O$ Secondary amine *N*-Nitroso amine 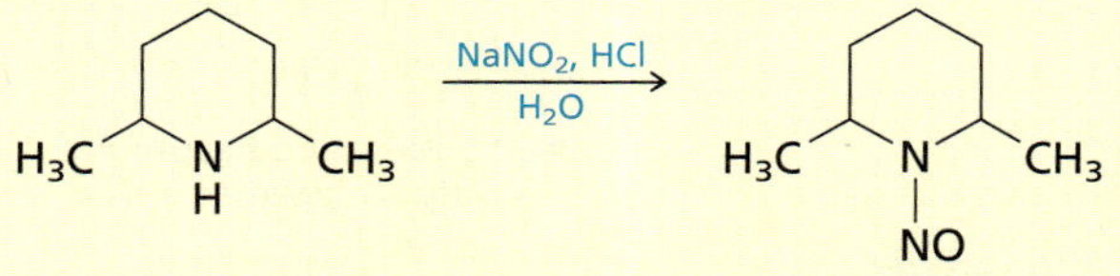 2,6-Dimethylpiperidine $\xrightarrow[\text{H}_2\text{O}]{NaNO_2,\ HCl}$ 2,6-Dimethyl-*N*-nitrosopiperidine (72%)
Tertiary alkylamines illustrate no useful chemistry on nitrosation. *Tertiary arylamines* undergo nitrosation of the ring by electrophilic aromatic substitution.	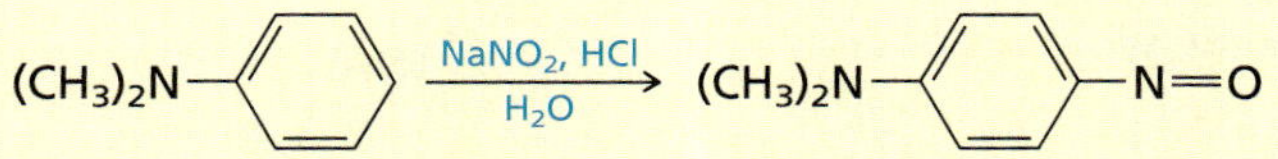 $(CH_3)_2N{-}$⟨ring⟩ $\xrightarrow[\text{H}_2\text{O}]{NaNO_2,\ HCl}$ $(CH_3)_2N{-}$⟨ring⟩$-N{=}O$ *N,N*-Dimethylaniline *N,N*-Dimethyl-4-nitrosoaniline (80–89%)

TABLE 22.7   Synthetically Useful Transformations Involving Aryl Diazonium Ions

Reaction and comments	General equation and specific example

Preparation of phenols Heating its aqueous acidic solution converts a diazonium salt to a phenol. This is the most general method for the synthesis of phenols.

$$ArNH_2 \xrightarrow[\text{2. } H_2O, \text{ heat}]{\text{1. } NaNO_2, H_2SO_4, H_2O} ArOH$$

Primary arylamine → Phenol

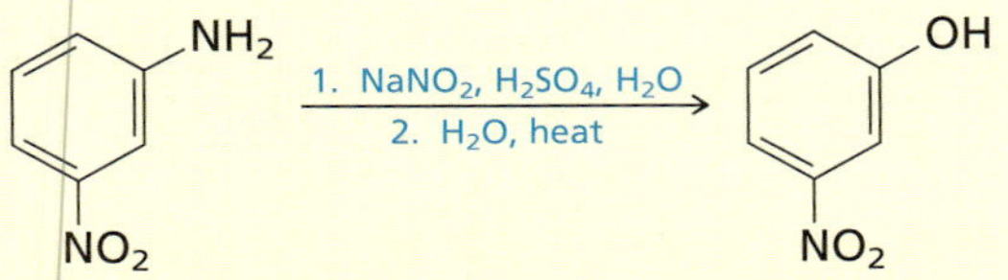

m-Nitroaniline → m-Nitrophenol (81–86%)

1. $NaNO_2$, H_2SO_4, H_2O
2. H_2O, heat

Preparation of aryl fluorides Addition of fluoroboric acid to a solution of a diazonium salt causes the precipitation of an aryl diazonium fluoroborate. When the dry aryl diazonium fluoroborate is heated, an aryl fluoride results. This is the Schiemann reaction; it is the most general method for the preparation of aryl fluorides.

$$ArNH_2 \xrightarrow[\text{2. } HBF_4]{\text{1. } NaNO_2, H_3O^+} Ar\overset{+}{N}\equiv N\text{:} \ \overset{-}{B}F_4 \xrightarrow{\text{heat}} ArF$$

Primary arylamine → Aryl diazonium fluoroborate → Aryl fluoride

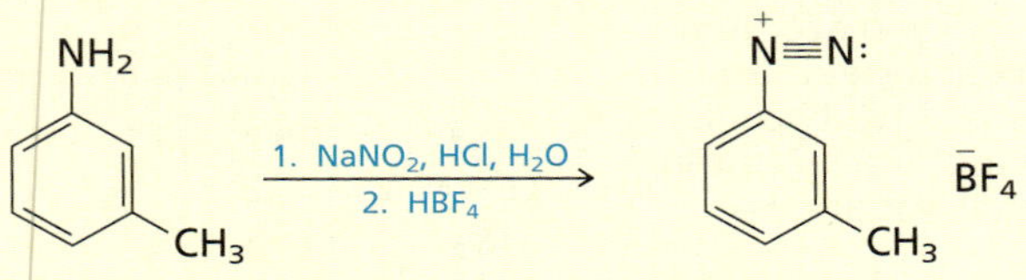

m-Toluidine

1. $NaNO_2$, HCl, H_2O
2. HBF_4

→ m-Methylbenzenediazonium fluoroborate (76–84%)

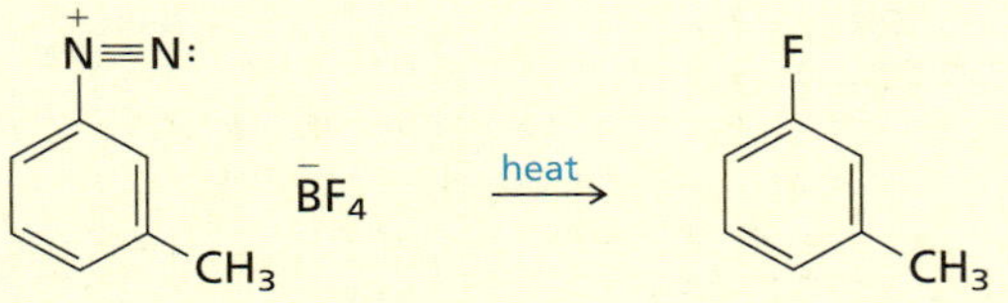

m-Methylbenzenediazonium fluoroborate $\xrightarrow{\text{heat}}$ m-Fluorotoluene (89%)

Preparation of aryl iodides Aryl diazonium salts react with sodium or potassium iodide to form aryl iodides. This is the most general method for the synthesis of aryl iodides.

$$ArNH_2 \xrightarrow[\text{2. NaI or KI}]{\text{1. } NaNO_2, H_3O^+} ArI$$

Primary arylamine → Aryl iodide

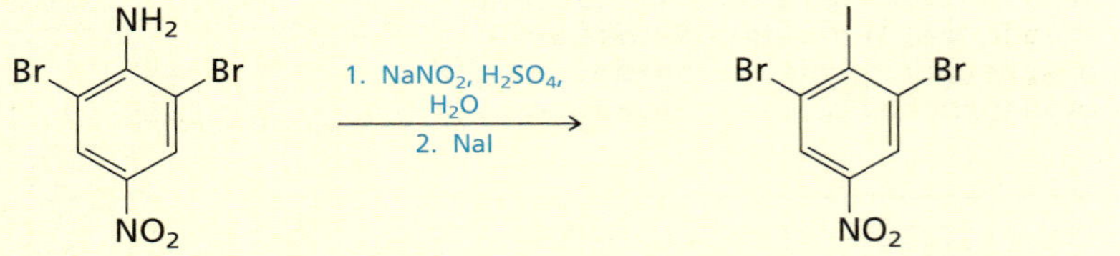

2,6-Dibromo-4-nitroaniline

1. $NaNO_2$, H_2SO_4, H_2O
2. NaI

→ 1,3-Dibromo-2-iodo-5-nitrobenzene (84–88%)

(Continued)

TABLE 22.7   Synthetically Useful Transformations Involving Aryl Diazonium Ions (Continued)

Reaction and comments	General equation and specific example
Preparation of aryl chlorides In the Sandmeyer reaction a solution containing an aryl diazonium salt is treated with copper(I) chloride to give an aryl chloride.	$ArNH_2 \xrightarrow[\text{2. CuCl}]{\text{1. NaNO}_2,\ \text{HCl, H}_2\text{O}} ArCl$ Primary arylamine → Aryl chloride

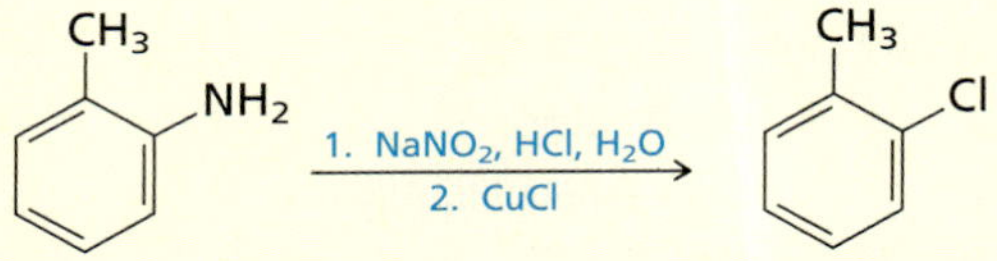

o-Toluidine → o-Chlorotoluene (74–79%)

| **Preparation of aryl bromides** The Sandmeyer reaction using copper(I) bromide is applicable to the conversion of primary arylamines to aryl bromides. | $ArNH_2 \xrightarrow[\text{2. CuBr}]{\text{1. NaNO}_2,\ \text{HBr, H}_2\text{O}} ArBr$
 Primary arylamine → Aryl bromide |

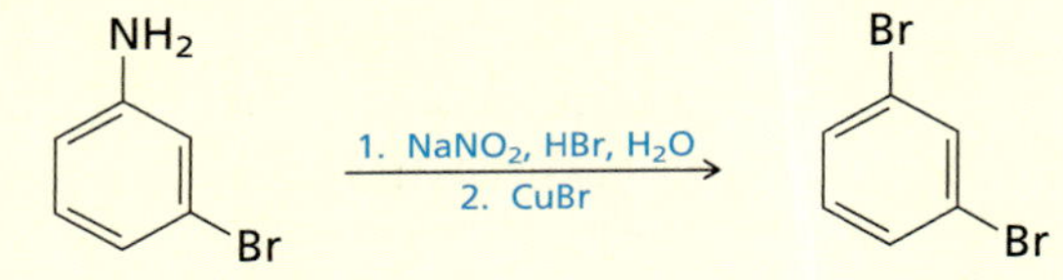

m-Bromoaniline → m-Dibromobenzene (80–87%)

| **Preparation of aryl nitriles** Copper(I) cyanide converts aryl diazonium salts to aryl nitriles. | $ArNH_2 \xrightarrow[\text{2. CuCN}]{\text{1. NaNO}_2,\ \text{H}_2\text{O}} ArCN$
 Primary arylamine → Aryl nitrile |

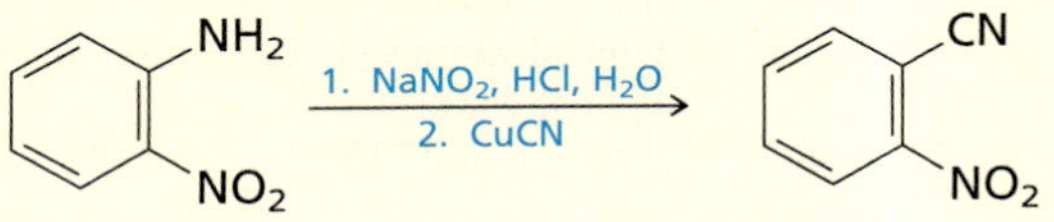

o-Nitroaniline → o-Nitrobenzonitrile (87%)

| **Reductive deamination of primary arylamines** The amino substituent of an arylamine can be replaced by hydrogen by treatment of its derived diazonium salt with ethanol or with hypophosphorous acid. | $ArNH_2 \xrightarrow[\text{2. CH}_3\text{CH}_2\text{OH or H}_3\text{PO}_2]{\text{1. NaNO}_2,\ \text{H}_3\text{O}^+} ArH$
 Primary arylamine → Arene |

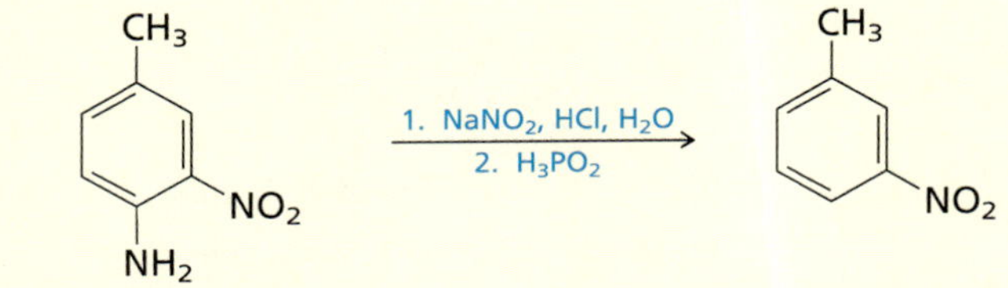

4-Methyl-2-nitroaniline → m-Nitrotoluene (80%)

PROBLEMS

22.21 Write structural formulas or build molecular models for all the amines of molecular formula $C_4H_{11}N$. Give an acceptable name for each one, and classify it as a primary, secondary, or tertiary amine.

22.22 Provide a structural formula for each of the following compounds:

(a) 2-Ethyl-1-butanamine

(b) *N*-Ethyl-1-butanamine

(c) Dibenzylamine

(d) Tribenzylamine

(e) Tetraethylammonium hydroxide

(f) *N*-Allylcyclohexylamine

(g) *N*-Allylpiperidine

(h) Benzyl 2-aminopropanoate

(i) 4-(*N,N*-Dimethylamino)cyclohexanone

(j) 2,2-Dimethyl-1,3-propanediamine

22.23 Many naturally occurring nitrogen compounds and many nitrogen-containing drugs are better known by common names than by their systematic names. A few of these follow. Write a structural formula for each one.

(a) *trans*-2-Phenylcyclopropylamine, better known as *tranylcypromine:* an antidepressant drug

(b) *N*-Benzyl-*N*-methyl-2-propynylamine, better known as *pargyline:* a drug used to treat high blood pressure

(c) 1-Phenyl-2-propanamine, better known as *amphetamine:* a stimulant

(d) 1-(*m*-Hydroxyphenyl)-2-(methylamino)ethanol: better known as *phenylephrine:* a nasal decongestant

22.24 (a) Give the structures or build molecular models and provide an acceptable name for all the isomers of molecular formula C_7H_9N that contain a benzene ring.

(b) Which one of these isomers is the strongest base?

(c) Which, if any, of these isomers yield an *N*-nitroso amine on treatment with sodium nitrite and hydrochloric acid?

(d) Which, if any, of these isomers undergo nitrosation of their benzene ring on treatment with sodium nitrite and hydrochloric acid?

22.25 Arrange the following compounds or anions in each group in order of decreasing basicity:

(a) H_3C^-, H_2N^-, HO^-, F^-

(b) H_2O, NH_3, HO^-, H_2N^-

(c) HO^-, H_2N^-, $:C{\equiv}N:$, NO_3^-

(d)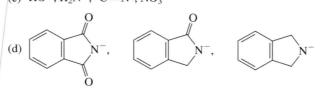

22.26 Arrange the members of each group in order of decreasing basicity:

(a) Ammonia, aniline, methylamine

(b) Acetanilide, aniline, *N*-methylaniline

(c) 2,4-Dichloroaniline, 2,4-dimethylaniline, 2,4-dinitroaniline

(d) 3,4-Dichloroaniline, 4-chloro-2-nitroaniline, 4-chloro-3-nitroaniline

(e) Dimethylamine, diphenylamine, *N*-methylaniline

22.27 *Physostigmine,* an alkaloid obtained from a West African plant, is used in the treatment of glaucoma. Treatment of physostigmine with methyl iodide gives a quaternary ammonium salt. What is the structure of this salt?

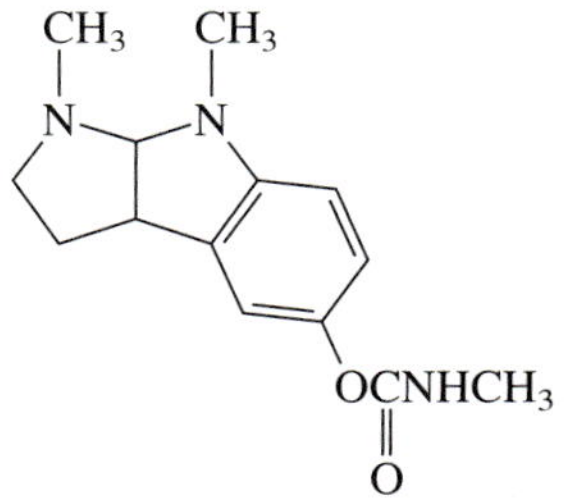

Physostigmine

22.28 Describe procedures for preparing each of the following compounds, using ethanol as the source of all their carbon atoms. Once you prepare a compound, you need not repeat its synthesis in a subsequent part of this problem.

(a) Ethylamine

(b) *N*-Ethylacetamide

(c) Diethylamine

(d) *N,N*-Diethylacetamide

(e) Triethylamine

(f) Tetraethylammonium bromide

22.29 Show by writing the appropriate sequence of equations how you could carry out each of the following transformations:

(a) 1-Butanol to 1-pentanamine

(b) *tert*-Butyl chloride to 2,2-dimethyl-1-propanamine

(c) Cyclohexanol to *N*-methylcyclohexylamine

(d) Isopropyl alcohol to 1-amino-2-methyl-2-propanol

(e) Isopropyl alcohol to 1-amino-2-propanol

(f) Isopropyl alcohol to 1-(*N,N*-dimethylamino)-2-propanol

(g) 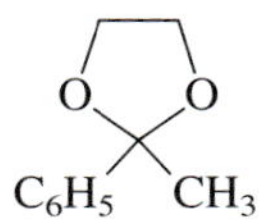 to

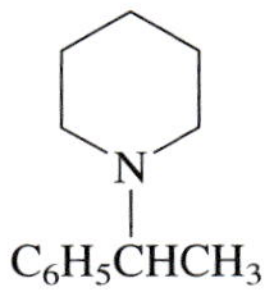

22.30 Each of the following dihaloalkanes gives an *N*-(haloalkyl)phthalimide on reaction with one equivalent of the potassium salt of phthalimide. Write the structure of the phthalimide derivative formed in each case and explain the basis for your answer.

(a) FCH_2CH_2Br

(b) $BrCH_2CH_2CH_2CHCH_3$ with Br substituent

(c) $BrCH_2CCH_2CH_2Br$ with two CH_3 substituents

22.31 Give the structure of the expected product formed when benzylamine reacts with each of the following reagents:

(a) Hydrogen bromide

(b) Sulfuric acid

(c) Acetic acid

(d) Acetyl chloride

(e) Acetic anhydride

(f) Acetone

(g) Acetone and hydrogen (nickel catalyst)

(h) Ethylene oxide

(i) 1,2-Epoxypropane

(j) Excess methyl iodide

(k) Sodium nitrite in dilute hydrochloric acid

22.32 Write the structure of the product formed on reaction of aniline with each of the following:

(a) Hydrogen bromide

(b) Excess methyl iodide

(c) Acetaldehyde

(d) Acetaldehyde and hydrogen (nickel catalyst)

(e) Acetic anhydride

(f) Benzoyl chloride

(g) Sodium nitrite, aqueous sulfuric acid, 0–5°C

(h) Product of part (g), heated in aqueous acid

(i) Product of part (g), treated with copper(I) chloride

(j) Product of part (g), treated with copper(I) bromide

(k) Product of part (g), treated with copper(I) cyanide

(l) Product of part (g), treated with hypophosphorous acid

(m) Product of part (g), treated with potassium iodide

(n) Product of part (g), treated with fluoroboric acid, then heated

(o) Product of part (g), treated with phenol

(p) Product of part (g), treated with N,N-dimethylaniline

22.33 Write the structure of the product formed on reaction of acetanilide with each of the following:

(a) Lithium aluminum hydride

(b) Nitric acid and sulfuric acid

(c) Sulfur trioxide and sulfuric acid

(d) Bromine in acetic acid

(e) tert-Butyl chloride, aluminum chloride

(f) Acetyl chloride, aluminum chloride

(g) 6 M hydrochloric acid, reflux

(h) Aqueous sodium hydroxide, reflux

22.34 Identify the principal organic products of each of the following reactions:

(a) Cyclohexanone + cyclohexylamine $\xrightarrow{H_2, Ni}$

(b) 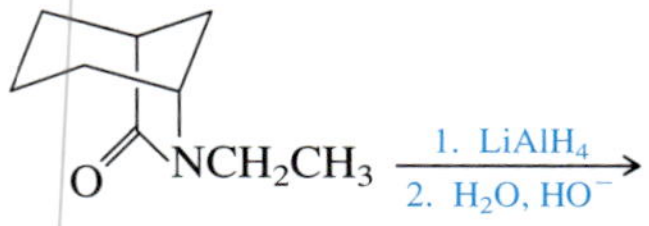 [structure] $\xrightarrow[\text{2. H}_2\text{O, HO}^-]{\text{1. LiAlH}_4}$

(c) $C_6H_5CH_2CH_2CH_2OH \xrightarrow[\text{2. (CH}_3)_2\text{NH (excess)}]{\text{1. } p\text{-toluenesulfonyl chloride, pyridine}}$

(d) $(CH_3)_2CHNH_2 +$ [structure] $\longrightarrow$

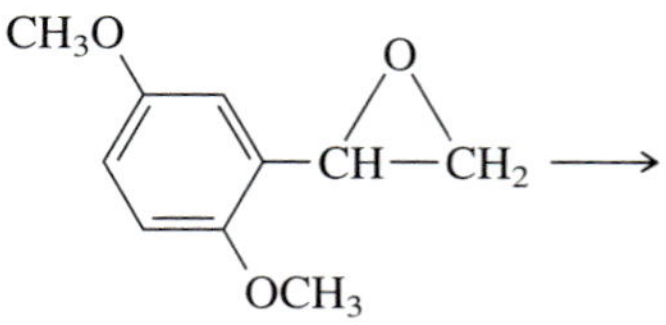

(e) $(C_6H_5CH_2)_2NH + CH_3CCH_2Cl \xrightarrow[\text{THF}]{\text{triethylamine}}$

(f) 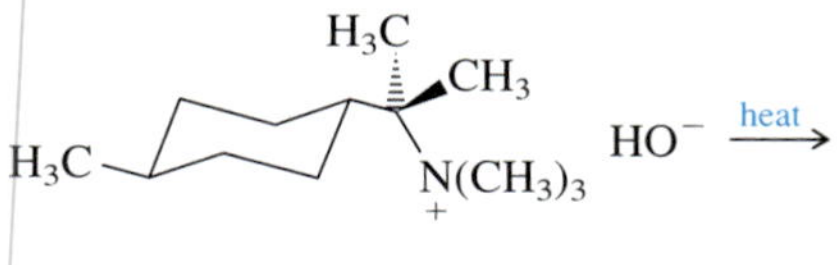 [structure] $\xrightarrow[\text{}]{\text{HO}^- \text{ heat}}$

(g) $(CH_3)_2CHNHCH(CH_3)_2 \xrightarrow[\text{HCl, H}_2\text{O}]{\text{NaNO}_2}$

22.35 Each of the following reactions has been reported in the chemical literature and proceeds in good yield. Identify the principal organic product of each reaction.

(a) 1,2-Diethyl-4-nitrobenzene $\xrightarrow[\text{ethanol}]{\text{H}_2,\text{ Pt}}$

(b) 1,3-Dimethyl-2-nitrobenzene $\xrightarrow[\text{2. HO}^-]{\text{1. SnCl}_2,\text{ HCl}}$

(c) Product of part (b) + $\text{ClCH}_2\overset{\overset{\displaystyle O}{\|}}{\text{C}}\text{Cl} \longrightarrow$

(d) Product of part (c) + $(CH_3CH_2)_2NH \longrightarrow$

(e) Product of part (d) + HCl $\longrightarrow$

(f) $C_6H_5NH\overset{\overset{\displaystyle O}{\|}}{\text{C}}CH_2CH_2CH_3 \xrightarrow[\text{2. H}_2\text{O}]{\text{1. LiAlH}_4}$

(g) Aniline + heptanal $\xrightarrow{\text{H}_2,\text{ Ni}}$

(h) Acetanilide + $\text{ClCH}_2\overset{\overset{\displaystyle O}{\|}}{\text{C}}\text{Cl} \xrightarrow{\text{AlCl}_3}$

(i) 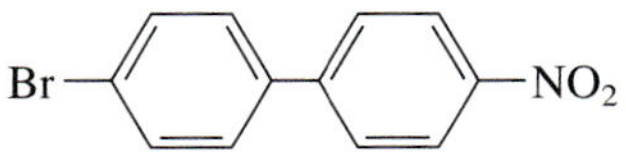 Br—⟨benzene⟩—⟨benzene⟩—$\text{NO}_2 \xrightarrow[\text{2. HO}^-]{\text{1. Fe, HCl}}$

(j) Product of part (i) $\xrightarrow[\text{2. H}_2\text{O, heat}]{\text{1. NaNO}_2,\text{ H}_2\text{SO}_4,\text{ H}_2\text{O}}$

(k) 2,6-Dinitroaniline $\xrightarrow[\text{2. CuCl}]{\text{1. NaNO}_2,\text{ H}_2\text{SO}_4,\text{ H}_2\text{O}}$

(l) *m*-Bromoaniline $\xrightarrow[\text{2. CuBr}]{\text{1. NaNO}_2,\text{ HBr, H}_2\text{O}}$

(m) *o*-Nitroaniline $\xrightarrow[\text{2. CuCN}]{\text{1. NaNO}_2,\text{ HCl, H}_2\text{O}}$

(n) 2,6-Diiodo-4-nitroaniline $\xrightarrow[\text{2. KI}]{\text{1. NaNO}_2,\text{ H}_2\text{SO}_4,\text{ H}_2\text{O}}$

(o) 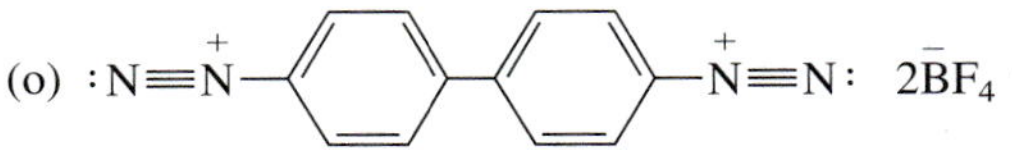 $:N\equiv\overset{+}{N}$—⟨benzene⟩—⟨benzene⟩—$\overset{+}{N}\equiv N: \quad 2\bar{B}F_4 \xrightarrow{\text{heat}}$

(p) 2,4,6-Trinitroaniline $\xrightarrow[\text{H}_2\text{O, H}_3\text{PO}_2]{\text{NaNO}_2,\text{ H}_2\text{SO}_4}$

(q) 2-Amino-5-iodobenzoic acid $\xrightarrow[\text{2. CH}_3\text{CH}_2\text{OH}]{\text{1. NaNO}_2,\text{ HCl, H}_2\text{O}}$

(r) Aniline $\xrightarrow[\text{2. 2,3,6-trimethylphenol}]{\text{1. NaNO}_2,\text{ H}_2\text{SO}_4,\text{ H}_2\text{O}}$

(s) 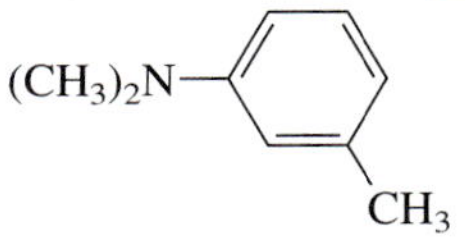 $(CH_3)_2N$—⟨benzene with CH_3⟩ $\xrightarrow[\text{2. HO}^-]{\text{1. NaNO}_2,\text{ HCl, H}_2\text{O}}$

22.36 Provide a reasonable explanation for each of the following observations:

(a) 4-Methylpiperidine has a higher boiling point than *N*-methylpiperidine.

$$HN\langle\rangle-CH_3 \qquad CH_3N\langle\rangle$$

4-Methylpiperidine   *N*-Methylpiperidine
(bp 129°C)   (bp 106°C)

(b) Two isomeric quaternary ammonium salts are formed in comparable amounts when 4-*tert*-butyl-*N*-methylpiperidine is treated with benzyl chloride. (*Hint:* Building a molecular model will help.)

$$CH_3N\langle\rangle-C(CH_3)_3$$

4-*tert*-Butyl-*N*-methylpiperidine

(c) When tetramethylammonium hydroxide is heated at 130°C, trimethylamine and methanol are formed.

(d) The major product formed on treatment of 1-propanamine with sodium nitrite in dilute hydrochloric acid is 2-propanol.

22.37 Give the structures, including stereochemistry, of compounds A through C.

$$(S)\text{-2-Octanol} + H_3C-\langle\rangle-SO_2Cl \xrightarrow{\text{pyridine}} \text{Compound A}$$

$$\downarrow \begin{array}{l} \text{NaN}_3, \\ \text{methanol–water} \end{array}$$

$$\text{Compound C} \xleftarrow[\text{2. H}_2\text{O}]{\text{1. LiAlH}_4} \text{Compound B}$$

22.38 Devise efficient syntheses of each of the following compounds from the designated starting materials. You may also use any necessary organic or inorganic reagents.

(a) 3,3-Dimethyl-1-butanamine from 1-bromo-2,2-dimethylpropane

(b) $H_2C{=}CH(CH_2)_8CH_2{-}N\langle\rangle$ from 10-undecenoic acid and pyrrolidine

(c) $\langle\rangle NH_2$ from $\langle\rangle$
     C₆H₅O   C₆H₅O   OH

(d) $C_6H_5CH_2NCH_2CH_2CH_2CH_2NH_2$ from $C_6H_5CH_2NHCH_3$ and $BrCH_2CH_2CH_2CN$
         |
         CH₃

(e) $NC-\langle\rangle-CH_2N(CH_3)_2$ from $NC-\langle\rangle-CH_3$

22.39 Each of the following compounds has been prepared from *p*-nitroaniline. Outline a reasonable series of steps leading to each one.

(a) *p*-Nitrobenzonitrile   (d) 3,5-Dibromoaniline

(b) 3,4,5-Trichloroaniline   (e) *p*-Acetamidophenol (*acetaminophen*)

(c) 1,3-Dibromo-5-nitrobenzene

22.40 Each of the following compounds has been prepared from *o*-anisidine (*o*-methoxyaniline). Outline a series of steps leading to each one.

(a) *o*-Bromoanisole

(b) *o*-Fluoroanisole

(c) 3-Fluoro-4-methoxyacetophenone

(d) 3-Fluoro-4-methoxybenzonitrile

(e) 3-Fluoro-4-methoxyphenol

22.41 Design syntheses of each of the following compounds from the indicated starting material and any necessary organic or inorganic reagents:

(a) *p*-Aminobenzoic acid from *p*-methylaniline

(b) $p\text{-}FC_6H_4\overset{\displaystyle O}{\overset{\displaystyle \|}{C}}CH_2CH_3$ from benzene

(c) 1-Bromo-2-fluoro-3,5-dimethylbenzene from *m*-xylene

(d)

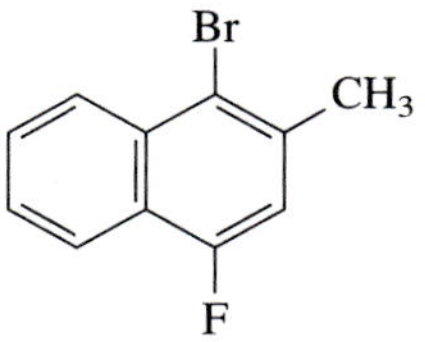

from

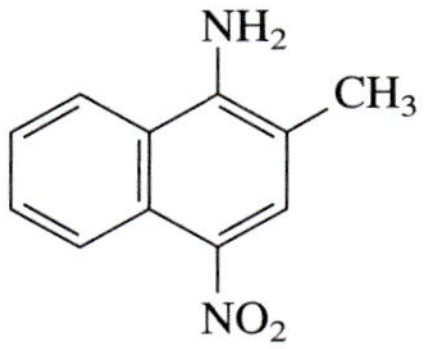

(e) $o\text{-}BrC_6H_4C(CH_3)_3$ from $p\text{-}O_2NC_6H_4C(CH_3)_3$

(f) $m\text{-}ClC_6H_4C(CH_3)_3$ from $p\text{-}O_2NC_6H_4C(CH_3)_3$

(g) 1-Bromo-3,5-diethylbenzene from *m*-diethylbenzene

(h)

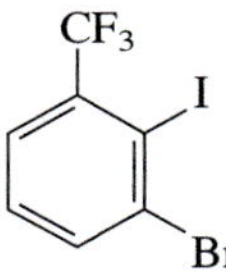

from

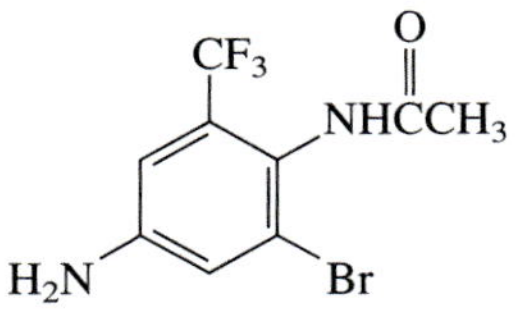

(i)

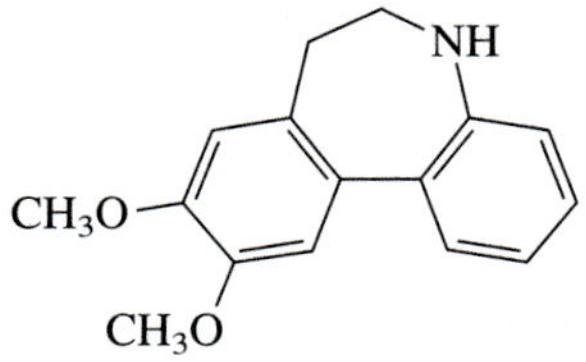

from

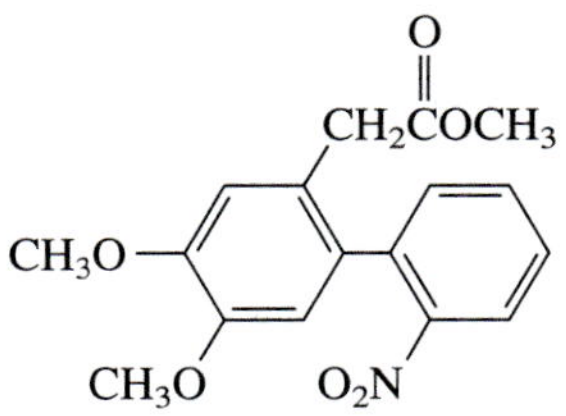

22.42 Ammonia and amines undergo conjugate addition to α,β-unsaturated carbonyl compounds (Section 18.12). On the basis of this information, predict the principal organic product of each of the following reactions:

(a) $(CH_3)_2C{=}CH\overset{\displaystyle O}{\overset{\displaystyle \|}{C}}CH_3 \ + \ NH_3 \longrightarrow$

(b)

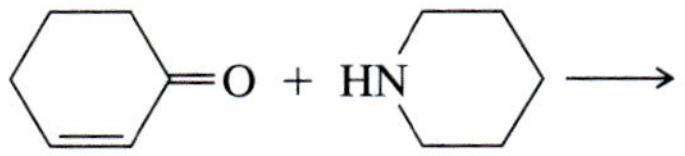

(c) 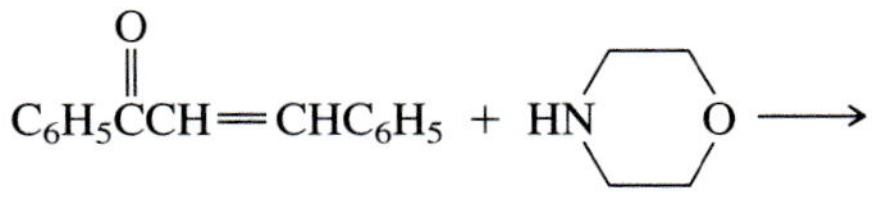 $C_6H_5\overset{\displaystyle O}{\overset{\displaystyle \|}{C}}CH{=}CHC_6H_5 \ + \ HN\overset{\frown}{\underset{\smile}{}}O \longrightarrow$

(d)

$$\xrightarrow{\text{spontaneous}} \text{C}_{15}\text{H}_{27}\text{NO}$$

(CH$_2$)$_3$CH(CH$_2$)$_4$CH$_3$

NH$_2$

22.43 A number of compounds of the type represented by compound A were prepared for evaluation as potential analgesic drugs. Their preparation is described in a retrosynthetic format as shown.

R′

R′

O

N

N

N

R

R

R

Compound A

RNH$_2$ + H$_2$C=CHCO$_2$CH$_2$CH$_3$ ⟸ RN(CH$_2$CH$_2$CO$_2$CH$_2$CH$_3$)$_2$

On the basis of this retrosynthetic analysis, design a synthesis of *N*-methyl-4-phenylpiperidine (compound A, where R = CH$_3$, R′ = C$_6$H$_5$). Present your answer as a series of equations, showing all necessary reagents and isolated intermediates.

22.44 *Mescaline,* a hallucinogenic amine obtained from the peyote cactus, has been synthesized in two steps from 3,4,5-trimethoxybenzyl bromide. The first step is nucleophilic substitution by sodium cyanide. The second step is a lithium aluminum hydride reduction. What is the structure of mescaline?

22.45 *Methamphetamine* is a notorious street drug. One synthesis involves reductive amination of benzyl methyl ketone with methylamine. What is the structure of methamphetamine?

 22.46 *N,N*-Dimethylaniline and pyridine are similar in basicity, whereas 4-(*N,N*-dimethylamino)-pyridine is considerably more basic than either.

N(CH$_3$)$_2$

N(CH$_3$)$_2$

N

N

N,N-Dimethylaniline
pK_a of conjugate
acid = 5.1

Pyridine
pK_a of conjugate
acid = 5.3

4-(*N,N*-Dimethylamino)pyridine
pK_a of conjugate
acid = 9.7

Identify the more basic of the two nitrogens of 4-(*N,N*-dimethylamino)pyridine, and suggest an explanation for its enhanced basicity as compared with pyridine and *N,N*-dimethylaniline. Refer to *Learning By Modeling* and compare your prediction to one based on the calculated charge and electrostatic potential of each nitrogen.

22.47 Compounds A and B are isomeric amines of molecular formula C$_8$H$_{11}$N. Identify each isomer on the basis of the ^{1}H NMR spectra given in Figure 22.9.

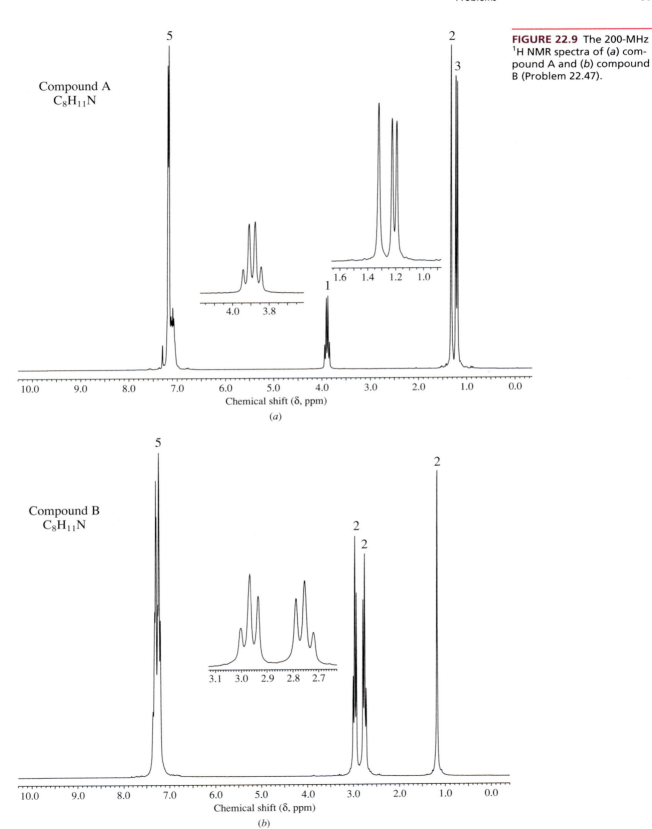

FIGURE 22.9 The 200-MHz ^{1}H NMR spectra of (a) compound A and (b) compound B (Problem 22.47).

Compound A
$C_8H_{11}N$

5

2

3

1

1.6   1.4   1.2   1.0

4.0   3.8

10.0   9.0   8.0   7.0   6.0   5.0   4.0   3.0   2.0   1.0   0.0
Chemical shift (δ, ppm)

(a)

Compound B
$C_8H_{11}N$

5

2

2

2

3.1   3.0   2.9   2.8   2.7

10.0   9.0   8.0   7.0   6.0   5.0   4.0   3.0   2.0   1.0   0.0
Chemical shift (δ, ppm)

(b)

22.48 The compound shown is a somewhat stronger base than ammonia. Which nitrogen do you think is protonated when it is treated with an acid? Write a structural formula for the species that results.

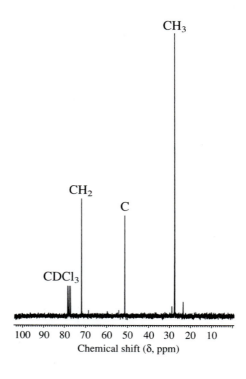

5-Methyl-γ-carboline
pK_a of conjugate acid = 10.5

Refer to *Learning By Modeling*, and compare your prediction to one based on the calculated charge and electrostatic potential of each nitrogen.

22.49 Does the ^{13}C NMR spectrum shown in Figure 22.10 correspond to that of 1-amino-2-methyl-2-propanol or to 2-amino-2-methyl-1-propanol? Could this compound be prepared by reaction of an epoxide with ammonia?

FIGURE 22.10 The ^{13}C NMR spectrum of the compound described in Problem 22.49.

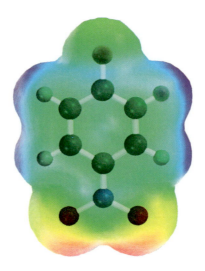

CHAPTER 23

ARYL HALIDES

The value of *alkyl halides* as starting materials for the preparation of a variety of organic functional groups has been stressed many times. In our earlier discussions, we noted that *aryl halides* are normally much less reactive than alkyl halides in reactions that involve carbon–halogen bond cleavage. In the present chapter you will see that aryl halides can exhibit their own patterns of chemical reactivity, and that these reactions are novel, useful, and mechanistically interesting.

23.1 BONDING IN ARYL HALIDES

Aryl halides are compounds in which a halogen substituent is attached directly to an aromatic ring. Representative aryl halides include

| Fluorobenzene | 1-Chloro-2-nitrobenzene | 1-Bromonaphthalene | p-Iodobenzyl alcohol |

Halogen-containing organic compounds in which the halogen is not directly bonded to an aromatic ring, even though an aromatic ring may be present, are not aryl halides. Benzyl chloride ($C_6H_5CH_2Cl$), for example, is not an aryl halide.

The carbon–halogen bonds of aryl halides are both shorter and stronger than the carbon–halogen bonds of alkyl halides. In this respect, as well as in their chemical behavior, they resemble vinyl halides more than alkyl halides. A hybridization effect seems to be responsible because, as the data in Table 23.1 indicate, similar patterns are seen for both carbon–hydrogen bonds and carbon–halogen bonds. An increase in *s*

TABLE 23.1	Carbon–Hydrogen and Carbon–Chlorine Bond Dissociation Energies of Selected Compounds		

Compound	Hybridization of carbon to which X is attached	Bond energy, kJ/mol (kcal/mol)	
		X = H	X = Cl
CH_3CH_2X	sp^3	410 (98)	339 (81)
$H_2C=CHX$	sp^2	452 (108)	368 (88)
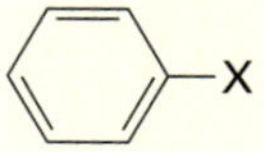 ⬡—X	sp^2	469 (112)	406 (97)

character of carbon from 25% (sp^3 hybridization) to 33.3% s character (sp^2 hybridization) increases its tendency to attract electrons and strengthens the bond.

> **PROBLEM 23.1** Consider all the isomers of C_7H_7Cl containing a benzene ring, and write the structure of the one that has the weakest carbon–chlorine bond as measured by its bond dissociation energy.

The strength of their carbon–halogen bonds causes aryl halides to react very slowly in reactions in which carbon–halogen bond cleavage is rate-determining, as in nucleophilic substitution, for example. Later in this chapter we will see examples of such reactions that do take place at reasonable rates but proceed by mechanisms distinctly different from the classical S_N1 and S_N2 pathways.

23.2 SOURCES OF ARYL HALIDES

The two main methods for the preparation of aryl halides, halogenation of arenes by electrophilic aromatic substitution and preparation by way of aryl diazonium salts, were described earlier and are reviewed in Table 23.2. A number of aryl halides occur naturally, some of which are shown in Figure 23.1.

23.3 PHYSICAL PROPERTIES OF ARYL HALIDES

Melting points and boiling points for some representative aryl halides are listed in Appendix 1.

Aryl halides resemble alkyl halides in many of their physical properties. All are practically insoluble in water and most are denser than water.

Aryl halides are polar molecules but are less polar than alkyl halides.

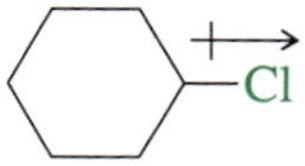

Compare the electronic charges at chlorine in chlorocyclohexane and chlorobenzene on *Learning By Modeling* to verify that the C—Cl bond is more polar in chlorocyclohexane.

Chlorocyclohexane
μ 2.2 D

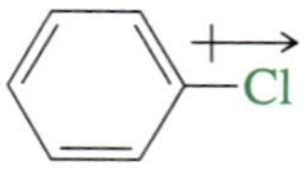

Chlorobenzene
μ 1.7 D

Because carbon is sp^2-hybridized in chlorobenzene, it is more electronegative than the sp^3-hybridized carbon of chlorocyclohexane. Consequently, the withdrawal of electron density away from carbon by chlorine is less pronounced in aryl halides than in alkyl halides, and the molecular dipole moment is smaller.

TABLE 23.2	Summary of Reactions Discussed in Earlier Chapters That Yield Aryl Halides

Reaction (section) and comments	General equation and specific example
Halogenation of arenes (Section 12.5) Aryl chlorides and bromides are conveniently prepared by electrophilic aromatic substitution. The reaction is limited to chlorination and bromination. Fluorination is difficult to control; iodination is too slow to be useful.	ArH + X_2 $\xrightarrow[\text{or FeX}_3]{\text{Fe}}$ ArX + HX Arene Halogen Aryl Hydrogen halide halide $O_2N-\!\!\bigcirc$ + Br_2 $\xrightarrow{\text{Fe}}$ $O_2N-\!\!\bigcirc\!\!-Br$ Nitrobenzene Bromine *m*-Bromonitrobenzene (85%)
The Sandmeyer reaction (Section 22.17) Diazotization of a primary arylamine followed by treatment of the diazonium salt with copper(I) bromide or copper(I) chloride yields the corresponding aryl bromide or aryl chloride.	ArNH_2 $\xrightarrow[\text{2. CuX}]{\text{1. NaNO}_2,\ \text{H}_3\text{O}^+}$ ArX Primary arylamine Aryl halide 1-Amino-8-chloronaphthalene $\xrightarrow[\text{2. CuBr}]{\text{1. NaNO}_2,\ \text{HBr}}$ 1-Bromo-8-chloronaphthalene (62%)
The Schiemann reaction (Section 22.17) Diazotization of an arylamine followed by treatment with fluoroboric acid gives an aryl diazonium fluoroborate salt. Heating this salt converts it to an aryl fluoride.	ArNH_2 $\xrightarrow[\text{2. HBF}_4]{\text{1. NaNO}_2,\ \text{H}_3\text{O}^+}$ $\text{Ar}\overset{+}{\text{N}}\!\!\equiv\!\!\text{N}\!: \ \bar{\text{B}}\text{F}_4$ $\xrightarrow{\text{heat}}$ ArF Primary Aryl diazonium Aryl arylamine fluoroborate fluoride $C_6H_5NH_2$ $\xrightarrow[\substack{\text{2. HBF}_4 \\ \text{3. heat}}]{\text{1. NaNO}_2,\ \text{H}_2\text{O, HCl}}$ C_6H_5F Aniline Fluorobenzene (51–57%)
Reaction of aryl diazonium salts with iodide ion (Section 22.17) Adding potassium iodide to a solution of an aryl diazonium ion leads to the formation of an aryl iodide.	ArNH_2 $\xrightarrow[\text{2. KI}]{\text{1. NaNO}_2,\ \text{H}_3\text{O}^+}$ ArI Primary arylamine Aryl iodide $C_6H_5NH_2$ $\xrightarrow[\text{2. KI}]{\text{1. NaNO}_2,\ \text{HCl, H}_2\text{O}}$ C_6H_5I Aniline Iodobenzene (74–76%)

23.4   REACTIONS OF ARYL HALIDES: A REVIEW AND A PREVIEW

Table 23.3 summarizes the reactions of aryl halides that we have encountered to this point.

Noticeably absent from Table 23.3 are nucleophilic substitutions. We have, so far, seen no nucleophilic substitution reactions of aryl halides in this text. Chlorobenzene, for example, is essentially inert to aqueous sodium hydroxide at room temperature. Reaction temperatures over 300°C are required for nucleophilic substitution to proceed at a reasonable rate.

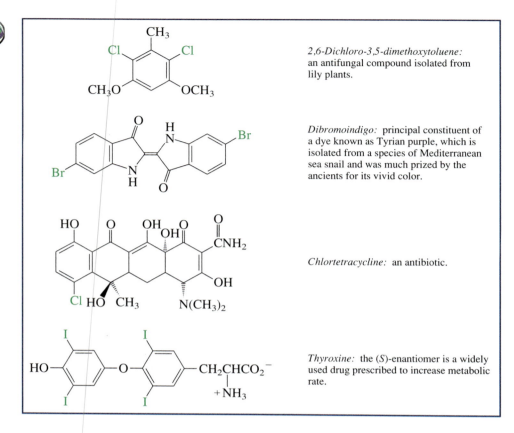

2,6-Dichloro-3,5-dimethoxytoluene: an antifungal compound isolated from lily plants.

Dibromoindigo: principal constituent of a dye known as Tyrian purple, which is isolated from a species of Mediterranean sea snail and was much prized by the ancients for its vivid color.

Chlortetracycline: an antibiotic.

Thyroxine: the (S)-enantiomer is a widely used drug prescribed to increase metabolic rate.

TABLE 23.3   Summary of Reactions of Aryl Halides Discussed in Earlier Chapters

Reaction (section) and comments	General equation and specific example
Electrophilic aromatic substitution (Section 12.14) Halogen substituents are slightly deactivating and ortho, para-directing.	Br—⬡ $\xrightarrow[\text{AlCl}_3]{\text{CH}_3\text{COCCH}_3}$ Br—⬡—$\overset{\text{O}}{\overset{\|}{\text{C}}}$CH₃ Bromobenzene p-Bromoacetophenone (69–79%)
Formation of aryl Grignard reagents (Section 14.4) Aryl halides react with magnesium to form the corresponding arylmagnesium halide. Aryl iodides are the most reactive, aryl fluorides the least. A similar reaction occurs with lithium to give aryllithium reagents (Section 14.3).	ArX + Mg $\xrightarrow{\text{diethyl ether}}$ ArMgX Aryl halide Magnesium Arylmagnesium halide ⬡—Br + Mg $\xrightarrow{\text{diethyl ether}}$ ⬡—MgBr Bromobenzene Magnesium Phenylmagnesium bromide (95%)

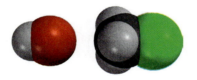

Chlorobenzene   Phenol (97%)

The mechanism of this reaction is discussed in Section 23.8.

Aryl halides are much less reactive than alkyl halides in nucleophilic substitution reactions. The carbon–halogen bonds of aryl halides are too strong, and aryl cations are too high in energy, to permit aryl halides to ionize readily in S_N1-type processes. Furthermore, as Figure 23.2 depicts, the optimal transition-state geometry required for S_N2 processes cannot be achieved. Nucleophilic attack from the side opposite the carbon–halogen bond is blocked by the aromatic ring.

23.5 NUCLEOPHILIC SUBSTITUTION IN NITRO-SUBSTITUTED ARYL HALIDES

One group of aryl halides that do undergo nucleophilic substitution readily consists of those that bear a nitro group ortho or para to the halogen.

p-Chloronitrobenzene   Sodium methoxide   p-Nitroanisole   Sodium chloride
                                                     (92%)

(a) Hydroxide ion + chloromethane

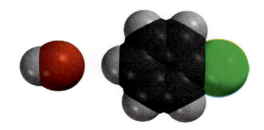

(b) Hydroxide ion + chlorobenzene

FIGURE 23.2 Nucleophilic substitution, with inversion of configuration, is blocked by the benzene ring of an aryl halide. (a) *Alkyl halide:* The new bond is formed by attack of the nucleophile at carbon from the side opposite the bond to the leaving group. Inversion of configuration is observed. (b) *Aryl halide:* The aromatic ring blocks the approach of the nucleophile to carbon at the side opposite the bond to the leaving group. Inversion of configuration is impossible.

An *ortho*-nitro group exerts a comparable rate-enhancing effect. *m*-Chloronitrobenzene, although much more reactive than chlorobenzene itself, is thousands of times less reactive than either *o*- or *p*-chloronitrobenzene.

The effect of *o*- and *p*-nitro substituents is cumulative, as the following rate data demonstrate:

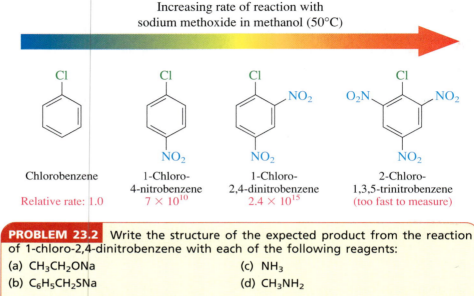

Increasing rate of reaction with
sodium methoxide in methanol (50°C)

Chlorobenzene — 1-Chloro-4-nitrobenzene — 1-Chloro-2,4-dinitrobenzene — 2-Chloro-1,3,5-trinitrobenzene
Relative rate: 1.0 — 7×10^{10} — 2.4×10^{15} — (too fast to measure)

PROBLEM 23.2 Write the structure of the expected product from the reaction of 1-chloro-2,4-dinitrobenzene with each of the following reagents:
(a) CH₃CH₂ONa
(b) C₆H₅CH₂SNa
(c) NH₃
(d) CH₃NH₂

SAMPLE SOLUTION (a) Sodium ethoxide is a source of the nucleophile $CH_3CH_2O^-$, which displaces chloride from 1-chloro-2,4-dinitrobenzene.

1-Chloro-2,4-dinitrobenzene + $CH_3CH_2O^-$ ⟶ 1-Ethoxy-2,4-dinitrobenzene + Cl^-
Ethoxide anion

In contrast to nucleophilic substitution in alkyl halides, where *alkyl fluorides* are exceedingly unreactive, *aryl fluorides* undergo nucleophilic substitution readily when the ring bears an *o*- or a *p*-nitro group.

The compound 1-fluoro-2,4-dinitrobenzene is exceedingly reactive toward nucleophilic aromatic substitution and was used in an imaginative way by Frederick Sanger (Section 27.11) in his determination of the structure of insulin.

p-Fluoronitrobenzene + KOCH₃ →(CH₃OH, 85°C) *p*-Nitroanisole (93%) + KF
Potassium methoxide — Potassium fluoride

Indeed, the order of leaving-group reactivity in nucleophilic aromatic substitution is the opposite of that seen in aliphatic substitution. *Fluoride is the most reactive leaving group in nucleophilic aromatic substitution, iodide the least reactive.*

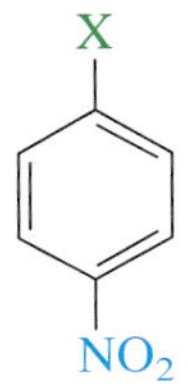

Relative reactivity
toward sodium
methoxide
in methanol (50°C):

X = F	312
X = Cl	1.0
X = Br	0.8
X = I	0.4

Kinetic studies of these reactions reveal that they follow a second-order rate law:

$$\text{Rate} = k[\text{Aryl halide}][\text{Nucleophile}]$$

Second-order kinetics is usually interpreted in terms of a bimolecular rate-determining step. In this case, then, we look for a mechanism in which both the aryl halide and the nucleophile are involved in the slowest step. Such a mechanism is described in the following section.

23.6 THE ADDITION–ELIMINATION MECHANISM OF NUCLEOPHILIC AROMATIC SUBSTITUTION

The generally accepted mechanism for nucleophilic aromatic substitution in nitro-substituted aryl halides, illustrated for the reaction of *p*-fluoronitrobenzene with sodium methoxide, is outlined in Figure 23.3. It is a two-step **addition–elimination mechanism,** in which addition of the nucleophile to the aryl halide is followed by elimination of the halide leaving group. Figure 23.4 shows the structure of the key intermediate. The mechanism is consistent with the following experimental observations:

This mechanism is sometimes called S_NAr (*substitution-nucleophilic-aromatic*).

1. *Kinetics:* As the observation of second-order kinetics requires, the rate-determining step (step 1) involves both the aryl halide and the nucleophile.

2. *Rate-enhancing effect of the nitro group:* The nucleophilic addition step is rate-determining because the aromatic character of the ring must be sacrificed to form the cyclohexadienyl anion intermediate. Only when the anionic intermediate is stabilized by the presence of a strong electron-withdrawing substituent ortho or para to the leaving group will the activation energy for its formation be low enough to provide a reasonable reaction rate. We can illustrate the stabilization that a *p*-nitro group provides by examining the resonance structures for the cyclohexadienyl anion formed from methoxide and *p*-fluoronitrobenzene:

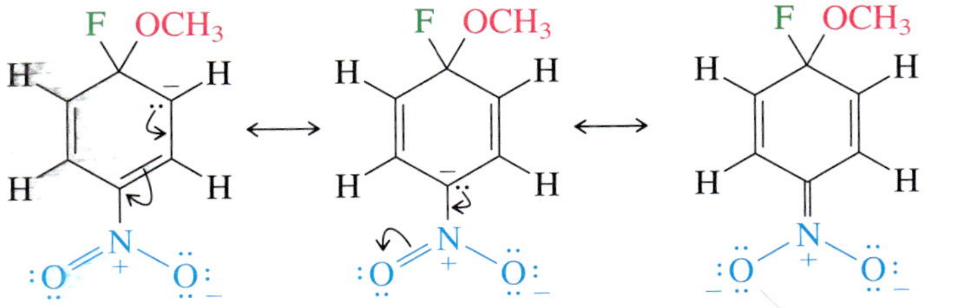

Most stable resonance
structure; negative
charge is on oxygen

Overall reaction:

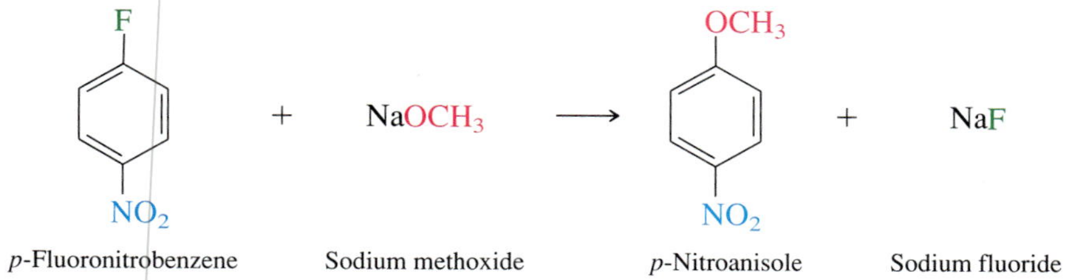

p-Fluoronitrobenzene   Sodium methoxide   p-Nitroanisole   Sodium fluoride

Step 1: Addition stage. The nucleophile, in this case methoxide ion, adds to the carbon atom that bears the leaving group to give a cyclohexadienyl anion intermediate.

p-Fluoronitrobenzene   Methoxide ion   Cyclohexadienyl anion intermediate

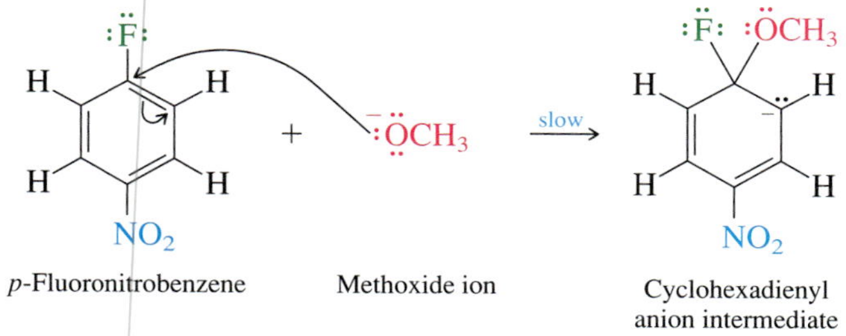

Step 2: Elimination stage. Loss of halide from the cyclohexadienyl intermediate restores the aromaticity of the ring and gives the product of nucleophilic aromatic substitution.

Cyclohexadienyl anion intermediate   p-Nitroanisole   Fluoride ion

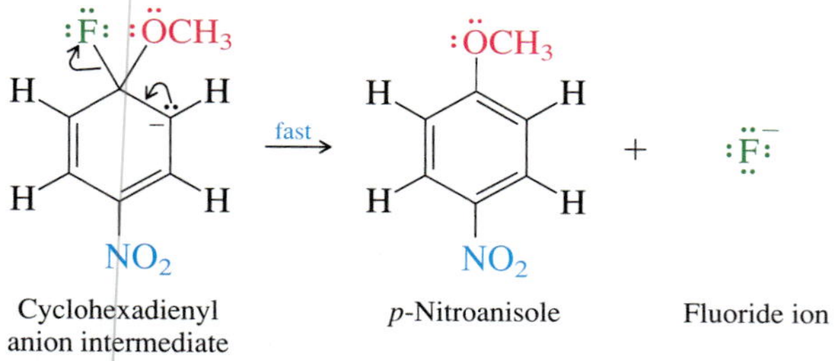

FIGURE 23.3 The addition–elimination mechanism of nucleophilic aromatic substitution.

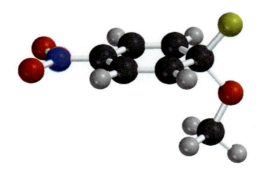

FIGURE 23.4 Structure of the rate-determining intermediate in the reaction of p-fluoronitrobenzene with methoxide ion.

PROBLEM 23.3 Write the most stable resonance structure for the cyclohexa-dienyl anion formed by reaction of methoxide ion with o-fluoronitrobenzene.

m-Fluoronitrobenzene reacts with sodium methoxide 10^5 times more slowly than its ortho and para isomers. According to the resonance description, direct conjugation of the negatively charged carbon with the nitro group is not possible in the cyclohexadienyl anion intermediate from *m*-fluoronitrobenzene, and the decreased reaction rate reflects the decreased stabilization afforded this intermediate.

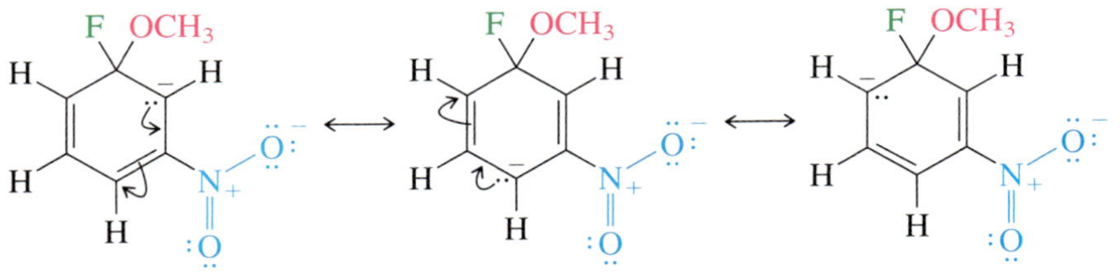

(Negative charge is restricted to carbon in all resonance forms)

PROBLEM 23.4 Reaction of 1,2,3-tribromo-5-nitrobenzene with sodium ethox-ide in ethanol gave a single product, $C_8H_7Br_2NO_3$, in quantitative yield. Suggest a reasonable structure for this compound.

3. *Leaving-group effects:* Because aryl fluorides have the strongest carbon–halogen bond and react fastest, the rate-determining step cannot involve carbon–halogen bond cleavage. According to the mechanism in Figure 23.3 the carbon–halogen bond breaks in the rapid elimination step that follows the rate-determining addition step. The unusually high reactivity of aryl fluorides arises because fluorine is the most electronegative of the halogens, and its greater ability to attract electrons increases the rate of formation of the cyclohexadienyl anion intermediate in the first step of the mechanism.

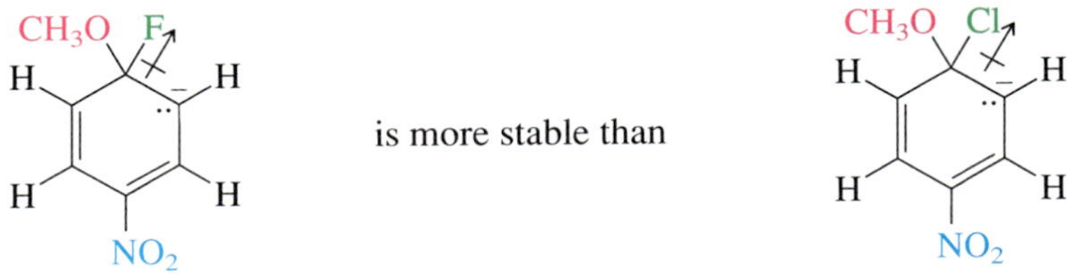

is more stable than

Fluorine stabilizes cyclohexadienyl anion by withdrawing electrons.

Chlorine is less electronegative than fluorine and does not stabilize cyclohexadienyl anion to as great an extent.

Before leaving this mechanistic discussion, we should mention that the addition–elimination mechanism for nucleophilic aromatic substitution illustrates a prin-ciple worth remembering. The words *activating* and *deactivating* as applied to substituent effects in organic chemistry are without meaning when they stand alone. When we say that a group is activating or deactivating, we need to specify the reaction type that is being considered. A nitro group is a strongly *deactivating* substituent in *electrophilic* aromatic substitution, where it markedly destabilizes the key cyclohexadienyl cation intermediate:

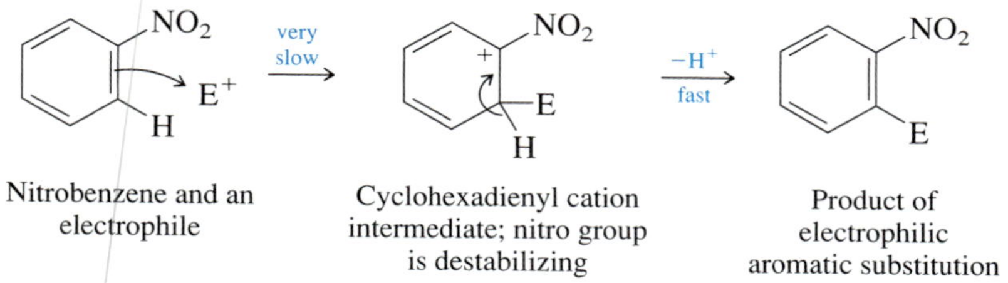

Nitrobenzene and an
electrophile

Cyclohexadienyl cation
intermediate; nitro group
is destabilizing

Product of
electrophilic
aromatic substitution

A nitro group is a strongly *activating* substituent in *nucleophilic* aromatic substitution, where it stabilizes the key cyclohexadienyl anion intermediate:

o-Halonitrobenzene
(X = F, Cl, Br, or I)
and a nucleophile

Cyclohexadienyl anion
intermediate; nitro group
is stabilizing

Product of
nucleophilic
aromatic substitution

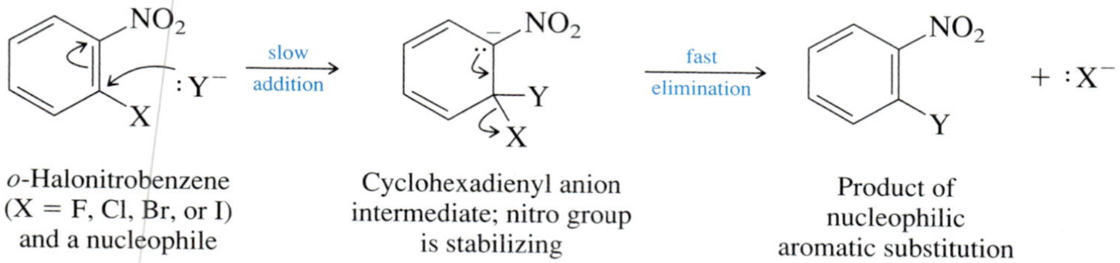

A nitro group behaves the same way in both reactions: it attracts electrons. Reaction is retarded when electrons flow from the aromatic ring to the attacking species (electrophilic aromatic substitution). Reaction is facilitated when electrons flow from the attacking species to the aromatic ring (nucleophilic aromatic substitution). By being aware of the connection between reactivity and substituent effects, you will sharpen your appreciation of how chemical reactions occur.

23.7   RELATED NUCLEOPHILIC AROMATIC SUBSTITUTION REACTIONS

The most common types of aryl halides in nucleophilic aromatic substitutions are those that bear *o*- or *p*-nitro substituents. Among other classes of reactive aryl halides, a few merit special consideration. One class includes highly fluorinated aromatic compounds such as hexafluorobenzene, which undergoes substitution of one of its fluorines on reaction with nucleophiles such as sodium methoxide.

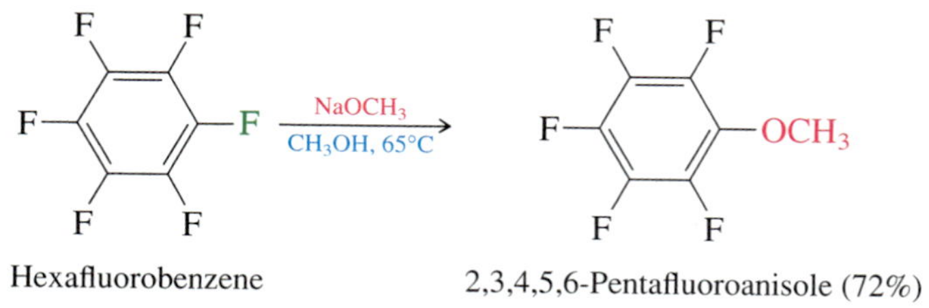

Hexafluorobenzene

2,3,4,5,6-Pentafluoroanisole (72%)

Here it is the combined electron-attracting effects of the six fluorine substituents that stabilize the cyclohexadienyl anion intermediate and permit the reaction to proceed so readily.

PROBLEM 23.5 Write equations describing the addition–elimination mechanism for the reaction of hexafluorobenzene with sodium methoxide, clearly showing the structure of the rate-determining intermediate.

Halides derived from certain heterocyclic aromatic compounds are often quite reactive toward nucleophiles. 2-Chloropyridine, for example, reacts with sodium methoxide some 230 million times faster than chlorobenzene at 50°C.

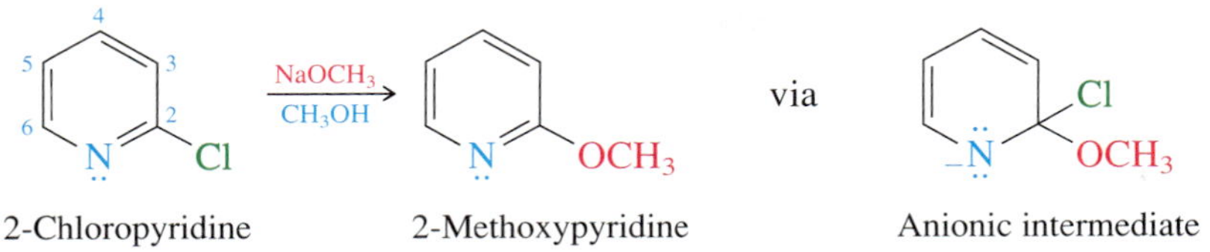

| 2-Chloropyridine | 2-Methoxypyridine | Anionic intermediate |

Again, rapid reaction is attributed to the stability of the intermediate formed in the addition step. In contrast to chlorobenzene, where the negative charge of the intermediate must be borne by carbon, the anionic intermediate in the case of 2-chloropyridine has its negative charge on nitrogen. Because nitrogen is more electronegative than carbon, the intermediate is more stable and is formed faster than the one from chlorobenzene.

PROBLEM 23.6 Offer an explanation for the observation that 4-chloropyridine is more reactive toward nucleophiles than 3-chloropyridine.

Another type of nucleophilic aromatic substitution occurs under quite different reaction conditions from those discussed to this point and proceeds by a different and rather surprising mechanism. It is described in the following section.

23.8 THE ELIMINATION–ADDITION MECHANISM OF NUCLEOPHILIC AROMATIC SUBSTITUTION: BENZYNE

Very strong bases such as sodium or potassium amide react readily with aryl halides, even those without electron-withdrawing substituents, to give products corresponding to nucleophilic substitution of halide by the base.

Comparing the pK_a of ammonia (36) and water (16) tells us that NH_2^- is 10^{20} times more basic than OH^-.

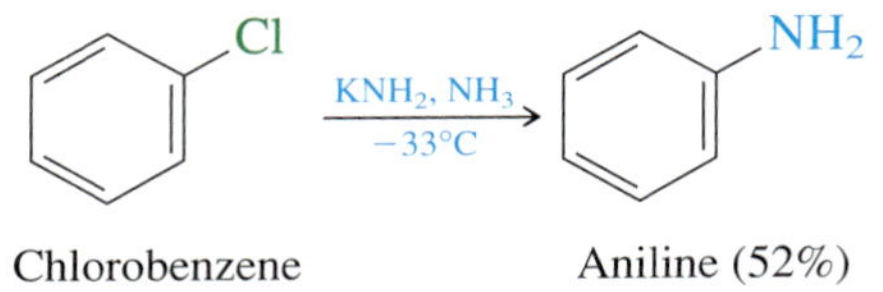

| Chlorobenzene | Aniline (52%) |

For a long time, observations concerning the regiochemistry of these reactions presented organic chemists with a puzzle. Substitution did not occur exclusively at the carbon from which the halide leaving group departed. Rather, a mixture of regioisomers was obtained in which the amine group was either on the carbon that originally bore the leaving group or on one of the carbons adjacent to it. Thus o-bromotoluene gave a mixture of o-methylaniline and m-methylaniline; p-bromotoluene gave m-methylaniline and p-methylaniline.

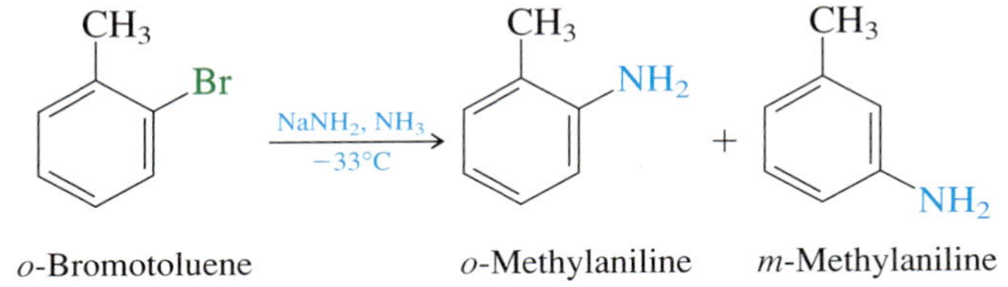

| o-Bromotoluene | o-Methylaniline | m-Methylaniline |

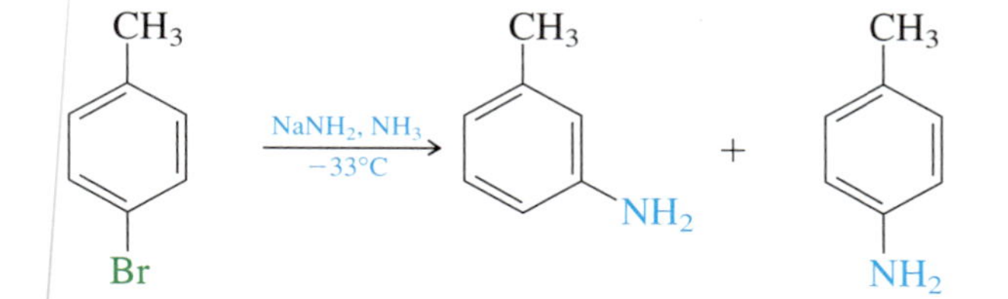

p-Bromotoluene   m-Methylaniline   p-Methylaniline

Three regioisomers (o-, m-, and p-methylaniline) were formed from m-bromotoluene.

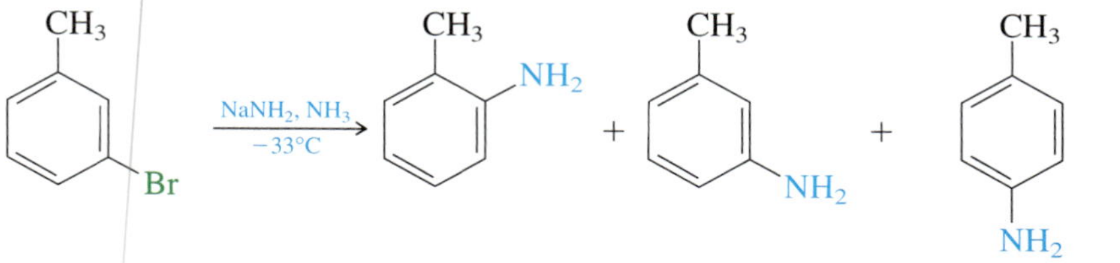

m-Bromotoluene   o-Methylaniline   m-Methylaniline   p-Methylaniline

These results rule out substitution by addition–elimination because that mechanism requires the nucleophile to attach itself to the carbon from which the leaving group departs.

A solution to the question of the mechanism of these reactions was provided by John D. Roberts in 1953 on the basis of an imaginative experiment. Roberts prepared a sample of chlorobenzene in which one of the carbons, the one bearing the chlorine, was the radioactive mass-14 isotope of carbon. Reaction with potassium amide in liquid ammonia yielded aniline containing almost exactly half of its ^{14}C label at C-1 and half at C-2:

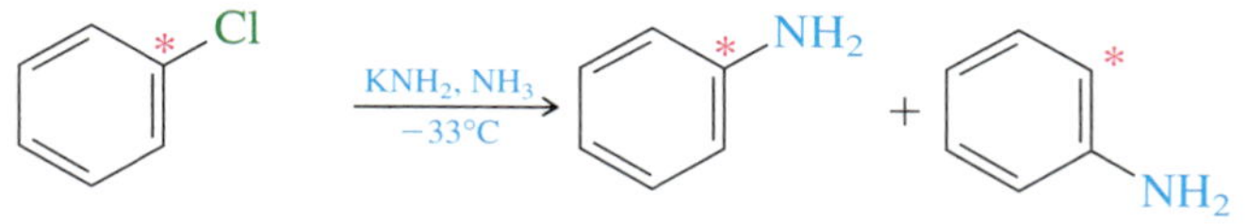

This work was done while Roberts was at MIT. He later moved to the California Institute of Technology, where he became a leader in applying NMR spectroscopy to nuclei other than protons, especially ^{13}C and ^{15}N.

Chlorobenzene-1-^{14}C   Aniline-1-^{14}C   Aniline-2-^{14}C
(* = ^{14}C)   (48%)   (52%)

The mechanism most consistent with the observations of this isotopic labeling experiment is the **elimination–addition mechanism** outlined in Figure 23.5. The first stage in this mechanism is a base-promoted dehydrohalogenation of chlorobenzene. The intermediate formed in this step contains a triple bond in an aromatic ring and is called **benzyne.** Aromatic compounds related to benzyne are known as **arynes.** The triple bond in benzyne is somewhat different from the usual triple bond of an alkyne, however. In benzyne one of the π components of the triple bond is part of the delocalized π system of the aromatic ring. The second π component results from overlapping sp^2-hybridized orbitals (*not* p-p overlap), lies in the plane of the ring, and does not interact with the aromatic π system. This π bond is relatively weak, because, as illustrated in Figure 23.6, its contributing sp^2 orbitals are not oriented properly for effective overlap.

Because the ring prevents linearity of the C—C≡C—C unit and π bonding in that unit is weak, benzyne is strained and highly reactive. This enhanced reactivity is

Overall reaction:

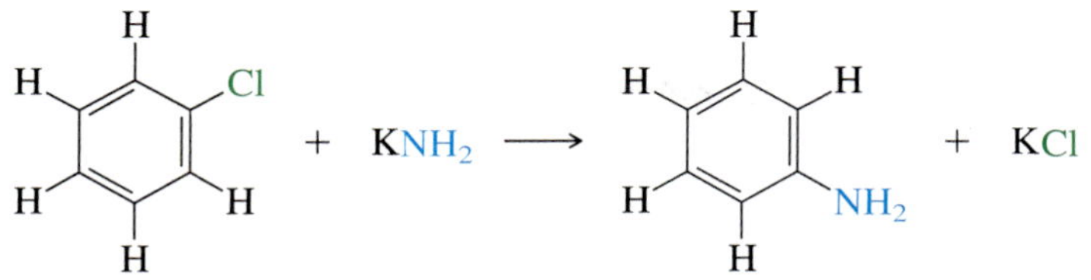

Chlorobenzene   Aniline

FIGURE 23.5 The elimination–addition mechanism of nucleophilic aromatic substitution.

Step 1: Elimination stage. Amide ion is a very strong base and brings about the dehydrohalogenation of chlorobenzene by abstracting a proton from the carbon adjacent to the one that bears the leaving group. The product of this step is an unstable intermediate called *benzyne*.

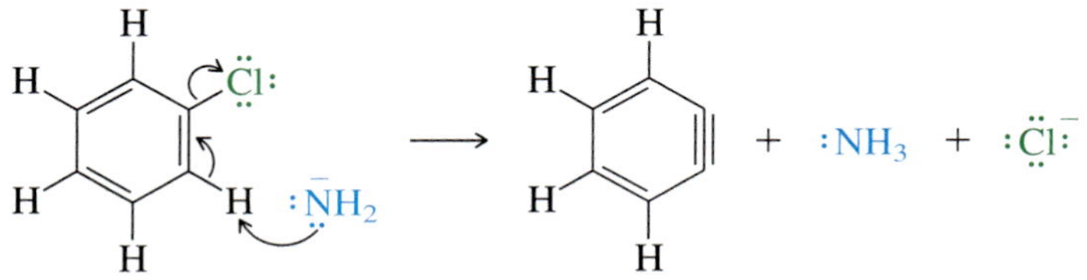

Chlorobenzene   Benzyne

Step 2: Beginning of addition phase. Amide ion acts as a nucleophile and adds to one of the carbons of the triple bond. The product of this step is a carbanion.

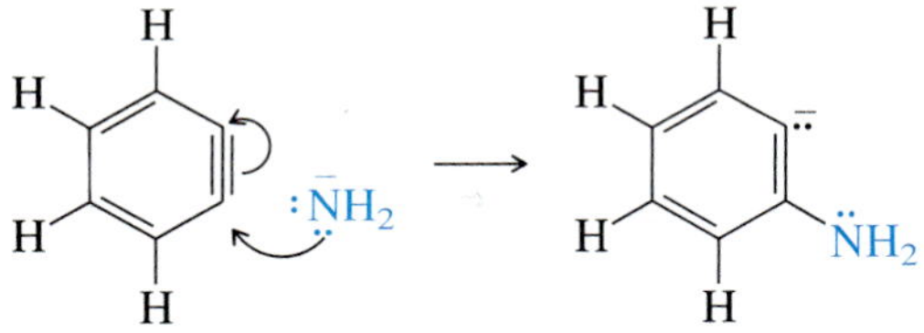

Benzyne   Aryl anion

Step 3: Completion of addition phase. The aryl anion abstracts a proton from the ammonia used as the solvent in the reaction.

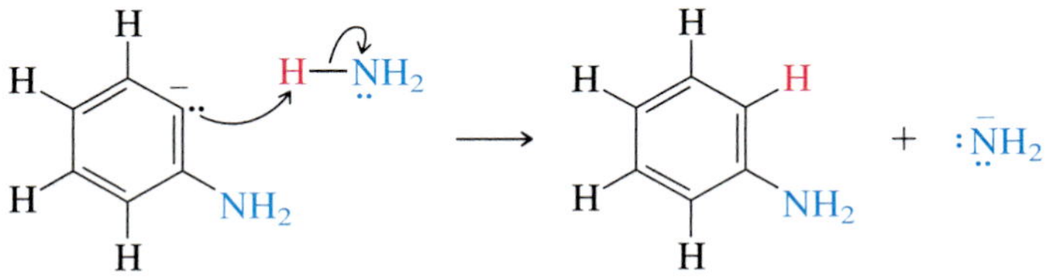

Aryl anion   Aniline

FIGURE 23.6 (*a*) The sp^2 orbitals in the plane of the ring in benzyne are not properly aligned for good overlap, and π bonding is weak. (*b*) The electrostatic potential map shows a region of high electron density associated with the "triple bond."

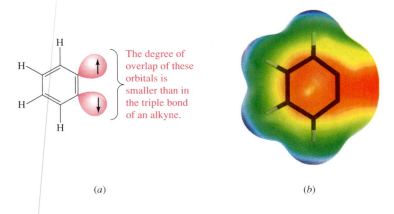

The degree of overlap of these orbitals is smaller than in the triple bond of an alkyne.

(*a*)   (*b*)

evident in the second stage of the elimination–addition mechanism as shown in steps 2 and 3 of Figure 23.5. In this stage the base acts as a nucleophile and adds to the strained bond of benzyne to form a carbanion. The carbanion, an *aryl anion,* then abstracts a proton from ammonia to yield the observed product.

The carbon that bears the leaving group and a carbon ortho to it become equivalent in the benzyne intermediate. Thus when chlorobenzene-1-^{14}C is the substrate, the amino group may be introduced with equal likelihood at either position.

> **PROBLEM 23.7** 2-Bromo-1,3-dimethylbenzene is inert to nucleophilic aromatic substitution on treatment with sodium amide in liquid ammonia. It is recovered unchanged even after extended contact with the reagent. Suggest an explanation for this lack of reactivity.

Once the intermediacy of an aryne intermediate was established, the reason for the observed regioselectivity of substitution in *o-, m-,* and *p*-chlorotoluene became evident. Only a single aryne intermediate may be formed from *o*-chlorotoluene, but this aryne yields a mixture containing comparable amounts of *o-* and *m*-methylaniline.

CH₃ / Cl $\xrightarrow[\text{NH}_3]{\text{KNH}_2}$ CH₃ (3-Methylbenzyne) $\xrightarrow[\text{NH}_3]{\text{KNH}_2}$ CH₃ / NH₂   +   CH₃ / NH₂

o-Chlorotoluene   3-Methylbenzyne   *o*-Methylaniline   *m*-Methylaniline

Similarly, *p*-chlorotoluene gives a single aryne, and this aryne gives a mixture of *m-* and *p*-methylaniline.

CH₃ / Cl $\xrightarrow[\text{NH}_3]{\text{KNH}_2}$ CH₃ (4-Methylbenzyne) $\xrightarrow[\text{NH}_3]{\text{KNH}_2}$ CH₃ / H₂N   +   CH₃ / NH₂

p-Chlorotoluene   4-Methylbenzyne   *m*-Methylaniline   *p*-Methylaniline

Two isomeric arynes give the three isomeric substitution products formed from *m*-chlorotoluene:

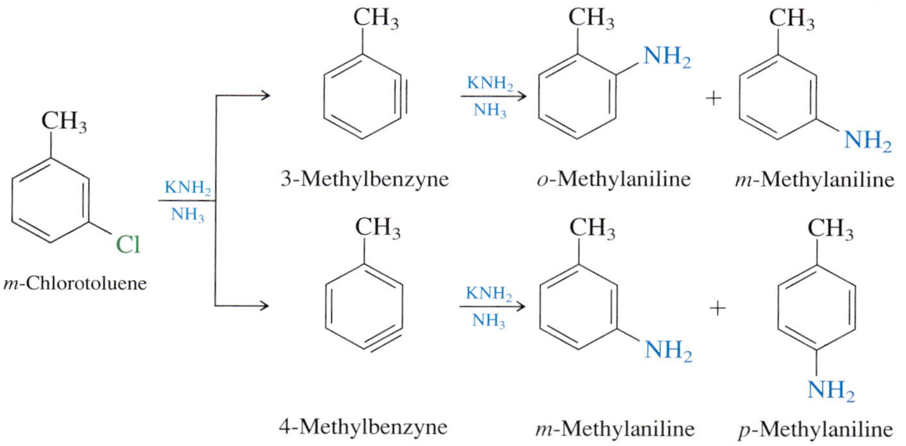

Although nucleophilic aromatic substitution by the elimination–addition mechanism is most commonly seen with very strong amide bases, it also occurs with bases such as hydroxide ion at high temperatures. A ^{14}C-labeling study revealed that hydrolysis of chlorobenzene proceeds by way of a benzyne intermediate.

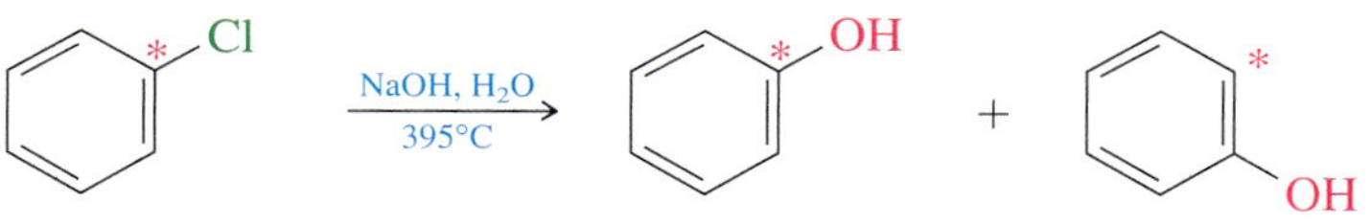

Chlorobenzene-1-^{14}C Phenol-1-^{14}C (54%) Phenol-2-^{14}C (43%)

PROBLEM 23.8 Two isomeric phenols are obtained in comparable amounts on hydrolysis of *p*-iodotoluene with 1 M sodium hydroxide at 300°C. Suggest reasonable structures for these two products.

23.9 DIELS–ALDER REACTIONS OF BENZYNE

Alternative methods for its generation have made it possible to use benzyne as an intermediate in a number of synthetic applications. One such method involves treating *o*-bromofluorobenzene with magnesium, usually in tetrahydrofuran as the solvent.

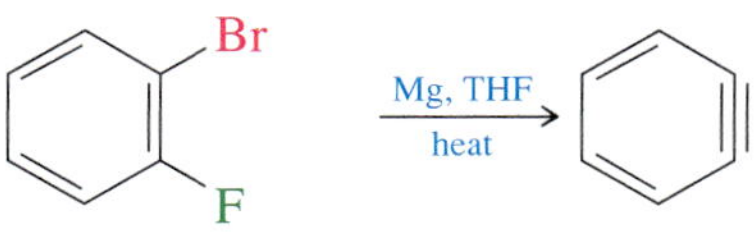

o-Bromofluorobenzene Benzyne

The reaction proceeds by formation of the Grignard reagent from *o*-bromofluorobenzene. Because the order of reactivity of magnesium with aryl halides is ArI > ArBr > ArCl > ArF, the Grignard reagent has the structure shown and forms benzyne by loss of the salt FMgBr:

o-Fluorophenylmagnesium bromide   Benzyne

Its strained triple bond makes benzyne a relatively good dienophile, and when benzyne is generated in the presence of a conjugated diene, Diels–Alder cycloaddition occurs.

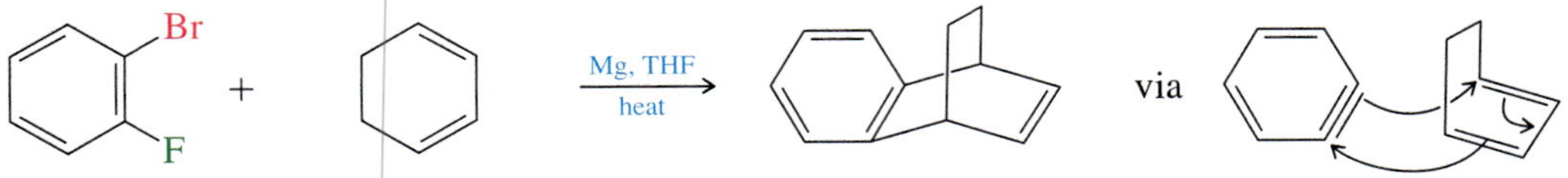

o-Bromo-   1,3-Cyclohexadiene   5,6-Benzobicyclo[2.2.2]-
fluorobenzene   octa-2,5-diene (46%)

> **PROBLEM 23.9** Give the structure of the cycloaddition product formed when benzyne is generated in the presence of furan. (See Section 11.22, if necessary, to remind yourself of the structure of furan.)

Benzyne may also be generated by treating o-bromofluorobenzene with lithium. In this case, o-fluorophenyllithium is formed, which then loses lithium fluoride to form benzyne.

23.10 SUMMARY

Section 23.1 Aryl halides are compounds of the type Ar—X where X = F, Cl, Br, or I. The carbon–halogen bond is stronger in ArX than in an alkyl halide (RX).

Section 23.2 Some aryl halides occur naturally, but most are the products of organic synthesis. The methods by which aryl halides are prepared were recalled in Table 23.2

Section 23.3 Aryl halides are less polar than alkyl halides.

Section 23.4 Aryl halides are less reactive than alkyl halides in reactions in which C—X bond breaking is rate-determining, especially in nucleophilic substitution reactions.

Section 23.5 Nucleophilic substitution in ArX is facilitated by the presence of a strong electron-withdrawing group, such as NO_2, ortho or para to the halogen.

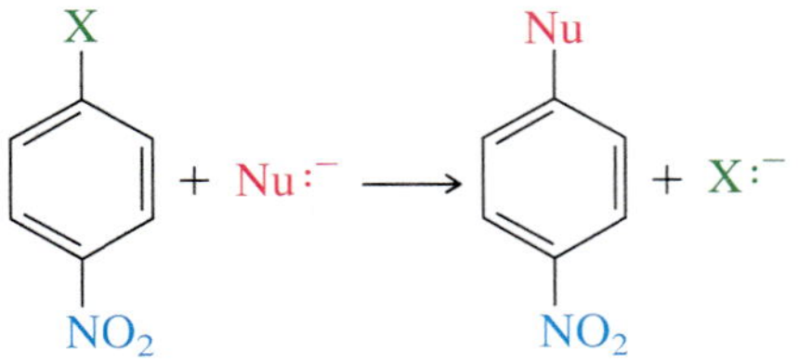

In reactions of this type, fluoride is the best leaving group of the halogens and iodide the poorest.

Section 23.6 Nucleophilic aromatic substitutions of the type just shown follow an **addition–elimination mechanism.**

Nitro-substituted aryl halide

Cyclohexadienyl anion intermediate

Product of nucleophilic aromatic substitution

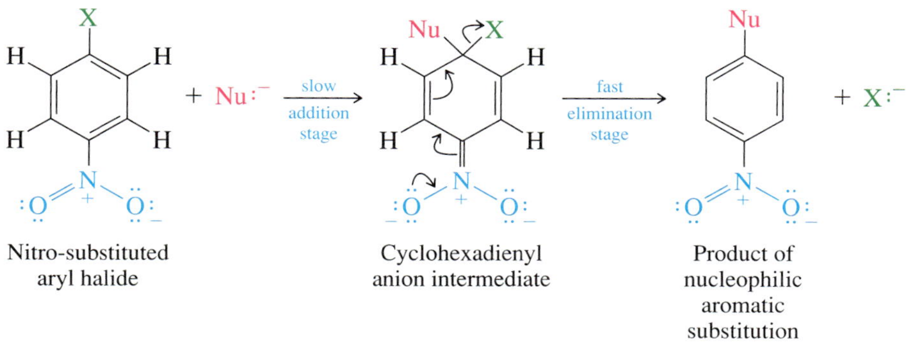

The rate-determining intermediate is a cyclohexadienyl anion and is stabilized by electron-withdrawing substituents.

Section 23.7 Other aryl halides that give stabilized anions can undergo nucleophilic aromatic substitution by the addition–elimination mechanism. Two examples are hexafluorobenzene and 2-chloropyridine.

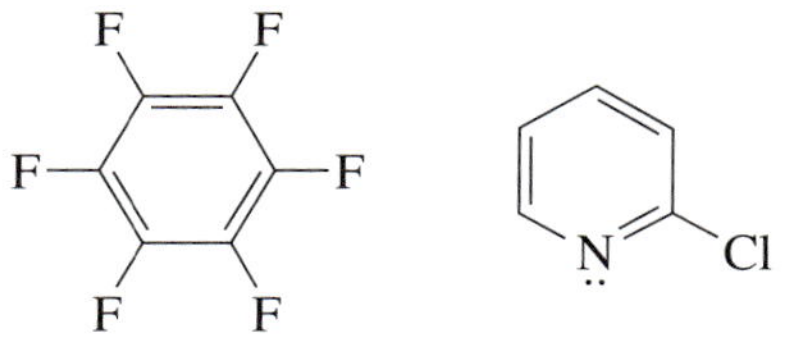

Hexafluorobenzene   2-Chloropyridine

Section 23.8 Nucleophilic aromatic substitution can also occur by an **elimination–addition mechanism.** This pathway is followed when the nucleophile is an exceptionally strong base such as amide ion in the form of sodium amide ($NaNH_2$) or potassium amide (KNH_2). **Benzyne** and related **arynes** are intermediates in nucleophilic aromatic substitutions that proceed by the elimination–addition mechanism.

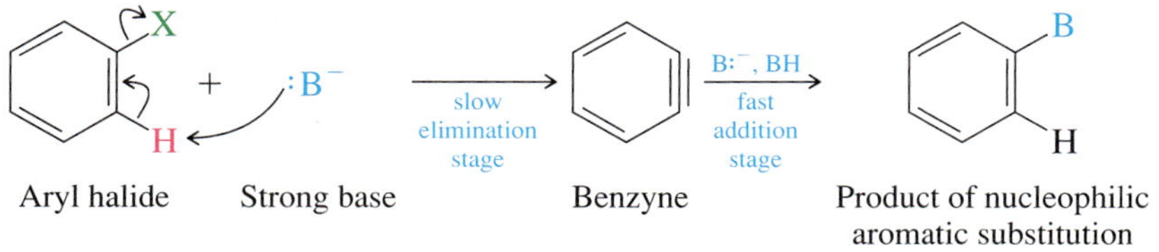

Aryl halide   Strong base   Benzyne   Product of nucleophilic aromatic substitution

Nucleophilic aromatic substitution by the elimination–addition mechanism can lead to substitution on the same carbon that bore the leaving group or on an adjacent carbon.

Section 23.9 Benzyne is a reactive dienophile and gives Diels–Alder products when generated in the presence of dienes. In these cases it is convenient to form benzyne by dissociation of the Grignard reagent of *o*-bromofluorobenzene.

PROBLEMS

23.10 Write a structural formula for each of the following:

(a) *m*-Chlorotoluene

(b) 2,6-Dibromoanisole

(c) *p*-Fluorostyrene

(d) 4,4′-Diiodobiphenyl

(e) 2-Bromo-1-chloro-4-nitrobenzene

(f) 1-Chloro-1-phenylethane

(g) *p*-Bromobenzyl chloride

(h) 2-Chloronaphthalene

(i) 1,8-Dichloronaphthalene

(j) 9-Fluorophenanthrene

23.11 Identify the major organic product of each of the following reactions. If two regioisomers are formed in appreciable amounts, show them both.

(a) Chlorobenzene + acetyl chloride $\xrightarrow{\text{AlCl}_3}$

(b) Bromobenzene + magnesium $\xrightarrow{\text{diethyl ether}}$

(c) Product of part (b) + dilute hydrochloric acid $\longrightarrow$

(d) Iodobenzene + lithium $\xrightarrow{\text{diethyl ether}}$

(e) Bromobenzene + sodium amide $\xrightarrow{\text{liquid ammonia, } -33°\text{C}}$

(f) *p*-Bromotoluene + sodium amide $\xrightarrow{\text{liquid ammonia, } -33°\text{C}}$

(g) 1-Bromo-4-nitrobenzene + ammonia $\longrightarrow$

(h) *p*-Bromobenzyl bromide + sodium cyanide $\longrightarrow$

(i) *p*-Chlorobenzenediazonium chloride + *N,N*-dimethylaniline $\longrightarrow$

(j) Hexafluorobenzene + sodium hydrogen sulfide $\longrightarrow$

23.12 Potassium *tert*-butoxide reacts with halobenzenes on heating in dimethyl sulfoxide to give *tert*-butyl phenyl ether.

(a) *o*-Fluorotoluene yields *tert*-butyl *o*-methylphenyl ether almost exclusively under these conditions. By which mechanism (addition–elimination or elimination–addition) do aryl fluorides react with potassium *tert*-butoxide in dimethyl sulfoxide?

(b) At 100°C, bromobenzene reacts over 20 times faster than fluorobenzene. By which mechanism do aryl bromides react?

23.13 Predict the products formed when each of the following isotopically substituted derivatives of chlorobenzene is treated with sodium amide in liquid ammonia. Estimate as quantitatively as possible the composition of the product mixture. The asterisk (*) in part (a) designates ^{14}C, and D in part (b) is ^{2}H.

(a) 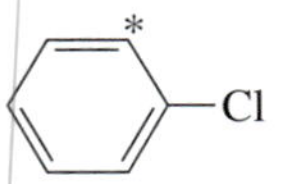   (b)

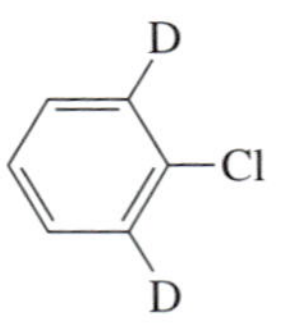

23.14 Choose the compound in each of the following pairs that reacts faster with sodium methoxide in methanol at 50°C:

(a) Chlorobenzene or *o*-chloronitrobenzene

(b) *o*-Chloronitrobenzene or *m*-chloronitrobenzene

(c) 4-Chloro-3-nitroacetophenone or 4-chloro-3-nitrotoluene

(d) 2-Fluoro-1,3-dinitrobenzene or 1-fluoro-3,5-dinitrobenzene

(e) 1,4-Dibromo-2-nitrobenzene or 1-bromo-2,4-dinitrobenzene

23.15 In each of the following reactions, an amine or a lithium amide derivative reacts with an aryl halide. Give the structure of the expected product, and specify the mechanism by which it is formed.

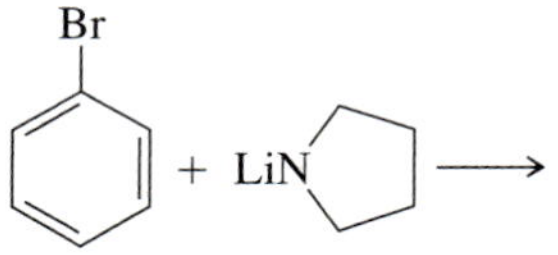

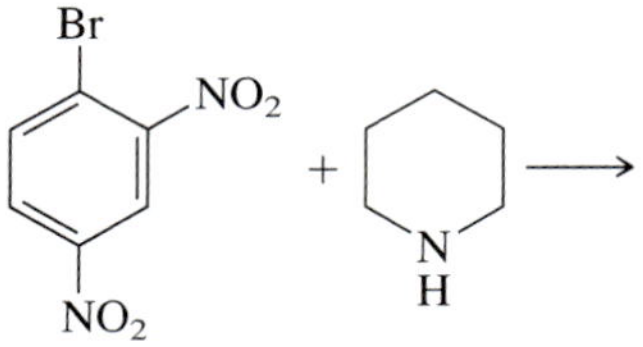

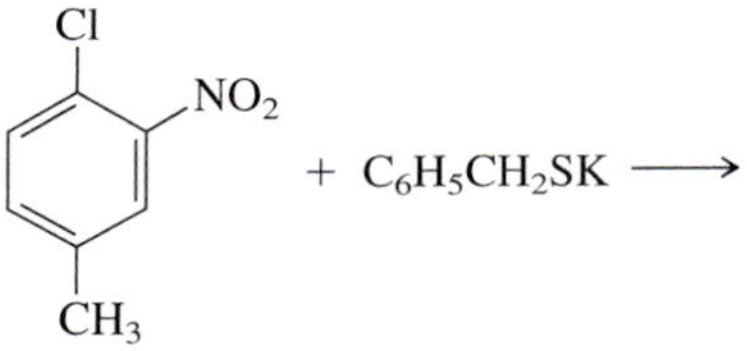

23.16 Piperidine, the amine reactant in parts (b) and (c) of the preceding problem, reacts with 1-bromonaphthalene on heating at 230°C to give a single product, compound A ($C_{15}H_{17}N$), as a noncrystallizable liquid. The same reaction using 2-bromonaphthalene yielded an isomeric product, compound B, a solid melting at 50–53°C. Mixtures of A and B were formed when either 1- or 2-bromonaphthalene was allowed to react with sodium piperidide in piperidine. Suggest reasonable structures for compounds A and B and offer an explanation for their formation under each set of reaction conditions.

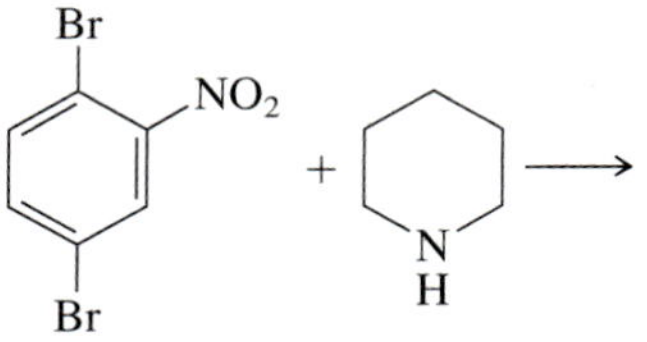

23.17 1,2,3,4,5-Pentafluoro-6-nitrobenzene reacts readily with sodium methoxide in methanol at room temperature to yield two major products, each having the molecular formula $C_7H_3F_4NO_3$. Suggest reasonable structures for these two compounds.

23.18 Predict the major organic product in each of the following reactions:

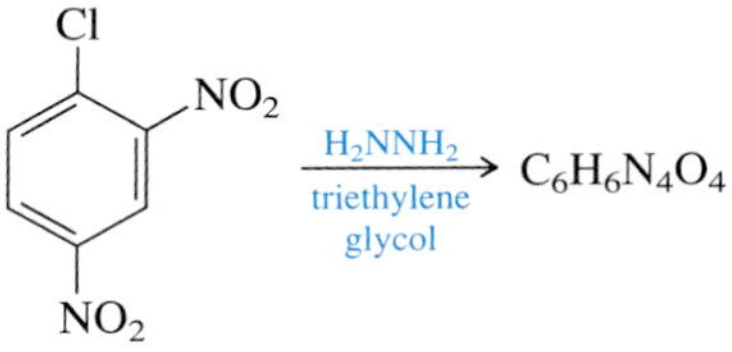

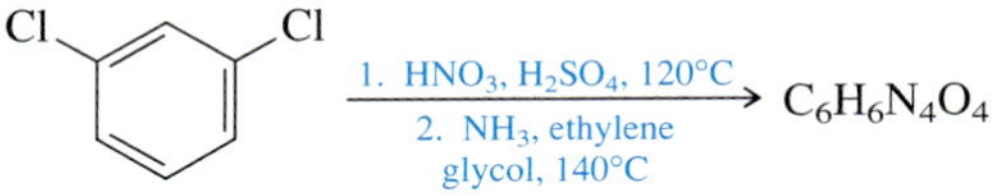

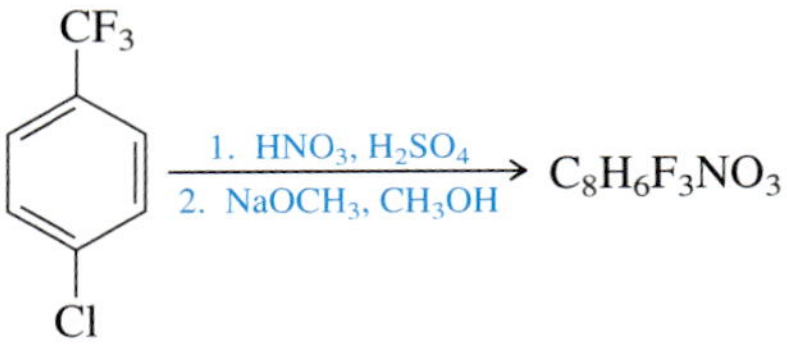

(e) 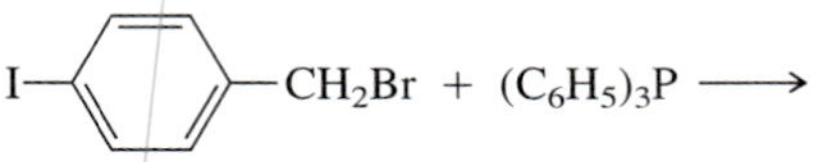 $I-\langle\!\!\!\bigcirc\!\!\!\rangle-CH_2Br + (C_6H_5)_3P \longrightarrow$

(f) 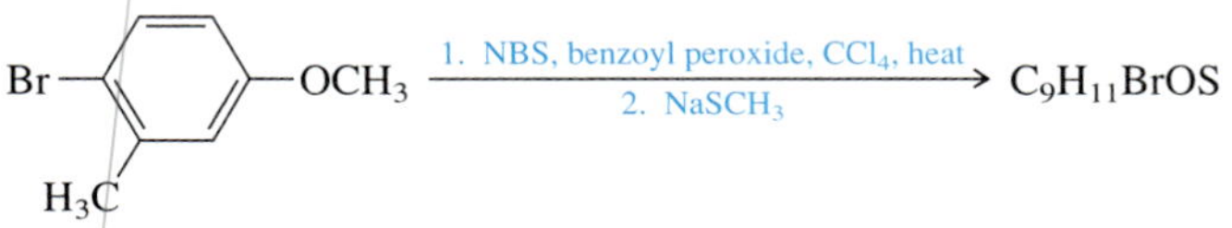 $Br-\langle\!\!\!\bigcirc\!\!\!\rangle-OCH_3$ $\xrightarrow[\text{2. NaSCH}_3]{\text{1. NBS, benzoyl peroxide, CCl}_4,\text{ heat}}$ $C_9H_{11}BrOS$

with H_3C substituent

23.19 The hydrolysis of *p*-bromotoluene with aqueous sodium hydroxide at 300°C yields *m*-methylphenol and *p*-methylphenol in a 5:4 ratio. What is the meta–para ratio for the same reaction carried out on *p*-chlorotoluene?

23.20 The herbicide *trifluralin* is prepared by the following sequence of reactions. Identify compound A and deduce the structure of trifluralin.

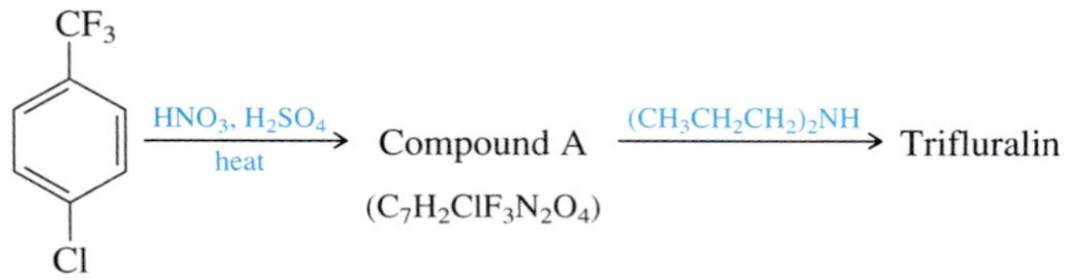

$$\underset{\text{Cl}}{\overset{\text{CF}_3}{\underset{}{\bigcirc}}} \xrightarrow[\text{heat}]{\text{HNO}_3,\text{ H}_2\text{SO}_4} \underset{(C_7H_2ClF_3N_2O_4)}{\text{Compound A}} \xrightarrow{(CH_3CH_2CH_2)_2NH} \text{Trifluralin}$$

23.21 *Chlorbenside* is a pesticide used to control red spider mites. It is prepared by the sequence shown. Identify compounds A and B in this sequence. What is the structure of chlorbenside?

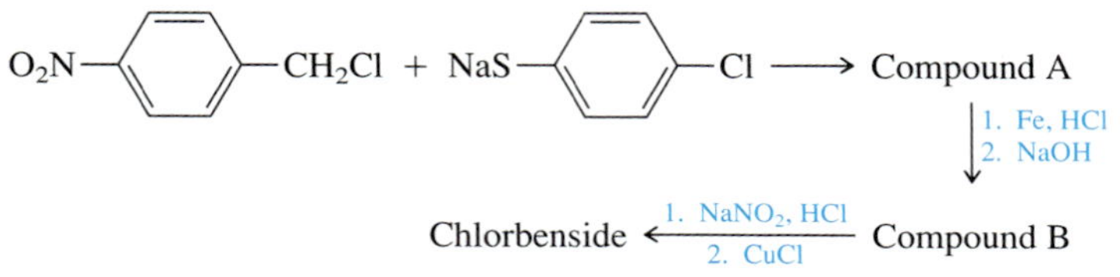

$$O_2N-\langle\!\!\!\bigcirc\!\!\!\rangle-CH_2Cl + NaS-\langle\!\!\!\bigcirc\!\!\!\rangle-Cl \longrightarrow \text{Compound A}$$

$$\downarrow \begin{array}{l}\text{1. Fe, HCl}\\ \text{2. NaOH}\end{array}$$

$$\text{Chlorbenside} \xleftarrow[\text{2. CuCl}]{\text{1. NaNO}_2,\text{ HCl}} \text{Compound B}$$

23.22 An article in the October 1998 issue of the *Journal of Chemical Education* (p. 1266) describes the following reaction.

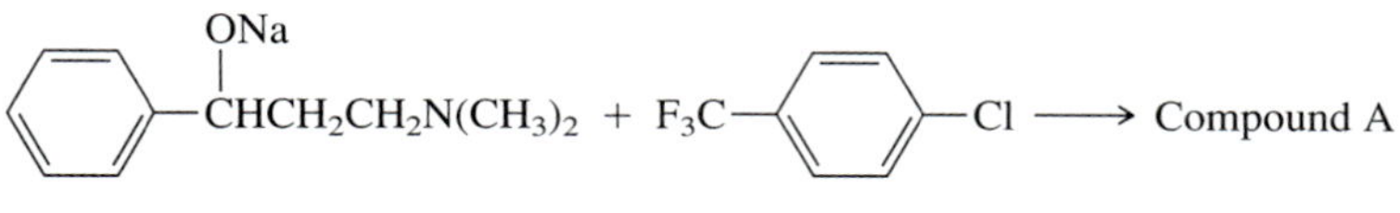

$$\langle\!\!\!\bigcirc\!\!\!\rangle-\underset{\underset{\text{ONa}}{|}}{\text{CH}}CH_2CH_2N(CH_3)_2 + F_3C-\langle\!\!\!\bigcirc\!\!\!\rangle-Cl \longrightarrow \text{Compound A}$$

Fluoxetine hydrochloride (Prozac) is a widely prescribed antidepressant drug introduced by Eli Lilly & Co. in 1986. It differs from Compound A in having an —NHCH₃ group in place of —N(CH₃)₂. What is the structure of Prozac?

23.23 A method for the generation of benzyne involves heating the diazonium salt from *o*-aminobenzoic acid (benzenediazonium-2-carboxylate). Using curved arrows, show how this substance forms benzyne. What two inorganic compounds are formed in this reaction?

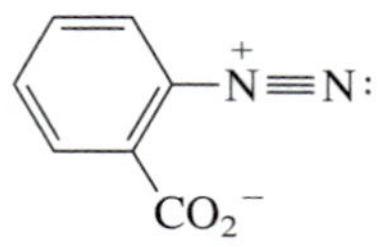

$$\underset{\text{CO}_2{}^-}{\overset{}{\bigcirc}}-\overset{+}{N}\!\!\equiv\!\!N\!:$$

Benzenediazonium-2-carboxylate

23.24 The compound *triptycene* may be prepared as shown. What is compound A?

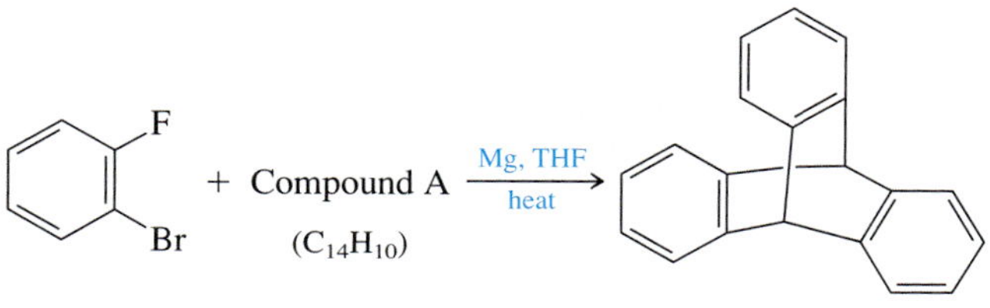

Triptycene

23.25 Nitro-substituted aromatic compounds that do not bear halide leaving groups react with nucleophiles according to the equation

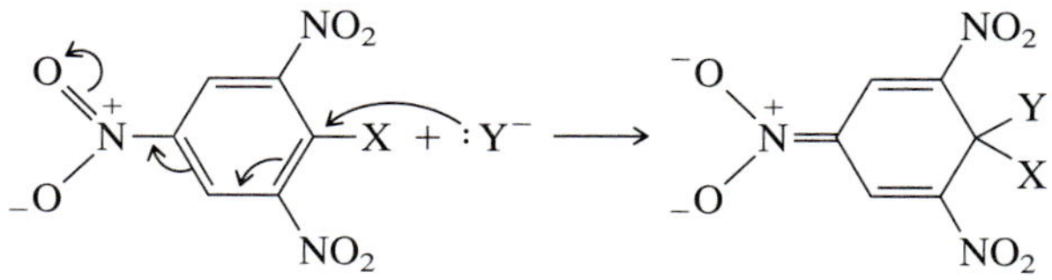

The product of this reaction, as its sodium salt, is called a *Meisenheimer complex* after the German chemist Jacob Meisenheimer, who reported on their formation and reactions in 1902. A Meisenheimer complex corresponds to the product of the nucleophilic addition stage in the addition–elimination mechanism for nucleophilic aromatic substitution.

(a) Give the structure of the Meisenheimer complex formed by addition of sodium ethoxide to 2,4,6-trinitroanisole.

(b) What other combination of reactants yields the same Meisenheimer complex as that of part (a)?

23.26 A careful study of the reaction of 2,4,6-trinitroanisole with sodium methoxide revealed that two different Meisenheimer complexes were present. Suggest reasonable structures for these two complexes.

23.27 Suggest a reasonable mechanism for each of the following reactions:

(a) $C_6H_5Br + CH_2(COOCH_2CH_3)_2 \xrightarrow[\text{2. } H_3O^+]{\text{1. excess NaNH}_2, \text{NH}_3} C_6H_5CH(COOCH_2CH_3)_2$

(b)

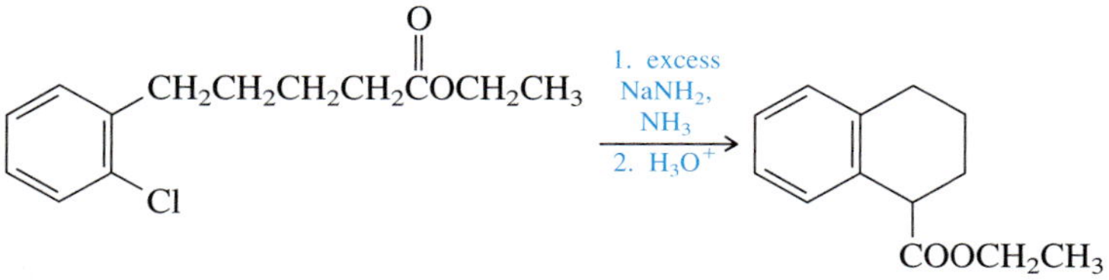

(c)

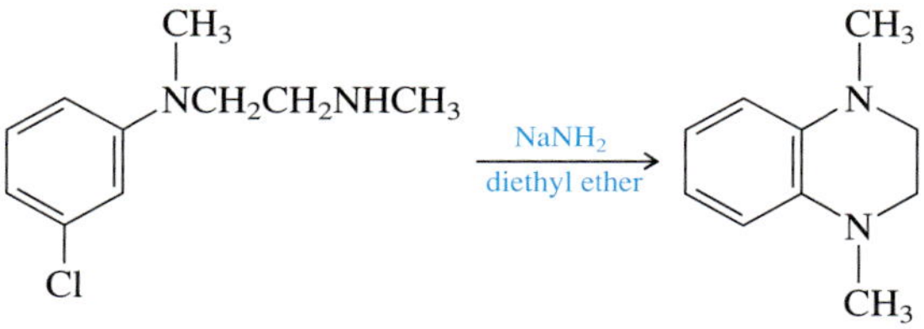

(d)

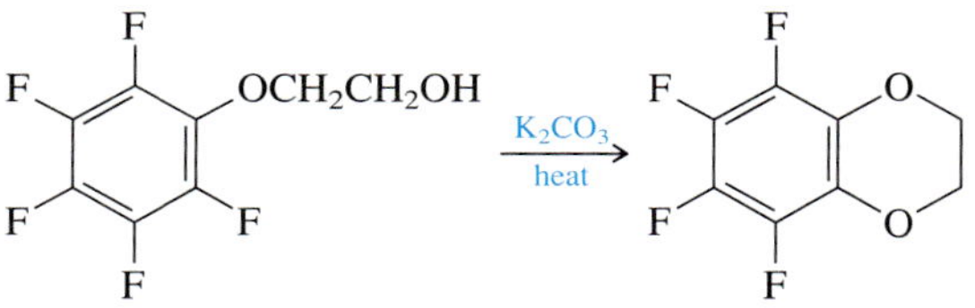

23.28 Mixtures of chlorinated derivatives of biphenyl, called *polychlorinated biphenyls,* or *PCBs,* were once prepared industrially on a large scale as insulating materials in electrical equipment. As equipment containing PCBs was discarded, the PCBs entered the environment at a rate that reached an estimated 25,000 lb/year. PCBs are very stable and accumulate in the fatty tissue of fish, birds, and mammals. They have been shown to be *teratogenic,* meaning that they induce mutations in the offspring of affected individuals. Some countries have banned the use of PCBs. A large number of chlorinated biphenyls are possible, and the commercially produced material is a mixture of many compounds.

(a) How many monochloro derivatives of biphenyl are possible?

(b) How many dichloro derivatives are possible?

(c) How many octachloro derivatives are possible?

(d) How many nonachloro derivatives are possible?

23.29 DDT-resistant insects have the ability to convert DDT to a less toxic substance called DDE. The mass spectrum of DDE shows a cluster of peaks for the molecular ion at *m/z* 316, 318, 320, 322, and 324. Suggest a reasonable structure for DDE.

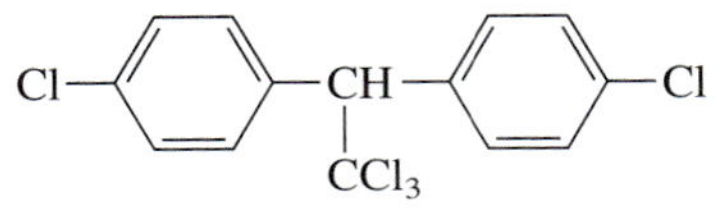

DDT (*d*ichloro*d*iphenyl*t*richloroethane)

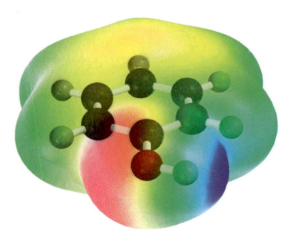

CHAPTER 24

PHENOLS

Phenols are compounds that have a hydroxyl group bonded directly to a benzene or benzenoid ring. The parent compound of this group, C_6H_5OH, called simply *phenol,* is an important industrial chemical. Many of the properties of phenols are analogous to those of alcohols, but this similarity is something of an oversimplification. Like arylamines, phenols are difunctional compounds; the hydroxyl group and the aromatic ring interact strongly, affecting each other's reactivity. This interaction leads to some novel and useful properties of phenols. A key step in the synthesis of aspirin, for example, is without parallel in the reactions of either alcohols or arenes. With periodic reminders of the ways in which phenols resemble alcohols and arenes, this chapter emphasizes the ways in which phenols are unique.

24.1 NOMENCLATURE

An old name for benzene was *phene,* and its hydroxyl derivative came to be called *phenol.** This, like many other entrenched common names, is an acceptable IUPAC name. Likewise, *o-, m-,* and *p*-cresol are acceptable names for the various ring-substituted hydroxyl derivatives of toluene. More highly substituted compounds are named as derivatives of phenol. Numbering of the ring begins at the hydroxyl-substituted carbon and proceeds in the direction that gives the lower number to the next substituted carbon. Substituents are cited in alphabetical order.

Phenol   *m*-Cresol   5-Chloro-2-methylphenol

*The systematic name for phenol is *benzenol.*

The three dihydroxy derivatives of benzene may be named as 1,2-, 1,3-, and 1,4-benzenediol, respectively, but each is more familiarly known by the common name indicated in parentheses below the structures shown here. These common names are permissible IUPAC names.

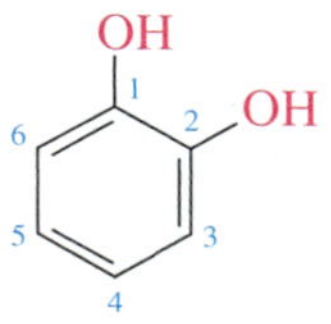

1,2-Benzenediol
(pyrocatechol)

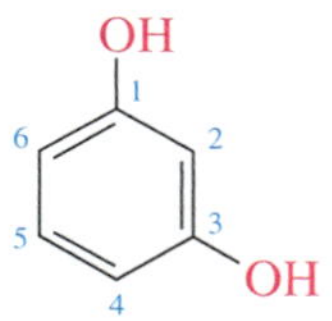

1,3-Benzenediol
(resorcinol)

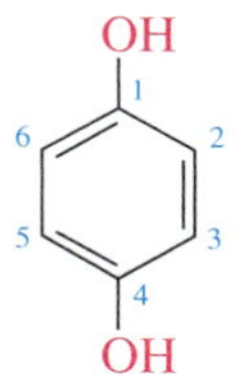

1,4-Benzenediol
(hydroquinone)

Pyrocatechol is often called *catechol.*

The common names for the two hydroxy derivatives of naphthalene are 1-naphthol and 2-naphthol. These are also acceptable IUPAC names.

PROBLEM 24.1   Write structural formulas for each of the following compounds:

(a) Pyrogallol (1,2,3-benzenetriol)   (c) 3-Nitro-1-naphthol

(b) *o*-Benzylphenol   (d) 4-Chlororesorcinol

SAMPLE SOLUTION   (a) Like the dihydroxybenzenes, the isomeric trihydroxybenzenes have unique names. Pyrogallol, used as a developer of photographic film, is 1,2,3-benzenetriol. The three hydroxyl groups occupy adjacent positions on a benzene ring.

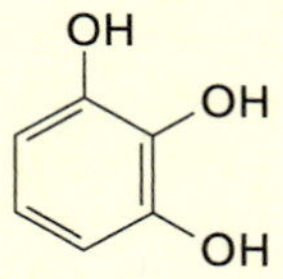

Pyrogallol
(1,2,3-benzenetriol)

Carboxyl and acyl groups take precedence over the phenolic hydroxyl in determining the base name. The hydroxyl is treated as a substituent in these cases.

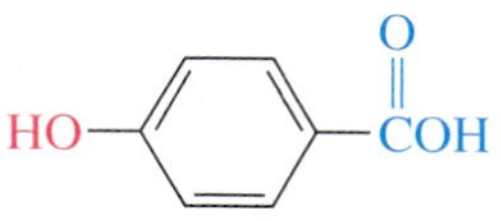

p-Hydroxybenzoic acid

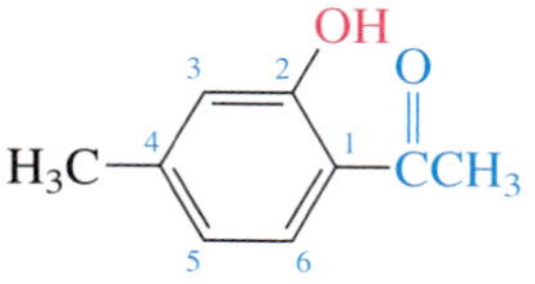

2-Hydroxy-4-methylacetophenone

24.2   STRUCTURE AND BONDING

Phenol is planar, with a C—O—H angle of 109°, almost the same as the tetrahedral angle and not much different from the 108.5° C—O—H angle of methanol:

The graphic that opened this chapter is a molecular model of phenol that shows its planar structure and electrostatic potential.

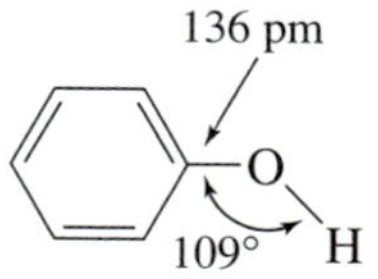

Phenol

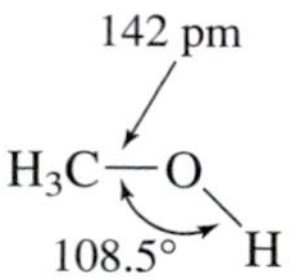

Methanol

As we've seen on a number of occasions, bonds to sp^2-hybridized carbon are shorter than those to sp^3-hybridized carbon, and the case of phenols is no exception. The carbon–oxygen bond distance in phenol is slightly less than that in methanol.

In resonance terms, the shorter carbon–oxygen bond distance in phenol is attributed to the partial double-bond character that results from conjugation of the unshared electron pair of oxygen with the aromatic ring.

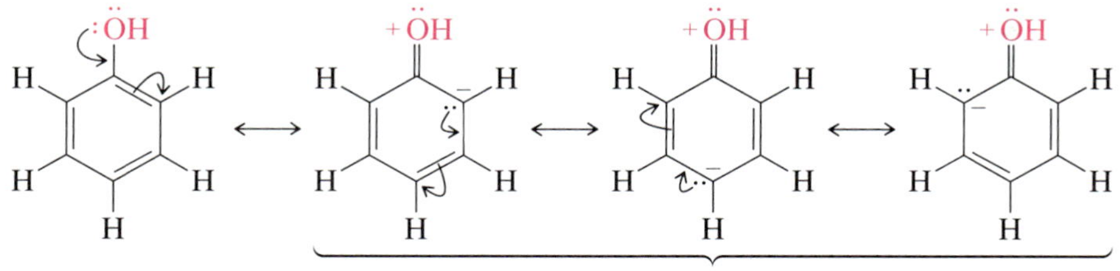

Most stable Lewis structure for phenol

Dipolar resonance forms of phenol

Many of the properties of phenols reflect the polarization implied by the resonance description. The hydroxyl oxygen is less basic, and the hydroxyl proton more acidic, in phenols than in alcohols. Electrophiles attack the aromatic ring of phenols much faster than they attack benzene, indicating that the ring, especially at the positions ortho and para to the hydroxyl group, is relatively "electron-rich."

24.3 PHYSICAL PROPERTIES

The physical properties of phenols are strongly influenced by the hydroxyl group, which permits phenols to form hydrogen bonds with other phenol molecules (Figure 24.1a) and with water (Figure 24.1b). Thus, phenols have higher melting points and boiling points and are more soluble in water than arenes and aryl halides of comparable molecular weight. Table 24.1 compares phenol, toluene, and fluorobenzene with regard to these physical properties.

Some ortho-substituted phenols, such as *o*-nitrophenol, have significantly lower boiling points than those of the meta and para isomers. This is because the *intramolecular* hydrogen bond that forms between the hydroxyl group and the substituent partially compensates for the energy required to go from the liquid state to the vapor.

The physical properties of some representative phenols are collected in Appendix 1.

| TABLE 24.1 | Comparison of Physical Properties of an Arene, a Phenol, and an Aryl Halide |

	Compound		
Physical property	Toluene, $C_6H_5CH_3$	Phenol, C_6H_5OH	Fluorobenzene, C_6H_5F
Molecular weight	92	94	96
Melting point	−95°C	43°C	−41°C
Boiling point (1 atm)	111°C	132°C	85°C
Solubility in water (25°C)	0.05 g/100 mL	8.2 g/100 mL	0.2 g/100 mL

FIGURE 24.1 (*a*) A hydrogen bond between two phenol molecules; (*b*) hydrogen bonds between water and phenol molecules.

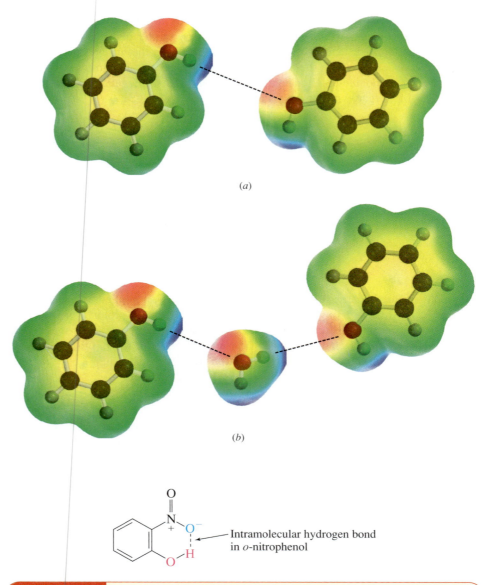

(*a*)

(*b*)

Intramolecular hydrogen bond in *o*-nitrophenol

> **PROBLEM 24.2** One of the hydroxybenzoic acids is known by the common name *salicylic acid*. Its methyl ester, methyl salicylate, occurs in oil of wintergreen. Methyl salicylate boils over 50°C lower than either of the other two methyl hydroxybenzoates. What is the structure of methyl salicylate? Why is its boiling point so much lower than that of either of its regioisomers?

Because of its acidity, phenol was known as *carbolic acid* when Joseph Lister introduced it as an antiseptic in 1865 to prevent postoperative bacterial infections that were then a life-threatening hazard in even minor surgical procedures.

24.4 ACIDITY OF PHENOLS

The most characteristic property of phenols is their acidity. Phenols are more acidic than alcohols but less acidic than carboxylic acids. Recall that carboxylic acids have pK_a's of approximately 5, whereas the pK_a's of alcohols are in the 16–20 range. The pK_a for most phenols is about 10.

To help us understand why phenols are more acidic than alcohols, let's compare the ionization equilibria for phenol and ethanol. In particular, consider the differences in charge delocalization in ethoxide ion and in phenoxide ion. The negative charge in ethoxide ion is localized on oxygen and is stabilized only by solvation forces.

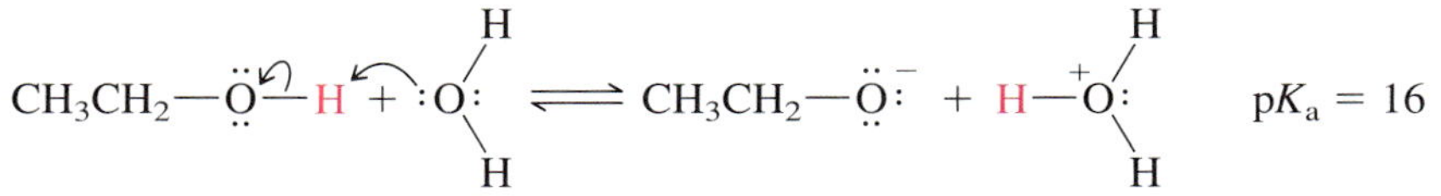

$$CH_3CH_2-\overset{..}{\underset{..}{O}}-H + :\overset{H}{\underset{H}{O}} \rightleftharpoons CH_3CH_2-\overset{..}{\underset{..}{O}}:^- + H-\overset{+}{\underset{H}{O}} \qquad pK_a = 16$$

The negative charge in phenoxide ion is stabilized both by solvation and by electron delocalization into the ring.

The negative charge in phenoxide ion is stabilized both by solvation and by electron delocalization into the ring.

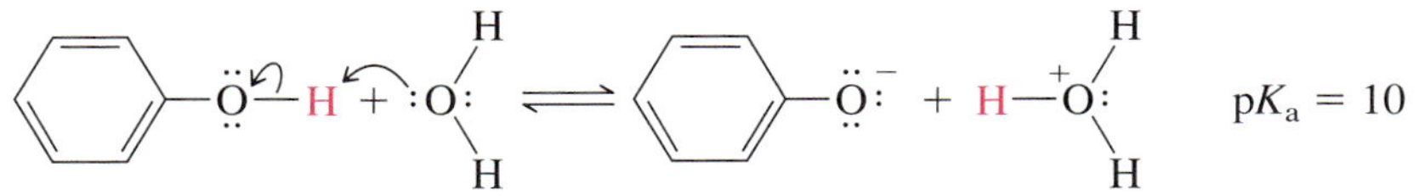

Electron delocalization in phenoxide is represented by resonance among the structures:

The negative charge in phenoxide ion is shared by the oxygen and the carbons that are ortho and para to it. Delocalization of its negative charge strongly stabilizes phenoxide ion.

To place the acidity of phenol in perspective, note that although phenol is more than a million times more acidic than ethanol, it is over a hundred thousand times weaker than acetic acid. Thus, phenols can be separated from alcohols because they are more acidic, and from carboxylic acids because they are less acidic. On shaking an ether solution containing both an alcohol and a phenol with dilute sodium hydroxide, the phenol is converted quantitatively to its sodium salt, which is extracted into the aqueous phase. The alcohol remains in the ether phase.

Phenol	Hydroxide ion	Phenoxide ion	Water
(stronger acid)	(stronger base)	(weaker base)	(weaker acid)

The electrostatic potential map of phenoxide ion on *Learning By Modeling* displays the delocalization of electrons into the ring.

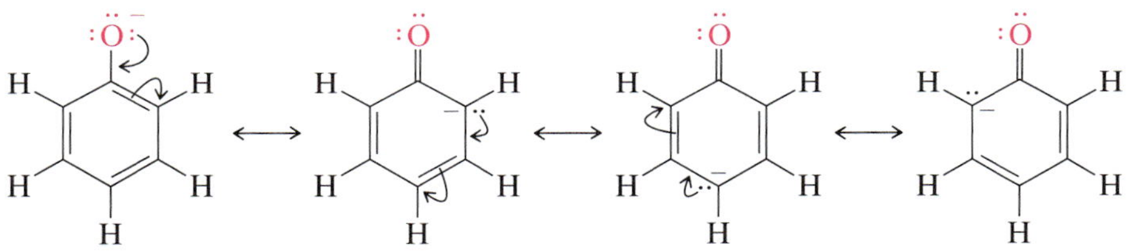

How do we know that water is a weaker acid than phenol? What are their respective pK_a values?

On shaking an ether solution of a phenol and a carboxylic acid with dilute sodium bicarbonate, the carboxylic acid is converted quantitatively to its sodium salt and extracted into the aqueous phase. The phenol remains in the ether phase.

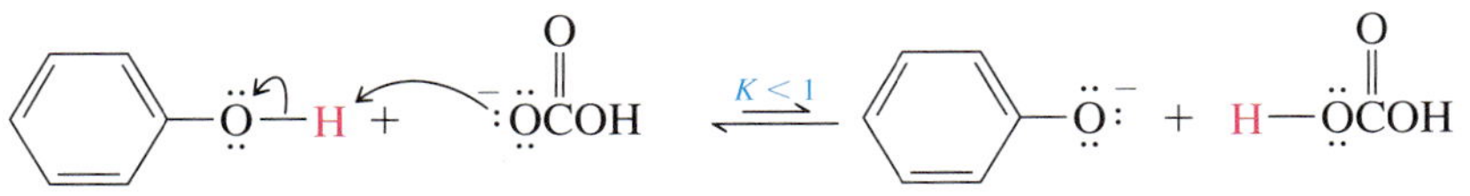

Phenol	Bicarbonate ion	Phenoxide ion	Carbonic acid
(weaker acid)	(weaker base)	(stronger base)	(stronger acid)

How do we know that carbonic acid is a stronger acid than phenol? What are their respective pK_a values?

It is necessary to keep the acidity of phenols in mind when we discuss preparation and reactions. Reactions that produce phenols, when carried out in basic solution, require an acidification step to convert the phenoxide ion to the neutral form of the phenol.

Phenoxide ion	Hydronium ion	Phenol	Water
(stronger base)	(stronger acid)	(weaker acid)	(weaker base)

How do we know that hydronium ion is a stronger acid than phenol? What are their respective pKₐ values?

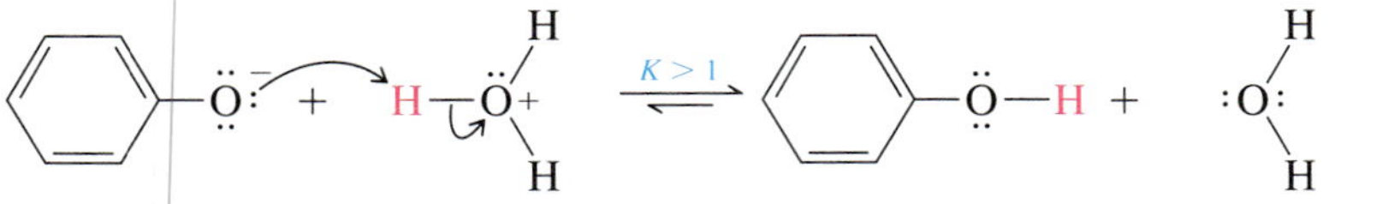

Many synthetic reactions involving phenols as nucleophiles are carried out in the presence of sodium or potassium hydroxide. Under these conditions the phenol is converted to the corresponding phenoxide ion, which is a far better nucleophile.

24.5   SUBSTITUENT EFFECTS ON THE ACIDITY OF PHENOLS

As Table 24.2 shows, most phenols have ionization constants similar to that of phenol itself. Substituent effects, in general, are small.

Alkyl substitution produces negligible changes in acidities, as do weakly electronegative groups attached to the ring.

Only when the substituent is strongly electron-withdrawing, as is a nitro group, is a substantial change in acidity noted. The ionization constants of o- and p-nitrophenol are several hundred times greater than that of phenol. An ortho- or para-nitro group greatly stabilizes the phenoxide ion by permitting a portion of the negative charge to be carried by its own oxygens.

Recall from Section 24.1 that cresols are methyl-substituted derivatives of phenol.

TABLE 24.2   Acidities of Some Phenols

Compound name	pK_a	Compound name	pK_a
Monosubstituted phenols			
Phenol	10.0	o-Methoxyphenol	10.0
o-Cresol	10.3	m-Methoxyphenol	9.6
m-Cresol	10.1	p-Methoxyphenol	10.2
p-Cresol	10.3	o-Nitrophenol	7.2
o-Chlorophenol	8.6	m-Nitrophenol	8.4
m-Chlorophenol	9.1	p-Nitrophenol	7.2
p-Chlorophenol	9.4		
Di- and trinitrophenols			
2,4-Dinitrophenol	4.0	2,4,6-Trinitrophenol	0.4
3,5-Dinitrophenol	6.7		
Naphthols			
1-Naphthol	9.2	2-Naphthol	9.5

Electron delocalization in **o**-*nitrophenoxide ion*

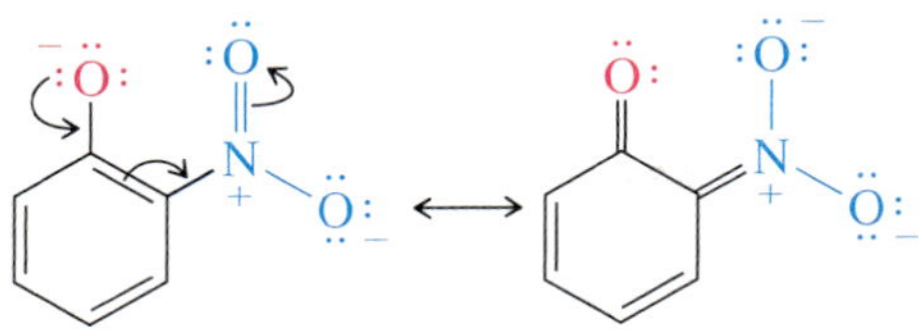

Electron delocalization in **p**-*nitrophenoxide ion*

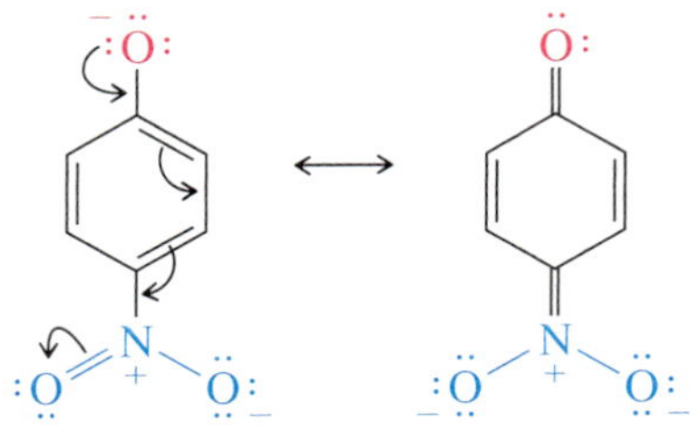

A meta-nitro group is not directly conjugated to the phenoxide oxygen and thus stabilizes a phenoxide ion to a smaller extent. *m*-Nitrophenol is more acidic than phenol but less acidic than either *o*- or *p*-nitrophenol.

PROBLEM 24.3   Which is the stronger acid in each of the following pairs? Explain your reasoning.

(a) Phenol or *p*-hydroxybenzaldehyde

(b) *m*-Cyanophenol or *p*-cyanophenol

(c) *o*-Fluorophenol or *p*-fluorophenol

SAMPLE SOLUTION   (a) The best approach when comparing the acidities of different phenols is to assess opportunities for stabilization of negative charge in their anions. Electron delocalization in the anion of *p*-hydroxybenzaldehyde is very effective because of conjugation.

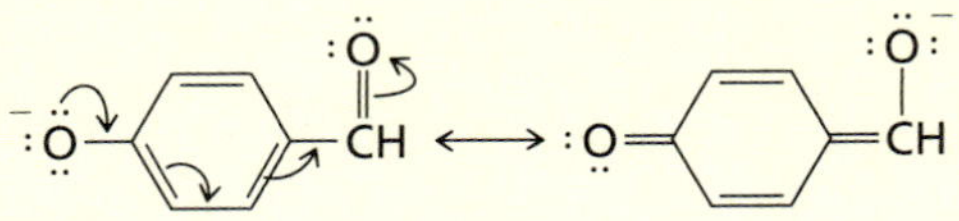

A carbonyl group is strongly electron-withdrawing and acid-strengthening, especially when ortho or para to the hydroxyl group. *p*-Hydroxybenzaldehyde is a stronger acid than phenol.

Multiple substitution by strongly electron-withdrawing groups greatly increases the acidity of phenols, as the pK_a values for 2,4-dinitrophenol (4.0) and 2,4,6-trinitrophenol (0.4) in Table 24.2 attest.

24.6   SOURCES OF PHENOLS

Phenol was first isolated in the early nineteenth century from coal tar, and a small portion of the more than 4 billion lb of phenol produced in the United States each year comes from this source. Although significant quantities of phenol are used to prepare aspirin and dyes, most of it is converted to phenolic resins used in adhesives and plastics.

Almost all the phenol produced commercially is synthetic, with several different processes in current use. These are summarized in Table 24.3.

The reaction of benzenesulfonic acid with sodium hydroxide (first entry in Table 24.3) proceeds by the addition–elimination mechanism of nucleophilic aromatic substitution (Section 23.6). Hydroxide replaces sulfite ion (SO_3^{2-}) at the carbon atom that bears the leaving group. Thus, p-toluenesulfonic acid is converted exclusively to p-cresol by an analogous reaction:

Can you recall how to prepare p-toluenesulfonic acid?

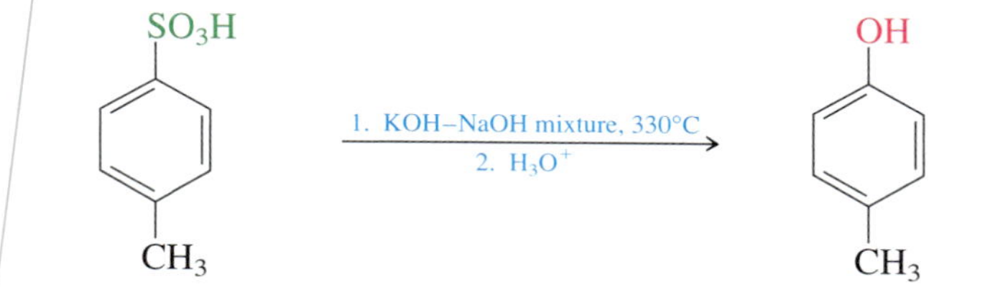

p-Toluenesulfonic acid → (1. KOH–NaOH mixture, 330°C; 2. H_3O^+) → p-Cresol (63–72%)

PROBLEM 24.4 Write a stepwise mechanism for the conversion of p-toluenesulfonic acid to p-cresol under the conditions shown in the preceding equation.

Can you recall how to prepare chlorobenzene?

On the other hand, ^{14}C-labeling studies have shown that the base-promoted hydrolysis of chlorobenzene (second entry in Table 24.3) proceeds by the elimination–addition mechanism and involves benzyne as an intermediate.

TABLE 24.3 Industrial Syntheses of Phenol

Reaction and comments	Chemical equation
Reaction of benzenesulfonic acid with sodium hydroxide This is the oldest method for the preparation of phenol. Benzene is sulfonated and the benzenesulfonic acid heated with molten sodium hydroxide. Acidification of the reaction mixture gives phenol.	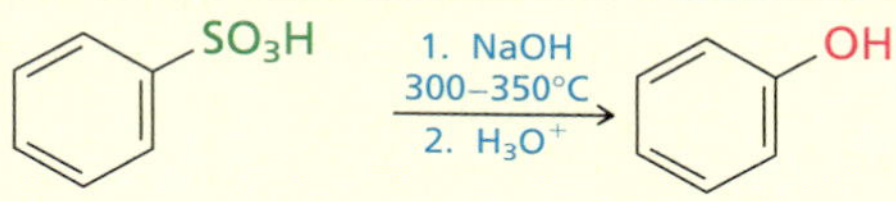 Benzenesulfonic acid → (1. NaOH 300–350°C; 2. H_3O^+) → Phenol
Hydrolysis of chlorobenzene Heating chlorobenzene with aqueous sodium hydroxide at high pressure gives phenol after acidification.	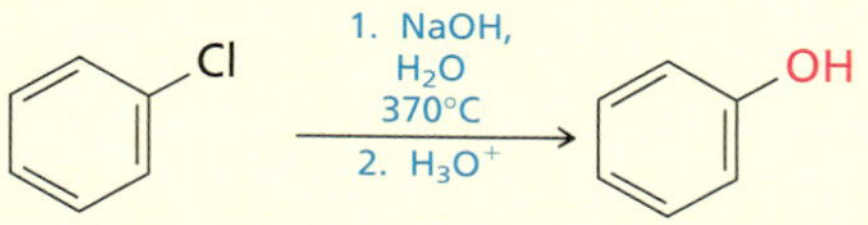 Chlorobenzene → (1. NaOH, H_2O 370°C; 2. H_3O^+) → Phenol
From cumene Almost all the phenol produced in the United States is prepared by this method. Oxidation of cumene takes place at the benzylic position to give a hydroperoxide. On treatment with dilute sulfuric acid, this hydroperoxide is converted to phenol and acetone.	Isopropylbenzene (cumene) —CH(CH$_3$)$_2$ → (O_2) → 1-Methyl-1-phenylethyl hydroperoxide —C(CH$_3$)$_2$ OOH
	1-Methyl-1-phenylethyl hydroperoxide —C(CH$_3$)$_2$ OOH → (H_2O / H_2SO_4) → Phenol —OH + Acetone $(CH_3)_2C$=O

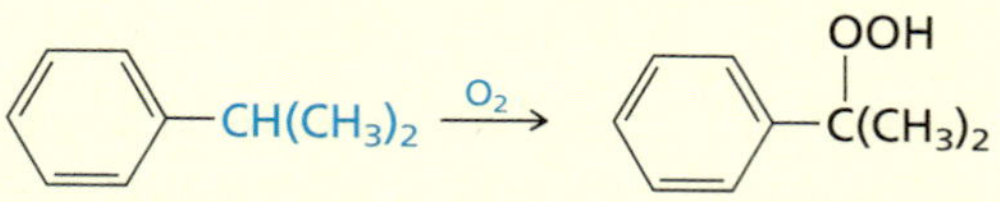

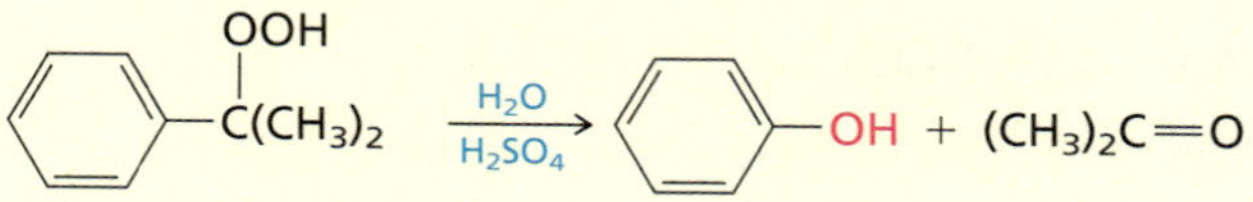

PROBLEM 24.5   Write a stepwise mechanism for the hydrolysis of chlorobenzene under the conditions shown in Table 24.3.

The most widely used industrial synthesis of phenol is based on isopropylbenzene (cumene) as the starting material and is shown in the third entry of Table 24.3. The economically attractive features of this process are its use of cheap reagents (oxygen and sulfuric acid) and the fact that it yields two high-volume industrial chemicals: phenol and acetone. The mechanism of this novel synthesis forms the basis of Problem 24.29 at the end of this chapter.

Can you recall how to prepare isopropylbenzene?

The most important synthesis of phenols in the laboratory is from amines by hydrolysis of their corresponding diazonium salts, as described in Section 22.17:

$$H_2N-\overset{\displaystyle}{\underset{NO_2}{\bigcirc}} \quad \xrightarrow[\text{2. H}_2\text{O, heat}]{\substack{\text{1. NaNO}_2\text{, H}_2\text{SO}_4 \\ \text{H}_2\text{O}}} \quad HO-\overset{\displaystyle}{\underset{NO_2}{\bigcirc}}$$

m-Nitroaniline   *m*-Nitrophenol (81–86%)

24.7   NATURALLY OCCURRING PHENOLS

Phenolic compounds are commonplace natural products. Figure 24.2 presents a sampling of some naturally occurring phenols. Phenolic natural products can arise by a number of different biosynthetic pathways. In animals, aromatic rings are hydroxylated by way of arene oxide intermediates formed by the enzyme-catalyzed reaction between an aromatic ring and molecular oxygen:

Thymol
(major constituent of oil of thyme)

2,5-Dichlorophenol
(isolated from defensive secretion
of a species of grasshopper)

Δ^9-Tetrahydrocannabinol
(active component of marijuana)

Catechin
(one of several flavonoids present
in wine; believed to reduce
risk of atherosclerosis)

FIGURE 24.2 Some naturally occurring phenols.

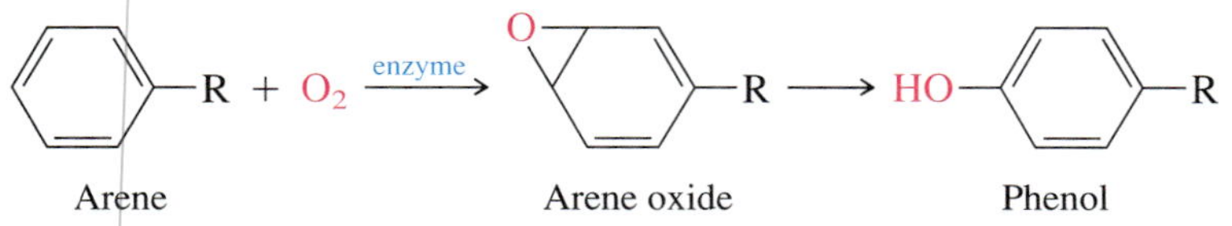

Arene   Arene oxide   Phenol

In plants, phenol biosynthesis proceeds by building the aromatic ring from carbohydrate precursors that already contain the required hydroxyl group.

24.8   REACTIONS OF PHENOLS: ELECTROPHILIC AROMATIC SUBSTITUTION

In most of their reactions phenols behave as nucleophiles, and the reagents that act on them are electrophiles. Either the hydroxyl oxygen or the aromatic ring may be the site of nucleophilic reactivity in a phenol. Reactions that take place on the ring lead to electrophilic aromatic substitution; Table 24.4 summarizes the behavior of phenols in reactions of this type.

A hydroxyl group is a very powerful activating substituent, and electrophilic aromatic substitution in phenols occurs far faster, and under milder conditions, than in benzene. The first entry in Table 24.4, for example, shows the monobromination of phenol in high yield at low temperature and in the absence of any catalyst. In this case, the reaction was carried out in the nonpolar solvent 1,2-dichloroethane. In polar solvents such as water it is difficult to limit the bromination of phenols to monosubstitution. In the following example, all three positions that are ortho or para to the hydroxyl undergo rapid substitution:

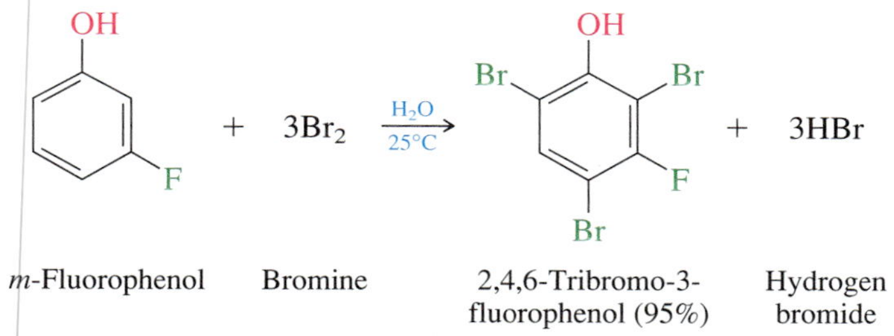

m-Fluorophenol   Bromine   2,4,6-Tribromo-3-   Hydrogen
                                fluorophenol (95%)   bromide

Other typical electrophilic aromatic substitution reactions—nitration (second entry), sulfonation (fourth entry), and Friedel–Crafts alkylation and acylation (fifth and sixth entries)—take place readily and are synthetically useful. Phenols also undergo electrophilic substitution reactions that are limited to only the most active aromatic compounds; these include nitrosation (third entry) and coupling with diazonium salts (seventh entry).

PROBLEM 24.6   Each of the following reactions has been reported in the chemical literature and gives a single organic product in high yield. Identify the product in each case.

(a) 3-Benzyl-2,6-dimethylphenol treated with bromine in chloroform

(b) 4-Bromo-2-methylphenol treated with 2-methylpropene and sulfuric acid

(c) 2-Isopropyl-5-methylphenol (thymol) treated with sodium nitrite and dilute hydrochloric acid

(d) p-Cresol treated with propanoyl chloride and aluminum chloride

TABLE 24.4   Electrophilic Aromatic Substitution Reactions of Phenols

Reaction and comments	Specific example
Halogenation Bromination and chlorination of phenols occur readily even in the absence of a catalyst. Substitution occurs primarily at the position para to the hydroxyl group. When the para position is blocked, ortho substitution is observed.	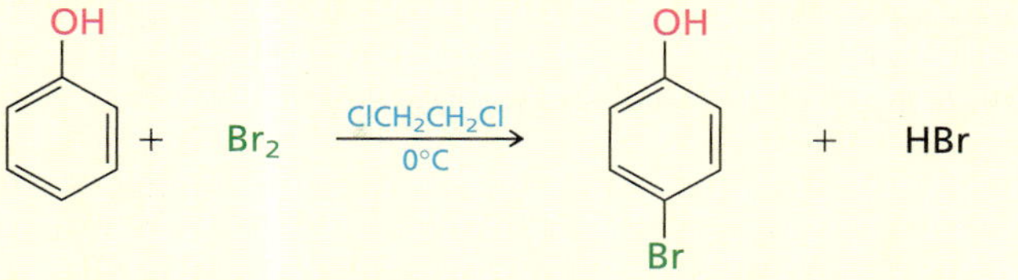 Phenol Bromine p-Bromophenol (93%) Hydrogen bromide
Nitration Phenols are nitrated on treatment with a dilute solution of nitric acid in either water or acetic acid. It is not necessary to use mixtures of nitric and sulfuric acids, because of the high reactivity of phenols.	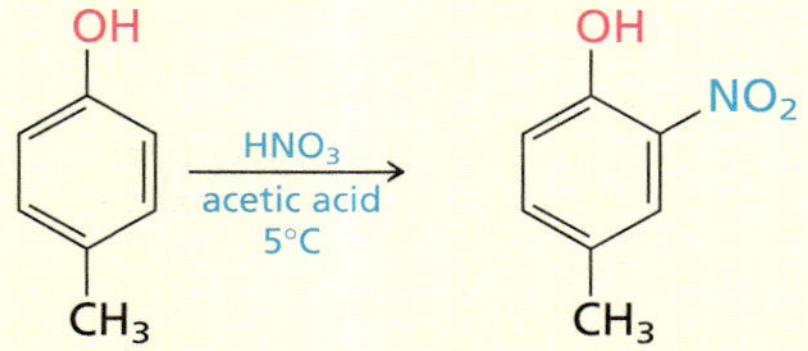 p-Cresol 4-Methyl-2-nitrophenol (73–77%)
Nitrosation On acidification of aqueous solutions of sodium nitrite, the nitrosyl cation ($:N{\equiv}O:$) is formed, which is a weak electrophile and attacks the strongly activated ring of a phenol. The product is a nitroso phenol.	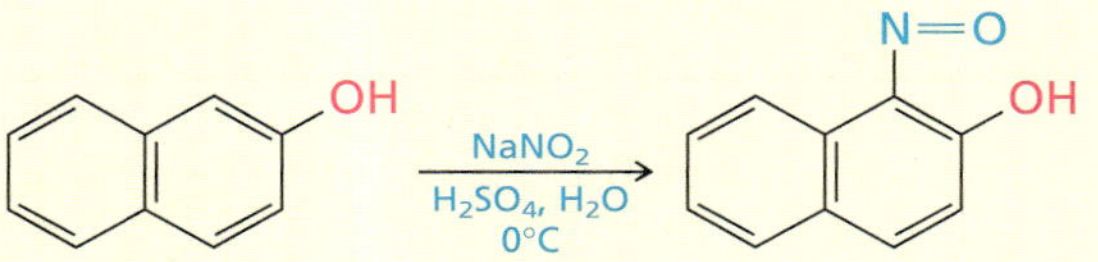 2-Naphthol 1-Nitroso-2-naphthol (99%)
Sulfonation Heating a phenol with concentrated sulfuric acid causes sulfonation of the ring.	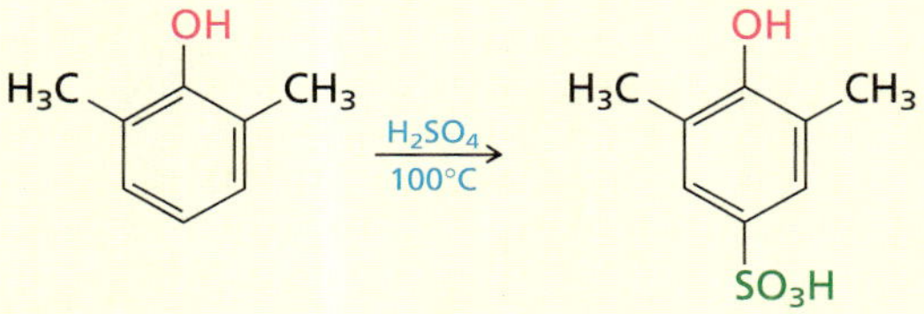 2,6-Dimethylphenol 4-Hydroxy-3,5-dimethylbenzenesulfonic acid (69%)
Friedel–Crafts alkylation Alcohols in combination with acids serve as sources of carbocations. Attack of a carbocation on the electron-rich ring of a phenol brings about its alkylation.	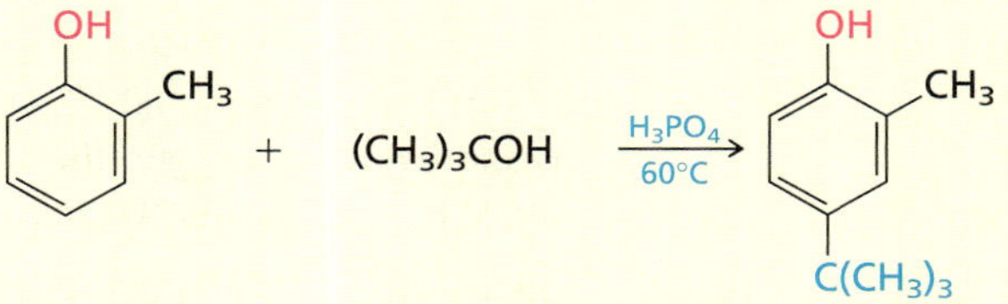 o-Cresol tert-Butyl alcohol 4-tert-Butyl-2-methylphenol (63%)

(Continued)

TABLE 24.4   Electrophilic Aromatic Substitution Reactions of Phenols (Continued)

Reaction and comments	Specific example
Friedel–Crafts acylation In the presence of aluminum chloride, acyl chlorides and carboxylic acid anhydrides acylate the aromatic ring of phenols.	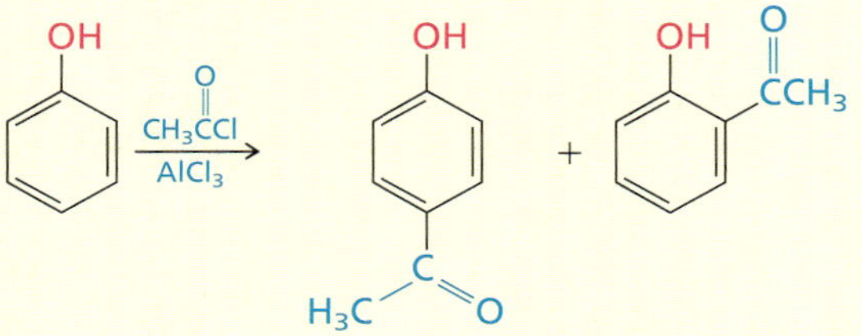 Phenol *p*-Hydroxyaceto-phenone (74%) *o*-Hydroxyaceto-phenone (16%)
Reaction with arenediazonium salts Adding a phenol to a solution of a diazonium salt formed from a primary aromatic amine leads to formation of an azo compound. The reaction is carried out at a pH such that a significant portion of the phenol is present as its phenoxide ion. The diazonium ion acts as an electrophile toward the strongly activated ring of the phenoxide ion.	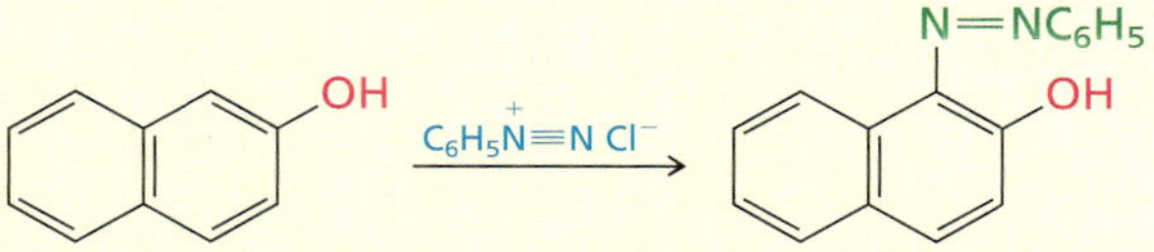 2-Naphthol 1-Phenylazo-2-naphthol (48%)

SAMPLE SOLUTION (a) The ring that bears the hydroxyl group is much more reactive than the other ring. In electrophilic aromatic substitution reactions of rings that bear several substituents, it is the most activating substituent that controls the orientation. Bromination occurs para to the hydroxyl group.

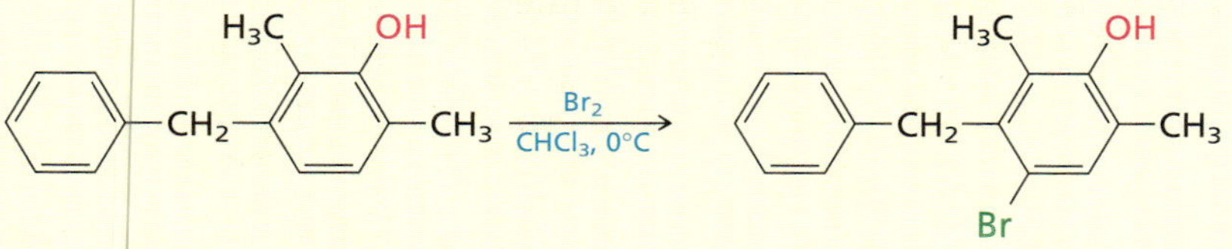

3-Benzyl-2,6-dimethylphenol   3-Benzyl-4-bromo-2,6-dimethylphenol (isolated in 100% yield)

The aromatic ring of a phenol, like that of an arylamine, is seen as an electron-rich functional unit and is capable of a variety of reactions. In some cases, however, it is the hydroxyl oxygen that reacts instead. An example of this kind of chemical reactivity is described in the following section.

24.9   ACYLATION OF PHENOLS

Acylating agents, such as acyl chlorides and carboxylic acid anhydrides, can react with phenols either at the aromatic ring (C-acylation) or at the hydroxyl oxygen (O-acylation):

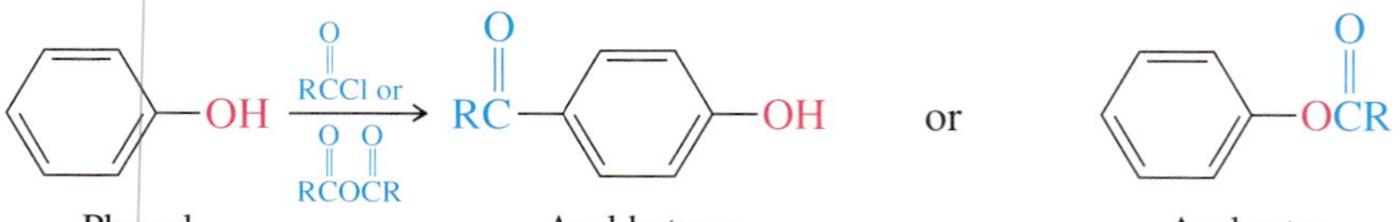

Phenol   Aryl ketone (product of C-acylation)   or   Aryl ester (product of O-acylation)

As shown in the sixth entry of Table 24.4, C-acylation of phenols is observed under the customary conditions of the Friedel–Crafts reaction (treatment with an acyl chloride or acid anhydride in the presence of aluminum chloride). In the absence of aluminum chloride, however, O-acylation occurs instead.

$$
\underset{\text{Phenol}}{\text{C}_6\text{H}_5\text{—OH}} + \underset{\substack{\text{Octanoyl chloride}}}{\text{CH}_3(\text{CH}_2)_6\overset{\overset{\displaystyle O}{\|}}{C}\text{Cl}} \longrightarrow \underset{\substack{\text{Phenyl octanoate}\\(95\%)}}{\text{C}_6\text{H}_5\text{—O}\overset{\overset{\displaystyle O}{\|}}{C}(\text{CH}_2)_6\text{CH}_3} + \underset{\substack{\text{Hydrogen}\\\text{chloride}}}{\text{HCl}}
$$

The O-acylation of phenols with carboxylic acid anhydrides can be conveniently catalyzed in either of two ways. One method involves converting the acid anhydride to a more powerful acylating agent by protonation of one of its carbonyl oxygens. Addition of a few drops of sulfuric acid is usually sufficient.

$$
\underset{\substack{p\text{-Fluorophenol}}}{\text{F}\text{—C}_6\text{H}_4\text{—OH}} + \underset{\substack{\text{Acetic}\\\text{anhydride}}}{\text{CH}_3\overset{\overset{\displaystyle O}{\|}}{C}\text{O}\overset{\overset{\displaystyle O}{\|}}{C}\text{CH}_3} \xrightarrow{\text{H}_2\text{SO}_4} \underset{\substack{p\text{-Fluorophenyl acetate}\\(81\%)}}{\text{F}\text{—C}_6\text{H}_4\text{—O}\overset{\overset{\displaystyle O}{\|}}{C}\text{CH}_3} + \underset{\substack{\text{Acetic}\\\text{acid}}}{\text{CH}_3\overset{\overset{\displaystyle O}{\|}}{C}\text{OH}}
$$

An alternative approach is to increase the nucleophilicity of the phenol by converting it to its phenoxide anion in basic solution:

$$
\underset{\substack{\text{Resorcinol}}}{\text{HO}\text{—C}_6\text{H}_4\text{—OH}} + \underset{\substack{\text{Acetic}\\\text{anhydride}}}{2\text{CH}_3\overset{\overset{\displaystyle O}{\|}}{C}\text{O}\overset{\overset{\displaystyle O}{\|}}{C}\text{CH}_3} \xrightarrow[\text{H}_2\text{O}]{\text{NaOH}} \underset{\substack{\text{1,3-Diacetoxybenzene}\\(93\%)}}{\text{C}_6\text{H}_4(\text{OCOCH}_3)_2} + \underset{\substack{\text{Sodium}\\\text{acetate}}}{2\text{CH}_3\overset{\overset{\displaystyle O}{\|}}{C}\text{ONa}}
$$

PROBLEM 24.7 Write chemical equations expressing each of the following:

(a) Preparation of o-nitrophenyl acetate by sulfuric acid catalysis of the reaction between a phenol and a carboxylic acid anhydride.

(b) Esterification of 2-naphthol with acetic anhydride in aqueous sodium hydroxide

(c) Reaction of phenol with benzoyl chloride

SAMPLE SOLUTION (a) The problem specifies that an acid anhydride be used; therefore, use acetic anhydride to prepare the acetate ester of o-nitrophenol:

$$
\underset{\substack{o\text{-Nitrophenol}}}{\text{C}_6\text{H}_4(\text{NO}_2)\text{—OH}} + \underset{\substack{\text{Acetic anhydride}}}{\text{CH}_3\overset{\overset{\displaystyle O}{\|}}{C}\text{O}\overset{\overset{\displaystyle O}{\|}}{C}\text{CH}_3} \xrightarrow{\text{H}_2\text{SO}_4} \underset{\substack{o\text{-Nitrophenyl acetate}\\(\text{isolated in 93\% yield by}\\\text{this method})}}{\text{C}_6\text{H}_4(\text{NO}_2)\text{—O}\overset{\overset{\displaystyle O}{\|}}{C}\text{CH}_3} + \underset{\substack{\text{Acetic acid}}}{\text{CH}_3\overset{\overset{\displaystyle O}{\|}}{C}\text{OH}}
$$

The preference for O-acylation of phenols arises because these reactions are *kinetically controlled*. O-acylation is faster than C-acylation. The C-acyl isomers are more stable, however, and it is known that aluminum chloride is a very effective catalyst for the conversion of aryl esters to aryl ketones. This isomerization is called the **Fries rearrangement.**

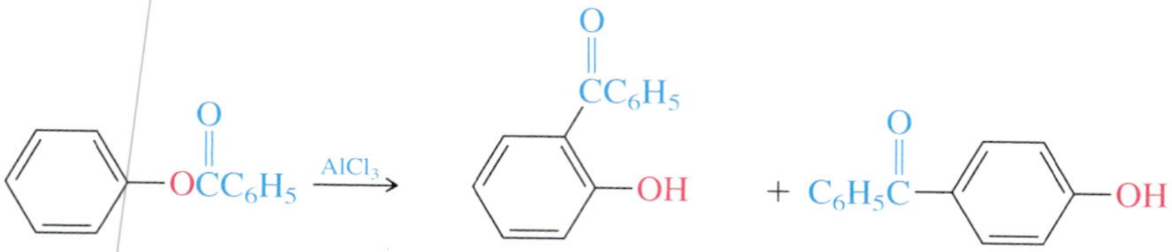

Phenyl benzoate   *o*-Hydroxybenzophenone   *p*-Hydroxybenzophenone
                                  (9%)   (64%)

Thus, ring acylation of phenols is observed under Friedel–Crafts conditions because the presence of aluminum chloride causes that reaction to be subject to *thermodynamic (equilibrium) control*.

Fischer esterification, in which a phenol and a carboxylic acid condense in the presence of an acid catalyst, is not used to prepare aryl esters.

24.10 CARBOXYLATION OF PHENOLS: ASPIRIN AND THE KOLBE–SCHMITT REACTION

The best known aryl ester is *O*-acetylsalicylic acid, better known as *aspirin*. It is prepared by acetylation of the phenolic hydroxyl group of salicylic acid:

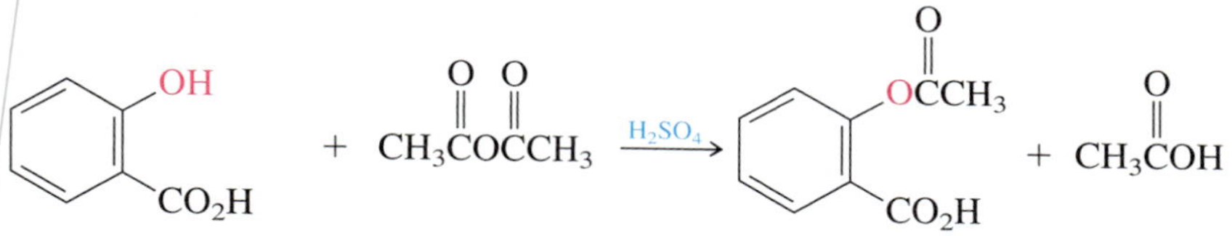

Salicylic acid   Acetic anhydride   *O*-Acetylsalicylic   Acetic acid
(*o*-hydroxybenzoic acid)   acid (aspirin)

An entertaining account of the history of aspirin can be found in the 1991 book *The Aspirin Wars: Money, Medicine, and 100 Years of Rampant Competition*, by Charles C. Mann.

Aspirin possesses a number of properties that make it an often-recommended drug. It is an analgesic, effective in relieving headache pain. It is also an antiinflammatory agent, providing some relief from the swelling associated with arthritis and minor injuries. Aspirin is an antipyretic compound; that is, it reduces fever. How aspirin does all this was once a mystery but is now better understood and will be discussed in Section 26.6. Each year, more than 40 million lb of aspirin is produced in the United States, a rate equal to 300 tablets per year for every man, woman, and child.

The key compound in the synthesis of aspirin, salicylic acid, is prepared from phenol by a process discovered in the nineteenth century by the German chemist Hermann Kolbe. In the Kolbe synthesis, also known as the **Kolbe–Schmitt reaction,** sodium phenoxide is heated with carbon dioxide under pressure, and the reaction mixture is subsequently acidified to yield salicylic acid:

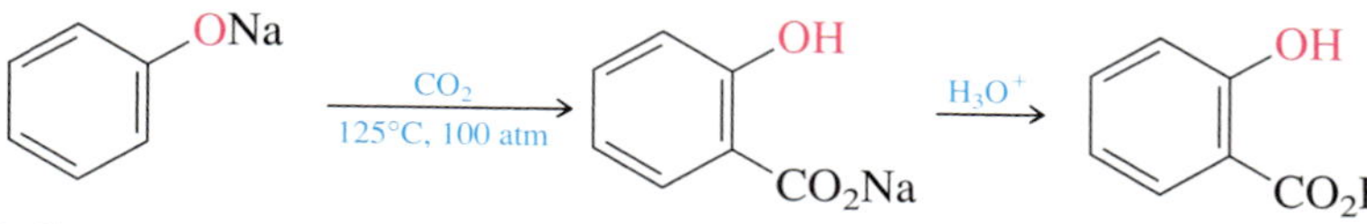

Sodium phenoxide   Sodium salicylate   Salicylic acid (79%)

Although a hydroxyl group strongly activates an aromatic ring toward electrophilic attack, an oxyanion substituent is an even more powerful activator. Electron delocalization in phenoxide anion leads to increased electron density at the positions ortho and para to oxygen.

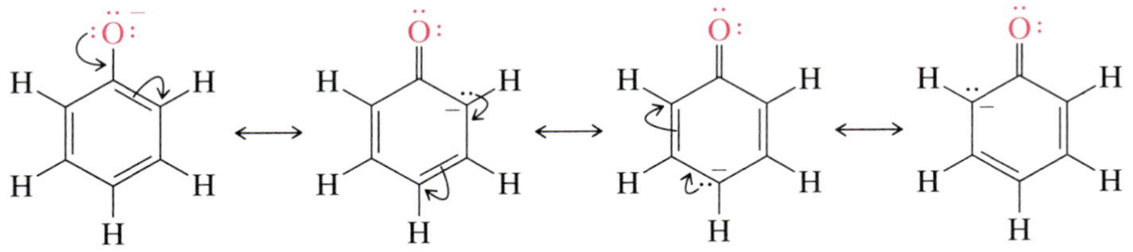

This is the same resonance description shown in Section 24.4.

The increased nucleophilicity of the ring permits it to react with carbon dioxide. An intermediate is formed that is simply the keto form of salicylate anion:

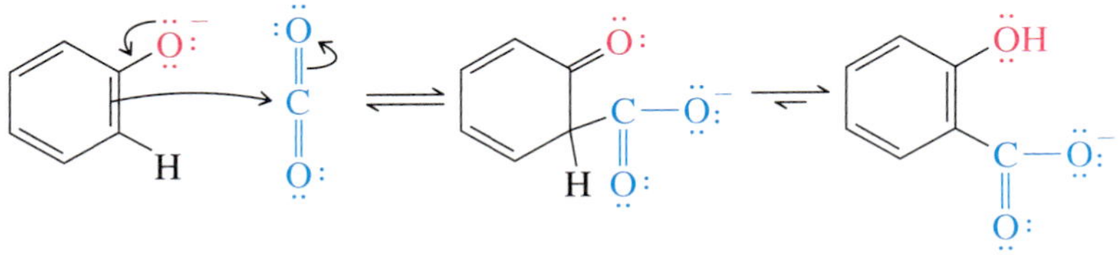

| Phenoxide anion (stronger base) | Carbon dioxide | Cyclohexadienone intermediate | Salicylate anion (weaker base) |

The Kolbe–Schmitt reaction is an equilibrium process governed by thermodynamic control. The position of equilibrium favors formation of the weaker base (salicylate ion) at the expense of the stronger one (phenoxide ion). Thermodynamic control is also responsible for the pronounced bias toward ortho over para substitution. Salicylate anion is a weaker base than *p*-hydroxybenzoate and predominates at equilibrium.

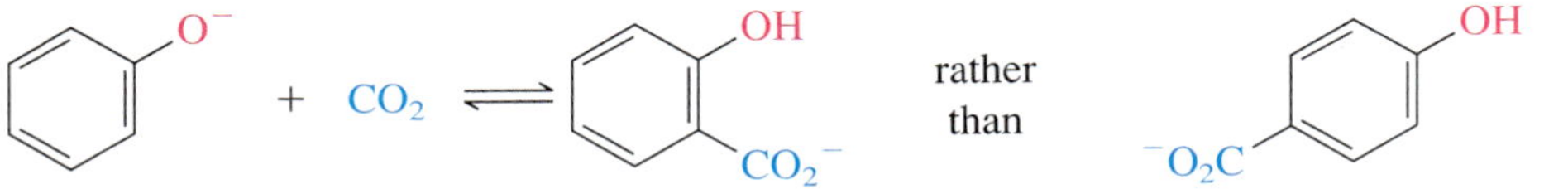

Phenoxide ion (strongest base; pK_a of conjugate acid, 10)   Carbon dioxide   Salicylate anion (weakest base; pK_a of conjugate acid, 3)   rather than   *p*-Hydroxybenzoate anion (pK_a of conjugate acid, 4.5)

Salicylate anion is a weaker base than *p*-hydroxybenzoate because it is stabilized by intramolecular hydrogen bonding.

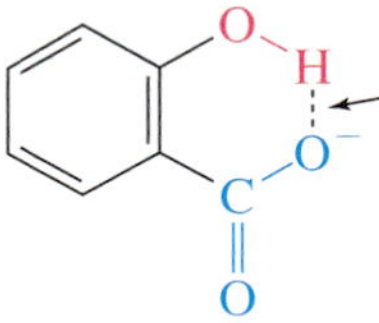

Intramolecular hydrogen bonding in salicylate anion

The Kolbe–Schmitt reaction has been applied to the preparation of other *o*-hydroxybenzoic acids. Alkyl derivatives of phenol behave very much like phenol itself.

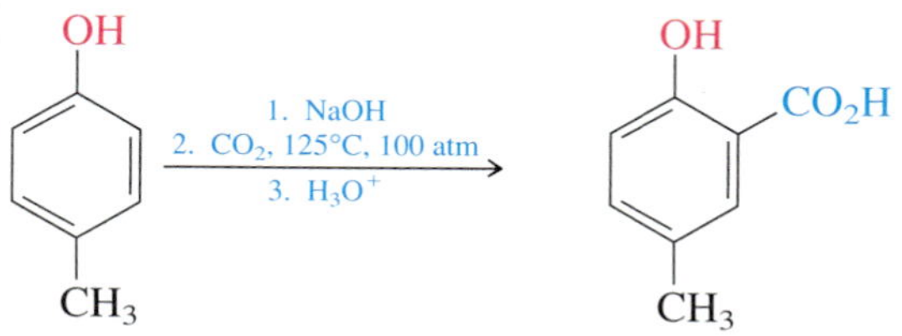

p-Cresol

2-Hydroxy-5-methylbenzoic
acid (78%)

Phenols that bear strongly electron-withdrawing substituents usually give low yields of carboxylated products; their derived phenoxide anions are less basic, and the equilibrium constants for their carboxylation are smaller.

24.11 PREPARATION OF ARYL ETHERS

Aryl ethers are best prepared by the Williamson method (Section 16.6). Alkylation of the hydroxyl oxygen of a phenol takes place readily when a phenoxide anion reacts with an alkyl halide.

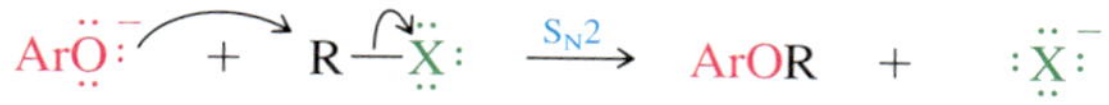

| Phenoxide anion | Alkyl halide | | Alkyl aryl ether | Halide anion |

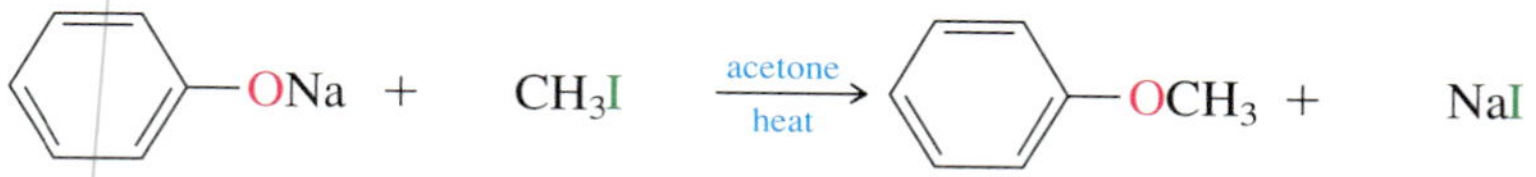

Sodium phenoxide   Iodomethane   Anisole (95%)   Sodium iodide

As the synthesis is normally performed, a solution of the phenol and alkyl halide is simply heated in the presence of a suitable base such as potassium carbonate:

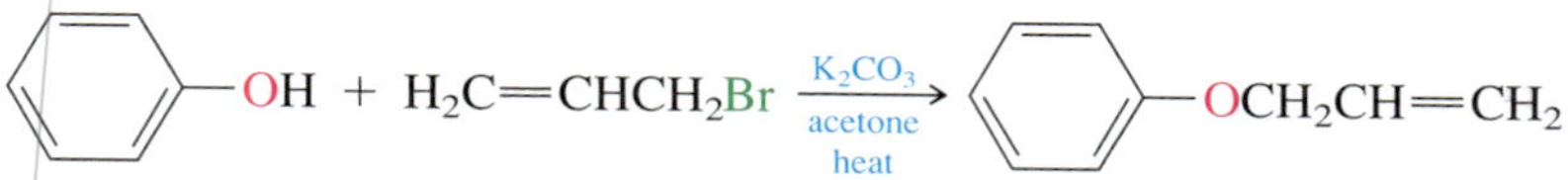

Phenol   Allyl bromide   Allyl phenyl ether (86%)

This is an example of an S$_N$2 reaction in a polar aprotic solvent.

The alkyl halide must be one that reacts readily by an S$_N$2 mechanism. Thus, methyl and primary alkyl halides are the most effective alkylating agents. Elimination competes with substitution when secondary alkyl halides are used and is the only reaction observed with tertiary alkyl halides.

PROBLEM 24.8 Reaction of phenol with 1,2-epoxypropane in aqueous sodium hydroxide at 150°C gives a single product, $C_9H_{12}O_2$, in 90% yield. Suggest a reasonable structure for this compound.

The reaction between an alkoxide ion and an aryl halide can be used to prepare alkyl aryl ethers only when the aryl halide is one that reacts rapidly by the addition–elimination mechanism of nucleophilic aromatic substitution (Section 23.6).

AGENT ORANGE AND DIOXIN

The once widely used herbicide 2,4,5-trichlorophenoxyacetic acid (2,4,5-T) is prepared by reaction of the sodium salt of 2,4,5-trichlorophenol with chloroacetic acid:

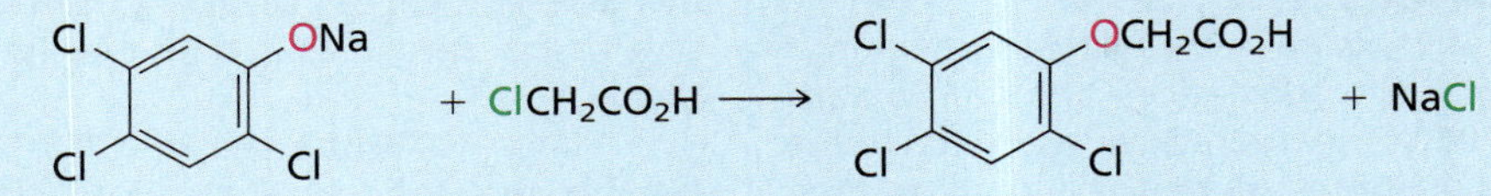

| Sodium 2,4,5-trichlorophenolate | Chloroacetic acid | | 2,4,5-Trichlorophenoxyacetic acid (2,4,5-T) | |

The starting material for this process, 2,4,5-trichlorophenol, is made by treating 1,2,4,5-tetrachlorobenzene with aqueous base. Nucleophilic aromatic substitution of one of the chlorines by an addition–elimination mechanism yields 2,4,5-trichlorophenol:

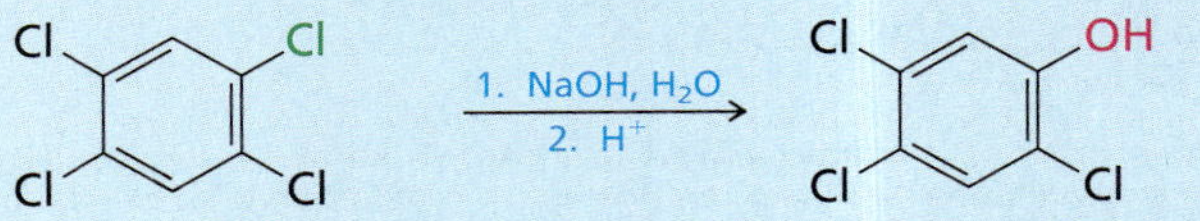

1,2,4,5-Tetrachlorobenzene   2,4,5-Trichlorophenol

In the course of making 2,4,5-trichlorophenol, it almost always becomes contaminated with small amounts of 2,3,7,8-tetrachlorodibenzo-*p*-dioxin, better known as *dioxin*.

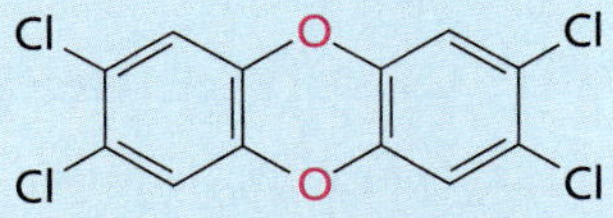

2,3,7,8-Tetrachlorodibenzo-*p*-dioxin
(dioxin)

—Cont.

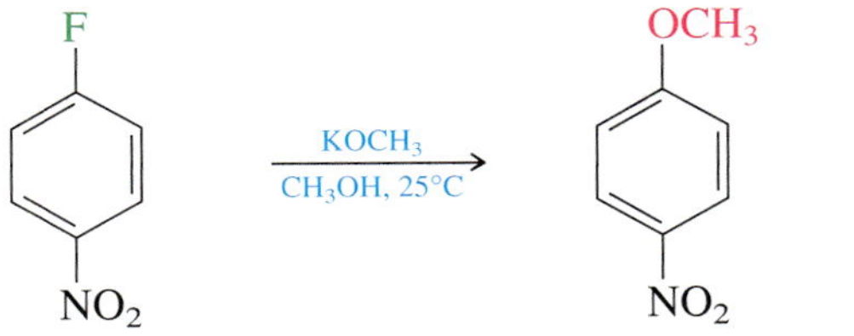

p-Fluoronitrobenzene   *p*-Nitroanisole (93%)

PROBLEM 24.9 Which of the following two combinations of reactants is more appropriate for the preparation of *p*-nitrophenyl phenyl ether?

Fluorobenzene and sodium *p*-nitrophenoxide

p-Fluoronitrobenzene and sodium phenoxide

Dioxin is carried along when 2,4,5-trichlorophenol is converted to 2,4,5-T and enters the environment when 2,4,5-T is sprayed on vegetation. Typically, the amount of dioxin present in 2,4,5-T is very small. *Agent Orange,* a 2,4,5-T-based defoliant used on a large scale in the Vietnam War, contained about 2 ppm of dioxin.

Tests with animals have revealed that dioxin is one of the most toxic substances known. Toward mice it is about 2000 times more toxic than strychnine and about 150,000 times more toxic than sodium cyanide. Fortunately, however, available evidence indicates that humans are far more resistant to dioxin than are test animals, and so far there have been no human fatalities directly attributable to dioxin. The most prominent short-term symptom seen so far has been a severe skin disorder known as *chloracne.* Yet to be determined is the answer to the question of

long-term effects. A 1991 study of the health records of over 5000 workers who were exposed to dioxin-contaminated chemicals indicated a 15% increase in incidences of cancer compared with those of a control group. Workers who were exposed to higher dioxin levels for prolonged periods exhibited a 50% increase in their risk of dying from cancer, especially soft-tissue sarcomas, compared with the control group.*

Since 1979, the use of 2,4,5-T has been regulated in the United States. It is likely that the United States Environmental Protection Agency will classify some dioxins as "known" and others as "probable" human carcinogens and recommend further controls be placed on processes that produce them. It appears, from decreasing dioxin levels in some soils, that existing regulations are having some effect.†

*The biological properties of dioxin include an ability to bind to a protein known as the AH (aromatic hydrocarbon) receptor. Dioxin is not a hydrocarbon, but it shares a certain structural property with aromatic hydrocarbons. Try constructing molecular models of dioxin and anthracene to see these similarities.

†For a detailed discussion of the sources and biological effects of doxins, see the article "Dioxins, Not Doomsday" in the December 1999 issue of the *Journal of Chemical Education,* pp. 1662–1666.

24.12 CLEAVAGE OF ARYL ETHERS BY HYDROGEN HALIDES

The cleavage of *dialkyl ethers* by hydrogen halides was discussed in Section 16.8, where it was noted that the same pair of alkyl halides results, irrespective of the order in which the carbon–oxygen bonds of the ether are broken.

$$\text{ROR}' + 2\text{HX} \longrightarrow \text{RX} + \text{R}'\text{X} + \text{H}_2\text{O}$$

Dialkyl ether   Hydrogen halide   Two alkyl halides   Water

Cleavage of *alkyl aryl ethers* by hydrogen halides always proceeds so that the alkyl–oxygen bond is broken and yields an alkyl halide and a phenol as the *final* products. Either hydrogen bromide or hydrogen iodide is normally used.

$$\text{ArOR} + \text{HX} \longrightarrow \text{ArOH} + \text{RX}$$

Alkyl aryl ether   Hydrogen halide   Phenol   Alkyl halide

Because phenols are not converted to aryl halides by reaction with hydrogen halides, reaction proceeds no further than shown in the preceding general equation. For example,

Guaiacol → Pyrocatechol (85–87%) + CH₃Br Methyl bromide (57–72%)

Guaiacol is obtained by chemical treatment of *lignum vitae,* the wood from a species of tree that grows in warm climates. It is sometimes used as an expectorant to help relieve bronchial congestion.

The first step in the reaction of an alkyl aryl ether with a hydrogen halide is protonation of oxygen to form an alkylaryloxonium ion:

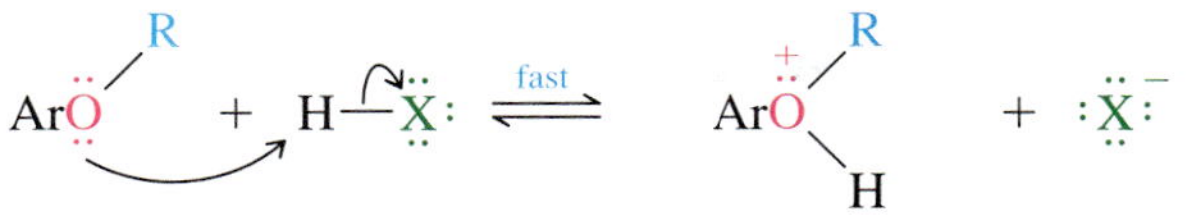

| Alkyl aryl ether | Hydrogen halide | Alkylaryloxonium ion | Halide ion |

This is followed by a nucleophilic substitution step:

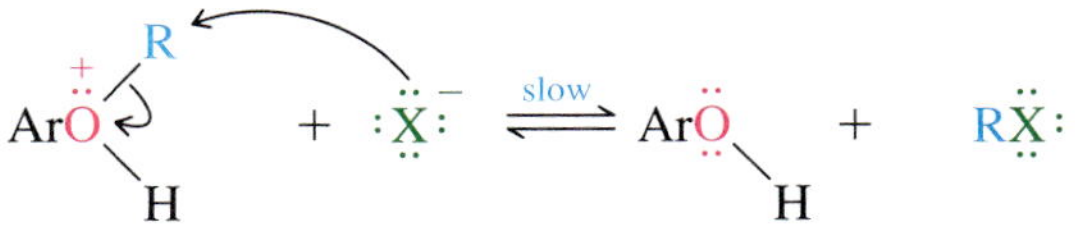

| Alkylaryloxonium ion | Halide ion | Phenol | Alkyl halide |

Attack by the halide nucleophile at the sp^3-hybridized carbon of the alkyl group is analogous to what takes place in the cleavage of dialkyl ethers. Attack at the sp^2-hybridized carbon of the aromatic ring is much slower. Indeed, nucleophilic aromatic substitution does not occur at all under these conditions.

24.13 CLAISEN REARRANGEMENT OF ALLYL ARYL ETHERS

Allyl aryl ethers undergo an interesting reaction, called the **Claisen rearrangement,** on being heated. The allyl group migrates from oxygen to the ring carbon ortho to it.

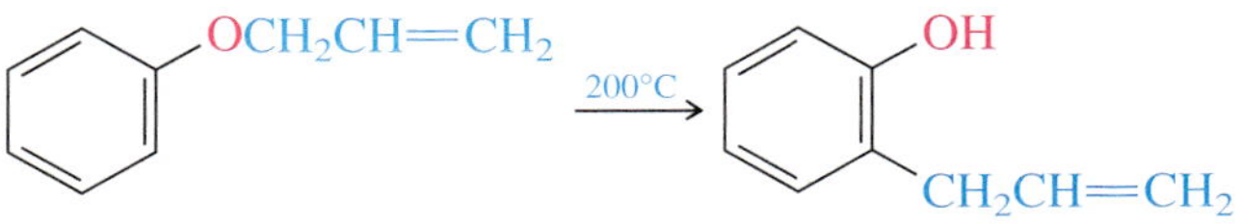

Allyl phenyl ether o-Allylphenol (73%)

Allyl phenyl ether is prepared by the reaction of phenol with allyl bromide, as described in Section 24.11.

Carbon-14 labeling of the allyl group reveals that the terminal carbon of the allyl group is the one that becomes bonded to the ring and suggests a mechanism involving a concerted electron reorganization in the first step. This step is followed by enolization of the resulting cyclohexadienone to regenerate the aromatic ring.

$* = {}^{14}C$

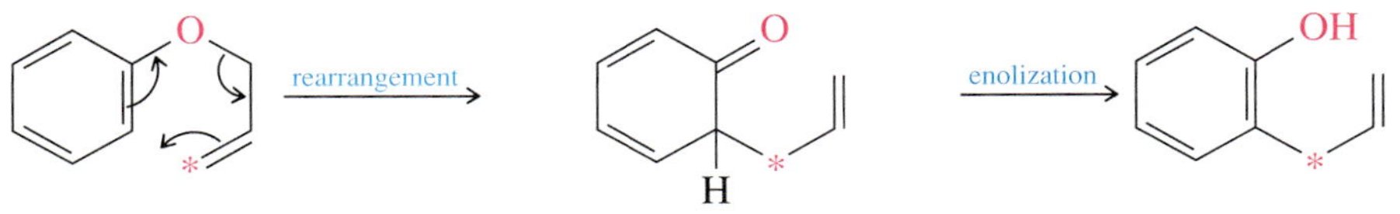

Allyl phenyl ether   6-Allyl-2,4-cyclohexadienone o-Allylphenol

PROBLEM 24.10 The mechanism of the Claisen rearrangement of other allylic ethers of phenol is analogous to that of allyl phenyl ether. What is the product of the Claisen rearrangement of $C_6H_5OCH_2CH=CHCH_3$?

The transition state for the first step of the Claisen rearrangement bears much in common with the transition state for the Diels–Alder cycloaddition. Both involve a concerted six-electron reorganization.

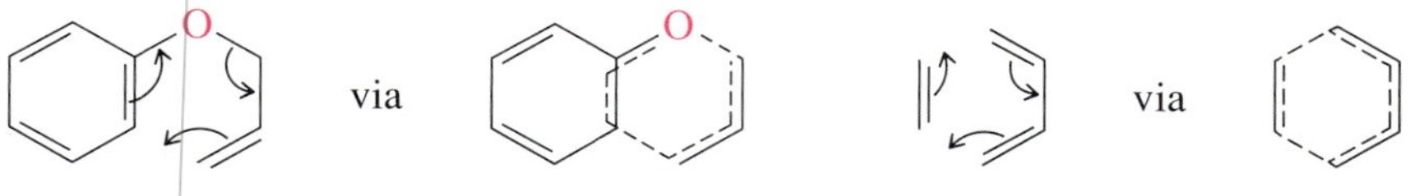

Claisen rearrangement   via   Diels–Alder cycloaddition

The Claisen rearrangement is an example of a **sigmatropic rearrangement.** A sigmatropic rearrangement is characterized by a transition state in which a σ bond migrates from one end of a conjugated π electron system to the other. In this case the σ bond to oxygen at one end of an allyl unit is broken and replaced by a σ bond to the ring carbon at the other end.

24.14 OXIDATION OF PHENOLS: QUINONES

Phenols are more easily oxidized than alcohols, and a large number of inorganic oxidizing agents have been used for this purpose. The phenol oxidations that are of the most use to the organic chemist are those involving derivatives of 1,2-benzenediol (pyrocatechol) and 1,4-benzenediol (hydroquinone). Oxidation of compounds of this type with silver oxide or with chromic acid yields conjugated dicarbonyl compounds called **quinones.**

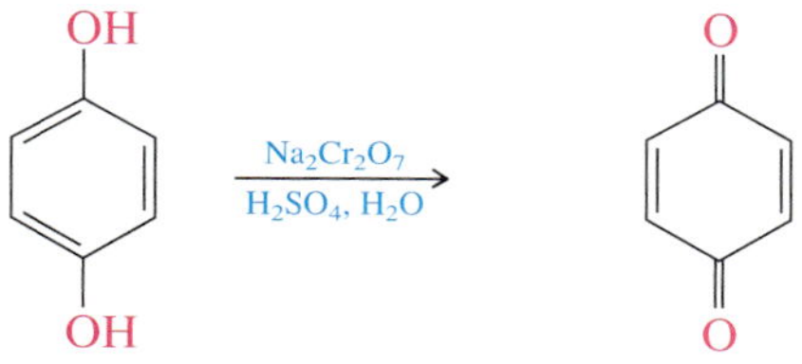

Hydroquinone   p-Benzoquinone (76–81%)

> Silver oxide is a weak oxidizing agent.

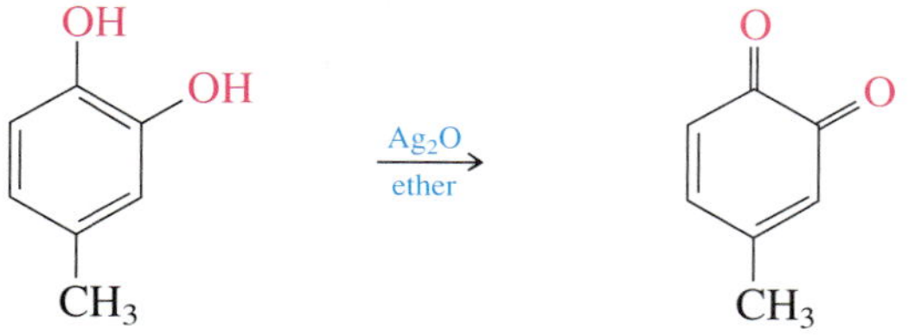

4-Methylpyrocatechol   4-Methyl-1,2-benzoquinone (68%)
(4-methyl-1,2-benzenediol)

Quinones are colored; p-benzoquinone, for example, is yellow. Many occur naturally and have been used as dyes. *Alizarin* is a red pigment extracted from the roots of the madder plant. Its preparation from anthracene, a coal tar derivative, in 1868 was a significant step in the development of the synthetic dyestuff industry.

> Quinones that are based on the anthracene ring system are called *anthraquinones.* Alizarin is one example of an *anthraquinone dye.*

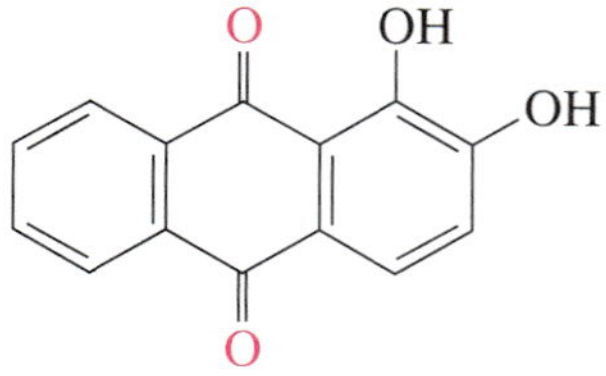

Alizarin

The oxidation–reduction process that connects hydroquinone and benzoquinone involves two 1-electron transfers:

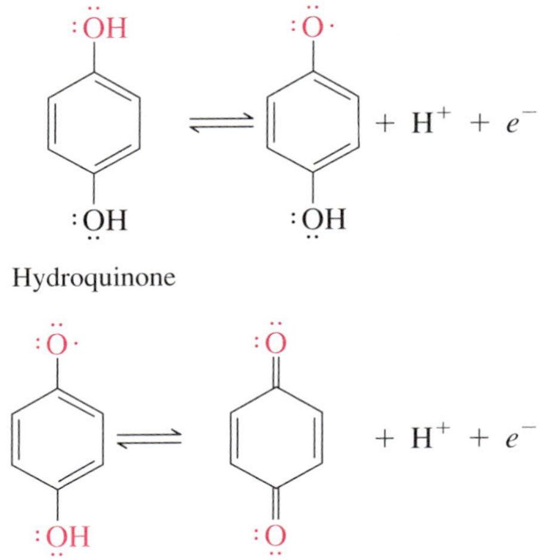

Hydroquinone

Benzoquinone

The ready reversibility of this reaction is essential to the role that quinones play in cellular respiration, the process by which an organism uses molecular oxygen to convert its food to carbon dioxide, water, and energy. Electrons are not transferred directly from the substrate molecule to oxygen but instead are transferred by way of an *electron transport chain* involving a succession of oxidation–reduction reactions. A key component of this electron transport chain is the substance known as *ubiquinone,* or coenzyme Q:

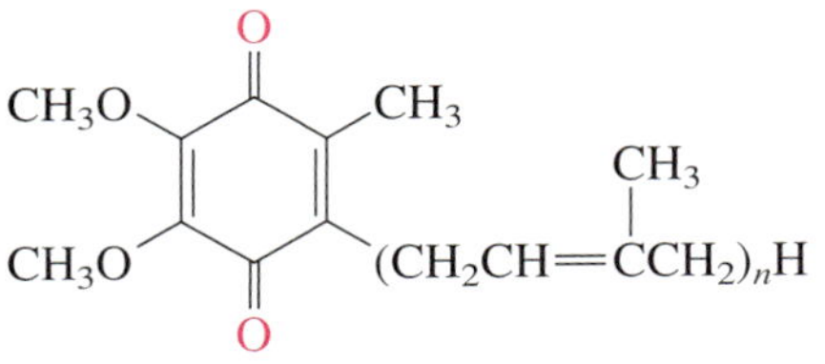

Ubiquinone (coenzyme Q)

The name *ubiquinone* is a shortened form of *ubiquitous quinone,* a term coined to describe the observation that this substance can be found in all cells. The length of its side chain varies among different organisms; the most common form in vertebrates has $n = 10$, and ubiquinones in which $n = 6$ to 9 are found in yeasts and plants.

Another physiologically important quinone is vitamin K. Here "K" stands for *koagulation* (Danish) because this substance was first identified as essential for the normal clotting of blood.

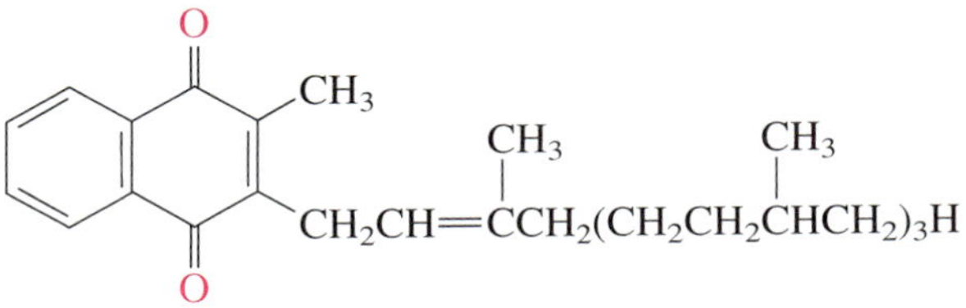

Vitamin K

Intestinal flora is a general term for the bacteria, yeast, and fungi that live in the large intestine.

Some vitamin K is provided in the normal diet, but a large proportion of that required by humans is produced by their intestinal flora.

24.15 SPECTROSCOPIC ANALYSIS OF PHENOLS

Infrared: The IR spectra of phenols combine features of those of alcohols and aromatic compounds. Hydroxyl absorbances resulting from O—H stretching are found in the 3600-cm^{-1} region, and the peak due to C—O stretching appears around 1200–1250 cm^{-1}. These features can be seen in the IR spectrum of *p*-cresol, shown in Figure 24.3.

^{1}H NMR: The ^{1}H NMR signals for the hydroxyl protons of phenols are often broad, and their chemical shift, like their acidity, lies between alcohols and carboxylic acids. The range is δ 4–12, with the exact chemical shift depending on the concentration, the solvent, and the temperature. The phenolic proton in the ^{1}H NMR spectrum shown for *p*-cresol, for example, appears at δ 5.1 (Figure 24.4).

^{13}C NMR: Compared with C—H, the carbon of C—O in a phenol is deshielded by about 25 ppm. In the case of *m*-cresol, for example, the C—O carbon gives the signal at lowest field.

The ^{13}C NMR spectrum of m-cresol appeared in Chapter 13 (Figure 13.21).

OH
155.1

112.3   116.1

129.4   139.8

121.7   CH$_3$
         21.3

^{13}C chemical shifts δ
in *m*-cresol (ppm)

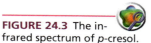

FIGURE 24.3 The infrared spectrum of *p*-cresol.

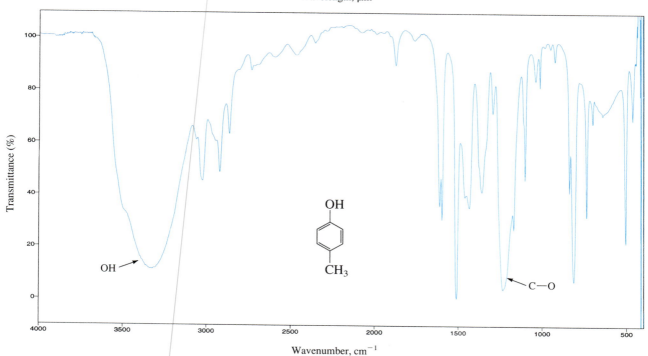

Notice, too, that the most shielded carbons of the aromatic ring are the ones that are ortho and para to the hydroxyl group in keeping with our experience that the OH group donates electrons preferentially to these positions.

UV-VIS: Just as with arylamines (Section 22.20), it is informative to look at the UV-VIS behavior of phenols in terms of how the OH group affects the benzene chromophore.

	X	λ_{max} (nm)
Benzene	H	204, 256
Aniline	NH_2	230, 280
Anilinium ion	NH_3^+	203, 254
Phenol	OH	210, 270
Phenoxide ion	O^-	235, 287

An OH group affects the UV-VIS spectrum of benzene in a way similar to that of an NH_2 group, but to a smaller extent. In basic solution, in which OH is converted to O^-, however, the shift to longer wavelengths exceeds that of an NH_2 group.

Mass Spectrometry: A peak for the molecular ion is usually quite prominent in the mass spectra of phenols. It is, for example, the most intense peak in phenol.

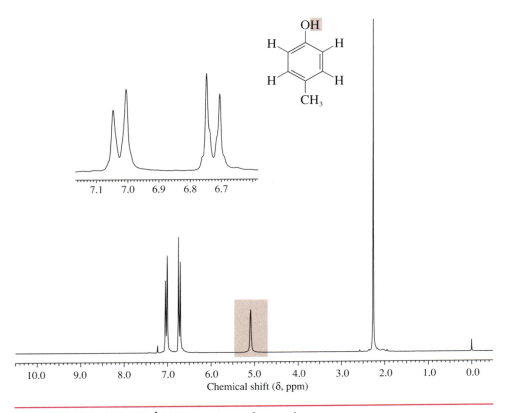

FIGURE 24.4 The 200-MHz ^{1}H NMR spectrum of *p*-cresol.

24.16 SUMMARY

Section 24.1   Phenol is both an important industrial chemical and the parent of a large class of compounds widely distributed as natural products. Although *benzenol* is the systematic name for C_6H_5OH, the IUPAC rules permit *phenol* to be used instead. Substituted derivatives are named on the basis of phenol as the parent compound.

Section 24.2   Phenols are polar compounds, but less polar than alcohols. They resemble arylamines in having an electron-rich aromatic ring.

Section 24.3   The —OH group of phenols makes it possible for them to participate in hydrogen bonding. This contributes to the higher boiling points and greater water-solubility of phenolic compounds compared with arenes and aryl halides.

Section 24.4   With pK_a's of approximately 10, phenols are stronger acids than alcohols, but weaker than carboxylic acids. They are converted quantitatively to phenoxide anions on treatment with aqueous sodium hydroxide.

$$ArOH + NaOH \longrightarrow ArONa + H_2O$$

Section 24.5   Electron-releasing substituents attached to the ring have a negligible effect on the acidity of phenols. Strongly electron-withdrawing groups increase the acidity. The compound 4-nitro-3-(trifluoromethyl)phenol, for example, is 10,000 times more acidic than phenol.

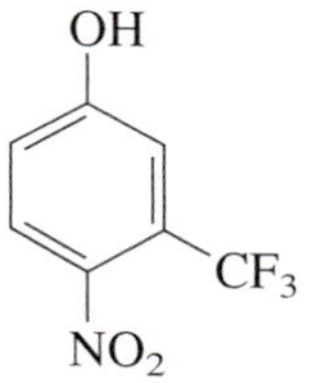

4-Nitro-3-(trifluoromethyl)phenol:
$pK_a = 6.0$

Section 24.6   Table 24.3 listed the main industrial methods for the preparation of phenol. Laboratory syntheses of phenols are usually carried out by hydrolysis of aryl diazonium salts.

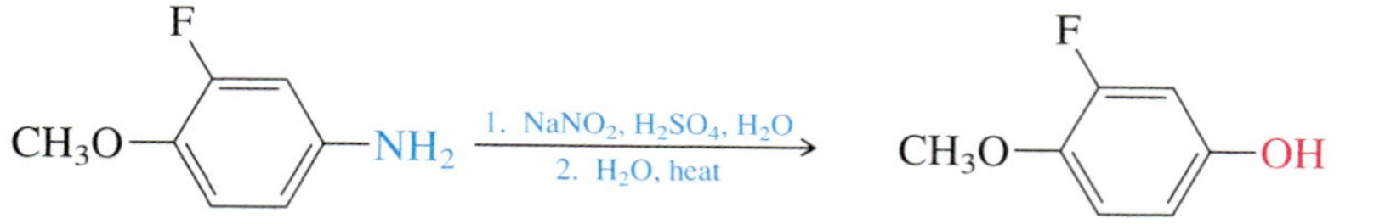

3-Fluoro-4-methoxyaniline

3-Fluoro-4-methoxyphenol (70%)

Section 24.7   Many phenols occur naturally.

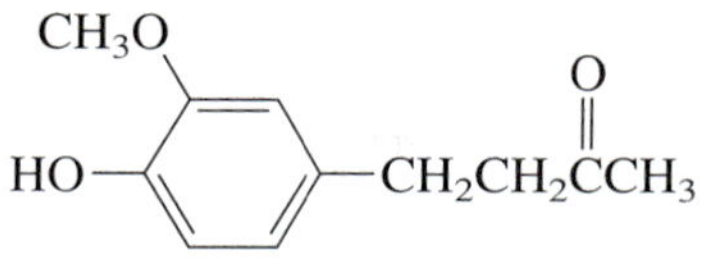

Zingerone
(responsible for spicy taste of ginger)

Phenol biosynthesis in plants proceeds from carbohydrate precursors, whereas the pathway in animals involves oxidation of aromatic rings.

Section 24.8   The hydroxyl group of a phenol is a strongly activating substituent, and electrophilic aromatic substitution occurs readily in phenol and its derivatives. Typical examples were presented in Table 24.4.

Section 24.9   On reaction with acyl chlorides and acid anhydrides, phenols may undergo either acylation of the hydroxyl group (O-acylation) or acylation of the ring (C-acylation). The product of C-acylation is more stable and predominates under conditions of thermodynamic control when aluminum chloride is present (see entry 6 in Table 24.4, Section 24.8). O-acylation is faster than C-acylation, and aryl esters are formed under conditions of kinetic control.

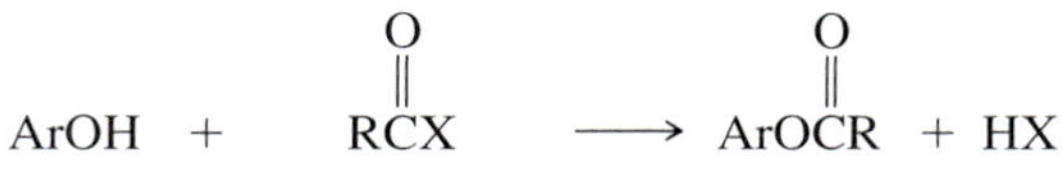

ArOH   +   RCX $\longrightarrow$ ArOCR   + HX

A phenol   Acylating agent   Aryl ester

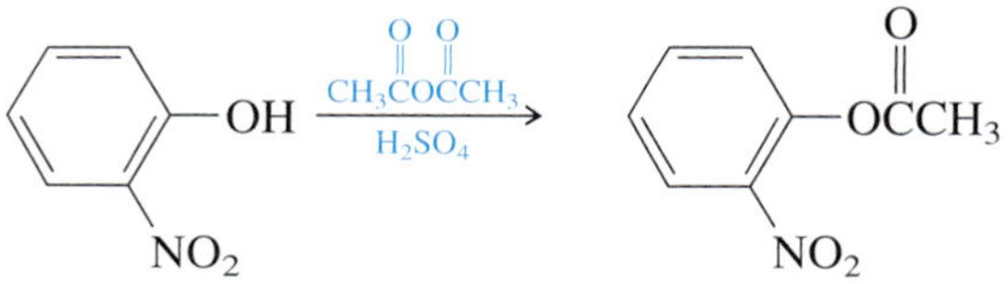

o-Nitrophenol   o-Nitrophenyl acetate (93%)

Section 24.10   The **Kolbe–Schmitt synthesis** of salicylic acid is a vital step in the preparation of aspirin. Phenols, as their sodium salts, undergo highly regioselective ortho carboxylation on treatment with carbon dioxide at elevated temperature and pressure.

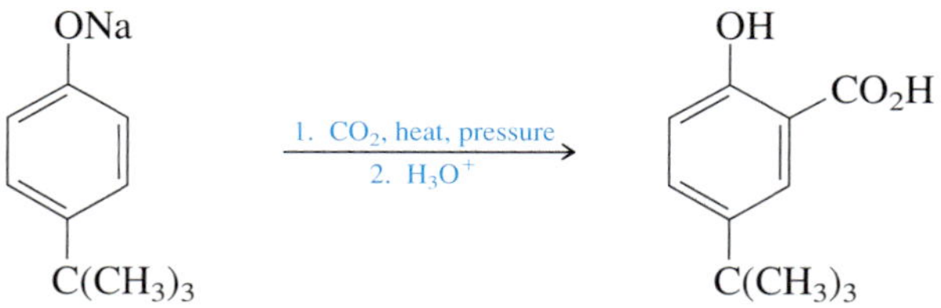

Sodium   5-tert-Butyl-2-
p-tert-butylphenoxide   hydroxybenzoic acid (74%)

Section 24.11 Phenoxide anions are nucleophilic toward alkyl halides, and the preparation of alkyl aryl ethers is easily achieved under S_N2 conditions.

$$ArO^- \quad + \quad RX \quad \longrightarrow \quad ArOR \quad + \quad X^-$$

| Phenoxide anion | Alkyl halide | Alkyl aryl ether | Halide anion |

o-Nitrophenol   Butyl *o*-nitrophenyl ether (75/80%)

Section 24.12 The cleavage of alkyl aryl ethers by hydrogen halides yields a phenol and an alkyl halide.

$$ArOR \quad + \quad HX \quad \xrightarrow{\text{heat}} \quad ArOH \quad + \quad RX$$

| Alkyl aryl ether | Hydrogen halide | A phenol | Alkyl halide |

m-Methoxyphenylacetic acid   *m*-Hydroxyphenylacetic acid (72%)   Methyl iodide

Section 24.13 On being heated, allyl aryl ethers undergo a **Claisen rearrangement** to form *o*-allylphenols. A cyclohexadienone, formed by a concerted six-π-electron reorganization, is an intermediate.

Allyl *o*-bromophenyl ether   2-Allyl-6-bromophenol (82%)   via   6-Allyl-2-bromo-2,4-cyclohexadienone

Section 24.14 Oxidation of 1,2- and 1,4-benzenediols gives colored compounds known as **quinones.**

3,4,5,6-Tetramethyl-1,2-benzenediol   3,4,5,6-Tetramethyl-1,2-benzoquinone (81%)

Section 24.15 The IR and ^{1}H NMR spectra of phenols are similar to those for alcohols, except that the OH proton is somewhat less shielded in a phenol than in an alcohol. In ^{13}C NMR, an OH group deshields the carbon of an aromatic ring to which it is attached. An OH group causes a shift in the UV-VIS spectrum of benzene to longer wavelengths. The effect is quite large in basic solution because of conversion of OH to O^-.

PROBLEMS

24.11 The IUPAC rules permit the use of common names for a number of familiar phenols and aryl ethers. These common names are listed here along with their systematic names. Write the structure of each compound.

  (a) *Vanillin* (4-hydroxy-3-methoxybenzaldehyde): a component of vanilla bean oil, which contributes to its characteristic flavor

  (b) *Thymol* (2-isopropyl-5-methylphenol): obtained from oil of thyme

  (c) *Carvacrol* (5-isopropyl-2-methylphenol): present in oil of thyme and marjoram

  (d) *Eugenol* (4-allyl-2-methoxyphenol): obtained from oil of cloves

  (e) *Gallic acid* (3,4,5-trihydroxybenzoic acid): prepared by hydrolysis of tannins derived from plants

  (f) *Salicyl alcohol* (*o*-hydroxybenzyl alcohol): obtained from bark of poplar and willow trees

24.12 Name each of the following compounds:

(a)

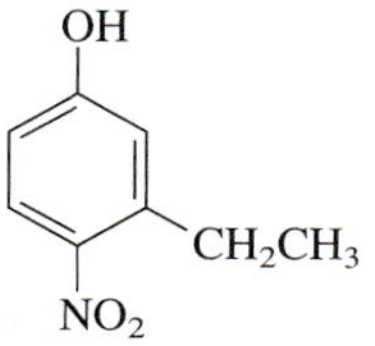

(b)

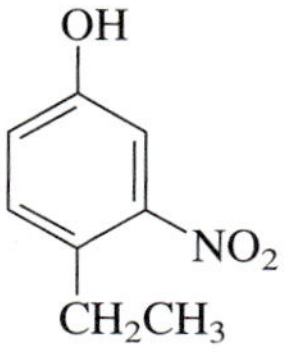

(c)

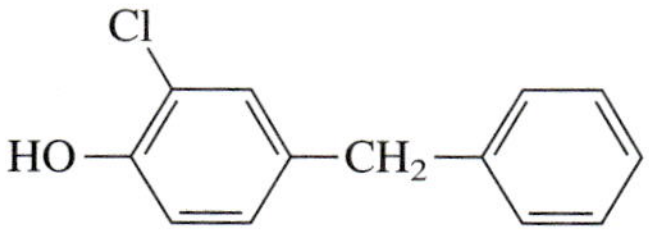

(d)

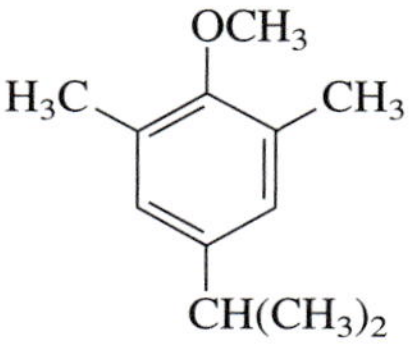

(e)

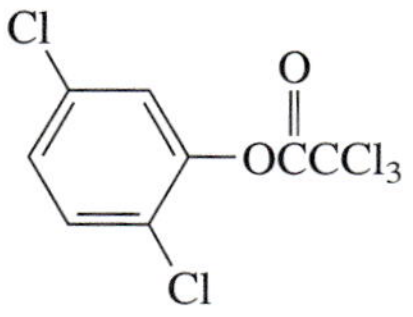

24.13 Write a balanced chemical equation for each of the following reactions:

  (a) Phenol + sodium hydroxide

  (b) Product of part (a) + ethyl bromide

  (c) Product of part (a) + butyl *p*-toluenesulfonate

  (d) Product of part (a) + acetic anhydride

  (e) *o*-Cresol + benzoyl chloride

  (f) *m*-Cresol + ethylene oxide

  (g) 2,6-Dichlorophenol + bromine

  (h) *p*-Cresol + excess aqueous bromine

  (i) Isopropyl phenyl ether + excess hydrogen bromide + heat

24.14 Which phenol in each of the following pairs is more acidic? Justify your choice.

(a) 2,4,6-Trimethylphenol or 2,4,6-trinitrophenol

(b) 2,6-Dichlorophenol or 3,5-dichlorophenol

(c) 3-Nitrophenol or 4-nitrophenol

(d) Phenol or 4-cyanophenol

(e) 2,5-Dinitrophenol or 2,6-dinitrophenol

24.15 Choose the reaction in each of the following pairs that proceeds at the faster rate. Explain your reasoning.

(a) Basic hydrolysis of phenyl acetate or *m*-nitrophenyl acetate

(b) Basic hydrolysis of *m*-nitrophenyl acetate or *p*-nitrophenyl acetate

(c) Reaction of ethyl bromide with phenol or with the sodium salt of phenol

(d) Reaction of ethylene oxide with the sodium salt of phenol or with the sodium salt of *p*-nitrophenol

(e) Bromination of phenol or phenyl acetate

24.16 Pentafluorophenol is readily prepared by heating hexafluorobenzene with potassium hydroxide in *tert*-butyl alcohol:

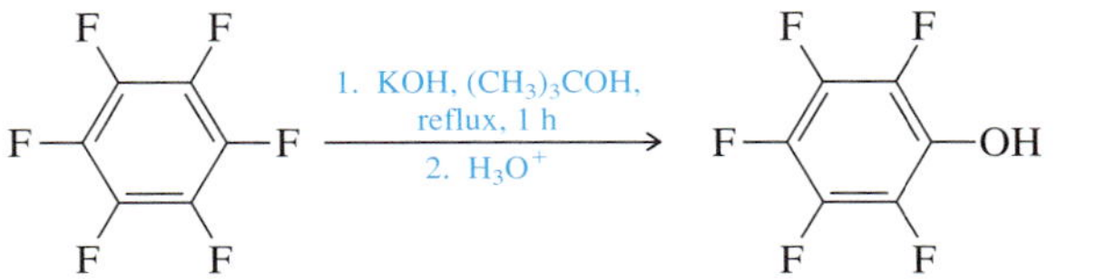

Hexafluorobenzene　　　　　　　　Pentafluorophenol (71%)

What is the most reasonable mechanism for this reaction? Comment on the comparative ease with which this conversion occurs.

24.17 Each of the following reactions has been reported in the chemical literature and proceeds cleanly in good yield. Identify the principal organic product in each case.

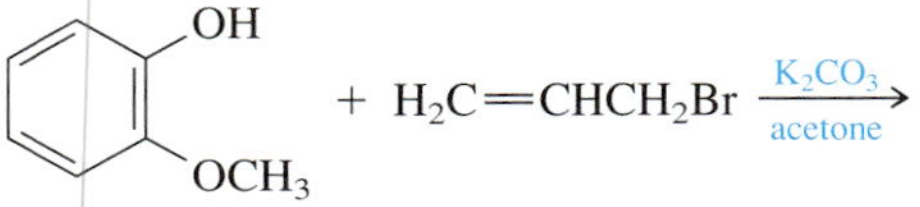

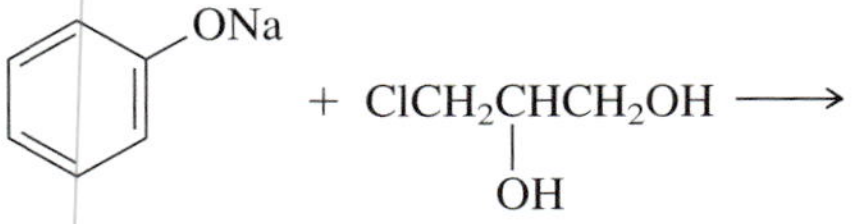

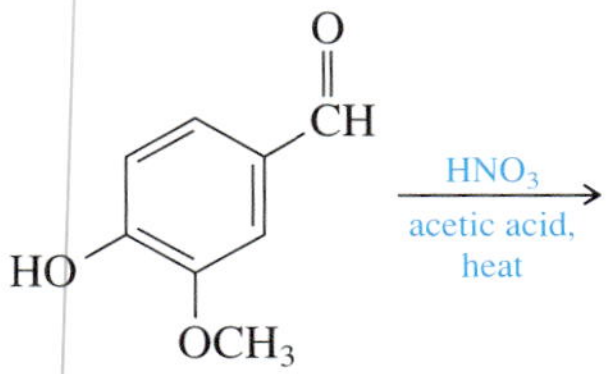

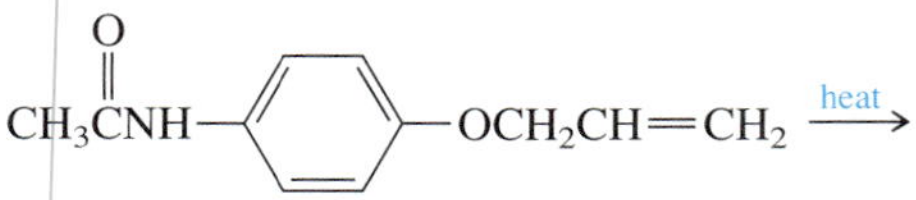

(e) 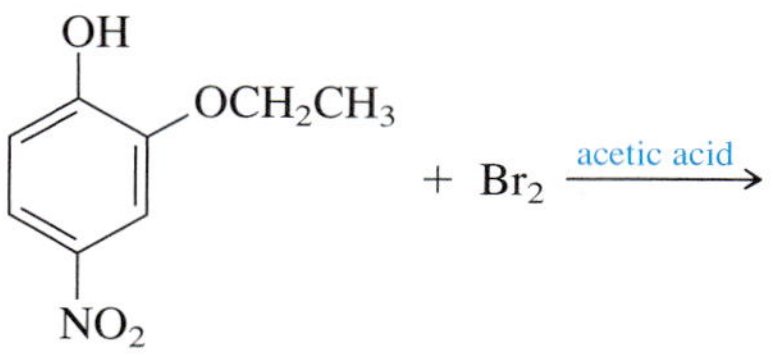 + Br₂ —acetic acid→

(f) 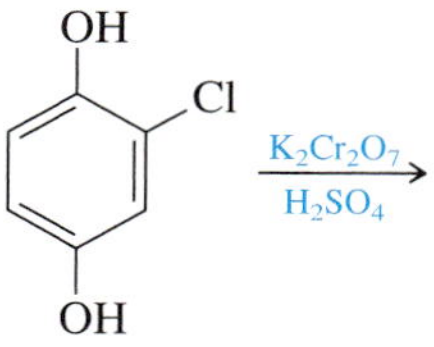 —K₂Cr₂O₇ / H₂SO₄→

(g) 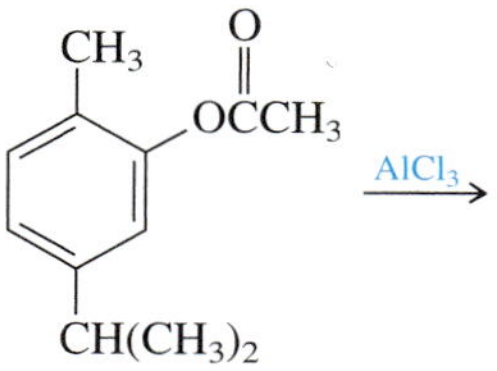 —AlCl₃→

(h) 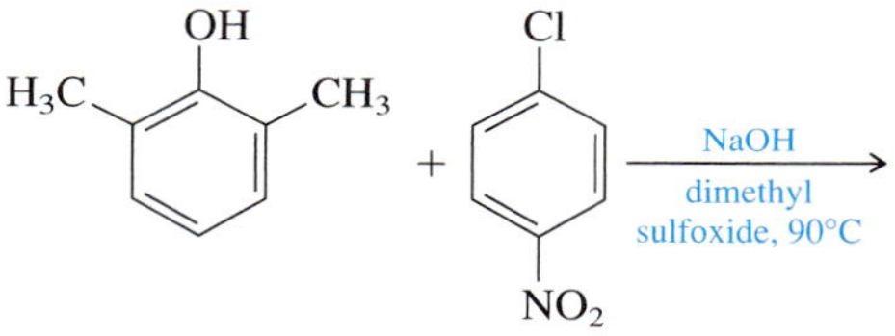 + —NaOH / dimethyl sulfoxide, 90°C→

(i) 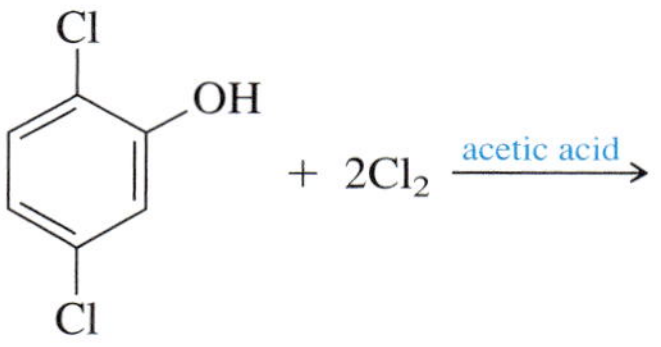 + 2Cl₂ —acetic acid→

(j) 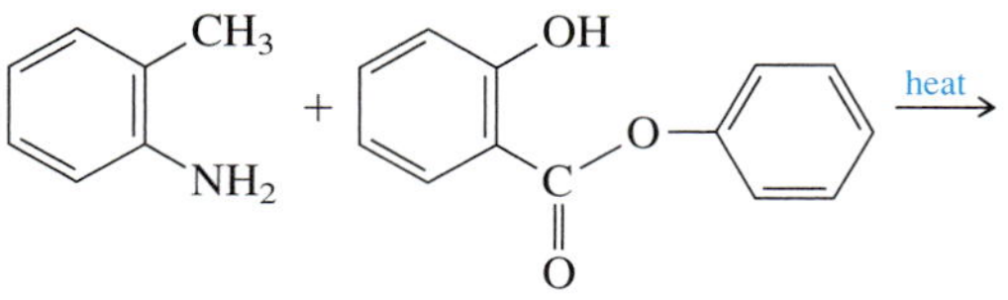 + —heat→

(k) 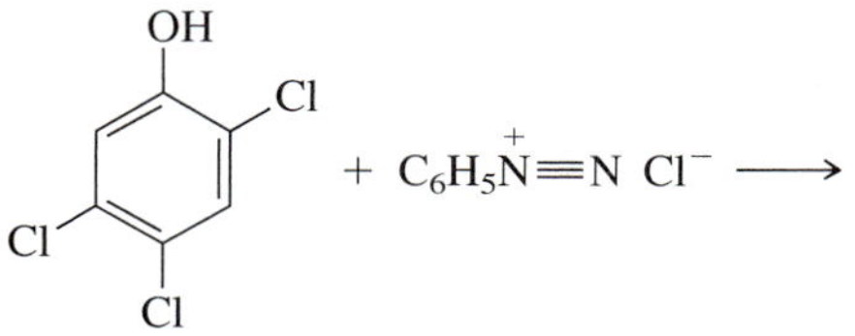 + C₆H₅N⁺≡N Cl⁻ ⟶

24.18 A synthesis of the analgesic substance *phenacetin* is outlined in the following equation. What is the structure of phenacetin?

$$\textit{p}\text{-Nitrophenol} \xrightarrow[\substack{\text{2. Fe, HCl; then HO}^-}]{\substack{\text{1. CH}_3\text{CH}_2\text{Br, NaOH}}} \text{phenacetin}$$

3. $CH_3\overset{O}{\overset{\|}{C}}O\overset{O}{\overset{\|}{C}}CH_3$

24.19 Identify compounds A through C in the synthetic sequence represented by equations (a) through (c).

(a) Phenol + H_2SO_4 $\xrightarrow{\text{heat}}$ Compound A ($C_6H_6O_7S_2$)

(b) Compound A + Br_2 $\xrightarrow[\text{2. H}^+]{\text{1. HO}^-}$ Compound B ($C_6H_5BrO_7S_2$)

(c) Compound B + H_2O $\xrightarrow[\text{heat}]{\text{H}^+}$ Compound C (C_6H_5BrO)

24.20 Treatment of 3,5-dimethylphenol with dilute nitric acid, followed by steam distillation of the reaction mixture, gave a compound A ($C_8H_9NO_3$, mp 66°C) in 36% yield. The nonvolatile residue from the steam distillation gave a compound B ($C_8H_9NO_3$, mp 108°C) in 25% yield on extraction with chloroform. Identify compounds A and B.

24.21 Outline a reasonable synthesis of 4-nitrophenyl phenyl ether from chlorobenzene and phenol.

24.22 As an allergen for testing purposes, synthetic 3-pentadecylcatechol is more useful than natural poison ivy extracts (of which it is one component). A stable crystalline solid, it is efficiently prepared in pure form from readily available starting materials. Outline a reasonable synthesis of this compound from 2,3-dimethoxybenzaldehyde and any necessary organic or inorganic reagents.

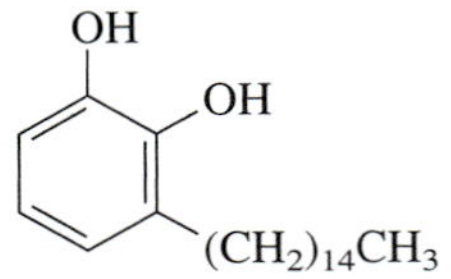

3-Pentadecylcatechol

24.23 Describe a scheme for carrying out the following synthesis. (In the synthesis reported in the literature, four separate operations were required.)

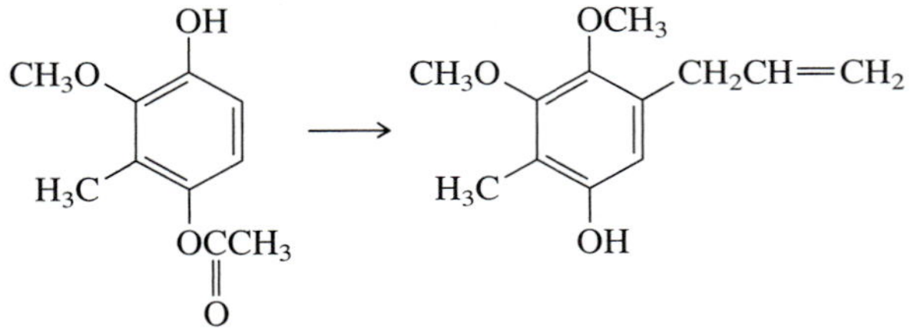

24.24 In a general reaction known as the *cyclohexadienone-phenol rearrangement,* cyclohexadienones are converted to phenols under conditions of acid catalysis. An example is

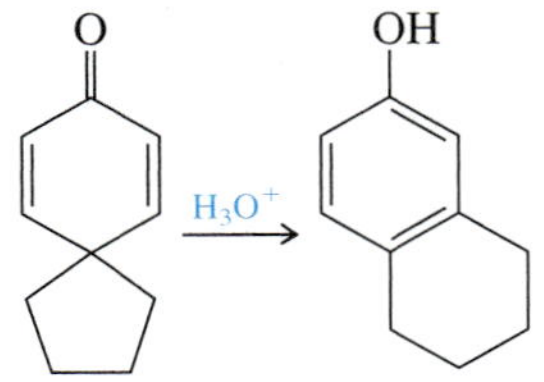

(100%)

Write a reasonable mechanism for this reaction.

24.25 Treatment of *p*-hydroxybenzoic acid with aqueous bromine leads to the evolution of carbon dioxide and the formation of 2,4,6-tribromophenol. Explain.

24.26 Treatment of phenol with excess aqueous bromine is actually more complicated than expected. A white precipitate forms rapidly, which on closer examination is not 2,4,6-tribromophenol but is instead 2,4,4,6-tetrabromocyclohexadienone. Explain the formation of this product.

24.27 Treatment of 2,4,6-tri-*tert*-butylphenol with bromine in cold acetic acid gives the compound $C_{18}H_{29}BrO$ in quantitative yield. The infrared spectrum of this compound contains absorptions at 1630 and 1655 cm^{-1}. Its ^{1}H NMR spectrum shows only three peaks (all singlets), at δ 1.2, 1.3, and 6.9, in the ratio 9:18:2. What is a reasonable structure for the compound?

24.28 Compound A undergoes hydrolysis of its acetal function in dilute sulfuric acid to yield 1,2-ethanediol and compound B ($C_6H_6O_2$), mp 54°C. Compound B exhibits a carbonyl stretching band in the infrared at 1690 cm^{-1} and has two singlets in its ^{1}H NMR spectrum, at δ 2.9 and 6.7, in the ratio 2:1. On standing in water or ethanol, compound B is converted cleanly to an isomeric substance, compound C, mp 172–173°C. Compound C has no peaks attributable to carbonyl groups in its infrared spectrum. Identify compounds B and C.

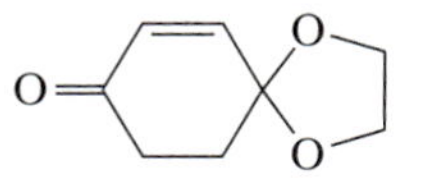

Compound A

24.29 One of the industrial processes for the preparation of phenol, discussed in Section 24.6, includes an acid-catalyzed rearrangement of cumene hydroperoxide as a key step. This reaction proceeds by way of an intermediate hemiacetal:

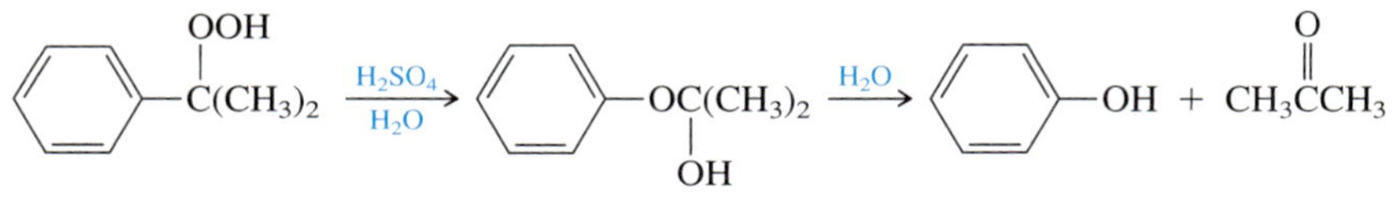

| Cumene hydroperoxide | Hemiacetal | Phenol | Acetone |

You learned in Section 17.8 of the relationship among hemiacetals, ketones, and alcohols; the formation of phenol and acetone is simply an example of hemiacetal hydrolysis. The formation of the hemiacetal intermediate is a key step in the synthetic procedure; it is the step in which the aryl–oxygen bond is generated. Can you suggest a reasonable mechanism for this step?

24.30 Identify the following compounds on the basis of the information provided:

    (a) $C_9H_{12}O$: Its IR and ^{13}C NMR spectra are shown in Figure 24.5.

    (b) $C_9H_{11}BrO$: Its IR and ^{13}C NMR spectra are shown in Figure 24.6.

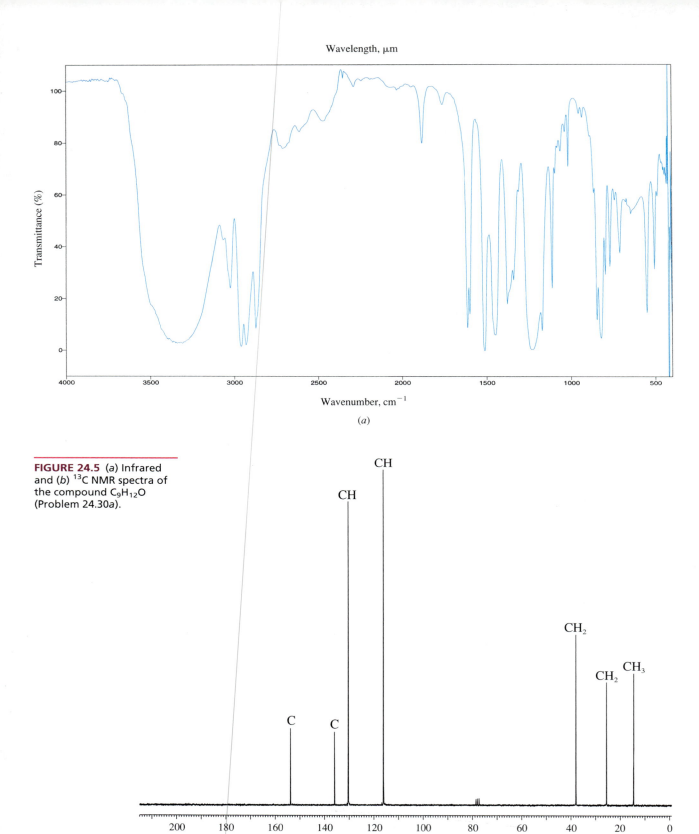

Wavelength, μm

Transmittance (%)

Wavenumber, cm^{-1}

(a)

FIGURE 24.5 (a) Infrared and (b) ^{13}C NMR spectra of the compound $C_9H_{12}O$ (Problem 24.30a).

CH

CH

CH$_2$

CH$_2$ CH$_3$

C   C

Chemical shift (δ, ppm)

(b)

Wavelength, μm

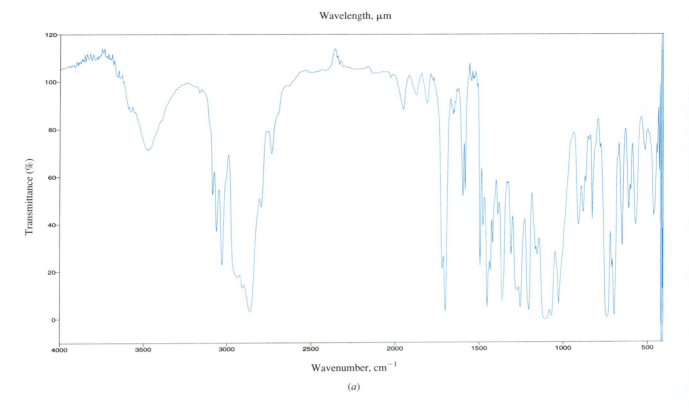

(a)

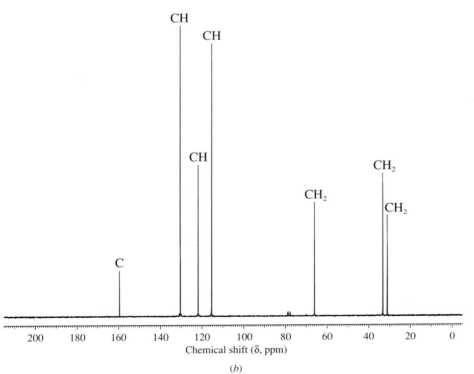

(b)

FIGURE 24.6 (a) Infrared and (b) ^{13}C NMR spectra of the compound $C_9H_{11}BrO$ (Problem 24.30b).

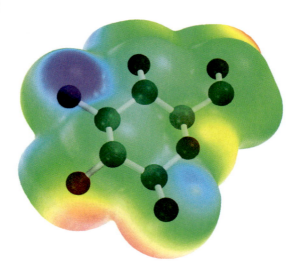

CHAPTER 25

CARBOHYDRATES

The major classes of organic compounds common to living systems are *lipids, proteins, nucleic acids,* and *carbohydrates.* Carbohydrates are very familiar to us—we call many of them "sugars." They make up a substantial portion of the food we eat and provide most of the energy that keeps the human engine running. Carbohydrates are structural components of the walls of plant cells and the wood of trees. Genetic information is stored and transferred by way of nucleic acids, specialized derivatives of carbohydrates, which we'll examine in more detail in Chapter 28.

Historically, carbohydrates were once considered to be "hydrates of carbon" because their molecular formulas in many (but not all) cases correspond to $C_n(H_2O)_m$. It is more realistic to define a carbohydrate as a *polyhydroxy aldehyde* or *polyhydroxy ketone,* a point of view closer to structural reality and more suggestive of chemical reactivity.

This chapter is divided into two parts. The first, and major, portion is devoted to carbohydrate *structure.* You will see how the principles of stereochemistry and conformational analysis combine to aid our understanding of this complex subject. The remainder of the chapter describes chemical *reactions* of carbohydrates. Most of these reactions are simply extensions of what you have already learned concerning alcohols, aldehydes, ketones, and acetals.

25.1 CLASSIFICATION OF CARBOHYDRATES

The Latin word for *sugar** is *saccharum,* and the derived term *saccharide* is the basis of a system of carbohydrate classification. A **monosaccharide** is a simple carbohydrate, one that on attempted hydrolysis is not cleaved to smaller carbohydrates. *Glucose* ($C_6H_{12}O_6$),

---

*Sugar is a combination of the Sanskrit words *su* (sweet) and *gar* (sand). Thus, its literal meaning is "sweet sand."

for example, is a monosaccharide. A **disaccharide** on hydrolysis is cleaved to two mono-saccharides, which may be the same or different. *Sucrose*—common table sugar—is a disaccharide that yields one molecule of glucose and one of fructose on hydrolysis.

Sucrose ($C_{12}H_{22}O_{11}$) + H_2O ⟶ glucose ($C_6H_{12}O_6$) + fructose ($C_6H_{12}O_6$)

An **oligosaccharide** (*oligos* is a Greek word that in its plural form means "few") yields two or more monosaccharides on hydrolysis. Thus, the IUPAC classifies disaccharides, trisaccharides, and so on as subcategories of oligosaccharides. **Polysaccharides** are hydrolyzed to "many" monosaccharides. The IUPAC has chosen not to specify the number of monosaccharide components that separates oligosaccharides from polysaccharides. The standard is a more practical one; it notes that an oligosaccharide is homogeneous. Each molecule of a particular oligosaccharide has the same number of monosaccharide units joined together in the same order as every other molecule of the same oligosaccharide. Polysaccharides are almost always mixtures of molecules having similar, but not necessarily the same, chain length. *Cellulose,* for example, is a polysaccharide that gives thousands of glucose molecules on hydrolysis but only a small fraction of the cellulose chains contain exactly the same number of glucose molecules.

Over 200 different monosaccharides are known. They can be grouped according to the number of carbon atoms they contain and whether they are polyhydroxy aldehydes or polyhydroxy ketones. Monosaccharides that are polyhydroxy aldehydes are called **aldoses;** those that are polyhydroxy ketones are **ketoses.** Aldoses and ketoses are further classified according to the number of carbon atoms in the main chain. Table 25.1 lists the terms applied to monosaccharides having four to eight carbon atoms.

25.2   FISCHER PROJECTIONS AND D–L NOTATION

Stereochemistry is the key to understanding carbohydrate structure, a fact that was clearly appreciated by the German chemist Emil Fischer. The projection formulas used by Fischer to represent stereochemistry in chiral molecules (Section 7.7) are particularly well-suited to studying carbohydrates. Figure 25.1 illustrates their application to the enantiomers of *glyceraldehyde* (2,3-dihydroxypropanal), a fundamental molecule in carbohydrate stereochemistry. When the Fischer projection is oriented as shown in the figure, with the carbon chain vertical and the aldehyde carbon at the top, the C-2 hydroxyl group points to the right in (+)-glyceraldehyde and to the left in (−)-glyceraldehyde.

Techniques for determining the absolute configuration of chiral molecules were not developed until the 1950s, and so it was not possible for Fischer and his contemporaries to relate the sign of rotation of any substance to its absolute configuration. A system evolved based on the arbitrary assumption, later shown to be correct, that the enantiomers

Fischer determined the structure of glucose in 1900 and won the Nobel Prize in chemistry in 1902.

TABLE 25.1	Some Classes of Monosaccharides	
Number of carbon atoms	Aldose	Ketose
Four	Aldotetrose	Ketotetrose
Five	Aldopentose	Ketopentose
Six	Aldohexose	Ketohexose
Seven	Aldoheptose	Ketoheptose
Eight	Aldooctose	Ketooctose

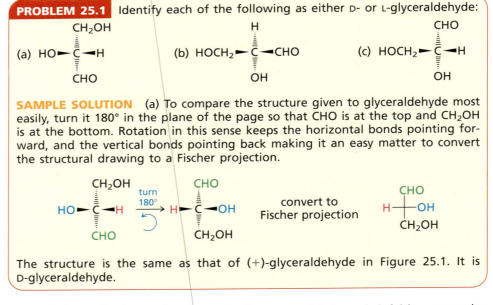

FIGURE 25.1 Three-dimensional representations and Fischer projections of the enantiomers of glyceraldehyde.

Adopting the enantiomers of glyceraldehyde as stereochemical reference compounds originated with proposals made in 1906 by M. A. Rosanoff, a chemist at New York University.

of glyceraldehyde have the signs of rotation and absolute configurations shown in Figure 25.1. Two stereochemical descriptors were defined: D and L. The absolute configuration of (+)-glyceraldehyde, as depicted in the figure, was said to be D and that of its enantiomer, (−)-glyceraldehyde, L. Compounds that had a spatial arrangement of substituents analogous to D-(+)- and L-(−)-glyceraldehyde were said to have the D and L configurations, respectively.

PROBLEM 25.1 Identify each of the following as either D- or L-glyceraldehyde:

SAMPLE SOLUTION (a) To compare the structure given to glyceraldehyde most easily, turn it 180° in the plane of the page so that CHO is at the top and CH₂OH is at the bottom. Rotation in this sense keeps the horizontal bonds pointing forward, and the vertical bonds pointing back making it an easy matter to convert the structural drawing to a Fischer projection.

The structure is the same as that of (+)-glyceraldehyde in Figure 25.1. It is D-glyceraldehyde.

Fischer projections and D–L notation have proved to be so helpful in representing carbohydrate stereochemistry that the chemical and biochemical literature is replete with their use. To read that literature you need to be acquainted with these devices, as well as the more modern Cahn–Ingold–Prelog *R,S* system.

25.3   THE ALDOTETROSES

Glyceraldehyde can be considered to be the simplest chiral carbohydrate. It is an **aldotriose** and, because it contains one chirality center, exists in two stereoisomeric forms: the D and L enantiomers. Moving up the scale in complexity, next come the **aldotetroses.** Examining their structures illustrates the application of the Fischer system to compounds that contain more than one chirality center.

The aldotetroses are the four stereoisomers of 2,3,4-trihydroxybutanal. Fischer projections are constructed by orienting the molecule in an eclipsed conformation with the aldehyde group at the top. The four carbon atoms define the main chain of the Fischer projection and are arranged vertically. Horizontal bonds are directed outward, vertical bonds back.

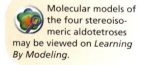

Molecular models of the four stereoisomeric aldotetroses may be viewed on *Learning By Modeling.*

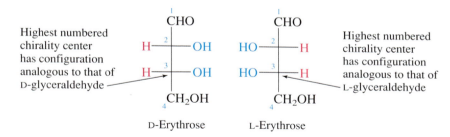

The particular aldotetrose just shown is called D-*erythrose*. The prefix D tells us that the configuration at the *highest numbered chirality center* is analogous to that of D-(+)-glyceraldehyde. Its mirror image is L-erythrose.

For a first-person account of the development of systematic carbohydrate nomenclature see C. D. Hurd's article in the December 1989 issue of the *Journal of Chemical Education*, pp. 984–988.

Highest numbered chirality center has configuration analogous to that of D-glyceraldehyde

Highest numbered chirality center has configuration analogous to that of L-glyceraldehyde

D-Erythrose   L-Erythrose

Relative to each other, both hydroxyl groups are on the same side in Fischer projections of the erythrose enantiomers. The remaining two stereoisomers have hydroxyl groups on opposite sides in their Fischer projections. They are diastereomers of D- and L-erythrose and are called D- and L-*threose.* The D and L prefixes again specify the configuration of the highest numbered chirality center. D-Threose and L-threose are enantiomers of each other:

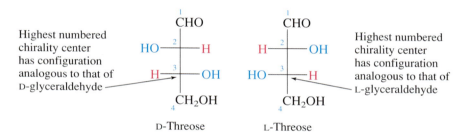

Highest numbered chirality center has configuration analogous to that of D-glyceraldehyde

Highest numbered chirality center has configuration analogous to that of L-glyceraldehyde

D-Threose   L-Threose

PROBLEM 25.2 Which aldotetrose is the structure shown? Is it D-erythrose, D-threose, L-erythrose, or L-threose? (Be careful! The conformation given is not the same as that used to generate a Fischer projection.)

As shown for the aldotetroses, an aldose belongs to the D or the L series according to the configuration of the chirality center farthest removed from the aldehyde function. Individual names, such as erythrose and threose, specify the particular arrangement of chirality centers within the molecule relative to each other. Optical activities cannot be determined directly from the D and L prefixes. As it turns out, both D-erythrose and D-threose are levorotatory, but D-glyceraldehyde is dextrorotatory.

25.4   ALDOPENTOSES AND ALDOHEXOSES

$2^3 = 8$

Aldopentoses have *three* chirality centers. The *eight stereoisomers* are divided into a set of four D-aldopentoses and an enantiomeric set of four L-aldopentoses. The aldopentoses are named *ribose, arabinose, xylose,* and *lyxose.* Fischer projections of the D stereoisomers of the aldopentoses are given in Figure 25.2. Notice that all these diastereomers have the same configuration at C-4 and that this configuration is analogous to that of D-(+)-glyceraldehyde.

PROBLEM 25.3 L-(+)-Arabinose is a naturally occurring L sugar. It is obtained by acid hydrolysis of the polysaccharide present in mesquite gum. Write a Fischer projection for L-(+)-arabinose.

Among the aldopentoses, D-ribose is a component of many biologically important substances, most notably the ribonucleic acids, and D-xylose is very abundant and is isolated by hydrolysis of the polysaccharides present in corncobs and the wood of trees.

The aldohexoses include some of the most familiar of the monosaccharides, as well as one of the most abundant organic compounds on earth, D-(+)-glucose. With *four* chirality centers, *16* stereoisomeric aldohexoses are possible; 8 belong to the D series and 8 to the L series. All are known, either as naturally occurring substances or as the products of synthesis. The eight D-aldohexoses are given in Figure 25.2; the spatial arrangement at C-5, hydrogen to the left in a Fischer projection and hydroxyl to the right, identifies them as carbohydrates of the D series.

$2^4 = 16$

PROBLEM 25.4 Use Figure 25.2 as a guide to help you name the aldose shown. What is its sign of rotation?

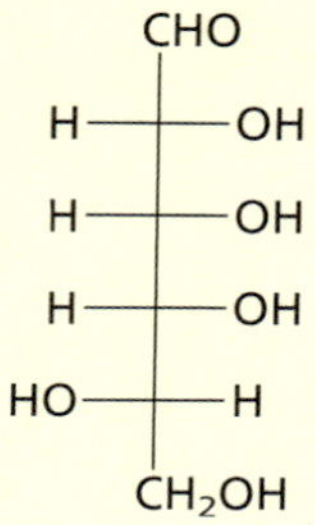

$$
\begin{array}{c}
\text{CHO} \\
\text{H}\!-\!\!\!-\!\text{OH} \\
\text{H}\!-\!\!\!-\!\text{OH} \\
\text{H}\!-\!\!\!-\!\text{OH} \\
\text{HO}\!-\!\!\!-\!\text{H} \\
\text{CH}_2\text{OH}
\end{array}
$$

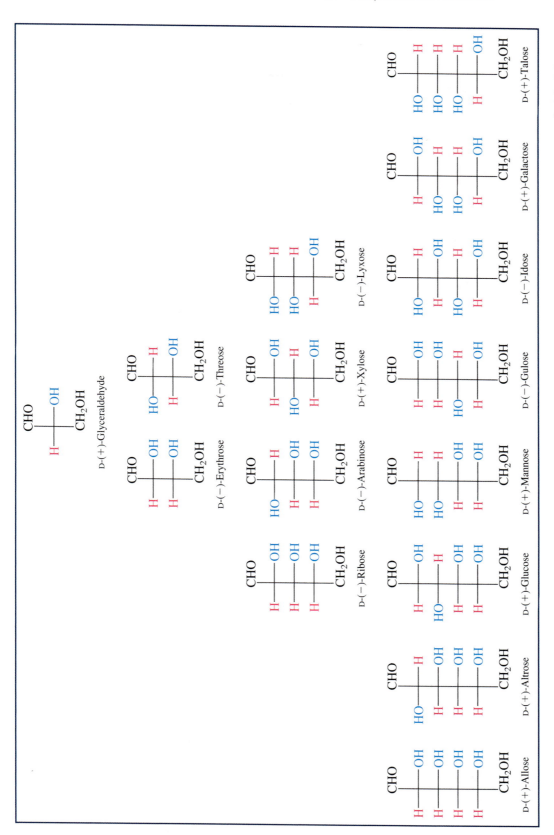

FIGURE 25.2
Configurations of the D series of aldoses containing three through six carbon atoms.

Cellulose is more abundant than glucose, but each cellulose molecule is a polysaccharide composed of thousands of glucose units (see Section 25.15). Methane may also be more abundant, but most of the methane comes from glucose.

Of all the monosaccharides, D-(+)-*glucose* is the best known, most important, and most abundant. Its formation from carbon dioxide, water, and sunlight is the central theme of photosynthesis. Carbohydrate formation by photosynthesis is estimated to be on the order of 10^{11} tons per year, a source of stored energy utilized, directly or indirectly, by all higher forms of life on the planet. Glucose was isolated from raisins in 1747 and by hydrolysis of starch in 1811. Its structure was determined, in work culminating in 1900, by Emil Fischer.

D-(+)-*Galactose* is a constituent of numerous polysaccharides. It is best obtained by acid hydrolysis of lactose (milk sugar), a disaccharide of D-glucose and D-galactose. L-(−)-Galactose also occurs naturally and can be prepared by hydrolysis of flaxseed gum and agar. The principal source of D-(+)-*mannose* is hydrolysis of the polysaccharide of the ivory nut, a large, nut-like seed obtained from a South American palm.

25.5   A MNEMONIC FOR CARBOHYDRATE CONFIGURATIONS

See, for example, the November 1955 issue of the *Journal of Chemical Education* (p. 584). An article giving references to a variety of chemistry mnemonics appears in the July 1960 issue of the *Journal of Chemical Education* (p. 366).

The task of relating carbohydrate configurations to names requires either a world-class memory or an easily recalled mnemonic. A mnemonic that serves us well here was popularized by the husband–wife team of Louis F. Fieser and Mary Fieser of Harvard University in their 1956 textbook, *Organic Chemistry.* As with many mnemonics, it's not clear who actually invented it, and references to this particular one appeared in the chemical education literature before publication of the Fiesers' text. The mnemonic has two features: (1) a system for setting down all the stereoisomeric D-aldohexoses in a logical order; and (2) a way to assign the correct name to each one.

A systematic way to set down all the D-aldohexoses (as in Figure 25.2) is to draw skeletons of the necessary eight Fischer projections, placing the C-5 hydroxyl group to the right in each so as to guarantee that they all belong to the D series. Working up the carbon chain, place the C-4 hydroxyl group to the right in the first four structures, and to the left in the next four. In each of these two sets of four, place the C-3 hydroxyl group to the right in the first two and to the left in the next two; in each of the resulting four sets of two, place the C-2 hydroxyl group to the right in the first one and to the left in the second.

Once the eight Fischer projections have been written, they are named in order with the aid of the sentence: "All altruists gladly make gum in gallon tanks." The words of the sentence stand for *allose, altrose, glucose, mannose, gulose, idose, galactose, talose.*

An analogous pattern of configurations can be seen in the aldopentoses when they are arranged in the order *ribose, arabinose, xylose, lyxose.* (RAXL is an easily remembered nonsense word that gives the correct sequence.) This pattern is discernible even in the aldotetroses erythrose and threose.

25.6   CYCLIC FORMS OF CARBOHYDRATES: FURANOSE FORMS

Aldoses incorporate two functional groups, C=O and OH, which are capable of reacting with each other. We saw in Section 17.8 that nucleophilic addition of an alcohol function to a carbonyl group gives a hemiacetal. When the hydroxyl and carbonyl groups are part of the same molecule, a *cyclic hemiacetal* results, as illustrated in Figure 25.3.

Cyclic hemiacetal formation is most common when the ring that results is five- or six-membered. Five-membered cyclic hemiacetals of carbohydrates are called **furanose** forms; six-membered ones are called **pyranose** forms. The ring carbon that is derived

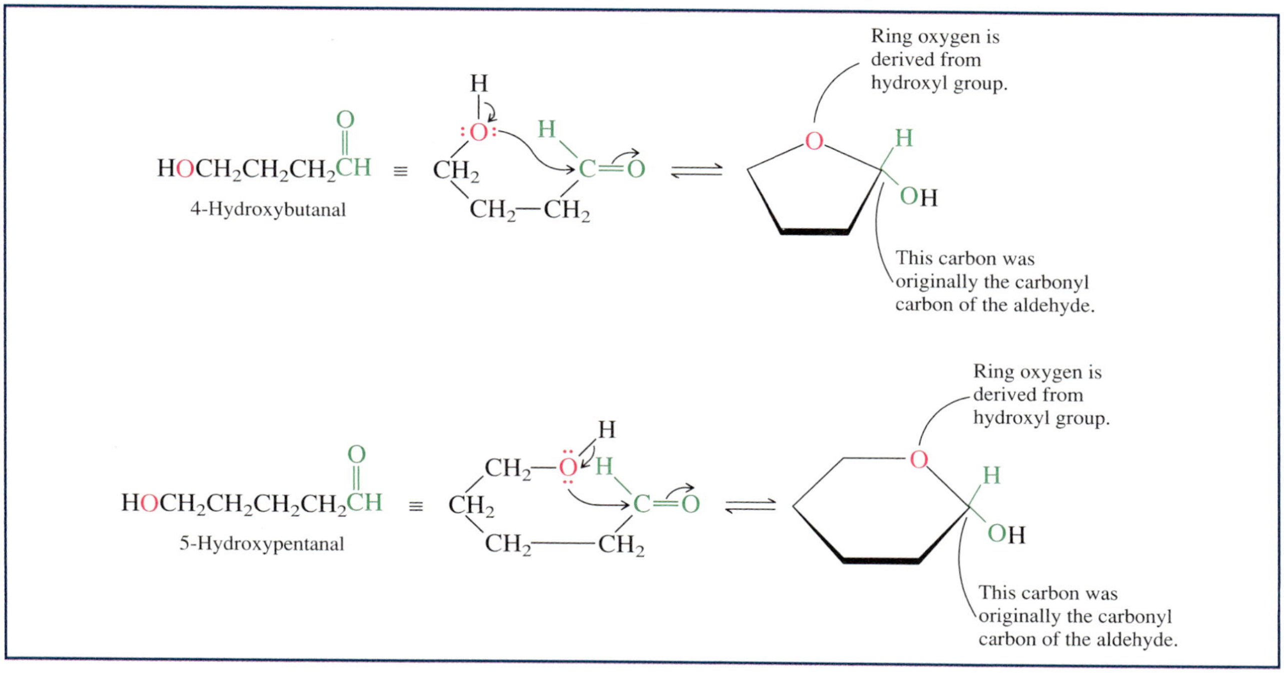

FIGURE 25.3 Cyclic hemiacetal formation in 4-hydroxybutanal and 5-hydroxypentanal.

from the carbonyl group, the one that bears two oxygen substituents, is called the **anomeric** carbon.

Aldoses exist almost exclusively as their cyclic hemiacetals; very little of the open-chain form is present at equilibrium. To understand their structures and chemical reactions, we need to be able to translate Fischer projections of carbohydrates into their cyclic hemiacetal forms. Consider first cyclic hemiacetal formation in D-erythrose. To visualize furanose ring formation more clearly, redraw the Fischer projection in a form more suited to cyclization, being careful to maintain the stereochemistry at each chirality center.

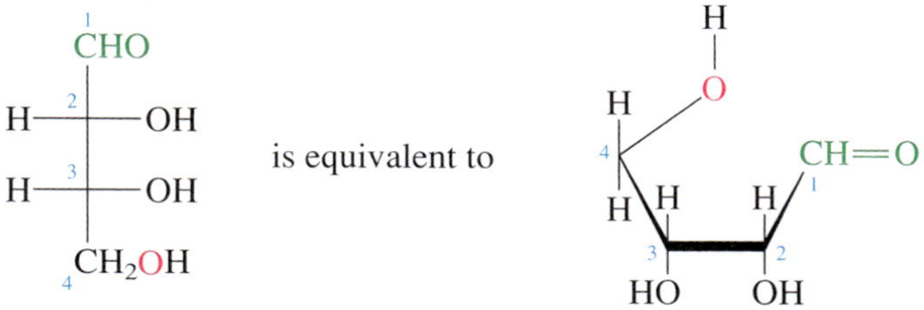

is equivalent to

D-Erythrose

Eclipsed conformation of
D-erythrose showing C-4
hydroxyl group in position to
add to carbonyl group

Making a molecular
model can help you
to visualize this
relationship.

Hemiacetal formation between the carbonyl group and the C-4 hydroxyl yields the five-membered furanose ring form. The anomeric carbon is a new chirality center; its hydroxyl group can be either cis or trans to the other hydroxyl groups of the molecule.

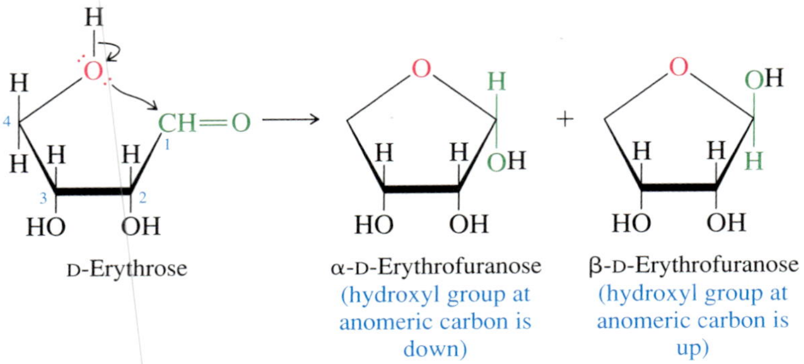

D-Erythrose → α-D-Erythrofuranose (hydroxyl group at anomeric carbon is down) + β-D-Erythrofuranose (hydroxyl group at anomeric carbon is up)

Structural drawings of carbohydrates of this type are called **Haworth formulas,** after the British chemist Sir Walter Norman Haworth (St. Andrew's University and the University of Birmingham). Early in his career Haworth contributed to the discovery that carbohydrates exist as cyclic hemiacetals rather than in open-chain forms. Later he collaborated on an efficient synthesis of vitamin C from carbohydrate precursors. This was the first chemical synthesis of a vitamin and provided an inexpensive route to its preparation on a commercial scale. Haworth was a corecipient of the Nobel Prize for chemistry in 1937.

The two stereoisomeric furanose forms of D-erythrose are named α-D-erythrofuranose and β-D-erythrofuranose. The prefixes α and β describe the *relative configuration* of the anomeric carbon. The configuration of the anomeric carbon is compared with that of the highest numbered chirality center in the molecule—the one that determines whether the carbohydrate is D or L. Chemists use a simplified, informal version of the IUPAC rules for assigning α and β that holds for carbohydrates up to and including hexoses.

The formal IUPAC rules for α and β notation in carbohydrates are more detailed and less easily understood than most purposes require. These rules can be accessed at http://www.chem.qmw.ac. uk/iupac/2carb/06n07.html.

1. Orient the Haworth formula of the carbohydrate with the ring oxygen at the back and the anomeric carbon at the right.

2. For carbohydrates of the D series, the configuration of the anomeric carbon is α if its hydroxyl group is *down,* β if the hydroxyl group at the anomeric carbon is *up.*

3. For carbohydrates of the L series, the configuration of the anomeric carbon is α if its hydroxyl group is *up,* β if the hydroxyl group at the anomeric carbon is *down.* This is exactly the reverse of the rule for the D series.

Substituents that are to the right in a Fischer projection are "down" in the corresponding Haworth formula; those to the left are "up."

PROBLEM 25.5 The structures shown are the four stereoisomeric threofuranoses. Assign the proper D, L and α, β stereochemical descriptors to each.

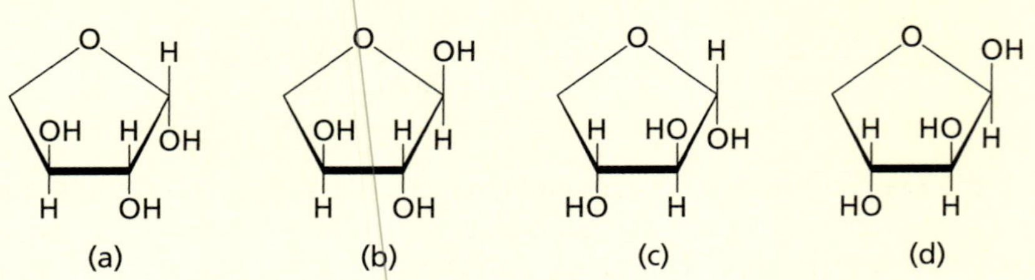

(a)   (b)   (c)   (d)

SAMPLE SOLUTION (a) The —OH group at the highest-numbered chirality center (C-3) is up, which places it to the left in the Fischer projection of the open-chain form. The stereoisomer belongs to the L series. The —OH group at the anomeric carbon (C-1) is down, making this the β-furanose form.

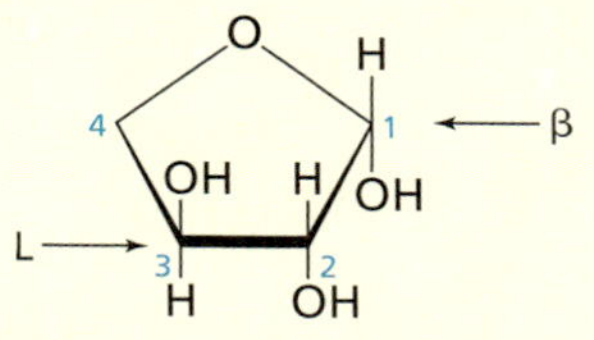

β-L-Threofuranose

Generating Haworth formulas to show stereochemistry in furanose forms of higher aldoses is slightly more complicated and requires an additional operation. Furanose forms of D-ribose are frequently encountered building blocks in biologically important organic molecules. They result from hemiacetal formation between the aldehyde group and the C-4 hydroxyl:

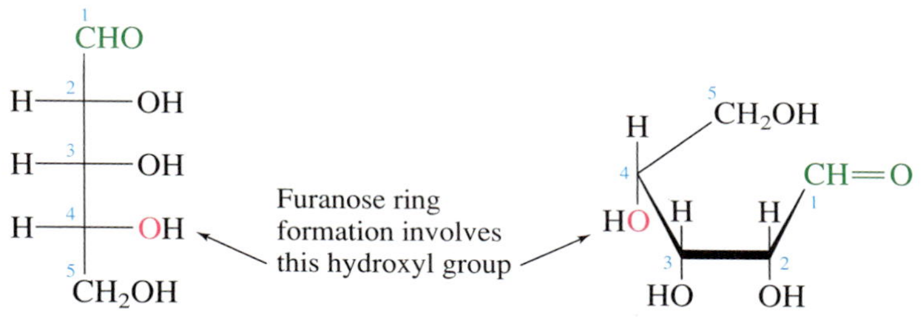

D-Ribose   Furanose ring formation involves this hydroxyl group   Eclipsed conformation of D-ribose

Notice that the eclipsed conformation of D-ribose derived directly from the Fischer projection does not have its C-4 hydroxyl group properly oriented for furanose ring formation. We must redraw it in a conformation that permits the five-membered cyclic hemiacetal to form. This is accomplished by rotation about the C(3)—C(4) bond, taking care that the configuration at C-4 is not changed.

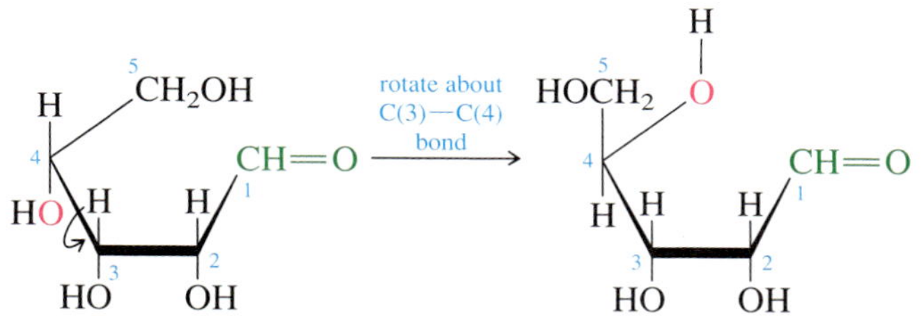

rotate about C(3)—C(4) bond

Conformation of D-ribose suitable for furanose ring formation

Try using a molecular model to help see this.

As viewed in the drawing, a 120° counterclockwise rotation of C-4 places its hydroxyl group in the proper position. At the same time, this rotation moves the CH_2OH group to a position such that it will become a substituent that is "up" on the five-membered ring. The hydrogen at C-4 then will be "down" in the furanose form.

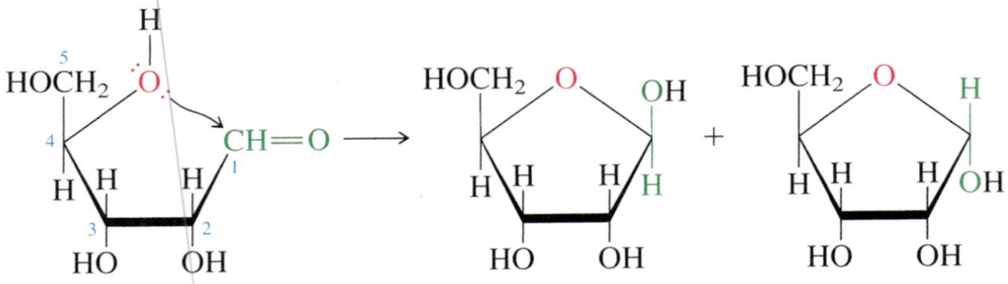

β-D-Ribofuranose   α-D-Ribofuranose

PROBLEM 25.6 Write Haworth formulas corresponding to the furanose forms of each of the following carbohydrates:

(a) D-Xylose

(c) L-Arabinose

(b) D-Arabinose

SAMPLE SOLUTION (a) The Fischer projection of D-xylose is given in Figure 25.2.

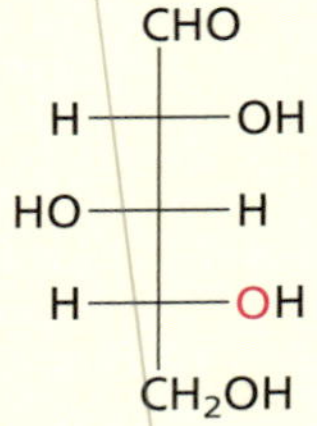

D-Xylose 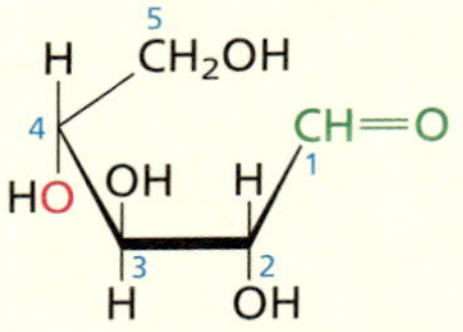   Eclipsed conformation of D-xylose

Carbon-4 of D-xylose must be rotated in a counterclockwise sense to bring its hydroxyl group into the proper orientation for furanose ring formation.

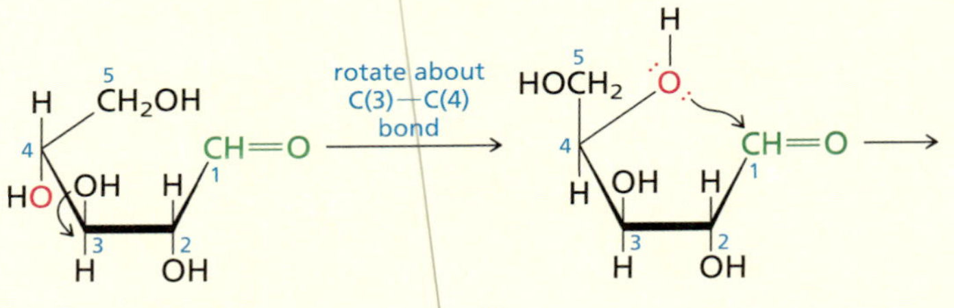

D-Xylose

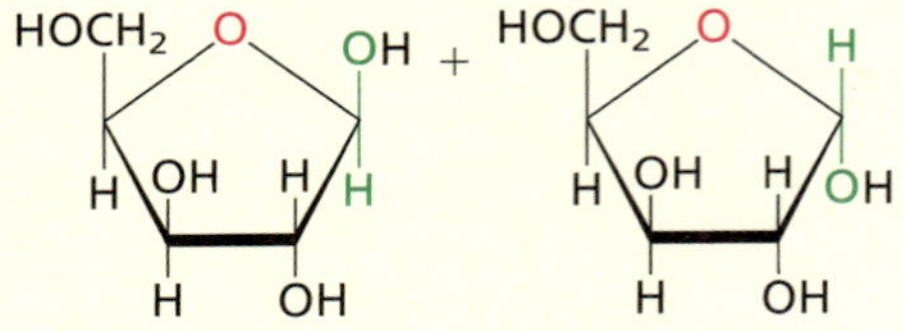

β-D-Xylofuranose   α-D-Xylofuranose

25.7 CYCLIC FORMS OF CARBOHYDRATES: PYRANOSE FORMS

During the discussion of hemiacetal formation in D-ribose in the preceding section, you may have noticed that aldopentoses have the potential of forming a six-membered cyclic hemiacetal via addition of the C-5 hydroxyl to the carbonyl group. This mode of ring closure leads to α- and β-*pyranose* forms:

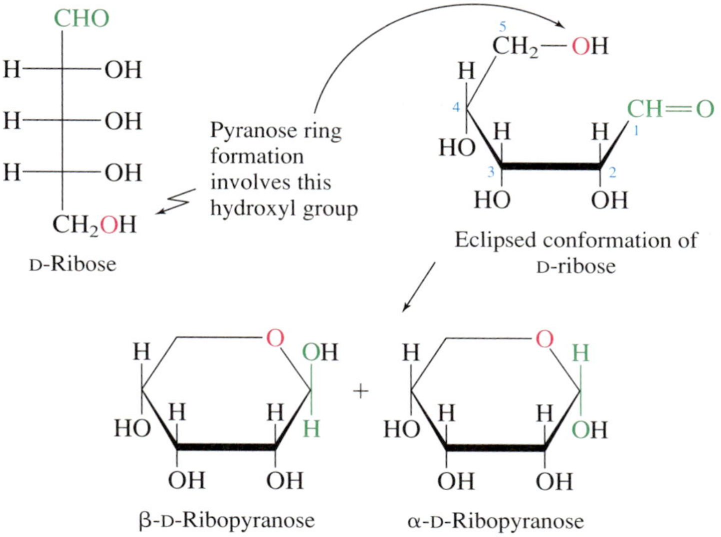

CHO

D-Ribose

Pyranose ring formation involves this hydroxyl group

Eclipsed conformation of D-ribose

β-D-Ribopyranose   +   α-D-Ribopyranose

Like aldopentoses, aldohexoses such as D-glucose are capable of forming two furanose forms (α and β) and two pyranose forms (α and β). The Haworth representations of the pyranose forms of D-glucose are constructed as shown in Figure 25.4; each has a CH_2OH group as a substituent on the six-membered ring.

FIGURE 25.4 Haworth formulas for α- and β-pyranose forms of D-glucose.

CHO

D-Glucose (hydroxyl group at C-5 is involved in pyranose ring formation)

Eclipsed conformation of D-Glucose; hydroxyl at C-5 is not properly oriented for ring formation

rotate about C-4–C-5 bond in counterclockwise direction

Eclipsed conformation of D-glucose in proper orientation for pyranose ring formation

β-D-Glucopyranose   +   α-D-Glucopyranose

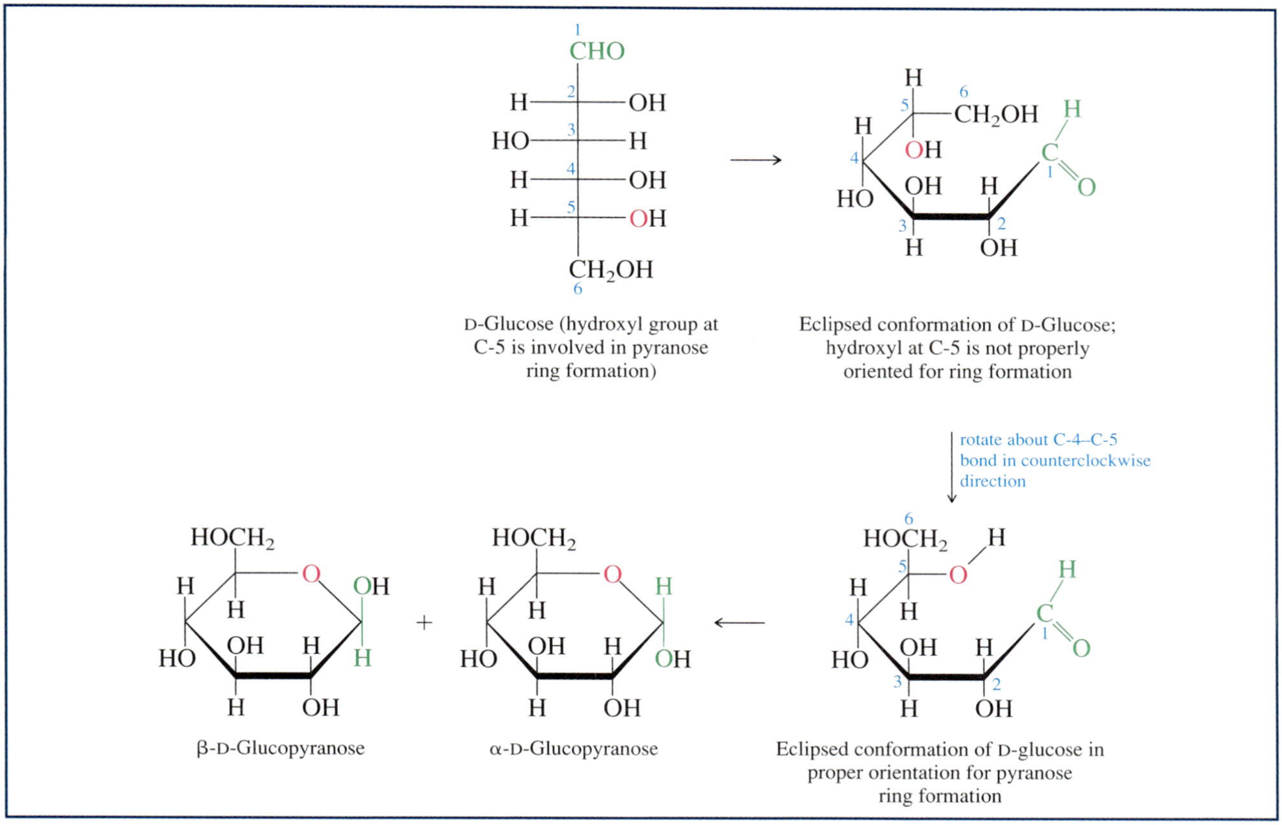

Haworth formulas are satisfactory for representing *configurational* relationships in pyranose forms but are uninformative as to carbohydrate *conformations*. X-ray crystallographic studies of a large number of carbohydrates reveal that the six-membered pyranose ring of D-glucose adopts a chair conformation:

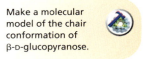

Make a molecular model of the chair conformation of β-D-glucopyranose.

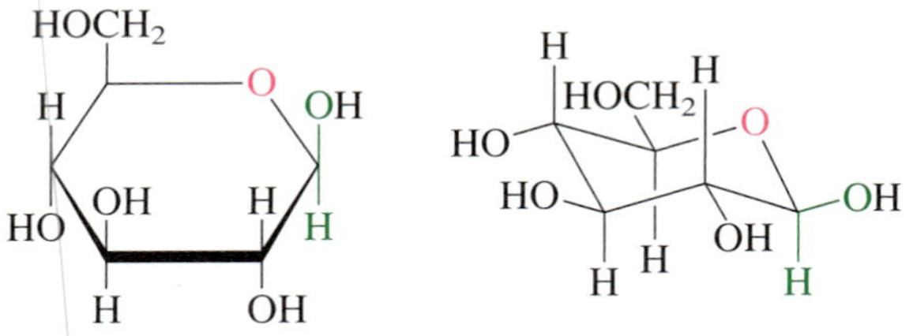

β-D-Glucopyranose

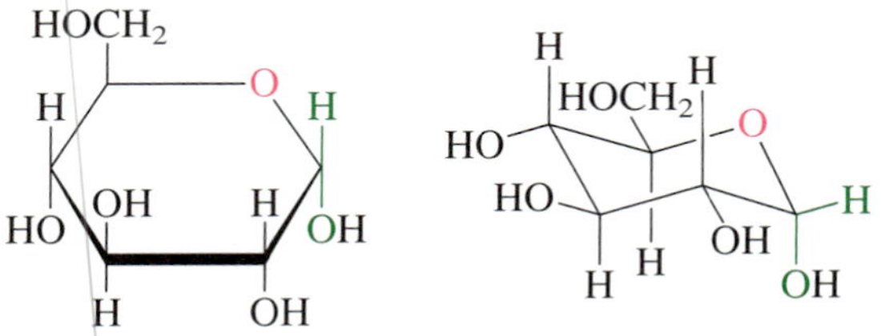

α-D-Glucopyranose

All the ring substituents in β-D-glucopyranose are equatorial in the most stable chair conformation. Only the anomeric hydroxyl group is axial in the α isomer; all the other substituents are equatorial.

Other aldohexoses behave similarly in adopting chair conformations that permit the CH_2OH substituent to occupy an equatorial orientation. Normally the CH_2OH group is the bulkiest, most conformationally demanding substituent in the pyranose form of a hexose.

PROBLEM 25.7   Clearly represent the most stable conformation of the β-pyranose form of each of the following sugars:

(a) D-Galactose

(c) L-Mannose

(b) D-Mannose

(d) L-Ribose

SAMPLE SOLUTION   (a) By analogy with the procedure outlined for D-glucose in Figure 25.4, first generate a Haworth formula for β-D-galactopyranose:

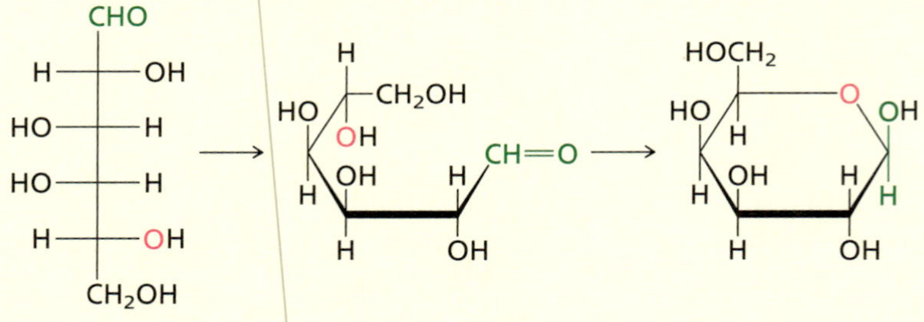

D-Galactose

β-D-Galactopyranose
(Haworth formula)

Next, convert the Haworth formula to the chair conformation that has the CH_2OH group equatorial.

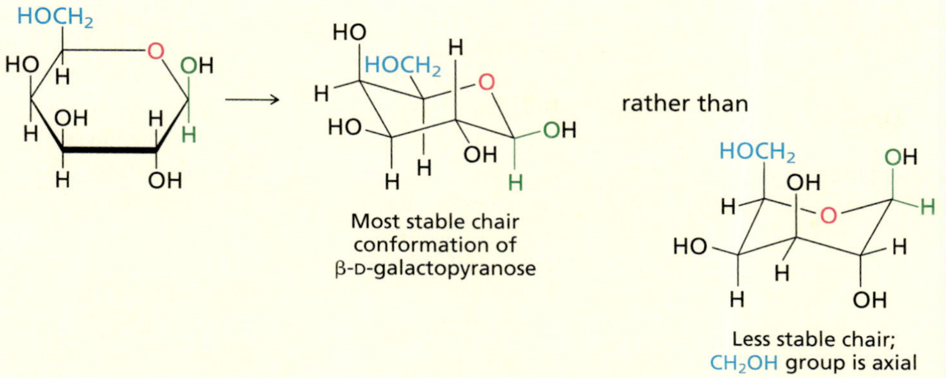

Most stable chair
conformation of
β-D-galactopyranose

rather than

Less stable chair;
CH₂OH group is axial

Galactose differs from glucose in configuration at C-4. The C-4 hydroxyl is axial in
β-D-galactopyranose, but it is equatorial in β-D-glucopyranose.

Because six-membered rings are normally less strained than five-membered ones,
pyranose forms are usually present in greater amounts than furanose forms at equilibrium, and the concentration of the open-chain form is quite small. The distribution of
carbohydrates among their various hemiacetal forms has been examined by using ^{1}H and
^{13}C NMR spectroscopy. In aqueous solution, for example, D-ribose is found to contain
the various α- and β-furanose and pyranose forms in the amounts shown in Figure 25.5.
The concentration of the open-chain form at equilibrium is too small to measure directly.
Nevertheless, it occupies a central position, in that interconversions of α and β anomers
and furanose and pyranose forms take place by way of the open-chain form as an intermediate. As will be seen later, certain chemical reactions also proceed by way of the
open-chain form.

β-D-Ribopyranose (56%)

Open-chain form
of D-ribose
(less than 1%)

β-D-Ribofuranose (18%)

α-D-Ribopyranose (20%)

α-D-Ribofuranose (6%)

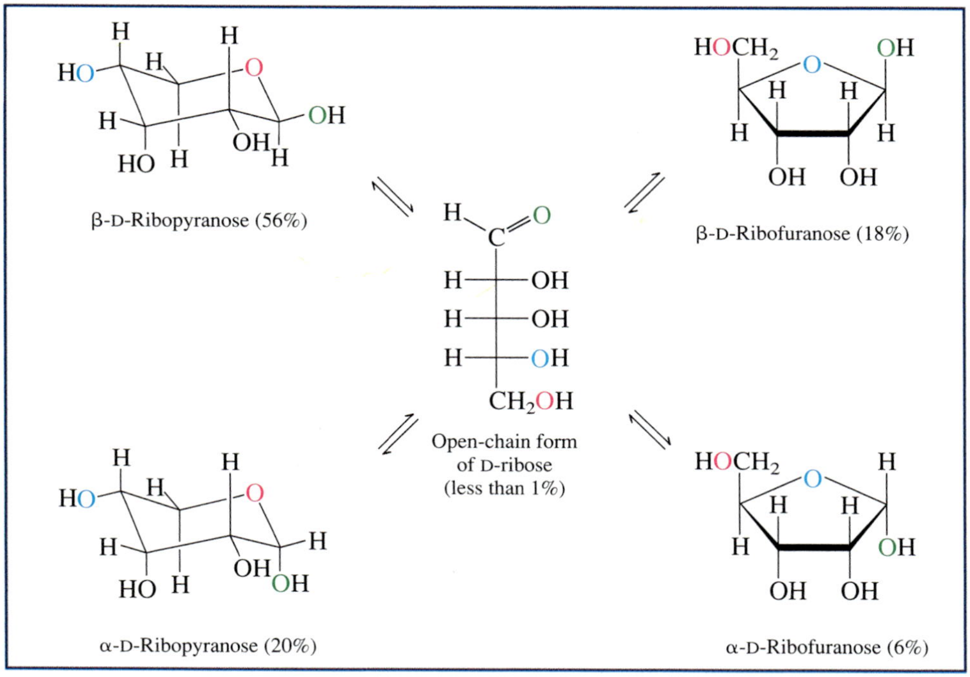

FIGURE 25.5 Distribution
of furanose, pyranose, and
open-chain forms of D-ribose
in aqueous solution as measured by ^{1}H and ^{13}C NMR
spectroscopy.

25.8 MUTAROTATION

In spite of their easy interconversion in solution, α and β forms of carbohydrates are capable of independent existence, and many have been isolated in pure form as crystalline solids. When crystallized from ethanol, D-glucose yields α-D-glucopyranose, mp 146°C, $[\alpha]_D$ +112.2°. Crystallization from a water–ethanol mixture produces β-D-glucopyranose, mp 148–155°C, $[\alpha]_D$ +18.7°. In the solid state the two forms do not interconvert and are stable indefinitely. Their structures have been unambiguously confirmed by X-ray crystallography.

The optical rotations just cited for each isomer are those measured immediately after each one is dissolved in water. On standing, the rotation of the solution containing the α isomer decreases from +112.2° to +52.5°; the rotation of the solution of the β isomer increases from +18.7° to the same value of +52.5°. This phenomenon is called **mutarotation.** What is happening is that each solution, initially containing only one anomeric form, undergoes equilibration to the same mixture of α- and β-pyranose forms. The open-chain form is an intermediate in the process.

α-D-Glucopyranose
(mp 146°C;
$[\alpha]_D$ +112.2°)

Open-chain form
of D-glucose

β-D-Glucopyranose
(mp 148–155°C;
$[\alpha]_D$ +18.7°)

A ^{13}C NMR study of D-glucose in water detected five species: the α-pyranose (38.8%), β-pyranose (60.9%), α-furanose (0.14%), and β-furanose (0.15%) forms, and the hydrate of the open-chain form (0.0045%).

The distribution between the α and β anomeric forms at equilibrium is readily calculated from the optical rotations of the pure isomers and the final optical rotation of the solution, and is determined to be 36% α to 64% β. Independent measurements have established that only the pyranose forms of D-glucose are present in significant quantities at equilibrium.

> **PROBLEM 25.8** The specific optical rotations of pure α- and β-D-mannopyranose are +29.3° and −17.0°, respectively. When either form is dissolved in water, mutarotation occurs, and the observed rotation of the solution changes until a final rotation of +14.2° is observed. Assuming that only α- and β-pyranose forms are present, calculate the percent of each isomer at equilibrium.

It is not possible to tell by inspection whether the α- or β-pyranose form of a particular carbohydrate predominates at equilibrium. As just described, the β-pyranose form is the major species present in an aqueous solution of D-glucose, whereas the α-pyranose form predominates in a solution of D-mannose (Problem 25.8). The relative abundance of α- and β-pyranose forms in solution depends on two factors. The first is solvation of the anomeric hydroxyl group. An equatorial OH is less crowded and better solvated by water than an axial one. This effect stabilizes the β-pyranose form in aqueous solution. The

The anomeric effect is best explained by a molecular orbital analysis that is beyond the scope of this text.

other factor, called the **anomeric effect,** involves an electronic interaction between the ring oxygen and the anomeric substituent and preferentially stabilizes the axial OH of the α-pyranose form. Because the two effects operate in different directions but are comparable in magnitude in aqueous solution, the α-pyranose form is more abundant for some carbohydrates and the β-pyranose form for others.

25.9 KETOSES

Up to this point all our attention has been directed toward aldoses, carbohydrates having an aldehyde function in their open-chain form. Aldoses are more common than ketoses, and their role in biological processes has been more thoroughly studied. Nevertheless, a large number of ketoses are known, and several of them are pivotal intermediates in carbohydrate biosynthesis and metabolism. Examples of some ketoses include D-*ribulose*, L-*xylulose*, and D-*fructose:*

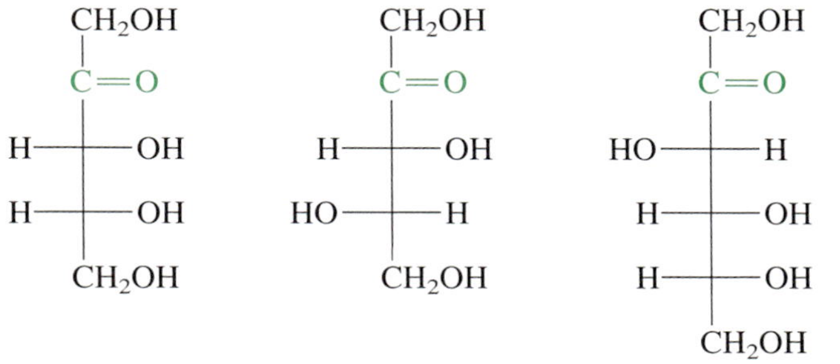

D-Ribulose
(a 2-ketopentose that is a key compound in photosynthesis)

L-Xylulose
(a 2-ketopentose excreted in excessive amounts in the urine of persons afflicted with the mild genetic disorder pentosuria)

D-Fructose
(a 2-ketohexose also known as *levulose*; it is found in honey and is signficantly sweeter than table sugar)

In these three examples the carbonyl group is located at C-2, which is the most common location for the carbonyl function in naturally occurring ketoses.

PROBLEM 25.9 How many ketotetroses are possible? Write Fischer projections for each.

Ketoses, like aldoses, exist mainly as cyclic hemiacetals. In the case of D-ribulose, furanose forms result from addition of the C-5 hydroxyl to the carbonyl group.

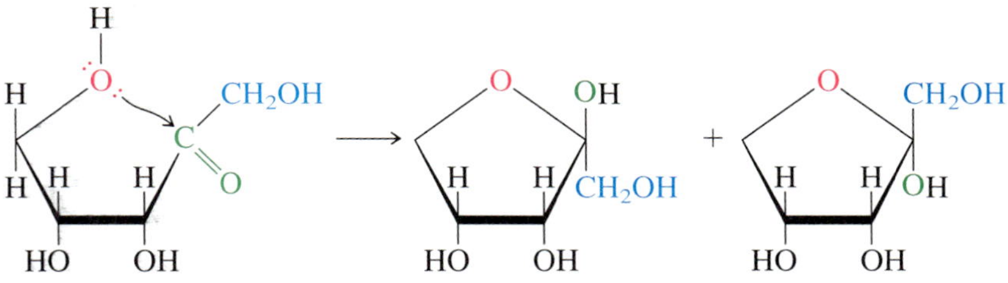

Eclipsed conformation of D-ribulose

β-D-Ribulofuranose

α-D-Ribulofuranose

The anomeric carbon of a furanose or pyranose form of a ketose bears both a hydroxyl group and a carbon substituent. In the case of 2-ketoses, this substituent is a CH_2OH group. As with aldoses, the anomeric carbon of a cyclic hemiacetal is readily identifiable because it is bonded to two oxygens.

25.10 DEOXY SUGARS

A commonplace variation on the general pattern seen in carbohydrate structure is the replacement of one or more of the hydroxyl substituents by some other atom or group. In **deoxy sugars** the hydroxyl group is replaced by hydrogen. Two examples of deoxy sugars are 2-deoxy-D-ribose and L-rhamnose:

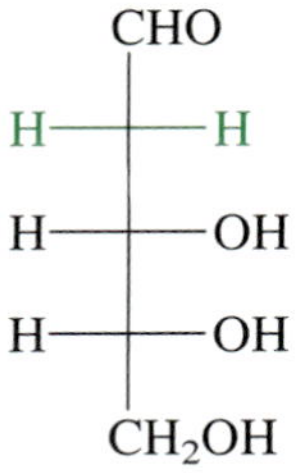

$$
\begin{array}{cc}
\text{CHO} & \text{CHO} \\
\text{H}\!-\!\!-\!\!\text{H} & \text{H}\!-\!\!-\!\!\text{OH} \\
\text{H}\!-\!\!-\!\!\text{OH} & \text{H}\!-\!\!-\!\!\text{OH} \\
\text{H}\!-\!\!-\!\!\text{OH} & \text{HO}\!-\!\!-\!\!\text{H} \\
\text{CH}_2\text{OH} & \text{HO}\!-\!\!-\!\!\text{H} \\
 & \text{CH}_3
\end{array}
$$

2-Deoxy-D-ribose	L-Rhamnose
	(6-deoxy-L-mannose)

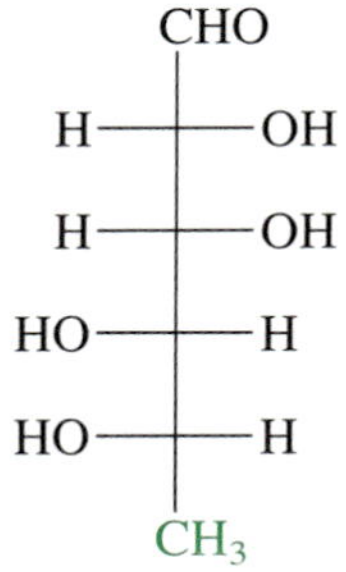

The hydroxyl at C-2 in D-ribose is absent in 2-deoxy-D-ribose. In Chapter 28 we shall see how derivatives of 2-deoxy-D-ribose, called *deoxyribonucleotides,* are the fundamental building blocks of deoxyribonucleic acid (DNA), the material responsible for storing genetic information. L-Rhamnose is a compound isolated from a number of plants. Its carbon chain terminates in a methyl rather than a CH₂OH group.

PROBLEM 25.10   Write Fischer projections of

(a) *Cordycepose* (3-deoxy-D-ribose): a deoxy sugar isolated by hydrolysis of the antibiotic substance *cordycepin*

(b) L-*Fucose* (6-deoxy-L-galactose): obtained from seaweed

SAMPLE SOLUTION   (a) The hydroxyl group at C-3 in D-ribose is replaced by hydrogen in 3-deoxy-D-ribose.

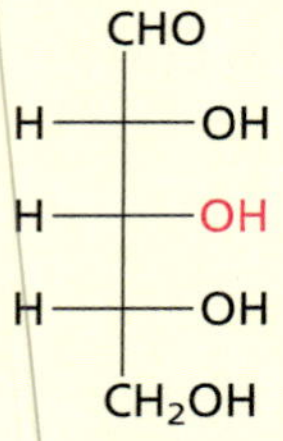

$$
\begin{array}{cc}
\text{CHO} & \text{CHO} \\
\text{H}\!-\!\!-\!\!\text{OH} & \text{H}\!-\!\!-\!\!\text{OH} \\
\text{H}\!-\!\!-\!\!\text{OH} & \text{H}\!-\!\!-\!\!\text{H} \\
\text{H}\!-\!\!-\!\!\text{OH} & \text{H}\!-\!\!-\!\!\text{OH} \\
\text{CH}_2\text{OH} & \text{CH}_2\text{OH}
\end{array}
$$

D-Ribose	3-Deoxy-D-ribose
(from Figure 25.2)	(cordycepose)

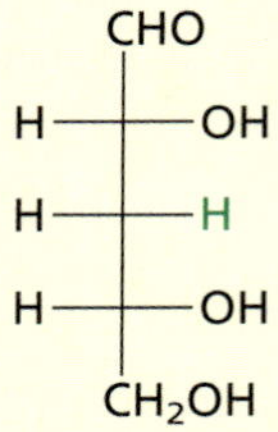

25.11 AMINO SUGARS

Another structural variation is the replacement of a hydroxyl group in a carbohydrate by an amino group to give an **amino sugar.** The most abundant amino sugar is one of the oldest and most abundant organic compounds on earth. *N*-Acetyl-D-glucosamine is the main component of the polysaccharide in *chitin,* the substance that makes up the tough outer skeleton of arthropods and insects. Chitin has been isolated from a 25-million-year-old beetle fossil, and more than 10^9 tons of chitin is produced in the biosphere each year. Lobster shells, for example, are mainly chitin. More than 60 amino sugars are known, many of them having been isolated and identified only recently as components of antibiotics. The anticancer drug doxorubicin hydrochloride (Adriamycin), for example, contains the amino sugar L-daunosamine as one of its structural units.

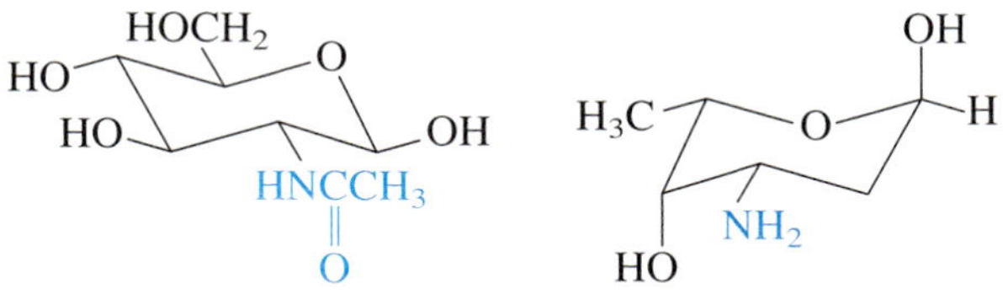

N-Acetyl-D-glucosamine   L-Daunosamine

25.12 BRANCHED-CHAIN CARBOHYDRATES

Carbohydrates that have a carbon substituent attached to the main chain are said to have a **branched chain.** D-Apiose and L-vancosamine are representative branched-chain carbohydrates:

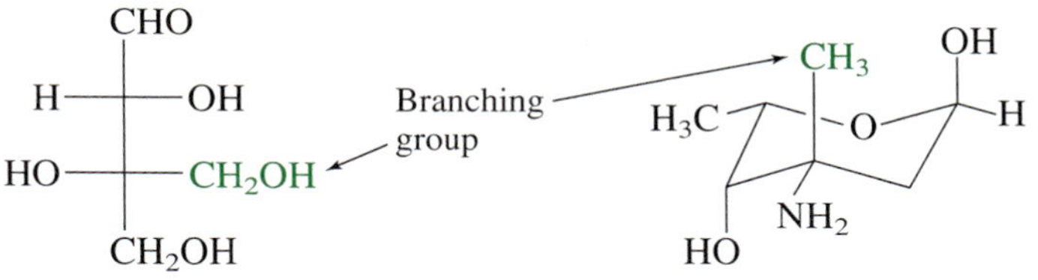

D-Apiose   L-Vancosamine

D-Apiose can be isolated from parsley and is a component of the cell wall polysaccharide of various marine plants. Among its novel structural features is the presence of only a single chirality center. L-Vancosamine is but one portion of vancomycin, a powerful antibiotic that has emerged as one of only a few antibiotics that are effective against drug-resistant bacteria. L-Vancosamine is not only a branched-chain carbohydrate, it is a deoxy sugar and an amino sugar as well.

25.13 GLYCOSIDES

Glycosides are a large and very important class of carbohydrate derivatives characterized by the replacement of the anomeric hydroxyl group by some other substituent. Glycosides are termed *O*-glycosides, *N*-glycosides, *S*-glycosides, and so on, according to the atom attached to the anomeric carbon.

For a review of the isolation of chitin from natural sources and some of its uses, see the November 1990 issue of the *Journal of Chemical Education* (pp. 938–942).

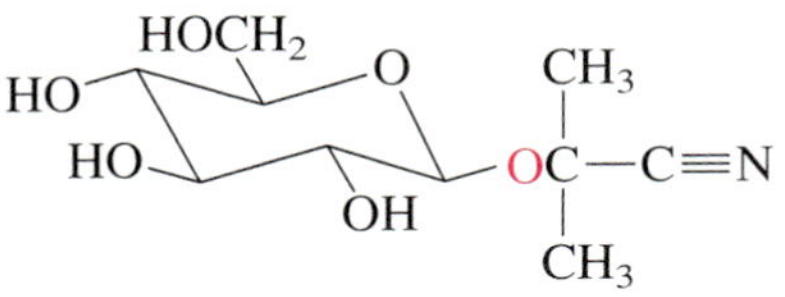

Linamarin
(an *O*-glycoside: obtained from manioc, a type of yam widely distributed in southeast Asia)

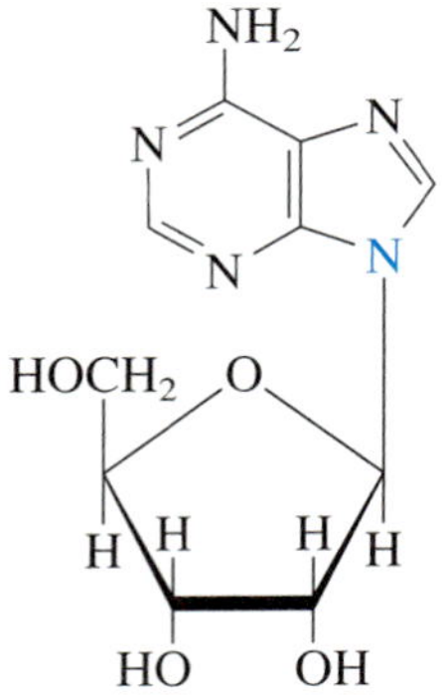

Adenosine
(an *N*-glycoside: also known as a nucleoside; adenosine is one of the fundamental molecules of biochemistry)

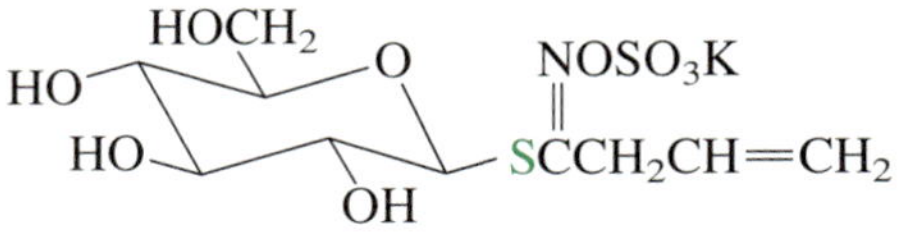

Sinigrin
(an *S*-glycoside: contributes to the characteristic flavor of mustard and horseradish)

Usually, the term *glycoside* without a prefix is taken to mean an *O*-glycoside and will be used that way in this chapter. Glycosides are classified as α or β in the customary way, according to the configuration at the anomeric carbon. All three of the glycosides just shown are β-glycosides. Linamarin and sinigrin are glycosides of D-glucose; adenosine is a glycoside of D-ribose.

Structurally, *O*-glycosides are mixed acetals that involve the anomeric position of furanose and pyranose forms of carbohydrates. Recall the sequence of intermediates in acetal formation (Section 17.8):

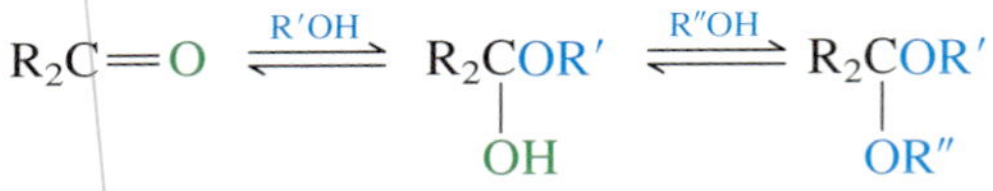

Aldehyde or ketone | Hemiacetal | Acetal

When this sequence is applied to carbohydrates, the first step takes place *intramolecularly* to yield a cyclic hemiacetal. The second step is *intermolecular,* requires an alcohol R″OH as a reactant, and proceeds readily only in the presence of an acid catalyst. An oxygen-stabilized carbocation is an intermediate.

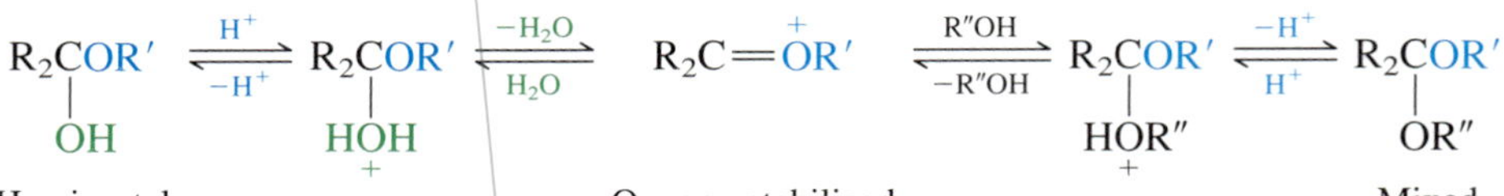

Hemiacetal | Oxygen-stabilized carbocation | Mixed acetal

The preparation of glycosides in the laboratory is carried out by simply allowing a carbohydrate to react with an alcohol in the presence of an acid catalyst:

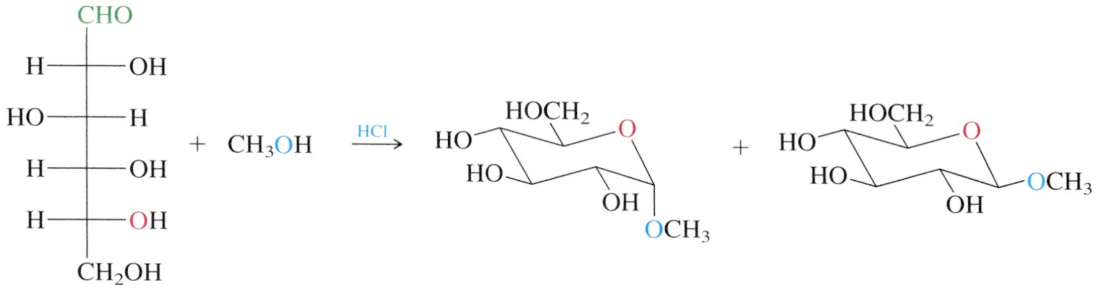

D-Glucose + Methanol → Methyl α-D-glucopyranoside (major product; isolated in 49% yield) + Methyl β-D-glucopyranoside (minor product)

PROBLEM 25.11 Write structural formulas for the α- and β-methyl pyranosides formed by reaction of D-galactose with methanol in the presence of hydrogen chloride.

A point to be emphasized about glycoside formation is that, despite the presence of a number of other hydroxyl groups in the carbohydrate, *only the anomeric hydroxyl group is replaced*. This is because a carbocation at the anomeric position is stabilized by the ring oxygen and is the only one capable of being formed under the reaction conditions.

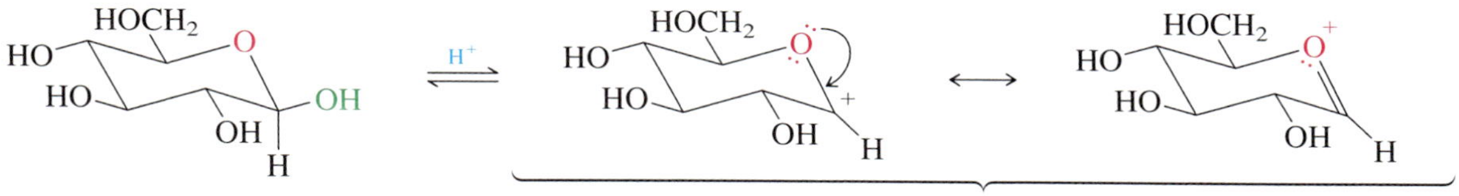

D-Glucose (shown in β-pyranose form)

Electron pair on ring oxygen can stabilize carbocation at anomeric position only.

Once the carbocation is formed, it is captured by the alcohol acting as a nucleophile. Attack can occur at either the α or β face of the carbocation.

Attack at the α face gives methyl α-D-glucopyranoside:

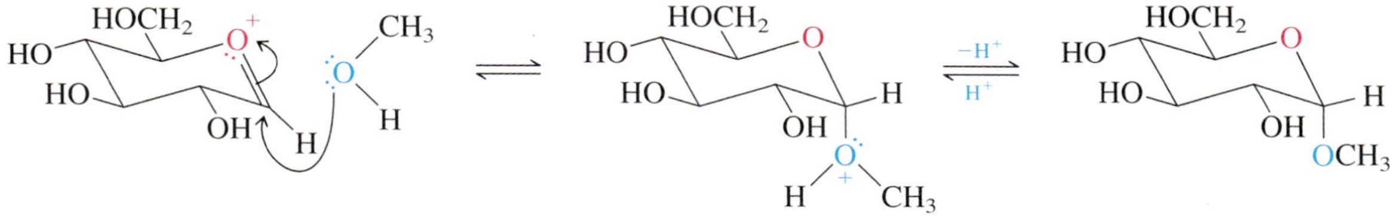

Carbocation intermediate + Methanol

Methyl α-D-glucopyranoside

Attack at the β face gives methyl β-D-glucopyranoside:

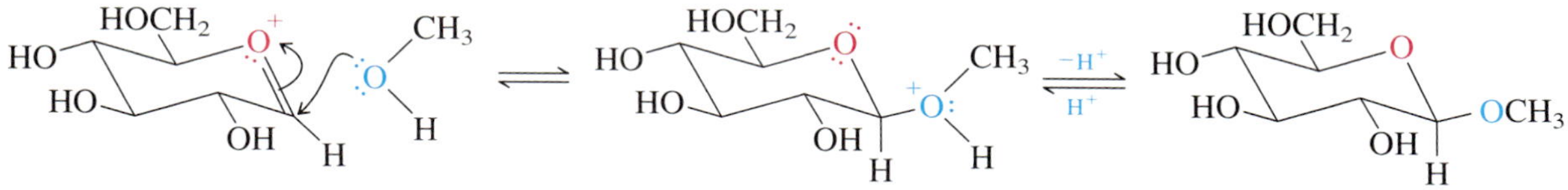

Carbocation intermediate + Methanol

Methyl β-D-glucopyranoside

All of the reactions, from D-glucose to the methyl glycosides via the carbocation, are reversible. The overall reaction is *thermodynamically controlled* and gives the same mixture of glycosides irrespective of which stereoisomeric pyranose form of D-glucose we start with. Nor does it matter whether we start with a pyranose form or a furanose form of D-glucose. Glucopyranosides are more stable than glucofuranosides and predominate at equilibrium.

PROBLEM 25.12 Methyl glycosides of 2-deoxy sugars have been prepared by the acid-catalyzed addition of methanol to unsaturated sugars known as *glycals.*

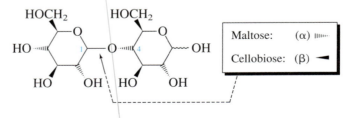

Galactal → Methyl 2-deoxy-α-D-lyxohexopyranoside (38%) + Methyl 2-deoxy-β-D-lyxohexopyranoside (36%)

CH_3OH / HCl

Suggest a reasonable mechanism for this reaction.

Under neutral or basic conditions glycosides are configurationally stable; unlike the free sugars from which they are derived, glycosides do not exhibit mutarotation. Converting the anomeric hydroxyl group to an ether function (hemiacetal ⟶ acetal) prevents its reversion to the open-chain form in neutral or basic media. In aqueous acid, acetal formation can be reversed and the glycoside hydrolyzed to an alcohol and the free sugar.

25.14 DISACCHARIDES

Disaccharides are carbohydrates that yield two monosaccharide molecules on hydrolysis. Structurally, disaccharides are *glycosides* in which the alkoxy group attached to the anomeric carbon is derived from a second sugar molecule.

Maltose, obtained by the hydrolysis of starch, and *cellobiose,* by the hydrolysis of cellulose, are isomeric disaccharides. In both maltose and cellobiose two D-glucopyranose units are joined by a glycosidic bond between C-1 of one unit and C-4 of the other. The two are diastereomers, differing only in the stereochemistry at the anomeric carbon of the glycoside bond; maltose is an α-glycoside, cellobiose is a β-glycoside.

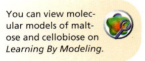

You can view molecular models of maltose and cellobiose on *Learning By Modeling.*

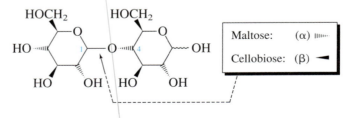

| Maltose: | (α) |
| Cellobiose: | (β) |

The stereochemistry and points of connection of glycosidic bonds are commonly designated by symbols such as α(1,4) for maltose and β(1,4) for cellobiose; α and β designate the stereochemistry at the anomeric position; the numerals specify the ring carbons involved.

Maltose

Cellobiose

FIGURE 25.6 Molecular models of the disaccharides maltose and cellobiose. Two D-glucopyranose units are connected by a glycoside linkage between C-1 and C-4. The glycosidic bond has the α orientation in maltose and is β in cellobiose. Maltose and cellobiose are diastereomers.

Both maltose and cellobiose have a free anomeric hydroxyl group that is not involved in a glycoside bond. The configuration at the free anomeric center is variable and may be either α or β. Indeed, two stereoisomeric forms of maltose have been isolated: one has its anomeric hydroxyl group in an equatorial orientation; the other has an axial anomeric hydroxyl.

The free anomeric hydroxyl group is the one shown at the far right of the preceding structural formula. The symbol ∿∿ is used to represent a bond of variable stereochemistry.

PROBLEM 25.13 The two stereoisomeric forms of maltose just mentioned undergo mutarotation when dissolved in water. What is the structure of the key intermediate in this process?

The single difference in their structures, the stereochemistry of the glycosidic bond, causes maltose and cellobiose to differ significantly in their three-dimensional shape, as the molecular models of Figure 25.6 illustrate. This difference in shape affects the way in which maltose and cellobiose interact with other chiral molecules such as proteins, and they behave much differently toward enzyme-catalyzed hydrolysis. An enzyme known as *maltase* catalyzes the hydrolytic cleavage of the α-glycosidic bond of maltose but is without effect in promoting the hydrolysis of the β-glycosidic bond of cellobiose. A different enzyme, *emulsin,* produces the opposite result: emulsin catalyzes the hydrolysis of cellobiose but not of maltose. The behavior of each enzyme is general for glucosides (glycosides of glucose). Maltase catalyzes the hydrolysis of α-glucosides and is also known as *α-glucosidase,* whereas emulsin catalyzes the hydrolysis of β-glucosides and is known as *β-glucosidase.* The specificity of these enzymes offers a useful tool for structure determination because it allows the stereochemistry of glycosidic linkages to be assigned.

Lactose is a disaccharide constituting 2–6% of milk and is known as *milk sugar.* It differs from maltose and cellobiose in that only one of its monosaccharide units is D-glucose. The other monosaccharide unit, the one that contributes its anomeric carbon to the glycoside bond, is D-galactose. Like cellobiose, lactose is a β-glycoside.

You can view molecular models of cellobiose and lactose on *Learning By Modeling.*

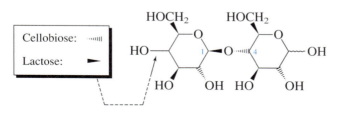

FIGURE 25.7 The structure of sucrose.

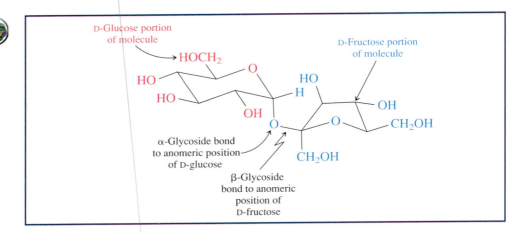

Digestion of lactose is facilitated by the β-glycosidase *lactase*. A deficiency of this enzyme makes it difficult to digest lactose and causes abdominal discomfort. Lactose intolerance is a genetic trait; it is treatable through over-the-counter formulations of lactase and by limiting the amount of milk in the diet.

The most familiar of all the carbohydrates is *sucrose*—common table sugar. Sucrose is a disaccharide in which D-glucose and D-fructose are joined at their anomeric carbons by a glycosidic bond (Figure 25.7). Its chemical composition is the same irrespective of its source; sucrose from cane and sucrose from sugar beets are chemically identical. Because sucrose does not have a free anomeric hydroxyl group, it does not undergo mutarotation.

25.15 POLYSACCHARIDES

Cellulose is the principal structural component of vegetable matter. Wood is 30–40% cellulose, cotton over 90%. Photosynthesis in plants is responsible for the formation of 10^9 tons per year of cellulose. Structurally, cellulose is a polysaccharide composed of several thousand D-glucose units joined by β(1,4)-glycosidic linkages (Figure 25.8). Complete hydrolysis of all the glycosidic bonds of cellulose yields D-glucose. The disaccharide fraction that results from partial hydrolysis is cellobiose.

As Figure 25.8 shows, the glucose units of cellulose are turned with respect to each other. The overall shape of the chain, however, is close to linear. Consequently, neighboring chains can pack together in bundles where networks of hydrogen bonds stabilize the structure and impart strength to cellulose fibers.

Animals lack the enzymes necessary to catalyze the hydrolysis of cellulose and so can't digest it. Cattle and other ruminants use cellulose as a food source indirectly.

FIGURE 25.8 Cellulose is a polysaccharide in which D-glucose units are connected by β(1,4)-glycoside linkages analogous to cellobiose. Hydrogen bonding, especially between the C-2 and C-6 hydroxyl groups, causes adjacent glucose units to tilt at an angle of 180° with each other.

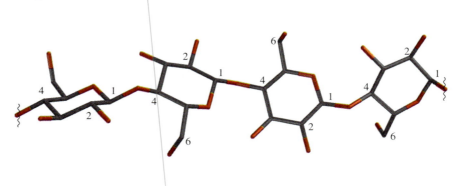

Colonies of bacteria that live in their digestive tract consume cellulose and in the process convert it to other substances that the animal can digest.

A more direct source of energy for animals is provided by the starches found in many plants. Starch is a mixture containing about 20% of a water-dispersible fraction called *amylose* and 80% of a second component, *amylopectin*.

Like cellulose, amylose is a polysaccharide of D-glucose. However, unlike cellulose in which all of the glycosidic linkages are β, all of the linkages in amylose are α. Amylopectin resembles amylose in being a polysaccharide built on a framework of α(1,4)-linked D-glucose units. In addition to this main framework, however, amylose incorporates polysaccharide branches of 24–30 glucose units joined by α(1,4)-glycosidic bonds. These branches sprout from C-6 of glucose units at various points along the main framework, connected to it by α(1,6)-glycosidic bonds.

The small change in stereochemistry between cellulose and amylose creates a large difference in their overall shape and in their properties. Some of this difference can be seen in the structure of a short portion of amylose in Figure 25.9. The presence of the α-glycosidic linkages imparts a twist to the amylose chain. Where the main chain is roughly linear in cellulose, it is helical in amylose. Attractive forces *between* chains are weaker in amylose, and amylose does not form the same kind of strong fibers that cellulose does.

Certain enzymes that cleave amylose produce unusual structures along the way. One is the cycloamylose shown in Figure 25.10. This oligosaccharide contains 26 α(1,4)-linked glucose molecules marked by the additional feature of joined ends. The result is a figure-8-shaped molecule. The left-handed helical nature of each half is clearly evident in part (*a*) of the figure. When looked at from the top, as in part (*b*), the interior region of the cycloamylose is seen to be a pair of hollow cylinders. The —OH groups of the glucose units are on the outside where they form hydrogen bonds with other —OH groups of the cycloamylose and with water molecules. The cylindrical inner channels are rich in C—H bonds and are hydrophobic.

A related, albeit simpler structural motif occurs in cyclodextrins. *Dextrins* are gummy, polysaccharide fragments formed when starch is degraded by heating or partial hydrolysis. *Cyclodextrins* are formed when starch is hydrolyzed in the presence of enzymes called *glucanotransferases*. These cyclodextrins contain six to eight glucose units, not nearly enough to form the kind of figure-8 shown in Figure 25.10. Instead, they simply close to form a structure that resembles a styrofoam cup. The hydrophobic interior of a cyclodextrin has a diameter of about 500 pm and can accommodate small organic molecules in a host-guest relationship.

One of the most important differences between cellulose and starch is that animals can digest starch. Because the glycosidic linkages in starch are α, an animal's α-glycosidase enzymes can catalyze their hydrolysis to glucose. When more glucose is

For more about starch, see "The Other Double Helix—The Fascinating Chemistry of Starch" in the August, 2000 issue of the *Journal of Chemical Education*, pp. 988–992.

Professor Ronald Breslow of Columbia University has carried out a number of organic reactions in the presence of cyclodextrins to study the effect of a molecule's environment on its chemical reactivity.

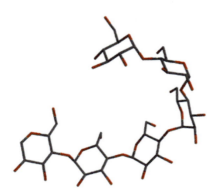

FIGURE 25.9 Amylose is a polysaccharide in which D-glucose units are connected by α(1,4)-glycoside linkages analogous to maltose. The geometry of the glycoside linkage is responsible for the left-hand helical twist of the chain.

(a)   (b)

FIGURE 25.10 A cy-
cloamylose containing 26
glucose units joined by
α(1,4)-glycoside bonds. (a)
The helical structure of amylose makes it possible for
two ends of amylose frag-
ments to bind together via
glycoside bonds. (b) Top
view showing interior of
each loop.

available than is needed as fuel, animals store some of it as glycogen. Glycogen resembles amylopectin in that it is a branched polysaccharide of α(1,4)-linked D-glucose units
with branches connected to C-6 of the main chain. The frequency of such branches is
greater in glycogen than in amylopectin.

25.16 CELL-SURFACE GLYCOPROTEINS

That carbohydrates play an informational role in biological interactions is a recent rev-
elation of great importance. *Glycoproteins,* protein molecules covalently bound to carbohydrates, are often the principal species involved. When a cell is attacked by a virus
or bacterium or when it interacts with another cell, the drama begins when the foreign
particle attaches itself to the surface of the host cell. The invader recognizes the host by
the glycoproteins on the cell surface. More specifically, it recognizes particular carbohydrate sequences at the end of the glycoprotein. For example, the receptor on the cell
surface to which an influenza virus attaches itself has been identified as a glycoprotein
terminating in a disaccharide of *N*-acetylgalactosamine and *N*-acetylneuraminic acid
(Figure 25.11). Because attachment of the invader to the surface of the host cell is the
first step in infection, one approach to disease prevention is to selectively inhibit this
"host-guest" interaction. Identifying the precise nature of the interaction is the first step
in the rational design of drugs that prevent it.

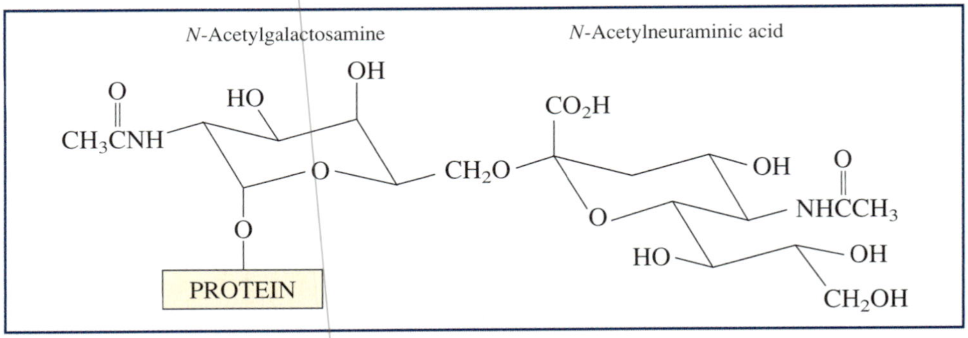

FIGURE 25.11 Diagram of a
cell-surface glycoprotein,
showing the disaccharide
unit that is recognized by an
invading influenza virus.

HOW SWEET IT IS!

How sweet is it? There is no shortage of compounds, natural or synthetic, that taste sweet. The most familiar are naturally occurring sugars, especially sucrose, glucose, and fructose. All occur naturally, with worldwide production of sucrose from cane and sugar beets exceeding 100 million tons per year. Glucose is prepared by the enzymatic hydrolysis of starch, and fructose is made by the isomerization of glucose.

Starch + H$_2$O $\longrightarrow$

D-(+)-Glucose   D-(−)-Fructose

Among sucrose, glucose, and fructose, fructose is the sweetest. Honey is sweeter than table sugar because it contains fructose formed by the isomerization of glucose as shown in the equation.

You may have noticed that most soft drinks contain "high-fructose corn syrup." Corn starch is hydrolyzed to glucose, which is then treated with glucose isomerase to produce a fructose-rich mixture. The enhanced sweetness permits less to be used, reducing the cost of production. Using less carbohydrate-based sweetener also reduces the number of calories.

Artificial sweeteners are a billion-dollar-per-year industry. The primary goal is, of course, to maximize sweetness and minimize calories. We'll look at the following three sweeteners to give us an overview of the field.

Saccharin   Sucralose   Aspartame

All three of these are hundreds of times sweeter than sucrose and variously described as "low-calorie" or "nonnutritive" sweeteners.

Saccharin was discovered at Johns Hopkins University in 1879 in the course of research on coal-tar derivatives and is the oldest artificial sweetener. In spite of its name, which comes from the Latin word for sugar, saccharin bears no structural relationship to any sugar. Nor is saccharin itself very soluble in water. The proton bonded to nitrogen, however, is fairly acidic and saccharin is normally marketed as its water-soluble sodium or calcium salt. Its earliest applications were not in weight control, but as a replacement for sugar in the diet of diabetics before insulin became widely available.

Sucralose has the structure most similar to sucrose. Galactose replaces the glucose unit of sucrose, and chlorines replace three of the hydroxyl groups. Sucralose is the newest artificial sweetener, having been approved by the U.S. Food and Drug Administration in 1998. The three chlorine substituents do not diminish sweetness, but do interfere with the ability of the body to metabolize sucralose. It, therefore, has no food value and is "noncaloric."

—*Cont.*

Aspartame is the market leader among artificial sweeteners. It is a methyl ester of a dipeptide, unrelated to any carbohydrate. It was discovered in the course of research directed toward developing drugs to relieve indigestion.

Saccharin, sucralose, and aspartame illustrate the diversity of structural types that taste sweet, and the vitality and continuing development of the industry of which they are a part.*

*For more information, including theories of structure–taste relationships, see the symposium "Sweeteners and Sweetness Theory" in the August, 1995 issue of the *Journal of Chemical Education*, pp. 671–683.

25.17 CARBOHYDRATE STRUCTURE DETERMINATION

Present-day techniques for structure determination in carbohydrate chemistry are substantially the same as those for any other type of compound. The full range of modern instrumental methods, including mass spectrometry and infrared and nuclear magnetic resonance spectroscopy, is brought to bear on the problem. If the unknown substance is crystalline, X-ray diffraction can provide precise structural information that in the best cases is equivalent to taking a three-dimensional photograph of the molecule.

Before the widespread availability of instrumental methods, the major approach to structure determination relied on a battery of chemical reactions and tests. The response of an unknown substance to various reagents and procedures provided a body of data from which the structure could be deduced. Some of these procedures are still used to supplement the information obtained by instrumental methods. To better understand the scope and limitations of these tests, a brief survey of the chemical reactions of carbohydrates is in order. In many cases these reactions are simply applications of chemistry you have already learned. Certain of the transformations, however, are unique to carbohydrates.

The classical approach to structure determination in carbohydrate chemistry is best exemplified by Fischer's work with D-glucose. A detailed account of this study appears in the August 1941 issue of the *Journal of Chemical Education* (pp. 353–357).

25.18 REDUCTION OF CARBOHYDRATES

Although carbohydrates exist almost entirely as cyclic hemiacetals in aqueous solution, they are in rapid equilibrium with their open-chain forms, and most of the reagents that react with simple aldehydes and ketones react in an analogous way with the carbonyl functional groups of carbohydrates.

The carbonyl group of carbohydrates can be reduced to an alcohol function. Typical procedures include catalytic hydrogenation and sodium borohydride reduction. Lithium aluminum hydride is not suitable, because it is not compatible with the solvents (water, alcohols) that are required to dissolve carbohydrates. The products of carbohydrate reduction are called **alditols.** Because these alditols lack a carbonyl group, they are, of course, incapable of forming cyclic hemiacetals and exist exclusively in noncyclic forms.

α-D-Galactofuranose, or
β-D-Galactofuranose, or
α-D-Galactopyranose, or
β-D-Galactopyranose

$\xrightleftharpoons{\qquad}$

$$\begin{array}{c} \text{CHO} \\ \text{H}\!-\!\!-\!\text{OH} \\ \text{HO}\!-\!\!-\!\text{H} \\ \text{HO}\!-\!\!-\!\text{H} \\ \text{H}\!-\!\!-\!\text{OH} \\ \text{CH}_2\text{OH} \end{array}$$

D-Galactose

$\xrightarrow{\text{NaBH}_4 \atop \text{H}_2\text{O}}$

$$\begin{array}{c} \text{CH}_2\text{OH} \\ \text{H}\!-\!\!-\!\text{OH} \\ \text{HO}\!-\!\!-\!\text{H} \\ \text{HO}\!-\!\!-\!\text{H} \\ \text{H}\!-\!\!-\!\text{OH} \\ \text{CH}_2\text{OH} \end{array}$$

D-Galactitol (90%)

Another name for glucitol, obtained by reduction of D-glucose, is *sorbitol;* it is used as a sweetener, especially in special diets required to be low in sugar. Reduction of D-fructose yields a mixture of glucitol and mannitol, corresponding to the two possible configurations at the newly generated chirality center at C-2.

25.19 OXIDATION OF CARBOHYDRATES

A characteristic property of an aldehyde function is its sensitivity to oxidation. A solution of copper(II) sulfate as its citrate complex **(Benedict's reagent)** is capable of oxidizing aliphatic aldehydes to the corresponding carboxylic acid.

$$\underset{\substack{\text{Aldehyde}}}{\overset{\displaystyle O \atop \displaystyle \|}{RCH}} \; + \; \underset{\substack{\text{From copper(II)} \\ \text{sulfate}}}{2Cu^{2+}} \; + \; \underset{\substack{\text{Hydroxide} \\ \text{ion}}}{5HO^-} \; \longrightarrow \; \underset{\substack{\text{Carboxylate} \\ \text{anion}}}{\overset{\displaystyle O \atop \displaystyle \|}{RCO^-}} \; + \; \underset{\substack{\text{Copper(I)} \\ \text{oxide}}}{Cu_2O} \; + \; \underset{\substack{\text{Water}}}{3H_2O}$$

The formation of a red precipitate of copper(I) oxide by reduction of Cu(II) is taken as a positive test for an aldehyde. Carbohydrates that give positive tests with Benedict's reagent are termed **reducing sugars.**

Aldoses are reducing sugars because they possess an aldehyde function in their open-chain form. Ketoses are also reducing sugars. Under the conditions of the test, ketoses equilibrate with aldoses by way of *enediol intermediates,* and the aldoses are oxidized by the reagent.

Benedict's reagent is the key material in a test kit available from drugstores that permits individuals to monitor the glucose levels in their urine.

$$\underset{\substack{\text{Ketose}}}{\overset{\displaystyle CH_2OH \atop \substack{\displaystyle | \\ \displaystyle C=O \\ \displaystyle | \\ \displaystyle R}}{}} \; \rightleftharpoons \; \underset{\substack{\text{Enediol}}}{\overset{\displaystyle CHOH \atop \substack{\displaystyle \| \\ \displaystyle C-OH \\ \displaystyle | \\ \displaystyle R}}{}} \; \rightleftharpoons \; \underset{\substack{\text{Aldose}}}{\overset{\displaystyle CH=O \atop \substack{\displaystyle | \\ \displaystyle CHOH \\ \displaystyle | \\ \displaystyle R}}{}} \; \xrightarrow{Cu^{2+}} \; \begin{array}{l}\text{positive test} \\ (Cu_2O \text{ formed})\end{array}$$

The same kind of equilibrium is available to α-hydroxy ketones generally; such compounds give a positive test with Benedict's reagent. Any carbohydrate that contains a free hemiacetal function is a reducing sugar. The free hemiacetal is in equilibrium with the open-chain form and through it is susceptible to oxidation. Maltose, for example, gives a positive test with Benedict's reagent.

Maltose   ⇌   Open-chain form of maltose $\xrightarrow{Cu^{2+}}$ positive test (Cu_2O formed)

Glycosides, in which the anomeric carbon is part of an acetal function, are not reducing sugars and do not give a positive test.

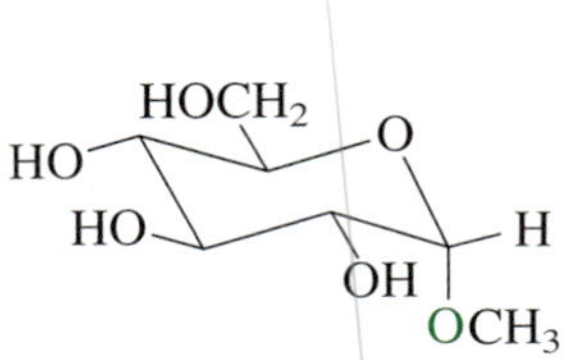

Methyl α-D-glucopyranoside:
not a reducing sugar

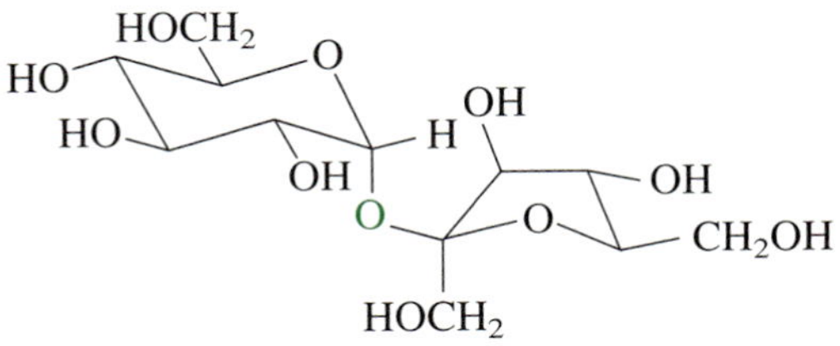

Sucrose: not a reducing sugar

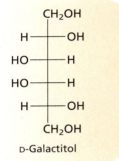

D-Galactitol

PROBLEM 25.15 Which of the following would be expected to give a positive test with Benedict's reagent? Why?

(a) D-Galactitol (see structure in margin)

(b) L-Arabinose

(c) 1,3-Dihydroxyacetone

(d) D-Fructose

(e) Lactose

(f) Amylose

SAMPLE SOLUTION (a) D-Galactitol lacks an aldehyde, an α-hydroxy ketone, or a hemiacetal function, so cannot be oxidized by Cu^{2+} and will not give a positive test with Benedict's reagent.

Fehling's solution, a tartrate complex of copper(II) sulfate, has also been used as a test for reducing sugars.

Derivatives of aldoses in which the terminal aldehyde function is oxidized to a carboxylic acid are called **aldonic acids.** Aldonic acids are named by replacing the *-ose* ending of the aldose by *-onic acid.* Oxidation of aldoses with bromine is the most commonly used method for the preparation of aldonic acids and involves the furanose or pyranose form of the carbohydrate.

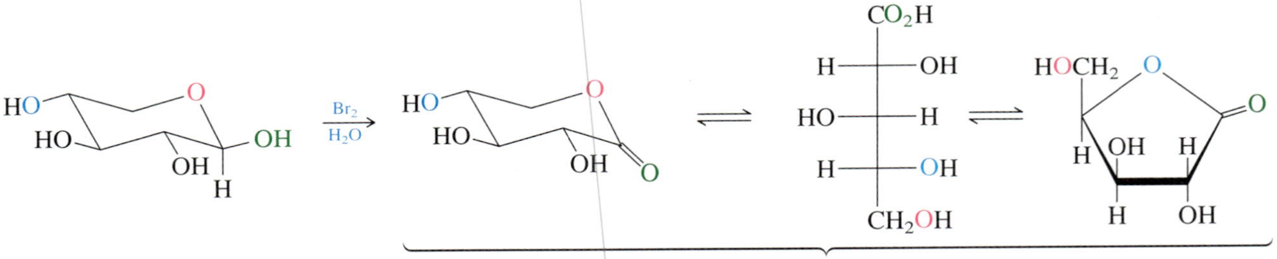

β-D-Xylopyranose

D-Xylonic acid
(90%)

Aldonic acids exist in equilibrium with their five- or six-membered lactones. They can be isolated as carboxylate salts of their open-chain forms on treatment with base.

The reaction of aldoses with nitric acid leads to the formation of **aldaric acids** by oxidation of both the aldehyde and the terminal primary alcohol function to carboxylic acid groups. Aldaric acids are also known as *saccharic acids* and are named by substituting *-aric acid* for the *-ose* ending of the corresponding carbohydrate.

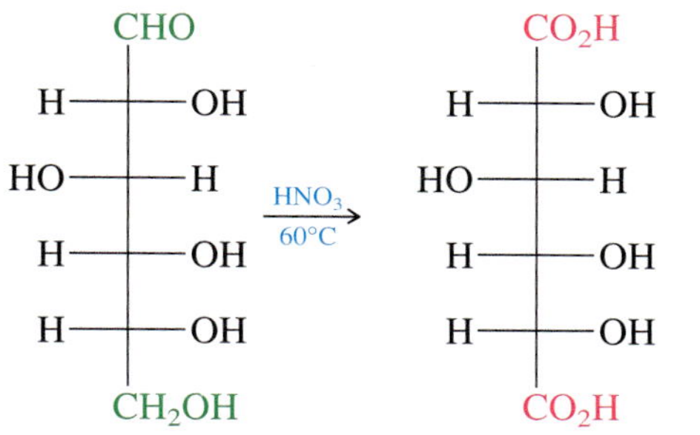

D-Glucose → D-Glucaric acid (41%)

Like aldonic acids, aldaric acids exist mainly as lactones.

PROBLEM 25.16 Another hexose gives the same aldaric acid on oxidation as does D-glucose. Which one?

Uronic acids occupy an oxidation state between aldonic and aldaric acids. They have an aldehyde function at one end of their carbon chain and a carboxylic acid group at the other.

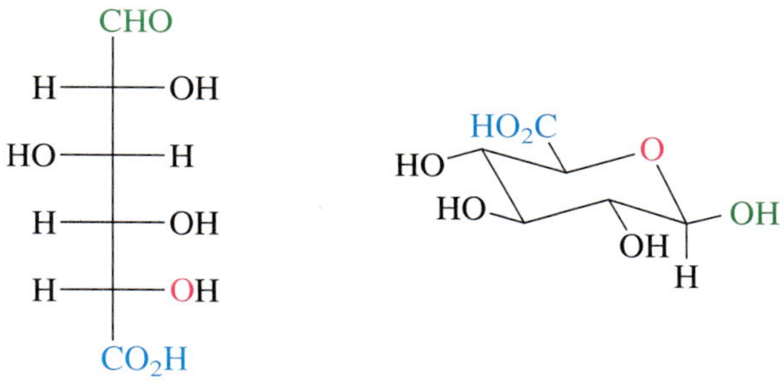

Fischer projection of D-glucuronic acid

β-Pyranose form of D-glucuronic acid

Uronic acids are biosynthetic intermediates in various metabolic processes; ascorbic acid (vitamin C), for example, is biosynthesized by way of glucuronic acid. Many metabolic waste products are excreted in the urine as their glucuronate salts.

25.20 CYANOHYDRIN FORMATION AND CHAIN EXTENSION

The presence of an aldehyde function in their open-chain forms makes aldoses reactive toward nucleophilic addition of hydrogen cyanide. Addition yields a mixture of diastereomeric cyanohydrins.

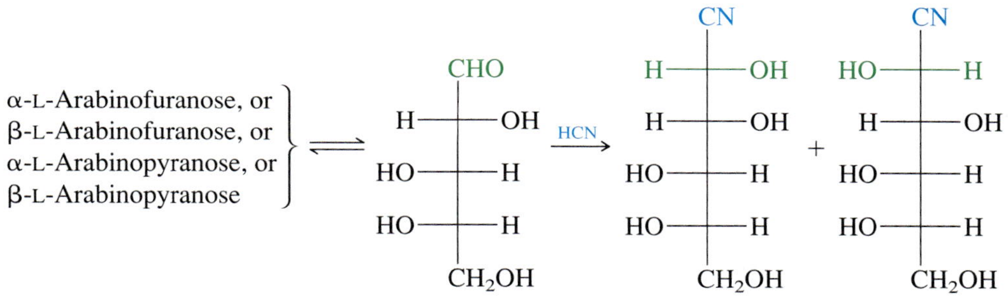

α-L-Arabinofuranose, or
β-L-Arabinofuranose, or
α-L-Arabinopyranose, or
β-L-Arabinopyranose

L-Arabinose → L-Mannononitrile + L-Gluconitrile

The reaction is used for the chain extension of aldoses in the synthesis of new or unusual sugars. In this case, the starting material, L-arabinose, is an abundant natural product and possesses the correct configurations at its three chirality centers for elaboration to the relatively rare L-enantiomers of glucose and mannose. After cyanohydrin formation, the cyano groups are converted to aldehyde functions by hydrogenation in aqueous solution. Under these conditions, $-C{\equiv}N$ is reduced to $-CH{=}NH$ and hydrolyzes rapidly to $-CH{=}O$. Use of a poisoned palladium-on-barium sulfate catalyst prevents further reduction to the alditols.

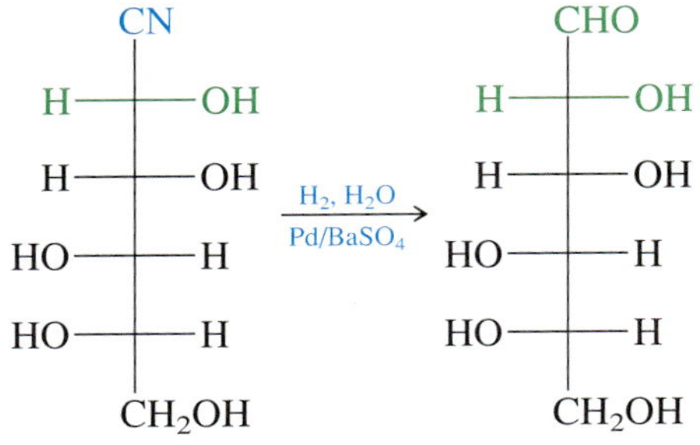

CN
H——OH
H——OH
HO——H
HO——H
CH₂OH

$\xrightarrow{\text{H}_2,\ \text{H}_2\text{O}\\ \text{Pd/BaSO}_4}$

CHO
H——OH
H——OH
HO——H
HO——H
CH₂OH

L-Mannononitrile · L-Mannose (56% yield from L-arabinose)

(Similarly, L-glucononitrile has been reduced to L-glucose; its yield was 26% from L-arabinose.)

An older version of this sequence is called the **Kiliani–Fischer synthesis.** It, too, proceeds through a cyanohydrin, but it uses a less efficient method for converting the cyano group to the required aldehyde.

25.21 EPIMERIZATION, ISOMERIZATION, AND RETRO-ALDOL CLEAVAGE

Carbohydrates undergo a number of isomerization and degradation reactions under both laboratory and physiological conditions. For example, a mixture of glucose, fructose, and mannose results when any one of them is treated with aqueous base. This reaction can be understood by examining the consequences of enolization of glucose:

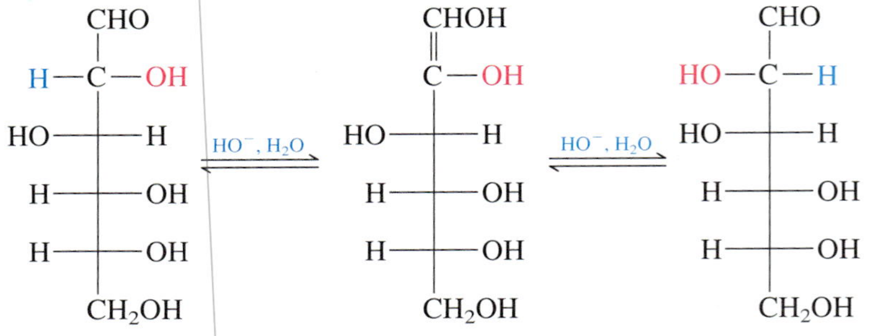

CHO
H—C—OH
HO——H
H——OH
H——OH
CH₂OH
D-Glucose (R configuration at C-2)

$\xrightleftharpoons{\text{HO}^-,\ \text{H}_2\text{O}}$

CHOH
‖
C—OH
HO——H
H——OH
H——OH
CH₂OH
Enediol (C-2 not a chirality center)

$\xrightleftharpoons{\text{HO}^-,\ \text{H}_2\text{O}}$

CHO
HO—C—H
HO——H
H——OH
H——OH
CH₂OH
D-Mannose (S configuration at C-2)

Because the configuration at C-2 is lost on enolization, the enediol intermediate can revert either to D-glucose or to D-mannose. Two stereoisomers that have multiple chirality centers but differ in configuration at only one of them are referred to as

epimers. Glucose and mannose are epimeric at C-2. Under these conditions epimerization occurs only at C-2 because it alone is α to the carbonyl group.

Epimers are a type of diastereomer.

There is another reaction available to the enediol intermediate. Proton transfer from water to C-1 converts the enediol not to an aldose but to the ketose D-fructose:

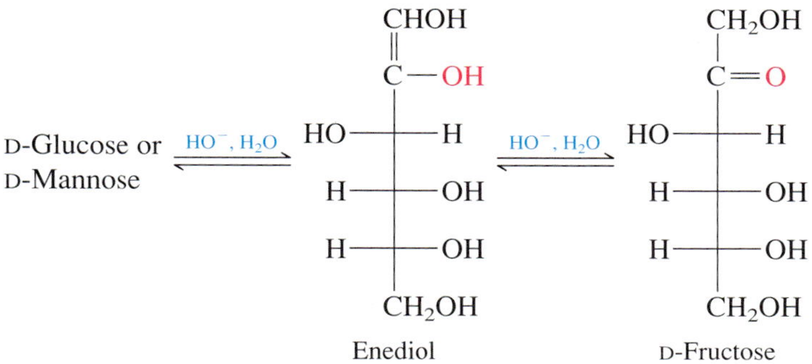

D-Glucose or D-Mannose $\xrightarrow{HO^-, H_2O}$ Enediol $\xrightarrow{HO^-, H_2O}$ D-Fructose

See the boxed essay "How Sweet It Is!" for more on this process.

The isomerization of D-glucose to D-fructose by way of an enediol intermediate is an important step in **glycolysis,** a complex process (11 steps) by which an organism converts glucose to chemical energy. The substrate is not glucose itself but its 6-phosphate ester. The enzyme that catalyzes the isomerization is called *phosphoglucose isomerase.*

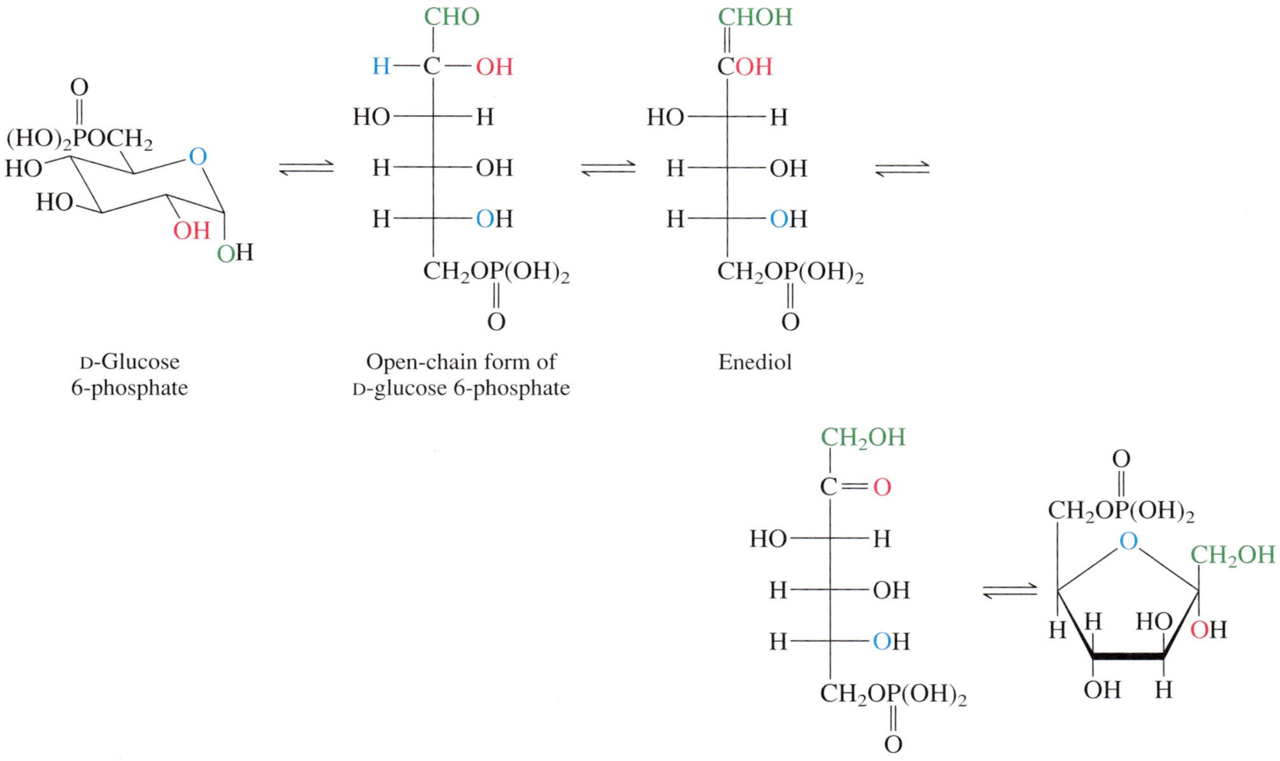

D-Glucose 6-phosphate   Open-chain form of D-glucose 6-phosphate   Enediol

Open-chain form of D-Fructose 6-phosphate   D-Fructose 6-phosphate

Following its formation, D-fructose 6-phosphate is converted to its corresponding 1,6-phosphate diester, which is then cleaved to two 3-carbon fragments under the influence of the enzyme *aldolase:*

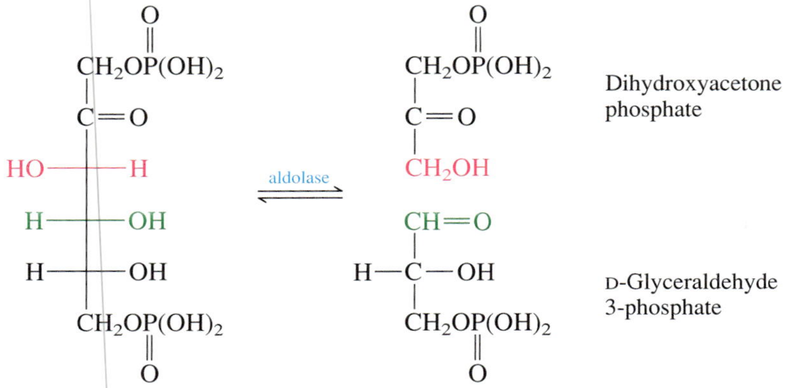

D-Fructose 1,6-diphosphate

This cleavage is a *retro-aldol* reaction. It is the reverse of the process by which D-fructose 1,6-diphosphate would be formed by aldol addition of the enolate of dihydroxyacetone phosphate to D-glyceraldehyde 3-phosphate. The enzyme aldolase catalyzes both the aldol addition of the two components and, in glycolysis, the retro-aldol cleavage of D-fructose 1,6-diphosphate.

Further steps in glycolysis use the D-glyceraldehyde 3-phosphate formed in the aldolase-catalyzed cleavage reaction as a substrate. Its coproduct, dihydroxyacetone phosphate, is not wasted, however. The enzyme *triose phosphate isomerase* converts dihydroxyacetone phosphate to D-glyceraldehyde 3-phosphate, which enters the glycolysis pathway for further transformations.

PROBLEM 25.17 Suggest a reasonable structure for the intermediate in the conversion of dihydroxyacetone phosphate to D-glyceraldehyde 3-phosphate.

Cleavage reactions of carbohydrates also occur on treatment with aqueous base for prolonged periods as a consequence of base-catalyzed retro-aldol reactions. As pointed out in Section 18.9, aldol addition is a reversible process, and β-hydroxy carbonyl compounds can be cleaved to an enolate and either an aldehyde or a ketone.

25.22 ACYLATION AND ALKYLATION OF HYDROXYL GROUPS IN CARBOHYDRATES

The alcohol groups of carbohydrates undergo chemical reactions typical of hydroxyl functions. They are converted to esters by reaction with acyl chlorides and carboxylic acid anhydrides.

α-D-Glucopyranose

Acetic anhydride

1,2,3,4,6-Penta-*O*-acetyl-α-D-glucopyranose (88%)

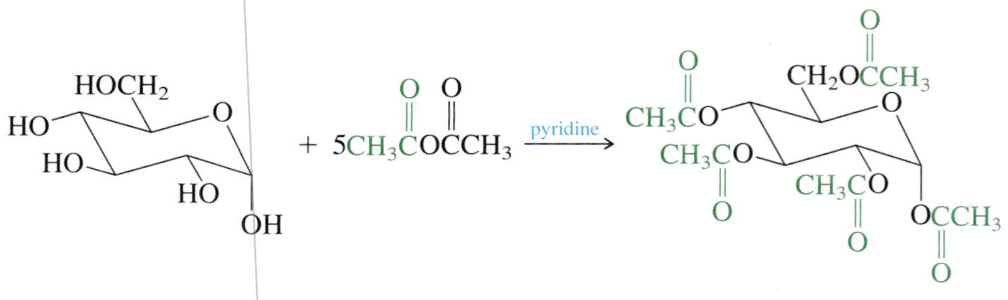

Ethers are formed under conditions of the Williamson ether synthesis. Methyl ethers of carbohydrates are efficiently prepared by alkylation with methyl iodide in the presence of silver oxide.

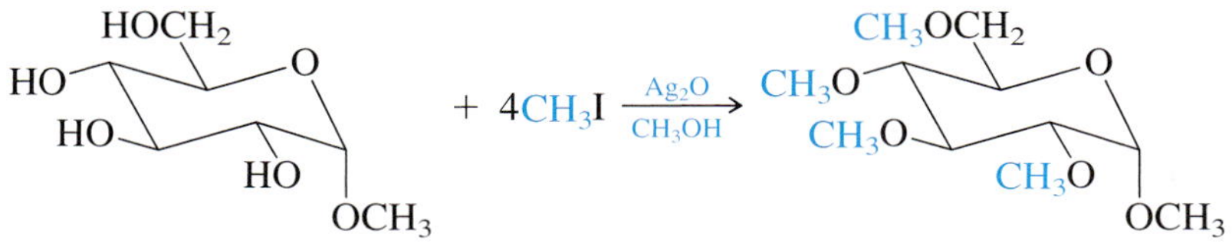

| Methyl α-D-glucopyranoside | Methyl iodide | Methyl 2,3,4,6-tetra-O-methyl-α-D-glucopyranoside (97%) |

This reaction has been used in an imaginative way to determine the ring size of glycosides. Once all the free hydroxyl groups of a glycoside have been methylated, the glycoside is subjected to acid-catalyzed hydrolysis. Only the anomeric methoxy group is hydrolyzed under these conditions—another example of the ease of carbocation formation at the anomeric position.

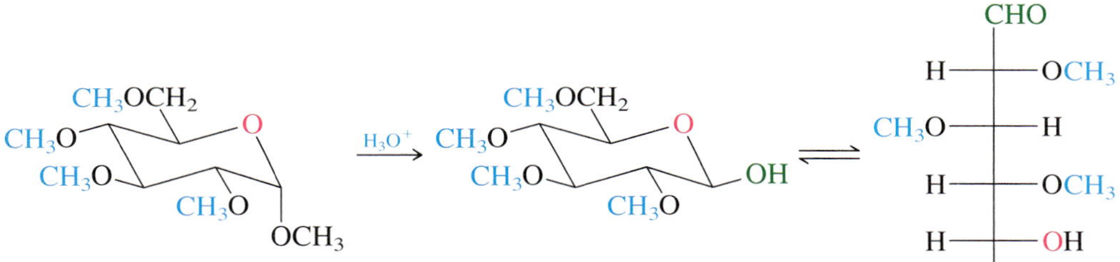

Methyl 2,3,4,6-tetra-O-methyl-α-D-glucopyranoside   2,3,4,6-Tetra-O-methyl-D-glucose

Notice that all the hydroxyl groups in the free sugar except C-5 are methylated. Carbon-5 is not methylated, because it was originally the site of the ring oxygen in the methyl glycoside. Once the position of the hydroxyl group in the free sugar has been determined, either by spectroscopy or by converting the sugar to a known compound, the ring size stands revealed.

25.23 PERIODIC ACID OXIDATION OF CARBOHYDRATES

Periodic acid oxidation (Section 15.12) finds extensive use as an analytical method in carbohydrate chemistry. Structural information is obtained by measuring the number of equivalents of periodic acid that react with a given compound and by identifying the reaction products. A vicinal diol consumes one equivalent of periodate and is cleaved to two carbonyl compounds:

$$R_2C-CR_2' + HIO_4 \longrightarrow R_2C=O + R_2'C=O + HIO_3 + H_2O$$
$$\quad | \quad\; |$$
$$\;HO \;\; OH$$

| Vicinal diol | Periodic acid | Two carbonyl compounds | Iodic acid | Water |

α-Hydroxy carbonyl compounds are cleaved to a carboxylic acid and a carbonyl compound:

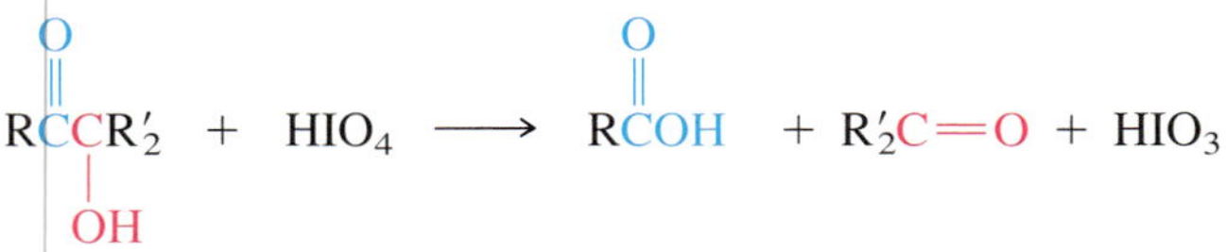

$$\underset{\substack{\text{OH}}}{\text{RCCR}'_2} + \text{HIO}_4 \longrightarrow \text{RCOH} + \text{R}'_2\text{C}{=}\text{O} + \text{HIO}_3$$

| α-Hydroxy carbonyl compound | Periodic acid | Carboxylic acid | Carbonyl compound | Iodic acid |

When three contiguous carbons bear hydroxyl groups, two moles of periodate is consumed per mole of carbohydrate and the central carbon is oxidized to a molecule of formic acid:

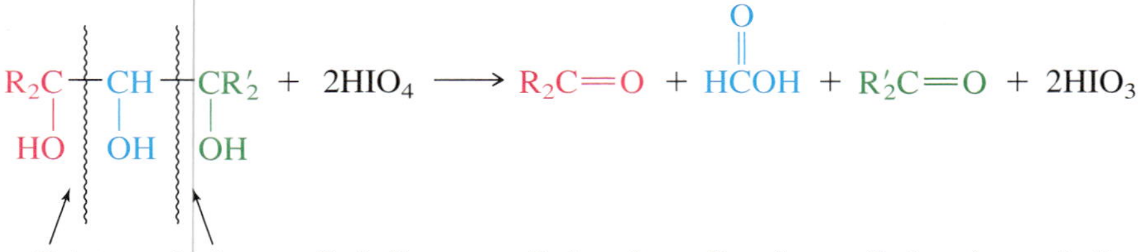

$$\text{R}_2\text{C}{-}\text{CH}{-}\text{CR}'_2 + 2\text{HIO}_4 \longrightarrow \text{R}_2\text{C}{=}\text{O} + \text{HCOH} + \text{R}'_2\text{C}{=}\text{O} + 2\text{HIO}_3$$

| Points at which cleavage occurs | Periodic acid | Carbonyl compound | Formic acid | Carbonyl compound | Iodic acid |

Ether and acetal functions are not affected by the reagent.

The use of periodic acid oxidation in structure determination can be illustrated by a case in which a previously unknown methyl glycoside was obtained by the reaction of D-arabinose with methanol and hydrogen chloride. The size of the ring was identified as five-membered because only one mole of periodic acid was consumed per mole of glycoside and no formic acid was produced. Were the ring six-membered, two moles of periodic acid would be required per mole of glycoside and one mole of formic acid would be produced.

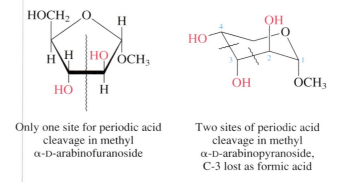

| Only one site for periodic acid cleavage in methyl α-D-arabinofuranoside | Two sites of periodic acid cleavage in methyl α-D-arabinopyranoside, C-3 lost as formic acid |

PROBLEM 25.18 Give the products of periodic acid oxidation of each of the following. How many moles of reagent will be consumed per mole of substrate in each case?

(a) D-Arabinose

(b) D-Ribose

(c) Methyl β-D-glucopyranoside

(d)

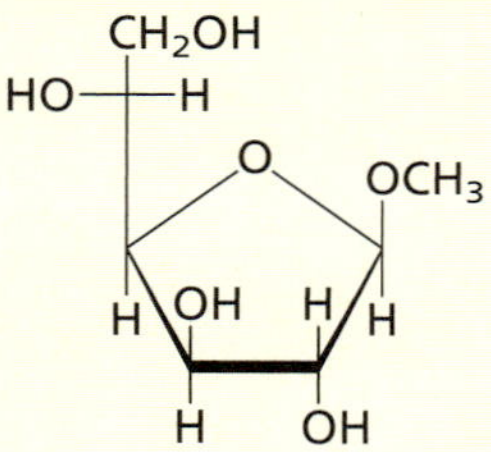

SAMPLE SOLUTION (a) The α-hydroxy aldehyde unit at the end of the sugar chain is cleaved, as well as all the vicinal diol functions. Four moles of periodic acid are required per mole of D-arabinose. Four moles of formic acid and one mole of formaldehyde are produced.

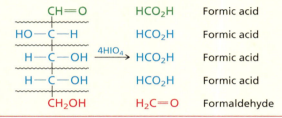

D-Arabinose, showing points of cleavage by periodic acid; each cleavage requires one equivalent of HIO_4.

25.24 SUMMARY

Section 25.1 Carbohydrates are marvelous molecules! In most of them, every carbon bears a functional group, and the nature of the functional groups changes as the molecule interconverts between open-chain and cyclic hemiacetal forms. Any approach to understanding carbohydrates must begin with structure.

Carbohydrates are polyhydroxy aldehydes and ketones. Those derived from aldehydes are classified as **aldoses;** those derived from ketones are **ketoses.**

Section 25.2 Fischer projections and D–L notation are commonly used to describe carbohydrate stereochemistry. The standards are the enantiomers of glyceraldehyde.

D-(+)-Glyceraldehyde   L-(−)-Glyceraldehyde

Section 25.3 Aldotetroses have two chirality centers, so four stereoisomers are possible. They are assigned to the D or the L series according to whether the configuration at their highest numbered chirality center is analogous to D- or L-glyceraldehyde, respectively. Both hydroxyl groups are on the same side of the Fischer projection in erythrose, but on opposite sides in threose. The Fischer projections of D-erythrose and D-threose are shown in Figure 25.2.

Section 25.4 Of the eight stereoisomeric aldopentoses, Figure 25.2 shows the Fischer projections of the D-enantiomers (D-ribose, D-arabinose, D-xylose, and D-lyxose). Likewise, Figure 25.2 gives the Fischer projections of the eight D-aldohexoses.

Section 25.5 The aldohexoses are allose, altrose, glucose, mannose, gulose, idose, galactose, and talose. The mnemonic "All altruists gladly make gum in gallon tanks" is helpful in writing the correct Fischer projection for each one.

Sections 25.6–25.7	Most carbohydrates exist as cyclic hemiacetals. Those with five-membered rings are called **furanose** forms; those with six-membered rings are called **pyranose** forms.

α-D-Ribofuranose   β-D-Glucopyranose

The **anomeric carbon** in a cyclic acetal is the one attached to *two* oxygens. It is the carbon that corresponds to the carbonyl carbon in the open-chain form. The symbols α and β refer to the configuration at the anomeric carbon.

Section 25.8	A particular carbohydrate can interconvert between furanose and pyranose forms and between the α and β configuration of each form. The change from one form to an equilibrium mixture of all the possible hemiacetals causes a change in optical rotation called **mutarotation.**

Section 25.9	Ketoses are characterized by the ending *-ulose* in their name. Most naturally occurring ketoses have their carbonyl group located at C-2. Like aldoses, ketoses cyclize to hemiacetals and exist as furanose or pyranose forms.

Sections 25.10 –25.12	Structurally modified carbohydrates include **deoxy sugars, amino sugars,** and **branched-chain carbohydrates.**

Section 25.13	Glycosides are acetals, compounds in which the anomeric hydroxyl group has been replaced by an alkoxy group. Glycosides are easily prepared by allowing a carbohydrate and an alcohol to stand in the presence of an acid catalyst.

D-Glucose + ROH $\xrightarrow{\text{H}^+}$ A glycoside + H_2O

A glycoside

Sections 25.14–25.15	**Disaccharides** are carbohydrates in which two monosaccharides are joined by a glycoside bond. **Polysaccharides** have many monosaccharide units connected through glycosidic linkages. Complete hydrolysis of disaccharides and polysaccharides cleaves the glycoside bonds, yielding the free monosaccharide components.

Section 25.16	Carbohydrates and proteins that are connected by a chemical bond are called **glycoproteins** and often occur on the surfaces of cells. They play an important role in the recognition events connected with the immune response.

Sections 25.17–25.24	Carbohydrates undergo chemical reactions characteristic of aldehydes and ketones, alcohols, diols, and other classes of compounds, depending on their structure. A review of the reactions described in this chapter is presented in Table 25.2. Although some of the reactions have synthetic value, many of them are used in analysis and structure determination.

Reaction (section) and comments	Example

Transformations of the carbonyl group

Reduction (Section 25.18) The carbonyl group of aldoses and ketoses is reduced by sodium borohydride or by catalytic hydrogenation. The products are called *alditols*.

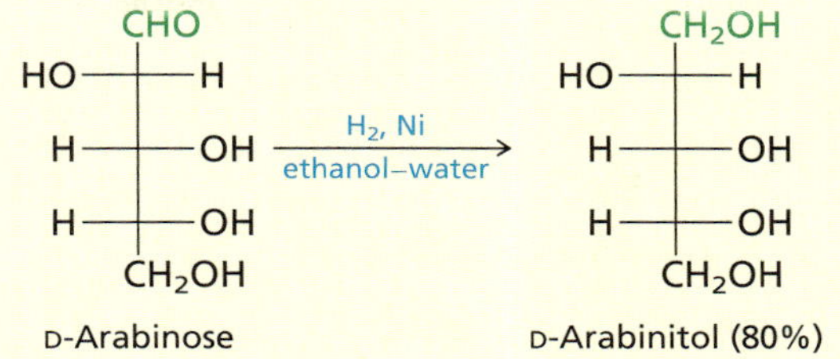

D-Arabinose $\xrightarrow[\text{ethanol–water}]{\text{H}_2,\ \text{Ni}}$ D-Arabinitol (80%)

Oxidation with Benedict's reagent (Section 25.19) Sugars that contain a free hemiacetal function are called reducing sugars. They react with copper(II) sulfate in a sodium citrate/sodium carbonate buffer (Benedict's reagent) to form a red precipitate of copper(I) oxide. Used as a qualitative test for reducing sugars.

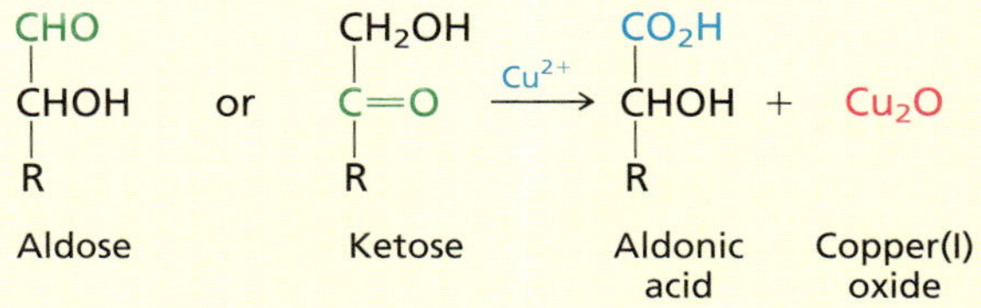

Aldose   or   Ketose $\xrightarrow{\text{Cu}^{2+}}$ Aldonic acid + Copper(I) oxide

Oxidation with bromine (Section 25.19) When a preparative method for an aldonic acid is required, bromine oxidation is used. The aldonic acid is formed as its lactone. More properly described as a reaction of the anomeric hydroxyl group than of a free aldehyde.

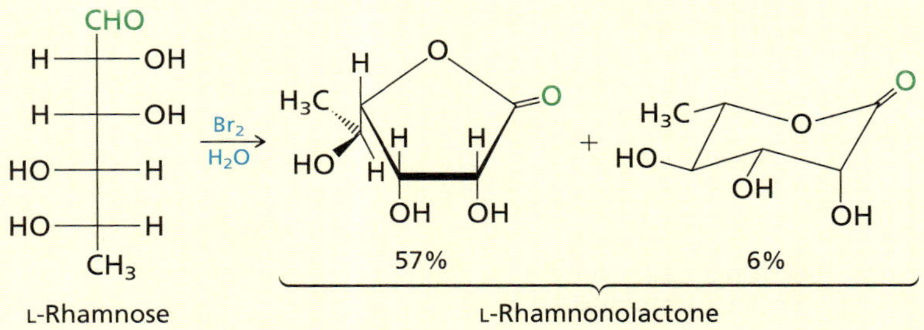

L-Rhamnose $\xrightarrow[\text{H}_2\text{O}]{\text{Br}_2}$ L-Rhamnonolactone (57% + 6%)

Chain extension by way of cyanohydrin formation (Section 25.20) The Kiliani–Fischer synthesis proceeds by nucleophilic addition of HCN to an aldose, followed by conversion of the cyano group to an aldehyde. A mixture of stereoisomers results; the two aldoses are epimeric at C-2. Section 25.20 describes the modern version of the Kiliani–Fischer synthesis. The example at the right illustrates the classical version.

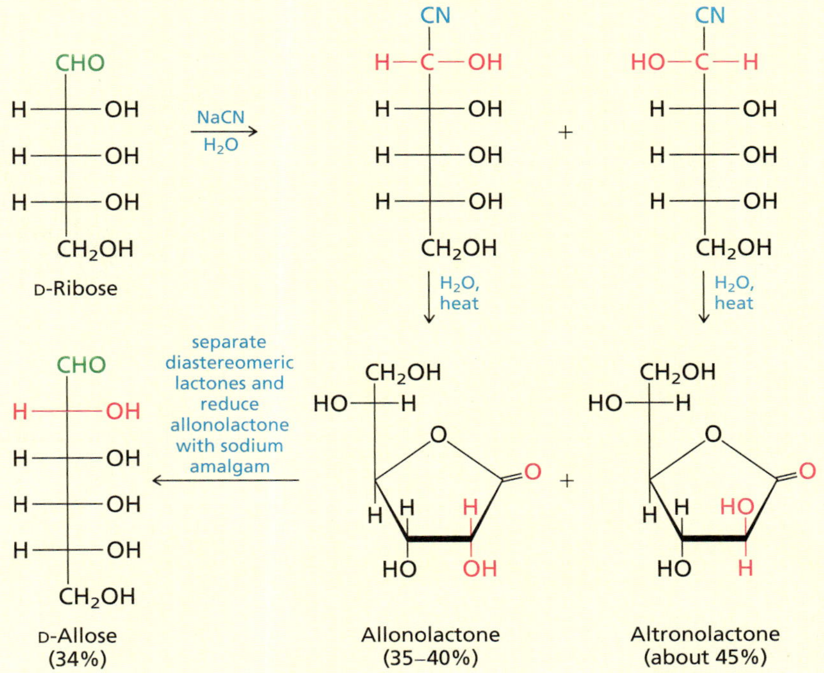

D-Ribose $\xrightarrow[\text{H}_2\text{O}]{\text{NaCN}}$

D-Allose (34%)   separate diastereomeric lactones and reduce allonolactone with sodium amalgam   Allonolactone (35–40%) + Altronolactone (about 45%)

TABLE 25.2   Summary of Reactions of Carbohydrates (Continued)

Reaction (section) and comments	Example

Enediol formation (Section 25.21) Enolization of an aldose or a ketose gives an enediol. Enediols can revert to aldoses or ketoses with loss of stereochemical integrity at the α-carbon atom.

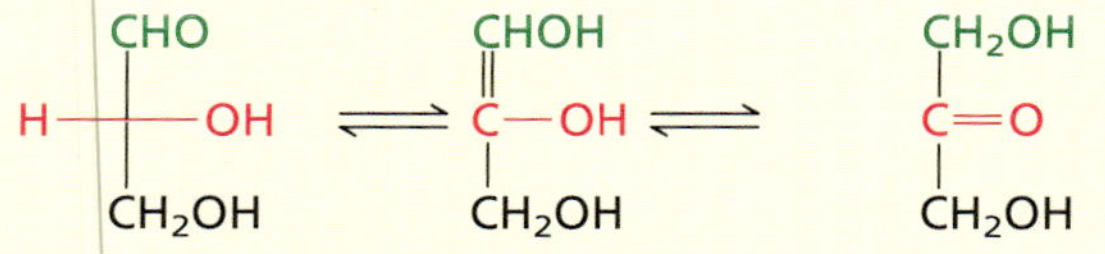

D-Glyceraldehyde   Enediol   1,3-Dihydroxyacetone

Reactions of the hydroxyl group

Acylation (Section 25.22) Esterification of the available hydroxyl groups occurs when carbohydrates are treated with acylating agents.

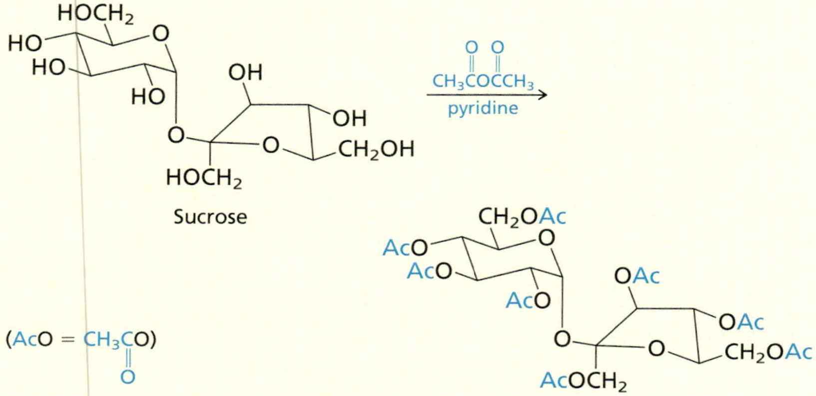

Sucrose

$(AcO = CH_3CO)$

Sucrose octaacetate (66%)

Alkylation (Section 25.22) Alkyl halides react with carbohydrates to form ethers at the available hydroxyl groups. An application of the Williamson ether synthesis to carbohydrates.

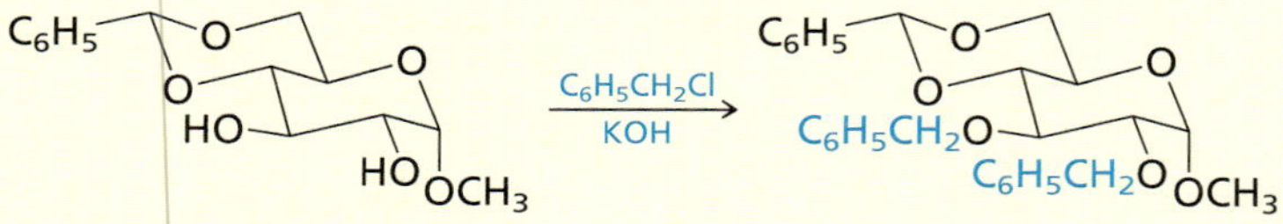

Methyl 4,6-O-benzylidene-α-D-glucopyranoside

Methyl 2,3-di-O-benzyl-4,6-O-benzylidene-α-D-glucopyranoside (92%)

Periodic acid oxidation (Section 25.23) Vicinal diol and α-hydroxy carbonyl functions in carbohydrates are cleaved by periodic acid. Used analytically as a tool for structure determination.

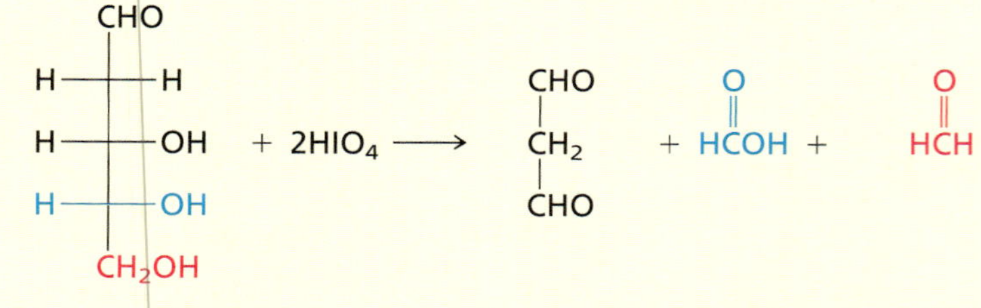

2-Deoxy-D-ribose   Propanedial   Formic acid   Formaldehyde

PROBLEMS

25.19 Refer to the Fischer projection of D-(+)-xylose in Figure 25.2 (Section 25.4) and give structural formulas for

   (a) (−)-Xylose (Fischer projection)

   (b) D-Xylitol

   (c) β-D-Xylopyranose

   (d) α-L-Xylofuranose

   (e) Methyl α-L-xylofuranoside

   (f) D-Xylonic acid (open-chain Fischer projection)

   (g) δ-Lactone of D-xylonic acid

   (h) γ-Lactone of D-xylonic acid

   (i) D-Xylaric acid (open-chain Fischer projection)

25.20 From among the carbohydrates shown in Figure 25.2, choose the D-aldohexoses that yield

   (a) An optically inactive product on reduction with sodium borohydride

   (b) An optically inactive product on oxidation with bromine

   (c) An optically inactive product on oxidation with nitric acid

   (d) The same enediol

25.21 Write the Fischer projection of the open-chain form of each of the following:

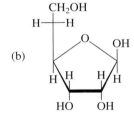

(a)

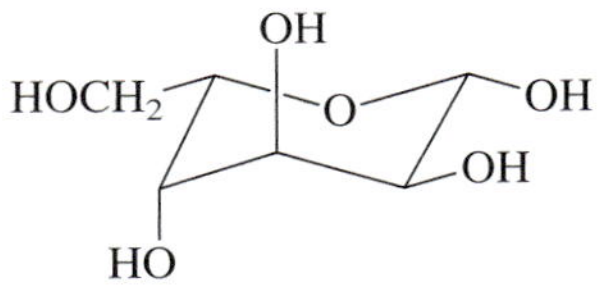

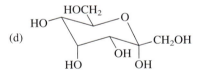

(c)

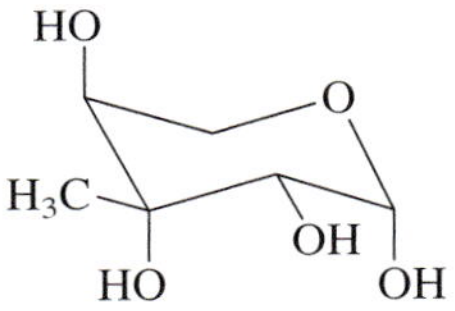

(b)

(d)

25.22 What are the R,S configurations of the three chirality centers in D-ribose? (A molecular model will be helpful here.)

25.23 From among the carbohydrates shown in Problem 25.21 choose the one(s) that

   (a) Belong to the L series

   (b) Are deoxy sugars

   (c) Are branched-chain sugars

   (d) Are ketoses

   (e) Are furanose forms

   (f) Have the α configuration at their anomeric carbon

25.24 How many pentuloses are possible? Write their Fischer projections.

25.25 The Fischer projection of the branched-chain carbohydrate D-apiose has been presented in Section 25.12.

(a) How many chirality centers are in the open-chain form of D-apiose?

(b) Does D-apiose form an optically active alditol on reduction?

(c) How many chirality centers are in the furanose forms of D-apiose?

(d) How many stereoisomeric furanose forms of D-apiose are possible? Write their Haworth formulas.

25.26 Treatment of D-mannose with methanol in the presence of an acid catalyst yields four isomeric products having the molecular formula $C_7H_{14}O_6$. What are these four products?

25.27 Maltose and cellobiose (Section 25.14) are examples of disaccharides derived from D-glucopyranosyl units.

(a) How many other disaccharides are possible that meet this structural requirement?

(b) How many of these are reducing sugars?

25.28 Gentiobiose has the molecular formula $C_{12}H_{22}O_{11}$ and has been isolated from gentian root and by hydrolysis of amygdalin. Gentiobiose exists in two different forms, one melting at 86°C and the other at 190°C. The lower melting form is dextrorotatory ($[\alpha]_D^{22} +16°$), the higher melting one is levorotatory ($[\alpha]_D^{22} -6°$). The rotation of an aqueous solution of either form, however, gradually changes until a final value of $[\alpha]_D^{22} +9.6°$ is observed. Hydrolysis of gentiobiose is efficiently catalyzed by emulsin and produces two moles of D-glucose per mole of gentiobiose. Gentiobiose forms an octamethyl ether, which on hydrolysis in dilute acid yields 2,3,4,6-tetra-O-methyl-D-glucose and 2,3,4-tri-O-methyl-D-glucose. What is the structure of gentiobiose?

25.29 *Cyanogenic glycosides* are potentially toxic because they liberate hydrogen cyanide on enzyme-catalyzed or acidic hydrolysis. Give a mechanistic explanation for this behavior for the specific cases of

(a) Linamarin

(b) Amygdalin

25.30 The following are the more stable anomers of the pyranose forms of D-glucose, D-mannose, and D-galactose:

β-D-Glucopyranose
(64% at equilibrium)

α-D-Mannopyranose
(68% at equilibrium)

β-D-Galactopyranose
(64% at equilibrium)

On the basis of these empirical observations and your own knowledge of steric effects in six-membered rings, predict the preferred form (α- or β-pyranose) at equilibrium in aqueous solution for each of the following:

(a) D-Gulose

(b) D-Talose

(c) D-Xylose

(d) D-Lyxose

25.31 Basing your answers on the general mechanism for the first stage of acid-catalyzed acetal hydrolysis

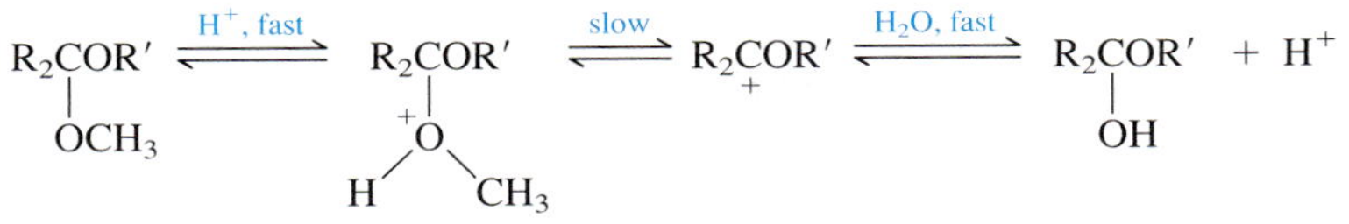

suggest reasonable explanations for the following observations:

(a) Methyl α-D-fructofuranoside (compound A) undergoes acid-catalyzed hydrolysis some 10^5 times faster than methyl α-D-glucofuranoside (compound B).

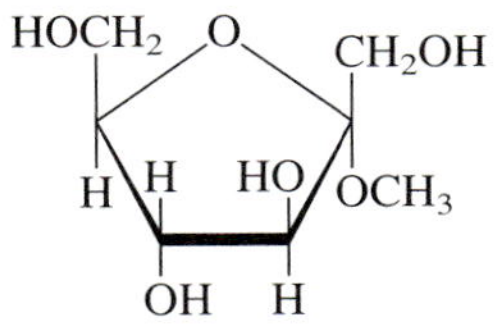

Compound A 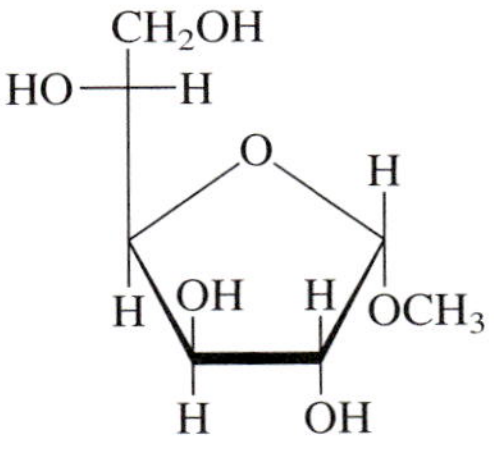   Compound B

(b) The β-methyl glucopyranoside of 2-deoxy-D-glucose (compound C) undergoes hydrolysis several thousand times faster than that of D-glucose (compound D).

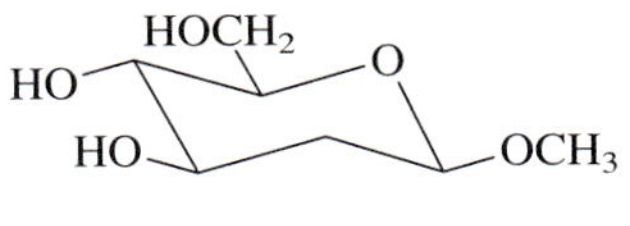

Compound C 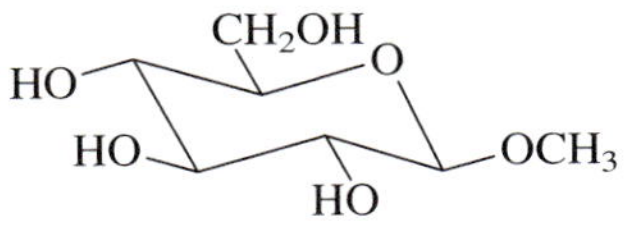   Compound D

25.32 D-Altrosan is converted to D-altrose by dilute aqueous acid. Suggest a reasonable mechanism for this reaction.

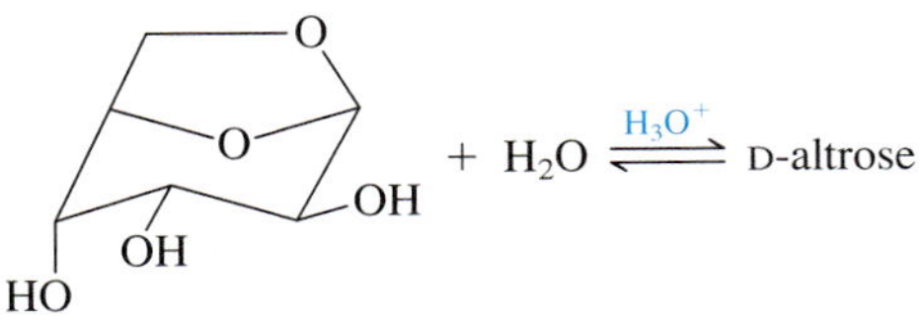

D-Altrosan

25.33 When D-galactose was heated at 165°C, a small amount of compound A was isolated:

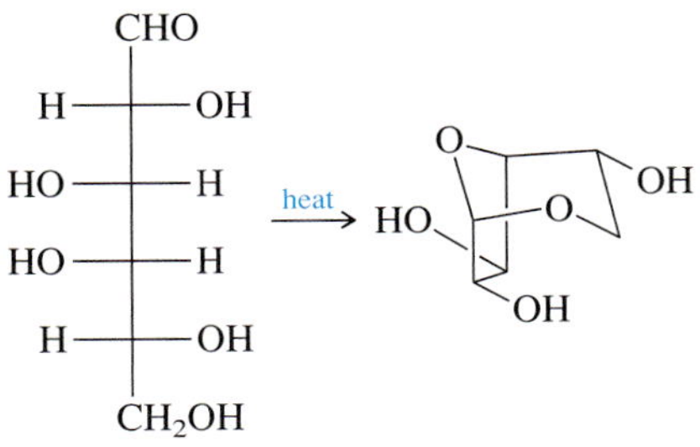

D-Galactose   Compound A

The structure of compound A was established, in part, by converting it to known compounds. Treatment of A with excess methyl iodide in the presence of silver oxide, followed by hydrolysis with dilute hydrochloric acid, gave a trimethyl ether of D-galactose. Comparing this trimethyl ether with known trimethyl ethers of D-galactose allowed the structure of compound A to be deduced.

How many trimethyl ethers of D-galactose are there? Which one is the same as the product derived from compound A?

25.34 Phlorizin is obtained from the root bark of apple, pear, cherry, and plum trees. It has the molecular formula $C_{21}H_{24}O_{10}$ and yields a compound A and D-glucose on hydrolysis in the presence of emulsin. When phlorizin is treated with excess methyl iodide in the presence of potassium carbonate and then subjected to acid-catalyzed hydrolysis, a compound B is obtained. Deduce the structure of phlorizin from this information.

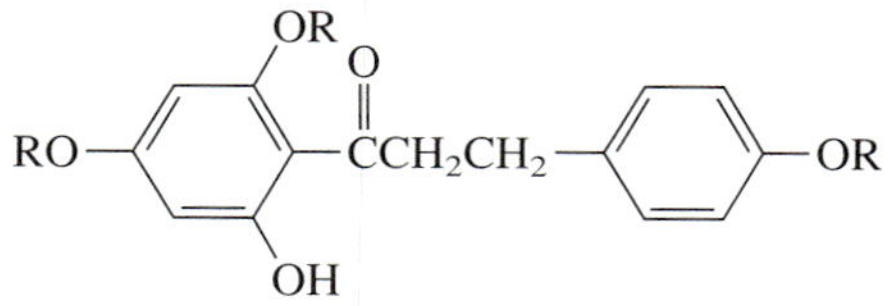

Compound A: R = H
Compound B: R = CH₃

25.35 Emil Fischer's determination of the structure of glucose was carried out as the nineteenth century ended and the twentieth began. The structure of no other sugar was known at that time, and none of the spectroscopic techniques that aid organic analysis were then available. All Fischer had was information from chemical transformations, polarimetry, and his own intellect. Fischer realized that (+)-glucose could be represented by 16 possible stereostructures. By arbitrarily assigning a particular configuration to the chirality center at C-5, the configurations of C-2, C-3, and C-4 could be determined relative to it. This reduces the number of structural possibilities to eight. Thus, he started with a structural representation shown as follows, in which C-5 of (+)-glucose has what is now known as the D configuration.

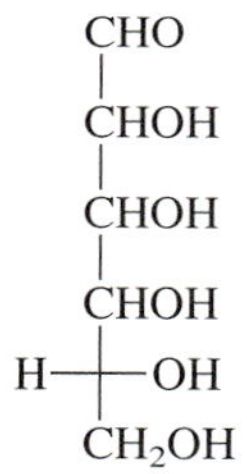

Eventually, Fischer's arbitrary assumption proved to be correct, and the structure he proposed for (+)-glucose is correct in an absolute as well as a relative sense. The following exercise uses information available to Fischer and leads you through a reasoning process similar to that employed in his determination of the structure of (+)-glucose. See if you can work out the configuration of (+)-glucose from the information provided, assuming the configuration of C-5 as shown here.

1. Chain extension of the aldopentose (−)-arabinose by way of the derived cyanohydrin gave a mixture of (+)-glucose and (+)-mannose.

2. Oxidation of (−)-arabinose with warm nitric acid gave an optically active aldaric acid.

3. Both (+)-glucose and (+)-mannose were oxidized to optically active aldaric acids with nitric acid.

4. There is another sugar, (+)-gulose, that gives the same aldaric acid on oxidation as does (+)-glucose.

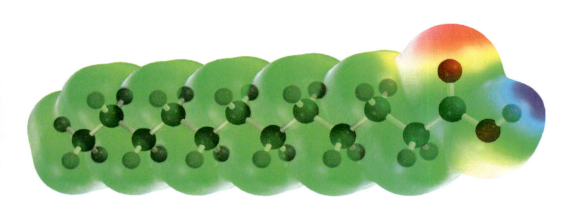

CHAPTER 26

LIPIDS

L ipids differ from the other classes of naturally occurring biomolecules (carbohydrates, proteins, and nucleic acids) in that they are more soluble in nonpolar to weakly polar solvents (diethyl ether, hexane, dichloromethane) than they are in water. They include a variety of structural types, a collection of which is introduced in this chapter.

In spite of the number of different structural types, lipids share a common biosynthetic origin in that they are ultimately derived from glucose. During one stage of carbohydrate metabolism, called *glycolysis,* glucose is converted to lactic acid. Pyruvic acid is an intermediate.

$$C_6H_{12}O_6 \longrightarrow \underset{\text{Pyruvic acid}}{CH_3\overset{\overset{\displaystyle O}{\|}}{C}CO_2H} \longrightarrow \underset{\text{Lactic acid}}{CH_3\overset{\overset{\displaystyle OH}{|}}{C}HCO_2H}$$

$\underset{\text{Glucose}}{}$

In most biochemical reactions the pH of the medium is close to 7. At this pH, carboxylic acids are nearly completely converted to their conjugate bases. Thus, it is common practice in biological chemistry to specify the derived carboxylate anion rather than the carboxylic acid itself. For example, we say that glycolysis leads to *lactate* by way of *pyruvate.*

Pyruvate is used by living systems in a number of different ways. One pathway, the one leading to lactate and beyond, is concerned with energy storage and production. This is not the only pathway available to pyruvate, however. A significant fraction of it is converted to acetate for use as a starting material in the biosynthesis of more complex substances, especially lipids. By far the major source of lipids is *biosynthesis* via acetate and this chapter is organized around that theme. We'll begin by looking at the reaction in which acetate (two carbons) is formed from pyruvate (three carbons).

26.1　ACETYL COENZYME A

The form in which acetate is used in most of its important biochemical reactions is **acetyl coenzyme A** (Figure 26.1a). Acetyl coenzyme A is a *thioester* (Section 20.13). Its formation from pyruvate involves several steps and is summarized in the overall equation:

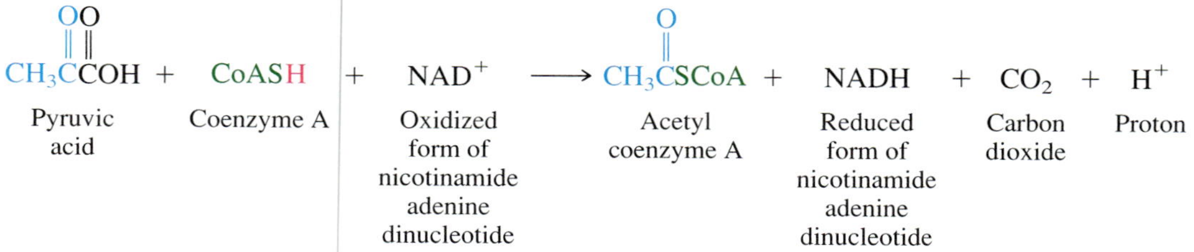

$$\underset{\substack{\text{Pyruvic}\\\text{acid}}}{CH_3\overset{O}{\underset{\|}{C}}\overset{O}{\underset{\|}{C}}OH} + \underset{\text{Coenzyme A}}{CoASH} + \underset{\substack{\text{Oxidized}\\\text{form of}\\\text{nicotinamide}\\\text{adenine}\\\text{dinucleotide}}}{NAD^+} \longrightarrow \underset{\substack{\text{Acetyl}\\\text{coenzyme A}}}{CH_3\overset{O}{\underset{\|}{C}}SCoA} + \underset{\substack{\text{Reduced}\\\text{form of}\\\text{nicotinamide}\\\text{adenine}\\\text{dinucleotide}}}{NADH} + \underset{\substack{\text{Carbon}\\\text{dioxide}}}{CO_2} + \underset{\text{Proton}}{H^+}$$

All the individual steps are catalyzed by enzymes. NAD^+ (Section 15.11) is required as an oxidizing agent, and coenzyme A (Figure 26.1b) is the acetyl group acceptor. Coenzyme A is a *thiol;* its chain terminates in a *sulfhydryl* (—SH) group. Acetylation of the sulfhydryl group of coenzyme A gives acetyl coenzyme A.

As we saw in Chapter 20, thioesters are more reactive than ordinary esters toward nucleophilic acyl substitution. They also contain a greater proportion of enol at equilibrium. Both properties are apparent in the properties of acetyl coenzyme A. In some reactions it is the carbonyl group of acetyl coenzyme A that reacts; in others it is the α-carbon atom.

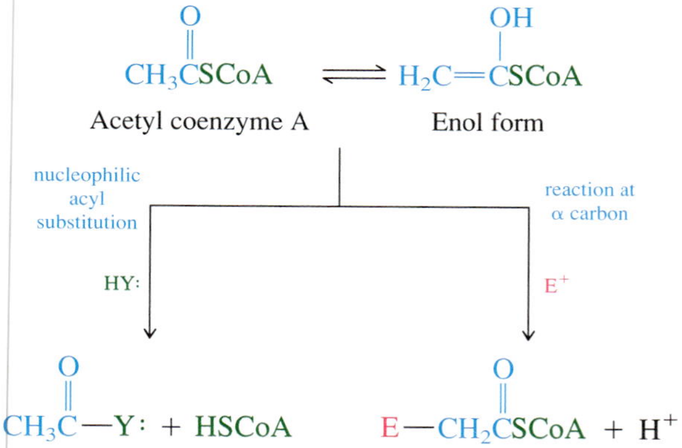

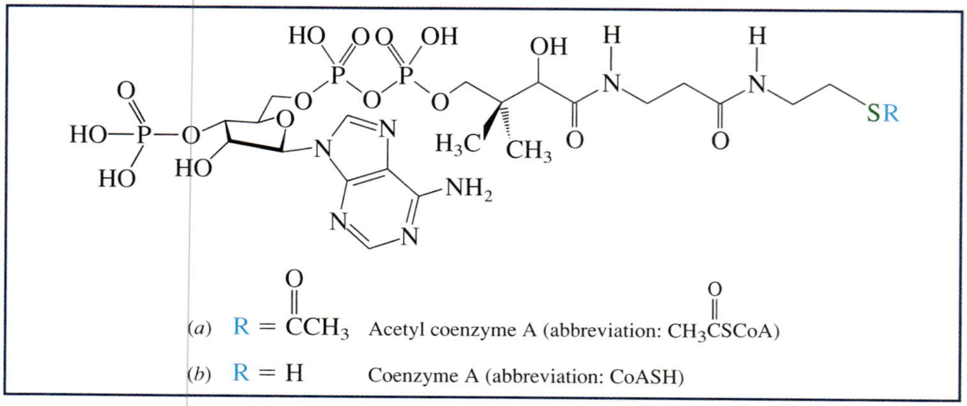

Coenzyme A was isolated and identified by Fritz Lipmann, an American biochemist. Lipmann shared the 1953 Nobel Prize in physiology or medicine for this work.

FIGURE 26.1 Structures of (a) acetyl coenzyme A and (b) coenzyme A.

(a)　R = $\overset{O}{\underset{\|}{C}}CH_3$　Acetyl coenzyme A (abbreviation: $CH_3\overset{O}{\underset{\|}{C}}SCoA$)

(b)　R = H　　　Coenzyme A (abbreviation: CoASH)

Biochemical examples of these two modes of reactivity are

Nucleophilic acyl substitution

$$\underset{\substack{\text{Acetyl coenzyme A}}}{CH_3\overset{\overset{\displaystyle O}{\|}}{C}SCoA} \;+\; \underset{\substack{\text{Choline}}}{HOCH_2CH_2\overset{+}{N}(CH_3)_3} \;\xrightarrow{\substack{\text{choline}\\ \text{acetylase}}}\; \underset{\substack{\text{Acetylcholine}}}{CH_3\overset{\overset{\displaystyle O}{\|}}{C}OCH_2CH_2\overset{+}{N}(CH_3)_3} \;+\; \underset{\substack{\text{Coenzyme A}}}{CoASH}$$

Reaction at α carbon

$$\underset{\substack{\text{Acetyl}\\ \text{coenzyme A}}}{CH_3\overset{\overset{\displaystyle O}{\|}}{C}SCoA} \;+\; \underset{\substack{\text{Carbon}\\ \text{dioxide}}}{CO_2} \;\xrightarrow{\substack{\text{acetyl-CoA}\\ \text{carboxylase}\\ \text{biotin, ATP}}}\; \underset{\substack{\text{Malonyl}\\ \text{coenzyme A}}}{HO_2CCH_2\overset{\overset{\displaystyle O}{\|}}{C}SCoA}$$

We'll see numerous examples of both reaction types in the following sections. Keep in mind that in vivo reactions (reactions in living systems) are enzyme-catalyzed and occur at far greater rates than those for the same transformations carried out in vitro ("in glass") in the absence of enzymes. In spite of the rapidity with which enzyme-catalyzed reactions take place, the nature of these transformations is essentially the same as the fundamental processes of organic chemistry described throughout this text.

Fats are one type of lipid. They have a number of functions in living systems, including that of energy storage. Although carbohydrates serve as a source of readily available energy, an equal weight of fat delivers over twice the amount of energy. It is more efficient for an organism to store energy in the form of fat because it requires less mass than storing the same amount of energy in carbohydrates or proteins.

How living systems convert acetate to fats is an exceedingly complex story, one that is well understood in broad outline and becoming increasingly clear in detail as well. We will examine several aspects of this topic in the next few sections, focusing mostly on its structural and chemical features.

26.2 FATS, OILS, AND FATTY ACIDS

Fats and oils are naturally occurring mixtures of *triacylglycerols*, also called *triglycerides*. They differ in that fats are solids at room temperature and oils are liquids. We generally ignore this distinction and refer to both groups as fats.

Triacylglycerols are built on a glycerol (1,2,3-propanetriol) framework.

> An experiment describing the analysis of the triglyceride composition of several vegetable oils is described in the May 1988 issue of the *Journal of Chemical Education* (pp. 464–466).

$$\underset{\text{Glycerol}}{\underset{\substack{|\\ \displaystyle OH}}{HOCH_2CHCH_2OH}} \qquad\qquad \underset{\text{A triacylglycerol}}{\underset{\substack{|\\ \displaystyle OCR''\\ \|\\ \displaystyle O}}{R\overset{\overset{\displaystyle O}{\|}}{C}OCH_2CHCH_2O\overset{\overset{\displaystyle O}{\|}}{C}R'}}$$

All three acyl groups in a triacylglycerol may be the same, all three may be different, or one may be different from the other two.

Figure 26.2 shows the structures of two typical triacylglycerols, 2-oleyl-1,3-distearylglycerol (Figure 26.2*a*) and tristearin (Figure 26.2*b*). Both occur naturally—in cocoa butter, for example. All three acyl groups in tristearin are stearyl (octadecanoyl) groups. In 2-oleyl-1,3-distearylglycerol, two of the acyl groups are stearyl, but the one

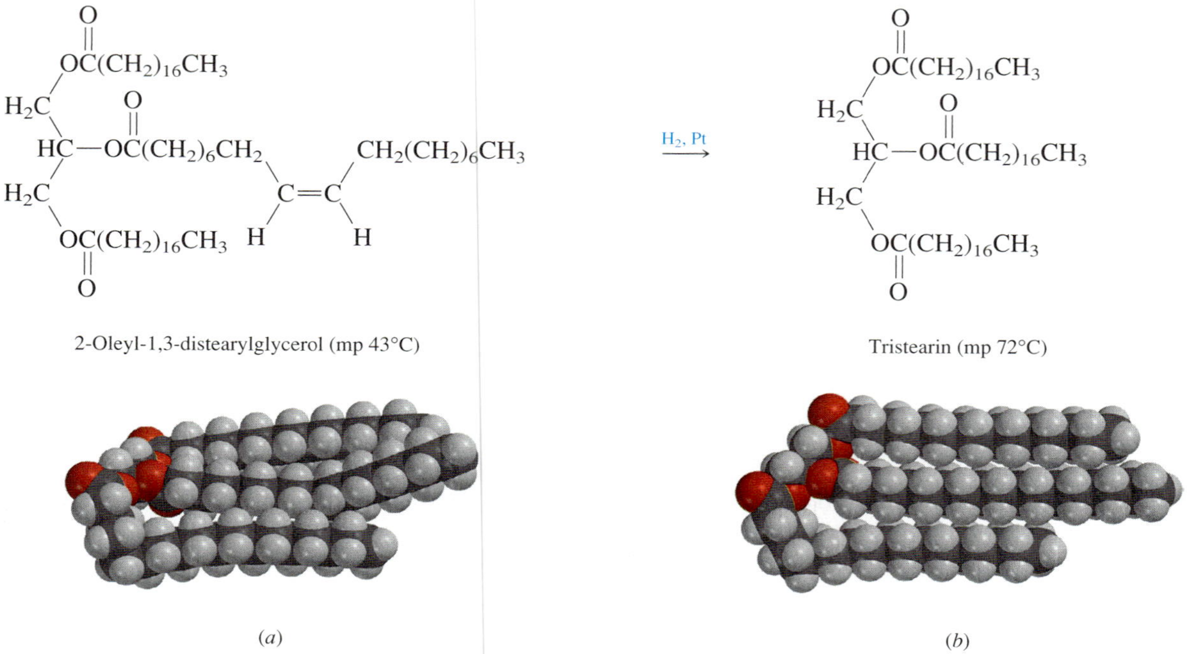

2-Oleyl-1,3-distearylglycerol (mp 43°C)

Tristearin (mp 72°C)

(a)

(b)

FIGURE 26.2 The structures of two typical triacylglycerols. (a) 2-Oleyl-1,3-distearylglycerol is a naturally occurring triacylglycerol found in cocoa butter. The cis double bond of its oleyl group gives the molecule a shape that interferes with efficient crystal packing. (b) Catalytic hydrogenation converts 2-oleyl-1,3-distearylglycerol to tristearin. Tristearin has a higher melting point than 2-oleyl-1,3-distearylglycerol.

in the middle is oleyl (*cis*-9-octadecenoyl). As the figure shows, tristearin can be prepared by catalytic hydrogenation of the carbon–carbon double bond of 2-oleyl-1,3-distearylglycerol. Hydrogenation raises the melting point from 43°C in 2-oleyl-1,3-distearylglycerol to 72°C in tristearin and is a standard technique in the food industry for converting liquid vegetable oils to solid "shortenings." The space-filling models of the two show the flatter structure of tristearin, which allows it to pack better in a crystal lattice than the more irregular shape of 2-oleyl-1,3-distearylglycerol permits. This irregular shape is a direct result of the cis double bond in the side chain.

Hydrolysis of fats yields glycerol and long-chain **fatty acids.** Thus, tristearin gives glycerol and three molecules of stearic acid on hydrolysis. Table 26.1 lists a few representative fatty acids. As these examples indicate, most naturally occurring fatty acids possess an even number of carbon atoms and an unbranched carbon chain. The carbon chain may be saturated or it can contain one or more double bonds. When double bonds are present, they are almost always cis. Acyl groups containing 14–20 carbon atoms are the most abundant in triacylglycerols.

Strictly speaking, the term *fatty acid* is restricted to those carboxylic acids that occur naturally in triacylglycerols. Many chemists and biochemists, however, refer to all unbranched carboxylic acids, irrespective of their origin and chain length, as fatty acids.

PROBLEM 26.1 What fatty acids are produced on hydrolysis of 2-oleyl-1,3-distearylglycerol? What other triacylglycerol gives the same fatty acids and in the same proportions as 2-oleyl-1,3-distearylglycerol?

A few fatty acids with trans double bonds (trans fatty acids) occur naturally, but the major source of trans fats comes from partial hydrogenation of vegetable oils in, for example, the preparation of margarine. However, the same catalysts that catalyze the

TABLE 26.1   Some Representative Fatty Acids

Number of carbons	Common name	Systematic name	Structural formula	Melting point, °C
Saturated fatty acids				
12	Lauric acid	Dodecanoic acid	$CH_3(CH_2)_{10}CO_2H$	44
14	Myristic acid	Tetradecanoic acid	$CH_3(CH_2)_{12}CO_2H$	58.5
16	Palmitic acid	Hexadecanoic acid	$CH_3(CH_2)_{14}CO_2H$	63
18	Stearic acid	Octadecanoic acid	$CH_3(CH_2)_{16}CO_2H$	69
20	Arachidic acid	Icosanoic acid	$CH_3(CH_2)_{18}CO_2H$	75
Unsaturated fatty acids				
18	Oleic acid	cis-9-Octadecenoic acid	(structure)	4
18	Linoleic acid	cis,cis-9,12-Octadecadienoic acid	(structure)	−12
18	Linolenic acid	cis,cis,cis-9,12,15-Octadecatrienoic acid	(structure)	—
20	Arachidonic acid	cis,cis,cis,cis-5,8,11,14-Icosatetraenoic acid	(structure)	−49

hydrogenation of the double bonds in a triacylglycerol also catalyze their stereoisomerization. The mechanism for conversion of a cis to a trans double bond follows directly from the mechanism of catalytic hydrogenation (Section 6.1) once one realizes that all of the steps in the mechanism are reversible.

Ester of *cis*-9-octadecenoic acid   Hydrogenated metal catalyst surface   Intermediate in catalytic hydrogenation

Ester of *trans*-9-octadecenoic acid   Hydrogenated metal catalyst surface

The intermediate in hydrogenation, formed by reaction of the unsaturated ester with the hydrogenated surface of the metal catalyst, not only can proceed to the saturated fatty acid ester, but also can dissociate to the original ester having a cis double bond or to its trans stereoisomer. Unlike polyunsaturated vegetable oils, which tend to reduce serum cholesterol levels, the trans fats produced by partial hydrogenation have cholesterol-raising effects similar to those of saturated fats.

Fatty acids occur naturally in forms other than as glyceryl triesters, and we'll see numerous examples as we go through the chapter. One recently discovered fatty acid derivative is *anandamide*.

Anandamide

The September 1997 issue of the *Journal of Chemical Education* (pp. 1030–1032) contains an article entitled "Trans Fatty Acids."

Anandamide is an ethanolamine ($H_2NCH_2CH_2OH$) amide of arachidonic acid (see Table 26.1). It was isolated from pig's brain in 1992 and identified as the substance that normally binds to the "cannabinoid receptor." The active component of marijuana, Δ^9-tetrahydrocannabinol (THC), must exert its effect by binding to a receptor, and scientists had long wondered what compound in the body was the natural substrate for this binding site. Anandamide is that compound, and it is now probably more appropriate to speak of cannabinoids binding to the anandamide receptor instead of vice versa. Anandamide seems to be involved in moderating pain. Once the identity of the "endogenous cannabinoid" was known, scientists looked specifically for it and found it in some surprising places—chocolate, for example.

Fatty acids are biosynthesized by way of acetyl coenzyme A. The following section outlines the mechanism of fatty acid biosynthesis.

Other than that both are lipids, there are no obvious structural similarities between anandamide and THC.

26.3   FATTY ACID BIOSYNTHESIS

We can describe the major elements of fatty acid biosynthesis by considering the formation of butanoic acid from two molecules of acetyl coenzyme A. The "machinery" responsible for accomplishing this conversion is a complex of enzymes known as **fatty acid synthetase.** Certain portions of this complex, referred to as **acyl carrier protein (ACP),** bear a side chain that is structurally similar to coenzyme A. An important early step in fatty acid biosynthesis is the transfer of the acetyl group from a molecule of acetyl coenzyme A to the sulfhydryl group of acyl carrier protein.

$$CH_3\overset{O}{\overset{\|}{C}}SCoA \ + \ HS-ACP \ \longrightarrow \ CH_3\overset{O}{\overset{\|}{C}}S-ACP \ + \ HSCoA$$

Acetyl   Acyl carrier   *S*-Acetyl acyl   Coenzyme A
coenzyme A   protein   carrier protein

> **PROBLEM 26.2**   Using HSCoA and HS—ACP as abbreviations for coenzyme A and acyl carrier protein, respectively, write a structural formula for the tetrahedral intermediate in the preceding reaction.

A second molecule of acetyl coenzyme A reacts with carbon dioxide (actually bicarbonate ion at biological pH) to give malonyl coenzyme A:

$$CH_3\overset{O}{\overset{\|}{C}}SCoA \ + \ HCO_3^- \ \longrightarrow \ {}^-O\overset{O}{\overset{\|}{C}}CH_2\overset{O}{\overset{\|}{C}}SCoA \ + \ H_2O$$

Acetyl   Bicarbonate   Malonyl   Water
coenzyme A   coenzyme A

Formation of malonyl coenzyme A is followed by a nucleophilic acyl substitution, which transfers the malonyl group to the acyl carrier protein as a thioester.

$${}^-O\overset{O}{\overset{\|}{C}}CH_2\overset{O}{\overset{\|}{C}}SCoA \ + \ HS-ACP \ \longrightarrow \ {}^-O\overset{O}{\overset{\|}{C}}CH_2\overset{O}{\overset{\|}{C}}S-ACP \ + \ HSCoA$$

Malonyl   Acyl carrier   *S*-Malonyl acyl   Coenzyme A
coenzyme A   protein   carrier protein

When both building block units are in place on the acyl carrier protein, carbon–carbon bond formation occurs between the α-carbon atom of the malonyl group and the carbonyl carbon of the acetyl group. This is shown in step 1 of Figure 26.3. Carbon–carbon bond formation is accompanied by decarboxylation and produces a four-carbon acetoacetyl (3-oxobutanoyl) group bound to acyl carrier protein.

The acetoacetyl group is then transformed to a butanoyl group by the reaction sequence illustrated in steps 2 to 4.

The four carbon atoms of the butanoyl group originate in two molecules of acetyl coenzyme A. Carbon dioxide assists the reaction but is not incorporated into the product. The same carbon dioxide that is used to convert one molecule of acetyl coenzyme A to malonyl coenzyme A is regenerated in the decarboxylation step that accompanies carbon–carbon bond formation.

Successive repetitions of the steps shown in Figure 26.3 give unbranched acyl groups having 6, 8, 10, 12, 14, and 16 carbon atoms. In each case, chain extension occurs

Step 1: An acetyl group is transferred to the α carbon atom of the malonyl group with evolution of carbon dioxide. Presumably decarboxylation gives an enol, which attacks the acetyl group.

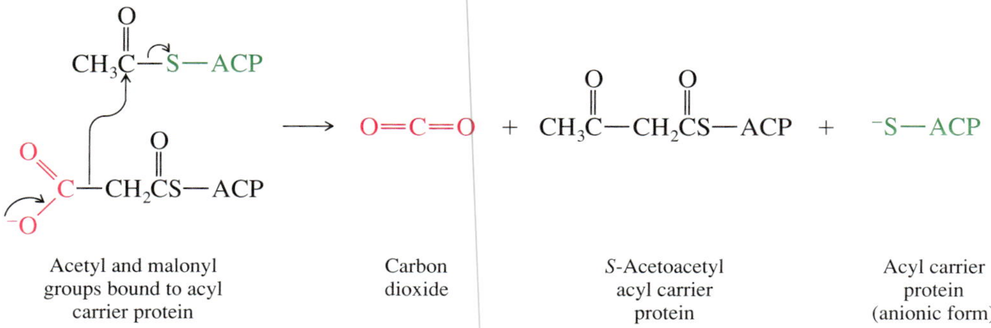

| Acetyl and malonyl groups bound to acyl carrier protein | Carbon dioxide | S-Acetoacetyl acyl carrier protein | Acyl carrier protein (anionic form) |

Step 2: The ketone carbonyl of the acetoacetyl group is reduced to an alcohol function. This reduction requires NADPH as a coenzyme. (NADPH is the phosphate ester of NADH and reacts similarly to it.)

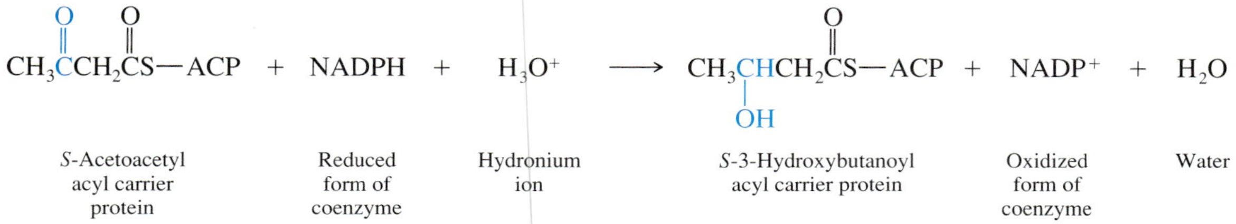

$$CH_3CCH_2CS—ACP + NADPH + H_3O^+ \longrightarrow CH_3CHCH_2CS—ACP + NADP^+ + H_2O$$

| S-Acetoacetyl acyl carrier protein | Reduced form of coenzyme | Hydronium ion | S-3-Hydroxybutanoyl acyl carrier protein | Oxidized form of coenzyme | Water |

Step 3: Dehydration of the β-hydroxy acyl group.

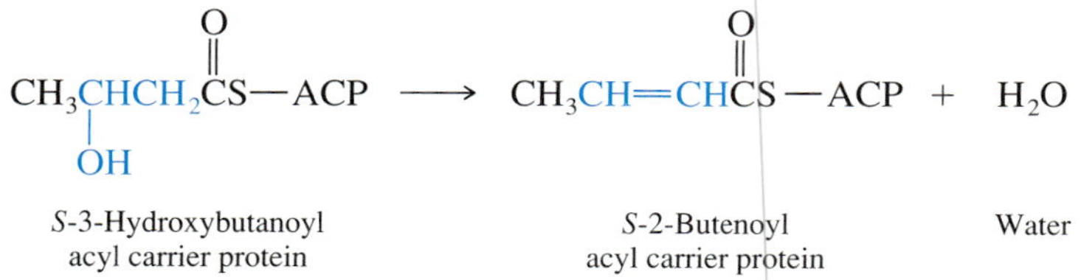

$$CH_3CHCH_2CS—ACP \longrightarrow CH_3CH=CHCS—ACP + H_2O$$

| S-3-Hydroxybutanoyl acyl carrier protein | S-2-Butenoyl acyl carrier protein | Water |

Step 4: Reduction of the double bond of the α, β-unsaturated acyl group. This step requires NADPH as a coenzyme.

$$CH_3CH=CHCS—ACP + NADPH + H_3O^+ \longrightarrow CH_3CH_2CH_2CS—ACP + NADP^+ + H_2O$$

| S-2-Butenoyl acyl carrier protein | Reduced form of coenzyme | Hydronium ion | S-Butanoyl acyl carrier protein | Oxidized form of coenzyme | Water |

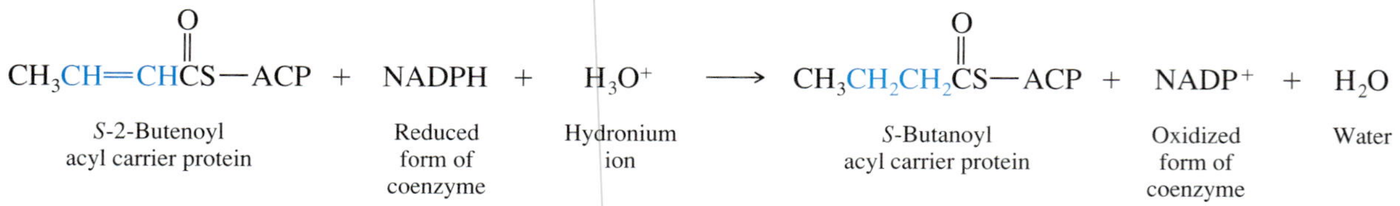

FIGURE 26.3 Mechanism of biosynthesis of a butanoyl group from acetyl and malonyl building blocks.

by reaction with a malonyl group bound to the acyl carrier protein. Thus, the biosynthesis of the 16-carbon acyl group of hexadecanoic (palmitic) acid can be represented by the overall equation:

$$CH_3\overset{O}{\underset{\|}{C}}S-ACP + 7HO\overset{O}{\underset{\|}{C}}CH_2\overset{O}{\underset{\|}{C}}S-ACP + 14NADPH + 14H_3O^+ \longrightarrow$$

| S-Acetyl acyl carrier protein | S-Malonyl acyl carrier protein | Reduced form of coenzyme | Hydronium ion |

$$CH_3(CH_2)_{14}\overset{O}{\underset{\|}{C}}S-ACP + 7CO_2 + 7HS-ACP + 14NADP^+ + 21H_2O$$

| S-Hexadecanoyl acyl carrier protein | Carbon dioxide | Acyl carrier protein | Oxidized form of coenzyme | Water |

PROBLEM 26.3 By analogy to the intermediates given in steps 1–4 of Figure 26.3, write the sequence of acyl groups that are attached to the acyl carrier protein in the conversion of

$$CH_3(CH_2)_{12}\overset{O}{\underset{\|}{C}}S-ACP \quad to \quad CH_3(CH_2)_{14}\overset{O}{\underset{\|}{C}}S-ACP$$

This phase of fatty acid biosynthesis concludes with the transfer of the acyl group from acyl carrier protein to coenzyme A. The resulting acyl coenzyme A molecules can then undergo a number of subsequent biological transformations. One such transformation is chain extension, leading to acyl groups with more than 16 carbons. Another is the introduction of one or more carbon–carbon double bonds. A third is acyl transfer from sulfur to oxygen to form esters such as triacylglycerols. The process by which acyl coenzyme A molecules are converted to triacylglycerols involves a type of intermediate called a *phospholipid* and is discussed in the following section.

26.4 PHOSPHOLIPIDS

Triacylglycerols arise, not by acylation of glycerol itself, but by a sequence of steps in which the first stage is acyl transfer to L-glycerol 3-phosphate (from reduction of dihydroxyacetone 3-phosphate, formed as described in Section 25.21). The product of this stage is called a **phosphatidic acid.**

$$\begin{array}{c} CH_2OH \\ HO-\!\!\!\!\!-H \\ CH_2OPO_3H_2 \end{array} + RC\overset{O}{\underset{\|}{S}}CoA + R'C\overset{O}{\underset{\|}{S}}CoA \longrightarrow \begin{array}{c} \overset{O}{\underset{\|}{C}}H_2OCR \\ R'C\overset{O}{\underset{\|}{O}}-H \\ CH_2OPO_3H_2 \end{array} + 2HSCoA$$

| L-Glycerol 3-phosphate | Two acyl coenzyme A molecules (R and R′ may be the same or they may be different) | Phosphatidic acid | Coenzyme A |

PROBLEM 26.4 What is the absolute configuration (R or S) of L-glycerol 3-phosphate? What must be the absolute configuration of the naturally occurring phosphatidic acids biosynthesized from it?

Hydrolysis of the phosphate ester function of the phosphatidic acid gives a diacylglycerol, which then reacts with a third acyl coenzyme A molecule to produce a triacylglycerol.

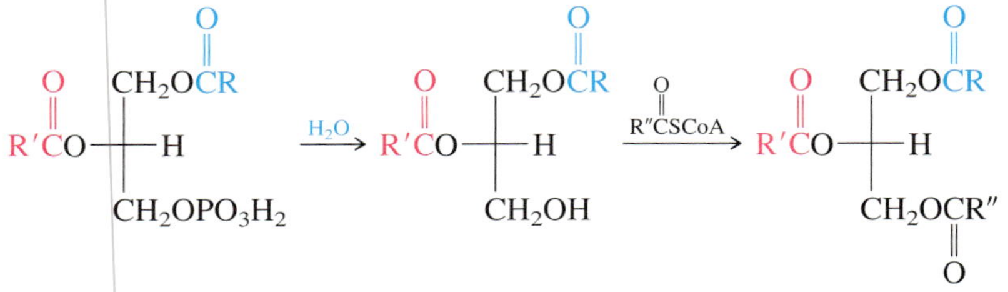

Phosphatidic acid Diacylglycerol Triacylglycerol

Phosphatidic acids not only are intermediates in the biosynthesis of triacylglycerols but also are biosynthetic precursors of other members of a group of compounds called **phosphoglycerides** or **glycerol phosphatides.** Phosphorus-containing derivatives of lipids are known as **phospholipids,** and phosphoglycerides are one type of phospholipid.

Lecithin is added to foods such as mayonnaise as an emulsifying agent to prevent the fat and water from separating into two layers.

One important phospholipid is **phosphatidylcholine,** also called *lecithin*. Phosphatidylcholine is a mixture of diesters of phosphoric acid. One ester function is derived from a diacylglycerol, whereas the other is a choline [$-OCH_2CH_2\overset{+}{N}(CH_3)_3$] unit.

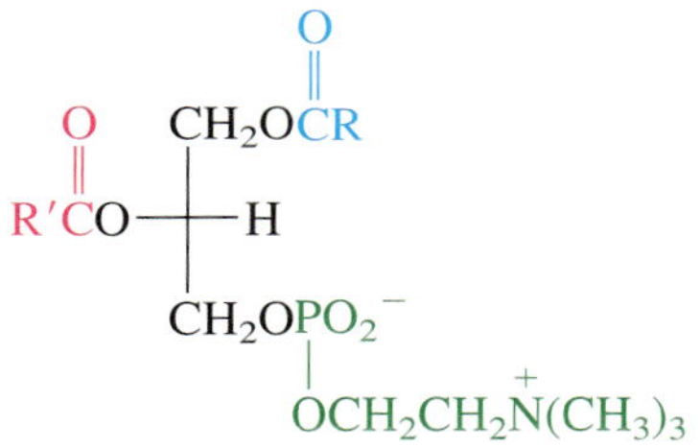

Phosphatidylcholine (R and R′ are usually different)

Phosphatidylcholine possesses a hydrophilic polar "head group" (the positively charged choline and negatively charged phosphate units) and two lipophilic nonpolar "tails" (the acyl groups). Under certain conditions, such as at the interface of two aqueous phases, phosphatidylcholine forms what is called a *lipid bilayer,* as shown in Figure 26.4. Because there are two long-chain acyl groups in each molecule, the most stable assembly has the polar groups solvated by water molecules at the top and bottom surfaces and the lipophilic acyl groups directed toward the interior of the bilayer.

Phosphatidylcholine is an important component of cell membranes, but cell membranes are more than simply lipid bilayers. Although their composition varies with their source, a typical membrane contains about equal amounts of lipid and protein, and the amount of cholesterol in the lipid fraction can approximate that of phosphatidylcholine.

The lipid fraction is responsible for the structure of the membrane. Phosphatidylcholine provides the bilayer that is the barrier between what is inside the cell and what is outside. Cholesterol intermingles with the phosphatidylcholine to confer an extra measure of rigidity to the membrane.

The protein fraction is responsible for a major part of membrane function. Nonpolar materials can diffuse through the bilayer from one side to the other relatively easily, but polar materials, particularly metal ions such as Na^+, K^+, and Ca^{2+} cannot. The

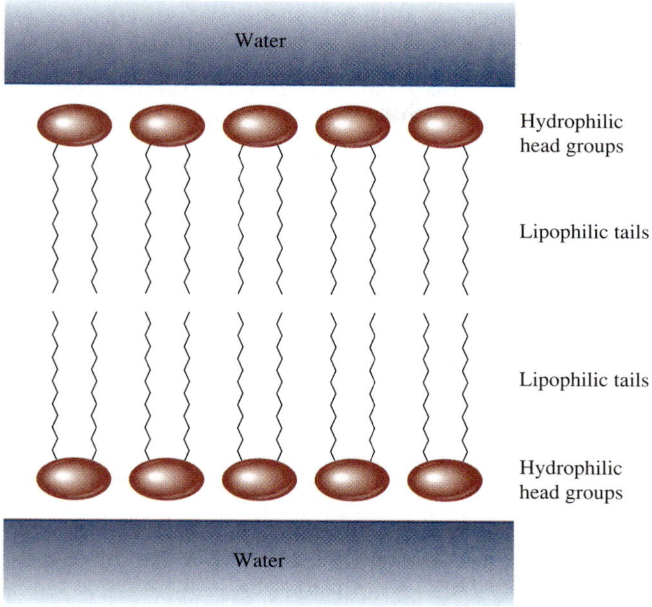

FIGURE 26.4 Cross section of a phospholipid bilayer.

transport of metal ions is assisted by the membrane proteins. These proteins pick up a metal ion from the aqueous phase on one side of the membrane and shield it from the hydrophobic environment within the membrane while transporting it to the aqueous phase on the other side of the membrane. Ionophore antibiotics such as monensin (Section 16.4) disrupt the normal functioning of cells by facilitating metal ion transport across cell membranes.

26.5   WAXES

Waxes are water-repelling solids that are part of the protective coatings of a number of living things, including the leaves of plants, the fur of animals, and the feathers of birds. They are usually mixtures of esters in which both the alkyl and acyl group are unbranched and contain a dozen or more carbon atoms. Beeswax, for example, contains the ester triacontyl hexadecanoate as one component of a complex mixture of hydrocarbons, alcohols, and esters.

$$CH_3(CH_2)_{14}\overset{\overset{\displaystyle O}{\|}}{C}OCH_2(CH_2)_{28}CH_3$$

Triacontyl hexadecanoate

PROBLEM 26.5 Spermaceti is a wax obtained from the sperm whale. It contains, among other materials, an ester known as *cetyl palmitate*, which is used as an emollient in a number of soaps and cosmetics. The systematic name for cetyl palmitate is *hexadecyl hexadecanoate*. Write a structural formula for this substance.

Fatty acids normally occur naturally as esters; fats, oils, phospholipids, and waxes all are unique types of fatty acid esters. There is, however, an important class of fatty acid derivatives that exists and carries out its biological role in the form of the free acid. This class of fatty acid derivatives is described in the following section.

26.6   PROSTAGLANDINS

Research in physiology carried out in the 1930s established that the lipid fraction of semen contains small amounts of substances that exert powerful effects on smooth muscle. Sheep prostate glands proved to be a convenient source of this material and yielded a mixture of structurally related substances referred to collectively as **prostaglandins.** We now know that prostaglandins are present in almost all animal tissues, where they carry out a variety of regulatory functions.

Prostaglandins are extremely potent substances and exert their physiological effects at very small concentrations. Because of this, their isolation was difficult, and it was not until 1960 that the first members of this class, designated PGE_1 and $PGF_{1\alpha}$ (Figure 26.5), were obtained as pure compounds. More than a dozen structurally related prostaglandins have since been isolated and identified. All the prostaglandins are 20-carbon carboxylic acids and contain a cyclopentane ring. All have hydroxyl groups at C-11 and C-15 (for the numbering of the positions in prostaglandins, see Figure 26.5). Prostaglandins belonging to the F series have an additional hydroxyl group at C-9, and a carbonyl function is present at this position in the various PGEs. The subscript numerals in their abbreviated names indicate the number of double bonds.

Physiological responses to prostaglandins encompass a variety of effects. Some prostaglandins relax bronchial muscle, others contract it. Some stimulate uterine contractions and have been used to induce therapeutic abortions. PGE_1 dilates blood vessels and lowers blood pressure; it inhibits the aggregation of platelets and offers promise as a drug to reduce the formation of blood clots.

Prostaglandins arise from unsaturated C_{20}-carboxylic acids such as arachidonic acid (see Table 26.1). Mammals cannot biosynthesize arachidonic acid directly. They obtain linoleic acid (Table 26.1) from vegetable oils in their diet and extend the carbon chain of linoleic acid from 18 to 20 carbons while introducing two more double bonds. Linoleic acid is said to be an **essential fatty acid,** forming part of the dietary requirement of mammals. Animals fed on diets that are deficient in linoleic acid grow poorly and suffer a number of other disorders, some of which are reversed on feeding them vegetable oils rich in linoleic acid and other *polyunsaturated fatty acids.* One function of these substances is to provide the raw materials for prostaglandin biosynthesis.

Studies of the biosynthesis of PGE_2 from arachidonic acid have shown that all three oxygens come from O_2. The enzyme involved, *prostaglandin endoperoxide synthase,* has *cyclooxygenase* (COX) activity and catalyzes the reaction of arachidonic acids with O_2 to give an *endoperoxide* (PGG_2).

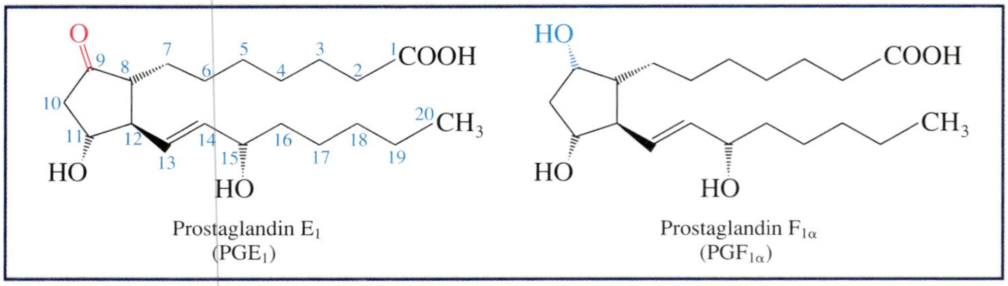

> Arachidonic acid gets its name from *arachidic acid,* the saturated C_{20} fatty acid isolated from peanut (*Arachis hypogaea*) oil.

FIGURE 26.5 Structures of two representative prostaglandins. The numbering scheme is illustrated in the structure of PGE_1.

Prostaglandin E_1
(PGE_1)

Prostaglandin $F_{1\alpha}$
($PGF_{1\alpha}$)

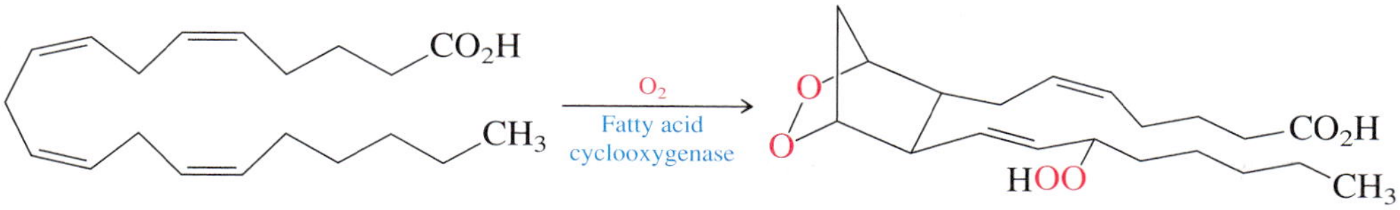

Arachidonic acid → PGG₂

In the next step, the —OOH group of PGG₂ is reduced to an alcohol function. Again, prostaglandin endoperoxide synthase is the enzyme responsible. The product of this step is called PGH₂.

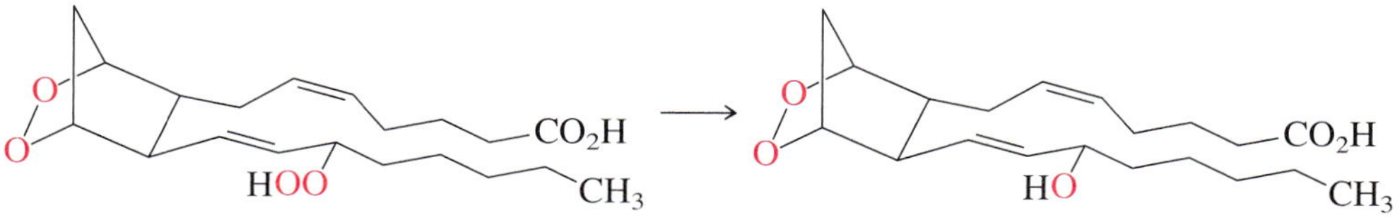

PGG₂ → PGH₂

PGH₂ is the precursor to a number of prostaglandins and related compounds, depending on the enzyme that acts on it. One of these cleaves the O—O bond of the endoperoxide and gives PGE₂.

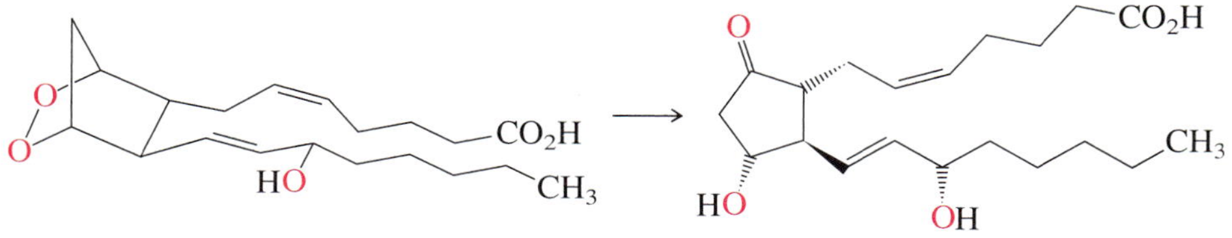

PGH₂ → PGE₂

Before leaving this biosynthetic scheme, notice that PGE₂ has four chirality centers. Even though arachidonic acid is achiral, only the stereoisomer shown in the equation is formed. Moreover, it is formed as a single enantiomer. The stereochemistry is controlled by the interaction of the substrate with the enzymes that act on it. Enzymes offer a chiral environment in which biochemical transformations occur and enzyme-catalyzed reactions almost always lead to a single stereoisomer. Many more examples will be seen in this chapter.

PROBLEM 26.6 Write the structural formula and give the IUPAC name for the fatty acid from which PGE₁ is biosynthesized. The structure of PGE₁ is shown in Figure 26.5

Prostaglandins belong to a group of compounds that, because they are related to icosanoic acid [$CH_3(CH_2)_{18}CO_2H$], are collectively known as *icosanoids*. The other icosanoids are *thromboxanes, prostacyclins,* and *leukotrienes.*

Thromboxane A₂ (TXA₂) promotes platelet aggregation and blood clotting. The biosynthetic pathway to TXA₂ is the same as that of PGE₂ up to PGH₂. At that point separate pathways lead to PGE₂ and TXA₂.

Older versions of the IUPAC rules called the unbranched carboxylic acid with 20 carbon atoms *eicosanoic acid.* Consequently, icosanoids are often referred to as *eicosanoids.*

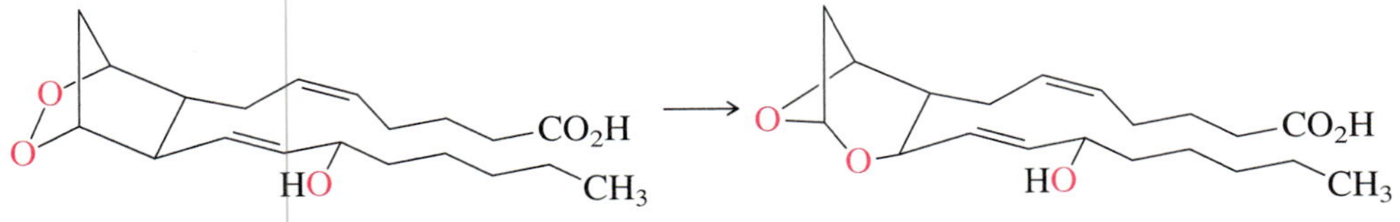

PGH$_2$ TXA$_2$

Prostacyclin I$_2$ (PGI$_2$) inhibits platelet aggregation and relaxes coronary arteries. Like PGE$_2$ and TXA$_2$, it is formed from arachidonic acid via PGH$_2$.

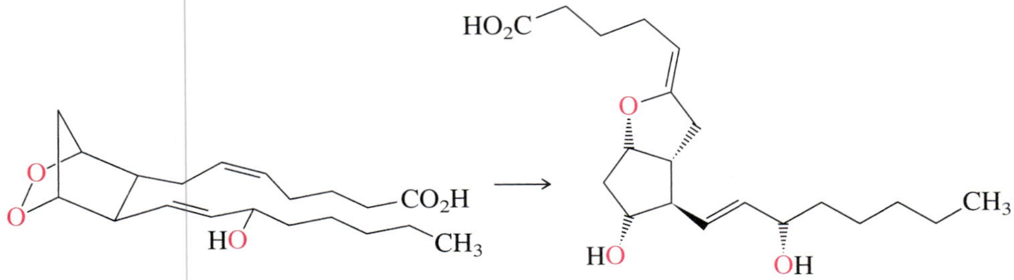

PGH$_2$ PGI$_2$

Leukotrienes are the substances most responsible for the constriction of bronchial passages during asthma attacks. They arise from arachidonic acid by a pathway different from the one that leads to prostaglandins and related compounds. The pathway to leukotrienes does not involve cyclooxygenation. Instead, oxidation simply introduces —OH groups at specific carbons along the chain. Allylic radicals are involved and some of the double bonds in the product are in different locations than those in arachidonic acid. The enzymes involved are known as *lipoxygenases* and are differentiated according to the carbon of the chain that is oxidized. The biosynthesis of the leukotriene shown begins with a 5-lipoxygenase-catalyzed oxidation of arachidonic acid.

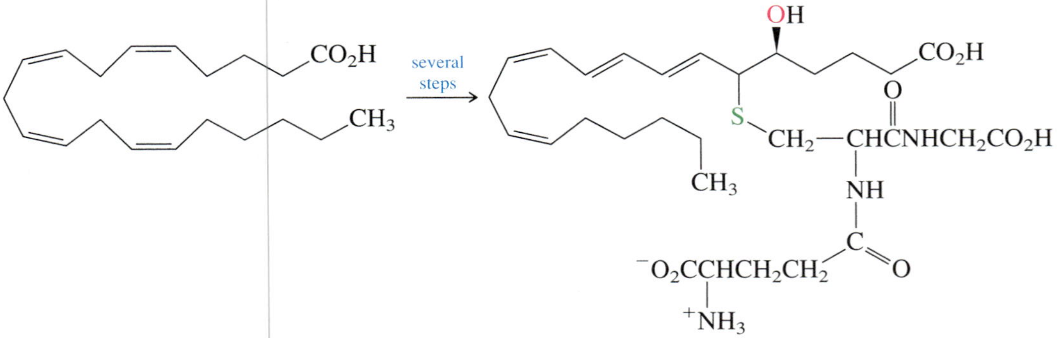

Arachidonic acid   Leukotriene C$_4$ (LTC$_4$)

PROBLEM 26.7 The carbon-sulfur bond in LTC$_4$ is formed by the reaction of glutathione (Section 15.13) with leukotriene A$_4$ (LTA$_4$). LTA$_4$ is an epoxide. Suggest a reasonable structure for LTA$_4$.

Most of the drugs such as epinephrine and albuterol used to treat asthma attacks are *bronchodilators*—substances that expand the bronchial passages. Newer drugs are designed to either inhibit the enzyme 5-lipoxygenase, which acts on arachidonic acid in the first stage of leukotriene biosynthesis, or to block leukotriene receptors.

NONSTEROIDAL ANTIINFLAMMATORY DRUGS (NSAIDs) AND COX-2 INHIBITORS

An injection of the steroid cortisone (Section 26.14) is often effective for reducing the pain and inflammation that comes from an injury. But chronic pain and inflammation, such as occurs with arthritis, is better managed with an orally administered remedy. Enter nonsteroidal antiinflammatory drugs (NSAIDs).

Aspirin (Section 24.10) is the oldest and best known NSAID. Over the years it has been joined by many others, a few of which are:

Ibuprofen   Naproxen   Diclofenac

The long-standing question of how aspirin works has been answered in terms of its effect on prostaglandin biosynthesis. Prostaglandins are made continuously in all mammalian cells and serve a variety of functions. They are biosynthesized in greater amounts at sites of tissue damage, and it is they that cause the pain and inflammation we feel. Cells contain two forms of the cyclooxygenase enzyme, COX-1 and COX-2; both catalyze prostaglandin biosynthesis. Some of the prostaglandins produced with the aid of COX-1 are involved in protecting the stomach and kidneys. COX-2 is concentrated in injured tissue where it works to catalyze the biosynthesis of the prostaglandins responsible for inflammation. Aspirin inhibits prostaglandin biosynthesis by inactivating both COX-1 and COX-2. Although inhibition of COX-2 has the desired effect of relieving pain and inflammation, inhibition of COX-1 causes irritation of the stomach lining.

A good antiinflammatory drug, therefore, will selectively inactivate COX-2 while leaving COX-1 untouched. Aspirin fails this test. In fact, aspirin is about ten times more effective toward inactivating the "wrong" COX. Beginning in the late 1990s, new antiinflammatory drugs became available that selectively inactivate the right one, COX-2. Two of these COX-2 inhibitors are *rofecoxib* and *celecoxib*.

Rofecoxib   Celecoxib
(Vioxx)   (Celebrex)

None of the NSAIDs mentioned so far has a structure that bears any resemblance to a typical lipid, yet all interact with enzymes that have lipids as their substrates. The classical period of drug development emphasized testing a large number of unrelated compounds for biological activity, identifying the structural features believed to be associated with the desired activity, then synthesizing and testing numerous analogs. The most recently developed NSAIDs, the COX-2 inhibitors rofecoxib and celecoxib, were developed with the intent of targeting the COX-2 enzyme for inactivation. They emerged by a combination of the classical "prepare and test" strategy and molecular modeling. Models of the three-dimensional structures of COX-1 and COX-2 were examined to guide thinking about the kinds of structural units that a drug should have to selectively inactivate COX-2. Although developed independently by different pharmaceutical companies, rofecoxib and celecoxib have several structural features in common, suggesting parallel strategies of drug discovery.

Much of the fundamental work on prostaglandins and related compounds was carried out by Sune Bergström and Bengt Samuelsson of the Karolinska Institute (Sweden) and by Sir John Vane of the Wellcome Foundation (Great Britain). These three shared the Nobel Prize for physiology or medicine in 1982.

26.7   TERPENES: THE ISOPRENE RULE

The word *essential* as applied to naturally occurring organic substances can have two different meanings. For example, as used in the previous section with respect to fatty acids, *essential* means "necessary." Linoleic acid is an "essential" fatty acid; it must be included in the diet for animals to grow properly because they lack the ability to biosynthesize it directly.

Essential is also used as the adjective form of the noun *essence*. The mixtures of substances that make up the fragrant material of plants are called *essential oils* because they contain the essence, that is, the odor, of the plant. The study of the composition of essential oils ranks as one of the oldest areas of organic chemical research. Very often, the principal volatile component of an essential oil belongs to a class of chemical substances called the **terpenes.**

Myrcene, a hydrocarbon isolated from bayberry oil, is a typical terpene:

Myrcene

The structural feature that distinguishes terpenes from other natural products is the **isoprene unit.** The carbon skeleton of myrcene (exclusive of its double bonds) corresponds to the head-to-tail union of two isoprene units.

Isoprene
(2-methyl-1,3-butadiene)

Two isoprene units
linked head-to-tail

The German chemist Otto Wallach (Nobel Prize in chemistry, 1910) determined the structures of many terpenes and is credited with setting forth the **isoprene rule:** terpenes are repeating assemblies of isoprene units, normally joined head-to-tail.

Terpenes are often referred to as *isoprenoid* compounds and are classified according to the number of isoprene units they contain (Table 26.2).

Although the term *terpene* once referred only to hydrocarbons, current usage includes functionally substituted derivatives as well, grouped together under the general term *isoprenoids*. Figure 26.6 (page 1086) presents the structural formulas for a number of representative examples. The isoprene units in some of these are relatively easy to identify. The three isoprene units in the sesquiterpene **farnesol,** for example, are indicated as follows in color. They are joined in a head-to-tail fashion.

There are more than 23,000 known isoprenoid compounds.

Isoprene units in farnesol

TABLE 26.2	Classification of Terpenes	
Class	**Number of isoprene units**	**Number of carbon atoms**
Monoterpene	2	10
Sesquiterpene	3	15
Diterpene	4	20
Sesterpene	5	25
Triterpene	6	30
Tetraterpene	8	40

Many terpenes contain one or more rings, but these also can be viewed as collections of isoprene units. An example is α-selinene. Like farnesol, it is made up of three isoprene units linked head-to-tail.

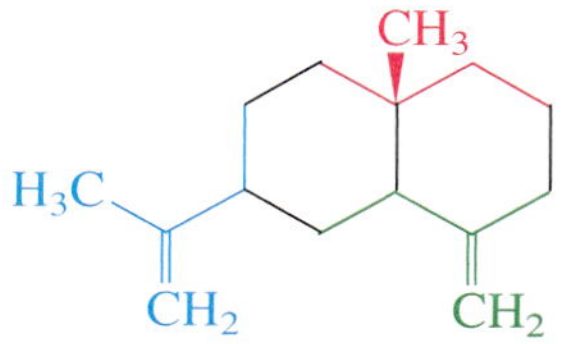

Isoprene units in α-selinene

PROBLEM 26.8 Locate the isoprene units in each of the monoterpenes, sesquiterpenes, and diterpenes shown in Figure 26.6. (In some cases there are two equally correct arrangements.)

Tail-to-tail linkages of isoprene units sometimes occur, especially in the higher terpenes. The C(12)—C(13) bond of squalene unites two C_{15} units in a tail-to-tail manner. Notice, however, that isoprene units are joined head-to-tail within each C_{15} unit of squalene.

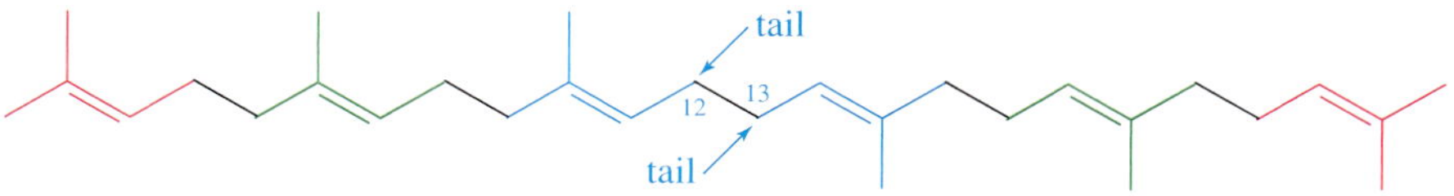

Isoprene units in squalene

PROBLEM 26.9 Identify the isoprene units in β-carotene (see Figure 26.6). Which carbons are joined by a tail-to-tail link between isoprene units?

Over time, Wallach's original isoprene rule was refined, most notably by Leopold Ruzicka of the Swiss Federal Institute of Technology (Zürich). Ruzicka, a corecipient of the 1939 Nobel Prize in chemistry, put forward a *biological isoprene rule* in which he connected the various classes of terpenes according to their biological precursors. Thus arose the idea of the *biological isoprene unit*. Isoprene is the fundamental structural unit of terpenes and related compounds, but isoprene does not occur naturally—at least in places where biosynthesis is going on. What then is the biological isoprene unit, how is this unit itself biosynthesized, and how do individual isoprene units combine to give terpenes?

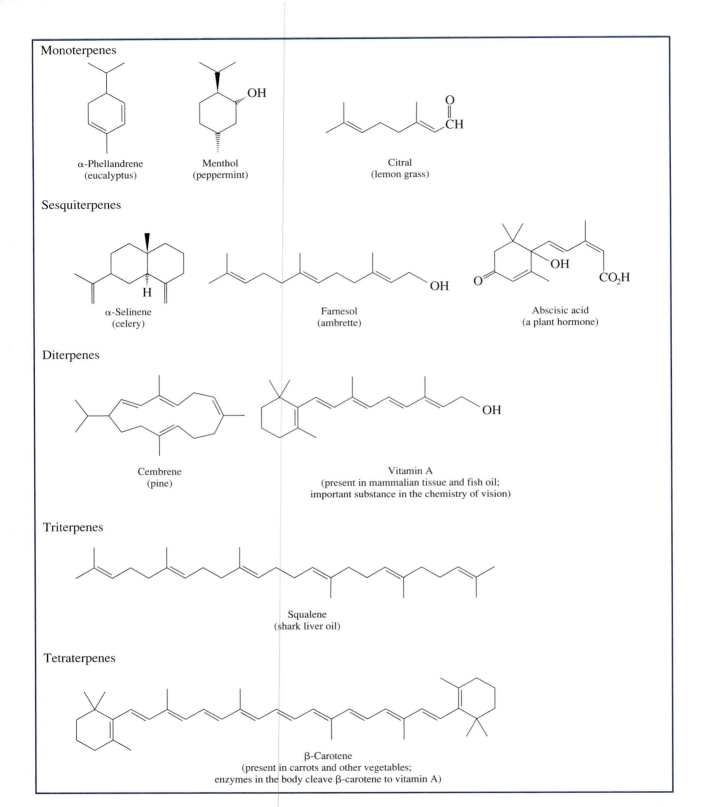

FIGURE 26.6 Some representative terpenes and related natural products. Structures are customarily depicted as carbon skeleton formulas when describing compounds of isoprenoid origin.

26.8 ISOPENTENYL PYROPHOSPHATE: THE BIOLOGICAL ISOPRENE UNIT

Isoprenoid compounds are biosynthesized from acetate by a process that involves several stages. The first stage is the formation of *mevalonic acid* from three molecules of acetic acid:

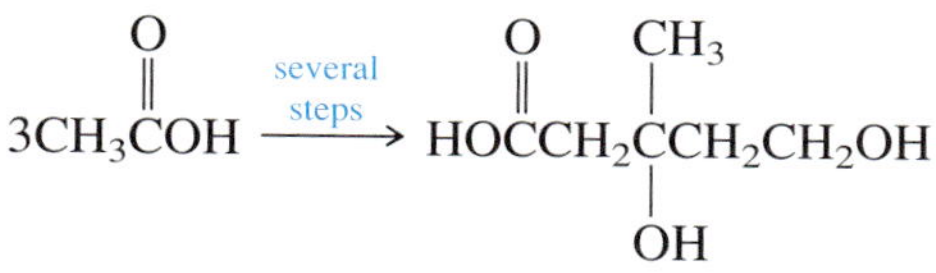

Acetic acid   Mevalonic acid

In the second stage, mevalonic acid is converted to *3-methyl-3-butenyl pyrophosphate (isopentenyl pyrophosphate):*

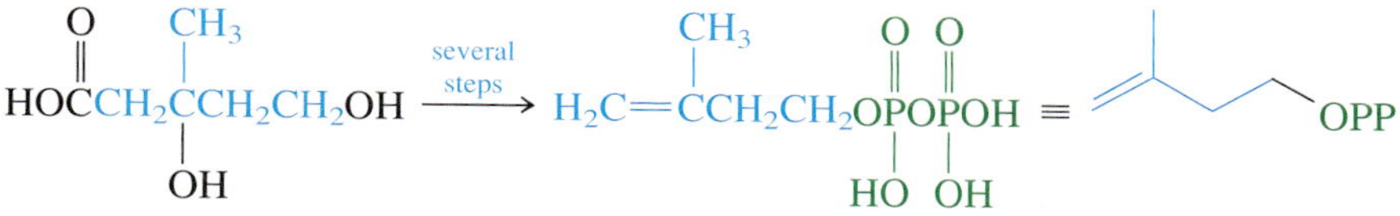

Mevalonic acid   Isopentenyl pyrophosphate

It is convenient to use the symbol —OPP to represent the pyrophosphate group.

Isopentenyl pyrophosphate is the biological isoprene unit; it contains five carbon atoms connected in the same order as in isoprene.

Isopentenyl pyrophosphate undergoes an enzyme-catalyzed reaction that converts it, in an equilibrium process, to *3-methyl-2-butenyl pyrophosphate (dimethylallyl pyrophosphate):*

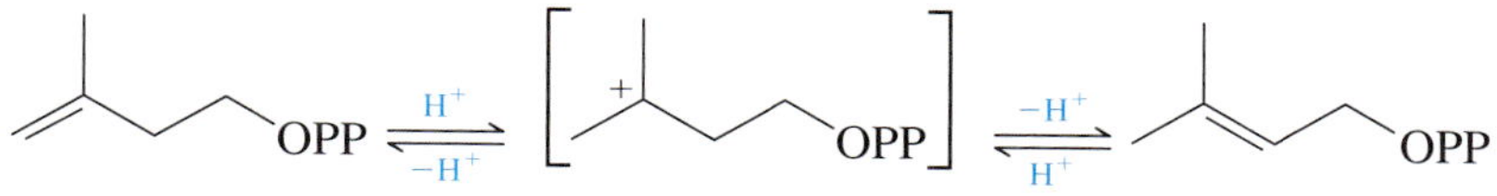

Isopentenyl pyrophosphate   Carbocation intermediate   Dimethylallyl pyrophosphate

Isopentenyl pyrophosphate and dimethylallyl pyrophosphate are structurally similar—both contain a double bond and a pyrophosphate ester unit—but the chemical reactivity expressed by each is different. The principal site of reaction in dimethylallyl pyrophosphate is the carbon that bears the pyrophosphate group. Pyrophosphate is a reasonably good leaving group in nucleophilic substitution reactions, especially when, as in dimethylallyl pyrophosphate, it is located at an allylic carbon. Isopentenyl pyrophosphate, on the other hand, does not have its leaving group attached to an allylic carbon and is far less reactive than dimethylallyl pyrophosphate toward nucleophilic reagents. The principal site of reaction in isopentenyl pyrophosphate is the carbon–carbon double bond, which, like the double bonds of simple alkenes, is reactive toward electrophiles.

26.9 CARBON–CARBON BOND FORMATION IN TERPENE BIOSYNTHESIS

The chemical properties of isopentenyl pyrophosphate and dimethylallyl pyrophosphate are complementary in a way that permits them to react with each other to form a carbon–carbon bond that unites two isoprene units. Using the π electrons of its double bond, isopentenyl pyrophosphate acts as a nucleophile and displaces pyrophosphate from dimethylallyl pyrophosphate.

Dimethylallyl   Isopentenyl   Ten-carbon carbocation
pyrophosphate   pyrophosphate

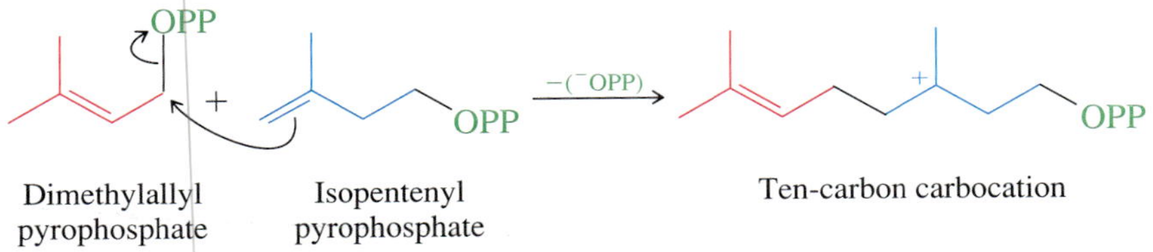

The tertiary carbocation formed in this step can react according to any of the various reaction pathways available to carbocations. One of these is loss of a proton to give a double bond.

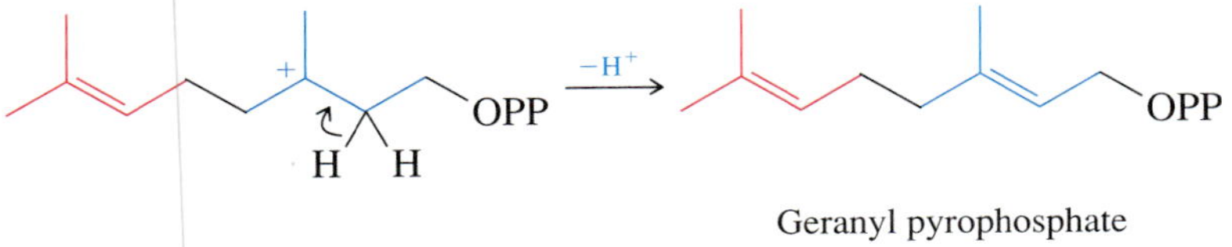

Geranyl pyrophosphate

The product of this reaction is *geranyl pyrophosphate.* Hydrolysis of the pyrophosphate ester group gives *geraniol,* a naturally occurring monoterpene found in rose oil.

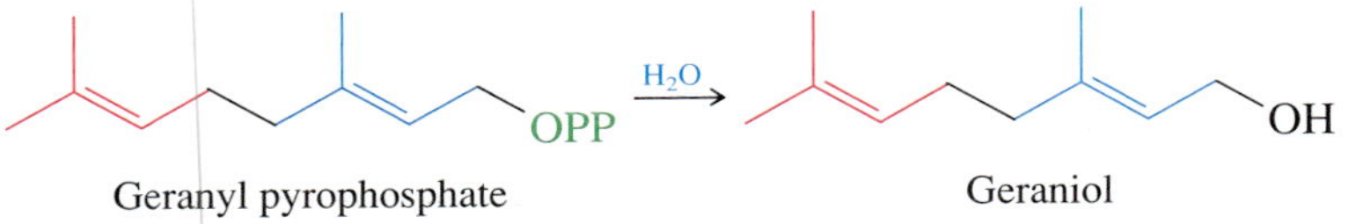

Geranyl pyrophosphate   Geraniol

Geranyl pyrophosphate is an allylic pyrophosphate and, like dimethylallyl pyrophosphate, can act as an alkylating agent toward a molecule of isopentenyl pyrophosphate. A 15-carbon carbocation is formed, which, on deprotonation, gives *farnesyl pyrophosphate.*

Geranyl pyrophosphate   Isopentenyl pyrophosphate

Farnesyl pyrophosphate

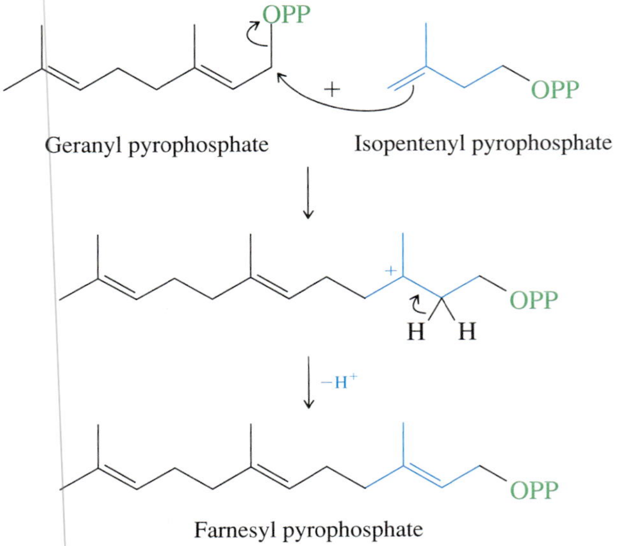

Hydrolysis of the pyrophosphate ester group converts farnesyl pyrophosphate to the corresponding alcohol *farnesol* (see Figure 26.6 for the structure of farnesol).

A repetition of the process just shown produces the diterpene geranylgeraniol from farnesyl pyrophosphate.

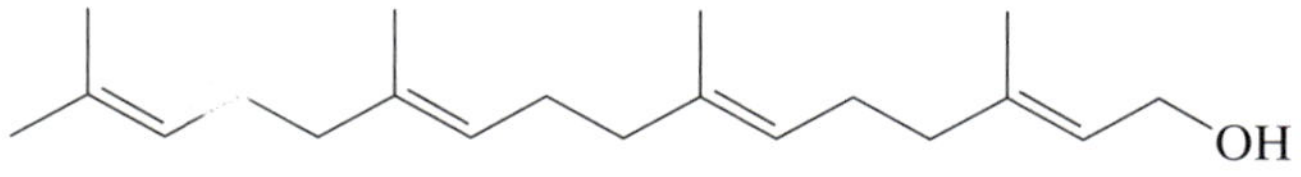

Geranylgeraniol

PROBLEM 26.10   Write a sequence of reactions that describes the formation of geranylgeraniol from farnesyl pyrophosphate.

The higher terpenes are formed not by successive addition of C_5 units but by the coupling of simpler terpenes. Thus, the triterpenes (C_{30}) are derived from two molecules of farnesyl pyrophosphate, and the tetraterpenes (C_{40}) from two molecules of geranylgeranyl pyrophosphate. These carbon–carbon bond-forming processes involve tail-to-tail couplings and proceed by a more complicated mechanism than that just described.

The enzyme-catalyzed reactions that lead to geraniol and farnesol (as their pyrophosphate esters) are mechanistically related to the acid-catalyzed dimerization of alkenes discussed in Section 6.21. The reaction of an allylic pyrophosphate or a carbocation with a source of π electrons is a recurring theme in terpene biosynthesis and is invoked to explain the origin of more complicated structural types. Consider, for example, the formation of cyclic monoterpenes. *Neryl pyrophosphate,* formed by an enzyme-catalyzed isomerization of the *E* double bond in geranyl pyrophosphate, has the proper geometry to form a six-membered ring via intramolecular attack of the double bond on the allylic pyrophosphate unit.

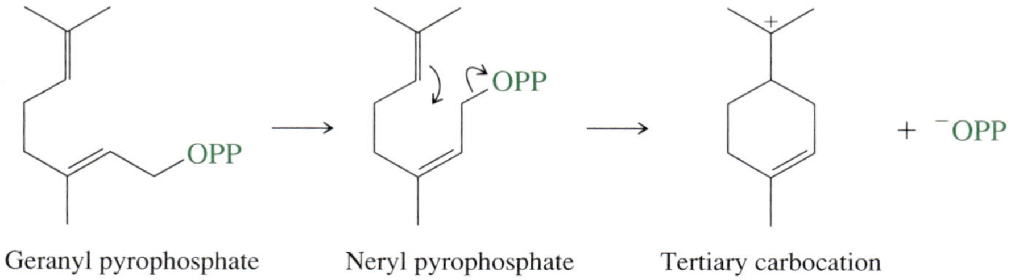

Geranyl pyrophosphate   Neryl pyrophosphate   Tertiary carbocation

Loss of a proton from the tertiary carbocation formed in this step gives *limonene,* an abundant natural product found in many citrus fruits. Capture of the carbocation by water gives *α-terpineol,* also a known natural product.

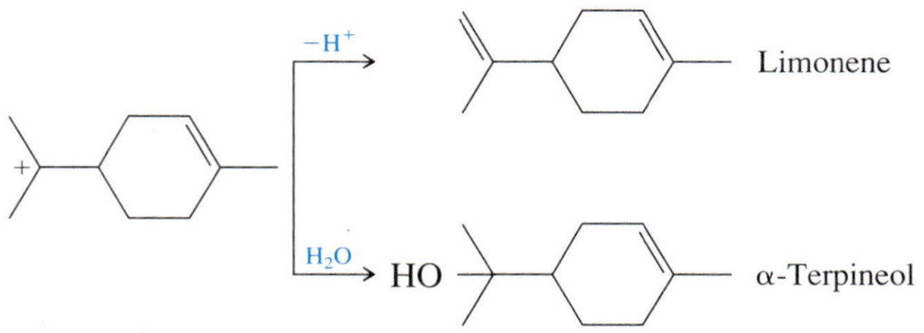

The same tertiary carbocation serves as the precursor to numerous bicyclic monoterpenes. A carbocation having a bicyclic skeleton is formed by intramolecular attack of the π electrons of the double bond on the positively charged carbon.

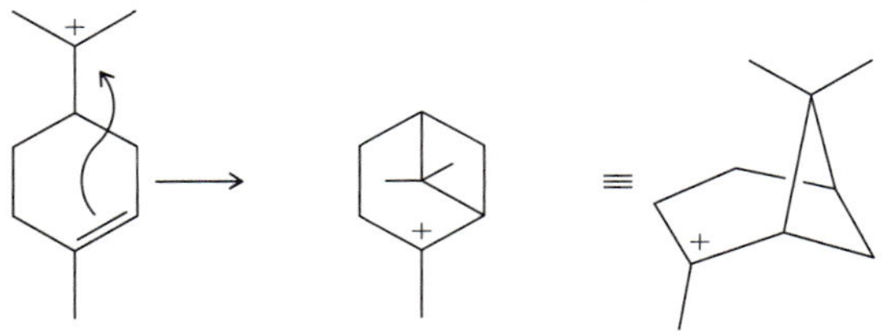

Bicyclic carbocation

This bicyclic carbocation then undergoes many reactions typical of carbocation intermediates to provide a variety of bicyclic monoterpenes, as outlined in Figure 26.7.

> **PROBLEM 26.11** The structure of the bicyclic monoterpene borneol is shown in Figure 26.7. Isoborneol, a stereoisomer of borneol, can be prepared in the laboratory by a two-step sequence. In the first step, borneol is oxidized to camphor by treatment with chromic acid. In the second step, camphor is reduced with sodium borohydride to a mixture of 85% isoborneol and 15% borneol. On the basis of these transformations, deduce structural formulas for isoborneol and camphor.

Analogous processes involving cyclizations and rearrangements of carbocations derived from farnesyl pyrophosphate produce a rich variety of structural types in the

A. Loss of a proton from the bicyclic carbocation yields α-pinene and β-pinene. The pinenes are the most abundant of the monoterpenes. They are the main constituents of turpentine.

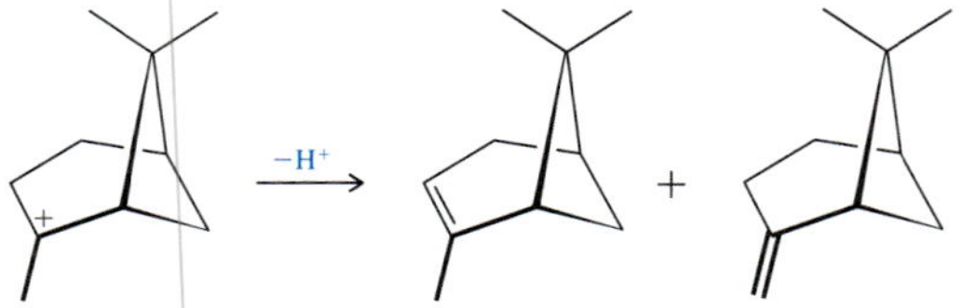

α-Pinene β-Pinene

B. Capture of the carbocation by water, accompanied by rearrangement of the bicyclo-[3.1.1] carbon skeleton to a bicyclo[2.2.1] unit, yields borneol. Borneol is found in the essential oil of certain trees that grow in Indonesia.

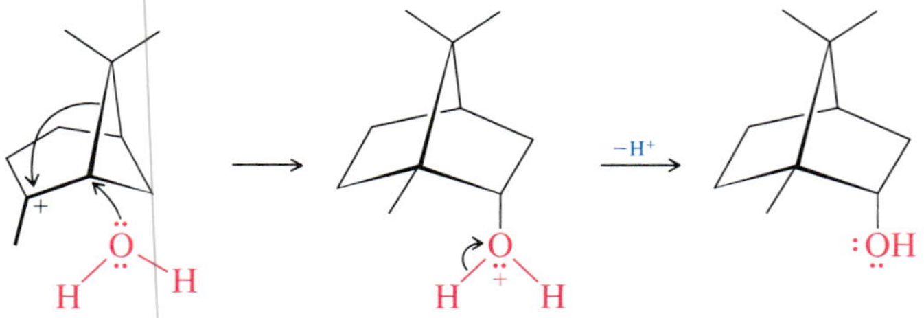

Borneol

FIGURE 26.7 Two of the reaction pathways available to the C_{10} bicyclic carbocation formed from neryl pyrophosphate. The same carbocation can lead to monoterpenes based on either the bicyclo[3.1.1] or the bicyclo[2.2.1] carbon skeleton.

sesquiterpene series. We will have more to say about the chemistry of higher terpenes, especially the triterpenes, later in this chapter. For the moment, however, let's return to smaller molecules to complete the picture of how isoprenoid compounds arise from acetate.

26.10 THE PATHWAY FROM ACETATE TO ISOPENTENYL PYROPHOSPHATE

The introduction to Section 26.8 pointed out that mevalonic acid is the biosynthetic precursor of isopentenyl pyrophosphate. The early steps in the biosynthesis of mevalonate from three molecules of acetic acid are analogous to those in fatty acid biosynthesis (Section 26.3) except that they do not involve acyl carrier protein. Thus, the reaction of acetyl coenzyme A with malonyl coenzyme A yields a molecule of acetoacetyl coenzyme A.

$$CH_3\overset{\overset{\displaystyle O}{\|}}{C}SCoA \;+\; {}^-O_2CCH_2\overset{\overset{\displaystyle O}{\|}}{C}SCoA \longrightarrow CH_3\overset{\overset{\displaystyle O}{\|}}{C}CH_2\overset{\overset{\displaystyle O}{\|}}{C}SCoA \;+\; CO_2$$

| Acetyl coenzyme A | Malonyl coenzyme A | Acetoacetyl coenzyme A | Carbon dioxide |

Carbon–carbon bond formation then occurs between the ketone carbonyl of acetoacetyl coenzyme A and the α carbon of a molecule of acetyl coenzyme A.

$$CH_3\overset{\overset{\displaystyle O}{\|}}{C}CH_2\overset{\overset{\displaystyle O}{\|}}{C}SCoA \;+\; CH_3\overset{\overset{\displaystyle O}{\|}}{C}SCoA \longrightarrow$$

3-Hydroxy-3-methylglutaryl coenzyme A (HMG CoA) + CoASH

| Acetoacetyl coenzyme A | Acetyl coenzyme A | 3-Hydroxy-3-methylglutaryl coenzyme A (HMG CoA) | Coenzyme A |

The product of this reaction, 3-hydroxy-3-methylglutaryl coenzyme A (HMG CoA), has the carbon skeleton of mevalonic acid and is converted to it by enzymatic reduction.

$$CH_3\overset{\displaystyle OH}{\underset{\displaystyle CH_2COH}{C}}CH_2\overset{\overset{\displaystyle O}{\|}}{C}SCoA \longrightarrow CH_3\overset{\displaystyle OH}{\underset{\displaystyle CH_2COH}{C}}CH_2CH_2OH$$

| 3-Hydroxy-3-methylglutaryl coenzyme A (HMG CoA) | Mevalonic acid |

Some of the most effective cholesterol-lowering drugs act by inhibiting the enzyme that catalyzes this reaction.

In keeping with its biogenetic origin in three molecules of acetic acid, mevalonic acid has six carbon atoms. The conversion of mevalonate to isopentenyl pyrophosphate involves loss of the "extra" carbon as carbon dioxide. First, the alcohol hydroxyl groups of mevalonate are converted to phosphate ester functions—they are enzymatically *phosphorylated,* with introduction of a simple phosphate at the tertiary site and a pyrophosphate at the primary site. Decarboxylation, in concert with loss of the tertiary phosphate, introduces a carbon–carbon double bond and gives isopentenyl pyrophosphate, the fundamental building block for formation of isoprenoid natural products.

Some bacteria, algae, and plants make isopentenyl pyrophosphate by a different route.

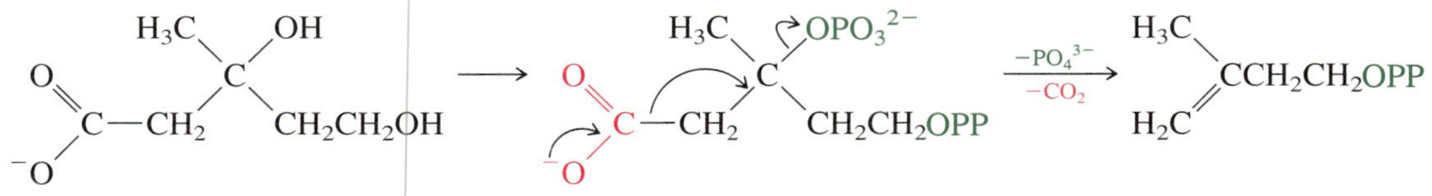

Mevalonate → (Unstable; undergoes rapid decarboxylation with loss of phosphate) → Isopentenyl pyrophosphate

Much of what we know concerning the pathway from acetate to mevalonate to isopentenyl pyrophosphate to terpenes comes from "feeding" experiments, in which plants are grown in the presence of radioactively labeled organic substances and the distribution of the radioactive label is determined in the products of biosynthesis. To illustrate, eucalyptus plants were allowed to grow in a medium containing acetic acid enriched with ^{14}C in its methyl group. *Citronellal* was isolated from the mixture of monoterpenes produced by the plants and shown, by a series of chemical degradations, to contain the radioactive ^{14}C label at carbons 2, 4, 6, and 8, as well as at the carbons of both branching methyl groups.

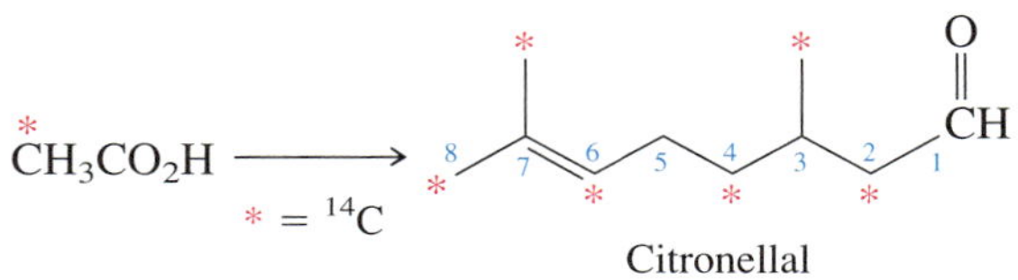

> Citronellal occurs naturally as the principal component of citronella oil and is used as an insect repellent.

Figure 26.8 traces the ^{14}C label from its origin in acetic acid to its experimentally determined distribution in citronellal.

PROBLEM 26.12 How many carbon atoms of citronellal would be radioactively labeled if the acetic acid used in the experiment were enriched with ^{14}C at C-1 instead of at C-2? Identify these carbon atoms.

A more recent experimental technique employs ^{13}C as the isotopic label. Instead of locating the position of a ^{14}C label by a laborious degradation procedure, the ^{13}C NMR spectrum of the natural product is recorded. The signals for the carbons that are enriched in ^{13}C are far more intense than those corresponding to carbons in which ^{13}C is present only at the natural abundance level.

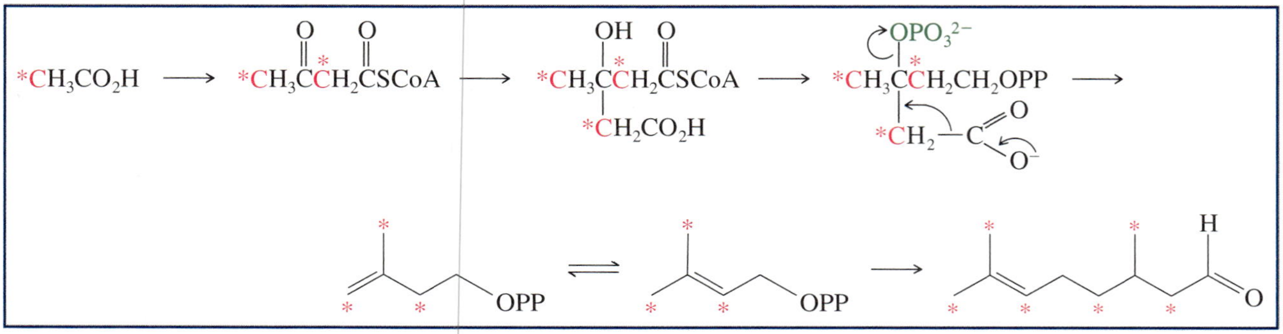

FIGURE 26.8 The distribution of the ^{14}C label in citronellal biosynthesized from acetate in which the methyl carbon was isotopically enriched with ^{14}C.

Isotope incorporation experiments have demonstrated the essential correctness of the scheme presented in this and preceding sections for terpene biosynthesis. Considerable effort has been expended toward its detailed elaboration because of the common biosynthetic origin of terpenes and another class of acetate-derived natural products, the steroids.

26.11 STEROIDS: CHOLESTEROL

Cholesterol is the central compound in any discussion of steroids. Its name is a combination of the Greek words for "bile" *(chole)* and "solid" *(stereos)* preceding the characteristic alcohol suffix *-ol*. It is the most abundant steroid present in humans and the most important one as well because all other steroids arise from it. An average adult has over 200 g of cholesterol; it is found in almost all body tissues, with relatively large amounts present in the brain and spinal cord and in gallstones. Cholesterol is the chief constituent of the plaque that builds up on the walls of arteries in atherosclerosis.

Cholesterol was isolated in the eighteenth century, but its structure is so complex that its correct constitution was not determined until 1932 and its stereochemistry not verified until 1955. Steroids are characterized by the tetracyclic ring system shown in Figure 26.9a. As shown in Figure 26.9b, cholesterol contains this tetracyclic skeleton modified to include an alcohol function at C-3, a double bond at C-5, methyl groups at C-10 and C-13, and a C_8H_{17} side chain at C-17. Isoprene units may be discerned in various portions of the cholesterol molecule, but the overall correspondence with the isoprene rule is far from perfect. Indeed, cholesterol has only 27 carbon atoms, three too few for it to be classed as a triterpene.

Animals accumulate cholesterol from their diet, but are also able to biosynthesize it from acetate. The pioneering work that identified the key intermediates in the complicated pathway of cholesterol biosynthesis was carried out by Konrad Bloch (Harvard) and Feodor Lynen (Munich), corecipients of the 1964 Nobel Prize for physiology or

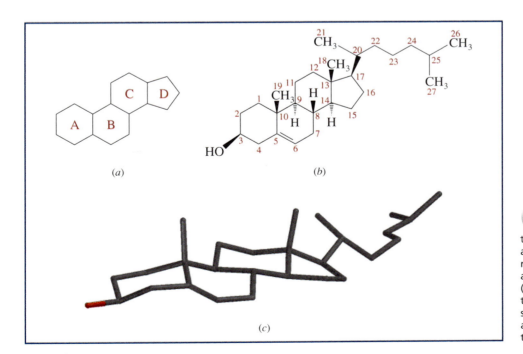

(a)

(b)

(c)

FIGURE 26.9 (a) The tetracyclic ring system characteristic of steroids. The rings are designated A, B, C, and D as shown. (b) and (c) The structure of cholesterol. A unique numbering system is used for steroids and is indicated in the structural formula.

Step 1: An electrophilic species, shown here as Enz—H$^+$, catalyzes ring-opening of squalene 2,3-epoxide. Ring opening is accompanied by cyclization to give a tricyclic tertiary carbocation. It is not known whether formation of the three new carbon–carbon bonds occurs in a single step or a series of steps.

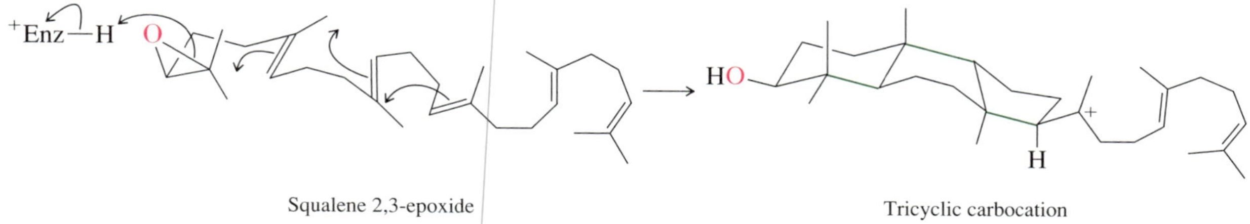

Squalene 2,3-epoxide   Tricyclic carbocation

Step 2: Ring expansion converts the five-membered ring of the carbocation formed in step 1 to a six-membered ring.

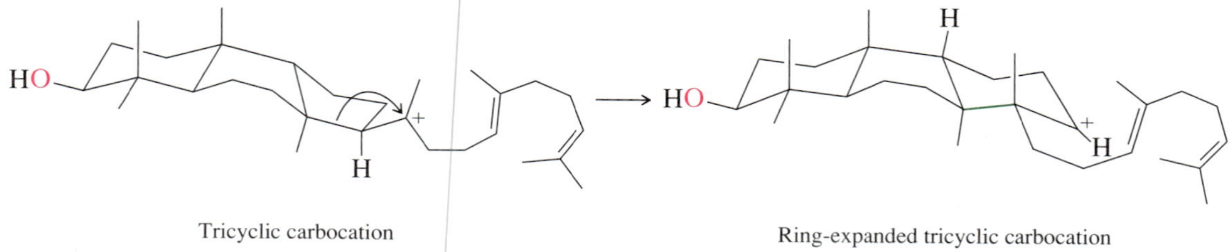

Tricyclic carbocation   Ring-expanded tricyclic carbocation

Step 3: Cyclization of the carbocation formed in step 2 gives a tetracyclic carbocation (*protosteryl cation*).

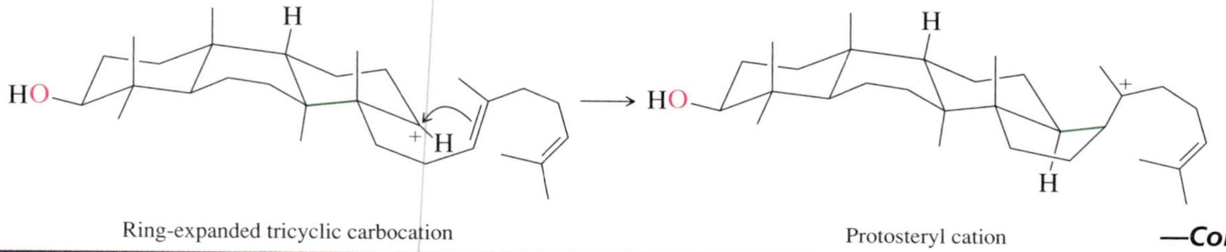

Ring-expanded tricyclic carbocation   Protosteryl cation   **—Cont.**

FIGURE 26.10 The biosynthetic conversion of squalene to cholesterol proceeds through lanosterol. Lanosterol is formed by enzyme-catalyzed cyclization of the 2,3-epoxide of squalene.

Lanosterol is one component of lanolin, a mixture of many substances that coats the wool of sheep.

medicine. An important discovery was that the triterpene *squalene* (see Figure 26.6) is an intermediate in the formation of cholesterol from acetate. Thus, *the early stages of cholesterol biosynthesis are the same as those of terpene biosynthesis* described in Sections 26.8–26.10. In fact, a significant fraction of our knowledge of terpene biosynthesis is a direct result of experiments carried out in the area of steroid biosynthesis.

How does the tetracyclic steroid cholesterol arise from the acyclic triterpene squalene? It begins with the epoxidation of squalene described earlier in Section 16.14 and continues from that point in Figure 26.10. Step 1 is an enzyme-catalyzed electrophilic ring opening of squalene 2,3-epoxide. Epoxide ring-opening triggers a series of carbocation reactions. These carbocation processes involve cyclization via carbon–carbon bond formation (step 1), ring-expansion via a carbocation rearrangement (step 2), another cyclization (step 3), followed by a cascade of methyl group migrations and hydride shifts (step 4). The result of all these steps is the tetracyclic triterpene *lanosterol*. Step 5 of Figure 26.10 summarizes the numerous remaining transformations by which lanosterol is converted to cholesterol.

Step 4: Rearrangement and deprotonation of protosteryl cation gives the tetracyclic triterpene lanosterol.

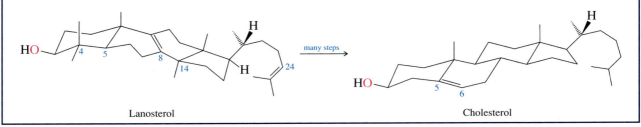

Protosteryl cation   Lanosterol

Step 5: A series of enzyme-catalyzed reactions converts lanosterol to cholesterol. The methyl groups at C-4 and C-14 are lost, the C-8 and C-24 double bonds are reduced, and a new double bond is introduced at C-5.

Lanosterol   Cholesterol

FIGURE 26.10 *(Continued)*

PROBLEM 26.13 The biosynthesis of cholesterol as outlined in Figure 26.10 is admittedly quite complicated. It will aid your understanding of the process if you consider the following questions:

(a) Which carbon atoms of squalene 2,3-epoxide correspond to the doubly bonded carbons of cholesterol?

(b) Which two hydrogen atoms of squalene 2,3-epoxide are the ones that migrate in step 3?

(c) Which methyl group of squalene 2,3-epoxide becomes the methyl group at the C,D ring junction of cholesterol?

(d) What three methyl groups of squalene 2,3-epoxide are lost during the conversion of lanosterol to cholesterol?

SAMPLE SOLUTION (a) As the structural formula in step 5 of Figure 26.10 indicates, the double bond of cholesterol unites C-5 and C-6 (steroid numbering). The corresponding carbons in the cyclization reaction of step 1 in the figure may be identified as C-7 and C-8 of squalene 2,3-epoxide (systematic IUPAC numbering).

Coiled form of squalene 2,3-epoxide

The conversion of lanosterol to cholesterol involves 19 steps and is described in the article "Cholesterol Biosynthesis: Lanosterol to Cholesterol" on pp. 377–384 in the March, 2002 issue of the *Journal of Chemical Education.*

GOOD CHOLESTEROL? BAD CHOLESTEROL? WHAT'S THE DIFFERENCE?

Cholesterol is biosynthesized in the liver, transported throughout the body to be used in a variety of ways, and returned to the liver where it serves as the biosynthetic precursor to other steroids. But cholesterol is a lipid and isn't soluble in water. How can it move through the blood if it doesn't dissolve in it? The answer is that it doesn't dissolve, but is instead carried through the blood and tissues as part of a *lipoprotein* (lipid + protein = lipoprotein).

The proteins that carry cholesterol from the liver are called low-density lipoproteins, or LDLs; those that return it to the liver are the *high-density lipoproteins,* or HDLs. If too much cholesterol is being transported by LDL, or too little by HDL, the extra cholesterol builds up on the walls of the arteries causing atherosclerosis. A thorough physical examination nowadays measures not only total cholesterol concentration but also the distribution between LDL and HDL cholesterol. An elevated level of LDL cholesterol is a risk factor for heart disease. LDL cholesterol is "bad" cholesterol. HDLs, on the other hand, remove excess cholesterol and are protective. HDL cholesterol is "good" cholesterol.

The distribution between LDL and HDL cholesterol depends mainly on genetic factors, but can be altered. Regular exercise increases HDL and reduces LDL cholesterol, as does limiting the amount of saturated fat in the diet. Much progress has been made in developing new drugs to lower cholesterol. The *statin* class, beginning with lovastatin in 1988 followed by simvastatin in 1991 have proven especially effective.

Simvastatin

The statins lower cholesterol by inhibiting the enzyme 3-hydroxy-3-methylglutaryl coenzyme A reductase, which is required for the biosynthesis of mevalonic acid (see Section 26.10). Mevalonic acid is an obligatory precursor to cholesterol, so less mevalonic acid translates into less cholesterol.

PROBLEM 26.14 The biosynthetic pathway shown in Figure 26.10 was developed with the aid of isotopic labeling experiments. Which carbon atoms of cholesterol would you expect to be labeled when acetate enriched with ^{14}C in its methyl group ($^{14}CH_3COOH$) is used as the carbon source?

Once formed, cholesterol undergoes a number of biochemical transformations. A very common one is acylation of its C-3 hydroxyl group by reaction with coenzyme A derivatives of fatty acids. Other processes convert cholesterol to the biologically important steroids described in the following sections.

26.12 VITAMIN D

A steroid very closely related structurally to cholesterol is its 7-dehydro derivative. 7-Dehydrocholesterol is formed by enzymatic oxidation of cholesterol and has a conjugated diene unit in its B ring. 7-Dehydrocholesterol is present in the tissues of the skin, where it is transformed to vitamin D_3 by a sunlight-induced photochemical reaction.

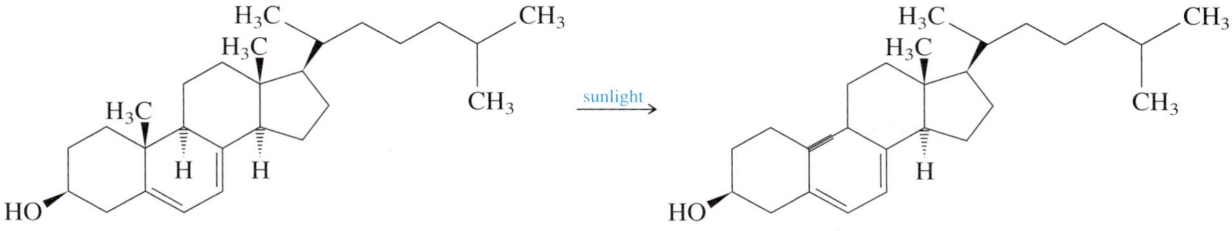

7-Dehydrocholesterol   Vitamin D₃

Vitamin D_3 is a key compound in the process by which Ca^{2+} is absorbed from the intestine. Low levels of vitamin D_3 lead to Ca^{2+} concentrations in the body that are insufficient to support proper bone growth, resulting in the bone disease called *rickets.*

Rickets was once more widespread than it is now. It was thought to be a dietary deficiency disease because it could be prevented in children by feeding them fish liver oil. Actually, rickets is an environmental disease brought about by a deficiency of sunlight. Where the winter sun is weak, children may not be exposed to enough of its light to convert the 7-dehydrocholesterol in their skin to vitamin D_3 at levels sufficient to promote the growth of strong bones. Fish have adapted to an environment that screens them from sunlight, and so they are not directly dependent on photochemistry for their vitamin D_3 and accumulate it by a different process. Although fish liver oil is a good source of vitamin D_3, it is not very palatable. Synthetic vitamin D_3, prepared from cholesterol, is often added to milk and other foods to ensure that children receive enough of the vitamin for their bones to develop properly. *Irradiated ergosterol* is another dietary supplement added to milk and other foods for the same purpose. Ergosterol, a steroid obtained from yeast, is structurally similar to 7-dehydrocholesterol and, on irradiation with sunlight or artificial light, is converted to vitamin D_2, a substance analogous to vitamin D_3 and comparable with its ability to support bone growth.

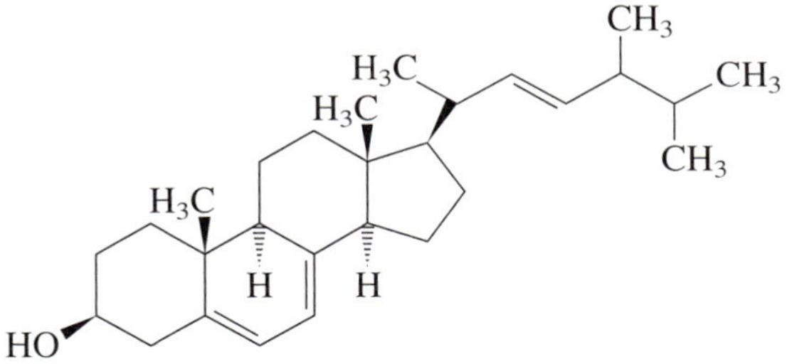

Ergosterol

PROBLEM 26.15   Suggest a reasonable structure for vitamin D_2.

26.13 BILE ACIDS

A significant fraction of the body's cholesterol is used to form **bile acids.** Oxidation in the liver removes a portion of the C_8H_{17} side chain, and additional hydroxyl groups are introduced at various positions on the steroid nucleus. *Cholic acid* is the most abundant of the bile acids. In the form of certain amide derivatives called **bile salts,** of which *sodium taurocholate* is one example, bile acids act as emulsifying agents to aid the digestion of fats.

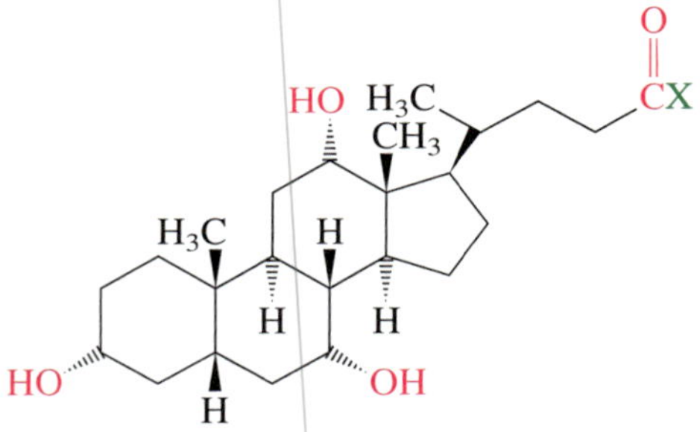

X = OH; cholic acid
X = NHCH$_2$CH$_2$SO$_3$Na;
sodium taurocholate

Cholic acid

The structure of cholic acid helps us understand how bile salts such as sodium taurocholate promote the transport of lipids through a water-rich environment. The bottom face of the molecule bears all of the polar groups, and the top face is exclusively hydrocarbon-like. Bile salts emulsify fats by forming micelles in which the fats are on the inside and the bile salts are on the outside. The hydrophobic face of the bile salt associates with the fat that is inside the micelle; the hydrophilic face is in contact with water on the outside.

26.14 CORTICOSTEROIDS

The outer layer, or *cortex,* of the adrenal gland is the source of a large group of substances known as **corticosteroids.** Like the bile acids, they are derived from cholesterol by oxidation, with cleavage of a portion of the alkyl substituent on the D ring. *Cortisol* is the most abundant of the corticosteroids, but *cortisone* is probably the best known. Cortisone is commonly prescribed as an antiinflammatory drug, especially in the treatment of rheumatoid arthritis.

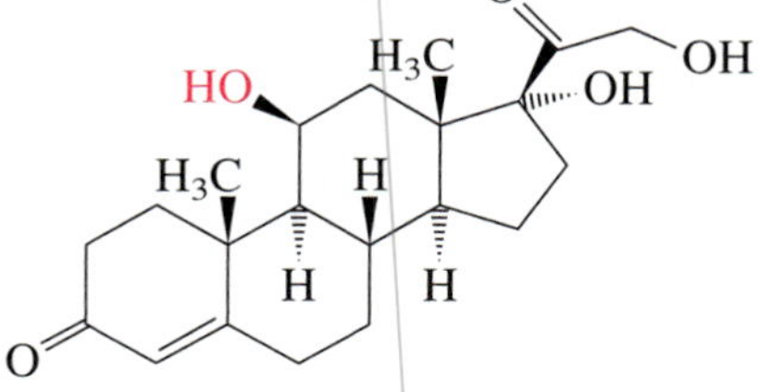

Cortisol

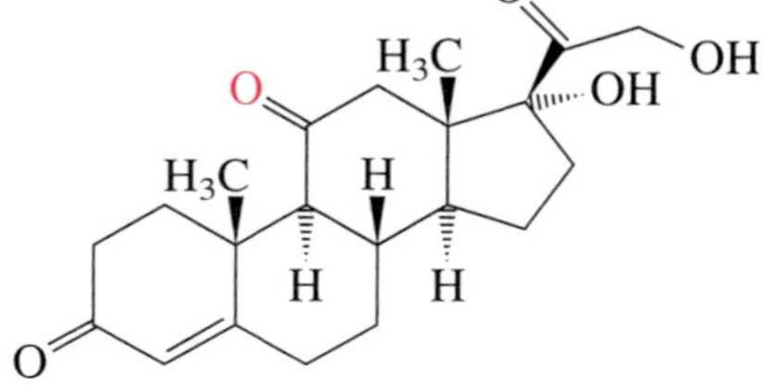

Cortisone

Corticosteroids exhibit a wide range of physiological effects. One important function is to assist in maintaining the proper electrolyte balance in body fluids. They also play a vital regulatory role in the metabolism of carbohydrates and in mediating the allergic response.

Many antiitch remedies contain dihydrocortisone.

26.15 SEX HORMONES

Hormones are the chemical messengers of the body; they are secreted by the endocrine glands and regulate biological processes. Corticosteroids, described in the preceding section, are hormones produced by the adrenal glands. The sex glands—testes in males, ovaries in females—secrete a number of hormones that are involved in sexual development and reproduction. *Testosterone* is the principal male sex hormone; it is an **androgen.**

ANABOLIC STEROIDS

As we have seen in this chapter, steroids have a number of functions in human physiology. Cholesterol is a component part of cell membranes and is found in large amounts in the brain. Derivatives of cholic acid assist the digestion of fats in the small intestine. Cortisone and its derivatives are involved in maintaining the electrolyte balance in body fluids. The sex hormones responsible for masculine and feminine characteristics as well as numerous aspects of pregnancy from conception to birth are steroids.

In addition to being an androgen, the principal male sex hormone testosterone promotes muscle growth and is classified as an anabolic steroid hormone. Biological chemists distinguish between two major classes of metabolism: catabolic and anabolic processes. **Catabolic** processes are degradative pathways in which larger molecules are broken down to smaller ones. **Anabolic** processes are the reverse; larger molecules are synthesized from smaller ones. Although the body mainly stores energy from food in the form of fat, a portion of that energy goes toward producing muscle from protein. An increase in the amount of testosterone, accompanied by an increase in the amount of food consumed, will cause an increase in the body's muscle mass.

Androstenedione, a close relative of testosterone, reached the public's attention in connection with Mark McGwire's successful bid to break Roger Maris' home run record in the summer of 1998. Androstenedione differs from testosterone in having a carbonyl group in the D ring where testosterone has a hydroxyl group. McGwire admitted to taking androstenedione, which is available as a nutritional supplement in health food stores and doesn't violate any of the rules of Major League Baseball. A controversy ensued as to the wisdom of androstenedione being sold without a prescription and the fairness of its use by athletes. Although the effectiveness of androstenedione as an anabolic steroid has not been established, it is clearly not nearly as potent as some others.

Androstenedione

The pharmaceutical industry has developed and studied a number of anabolic steroids for use in veterinary medicine and in rehabilitation from injuries that are accompanied by deterioration of muscles. The ideal agent would be one that possessed the anabolic properties of testosterone without its androgenic (masculinizing) effects. Methandrostenolone *(Dianabol)* and *stanozolol* are among the many synthetic anabolic steroids that require a prescription.

Dianabol

Stanozolol

Some scientific studies indicate that the gain in performance obtained through the use of anabolic steroids is small. This may be a case, though, in which the anecdotal evidence of the athletes may be closer to the mark than the scientific studies. The scientific studies are done under ethical conditions in which patients are treated with "prescription-level" doses of steroids. A 240-pound offensive tackle ("too small" by today's standards) may take several anabolic steroids at a time at 10–20 times their prescribed doses in order to weigh the 280 pounds he (or his coach) feels is necessary. The price athletes pay for gains in size and strength can be enormous. This price includes emotional costs (friendships lost because of heightened aggressiveness), sterility, testicular atrophy (the testes cease to function once the body starts to obtain a sufficient supply of testosterone-like steroids from outside), and increased risk of premature death from liver cancer or heart disease.

Testosterone promotes muscle growth, deepening of the voice, the growth of body hair, and other male secondary sex characteristics. Testosterone is formed from cholesterol and is the biosynthetic precursor of estradiol, the principal female sex hormone, or **estrogen.** *Estradiol* is a key substance in the regulation of the menstrual cycle and the reproductive process. It is the hormone most responsible for the development of female secondary sex characteristics.

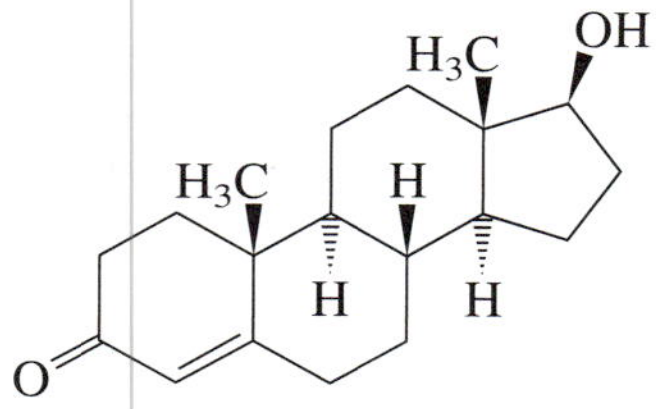

Testosterone 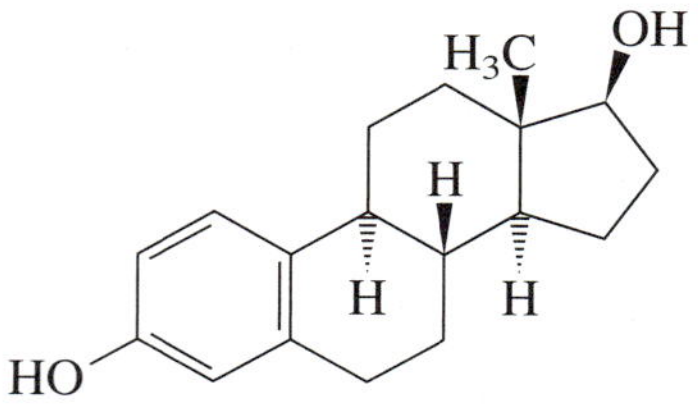   Estradiol

Testosterone and estradiol are present in the body in only minute amounts, and their isolation and identification required heroic efforts. In order to obtain 0.012 g of estradiol for study, for example, 4 tons of sow ovaries had to be extracted!

A separate biosynthetic pathway leads from cholesterol to *progesterone,* a female sex hormone. One function of progesterone is to suppress ovulation at certain stages of the menstrual cycle and during pregnancy. Synthetic substances, such as *norethindrone,* have been developed that are superior to progesterone when taken orally to "turn off" ovulation. By inducing temporary infertility, they form the basis of most oral contraceptive agents.

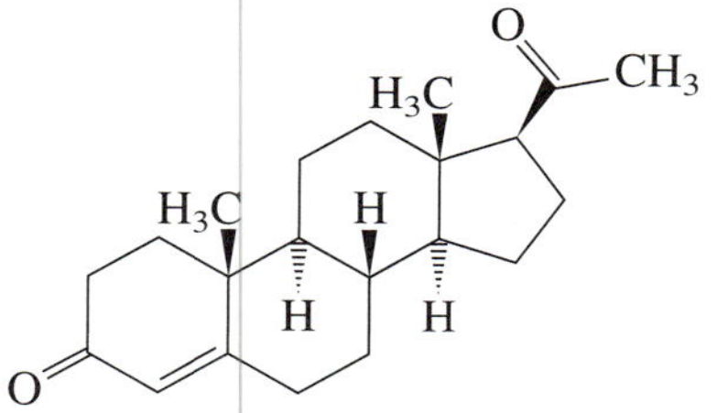

Progesterone 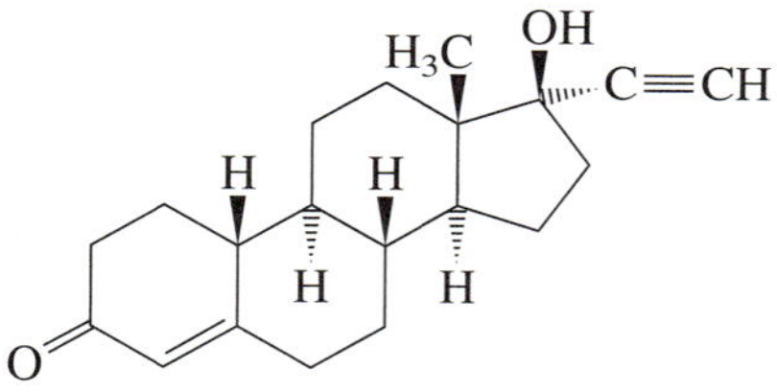   Norethindrone

26.16 CAROTENOIDS

Carotenoids are natural pigments characterized by a tail-to-tail linkage between two C_{20} units and an extended conjugated system of double bonds. They are the most widely distributed of the substances that give color to our world and occur in flowers, fruits, plants, insects, and animals. It has been estimated that biosynthesis from acetate produces approximately a hundred million tons of carotenoids per year. The most familiar carotenoids are lycopene and β-carotene, pigments found in numerous plants and easily isolable from ripe tomatoes and carrots, respectively.

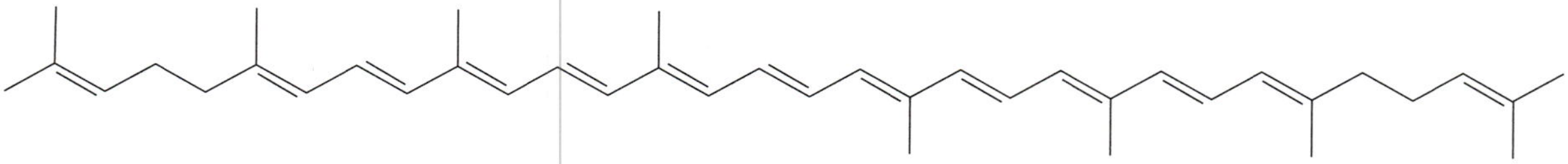

Lycopene (tomatoes)

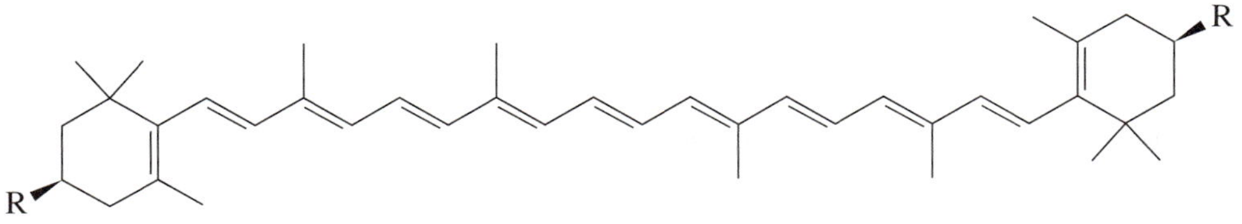

R = H; β-Carotene (carrots)
R = OH; Zeaxanthyn (yellow corn)

Not all carotenoids are hydrocarbons. Oxygen-containing carotenes called *xantho-phylls,* which are often the pigments responsible for the yellow color of flowers, are especially abundant.

Carotenoids absorb visible light (Section 13.21) and dissipate its energy as heat, thereby protecting the organism from any potentially harmful effects associated with sunlight-induced photochemistry. They are also indirectly involved in the chemistry of vision, owing to the fact that β-carotene is the biosynthetic precursor of vitamin A, also known as retinol, a key substance in the visual process.

The structural chemistry of the visual process, beginning with β-carotene, was described in the boxed essay entitled "Imines in Biological Chemistry" in Chapter 17.

26.17 SUMMARY

Section 26.1 Chemists and biochemists find it convenient to divide the principal organic substances present in cells into four main groups: *carbohydrates, proteins, nucleic acids,* and **lipids.** Structural differences separate carbohydrates from proteins, and both of these are structurally distinct from nucleic acids. Lipids, on the other hand, are characterized by a *physical property,* their solubility in nonpolar solvents, rather than by their structure. In this chapter we have examined lipid molecules that share a common biosynthetic origin in that all their carbons are derived from acetic acid (acetate). The form in which acetate occurs in many of these processes is a thioester called acetyl coenzyme A.

$$\overset{\overset{\displaystyle O}{\parallel}}{CH_3CSCoA}$$

Abbreviation for acetyl coenzyme A
(for complete structure, see Figure 26.1)

Section 26.2 Acetyl coenzyme A is the biosynthetic precursor to the **fatty acids,** which most often occur naturally as esters. **Fats** and **oils** are glycerol esters of long-chain carboxylic acids. Typically, these chains are unbranched and contain even numbers of carbon atoms.

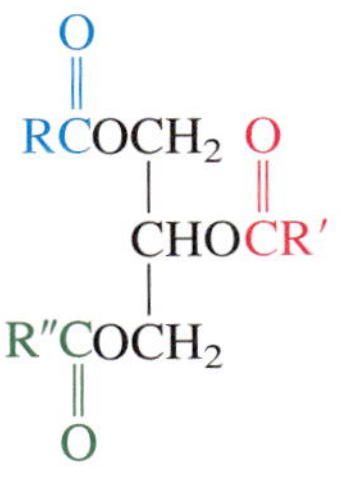

$$RCOCH_2$$
$$\mid$$
$$CHOCR'$$
$$\mid$$
$$R''COCH_2$$

Triacylglycerol
(R, R', and R'' may be the same or different)

Section 26.3 The biosynthesis of fatty acids follows the pathway outlined in Figure 26.3. Malonyl coenzyme A is a key intermediate.

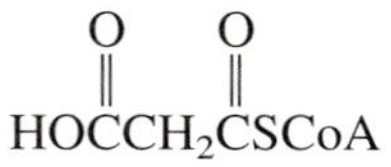

$$HOCCH_2CSCoA$$

Malonyl coenzyme A

Section 26.4 **Phospholipids** are intermediates in the biosynthesis of triacylglycerols from fatty acids and are the principal constituents of the lipid bilayer component of cell membranes.

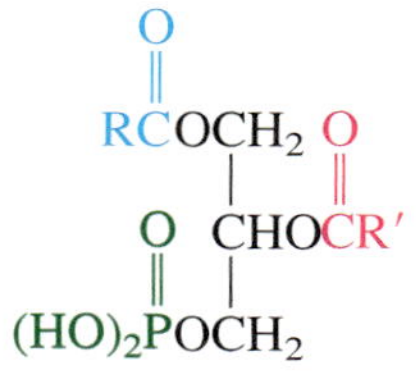

A phospholipid

Section 26.5 **Waxes** are mixtures of substances that usually contain esters of fatty acids and long-chain alcohols.

Section 26.6 **Icosanoids** are a group of naturally occurring compounds derived from unsaturated C_{20} carboxylic acids. Icosanoids include **prostaglandins, prostacyclins, thromboxanes,** and **leukotrienes.** Although present in very small amounts, icosanoids play regulatory roles in a very large number of biological processes.

Section 26.7 **Terpenes** are said to have structures that follow the isoprene rule in that they can be viewed as collections of isoprene units.

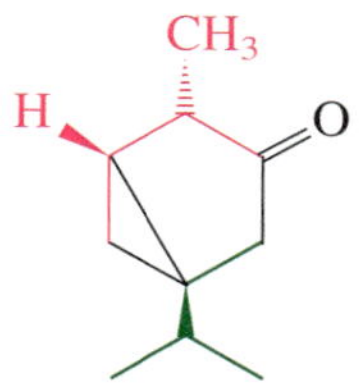

β-Thujone: a toxic monoterpene
present in absinthe

Section 26.8 Terpenes and related *isoprenoid* compounds are biosynthesized from *isopentenyl pyrophosphate.*

Isopentenyl pyrophosphate is
the "biological isoprene unit."

Section 26.9   Carbon–carbon bond formation between isoprene units can be understood on the basis of nucleophilic attack of the π electrons of a double bond on a carbocation or an allylic carbon that bears a pyrophosphate leaving group.

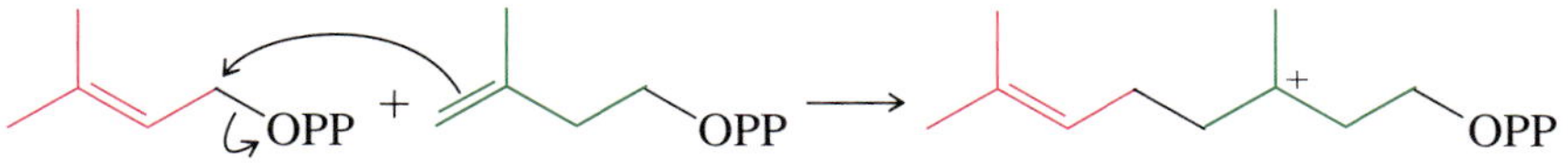

Section 26.10   The biosynthesis of isopentenyl pyrophosphate begins with acetate and proceeds by way of *mevalonic acid.*

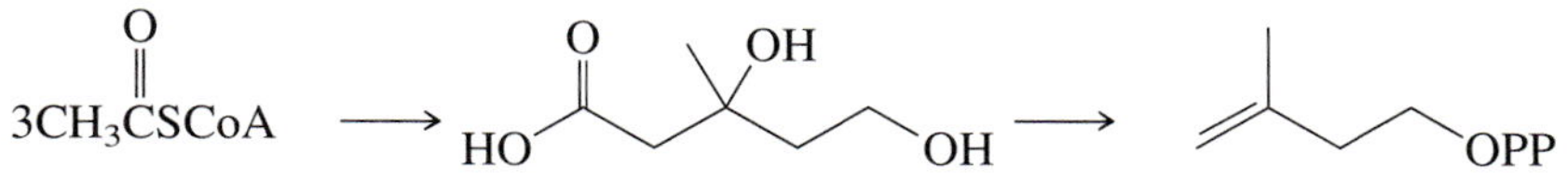

| Acetyl coenzyme A | Mevalonic acid | Isopentenyl pyrophosphate |

Section 26.11   The triterpene *squalene* is the biosynthetic precursor to cholesterol by the pathway shown in Figure 26.10.

Sections 26.12–26.15   Most of the steroids in animals are formed by biological transformations of cholesterol.

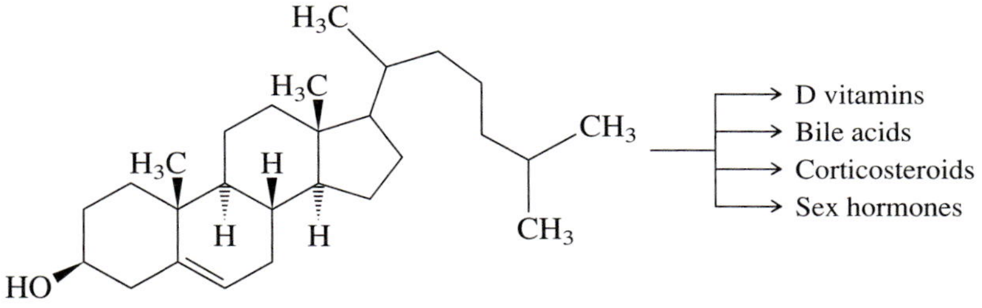

Cholesterol

Section 26.16   **Carotenoids** are tetraterpenes. They have 40 carbons and numerous double bonds. Many of the double bonds are conjugated, causing carotenes to absorb visible light and be brightly colored. They are often plant pigments.

PROBLEMS

26.16   The structures of each of the following are given within the chapter. Identify the carbon atoms expected to be labeled with ^{14}C when each is biosynthesized from acetate enriched with ^{14}C in its methyl group.

(a) Palmitic acid

(b) PGE_2

(c) PGI_2

(d) Limonene

(e) β-Carotene

26.17 Identify the isoprene units in each of the following naturally occurring substances:

(a) *Ascaridole*, a naturally occurring peroxide present in chenopodium oil:

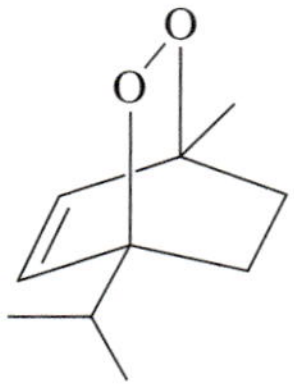

(b) *Dendrolasin*, a constituent of the defense secretion of a species of ant:

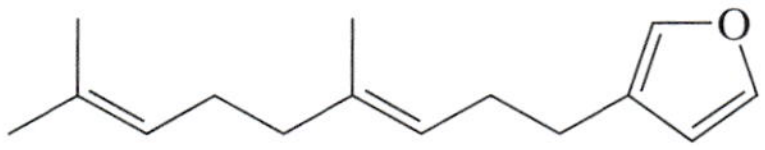

(c) *γ-Bisabolene*, a sesquiterpene found in the essential oils of a large number of plants:

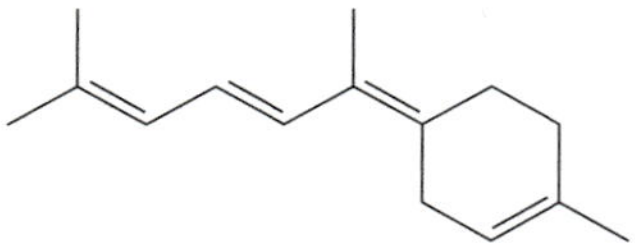

(d) *α-Santonin*, a lactone found in artemisia flowers:

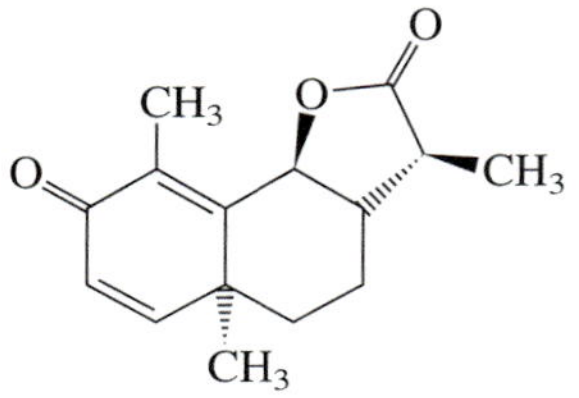

(e) *Tetrahymanol*, a pentacyclic triterpene isolated from a species of protozoans:

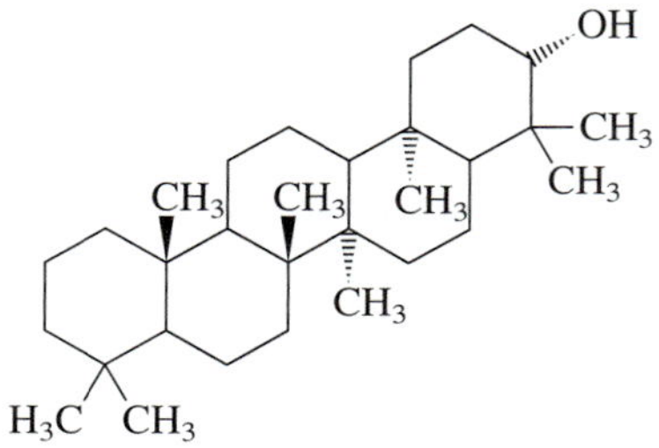

26.18 *Cubitene* is a diterpene present in the defense secretion of a species of African termite. What unusual feature characterizes the joining of isoprene units in cubitene?

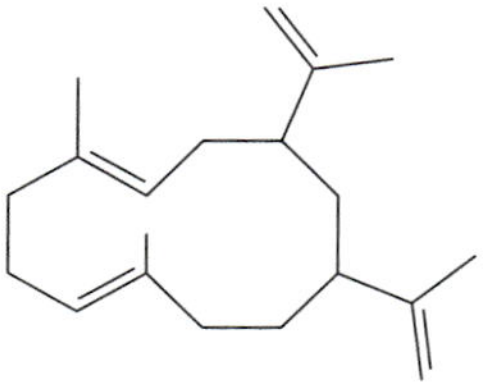

26.19 *Pyrethrins* are a group of naturally occurring insecticidal substances found in the flowers of various plants of the chrysanthemum family. The following is the structure of a typical pyrethrin, *cinerin I* (exclusive of stereochemistry):

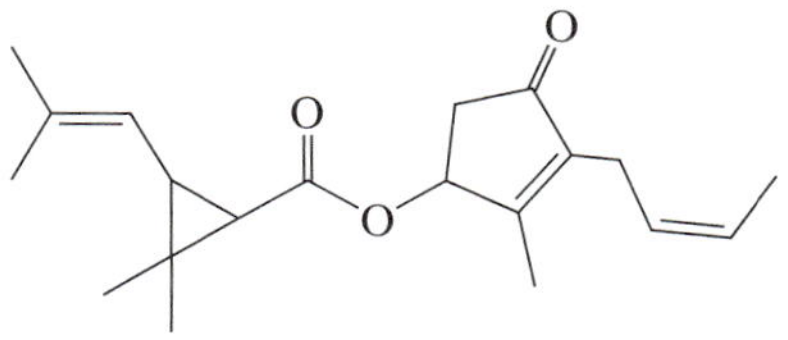

(a) Locate any isoprene units present in cinerin I.

(b) Hydrolysis of cinerin I gives an optically active carboxylic acid, (+)-chrysanthemic acid. Ozonolysis of (+)-chrysanthemic acid, followed by oxidation, gives acetone and an optically active dicarboxylic acid, (−)-caronic acid ($C_7H_{10}O_4$). What is the structure of (−)-caronic acid? Are the two carboxyl groups cis or trans to each other? What does this information tell you about the structure of (+)-chrysanthemic acid?

26.20 *Cerebrosides* are found in the brain and in the myelin sheath of nerve tissue. The structure of the cerebroside *phrenosine* is

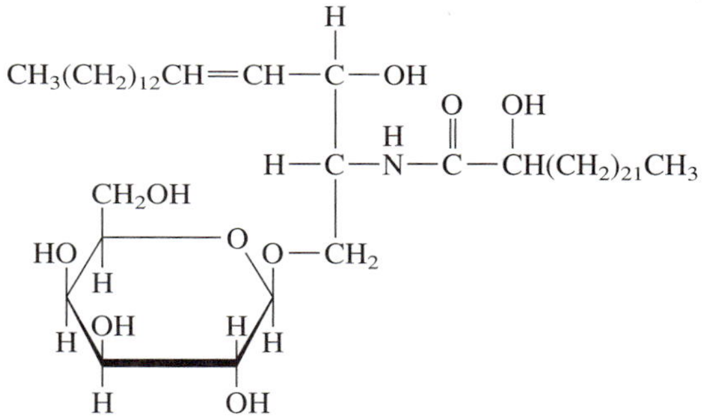

(a) What hexose is formed on hydrolysis of the glycoside bond of phrenosine? Is phrenosine an α- or a β-glycoside?

(b) Hydrolysis of phrenosine gives, in addition to the hexose in part (*a*), a fatty acid called *cerebronic acid,* along with a third substance called *sphingosine*. Write structural formulas for both cerebronic acid and sphingosine.

26.21 Each of the following reactions has been reported in the chemical literature and proceeds in good yield. What are the principal organic products of each reaction? In some of the exercises more than one diastereomer may be theoretically possible, but in such instances one diastereomer is either the major product or the only product. For those reactions in which one diastereomer is formed preferentially, indicate its expected stereochemistry.

(a) $CH_3(CH_2)_7C\equiv C(CH_2)_7COOH + H_2 \xrightarrow{\text{Lindlar Pd}}$

(b) $CH_3(CH_2)_7C\equiv C(CH_2)_7COOH \xrightarrow[\text{2. H}^+]{\text{1. Li, NH}_3}$

(c) $(Z)\text{-}CH_3(CH_2)_7CH=CH(CH_2)_7\overset{\displaystyle O}{\overset{\|}{C}}OCH_2CH_3 + H_2 \xrightarrow{\text{Pt}}$

(d) $(Z)\text{-}CH_3(CH_2)_5\underset{\underset{OH}{|}}{C}HCH_2CH=CH(CH_2)_7\overset{\displaystyle O}{\overset{\|}{C}}OCH_3 \xrightarrow[\text{2. H}_2\text{O}]{\text{1. LiAlH}_4}$

(e) (Z)-$CH_3(CH_2)_7CH{=}CH(CH_2)_7COOH + C_6H_5\overset{\overset{\displaystyle O}{\displaystyle \|}}{C}OOH \longrightarrow$

(f) Product of part (e) + $H_3O^+ \longrightarrow$

(g) (Z)-$CH_3(CH_2)_7CH{=}CH(CH_2)_7COOH \xrightarrow[\text{2. } H^+]{\text{1. } OsO_4, (CH_3)_3COOH, HO^-}$

(h) 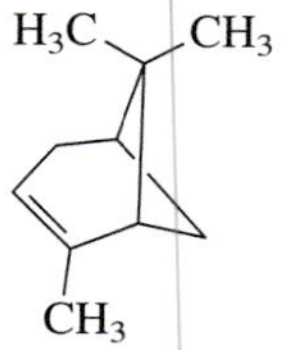 $\xrightarrow[\text{2. } H_2O_2, HO^-]{\text{1. } B_2H_6, \text{diglyme}}$

(i) 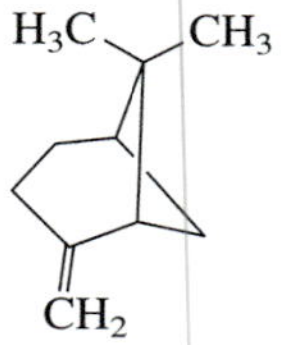 $\xrightarrow[\text{2. } H_2O_2, HO^-]{\text{1. } B_2H_6, \text{diglyme}}$

(j) 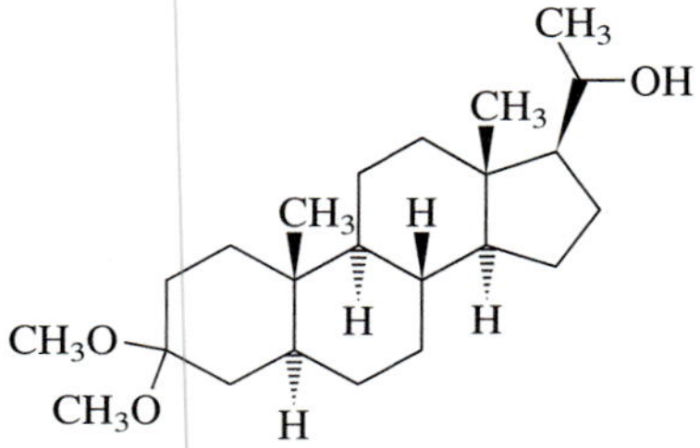 $\xrightarrow{HCl, H_2O} C_{21}H_{34}O_2$

26.22 Describe an efficient synthesis of each of the following compounds from octadecanoic (stearic) acid using any necessary organic or inorganic reagents:

(a) Octadecane

(b) 1-Phenyloctadecane

(c) 3-Ethylicosane

(d) Icosanoic acid

(e) 1-Octadecanamine

(f) 1-Nonadecanamine

26.23 A synthesis of triacylglycerols has been described that begins with the substance shown.

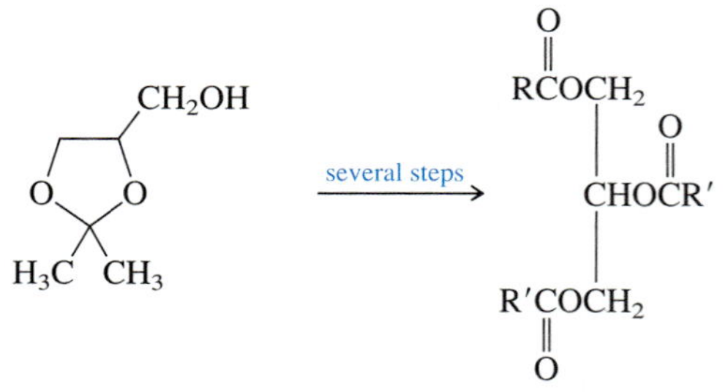

4-(Hydroxymethyl)-
2,2-dimethyl-1,3-dioxolane

Triacylglycerol

Outline a series of reactions suitable for the preparation of a triacylglycerol of the type illustrated in the equation, where R and R' are different.

26.24 The isoprenoid compound shown is a scent marker present in the urine of the red fox. Suggest a reasonable synthesis for this substance from 3-methyl-3-buten-1-ol and any necessary organic or inorganic reagents.

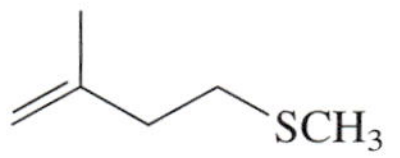

26.25 *Sabinene* is a monoterpene found in the oil of citrus fruits and plants. It has been synthesized from 6-methyl-2,5-heptanedione by the sequence that follows. Suggest reagents suitable for carrying out each of the indicated transformations.

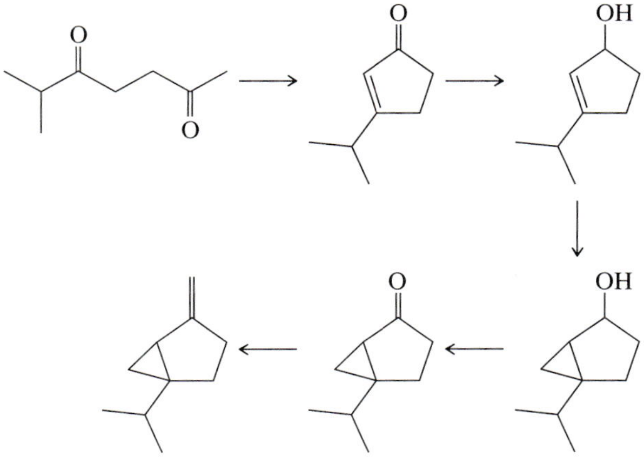

Sabinene

26.26 Isoprene has sometimes been used as a starting material in the laboratory synthesis of terpenes. In one such synthesis, the first step is the electrophilic addition of 2 moles of hydrogen bromide to isoprene to give 1,3-dibromo-3-methylbutane.

$$H_2C=CCH=CH_2 \ + \ 2HBr \longrightarrow (CH_3)_2CCH_2CH_2Br$$

2-Methyl-1,3-butadiene (isoprene)   Hydrogen bromide   1,3-Dibromo-3-methylbutane

Write a series of equations describing the mechanism of this reaction.

26.27 The ionones are fragrant substances present in the scent of iris and are used in perfume. A mixture of α- and β-ionone can be prepared by treatment of pseudoionone with sulfuric acid.

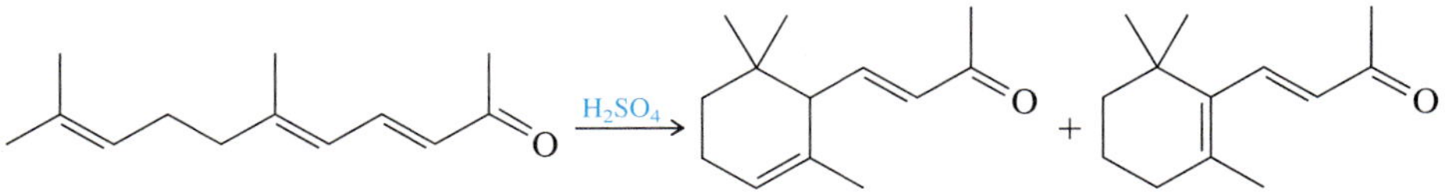

Pseudoionone   α-Ionone   β-Ionone

Write a stepwise mechanism for this reaction.

26.28 β,γ-Unsaturated steroidal ketones represented by the partial structure shown here are readily converted in acid to their α,β-unsaturated isomers. Write a stepwise mechanism for this reaction.

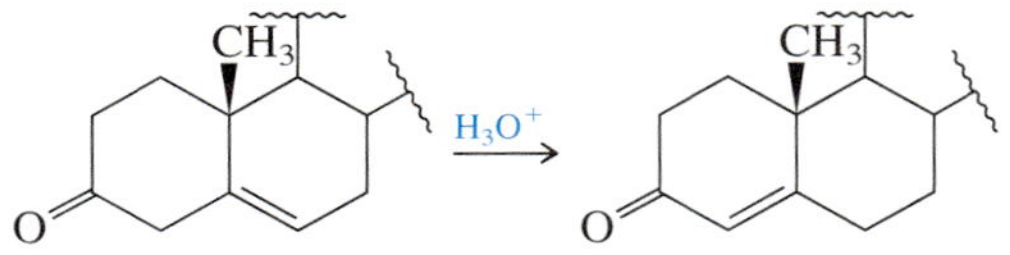

$$\xrightarrow{\text{H}_3\text{O}^+}$$

26.29 (a) Suggest a mechanism for the following reaction.

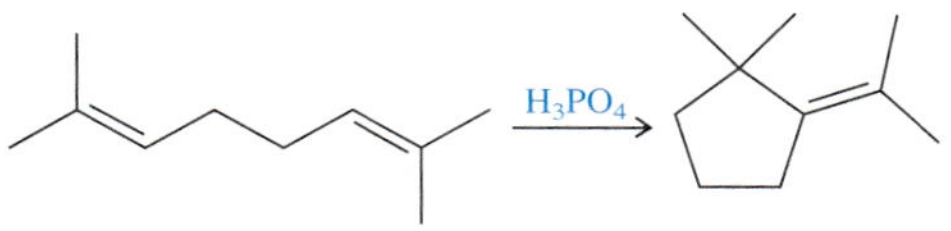

$$\xrightarrow{\text{H}_3\text{PO}_4}$$

(b) The following two compounds are also formed in the reaction given in part (a). How are these two products formed?

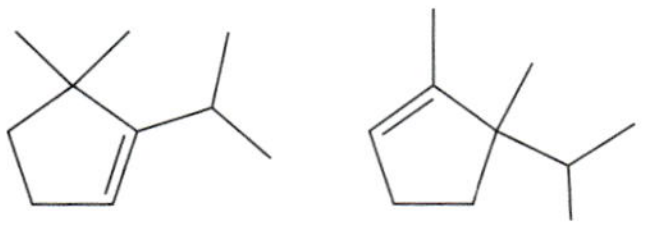

(*Note:* The solution to this problem is not given in the *Solutions Manual.* It is discussed in detail, however, in a very interesting article on pages 541–542 of the June 1995 issue of the *Journal of Chemical Education.*)

26.30 The compound shown is *diethylstilbestrol* (DES); it has a number of therapeutic uses in estrogen-replacement therapy. DES is not a steroid, but can adopt a shape that allows it to mimic estrogens such as estradiol (p. 1100) and bind to the same receptor sites. Construct molecular models of DES and estradiol that illustrate this similarity in molecular size, shape, and location of polar groups.

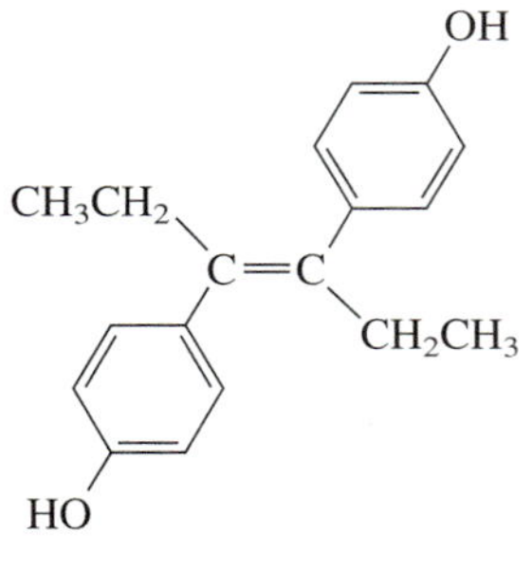

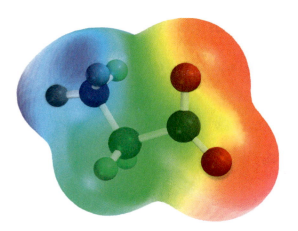

CHAPTER 27

AMINO ACIDS, PEPTIDES, AND PROTEINS

The relationship between structure and function reaches its ultimate expression in the chemistry of amino acids, peptides, and proteins.

Amino acids are carboxylic acids that contain an amine function. An amide bond between the carboxylic acid function of one amino acid and the amino nitrogen of another is called a **peptide bond.**

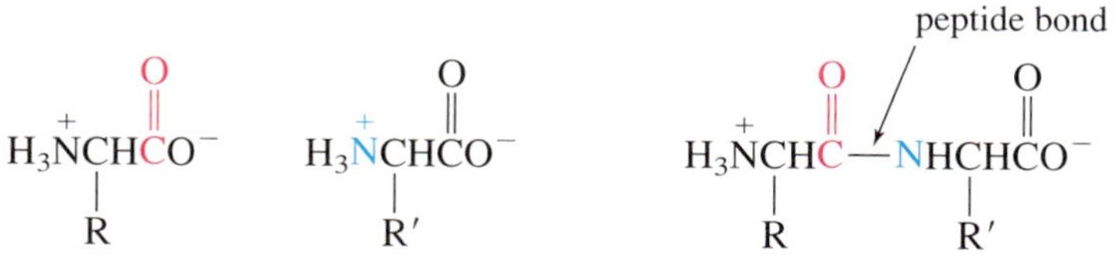

Two α-amino acids

Dipeptide

peptide bond

A **dipeptide** is a molecule consisting of two amino acids joined by a peptide bond. A **tripeptide** has three amino acids joined by two peptide bonds, a **tetrapeptide** has four amino acids, and so on. Peptides with more than 30–50 amino acids are **polypeptides. Proteins** are polypeptides that have some biological function.

The most striking thing about proteins is the diversity of their roles in living systems: silk is a protein, skin and hair are mostly proteins, many hormones are proteins, a protein carries oxygen from the lungs to the tissues where it is stored by another protein, and all enzymes are proteins.

As in most aspects of chemistry and biochemistry, structure is the key to function. We'll explore the structure of proteins by first concentrating on their fundamental building block units, the α-amino acids. Then, after developing the principles of peptide structure, we'll see how the insights gained from these smaller molecules aid our understanding of proteins.

27.1   CLASSIFICATION OF AMINO ACIDS

Amino acids are classified as α, β, γ, and so on, according to the location of the amine group on the carbon chain that contains the carboxylic acid function.

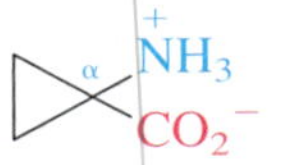

1-Aminocyclopropanecarboxylic acid: an α-amino acid that is the biological precursor to ethylene in plants

$$\overset{+}{H_3}NCH_2CH_2CO_2^{-}$$

3-Aminopropanoic acid: known as β-alanine, it is a β-amino acid that makes up one of the structural units of coenzyme A

$$\overset{+}{H_3}NCH_2CH_2CH_2CO_2^{-}$$

4-Aminobutanoic acid: known as γ-aminobutyric acid (GABA), it is a γ-amino acid and is involved in the transmission of nerve impulses

Although more than 700 different amino acids are known to occur naturally, a group of 20 of them commands special attention. These 20 are the amino acids that are normally present in proteins and are listed in Table 27.1. All the amino acids from which proteins are derived are α-amino acids, and all but one of these contain a primary amino function and conform to the general structure

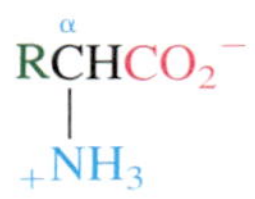

$$\overset{\alpha}{R}CHCO_2^{-}$$
$$\underset{+NH_3}{|}$$

The one exception is proline, a secondary amine in which the amino nitrogen is incorporated into a five-membered ring.

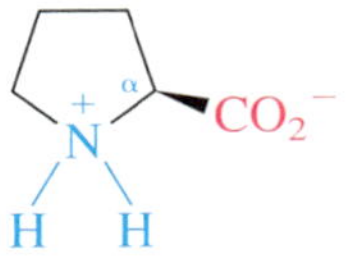

Proline

Table 27.1 includes three-letter and one-letter abbreviations for the amino acids. Both enjoy wide use.

Our bodies can make some of the amino acids shown in the table. The others, which are called **essential amino acids,** we have to get from what we eat.

The most important aspect of Table 27.1 is that the 20 amino acids that occur in proteins share the common feature of being α-amino acids, and the differences among them are in their side chains. Peptide bonds linking carboxyl and α-amino groups characterize the *structure* of proteins, but it is the side chains that are mainly responsible for their *properties.* The side chains of the 20 commonly occurring amino acids encompass both large and small differences. The major differences between amino acid side chains concern:

1. Size and shape
2. Electronic characteristics and their effects on the ability of side chains to engage in ionic bonding, covalent bonding, hydrogen bonding, van der Waals forces, and acid–base chemistry

TABLE 27.1 α-Amino Acids Found in Proteins

Learning By Modeling contains electrostatic potential maps of all the amino acids in this table.

Name	Abbreviation	Structural formula*
Amino acids with nonpolar side chains		
Glycine	Gly (G)	$\overset{\overset{+}{N}H_3}{H-CHCO_2^-}$
Alanine	Ala (A)	$H_3C-\overset{\overset{+}{N}H_3}{CHCO_2^-}$
Valine[†]	Val (V)	$(CH_3)_2CH-\overset{\overset{+}{N}H_3}{CHCO_2^-}$
Leucine[†]	Leu (L)	$(CH_3)_2CHCH_2-\overset{\overset{+}{N}H_3}{CHCO_2^-}$
Isoleucine[†]	Ile (I)	$CH_3CH_2\overset{CH_3}{CH}-\overset{\overset{+}{N}H_3}{CHCO_2^-}$
Methionine[†]	Met (M)	$CH_3SCH_2CH_2-\overset{\overset{+}{N}H_3}{CHCO_2^-}$
Proline	Pro (P)	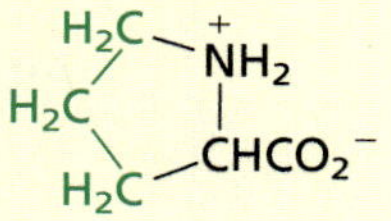 (cyclic structure)
Phenylalanine[†]	Phe (F)	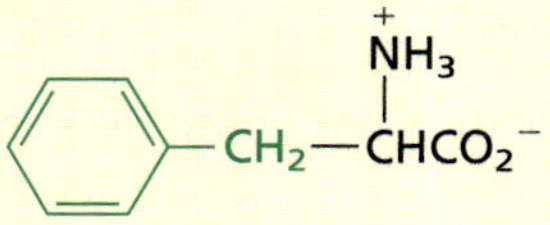 $C_6H_5-CH_2-\overset{\overset{+}{N}H_3}{CHCO_2^-}$
Tryptophan[†]	Trp (W)	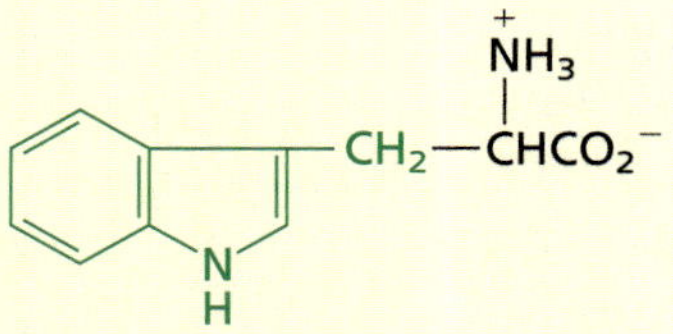 (indole)$-CH_2-\overset{\overset{+}{N}H_3}{CHCO_2^-}$
Amino acids with polar but nonionized side chains		
Asparagine	Asn (N)	$H_2N\overset{O}{\overset{\|}{C}}CH_2-\overset{\overset{+}{N}H_3}{CHCO_2^-}$

*All amino acids are shown in the form present in greatest concentration at pH 7.
[†]An essential amino acid, which must be present in the diet of animals to ensure normal growth.

(Continued)

TABLE 27.1   α-Amino Acids Found in Proteins *(Continued)*

Name	Abbreviation	Structural formula*
Amino acids with polar but nonionized side chains		
Glutamine	Gln (Q)	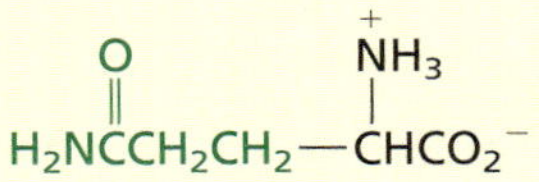 $\overset{\overset{\displaystyle O}{\|\|}}{H_2NCCH_2CH_2}\!-\!\overset{\overset{\displaystyle \overset{+}{N}H_3}{\|}}{CHCO_2^{-}}$
Serine	Ser (S)	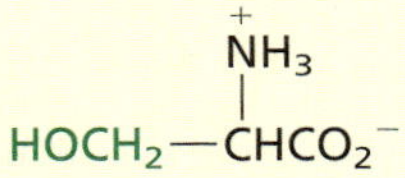 $HOCH_2\!-\!\overset{\overset{\displaystyle \overset{+}{N}H_3}{\|}}{CHCO_2^{-}}$
Threonine[†]	Thr (T)	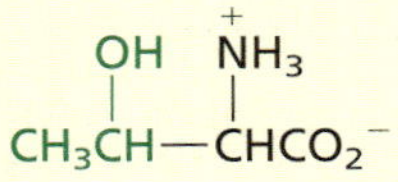 $CH_3\overset{\overset{\displaystyle OH}{\|}}{CH}\!-\!\overset{\overset{\displaystyle \overset{+}{N}H_3}{\|}}{CHCO_2^{-}}$
Tyrosine	Tyr (Y)	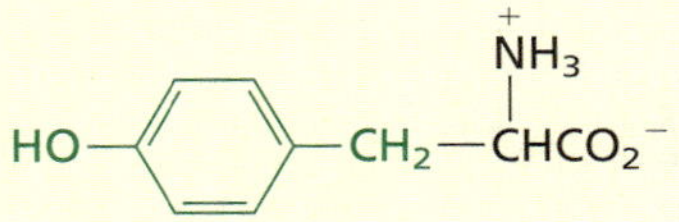 $HO\!-\!\!\bigcirc\!\!-\!CH_2\!-\!\overset{\overset{\displaystyle \overset{+}{N}H_3}{\|}}{CHCO_2^{-}}$
Cysteine	Cys (C)	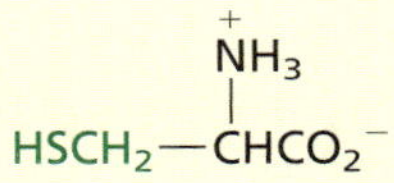 $HSCH_2\!-\!\overset{\overset{\displaystyle \overset{+}{N}H_3}{\|}}{CHCO_2^{-}}$
Amino acids with acidic side chains		
Aspartic acid	Asp (D)	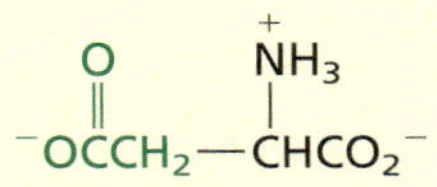 $\overset{\overset{\displaystyle O}{\|\|}}{{}^{-}OCCH_2}\!-\!\overset{\overset{\displaystyle \overset{+}{N}H_3}{\|}}{CHCO_2^{-}}$
Glutamic acid	Glu (E)	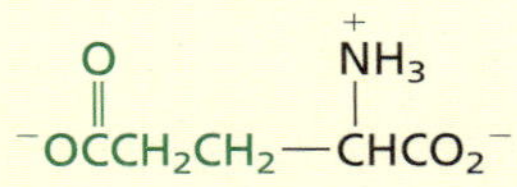 $\overset{\overset{\displaystyle O}{\|\|}}{{}^{-}OCCH_2CH_2}\!-\!\overset{\overset{\displaystyle \overset{+}{N}H_3}{\|}}{CHCO_2^{-}}$
Amino acids with basic side chains		
Lysine[†]	Lys (K)	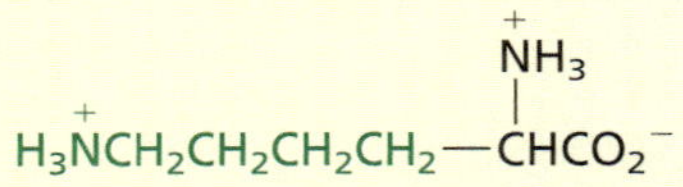 $\overset{+}{H_3}NCH_2CH_2CH_2CH_2\!-\!\overset{\overset{\displaystyle \overset{+}{N}H_3}{\|}}{CHCO_2^{-}}$
Arginine[†]	Arg (R)	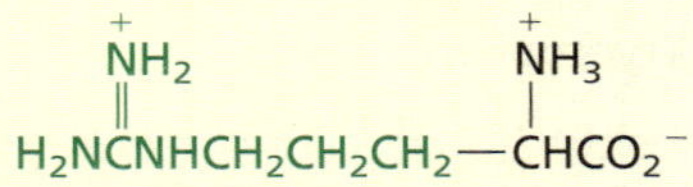 $H_2N\overset{\overset{\displaystyle \overset{+}{N}H_2}{\|\|}}{C}NHCH_2CH_2CH_2\!-\!\overset{\overset{\displaystyle \overset{+}{N}H_3}{\|}}{CHCO_2^{-}}$
Histidine[†]	His (H)	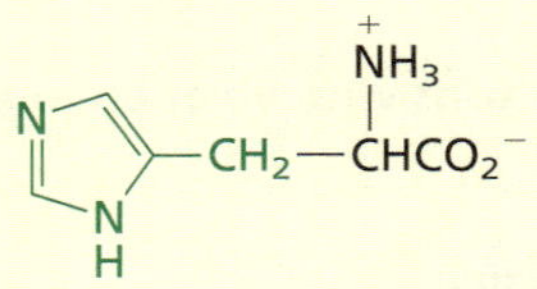 $\underset{\underset{H}{N}}{N}\!\!\diagup\!\!-\!CH_2\!-\!\overset{\overset{\displaystyle \overset{+}{N}H_3}{\|}}{CHCO_2^{-}}$

In addition to Table 27.1, we'll use the electrostatic potential maps in Figure 27.1 to survey these differences. Figure 27.1 shows the amino acids in the form in which they exist at a pH of 7; amine groups as positively charged ammonium ions, and carboxylic acid groups as negatively charged carboxylates. Because the regions of greatest positive and negative charge are associated with these two functional groups and are virtually the same in all the amino acids, the color range for the side chains is similar throughout the group and allows direct comparison. The electrostatic potential is mapped on the van der Waals surface of the molecule and so displays the charge distribution, size, and shape at the same time.

Nonpolar side chains: *Glycine* is the smallest amino acid because it has no side chain. The main service it offers is to the polypeptide chain itself. It can add length and flexibility to a polypeptide without sacrificing strength or making spatial demands of its own.

After glycine, the next four amino acids in the figure all have alkyl groups (R) as side chains: *alanine* (R = methyl), *valine* (R = isopropyl), *leucine* (R = isobutyl), and *isoleucine* (R = *sec*-butyl). All are hydrophobic side chains and although electronically similar, they differ in size. Alanine is slightly larger than glycine, valine slightly larger than alanine, leucine slightly larger than valine, and isoleucine somewhat more spherical than leucine.

Compared with the alkyl groups of the four just mentioned, the presence of sulfur makes the side chain of *methionine* (R = —$CH_2CH_2SCH_3$) somewhat more polarizable. Greater polarizability leads to stronger dispersion forces.

Proline is relatively compact because of the cyclic nature of its side chain. It has less conformational flexibility than the other amino acids, and the presence of proline affects the shape of a peptide more than other amino acids.

Phenylalanine and *tryptophan* have side chains that incorporate aromatic rings, which are large and hydrophobic. The aromatic portion of tryptophan is bicyclic, which makes it larger than phenylalanine. Tryptophan also has a more electron-rich aromatic ring and is more polarizable than phenylalanine. Its role is more specialized, and it is less abundant in proteins than most of the other amino acids.

Amino acids with polar but nonionized side chains: Among amino acids with polar side chains, *serine* is the smallest; it is not much larger than alanine. With a —CH_2OH side chain, serine participates well in hydrogen bonding and often occurs in regions of a peptide that are exposed to water.

Threonine has a methyl group in place of one of the hydrogens of the —CH_2OH group of serine, sterically hindering the OH group and making it less effective in hydrogen bonding.

Cysteine is related to serine in that its side chain is —CH_2SH rather than —CH_2OH. The number of cysteines in a protein is often relatively small, but their effect on its three-dimensional shape is substantial. Oxidation of two cysteines converts their —CH_2SH side chains to a —CH_2S—SCH_2— bridge between them (Section 15.13). This ties together two often remote amino acids and helps guide the folding of the protein into a specific shape.

As a *p*-hydroxy derivative of phenylalanine, *tyrosine* has properties similar to those of phenylalanine plus the ability to engage in hydrogen bonding via its —OH group.

Asparagine and *glutamine* are not amines; they are amides. The side chains of both

$$\overset{\displaystyle O}{\underset{\displaystyle \|}{}}$$

terminate in —$\overset{\text{O}}{\overset{\|}{C}}$$NH_2$ and differ by only a single CH_2 group. Amide functions are quite polar and interact strongly with water molecules by hydrogen bonding. Like serine, asparagine and glutamine are often found in regions of a peptide that are in contact with water.

Amino acids with nonpolar side chains

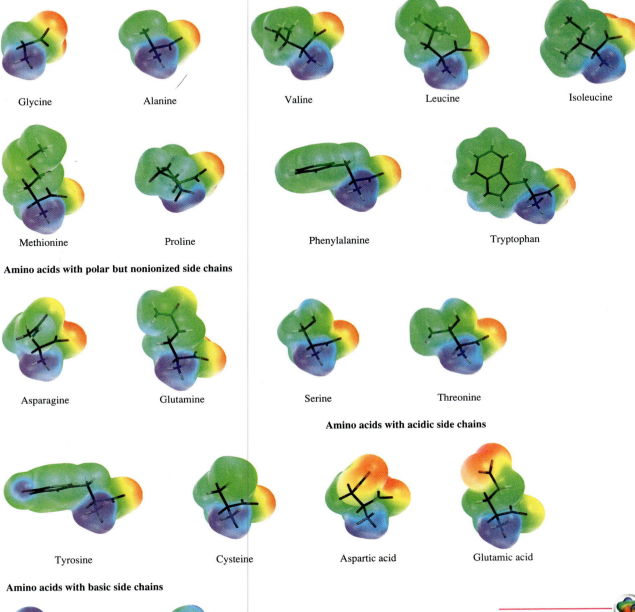

Glycine   Alanine   Valine   Leucine   Isoleucine

Methionine   Proline   Phenylalanine   Tryptophan

Amino acids with polar but nonionized side chains

Asparagine   Glutamine   Serine   Threonine

Amino acids with acidic side chains

Tyrosine   Cysteine   Aspartic acid   Glutamic acid

Amino acids with basic side chains

Lysine   Arginine   Histidine

FIGURE 27.1 Electrostatic potential maps of the 20 common amino acids listed in Table 27.1. Each amino acid is oriented so that its side chain is in the upper left corner. The side chains affect the shape and properties of the amino acids.

Amino acids with acidic side chains: The electrostatic potential maps of *aspartic acid* and *glutamic acid* are the most prominent ones in Figure 27.1. The —CO_2H side chains of both aspartic and glutamic acid are almost completely deprotonated to —CO_2^- at biological pH, giving these species the most electron-rich units of all of the common amino acids. Their most important function is in ionic bonding to positively charged species. These ionic bonds can be to metal ions and —$\overset{+}{\underset{|}{\overset{|}{N}}}$— species among others.

Amino acids with basic side chains: Basic amino acids are the opposite of acidic amino acids. Their most important role is to form ionic bonds to negative ions—phosphate and the like. Lysine is a simple example. The side chain contains four CH_2 groups and terminates in —NH_3^+. Arginine has an even more basic, if somewhat more complicated and larger, side chain. Conversely, the side chain of histidine is not as basic as that of lysine and the concentrations of the unprotonated and protonated forms of histidine are almost equal at biological pH. Strong Lewis acid-Lewis base complexes between the unprotonated form of histidine and metal ions is very common in proteins. Histidine side chains are also involved in moving protons from one atom to another.

What is remarkable is not only the range of properties that are covered with just 20 amino acids, but also the fine tuning with respect to a particular property.

27.2 STEREOCHEMISTRY OF AMINO ACIDS

Glycine is the simplest amino acid and the only one in Table 27.1 that is achiral. The α-carbon atom is a chirality center in all the others. Configurations in amino acids are normally specified by the D, L notational system. All the chiral amino acids obtained from proteins have the L configuration at their α-carbon atom, meaning that the amine group is at the left when a Fischer projection is arranged so the carboxyl group is at the top.

Glycine
(achiral)

Fischer projection
of an L-amino acid

PROBLEM 27.1 What is the absolute configuration (*R* or *S*) at the α-carbon atom in each of the following L-amino acids?

(a) $H_3\overset{+}{N}\text{—}\overset{\displaystyle CO_2^-}{\underset{\displaystyle CH_2OH}{\rule{0pt}{1em}}}\text{—}H$

L-Serine

(b) $H_3\overset{+}{N}\text{—}\overset{\displaystyle CO_2^-}{\underset{\displaystyle CH_2SH}{\rule{0pt}{1em}}}\text{—}H$

L-Cysteine

(c) $H_3\overset{+}{N}\text{—}\overset{\displaystyle CO_2^-}{\underset{\displaystyle CH_2CH_2SCH_3}{\rule{0pt}{1em}}}\text{—}H$

L-Methionine

SAMPLE SOLUTION (a) First identify the four groups attached directly to the chirality center, and rank them in order of decreasing sequence rule precedence. For L-serine these groups are

$$H_3\overset{+}{N}\text{—} \quad > \quad \text{—}CO_2^- \quad > \quad \text{—}CH_2OH \quad > \quad H$$

Highest ranked   Lowest ranked

Next, translate the Fischer projection of L-serine to a three-dimensional representation, and orient it so that the lowest ranked substituent at the chirality center is directed away from you.

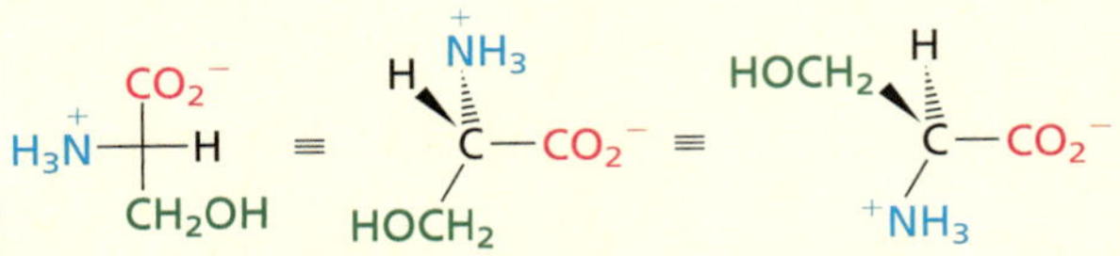

In order of decreasing precedence the three highest ranked groups trace a counterclockwise path.

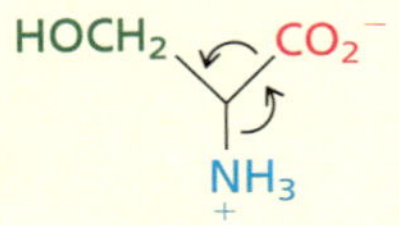

The absolute configuration of L-serine is S.

PROBLEM 27.2 Which of the amino acids in Table 27.1 have more than one chirality center?

Although all the chiral amino acids obtained from proteins have the L configuration at their α carbon, that should not be taken to mean that D-amino acids are unknown. In fact, quite a number of D-amino acids occur naturally. D-Alanine, for example, is a constituent of bacterial cell walls and D-serine occurs in brain tissue. The point is that D-amino acids are not constituents of proteins.

A novel technique for dating archaeological samples called *amino acid racemization* (AAR) is based on the stereochemistry of amino acids. Over time, the configuration at the α-carbon atom of a protein's amino acids is lost in a reaction that follows first-order kinetics. When the α carbon is the only chirality center, this process corresponds to racemization. For an amino acid with two chirality centers, changing the configuration of the α carbon from L to D gives a diastereomer. In the case of isoleucine, for example, the diastereomer is an amino acid not normally present in proteins, called *alloisoleucine*.

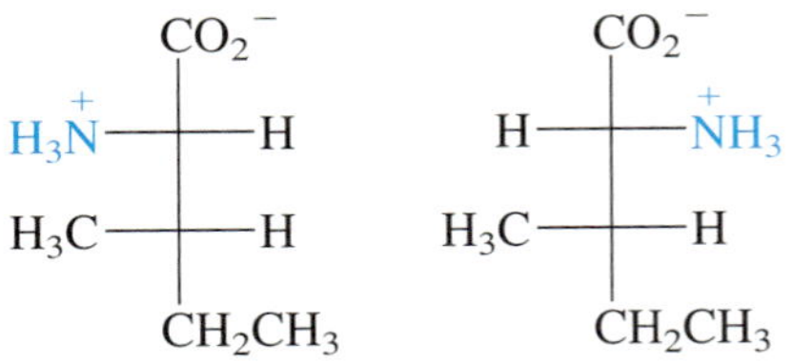

L-Isoleucine   D-Alloisoleucine

By measuring the L-isoleucine/D-alloisoleucine ratio in the protein isolated from the eggshells of an extinct Australian bird, a team of scientists recently determined that this bird lived approximately 50,000 years ago. Radiocarbon (^{14}C) dating is not accurate for samples older than about 35,000 years, so AAR is a useful addition to the tools available to paleontologists.

27.3 ACID–BASE BEHAVIOR OF AMINO ACIDS

The physical properties of a typical amino acid such as glycine suggest that it is a very polar substance, much more polar than would be expected on the basis of its formulation as $H_2NCH_2CO_2H$. Glycine is a crystalline solid; it does not melt, but on being heated it eventually decomposes at 233°C. It is very soluble in water but practically insoluble in nonpolar organic solvents. These properties are attributed to the fact that the stable form of glycine is a **zwitterion,** or **inner salt.**

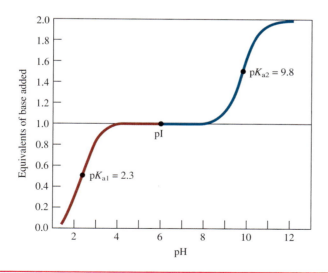

Zwitterionic form of glycine

> The zwitterion is also often referred to as a *dipolar ion.* Note, however, that it is not an ion, but a neutral molecule.

The equilibrium expressed by the preceding equation lies overwhelmingly to the side of the zwitterion.

Glycine, as well as other amino acids, is *amphoteric,* meaning it contains an acidic functional group and a basic functional group. The acidic functional group is the ammonium ion $\overset{+}{H_3N}$—; the basic functional group is the carboxylate ion —CO_2^-. How do we know this? Aside from its physical properties, the acid–base properties of glycine, as illustrated by the titration curve in Figure 27.2, require it. In a strongly acidic medium the species present is $\overset{+}{H_3N}CH_2CO_2H$. As the pH is raised, a proton is removed from this species. Is the proton removed from the positively charged nitrogen or from the carboxyl group? We know what to expect for the relative acid strengths of $R\overset{+}{N}H_3$ and RCO_2H. A typical ammonium ion has $pK_a \approx 9$, and a typical carboxylic acid has $pK_a \approx 5$. The measured pK_a for the conjugate acid of glycine is 2.35, a value closer to that expected for deprotonation of the carboxyl group. As the pH is raised, a second deprotonation step,

FIGURE 27.2 The titration curve of glycine. At pH values less than pK_{a1}, $\overset{+}{H_3N}CH_2CO_2H$ is the major species present. At pH values between pK_{a1} and pK_{a2}, the principal species is the zwitterion $\overset{+}{H_3N}CH_2CO_2^-$. The concentration of the zwitterion is a maximum at the isoelectric point pI. At pH values greater than pK_{a2}, $H_2NCH_2CO_2^-$ is the species present in greatest concentration.

corresponding to removal of a proton from nitrogen of the zwitterion, is observed. The pK_a associated with this step is 9.78, much like that of typical alkylammonium ions.

Species present in strong acid — Zwitterion — Species present in strong base

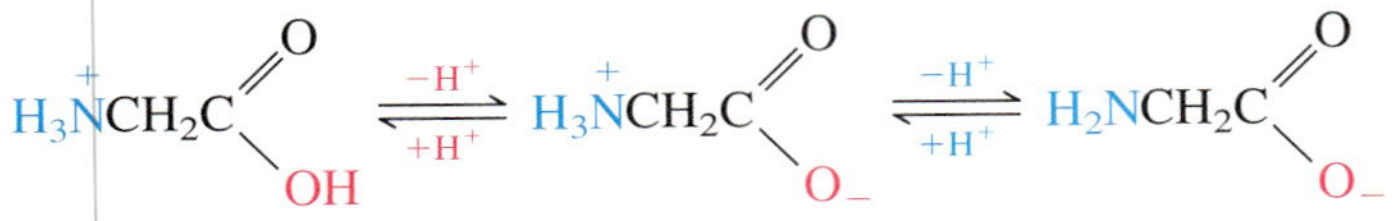

Thus, glycine is characterized by two pK_a values: the one corresponding to the more acidic site is designated pK_{a1}, the one corresponding to the less acidic site is designated pK_{a2}. Table 27.2 lists pK_{a1} and pK_{a2} values for the α-amino acids that have neutral side chains, which are the first two groups of amino acids given in Table 27.1. In all cases their pK_a values are similar to those of glycine.

Table 27.2 includes a column labeled pI, which is *the isoelectric point* of the amino acid. The **isoelectric point,** also called the **isoionic point,** is the pH at which the amino acid has no net charge. It is the pH at which the concentration of the zwitterion is a maximum. At a pH lower than pI, the amino acid is positively charged; at a pH higher than pI, the amino acid is negatively charged. For the amino acids in Table 27.2, pI is the average of pK_{a1} and pK_{a2} and lies slightly to the acid side of neutrality.

Some amino acids have side chains that bear acidic or basic groups. As Table 27.3 indicates, these amino acids are characterized by three pK_a values. The third pK_a reflects the nature of the side chain. Acidic amino acids (aspartic and glutamic acid) have acidic side chains; basic amino acids (lysine, arginine, and histidine) have basic side chains.

The isoelectric points of the amino acids in Table 27.3 are midway between the pK_a values of the zwitterion and its conjugate acid. Take two examples: aspartic acid and lysine. Aspartic acid has an acidic side chain and a pI of 2.77. Lysine has a basic side chain and a pI of 9.74.

Aspartic acid:

Conjugate acid of zwitterion — Zwitterion

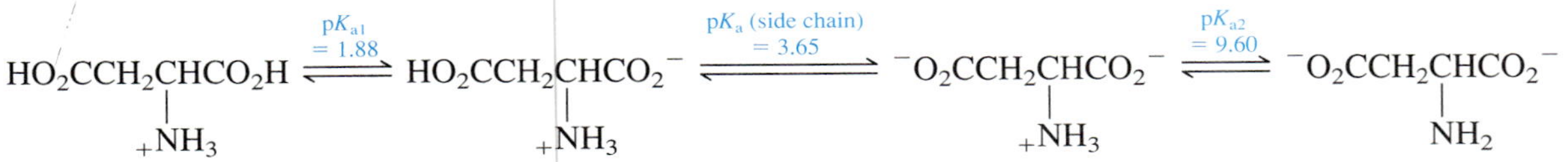

The pI of aspartic acid is the average of pK_{a1} (1.88) and the pK_a of the side chain (3.65) or 2.77.

Lysine:

Conjugate acid of zwitterion — Zwitterion

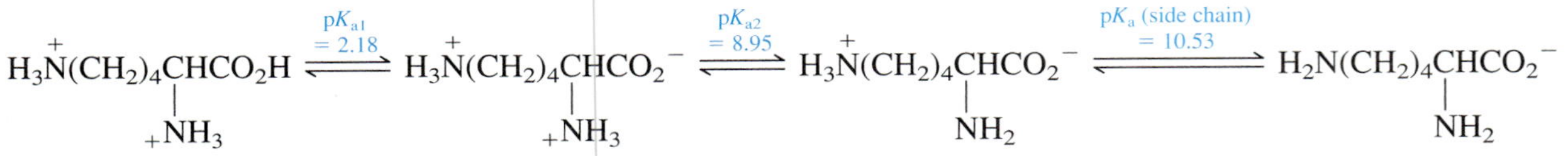

The pI of lysine is the average of pK_{a2} (8.95) and the pK_a of the side chain (10.53) or 9.74.

TABLE 27.2	Acid–Base Properties of Amino Acids with Neutral Side Chains		
Amino acid	**pK_{a1}***	**pK_{a2}***	**pI**
Glycine	2.34	9.60	5.97
Alanine	2.34	9.69	6.00
Valine	2.32	9.62	5.96
Leucine	2.36	9.60	5.98
Isoleucine	2.36	9.60	6.02
Methionine	2.28	9.21	5.74
Proline	1.99	10.60	6.30
Phenylalanine	1.83	9.13	5.48
Tryptophan	2.83	9.39	5.89
Asparagine	2.02	8.80	5.41
Glutamine	2.17	9.13	5.65
Serine	2.21	9.15	5.68
Threonine	2.09	9.10	5.60
Tyrosine	2.20	9.11	5.66

*In all cases pK_{a1} corresponds to ionization of the carboxyl group; pK_{a2} corresponds to deprotonation of the ammonium ion.

TABLE 27.3	Acid–Base Properties of Amino Acids with Ionizable Side Chains			
Amino acid	**pK_{a1}***	**pK_{a2}**	**pK_a of side chain**	**pI**
Aspartic acid	1.88	9.60	3.65	2.77
Glutamic acid	2.19	9.67	4.25	3.22
Lysine	2.18	8.95	10.53	9.74
Arginine	2.17	9.04	12.48	10.76
Histidine	1.82	9.17	6.00	7.59

*In all cases pK_{a1} corresponds to ionization of the carboxyl group of $RCHCO_2H$, and pK_{a2} to ionization of

$$\overset{|}{\underset{+}{N}H_3}$$

the ammonium ion.

PROBLEM 27.3 Cysteine has $pK_{a1} = 1.96$ and $pK_{a2} = 10.28$. The pK_a for ionization of the —SH group of the side chain is 8.18. What is the isoelectric point of cysteine?

PROBLEM 27.4 Above a pH of about 10, the major species present in a solution of tyrosine has a net charge of −2. Suggest a reasonable structure for this species.

Individual amino acids differ in their acid–base properties. This is important in peptides and proteins, where the properties of the substance depend on its amino acid constituents, especially on the nature of the side chains. It is also important in analyses in which a complex mixture of amino acids is separated into its components by taking advantage of the differences in their proton-donating and accepting power.

ELECTROPHORESIS

Electrophoresis is a method for separation and purification that depends on the movement of charged particles in an electric field. Its principles can be introduced by considering the electrophoretic behavior of some representative amino acids. The medium is a cellulose acetate strip that is moistened with an aqueous solution buffered at a particular pH. The opposite ends of the strip are placed in separate compartments containing the buffer, and each compartment is connected to a source of direct electric current (Figure 27.3a). If the buffer solution is more acidic than the isoelectric point (pI) of the amino acid, the amino acid has a net positive charge and migrates toward the negatively charged electrode. Conversely, when the buffer is more basic than the pI of the amino acid, the amino acid has a net negative charge and migrates toward the positively charged

electrode. When the pH of the buffer corresponds to the pI, the amino acid has no net charge and does not migrate from the origin.

Thus if a mixture containing alanine, aspartic acid, and lysine is subjected to electrophoresis in a buffer that matches the isoelectric point of alanine (pH 6.0), aspartic acid (pI = 2.8) migrates toward the positive electrode, alanine remains at the origin, and lysine (pI = 9.7) migrates toward the negative electrode (Figure 27.3b).

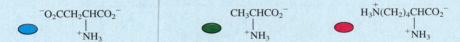

$$^-O_2CCH_2CHCO_2^- \qquad CH_3CHCO_2^- \qquad \overset{+}{H_3N}(CH_2)_4CHCO_2^-$$
$$\underset{+NH_3}{|} \qquad \underset{+NH_3}{|} \qquad \underset{+NH_3}{|}$$

Aspartic acid (−1 ion)   Alanine (neutral)   Lysine (+1 ion)

—Cont.

A mixture of amino acids

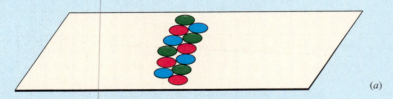

$$^-O_2CCH_2CHCO_2^- \qquad CH_3CHCO_2^- \qquad \overset{+}{H_3N}(CH_2)_4CHCO_2^-$$
$$\underset{+NH_3}{|} \qquad \underset{+NH_3}{|} \qquad \underset{+NH_3}{|}$$

is placed at the center of a sheet of cellulose acetate. The sheet is soaked with an aqueous solution buffered at a pH of 6.0. At this pH aspartic acid ⬤ exists as its −1 ion, alanine ⬤ as its zwitterion, and lysine ⬤ as its +1 ion.

(a)

Application of an electric current causes the negatively charged ions to migrate to the + electrode, and the positively charged ions to migrate to the − electrode. The zwitterion, with a net charge of zero, remains at its original position.

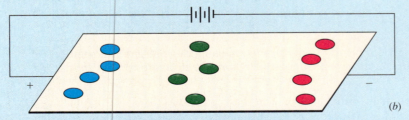

(b)

FIGURE 27.3 Application of electrophoresis to the separation of aspartic acid, alanine, and lysine according to their charge type at a pH corresponding to the isoelectric point (pI) of alanine.

Electrophoresis is used primarily to analyze mixtures of peptides and proteins, rather than individual amino acids, but analogous principles apply. Because they incorporate different numbers of amino acids and because their side chains are different, two peptides will have slightly different acid–base properties and slightly different net charges at a particular pH. Thus, their mobilities in an electric field will be different, and electrophoresis can be used to separate them. The medium used to separate peptides and proteins is typically a polyacrylamide gel, leading to the term **gel electrophoresis** for this technique.

A second factor that governs the rate of migration during electrophoresis is the size (length and shape) of the peptide or protein. Larger molecules move through the polyacrylamide gel more slowly than smaller ones. In current practice, the experiment is modified to exploit differences in size more than differences in net charge, especially in the **SDS gel electrophoresis** of proteins. Approximately 1.5 g of the detergent *sodium dodecyl sulfate* (SDS, page 800)

per gram of protein is added to the aqueous buffer. SDS binds to the protein, causing the protein to unfold so that it is roughly rod-shaped with the $CH_3(CH_2)_{10}CH_2-$ groups of SDS associated with the lipophilic portions of the protein. The negatively charged sulfate groups are exposed to the water. The SDS molecules that they carry ensure that all the protein molecules are negatively charged and migrate toward the positive electrode. Furthermore, all the proteins in the mixture now have similar shapes and tend to travel at rates proportional to their chain length. Thus, when carried out on a preparative scale, SDS gel electrophoresis permits proteins in a mixture to be separated according to their molecular weight. On an analytical scale, it is used to estimate the molecular weight of a protein by comparing its electrophoretic mobility with that of proteins of known molecular weight.

Later, in Chapter 28, we will see how gel electrophoresis is used in nucleic acid chemistry.

27.4   SYNTHESIS OF AMINO ACIDS

One of the oldest methods for the synthesis of amino acids dates back to the nineteenth century and is simply a nucleophilic substitution in which ammonia reacts with an α-halo carboxylic acid.

$$CH_3CHCO_2H \underset{Br}{} \quad + \quad 2NH_3 \xrightarrow{H_2O} \quad CH_3CHCO_2^- \underset{\overset{+}{N}H_3}{} \quad + \quad NH_4Br$$

2-Bromopropanoic acid   Ammonia   Alanine (65–70%)   Ammonium bromide

The α-halo acid is normally prepared by the Hell–Volhard–Zelinsky reaction (see Section 19.16).

> **PROBLEM 27.5**   Outline the steps in a synthesis of valine from 3-methylbutanoic acid.

In the **Strecker synthesis** an aldehyde is converted to an α-amino acid with one more carbon atom by a two-stage procedure in which an α-amino nitrile is an intermediate. The α-amino nitrile is formed by reaction of the aldehyde with ammonia or an ammonium salt and a source of cyanide ion. Hydrolysis of the nitrile group to a carboxylic acid function completes the synthesis.

$$\overset{O}{\overset{\|}{CH_3CH}} \xrightarrow[NaCN]{NH_4Cl} \quad CH_3\underset{NH_2}{CH}C\equiv N \xrightarrow[\text{2. HO}^-]{\text{1. } H_2O, \text{ HCl, heat}} \quad CH_3\underset{\overset{+}{N}H_3}{CH}CO_2^-$$

Acetaldehyde   2-Aminopropanenitrile   Alanine (52–60%)

The synthesis of alanine was described by Adolf Strecker of the University of Würzburg (Germany) in a paper published in 1850.

PROBLEM 27.6 Outline the steps in the preparation of valine by the Strecker synthesis.

The most widely used method for the laboratory synthesis of α-amino acids is a modification of the malonic ester synthesis (Section 21.7). The key reagent is *diethyl acetamidomalonate*, a derivative of malonic ester that already has the critical nitrogen substituent in place at the α-carbon atom. The side chain is introduced by alkylating diethyl acetamidomalonate in the same way as diethyl malonate itself is alkylated.

$$
\underset{\substack{\text{Diethyl}\\\text{acetamidomalonate}}}{CH_3\overset{O}{\overset{\|}{C}}NHCH(CO_2CH_2CH_3)_2}
\xrightarrow[\text{CH}_3\text{CH}_2\text{OH}]{\text{NaOCH}_2\text{CH}_3}
\underset{\substack{\text{Na}^+\\\text{Sodium salt of}\\\text{diethyl acetamidomalonate}}}{CH_3\overset{O}{\overset{\|}{C}}NH\overset{..}{\overset{-}{C}}(CO_2CH_2CH_3)_2}
\xrightarrow{C_6H_5CH_2Cl}
\underset{\substack{CH_2C_6H_5\\\text{Diethyl}\\\text{acetamidobenzylmalonate}\\(90\%)}}{CH_3\overset{O}{\overset{\|}{C}}NHC(CO_2CH_2CH_3)_2}
$$

Hydrolysis removes the acetyl group from nitrogen and converts the two ester functions to carboxyl groups. Decarboxylation gives the desired product.

$$
\underset{\substack{CH_2C_6H_5\\\text{Diethyl}\\\text{acetamidobenzylmalonate}}}{CH_3\overset{O}{\overset{\|}{C}}NHC(CO_2CH_2CH_3)_2}
\xrightarrow[\text{H}_2\text{O, heat}]{\text{HBr}}
\underset{\substack{CH_2C_6H_5\\\text{(not isolated)}}}{H_3\overset{+}{N}C(CO_2H)_2}
\xrightarrow[-\text{CO}_2]{\text{heat}}
\underset{\substack{NH_3\\\overset{+}{}\\\text{Phenylalanine}\\(65\%)}}{C_6H_5CH_2CHCO_2^-}
$$

PROBLEM 27.7 Outline the steps in the synthesis of valine from diethyl acetamidomalonate. The overall yield of valine by this method is reported to be rather low (31%). Can you think of a reason why this synthesis is not very efficient?

Unless a resolution step is included, the α-amino acids prepared by the synthetic methods just described are racemic. Optically active amino acids, when desired, may be obtained by resolving a racemic mixture or by **enantioselective synthesis.** A synthesis is described as enantioselective if it produces one enantiomer of a chiral compound in an amount greater than its mirror image. Recall from Section 7.9 that optically inactive reactants cannot give optically active products. Enantioselective syntheses of amino acids therefore require an enantiomerically enriched chiral reagent or catalyst at some point in the process. If the chiral reagent or catalyst is a single enantiomer and if the reaction sequence is completely enantioselective, an optically pure amino acid is obtained. Chemists have succeeded in preparing α-amino acids by techniques that are more than 95% enantioselective. Although this is an impressive feat, we must not lose sight of the fact that the enzyme catalyzed reactions that produce amino acids in living systems do so with 100% enantioselectivity.

27.5   REACTIONS OF AMINO ACIDS

Amino acids undergo reactions characteristic of both their amine and carboxylic acid functional groups. Acylation is a typical reaction of the amino group.

$$\overset{+}{H_3}NCH_2CO_2^- + CH_3\overset{O}{\overset{\|}{C}}O\overset{O}{\overset{\|}{C}}CH_3 \longrightarrow CH_3\overset{O}{\overset{\|}{C}}NHCH_2CO_2H + CH_3CO_2H$$

| Glycine | Acetic anhydride | *N*-Acetylglycine (89–92%) | Acetic acid |

Ester formation is a typical reaction of the carboxyl group.

$$CH_3\underset{\underset{+}{\overset{|}{NH_3}}}{\overset{|}{C}H}CO_2^- + CH_3CH_2OH \xrightarrow{HCl} CH_3\underset{\underset{+}{\overset{|}{NH_3}}}{\overset{|}{C}H}\overset{O}{\overset{\|}{C}}OCH_2CH_3 \; Cl^-$$

| Alanine | Ethanol | Hydrochloride salt of alanine ethyl ester (90–95%) |

The presence of amino acids can be detected by the formation of a purple color on treatment with *ninhydrin*. The same compound responsible for the purple color is formed from all amino acids in which the α-amino group is primary.

Ninhydrin is used to detect fingerprints.

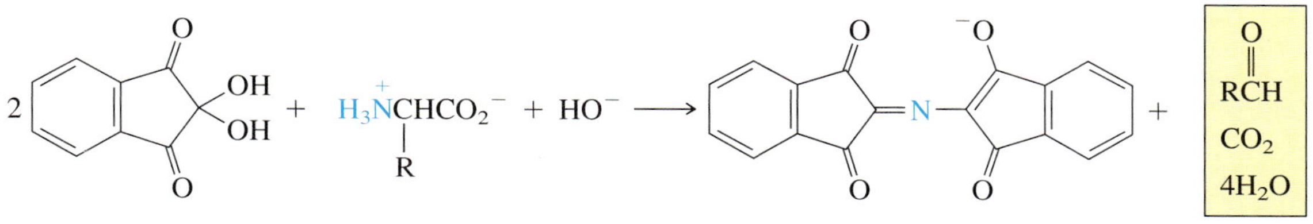

| Ninhydrin | | Violet dye ("Ruhemann's purple") | (Formed, but not normally isolated) |

Proline, in which the α-amino group is secondary, gives an orange compound on reaction with ninhydrin.

PROBLEM 27.8 Suggest a reasonable mechanism for the reaction of an α-amino acid with ninhydrin.

27.6   SOME BIOCHEMICAL REACTIONS OF AMINO ACIDS

The 20 amino acids listed in Table 27.1 are biosynthesized by a number of different pathways, and we will touch on only a few of them in an introductory way. We will examine the biosynthesis of glutamic acid first because it illustrates a biochemical process analogous to a reaction we discussed earlier in the context of amine synthesis, *reductive amination* (Section 22.10).

Glutamic acid is formed in most organisms from ammonia and α-ketoglutaric acid. α-Ketoglutaric acid is one of the intermediates in the **tricarboxylic acid cycle** (also called the **Krebs cycle**) and arises via metabolic breakdown of food sources: carbohydrates, fats, and proteins.

The August 1986 issue of the *Journal of Chemical Education* (pp. 673–677) contains a review of the Krebs cycle.

$$\text{HO}_2\text{CCH}_2\text{CH}_2\overset{\overset{\text{O}}{\|}}{\text{C}}\text{CO}_2\text{H} \ + \ \text{NH}_3 \ \xrightarrow[\text{reducing agents}]{\text{enzymes}} \ \text{HO}_2\text{CCH}_2\text{CH}_2\underset{\underset{+}{\overset{|}{\text{NH}_3}}}{\text{CHCO}_2^-}$$

α-Ketoglutaric acid   Ammonia   L-Glutamic acid

Ammonia reacts with the ketone carbonyl group to give an imine (C=NH), which is then reduced to the amine function of the α-amino acid. Both imine formation and reduction are enzyme-catalyzed. The reduced form of nicotinamide adenine diphosphonucleotide (NADPH) is a coenzyme and acts as a reducing agent. The step in which the imine is reduced is the one in which the chirality center is introduced and gives only L-glutamic acid.

L-Glutamic acid is not an essential amino acid. It need not be present in the diet because animals can biosynthesize it from sources of α-ketoglutaric acid. It is, however, a key intermediate in the biosynthesis of other amino acids by a process known as **transamination.** L-Alanine, for example, is formed from pyruvic acid by transamination from L-glutamic acid.

$$\overset{\overset{\text{O}}{\|}}{\text{CH}_3\text{C}}\text{CO}_2\text{H} \ + \ \text{HO}_2\text{CCH}_2\text{CH}_2\underset{\underset{+}{\overset{|}{\text{NH}_3}}}{\text{CHCO}_2^-} \ \xrightarrow{\text{enzymes}} \ \text{CH}_3\underset{\underset{+}{\overset{|}{\text{NH}_3}}}{\text{CHCO}_2^-} \ + \ \text{HO}_2\text{CCH}_2\text{CH}_2\overset{\overset{\text{O}}{\|}}{\text{C}}\text{CO}_2\text{H}$$

Pyruvic acid   L-Glutamic acid   L-Alanine   α-Ketoglutaric acid

In transamination an amine group is transferred from L-glutamic acid to pyruvic acid. An outline of the mechanism of transamination is presented in Figure 27.4.

One amino acid often serves as the biological precursor to another. L-Phenylalanine is classified as an essential amino acid, whereas its *p*-hydroxy derivative, L-tyrosine, is not. This is because animals can convert L-phenylalanine to L-tyrosine by hydroxylation of the aromatic ring. An *arene oxide* (Section 24.7) is an intermediate.

L-Phenylalanine   Arene oxide intermediate   L-Tyrosine

Some people lack the enzymes necessary to convert L-phenylalanine to L-tyrosine. Any L-phenylalanine that they obtain from their diet is diverted along a different metabolic pathway, giving phenylpyruvic acid:

L-Phenylalanine   Phenylpyruvic acid

Step 1: The amine function of L-glutamate reacts with the ketone carbonyl of pyruvate to form an imine.

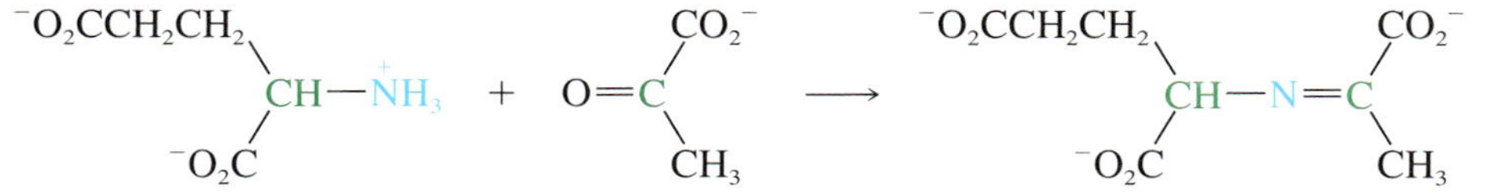

$$^-O_2CCH_2CH_2\!-\!\underset{^-O_2C}{\overset{}{CH}}\!-\!\overset{+}{N}H_3 \;+\; O\!=\!\underset{CH_3}{\overset{CO_2^-}{C}} \;\longrightarrow\; ^-O_2CCH_2CH_2\!-\!\underset{^-O_2C}{\overset{}{CH}}\!-\!N\!=\!\underset{CH_3}{\overset{CO_2^-}{C}}$$

L-Glutamate   Pyruvate   Imine

Step 2: Enzyme-catalyzed proton-transfer steps cause migration of the double bond, converting the imine formed in step 1 to an isomeric imine.

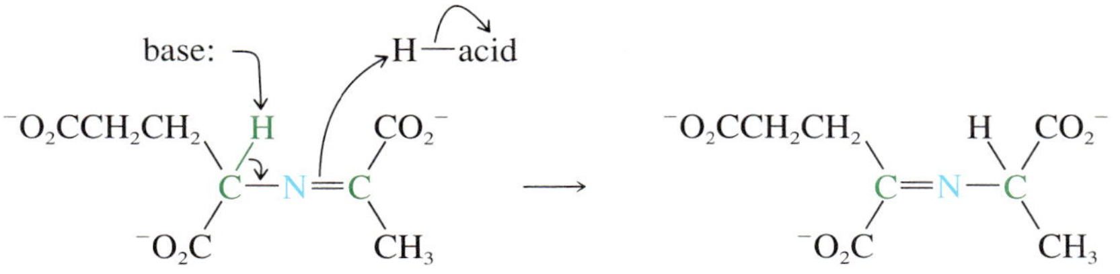

base: ⌐   H—acid

$$^-O_2CCH_2CH_2\!-\!\underset{^-O_2C}{\overset{H}{C}}\!-\!N\!=\!\underset{CH_3}{\overset{CO_2^-}{C}} \;\longrightarrow\; ^-O_2CCH_2CH_2\!-\!\underset{^-O_2C}{\overset{}{C}}\!=\!N\!-\!\underset{CH_3}{\overset{H}{C}}CO_2^-$$

Imine from step 1   Rearranged imine

Step 3: Hydrolysis of the rearranged imine gives L-alanine and α-ketoglutarate.

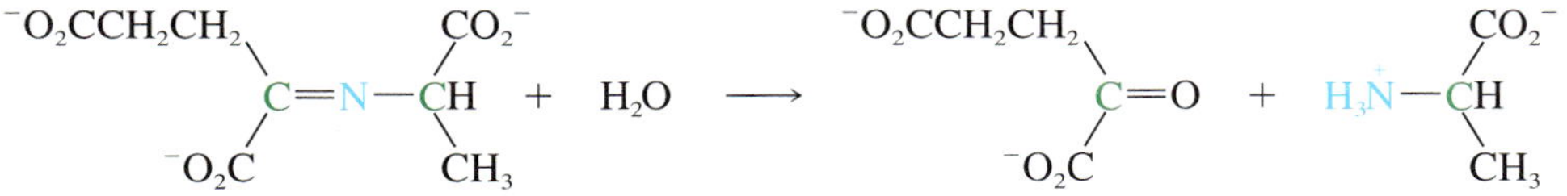

$$^-O_2CCH_2CH_2\!-\!\underset{^-O_2C}{\overset{}{C}}\!=\!N\!-\!\underset{CH_3}{\overset{CO_2^-}{CH}} \;+\; H_2O \;\longrightarrow\; ^-O_2CCH_2CH_2\!-\!\underset{^-O_2C}{\overset{}{C}}\!=\!O \;+\; \overset{+}{H_3}N\!-\!\underset{CH_3}{\overset{CO_2^-}{CH}}$$

Rearranged imine   Water   α-Ketoglutarate   L-Alanine

FIGURE 27.4 The mechanism of transamination. All the steps are enzyme-catalyzed.

Foods sweetened with Aspartame (page 1051) contain a PKU warning. Can you see why?

Phenylpyruvic acid can cause mental retardation in infants who are deficient in the enzymes necessary to convert L-phenylalanine to L-tyrosine. This disorder is called **phenylketonuria,** or **PKU disease.** PKU disease can be detected by a simple test routinely administered to newborns. It cannot be cured, but is controlled by restricting the dietary intake of L-phenylalanine. In practice this means avoiding foods such as meat that are rich in L-phenylalanine.

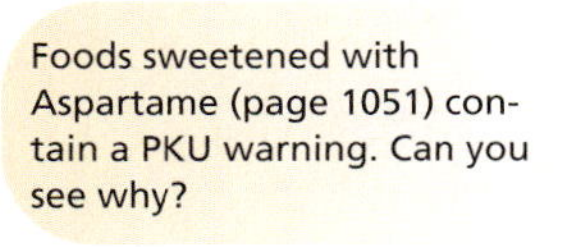

Among the biochemical reactions that amino acids undergo is **decarboxylation** to amines. Decarboxylation of histidine, for example, gives histamine, a powerful vasodilator normally present in tissue and formed in excessive amounts under conditions of traumatic shock.

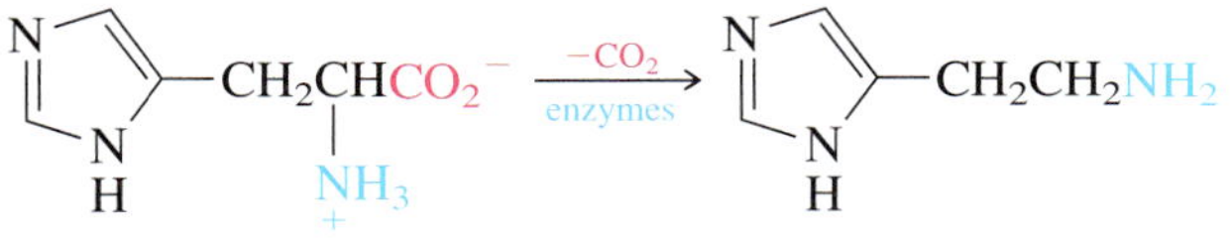

$$\text{Histidine} \xrightarrow[\text{enzymes}]{-CO_2} \text{Histamine}$$

Histidine: imidazole—CH_2CHCO_2^- with NH_3^+ ; Histamine: imidazole—CH_2CH_2NH_2

Histidine   Histamine

Histamine is responsible for many of the symptoms associated with hay fever and other allergies. An antihistamine relieves these symptoms by blocking the action of histamine.

FIGURE 27.5 Tyrosine is the biosynthetic precursor to a number of neurotransmitters. Each transformation is enzyme-catalyzed. Hydroxylation of the aromatic ring of tyrosine converts it to 3,4-dihydroxyphenylalanine (L-dopa), decarboxylation of which gives dopamine. Hydroxylation of the benzylic carbon of dopamine converts it to norepinephrine (noradrenaline), and methylation of the amino group of norepinephrine yields epinephrine (adrenaline).

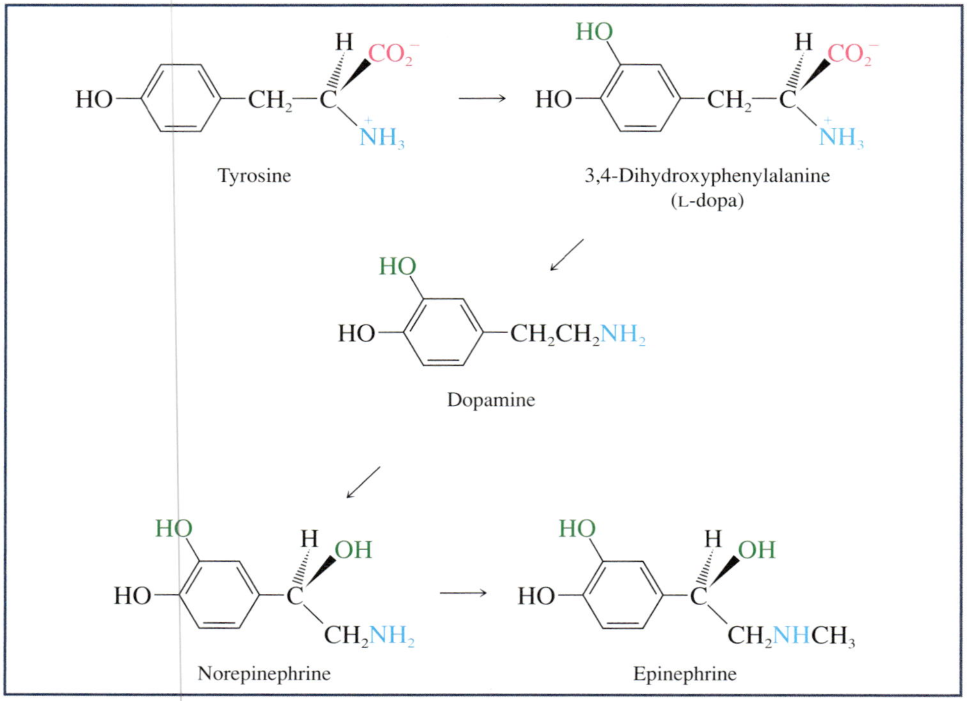

Tyrosine

3,4-Dihydroxyphenylalanine
(L-dopa)

Dopamine

Norepinephrine

Epinephrine

PROBLEM 27.9   One of the amino acids in Table 27.1 is the biological precursor to γ-aminobutyric acid (4-aminobutanoic acid), which it forms by a decarboxylation reaction. Which amino acid is this?

For a review of neurotransmitters, see the February 1988 issue of the *Journal of Chemical Education* (pp. 108–111).

The chemistry of the brain and central nervous system is affected by a group of substances called **neurotransmitters,** substances that carry messages across a synapse from one neuron to another. Several of these neurotransmitters arise from L-tyrosine by structural modification and decarboxylation, as outlined in Figure 27.5.

27.7   PEPTIDES

A key biochemical reaction of amino acids is their conversion to peptides, polypeptides, and proteins. In all these substances amino acids are linked together by amide bonds. The amide bond between the amino group of one amino acid and the carboxyl of another is called a **peptide bond.** Alanylglycine is a representative dipeptide.

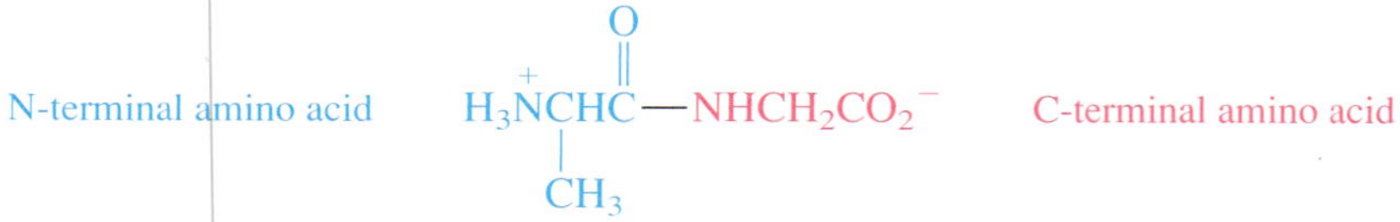

N-terminal amino acid $\overset{+}{H_3}NCHC-NHCH_2CO_2^-$ C-terminal amino acid

$\underset{CH_3}{|}$

Alanylglycine
(Ala-Gly or AG)

It is understood that α-amino acids occur as their L stereoisomers unless otherwise indicated. The D notation is explicitly shown when a D amino acid is present, and a racemic amino acid is identified by the prefix DL.

By agreement, peptide structures are written so that the amino group (as $\overset{+}{H_3}N$— or H_2N—) is at the left and the carboxyl group (as CO_2^- or CO_2H) is at the right. The

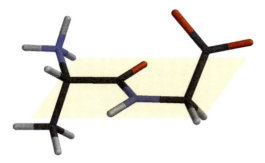

FIGURE 27.6 Structural features of the dipeptide L-alanylglycine as determined by X-ray crystallography.

left and right ends of the peptide are referred to as the **N terminus** (or amino terminus) and the **C terminus** (or carboxyl terminus), respectively. Alanine is the N-terminal amino acid in alanylglycine; glycine is the C-terminal amino acid. A dipeptide is named as an acyl derivative of the C-terminal amino acid. We call the precise order of bonding in a peptide its amino acid **sequence.** The amino acid sequence is conveniently specified by using the three-letter amino acid abbreviations for the respective amino acids and connecting them by hyphens. One-letter abbreviations may also be used. Individual amino acid components of peptides are often referred to as amino acid **residues.**

PROBLEM 27.10 Write structural formulas showing the constitution of each of the following dipeptides. Rewrite each sequence using one-letter abbreviations for the amino acids.

(a) Gly-Ala   (d) Gly-Glu

(b) Ala-Phe   (e) Lys-Gly

(c) Phe-Ala   (f) D-Ala-D-Ala

SAMPLE SOLUTION (a) Gly-Ala is a constitutional isomer of Ala-Gly. Glycine is the N-terminal amino acid in Gly-Ala; alanine is the C-terminal amino acid.

$$\text{N-terminal amino acid} \quad \overset{+}{H_3}NCH_2\overset{\displaystyle O}{\overset{\|}{C}}-NHCHCO_2^{-} \quad \text{C-terminal amino acid}$$
$$\underset{\displaystyle CH_3}{|}$$

Glycylalanine (GA)

Figure 27.6 shows the structure of Ala-Gly as determined by X-ray crystallography. An important feature is the planar geometry associated with the peptide bond, and the most stable conformation with respect to this bond has the two α-carbon atoms anti to each other. Rotation about the amide linkage is slow because delocalization of the unshared electron pair of nitrogen into the carbonyl group gives partial double-bond character to the carbon–nitrogen bond.

PROBLEM 27.11 Expand your answer to Problem 27.10 by showing the structural formula for each dipeptide in a manner that reveals the stereochemistry at the α-carbon atom.

SAMPLE SOLUTION   (a) Glycine is achiral, and so Gly-Ala has only one chirality center, the α-carbon atom of the L-alanine residue. When the carbon chain is drawn in an extended zigzag fashion and L-alanine is the C terminus, its structure is as shown:

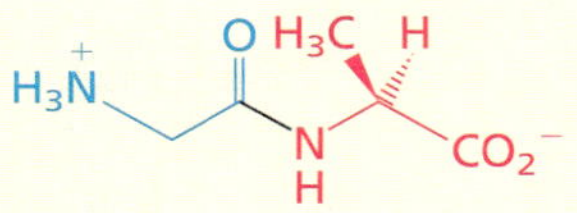

Glycyl-L-alanine (Gly-Ala)

The structures of higher peptides follow in an analogous fashion. Figure 27.7 gives the structural formula and amino acid sequence of a naturally occurring pentapeptide

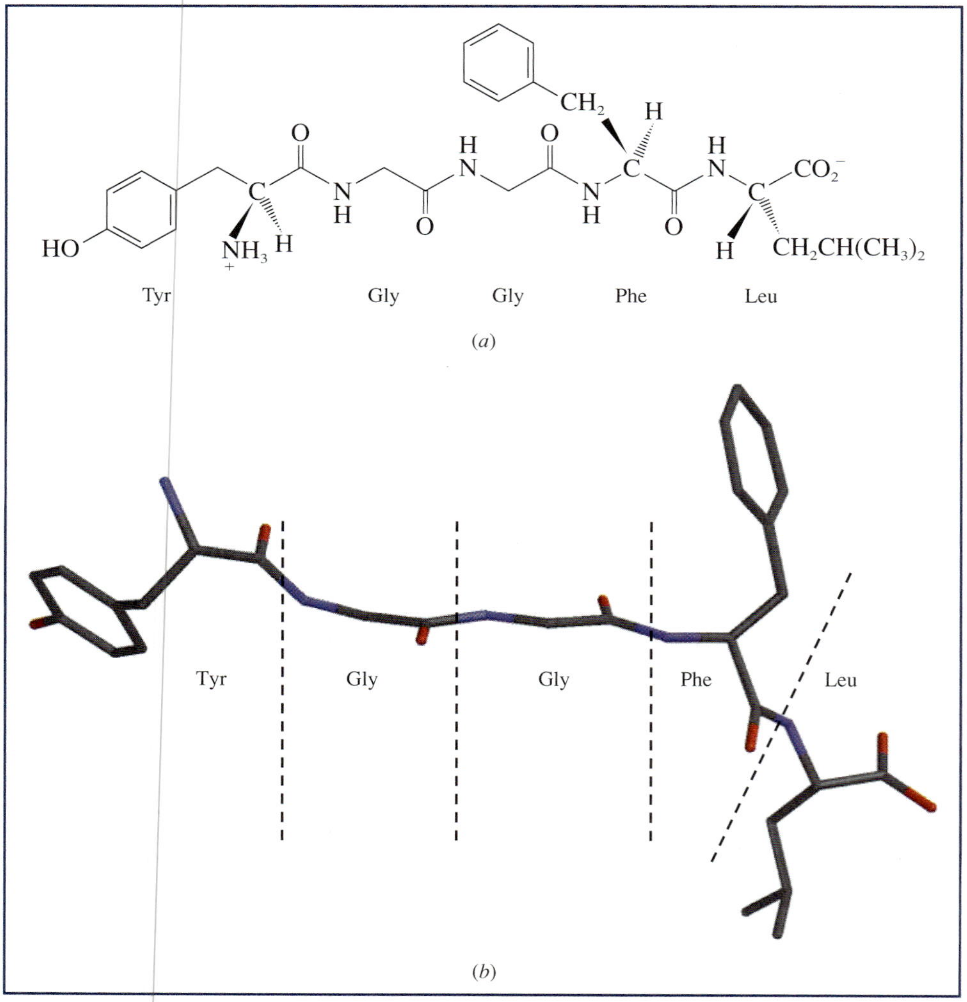

FIGURE 27.7 The structure of the pentapeptide leucine enkephalin shown as (a) a structural drawing and (b) as a molecular model. The shape of the molecular model was determined by X-ray crystallography. Hydrogens have been omitted for clarity.

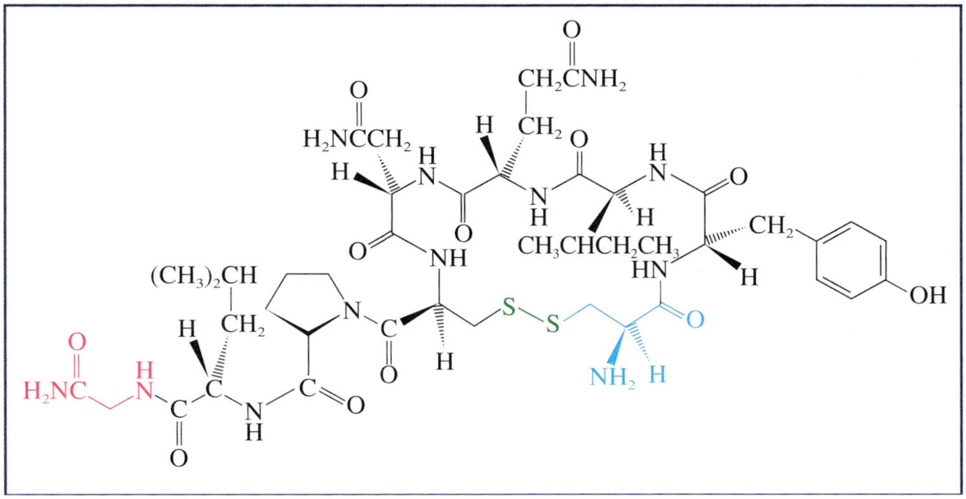

FIGURE 27.8 The structure of oxytocin, a nonapeptide containing a disulfide bond (green) between two cysteine residues. One of these cysteines is the N-terminal amino acid; it is shown in blue. The C-terminal amino acid is the amide of glycine and is shown in red. There are no free carboxyl groups in the molecule; all exist in the form of amides.

known as *leucine enkephalin*. Enkephalins are pentapeptide components of **endorphins,** polypeptides present in the brain that act as the body's own painkillers. A second substance, known as *methionine enkephalin,* is also present in endorphins. Methionine enkephalin is about 20 times more potent than leucine enkephalin. It differs from leucine enkephalin only in having methionine instead of leucine as its C-terminal amino acid.

> **PROBLEM 27.12** What is the amino acid sequence (using three-letter abbreviations) of methionine enkephalin? Also show it using one-letter abbreviations.

Peptides having structures slightly different from those described to this point are known. One such variation is seen in the nonapeptide *oxytocin,* shown in Figure 27.8. Oxytocin is a hormone secreted by the pituitary gland that stimulates uterine contractions during childbirth. Rather than terminating in a carboxyl group, the C-terminal glycine residue in oxytocin has been modified so that it exists as the corresponding amide. Two cysteine units, one of them the N-terminal amino acid, are joined by the sulfur–sulfur bond of a large-ring cyclic disulfide unit. This is a common structural modification in polypeptides and proteins that contain cysteine residues. It provides a covalent bond between regions of peptide chains that may be many amino acid residues removed from each other.

> Recall from Section 15.13 that compounds of the type RSH are readily oxidized to RSSR.

27.8 INTRODUCTION TO PEPTIDE STRUCTURE DETERMINATION

There are several levels of peptide structure. The **primary structure** is the amino acid sequence plus any disulfide links. With the 20 amino acids of Table 27.1 as building blocks, 20^2 dipeptides, 20^3 tripeptides, 20^4 tetrapeptides, and so on, are possible. Given a peptide of unknown structure, how do we determine its amino acid sequence?

We'll describe peptide structure determination by first looking at one of the great achievements of biochemistry, the determination of the amino acid sequence of insulin by Frederick Sanger of Cambridge University (England). Sanger was awarded the 1958 Nobel Prize in chemistry for this work, which he began in 1944 and completed 10 years later. The methods used by Sanger and his coworkers are, of course, dated by now, but the overall strategy hasn't changed very much. We'll use Sanger's insulin work to orient us with respect to strategy, then show how current methods of protein sequencing have evolved from it.

> Sanger was a corecipient of a second Nobel Prize in 1980 for devising methods for sequencing nucleic acids. Sanger's strategy for nucleic acid sequencing will be described in Section 28.14.

Sanger's strategy can be outlined as follows:

1. Determine what amino acids are present and their molar ratios.
2. Cleave the peptide into smaller fragments, separate these fragments, and determine the amino acid composition of the fragments.
3. Identify the N-terminal and the C-terminal amino acid in the original peptide and in each fragment.
4. Organize the information so that the amino acid sequences of small fragments can be overlapped to reveal the full sequence.

27.9 AMINO ACID ANALYSIS

The chemistry behind amino acid analysis is nothing more than acid-catalyzed hydrolysis of amide (peptide) bonds. The peptide is hydrolyzed by heating in 6 M hydrochloric acid for about 24 h to give a solution that contains all the amino acids. This mixture is then separated by **ion-exchange chromatography,** which separates the amino acids mainly according to their acid–base properties. As the amino acids leave the chromatography column, they are mixed with ninhydrin and the intensity of the ninhydrin color monitored electronically. The amino acids are identified by comparing their chromatographic behavior with that of known samples, and their relative amounts from peak areas as recorded on a strip chart.

The entire operation is carried out automatically using an **amino acid analyzer** and is so sensitive that as little as 10^{-5}–10^{-7} g of the peptide is required.

> **PROBLEM 27.13** Amino acid analysis of a certain tetrapeptide gave alanine, glycine, phenylalanine, and valine in equimolar amounts. What amino acid sequences are possible for this tetrapeptide?

27.10 PARTIAL HYDROLYSIS OF PEPTIDES

Whereas acid-catalyzed hydrolysis of peptides cleaves amide bonds indiscriminately and eventually breaks all of them, enzymatic hydrolysis is much more selective and is the method used to convert a peptide into smaller fragments.

The enzymes that catalyze the hydrolysis of peptides are called **peptidases,** **proteases,** or **proteolytic enzymes.** *Trypsin,* a digestive enzyme present in the intestine, catalyzes only the hydrolysis of peptide bonds involving the carboxyl group of a lysine or arginine residue. *Chymotrypsin,* another digestive enzyme, is selective for peptide bonds involving the carboxyl group of amino acids with aromatic side chains (phenylalanine, tryosine, tryptophan). One group of pancreatic enzymes, known as *carboxypeptidases,* catalyzes only the hydrolysis of the peptide bond to the C-terminal amino acid. In addition to these, many other digestive enzymes are known and their selectivity exploited in the selective hydrolysis of peptides.

Papain, the active component of most meat tenderizers, is a proteolytic enzyme.

Site of chymotrypsin-catalyzed
hydrolysis when R′ is an
aromatic side chain

PROBLEM 27.14 Digestion of the tetrapeptide of Problem 27.13 with chymotrypsin gave a dipeptide that on amino acid analysis gave phenylalanine and valine in equimolar amounts. What amino acid sequences are possible for the tetrapeptide?

27.11 END GROUP ANALYSIS

An amino acid sequence is ambiguous unless we know the direction in which to read it—left to right, or right to left. We need to know which end is the N terminus and which is the C terminus. As we saw in the preceding section, carboxypeptidase-catalyzed hydrolysis cleaves the C-terminal amino acid and so can be used to identify it. What about the N terminus?

Several chemical methods have been devised for identifying the N-terminal amino acid. They all take advantage of the fact that the N-terminal amino group is free and can act as a nucleophile. The α-amino groups of all the other amino acids are part of amide linkages, are not free, and are much less nucleophilic. Sanger's method for N-terminal residue analysis involves treating a peptide with 1-fluoro-2,4-dinitrobenzene, which is very reactive toward nucleophilic aromatic substitution (Chapter 23).

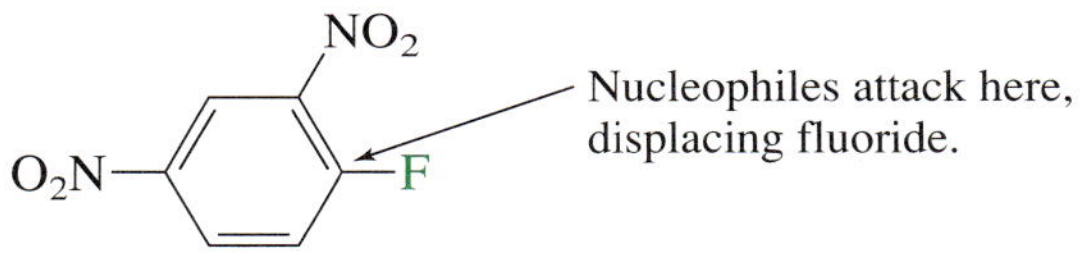

NO$_2$

O$_2$N—F   Nucleophiles attack here, displacing fluoride.

1-Fluoro-2,4-dinitrobenzene

1-Fluoro-2,4-dinitrobenzene is commonly referred to as *Sanger's reagent.*

The amino group of the N-terminal amino acid displaces fluoride from 1-fluoro-2,4-dinitrobenzene and gives a peptide in which the N-terminal nitrogen is labeled with a 2,4-dinitrophenyl (DNP) group. This is shown for the case of Val-Phe-Gly-Ala in Figure 27.9. The 2,4-dinitrophenyl-labeled peptide DNP-Val-Phe-Gly-Ala is isolated and subjected to hydrolysis, after which the 2,4-dinitrophenyl derivative of the N-terminal amino acid is isolated and identified as DNP-Val by comparing its chromatographic behavior with that of standard samples of 2,4-dinitrophenyl-labeled amino acids. None of the other amino acid residues bear a 2,4-dinitrophenyl group; they appear in the hydrolysis product as the free amino acids.

Labeling the N-terminal amino acid as its DNP derivative is mainly of historical interest and has been replaced by other methods. We'll discuss one of these—the Edman degradation—in Section 27.13. First, though, we'll complete our review of the general strategy for peptide sequencing by seeing how Sanger tied all of the information together into a structure for insulin.

27.12 INSULIN

Insulin has 51 amino acids, divided between two chains. One of these, the A chain, has 21 amino acids; the other, the B chain, has 30. The A and B chains are joined by disulfide bonds between cysteine residues (Cys-Cys). Figure 27.10 shows some of the information that defines the amino acid sequence of the B chain.

- Reaction of the B chain peptide with 1-fluoro-2,4-dinitrobenzene established that phenylalanine is the N terminus.

The reaction is carried out by mixing the peptide and 1-fluoro-2,4-dinitrobenzene in the presence of a weak base such as sodium carbonate. In the first step the base abstracts a proton from the terminal $H_3\overset{+}{N}$ group to give a free amino function. The nucleophilic amino group attacks 1-fluoro-2,4-dinitrobenzene, displacing fluoride.

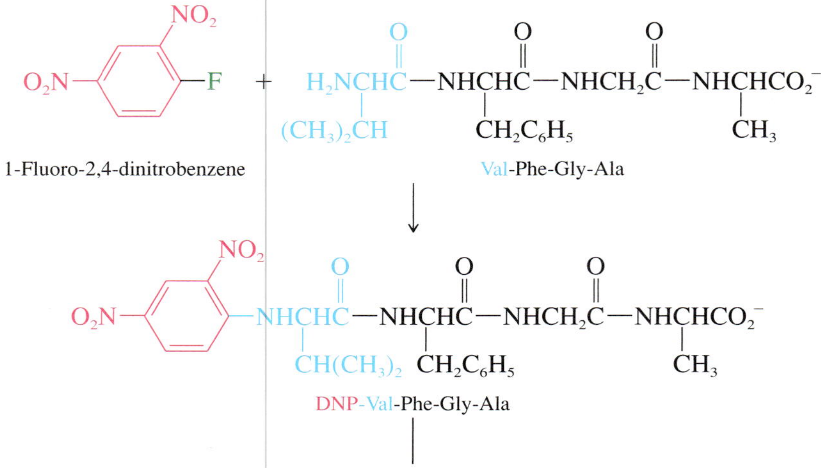

Acid hydrolysis cleaves the amide bonds of the 2,4-dinitrophenyl-labeled peptide, giving the 2,4-dinitrophenyl-labeled N-terminal amino acid and a mixture of unlabeled amino acids.

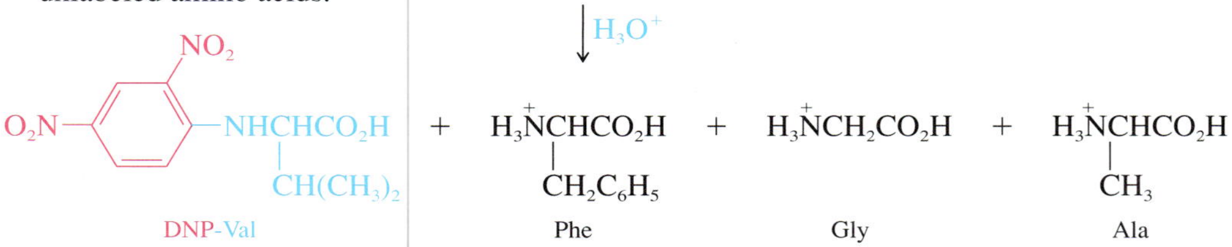

FIGURE 27.9 Use of 1-fluoro-2,4-dinitrobenzene to identify the N-terminal amino acid of a peptide.

- Pepsin-catalyzed hydrolysis gave the four peptides shown in blue in Figure 27.10. (Their sequences were determined in separate experiments.) These four peptides contain 27 of the 30 amino acids in the B chain, but there are no points of overlap between them.
- The sequences of the four tetrapeptides shown in red in Figure 27.10 bridge the gaps between three of the four "blue" peptides to give an unbroken sequence from 1 through 24.
- The peptide shown in green was isolated by trypsin-catalyzed hydrolysis and has an amino acid sequence that completes the remaining overlaps.

Sanger also determined the sequence of the A chain and identified the cysteine residues involved in disulfide bonds between the A and B chains as well as in the disulfide linkage within the A chain. The complete insulin structure is shown in Figure 27.11. The structure shown is that of bovine insulin (from cattle). The A chains of human insulin and bovine insulin differ in only two amino acid residues; their B chains are identical except for the amino acid at the C terminus.

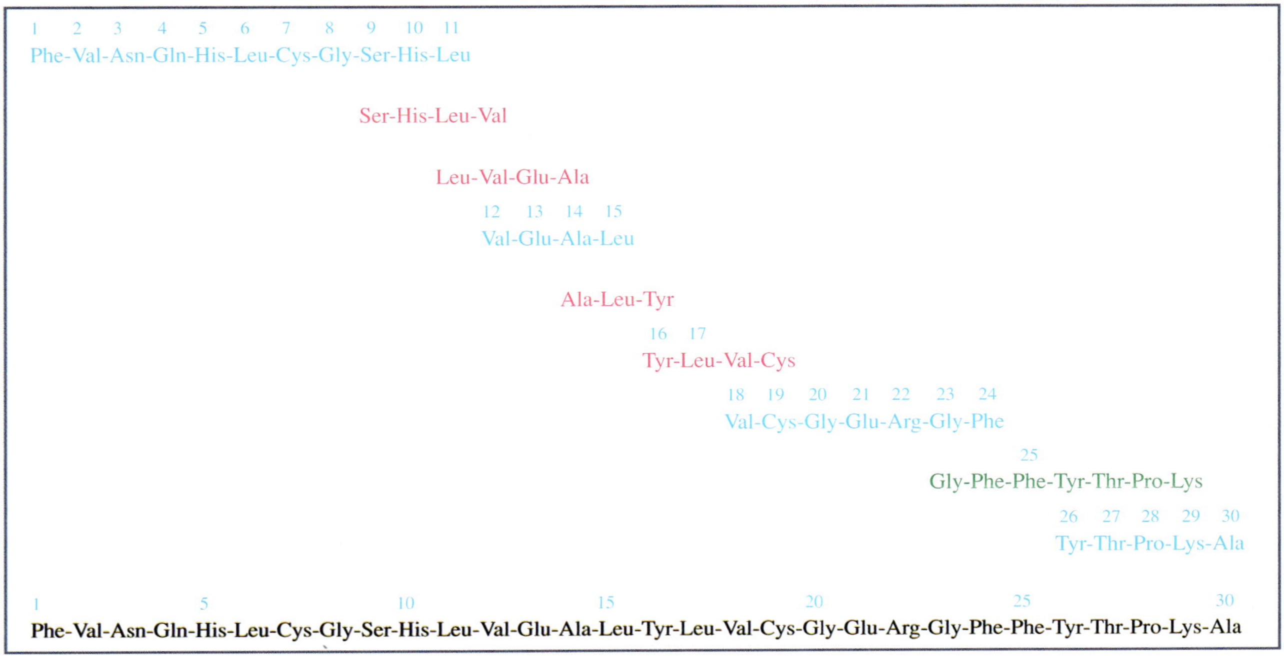

1 2 3 4 5 6 7 8 9 10 11
Phe-Val-Asn-Gln-His-Leu-Cys-Gly-Ser-His-Leu
Ser-His-Leu-Val
Leu-Val-Glu-Ala
12 13 14 15
Val-Glu-Ala-Leu
Ala-Leu-Tyr
16 17
Tyr-Leu-Val-Cys
18 19 20 21 22 23 24
Val-Cys-Gly-Glu-Arg-Gly-Phe
25
Gly-Phe-Phe-Tyr-Thr-Pro-Lys
26 27 28 29 30
Tyr-Thr-Pro-Lys-Ala
1 5 10 15 20 25 30
Phe-Val-Asn-Gln-His-Leu-Cys-Gly-Ser-His-Leu-Val-Glu-Ala-Leu-Tyr-Leu-Val-Cys-Gly-Glu-Arg-Gly-Phe-Phe-Tyr-Thr-Pro-Lys-Ala

FIGURE 27.10 Diagram showing how the amino acid sequence of the B chain of bovine insulin can be determined by overlap of peptide fragments. Pepsin-catalyzed hydrolysis produced the fragments shown in blue, trypsin produced the one shown in green, and acid-catalyzed hydrolysis gave many fragments, including the four shown in red.

N terminus of A chain
C terminus of A chain
N terminus of B chain
C terminus of B chain
S — S
Gly Ile Val Glu Gln Cys Cys Ala Ser Val Cys Ser Leu Tyr Gln Leu Glu Asn Tyr Cys Asn
5
10
15
20
S
S
Phe Val Asn Gln His Leu Cys Gly Ser His Leu Val Glu Ala Leu Tyr Leu Val Cys Gly Glu Arg Gly Phe Phe Tyr Thr Pro Lys Ala
5
10
15
20
25
30

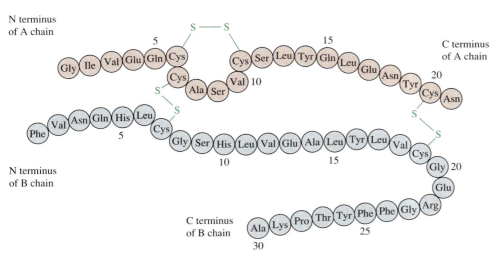

FIGURE 27.11 The amino acid sequence in bovine insulin. The A chain is shown in red and the B chain in blue. The A chain is joined to the B chain by two disulfide units (shown in green). There is also a disulfide bond linking cysteines 6 and 11 in the A chain. Human insulin has threonine and isoleucine at residues 8 and 10, respectively, in the A chain and threonine as the C-terminal amino acid in the B chain.

27.13 THE EDMAN DEGRADATION AND AUTOMATED SEQUENCING OF PEPTIDES

The years that have passed since Sanger determined the structure of insulin have seen refinements in technique while retaining the same overall strategy. Enzyme-catalyzed hydrolysis to convert a large peptide to smaller fragments remains an important component, as does searching for overlaps among these smaller fragments. The method for N-terminal residue analysis, however, has been improved so that much smaller amounts of peptide are required, and the analysis has been automated.

When Sanger's method for N-terminal residue analysis was discussed, you may have wondered why it was not done sequentially. Simply start at the N terminus and work steadily back to the C terminus identifying one amino acid after another. The idea is fine, but it just doesn't work well in practice, at least with 1-fluoro-2,4-dinitrobenzene.

A major advance was devised by Pehr Edman (University of Lund, Sweden) that has become the standard method for N-terminal residue analysis. The **Edman degradation** is based on the chemistry shown in Figure 27.12. A peptide reacts with phenyl isothiocyanate to give a *phenylthiocarbamoyl* (PTC) derivative, as shown in the first step. This PTC derivative is then treated with an acid in an *anhydrous* medium (Edman used nitromethane saturated with hydrogen chloride) to cleave the amide bond between the N-terminal amino acid and the remainder of the peptide. No other peptide bonds are cleaved in this step as amide bond hydrolysis requires water. When the PTC derivative is treated with acid in an anhydrous medium, the sulfur atom of the C=S unit acts as

FIGURE 27.12 Identification of the N-terminal amino acid of a peptide by Edman degradation.

Step 1: A peptide is treated with phenyl isothiocyanate to give a phenylthiocarbamoyl (PTC) derivative.

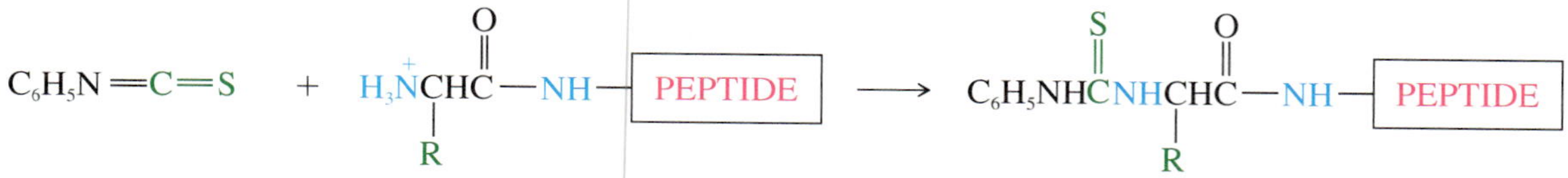

Phenyl isothiocyanate

PTC derivative

Step 2: On reaction with hydrogen chloride in an anhydrous solvent, the thiocarbonyl sulfur of the PTC derivative attacks the carbonyl carbon of the N-terminal amino acid. The N-terminal amino acid is cleaved as a thiazolone derivative from the remainder of the peptide.

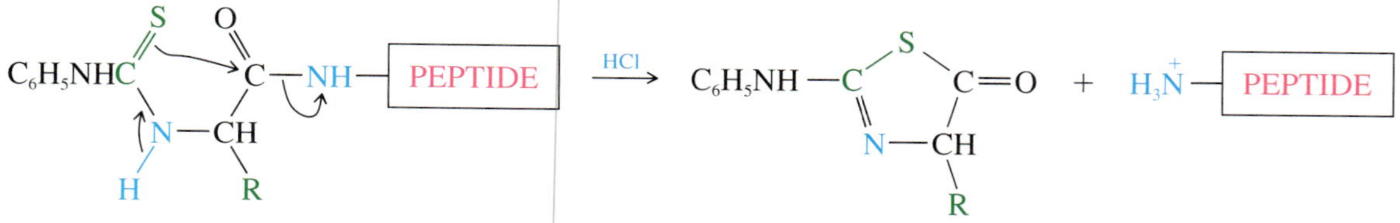

PTC derivative

Thiazolone

Remainder of peptide

Step 3: Once formed, the thiazolone derivative isomerizes to a more stable phenylthiohydantoin (PTH) derivative, which is isolated and characterized, thereby providing identification of the N-terminal amino acid. The remainder of the peptide (formed in step 2) can be isolated and subjected to a second Edman degradation.

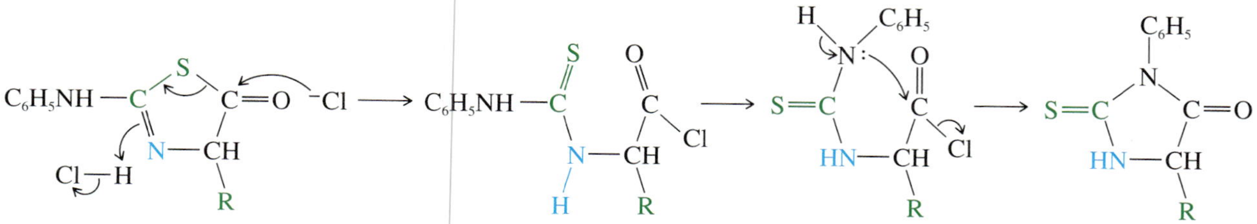

Thiazolone

PTH derivative

an internal nucleophile, and the only amide bond cleaved under these conditions is the one to the N-terminal amino acid. The product of this cleavage, called a *thiazolone*, is unstable under the conditions of its formation and rearranges to a *phenylthiohydantoin* (PTH), which is isolated and identified by comparing it with standard samples of PTH derivatives of known amino acids. This is normally done by chromatographic methods, but mass spectrometry has also been used.

Only the N-terminal amide bond is broken in the Edman degradation; the rest of the peptide chain remains intact. It can be isolated and subjected to a second Edman procedure to determine its new N terminus. We can proceed along a peptide chain by beginning with the N terminus and determining each amino acid in order. The sequence is given directly by the structure of the PTH derivative formed in each successive degradation.

> **PROBLEM 27.15**   Give the structure of the PTH derivative isolated in the second Edman cycle of the tetrapeptide Val-Phe-Gly-Ala.

Ideally, one could determine the primary structure of even the largest protein by repeating the Edman procedure. Because anything less than 100% conversion in any single Edman degradation gives a mixture containing some of the original peptide along with the degraded one, two different PTH derivatives are formed in the next Edman cycle, and the ideal is not realized in practice. Nevertheless, some impressive results have been achieved. It is a fairly routine matter to sequence the first 20 amino acids from the N terminus by repetitive Edman cycles, and even 60 residues have been determined on a single sample of the protein myoglobin. The entire procedure has been automated and incorporated into a device called an **Edman sequenator,** which carries out all the operations under computer control.

The amount of sample required is quite small; as little as 10^{-10} mole is typical. So many peptides and proteins have been sequenced now that it is impossible to give an accurate count. What was Nobel Prize-winning work in 1958 is routine today. Nor has the story ended. Sequencing of *nucleic acids* has advanced so dramatically that it is possible to clone the gene that codes for a particular protein, sequence its DNA, and deduce the structure of the protein from the nucleotide sequence of the DNA. We'll have more to say about DNA sequencing in the next chapter.

27.14 THE STRATEGY OF PEPTIDE SYNTHESIS

One way to confirm the structure proposed for a peptide is to synthesize a peptide having a specific sequence of amino acids and compare the two. This was done, for example, in the case of *bradykinin,* a peptide present in blood that acts to lower blood pressure. Excess bradykinin, formed as a response to the sting of wasps and other insects containing substances in their venom that stimulate bradykinin release, causes severe local pain. Bradykinin was originally believed to be an octapeptide containing two proline residues; however, a nonapeptide containing three prolines in the following sequence was synthesized and determined to be identical with natural bradykinin in every respect, including biological activity:

Arg-Pro-Pro-Gly-Phe-Ser-Pro-Phe-Arg

Bradykinin

A reevaluation of the original sequence data established that natural bradykinin was indeed the nonapeptide shown. Here the synthesis of a peptide did more than confirm structure; synthesis was instrumental in determining structure.

Chemists and biochemists also synthesize peptides in order to better understand how they act. By systematically altering the sequence, it's sometimes possible to find out which amino acids are essential to the reactions that involve a particular peptide.

Only a few prescription drugs are peptides, the most prominent being insulin and calcitonin. The insulin required for the treatment of diabetes used to be obtained by extraction from the pancreas glands of cows and pigs. Since the early 1980s, this "natural" insulin has been replaced by "synthetic" human insulin prepared by recombinant DNA technology. Synthetic insulin is not only identical to human insulin, it is less expensive than insulin obtained from animals. A somewhat smaller polypeptide, *calcitonin* with 32 amino acids, is prepared by more traditional methods of synthetic organic chemistry. Synthetic calcitonin is identical to that obtained from salmon and is widely used for the treatment of osteoporosis. How calcitonin acts remains uncertain, but one possibility is that it maintains bone mass, not by increasing the rate of bone growth, but by decreasing the rate of bone loss.

Other than the biochemical methods typified by the synthesis of insulin, there are two major approaches to peptide synthesis:

1. Solution phase

2. Solid phase

Although the two approaches differ in respect to the phase in which the synthesis is carried out, the overall strategy is the same in both.

The objective in peptide synthesis may be simply stated: to connect amino acids in a prescribed sequence by amide bond formation between them. A number of very effective methods and reagents have been designed for peptide bond formation, so that the joining together of amino acids by amide linkages is not difficult. The real difficulty lies in ensuring that the correct sequence is obtained. This can be illustrated by considering the synthesis of a representative dipeptide, Phe-Gly. Random peptide bond formation in a mixture containing phenylalanine and glycine would be expected to lead to four dipeptides:

$$\underset{\underset{\text{Phenylalanine}}{\overset{|}{\underset{}{\text{CH}_2\text{C}_6\text{H}_5}}}}{\overset{+}{\text{H}_3\text{NCHCO}_2^-}} \;+\; \underset{\text{Glycine}}{\overset{+}{\text{H}_3\text{NCH}_2\text{CO}_2^-}} \longrightarrow \text{Phe-Gly} \;+\; \text{Phe-Phe} \;+\; \text{Gly-Phe} \;+\; \text{Gly-Gly}$$

To direct the synthesis so that only Phe-Gly is formed, the amino group of phenylalanine and the carboxyl group of glycine must be protected so that they cannot react under the conditions of peptide bond formation. We can represent the peptide bond formation step by the following equation, where X and Y are amine- and carboxyl-protecting groups, respectively:

$$\underset{\substack{| \\ \text{CH}_2\text{C}_6\text{H}_5 \\ \text{N-Protected} \\ \text{phenylalanine}}}{\text{X}-\text{NHCHCOH}} \;+\; \underset{\substack{\text{C-Protected} \\ \text{glycine}}}{\text{H}_2\text{NCH}_2\overset{\text{O}}{\overset{\|}{\text{C}}}-\text{Y}} \;\xrightarrow{\text{couple}}\; \underset{\substack{| \\ \text{CH}_2\text{C}_6\text{H}_5 \\ \text{Protected Phe-Gly}}}{\text{X}-\text{NHCHC}-\text{NHCH}_2\text{C}-\text{Y}} \;\xrightarrow{\text{deprotect}}\; \underset{\substack{| \\ \text{CH}_2\text{C}_6\text{H}_5 \\ \text{Phe-Gly}}}{\overset{+}{\text{H}_3\text{NCHC}}-\text{NHCH}_2\text{CO}^-}$$

Thus, the synthesis of a dipeptide of prescribed sequence requires at least three operations:

1. *Protect* the amino group of the N-terminal amino acid and the carboxyl group of the C-terminal amino acid.

2. *Couple* the two protected amino acids by amide bond formation between them.

3. *Deprotect* the amino group at the N terminus and the carboxyl group at the C terminus.

Higher peptides are prepared in an analogous way by a direct extension of the logic just outlined for the synthesis of dipeptides.

Sections 27.15 through 27.17 describe the chemistry associated with the protection and deprotection of amino and carboxyl functions, along with methods for peptide bond formation. The focus in those sections is on solution-phase peptide synthesis. Section 27.18 shows how these methods are adapted to solid-phase synthesis.

27.15 AMINO GROUP PROTECTION

The reactivity of an amino group is suppressed by converting it to an amide, and amino groups are most often protected by acylation. The benzyloxycarbonyl group

$(C_6H_5CH_2O\overset{\overset{\displaystyle O}{\|}}{C}-)$ is one of the most often used amino-protecting groups. It is attached by acylation of an amino acid with benzyloxycarbonyl chloride.

> Another name for the benzyloxycarbonyl group is *carbobenzoxy*. This name, and its abbreviation *Cbz*, are often found in the older literature, but are no longer a part of IUPAC nomenclature.

| Benzyloxycarbonyl chloride | Phenylalanine | *N*-Benzyloxycarbonylphenylalanine (82–87%) |

PROBLEM 27.16 Lysine reacts with two equivalents of benzyloxycarbonyl chloride to give a derivative containing two benzyloxycarbonyl groups. What is the structure of this compound?

Just as it is customary to identify individual amino acids by abbreviations, so too with protected amino acids. The approved abbreviation for a benzyloxycarbonyl group is the letter Z. Thus, *N*-benzyloxycarbonylphenylalanine is represented as

$$\underset{\underset{\displaystyle CH_2C_6H_5}{|}}{ZNHCHCO_2H} \qquad \text{or more simply as} \qquad \text{Z-Phe}$$

The value of the benzyloxycarbonyl protecting group is that it is easily removed by reactions other than hydrolysis. In peptide synthesis, amide bonds are formed. We protect the N terminus as an amide but need to remove the protecting group without cleaving the very amide bonds we labored so hard to construct. Removing the protecting group by hydrolysis would surely bring about cleavage of peptide bonds as well. One advantage that the benzyloxycarbonyl protecting group enjoys over more familiar acyl groups such as acetyl is that it can be removed by *hydrogenolysis* in the presence of palladium. The following equation illustrates this for the removal of the benzyloxycarbonyl protecting group from the ethyl ester of Z-Phe-Gly:

> *Hydrogenolysis* refers to the cleavage of a molecule under conditions of catalytic hydrogenation.

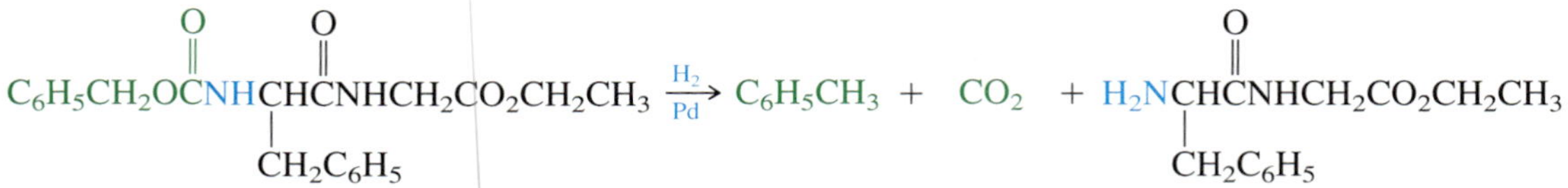

$$C_6H_5CH_2OCNHCHCNHCH_2CO_2CH_2CH_3 \xrightarrow[\text{Pd}]{H_2} C_6H_5CH_3 + CO_2 + H_2NCHCNHCH_2CO_2CH_2CH_3$$

with structural detail:

N-Benzyloxycarbonylphenylalanylglycine ethyl ester   Toluene   Carbon dioxide   Phenylalanylglycine ethyl ester (100%)

Alternatively, the benzyloxycarbonyl protecting group may be removed by treatment with hydrogen bromide in acetic acid:

$$C_6H_5CH_2OCNHCHCNHCH_2CO_2CH_2CH_3 \xrightarrow{HBr} C_6H_5CH_2Br + CO_2 + \overset{+}{H_3}NCHCNHCH_2CO_2CH_2CH_3 \; Br^-$$

N-Benzyloxycarbonylphenylalanylglycine ethyl ester   Benzyl bromide   Carbon dioxide   Phenylalanylglycine ethyl ester hydrobromide (82%)

Deprotection by this method rests on the ease with which benzyl esters are cleaved by nucleophilic attack at the benzylic carbon in the presence of strong acids. Bromide ion is the nucleophile.

A related N-terminal-protecting group is *tert*-butoxycarbonyl, abbreviated *Boc:*

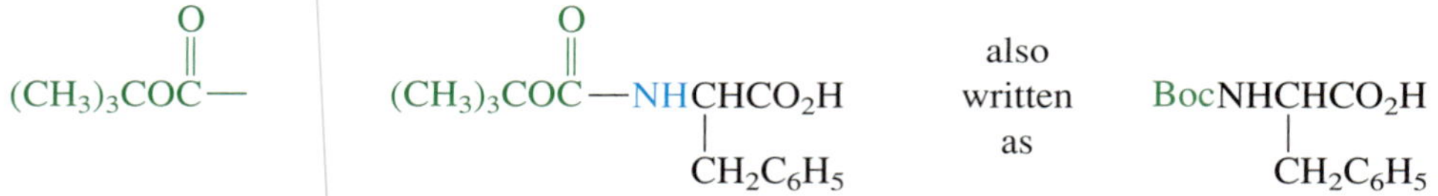

$(CH_3)_3COC-$ $(CH_3)_3COC-NHCHCO_2H$ also written as $BocNHCHCO_2H$

tert-Butoxycarbonyl (Boc-)   N-*tert*-Butoxycarbonylphenylalanine   Boc-Phe

Like the benzyloxycarbonyl protecting group, the Boc group may be removed by treatment with hydrogen bromide (it is stable to hydrogenolysis, however):

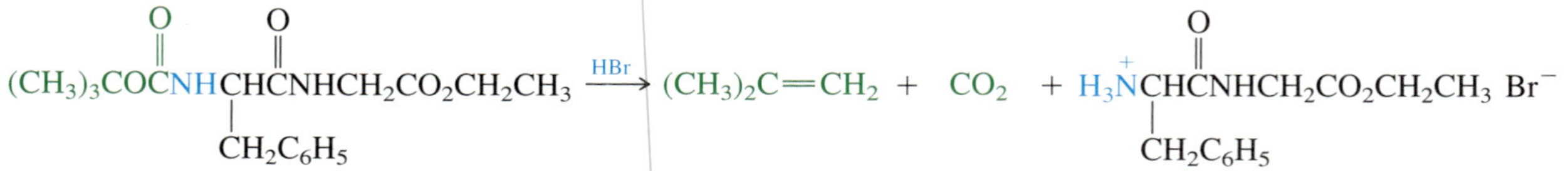

$$(CH_3)_3COCNHCHCNHCH_2CO_2CH_2CH_3 \xrightarrow{HBr} (CH_3)_2C=CH_2 + CO_2 + \overset{+}{H_3}NCHCNHCH_2CO_2CH_2CH_3 \; Br^-$$

N-*tert*-Butoxycarbonylphenylalanylglycine ethyl ester   2-Methylpropene   Carbon dioxide   Phenylalanylglycine ethyl ester hydrobromide (86%)

An experiment using Boc protection in the synthesis of a dipeptide can be found in the November 1989 issue of the *Journal of Chemical Education*, pp. 965–967.

The *tert*-butyl group is cleaved as the corresponding carbocation. Loss of a proton from *tert*-butyl cation converts it to 2-methylpropene. Because of the ease with which a *tert*-butyl group is cleaved as a carbocation, other acidic reagents, such as trifluoroacetic acid, may also be used.

27.16 CARBOXYL GROUP PROTECTION

Carboxyl groups of amino acids and peptides are normally protected as esters. Methyl and ethyl esters are prepared by Fischer esterification. Deprotection of methyl and ethyl esters is accomplished by hydrolysis in base. Benzyl esters are a popular choice because they can also be removed by hydrogenolysis. Thus a synthetic peptide, protected at both its N terminus with a Z group and at its C terminus as a benzyl ester, can be completely deprotected in a single operation.

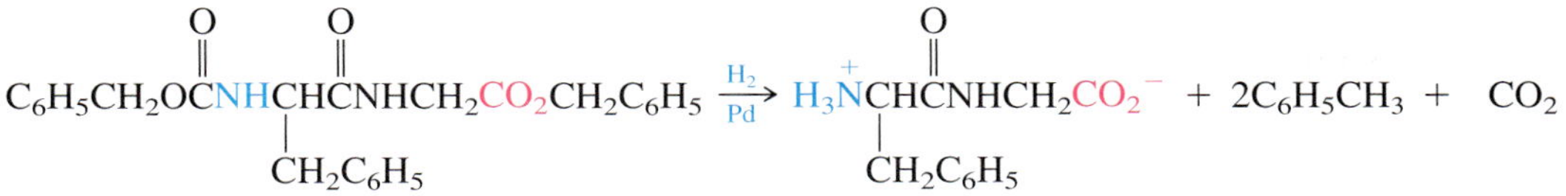

$$
C_6H_5CH_2O\overset{\displaystyle O}{\overset{\|}{C}}NHCHC\overset{\displaystyle O}{\overset{\|}{C}}NHCH_2CO_2CH_2C_6H_5 \xrightarrow[\text{Pd}]{H_2} H_3\overset{+}{N}CHC\overset{\displaystyle O}{\overset{\|}{C}}NHCH_2CO_2^- + 2C_6H_5CH_3 + CO_2
$$

$\underset{CH_2C_6H_5}{}$		$\underset{CH_2C_6H_5}{}$		
N-Benzyloxycarbonylphenylalanylglycine benzyl ester		Phenylalanylglycine (87%)	Toluene	Carbon dioxide

Several of the amino acids listed in Table 27.1 bear side-chain functional groups, which must also be protected during peptide synthesis. In most cases, protecting groups are available that can be removed by hydrogenolysis.

27.17 PEPTIDE BOND FORMATION

To form a peptide bond between two suitably protected amino acids, the free carboxyl group of one of them must be *activated* so that it is a reactive acylating agent. The most familiar acylating agents are acyl chlorides, and they were once extensively used to couple amino acids. Certain drawbacks to this approach, however, led chemists to seek alternative methods.

In one method, treatment of a solution containing the N-protected and the C-protected amino acids with *N,N'*-dicyclohexylcarbodiimide (DCCI) leads directly to peptide bond formation:

$$
\underset{\underset{CH_2C_6H_5}{|}}{ZNHCHC\overset{\displaystyle O}{\overset{\|}{}}OH} + H_2NCH_2\overset{\displaystyle O}{\overset{\|}{C}}OCH_2CH_3 \xrightarrow[\text{chloroform}]{DCCI} \underset{\underset{CH_2C_6H_5}{|}}{ZNHCHC\overset{\displaystyle O}{\overset{\|}{}}}-NHCH_2\overset{\displaystyle O}{\overset{\|}{C}}OCH_2CH_3
$$

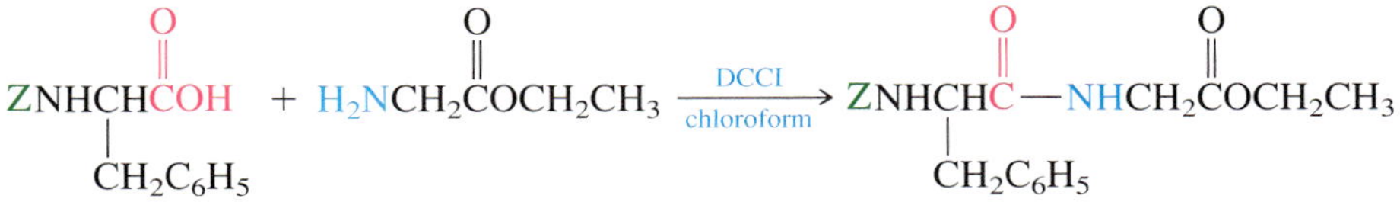

Z-Protected phenylalanine	Glycine ethyl ester	Z-Protected Phe-Gly ethyl ester (83%)

N,N'-Dicyclohexylcarbodiimide has the structure shown:

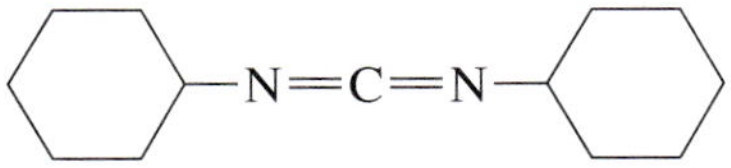

⬡—N=C=N—⬡

N,N'-Dicyclohexylcarbodiimide (DCCI)

The mechanism by which DCCI promotes the condensation of an amine and a carboxylic acid to give an amide is outlined in Figure 27.13.

> **PROBLEM 27.17** Show the steps involved in the synthesis of Ala-Leu from alanine and leucine using benzyloxycarbonyl and benzyl ester protecting groups and DCCI-promoted peptide bond formation.

In the second major method of peptide synthesis the carboxyl group is activated by converting it to an *active ester*, usually a *p*-nitrophenyl ester. Recall from Section 20.12 that esters react with ammonia and amines to give amides. *p*-Nitrophenyl esters are much more reactive than methyl and ethyl esters in these reactions because *p*-nitrophenoxide is a better (less basic) leaving group than methoxide and ethoxide. Simply allowing the active ester and a C-protected amino acid to stand in a suitable solvent is sufficient to bring about peptide bond formation by nucleophilic acyl substitution.

Overall reaction:

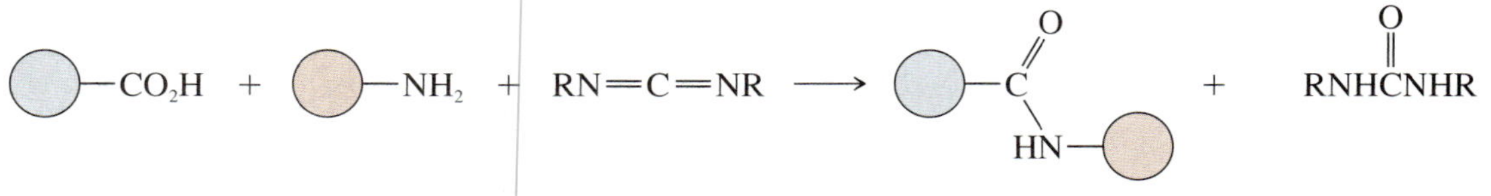

| Carboxylic acid | Amine | DCCI | Amide | N,N'-Dicyclohexylurea |

DCCI = N,N'-dicyclohexylcarbodiimide; R = cyclohexyl

Mechanism:

Step 1: In the first stage of the reaction, the carboxylic acid adds to one of the double bonds of DCCI to give an O-acylisourea.

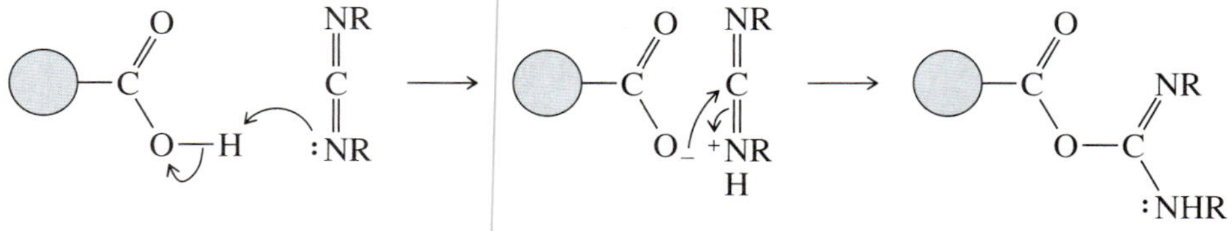

Carboxylic acid   DCCI   O-Acylisourea

Step 2: Structurally, O-acylisoureas resemble carboxylic acid anhydrides and are powerful acylating agents. In the reaction's second stage the amine adds to the carbonyl group of the O-acylisourea to give a tetrahedral intermediate.

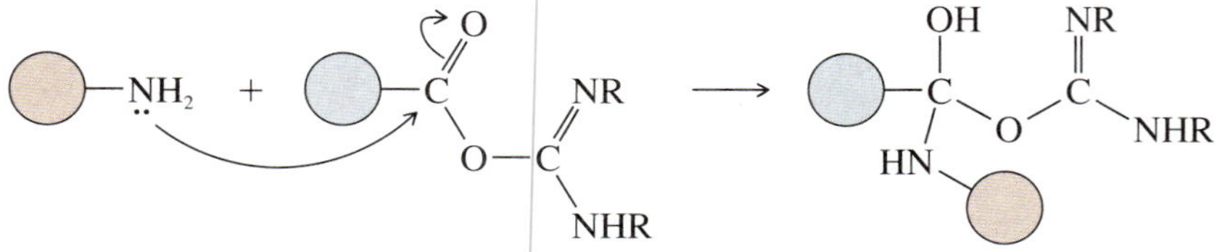

Amine   O-Acylisourea   Tetrahedral intermediate

Step 3: The tetrahedral intermediate dissociates to an amide and N,N'-dicyclohexylurea.

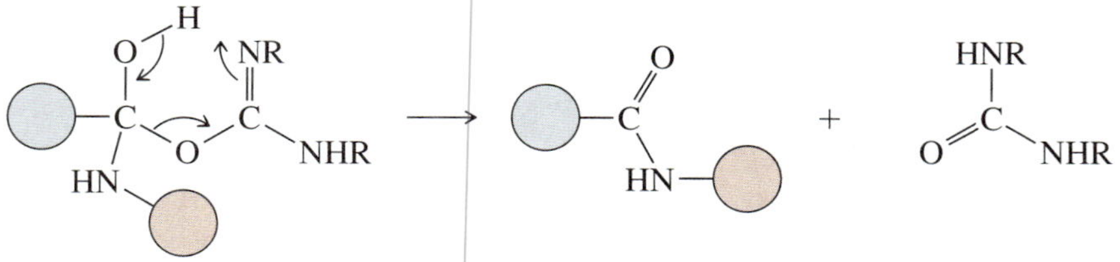

Tetrahedral intermediate   Amide   N,N'-Dicyclohexylurea

FIGURE 27.13 The mechanism of amide bond formation by N,N'-dicyclohexylcarbodiimide-promoted condensation of a carboxylic acid and an amine.

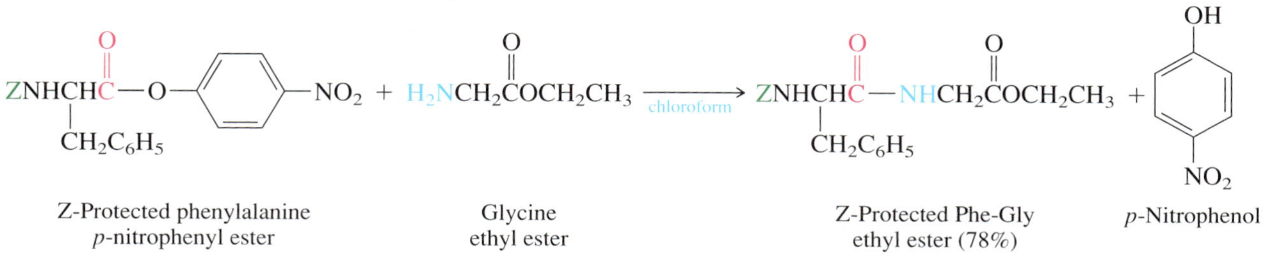

The chemical reaction scheme shows:

Z-Protected phenylalanine *p*-nitrophenyl ester + Glycine ethyl ester $\xrightarrow{\text{chloroform}}$ Z-Protected Phe-Gly ethyl ester (78%) + *p*-Nitrophenol

The *p*-nitrophenol formed as a byproduct in this reaction is easily removed by extraction with dilute aqueous base. Unlike free amino acids and peptides, protected peptides are not zwitterionic and are more soluble in organic solvents than in water.

> **PROBLEM 27.18** *p*-Nitrophenyl esters are made from Z-protected amino acids by reaction with *p*-nitrophenol in the presence of *N,N'*-dicyclohexylcarbodiimide. Suggest a reasonable mechanism for this reaction.

> **PROBLEM 27.19** Show how you could convert the ethyl ester of Z-Phe-Gly to Leu-Phe-Gly (as its ethyl ester) by the active ester method.

Higher peptides are prepared either by stepwise extension of peptide chains, one amino acid at a time, or by coupling of fragments containing several residues (the **fragment condensation** approach). Human pituitary adrenocorticotropic hormone (ACTH), for example, has 39 amino acids and was synthesized by coupling of smaller peptides containing residues 1–10, 11–16, 17–24, and 25–39. An attractive feature of this approach is that the various protected peptide fragments may be individually purified, which simplifies the purification of the final product. Among the substances that have been synthesized by fragment condensation are insulin (51 amino acids) and the protein ribonuclease A (124 amino acids). In the stepwise extension approach, the starting peptide in a particular step differs from the coupling product by only one amino acid residue and the properties of the two peptides may be so similar as to make purification by conventional techniques all but impossible. The solid-phase method described below overcomes many of the difficulties involved in the purification of intermediates.

27.18 SOLID-PHASE PEPTIDE SYNTHESIS: THE MERRIFIELD METHOD

In 1962, R. Bruce Merrifield of Rockefeller University reported the synthesis of the nonapeptide bradykinin by a novel method. In Merrifield's method, peptide coupling and deprotection are carried out not in homogeneous solution but at the surface of an insoluble polymer, or *solid support*. Beads of a copolymer prepared from styrene containing about 2% divinylbenzene are treated with chloromethyl methyl ether and tin(IV) chloride to give a resin in which about 10% of the aromatic rings bear —CH₂Cl groups (Figure 27.14). The growing peptide is anchored to this polymer, and excess reagents, impurities, and byproducts are removed by thorough washing after each operation. This greatly simplifies the purification of intermediates.

The actual process of solid-phase peptide synthesis, outlined in Figure 27.15, begins with the attachment of the C-terminal amino acid to the chloromethylated polymer in step 1. Nucleophilic substitution by the carboxylate anion of an *N*-Boc-protected C-terminal

Merrifield was awarded the 1984 Nobel Prize in chemistry for developing the solid-phase method of peptide synthesis.

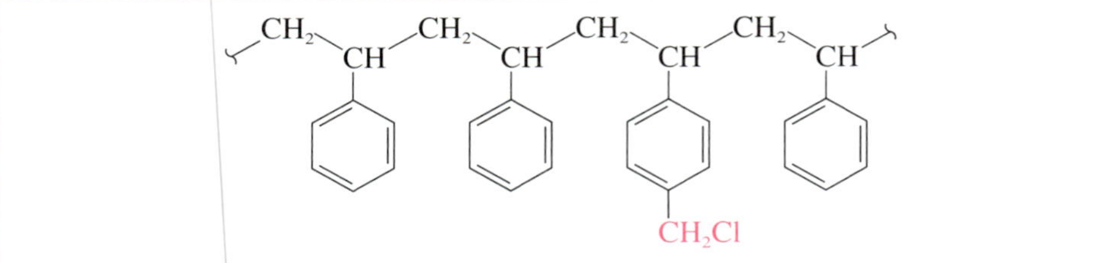

FIGURE 27.14 A section of polystyrene showing one of the benzene rings modified by chloromethylation. Individual polystyrene chains in the resin used in solid-phase peptide synthesis are connected to one another at various points (cross-linked) by adding a small amount of *p*-divinylbenzene to the styrene monomer. The chloromethylation step is carried out under conditions such that only about 10% of the benzene rings bear —CH₂Cl groups.

amino acid displaces chloride from the chloromethyl group of the polymer to form an ester, protecting the C terminus while anchoring it to a solid support. Next, the Boc group is removed by treatment with acid (step 2), and the polymer containing the unmasked N terminus is washed with a series of organic solvents. Byproducts are removed, and only the polymer and its attached C-terminal amino acid residue remain. Next (step 3), a peptide bond to an *N*-Boc-protected amino acid is formed by condensation in the presence of *N,N'*-dicyclohexylcarbodiimide. Again, the polymer is washed thoroughly. The Boc-protecting group is then removed by acid treatment (step 4), and after washing, the polymer is now ready for the addition of another amino acid residue by a repetition of the cycle. When all the amino acids have been added, the synthetic peptide is removed from the polymeric support by treatment with hydrogen bromide in trifluoroacetic acid.

By successively adding amino acid residues to the C-terminal amino acid, it took Merrifield only eight days to synthesize bradykinin in 68% yield. The biological activity of synthetic bradykinin was identical with that of natural material.

PROBLEM 27.20 Starting with phenylalanine and glycine, outline the steps in the preparation of Phe-Gly by the Merrifield method.

Merrifield successfully automated all the steps in solid-phase peptide synthesis, and computer-controlled equipment is now commercially available to perform this synthesis. Using an early version of his "peptide synthesizer," in collaboration with coworker Bernd Gutte, Merrifield reported the synthesis of the enzyme ribonuclease in 1969. It took them only six weeks to perform the 369 reactions and 11,391 steps necessary to assemble the sequence of 124 amino acids of ribonuclease.

Solid-phase peptide synthesis does not solve all purification problems, however. Even if every coupling step in the ribonuclease synthesis proceeded in 99% yield, the product would be contaminated with many different peptides containing 123 amino acids, 122 amino acids, and so on. Thus, Merrifield and Gutte's six weeks of synthesis was followed by four months spent in purifying the final product. The technique has since been refined to the point that yields at the 99% level and greater are achieved with current instrumentation, and thousands of peptides and peptide analogs have been prepared by the solid-phase method.

Merrifield's concept of a solid-phase method for peptide synthesis and his development of methods for carrying it out set the stage for an entirely new way to do chemical reactions. Solid-phase synthesis has been extended to include numerous other classes of compounds and has helped spawn a whole new field called **combinatorial chemistry.** Combinatorial synthesis allows a chemist, using solid-phase techniques, to prepare hundreds of related compounds (called *libraries*) at a time. It is one of the most active areas of organic synthesis, especially in the pharmaceutical industry.

Step 1: The Boc-protected amino acid is anchored to the resin. Nucleophilic substitution of the benzylic chloride by the carboxylate anion gives an ester.

$$BocNHCHC\underset{R}{\underset{|}{}}\overset{O}{\overset{\|}{}}\underset{O^-}{} \quad \overset{Cl}{\underset{CH_2}{}} - \boxed{Resin}$$

Step 2: The Boc protecting group is removed by treatment with hydrochloric acid in dilute acetic acid. After the resin has been washed, the C-terminal amino acid is ready for coupling.

$$BocNHCHCO\underset{R}{\underset{|}{}}\overset{O}{\overset{\|}{}}-CH_2-\boxed{Resin}$$

HCl ↓

$$H_2NCHCO\underset{R}{\underset{|}{}}\overset{O}{\overset{\|}{}}-CH_2-\boxed{Resin}$$

Step 3: The resin-bound C-terminal amino acid is coupled to an N-protected amino acid by using N,N'-dicyclohexylcarbodiimide. Excess reagent and N,N'-dicyclohexylurea are washed away from the resin after coupling is complete.

$$BocNHCHCO_2H\underset{R'}{\underset{|}{}} \quad DCCI ↓$$

$$BocNHCHC\underset{R'}{\underset{|}{}}\overset{O}{\overset{\|}{}}-NHCHCO\underset{R}{\underset{|}{}}\overset{O}{\overset{\|}{}}-CH_2-\boxed{Resin}$$

Step 4: The Boc protecting group is removed as in step 2. If desired, steps 3 and 4 may be repeated to introduce as many amino acid residues as desired.

HCl ↓

$$H_2NCHCNHCHCO\underset{R'}{\underset{|}{}}\overset{O}{\overset{\|}{}}\underset{R}{\underset{|}{}}\overset{O}{\overset{\|}{}}-CH_2-\boxed{Resin}$$

Step n: When the peptide is completely assembled, it is removed from the resin by treatment with hydrogen bromide in trifluoroacetic acid.

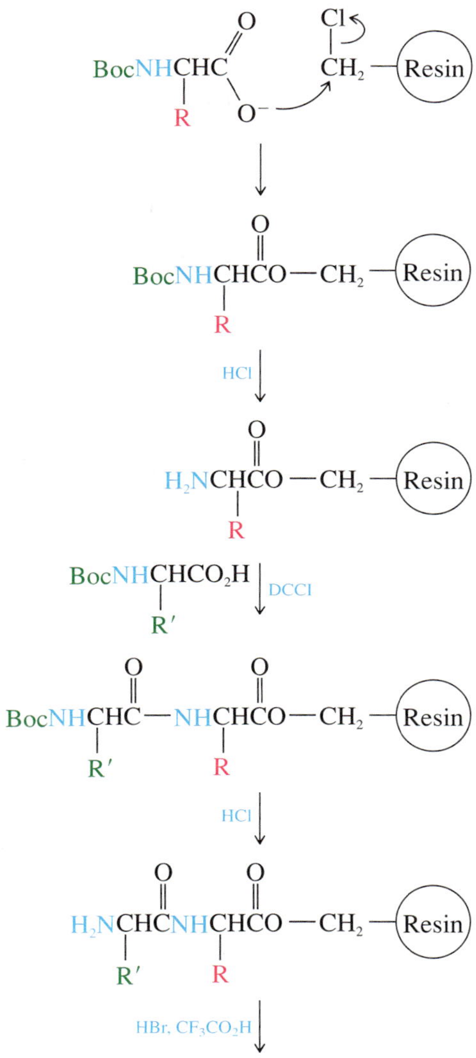

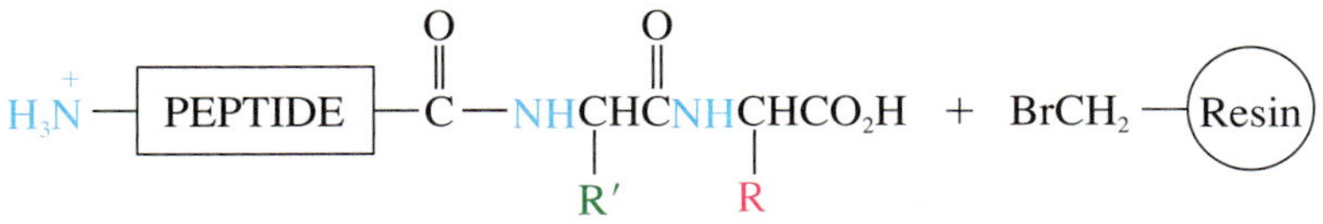

HBr, CF₃CO₂H ↓

$$\overset{+}{H_3N}-\boxed{PEPTIDE}-\overset{O}{\overset{\|}{C}}-NHCHCNHCHCO_2H\underset{R'}{\underset{|}{}}\overset{O}{\overset{\|}{}}\underset{R}{\underset{|}{}} \quad + \quad BrCH_2-\boxed{Resin}$$

27.19   SECONDARY STRUCTURES OF PEPTIDES AND PROTEINS

FIGURE 27.15 Peptide synthesis by the solid-phase method. Amino acid residues are attached sequentially beginning at the C terminus.

The primary structure of a peptide is its amino acid sequence. We also speak of the **secondary structure** of a peptide, that is, the conformational relationship of nearest neighbor amino acids with respect to each other. On the basis of X-ray crystallographic studies and careful examination of molecular models, Linus Pauling and Robert B. Corey of the California Institute of Technology showed that certain peptide conformations were more stable than others. Two arrangements, the **α helix** and the **β sheet**, stand out as secondary structural units that are both particularly stable and commonly encountered. Both of these incorporate two important features:

1. The geometry of the peptide bond is planar and the main chain is arranged in an anti conformation (Section 27.7).

2. Hydrogen bonding can occur when the N—H group of one amino acid unit and the C=O group of another are close in space; conformations that maximize the number of these hydrogen bonds are stabilized by them.

Figure 27.16 illustrates a β sheet structure for a protein composed of alternating glycine and alanine residues. There are hydrogen bonds between the C=O and H—N groups of adjacent chains. Van der Waals repulsions between the α hydrogens of glycine and the methyl groups of alanine cause the chains to rotate with respect to one another to give a rippled effect. Hence the name *pleated β sheet*. The pleated β sheet is an important secondary structure, especially in proteins that are rich in amino acids with small side chains, such as H (glycine), CH_3 (alanine), and CH_2OH (serine). *Fibroin,* the major protein of most silk fibers, is almost entirely pleated β sheet, and over 80% of it is a repeating sequence of the six-residue unit -Gly-Ser-Gly-Ala-Gly-Ala-. The pleated β sheet is flexible, but because the peptide chains are nearly in an extended conformation, it resists stretching.

Unlike the pleated β sheet, in which hydrogen bonds are formed *between* two chains, the α *helix* is stabilized by hydrogen bonds *within* a single chain. Figure 27.17 illustrates a section of peptide α helix constructed from L-alanine. A right-handed helical conformation with about 3.6 amino acids per turn permits each carbonyl oxygen to be hydrogen-bonded to an amide proton and vice versa. The α helix is found in many proteins; the principal protein components of muscle *(myosin)* and wool *(α-keratin)*, for example, contain high percentages of α helix. When wool fibers are stretched, these helical regions are elongated by the breaking of hydrogen bonds. Disulfide bonds between cysteine residues of neighboring α-keratin chains are too strong to be broken during stretching, however, and they limit the extent of distortion. After the stretching force is removed, the hydrogen bonds re-form spontaneously, and the wool fiber returns to its original shape. Wool has properties that are different from those of silk because the secondary structures of the two fibers are different, and their secondary structures are different because the primary structures are different.

Proline is the only amino acid in Table 27.1 that is a secondary amine, and its presence in a peptide chain introduces an amide nitrogen that has no hydrogen available for hydrogen bonding. This disrupts the network of hydrogen bonds and divides the peptide into two separate regions of α helix. The presence of proline is often associated with a bend in the peptide chain.

FIGURE 27.16 The β-sheet secondary structure of a protein, composed of alternating glycine and alanine residues. Hydrogen bonding occurs between the amide N—H of one chain and the carbonyl oxygen of another. Van der Waals repulsions between substituents at the α carbon atoms, shown here as vertical methyl groups, introduces creases in the sheet. The structure of the pleated β sheet is seen more clearly on *Learning By Modeling* by rotating it in three dimensions.

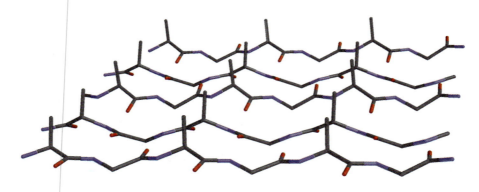

FIGURE 27.17 An α helix of a portion of a protein in which all of the amino acids are alanine. The helix is stabilized by hydrogen bonds between the N—H proton of one amide group and the carbonyl oxygen of another. The methyl groups at the α carbon project away from the outer surface of the helix. When viewed along the helical axis, the chain turns in a clockwise direction (a right-handed helix). The structure of the α helix is seen more clearly by examining the molecular model on *Learning By Modeling* and rotating it in three dimensions.

Proteins, or sections of proteins, sometimes exist as **random coils,** an arrangement that lacks the regularity of the α helix or pleated β sheet.

27.20 TERTIARY STRUCTURE OF PEPTIDES AND PROTEINS

The **tertiary structure** of a peptide or protein refers to the folding of the chain. The way the chain is folded affects both the physical properties of a protein and its biological function. Structural proteins, such as those present in skin, hair, tendons, wool, and silk, may have either helical or pleated-sheet secondary structures, but in general are elongated in shape, with a chain length many times the chain diameter. They are classed as *fibrous* proteins and, as befits their structural role, tend to be insoluble in water. Many other proteins, including most enzymes, operate in aqueous media; some are soluble, but most are dispersed as colloids. Proteins of this type are called *globular* proteins. Globular proteins are approximately spherical. Figure 27.18 shows carboxypeptidase A, a globular protein containing 307 amino acids. A typical protein such as carboxypeptidase A

FIGURE 27.18 The structure of carboxypeptidase A displayed as (a) a tube model and (b) a ribbon diagram. The tube model shows all of the amino acids and their side chains. The most evident feature illustrated by (a) is the globular shape of the enzyme. The ribbon diagram emphasizes the folding of the chain and the helical regions. As can be seen in (b), a substantial portion of the protein, the sections colored gray, is not helical but is random coil. The orientation of the protein and the color-coding are the same in both views.

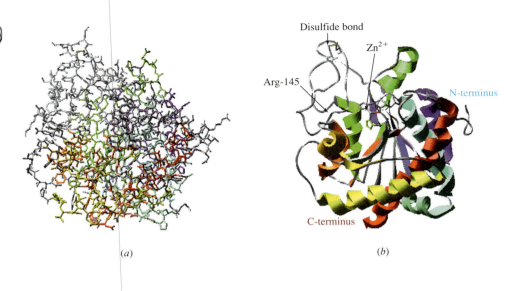

(a)

(b)

Disulfide bond

Zn^{2+}

Arg-145

N-terminus

C-terminus

For their work on myoglobin and hemoglobin, respectively, John C. Kendrew and Max F. Perutz were awarded the 1962 Nobel Prize in chemistry.

incorporates elements of a number of secondary structures: some segments are helical; others, pleated sheet; and still others correspond to no simple description.

The shape of a large protein is influenced by many factors, including, of course, its primary and secondary structure. The disulfide bond shown in Figure 27.18 links Cys-138 of carboxypeptidase A to Cys-161 and contributes to the tertiary structure. Carboxypeptidase A contains a Zn^{2+} ion, which is essential to the catalytic activity of the enzyme, and its presence influences the tertiary structure. The Zn^{2+} ion lies near the center of the enzyme, where it is coordinated to the imidazole nitrogens of two histidine residues (His-69, His-196) and to the carboxylate side chain of Glu-72.

Protein tertiary structure is also influenced by the environment. In water a globular protein usually adopts a shape that places its hydrophobic groups toward the interior, with its polar groups on the surface, where they are solvated by water molecules. About 65% of the mass of most cells is water, and the proteins present in cells are said to be in their *native state*—the tertiary structure in which they express their biological activity. When the tertiary structure of a protein is disrupted by adding substances that cause the protein chain to unfold, the protein becomes *denatured* and loses most, if not all, of its activity. Evidence that supports the view that the tertiary structure is dictated by the primary structure includes experiments in which proteins are denatured and allowed to stand, whereupon they are observed to spontaneously readopt their native-state conformation with full recovery of biological activity.

Most protein tertiary structures are determined by X-ray crystallography. The first, myoglobin, the oxygen storage protein of muscle, was determined in 1957. Since then thousands more have been determined. The data are deposited as a table of crystallographic coordinates in the **Protein Data Bank** and are freely available. The three-dimensional structure of carboxypeptidase A in Figure 27.18, for example, was produced by downloading the coordinates from the Protein Data Bank and converting them to a molecular model. At present, the Protein Data Bank averages about one new protein structure per day.

Knowing how the protein chain is folded is a key ingredient in understanding the mechanism by which an enzyme catalyzes a reaction. Take carboxypeptidase A for example. This enzyme catalyzes the hydrolysis of the peptide bond at the C terminus. It is

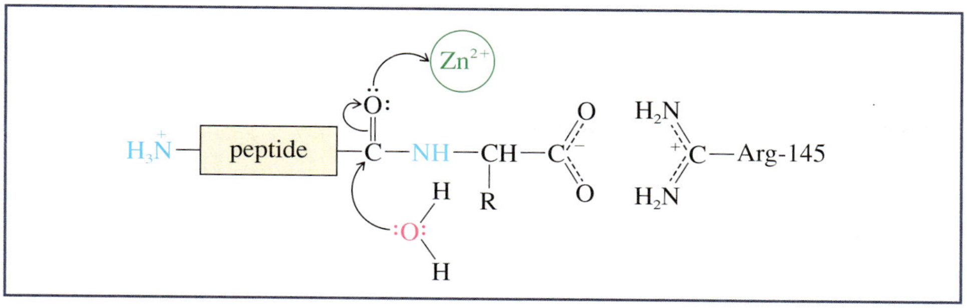

FIGURE 27.19 Proposed mechanism of hydrolysis of a peptide catalyzed by carboxypeptidase A. The peptide is bound at the active site by an ionic bond between its C-terminal amino acid and the positively charged side chain of arginine-145. Coordination of Zn^{2+} to oxygen makes the carbon of the carbonyl group more positive and increases the rate of nucleophilic attack by water.

believed that an ionic bond between the positively charged side chain of an arginine residue (Arg-145) of the enzyme and the negatively charged carboxylate group of the substrate's C-terminal amino acid binds the peptide at the **active site,** a cavity on the enzyme's surface where the catalytically important functional groups are located. There, the Zn^{2+} ion acts as a Lewis acid toward the carbonyl oxygen of the peptide substrate, increasing its susceptibility to attack by a water molecule (Figure 27.19).

Living systems contain thousands of different enzymes. As we have seen, all are structurally quite complex, and no sweeping generalizations can be made to include all aspects of enzymic catalysis. The case of carboxypeptidase A illustrates one mode of enzyme action, the bringing together of reactants and catalytically active functions at the active site.

27.21 COENZYMES

The number of chemical processes that protein side chains can engage in is rather limited. Most prominent among them are proton donation, proton abstraction, and nucleophilic addition to carbonyl groups. In many biological processes a richer variety of reactivity is required, and proteins often act in combination with substances other than proteins to carry out the necessary chemistry. These substances are called **cofactors,** and they can be organic or inorganic and strongly or weakly bound to the enzyme. Among cofactors that are organic molecules, the term **coenzyme** is applied to those that are not covalently bound to the enzyme, and **prosthetic group** to those that are. Acting alone, for example, proteins lack the necessary functionality to be effective oxidizing or reducing agents. They can catalyze biological oxidations and reductions, however, in the presence of a suitable coenzyme. In earlier sections we saw numerous examples of these reactions in which the coenzyme NAD^+ acted as an oxidizing agent, and others in which NADH acted as a reducing agent.

Heme (Figure 27.20) is an important prosthetic group in which iron(II) is coordinated with the four nitrogen atoms of a type of tetracyclic aromatic substance known as a *porphyrin*. The oxygen-storing protein of muscle, myoglobin, represented schematically in Figure 27.21, consists of a heme group surrounded by a protein of 153 amino acids. Four of the six available coordination sites of Fe^{2+} are taken up by the nitrogens of the

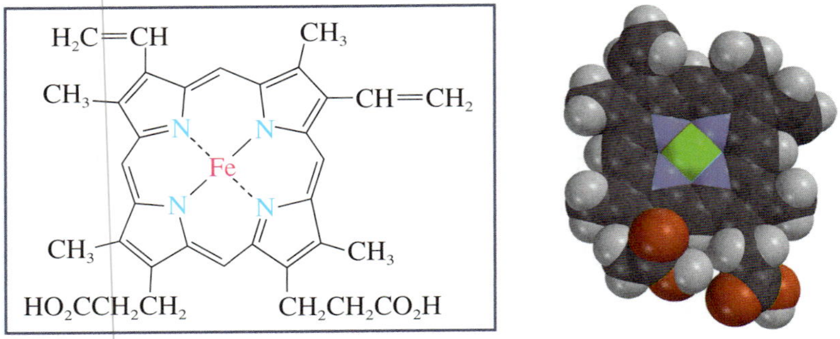

FIGURE 27.20 Heme shown as (a) a structural drawing and as (b) a space-filling model. The space-filling model shows the coplanar arrangement of the groups surrounding iron.

$$H_2C=CH \quad CH_3$$

Fe

$$CH_3 \quad CH=CH_2$$

$$CH_3 \quad CH_3$$

$$HO_2CCH_2CH_2 \quad CH_2CH_2CO_2H$$

(a)

(b)

porphyrin, one by a histidine residue of the protein, and the last by a water molecule. Myoglobin stores oxygen obtained from the blood by formation of an $Fe-O_2$ complex. The oxygen displaces water as the sixth ligand on iron and is held there until needed. The protein serves as a container for the heme and prevents oxidation of Fe^{2+} to Fe^{3+}, an oxidation state in which iron lacks the ability to bind oxygen. Separately, neither heme nor the protein binds oxygen in aqueous solution; together, they do it very well.

27.22 PROTEIN QUATERNARY STRUCTURE: HEMOGLOBIN

Rather than existing as a single polypeptide chain, some proteins are assemblies of two or more chains. The manner in which these subunits are organized is called the **quaternary structure** of the protein.

Hemoglobin is the oxygen-carrying protein of blood. It binds oxygen at the lungs and transports it to the muscles, where it is stored by myoglobin. Hemoglobin binds oxygen in very much the same way as myoglobin, using heme as the prosthetic group. Hemoglobin is much larger than myoglobin, however, having a molecular weight of 64,500, whereas that of myoglobin is 17,500; hemoglobin contains four heme units, myoglobin only one. Hemoglobin is an assembly of four hemes and four protein chains, including two identical chains called the *alpha chains* and two identical chains called the *beta chains*.

An article entitled "Hemoglobin: Its Occurrence, Structure, and Adaptation" appeared in the March 1982 issue of the *Journal of Chemical Education* (pp. 173–178).

FIGURE 27.21 The structure of sperm-whale myoglobin displayed as (a) a tube model and (b) a ribbon diagram. The tube model shows all of the amino acids in the chain; the ribbon diagram shows the folding of the chain. There are five separate regions of α helix in myoglobin, which are shown in different colors to distinguish them more clearly. The heme portion is included in both drawings, but is easier to locate in the ribbon diagram, as is the histidine side chain that is attached to the iron of heme.

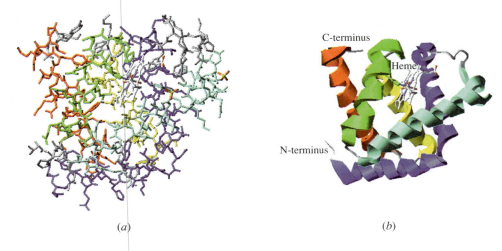

C-terminus

Heme

N-terminus

(a)

(b)

OH NO! IT'S INORGANIC!

The amino acid L-arginine undergoes an interesting biochemical conversion.

$$\underset{\text{L-Arginine}}{\overset{\displaystyle\overset{+}{\underset{\|}{N}}H_2}{H_2NCNHCH_2CH_2CH_2\underset{\overset{|}{+}NH_3}{CHCO^-}}} \longrightarrow$$

$$\underset{\text{L-Citrulline}}{\overset{\displaystyle\overset{O}{\|}}{H_2NCNHCH_2CH_2CH_2\underset{\overset{|}{+}NH_3}{\overset{\overset{O}{\|}}{CHCO^-}}}} + \quad \underset{\text{Nitric oxide}}{NO}$$

Our experience conditions us to focus on the organic components of the reaction—L-arginine and L-citrulline—and to give less attention to the inorganic one—nitric oxide (nitrogen monoxide, NO). To do so, however, would lead us to overlook one of the most important discoveries in biology in the last quarter of the twentieth century.

Our story starts with the long-standing use of nitroglycerin to treat the chest pain that characterizes angina, a condition in diseases such as atherosclerosis in which restricted blood flow to the heart muscle itself causes it to receive an insufficient amount of oxygen. Placing a nitroglycerin tablet under the tongue provides rapid relief by expanding the blood vessels feeding the heart. A number of other nitrogen-containing compounds such as amyl nitrite and sodium nitroprusside exert a similar effect.

$$\underset{\text{Nitroglycerin}}{\overset{\displaystyle O_2NCH_2CHCH_2ONO_2}{\underset{\overset{|}{ONO_2}}{}}} \qquad \underset{\text{Amyl nitrite}}{(CH_3)_2CHCH_2CH_2ONO}$$

$$\underset{\text{Sodium nitroprusside}}{Na_2[O{=}N{-}Fe(CN)_5]}$$

A chemical basis for their action was proposed in 1977 by Ferid Murad who showed that all were sources of NO, thereby implicating it as the active agent.

Three years later, Robert F. Furchgott discovered that the relaxing of smooth muscles, such as blood vessel walls, was stimulated by an unknown substance produced in the lining of the blood vessels (the *endothelium*). He called this substance the *endothelium-dependent relaxing factor*, or EDRF and, in 1986, showed that EDRF was NO. Louis J. Ignarro reached the same conclusion at about the same time. Further support was provided by Salvador Moncada who showed that endothelial cells did indeed produce NO and that the L-arginine-to-L-citrulline conversion was responsible.

The initial skepticism that greeted the idea that NO, which is (a) a gas, (b) toxic, (c) inorganic, and (d) a free radical, could be a biochemical messenger was quickly overcome. An avalanche of results confirmed not only NO's role in smooth-muscle relaxation, but added more and more examples to an ever-expanding list of NO-stimulated biochemical processes. Digestion is facilitated by the action of NO on intestinal muscles. The drug Viagra (*sildenafil*), prescribed to treat erectile dysfunction, works by increasing the concentration of a hormone, the release of which is signaled by NO. A theory that NO is involved in long-term memory receives support from the fact the brain is a rich source of the enzyme *nitric oxide synthase* (NOS), which catalyzes the formation of NO from L-arginine. NO even mediates the glow of fireflies. They glow nonstop when placed in a jar containing NO, but not at all when measures are taken to absorb NO.

Identifying NO as a signaling molecule in biological processes clearly justified a Nobel Prize. The only mystery was who would get it. Nobel Prizes are often shared, but never among more than three persons. Although four scientists—Murad, Furchgott, Ignarro, and Moncada—made important contributions, the Nobel committee followed tradition and recognized only the first three of them with the 1998 Nobel Prize in medicine or physiology.

Some substances, such as CO, form strong bonds to the iron of heme, strong enough to displace O_2 from it. Carbon monoxide binds 30–50 times more effectively than oxygen to myoglobin and hundreds of times better than oxygen to hemoglobin. Strong binding of CO at the active site interferes with the ability of heme to perform its biological task of transporting and storing oxygen, with potentially lethal results.

How function depends on structure can be seen in the case of the genetic disorder *sickle cell anemia.* This is a debilitating, sometimes fatal, disease in which red blood cells become distorted ("sickle-shaped") and interfere with the flow of blood through the capillaries. This condition results from the presence of an abnormal hemoglobin in affected people. The primary structures of the beta chain of normal and sickle cell hemoglobin differ by a single amino acid out of 146; sickle cell hemoglobin has valine in place of glutamic acid as the sixth residue from the N terminus of the β chain. A tiny change in amino acid sequence can produce a life-threatening result! This modification is genetically controlled and probably became established in the gene pool because bearers of the trait have an increased resistance to malaria.

27.23 SUMMARY

This chapter revolves around **proteins.** The first half describes the building blocks of proteins, progressing through **amino acids** and **peptides.** The second half deals with proteins themselves.

Section 27.1	A group of 20 amino acids, listed in Table 27.1, regularly appears as the hydrolysis products of proteins. All are α-amino acids.
Section 27.2	Except for glycine, which is achiral, all of the α-amino acids present in proteins are chiral and have the L configuration at the α carbon.
Section 27.3	The most stable structure of a neutral amino acid is a **zwitterion.** The pH of an aqueous solution at which the concentration of the zwitterion is a maximum is called the isoelectric point (pI).

$$\underset{\text{CH(CH}_3)_2}{\overset{\text{CO}_2^-}{\overset{|}{\underset{|}{H_3\overset{+}{N}\text{———}H}}}}$$

Fischer projection of
L-valine in its zwitterionic form

Section 27.4	Amino acids are synthesized in the laboratory from

1. α-Halo acids by reaction with ammonia

2. Aldehydes by reaction with ammonia and cyanide ion (the Strecker synthesis)

3. Alkyl halides by reaction with the enolate anion derived from diethyl acetamidomalonate

The amino acids prepared by these methods are formed as racemic mixtures and are optically inactive.

Section 27.5	Amino acids undergo reactions characteristic of the amino group (e.g., amide formation) and the carboxyl group (e.g., esterification). Amino acid side chains undergo reactions characteristic of the functional groups they contain.

Section 27.6   The reactions that amino acids undergo in living systems include **transamination** and **decarboxylation.**

Section 27.7   An amide linkage between two α-amino acids is called a **peptide bond.** By convention, peptides are named and written beginning at the N terminus.

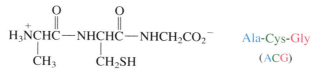

Alanylcysteinylglycine

Ala-Cys-Gly

(ACG)

Section 27.8   The **primary structure** of a peptide is given by its amino acid sequence plus any disulfide bonds between two cysteine residues. The primary structure is determined by a systematic approach in which the protein is cleaved to smaller fragments, even individual amino acids. The smaller fragments are sequenced and the main sequence deduced by finding regions of overlap among the smaller peptides.

Section 27.9   Complete hydrolysis of a peptide gives a mixture of amino acids. An **amino acid analyzer** identifies the individual amino acids and determines their molar ratios.

Section 27.10   Selective hydrolysis can be accomplished by using enzymes to catalyze cleavage at specific peptide bonds.

Section 27.11   Carboxypeptidase-catalyzed hydrolysis can be used to identify the C-terminal amino acid. The N terminus is determined by chemical means. One reagent used for this purpose is Sanger's reagent, 1-fluoro-2,4-dinitrobenzene (see Figure 27.9).

Section 27.12   The procedure described in Sections 27.8–27.11 was used to determine the amino acid sequence of insulin.

Section 27.13   Modern methods of peptide sequencing follow a strategy similar to that used to sequence insulin, but are automated and can be carried out on a small scale. A key feature is repetitive N-terminal identification using the **Edman degradation.**

Section 27.14   Synthesis of a peptide of prescribed sequence requires the use of protecting groups to minimize the number of possible reactions.

Section 27.15   Amino-protecting groups include *benzyloxycarbonyl* (Z) and *tert-butoxycarbonyl* (Boc).

$$C_6H_5CH_2OC\!-\!NHCHCO_2H \qquad (CH_3)_3COC\!-\!NHCHCO_2H$$

Benzyloxycarbonyl-protected   *tert*-Butoxycarbonyl-protected
amino acid   amino acid

Hydrogen bromide may be used to remove either the benzyloxycarbonyl or *tert*-butoxycarbonyl protecting group. The benzyloxycarbonyl protecting group may also be removed by catalytic hydrogenolysis.

Section 27.16   Carboxyl groups are normally protected as benzyl, methyl, or ethyl esters. Hydrolysis in dilute base is normally used to deprotect methyl and ethyl esters. Benzyl protecting groups are removed by hydrogenolysis.

Section 27.17   Peptide bond formation between a protected amino acid having a free carboxyl group and a protected amino acid having a free amino group can be accomplished with the aid of *N,N'*-dicyclohexylcarbodiimide (DCCI).

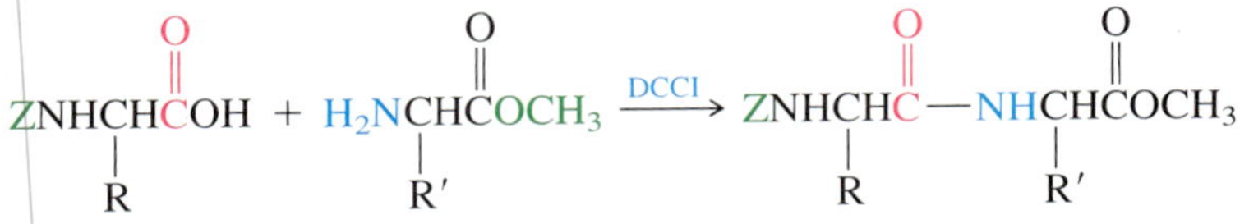

$$\underset{\substack{|\\R}}{ZNHCHCOH} + \underset{\substack{|\\R'}}{H_2NCHCOCH_3} \xrightarrow{\text{DCCI}} \underset{\substack{|\\R}}{ZNHCHC}-\underset{\substack{|\\R'}}{NHCHCOCH_3}$$

Section 27.18   In the Merrifield method the carboxyl group of an amino acid is anchored to a solid support and the chain extended one amino acid at a time. When all the amino acid residues have been added, the polypeptide is removed from the solid support.

Section 27.19   Two **secondary structures** of proteins are particularly prominent. The *pleated β sheet* is stabilized by hydrogen bonds between N—H and C=O groups of adjacent chains. The α *helix* is stabilized by hydrogen bonds within a single polypeptide chain.

Section 27.20   The folding of a peptide chain is its **tertiary structure.** The tertiary structure has a tremendous influence on the properties of the peptide and the biological role it plays. The tertiary structure is normally determined by X-ray crystallography.

Many globular proteins are enzymes. They accelerate the rates of chemical reactions in biological systems, but the kinds of reactions that take place are the fundamental reactions of organic chemistry. One way in which enzymes accelerate these reactions is by bringing reactive functions together in the presence of catalytically active functions of the protein.

Section 27.21   Often the catalytically active functions of an enzyme are nothing more than proton donors and proton acceptors. In many cases a protein acts in cooperation with a **coenzyme,** a small molecule having the proper functionality to carry out a chemical change not otherwise available to the protein itself.

Section 27.22   Many proteins consist of two or more chains, and the way in which the various units are assembled in the native state of the protein is called its **quaternary structure.**

PROBLEMS

27.21 The imidazole ring of the histidine side chain acts as a proton acceptor in certain enzyme-catalyzed reactions. Which is the more stable protonated form of the histidine residue, A or B? Why?

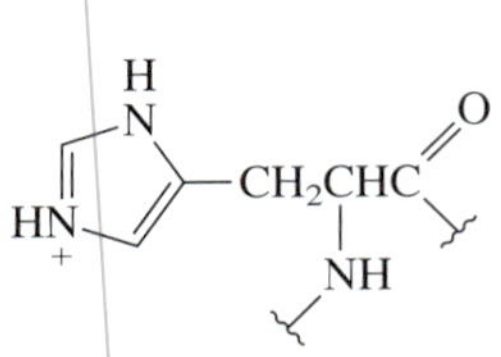

A 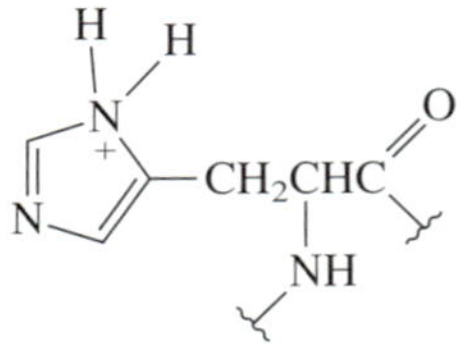   B

27.22 Acrylonitrile ($H_2C=CHC\equiv N$) readily undergoes conjugate addition when treated with nucleophilic reagents. Describe a synthesis of β-alanine ($H_3\overset{+}{N}CH_2CH_2CO_2{}^-$) that takes advantage of this fact.

27.23 (a) Isoleucine has been prepared by the following sequence of reactions. Give the structure of compounds A through D isolated as intermediates in this synthesis.

$$CH_3CH_2\underset{\underset{Br}{|}}{CH}CH_3 \xrightarrow[\text{sodium ethoxide}]{\text{diethyl malonate}} A \xrightarrow[\text{2. HCl}]{\text{1. KOH}} B\ (C_7H_{12}O_4)$$

$$B \xrightarrow{Br_2} C\ (C_7H_{11}BrO_4) \xrightarrow{\text{heat}} D \xrightarrow[H_2O]{NH_3} \text{isoleucine (racemic)}$$

(b) An analogous procedure has been used to prepare phenylalanine. What alkyl halide would you choose as the starting material for this synthesis?

27.24 Hydrolysis of the following compound in concentrated hydrochloric acid for several hours at 100°C gives one of the amino acids in Table 27.1. Which one? Is it optically active?

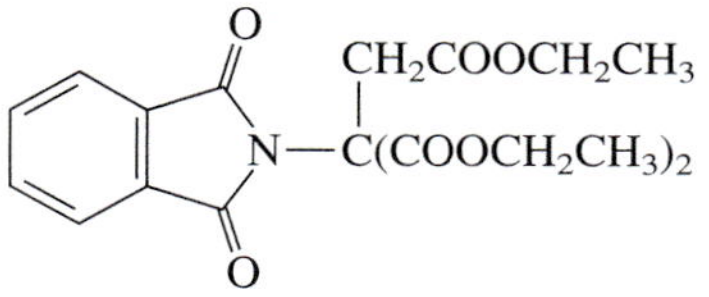

27.25 If you synthesized the tripeptide Leu-Phe-Ser from amino acids prepared by the Strecker synthesis, how many stereoisomers would you expect to be formed?

27.26 How many peaks would you expect to see on the strip chart after amino acid analysis of bradykinin?

Arg-Pro-Pro-Gly-Phe-Ser-Pro-Phe-Arg

Bradykinin

27.27 Automated amino acid analysis of peptides containing asparagine (Asn) and glutamine (Gln) residues gives a peak corresponding to ammonia. Why?

27.28 What are the products of each of the following reactions? Your answer should account for all the amino acid residues in the starting peptides.

(a) Reaction of Leu-Gly-Ser with 1-fluoro-2,4-dinitrobenzene

(b) Hydrolysis of the compound in part (a) in concentrated hydrochloric acid (100°C)

(c) Treatment of Ile-Glu-Phe with $C_6H_5N=C=S$, followed by hydrogen bromide in nitromethane

(d) Reaction of Asn-Ser-Ala with benzyloxycarbonyl chloride

(e) Reaction of the product of part (d) with *p*-nitrophenol and *N,N′*-dicyclohexylcarbodiimide

(f) Reaction of the product of part (e) with the ethyl ester of valine

(g) Hydrogenolysis of the product of part (f) by reaction with H_2 over palladium

27.29 Hydrazine cleaves amide bonds to form *acylhydrazides* according to the general mechanism of nucleophilic acyl substitution discussed in Chapter 20:

$$
\underset{\text{Amide}}{\overset{\overset{\displaystyle O}{\parallel}}{RCNHR'}} + \underset{\text{Hydrazine}}{H_2NNH_2} \longrightarrow \underset{\text{Acylhydrazide}}{\overset{\overset{\displaystyle O}{\parallel}}{RCNHNH_2}} + \underset{\text{Amine}}{R'NH_2}
$$

This reaction forms the basis of one method of terminal residue analysis. A peptide is treated with excess hydrazine in order to cleave all the peptide linkages. One of the terminal amino acids is cleaved as the free amino acid and identified; all the other amino acid residues are converted to acylhydrazides. Which amino acid is identified by *hydrazinolysis*, the N terminus or the C terminus?

27.30 *Somatostatin* is a tetradecapeptide of the hypothalamus that inhibits the release of pituitary growth hormone. Its amino acid sequence has been determined by a combination of Edman degradations and enzymic hydrolysis experiments. On the basis of the following data, deduce the primary structure of somatostatin:

1. Edman degradation gave PTH-Ala.
2. Selective hydrolysis gave peptides having the following indicated sequences:

   Phe-Trp

   Thr-Ser-Cys

   Lys-Thr-Phe

   Thr-Phe-Thr-Ser-Cys

   Asn-Phe-Phe-Trp-Lys

   Ala-Gly-Cys-Lys-Asn-Phe
3. Somatostatin has a disulfide bridge.

27.31 What protected amino acid would you anchor to the solid support in the first step of a synthesis of oxytocin (see Figure 27.8) by the Merrifield method?

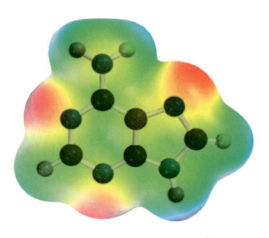

CHAPTER 28

NUCLEOSIDES, NUCLEOTIDES, AND NUCLEIC ACIDS

I n Chapter 1 we saw that a major achievement of the first half of the twentieth century was the picture of atomic and molecular structure revealed by quantum mechanics. In this, the last chapter, we examine the major achievement of the second half of that century—a molecular view of genetics based on the structure and biochemistry of nucleic acids.

Nucleic acids are acidic substances present in the nuclei of cells and were known long before anyone suspected they were the primary substances involved in the storage, transmission, and processing of genetic information. There are two kinds of nucleic acids: ribonucleic acid (RNA) and deoxyribonucleic acid (DNA). Both are complicated biopolymers, based on three structural units: a carbohydrate, a phosphate ester linkage between carbohydrates, and a heterocyclic aromatic compound. The heterocyclic aromatic compounds are referred to as purine and pyrimidine bases. We'll begin with them and follow the structural thread:

Purine and pyrimidine bases → Nucleosides → Nucleotides → Nucleic acids

There will be a few pauses along the way as we stop to examine some biochemical roles played by these compounds independent of their genetic one.

28.1 PYRIMIDINES AND PURINES

Two nitrogen-containing heterocyclic aromatic compounds—**pyrimidine** and **purine**—are the parents of the "bases" that constitute a key structural unit of nucleic acids.

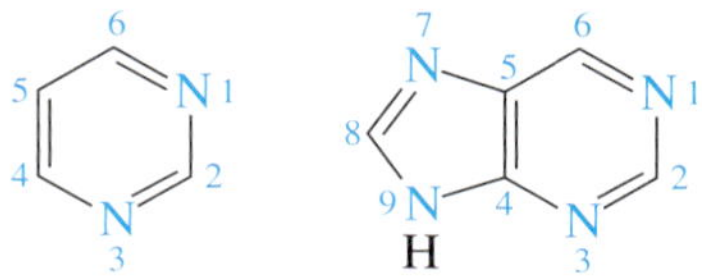

Pyrimidine

Purine

Both pyrimidine and purine are planar. You will see how important this flat shape is when we consider the structure of nucleic acids. In terms of their chemistry, pyrimidine and purine resemble pyridine. They are weak bases and relatively unreactive toward electrophilic aromatic substitution.

Pyrimidine and purine themselves do not occur naturally, but many of their derivatives do. Before going too far, we need to point out an important structural difference between derivatives that bear —OH groups and those with —NH₂ groups. The structure of a pyrimidine or purine that bears an —NH₂ group follows directly from the structure of the parent ring system.

6-Aminopurine is *adenine* and will appear numerous times in this chapter.

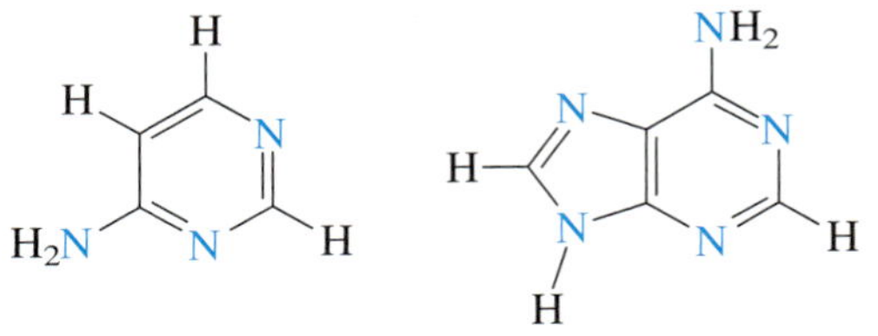

4-Aminopyrimidine   6-Aminopurine

However, the corresponding compounds that have an —OH group resemble enols:

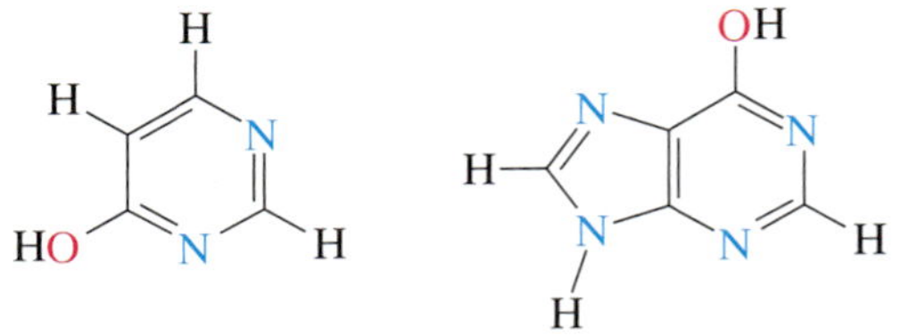

4-Hydroxypyrimidine   6-Hydroxypurine

and exist instead in their keto forms.

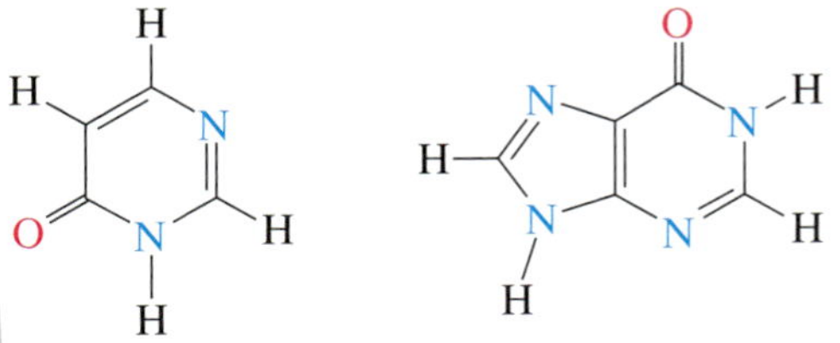

Keto form of   Keto form of
4-hydroxypyrimidine   6-hydroxypurine

By analogy to phenols, we would expect the isomers with —OH groups on benzene-like rings to be more stable. This turns out not to be true because the keto forms are also aromatic owing to amide resonance.

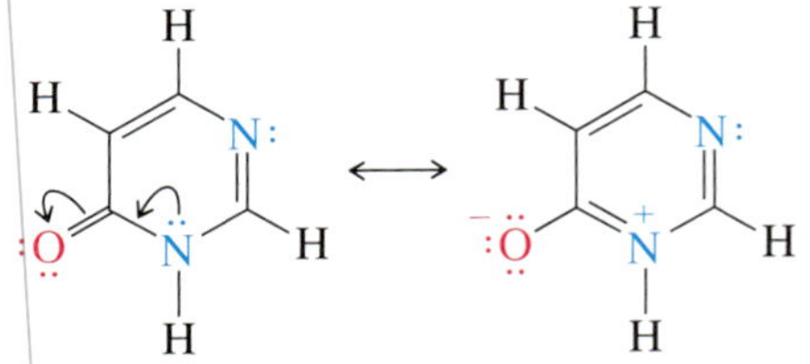

Resonance in keto form of 4-hydroxypyrimidine

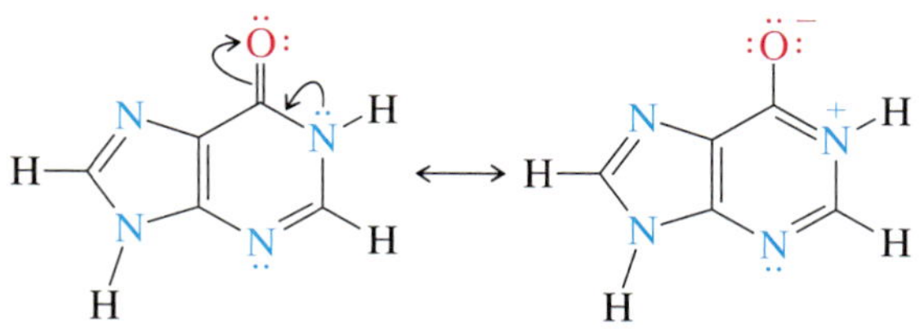

Resonance in keto form of 6-hydroxypurine

These relationships are general. Hydroxyl-substituted purines and pyrimidines exist in their keto forms; amino-substituted ones retain structures with an amino group on the ring. The pyrimidine and purine bases in DNA and RNA listed in Table 28.1 follow this general rule. Beginning in Section 28.7 we'll see how critical it is that we know the correct tautomeric forms of the nucleic acid bases.

TABLE 28.1 Pyrimidines and Purines That Occur in DNA and/or RNA

Name	Structure	Occurrence
Pyrimidines		
Cytosine	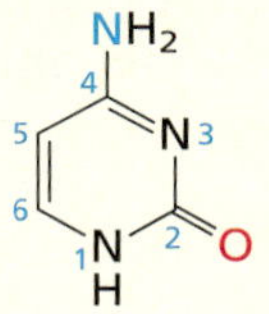	DNA and RNA
Thymine	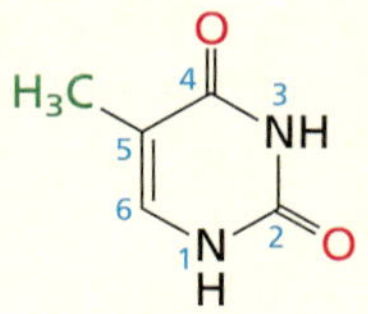	DNA
Uracil	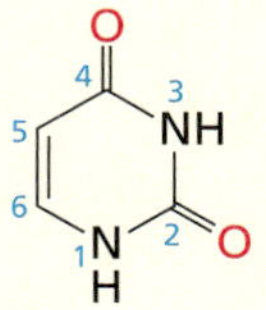	RNA
Purines		
Adenine	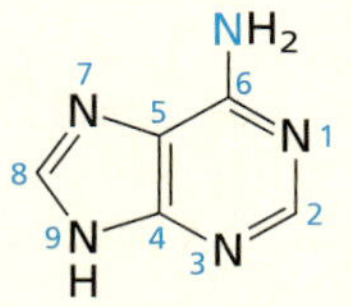	DNA and RNA
Guanine	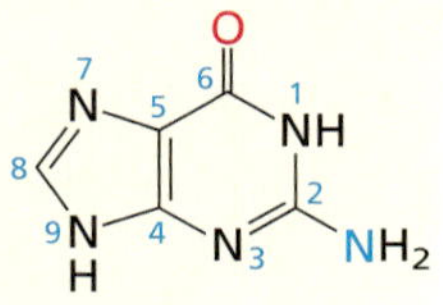	DNA and RNA

PROBLEM 28.1 Write a structural formula for the enol tautomer of cytosine.

PROBLEM 28.2 Write a resonance form for guanine in which the six-membered ring has an electronic structure analogous to benzene. Show all unshared pairs and don't forget to include formal charges.

Pyrimidines and purines occur naturally in substances other than nucleic acids. Coffee, for example, is a familiar source of caffeine. Tea contains both caffeine and theobromine.

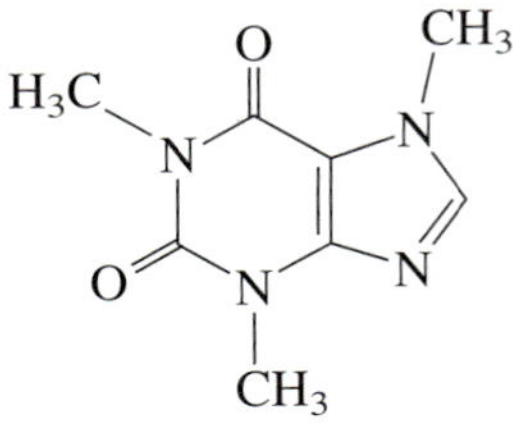

Caffeine 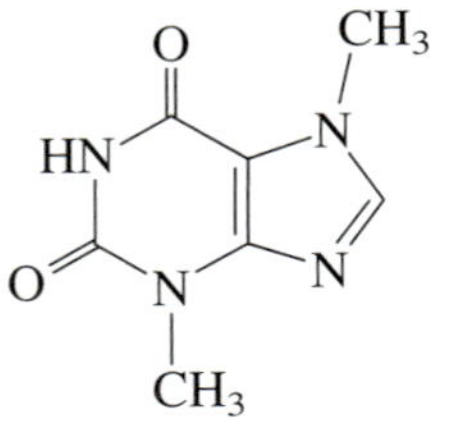   Theobromine

PROBLEM 28.3 Classify caffeine and theobromine according to whether each is a pyrimidine or a purine. One of these cannot isomerize to an enolic form; two different enols are possible for the other. Explain and write structural formulas for the possible enols.

Several synthetic pyrimidines and purines are useful drugs. *Acyclovir* was the first effective antiviral compound and is used to treat herpes infections. *6-Mercaptopurine* is one of the drugs used to treat childhood leukemia, which has become a very treatable form of cancer with a cure rate approaching 80%.

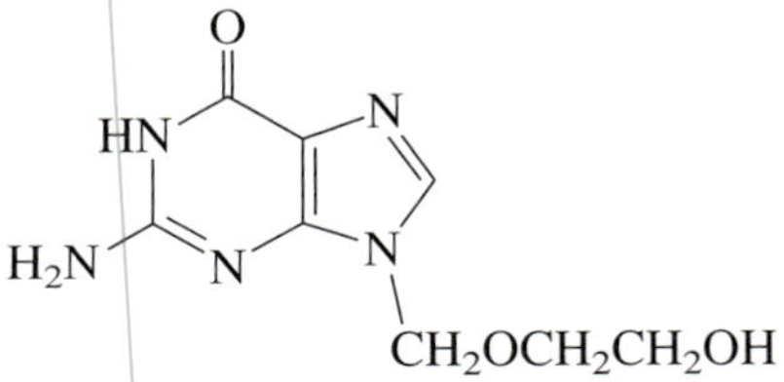

Acyclovir 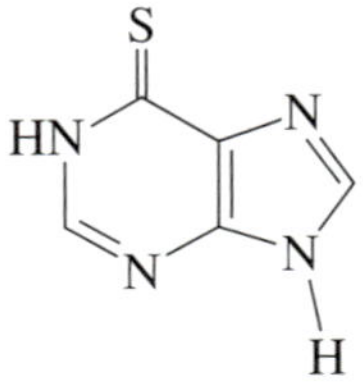   6-Mercaptopurine

28.2 NUCLEOSIDES

The most important derivatives of pyrimidines and purines are nucleosides. **Nucleosides** are *N*-glycosides in which a pyrimidine or purine nitrogen is bonded to the anomeric carbon of a carbohydrate. The nucleosides listed in Table 28.2 are the main building blocks of nucleic acids. In RNA the carbohydrate component is D-ribofuranose; in DNA it is 2-deoxy-D-ribofuranose.

Among the points to be made concerning Table 28.2 are the following:

1. Three of the bases (cytosine, adenine, and guanine) occur in both RNA and DNA.
2. Uracil occurs only in RNA; thymine occurs only in DNA.

TABLE 28.2 — The Major Pyrimidine and Purine Nucleosides in RNA and DNA

Name	Pyrimidines			Purines	
	Cytidine	Thymidine	Uridine	Adenosine	Guanosine
Abbreviation*	C	T	U	A	G
Systematic name	1-β-D-Ribo-furanosylcytosine	2'-Deoxy-1-β-D-ribo-furanosylthymine	1-β-D-Ribo-furanosyluracil	9-β-D-Ribo-furanosyladenine	9-β-D-Ribo-furanosylguanine
Structural formula					
Molecular model					
Found in	RNA 2'-Deoxy analog in DNA	DNA	RNA	RNA 2'-Deoxy analog in DNA	RNA 2'-Deoxy analog in DNA

*Sometimes the abbreviation applies to the pyrimidine or purine base, sometimes to the nucleoside. Though this may seem confusing, it is normally clear from the context what is intended and causes no confusion in practice.

3. The anomeric carbon is attached to N-1 in pyrimidine nucleosides and to N-9 in purines.

4. The pyrimidine and purine bases are cis to the —CH₂OH group of the furanose ring (β stereochemistry).

5. Potential hydrogen-bonding groups (—NH₂ and C=O) point away from the furanose ring.

The numbering scheme used for nucleosides maintains the independence of the two structural units. The pyrimidine or purine is numbered in the usual way. So is the carbohydrate, except that a prime symbol (′) follows each locant. Thus adenosine is a nucleoside of D-ribose, and 2′-deoxyadenosine is a nucleoside of 2-deoxy-D-ribose.

PROBLEM 28.4 The nucleoside *cordycepin* was isolated from cultures of *Cordyceps militaris* and found to be 3′-deoxyadenosine. Write its structural formula.

Table 28.2 doesn't include all of the nucleoside components of nucleic acids. The presence of methyl groups on pyrimidine and purine rings is a common, and often important, variation on the general theme.

Although the term *nucleoside* was once limited to the compounds in Table 28.2 and a few others, current use is more permissive. Pyrimidine derivatives of D-arabinose, for example, occur in the free state in certain sponges and are called *spongonucleosides*. The powerful antiviral drug ribavirin, used to treat hepatitis C and Lassa fever, is a synthetic nucleoside analog in which the base, rather than being a pyrimidine or purine, is a *triazole*.

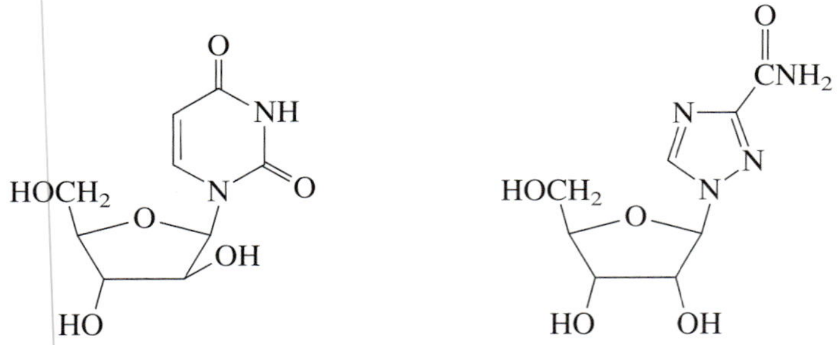

1-β-D-Arabinofuranosyluracil ("spongouridine")   Ribavirin

28.3 NUCLEOTIDES

Nucleotides are phosphoric acid esters of nucleosides. Those derived from adenosine, of which *adenosine 5′-monophosphate* (AMP) is but one example, are especially prominent. AMP is a weak diprotic acid with pK_a's for ionization of 3.8 and 6.2, respectively. In aqueous solution at pH 7, both OH groups of the $P(O)(OH)_2$ unit are ionized.

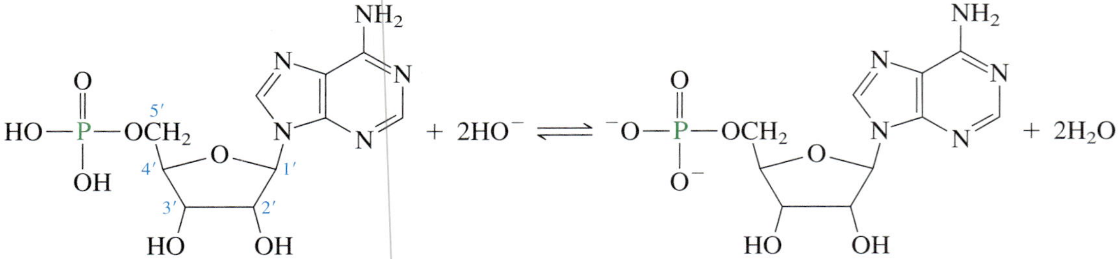

Adenosine 5′-monophosphate (AMP)   Major species at pH 7

PROBLEM 28.5 Write a structural formula for 2'-deoxycytidine 3'-monophosphate. You may wish to refer to Table 28.2 for the structure of cytidine.

Other important 5'-nucleotides of adenosine include *adenosine 5'-diphosphate* (ADP) and *adenosine 5'-triphosphate* (ATP):

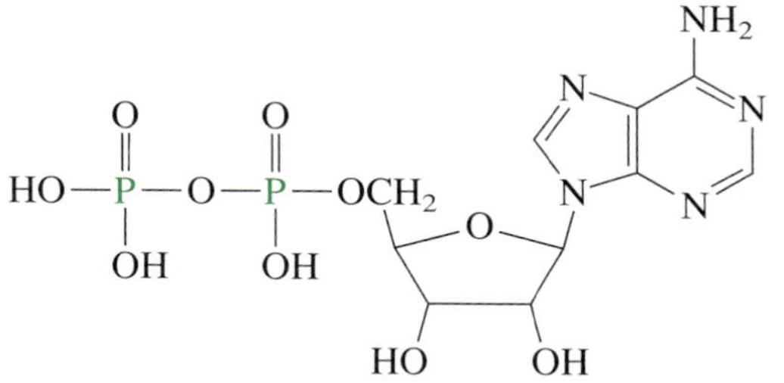

Adenosine 5'-diphosphate (ADP)

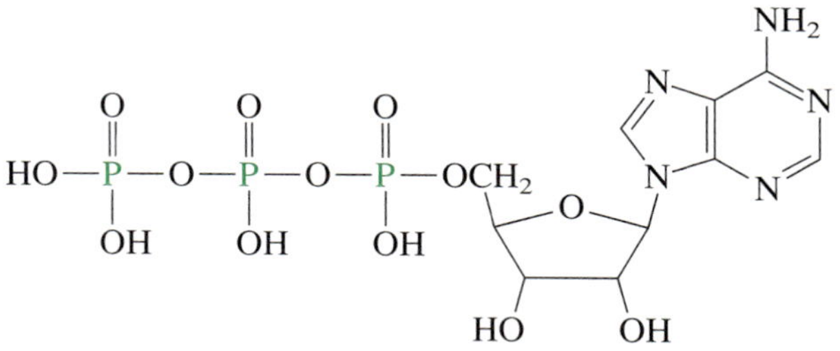

Adenosine 5'-triphosphate (ATP)

ATP is the main energy-storing molecule for practically every form of life on earth. We often speak of ATP as a "high-energy compound" and its P—O bonds as "high-energy bonds." This topic is discussed in more detail in Sections 28.4 and 28.5.

The biological transformations that involve ATP are both numerous and fundamental. They include, for example, many *phosphorylation* reactions in which ATP transfers one of its phosphate units to the —OH of another molecule. These phosphorylations are catalyzed by enzymes called *kinases*. An example is the first step in the metabolism of glucose:

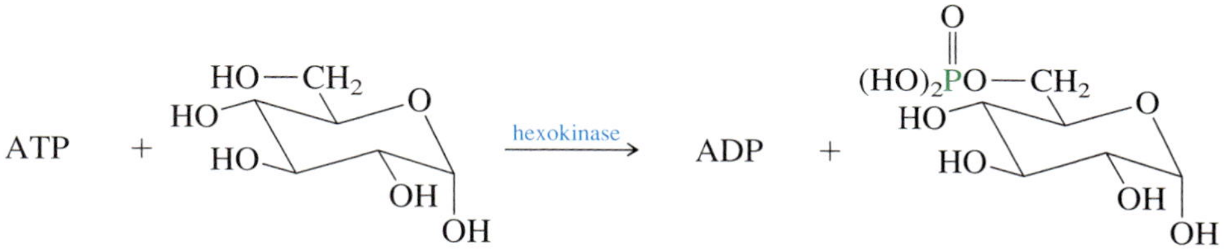

ATP + α-D-Glucopyranose → (hexokinase) → ADP + α-D-Glucopyranose 6-phosphate

Adenosine triphosphate

α-D-Glucopyranose

Adenosine diphosphate

α-D-Glucopyranose 6-phosphate

Both adenosine and guanosine form cyclic monophosphates (*cyclic-AMP* or *cAMP* and *cyclic-GMP* or *cGMP*, respectively) that are involved in a large number of biological processes as "second messengers." Many hormones (the "first messengers") act by first stimulating the formation of cAMP or cGMP on a cell surface, which triggers a series of events characteristic of the organism's response to the hormone.

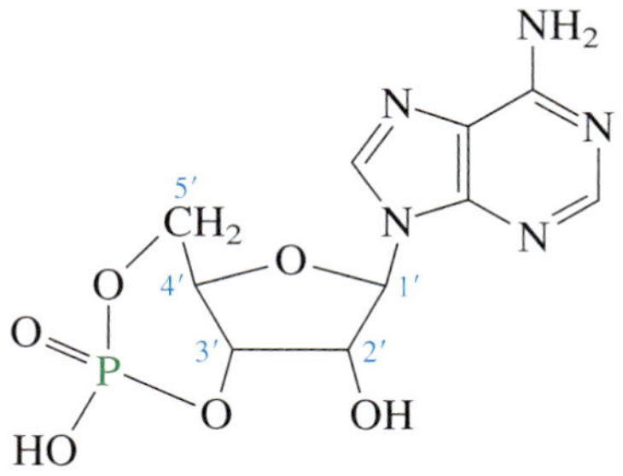

Earl Sutherland of Vanderbilt University won the 1971 Nobel Prize in physiology or medicine for uncovering the role of cAMP as a second messenger in connection with his studies of the "fight or flight" hormone epinephrine (Section 27.6).

Adenosine 3',5'-cyclic monophosphate (cAMP)

> **PROBLEM 28.6** Cyclic-AMP is formed from ATP in a reaction catalyzed by the enzyme *adenylate cyclase*. Assume that adenylate cyclase acts as a base to remove a proton from the 3′-hydroxyl group of ATP and write a mechanism for the formation of cAMP.

28.4 BIOENERGETICS

Bioenergetics is the study of the thermodynamics of biological processes, especially those that are important in energy storage and transfer. Some of its conventions are slightly different from those we are accustomed to. First, it is customary to focus on changes in free energy (ΔG) rather than changes in enthalpy (ΔH). Consider the reaction

> Recall that free energy is the energy available to do work. By focusing on free energy, we concern ourselves more directly with what is important to a living organism.

$$mA(aq) \rightleftharpoons nB(aq)$$

where (aq) indicates that both A and B are in aqueous solution. The reaction is spontaneous in the direction written when ΔG is negative, nonspontaneous when ΔG is positive.

But spontaneity depends on the concentrations of reactants and products. If the ratio $[B]^n/[A]^m$ is less than a certain value, the reaction is spontaneous in the forward direction; if $[B]^n/[A]^m$ exceeds this value, the reaction is spontaneous in the reverse direction. Therefore, it is useful to define a **standard free-energy change ($\Delta G°$)** which applies to a standard state where $[A] = [B] = 1$ M.

$$mA(aq) \rightleftharpoons nB(aq)$$
$$\text{standard state:} \quad 1\text{ M} \qquad 1\text{ M}$$

Reactions are classified as **exergonic** or **endergonic** according to the sign of $\Delta G°$. An exergonic reaction is one in which $\Delta G°$ is negative, an endergonic reaction has a positive value of $\Delta G°$.

Thus, ΔG tells us about the *reaction* with respect to the substances present and their concentrations. $\Delta G°$ focuses more clearly on the differences in free energy between the reactants and products by removing their concentrations from consideration.

The next point takes the standard-state idea and makes it more suitable for biological processes by defining a new $\Delta G°$, called $\Delta G°'$. This new standard state is one with a pH of 7. This is the standard state used most of the time for biochemical reactions and is the one we will use. Not only does it make a big difference in reactions in which H^+ is consumed or produced, it also requires us to be aware of the form in which various species exist at a pH of 7. A reaction that is endergonic at $[H^+] = 1$ M can easily become exergonic at $[H^+] = 10^{-7}$ M (pH = 7) and vice versa.

28.5 ATP AND BIOENERGETICS

The key reaction in bioenergetics is the interconversion of ATP and ADP, usually expressed in terms of the hydrolysis of ATP.

> HPO_4^{2-} is often referred to as "inorganic phosphate" and abbreviated P_i.

$$\text{ATP} \;+\; \text{H}_2\text{O} \longrightarrow \text{ADP} \;+\; \text{HPO}_4^{2-} \qquad \Delta G°' = -31 \text{ kJ } (-7.4 \text{ kcal})$$

| Adenosine triphosphate | Water | Adenosine diphosphate | Hydrogen phosphate |

As written, the reaction is exergonic at pH = 7. The reverse process—conversion of ADP to ATP—is endergonic. Relative to ADP + HPO_4^{2-}, ATP is a "high-energy compound."

When coupled to some other process, the conversion of ATP to ADP can provide the free energy to transform an otherwise endergonic process to an exergonic one. Take, for example, the conversion of glutamic acid to glutamine at pH = 7.

Equation 1:

$$\overset{O}{\underset{}{\overset{\|}{C}}}\text{OCCH}_2\text{CH}_2\underset{\underset{+NH_3}{|}}{\text{CHCO}^-} + \text{NH}_4^+ \longrightarrow \text{H}_2\text{NCCH}_2\text{CH}_2\underset{\underset{+NH_3}{|}}{\text{CHCO}^-} + \text{H}_2\text{O}$$

Glutamic acid   Glutamine

Equation 1 has $\Delta G^{\circ\prime} = +14$ kJ and is endergonic. The main reason for this is that one of the very stable carboxylate groups of glutamic acid is converted to a less-stable amide function.

Nevertheless, the biosynthesis of glutamine proceeds from glutamic acid. The difference is that the endergonic process in Equation 1 is coupled with the strongly exergonic hydrolysis of ATP.

Equation 2:

$$^-\text{OCCH}_2\text{CH}_2\underset{\underset{+NH_3}{|}}{\text{CHCO}^-} + \text{NH}_4^+ + \text{ATP} \longrightarrow \text{H}_2\text{NCCH}_2\text{CH}_2\underset{\underset{+NH_3}{|}}{\text{CHCO}^-} + \text{P}_i + \text{ADP}$$

Glutamic acid   Glutamine

Adding the value of $\Delta G^{\circ\prime}$ for the hydrolysis of ATP (-31 kJ) to that of Equation 1 ($+14$ kJ) gives $\Delta G^{\circ\prime} = -17$ kJ for Equation 2. The biosynthesis of glutamine from glutamic acid is exergonic because it is coupled to the hydrolysis of ATP.

PROBLEM 28.7 Verify that Equation 2 is obtained by adding Equation 1 to the equation for the hydrolysis of ATP.

There is an important qualification to the idea that ATP can serve as a free-energy source for otherwise endergonic processes. There must be some mechanism by which ATP reacts with one or more species along the reaction pathway. Simply being present and undergoing independent hydrolysis isn't enough. More often than not, the mechanism involves transfer of a phosphate unit from ATP to some nucleophilic site. In the case of glutamine synthesis, this step is phosphate transfer to glutamic acid to give γ-glutamyl phosphate as a reactive intermediate.

$$^-\text{OCCH}_2\text{CH}_2\underset{\underset{+NH_3}{|}}{\text{CHCO}^-} + \text{ATP} \longrightarrow {}^-\text{O}-\underset{\underset{-O}{|}}{\overset{O}{\overset{\|}{P}}}-\text{OCCH}_2\text{CH}_2\underset{\underset{+NH_3}{|}}{\text{CHCO}^-} + \text{ADP}$$

Glutamic acid   γ-Glutamyl phosphate

The γ-glutamyl phosphate formed in this step is a mixed anhydride of glutamic acid and phosphoric acid. It is activated toward nucleophilic acyl substitution and gives glutamine when attacked by ammonia.

PROBLEM 28.8 Write a stepwise mechanism for the formation of glutamine by attack of NH_3 on γ-glutamyl phosphate.

If free energy is stored and transferred by way of ATP, where does the ATP come from? It comes from ADP by the endergonic reaction

$$\text{ADP} \quad + \quad \text{HPO}_4{}^{2-} \quad \longrightarrow \quad \text{ATP} \quad + \quad \text{H}_2\text{O} \qquad \Delta G^{\circ\prime} = +31 \text{ kJ } (+7.4 \text{ kcal})$$

| Adenosine diphosphate | Hydrogen phosphate | Adenosine triphosphate | Water |

which you recognize as the reverse of the exergonic hydrolysis of ATP. The free energy to drive this endergonic reaction comes from the metabolism of energy sources such as fats and carbohydrates. In the metabolism of glucose during glycolysis (Section 25.21), for example, about one third of the free energy produced is used to convert ADP to ATP.

As important as nucleotides of adenosine are to bioenergetics, that is not the only indispensable part they play in biology. The remainder of this chapter describes how these and related nucleotides are the key compounds in storing and expressing genetic information.

28.6   PHOSPHODIESTERS, OLIGONUCLEOTIDES, AND POLYNUCLEOTIDES

Just as amino acids can join together to give dipeptides, tripeptides, and so on up to polypeptides and proteins, so too can nucleotides join to form larger molecules. Analogous to the "peptide bond" that connects two amino acids, a *phosphodiester* joins two nucleosides. Figure 28.1 shows the structure and highlights the two phosphodiester units of a trinucleotide of 2′-deoxy-D-ribose in which the bases are adenine (A), thymine (T), and guanine (G). Phosphodiester units connect the 3′-oxygen of one nucleoside to the 5′-oxygen of the next. Nucleotide sequences are written with the free 5′ end at the left and the free 3′ end at the right. Thus, the trinucleotide sequence shown in Figure 28.1 is written as ATG.

The same kind of 5′→3′ phosphodiester units that join the 2′-deoxy-D-ribose units in Figure 28.1 are also responsible for connecting nucleosides of D-ribose.

> **PROBLEM 28.9**   How would the structures of the trinucleotides AUG and GUA in which all of the pentoses are D-ribose differ from the trinucleotide in Figure 28.1?

Adding nucleotides to the 3′-oxygen of an existing structure is called *elongation* and leads ultimately to a *polynucleotide*. The most important polynucleotides are ribonucleic acid (RNA) and deoxyribonucleic acid (DNA). As we shall see in later sections, the polynucleotide chains of DNA and some RNAs are quite long and contain hundreds of thousands of bases.

Polynucleotides of modest chain length, say 50 or fewer, are called **oligonucleotides.** With the growth of the biotechnology industry, the chemical synthesis of oligonucleotides has become a thriving business with hundreds of companies offering custom syntheses of oligonucleotides of prescribed sequences. Such oligonucleotides are required as "primers" in the polymerase chain reaction (Section 28.16) and as "probes" in DNA cloning and genetic engineering. Their synthesis is modeled after the Merrifield solid-phase method and, like it, is automated. The synthesis of a typical oligonucleotide containing 20–50 bases can be accomplished in a few hours.

Oligonucleotide synthesis involves specialized blocking and coupling reactions the chemistry of which is beyond the scope of a typical introductory course. The interested reader is referred to http://www.bi.umist.ac.uk/users/dberrisford/1MBL/nucleicacid3.html.

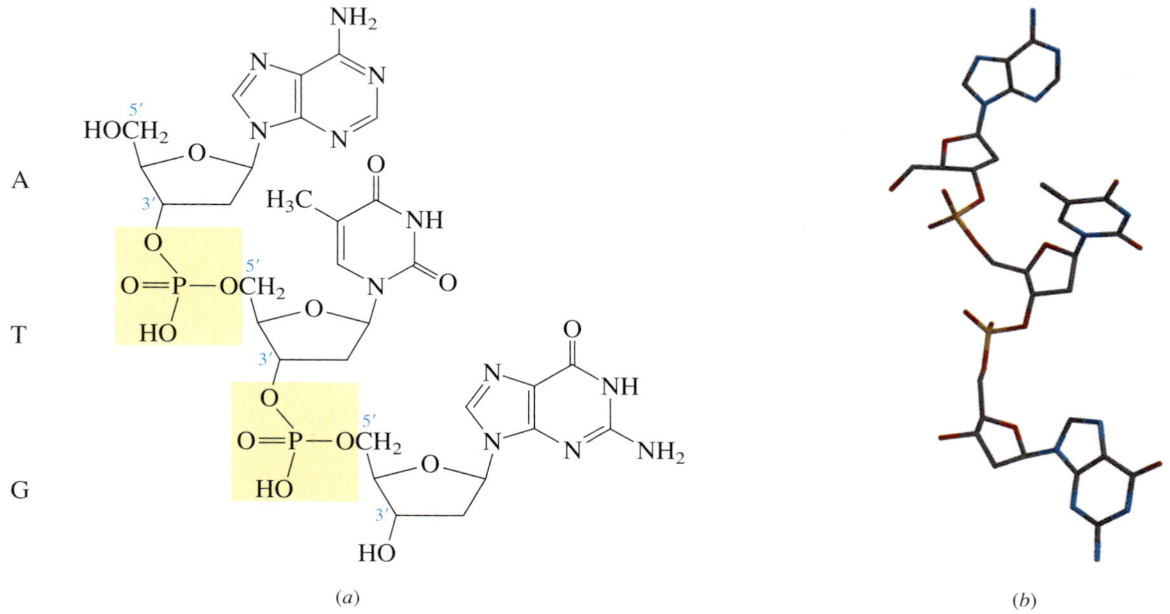

FIGURE 28.1 (a) Structural formula and (b) molecular model of the trinucleotide ATG. The phosphodiester units highlighted in yellow in (a) join the oxygens at 3' of one nucleoside to 5' of the next. By convention, the sequence is read in the direction that starts at the free CH₂OH group (5') and proceeds toward the free 3' OH group at other end.

28.7 NUCLEIC ACIDS

The nineteenth century saw three things happen that, taken together, prepared the way for our present understanding of genetics. In 1854, an Augustinian monk named Gregor Mendel began growing peas in a Moravian monastery. He soon discovered some fundamental relationships about their inherited characteristics. He described these at a scientific meeting in 1865 and sent copies of a paper describing his work to a number of prominent scientists. At about the same time (1859), Charles Darwin published his book *On the Origin of Species by means of Natural Selection.* Mendel's work was ignored until it was rediscovered in 1900; Darwin's was widely known and vigorously debated. The third event occurred in 1869 when Johann Miescher isolated a material he called *nuclein* from the nuclei of white blood cells harvested from the pus of surgical bandages. Miescher's nuclein contained both a protein and an acidic, phosphorus-rich substance that, when eventually separated from the protein, was given the name *nucleic acid.*

After 1900, genetic research—but not research on nucleic acids—blossomed. Nucleic acids were difficult to work with, hard to purify, and, even though they were present in all cells, did not seem to be very interesting. Early analyses, later shown to be incorrect, were interpreted to mean that nucleic acids were polymers consisting of repeats of some sequence of adenine (A), thymine (T), guanine (G), and cytosine (C) in a 1:1:1:1 ratio. Nucleic acids didn't seem to offer a rich enough alphabet from which to build a genetic dictionary. Most workers in the field believed proteins to be better candidates.

The bacterium was *Strepto-coccus pneumoniae*, also called *Pneumococcus*.

More attention began to be paid to nucleic acids in 1945 when Oswald Avery of the Rockefeller Institute for Medical Research found that he could cause a nonvirulent strain of bacterium to produce virulent offspring by incubating them with a substance isolated from a virulent strain. What was especially important was that this virulence was passed on to succeeding generations and could only result from a permanent change in the genetic makeup—what we now call the *genome*—of the bacterium. Avery established that the substance responsible was DNA and in a letter to his brother speculated that it "may be a gene."

Avery's paper prompted other biochemists to rethink their ideas about DNA. One of them, Erwin Chargaff of Columbia University, soon discovered that the distribution of adenine, thymine, cytosine, and guanine differed from species to species, but was the same within a species and within all the cells of a species. Perhaps DNA did have the capacity to carry genetic information after all. Chargaff also found that regardless of the source of the DNA, half the bases were purines and the other half were pyrimidines. Significantly, the ratio of the purine adenine (A) to the pyrimidine thymine (T) was always close to 1:1. Likewise, the ratio of the purine guanine (G) to the pyrimidine cytosine (C) was also close to 1:1. For human DNA the values are:

Purine	Pyrimidine	Base ratio
Adenine (A) 30.3%	Thymine (T) 30.3%	A/T = 1.00
Guanine (G) 19.5%	Cytosine (C) 19.9%	G/C = 0.98
Total purines 49.8%	Total pyrimidines 50.1%	

PROBLEM 28.10 Estimate the guanine content in turtle DNA if adenine = 28.7% and cytosine = 21.3%.

Avery's studies shed light on the *function* of DNA. Chargaff's touched on *structure* in that knowing the distribution of A, T, G, and C in DNA is analogous to knowing the amino acid composition of a protein, but not its sequence or three-dimensional shape.

The breakthrough came in 1953 when James D. Watson and Francis H. C. Crick proposed a structure for DNA. The Watson–Crick proposal ranks as one of the most important in all of science and has spurred a revolution in our understanding of genetics. The structure of DNA is detailed in the next section. The boxed essay *"It Has Not Escaped Our Notice . . ."* describes how it came about.

28.8   SECONDARY STRUCTURE OF DNA: THE DOUBLE HELIX

Watson and Crick shared the 1962 Nobel Prize in physiology or medicine with Maurice Wilkins who, with Rosalind Franklin, was responsible for the X-ray crystallographic work.

Watson and Crick relied on molecular modeling to guide their thinking about the structure of DNA. Because X-ray crystallographic evidence suggested that DNA was composed of two polynucleotide chains running in opposite directions, they focused on the forces holding the two chains together. Hydrogen bonding between bases seemed the most likely candidate. After exploring a number of possibilities, Watson and Crick hit on the arrangement shown in Figure 28.3 in which adenine and thymine comprise one complementary *base pair* and guanine and cytosine another. This base-pairing scheme has several desirable features.

Our text began with an application of physics to chemistry when we described the electronic structure of atoms. We saw then that Erwin Schödinger's introduction of wave mechanics figured prominently in developing the theories that form the basis for our present understanding. As we near the end of our text, we see applications of chemistry to areas of biology that are fundamental to life itself. Remarkably, Schrödinger appears again, albeit less directly. His 1944 book *What Is Life?* made the case for studying genes, their structure, and function.

Schrödinger's book inspired a number of physicists to change fields and undertake research in biology from a physics perspective. One of these was Francis Crick who, after earning an undergraduate degree in physics from University College, London, and while employed in defense work for the British government, decided that the most interesting scientific questions belonged to biology. Crick entered Cambridge University in 1949 as a 30-year-old graduate student, eventually settling on a research problem involving X-ray crystallography of proteins.

One year later, 22-year-old James Watson completed his Ph.D. studies on bacterial viruses at Indiana University and began postdoctoral research in biochemistry in Copenhagen. After a year at Copenhagen, Watson decided Cambridge was the place to be.

Thus it was that the paths of James Watson and Francis Crick crossed in the fall of 1951. One was a physicist, the other a biologist. Both were ambitious in the sense of wanting to do great things and shared a belief that the chemical structure of DNA was the most important scientific question of the time. At first, Watson and Crick talked about DNA in their spare time because each was working on another project. Soon, however, it became their major effort. Their sense of urgency grew when they learned that Linus Pauling, fresh from his proposal of helical protein structures, had turned his attention to DNA. Indeed, Watson and Crick were using the Pauling approach to structure—take what is known about the structure of small molecules, couple it to structural information about larger ones, and build molecular models consistent with the data.

At the same time, Maurice Wilkins and Rosalind Franklin at King's College, Cambridge, were beginning to obtain high quality X-ray crystallographic data of DNA. Some of their results were presented in a seminar at King's attended by Watson, and even more were disclosed in a progress report to the Medical Research Council of the U.K. Armed with Chargaff's A = T and G = C relationships and Franklin's

FIGURE 28.2 Molecular modeling—1953 style. James Watson *(left)* and Francis Crick *(right)* with their DNA model. © A. Barrington Brown/Science Source Photo Researchers, Inc.

X-ray data, Watson and Crick began their model building. A key moment came when Jerry Donohue, a postdoctoral colleague from the United States, noticed that they were using the wrong structures for the pyrimidine and purine bases. Watson and Crick were using models of the enol forms of thymine, cytosine, and guanine, rather than the correct keto forms (recall Section 28.1). Once they fixed this error, the now-familiar model shown in Figure 28.2 emerged fairly quickly and they had the structure of DNA.

Watson and Crick published their work in a paper entitled "A Structure for Deoxyribose Nucleic Acid" in the British journal *Nature* on April 25, 1953. In addition to being one of the most important papers of the twentieth century, it is also remembered for one brief sentence appearing near the end.

> *"It has not escaped our notice that the specific pairing we have postulated immediately suggests a possible copying mechanism for the genetic material."*

True to their word, Watson and Crick followed up their April 25 paper with another on May 30. This second paper, "Genetical Implications of the Structure of Deoxyribonucleic Acid," outlines a mechanism for DNA replication that is still accepted as essentially correct.

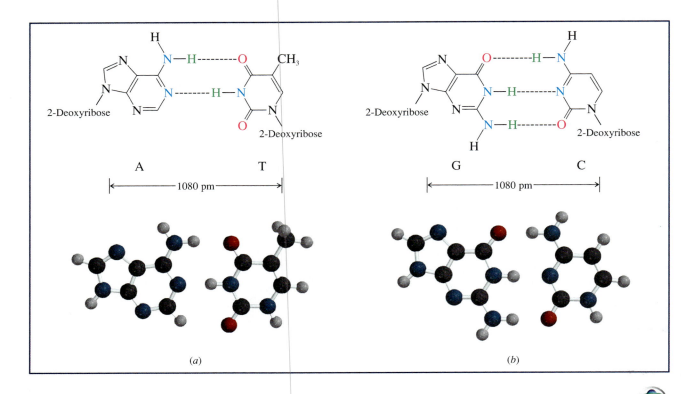

FIGURE 28.3 Hydrogen bonding between DNA bases shown as structural drawings of nucleosides (top) and as molecular models (bottom) of (*a*) adenine and thymine and (*b*) guanine and cytosine.

1. Pairing A with T and G with C gives the proper Chargaff ratios (A = T and G = C).

2. Each pair contains one purine and one pyrimidine base. This makes the A---T and G---C pairs approximately the same size and ensures a consistent distance between the two DNA strands.

3. Complementarity between A and T and G and C suggests a mechanism for copying DNA. This is called replication and is discussed in Section 28.10.

Figure 28.4 supplements Figure 28.3 by showing portions of two DNA strands arranged side by side with the base pairs in the middle.

Hydrogen bonding between complementary bases is responsible for association between the strands, whereas conformational features of its carbohydrate–phosphate backbone and the orientation of the bases with respect to the furanose rings govern the overall shape of each strand. Using the X-ray crystallographic data available to them, Watson and Crick built a molecular model in which each strand took the shape of a right-handed helix. Joining two antiparallel strands by appropriate hydrogen bonds produced the double helix shown in the photograph (Figure 28.2). Figure 28.5 shows two modern renderings of DNA models.

In addition to hydrogen bonding between the two polynucleotide chains, the double-helical arrangement is stabilized by having its negatively charged phosphate groups on the outside where they are in contact with water and various cations, Na^+, Mg^{2+}, and ammonium ions, for example. Attractive van der Waals forces between the

A helical structure for DNA strands had been suggested in 1949 by Sven Furberg in his Ph.D. dissertation at the University of London.

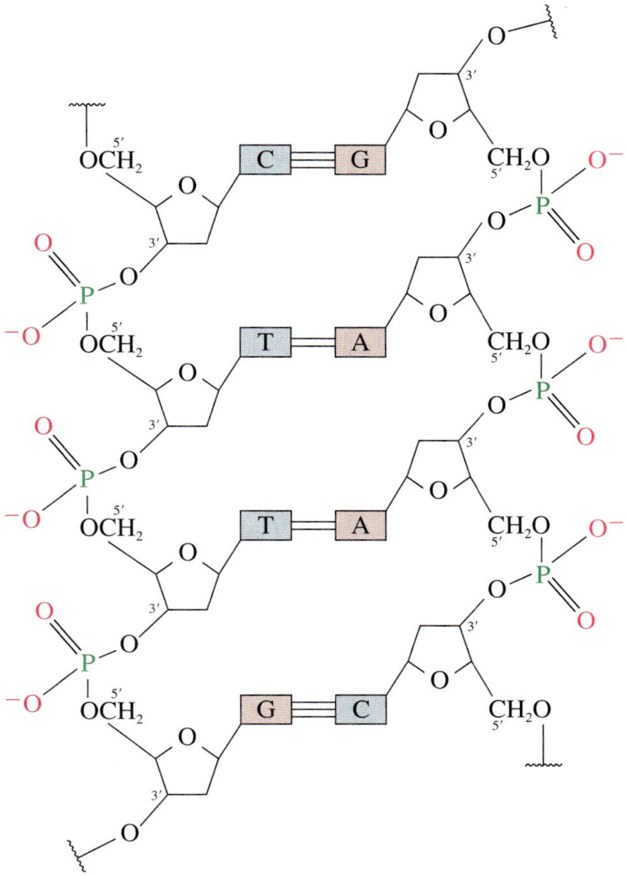

FIGURE 28.4 Hydrogen bonding between complementary bases (A and T, and G and C) permit pairing of two DNA strands. The strands are antiparallel; the 5′ end of the left strand is at the top, and the 5′ end of the right strand is at the bottom.

aromatic pyrimidine and purine rings, called *π-stacking,* stabilize the layered arrangement of the bases on the inside. Even though the bases are on the inside, they are accessible to other substances through two grooves that run along the axis of the double helix. They are more accessible via the *major groove,* which is almost twice as wide as the *minor groove.* The grooves differ in size because of the way the bases are tilted with respect to the furanose ring.

The structure proposed by Watson and Crick was modeled to fit crystallographic data obtained on a sample of the most common form of DNA called B-DNA. Other forms include A-DNA, which is similar to, but more compact than B-DNA, and Z-DNA, which is a left-handed double helix.

By analogy to the levels of structure of proteins, the **primary structure** of DNA is the sequence of bases along the polynucleotide chain, and the A-DNA, B-DNA, and Z-DNA helices are varieties of **secondary structures.**

Not all DNAs are double helices (*duplex DNA*). Some types of viral DNA are single-stranded, and even a few DNA triple and quadruple helices are known.

FIGURE 28.5 (a) Tube and (b) space-filling models of a DNA double helix. The carbohydrate–phosphate "backbone" is on the outside and can be roughly traced in (b) by the red oxygen atoms. The blue atoms belong to the purine and pyrimidine bases and lie on the inside. The base-pairing is more clearly seen in (a).

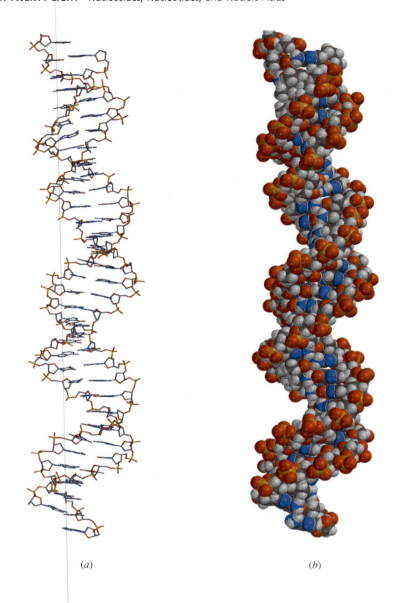

(a)   (b)

28.9   TERTIARY STRUCTURE OF DNA: SUPERCOILS

We have, so far, described the structure of DNA as an extended double helix. The crystallographic evidence that gave rise to this picture was obtained on a sample of DNA removed from the cell that contained it. Within a cell—its native state—DNA almost always adopts some shape other than an extended chain. We can understand why by doing a little arithmetic. Each helix of B-DNA makes a complete turn every 3.4×10^{-9} m and there are about 10 base pairs per turn. A typical human DNA contains 10^8 base pairs. Therefore,

$$\text{Length of DNA chain} = \frac{3.4 \times 10^{-9} \text{ m/turn}}{10 \text{ base pairs/turn}} \times 10^8 \text{ base pairs}$$

$$\text{Length of DNA chain} = 3.4 \times 10^{-2} \text{ m} = 3.4 \text{ cm}$$

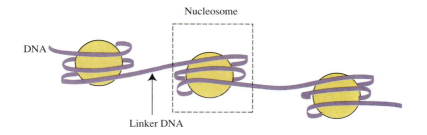

Nucleosome

DNA

Linker DNA

FIGURE 28.6 The effective length of DNA is reduced by coiling around the surface of histones to form nucleosomes. The histone proteins are represented by the spheres and the DNA double helix by the ribbon.

For a 3-cm-long molecule of DNA to fit inside a cell so tiny that we can only see it with a microscope, the polynucleotide chain must be folded into a more compact form. Not only must the DNA be compacted, it must be folded in a way that allows it to carry out its main functions. The way the chain is folded defines the **tertiary structure** of a nucleic acid.

The compacting mechanism is a marvel of cellular engineering. A twisted tangle of indefinite shape would present serious problems as a vessel for storing genetic information. Coiling the duplex, however, reduces its length without blocking access to important parts of its structure. Remember though, that DNA is negatively charged at biological pH. Thus, the tighter the coil, the closer together are the negatively charged phosphate units and the less stable the coil. Nature solves this puzzle for chromosomes by wrapping short sections of the DNA around proteins called histones (Figure 28.6). **Histones** are a family of five proteins rich in basic amino acids such as arginine and lysine, which are positively charged at biological pH. The positively charged histones stabilize the coiled form of the negatively charged DNA. The species formed between a section of DNA and histones is called a **nucleosome.** Each nucleosome contains about one and three quarters turns of coil comprising 146 base pairs of DNA and is separated from the next nucleosome by a "linker" of about 50 base pairs of DNA. Figure 28.7 shows a molecular model of a single nucleosome.

PROBLEM 28.11   Approximately how many nucleosomes are in a gene with 10,000 base pairs?

FIGURE 28.7 Molecular models of a nucleosome and its components. The nucleosome has a protein core around which is wound a supercoil of duplex DNA.

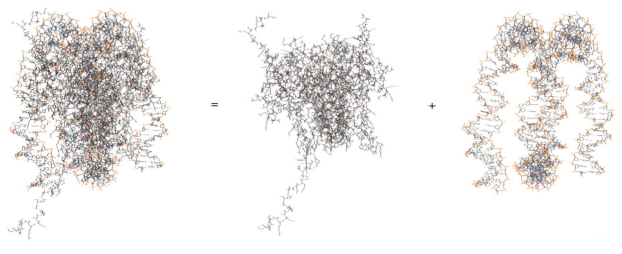

Nucleosome   Histone proteins   Supercoiled DNA

A single helix is a coil; a double helix is two nested coils. The tertiary structure of DNA in a nucleosome is a coiled coil. Coiled coils are referred to as **supercoils** and are quite common.

A coiled α-helix in a protein is another example of a supercoil.

28.10 REPLICATION OF DNA

Every time a cell divides, its DNA is duplicated so that the DNA in the new cell is identical to that in the original one. As Figure 28.8 shows, Watson–Crick base-pairing provides the key to understanding this process of DNA **replication.** During cell division the DNA double helix begins to unwind, generating a **replication fork** separating the two strands. Each strand serves as a template on which a new DNA strand is constructed. The A---T, G---C base-pairing requirement ensures that each new strand is the precise complement of its template strand. Each of the two new duplex DNA molecules contains one original and one new strand.

Both new chains grow in their $5' \rightarrow 3'$ direction. Because of this, one grows toward the replication fork (the **leading strand**) and the other away from it (the **lagging strand**), making the details of chain extension somewhat different for the two. The fundamental chemistry, however, is straightforward (Figure 28.9). The hydroxyl group at the $3'$ end of the growing polynucleotide chain acts as a nucleophile, attacking the $5'$-triphosphate of $2'$-deoxyadenosine, $2'$-deoxyguanosine, $2'$-deoxycytidine, or thymidine to form the new phosphodiester linkage. The enzyme that catalyzes phosphodiester bond formation is called *DNA polymerase;* different DNA polymerases operate on the leading strand and the lagging strand.

All of the steps, from the unwinding of the original DNA double helix to the supercoiling of the new DNAs, are catalyzed by enzymes.

Genes are DNA and carry the inheritable characteristics of an organism and these characteristics are normally **expressed** at the molecular level via protein synthesis. Gene expression consists of two stages, **transcription** and **translation,** both of which involve RNAs. Sections 28.11 and 28.12 describe these RNAs and their roles in transcription and translation.

28.11 RIBONUCLEIC ACIDS

Unlike DNA, most of which is in the nucleus, RNA is found mostly in the cell's main compartment, the cytoplasm. There are three different kinds of RNA, which differ substantially from one another in both structure and function:

1. Messenger RNA *(mRNA)*
2. Transfer RNA *(tRNA)*
3. Ribosomal RNA *(rRNA)*

All are important in the biosynthesis of proteins.

Messenger RNA (mRNA): According to Crick, the so-called central dogma of molecular biology is "DNA makes RNA makes protein." The first part can be restated more exactly as "DNA makes mRNA." This is what transcription is—transcribing the message of DNA to a complementary RNA, in this case messenger RNA. mRNA is the least abundant of the RNAs and is the only one that is synthesized in the cell's nucleus. This transcription process is illustrated in Figure 28.10. Transcription resembles DNA replication in that a DNA strand serves as the template for construction of, in this case, a ribonucleic acid. mRNA synthesis begins at its $5'$ end, and ribonucleotides complementary to the DNA strand being copied are added. The phosphodiester linkages are formed

1. The DNA to be copied is a double helix, shown here as flat for clarity.

2. The two strands begin to unwind. Each strand will become a template for construction of its complement.

3. As the strands unwind, the pyrimidine and purine bases become exposed. Notice that the bases are exposed in the $3' \rightarrow 5'$ direction in one strand, and in the $5' \rightarrow 3'$ direction in the other.

4. Two new strands form as nucleotides that are complementary to those of the original strands are joined by phosphodiester linkages. The sources of the new bases are dATP, dGTP, dCTP, and dTTP already present in the cell.

5. Because nucleotides are added in the $5' \rightarrow 3'$ direction, the processes by which the two new chains grow are different. Chain growth can be continuous in the leading strand, but not in the lagging strand.

6. Two duplex DNA molecules result, each of which is identical to the original DNA.

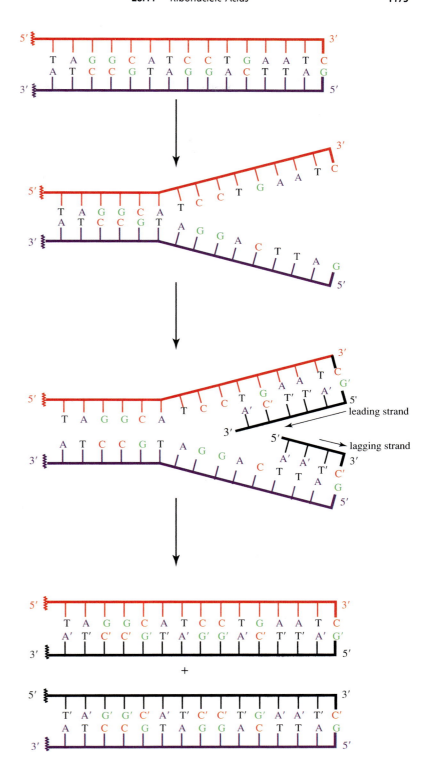

FIGURE 28.8 Outline of DNA replication. The original strands are shown in red and blue and are the templates from which the new strands, shown in black, are copied.

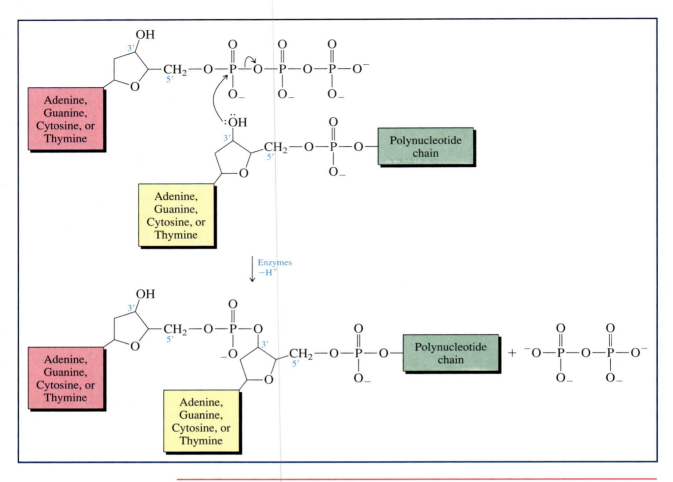

FIGURE 28.9 The new polynucleotide chain grows by reaction of its free 3′-OH group with the 5′-triphosphate of an appropriate 2′-deoxyribonucleoside.

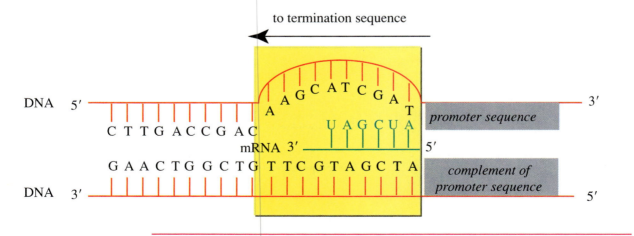

FIGURE 28.10 During transcription a molecule of mRNA is assembled from a DNA template. Transcription begins at a promoter sequence and proceeds in the 5′→3′ direction of the mRNA until a termination sequence of the DNA is reached. Only a region of about 10 base pairs is unwound at any time.

by reaction of the free 3′-OH group of the growing mRNA with ATP, GTP, CTP, or UTP (recall that uracil, not thymine, is the complement of adenine in RNA). The enzyme that catalyzes this reaction is *RNA polymerase*. Only a small section of about 10 base pairs of the DNA template is exposed at a time. As the synthesis zone moves down the DNA chain, restoration of hydrogen bonds between the two original DNA strands displaces the newly synthesized single-stranded mRNA. The entire DNA molecule is not transcribed as a single mRNA. Transcription begins at a prescribed sequence of bases (the *promoter sequence*) and ends at a *termination sequence*. Thus, one DNA molecule can give rise to many different mRNAs and code for many different proteins. There are thousands of mRNAs and they vary in length from about 500 to 6000 nucleotides.

The **genetic code** (Table 28.3) is the message carried *by* mRNA. It is made up of triplets of adjacent nucleotide bases called **codons**. Because mRNA has only four different bases and 20 amino acids must be coded for, codes using either one or two nucleotides per amino acid are inadequate. If nucleotides are read in sets of three, however, the four mRNA bases generate 64 possible "words," more than suffecient to code for 20 amino acids.

In addition to codons for amino acids, there are *start* and *stop* codons. Protein biosynthesis begins at a start codon and ends at a stop codon of mRNA. The start codon is the nucleotide triplet AUG, which is also the codon for methionine. The stop codons are UAA, UAG, and UGA.

Transfer RNA (tRNA):   Transfer-RNAs are relatively small nucleic acids, containing only about 70 nucleotides. They get their name because they transfer amino acids to the ribosome for incorporation into a polypeptide. Although 20 amino acids need to be transferred, there are 50–60 tRNAs, some of which transfer the same amino acids. Figure 28.11 shows the structure of phenylalanine tRNA (tRNA^Phe). Like all tRNAs it is composed of a single strand, with a characteristic shape that results from the presence of paired bases in some regions and their absence in others.

The 1968 Nobel Prize in physiology or medicine was shared by Robert W. Holley of Cornell University for determining the nucleotide sequence of phenylalanine transfer RNA.

TABLE 28.3   The Genetic Code (Messenger RNA Codons)

		Second Position				
		U	**C**	**A**	**G**	
First Position (5′ end)	**U**	UUU Phe UUC Phe UUA Leu UUG Leu	UCU Ser UCC Ser UCA Ser UCG Ser	UAU Tyr UAC Tyr UAA Stop UAG Stop	UGU Cys UGC Cys UGA Stop UGG Trp	U C A G
	C	CUU Leu CUC Leu CUA Leu CUG Leu	CCU Pro CCC Pro CCA Pro CCG Pro	CAU His CAC His CAA Gln CAG Gln	CGU Arg CGC Arg CGA Arg CGG Arg	U C A G
	A	AUU Ile AUC Ile AUA Ile AUG Met	ACU Thr ACC Thr ACA Thr ACG Thr	AAU Asn AAC Asn AAA Lys AAG Lys	AGU Ser AGC Ser AGA Arg AGG Arg	U C A G
	G	GUU Val GUC Val GUA Val GUG Val	GCU Ala GCC Ala GCA Ala GCG Ala	GAU Asp GAC Asp GAA Glu GAG Glu	GGU Gly GGC Gly GGA Gly GGG Gly	U C A G

Third Position (3′ end)

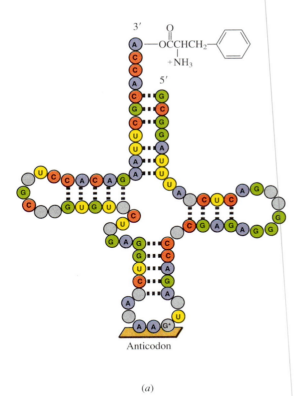

(a)

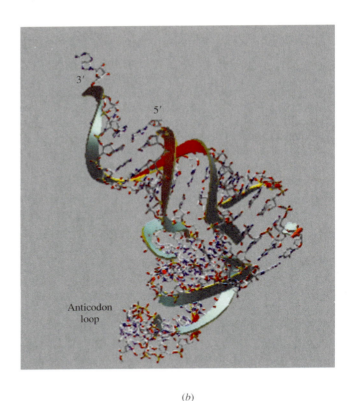

(b)

FIGURE 28.11 Phenylalanine tRNA from yeast. (a) A schematic drawing showing the sequence of bases. Transfer RNAs usually contain a number of modified bases (gray circles). One of these is a modified guanosine (G*) in the anticodon. Hydrogen bonds, where present, are shown as dashed lines. (b) The structure of yeast tRNA^Phe as determined by X-ray crystallography.

Among the 76 nucleotides of tRNA^Phe are two sets of three that are especially important. The first is a group of three bases called the **anticodon,** which is complementary to the mRNA codon for the amino acid being transferred. Table 28.3 lists two mRNA codons for phenylalanine, UUU and UUC (reading in the 5′→3′ direction). Because base-pairing requires the mRNA and tRNA to be antiparallel, the two anticodons are read in the 3′→5′ direction as AAA and AAG.

$$3'\ \{-A-A-G-\}\ 5'\qquad \text{tRNA anticodon}$$

$$5'\ \{-U-U-C-\}\ 3'\qquad \text{mRNA codon}$$

The other important sequence is the CCA triplet at the 3′ end. The amino acid that is to be transferred is attached through an ester linkage to the terminal 3′-oxygen of this sequence. *All tRNAs have a CCA sequence at their 3′ end.*

Transfer RNAs normally contain some bases other than A, U, G, and C. Of the 76 bases in tRNA^Phe, for example, 13 are of the modified variety. One of these, marked G* in Figure 28.11, is a modified guanosine in the anticodon. Many of the modified bases, including G*, are methylated derivatives of the customary RNA bases.

Ribosomal RNA (rRNA):   Ribosomes, which are about two thirds nucleic acid and about one third protein, constitute about 90% of a cell's RNA. A ribosome is made up of two subunits. The larger one contains two rRNAs, one with 122 nucleotides and the other with 2923; the smaller subunit contains one rRNA which has 1500 nucleotides.

The ribosome is where the message carried by the mRNA is **translated** into the amino sequence of a protein. How it occurs is described in the next section. One of its most noteworthy aspects was discovered only recently. It was formerly believed that the RNA part of the ribosome was a structural component and the protein part was the catalyst for protein biosynthesis. Present thinking tilts toward reversing these two functions by ascribing the structural role to the protein and the catalytic one to rRNA. RNAs that catalyze biological processes are called **ribozymes.** Catalysis by RNA is an important element in origins of life theories as outlined in the accompanying boxed essay *"RNA World."*

Sidney Altman (Yale University) and Thomas Cech (University of Colorado) shared the 1989 Nobel Prize in chemistry for showing that RNAs could function as biological catalysts.

RNA WORLD

Once the broad outlines of DNA replication and protein biosynthesis were established, scientists speculated about how these outlines affected various "origins of life" scenarios. A key question concerned the fact that proteins are required for the synthesis of DNA, yet the synthesis of these proteins is coded for by DNA. Which came first, DNA or proteins? How could DNA store genetic information if there were no enzymes to catalyze the polymerization of its nucleotide components? How could there be proteins if there were no DNA to code for them?

Molecular biologists entertained a number of possibilities but identified weaknesses in all of them. The simplest hypothesis, the chance combination of a polynucleotide and a polypeptide to produce even a primitive self-replicating system, seemed statistically improbable. Other suggestions assigned both the catalytic and storage of genetic information tasks to the same component, either a polynucleotide or a polypeptide. Once a molecule became self-replicating, natural selection could favor dividing the catalytic and genetic tasks—adding a polynucleotide component to a self-replicating polypeptide, or vice versa. The problem with this hypothesis was that there was no precedent for either polypeptides serving as reservoirs for storing genetic information, or for polynucleotides acting as catalysts.

There was a sense, however, that the sequence that later came to be called "RNA World" was the most reasonable one. RNA World denotes an early period in the development of self-replicating systems in which RNA assumes both the catalytic and information-storing roles. DNA and proteins don't arrive on the stage as functioning participants until much later.

The discovery of ribozymes (Section 28.11) in the late 1970s and early 1980s by Sidney Altman of Yale University and Thomas Cech of the University of Colorado placed the RNA World idea on a more solid footing. Altman and Cech independently discovered that RNA can catalyze the formation and cleavage of phosphodiester bonds—exactly the kinds of bonds that unite individual ribonucleotides in RNA. That, plus the recent discovery that ribosomal RNA catalyzes the addition of amino acids to the growing peptide chain in protein biosynthesis, takes care of the most serious deficiencies in the RNA World model by providing precedents for the catalysis of biological processes by RNA.

Even if it could be shown that RNA preceded both DNA and proteins in the march toward living things, that doesn't automatically make RNA the first self-replicating molecule. Another possibility is that a self-replicating polynucleotide based on some carbohydrate other than D-ribose was a precursor to RNA. Over many generations, natural selection could have led to the replacement of the other carbohydrate by D-ribose, giving RNA. Recent research on unnatural polynucleotides by Professor Albert Eschenmoser of the Swiss Federal Institute of Technology (Zürich) has shown, for example, that nucleic acids based on L-threose possess many of the properties of RNA and DNA.

28.12 PROTEIN BIOSYNTHESIS

As described in the preceding sections, protein synthesis involves transcription of the DNA to mRNA, followed by translation of the mRNA as an amino acid sequence. In addition to outlining the mechanics of transcription, we have described the relationship among mRNA codons, tRNA anticodons, and amino acids.

During translation the protein is synthesized beginning at its N-terminus (Figure 28.12). The mRNA is read in its 5'→3' direction beginning at the start codon AUG and ending at a stop codon (UAA, UAG, or UGA). Because the start codon is always AUG, the N-terminal amino acid is always methionine (as its *N*-formyl derivative). However, this *N*-formylmethionine residue is normally lost in a subsequent process and the N-terminus of the expressed protein is therefore determined by the second mRNA codon. The portion of the mRNA between the start and stop codons is called the coding sequence and is flanked on either side by noncoding regions.

In addition to illustrating the mechanics of translation, Figure 28.12 is important in that it shows the mechanism of peptide bond formation as a straightforward nucleophilic acyl substitution. Both methionine and alanine are attached to their respective tRNAs as esters. The amino group of alanine attacks the methionine carbonyl, displacing methionine from its tRNA and converting the carbonyl group of methionine from an ester to an amide function.

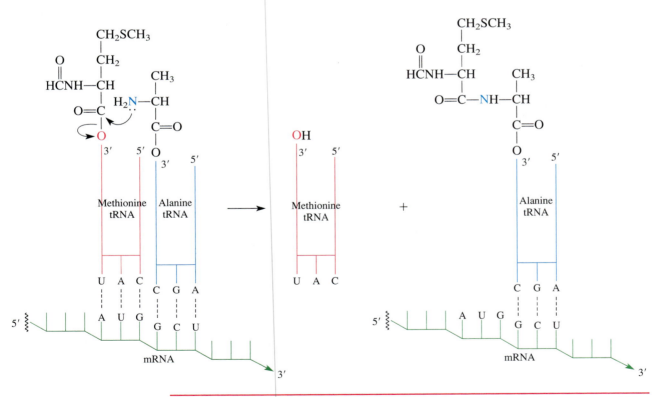

FIGURE 28.12 Translation of mRNA to an amino acid sequence of a protein starts at an mRNA codon for methionine. Nucleophilic acyl substitution transfers the *N*-formylmethionine residue from its tRNA to the amino group of the next amino acid (shown here as alanine). The process converts an ester to an amide.

PROBLEM 28.12 Modify Figure 28.12 so that it corresponds to translation of an mRNA in which the sequence of the first six bases of the coding sequence are AUGUCU.

28.13 AIDS

The explosive growth of our knowledge of nucleic acid chemistry and its role in molecular biology in the 1980s coincided with the emergence of AIDS (acquired immune deficiency syndrome) as a major public health threat. In AIDS, a virus devastates the body's defenses to the extent that its victims can die from infections that are normally held in check by a healthy immune system. In the short time since its discovery in the early 1980s, AIDS has claimed the lives of over 22 million people, and current estimates place the number of those infected at more than 36 million. According to the World Health Organization (WHO), AIDS is now the fourth leading cause of death worldwide and the leading cause of death in Africa.

The viruses responsible for AIDS are human immunodeficiency virus 1 and 2 (HIV-1 and HIV-2). Both are **retroviruses,** meaning that their genetic material is RNA rather than DNA. HIVs require a host cell to reproduce, and the hosts in humans are the T4 lymphocytes, which are the cells primarily responsible for inducing the immune system to respond when provoked. The HIV penetrates the cell wall of a T4 lymphocyte and deposits both its RNA and an enzyme called *reverse transcriptase* inside. There, the reverse transcriptase catalyzes the formation of a DNA strand that is complementary to the viral RNA. The transcribed DNA then serves as the template from which the host lymphocyte produces copies of the virus, which then leave the host to infect other T4 cells. In the course of HIV reproduction, the ability of the T4 lymphocyte to reproduce itself is compromised. As the number of T4 cells decrease, so does the body's ability to combat infections.

PROBLEM 28.13 When the RNA of a retrovirus is transcribed, what DNA base is the complement of the uracil in the viral RNA?

Although there is no known cure for AIDS, progress is being made in delaying the onset of symptoms and prolonging the lives of those infected with HIV. The first advance in treatment came with drugs such as the nucleoside *zidovudine,* also known as azidothymine, or AZT. During reverse transcription, AZT replaces thymidine in the DNA being copied from the viral RNA. AZT has a 5′-OH group, so can be incorporated into a growing polynucleotide chain. But because it lacks a 3′-OH group, the chain cannot be extended beyond it and synthesis of the viral DNA stops before the chain is complete.

Zidovudine (AZT)   2′,3′-Dideoxyinosine (ddI)

Other nucleosides such as 2′,3′-dideoxyinosine (ddI) also block the action of reverse transcriptase and are often combined with AZT in "drug cocktails." Using a mixture of drugs makes it more difficult for a virus to develop resistance than using a single drug.

The most recent advance in treating HIV infections has been to simultaneously attack the virus on a second front using a *protease inhibitor*. Recall from Section 27.10 that proteases are enzymes that catalyze the hydrolysis of proteins at specific points. When HIV uses a cell's DNA to synthesize its own proteins, the initial product is a long polypeptide that contains several different proteins joined together. To be useful, the individual proteins must be separated from the aggregate by protease-catalyzed hydrolysis of peptide bonds. Protease inhibitors prevent this hydrolysis and, in combination with reverse transcriptase inhibitors, slow the reproduction of HIV. Dramatic reductions in the "viral load" in HIV-infected patients have been achieved with this approach.

28.14 DNA SEQUENCING

Once the Watson–Crick structure was proposed, determining the nucleotide sequence of DNA emerged as an important area of research. Some difficulties were apparent from the beginning, especially if one draws comparisons to protein sequencing. First, most DNAs are much larger biopolymers than proteins. Not only does it take three nucleotides to code for a single amino acid, but vast regions of DNA don't seem to code for anything at all. A less obvious problem is that the DNA alphabet contains only four letters (A, G, C, and T) compared with the 20 amino acids from which proteins are built. Recall too that protein sequencing benefits from having proteases available that cleave the chain at specific amino acids. Not only are there no enzymes that cleave nucleic acids at specific bases but, with only four bases to work with, the resulting fragments would be too small to give useful information. In spite of this, DNA sequencing not only developed very quickly, but also has turned out to be much easier to do than protein sequencing.

To explain how DNA sequencing works, we must first mention **restriction enzymes.** Like all organisms, bacteria are subject to infection by external invaders (e.g., viruses and other bacteria) and possess defenses in the form of restriction enzymes that destroy the invader by cleaving its DNA. About 200 different restriction enzymes are known. Unlike proteases, which recognize a single amino acid, restriction enzymes recognize specific nucleotide *sequences*. Cleavage of the DNA at prescribed sequences gives fragments small enough to be sequenced conveniently. These smaller DNA fragments are separated and purified by gel electrophoresis. At a pH of 7.4, each phosphate link between adjacent nucleotides is ionized, giving the DNA fragments a negative charge and causing them to migrate to the positively charged electrode. Separation is size-dependent. Larger polynucleotides move more slowly through the polyacrylamide gel than smaller ones. The technique is so sensitive that two polynucleotides differing in length by only a single nucleotide can be separated from each other on polyacrylamide gels.

Once the DNA is separated into smaller fragments, each fragment is sequenced independently. Again, gel electrophoresis is used, this time as an analytical tool. In the technique devised by Frederick Sanger, the two strands of a sample of a small fragment of DNA, 100–200 base pairs in length, are separated and one strand is used as a template to create complements of itself. The single-stranded sample is divided among four test tubes, each of which contains the materials necessary for DNA synthesis. These materials include the four nucleosides present in DNA, 2′-deoxyadenosine (dA), 2′-deoxythymidine (dT), 2′-deoxyguanosine (dG), and 2′-deoxycytidine (dC) as their triphosphates dATP, dTTP, dGTP, and dCTP.

OH   OH   OH
| | |
HO—P—O—P—O—P—O—CH$_2$ base
‖   ‖   ‖   O
O   O   O

X = OH   X = H
dATP   ddATP
dTTP   ddTTP
dGTP   ddGTP
dCTP   ddCTP

Also present in the first test tube is a synthetic analog of ATP in which both the 2′- and 3′-hydroxyl groups have been replaced by hydrogens. This compound is called 2′,3′-dideoxyadenosine triphosphate (ddATP). Similarly, ddTTP is added to the second tube, ddGTP to the third, and ddCTP to the fourth. Each tube also contains a "primer." The primer is a short section of the complementary DNA strand, which has been labeled with a radioactive isotope of phosphorus (^{32}P). When the electrophoresis gel is examined at the end of the experiment, the positions of the DNAs formed by chain extension of the primer are located by a technique called *autoradiography,* which detects the particles emitted by the ^{32}P isotope.

As DNA synthesis proceeds, nucleotides from the solution are added to the growing polynucleotide chain. Chain extension takes place without complication as long as the incorporated nucleotides are derived from dATP, dTTP, dGTP, and dCTP. If, however, the incorporated species is derived from a dideoxy analog, chain extension stops. Because the dideoxy species ddA, ddT, ddG, and ddC lack hydroxyl groups at 3′, they cannot engage in the 3′→5′ phosphodiester linkage necessary for chain extension. Thus, the first tube—the one containing ddATP—contains a mixture of DNA fragments of different length, *all of which terminate in ddA.* Similarly, all the polynucleotides in the second tube terminate in ddT, those in the third tube terminate in ddG, and those in the fourth terminate in ddC.

The contents of each tube are then subjected to electrophoresis in separate lanes on the same sheet of polyacrylamide gel and the DNAs located by autoradiography. A typical electrophoresis gel of a DNA fragment containing 50 nucleotides will exhibit a pattern of 50 bands distributed among the four lanes with no overlaps. Each band corresponds to a polynucleotide that is one nucleotide longer than the one that precedes it (which may be in a different lane). One then simply "reads" the nucleotide sequence according to the lane in which each succeeding band appears.

The Sanger method for DNA sequencing is summarized in Figure 28.13. This work produced a second Nobel Prize for Sanger. (His first was for protein sequencing in 1958.) Sanger shared the 1980 chemistry prize with Walter Gilbert of Harvard University, who developed a chemical method for DNA sequencing (the Maxam–Gilbert method), and with Paul Berg of Stanford University, who was responsible for many of the most important techniques in nucleic acid chemistry and biology.

A recent modification of Sanger's method has resulted in the commercial availability of automated **DNA sequenators** based on Sanger's use of dideoxy analogs of nucleotides. Instead, however, of tagging a primer with ^{32}P, the purine and pyrimidine base portions of the dideoxynucleotides are each modified to contain a side chain that bears a different fluorescent dye, and all the dideoxy analogs are present in the same reaction. After electrophoretic separation of the products in a single lane, the gel is read by argon-laser irradiation at four different wavelengths. One wavelength causes the modified ddA-containing polynucleotides to fluoresce, another causes modified-ddT fluorescence, and so on. The data are stored and analyzed in a computer and printed out as the DNA sequence. A single instrument can sequence about 10,000 bases per day.

FIGURE 28.13 Sequencing of a short strand of DNA (10 bases) by Sanger's method using dideoxynucleotides to halt polynucleotide chain extension. Double-stranded DNA is separated and one of the strands used to produce complements of itself in four different tubes. All of the tubes contain a primer tagged with ^{32}P, dATP, dTTP, dGTP, and dCTP (see text for abbreviations). The first tube also contains ddATP, the second ddTTP, the third, ddGTP, and the fourth ddCTP. All of the DNA fragments in the first tube terminate in A, those in the second terminate in T, those in the third terminate in G, and those in the fourth terminate in C. Location of the zones by autoradiographic detection of ^{32}P identifies the terminal nucleoside. The original DNA strand is its complement.

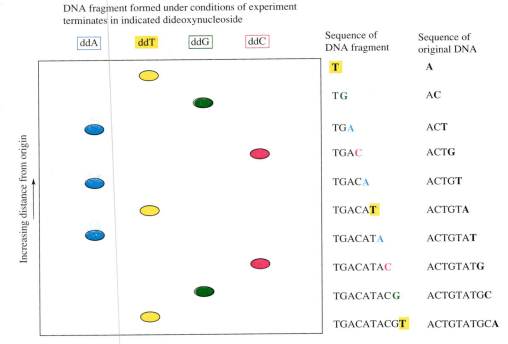

DNA fragment formed under conditions of experiment
terminates in indicated dideoxynucleoside

Sequence of DNA fragment	Sequence of original DNA
T	A
T G	A C
T G A	A C T
T G A C	A C T G
T G A C A	A C T G T
T G A C A T	A C T G T A
T G A C A T A	A C T G T A T
T G A C A T A C	A C T G T A T G
T G A C A T A C G	A C T G T A T G C
T G A C A T A C G T	A C T G T A T G C A

In addition to sequencing bits of DNA or individual genes, DNA sequencing has become so powerful a technique that the entire genomes of more than a thousand organisms have been sequenced. The first and largest number of these organisms were viruses—organisms with relatively small genomes. Then came a bacterium with 1.8 million base pairs, then baker's yeast with 12 million base pairs, followed by a roundworm with 97 million. The year 2000 brought announcements of the sequences of the 100 million base-pair genome of the wild mustard plant and the 180 million base-pair genome of the fruit fly. On the horizon was the 3-billion-base-pair human genome.

28.15 THE HUMAN GENOME PROJECT

In 1988, the National Research Council (NRC) recommended that the United States mount a program to map and then sequence the human genome. Shortly thereafter, the U.S. Congress authorized the first allocation of funds for what became a 15-year $3-billion-dollar project. Most of the NRC's recommendations for carrying out the project were adopted, including a strategy emphasizing technology development in the early stages followed by the sequencing of model organisms before attacking the human genome. The NRC's recommendation that the United States collaborate with other countries was also realized with the participation of teams from the United Kingdom, Japan, France, Germany, and China.

What was not anticipated was that in 1998 Celera Genomics of Rockville, Maryland, would undertake its own privately funded program toward the same goal. By 2000, the two groups agreed to some coordination of their efforts and published draft sequences in 2001 with final versions expected in 2003.

Because a fruit fly, for example, has about 13,000 genes, scientists expected humans to have on the order of 100,000 genes. The first surprise to emerge from the

The International Human Genome Sequencing Consortium was headed by Francis S. Collins of the U.S. National Institutes of Health. J. Craig Venter led the Celera effort.

human genome sequence is that we have far fewer genes than we thought—only about 35,000. Because human DNA has more proteins to code for than fruit-fly DNA, gene expression must be more complicated than the phrase "one gene–one protein" suggests. Puzzles such as this belong to the new research field of **genomics**—the study of genome sequences and their function.

The human genome sequence has been called "the book of life" and, more modestly, a "tool box" and an "instruction manual." Regardless of what we call it, it promises a future characterized by a fuller understanding of human biology and medical science.

28.16 DNA PROFILING AND THE POLYMERASE CHAIN REACTION

DNA sequencing and DNA profiling are different. The former, as we have seen, applies to procedures for determining the sequence of nucleotides in DNA. The latter is also a familiar term, usually encountered in connection with evidence in legal proceedings. In DNA profiling, the genes themselves are of little interest because their role in coding for proteins demands that they differ little, if at all, between individuals. But less than 2% of the human genome codes for proteins. Most of it lies in noncoding regions and this DNA does vary between individuals. Enzymatic cleavage of DNA produces a mixture of fragments that can be separated by electrophoresis to give a pattern of bands more likely to belong to one individual than others. Repeating the process with other cleaving enzymes gives a different pattern of bonds and increases the probability that the identification is correct. Until the 1980s, the limiting factor in both DNA profiling and sequencing was often the small amount of sample that was available. A major advance, called the **polymerase chain reaction** (PCR), effectively overcomes this obstacle and was recognized with the award of the 1993 Nobel Prize in chemistry to its inventor Kary B. Mullis.

The main use of PCR is to "amplify," or make hundreds of thousands—even millions—of copies of a portion of the polynucleotide sequence in a sample of DNA. Suppose, for example, we wish to copy a 500-base-pair region of a DNA that contains a total of 1 million base pairs. We would begin as described in Section 28.14 by cleaving the DNA into smaller fragments using restriction enzymes, then use PCR to make copies of the desired fragment.

Figure 28.14 illustrates how PCR works. In general, it involves multiple cycles of a three-step sequence. In working through Figure 28.14, be alert to the fact that the material we want does not arise until after the third cycle. After that, its contribution to the mixture of DNA fragments increases disproportionately. Repetitive PCR cycling increases both the amount of material and its homogeneity (Table 28.4 on page 1185). If every step proceeds in 100% yield, a greater than 1-billionfold amplification is possible after 30 cycles.

Each cycle incorporates three steps:

1. Denaturation
2. Annealing (also called priming)
3. Synthesis (also called extension or elongation)

All of the substances necessary for PCR are present throughout, and proceeding from one cycle to the next requires only changing the temperature after suitable time intervals. The entire process is carried out automatically, and 30 cycles can be completed within a few hours.

The double-stranded DNA shown in Figure 28.14(*a*) contains the polynucleotide sequence (the target region) we wish to amplify. The DNA is denatured by heating to

PCR is reviewed in the April 1993 issue of the *Journal of Chemical Education*, pp. 273–280. A PCR experiment suitable for undergraduate laboratories appears in the April 1994 issue, pp. 340–341.

FIGURE 28.14 The polymerase chain reaction (PCR). Three cycles are shown; the target region appears after the third cycle. Additional cycles lead to amplification of the target region.

(*a*) Consider double-stranded DNA containing a polynucleotide sequence (the **target region**) that you wish to amplify (make millions of copies of).

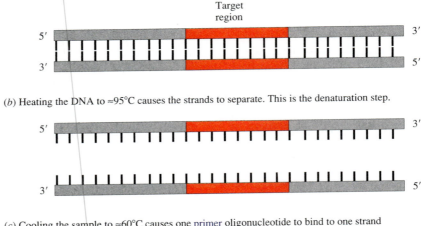

(*b*) Heating the DNA to ≈95°C causes the strands to separate. This is the denaturation step.

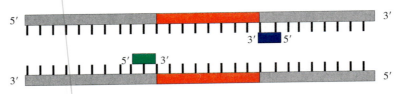

(*c*) Cooling the sample to ≈60°C causes one primer oligonucleotide to bind to one strand and the other primer to the other strand. This is the annealing step.

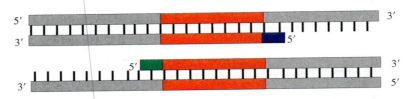

(*d*) In the presence of the four DNA nucleotides and the enzyme DNA polymerase, the primer is extended in its 3′ direction as it adds nucleotides that are complementary to the original DNA strand. This is the synthesis step and is carried out at ≈72°C.

(*e*) Steps (*a*)–(*d*) constitute one cycle of the polymerase chain reaction and produce two double-stranded DNA molecules from one. Denaturing the two DNAs and priming the four strands gives:

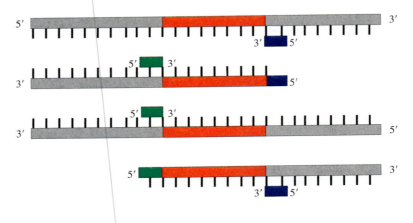

(*f*) Elongation of the primed polynucleotide fragments completes the second cycle and gives four DNAs.

FIGURE 28.14 *(Continued)*

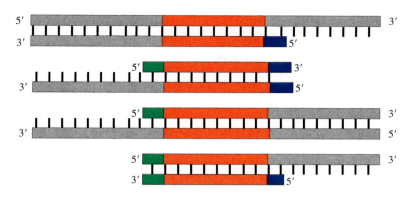

(*g*) Among the eight DNAs formed in the third cycle are two having the structure shown. This is the structure that increases disproportionately in the succeeding cycles.

≈95°C which causes the strands to separate by breaking the hydrogen bonds between them [Figure 28.14(*b*)].

The solution is then cooled to ≈60°C, allowing new hydrogen bonds to form [Figure 28.14(*c*)]. However, the reaction mixture contains much larger concentrations of two primer molecules than DNA, and the new hydrogen bonds are between the separated DNA strands and the primers rather than between the two strands.

Each primer is a synthetic oligonucleotide of about 20 bases, prepared so that their sequences are complementary to the (previously determined) sequences that flank the target regions on opposite strands. Thus, one primer is annealed to one strand, the other to the other strand. The 3′-hydroxyl end of each primer points toward the target region.

The stage is now set for DNA synthesis to proceed from the 3′ end of each primer [Figure 28.14(*d*)]. The solution contains a DNA polymerase and Mg^{2+} in addition to the

TABLE 28.4   Distribution of DNAs with Increasing Number of PCR Cycles

Cycle Number	Total Number of DNAs*	Number of DNAs Containing Only the Target Region
0 (start)	1	0
1	2	0
2	4	0
3	8	2
4	16	8
5	32	22
10	1,024	1,004
20	1,048,566	1,048,526
30	1,073,741,824	1,073,741,764

*Total number of DNAs is 2^n, where n = number of cycles

Taq polymerase was first found in a bacterium (*Thermus aquaticus*) that lives in hot springs in Yellowstone National Park. Bacteria of this type are called thermophiles because they thrive in warm environments.

deoxynucleoside triphosphates dATP, dTTP, dGTP, and dCTP. The particular DNA polymerase used is one called *Taq polymerase* that is stable and active at the temperature at which the third step of the cycle is carried out (72°C).

The products of the first cycle are two DNAs, each of which is composed of a longer and a shorter strand. These products are subjected to a second three-step cycle [Figure 28.14(e)–(f)] to give four DNAs. Two of these four contain a "strand" which is nothing more than the target region flanked by primers. In the third cycle, these two ultrashort "strands" produce two DNAs of the kind shown in Figure 28.14(g). This product contains only the target region plus the primers and is the one that increases disproportionately in subsequent cycles.

Since its introduction in 1985, PCR has been applied to practically every type of study that requires samples of DNA. These include screening for genetic traits such as sickle cell anemia, Huntington's disease, and cystic fibrosis. PCR can detect HIV infection when the virus is present in such small concentrations that no AIDS symptoms have as yet appeared. In forensic science, analysis of PCR-amplified DNA from tiny amounts of blood or semen have helped convict the guilty and free the innocent. Anthropologists increasingly use information from DNA analysis to trace the origins of racial and ethnic groups but sometimes find it difficult, for cultural reasons, to convince individuals to volunteer blood samples. Thanks to PCR, a strand of hair is now sufficient.

Within a few weeks of being brought into a case, scientists at the U.S. Centers for Disease Control (CDC) used PCR to help identify the infectious agent responsible for an outbreak of an especially dangerous hemorrhagic fever that struck the U.S. southwest in 1993. By annealing with synthetic oligonucleotide primers having sequences complementary to known hantaviruses, portions of the viral DNA obtained from those infected with the disease could be sucessfully amplified. Not only did this provide material for analysis, it also suggested that the new viral DNA had stretches where its sequence was the same as already known hantaviruses. Thus, the "Four Corners virus" was found to be a new strain of hantavirus and diagnostic procedures were developed specific for it.

More recently, PCR proved to be a valuable detection and analytical tool during the terrorist-inspired anthrax outbreak in the fall of 2001.

Although not an official name, "Four Corners" succinctly describes where the virus was first discovered. It is the region where Arizona, New Mexico, Colorado, and Utah meet.

28.17 SUMMARY

Section 28.1   Many biologically important compounds are related to the heterocyclic aromatic compounds pyrimidine and purine.

Pyrimidine   Purine

The structure of guanine illustrates an important feature of substituted pyrimidines and purines. Oxygen substitution on the ring favors the keto form rather than the enol. Amino substitution does not.

Guanine

Section 28.2   **Nucleosides** are carbohydrate derivatives of pyrimidine and purine bases. The most important nucleosides are derived from D-ribose and 2-deoxy-D-ribose.

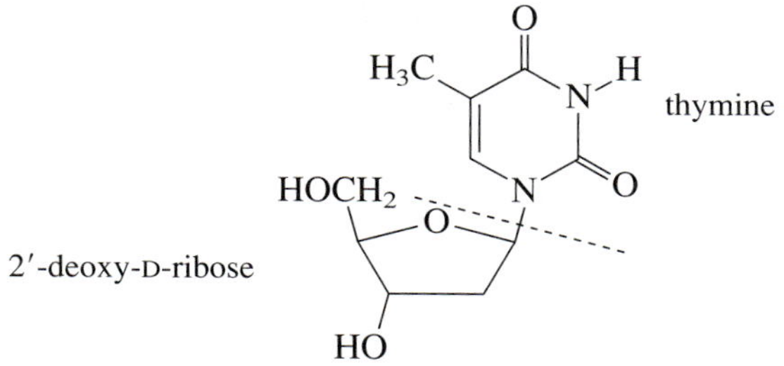

Thymidine

Section 28.3   **Nucleotides** are phosphoric acid esters of nucleosides.

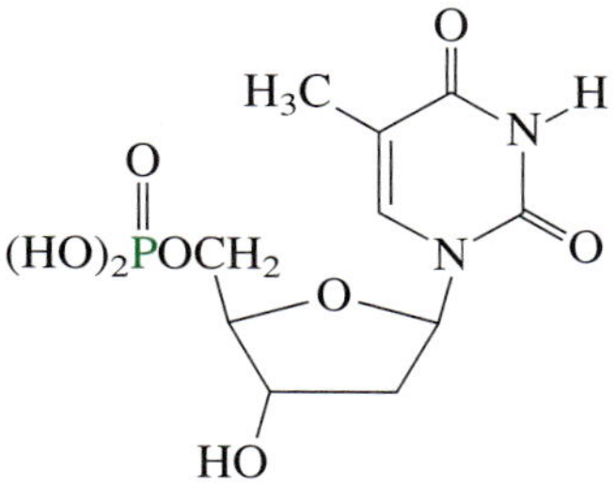

Thymidine 5′-monophosphate

In the example shown, the 5′-OH group is phosphorylated. Nucleotides are also possible in which some other OH group bears the phosphate ester function. Cyclic phosphates are common and important as biochemical messengers.

Section 28.4   **Bioenergetics** is concerned with the thermodynamics of biological processes. Particular attention is paid to $\Delta G^{\circ\prime}$, the standard free-energy change of reactions at pH = 7. When the sign of $\Delta G^{\circ\prime}$ is +, the reaction is **endergonic;** when the sign of $\Delta G^{\circ\prime}$ is −, the reaction is **exergonic.**

Section 28.5   **Adenosine triphosphate (ATP)** is a key compound in biological energy storage and delivery.

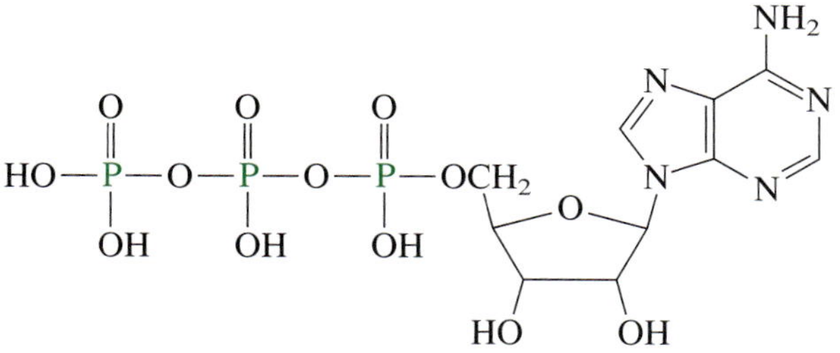

Adenosine triphosphate (ATP)

The hydrolysis of ATP to ADP and $HPO_4{}^{2-}$ is exergonic.

$$ATP + H_2O \longrightarrow ADP + HPO_4{}^{2-} \qquad \Delta G^{\circ\prime} = -31 \text{ kJ } (-7.4 \text{ kcal})$$

Many formally endergonic biochemical processes become exergonic when they are coupled mechanistically to the hydrolysis of ATP.

Section 28.6   Many important compounds contain two or more nucleotides joined together by a **phosphodiester** linkage. The best known are those in which the phosphodiester joins the 5'-oxygen of one nucleotide to the 3'-oxygen of the other.

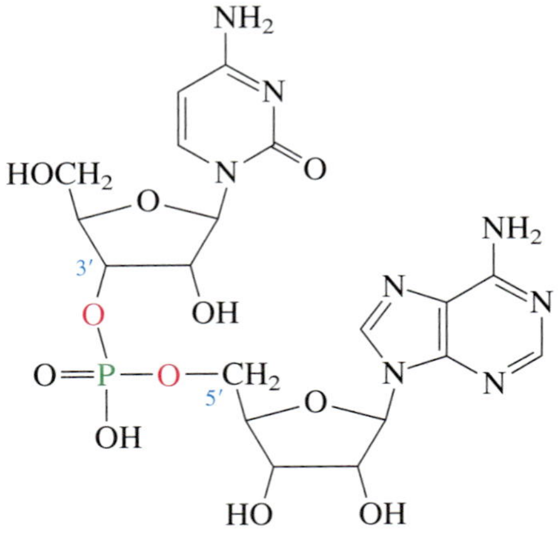

Oligonucleotides contain about 50 or fewer nucleotides held together by phosphodiester links; **polynucleotides** can contain thousands of nucleotides.

Section 28.7   **Nucleic acids** are polynucleotides present in cells. The carbohydrate component is D-ribose in ribonucleic acid (RNA) and 2-deoxy-D-ribose in deoxyribonucleic acid (DNA).

Section 28.8   The most common form of DNA is B-DNA, which exists as a right-handed double helix. The carbohydrate–phosphate backbone lies on the outside, the purine and pyrimidine bases on the inside. The double helix is stabilized by complementary hydrogen bonding (base pairing) between adenine (A) and thymine (T), and guanine (G) and cytosine (C).

Section 28.9   Within the cell nucleus, double-helical DNA adopts a **supercoiled** terti-ary structure in which short sections are wound around proteins called **histones.** This reduces the effective length of the DNA and maintains it in an ordered arrangement.

Section 28.10   During DNA replication the two strands of the double helix begin to unwind, exposing the pyrimidine and purine bases in the interior. Nucleotides with complementary bases hydrogen bond to the original strands and are joined together by phosphodiester linkages with the aid of DNA polymerase. Each new strand grows in its $5' \rightarrow 3'$ direction.

Section 28.11   Three RNAs are involved in gene expression. In the **transcription** phase, a strand of **messenger RNA (mRNA)** is synthesized from a DNA tem-plate. The four bases A, G, C, and U, taken three at a time, generate 64 possible combinations called **codons.** These 64 codons comprise the **genetic code** and code for the 20 amino acids found in proteins plus start and stop signals. The mRNA sequence is **translated** into a prescribed protein sequence at the ribosomes. There, small polynucleotides called transfer RNA (tRNA), each of which contains an **anticodon** complementary to an mRNA codon carries the correct amino acid for incorporation into the growing protein. **Ribosomal RNA (rRNA)** is the main constituent of ribosomes and appears to catalyze protein biosynthesis.

Section 28.12 The start codon for protein biosynthesis is AUG, which is the same as the codon for methionine. Thus, all proteins initially have methionine as their N-terminal amino acid, but lose it subsequent to their formation. The reaction responsible for extending the protein chain is nucleophilic acyl substitution.

Section 28.13 HIV, which causes AIDS, is a retrovirus. Its genetic material is RNA instead of DNA. HIV contains an enzyme called reverse transcriptase that allows its RNA to serve as a template for DNA synthesis in the host cell.

Section 28.14 The nucleotide sequence of DNA can be determined by a technique in which a short section of single-stranded DNA is allowed to produce its complement in the presence of dideoxy analogs of ATP, TTP, GTP, and CTP. DNA formation terminates when a dideoxy analog is incorporated into the growing polynucleotide chain. A mixture of polynucleotides differing from one another by an incremental nucleoside is produced and analyzed by electrophoresis. From the observed sequence of the complementary chain, the sequence of the original DNA is deduced.

Section 28.15 A rough draft of the sequence of nucleotides that make up the human genome has been completed. There is every reason to believe that the increased knowledge of human biology it offers will dramatically affect the practice of medicine.

Section 28.16 In **DNA profiling** the noncoding regions are cut into smaller fragments using enzymes that recognize specific sequences, and these smaller bits of DNA are then separated by electrophoresis. The observed pattern of DNA fragments is believed to be highly specific for the source of the DNA. Using the **polymerase chain reaction (PCR),** millions of copies of minute amounts of DNA can be produced in a relatively short time.

PROBLEMS

28.14 *5-Fluorouracil* is one component of a mixture of three drugs used in breast-cancer chemotherapy. What is its structure?

28.15 (a) Which isomer, the keto or enol form of cytosine, is the stronger acid?

keto   enol

(b) What is the relationship between the conjugate base of the keto form and the conjugate base of the enol form?

28.16 Birds excrete nitrogen as *uric acid*. Uric acid is a purine having the molecular formula $C_5H_4N_4O_3$; it has no C—H bonds. Write a structural formula for uric acid.

28.17 *Nebularine* is a toxic nucleoside isolated from a species of mushroom. Its systematic name is 9-β-D-ribofuranosylpurine. Write a structural formula for nebularine.

28.18 The D-arabinose analog of adenosine is an anitviral agent (vidarabine) used to treat conjunctivitis and shingles. Write a structural formula for this compound.

28.19 Adenine is a weak base. Which one of the three nitrogens designated by arrows in the structural formula shown is protonated in acidic solution? A resonance evaluation of the three protonated forms will tell you which one is the most stable.

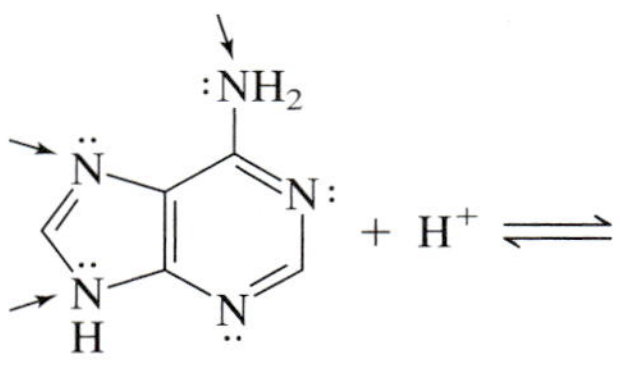

28.20 When 6-chloropurine is heated with aqueous sodium hydroxide, it is quantitatively converted to *hypoxanthine*. Suggest a reasonable mechanism for this reaction.

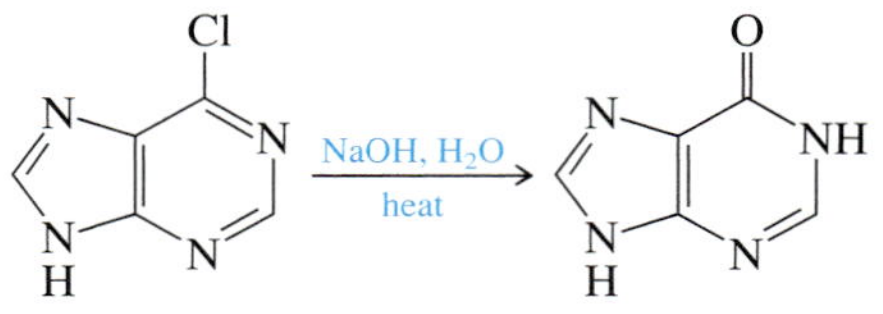

6-Chloropurine   Hypoxanthine

28.21 Treatment of adenosine with nitrous acid gives a nucleoside known as *inosine*. Suggest a reasonable mechanism for this reaction.

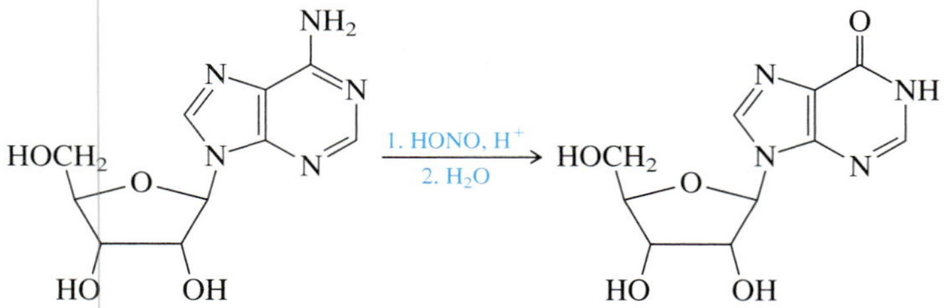

Adenosine   Inosine

28.22 The 5'-nucleotide of inosine, *inosinic acid* ($C_{10}H_{13}N_4O_8P$) is added to foods as a flavor enhancer. What is the structure of inosinic acid? (The structure of inosine is given in Problem 28.21).

28.23 The phosphorylation of α-D-glucopyranose by ATP (Section 28.3) has $\Delta G^{\circ\prime} = -23$ kJ at 298 K.

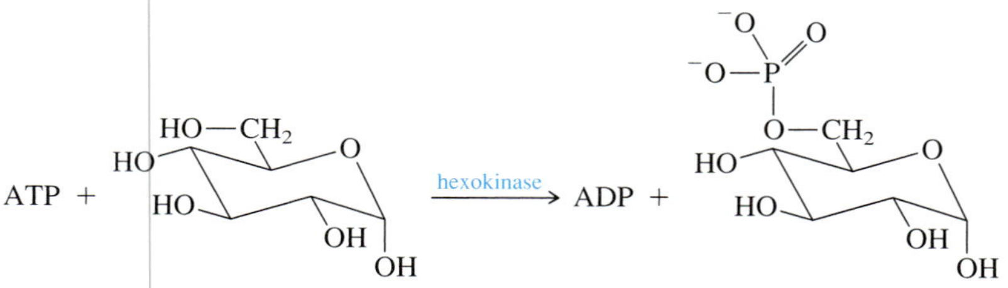

(a) Is this reaction exergonic or endergonic?

(b) How would the value of $\Delta G^{\circ\prime}$ change in the absence of the enzyme hexokinase? Would it become more positive, more negative, or would it stay the same? Why?

(c) Use the value for the hydrolysis of ATP to ADP (Section 28.5) to calculate $\Delta G^{\circ\prime}$ for the reaction of α-D-glucopyranose with inorganic phosphate. Is this reaction exergonic or endergonic?

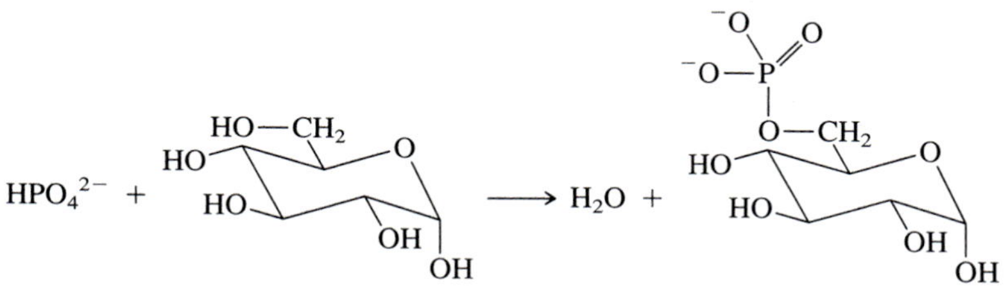

28.24 In one of the early experiments designed to elucidate the genetic code, Marshall Nirenberg of the U.S. National Institutes of Health (Nobel Prize in physiology or medicine, 1968) prepared a synthetic mRNA in which all the bases were uracil. He added this poly(U) to a cell-free system containing all the necessary materials for protein biosynthesis. A polymer of a single amino acid was obtained. What amino acid was polymerized?

PHYSICAL PROPERTIES

TABLE A Selected Physical Properties of Representative Hydrocarbons

Compound name	Molecular formula	Structural formula	Melting point, °C	Boiling point, °C (1 atm)
Alkanes				
Methane	CH_4	CH_4	−182.5	−160
Ethane	C_2H_6	CH_3CH_3	−183.6	−88.7
Propane	C_3H_8	$CH_3CH_2CH_3$	−187.6	−42.2
Butane	C_4H_{10}	$CH_3CH_2CH_2CH_3$	−139.0	−0.4
2-Methylpropane	C_4H_{10}	$(CH_3)_3CH$	−160.9	−10.2
Pentane	C_5H_{12}	$CH_3(CH_2)_3CH_3$	−129.9	36.0
2-Methylbutane	C_5H_{12}	$(CH_3)_2CHCH_2CH_3$	−160.5	27.9
2,2-Dimethylpropane	C_5H_{12}	$(CH_3)_4C$	−16.6	9.6
Hexane	C_6H_{14}	$CH_3(CH_2)_4CH_3$	−94.5	68.8
Heptane	C_7H_{16}	$CH_3(CH_2)_5CH_3$	−90.6	98.4
Octane	C_8H_{18}	$CH_3(CH_2)_6CH_3$	−56.9	125.6
Nonane	C_9H_{20}	$CH_3(CH_2)_7CH_3$	−53.6	150.7
Decane	$C_{10}H_{22}$	$CH_3(CH_2)_8CH_3$	−29.7	174.0
Dodecane	$C_{12}H_{26}$	$CH_3(CH_2)_{10}CH_3$	−9.7	216.2
Pentadecane	$C_{15}H_{32}$	$CH_3(CH_2)_{13}CH_3$	10.0	272.7
Icosane	$C_{20}H_{42}$	$CH_3(CH_2)_{18}CH_3$	36.7	205 (15 mm)
Hectane	$C_{100}H_{202}$	$CH_3(CH_2)_{98}CH_3$	115.1	
Cycloalkanes				
Cyclopropane	C_3H_6		−127.0	−32.9
Cyclobutane	C_4H_8			13.0
Cyclopentane	C_5H_{10}		−94.0	49.5
Cyclohexane	C_6H_{12}		6.5	80.8
Cycloheptane	C_7H_{14}		−13.0	119.0
Cyclooctane	C_8H_{16}		13.5	149.0
Cyclononane	C_9H_{18}			171
Cyclodecane	$C_{10}H_{20}$		9.6	201
Cyclopentadecane	$C_{15}H_{30}$		60.5	112.5 (1 mm)
Alkenes and cycloalkenes				
Ethene (ethylene)	C_2H_4	$H_2C{=}CH_2$	−169.1	−103.7
Propene	C_3H_6	$CH_3CH{=}CH_2$	−185.0	−47.6
1-Butene	C_4H_8	$CH_3CH_2CH{=}CH_2$	−185	−6.1
2-Methylpropene	C_4H_8	$(CH_3)_2C{=}CH_2$	−140	−6.6
Cyclopentene	C_5H_8		−98.3	44.1

(Continued)

TABLE A Selected Physical Properties of Representative Hydrocarbons *(Continued)*

Compound name	Molecular formula	Structural formula	Melting point, °C	Boiling point, °C (1 atm)
1-Pentene	C_5H_{10}	$CH_3CH_2CH_2CH{=}CH_2$	−138.0	30.2
2-Methyl-2-butene	C_5H_{10}	$(CH_3)_2C{=}CHCH_3$	−134.1	38.4
Cyclohexene	C_6H_{10}	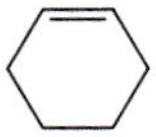	−104.0	83.1
1-Hexene	C_6H_{12}	$CH_3CH_2CH_2CH_2CH{=}CH_2$	−138.0	63.5
2,3-Dimethyl-2-butene	C_6H_{12}	$(CH_3)_2C{=}C(CH_3)_2$	−74.6	73.5
1-Heptene	C_7H_{14}	$CH_3(CH_2)_4CH{=}CH_2$	−119.7	94.9
1-Octene	C_8H_{16}	$CH_3(CH_2)_5CH{=}CH_2$	−104	119.2
1-Decene	$C_{10}H_{20}$	$CH_3(CH_2)_7CH{=}CH_2$	−80.0	172.0
Alkynes				
Ethyne (acetylene)	C_2H_2	$HC{\equiv}CH$	−81.8	−84.0
Propyne	C_3H_4	$CH_3C{\equiv}CH$	−101.5	−23.2
1-Butyne	C_4H_6	$CH_3CH_2C{\equiv}CH$	−125.9	8.1
2-Butyne	C_4H_6	$CH_3C{\equiv}CCH_3$	−32.3	27.0
1-Hexyne	C_6H_{10}	$CH_3(CH_2)_3C{\equiv}CH$	−132.4	71.4
3,3-Dimethyl-1-butyne	C_6H_{10}	$(CH_3)_3CC{\equiv}CH$	−78.2	37.7
1-Octyne	C_8H_{14}	$CH_3(CH_2)_5C{\equiv}CH$	−79.6	126.2
1-Nonyne	C_9H_{16}	$CH_3(CH_2)_6C{\equiv}CH$	−36.0	160.6
1-Decyne	$C_{10}H_{18}$	$CH_3(CH_2)_7C{\equiv}CH$	−40.0	182.2
Arenes				
Benzene	C_6H_6		5.5	80.1
Toluene	C_7H_8	 —CH_3	−95	110.6
Styrene	C_8H_8	 —$CH{=}CH_2$	−33	145
p-Xylene	C_8H_{10}	H_3C—  —CH_3	−13	138
Ethylbenzene	C_8H_{10}	 —CH_2CH_3	−94	136.2
Naphthalene	$C_{10}H_8$	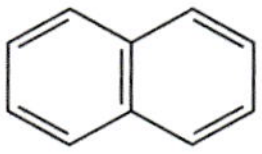	80.3	218
Diphenylmethane	$C_{13}H_{12}$	$(C_6H_5)_2CH_2$	26	261
Triphenylmethane	$C_{19}H_{16}$	$(C_6H_5)_3CH$	94	

TABLE B Selected Physical Properties of Representative Organic Halogen Compounds

Alkyl Halides

Compound name	Structural formula	Boiling point, °C (1 atm)				Density, g/mL (20°C)		
		Fluoride	Chloride	Bromide	Iodide	Chloride	Bromide	Iodide
Halomethane	CH_3X	−78	−24	3	42			2.279
Haloethane	CH_3CH_2X	−32	12	38	72	0.903	1.460	1.933
1-Halopropane	$CH_3CH_2CH_2X$	−3	47	71	103	0.890	1.353	1.739
2-Halopropane	$(CH_3)_2CHX$	−11	35	59	90	0.859	1.310	1.714
1-Halobutane	$CH_3CH_2CH_2CH_2X$		78	102	130	0.887	1.276	1.615
2-Halobutane	$CH_3CHCH_2CH_3$ $\quad\;$ | $\quad\;$ X		68	91	120	0.873	1.261	1.597
1-Halo-2-methylpropane	$(CH_3)_2CHCH_2X$	16	68	91	121	0.878	1.264	1.603
2-Halo-2-methylpropane	$(CH_3)_3CX$		51	73	99	0.847	1.220	1.570
1-Halopentane	$CH_3(CH_2)_3CH_2X$	65	108	129	157	0.884	1.216	1.516
1-Halohexane	$CH_3(CH_2)_4CH_2X$	92	134	155	180	0.879	1.175	1.439
1-Halooctane	$CH_3(CH_2)_6CH_2X$	143	183	202	226	0.892	1.118	1.336
Halocyclopentane	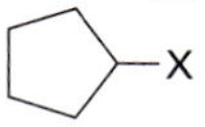 ⬠—X		114	138	166	1.005	1.388	1.694
Halocyclohexane	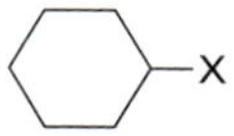 ⬡—X		142	167	192	0.977	1.324	1.626

Aryl Halides

Compound	Halogen substituent (X)*							
	Fluorine		Chlorine		Bromine		Iodine	
	mp	bp	mp	bp	mp	bp	mp	bp
C_6H_5X	−41	85	−45	132	−31	156	−31	188
$o\text{-}C_6H_4X_2$	−34	91	−17	180	7	225	27	286
$m\text{-}C_6H_4X_2$	−59	83	−25	173	−7	218	35	285
$p\text{-}C_6H_4X_2$	−13	89	53	174	87	218	129	285
$1,3,5\text{-}C_6H_3X_3$	−5	76	63	208	121	271	184	
C_6X_6	5	80	230	322	327		350	

*All boiling points and melting points cited are in degrees Celsius.

TABLE C Selected Physical Properties of Representative Alcohols, Ethers, and Phenols

Compound name	Structural formula	Melting point, °C	Boiling point, °C (1 atm)	Solubility, g/100 mL H_2O
Alcohols				
Methanol	CH_3OH	−94	65	∞
Ethanol	CH_3CH_2OH	−117	78	∞
1-Propanol	$CH_3CH_2CH_2OH$	−127	97	∞
2-Propanol	$(CH_3)_2CHOH$	−90	82	∞
1-Butanol	$CH_3CH_2CH_2CH_2OH$	−90	117	9
2-Butanol	$CH_3CHCH_2CH_3$ 　　$\mid$ 　　OH	−115	100	26
2-Methyl-1-propanol	$(CH_3)_2CHCH_2OH$	−108	108	10
2-Methyl-2-propanol	$(CH_3)_3COH$	26	83	∞
1-Pentanol	$CH_3(CH_2)_3CH_2OH$	−79	138	
1-Hexanol	$CH_3(CH_2)_4CH_2OH$	−52	157	0.6
1-Dodecanol	$CH_3(CH_2)_{10}CH_2OH$	26	259	Insoluble
Cyclohexanol	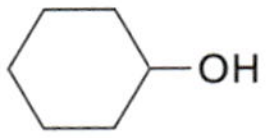 ⬡—OH	25	161	3.6
Ethers				
Dimethyl ether	CH_3OCH_3	−138.5	−24	Very soluble
Diethyl ether	$CH_3CH_2OCH_2CH_3$	−116.3	34.6	7.5
Dipropyl ether	$CH_3CH_2CH_2OCH_2CH_2CH_3$	−122	90.1	Slight
Diisopropyl ether	$(CH_3)_2CHOCH(CH_3)_2$	−60	68.5	0.2
1,2-Dimethoxyethane	$CH_3OCH_2CH_2OCH_3$		83	∞
Diethylene glycol dimethyl ether (diglyme)	$CH_3OCH_2CH_2OCH_2CH_2OCH_3$		161	∞
Ethylene oxide	▽O	−111.7	10.7	∞
Tetrahydrofuran	 ⬠O	−108.5	65	∞
Phenols				
Phenol		43	182	8.2
o-Cresol		31	191	2.5
m-Cresol		12	203	0.5
p-Cresol		35	202	1.8
o-Chlorophenol		7	175	2.8
m-Chlorophenol		32	214	2.6
p-Chlorophenol		42	217	2.7
o-Nitrophenol		45	217	0.2
m-Nitrophenol		96		1.3
p-Nitrophenol		114	279	1.6
1-Naphthol		96	279	Slight
2-Naphthol		122	285	0.1
Pyrocatechol		105	246	45.1
Resorcinol		110	276	147.3
Hydroquinone		170	285	6

TABLE D Selected Physical Properties of Representative Aldehydes and Ketones

Compound name	Structural formula	Melting point, °C	Boiling point, °C (1 atm)	Solubility, g/100 mL H$_2$O
Aldehydes				
Formaldehyde	$\overset{\text{O}}{\overset{\|}{\text{HCH}}}$	−92	−21	Very soluble
Acetaldehyde	$\overset{\text{O}}{\overset{\|}{\text{CH}_3\text{CH}}}$	−123.5	20.2	∞
Propanal	$\overset{\text{O}}{\overset{\|}{\text{CH}_3\text{CH}_2\text{CH}}}$	−81	49.5	20
Butanal	$\overset{\text{O}}{\overset{\|}{\text{CH}_3\text{CH}_2\text{CH}_2\text{CH}}}$	−99	75.7	4
Benzaldehyde	$\overset{\text{O}}{\overset{\|}{\text{C}_6\text{H}_5\text{CH}}}$	−26	178	0.3
Ketones				
Acetone	$\overset{\text{O}}{\overset{\|}{\text{CH}_3\text{CCH}_3}}$	−94.8	56.2	∞
2-Butanone	$\overset{\text{O}}{\overset{\|}{\text{CH}_3\text{CCH}_2\text{CH}_3}}$	−86.9	79.6	37
2-Pentanone	$\overset{\text{O}}{\overset{\|}{\text{CH}_3\text{CCH}_2\text{CH}_2\text{CH}_3}}$	−77.8	102.4	Slight
3-Pentanone	$\overset{\text{O}}{\overset{\|}{\text{CH}_3\text{CH}_2\text{CCH}_2\text{CH}_3}}$	−39.9	102.0	4.7
Cyclopentanone	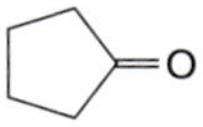 ⬠=O	−51.3	130.7	43.3
Cyclohexanone	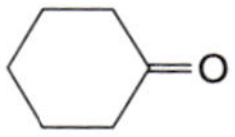 ⬡=O	−45	155	
Acetophenone	$\overset{\text{O}}{\overset{\|}{\text{C}_6\text{H}_5\text{CCH}_3}}$	21	202	Insoluble
Benzophenone	$\overset{\text{O}}{\overset{\|}{\text{C}_6\text{H}_5\text{CC}_6\text{H}_5}}$	48	306	Insoluble

TABLE E Selected Physical Properties of Representative Carboxylic Acids and Dicarboxylic Acids

Compound name	Structural formula	Melting point, °C	Boiling point, °C (1 atm)	Solubility, g/100 mL H_2O
Carboxylic acids				
Formic acid	HCO_2H	8.4	101	∞
Acetic acid	CH_3CO_2H	16.6	118	∞
Propanoic acid	$CH_3CH_2CO_2H$	−20.8	141	∞
Butanoic acid	$CH_3CH_2CH_2CO_2H$	−5.5	164	∞
Pentanoic acid	$CH_3(CH_2)_3CO_2H$	−34.5	186	3.3 (16°C)
Decanoic acid	$CH_3(CH_2)_8CO_2H$	31.4	269	0.003 (15°C)
Benzoic acid	$C_6H_5CO_2H$	122.4	250	0.21 (17°C)
Dicarboxylic acids				
Oxalic acid	HO_2CCO_2H	186	Sublimes	10 (20°C)
Malonic acid	$HO_2CCH_2CO_2H$	130–135	Decomposes	138 (16°C)
Succinic acid	$HO_2CCH_2CH_2CO_2H$	189	235	6.8 (20°C)
Glutaric acid	$HO_2CCH_2CH_2CH_2CO_2H$	97.5		63.9 (20°C)

TABLE F Selected Physical Properties of Representative Amines

Alkylamines

Compound name	Structural formula	Melting point, °C	Boiling point, °C	Solubility, g/100 mL H_2O
Primary amines				
Methylamine	CH_3NH_2	−92.5	−6.7	Very high
Ethylamine	$CH_3CH_2NH_2$	−80.6	16.6	∞
Butylamine	$CH_3CH_2CH_2CH_2NH_2$	−50	77.8	∞
Isobutylamine	$(CH_3)_2CHCH_2NH_2$	−85	68	∞
sec-Butylamine	$CH_3CH_2CHNH_2$ $\vert$ CH_3	−104	66	∞
tert-Butylamine	$(CH_3)_3CNH_2$	−67.5	45.2	
Hexylamine	$CH_3(CH_2)_5NH_2$	−19	129	Slightly soluble
Cyclohexylamine	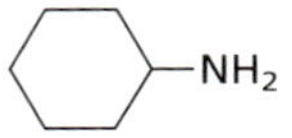 ⬡—NH_2	−18	134.5	∞
Benzylamine	$C_6H_5CH_2NH_2$	10	184.5	∞
Secondary amines				
Dimethylamine	$(CH_3)_2NH$	−92.2	6.9	Very soluble
Diethylamine	$(CH_3CH_2)_2NH$	−50	55.5	Very soluble
N-Methylpropylamine	$CH_3NHCH_2CH_2CH_3$		62.4	Soluble
Piperidine	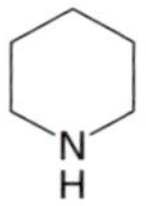 ⬡ with N—H	−10.5	106.4	∞

(Continued)

TABLE F Selected Physical Properties of Representative Amines *(Continued)*

Alkylamines

Compound name	Structural formula	Melting point, °C	Boiling point, °C	Solubility, g/100 mL H_2O
Tertiary amines				
Trimethylamine	$(CH_3)_3N$	−117.1	2.9	41
Triethylamine	$(CH_3CH_2)_3N$	−114.7	89.4	∞
N-Methylpiperidine	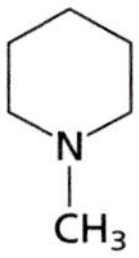	3	107	

Arylamines

Primary amines

Aniline	−6.3	184	
o-Toluidine	−14.7	200	
m-Toluidine	−30.4	203	
p-Toluidine	44	200	
o-Chloroaniline	−14	209	
m-Chloroaniline	−10	230	
p-Chloroaniline	72.5	232	
o-Nitroaniline	71.5	284	
m-Nitroaniline	114	306	
p-Nitroaniline	148	332	

Secondary amines

N-Methylaniline	−57	196
N-Ethylaniline	−63	205

Tertiary amines

N,N-Dimethylaniline	2.4	194
Triphenylamine	127	365

APPENDIX 2

ANSWERS TO IN-TEXT PROBLEMS

Problems are of two types: in-text problems that appear within the body of each chapter, and end-of-chapter problems. This appendix gives brief answers to all the in-text problems. More detailed discussions of in-text problems as well as detailed solutions to all the end-of-chapter problems are provided in a separate *Solutions Manual*. Answers to part (a) of those in-text problems with multiple parts have been provided in the form of a sample solution within each chapter and are not repeated here.

CHAPTER 1

1.1 4

1.2 All the third-row elements have a neon core containing ten electrons ($1s^2 2s^2 2p^6$). The elements in the third row, their atomic numbers Z, and their electron configurations beyond the neon core are Na($Z = 11$)$3s^1$; Mg($Z = 12$)$3s^2$; Al($Z = 13$) $3s^2 3p_x^1$; Si($Z = 14$) $3s^2 3p_x^1 3p_y^1$; P ($Z = 15$) $3s^2 3p_x^1 3p_y^1 3p_z^1$; S ($Z = 16$) $3s^2 3p_x^2 3p_y^1 3p_z^1$; Cl ($Z = 17$) $3s^2 3p_x^2 3p_y^2 3p_z^1$; Ar ($Z = 18$) $3s^2 3p_x^2 3p_y^2 3p_z^2$.

1.3 Those ions that possess a noble gas electron configuration are (a) K^+; (c) H^-; (e) F^-; and (f) Ca^{2+}.

1.4 Electron configuration of C^+ is $1s^2 2s^2 2p^1$; electron configuration of C^- is $1s^2 2s^2 2p^3$. Neither C^+ nor C^- possesses a noble gas electron configuration.

1.5 H:F̈:

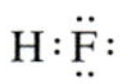

1.6
```
     H H
H:C:C:H
     H H
```

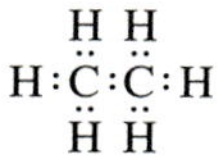

1.7 (b) 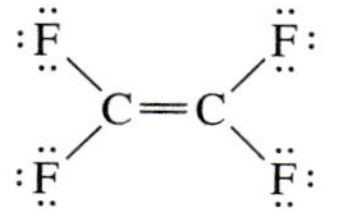
(c)

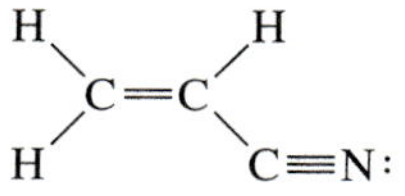

1.8 Carbon bears a partial positive charge in CH_3Cl. It is partially negative in both CH_4 and CH_3Li, but the degree of negative charge is greater in CH_3Li.

1.9 (b) Sulfur has a formal charge of $+2$ in the Lewis structure given for sulfuric acid, the two oxygens bonded only to sulfur each have a formal charge of -1, and the oxygens and hydrogens of the two OH groups have no formal charge; (c) none of the atoms has a formal charge in the Lewis structure given for nitrous acid.

1.10 The electron counts of nitrogen in ammonium ion and boron in borohydride ion are both 4 (half of eight electrons in covalent bonds). Because a neutral nitrogen has five electrons in its valence shell, an electron count of 4 gives it a formal charge of $+1$. A neutral boron has three valence electrons, so that an electron count of 4 in borohydride ion corresponds to a formal charge of -1.

1.11 $\overset{\delta+}{H}-\overset{\delta+}{\underset{\delta+}{N}}-\overset{\delta+}{H}$ (with H above, δ+, and H below δ+)   and $\overset{\delta-}{H}-\overset{\delta-}{\underset{\delta-}{B}}-\overset{\delta-}{H}$ (with H above δ−, and H below δ−)

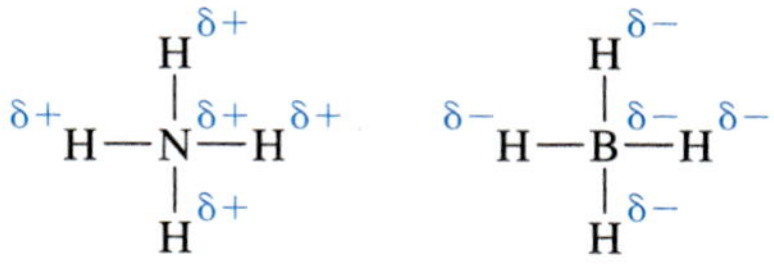

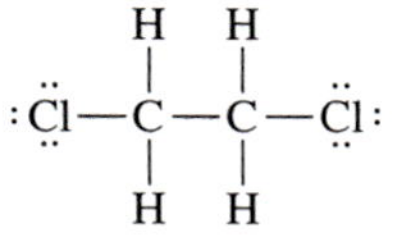

1.12 (b) 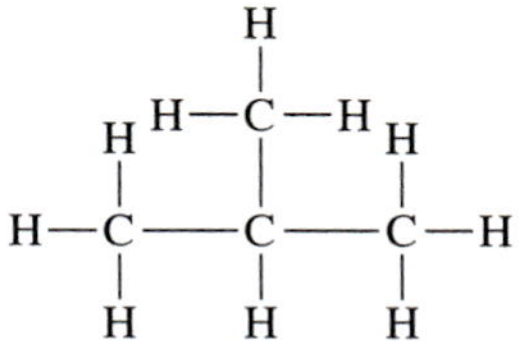
```
            H
            |
        H—C—H
      H     H     H
      |     |     |
  H—C———C———C—H
      |     |     |
      H     H     H
```
(c)
```
     H   H
     |   |
 :Cl—C———C—Cl:
     |   |
     H   H
```
(d) 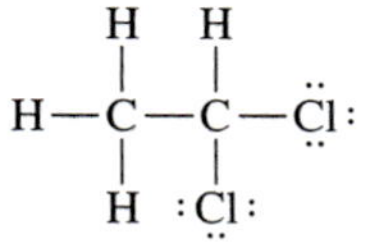
```
     H   H
     |   |
 H—C———C—Cl:
     |   |
     H  :Cl:
```

(e) 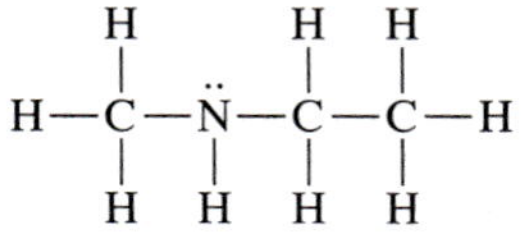
```
     H       H   H
     |       |   |
 H—C———N———C———C—H
     |   |   |   |
     H   H   H   H
```
(f) 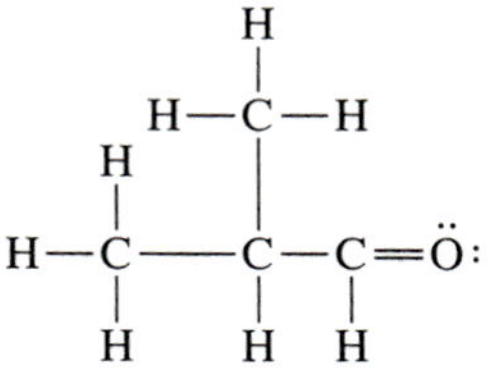
```
              H
              |
          H—C—H
      H       |
      |       |
  H—C———C———C=O:
      |       |
      H       H   H
```

1.13 (b) $(CH_3)_2CHCH(CH_3)_2$ (c) 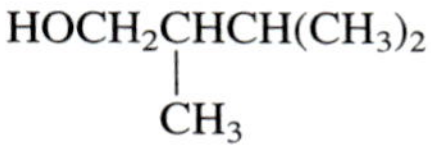 $HOCH_2CHCH(CH_3)_2$ with CH_3 below   (d) 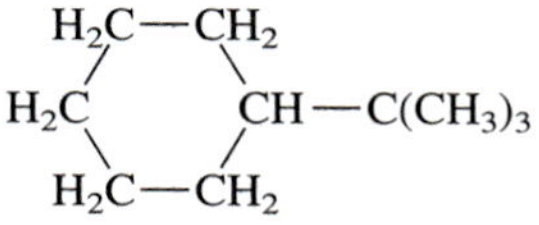
```
        H2C—CH2
    H2C         CH—C(CH3)3
        H2C—CH2
```

1.14
```
         :O:
         ‖
 H—N—C—O—H
     |
     H
```

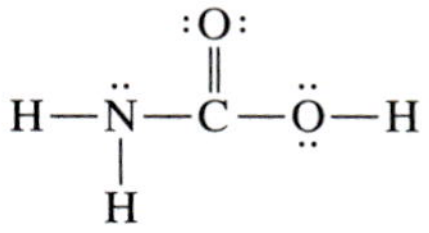

1.15 (b) $CH_3CH_2CH_2OH$, $(CH_3)_2CHOH$, and $CH_3CH_2OCH_3$. (c) There are seven isomers of $C_4H_{10}O$. Four have —OH groups: $CH_3CH_2CH_2CH_2OH$, $(CH_3)_2CHCH_2OH$, $(CH_3)_3COH$, and $CH_3CHCH_2CH_3$ (with OH below). Three have C—O—C units: $CH_3OCH_2CH_2CH_3$, $CH_3CH_2OCH_2CH_3$, and $(CH_3)_2CHOCH_3$

1.16 (b) 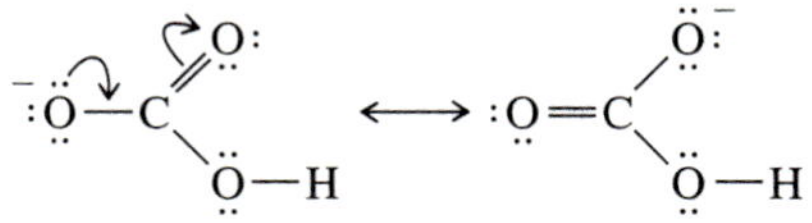   [resonance structures of carboxylate/carbonate type]

(c) 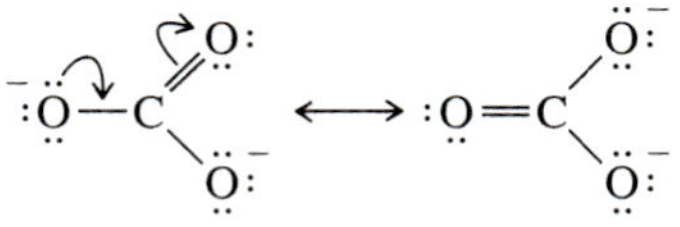   [resonance structures]

and

[resonance structures]

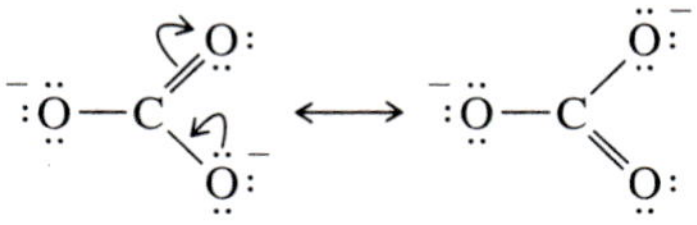

(d)

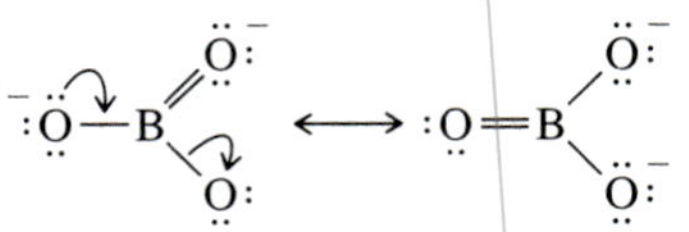

and

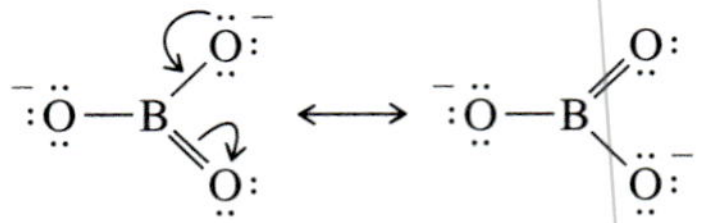

1.17 The H—B—H angles in BH_4^- are 109.5° (tetrahedral).

1.18 (b) Tetrahedral; (c) linear; (d) trigonal planar

1.19 (b) Oxygen is negative end of dipole moment directed along bisector of H—O—H angle; (c) no dipole moment; (d) dipole moment directed along axis of C—Cl bond, with chlorine at negative end, and carbon and hydrogens partially positive; (e) dipole moment directed along bisector of H—C—H angle, with oxygen at negative end; (f) dipole moment aligned with axis of linear molecule, with nitrogen at negative end.

1.20 $pK_a = 2.97$

1.21 $K_a = 7.9 \times 10^{-10}$

1.22 (b)

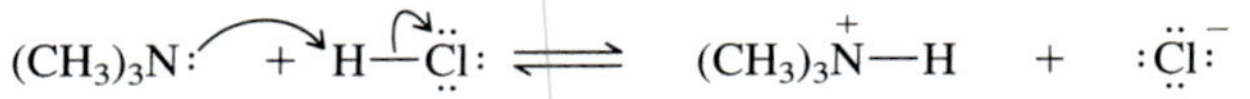

Trimethylamine	Hydrogen chloride	Trimethylammonium ion	Chloride ion
(Base)	(Acid)	(Conjugate acid)	(Conjugate base)

1.23 The conjugate acid of hydride ion is H_2.

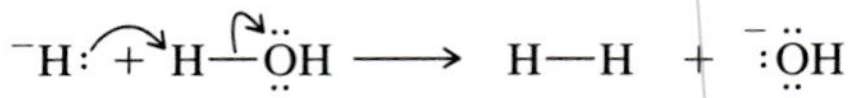

1.24 (b) Sodium amide; (c) sodium acetylide

1.25 H_2S is a stronger acid than H_2O. HO^- is a stronger base than HS^-.

1.26 (b) $(CH_3)_3N$ is a stronger base than $(CH_3)_2O$.

1.27 Each oxygen has two-thirds of a negative charge (-0.67).

1.28 The conjugate bases A and B are resonance forms of each other.

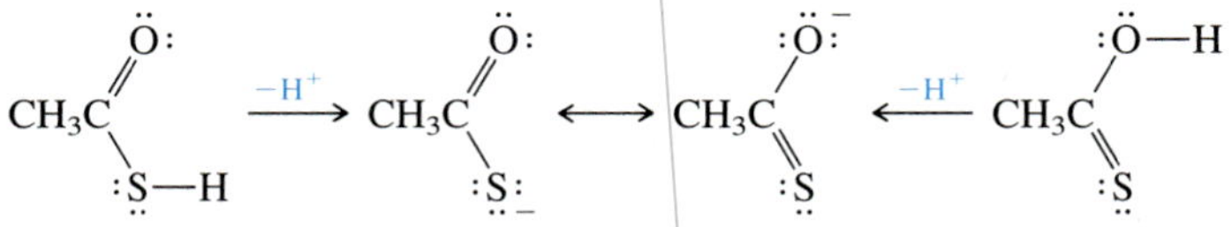

The structure of the conjugate base is more like resonance structure B than A because the negative charge is on the more electronegative atom (O versus S).

1.29 (b) K is less than 1 for the reaction of fluoride ion with acetic acid. The conjugate acid of fluoride ion (HF) is stronger than acetic acid.

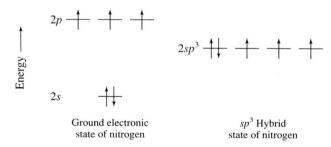

| Fluoride ion | Acetic acid $pK_a = 4.7$ weaker acid | Hydrogen fluoride $pK_a = 3.1$ stronger acid | Acetate ion |

1.30 $[H_3O^+] > [HSO_4^-] > [SO_4^{2-}] > [H_2SO_4]$

1.31 In the reaction of phenol with hydroxide ion, the stronger acid (phenol) is on the left side of the equation and the weaker acid (water) is on the right.

$$C_6H_5-\ddot{O}-H + \ :\ddot{O}-H \ \overset{K>1}{\rightleftharpoons} \ C_6H_5-\ddot{O}:^- \ + \ H-\ddot{O}-H$$

Stronger acid $pK_a = 10$ 　　　　　　　　　 Weaker acid $pK_a = 15.7$

Phenol is a weaker acid than carbonic acid. The equilibrium shown lies to the left.

$$C_6H_5-\ddot{O}-H + \ :\ddot{O}-\overset{O}{\overset{\|}{C}}OH \ \overset{K<1}{\rightleftharpoons} \ C_6H_5-\ddot{O}:^- \ + \ H-\ddot{O}-\overset{O}{\overset{\|}{C}}OH$$

Weaker acid $pK_a = 10$ 　　　　　　　　　 Stronger acid $pK_a = 6.4$

1.32 $F_3B + \ :\overset{\displaystyle CH_3}{\underset{\displaystyle CH_3}{S}}: \ \rightleftharpoons \ F_3\bar{B}-\overset{\displaystyle CH_3}{\underset{\displaystyle CH_3}{\overset{+}{S}}}:$

CHAPTER 2

2.1 The sp^3 hybrid state of nitrogen is just like that of carbon except nitrogen has one more electron. Each N—H bond in NH_3 involves overlap of an sp^3 hybrid orbital of N with a $1s$ orbital of hydrogen. The unshared pair of NH_3 occupies an sp^3 orbital.

Energy →

2p ↿ ↿ ↿

$2sp^3$ ⇅ ↿ ↿ ↿

2s ⇅

Ground electronic state of nitrogen

sp^3 Hybrid state of nitrogen

2.2 Each carbon in propane is bonded to four atoms and is sp^3-hybridized. The C—C bonds are σ bonds involving overlap of a half-filled sp^3 hybrid orbital of one carbon with a half-filled sp^3 hybrid orbital of the other. The C—H bonds are σ bonds involving overlap of a half-filled sp^3 hybrid oribital of carbon with a half-filled hydrogen $1s$ orbital.

2.3 $CH_3(CH_2)_{26}CH_3$

2.4 The molecular formula is $C_{11}H_{24}$; the condensed structural formula is $CH_3(CH_2)_9CH_3$.

2.5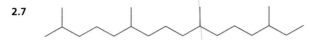

2.6 (b) $CH_3(CH_2)_{26}CH_3$; (c) undecane

2.7

2.8 (b) $CH_3CH_2CH_2CH_2CH_3$ (pentane), $(CH_3)_2CHCH_2CH_3$ (2-methylbutane), $(CH_3)_4C$ (2,2-dimethylpropane); (c) 2,2,4-trimethylpentane; (d) 2,2,3,3-tetramethylbutane

2.9 $CH_3CH_2CH_2CH_2CH_2-$ (pentyl, primary); $CH_3CH_2CH_2\overset{|}{C}HCH_3$ (1-methylbutyl, secondary); $CH_3CH_2\overset{|}{C}HCH_2CH_3$ (1-ethylpropyl, secondary); $(CH_3)_2CHCH_2CH_2-$ (3-methylbutyl, primary); $CH_3CH_2CH(CH_3)CH_2-$ (2-methylbutyl, primary); $(CH_3)_2\overset{|}{C}CH_2CH_3$ (1,1-dimethyl propyl, tertiary); and $(CH_3)_2CH\overset{|}{C}HCH_3$ (1,2-dimethylpropyl, secondary)

2.10 (b) 4-Ethyl-2-methylhexane; (c) 8-ethyl-4-isopropyl-2,6-dimethyldecane

2.11 (b) 4-Isopropyl-1,1-dimethylcyclodecane; (c) cyclohexylcyclohexane

2.12 2,2,3,3-Tetramethylbutane (106°C); 2-methylheptane (116°C); octane (126°C); nonane (151°C)

2.13 $\hexagon$ $+ 9O_2 \longrightarrow 6CO_2 + 6H_2O$

2.14 13,313 kJ/mol

2.15 Hexane $(CH_3CH_2CH_2CH_2CH_2CH_3)$ > pentane $(CH_3CH_2CH_2CH_2CH_3)$ > isopentane $[(CH_3)_2CHCH_2CH_3]$ > neopentane $[(CH_3)_4C]$

2.16 When calculating an oxidation number from a molecular formula, you obtain the average oxidation number of all of the atoms of a particular element. There are two different types of carbon in ethanol (CH_3CH_2OH). One carbon is bonded to three hydrogens and one carbon; the other is bonded to two hydrogens, one carbon, and an oxygen. One carbon has an oxidation number of -3; the other has an oxidation number of -1. Their average oxidation number is -2, which is what is obtained if you calculate the oxidation number directly from the molecular formula (C_2H_6O) where no distinction is made between the two carbons.

2.17 The first reaction:

$$(CH_3)_3COH + HCl \longrightarrow (CH_3)_3CCl + H_2O$$

is not oxidation–reduction. The second reaction is oxidation–reduction.

$$(CH_3)_3CH + Br_2 \longrightarrow (CH_3)_3CBr + HBr$$

2.18 The indicated bond is a σ bond. The carbon of the CH_3 group is sp^3-hybridized. The carbon to which the CH_3 group is attached is sp^2-hybridized.

2.19 The C—C single bond in vinylacetylene is a σ bond generated by overlap of an sp^2-hybridized orbital on one carbon with an sp-hybridized orbital on the other. Vinylacetylene has three σ bonds and three π bonds.

CHAPTER 3

3.2 Red circles gauche: 60° and 300°. Red circles anti: 180°. Gauche and anti relationships occur only in staggered conformations; therefore, ignore the eclipsed conformations (0°, 120°, 240°, 360°).

3.3 Shape of potential energy diagram is identical with that for ethane (Figure 3.4). Activation energy for rotation about the C—C bond is higher than that of ethane, lower than that of butane.

3.4 Ethylcyclopropane: 3384 kJ/mol (808.8 kcal/mol); methylcyclobutane: 3352 kJ/mol (801.2 kcal/mol)

3.5 (b)
X
A
(c)
X
A
(d)
X
A

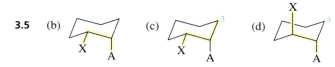

3.6 (b) Less stable; (c) methyl is equatorial and down

3.7
CH3
C(CH3)3

3.8 1,1-Dimethylcyclopropane, ethylcyclopropane, methylcyclobutane, and cyclopentane

3.9 *cis*-1,3,5-Trimethylcyclohexane is more stable.

3.10 (b)
H3C
C(CH3)3
H
H
(c)
H
H3C
C(CH3)3
H

(d)
CH3
H
C(CH3)3
H

3.11
CH=CH2
and
=CH2

3.12

CH3
CH3
CH2
CH3
CH3
CH2
CH3
CH3
CH2

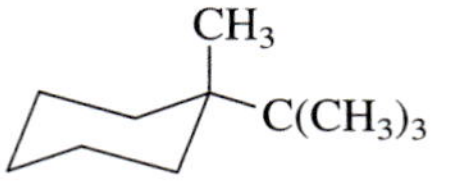

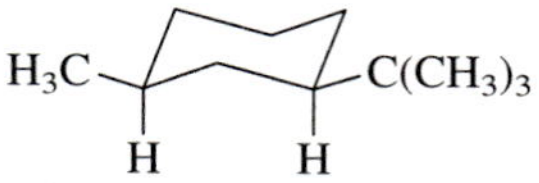

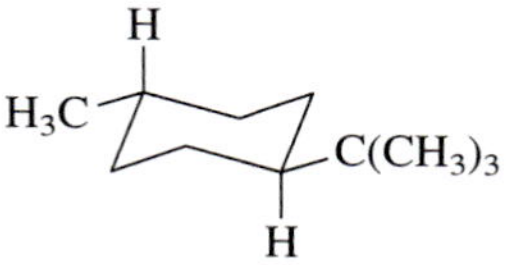

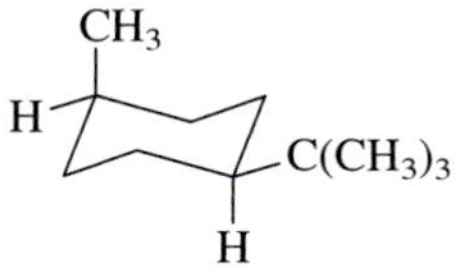

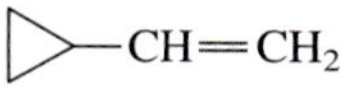

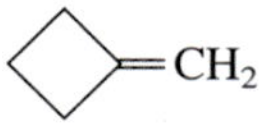

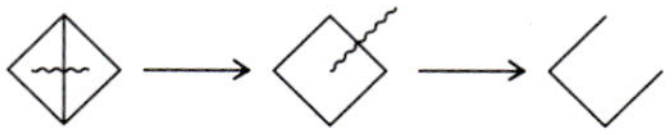

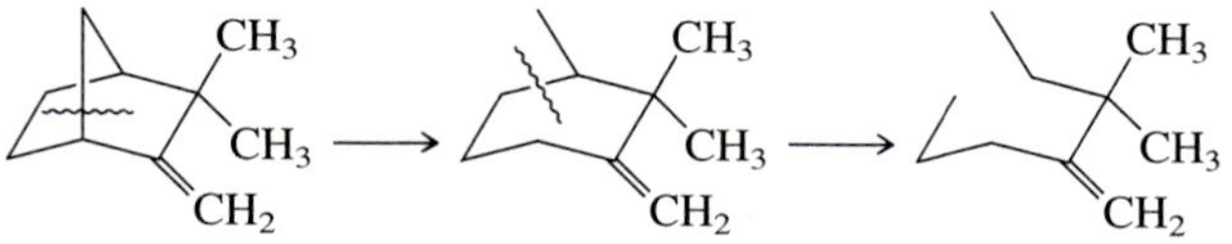

Other pairs of bond cleavages are also possible.

3.13 (b)
(c)
(d)

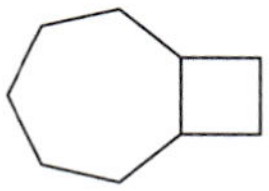

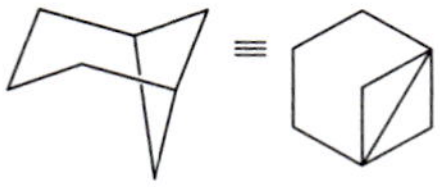

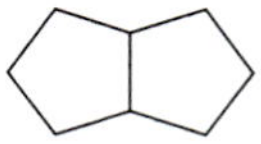

3.14
N
CH3

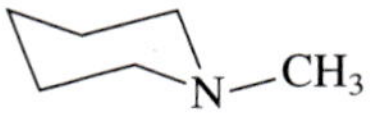

CHAPTER 4

4.1   (b) $CH_3CH_2CH_2SH$ and $(CH_3)_2CHSH$

4.2

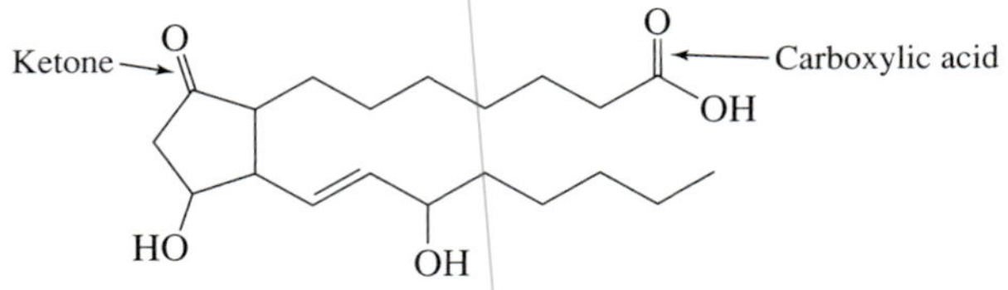

The most acidic proton is the one of the carboxylic acid group and should have a pK_a of approximately 5.

4.3

	$CH_3CH_2CH_2CH_2Cl$	$CH_3CHCH_2CH_3$
		         |
		         Cl

Substitutive name:	1-Chlorobutane	2-Chlorobutane
Functional class names:	*n*-Butyl chloride or butyl chloride	*sec*-Butyl chloride or 1-methylpropyl chloride

	$(CH_3)_2CHCH_2Cl$	$(CH_3)_3CCl$
	1-Chloro-2-methylpropane Isobutyl chloride or 2-methylpropyl chloride	2-Chloro-2-methylpropane *tert*-Butyl chloride or 1,1-dimethylethyl chloride

4.4

	$CH_3CH_2CH_2CH_2OH$	$CH_3CHCH_2CH_3$
		         |
		         OH

Substitutive name:	1-Butanol	2-Butanol
Functional class names:	*n*-Butyl alcohol or butyl alcohol	*sec*-Butyl alcohol or 1-methylpropyl alcohol

	$(CH_3)_2CHCH_2OH$	$(CH_3)_3COH$
	2-Methyl-1-propanol Isobutyl alcohol or 2-methylpropyl alcohol	2-Methyl-2-propanol *tert*-Butyl alcohol or 1,1-dimethylethyl alcohol

4.5 $CH_3CH_2CH_2CH_2OH$ $CH_3CHCH_2CH_3$ $(CH_3)_2CHCH_2OH$ $(CH_3)_3COH$
                                                       |
                                                       OH

            Primary   Secondary   Primary   Tertiary

4.6   The carbon–bromine bond is longer than the carbon–chlorine bond; therefore, although the charge e in the dipole moment expression $\mu = e \cdot d$ is smaller for the bromine than for the chlorine compound, the distance d is greater.

4.7   Hydrogen bonding in ethanol (CH_3CH_2OH) makes its boiling point higher than that of dimethyl ether (CH_3OCH_3), in which hydrogen bonding is absent.

4.8   (b) $(CH_3CH_2)_3COH + HCl \longrightarrow (CH_3CH_2)_3CCl + H_2O$
          (c) $CH_3(CH_2)_{12}CH_2OH + HBr \longrightarrow CH_3(CH_2)_{12}CH_2Br + H_2O$

4.9 $(CH_3)_2\overset{+}{C}CH_2CH_3$

4.10   Electron pairs in bonds β to C$^+$ stabilize the carbocation.

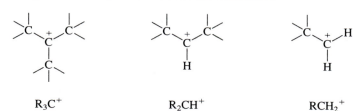

R_3C^+	R_2CH^+	$RCH_2{}^+$
Nine electron pairs β to C$^+$	Six electron pairs β to C$^+$	Three electron pairs β to C$^+$

4.11

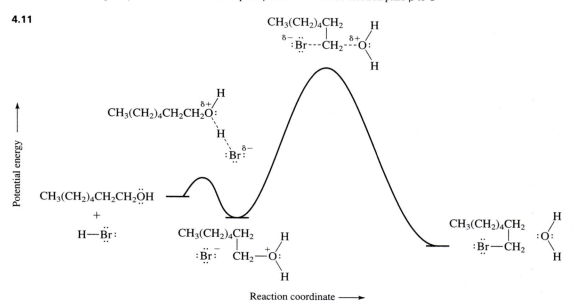

Reaction coordinate ⟶

4.12   1-Butanol: Rate-determining step is bimolecular; therefore, S$_N$2.

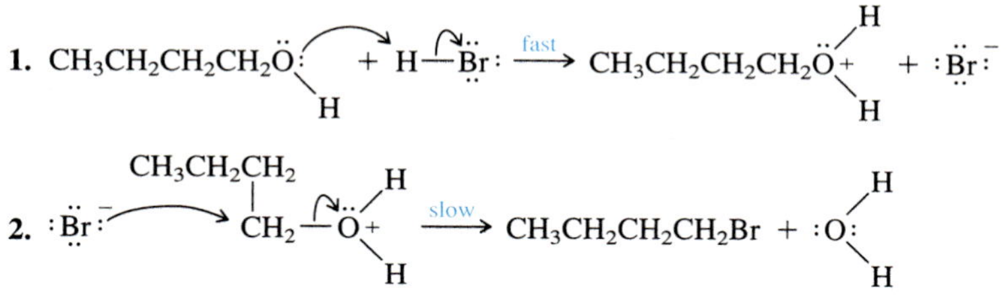

1. $CH_3CH_2CH_2CH_2\ddot{O}$: + H—$\ddot{B}r$: $\xrightarrow{\text{fast}}$ $CH_3CH_2CH_2CH_2\overset{+}{\ddot{O}}$ + :$\ddot{B}r$:$^-$

2. :$\ddot{B}r$:$^-$ $CH_3CH_2CH_2$—CH_2—$\overset{+}{\ddot{O}}$ $\xrightarrow{\text{slow}}$ $CH_3CH_2CH_2CH_2Br$ + :$\ddot{O}$

2-Butanol: Rate-determining step is unimolecular, therefore, S$_N$1.

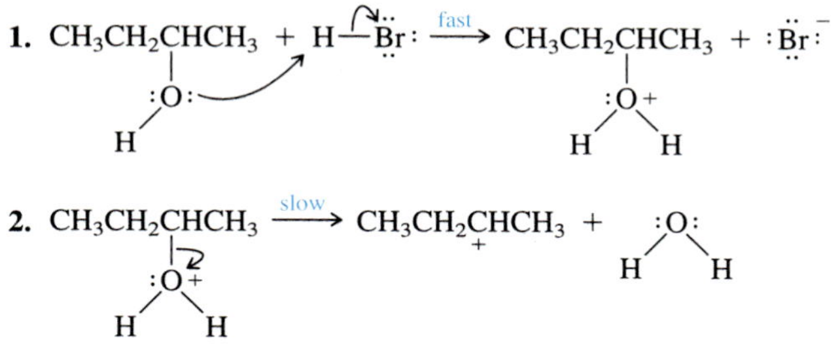

1. $CH_3CH_2CHCH_3$ + H—$\ddot{B}r$: $\xrightarrow{\text{fast}}$ $CH_3CH_2CHCH_3$ + :$\ddot{B}r$:$^-$

2. $CH_3CH_2CHCH_3$ $\xrightarrow{\text{slow}}$ $CH_3CH_2\overset{+}{C}HCH_3$ + :$\ddot{O}$

3. 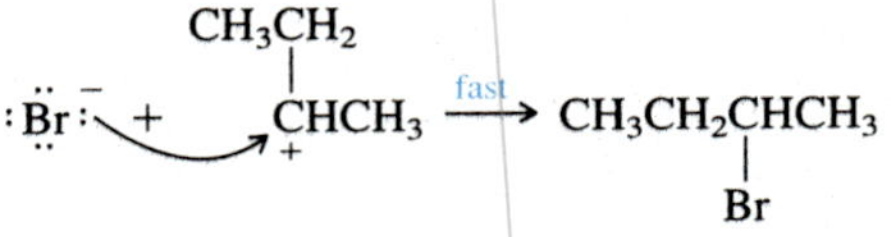 $:\overset{..}{\underset{..}{Br}}:^{-} +$ $\overset{\displaystyle CH_3CH_2}{\underset{\displaystyle +}{|}}$ $\overset{}{C}HCH_3$ $\xrightarrow{\text{fast}}$ $CH_3CH_2CHCH_3$

$\underset{\displaystyle Br}{|}$

4.13 $(CH_3)_2\overset{\cdot}{C}CH_2CH_3$

4.14 (b) The carbon–carbon bond dissociation energy is lower for 2-methylpropane because it yields a more stable (secondary) radical; propane yields a primary radical. (c) The carbon–carbon bond dissociation energy is lower for 2,2-dimethylpropane because it yields a still more stable tertiary radical.

4.15 Initiation: $:\overset{..}{\underset{..}{Cl}}-\overset{..}{\underset{..}{Cl}}:$ $\longrightarrow$ $:\overset{..}{\underset{..}{Cl}}\cdot + \cdot\overset{..}{\underset{..}{Cl}}:$

Chlorine   2 Chlorine atoms

Propagation:

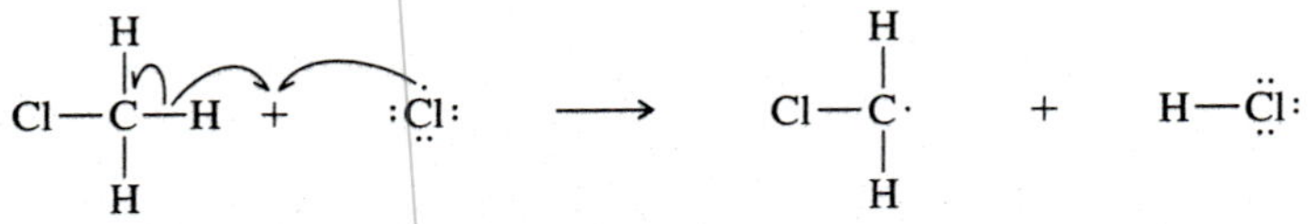

Cl—C—H + :Cl: $\longrightarrow$ Cl—C· + H—Cl:

Chloromethane   Chlorine atom   Chloromethyl radical   Hydrogen chloride

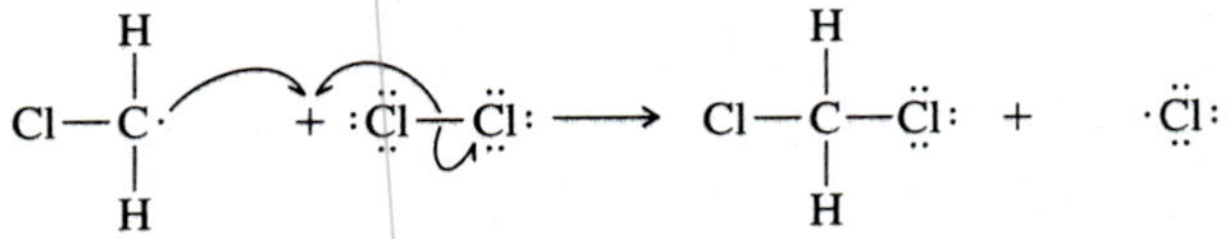

Cl—C· + :Cl—Cl: $\longrightarrow$ Cl—C—Cl: + ·Cl:

Chloromethyl radical   Chlorine   Dichloromethane   Chlorine atom

4.16 CH_3CHCl_2 and $ClCH_2CH_2Cl$

4.17 1-Chloropropane (43%); 2-chloropropane (57%)

4.18 (b)

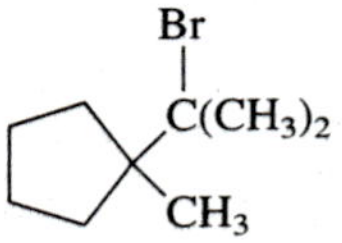

$\underset{\displaystyle CH_3}{\overset{\displaystyle Br}{\overset{|}{\underset{|}{C(CH_3)_2}}}}$ (c)

CHAPTER 5

5.1 (b) 3,3-Dimethyl-1-butene; (c) 2-methyl-2-hexene; (d) 4-chloro-1-pentene; (e) 4-penten-2-ol

5.2

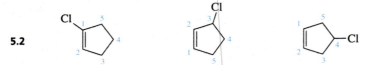

1-Chlorocyclopentene   3-Chlorocyclopentene   4-Chlorocyclopentene

5.3 (b) 3-Ethyl-3-hexene; (c) two carbons are sp^2-hybridized, six are sp^3-hybridized; (d) there are three sp^2–sp^3 σ bonds and three sp^3–sp^3 σ bonds.

5.4

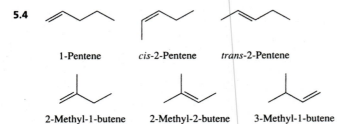

1-Pentene   cis-2-Pentene   trans-2-Pentene

2-Methyl-1-butene   2-Methyl-2-butene   3-Methyl-1-butene

5.5

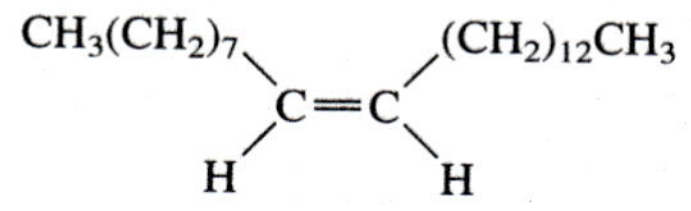

5.6 (b) *Z;* (c) *E;* (d) *E*

5.7 Molecular modeling problem. See *Learning By Modeling.*

5.8

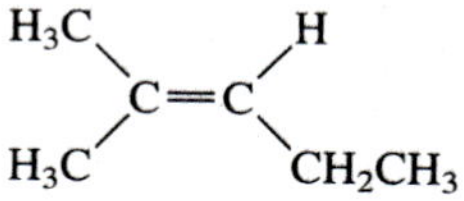

2-Methyl-2-pentene 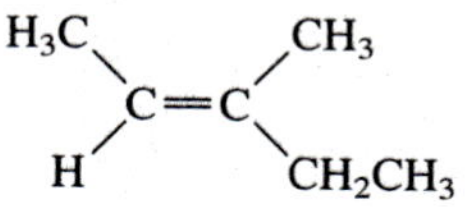   (E)-3-Methyl-2-pentene 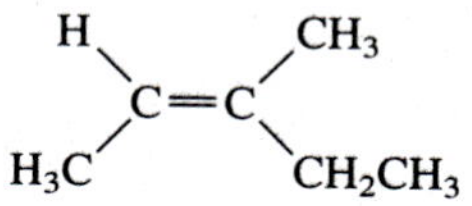   (Z)-3-Methyl-2-pentene

5.9 $(CH_3)_2C{=}C(CH_3)_2$

5.10 2-Methyl-2-butene (most stable) > (E)-2-pentene > (Z)-2-pentene > 1-pentene (least stable)

5.11 Bulky *tert*-butyl groups are cis to one another on each side of the double bond and van der Waals strain destabilizes the alkene.

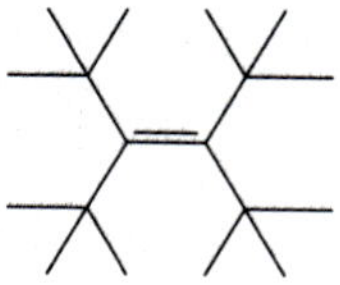

5.12 (c)

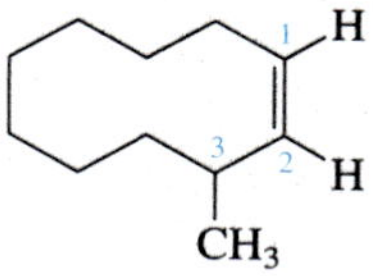

(d)

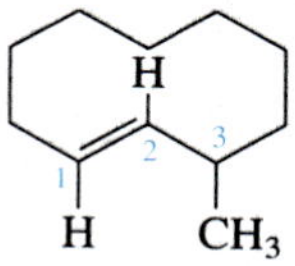

(e)

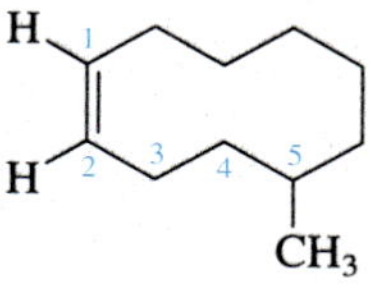

(f)

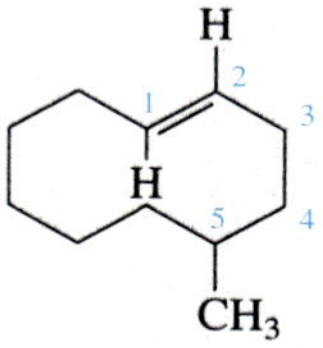

5.13 (b) Propene; (c) propene; (d) 2,3,3-trimethyl-1-butene

5.14 (b)

Major   and 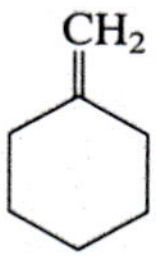   Minor

(c)

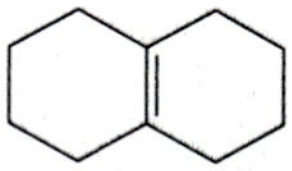

and

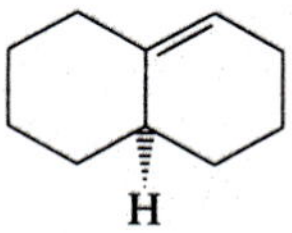

Major   Minor

5.15 1-Pentene, *cis*-2-pentene, and *trans*-2-pentene

5.16 (b)

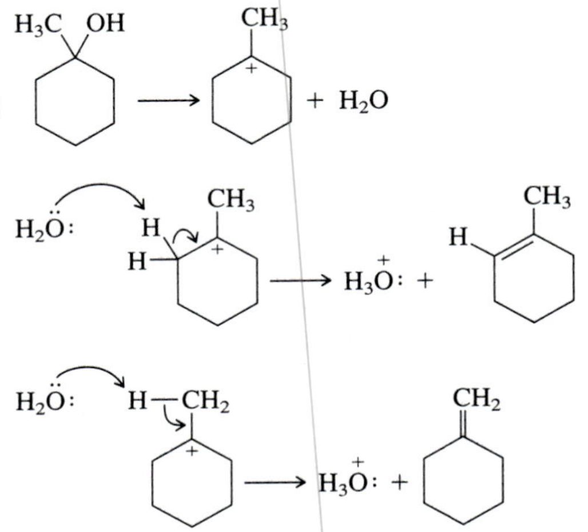

and

(c)

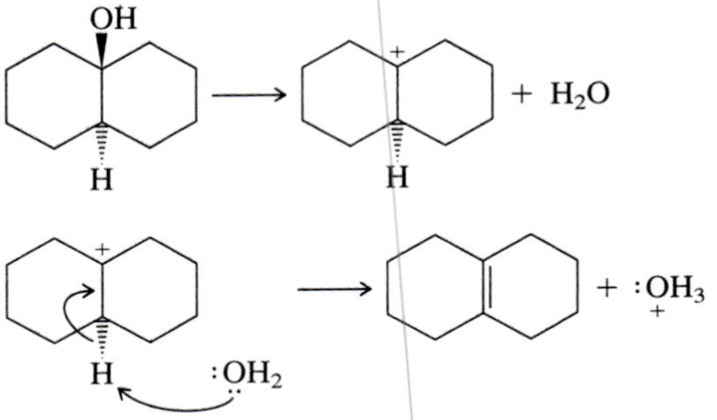

and

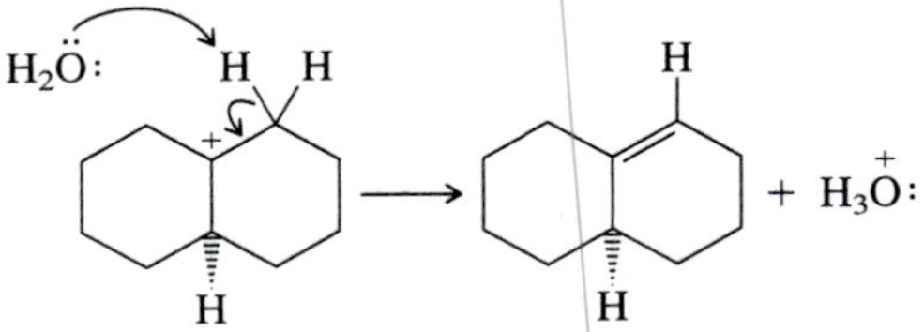

5.17

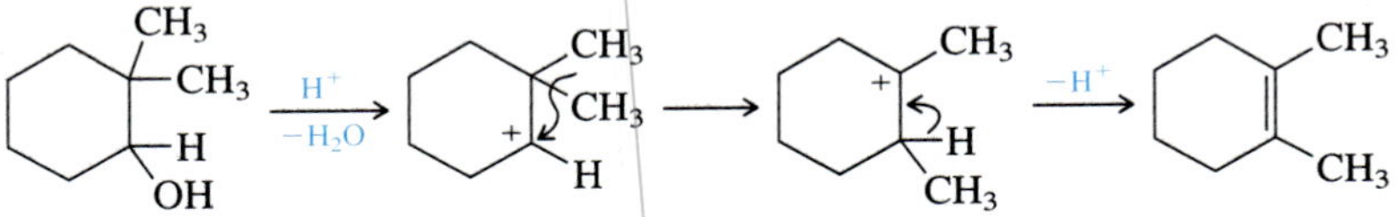

5.18 (b) $(CH_3)_2C=CH_2$;   (c) $CH_3CH=C(CH_2CH_3)_2$;   (d) $CH_3CH=C(CH_3)_2$ (major) and
$H_2C=CHCH(CH_3)_2$ (minor);   (e) $H_2C=CHCH(CH_3)_2$;   (f) 1-methylcyclohexene (major) and
methylenecyclohexane (minor)

5.19 $H_2C=CHCH_2CH_3$, cis-$CH_3CH=CHCH_3$, and $trans$-$CH_3CH=CHCH_3$.

5.20

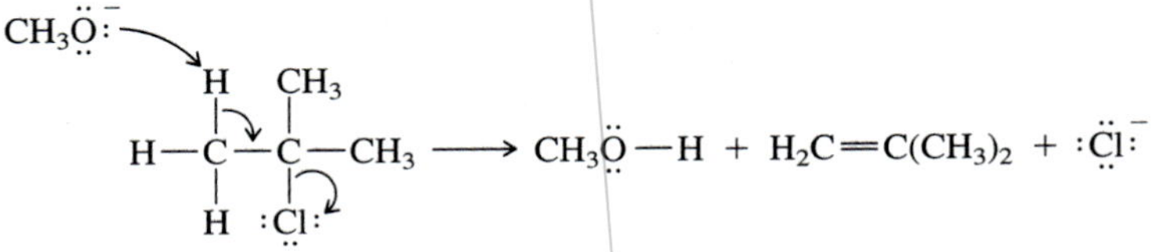

5.21 $(CH_3)_3C$—

$(CH_3)_3C$—$\ddot{O}:^-$

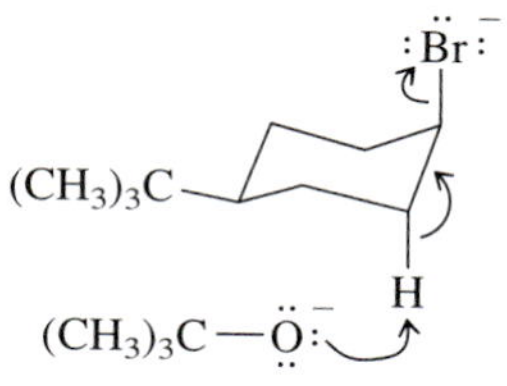

CHAPTER 6

6.1 2-Methyl-1-butene, 2-methyl-2-butene, and 3-methyl-1-butene

6.2 2-Methyl-2-butene (112 kJ/mol, 26.7 kcal/mol), 2-methyl-1-butene (118 kJ/mol, 28.2 kcal/mol), and 3-methyl-1-butene (126 kJ/mol, 30.2 kcal/mol)

6.3 (b) $(CH_3)_2CCH_2CH_3$ with Cl   (c) $CH_3CHCH_2CH_3$ with Cl   (d) CH_3CH_2 with Cl on cyclohexane

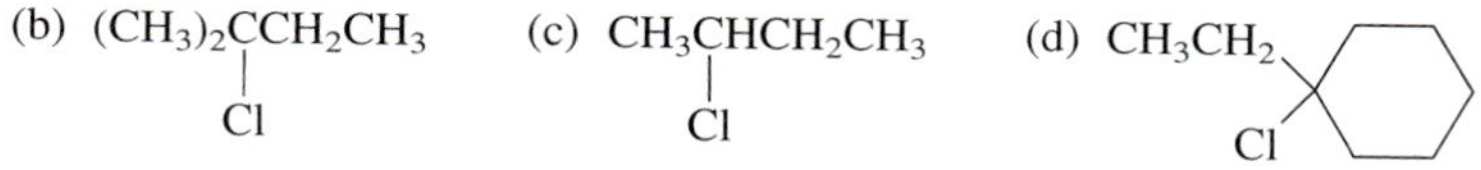

6.4 (b) $(CH_3)_2\overset{+}{C}CH_2CH_3$ (c) $CH_3\overset{+}{C}HCH_2CH_3$ (d) CH_3CH_2—$\overset{+}{}$ cyclohexyl

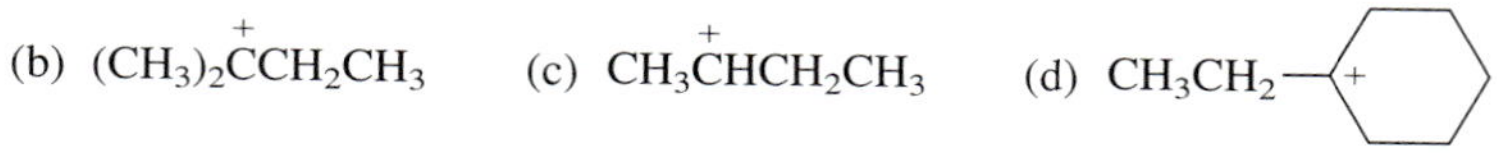

6.5
$CH_3\overset{CH_3}{\underset{CH_3}{C}}$—CH=CH$_2$ $\xrightarrow{HCl}$ $CH_3\overset{CH_3}{\underset{CH_3}{C}}$—$\overset{+}{C}HCH_3$ $\xrightarrow{CH_3 \text{ shift}}$ $CH_3\overset{CH_3}{\underset{CH_3}{\overset{+}{C}}}$—CHCH$_3$

$\downarrow Cl^-$ $\downarrow Cl^-$

$CH_3\overset{CH_3}{\underset{CH_3 Cl}{C}}$—CHCH$_3$ $CH_3\overset{Cl}{\underset{CH_3}{C}}$—$\overset{CH_3}{C}HCH_3$

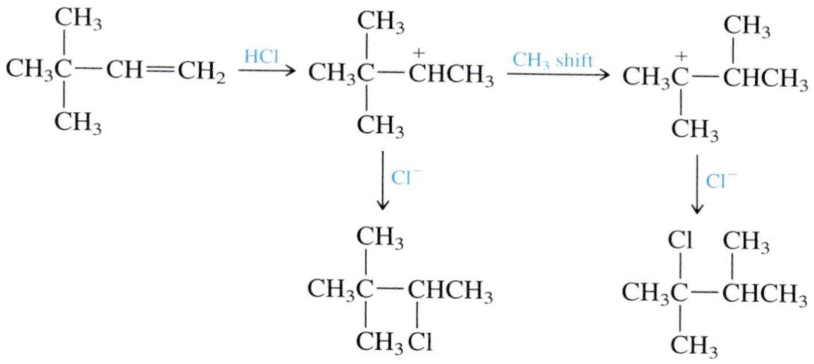

6.6 Addition in accordance with Markovnikov's rule gives 1,2-dibromopropane. Addition opposite to Markovnikov's rule gives 1,3-dibromopropane.

6.7 Absence of peroxides: (b) 2-bromo-2-methylbutane; (c) 2-bromobutane; (d) 1-bromo-1-ethylcyclohexane. Presence of peroxides: (b) 1-bromo-2-methylbutane; (c) 2-bromobutane; (d) (1-bromoethyl)cyclohexane.

6.8 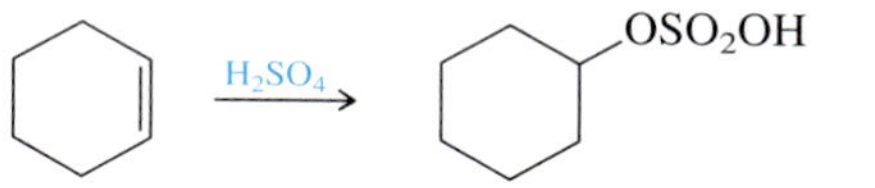 cyclohexene $\xrightarrow{H_2SO_4}$ cyclohexyl—OSO$_2$OH

Cyclohexene   Cyclohexyl hydrogen sulfate

6.9 The concentration of hydroxide ion is too small in acid solution to be chemically significant.

6.10 cyclopropyl—$\overset{CH_3}{C}$=CH$_2$ is more reactive, because it gives a tertiary carbocation cyclopropyl—$\overset{CH_3}{\underset{CH_3}{C}}^+$ when it is protonated in acid solution.

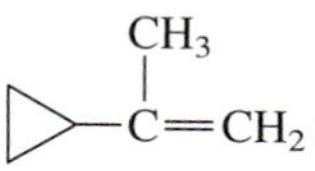

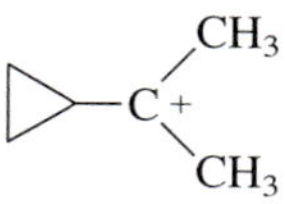

6.11 E1

6.12 (b) 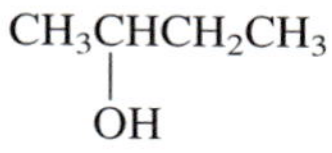 $CH_3CHCH_2CH_3$ with OH   (c) 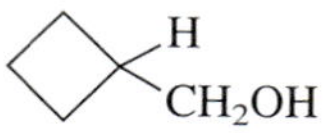 cyclobutyl with H and CH$_2$OH   (d) 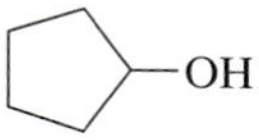 cyclopentyl—OH

(e) CH$_3$CHCH(CH$_2$CH$_3$)$_2$
    |
    OH

(f) HOCH$_2$CH$_2$CH(CH$_2$CH$_3$)$_2$

6.13

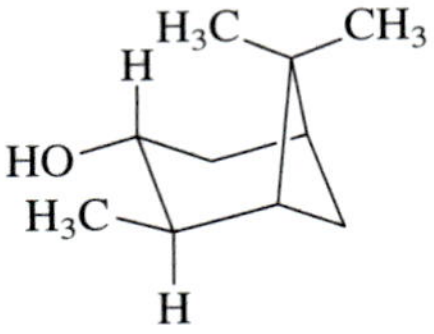

6.14

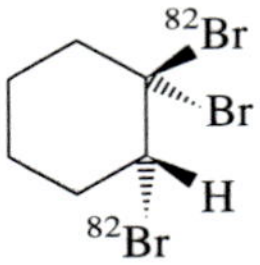

6.15 2-Methyl-2-butene (most reactive) > 2-methyl-1-butene > 3-methyl-1-butene (least reactive)

6.16 (b) (CH$_3$)$_2$C—CHCH$_3$
         | |
        OH Br

(c) BrCH$_2$CHCH(CH$_3$)$_2$
       |
      OH

(d)

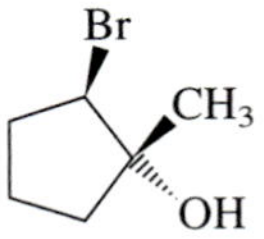

6.17 *cis*-2-Methyl-7,8-epoxyoctadecane

6.18 *cis*-(CH$_3$)$_2$CHCH$_2$CH$_2$CH$_2$CH$_2$CH=CH(CH$_2$)$_9$CH$_3$

6.19 2,4,4-Trimethyl-1-pentene

6.20 (CH$_3$)$_3$CBr $\xrightarrow[\text{heat}]{\text{NaOCH}_2\text{CH}_3}$ (CH$_3$)$_2$C=CH$_2$ $\xrightarrow[\text{H}_2\text{O}]{\text{Br}_2}$ (CH$_3$)$_2$C—CH$_2$Br
                                                   |
                                                   OH

6.21 Hydrogenation over a metal catalyst such as platinum, palladium, or nickel

CHAPTER 7

7.1 (c) C-2 is a chirality center; (d) no chirality centers.

7.2 (b) C-2 is a chirality center in 1,1,2-trimethylcyclobutane. 1,1,3-Trimethylcyclobutane does not contain a chirality center.

7.3 (b) (Z)-1,2-Dichloroethene is achiral. The plane of the molecule is a plane of symmetry. A second plane of symmetry is perpendicular to the plane of the molecule and bisects the carbon–carbon bond.

(c) *cis*-1,2-Dichlorocyclopropane is achiral. It has a plane of symmetry that bisects the C-1—C-2 bond and passes through C-3.

(d) *trans*-1,2-Dichlorocyclopropane is chiral. It has neither a plane of symmetry nor a center of symmetry.

7.4 $[\alpha]_D$ − 39°

7.5 Two-thirds (66.7%)

7.6 (+)-2-Butanol

7.7 (b) *R*; (c) *S*; (d) *S*

7.8 (b)

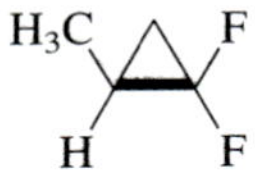

7.9 *R*

7.10 (b) Change *R* to *S*; (c) no change; (d) no change; (e) no change

7.11 Equal amounts of (*R*)- and (*S*)-3-methylhexane

7.12 The carboxyl groups of the two —CH_2CO_2H groups are enantiotopic.

7.13 *S*

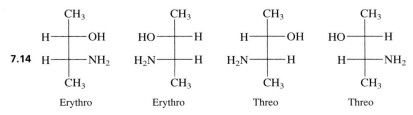

7.14

Erythro   Erythro   Threo   Threo

7.15 2*S*,3*R*

7.16 2,4-Dibromopentane

7.17 *cis*-1,3-Dimethylcyclohexane

7.18 *RRR RRS RSR SRR SSS SSR SRS RSS*

7.19 Eight

7.20 Epoxidation of *cis*-2-butene gives *meso*-2,3-epoxybutane; *trans*-2-butene gives a racemic mixture of (2*R*,3*R*) and (2*S*,3*S*)-2,3-epoxybutane.

7.21 No. The major product *cis*-1,2-dimethylcyclohexane is less stable than the minor product *trans*-1,2-dimethylcyclohexane.

7.22

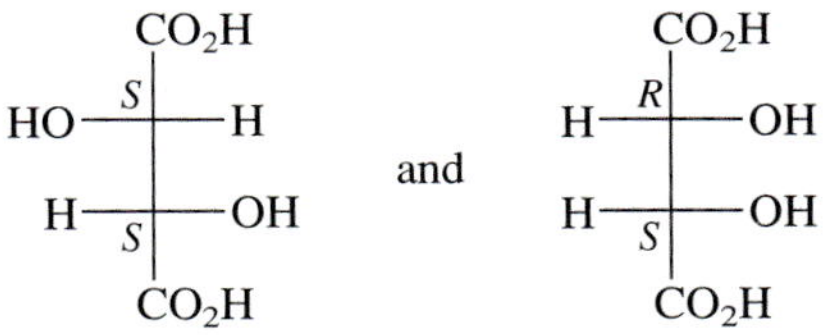

$$\begin{array}{ccc}
\text{CO}_2\text{H} & & \text{CO}_2\text{H} \\
\text{HO}\underset{S}{\overset{S}{\rule{0pt}{0pt}}}\text{H} & \text{and} & \text{H}\overset{R}{\rule{0pt}{0pt}}\text{OH} \\
\text{H}\underset{S}{\rule{0pt}{0pt}}\text{OH} & & \text{H}\underset{S}{\rule{0pt}{0pt}}\text{OH} \\
\text{CO}_2\text{H} & & \text{CO}_2\text{H}
\end{array}$$

7.23 No

7.24 (*S*)-1-Phenylethylammonium (*S*)-malate

CHAPTER 8

8.1 (b) $CH_3OCH_2CH_3$ (c) 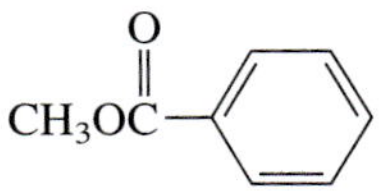 $CH_3O\overset{\overset{\displaystyle O}{\|}}{C}-$⟨benzene ring⟩ (d) $CH_3\overset{+}{N}{=}N{=}\overset{-}{N}:$

(e) $CH_3C{\equiv}N$ (f) CH_3SH (g) CH_3I

8.2 $ClCH_2CH_2CH_2C{\equiv}N$

8.3 No

8.4

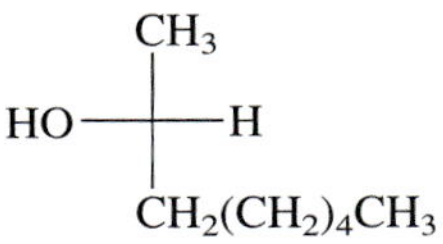

$$\begin{array}{c}
\text{CH}_3 \\
\text{HO}\!-\!\!\!\overline{\rule{0pt}{0pt}}\!\!\!-\!\text{H} \\
\text{CH}_2(\text{CH}_2)_4\text{CH}_3
\end{array}$$

8.5 Hydrolysis of (*R*)-(−)-2-bromooctane by the S_N2 mechanism yields optically active (*S*)-(+)-2-octanol. The 2-octanol obtained by hydrolysis of racemic 2-bromooctane is not optically active.

8.6   (b) 1-Bromopentane; (c) 2-chloropentane; (d) 2-bromo-5-methylhexane; (e) 1-bromodecane

8.7   **Step 1:**

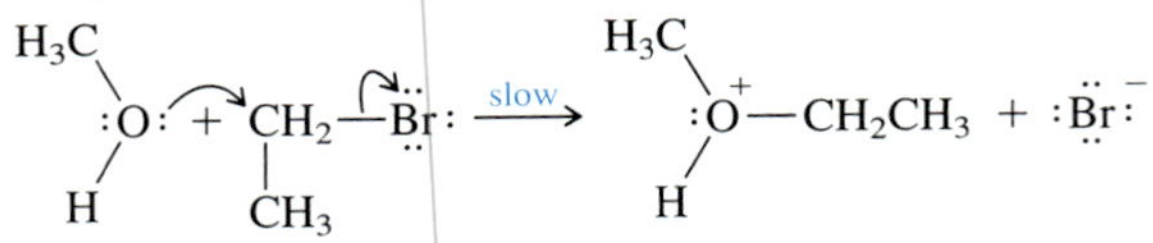

Step 2:

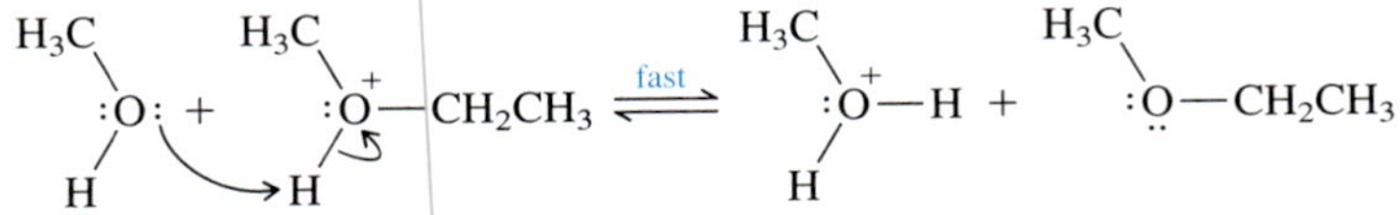

8.8   Product is $(CH_3)_3COCH_3$. The mechanism of solvolysis is S_N1.

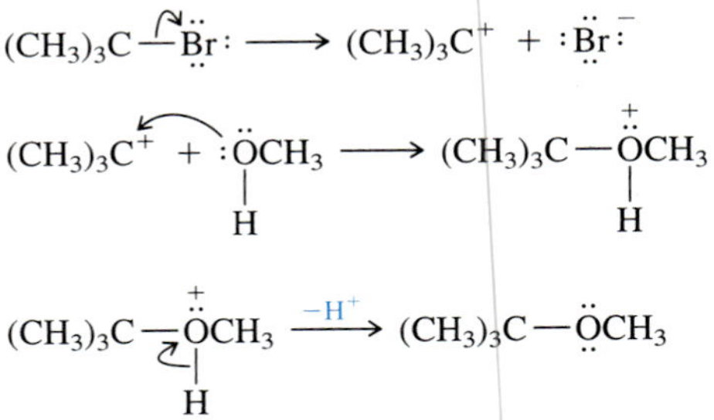

8.9   (b) 1-Methylcyclopentyl iodide; (c) cyclopentyl bromide; (d) *tert*-butyl iodide

8.10   Both *cis*- and *trans*-1,4-dimethylcyclohexanol are formed in the hydrolysis of either *cis*- or *trans*-1,4-dimethylcyclohexyl bromide.

8.11   A hydride shift produces a tertiary carbocation; a methyl shift produces a secondary carbocation.

8.12   (b)

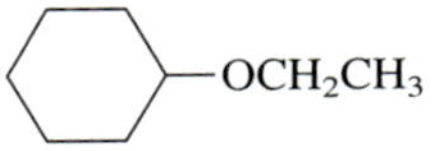

(c) 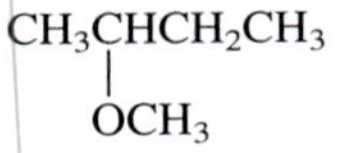 $CH_3CHCH_2CH_3$
                                    $|$
                                    OCH_3

(d) *cis*- and *trans*-$CH_3CH{=}CHCH_3$ and $H_2C{=}CHCH_2CH_3$

8.13 $CH_3(CH_2)_{16}CH_2OH$ +

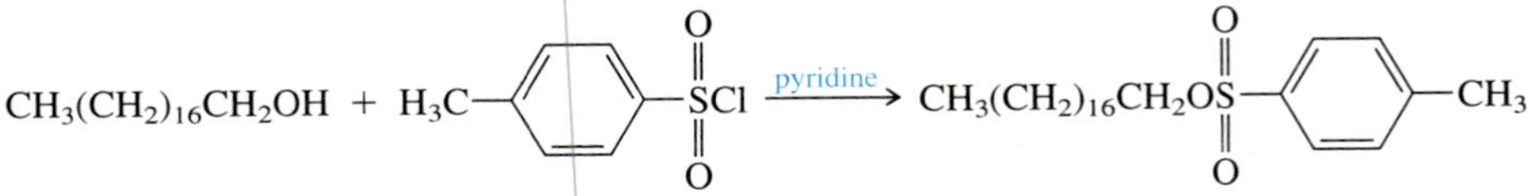

8.14   (b) $CH_3(CH_2)_{16}CH_2I$; (c) $CH_3(CH_2)_{16}CH_2C{\equiv}N$; (d) $CH_3(CH_2)_{16}CH_2SH$;
           (e) $CH_3(CH_2)_{16}CH_2SCH_2CH_2CH_2CH_3$

8.15   The product has the R configuration and a specific rotation $[\alpha]_D$ of $-9.9°$.

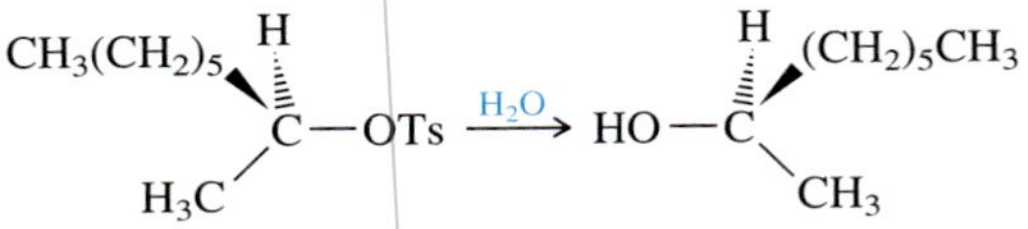

8.16 $CH_3CH_2C(CH_3)_2$
           $\quad\quad\ \ |$
           $\quad\quad\ \ Cl$

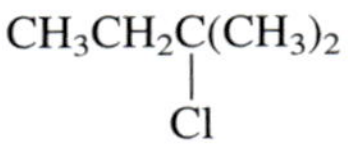

CHAPTER 9

9.1

$$:\overset{-}{C}\equiv\overset{-}{C}: \ + \ H\!-\!\overset{..}{\underset{..}{O}}\!-\!H \longrightarrow :\overset{-}{C}\equiv C\!-\!H \ + \ :\overset{..}{\underset{..}{O}}\!-\!H$$

Carbide ion   Water   Acetylide ion   Hydroxide ion

$$H\!-\!\overset{..}{\underset{..}{O}}\!-\!H \ + \ :\overset{-}{C}\equiv C\!-\!H \longrightarrow H\!-\!\overset{..}{\underset{..}{O}}:^{-} \ + \ H\!-\!C\equiv C\!-\!H$$

Water   Acetylide ion   Hydroxide ion   Acetylene

9.2 $CH_3CH_2CH_2C\equiv CH$ (1-pentyne), $CH_3CH_2C\equiv CCH_3$ (2-pentyne), $(CH_3)_2CHC\equiv CH$ (3-methyl-1-butyne)

9.3 The bonds become shorter and stronger in the series as the electronegativity increases; N—H longest and weakest, H—F shortest and strongest.

9.4 (b)
$$HC\equiv C\!-\!H \ + \ :\overset{-}{C}H_2CH_3 \ \underset{}{\overset{K\gg1}{\rightleftharpoons}} \ HC\equiv \overset{-}{C}: \ + \ CH_3CH_3$$

Acetylene   Ethyl anion   Acetylide ion   Ethane
(stronger acid)   (stronger base)   (weaker base)   (weaker acid)

(c)
$$H_2C\!=\!CH\!-\!H \ + \ :\overset{-}{N}H_2 \ \overset{K\ll1}{\rightleftharpoons} \ H_2C\!=\!\overset{-}{C}H \ + \ :NH_3$$

Ethylene   Amide ion   Vinyl anion   Ammonia
(weaker acid)   (weaker base)   (stronger base)   (stronger acid)

(d)
$$CH_3C\equiv CCH_2\overset{..}{\underset{..}{O}}\!-\!H \ + \ :\overset{-}{N}H_2 \ \overset{K\gg1}{\rightleftharpoons} \ CH_3C\equiv CCH_2\overset{..}{\underset{..}{O}}:^{-} \ + \ :NH_3$$

2-Butyn-1-ol   Amide ion   2-Butyn-1-olate anion   Ammonia
(stronger acid)   (stronger base)   (weaker base)   (weaker acid)

9.5 (b) $HC\equiv CH \ \xrightarrow[\text{2. } CH_3Br]{\text{1. } NaNH_2,\ NH_3} \ CH_3C\equiv CH \ \xrightarrow[\text{2. } CH_3CH_2CH_2Br]{\text{1. } NaNH_2,\ NH_3} \ CH_3C\equiv CCH_2CH_2CH_2CH_3$

(c) $HC\equiv CH \ \xrightarrow[\text{2. } CH_3CH_2CH_2Br]{\text{1. } NaNH_2,\ NH_3} \ CH_3CH_2CH_2C\equiv CH \ \xrightarrow[\text{2. } CH_3CH_2Br]{\text{1. } NaNH_2,\ NH_3} \ CH_3CH_2CH_2C\equiv CCH_2CH_3$

9.6 Both $CH_3CH_2CH_2C\equiv CH$ and $CH_3CH_2C\equiv CCH_3$ can be prepared by alkylation of acetylene. The alkyne $(CH_3)_2CHC\equiv CH$ cannot be prepared by alkylation of acetylene because the required alkyl halide, $(CH_3)_2CHBr$, is secondary and will react with the strongly basic acetylide ion by elimination.

9.7
$$(CH_3)_3\overset{\overset{\displaystyle Br}{|}}{\underset{\underset{\displaystyle Br}{|}}{C}}CCH_3 \quad \text{or} \quad (CH_3)_3CCH_2CHBr_2 \quad \text{or} \quad (CH_3)_3CCH\overset{}{\underset{\underset{\displaystyle Br}{|}}{C}}H_2Br$$

9.8 (b) $CH_3CH_2CH_2OH \ \xrightarrow[\text{heat}]{H_2SO_4} \ CH_3CH\!=\!CH_2 \ \xrightarrow{Br_2} \ CH_3\overset{\overset{\displaystyle Br}{|}}{C}HCH_2Br \ \xrightarrow[\text{2. } H_2O]{\text{1. } NaNH_2} \ CH_3C\equiv CH$

(c) $(CH_3)_2CHBr \ \xrightarrow{NaOCH_2CH_3} \ CH_3CH\!=\!CH_2$; then proceed as in parts (a) and (b).

(d) $CH_3CHCl_2 \ \xrightarrow[\text{2. } H_2O]{\text{1. } NaNH_2} \ HC\equiv CH \ \xrightarrow[\text{2. } CH_3Br]{\text{1. } NaNH_2} \ CH_3C\equiv CH$

(e) $CH_3CH_2OH \ \xrightarrow[\text{heat}]{H_2SO_4} \ H_2C\!=\!CH_2 \ \xrightarrow{Br_2} \ BrCH_2CH_2Br \ \xrightarrow[\text{2. } H_2O]{\text{1. } NaNH_2} \ HC\equiv CH$; then proceed as in part (d).

9.9 $HC\equiv CH$ $\xrightarrow[\text{2. } CH_3CH_2CH_2Br]{\text{1. } NaNH_2,\ NH_3}$ $CH_3CH_2CH_2C\equiv CH$ $\xrightarrow[\text{2. } CH_3CH_2CH_2Br]{\text{1. } NaNH_2,\ NH_3}$

$CH_3CH_2CH_2C\equiv CCH_2CH_2CH_3$ $\xrightarrow[\text{Pt}]{H_2}$ $CH_3(CH_2)_6CH_3$

or $HC\equiv CH$ $\xrightarrow[\text{2. } CH_3(CH_2)_5Br]{\text{1. } NaNH_2,\ NH_3}$ $CH_3(CH_2)_5C\equiv CH$ $\xrightarrow[\text{Pt}]{H_2}$ $CH_3(CH_2)_6CH_3$

9.10 $HC\equiv CH$ $\xrightarrow[NH_3]{NaNH_2}$ $HC\equiv CNa$ $\xrightarrow{CH_3CH_2CH_2CH_2Br}$ $CH_3CH_2CH_2CH_2C\equiv CH$

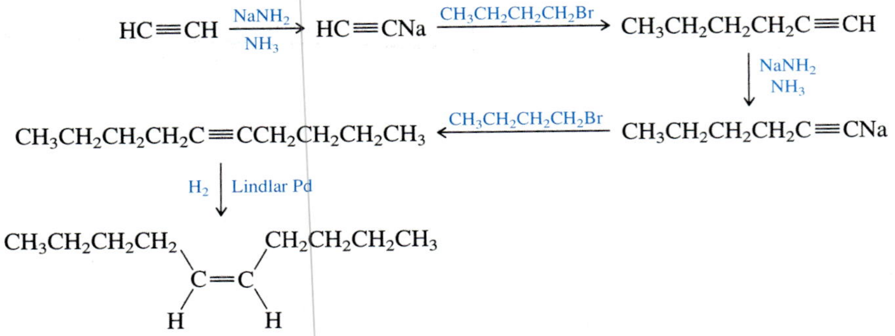

$\Big\downarrow \begin{array}{c} NaNH_2 \\ NH_3 \end{array}$

$CH_3CH_2CH_2CH_2C\equiv CCH_2CH_2CH_2CH_3$ $\xleftarrow{CH_3CH_2CH_2CH_2Br}$ $CH_3CH_2CH_2CH_2C\equiv CNa$

$H_2 \Big\downarrow$ Lindlar Pd

$CH_3CH_2CH_2CH_2$ $CH_2CH_2CH_2CH_3$

$\hspace{4em} C=C$

$\hspace{3em} H \hspace{3em} H$

9.11 $CH_3C\equiv CH$ $\xrightarrow[\text{2. } CH_3CH_2CH_2CH_2Br]{\text{1. } NaNH_2,\ NH_3}$ $CH_3C\equiv CCH_2CH_2CH_2CH_3$ $\xrightarrow[NH_3]{Li}$

$H_3C \hspace{4em} H$

$\hspace{3em} C=C$

$H \hspace{4em} CH_2CH_2CH_2CH_3$

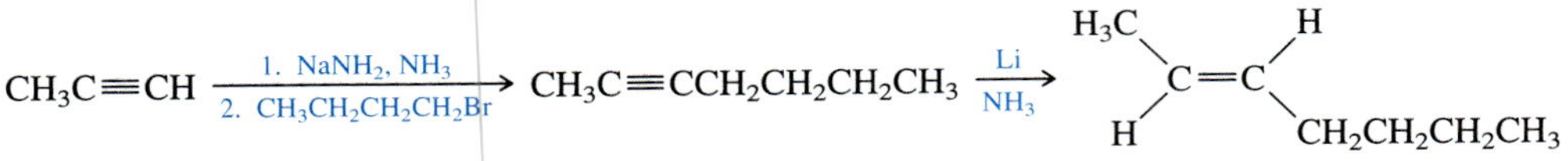

9.12   (b) $H_2C=CHCl$ $\xrightarrow{HCl}$ CH_3CHCl_2

(c) CH_3CHBr_2 $\xrightarrow[\text{2. } H_2O]{\text{1. } NaNH_2,\ NH_3}$ $HC\equiv CH$ $\xrightarrow{2HCl}$ CH_3CHCl_2

9.13 $CH_3C\equiv CCH_3$ $\xrightarrow[H_2SO_4]{H_2O,\ Hg^{2+}}$ $\left[\begin{array}{c} CH_3C=CHCH_3 \\ | \\ OH \end{array} \right]$ $\longrightarrow$ $CH_3\overset{\overset{\displaystyle O}{\|}}{C}CH_2CH_3$

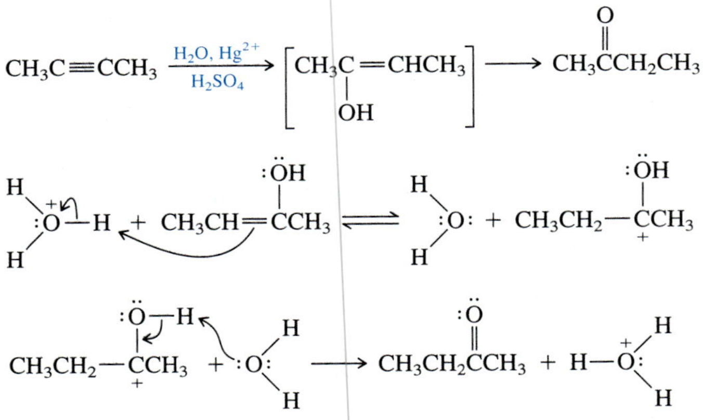

9.14   2-Octanone is prepared as shown:

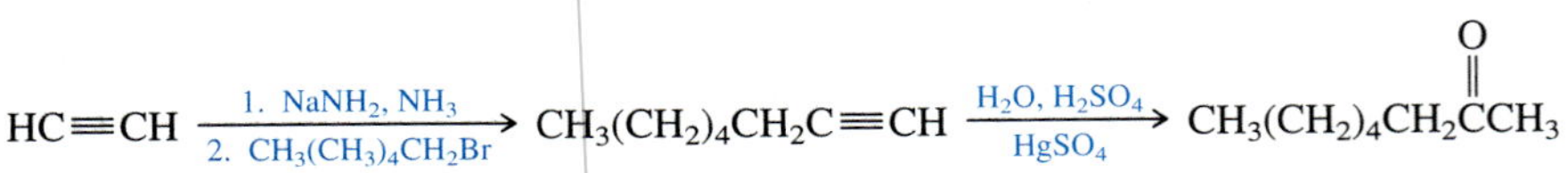

$HC\equiv CH$ $\xrightarrow[\text{2. } CH_3(CH_3)_4CH_2Br]{\text{1. } NaNH_2,\ NH_3}$ $CH_3(CH_2)_4CH_2C\equiv CH$ $\xrightarrow[HgSO_4]{H_2O,\ H_2SO_4}$ $CH_3(CH_2)_4CH_2\overset{\overset{\displaystyle O}{\|}}{C}CH_3$

4-Octyne is prepared as described in Problem 9.9 and converted to 4-octanone by hydration with H_2O, H_2SO_4, and $HgSO_4$.

9.15 $CH_3(CH_2)_4C\equiv CCH_2CH_2C\equiv C(CH_2)_4CH_3$

CHAPTER 10

10.1 (b)

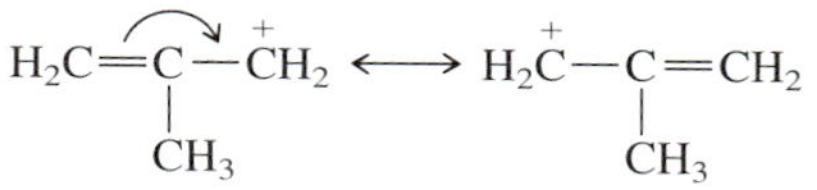

(c)

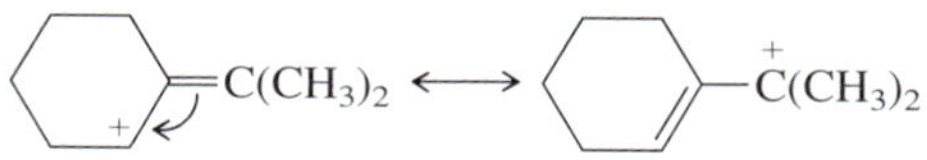

10.2 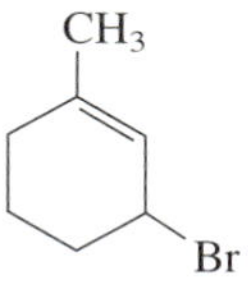 and

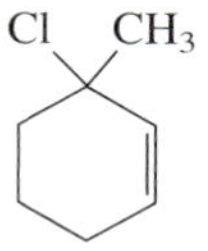

10.3 (b) 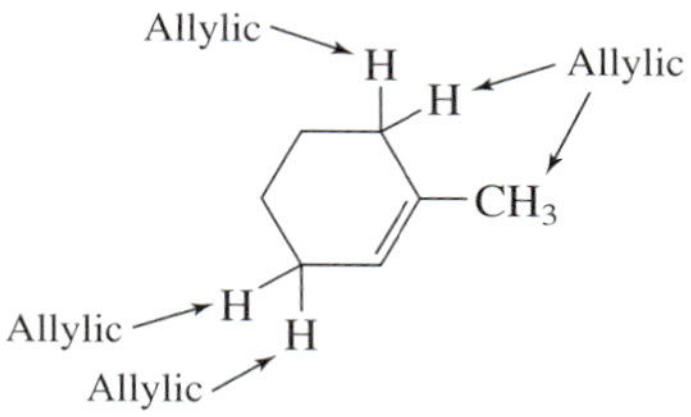 (c)

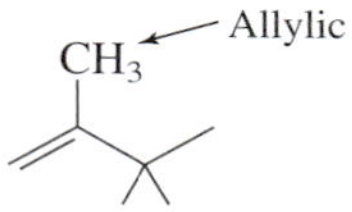

(d)

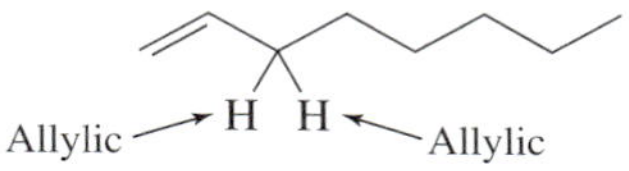

10.4 (Propagation step 1)

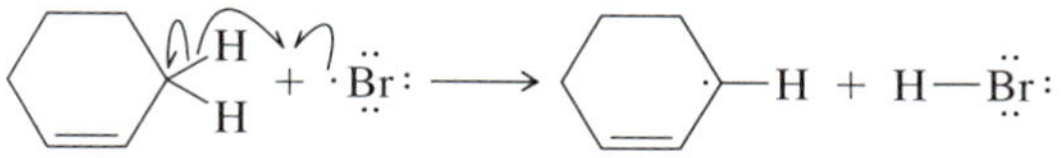

(Propagation step 2)

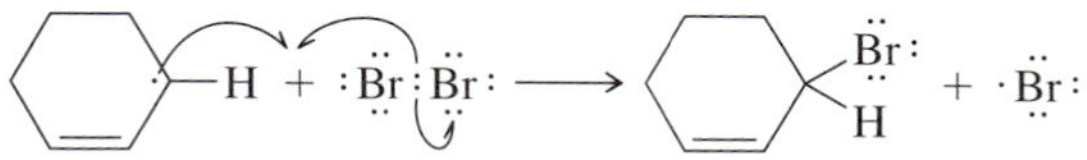

10.5 2,3,3-Trimethyl-1-butene gives only $(CH_3)_3CC=CH_2$ 1-Octene gives a mixture of CH_2Br

$H_2C=CHCH(CH_2)_4CH_3$ as well as the cis and trans stereoisomers of $BrCH_2CH=CH(CH_2)_4CH_3$. Br

10.6 Isolated

10.7 (b) Two of the double bonds in cembrene are conjugated to each other but isolated from the remaining double bonds in the molecule. (c) The $CH=C=CH$ unit is a cumulated double bond; it is conjugated to the double bond at C-2.

10.8 1,2-Pentadiene (3251 kJ/mol); (*E*)-1,3-pentadiene (3186 kJ/mol); 1,4-pentadiene (3217 kJ/mol)

10.9 2-Methyl-2,3-pentadiene is achiral. 2-Chloro-2,3-pentadiene is chiral.

10.10 $H_2C=CHCH_2\overset{\overset{\displaystyle CH_3}{|}}{C}=CHCH_3$ (*cis* + *trans*)   and $H_2C=CHCH_2\overset{\overset{\displaystyle CH_2}{\|}}{C}CH_2CH_3$

10.11 $(CH_3)_2CCH\!=\!CH_2$
　　　　　　　　|
　　　　　　　Cl

10.12 3,4-Dibromo-3-methyl-1-butene; 3,4-dibromo-2-methyl-1-butene; and 1,4-dibromo-2-methyl-2-butene

10.13

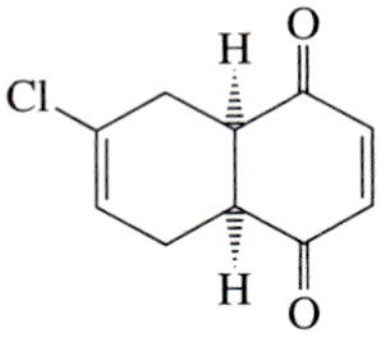

10.14 (b) $H_2C\!=\!CHCH\!=\!CH_2 + cis\text{-}N\!\equiv\!CCH\!=\!CHC\!\equiv\!N$

(c) $CH_3CH\!=\!CHCH\!=\!CH_2 +$

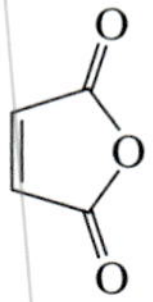

10.15

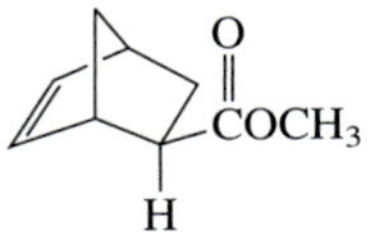

and

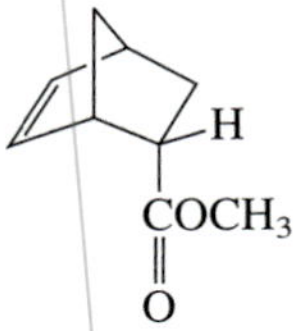

10.16 π

10.17 There is a mismatch between the ends of the HOMO of one 1,3-butadiene molecule and the LUMO of the other (Fig. 10.10). The reaction is forbidden.

CHAPTER 11

11.1 (a)

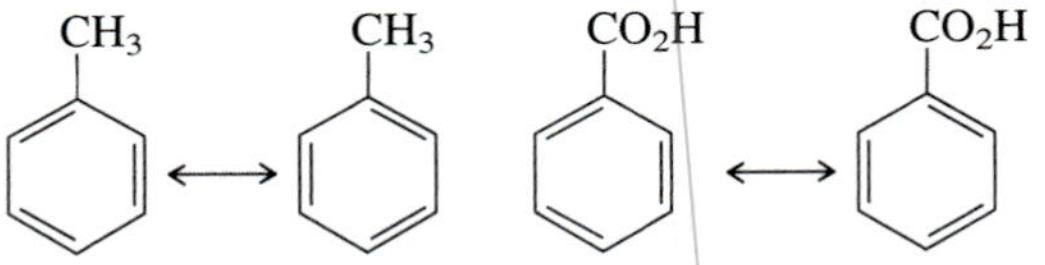

(b)

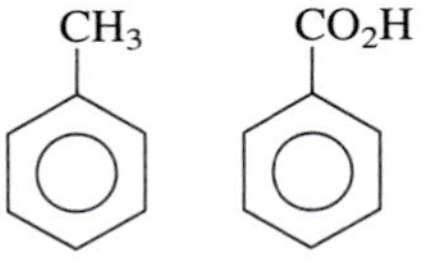

11.2 1,3,5-Cycloheptatriene resonance energy = 25 kJ/mol (5.9 kcal/mol). It is about six times smaller than the resonance energy of benzene.

11.3 (b)

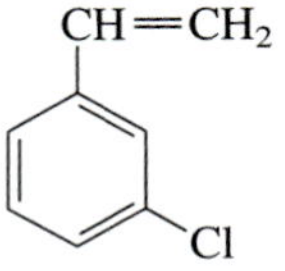

(c)

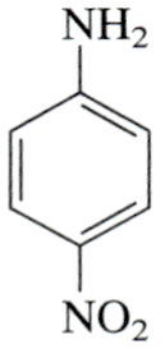

11.4

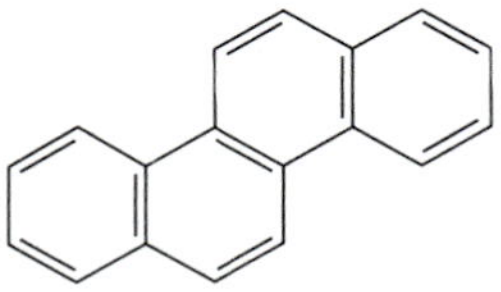

11.5 H_3C——⬡——CH_3

11.6   The aromaticity of the ring is retained when chlorine bonds to the benzylic carbon; aromaticity is lost when chlorine bonds to one of the ring carbons.

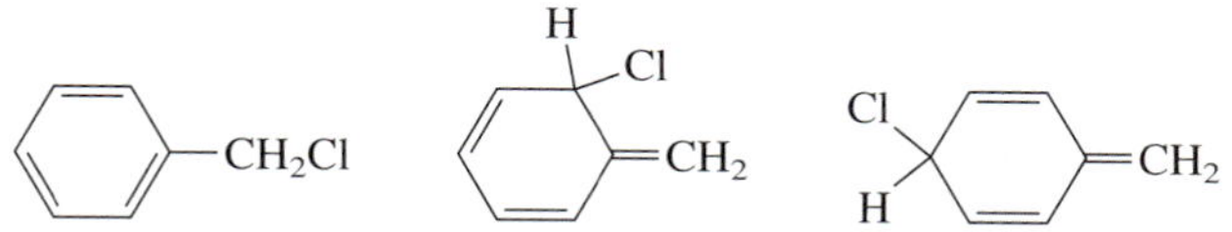

⬡——CH_2Cl

Product from
benzylic chlorination

⬡——CH_2 (with H, Cl)

Product from
ortho chlorination

⬡——CH_2 (with Cl, H)

Product from
para chlorination

11.7   (b) 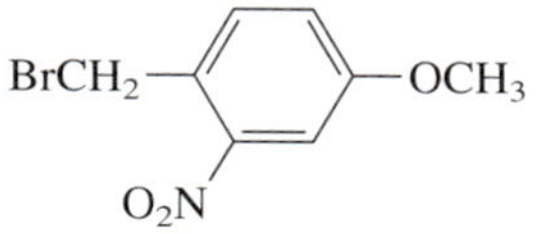 $BrCH_2$——⬡——OCH_3 (with O_2N)

11.8 $(CH_3)_3C$——⬡——CO_2H (with CO_2H)

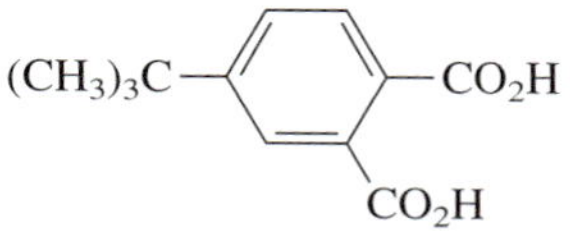

11.9   (b) $C_6H_5CH_2OC(CH_3)_3$ (c) $C_6H_5CH_2\overset{..}{N}{=}\overset{+}{N}{=}\overset{..}{N}{:}^{-}$ (d) $C_6H_5CH_2SH$
(e) $C_6H_5CH_2I$

11.10   1,2-Dihydronaphthalene, 101 kJ/mol (24.1 kcal/mol); 1,4-dihydronaphthalene, 113 kJ/mol (27.1 kcal/mol)

11.11   (b) $C_6H_5CHCH_2OH$ (with CH_3)   (c) $C_6H_5CHCH_2Br$ (with OH)   (d) $C_6H_5CH{-}CH_2$ (with O) $+ C_6H_5CO_2H$

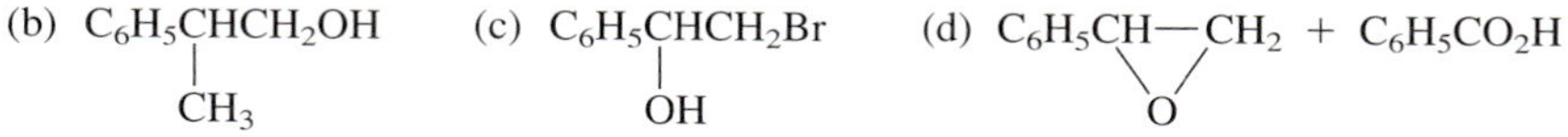

11.12   Styrene, 4393 kJ/mol (1050 kcal/mol); cyclooctatetraene, 4543 kJ/mol (1086 kcal/mol)

11.13   Diels–Alder reaction

11.14   (b) Cyclic conjugation is necessary for aromaticity; one of the double bonds is not part of the ring. (c) Two benzene rings are connected by a single bond; each ring is aromatic, making the molecule aromatic.

11.15   (b) Two of the π electrons of [12]annulene are unpaired and occupy nonbonding orbitals. [12]Annulene is not aromatic.

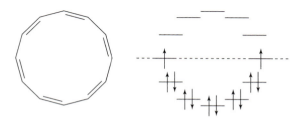

11.16   Divide the heats of combustion by the number of carbons. The two aromatic hydrocarbons (benzene and [18]annulene) have heats of combustion per carbon that are less than those of the nonaromatic hydrocarbons (cyclooctatetraene and [16]annulene). On a per carbon basis, the aromatic hydrocarbons have lower potential energy (are more stable) than the nonaromatic hydrocarbons.

11.17

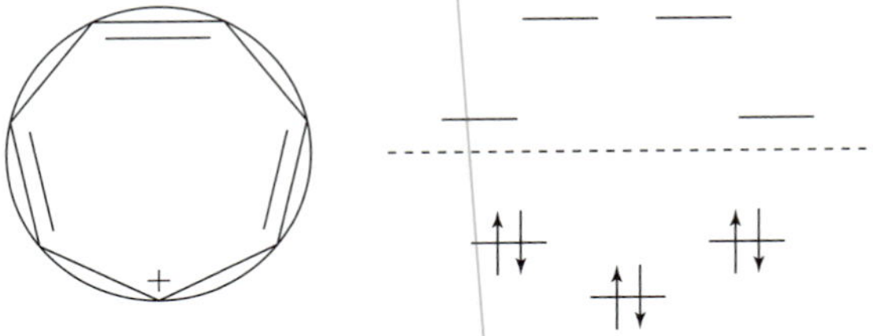

11.18 Cycloheptatrienyl radical has seven π electrons. Therefore, it does not satisfy the Hückel $4n + 2$ rule and is not aromatic.

11.19

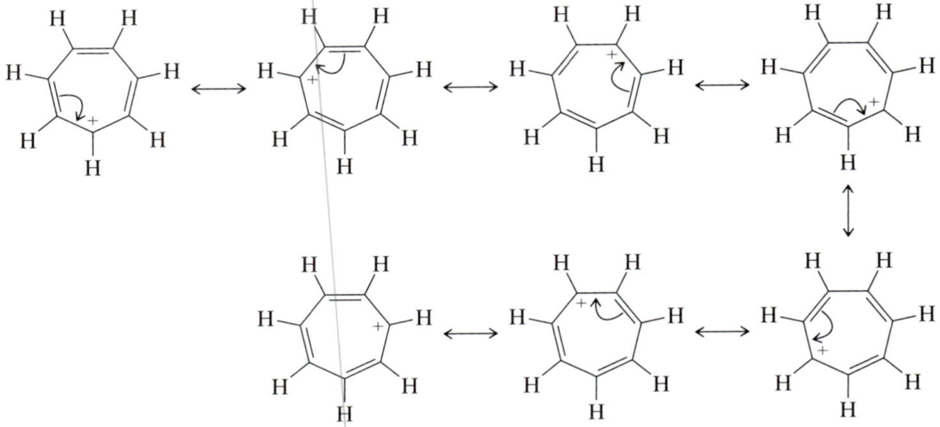

11.20

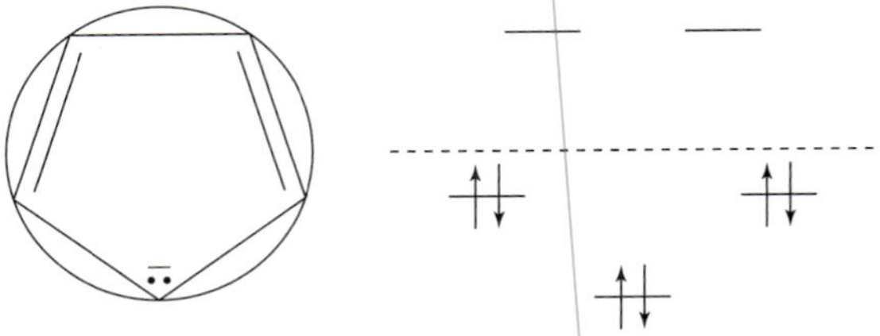

11.21

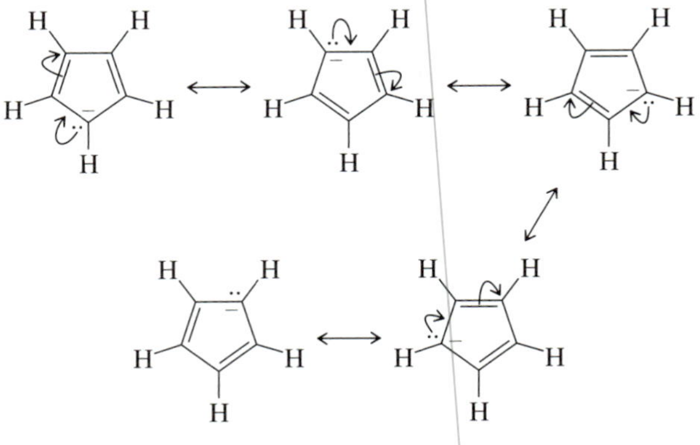

11.22 (b) Cyclononatetraenide anion is aromatic.

11.23 Indole is more stable than isoindole.

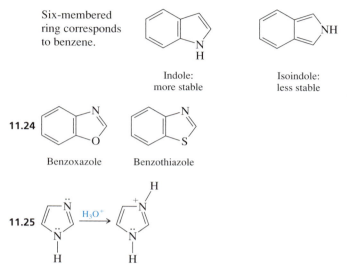

Six-membered ring corresponds to benzene.

Indole: more stable

Six-membered ring does not have same pattern of bonds as benzene.

Isoindole: less stable

11.24

Benzoxazole   Benzothiazole

11.25

CHAPTER 12

12.1 The positive charge is shared by the three carbons indicated in the three most stable resonance structures:

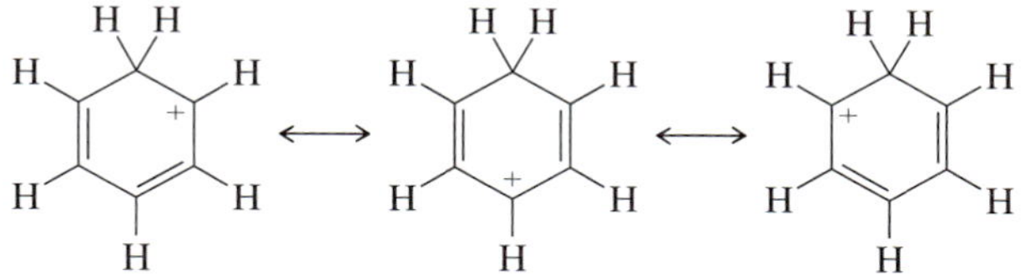

Provided that these structures contribute equally, the resonance picture coincides with the MO treatment in assigning one third of a positive charge ($+0.33$) to each of the indicated carbons.

12.2

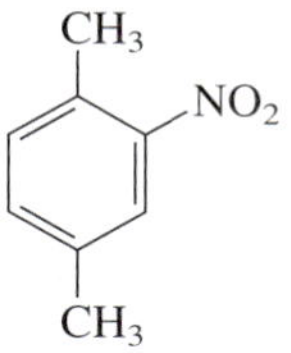

12.3

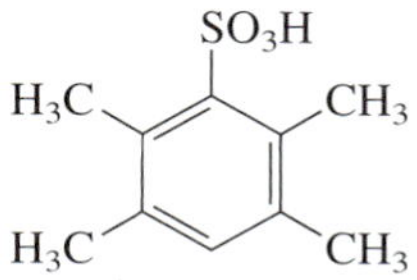

12.4 The major product is isopropylbenzene. Ionization of 1-chloropropane is accompanied by a hydride shift to give $CH_3\overset{+}{C}HCH_3$, which then attacks benzene.

12.5

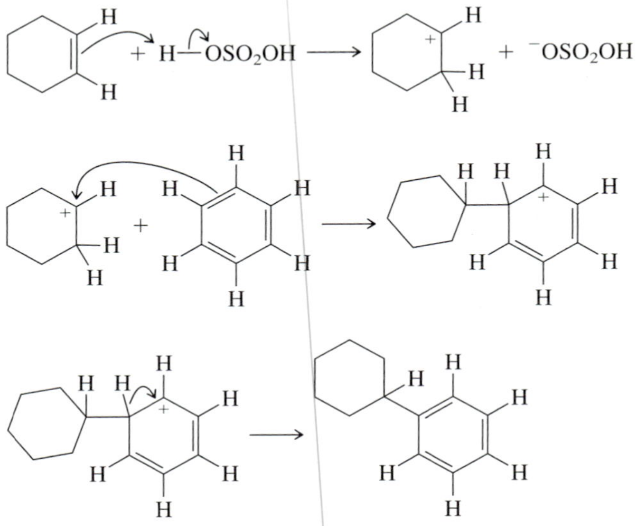

12.6

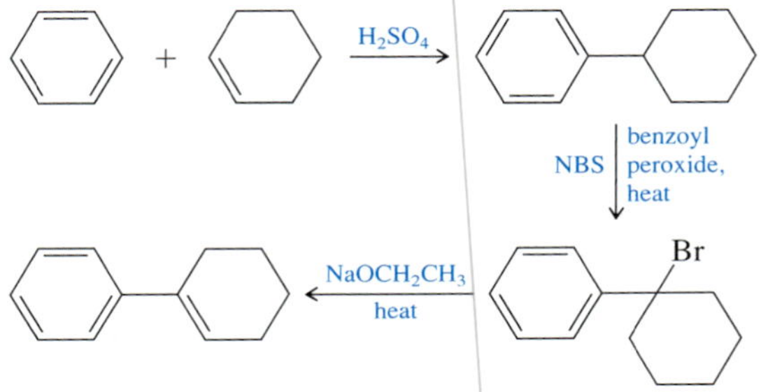

12.7 CH_3O—

with substituents OCH_3, OCH_3, and $\overset{O}{\overset{\|}{C}}CH_2CH(CH_3)_2$

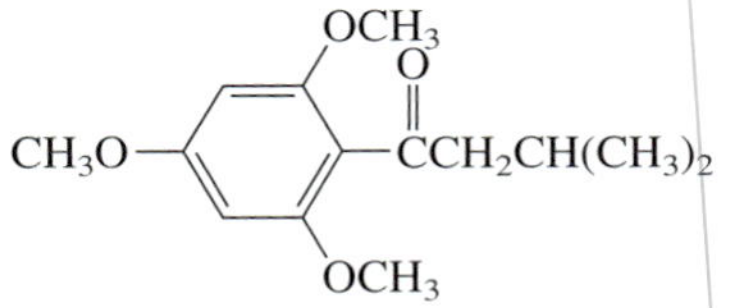

12.8

$\overset{O}{\overset{\|}{C}}CH_2CH_2\overset{O}{\overset{\|}{C}}OH$

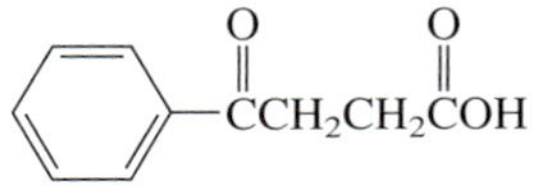

12.9 (b) Friedel– Crafts acylation of benzene with $(CH_3)_3\overset{O}{\overset{\|}{C}}CCl$, followed by reduction with Zn(Hg) and hydrochloric acid

12.10 (b) Toluene is 1.7 times more reactive than *tert*-butylbenzene. (c) Ortho (10%), meta (6.7%), para (83.3%)

12.11

—CH_2Cl	—$CHCl_2$	—CCl_3
Deactivating ortho, para-directing	Deactivating ortho, para-directing	Deactivating meta-directing

12.12 (b)

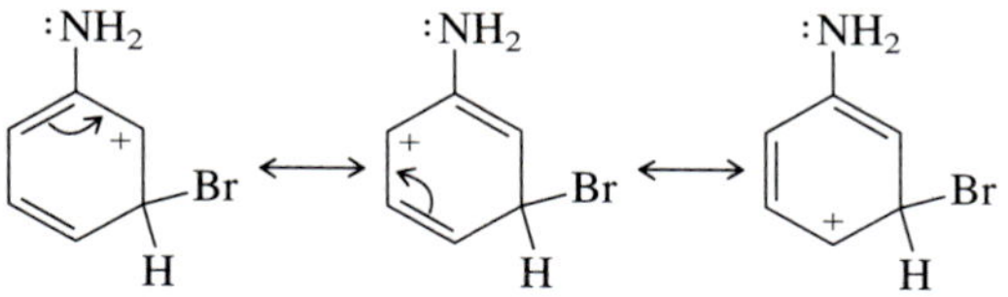

(c)

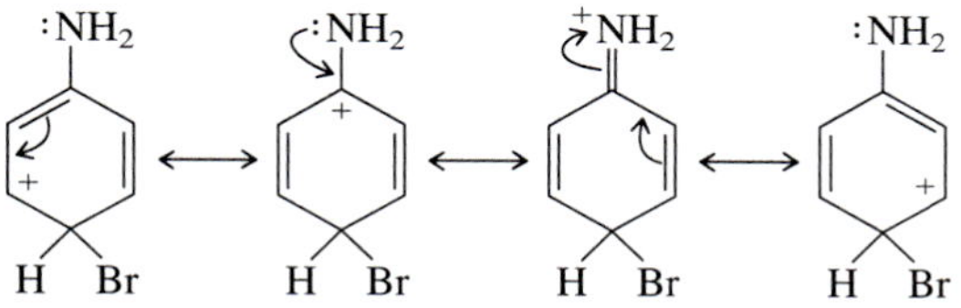

12.13

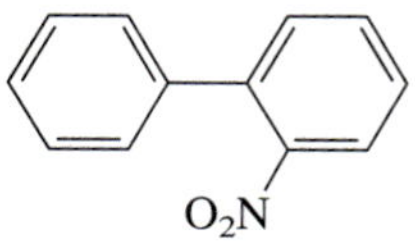

and

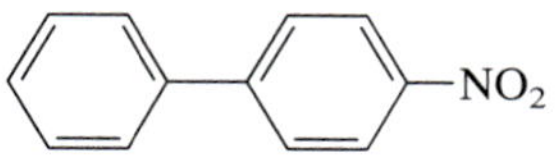

12.14 (b)

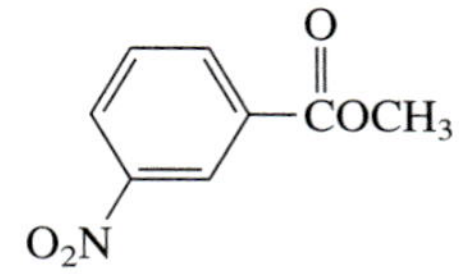

(c)

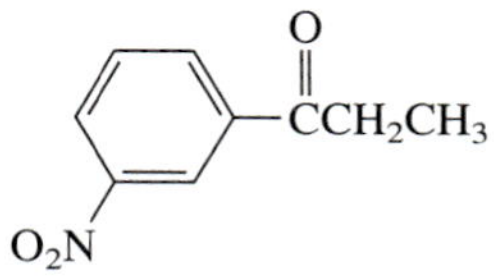

12.15 The group $-\overset{+}{N}(CH_3)_3$ is strongly deactivating and meta-directing. Its positively charged nitrogen makes it a powerful electron-withdrawing substituent. It resembles a nitro group.

12.16

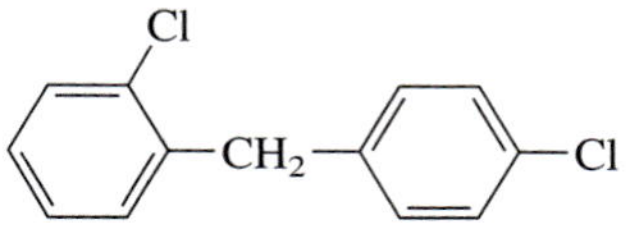

and

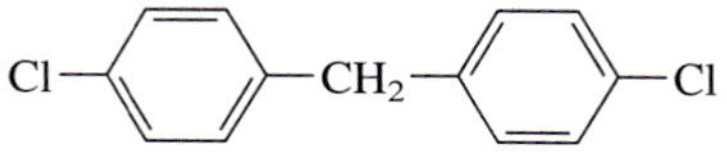

12.17 (b)

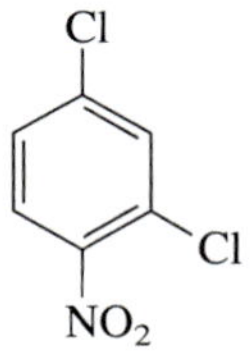

(c)

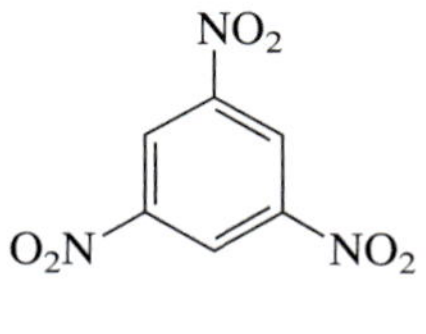

(d) 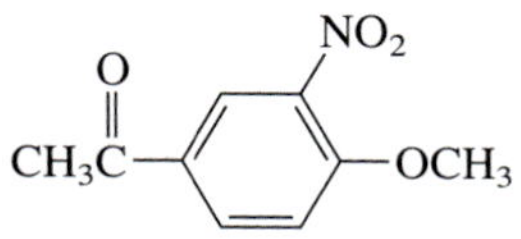 $CH_3\overset{O}{\overset{\|}{C}}$

(e) 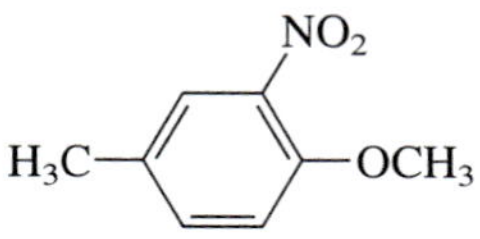 H_3C- (f)

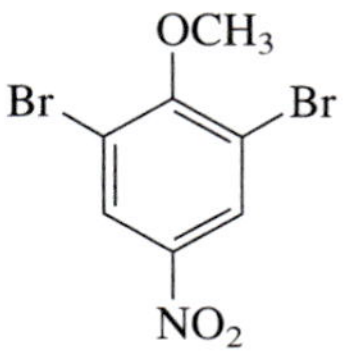

12.18 *m*-Bromonitrobenzene:

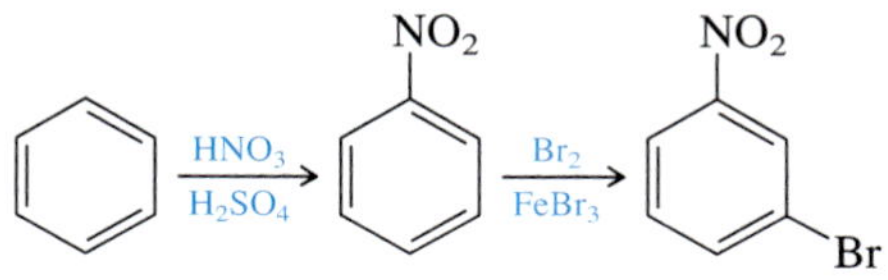

p-Bromonitrobenzene:

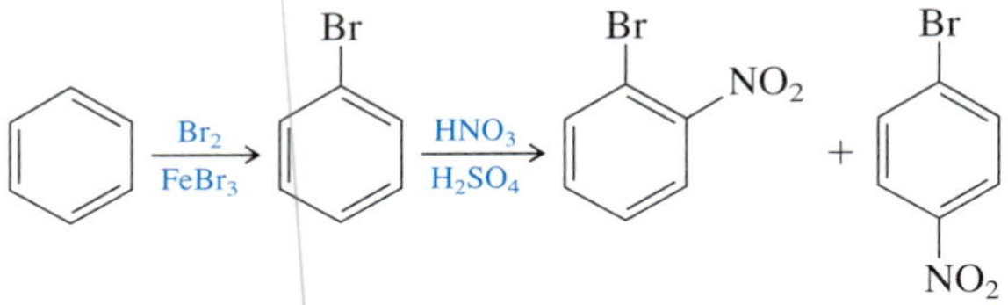

12.19

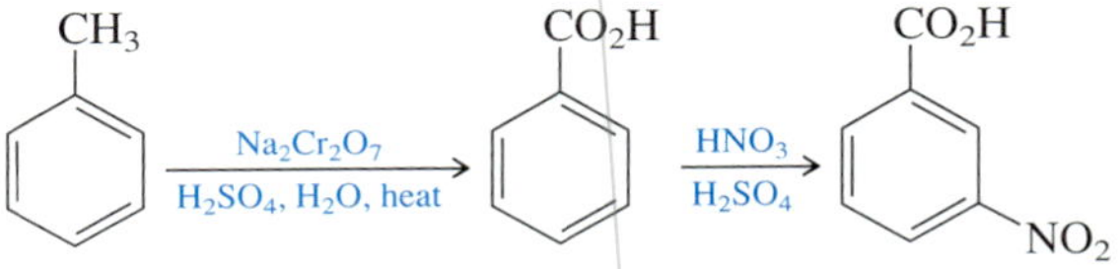

12.20

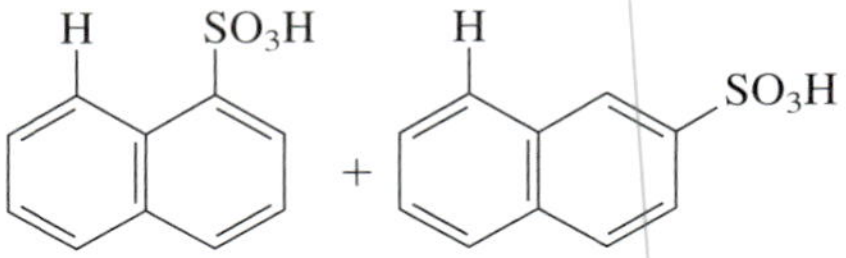

Formed faster   More stable

The hydrogen at C-8 (the one shown in the structural formulas) crowds the —SO₃H group in the less stable isomer.

12.21

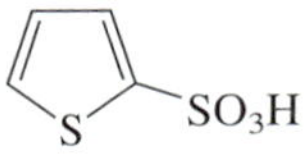

CHAPTER 13

13.1 1.41 T

13.2 25.2 MHz

13.3 (a) 6.88 ppm; (b) higher field; more shielded

13.4 (b) Cl_3C—CH_2—CH_2—CH_3

        least shielded   most shielded

(c) most shielded →   ← most shielded

    least shielded →   ← least shielded

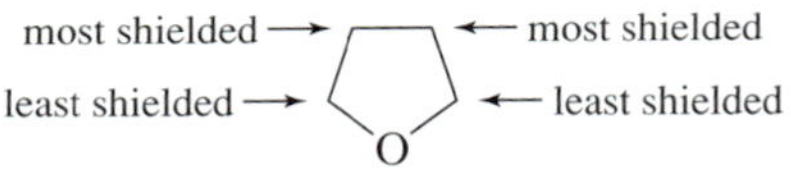

13.5

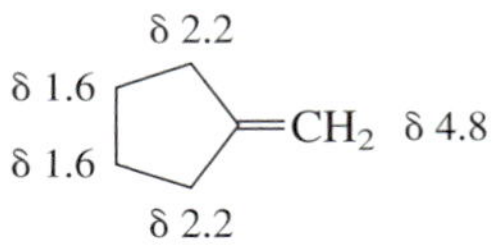

13.6

H_3C—$\overset{\overset{\textstyle O}{\|}}{C}$—$CH_2$—$CH_2$—$CH_3$

δ 2.0   δ 2.3   δ 1.7   δ 1.1

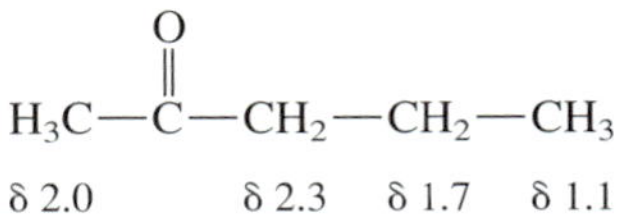

13.7

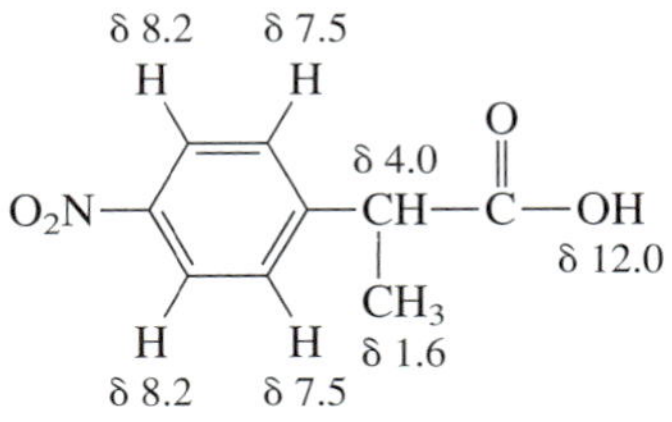

13.8 The chemical shift of the methyl protons is δ 2.2. The chemical shift of the protons attached to the aromatic ring is δ 7.0.

13.9 (b) Five; (c) two; (d) two; (e) three; (f) one; (g) four; (h) three

13.10 (b) One; (c) one; (d) one; (e) four; (f) four

13.11 (b) One signal (singlet); (c) two signals (doublet and triplet); (d) two signals (both singlets); (e) two signals (doublet and quartet)

13.12 (b) Three signals (singlet, triplet, and quartet); (c) two signals (triplet and quartet); (d) three signals (singlet, triplet, and quartet); (e) four signals (three triplets and quartet)

13.13 Both H_b and H_c appear as doublets of doublets:

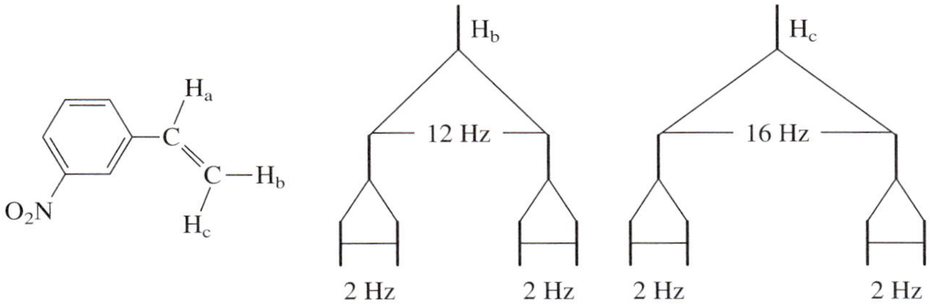

13.14 (b) The signal for the proton at C-2 is split into a quartet by the methyl protons, and each line of this quartet is split into a doublet by the aldehyde proton. It appears as a doublet of quartets.

13.15 (b) Six; (c) six; (d) nine; (e) three

13.16 $Br—CH_2—CH_2—CH_2—Cl$

δ 35 δ 30 δ 43

13.17

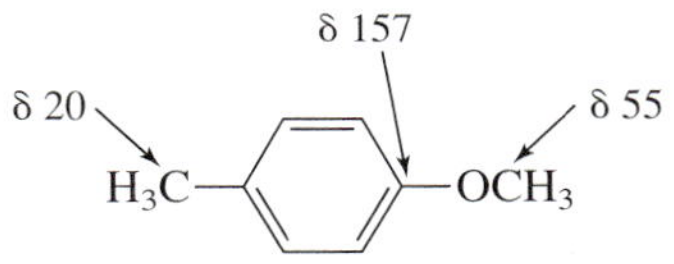

13.18 Hydrogen is less electronegative than carbon, so the carbonyl carbon of —CH=O is more shielded than the carbonyl carbon of a ketone.

13.19 1,2,4-Trimethylbenzene

13.20 Benzyl alcohol. Infrared spectrum has peaks for O—H and sp^3 C—H; lacks peak for C=O.

13.21 HOMO–LUMO energy difference in ethylene is greater than that of *cis,trans*-1,3-cyclooctadiene.

13.22 2-Methyl-1,3-butadiene

13.23 (b) Three peaks (*m/z* 146, 148, and 150); (c) three peaks (*m/z* 234, 236, and 238); (d) three peaks (*m/z* 190, 192, and 194)

13.24

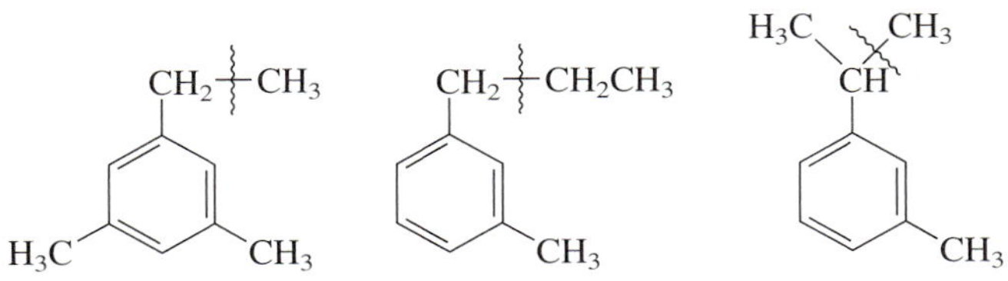

Base peak $C_9H_{11}^+$ (*m/z* 119) Base peak $C_8H_9^+$ (*m/z* 105) Base peak $C_9H_{11}^+$ (*m/z* 119)

13.25 (b) 3; (c) 2; (d) 3; (e) 2; (f) 2

CHAPTER 14

14.1 (b) Cyclohexylmagnesium chloride

14.2 (b) $CH_3CHCH_2CH_3 + 2Li \longrightarrow CH_3CHCH_2CH_3 + LiBr$
　　　　　　|　　　　　　　　　　　　　　　　|
　　　　　　Br　　　　　　　　　　　　　　　Li

14.3 (b) $H_2C=CHCH_2MgCl$　(c) 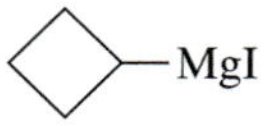 ⬦—MgI　(d) 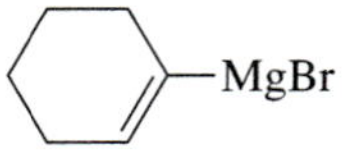 ⬡—MgBr

14.4 $Li-CH_2CH_2CH_2CH_3$　and　$BrMg-C_6H_5$

$\qquad\qquad \searrow H-\overset{..}{\underset{..}{O}}H \qquad\qquad\qquad \searrow H-\overset{..}{\underset{..}{O}}CH_3$

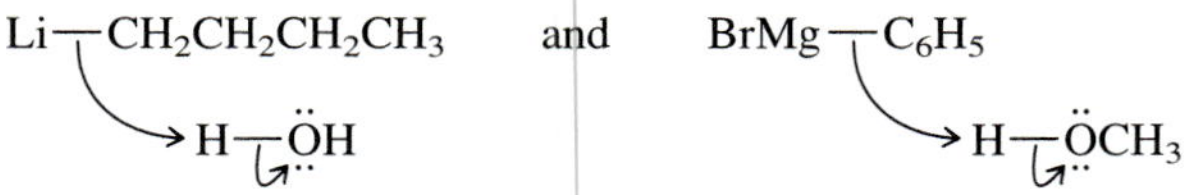

14.5 (b) $CH_3(CH_2)_4CH_2OH + CH_3CH_2CH_2CH_2Li \longrightarrow CH_3CH_2CH_2CH_3 + CH_3(CH_2)_4CH_2OLi$
(c) $C_6H_5SH + CH_3CH_2CH_2CH_2Li \longrightarrow CH_3CH_2CH_2CH_3 + C_6H_5SLi$

14.6 (b) $C_6H_5\underset{\underset{\textstyle OH}{|}}{C}HCH_2CH_2CH_3$　(c) 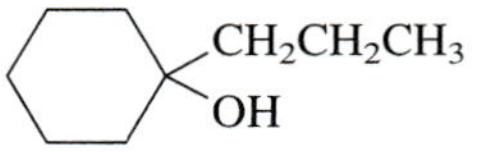 ⬡ with $\underset{\underset{\textstyle OH}{|}}{}CH_2CH_2CH_3$　(d) $CH_3CH_2CH_2\underset{\underset{\textstyle CH_3CH_2}{|}}{\overset{\overset{\textstyle CH_3}{|}}{C}}OH$

14.7 $CH_3\overset{..}{C}H_2^- + H-\overset{\frown}{C}\equiv CCH_2CH_2CH_2CH_3 \longrightarrow CH_3CH_3 + \ :\bar{C}\equiv CCH_2CH_2CH_2CH_3$
　　Ethyl anion　　　　　1-Hexyne　　　　　　　　Ethane　　　Conjugate base of 1-hexyne

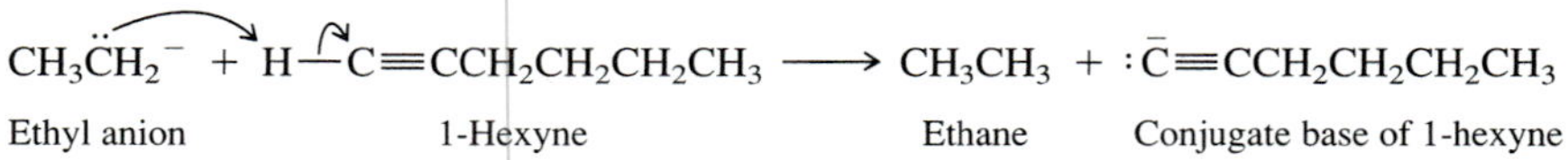

14.8 (b) $CH_3MgI + C_6H_5\overset{\overset{\textstyle O}{||}}{C}CH_3 \xrightarrow[\text{2. }H_3O^+]{\text{1. diethyl ether}} C_6H_5\underset{\underset{\textstyle OH}{|}}{\overset{\overset{\textstyle CH_3}{|}}{C}}CH_3$

and　$C_6H_5MgBr + CH_3\overset{\overset{\textstyle O}{||}}{C}CH_3 \xrightarrow[\text{2. }H_3O^+]{\text{1. diethyl ether}} C_6H_5\underset{\underset{\textstyle OH}{|}}{\overset{\overset{\textstyle CH_3}{|}}{C}}CH_3$

14.9 (b) $2C_6H_5MgBr + $ ▷—$\overset{\overset{\textstyle O}{||}}{C}OCH_2CH_3$

14.10 (b) $LiCu(CH_3)_2 + $ (cyclopentene with Br and two CH_3 groups)

14.11 ⬦$=CH_2$

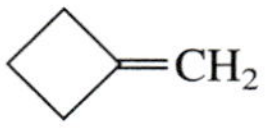

14.12 *cis*-2-Butene $\longrightarrow$ (dibromocyclopropane, H_3C, CH_3 cis; H, H; Br, Br)　　*trans*-2-Butene $\longrightarrow$ (dibromocyclopropane, H_3C, H; H, CH_3; Br, Br)

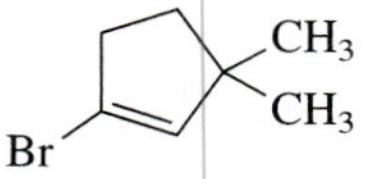

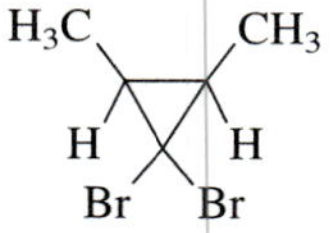

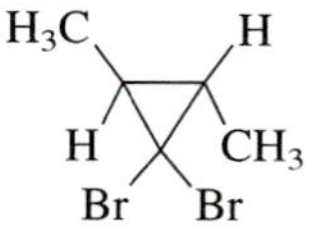

14.13 $Fe(CO)_5$

CHAPTER 15

15.1 The primary alcohols $CH_3CH_2CH_2CH_2OH$ and $(CH_3)_2CHCH_2OH$ can each be prepared by hydrogenation of an aldehyde. The secondary alcohol $CH_3\overset{\underset{|}{OH}}{CH}CH_2CH_3$ can be prepared by hydrogenation of a ketone. The tertiary alcohol $(CH_3)_3COH$ cannot be prepared by hydrogenation.

15.2 (b) $CH_3\overset{\overset{D}{|}}{\underset{\underset{OD}{|}}{C}}CH_3$ (c) $C_6H_5\overset{\overset{D}{|}}{\underset{\underset{H}{|}}{C}}OH$ (d) DCH_2OD

15.3 $CH_3CH_2\overset{\overset{O}{||}}{C}OCH(CH_3)_2$

15.4

15.5 $HO\overset{\overset{O}{||}}{C}CH_2\overset{\underset{\underset{CH_3}{|}}{}}{CH}CH_2\overset{\overset{O}{||}}{C}OH \xrightarrow[\text{2. H}_2\text{O}]{\text{1. LiAlH}_4} HOCH_2CH_2\overset{\underset{\underset{CH_3}{|}}{}}{CH}CH_2CH_2OH$

$CH_3O\overset{\overset{O}{||}}{C}CH_2\overset{\underset{\underset{CH_3}{|}}{}}{CH}CH_2\overset{\overset{O}{||}}{C}OCH_3 \xrightarrow[\text{2. H}_2\text{O}]{\text{1. LiAlH}_4} HOCH_2CH_2\overset{\underset{\underset{CH_3}{|}}{}}{CH}CH_2CH_2OH + 2CH_3OH$

15.6 *cis*-2-Butene yields the meso stereoisomer of 2,3-butanediol:

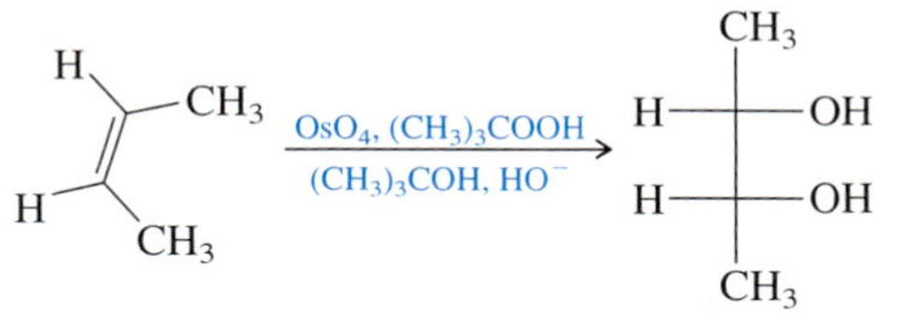

trans-2-Butene gives equal quantities of the two enantiomers of the chiral diol:

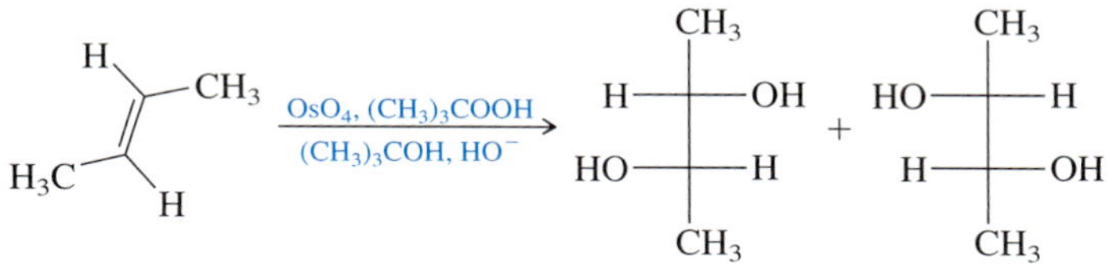

15.7 Step 1:

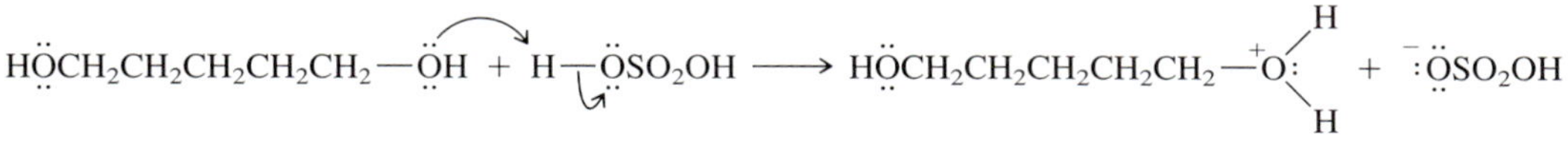

Step 2:

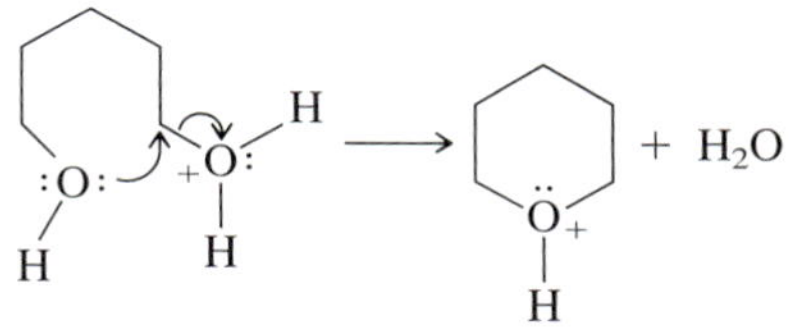

Step 3:

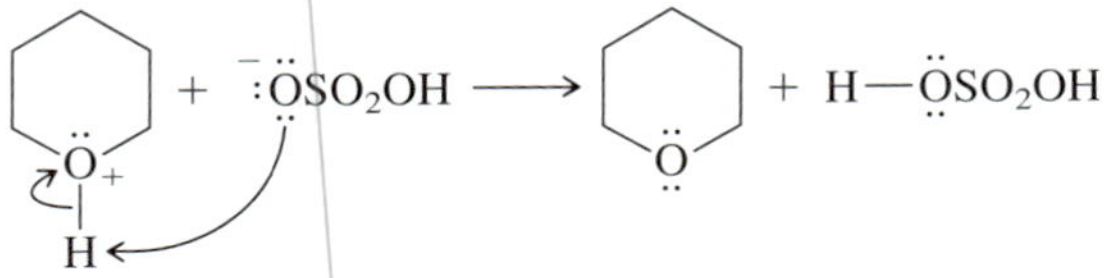

15.8 (b) 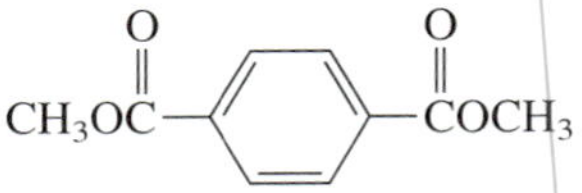 $CH_3O\overset{O}{\overset{\|}{C}}$—[benzene ring]—$\overset{O}{\overset{\|}{C}}OCH_3$

15.9

$(CH_3)_3C$—[cyclohexane]—OH $\xrightarrow[\text{anhydride}]{\text{acetic}}$ $(CH_3)_3C$—[cyclohexane]—$O\overset{O}{\overset{\|}{C}}CH_3$

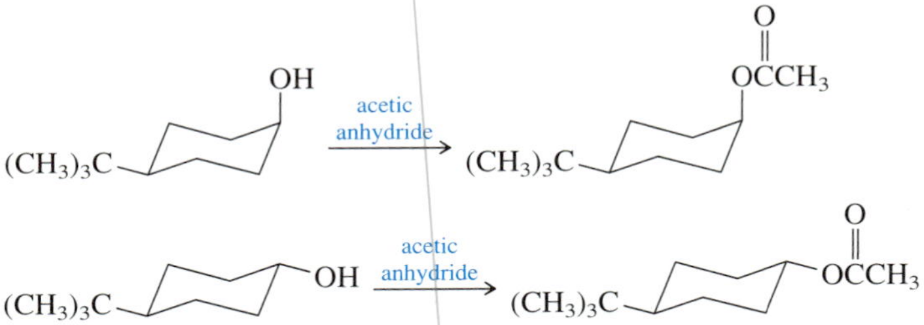

15.10 $O_2NOCH_2CHCH_2ONO_2$
        |
        ONO_2

15.11 (b) $CH_3\overset{O}{\overset{\|}{C}}(CH_2)_5CH_3$　　(c) $CH_3(CH_2)_5\overset{O}{\overset{\|}{C}}H$

15.12 (b) One; (c) none

15.13 (b) $(CH_3)_2CHCH_2\overset{O}{\overset{\|}{C}}H$ + $C_6H_5CH_2\overset{O}{\overset{\|}{C}}H$　　(c) 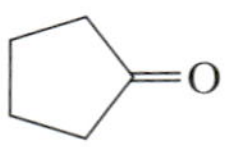 [cyclopentane]=O + $H\overset{O}{\overset{\|}{C}}H$

15.14 $CH_3CHCH_2CH_2SH$
        |
        CH_3

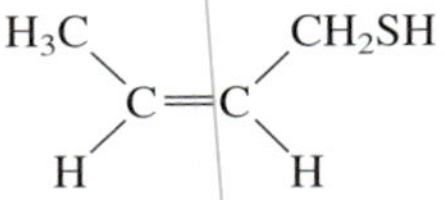

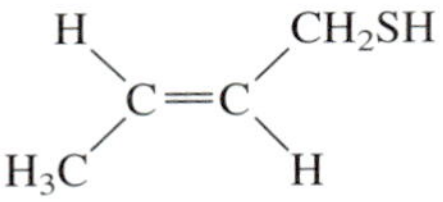

3-Methyl-1-butanethiol

cis-2-Butene-1-thiol

trans-2-Butene-1-thiol

15.15 $CH_3(CH_2)_4CH_2OH \xrightarrow[\text{heat}]{\text{HBr}} CH_3(CH_2)_4CH_2Br \xrightarrow[\text{ethanol}]{\text{KSH}} CH_3(CH_2)_4CH_2SH$

1-Hexanol　　　　　　1-Bromohexane　　　　　　1-Hexanethiol

15.16 The peak at m/z 70 corresponds to loss of water from the molecular ion. The peaks at m/z 59 and 73 correspond to the cleavages indicated:

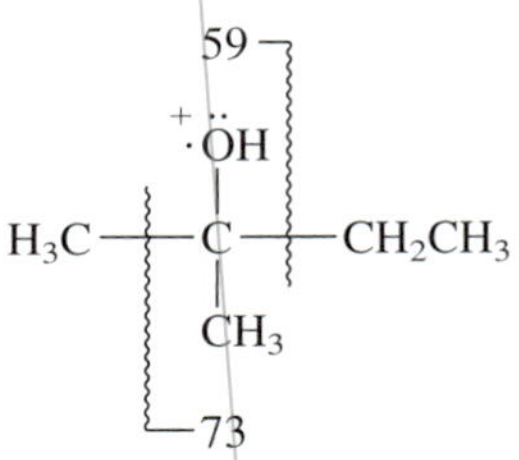

CHAPTER 16

16.1 (b) 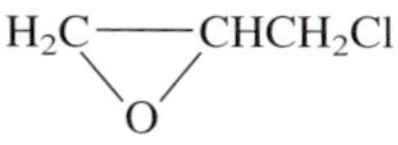 $H_2C\overset{O}{\underset{}{-}}CHCH_2Cl$ (c) 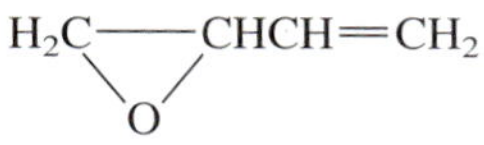 $H_2C\overset{O}{\underset{}{-}}CHCH{=}CH_2$

16.2 1,2-Epoxybutane, 2546 kJ/mol; tetrahydrofuran, 2499 kJ/mol

16.3

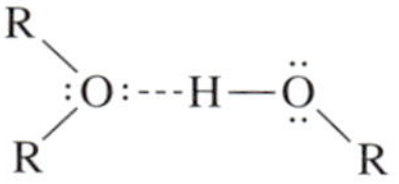

$$\overset{R}{\underset{R}{\diagdown}}\ddot{O}{:}{-}{-}{-}H{-}\overset{\cdot\cdot}{\ddot{O}}\underset{R}{\diagdown}$$

16.4 1,4-Dioxane

16.5 $(CH_3)_2C{=}CH_2 \xrightarrow{H^+} (CH_3)_2\overset{+}{C}{-}CH_3 \xrightarrow{H\ddot{O}CH_3} (CH_3)_3C{-}\overset{+}{\underset{H}{\ddot{O}CH_3}} \xrightarrow{-H^+} (CH_3)_3C\ddot{O}CH_3$

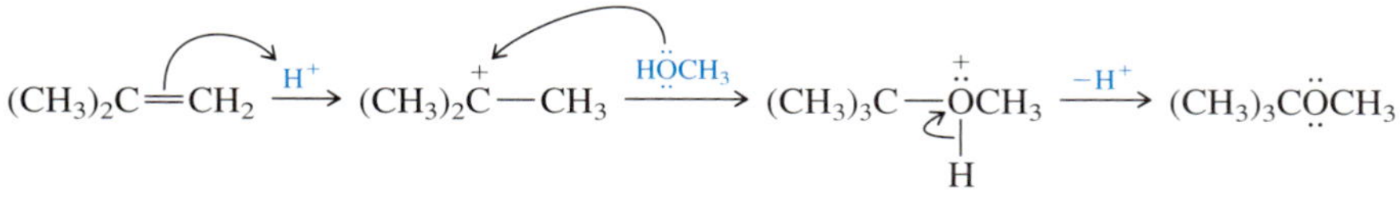

16.6 $C_6H_5CH_2ONa + CH_3CH_2Br \longrightarrow C_6H_5CH_2OCH_2CH_3 + NaBr$
and $CH_2CH_2ONa + C_6H_5CH_2Br \longrightarrow C_6H_5CH_2OCH_2CH_3 + NaBr$

16.7 (b) $(CH_3)_2CHONa + H_2C{=}CHCH_2Br \longrightarrow H_2C{=}CHCH_2OCH(CH_3)_2 + NaBr$
(c) $(CH_3)_3COK + C_6H_5CH_2Br \longrightarrow (CH_3)_3COCH_2C_6H_5 + KBr$

16.8 $CH_3CH_2OCH_2CH_3 + 6O_2 \longrightarrow 4CO_2 + 5H_2O$

16.9 (b) $C_6H_5CH_2OCH_2C_6H_5$ (c) 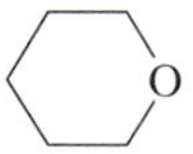 [structure]

16.10 [mechanism structures]

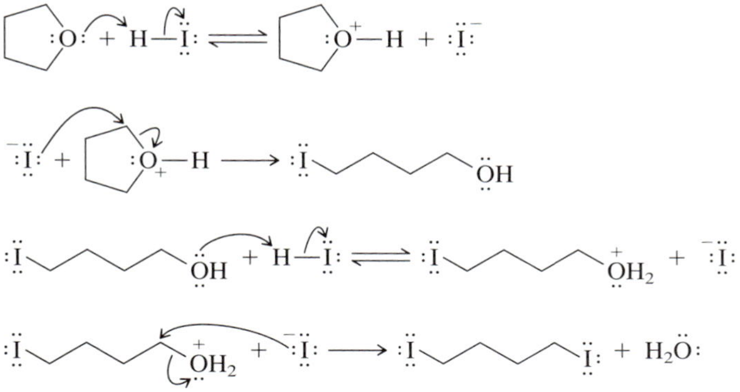

16.11 Only the trans epoxide is chiral. As formed in this reaction, neither product is optically active.

16.12 (b) $N_3CH_2CH_2OH$ (c) $HOCH_2CH_2OH$ (d) $C_6H_5CH_2CH_2OH$
(e) $CH_3CH_2C{\equiv}CCH_2CH_2OH$

16.13 Compound B

16.14 Compound A

16.15 *trans*-2-Butene gives *meso*-2,3-butanediol on epoxidation followed by acid-catalyzed hydrolysis. *cis*-2-Butene gives *meso*-2,3-butanediol on osmium tetraoxide hydroxylation.

16.16 The product has the S configuration.

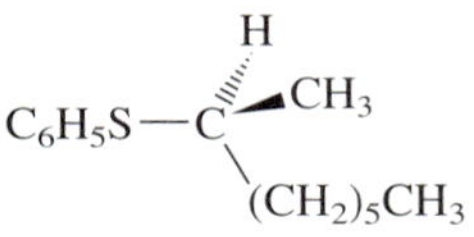

[structure: $C_6H_5S{-}C$ with H, CH_3, $(CH_2)_5CH_3$]

16.17 Phenyl vinyl sulfoxide is chiral. Phenyl vinyl sulfone is achiral.

16.18 $CH_3SCH_3 + CH_3(CH_2)_{10}CH_2I$ will yield the same sulfonium salt. This combination is not as effective as $CH_3I + CH_3(CH_2)_{10}CH_2SCH_3$, because the reaction mechanism is S_N2 and CH_3I is more reactive than $CH_3(CH_2)_{10}CH_2I$ in reactions of this type because it is less crowded.

16.19 $H_2C{=}\overset{+}{\ddot{O}}CH(CH_3)CH_2CH_3$

CHAPTER 17

17.1 (b) Pentanedial; (c) 2,3-dihydroxypropanal; (d) 4-hydroxy-3-methoxybenzaldehyde

17.2 (b) 2-Methyl-3-pentanone; (c) 4,4-dimethyl-2-pentanone; (d) 4-penten-2-one

17.3

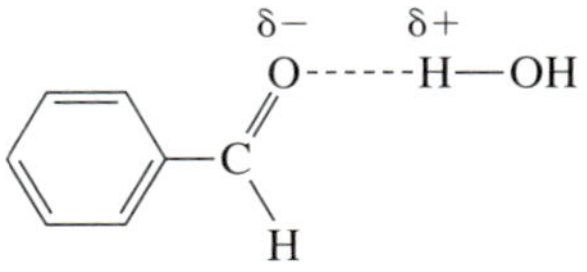

17.4 No. Carboxylic acids are inert to catalytic hydrogenation.

17.5

$$CH_3\overset{O}{\overset{\|}{C}}OH \xrightarrow[\text{2. } H_2O]{\text{1. } LiAlH_4} CH_3CH_2OH \xrightarrow[CH_2Cl_2]{PCC} CH_3\overset{O}{\overset{\|}{C}}H$$

$$CH_3CH_2OH \xrightarrow[\text{heat}]{HBr} CH_3CH_2Br \xrightarrow[\text{diethyl ether}]{Mg} CH_3CH_2MgBr$$

$$CH_3CH_2MgBr + CH_3\overset{O}{\overset{\|}{C}}H \xrightarrow[\text{2. } H_3O^+]{\text{1. diethyl ether}} CH_3\overset{OH}{\overset{|}{C}}HCH_2CH_3 \xrightarrow[CH_2Cl_2]{PCC} CH_3\overset{O}{\overset{\|}{C}}CH_2CH_3$$

17.6 $Cl_3CCH(OH)_2$

17.7 $H_2C{=}C(CH_3)C{\equiv}N$

17.8

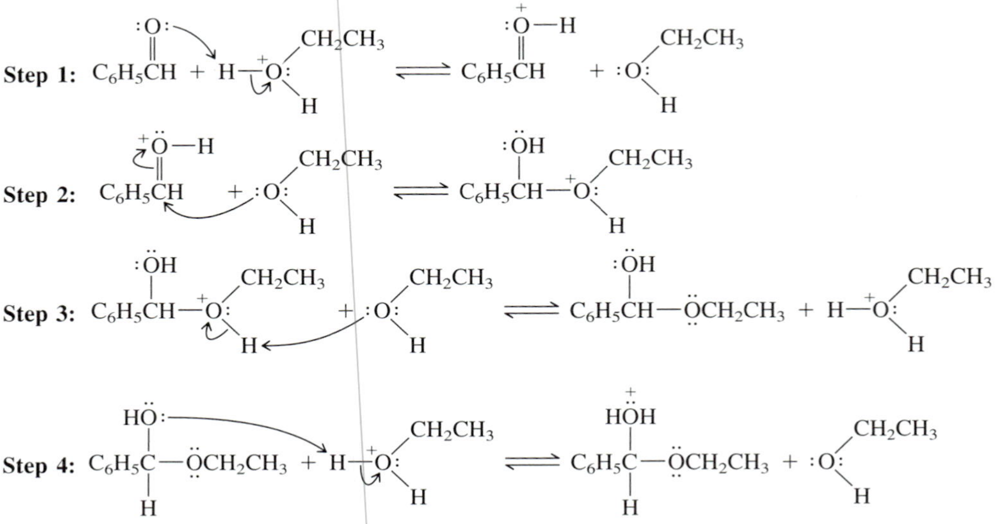
Step 1:
Step 2:
Step 3:
Step 4:

Step 5: $C_6H_5CH-OCH_2CH_3 \rightleftharpoons C_6H_5\overset{+}{C}H-OCH_2CH_3 + H-O-H$

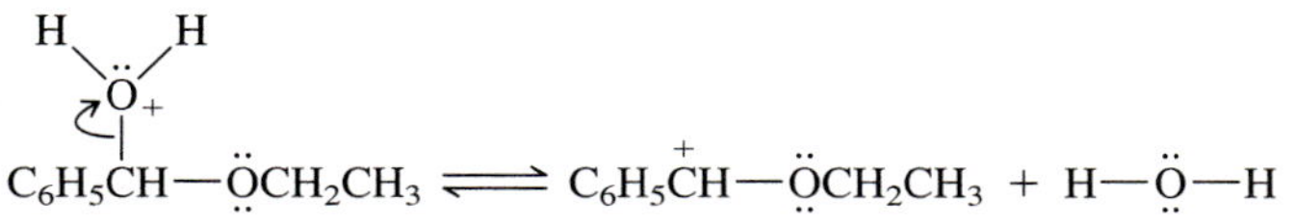

17.9 (b)   (c)   (d)

C_6H_5 H $(CH_3)_2CHCH_2$ CH_3 $(CH_3)_2CHCH_2$ CH_3

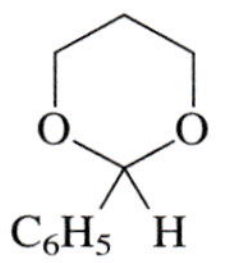

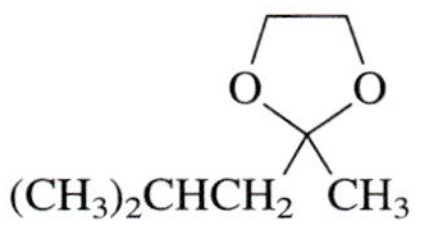

17.10 Step 1: $C_6H_5CH-OCH_2CH_3 + H-\overset{+}{O} \rightleftharpoons C_6H_5CH-OCH_2CH_3 + :O:$

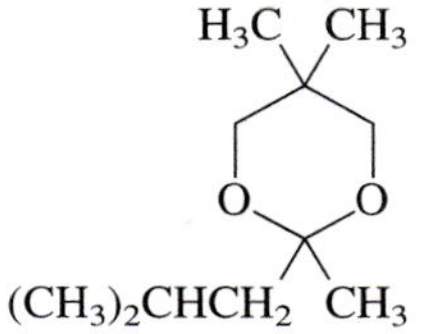

$:OCH_2CH_3$ $CH_3CH_2 \quad H$

Step 2: $C_6H_5CH-OCH_2CH_3 \rightleftharpoons C_6H_5\overset{+}{C}H-OCH_2CH_3 + :O:$

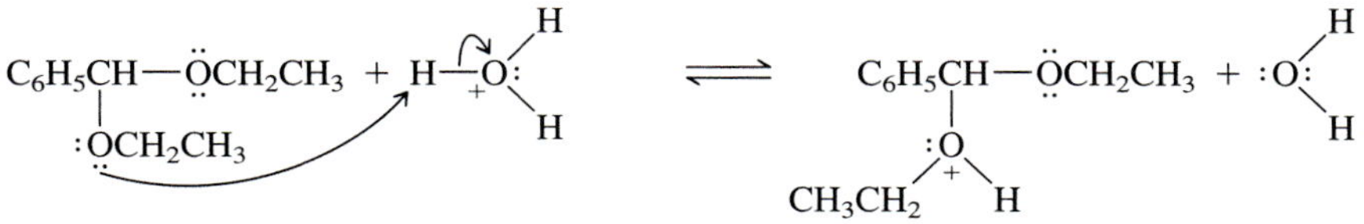

$CH_3CH_2 \quad H$ CH_2CH_3 H

Step 3: $C_6H_5\overset{+}{C}H-OCH_2CH_3 + :O: \rightleftharpoons C_6H_5CH-OCH_2CH_3$

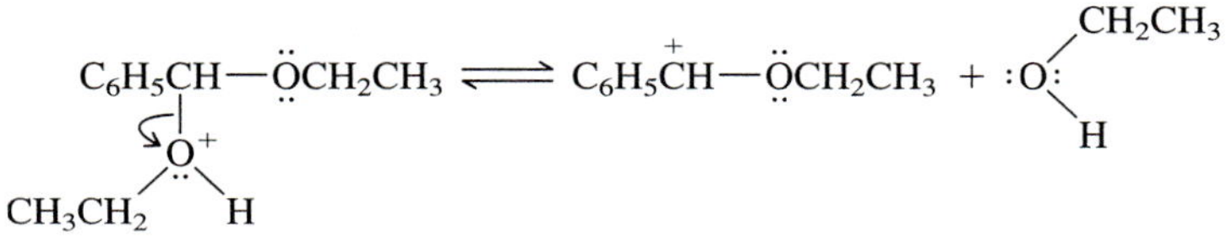

H

Step 4: $C_6H_5\overset{+}{C}-OCH_2CH_3 + :O: \rightleftharpoons C_6H_5C-OCH_2CH_3 + H-\overset{+}{O}:$

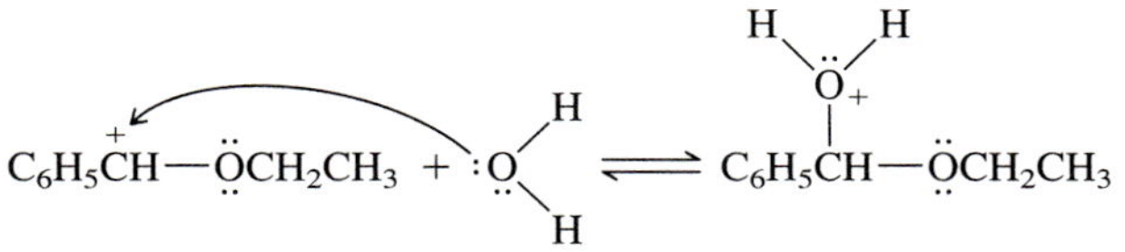

H H $HO:$ H

Step 5: $C_6H_5CH-OCH_2CH_3 + H-\overset{+}{O}: \rightleftharpoons C_6H_5CH-\overset{+}{O} \quad CH_2CH_3 + :O:$

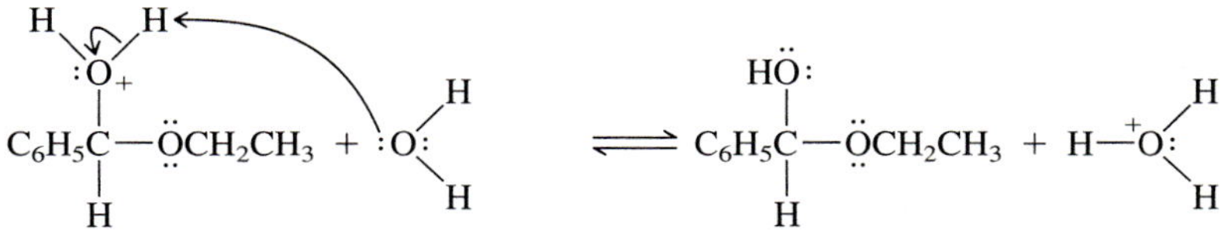

$:OH$ H $:OH$ H

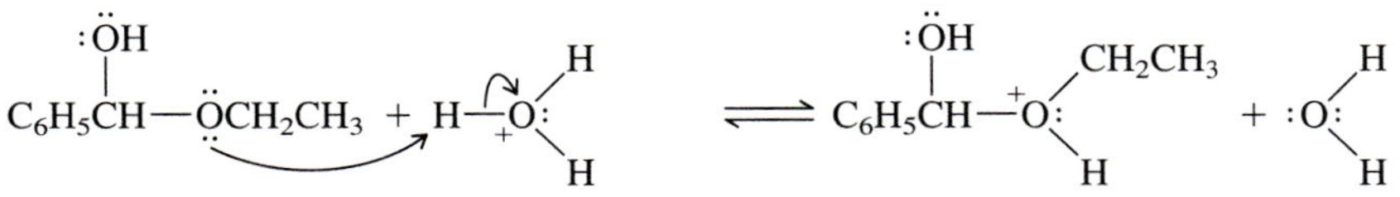

Step 6: $C_6H_5CH-\overset{+}{O}: \rightleftharpoons C_6H_5\overset{+}{C}H + :O:$

$:OH \quad CH_2CH_3$ $\overset{+}{O}-H \quad CH_2CH_3$

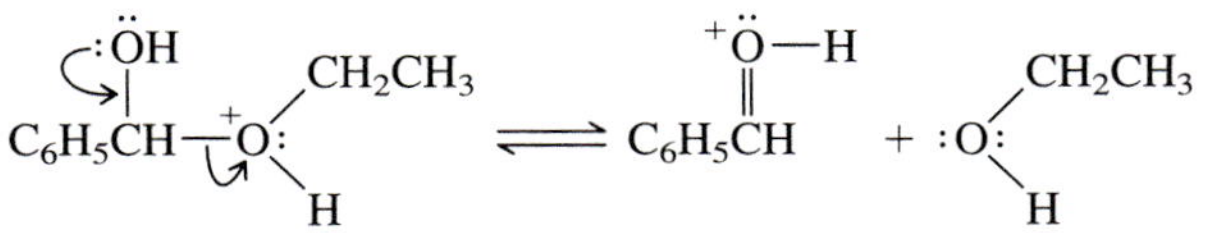

H H

Step 7: $C_6H_5\overset{+}{C}H + :O: \rightleftharpoons C_6H_5CH + H-\overset{+}{O}:$

$:O-H$ H $:O:$ H

H

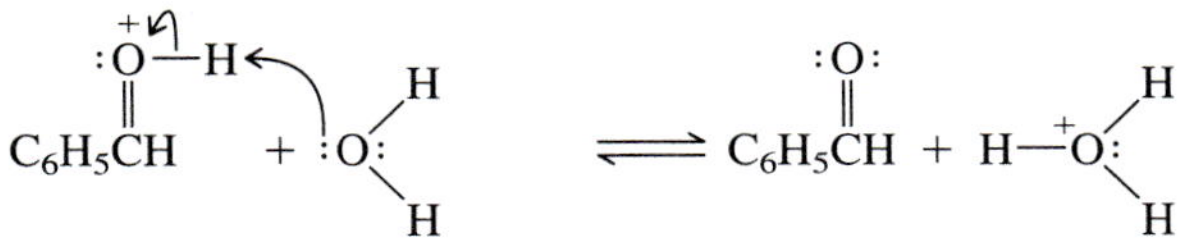

17.11

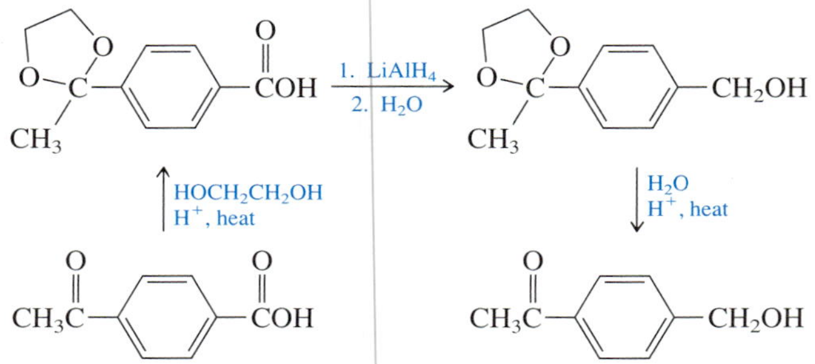

$$\xrightarrow[\text{2. } H_2O]{\text{1. } LiAlH_4}$$

$$\underset{H^+, \text{ heat}}{\overset{HOCH_2CH_2OH}{\uparrow}}$$

$$\underset{H^+, \text{ heat}}{\overset{H_2O}{\downarrow}}$$

17.12 (b) $\underset{\overset{|}{OH}}{C_6H_5CHNHCH_2CH_2CH_2CH_3} \longrightarrow C_6H_5CH{=}NCH_2CH_2CH_2CH_3$

(c)

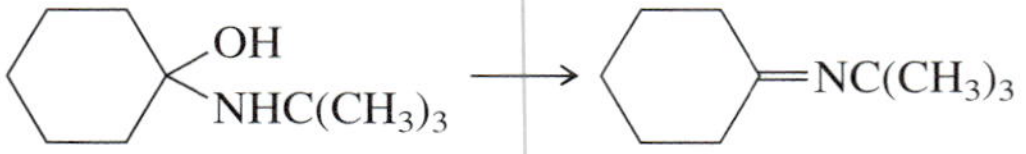

(d) 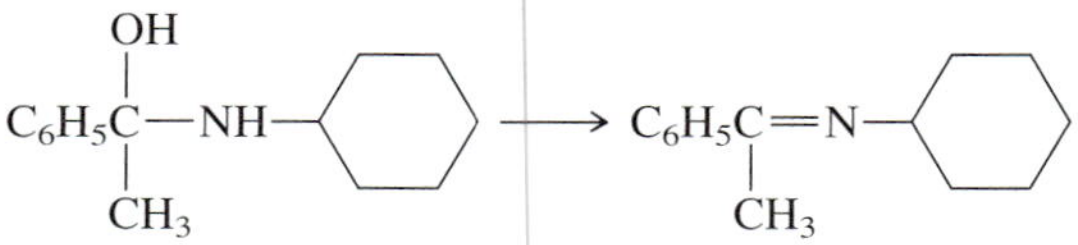 $\underset{\overset{|}{CH_3}}{\overset{\overset{OH}{|}}{C_6H_5C}{-}NH}{-}$ $\longrightarrow C_6H_5\underset{\overset{|}{CH_3}}{C}{=}N{-}$

17.13 (b) 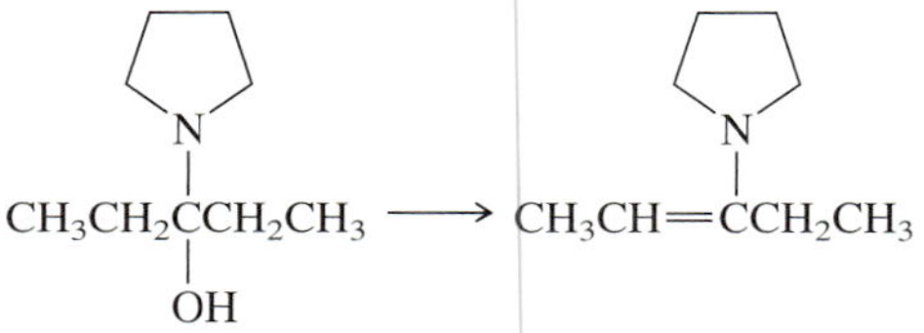 $\underset{\overset{|}{OH}}{CH_3CH_2CCH_2CH_3} \longrightarrow CH_3CH{=}CCH_2CH_3$

(c) 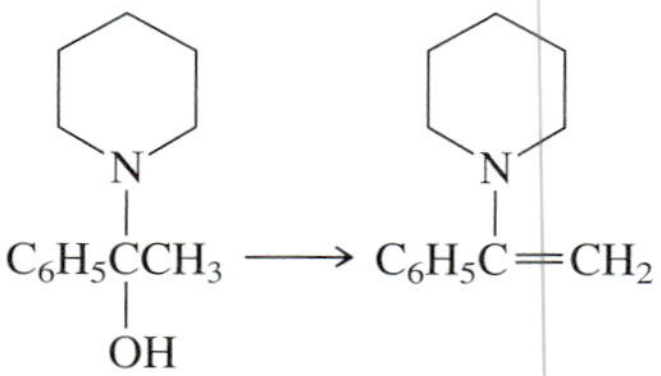 $\underset{\overset{|}{OH}}{C_6H_5CCH_3} \longrightarrow C_6H_5C{=}CH_2$

17.14 (b) $CH_3CH_2CH_2CH{=}CHCH{=}CH_2$ (c)

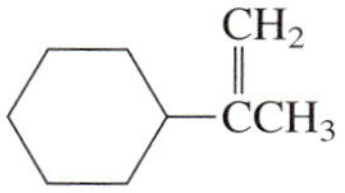

17.15 $(C_6H_5)_3P{=}CH_2$

17.16 (b) $CH_3CH_2CH_2\overset{\overset{O}{||}}{CH} + (C_6H_5)_3\overset{+}{P}{-}\overset{..}{\overset{..}{C}}H_2$ or $H\overset{\overset{O}{||}}{CH} + CH_3CH_2CH_2\overset{..}{C}H{-}\overset{+}{P}(C_6H_5)_3$

17.17 $\underset{\overset{|}{Br}}{CH_3CHCH_2CH_3} + (C_6H_5)_3P{:} \longrightarrow \underset{\overset{+}{\underset{|}{P(C_6H_5)_3}}}{CH_3CHCH_2CH_3}\ Br^-$

$\underset{\overset{+}{\underset{|}{P(C_6H_5)_3}}}{CH_3CHCH_2CH_3}\ Br^- \xrightarrow[\text{DMSO}]{NaCH_2\overset{\overset{O}{||}}{S}CH_3} \underset{\overset{+}{\underset{|}{P(C_6H_5)_3}}}{CH_3\overset{..}{C}CH_2CH_3}$

17.18 Diastereomers

17.19

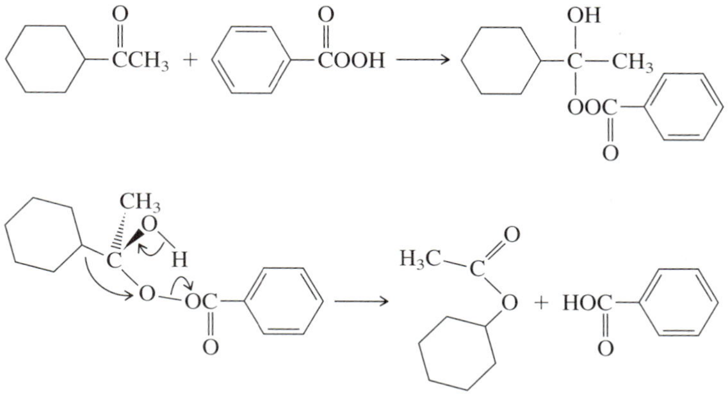

CHAPTER 18

18.1 (b) Zero; (c) five; (d) four

18.2 $ClCH_2CCH_2CH_3$ and CH_3CCHCH_3

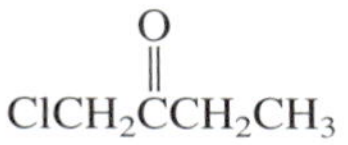

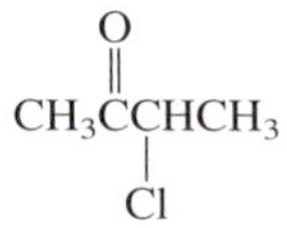

18.3 $H_2C{=}CCH_2CH_3 \xrightarrow{Cl_2} ClCH_2CCH_2CH_3$ $CH_3C{=}CHCH_3 \xrightarrow{Cl_2} CH_3CCHCH_3$

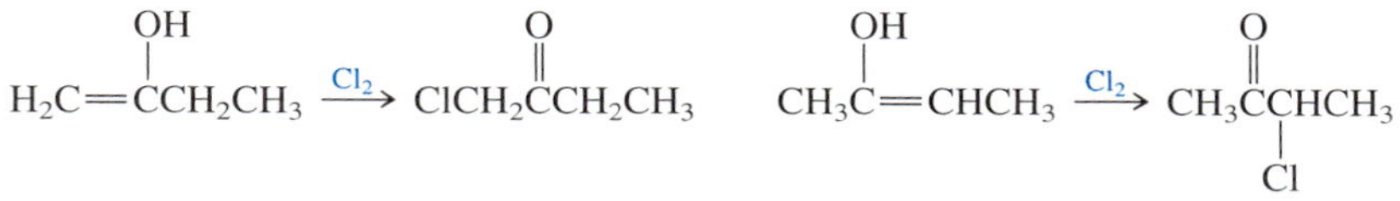

18.4

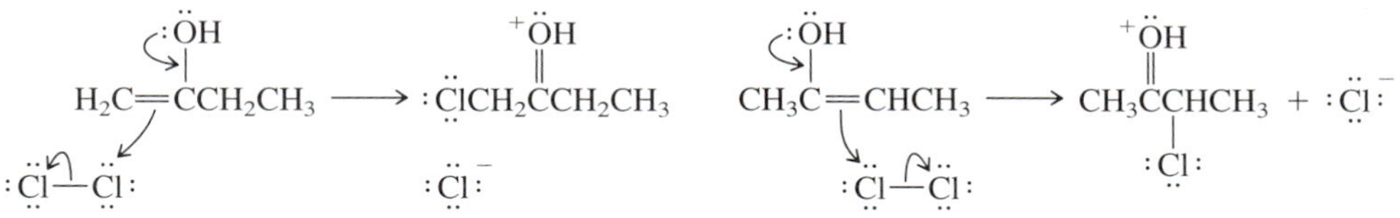

18.5 (b) 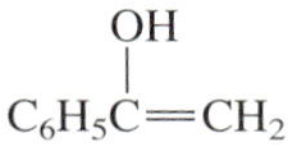 $C_6H_5C{=}CH_2$ (c)

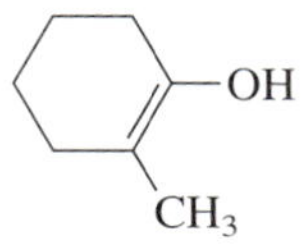

and

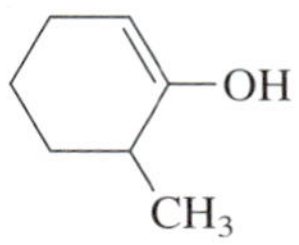

18.6 Conjugation is more important. 1,3-Cyclohexanedione exists mainly in its enol form in spite of the fact that intramolecular hydrogen bonding is impossible due to the distance between the carbonyl group and the enolic —OH group.

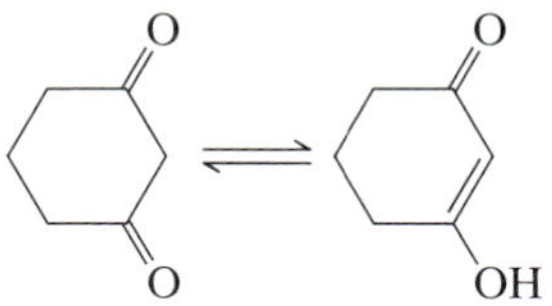

18.7 (b) 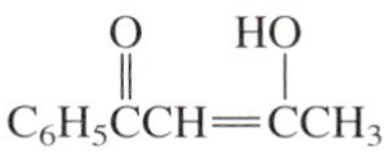 $C_6H_5CCH{=}CCH_3$ and $C_6H_5C{=}CHCCH_3$

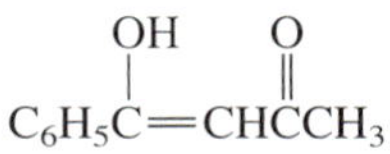

18.8 (b) 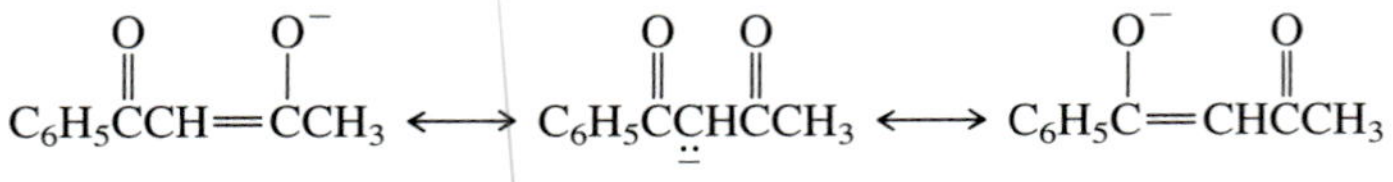 $C_6H_5\overset{O}{\underset{||}{C}}CH=\overset{O^-}{\underset{|}{C}}CH_3 \longleftrightarrow C_6H_5\overset{O}{\underset{||}{C}}\overset{O}{\underset{||}{C}}\ddot{C}HCCH_3 \longleftrightarrow C_6H_5\overset{O^-}{\underset{|}{C}}=CH\overset{O}{\underset{||}{C}}CH_3$

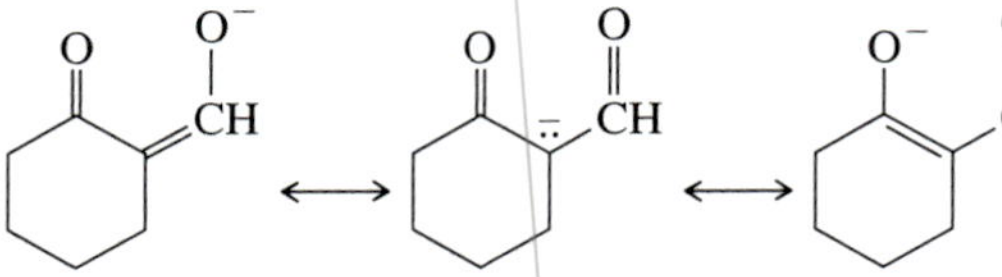

(c) (three cyclohexanone enolate resonance structures with —CH groups)

18.9 Hydrogen–deuterium exchange at α carbons via enolate:

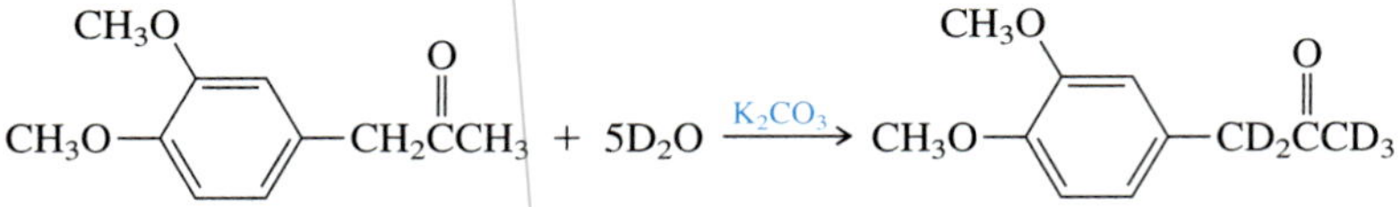

CH_3O— (dimethoxyphenyl)—$CH_2\overset{O}{\underset{||}{C}}CH_3$ + $5D_2O$ $\xrightarrow{K_2CO_3}$ CH_3O—(dimethoxyphenyl)—$CD_2\overset{O}{\underset{||}{C}}CD_3$

18.10 The product is chiral, but is formed as a racemic mixture because it arises from an achiral intermediate (the enol); it is therefore not optically active.

18.11 (b) 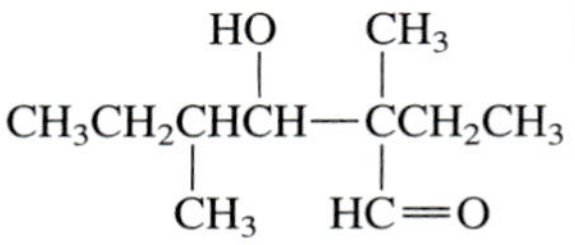 $CH_3CH_2\underset{\underset{CH_3}{|}}{CH}\underset{\underset{HC=O}{|}}{\overset{\overset{HO}{|}}{CH}}-\underset{\underset{HC=O}{|}}{\overset{\overset{CH_3}{|}}{C}}CH_2CH_3$ (c) $(CH_3)_2CHCH_2\underset{\underset{HC=O}{|}}{\overset{\overset{OH}{|}}{CH}}-CHCH(CH_3)_2$

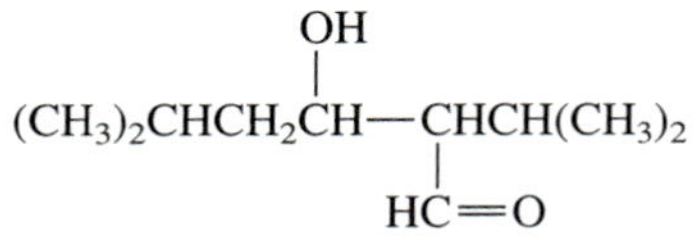

18.12 (b) 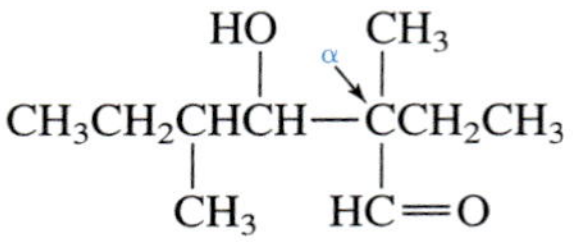 $CH_3CH_2\underset{\underset{CH_3}{|}}{CH}\overset{\overset{HO \quad {}^{\alpha}CH_3}{|\quad|}}{CH}-\underset{\underset{HC=O}{|}}{\overset{\overset{CH_3}{|}}{C}}CH_2CH_3$ (c) $(CH_3)_2CHCH_2CH=\underset{\underset{HC=O}{|}}{C}CH(CH_3)_2$

Cannot dehydrate; no protons on α-carbon atom

18.13 $CH_3CH_2CH_2\overset{O}{\underset{||}{C}}H \xrightarrow[\text{H}_2\text{O, heat}]{\text{NaOH}} CH_3CH_2CH_2CH=\underset{\underset{CH_2CH_3}{|}}{\overset{\overset{O}{||}}{C}}CH \xrightarrow[\text{Pt}]{\text{H}_2} CH_3CH_2CH_2CH_2\underset{\underset{CH_2CH_3}{|}}{CH}CH_2OH$

18.14 $CH_3\overset{:\ddot{O}:}{\underset{||}{C}}H \quad + \quad H_2C=\overset{:\ddot{O}:}{\underset{||}{C}}H \longrightarrow CH_3\overset{:\ddot{O}:}{\underset{|}{C}}H-CH_2\overset{\ddot{O}:}{\underset{||}{C}}H$

$CH_3\overset{:\ddot{O}:}{\underset{||}{C}}H \; + \; CH_3CH=\overset{:\ddot{O}:}{\underset{|}{C}}H \longrightarrow CH_3\overset{:\ddot{O}:}{\underset{|}{C}}H-\underset{\underset{CH_3}{|}}{CH}\overset{\ddot{O}:}{\underset{||}{C}}H$

$CH_3CH_2\overset{:\ddot{O}:}{\underset{||}{C}}H \quad + \quad H_2C=\overset{:\ddot{O}:}{\underset{||}{C}}H \longrightarrow CH_3CH_2\overset{:\ddot{O}:}{\underset{|}{C}}H-CH_2\overset{\ddot{O}:}{\underset{||}{C}}H$

$CH_3CH_2\overset{:\ddot{O}:}{\underset{||}{C}}H \; + \; CH_3CH=\overset{:\ddot{O}:}{\underset{|}{C}}H \longrightarrow CH_3CH_2\overset{:\ddot{O}:}{\underset{|}{C}}H-\underset{\underset{CH_3}{|}}{CH}\overset{\ddot{O}:}{\underset{||}{C}}H$

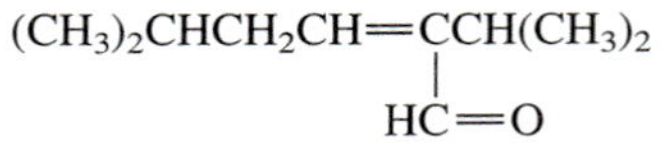

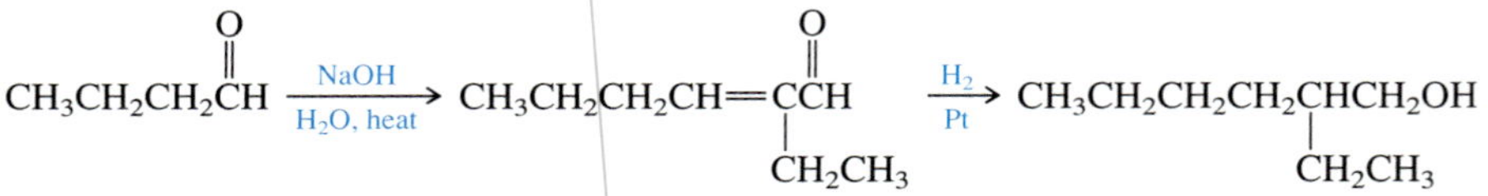

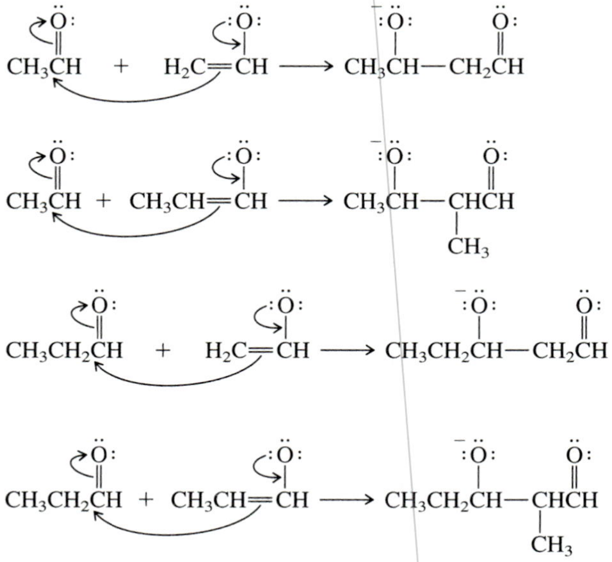

18.15 (b) $C_6H_5CH=CHCC(CH_3)_3$ (with O double bond on the carbonyl) (c) 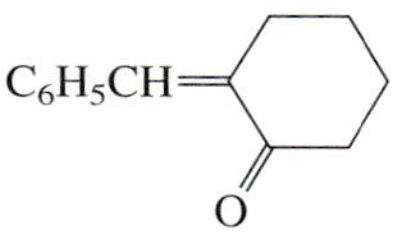 $C_6H_5CH=$ (cyclohexanone ring with O)

18.16 $CH_3CCH_2CCH_3$ (with O above first carbonyl and CH_2 below second carbon as $=CH_2$)

18.17 Acrolein ($H_2C=CHCH=O$) undergoes conjugate addition with sodium azide in aqueous solution to give $N_3CH_2CH_2CH=O$. Propanal is not an α,β-unsaturated carbonyl compound and cannot undergo conjugate addition.

18.18 $C_6H_5CH_2CCHC_6H_5$ with $CH_2CH_2CCH_3$ substituent (carbonyl groups marked with O)   and   (cyclohexanone structure with H, C_6H_5, H_3C, HO, and C_6H_5 substituents)

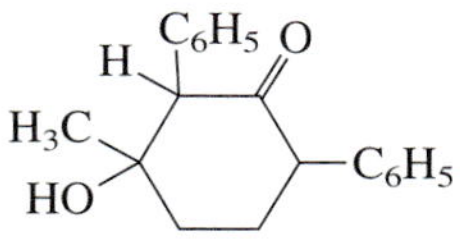

18.19 $CH_3CH_2CH_2CH_2CH=CHCCH_3 + LiCu(CH_3)_2$ (carbonyl with O)

CHAPTER 19

19.1 (b) (E)-2-butenoic acid; (c) ethanedioic acid; (d) p-methylbenzoic acid or 4-methylbenzoic acid.

19.2 The negative charge in CH_3COO^- cannot be delocalized into the carbonyl group.

19.3 (b) $CH_3CO_2H + (CH_3)_3CO^- \rightleftharpoons CH_3CO_2^- + (CH_3)_3COH$
(The position of equilibrium lies to the right.)
    (c) $CH_3CO_2H + Br^- \rightleftharpoons CH_3CO_2^- + HBr$
(The position of equilibrium lies to the left.)
    (d) $CH_3CO_2H + HC\equiv C:^- \rightleftharpoons CH_3CO_2^- + HC\equiv CH$
(The position of equilibrium lies to the right.)
    (e) $CH_3CO_2H + NO_3^- \rightleftharpoons CH_3CO_2^- + HNO_3$
(The position of equilibrium lies to the left.)
    (f) $CH_3CO_2H + H_2N^- \rightleftharpoons CH_3CO_2^- + NH_3$
(The position of equilibrium lies to the right.)

19.4 (b) CH_3CHCO_2H with OH substituent   (c) CH_3CCO_2H (with O)   (d) $CH_3SCH_2CO_2H$ (with two O)

19.5 $HC\equiv CCO_2H$

19.6 The "true K_1" for carbonic acid is 1.4×10^{-4}.

19.7 (b) The conversion proceeding by way of the nitrile is satisfactory.

$$HOCH_2CH_2Cl \xrightarrow{\text{NaCN}} HOCH_2CH_2CN \xrightarrow{\text{hydrolysis}} HOCH_2CH_2CO_2H$$

Because 2-chloroethanol has a proton bonded to oxygen, it is not an appropriate substrate for conversion to a stable Grignard reagent.

(c) The procedure involving a Grignard reagent is satisfactory.

$$(CH_3)_3CCl \xrightarrow{Mg} (CH_3)_3CMgCl \xrightarrow[\text{2. } H_3O^+]{\text{1. } CO_2} (CH_3)_3CCO_2H$$

The reaction of *tert*-butyl chloride with cyanide ion proceeds by elimination rather than substitution.

19.8 Water labeled with ^{18}O adds to benzoic acid to give the tetrahedral intermediate shown. This intermediate can lose unlabeled H_2O to give benzoic acid containing ^{18}O.

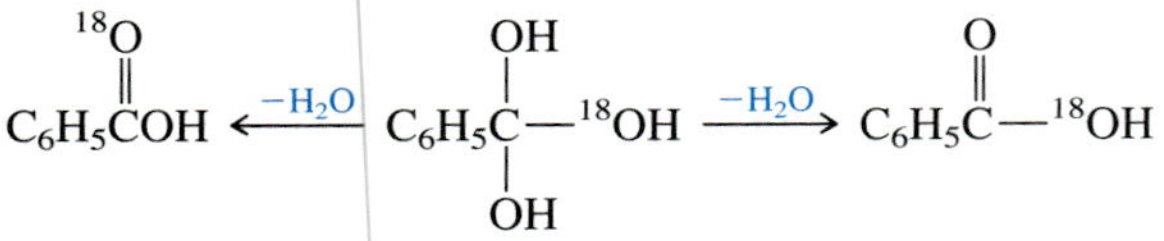

19.9 (b) $HOCH_2(CH_2)_{13}CO_2H$;　(c)

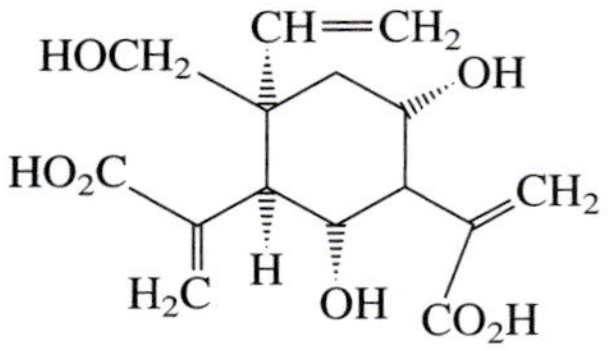

19.10 $CH_3(CH_2)_{15}CH_2CO_2H \xrightarrow[PCl_3]{Br_2} CH_3(CH_2)_{15}\underset{Br}{CH}CO_2H \xrightarrow[\text{acetone}]{NaI} CH_3(CH_2)_{15}\underset{I}{CH}CO_2H$

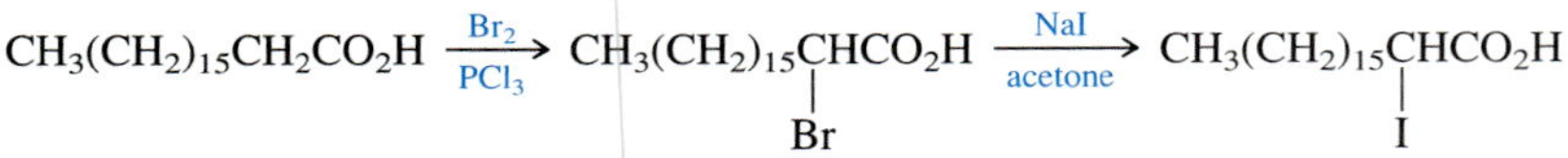

19.11 (b) $CH_3(CH_2)_6CH_2CO_2H$　via

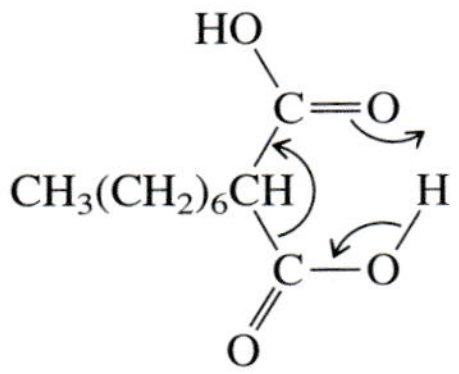

(c) $C_6H_5\underset{CH_3}{CH}CO_2H$　via

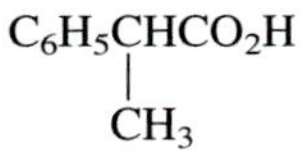

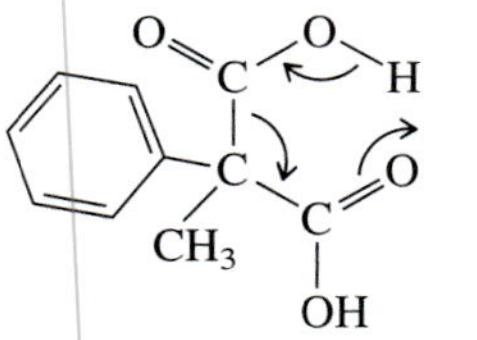

19.12 (b)

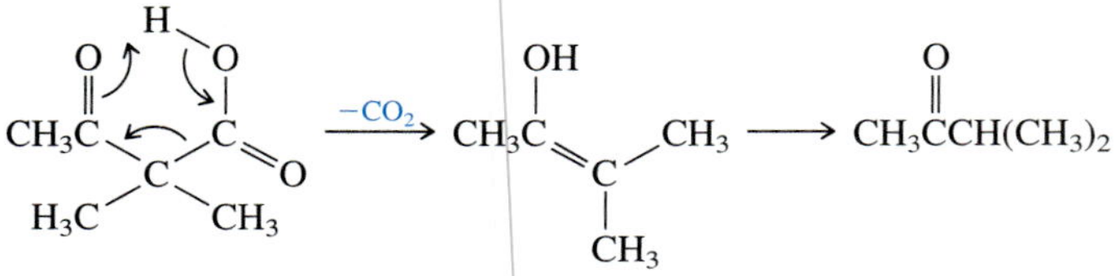

CHAPTER 20

20.1 (b)

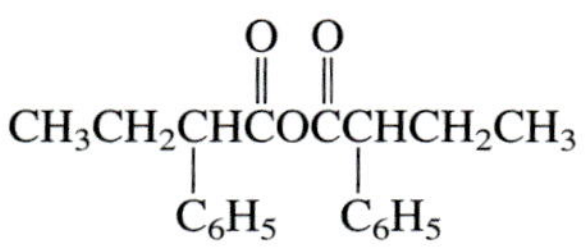

(c) $CH_3CH_2\underset{C_6H_5}{CH}\overset{O}{\overset{\|}{C}}OCH_2CH_2CH_2CH_3$

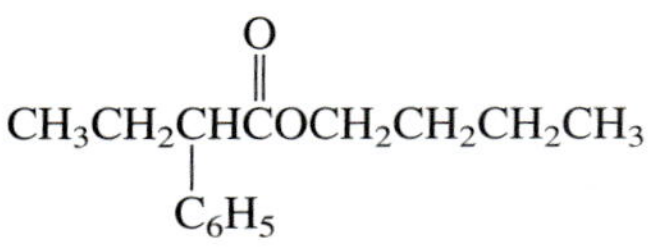

(d) $CH_3CH_2CH_2\overset{\overset{O}{\|}}{C}OCH_2\underset{\underset{C_6H_5}{|}}{C}HCH_2CH_3$ (e) $CH_3CH_2\underset{\underset{C_6H_5}{|}}{C}H\overset{\overset{O}{\|}}{C}NH_2$

(f) $CH_3CH_2\underset{\underset{C_6H_5}{|}}{C}H\overset{\overset{O}{\|}}{C}NHCH_2CH_3$ (g) $CH_3CH_2\underset{\underset{C_6H_5}{|}}{C}HC\equiv N$

20.2 Rotation about the carbon–nitrogen bond is slow in amides. The methyl groups of *N,N*-dimethylformamide are nonequivalent because one is cis to oxygen, the other cis to hydrogen.

20.3 (b) $C_6H_5\overset{\overset{O}{\|}}{C}O\overset{\overset{O}{\|}}{C}C_6H_5$ (c) $C_6H_5\overset{\overset{O}{\|}}{C}OCH_2CH_3$ (d) $C_6H_5\overset{\overset{O}{\|}}{C}NHCH_3$

(e) $C_6H_5\overset{\overset{O}{\|}}{C}N(CH_3)_2$ (f) $C_6H_5\overset{\overset{O}{\|}}{C}OH$

20.4 (b)
$$B:\curvearrowright H \quad C_6H_5\overset{\overset{O}{\|}}{\underset{\underset{Cl}{|}}{C}}O\overset{\overset{O}{\|}}{C}C_6H_5 \longrightarrow C_6H_5\overset{\overset{O}{\|}}{C}O\overset{\overset{O}{\|}}{C}C_6H_5 + \overset{+}{B}{-}H + Cl^-$$

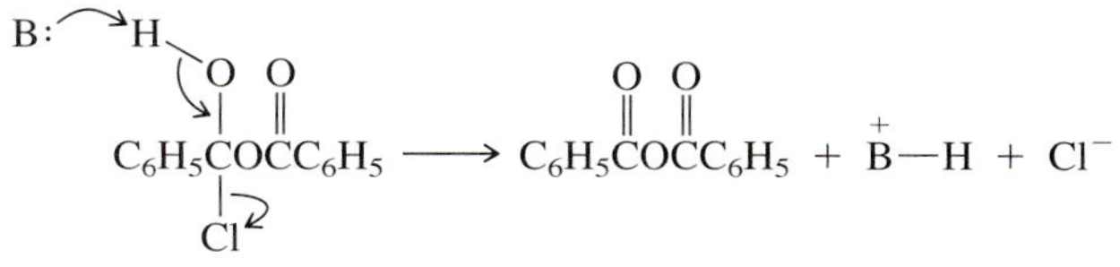

(c)
$$B:\curvearrowright H \quad C_6H_5\overset{\overset{O}{\|}}{\underset{\underset{Cl}{|}}{C}}OCH_2CH_3 \longrightarrow C_6H_5\overset{\overset{O}{\|}}{C}OCH_2CH_3 + \overset{+}{B}{-}H + Cl^-$$

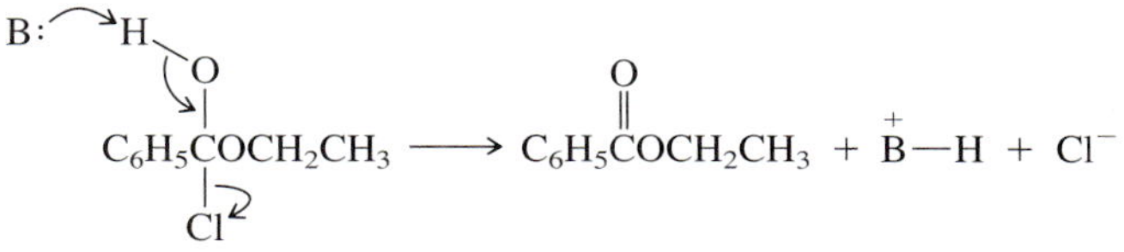

(d)
$$B:\curvearrowright H \quad C_6H_5\overset{\overset{O}{\|}}{\underset{\underset{Cl}{|}}{C}}NHCH_3 \longrightarrow C_6H_5\overset{\overset{O}{\|}}{C}NHCH_3 + \overset{+}{B}{-}H + Cl^-$$

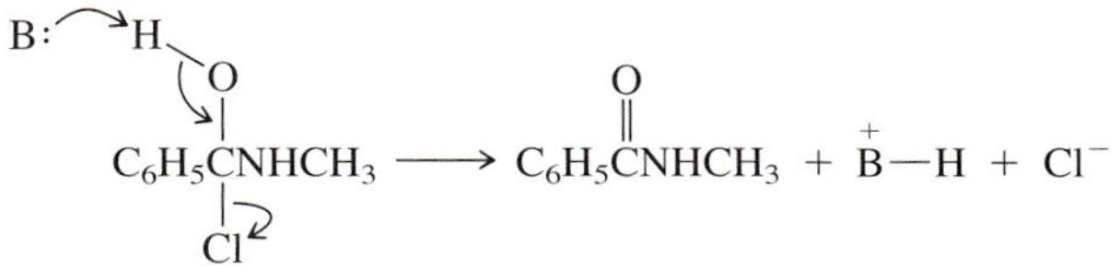

(e)
$$B:\curvearrowright H \quad C_6H_5\overset{\overset{O}{\|}}{\underset{\underset{Cl}{|}}{C}}N(CH_3)_2 \longrightarrow C_6H_5\overset{\overset{O}{\|}}{C}N(CH_3)_2 + \overset{+}{B}{-}H + Cl^-$$

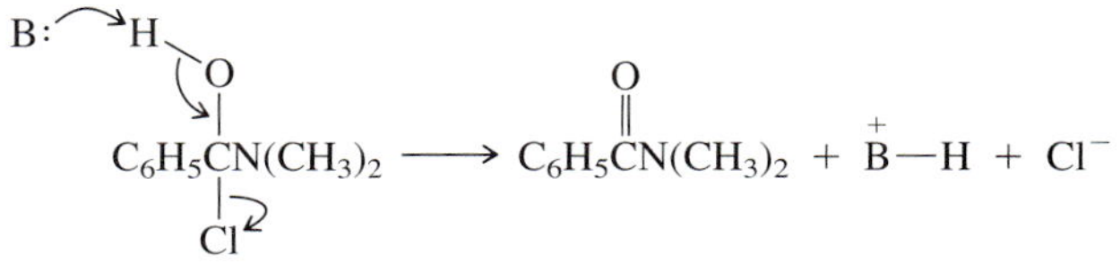

(f)
$$B:\curvearrowright H \quad C_6H_5\overset{\overset{O}{\|}}{\underset{\underset{Cl}{|}}{C}}OH \longrightarrow C_6H_5\overset{\overset{O}{\|}}{C}OH + \overset{+}{B}{-}H + Cl^-$$

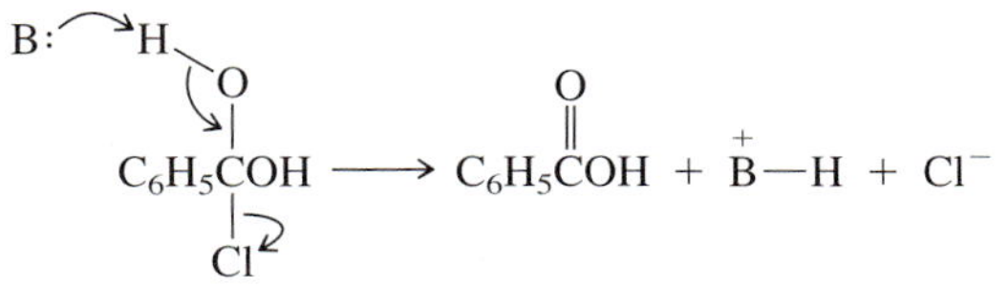

20.5 $C_6H_5\overset{\overset{O}{\|}}{C}Cl + H_2O \longrightarrow C_6H_5\overset{\overset{O}{\|}}{C}OH + HCl$

$C_6H_5\overset{\overset{O}{\|}}{C}Cl + C_6H_5\overset{\overset{O}{\|}}{C}OH \longrightarrow C_6H_5\overset{\overset{O}{\|}}{C}O\overset{\overset{O}{\|}}{C}C_6H_5 + HCl$

20.6 $CH_3\overset{+}{C}\!=\!\ddot{O}: \longleftrightarrow CH_3C\!\equiv\!\overset{+}{O}:$

20.7 (b) $\underset{\underset{\displaystyle O}{\|}}{CH_3C}NH_2 + CH_3CO_2^-\ \overset{+}{N}H_4$ (c)

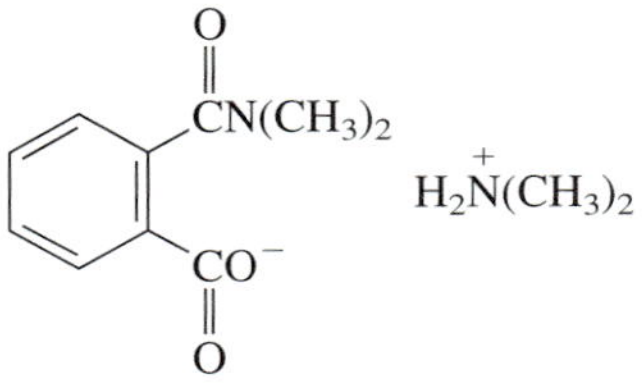

$H_2\overset{+}{N}(CH_3)_2$

(d)

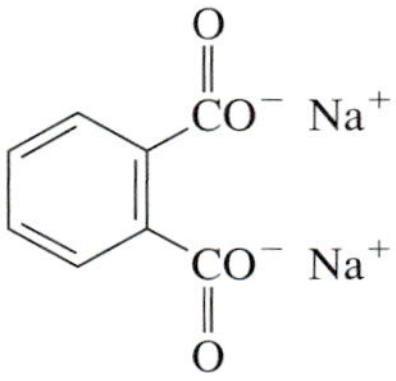

20.8 (b) $\underset{\underset{\displaystyle NH_2}{|}}{CH_3\overset{\overset{\displaystyle OH}{|}}{C}}\!-\!\underset{\displaystyle O}{\overset{\displaystyle \|}{O}}\!CCH_3$

(c)

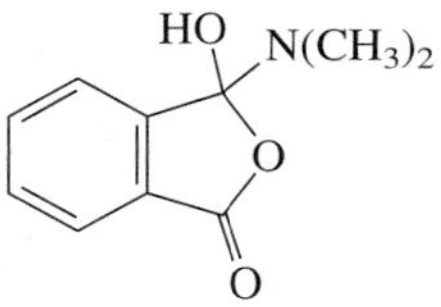

(d)

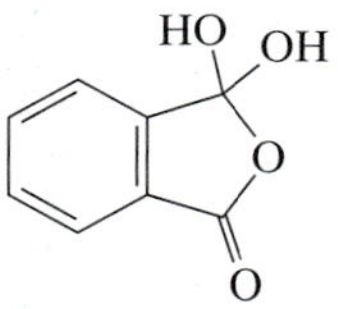

20.9 $HOCH_2\underset{\underset{\displaystyle OH}{|}}{CH}CH_2CH_2CH_2OH\ (C_5H_{12}O_3)$ and CH_3CO_2H

20.10 Step 1: Protonation of the carbonyl oxygen

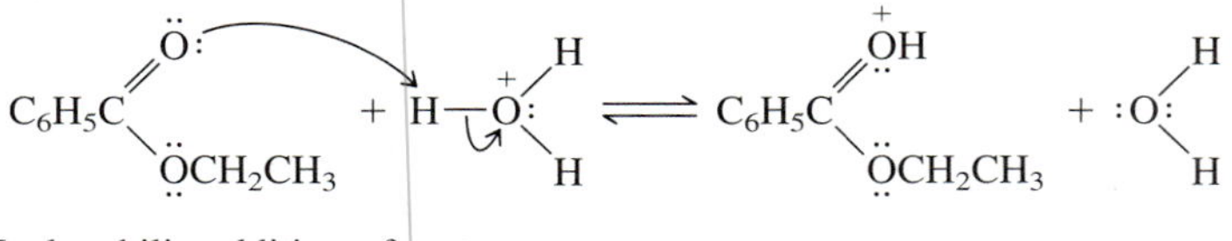

$C_6H_5C\underset{\ddot{O}CH_2CH_3}{\overset{\ddot{O}:}{\diagup\!\!\!\diagdown}} + H\!-\!\overset{+}{\underset{H}{\ddot{O}}}{-}H \rightleftharpoons C_6H_5C\underset{\ddot{O}CH_2CH_3}{\overset{\overset{+}{O}H}{\diagup\!\!\!\diagdown}} + :\overset{H}{\underset{H}{\ddot{O}}}$

Step 2: Nucleophilic addition of water

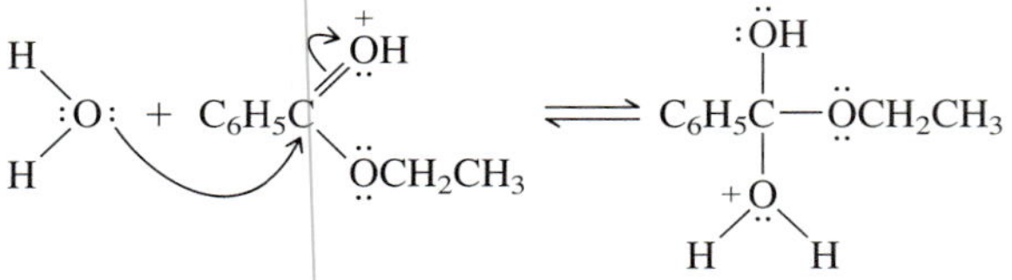

$\underset{H}{\overset{H}{\ddot{O}:}} + C_6H_5C\underset{\ddot{O}CH_2CH_3}{\overset{\overset{+}{O}H}{\diagup\!\!\!\diagdown}} \rightleftharpoons C_6H_5\underset{\underset{H\ \ H}{\overset{+}{O}}}{\overset{:\ddot{O}H}{C}}\!-\!\ddot{O}CH_2CH_3$

Step 3: Deprotonation of oxonium ion to give neutral form of tetrahedral intermediate

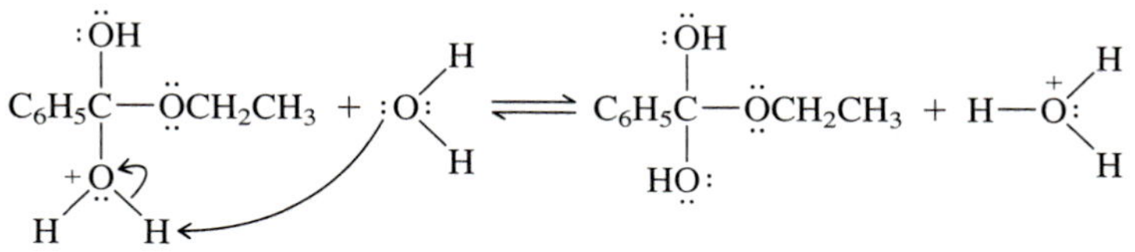

Step 4: Protonation of ethoxy oxygen

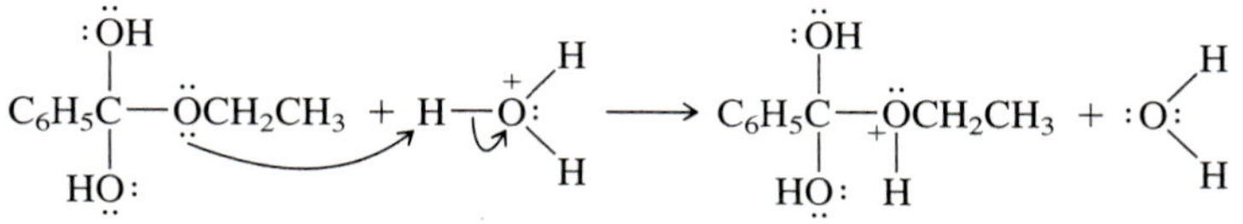

Step 5: Dissociation of protonated form of tetrahedral intermediate

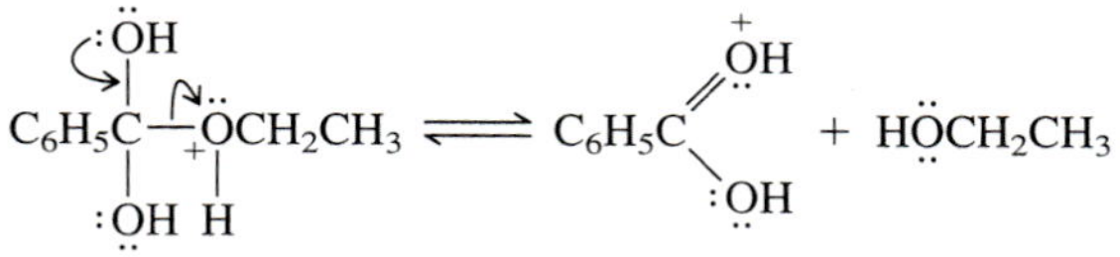

Step 6: Deprotonation of protonated form of benzoic acid

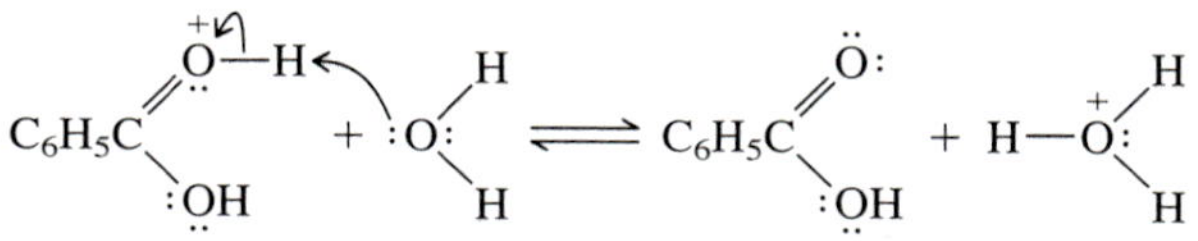

20.11 The carbonyl oxygen of the lactone became labeled with ^{18}O.

20.12 $CH_3(CH_2)_{12}CO$ ⸺ $OC(CH_2)_{12}CH_3$

$OC(CH_2)_{12}CH_3$

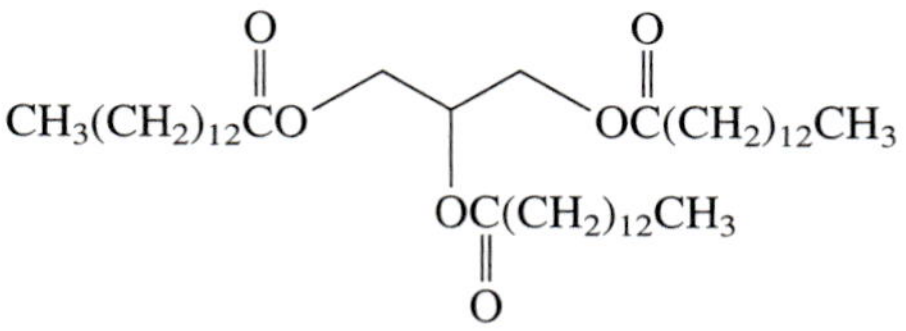

20.13 The isotopic label appeared in the acetate ion.

20.14 Step 1: Nucleophilic addition of hydroxide ion to the carbonyl group

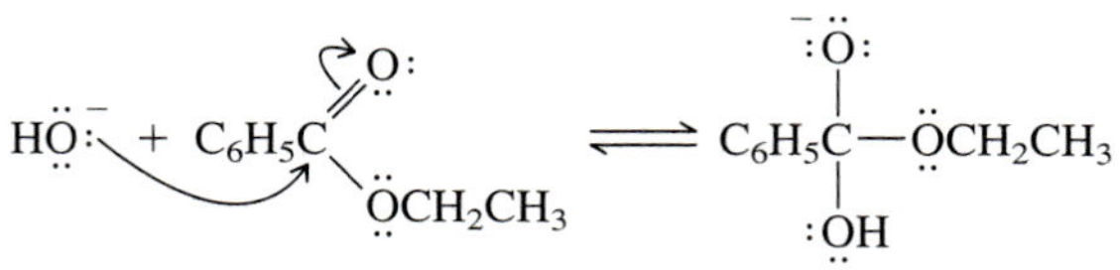

Step 2: Proton transfer from water to give neutral form of tetrahedral intermediate

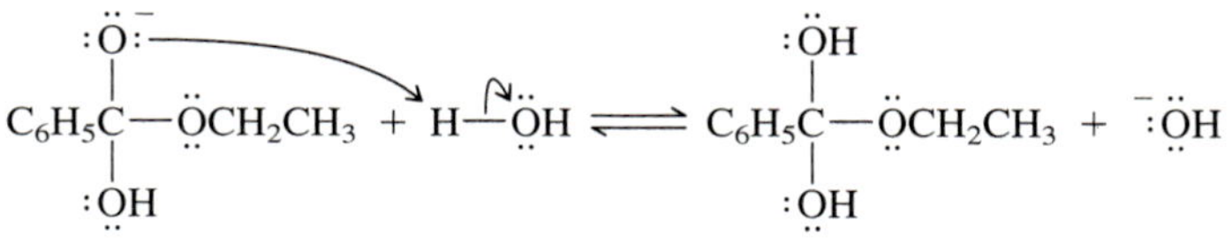

Step 3: Hydroxide ion-promoted dissociation of tetrahedral intermediate

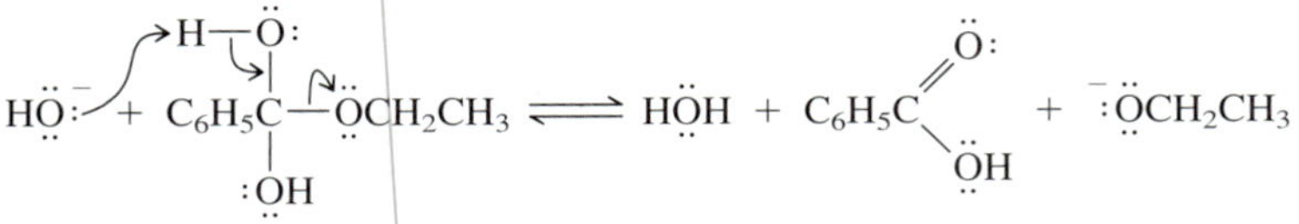

$$HO^{-} + C_6H_5C \overset{\underset{|}{:OH}}{\underset{}{\overset{}{}}} OCH_2CH_3 \;\rightleftharpoons\; H\ddot{O}H + C_6H_5C \overset{\ddot{O}:}{\underset{\ddot{O}H}{}} + \; {^{-}}\!\ddot{O}CH_2CH_3$$

Step 4: Proton transfer steps to yield ethanol and benzoate ion

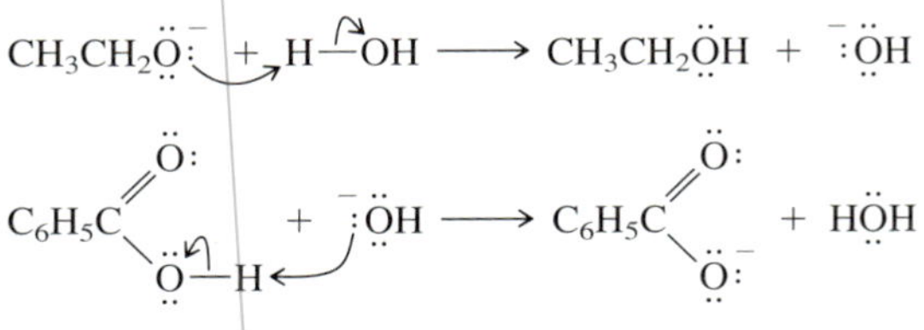

$$CH_3CH_2\ddot{O}:^{-} + H\!-\!OH \longrightarrow CH_3CH_2\ddot{O}H + \; ^{-}\!:\ddot{O}H$$

$$C_6H_5C \overset{\ddot{O}:}{\underset{\ddot{O}\!-\!H}{}} + \; ^{-}:\ddot{O}H \longrightarrow C_6H_5C \overset{\ddot{O}:}{\underset{\ddot{O}:^{-}}{}} + \; H\ddot{O}H$$

20.15
$$CH_3NH\overset{O}{\overset{\|}{C}}CH_2CH_2\overset{}{\underset{OH}{C}}HCH_3$$

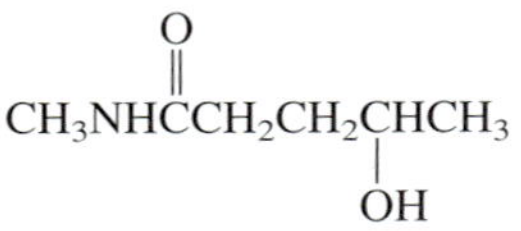

20.16
$$CH_3\overset{OH}{\underset{OCH_3}{C}}SCH_2CH_2OC_6H_5$$

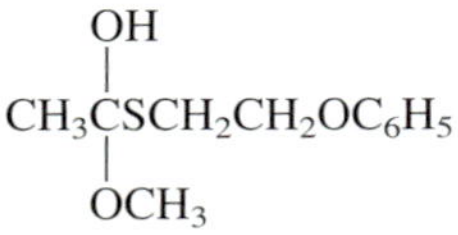

20.17 (b) 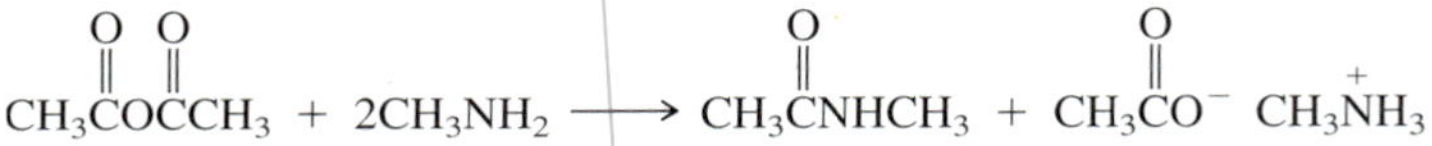
$$CH_3\overset{O}{\overset{\|}{C}}O\overset{O}{\overset{\|}{C}}CH_3 + 2CH_3NH_2 \longrightarrow CH_3\overset{O}{\overset{\|}{C}}NHCH_3 + CH_3\overset{O}{\overset{\|}{C}}O^{-}\; CH_3\overset{+}{N}H_3$$

(c)
$$H\overset{O}{\overset{\|}{C}}OCH_3 + HN(CH_3)_2 \longrightarrow H\overset{O}{\overset{\|}{C}}N(CH_3)_2 + CH_3OH$$

20.18

A benzene ring bearing ortho substituents: $\overset{O}{\overset{\|}{C}}NH_2$ and $\overset{}{\underset{O}{\overset{\|}{C}}}O^{-}\; \overset{+}{N}H_4$

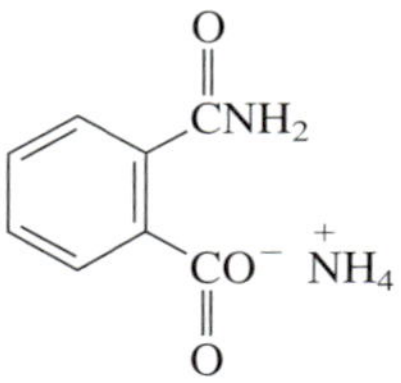

20.19 Step 1: Protonation of the carbonyl oxygen

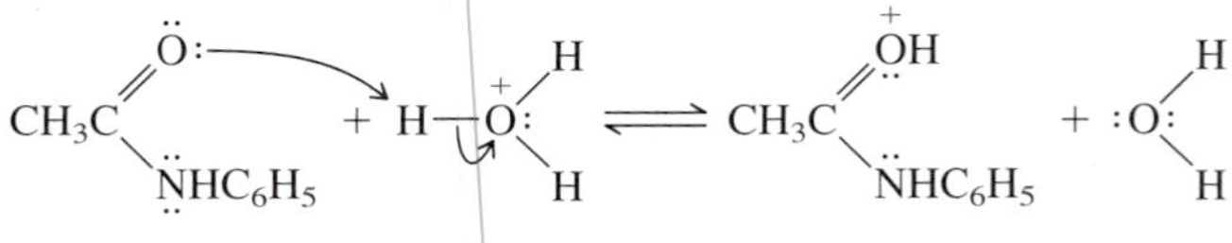

$$CH_3C \overset{\ddot{O}:}{\underset{\ddot{N}HC_6H_5}{}} + H\!-\!\overset{+}{\ddot{O}}\overset{H}{\underset{H}{}} \;\rightleftharpoons\; CH_3C \overset{\overset{+}{O}H}{\underset{\ddot{N}HC_6H_5}{}} + \; :\ddot{O}\overset{H}{\underset{H}{}}$$

Step 2: Nucleophilic addition of water

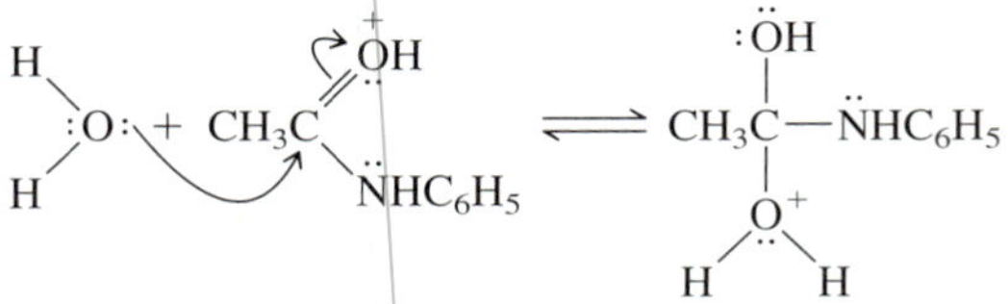

$$\overset{H}{\underset{H}{}}\ddot{O}: + CH_3C \overset{\overset{+}{O}H}{\underset{\ddot{N}HC_6H_5}{}} \;\rightleftharpoons\; CH_3\overset{:\ddot{O}H}{\underset{\overset{+}{O}}{\overset{|}{C}}}\!-\!\ddot{N}HC_6H_5 \; \underset{H \quad H}{}$$

Step 3: Deprotonation of oxonium ion to give neutral form of tetrahedral intermediate

$$CH_3\overset{\overset{\displaystyle :\overset{..}{O}H}{|}}{\underset{\underset{\displaystyle H \quad H}{+\overset{..}{O}}}{C}}\!-\!\overset{..}{N}HC_6H_5 \ + \ :\overset{..}{\underset{\underset{\displaystyle H}{|}}{O}}: \ \rightleftharpoons \ CH_3\overset{\overset{\displaystyle :\overset{..}{O}H}{|}}{\underset{\underset{\displaystyle :\overset{..}{O}H}{|}}{C}}\!-\!\overset{..}{N}HC_6H_5 \ + \ H\!-\!\overset{+}{\underset{\underset{\displaystyle H}{|}}{O}}\diagdown^H$$

Step 4: Protonation of amino group of tetrahedral intermediate

$$CH_3\overset{\overset{\displaystyle :\overset{..}{O}H}{|}}{\underset{\underset{\displaystyle :\overset{..}{O}H}{|}}{C}}\!-\!\overset{..}{N}HC_6H_5 \ + \ H\!-\!\overset{+}{\underset{\underset{\displaystyle H}{|}}{O}}: \ \rightleftharpoons \ CH_3\overset{\overset{\displaystyle :\overset{..}{O}H \ \ H}{|}}{\underset{\underset{\displaystyle :\overset{..}{O}H \ \ H}{|}}{C}}\!-\!\overset{+}{N}C_6H_5 \ + \ :\overset{..}{\underset{\underset{\displaystyle H}{|}}{O}}\diagdown^H$$

Step 5: Dissociation of N-protonated form of tetrahedral intermediate

$$CH_3\overset{\overset{\displaystyle :\overset{..}{O}H \ \ H}{|}}{\underset{\underset{\displaystyle :\overset{..}{O}H \ \ H}{|}}{C}}\!-\!\overset{+}{N}C_6H_5 \ \rightleftharpoons \ CH_3C\overset{\overset{\displaystyle +}{\diagup}OH}{\diagdown_{\overset{..}{O}H}} \ + \ H_2\overset{..}{N}C_6H_5$$

Step 6: Proton-transfer processes

$$\overset{H}{\underset{H}{\diagdown}}\overset{+}{\overset{..}{O}}\!-\!H \ + \ H_2\overset{..}{N}C_6H_5 \ \rightleftharpoons \ \overset{H}{\underset{H}{\diagdown}}\overset{..}{O}: \ + \ H_3\overset{+}{N}C_6H_5$$

$$CH_3C\overset{\overset{\displaystyle +}{\diagup}\overset{..}{O}\!-\!H}{\diagdown_{\overset{..}{O}H}} \ + \ :\overset{..}{\underset{\underset{\displaystyle H}{|}}{O}}\diagup^H \ \rightleftharpoons \ CH_3C\overset{\diagup:O:}{\diagdown_{\overset{..}{O}H}} \ + \ H\!-\!\overset{+}{\overset{..}{O}}\diagdown_H^H$$

20.20 Step 1: Nucleophilic addition of hydroxide ion to the carbonyl group

$$H\overset{..}{\underset{..}{O}}:^- \ + \ H\overset{\overset{\displaystyle :\overset{..}{O}}{\|}}{C}\overset{..}{N}(CH_3)_2 \ \longrightarrow \ H\overset{\overset{\displaystyle :\overset{..}{O}:^-}{|}}{\underset{\underset{\displaystyle :\overset{..}{O}H}{|}}{C}}\!-\!\overset{..}{N}(CH_3)_2$$

Step 2: Proton transfer to give neutral form of tetrahedral intermediate

$$H\overset{\overset{\displaystyle :\overset{..}{O}:^-}{|}}{\underset{\underset{\displaystyle :\overset{..}{O}H}{|}}{C}}\!-\!\overset{..}{N}(CH_3)_2 \ + \ H\!-\!\overset{..}{O}H \ \longrightarrow \ H\overset{\overset{\displaystyle :\overset{..}{O}H}{|}}{\underset{\underset{\displaystyle :\overset{..}{O}H}{|}}{C}}\!-\!\overset{..}{N}(CH_3)_2 \ + \ ^-:\overset{..}{O}H$$

Step 3: Proton transfer from water to nitrogen of tetrahedral intermediate

$$H\overset{\overset{\displaystyle :\overset{..}{O}H}{|}}{\underset{\underset{\displaystyle :\overset{..}{O}H}{|}}{C}}\!-\!\overset{..}{N}(CH_3)_2 \ + \ H\!-\!\overset{..}{O}H \ \rightleftharpoons \ H\overset{\overset{\displaystyle :\overset{..}{O}H}{|}}{\underset{\underset{\displaystyle :\overset{..}{O}H}{|}}{C}}\!-\!\overset{+}{N}H(CH_3)_2 \ + \ ^-:\overset{..}{O}H$$

Step 4: Dissociation of N-protonated form of tetrahedral intermediate

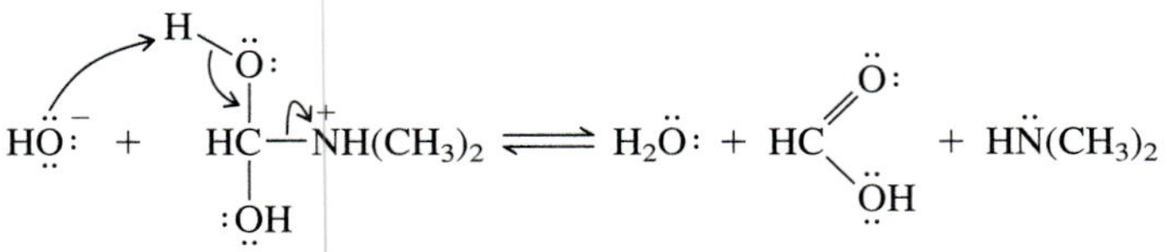

Step 5: Irreversible formation of formate ion

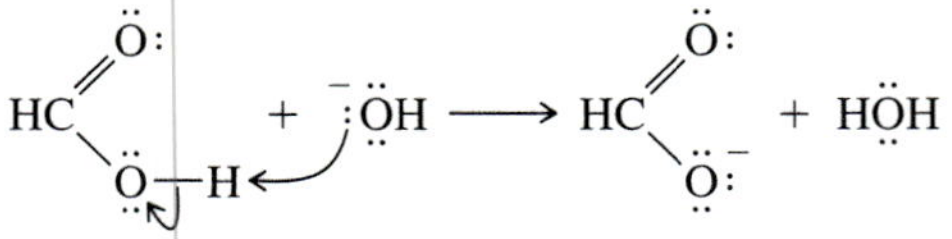

20.21 $CH_3CH_2OH \xrightarrow[H_2SO_4, \text{ heat}]{Na_2Cr_2O_7, H_2O} CH_3COH \xrightarrow[\text{2. } NH_3]{\text{1. } SOCl_2} CH_3CNH_2 \xrightarrow{P_4O_{10}} CH_3C{\equiv}N$

$CH_3CH_2OH \xrightarrow[\text{or HBr}]{PBr_3} CH_3CH_2Br \xrightarrow{NaCN} CH_3CH_2CN$

20.22 In acid, the nitrile is protonated on nitrogen. Nucleophilic addition of water yields an imino acid.

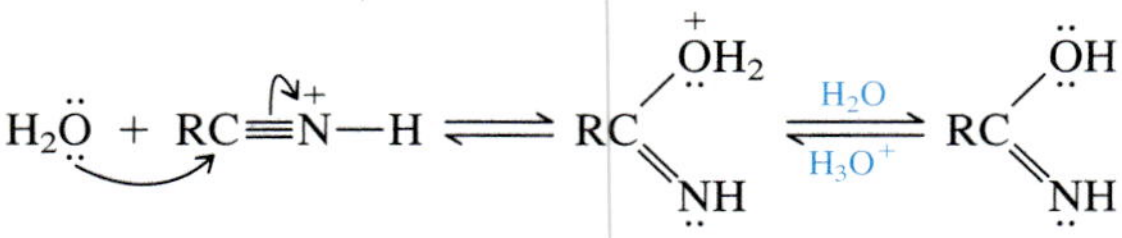

A series of proton transfers converts the imino acid to an amide.

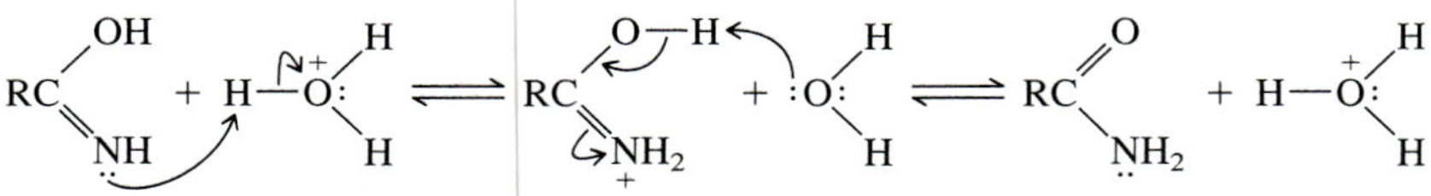

20.23 $CH_3CH_2CN + C_6H_5MgBr \xrightarrow[\text{2. } H_3O^+, \text{ heat}]{\text{1. diethyl ether}} C_6H_5CCH_2CH_3$

The imine intermediate is $C_6H_5\overset{NH}{\overset{\|}{C}}CH_2CH_3$.

CHAPTER 21

21.1 Ethyl benzoate cannot undergo the Claisen condensation.

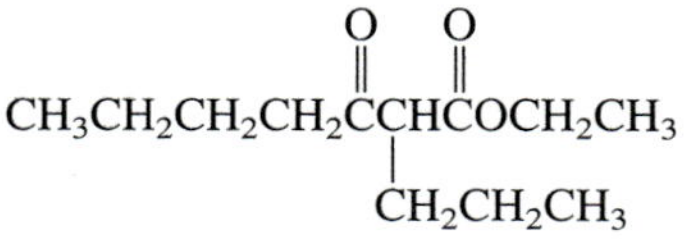

$$CH_3CH_2CH_2CH_2\overset{O}{\overset{\|}{C}}\overset{O}{\overset{\|}{C}}HCOCH_2CH_3$$
$$\underset{CH_2CH_2CH_3}{|}$$

Claisen condensation product of
ethyl pentanoate

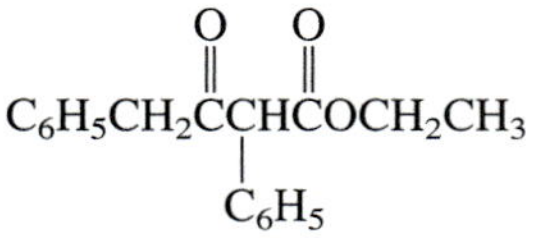

$$C_6H_5CH_2\overset{O}{\overset{\|}{C}}\overset{O}{\overset{\|}{C}}HCOCH_2CH_3$$
$$\underset{C_6H_5}{|}$$

Claisen condensation product of
ethyl phenylacetate

21.2 (b) (c)

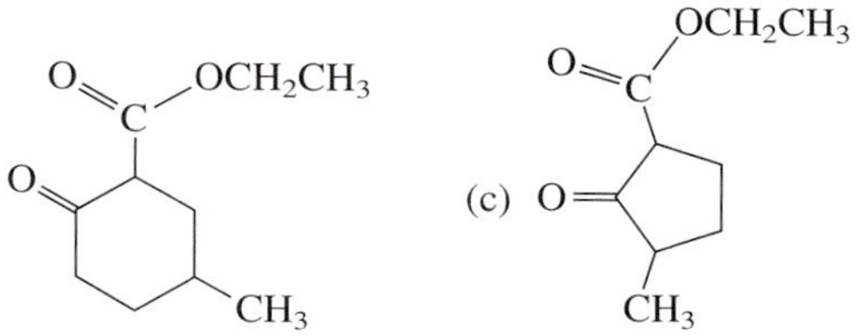

21.3 (b) $C_6H_5CHCCOCH_2CH_3$ with two C=O, and $COCH_2CH_3$ group below

(c) C_6H_5CHCH with $COCH_2CH_3$ group below

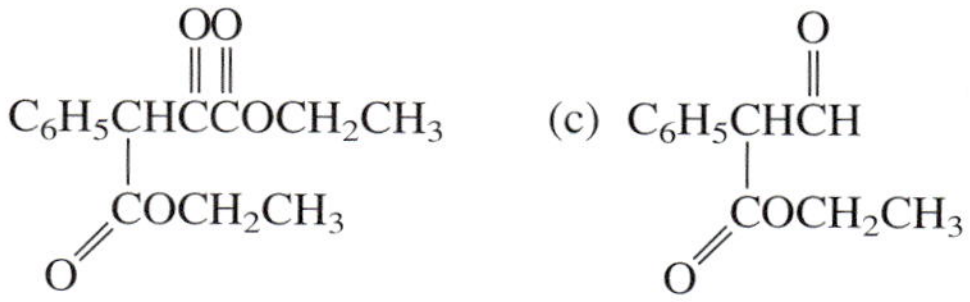

21.4

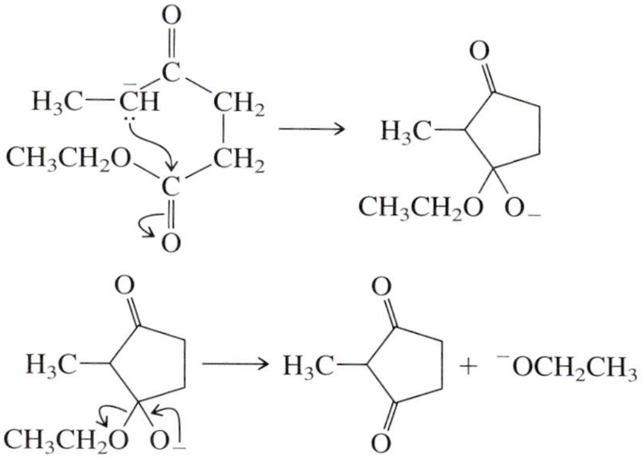

21.5 $CH_3CH_2OCCH_2CH_2CH_2CH_2COCH_2CH_3$ $\xrightarrow[\text{2. } H_3O^+]{\text{1. NaOCH}_2CH_3}$ $\xrightarrow[\substack{\text{2. } H_3O^+ \\ \text{3. heat}}]{\text{1. } HO^-, H_2O}$

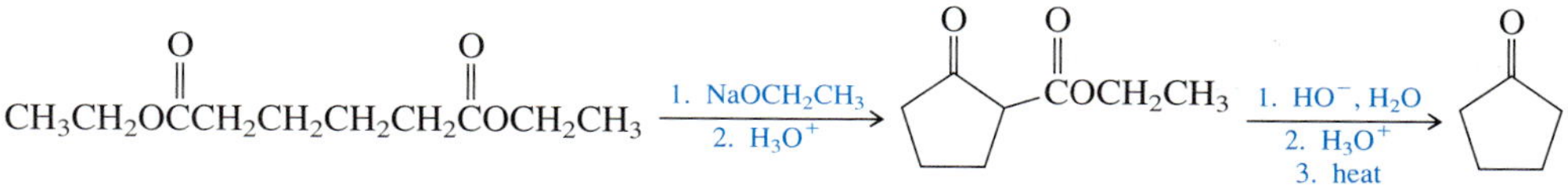

21.6 (b) $C_6H_5CH_2Br + CH_3CCH_2COCH_2CH_3$ $\xrightarrow[\substack{\text{2. } HO^-, H_2O \\ \text{3. } H_3O^+ \\ \text{4. heat}}]{\text{1. NaOCH}_2CH_3}$ $C_6H_5CH_2CH_2CCH_3$

(c) $H_2C=CHCH_2Br + CH_3CCH_2COCH_2CH_3$ $\xrightarrow[\substack{\text{2. } HO^-, H_2O \\ \text{3. } H_3O^+ \\ \text{4. heat}}]{\text{1. NaOCH}_2CH_3}$ $H_2C=CHCH_2CH_2CCH_3$

21.7 (b) $CH_3(CH_2)_5CH_2Br + CH_2(COOCH_2CH_3)_2$ $\xrightarrow[\text{ethanol}]{\text{NaOCH}_2CH_3}$ $CH_3(CH_2)_5CH_2CH(COOCH_2CH_3)_2$

$\downarrow \substack{\text{1. } HO^-, H_2O \\ \text{2. } H_3O^+ \\ \text{3. heat}}$

$CH_3(CH_2)_5CH_2CH_2COH$

(c) $CH_3CH_2CHCH_2Br$ + $CH_2(COOCH_2CH_3)_2$ $\xrightarrow[\text{ethanol}]{NaOCH_2CH_3}$ $CH_3CH_2CHCH_2CH(COOCH_2CH_3)_2$
　　　　　|　　　　　　　　　　　　　　　　　　　　　　　　　　　　　　　　　　|
　　　　CH_3　　　　　　　　　　　　　　　　　　　　　　　　　　　　　　CH_3

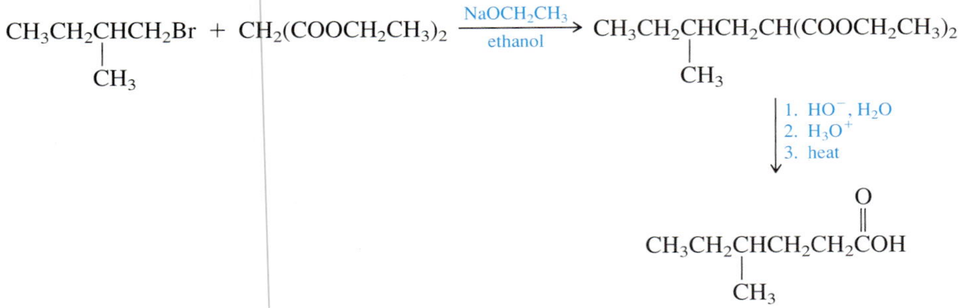

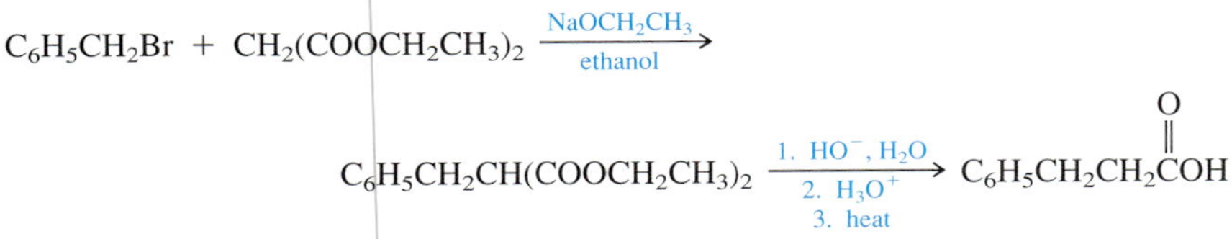

$$\xrightarrow[\substack{1.\ HO^-,\ H_2O \\ 2.\ H_3O^+ \\ 3.\ heat}]{}$$

$$CH_3CH_2CHCH_2CH_2\overset{\displaystyle O}{\overset{\|}{C}}OH$$
　　　　　　　|
　　　　　　CH_3

(d) $C_6H_5CH_2Br$ + $CH_2(COOCH_2CH_3)_2$ $\xrightarrow[\text{ethanol}]{NaOCH_2CH_3}$

$C_6H_5CH_2CH(COOCH_2CH_3)_2$ $\xrightarrow[\substack{2.\ H_3O^+ \\ 3.\ heat}]{1.\ HO^-,\ H_2O}$ $C_6H_5CH_2CH_2\overset{\displaystyle O}{\overset{\|}{C}}OH$

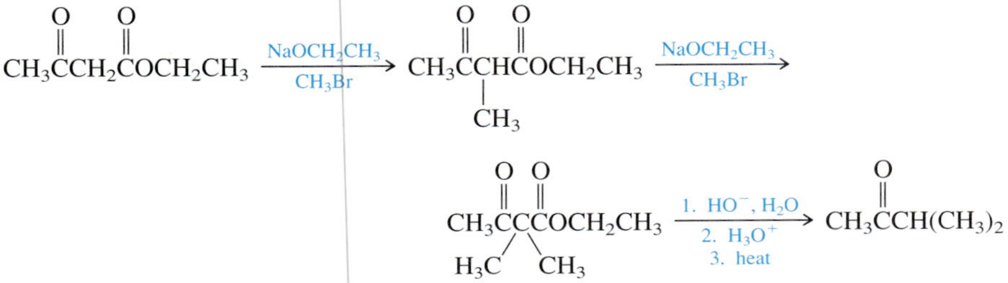

21.8 $CH_3\overset{\displaystyle O}{\overset{\|}{C}}CH_2\overset{\displaystyle O}{\overset{\|}{C}}OCH_2CH_3$ $\xrightarrow[CH_3Br]{NaOCH_2CH_3}$ $CH_3\overset{\displaystyle O}{\overset{\|}{C}}CHC\overset{\displaystyle O}{\overset{\|}{}}OCH_2CH_3$ $\xrightarrow[CH_3Br]{NaOCH_2CH_3}$
　　　　　　　　　　　　　　　　　　　　　　　　　　　　　　　|
　　　　　　　　　　　　　　　　　　　　　　　　　　　　　CH_3

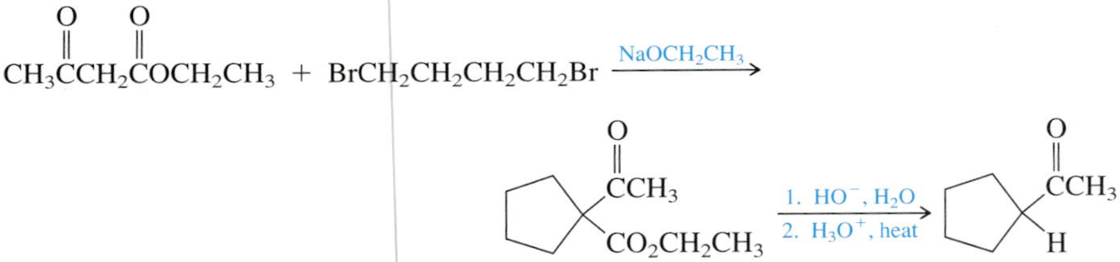

$CH_3\overset{\displaystyle O}{\overset{\|}{C}}\overset{}{C}\overset{\displaystyle O}{\overset{\|}{C}}OCH_2CH_3$ $\xrightarrow[\substack{2.\ H_3O^+ \\ 3.\ heat}]{1.\ HO^-,\ H_2O}$ $CH_3\overset{\displaystyle O}{\overset{\|}{C}}CH(CH_3)_2$
　　　　|　　|
　　H_3C　CH_3

21.9 $CH_3\overset{\displaystyle O}{\overset{\|}{C}}CH_2\overset{\displaystyle O}{\overset{\|}{C}}OCH_2CH_3$ + $BrCH_2CH_2CH_2CH_2Br$ $\xrightarrow{NaOCH_2CH_3}$

$\xrightarrow[\text{2. }H_3O^+,\ heat]{\text{1. }HO^-,\ H_2O}$

21.10 $CH_2(COOCH_2CH_3)_2$ $\xrightarrow[\text{2. }NaOCH_2CH_3,\ CH_3CH_2Br]{\text{1. }NaOCH_2CH_3,\ CH_3CH_2CH_2\overset{Br}{\overset{|}{C}}HCH_3}$

$$\substack{CH_3 \\ | \\ CH_3CH_2CH_2CH} \overset{}{\underset{\substack{| \\ CH_3CH_2}}{C}}(COOCH_2CH_3)_2$$

$$\substack{CH_3 \\ | \\ CH_3CH_2CH_2CH} \overset{\overset{\displaystyle O}{\|}}{\underset{\substack{| \\ CH_3CH_2 \quad \underset{\displaystyle O}{\|}}}{\overset{COCH_2CH_3}{\underset{COCH_2CH_3}{C}}}}$$ + $H_2N\overset{\displaystyle O}{\overset{\|}{C}}NH_2$ ⟶

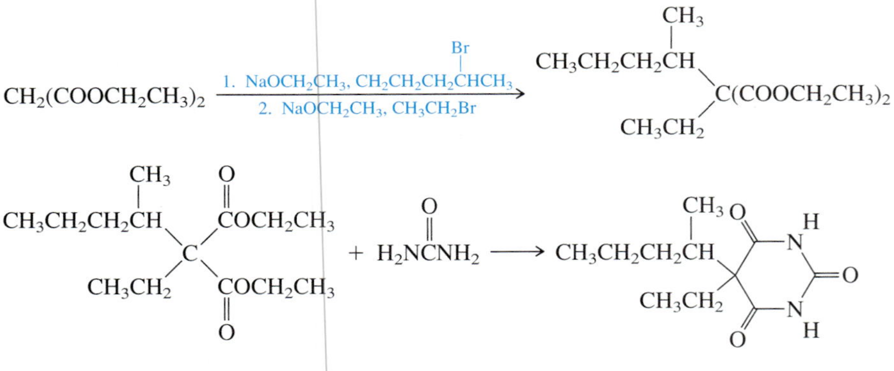

21.11

⟷

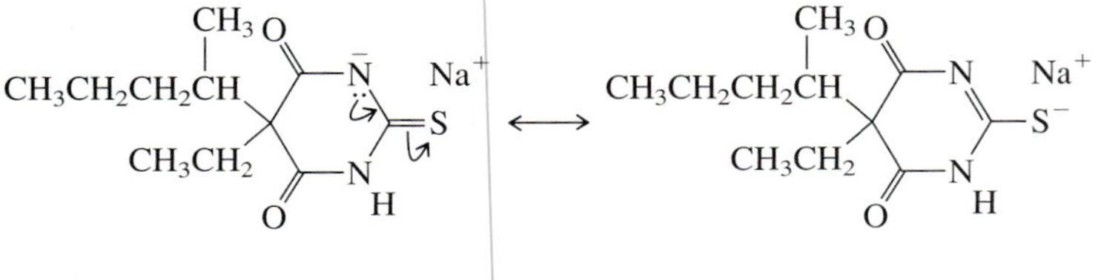

21.12 $C_6H_5CH_2\overset{\overset{\displaystyle O}{\|}}{C}OCH_2CH_3$ + $CH_3CH_2O\overset{\overset{\displaystyle O}{\|}}{C}OCH_2CH_3$ $\xrightarrow{\text{NaOCH}_2\text{CH}_3}$ $C_6H_5CH(COOCH_2CH_3)_2$

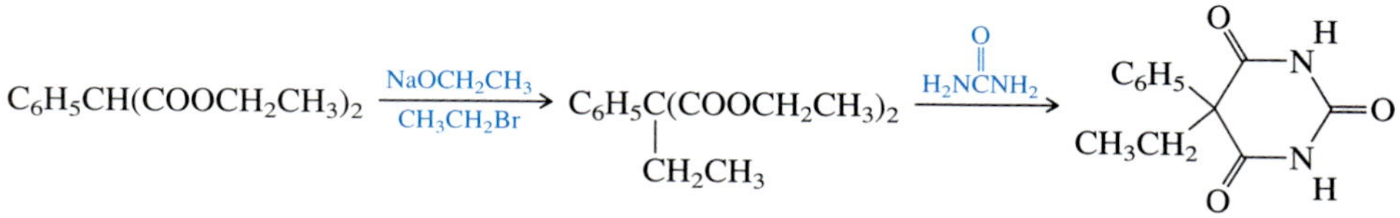

$C_6H_5CH(COOCH_2CH_3)_2$ $\xrightarrow[\text{CH}_3\text{CH}_2\text{Br}]{\text{NaOCH}_2\text{CH}_3}$ $C_6H_5\underset{\underset{\displaystyle CH_2CH_3}{|}}{C}(COOCH_2CH_3)_2$ $\xrightarrow{\overset{\overset{\displaystyle O}{\|}}{H_2N}CNH_2}$

21.13

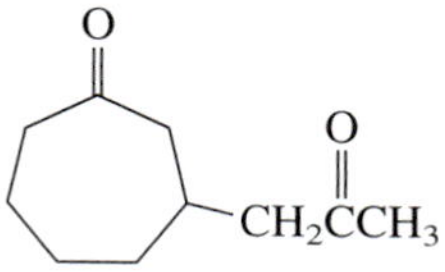

21.14 (b) $C_6H_5CH_2CO_2CH_3$ $\xrightarrow[\text{2. CH}_3\text{I}]{\text{1. LDA, THF}}$ $C_6H_5\underset{\underset{\displaystyle CH_3}{|}}{C}HCO_2CH_3$

(c)

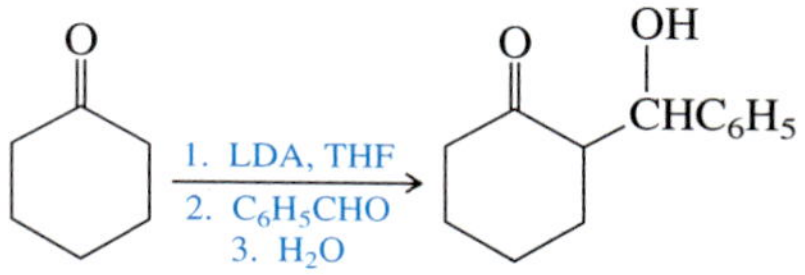

(d) 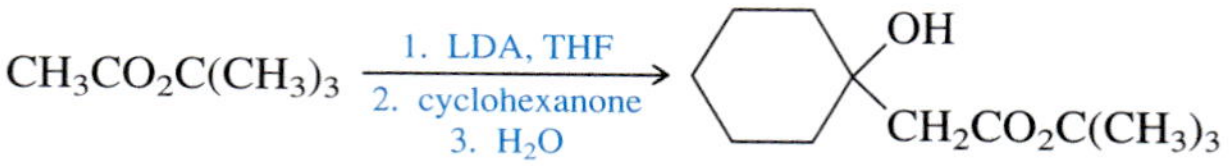 $CH_3CO_2C(CH_3)_3$ $\xrightarrow[\substack{\text{2. cyclohexanone} \\ \text{3. H}_2\text{O}}]{\text{1. LDA, THF}}$

CHAPTER 22

22.1 (b) 1-Phenylethanamine or 1-phenylethylamine; (c) 2-propen-1-amine or allylamine

22.2 *N,N*-Dimethylcycloheptanamine

22.3 Tertiary amine; *N*-ethyl-4-isopropyl-*N*-methylaniline

22.4

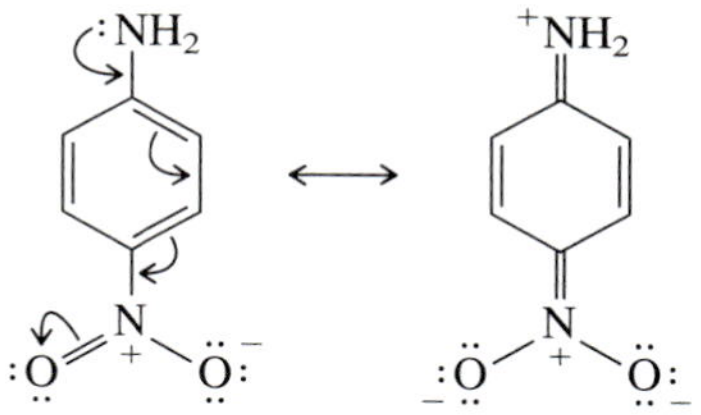

22.5 $\log (CH_3NH_3^+/CH_3NH_2) = 10.7 - 7 = 3.7$; $(CH_3NH_3^+/CH_3NH_2) = 10^{3.7} = 5000$

22.6 Tetrahydroisoquinoline is a stronger base than tetrahydroquinoline. The unshared electron pair of tetrahydroquinoline is delocalized into the aromatic ring, and this substance resembles aniline in its basicity, whereas tetrahydroisoquinoline resembles an alkylamine.

22.7 (b) The lone pair of nitrogen is delocalized into the carbonyl group by amide resonance.

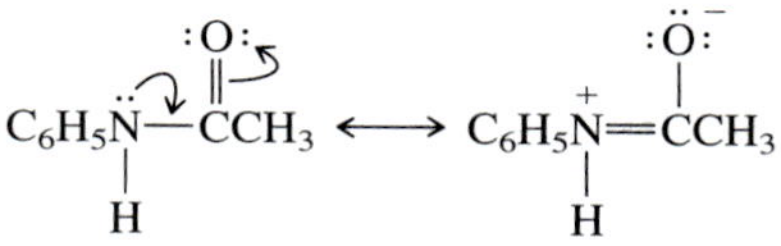

(c) The amino group is conjugated to the carbonyl group through the aromatic ring.

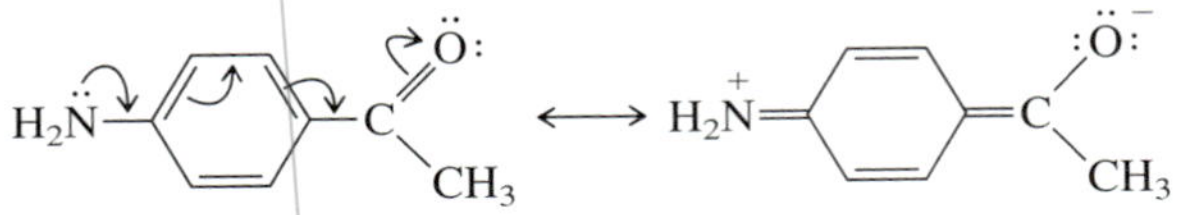

22.8 A 1 M solution of imidazolium chloride is acidic. A 1 M solution of imidazole is basic. A solution containing equal-molar quantities of imidazole and imidazolium chloride has a pH of 7.

22.9 $H_2C{=}CHCH_3 \xrightarrow[400°C]{Cl_2} H_2C{=}CHCH_2Cl \xrightarrow{NH_3} H_2C{=}CHCH_2NH_2$

22.10 Isobutylamine and 2-phenylethylamine can be prepared by the Gabriel synthesis; *tert*-butyl-amine, *N*-methylbenzylamine, and aniline cannot.

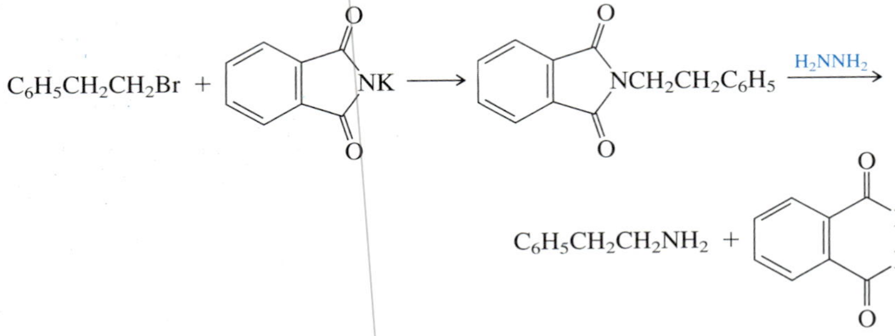

22.11 (b) Prepare *p*-isopropylnitrobenzene as in part (a); then reduce with H_2, Ni (or Fe + HCl or Sn + HCl, followed by base). (c) Prepare isopropylbenzene as in part (a); then dinitrate with HNO_3 + H_2SO_4; then reduce both nitro groups. (d) Chlorinate benzene with Cl_2 + $FeCl_3$; then nitrate (HNO_3, H_2SO_4), separate the desired para isomer from the unwanted ortho isomer, and reduce. (e) Acetylate benzene by a Friedel–Crafts reaction (acetyl chloride + $AlCl_3$); then nitrate (HNO_3, H_2SO_4); then reduce the nitro group.

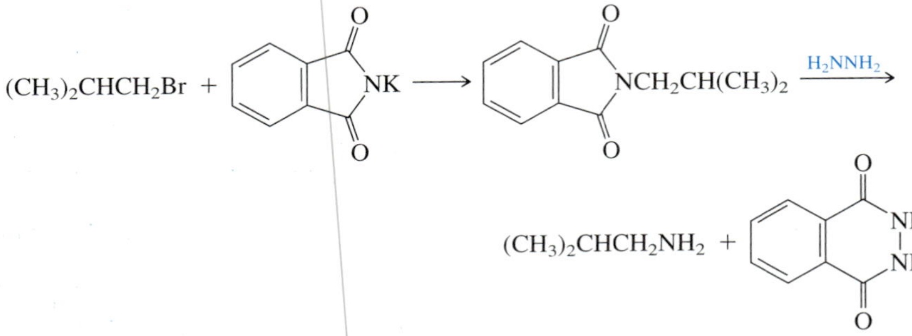

22.12 (b) $C_6H_5\overset{O}{\overset{\|}{C}}H + C_6H_5CH_2NH_2 \xrightarrow{H_2,\ Ni} C_6H_5CH_2NHCH_2C_6H_5$

(c) $C_6H_5\overset{O}{\overset{\|}{C}}H + (CH_3)_2NH \xrightarrow{H_2,\ Ni} C_6H_5CH_2N(CH_3)_2$

(d) 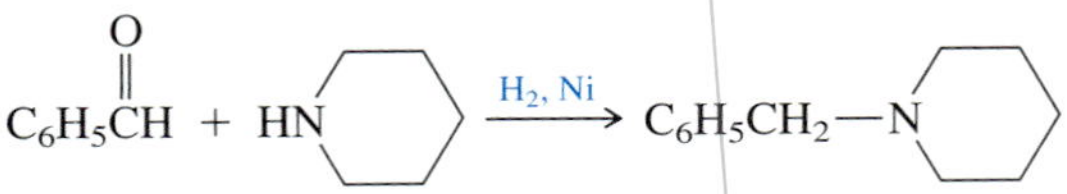 $C_6H_5\overset{O}{\overset{\|}{C}}H + HN\bigcirc \xrightarrow{H_2,\ Ni} C_6H_5CH_2{-}N\bigcirc$

22.13 (b) 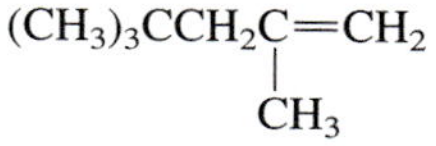 $(CH_3)_3CCH_2\underset{\underset{CH_3}{|}}{C}=CH_2$ (c) $H_2C=CH_2$

22.14 (b) Prepare acetanilide as in part (a); dinitrate (HNO_3, H_2SO_4); then hydrolyze the amide in either acid or base. (c) Prepare p-nitroacetanilide as in part (a); then reduce the nitro group with H_2 (or Fe + HCl or Sn + HCl, followed by base).

22.15

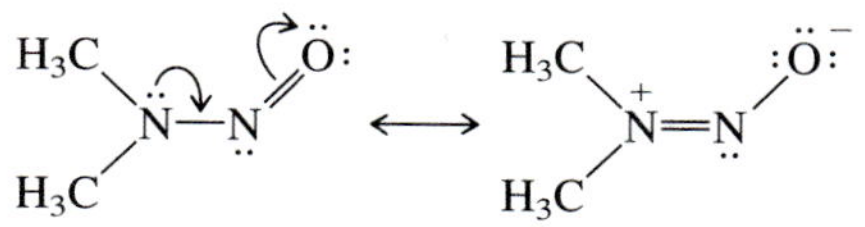

22.16 The diazonium ion from 2,2-dimethylpropylamine rearranges via a methyl shift on loss of nitrogen to give 1,1-dimethylpropyl cation.

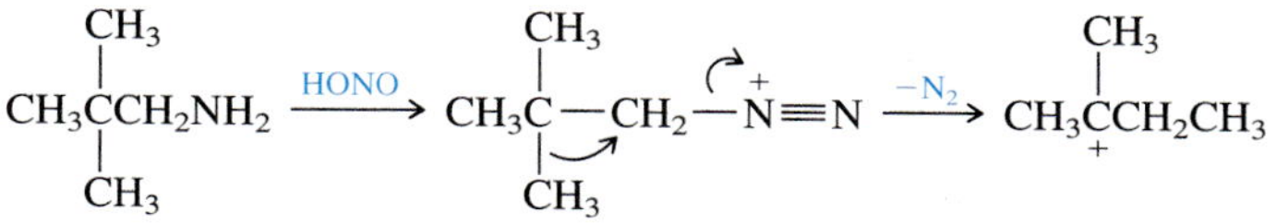

22.17 Intermediates: benzene to nitrobenzene to m-bromonitrobenzene to m-bromoaniline to m-bromophenol. Reagents: HNO_3, H_2SO_4; Br_2, $FeBr_3$; Fe, HCl then HO^-; $NaNO_2$, H_2SO_4, H_2O, then heat in H_2O.

22.18 Prepare m-bromoaniline as in Problem 22.17; then $NaNO_2$, HCl, H_2O followed by KI.

22.19 Intermediates: benzene to ethyl phenyl ketone to ethyl m-nitrophenyl ketone to m-aminophenyl ethyl ketone to ethyl m-fluorophenyl ketone. Reagents: propanoyl chloride, $AlCl_3$; HNO_3, H_2SO_4; Fe, HCl, then HO^-; $NaNO_2$, H_2O, HCl, then HBF_4, then heat.

22.20 Intermediates: isopropylbenzene to p-isopropylnitrobenzene to p-isopropylaniline to p-isopropylacetanilide to 4-isopropyl-2-nitroacetanilide to 4-isopropyl-2-nitroaniline to m-isopropylnitrobenzene. Reagents: HNO_3, H_2SO_4; Fe, HCl, then HO^-; acetyl chloride; HNO_3, H_2SO_4; acid or base hydrolysis; $NaNO_2$, HCl, H_2O, and CH_3CH_2OH or H_3PO_2.

CHAPTER 23

23.1 $C_6H_5CH_2Cl$

23.2 (b) 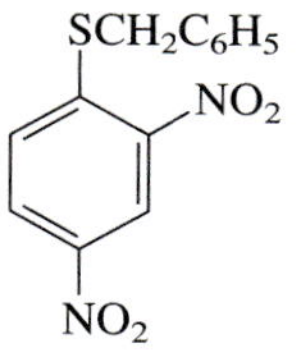   (c) 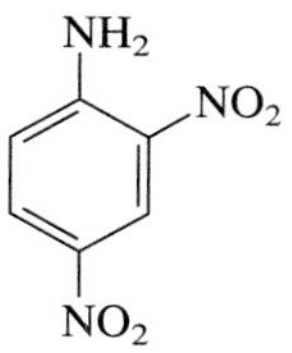   (d)

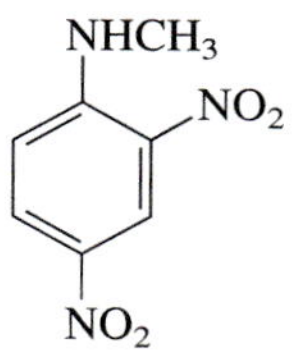

23.3

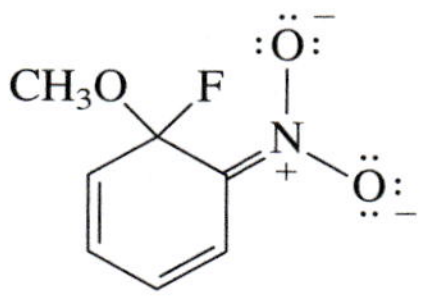

23.4

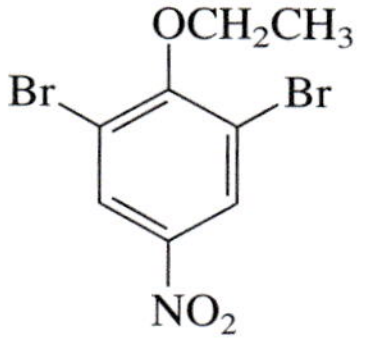

23.5

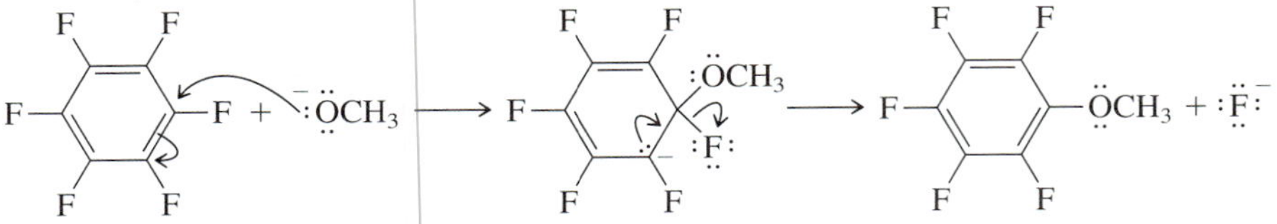

23.6 Nitrogen bears a portion of the negative charge in the anionic intermediate formed in the nucleophilic addition step in 4-chloropyridine, but not in 3-chloropyridine.

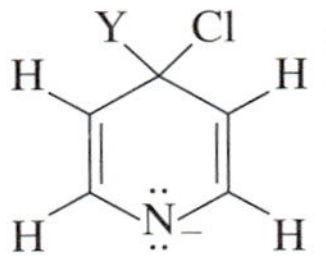

is more stable and formed faster than

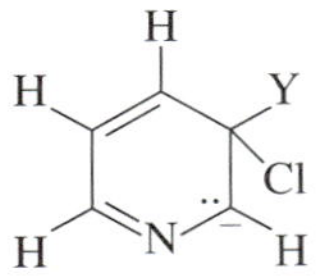

23.7 A benzyne intermediate is impossible because neither of the carbons ortho to the intended leaving group bears a proton.

23.8 3-Methylphenol and 4-methylphenol (*m*-cresol and *p*-cresol)

23.9

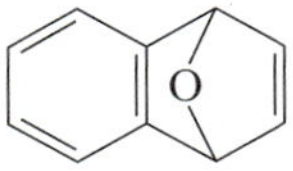

CHAPTER 24

24.1 (b) 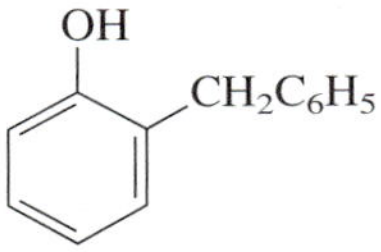   (c) 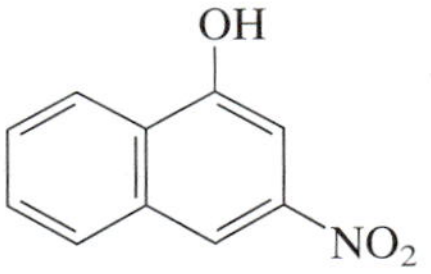   (d)

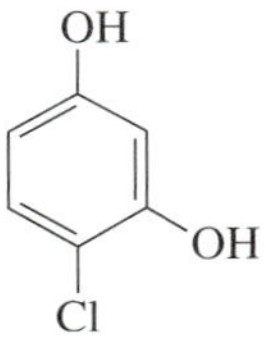

24.2 Methyl salicylate is the methyl ester of *o*-hydroxybenzoic acid. Intramolecular (rather than intermolecular) hydrogen bonding is responsible for its relatively low boiling point.

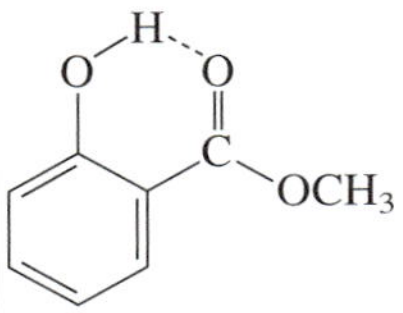

24.3 (b) *p*-Cyanophenol is the stronger acid because of conjugation of the cyano group with the phenoxide oxygen. (c) *o*-Fluorophenol is the stronger acid because the electronegative fluorine substituent can stabilize a negative charge better when fewer bonds intervene between it and the phenoxide oxygen.

24.4

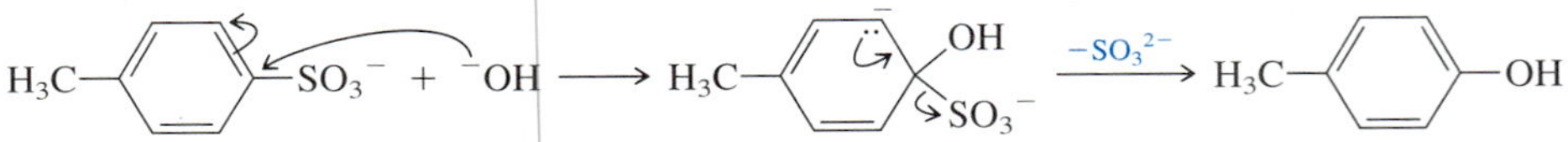

24.5

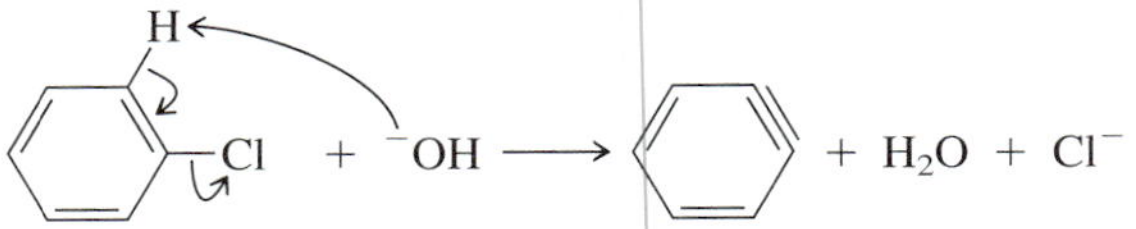

then

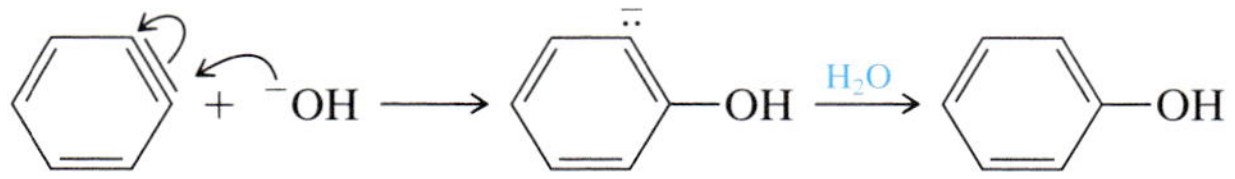

24.6 (b) 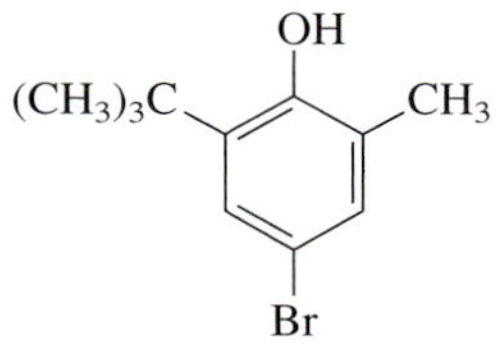 (c)

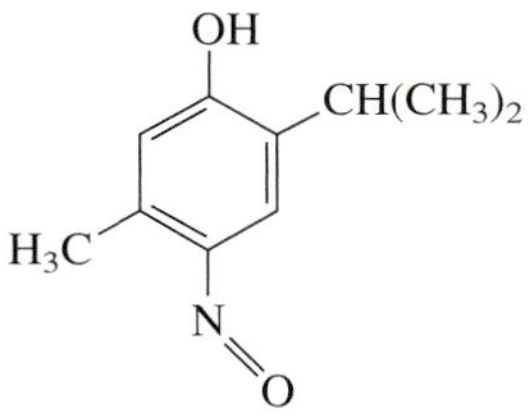

(d)

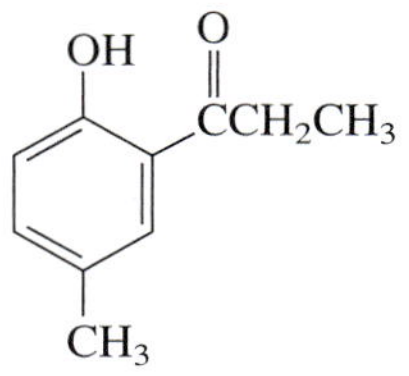

24.7 (b)

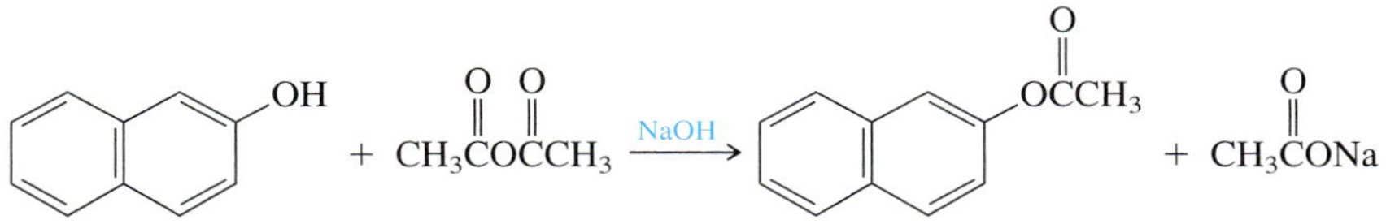

(c) $C_6H_5OH + C_6H_5\overset{O}{\overset{\|}{C}}Cl \longrightarrow C_6H_5O\overset{O}{\overset{\|}{C}}C_6H_5 + HCl$

24.8 $C_6H_5OCH_2\underset{\underset{OH}{|}}{C}HCH_3$

24.9 *p*-Fluoronitrobenzene and phenol (as its sodium or potassium salt)

24.10

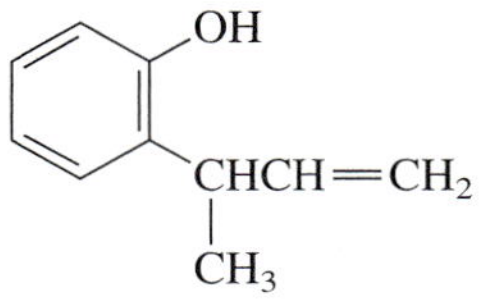

CHAPTER 25

25.1 (b) L-Glyceraldehyde; (c) D-glyceraldehyde

25.2 L-Erythrose

25.3

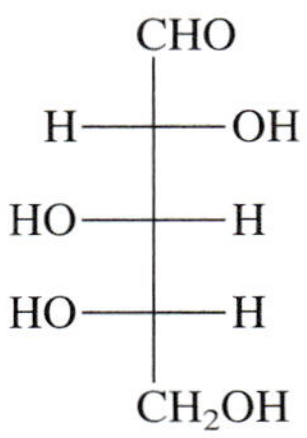

25.4   L-(−)-Talose

25.5   (b) α-L; (c) α-D; (d) β-D

25.6   (b)

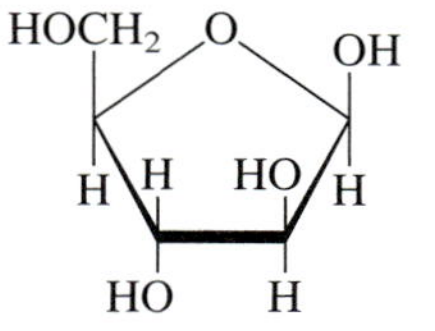

and

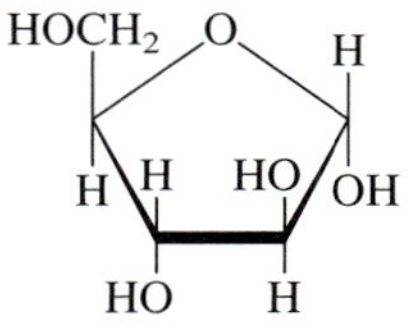

      (c)

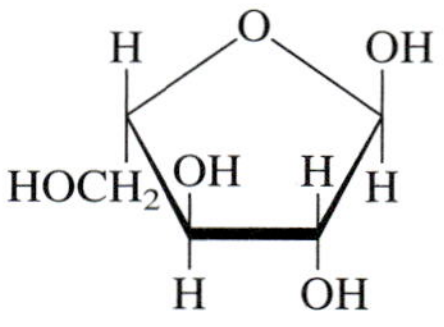

and

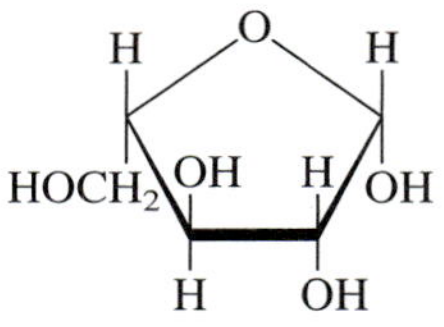

25.7   (b)

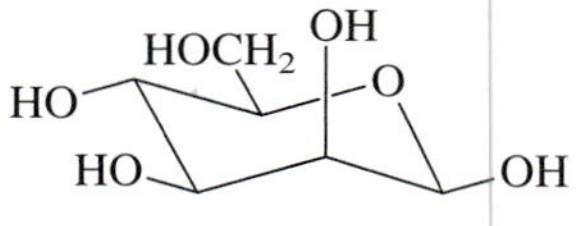

      (c)

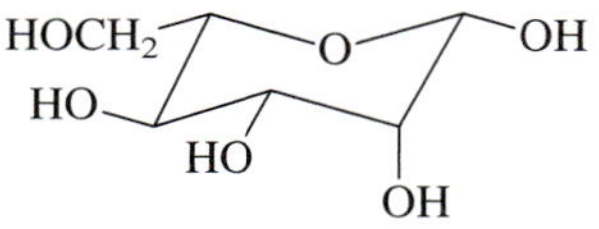

      (d)

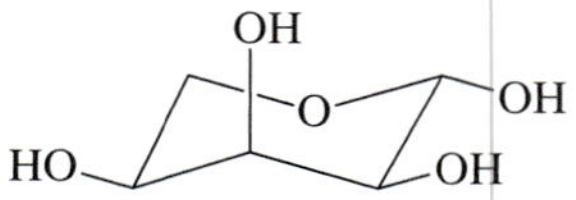

25.8   67% α, 33% β

25.9

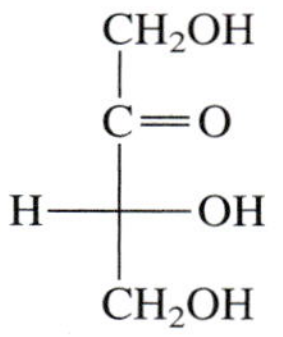

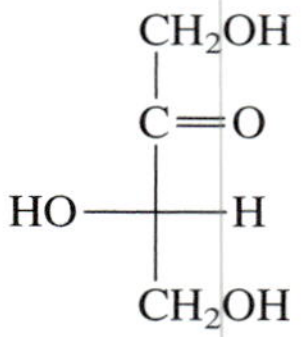

25.10   (b)

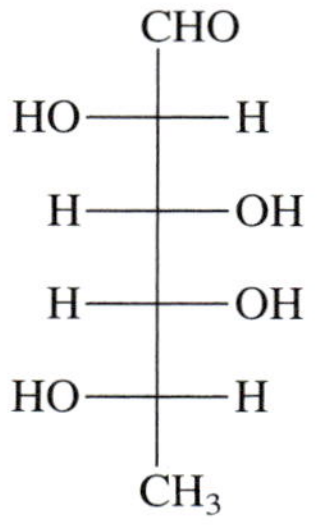

25.11

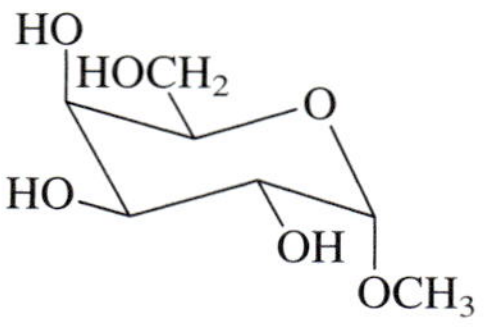

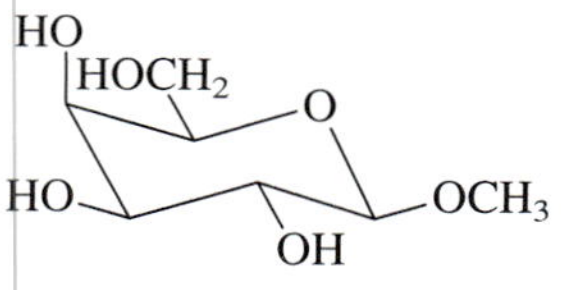

         α   β

25.12 The mechanism for formation of the β-methyl glycoside is shown. The mechanism for formation of the α isomer is the same except that methanol approaches the carbocation from the axial direction.

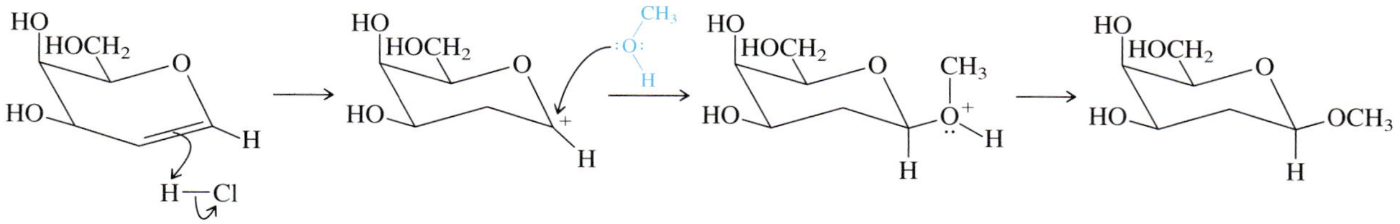

25.13

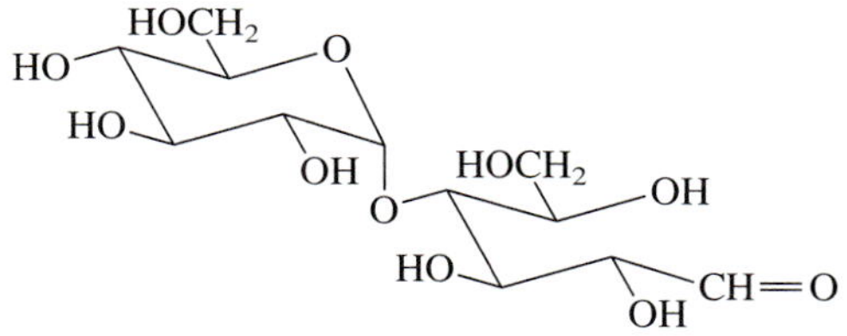

25.14 No. The product is a meso form.

25.15 All [(b) through (f)] will give positive tests.

25.16 L-Gulose

25.17 The intermediate is an enediol, $HOCH=CCH_2OP(OH)_2$ with $\overset{O}{\underset{OH}{\|}}$

25.18 (b) Four equivalents of periodic acid are required. One molecule of formaldehyde and four molecules of formic acid are formed from each molecule of D-ribose.
(c) Two equivalents

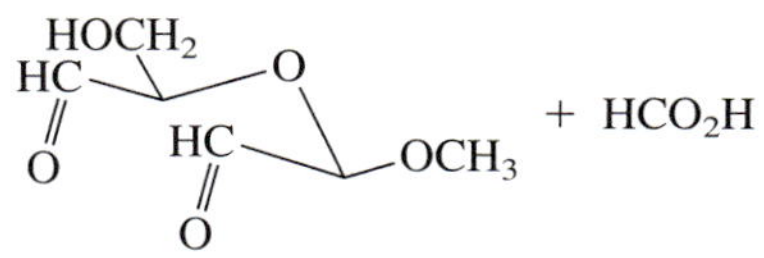

(d) Two equivalents

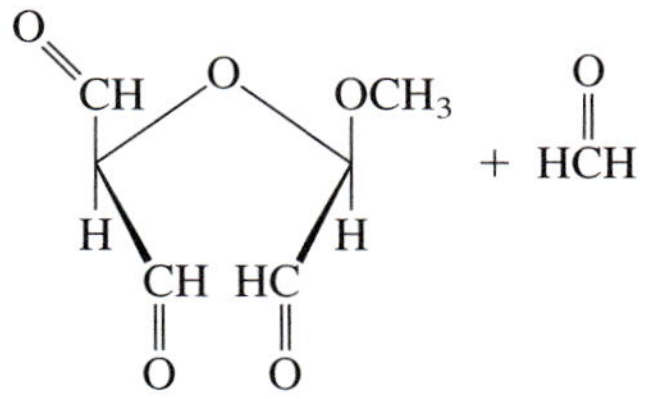

CHAPTER 26

26.1 Hydrolysis gives $CH_3(CH_2)_{16}CO_2H$ (2 mol) and (Z)-$CH_3(CH_2)_7CH=CH(CH_2)_7CO_2H$ (1 mol). The same mixture of products is formed from 1-oleyl-2,3-distearylglycerol.

26.2

$$CH_3\overset{\displaystyle OH}{\underset{\displaystyle S-ACP}{\overset{|}{\underset{|}{C}}}}-SCoA$$

26.3 $\mathrm{CH_3(CH_2)_{12}\overset{O}{\overset{\|}{C}}S{-}ACP \longrightarrow CH_3(CH_2)_{12}\overset{O}{\overset{\|}{C}}CH_2\overset{O}{\overset{\|}{C}}{-}ACP \longrightarrow}$

$\mathrm{CH_3(CH_2)_{12}\underset{OH}{\underset{|}{C}}HCH_2\overset{O}{\overset{\|}{C}}S{-}ACP \longrightarrow CH_3(CH_2)_{12}CH{=}CH\overset{O}{\overset{\|}{C}}S{-}ACP \longrightarrow CH_3(CH_2)_{12}CH_2CH_2\overset{O}{\overset{\|}{C}}S{-}ACP}$

26.4 *R* in both cases

26.5 $\mathrm{CH_3(CH_2)_{14}\overset{O}{\overset{\|}{C}}O(CH_2)_{15}CH_3}$

26.6 The biosynthetic precursor of PGE_1 is *cis,cis,cis*-icosa-8,11,14-trienoic acid.

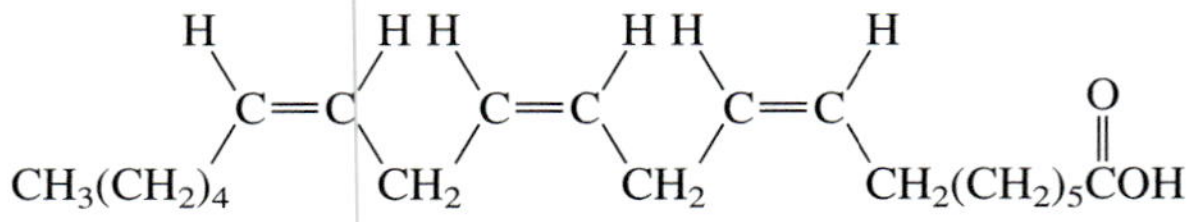
H
H H
H H
H
C=C
C=C
C=C
O
CH3(CH2)4
CH2
CH2
CH2(CH2)5COH

26.7

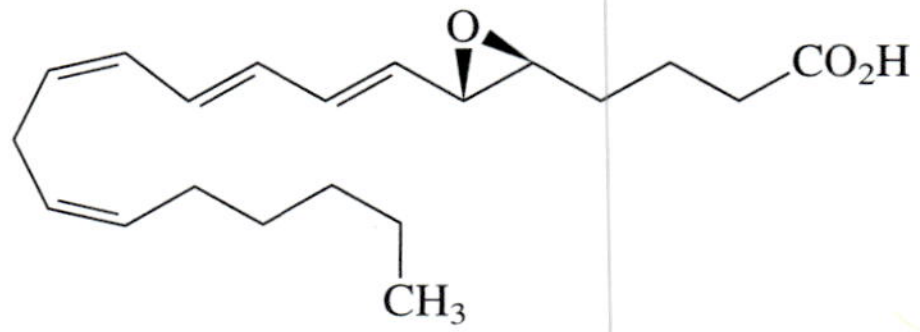
O
CO2H
CH3

26.8

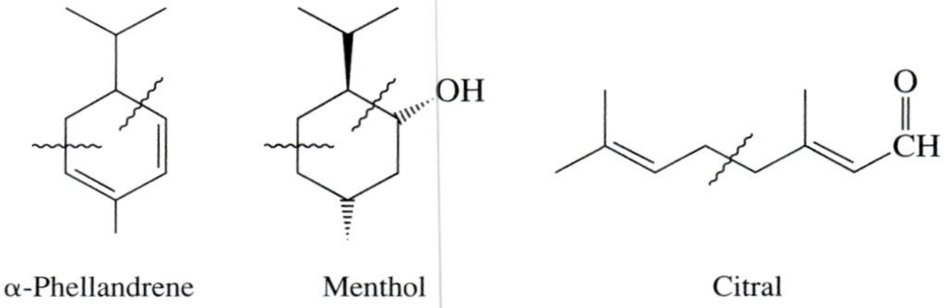
OH
O
CH
α-Phellandrene
Menthol
Citral

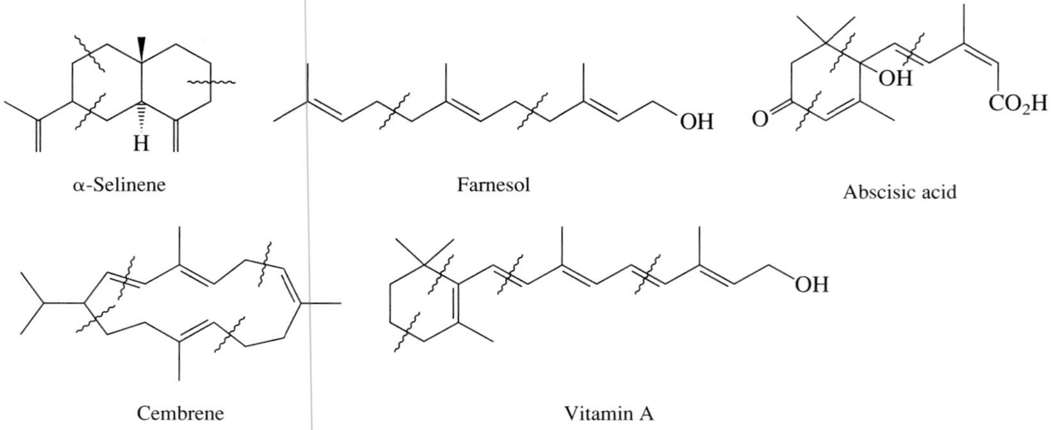
H
OH
OH
O
CO2H
α-Selinene
Farnesol
Abscisic acid
OH
Cembrene
Vitamin A

26.9

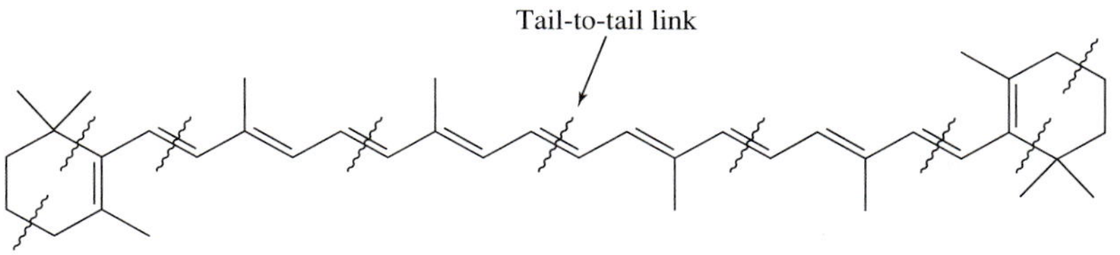

Tail-to-tail link

26.10

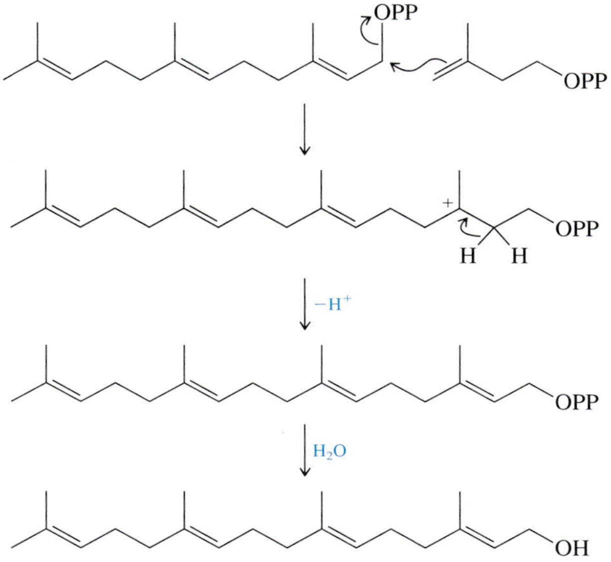

$-H^+$

H_2O

26.11

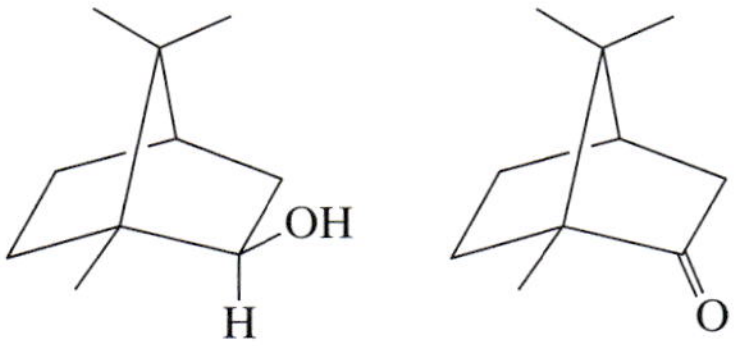

Isoborneol   Camphor

26.12 Four carbons would be labeled with ^{14}C; they are C-1, C-3, C-5, and C-7.

26.13 (b) Hydrogens that migrate in step 4 are those attached to C-14 and C-18 of squalene 2,3-epoxide; (c) carbons at the C,D ring junction of cholesterol are C-14 and C-15 of squalene 2,3-epoxide; (d) both methyl groups at C-2 are lost, as well as the methyl group originally attached to C-10 of squalene 2,3-epoxide.

26.14 Labeled carbons of squalene 2,3-epoxide: all methyl groups plus C-3, C-5, C-7, C-9, C-11, C-14, C-16, C-18, C-20, and C-22. Corresponding carbons of cholesterol (steroid numbering): C-1, C-3, C-5, C-7, C-9, C-13, C-15, C-17, C-18, C-19, C-21, C-22, C-24, C-26, C-27.

26.15 The structure of vitamin D_2 is the same as that of vitamin D_3 except that vitamin D_2 has a double bond between C-22 and C-23 and a methyl substituent at C-24.

CHAPTER 27

27.1 (b) R; (c) S

27.2 Isoleucine and threonine

27.3 5.07

27.4 ^-O—(benzene ring)—$CH_2CHCO_2^-$ with NH_2 substituent

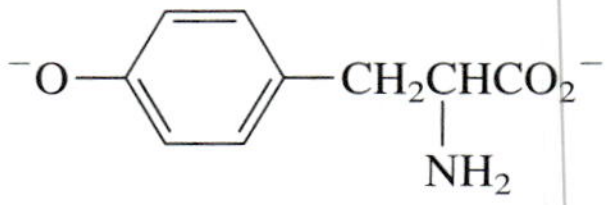

27.5 $(CH_3)_2CHCH_2CO_2H \xrightarrow[P]{Br_2} (CH_3)_2CHCHCO_2H$ (with Br) $\xrightarrow{NH_3} (CH_3)_2CHCHCO_2^-$ (with $^+NH_3$)

27.6 $(CH_3)_2CHCHO \xrightarrow[NaCN]{NH_4Cl} (CH_3)_2CHCHCN$ (with NH_2) $\xrightarrow[\text{2. } HO^-]{\text{1. } H_2O, HCl, heat} (CH_3)_2CHCHCO_2^-$ (with $^+NH_3$)

27.7 Treat the sodium salt of diethyl acetamidomalonate with isopropyl bromide. Remove the amide and ester functions by hydrolysis in aqueous acid; then heat to cause $(CH_3)_2CHC(CO_2H)_2$ (with $^+NH_3$)

to decarboxylate to give valine. The yield is low because isopropyl bromide is a secondary alkyl halide, because it is sterically hindered to nucleophilic attack, and because elimination competes with substitution.

27.8

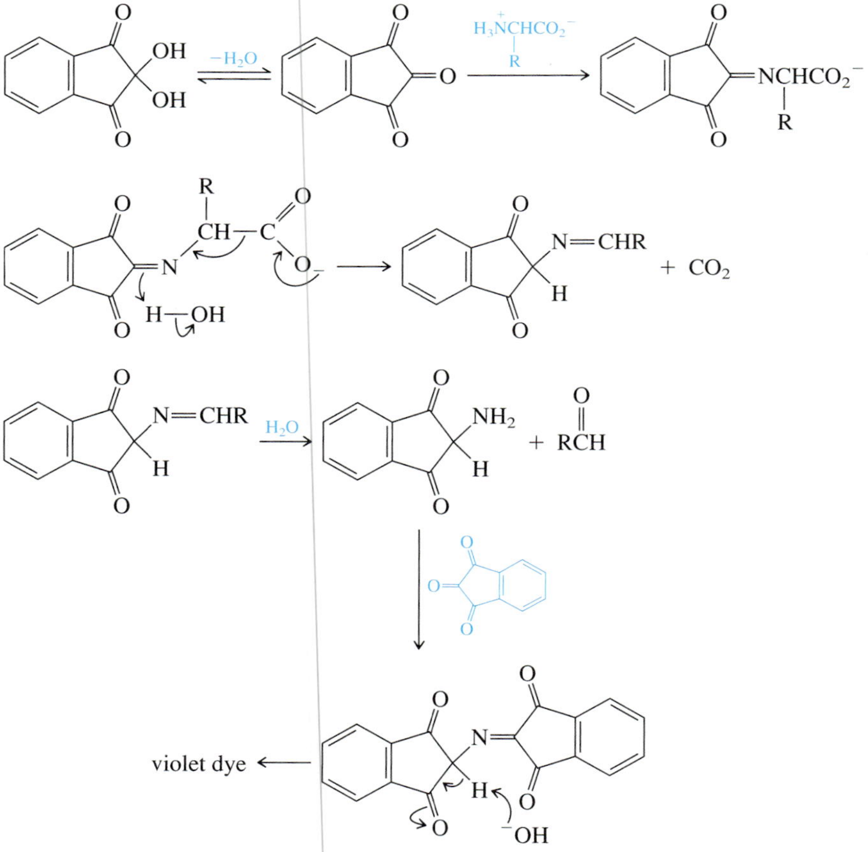

(reaction scheme showing ninhydrin hydrate $\xrightarrow{-H_2O}$ ninhydrin ketone, then reaction with $H_3\overset{+}{N}CHCO_2^-$ (R) to form the N=CHCO₂⁻ imine derivative)

(mechanism scheme showing decarboxylation to N=CHR with + CO₂)

(scheme: N=CHR $\xrightarrow{H_2O}$ NH₂ + RCHO)

(final scheme producing violet dye)

27.9 Glutamic acid

27.10 (b) 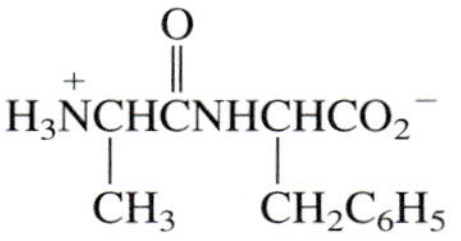 $\overset{+}{H_3}NCHCNHCHCO_2^-$ with $\overset{O}{\parallel}$ above, CH_3 and $CH_2C_6H_5$ substituents

(c) 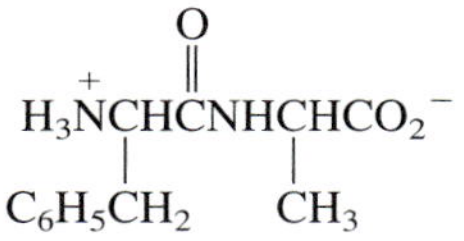 $\overset{+}{H_3}NCHCNHCHCO_2^-$ with $\overset{O}{\parallel}$ above, $C_6H_5CH_2$ and CH_3 substituents

(d) 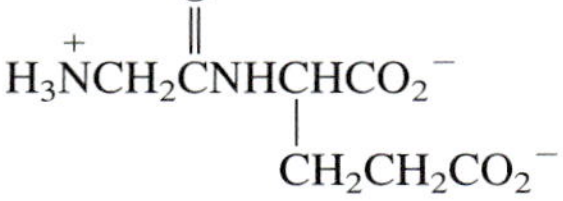 $\overset{+}{H_3}NCH_2CNHCHCO_2^-$ with $\overset{O}{\parallel}$ above, $CH_2CH_2CO_2^-$ substituent

(e) 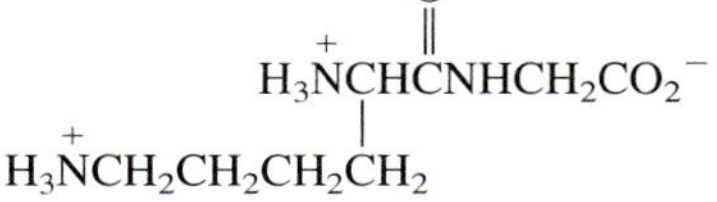 $\overset{+}{H_3}NCHCNHCH_2CO_2^-$ with $\overset{O}{\parallel}$ above, $\overset{+}{H_3}NCH_2CH_2CH_2CH_2$ substituent

(f) 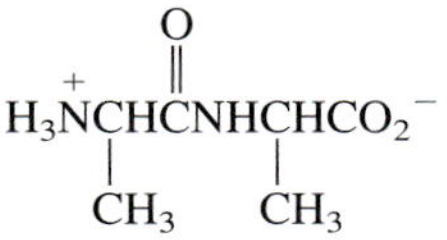 $\overset{+}{H_3}NCHCNHCHCO_2^-$ with $\overset{O}{\parallel}$ above, CH_3 and CH_3 substituents

One-letter abbreviations: (b) AF; (c) FA; (d) GE; (e) KG; (f) D-A-D-A

27.11 (b), 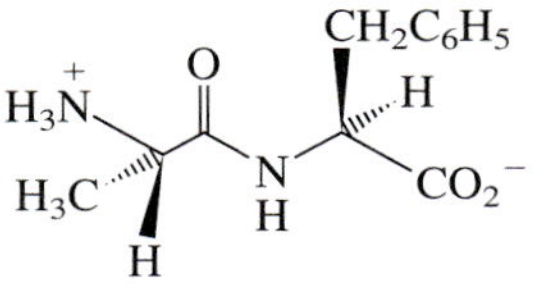 (c), 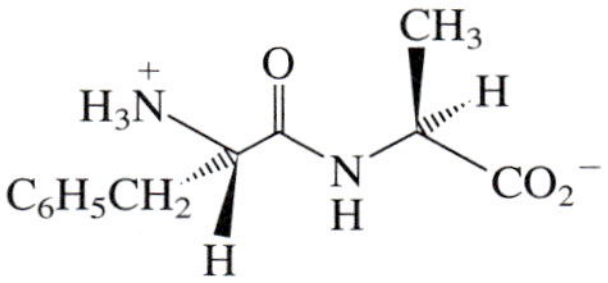 (d), 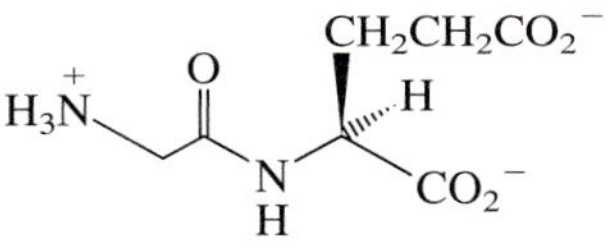 (e), 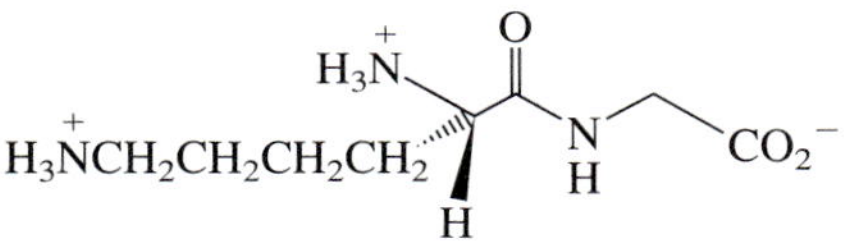 (f) 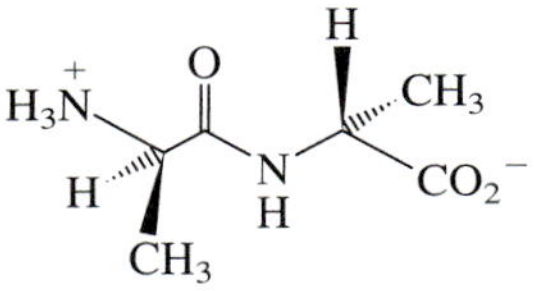 stereochemical structures

(b) $H_3\overset{+}{N}$, H_3C, H, with $\overset{O}{\parallel}$, N–H, $CH_2C_6H_5$, H, CO_2^-

(c) $H_3\overset{+}{N}$, $C_6H_5CH_2$, H, with $\overset{O}{\parallel}$, N–H, CH_3, H, CO_2^-

(d) $H_3\overset{+}{N}$, with $\overset{O}{\parallel}$, N–H, $CH_2CH_2CO_2^-$, H, CO_2^-

(e) $\overset{+}{H_3}NCH_2CH_2CH_2CH_2$, H, $H_3\overset{+}{N}$, with $\overset{O}{\parallel}$, N–H, CO_2^-

(f) $H_3\overset{+}{N}$, H, CH_3, with $\overset{O}{\parallel}$, N–H, H, CH_3, CO_2^-

27.12 Tyr-Gly-Gly-Phe-Met; YGGFM

27.13

Ala-Gly-Phe-Val	Gly-Ala-Phe-Val	Phe-Gly-Ala-Val	Val-Gly-Phe-Ala
Ala-Gly-Val-Phe	Gly-Ala-Val-Phe	Phe-Gly-Val-Ala	Val-Gly-Ala-Phe
Ala-Phe-Gly-Val	Gly-Phe-Ala-Val	Phe-Ala-Gly-Val	Val-Phe-Gly-Ala
Ala-Phe-Val-Gly	Gly-Phe-Val-Ala	Phe-Ala-Val-Gly	Val-Phe-Ala-Gly
Ala-Val-Gly-Phe	Gly-Val-Ala-Phe	Phe-Val-Gly-Ala	Val-Ala-Gly-Phe
Ala-Val-Phe-Gly	Gly-Val-Phe-Ala	Phe-Val-Ala-Gly	Val-Ala-Phe-Gly

27.14 Val-Phe-Gly-Ala   Val-Phe-Ala-Gly

27.15 Hydantoin structure with C_6H_5 on N, $S{=}$, HN, $=O$, $CH_2C_6H_5$

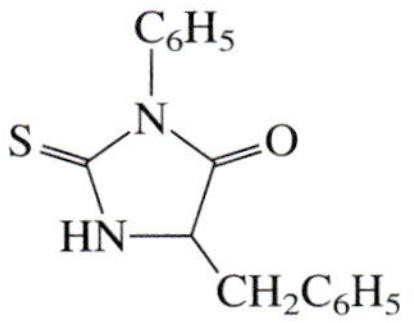

27.16 $C_6H_5CH_2OCNHCHCO_2H$ with $\overset{O}{\parallel}$ and $CH_2C_6H_5$ substituent; $C_6H_5CH_2OCNHCH_2CH_2CH_2CH_2$ with $\overset{\parallel}{O}$

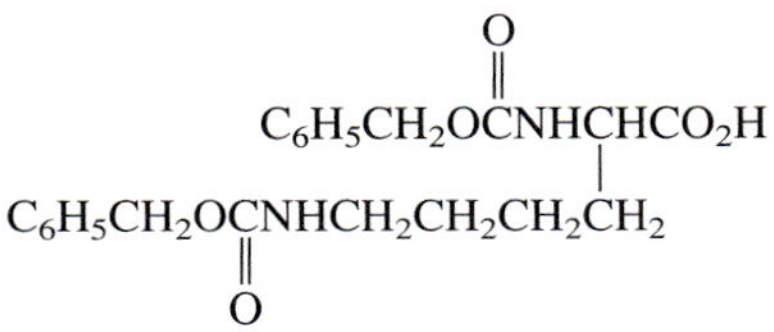

27.17 $\overset{+}{H_3}NCHCO_2^- + C_6H_5CH_2O\overset{O}{\overset{\|}{C}}Cl \longrightarrow C_6H_5CH_2O\overset{O}{\overset{\|}{C}}NHCHCO_2H$
$\quad\quad\quad\quad\;\; |$ $\quad\quad\quad\quad\quad\quad\quad\quad\quad\quad\quad\quad\quad\quad\quad\quad\quad |$
$\quad\quad\quad\quad\; CH_3$ $\quad\quad\quad\quad\quad\quad\quad\quad\quad\quad\quad\quad\quad\quad\quad\quad\; CH_3$

$\overset{+}{H_3}NCHCO_2^- + C_6H_5CH_2OH \xrightarrow[\text{2. HO}^-]{\text{1. H}^+\text{, heat}} H_2NCHCO_2CH_2C_6H_5$
$\quad\quad\; |$ $\quad\quad\quad\quad\quad\quad\quad\quad\quad\quad\quad\quad\quad\quad\quad\quad |$
$(CH_3)_2CHCH_2$ $\quad\quad\quad\quad\quad\quad\quad\quad\quad\quad\quad (CH_3)_2CHCH_2$

$C_6H_5CH_2O\overset{O}{\overset{\|}{C}}NHCHCO_2H + H_2NCH\overset{O}{\overset{\|}{C}}OCH_2C_6H_5 \xrightarrow{\text{DCCI}} C_6H_5CH_2O\overset{O}{\overset{\|}{C}}NHCHCH\overset{O}{\overset{\|}{C}}NHCH\overset{O}{\overset{\|}{C}}OCH_2C_6H_5$
$\quad\quad\quad\quad\quad\quad |$ $\quad\quad\quad\quad |$ $\quad\quad\quad\quad\quad\quad\quad\quad\quad\quad\quad\quad\quad\quad\quad\quad |\quad\quad\quad\quad\quad |$
$\quad\quad\quad\quad\quad\; CH_3$ $\quad (CH_3)_2CHCH_2$ $\quad\quad\quad\quad\quad\quad\quad\quad\quad\quad\quad\quad\quad\quad\; CH_3\quad\quad CH_2CH(CH_3)_2$

$C_6H_5CH_2O\overset{O}{\overset{\|}{C}}NHCHCH\overset{O}{\overset{\|}{C}}NHCH\overset{O}{\overset{\|}{C}}OCH_2C_6H_5 \xrightarrow[\text{Pd}]{\text{H}_2} \text{Ala-Leu}$
$\quad\quad\quad\quad\quad\quad |\quad\quad\quad\quad\quad |$
$\quad\quad\quad\quad\quad CH_3\quad\quad CH_2CH(CH_3)_2$

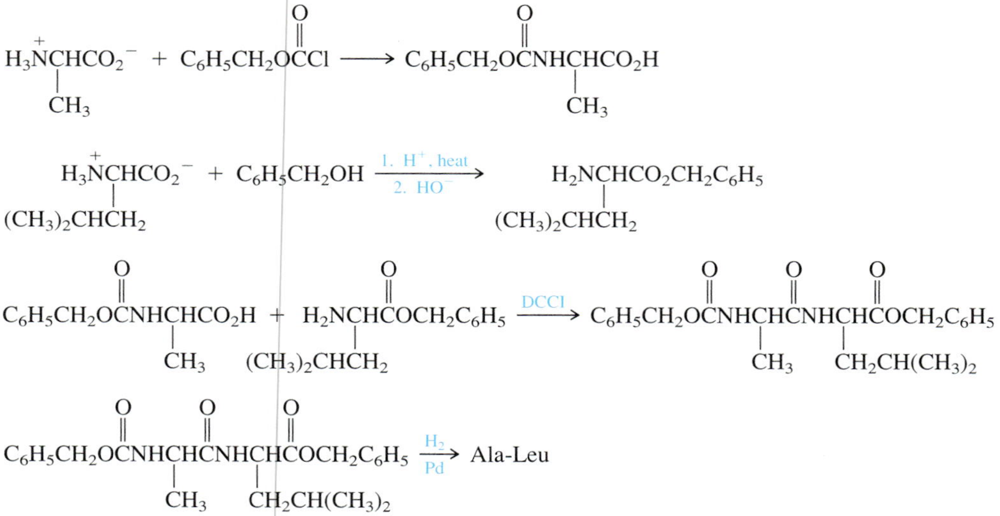

27.18 An *O*-acylisourea is formed by addition of the Z-protected amino acid to *N,N*′-dicyclo-hexylcarbodiimide, as shown in Figure 27.13. This *O*-acylisourea is attacked by *p*-nitrophenol.

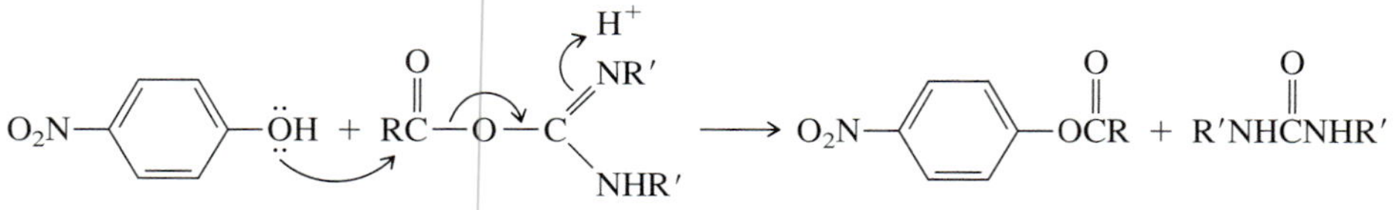

27.19 Remove the Z protecting group from the ethyl ester of Z-Phe-Gly by hydrogenolysis. Couple with the *p*-nitrophenyl ester of Z-Leu; then remove the Z group of the ethyl ester of Z-Leu-Phe-Gly.

27.20 Protect glycine as its Boc derivative and anchor this to the solid support. Remove the protecting group and treat with Boc-protected phenylalanine and DCCI. Remove the Boc group with HCl; then treat with HBr in trifluoroacetic acid to cleave Phe-Gly from the solid support.

CHAPTER 28

28.1

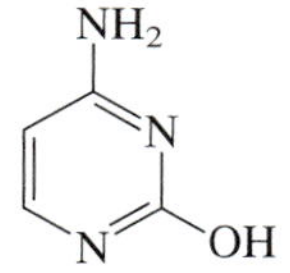

28.2

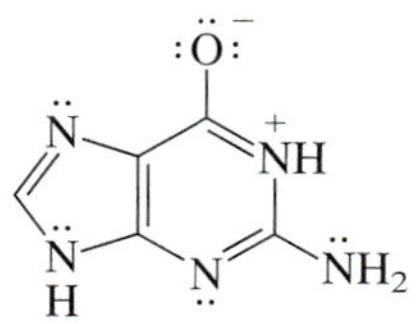

28.3 Caffeine and theobromine are both purines. Caffeine lacks $H—N—C=O$ units so cannot enolize. Two consitutionally isomeric enols are possible for theobromine.

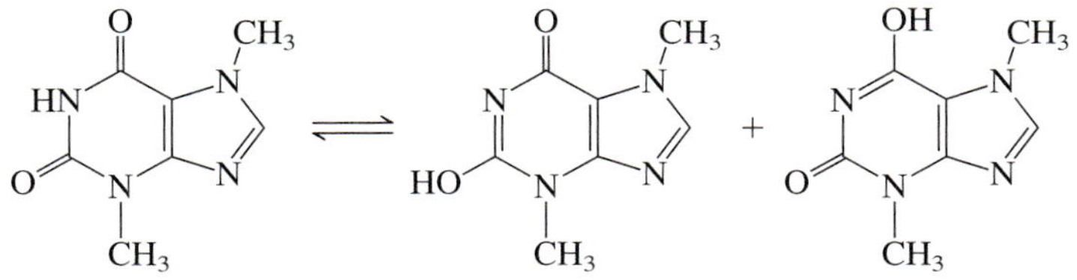

28.4

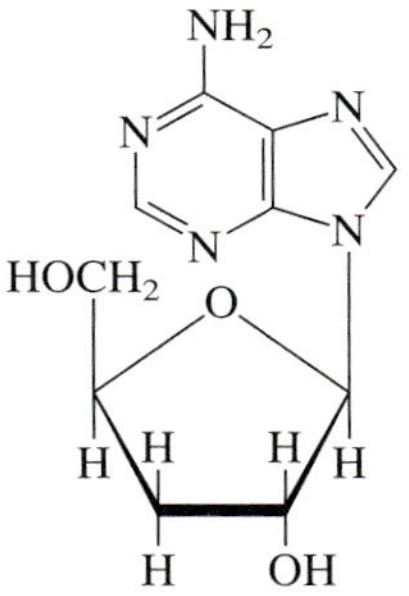

28.5

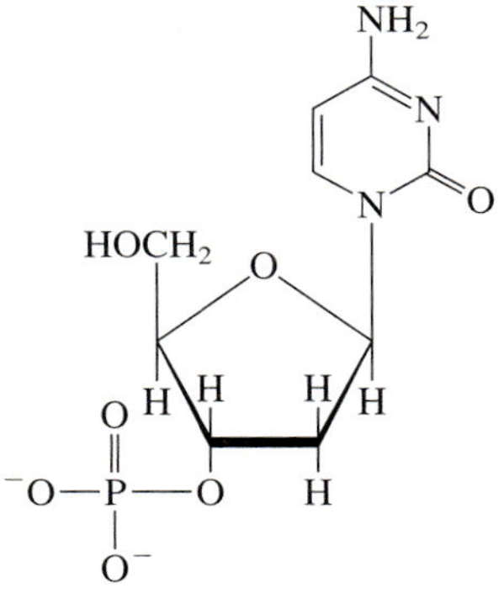

28.6

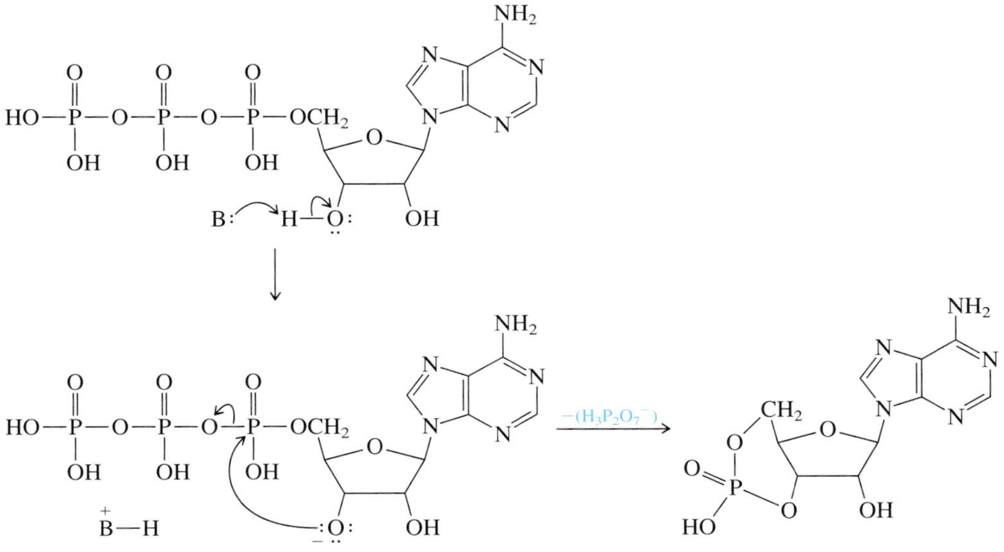

28.7
$$ATP + H_2O \longrightarrow ADP + HPO_4^{2-}$$

$$^-OCCH_2CH_2CHCO^- + NH_4^+ \longrightarrow H_2NCCH_2CH_2CHCO^- + H_2O$$

(with O above carbonyls and $_+NH_3$ below central carbon)

$$^-OCCH_2CH_2CHCO^- + NH_4^+ + ATP \longrightarrow H_2NCCH_2CH_2CHCO^- + HPO_4^{2-} + ADP$$

(with $_+NH_3$ below central carbon)

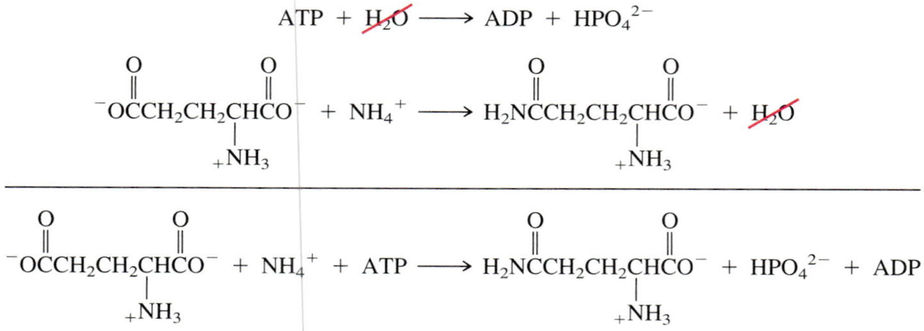

28.8

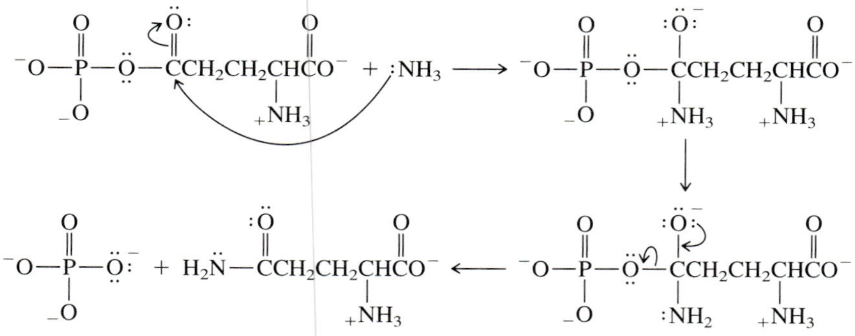

28.9

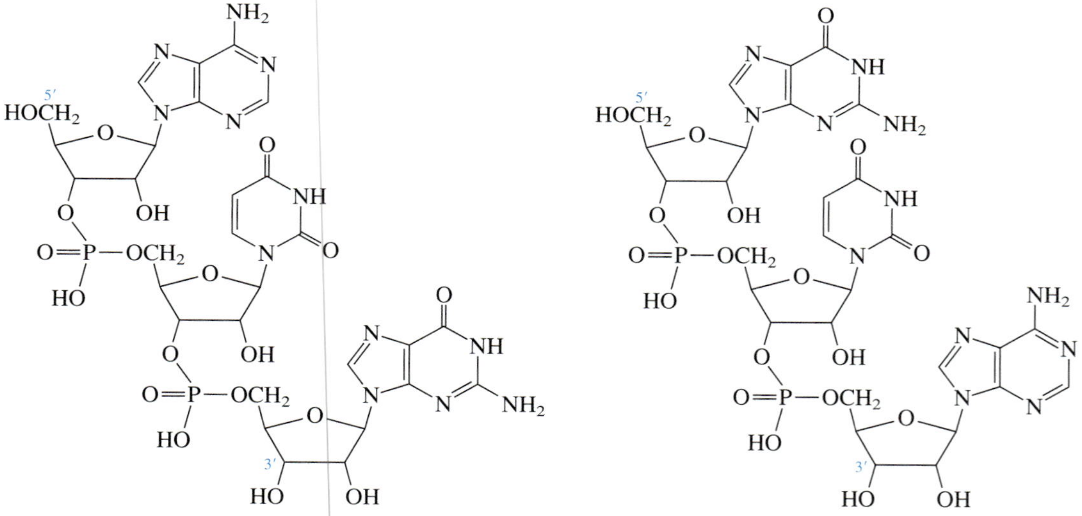

AUG   GUA

28.10 G = C = 21.3%

28.11 50

28.12 The amino acid transferred to methionine is serine instead of alanine. The serine tRNA sequence that is complementary to the UCU sequence of mRNA is AGA.

28.13 Adenine

A P P E N D I X 3

LEARNING CHEMISTRY WITH MOLECULAR MODELS: USING *SpartanBuild* AND *SpartanView*

Alan J. Shusterman, Department of Chemistry, Reed College, Portland, OR
Warren J. Hehre, Wavefunction, Inc., Irvine, CA

SpartanBuild: AN ELECTRONIC MODEL KIT

SpartanBuild is a program for building and displaying molecular models. It gives detailed information about molecular geometry (bond lengths and angles) and stability (strain energy). The program is located on the CD *Learning By Modeling* included with your text and may be run on any Windows (95/98/NT) or Power Macintosh computer.

SpartanBuild is intended both to assist you in solving problems in the text (these problems are matched with the following icon)

and more generally as a "replacement" to the plastic "model kits" that have been a mainstay in organic chemistry courses.

The tutorials that follow contain instructions for using *SpartanBuild*. Each tutorial gives instructions for a related group of tasks (install software, change model display, etc.). Computer instructions are listed in the left-hand column, and comments are listed in the right-hand column. Please perform these instructions on your computer as you read along.

Installing *SpartanBuild*

1. Insert *Learning By Modeling* CD.
2. Double-click on the CD's icon.

SpartanBuild is "CD-protected." The CD must remain in the drive at all times.

Starting *SpartanBuild*

3. Double-click on the *SpartanBuild* icon.

Starting the program opens a large *SpartanBuild* window (blank initially), a model kit, and a tool bar. Models are assembled in the window.

Quitting *SpartanBuild*

4. Select **Quit** from the **File** menu.

Restart *SpartanBuild* to continue.

BUILDING A MODEL WITH ATOMS

One way to build a model is to start with one atom and then add atoms one at a time as needed. For example, propanal, $CH_3CH_2CH{=}O$, can be assembled from four "atoms" (sp^3 C, sp^3 C, sp^2 C, and sp^2 O).

Starting to build propanal, $CH_3CH_2CH{=}O$
If necessary, start *SpartanBuild*.

1. Click on ⟩C– in the model kit.   The ⟩C– button becomes highlighted.

2. Click anywhere in the window.   A carbon atom with four unfilled va-
                                          lences (white) appears in the *Spartan-*
                                          Build window as a ball-and-wire model.

You start building propanal using an sp^3 C from the model kit. Note that five different types of carbon are available. Each is defined by a particular number of *unfilled valences* (these are used to make bonds) and a particular "idealized geometry." Valences that are not used for bonds are automatically turned into hydrogen atoms, so it is normally unnecessary to build hydrogens into a model.

Atom button	⟩C—	⟩C=	—C≡	⟩C—	⟩C⁺—
Atom label	sp^3 C	sp^2 C	sp C	delocalized C	trigonal C
Unfilled valences	4 single	2 single 1 double	1 single 1 triple	1 single 2 partial double	3 single
Ideal bond angles	109.5°	120°	180°	120°	120°

You can rotate a model (in this case, just an sp^3 C), move it around the screen, and change its size using the mouse in conjunction with the keyboard (see the following table). Try these operations now.

Operation	PC	Mac
Rotate	Move mouse with left button depressed.	Move mouse with button depressed.
Translate	Move mouse with right button depressed.	Press **option** key, and move mouse with button depressed.
Scale	Press **shift** key, and move mouse with right button depressed.	Simultaneously press **option** and **control** keys, and move mouse with button depressed.

To finish building propanal, you need to add two carbons and an oxygen. Start by adding another sp^3 C (it should still be selected), and continue by adding an sp^2 C and an sp^2 O. Atoms are added by clicking on unfilled valences in the model (the valences turn into bonds).

If you make a mistake at any point, you can undo the last operation by selecting **Undo** from the **Edit** menu, or you can start over by selecting **Clear** from the **Edit** menu.

To finish building propanal, $CH_3CH_2CH{=}O$

3. If necessary, click on sp^3 C in the   This selects the carbon atom with four
   model kit.   single valences.

4. Click on the tip of any unfilled valence in the window.	This makes a carbon–carbon single bond (the new bond appears as a dashed line).
5. Click on *sp²* C in the model kit.	This selects the carbon atom with one double and two single valences.
6. Click on the tip of any unfilled valence in the window.	This makes a carbon–carbon single bond. Bonds can only be made between valences of the same type (single + single, double + double, etc.).
7. Click on *sp²* O in the model kit.	This selects the oxygen atom with one double valence.
8. Click on the tip of the double unfilled valence in the window.	This makes a carbon–oxygen double bond. *Note:* If you cannot see which valence is the double valence, then rotate the model first.

MEASURING MOLECULAR GEOMETRY

Three types of geometry measurements can be made using *SpartanBuild:* distances between pairs of atoms, angles involving any three atoms, and dihedral angles involving any four atoms. These are accessible from the **Geometry** menu and from the toolbar. Try these operations now.

Geometry Menu	PC	Mac
Distance		
Angle		
Dihedral		

CHANGING MODEL DISPLAY

The ball-and-wire display is used for model building. Although it is convenient for this purpose, other model displays show three-dimensional molecular structure more clearly and may be preferred. The space-filling display is unique in that it portrays a molecule as a set of atom-centered spheres. The individual sphere radii are taken from experimental data and roughly correspond to the size of atomic electron clouds. Thus, the space-filling display attempts to show how much space a molecule takes up.

Changing the Model Display

1. One after the other, select **Wire, Tube, Ball and Spoke,** and **Space Filling** from the **Model** menu.

BUILDING A MODEL USING GROUPS

Organic chemistry is organized around "functional groups," collections of atoms that display similar structures and properties in many different molecules. *SpartanBuild* simplifies the construction of molecular models that contain functional groups by providing a small library of prebuilt groups. For example, malonic acid, $HO_2C\!-\!CH_2\!-\!CO_2H$, is easily built using the **Carboxylic Acid** group.

Building malonic acid, $HO_2C\!-\!CH_2\!-\!CO_2H$

1. Select **Clear** from the **Edit** menu.	This removes the existing model from the *SpartanBuild* window.
2. Click on *sp*^{3} C in the model kit, then click in the *SpartanBuild* window.	
3. Click on the **Groups** button in the model kit.	This indicates that a functional group is to be selected.
4. Select **Carboxylic Acid** from the **Groups** menu.	This makes this group appear in the model kit.
5. Examine the unfilled valences of the carboxylic acid group, and find the one marked by a small circle. If necessary, click on the group to make this circle move to the valence on carbon.	The carboxylic acid group has two structurally distinct valences that can be used to connect this group to the model. The "active" valence is marked by a small circle and can be changed by clicking anywhere on the group.
6. Click on the tip of any unfilled valence in the window.	A new carbon–carbon bond forms and an entire carboxylic acid group is added to the model.
7. Click on the tip of any unfilled valence on carbon.	This adds a second carboxylic acid group to the model.

BUILDING A MODEL USING RINGS

Many organic molecules contain one or more rings. *SpartanBuild* contains a small library of prebuilt structures representing some of the most common rings. For example, *trans*-1,4-diphenylcyclohexane can be constructed most easily using **Benzene** and **Cyclohexane** rings.

trans-1,4-Diphenylcyclohexane

Building *trans*-1,4-phenylcyclohexane

1. Select **Clear** from the **Edit** menu.	This removes the existing model from the *SpartanBuild* window.
2. Click on the **Rings** button.	This indicates that a ring is to be selected.
3. Select **Cyclohexane** from the **Rings** menu.	This makes this ring appear in the model kit.
4. Click anywhere in the *SpartanBuild* window.	This places an entire cyclohexane ring in the window.

5. Select **Benzene** from the **Rings** menu.	This makes this ring appear in the model kit.
6. Click on the tip of any equatorial unfilled valence.	This adds an entire benzene ring to the model.
7. Click on the tip of the equatorial unfilled valence directly across the ring (the valence on C-4).	This adds a second benzene ring to the model.

ADDITIONAL TOOLS

Many models can be built with the tools that have already been described. Some models, however, require special techniques (or are more easily built) using some of the *SpartanBuild* tools described in the following table.

Tool	PC	Mac	Use	Example
Make Bond			Click on two unfilled valences. The valences are replaced by a bond.	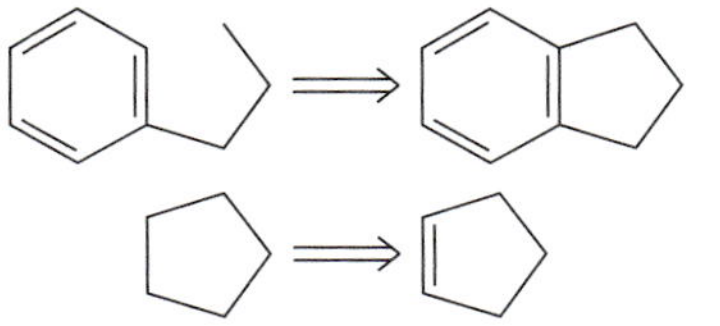
Break Bond		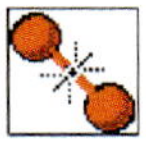	Click on bond. The bond is replaced by two unfilled valences.	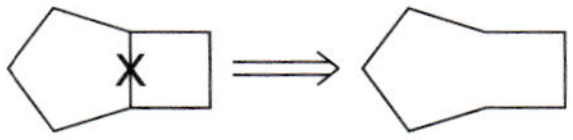
Delete		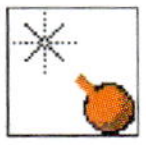	Click on atom or unfilled valence. Deleting an atom removes all unfilled valences associated with atom.	
Internal Rotation			Click on bond to select it for rotation. Press **Alt** key (PC) or **space bar** (Mac), and move mouse with button depressed (left button on PC). One part of the model rotates about the selected bond relative to other part.	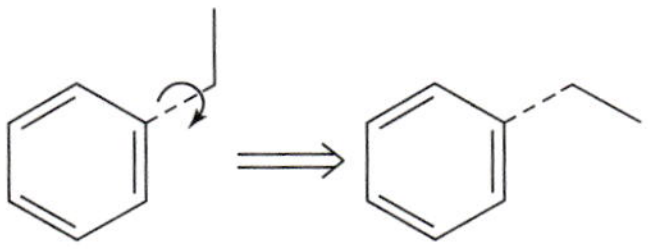
Atom Replacement			Select atom from model kit, then double-click on atom in model. Valences on the new atom must match bonds in the model or replacement will not occur.	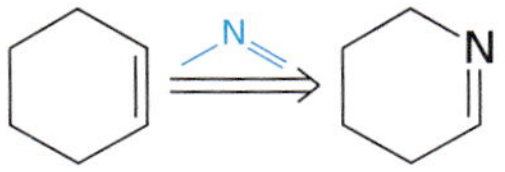

MINIMIZE: GENERATING REALISTIC STRUCTURES AND STRAIN ENERGY

In some cases, the model that results from building may be severely distorted. For example, using **Make Bond** to transform *axial* methylcyclohexane into bicyclo[2.2.1]heptane (norbornane) gives a highly distorted model (the new bond is too long and the ring has the wrong conformation).

The distorted structure can be replaced by a "more reasonable" structure using an empirical "molecular mechanics" calculation. This calculation, which is invoked in *SpartanBuild* by clicking on **Minimize,** automatically finds the structure with the smallest strain energy (in this case, a structure with "realistic" bond distances and a boat conformation for the six-membered ring).

It is difficult to tell which models contain structural distortions. You should "minimize" all models after you finish building them.

Molecular mechanics strain energies have another use. They can also be used to compare the energies of models that share the same molecular formula, that is, models that are either stereoisomers or different conformations of a single molecule (allowed comparisons are shown here).

Anti   Gauche

SpartanBuild reports strain energies in kilocalories per mole (1 kcal/mol = 4.184 kJ/mol) in the lower left-hand corner of the *SpartanBuild* window.

SpartanView: VIEWING AND INTERPRETING MOLECULAR-MODELING DATA

Learning By Modeling contains a program, *SpartanView*, which displays preassembled molecular models, and also a library of *SpartanView* models to which you can refer. These models differ in two respects from the models that you can build with *SpartanBuild*. Some models are animations that show how a molecule changes its shape during a chemical reaction, vibration, or conformation change. Others contain information about electron distribution and energy that can only be obtained from sophisticated quantum chemical calculations. The following sections describe how to use *SpartanView*.

SpartanView models are intended to give you a "molecule's eye view" of chemical processes and to help you solve certain text problems. The text uses the following icon to alert you to corresponding models on the CD.

Each icon corresponds to a model or a group of models on the CD. All of the models for a given chapter are grouped together in the same folder. For example, the models for this appendix are grouped together in a folder named "Appendix." The location of models within each folder can be determined by paying attention to the context of the icon. When an icon accompanies a numbered figure or problem, the figure or problem number is used to identify the model on the CD. When an icon appears next to an unnumbered figure, the name of the model is listed next to the icon.

Some *SpartanView* procedures are identical to *SpartanBuild* procedures and are not described in detail. In particular, the same mouse button-keyboard combinations are used to rotate, translate, and scale models. Also, the same menu commands are used to change the model display and obtain geometry data. Please refer back to the *SpartanBuild* instructions for help with these operations.

START *SpartanView*, OPEN AND CLOSE MODELS, SELECT AND MOVE "ACTIVE" MODEL

One difference between *SpartanView* and *SpartanBuild* is the number of models that the two programs can display. *SpartanBuild* can display only a single model, but *SpartanView* allows the simultaneous display of several models. Only one *SpartanView* model can be "active" at any time, and most mouse and menu operations affect only the "active" model.

The following tutorials contain instructions for using *SpartanView*. Please perform these operations on your computer as you read along.

Installing *Spartan View*

1. Insert *SpartanView* CD.
2. Double-click on the CD's icon.

SpartanView and *SpartanBuild* are located on *Learning By Modeling*. Both programs are "CD-protected."

Starting *SpartanView*

3. Double-click on the *SpartanView* icon.

This causes the *SpartanView* window to open. The window is blank initially.

Opening models

4. Select **Open** from the **File** menu.
5. Double-click on "Appendix," then double-click on "Appendix A."

"Appendix A" in the Appendix folder contains three models: water, methanol, and hydrogen chloride.

Making hydrogen chloride, HCl, the "active" model

6. Move the cursor to any part of the hydrogen chloride model, and click on it.

This makes hydrogen chloride the active model. The name of the active model is displayed at the top of the *SpartanView* window. Only one model can be active at any time.

Moving a model

7. Rotate, translate, and scale the active model using the same mouse and keyboard operations as those used with *SpartanBuild*.

Rotation and translation affect only the active model, but scaling affects all models on the screen.

Closing model

8. Select **Close** from the **File** menu.

Close affects only the active model.

QUANTUM MECHANICAL MODELS

Most of the *SpartanView* models on the CD have been constructed using quantum mechanical calculations, although some simplifications have been used to accelerate the calculations. This means that the models, although closely resembling real molecules, never precisely duplicate the properties of real molecules. Even so, the models are sufficiently similar to real molecules that they can usually be treated as equivalent. This is important because models can contain more types of information, and models can be constructed for molecules that cannot be studied in the laboratory. Also, models can be joined together to make "animations" that show how molecules move.

MEASURING AND USING MOLECULAR PROPERTIES

SpartanView models provide information about molecular energy, dipole moment, atomic charges, and vibrational frequencies (these data are accessed from the **Properties** menu). Energies and charges are available for all quantum mechanical models, whereas dipole moments and vibrational frequencies are provided for selected models only.

Energy is the most useful molecular property because changes in energy indicate whether or not a chemical reaction is favorable and how fast it can occur. *SpartanView* reports energies in "atomic units," or au (1 au = 2625.5 kJ/mol). The energy of any system made up of infinitely separated (and stationary) nuclei and electrons is exactly 0 au. A molecule's energy can therefore be thought of as the energy change that occurs when its component nuclei and electrons are brought together to make the molecule. The "assembly" process releases a vast amount of energy, so molecular energies are always large and negative.

The energies of two molecules (or two groups of molecules) can be compared as long as they contain exactly the same nuclei and exactly the same number of electrons, a condition that is satisfied by isomers. It is also satisfied by the reactants and products of a balanced chemical reaction. For example, the energy change, ΔE, for a chemical reaction, A + B → C + D, is obtained by subtracting the energies of the reactant molecules from the energies of the product molecules: $\Delta E = E_C + E_D - E_A - E_B$. ΔE is roughly equivalent to the reaction enthalpy, $\Delta H°$. The same type of computation is used to calculate the activation energy, E_{act}. This energy is obtained by subtracting the energies of the reactant molecules from that of the transition state.

Making water the active model

1. Move the cursor to any part of the water model, and click on it.

Measuring the calculated energy

2. Select **Energy** from the **Properties** menu.

3. Click on **Done** when finished.

The calculated energy of water (−75.5860 au) is displayed at the bottom of the screen.

Measuring the dipole moment

4. Select **Dipole Moment** from the **Properties** menu.

5. Click on **Done** when finished.

The calculated magnitude of the dipole moment of water (2.39 D) is displayed at the bottom of the screen. The calculated direction is indicated by a yellow arrow.

DISPLAYING MOLECULAR VIBRATIONS AND MEASURING VIBRATIONAL FREQUENCIES

Molecular vibrations are the basis of infrared (IR) spectroscopy. Certain groups of atoms vibrate at characteristic frequencies and these frequencies can be used to detect the presence of these groups in a molecule.

SpartanView displays calculated vibrations and frequencies for selected models. Calculated frequencies are listed in wavenumbers (cm^{-1}) and are consistently larger than observed frequencies (observed frequency $= 0.9 \times$ calculated frequency is a good rule of thumb).

Displaying a list of vibrational frequencies for water

1. Select **Frequencies** from the **Properties** menu.

Frequencies (in cm^{-1}) are listed in numerical order from smallest (or imaginary) at the top to largest at the bottom.

Displaying a vibration

2. *Double-click* on a frequency to make it active.
3. Click on **OK** to close the window.

A checkmark indicates the active vibration (only one vibration can be displayed at a time). Atom motions are exaggerated to make them easier to see.

4. Select **Ball and Spoke** from the **Model** menu.

Vibrations appear most clearly when a molecule is displayed as a ball-and-spoke model.

Stopping the display of a vibration

5. Repeat step 1, double-click on the active vibration, and click on **OK**.

Double-clicking on an active vibration deactivates it.

DISPLAYING ELECTROSTATIC POTENTIAL MAPS

One of the most important uses of models is to show how electrons are distributed inside molecules. The "laws" of quantum mechanics state that an electron's spatial location cannot be precisely specified, but the likelihood of detecting an electron at a particular location can be calculated (and measured). This likelihood is called the "electron density" (see Chapter 1), and *SpartanView* can display three-dimensional graphs that show regions of high and low electron density inside a molecule.

The electron density at a given location is equivalent to the amount of negative charge at that location. Thus, a hydrogen atom, which consists of a proton and an electron, can be thought of as a proton embedded in a "cloud" of negative charge. The total amount of charge in the cloud exactly equals the charge on a single electron, but the charge at any given point in the cloud is considerably smaller and varies as shown in the following graph.

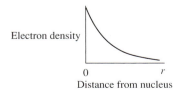

The graph shows that negative charge (or electron density) falls off as one goes farther away from the nucleus. It also shows that the charge cloud lacks a sharp boundary, or "edge." The apparent lack of an edge is problematic because we know from experimental observations that molecules do, in fact, possess a characteristic size and shape. *SpartanView* models solve this problem by using an arbitrarily selected value of the electron density to define the edge of a molecule's electron cloud. The program searches for all of the locations where the electron density takes on this edge value. Then it connects these locations together to make a smooth surface called a "size density surface," or more simply, a "density surface." Such density surfaces can be used as quantum mechanical "space-filling" models. The size and shape of density surfaces are in good agreement with the size and shape of empirical space-filling models, and the amount of electron density that lies outside the density surface is usually inconsequential.

A density surface marks the edge of a charge cloud, but it does not tell us how electron density is distributed inside the cloud. We can get a feel for the latter by calculating the electrostatic potential at different points on the density surface. The electrostatic potential at any point *(x, y, z)* on the density surface is defined as the change in energy that occurs when a "probe" particle with +1 charge is brought to this point starting from another point that is infinitely far removed from the molecule (see figure). If the energy rises (positive potential), the probe is repelled by the molecule at point *(x, y, z)*. If the energy falls (negative potential), the probe is attracted by the molecule.

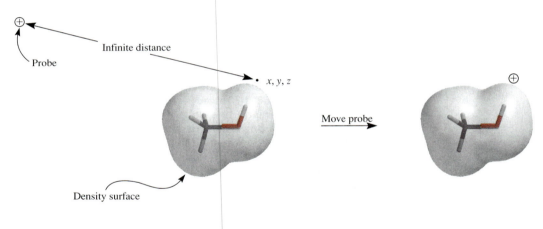

Probe

Infinite distance

• *x, y, z*

Move probe

Density surface

The electrostatic potential gives us information about the distribution of electron density in the molecule because the potential at point *(x, y, z)* is usually influenced most by the atom closest to this point. For example, if a molecule is neutral and the potential at point *(x, y, z)* is positive, then it is likely that the atom closest to this point has a net positive charge. If the potential at *(x, y, z)* is negative, then it is likely that the closest atom has a net negative charge. The size of the potential is also useful. The larger the potential at a given point, the larger the charge on the nearest atom.

These rules for assigning atomic charges work well for most neutral molecules, but they do not work for ions. This is because an ion's overall charge dominates the potential near the ion. For example, positive ions generate a positive potential everywhere around the ion. The rules also fail for atoms with highly distorted electron clouds. In such cases, positive and negative potentials are both found near the atom, and the charge is ambiguous.

SpartanView uses color to display the value of the electrostatic potential on the density surface. These colored diagrams are called "electrostatic potential maps" or just "potential maps." Different potentials are assigned different colors as follows: red (most negative potential on the map) < orange < yellow < (green) < blue (most positive potential on the map). The following potential map of water shows how this works (refer to the ball-and-spoke model for the molecule's orientation). The most negative potential (red) is found near oxygen, and the most positive potentials (blue) are found near the hydrogens. Thus, we can assign a partial negative charge to oxygen and partial positive charges to the hydrogens.

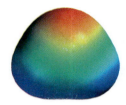

The potential map of water tells us the relative charges on oxygen and hydrogen, but it does not tell us if these charges are large or small. To discover this, we need to know the *magnitude* of the potentials. As it turns out, the most positive potentials (the blue regions) on this map are about 250 kJ/mol—a large value for a neutral molecule—so the atomic charges must be fairly large.

Potential maps can be used to compare electron distributions in different molecules providing *all of the maps assign the same color to the same potential*, that is, the maps all use the same color–potential scale. A "normal" potential map for methane (CH_4) is shown on the left (by "normal" we mean that the map displays the most negative potential as red and the most positive potential as blue). This map tells us that carbon carries a partial negative charge and the hydrogens carry partial positive charges. But, just like before, the map does not tell us the magnitude of these charges. One way to get at this information is to reassign the colors using the color–potential scale that was previously used to make water's potential map (see preceding discussion). This gives a new map that looks more or less green everywhere. This fact, along with the total absence of red and blue, tells us that the potentials, and the atomic charges, in methane are much smaller than those in water. (The most positive potential on methane's map is only 50 kJ/mol.)

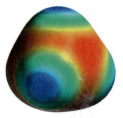

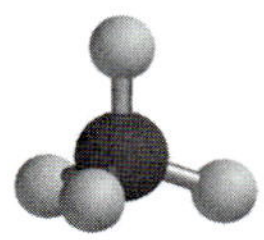

normal color assignments   color assignments based on water molecule's potential map (see above)

Size density surface *(top left)*, space-filling model *(top right)*, potential map *(bottom left)*, and tube model *(bottom right)* for methanol.

Making methanol the active model

1. Move the cursor to any part of the methanol model, and click on it.

Displaying a size density surface

2. Select **Density** from the **Surfaces** menu, then select **Transparent** from the sub-menu.

SpartanView uses the word "density" to identify size density surfaces. The size density surface is similar in size and shape to a space-filling model.

Stopping the display of a surface

3. Select **Density** from the **Surfaces** menu, then select **None** from the sub-menu.

This removes the size density surface.

Displaying an electrostatic potential map

4. Select **Potential Map** from the **Surfaces** menu, then select **Solid** from the sub-menu.

The red part of the map identifies oxygen as a negatively charged atom, and the blue part identifies the most positively charged hydrogen atom.

Closing all of the models

5. Select **Close All** from the **File** menu.

CHEMICAL APPLICATIONS OF ELECTROSTATIC POTENTIAL MAPS

Potential maps are a very powerful tool for thinking about a variety of chemical and physical phenomena. For example, water's potential map suggests that two water molecules will be attracted to each other in a way that brings a positive hydrogen in one molecule close to the negative oxygen in the other molecule (see following figure). This type of intermolecular bonding is called a "hydrogen bond." Significant hydrogen bonding does not occur between methane molecules because methane molecules create much smaller potentials.

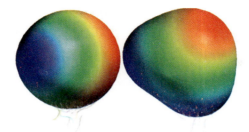

Potential maps can also be useful predictors of chemical reactivity. For example, the nitrogen atoms in ethylamine, $CH_3CH_2NH_2$, and in formamide, $O{=}CHNH_2$, appear to be identical, and we might therefore predict similar chemical reactivity patterns, but the potential maps of these compounds tell a different story. The potential map of ethylamine (see following figure, *left*) shows a region of negative potential that coincides with the location of the lone-pair electron density. This nitrogen is a good electron donor and can act as a base or nucleophile. Formamide's map (see figure, *right*), on the other hand, shows that the oxygen atom might act as an electron donor, but not the nitrogen atom. The nitrogen atoms in these compounds are very different, and they will display different chemical behavior as well.

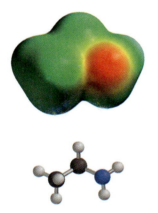

The same kinds of comparisons can also be applied to the short-lived (and therefore hard-to-observe) molecules that form during a chemical reaction. The potential maps of *n*-butyl cation, $CH_3CH_2CH_2\overset{+}{C}H_2$, and *tert*-butyl cation, $(CH_3)_3C^+$, show us that these highly reactive species differ in significant ways. The electrostatic potentials for *n*-butyl cation vary over a wider range, and the positive charge is clearly associated with the end carbon (see following figure, *left*). *tert*-Butyl cation's map, by comparison, shows a much smaller range of potentials (see figure, *right*). The central carbon is positively charged, but the potential never becomes as positive as those found in *n*-butyl cation. This tells us that some of the electron density normally associated with the methyl groups has been transferred to the central carbon.

$$CH_3CH_2CH_2\overset{+}{C}H_2 \qquad\qquad (CH_3)_3C^+$$

As a final example, we compare potential maps of the reactants, transition state, and products for an S_N2 reaction, $Cl^- + CH_3Br \rightarrow ClCH_3 + Br^-$. The reactant and product maps show negatively charged chloride and bromide ions, respectively; therefore, this reaction causes electron density to shift from one atom to another. The transition state map is distinctive in that it shows *partial* negative charges on both Cl and Br, that is, the negative charge is delocalized over Cl and Br in the transition state.

$Cl^- + CH_3—Br$

$[Cl---CH_3---Br]^-$

$Cl—CH_3 + Br^-$

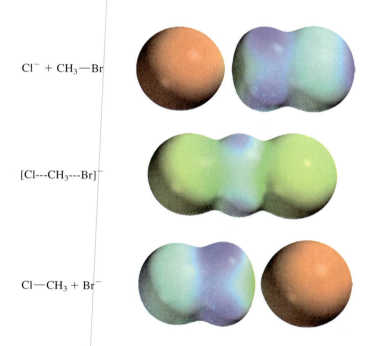

DISPLAYING MOLECULAR ORBITAL SURFACES

SpartanView displays molecular orbitals as colored surfaces. An orbital surface connects points in space where the selected orbital has a particular numerical *magnitude,* and different colors are used to indicate surfaces corresponding to negative and positive values of the orbital.

The most important molecular orbitals are the so-called frontier molecular orbitals. These are the highest (energy) occupied molecular orbital (HOMO), and lowest (energy) unoccupied molecular orbital (LUMO). The following picture shows the LUMO surface for the hydrogen molecule, H_2. The LUMO consists of two separate surfaces, a red surface surrounding one hydrogen and a blue surface surrounding the other. The colors tell us that the orbital's value is negative near one hydrogen, and positive near the other. We can also tell from this that the orbital's value must pass through zero somewhere in the empty space between the two surfaces (the "zero" region is called a "node"). Any node that crosses the bonding region makes an orbital "antibonding" and raises the orbital's energy. As a rule, electrons are only found in low-energy bonding orbitals, but this can change during a chemical reaction.

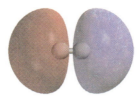

Molecular orbitals are useful tools for identifying reactive sites in a molecule. For example, the positive charge in allyl cation is delocalized over the two terminal carbon atoms, and both atoms can act as electron acceptors. This is normally shown using two resonance structures, but a more "compact" way to see this is to look at the shape of the ion's LUMO (the LUMO is a molecule's electron-acceptor orbital). Allyl cation's LUMO appears as four surfaces. Two surfaces are positioned near each of the terminal carbon atoms, and they identify allyl cation's electron-acceptor sites.

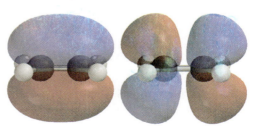

Moving into "Appendix B" and making ethylene the active model

1. Select **Open** from the **File** menu and double click on "Appendix B." Move the cursor to any part of the ethylene model, and click on it.

Appendix B contains two models: ethylene and butane.

The HOMO *(left)* and LUMO *(right)* of ethylene.

Displaying an orbital surface

2. Select **LUMO** from the **Surfaces** menu, then select **Transparent** from the sub-menu.

This displays the LUMO of ethylene. This is an unoccupied antibonding molecular orbital.

Stopping the display of an orbital surface

3. Select **LUMO** again from the **Surfaces** menu, then select **None** from the sub-menu.

The orbital is no longer displayed.

4. Select **HOMO** from the **Surfaces** menu, then select **Transparent** from the sub-menu.

This displays the HOMO of ethylene. This is an occupied bonding molecular orbital.

DISPLAYING *SpartanView* SEQUENCES (ANIMATIONS)

SpartanView can display atom motions that occur during a conformational change or chemical reaction.

Making butane the active model

1. Move the cursor to any part of the butane model, and click on it.

Animating a sequence

2. Click on the "arrow" button in the lower left-hand corner of the window.

The scroll bar slides back and forth, and the "step" label is updated during the animation. You can rotate, translate, and scale the model at any point during the animation.

Stopping the animation

3. Click on the "double bar" button in the lower left-hand corner of the window.

The animation and the scroll bar stop at the current step in the sequence.

Stepping through a sequence

4. Click on the "bar-arrows" at the right end of the scroll bar.

The scroll bar jumps to a new position, and the step label is updated, to show the current location in the sequence.

Measuring a property for a sequence

5. Select **Energy** from the Properties menu.

6. Repeat step 4 to see other energies.

All properties (energy, dipole moment, atomic charges) and geometry parameters (distance, angle, dihedral angle) can be animated or stepped through.

Quitting *SpartanView*

7. Select **Quit** from the **File** menu.

GLOSSARY

Absolute configuration (Section 7.5): The three-dimensional arrangement of atoms or groups at a chirality center.

Acetal (Section 17.8): Product of the reaction of an aldehyde or a ketone with two moles of an alcohol according to the equation

$$\underset{\substack{\| \\ O}}{RCR'} + 2R''OH \xrightarrow{H^+} \underset{\substack{| \\ OR''}}{\overset{OR''}{RCR'}} + H_2O$$

Acetoacetic ester synthesis (Section 21.6): A synthetic method for the preparation of ketones in which alkylation of the enolate of ethyl acetoacetate

$$\underset{\substack{\| \\ O}}{CH_3CCH_2} \underset{\substack{\| \\ O}}{COCH_2CH_3}$$

is the key carbon–carbon bond-forming step.

Acetyl coenzyme A (Section 26.1): A thiol ester abbreviated as

$$\underset{\substack{\| \\ O}}{CH_3CSCoA}$$

that acts as the source of acetyl groups in biosynthetic processes involving acetate.

Acetylene (Sections 2.21 and 9.1): The simplest alkyne, $HC\equiv CH$.

Achiral (Section 7.1): Opposite of *chiral.* An achiral object is superimposable on its mirror image.

Acid: According to the Arrhenius definition (Section 1.12), a substance that ionizes in water to produce protons. According to the Brønsted–Lowry definition (Section 1.13), a substance that donates a proton to some other substance. According to the Lewis definition (Section 1.17), an electron-pair acceptor.

Acid anhydride (Sections 4.1 and 20.1): Compound of the type

$$\underset{\substack{\| \ \| \\ O \ \ O}}{RCOCR}$$

Both R groups are usually the same, although they need not always be.

Acid dissociation constant K_a (Section 1.12): Equilibrium constant for dissociation of an acid:

$$K_a = \frac{[H^+][A^-]}{[HA]}$$

Activating substituent (Sections 12.10 and 12.12): A group that when present in place of a hydrogen causes a particular reaction to occur faster. Term is most often applied to substituents that increase the rate of electrophilic aromatic substitution.

Active site (Section 27.20): The region of an enzyme at which the substrate is bound.

Acylation (Section 12.7 and Chapter 20): Reaction in which an acyl group becomes attached to some structural unit in a molecule. Examples include the Friedel–Crafts acylation and the conversion of amines to amides.

Acyl chloride (Sections 4.1 and 20.1): Compound of the type

$$\underset{\substack{\| \\ O}}{RCCl}$$

R may be alkyl or aryl.

Acyl group (Sections 12.7 and 20.1): The group

$$\underset{\substack{\| \\ O}}{RC-}$$

R may be alkyl or aryl.

Acylium ion (Section 12.7): The cation $R-\overset{+}{C}\equiv\overset{..}{O}:$

Acyl transfer (Section 20.3): A nucleophilic acyl substitution. A reaction in which one type of carboxylic acid derivative is converted to another.

Addition (Section 6.1): Reaction in which a reagent X—Y adds to a multiple bond so that X becomes attached to one of the carbons of the multiple bond and Y to the other.

1,2 Addition (Section 10.10): Addition of reagents of the type X—Y to conjugated dienes in which X and Y add to adjacent doubly bonded carbons:

$$R_2C=CH-CH=CR_2 \xrightarrow{X-Y} \underset{\substack{| \ \ | \\ X \ \ Y}}{R_2C-CH-CH=CR_2}$$

1,4 Addition (Section 10.10): Addition of reagents of the type X—Y to conjugated dienes in which X and Y add to the termini of the diene system:

$$R_2C=CH-CH=CR_2 \xrightarrow{X-Y} \underset{\substack{| \ \ \ \ \ \ \ \ \ | \\ X \ \ \ \ \ \ \ \ \ Y}}{R_2C-CH=CH-CR_2}$$

Addition–elimination mechanism (Section 23.6): Two-stage mechanism for nucleophilic aromatic substitution. In the addition stage, the nucleophile adds to the carbon that bears the leaving group. In the elimination stage, the leaving group is expelled.

Alcohol (Section 4.1): Compound of the type ROH.

Alcohol dehydrogenase (Section 15.11): Enzyme in the liver that catalyzes the oxidation of alcohols to aldehydes and ketones.

Aldaric acid (Section 25.19): Carbohydrate in which carboxylic acid functions are present at both ends of the chain. Aldaric acids are typically prepared by oxidation of aldoses with nitric acid.

Aldehyde (Sections 4.1 and 17.1): Compound of the type

$$\underset{RCH}{\overset{O}{\parallel}} \quad \text{or} \quad \underset{ArCH}{\overset{O}{\parallel}}$$

Alditol (Section 25.18): The polyol obtained on reduction of the carbonyl group of a carbohydrate.

Aldol addition (Section 18.9): Nucleophilic addition of an aldehyde or ketone enolate to the carbonyl group of an aldehyde or a ketone. The most typical case involves two molecules of an aldehyde, and is usually catalyzed by bases.

$$2RCH_2\overset{O}{\overset{\parallel}{C}H} \xrightarrow{HO^-} \underset{\underset{CH=O}{|}}{RCH_2\overset{OH}{\overset{|}{C}H}CHR}$$

Aldol condensation (Sections 18.9–18.10): When an aldol addition is carried out so that the β-hydroxy aldehyde or ketone dehydrates under the conditions of its formation, the product is described as arising by an aldol condensation.

$$2RCH_2\overset{O}{\overset{\parallel}{C}H} \xrightarrow[heat]{HO^-} \underset{\underset{CH=O}{|}}{RCH_2CH=CR} + H_2O$$

Aldonic acid (Section 25.19): Carboxylic acid obtained by oxidation of the aldehyde function of an aldose.

Aldose (Section 25.1): Carbohydrate that contains an aldehyde carbonyl group in its open-chain form.

Alicyclic (Section 2.15): Term describing an *aliphatic cyclic* structural unit.

Aliphatic (Section 2.1): Term applied to compounds that do not contain benzene or benzene-like rings as structural units. (Historically, *aliphatic* was used to describe compounds derived from fats and oils.)

Alkadiene (Section 10.5): Hydrocarbon that contains two carbon–carbon double bonds; commonly referred to as a *diene*.

Alkaloid (Section 22.4): Amine that occurs naturally in plants. The name derives from the fact that such compounds are weak bases.

Alkane (Section 2.1): Hydrocarbon in which all the bonds are single bonds. Alkanes have the general formula C_nH_{2n+2}.

Alkene (Section 2.1): Hydrocarbon that contains a carbon–carbon double bond (C=C); also known by the older name *olefin*.

Alkoxide ion (Section 5.14): Conjugate base of an alcohol; a species of the type $R-\ddot{O}:^-$.

Alkylamine (Section 22.1): Amine in which the organic groups attached to nitrogen are alkyl groups.

Alkylation (Section 9.6): Reaction in which an alkyl group is attached to some structural unit in a molecule.

Alkyl group (Section 2.13): Structural unit related to an alkane by replacing one of the hydrogens by a potential point of attachment to some other atom or group. The general symbol for an alkyl group is R—.

Alkyl halide (Section 4.1): Compound of the type RX, in which X is a halogen substituent (F, Cl, Br, I).

Alkyloxonium ion (Section 4.7): Positive ion of the type ROH_2^+.

Alkyne (Section 2.1): Hydrocarbon that contains a carbon–carbon triple bond.

Allene (Section 10.5): The compound $H_2C=C=CH_2$.

Allyl cation (Section 10.2): The carbocation

$$H_2C=CH\overset{+}{C}H_2$$

The carbocation is stabilized by delocalization of the π electrons of the double bond, and the positive charge is shared by the two CH_2 groups. Substituted analogs of allyl cation are called *allylic carbocations*.

Allyl group (Sections 5.1, 10.1): The group

$$H_2C=CHCH_2-$$

Allylic rearrangement (Section 10.2): Functional group transformation in which double-bond migration has converted one allylic structural unit to another, as in:

$$R_2C=CHCH_2X \longrightarrow \underset{\underset{Y}{|}}{R_2CCH=CH_2}$$

Amide (Sections 4.1 and 20.1): Compound of the type $\overset{O}{\overset{\parallel}{RCNR'_2}}$

Amine (Chapter 22): Molecule in which a nitrogen-containing group of the type $-NH_2$, $-NHR$, or $-NR_2$ is attached to an alkyl or aryl group.

α-Amino acid (Section 27.1): A carboxylic acid that contains an amino group at the α-carbon atom. α-Amino acids are the building blocks of peptides and proteins. An α-amino acid normally exists as a *zwitterion*.

$$\underset{\underset{+NH_3}{|}}{RCHCO_2^-}$$

L-Amino acid (Section 27.2): A description of the stereochemistry at the α-carbon atom of a chiral amino acid. The Fischer projection of an α-amino acid has the amino group on the left when the carbon chain is vertical with the carboxyl group at the top.

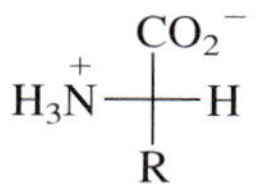

$$\overset{\underset{|}{R}}{\overset{CO_2^-}{\underset{|}{H_3\overset{+}{N}{-}}}}{-}H$$

Amino acid racemization (Section 27.2): A method for dating archeological samples based on the rate at which the stereochemistry at the α carbon of amino acid components is randomized. It is useful for samples too old to be reliably dated by ^{14}C decay.

Amino acid residues (Section 27.7): Individual amino acid components of a peptide or protein.

Amino sugar (Section 25.11): Carbohydrate in which one of the hydroxyl groups has been replaced by an amino group.

Amphiphilic (Section 19.5): Possessing both hydrophilic and lipophilic properties within the same species.

Amylopectin (Section 25.15): A polysaccharide present in starch. Amylopectin is a polymer of α(1,4)-linked glucose units, as is amylose (see *amylose*). Unlike amylose, amylopectin contains branches of 24–30 glucose units connected to the main chain by an α(1,6) linkage.

Amylose (Section 25.15): The water-dispersible component of starch. It is a polymer of α(1,4)-linked glucose units.

Anabolic steroid (Section 26.15): A steroid that promotes muscle growth.

Androgen (Section 26.15): A male sex hormone.

Angle strain (Section 3.4): The strain a molecule possesses because its bond angles are distorted from their normal values.

Anion (Section 1.2): Negatively charged ion.

Annulene (Section 11.20): Monocyclic hydrocarbon characterized by a completely conjugated system of double bonds. Annulenes may or may not be aromatic.

Anomeric carbon (Section 25.6): The carbon atom in a furanose or pyranose form that is derived from the carbonyl carbon of the open-chain form. It is the ring carbon that is bonded to two oxygens.

Anomeric effect (Section 25.8): The preference for an electronegative substituent, especially a hydroxyl group, to occupy an axial orientation when bonded to the anomeric carbon in the pyranose form of a carbohydrate.

Anti (Section 3.1): Term describing relative position of two substituents on adjacent atoms when the angle between their bonds is on the order of 180°. Atoms X and Y in the structure shown are anti to each other.

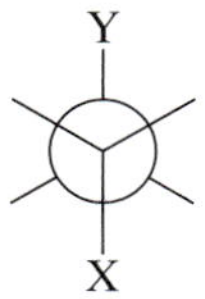

Anti addition (Section 6.3): Addition reaction in which the two portions of the attacking reagent X—Y add to opposite faces of the double bond.

Antiaromatic (Section 11.18): The quality of being destabilized by electron delocalization.

Antibonding orbital (Section 2.4): An orbital in a molecule in which an electron is less stable than when localized on an isolated atom.

Anticodon (Section 27.28): Sequence of three bases in a molecule of tRNA that is complementary to the codon of mRNA for a particular amino acid.

Anti-Markovnikov addition (Sections 6.8, 6.11): Addition reaction for which the regioselectivity is opposite to that predicted on the basis of Markovnikov's rule.

Aprotic solvent (Section 8.12): A solvent that does not have easily exchangeable protons such as those bonded to oxygen of hydroxyl groups.

Arene (Section 2.1): Aromatic hydrocarbon. Often abbreviated ArH.

Arenium ion (Section 12.2): The carbocation intermediate formed by attack of an electrophile on an aromatic substrate in electrophilic aromatic substitution. See *cyclohexadienyl cation*.

Aromatic compound (Section 11.3): An electron-delocalized species that is much more stable than any structure written for it in which all the electrons are localized either in covalent bonds or as unshared electron pairs.

Aromaticity (Section 11.4): Special stability associated with aromatic compounds.

Arylamine (Section 22.1): An amine that has an aryl group attached to the amine nitrogen.

Aryne (Section 23.8): A species that contains a triple bond within an aromatic ring (see *benzyne*).

Asymmetric (Section 7.1): Lacking all significant symmetry elements; an asymmetric object does not have a plane, axis, or center of symmetry.

Asymmetric center (Section 7.2): Obsolete name for a *chirality center*.

Atactic polymer (Section 7.15): Polymer characterized by random stereochemistry at its chirality centers. An atactic polymer, unlike an isotactic or a syndiotactic polymer, is not a stereoregular polymer.

Atomic number (Section 1.1): The number of protons in the nucleus of a particular atom. The symbol for atomic number is Z, and each element has a unique atomic number.

Axial bond (Section 3.8): A bond to a carbon in the chair conformation of cyclohexane oriented like the six "up-and-down" bonds in the following:

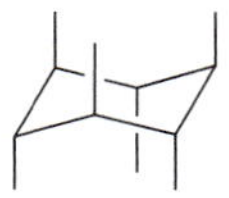

Azo coupling (Section 22.18): Formation of a compound of the type $ArN{=}NAr'$ by reaction of an aryl diazonium salt with an arene. The arene must be strongly activated toward electrophilic aromatic substitution; that is, it must bear a powerful electron-releasing substituent such as —OH or —NR₂.

Baeyer strain theory (Section 3.4): Incorrect nineteenth-century theory that considered the rings of cycloalkanes to be planar and assessed their stabilities according to how much the angles of a corresponding regular polygon deviated from the tetrahedral value of 109.5°.

Baeyer–Villiger oxidation (Section 17.16): Oxidation of an aldehyde or, more commonly, a ketone with a peroxy acid. The product of Baeyer–Villiger oxidation of a ketone is an ester.

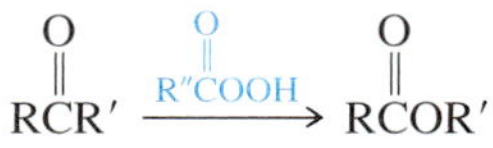

$$\underset{RCR'}{\overset{O}{\|}} \quad \xrightarrow{R''COOH} \quad \underset{RCOR'}{\overset{O}{\|}}$$

Ball-and-stick model (Section 1.10): Type of molecular model in which balls representing atoms are connected by sticks representing bonds. Similar to ball-and-spoke models of *Learning By Modeling.*

Base: According to the Arrhenius definition (Section 1.12), a substance that ionizes in water to produce hydroxide ions. According to the Brønsted–Lowry definition (Section 1.13), a substance that accepts a proton from some suitable donor. According to the Lewis definition (Section 1.17), an electron-pair donor.

Base pair (Section 28.7): Term given to the purine of a nucleotide and its complementary pyrimidine. Adenine (A) is complementary to thymine (T), and guanine (G) is complementary to cytosine (C).

Base peak (Section 13.22): The most intense peak in a mass spectrum. The base peak is assigned a relative intensity of 100, and the intensities of all other peaks are cited as a percentage of the base peak.

Basicity constant K_b (Section 1.14): A measure of base strength, especially of amines.

$$K_b = \frac{[R_3NH^+][HO^-]}{[R_3N]}$$

Bending vibration (Section 13.20): The regular, repetitive motion of an atom or a group along an arc the radius of which is the bond connecting the atom or group to the rest of the molecule. Bending vibrations are one type of molecular motion that gives rise to a peak in the infrared spectrum.

Benedict's reagent (Section 25.19): A solution containing the citrate complex of $CuSO_4$. It is used to test for the presence of reducing sugars.

Benzene (Section 11.1): The most typical aromatic hydrocarbon:

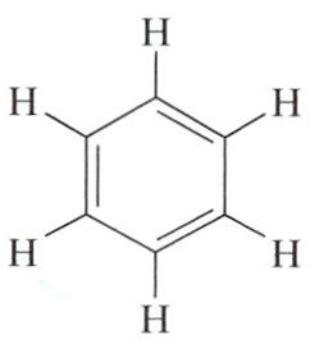

Benzyl group (Section 11.7): The group $C_6H_5CH_2$—.

Benzylic carbon (Section 11.10): A carbon directly attached to a benzene ring. A hydrogen attached to a benzylic carbon is a benzylic hydrogen. A carbocation in which the benzylic carbon is positively charged is a benzylic carbocation. A free radical in which the benzylic carbon bears the unpaired electron is a benzylic radical.

Benzyne (Section 23.8): The compound

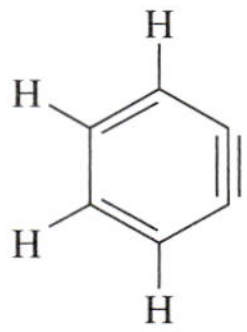

Benzyne is formed as a reactive intermediate in the reaction of aryl halides with very strong bases such as potassium amide.

Bile acids (Section 26.13): Steroid derivatives biosynthesized in the liver that aid digestion by emulsifying fats.

Bimolecular (Section 4.8): A process in which two particles react in the same elementary step.

Biological isoprene unit (Section 26.8): Isopentenyl pyrophosphate, the biological precursor to terpenes and steroids:

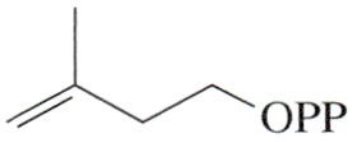

Birch reduction (Section 11.11): Reduction of an aromatic ring to a 1,4-cyclohexadiene on treatment with a group I metal (Li, Na, K) and an alcohol in liquid ammonia.

Boat conformation (Section 3.7): An unstable conformation of cyclohexane, depicted as

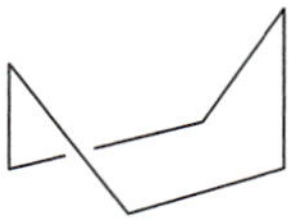

π bond (Section 2.20): In alkenes, a bond formed by overlap of *p* orbitals in a side-by-side manner. A π bond is weaker than a σ bond. The carbon–carbon double bond in alkenes consists of two sp^2-hybridized carbons joined by a σ bond and a π bond.

σ bond (Section 2.3): A connection between two atoms in which the electron probability distribution has rotational symmetry along the internuclear axis. A cross section perpendicular to the internuclear axis is a circle.

Bond dissociation energy (Sections 1.3 and 4.16): For a substance A:B, the energy required to break the bond between A and B so that each retains one of the electrons in the bond. Table 4.3 gives bond dissociation energies for some representative compounds.

Bonding orbital (Section 2.4): An orbital in a molecule in which an electron is more stable than when localized on an isolated atom. All the bonding orbitals are normally doubly occupied in stable neutral molecules.

Bond-line formula (Section 1.7): Formula in which connections between carbons are shown but individual carbons and hydrogens are not. The bond-line formula

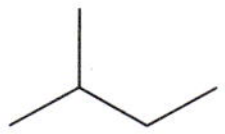

represents the compound (CH₃)₂CHCH₂CH₃.

Boundary surface (Section 1.1): The surface that encloses the region where the probability of finding an electron is high (90–95%).

Branched-chain carbohydrate (Section 25.12): Carbohydrate in which the main carbon chain bears a carbon substituent in place of a hydrogen or hydroxyl group.

Bromohydrin (Section 6.17): A halohydrin in which the halogen is bromine (see *halohydrin*).

Bromonium Ion (Section 6.16): A halonium ion in which the halogen is bromine (see *halonium ion*).

Brønsted acid See *acid*.

Brønsted base See *base*.

Buckminsterfullerene (Chapter 11, essay, "Carbon Clusters, Fullerenes, and Nanotubes"): Name given to the C₆₀ cluster with structure resembling the geodesic domes of R. Buckminster Fuller; see front cover.

n-Butane (Section 2.8): Common name for butane CH₃CH₂CH₂CH₃.

n-Butyl group (Section 2.13): The group CH₃CH₂CH₂CH₂—.

sec-Butyl group (Section 2.13): The group

$$CH_3CH_2\underset{|}{C}HCH_3$$

tert-Butyl group (Section 2.13): The group (CH₃)₃C—.

Cahn–Ingold–Prelog notation (Section 7.6): System for specifying absolute configuration as *R* or *S* on the basis of the order in which atoms or groups are attached to a chirality center. Groups are ranked in order of precedence according to rules based on atomic number.

Carbanion (Section 9.5): Anion in which the negative charge is borne by carbon. An example is acetylide ion.

Carbene (Section 14.13): A neutral species in which one of the carbon atoms is associated with six valence electrons.

Carbinolamine (Section 17.10): Compound of the type

$$HO\underset{|}{\overset{|}{-C-}}NR_2$$

Carbinolamines are formed by nucleophilic addition of an amine to a carbonyl group and are intermediates in the formation of imines and enamines.

Carbocation (Section 4.8): Positive ion in which the charge resides on carbon. An example is *tert*-butyl cation, (CH₃)₃C⁺. Carbocations are unstable species that, though they cannot normally be isolated, are believed to be intermediates in certain reactions.

Carboxylate ion (Section 19.5): The conjugate base of a carboxylic acid, an ion of the type RCO₂⁻.

Carboxylation (Section 19.11): In the preparation of a carboxylic acid, the reaction of a carbanion with carbon dioxide. Typically, the carbanion source is a Grignard reagent.

Carboxylic acid (Sections 2.1 and 19.1): Compound of the type

$$R\overset{O}{\overset{\|}{C}}OH$$, also written as RCO₂H.

Carboxylic acid derivative (Section 20.1): Compound that yields a carboxylic acid on hydrolysis. Carboxylic acid derivatives include acyl chlorides, acid anhydrides, esters, and amides.

Carotenoids (Section 26.16): Naturally occurring tetraterpenoid plant pigments.

Cation (Section 1.2): Positively charged ion.

Cellobiose (Section 25.14): A disaccharide in which two glucose units are joined by a β(1,4) linkage. Cellobiose is obtained by the hydrolysis of cellulose.

Cellulose (Section 25.15): A polysaccharide in which thousands of glucose units are joined by β(1,4) linkages.

Center of symmetry (Section 7.3): A point in the center of a structure located so that a line drawn from it to any element of the structure, when extended an equal distance in the opposite direction, encounters an identical element. Benzene, for example, has a center of symmetry.

Chain reaction (Section 4.17): Reaction mechanism in which a sequence of individual steps repeats itself many times, usually because a reactive intermediate consumed in one step is regenerated in a subsequent step. The halogenation of alkanes is a chain reaction proceeding via free-radical intermediates.

Chair conformation (Section 3.7): The most stable conformation of cyclohexane:

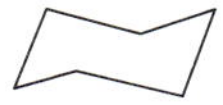

Chemical shift (Section 13.4): A measure of how shielded the nucleus of a particular atom is. Nuclei of different atoms have different chemical shifts, and nuclei of the same atom have chemical shifts that are sensitive to their molecular environment. In proton and carbon-13 NMR, chemical shifts are cited as δ, or parts per million (ppm), from the hydrogens or carbons, respectively, of tetramethylsilane.

Chiral (Section 7.1): Term describing an object that is not superimposable on its mirror image.

Chirality axis (Section 10.8): Line drawn through a molecule that is analogous to the long axis of a right-handed or left-handed screw or helix.

Chirality center (Section 7.2): An atom that has four nonequivalent atoms or groups attached to it. At various times chirality centers have been called *asymmetric centers* or *stereogenic centers*.

Chlorohydrin (Section 6.17): A halohydrin in which the halogen is chlorine (see *halohydrin*).

Chloronium ion (Section 6.16): A halonium ion in which the halogen is chlorine (see *halonium ion*).

Cholesterol (Section 26.11): The most abundant steroid in animals and the biological precursor to other naturally occurring ring steroids, including the bile acids, sex hormones, and corticosteroids.

Chromatography (Section 13.22): A method for separation and analysis of mixtures based on the different rates at which different compounds are removed from a stationary phase by a moving phase.

Chromophore (Section 13.21): The structural unit of a molecule principally responsible for absorption of radiation of a particular frequency; a term usually applied to ultraviolet-visible spectroscopy.

Chymotrypsin (Section 27.10): A digestive enzyme that catalyzes the hydrolysis of proteins. Chymotrypsin selectively catalyzes the cleavage of the peptide bond between the carboxyl group of phenylalanine, tyrosine, or tryptophan and some other amino acid.

cis- (Section 3.11): Stereochemical prefix indicating that two substituents are on the same side of a ring or double bond. (Contrast with the prefix *trans-*.)

Claisen condensation (Section 21.1): Reaction in which a β-keto ester is formed by condensation of two moles of an ester in base:

$$RCH_2COR' \xrightarrow[\text{2. } H_3O^+]{\text{1. } NaOR'} RCH_2CCHCOR' + R'OH$$
$$|$$
$$R$$

Claisen rearrangement (Section 24.13): Thermal conversion of an allyl phenyl ether to an *o*-allyl phenol. The rearrangement proceeds via a cyclohexadienone intermediate.

Claisen–Schmidt condensation (Section 18.10): A mixed aldol condensation involving a ketone enolate and an aromatic aldehyde or ketone.

Clathrate (Section 2.5): A mixture of two substances in which molecules of the minor component are held by van der Waals forces within a framework of molecules of the major component.

Clemmensen reduction (Section 12.8): Method for reducing the carbonyl group of aldehydes and ketones to a methylene group ($C{=}O \longrightarrow CH_2$) by treatment with zinc amalgam [Zn(Hg)] in concentrated hydrochloric acid.

Closed-shell electron configuration (Sections 1.1 and 11.6): Stable electron configuration in which all the lowest energy orbitals of an atom (in the case of the noble gases), an ion (e.g., Na^+), or a molecule (e.g., benzene) are filled.

^{13}C NMR (Section 13.14): Nuclear magnetic resonance spectroscopy in which the environments of individual carbon atoms are examined via their mass 13 isotope.

Codon (Section 28.11): Set of three successive nucleotides in mRNA that is unique for a particular amino acid. The 64 codons possible from combinations of A, T, G, and C code for the 20 amino acids from which proteins are constructed.

Coenzyme (Section 27.21): Molecule that acts in combination with an enzyme to bring about a reaction.

Coenzyme Q (Section 24.14): Naturally occurring group of related quinones involved in the chemistry of cellular respiration. Also known as *ubiquinone*.

Combinatorial chemistry (Section 27.18): A method for carrying out a large number of reactions on a small scale in the solid phase so as to generate a "library" of related compounds for further study, such as biological testing.

Combustion (Section 2.18): Burning of a substance in the presence of oxygen. All hydrocarbons yield carbon dioxide and water when they undergo combustion.

Common nomenclature (Section 2.11): Names given to compounds on some basis other than a comprehensive, systematic set of rules.

Concerted reaction (Section 4.8): Reaction that occurs in a single elementary step.

Condensation polymer (Section 20.17): Polymer in which the bonds that connect the monomers are formed by condensation reactions. Typical condensation polymers include polyesters and polyamides.

Condensation reaction (Section 15.7): Reaction in which two molecules combine to give a product accompanied by the expulsion of some small stable molecule (such as water). An example is acid-catalyzed ether formation:

$$2ROH \xrightarrow{H_2SO_4} ROR + H_2O$$

Condensed structural formula (Section 1.7): A standard way of representing structural formulas in which subscripts are used to indicate replicated atoms or groups, as in $(CH_3)_2CHCH_2CH_3$.

Conformational analysis (Section 3.1): Study of the conformations available to a molecule, their relative stability, and the role they play in defining the properties of the molecule.

Conformations (Section 3.1): Nonidentical representations of a molecule generated by rotation about single bonds.

Conformers (Section 3.1): Different conformations of a single molecule.

Conjugate acid (Section 1.13): The species formed from a Brønsted base after it has accepted a proton.

Conjugate addition (Sections 10.10 and 18.12): Addition reaction in which the reagent adds to the termini of the conjugated system with migration of the double bond; synonymous with 1,4 addition. The most common examples include conjugate addition to 1,3-dienes and to α,β-unsaturated carbonyl compounds.

Conjugate base (Section 1.13): The species formed from a Brønsted acid after it has donated a proton.

Conjugated diene (Section 10.5): System of the type C=C—C=C, in which two pairs of doubly bonded carbons are joined by a single bond. The π electrons are delocalized over the unit of four consecutive sp^2-hybridized carbons.

Connectivity (Section 1.6): Order in which a molecule's atoms are connected. Synonymous with *constitution*.

Constitution (Section 1.6): Order of atomic connections that defines a molecule.

Constitutional isomers (Section 1.8): Isomers that differ in respect to the order in which the atoms are connected. Butane ($CH_3CH_2CH_2CH_3$) and isobutane [$(CH_3)_3CH$] are constitutional isomers.

Copolymer (Section 10.11): Polymer formed from two or more different monomers.

COSY (Section 13.19): A 2D NMR technique that correlates the chemical shifts of spin-coupled nuclei. COSY stands for *correlated spectroscopy.*

Coupling constant J (Section 13.7): A measure of the extent to which two nuclear spins are coupled. In the simplest cases, it is equal to the distance between adjacent peaks in a split NMR signal.

Covalent bond (Section 1.3): Chemical bond between two atoms that results from their sharing of two electrons.

COX-2 (Section 26.6): Cyclooxygenase-2, an enzyme that catalyzes the biosynthesis of prostaglandins. COX-2 inhibitors reduce pain and inflammation by blocking the activity of this enzyme.

Cracking (Section 2.16): A key step in petroleum refining in which high-molecular-weight hydrocarbons are converted to lower molecular-weight ones by thermal or catalytic carbon–carbon bond cleavage.

Critical micelle concentration (Section 19.5): Concentration above which substances such as salts of fatty acids aggregate to form micelles in aqueous solution.

Crown ether (Section 16.4): A cyclic polyether that, via ion–dipole attractive forces, forms stable complexes with metal ions. Such complexes, along with their accompanying anion, are soluble in nonpolar solvents.

C terminus (Section 27.7): The amino acid at the end of a peptide or protein chain that has its carboxyl group intact—that is, in which the carboxyl group is not part of a peptide bond.

Cumulated diene (Section 10.5): Diene of the type C=C=C, in which a single carbon atom participates in double bonds with two others.

Cyanohydrin (Section 17.7): Compound of the type

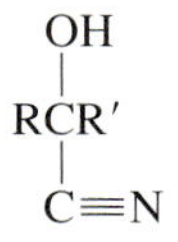

$$\begin{array}{c} OH \\ | \\ R\overset{|}{C}R' \\ | \\ C \equiv N \end{array}$$

Cyanohydrins are formed by nucleophilic addition of HCN to the carbonyl group of an aldehyde or a ketone.

Cycloaddition (Section 10.12): Addition, such as the Diels–Alder reaction, in which a ring is formed via a cyclic transition state.

Cycloalkane (Section 2.15): An alkane in which a ring of carbon atoms is present.

Cycloalkene (Section 5.1): A cyclic hydrocarbon characterized by a double bond between two of the ring carbons.

Cycloalkyne (Section 9.4): A cyclic hydrocarbon characterized by a triple bond between two of the ring carbons.

Cyclohexadienyl anion (Section 23.6): The key intermediate in nucleophilic aromatic substitution by the addition–elimination mechanism. It is represented by the general structure shown, where Y is the nucleophile and X is the leaving group.

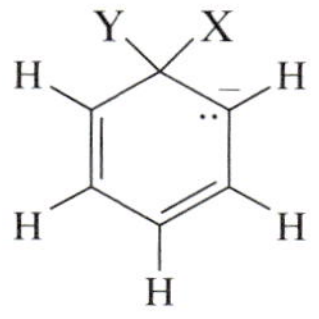

Cyclohexadienyl cation (Section 12.2): The key intermediate in electrophilic aromatic substitution reactions. It is represented by the general structure

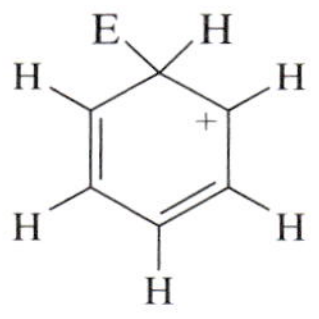

where E is derived from the electrophile that attacks the ring.

Deactivating substituent (Sections 12.11 and 12.13): A group that when present in place of hydrogen causes a particular reaction to occur more slowly. The term is most often applied to the effect of substituents on the rate of electrophilic aromatic substitution.

Debye unit (D) (Section 1.5): Unit customarily used for measuring dipole moments:

$$1\ D = 1 \times 10^{-18}\ esu \cdot cm$$

Decarboxylation (Section 19.17): Reaction of the type $RCO_2H \longrightarrow RH + CO_2$, in which carbon dioxide is lost from a carboxylic acid. Decarboxylation normally occurs readily only when the carboxylic acid is a 1,3-dicarboxylic acid or a β-keto acid.

Decoupling (Section 13.17): In NMR spectroscopy, any process that destroys the coupling of nuclear spins between two nuclei. Two types of decoupling are employed in ^{13}C NMR spectroscopy. *Broadband decoupling* removes all the $^1H-^{13}C$ couplings; *off-resonance decoupling* removes all $^1H-^{13}C$ couplings except those between directly bonded atoms.

Dehydration (Section 5.9): Removal of H and OH from adjacent atoms. The term is most commonly employed in the preparation of alkenes by heating alcohols in the presence of an acid catalyst.

Dehydrogenation (Section 5.1): Elimination in which H_2 is lost from adjacent atoms. The term is most commonly encountered in the industrial preparation of ethylene from ethane, propene from propane, 1,3-butadiene from butane, and styrene from ethylbenzene.

Dehydrohalogenation (Section 5.14): Reaction in which an alkyl halide, on being treated with a base such as sodium ethoxide, is converted to an alkene by loss of a proton from one carbon and the halogen from the adjacent carbon.

Delocalization (Section 1.9): Association of an electron with more than one atom. The simplest example is the shared electron pair (covalent) bond. Delocalization is important in conjugated π electron systems, where an electron may be associated with several carbon atoms.

Deoxy sugar (Section 25.10): A carbohydrate in which one of the hydroxyl groups has been replaced by a hydrogen.

DEPT (Section 13.18): Abbreviation for *d*istortionless *e*nhancement of *p*olarization *t*ransfer. DEPT is an NMR technique that reveals the number of hydrogens directly attached to a carbon responsible for a particular signal.

Detergents (Section 19.5): Substances that clean by micellar action. Although the term usually refers to a synthetic detergent, soaps are also detergents.

Diastereomers (Section 7.10): Stereoisomers that are not enantiomers—stereoisomers that are not mirror images of one another.

Diastereotopic (Section 7.13): Describing two atoms or groups in a molecule that are attached to the same atom but are in stereochemically different environments that are not mirror images of each other. The two protons shown in bold in $H_2C=CHCl$, for example, are diastereotopic. One is cis to chlorine, the other is trans.

1,3-Diaxial repulsion (Section 3.10): Repulsive forces between axial substituents on the same side of a cyclohexane ring.

Diazonium ion (Sections 22.15–22.16): Ion of the type $R-\overset{+}{N}\equiv N:$. Aryl diazonium ions are formed by treatment of primary aromatic amines with nitrous acid. They are extremely useful in the preparation of aryl halides, phenols, and aryl cyanides.

Diazotization (Section 22.16): The reaction by which a primary arylamine is converted to the corresponding diazonium ion by nitrosation.

Dieckmann reaction (Section 21.2): An intramolecular version of the Claisen condensation.

Dielectric constant (Section 8.12): A measure of the ability of a material to disperse the force of attraction between oppositely charged particles. The symbol for dielectric constant is ε.

Diels–Alder reaction (Section 10.12): Conjugate addition of an alkene to a conjugated diene to give a cyclohexene derivative. Diels–Alder reactions are extremely useful in synthesis.

Dienophile (Section 10.12): The alkene that adds to the diene in a Diels–Alder reaction.

β-Diketone (Section 18.5): Compound of the type

also referred to as a 1,3-diketone.

Dimer (Section 6.21): Molecule formed by the combination of two identical molecules.

Dipeptide (Section 27.7): A compound in which two α-amino acids are linked by an amide bond between the amino group of one and the carboxyl group of the other:

Dipole–dipole attraction (Section 2.17): A force of attraction between oppositely polarized atoms.

Dipole/induced-dipole attraction (Section 4.6): A force of attraction that results when a species with a permanent dipole induces a complementary dipole in a second species.

Dipole moment (Section 1.5): Product of the attractive force between two opposite charges and the distance between them. Dipole moment has the symbol μ and is measured in Debye units (D).

Disaccharide (Sections 25.1 and 25.14): A carbohydrate that yields two monosaccharide units (which may be the same or different) on hydrolysis.

Dispersion force (Section 2.17): Attractive force that involves induced dipoles.

Disubstituted alkene (Section 5.6): Alkene of the type $R_2C=CH_2$ or $RCH=CHR$. The groups R may be the same or different, they may be any length, and they may be branched or unbranched. The significant point is that there are two carbons *directly* bonded to the carbons of the double bond.

Disulfide bridge (Section 27.7): An S—S bond between the sulfur atoms of two cysteine residues in a peptide or protein.

DNA (deoxyribonucleic acid) (Section 28.7): A polynucleotide of 2′-deoxyribose present in the nuclei of cells that serves to store and replicate genetic information. Genes are DNA.

Double bond (Section 1.4): Bond formed by the sharing of four electrons between two atoms.

Double dehydrohalogenation (Section 9.7): Reaction in which a geminal dihalide or vicinal dihalide, on being treated with a very strong base such as sodium amide, is converted to an alkyne by loss of two protons and the two halogen substituents.

Double helix (Section 28.8): The form in which DNA normally occurs in living systems. Two complementary strands of DNA are associated with each other by hydrogen bonds between their base pairs, and each DNA strand adopts a helical shape.

Downfield (Section 13.4): The low-field region of an NMR spectrum. A signal that is downfield with respect to another lies to its left in the spectrum.

Eclipsed conformation (Section 3.1): Conformation in which bonds on adjacent atoms are aligned with one another. For example, the C—H bonds indicated in the structure shown are eclipsed.

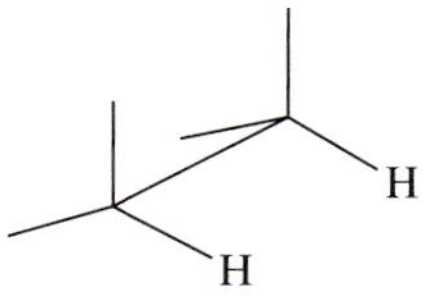

Edman degradation (Section 27.13): Method for determining the N-terminal amino acid of a peptide or protein. It involves treating the material with phenyl isothiocyanate ($C_6H_5N=C=S$), cleaving with acid, and then identifying the phenylthiohydantoin (PTH derivative) produced.

Elastomer (Section 10.11): A synthetic polymer that possesses elasticity.

Electromagnetic radiation (Section 13.1): Various forms of radiation propagated at the speed of light. Electromagnetic radiation includes (among others) visible light; infrared, ultraviolet, and microwave radiation; and radio waves, cosmic rays, and X-rays.

Electron affinity (Section 1.2): Energy change associated with the capture of an electron by an atom.

Electronegativity (Section 1.5): A measure of the ability of an atom to attract the electrons in a covalent bond toward itself. Fluorine is the most electronegative element.

Electronic effect (Section 5.6): An effect on structure or reactivity that is attributed to the change in electron distribution that a substituent causes in a molecule.

Electron impact (Section 13.22): Method for producing positive ions in mass spectrometry whereby a molecule is bombarded by high-energy electrons.

18-Electron rule (Section 14.14): The number of ligands that can be attached to a transition metal are such that the sum of the electrons brought by the ligands plus the valence electrons of the metal equals 18.

Electrophile (Section 4.8): A species (ion or compound) that can act as a Lewis acid, or electron pair acceptor; an "electron seeker." Carbocations are one type of electrophile.

Electrophilic addition (Section 6.4): Mechanism of addition in which the species that first attacks the multiple bond is an electrophile ("electron seeker").

Electrophilic aromatic substitution (Section 12.1): Fundamental reaction type exhibited by aromatic compounds. An electrophilic species (E^+) attacks an aromatic ring and replaces one of the hydrogens.

$$Ar—H + E—Y \longrightarrow Ar—E + H—Y$$

Electrophoresis (Section 27.3): Method for separating substances on the basis of their tendency to migrate to a positively or negatively charged electrode at a particular pH.

Electrostatic attraction (Section 1.2): Force of attraction between oppositely charged particles.

Electrostatic potential (Section 1.10): The energy of interaction between a point positive charge and the charge field of a molecule.

Elementary step (Section 4.8): A step in a reaction mechanism in which each species shown in the equation for this step participates in the same transition state. An elementary step is characterized by a single transition state.

Elements of unsaturation: See *index of hydrogen deficiency.*

β Elimination (Section 5.8): Reaction in which a double or triple bond is formed by loss of atoms or groups from adjacent atoms. (See *dehydration, dehydrogenation, dehydrohalogenation,* and *double dehydrohalogenation.*)

Elimination–addition mechanism (Section 23.8): Two-stage mechanism for nucleophilic aromatic substitution. In the first stage, an aryl halide undergoes elimination to form an aryne intermediate. In the second stage, nucleophilic addition to the aryne yields the product of the reaction.

Elimination bimolecular (E2) mechanism (Section 5.15): Mechanism for elimination of alkyl halides characterized by a transition state in which the attacking base removes a proton at the same time that the bond to the halide leaving group is broken.

Elimination unimolecular (E1) mechanism (Section 5.17): Mechanism for elimination characterized by the slow formation of a carbocation intermediate followed by rapid loss of a proton from the carbocation to form the alkene.

Enamine (Section 17.11): Product of the reaction of a secondary amine and an aldehyde or a ketone. Enamines are characterized by the general structure

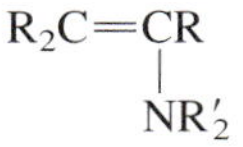

$$R_2C=CR$$
$$\underset{NR_2'}{|}$$

Enantiomeric excess (Section 7.4): Difference between the percentage of the major enantiomer present in a mixture and the percentage of its mirror image. An optically pure material has an enantiomeric excess of 100%. A racemic mixture has an enantiomeric excess of zero.

Enantiomers (Section 7.1): Stereoisomers that are related as an object and its nonsuperimposable mirror image.

Enantioselective synthesis (Section 27.4): Reaction that converts an achiral or racemic starting material to a chiral product in which one enantiomer is present in excess of the other.

Enantiotopic (Section 7.9): Describing two atoms or groups in a molecule whose environments are nonsuperimposable mirror images of each other. The two protons shown in bold in CH_3CH_2Cl, for example, are enantiotopic. Replacement of first one, then the other, by some arbitrary test group yields compounds that are enantiomers of each other.

Endergonic (Section 28.4): A process in which $\Delta G°$ is positive.

Endothermic (Section 1.2): Term describing a process or reaction that absorbs heat.

Enediyne antibiotics (Section 9.4): A family of tumor-inhibiting substances that is characterized by the presence of a C≡C—C=C—C≡C unit as part of a nine- or ten-membered ring.

Energy of activation (Section 3.2): Minimum energy that a reacting system must possess above its most stable state in order to undergo a chemical or structural change.

Enol (Section 9.12): Compound of the type

$$\underset{\displaystyle RC=CR_2}{\overset{\displaystyle OH}{|}}$$

Enols are in equilibrium with an isomeric aldehyde or ketone, but are normally much less stable than aldehydes and ketones.

Enolate ion (Section 18.6): The conjugate base of an enol. Enolate ions are stabilized by electron delocalization.

$$\underset{\displaystyle RC=CR_2}{\overset{\displaystyle :\ddot{O}:}{|}} \longleftrightarrow \underset{\displaystyle RC-\bar{C}R_2}{\overset{\displaystyle \ddot{O}:}{\|}}$$

Enthalpy (Section 2.18): The heat content of a substance; symbol, H.

Envelope (Section 3.6): One of the two most stable conformations of cyclopentane. Four of the carbons in the envelope conformation are coplanar; the fifth carbon lies above or below this plane.

Enzymatic resolution (Section 7.13): Resolution of a mixture of enantiomers based on the selective reaction of one of them under conditions of enzyme catalysis.

Enzyme (Section 27.20): A protein that catalyzes a chemical reaction in a living system.

Epimers (Section 25.21): Diastereomers that differ in configuration at only one of their stereogenic centers.

Epoxidation (Section 6.18): Conversion of an alkene to an epoxide by treatment with a peroxy acid.

Epoxide (Section 6.18): Compound of the type

$$\underset{\displaystyle O}{R_2C \text{——} CR_2}$$

Equatorial bond (Section 3.8): A bond to a carbon in the chair conformation of cyclohexane oriented approximately along the equator of the molecule.

Erythro (Section 7.11): Term applied to the relative configuration of two chirality centers within a molecule. The erythro stereoisomer has like substituents on the same side of a Fischer projection.

Essential amino acids (Section 27.1): Amino acids that must be present in the diet for normal growth and good health.

Essential fatty acids (Section 26.6): Fatty acids that must be present in the diet for normal growth and good health.

Essential oils (Section 26.7): Pleasant-smelling oils of plants consisting of mixtures of terpenes, esters, alcohols, and other volatile organic substances.

Ester (Sections 4.1 and 20.1): Compound of the type

Estrogen (Section 26.15): A female sex hormone.

Ethene (Section 5.1): IUPAC name for $H_2C=CH_2$. The common name *ethylene*, however, is used far more often, and the IUPAC rules permit its use.

Ether (Section 16.1): Molecule that contains a C—O—C unit such as ROR′, ROAr, or ArOAr.

Ethylene (Section 5.1): $H_2C=CH_2$, the simplest alkene and the most important industrial organic chemical.

Ethyl group (Section 2.13): The group CH_3CH_2-.

Exergonic (Section 28.4): A process in which $\Delta G°$ is negative.

Exothermic (Section 1.2): Term describing a reaction or process that gives off heat.

Extinction coefficient: See *molar absorptivity.*

E–Z notation for alkenes (Section 5.4): System for specifying double-bond configuration that is an alternative to cis–trans notation. When higher ranked substituents are on the same side of the double bond, the configuration is Z. When higher ranked substituents are on opposite sides, the configuration is E. Rank is determined by the Cahn–Ingold–Prelog system.

Fats and oils (Section 26.2): Triesters of glycerol. Fats are solids at room temperature, oils are liquids.

Fatty acid (Section 26.2): Carboxylic acids obtained by hydrolysis of fats and oils. Fatty acids typically have unbranched chains and contain an even number of carbon atoms in the range of 12–20 carbons. They may include one or more double bonds.

Fatty acid synthetase (Section 26.3): Complex of enzymes that catalyzes the biosynthesis of fatty acids from acetate.

Field effect (Section 19.6): An electronic effect in a molecule that is transmitted from a substituent to a reaction site via the medium (e.g., solvent).

Fingerprint region (Section 13.20): The region 1400–625 cm^{-1} of an infrared spectrum. This region is less characteristic of functional groups than others, but varies so much from one molecule to another that it can be used to determine whether two substances are identical or not.

Fischer esterification (Sections 15.8 and 19.14): Acid-catalyzed ester formation between an alcohol and a carboxylic acid:

$$\underset{\displaystyle RCOH}{\overset{\displaystyle O}{\|}} + R'OH \overset{H^+}{\longrightarrow} \underset{\displaystyle RCOR'}{\overset{\displaystyle O}{\|}} + H_2O$$

Fischer projection (Section 7.7): Method for representing stereochemical relationships. The four bonds to a chirality carbon are represented by a cross. The horizontal bonds are understood to project toward the viewer and the vertical bonds away from the viewer.

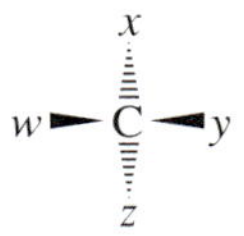

is represented in a Fischer projection as

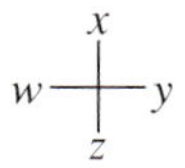

Formal charge (Section 1.6): The charge, either positive or negative, on an atom calculated by subtracting from the number of valence electrons in the neutral atom a number equal to the sum of its unshared electrons plus half the electrons in its covalent bonds.

Fragmentation pattern (Section 13.22): In mass spectrometry, the ions produced by dissociation of the molecular ion.

Free energy (Section 3.10): The available energy of a system; symbol, G.

Free radical (Section 4.16): Neutral species in which one of the electrons in the valence shell of carbon is unpaired. An example is methyl radical, $\cdot CH_3$.

Frequency (Section 13.1): Number of waves per unit time. Although often expressed in hertz (Hz), or cycles per second, the SI unit for frequency is s^{-1}.

Friedel–Crafts acylation (Section 12.7): An electrophilic aromatic substitution in which an aromatic compound reacts with an acyl chloride or a carboxylic acid anhydride in the presence of aluminum chloride. An acyl group becomes bonded to the ring.

Friedel–Crafts alkylation (Section 12.6): An electrophilic aromatic substitution in which an aromatic compound reacts with an alkyl halide in the presence of aluminum chloride. An alkyl group becomes bonded to the ring.

Fries rearrangement (Section 24.9): Aluminum chloride-promoted rearrangement of an aryl ester to a ring-acylated derivative of phenol.

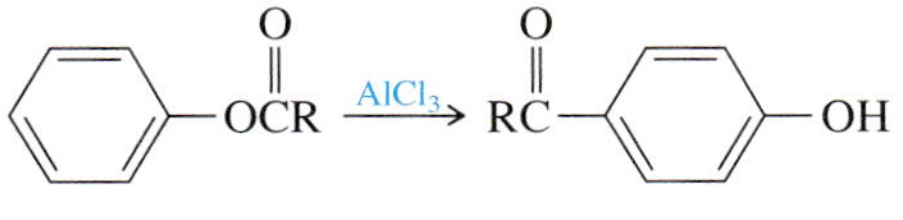

Frontier orbitals (Section 10.14): Orbitals involved in a chemical reaction, usually the highest occupied molecular orbital of one reactant and the lowest unoccupied molecular orbital of the other.

Frost's circle (Section 11.19): A mnemonic that gives the Hückel π MOs for cyclic conjugated molecules and ions.

Functional class nomenclature (Section 4.2): Type of IUPAC nomenclature in which compounds are named according to functional group families. The last word in the name identifies the functional group; the first word designates the alkyl or aryl group that bears the functional group. *Methyl bromide, ethyl alcohol,* and *diethyl ether* are examples of functional class names.

Functional group (Section 4.1): An atom or a group of atoms in a molecule responsible for its reactivity under a given set of conditions.

Furanose form (Section 25.6): Five-membered ring arising via cyclic hemiacetal formation between the carbonyl group and a hydroxyl group of a carbohydrate.

Gabriel synthesis (Section 22.8): Method for the synthesis of primary alkylamines in which a key step is the formation of a carbon–nitrogen bond by alkylation of the potassium salt of phthalimide.

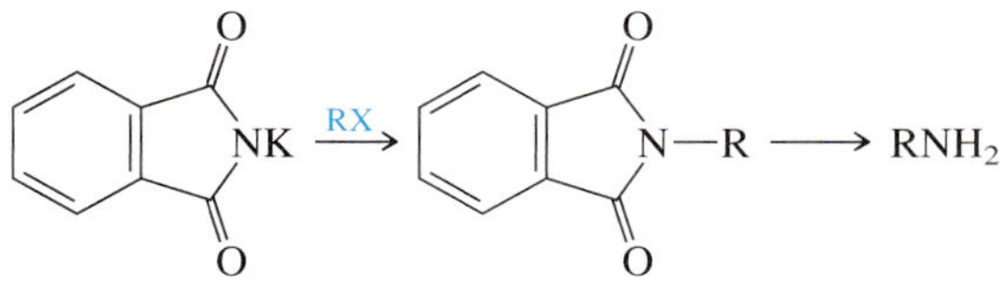

Gauche (Section 3.1): Term describing the position relative to each other of two substituents on adjacent atoms when the angle between their bonds is on the order of 60°. Atoms X and Y in the structure shown are gauche to each other.

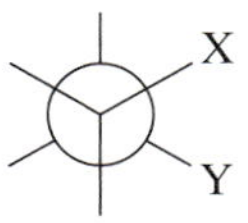

Geminal dihalide (Section 9.7): A dihalide of the form R_2CX_2, in which the two halogen substituents are located on the same carbon.

Geminal diol (Section 17.6): The hydrate $R_2C(OH)_2$ of an aldehyde or a ketone.

Genome (Section 28.7): The aggregate of all the genes that determine what an organism becomes.

Genomics (section 28.15): The study of genome sequences and their function.

Globular protein (Section 27.20): An approximately spherically shaped protein that forms a colloidal dispersion in water. Most enzymes are globular proteins.

Glycogen (Section 25.15): A polysaccharide present in animals that is derived from glucose. Similar in structure to amylopectin.

Glycolysis (Section 25.21): Biochemical process in which glucose is converted to pyruvate with release of energy.

Glycoprotein (Section 25.16): A protein to which carbohydrate molecules are attached by covalent bonds.

Glycoside (Section 25.13): A carbohydrate derivative in which the hydroxyl group at the anomeric position has been replaced by some other group. An *O*-glycoside is an ether of a carbohydrate in which the anomeric position bears an alkoxy group.

Grain alcohol (Section 4.3): A common name for ethanol (CH_3CH_2OH).

Grignard reagent (Section 14.4): An organomagnesium compound of the type RMgX formed by the reaction of magnesium with an alkyl or aryl halide.

Half-chair (Section 3.6): One of the two most stable conformations of cyclopentane. Three consecutive carbons in the half-chair conformation are coplanar. The fourth and fifth carbon lie, respectively, above and below the plane.

Haloform reaction (Section 18.7): The formation of CHX_3 ($X = Br$, Cl, or I) brought about by cleavage of a methyl ketone on treatment with Br_2, Cl_2, or I_2 in aqueous base.

$$\underset{RCCH_3}{\overset{O}{\|}} \xrightarrow[HO^-]{X_2} \underset{RCO^-}{\overset{O}{\|}} + CHX_3$$

Halogenation (Sections 4.14 and 12.5): Replacement of a hydrogen by a halogen. The most frequently encountered examples are the free-radical halogenation of alkanes and the halogenation of arenes by electrophilic aromatic substitution.

Halohydrin (Section 6.17): A compound that contains both a halogen atom and a hydroxyl group. The term is most often used for compounds in which the halogen and the hydroxyl group are on adjacent atoms (*vicinal halohydrins*). The most commonly encountered halohydrins are *chlorohydrins* and *bromohydrins*.

Halonium ion (Section 6.16): A species that incorporates a positively charged halogen. Bridged halonium ions are intermediates in the addition of halogens to the double bond of an alkene.

Hammond's postulate (Section 4.8): Principle used to deduce the approximate structure of a transition state. If two states, such as a transition state and an unstable intermediate derived from it, are similar in energy, they are believed to be similar in structure.

Haworth formulas (Section 25.6): Planar representations of furanose and pyranose forms of carbohydrates.

Heat of combustion (Section 2.18): Heat evolved on combustion of a substance. It is the value of $-\Delta H°$ for the combustion reaction.

Heat of formation (Section 2.18): The value of $\Delta H°$ for formation of a substance from its elements.

Heat of hydrogenation (Section 6.1): Heat evolved on hydrogenation of a substance. It is the value of $-\Delta H°$ for the addition of H_2 to a multiple bond.

α Helix (Section 27.19): One type of protein secondary structure. It is a right-handed helix characterized by hydrogen bonds between NH and $C=O$ groups. It contains approximately 3.6 amino acids per turn.

Hell–Volhard–Zelinsky reaction (Section 19.16): The phosphorus trihalide-catalyzed α halogenation of a carboxylic acid:

$$R_2CHCO_2H + X_2 \xrightarrow[\text{or } PX_3]{P} \underset{X}{\overset{|}{R_2CCO_2H}} + HX$$

Hemiacetal (Section 17.8): Product of nucleophilic addition of one molecule of an alcohol to an aldehyde or a ketone. Hemiacetals are compounds of the type

$$\underset{R_2C-OR'}{\overset{\overset{\displaystyle OH}{|}}{}}$$

Hemiketal (Section 17.8): A hemiacetal derived from a ketone.

Henderson–Hasselbalch equation (Section 19.4): An equation that relates degree of dissociation of an acid at a particular pH to its pK_a.

$$pH = pK_a + \log \frac{[\text{conjugate base}]}{[\text{acid}]}$$

HETCOR (Section 13.19): A 2D NMR technique that correlates the ^1H chemical shift of a proton to the ^{13}C chemical shift of the carbon to which it is attached. HETCOR stands for *heteronuclear chemical shift correlation*.

Heteroatom (Section 1.7): An atom in an organic molecule that is neither carbon nor hydrogen.

Heterocyclic compound (Section 3.15): Cyclic compound in which one or more of the atoms in the ring are elements other than carbon. Heterocyclic compounds may or may not be aromatic.

Heterogeneous reaction (Section 6.1): A reaction involving two or more substances present in different phases. Hydrogenation of alkenes is a heterogeneous reaction that takes place on the surface of an insoluble metal catalyst.

Heterolytic cleavage (Section 4.16): Dissociation of a two-electron covalent bond in such a way that both electrons are retained by one of the initially bonded atoms.

Hexose (Section 25.4): A carbohydrate with six carbon atoms.

High-density lipoprotein (HDL) (Section 26.11): A protein that carries cholesterol from the tissues to the liver where it is metabolized. HDL is often called "good cholesterol."

Histones (Section 28.9): Proteins that are associated with DNA in nucleosomes.

Hofmann elimination (Section 22.14): Conversion of a quaternary ammonium hydroxide, especially an alkyltrimethylammonium hydroxide, to an alkene on heating. Elimination occurs in the direction that gives the less substituted double bond.

$$\underset{\overset{|}{\overset{+}{N}(CH_3)_3}}{R_2CH-CR_2'} \; HO^- \xrightarrow{\text{heat}} R_2C=CR_2' + N(CH_3)_3 + H_2O$$

HOMO (Section 10.13): Highest occupied molecular orbital (the orbital of highest energy that contains at least one of a molecule's electrons).

Homologous series (Section 2.9): Group of structurally related substances in which successive members differ by a CH_2 group.

Homolytic cleavage (Section 4.16): Dissociation of a two-electron covalent bond in such a way that one electron is retained by each of the initially bonded atoms.

Hückel's rule (Section 11.19): Completely conjugated planar monocyclic hydrocarbons possess special stability when the number of their π electrons $= 4n + 2$, where n is an integer.

Hund's rule (Section 1.1): When two orbitals are of equal energy, they are populated by electrons so that each is half-filled before either one is doubly occupied.

Hybrid orbital (Section 2.6): An atomic orbital represented as a mixture of various contributions of that atom's s, p, d, etc. orbitals.

Hydration (Section 6.10): Addition of the elements of water (H, OH) to a multiple bond.

Hydride shift (Section 5.13): Migration of a hydrogen with a pair of electrons (H:) from one atom to another. Hydride shifts are most commonly seen in carbocation rearrangements.

Hydroboration–oxidation (Section 6.11): Reaction sequence involving a separate hydroboration stage and oxidation stage. In the hydroboration stage, diborane adds to an alkene to give an alkylborane. In the oxidation stage, the alkylborane is oxidized with hydrogen peroxide to give an alcohol. The reaction product is an alcohol corresponding to the anti-Markovnikov, syn hydration of an alkene.

Hydrocarbon (Section 2.1): A compound that contains only carbon and hydrogen.

Hydroformylation (Section 17.5): An industrial process for preparing aldehydes ($RCH_2CH_2CH{=}O$) by the reaction of terminal alkenes ($RCH{=}CH_2$) with carbon monoxide.

Hydrogenation (Section 6.1): Addition of H_2 to a multiple bond.

Hydrogen bonding (Section 4.6): Type of dipole–dipole attractive force in which a positively polarized hydrogen of one molecule is weakly bonded to a negatively polarized atom of an adjacent molecule. Hydrogen bonds typically involve the hydrogen of one —OH or —NH group and the oxygen or nitrogen of another.

Hydrolysis (Section 6.9): Water-induced cleavage of a bond.

Hydronium ion (Section 1.13): The species H_3O^+.

Hydrophilic (Section 19.5): Literally, "water-loving"; a term applied to substances that are soluble in water, usually because of their ability to form hydrogen bonds with water.

Hydrophobic (Section 19.5): Literally, "water-hating"; a term applied to substances that are not soluble in water, but are soluble in nonpolar, hydrocarbon-like media.

Hydroxylation (Section 15.5): Reaction or sequence of reactions in which an alkene is converted to a vicinal diol.

Hyperconjugation (Section 4.10): Delocalization of σ electrons.

Icosanoids (Section 26.6): A group of naturally occurring compounds derived from unsaturated C_{20} carboxylic acids.

Imide (Section 20.16): Compound of the type

$$\overset{\displaystyle O}{\overset{\displaystyle \|}{RN(CR')_2}}$$

in which two acyl groups are bonded to the same nitrogen.

Imine (Section 17.10): Compound of the type $R_2C{=}NR'$ formed by the reaction of an aldehyde or a ketone with a

primary amine ($R'NH_2$). Imines are sometimes called *Schiff's bases.*

Index of hydrogen deficiency (Section 13.23): A measure of the total double bonds and rings a molecule contains. It is determined by comparing the molecular formula C_nH_x of the compound to that of an alkane that has the same number of carbons according to the equation:

$$\text{Index of hydrogen deficiency} = \frac{1}{2}(C_nH_{2n+2} - C_nH_x)$$

Induced-dipole/induced-dipole attraction (Section 2.17): Force of attraction resulting from a mutual and complementary polarization of one molecule by another. Also referred to as *London forces* or *dispersion forces.*

Inductive effect (Section 1.15): An electronic effect transmitted by successive polarization of the σ bonds within a molecule or an ion.

Infrared (IR) spectroscopy (Section 13.20): Analytical technique based on energy absorbed by a molecule as it vibrates by stretching and bending bonds. Infrared spectroscopy is useful for analyzing the functional groups in a molecule.

Initiation step (Section 4.17): A process which causes a reaction, usually a free-radical reaction, to begin but which by itself is not the principal source of products. The initiation step in the halogenation of an alkane is the dissociation of a halogen molecule to two halogen atoms.

Integrated area (Section 13.6): The relative area of a signal in an NMR spectrum. Areas are proportional to the number of equivalent protons responsible for the peak.

Intermediate (Section 3.9): Transient species formed during a chemical reaction. Typically, an intermediate is not stable under the conditions of its formation and proceeds further to form the product. Unlike a transition state, which corresponds to a maximum along a potential energy surface, an intermediate lies at a potential energy minimum.

Intermolecular forces (Section 2.17): Forces, either attractive or repulsive, between two atoms or groups in *separate* molecules.

Intramolecular forces (Section 2.18): Forces, either attractive or repulsive, between two atoms or groups *within* the same molecule.

Inversion of configuration (Section 8.4): Reversal of the three-dimensional arrangement of the four bonds to sp^3-hybridized carbon. The representation shown illustrates inversion of configuration in a nucleophilic substitution where LG is the leaving group and Nu is the nucleophile.

Ionic bond (Section 1.2): Chemical bond between oppositely charged particles that results from the electrostatic attraction between them.

Ionization energy (Section 1.2): Amount of energy required to remove an electron from some species.

Isobutane (Section 2.8): The common name for 2-methylpropane, $(CH_3)_3CH$.

Isobutyl group (Section 2.13): The group $(CH_3)_2CHCH_2—$.

Isoelectric point (Section 27.3): pH at which the concentration of the zwitterionic form of an amino acid is a maximum. At a pH below the isoelectric point the dominant species is a cation. At higher pH, an anion predominates. At the isoelectric point the amino acid has no net charge.

Isolated diene (Section 10.5): Diene of the type

$$C=C—(C)_x—C=C$$

in which the two double bonds are separated by one or more sp^3-hybridized carbons. Isolated dienes are slightly less stable than isomeric conjugated dienes.

Isomers (Section 1.8): Different compounds that have the same molecular formula. Isomers may be either constitutional isomers or stereoisomers.

Isopentane (Section 2.10): The common name for 2-methylbutane, $(CH_3)_2CHCH_2CH_3$.

Isoprene unit (Section 26.7): The characteristic five-carbon structural unit found in terpenes:

Isopropyl group (Section 2.13): The group $(CH_3)_2CH—$.

Isotactic polymer (Section 7.15): A stereoregular polymer in which the substituent at each successive chirality center is on the same side of the zigzag carbon chain.

Isotopic cluster (Section 13.22): In mass spectrometry, a group of peaks that differ in m/z because they incorporate different isotopes of their component elements.

IUPAC nomenclature (Section 2.11): The most widely used method of naming organic compounds. It uses a set of rules proposed and periodically revised by the International Union of Pure and Applied Chemistry.

Kekulé structure (Section 11.2): Structural formula for an aromatic compound that satisfies the customary rules of bonding and is usually characterized by a pattern of alternating single and double bonds. There are two Kekulé formulations for benzene:

A single Kekulé structure does not completely describe the actual bonding in the molecule.

Ketal (Section 17.8): An acetal derived from a ketone.

Keto−enol tautomerism (Section 18.4): Process by which an aldehyde or a ketone and its enol equilibrate:

$$\underset{\text{O}}{\overset{\text{O}}{\underset{\|}{RC}}}—CHR_2 \rightleftharpoons \underset{\text{OH}}{\overset{\text{OH}}{\underset{|}{RC}}}=CR_2$$

β-Keto ester (Section 21.1): A compound of the type

Ketone (Sections 4.1 and 17.1): A member of the family of compounds in which both atoms attached to a carbonyl group (C=O) are carbon, as in

$$\underset{RCR}{\overset{O}{\|}} \qquad \underset{RCAr}{\overset{O}{\|}} \qquad \underset{ArCAr}{\overset{O}{\|}}$$

Ketose (Section 25.1): A carbohydrate that contains a ketone carbonyl group in its open-chain form.

Kiliani−Fischer synthesis (Section 25.20): A synthetic method for carbohydrate chain extension. The new carbon−carbon bond is formed by converting an aldose to its cyanohydrin. Reduction of the cyano group to an aldehyde function completes the synthesis.

Kinases (Section 28.3): Enzymes that catalyze the transfer of phosphate from ATP to some other molecule.

Kinetically controlled reaction (Section 10.10): Reaction in which the major product is the one that is formed at the fastest rate.

Kolbe−Schmitt reaction (Section 24.10): The high-pressure reaction of the sodium salt of a phenol with carbon dioxide to give an o-hydroxybenzoic acid. The Kolbe−Schmitt reaction is used to prepare salicylic acid in the synthesis of aspirin.

Lactam (Section 20.15): A cyclic amide.

Lactone (Section 19.15): A cyclic ester.

Lactose (Section 25.14): Milk sugar; a disaccharide formed by a β-glycosidic linkage between C-4 of glucose and C-1 of galactose.

Lagging strand (Section 28.10): In DNA replication, the strand that grows away from the replication fork.

LDA (Section 21.10): Abbreviation for lithium diisopropylamide $LiN[CH(CH_3)_2]$. LDA is a strong, sterically hindered base, used to convert esters to their enolates.

Leading strand (Section 28.10): In DNA replication, the strand that grows toward the replication fork.

Leaving group (Section 5.15): The group, normally a halide ion, that is lost from carbon in a nucleophilic substitution or elimination.

Le Châtelier's principle (Section 6.10): A reaction at equilibrium responds to any stress imposed on it by shifting the equilibrium in the direction that minimizes the stress.

Lewis acid: See *acid.*

Lewis base: See *base.*

Lewis structure (Section 1.3): A chemical formula in which electrons are represented by dots. Two dots (or a line) between two atoms represent a covalent bond in a Lewis structure. Unshared electrons are explicitly shown, and stable Lewis structures are those in which the octet rule is satisfied.

Lindlar catalyst (Section 9.9): A catalyst for the hydrogenation of alkynes to *cis*-alkenes. It is composed of palladium, which has been "poisoned" with lead(II) acetate and quinoline, supported on calcium carbonate.

Lipid bilayer (Section 26.4): Arrangement of two layers of phospholipids that constitutes cell membranes. The polar termini are located at the inner and outer membrane–water interfaces, and the lipophilic hydrocarbon tails cluster on the inside.

Lipids (Section 26.1): Biologically important natural products characterized by high solubility in nonpolar organic solvents.

Lipophilic (Section 19.5): Literally, "fat-loving"; synonymous in practice with *hydrophobic*.

Locant (Section 2.12): In IUPAC nomenclature, a prefix that designates the atom that is associated with a particular structural unit. The locant is most often a number, and the structural unit is usually an attached substituent as in *2-chlorobutane*.

London force (Section 2.17): See *induced-dipole/induced-dipole attraction*.

Low-density lipoprotein (LDL) (Section 26.11): A protein which carries cholesterol from the liver through the blood to the tissues. Elevated LDL levels are a risk factor for heart disease; LDL is often called "bad cholesterol."

LUMO (Section 10.13): The orbital of lowest energy that contains none of a molecule's electrons; the lowest unoccupied molecular orbital.

Magnetic resonance imaging (MRI) (Section 13.13): A diagnostic method in medicine in which tissues are examined by NMR.

Malonic ester synthesis (Section 21.7): Synthetic method for the preparation of carboxylic acids involving alkylation of the enolate of diethyl malonate

$$\underset{\displaystyle CH_3CH_2OCCH_2COCH_2CH_3}{\overset{\displaystyle O \qquad\quad O}{\underset{\|}{} \qquad \underset{\|}{}}}$$

as the key carbon–carbon bond-forming step.

Maltose (Section 25.14): A disaccharide obtained from starch in which two glucose units are joined by an $\alpha(1,4)$-glycosidic link.

Markovnikov's rule (Section 6.5): An unsymmetrical reagent adds to an unsymmetrical double bond in the direction that places the positive part of the reagent on the carbon of the double bond that has the greater number of hydrogens.

Mass spectrometry (Section 13.22): Analytical method in which a molecule is ionized and the various ions are examined on the basis of their mass-to-charge ratio.

Mechanism (Section 4.8): The sequence of steps that describes how a chemical reaction occurs; a description of the intermediates and transition states that are involved during the transformation of reactants to products.

Mercaptan (Section 15.13): An old name for the class of compounds now known as *thiols*.

Merrifield method: See *solid-phase peptide synthesis*.

Meso stereoisomer (Section 7.11): An achiral molecule that has chirality centers. The most common kind of meso compound is a molecule with two chirality centers and a plane of symmetry.

Messenger RNA (mRNA) (Section 28.11): A polynucleotide of ribose that "reads" the sequence of bases in DNA and interacts with tRNAs in the ribosomes to promote protein biosynthesis.

Meta (Section 11.7): Term describing a 1,3 relationship between substituents on a benzene ring.

Meta director (Section 12.9): A group that when present on a benzene ring directs an incoming electrophile to a position meta to itself.

Metallocene (Section 14.14): A transition metal complex that bears a cyclopentadienyl ligand.

Metalloenzyme (Section 27.20): An enzyme in which a metal ion at the active site contributes in a chemically significant way to the catalytic activity.

Methanogen (Section 2.5): An organism that produces methane.

Methine group (Section 2.8): The group CH.

Methylene group (Section 2.8): The group $—CH_2—$.

Methyl group (Section 2.7): The group $—CH_3$.

Mevalonic acid (Section 26.10): An intermediate in the biosynthesis of steroids from acetyl coenzyme A.

Micelle (Section 19.5): A spherical aggregate of species such as carboxylate salts of fatty acids that contain a lipophilic end and a hydrophilic end. Micelles containing 50–100 carboxylate salts of fatty acids are soaps.

Michael addition (Sections 18.13 and 21.9): The conjugate addition of a carbanion (usually an enolate) to an α,β-unsaturated carbonyl compound.

Microscopic reversibility (Section 6.10): The principle that the intermediates and transition states in the forward and backward stages of a reversible reaction are identical, but are encountered in the reverse order.

Molar absorptivity (Section 13.21): A measure of the intensity of a peak, usually in UV-VIS spectroscopy.

Molecular dipole moment (Section 1.11): The overall measured dipole moment of a molecule. It can be calculated as the resultant (or vector sum) of all the individual bond dipole moments.

Molecular formula (Section 1.7): Chemical formula in which subscripts are used to indicate the number of atoms of each element present in one molecule. In organic compounds, carbon is cited first, hydrogen second, and the remaining elements in alphabetical order.

Molecular ion (Section 13.22): In mass spectrometry, the species formed by loss of an electron from a molecule.

Molecular orbital theory (Section 2.4): Theory of chemical bonding in which electrons are assumed to occupy orbitals in molecules much as they occupy orbitals in atoms. The molecular orbitals are described as combinations of the orbitals of all of the atoms that make up the molecule.

Molecularity (Section 4.8): The number of species that react together in the same elementary step of a reaction mechanism.

Monomer (Section 6.21): The simplest stable molecule from which a particular polymer may be prepared.

Monosaccharide (Section 25.1): A carbohydrate that cannot be hydrolyzed further to yield a simpler carbohydrate.

Monosubstituted alkene (Section 5.6): An alkene of the type $RCH=CH_2$, in which there is only one carbon *directly* bonded to the carbons of the double bond.

Multiplicity (Section 13.7): The number of peaks into which a signal is split in nuclear magnetic resonance spectroscopy. Signals are described as *singlets, doublets, triplets,* and so on, according to the number of peaks into which they are split.

Mutarotation (Section 25.8): The change in optical rotation that occurs when a single form of a carbohydrate is allowed to equilibrate to a mixture of isomeric hemiacetals.

Nanotube (Section 11.8): A form of elemental carbon composed of a cylindrical cluster of carbon atoms.

Neopentane (Section 2.10): The common name for 2,2-dimethylpropane, $(CH_3)_4C$.

Neurotransmitter (Section 22.4): Substance, usually a naturally occurring amine, that mediates the transmission of nerve impulses.

Newman projection (Section 3.1): Method for depicting conformations in which one sights down a carbon–carbon bond and represents the front carbon by a point and the back carbon by a circle.

Nitration (Section 12.3): Replacement of a hydrogen by an —NO_2 group. The term is usually used in connection with electrophilic aromatic substitution.

$$Ar-H \xrightarrow[H_2SO_4]{HNO_3} Ar-NO_2$$

Nitrile (Section 20.1): A compound of the type $RC{\equiv}N$. R may be alkyl or aryl. Also known as *alkyl* or *aryl cyanides.*

Nitrogen rule (Section 13.23): The molecular weight of a substance that contains C, H, O, and N is odd if the number of nitrogens is odd. The molecular weight is even if the number of nitrogens is even.

Nitrosamine: See *N-nitroso amine.*

Nitrosation (Section 22.15): The reaction of a substance, usually an amine, with nitrous acid. Primary amines yield diazonium ions; secondary amines yield *N*-nitroso amines. Tertiary aromatic amines undergo nitrosation of their aromatic ring.

N-Nitroso amine (Section 22.15): A compound of the type $R_2N-N=O$. R may be alkyl or aryl groups, which may be the same or different. *N*-Nitroso amines are formed by nitrosation of secondary amines.

Noble gases (Section 1.1): The elements in group VIIIA of the periodic table (helium, neon, argon, krypton, xenon, radon).

Also known as the *rare gases,* they are, with few exceptions, chemically inert.

Nodal surface (Section 1.1): A plane drawn through an orbital where the algebraic sign of a wave function changes. The probability of finding an electron at a node is zero.

NSAIDs (Section 26.6): Nonsteroidal antiinflammatory drugs.

N terminus (Section 27.7): The amino acid at the end of a peptide or protein chain that has its α-amino group intact; that is, the α-amino group is not part of a peptide bond.

Nuclear magnetic resonance (NMR) spectroscopy (Section 13.3): A method for structure determination based on the effect of molecular environment on the energy required to promote a given nucleus from a lower energy spin state to a higher energy state.

Nucleic acid (Section 28.7): A polynucleotide present in the nuclei of cells.

Nucleophile (Section 4.8): An atom or ion that has an unshared electron pair which can be used to form a bond to carbon. Nucleophiles are Lewis bases.

Nucleophilic acyl substitution (Section 20.3): Nucleophilic substitution at the carbon atom of an acyl group.

Nucleophilic addition (Section 17.6): The characteristic reaction of an aldehyde or a ketone. An atom possessing an unshared electron pair bonds to the carbon of the $C=O$ group, and some other species (normally hydrogen) bonds to the oxygen.

$$\underset{\underset{R'}{|}}{\overset{\overset{O}{\|}}{R C}} R' + H-Y: \longrightarrow \underset{\underset{R'}{|}}{\overset{\overset{OH}{|}}{R C}}-Y:$$

Nucleophilic aliphatic substitution (Chapter 8): Reaction in which a nucleophile replaces a leaving group, usually a halide ion, from sp^3-hybridized carbon. Nucleophilic aliphatic substitution may proceed by either an S_N1 or an S_N2 mechanism.

Nucleophilic aromatic substitution (Chapter 23): A reaction in which a nucleophile replaces a leaving group as a substituent on an aromatic ring. Substitution may proceed by an addition–elimination mechanism or an elimination–addition mechanism.

Nucleophilicity (Section 8.7): A measure of the reactivity of a Lewis base in a nucleophilic substitution reaction.

Nucleoside (Section 28.2): The combination of a purine or pyrimidine base and a carbohydrate, usually ribose or 2-deoxyribose.

Nucleosome (Section 28.9): A DNA–protein complex by which DNA is stored in cells.

Nucleotide (Section 28.3): The phosphate ester of a nucleoside.

Octane rating (Section 2.16): The capacity of a sample of gasoline to resist "knocking," expressed as a number equal to the percentage of 2,2,4-trimethylpentane ("isooctane") in an isooctane–heptane mixture that has the same knocking characteristics.

Octet rule (Section 1.3): When forming compounds, atoms gain, lose, or share electrons so that the number of their valence electrons is the same as that of the nearest noble gas. For the elements carbon, nitrogen, oxygen, and the halogens, this number is 8.

Oligomer (Section 14.15): A molecule composed of too few monomer units for it to be classified as a polymer, but more than in a dimer, trimer, tetramer, etc.

Oligonucleotide (Section 28.6): A polynucleotide containing a relatively small number of bases.

Oligosaccharide (Section 25.1): A carbohydrate that gives three to ten monosaccharides on hydrolysis.

Optical activity (Section 7.4): Ability of a substance to rotate the plane of polarized light. To be optically active, a substance must be chiral, and one enantiomer must be present in excess of the other.

Optically pure (Section 7.4): Describing a chiral substance in which only a single enantiomer is present.

Orbital (Section 1.1): Strictly speaking, a wave function ψ. It is convenient, however, to think of an orbital in terms of the probability ψ^2 of finding an electron at some point relative to the nucleus, as the volume inside the boundary surface of an atom, or the region in space where the probability of finding an electron is high.

σ Orbital (Section 2.4): A bonding orbital characterized by rotational symmetry.

σ* Orbital (Section 2.4): An antibonding orbital characterized by rotational symmetry.

Organometallic compound (Section 14.1): A compound that contains a carbon-to-metal bond.

Ortho (Section 11.7): Term describing a 1,2 relationship between substituents on a benzene ring.

Ortho, para director (Section 12.9): A group that when present on a benzene ring directs an incoming electrophile to the positions ortho and para to itself.

Oxidation (Section 2.19): A decrease in the number of electrons associated with an atom. In organic chemistry, oxidation of carbon occurs when a bond between carbon and an atom that is less electronegative than carbon is replaced by a bond to an atom that is more electronegative than carbon.

Oxidation number (Section 2.19): The formal charge an atom has when the atoms in its covalent bonds are assigned to the more electronegative partner.

Oxidation state: See *oxidation number.*

Oxime (Section 17.10): A compound of the type $R_2C=NOH$, formed by the reaction of hydroxylamine (NH_2OH) with an aldehyde or a ketone.

Oxonium ion (Section 1.13): The species H_3O^+ (also called *hydronium ion*).

Ozonolysis (Section 6.19): Ozone-induced cleavage of a carbon–carbon double or triple bond.

Para (Section 11.7): Term describing a 1,4 relationship between substituents on a benzene ring.

Paraffin hydrocarbons (Section 2.18): An old name for alkanes and cycloalkanes.

Partial rate factor (Section 12.10): In electrophilic aromatic substitution, a number that compares the rate of attack at a particular ring carbon with the rate of attack at a single position of benzene.

Pauli exclusion principle (Section 1.1): No two electrons can have the same set of four quantum numbers. An equivalent expression is that only two electrons can occupy the same orbital, and then only when they have opposite spins.

PCC (Section 15.10): Abbreviation for pyridinium chlorochromate $C_5H_5NH^+$ $ClCrO_3^-$. When used in an anhydrous medium, PCC oxidizes primary alcohols to aldehydes and secondary alcohols to ketones.

PDC (Section 15.10): Abbreviation for pyridinium dichromate $(C_5H_5NH)_2^{2+}$ $Cr_2O_7^{2-}$. Used in same manner and for same purposes as PCC (see preceding entry).

n-Pentane (Section 2.10): The common name for pentane, $CH_3CH_2CH_2CH_2CH_3$.

Pentose (Section 25.4): A carbohydrate with five carbon atoms.

Peptide (Section 27.7): Structurally, a molecule composed of two or more α-amino acids joined by peptide bonds.

Peptide bond (Section 27.7): An amide bond between the carboxyl group of one α-amino acid and the amino group of another.

(The bond highlighted in yellow is the peptide bond.)

Pericyclic reaction (Section 10.12): A reaction that proceeds through a cyclic transition state.

Period (Section 1.1): A horizontal row of the periodic table.

Peroxide (Section 6.8): A compound of the type ROOR.

Peroxide effect (Section 6.8): Reversal of regioselectivity observed in the addition of hydrogen bromide to alkenes brought about by the presence of peroxides in the reaction mixture.

Phase-transfer catalysis (Section 22.5): Method for increasing the rate of a chemical reaction by transporting an ionic reactant from an aqueous phase where it is solvated and less reactive to an organic phase where it is not solvated and is more reactive. Typically, the reactant is an anion that is carried to the organic phase as its quaternary ammonium salt.

Phenols (Section 24.1): Family of compounds characterized by a hydroxyl substituent on an aromatic ring as in ArOH. *Phenol* is also the name of the parent compound, C_6H_5OH.

Phenyl group (Section 11.7): The group

It is often abbreviated C_6H_5—.

Phosphodiester (Section 28.6): Compound of the type shown, especially when R and R′ are D-ribose or 2-deoxy-D-ribose.

$$R-O-\overset{\overset{\displaystyle O}{\|}}{\underset{\underset{\displaystyle OH}{|}}{P}}-O-R'$$

Phospholipid (Section 26.4): A diacylglycerol bearing a choline-phosphate "head group." Also known as *phosphatidylcholine.*

Photochemical reaction (Section 4.18): A chemical reaction that occurs when light is absorbed by a substance.

Photon (Section 13.1): Term for an individual "bundle" of energy, or particle, of electromagnetic radiation.

pK_a (Section 1.12): A measure of acid strength defined as $-\log K_a$. The stronger the acid, the smaller the value of pK_a.

Planck's constant (Section 13.1): Constant of proportionality (h) in the equation $E = h\nu$, which relates the energy (E) to the frequency (ν) of electromagnetic radiation.

Plane of symmetry (Section 7.3): A plane that bisects an object, such as a molecule, into two mirror-image halves; also called a *mirror plane*. When a line is drawn from any element in the object perpendicular to such a plane and extended an equal distance in the opposite direction, a duplicate of the element is encountered.

Pleated β sheet (Section 27.19): Type of protein secondary structure characterized by hydrogen bonds between NH and C=O groups of adjacent parallel peptide chains. The individual chains are in an extended zigzag conformation.

Polar covalent bond (Section 1.5): A shared electron pair bond in which the electrons are drawn more closely to one of the bonded atoms than the other.

Polarimeter (Section 7.4): An instrument used to measure optical activity.

Polarizability (Section 4.6): A measure of the ease of distortion of the electric field associated with an atom or a group. A fluorine atom in a molecule, for example, holds its electrons tightly and is very nonpolarizable. Iodine is very polarizable.

Polarized light (Section 7.4): Light in which the electric field vectors vibrate in a single plane. Polarized light is used in measuring optical activity.

Polyamide (Section 20.17): A polymer in which individual structural units are joined by amide bonds. Nylon is a synthetic polyamide; proteins are naturally occurring polyamides.

Polyamine (Section 22.4): A compound that contains many amino groups. The term is usually applied to a group of naturally occurring substances, including spermine, spermidine, and putrescine, that are believed to be involved in cell differentiation and proliferation.

Polycyclic aromatic hydrocarbon (Section 11.8): An aromatic hydrocarbon characterized by the presence of two or more fused benzene rings.

Polycyclic hydrocarbon (Section 3.14): A hydrocarbon in which two carbons are common to two or more rings.

Polyester (Section 20.17): A polymer in which individual structural units are joined by ester bonds.

Polyether (Section 16.4): A molecule that contains many ether linkages. Polyethers occur naturally in a number of antibiotic substances.

Polyethylene (Section 6.21): A polymer of ethylene.

Polymer (Section 6.21): Large molecule formed by the repetitive combination of many smaller molecules (monomers).

Polymerase chain reaction (Section 28.16): A laboratory method for making multiple copies of DNA.

Polymerization (Section 6.21): Process by which a polymer is prepared. The principal processes include *free-radical, cationic, coordination,* and *condensation polymerization.*

Polypeptide (Section 27.1): A polymer made up of "many" (more than eight to ten) amino acid residues.

Polypropylene (Section 6.21): A polymer of propene.

Polysaccharide (Sections 25.1 and 25.15): A carbohydrate that yields "many" monosaccharide units on hydrolysis.

Potential energy (Section 2.18): The energy a system has exclusive of its kinetic energy.

Potential energy diagram (Section 4.8): Plot of potential energy versus some arbitrary measure of the degree to which a reaction has proceeded (the reaction coordinate). The point of maximum potential energy is the transition state.

Primary alkyl group (Section 2.13): Structural unit of the type RCH_2-, in which the point of attachment is to a primary carbon.

Primary amine (Section 22.1): An amine with a single alkyl or aryl substituent and two hydrogens: an amine of the type RNH_2 (primary alkylamine) or $ArNH_2$ (primary arylamine).

Primary carbon (Section 2.13): A carbon that is directly attached to only one other carbon.

Primary structure (Section 27.8): The sequence of amino acids in a peptide or protein.

Principal quantum number (Section 1.1): The quantum number (n) of an electron that describes its energy level. An electron with $n = 1$ must be an s electron; one with $n = 2$ has s and p states available.

Prochiral (Section 7.9): The capacity of an achiral molecule to become chiral by replacement of an existing atom or group by a different one.

Propagation steps (Section 4.17): Elementary steps that repeat over and over again in a chain reaction. Almost all of the products in a chain reaction arise from the propagation steps.

Protease inhibitor (Section 28.13): A substance that interferes with enzyme-catalyzed hydrolysis of peptide bonds.

Protecting group (Section 17.9): A temporary alteration in the nature of a functional group so that it is rendered inert under the conditions in which reaction occurs somewhere else in the molecule. To be synthetically useful, a protecting group must be stable under a prescribed set of reaction conditions, yet be easily introduced and removed.

Protein (Chapter 27): A naturally occurring polypeptide that has a biological function.

Protein Data Bank (Section 27.20): A central repository in which crystallographic coordinates for biological molecules, especially proteins, are stored. The data are accessible via the Worldwide Web and can be transformed into three-dimensional images with appropriate molecular-modeling software.

Protic solvent (Section 8.12): A solvent that has easily exchangeable protons, especially protons bonded to oxygen as in hydroxyl groups.

Purine (Section 28.1): The heterocyclic aromatic compound.

Pyranose form (Section 25.7): Six-membered ring arising via cyclic hemiacetal formation between the carbonyl group and a hydroxyl group of a carbohydrate.

Pyrimidine (Section 28.1): The heterocyclic aromatic compound.

Quantum (Section 13.1): The energy associated with a photon.

Quaternary ammonium salt (Section 22.1): Salt of the type $R_4N^+ X^-$. The positively charged ion contains a nitrogen with a total of four organic substituents (any combination of alkyl and aryl groups).

Quaternary carbon (Section 2.13): A carbon that is directly attached to four other carbons.

Quaternary structure (Section 27.22): Description of the way in which two or more protein chains, not connected by chemical bonds, are organized in a larger protein.

Quinone (Section 24.14): The product of oxidation of an ortho or para dihydroxybenzene derivative. Examples of quinones include

R (Section 4.1): Symbol for an alkyl group.

Racemic mixture (Section 7.4): Mixture containing equal quantities of enantiomers.

Rate-determining step (Section 4.9): Slowest step of a multistep reaction mechanism. The overall rate of a reaction can be no faster than its slowest step.

Rearrangement (Section 5.13): Intramolecular migration of an atom, a group, or a bond from one atom to another.

Reducing sugar (Section 25.19): A carbohydrate that can be oxidized with substances such as Benedict's reagent. In general, a carbohydrate with a free hydroxyl group at the anomeric position.

Reduction (Section 2.19): Gain in the number of electrons associated with an atom. In organic chemistry, reduction of carbon occurs when a bond between carbon and an atom which is more electronegative than carbon is replaced by a bond to an atom which is less electronegative than carbon.

Reductive amination (Section 22.10): Method for the preparation of amines in which an aldehyde or a ketone is treated with ammonia or an amine under conditions of catalytic hydrogenation.

Refining (Section 2.16): Conversion of crude oil to useful materials, especially gasoline.

Reforming (Section 2.16): Step in oil refining in which the proportion of aromatic and branched-chain hydrocarbons in petroleum is increased so as to improve the octane rating of gasoline.

Regioselective (Section 5.10): Term describing a reaction that can produce two (or more) constitutional isomers but gives one of them in greater amounts than the other. A reaction that is 100% regioselective is termed regiospecific.

Relative configuration (Section 7.5): Stereochemical configuration on a comparative, rather than an absolute, basis. Terms such as D, L, erythro, threo, α, and β describe relative configuration.

Replication fork (Section 28.9): Point at which strands of double-helical DNA separate.

Resolution (Section 7.14): Separation of a racemic mixture into its enantiomers.

Resonance (Section 1.9): Method by which electron delocalization may be shown using Lewis structures. The true electron distribution in a molecule is regarded as a hybrid of the various Lewis structures that can be written for a molecule.

Resonance energy (Section 10.6): Extent to which a substance is stabilized by electron delocalization. It is the difference in energy between the substance and a hypothetical model in which the electrons are localized.

Restriction enzymes (Section 28.14): Enzymes that catalyze the cleavage of DNA at specific sites.

Retention of configuration (Section 6.13): Stereochemical pathway observed when a new bond is made that has the same spatial orientation as the bond that was broken.

Retrosynthetic analysis (Section 14.9): Technique for synthetic planning based on reasoning backward from the target molecule to appropriate starting materials. An arrow of the type ⇨ designates a retrosynthetic step.

Retrovirus (Section 28.13): A virus for which the genetic material is RNA rather than DNA.

Reverse transcriptase (Section 28.13): Enzyme that catalyzes the transcription of RNA to DNA.

Ribozyme (Section 28.11): A polynucleotide that has catalytic activity.

Ring current (Section 13.5): Electric field associated with circulating system of π electrons.

Ring inversion (Section 3.9): Process by which a chair conformation of cyclohexane is converted to a mirror-image chair. All of the equatorial substituents become axial, and vice versa. Also called *ring flipping,* or *chair–chair interconversion.*

RNA (ribonucleic acid) (Section 28.11): A polynucleotide of ribose.

Robinson annulation (Section 18.13): The combination of a Michael addition and an intramolecular aldol condensation used as a synthetic method for ring formation.

Rotamer (Section 3.1): Synonymous with *conformer.*

Sandmeyer reaction (Section 22.17): Reaction of an aryl diazonium ion with CuCl, CuBr, or CuCN to give, respectively, an aryl chloride, aryl bromide, or aryl cyanide (nitrile).

Sanger's reagent (Section 27.11): The compound 1-fluoro-2,4-dinitrobenzene, used in N-terminal amino acid identification.

Saponification (Section 20.11): Hydrolysis of esters in basic solution. The products are an alcohol and a carboxylate salt. The term means "soap making" and derives from the process whereby animal fats were converted to soap by heating with wood ashes.

Saturated hydrocarbon (Section 6.1): A hydrocarbon in which there are no multiple bonds.

Sawhorse formula (Section 3.1): A representation of the three-dimensional arrangement of bonds in a molecule by a drawing of the type shown.

Schiemann reaction (Section 22.17): Preparation of an aryl fluoride by heating the diazonium fluoroborate formed by addition of tetrafluoroboric acid (HBF$_4$) to a diazonium ion.

Schiff's base (Section 17.10): Another name for an imine; a compound of the type R$_2$C=NR′.

Scientific method (Section 6.6): A systematic approach to establishing new knowledge in which observations lead to laws, laws to theories, theories to testable hypotheses, and hypotheses to experiments.

Secondary alkyl group (Section 2.13): Structural unit of the type R$_2$CH—, in which the point of attachment is to a secondary carbon.

Secondary amine (Section 22.1): An amine with any combination of two alkyl or aryl substituents and one hydrogen on nitrogen; an amine of the type

RNHR′   or   RNHAr   or   ArNHAr′

Secondary carbon (Section 2.13): A carbon that is directly attached to two other carbons.

Secondary structure (Section 27.19): The conformation with respect to nearest neighbor amino acids in a peptide or protein. The α helix and the pleated β sheet are examples of protein secondary structures.

Sequence rule (Section 7.6): Foundation of the Cahn–Ingold–Prelog system. It is a procedure for ranking substituents on the basis of atomic number.

Shielding (Section 13.4): Effect of a molecule's electrons that decreases the strength of an external magnetic field felt by a proton or another nucleus.

Sigmatropic rearrangement (Section 24.13): Migration of a σ bond from one end of a conjugated π electron system to the other. The Claisen rearrangement is an example.

Simmons–Smith reaction (Section 14.12): Reaction of an alkene with iodomethylzinc iodide to form a cyclopropane derivative.

Skew boat (Section 3.7): An unstable conformation of cyclohexane. It is slightly more stable than the boat conformation.

Soaps (Section 19.5): Cleansing substances obtained by the hydrolysis of fats in aqueous base. Soaps are sodium or potassium salts of unbranched carboxylic acids having 12–18 carbon atoms.

Solid-phase peptide synthesis (Section 27.18): Method for peptide synthesis in which the C-terminal amino acid is covalently attached to an inert solid support and successive amino acids are attached via peptide bond formation. At the completion of the synthesis the polypeptide is removed from the support.

Solvolysis reaction (Section 8.7): Nucleophilic substitution in a medium in which the only nucleophiles present are the solvent and its conjugate base.

Space-filling model (Section 1.9): A type of molecular model that attempts to represent the volume occupied by the atoms.

Specific rotation (Section 7.4): Optical activity of a substance per unit concentration per unit path length:

$$[\alpha] = \frac{100\alpha}{cl}$$

where α is the observed rotation in degrees, c is the concentration in g/100 mL, and l is the path length in decimeters.

Spectrometer (Section 13.1): Device designed to measure absorption of electromagnetic radiation by a sample.

Spectrum (Section 13.2): Output, usually in chart form, of a spectrometer. Analysis of a spectrum provides information about molecular structure.

***sp* Hybridization** (Section 2.21): Hybridization state adopted by carbon when it bonds to two other atoms as, for example, in alkynes. The *s* orbital and one of the 2*p* orbitals mix to form two equivalent *sp*-hybridized orbitals. A linear geometry is characteristic of *sp* hybridization.

***sp*^{2} Hybridization** (Section 2.20): A model to describe the bonding of a carbon attached to three other atoms or groups. The carbon 2*s* orbital and the two 2*p* orbitals are combined to give a set of three equivalent *sp*^{2} orbitals having 33.3% *s* character and 66.7% *p* character. One *p* orbital remains unhybridized. A trigonal planar geometry is characteristic of *sp²* hybridization.

sp³ Hybridization (Section 2.6): A model to describe the bonding of a carbon attached to four other atoms or groups. The carbon 2*s* orbital and the three 2*p* orbitals are combined to give a set of four equivalent orbitals having 25% *s* character and 75% *p* character. These orbitals are directed toward the corners of a tetrahedron.

Spin density (Section 10.3): A measure of the unpaired electron distribution at the various atoms in a molecule.

Spin quantum number (Section 1.1): One of the four quantum numbers that describe an electron. An electron may have either of two different spin quantum numbers, $+\frac{1}{2}$ or $-\frac{1}{2}$.

Spin–spin coupling (Section 13.7): The communication of nuclear spin information between two nuclei.

Spin–spin splitting (Section 13.7): The splitting of NMR signals caused by the coupling of nuclear spins. Only nonequivalent nuclei (such as protons with different chemical shifts) can split one another's signals.

Spirocyclic hydrocarbon (Section 3.14): A hydrocarbon in which a single carbon is common to two rings.

Squalene (Section 26.11): A naturally occurring triterpene from which steroids are biosynthesized.

Staggered conformation (Section 3.1): Conformation of the type shown, in which the bonds on adjacent carbons are as far away from one another as possible.

Stereochemistry (Chapter 7): Chemistry in three dimensions; the relationship of physical and chemical properties to the spatial arrangement of the atoms in a molecule.

Stereoelectronic effect (Section 5.16): An electronic effect that depends on the spatial arrangement between the orbitals of the electron donor and acceptor.

Stereoisomers (Section 3.11): Isomers with the same constitution but that differ in respect to the arrangement of their atoms in space. Stereoisomers may be either *enantiomers* or *diastereomers.*

Stereoregular polymer (Section 7.15): Polymer containing chirality centers according to a regular repeating pattern. Syndiotactic and isotactic polymers are stereoregular.

Stereoselective reaction (Sections 5.11 and 6.3): Reaction in which a single starting material has the capacity to form two or more stereoisomeric products but forms one of them in greater amounts than any of its stereoisomers. Terms such as *addition to the less hindered side* describe stereoselectivity.

Stereospecific reaction (Section 7.13): Reaction in which stereoisomeric starting materials give stereoisomeric products. Terms such as *syn addition, anti elimination,* and *inversion of configuration* describe stereospecific reactions.

Steric hindrance (Sections 3.2, 6.3, and 8.6): An effect on structure or reactivity that depends on van der Waals repulsive forces.

Steric strain (Section 3.2): Destabilization of a molecule as a result of van der Waals repulsion, distorted bond distances, bond angles, or torsion angles.

Steroid (Section 26.11): Type of lipid present in both plants and animals characterized by a nucleus of four fused rings (three are six-membered, one is five-membered). Cholesterol is the most abundant steroid in animals.

Strecker synthesis (Section 27.4): Method for preparing amino acids in which the first step is reaction of an aldehyde with ammonia and hydrogen cyanide to give an amino nitrile, which is then hydrolyzed.

$$\overset{O}{\overset{\|}{RCH}} \xrightarrow[\text{HCN}]{\text{NH}_3} \underset{\underset{NH_2}{|}}{RCHC} \equiv N \xrightarrow{\text{hydrolysis}} \underset{\underset{^+NH_3}{|}}{RCHCO_2^-}$$

Stretching vibration (Section 13.20): A regular, repetitive motion of two atoms or groups along the bond that connects them.

Strong acid (Section 1.16): An acid that is stronger than H_3O^+.

Strong base (Section 1.16): A base that is stronger than HO^-.

Structural isomer (Section 1.8): Synonymous with *constitutional isomer.*

Substitution nucleophilic bimolecular (S$_N$2) mechanism (Sections 4.12 and 8.3): Concerted mechanism for nucleophilic substitution in which the nucleophile attacks carbon from the side opposite the bond to the leaving group and assists the departure of the leaving group.

Substitution nucleophilic unimolecular (S$_N$1) mechanism (Sections 4.9 and 8.8): Mechanism for nucleophilic substitution characterized by a two-step process. The first step is rate-determining and is the ionization of an alkyl halide to a carbocation and a halide ion.

Substitution reaction (Section 1.17): Chemical reaction in which an atom or a group of a molecule is replaced by a different atom or group.

Substitutive nomenclature (Section 4.2): Type of IUPAC nomenclature in which a substance is identified by a name ending in a suffix characteristic of the type of compound. 2-Methylbutanol, 3-pentanone, and 2-phenylpropanoic acid are examples of substitutive names.

Sucrose (Section 25.14): A disaccharide of glucose and fructose in which the two monosaccharides are joined at their anomeric positions.

Sulfide (Section 16.1): A compound of the type RSR′. Sulfides are the sulfur analogs of ethers.

Sulfonation (Section 12.4): Replacement of a hydrogen by an —SO$_3$H group. The term is usually used in connection with electrophilic aromatic substitution.

$$Ar-H \xrightarrow[\text{H}_2\text{SO}_4]{\text{SO}_3} Ar-SO_3H$$

Sulfone (Section 16.16): Compound of the type

$$R-\overset{\overset{:\ddot{O}:^-}{|}}{\underset{\underset{:\ddot{O}:^-}{|}}{S^{2+}}}-R$$

Sulfoxide (Section 16.16): Compound of the type

$$R-\overset{\overset{:\ddot{O}:^-}{|}}{\underset{}{S^+}}-R$$

Supercoil (Section 28.9): Coiled DNA helices.

Symmetry-allowed reaction (Section 10.14): Concerted reaction in which the orbitals involved overlap in phase at all stages of the process. The conrotatory ring opening of cyclobutene to 1,3-butadiene is a symmetry-allowed reaction.

Symmetry-forbidden reaction (Section 10.14): Concerted reaction in which the orbitals involved do not overlap in phase at all stages of the process. The disrotatory ring opening of cyclobutene to 1,3-butadiene is a symmetry-forbidden reaction.

Syn addition (Section 6.3): Addition reaction in which the two portions of the reagent that add to a multiple bond add from the same side.

Syndiotactic polymer (Section 7.15): Stereoregular polymer in which the configuration of successive chirality centers alternates along the chain.

Synthon (Section 21.6): A structural unit in a molecule that is related to a synthetic operation.

Systematic nomenclature (Section 2.11): Names for chemical compounds that are developed on the basis of a prescribed set of rules. Usually the IUPAC system is meant when the term *systematic nomenclature* is used.

Tautomerism (Sections 9.12 and 18.4): Process by which two isomers are interconverted by the movement of an atom or a group. Enolization is a form of tautomerism.

$$\overset{\overset{O}{\|}}{RC}-CHR_2 \rightleftharpoons RC=CR_2\overset{OH}{\underset{|}{}}$$

Terminal alkyne (Section 9.1): Alkyne of the type $RC\equiv CH$, in which the triple bond appears at the end of the chain.

Termination steps (Section 4.17): Reactions that halt a chain reaction. In a free-radical chain reaction, termination steps consume free radicals without generating new radicals to continue the chain.

Terpenes (Section 26.7): Compounds that can be analyzed as clusters of isoprene units. Terpenes with 10 carbons are classified as *monoterpenes,* those with 15 are *sesquiterpenes,* those with 20 are *diterpenes,* and those with 30 are *triterpenes.*

Tertiary alkyl group (Section 2.13): Structural unit of the type R_3C-, in which the point of attachment is to a tertiary carbon.

Tertiary amine (Section 22.1): Amine of the type R_3N with any combination of three alkyl or aryl substituents on nitrogen.

Tertiary carbon (Section 2.13): A carbon that is directly attached to three other carbons.

Tertiary structure (Section 27.20): A description of how a protein chain is folded.

Tesla (Section 13.3): SI unit for magnetic field strength.

Tetrahedral intermediate (Section 19.14 and Chapter 20): The key intermediate in nucleophilic acyl substitution. Formed by nucleophilic addition to the carbonyl group of a carboxylic acid derivative.

Tetramethylsilane (TMS) (Section 13.4): The molecule $(CH_3)_4Si$, used as a standard to calibrate proton and carbon-13 NMR spectra.

Tetrasubstituted alkene (Section 5.6): Alkene of the type $R_2C=CR_2$, in which there are four carbons *directly* bonded to the carbons of the double bond. (The R groups may be the same or different.)

Tetrose (Section 25.3): A carbohydrate with four carbon atoms.

Thermochemistry (Section 2.18): The study of heat changes that accompany chemical processes.

Thermodynamically controlled reaction (Section 10.10): Reaction in which the reaction conditions permit two or more products to equilibrate, giving a predominance of the most stable product.

Thioester (Section 20.13): An *S*-acyl derivative of a thiol; a compound of the type

Thiol (Section 15.13): Compound of the type RSH or ArSH.

Threo (Section 7.11): Term applied to the relative configuration of two stereogenic centers within a molecule. The threo stereoisomer has like substituents on opposite sides of a Fischer projection.

Torsional strain (Section 3.1): Decreased stability of a molecule associated with eclipsed bonds.

trans- (Section 3.11): Stereochemical prefix indicating that two substituents are on opposite sides of a ring or a double bond. (Contrast with the prefix *cis-*.)

Transcription (Section 28.11): Construction of a strand of mRNA complementary to a DNA template.

Transfer RNA (tRNA) (Section 28.11): A polynucleotide of ribose that is bound at one end to a unique amino acid. This amino acid is incorporated into a growing peptide chain.

Transition state (Section 3.1): The point of maximum energy in an elementary step of a reaction mechanism.

Translation (Section 28.12): The "reading" of mRNA by various tRNAs, each one of which is unique for a particular amino acid.

Triacylglycerol (Section 26.2): A derivative of glycerol (1,2,3-propanetriol) in which the three oxygens bear acyl groups derived from fatty acids.

Tripeptide (Section 27.1): A compound in which three α-amino acids are linked by peptide bonds.

Triple bond (Section 1.4): Bond formed by the sharing of six electrons between two atoms.

Trisubstituted alkene (Section 5.6): Alkene of the type $R_2C{=}CHR$, in which there are three carbons *directly* bonded to the carbons of the double bond. (The R groups may be the same or different.)

Trivial nomenclature (Section 2.11): Term synonymous with *common nomenclature.*

Trypsin (Section 27.10): A digestive enzyme that catalyzes the hydrolysis of proteins. Trypsin selectively catalyzes the cleavage of the peptide bond between the carboxyl group of lysine or arginine and some other amino acid.

Twist boat (Section 3.7): Synonymous with *skew boat.*

Ultraviolet-visible (UV-VIS) spectroscopy (Section 13.21): Analytical method based on transitions between electronic energy states in molecules. Useful in studying conjugated systems such as polyenes.

Unimolecular (Section 4.8): Describing a step in a reaction mechanism in which only one particle undergoes a chemical change at the transition state.

α, β-Unsaturated aldehyde or ketone (Section 18.11): Aldehyde or ketone that bears a double bond between its α and β carbons as in

$$\underset{R_2C=CHCR'}{\overset{\displaystyle O}{\|}}$$

Unsaturated hydrocarbon (Section 6.1): A hydrocarbon that can undergo addition reactions; that is, one that contains multiple bonds.

Upfield (Section 13.4): The high-field region of an NMR spectrum. A signal that is upfield with respect to another lies to its right on the spectrum.

Uronic acids (Section 25.19): Carbohydrates that have an aldehyde function at one end of their carbon chain and a carboxylic acid group at the other.

Valence bond theory (Section 2.3): Theory of chemical bonding based on overlap of half-filled atomic orbitals between two atoms. Orbital hybridization is an important element of valence bond theory.

Valence electrons (Section 1.1): The outermost electrons of an atom. For second-row elements these are the $2s$ and $2p$ electrons.

Valence shell electron-pair repulsion (VSEPR) model (Section 1.10): Method for predicting the shape of a molecule based on the notion that electron pairs surrounding a central atom repel one another. Four electron pairs will arrange themselves in a tetrahedral geometry, three will assume a trigonal planar geometry, and two electron pairs will adopt a linear arrangement.

Van der Waals forces (Section 2.17): Intermolecular forces that do not involve ions (dipole–dipole, dipole/induced-dipole, and induced-dipole/induced-dipole forces).

Van der Waals radius (Section 2.17): A measure of the effective size of an atom or a group. The repulsive force between two atoms increases rapidly when they approach each other at distances less than the sum of their van der Waals radii.

Van der Waals strain (Section 3.2): Destabilization that results when two atoms or groups approach each other too closely. Also known as *van der Waals repulsion.*

Vicinal (Section 6.14): Describing two atoms or groups attached to adjacent atoms.

Vicinal coupling (Section 13.7): Coupling of the nuclear spins of atoms X and Y on adjacent atoms as in X—A—B—Y. Vicinal coupling is the most common cause of spin–spin splitting in ^{1}H NMR spectroscopy.

Vicinal diol (Section 15.5): Compound that has two hydroxyl (—OH) groups on adjacent sp^3-hybridized carbons.

Vinyl group (Section 5.1): The group $H_2C{=}CH{-}$.

Vitalism (Introduction): A nineteenth-century theory that divided chemical substances into two main classes, organic and inorganic, according to whether they originated in living (animal or vegetable) or nonliving (mineral) matter, respectively. Vitalist doctrine held that the conversion of inorganic substances to organic ones could be accomplished only through the action of some "vital force."

Walden inversion (Section 8.4): Originally, a reaction sequence developed by Paul Walden whereby a chiral starting material was transformed to its enantiomer by a series of stereospecific reactions. Current usage is more general and refers to the inversion of configuration that attends any bimolecular nucleophilic substitution.

Wave functions (Section 1.1): The solutions to arithmetic expressions that express the energy of an electron in an atom.

Wavelength (Section 13.1): Distance between two successive maxima (peaks) or two successive minima (troughs) of a wave.

Wavenumbers (Section 13.20): Conventional units in infrared spectroscopy that are proportional to frequency. Wavenumbers are cited in reciprocal centimeters (cm^{-1}).

Wax (Section 26.5): A mixture of water-repellent substances that form a protective coating on the leaves of plants, the fur of animals, and the feathers of birds, among other things. A principal component of a wax is often an ester in which both the acyl portion and the alkyl portion are characterized by long carbon chains.

Weak acid (Section 1.16): An acid that is weaker than H_3O^+.

Weak base (Section 1.16): A base that is weaker than HO^-.

Williamson ether synthesis (Section 16.6): Method for the preparation of ethers involving an S_N2 reaction between an alkoxide ion and a primary alkyl halide:

$$RONa + R'CH_2Br \longrightarrow R'CH_2OR + NaBr$$

Wittig reaction (Section 17.12): Method for the synthesis of alkenes by the reaction of an aldehyde or a ketone with a phosphorus ylide.

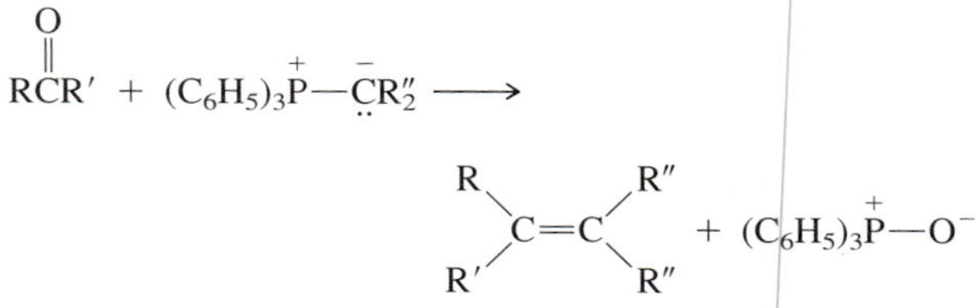

$$\overset{O}{\overset{\|}{R}}\overset{}{C}R' \; + \; (C_6H_5)_3\overset{+}{P}-\overset{-}{C}R''_2 \longrightarrow$$

$$\begin{array}{c} R \\ \diagdown \\ \diagup \\ R' \end{array} C=C \begin{array}{c} R'' \\ \diagup \\ \diagdown \\ R'' \end{array} \; + \; (C_6H_5)_3\overset{+}{P}-O^-$$

Wolff–Kishner reduction (Section 12.8): Method for reducing the carbonyl group of aldehydes and ketones to a methylene group $(C=O \longrightarrow CH_2)$ by treatment with hydrazine (H_2NNH_2) and base (KOH) in a high-boiling alcohol solvent.

Wood alcohol (Section 4.3): A common name for methanol, CH_3OH.

Ylide (Section 17.12): A neutral molecule in which two oppositely charged atoms, each with an octet of electrons, are directly bonded to each other. The compound

$$(C_6H_5)_3\overset{+}{P}-\overset{-}{C}H_2$$

is an example of an ylide.

Zaitsev's rule (Section 5.10): When two or more alkenes are capable of being formed by an elimination reaction, the one with the more highly substituted double bond (the more stable alkene) is the major product.

Zwitterion (Section 27.3): The form in which neutral amino acids actually exist. The amino group is in its protonated form and the carboxyl group is present as a carboxylate

$$\overset{}{\underset{\overset{|}{^+NH_3}}{R}}CHCO_2^-$$

CREDITS

INTRODUCTION

Pages 2, 3, 4, 5 Stamps are courtesy of James O. Schreck, Professor of Chemistry, University of Northern Colorado.

CHAPTER 11

Page 436 (Figure 11.5) was generated using crystallographic coordinates obtained from the Center for Computational Materials Science at the United States Naval Research Laboratory via http://cst-www.nrl.navy.mil/lattice/struk/a9.html.
Page 437 (Figure 11.7) was obtained from the Center for Nanoscale Science and Technology at Rice University via http://cnst.rice.edu/images/Tube1010a.tif.

CHAPTER 13

Page 520 (Figure 13.1) is from M. Silberberg, *Chemistry,* 2nd ed., p. 260. McGraw-Hill, New York, 2000.
Page 546 (Figure 13.24) is courtesy of Simon Fraser/Science Photo Library/Photo Researchers, Inc. Newcastle upon Tyne.
Page 565 (Figure 13.37) is adapted from R. Isaksson, J. Roschester, J. Sandstrom, and L. G. Wistrand, *Journal of the American Chemical Society,* 1985, 107, 4074–4075 with permission of the American Chemical Society.
Page 568 (Figure 13.39) is from M. Silberberg, *Chemistry,* 2nd ed., p. 56. McGraw-Hill, New York, 2000.
Page 572 (Figure 13.44) is adapted from H. D. Durst and G. W. Gokel, *Experimental Organic Chemistry,* 2nd ed., McGraw-Hill, New York, 1987.
Mass spectra are reproduced with permission from "EPA/NIH Mass Spectral Data Base," Supplement I, S. R. Heller and G. W. A. Milne, National Bureau of Standards, 1980.

CHAPTER 15

Page 646 (Figure 15.5) is adapted from crystallographic coordinates deposited with The Protein Data Bank, PDB ID: 1A71. Colby, T. D., Bahnson, B. J., Chin, J. K., Klinman, J. P., Goldstein, B. M., Active Site Modifications in a Double Mutant of Liver Alcohol Dehydrogenase: Structural Studies of Two Enzyme-Ligand Complexes. To be published.

CHAPTER 16

Page 670 (Figure 16.3) is adapted from crystallographic coordinates deposited with The Cambridge Crystallographic Data Centre, CCDC ID: NAMNSB. Duax, W. L., Smith, G. D., Strong, P. D., *Journal of the American Chemical Society,* 1980, 102, 6725.

CHAPTER 25

Page 1048 (Figure 25.8) is adapted from crystallographic coordinates deposited with the Protein Data Bank. PDB ID: 4TF4. Sakon, J., Irwin, D., Wilson, D. B., Karplus, P. A., Structure and Mechanism of Endo/Exocellulase E4 from Thermomonospora Fusca. To be published.
Pages 1049 (Figure 25.9) and **1050** (Figure 25.10) are adapted from crystallographic coordinates deposited with The Protein Data Bank, PDB ID: 1C58. Gessler, K, Uson, I., Takahan, T., Krauss, N., Smith, S. M., Okada, G. M., Sheldrick, G. M., Saenger, W., V-Amylose at Atomic Resolution: X-Ray Structure of a Cycloamylose with 26 Glucose Residues (Cyclomaltohexaicosaose). *Proc.Nat.Acad.Sci.USA,* 1999, 96, 4246.

CHAPTER 26

Page 1093 (Figure 26.9c) is adapted from crystallographic coordinates deposited with the Protein Data Bank. PDB ID: 1CLE. Ghosh, D., Wawrzak, Z., Pletnev, V. Z., Li, N., Kaiser, R., Pangborn, W., Jornvall, H., Erman, M., Duax, W. L., Structure of Uncomplexed and Linoleate-Bound Candida Cholesterol Esterase. To be published.

CHAPTER 27

Page 1144 (Figure 27.16) is adapted from crystallographic coordinates deposited with the Protein Data Bank. PDB ID: 2SLK. Fossey S. A., Nemethy, G., Gibson, K. D., Scheraga, H. A., Conformational Energy Studies of Beta-Sheets of Model Silk Fibroin Peptides. I. Sheets of Poly(Ala-Gly) Chains. Biopolymers 31, p. 1529 (1991).
Page 1146 (Figure 27.18) is adapted from crystallographic coordinates deposited with the Protein Data Bank. PDB ID: 2CTB. Teplyakov, A., Wilson, K. S., Orioli, P., Mangani S., The High Resolution Structure of the Complex between Carboxypeptidase A and L-Phenyl Lactate. To be published.
Page 1148 (Figure 27.21) is adapted from crystallographic coordinates deposited with the Protein Data Bank. PDB ID: 1VXH. Yang, F., Phillips Jr., G. N., Structures of Co-, Deoxy- and met-Myoglobins at Various pH Values. To be published.

CHAPTER 28

Page 1170 (Figure 28.5) is adapted from crystallographic coordinates deposited with the Protein Data Bank. PDB ID: 1DDN. White, A., Ding, X., Vanderspek, J. C., Murphy J. R., Ringe, D., Structure of the Metal-Ion-Activated Diphtheria Toxin Repressor/Tox Operator Complex. *Nature* 394, p. 502, (1998).
Page 1171 (Figure 28.7) is adapted from crystallographic coordinates deposited with The Protein Data Bank, PDB ID: 1A01. Luger, A., Mader, W. Richmond, R. K., Sargent, D. F., Richmond, T. J., Crystal Structure of the Nucleosome Core Particle at 2.8 A Resolution. *Nature,* 1997, V. 389, 251.
Page 1176 (Figure 28.11) is adapted from crystallographic coordinates deposited with the Protein Data Bank. PDB ID: 6TNA. Sussman, J. L., Holbrook, S. R., Warrant, R. W., Church, G. M., Kim, S. H., Crystal Structure of Yeast Phenylalanine tRNA. I. Crystallographic Refinement, *J. Mol. Biol.,* 1978, 126, 607. (1978).

INDEX

WHERE TO FIND IT
A GUIDE TO FREQUENTLY CONSULTED TABLES AND FIGURES

Acids and Bases

    Dissociation Constants for Selected Brønsted Acids (Table 1.7, pp. 36-37)

    Acidities of Hydrocarbons (Table 14.2, p. 593)

    Acidities of Carboxylic Acids (Table19.2, p.801)

    Acidities of Phenols (Table 24.2, p.998)

    Acidities of Substituted Benzoic Acids (Table 19.3, p.804)

    Acid-Base Properties of Amino Acids (Tables 27.2 and 27.3, p. 1119)

    Basicities of Alkylamines (Table 22.1, p. 920)

    Basicities of Arylamines (Table 22.2, p.921)

Classification of Isomers (Table 7.2, p. 315)

Free-Energy Difference-Composition Relationship in an Equilibrium Mixture (Figure 3.19, p.123)

IUPAC Nomenclature

    Names of Unbranched Alkanes (Table 2.2, p. 71)

    Rules for Alkanes and Cycloalkanes (Table 2.6, pp. 96-97)

    Rules for Alkyl Groups (Table 2.7, p. 98)

Reactivity

    Nucleophilicity of Some Common Nucleophiles (Table 8.4, p. 338)

    Leaving Groups in Nucleophilic Substitution (Table 8.8, p. 352)

    Substituent Effects in Electrophilic Aromatic Substitution (Table 12.2, p. 495)

Spectroscopy Correlation Tables

    Proton Chemical Shifts (Table 13.1, p. 528)

    [13] Chemical Shifts (Table 13.3, p. 549)

    Infrared Absorption Frequencies (Table 13.4, p. 561)

Stereochemistry

    Cahn-Ingold-Prelog Priority Rules (Table 5.1, p. 195)

    Absolute Configuration Using Cahn-Ingold-Prelog Notation (Table 7.1, p. 291)

    Fisher Projections of D-Aldoses (Figure 25.2, p. 1031)

Structure and Bonding

    Electronegativities of Selected Atoms (Table 1.2, p. 15; Table 14.1, p.588)

    How to Write Lewis Structures (Table 1.4, p. 20)

    Rules of Resonance (Table 1.5, pp. 26-27)

    Bond Dissociation Energies (Table 4.3, p. 170)

    Bond Distances, Bond Angles, and Bond Energies in Ethane, Ethene, and Ethyne (Table 9.1, p.367)

    Structures of α-Amino Acids (Table 27.1, pp.1111-1112)

Periodic Table of the Elements

MAIN-GROUP ELEMENTS

MAIN-GROUP ELEMENTS

Period

Color key:
- Metals (main-group)
- Metals (transition)
- Metals (inner transition)
- Metalloids
- Nonmetals

TRANSITION ELEMENTS

INNER TRANSITION ELEMENTS

As of mid-1999, elements 110 through 112 have not yet been named.

Period	1A (1)	2A (2)	3B (3)	4B (4)	5B (5)	6B (6)	7B (7)	8B (8)	8B (9)	8B (10)	1B (11)	2B (12)	3A (13)	4A (14)	5A (15)	6A (16)	7A (17)	8A (18)
1	1 **H** 1.008																	2 **He** 4.003
2	3 **Li** 6.941	4 **Be** 9.012											5 **B** 10.81	6 **C** 12.01	7 **N** 14.01	8 **O** 16.00	9 **F** 19.00	10 **Ne** 20.18
3	11 **Na** 22.99	12 **Mg** 24.31											13 **Al** 26.98	14 **Si** 28.09	15 **P** 30.97	16 **S** 32.07	17 **Cl** 35.45	18 **Ar** 39.95
4	19 **K** 39.10	20 **Ca** 40.08	21 **Sc** 44.96	22 **Ti** 47.88	23 **V** 50.94	24 **Cr** 52.00	25 **Mn** 54.94	26 **Fe** 55.85	27 **Co** 58.93	28 **Ni** 58.69	29 **Cu** 63.55	30 **Zn** 65.39	31 **Ga** 69.72	32 **Ge** 72.61	33 **As** 74.92	34 **Se** 78.96	35 **Br** 79.90	36 **Kr** 83.80
5	37 **Rb** 85.47	38 **Sr** 87.62	39 **Y** 88.91	40 **Zr** 91.22	41 **Nb** 92.91	42 **Mo** 95.94	43 **Tc** (98)	44 **Ru** 101.1	45 **Rh** 102.9	46 **Pd** 106.4	47 **Ag** 107.9	48 **Cd** 112.4	49 **In** 114.8	50 **Sn** 118.7	51 **Sb** 121.8	52 **Te** 127.6	53 **I** 126.9	54 **Xe** 131.3
6	55 **Cs** 132.9	56 **Ba** 137.3	57 **La** 138.9	72 **Hf** 178.5	73 **Ta** 180.9	74 **W** 183.9	75 **Re** 186.2	76 **Os** 190.2	77 **Ir** 192.2	78 **Pt** 195.1	79 **Au** 197.0	80 **Hg** 200.6	81 **Tl** 204.4	82 **Pb** 207.2	83 **Bi** 209.0	84 **Po** (209)	85 **At** (210)	86 **Rn** (222)
7	87 **Fr** (223)	88 **Ra** (226)	89 **Ac** (227)	104 **Rf** (261)	105 **Db** (262)	106 **Sg** (266)	107 **Bh** (262)	108 **Hs** (265)	109 **Mt** (266)	110 (269)	111 (272)	112 (277)						

Lanthanides (6):

58 **Ce** 140.1	59 **Pr** 140.9	60 **Nd** 144.2	61 **Pm** (145)	62 **Sm** 150.4	63 **Eu** 152.0	64 **Gd** 157.3	65 **Tb** 158.9	66 **Dy** 162.5	67 **Ho** 164.9	68 **Er** 167.3	69 **Tm** 168.9	70 **Yb** 173.0	71 **Lu** 175.0

Actinides (7):

90 **Th** 232.0	91 **Pa** (231)	92 **U** 238.0	93 **Np** (237)	94 **Pu** (242)	95 **Am** (243)	96 **Cm** (247)	97 **Bk** (247)	98 **Cf** (251)	99 **Es** (252)	100 **Fm** (257)	101 **Md** (258)	102 **No** (259)	103 **Lr** (260)